COMPOUND AMOUNT

If P dollars is deposited for n time periods at a compound interest rate i per period, the **compound amount** A is

$$A = P(1 + i)^n.$$

Consider an ordinary annuity of n payments of R dollars each at the end of consecutive interest periods with interest compounded at a rate of i per period. The **future value** S of the annuity is:

$$S = R\left[\frac{(1 + i)^n - 1}{i}\right].$$

The **present value** P of the annuity is:

$$P = R\left[\frac{(1 - (1 + i)^{-n}}{i}\right].$$

The **matrix** version of the elimination method uses the following **matrix row operations** to obtain the augmented matrix of an equivalent system. They correspond to using elementary row operations on a system of equations.

1. Interchange any two rows.
2. Multiply each element of a row by a nonzero constant.
3. Replace a row by the sum of itself and a constant multiple of another row in the matrix.

UNION RULE

For any events E and F from a sample space S.

$$P(E \cup F) = P(E) + P(F) - P(E \cap F).$$

PROPERTIES OF PROBABILITY

Let S be a sample space consisting of n distinct outcomes $s_1, s_2, \ldots, s_n$. An acceptable probability assignment consists of assigning to each outcome s_1 a number p_1 (the probability of s_1) according to these rules.

1. The probability of each outcome is a number between 0 and 1.

$$0 \le p_1 \le 1, \quad 0 \le p_2 \le 1, \quad \ldots, \quad 0 \le p_n \le 1.$$

2. The sum of the probabilities of all possible outcomes is 1.

$$p_1 + p_2 + p_3 + \cdots + p_n = 1.$$

BAYES' FORMULA

For any events E and $F_1, F_2, \ldots, F_n$, from a sample space S, where $F_1 \cup F_2 \ldots \cup F_n = S$,

$$P(F_i|E) = \frac{P(F_i) \cdot P(E|F_i)}{P(F_1) \cdot P(E|F_1) + \cdots + P(F_n) \cdot P(E|F_n)}.$$

TENTH EDITION

Mathematics with Applications

In the Management, Natural, and Social Sciences

Margaret L. Lial
American River College

Thomas W. Hungerford
Saint Louis University

John P. Holcomb, Jr.
Cleveland State University

Addison-Wesley

Boston Columbus Indianapolis New York San Francisco Upper Saddle River
Amsterdam Cape Town Dubai London Madrid Milan Munich Paris Montreal Toronto
Delhi Mexico City Sao Paulo Sydney Hong Kong Seoul Singapore Taipei Tokyo

Editor-in-Chief: Deirdre Lynch
Executive Acquisitions Editor: Jennifer Crum
Project Editor: Elizabeth Bernardi
Editorial Supervisor: Joanne Wendelken
Managing Editor: Karen Wernholm
Associate Managing Editor: Tamela Ambush
Senior Production Project Manager: Sheila Spinney
Executive Marketing Manager: Jeff Weidenaar
Marketing Assistant: Kendra Bassi
Associate Media Producer: Jean Choe
Software Editor/MathXL: Janet McHugh
Software Editor/TestGen: Mary Durnwald
Manufacturing Manager: Evelyn Beaton
Senior Designer: Beth Paquin
Cover and Interior Design: Nancy Goulet, Studio;Wink
Production Coordination, Composition, and Illustrations: Aptara Corporation
Cover Photo: Carlos Dominquez/Corbis

Photo Credits:
Page 1 © iStockphoto; **page 11** © Photos.com; **page 73** © Photos.com; **page 75** © iStockphoto; **page 107** © Shutterstock; **page 133** © iStockphoto; **page 165** © iStockphoto; **page 213** © Shutterstock; **page 222** © Shutterstock; **page 242** © NOAA; **page 253** © iStockphoto; **page 257** © NOAA; **page 259** © iStockphoto; **page 301** © iStockphoto; **page 306** © University of Southern Mississippi; **page 306** © iStockphoto; **page 311** © Shutterstock; **page 317** © Shutterstock; **page 349** © iStockphoto; **page 365** © Photos.com; **page 387** © Shutterstock; **page 405** © Shutterstock; **page 431** © Shutterstock; **page 459** © Shutterstock; **page 463** © Shutterstock; **page 465** © Photos.com; **page 473** © iStockphoto; **page 486** © Shutterstock; **page 496** © Photos.com; **page 503** © Shutterstock; **page 533** © Photos.com; **page 535** © iStockphoto; **page 545** © iStockphoto; **page 566** © iStockphoto; **page 595** © Shutterstock; **page 597** © Shutterstock; **page 611** © iStockphoto; **page 639** © Shutterstock; **page 651** © iStockphoto; **page 653** © iStockphoto; **page 717** © iStockphoto; **page 727** © Photos.com; **page 739** © Shutterstock; **page 766** © Shutterstock; **page 769** © iStockphoto; **page 775** © Shutterstock; **page 784** © Natural Resources Conservation Service/USDA; **page 810** © Shutterstock; **page 825** © Photos.com; **page 827** © Photos.com; **page 837** © Photos.com; **page 870** © Shutterstock; **page 897** © iStockphoto; **page 899** © Shutterstock; **page 901** © U.S. Navy photo; **page 913** © Photos.com; **page 926** © iStockphoto; **page 947** © iStockphoto; **page 948** © iStockphoto

Library of Congress Cataloging-in-Publication Data
Lial, Margaret L.
Mathematics with applications in the management, natural, and social sciences/Margaret L.—10th ed./ Thomas W. Hungerford, John Holcomb, Jr.
p. cm.
Includes index.
ISBN 0-321-64553-7 (hardcover)
1. Mathematics—Textbooks. 2. Social sciences—Mathematics—Textbooks. 3. Medical sciences—Mathematics—Textbooks. I. Hungerford, Thomas W. II. Holcomb, John P. III. Title.
QA37.3.L55 2010
510—dc22 2009025393

2 3 4 5 6 7 8 9 10–DOW–10

Addison-Wesley
is an imprint of

www.pearsonhighered.com

ISBN 10: 0-321-64553-7
ISBN 13: 978-0-321-64553-1

Contents

Preface

Mathematics with Applications is an applications-oriented text for students in business, management, and natural and social sciences. The text can be used for a variety of different courses, and the only prerequisite is a course in algebra. Chapter 1 provides a thorough review of basic algebra for those students who need it.

It has been our primary goal to present sound mathematics in an understandable manner, proceeding from the familiar to new material and from concrete examples to general rules and formulas. There is an ongoing focus on real-world problem solving, and almost every section includes relevant, contemporary applications.

New to This Edition

This edition covers the same topics as the previous one, but with the following significant additions and changes to both content and presentation:

- Chapter 5 (Mathematics of Finance) has been substantially rewritten to introduce material that plays a significant role in today's business environment. It includes new material on treasury bills, corporate bonds, zero-coupon bonds, pension distributions, bond pricing, and annuities due. The discussion of the present value of an annuity has been completely redone so that the topic is much more understandable.
- The first two sections of Chapter 6 (Systems of Linear Equations and Matrices) have been expanded to three. Systems of two equations in two variables are now in a section by themselves—to satisfy both those instructors who want to omit this discussion and proceed directly to larger systems, and those who requested expanded coverage of this topic. Both the elimination method and the Gauss–Jordan method for solving large systems of equations are now in the same section, since many instructors prefer to get to the Gauss–Jordan method as soon as possible. Finally, an entire section is devoted to applications of systems of equations.
- At the request of a number of users, the discussion of frequency probability has been moved from Section 8.4 to 8.3.
- Some requested changes have also been made in Chapter 10 (Introduction to Statistics). The first section has been split into two to give separate treatment to two important topics. Section 10.1 has an expanded discussion of frequency distributions, and Section 10.2 covers measures of central tendency.
- Three of the end-of-chapter Cases are new and several others have been updated.
- Approximately 20% of the real data examples have been updated. Updates were focused on situations in which data could be updated to reflect current conditions.
- Approximately 18% of the 5,288 exercises in the book are new and 11% more have been updated. 15% of the 623 examples in the text are new and 24% have been updated.

- The margin exercises that have always been a part of this book have been renamed *Checkpoints,* to more accurately reflect their pedagogical purpose. The checkpoint icons (such as 3✓) within the body of the text indicate when it's appropriate to work the Checkpoint exercise. Their answers have been moved to the end of the section to provide a more open and inviting page layout and to encourage students to work the problems before looking at the answers.
- In the Examples, real-world applications now carry titles corresponding to the ones used in the Exercises (Business, Finance, Health, Social Science, etc.).

Continuing Pedagogical Features

- *Balanced Approach* Multiple representations of a topic (symbolic, numerical, graphical, verbal) are given when appropriate. However, we do not believe that all representations are useful for all topics, so effective alternatives are discussed only when they are likely to increase student understanding.
- *Real-Data Examples and Explanations* Real-data exercises have long been a popular feature of this book. A significant number of new real-data examples have also been introduced into the text.
- *Cautions* highlight common student difficulties or warn against frequently made mistakes.
- *Exercises* In addition to the drill, conceptual, and application-based exercises, as well as questions from past CPA exams, there are some specially-marked exercises:

  Writing Exercises (see page 241)

  Connection Exercises that relate current topics to earlier sections (see page 317); and exercises that require technology (see page 152).
- *Selected exercises* include a reference back to related example(s) within the section (See Examples 6 and 7). This helps students as they get started with homework.
- *End-of-Chapter* materials include a summary of key terms and symbols and key concepts, as well as a set of chapter-review exercises.
- *Cases* appear at the end of each chapter and offer contemporary real-world applications of some of the mathematics presented in the chapter. Not only do these provide an opportunity for students to see the mathematics they are learning in action, but they also provide at least a partial answer to the question, "What's this stuff good for?"

Technology

It is assumed that all students have a calculator that will handle exponential and logarithmic functions. Beyond that, however, *the use of technology in this text is optional.* Examples and exercises that definitely require some sort of technology (graphing calculators, spreadsheets, or other computer programs) are marked with the icon , so instructors who want to omit these discussions and exercises can easily do so.

Instructors who routinely use technology in their courses will find more than enough material here to satisfy their needs. Here are some of the features they may want to incorporate into their courses:

- *Examples and Exercises marked with* A number of examples show students how various features of graphing calculators and spreadsheets can be applied to the topics in this book.
- *Technology Tips* These are placed at appropriate points in the text to inform students of various features of their graphing calculator, spreadsheet, or other computer programs. Note that

Technology Tips for	**Also apply to**
TI-84+	TI-83+, TI-Nspire, and usually TI-83
Casio	Casio 9750 GII and 9860 GII, and usually other models.

- *Appendix A: Graphing Calculators* This consists of a brief introduction to the relevant features of the latest Texas Instruments and Casio graphing calculators. An outline of the appendix is on page 951, and the full appendix is available in MyMathLab or at pearsonhighered.com/MWA10e. Additionally, the Web sites contain downloadable programs for both TI and Casio calculators that add functionality that is specifically relevant to this course.
- A separate Graphing Calculator and Excel Spreadsheet Manual is also available. This manual, correlated to examples in the textbook, provides students with the support they need to make use of graphing calculators and Excel.

Course Flexibility

The content of the text is divided into three parts:

- College Algebra (Chapters 1–4),
- Finite Mathematics (Chapters 5–10),
- Calculus (Chapters 11–14)

This coverage of the material offers flexibility, making the book appropriate for a variety of courses, including:

Finite Mathematics and Calculus (one year or less). Use the entire book; cover topics from Chapters 1–4 as needed before proceeding to further topics.

Finite Mathematics (one semester or two quarters). Use as much of Chapters 1–4 as needed, and then go into Chapters 5–10 as time permits and local needs require.

Calculus (one semester or quarter). Cover the precalculus topics in Chapters 1–4 as necessary, and then use Chapters 11–14.

College Algebra with Applications (one semester or quarter). Use Chapters 1–8, with Chapters 7 and 8 being optional.

Chapter interdependence is as follows:

	Chapter	*Prerequisite*
1	Algebra and Equations	None
2	Graphs, Lines, and Inequalities	Chapter 1
3	Functions and Graphs	Chapters 1 and 2
4	Exponential and Logarithmic Functions	Chapter 3
5	Mathematics of Finance	Chapter 4
6	Systems of Linear Equations and Matrices	Chapters 1 and 2
7	Linear Programming	Chapters 3 and 6
8	Sets and Probability	None
9	Counting, Probability Distributions, and Further Topics in Probability	Chapter 8
10	Introduction to Statistics	Chapter 8
11	Differential Calculus	Chapters 1–4
12	Applications of the Derivative	Chapter 11
13	Integral Calculus	Chapters 11 and 12
14	Multivariate Calculus	Chapters 11–13

Contact your local Pearson Education sales representative to order a customized version of this text.

Student Supplements

Student's Solutions Manual

- By Beverly Fusfield and James J. Ball, Indiana State University
- This manual contains detailed, carefully worked out solutions to all odd-numbered section exercises and all Chapter Review and Case exercises.
  ISBN 13: 978-0-321-64582-1;
  ISBN 10: 0-321-64582-0

Graphing Calculator and Excel Spreadsheet Manual

- By Victoria Baker, Nicholls State University and Stela Pudar-Hozo, Indiana University—Northwest
- The Graphing Calculator portion contains detailed instruction for using the TI-83/TI-83+/TI-84+, TI-Nspire, and Casio 9750 GII and Casio 9860 GII with this textbook. Instructions are organized by chapter and section.
- The Excel spreadsheet portion contains detailed instructions for using Excel with this textbook.
- Available in MyMathLab or through www.coursesmart.com
  ISBN-13: 978-0-321-65513-3
  ISBN-10: 0-321-65513-4

Instructor Supplements

Instructor's Edition

- This book contains answers to all exercises in the text.
  ISBN 13: 978-0-321-64632-3;
  ISBN 10: 0-321-64632-0

Online Instructor's Solutions Manual (downloadable only)

- By Beverly Fusfield and James J. Ball, Indiana State University
- This manual contains detailed solutions to all text exercises, suggested course outlines, and a chapter interdependence chart.
  Go to www.pearsonhighered.com/educator or MyMathLab.

Online Test Bank (downloadable only)

- By David Bridge, University of Central Oklahoma
- This test bank includes four alternate tests per chapter that parallel the text's Chapter Tests.
- Available through http://www.pearsonhighered.com/educator or in MyMathLab,

(*continued*)

Instructor Supplements (*continued*)

TestGen®

- TestGen enables instructors to build, edit, print, and administer tests using a computerized bank of questions developed to cover all the objectives of the text. TestGen is algorithmically based, allowing instructors to create multiple equivalent versions of the same question or test with the click of a button. Instructors also can modify test bank questions or add new questions. Tests can be printed or administered online. Available for download through http://www.pearsonhighered.com/educator.

PowerPoint® Lecture Slides

- These slides present key concepts and definitions from the text. They are available in MyMathLab or at http://www.pearsonhighered.com/educator.

Media Supplements

MyMathLab® Online Course (access code required)
MyMathLab is a text-specific, easily customizable online course that integrates interactive multimedia instruction with textbook content. MyMathLab gives you the tools you need to deliver all or a portion of your course online, whether your students are in a lab setting or working from home.

- **Interactive homework exercises,** correlated to your textbook at the objective level, are algorithmically generated for unlimited practice and mastery. Most exercises are free-response and provide guided solutions, sample problems, and learning aids for extra help.
- **Personalized Study Plan,** generated when students complete a test or quiz, indicates which topics have been mastered and links to tutorial exercises for topics students have not mastered.
- **Multimedia learning aids,** such as video lectures, animations, and complete multimedia textbook, help students independently improve their understanding and performance.
- **Assessment Manager** lets you create online homework, quizzes, and tests that are automatically graded. Select just the right mix of questions from the MyMathLab exercise bank, instructor-created custom exercises, and/or TestGen test items.
- **Gradebook,** designed specifically for mathematics and statistics, automatically tracks students' results and give you control over how to calculate final grades. You can also add offline (paper-and-pencil) grades to the gradebook.
- **MathXL® Exercise Builder** allows you to create static and algorithmic exercises for your online assignments. You can use the library of sample exercises as an easy starting point.
- **Pearson Tutor Center** (www.pearsontutorsercvices.com) access is automatically included with MyMathLab. The Tutor Center is staffed by qualified math instructors who provide textbook-specific tutoring for students via toll-free phone, fax, email, and interactive Web sessions.

MyMathLab is powered by CourseCompass™, Pearson Education's online teaching and learning environment, and by MathXL, our online homework, tutorial, and assessment system. MyMathLab is available to qualified adopters. For more information, visit www.mymathlab.com or contact your Pearson sales representative.

MathXL Online Course (access code required)

MathXL is an online homework, tutorial, and assessment system that accompanies Pearson's textbooks in mathematics or statistics.

- **Interactive homework exercises,** correlated to your textbook at the objective level, are algorithmically generated for unlimited practice and mastery. Most exercises are free-response and provide guided solutions, sample problems, and learning aids for extra help.
- **Personalized Study Plan,** generated when students complete a test or quiz, indicates which topics have been mastered and links to tutorial exercises for topics students have not mastered.
- **Multimedia learning aids,** such as video lectures and animations, help students independently improve their understanding and performance.
- **Gradebook,** designed specifically for mathematics and statistics, automatically tracks students' results and gives you control over how to calculate final grades.
- **MathXL Exercise Builder** allows you to create static and algorithmic exercises for your online assignments. You can use the library of sample exercises as an easy starting point.
- **Assessment Manager** lets you create online homework, quizzes, and tests that are automatically graded. Select just the right mix of questions from the MathXL exercise bank, instructor-created custom exercises, and/or TestGen test items.

MathXL is available to qualified adopters. For more information, visit our Web site www.mathxl.com, or contact your Pearson sales representative.

New! ***Video Lectures on DVD with Optional Subtitles***

The video lectures for this text are available on DVD, making it easy and convenient for students to watch the videos from a computer at home or on campus. The videos feature engaging chapter summaries and worked-out examples. This format provides distance-learning students with critical video instruction. The videos have optional English subtitles; they can easily be turned on or off for individual student needs.
ISBN-13: 978-0-321-64577-7; ISBN 10: 0-321-64577-4

Acknowledgments

The authors wish to thank the following reviewers for their helpful comments and suggestions for this and previous editions of the text (reviewers of the 10th edition are noted with an asterisk):

Erol Barbut, University of Idaho
Bob Beul, St. Louis University–Metropolitan College
Richard Bieberich, Ball State University
Chris Boldt, Eastfield College
Michael J. Bradley, Merrimack College
James F. Brown, Midland College
Tarek Buhagiar, University of Central Florida*

James E. Carpenter, Iona College
Jesus Carreon, Mesa Community College
Faith Y. Chao, Golden Gate University
Jan S. Collins, Embry-Riddle University
Jerry Currence, University of South Carolina–Lancaster
Juli D'Ann Ratheal, Western Texas A & M University
Frederick Davidson, Old Dominion University
Jean Davis, Southwest Texas State University
Duane E. Deal, Ball State University
Richard D. Derderian, Providence College
Carol E. DeVille, Louisiana Tech University
Wayne Ehler, Anne Arundel Community College
George A. Emerson, National University
Garret Etgen, University of Houston
George Evanovich, Iona College
Richard Fast, Mesa Community College
Gordon Feathers, Iona College
J. Franklin Fitzgerald, Boston University
Leland J. Fry, Kirkwood Community College
Dauhrice K. Gibson, Gulf Coast Community College
Robert E. Goad, Sam Houston State University
Mark Goldstein, West Virginia Northern Community College*
Richard E. Goodrick, University of Washington
Kim Gregor, Delaware Technical and Community College
Kay Gura, Ramapo College of New Jersey
Joseph A. Guthrie, University of Texas at El Paso
Patricia Hirschy, Delaware Technical and Community College
Arthur M. Hobbs, Texas A & M University
Irene Hollman, Southwestern College
Miles Hubbard, St. Cloud State University
Katherine J. Huppler, St. Cloud State University
Carol M. Hurwitz, Manhattan College
Donald R. Ignatz, Lorain County Community College
Alec Ingraham, New Hampshire College
Robert H. Johnston, Virginia Commonwealth University
June Jones, Macon College
Paul Kaczur, Phoenix College
Michael J. Kallaher, Washington State University
Akihiro Kanamori, Boston University
Terence J. Keegan, Providence College
Hubert C. Kennedy, Providence College
Raja Khoury, Collin County Community College
Clint Kolaski, University of Texas at San Antonio
Archille J. Laferriere, Boston College
Steve Laroe, University of Alaska–Fairbanks
Jeffrey Lee, Texas Tech University
Arthur M. Lieberman, Cleveland State University
Norman Lindquist, Western Washington University
Laurence P. Maher, Jr., North Texas State University
Norman R. Martin, Northern Arizona University
Donald Mason, Elmhurst College
James Mazzarella, Holy Family College
Walter S. McVoy, Illinois State University
C. G. Mendez, Metropolitan State College of Denver
Shannon Michaux, University of Colorado at Colorado Springs
W. W. Mitchell, Jr., Phoenix College
Robert A. Moreland, Texas Tech University
Ruth M. Murray, College of DuPage
Kandasamy Muthuvel, University of Wisconsin–Oshkosh
Javad Namazi, Fairleigh Dickinson University*
Carol Nessmith, Georgia Southern College
Peter Nicholls, Northern Illinois University
Ann O'Connell, Providence College
Kathy O'Dell, University of Alabama at Huntsville
Charles Odion, Houston Community College
Thomas J. Ordoyne, University of South Carolina at Spartanburg

Marian Paysinger, University of Texas at Arlington
Julienne K. Pendleton, Brookhaven College
Sandra Peskin, Queensborough Community College
S. Pierce, West Coast University
John M. Plachy, Metropolitan State College
Elizabeth Polenzani, Pasadena City College
Donald G. Poulson, Mesa Community College
Wayne B. Powell, Oklahoma State University
Michael I. Ratliff, Northern Arizona University
Clark P. Rhoades, Loyola University at New Orleans
Sarah Sabinson, Queensborough Community College
Leon Sagan, Anne Arundel Community College
C. Edward Sandifer, Western Connecticut State University*
Warren Sargent, College of the Sequoias*
Subhash C. Saxena, University of South Carolina, Coastal Carolina College
Harold Schachter, Queensborough Community College
Steven A. Schonefeld, Tri-State University
Robert Seaver, Lorain County Community College
Surinder Sehgal, Ohio State University
Gordon Shilling, University of Texas at Arlington
Calvin Shipley, Henderson State University
Pradeep Shukla, Suffolk University
James L. Southam, San Francisco State University
John Spellman, Southwest Texas State University
Joan M. Spetich, Baldwin-Wallace College
William D. Stark, Navarro College
Jo Steig, Phoenix College
David Stoneman, University of Wisconsin–Whitewater
David P. Sumner, University of South Carolina
Daniel F. Symancyk, Anne Arundel Community College
Giovanni Viglino, Ramapo College of New Jersey
Deborah A. Vrooman, Coastal Carolina College
H. J. Wellenzohn, Niagara University
Stephen H. West, Coastal Carolina College
Thelma West, University of Louisiana at Lafayette
Richard J. Wilders, North Central College
John L. Wisthoff, Anne Arundel Community College
Hing-Sing Yu, University of Texas at San Antonio
Cathy Zucco-Teveloff, Trinity College

We also wish to thank our accuracy checkers, who did an excellent job of checking both text and exercise answers: Paul Lorczak, Debra McGivney, John and Ann Corbeil. Thanks also to the supplements authors: Jim Ball, Beverly Fusfield, David Bridge, Victoria Baker and Stela Pudar-Hozo. Special thanks to Becky Troutman, who carefully compiled the Index of Applications.

We want to thank the staff of Pearson Education for their assistance with, and contributions to, this book, particularly Greg Tobin, Deirdre Lynch, Jennifer Crum, Jeff Weidenaar, Sheila Spinney, Elizabeth Bernardi, Joanne Wendelken, Kendra Bassi, and Jean Choe. Finally, we wish to express our appreciation to Denise Showers of Aptara Corporation, who was a pleasure to work with.

Margaret L. Lial
Thomas W. Hungerford
John P. Holcomb, Jr.

To the Student

This book has several features designed to help you understand and apply the mathematics presented here.

Checkpoints These problems appear in the margin and are keyed to the discussion in the text. When you see the symbol ✓ in the text, you are advised to work the corresponding checkpoint problem in the margin. Doing so will help you to understand and apply new concepts.

Technology Examples and exercises that require graphing calculators, spreadsheets, or other technology are marked with the icon. Some instructors may make this material part of the course, whereas others will not require you to use technology other than a calculator that can handle exponential functions and logarithms.

For those who use technology, there is a *Graphing Calculator Appendix* that covers the basics of calculator use and provides a number of helpful programs for working with some of the topics in the text. An outline of the appendix is on page 951 and the full appendix is available in MyMathLab or at http://www.pearsonhighered.com/MWA10e.

Finally, there are *Technology Tips* throughout the text that describe the proper menu or keys to use for various procedures on a graphing calculator. Note that

Technology Tips for	**Also apply to**
TI-84+	TI-83+, TI-Nspire, and usually TI-83
Casio	Casio 9750 GII and 9860 GII, and usually other models.

The key to succeeding in this course is to remember that

mathematics is not a spectator sport.

You can't expect to learn mathematics without *doing* mathematics any more than you could learn to swim without getting wet. You have to take an active role, making use of all the resources at your disposal: your instructor, your fellow students, and this book. To get the most out of the book, you need to read it regularly (preferably *before* starting the exercises).

Finally, remember the words of the great Hillel: "The bashful do not learn." There is no such thing as a "dumb question" (assuming, of course, that you have read the book and your class notes and attempted the homework). Your instructor will welcome questions that arise from a serious effort on your part. So get your money's worth: Ask questions.

Algebra and Equations

Mathematics is widely used in business, finance, and the biological, social, and physical sciences, from developing efficient production schedules for a factory to mapping the human genome. Mathematics also plays a role in determining interest on a loan from a bank and the cancer risk from a pollutant, as well as in the study of falling objects. See Exercises 61–63 on page 58 and Exercise 63 on page 68.

CASE 1: Consumers Often Defy Common Sense

Algebra and equations are the basic mathematical tools for handling many applications. Your success in this course will depend on your having the algebraic skills presented in this chapter.

1.1 The Real Numbers

Only real numbers will be used in this book.* The names of the most common types of real numbers are as follows.

The Real Numbers

Natural (counting) numbers	$1, 2, 3, 4, \ldots$
Whole numbers	$0, 1, 2, 3, 4, \ldots$
Integers	$\ldots, -3, -2, -1, 0, 1, 2, 3, \ldots$
Rational numbers	All numbers that can be written in the form p/q, where p and q are integers and $q \neq 0$
Irrational numbers	Real numbers that are not rational

As you can see, every natural number is a whole number, and every whole number is an integer. Furthermore, every integer is a rational number. For instance, the integer 7 can be written as the fraction $\frac{7}{1}$ and is therefore a rational number.

One example of an irrational number is π, the ratio of the circumference of a circle to its diameter. The number π can be approximated as $\pi \approx 3.14159$ ($\approx$ means "is approximately equal to"), but there is no rational number that is exactly equal to π.

Example 1 What kind of number is each of the following?

(a) 6

Solution The number 6 is a natural number, a whole number, an integer, a rational number, and a real number.

(b) $\frac{3}{4}$

Solution This number is rational and real.

(c) 3π

Solution Because π is not a rational number, 3π is irrational and real. 

✓Checkpoint 1

Name all the types of numbers that apply to the following.

(a) -2

(b) $-5/8$

(c) $\pi/5$

Answers: See page 12.

All real numbers can be written in decimal form. A rational number, when written in decimal form, is either a terminating decimal, such as .5 or .128, or a repeating decimal, in which some block of digits eventually repeats forever, such as 1.3333... or 4.7234234234....‡ Irrational numbers are decimals that neither terminate nor repeat.

*Not all numbers are real numbers. For example, $\sqrt{-1}$ is a number that is *not* a real number.

†The use of Checkpoint problems is explained in the "To the Student" section preceding this chapter.

‡Some graphing calculators have a FRAC key that automatically converts some repeating decimals to fraction form. FRAC programs for other graphing calculators are in the Program Appendix.

When a calculator is used for computations, the answers it produces are often decimal *approximations* of the actual answers; they are accurate enough for most applications. To ensure that your final answer is as accurate as possible,

you should not round off any numbers during long calculator computations.

It is usually OK to round off the final answer to a reasonable number of decimal places once the computation is finished.

The important basic properties of the real numbers are as follows.

Properties of the Real Numbers

For all real numbers, a, b, and c, the following properties hold true:

Commutative properties $\quad a + b = b + a \qquad ab = ba$

Associative properties $\quad (a + b) + c = a + (b + c) \qquad (ab)c = a(bc)$

Identity properties There exists a unique real number 0, called the **additive identity**, such that

$$a + 0 = a \quad \text{and} \quad 0 + a = a.$$

There exists a unique real number 1, called the **multiplicative identity**, such that

$$a \cdot 1 = a \quad \text{and} \quad 1 \cdot a = a.$$

Inverse properties For each real number a, there exists a unique real number $-a$, called the **additive inverse** of a, such that

$$a + (-a) = 0 \quad \text{and} \quad (-a) + a = 0.$$

If $a \neq 0$, there exists a unique real number $1/a$, called the **multiplicative inverse** of a, such that

$$a \cdot \frac{1}{a} = 1 \quad \text{and} \quad \frac{1}{a} \cdot a = 1.$$

Distributive property $\quad a(b + c) = ab + ac$ and $(b + c)a = ba + ca$.

The next five examples illustrate the properties listed in the preceding box.

Example 2 The commutative property says that the order in which you add or multiply two quantities doesn't matter.

(a) $(6 + x) + 9 = 9 + (6 + x) = 9 + (x + 6)$ **(b)** $5 \cdot (9 \cdot 8) = (9 \cdot 8) \cdot 5$

Example 3 When the associative property is used, the order of the numbers does not change, but the placement of parentheses does.

(a) $4 + (9 + 8) = (4 + 9) + 8$ **(b)** $3(9x) = (3 \cdot 9)x$ 2✓

✓**Checkpoint 2**

Name the property illustrated in each of the following examples.

(a) $(2 + 3) + 9$ $= (3 + 2) + 9$

(b) $(2 + 3) + 9$ $= 2 + (3 + 9)$

(c) $(2 + 3) + 9$ $= 9 + (2 + 3)$

(d) $(4 \cdot 6)p = (6 \cdot 4)p$

(e) $4(6p) = (4 \cdot 6)p$

Answers: See page 12.

Example 4 By the identity properties,

(a) $-8 + 0 = -8$ **(b)** $(-9) \cdot 1 = -9.$

TECHNOLOGY TIP To enter -8 on a calculator, use the negation key (labeled (−) or +/−), *not* the subtraction key. On most one-line scientific calculators, key in 8 +/−. On graphing calculators or two-line scientific calculators, key in either (−) 8 or +/− 8.

Example 5 By the inverse properties, the statements in parts (a) through (d) are true.

(a) $9 + (-9) = 0$ **(b)** $-15 + 15 = 0$

(c) $-8 \cdot \left(\frac{1}{-8}\right) = 1$ **(d)** $\frac{1}{\sqrt{5}} \cdot \sqrt{5} = 1$

✓**Checkpoint 3**

Name the property illustrated in each of the following examples.

(a) $2 + 0 = 2$

(b) $-\frac{1}{4} \cdot (-4) = 1$

(c) $-\frac{1}{4} + \frac{1}{4} = 0$

(d) $1 \cdot \frac{2}{3} = \frac{2}{3}$

Answers: See page 12.

NOTE There is no real number x such that $0 \cdot x = 1$, so 0 has no multiplicative inverse. 3✓

Example 6 By the distributive property,

(a) $9(6 + 4) = 9 \cdot 6 + 9 \cdot 4$

(b) $3(x + y) = 3x + 3y$

(c) $-8(m + 2) = (-8)(m) + (-8)(2) = -8m - 16$

(d) $(5 + x)y = 5y + xy.$ 4✓

✓**Checkpoint 4**

Use the distributive property to complete each of the following.

(a) $4(-2 + 5)$

(b) $2(a + b)$

(c) $-3(p + 1)$

(d) $(8 - k)m$

(e) $5x + 3x$

Answers: See page 12.

ORDER OF OPERATIONS

Some complicated expressions may contain many sets of parentheses. To avoid ambiguity, the following procedure should be used.

Parentheses

Work separately above and below any fraction bar. Within each set of parentheses or square brackets, start with the innermost set and work outward.

Example 7 Simplify: $[(3 + 2) - 7]5 + 2([6 \cdot 3] - 13)$.

Solution On each segment, work from the inside out:

$$\begin{aligned}[(3 + 2) - 7]5 - 2([6 \cdot 3] - 13) \\ = [5 - 7]5 + 2(18 - 13) \\ = [-2]5 + 2(5) \\ = -10 + 10 = 0.\end{aligned}$$

Does the expression $2 + 4 \times 3$ mean

$$(2 + 4) \times 3 = 6 \times 3 = 18?$$

Or does it mean

$$2 + (4 \times 3) = 2 + 12 = 14?$$

To avoid this ambiguity, mathematicians have adopted the following rules (which are also followed by almost all scientific and graphing calculators).

Order of Operations

1. Find all powers and roots, working from left to right.
2. Do any multiplications or divisions in the order in which they occur, working from left to right.
3. Finally, do any additions or subtractions in the order in which they occur, working from left to right.

If sets of parentheses or square brackets are present, use the rules in the preceding box within each set, working from the innermost set outward.

According to these rules, multiplication is done *before* addition, so $2 + 4 \times 3 = 2 + 12 = 14$. Here are some additional examples.

Example 8 Use the order of operations to evaluate each expression if $x = -2$, $y = 5$, and $z = -3$.

(a) $-4x^2 - 7y + 4z$

Solution Use parentheses when replacing letters with numbers:

$$\begin{aligned}-4x^2 - 7y + 4z &= -4(-2)^2 - 7(5) + 4(-3) \\ &= -4(4) - 7(5) + 4(-3) = -16 - 35 - 12 = -63.\end{aligned}$$

(b) $$\begin{aligned}\frac{2(x - 5)^2 + 4y}{z + 4} &= \frac{2(-2 - 5)^2 + 4(5)}{-3 + 4} \\ &= \frac{2(-7)^2 + 20}{1} \\ &= 2(49) + 20 = 118.\end{aligned}$$ 5 ✓

✓ Checkpoint 5

Evaluate the following if $m = -5$ and $n = 8$.

(a) $-2mn - 2m^2$

(b) $\frac{4(n - 5)^2 - m}{m + n}$

Answers: See page 12.

Example 9 Use a calculator to evaluate

$$\frac{-9(-3) + (-5)}{3(-4) - 5(2)}.$$

Solution Use extra parentheses (shown here in blue) around the numerator and denominator when you enter the number in your calculator, and be careful to distinguish the negation key from the subtraction key.

$$\overbrace{(-9\,(-3) + (-5))}^{\text{numerator}}/\overbrace{(3\,(-4) - 5(2))}^{\text{denominator}}$$

Negation key   Subtraction key

If you don't get -1 as the answer, then you are entering something incorrectly. 6

Checkpoint 6

Use a calculator to evaluate the following.

(a) $4^2 \div 8 + 3^2 \div 3$

(b) $[-7 + (-9)]\cdot(-4) - 8(3)$

(c) $\dfrac{-11 - (-12) - 4\cdot 5}{4(-2) - (-6)(-5)}$

(d) $\dfrac{36 \div 4\cdot 3 \div 9 + 1}{9 \div (-6)\cdot 8 - 4}$

Answers: See page 12.

SQUARE ROOTS

There are two numbers whose square is 16, namely, 4 and -4. The positive one, 4, is called the **square root** of 16. Similarly, the square root of a nonnegative number d is defined to be the *nonnegative* number whose square is d; this number is denoted $\sqrt{d}$. For instance,

$$\sqrt{36} = 6 \text{ because } 6^2 = 36, \quad \sqrt{0} = 0 \text{ because } 0^2 = 0, \text{ and}$$
$$\sqrt{1.44} = 1.2 \text{ because } (1.2)^2 = 1.44.$$

No negative number has a square root that is a real number. For instance, there is no real number whose square is -4, so -4 has no square root.

Every nonnegative real number has a square root. Unless an integer is a perfect square (such as $64 = 8^2$), its square root is an irrational number. A calculator can be used to obtain a rational approximation of these square roots.

TECHNOLOGY TIP On one-line scientific calculators, $\sqrt{40}$ is entered as 40 $\sqrt{\ }$. On graphing calculators and two-line scientific calculators, key in $\sqrt{\ }$ 40 ENTER (or EXE).

Example 10 Estimate each of the given quantities. Verify your estimates with a calculator.

(a) $\sqrt{40}$

Solution Since $6^2 = 36$ and $7^2 = 49$, $\sqrt{40}$ must be a number between 6 and 7. A typical calculator shows that $\sqrt{40} \approx 6.32455532$.

(b) $5\sqrt{7}$

Solution $\sqrt{7}$ is between 2 and 3 because $2^2 = 4$ and $3^2 = 9$, so $5\sqrt{7}$ must be a number between $5\cdot 2 = 10$ and $5\cdot 3 = 15$. A calculator shows that $5\sqrt{7} \approx 13.22875656$. 7

Checkpoint 7

Estimate each of the following.

(a) $\sqrt{73}$

(b) $\sqrt{22} + 3$

(c) Confirm your estimates in parts (a) and (b) with a calculator.

Answers: See page 12.

CAUTION If c and d are positive real numbers, then $\sqrt{c + d}$ is *not* equal to $\sqrt{c} + \sqrt{d}$. For example, $\sqrt{9 + 16} = \sqrt{25} = 5$, but $\sqrt{9} + \sqrt{16} = 3 + 4 = 7$.

THE NUMBER LINE

The real numbers can be illustrated geometrically with a diagram called a **number line**. Each real number corresponds to exactly one point on the line and vice versa. A number line with several sample numbers located (or **graphed**) on it is shown in Figure 1.1. 8✓

✓Checkpoint 8

Draw a number line, and graph the numbers -4, -1, 0, 1, 2.5, and 13/4 on it.

Answer: See page 12.

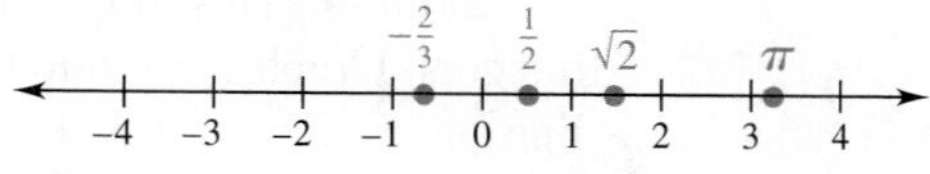

FIGURE 1.1

When comparing the sizes of two real numbers, the following symbols are used.

Symbol	Read	Meaning
$a < b$	a is less than b.	a lies to the *left* of b on the number line.
$b > a$	b is greater than a.	b lies to the *right* of a on the number line.

Note that $a < b$ means the same thing as $b > a$. The inequality symbols are sometimes joined with the equals sign, as follows.

Symbol	Read	Meaning
$a \le b$	a is less than or equal to b.	either $a < b$ or $a = b$
$b \ge a$	b is greater than or equal to a.	either $b > a$ or $b = a$

TECHNOLOGY TIP If your graphing calculator has inequality symbols (usually located on the TEST menu), you can key in statements such as "5 < 12" or "$-2 \ge 3$." When you press ENTER, the calculator will display 1 if the statement is true and 0 if it is false.

Only one part of an "either . . . or" statement needs to be true for the entire statement to be considered true. So the statement $3 \le 7$ is true because $3 < 7$, and the statement $3 \le 3$ is true because $3 = 3$.

Example 11 Write *true* or *false* for each of the following.

(a) $8 < 12$

Solution This statement says that 8 is less than 12, which is true.

(b) $-6 > -3$

Solution The graph in Figure 1.2 shows that -6 is to the *left* of -3. Thus, $-6 < -3$, and the given statement is false.

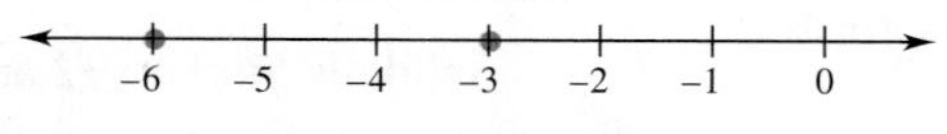

FIGURE 1.2

(c) $-2 \le -2$

Solution Because $-2 = -2$, this statement is true. 

✓Checkpoint 9

Write *true* or *false* for the following.

(a) $-9 \le -2$

(b) $8 > -3$

(c) $-14 \le -20$

Answers: See page 12.

A number line can be used to draw the graph of a set of numbers, as shown in the next few examples.

Example 12 Graph all real numbers x such that $1 < x < 5$.

Solution This graph includes all the real numbers between 1 and 5, not just the integers. Graph these numbers by drawing a heavy line from 1 to 5 on the number line, as in Figure 1.3. Parentheses at 1 and 5 show that neither of these points belongs to the graph. 10✓

✓Checkpoint 10

Graph all real numbers x such that

(a) $-5 < x < 1$

(b) $4 < x < 7$.

Answers: See page 12.

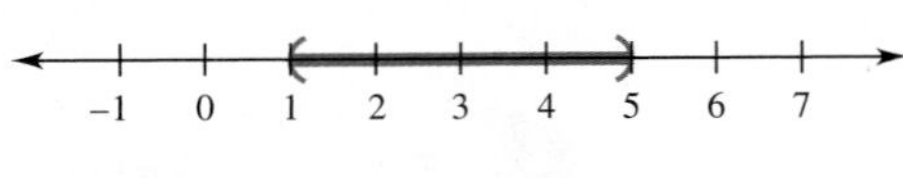

FIGURE 1.3

A set that consists of all the real numbers between two points, such as $1 < x < 5$ in Example 12, is called an **interval**. A special notation called **interval notation** is used to indicate an interval on the number line. For example, the interval including all numbers x such that $-2 < x < 3$ is written as $(-2, 3)$. The parentheses indicate that the numbers -2 and 3 are *not* included. If -2 and 3 are to be included in the interval, square brackets are used, as in $[-2, 3]$. The following chart shows several typical intervals, where $a < b$.

Intervals

Inequality	Interval Notation	Explanation
$a \le x \le b$	$[a, b]$	Both a and b are included.
$a \le x < b$	$[a, b)$	a is included; b is not.
$a < x \le b$	$(a, b]$	b is included; a is not.
$a < x < b$	(a, b)	Neither a nor b is included.

Interval notation is also used to describe sets such as the set of all numbers x such that $x \ge -2$. This interval is written $[-2, \infty)$. The set of all real numbers is written $(-\infty, \infty)$ in interval notation.

Example 13 Graph the interval $[-2, \infty)$.

Solution Start at -2 and draw a heavy line to the right, as in Figure 1.4. Use a square bracket at -2 to show that -2 itself is part of the graph. The symbol ∞, read "infinity," *does not* represent a number. This notation simply indicates that *all* numbers greater than -2 are in the interval. Similarly, the notation $(-\infty, 2)$ indicates the set of all numbers x such that $x < 2$. 11✓

✓Checkpoint 11

Graph all real numbers x in the given interval.

(a) $(-\infty, 4]$

(b) $[-2, 1]$

Answers: See page 12.

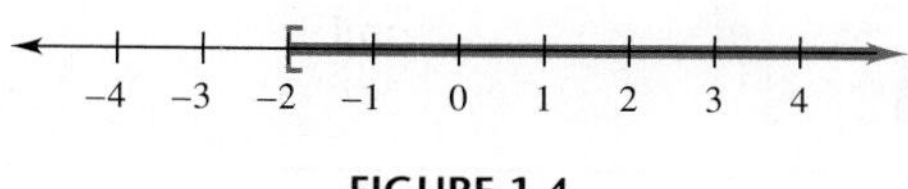

FIGURE 1.4

ABSOLUTE VALUE

The **absolute value** of a real number a is the distance from a to 0 on the number line and is written $|a|$. For example, Figure 1.5 shows that the distance from 9 to 0 on the number line is 9, so we have $|9| = 9$. The figure also shows that $|-9| = 9$, because the distance from -9 to 0 is also 9.

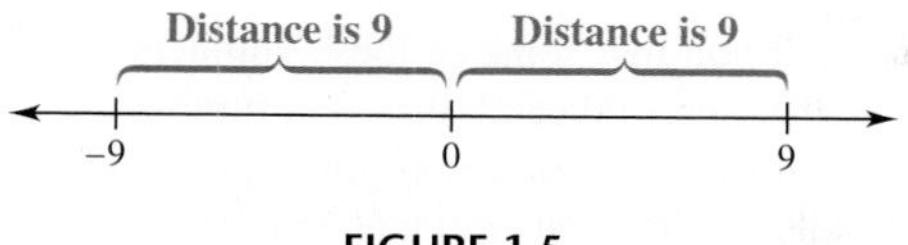

FIGURE 1.5

The facts that $|9| = 9$ and $|-9| = 9 = -(-9)$ suggest the following algebraic definition of absolute value.

Absolute Value

For any real number a,

$$|a| = a \quad \text{if } a \geq 0$$
$$|a| = -a \quad \text{if } a < 0.$$

The first part of the definition shows that $|0| = 0$ (because $0 \geq 0$). It also shows that the absolute value of any positive number a is the number itself, so $|a|$ is positive in such cases. The second part of the definition says that the absolute value of a negative number a is the *negative* of a. For instance, if $a = -5$, then $|-5| = -(-5) = 5$. So $|-5|$ is positive. The same thing works for any negative number—that is, its absolute value (the negative of a negative number) is positive. Thus, we can state the following:

For every nonzero real number a, the number $|a|$ is positive.

✔ Checkpoint 12

Find the following.

(a) $|-6|$

(b) $-|7|$

(c) $-|-2|$

(d) $|-3 - 4|$

(e) $|2 - 7|$

Answers: See page 12.

Example 14 Evaluate $|8 - 9|$.

Solution First, simplify the expression within the absolute-value bars:

$$|8 - 9| = |-1| = 1.$$

Similarly, $-|-5 - 8| = -|-13| = -13$. 12✔

1.1 ▸ Exercises

In Exercises 1 and 2, label the statement true or false.

1. Every integer is a rational number.
2. Every real number is an irrational number.
3. The decimal expansion of the irrational number π begins 3.141592653589793 . . . . Use your calculator to determine which of the following rational numbers is the best approximation for the irrational number π:

$$\frac{22}{7}, \quad \frac{355}{113}, \quad \frac{103{,}993}{33{,}102}, \quad \frac{2{,}508{,}429{,}787}{798{,}458{,}000}.$$

Your calculator may tell you that some of these numbers are equal to π, but that just indicates that the number agrees with π for as many decimal places as your calculator can handle (usually 10–14). No rational number is exactly equal to π.

Identify the properties that are illustrated in each of the following. (See Examples 2–6.)

4. $-5 + 0 = -5$
5. $6(t + 4) = 6t + 6 \cdot 4$
6. $3 + (-3) = (-3) + 3$
7. $0 + (-7) = -7 + 0$
8. $8 + (12 + 6) = (8 + 12) + 6$
9. How is the additive inverse property related to the additive identity property? the multiplicative inverse property to the multiplicative identity property?
10. Explain the distinction between the commutative and associative properties.

Evaluate each of the following if $p = -2$, $q = 3$, and $r = -5$. (See Examples 7–9.)

11. $-3(p + 5q)$
12. $2(q - r)$
13. $\dfrac{q + r}{q + p}$
14. $\dfrac{3q}{3p - 2r}$

Business *Lenders are required to state the annual percentage rate (APR) for every loan, using the formula APR = 12r, where r is the monthly interest rate. Find the APR when*

15. $r = 1.5$
16. $r = 1.67$

Find the monthly interest rate when

17. $APR = 9$
18. $APR = 19.5$

Evaluate each expression, using the order of operations given in the text. (See Examples 7–9.)

19. $3 - 4 \cdot 5 + 5$
20. $(4 - 5) \cdot 6 + 6$
21. $8 - 4^2 - (-12)$
22. $8 - (-4)^2 - (-12)$
23. $-(3 - 5) - [2 - (3^2 - 13)]$
24. $\dfrac{2(3 - 7) + 4(8)}{4(-3) + (-3)(-2)}$
25. $\dfrac{2(-3) + 3/(-2) - 2/(-\sqrt{16})}{\sqrt{64} - 1}$
26. $\dfrac{6^2 - 3\sqrt{25}}{\sqrt{6^2 + 13}}$

Use a calculator to help you list the given numbers in order from smallest to largest.

27. $\dfrac{189}{37}, \dfrac{4587}{691}, \sqrt{47}, 6.735, \sqrt{27}, \dfrac{2040}{523}$
28. $\dfrac{385}{117}, \sqrt{10}, \dfrac{187}{63}, \pi, \sqrt{\sqrt{85}}, 2.9884$

Express each of the following statements in symbols, using $<, >, \leq$, or $\geq$.

29. 12 is less than 18.5.
30. -2 is greater than -20.
31. x is greater than or equal to 5.7.
32. y is less than or equal to -5.
33. z is at most 7.5.
34. w is negative.

Fill the blank with $<$, $=$, or $>$ so that the resulting statement is true.

35. -6 ___ -2
36. $3/4$ ___ $.75$
37. 3.14 ___ π
38. $1/3$ ___ $.33$

Fill the blank so as to produce two equivalent statements. For example, the arithmetic statement "a is negative" is equivalent to the geometric statement "the point a lies to the left of the point 0."

	Arithmetic Statement	***Geometric Statement***
39.	$a \geq b$	________
40.	________	a lies c units to the right of b
41.	________	a lies between b and c, and to the right of c
42.	a is positive	________

Graph the given intervals on a number line. (See Examples 12 and 13.)

43. $(-8, -1)$
44. $[-1, 10]$
45. $[-2, 2)$
46. $(-2, 3]$
47. $(-2, \infty)$
48. $(-\infty, -2]$

Business *The Consumer Price Index (CPI) tracks that cost of a typical sample of consumer goods. The following table shows the percentage increase in the CPI for each year in a 10-year period.**

*U.S. Bureau of Labor Statistics.

Year	1998	1999	2000	2001	2002
% Increase in CPI	1.6	2.2	3.4	2.8	1.6

Year	2003	2004	2005	2006	2007
% Increase in CPI	2.3	2.7	2.5	3.2	4.1

Let r denote the yearly percentage increase in the CPI. For each of the following inequalities, find the number *of years during the given period that r satisfied the inequality.*

49. $r > 2.8$ **50.** $r < 2.8$

51. $r \le 2.8$ **52.** $r \ge 3$

53. $r > 4.1$ **54.** $r \le 2.5$

Health *The* body mass index *(BMI), a number B that measures the relationship between a person's height H (in inches) and weight W (in pounds), is given by the formula*

$$B = \frac{.455W}{(.0254H)^2}.*$$

Federal guidelines suggest that the desirable range for B is $19 \le B \le 25$.

(a) *Find the BMI of the given athletes.*

(b) *Determine whether each athlete's BMI falls in the desirable range or not.*

55. Steffi Graf (119 pounds; 5 ft, 9 in.)

56. Jackie Joyner-Kersee (153 pounds; 5 ft, 10 in.)

57. Tiger Woods (180 pounds; 6 ft, 2 in.)

58. Shaquille O'Neal (300 pounds; 7 ft, 1 in.)

*Washington Post.

Physical Science *The windchill factor is a measure of the cooling effect that the wind has on a person's skin. It calculates the equivalent cooling temperature if there were no wind.**

Temperature (°F)	Wind (mph)								
	Calm	5	10	15	20	25	30	35	40
	40	36	34	32	30	29	28	28	27
	30	25	21	19	17	16	15	14	13
	20	13	9	6	4	3	1	0	−1
	10	1	−4	−7	−9	−11	−12	−14	−15
	0	−11	−16	−19	−22	−24	−26	−27	−29
	−10	−22	−28	−32	−35	−37	−39	−41	−43
	−20	−34	−41	−45	−48	−51	−53	−55	−57
	−30	−46	−53	−58	−61	−64	−67	−69	−71
	−40	−57	−66	−71	−74	−78	−80	−82	−84

Suppose that we wish to determine the difference between two of the entries in the foregoing table and we are interested only in the magnitude, or absolute value, of this difference. Then we subtract the two entries and find the absolute value. For example, the difference in windchill factors for wind at 20 miles per hour with a 20° air temperature and wind at 30 miles per hour with a 10° air temperature is $|-12° - 4°| = 16°$, *or equivalently,* $|4° - (-12°)| = 16°$.

Find the absolute value of the difference between the two indicated windchill factors.

59. Wind at 10 miles per hour with a 30° air temperature and wind at 30 miles per hour with a −10° air temperature

60. Wind at 20 miles per hour with a −20° air temperature and wind at 5 miles per hour with a 30° air temperature

61. Wind at 25 miles per hour with a −30° air temperature and wind at 15 miles per hour with a −30° air temperature

62. Wind at 40 miles per hour with a 40° air temperature and wind at 25 miles per hour with a −30° air temperature

Evaluate each of the following expressions (see Example 14).

63. $|8| - |-4|$ **64.** $|-9| - |-12|$

65. $-|-4| - |-1 - 14|$ **66.** $-|6| - |-12 - 4|$

In each of the following problems, fill in the blank with either =, <, *or* >, *so that the resulting statement is true.*

67. $|5|$ ______ $|-5|$

68. $-|-4|$ ______ $|4|$

69. $|10 - 3|$ ______ $|3 - 10|$

70. $|6 - (-4)|$ ______ $|-4 - 6|$

71. $|-2 + 8|$ ______ $|2 - 8|$

72. $|3| \cdot |-5|$ ______ $|3(-5)|$

*Table from the Joint Action Group for Temperature Indices, 2001.

73. $|3-5|$ ______ $|3|-|5|$

74. $|-5+1|$ ______ $|-5|+|1|$

Write the expression without using absolute-value notation.

75. $|a-7|$ if $a<7$

76. $|b-c|$ if $b \geq c$

77. If a and b are any real numbers, is it always true that $|a+b|=|a|+|b|$? Explain your answer.

78. If a and b are any two real numbers, is it always true that $|a-b|=|b-a|$? Explain your answer.

79. For which real numbers b does $|2-b|=|2+b|$? Explain your answer.

80. According to data from the Center for Science in the Public Interest, the healthy weight range for a person depends on the person's height. For example,

Height	Healthy Weight Range (lb)
5 ft 8 in.	143 ± 21
6 ft 0 in.	163 ± 26

Express each of these ranges as an absolute-value inequality in which x is the weight of the preson.

81. The following graph shows the number of Schedule C and C-EZ forms (in millions) that were filed with the IRS over a six-year period.*

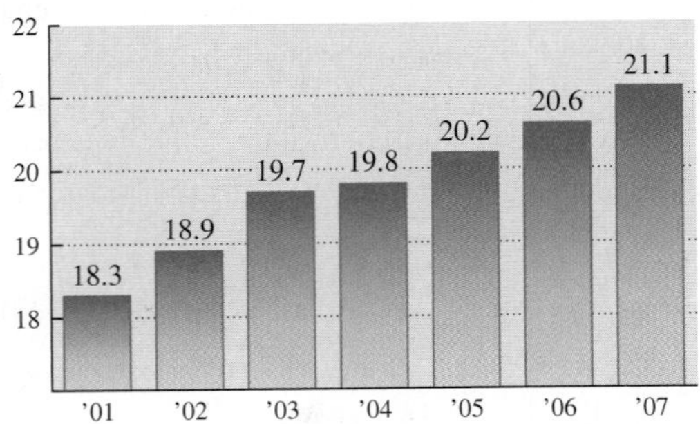

In what years was the following statement true:

$$|x-20{,}000{,}000| > 600{,}000,$$

where x is the number of Schedule C and C-EZ forms in that year?

82. At Ergonomics, Inc., each department's expenses are reviewed monthly. A department can fail to pass the budget variance test in a category if either (i) the absolute value of the difference between actual expenses and the budget is more than \$500 or (ii) the absolute value of the difference between the actual expenses and the budget is more than 5% of the budgeted amount. Which of the following items fail the budget variance test? Explain your answers.

Item	Budgeted Expense (\$)	Actual Expense (\$)
Wages	220,750	221,239
Overtime	10,500	11,018
Shipping and Postage	530	589

✓ Checkpoint Answers

1. **(a)** Integer, rational, real
(b) Rational, real
(c) Irrational, real

2. **(a)** Commutative property
(b) Associative property
(c) Commutative property
(d) Commutative property
(e) Associative property

3. **(a)** Additive identity property
(b) Multiplicative inverse property
(c) Additive inverse property
(d) Multiplicative identity property

4. **(a)** $4(-2)+4(5)=12$ **(b)** $2a+2b$
(c) $-3p-3$ **(d)** $8m-km$
(e) $(5+3)x=8x$

5. **(a)** 30 **(b)** $\frac{41}{3}$

6. **(a)** 5 **(b)** 40
(c) $\frac{19}{38}=\frac{1}{2}=.5$ **(d)** $-\frac{1}{4}=-.25$

7. **(a)** Between 8 and 9 **(b)** Between 7 and 8
(c) 8.5440; 7.6904

8.

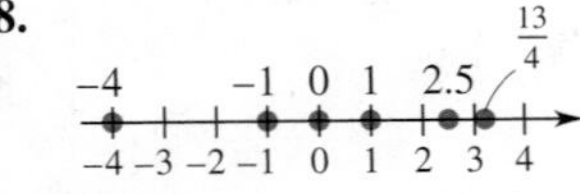

9. **(a)** True **(b)** True **(c)** False

10. **(a)**

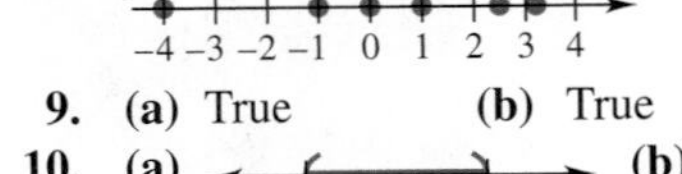

(b) 4 7

11. **(a)** **(b)**

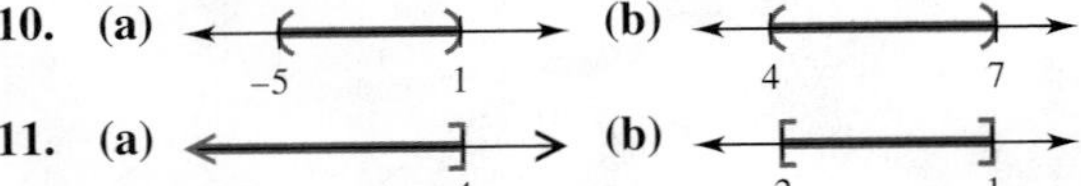

12. **(a)** 6 **(b)** -7 **(c)** -2
(d) 7 **(e)** 5

*Internal Revenue Service. These schedules are for self-employed individuals.

1.2 ▶ Polynomials

Polynomials are the fundamental tools of algebra and will play a central role in this course. In order to do polynomial arithmetic, you must first understand exponents. So we begin with them. You are familiar with the usual notation for squares and cubes, such as:

$$5^2 = 5 \cdot 5 \quad \text{and} \quad 6^3 = 6 \cdot 6 \cdot 6.$$

We now extend this convenient notation to other cases.

If n is a natural number and a is any real number, then

a^n denotes the product $a \cdot a \cdot a \cdots a$ (n factors).

The number a is the **base**, and the number n is the **exponent**.

Example 1 4^6, which is read "four to the sixth," or "four to the sixth power," is the number

$$4 \cdot 4 \cdot 4 \cdot 4 \cdot 4 \cdot 4 = 4096.$$

Similarly, $(-5)^3 = (-5)(-5)(-5) = -125$, and

$$\left(\frac{3}{2}\right)^4 = \frac{3}{2} \cdot \frac{3}{2} \cdot \frac{3}{2} \cdot \frac{3}{2} = \frac{81}{16}.$$

Example 2 Use a calculator to approximate the given expressions.

(a) $(1.2)^8$

Solution Key in 1.2 and then use the x^y key (labeled ∧ on some calculators); finally, key in the exponent 8. The calculator displays the (exact) answer 4.29981696.

(b) $\left(\frac{12}{7}\right)^{23}$

Solution Don't compute 12/7 separately. Use parentheses and key in (12/7), followed by the x^y key and the exponent 23 to obtain the approximate answer 242,054.822. 1 ✓

✓Checkpoint 1

Evaluate the following.

(a) 6^3

(b) 5^{12}

(c) 1^9

(d) $\left(\frac{7}{5}\right)^8$

Answers: See page 20.

CAUTION A common error in using exponents occurs with expressions such as $4 \cdot 3^2$. The exponent of 2 applies only to the base 3, so that

$$4 \cdot 3^2 = 4 \cdot 3 \cdot 3 = 36.$$

On the other hand,

$$(4 \cdot 3)^2 = (4 \cdot 3)(4 \cdot 3) = 12 \cdot 12 = 144,$$

so

$$4 \cdot 3^2 \neq (4 \cdot 3)^2.$$

Be careful to distinguish between expressions like -2^4 and $(-2)^4$:

$$-2^4 = -(2^4) = -(2 \cdot 2 \cdot 2 \cdot 2) = -16$$
$$(-2)^4 = (-2)(-2)(-2)(-2) = 16,$$

so

$$-2^4 \neq (-2)^4. \quad 2 ✓$$

✓Checkpoint 2

Evaluate the following.

(a) $3 \cdot 6^2$

(b) $5 \cdot 4^3$

(c) -3^6

(d) $(-3)^6$

(e) $-2 \cdot (-3)^5$

Answers: See page 20.

By the definition of an exponent,

$$3^4 \cdot 3^2 = (3 \cdot 3 \cdot 3 \cdot 3)(3 \cdot 3) = 3^6.$$

Since $6 = 4 + 2$, we can write the preceding equation as $3^4 \cdot 3^2 = 3^{4+2}$. This result suggests the following fact, which applies to any real number a and natural numbers m and n.

Multiplication with Exponents

To multiply a^m by a^n, *add* the exponents:

$$a^m \cdot a^n = a^{m+n}.$$

Example 3 Verify each of the following simplifications.

(a) $7^4 \cdot 7^6 = 7^{4+6} = 7^{10}$

(b) $(-2)^3 \cdot (-2)^5 = (-2)^{3+5} = (-2)^8$

(c) $(3k)^2 \cdot (3k)^3 = (3k)^5$

(d) $(m + n)^2 \cdot (m + n)^5 = (m + n)^7$ 3 ✓

✓Checkpoint 3

Simplify the following.

(a) $5^3 \cdot 5^6$

(b) $(-3)^4 \cdot (-3)^{10}$

(c) $(5p)^2 \cdot (5p)^8$

Answers: See page 20.

The multiplication property of exponents has a convenient consequence. By definition,

$$(5^2)^3 = 5^2 \cdot 5^2 \cdot 5^2 = 5^{2+2+2} = 5^6.$$

Note that $2 + 2 + 2$ is $3 \cdot 2 = 6$. This is an example of a more general fact about any real number a and natural numbers m and n.

Power of a Power

To find a power of a power, $(a^m)^n$, *multiply* the exponents:

$$(a^m)^n = a^{mn}.$$

Example 4 Verify the following computations.

(a) $(x^3)^4 = x^{3 \cdot 4} = x^{12}$.

(b) $[(-3)^5]^3 = (-3)^{5 \cdot 3} = (-3)^{15}$.

(c) $[(6z)^4]^4 = (6z)^{4 \cdot 4} = (6z)^{16}$. 4 ✓

✓Checkpoint 4

Compute the following.

(a) $(6^3)^7$

(b) $[(4k)^5]^6$

Answers: See page 20.

It will be convenient to give a zero exponent a meaning. If the multiplication property of exponents is to remain valid, we should have, for example, $3^5 \cdot 3^0 = 3^{5+0} = 3^5$. But this will be true only when $3^0 = 1$. So we make the following definition.

Zero Exponent

If a is any nonzero real number, then

$$a^0 = 1.$$

✓Checkpoint 5

Evaluate the following.

(a) 17^0

(b) 30^0

(c) $(-10)^0$

(d) $-(12)^0$

Answers: *See page 20.*

For example, $6^0 = 1$ and $(-9)^0 = 1$. Note that 0^0 is *not* defined. 5✓

POLYNOMIALS

A **polynomial** is an algebraic expression such as

$$5x^4 + 2x^3 + 6x, \qquad 8m^3 + 9m^2 + \frac{3}{2}m + 3, \qquad -10p, \qquad \text{or} \qquad 8.$$

The letter used is called a **variable**, and a polynomial is a sum of **terms** of the form

(constant) × (nonnegative integer power of the variable).

We assume that $x^0 = 1$, $m^0 = 1$, etc., so terms such as 3 or 8 may be thought of as $3x^0$ and $8x^0$, respectively. The constants that appear in each term of a polynomial are called the **coefficients** of the polynomial. The coefficient of x^0 is called the **constant term.**

Example 5 Identify the coefficients and the constant term of the given polynomials.

(a) $5x^2 - x + 12$

Solution The coefficients are 5, -1, and 12, and the constant term is 12.

(b) $7x^3 + 2x - 4$

Solution The coefficients are 7, 0, 2, and -4, because the polynomial can be written $7x^3 + 0x^2 + 2x - 4$. The constant term is -4.

A polynomial that consists only of a constant term, such as 15, is called a **constant polynomial**. The **zero polynomial** is the constant polynomial 0. The **degree** of a polynomial is the *exponent* of the highest power of x that appears with a *nonzero* coefficient, and the nonzero coefficient of this highest power of x is the **leading coefficient** of the polynomial. For example,

Polynomial	Degree	Leading Coefficient	Constant Term
$6x^7 + 4x^3 + 5x^2 - 7x + 10$	7	6	10
$-x^4 + 2x^3 + \frac{1}{2}$	4	-1	$\frac{1}{2}$
x^3	3	1	0
12	0	12	12

✓Checkpoint 6

Find the degree of each polynomial.

(a) $x^4 - x^2 + x + 5$

(b) $7x^5 + 6x^3 - 3x^8 + 2$

(c) 17

(d) 0

Answers: See page 20.

The degree of the zero polynomial is *not defined*, since no exponent of x occurs with nonzero coefficient. First-degree polynomials are often called **linear polynomials**. Second- and third-degree polynomials are called **quadratics** and **cubics**, respectively. 6✓

ADDITION AND SUBTRACTION

Two terms having the same variable with the same exponent are called **like terms**; other terms are called **unlike terms**. Polynomials can be added or subtracted by using the distributive property to combine like terms. Only like terms can be combined. For example,

$$12y^4 + 6y^4 = (12 + 6)y^4 = 18y^4$$

and

$$-2m^2 + 8m^2 = (-2 + 8)m^2 = 6m^2.$$

The polynomial $8y^4 + 2y^5$ has unlike terms, so it cannot be further simplified.

In more complicated cases of addition, you may have to eliminate parentheses, use the commutative and associative laws to regroup like terms, and then combine them.

Example 6 Add the following polynomials.

(a) $(8x^3 - 4x^2 + 6x) + (3x^3 + 5x^2 - 9x + 8)$

Solution

$8x^3 - 4x^2 + 6x + 3x^3 + 5x^2 - 9x + 8$ Eliminate parentheses.

$= (8x^3 + 3x^3) + (-4x^2 + 5x^2) + (6x - 9x) + 8$ Group like terms.

$= 11x^3 + x^2 - 3x + 8$ Combine like terms.

(b) $(-4x^4 + 6x^3 - 9x^2 - 12) + (-3x^3 + 8x^2 - 11x + 7)$

Solution

$-4x^4 + 6x^3 - 9x^2 - 12 - 3x^3 + 8x^2 - 11x + 7$ Eliminate parentheses.

$= -4x^4 + (6x^3 - 3x^3) + (-9x^2 + 8x^2) - 11x + (-12 + 7)$ Group like terms.

$= -4x^4 + 3x^3 - x^2 - 11x - 5$ Combine like terms.

Care must be used when parentheses are preceded by a minus sign. For example, we know that

$$-(4 + 3) = -(7) = -7 = -4 - 3.$$

If you simply delete the parentheses in $-(4 + 3)$, you obtain $-4 + 3 = -1$, which is the wrong answer. This fact and the preceding examples illustrate the following rules.

Rules for Eliminating Parentheses

Parentheses preceded by a plus sign (or no sign) may be deleted.

Parentheses preceded by a minus sign may be deleted *provided that* the sign of every term within the parentheses is changed.

Example 7 Subtract: $(2x^2 - 11x + 8) - (7x^2 - 6x + 2)$.

Solution $2x^2 - 11x + 8 - 7x^2 + 6x - 2$ Eliminate parentheses.

$= (2x^2 - 7x^2) + (-11x + 6x) + (8 - 2)$ Group like terms.

$= -5x^2 - 5x + 6$ Combine like terms. 7 ✓

✓Checkpoint 7

Add or subtract as indicated.

(a) $(-2x^2 + 7x + 9) + (3x^2 + 2x - 7)$

(b) $(4x + 6) - (13x - 9)$

(c) $(9x^3 - 8x^2 + 2x) - (9x^3 - 2x^2 - 10)$

Answers: *See page 20.*

MULTIPLICATION

The distributive property is also used to multiply polynomials. For example, the product of $8x$ and $6x - 4$ is found as follows:

$$8x(6x - 4) = 8x(6x) - 8x(4) \quad \text{Distributive property}$$
$$= 48x^2 - 32x \quad x \cdot x = x^2$$

Example 8 Use the distributive property to find each product.

(a) $2p^3(3p^2 - 2p + 5) = 2p^3(3p^2) + 2p^3(-2p) + 2p^3(5)$
$= 6p^5 - 4p^4 + 10p^3$

(b) $(3k - 2)(k^2 + 5k - 4) = 3k(k^2 + 5k - 4) - 2(k^2 + 5k - 4)$
$= 3k^3 + 15k^2 - 12k - 2k^2 - 10k + 8$
$= 3k^3 + 13k^2 - 22k + 8$ 8 ✓

✓Checkpoint 8

Find the following products.

(a) $-6r(2r - 5)$

(b) $(8m + 3) \cdot (m^4 - 2m^2 + 6m)$

Answers: *See page 20.*

Example 9 The product $(2x - 5)(3x + 4)$ can be found by using the distributive property twice:

$$(2x - 5)(3x + 4) = 2x(3x + 4) - 5(3x + 4)$$
$$= 2x \cdot 3x + 2x \cdot 4 + (-5) \cdot 3x + (-5) \cdot 4$$
$$= 6x^2 + 8x - 15x - 20$$
$$= 6x^2 - 7x - 20$$

Observe the pattern in the second line of Example 8 and its relationship to the terms being multiplied:

$$(2x - 5)(3x + 4) = 2x \cdot 3x + 2x \cdot 4 + (-5) \cdot 3x + (-5) \cdot 4$$

$(2x - 5)(3x + 4)$ First terms

$(2x - 5)(3x + 4)$ Outside terms

$(2x - 5)(3x + 4)$ Inside terms

$(2x - 5)(3x + 4)$ Last terms

This pattern is easy to remember by using the acronym **FOIL** (**F**irst, **O**utside, **I**nside, **L**ast). The FOIL method makes it easy to find products such as this one mentally, without the necessity of writing out the intermediate steps.

Example 10 Use FOIL to find the product of the given polynomials.

(a) $(3x + 2)(x + 5) = 3x^2 + 15x + 2x + 10 = 3x^2 + 17x + 10$

First ↑ Outside ↑ Inside ↑ Last ↑

(b) $(x + 3)^2 = (x + 3)(x + 3) = x^2 + 3x + 3x + 9 = x^2 + 6x + 9$

(c) $(2x + 1)(2x - 1) = 4x^2 - 2x + 2x - 1 = 4x^2 - 1$ ✓9

✓Checkpoint 9

Use FOIL to find these products.

(a) $(5k - 1)(2k + 3)$

(b) $(7z - 3)(2z + 5)$

Answers: See page 20.

APPLICATIONS

In business, the *revenue* from the sales of an item is given by

Revenue = (price per item) × (number of items sold).

The *cost* to manufacture and sell these items is given by

Cost = Fixed Costs + Variable Costs,

where the fixed costs include such things as buildings and machinery (which do not depend on how many items are made) and variable costs include such things as labor and materials (which vary, depending on how many items are made). Then

Profit = Revenue − Cost.

Example 11 Hot Rocks Movies sells DVDs for \$8 each (wholesale) and can produce a maximum of 200,000 DVDs. The variable cost of producing x thousand DVDs is $3550x - 8x^2$ dollars, and the fixed costs for the manufacturing operation are \$215,000. If x thousand DVDs are manufactured and sold, find expressions for the revenue, cost, and profit.

Solution If x thousand DVDs are sold at \$8 each, then

$$\text{Revenue} = (\text{price per item}) \times (\text{number of items sold})$$
$$R = 8 \times 1000x = 8000x,$$

where $x \le 200$ (because only 200,000 DVDs can be made). The variable cost of making x thousand DVDs is $3550x - 8x^2$, so that

$$\text{Cost} = \text{Fixed Costs} + \text{Variable Costs},$$
$$C = 215{,}000 + (3550x - 8x^2) \qquad (x \le 200).$$

Therefore, the profit is given by

$$P = R - C = 8000x - (215{,}000 + 3550x - 8x^2)$$
$$= 8000x - 215{,}000 - 3550x + 8x^2$$
$$P = 8x^2 + 4450x - 215{,}000 \qquad (x \le 200).$$ ✓10

✓Checkpoint 10

Suppose revenue is given by $7x^2 - 3x$, fixed costs are \$500, and variable costs are given by $3x^2 + 5x - 25$. Write an expression for

(a) Cost

(b) Profit

Answers: See page 20.

1.2 ▸ Exercises

Use a calculator to approximate these numbers. (See Examples 1 and 2.)

1. 11.2^6 **2.** $(-6.54)^{11}$

3. $(-18/7)^6$ **4.** $(5/9)^7$

5. Explain how the value of -3^2 differs from $(-3)^2$. Do -3^3 and $(-3)^3$ differ in the same way? Why or why not?

6. Describe the steps used to multiply 4^3 and 4^5. Is the product of 4^3 and 3^4 found in the same way? Explain.

Simplify each of the given expressions. Leave your answers in exponential notation. (See Examples 3 and 4.)

7. $4^2 \cdot 4^3$ **8.** $(-4)^4 \cdot (-4)^6$

9. $(-6)^2 \cdot (-6)^5$ **10.** $(2z)^5 \cdot (2z)^6$

11. $[(5u)^4]^7$ **12.** $(6y)^3 \cdot [(6y)^5]^4$

List the degree of the given polynomial, its coefficients, and its constant term. (See Example 5.)

13. $6.2x^4 - 5x^3 + 4x^2 - 3x + 3.7$

14. $6x^7 + 4x^6 - x^3 + x$

State the degree of the given polynomial.

15. $1 + x + 2x^2 + 3x^3$

16. $5x^4 - 4x^5 - 6x^3 + 7x^4 - 2x + 8$

Add or subtract as indicated. (See Examples 6 and 7.)

17. $(3x^3 + 2x^2 - 5x) + (-4x^3 - x^2 - 8x)$

18. $(-2p^3 - 5p + 7) + (-4p^2 + 8p + 2)$

19. $(-4y^2 - 3y + 8) - (2y^2 - 6y + 2)$

20. $(7b^2 + 2b - 5) - (3b^2 + 2b - 6)$

21. $(2x^3 + 2x^2 + 4x - 3) - (2x^3 + 8x^2 + 1)$

22. $(3y^3 + 9y^2 - 11y + 8) - (-4y^2 + 10y - 6)$

Find each of the given products. (See Examples 8–10.)

23. $-9m(2m^2 + 6m - 1)$

24. $2a(4a^2 - 6a + 8)$

25. $(3z + 5)(4z^2 - 2z + 1)$

26. $(2k + 3)(4k^3 - 3k^2 + k)$

27. $(6k - 1)(2k + 3)$

28. $(8r + 3)(r - 1)$

29. $(3y + 5)(2y + 1)$

30. $(5r - 3s)(5r - 4s)$

31. $(9k + q)(2k - q)$

32. $(.012x - .17)(.3x + .54)$

33. $(6.2m - 3.4)(.7m + 1.3)$

34. $2p - 3[4p - (8p + 1)]$

35. $5k - [k + (-3 + 5k)]$

36. $(3x - 1)(x + 2) - (2x + 5)^2$

Business *Find expressions for the revenue, cost, and profit from selling x thousand items. (See Example 11.)*

	Item Price	*Fixed Costs*	*Variable Costs*
37.	\$5.00	\$200,000	$1800x$
38.	\$8.50	\$225,000	$4200x$

39. Business Beauty Works sells its cologne wholesale for \$9.75 per bottle. The variable costs of producing x thousand bottles is $-3x^2 + 3480x - 325$ dollars, and the fixed costs of manufacturing are \$260,000. Find expressions for the revenue, cost, and profit from selling x thousand items.

40. Business A well-known golfer sells his book *Help for Hackers* for \$13.25 per copy. His fixed costs are \$125,000 and he estimates that the variable cost of printing, binding, and distribution are given by $-4x^2 + 2880x - 295$ dollars. Find expressions for the revenue, cost, and profit from selling x thousand copies of the book.

Business *The accompanying bar graph shows the net earnings (in millions of dollars) of the Starbucks Corporation, as given in its annual reports. The polynomial*

$$.875x^4 - 16.47x^3 + 110.35x^2 - 212.6x + 300$$

gives a good approximation of Starbucks's net earnings in year x, where x = 1 corresponds to 2001, x = 2 to 2002, and so on (1 ≤ x ≤ 7). For each of the given years,

(a) *use the bar graph to determine the net earnings;*

(b) *use the polynomial to determine the net earnings.*

41. 2001 **42.** 2004 **43.** 2005 **44.** 2007

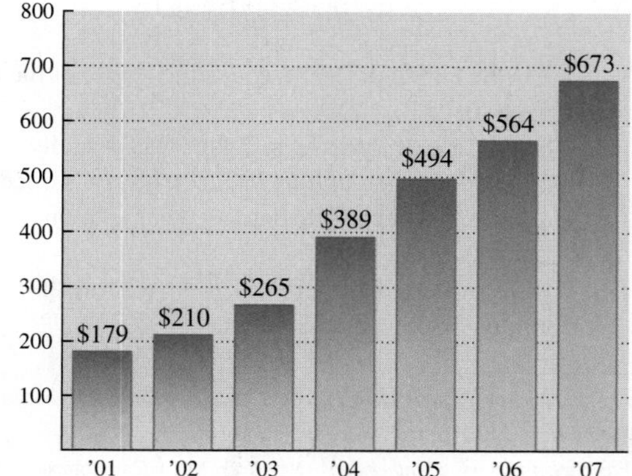

Assuming that the polynomial approximation in Exercises 41–44 remains accurate in later years, use it to estimate Starbucks's net earnings in each of the following years.

45. 2008 **46.** 2009 **47.** 2010

48. Do the estimates in Exercises 45–47 seem plausible? Explain.

Health *The number of knee implants (in millions) in year x is approximated by the polynomial* $.002722x^2 + .003x + .37$*, where* $x = 0$ *corresponds to the year 2000.* Assuming the polynomial approximation is accurate through the year 2021, use it to determine whether each of the given statements is true or false.*

49. There were about 370,000 knee implants in 2000.

50. There will be more than 750,000 knee implants in 2011.

51. There will be more than a million knee implants in 2014.

52. There will be at least a million and a half knee implants in 2021.

Health *According to data from a leading insurance company, if a person is 65 years old, the probability that he or she will live for another x years is approximated by the polynomial*

$$1 - .0058x - .00076x^2.†$$

Find the probability that a 65-year-old person will live to the following ages.

53. 75 (that is, 10 years past 65) **54.** 80

55. 87 **56.** 95

57. Physical Science One of the most amazing formulas in all of ancient mathematics is the formula discovered by the Egyptians to find the volume of the frustum of a square pyramid, as shown in the following figure:

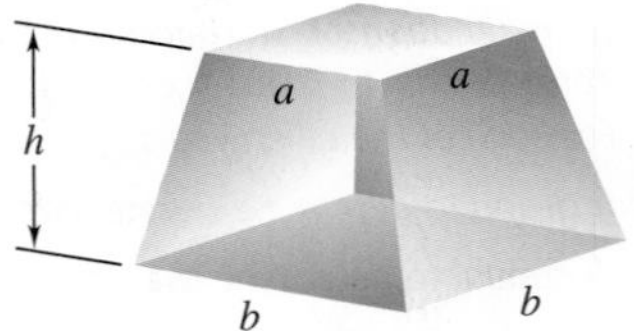

The volume of this pyramid is given by

$$(1/3)h \cdot (a^2 + ab + b^2),$$

where b is the length of the base, a is the length of the top, and h is the height.‡

(a) When the Great Pyramid in Egypt was partially completed to a height h of 200 feet, b was 756 feet and a was 314 feet. Calculate its volume at this stage of construction.

(b) Try to visualize the figure if $a = b$. What is the resulting shape? Find its volume.

(c) Let $a = b$ in the Egyptian formula and simplify. Are the results the same?

58. Physical Science Refer to the formula and the discussion in Exercise 57.

(a) Use the expression $(1/3)h(a^2 + ab + b^2)$ to determine a formula for the volume of a pyramid with a square base b and height h by letting $a = 0$.

(b) The Great Pyramid in Egypt had a square base of 756 feet and a height of 481 feet. Find the volume of the Great Pyramid. Compare it with the volume of the 273-foot-tall Louisiana Superdome, which has an approximate volume of 125 million cubic feet.*

(c) The Superdome covers an area of 13 acres. How many acres does the Great Pyramid cover? (*Hint:* 1 acre = 43,560 ft^2.)

59. Suppose one polynomial has degree 3 and another also has degree 3. Find all possible values for the degree of their

(a) sum;

(b) difference;

(c) product.

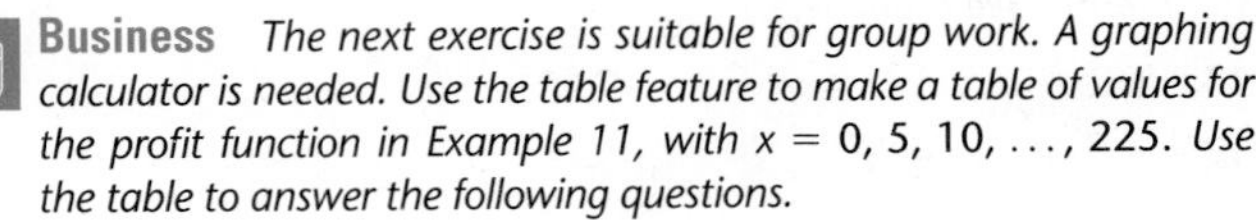

Business *The next exercise is suitable for group work. A graphing calculator is needed. Use the table feature to make a table of values for the profit function in Example 11, with* $x = 0, 5, 10, \ldots, 225$*. Use the table to answer the following questions.*

60. (a) What is the profit or loss (= negative profit) when 20,000 DVDs are sold? when 40,000 are sold? Explain these answers.

(b) Approximately how many DVDs must be sold in order for the company to make a profit?

(c) What is the profit from selling 100,000 DVDs? from 150,000 DVDs? from 200,000 DVDs?

(d) Explain why the profit amounts shown in the table for 205,000 DVDs and beyond are meaningless.

✓ Checkpoint Answers

1. (a) 216 (b) 244,140,625 (c) 1 (d) 14.75789056
2. (a) 108 (b) 320 (c) -729 (d) 729 (e) 486
3. (a) 5^9 (b) $(-3)^{14}$ (c) $(5p)^{10}$
4. (a) 6^{21} (b) $(4k)^{30}$
5. (a) 1 (b) 1 (c) 1 (d) -1
6. (a) 4 (b) 8 (c) 0 (d) Not defined
7. (a) $x^2 + 9x + 2$ (b) $-9x + 15$ (c) $-6x^2 + 2x + 10$
8. (a) $-12r^2 + 30r$ (b) $8m^5 + 3m^4 - 16m^3 + 42m^2 + 18m$
9. (a) $10k^2 + 13k - 3$ (b) $14z^2 + 29z - 15$
10. (a) $C = 3x^2 + 5x + 475$ (b) $P = 4x^2 - 8x - 475$

*Based on data and projections from Exponent, Inc., as published in *Newsweek*.

†Provided by Ralph DeMarr, University of New Mexico.

‡H. A. Freebury, *A History of Mathematics*. (New York: MacMillan Company, 1968).

*Louisiana Superdome (www.superdome.com).

1.3 ▸ Factoring

The number 18 can be written as a product in several ways: $9 \cdot 2$, $(-3)(-6)$, $1 \cdot 18$, etc. The numbers in each product (9, 2, −3, etc.) are called **factors**, and the process of writing 18 as a product of factors is called **factoring**. Thus, factoring is the reverse of multiplication.

Factoring of polynomials is a means of simplifying many expressions and of solving certain types of equations. As is the usual custom, factoring of polynomials in this book will be restricted to finding factors with *integer* coefficients (otherwise there may be an infinite number of possible factors).

GREATEST COMMON FACTOR

The algebraic expression $15m + 45$ is made up of two terms: $15m$ and 45. Each of these terms has 15 as a factor. In fact, $15m = 15 \cdot m$ and $45 = 15 \cdot 3$. By the distributive property,

$$15m + 45 = 15 \cdot m + 15 \cdot 3 = 15(m + 3).$$

Both 15 and $m + 3$ are factors of $15m + 45$. Since 15 divides evenly into all terms of $15m + 45$ and is the largest number that will do so, it is called the **greatest common factor** for the polynomial $15m + 45$. The process of writing $15m + 45$ as $15(m + 3)$ is called **factoring out** the greatest common factor.

Example 1 Factor out the greatest common factor.

(a) $12p - 18q$

Solution Both $12p$ and $18q$ are divisible by 6, and

$$\begin{aligned} 12p - 18q &= 6 \cdot 2p - 6 \cdot 3q \\ &= 6(2p - 3q). \end{aligned}$$

(b) $8x^3 - 9x^2 + 15x$

Solution Each of these terms is divisible by x:

$$\begin{aligned} 8x^3 - 9x^2 + 15x &= (8x^2) \cdot x - (9x) \cdot x + 15 \cdot x \\ &= x(8x^2 - 9x + 15) \end{aligned}$$

(c) $5(4x - 3)^3 + 2(4x - 3)^2$

Solution The quantity $(4x - 3)^2$ is a common factor. Factoring it out gives

$$\begin{aligned} 5(4x - 3)^3 + 2(4x - 3)^2 &= (4x - 3)^2[5(4x - 3) + 2] \\ &= (4x - 3)^2(20x - 15 + 2) \\ &= (4x - 3)^2(20x - 13). \end{aligned}$$

✓1

✓Checkpoint 1

Factor out the greatest common factor.

(a) $12r + 9k$

(b) $75m^2 + 100n^2$

(c) $6m^4 - 9m^3 + 12m^2$

(d) $3(2k + 1)^3 + 4(2k + 1)^4$

Answers: See page 28.

FACTORING QUADRATICS

If we multiply two first-degree polynomials, the result is a quadratic. For instance, using FOIL, we see that $(x + 1)(x - 2) = x^2 - x - 2$. Since factoring is the reverse of multiplication, factoring quadratics requires using FOIL backwards.

Example 2 Factor $x^2 + 9x + 18$.

Solution We must find integers b and d such that

$$\begin{aligned} x^2 + 9x + 18 &= (x + b)(x + d) \\ &= x^2 + dx + bx + bd \\ x^2 + 9x + 18 &= x^2 + (b + d)x + bd. \end{aligned}$$

Since the constant coefficients on each side of the equation must be equal, we must have $bd = 18$; that is, b and d are factors of 18. Similarly, the coefficients of x must be the same, so that $b + d = 9$. The possibilities are summarized in this table:

Factors b, d of 18	Sum $b + d$
$18 \cdot 1$	$18 + 1 = 19$
$9 \cdot 2$	$9 + 2 = 11$
$6 \cdot 3$	$6 + 3 = 9$

There is no need to list negative factors, such as $(-3)(-6)$, because their sum is negative. The table suggests that 6 and 3 will work. Verify that

$$(x + 6)(x + 3) = x^2 + 9x + 18. \quad 2✓$$

✓Checkpoint 2

Factor the following.

(a) $r^2 + 7r + 10$

(b) $x^2 + 4x + 3$

(c) $y^2 + 6y + 8$

Answers: See page 28.

Example 3 Factor $x^2 + 3x - 10$.

Solution As in Example 2, we must find factors b and d whose product is -10 (the constant term) and whose sum is 3 (the coefficient of x). The following table shows the possibilities.

Factors b, d of -10	Sum $b + d$
$1(-10)$	$1 + (-10) = -9$
$(-1)10$	$-1 + 10 = 9$
$2(-5)$	$2 + (-5) = -3$
$(-2)5$	$-2 + 5 = 3$

The only factors with product -10 and sum 3 are -2 and 5. So the correct factorization is

$$x^2 + 3x - 10 = (x - 2)(x + 5),$$

as you can readily verify.

It is usually not necessary to construct tables as was done in Examples 2 and 3—you can just mentally check the various possibilities. The approach used in Examples 2 and 3 (with minor modifications) also works for factoring quadratic polynomials whose leading coefficient is not 1.

Example 4 Factor $4y^2 - 11y + 6$.

Solution We must find integers a, b, c, and d such that

$$\begin{aligned} 4y^2 - 11y + 6 &= (ay + b)(cy + d) \\ &= acy^2 + ady + bcy + bd \\ 4y^2 - 11y + 6 &= acy^2 + (ad + bc)y + bd. \end{aligned}$$

Since the coefficients of y^2 must be the same on both sides, we see that $ac = 4$. Similarly, the constant terms show that $bd = 6$. The positive factors of 4 are 4 and 1 or 2 and 2. Since the middle term is negative, we consider only negative factors of 6. The possibilities are -2 and -3 or -1 and -6. Now we try various arrangements of these factors until we find one that gives the correct coefficient of y:

$$(2y - 1)(2y - 6) = 4y^2 - \mathbf{14y} + 6 \quad \text{Incorrect}$$
$$(2y - 2)(2y - 3) = 4y^2 - \mathbf{10y} + 6 \quad \text{Incorrect}$$
$$(y - 2)(4y - 3) = 4y^2 - \mathbf{11y} + 6 \quad \text{Correct}$$

The last trial gives the correct factorization. ✓3

✓Checkpoint 3

Factor the following.

(a) $x^2 - 4x + 3$

(b) $2y^2 - 5y + 2$

(c) $6z^2 - 13z + 6$

Answers: See page 28.

Example 5 Factor $6p^2 - 7pq - 5q^2$.

Solution Again, we try various possibilities. The positive factors of 6 could be 2 and 3 or 1 and 6. As factors of -5, we have only -1 and 5 or -5 and 1. Try different combinations of these factors until the correct one is found:

$$(2p - 5q)(3p + q) = 6p^2 - \mathbf{13pq} - 5q^2 \quad \text{Incorrect}$$
$$(3p - 5q)(2p + q) = 6p^2 - \mathbf{7pq} - 5q^2 \quad \text{Correct}$$

So $6p^2 - 7pq - 5q^2$ factors as $(3p - 5q)(2p + q)$. ✓4

✓Checkpoint 4

Factor the following.

(a) $r^2 - 5r - 14$

(b) $3m^2 + 5m - 2$

(c) $6p^2 + 13pq - 5q^2$

Answers: See page 28.

NOTE In Examples 2–4, we chose positive factors of the positive first term. Of course, we could have used two negative factors, but the work is easier if positive factors are used.

Example 6 Factor $x^2 + x + 3$.

Solution There are only two ways to factor 3, namely, $3 = 1 \cdot 3$ and $3 = (-1)(-3)$. They lead to these products:

$$(x + 1)(x + 3) = x^2 + 4x + 3 \quad \text{Incorrect}$$
$$(x - 1)(x - 3) = x^2 - 4x + 3. \quad \text{Incorrect}$$

Therefore, this polynomial cannot be factored.

FACTORING PATTERNS

In some cases, you can factor a polynomial with a minimum amount of guesswork by recognizing common patterns. The easiest pattern to recognize is the *difference of squares*.

$$x^2 - y^2 = (x + y)(x - y).$$ Difference of squares

To verify the accuracy of the preceding equation, multiply out the right side.

Example 7 Factor each of the following.

(a) $4m^2 - 9$

Solution Notice that $4m^2 - 9$ is the difference of two squares, since $4m^2 = (2m)^2$ and $9 = 3^2$. Use the pattern for the difference of two squares, letting $2m$ replace x and 3 replace y. Then the pattern $x^2 - y^2 = (x + y)(x - y)$ becomes

$$\begin{aligned} 4m^2 - 9 &= (2m)^2 - 3^2 \\ &= (2m + 3)(2m - 3). \end{aligned}$$

(b) $128p^2 - 98q^2$

Solution First factor out the common factor of 2:

$$\begin{aligned} 128p^2 - 98q^2 &= 2(64p^2 - 49q^2) \\ &= 2[(8p)^2 - (7q)^2] \\ &= 2(8p + 7q)(8p - 7q). \end{aligned}$$

(c) $x^2 + 36$

Solution The *sum* of two squares cannot be factored. To convince yourself of this, check some possibilities:

$$(x + 6)(x + 6) = (x + 6)^2 = x^2 + 12x + 36;$$
$$(x + 4)(x + 9) = x^2 + 13x + 36.$$

(d) $(x - 2)^2 - 49$

Solution Since $49 = 7^2$, this is a difference of two squares. So it factors as follows:

$$\begin{aligned} (x - 2)^2 - 49 &= (x - 2)^2 - 7^2 \\ &= [(x - 2) + 7][(x - 2) - 7] \\ &= (x + 5)(x - 9). \end{aligned}$$ 5

✓Checkpoint 5

Factor the following.

(a) $9p^2 - 49$

(b) $y^2 + 100$

(c) $(x + 3)^2 - 64$

Answers: See page 28.

Another common pattern is the *perfect square*. Verify each of the following factorizations by multiplying out the right side.

$$x^2 + 2xy + y^2 = (x + y)^2$$
$$x^2 - 2xy + y^2 = (x - y)^2$$
Perfect Squares

Whenever you have a quadratic whose first and last terms are squares, it *may* factor as a perfect square. The key is to look at the middle term. To have a perfect square whose first and last terms are x^2 and y^2, the middle term must be $\pm 2xy$. To avoid errors, always check this.

Example 8 Factor each polynomial, if possible.

(a) $16p^2 - 40pq + 25q^2$

Solution The first and last terms are squares, namely, $16p^2 = (4p)^2$ and $25q^2 = (5q)^2$. So the second perfect-square pattern, with $x = 4p$ and $y = 5q$, might work. To have a perfect square, the middle term $-40pq$ must equal $-2(4p)(5q)$, which it does. So the polynomial factors as

$$16p^2 - 40pq + 25q^2 = (4p - 5q)(4p - 5q),$$

as you can easily verify.

(b) $9u^2 + 5u + 1$

Solution Again, the first and last terms are squares: $9u^2 = (3u)^2$ and $1 = 1^2$. The middle term is positive, so the first perfect-square pattern might work, with $x = 3u$ and $y = 1$. To have a perfect square, however, the middle term would have to be $2(3u) \cdot 1 = 6u$, which is *not* the middle term of the given polynomial. So it is not a perfect square—in fact, it cannot be factored.

(c) $169x^2 + 104xy^2 + 16y^4$

Solution This polynomial may be factored as $(13x + 4y^2)^2$, since $169x^2 = (13x)^2$, $16y^4 = (4y^2)^2$, and $2(13x)(4y^2) = 104xy^2$. 6✓

✓Checkpoint 6

Factor.

(a) $4m^2 + 4m + 1$

(b) $25z^2 - 80zt + 64t^2$

(c) $9x^2 + 15x + 25$

Answers: See page 28.

Example 9 Factor each of the given polynomials.

(a) $12x^2 - 26x - 10$

Solution Look first for a greatest common factor. Here, the greatest common factor is 2: $12x^2 - 26x - 10 = 2(6x^2 - 13x - 5)$. Now try to factor $6x^2 - 13x - 5$. Possible factors of 6 are 3 and 2 or 6 and 1. The only factors of -5 are -5 and 1 or 5 and -1. Try various combinations. You should find that the quadratic factors as $(3x + 1)(2x - 5)$. Thus,

$$12x^2 - 26x - 10 = 2(3x + 1)(2x - 5).$$

(b) $4z^2 + 12z + 9 - w^2$

Solution There is no common factor here, but notice that the first three terms can be factored as a perfect square:

$$4z^2 + 12z + 9 - w^2 = (2z + 3)^2 - w^2$$

Written in this form, the expression is the difference of squares, which can be factored as follows:

$$\begin{aligned}(2z + 3)^2 - w^2 &= [(2z + 3) + w][(2z + 3) - w] \\ &= (2z + 3 + w)(2z + 3 - w).\end{aligned}$$

(c) $16a^2 - 100 - 48ac + 36c^2$

Solution Factor out the greatest common factor of 4 first:

$$\begin{aligned} 16a^2 - 100 - 48ac + 36c^2 &= 4[4a^2 - 25 - 12ac + 9c^2] \\ &= 4[(4a^2 - 12ac + 9c^2) - 25] && \text{Rearrange terms and group.} \\ &= 4[(2a - 3c)^2 - 25] && \text{Factor.} \\ &= 4(2a - 3c + 5)(2a - 3c - 5) && \text{Factor the difference of squares.} \end{aligned}$$

✓Checkpoint 7

Factor the following.

(a) $6x^2 - 27x - 15$

(b) $9r^2 + 12r + 4 - t^2$

(c) $18 - 8xy - 2y^2 - 8x^2$

Answers: See page 28.

CAUTION Remember always to look first for a greatest common factor.

HIGHER DEGREE POLYNOMIALS

Polynomials of degree greater than 2 are often difficult to factor. However, factoring is relatively easy in two cases: *the difference and the sum of cubes.* By multiplying out the right side, you can readily verify each of the following factorizations.

$$x^3 - y^3 = (x - y)(x^2 + xy + y^2) \quad \text{Difference of cubes}$$
$$x^3 + y^3 = (x + y)(x^2 - xy + y^2) \quad \text{Sum of cubes}$$

Example 10 Factor each of the following polynomials.

(a) $k^3 - 8$

Solution Since $8 = 2^3$, use the pattern for the difference of two cubes to obtain

$$k^3 - 8 = k^3 - 2^3 = (k - 2)(k^2 + 2k + 4).$$

(b) $m^3 + 125$

Solution

$$m^3 + 125 = m^3 + 5^3 = (m + 5)(m^2 - 5m + 25)$$

(c) $8k^2 - 27z^3$

Solution

$$8k^3 - 27z^3 = (2k)^3 - (3z)^3 = (2k - 3z)(4k^2 + 6kz + 9z^2)$$

✓Checkpoint 8

Factor the following.

(a) $a^3 + 1000$

(b) $z^3 - 64$

(c) $1000m^3 - 27z^3$

Answers: See page 28.

Substitution and appropriate factoring patterns can sometimes be used to factor higher degree expressions.

Example 11 Factor the following polynomials.

(a) $x^8 + 4x^4 + 3$

Solution The idea is to make a substitution that reduces the polynomial to a quadratic or cubic that we can deal with. Note that $x^8 = (x^4)^2$. Let $u = x^4$. Then

$$\begin{aligned} x^8 + 4x^4 + 3 &= (x^4)^2 + 4x^4 + 3 && \text{Power of a power} \\ &= u^2 + 4u + 3 && \text{Substitute } x^4 = u. \\ &= (u + 3)(u + 1) && \text{Factor.} \\ &= (x^4 + 3)(x^4 + 1). && \text{Substitute } u = x^4. \end{aligned}$$

(b) $x^4 - y^4$

Solution Note that $x^4 = (x^2)^2$, and similarly for the y term. Let $u = x^2$ and $v = y^2$. Then

$$\begin{aligned} x^4 - y^4 &= (x^2)^2 - (y^2)^2 && \text{Power of a power} \\ &= u^2 - v^2 && \text{Substitute } x^2 = u \text{ and } y^2 = v. \\ &= (u + v)(u - v) && \text{Difference of squares} \\ &= (x^2 + y^2)(x^2 - y^2) && \text{Substitute } u = x^2 \text{ and } v = y^2. \\ &= (x^2 + y^2)(x + y)(x - y). && \text{Difference of squares} \end{aligned}$$

9✓

✓Checkpoint 9

Factor each of the following.

(a) $2x^4 + 5x^2 + 2$

(b) $3x^4 - x^2 - 2$

Answers: See page 28.

Once you understand Example 11, you can often factor without making explicit substitutions.

Example 12 Factor $256k^4 - 625m^4$.

Solution Use the difference of squares twice, as follows:

$$\begin{aligned} 256k^4 - 625m^4 &= (16k^2)^2 - (25m^2)^2 \\ &= (16k^2 + 25m^2)(16k^2 - 25m^2) \\ &= (16k^2 + 25m^2)(4k + 5m)(4k - 5m). \end{aligned}$$

10✓

✓Checkpoint 10

Factor $81x^4 - 16y^4$.

Answer: See page 28.

1.3 ▸ Exercises

Factor out the greatest common factor in each of the given polynomials. (See Example 1.)

1. $12x^2 - 24x$
2. $5y - 65xy$
3. $r^3 - 5r^2 + r$
4. $t^3 + 3t^2 + 8t$
5. $6z^3 - 12z^2 + 18z$
6. $5x^3 + 55x^2 + 10x$
7. $3(2y - 1)^2 + 7(2y - 1)^3$
8. $(3x + 7)^5 - 4(3x + 7)^3$
9. $3(x + 5)^4 + (x + 5)^6$
10. $3(x + 6)^2 + 6(x + 6)^4$

Factor the polynomial. (See Examples 2 and 3.)

11. $x^2 + 5x + 4$
12. $u^2 + 7u + 6$
13. $x^2 + 7x + 12$
14. $y^2 + 8y + 12$
15. $x^2 + x - 6$
16. $x^2 + 4x - 5$
17. $x^2 + 2x - 3$
18. $y^2 + y - 12$
19. $x^2 - 3x - 4$
20. $u^2 - 2u - 8$
21. $z^2 - 9z + 14$
22. $w^2 - 6w - 16$
23. $z^2 + 10z + 24$
24. $r^2 + 16r + 60$

Factor the polynomial. (See Examples 4–6.)

25. $2x^2 - 9x + 4$
26. $3w^2 - 8w + 4$
27. $15p^2 - 23p + 4$
28. $8x^2 - 14x + 3$
29. $4z^2 - 16z + 15$
30. $12y^2 - 29y + 15$
31. $6x^2 - 5x - 4$
32. $12z^2 + z - 1$
33. $10y^2 + 21y - 10$
34. $15u^2 + 4u - 4$
35. $6x^2 + 5x - 4$
36. $12y^2 + 7y - 10$

Factor each polynomial completely. Factor out the greatest common factor as necessary. (See Examples 2–9.)

37. $3a^2 + 2a - 5$
38. $6a^2 - 48a - 120$
39. $x^2 - 81$
40. $x^2 + 17xy + 72y^2$
41. $9p^2 - 12p + 4$
42. $3r^2 - r - 2$
43. $r^2 + 3rt - 10t^2$
44. $2a^2 + ab - 6b^2$
45. $m^2 - 8mn + 16n^2$
46. $8k^2 - 16k - 10$
47. $4u^2 + 12u + 9$
48. $9p^2 - 16$
49. $25p^2 - 10p + 4$
50. $10x^2 - 17x + 3$

51. $4r^2 - 9v^2$
52. $x^2 + 3xy - 28y^2$
53. $x^2 + 4xy + 4y^2$
54. $16u^2 + 12u - 18$
55. $3a^2 - 13a - 30$
56. $3k^2 + 2k - 8$
57. $21m^2 + 13mn + 2n^2$
58. $81y^2 - 100$
59. $y^2 - 4yz - 21z^2$
60. $49a^2 + 9$
61. $121x^2 - 64$
62. $4z^2 + 56zy + 196y^2$

Factor each of these polynomials. (See Example 10.)

63. $a^3 - 64$
64. $b^3 + 216$
65. $8r^3 - 27s^3$
66. $1000p^3 + 27q^3$
67. $64m^3 + 125$
68. $216y^3 - 343$
69. $1000y^3 - z^3$
70. $125p^3 + 8q^3$

Factor each of these polynomials. (See Examples 11 and 12.)

71. $x^4 + 5x^2 + 6$
72. $y^4 + 7y^2 + 10$
73. $b^4 - b^2$
74. $z^4 - 3z^2 - 4$
75. $x^4 - x^2 - 12$
76. $4x^4 + 27x^2 - 81$
77. $16a^4 - 81b^4$
78. $x^6 - y^6$
79. $x^8 + 8x^2$
80. $x^9 - 64x^3$
81. When asked to factor $6x^4 - 3x^2 - 3$ completely, a student gave the following result:

$$6x^4 - 3x^2 - 3 = (2x^2 + 1)(3x^2 - 3).$$

Is this answer correct? Explain why.

82. When can the sum of two squares be factored? Give examples.
83. Explain why $(x + 2)^3$ is not the correct factorization of $x^3 + 8$, and give the correct factorization.
84. Describe how factoring and multiplication are related. Give examples.

✓ Checkpoint Answers

1. **(a)** $3(4r + 3k)$ **(b)** $25(3m^2 + 4n^2)$ **(c)** $3m^2(2m^2 - 3m + 4)$ **(d)** $(2k + 1)^3(7 + 8k)$
2. **(a)** $(r + 2)(r + 5)$ **(b)** $(x + 3)(x + 1)$ **(c)** $(y + 2)(y + 4)$
3. **(a)** $(x - 3)(x - 1)$ **(b)** $(2y - 1)(y - 2)$ **(c)** $(3z - 2)(2z - 3)$
4. **(a)** $(r - 7)(r + 2)$ **(b)** $(3m - 1)(m + 2)$ **(c)** $(2p + 5q)(3p - q)$
5. **(a)** $(3p + 7)(3p - 7)$ **(b)** Cannot be factored **(c)** $(x + 11)(x - 5)$
6. **(a)** $(2m + 1)^2$ **(b)** $(5z - 8t)^2$ **(c)** Does not factor
7. **(a)** $3(2x + 1)(x - 5)$ **(b)** $(3r + 2 + t)(3r + 2 - t)$ **(c)** $2(3 - 2x - y)(3 + 2x + y)$
8. **(a)** $(a + 10)(a^2 - 10a + 100)$ **(b)** $(z - 4)(z^2 + 4z + 16)$ **(c)** $(10m - 3z)(100m^2 + 30mz + 9z^2)$
9. **(a)** $(2x^2 + 1)(x^2 + 2)$ **(b)** $(3x^2 + 2)(x + 1)(x - 1)$
10. $(9x^2 + 4y^2)(3x + 2y)(3x - 2y)$

1.4 ▸ Rational Expressions

A **rational expression** is an expression that can be written as the quotient of two polynomials, such as

$$\frac{8}{x - 1}, \quad \frac{3x^2 + 4x}{5x - 6}, \quad \text{and} \quad \frac{2y + 1}{y^4 + 8}.$$

It is sometimes important to know the values of the variable that make the denominator 0 (in which case the quotient is not defined). For example, 1 cannot be used as a replacement for x in the first expression above and 6/5 cannot be used in the second one, since these values make the respective denominators equal 0. *Throughout this section, we assume that all denominators are nonzero*, which means that some replacement values for the variables may have to be excluded. 1✓

✓ Checkpoint 1

What value of the variable makes each denominator equal 0?

(a) $\dfrac{5}{x - 3}$

(b) $\dfrac{2x - 3}{4x - 1}$

(c) $\dfrac{x + 2}{x}$

(d) Why do we need to determine these values?

Answers: See page 35.

SIMPLIFYING RATIONAL EXPRESSIONS

A key tool for simplification is the following fact.

For all expressions P, Q, R, and S, with $Q \neq 0$ and $S \neq 0$,

$$\frac{PS}{QS} = \frac{P}{Q}. \quad \text{Cancellation Property}$$

Example 1 Write each of the following rational expressions in lowest terms (so that the numerator and denominator have no common factor with integer coefficients except 1 or -1).

(a) $\frac{12m}{-18}$

Solution Both $12m$ and -18 are divisible by 6. By the cancellation property,

$$\frac{12m}{-18} = \frac{2m \cdot 6}{-3 \cdot 6} = \frac{2m}{-3} = -\frac{2m}{3}.$$

(b) $\frac{8x + 16}{4}$

Solution Factor the numerator and cancel:

$$\frac{8x + 16}{4} = \frac{8(x + 2)}{4} = \frac{4 \cdot 2(x + 2)}{4} = \frac{2(x + 2)}{1} = 2(x + 2).$$

The answer could also be written as $2x + 4$ if desired.

(c) $\frac{k^2 + 7k + 12}{k^2 + 2k - 3}$

Solution Factor the numerator and denominator and cancel:

$$\frac{k^2 + 7k + 12}{k^2 + 2k - 3} = \frac{(k + 4)(k + 3)}{(k - 1)(k + 3)} = \frac{k + 4}{k - 1}.$$ ✓2

✓Checkpoint 2

Write each of the following in lowest terms.

(a) $\frac{12k + 36}{18}$

(b) $\frac{15m + 30m^2}{5m}$

(c) $\frac{2p^2 + 3p + 1}{p^2 + 3p + 2}$

Answers: See page 35.

MULTIPLICATION AND DIVISION

The rules for multiplying and dividing rational expressions are the same fraction rules you learned in arithmetic.

For all expressions P, Q, R, and S, with $Q \neq 0$ and $S \neq 0$,

$$\frac{P}{Q} \cdot \frac{R}{S} = \frac{PR}{QS}$$ Multiplication Rule

and

$$\frac{P}{Q} \div \frac{R}{S} = \frac{P}{Q} \cdot \frac{S}{R} \quad (R \neq 0).$$ Division Rule

Example 2

(a) Multiply $\frac{2}{3} \cdot \frac{y}{5}$.

Solution Use the multiplication rule. Multiply the numerators and then the denominators:

$$\frac{2}{3} \cdot \frac{y}{5} = \frac{2 \cdot y}{3 \cdot 5} = \frac{2y}{15}$$

The result, $2y/15$, is in lowest terms.

(b) Multiply $\frac{3y + 9}{6} \cdot \frac{18}{5y + 15}$.

Solution Factor where possible:

$$\frac{3y + 9}{6} \cdot \frac{18}{5y + 15} = \frac{3(y + 3)}{6} \cdot \frac{18}{5(y + 3)}$$

$$= \frac{3 \cdot 18(y + 3)}{6 \cdot 5(y + 3)} \quad \text{Multiply numerators and denominators.}$$

$$= \frac{3 \cdot 6 \cdot 3(y + 3)}{6 \cdot 5(y + 3)} \quad 18 = 6 \cdot 3$$

$$= \frac{3 \cdot 3}{5} \quad \text{Write in lowest terms.}$$

$$= \frac{9}{5}.$$

(c) Multiply $\frac{m^2 + 5m + 6}{m + 3} \cdot \frac{m^2 + m - 6}{m^2 + 3m + 2}$.

Solution Factor numerators and denominators:

$$\frac{(m + 2)(m + 3)}{m + 3} \cdot \frac{(m - 2)(m + 3)}{(m + 2)(m + 1)} \quad \text{Factor.}$$

$$= \frac{(m + 2)(m + 3)(m - 2)(m + 3)}{(m + 3)(m + 2)(m + 1)} \quad \text{Multiply.}$$

$$= \frac{(m - 2)(m + 3)}{m + 1} \quad \text{Lowest terms}$$

$$= \frac{m^2 + m - 6}{m + 1}.$$

✓3

✓Checkpoint 3

Multiply.

(a) $\frac{3r^2}{5} \cdot \frac{20}{9r}$

(b) $\frac{y - 4}{y^2 - 2y - 8} \cdot \frac{y^2 - 4}{3y}$

Answers: See page 35.

Example 3

(a) Divide $\frac{8x}{5} \div \frac{11x^2}{20}$.

Solution Invert the second expression and multiply (division rule):

$$\frac{8x}{5} \div \frac{11x^2}{20} = \frac{8x}{5} \cdot \frac{20}{11x^2} \quad \text{Invert and multiply.}$$

$$= \frac{8x \cdot 20}{5 \cdot 11x^2} \quad \text{Multiply.}$$

$$= \frac{32}{11x}. \quad \text{Lowest terms}$$

(b) Divide $\frac{9p - 36}{12} \div \frac{5(p - 4)}{18}$.

Solution We have

$$\frac{9p - 36}{12} \cdot \frac{18}{5(p - 4)} \quad \text{Invert and multiply.}$$

$$= \frac{9(p - 4)}{12} \cdot \frac{18}{5(p - 4)} \quad \text{Factor.}$$

$$= \frac{27}{10}. \quad \text{Cancel, multiply, and write in lowest terms.}$$

4 ✓

✓Checkpoint 4

Divide.

(a) $\frac{5m}{16} \div \frac{m^2}{10}$

(b) $\frac{2y - 8}{6} \div \frac{5y - 20}{3}$

(c) $\frac{m^2 - 2m - 3}{m(m + 1)} \div \frac{m + 4}{5m}$

Answers: See page 35.

ADDITION AND SUBTRACTION

As you know, when two numerical fractions have the same denominator, they can be added or subtracted. The same rules apply to rational expressions.

For all expressions P, Q, R, with $Q \neq 0$,

$$\frac{P}{Q} + \frac{R}{Q} = \frac{P + R}{Q} \quad \text{Addition Rule}$$

and

$$\frac{P}{Q} - \frac{R}{Q} = \frac{P - R}{Q}. \quad \text{Subtraction Rule}$$

Example 4 Add or subtract as indicated.

(a) $\frac{4}{5k} + \frac{11}{5k}$

Solution Since the denominators are the same, we add the numerators:

$$\frac{4}{5k} + \frac{11}{5k} = \frac{4 + 11}{5k} = \frac{15}{5k} \quad \text{Addition rule}$$

$$= \frac{3}{k}. \quad \text{Lowest terms}$$

(b) $\dfrac{2x^2 + 3x + 1}{x^5 + 1} - \dfrac{x^2 - 7x}{x^5 + 1}$

Solution The denominators are the same, so we subtract numerators, paying careful attention to parentheses:

$$\frac{2x^2 + 3x + 1}{x^5 + 1} - \frac{x^2 - 7x}{x^5 + 1} = \frac{(2x^2 + 3x + 1) - (x^2 - 7x)}{x^5 + 1} \quad \text{Subtraction rule}$$

$$= \frac{2x^2 + 3x + 1 - x^2 - (-7x)}{x^5 + 1} \quad \text{Subtract numerators.}$$

$$= \frac{2x^2 + 3x + 1 - x^2 + 7x}{x^5 + 1}$$

$$= \frac{x^2 + 10x + 1}{x^5 + 1}. \quad \text{Simplify the numerator.}$$

When fractions do not have the same denominator, you must first find a common denominator before you can add or subtract. A common denominator is a denominator that has each fraction's denominator as a factor.

Example 5 Add or subtract as indicated.

(a) $\dfrac{7}{p^2} + \dfrac{9}{2p} + \dfrac{1}{3p^2}$

Solution These three denominators are different, so we must find a common denominator that has each of p^2, $2p$, and $3p^2$ as factors. Observe that $6p^2$ satisfies these requirements. Use the cancellation property to rewrite each fraction as one that has $6p^2$ as its denominator and then add them:

$$\frac{7}{p^2} + \frac{9}{2p} + \frac{1}{3p^2} = \frac{6 \cdot 7}{6 \cdot p^2} + \frac{3p \cdot 9}{3p \cdot 2p} + \frac{2 \cdot 1}{2 \cdot 3p^2} \quad \text{Cancellation property}$$

$$= \frac{42}{6p^2} + \frac{27p}{6p^2} + \frac{2}{6p^2}$$

$$= \frac{42 + 27p + 2}{6p^2} \quad \text{Addition rule}$$

$$= \frac{27p + 44}{6p^2}. \quad \text{Simplify.}$$

(b) $\dfrac{k^2}{k^2 - 1} - \dfrac{2k^2 - k - 3}{k^2 + 3k + 2}$

Solution Factor the denominators to find a common denominator:

$$\frac{k^2}{k^2 - 1} - \frac{2k^2 - k - 3}{k^2 + 3k + 2} = \frac{k^2}{(k + 1)(k - 1)} - \frac{2k^2 - k - 3}{(k + 1)(k + 2)}$$

A common denominator here is $(k + 1)(k - 1)(k + 2)$, because each of the preceding denominators is a factor of this common denominator. Write each fraction with the common denominator:

$$\frac{k^2}{(k+1)(k-1)} - \frac{2k^2-k-3}{(k+1)(k+2)}$$

$$= \frac{k^2(k+2)}{(k+1)(k-1)(k+2)} - \frac{(2k^2-k-3)(k-1)}{(k+1)(k-1)(k+2)}$$

$$= \frac{k^3+2k^2-(2k^2-k-3)(k-1)}{(k+1)(k-1)(k+2)}$$ Subtract fractions.

$$= \frac{k^3+2k^2-(2k^3-3k^2-2k+3)}{(k+1)(k-1)(k+2)}$$ Multiply $(2k^2-k-3)(k-1)$.

$$= \frac{k^3+2k^2-2k^3+3k^2+2k-3}{(k+1)(k-1)(k+2)}$$ Polynomial subtraction

$$= \frac{-k^3+5k^2+2k-3}{(k+1)(k-1)(k+2)}.$$ Combine terms. 5✓

✓Checkpoint 5

Add or subtract.

(a) $\frac{3}{4r} + \frac{8}{3r}$

(b) $\frac{1}{m-2} - \frac{3}{2(m-2)}$

(c) $\frac{p+1}{p^2-p} - \frac{p^2-1}{p^2+p-2}$

Answers: *See page 35.*

COMPLEX FRACTIONS

Any quotient of rational expressions is called a **complex fraction**. Complex fractions are simplified as demonstrated in Example 6.

Example 6 Simplify the complex fraction

$$\frac{6-\frac{5}{k}}{1+\frac{5}{k}}.$$

Solution Multiply both numerator and denominator by the common denominator k:

$$\frac{6-\frac{5}{k}}{1+\frac{5}{k}} = \frac{k\left(6-\frac{5}{k}\right)}{k\left(1+\frac{5}{k}\right)}$$ Multiply by $\frac{k}{k}$.

$$= \frac{6k-k\left(\frac{5}{k}\right)}{k+k\left(\frac{5}{k}\right)}$$ Distributive property

$$= \frac{6k-5}{k+5}.$$ Simplify.

1.4 Exercises

Write each of the given expressions in lowest terms. Factor as necessary. (See Example 1.)

1. $\frac{8x^2}{56x}$
2. $\frac{27m}{81m^3}$
3. $\frac{25p^2}{35p^3}$
4. $\frac{18y^4}{24y^2}$
5. $\frac{5m+15}{4m+12}$
6. $\frac{10z+5}{20z+10}$
7. $\frac{4(w-3)}{(w-3)(w+6)}$
8. $\frac{-6(x+2)}{(x+4)(x+2)}$
9. $\frac{3y^2-12y}{9y^3}$
10. $\frac{15k^2+45k}{9k^2}$
11. $\frac{m^2-4m+4}{m^2+m-6}$
12. $\frac{r^2-r-6}{r^2+r-12}$
13. $\frac{x^2+2x-3}{x^2-1}$
14. $\frac{z^2+4z+4}{z^2-4}$

Multiply or divide as indicated in each of the exercises. Write all answers in lowest terms. (See Examples 2 and 3.)

15. $\frac{3a^2}{64}\cdot\frac{8}{2a^3}$
16. $\frac{2u^2}{8u^4}\cdot\frac{10u^3}{9u}$
17. $\frac{7x}{11}\div\frac{14x^3}{66y}$
18. $\frac{6x^2y}{2x}\div\frac{21xy}{y}$
19. $\frac{2a+b}{3c}\cdot\frac{15}{4(2a+b)}$
20. $\frac{4(x+2)}{w}\cdot\frac{3w^2}{8(x+2)}$
21. $\frac{15p-3}{6}\div\frac{10p-2}{3}$
22. $\frac{2k+8}{6}\div\frac{3k+12}{3}$
23. $\frac{9y-18}{6y+12}\cdot\frac{3y+6}{15y-30}$
24. $\frac{12r+24}{36r-36}\div\frac{6r+12}{8r-8}$
25. $\frac{4a+12}{2a-10}\div\frac{a^2-9}{a^2-a-20}$
26. $\frac{6r-18}{9r^2+6r-24}\cdot\frac{12r-16}{4r-12}$
27. $\frac{k^2-k-6}{k^2+k-12}\cdot\frac{k^2+3k-4}{k^2+2k-3}$
28. $\frac{n^2-n-6}{n^2-2n-8}\div\frac{n^2-9}{n^2+7n+12}$
29. In your own words, explain how to find the least common denominator of two fractions.
30. Describe the steps required to add three rational expressions. You may use an example to illustrate.

Add or subtract as indicated in each of the following. Write all answers in lowest terms. (See Example 4.)

31. $\frac{2}{7z}-\frac{1}{5z}$
32. $\frac{4}{3z}-\frac{5}{4z}$
33. $\frac{r+2}{3}-\frac{r-2}{3}$
34. $\frac{3y-1}{8}-\frac{3y+1}{8}$
35. $\frac{4}{x}+\frac{1}{5}$
36. $\frac{6}{r}-\frac{3}{4}$
37. $\frac{1}{m-1}+\frac{2}{m}$
38. $\frac{8}{y+2}-\frac{3}{y}$
39. $\frac{7}{b+2}+\frac{2}{5(b+2)}$
40. $\frac{4}{3(k+1)}+\frac{3}{k+1}$
41. $\frac{2}{5(k-2)}+\frac{5}{4(k-2)}$
42. $\frac{11}{3(p+4)}-\frac{5}{6(p+4)}$
43. $\frac{2}{x^2-4x+3}+\frac{5}{x^2-x-6}$
44. $\frac{3}{m^2-3m-10}+\frac{7}{m^2-m-20}$
45. $\frac{2y}{y^2+7y+12}-\frac{y}{y^2+5y+6}$
46. $\frac{-r}{r^2-10r+16}-\frac{3r}{r^2+2r-8}$

In each of the exercises in the next set, simplify the complex fraction. (See Example 5.)

47. $\frac{1+\frac{1}{x}}{1-\frac{1}{x}}$
48. $\frac{2-\frac{2}{y}}{2+\frac{2}{y}}$
49. $\frac{\frac{1}{x+h}-\frac{1}{x}}{h}$
50. $\frac{\frac{1}{(x+h)^2}-\frac{1}{x^2}}{h}$

Geometry *Each figure in the following exercises is a dartboard. The probability that a dart which hits the board lands in the shaded area is the fraction*

$$\frac{\text{area of the shaded region}}{\text{area of the dartboard}}.$$

(a) *Express the probability as a rational expression in x. (Hint: Area formulas are given in Appendix B.)*

(b) *Then reduce the expression to lowest terms.*

51.

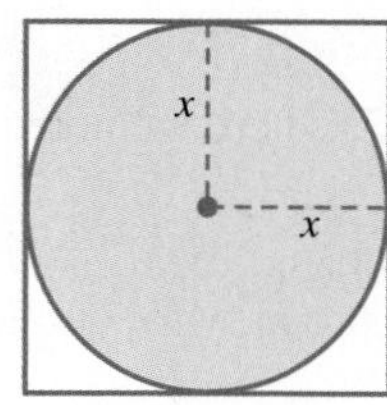

52.

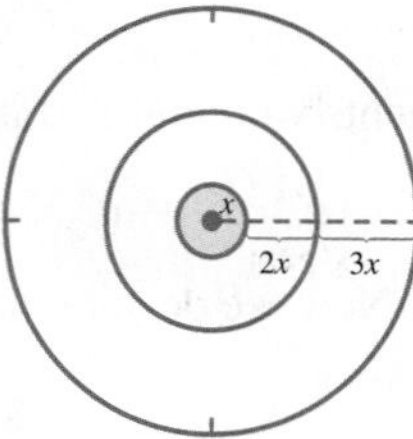

53.

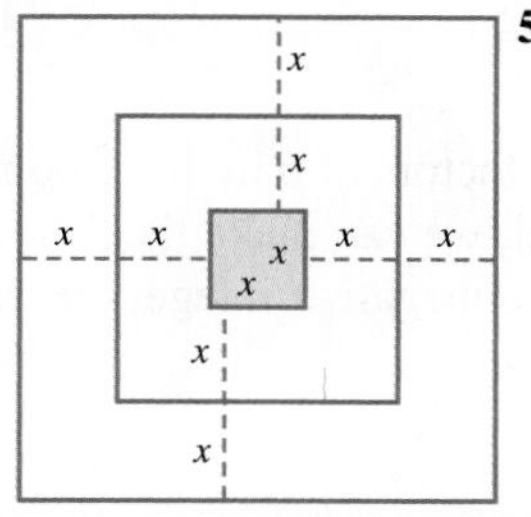

54.

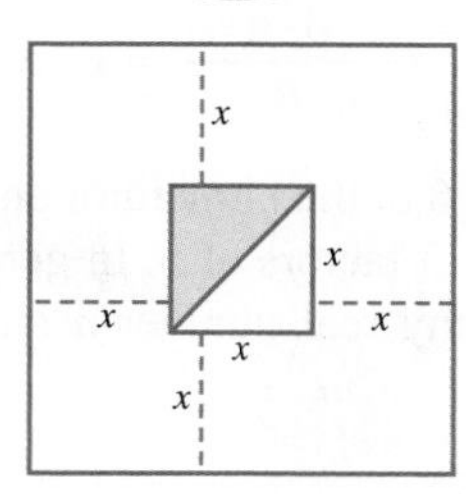

55. **Social Sciences** When 10 cars per minute, on average, are arriving at the entrance to an amusement park and the average number being admitted per minute is x (with $x > 10$), then the average waiting time in minutes for each car is given by

$$\frac{x^2 - 10x + 25}{x^3 - 15x^2 + 50x}.*$$

(a) Reduce this fraction to lowest terms.
(b) Use the reduced fraction to determine the average waiting time *in seconds* when $x = 11$, 15, and 20.

56. **Business** In Example 11 of Section 1.2, we saw that the cost C of producing x thousand DVDs is given by

$$C = -8x^2 + 3550x + 215{,}000 \qquad (x \le 200).$$

(a) Write a rational expression in lowest terms that gives the average cost per DVD when x thousand are produced. (*Hint:* The average cost is the total cost C divided by the number of DVDs produced.)
(b) Find the average cost per DVD for each of these production levels: 20,000, 50,000, and 150,000.

57. **Business** The cost (in millions of dollars) for a 30-second ad during the TV broadcast of the Superbowl can be approximated by

$$\frac{.072x^2 + .744x + 1.2}{x + 2},$$

where $x = 0$ corresponds to the year 1980.*
(a) How much did an ad cost in 2005?
(b) If this trend continues, will the cost of an ad reach \$4 million by 2020?

58. **Health** The average company cost per hour of an employee's health insurance in year x is approximated by

$$\frac{.07x^2 + 1.15x + 1.08}{x + 1},$$

where $x = 0$ corresponds to the year 1998.†
(a) What was the hourly health insurance cost in 2004?
(b) Assuming that this model remains accurate and that an employee works 2100 hours per year, what is the annual company cost of her health insurance in 2008? Will annual costs reach \$4200 by 2011?

✓ Checkpoint Answers

1. (a) 3 (b) 1/4 (c) 0
(d) Because division by 0 is undefined
2. (a) $\frac{2(k+3)}{3}$ or $\frac{2k+6}{3}$
(b) $3(1 + 2m)$ or $3 + 6m$ (c) $\frac{2p+1}{p+2}$
3. (a) $\frac{4r}{3}$ (b) $\frac{y-2}{3y}$
4. (a) $\frac{25}{8m}$ (b) $\frac{1}{5}$ (c) $\frac{5(m-3)}{m+4}$
5. (a) $\frac{41}{12r}$ (b) $\frac{-1}{2(m-2)}$ (c) $\frac{-p^3 + p^2 + 4p + 2}{p(p-1)(p+2)}$

*L. Haefner, *Introduction to Transportation Systems.* (New York: Holt, Rinehart and Winston, 1986).

*Based on data from *Ad Age* and the *St. Louis Post–Dispatch.*
†Based on data from the U.S. Department of Labor.

1.5 ▸ Exponents and Radicals

Exponents were introduced in Section 1.2. In this section, the definition of exponents will be extended to include negative exponents and rational-number exponents such as 1/2 and 7/3.

INTEGER EXPONENTS

Positive-integer and zero exponents were defined in Section 1.2, where we noted that

$$a^m \cdot a^n = a^{m+n}$$

for nonegative integers m and n. Now we develop an analogous property for quotients. By definition,

$$\frac{6^5}{6^2} = \frac{6 \cdot 6 \cdot 6 \cdot \not{6} \cdot \not{6}}{\not{6} \cdot \not{6}} = 6 \cdot 6 \cdot 6 = 6^3.$$

Because there are 5 factors of 6 in the numerator and 2 factors of 6 in the denominator, the quotient has $5 - 2 = 3$ factors of 6. In general, we can make the following statement, which applies to any real number a and nonnegative integers m and n with $m > n$.

Division with Exponents

To divide a^m by a^n, *subtract* the exponents:

$$\frac{a^m}{a^n} = a^{m-n}.$$

Example 1 Compute each of the following.

(a) $\frac{5^7}{5^4} = 5^{7-4} = 5^3.$

(b) $\frac{(-8)^{10}}{(-8)^5} = (-8)^{10-5} = (-8)^5.$

(c) $\frac{(3c)^9}{(3c)^3} = (3c)^{9-3} = (3c)^6.$ ✓1

✓Checkpoint 1

Evaluate each of the following.

(a) $\frac{2^{14}}{2^5}$

(b) $\frac{(-5)^9}{(-5)^5}$

(c) $\frac{(xy)^{17}}{(xy)^{12}}$

Answers: See page 48.

When an exponent applies to the product of two numbers, such as $(7 \cdot 19)^3$, use the definitions carefully. For instance,

$$(7 \cdot 19)^3 = (7 \cdot 19)(7 \cdot 19)(7 \cdot 19) = 7 \cdot 7 \cdot 7 \cdot 19 \cdot 19 \cdot 19 = 7^3 \cdot 19^3.$$

In other words, $(7 \cdot 19)^3 = 7^3 \cdot 19^3$. This is an example of the following fact, which applies to any real numbers a and b and any nonnegative-integer exponent n.

Product to a Power

To find $(ab)^n$, apply the exponent to *every* term inside the parentheses:

$$(ab)^n = a^n b^n.$$

CAUTION A common mistake is to write an expression such as $(2x)^5$ as $2x^5$, rather than the correct answer $(2x)^5 = 2^5x^5 = 32x^5$

Analogous conclusions are valid for quotients (where a and b are any real numbers with $b \neq 0$ and n is a nonnegative-integer exponent).

Quotient to a Power

To find $\left(\frac{a}{b}\right)^n$, apply the exponent to both numerator and denominator:

$$\left(\frac{a}{b}\right)^n = \frac{a^n}{b^n}.$$

Example 2 Compute each of the following.

(a) $(5y)^3 = 5^3y^3 = 125y^3$ Product to a power

(b) $(c^2d^3)^4 = (c^2)^4(d^3)^4$ Product to a power

$= c^8d^{12}$ Power of a power

(c) $\left(\frac{x}{2}\right)^6 = \frac{x^6}{2^6} = \frac{x^6}{64}$ Quotient to a power

(d) $\left(\frac{a^4}{b^3}\right)^3 = \frac{(a^4)^3}{(b^3)^3}$ Quotient to a power

$= \frac{a^{12}}{b^9}$ Power of a power

(e) $\left(\frac{(rs)^3}{r^4}\right)^2$

Solution Use several of the preceding properties in succession:

$\left(\frac{(rs)^3}{r^4}\right)^2 = \left(\frac{r^3s^3}{r^4}\right)^2$ Product to a power in numerator

$= \left(\frac{s^3}{r}\right)^2$ Cancel.

$= \frac{(s^3)^2}{r^2}$ Quotient to a power

$= \frac{s^6}{r^2}.$ Power of a power in numerator

As is often the case, there is another way to reach the last expression. You should be able to supply the reasons for each of the following steps:

$$\left(\frac{(rs)^3}{r^4}\right)^2 = \frac{[(rs)^3]^2}{(r^4)^2} = \frac{(rs)^6}{r^8} = \frac{r^6s^6}{r^8} = \frac{s^6}{r^2}.$$

✓2

✓Checkpoint 2

Compute each of the following.

(a) $(3x)^4$

(b) $(r^2s^5)^6$

(c) $\left(\frac{2}{z}\right)^5$

(d) $\left(\frac{3a^5}{(ab)^3}\right)^2$

Answers: See page 48.

NEGATIVE EXPONENTS

The next step is to define negative-integer exponents. If they are to be defined in such a way that the quotient rule for exponents remains valid, then we must have, for example,

$$\frac{3^2}{3^4} = 3^{2-4} = 3^{-2}.$$

However,

$$\frac{3^2}{3^4} = \frac{\cancel{3} \cdot \cancel{3}}{\cancel{3} \cdot \cancel{3} \cdot 3 \cdot 3} = \frac{1}{3^2},$$

which suggests that 3^{-2} should be defined to be $1/3^2$. Thus, we have the following definition of a negative exponent.

Negative Exponent

If n is a natural number, and if $a \neq 0$, then

$$a^{-n} = \frac{1}{a^n}.$$

Example 3 Evaluate the following.

(a) $3^{-2} = \frac{1}{3^2} = \frac{1}{9}.$

(b) $5^{-4} = \frac{1}{5^4} = \frac{1}{625}.$

(c) $x^{-1} = \frac{1}{x^1} = \frac{1}{x}.$

(d) $-4^{-2} = -\frac{1}{4^2} = -\frac{1}{16}.$

(e) $\left(\frac{3}{4}\right)^{-1} = \frac{1}{\left(\frac{3}{4}\right)^1} = \frac{1}{\frac{3}{4}} = \frac{4}{3}.$ 3✓

✓Checkpoint 3

Evaluate the following.

(a) 6^{-2}

(b) -6^{-3}

(c) -3^{-4}

(d) $\left(\frac{5}{8}\right)^{-1}$

Answers: See page 48.

There is a useful property that makes it easy to raise a fraction to a negative exponent. Consider, for example,

$$\left(\frac{2}{3}\right)^{-4} = \frac{1}{\left(\frac{2}{3}\right)^4} = \frac{1}{\left(\frac{2^4}{3^4}\right)} = 1 \cdot \frac{3^4}{2^4} = \left(\frac{3}{2}\right)^4.$$

This example is easily generalized to the following property (in which a/b is a nonzero fraction and n a positive integer).

Inversion Property

$$\left(\frac{a}{b}\right)^{-n} = \left(\frac{b}{a}\right)^n.$$

Checkpoint 4

Compute each of the following.

(a) $\left(\frac{5}{8}\right)^{-1}$

(b) $\left(\frac{1}{2}\right)^{-5}$

(c) $\left(\frac{a^2}{b}\right)^{-3}$

Answers: See page 48.

Example 4 Use the inversion property to compute each of the following.

(a) $\left(\frac{2}{5}\right)^{-3} = \left(\frac{5}{2}\right)^{3} = \frac{5^3}{2^3} = \frac{125}{8}.$

(b) $\left(\frac{3}{x}\right)^{-5} = \left(\frac{x}{3}\right)^{5} = \frac{x^5}{3^5} = \frac{x^5}{243}.$ ✓4

When keying in negative exponents on a calculator, be sure to use the negation key (labeled (−) or +/−), not the subtraction key. Calculators normally display answers as decimals, as shown in Figure 1.6. Some graphing calculators have a FRAC key that converts these decimals to fractions, as shown in Figure 1.7.

TECHNOLOGY TIP The FRAC key is in the MATH menu of TI graphing calculators. A FRAC program for other graphing calculators is in the Program Appendix. Fractions can be displayed on some graphing calculators by changing the number display format (in the MODES menu) to "fraction" or "exact."

```
3^-2
          .1111111111
5^-4
                .0016
(2/3)^-3
                3.375
```

FIGURE 1.6

```
3^-2▸Frac
                  1/9
5^-4▸Frac
                1/625
(2/3)^-3▸Frac
                 27/8
```

FIGURE 1.7

ROOTS AND RATIONAL EXPONENTS

There are two numbers whose square is 16: 4 and −4. As we saw in Section 1.1, the positive one, 4, is called the *square root* (or second root) of 16. Similarly, there are two numbers whose fourth power is 16: 2 and −2. We call 2 the **fourth root** of 16. This suggests the following generalization.

> If n is even, the ***n*th root of *a*** is the positive real number whose nth power is a.

All nonnegative numbers have nth roots for every natural number n, but *no negative number has a real, even nth root.* For example, there is no real number whose square is −16, so −16 has no square root.

We say that the **cube root** (or third root) of 8 is 2 because $2^3 = 8$. Similarly, since $(-2)^3 = -8$, we say that −2 is the cube root of −8. Again, we can make the following generalization.

> If n is odd, the ***n*th root of *a*** is the real number whose nth power is a.

Every real number has an nth root for every *odd* natural number n.

We can now define rational exponents. If they are to have the same properties as integer exponents, we want $a^{1/2}$ to be a number such that

$$(a^{1/2})^2 = a^{1/2} \cdot a^{1/2} = a^{1/2+1/2} = a^1 = a.$$

Thus, $a^{1/2}$ should be a number whose square is a, and it is reasonable to *define* $a^{1/2}$ to be the square root of a (if it exists). Similarly, $a^{1/3}$ is defined to be the cube root of a, and we have the following definition.

> If a is a real number and n is a positive integer, then
>
> $a^{1/n}$ is defined to be the nth root of a (if it exists).

Example 5 Examine the reasoning used to evaluate the following roots.

(a) $36^{1/2} = 6$ because $6^2 = 36$.

(b) $100^{1/2} = 10$ because $10^2 = 100$.

(c) $-(225^{1/2}) = -15$ because $15^2 = 225$.

(d) $625^{1/4} = 5$ because $5^4 = 625$.

(e) $(-1296)^{1/4}$ is not a real number.

(f) $-1296^{1/4} = -6$ because $6^4 = 1296$.

(g) $(-27)^{1/3} = -3$ because $(-3)^3 = -27$.

(h) $-32^{1/5} = -2$ because $2^5 = 32$.

Evaluate the following.

(a) $16^{1/2}$

(b) $16^{1/4}$

(c) $-256^{1/2}$

(d) $(-256)^{1/2}$

(e) $-8^{1/3}$

(f) $243^{1/5}$

Answers: See page 48.

A calculator can be used to evaluate expressions with fractional exponents. Whenever it is easy to do so, enter the fractional exponents in their equivalent decimal form. For instance, to find $625^{1/4}$, enter $625^{.25}$ into the calculator. When the decimal equivalent of a fraction is an infinitely repeating decimal, however, it is best to enter the fractional exponent directly, using parentheses, as in Figure 1.8. If you omit the parentheses or use a shortened decimal approximation (such as .333 for 1/3), you will not get the correct answers. Compare the incorrect answers in Figure 1.9 with the correct ones in Figure 1.8.

FIGURE 1.8

FIGURE 1.9

For other rational exponents, the symbol $a^{m/n}$ should be defined so that the properties for exponents still hold. For example, by the product property, we want

$$(a^{1/3})^2 = a^{1/3} \cdot a^{1/3} = a^{1/3+1/3} = a^{2/3}.$$

This result suggests the following definition.

For all integers m and all positive integers n, and for all real numbers a for which $a^{1/n}$ is a real number,

$$a^{m/n} = (a^{1/n})^m.$$

Example 6 Verify each of the following calculations.

(a) $27^{2/3} = (27^{1/3})^2 = 3^2 = 9.$

(b) $32^{2/5} = (32^{1/5})^2 = 2^2 = 4.$

(c) $64^{4/3} = (64^{1/3})^4 = 4^4 = 256.$

(d) $25^{3/2} = (25^{1/2})^3 = 5^3 = 125.$ 6

Checkpoint 6

Evaluate the following.

(a) $16^{3/4}$

(b) $25^{5/2}$

(c) $32^{7/5}$

(d) $100^{3/2}$

Answers: See page 48.

CAUTION When the base is negative, as in $(-8)^{2/3}$, some calculators produce an error message. On such calculators, you should first compute $(-8)^{1/3}$ and then square the result; that is, compute $[(-8)^{1/3}]^2$.

Since every terminating decimal is a rational number, decimal exponents now have a meaning. For instance, $5.24 = \frac{524}{100}$, so $3^{5.24} = 3^{524/100}$, which is easily approximated by a calculator (Figure 1.10).

Rational exponents were defined so that one of the familiar properties of exponents remains valid. In fact, it can be proved that *all* of the rules developed earlier for integer exponents are valid for rational exponents. The following box summarizes these rules, which are illustrated in Examples 7–9.

FIGURE 1.10

Properties of Exponents

For any rational numbers m and n, and for any real numbers a and b for which the following exist,

(a) $a^m \cdot a^n = a^{m+n}$ Product property

(b) $\dfrac{a^m}{a^n} = a^{m-n}$ Quotient property

(c) $(a^m)^n = a^{mn}$ Power of a power

(d) $(ab)^m = a^m \cdot b^m$ Product to a power

(e) $\left(\dfrac{a}{b}\right)^m = \dfrac{a^m}{b^m}$ Quotient to a power

(f) $a^0 = 1$ Zero exponent

(g) $a^{-n} = \dfrac{1}{a^n}$ Negative exponent

(h) $\left(\dfrac{a}{b}\right)^{-n} = \left(\dfrac{b}{a}\right)^n.$ Inversion property

The power-of-a-power property provides another way to compute $a^{m/n}$ (when it exists):

(1) $$a^{m/n} = a^{m(1/n)} = (a^m)^{1/n}.$$

For example, we can now find $4^{3/2}$ in two ways:

$$4^{3/2} = (4^{1/2})^3 = 2^3 = 8 \quad \text{or} \quad 4^{3/2} = (4^3)^{1/2} = 64^{1/2} = 8.$$

Definition of $a^{m/n}$ — Statement (1)

Example 7 Simplify each of the following expressions.

(a) $7^{-4} \cdot 7^6 = 7^{-4+6} = 7^2 = 49.$ Product property

(b) $5x^{2/3} \cdot 2x^{1/4} = 10x^{2/3}x^{1/4}$

$= 10x^{2/3+1/4}$ Product property

$= 10x^{11/12}.$ $\frac{2}{3}+\frac{1}{4}=\frac{8}{12}+\frac{3}{12}=\frac{11}{12}$

(c) $\dfrac{9^{14}}{9^{-6}} = 9^{14-(-6)} = 9^{20}.$ Quotient property

(d) $\dfrac{c^5}{2c^{4/3}} = \dfrac{1}{2} \cdot \dfrac{c^5}{c^{4/3}}$

$= \dfrac{1}{2}c^{5-4/3}$ Quotient property

$= \dfrac{1}{2}c^{11/3} = \dfrac{c^{11/3}}{2}.$ $5-\frac{4}{3}=\frac{15}{3}-\frac{4}{3}=\frac{11}{3}$

(e) $\dfrac{27^{1/3} \cdot 27^{5/3}}{27^3} = \dfrac{27^{1/3+5/3}}{27^3}$ Product property

$= \dfrac{27^2}{27^3} = 27^{2-3}$ Quotient property

$= 27^{-1} = \dfrac{1}{27}.$ Definition of negative exponent 7✓

✓Checkpoint 7

Simplify each of the following.

(a) $9^7 \cdot 9^{-5}$

(b) $3x^{1/4} \cdot 5x^{5/4}$

(c) $\dfrac{8^7}{8^{-3}}$

(d) $\dfrac{5^{2/3} \cdot 5^{-4/3}}{5^2}$

Answers: See page 48.

FIGURE 1.11

You can use a calculator to check numerical computations, such as those in Example 7, by computing the left and right sides separately and confirming that the answers are the same in each case. Figure 1.11 shows this technique for part (e) of Example 7.

Example 8 Perform the indicated operations.

(a) $(2^{-3})^{-4/7} = 2^{(-3)(-4/7)} = 2^{12/7}.$ Power of a power

(b) $\left(\dfrac{3m^{5/6}}{y^{3/4}}\right)^2 = \dfrac{(3m^{5/6})^2}{(y^{3/4})^2}$ Quotient to a power

$= \dfrac{3^2(m^{5/6})^2}{(y^{3/4})^2}$ Product to a power

$$= \frac{9m^{(5/6)2}}{y^{(3/4)2}}$$ Power of a power

$$= \frac{9m^{5/3}}{y^{3/2}}.$$ $\frac{5}{6}\cdot 2 = \frac{10}{6} = \frac{5}{3}$ and $\frac{3}{4}\cdot 2 = \frac{3}{2}$

(c) $m^{2/3}(m^{7/3} + 2m^{1/3}) = m^{2/3}m^{7/3} + m^{2/3}2m^{1/3}$ Distributive property

$= m^{2/3+7/3} + 2m^{2/3+1/3} = m^3 + 2m.$ Product rule 8✓

✓Checkpoint 8

Simplify each of the following.

(a) $(7^{-4})^{-2} \cdot (7^4)^{-2}$

(b) $\frac{c^4c^{-1/2}}{c^{3/2}d^{1/2}}$

(c) $a^{5/8}(2a^{3/8} + a^{-1/8})$

Answers: See page 48.

Example 9 Simplify each expression in parts (a)–(c). Give answers with only positive exponents.

(a) $\frac{(m^3)^{-2}}{m^4} = \frac{m^{-6}}{m^4} = m^{-6-4} = m^{-10} = \frac{1}{m^{10}}.$

(b) $6y^{2/3} \cdot 2y^{-1/2} = 12y^{2/3-1/2} = 12y^{1/6}.$

(c) $\frac{x^{1/2}(x-2)^{-3}}{5(x-2)} = \frac{x^{1/2}}{5}\cdot\frac{(x-2)^{-3}}{x-2} = \frac{x^{1/2}}{5}\cdot(x-2)^{-3-1}$

$$= \frac{x^{1/2}}{5}\cdot\frac{1}{(x-2)^4} = \frac{x^{1/2}}{5(x-2)^4}.$$

(d) Write $a^{-1} + b^{-1}$ as a single quotient.

Solution Be careful here. $a^{-1} + b^{-1}$ does *not* equal $(a+b)^{-1}$; the exponent properties deal only with products and quotients, not with sums. However, using the definition of negative exponents and addition of fractions, we have

$$a^{-1} + b^{-1} = \frac{1}{a} + \frac{1}{b} = \frac{b+a}{ab}.$$ 9✓

✓Checkpoint 9

Simplify the given expressions. Give answers with only positive exponents.

(a) $(3x^{2/3})(2x^{-1})(y^{-1/3})^2$

(b) $\frac{(t^{-1})^2}{t^{-5}}$

(c) $\left(\frac{2k^{1/3}}{p^{5/4}}\right)^2 \cdot \left(\frac{4k^{-2}}{p^5}\right)^{3/2}$

(d) $x^{-1} - y^{-2}$

Answers: See page 48.

RADICALS

Earlier, we denoted the *n*th root of *a* as $a^{1/n}$. An alternative notation for *n*th roots uses the radical symbol $\sqrt[n]{\ }$.

> If *n* is an even natural number and $a \geq 0$, or if *n* is an odd natural number,
>
> $$\sqrt[n]{a} = a^{1/n}.$$

In the radical expression $\sqrt[n]{a}$, *a* is called the *radicand* and *n* is called the *index*. When $n = 2$, the familiar square-root symbol $\sqrt{a}$ is used instead of $\sqrt[2]{a}$.

✓Checkpoint 10

Simplify.

(a) $\sqrt[3]{27}$

(b) $\sqrt[4]{625}$

(c) $\sqrt[6]{64}$

(d) $\sqrt[3]{\frac{64}{125}}$

Answers: See page 48.

Example 10 Simplify the following radicals.

(a) $\sqrt[4]{16} = 16^{1/4} = 2.$

(b) $\sqrt[5]{-32} = -2.$

(c) $\sqrt[3]{1000} = 10.$

(d) $\sqrt[6]{\frac{64}{729}} = \left(\frac{64}{729}\right)^{1/6} = \frac{64^{1/6}}{729^{1/6}} = \frac{2}{3}.$ 10✓

Recall that $a^{m/n} = (a^{1/n})^m$ by definition and $a^{m/n} = (a^m)^{1/n}$ by statement **(1)** on page 42 (provided that all terms are defined). We translate these facts into radical notation as follows.

For all rational numbers m/n and all real numbers a for which $\sqrt[n]{a}$ exists,

$$a^{m/n} = (\sqrt[n]{a})^m \quad \text{or} \quad a^{m/n} = \sqrt[n]{a^m}.$$

Notice that $\sqrt[n]{x^n}$ cannot be written simply as x when n is even. For example, if $x = -5$, then

$$\sqrt{x^2} = \sqrt{(-5)^2} = \sqrt{25} = 5 \neq x.$$

However, $|-5| = 5$, so that $\sqrt{x^2} = |x|$ when x is -5. This relationship is true in general.

For any real number a and any natural number n,

$$\sqrt[n]{a^n} = |a| \qquad \textbf{if } n \textbf{ is even}$$

and

$$\sqrt[n]{a^n} = a \qquad \textbf{if } n \textbf{ is odd.}$$

To avoid the difficulty that $\sqrt[n]{a^n}$ is not necessarily equal to a, we shall assume that all variables in radicands represent only nonnegative numbers, as they usually do in applications.

The properties of exponents can be written with radicals as follows.

For all real numbers a and b, and for positive integers n for which all indicated roots exist,

(a) $\sqrt[n]{a} \cdot \sqrt[n]{b} = \sqrt[n]{ab}$ and

(b) $\dfrac{\sqrt[n]{a}}{\sqrt[n]{b}} = \sqrt[n]{\dfrac{a}{b}} \quad (b \neq 0).$

Example 11 Simplify the following expressions.

(a) $\sqrt{6} \cdot \sqrt{54} = \sqrt{6 \cdot 54} = \sqrt{324} = 18.$

Alternatively, simplify $\sqrt{54}$ first:

$$\begin{aligned}\sqrt{6} \cdot \sqrt{54} &= \sqrt{6} \cdot \sqrt{9 \cdot 6} \\ &= \sqrt{6} \cdot 3\sqrt{6} = 3 \cdot 6 = 18.\end{aligned}$$

(b) $\sqrt{\dfrac{7}{64}} = \dfrac{\sqrt{7}}{\sqrt{64}} = \dfrac{\sqrt{7}}{8}.$

(c) $\sqrt{75} - \sqrt{12}$

Solution Note that $12 = 4 \cdot 3$ and that 4 is a perfect square. Similarly, $75 = 25 \cdot 3$ and 25 is a perfect square. Consequently,

$$\begin{aligned} \sqrt{75} - \sqrt{12} &= \sqrt{25 \cdot 3} - \sqrt{4 \cdot 3} && \text{Factor.} \\ &= \sqrt{25}\sqrt{3} - \sqrt{4}\sqrt{3} && \text{Property (a)} \\ &= 5\sqrt{3} - 2\sqrt{3} = 3\sqrt{3}. && \text{Simplify.} \end{aligned}$$

11 ✓

✓Checkpoint 11

Simplify.

(a) $\sqrt{3} \cdot \sqrt{27}$

(b) $\sqrt{\dfrac{3}{49}}$

(c) $\sqrt{50} + \sqrt{72}$

Answers: See page 48.

CAUTION When a and b are nonzero real numbers,

$\sqrt[n]{a+b}$ is **NOT** equal to $\sqrt[n]{a} + \sqrt[n]{b}$.

For example,

$$\sqrt{9+16} = \sqrt{25} = 5, \text{ but } \sqrt{9} + \sqrt{16} = 3 + 4 = 7,$$

so $\sqrt{9+16} \neq \sqrt{9} + \sqrt{16}$.

Multiplying radical expressions is much like multiplying polynomials.

Example 12 Perform the following multiplications.

(a) $$\begin{aligned} (\sqrt{2} + 3)(\sqrt{8} - 5) &= \sqrt{2}(\sqrt{8}) - \sqrt{2}(5) + 3\sqrt{8} - 3(5) && \text{FOIL} \\ &= \sqrt{16} - 5\sqrt{2} + 3(2\sqrt{2}) - 15 \\ &= 4 - 5\sqrt{2} + 6\sqrt{2} - 15 \\ &= -11 + \sqrt{2}. \end{aligned}$$

(b) $$\begin{aligned} (\sqrt{7} - \sqrt{10})(\sqrt{7} + \sqrt{10}) &= (\sqrt{7})^2 - (\sqrt{10})^2 \\ &= 7 - 10 = -3. \end{aligned}$$

12 ✓

✓Checkpoint 12

Multiply.

(a) $(\sqrt{5} - \sqrt{2})(3 + \sqrt{2})$

(b) $(\sqrt{3} + \sqrt{7})(\sqrt{3} - \sqrt{7})$

Answers: See page 48.

RATIONALIZING DENOMINATORS AND NUMERATORS

Before the invention of calculators, it was customary to **rationalize the denominators** of fractions (that is, write equivalent fractions with no radicals in the denominator), because this made many computations easier. Although there is no longer a computational reason to do so, rationalization of denominators (and sometimes numerators) is still used today to simplify expressions and to derive useful formulas.

Example 13 Rationalize each denominator.

(a) $\dfrac{4}{\sqrt{3}}$

Solution The key is to multiply by 1, with 1 written as a radical fraction:

$$\frac{4}{\sqrt{3}} = \frac{4}{\sqrt{3}} \cdot 1 = \frac{4}{\sqrt{3}} \cdot \frac{\sqrt{3}}{\sqrt{3}} = \frac{4\sqrt{3}}{3}.$$

(b) $\dfrac{1}{3-\sqrt{2}}$

Solution The same technique works here, using $1 = \dfrac{3+\sqrt{2}}{3+\sqrt{2}}$:

$$\frac{1}{3-\sqrt{2}} = \frac{1}{3-\sqrt{2}}\cdot 1 = \frac{1}{3-\sqrt{2}}\cdot\frac{3+\sqrt{2}}{3+\sqrt{2}} = \frac{3+\sqrt{2}}{(3-\sqrt{2})(3+\sqrt{2})}$$
$$= \frac{3+\sqrt{2}}{9-2} = \frac{3+\sqrt{2}}{7}.$$

13✓

✓Checkpoint 13

Rationalize the denominator.

(a) $\dfrac{2}{\sqrt{5}}$

(b) $\dfrac{1}{2+\sqrt{3}}$

Answers: See page 48.

Example 14 Rationalize the numerator of $\dfrac{2+\sqrt{5}}{1+\sqrt{3}}$.

Solution As in Example 13 (b), we must write 1 as a suitable fraction. Since we want to rationalize the numerator here, we multiply by the fraction $1 = \dfrac{2-\sqrt{5}}{2-\sqrt{5}}$:

$$\frac{2+\sqrt{5}}{1+\sqrt{3}} = \frac{2+\sqrt{5}}{1+\sqrt{3}}\cdot\frac{2-\sqrt{5}}{2-\sqrt{5}} = \frac{4-5}{2-\sqrt{5}+2\sqrt{3}-\sqrt{3}\sqrt{5}}$$
$$= \frac{-1}{2-\sqrt{5}+2\sqrt{3}-\sqrt{15}}.$$

1.5 ▶ Exercises

Perform the indicated operations and simplify your answer. (See Examples 1 and 2.)

1. $\dfrac{7^5}{7^3}$
2. $\dfrac{(-6)^{14}}{(-6)^6}$
3. $(4c)^2$
4. $(-2x)^4$
5. $\left(\dfrac{2}{x}\right)^5$
6. $\left(\dfrac{5}{xy}\right)^3$
7. $(3u^2)^3(2u^3)^2$
8. $\dfrac{(5v^2)^3}{(2v)^4}$

Perform the indicated operations and simplify your answer, which should not have any negative exponents. (See Examples 3 and 4.)

9. 7^{-1}
10. 10^{-3}
11. 2^{-5}
12. -5^{-3}
13. -6^{-5}
14. $(-x)^{-4}$
15. $(-y)^{-3}$
16. $\left(\dfrac{1}{6}\right)^{-2}$
17. $\left(\dfrac{1}{7}\right)^{-3}$
18. $\left(\dfrac{3}{4}\right)^{-5}$
19. $\left(\dfrac{4}{3}\right)^{-2}$
20. $\left(\dfrac{x}{y^2}\right)^{-2}$
21. $\left(\dfrac{a}{b^3}\right)^{-1}$
22. Explain why $-2^{-4} = -1/16$, but $(-2)^{-4} = 1/16$.

Evaluate each expression. Write all answers without exponents. Round decimal answers to two places. (See Examples 5 and 6.)

23. $49^{1/2}$
24. $8^{1/3}$
25. $(5.71)^{1/4}$
26. $(93.68)^{1/5}$
27. $27^{2/3}$
28. $12^{5/2}$
29. $-64^{2/3}$
30. $-64^{3/2}$
31. $(8/27)^{-4/3}$
32. $(27/64)^{-1/3}$

Simplify each expression. Write all answers using only positive exponents. (See Example 7.)

33. $\dfrac{5^{-3}}{4^{-2}}$
34. $\dfrac{7^{-4}}{7^{-3}}$
35. $4^{-3}\cdot 4^6$
36. $9^{-9}\cdot 9^{10}$
37. $8^{2/3}\cdot 8^{-1/3}$
38. $12^{-3/4}\cdot 12^{1/4}$
39. $\dfrac{4^{10}\cdot 4^{-6}}{4^{-4}}$
40. $\dfrac{5^{-4}\cdot 5^6}{5^{-1}}$
41. $\dfrac{9^{-5/3}}{9^{2/3}\cdot 9^{-1/5}}$
42. $\dfrac{3^{5/3}\cdot 3^{-3/4}}{3^{-1/4}}$

Simplify each expression. Assume all variables represent positive real numbers. Write answers with only positive exponents. (See Examples 8 and 9.)

43. $\frac{z^6 \cdot z^2}{z^5}$ **44.** $\frac{k^6 \cdot k^9}{k^{12}}$

45. $\frac{3^{-1}(p^{-2})^3}{3p^{-7}}$ **46.** $\frac{(5x^3)^{-2}}{x^4}$

47. $(q^{-5}r^3)^{-1}$ **48.** $(2y^2z^{-2})^{-3}$

49. $(2p^{-1})^3 \cdot (5p^2)^{-2}$ **50.** $(4^{-1}x^3)^{-2} \cdot (3x^{-3})^4$

51. $(2p)^{1/2} \cdot (2p^3)^{1/3}$ **52.** $(5k^2)^{3/2} \cdot (5k^{1/3})^{3/4}$

53. $p^{2/3}(2p^{1/3} + 5p)$ **54.** $3x^{3/2}(2x^{-3/2} + x^{3/2})$

55. $\frac{(x^2)^{1/3}(y^2)^{2/3}}{3x^{2/3}y^2}$ **56.** $\frac{(c^{1/2})^3(d^3)^{1/2}}{(c^3)^{1/4}(d^{1/4})^3}$

57. $\frac{(7a)^2(5b)^{3/2}}{(5a)^{3/2}(7b)^4}$ **58.** $\frac{(4x)^{1/2}\sqrt{xy}}{x^{3/2}y^2}$

59. $x^{1/2}(x^{2/3} - x^{4/3})$ **60.** $x^{1/2}(3x^{3/2} + 2x^{-1/2})$

61. $(x^{1/2} + y^{1/2})(x^{1/2} - y^{1/2})$

62. $(x^{1/3} + y^{1/2})(2x^{1/3} - y^{3/2})$

Match the rational-exponent expression in Column I with the equivalent radical expression in Column II. Assume that x is not zero.

I	II
63. $(-3x)^{1/3}$	**(a)** $\frac{3}{\sqrt[3]{x}}$
64. $-3x^{1/3}$	**(b)** $-3\sqrt[3]{x}$
65. $(-3x)^{-1/3}$	**(c)** $\frac{1}{\sqrt[3]{3x}}$
66. $-3x^{-1/3}$	**(d)** $\frac{-3}{\sqrt[3]{x}}$
67. $(3x)^{1/3}$	**(e)** $3\sqrt[3]{x}$
68. $3x^{-1/3}$	**(f)** $\sqrt[3]{-3x}$
69. $(3x)^{-1/3}$	**(g)** $\sqrt[3]{3x}$
70. $3x^{1/3}$	**(h)** $\frac{1}{\sqrt[3]{-3x}}$

Simplify each of the given radical expressions. (See Examples 10–12.)

71. $\sqrt[3]{125}$ **72.** $\sqrt[6]{64}$ **73.** $\sqrt[4]{625}$

74. $\sqrt[5]{-243}$ **75.** $\sqrt[7]{-128}$ **76.** $\sqrt{63}\sqrt{7}$

77. $\sqrt[3]{81} \cdot \sqrt[3]{9}$ **78.** $\sqrt{49 - 16}$ **79.** $\sqrt{81 - 4}$

80. $\sqrt{49} - \sqrt{16}$ **81.** $\sqrt{81} - \sqrt{4}$

82. $\sqrt{5}\sqrt{15}$ **83.** $\sqrt{8}\sqrt{96}$

84. $\sqrt{50} - \sqrt{72}$ **85.** $\sqrt{75} + \sqrt{192}$

86. $5\sqrt{20} - \sqrt{45} + 2\sqrt{80}$

87. $(\sqrt{3} + 2)(\sqrt{3} - 2)$

88. $(\sqrt{5} + \sqrt{2})(\sqrt{5} - \sqrt{2})$

89. $(\sqrt{3} + 4)(\sqrt{5} - 4)$

90. What is wrong with the statement $\sqrt[3]{4} \cdot \sqrt[3]{4} = 4$?

Rationalize the denominator of each of the given expressions. (See Example 13.)

91. $\frac{3}{1 - \sqrt{2}}$ **92.** $\frac{2}{1 + \sqrt{5}}$

93. $\frac{9 - \sqrt{3}}{3 - \sqrt{3}}$ **94.** $\frac{\sqrt{3} - 1}{\sqrt{3} - 2}$

Rationalize the numerator of each of the given expressions. (See Example 14.)

95. $\frac{3 - \sqrt{2}}{3 + \sqrt{2}}$ **96.** $\frac{1 + \sqrt{7}}{2 - \sqrt{3}}$

The following exercises are applications of exponentiation and radicals.

97. Business The theory of economic lot size shows that, under certain conditions, the number of units to order to minimize total cost is

$$x = \sqrt{\frac{kM}{f}},$$

where k is the cost to store one unit for one year, f is the (constant) setup cost to manufacture the product, and M is the total number of units produced annually. Find x for the following values of f, k, and M.

(a) $k = \$1, f = \$500, M = 100{,}000$
(b) $k = \$3, f = \$7, M = 16{,}700$
(c) $k = \$1, f = \$5, M = 16{,}800$

98. Health The threshold weight T for a person is the weight above which the risk of death increases greatly. One researcher found that the threshold weight in pounds for men aged 40–49 is related to height in inches by the equation $h = 12.3T^{1/3}$. What height corresponds to a threshold of 216 pounds for a man in this age group?

Business *According to data in the* New York Times *the domestic sales of DVDs and videotapes (in billions of dollars) can be approximated by*

$$11.68x^{.1954} \quad (x \geq 3),$$

where $x = 3$ corresponds to the year 2003. Assuming the model remains accurate, approximate the sales in the following years.

99. 2005 **100.** 2007 **101.** 2011 **102.** 2013

Health *The number of kidney transplants in year x can be approximated by*

$$5274.7x^{.4159} \quad (x \geq 7),$$

where $x = 7$ corresponds to the year 1997. Find the approximate number of kidney transplants in each of the following years.*

103. 2001 **104.** 2003 **105.** 2006 **106.** 2010

107. In late 2008, there were approximately 76,300 people waiting for a kidney transplant. About how many of these people did *not* get a kidney transplant?

Health *The average life expectancy of someone of age x is approximately*

$$\frac{92.7}{1 + .237 \times 10^{.0189x}}.†$$

Find the life expectancy of a person whose age is

108. 17 **109.** 22 **110.** 35 **111.** Your age

*Based on data from the United Network for Organ Sharing.

†Based on data from the U.S. National Center for Health Statistics.

✓ Checkpoint Answers

1. (a) 2^9 (b) $(-5)^4$ (c) $(xy)^5$
2. (a) $81x^4$ (b) $r^{12}s^{30}$ (c) $\frac{32}{z^5}$ (d) $\frac{9a^4}{b^6}$
3. (a) 1/36 (b) −1/216 (c) −1/81 (d) 8/5
4. (a) 8/5 (b) 32 (c) b^3/a^6
5. (a) 4 (b) 2 (c) −16 (d) Not a real number (e) −2 (f) 3
6. (a) 8 (b) 3125 (c) 128 (d) 1000
7. (a) 81 (b) $15x^{3/2}$ (c) 8^{10} (d) $5^{-8/3}$ or $1/5^{8/3}$
8. (a) 1 (b) $c^2/d^{1/2}$ (c) $2a + a^{1/2}$
9. (a) $\frac{6}{x^{1/3}y^{2/3}}$ (b) t^3 (c) $32/(p^{10}k^{7/3})$ (d) $\frac{y^2 - x}{xy^2}$
10. (a) 3 (b) 5 (c) 2 (d) 4/5
11. (a) 9 (b) $\frac{\sqrt{3}}{7}$ (c) $11\sqrt{2}$
12. (a) $3\sqrt{5} + \sqrt{10} - 3\sqrt{2} - 2$ (b) −4
13. (a) $\frac{2\sqrt{5}}{5}$ (b) $2 - \sqrt{3}$

1.6 ▶ First-Degree Equations

An **equation** is a statement that two mathematical expressions are equal; for example,

$$5x - 3 = 13, \qquad 8y = 4, \qquad \text{and} \qquad -3p + 5 = 4p - 8$$

are equations.

The letter in each equation is called the variable. This section concentrates on **first-degree equations**, which are equations that involve only constants and the first power of the variable. All of the equations displayed above are first-degree equations, but neither of the following equations is of first degree:

$$2x^2 = 5x + 6 \quad \text{(the variable has an exponent greater than 1);}$$
$$\sqrt{x + 2} = 4 \quad \text{(the variable is under the radical).}$$

A **solution** of an equation is a number that can be substituted for the variable in the equation to produce a true statement. For example, substituting the number 9 for x in the equation $2x + 1 = 19$ gives

$$2x + 1 = 19$$
$$2(9) + 1 \stackrel{?}{=} 19 \quad \text{Let } x = 9.$$
$$18 + 1 = 19. \quad \text{True}$$

This true statement indicates that 9 is a solution of $2x + 1 = 19$. ✓1

✓ Checkpoint 1

Is −4 a solution of the equations in parts (a) and (b)?

(a) $3x + 5 = -7$

(b) $2x - 3 = 5$

(c) Is there more than one solution of the equation in part (a)?

Answers: See page 59.

The following properties are used to solve equations.

Properties of Equality

1. The same number may be added to or subtracted from both sides of an equation:

   If $a = b$, then $a + c = b + c$ and $a - c = b - c$.

2. Both sides of an equation may be multiplied or divided by the same nonzero number:

   If $a = b$ and $c \neq 0$, then $ac = bc$ and $\frac{a}{c} = \frac{b}{c}$.

Example 1 Solve the equation $5x - 3 = 12$.

Solution Using the first property of equality, add 3 to both sides. This isolates the term containing the variable on one side of the equation:

$$5x - 3 = 12$$
$$5x - 3 + 3 = 12 + 3 \quad \text{Add 3 to both sides.}$$
$$5x = 15.$$

Now arrange for the coefficient of x to be 1 by using the second property of equality:

$$5x = 15$$
$$\frac{5x}{5} = \frac{15}{5} \quad \text{Divide both sides by 5.}$$
$$x = 3.$$

The solution of the original equation, $5x - 3 = 12$, is 3. Check the solution by substituting 3 for x in the original equation. 2✓

✓Checkpoint 2

Solve the following.

(a) $3p - 5 = 19$

(b) $4y + 3 = -5$

(c) $-2k + 6 = 2$

Answers: See page 59.

Example 2 Solve $2k + 3(k - 4) = 2(k - 3)$.

Solution First, simplify the equation by using the distributive property on the left-side term $3(k - 4)$ and right-side term $2(k - 3)$:

$$2k + 3(k - 4) = 2(k - 3)$$
$$2k + 3k - 12 = 2(k - 3) \quad \text{Distributive property}$$
$$2k + 3k - 12 = 2k - 6 \quad \text{Distributive property}$$
$$5k - 12 = 2k - 6. \quad \text{Collect like terms on left side.}$$

One way to proceed is to add $-2k$ to both sides:

$$5k - 12 + (-2k) = 2k - 6 + (-2k) \quad \text{Add } -2k \text{ to both sides.}$$
$$3k - 12 = -6$$
$$3k - 12 + 12 = -6 + 12 \quad \text{Add 12 to both sides.}$$
$$3k = 6$$

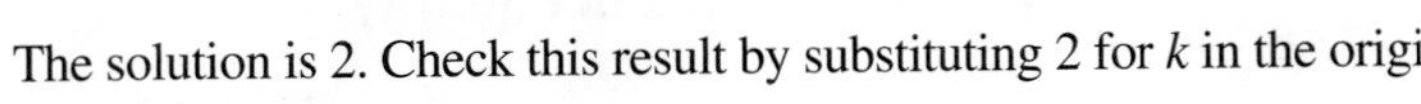

$$\frac{1}{3}(3k) = \frac{1}{3}(6) \qquad \text{Multiply both sides by } \frac{1}{3}.$$

$$k = 2.$$

The solution is 2. Check this result by substituting 2 for k in the original equation. [3] ✓

✓Checkpoint 3

Solve the following.

(a) $3(m - 6) + 2(m + 4) = 4m - 2$

(b) $-2(y + 3) + 4y = 3(y + 1) - 6$

Answers: See page 59.

Example 3 **Business** The percentage y of workers in private industry who participate in a "defined contribution" retirement plan (such as a 401(k)) in year x is approximated by the equation

$$.08(x - 1992) = 14y - 4.9.^*$$

Assuming this equation remains valid, use a calculator to determine when 51% of workers will participate in such a plan.

Solution Since $51\% = .51$, let $y = .51$ in the equation and solve for x. To avoid any rounding errors in the intermediate steps, it is often a good idea to do all the algebra first, before using the calculator:

$$\begin{aligned}
.08(x - 1992) &= 14y - 4.9 \\
.08(x - 1992) &= 14 \cdot .51 - 4.9 && \text{Substitute } y = .51. \\
.08x - .08 \cdot 1992 &= 14 \cdot .51 - 4.9 && \text{Distributive property} \\
.08x &= 14 \cdot .51 - 4.9 + .08 \cdot 1992 && \text{Add (.08) 1992 to both sides.} \\
x &= \frac{14 \cdot .51 - 4.9 + .08 \cdot 1992}{.08}. && \text{Divide both sides by .08.}
\end{aligned}$$

Now use a calculator (inserting parentheses around the numerator) to determine that $x = 2020$, as shown in Figure 1.12. So, 51% of workers will participate in a defined contribution plan in 2020. [4] ✓

✓Checkpoint 4

In Example 3, when were 40% of workers covered by a defined contribution plan?

Answer: See page 59.

```
(14*.51-4.9+.08*
1992)/.08
                2020
```

FIGURE 1.12

The next three examples show how to simplify the solution of first-degree equations involving fractions. We solve these equations by multiplying both sides of the equation by a *common denominator.* This step will eliminate the fractions.

*Based on data from the U.S. Bureau of Labor Statistics.

Example 4 Solve $\frac{r}{10} - \frac{2}{15} = \frac{3r}{20} - \frac{1}{5}$.

Solution Here, the denominators are 10, 15, 20, and 5. Each of these numbers can be divided into 60; therefore, 60 is a common denominator. Multiply both sides of the equation by 60:

$$60\left(\frac{r}{10} - \frac{2}{15}\right) = 60\left(\frac{3r}{20} - \frac{1}{5}\right)$$

$$60\left(\frac{r}{10}\right) - 60\left(\frac{2}{15}\right) = 60\left(\frac{3r}{20}\right) - 60\left(\frac{1}{5}\right) \quad \text{Distributive property}$$

$$6r - 8 = 9r - 12$$

$$6r - 8 + (-6r) + 12 = 9r - 12 + (-6r) + 12 \quad \text{Add } -6r \text{ and 12 to both sides.}$$

$$4 = 3r$$

$$r = \frac{4}{3}. \quad \text{Multiply both sides by 1/3.}$$

Check this solution in the original equation. 5 ✓

✓Checkpoint 5

Solve the following.

(a) $\frac{x}{2} - \frac{x}{4} = 6$

(b) $\frac{2x}{3} + \frac{1}{2} = \frac{x}{4} - \frac{9}{2}$

Answers: See page 59.

CAUTION Multiplying *both* sides of an *equation* by a number to eliminate fractions is valid. But multiplying a single fraction by a number to simplify it is not valid. For instance, multiplying $\frac{3x}{8}$ by 8 *changes* it to $3x$, which is *not equal* to $\frac{3x}{8}$.

The second property of equality (page 49) applies only to *nonzero* quantities. Multiplying or dividing both sides of an equation by a quantity involving the variable (which might be zero for some values) may lead to an **extraneous solution**—that is, a number that does not satisfy the original equation. To avoid errors in such situations, always *check your solutions in the original equation.*

Example 5 Solve $\frac{4}{3(k+2)} - \frac{k}{3(k+2)} = \frac{5}{3}$.

Solution Multiply both sides of the equation by the common denominator $3(k + 2)$. Here, $k \neq -2$, since $k = -2$ would give a 0 denominator, making the fraction undefined. So, we have

$$3(k+2)\cdot\frac{4}{3(k+2)} - 3(k+2)\cdot\frac{k}{3(k+2)} = 3(k+2)\cdot\frac{5}{3}.$$

Simplify each side and solve for k:

$$4 - k = 5(k + 2)$$

$$4 - k = 5k + 10 \quad \text{Distributive property}$$

$$4 - k + k = 5k + 10 + k \quad \text{Add } k \text{ to both sides.}$$

$$4 = 6k + 10$$

$$4 + (-10) = 6k + 10 + (-10) \quad \text{Add } -10 \text{ to both sides.}$$

$$-6 = 6k$$

$$-1 = k. \quad \text{Multiply both sides by } \frac{1}{6}.$$

The solution is -1. Substitute -1 for k in the original equation as a check:

$$\frac{4}{3(-1+2)} - \frac{-1}{3(-1+2)} \stackrel{?}{=} \frac{5}{3}$$

$$\frac{4}{3} - \frac{-1}{3} \stackrel{?}{=} \frac{5}{3}$$

$$\frac{5}{3} = \frac{5}{3}.$$

The check shows that -1 is the solution. 6✓

✓Checkpoint 6

Solve the equation

$$\frac{5p+1}{3(p+1)} = \frac{3p-3}{3(p+1)} + \frac{9p-3}{3(p+1)}.$$

Answer: See page 59.

Example 6 Solve $\dfrac{3x-4}{x-2} = \dfrac{x}{x-2}$.

Solution Multiplying both sides by $x - 2$ produces

$$3x - 4 = x$$

$2x - 4 = 0$ Subtract x from both sides.

$2x = 4$ Add 4 to both sides.

$x = 2.$ Divide both sides by 2.

Substituting 2 for x in the original equation produces fractions with 0 denominators. Since division by 0 is not defined, $x = 2$ is an extraneous solution. So the original equation has no solution. 7✓

✓Checkpoint 7

Solve each equation.

(a) $\dfrac{3p}{p+1} = 1 - \dfrac{3}{p+1}$

(b) $\dfrac{8y}{y-4} = \dfrac{32}{y-4} - 3$

Answers: See page 59.

Sometimes an equation with several variables must be solved for one of the variables. This process is called **solving for a specified variable**.

Example 7 Solve for x: $3(ax - 5a) + 4b = 4x - 2$.

Solution Use the distributive property to get

$$3ax - 15a + 4b = 4x - 2.$$

Treat x as the variable and the other letters as constants. Get all terms with x on one side of the equation and all terms without x on the other side:

$3ax - 4x = 15a - 4b - 2$ Isolate terms with x on the left.

$(3a - 4)x = 15a - 4b - 2$ Distributive property

$x = \dfrac{15a - 4b - 2}{3a - 4}$ Multiply both sides by $\dfrac{1}{3a-4}$.

The final equation is solved for x, as required. 8✓

✓Checkpoint 8

Solve for x.

(a) $2x - 7y = 3xk$

(b) $8(4 - x) + 6p = -5k - 11yx$

Answers: See page 59.

ABSOLUTE-VALUE EQUATIONS

Recall from Section 1.1 that the absolute value of a number a is either a or $-a$, whichever one is positive. For instance, $|4| = 4$ and $|-7| = -(-7) = 7$.

Example 8 Solve $|x| = 3$.

Solution Since $|x|$ is either x or $-x$, the equation says that

$$x = 3 \quad \text{or} \quad -x = 3$$
$$x = -3.$$

The solutions of $|x| = 3$ are 3 and -3.

Example 9 Solve $|p - 4| = 2$.

Solution Since $|p - 4|$ is either $p - 4$ or $-(p - 4)$, we have

$$p - 4 = 2 \quad \text{or} \quad -(p - 4) = 2$$
$$p = 6 \qquad -p + 4 = 2$$
$$-p = -2$$
$$p = 2,$$

so that 6 and 2 are possible solutions. Checking them in the original equation shows that both are solutions. 9✓

✓Checkpoint 9

Solve each equation.

(a) $|y| = 9$

(b) $|r + 3| = 1$

(c) $|2k - 3| = 7$

Answers: See page 59.

Example 10 Solve $|4m - 3| = |m + 6|$.

Solution To satisfy the equation, the quantities in absolute-value bars must either be equal or be negatives of one another. That is,

$$4m - 3 = m + 6 \quad \text{or} \quad 4m - 3 = -(m + 6)$$
$$3m = 9 \qquad 4m - 3 = -m - 6$$
$$m = 3 \qquad 5m = -3$$
$$m = -\frac{3}{5}.$$

Check that the solutions for the original equation are 3 and $-3/5$. 10✓

✓Checkpoint 10

Solve each equation.

(a) $|r + 6| = |2r + 1|$

(b) $|5k - 7| = |10k - 2|$

Answers: See page 59.

APPLICATIONS

One of the main reasons for learning mathematics is to be able to use it to solve practical problems. There are no hard-and-fast rules for dealing with real-world applications, except perhaps to use common sense. However, you will find it much easier to deal with such problems if you do not try to do everything at once. After reading the problem carefully, attack it in stages, as suggested in the following guidelines.

Solving Applied Problems

Step 1 Read the problem carefully, focusing on the facts you are given and the unknown values you are asked to find. Look up any words you do not understand. You may have to read the problem more than once, until you understand exactly what you are being asked to do.

(continued on next page)

Step 2 Identify the unknown. (If there is more than one, choose one of them, and see Step 3 for what to do with the others.) Name the unknown with some variable that you *write down.* Many students try to skip this step. They are eager to get on with the writing of the equation. But this is an important step. If you do not know what the variable represents, how can you write a meaningful equation or interpret a result?

Step 3 Decide on a variable expression to represent any other unknowns in the problem. For example, if x represents the width of a rectangle, and you know that the length is one more than twice the width, then *write down* the fact that the length is $1 + 2x$.

Step 4 Draw a sketch or make a chart, if appropriate, showing the information given in the problem.

Step 5 Using the results of Steps 1–4, write an equation that expresses a condition that must be satisfied.

Step 6 Solve the equation.

Step 7 Check the solution in the words of the *original problem,* not just in the equation you have written.

Example 11 **Finance** A financial manager has $14,000 to invest for her company. She plans to invest part of the money in tax-free bonds at 6% interest and the remainder in taxable bonds at 9%. She wants to earn $1005 per year in interest from the investments. Find the amount she should invest at each rate.

Solution

Step 1 We are asked to find how much of the $14,000 should be invested at 6% and how much at 9%, in order to earn the required interest.

Step 2 Let x represent the amount to be invested at 6%.

Step 3 After x dollars are invested, the remaining amount is $14{,}000 - x$ dollars, which is to be invested at 9%.

Step 4 Interest for one year is given by rate $\times$ amount invested. For instance, 6% of x dollars is $.06x$. The given information is summarized in the following chart.

Investment	Amount Invested	Interest Rate	Interest Earned in 1 Year
Tax-free bonds	x	$6\% = .06$	$.06x$
Taxable bonds	$14{,}000 - x$	$9\% = .09$	$.09(14{,}000 - x)$
Totals	14,000		1005

Step 5 Because the total interest is to be $1005, the last column of the tables shows that

$$.06x + .09(14{,}000 - x) = 1005.$$

Step 6 Solve the preceding equation as follows:

$$.06x + .09(14{,}000 - x) = 1005$$
$$.06x + .09(14{,}000) - .09x = 1005$$
$$.06x + 1260 - .09x = 1005$$
$$-.03x = -255$$
$$x = 8500.$$

The manager should invest \$8500 at 6% and \$14,000 − \$8500 = \$5500 at 9%.

Step 7 Check these results in the original problem. If \$8500 is invested at 6%, the interest is .06(8500) = \$510. If \$5500 is invested at 9%, the interest is .09(5500) = \$495. So the total interest is \$510 + \$495 = \$1005, as required. 11 ✓

✓Checkpoint 11

An investor owns two pieces of property. One, worth twice as much as the other, returns 6% in annual interest, while the other returns 4%. Find the value of each piece of property if the total annual interest earned is \$8000.

Answer: See page 59.

Example 12 **Travel** Chuck travels 80 kilometers in the same time that Mary travels 180 kilometers. Mary travels 50 kilometers per hour faster than Chuck. Find the speed of each person.

Solution

Step 1 We must find Chuck's speed and Mary's speed.

Step 2 Let x represent Chuck's speed.

Step 3 Since Mary travels 50 kilometers per hour faster than Chuck, her speed is $x + 50$.

Step 4 Constant-rate problems of this kind require the distance formula

$$d = rt,$$

where d is the distance traveled in t hours at a constant rate of speed, r. The distance traveled by each person is given, along with the fact that the time traveled by each person is the same. Solve the formula $d = rt$ for t:

$$d = rt$$
$$\frac{1}{r} \cdot d = \frac{1}{r} \cdot rt$$
$$\frac{d}{r} = t.$$

For Chuck, $d = 80$ and $r = x$, giving $t = 80/x$. For Mary, $d = 180$, $r = x + 50$, and $t = 180/(x + 50)$. Use these facts to complete a chart, which organizes the information given in the problem.

	d	r	t
Chuck	80	x	$\frac{80}{x}$
Mary	180	$x + 50$	$\frac{180}{x + 50}$

Step 5 Because both people traveled for the *same amount of time*, the equation is

$$\frac{80}{x} = \frac{180}{x + 50}.$$

Step 6 Multiply both sides of the equation by $x(x + 50)$:

$$\begin{aligned} x(x + 50)\frac{80}{x} &= x(x + 50)\frac{180}{x + 50} \\ 80(x + 50) &= 180x \\ 80x + 4000 &= 180x \\ 4000 &= 100x \\ 40 &= x. \end{aligned}$$

Step 7 Since x represents Chuck's speed, Chuck went 40 kilometers per hour. Mary's speed is $x + 50$, or $40 + 50 = 90$ kilometers per hour. Check these results in the words of the original problem. 12

✓ **Checkpoint 12**

(a) Tom and Tyrone enter a run for charity. Tom runs at 7 mph and Tyrone runs at 5 mph. If they start at the same time, how long will it be until they are $\frac{1}{2}$ mile apart?

(b) In part (a), suppose the run has a staggered start. If Tyrone starts first, and Tom starts 10 minutes later, how long will it be until they are neck and neck?

Answers: See page 59.

Example 13 **Business** An oil company distributor needs to fill orders for 89-octane gas, but has only 87- and 93-octane gas on hand. How much of each type should be mixed together to produce 54,000 gallons of 89-octane gas?

Solution

Step 1 We must find how much 87-octane gas and how much 93-octane gas are needed for the 54,000 gallon mixture.

Step 2 Let x be the amount of 87-octane gas.

Step 3 Then $54{,}000 - x$ is the amount of 93-octane gas.

Step 4 The octane rating of a gasoline indicates that it has the same anti-knock qualities as a standard fuel made of heptane and isooctane. The octane rating is the percentage of isooctane in the standard fuel.* So we assume that standard fuels are being mixed. We can summarize the relevant information in a chart.

Type of Gas	Quantity	% Isooctane	Amount of Isooctane
87-octane	x	87%	$.87x$
93-octane	$54{,}000 - x$	93%	$.93(54{,}000 - x)$
Mixture	54,000	89%	$.89(54{,}000)$

Step 5 The amount of isooctane satisfies this equation:

(Amount in 87-octane) + (Amount in 93-octane) = Amount in mixture

$$.87x + .93(54{,}000 - x) = .89(54{,}000).$$

*This is one of several possible ways of determining octane ratings.

Step 6 To solve the equation, use the distributive property on the left side and multiply out the right side:

$$.87x + 50{,}220 - .93x = 48{,}060$$
$$-.06x = -2160$$
$$x = \frac{-2160}{-.06} = 36{,}000.$$

Step 7 So the distributor should mix 36,000 gallons of 87-octane gas with $54{,}000 - 36{,}000 = 18{,}000$ gallons of 93-octane gas. Then the amount of isooctane in the mixture is

$$.87(36{,}000) + .93(18{,}000)$$
$$= 31{,}320 + 16{,}740$$
$$= 48{,}060.$$

Hence, the octane rating of the mixture is $\frac{48{,}060}{54{,}000} = .89$, as required. 13 ✓

✓Checkpoint 13

How much 89-octane gas and how much 94-octane gas are needed to produce 1000 gallons of 91-octane gas?

Answer: *See page 59.*

1.6 ▶ Exercises

Solve each equation. (See Examples 1–6.)

1. $3x + 8 = 20$ **2.** $4 - 5y = 19$

3. $.6k - .3 = .5k + .4$

4. $2.5 + 5.04m = 8.5 - .06m$

5. $2a - 1 = 4(a + 1) + 7a + 5$

6. $3(k - 2) - 6 = 4k - (3k - 1)$

7. $2[x - (3 + 2x) + 9] = 3x - 8$

8. $-2[4(k + 2) - 3(k + 1)] = 14 + 2k$

9. $\frac{3x}{5} - \frac{4}{5}(x + 1) = 2 - \frac{3}{10}(3x - 4)$

10. $\frac{4}{3}(x - 2) - \frac{1}{2} = 2\left(\frac{3}{4}x - 1\right)$

11. $\frac{5y}{6} - 8 = 5 - \frac{2y}{3}$ **12.** $\frac{x}{2} - 3 = \frac{3x}{5} + 1$

13. $\frac{m}{2} - \frac{1}{m} = \frac{6m + 5}{12}$

14. $-\frac{3k}{2} + \frac{9k - 5}{6} = \frac{11k + 8}{k}$

15. $\frac{4}{x - 3} - \frac{8}{2x + 5} + \frac{3}{x - 3} = 0$

16. $\frac{5}{2p + 3} - \frac{3}{p - 2} = \frac{4}{2p + 3}$

17. $\frac{3}{2m + 4} = \frac{1}{m + 2} - 2$ **18.** $\frac{8}{3k - 9} - \frac{5}{k - 3} = 4$

Use a calculator to solve each equation. Round your answer to the nearest hundredth. (See Example 3.)

19. $9.06x + 3.59(8x - 5) = 12.07x + .5612$

20. $-5.74(3.1 - 2.7p) = 1.09p + 5.2588$

21. $\frac{2.63r - 8.99}{1.25} - \frac{3.90r - 1.77}{2.45} = r$

22. $\frac{8.19m + 2.55}{4.34} - \frac{8.17m - 9.94}{1.04} = 4m$

Solve each equation for x. (See Example 7.)

23. $4(a + x) = b - a + 2x$

24. $(3a - b) - bx = a(x - 2) \quad (a \neq -b)$

25. $5(b - x) = 2b + ax \quad (a \neq -5)$

26. $bx - 2b = 2a - ax$

Solve each equation for the specified variable. Assume that all denominators are nonzero. (See Example 7.)

27. $PV = k$ for V **28.** $i = prt$ for p

29. $V = V_0 + gt$ for g **30.** $S = S_0 + gt^2 + k$ for g

31. $A = \frac{1}{2}(B + b)h$ for B **32.** $C = \frac{5}{9}(F - 32)$ for F

Solve each equation. (See Examples 8–10.)

33. $|2h - 1| = 5$ **34.** $|4m - 3| = 12$

35. $|6 + 2p| = 10$ **36.** $|-5x + 7| = 15$

37. $\left|\frac{5}{r-3}\right| = 10$ **38.** $\left|\frac{3}{2h-1}\right| = 4$

Solve the following applied problems.

Health *According to the American Heart Association, the number y of brain neurons (in billions) that are lost in a stroke lasting x hours is given by* $y = \frac{x}{8}$. *Find the length of the stroke for the given number of neurons lost.*

39. 1,250,000,000 **40.** 2,400,000,000

Natural Science *The equation that relates Fahrenheit temperature F to Celsius temperature C is*

$$C = \frac{5}{9}(F - 32).$$

Find the Fahrenheit temperature corresponding to these Celsius temperatures.

41. -5 **42.** -15

43. 22 **44.** 36

Finance *The gross federal debt y (in trillions of dollars) in year x is approximated by*

$$y = .5x + 5.33,$$

where x is the number of years after 2000. In what year will the federal debt be the given amount?*

45. \$8.83 trillion **46.** \$10.33 trillion

47. \$12.83 trillion **48.** \$14.83 trillion

Health *The total health care expenditures E in the United States (in billions of dollars) can be approximated by*

$$E = 86x + 1648,$$

where x is the number of years since 2000.† Determine the year in which health care expenditures are at the given level.

49. \$2250 billion **50.** \$2422 billion

51. \$2680 billion **52.** \$2938 billion

Business *The percentage of workers participating in defined contribution retirement plans was discussed in Example 3. The percentage y of workers who are covered by a traditional pension in year x is given by the equation*

$$-.12(x - 2006) = 14y - 2.8.‡$$

Find the year in which the given percentage of workers covered by a traditional pension plan is the given amount:

53. 26% **54.** 18% **55.** 14%

56. Find the year in which the percentage of workers participating in a traditional pension was the same as the percentage participating in a defined contribution plan. What was that percentage? (*Hint:* Solve both the defined-benefits equation in Example 3 and the preceding traditional-pension equation for y. Set the results equal to each other and solve for x. Use this value of x to determine y.)

Social Science *The population P of the United States (in millions) is approximated by*

$$P = 2.67x + 282.62,$$

where x is the number of years since 2000. Determine the year in which the United States will have the given population.*

57. 320,000,000 **58.** 330,680,000

59. 346,700,000 **60.** 373,400,000

Business *When a loan is paid off early, a portion of the finance charge must be returned to the borrower. By one method of calculating the finance charge (called the rule of 78), the amount of unearned interest (finance charge to be returned) is given by*

$$u = f\cdot\frac{n(n+1)}{q(q+1)},$$

where u represents unearned interest, f is the original finance charge, n is the number of payments remaining when the loan is paid off, and q is the original number of payments. Find the amount of the unearned interest in each of the given cases.

61. Original finance charge = \$800, loan scheduled to run 36 months, paid off with 18 payments remaining

62. Original finance charge = \$1400, loan scheduled to run 48 months, paid off with 12 payments remaining

63. **Natural Science** The excess lifetime cancer risk R is a measure of the likelihood that an individual will develop cancer from a particular pollutant. For example, if $R = .01$, then a person has a 1% increased chance of developing cancer during a lifetime. The value of R for formaldehyde can be calculated from the equation $R = kd$, where k is a constant and d is the daily dose in parts per million. The constant k for formaldehyde can be calculated from the formula $k = .132(\frac{B}{W})$, where B is the total number of cubic meters of air a person breathes in one day and W is a person's weight in kilograms.†

(a) Find k for a person who breathes in 20 cubic meters of air per day and weighs 75 kg.

(b) Mobile homes in Minnesota were found to have a mean daily dose d of .42 parts per million.‡ Calculate R.

(c) For every 5000 people, how many cases of cancer could be expected each year from the preceding levels of formaldehyde? Assume an average life expectancy of 72 years.

---

*Based on data from the U.S. Treasury.

†Based on data and projections from the U.S. Centers for Medicare and Medicaid Services.

‡Based on data from the U.S. Bureau of Labor Statistics.

*Based on data and projections from the U.S. Census Bureau.

†A. Hines, T. Ghosh, S. Layalka, and R. Warder, *Indoor Air Quality and Control* (Upper Saddle River, NJ: Prentice Hall, 1993), (TD 883.I.1476 1993).

‡I. Ritchie and R. Lehnen, "An Analysis of Formaldehyde Concentration in Mobile and Conventional Homes," *Journal of Environmental Health* 47:300–305.

Business *Solve the following investment problems. (See Example 11.)*

64. Joe Gonzalez received \$52,000 profit from the sale of some land. He invested part at 5% interest and the rest at 4% interest. He earned a total of \$2290 interest per year. How much did he invest at 5%?

65. Weijen Luan invests \$20,000 received from an insurance settlement in two ways: some at 6% and some at 4%. Altogether, she makes \$1040 per year in interest. How much is invested at 4%?

66. Maria Martinelli bought two plots of land for a total of \$120,000. On the first plot, she made a profit of 15%. On the second, she lost 10%. Her total profit was \$5500. How much did she pay for each piece of land?

67. Suppose \$20,000 is invested at 5%. How much additional money must be invested at 4% to produce a yield of 4.8% on the entire amount invested?

Solve the given applied problems. (See Example 12.)

68. A plane flies nonstop from New York to London, cities that are about 3500 miles apart. After 1 hour and 6 minutes in the air, the plane passes over Halifax, Nova Scotia, which is 600 miles from New York. Estimate the flying time from New York to London.

69. On vacation, Le Hong averaged 50 mph traveling from Denver to Minneapolis. Returning by a different route that covered the same number of miles, he averaged 55 mph. What is the distance between the two cities if his total traveling time was 32 hours?

70. Russ and Janet are running in the Apple Hill Fun Run. Russ runs at 7 mph, Janet at 5 mph. If they start at the same time, how long will it be before they are $\frac{2}{3}$ mile apart?

Natural Science *Using the same assumptions about octane ratings as in Example 13, solve the following problems.*

71. How many liters of 94-octane gasoline should be mixed with 200 liters of 99-octane gasoline to get a mixture that is 97-octane gasoline?

72. A service station has 92-octane and 98-octane gasoline. How many liters of each gasoline should be mixed to provide 12 liters of 96-octane gasoline for a chemistry experiment?

Solve the following applied problems.

73. Business A major car rental firm charges \$45.56 a day with unlimited mileage. A discount firm offers a similar car for \$20 a day plus 18 cents per mile. How far must you drive in a day in order for the cost to be the same at both firms?

74. Natural Science A car radiator contains 8 quarts of fluid, 40% of which is antifreeze. How much fluid should be drained and replaced with pure (100%) antifreeze in order that the new mixture be 60% antifreeze?

75. Transportation In Massachusetts, speeding fines are determined by the equation

$$y = 10(x - 65) + 50, \qquad x \geq 65,$$

where y is the amount of the fine (in dollars) for driving x miles per hour. If Paul was fined \$100 for speeding, how fast was he driving?

76. Finance Jack borrowed his father's luxury car and promised to return it with a full tank of premium gas, which costs \$2.50 per gallon. From experience, he knows that he needs 15.5 gallons. But he has only \$38 (and no credit card), which isn't enough. He decides to get as much premium as possible and fill the remainder of the tank with regular gas, which costs \$2.30 per gallon. How much of each type of gas should he get?

Geometry *Solve each of these geometry problems. (Hint: In each case, draw an appropriate figure and label its parts.)*

77. The length of a rectangular label is 3 centimeters less than twice the width. The perimeter is 54 centimeters. Find the width.

78. A puzzle piece in the shape of a triangle has a perimeter of 30 centimeters. Two sides of the triangle are each twice as long as the shortest side. Find the length of the shortest side.

79. A triangle has a perimeter of 27 centimeters. One side is twice as long as the shortest side. The third side is 7 centimeters longer than the shortest side. Find the length of the shortest side.

80. A closed recycling bin is in the shape of a rectangular box. Find the height of the bin if its length is 18 feet, its width is 8 feet, and its surface area is 496 square feet.

✓ Checkpoint Answers

1. **(a)** Yes **(b)** No **(c)** No
2. **(a)** 8 **(b)** -2 **(c)** 2
3. **(a)** 8 **(b)** -3
4. About late 2000 ($x \approx 2000.75$)
5. **(a)** 24 **(b)** -12
6. 1
7. Neither equation has a solution.
8. **(a)** $x = \dfrac{7y}{2 - 3k}$ **(b)** $x = \dfrac{5k + 32 + 6p}{8 - 11y}$
9. **(a)** $9, -9$ **(b)** $-2, -4$ **(c)** $5, -2$
10. **(a)** $5, -7/3$ **(b)** $-1, 3/5$
11. 6% return: \$100,000; 4% return: \$50,000
12. **(a)** 15 minutes ($\frac{1}{4}$ hour)
(b) After Tom has run 25 minutes
13. 600 gallons of 89-octane gas; 400 gallons of 94-octane gas

1.7 ▶ Quadratic Equations

An equation that can be written in the form

$$ax^2 + bx + c = 0,$$

where a, b, and c are real numbers with $a \neq 0$, is called a **quadratic equation**. For example, each of

$$2x^2 + 3x + 4 = 0, \quad x^2 = 6x - 9, \quad 3x^2 + x = 6, \quad \text{and} \quad x^2 = 5$$

is a quadratic equation. A solution of an equation that is a real number is said to be a **real solution** of the equation.

One method of solving quadratic equations is based on the following property of real numbers.

Zero-Factor Property

If a and b are real numbers, with $ab = 0$, then $a = 0$ or $b = 0$ or both.

Example 1 Solve the equation $(x - 4)(3x + 7) = 0$.

Solution By the zero-factor property, the product $(x - 4)(3x + 7)$ can equal 0 only if at least one of the factors equals 0. That is, the product equals zero only if $x - 4 = 0$ or $3x + 7 = 0$. Solving each of these equations separately will give the solutions of the original equation:

$$\begin{aligned} x - 4 &= 0 \quad \text{or} \quad & 3x + 7 &= 0 \\ x &= 4 \quad \text{or} \quad & 3x &= -7 \\ & & x &= -\frac{7}{3}. \end{aligned}$$

The solutions of the equation $(x - 4)(3x + 7) = 0$ are 4 and $-7/3$. Check these solutions by substituting them into the original equation. ✓1

✓Checkpoint 1

Solve the following equations.

(a) $(y - 6)(y + 2) = 0$

(b) $(5k - 3)(k + 5) = 0$

(c) $(2r - 9)(3r + 5) \cdot (r + 3) = 0$

Answers: See page 69.

Example 2 Solve $6r^2 + 7r = 3$.

Solution Rewrite the equation as

$$6r^2 + 7r - 3 = 0.$$

Now factor $6r^2 + 7r - 3$ to get

$$(3r - 1)(2r + 3) = 0.$$

By the zero-factor property, the product $(3r - 1)(2r + 3)$ can equal 0 only if

$$3r - 1 = 0 \quad \text{or} \quad 2r + 3 = 0.$$

Solving each of these equations separately gives the solutions of the original equation:

$$3r = 1 \quad \text{or} \quad 2r = -3$$
$$r = \frac{1}{3} \qquad r = -\frac{3}{2}.$$

Verify that both $1/3$ and $-3/2$ are solutions by substituting them into the original equation. ✓2

✓Checkpoint 2

Solve each equation by factoring.

(a) $y^2 + 3y = 10$

(b) $2r^2 + 9r = 5$

(c) $4k^2 = 9k$

Answers: See page 69.

An equation such as $x^2 = 5$ has two solutions: $\sqrt{5}$ and $-\sqrt{5}$. This fact is true in general.

Square-Root Property

If $b > 0$, then the solutions of $\boldsymbol{x^2 = b}$ are $\boldsymbol{\sqrt{b}}$ and $\boldsymbol{-\sqrt{b}}$.

The two solutions are sometimes abbreviated $\pm\sqrt{b}$.

Example 3 Solve each equation.

(a) $m^2 = 17$

Solution By the square-root property, the solutions are $\sqrt{17}$ and $-\sqrt{17}$, abbreviated $\pm\sqrt{17}$.

(b) $(y - 4)^2 = 11$

Solution Use a generalization of the square-root property, we work as follows.

$$(y - 4)^2 = 11$$
$$y - 4 = \sqrt{11} \quad \text{or} \quad y - 4 = -\sqrt{11}$$
$$y = 4 + \sqrt{11} \qquad y = 4 - \sqrt{11}.$$

Abbreviate the solutions as $4 \pm \sqrt{11}$. ✓3

✓Checkpoint 3

Solve each equation by using the square-root property.

(a) $p^2 = 21$

(b) $(m + 7)^2 = 15$

(c) $(2k - 3)^2 = 5$

Answers: See page 69.

When a quadratic equation cannot be easily factored, it can be solved by using the following formula, which you should memorize.*

Quadratic Formula

The solutions of the quadratic equation $ax^2 + bx + c = 0$, where $a \neq 0$, are given by

$$\boldsymbol{x = \frac{-b \pm \sqrt{b^2 - 4ac}}{2a}}.$$

*A proof of the quadratic formula can be found in many College Algebra books.

CAUTION When using the quadratic formula, remember that the equation must be in the form $ax^2 + bx + c = 0$. Also, notice that the fraction bar in the quadratic formula extends under *both* terms in the numerator. Be sure to add $-b$ to $\pm\sqrt{b^2 - 4ac}$ *before* dividing by $2a$.

Example 4 Solve $x^2 + 1 = 4x$.

Solution First add $-4x$ to both sides to get 0 alone on the right side:

$$x^2 - 4x + 1 = 0.$$

Now identify the values of a, b, and c. Here, $a = 1$, $b = -4$, and $c = 1$. Substitute these numbers into the quadratic formula to obtain

$$\begin{aligned} x &= \frac{-(-4) \pm \sqrt{(-4)^2 - 4(1)(1)}}{2(1)} \\ &= \frac{4 \pm \sqrt{16 - 4}}{2} \\ &= \frac{4 \pm \sqrt{12}}{2} \\ &= \frac{4 \pm 2\sqrt{3}}{2} && \sqrt{12} = \sqrt{4 \cdot 3} = \sqrt{4} \cdot \sqrt{3} = 2\sqrt{3} \\ &= \frac{2(2 \pm \sqrt{3})}{2} && \text{Factor } 4 \pm 2\sqrt{3}. \\ x &= 2 \pm \sqrt{3}. && \text{Cancel 2.} \end{aligned}$$

The $\pm$ sign represents the two solutions of the equation. First use + and then use − to find each of the solutions: $2 + \sqrt{3}$ and $2 - \sqrt{3}$. 4

✓Checkpoint 4

Use the quadratic formula to solve each equation.

(a) $x^2 - 2x = 2$

(b) $u^2 - 6u + 4 = 0$

Answers: See page 69.

Example 4 shows that the quadratic formula produces exact solutions. In many real-world applications, however, you must use a calculator to find decimal approximations of the solutions. The approximate solutions in Example 4 are

$$x = 2 + \sqrt{3} \approx 3.732050808 \quad \text{and} \quad x = 2 - \sqrt{3} \approx .2679491924,$$

as shown in Figure 1.13.

```
2+√(3)
          3.732050808
2-√(3)
          .2679491924
```

FIGURE 1.13

Example 5 **Social Science** The yearly amount y of food distributed by the St. Louis Area Foodbank is approximated by the equation

$$.027x^2 - .037x + 7.16 = y \qquad (x \geq 0),$$

where $x = 0$ corresponds to the year 1990 and y is measured in millions of pounds. Use the quadratic formula and a calculator to find the year in which 17,000,000 pounds of food were distributed.

Solution To find the year x, solve the equation above when $y = 17$ (because y is measured in millions):

$$.027x^2 - .037x + 7.16 = y$$
$$.027x^2 - .037x + 7.16 = 17 \qquad \text{Substitute } y = 17.$$
$$.027x^2 - .037x - 9.84 = 0. \qquad \text{Subtract 17 from both sides.}$$

To apply the quadratic formula, first compute the radical part:

$$\sqrt{b^2 - 4ac} = \sqrt{(-.037)^2 - 4(.027)(-9.84)} = \sqrt{1.064089}.$$

Then store $\sqrt{1.064089}$ (which we denote by M) in the calculator memory. (Check your instruction manual for how to store and recall numbers.) By the quadratic formula, the exact solutions of the equation are

FIGURE 1.14

$$x = \frac{-b + \sqrt{b^2 - 4ac}}{2a} = \frac{-b \pm M}{2a} = \frac{-(-.037) \pm M}{2(.027)} = \frac{.037 \pm M}{.054}.$$

Figure 1.14 shows that the approximate solutions are

$$\frac{.037 + M}{.054} \approx 19.7879 \quad \text{and} \quad \frac{.037 - M}{.054} \approx -18.4175.$$

Since we are given that $x \geq 0$, the only applicable solution here is $x \approx 19.7879$, which corresponds to late 2009. 5✓

✓Checkpoint 5

Use the method in Example 5 to find approximate solutions for $5.1x^2 - 3.3x - 240.624 = 0$.

Answer: See page 69.

Example 6 Solve $9x^2 - 30x + 25 = 0$.

Solution Applying the quadratic formula with $a = 9$, $b = -30$, and $c = 25$, we have

$$x = \frac{-(-30) \pm \sqrt{(-30)^2 - 4(9)(25)}}{2(9)}$$
$$= \frac{30 \pm \sqrt{900 - 900}}{18} = \frac{30 \pm 0}{18} = \frac{30}{18} = \frac{5}{3}.$$

Therefore, the given equation has only one real solution. The fact that the solution is a rational number indicates that this equation could have been solved by factoring.

TECHNOLOGY TIP You can approximate the solutions of quadratic equations on a graphing calculator by using a quadratic formula program (see the Program Appendix) or using a built-in quadratic equation solver if your calculator has one. Then you need only enter the values of the coefficients a, b, and c to obtain the approximate solutions.

Example 7 Solve $x^2 - 6x + 10 = 0$.

Solution Apply the quadratic formula with $a = 1$, $b = -6$, and $c = 10$:

$$x = \frac{-(-6) \pm \sqrt{(-6)^2 - 4(1)(10)}}{2(1)}$$
$$= \frac{6 \pm \sqrt{36 - 40}}{2}$$
$$= \frac{6 \pm \sqrt{-4}}{2}.$$

Since no negative number has a square root in the real-number system, $\sqrt{-4}$ is not a real number. Hence, the equation has no real solutions. 6✓

Checkpoint 6

Solve each equation.

(a) $9k^2 - 6k + 1 = 0$

(b) $4m^2 + 28m + 49 = 0$

(c) $2x^2 - 5x + 5 = 0$

Answers: See page 69.

Examples 4–7 show that the number of solutions of the quadratic equation $ax^2 + bx + c = 0$ is determined by $b^2 - 4ac$, the quantity under the radical, which is called the **discriminant** of the equation. 7✓

✓Checkpoint 7

Use the discriminant to determine the number of real solutions of each equation.

(a) $x^2 + 8x + 3 = 0$

(b) $2x^2 + x + 3 = 0$

(c) $x^2 - 194x + 9409 = 0$

Answers: See page 69.

The Discriminant

The equation $\boldsymbol{ax^2 + bx + c = 0}$ has either two, or one, or no real solutions.

If $\boldsymbol{b^2 - 4ac > 0}$, there are two real solutions. (*Examples 4 and 5*)

If $\boldsymbol{b^2 - 4ac = 0}$, there is one real solution. (*Example 6*)

If $\boldsymbol{b^2 - 4ac < 0}$, there are no real solutions. (*Example 7*)

APPLICATIONS

Quadratic equations arise in a variety of settings, as illustrated in the next set of examples. Example 8 depends on the following useful fact from geometry.

The Pythagorean Theorem

In a right triangle with legs of lengths a and b and hypotenuse of length c,

$$a^2 + b^2 = c^2.$$

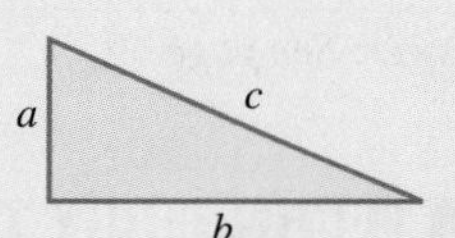

Example 8 The size of a computer monitor is the diagonal measurement of its screen. The height of the screen is approximately 3/4 of its width. Kathy claims that John's 14-inch monitor has less than half the viewing area of her 21-inch monitor. John says, "No way! 14/21 is 2/3, so my monitor has 2/3 of the viewing area of yours." Who is right?

Solution First, find the area of Kathy's screen. Let x be its width. Then its height is 3/4 of x (that is, $.75x$, as shown in Figure 1.15).

By the Pythagorean theorem,

$$\begin{aligned} x^2 + (.75x)^2 &= 21^2 \\ x^2 + .5625x^2 &= 441 && \text{Expand } (.75x)^2. \\ (1 + .5625)x^2 &= 441 && \text{Distributive property} \\ 1.5625x^2 &= 441 \\ x^2 &= 282.24 && \text{Divide both sides by 1.5625.} \\ x &= \pm\sqrt{282.24} && \text{Square-root property} \\ x &= \pm 16.8. \end{aligned}$$

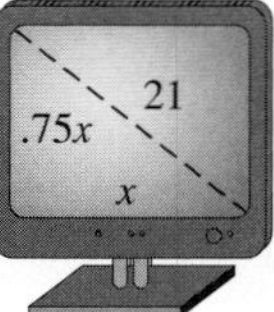

FIGURE 1.15

We can ignore the negative solution, since x is a width. Thus, the width is 16.8 inches and the height is $.75x = .75(16.8) = 12.6$ inches, so the area is

$$\text{Area} = \text{width} \times \text{height} = 16.8 \times 12.6 = 211.68 \text{ square inches.}$$

Next, find the area of John's screen by working Checkpoint 8, which shows that John's screen has an area of 94.08 square inches. So who is right? 8

✓Checkpoint 8

Let x be the width of John's screen in Example 8.

(a) Write an equation that expresses the relationship between the width, height, and diagonal of the screen.

(b) Find the dimensions of the screen.

(c) Find its area.

Answers: See page 69.

Example 9 **Business** A landscape architect wants to make an exposed gravel border of uniform width around a small shed behind a company plant. The shed is 10 feet by 6 feet. He has enough gravel to cover 36 square feet. How wide should the border be?

Solution A sketch of the shed with border is given in Figure 1.16. Let x represent the width of the border. Then the width of the large rectangle is $6 + 2x$, and its length is $10 + 2x$.

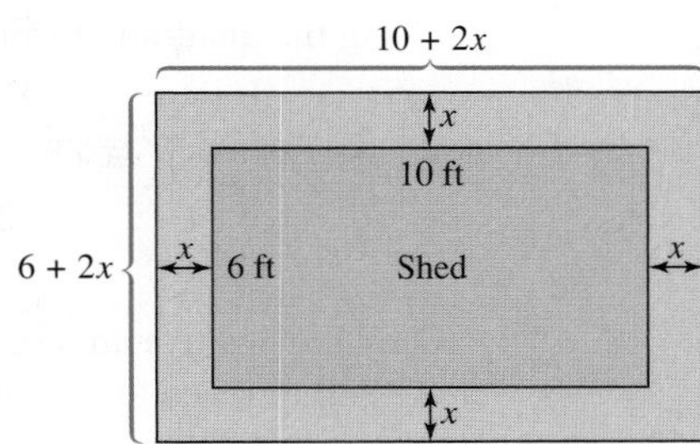

FIGURE 1.16

We must write an equation relating the given areas and dimensions. The area of the large rectangle is $(6 + 2x)(10 + 2x)$. The area occupied by the shed is $6 \cdot 10 = 60$. The area of the border is found by subtracting the area of the shed from the area of the large rectangle. This difference should be 36 square feet, giving the equation

$$(6 + 2x)(10 + 2x) - 60 = 36.$$

Solve this equation with the following sequence of steps:

$$60 + 32x + 4x^2 - 60 = 36 \quad \text{Multiply out left side.}$$
$$4x^2 + 32x - 36 = 0 \quad \text{Simplify.}$$
$$x^2 + 8x - 9 = 0 \quad \text{Divide both sides by 4.}$$
$$(x + 9)(x - 1) = 0 \quad \text{Factor.}$$
$$x + 9 = 0 \quad \text{or} \quad x - 1 = 0 \quad \text{Zero-factor property}$$
$$x = -9 \quad \text{or} \quad x = 1.$$

The number -9 cannot be the width of the border, so the solution is to make the border 1 foot wide. 9

✓Checkpoint 9

The length of a picture is 2 inches more than the width. It is mounted on a mat that extends 2 inches beyond the picture on all sides. What are the dimensions of the picture if the area of the mat is 99 square inches?

Answer: See page 69.

Example 10 **Physical Science** If an object is thrown upward, dropped, or thrown downward and travels in a straight line subject only to gravity (with wind resistance ignored), the height h of the object above the ground (in feet) after t seconds is given by

$$h = -16t^2 + v_0 t + h_0,$$

where h_0 is the height of the object when $t = 0$ and v_0 is the initial velocity at time $t = 0$. The value of v_0 is taken to be positive if the object moves upward and negative if it moves downward. Suppose that a golf ball is thrown downward from the top of a 625-foot-high building with an initial velocity of 65 feet per second. How long does it take to reach the ground?

Solution In this case, $h_0 = 625$ (the height of the building) and $v_0 = -65$ (negative because the ball is thrown downward). The object is on the ground when $h = 0$, so we must solve the equation

$$h = -16t^2 + v_0 t + h_0$$

$$0 = -16t^2 - 65t + 625. \quad \text{Let } h = 0,\ v_0 = -65, \text{ and } h_0 = 625.$$

Using the quadratic formula and a calculator, we see that

$$t = \frac{-(-65) \pm \sqrt{(-65)^2 - 4(-16)(625)}}{2(-16)} = \frac{65 \pm \sqrt{44{,}225}}{-32} \approx \begin{cases} -8.60 \\ \text{or} \\ 4.54 \end{cases}$$

Only the positive answer makes sense in this case. So it takes about 4.54 seconds for the ball to reach the ground.

In some applications, it may be necessary to solve an equation in several variables for a specific variable.

Example 11 Solve $v = mx^2 + x$ for x. (Assume that m and v are positive.)

Solution The equation is quadratic in x because of the x^2 term. Before we use the quadratic formula, we write the equation in standard form:

$$v = mx^2 + x$$

$$0 = mx^2 + x - v.$$

Let $a = m$, $b = 1$, and $c = -v$. Then the quadratic formula gives

$$x = \frac{-1 \pm \sqrt{1^2 - 4(m)(-v)}}{2m}$$

$$x = \frac{-1 \pm \sqrt{1 + 4mv}}{2m}. \quad 10✓$$

✓Checkpoint 10

Solve each of the given equations for the indicated variable. Assume that all variables are positive.

(a) $k = mp^2 - bp$ for p

(b) $r = \dfrac{APk^2}{3}$ for k

Answers: See page 69.

1.7 ▶ Exercises

Use factoring to solve each equation. (See Examples 1 and 2.)

1. $(x + 4)(x - 14) = 0$
2. $(p - 16)(p - 5) = 0$
3. $x(x + 6) = 0$
4. $x^2 - 2x = 0$
5. $2z^2 = 4z$
6. $x^2 - 64 = 0$
7. $y^2 + 15y + 56 = 0$
8. $k^2 - 4k - 5 = 0$
9. $2x^2 = 7x - 3$
10. $2 = 15z^2 + z$
11. $6r^2 + r = 1$
12. $3y^2 = 16y - 5$
13. $2m^2 + 20 = 13m$
14. $6a^2 + 17a + 12 = 0$
15. $m(m + 7) = -10$
16. $z(2z + 7) = 4$
17. $9x^2 - 16 = 0$
18. $36y^2 - 49 = 0$
19. $16x^2 - 16x = 0$
20. $12y^2 - 48y = 0$

Solve each equation by using the square-root property. (See Example 3.)

21. $(r - 2)^2 = 7$
22. $(b + 4)^2 = 27$
23. $(4x - 1)^2 = 20$
24. $(3t + 5)^2 = 11$

Use the quadratic formula to solve each equation. If the solutions involve square roots, give both the exact and approximate solutions. (See Examples 4–7.)

25. $2x^2 + 7x + 1 = 0$
26. $3x^2 - x - 7 = 0$
27. $4k^2 + 2k = 1$
28. $r^2 = 3r + 5$
29. $5y^2 + 5y = 2$
30. $2z^2 + 3 = 8z$
31. $6x^2 + 6x + 4 = 0$
32. $3a^2 - 2a + 2 = 0$
33. $2r^2 + 3r - 5 = 0$
34. $8x^2 = 8x - 3$
35. $2x^2 - 7x + 30 = 0$
36. $3k^2 + k = 6$
37. $1 + \frac{7}{2a} = \frac{15}{2a^2}$
38. $5 - \frac{4}{k} - \frac{1}{k^2} = 0$

Use the discriminant to determine the number of real solutions of each equation. You need not solve the equations.

39. $25t^2 + 49 = 70t$
40. $9z^2 - 12z = 1$
41. $13x^2 + 24x - 5 = 0$
42. $20x^2 + 19x + 5 = 0$

Use a calculator and the quadratic formula to find approximate solutions of each equation. (See Example 5.)

43. $4.42x^2 - 10.14x + 3.79 = 0$
44. $3x^2 - 82.74x + 570.4923 = 0$
45. $7.63x^2 + 2.79x = 5.32$
46. $8.06x^2 + 25.8726x = 25.047256$

Solve the next two problems. (See Example 3.)

47. **Transportation** According to the Federal Aviation Administration, the maximum recommended taxiing speed x (in miles per hour) for a plane on a curved runway exit is given by $R = .5x^2$, where R is the radius of the curve (in feet). Find the maximum taxiing speed for planes on such exits when the radius of the exit is
(a) 450 ft **(b)** 615 ft **(c)** 970 ft

48. **Social Science** The enrollment E in public colleges and universities (in millions) is approximated by $E = .008x^2 + 10.845$, where x is the number of years since 1990.* Find the year when enrollment is
(a) 14,045,000 **(b)** 15,845,000

Solve the next two problems. (See Example 5.)

49. **Social Science** The number of drivers who die in automobile accidents is related to the ages of the drivers, with teenagers and the elderly having the worst records. According to data from the National Highway Traffic Safety Administration, the driver fatality rate D per 1000 licensed drivers every 100 million miles can be approximated by the equation $D = .0031x^2 - .291x + 7.1$, where x is the age of the driver.
(a) For what ages is the driver fatality rate about 1 death per 1000?
(b) For what ages is the rate three times greater than in part (a)?

50. **Finance** The total resources T (in billions of dollars) of the Pension Benefit Guaranty Corporation, the government agency that insures pensions, can be approximated by the equation $T = -.26x^2 + 3.62x + 30.18$, where x is the number of years after 2000.† Determine when the total resources are at the given level.
(a) $27 billion
(b) $30 billion
(c) When will the Corporation be out of money ($T = 0$)?

51. **Business** The revenues R of the Dell Corporation (in billions of dollars) can be approximated by

$$R = -.315x^2 + 8.44x + 14.27 \quad (2 \le x \le 8),$$

where $x = 2$ corresponds to the year 2002.‡
(a) When were revenues about $58 billion dollars?
(b) If this formula remains valid until 2015, when do revenues reach $67,170,000,000?

52. **Health** Total spending S on health care in the United States (in billions of dollars) can be approximated by

$$S = .78x^2 + 5.93x + 232.2,$$

*Based on data in the *Statistical Abstract of the United States: 2008.*

†Based on data from the Center on Federal Financial Institutions.

‡Based on data from Dell's Web site.

where x is the number of years since 1960.* Find the year in which total spending is
(a) \$2,351,000,000,000 **(b)** \$2,650,000,000,000

Solve the following problems. (See Examples 8 and 9).

53. A 13-foot-long ladder leans on a wall, as shown in the accompanying figure. The bottom of the ladder is 5 feet from the wall. If the bottom is pulled out 2 feet farther from the wall, how far does the top of the ladder move down the wall? [*Hint:* Draw pictures of the right triangle formed by the ladder, the ground, and the wall before and after the ladder is moved. In each case, use the Pythagorean theorem to find the distance from the top of the ladder to the ground.]

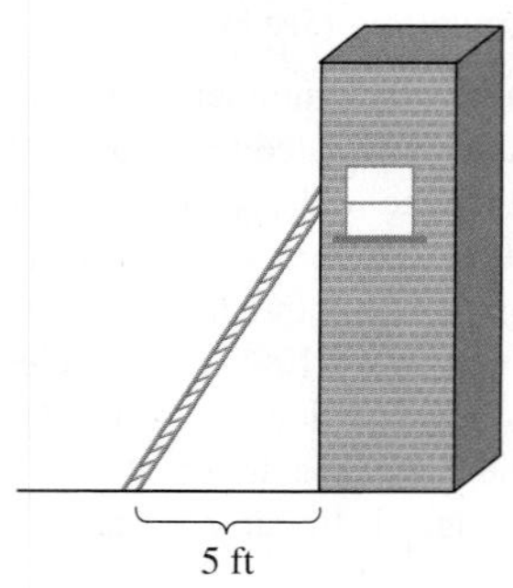

54. A 15-foot-long pole leans against a wall. The bottom is 9 feet from the wall. How much farther should the bottom be pulled away from the wall so that the top moves the same amount down the wall?

55. **Transportation** Two trains leave the same city at the same time, one going north and the other east. The eastbound train travels 20 mph faster than the northbound one. After 5 hours, the trains are 300 miles apart. Determine the speed of each train, using the following steps.
(a) Let x denote the speed of the northbound train. Express the speed of the eastbound train in terms of x.
(b) Write expressions that give the distance traveled by each train after 5 hours.
(c) Use part (b) and the fact that the trains are 300 miles apart after 5 hours to write an equation. (A diagram of the situation may help.)
(d) Solve the equation and determine the speeds of the trains.

56. Chris and Josh have received walkie-talkies for Christmas. If they leave from the same point at the same time, Chris walking north at 2.5 mph and Josh walking east at 3 mph, how long will they be able to talk to each other if the range of the walkie-talkies is 4 miles? Round your answer to the nearest minute.

57. An ecology center wants to set up an experimental garden. It has 300 meters of fencing to enclose a rectangular area of 5000 square meters. Find the length and width of the rectangle as follows.
(a) Let x = the length and write an expression for the width.
(b) Write an equation relating the length, width, and area, using the result of part (a).
(c) Solve the problem.

58. A landscape architect has included a rectangular flower bed measuring 9 feet by 5 feet in her plans for a new building. She wants to use two colors of flowers in the bed, one in the center and the other for a border of the same width on all four sides. If she can get just enough plants to cover 24 square feet for the border, how wide can the border be?

59. Joan wants to buy a rug for a room that is 12 feet by 15 feet. She wants to leave a uniform strip of floor around the rug. She can afford 108 square feet of carpeting. What dimensions should the rug have?

60. In 2008, Scott Dixon won the (500-mile) Indianapolis 500 race. His speed (rate) was 69 mph (to the nearest mph) faster than that of the 1911 winner, Ray Harroun. Dixon completed the race in 3.22 hours less time than Harroun. Find Harroun's and Dixon's rates to the nearest tenth.

Physical Science *Use the height formula in Example 10 to work the given problems. Note that an object that is dropped (rather than thrown downward) has initial velocity $v_0 = 0$.*

61. How long does it take a baseball to reach the ground if it is dropped from the top of a 625-foot-high building? Compare the answer with that in Example 10.

62. After the baseball in Exercise 61 is dropped, how long does it take for the ball to fall 196 feet? (*Hint:* How high is the ball at that time?)

63. You are standing on a cliff that is 200 feet high. How long will it take a rock to reach the ground if
(a) you drop it?
(b) you throw it downward at an initial velocity of 40 feet per second?
(c) How far does the rock fall in 2 seconds if you throw it downward with an initial velocity of 40 feet per second?

64. A rocket is fired straight up from ground level with an initial velocity of 800 feet per second.
(a) How long does it take the rocket to rise 3200 feet?
(b) When will the rocket hit the ground?

65. A ball is thrown upward from ground level with an initial velocity of 64 ft per second. In how many seconds will the ball reach the given height?
(a) 64 ft **(b)** 39 ft
(c) Why are two answers possible in part (b)?

66. **Health** The length L (in mm) of the mesiodistal crown of the first molar of a human fetus is approximated by

$$L = -.01t^2 + .788t - 7.048,$$

*Based on data from U.S. Department of Health and Human Services.

where t is the number of weeks since conception.*

(a) At how many weeks is the length about 4 mm?

(b) Explain why the second answer in part (a) does not apply here.

Solve each of the given equations for the indicated variable. Assume that all denominators are nonzero and that all variables represent positive real numbers. (See Example 11.)

67. $S = \frac{1}{2}gt^2$ for t

68. $a = \pi r^2$ for r

69. $L = \frac{d^4k}{h^2}$ for h

70. $F = \frac{kMv^2}{r}$ for v

71. $P = \frac{E^2R}{(r + R)^2}$ for R

72. $S = 2\pi rh + 2\pi r^2$ for r

73. Solve the equation $z^4 - 2z^2 = 15$ as follows.

(a) Let $x = z^2$ and write the equation in terms of x.

(b) Solve the new equation for x.

(c) Set z^2 equal to each positive answer in part (b) and solve the resulting equation.

Solve each of the given equations. (See Exercise 73.)

74. $6p^4 = p^2 + 2$

75. $2q^4 + 3q^2 - 9 = 0$

76. $4a^4 = 2 - 7a^2$

77. $z^4 - 3z^2 - 1 = 0$

78. $2r^4 - r^2 - 5 = 0$

*Based on E. F. Harris, J. D. Hicks, and B. D. Barcroft, "Tissue Contributions to Sex and Race: Differences in Tooth Crown Size of Deciduous Molars," *American Journal of Physical Anthropology*, Vol. 115, 2001, p. 223.

✓ Checkpoint Answers

1. **(a)** 6, −2 **(b)** 3/5, −5 **(c)** 9/2, −5/3, −3

2. **(a)** 2, −5 **(b)** 1/2, −5 **(c)** 9/4, 0

3. **(a)** $\pm\sqrt{21}$ **(b)** $-7 \pm \sqrt{15}$ **(c)** $(3 \pm \sqrt{5})/2$

4. **(a)** $x = 1 + \sqrt{3}$ or $1 - \sqrt{3}$

(b) $u = 3 + \sqrt{5}$ or $3 - \sqrt{5}$

5. $x \approx 7.2$ or $x \approx -6.5529$

6. **(a)** 1/3 **(b)** −7/2 **(c)** No real solutions

7. **(a)** 2 **(b)** 0 **(c)** 1

8. **(a)** $x^2 + (.75x)^2 = 14^2$

(b) 11.2 by 8.4 inches

(c) 94.08 square inches

9. 5 inches by 7 inches

10. **(a)** $p = \frac{b \pm \sqrt{b^2 + 4mk}}{2m}$

(b) $k = \pm\sqrt{\frac{3r}{AP}}$ or $\frac{\pm\sqrt{3rAP}}{AP}$

CHAPTER 1 ▸ Summary

KEY TERMS AND SYMBOLS

1.1 ▸ $\approx$ is approximately equal to
π pi
$|a|$ absolute value of a
real number
natural (counting) number
whole number
integer
rational number
irrational number
properties of real numbers
additive inverse
multiplicative inverse
order of operations
square roots
number line
interval
interval notation
absolute value

1.2 ▸ a^n a to the power n
exponent or power
multiplication with exponents
power of a power rule
zero exponent
base
polynomial
variable
coefficient
term
constant term
degree of a polynomial
zero polynomial
leading coefficient
quadratics
cubics
like terms
FOIL
revenue
fixed cost
variable cost
profit

1.3 ▸ factor
factoring
greatest common factor
difference of squares
perfect squares
sum and difference of cubes

1.4 ▸ rational expression
cancellation property
operations with rational expressions
complex fraction

1.5 ▸ $a^{1/n}$ nth root of a
$\sqrt{a}$ square root of a
$\sqrt[n]{a}$ nth root of a
properties of exponents
radical
radicand
index

rationalizing the denominator
rationalizing the numerator

1.6 first-degree equation
solution of an equation
properties of equality
extraneous solution

solving for a specified variable
absolute-value equations
solving applied problems

1.7 quadratic equation
real solution
zero-factor property
square-root property
quadratic formula
discriminant
Pythagorean theorem

CHAPTER 1 KEY CONCEPTS

Factoring $x^2 + 2xy + y^2 = (x + y)^2$

$x^2 - 2xy + y^2 = (x - y)^2$

$x^2 - y^2 = (x + y)(x - y)$

$x^3 - y^3 = (x - y)(x^2 + xy + y^2)$

$x^3 + y^3 = (x + y)(x^2 - xy + y^2)$

Properties of Radicals Let a and b be real numbers, n be a positive integer, and m be any integer for which the given relationships exist. Then

$a^{m/n} = \sqrt[n]{a^m} = (\sqrt[n]{a})^m$; $\quad \sqrt[n]{a^n} = |a|$ if n is even; $\quad \sqrt[n]{a^n} = a$ if n is odd;

$\sqrt[n]{a} \cdot \sqrt[n]{b} = \sqrt[n]{ab}$; $\quad \dfrac{\sqrt[n]{a}}{\sqrt[n]{b}} = \sqrt[n]{\dfrac{a}{b}} \quad (b \neq 0)$.

Properties of Exponents Let a, b, r, and s be any real numbers for which the following exist. Then

$a^{-r} = \dfrac{1}{a^r}$ $\quad a^0 = 1$ $\quad \left(\dfrac{a}{b}\right)^r = \dfrac{a^r}{b^r}$

$a^r \cdot a^s = a^{r+s}$ $\quad (a^r)^s = a^{rs}$ $\quad a^{1/r} = \sqrt[r]{a}$

$\dfrac{a^r}{a^s} = a^{r-s}$ $\quad (ab)^r = a^r b^r$ $\quad \left(\dfrac{a}{b}\right)^{-r} = \left(\dfrac{b}{a}\right)^r$

Absolute Value Assume that a and b are real numbers with $b > 0$.
The solutions of $|a| = b$ or $|a| = |b|$ are $a = b$ or $a = -b$.

Quadratic Equations Facts needed to solve quadratic equations (in which a, b, and c are real numbers):

Factoring If $ab = 0$, then $a = 0$ or $b = 0$ or both.

Square-Root Property If $b > 0$, then the solutions of $x^2 = b$ are $\sqrt{b}$ and $-\sqrt{b}$.

Quadratic Formula The solutions of $ax^2 + bx + c = 0$ (with $a \neq 0$) are

$$x = \frac{-b \pm \sqrt{b^2 - 4ac}}{2a}.$$

Discriminant There are two real solutions of $ax^2 + bx + c = 0$ if $b^2 - 4ac > 0$, one real solution if $b^2 - 4ac = 0$, and no real solutions if $b^2 - 4ac < 0$.

CHAPTER 1 REVIEW EXERCISES

Name the numbers from the list $-12, -6, -9/10, -\sqrt{7}, -\sqrt{4}, 0, 1/8, \pi/4, 6,$ *and* $\sqrt{11}$ *that are*

1. whole numbers
2. integers
3. rational numbers
4. irrational numbers

Identify the properties of real numbers that are illustrated in each of the following expressions.

5. $9[(-3)4] = 9[4(-3)]$
6. $7(4 + 5) = (4 + 5)7$
7. $6(x + y - 3) = 6x + 6y + 6(-3)$
8. $11 + (5 + 3) = (11 + 5) + 3$

Express each statement in symbols.

9. x is at least 9.
10. x is negative.

Write the following numbers in numerical order from smallest to largest.

11. $-7, -3, 8, \pi, -2, 0$

12. $\frac{5}{6}, \frac{1}{2}, -\frac{2}{3}, -\frac{5}{4}, -\frac{3}{8}$

13. $|6-4|, -|-2|, |8+1|, -|3-(-2)|$

14. $\sqrt{7}, -\sqrt{8}, -|\sqrt{16}|, |-\sqrt{12}|$

Write the following without absolute-value bars.

15. $-|-4| + |3|$
16. $|-6| + |-9|$
17. $7 - |-8|$
18. $|-3| - |-9 + 6|$

Graph each of the following on a number line.

19. $x \geq -3$
20. $-4 < x \leq 6$
21. $x < -2$
22. $x \leq 1$

Use the order of operations to simplify each of the following.

23. $(-6 + 3 \cdot 5)(-2)$
24. $-4(-8 - 9 \div 3)$
25. $\dfrac{-9 + (-6)(-3) \div 9}{6 - (-3)}$
26. $\dfrac{20 \div 4 \cdot 2 \div 5 - 1}{-9 - (-3) - 12 \div 3}$

Perform each of the indicated operations.

27. $(3x^4 - x^2 + 5x) - (-x^4 + 3x^2 - 6x)$
28. $(-8y^3 + 8y^2 - 5y) - (2y^3 + 4y^2 - 10)$
29. $-2(q^4 - 3q^3 + 4q^2) + 4(q^4 + 2q^3 + q^2)$
30. $5(3y^4 - 4y^5 + y^6) - 3(2y^4 + y^5 - 3y^6)$
31. $(4z + 2)(3z - 2)$
32. $(8p - 4)(5p + 2)$
33. $(5k - 2h)(5k + 2h)$
34. $(2r - 5y)(2r + 5y)$
35. $(3x + 4y)^2$
36. $(2a - 5b)^2$

Factor each of the following as completely as possible.

37. $2kh^2 - 4kh + 5k$
38. $2m^2n^2 + 6mn^2 + 16n^2$
39. $5a^4 + 12a^3 + 4a^2$
40. $24x^3 + 4x^2 - 4x$
41. $6y^2 - 13y + 6$
42. $8q^2 + 3m + 4qm + 6q$
43. $25a^2 - 20a + 4$
44. $36p^2 + 12p + 1$
45. $144p^2 - 169q^2$
46. $81z^2 - 25x^2$
47. $27y^3 - 1$
48. $125a^3 + 216$

Perform each operation.

49. $\dfrac{3x}{5} \cdot \dfrac{45x}{12}$

50. $\dfrac{5k^2}{24} - \dfrac{70k}{36}$

51. $\dfrac{c^2 - 3c + 2}{2c(c - 1)} \div \dfrac{c - 2}{8c}$

52. $\dfrac{p^3 - 2p^2 - 8p}{3p(p^2 - 16)} \div \dfrac{p^2 + 4p + 4}{9p^2}$

53. $\dfrac{2y - 6}{5y} \cdot \dfrac{20y - 15}{14}$

54. $\dfrac{m^2 - 2m}{15m^3} \cdot \dfrac{5}{m^2 - 4}$

55. $\dfrac{2m^2 - 4m + 2}{m^2 - 1} \div \dfrac{6m + 18}{m^2 + 2m - 3}$

56. $\dfrac{x^2 + 6x + 5}{4(x^2 + 1)} \cdot \dfrac{2x(x + 1)}{x^2 - 25}$

57. $\dfrac{6}{15z} + \dfrac{2}{3z} - \dfrac{9}{10z}$

58. $\dfrac{5}{y - 3} - \dfrac{4}{y}$

59. $\dfrac{2}{5q} + \dfrac{10}{7q}$

60. Give two ways to evaluate $125^{2/3}$ and then compare them. Which do you prefer? Why?

Simplify each of the given expressions. Write all answers without negative exponents. Assume that all variables represent positive real numbers.

61. 5^{-3}
62. 10^{-2}
63. -8^0
64. -5^{-1}
65. $\left(-\frac{5}{6}\right)^{-2}$
66. $\left(\frac{3}{2}\right)^{-3}$
67. $4^6 \cdot 4^{-3}$
68. $7^{-5} \cdot 7^{-2}$
69. $\dfrac{8^{-5}}{8^{-4}}$
70. $\dfrac{6^{-3}}{6^4}$
71. $\dfrac{9^4 \cdot 9^{-5}}{(9^{-2})^2}$
72. $\dfrac{k^4 \cdot k^{-3}}{(k^{-2})^{-3}}$
73. $5^{-1} + 2^{-1}$
74. $5^{-2} + 5^{-1}$
75. $125^{2/3}$
76. $128^{3/7}$
77. $8^{-5/3}$
78. $\left(\frac{144}{49}\right)^{-1/2}$
79. $\dfrac{5^{1/3} \cdot 5^{1/2}}{5^{3/2}}$
80. $\dfrac{2^{3/4} \cdot 2^{-1/2}}{2^{1/4}}$
81. $(3a^2)^{1/2} \cdot (3^2a)^{3/2}$
82. $(4p)^{2/3} \cdot (2p^3)^{3/2}$

Simplify each of the following expressions.

83. $\sqrt[3]{27}$
84. $\sqrt[6]{-64}$
85. $\sqrt{99}$
86. $\sqrt{63}$
87. $\sqrt[3]{54p^3q^5}$
88. $\sqrt[4]{64a^5b^3}$
89. $\sqrt{\dfrac{5n^2}{6m}}$
90. $\sqrt{\dfrac{3x^3}{2z}}$
91. $3\sqrt{3} - 12\sqrt{12}$
92. $8\sqrt{7} + 2\sqrt{63}$
93. $(\sqrt{8} - 1)(\sqrt{8} + 1)$
94. $(\sqrt{7} - \sqrt{3})(\sqrt{7} + \sqrt{3})$

Rationalize each denominator.

95. $\dfrac{\sqrt{3}}{1 + \sqrt{2}}$
96. $\dfrac{4 + \sqrt{2}}{4 - \sqrt{5}}$

Social Science In our system of government, the president is elected by the electoral college and not by individual voters. Because of this, smaller states have a greater voice in the selection of a president than they otherwise would have. Two political scientists have studied the problems of campaigning for president under the current system and have concluded that candidates should allot their money according to the formula

$$\text{Amount for large state} = \left(\frac{E_{\text{large}}}{E_{\text{small}}}\right)^{3/2} \times \text{amount for small state.}$$

Here, E_{large} represents the electoral vote of the large state, and E_{small} represents the electoral vote of the small state. Find the amount that should be spent in each of the following larger states if \$1,000,000 is spent in the small state and the following statements are true.

97. Florida has 27 electoral votes, and Vermont has 3.

98. New York has 31 electoral votes, and Kansas has 6.

99. Nevada has 5 electoral votes, and California has 55.

100. Wisconsin has 10 electoral votes, and Texas has 34.

Solve each equation.

101. $3x - 4(x - 2) = 2x + 9$

102. $4y + 9 = -3(1 - 2y) + 5$

103. $\frac{2z}{5} - \frac{4z - 3}{10} = \frac{-z + 1}{10}$

104. $\frac{p}{p + 2} - \frac{3}{4} = \frac{2}{p + 2}$

105. $\frac{2m}{m - 3} = \frac{6}{m - 3} + 4$

106. $\frac{15}{k + 5} = 4 - \frac{3k}{k + 5}$

Solve for x.

107. $8ax - 3 = 2x$

108. $6x - 5y = 4bx$

109. $\frac{2x}{3 - c} = ax + 1$

110. $b^2x - 2x = 4b^2$

Solve each equation.

111. $|m - 4| = 7$

112. $|5 - x| = 15$

113. $\left|\frac{2 - y}{5}\right| = 8$

114. $|4k + 1| = |6k - 3|$

115. **Energy** World oil demand D (in millions of barrels per day equivalent) is approximated by $D = 3.7x + 111$, where $x = 0$ corresponds to the year 1970.* Find the year in which demand reached or will reach
(a) 259 millions of barrels per day;
(b) 300 millions of barrels per day.

*Based on data and projections from Exxon Mobil.

116. **Business** A laser printer is on sale for 15% off. The sale price is \$306. What was the original price?

117. **Social Science** The number N of daily newspapers in the United States can be approximated by

$$N = -10.2x + 1801,$$

where x is the number of years after 1970.*
(a) In what year were there about 1444 daily newspapers?
(b) According to this model, how many daily newspapers are there this year?

118. **Finance** Ellen borrowed \$500 from a credit union at 12% annual interest and got \$250 in cash with her credit card at 18% annual interest. What single rate of interest on \$750 results in the same total amount of interest?

119. **Finance** A real estate firm invests \$100,000 in proceeds from a sale in two ways. The first portion is invested in a shopping center that provides an annual return of 8%. The rest is invested in a small apartment building with an annual return of 5%. The firm wants an annual income of \$6800 from these investments. How much should be put into each investment?

120. **Business** To make a special mix for Valentine's Day, the owner of a candy store wants to combine chocolate hearts that sell for \$5 per pound with candy kisses that sell for \$3.50 per pound. How many pounds of each kind should be used to get 30 pounds of a mix that can be sold for \$4.50 per pound?

Determine the number of real solutions of each quadratic equation.

121. $x^2 - 6x = 4$

122. $-3x^2 + 5x + 2 = 0$

123. $4x^2 - 12x + 9 = 0$

124. $5x^2 + 2x + 1 = 0$

125. $x^2 + 3x + 5 = 0$

Find all real solutions of each equation.

126. $(b + 7)^2 = 5$

127. $(2p + 1)^2 = 7$

128. $2p^2 + 3p = 2$

129. $2y^2 = 15 + y$

130. $x^2 - 2x = 2$

131. $r^2 + 4r = 1$

132. $2m^2 - 12m = 11$

133. $9k^2 + 6k = 2$

134. $2a^2 + a - 15 = 0$

135. $12x^2 = 8x - 1$

136. $2q^2 - 11q = 21$

137. $3x^2 + 2x = 16$

138. $6k^4 + k^2 = 1$

139. $21p^4 = 2 + p^2$

140. $2x^4 = 7x^2 + 15$

141. $3m^4 + 20m^2 = 7$

142. $3 = \frac{13}{z} + \frac{10}{z^2}$

Solve each equation for the specified variable.

143. $p = \frac{E^2R}{(r + R)^2}$ for r

144. $p = \frac{E^2R}{(r + R)^2}$ for E

145. $K = s(s - a)$ for s

146. $kz^2 - hz - t = 0$ for z

*Based on data in the *Statistical Abstract of the United States: 2008,* Table 1102.

147. Physical Science The atmospheric pressure a (in pounds per square foot) at height h thousand feet above sea level is approximately $a = .8315h^2 - 73.93h + 2116.1$.

(a) Find the atmospheric pressure at sea level and at the top of Mount Everest, the tallest mountain in the world (29,035 feet). Remember that h is measured in thousands.

(b) The atmospheric pressure at the top of Mount Rainier is 1223.43 pounds per square foot. How high is Mount Rainier?

148. Business A landscaper wants to put a cement walk of uniform width around a rectangular garden that measures 24 by 40 feet. She has enough cement to cover 740 square feet. To the nearest tenth of a foot, how wide should the walk be in order to use up all the cement?

149. Business A recreation director wants to fence off a rectangular playground beside an apartment building. The building forms one boundary, so she needs to fence only the other three sides. The area of the playground is to be 11,250 square meters. She has enough material to build 325 meters of fence. Find the length and width of the playground.

150. Transportation Two cars leave an intersection at the same time. One travels north, and the other heads west traveling 10 mph faster. After 1 hour, they are 50 miles apart. What were their speeds?

151. Physical Science A rocket loaded with fireworks is to be shot vertically upward from ground level with an initial velocity of 200 feet per second. When the rocket reaches a height of 400 feet on its upward trip, the fireworks will be detonated. How many seconds after liftoff will the detonation take place?

152. Geometry How long is piece x in the right triangle in the figure?

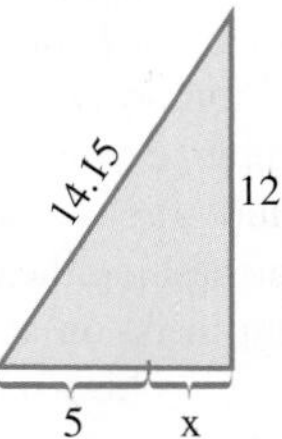

CASE 1

Consumers Often Defy Common Sense*

Imagine two refrigerators in the appliance section of a department store. One sells for \$700 and uses \$85 worth of electricity a year. The other is \$100 more expensive, but costs only \$25 a year to run. Given that either refrigerator should last at least 10 years without repair, consumers would overwhelmingly buy the second model, right?

Well, not exactly. Many studies by economists have shown that in a wide range of decisions about money—from paying taxes to buying major appliances—consumers consistently make decisions that defy common sense.

In some cases—as in the refrigerator example—this means that people are generally unwilling to pay a little more money up front to save a lot of money in the long run. At times,

*"Consumers' Choices About Money Consistently Defy Common Sense." Malcolm Gladwell, *The Washington Post*, February 12, 1990.

psychological studies have shown, consumers appear to assign entirely whimsical values to money, values that change depending on time and circumstances.

In recent years, these apparently irrational patterns of human behavior have become a subject of intense interest among economists and psychologists, both for what they say about the way the human mind works and because of their implications for public policy.

How, for example, can the United States move toward a more efficient use of electricity if so many consumers refuse to buy energy-efficient appliances even when such a move is in their own best interest?

At the heart of research into the economic behavior of consumers is a concept known as the discount rate. It is a measure of how consumers compare the value of a dollar received today with one received tomorrow.

Consider, for example, if you won $1000 in a lottery. How much more money would officials have to give you before you would agree to postpone cashing the check for a year?

Some people might insist on at least another $100, or 10 percent, since that is roughly how much it would take to make up for the combined effects of a year's worth of inflation and lost interest.

But the studies show that someone who wants immediate gratification might not be willing to postpone receiving the $1000 for 20 percent, 30 percent, or even 40 percent more money.

In the language of economists, this type of person has a high discount rate: He or she discounts the value of $1000 so much over a year that it would take hundreds of extra dollars to make waiting as attractive as getting the money immediately.

Of the two alternatives, waiting a year for more money is clearly more rational than taking the check now. Why would people turn down $1400 dollars next year in favor of $1000 today? Even if they needed the $1000 immediately, they would be better off borrowing it from a bank, even at 20 percent or even 30 percent interest. Then, a year later, they could pay off the loan—including the interest—with the $1400 and pocket the difference.

The fact is, however, that economists find numerous examples of such high discount rates implicit in consumer behavior.

While consumers were very much aware of savings to be made at the point of purchase, they so heavily discounted the value of monthly electrical costs that they would pay over the lifetime of their dryer or freezer that they were oblivious to the potential for greater savings.

Gas water heaters, for example, were found to carry an implicit discount rate of 100 percent. This means that in deciding which model was cheapest over the long run, consumers acted as if they valued a $100 gas bill for the first year as if it were really $50. Then, in the second year, they would value the next $100 gas bill as if it were really worth $25, and so on through the life of the appliance.

Few consumers actually make this formal calculation, of course. But there are clearly bizarre behavioral patterns in evidence.

Some experiments, for example, have shown that the way in which consumers make decisions about money depends a great deal on how much is at stake. Few people are willing to give up $10 now for $15 next year. But they are if the choice is between $100 now and $150 next year, a fact that would explain why consumers appear to care less about many small electricity bills—even if they add up to a lot—than one big initial outlay.

EXERCISES

1. Suppose a refrigerator that sells for $700 costs $85 a year for electricity. Write an expression for the cost to buy and run the refrigerator for x years.
2. Suppose another refrigerator sells for $1000 and costs $25 a year for electricity. Write an expression for the total cost for this refrigerator over x years.
3. Over 10 years, which refrigerator costs the most? By how much?
4. In how many years will the total costs for the two refrigerators be equal?

CHAPTER

2

Graphs, Lines, and Inequalities

Data from current and past events is often a useful tool in business and in the social and health sciences. Gathering data is the first step in developing mathematical models that can be used to analyze a situation and predict future performance. For examples of linear models in business, transportation, and health science, see Exercises 18, 21, and 22 on pages 110–111.

CASE 2: Using Extrapolation for Prediction

Graphical representations of data are commonly used in business and in the health and social sciences. Lines, equations, and inequalities play an important role in developing mathematical models from such data. This chapter presents both algebraic and graphical methods for dealing with these topics.

2.1 ▸ Graphs

Just as the number line associates the points on a line with real numbers, a similar construction in two dimensions associates points in a plane with *ordered pairs* of real numbers. A **Cartesian coordinate system**, as shown in Figure 2.1, consists of a horizontal number line (usually called the **x-axis**) and a vertical number line (usually called the **y-axis**). The point where the number lines meet is called the **origin**. Each point in a Cartesian coordinate system is labeled with an **ordered pair** of real numbers, such as $(-2, 4)$ or $(3, 2)$. Several points and their corresponding ordered pairs are shown in Figure 2.1.

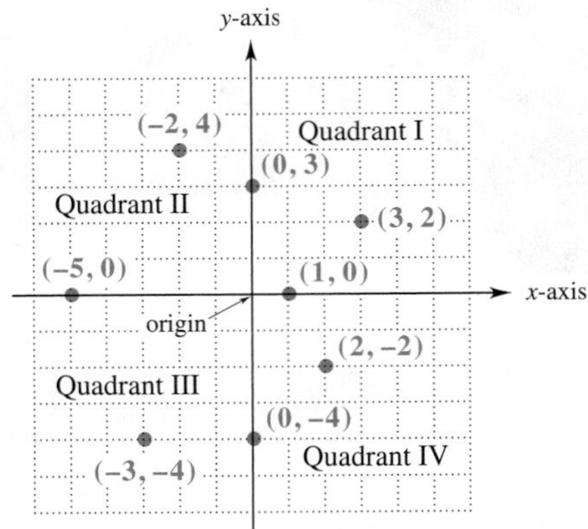

FIGURE 2.1

For the point labeled $(-2, 4)$, for example, -2 is the **x-coordinate** and 4 is the **y-coordinate**. You can think of these coordinates as directions telling you how to move to this point from the origin: You go 2 horizontal units to the left (x-coordinate) and 4 vertical units upward (y-coordinate). From now on, instead of referring to "the point labeled by the ordered pair $(-2, 4)$," we will say "the point $(-2, 4)$." ✓1

The x-axis and the y-axis divide the plane into four parts, or **quadrants**, which are numbered as shown in Figure 2.1. The points on the coordinate axes belong to no quadrant.

✓Checkpoint 1

Locate $(-1, 6)$, $(-3, -5)$, $(4, -3)$, $(0, 2)$, and $(-5, 0)$ on a coordinate system.

Answer: See page 86.

EQUATIONS AND GRAPHS

A **solution of an equation** in two variables, such as

$$y = -2x + 3$$

or

$$y = x^2 + 7x - 2,$$

is an ordered pair of numbers such that the substitution of the first number for x and the second number for y produces a true statement.

✓Checkpoint 2

Which of the following are solutions of

$$y = x^2 + 5x - 3?$$

(a) $(1, 3)$

(b) $(-2, -3)$

(c) $(-1, -7)$

Answer: See page 87.

Example 1 Which of the following are solutions of $y = -2x + 3$?

(a) $(2, -1)$

Solution This is a solution of $y = -2x + 3$ because "$-1 = -2 \cdot 2 + 3$" is a true statement.

(b) $(4, 7)$

Solution Since $-2 \cdot 4 + 3 = -5$, and not 7, the ordered pair $(4, 7)$ is not a solution of $y = -2x + 3$. ✓2

Equations in two variables, such as $y = -2x + 3$, typically have an infinite number of solutions. To find one, choose a number for x and then compute the value of y that produces a solution. For instance, if $x = 5$, then $y = -2 \cdot 5 + 3 = -7$, so that the pair $(5, -7)$ is a solution of $y = -2x + 3$. Similarly, if $x = 0$, then $y = -2 \cdot 0 + 3 = 3$, so that $(0, 3)$ is also a solution.

The **graph** of an equation in two variables is the set of points in the plane whose coordinates (ordered pairs) are solutions of the equation. Thus, the graph of an equation is a picture of its solutions. Since a typical equation has infinitely many solutions, its graph has infinitely many points.

Example 2 Sketch the graph of $y = -2x + 5$.

Solution Since we cannot plot infinitely many points, we construct a table of y-values for a reasonable number of x-values, plot the corresponding points, and make an "educated guess" about the rest. The table of values and points in Figure 2.2 suggests that the graph is a straight line, as shown in Figure 2.3. ✓3

✓Checkpoint 3

Graph $x = 5y$.

Answer: See page 87.

x	−1	0	2	4	5
$-2x + 5$	7	5	1	−3	−5

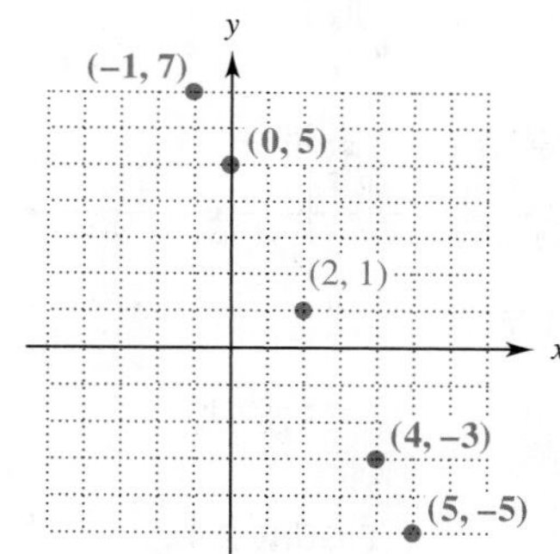

FIGURE 2.2

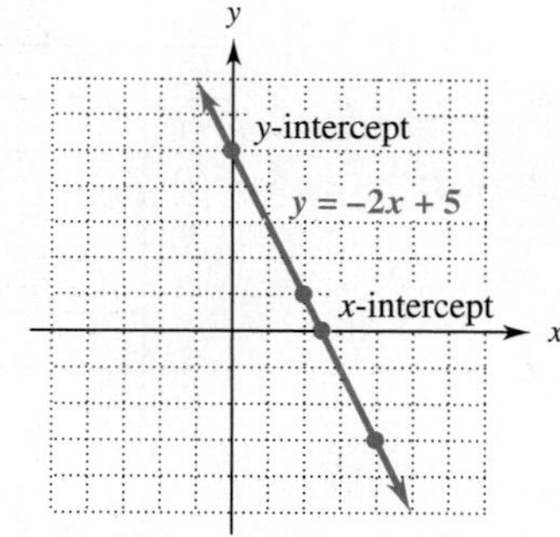

FIGURE 2.3

An ***x*-intercept** of a graph is the x-coordinate of a point where the graph intersects the x-axis. The y-coordinate of this point is 0, since it is on the axis. Consequently, to find the x-intercepts of the graph of an equation, set $y = 0$ and solve for x. For instance, in Example 2, the x-intercept of the graph of $y = -2x + 5$ (see Figure 2.3) is found by setting $y = 0$ and solving for x:

$$0 = -2x + 5$$
$$2x = 5$$
$$x = \frac{5}{2}.$$

Similarly, a ***y*-intercept** of a graph is the y-coordinate of a point where the graph intersects the y-axis. The x-coordinate of this point is 0. The y-intercepts are found by setting $x = 0$ and solving for y. For example, the graph of $y = -2x + 5$ in Figure 2.3 has y-intercept 5. ✓4

✓Checkpoint 4

Find the x- and y-intercepts of the graphs of these equations.

(a) $3x + 4y = 12$

(b) $5x - 2y = 8$

Answers: See page 87.

Example 3 Find the x- and y-intercepts of the graph of $y = x^2 - 2x - 8$, and sketch the graph.

Solution To find the y-intercept, set $x = 0$ and solve for y:

$$y = x^2 - 2x - 8 = 0^2 - 2 \cdot 0 - 8 = -8.$$

The y-intercept is -8. To find the x-intercept, set $y = 0$ and solve for x.

$$x^2 - 2x - 8 = y$$

$$x^2 - 2x - 8 = 0 \qquad \text{Set } y = 0.$$

$$(x + 2)(x - 4) = 0 \qquad \text{Factor.}$$

$$x + 2 = 0 \quad \text{or} \quad x - 4 = 0 \qquad \text{Zero-factor property}$$

$$x = -2 \quad \text{or} \quad x = 4$$

The x-intercepts are -2 and 4. Now make a table, using both positive and negative values for x, and plot the corresponding points, as in Figure 2.4. These points suggest that the entire graph looks like Figure 2.5.

x	$x^2 - 2x - 8$
−3	7
−2	0
−1	−5
0	−8
1	−9
3	−5
4	0
5	7

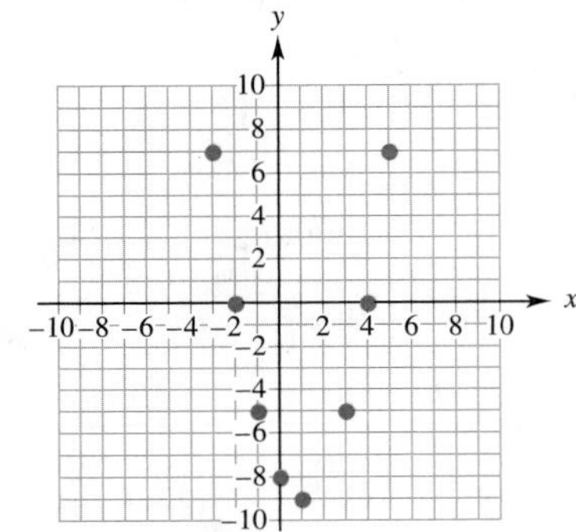

FIGURE 2.4

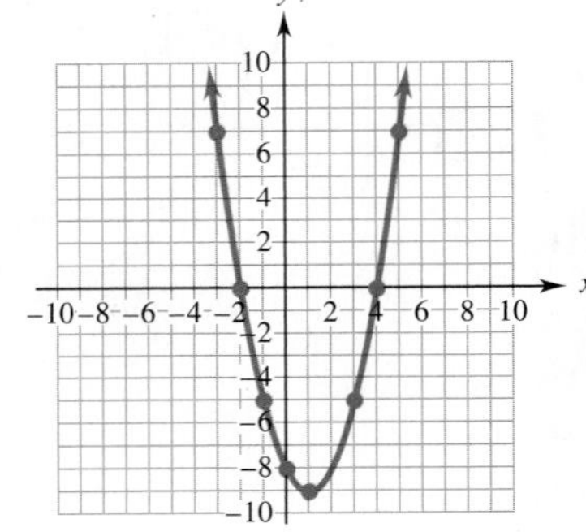

FIGURE 2.5

Example 4 Sketch the graph of $y = \sqrt{x + 2}$.

Solution Notice that $\sqrt{x + 2}$ is a real number only when $x + 2 \geq 0$—that is, when $x \geq -2$. Furthermore, $y = \sqrt{x + 2}$ is always nonnegative. Hence, all points on the graph lie on or above the x-axis and on or to the right of $x = -2$. Computing some typical values, we obtain the graph in Figure 2.6.

x	$\sqrt{x + 2}$
−2	0
0	$\sqrt{2} \approx 1.414$
2	2
5	$\sqrt{7} \approx 2.646$
7	3
9	$\sqrt{11} \approx 3.317$

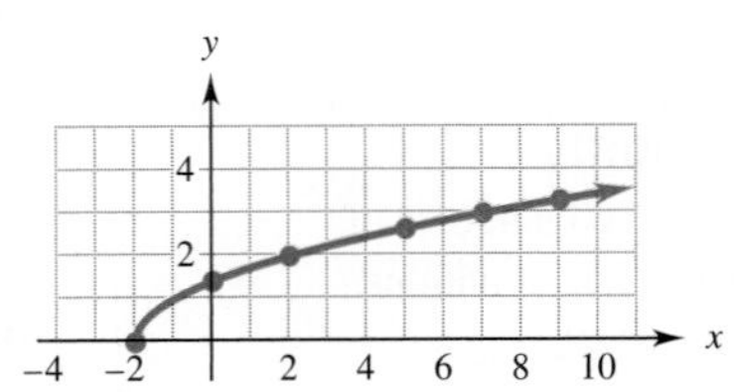

FIGURE 2.6

Example 2 shows that the solution of the equation $-2x + 5 = 0$ is the x-intercept of the graph of $y = -2x + 5$. Example 3 shows that the solutions of the equation $x^2 - 2x - 8 = 0$ are the x-intercepts of the graph $y = x^2 - 2x - 8$. Similar facts hold in the general case.

Intercepts and Equations

The real solutions of a one-variable equation of the form

expression in x = 0

are the x-intercepts of the graph of

y = same expression in x.

GRAPH READING

Information is often given in graphical form, so you must be able to read and interpret graphs—that is, translate graphical information into statements in English.

Example 5 **Natural Science** At various locations around the United States, the National Weather Service continuously records the temperature in graphical form. The results for July 2, 2008, in Salt Lake City, Utah, are displayed in Figure 2.7. The first coordinate of each point on the graph represents the time (measured in hours after midnight), and the second coordinate represents the temperature at that time.

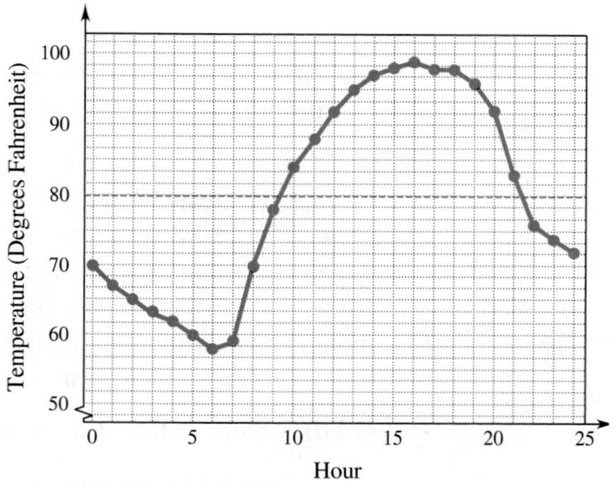

FIGURE 2.7

(a) What was the temperature at 8 AM and at 7 PM?

Solution The point (8, 70) is on the graph, which means that the temperature at 8 AM was 70° Fahrenheit. Now, 7 PM is 19 hours after midnight, and the point (19, 97) is on the graph. So the temperature at 7 PM was 97°.

(b) At what times during the day was the temperature below 80°?

Solution Look for the points whose second coordinates are less than 80—that is, points that lie below the horizontal line through 80° (shown in red in Figure 2.7). The first coordinates of these points are the times when the temperature was below 80°. The figure shows that these are the points whose first coordinates are less than approximately 9.5 or greater than about 21.5. Since 9.5 corresponds to 9:30 AM and 21.5 corresponds to 9:30 PM, the temperature was below 80° from midnight to 9:30 AM and 9:30 PM to midnight. ✓5

✓Checkpoint 5

In Example 5, what were the highest and lowest temperatures during the day? When did they occur?

Answers: See page 87.

The next example deals with the basic business relationship that was introduced in Section 1.2:

$$\text{Profit} = \text{Revenue} - \text{Cost}.$$

Example 6 **Business** Monthly revenue and costs for the Webster Cell Phone Company are determined by the number t of phones produced and sold, as shown in Figure 2.8.

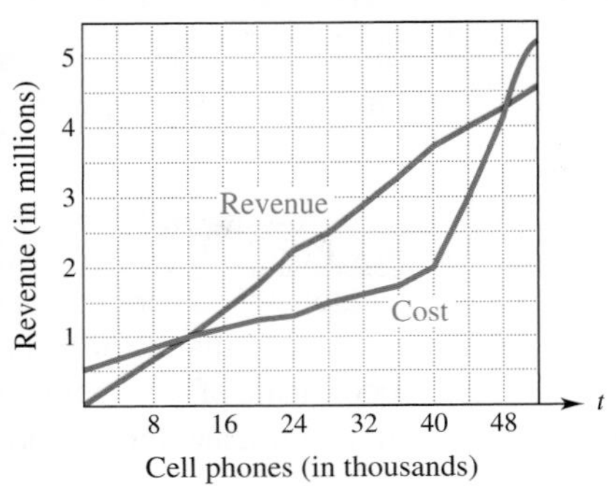

FIGURE 2.8

(a) How many phones should be produced each month if the company is to make a profit (assuming that all phones produced are sold)?

Solution Profit is revenue minus cost, so the company makes a profit whenever revenue is greater than cost—that is, when the revenue graph is above the cost graph. Figure 2.8 shows that this occurs between $t = 12$ and $t = 48$—that is, when 12,000 to 48,000 phones are produced. If the company makes fewer than 12,000 phones, it will lose money (because costs will be greater than revenue.) It also loses money by making more than 48,000 phones. (One reason might be that high production levels require large amounts of overtime pay, which drives costs up too much.)

(b) Is it more profitable to make 40,000 or 44,000 phones?

Solution On the revenue graph, the point with first coordinate 40 has second coordinate of approximately 3.7, meaning that the revenue from 40,000 phones is about 3.7 million dollars. The point with first coordinate 40 on the cost graph is (40, 2), meaning that the cost of producing 40,000 phones is 2 million dollars. Therefore, the profit on 40,000 phones is about $3.7 - 2 = 1.7$ million dollars. For 44,000 phones, we have the approximate points (44, 4) on the revenue graph and (44, 3) on the cost graph. So the profit on 44,000 phones is $4 - 3 = 1$ million dollars. Consequently, it is more profitable to make 40,000 phones. ✓6

✓Checkpoint 6

In Example 6, find the profit from making

(a) 32,000 phones;

(b) 4000 phones.

Answers: See page 87.

TECHNOLOGY AND GRAPHS

A graphing calculator or computer graphing program follows essentially the same procedure used when graphing by hand: The calculator selects a large number of x-values (95 or more), equally spaced along the x-axis, and plots the corresponding points, simultaneously connecting them with line segments. Calculator-generated graphs are generally quite accurate, although they may not appear as smooth as hand-drawn ones. The next example illustrates the basics of graphing on a graphing calculator. (Computer graphing software operates similarly.)

```
WINDOW
 Xmin=-9
 Xmax=9
 Xscl=2
 Ymin=-6
 Ymax=6
 Yscl=1
 Xres=1
```

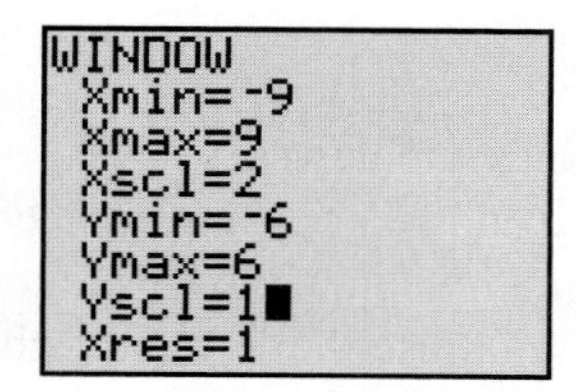

FIGURE 2.9

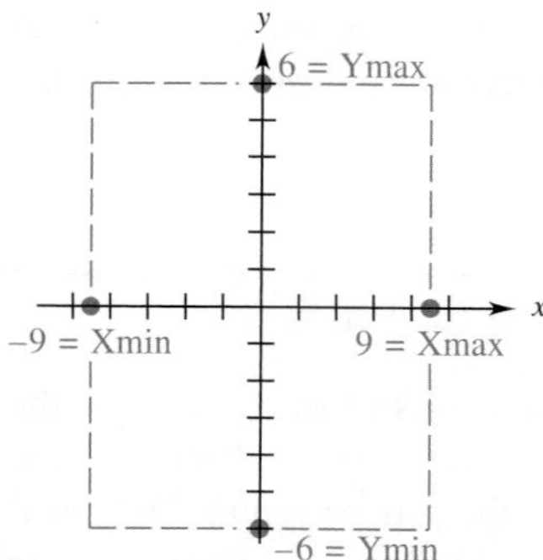

FIGURE 2.10

Example 7 Use a graphing calculator to sketch the graph of the equation $2x^3 - 2y - 10x + 2 = 0$.

Solution *First, set the* ***viewing window***—the portion of the coordinate plane that will appear on the screen. Press the WINDOW key (labeled RANGE or PLOT-SETUP on some calculators) and enter the appropriate numbers, as in Figure 2.9 (which shows the screen from a TI-84+; other calculators are similar). Then the calculator will display the portion of the plane inside the dashed lines shown in Figure 2.10—that is, the points (x, y) with $-9 \le x \le 9$ and $-6 \le y \le 6$.

In Figure 2.9, we have set Xscl = 2 and Yscl = 1, which means the **tick marks** on the x-axis are two units apart and the tick marks on the y-axis are one unit apart (as shown in Figure 2.10).

Second, enter the equation to be graphed in the equation memory. To do this, you must first solve the equation for y (because a calculator accepts only equations of the form y = expression in x):

$$2y = 2x^3 - 10x + 2$$
$$y = x^3 - 5x + 1.$$

Now press the Y = key (labeled SYMB on some calculators) and enter the equation, using the "variable key" for x. (This key is labeled X,T,θ,n or X,θ,T or x-VAR, depending on the calculator.) Figure 2.11 shows the equation entered on a TI-84+; other calculators are similar. Now press GRAPH (or PLOT or DRW on some calculators), and obtain Figure 2.12.

FIGURE 2.11

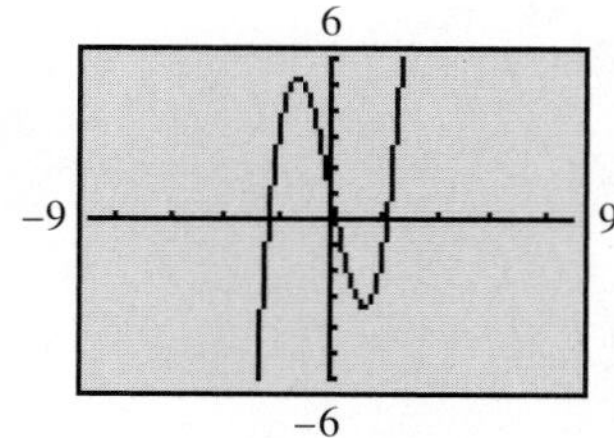

FIGURE 2.12

✓**Checkpoint 7**

Use a graphing calculator to graph $y = 18x - 3x^3$ in the following viewing windows:

(a) $-10 \le x \le 10$ and $-10 \le y \le 10$ with Xscl = 1, Yscl = 1;

(b) $-5 \le x \le 5$ and $-20 \le y \le 20$ with Xscl = 1, Yscl = 5.

Answers: See page 87.

Finally, if necessary, change the viewing window to obtain a more readable graph. It is difficult to see the y-intercept in Figure 2.12, so press WINDOW and change the viewing window (Figure 2.13) on the next page; then press GRAPH to obtain Figure 2.14, in which the y-intercept at $y = 1$ is clearly shown. (It isn't necessary to reenter the equation.) 7✓

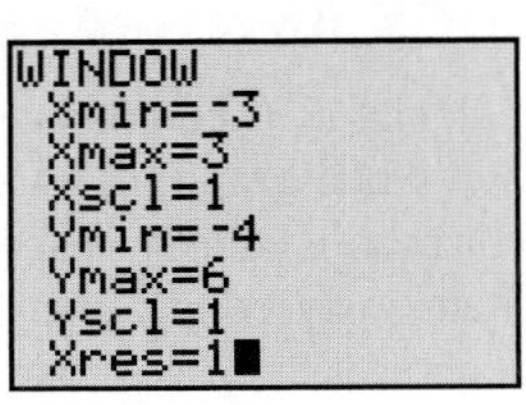

FIGURE 2.13

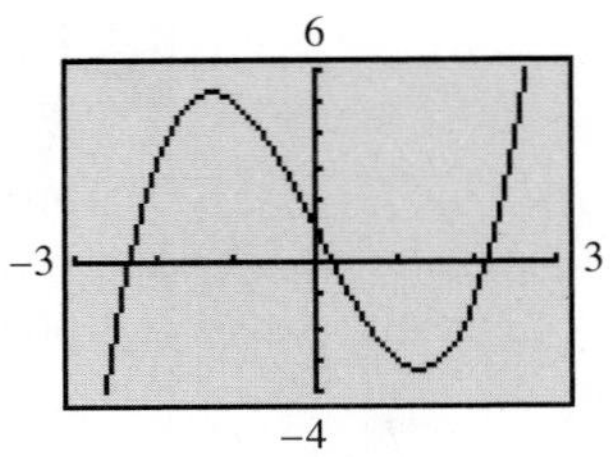

FIGURE 2.14

TECHNOLOGY TOOLS

In addition to graphing equations, graphing calculators (and graphing software) provide convenient tools for solving equations and reading graphs. For example, when you have graphed an equation, you can readily determine the points the calculator plotted. Press **trace** (a cursor will appear on the graph), and use the left and right arrow keys to move the cursor along the graph. The coordinates of the point the cursor is on appear at the bottom of the screen.

Recall that the solutions of an equation, such as $x^3 - 5x + 1 = 0$—are the x-intercepts of the graph of $y = x^3 - 5x + 1$. (See the box on page 79.) A **graphical root finder** enables you to find these x-intercepts and thus to solve the equation.

Checkpoint 8

Use a graphical root finder to approximate the third solution of the equation in Example 8.

Answer: See page 87.

Example 8 Use a graphical root finder to solve $x^3 - 5x + 1 = 0$.

Solution First, graph $y = x^3 - 5x + 1$. The x-intercepts of this graph are the solutions of the equation. To find these intercepts, look for "root" or "zero" in the appropriate menu.* Check your instruction manual for the proper syntax. A typical root finder (see Figure 2.15) shows that two of the solutions (x-intercepts) are $x \approx .2016$ and $x \approx 2.1284$. For the third solution, see the problem at the side.

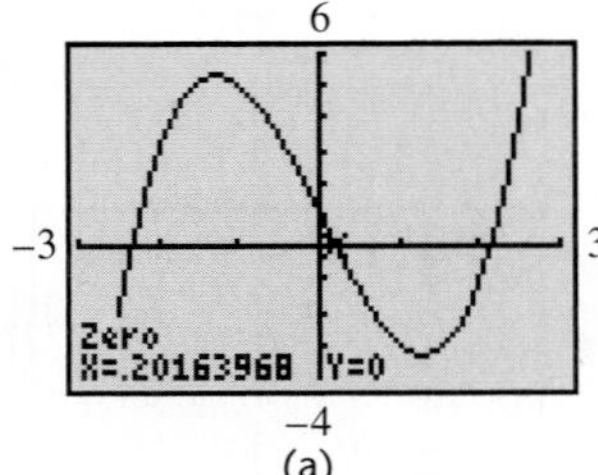

(a)

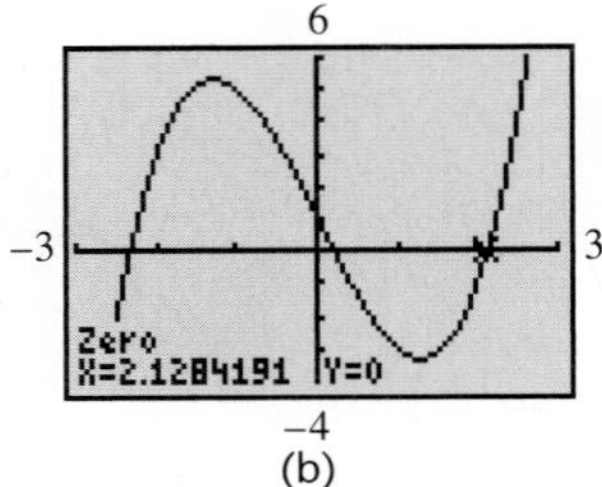

(b)

FIGURE 2.15

Many graphs have peaks and valleys (for instance, the graphs in Figure 2.15). A **maximum/minimum finder** provides accurate approximations of the locations of the "tops" of the peaks and the "bottoms" of the valleys.

*CALC on TI-84+, GRAPH MATH on TI-86, G-SOLV on Casio.

Example 9 **Health** The percentage y of 12th graders who have ever tried an illicit drug is approximated by

$$y = -.00374x^3 - .0591x^2 + 2.534x + 38.21,$$

where $x = 1$ corresponds to the year 1991.* In what year was the percentage of 12th graders who tried illicit drugs the greatest? What was that percentage?

Solution The graph of $y = -.00374x^3 - .0591x^2 + 2.534x + 38.21$ is shown in Figure 2.16. The highest percentage of 12th graders who tried illicit drugs corresponds to the highest point on this graph. To find this point, look for "maximum" or "max" or "extremum" in the same menu as the graphical root finder. Check your instruction manual for the proper syntax. A typical maximum finder (Figure 2.17) shows that the highest point has approximate coordinates (10.66, 53.98). The first coordinate is the year and the second is the percentage. So the largest percentage of 12th graders who tried illicit drugs (about 53.98%) occurred in mid-2000 ($x \approx 10.66$). ✓9

✓Checkpoint 9

Use a minimum finder to locate the approximate coordinates of the lowest point to the left of the y-axis on the graph of

$$y = 3x - 2x^3.$$

Answer: *See page 87.*

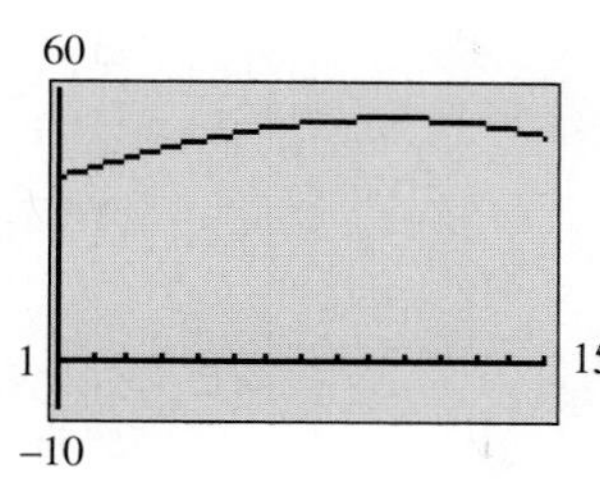

FIGURE 2.16

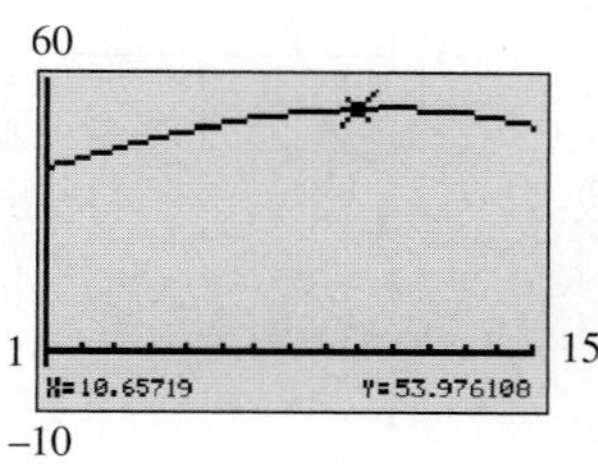

FIGURE 2.17

*L. D. Johnston, P. M. O'Malley, J. G. Bachman, and J. E. Schulenberg. "Teen smoking resumes decline." Ann Arbor, MI: University of Michigan News Service [Online]. Available: www.monitoringthefuture.org.

2.1 ▸ Exercises

State the quadrant in which each point lies.

1. $(1, -2), (-2, 1), (3, 4), (-5, -6)$
2. $(\pi, 2), (3, -\sqrt{2}), (4, 0), (-\sqrt{3}, \sqrt{3})$

Determine whether the given ordered pair is a solution of the given equation. (See Example 1.)

3. $(1, -3); 3x - y - 6 = 0$
4. $(2, -1); x^2 + y^2 - 6x + 8y = -15$
5. $(3, 4); (x - 2)^2 + (y + 2)^2 = 6$
6. $(1, -1); \frac{x^2}{2} + \frac{y^2}{3} = -4$

Sketch the graph of each of these equations. (See Example 2.)

7. $4y + 3x = 12$
8. $2x + 7y = 14$
9. $8x + 3y = 12$
10. $9y - 4x = 12$
11. $x = 2y + 3$
12. $x - 3y = 0$

List the x-intercepts and y-intercepts of each graph.

13.

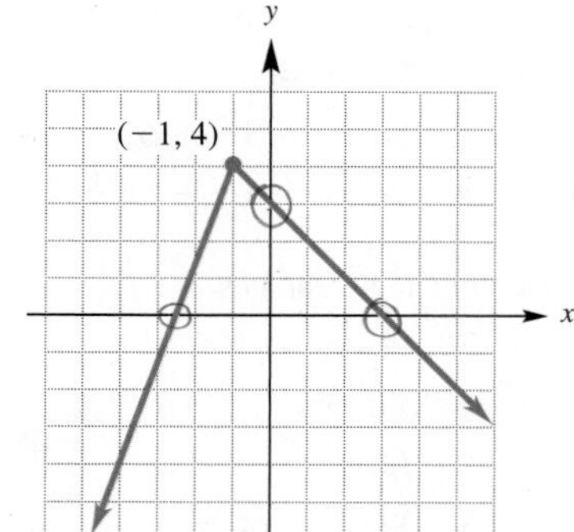

14.

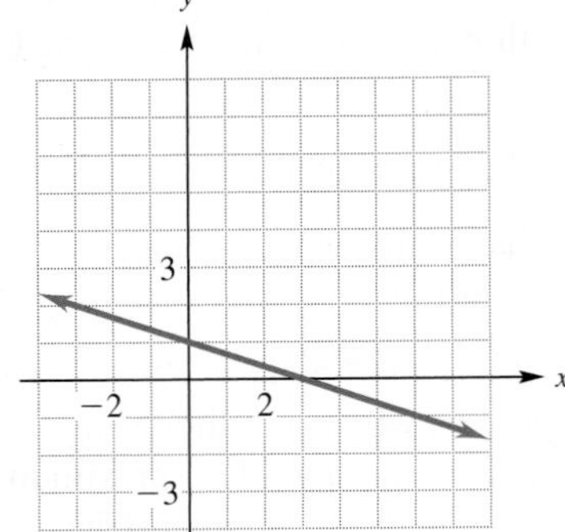

15.

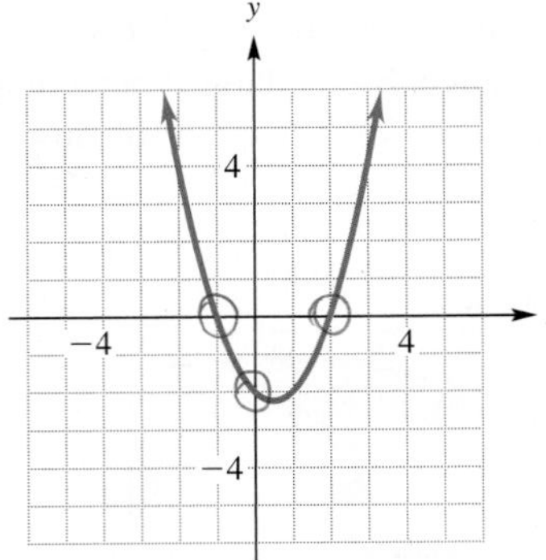

16.

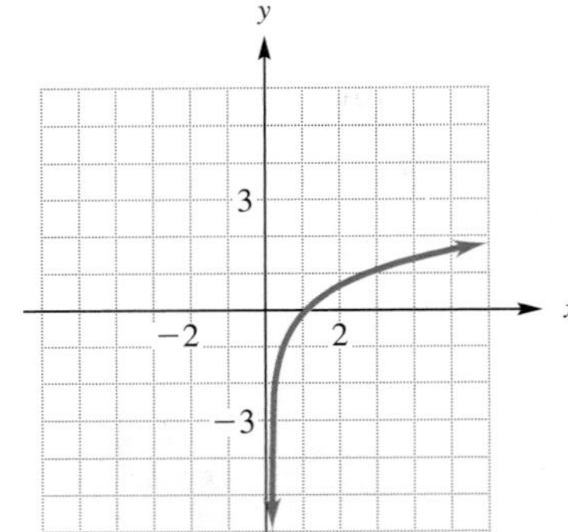

Find the x-intercepts and y-intercepts of the graph of each equation. You need not sketch the graph. (See Example 3.)

17. $3x + 4y = 12$ **18.** $x - 2y = 5$

19. $2x - 3y = 24$ **20.** $3x + y = 4$

21. $y = x^2 - 9$ **22.** $y = x^2 + 4$

23. $y = x^2 + x - 20$ **24.** $y = 5x^2 + 6x + 1$

25. $y = 2x^2 - 5x + 7$ **26.** $y = 3x^2 + 4x - 4$

Sketch the graph of the equation. (See Examples 2–4.)

27. $y = x^2$ **28.** $y = x^2 + 2$

29. $y = x^2 - 3$ **30.** $y = 2x^2$

31. $y = x^2 - 6x + 5$ **32.** $y = x^2 + 2x - 3$

33. $y = x^3$ **34.** $y = x^3 - 3$

35. $y = x^3 + 1$ **36.** $y = x^3/2$

37. $y = \sqrt{x + 4}$ **38.** $y = \sqrt{x - 2}$

39. $y = \sqrt{4 - x^2}$ **40.** $y = \sqrt{9 - x^2}$

Physical Science *The graphs for the hourly temperature of Portland, ME, and San Diego, CA, on July 8, 2008, are shown in the accompanying figure. Use them to do the next set of exercises. (See Examples 5 and 6.)*

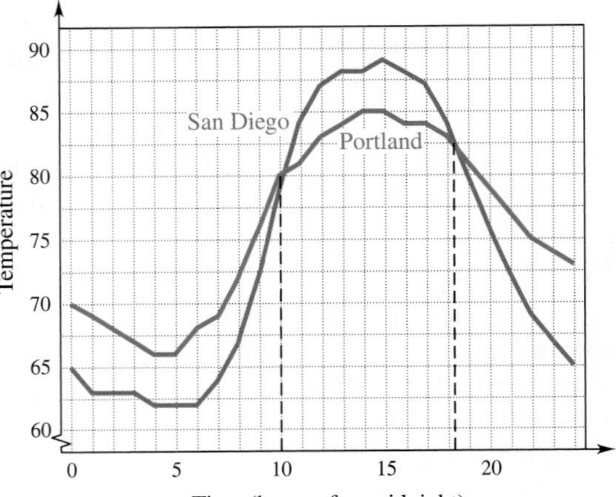

41. Approximately what time did the temperature first reach 75° in San Diego? in Portland?

42. At what times during the day did the two cities have the same temperature?

43. At what times during the day was it warmer in San Diego than Portland?

44. Was there any time when it was at least 10 degrees cooler in San Diego than Portland?

Business *Use the revenue and cost graphs for the Webster Cell Phone Company in Example 6 to do Exercises 45–48.*

45. Find the approximate cost of manufacturing the given number of phones.
(a) 20,000 **(b)** 36,000 **(c)** 48,000

46. Find the approximate revenue from selling the given number of phones.
(a) 12,000 **(b)** 24,000 **(c)** 36,000

47. Find the approximate profit from manufacturing the given number of phones.
(a) 20,000 **(b)** 28,000 **(c)** 36,000

48. The company must replace its aging machinery with better, but much more expensive, machines. In addition, raw material prices increase, so that monthly costs go up by $250,000. Owing to competitive pressure, phone prices cannot be increased, so revenue remains the same. Under these new circumstances, find the approximate profit from manufacturing the given number of phones.
(a) 20,000 **(b)** 36,000 **(c)** 40,000

Business *The graph below gives the annual per-person consumption of beef, chicken, and poultry (in pounds) from 1998 to 2006.* Use it to answer the given questions.*

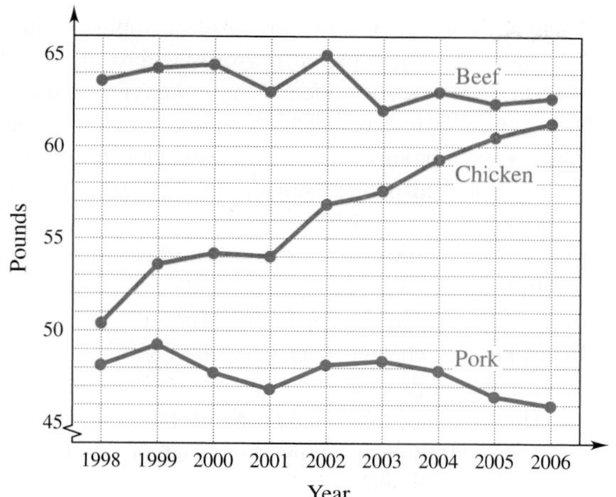

49. What is the approximate annual per-person consumption of beef, chicken, and pork in 2006?

50. In what year was annual per-person pork consumption the highest?

51. In what year was annual per-person beef consumption the highest?

52. How much has annual per-person chicken consumption increased between 1998 and 2006?

Business *The graph below shows the total spending on gasoline (in billions of dollars) within the United States from 2003 to 2012. (Years 2008–2012 are projections.)† Use the graph to answer the given questions.*

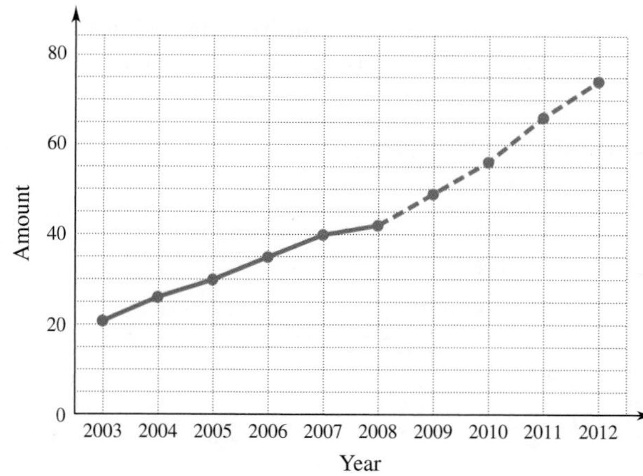

53. What was the total spending on gasoline in 2003?

54. In what years is spending below 50 billion dollars?

55. In what years is spending above 65 billion dollars?

56. How much is spending projected to increase from 2008 to 2012?

Business *The graph below shows the median weekly earnings for full-time employed women and men for four quarterly points during each year. (Median earnings mean that half the people earn more than this amount and half earn less.)* Use this graph to answer the given questions.*

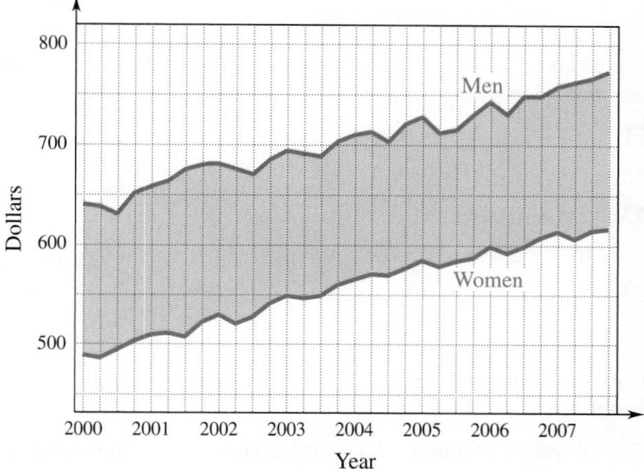

57. What were the approximate median earnings for men in 2006? for women?

58. What were the lowest median earnings for men, and in what year did this occur?

59. What were the highest median earnings for women, and in what year did this occur?

60. Was the difference between men's median earnings and women's median earnings ever less than $100 per week?

61. Was the difference between men's median earnings and women's median earnings ever more than $200 per week?

62. By how much did men's and women's median weekly earnings increase from 2000 to 2007?

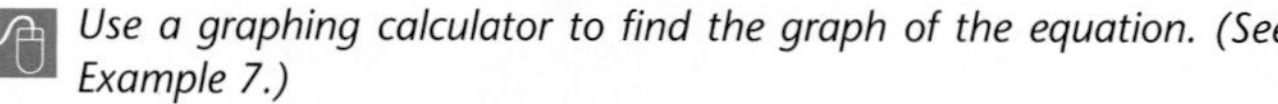
Use a graphing calculator to find the graph of the equation. (See Example 7.)

63. $y = x^2 + x + 1$

64. $y = 2 - x - x^2$

65. $y = (x - 3)^3$

66. $y = x^3 + 2x^2 + 2$

67. $y = x^3 - 3x^2 + x - 1$

68. $y = x^4 - 5x^2 - 2$

*U.S. Department of Agriculture.

†Data and projections from Stephen Brown, Federal Reserve Bank of Dallas, as published in *Newsweek*, July 21, 2008.

*Bureau of Labor Statistics.

Use a graphing calculator for Exercises 69–70.

69. Graph $y = x^4 - 2x^3 + 2x$ in a window with $-3 \le x \le 3$. Is the "flat" part of the graph near $x = 1$ really a horizontal line segment? (*Hint*: Use the trace feature to move along the "flat" part and watch the y-coordinates. Do they remain the same [as they should on a horizontal segment]?)

70. **(a)** Graph $y = x^4 - 2x^3 + 2x$ in the **standard window** (the one with $-10 \le x \le 10$ and $-10 \le y \le 10$). Use the trace feature to approximate the coordinates of the lowest point on the graph.

(b) Use a minimum finder to obtain an accurate approximation of the lowest point. How does this compare with your answer in part (a)?

Use a graphing calculator to approximate all real solutions of the equation. (See Example 8.)

71. $x^3 - 3x^2 + 5 = 0$

72. $x^3 + x - 1 = 0$

73. $2x^3 - 4x^2 + x - 3 = 0$

74. $6x^3 - 5x^2 + 3x - 2 = 0$

75. $x^5 - 6x + 6 = 0$

76. $x^3 - 3x^2 + x - 1 = 0$

Use a graphing calculator to work Exercises 77–82. (See Examples 8 and 9.)

77. **Physical Science** The surface area S of the right circular cone in the figure is given by $S = \pi r\sqrt{r^2 + h^2}$. What radius should be used to produce a cone of height 5 inches and surface area 100 square inches?

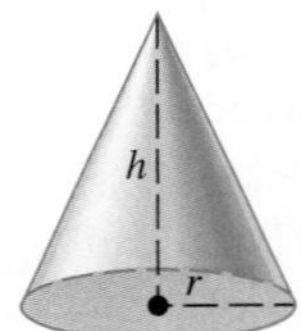

78. **Physical Science** The surface area of the right square pyramid in the figure is given by $S = b\sqrt{b^2 + 4h^2}$. If the pyramid has height 10 feet and surface area 100 square feet, what is the length of a side b of its base?

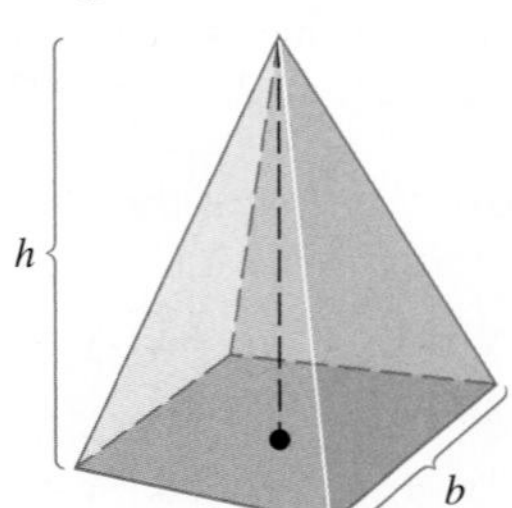

79. **Health** The number of weight-loss surgeries performed in the United States can be approximated by

$$y = 216.9x^4 - 1202.3x^3 + 3223.9x^2 + 2596.8x + 29{,}087.2,$$

where $x = 0$ corresponds to the year 1999.* According to this model, in what year did the number of weight-loss surgeries first reach 150,000?

80. **Natural Science** Carbon monoxide combines with the hemoglobin of the blood to form carboxyhemoglobin (COHb), which reduces the transport of oxygen to tissue. A 4%-to-6% COHb level in the blood (typical for smokers) can cause symptoms such as alterations in blood flow, visual impairment, and poorer vigilance. When $50 \le x \le 100$, the equation $T = .00787x^2 - 1.528x + 75.89$ approximates the exposure time T (in hours) necessary to reach this 4%-to-6% level, where x is the amount of carbon monoxide present in the air in parts per million (ppm).†

(a) A kerosene heater or a room full of smokers is capable of producing 50 ppm of carbon monoxide. How long would it take for a nonsmoking person to start feeling the aforementioned symptoms under such conditions?

(b) What carbon monoxide concentration will cause a person to reach a 4%-to-6% COHb level in three hours?

81. **Health** The number of doctors per 1000 residents in the United States from 1970 to 2020 can be approximated by

$$y = -.000018x^3 + .000488x^2 + .0366x + 1.6,$$

where $x = 0$ corresponds to the year 1970.‡ In what year is the number of doctors the highest? How many doctors are there per 1000 residents in that year?

82. **Business** From 2000 to 2007 the total net assets of U.S. equity mutual funds (in billions of dollars) could be approximated by

$$y = 331.35x^3 - 1099.6x^2 + 224.15x + 3962.3,$$

where $x = 0$ corresponds to the year 2000.§ What was the lowest amount of total assets, and when did this occur?

✓ Checkpoint Answers

1.

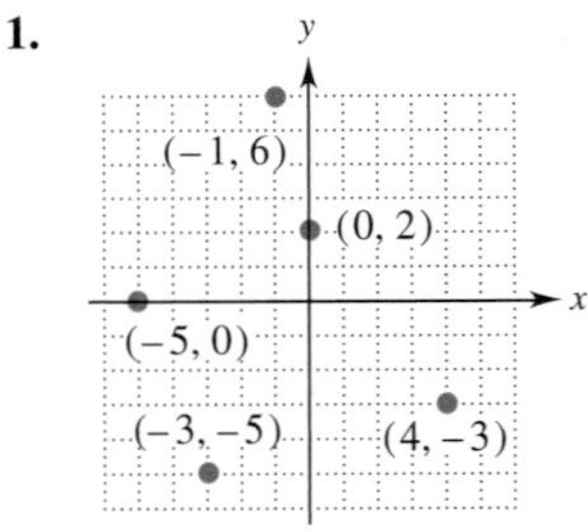

*Based on data from the American Society for Bariatric Surgery.

†U.S. Department of Energy.

‡Based on data and projections by *Health Affairs* and Dr. Richard Cooper, Institute of Health Policy at the Medical College of Wisconsin.

§Based on data from the Investment Company Institute: *Fund Fact Book* (2004).

2. (a) and (c)
3.

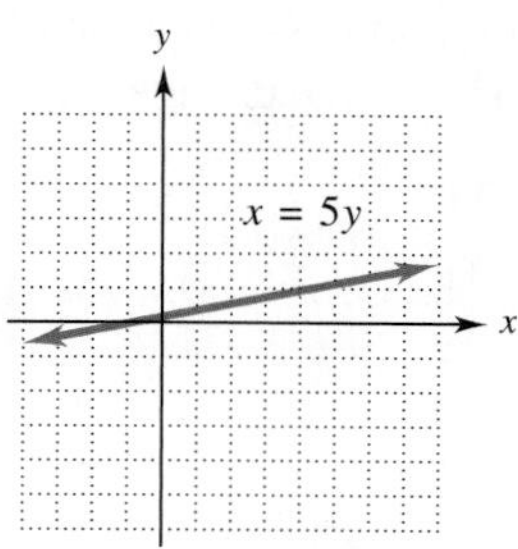

4. **(a)** x-intercept 4, y-intercept 3
   (b) x-intercept $8/5$, y-intercept -4
5. The highest was 99° at 4 PM.
   The lowest was 68° at 6 AM.
6. **(a)** About \$1,200,000 (rounded)
   (b) About $-$\$500,000
   (that is, a loss of \$500,000)
7. **(a)**

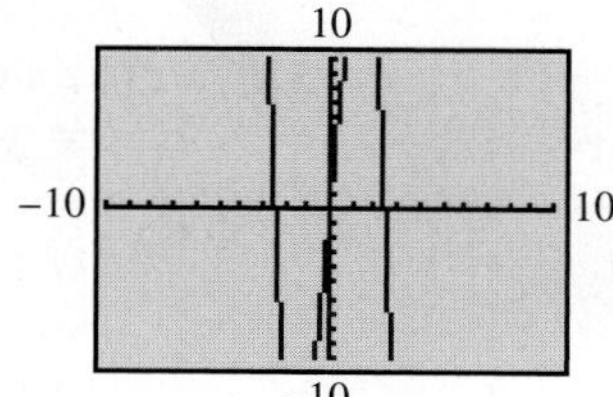

(b)

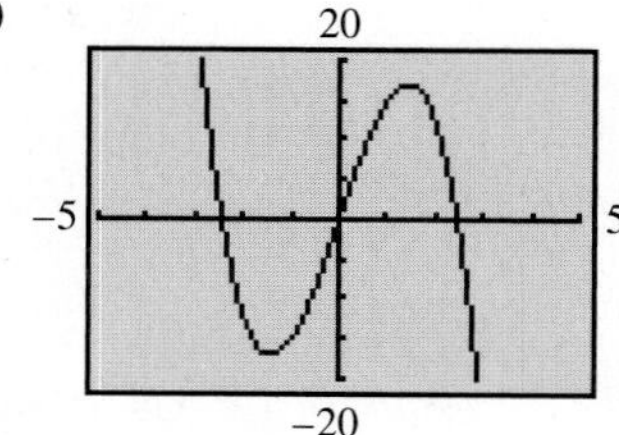

8. $x \approx -2.330059$
9. $(-.70, -1.41)$

2.2 ▶ Equations of Lines

Straight lines, which are the simplest graphs, play an important role in a wide variety of applications. They are considered here from both a geometric and an algebraic point of view.

The key geometric feature of a nonvertical straight line is how steeply it rises or falls as you move from left to right. The "steepness" of a line can be represented numerically by a number called the *slope* of the line.

To see how the slope is defined, start with Figure 2.18, which shows a line passing through the two different points $(x_1, y_1) = (-3, 5)$ and $(x_2, y_2) = (2, -4)$. The difference in the two x-values,

$$x_2 - x_1 = 2 - (-3) = 5,$$

is called the **change in *x***. Similarly the **change in *y*** is the difference in the two y-values:

$$y_2 - y_1 = -4 - 5 = -9.$$

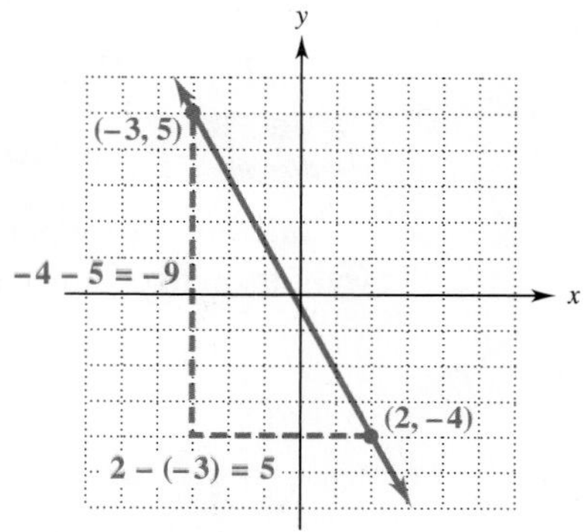

FIGURE 2.18

The **slope** of the line through the two points (x_1, y_1) and (x_2, y_2), where $x_1 \neq x_2$, is defined as the quotient of the change in y and the change in x:

$$\textbf{slope} = \frac{\textbf{change in } y}{\textbf{change in } x} = \frac{y_2 - y_1}{x_2 - x_1}.$$

The slope of the line in Figure 2.18 is

$$\text{slope} = \frac{-4 - 5}{2 - (-3)} = -\frac{9}{5}.$$

Using similar triangles from geometry, we can show that the slope is independent of the choice of points on the line. That is, the same value of the slope will be obtained for *any* choice of two different points on the line.

Example 1 Find the slope of the line through the points $(-6, 8)$ and $(5, 4)$.

Solution Let $(x_1, y_1) = (-6, 8)$ and $(x_2, y_2) = (5, 4)$. Use the definition of slope as follows:

$$\text{slope} = \frac{y_2 - y_1}{x_2 - x_1} = \frac{4 - 8}{5 - (-6)} = \frac{-4}{11} = -\frac{4}{11}.$$

The slope can also be found be letting $(x_1, y_1) = (5, 4)$ and $(x_2, y_2) = (-6, 8)$. In that case,

$$\text{slope} = \frac{y_2 - y_1}{x_2 - x_1} = \frac{8 - 4}{-6 - 5} = \frac{4}{-11} = -\frac{4}{11},$$

which is the same answer. 1 ✓

✓ Checkpoint 1

Find the slope of the line through the following pairs of points.

(a) $(5, 9), (-5, -3)$

(b) $(-4, 2), (-2, -7)$

Answers: See page 100.

CAUTION When finding the slope of a line, be careful to subtract the *x*-values and the *y*-values in the same order. For example, with the points (4, 3) and (2, 9), if you use 9 − 3 for the numerator, you must use 2 − 4 (*not* 4 − 2) for the denominator.

Example 2 Find the slope of the horizontal line in Figure 2.19.

Solution Every point on the line has the same *y*-coordinate, −5. Choose any two of them to compute the slope, say, $(x_1, y_1) = (-3, -5)$ and $(x_2, y_2) = (2, -5)$:

$$\begin{aligned}\text{slope} &= \frac{-5 - (-5)}{2 - (-3)} \\ &= \frac{0}{5} \\ &= 0.\end{aligned}$$

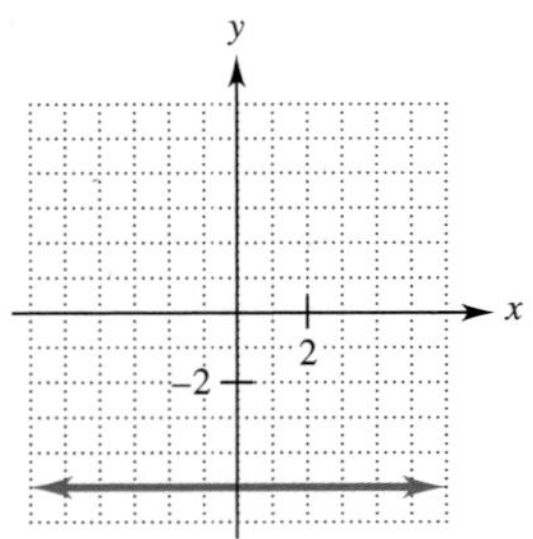

FIGURE 2.19

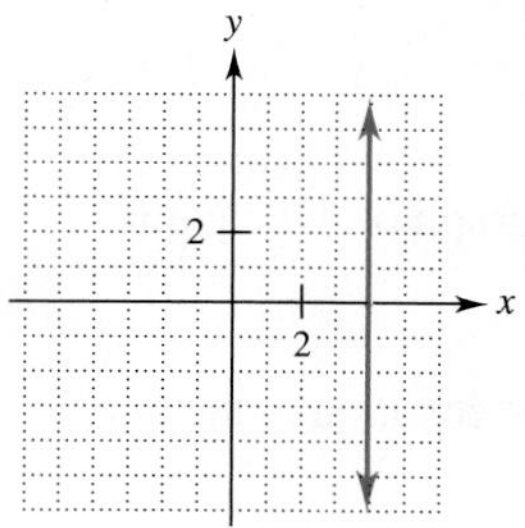

FIGURE 2.20

Example 3 What is the slope of the vertical line in Figure 2.20?

Solution Every point on the line has the same x-coordinate, 4. If we attempt to compute the slope with two of these points, say, $(x_1, y_1) = (4, -2)$ and $(x_2, y_2) = (4, 1)$, we obtain

$$\text{slope} = \frac{1 - (-2)}{4 - 4} = \frac{3}{0}.$$

Division by 0 is not defined, so the slope of this line is undefined.

The arguments used in Examples 2 and 3 work in the general case and lead to the following conclusion.

> The slope of every horizontal line is 0.
>
> The slope of every vertical line is undefined.

SLOPE–INTERCEPT FORM

The slope can be used to develop an algebraic description of nonvertical straight lines. Assume that a line with slope m has y-intercept b, so that it goes through the point $(0, b)$. (See Figure 2.21.) Let (x, y) be any point on the line other than $(0, b)$. Using the definition of slope with the points $(0, b)$ and (x, y) gives

$$m = \frac{y - b}{x - 0}$$

$$m = \frac{y - b}{x}$$

$$mx = y - b \qquad \text{Multiply both sides by } x.$$

$$y = mx + b. \qquad \text{Add } b \text{ to both sides. Reverse the equation.}$$

In other words, the coordinates of any point on the line satisfy the equation $y = mx + b$.

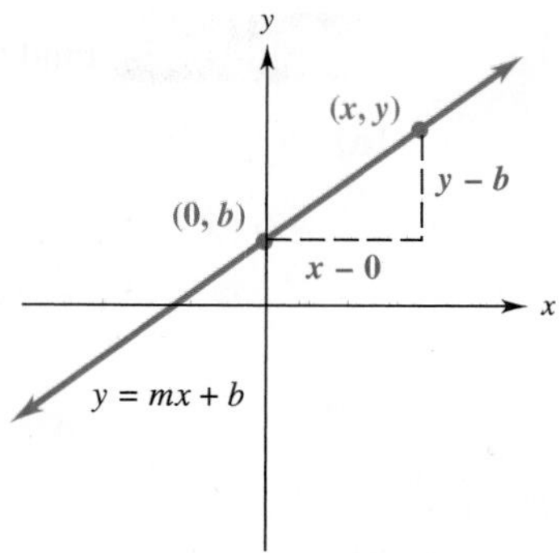

FIGURE 2.21

Slope–Intercept Form

If a line has slope m and y-intercept b, then it is the graph of the equation

$$y = mx + b.$$

This equation is called the **slope–intercept form** of the equation of the line.

Example 4 Find an equation for the line with y-intercept $7/2$ and slope $-5/2$.

Solution Use the slope–intercept form with $b = 7/2$ and $m = -5/2$:

$$y = mx + b$$
$$y = -\frac{5}{2}x + \frac{7}{2}. \quad ✓2$$

✓Checkpoint 2

Find an equation for the line with

(a) y-intercept -3 and slope $2/3$;

(b) y-intercept $1/4$ and slope $-3/2$.

Answers: See page 100.

Example 5 Find the equation of the horizontal line with y-intercept 3.

Solution The slope of the line is 0 (why?) and its y-intercept is 3, so its equation is

$$y = mx + b$$
$$y = 0x + 3$$
$$y = 3.$$

The argument in Example 5 also works in the general case.

If k is a constant, then the graph of the equation $y = k$ is the horizontal line with y-intercept k.

Example 6 Find the slope and y-intercept for each of the following lines.

(a) $5x - 3y = 1$

Solution Solve for y:

$$5x - 3y = 1$$
$$-3y = -5x + 1 \qquad \text{Subtract } 5x \text{ from both sides.}$$
$$y = \frac{5}{3}x - \frac{1}{3}. \qquad \text{Divide both sides by } -3.$$

This equation is in the form $y = mx + b$, with $m = 5/3$ and $b = -1/3$. So the slope is $5/3$ and the y-intercept is $-1/3$.

(b) $-9x + 6y = 2$

Solution Solve for y:

$$-9x + 6y = 2$$

$$6y = 9x + 2 \quad \text{Add } 9x \text{ to both sides.}$$

$$y = \frac{3}{2}x + \frac{1}{3}. \quad \text{Divide both sides by 6.}$$

The slope is 3/2 (the coefficient of x), and the y-intercept is 1/3. 3✓

✓Checkpoint 3

Find the slope and y-intercept of

(a) $x + 4y = 6$;

(b) $3x - 2y = 1$.

Answers: See page 100.

The slope–intercept form can be used to show how the slope measures the steepness of a line. Consider the straight lines A, B, C, and D given by the following equations, where each has y-intercept 0 and slope as indicated:

A: $y = .5x$;	B: $y = x$;	C: $y = 3x$;	D: $y = 7x$.
Slope .5	Slope 1	Slope 3	Slope 7

For these lines, Figure 2.22 shows that the bigger the slope, the more steeply the line rises from left to right. 4✓

✓Checkpoint 4

(a) List the slopes of the following lines:
E: $y = -.3x$; F: $y = -x$;
G: $y = -2x$; H: $y = -5x$.

(b) Graph all four lines on the same set of axes.

(c) How are the slopes of the lines related to their steepness?

Answers: See page 100.

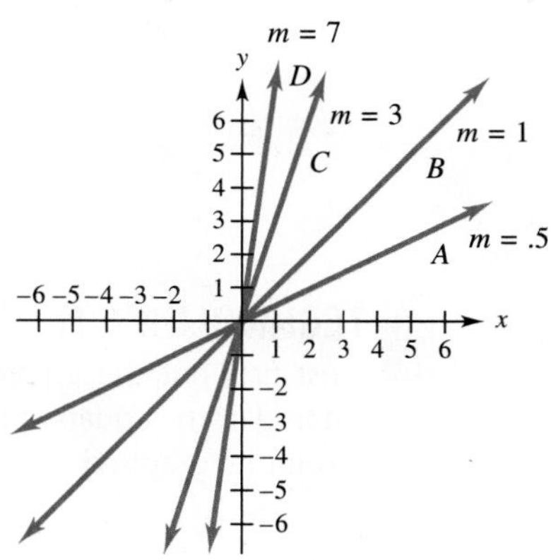

FIGURE 2.22

The preceding discussion and Checkpoint 4 may be summarized as follows.

Direction of Line (moving from left to right)	Slope
Upward	**Positive** (larger for steeper lines)
Horizontal	**0**
Downward	**Negative** (larger in absolute value for steeper lines)
Vertical	**Undefined**

Example 7 Sketch the graph of $x + 2y = 5$, and label the intercepts.

Solution Find the x-intercept by setting $y = 0$ and solving for x:

$$x + 2 \cdot 0 = 5$$
$$x = 5.$$

The x-intercept is 5, and (5, 0) is on the graph. The y-intercept is found similarly, by setting $x = 0$ and solving for y:

$$0 + 2y = 5$$
$$y = 5/2.$$

The y-intercept is $5/2$, and $(0, 5/2)$ is on the graph. The points (5, 0) and $(0, 5/2)$ can be used to sketch the graph (Figure 2.23). 5

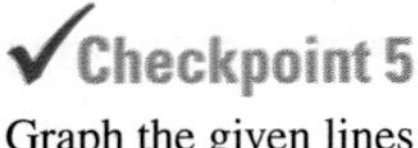

Checkpoint 5

Graph the given lines and label the intercepts.

(a) $3x + 4y = 12$

(b) $5x - 2y = 8$

Answers: See page 100.

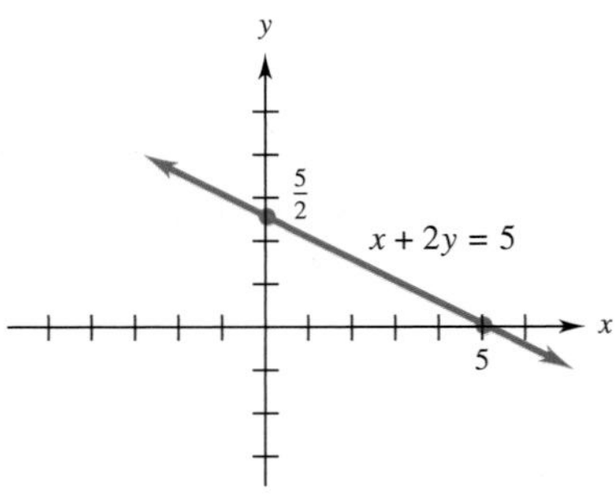

FIGURE 2.23

TECHNOLOGY TIP To graph a linear equation on a graphing calculator, you must first put the equation in slope–intercept form $y = mx + b$ so that it can be entered in the equation memory (called the Y=list on some calculators). Vertical lines cannot be graphed on most calculators.

SLOPES OF PARALLEL AND PERPENDICULAR LINES

We shall assume the following facts without proof. The first one is a consequence of the fact that the slope measures steepness and that parallel lines have the same steepness.

> Two nonvertical lines are parallel whenever they have the same slope.
>
> Two nonvertical lines are perpendicular whenever the product of their slopes is -1.

Example 8 Determine whether each of the given pairs of lines are *parallel*, *perpendicular*, or *neither*.

(a) $2x + 3y = 5$ and $4x + 5 = -6y$.

Solution Put each equation in slope–intercept form by solving for y:

$$3y = -2x + 5 \qquad\qquad -6y = 4x + 5$$

$$y = -\frac{2}{3}x + \frac{5}{3} \qquad\qquad y = -\frac{2}{3}x - \frac{5}{6}.$$

In each case, the slope (the coefficient of x) is $-2/3$, so the lines are parallel.

(b) $3x = y + 7$ and $x + 3y = 4$.

Solution Put each equation in slope–intercept form to determine the slope of the associated line:

$$3x = y + 7 \qquad\qquad 3y = -x + 4$$

$$y = 3x - 7 \qquad\qquad y = -\frac{1}{3}x + \frac{4}{3}$$

slope 3 slope $-1/3$.

Since $3(-1/3) = -1$, these lines are perpendicular.

(c) $x + y = 4$ and $x - 2y = 3$.

Solution Verify that the slope of the first line is -1 and the slope of the second is $1/2$. The slopes are not equal and their product is not -1, so the lines are neither parallel nor perpendicular. 6✓

✓Checkpoint 6

Tell whether the lines in each of the following pairs are *parallel*, *perpendicular*, or *neither*.

(a) $x - 2y = 6$ and $2x + y = 5$

(b) $3x + 4y = 8$ and $x + 3y = 2$

(c) $2x - y = 7$ and $2y = 4x - 5$

Answers: See page 100.

TECHNOLOGY TIP Perpendicular lines may not appear perpendicular on a graphing calculator unless you use a *square window*—a window in which a one-unit segment on the y-axis is the same length as a one-unit segment on the x-axis. To obtain such a window on most calculators, use a viewing window in which the y-axis is about two-thirds as long as the x-axis. The SQUARE (or ZSQUARE) key in the ZOOM menu will change the current window to a square window by automatically adjusting the length of one of the axes.

POINT–SLOPE FORM

The slope–intercept form of the equation of a line is usually the most convenient for graphing and for understanding how slopes and lines are related. However, it is not always the best way to *find* the equation of a line. In many situations (particularly in calculus), the slope and a point on the line are known and you must find the equation of the line. In such cases, the best method is to use the *point–slope form*, which we now explain.

Suppose that a line has slope m and that (x_1, y_1) is a point on the line. Let (x, y) represent any other point on the line. Since m is the slope, then, by the definition of slope,

$$\frac{y - y_1}{x - x_1} = m.$$

Multiplying both sides by $x - x_1$ yields

$$y - y_1 = m(x - x_1).$$

Point–Slope Form

If a line has slope m and passes through the point (x_1, y_1), then

$$y - y_1 = m(x - x_1)$$

is the **point–slope form** of the equation of the line.

Example 9 Find the equation of the line satisfying the given conditions.

(a) Slope 2; the point (5, 3) is on the line.

Solution Use the point–slope form with $m = 2$ and $(x_1, y_1) = (5, 3)$. Substitute $x_1 = 5$, $y_1 = 3$, and $m = 2$ into the point–slope form of the equation.

$$y - y_1 = m(x - x_1)$$
$$y - 3 = 2(x - 5).$$

For some purposes, this form of the equation is fine; in other cases, you may want to rewrite it in the slope–intercept form.

(b) Slope -3; the point $(-4, 1)$ is on the line.

Solution Use the point–slope form with $m = -3$ and $(x_1, y_1) = (-4, 1)$:

$$y - y_1 = m(x - x_1)$$
$$y - 1 = -3[x - (-4)]. \quad \text{Point–slope form}$$

Using algebra, we obtain the slope–intercept form of this equation:

$$y - 1 = -3(x + 4)$$
$$y - 1 = -3x - 12 \quad \text{Distributive property}$$
$$y = -3x - 11. \quad \text{Slope–intercept form}$$

7 ✓

✓Checkpoint 7

Find both the point–slope and the slope–intercept form of the equation of the line having the given slope and passing through the given point.

(a) $m = -3/5$, $(5, -2)$

(b) $m = 1/3$, $(6, 8)$

Answers: See page 100.

The point–slope form can also be used to find an equation of a line, given two different points on the line. The procedure is shown in the next example.

Example 10 Find an equation of the line through (5, 4) and $(-10, -2)$.

Solution Begin by using the definition of the slope to find the slope of the line that passes through the two points:

$$\text{slope} = m = \frac{-2 - 4}{-10 - 5} = \frac{-6}{-15} = \frac{2}{5}.$$

Use $m = 2/5$ and either of the given points in the point–slope form. If $(x_1, y_1) = (5, 4)$, then

$$y - y_1 = m(x - x_1)$$

$$y - 4 = \frac{2}{5}(x - 5) \quad \text{Let } y_1 = 4,\ m = \frac{2}{5},\ \text{and } x_1 = 5.$$

$$5(y - 4) = 2(x - 5) \quad \text{Multiply both sides by 5.}$$

$$5y - 20 = 2x - 10 \quad \text{Distributive property}$$

$$5y = 2x + 10.$$

Check that the result is the same when $(x_1, y_1) = (-10, -2)$. 8✓

✓Checkpoint 8

Find an equation of the line through

(a) (2, 3) and (−4, 6);

(b) (−8, 2) and (3, −6).

Answers: See page 100.

VERTICAL LINES

The equation forms we just developed do not apply to vertical lines, because the slope is not defined for such lines. However, vertical lines can easily be described as graphs of equations.

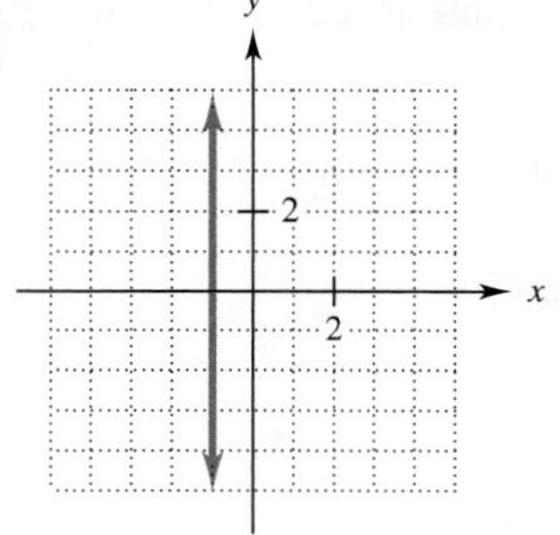

FIGURE 2.24

Example 11 Find the equation whose graph is the vertical line in Figure 2.24.

Solution Every point on the line has x-coordinate -1 and hence has the form $(-1, y)$. Thus, every point is a solution of the equation $x + 0y = -1$, which is usually written simply as $x = -1$. Note that -1 is the x-intercept of the line.

The argument in Example 11 also works in the general case.

> If k is a constant, then the graph of the equation $x = k$ is the vertical line with x-intercept k.

LINEAR EQUATIONS

An equation in two variables whose graph is a straight line is called a **linear equation**. Linear equations have a variety of forms, as summarized in the following table.

Equation	Description
$x = k$	**Vertical line**, x-intercept k, no y-intercept, undefined slope
$y = k$	**Horizontal line**, y-intercept k, no x-intercept, slope 0
$y = mx + b$	**Slope–intercept form**, slope m, y-intercept b
$y - y_1 = m(x - x_1)$	**Point–slope form**, slope m, the line passes through (x_1, y_1)
$ax + by = c$	**General form**. If $a \neq 0$ and $b \neq 0$, the line has x-intercept c/a, y-intercept c/b, and slope $-a/b$.

Note that every linear equation can be written in general form. For example, $y = 4x - 5$ can be written in general form as $4x - y = 5$, and $x = 6$ can be written in general form as $x + 0y = 6$.

APPLICATIONS

Many relationships are linear or almost linear, so that they can be approximated by linear equations.

Example 12 **Business** Each year, the Web site www.teammarketing.com posts data on the cost for a family of four to attend a major sporting event. Using data from the 2007–2008 National Football League (NFL) season, the linear equation

$$y = .39x + 59.40$$

was found to be a good approximation of the relationship between the price x for parking and the price y of four tickets.

(a) If the price of parking was \$20, what was the approximate price for four tickets to an NFL game?

Solution Substitute $x = 20$ in the equation and use a calculator to compute y:

$$y = .39x + 59.40$$
$$y = .39(20) + 59.40 = 67.20.$$

The price of four tickets was approximately \$67.20.

(b) When the price of tickets was \$76, how much was parking?

Solution Substitute $y = 76$ in the equation and solve for x:

$$76 = .39x + 59.40$$
$$76 - 59.40 = .39x$$
$$x = \frac{76 - 59.40}{.39} = 42.56.$$

The price of parking was approximately \$42.56.

Example 13 **Education** According to data from the National Center for Education Statistics, the average cost of tuition and fees in public four-year colleges was \$2987 in the fall of 1996 and grew in an approximately linear fashion to \$5685 in the fall of 2006.

(a) Find a linear equation for this data.

Solution Measure time along the x-axis and cost along the y-axis. Then the x-coordinate of each point is a year and the y-coordinate is the average cost of tuition and fees in that year. For convenience, let $x = 0$ correspond to 1990, so that $x = 6$ is 1996 and $x = 16$ is 2006. Then the given data points are (6, 2987) and (16, 5685). The slope of the line joining these points is

$$\frac{5685 - 2987}{16 - 6} = \frac{2698}{10} = 269.80.$$

Using the point (6, 2987), we obtain the equation of this line:

$$y - 2987 = 269.80(x - 6) \quad \text{Point–slope form}$$
$$y - 2987 = 269.80x - 1618.80 \quad \text{Distributive property}$$
$$y = 269.80x + 1368.20. \quad \text{Add 2987 to both sides.}$$

(b) Use this equation to estimate the average cost of tuition and fees in the fall of 2004.

Solution Since 2004 corresponds to $x = 14$, let $x = 14$ in the equation part of (a). Then

$$y = 269.80(14) + 1368.20 = \$5145.40.$$

(c) Assuming the equation remains valid beyond fall 2006, estimate the average cost of tuition and fees in the fall of 2014.

Solution 2014 corresponds to $x = 24$, so the average cost is

$$y = 269.80(24) + 1368.20 = \$7843.40. \quad 9✓$$

✓Checkpoint 9

The average cost of tuition and fees in private four-year colleges was \$12,881 in 1996 and \$20,492 in 2006.

(a) Let $x = 0$ correspond to 1990, and find a linear equation for the given data.

(b) Assuming that your equation remains accurate, estimate the average cost in 2012.

Answers: See page 100.

2.2 ▸ Exercises

Find the slope of the given line, if it is defined. (See Examples 1–3.)

1. The line through (2, 5) and (0, 8)
2. The line through (9, 0) and (12, 12)
3. The line through (−4, 14) and (3, 0)
4. The line through (−5, −2) and (−4, 11)
5. The line through the origin and (−4, 10)
6. The line through the origin and (8, −2)
7. The line through (−1, 4) and (−1, 6)
8. The line through (−3, 5) and (2, 5)

Find an equation of the line with the given y-intercept and slope m. (See Examples 4 and 5.)

9. $5, m = 4$
10. $-3, m = -7$
11. $1.5, m = -2.3$
12. $-4.5, m = 2.5$
13. $4, m = -3/4$
14. $-3, m = 4/3$

Find the slope m and the y-intercept b of the line whose equation is given. (See Example 6.)

15. $2x - y = 9$
16. $x + 2y = 7$
17. $6x = 2y + 4$
18. $4x + 3y = 24$
19. $6x - 9y = 16$
20. $4x + 2y = 0$
21. $2x - 3y = 0$
22. $y = 7$
23. $x = y - 5$
24. On one graph, sketch six straight lines that meet at a single point and satisfy this condition: one line has slope 0, two lines have positive slope, two lines have negative slope, and one line has undefined slope.
25. For which of the line segments in the figure is the slope
(a) largest? **(b)** smallest?
(c) largest in absolute value? **(d)** closest to 0?

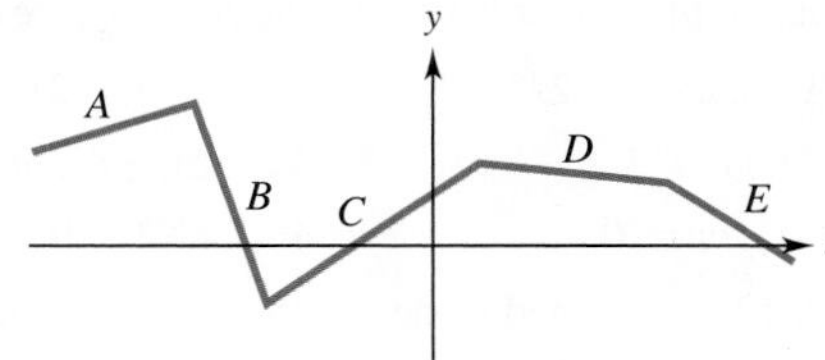

26. Match each equation with the line that most closely resembles its graph. (*Hint*: Consider the signs of m and b in the slope–intercept form.)
(a) $y = 3x + 2$ **(b)** $y = -3x + 2$
(c) $y = 3x - 2$ **(d)** $y = -3x - 2$

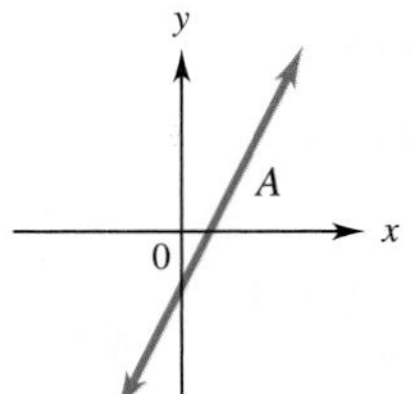

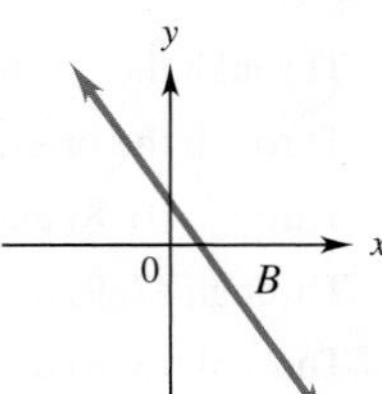

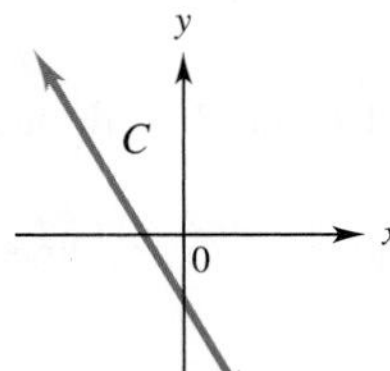

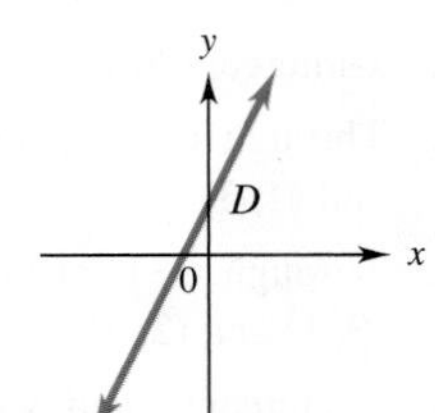

Sketch the graph of the given equation and label its intercepts. (See Example 7.)

27. $2x - y = -2$

28. $2y + x = 4$

29. $2x + 3y = 4$

30. $-5x + 4y = 3$

31. $4x - 5y = 2$

32. $3x + 2y = 8$

Determine whether each pair of lines is parallel, perpendicular, or neither. (See Example 8.)

33. $4x - 3y = 6$ and $3x + 4y = 8$

34. $2x - 5y = 7$ and $15y - 5 = 6x$

35. $3x + 2y = 8$ and $6y = 5 - 9x$

36. $x - 3y = 4$ and $y = 1 - 3x$

37. $4x = 2y + 3$ and $2y = 2x + 3$

38. $2x - y = 6$ and $x - 2y = 4$

39. **(a)** Find the slope of each side of the triangle with vertices $(9, 6)$, $(-1, 2)$, and $(1, -3)$.
(b) Is this triangle a right triangle? (*Hint*: Are two sides perpendicular?)

40. **(a)** Find the slope of each side of the quadrilateral with vertices $(-5, -2)$, $(-3, 1)$, $(3, 0)$, and $(1, -3)$.
(b) Is this quadrilateral a parallelogram? (*Hint*: Are opposite sides parallel?)

Find an equation of the line with slope m that passes through the given point. Put the answer in slope–intercept form. (See Example 10.)

41. $(-3, 2)$, $m = -2/3$

42. $(-5, -2)$, $m = 4/5$

43. $(2, 3)$, $m = 3$

44. $(3, -4)$, $m = -1/4$

45. $(10, 1)$, $m = 0$

46. $(-3, -9)$, $m = 0$

47. $(-2, 12)$, undefined slope

48. $(1, 1)$, undefined slope

Find an equation of the line that passes through the given points. (See Example 10.)

49. $(-1, 1)$ and $(2, 7)$

50. $(2, 5)$ and $(0, 6)$

51. $(1, 2)$ and $(3, 9)$

52. $(-1, -2)$ and $(2, -1)$

Find an equation of the line satisfying the given conditions.

53. Through the origin with slope 5

54. Through the origin and horizontal

55. Through $(6, 8)$ and vertical

56. Through $(7, 9)$ and parallel to $y = 6$

57. Through $(3, 4)$ and parallel to $4x - 2y = 5$

58. Through $(6, 8)$ and perpendicular to $y = 2x - 3$

59. x-intercept 6; y-intercept -6

60. Through $(-5, 2)$ and parallel to the line through $(1, 2)$ and $(4, 3)$

61. Through $(-1, 3)$ and perpendicular to the line through $(0, 1)$ and $(2, 3)$

62. y-intercept 3 and perpendicular to $2x - y + 6 = 0$

Business *The lost value of equipment over a period of time is called depreciation. The simplest method for calculating depreciation is straight-line depreciation. The annual straight-line depreciation D of an item that cost x dollars with a useful life of n years is $D = (1/n)x$. Find the depreciation for items with the given characteristics.*

63. Cost: \$15,965; life 12 yr

64. Cost: \$41,762; life 15 yr

65. Cost: \$201,457; life 30 yr

66. **Business** Ral Corp. has an incentive compensation plan under which a branch manager receives 10% of the branch's income after deduction of the bonus, but before deduction of income tax.* The income of a particular branch before the bonus and income tax was \$165,000. The tax rate was 30%. The bonus amounted to
(a) \$12,600
(b) \$15,000
(c) \$16,500
(d) \$18,000

67. **Health** According to data from the U.S. Census Bureau, the number of retail drug prescriptions (in millions) can be approximated by

$$y = 83.54x + 2920.94,$$

where $x = 0$ corresponds to the year 2000. Find the approximate number of prescriptions in the following years.
(a) 2004
(b) 2009
(c) Assuming that this model remains accurate, in what year will the number of prescriptions be 4,750,000,000?

68. **Business** Data from the National Football League (NFL) shows that the average yearly salary (in thousands) for a player can be approximated by the equation

$$y = 134.47x + 827.93,$$

where $x = 0$ corresponds to the year 2000. Assume the trend continues indefinitely.
(a) What is the average salary for a player in 2011?
(b) In what year will the average salary be \$2.8 million?

69. **Business** According to data from the Motion Picture Association of America, total box-office receipts (in millions of dollars) declined from 2002 to 2006. The receipts in year x can be approximated by

$$y = -60.30x + 9635.4,$$

where $x = 2$ corresponds to 2002. Assume this equation remains valid through 2012.
(a) Estimate the box-office receipts in 2006.
(b) In what year will the box-office receipts be \$9 billion?

---

*Uniform CPA Examination question.

70. Business The number of ski resorts in the United States has been declining and can be approximated with the equation

$$y = -3.14x + 502,$$

where $x = 2$ corresponds to the year 2002.*

(a) How many ski resorts were there in 2007?

(b) If the trend holds indefinitely, in what year are there 460 ski resorts?

71. Business In the United States, total alcoholic beverage sales were $119.1 billion in 2003 and $132.3 billion in 2005.†

(a) Let the x-axis denote time and the y-axis denote the amount of alcohol sales (in billions). Let $x = 0$ correspond to 2003. Fill in the blanks: The given data is represented by the points (____, 119.1) and (2, ____).

(b) Find the linear equation determined by the two points in part (a).

(c) Use the equation in part (b) to estimate the alcohol sales produced in 2004.

(d) If the model remains accurate, when did alcohol sales reach $150 billion?

72. Business According to the U.S. Bureau of Labor Statistics, there were approximately 16.3 million union workers in 2000 and 15.4 million union workers in 2006.

(a) If the change in the number of union workers is considered to be linear, write an equation expressing the number y of union workers in terms of the number x of years since 2000.

(b) Assuming that the equation in part (a) remains accurate, use it to predict the number of union workers in 2010.

73. Business In 1995, there were 41,235 shopping centers in the United States. In 2005, there were 48,695.‡

(a) Write an equation expressing the number y of shopping centers in terms of the number x of years after 1995.

(b) Assuming the future accuracy of the equation you found in part (a), predict the number of shopping centers in 2012.

(c) When will the number of shopping centers reach 60,000?

74. Social Science The percentage of people 25 years old and older who have completed four or more years of college was 21.3% in 1990 and 28.0% in 2006.§

(a) Find a linear equation that gives the percentage of people 25 and over who have completed four or more years of college in terms of time t, where t is the number of years since 1990. Assume that this equations remains valid in the future.

(b) What will the percentage be in 2012?

(c) When will 40% of those 25 and over have completed four or more years of college?

75. Business According to the U.S. Bureau of Labor Statistics, in 1990, 529,000 people worked in the air transportation industry. In 2006, the number was 487,000.

(a) Find a linear equation giving the number of employees in the air transportation industry in terms of x, the number of years since 1990.

(b) Assuming the equation remains valid in the future, in what year will there be 450,000 employees of the air transportation industry?

76. Business The U.S. Bureau of Labor Statistics estimated that in 1990, 1.1 million people worked in the truck transportation industry. In 2006, the number was 1.4 million.

(a) Find a linear equation giving the number of employees in the truck transportation industry in terms of x, the number of years since 1990.

(b) Assuming the equation remains valid in the future, in what year will there be 1.9 million employees in the truck transportation industry?

77. Physical Science The accompanying graph shows the winning time (in minutes) at the Olympic Games from 1952 to 2008 for the men's 5000-meter run, together with a linear approximation of the data:*

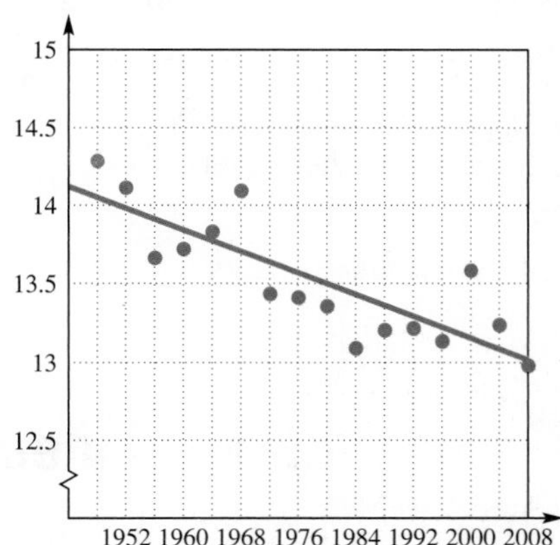

(a) The equation for the linear approximation is

$$y = -.01723x + 47.61.$$

What does the slope of this line represent? Why is the slope negative?

(b) Use the approximation to estimate the winning time in the 2012 Olympics. If possible, check this estimate against the actual time.

*_Statistical Abstract of the United States_: 2008.

†U.S. Department of Agriculture.

‡www.icsc.org/.

§U.S. Census Bureau.

*_The World Almanac and Book of Facts_: 2009.

78. Business The accompanying graph shows the number of FM commercial radio stations in the United States in selected years, along with a linear approximation of the data.*

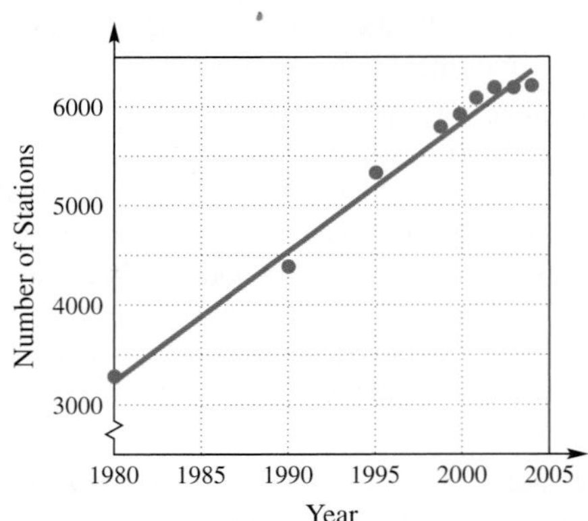

(a) Use the data points (1980, 3282) and (2000, 5892) to estimate the slope of the line shown. Interpret this number.

(b) Use part (a) to approximate the number of stations in 1996.

(c) Discuss the accuracy of the linear approximation.

✓ Checkpoint Answers

1. **(a)** 6/5 **(b)** −9/2

2. **(a)** $y = \frac{2}{3}x - 3$ **(b)** $y = -\frac{3}{2}x + \frac{1}{4}$

3. **(a)** Slope −1/4; y-intercept 3/2
   (b) Slope 3/2; y-intercept −1/2

4. **(a)** Slope of $E = -.3$; slope of $F = -1$; slope of $G = -2$; slope of $H = -5$.
   (b)

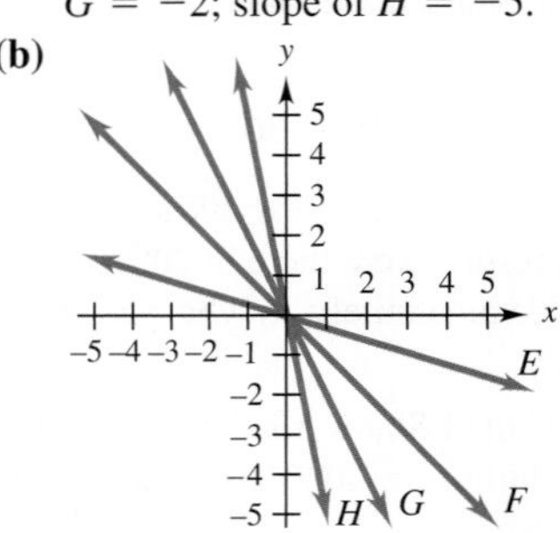

   (c) The larger the slope in absolute value, the more steeply the line falls from left to right.

5. **(a)**

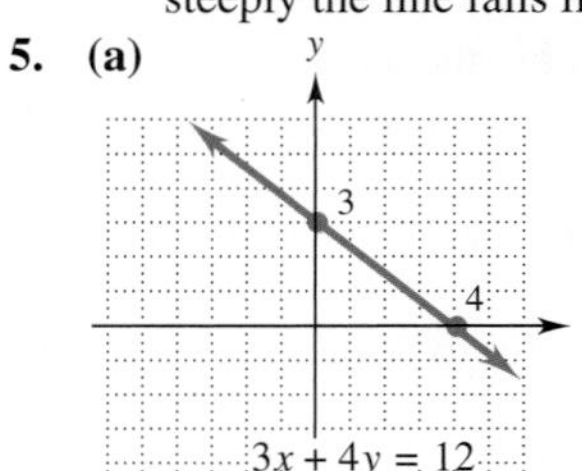

   (b)

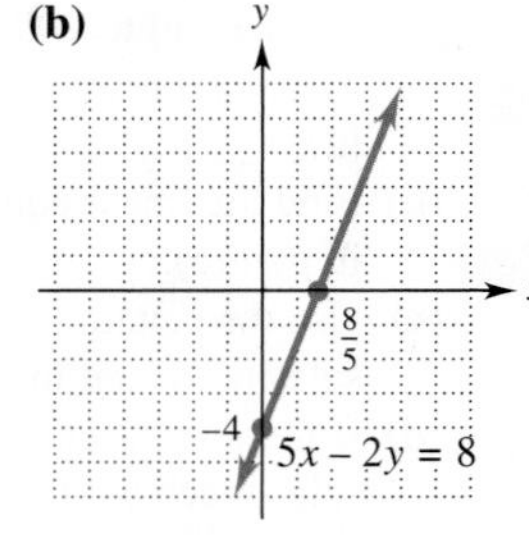

6. **(a)** Perpendicular **(b)** Neither **(c)** Parallel

7. **(a)** $y + 2 = -\frac{3}{5}(x - 5)$; $y = -\frac{3}{5}x + 1$.
   (b) $y - 8 = \frac{1}{3}(x - 6)$; $y = \frac{1}{3}x + 6$.

8. **(a)** $2y = -x + 8$ **(b)** $11y = -8x - 42$

9. **(a)** $y = 761.10x + 8314.40$ **(b)** \$25,058.60

*_Statistical Abstract of the United States_: 2008.

2.3 ▸ Linear Models

In business and science, it is often necessary to make judgments on the basis of data from the past. For instance, a stock analyst might use a company's profits in previous years to estimate the next year's profits. Or a life insurance company might look at life expectancies of people born in various years to predict how much money it should expect to pay out in the next year.

In such situations, the available data is used to construct a mathematical model, such as an equation or a graph, which is used to approximate the likely outcome in cases where complete data is not available. In this section, we consider applications in which the data can be modeled by a linear equation.

The simplest way to construct a linear model is to use the line determined by two of the data points, as illustrated in the following example.

Example 1 **Business** The profits of the General Electric Company (GE), in billions of dollars, over a five-year period are shown in the following table*:

Year	2003	2004	2005	2006	2007
Profit	15.0	16.6	16.4	20.8	22.2

(a) Let $x = 3$ correspond to 2003, and plot the points (x, y), where x is the year and y is the profit.

Solution The data points are (3, 15.0), (4, 16.6), (5, 16.4), (6, 20.8), and (7, 22.2), as shown in Figure 2.25.

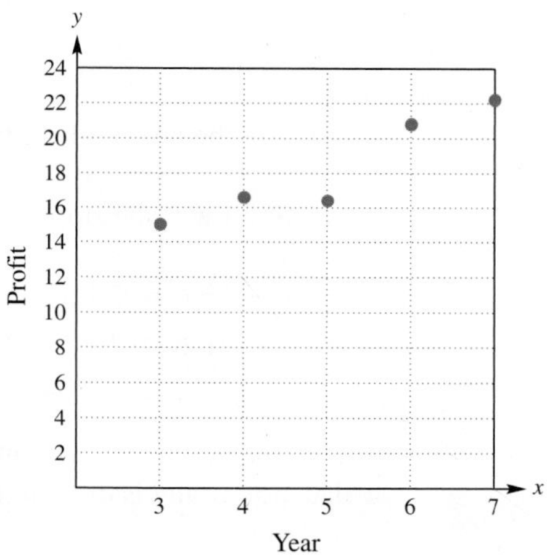

FIGURE 2.25

(b) Use the data points (3, 15.0) and (7, 22.2) to find a line that models the data.

Solution The slope of the line through (3, 15.0) and (7, 22.2) is $\frac{22.2 - 15.0}{7 - 3} = \frac{7.2}{4} = 1.8$. Using the point (3, 15) and the slope 1.8, we find that the equation of this line is

$$y - 15 = 1.8(x - 3) \quad \text{Point–slope form for the equation of a line}$$
$$y - 15 = 1.8x - 5.4 \quad \text{Distributive property}$$
$$y = 1.8x + 9.6. \quad \text{Slope–intercept form}$$

The line and the data points are shown in Figure 2.26 on the next page. Although the line fits the two endpoints perfectly, it overestimates profits in 2004 and 2005 while underestimating profits in 2006.

*Morningstar.com. The profit figures (net income) are in billions of dollars.

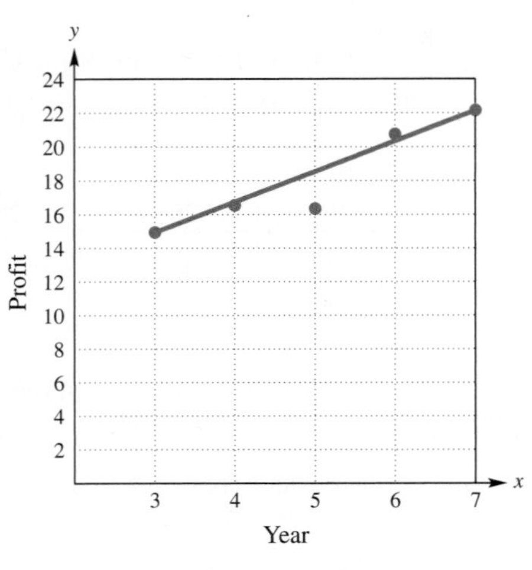

FIGURE 2.26

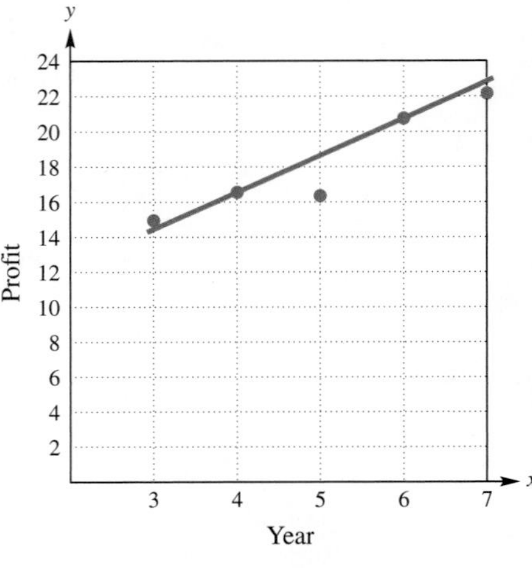

FIGURE 2.27

(c) Use the points (4, 16.6) and (6, 20.8) to find another line that models the data.

Solution The slope is $\frac{20.8 - 16.6}{6 - 4} = \frac{4.2}{2} = 2.1$, and the equation is

$$y - 16.6 = 2.1(x - 4) \qquad \text{Point–slope form, using (4, 16.6) and slope 2.1}$$
$$y - 16.6 = 2.1x - 8.4 \qquad \text{Distributive property}$$
$$y = 2.1x + 8.2. \qquad \text{Slope–intercept form}$$

The line and the data points are shown in Figure 2.27. This line passes through two data points and is close to two other points, but significantly overestimates the profit in 2005.

Checkpoint 1

Use the points (3, 15) and (6, 20.8) to find another linear model for the data in Example 1.

Answer: See page 112.

(d) Use the model in part (b) to estimate the profits in 2006.

Solution Substitute $x = 6$ (corresponding to 2006) into the equation:

$$y = 1.8x + 9.6$$
$$y = 1.8(6) + 9.6 = 20.4 \text{ billion.}$$

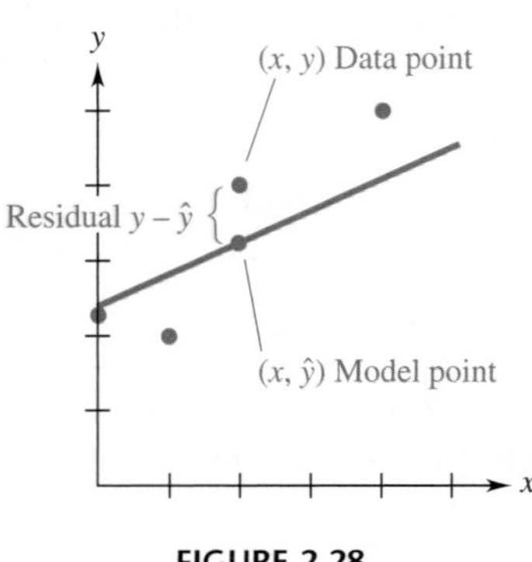

FIGURE 2.28

Opinions may vary as to which of the lines in Example 1 best fits the data. To make a decision, we might try to measure the amount of error in each model. One way to do this is to compute the difference between the actual profit y and the amount $\hat{y}$ given by the model. If the data point is (x, y) and the corresponding point on the line is $(x, \hat{y})$, then the difference $y - \hat{y}$ measures the error in the model for that particular value of x. The number $y - \hat{y}$ is called a **residual**. As shown in Figure 2.28, the residual $y - \hat{y}$ is the vertical distance from the data point to the line (positive when the data point is above the line, negative when it is below the line, and 0 when it is on the line).

One way to determine how well a line fits the data points is to compute the sum of its residuals—that is, the sum of the individual errors. Unfortunately, however, the sum of the residuals of two different lines might be equal, thwarting our effort to decide which is the better fit. Furthermore, the residuals may sum to 0, which doesn't mean that there is no error, but only that the positive and negative errors (which might be quite large) cancel each other out. (See Exercise 11 at the end of this section for an example.)

To avoid this difficulty, mathematicians use the sum of the *squares* of the residuals to measure how well a line fits the data points. When the sum of the squares is used, a smaller sum means a smaller overall error and hence a better fit. The error is 0 only when all the data points lie on the line (a perfect fit).

Example 2 **Business** Two linear models for GE's profits were constructed in Example 1:

$$y = 1.8x + 9.6 \quad \text{and} \quad y = 2.1x + 8.2.$$

For each model, determine the five residuals, the square of each residual, and the sum of the squares of the residuals.

Solution The information for each model is summarized in the following tables:

$$y = 1.8x + 9.6$$

Data Point (x, y)	Model Point $(x, \hat{y})$	Residual $y - \hat{y}$	Squared Residual $(y - \hat{y})^2$
(3, 15.0)	(3, 15.0)	0	0
(4, 16.6)	(4, 16.8)	−.2	.04
(5, 16.4)	(5, 18.6)	−2.2	4.84
(6, 20.8)	(6, 20.4)	.4	.16
(7, 22.2)	(7, 22.2)	0	0
			Sum = 5.04

$$y = 2.1x + 8.2$$

Data Point (x, y)	Model Point $(x, \hat{y})$	Residual $y - \hat{y}$	Squared Residual $(y - \hat{y})^2$
(3, 15.0)	(3, 14.5)	.5	.25
(4, 16.6)	(4, 16.6)	0	0
(5, 16.4)	(5, 18.7)	−2.3	5.29
(6, 20.8)	(6, 20.8)	0	0
(7, 22.2)	(7, 22.9)	−.7	.49
			Sum = 6.03

According to this measure of error, the line $y = 1.8x + 9.6$ is a better fit for the data because the sum of the squares of its residuals is smaller than the sum of the squares of the residuals for $y = 2.1x + 8.2$. 2✓

✓Checkpoint 2

Another model for the data in Example 1 is $y = 1.9x + 9.4$. Use this line to find

(a) the residuals and

(b) the sum of the squares of the residuals.

(c) Does this line fit the data better than the two lines in Examples 1 and 2?

Answers: See page 112.

LINEAR REGRESSION (OPTIONAL)*

Mathematical techniques from multivariable calculus can be used to prove the following result.

> For any set of data points, there is one, and only one, line for which the sum of the squares of the residuals is as small as possible.

This *line of best fit* is called the **least-squares regression line**, and the computational process for finding its equation is called **linear regression**. Linear-regression formulas are quite complicated and require a large amount of computation. Fortunately, most graphing calculators and spreadsheet programs can do linear regression quickly and easily.

Example 3 **Business** Recall that the profits of GE (in billions of dollars) were as follows.

Year	2003	2004	2005	2006	2007
Profit	15.0	16.6	16.4	20.8	22.2

Use a graphing calculator to do the following:

(a) Plot the data points, with $x = 3$ corresponding to 2003.†

Solution The data points are (3, 15.0), (4, 16.6), (5, 16.4), (6, 20.8), and (7, 22.2). Press STAT EDIT to bring up the statistics editor. Enter the x-coordinates as list L_1 and the corresponding y-coordinates as list L_2, as shown in Figure 2.29. To plot the data points, go to the STAT PLOT menu, choose a plot (here, it is Plot 1), choose ON, and enter the lists L_1 and L_2, as shown in Figure 2.30. Then set the viewing window as usual and press GRAPH to produce the plot in Figure 2.31.

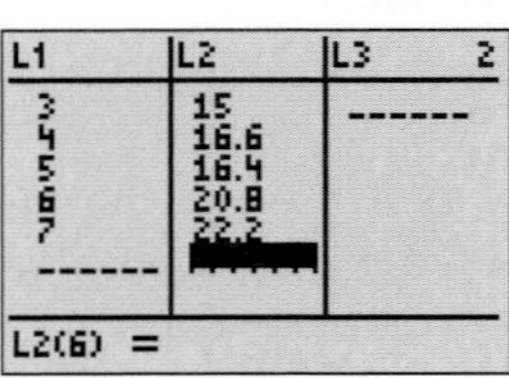

FIGURE 2.29

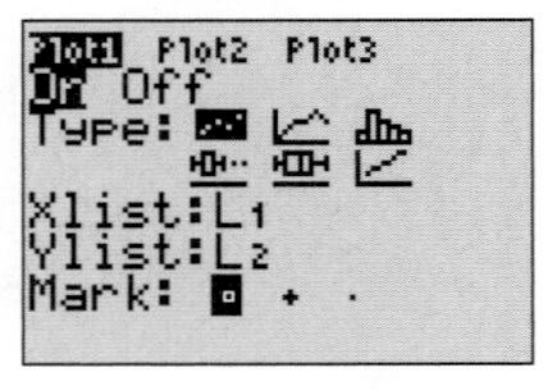

FIGURE 2.30

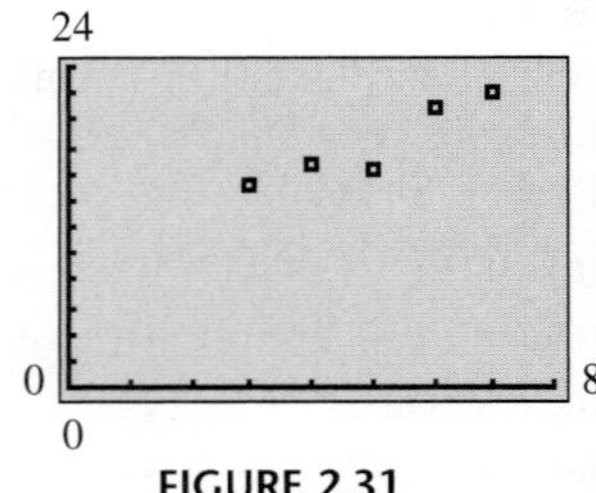

FIGURE 2.31

*Examples 3–6 require either a graphing calculator or a spreadsheet program.

†The process outlined here works for most TI graphing calculators. Other graphing calculators and spreadsheet programs operate similarly, but check your instruction manual or see the Graphing Calculator Appendix.

(b) Find the least-squares regression line for this data.

Solution Go to the STAT CALC menu and choose LIN REG, which returns you to the home screen. As shown in Figure 2.32, enter the list names and the place where the equation of the regression line should be stored (here, Y_1 is chosen; it is on the VARS Y-VARS FUNCTION menu); then press ENTER. Figure 2.33 shows that the equation of the regression line is

$$y = 1.86x + 8.9.$$

(The number r in Figure 2.33 will be discussed shortly.)

FIGURE 2.32

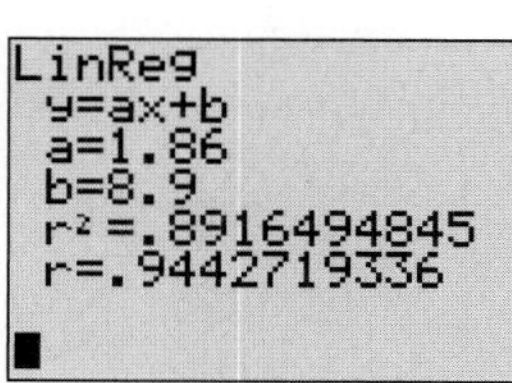

FIGURE 2.33

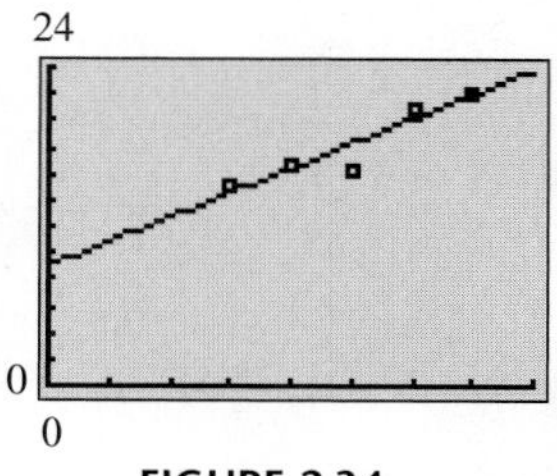

FIGURE 2.34

(c) Graph the data points and the regression line on the same screen.

Solution Press GRAPH to see the line plotted with the data points (Figure 2.34). 3

Checkpoint 3

Use the least-squares regression line $y = 1.86x + 8.9$ and the data points of Example 3 to find

(a) the residuals and

(b) the sum of the squares of the residuals.

(c) How does this line compare with those in Example 1 and side problem 2?

Answers: See page 112.

Example 4 The following table shows the revenue (in billions of dollars) for the wireless company Sprint Nextel Corporation from 1998 to 2007*:

Year	Revenue	Year	Revenue
1998	17.1	2003	26.2
1999	19.9	2004	27.4
2000	23.6	2005	34.7
2001	26.1	2006	41.0
2002	26.6	2007	40.1

(a) Let $x = 8$ correspond to 1998. Use a graphing calculator or spreadsheet program to find the least-squares regression line that models the data in the table.

Solution The data points are (8, 17.1), (9, 19.9), ..., (17, 40.1). Enter the x-coordinates as list L_1 and the corresponding y-coordinates as list L_2 in a graphing calculator and then find the regression line as in Figure 2.35 on the next page. Rounding the coefficients, the equation of the line is $y = 2.51x - 3.07$.

*Morningstar.com.

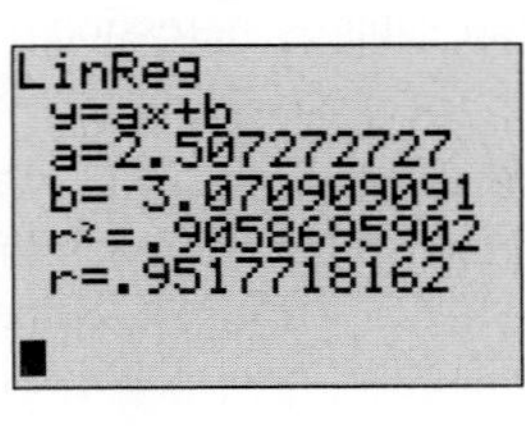

FIGURE 2.35

FIGURE 2.36

(b) Plot the data points and the regression line on the same screen.

Solution See Figure 2.36, which shows that the line is a reasonable model for the data.

(c) Assuming the linear trend continues, estimate the revenue in 2010.

Solution 2010 corresponds to $x = 20$. Substitute $x = 20$ into the regression-line equation:

$$y = 2.51(20) - 3.07 \approx 47.13.$$

This model estimates that Sprint Nextel Corporation will earn 47.13 billion dollars in revenue in 2010. 4✓

✓Checkpoint 4

Using only the data from 2003 and later in Example 4, find the equation of the least-squares regression line. Round the coefficients to two decimal places.

Answer: See page 112.

CORRELATION

Although the "best fit" line can always be found by linear regression, it may not be a good model. For instance, if the data points are widely scattered, no straight line will model the data accurately. When the linear-regression line was computed in Examples 3 and 4, the calculator also displayed a number r and its square (Figures 2.33 and 2.35). This number is called the **correlation coefficient**. It measures how closely the data points fit the regression line and thus indicates how good the regression line is for predictive purposes.

The correlation coefficient r is always between -1 and 1. When $r = \pm 1$, the data points all lie on the regression line (a perfect fit). When the absolute value of r is close to 1, the line fits the data quite well, and when r is close to 0, the line is a poor fit for the data (but some other curve might be a good fit). Figure 2.37 shows how the value of r varies, depending on the pattern of the data points.

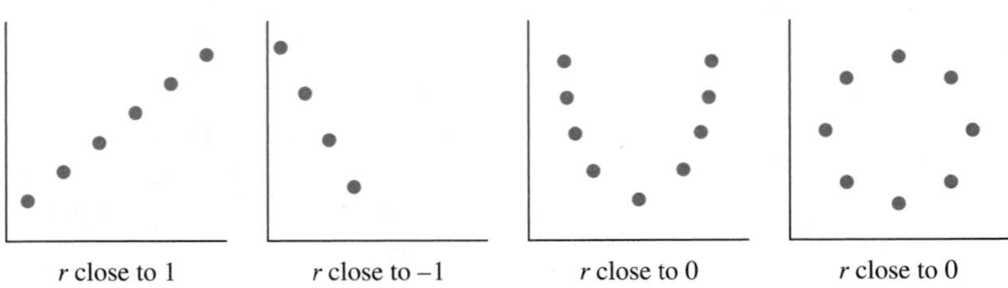

FIGURE 2.37

Example 5 **Business** The number of unemployed people in the U.S. labor force (in millions) in recent years is shown in the table.*

Year	Unemployed	Year	Unemployed	Year	Unemployed
1990	7.047	1996	7.236	2002	8.378
1991	8.628	1997	6.739	2003	8.774
1992	9.613	1998	6.210	2004	8.149
1993	8.940	1999	5.880	2005	7.591
1994	7.996	2000	5.692	2006	7.001
1995	7.404	2001	6.801	2007	7.078

Determine whether a linear equation is a good model for this data or not.

Solution Let $x = 0$ correspond to 1990, and plot the data points (0, 7.047), etc., either by hand or with a graphing calculator, as in Figure 2.38. They do not form a linear pattern (because unemployment tends to rise and fall). Alternatively, you could compute the regression equation for the data, as in Figure 2.39. The correlation coefficient is $r \approx -.256$, which is a relatively low value. This indicates that the regression line is a poor fit for the data. Therefore, a linear equation is not a good model for this data. 5✓

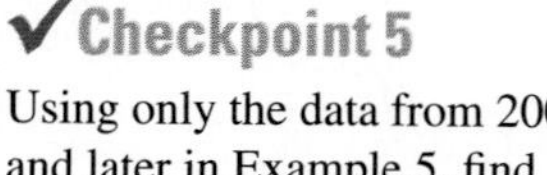

✓Checkpoint 5

Using only the data from 2002 and later in Example 5, find

(a) the equation of the least-squares regression line and

(b) the correlation coefficient.

(c) How well does this line fit the data?

Answers: See page 112.

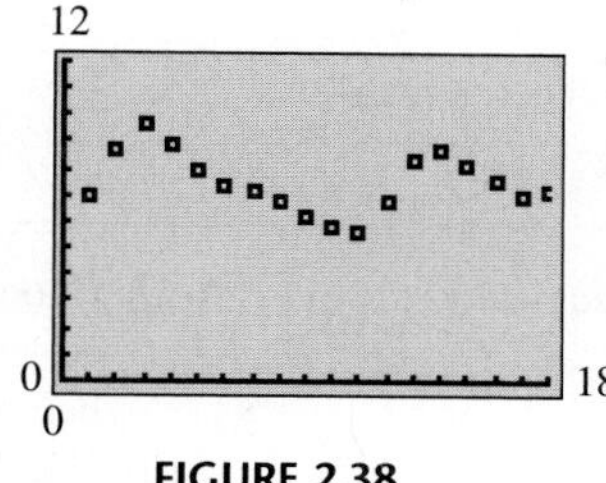

FIGURE 2.38

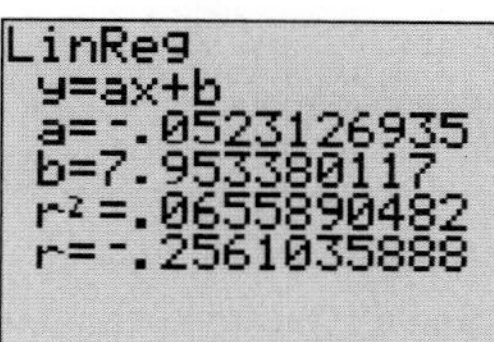

FIGURE 2.39

*U.S. Department of Labor, Bureau of Labor Statistics.

Example 6 **Education** Enrollment projections (in millions) for all U.S. colleges and universities in selected years are shown in the following table*:

Year	Enrollment	Year	Enrollment
2000	15.313	2005	16.679
2001	15.928	2006	16.887
2002	16.103	2007	17.958
2003	16.360	2008	18.264
2004	16.468		

(a) Let $x = 0$ correspond to 2000. Use a graphing calculator or spreadsheet program to find a linear model for the data and determine how well it fits the data points.

Solution The least-squares regression line (with coefficients rounded) is

$$y = .3297x + 15.3435,$$

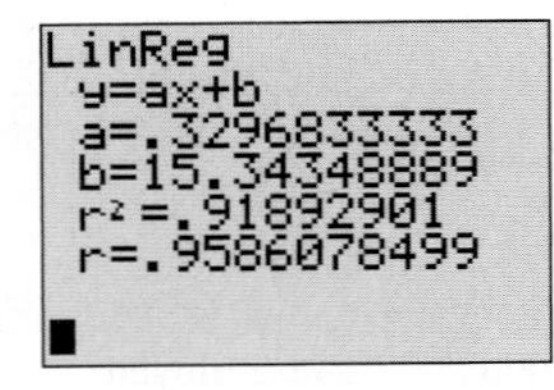

FIGURE 2.40

as shown in Figure 2.40. The correlation coefficient is $r \approx .96$, which is very close to 1, so this line fits the data well.

(b) Assuming the trend continues, predict the enrollment in 2014.

Solution Let $x = 14$ (corresponding to 2014) in the regression equation:

$$y = .3297(14) + 15.3435 = 19.9593.$$

Therefore, the enrollment in 2014 will be approximately 19,959,300 students.

(c) According to this model, in what year will enrollment reach 21 million?

Solution Let $y = 21$ and solve the regression equation for x:

$$y = .3297x + 15.3435$$

$$21 = .3297x + 15.3435 \quad \text{Let } y = 21.$$

$$5.6565 = .3297x \quad \text{Subtract 15.3435 from both sides.}$$

$$x \approx 17.16. \quad \text{Divide both sides by 0.3297.}$$

Since these enrollment figures change once a year, use the nearest integer value for x, namely, 17. So enrollment will reach 21 million in 2017.

*As of fall of each year; *Statistical Abstracts of the United States*: 2008.

2.3 ▶ Exercises

1. **Physical Science** The following table shows equivalent Fahrenheit and Celsius temperatures:

Degrees Fahrenheit	32	68	104	140	176	212
Degrees Celsius	0	20	40	60	80	100

(a) Choose any two data points and use them to construct a linear equation that models the data, with x being Fahrenheit and y being Celsius.

(b) Use the model in part (a) to find the Celsius temperature corresponding to

50° Fahrenheit and 75° Fahrenheit.

Physical Science *Use the linear equation derived in Exercise 1 to work the following problems.*

2. Convert each temperature.
 (a) 58°F to Celsius
 (b) 50°C to Fahrenheit
 (c) −10°C to Fahrenheit
 (d) −20°F to Celsius

3. According to the *World Almanac and Book of Facts*, 2008, Venus is the hottest planet, with a surface temperature of 867° Fahrenheit. What is this temperature in degrees Celsius?

4. Find the temperature at which Celsius and Fahrenheit temperatures are numerically equal.

In each of the next set of problems, assume that the data can be modeled by a straight line and that the trend continues indefinitely. Use two data points to find such a line and then answer the question. (See Example 1.)

5. **Business** The Consumer Price Index (CPI), which measures the cost of a typical package of consumer goods, stood at 168.8 in January 2000 and 211.1 in January 2008. Let $x = 0$ correspond to 2000, and estimate the CPI in 2005 and 2011.

6. **Finance** The approximate number of tax returns examined by the Internal Revenue Service in 1996 was 1.9 million. In 2006, it was 1.2 million.* Let $x = 0$ correspond to 1996, and estimate the number of examined returns in 2008.

7. **Business** Forrester Research, Inc., estimated the amount of online retail sales to be \$132.1 billion in 2006 and \$271.1 billion in 2011.† Let $x = 0$ correspond to 2006, and estimate the amount of online retail sales in 2010.

8. **Business** As in Exercise 7, Forrester Research estimated \$9.9 billion in sales of music and videos online in 2007 and \$15.6 billion in 2010. Let $x = 0$ correspond to 2007, and estimate the online sales of music and video in 2012.

9. **Physical Science** Suppose a baseball is thrown at 85 miles per hour. The ball will travel 320 feet when hit by a bat swung at 50 miles per hour and will travel 440 feet when hit by a bat swung at 80 miles per hour. Let y be the number of feet traveled by the ball when hit by a bat swung at x miles per hour. (*Note*: The preceding data are valid for $50 \le x \le 90$, where the bat is 35 inches long, weighs 32 ounces, and strikes a waist-high pitch so that the place of the swing lies at 10° from the diagonal.‡) How much farther will a ball travel for each mile-per-hour increase in the speed of the bat?

10. **Physical Science** Ski resorts require large amounts of water in order to make snow. Snowmass Ski Area in Colorado plans to pump at least 1120 gallons of water per minute for at least 12 hours a day from Snowmass Creek between mid-October and late December.* Environmentalists are concerned about the effects on the ecosystem. Find the minimum amount of water pumped in 30 days. (*Hint*: Let y be the total number of gallons pumped x days after pumping begins. Note that (0, 0) is on the graph of the equation.)

In each of the next two problems, two linear models are given for the data. For each model,
(a) *find the residuals and their sum;*
(b) *find the sum of the squares of the residuals;*
(c) *decide which model is the better fit. (See Example 2.)*

11. **Finance** The following table shows the amount of consumer credit outstanding (excluding real estate loans) in trillions of dollars†:

Year	1995	2000	2002	2004	2006
Credit	1.1	1.7	2.0	2.2	2.4

Let $x = 0$ correspond to 1995. Two equations that model the data are $y = .12x + 1.1$ and $y = .1x + 1.3$.

12. **Business** The weekly amount spent on advertising and the weekly sales revenue of a small store over a five-week period are shown in the following table:

Advertising Expenditure x (in hundreds of dollars)	1	2	3	4	5
Sales Revenue y (in thousands of dollars)	2	2	3	3	5

Two equations that model the data are $y = .5x + 1.5$ and $y = x$.

In each of the following problems, determine whether a straight line is a good model for the data. You may do this either visually, by plotting the data points, or analytically, by finding the correlation coefficient for the least-squares regression line. (See Examples 5 and 6.)

13. **Natural Science** The accompanying table shows the average monthly temperature (in degrees Fahrenheit) in Cleveland, Ohio, based on data from 1961 to 1990.‡ Let $x = 2$ correspond to February, $x = 4$ to April, etc.

Month	Feb	April	June	Aug	Oct	Dec
Temperature	27.3	47.5	67.5	70.3	52.7	30.9

*U.S. Internal Revenue Service, *IRS Data Book*, Publication 55B.

†Forrester Research, Inc., *U.S. eCommerce: Five Year Forecast and Data Overview.* (Cambridge, MA: 2006).

‡Robert K. Adair, *The Physics of Baseball* (HarperCollins, 1990).

*York Snow, Inc.

†Board of Governors of the Federal Reserve System, *Statistical Supplement to the Federal Reserve Bulletin.*

‡National Climatic Research Center.

14. **Health** The accompanying table shows the number of deaths per 100,000 people from heart disease in selected years.* Let $x = 0$ correspond to 1960.

Year	1960	1970	1980	1990	2000	2004
Deaths	559	483	412	322	258	217

In Exercises 15–18 find the required linear model as follows: If you do not have a graphing calculator or spreadsheet program, use the first and last data points to determine a line. If you do have a graphing calculator or spreadsheet program, find the least-squares regression line. (See Examples 1, 3, 4, and 6.)

15. **Health** Use the data on death from heart disease in Exercise 14.
   (a) Find a linear model for the data, with $x = 0$ corresponding to 1960.
   (b) Assuming the trend continues, estimate the number of deaths from heart disease in 2010.

16. **Social Science** The accompanying table shows how poverty-level income cutoffs (in dollars) for a family of four have changed over time (in large part because of inflation).†

Year	1980	1985	1990	1995	2000	2005
Income	8414	10,989	13,359	15,569	17,604	19,971

   (a) Find a linear model for the data, with $x = 0$ corresponding to 1980.
   (b) Assuming the trend continues, what is the approximate poverty-level income cutoff in 2009? in 2011?

17. **Physical Science** While shopping for an air conditioner, Mario Cekada consulted the following table giving a machine's BTUs and the square footage (ft^2) that it would cool‡:

ft^2 (x)	BTUs (y)
150	5000
175	5500
215	6000
250	6500
280	7000
310	7500
350	8000
370	8500
420	9000
450	9500

   (a) Find a linear model for the data.
   (b) To check the fit of the data to the line, use the results from part (a) to find the number of BTUs required to cool rooms of 150 ft^2, 280 ft^2, and 420 ft^2. How well do the actual data agree with the predicted values?
   (c) Suppose Adam's room measures 235 ft^2. Use the results from part (a) to decide how many BTUs it requires. If air conditioners are available only with the BTU choices in the table, which should Adam choose?

18. **Business** The accompanying table shows the daily newspaper circulation (in millions)*:

Year	Circulation
1990	62.3
1995	58.2
2000	55.8
2001	55.6
2002	55.2
2003	55.2
2004	54.5
2005	53.3

   (a) Find a linear model for the data, with $x = 0$ corresponding to 1990.
   (b) Assuming the trend continues indefinitely, estimate the circulation in 2010 and 2012.

Work these problems.

19. The accompanying graph shows Dell Computer's revenue (in billions) of dollars from 1999 to 2007.†

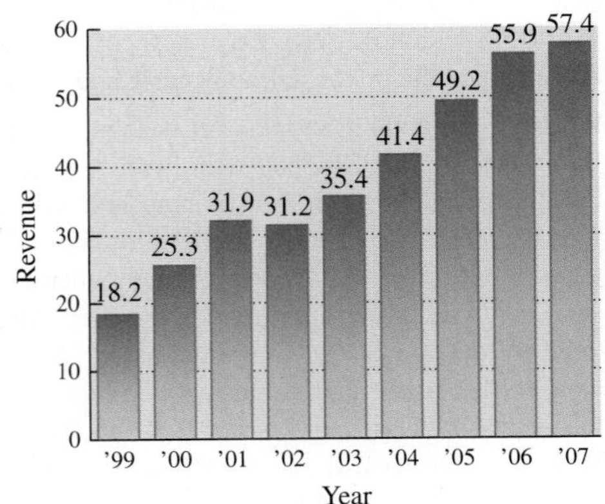

   (a) List the data points, with $x = 9$ corresponding to 1999.
   (b) Find the least-squares regression line that models the data.
   (c) If the trend shown continues, what will Dell earn in revenues in 2012?

*U.S. National Center for Health Statistics.
†U.S. Census Bureau, *Current Population Reports*.
‡Morris Carey and James Carey, "On the House," *Sacramento Bee*, July 29, 2000.

**Statistical Abstract of the United States*: 2008.
†Morningstar.com.

20. Natural Science Biologists have observed a linear relationship between the temperature and the frequency with which a cricket chirps. The following data were measured for the striped ground cricket*:

Temperature °F (x)	Chirps per second (y)
88.6	20.0
71.6	16.0
93.3	19.8
84.3	18.4
80.6	17.1
75.2	15.5
69.7	14.7
82.0	17.1
69.4	15.4
83.3	16.2
79.6	15.0
82.6	17.2
80.6	16.0
83.5	17.0
76.3	14.4

(a) Find the least-squares regression line that models the data.
(b) Use the results of part (a) to determine how many chirps per second you would expect to hear from the striped ground cricket if the temperature were 73°F.
(c) Use the results of part (a) to determine what the temperature is when the striped ground crickets are chirping at a rate of 18 times per second.
(d) Find the correlation coefficient.

21. Transportation The estimated number of scheduled passengers on U.S. commercial airlines (in billions) in selected years is shown in the following table†:

Year	Passengers
2002	.63
2003	.64
2004	.69
2005	.72
2006	.77
2007	.79
2008	.80
2009	.84

(a) Find the least-squares regression line that models these data, with $x = 2$ corresponding to 2002.
(b) Extimate the number of passengers in 2012 and in 2016.

22. Health The following table shows men's and women's life expectancy at birth (in years) for selected birth years in the United States*:

Birth Year	Life Expectancy	
	Men	Women
1970	67.1	74.7
1975	68.8	76.6
1980	70.0	77.4
1985	71.1	78.2
1990	71.8	78.8
1995	72.5	78.9
1998	73.8	79.5
2000	74.3	79.7
2001	74.4	79.8
2004	75.2	80.4
2010	75.6	81.4

(a) Find the least-squares regression line for the men's data, with $x = 70$ corresponding to 1970.
(b) Find the least-squares regression line for the women's data, with $x = 70$ corresponding to 1970.
(c) Suppose life expectancy continues to increase as predicted by the equations in parts (a) and (b). Will men's life expectancy ever be the same as women's? If so, in what birth year will this occur?

23. Finance The accompanying table gives the median household income (in thousands of dollars) for men living without a spouse (this means that 50% of males earn more than the median and 50% earn less than that amount) for various years†:

Year	1980	1990	1995	1997	1998	1999
Income	17.5	29.0	30.4	33.0	35.7	37.3

Year	2000	2001	2002	2003	2004	2005
Income	37.7	36.6	37.7	38.0	40.3	41.1

(a) Find a linear model for these data, with $x = 0$ corresponding to 1980.
(b) Interpret the meaning of the slope.
(c) If the model remains accurate in the future, what will be the median household income for men living without a spouse in 2010?

*Reprinted by permission of the publishers from *The Songs of Insects* by George W. Pierce. Cambridge, MA: Harvard University Press, copyright © 1948 by the President and Fellows of Harvard College.

†Data and projections are from the Federal Aviation Administration.

*U.S. Center for National Health Statistics.

†U.S. Census Bureau.

24. **Finance** The accompanying table gives the median household income (in thousands of dollars) for women living without a spouse (this means that 50% of females earn more than the median and 50% earn less than that amount) for various years*:

Year	1980	1990	1995	1997	1998	1999
Income	10.4	16.9	19.7	21.0	22.1	23.8

Year	2000	2001	2002	2003	2004	2005
Income	25.7	25.7	26.4	26.6	27.0	27.2

(a) Find a linear model for these data, with $x = 0$ corresponding to 1980.
(b) Interpret the meaning of the slope.
(c) Comparing your answer for (b) with your answer for part (b) in Exercise 23, are men or women gaining more in median income each year?

25. **Business** The following table shows total operating revenue (in billions of dollars) for the U.S. airline industry for several years†:

Year	2002	2003	2004	2005	2006
Operating Revenue	107	118	134	151	164

(a) Find the least-squares regression line, with $x = 2$ corresponding to 2002.
(b) Assuming the trend continues indefinitely, predict the operating revenue for 2010.
(c) Based on the model, when will operating revenue reach $180 billion?

*U.S. Census Bureau.

†Air Transportation Association of America.

✓ Checkpoint Answers

1. $y = 1.933x + 9.201$
2. (a) $-.1, -.4, -2.5, 0, -.5$
   (b) 6.67
   (c) No, because the sum of the squares of the residuals is higher.
3. (a) $.52, .26, -1.80, .74, .28$
   (b) 4.204
   (c) It fits the data best because the sum of the squares of its residuals is smallest.
4. $y = 4.14x - 28.22$
5. (a) $y = -.354x + 12.956$
   (b) $r \approx -.92$
   (c) It fits reasonably well because $|r|$ is close to 1.

2.4 Linear Inequalities

An **inequality** is a statement that one mathematical expression is greater than (or less than) another. Inequalities are very important in applications. For example, a company wants revenue to be *greater than* costs and must use *no more than* the total amount of capital or labor available.

Inequalities may be solved by algebraic or geometric methods. In this section, we shall concentrate on algebraic methods for solving **linear inequalities**, such as

$$4 - 3x \leq 7 + 2x \quad \text{and} \quad -2 < 5 + 3m < 20,$$

and absolute-value inequalities, such as $|x - 2| < 5$. The following properties are the basic algebraic tools for working with inequalities.

Properties of Inequality

For real numbers a, b, and c,

(a) if $a < b$, then $a + c < b + c$;
(b) if $a < b$, and if $c > 0$, then $ac < bc$;
(c) if $a < b$, and if $c < 0$, then $ac > bc$.

Throughout this section, definitions are given only for $<$, but they are equally valid for $>$, $\leq$, and $\geq$.

CAUTION Pay careful attention to part (c) in the previous box; if both sides of an inequality are multiplied by a negative number, the direction of the inequality symbol must be reversed. For example, starting with the true statement $-3 < 5$ and multiplying both sides by the positive number 2 gives

$$-3 \cdot \mathbf{2} < 5 \cdot \mathbf{2},$$

or

$$-6 < 10,$$

still a true statement. However, starting with $-3 < 5$ and multiplying both sides by the negative number -2 gives a true result only if the direction of the inequality symbol is reversed:

$$-3(-\mathbf{2}) > 5(-\mathbf{2})$$
$$6 > -10. \quad ✓1$$

✓Checkpoint 1

(a) First multiply both sides of $-6 < -1$ by 4, and then multiply both sides of $-6 < -1$ by -7.

(b) First multiply both sides of $9 \geq -4$ by 2, and then multiply both sides of $9 \geq -4$ by -5.

(c) First add 4 to both sides of $-3 < -1$, and then add -6 to both sides of $-3 < -1$.

Answers: *See page 119.*

Example 1 Solve $3x + 5 > 11$. Graph the solution.

Solution First add -5 to both sides:

$$3x + 5 + (-5) > 11 + (-5)$$
$$3x > 6.$$

Now multiply both sides by 1/3:

$$\frac{1}{3}(3x) > \frac{1}{3}(6)$$
$$x > 2.$$

(Why was the direction of the inequality symbol not changed?) In interval notation (introduced in Section 1.1), the solution is the interval $(2, \infty)$, which is graphed on the number line in Figure 2.41. The parenthesis at 2 shows that 2 is not included in the solution.

FIGURE 2.41

As a partial check, note that 0, which is not part of the solution, makes the inequality false, while 3, which is part of the solution, makes it true:

$$3(\mathbf{0}) + 5 \overset{?}{>} 11 \qquad\qquad 3(\mathbf{3}) + 5 \overset{?}{>} 11$$
$$5 > 11 \quad \text{False} \qquad\qquad 14 > 11. \quad \text{True} \quad ✓2$$

✓Checkpoint 2

Solve these inequalities. Graph each solution.

(a) $5z - 11 < 14$

(b) $-3k \leq -12$

(c) $-8y \geq 32$

Answers: *See page 119.*

Example 2 Solve $4 - 3x \le 7 + 2x$.

Solution Add -4 to both sides:

$$4 - 3x + (-4) \le 7 + 2x + (-4)$$
$$-3x \le 3 + 2x.$$

Add $-2x$ to both sides (remember that *adding* to both sides never changes the direction of the inequality symbol):

$$-3x + (-2x) \le 3 + 2x + (-2x)$$
$$-5x \le 3.$$

Multiply both sides by $-1/5$. Since $-1/5$ is negative, change the direction of the inequality symbol:

$$-\frac{1}{5}(-5x) \ge -\frac{1}{5}(3)$$
$$x \ge -\frac{3}{5}.$$

Figure 2.42 shows a graph of the solution, $[-3/5, \infty)$. The bracket in Figure 2.42 shows that $-3/5$ is included in the solution. 3✓

✓Checkpoint 3

Solve these inequalities. Graph each solution.

(a) $8 - 6t \ge 2t + 24$

(b) $-4r + 3(r + 1) < 2r$

Answers: See page 119.

FIGURE 2.42

Example 3 Solve $-2 < 5 + 3m < 20$. Graph the solution.

Solution The inequality $-2 < 5 + 3m < 20$ says that $5 + 3m$ is between -2 and 20. We can solve this inequality with an extension of the properties given at the beginning of this section. Work as follows, first adding -5 to each part:

$$-2 + (-5) < 5 + 3m + (-5) < 20 + (-5)$$
$$-7 < 3m < 15.$$

Now multiply each part by $1/3$:

$$-\frac{7}{3} < m < 5.$$

A graph of the solution, $(-7/3, 5)$, is given in Figure 2.43. 4✓

✓Checkpoint 4

Solve each of the given inequalities. Graph each solution.

(a) $9 < k + 5 < 13$

(b) $-6 \le 2z + 4 \le 12$

Answers: See page 119.

FIGURE 2.43

Example 4 The formula for converting from Celsius to Fahrenheit temperature is

$$F = \frac{9}{5}C + 32.$$

What Celsius temperature range corresponds to the range from 32°F to 77°F?

Solution The Fahrenheit temperature range is $32 < F < 77$. Since $F = (9/5)C + 32$, we have

$$32 < \frac{9}{5}C + 32 < 77.$$

Solve the inequality for C:

$$32 < \frac{9}{5}C + 32 < 77$$

$$0 < \frac{9}{5}C < 45 \qquad \text{Subtract 32 from each part.}$$

$$\frac{5}{9}\cdot 0 < \frac{5}{9}\cdot\frac{9}{5}C < \frac{5}{9}\cdot 45 \qquad \text{Multiply each part by } \frac{5}{9}.$$

$$0 < C < 25.$$

The corresponding Celsius temperature range is 0°C to 25°C. 5✓

✓Checkpoint 5

In Example 4, what Celsius temperatures correspond to the range from 5°F to 95°F?

Answer: See page 119.

A product will break even or produce a profit only if the revenue R from selling the product at least equals the cost C of producing it—that is, if $R \geq C$.

Example 5 A company analyst has determined that the cost to produce and sell x units of a certain product is $C = 20x + 1000$. The revenue for that product is $R = 70x$. Find the values of x for which the company will break even or make a profit on the product.

Solution Solve the inequality $R \geq C$:

$$R \geq C$$

$$70x \geq 20x + 1000 \qquad \text{Let } R = 70x \text{ and } C = 20x + 1000.$$

$$50x \geq 1000 \qquad \text{Subtract } 20x \text{ from both sides.}$$

$$x \geq 20. \qquad \text{Divide both sides by 50.}$$

The company must produce and sell 20 items to break even and more than 20 to make a profit.

Example 6 **Business** A pretzel manufacturer can sell a 6-ounce bag of pretzels to a wholesaler for \$.35 a bag. The variable cost of producing each bag is \$.20 per bag, and the fixed cost for the manufacturing operation is \$110,000.* How many bags of pretzels need to be sold in order to break even or earn a profit?

*Variable costs, fixed costs, and revenue were discussed on page 18.

Solution Let x be the number of bags produced. Then the revenue equation is

$$R = .35x,$$

and the cost is given by

$$\begin{aligned} \text{Cost} &= \text{Fixed Costs} + \text{Variable Costs} \\ C &= 110{,}000 + .20x. \end{aligned}$$

We now solve the inequality $R \geq C$:

$$\begin{aligned} R &\geq C \\ .35x &\geq 110{,}000 + .25x \\ .1x &\geq 110{,}000 \\ x &\geq 1{,}110{,}000. \end{aligned}$$

The manufacturer must produce and sell 1,110,000 bags of pretzels to break even and more than that to make a profit.

ABSOLUTE-VALUE INEQUALITIES

You may wish to review the definition of absolute value in Section 1.1 before reading the following examples, which show how to solve inequalities involving absolute values.

Example 7 Solve each inequality.

(a) $|x| < 5$

Solution Because absolute value gives the distance from a number to 0, the inequality $|x| < 5$ is true for all real numbers whose distance from 0 is less than 5. This includes all numbers between -5 and 5, or numbers in the interval $(-5, 5)$. A graph of the solution is shown in Figure 2.44.

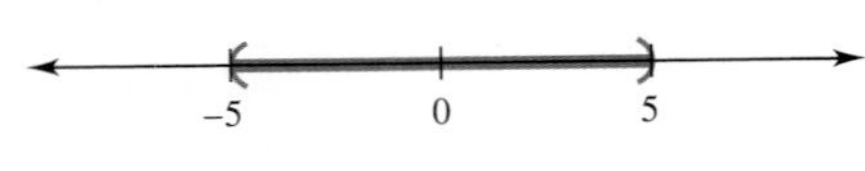

FIGURE 2.44

(b) $|x| > 5$

Solution The solution of $|x| > 5$ is given by all those numbers whose distance from 0 is *greater* than 5. This includes the numbers satisfying $x < -5$ or $x > 5$. A graph of the solution, all numbers in

$$(-\infty, -5) \quad \text{or} \quad (5, \infty),$$

is shown in Figure 2.45. 6✓

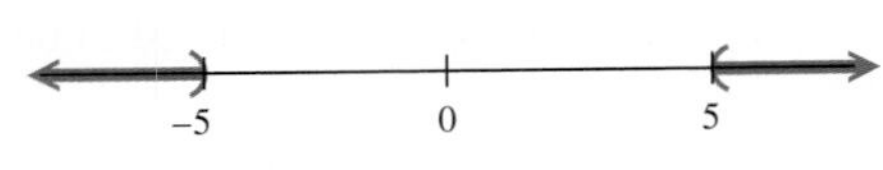

FIGURE 2.45

✓Checkpoint 6

Solve each inequality. Graph each solution.

(a) $|x| \leq 1$

(b) $|y| \geq 3$

Answers: See page 119.

The preceding examples suggest the following generalizations.

> Assume that a and b are real numbers and that b is positive.
> 1. Solve $|a| < b$ by solving $-b < a < b$.
> 2. Solve $|a| > b$ by solving $a < -b$ or $a > b$.

Example 8 Solve $|x - 2| < 5$.

Solution Replace a with $x - 2$ and b with 5 in property (1) in the box above. Now solve $|x - 2| < 5$ by solving the inequality

$$-5 < x - 2 < 5.$$

Add 2 to each part, getting the solution

$$-3 < x < 7,$$

which is graphed in Figure 2.46. 7✓

FIGURE 2.46

✓Checkpoint 7

Solve each inequality. Graph each solution.

(a) $|p + 3| < 4$

(b) $|2k - 1| \leq 7$

Answers: See page 120.

Example 9 Solve $|2 - 7m| - 1 > 4$.

Solution First add 1 to both sides:

$$|2 - 7m| > 5$$

Now use property (2) from the preceding box to solve $|2 - 7m| > 5$ by solving the inequality

$$2 - 7m < -5 \quad \text{or} \quad 2 - 7m > 5.$$

Solve each part separately:

$$-7m < -7 \quad \text{or} \quad -7m > 3$$

$$m > 1 \quad \text{or} \quad m < -\frac{3}{7}.$$

The solution, all numbers in $\left(-\infty, -\frac{3}{7}\right)$ or $(1, \infty)$, is graphed in Figure 2.47. 8✓

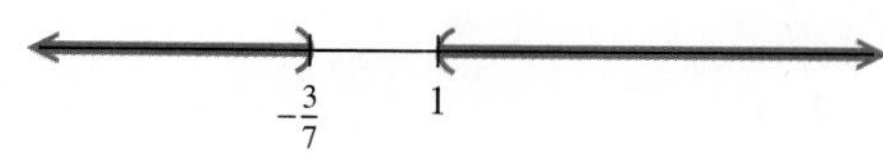

FIGURE 2.47

✓Checkpoint 8

Solve each inequality. Graph each solution.

(a) $|y - 2| > 5$

(b) $|3k - 1| \geq 2$

(c) $|2 + 5r| - 4 \geq 1$

Answers: See page 120.

✓ Checkpoint 9

Solve each inequality.

(a) $|5m - 3| > -10$

(b) $|6 + 5a| < -9$

(c) $|8 + 2r| > 0$

Answers: See page 120.

Example 10 Solve $|3 - 7x| \geq -8$.

Solution The absolute value of a number is always nonnegative. Therefore, $|3 - 7x| \geq -8$ is always true, so the solution is the set of all real numbers. Note that the inequality $|3 - 7x| \leq -8$ has no solution, because the absolute value of a quantity can never be less than a negative number. 9 ✓

2.4 ▶ Exercises

1. Explain how to determine whether a parenthesis or a bracket is used when graphing the solution of a linear inequality.

2. The three-part inequality $p < x < q$ means "p is less than x and x is less than q." Which one of the given inequalities is not satisfied by any real number x? Explain why.

(a) $-3 < x < 5$ **(b)** $0 < x < 4$

(c) $-7 < x < -10$ **(d)** $-3 < x < -2$

Solve each inequality and graph each solution. (See Examples 1–3.)

3. $-8k \leq 32$ **4.** $-4a \leq 36$

5. $-2b > 0$ **6.** $6 - 6z < 0$

7. $3x + 4 \leq 14$ **8.** $2y - 7 < 9$

9. $-5 - p \geq 3$ **10.** $5 - 3r \leq -4$

11. $7m - 5 < 2m + 10$ **12.** $6x - 2 > 4x - 10$

13. $m - (4 + 2m) + 3 < 2m + 2$

14. $2p - (3 - p) \leq -7p - 2$

15. $-2(3y - 8) \geq 5(4y - 2)$

16. $5r - (r + 2) \geq 3(r - 1) + 6$

17. $3p - 1 < 6p + 2(p - 1)$

18. $x + 5(x + 1) > 4(2 - x) + x$

19. $-7 < y - 2 < 5$ **20.** $-3 < m + 6 < 2$

21. $8 \leq 3r + 1 \leq 16$ **22.** $-6 < 2p - 3 \leq 5$

23. $-4 \leq \dfrac{2k - 1}{3} \leq 2$ **24.** $-1 \leq \dfrac{5y + 2}{3} \leq 4$

25. $\dfrac{3}{5}(2p + 3) \geq \dfrac{1}{10}(5p + 1)$

26. $\dfrac{8}{3}(z - 4) \leq \dfrac{2}{9}(3z + 2)$

In the following exercises, write a linear inequality that describes the given graph.

27.

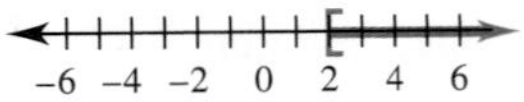

28.

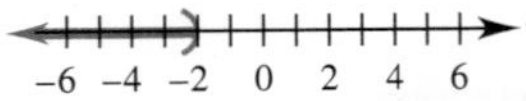

29.

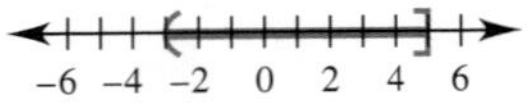

30.

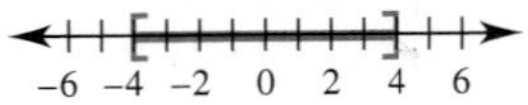

31. Natural Science Federal guidelines require drinking water to have less than .050 milligram per liter of lead. A test using 21 samples of water in a midwestern city found that the average amount of lead in the samples was .040 milligram per liter. All samples had lead content within 5% of the average.

(a) Select a variable and write down what it represents.

(b) Write a three-part inequality to express the results obtained from the sample.

(c) Did all the samples meet the federal requirement?

32. Social Science A Gallup poll found that among American alcohol consumers, between 39% and 45% prefer beer.* Let B represent the percentage of American alcohol consumers who prefer beer. Write the preceding information as an inequality.

33. Finance The following table shows the federal income tax for a single person in 2008:

If Taxable Income is Over	But Not Over	The Tax Is	Of Amount Over
\$0	\$8025	10%	\$0
\$8025	\$32,550	\$802.50 + 15%	\$8025
\$32,550	\$78,850	\$4481.25 + 25%	\$32,550
\$78,850	\$164,550	\$16,056.25 + 28%	\$78,850
\$164,550	\$357,700	\$40,052.25 + 33%	\$164,550
\$357,700	no limit	\$103,791.75 + 35%	\$357,700

(a) Let x denote the taxable income. Write each of the six income ranges in the table as an inequality.

(b) Let T denote the income tax. Write an inequality that gives the tax range in dollars for each of the six income ranges in the table.

Solve each inequality. Graph each solution. (See Examples 7–10.)

34. $|p| > 7$ **35.** $|m| < 2$ **36.** $|r| \leq 5$

37. $|a| < -2$ **38.** $|b| > -5$

39. $|2x + 5| < 1$ **40.** $\left|x - \dfrac{1}{2}\right| < 2$

*Gallup.com, "Beer back to double-digit lead over wine as favored drink," July 25, 2008.

41. $|3z + 1| \geq 4$

42. $|8b + 5| \geq 7$

43. $\left|5x + \frac{1}{2}\right| - 2 < 5$

44. $\left|x + \frac{2}{3}\right| + 1 < 4$

Physical Science *The given inequality describes the monthly average high daily temperature T in degrees Fahrenheit in the given location.* What range of temperatures corresponds to the inequality?*

45. $|T - 83| \leq 7$; Miami, Florida

46. $|T - 63| \leq 27$; Boise, Idaho

47. $|T - 61| \leq 21$; Flagstaff, Arizona

48. $|T - 43| \leq 22$; Anchorage, Alaska

49. Natural Science Human beings emit carbon dioxide when they breathe. In one study, the emission rates of carbon dioxide by college students were measured both during lectures and during exams. The average individual rate R_L (in grams per hour) during a lecture class satisfied the inequality $|R_L - 26.75| \leq 1.42$, whereas during an exam, the rate R_E satisfied the inequality $|R_E - 38.75| \leq 2.17$.†

(a) Find the range of values for R_L and R_E.

(b) A class had 225 students. If T_L and T_E represent the total amounts of carbon dioxide (in grams) emitted during a one-hour lecture and one-hour exam, respectively, write inequalities that describe the ranges for T_L and T_E.

50. Social Science When administering a standard intelligence quotient (IQ) test, we expect about one-third of the scores to be more than 12 units above 100 or more than 12 units below 100. Describe this situation by writing an absolute value inequality.

51. Business A model for data from Dell Computer obtained from Morningstar.com yields the equation $y = 4.89x - 25.13$, where $x = 9$ corresponds to 1999 and y is the annual revenue (in billions of dollars) in year x. During what years, according to the model, was revenue between 30 and 40 billion dollars?

52. Business A linear model for daily newspaper circulation y (in millions) is $y = -.54x + 61.69$, where $x = 0$ corresponds to 1990.‡ If the model remains accurate indefinitely, in what year will circulation be below 50 million?

53. Business The number of ski resorts y in the United States has declined and can be approximated by the equation $y = -3.14x + 502$, with $x = 2$ corresponding to the year 2002.§ In what years was the number of ski resorts between 480 and 470?

54. Business Data from the U.S. Department of Agriculture indicates that the model $y = 4.35x + 22.84$ is useful for predicting the average number of dollars earned per 100 pounds for beef, y, between 1999 and 2006 ($x = 9$ corresponds to 1999). Assuming this model holds for years before 1999 and years after 2006,

(a) in what years were the dollars earned per 100 pounds less than 72?

(b) in what years were the dollars earned per 100 pounds greater than 85?

55. Business The model for the average number of dollars earned per 100 pounds of hog meat, y, is $y = 2.38x + 12.46$ for the years between 1999 and 2006 ($x = 9$ corresponds to 1999). Assuming this model holds for the years before 1999 and the years after 2006,

(a) in what years were the dollars earned per 100 pounds less than 40?

(b) in what years were the dollars earned per 100 pounds greater than 50?

56. Transportation The cost of a taxi in New York City (as of 2008) is \$2.50, plus 40 cents for each 1/5 mile. How far could you travel for at least \$8.50, but not more than \$12.50?

Business *In Exercises 57–62, find all values of x for which the given products will at least break even. (See Examples 5 and 6.)*

57. The cost to produce x units of wire is $C = 50x + 6000$, while the revenue is $R = 65x$.

58. The cost to produce x units of squash is $C = 100x + 6000$, while the revenue is $R = 500x$.

59. $C = 85x + 1000$; $R = 105x$

60. $C = 70x + 500$; $R = 60x$

61. $C = 1000x + 5000$; $R = 900x$

62. $C = 25{,}000x + 21{,}700{,}000$; $R = 102{,}500x$

✓ Checkpoint Answers

1. (a) $-24 < -4$; $42 > 7$ (b) $18 \geq -8$; $-45 \leq 20$
(c) $1 < 3$; $-9 < -7$

2. (a) $z < 5$

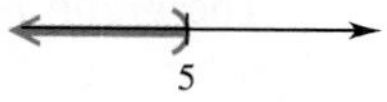

(b) $k \geq 4$

(c) $y \leq -4$

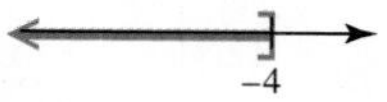

3. (a) $t \leq -2$

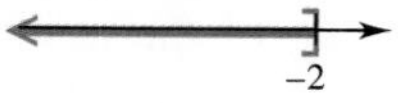

(b) $r > 1$

4. (a) $4 < k < 8$

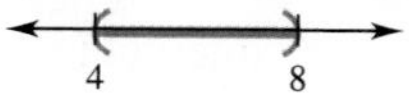

(b) $-5 \leq z \leq 4$

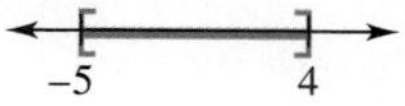

5. −15°C to 35°C

6. (a) $[-1, 1]$

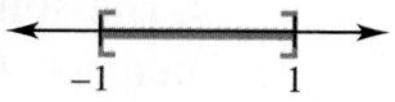

(b) All numbers in $(-\infty, -3]$ or $[3, \infty)$

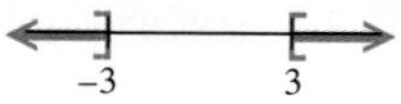

---

*Weatherbase.com.

†T.C. Wang, *ASHRAE Transactions* 81 (Part 1), 32 (1975).

‡*Statistical Abstract of the United States*: 2008.

§*Statistical Abstract of the United States*: 2008.

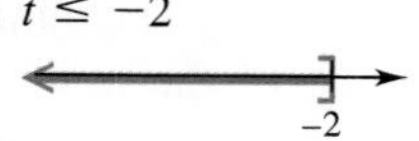

7. **(a)** $(-7, 1)$ **(b)** $[-3, 4]$

8. **(a)** All numbers in $(-\infty, -3)$ or $(7, \infty)$

(b) All numbers in $\left(-\infty, -\frac{1}{3}\right]$ or $[1, \infty)$

(c) All numbers in $\left(-\infty, -\frac{7}{5}\right]$ or $\left[\frac{3}{5}, \infty\right)$

9. **(a)** All real numbers **(b)** No solution
(c) All real numbers except -4

2.5 ▶ Polynomial and Rational Inequalities

This section deals with the solution of polynomial and rational inequalities, such as

$$r^2 + 3r - 4 \geq 0, \qquad x^3 - x \leq 0, \qquad \text{and} \qquad \frac{2x - 1}{3x + 4} < 5.$$

We shall concentrate on algebraic solution methods, but to understand why these methods work, we must first look at such inequalities from a graphical point of view.

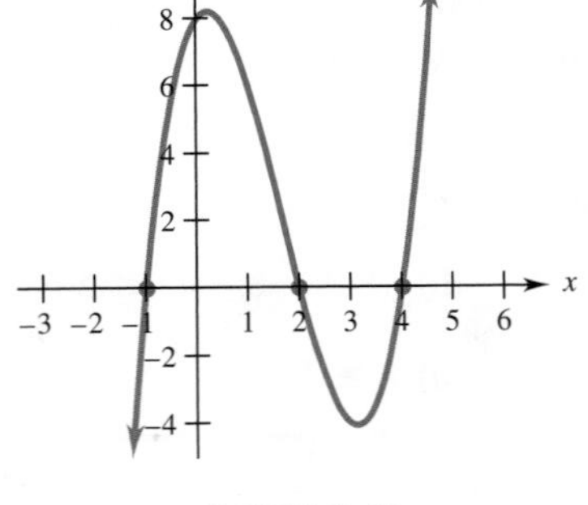

FIGURE 2.49

Example 1 Use the graph of $y = x^3 - 5x^2 + 2x + 8$ in Figure 2.49 to solve each of the given inequalities.

(a) $x^3 - 5x^2 + 2x + 8 > 0$

Solution Each point on the graph has coordinates of the form $(x, x^3 - 5x^2 + 2x + 8)$. The number x is a solution of the inequality exactly when the second coordinate of this point is positive—that is, when the point lies *above* the x-axis. So to solve the inequality, we need only find the first coordinates of points on the graph that are above the x-axis. This information can be read from Figure 2.49. The graph is above the x-axis when $-1 < x < 2$ and when $x > 4$. Therefore, the solutions of the inequality are all numbers x in the interval $(-1, 2)$ or the interval $(4, \infty)$.

(b) $x^3 - 5x^2 + 2x + 8 < 0$

The number x is a solution of the inequality exactly when the second coordinate of the point $(x, x^3 - 5x^2 + 2x + 8)$ on the graph is negative—that is, when the point lies *below* the x-axis. Figure 2.49 shows that the graph is below the x-axis when $x < -1$ and when $2 < x < 4$. Hence, the solutions are all numbers x in the interval $(-\infty, -1)$ or the interval $(2, 4)$.

The solution process in Example 1 depends only on knowing the graph and its x-intercepts (that is, the points where the graph intersects the x-axis). This information can often be obtained algebraically, without doing any graphing, as illustrated in the next example.

Example 2 Solve each of the given quadratic inequalities.

(a) $x^2 - x < 12$

Solution First rewrite the inequality as $x^2 - x - 12 < 0$. Now, we don't know what the graph of $y = x^2 - x - 12$ looks like, but we can still find its x-intercepts by setting $y = 0$ and solving for x:

$$x^2 - x - 12 = 0$$
$$(x + 3)(x - 4) = 0.$$
$$x + 3 = 0 \quad \text{or} \quad x - 4 = 0$$
$$x = -3 \qquad\qquad x = 4.$$

These numbers divide the x-axis (number line) into three regions, as indicated in Figure 2.50.

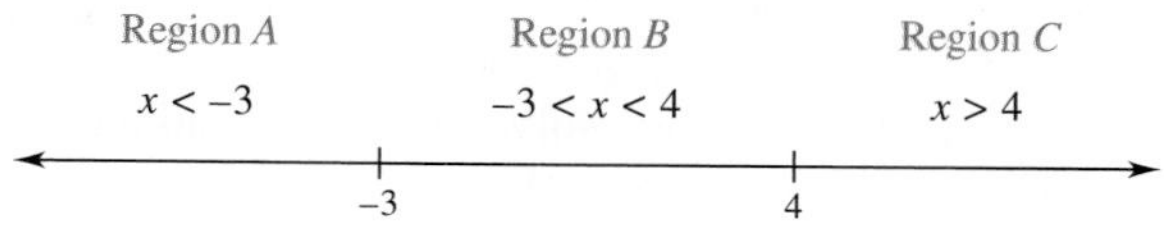

FIGURE 2.50

In each region, the graph of $y = x^2 - x - 12$ is an unbroken curve, so it will be entirely above or entirely below the axis. It can pass from above to below the x-axis only at the x-intercepts. To see whether the graph is above or below the x-axis when x is in region A, choose a value of x in region A, say, $x = -5$, and substitute it into the equation:

$$y = x^2 - x - 12 = (-5)^2 - (-5) - 12 = 18.$$

Therefore, the point $(-5, 18)$ is on the graph. Since its y-coordinate 18 is positive, this point lies above the x-axis; hence, the entire graph lies above the x-axis in region A.

Similarly, we can choose a value of x in region B, say, $x = 0$. Then

$$y = x^2 - x - 12 = 0^2 - 0 - 12 = -12,$$

so that $(0, -12)$ is on the graph. Since this point lies below the x-axis (why?), the entire graph in region B must be below the x-axis. Finally, in region C, let $x = 5$. Then $y = 5^2 - 5 - 12 = 8$, so that $(5, 8)$ is on the graph, and the entire graph in region C lies above the x-axis. We can summarize the results as follows:

Interval	$x < -3$	$-3 < x < 4$	$x > 4$
Test value in interval	−5	0	5
Value of $x^2 - x - 12$	18	−12	8
Graph	above x-axis	below x-axis	above x-axis
Conclusion	$x^2 - x - 12 > 0$	$x^2 - x - 12 < 0$	$x^2 - x - 12 > 0$

The last row shows that the only region where $x^2 - x - 12 < 0$ is region B, so the solutions of the inequality are all numbers x with $-3 < x < 4$—that is, the interval $(-3, 4)$, as shown in the number line graph in Figure 2.51 on the next page.

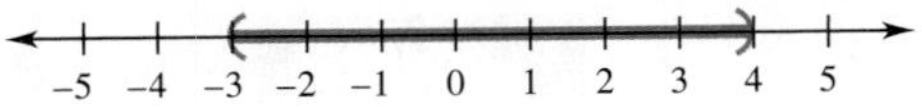

−5 −4 −3 −2 −1 0 1 2 3 4 5

FIGURE 2.51

(b) $x^2 - x - 12 > 0$

Solution Use the chart in part (a). The last row shows that $x^2 - x - 12 > 0$ only when x is in region A or region C. Hence, the solutions of the inequality are all numbers x with $x < -3$ or $x > 4$—that is, all numbers in the interval $(-\infty, -3)$ or the interval $(4, \infty)$. 1✓

✓Checkpoint 1

Solve each inequality. Graph the solution on the number line.

(a) $x^2 - 2x < 15$

(b) $2x^2 - 3x - 20 < 0$

Answers: See page 127.

Example 3 Solve the quadratic inequality $r^2 + 3r \geq 4$.

Solution First rewrite the inequality so that one side is 0:

$$r^2 + 3r \geq 4$$
$$r^2 + 3r - 4 \geq 0. \quad \text{Add } -4 \text{ to both sides.}$$

Now solve the corresponding equation (which amounts to finding the x-intercepts of $y = r^3 + 3r - 4$):

$$r^2 + 3r - 4 = 0$$
$$(r - 1)(r + 4) = 0$$
$$r = 1 \quad \text{or} \quad r = -4.$$

These numbers separate the number line into three regions, as shown in Figure 2.52. Test a number from each region:

Let $x = -5$ from region $\boldsymbol{A}$: $(-5)^2 + 3(-5) - 4 = 6 > 0$.
Let $x = 0$ from region $\boldsymbol{B}$: $(0)^2 + 3(0) - 4 = -4 < 0$.
Let $x = 2$ from region $\boldsymbol{C}$: $(2)^2 + 3(2) - 4 = 6 > 0$.

We want the inequality to be positive or 0. The solution includes numbers in region A and in region C, as well as -4 and 1, the endpoints. The solution, which includes all numbers in the interval $(-\infty, -4]$ or the interval $[1, \infty)$, is graphed in Figure 2.52. 2✓

Solve each inequality. Graph each solution.

(a) $k^2 + 2k - 15 \geq 0$

(b) $3m^2 + 7m \geq 6$

Answers: See page 127.

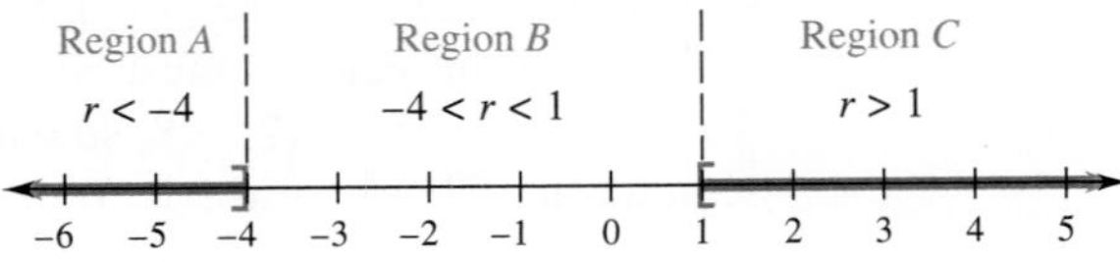

FIGURE 2.52

Example 4 Solve $q^3 - 4q > 0$.

Solution Solve the corresponding equation by factoring:

$$q^3 - 4q = 0$$
$$q(q^2 - 4) = 0$$
$$q(q + 2)(q - 2) = 0$$
$$q = 0 \quad \text{or} \quad q + 2 = 0 \quad \text{or} \quad q - 2 = 0$$
$$q = 0 \qquad q = -2 \qquad q = 2.$$

These three numbers separate the number line into the four regions shown in Figure 2.53. Test a number from each region:

$$\begin{aligned} A&: \text{ If } q = -3, (-3)^3 - 4(-3) = -15 < 0. \\ B&: \text{ If } q = -1, (-1)^3 - 4(-1) = 3 > 0. \\ C&: \text{ If } q = 1, (1)^3 - 4(1) = -3 < 0. \\ D&: \text{ If } q = 3, (3)^3 - 4(3) = 15 > 0. \end{aligned}$$

The numbers that make the polynomial positive are in the interval $(-2, 0)$ or the interval $(2, \infty)$, as graphed in Figure 2.53. 3 ✓

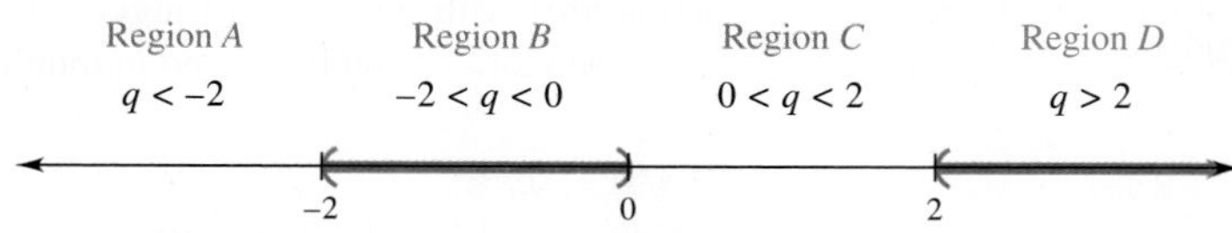

FIGURE 2.53

✓Checkpoint 3

Solve each inequality. Graph each solution.

(a) $m^3 - 9m > 0$

(b) $2k^3 - 50k \leq 0$

Answers: See page 127.

A graphing calculator can be used to solve inequalities without the need to evaluate at a test number in each interval. It is also useful for finding approximate solutions when the x-intercepts of the graph cannot be found algebraically.

Example 5 Use a graphing calculator to solve $x^3 - 5x^2 + x + 6 > 0$.

Solution Begin by graphing $y = x^3 - 5x^2 + x + 6$ (Figure 2.54). Find the x-intercepts by solving $x^3 - 5x^2 + x + 6 = 0$. Since this cannot readily be done algebraically, use the graphical root finder to determine that the solutions (x-intercepts) are approximately $-.9254$, 1.4481, and 4.4774.

The graph is above the x-axis when $-.9254 < x < 1.4481$ and when $x > 4.4774$. Therefore, the approximate solutions of the inequality are all numbers in the interval $(-.9254, 1.4481)$ or the interval $(4.4774, \infty)$. 4 ✓

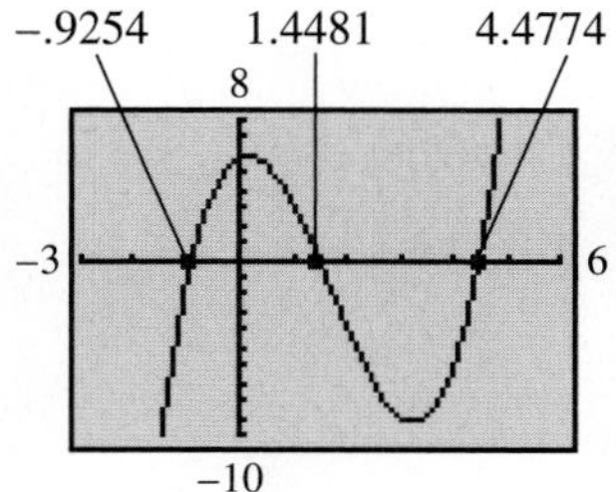

FIGURE 2.54

Use graphical methods to find approximate solutions of these inequalities.

(a) $x^2 - 6x + 2 > 0$

(b) $x^2 - 6x + 2 < 0$

Answers: See page 127.

Example 6 **Business** A wholesale DVD company sells DVDs for \$9 each. The variable cost of producing x thousand DVDs is $3x - 2x^2$ (in thousands of dollars), and the fixed cost is \$120 (in thousands). Find the values of x for which the company will break even or make a profit on the product.

Solution If x thousand DVDs are sold at \$9 each, then

$$\begin{aligned} \text{Revenue} &= (\text{price per item}) \times (\text{number of items sold}) \\ R &= 9 \times x = 9x. \end{aligned}$$

The cost function is (in thousands of dollars) is

$$\begin{aligned} \text{Cost} &= \text{Fixed Costs} + \text{Variable Costs} \\ C &= 120 + (3x - 2x^2). \end{aligned}$$

Therefore, to break even or earn a profit, we need

$$\begin{aligned} R &\geq C \\ 9x &\geq 120 + 3x - 2x^2 \\ 2x^2 + 6x - 120 &\geq 0 \\ x^2 + 3x - 60 &\geq 0. \end{aligned}$$

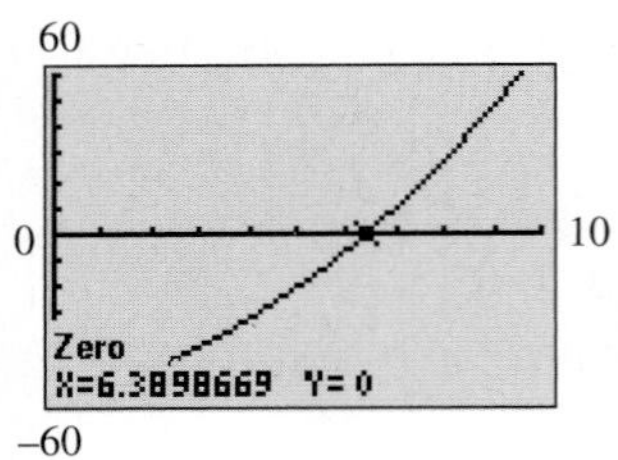

FIGURE 2.55

Now graph $y = x^2 + 3x - 60$. Since x has to be positive in this situation (why?), we need only look at the graph in the right half of the plane. Here, and in other cases, you may have to try several viewing windows before you find one that shows what you need. Once you find a suitable window, such as in Figure 2.55, use the graphical root finder to determine the relevant x-intercept. Figure 2.55 shows the intercept is approximately $x \approx 6.39$. Hence, the company must manufacture $6.39 \times 1000 = 6390$ DVDs to make a profit.

RATIONAL INEQUALITIES

Inequalities with quotients of algebraic expressions are called **rational inequalities**. These inequalities can be solved in much the same way as polynomial inequalities can.

Example 7 Solve the rational inequality $\dfrac{5}{x+4} \geq 1$.

Solution Write an equivalent inequality with one side equal to 0:

$$\frac{5}{x+4} \geq 1$$
$$\frac{5}{x+4} - 1 \geq 0.$$

Write the left side as a single fraction:

$$\frac{5}{x+4} - \frac{x+4}{x+4} \geq 0 \qquad \text{Get a common denominator.}$$
$$\frac{5-(x+4)}{x+4} \geq 0 \qquad \text{Subtract fractions.}$$
$$\frac{5-x-4}{x+4} \geq 0 \qquad \text{Distributive property}$$
$$\frac{1-x}{x+4} \geq 0.$$

The quotient can change sign only at places where the denominator is 0 or the numerator is 0. (In graphical terms, these are the only places where the graph of $y = \dfrac{1-x}{x+4}$ can change from above the x-axis to below.) This happens when

$$1 - x = 0 \quad \text{or} \quad x + 4 = 0$$
$$x = 1 \quad \text{or} \quad x = -4.$$

As in the earlier examples, the numbers -4 and 1 divide the x-axis into three regions. Test a number from each of these regions:

$$x < -4 \quad \text{Let } x = -5\text{:} \quad \frac{1-(-5)}{-5+4} = -6 < 0.$$
$$-4 < x < 1 \quad \text{Let } x = 0\text{:} \quad \frac{1-0}{0+4} = \frac{1}{4} > 0.$$
$$x > 1 \quad \text{Let } x = 2\text{:} \quad \frac{1-2}{2+4} = -\frac{1}{6} < 0.$$

The test shows that numbers in $(-4, 1)$ satisfy the inequality. With a quotient, the endpoints must be considered individually to make sure that no denominator is 0. In this inequality, -4 makes the denominator 0, while 1 satisfies the given inequality. Write the solution as $(-4, 1]$.

CAUTION As suggested by Example 6, be very careful with the endpoints of the intervals in the solution of rational inequalities. 5✓

Checkpoint 5

Solve each inequality.

(a) $\dfrac{3}{x-2} \geq 4$

(b) $\dfrac{p}{1-p} < 3$

(c) Why is 2 excluded from the solution in part (a)?

Answers: See page 127.

Example 8 Solve $\dfrac{2x-1}{3x+4} < 5$.

Solution Write an equivalent inequality with 0 on one side. Begin by subtracting 5 on both sides and combining the terms on the left into a single fraction:

$$\frac{2x-1}{3x+4} < 5$$

$$\frac{2x-1}{3x+4} - 5 < 0 \qquad \text{Get 0 on one side.}$$

$$\frac{2x-1-5(3x+4)}{3x+4} < 0 \qquad \text{Subtract.}$$

$$\frac{-13x-21}{3x+4} < 0. \qquad \text{Combine terms.}$$

Set the numerator and denominator each equal to 0 and solve the two equations:

$$-13x - 21 = 0 \quad \text{or} \quad 3x + 4 = 0$$

$$x = -\frac{21}{13} \quad \text{or} \quad x = -\frac{4}{3}.$$

Use the values $-21/13$ and $-4/3$ to divide the number line into three intervals. Test a number from each interval in the inequality. The quotient is negative for numbers in $(-\infty, -21/13)$ or $(-4/3, \infty)$. Neither endpoint satisfies the given inequality. 6✓

Checkpoint 6

Solve each rational inequality.

(a) $\dfrac{3y-2}{2y+5} < 1$

(b) $\dfrac{3c-4}{2-c} \geq -5$

Answers: See page 127.

CAUTION In problems like those in Examples 7 and 8, you should *not* begin by multiplying both sides by the denominator to simplify the inequality. Doing so will usually produce a wrong answer. For the reason, see Exercise 38. For an example, see Checkpoint 7. 7✓

Checkpoint 7

(a) Solve the inequality

$$\frac{10}{x+2} \geq 3$$

by first multiplying both sides by $x + 2$.

(b) Show that this method produces a wrong answer by testing $x = -3$.

Answer: See page 127.

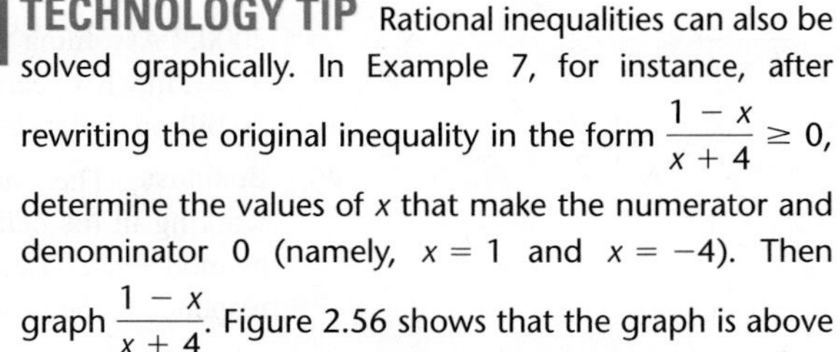

TECHNOLOGY TIP Rational inequalities can also be solved graphically. In Example 7, for instance, after rewriting the original inequality in the form $\dfrac{1-x}{x+4} \geq 0$, determine the values of x that make the numerator and denominator 0 (namely, $x = 1$ and $x = -4$). Then graph $\dfrac{1-x}{x+4}$. Figure 2.56 shows that the graph is above the x-axis when it is between the vertical asymptote at $x = -4$ and the x-intercept at $x = 1$. So the solution of the inequality is the interval $(-4, 1]$. (When the values that make the numerator and denominator 0 cannot be found algebraically, as they were here, you can use the root finder to approximate them.)

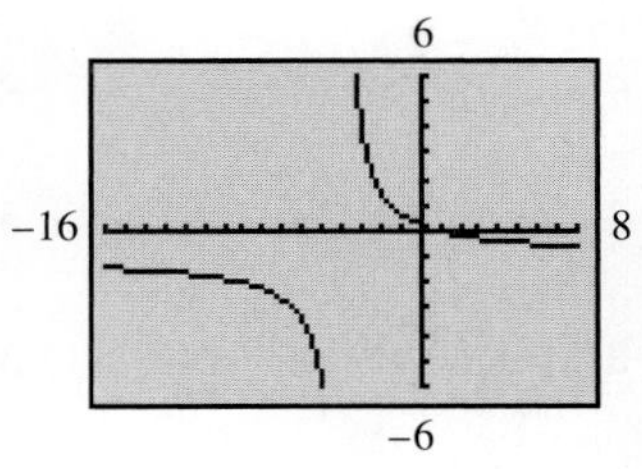

FIGURE 2.56

2.5 Exercises

Solve each of these quadratic inequalities. Graph the solutions on the number line. (See Examples 2 and 3.)

1. $(x + 4)(2x - 3) \le 0$
2. $(5y - 1)(y + 3) > 0$
3. $r^2 + 4r > -3$
4. $z^2 + 6z > -8$
5. $4m^2 + 7m - 2 \le 0$
6. $6p^2 - 11p + 3 \le 0$
7. $4x^2 + 3x - 1 > 0$
8. $3x^2 - 5x > 2$
9. $x^2 \le 36$
10. $y^2 \ge 9$
11. $p^2 - 16p > 0$
12. $r^2 - 9r < 0$

Solve these inequalities. (See Example 4.)

13. $x^3 - 9x \ge 0$
14. $p^3 - 25p \le 0$
15. $(x + 7)(x + 2)(x - 2) \ge 0$
16. $(2x + 4)(x^2 - 9) \le 0$
17. $(x + 5)(x^2 - 2x - 3) < 0$
18. $x^3 - 2x^2 - 3x \le 0$
19. $6k^3 - 5k^2 < 4k$
20. $2m^3 + 7m^2 > 4m$
21. A student solved the inequality $p^2 < 16$ by taking the square root of both sides to get $p < 4$. She wrote the solution as $(-\infty, 4)$. Is her solution correct?

Use a graphing calculator to solve these inequalities. (See Example 5.)

22. $6x + 7 < 2x^2$
23. $.5x^2 - 1.2x < .2$
24. $3.1x^2 - 7.4x + 3.2 > 0$
25. $x^3 - 2x^2 - 5x + 7 \ge 2x + 1$
26. $x^4 - 6x^3 + 2x^2 < 5x - 2$
27. $2x^4 + 3x^3 < 2x^2 + 4x - 2$
28. $x^5 + 5x^4 > 4x^3 - 3x^2 - 2$

Solve these rational inequalities. (See Examples 7 and 8.)

29. $\dfrac{r - 4}{r - 1} \ge 0$
30. $\dfrac{z + 6}{z + 4} > 1$
31. $\dfrac{a - 2}{a - 5} < -1$
32. $\dfrac{1}{3k - 5} < \dfrac{1}{3}$
33. $\dfrac{1}{p - 2} < \dfrac{1}{3}$
34. $\dfrac{7}{k + 2} \ge \dfrac{1}{k + 2}$
35. $\dfrac{5}{p + 1} > \dfrac{12}{p + 1}$
36. $\dfrac{x^2 - 4}{x} > 0$
37. $\dfrac{x^2 - x - 6}{x} < 0$
38. Determine whether $x + 4$ is positive or negative when
(a) $x > -4$; **(b)** $x < -4$.
(c) If you multiply both sides of the inequality $\dfrac{1 - x}{x + 4} \ge 0$ by $x + 4$, should you change the direction of the inequality sign? If so, when?
(d) Explain how you can use parts (a)–(c) to solve $\dfrac{1 - x}{x + 4} \ge 0$ correctly.

Use a graphing calculator to solve these inequalities. You may have to approximate the roots of the numerators or denominators.

39. $\dfrac{2x^2 + x - 1}{x^2 - 4x + 4} \le 0$
40. $\dfrac{x^3 - 3x^2 + 5x - 29}{x^2 - 7} > 3$
41. **Business** An analyst has found that her company's profits, in hundreds of thousands of dollars, are given by $P = 2x^2 - 12x - 32$, where x is the amount, in hundreds of dollars, spent on advertising. For what values of x does the company make a profit?
42. **Business** The commodities market is highly unstable; money can be made or lost quickly on investments in soy beans, wheat, and so on. Suppose that an investor kept track of his total profit P at time t, in months, after he began investing, and he found that $P = 4t^2 - 30t + 14$. Find the time intervals during which he has been ahead.
43. **Business** The manager of a large apartment complex has found that the profit is given by
$$P = -x^2 + 280x - 16{,}000,$$
where x is the number of apartments rented. For what values of x does the complex produce a profit?

Use a graphing calculator or other technology to complete Exercises 44–48. You may need a larger viewing window for some of these problems.

44. **Business** Annual revenue for Apple computer company (in billions of dollars) can be approximated by $.7x^2 - 2.4x + 7.6$, where $x = 0$ corresponds to the year 2000.* Assuming the model continues to be valid indefinitely, in what years since 2000 were revenues higher than 20 billion dollars?
45. **Business** The number of employees (in thousands) working in the cellular telecommunications industry can be modeled by $1.35x^2 + 3.7x + 183.2$, where $x = 0$ corresponds to the year 2000.† Assuming the model holds for

*Morningstar.com.

†CTIA—The Wireless Association, *Semi-annual Wireless Survey*.

the indefinite future, in what years does the model predict more than 250,000 employees?

46. **Social Science** According to data from the U.S. Department of Justice, the number of violent crimes (in thousands) can be approximated by $.8127x^3 - 17.62x^2 + 52.33x + 1863.8$, where $x = 0$ corresponds to the year 1990. During what years were there at least 1.7 million violent crimes? (*Hint*: 1.7 million is 1,700 thousand).

47. **Business** A manufacturing company can sell 100 cartons of ink pens for \$150. The costs associated with the manufacturing of the pens is a fixed cost (in thousands) of \$114 and a variable cost (in thousands) given by $8x - x^2$, where x represents the number of cartons (in hundreds). Find the values for x in which the manufacturer will break even or make a profit. (*Hint*: 150 = .15 thousand).

48. **Business** Use the information described in Exercise 47, but the fixed cost (in thousands) is now \$200. How many cartons of pens now need to be sold in order to make a profit?

✓ Checkpoint Answers

1. **(a)** $(-3, 5)$

(b) $(-5/2, 4)$

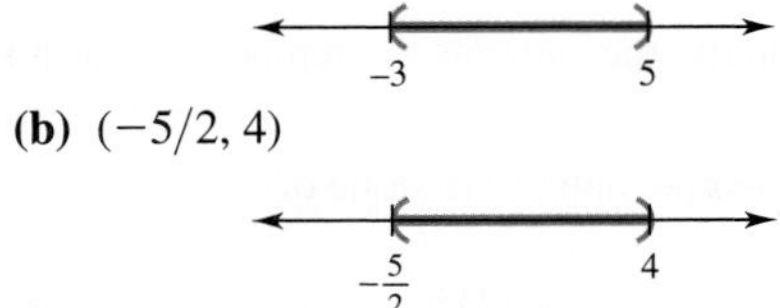

2. **(a)** All numbers in $(-\infty, -5]$ or $[3, \infty)$

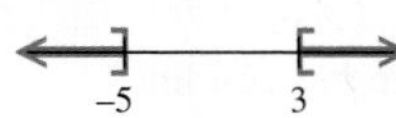

(b) All numbers in $(-\infty, -3]$ or $[2/3, \infty)$

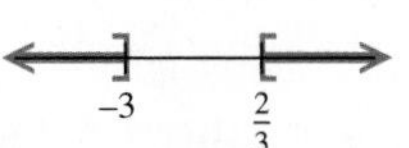

3. **(a)** All numbers in $(-3, 0)$ or $(3, \infty)$

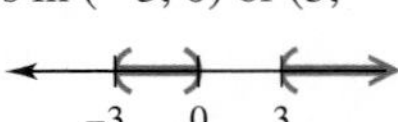

(b) All numbers in $(-\infty, -5]$ or $[0, 5]$

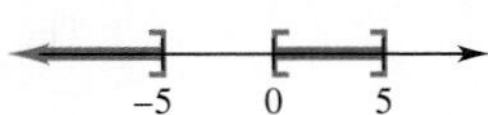

4. **(a)** All numbers in $(-\infty, .3542)$ or $(5.6458, \infty)$

(b) All numbers in $(.3542, 5.6458)$

5. **(a)** $(2, 11/4]$

(b) All numbers in $(-\infty, 3/4)$ or $(1, \infty)$

(c) When $x = 2$, the fraction is undefined.

6. **(a)** $(-5/2, 7)$

(b) All numbers in $(-\infty, 2)$ or $[3, \infty)$

7. **(a)** $x \le \frac{4}{3}$

(b) $x = -3$ is a solution of $x \le \frac{4}{3}$, but not of the original inequality $\frac{10}{x + 2} \ge 3$.

CHAPTER 2 ▸ Summary

KEY TERMS AND SYMBOLS*

2.1 ▸ Cartesian coordinate system
x-axis
y-axis
origin
ordered pair
x-coordinate
y-coordinate
quadrant
solution of an equation
graph
x-intercept
y-intercept
[viewing window]
[trace]
[graphical root finder]
[maximum and minimum finder]
graph reading

2.2 ▸ change in x
change in y
slope
slope–intercept form
parallel and perpendicular lines
point–slope form
linear equations
general form

*Terms in brackets deal with material in which a graphing calculator or other technology is used.

2.3 ▶ linear models
residual
[least-squares regression line]
[linear regression]
[correlation coefficient]

2.4 ▶ linear inequality
properties of inequality
absolute-value inequality

2.5 ▶ polynomial inequality
algebraic solution methods
[graphical solution methods]
rational inequality

CHAPTER 2 KEY CONCEPTS

The **slope** of the line through the points (x_1, y_1) and (x_2, y_2), where $x_1 \neq x_2$, is $m = \frac{y_2 - y_1}{x_2 - x_1}$.

The line with equation $\boldsymbol{y = mx + b}$ has slope m and y-intercept b.

The line with equation $\boldsymbol{y - y_1 = m(x - x_1)}$ has slope m and goes through (x_1, y_1).

The line with equation $\boldsymbol{ax + by = c}$ (with $a \neq 0, b \neq 0$) has x-intercept c/a and y-intercept c/b.

The line with equation $\boldsymbol{x = k}$ is vertical, with x-intercept k, no y-intercept, and undefined slope.

The line with equation $\boldsymbol{y = k}$ is horizontal, with y-intercept k, no x-intercept, and slope 0.

Nonvertical **parallel lines** have the same slope, and **perpendicular lines**, if neither is vertical, have slopes with a product of -1.

CHAPTER 2 REVIEW EXERCISES

Which of the ordered pairs $(-2, 3)$, $(0, -5)$, $(2, -3)$, $(3, -2)$, $(4, 3)$, and $(7, 2)$ are solutions of the given equation?

1. $y = x^2 - 2x - 5$ **2.** $x - y = 5$

Sketch the graph of each equation.

3. $5x - 3y = 15$ **4.** $2x + 7y - 21 = 0$

5. $y + 3 = 0$ **6.** $y - 2x = 0$

7. $y = .25x^2 + 1$ **8.** $y = \sqrt{x + 4}$

9. The following temperature graph was recorded in Bratenahl, Ohio:

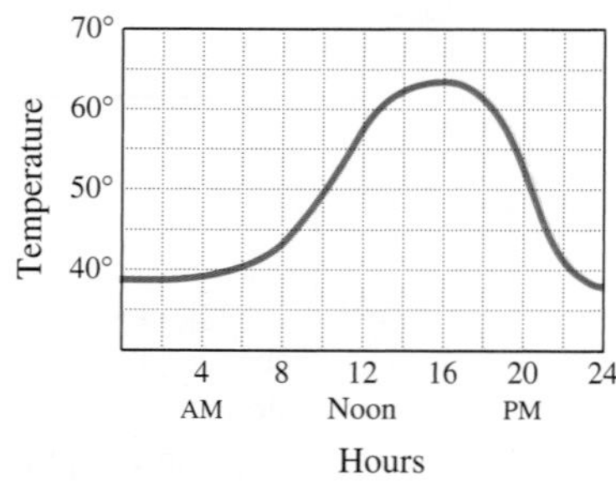

(a) At what times during the day was the temperature over 55°?

(b) When was the temperature below 40°?

10. Greenville, South Carolina, is 500 miles south of Bratenahl, Ohio, and its temperature is 7° higher all day long. (See the graph in Exercise 9). At what time was the temperature in Greenville the same as the temperature at noon in Bratenahl?

11. In your own words, define the slope of a line.

In Exercises 12–21, find the slope of the line defined by the given conditions.

12. Through $(-1, 3)$ and $(2, 6)$

13. Through $(4, -5)$ and $(1, 4)$

14. Through $(8, -3)$ and the origin

15. Through $(8, 2)$ and $(0, 4)$

16. $3x + 5y = 25$ **17.** $6x - 2y = 7$

18. $x - 2 = 0$ **19.** $y = -4$

20. Parallel to $3x + 8y = 0$

21. Perpendicular to $x = 3y$

22. Graph the line through $(0, 5)$ with slope $m = -2/3$.

23. Graph the line through $(-4, 1)$ with $m = 3$.

24. What information is needed to determine the equation of a line?

Find an equation for each of the following lines.

25. Through $(5, -1)$, slope $2/3$

26. Through $(8, 0)$, slope $-1/4$

27. Through $(5, -2)$ and $(1, 3)$

28. Through $(2, -3)$ and $(-3, 4)$

29. Undefined slope, through $(-1, 4)$

30. Slope 0, through $(-2, 5)$

31. x-intercept -3, y-intercept 5

32. Here is a sample SAT question: Which of the following is an equation of the line that has a y-intercept of 2 and an x-intercept of 3?
(a) $-2x + 3y = 4$ **(b)** $-2x + 3y = 6$
(c) $2x + 3y = 4$ **(d)** $2x + 3y = 6$
(e) $3x + 2y = 6$

33. Business According to the U.S. Department of Agriculture, in 1995, the United States exported 30.2 million metric tons of wheat. In 2005, that number was 27.2 million metric tons.
(a) Assuming the decline in wheat exports is linear, write an equation that gives the amount of wheat exported in year x, with $x = 0$ corresponding to 1995.
(b) Is the slope of the line positive or negative? Why?
(c) Assuming the linear trend continues, estimate the amount of wheat exports in 2011.

34. Business In 2000, Canada produced 3.0 million new motor vehicles. In 2006, it produced 2.5 million.*
(a) Assuming the decline in automobiles is linear, write an equation that gives the number of motor vehicles produced in year x, where $x = 0$ corresponds to 2000.
(b) Graph the equation for years 2000 through 2006.
(c) Assuming the linear trend continues, estimate the number of vehicles produced in 2011.

35. Business The following table gives the average hourly earnings of U.S. production workers†:

Year	2001	2002	2004	2005	2006
Earnings	14.54	14.97	15.69	16.13	16.76

(a) Use the first and last data points to find a linear model for the data, with $x = 1$ corresponding to 2001.
(b) If you have access to the necessary technology, find the least-squares regression line for the data.
(c) Use the models from part (a) and/or (b) to estimate the hourly earnings in 2003. The actual average in 2003 was \$15.37. How far off is the model?
(d) Use the model from part (a) and/or (b), and assume the trend continues, to estimate the hourly earnings in 2010.

36. Business If current trends continue, the weekly median wages that a nonunion male who is 16–24 years old can expect to earn can be given by the equation $y = 11x + 303$, where $x = 0$ corresponds to the year 1996. The equation for nonunion females in the same age group is $y = 11.1x + 280$.‡
(a) In what year will nonunion men in that age group earn \$450 a week?
(b) In what year will nonunion women in that age group earn \$450 a week?

37. Business The following table shows data provided by the National Center for Health Statistics on total national health expenditures (in billions of dollars) for the United States in various years:

Year	1980	1990	1995	2000	2002	2003	2004
Amount	254.9	717.3	1024.4	1358.5	1607.9	1740.6	1877.6

(a) Let $x = 0$ correspond to 1980, and find the least-squares regression line for the data.
(b) What is the correlation coefficient? Does it indicate that the line is a good fit?
(c) Assuming the trend continues, estimate the total health care expenditures for 2012.

38. Finance According to the U.S. Department of Education, the average financial aid award (in dollars) for a full-time student is given in the following table for various years:

Year	1995	2000	2003	2004	2005	2006	2007
Amount	2,596	2,925	3,208	3,324	3,407	3,458	3,629

(a) Let $x = 5$ correspond to 1995, and find the least-squares regression line for the data.
(b) What is the correlation coefficient? Does it indicate that the line is a good fit?
(c) Assuming the trend continues, estimate the average financial aid award for a student in 2011.

Solve each inequality.

39. $-6x + 3 < 2x$ **40.** $12z \geq 5z - 7$

41. $2(3 - 2m) \geq 8m + 3$ **42.** $6p - 5 > -(2p + 3)$

43. $-3 \leq 4x - 1 \leq 7$ **44.** $0 \leq 3 - 2a \leq 15$

45. $|b| \leq 8$ **46.** $|a| > 7$

47. $|2x - 7| \geq 3$ **48.** $|4m + 9| \leq 16$

49. $|5k + 2| - 3 \leq 4$ **50.** $|3z - 5| + 2 \geq 10$

51. Health Here is a sample SAT question: For pumpkin carving, Mr. Sephera will not use pumpkins that weigh less than 2 pounds or more than 10 pounds. If x represents the weight of a pumpkin (in pounds) that he will *not* use, which of the following inequalities represents all possible values of x?
(a) $|x - 2| > 10$ **(b)** $|x - 4| > 6$
(c) $|x - 5| > 5$ **(d)** $|x - 6| > 4$
(e) $|x - 10| > 4$

52. Business Prices at several retail outlets for a new 24-inch, 2-cycle snow thrower were all within \$55 of \$600. Write this information as an inequality, using absolute-value notation.

*Automotive News Data Center and R.L. Polk Marketing Systems GmbH.

†U.S. Bureau of Labor Statistics.

‡U.S. Bureau of Labor Statistics.

53. **Business** The number of barrels of crude oil (in millions) imported to the United States from Canada was 492 in 2000 and 651 in 2006.* Assume the amount imported is increasing linearly.
   (a) Find the linear equation that gives the number of barrels imported from Canada, with $x = 0$ corresponding to 2000.
   (b) Assume the linear trend continues indefinitely in either direction. Determine when the number of barrels exceeds 550 million.
   (c) When is the number of barrels higher than 700 million?

54. **Business** One car rental firm charges $125 for a weekend rental (Friday afternoon through Monday morning) and gives unlimited mileage. A second firm charges $95 plus $.20 a mile. For what range of miles driven is the second firm cheaper?

Solve each inequality.

55. $r^2 + r - 6 < 0$

56. $y^2 + 4y - 5 \geq 0$

57. $2z^2 + 7z \geq 15$

58. $3k^2 \leq k + 14$

59. $(x - 3)(x^2 + 7x + 10) \leq 0$

60. $(x + 4)(x^2 - 1) \geq 0$

61. $\frac{m + 2}{m} \leq 0$

62. $\frac{q - 4}{q + 3} > 0$

63. $\frac{5}{p + 1} > 2$

64. $\frac{6}{a - 2} \leq -3$

65. $\frac{2}{r + 5} \leq \frac{3}{r - 2}$

66. $\frac{1}{z - 1} > \frac{2}{z + 1}$

67. **Business** The revenue earned by Texas Instruments Corporation (in billions) can be modeled by the equation $y = 0.145x^2 - .27x + 10.0$, where $x = 0$ corresponds to the year 2000.* In what years between 2000 and 2006 did revenue exceed $11 billion?

Statistical Abstract of the United States: 2008.

*Morningstar.com.

CASE 2

Using Extrapolation for Prediction

One reason for developing a mathematical model is to make predictions. If your model is a least-squares line, you can predict the y-value corresponding to some new x-value by substituting that x-value into an equation of the form $\hat{y} = mx + b$. (We use $\hat{y}$ to remind us that we're getting a predicted value rather than an actual data value.) Data analysts distinguish two very different kinds of prediction: *interpolation* and *extrapolation*. An interpolation uses a new x inside the x-range of your original data. For example, if you have inflation data at five-year intervals from 1950 to 2010, estimating the rate of inflation in 1957 is an interpolation problem. But if you use the same data to estimate what the inflation rate was in 1920, or what it will be in 2020, you are extrapolating.

In general, interpolation is much safer than extrapolation, because data that are approximately linear over a short interval may be nonlinear over a larger interval. One way to detect nonlinearity is to look at *residuals*, which are the differences between the actual data values and the values predicted by the line of best fit. A simple example is shown in Figure 1.

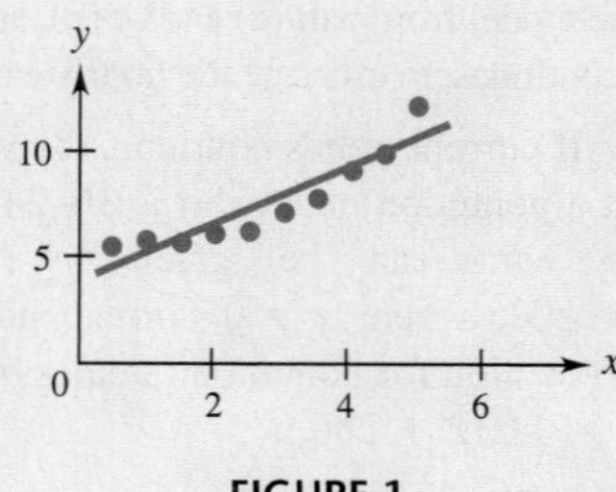

FIGURE 1

The regression equation for the best-fit line in Figure 1 is $\hat{y} = 3.431 + 1.334x$. Since the r value for this regression line is

.93, our linear model fits the data very well. But we might notice that the predictions are a bit low at the ends and high in the middle. We can get a better look at this pattern by plotting the residuals. To find them, we put each value of the independent variable into the regression equation, calculate the predicted value $\hat{y}$, and subtract it from the actual value of y. The residual plot in Figure 2

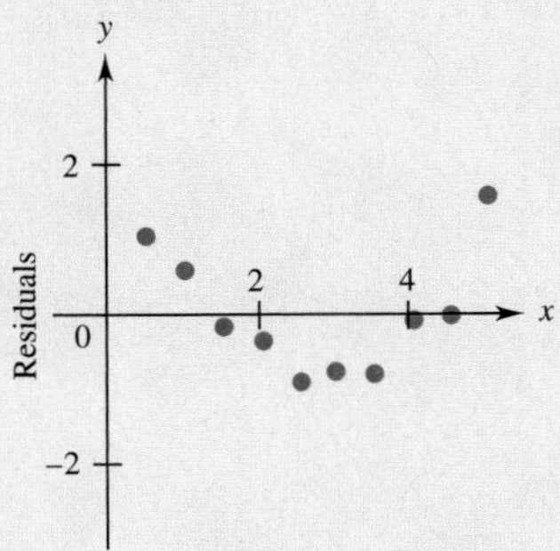

FIGURE 2

has the vertical axis rescaled to exaggerate the pattern. The residuals indicate that our data have a nonlinear U-shaped component that isn't captured by the linear fit. Extrapolating from this data set is probably not a good idea; our linear prediction for the value of y when x is 10 may be much too low.

EXERCISES

Business *The following table gives the average hourly earnings (in dollars) of U.S. production workers from 1970 to 2005*:*

Year	Average Hourly Wage (Dollars)
1970	3.40
1975	4.73
1980	6.85
1985	8.74
1990	10.20
1995	11.65
2000	14.02
2005	16.13

1. If you have appropriate technology, verify that the least-squares regression line that models these data (with coefficients rounded) is $y = .361x - 22.08$, where y is the average hourly wage in year x and $x = 0$ corresponds to 1900.
2. Use the model from Exercise 1 to interpolate a prediction of the hourly wage in 2002. The actual value was \$14.97. How close was your prediction?
3. Use the model from Exercise 1 to extrapolate to 1960 and predict the hourly average wage. Why is this prediction nonsensical?
4. Using the model from Exercise 1, find the average hourly wage for each year in the table, and subtract if from the actual value in the second column. This gives you a table of the residuals. Plot your residuals as points on a graph.
5. What will happen if you try linear regression on the *residuals*? If you're not sure, use technology such as a graphing calculator to find the regression equation for the residuals. Why does this result make sense?

*U.S. Bureau of Labor Statistics.

CHAPTER

3

Functions and Graphs

The modern world is overwhelmed with data—from the cost of college to mortgage rates, health care expenditures, and hundreds of other pieces of information. Functions enable us to construct mathematical models that can sometimes be used to estimate outcomes. Graphs of functions allow us to visualize a situation and to detect trends more easily. See Example 11 on page 149 and Exercises 57 and 58 on page 208.

CASE 3: Architectural Arches

Functions are an extremely useful way of describing many real-world situations in which the value of one quantity varies with, depends on, or determines the value of another. In this chapter, you will be introduced to functions, learn how to use functional notation, develop skills in constructing and interpreting the graphs of functions, and, finally, learn to apply this knowledge in a variety of situations.

3.1 Functions

To understand the origin of the concept of a function, we consider some "real-life" situations in which one numerical quantity depends on, corresponds to, or determines another.

Example 1 The amount of income tax you pay depends on the amount of your income. The way in which the income determines the tax is given by the tax law.

Example 2 The graph in Figure 3.1 shows the temperatures in Palm Springs, CA, on January 25, 2009, as recorded by the U.S. Weather Bureau. It displays the temperature that corresponds to each time during the day.

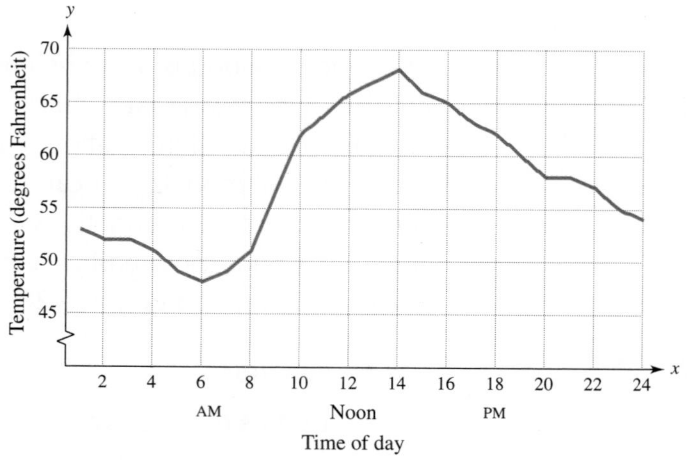

FIGURE 3.1

Example 3 Suppose a rock is dropped straight down from a high point. From physics, we know that the distance traveled by the rock in t seconds is $16t^2$ feet. So the distance depends on the time.

These examples share a couple of features. First, each involves two sets of numbers, which we can think of as inputs and outputs. Second, in each case, there is a rule by which each input determines an output, as summarized here:

	Set of Inputs	Set of Outputs	Rule
Example 1	All incomes	All tax amounts	The tax law
Example 2	Hours since midnight	Temperatures during the day	Time–temperature graph
Example 3	Seconds elapsed after dropping the rock	Distances the rock travels	Distance $= 16t^2$

Each of these examples may be mentally represented by an idealized calculator that has a single operation key: A number is entered [*input*], the rule key is pushed [*rule*], and an answer is displayed [*output*]. The formal definition of a function incorporates these same common features (input–rule–output), with a slight change in terminology.

A **function** consists of a set of inputs called the **domain**, a set of outputs called the **range**, and a rule by which each input determines *exactly one* output.

Example 4 Find the domains and ranges for the functions in Examples 1 and 2.

Solution In Example 1, the domain consists of all possible income amounts and the range consists of all possible tax amounts.

In Example 2, the domain is the set of hours in the day—that is, all real numbers in the interval [0, 24]. The range consists of the temperatures that actually occur during that day. Figure 3.1 suggests that these are all numbers in the interval [48, 68]. ✓1

✓Checkpoint 1

Find the domain and range of the function in Example 3, assuming (unrealistically) that the rock can fall forever.

Answer: See page 142.

Be sure that you understand the phrase "exactly one output" in the definition of the rule of a function. In Example 2, for instance, each time of day (input) determines exactly one temperature (output)—you can't have two different temperatures at the same time. However, it is quite possible to have the same temperature (output) at two different times (inputs). In other words, we can say the following:

In a function, each input produces a single output, but different inputs may produce the same output.

Example 5 Which of the following rules describe functions?

(a) Use the optical reader at the checkout counter of the supermarket to convert codes to prices.

Solution For each code, the reader produces exactly one price, so this is a function.

(b) Enter a number in a calculator and press the x^2 key.

Solution This is a function, because the calculator produces just one number x^2 for each number x that is entered.

(c) Assign to each number x the number y given by this table:

x	1	1	2	2	3	3
y	3	−3	−5	−5	8	−8

Solution Since $x = 1$ corresponds to more than one y-value (as does $x = 3$), this table does not define a function.

(d) Assign to each number x the number y given by the equation $y = 3x - 5$.

Solution Because the equation determines a unique value of y for each value of x, it defines a function. ✓2

✓Checkpoint 2

Do the following define functions?

(a) the correspondence defined by the rule $y = x^2 + 5$, where x is the input and y is the output

(b) the correspondence defined by entering a nonzero number in a calculator and pressing the $1/x$ key.

(c) the correspondence between a computer, x, and several users of the computer, y

Answers: See page 142.

The equation $y = 3x - 5$ in part (d) of Example 5 defines a function, with x as input and y as output, because each value of x determines a *unique* value of y. In such a case, the equation is said to **define y as a function of x**.

Example 6 Decide whether each of the following equations defines y as a function of x.

(a) $y = -4x + 11$

Solution For a given value of x, calculating $-4x + 11$ produces exactly one value of y. (For example, if $x = -7$, then $y = -4(-7) + 11 = 39$.) Because one value of the input variable leads to exactly one value of the output variable, $y = -4x + 11$ defines y as a function of x.

(b) $y^2 = x$

Solution Suppose $x = 36$. Then $y^2 = x$ becomes $y^2 = 36$, from which $y = 6$ or $y = -6$. Since one value of x can lead to two values of y, $y^2 = x$ does *not* define y as a function of x. ✓3

✓Checkpoint 3

Do the following define y as a function of x?

(a) $y = -6x + 1$

(b) $y = x^2$

(c) $x = y^2 - 1$

(d) $y < x + 2$

Answers: See page 142.

Almost all the functions in this book are defined by formulas or equations, as in part (a) of Example 6. The domain of such a function is determined by the following **agreement on domains**.

> Unless otherwise stated, assume that the domain of any function defined by a formula or an equation is the largest set of real numbers (inputs) that each produce a real number as output.

Example 7 Each of the given equations defines y as a function of x. Find the domain of each function.

(a) $y = x^4$

Solution Any number can be raised to the fourth power, so the domain is the set of all real numbers, which is sometimes written as $(-\infty, \infty)$.

(b) $y = \sqrt{x - 6}$

Solution For y to be a real number, $x - 6$ must be nonnegative. This happens only when $x - 6 \geq 0$—that is, when $x \geq 6$. So the domain is the interval $[6, \infty)$.

(c) $y = \sqrt{4 - x}$

Solution For y to be a real number here, we must have $4 - x \geq 0$, which is equivalent to $x \leq 4$. So the domain is the interval $(-\infty, 4]$.

(d) $y = \dfrac{1}{x + 3}$

Solution Because the denominator cannot be 0, $x \neq -3$ and the domain consists of all numbers in the intervals,

$$(-\infty, -3) \quad \text{or} \quad (-3, \infty).$$

(e) $y = \dfrac{\sqrt{x}}{x^2 - 3x + 2}$

Solution The numerator is defined only when $x \geq 0$. The domain cannot contain any numbers that make the denominator 0—that is, the numbers that are solutions of

$$x^2 - 3x + 2 = 0.$$

$$(x - 1)(x - 2) = 0 \qquad \text{Factor.}$$

$$x - 1 = 0 \quad \text{or} \quad x - 2 = 0 \qquad \text{Zero-factor property}$$

$$x = 1 \quad \text{or} \quad x = 2.$$

Therefore, the domain consists of all nonnegative real numbers except 1 and 2. ✓4

✓Checkpoint 4

Give the domain of each function.

(a) $y = 3x + 1$

(b) $y = x^2$

(c) $y = \sqrt{-x}$

(d) $y = \dfrac{3}{x^2 - 1}$

Answers: See page 142.

FUNCTIONAL NOTATION

In actual practice, functions are seldom presented in the style of domain–rule–range, as they have been here. Functions are usually denoted by a letter such as f. If x is an input, then $f(x)$ denotes the output number that the function f produces from the input x. The symbol $f(x)$ is read "f of x." The rule is usually given by a formula, such as $f(x) = \sqrt{x^2 + 1}$. This formula can be thought of as a set of directions:

Name of function — Input number

$$f(x) = \sqrt{x^2 + 1}$$

Output number — Directions that tell you what to do with input x in order to produce the corresponding output $f(x)$—namely, "square it, add 1, and take the square root of the result."

For example, to find $f(3)$ (the output number produced by the input 3), simply replace x by 3 in the formula:

$$f(3) = \sqrt{3^2 + 1} = \sqrt{10}.$$

Similarly, replacing x respectively by -5 and 0 shows that

$$f(-5) = \sqrt{(-5)^2 + 1} = \sqrt{26} \quad \text{and} \quad f(0) = \sqrt{0^2 + 1} = 1.$$

These directions can be applied to any quantities, such as $a + b$ or c^4 (where a, b, and c are real numbers). Thus, to compute $f(a + b)$, the output corresponding to input $a + b$, we square the input [obtaining $(a + b)^2$], add 1 [obtaining $(a + b)^2 + 1$], and take the square root of the result:

$$\begin{aligned} f(a + b) &= \sqrt{(a + b)^2 + 1} \\ &= \sqrt{a^2 + 2ab + b^2 + 1}. \end{aligned}$$

Similarly, the output $f(c^4)$, corresponding to the input c^4, is computed by squaring the input $[(c^4)^2]$, adding 1 $[(c^4)^2 + 1]$, and taking the square root of the result:

$$\begin{aligned} f(c^4) &= \sqrt{(c^4)^2 + 1} \\ &= \sqrt{c^8 + 1}. \end{aligned}$$

Example 8 Let $g(x) = -x^2 + 4x - 5$. Find each of the given outputs.

(a) $g(-2)$

Solution Replace x with -2:

$$\begin{aligned} g(-2) &= -(-2)^2 + 4(-2) - 5 \\ &= -4 - 8 - 5 \\ &= -17. \end{aligned}$$

(b) $g(x + h)$

Solution Replace x by the quantity $x + h$ in the rule of g:

$$\begin{aligned} g(x + h) &= -(x + h)^2 + 4(x + h) - 5 \\ &= -(x^2 + 2xh + h^2) + (4x + 4h) - 5 \\ &= -x^2 - 2xh - h^2 + 4x + 4h - 5. \end{aligned}$$

(c) $g(x + h) - g(x)$

Solution Use the result from part (b) and the rule for $g(x)$:

$$\begin{aligned} g(x + h) - g(x) &= \overbrace{(-x^2 - 2xh - h^2 + 4x + 4h - 5)}^{g(x+h)} - \overbrace{(-x^2 + 4x - 5)}^{g(x)} \\ &= -2xh - h^2 + 4h. \end{aligned}$$

(d) $\dfrac{g(x + h) - g(x)}{h}$ (assuming that $h \neq 0$)

Solution The numerator was found in part (c). Divide it by h as follows:

$$\begin{aligned} \frac{g(x + h) - g(x)}{h} &= \frac{-2xh - h^2 + 4h}{h} \\ &= \frac{h(-2x - h + 4)}{h} \\ &= -2x - h + 4. \end{aligned}$$

The quotient found in Example 8(d),

$$\frac{g(x + h) - g(x)}{h},$$

is called the **difference quotient** of the function g. Difference quotients are important in calculus. 5

✓Checkpoint 5

Let $f(x) = 5x^2 - 2x + 1$. Find the following.

(a) $f(1)$

(b) $f(3)$

(c) $f(1 + 3)$

(d) $f(1) + f(3)$

(e) $f(m)$

(f) $f(x + h) - f(x)$

(g) $\frac{f(x + h) - f(x)}{h}$ $(h \neq 0)$

Answers: See page 142.

CAUTION Functional notation is *not* the same as ordinary algebraic notation. You cannot simplify an expression such as $f(x + h)$ by writing $f(x) + f(h)$. To see why, consider the answers to Checkpoints 5(c) and (d), which show that

$$f(1 + 3) \neq f(1) + f(3).$$

APPLICATIONS

Example 9 **Finance** If you were a single person in Connecticut in 2008 with a taxable income of x dollars, then your state income tax T was determined by the rule

$$T(x) = \begin{cases} .03x & \text{if } 0 \leq x \leq 10{,}000 \\ 300 + .05(x - 10{,}000) & \text{if } x > 10{,}000. \end{cases}$$

Find the income tax paid by a single person with the given taxable income.

(a) \$9200

Solution We must find $T(9200)$. Since 9200 is less than 10,000, the first part of the rule applies:

$$T(x) = .03x$$
$$T(9200) = .03(9200) = \$276. \quad \text{Let } x = 9200.$$

(b) \$30,000

Solution Now we must find $T(30{,}000)$. Since 30,000 is greater than \$10,000, the second part of the rule applies:

$$\begin{aligned} T(x) &= 300 + .05(x - 10{,}000) \\ T(30{,}000) &= 300 + .05(30{,}000 - 10{,}000) && \text{Let } x = 30{,}000. \\ &= 300 + .05(20{,}000) && \text{Simplify.} \\ &= 300 + 1000 = \$1300. \end{aligned}$$ 6

✓Checkpoint 6

Use Example 9 to find the tax on each of these incomes.

(a) \$48,750

(b) \$7345

Answers: See page 142.

A function with a multipart rule, as in Example 9, is called a **piecewise-defined function.**

Example 10 **Business** Suppose the projected sales (in thousands of dollars) of a small company over the next 10 years are approximated by the function

$$S(x) = .07x^4 - .05x^3 + 2x^2 + 7x + 62.$$

✓Checkpoint 7

A developer estimates that the total cost of building x large apartment complexes in a year is approximated by

$$A(x) = x^2 + 80x + 60,$$

where $A(x)$ represents the cost in hundred thousands of dollars. Find the cost of building

(a) 4 complexes;

(b) 10 complexes.

Answers: See page 142.

(a) What are the projected sales for the current year?

Solution The current year corresponds to $x = 0$, and the sales for this year are given by $S(0)$. Substituting 0 for x in the rule for S, we see that $S(0) = 62$. So the current projected sales are \$62,000.

(b) What will sales be in four years?

Solution The sales in four years from now are given by $S(4)$, which can be computed by hand or with a calculator:

$$S(x) = .07x^4 - .05x^3 + 2x^2 + 7x + 62$$
$$S(4) = .07(4)^4 - .05(4)^3 + 2(4)^2 + 7(4) + 62 \quad \text{Let } x = 4.$$
$$= 136.72.$$

Thus, sales are projected to be \$136,720. 7✓

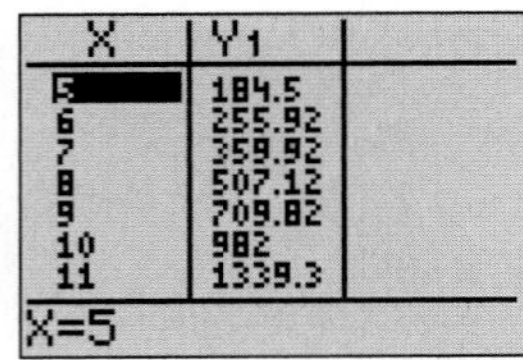

X	Y1
5	184.5
6	255.92
7	359.92
8	507.12
9	709.82
10	982
11	1339.3

X=5

FIGURE 3.2

Example 11 **Business** Use the table feature of the graphing calculator to find the projected sales of the company in Example 10 for years 5 through 10.

Enter the sales equation $y = .07x^4 - .05x^3 + 2x^2 + 7x + 62$ into the equation memory of the calculator (often called the Y = list.) Check your instruction manual for how to set the table to start at $x = 5$ and go at least through $x = 10$. Then display the table, as in Figure 3.2. The figure shows that sales are projected to rise from \$184,500 in year 5 to \$982,000 in year 10.

3.1 ▶ Exercises

For each of the following rules, state whether it defines y as a function of x or not. (See Examples 5 and 6.)

1.

x	3	2	1	0	−1	−2	−3
y	9	4	1	0	1	4	9

2.

x	9	4	1	0	1	4	9
y	3	2	1	0	−1	−2	−3

3. $y = x^3$

4. $y = \sqrt{x - 1}$

5. $x = |y + 2|$

6. $x = y^2 + 3$

7. $y = \dfrac{-1}{x - 1}$

8. $y = \dfrac{4}{2x + 3}$

State the domain of each function. (See Example 7.)

9. $f(x) = 4x - 1$

10. $f(x) = 2x + 7$

11. $f(x) = x^4 - 1$

12. $f(x) = (2x + 5)^2$

13. $f(x) = \sqrt{-x + 3}$

14. $f(x) = \sqrt{5 - x}$

15. $g(x) = \dfrac{1}{x - 2}$

16. $g(x) = \dfrac{x}{x^2 + x - 2}$

17. $g(x) = \dfrac{x^2 + 4}{x^2 - 4}$

18. $g(x) = \dfrac{x^2 - 1}{x^2 + 1}$

19. $h(x) = \dfrac{\sqrt{x + 4}}{x^2 + x - 12}$

20. $h(x) = |5 - 4x|$

21. $g(x) = \begin{cases} 1/x & \text{if } x < 0 \\ \sqrt{x^2 + 1} & \text{if } x \geq 0 \end{cases}$

22. $f(x) = \begin{cases} 2x + 3 & \text{if } x < 4 \\ x^2 - 1 & \text{if } 4 \leq x \leq 10 \end{cases}$

For each of the following functions, find

(a) $f(4)$; **(b)** $f(-3)$; **(c)** $f(2.7)$; **(d)** $f(-4.9)$.

(See Examples 8 and 9.)

23. $f(x) = 8$

24. $f(x) = 0$

25. $f(x) = 2x^2 + 4x$

26. $f(x) = x^2 - 2x$

27. $f(x) = \sqrt{x + 3}$

28. $f(x) = \sqrt{5 - x}$

29. $f(x) = |x^2 - 6x - 4|$

30. $f(x) = |x^3 - x^2 + x - 1|$

31. $f(x) = \frac{\sqrt{x - 1}}{x^2 - 1}$

32. $f(x) = \sqrt{-x} + \frac{2}{x + 1}$

33. $f(x) = \begin{cases} x^2 & \text{if } x < 2 \\ 5x - 7 & \text{if } x \geq 2 \end{cases}$

34. $f(x) = \begin{cases} -2x + 4 & \text{if } x \leq 1 \\ 3 & \text{if } 1 < x < 4 \\ x + 1 & \text{if } x \geq 4 \end{cases}$

For each of the following functions, find

(a) $f(p)$; **(b)** $f(-r)$; **(c)** $f(m + 3)$.

(See Example 8.)

35. $f(x) = 6 - x$

36. $f(x) = 3x + 5$

37. $f(x) = \sqrt{4 - x}$

38. $f(x) = \sqrt{-2x}$

39. $f(x) = x^3 + 1$

40. $f(x) = 3 - x^3$

41. $f(x) = \frac{3}{x - 1}$

42. $f(x) = \frac{-1}{5 + x}$

For each of the following functions, find the difference quotient

$$\frac{f(x + h) - f(x)}{h} \quad (h \neq 0).$$

(See Example 8.)

43. $f(x) = 2x - 4$

44. $f(x) = 2 + 4x$

45. $f(x) = x^2 + 1$

46. $f(x) = x^2 - x$

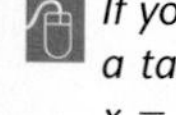

If you have a graphing calculator with table-making ability, display a table showing the (approximate) values of the given function at $x = 3.5, 3.9, 4.3, 4.7, 5.1$, and 5.5. (See Example 11.)

47. $g(x) = 3x^4 - x^3 + 2x$

48. $f(x) = \sqrt{x^2 - 2.4x + 8}$

Use a calculator to work these exercises. (See Examples 9 and 10.)

49. **Finance** The Minnesota state income tax for a single person in 2008 was determined by the rule

$$T(x) = \begin{cases} .0535x & \text{if } 0 \leq x \leq 21{,}310 \\ 1140 + .0705(x - 21{,}310) & \text{if } 21{,}310 < x \leq 69{,}990, \\ 4572 + .0785(x - 69{,}990) & \text{if } x > 69{,}990, \end{cases}$$

where x is the person's taxable income. Find the tax on each of these incomes.
(a) \$17,800 **(b)** \$58,872 **(c)** \$115,412

50. **Business** During the first half of this century, the gross domestic product (GDP) of the United States, which measures the overall size of our economy in trillions of dollars, is projected to be given by the function

$$f(x) = .052x^2 + .264x + 9.9,$$

where $x = 0$ corresponds to the year 2000.* Estimate the GDP in the given years.
(a) 2010 **(b)** 2015 **(c)** 2025

51. **Business** The revenue for Intel Corporation between 1998 and 2007 is approximated by

$$R(x) = -.042x^4 + 2.10x^3 - 38.2x^2 + 302.02x - 846.2,$$

where $x = 8$ corresponds to 1998 and $R(x)$ is in billions of dollars.†
(a) What was the revenue in 2000?
(b) What was the revenue in 2007?

52. **Business** The number of U.S. aircraft departures (in millions) for the years 2000–2006 can be approximated by the function

$$f(x) = .0154x^4 - .2618x^3 + 1.33x^2 - 1.54x + 9.1,$$

where $x = 0$ corresponds to the year 2000.‡ How many departures were there in 2006?

53. **Business** The number of cargo ton miles (in millions) transported by aircraft can be approximated by the function $g(x) = .011x^2 + 1.05x + 22.8$, where $x = 0$ corresponds to the year 2000.‡
(a) What was the number of cargo ton miles transported in 2005?
(b) If the current trend continues, predict the number of cargo ton miles that will be transported in 2012.

54. **Natural Science** High concentrations of zinc ions in water are lethal to rainbow trout. The function

$$f(x) = \left(\frac{x}{1960}\right)^{-.833}$$

gives the approximate average survival time (in minutes) for trout exposed to x milligrams per liter (mg/L) of zinc ions.§ Find the survival time (to the nearest minute) for the given concentrations of zinc ions.
(a) 110 **(b)** 525 **(c)** 1960 **(d)** 4500

55. **Transportation** The distance from Chicago to Sacramento, CA, is approximately 2050 miles. A plane flying directly to Sacramento passes over Chicago at noon. If the plane travels at 500 mph, find the rule of the function $f(t)$ that gives the distance of the plane from Sacramento at time t hours (with $t = 0$ corresponding to noon).

---

*U.S. Bureau of Economic Analysis.

†Morningstar.com.

‡Air Transportation of America.

§C. Mason, *Biology of Freshwater Pollution*, Addison: Wesley Longman, 1996.

56. **Business** The profits for durable-goods industries (in billions of dollars) in the United States are approximated by $P(x) = .0991x^4 - 2.82x^3 + 23.5x^2 - 41.8x + 34$, where $x = 0$ corresponds to the year 1990.*
   (a) Estimate the profits for the year 2005.
   (b) Assuming the trend remains valid, estimate the profits for the year 2014.

57. **Finance** In 2008, the state of Illinois had a flat income tax rate of 3% for a single person. Write a function to express the amount of state tax owed for x taxable dollars earned.

58. **Business** A pretzel factory has daily fixed costs of \$1800. In addition, it costs 50 cents to produce each bag of pretzels. A bag of pretzels sells for \$1.20.
   (a) Find the rule of the cost function $c(x)$ that gives the total daily cost of producing x bags of pretzels.
   (b) Find the rule of the revenue function $r(x)$ that gives the daily revenue from selling x bags of pretzels.
   (c) Find the rule of the profit function $p(x)$ that gives the daily profit from x bags of pretzels.

59. **Geometry** Find an equation that expresses the area y of a square as a function of its
   (a) side x; **(b)** diagonal d.

Use the table feature of a graphing calculator to do these exercises. (See Example 11.)

60. **Business** The sales for durable-goods industries (in billions of dollars) in the United States are approximated by $h(x) = .252x^3 - 7.53x^2 + 156x + 1202$, where $x = 0$ corresponds to the year 1990.†
   (a) Create a table that gives the amount of sales for the years 2000–2006.
   (b) Assuming the trend remains valid, make a table for the amount of sales in years 2010–2014.

**Statistical Abstract of the United States*: 2008.
†U.S. Census Bureau.

61. **Health** The number of emergency-room physicians (in thousands) working annually in the United States can be approximated by

$$f(x) = .0005456x^3 - .014x^2 + .932x + 5.703,$$

where x is the number of years since 1980.* Find the number of emergency-room physicians working in 2006.

62. **Health** The number of prescriptions (in millions) filled annually in the United States is approximated by

$$g(x) = 3.3788x^3 - 39.1082x^2 + 202.6667x + 2858.439,$$

where x is the number of years since 2000.*
   (a) Create a table that gives the number of prescriptions (in millions) at the beginning of each year from 2002 to 2007.
   (b) Create a table that gives the number of prescriptions (in millions) at the middle of each year from 2002 to 2007. (*Hint*: Mid-2002 corresponds to $x = 2.5$.)

✓ Checkpoint Answers

1. The domain consists of all possible times—that is, all nonnegative real numbers. The range consists of all possible distances; thus, the range is also the set of all nonnegative real numbers.
2. **(a)** Yes **(b)** Yes **(c)** No
3. **(a)** Yes **(b)** Yes **(c)** No **(d)** No
4. **(a)** $(-\infty, \infty)$ **(b)** $(-\infty, \infty)$
   (c) $(-\infty, 0]$ **(d)** All real numbers except 1 and -1
5. **(a)** 4 **(b)** 40 **(c)** 73 **(d)** 44
   (e) $5m^2 - 2m + 1$ **(f)** $10xh + 5h^2 - 2h$
   (g) $10x + 5h - 2$
6. **(a)** \$2237.50 **(b)** \$220.35
7. **(a)** \$39,600,000 **(b)** \$96,000,000

**Statistical Abstract of the United States*: 2009.

3.2 ▶ Graphs of Functions

The **graph** of a function $f(x)$ is defined to be the graph of the *equation* $y = f(x)$. It consists of all points $(x, f(x))$—that is, every point whose first coordinate is an input number from the domain of f and whose second coordinate is the corresponding output number.

Example 1 The graph of the function $g(x) = .5x - 3$ is the graph of the equation $y = .5x - 3$. So the graph is a straight line with slope .5 and y-intercept -3, as shown in Figure 3.3.

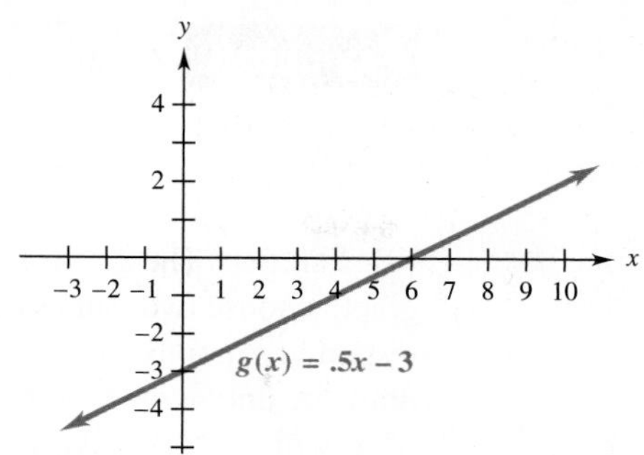

FIGURE 3.3

A function whose graph is a straight line, as in Example 1, is called a **linear function**. The rule of a linear function can always be put into the form

$$f(x) = ax + b$$

for some constants a and b.

PIECEWISE LINEAR FUNCTIONS

We now consider functions whose graphs consist of straight-line segments. Such functions are called **piecewise linear functions** and are typically defined with different equations for different parts of the domain.

Example 2 Graph the following function:

$$f(x) = \begin{cases} x + 1 & \text{if } x \le 2 \\ -2x + 7 & \text{if } x > 2. \end{cases}$$

Solution Consider the two parts of the rule of f. The graphs of $y = x + 1$ and $y = -2x + 7$ are straight lines. The graph of f consists of

the part of the line $y = x + 1$ with $x \le 2$ and
the part of the line $y = -2x + 7$ with $x > 2$.

Each of these line segments can be graphed by plotting two points in the appropriate interval, as shown in Figure 3.4.

$x \le 2$

x	0	2
$y = x + 1$	1	3

$x > 2$

x	3	4
$y = -2x + 7$	1	−1

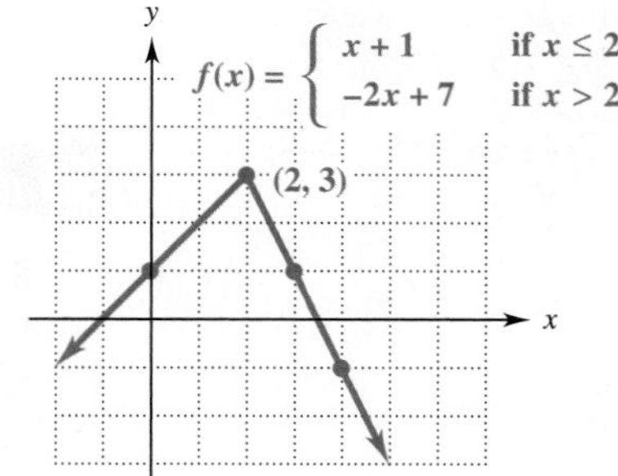

FIGURE 3.4

Note that the left and right parts of the graph each extend to the vertical line through $x = 2$, where the two halves of the graph meet at the point (2, 3). ✓1

✓Checkpoint 1

Graph

$$f(x) = \begin{cases} x + 2 & \text{if } x < 0 \\ 2 - x & \text{if } x \ge 0 \end{cases}$$

Answer: See page 155.

Example 3 Graph the function

$$f(x) = \begin{cases} x - 2 & \text{if } x \le 3 \\ -x + 8 & \text{if } x > 3. \end{cases}$$

Solution The graph consists of parts of two lines. To find the left side of the graph, choose two values of x with $x \le 3$, say, $x = 0$ and $x = 3$. Then find the corresponding points on $y = x - 2$, namely, $(0, -2)$ and $(3, 1)$. Use these points to draw the line segment to the left of $x = 3$, as in Figure 3.5. Next, choose two values of x with $x > 3$, say, $x = 4$ and $x = 6$, and find the corresponding points on $y = -x + 8$, namely, $(4, 4)$ and $(6, 2)$. Use these points to draw the line segment to the right of $x = 3$, as in Figure 3.5.

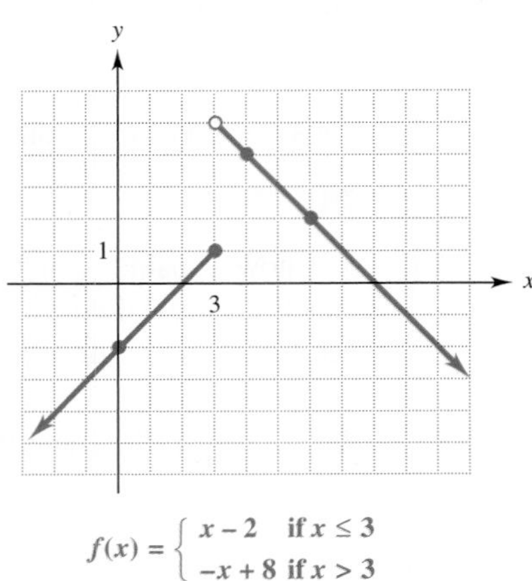

$$f(x) = \begin{cases} x - 2 & \text{if } x \le 3 \\ -x + 8 & \text{if } x > 3 \end{cases}$$

FIGURE 3.5

Note that both line segments of the graph of f extend to the vertical line through $x = 3$. The closed circle at $(3, 1)$ indicates that this point is on the graph of f, whereas the open circle at $(3, 5)$ indicates that this point is *not* on the graph of f (although it is on the graph of the line $y = -x + 8$). ✓2

✓Checkpoint 2

Graph

$$f(x) = \begin{cases} -2x - 3 & \text{if } x < 1 \\ x - 2 & \text{if } x \ge 1. \end{cases}$$

Answer: See page 155.

Example 4 Graph the **absolute-value function**, whose rule is $f(x) = |x|$.

Solution The definition of absolute value on page 9 shows that the rule of f can be written as

$$f(x) = \begin{cases} x & \text{if } x \ge 0 \\ -x & \text{if } x < 0. \end{cases}$$

So the right half of the graph (that is, where $x \ge 0$) will consist of a portion of the line $y = x$. It can be graphed by plotting two points, say, $(0, 0)$ and $(1, 1)$. The left half of the graph (where $x < 0$) will consist of a portion of the line $y = -x$, which can be graphed by plotting $(-2, 2)$ and $(-1, 1)$, as shown in Figure 3.6. ✓3

✓Checkpoint 3

Graph $f(x) = |3x - 4|$.

Answer: See page 155.

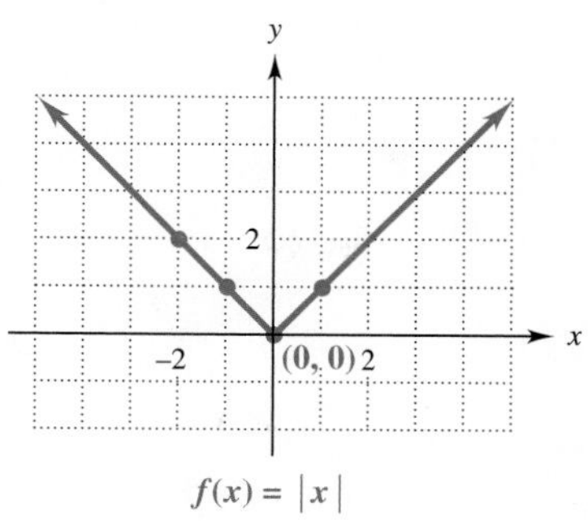

FIGURE 3.6

TECHNOLOGY TIP To graph most piecewise linear functions on a graphing calculator, you must use a special syntax. For example, on TI calculators, the best way to obtain the graph in Example 3 is to graph two separate equations on the same screen:

$$y_1 = (x - 2)/(x \leq 3) \quad \text{and} \quad y_2 = (-x + 8)/(x > 3).$$

The inequality symbols are in the TEST (or CHAR) menu. However, most calculators will graph absolute-value functions directly. To graph $f(x) = |x + 2|$, for instance, graph the equation $y = abs(x + 2)$. "Abs" (for absolute value) is on the keyboard or in the MATH menu.

STEP FUNCTIONS

The **greatest-integer function**, usually written $f(x) = [x]$, is defined by saying that $[x]$ denotes the largest integer that is less than or equal to x. For example, $[8] = 8$, $[7.45] = 7$, $[\pi] = 3$, $[-1] = -1$, $[-2.6] = -3$, and so on.

Example 5 Graph the greatest-integer function $f(x) = [x]$.

Solution Consider the values of the function between each two consecutive integers—for instance,

x	$-2 \leq x < -1$	$-1 \leq x < 0$	$0 \leq x < 1$	$1 \leq x < 2$	$2 \leq x < 3$
$[x]$	-2	-1	0	1	2

Thus, between $x = -2$ and $x = -1$, the value of $f(x) = [x]$ is always -2, so the graph there is a horizontal line segment, all of whose points have second coordinate -2. The rest of the graph is obtained similarly (Figure 3.7) on the next page. An open circle in that figure indicates that the endpoint of the segment is *not* on the graph, whereas a closed circle indicates that the endpoint *is* on the graph. ✓4

✓**Checkpoint 4**

Graph $y = [\frac{1}{2}x + 1]$.

Answer: See page 155.

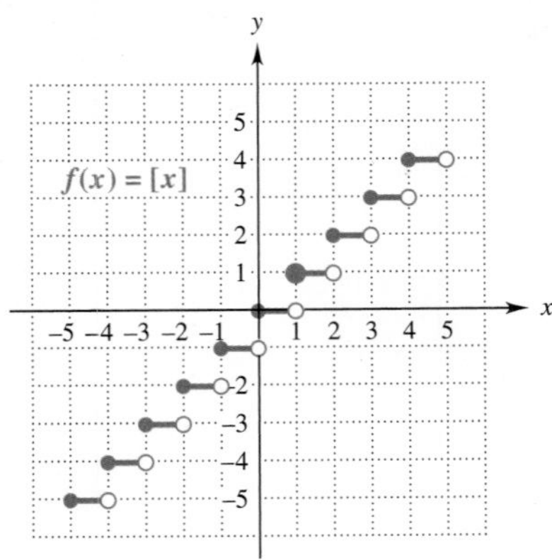

FIGURE 3.7

Functions whose graphs resemble the graph of the greatest-integer function are sometimes called **step functions**.

Example 6 **Business** An overnight delivery service charges \$30 for a package weighing up to 2 pounds. For each additional pound or fraction of a pound, there is an additional charge of \$3.50. Let $D(x)$ represent the cost to send a package weighing x pounds. Graph $D(x)$ for x in the interval $(0, 7]$.

Solution For x in the interval $(0, 2]$, $y = 30$. For x in $(2, 3]$, $y = 30 + 3.50$. For x in $(3, 4]$, $y = 33.50 + 3.50 = 37$, and so on. The graph, which is that of a step function, is shown in Figure 3.8. ✓5

✓Checkpoint 5

In 2008, the U.S. Post Office charged a fee of \$1.20 for up to the first ounce to mail a letter to France and then \$.80 for each additional ounce or fraction of an ounce. Graph the ordered pairs (ounces, cost).

Answer: See page 155.

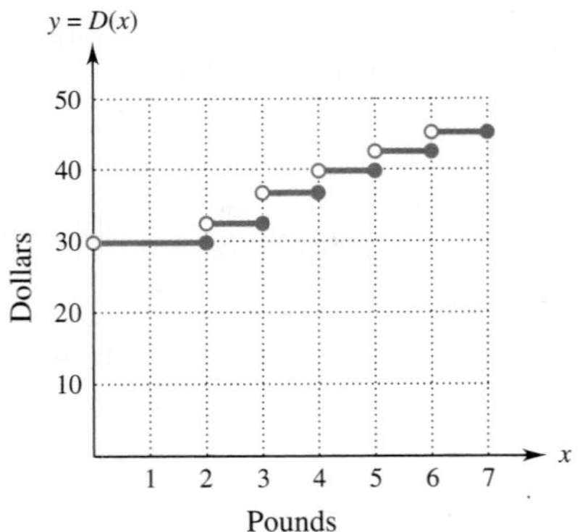

FIGURE 3.8

TECHNOLOGY TIP On most graphing calculators, the greatest-integer function is denoted INT or FLOOR. (Look on the MATH menu or its NUM submenu.) Casio calculators use INTG for the greatest-integer function and INT for a different function. When graphing these functions, put your calculator in "dot" graphing mode rather than the usual "connected" mode to avoid erroneous vertical line segments in the graph.

OTHER FUNCTIONS

The graphs of many functions do not consist only of straight-line segments. As a general rule when graphing functions by hand, you should follow the procedure introduced in Section 2.1 and summarized here.

Graphing a Function by Plotting Points

1. Determine the domain of the function.
2. Select a few numbers in the domain of f (include both negative and positive ones when possible), and compute the corresponding values of $f(x)$.
3. Plot the points $(x, f(x))$ computed in step 2. Use these points and any other information you may have about the function to make an educated guess about the shape of the entire graph.
4. Unless you have information to the contrary, assume that the graph is continuous (unbroken) wherever it is defined.

This method was used to find the graphs of the functions $f(x) = x^2 - 2x - 8$ and $g(x) = \sqrt{x+2}$ in Examples 3 and 4 of Section 2.1. Here are some more examples.

Example 7 Graph $g(x) = \sqrt{x-1}$.

Solution Because the rule of the function is defined only when $x - 1 \geq 0$ (that is, when $x \geq 1$), the domain of g is the interval $[1, \infty)$. Use a calculator to make a table of values such as the one in Figure 3.9. Plot the corresponding points and connect them to get the graph in Figure 3.9. 6 ✓

✓Checkpoint 6

Graph $f(x) = \sqrt{5-2x}$.

Answer: See page 155.

x	$g(x) = \sqrt{x-1}$
1	0
2	1
3	$\sqrt{2} \approx 1.44$
5	2
7	$\sqrt{6} \approx 2.449$
10	3

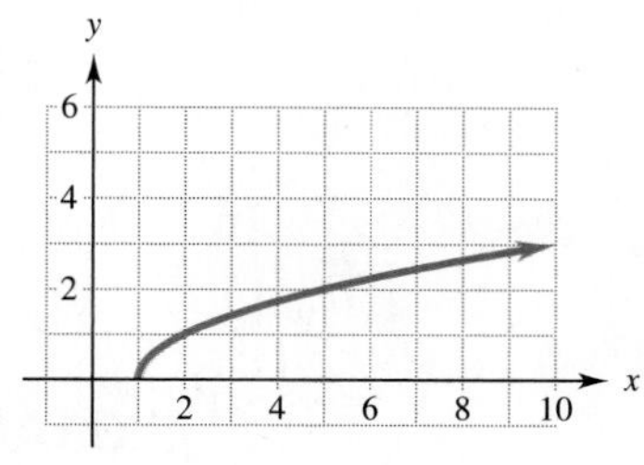

FIGURE 3.9

Example 8 Graph the function whose rule is $f(x) = 3 - \dfrac{x^3}{4}$.

Solution Make a table of values and plot the corresponding points. They suggest the graph in Figure 3.10.

x	$f(x) = 3 - \frac{x^3}{4}$
−4	19.0
−3	9.8
−2	5.0
−1	3.3
0	3.0
1	2.8
2	1.0
3	−3.8
4	−13.0

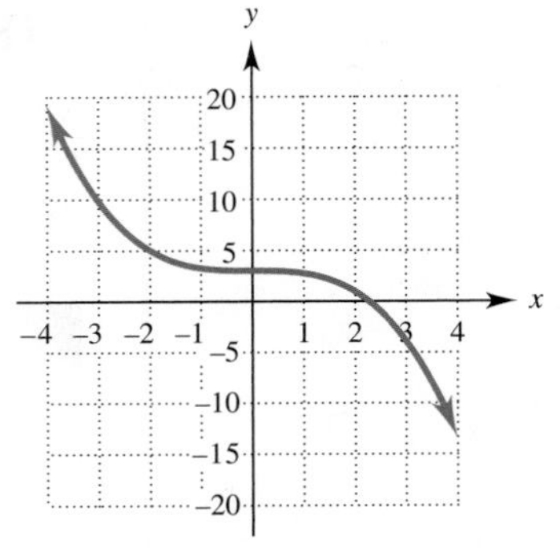

FIGURE 3.10

Example 9 Graph the piecewise defined function

$$f(x) = \begin{cases} x^2 & \text{if } x \leq 2 \\ \sqrt{x - 1} & \text{if } x > 2. \end{cases}$$

Solution When $x \leq 2$, the rule of the function is $f(x) = x^2$. Make a table of values such as the one in Figure 3.11. Plot the corresponding points and connect them to get the left half of the graph in Figure 3.9. When $x > 2$, the rule of the function is $f(x) = \sqrt{x - 1}$, whose graph is shown in Figure 3.9. In Example 7, the entire graph was given, beginning at $x = 1$. Here we use only the part of the graph to the right of $x = 2$, as shown in Figure 3.11. The open circle at (2, 1) indicates that this point is not part of the graph of f (why?).

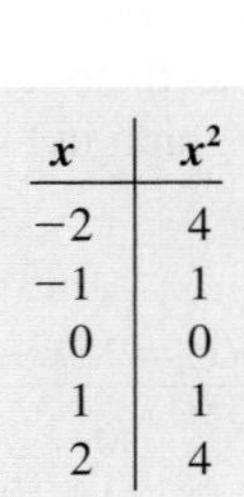

x	x^2
−2	4
−1	1
0	0
1	1
2	4

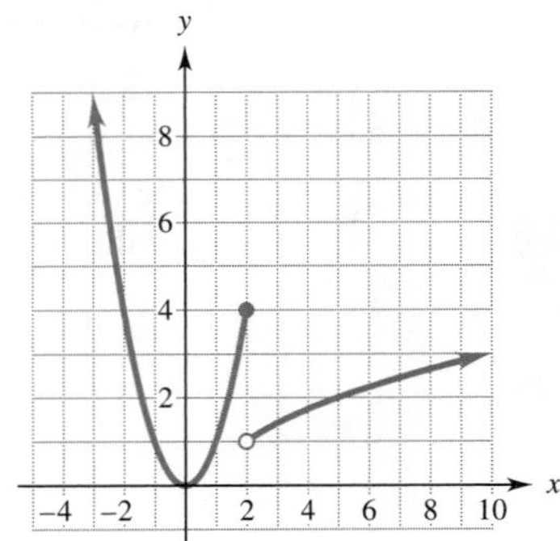

FIGURE 3.11

GRAPH READING

Graphs are often used in business and the social sciences to present data. It is just as important to know how to *read* such graphs as it is to construct them.

Example 10 **Business** The following graph shows the averages sales prices of new one-family houses by region of the United States (northeast, midwest, south, and west) from 2001 to 2006*:

The New York Times Almanac, 2008.

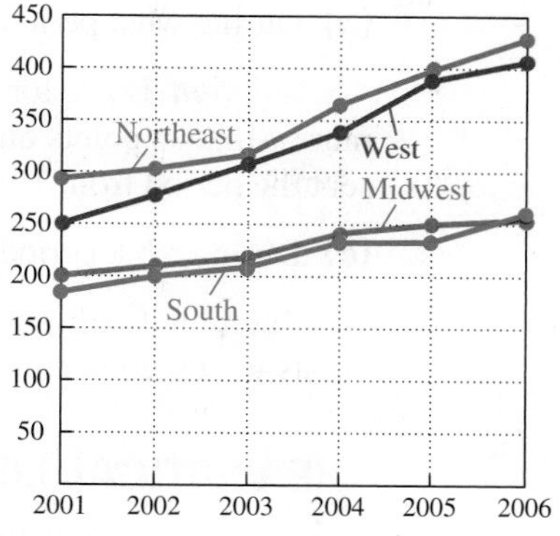

FIGURE 3.12

(a) How do the prices of homes in the west compare with those of homes in the midwest over the period shown in this graph?

Solution The prices of homes in the west were about \$50,000 higher in 2001 than those of homes in the midwest. Both regions saw the average price rise over the next few years. However, prices rose more steeply for homes in the west, and in 2006, the average price was approximately \$150,000 higher than in the midwest.

(b) Which regions typically have the highest and lowest average price of a new home?

Solution The northeast is highest for all years in the period, and the south is lowest.

Example 11 Figure 3.13 is the graph of the function f whose rule is $f(x) =$ average interest rate on a 30-year fixed-rate mortgage in year x.

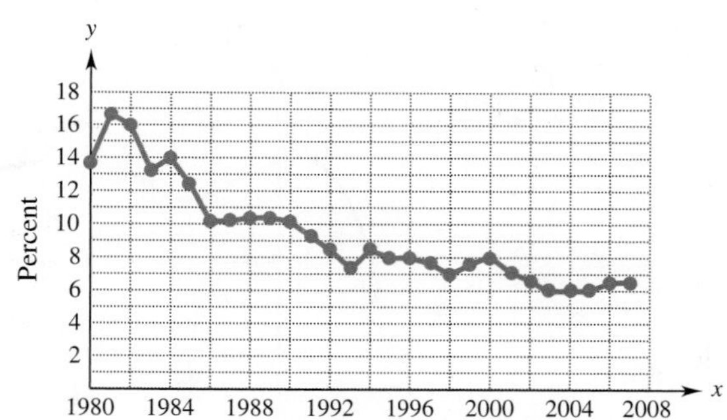

FIGURE 3.13

(a) Find the function values of $f(1984)$ and $f(2004)$.

Solution The point (1984, 14) is on the graph, which means that $f(1984) = 14$. Similarly, $f(2004) = 6$, because the point (2004, 6) is on the graph. These values tell us that average mortgage rates were 14% in 1984 and 6% in 2004.

(b) During what period were mortgage rates at or above 10%?

Solution Look for points on the graph whose second coordinates are 10 or more—that is, points on or above the horizontal line through 10. These points represent the period from 1980 to 1990.

(c) During what period were mortgage rates below 6%?

Solution Look for points that are below the horizontal line through 6—that is, points with second coordinates less than 6. They occur from 2003 to 2005.

THE VERTICAL-LINE TEST

The following fact distinguishes function graphs from other graphs.

> **Vertical-Line Test**
>
> No vertical line intersects the graph of a function $y = f(x)$ at more than one point.

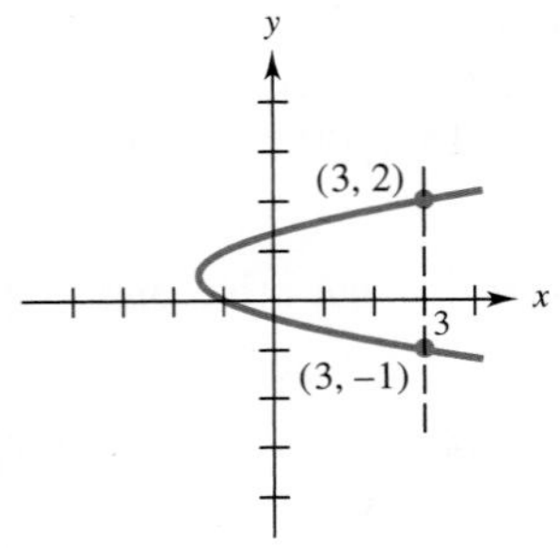

FIGURE 3.14

In other words, if a vertical line intersects a graph at more than one point, the graph is not the graph of a function. To see why this is true, consider the graph in Figure 3.14. The vertical line $x = 3$ intersects the graph at two points. If this were the graph of a function f, it would mean that $f(3) = 2$ (because $(3, 2)$ is on the graph) and that $f(3) = -1$ (because $(3, -1)$ is on the graph). This is impossible, because a *function* can have only one value when $x = 3$ (because each input determines exactly one output). Therefore, the graph in Figure 3.14 cannot be the graph of a function. A similar argument works in the general case.

Example 12 Use the vertical-line test to determine which of the graphs in Figure 3.15 are the graphs of functions.

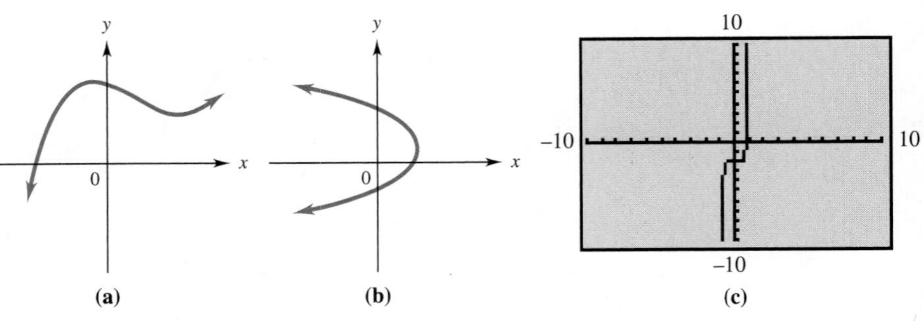

FIGURE 3.15

Solution To use the vertical-line test, imagine dragging a ruler held vertically across the graph from left to right. If the graph is that of a function, the edge of the ruler would hit the graph only once for every x-value. If you do this for graph (a), every vertical line intersects the graph in at most one point, so this graph is the graph of a function. Many vertical lines (including the y-axis) intersect graph (b) twice, so it is not the graph of a function.

Graph (c) appears to fail the vertical-line test near $x = 1$ and $x = -1$, indicating that it is not the graph of a function. But this appearance is misleading because of the low resolution of the calculator screen. The table in Figure 3.16 and the very narrow segment of the graph in Figure 3.17 show that the graph actually rises as it moves to the right. The same thing happens near $x = -1$. So this graph *does* pass the vertical-line test and *is* the graph of a function. (Its rule is $f(x) = 15x^{11} - 2$.) The moral of this story is that you can't always trust images produced by a graphing calculator. When in doubt, try other viewing windows or a table to see what is really going on. 7 ✓

✓Checkpoint 7

Find a viewing window that indicates the actual shape of the graph of the function $f(x) = 15x^{11} - 2$ of Example 12 near the point $(-1, -17)$.

Answer: See page 155.

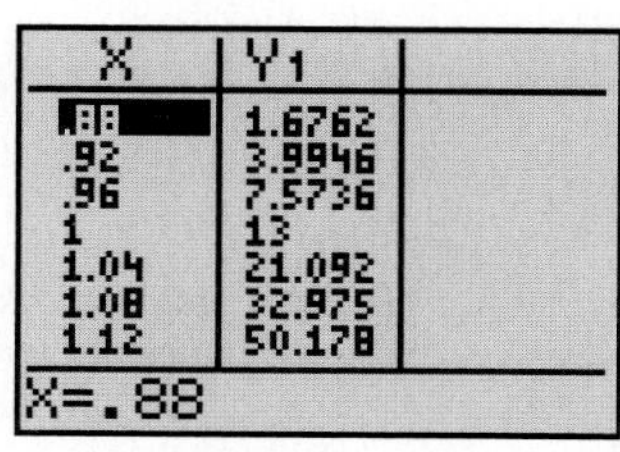

FIGURE 3.16

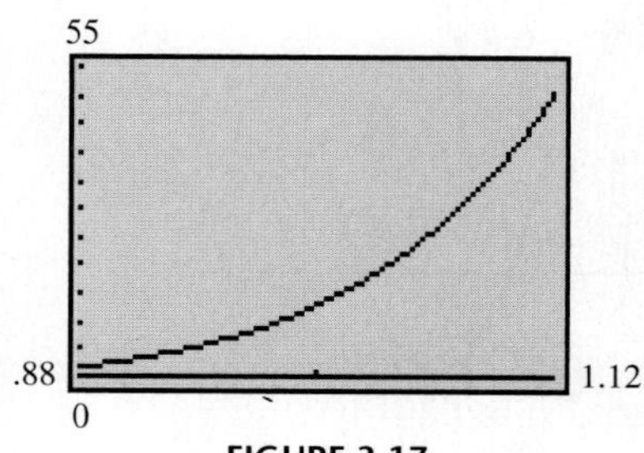

FIGURE 3.17

3.2 ▶ Exercises

Graph each function. (See Examples 1–4.)

1. $f(x) = -.5x + 2$
2. $g(x) = 3 - x$
3. $f(x) = \begin{cases} x + 3 & \text{if } x \le 1 \\ 4 & \text{if } x > 1 \end{cases}$
4. $g(x) = \begin{cases} 2x - 1 & \text{if } x < 0 \\ -1 & \text{if } x \ge 0 \end{cases}$
5. $y = \begin{cases} 4 - x & \text{if } x \le 0 \\ 3x + 4 & \text{if } x > 0 \end{cases}$
6. $y = \begin{cases} x + 5 & \text{if } x \le 1 \\ 2 - 3x & \text{if } x > 1 \end{cases}$
7. $f(x) = \begin{cases} |x| & \text{if } x < 2 \\ -2x & \text{if } x \ge 2 \end{cases}$
8. $g(x) = \begin{cases} -|x| & \text{if } x \le 1 \\ 2x & \text{if } x > 1 \end{cases}$
9. $f(x) = |x - 4|$
10. $g(x) = |4 - x|$
11. $f(x) = |3 - 3x|$
12. $g(x) = -|x|$
13. $y = -|x - 1|$
14. $f(x) = |x| - 2$
15. $y = |x - 2| + 3$
16. $|x| + |y| = 1$ (*Hint*: This is not the graph of a function, but is made up of four straight-line segments. Find them by using the definition of absolute value in these four cases: $x \ge 0$ and $y \ge 0$; $x \ge 0$ and $y < 0$; $x < 0$ and $y \ge 0$; $x < 0$ and $y < 0$).

Graph each function. (See Examples 5 and 6.)

17. $f(x) = [x - 3]$
18. $g(x) = [x + 3]$
19. $g(x) = [-x]$
20. $f(x) = -[x]$
21. $f(x) = [x] + [-x]$ (The graph contains horizontal segments, but is *not* a horizontal line.)
22. The accompanying table gives rates charged by the U.S. Postal Service for first-class letters in January 2009. Graph the function $f(x)$ that gives the price of mailing a first-class letter, where x represents the weight of the letter in ounces and $0 \le x \le 3.5$.

Weight Not Over	Price
1 ounce	\$0.42
2 ounces	\$0.59
3 ounces	\$0.76
3.5 ounces	\$0.93

Graph each function. (See Examples 7–9.)

23. $f(x) = 3 - 2x^2$

24. $g(x) = 2 - x^2$

25. $h(x) = x^3/10 + 2$

26. $f(x) = x^3/20 - 3$

27. $g(x) = \sqrt{-x}$

28. $h(x) = \sqrt{x} - 1$

29. $f(x) = \sqrt[3]{x}$

30. $g(x) = \sqrt[3]{x} - 4$

31. $f(x) = \begin{cases} x^2 & \text{if } x < 2 \\ -2x + 2 & \text{if } x \geq 2 \end{cases}$

32. $g(x) = \begin{cases} \sqrt{-x} & \text{if } x \leq -4 \\ \dfrac{x^2}{4} & \text{if } x > -4 \end{cases}$

Determine whether each graph is a graph of a function or not. (See Example 12.)

33.

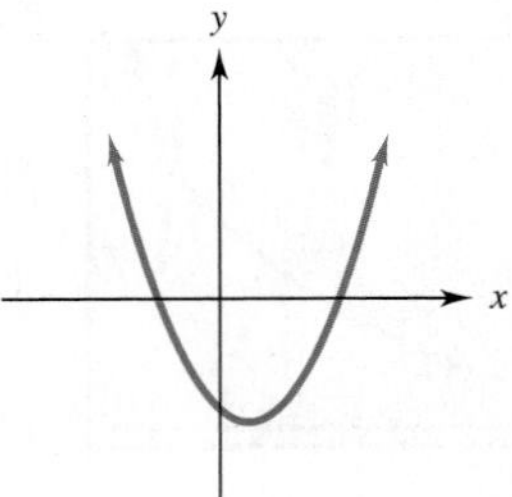

34.

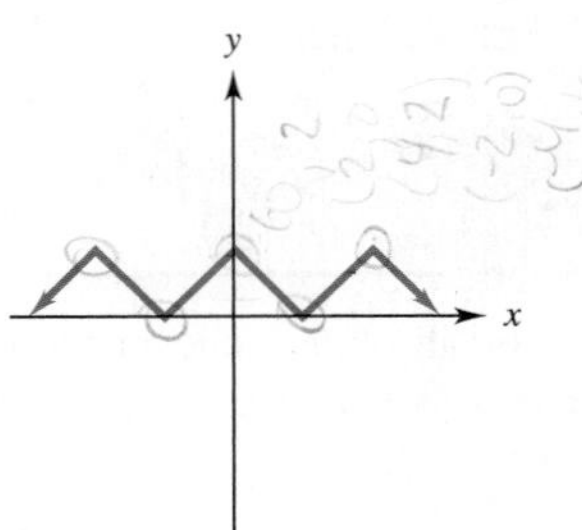

35.

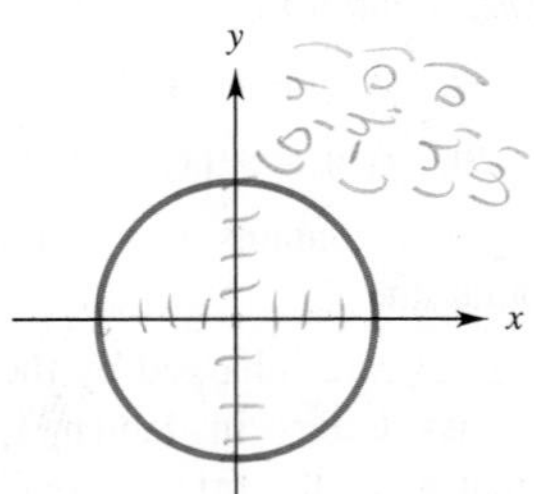

36.

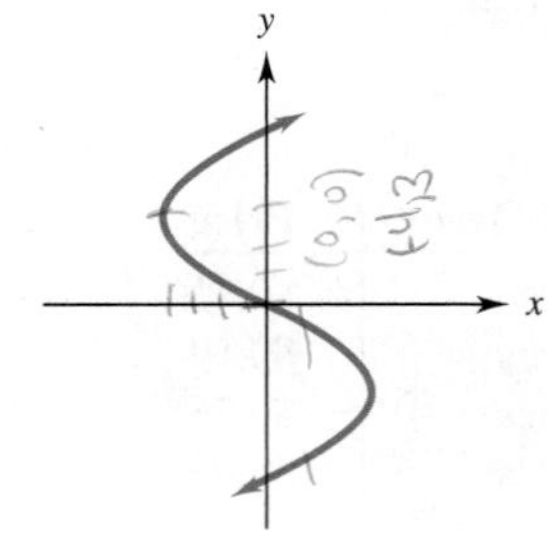

37.

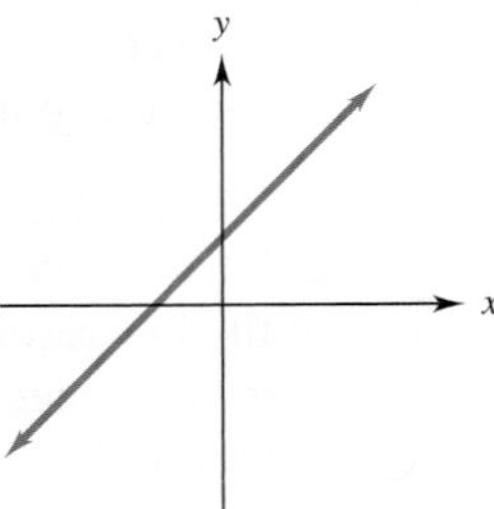

38.

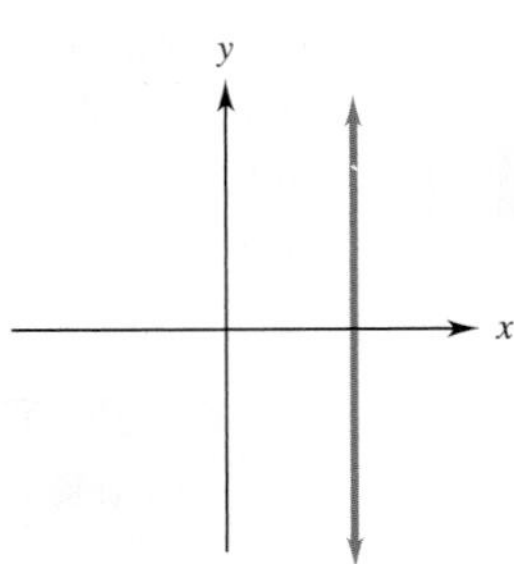

Use a graphing calculator or other technology to graph each of the given functions. If the graph has any endpoints, indicate whether they are part of the graph or not.

39. $f(x) = .2x^3 - .8x^2 - 4x + 9.6$

40. $g(x) = .1x^4 - .3x^3 - 1.3x^2 + 1.5x$

41. $g(x) = \begin{cases} 2x^2 + x & \text{if } x < 1 \\ x^3 - x - 1 & \text{if } x \geq 1 \end{cases}$ (*Hint*: See the Technology Tip on page 145.)

42. $f(x) = \begin{cases} x|x| & \text{if } x \leq 0 \\ -x^2|x| + 2 & \text{if } x > 0 \end{cases}$

Use a graphical root finder to determine the x-intercepts of the graph of

43. f in Exercise 39;

44. g in Exercise 40.

Use a maximum–minimum finder to determine the location of the peaks and valleys in the graph of

45. g in Exercise 40;

46. f in Exercise 39.

See Examples 2, 3, 10, and 11 as you do Exercises 47–52.

47. Finance The maximum allowable yearly contribution to an individual retirement account (IRA) was \$2000 from 1981 to 2001 and has increased steadily since then. When the maximum is adjusted for inflation, however, the picture is a little different. The approximate maximum IRA contribution in 1981 dollars in year x is given by the function

$$f(x) = \begin{cases} -50x + 2050 & \text{if } 1 \leq x < 22 \\ 84x - 348 & \text{if } 22 \leq x < 29, \end{cases}$$

where $x = 0$ corresponds to 1980.

(a) Graph this function.

(b) What does the graph say about inflation from 1981 to 2001?

48. Finance The Alabama state income tax for a single person in 2009 was determined by the rule

$$T(x) = \begin{cases} .02x & \text{if } 0 \le x \le 500 \\ 10 + .04(x - 500) & \text{if } 500 < x \le 3000 \\ 110 + .05(x - 3000) & \text{if } x > 3000, \end{cases}$$

where x is the person's taxable income in dollars. Graph the function $T(x)$ for taxable incomes between 0 and \$5000.

49. Finance The price of Starbucks stock in 2008 can be modeled by the function

$$f(x) = \begin{cases} 18.35 - .18x & \text{if } 0 \le x \le 22 \\ 12.19 + .10x & \text{if } 22 < x \le 44, \end{cases}$$

where x represents the number of business days past June 16, 2008.*

(a) Graph $f(x)$.

(b) What is the lowest stock price during the period defined by $f(x)$?

50. Natural Science The number of minutes of daylight in Washington, DC, can be modeled by the function

$$g(x) = \begin{cases} .0106x^2 + 1.24x + 565 & \text{if } 1 \le x < 84 \\ -.0202x^2 + 6.97x + 295 & \text{if } 84 \le x < 263 \\ .0148x^2 - 11.1x + 2632 & \text{if } 263 < x \le 365, \end{cases}$$

where x represents the number of days of the year, starting on January 1. Find the number of minutes of daylight on

(a) day 32. **(b)** day 90.

(c) day 200. **(d)** day 270.

(e) Graph $g(x)$.

(f) On what day is the number of minutes of daylight the greatest?

51. Health The following table shows the consumer price index (CPI) for medical care in selected years†:

Year	Medical CPI
1950	15.1
1975	47.5
2007	351.05

(a) Let $x = 0$ correspond to 1950. Find the rule of a piecewise linear function that models these data—that is, a piecewise linear function f with $f(0) = 15.1$, $f(25) = 47.5$, and $f(57) = 351.05$. Round all coefficients in your final answer to one decimal place.

(b) Graph the function f for $0 \le x \le 65$.

(c) Use the function to estimate the medical-care CPI in 2004.

(d) Assuming that this model remains accurate after 2007, estimate the medical-care CPI for 2012.

*Morningstar.com.

†U.S. Bureau of Labor Statistics.

52. Business The following graph from the U.S. Office of Management and Budget shows the federal debt from 1990 to 2006 (in billions of dollars), with $x = 0$ corresponding to 1990.

Find the rule of a linear function g that passes through the two points corresponding to 1990 and 2006. Draw the graph of g.

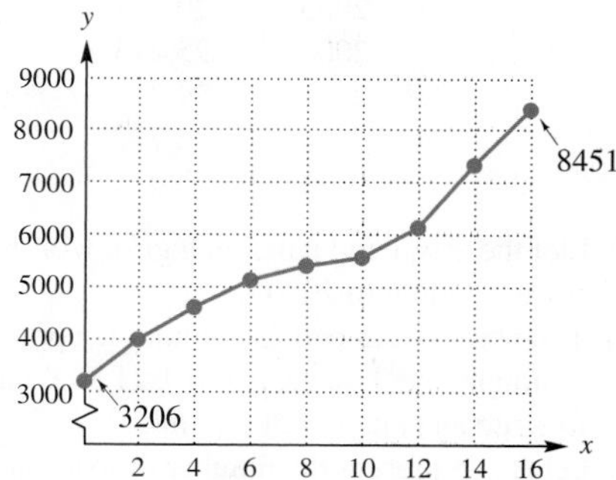

Finance *Use the accompanying graph to answer Exercises 53 and 54. Explain how you obtained your answer from the graph, which shows the annual percent change in various consumer price indexes (CPIs). (See Examples 10 and 11.)*

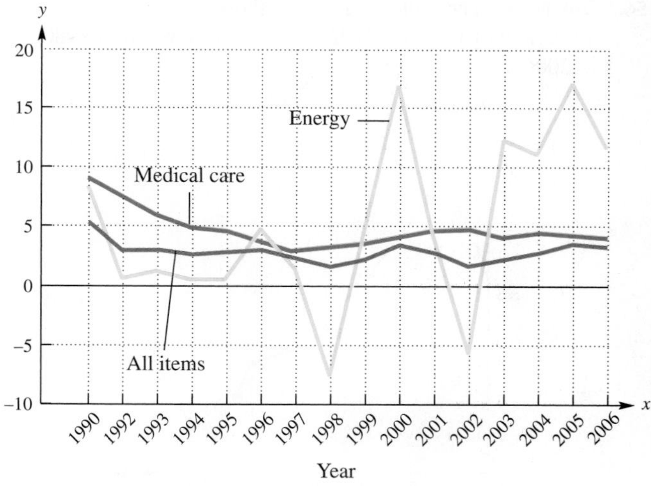

Source: U.S. Census Bureau.

53. **(a)** Was there any period between 1990 and 2006 when the CPI for all items was decreasing?

(b) During what years was the CPI for energy decreasing?

(c) During what years was the CPI for energy increasing?

54. **(a)** When was the energy CPI decreasing at the fastest rate?

(b) When was the energy CPI increasing at the fastest rate?

(c) Was there ever a time after 2000 when the medical-care CPI and the energy CPI were increasing at the same rate? How can you tell from the graph?

55. **Business** The following table gives the number of small-business loans to minority-owned small businesses from 2000 to 2006*:

Year	Number
2000	11,999
2003	20,184
2004	25,413
2005	29,722
2006	33,772

(a) Plot these ordered pairs on a grid, with $x = 0$ corresponding to 2000.
(b) The points from part (a) should lie approximately on a straight line. Use the pairs (0, 11,999) and (6, 33,772) to write an equation of the line.
(c) Let $f(x)$ represent the number of loans and x represent the year. Write the equation from part (b) as the rule that defines a function.
(d) Find $f(4)$. Does it agree fairly closely with the number in the table that corresponds to 2004? Do you think the expression from part (c) describes this function adequately?

56. **Business** The following graph shows the percent-per-year yield of the 10-year U.S. Treasury Bond from 1980 to 2006†:

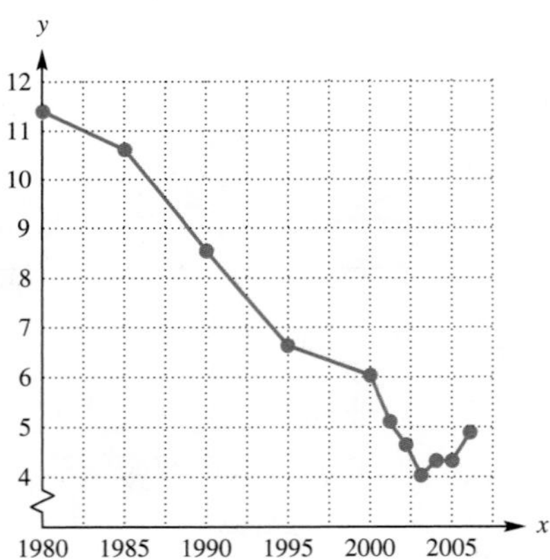

(a) Is this the graph of a function?
(b) What does the domain represent?
(c) Estimate the range.

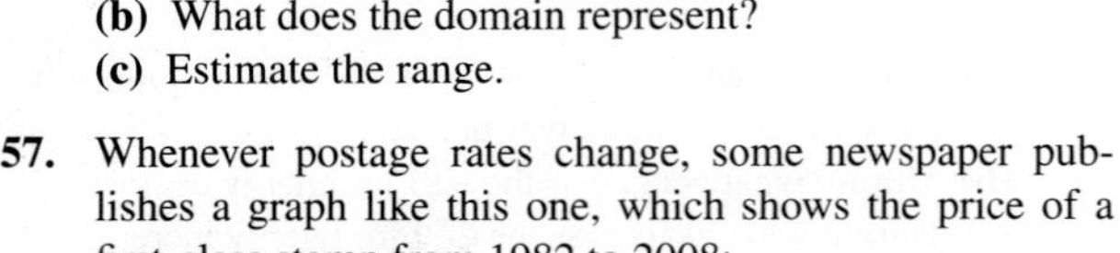
57. Whenever postage rates change, some newspaper publishes a graph like this one, which shows the price of a first-class stamp from 1982 to 2008:

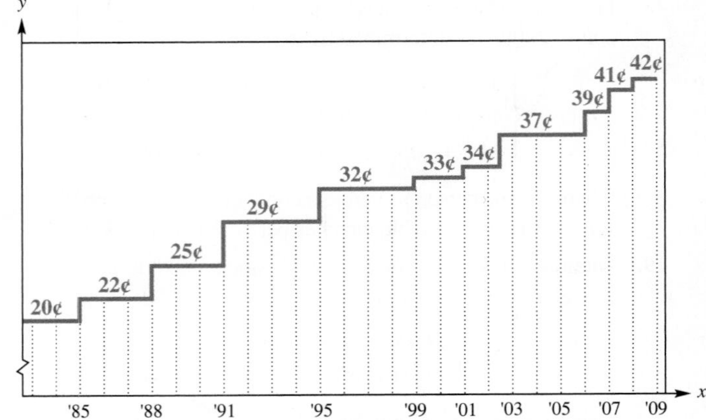

(a) Let f be the function whose rule is

$$f(x) = \text{cost of a first-class stamp in year } x.$$

Find $f(2000)$ and $f(2006)$.
(b) Explain why the graph in the figure is not the graph of the *function f*. What must be done to the figure to make it an accurate graph of the function f?

58. A chain-saw rental firm charges \$20 per day or fraction of a day to rent a saw, plus a fixed fee of \$7 for resharpening the blade. Let $S(x)$ represent the cost of renting a saw for x days. Find each of the following.

(a) $S\left(\frac{1}{2}\right)$ **(b)** $S(1)$ **(c)** $S\left(1\frac{2}{3}\right)$ **(d)** $S\left(4\frac{3}{4}\right)$

(e) What does it cost to rent for $5\frac{7}{8}$ days?
(f) A portion of the graph of $y = S(x)$ is shown here. Explain how the graph could be continued.

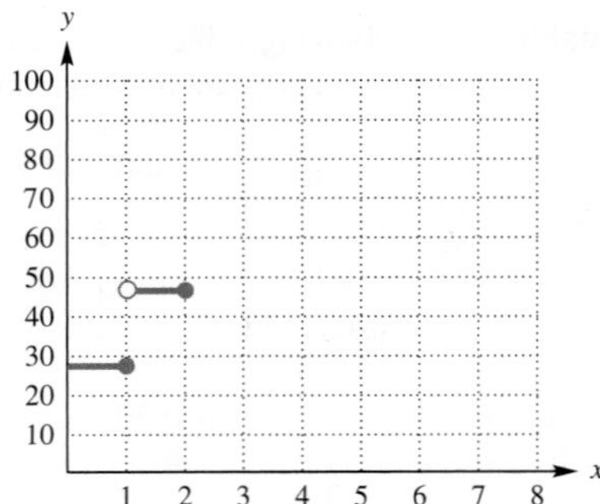

(g) What is the domain variable?
(h) What is the range variable?
(i) Write a sentence or two explaining what (c) and its answer represent.
(j) We have left $x = 0$ out of the graph. Discuss why it should or should not be included. If it were included, how would you define $S(0)$?

59. Sarah Hendrickson needs to rent a van to pick up a new couch she has purchased. The cost of the van is \$19.99 for the first 75 minutes and then an additional \$5 for

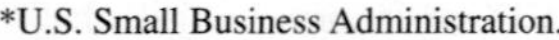

*U.S. Small Business Administration.

†Board of Governors of the Federal Reserve System.

each block of 15 minutes beyond 75. Find the cost to rent a van for

(a) 2 hours; **(b)** 1.5 hours;
(c) 3.5 hours; **(d)** 4 hours.
(e) Graph the ordered pairs (hours, cost).

60. A delivery company charges \$25 plus 60¢ per mile or part of a mile. Find the cost for a trip of

(a) 3 miles; **(b)** 4.2 miles;
(c) 5.9 miles; **(d)** 8 miles.
(e) Graph the ordered pairs (miles, cost).
(f) Is this a function?

Work these problems.

61. Natural Science A laboratory culture contains about 1 million bacteria at midnight. The culture grows very rapidly until noon, when a bactericide is introduced and the bacteria population plunges. By 4 PM, the bacteria have adapted to the bactericide and the culture slowly increases in population until 9 PM, when the culture is accidentally destroyed by the cleanup crew. Let $g(t)$ denote the bacteria population at time t (with $t = 0$ corresponding to midnight). Draw a plausible graph of the function g. (Many correct answers are possible.)

62. Transportation A plane flies from Austin, Texas, to Cleveland, Ohio, a distance of 1200 miles. Let f be the function whose rule is

$$f(t) = \text{distance (in miles) from Austin at time } t \text{ hours,}$$

with $t = 0$ corresponding to the 4 PM takeoff. In each part of this exercise, draw a plausible graph of f under the given circumstances. (There are many correct answers for each part.)

(a) The flight is nonstop and takes between 3.5 and 4 hours.
(b) Bad weather forces the plane to land in Dallas (about 200 miles from Austin) at 5 PM, remain overnight, and leave at 8 AM the next day, flying nonstop to Cleveland.
(c) The plane flies nonstop, but due to heavy traffic it must fly in a holding pattern for an hour over Cincinnati (about 200 miles from Cleveland) and then go on to Cleveland.

Checkpoint Answers

1.

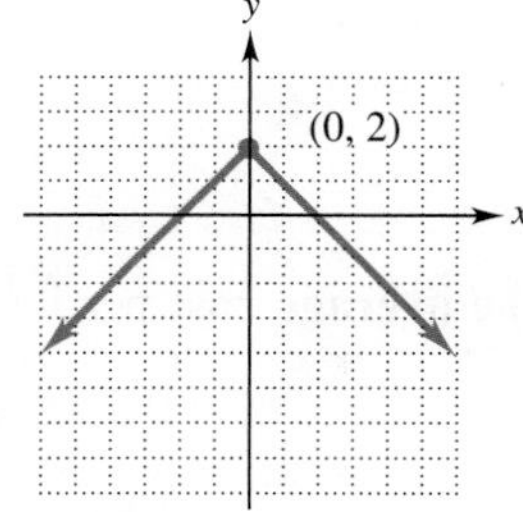

2.

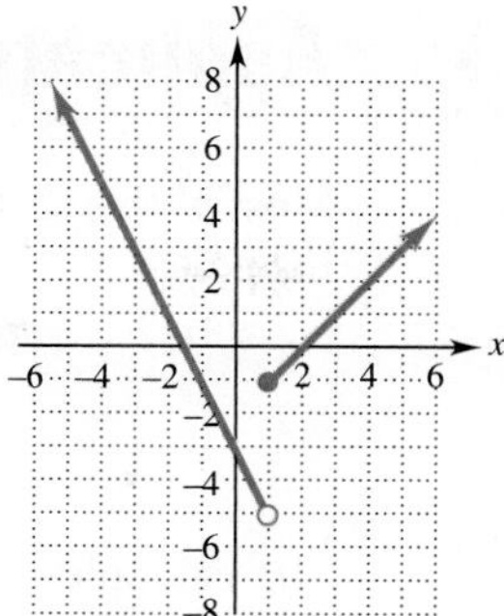

3.

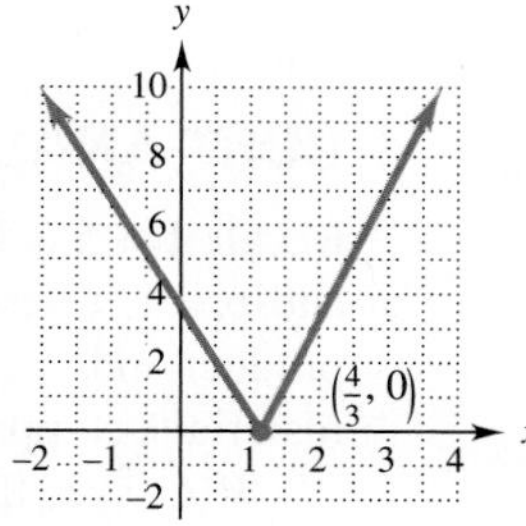

4.

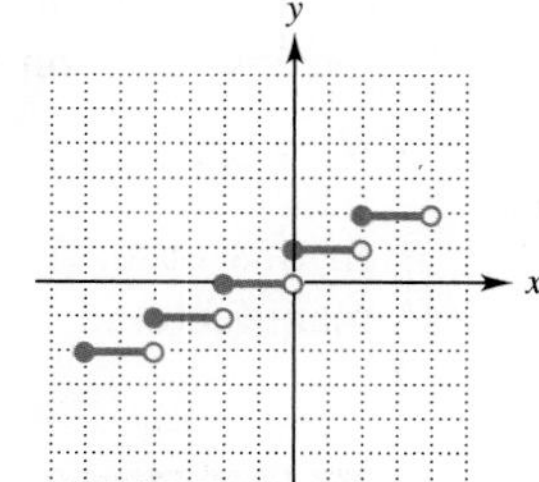

5.

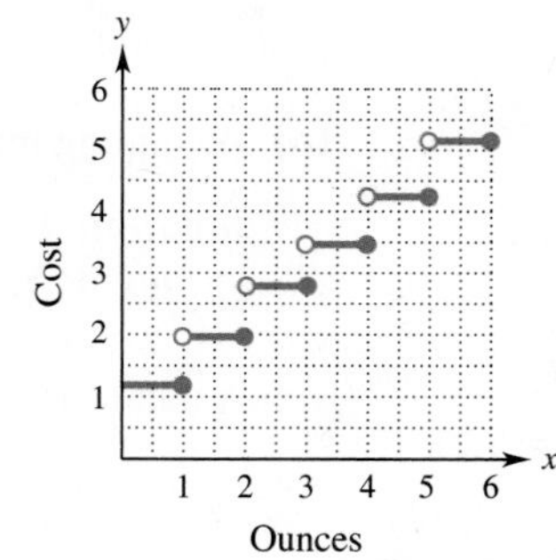

6.

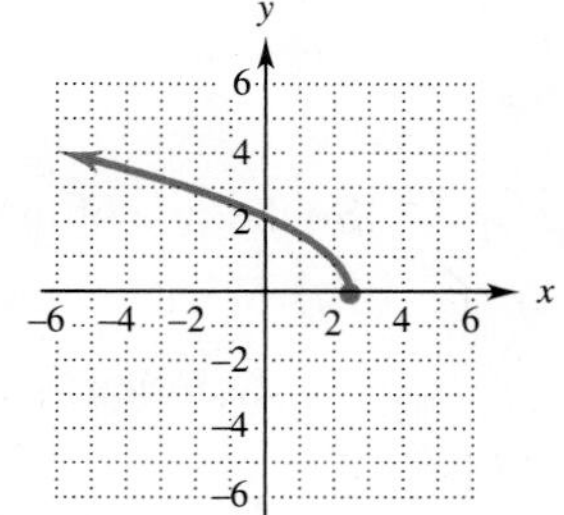

7. There are many correct answers, including, $-1.4 \le x \le -.6$ and $-30 \le y \le 0$.

3.3 Applications of Linear Functions

Most of this section deals with the basic business relationships that were introduced in Section 1.2:

Revenue = (Price per item) × (Number of items);

Cost = Fixed Costs + Variable Costs;

Profit = Revenue − Cost.

The examples will use only linear functions, but the methods presented here also apply to more complicated functions.

COST ANALYSIS

Recall that fixed costs are for such things as buildings, machinery, real-estate taxes, and product design. Within broad limits, the fixed cost is constant for a particular product and does not change as more items are made. Variable costs are for labor, materials, shipping, and so on, and depend on the number of items made.

If $C(x)$ is the cost of making x items, then the fixed cost (the cost that occurs even when no items are produced) can be found by letting $x = 0$. For example, for the cost function $C(x) = 45x + 250{,}000$, the fixed cost is

$$C(0) = 45(0) + 250{,}000 = \$250{,}000.$$

In this case, the variable cost of making x items is $45x$—that is, \$45 per item manufactured.

Example 1 An anticlot drug can be made for \$10 per unit. The total cost to produce 100 units is \$1500.

(a) Assuming that the cost function is linear, find its rule.

Solution Since the cost function $C(x)$ is linear, its rule is of the form $C(x) = mx + b$. We are given that m (the cost per item) is 10, so the rule is $C(x) = 10x + b$. To find b, use the fact that it costs \$1500 to produce 100 units, which means that

$$\begin{aligned} C(100) &= 1500 \\ 10(100) + b &= 1500 \qquad C(x) = 10x + b. \\ 1000 + b &= 1500 \\ b &= 500. \end{aligned}$$

So the rule of the cost function is $C(x) = 10x + 500$.

(b) What are the fixed costs?

Solution The fixed costs are $C(0) = 10(0) + 500 = \$500$. ✓1

✓Checkpoint 1

The total cost of producing 10 calculators is \$100. The variable costs per calculator are \$4. Find the rule of the linear cost function.

Answer: See page 167.

If $C(x)$ is the total cost to produce x items, then the **average cost** per item is given by

$$\overline{C}(x) = \frac{C(x)}{x}.$$

As more and more items are produced, the average cost per item typically decreases.

Example 2 Find the average cost of producing 100 and 1000 units of the anticlot drug in Example 1.

Solution The cost function is $C(x) = 10x + 500$, so the average cost of producing 100 units is

$$\overline{C}(100) = \frac{C(100)}{100} = \frac{10(100) + 500}{100} = \frac{1500}{100} = \$15.00 \text{ per unit.}$$

The average cost of producing 1000 units is

$$\overline{C}(1000) = \frac{C(1000)}{1000} = \frac{10(1000) + 500}{1000} = \frac{10{,}500}{1000} = \$10.50 \text{ per unit.}$$

✓ **Checkpoint 2**

In Checkpoint 1, find the average cost per calculator when 100 are produced.

Answer: See page 167.

RATES OF CHANGE

The rate at which a quantity (such as revenue or profit) is changing can be quite important. For instance, if a company determines that the rate of change of its revenue is decreasing, then sales growth is slowing down, a trend that may require a response.

The rate of change of a linear function is easily determined. For example, suppose $f(x) = 3x + 5$ and consider the table of values in the margin. The table shows that each time x changes by 1, the corresponding value of $f(x)$ changes by 3. Thus, the rate of change of $f(x) = 3x + 5$ with respect to x is 3, which is the slope of the line $y = 3x + 5$. The same thing happens for any linear function:

x	$f(x) = 3x + 5$
1	8
2	11
3	14
4	17
5	20

The rate of change of a linear function $f(x) = mx + b$ is the slope m.

In particular, the rate of change of a linear function is constant.

The value of a computer, or an automobile, or a machine *depreciates* (decreases) over time. **Linear depreciation** means that the value of the item at time x is given by a linear function $f(x) = mx + b$. The slope m of this line gives the rate of depreciation.

Example 3 **Business** According to the *Kelley Blue Book*, a Ford Mustang two-door convertible that is worth \$14,776 today will be worth \$10,600 in two years (if it is in excellent condition with average mileage).

(a) Assuming linear depreciation, find the depreciation function for this car.

Solution We know the car is worth \$14,776 now ($x = 0$) and will be worth \$10,600 in two years ($x = 2$). So the points (0, 14,776) and (2, 10,600) are on the graph of the linear-depreciation function and can be used to determine its slope:

$$m = \frac{10{,}600 - 14{,}776}{2} = \frac{-4176}{2} = -2088.$$

Using the point (0, 14,776), we find that the equation of the line is

$$y - 14{,}776 = -2088(x - 0) \qquad \text{Point–slope form}$$
$$y = -2088x + 14{,}776. \qquad \text{Slope–intercept form}$$

Therefore, the rule of the depreciation function is $f(x) = -2088x + 14{,}776$.

(b) What will the car be worth in 4 years?

Solution Evaluate f when $x = 4$:

$$f(x) = -2088x + 14{,}776$$
$$f(4) = -2088(4) + 14{,}776 = \$6424.$$

(c) At what rate is the car depreciating?

Solution The depreciation rate is given by the slope of $f(x) = -2088x + 14{,}776$, namely, -2088. This negative slope means that the car is decreasing in value an average of \$2088 a year. [3]

✓Checkpoint 3

Using the information from Example 3, determine what the car will be worth in 6 years.

Answer: See page 167.

In economics, the rate of change of the cost function is called the **marginal cost**. Marginal cost is important to management in making decisions in such areas as cost control, pricing, and production planning. When the cost function is linear, say, $C(x) = mx + b$, the marginal cost is the number m (the slope of the graph of C). Marginal cost can also be thought of as the cost of producing one more item, as the next example demonstrates.

Example 4 An electronics company manufactures handheld PCs. The cost function for one of its models is $C(x) = 160x + 750{,}000$.

(a) What are the fixed costs for this product?

Solution The fixed costs are $C(0) = 160(0) + 750{,}000 = \$750{,}000$.

(b) What is the marginal cost?

Solution The slope of $C(x) = 160x + 750{,}000$ is 160, so the marginal cost is \$160.

(c) After 50,000 units have been produced, what is the cost of producing one more?

Solution The cost of producing 50,000 is

$$C(50{,}000) = 160(50{,}000) + 750{,}000 = \$8{,}750{,}000.$$

The cost of 50,001 units is

$$C(50{,}001) = 160(50{,}001) + 750{,}000 = \$8{,}750{,}160.$$

The cost of the additional unit is the difference

$$C(50{,}001) - C(50{,}000) = 8{,}750{,}160 - 8{,}750{,}000 = \$160.$$

Thus, the cost of one more item is the marginal cost. [4]

✓Checkpoint 4

The cost in dollars to produce x kilograms of chocolate candy is given by $C(x) = 3.5x + 800$. Find each of the following.

(a) The fixed cost

(b) The total cost for 12 kilograms

(c) The marginal cost of the 40th kilogram

(d) The marginal cost per kilogram

Answers: See page 167.

Similarly, the rate of change of a revenue function is called the **marginal revenue**. When the revenue function is linear, the marginal revenue is the slope of the line, as well as the revenue from producing one more item.

Example 5 The energy company New York State Electric and Gas charges each residential customer a basic fee of \$13.11, plus \$0.0347 per kilowatt hour (kWh).

(a) Assuming there are 700,000 residential customers, find the company's revenue function.

Solution The monthly revenue from the basic fee is

$$13.11(700{,}000) = \$9{,}177{,}000.$$

If x is the total number of kilowatt hours used by all customers, then the revenue from electricity use is $.0347x$. So the monthly revenue function is given by

$$R(x) = .0347x + 9{,}177{,}000.$$

✓Checkpoint 5

Assume that the average customer in Example 5 uses 1600 kWh in a month.

(a) What is the total number of kWh used by all customers?

(b) What is the company's monthly revenue?

Answers: See page 167.

(b) What is the marginal revenue?

Solution The marginal revenue (the rate at which revenue is changing) is given by the slope of the rate function: \$0.0347 per kWh. 5✓

Examples 4 and 5 are typical of the general case, as summarized here.

In a **linear cost function** $C(x) = mx + b$, the marginal cost is m (the slope of the cost line) and the fixed cost is b (the y-intercept of the cost line). The marginal cost is the cost of producing one more item.

Similarly, in a **linear revenue function** $R(x) = kx + d$, the marginal revenue is k (the slope of the revenue line), which is the revenue from selling one more item.

BREAK-EVEN ANALYSIS

A typical company must analyze its costs and the potential market for its product to determine when (or even whether) it will make a profit.

Example 6 **Business** A company manufactures a 42-inch plasma HDTV that sells to retailers for \$550. The cost of making x of these TVs for a month is given by the cost function $C(x) = 250x + 213{,}000$.

(a) Find the function R that gives the revenue from selling x TVs.

Solution Since revenue is the product of the price per item and the number of items, $R(x) = 550x$.

(b) What is the revenue from selling 600 TVs?

Solution Evaluate the revenue function R at 600:

$$R(600) = 550(600) = \$330{,}000.$$

(c) Find the profit function P.

Solution Since Profit = Revenue − Cost,

$$P(x) = R(x) - C(x) = 550x - (250x + 213{,}000) = 300x - 213{,}000.$$

(d) What is the profit from selling 500 TVs?

Solution Evaluate the profit function at 500 to obtain

$$P(500) = 300(500) - 213{,}000 = -63{,}000,$$

that is, a loss of \$63,000.

A company can make a profit only if the revenue on a product exceeds the cost of manufacturing it. The number of units at which revenue equals cost (that is, profit is 0) is the **break-even point**.

Example 7 **Business** Find the break-even point for the company in Example 6.

Solution The company will break even when revenue equals cost—that is, when

$$R(x) = C(x)$$
$$550x = 250x + 213{,}000$$
$$300x = 213{,}000$$
$$x = 710.$$

The company breaks even by selling 710 TVs. The graphs of the revenue and cost functions and the break-even point (where $x = 710$) are shown in the Figure 3.18. The company must sell more than 710 TVs ($x > 710$) in order to make a profit. 6

Checkpoint 6

For a certain newsletter, the cost equation is $C(x) = 0.90x + 1500$, where x is the number of newsletters sold. The newsletter sells for \$1.25 per copy. Find the break-even point.

Answer: See page 167.

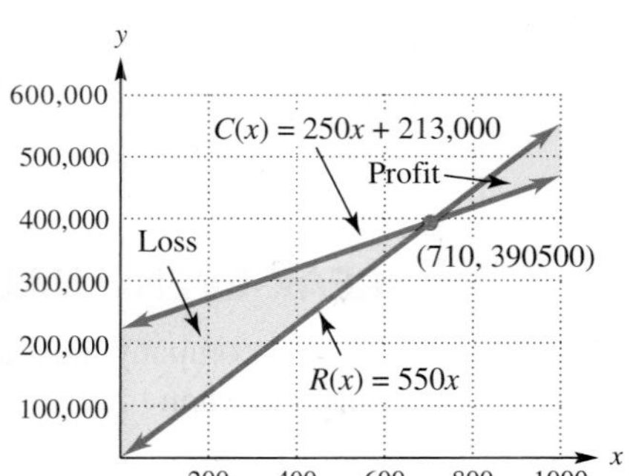

FIGURE 3.18

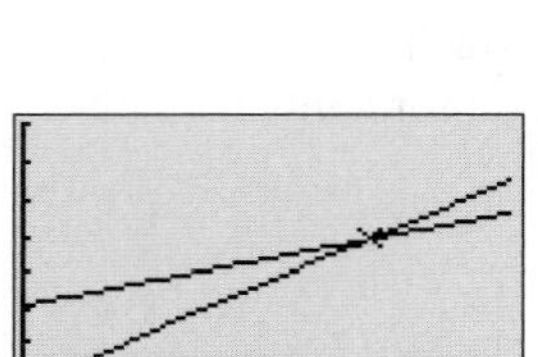

FIGURE 3.19

TECHNOLOGY TIP The break-even point in Example 7 can be found on a graphing calculator by graphing the cost and revenue functions on the same screen and using the calculator's intersection finder, as shown in Figure 3.19. Depending on the calculator, the intersection finder is in the CALC or G-SOLVE menu or in the MATH or FCN submenu of the GRAPH menu.

SUPPLY AND DEMAND

The supply of and demand for an item are usually related to its price. Producers will supply large numbers of the item at a high price, but consumer demand will be low. As the price of the item decreases, consumer demand increases, but producers are less willing to supply large numbers of the item. The curves showing the quantity that will be supplied at a given price and the quantity that will be demanded at a given price are called **supply and demand curves**, respectively. In supply-and-demand problems, we use p for price and q for quantity. We will discuss the economic concepts of supply and demand in more detail in later chapters.

Example 8 Bill Fullington, an economist, has studied the supply and demand for aluminum siding and has determined that the price per unit,* p, and the quantity demanded, q, are related by the linear equation

$$p = 60 - \frac{3}{4}q.$$

*An appropriate unit here might be, for example, one thousand square feet of siding.

(a) Find the demand at a price of $40 per unit.

Solution Let $p = 40$. Then we have

$$p = 60 - \frac{3}{4}q$$

$$40 = 60 - \frac{3}{4}q \quad \text{Let } p = 40.$$

$$-20 = -\frac{3}{4}q \quad \text{Add } -60 \text{ on both sides.}$$

$$\frac{80}{3} = q. \quad \text{Multiply both sides by } -\frac{4}{3}.$$

At a price of $40 per unit, 80/3 (or $26\frac{2}{3}$) units will be demanded.

(b) Find the price if the demand is 32 units.

Solution Let $q = 32$. Then we have

$$p = 60 - \frac{3}{4}q$$

$$p = 60 - \frac{3}{4}(32) \quad \text{Let } q = 32.$$

$$p = 60 - 24$$

$$p = 36.$$

With a demand of 32 units, the price is $36.

(c) Graph $p = 60 - \frac{3}{4}q$.

Solution It is customary to use the horizontal axis for the quantity q and the vertical axis for the price p. In part (a), we saw that 80/3 units would be demanded at a price of $40 per unit; this gives the ordered pair (80/3, 40). Part (b) shows that with a demand of 32 units, the price is $36, which gives the ordered pair (32, 36). Using the points (80/3, 40) and (32, 36) yields the demand graph depicted in Figure 3.20. Only the portion of the graph in Quadrant I is shown, because supply and demand are meaningful only for positive values of p and q. 7✓

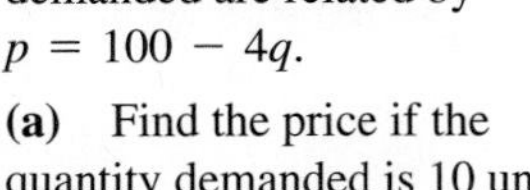

Suppose price and quantity demanded are related by $p = 100 - 4q$.

(a) Find the price if the quantity demanded is 10 units.

(b) Find the quantity demanded if the price is $80.

(c) Write the corresponding ordered pairs.

Answers: See page 167.

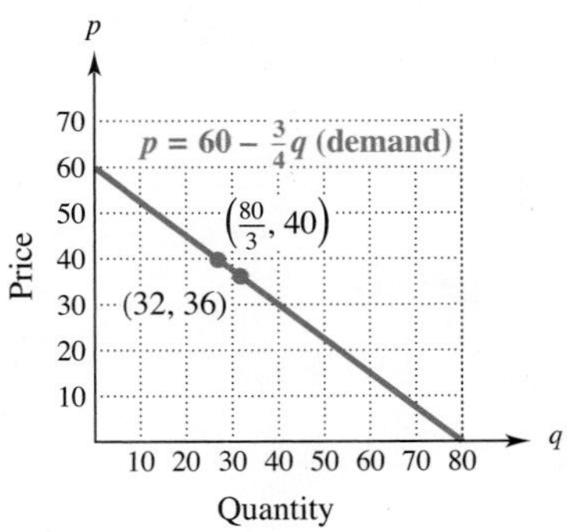

FIGURE 3.20

(d) From Figure 3.20, at a price of $30, what quantity is demanded?

Solution Price is located on the vertical axis. Look for 30 on the p-axis, and read across to where the line $p = 30$ crosses the demand graph. As the graph shows, this occurs where the quantity demanded is 40.

(e) At what price will 60 units be demanded?

Solution Quantity is located on the horizontal axis. Find 60 on the q-axis, and read up to where the vertical line $q = 60$ crosses the demand graph. This occurs where the price is about \$15 per unit.

(f) What quantity is demanded at a price of \$60 per unit?

Solution The point (0, 60) on the demand graph shows that the demand is 0 at a price of \$60 (that is, there is no demand at such a high price).

Example 9 Suppose the economist in Example 8 concludes that the supply q of siding is related to its price p by the equation

$$p = .85q.$$

(a) Find the supply if the price is \$51 per unit.

Solution

$$51 = .85q \quad \text{Let } p = 51.$$
$$60 = q. \quad \text{Divide both sides by .85}$$

If the price is \$51 per unit, then 60 units will be supplied to the marketplace.

(b) Find the price per unit if the supply is 20 units.

Solution $p = .85(20) = 17.$ Let $q = 20.$

If the supply is 20 units, then the price is \$17 per unit.

(c) Graph the supply equation $p = .85q$.

Solution As with demand, each point on the graph has quantity q as its first coordinate and the corresponding price p as its second coordinate. Part (a) shows that the ordered pair (60, 51) is on the graph of the supply equation, and part (b) shows that (20, 17) is on the graph. Using these points, we obtain the supply graph in Figure 3.21.

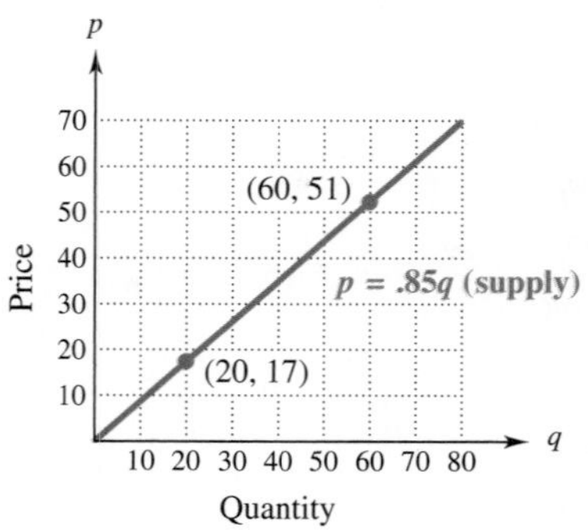

FIGURE 3.21

(d) Use the graph in Figure 3.21 to find the approximate price at which 35 units will be supplied. Then use algebra to find the exact price.

Solution The point on the graph with first coordinate $q = 35$ is approximately (35, 30). Therefore, 35 units will be supplied when the price is approximately \$30. To determine the exact price algebraically, substitute $q = 35$ into the supply equation:

$$p = .85q = .85(35) = \$29.75.$$

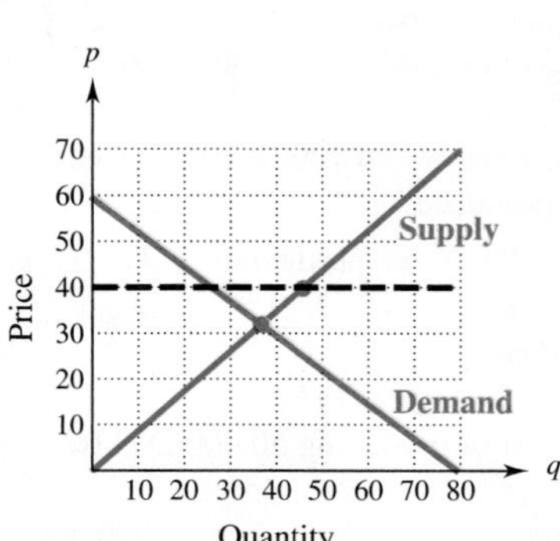

FIGURE 3.22

Example 10 The supply and demand curves of Examples 8 and 9 are shown in Figure 3.22. Determine graphically whether there is a surplus or a shortage of supply at a price of \$40 per unit.

Solution Find 40 on the vertical axis in Figure 3.22 and read across to the point where the horizontal line $p = 40$ crosses the supply graph (that is, the point corresponding to a price of \$40). This point lies above the demand graph, so supply is greater than demand at a price of \$40, and there is a surplus of supply.

Supply and demand are equal at the point where the supply curve intersects the demand curve. This is the **equilibrium point**. Its second coordinate is the **equilibrium price**, the price at which the same quantity will be supplied as is demanded. Its first coordinate is the quantity that will be demanded and supplied at the equilibrium price; this number is called the **equilibrium quantity**.

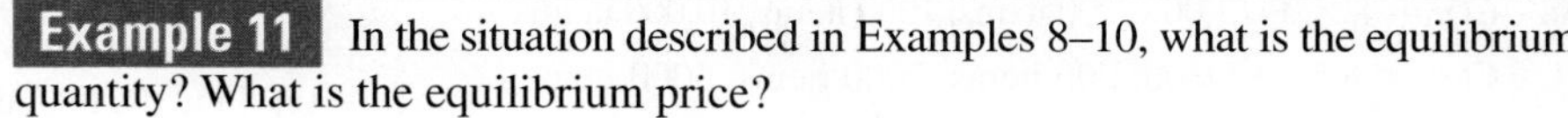

Example 11 In the situation described in Examples 8–10, what is the equilibrium quantity? What is the equilibrium price?

Solution The equilibrium point is where the supply and demand curves in Figure 3.22 intersect. To find the quantity q at which the price given by the demand equation $p = 60 - .75q$ (Example 8) is the same as that given by the supply equation $p = .85q$ (Example 9), set these two expressions for p equal to each other and solve the resulting equation:

$$\begin{aligned} 60 - .75q &= .85q \\ 60 &= 1.6q \\ 37.5 &= q. \end{aligned}$$

Therefore, the equilibrium quantity is 37.5 units, the number of units for which supply will equal demand. Substituting $q = 37.5$ into either the demand or supply equation shows that

$$p = 60 - .75(37.5) = 31.875 \quad \text{or} \quad p = .85(37.5) = 31.875.$$

So the equilibrium price is \$31.875 (or \$31.88, rounded). (To avoid error, it is a good idea to substitute into both equations, as we did here, to be sure that the same value of p results; if it does not, a mistake has been made.) In this case, the equilibrium point—the point whose coordinates are the equilibrium quantity and price—is (37.5, 31.875). 8✓

✓Checkpoint 8

The demand for a certain commodity is related to the price by $p = 80 - (2/3)q$. The supply is related to the price by $p = (4/3)q$. Find

(a) the equilibrium quantity;

(b) the equilibrium price.

Answers: See page 167.

TECHNOLOGY TIP The equilibrium point (37.5, 31.875) can be found on a graphing calculator by graphing the supply and demand curves on the same screen and using the calculator's intersection finder to locate their point of intersection.

3.3 Exercises

Business *Write a cost function for each of the given scenarios. Identify all variables used. (See Example 1.)*

1. A chain-saw rental firm charges $25, plus $5 per hour.
2. A trailer-hauling service charges $95, plus $8 per mile.
3. A parking garage charges $8.00, plus $2.50 per half hour.
4. For a 1-day rental, a car rental firm charges $65, plus 45¢ per mile.

Business *Assume that each of the given situations can be expressed as a linear cost function. Find the appropriate cost function in each case. (See Examples 1 and 4.)*

5. Fixed cost, $200; 50 items cost $2000 to produce.
6. Fixed cost, $2000; 40 items cost $5000 to produce.
7. Marginal cost, $120; 100 items cost $15,800 to produce.
8. Marginal cost, $90; 150 items cost $16,000 to produce.

Business *In Exercises 9–12, a cost function is given. Find the average cost per item when the required numbers of items are produced. (See Example 2.)*

9. $C(x) = 12x + 1800$; 50 items, 500 items, 1000 items
10. $C(x) = 80x + 12{,}000$; 100 items, 1000 items, 10,000 items
11. $C(x) = 6.5x + 9800$; 200 items, 2000 items, 5000 items
12. $C(x) = 8.75x + 16{,}500$; 1000 items, 10,000 items, 75,000 items

Business *Work these exercises. (See Example 3.)*

13. A Volkswagen Beetle convertible sedan is worth $16,615 now and is expected to be worth $8950 in 4 years.
    (a) Find a linear depreciation function for this car.
    (b) Estimate the value of the car 5 years from now.
    (c) At what rate is the car depreciating?

14. A computer that cost $1250 new is expected to depreciate linearly at a rate of $250 per year.
    (a) Find the depreciation function f.
    (b) Explain why the domain of f is $[0, 5]$.

15. A machine is now worth $120,000 and will be depreciated linearly over an 8-year period, at which time it will be worth $25,000 as scrap.
    (a) Find the rule of the depreciation function f.
    (b) What is the domain of f?
    (c) What will the machine be worth in 6 years?

16. A house increases in value in an approximately linear fashion from $222,000 to $300,000 in 6 years.
    (a) Find the *appreciation function* that gives the value of the house in year x.
    (b) If the house continues to appreciate at this rate, what will it be worth 12 years from now?

Business *Work these problems. (See Example 4.)*

17. The total cost (in dollars) of producing x college algebra books is $C(x) = 42.5x + 80{,}000$.
    (a) What are the fixed costs?
    (b) What is the marginal cost per book?
    (c) What is the total cost of producing 1000 books? 32,000 books?
    (d) What is the average cost when 1000 are produced? when 32,000 are produced?

18. The total cost (in dollars) of producing x DVDs is $C(x) = 6.80x + 450{,}000$.
    (a) What are the fixed costs?
    (b) What is the marginal cost per DVD?
    (c) What is the total cost of producing 50,000 DVDs? 600,000 DVDs?
    (d) What is the average cost per DVD when 50,000 are produced? when 500,000 are produced?

19. The manager of a restaurant found that the cost of producing 100 cups of coffee is $11.02, while the cost of producing 400 cups is $40.12. Assume that the cost $C(x)$ is a linear function of x, the number of cups produced.
    (a) Find a formula for $C(x)$.
    (b) Find the total cost of producing 1000 cups.
    (c) Find the total cost of producing 1001 cups.
    (d) Find the marginal cost of producing the 1000th cup.
    (e) What is the marginal cost of producing *any* cup?

20. In deciding whether to set up a new manufacturing plant, company analysts have determined that a linear function is a reasonable estimation for the total cost $C(x)$ in dollars of producing x items. They estimate the cost of producing 10,000 items as $547,500 and the cost of producing 50,000 items as $737,500.
    (a) Find a formula for $C(x)$.
    (b) Find the total cost of producing 100,000 items.
    (c) Find the marginal cost of the items to be produced in this plant.

Business *Work these problems. (See Example 5.)*

21. In the Irvine Ranch Water District in California, a customer with a $\frac{5}{8}''$ water meter pays $7.50 per month, plus $1.43 per 1000 gallons of water used.* If the District has 70,000 such customers, find its monthly revenue function $R(x)$, where the total number of gallons x is measured in thousands.

22. The Lacledge Gas Company in St. Louis charges a residential customer $15.93 per month, plus $1.75851 per therm for the first 30 therms of gas used and $1.11308 for each therm above 30.*

*Rates in October 2008.

(a) How much revenue does the company get from a customer who uses exactly 30 therms of gas in a month?

(b) Find the rule of the function $R(x)$ that gives the company's monthly revenue from one customer, where x is the number of therms of gas used. (*Hint*: $R(x)$ is a piecewise-defined function that has a two-part rule, one part for $x \leq 30$ and the other for $x > 30$.)

Business *Assume that each row of the accompanying table has a linear cost function. Find (a) the cost function; (b) the revenue function; (c) the profit function; (d) the profit on 100 items. (See Example 6.)*

	Fixed Cost	Marginal Cost per Item	Item Sells For
23.	\$750	\$10	\$35
24.	\$150	\$11	\$20
25.	\$300	\$18	\$28
26.	\$17,000	\$30	\$80
27.	\$20,000	\$12.50	\$30

28. **Business** In the following profit–volume chart, EF and GH represent the profit–volume graphs of a single-product company for, 2009 and 2010, respectively*:

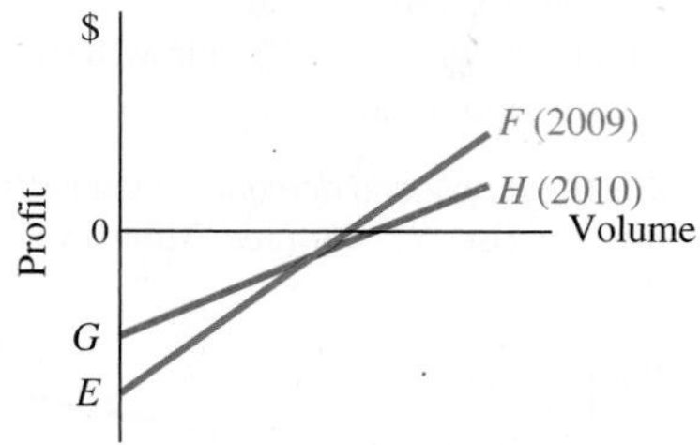

If the 2009 and 2010 unit sales prices are identical, how did the total fixed costs and unit variable costs of 2010 change compared with their values in 2009? Choose one:

	2009 Total Fixed Costs	2010 Unit Variable Costs
(a)	Decreased	Increased
(b)	Decreased	Decreased
(c)	Increased	Increased
(d)	Increased	Decreased

Use algebra to find the intersection points of the graphs of the given equations. (See Examples 7 and 11.)

29. $2x - y = 7$ and $y = 8 - 3x$

30. $6x - y = 2$ and $y = 4x + 7$

*Adapted from Uniform CPA Examination, American Institute of Certified Public Accountants.

31. $y = 3x - 7$ and $y = 7x + 4$

32. $y = 3x + 5$ and $y = 12 - 2x$

Business *Work the following problems. (See Example 7.)*

33. An insurance company claims that for x thousand policies, its monthly revenue in dollars is given by $R(x) = 125x$ and its monthly cost in dollars is given by $C(x) = 100x + 5000$.
(a) Find the break-even point.
(b) Graph the revenue and cost equations on the same axes.
(c) From the graph, estimate the revenue and cost when $x = 100$ (100,000 policies).

34. The owners of a parking lot have determined that their weekly revenue and cost in dollars are given by $R(x) = 80x$ and $C(x) = 50x + 2400$, where x is the number of long-term parkers.
(a) Find the break-even point.
(b) Graph $R(x)$ and $C(x)$ on the same axes.
(c) From the graph, estimate the revenue and cost when there are 60 long-term parkers.

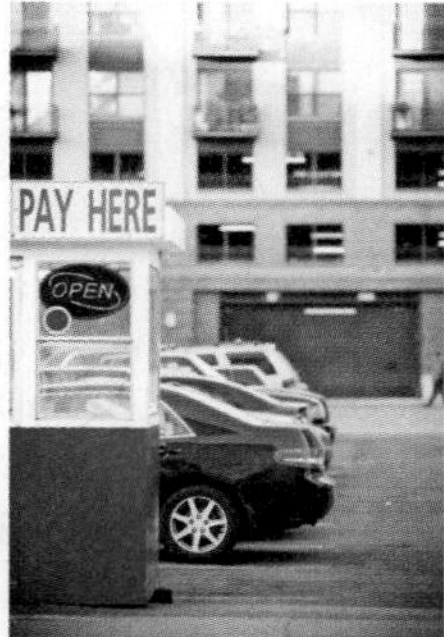

35. The revenue (in millions of dollars) from the sale of x units at a home supply outlet is given by $R(x) = .21x$. The profit (in millions of dollars) from the sale of x units is given by $P(x) = .084x - 1.5$.
(a) Find the cost equation.
(b) What is the cost of producing 7 units?
(c) What is the break-even point?

36. The profit (in millions of dollars) from the sale of x million units of Blue Glue is given by $P(x) = .7x - 25.5$. The cost is given by $C(x) = .9x + 25.5$.
(a) Find the revenue equation.
(b) What is the revenue from selling 10 million units?
(c) What is the break-even point?

Business *Suppose you are the manager of a firm. The accounting department has provided cost estimates, and the sales department sales estimates, on a new product. You must analyze the data they give you, determine what it will take to break even, and decide whether to go ahead with production of the new product. (See Example 7.)*

37. Cost is estimated by $C(x) = 80x + 7000$ and revenue is estimated by $R(x) = 95x$; no more than 400 units can be sold.

38. Cost is $C(x) = 140x + 3000$ and revenue is $R(x) = 125x$.

39. Cost is $C(x) = 125x + 42{,}000$ and revenue is $R(x) = 165.5x$; no more than 2000 units can be sold.

40. Cost is $C(x) = 1750x + 95{,}000$ and revenue is $R(x) = 1975x$; no more than 600 units can be sold.

41. **Business** The accompanying graph shows the hourly compensation (wages and benefits), in U.S. dollars, of production workers in Japan and the United States since 1999.* Estimate the break-even point (the point at which workers in both countries had the same compensation).

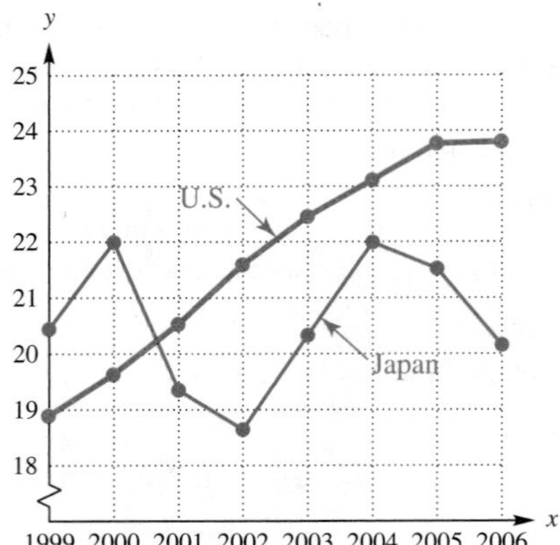

42. **Business** The accompanying graph shows U.S. exports to Iceland and U.S. imports from Iceland (in millions of dollars).† Estimate the year in which the break-even point (the point at which the values of exports and imports were the same) occurred.

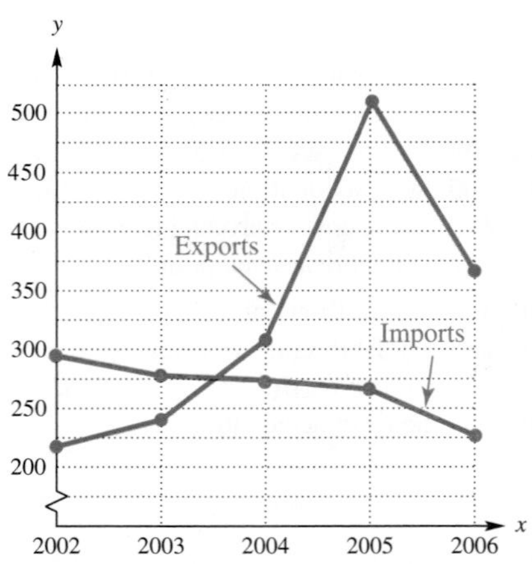

43. **Business** The amount of crude oil produced in the United States (domestic supply) and the amount of imported crude oil, in millions of barrels per day, in 1990 and 2007 are shown in the following table‡:

Year	1990	2007
Domestic	7.4	5.1
Imported	5.9	10.0

(a) Over the period from 1990 to 2007, domestic crude oil produced and crude oil imported each changed in an approximately linear fashion. If $x = 0$ corresponds to 1990, find two linear functions (one for domestic production, one for imports) that give the amount of oil for each year x.

(b) Graph the functions in part (a) on the same coordinate axes.

(c) Find the intersection point on the graphs and interpret your answer.

44. **Social Science** The population (in hundreds of thousands) of Florida from 2000 to 2007 can be approximated by the function $f(x) = 3.14x + 160.5$. Similarly, the population (in hundreds of thousands) of New York during the same period can be approximated by the function $g(x) = .43x + 190$.*

(a) Graph both functions on the same coordinate axes, with $x = 0$ corresponding to 2000 and $x = 7$ to 2007.

(b) Do the graphs intersect in this window?

(c) If trends continue at the same rate, will Florida overtake New York in population?

(d) Estimate in what year Florida will overtake New York in population

Business *Use the supply and demand curves in the accompanying graph to answer Exercises 45–48. (See Examples 8–11.)*

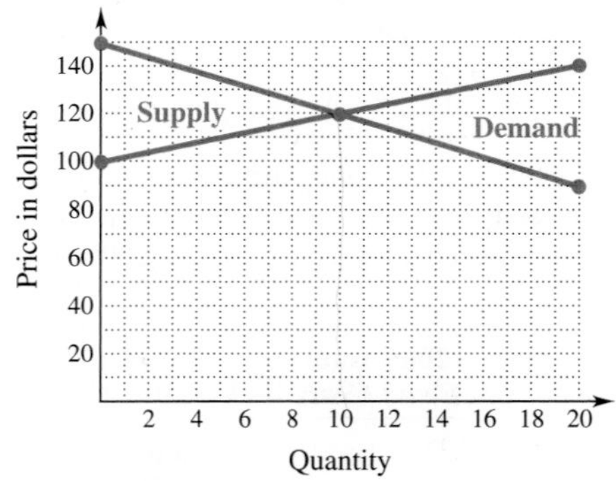

45. At what price are 20 items supplied?

46. At what price are 20 items demanded?

47. Find the equilibrium quantity.

48. Find the equilibrium price.

Business *Work the following exercises. (See Examples 8–11.)*

49. Suppose that the demand and price for a certain brand of shampoo are related by

$$p = 16 - \frac{5}{4}q,$$

*U.S. Bureau of Labor Statistics.

†U.S. Census Bureau.

‡U.S. Department of Energy.

*U.S. Census Bureau.

where p is price in dollars and q is demand. Find the price for a demand of
(a) 0 units; **(b)** 4 units; **(c)** 8 units.

Find the demand for the shampoo at a price of
(d) \$6; **(e)** \$11; **(f)** \$16.
(g) Graph $p = 16 - (5/4)q$.

Suppose the price and supply of the shampoo are related by

$$p = \frac{3}{4}q,$$

where q represents the supply and p the price. Find the supply when the price is
(h) \$0; **(i)** \$10; **(j)** \$20.
(k) Graph $p = (3/4)q$ on the same axes used for part (g).
(l) Find the equilibrium quantity.
(m) Find the equilibrium price.

50. Let the supply and demand for radial tires in dollars be given by

$$\text{supply: } p = \frac{3}{2}q; \quad \text{demand: } p = 81 - \frac{3}{4}q.$$

(a) Graph these equations on the same axes.
(b) Find the equilibrium quantity.
(c) Find the equilibrium price.

51. Let the supply and demand for bananas in cents per pound be given by

$$\text{supply: } p = \frac{2}{5}q; \quad \text{demand: } p = 100 - \frac{2}{5}q.$$

(a) Graph these equations on the same axes.
(b) Find the equilibrium quantity.
(c) Find the equilibrium price.
(d) On what interval does demand exceed supply?

52. Let the supply and demand for sugar be given by

$$\text{supply: } p = 1.4q - .6$$

and

$$\text{demand: } p = -2q + 3.2,$$

where p is in dollars.
(a) Graph these on the same axes.
(b) Find the equilibrium quantity.
(c) Find the equilibrium price.
(d) On what interval does supply exceed demand?

53. Explain why the graph of the (total) cost function is always above the x-axis and can never move downward as you go from left to right. Is the same thing true of the graph of the average cost function?

54. Explain why the graph of the profit function can rise or fall (as you go from left to right) and can be below the x-axis.

✓ Checkpoint Answers

1. $C(x) = 4x + 60$. **2.** \$4.60 **3.** \$2248
4. **(a)** \$800 **(b)** \$842 **(c)** \$3.50 **(d)** \$3.50
5. **(a)** 1,120,000,000 **(b)** \$48,041,000
6. 4286 newsletters
7. **(a)** \$60 **(b)** 5 units **(c)** (10, 60); (5, 80)
8. **(a)** 40 **(b)** $160/3 \approx \$53.33$

3.4 ▸ Quadratic Functions

A **quadratic function** is a function whose rule is given by a quadratic polynomial, such as

$$f(x) = x^2, \qquad g(x) = 3x^2 + 30x + 67, \qquad \text{and} \qquad h(x) = -x^2 + 4x.$$

Thus, a quadratic function is a function whose rule can be written in the form

$$\boldsymbol{f(x) = ax^2 + bx + c}$$

for some constants a, b, and c, with $a \neq 0$.

Example 1 Graph each of these quadratic functions:

$$f(x) = x^2; \qquad g(x) = 4x^2; \qquad h(x) = -.2x^2.$$

Solution In each case, choose several numbers (negative, positive, and 0) for x, find the values of the function at these numbers, and plot the corresponding points. Then connect the points with a smooth curve to obtain Figure 3.23 on the next page.

$f(x) = x^2$					
x	−2	−1	0	1	2
x^2	4	1	0	1	4

$g(x) = 4x^2$					
x	−2	−1	0	1	2
$4x^2$	16	4	0	4	16

$h(x) = -.2x^2$						
x	−5	−3	−1	0	2	4
$-.2x^2$	−5	−1.8	−.2	0	−.8	−3.2

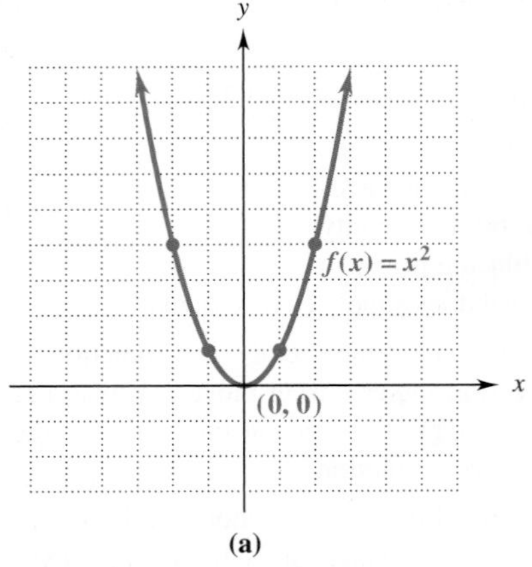

(a)

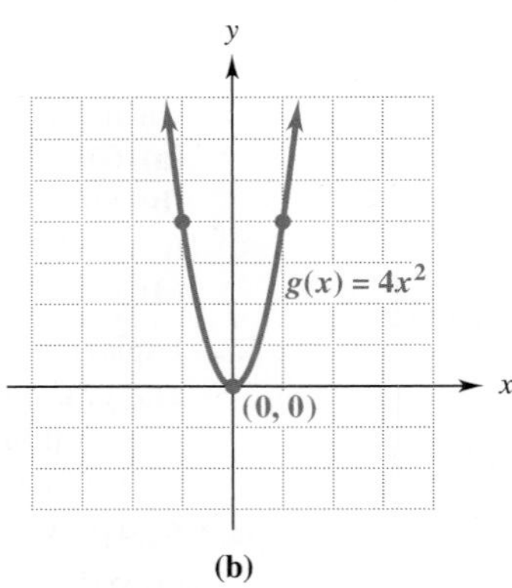

(b)

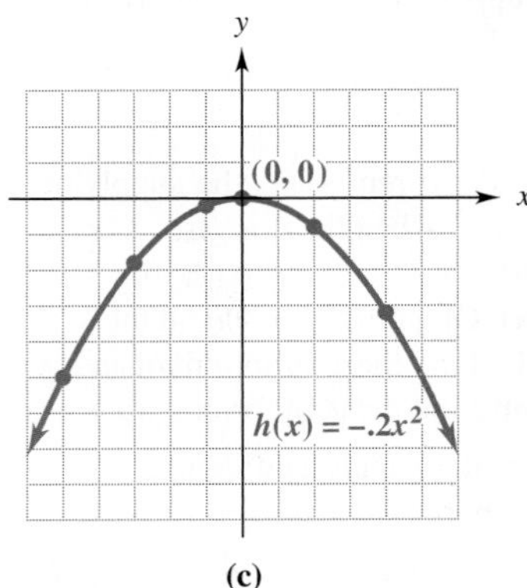

(c)

FIGURE 3.23

Each of the curves in Figure 3.23 is a **parabola**. It can be shown that the graph of every quadratic function is a parabola. Parabolas have many useful properties. Cross sections of radar dishes and spotlights form parabolas. Disks often visible on the sidelines of televised football games are microphones having reflectors with parabolic cross sections. These microphones are used by the television networks to pick up the signals shouted by the quarterbacks.

All parabolas have the same basic "cup" shape, although the cup may be broad or narrow and open upward or downward. The general shape of a parabola is determined by the coefficient of x^2 in its rule, as summarized here and illustrated in Example 1. 1✓

✓Checkpoint 1

Graph each quadratic function.

(a) $f(x) = x^2 - 4$

(b) $g(x) = -x^2 + 4$

Answers: See page 175.

The graph of a quadratic function $f(x) = ax^2 + bx + c$ is a parabola.

If $a > 0$, the parabola opens upward. [*Figure 3.23(a) and 3.23(b)*]

If $a < 0$, the parabola opens downward. [*Figure 3.23(c)*]

If $|a| < 1$, the parabola appears wider than the graph of $y = x^2$. [*Figure 3.23(c)*]

If $|a| > 1$, the parabola appears narrower than the graph of $y = x^2$. [*Figure 3.23(b)*]

When a parabola opens upward [as in Figure 3.23(a), (b)], its lowest point is called the **vertex**. When a parabola opens downward [as in Figure 3.23(c)], its highest point is called the **vertex**. The vertical line through the vertex of a parabola is called the **axis of the parabola**. For example, (0, 0) is the vertex of each of the parabolas in Figure 3.23, and the axis of each parabola is the y-axis. If you were to fold the graph of a parabola along its axis, the two halves of the parabola would match exactly. This means that a parabola is *symmetric* about its axis.

Although the vertex of a parabola can be approximated by a graphing calculator's maximum or minimum finder, its exact coordinates can be found algebraically, as in the following examples.

Example 2 Consider the function $g(x) = 2(x - 3)^2 + 1$.

(a) Show that g is a quadratic function.

Solution Multiply out the rule of g to show that it has the required form:

$$\begin{aligned} g(x) &= 2(x - 3)^2 + 1 \\ &= 2(x^2 - 6x + 9) + 1 \\ &= 2x^2 - 12x + 18 + 1 \\ g(x) &= 2x^2 - 12x + 19. \end{aligned}$$

According to the preceding box, the graph of g is somewhat narrow, upward-opening parabola.

(b) Show that the vertex of the graph of $g(x) = 2(x - 3)^2 + 1$ is $(3, 1)$.

Solution Since $g(3) = 2(3 - 3)^2 + 1 = 0 + 1 = 1$, the point $(3, 1)$ is on the graph. The vertex of an upward-opening parabola is the lowest point on the graph, so we must show that $(3, 1)$ is the lowest point. Let x be any number except 3 (so that $x - 3 \neq 0$. Then the quantity $2(x - 3)^2$ is positive, and hence

$$g(x) = 2(x - 3)^2 + 1 = (\text{a positive number}) + 1,$$

which means that $g(x) > 1$. Therefore, every point $(x, g(x))$ on the graph, where $x \neq 3$, has second coordinate $g(x)$ greater than 1. Hence $(x, g(x))$ lies *above* $(3, 1)$. In other words, $(3, 1)$ is the lowest point on the graph—the vertex of the parabola.

(c) Graph $g(x) = 2(x - 3)^2 + 1$.

Solution Plot some points on both sides of the vertex $(3, 1)$ to obtain the graph in Figure 3.24. The vertical line $x = 3$ through the vertex is the axis of the parabola.

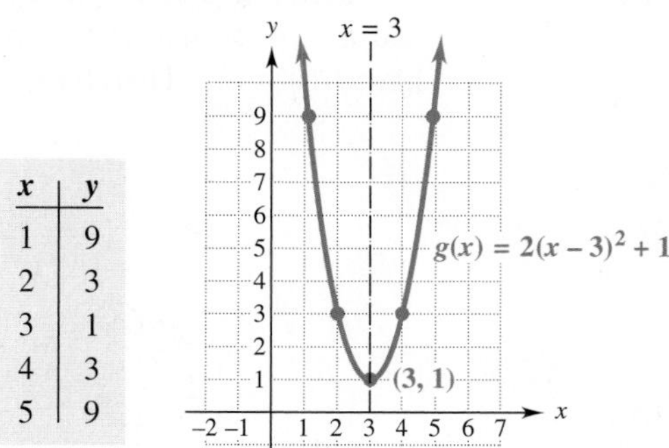

x	y
1	9
2	3
3	1
4	3
5	9

FIGURE 3.24

In Example 2, notice how the rule of the function g is related to the coordinates of the vertex:

$$g(x) = 2(x - 3)^2 + 1. \qquad (3, 1).$$

Arguments similar to those in Example 2 lead to the following fact.

The graph of the quadratic function $f(x) = a(x - h)^2 + k$ is a parabola with vertex (h, k). It opens upward when $a > 0$ and downward when $a < 0$.

Example 3 Determine algebraically whether the given parabola opens upward or downward, and find its vertex.

(a) $f(x) = -3(x - 4)^2 - 7$

Solution The rule of the function is in the form $f(x) = a(x - h)^2 + k$ (with $a = -3$, $h = 4$, and $k = -7$). The parabola opens downward ($a < 0$), and its vertex is $(h, k) = (4, -7)$.

(b) $g(x) = 2(x + 3)^2 + 5$

Solution Be careful here: The vertex is *not* (3, 5). To put the rule of $g(x)$ in the form $a(x - h)^2 + k$, we must rewrite it so that there is a minus sign inside the parentheses:

$$\begin{aligned} g(x) &= 2(x + 3)^2 + 5 \\ &= 2(x - (-3))^2 + 5. \end{aligned}$$

This is the required form, with $a = 2$, $h = -3$, and $k = 5$. The parabola opens upward, and its vertex is $(-3, 5)$. ✓2

✓ Checkpoint 2

Determine the vertex of each parabola, and graph the parabola.

(a) $f(x) = (x + 4)^2 - 3$

(b) $f(x) = -2(x - 3)^2 + 1$

Answers: See page 175.

Example 4 Find the rule of a quadratic function whose graph has vertex (3, 4) and passes through the point (6, 22).

Solution The graph of $f(x) = a(x - h)^2 + k$ has vertex (h, k). We want $h = 3$ and $k = 4$, so that $f(x) = a(x - 3)^2 + 4$. Since (6, 22) is on the graph, we must have $f(6) = 22$. Therefore,

$$\begin{aligned} f(x) &= a(x - 3)^2 + 4 \\ f(6) &= a(6 - 3)^2 + 4 \\ 22 &= a(3)^2 + 4 \\ 9a &= 18 \\ a &= 2. \end{aligned}$$

Thus, the graph of $f(x) = 2(x - 3)^2 + 4$ is a parabola with vertex (3, 4) that passes through (6, 22).

The vertex of each parabola in Examples 2 and 3 was easily determined because the rule of the function had the form

$$f(x) = a(x - h)^2 + k.$$

The rule of *any* quadratic function can be put in this form by using the technique of **completing the square**, which is illustrated in the next example.

Example 5 Determine the vertex of the graph of $f(x) = x^2 - 4x + 2$. Then graph the parabola.

Solution In order to get $f(x)$ in the form $a(x - h)^2 + k$, take half the coefficient of x, namely $\frac{1}{2}(-4) = -2$, and *square it:* $(-2)^2 = 4$. Then proceed as follows.

$$f(x) = x^2 - 4x + 2$$

$$f(x) = x^2 - 4x + 4 - 4 + 2 \quad \text{Add and subtract 4}$$

$$= (x^2 - 4x + 4) - 4 + 2 \quad \text{Insert parentheses.}$$

$$= (x - 2)^2 - 2 \quad \text{Factor expression in parentheses}$$

Adding and subtracting 4 did not change the rule of $f(x)$, but did make it possible to have a perfect square as part of its rule: $f(x) = (x - 2)^2 - 2$. Now we can see that the graph is an upward-opening parabola, as shown in Figure 3.25. 3✓

✓Checkpoint 3

Rewrite the rule of each function by completing the square, and use this form to find the vertex of the graph.

(a) $f(x) = x^2 + 6x + 5$ (*Hint*: add and subtract the square of half the coefficient of x.)

(b) $g(x) = x^2 - 12x + 33$

Answers: See page 175.

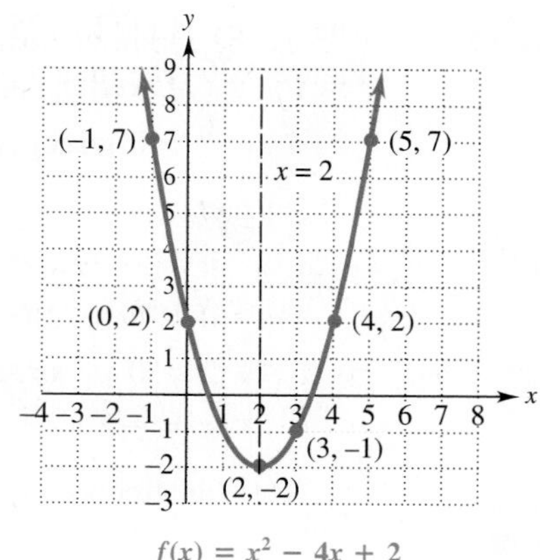

$f(x) = x^2 - 4x + 2$

$f(x) = (x - 2)^2 - 2$

FIGURE 3.25

CAUTION The technique of completing the square only works when the coefficient of x^2 is 1. To find the vertex of a quadratic function such as

$$f(x) = 2x^2 + 12x - 19,$$

you must first factor out the coefficient of x^2 and write the rule as

$$f(x) = 2(x^2 + 6x - \tfrac{19}{2}).$$

Now complete the square on the expression in parentheses by adding and subtracting 9 (the square of half the coefficient of x), and proceed as in Example 5.

The technique of completing the square can be used to rewrite the general equation $f(x) = ax^2 + bx + c$ in the form $f(x) = a(x - h)^2 + k$, as is shown in Exercise 54. When this is done, we obtain a formula for the coordinates of the vertex.

The graph of $f(x) = ax^2 + bx + c$ is a parabola with vertex (h, k), where

$$h = \frac{-b}{2a} \quad \text{and} \quad k = f(h).$$

Example 6 Find the vertex, the axis, and the x- and y-intercepts of the graph of $f(x) = x^2 - x - 6$.

Solution Since $a = 1$ and $b = -1$, the x-value of the vertex is

$$\frac{-b}{2a} = \frac{-(-1)}{2 \cdot 1} = \frac{1}{2}.$$

The y-value of the vertex is

$$f\left(\frac{1}{2}\right) = \left(\frac{1}{2}\right)^2 - \frac{1}{2} - 6 = -\frac{25}{4}.$$

The vertex is $(1/2, -25/4)$ and the axis of the parabola is $x = 1/2$, as shown in Figure 3.26. The intercepts are found by setting each variable equal to 0.

***x*-intercepts**

Set $f(x) = y = 0$:

$$0 = x^2 - x - 6$$

$$0 = (x + 2)(x - 3)$$

$$x + 2 = 0 \quad \text{or} \quad x - 3 = 0$$

$$x = -2 \qquad\qquad x = 3.$$

The x-intercepts are -2 and 3. 4✓

***y*-intercept**

Set $x = 0$:

$$f(x) = y = 0^2 - 0 - 6 = -6.$$

The y-intercept is -6.

✓Checkpoint 4

Find the vertex of the graph of the given equation.

(a) $f(x) = 2x^2 - 5x + 10$

(b) $k(x) = -.3x^2 + 1.8x + 7$

Answers: See page 175.

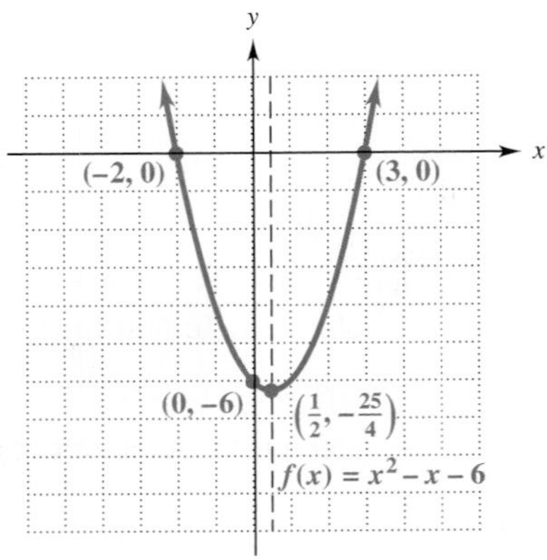

FIGURE 3.26

TECHNOLOGY TIP The maximum or minimum finder on a graphing calculator can approximate the vertex of a parabola with a high degree of accuracy. The max–min finder is in the CALC menu or in the MATH or FCN submenu of the GRAPH menu. Similarly, the calculator's graphical root finder can approximate the *x*-intercepts of a parabola.

3.4 ▶ Exercises

The graph of each of the functions in Exercises 1–6 is a parabola. Without graphing, determine whether the parabola opens upward or downward. (See Example 1.)

1. $f(x) = x^2 - 3x - 12$
2. $g(x) = -x^2 + 5x + 15$
3. $h(x) = -3x^2 + 14x + 1$
4. $f(x) = 6.5x^2 - 7.2x + 4$
5. $g(x) = 2.9x^2 - 12x - 5$
6. $h(x) = -4x^2 + 4.7x - 6$

Without graphing, determine the vertex of the parabola that is the graph of the given function. State whether the parabola opens upward or downward. (See Examples 2 and 3.)

7. $f(x) = -2(x - 5)^2 + 7$
8. $g(x) = -7(x - 8)^2 - 3$
9. $h(x) = 4(x + 1)^2 - 9$
10. $f(x) = -8(x + 12)^2 + 9$
11. $g(x) = 3.2(x - 4.5)^2 + 7.1$
12. $h(x) = 4.28(x + 5.12)^2 - 10$

Match each function with its graph, which is one of those shown at the bottom of this page or the top of page 174. (See Examples 1–3.)

13. $f(x) = x^2 + 2$
14. $g(x) = x^2 - 2$
15. $g(x) = (x - 2)^2$
16. $f(x) = -(x + 2)^2$
17. $f(x) = 2(x - 2)^2 + 2$
18. $g(x) = -2(x - 2)^2 - 2$
19. $g(x) = -2(x + 2)^2 + 2$
20. $f(x) = 2(x + 2)^2 - 2$

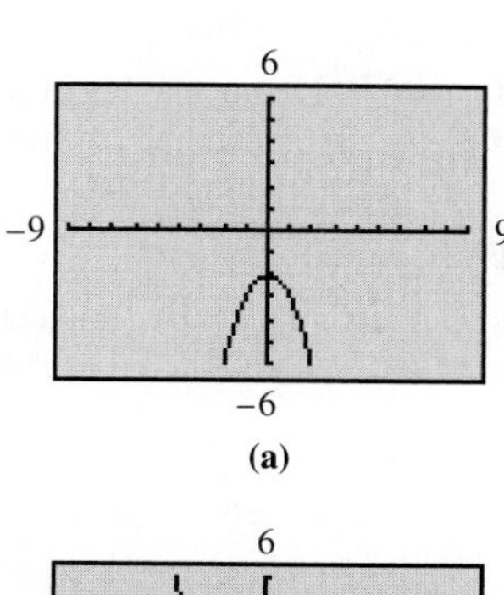

(a)

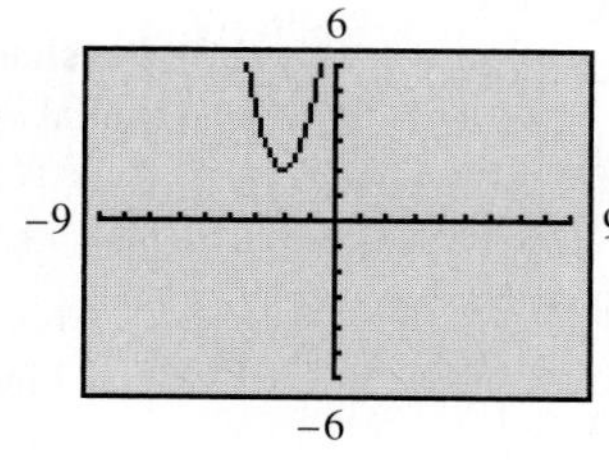

(b)

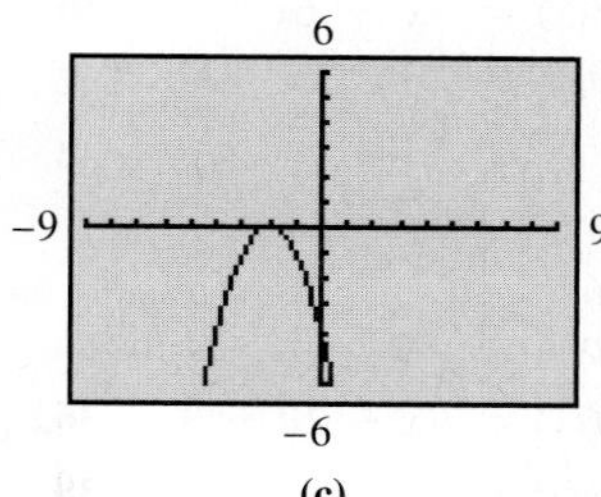

(c)

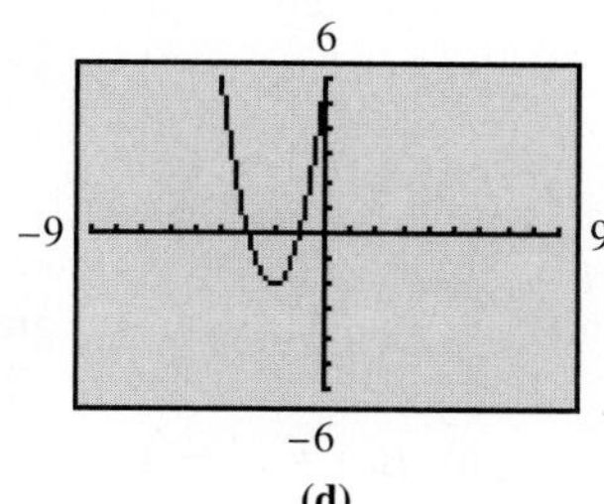

(d)

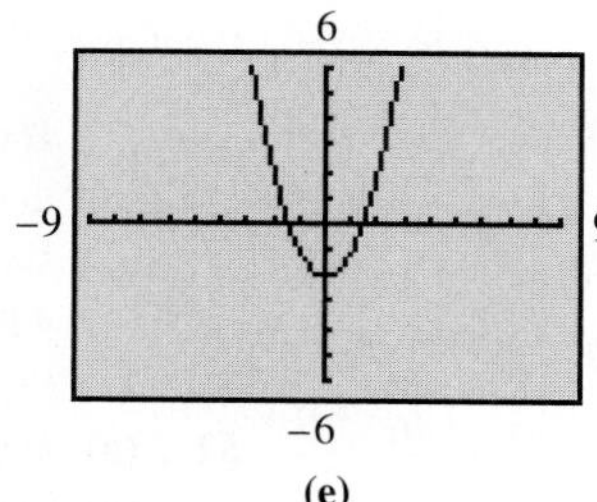

(e)

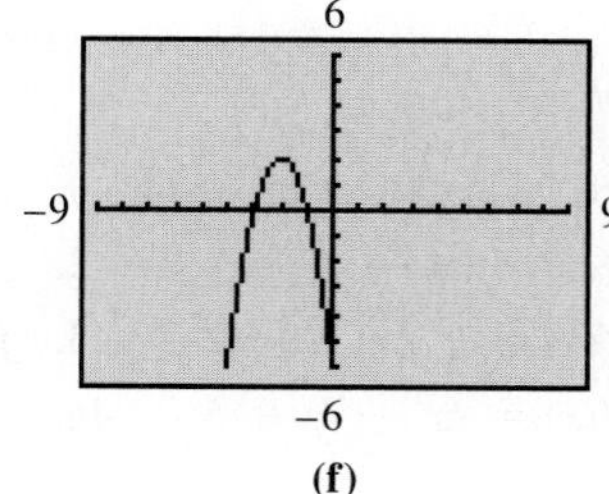

(f)

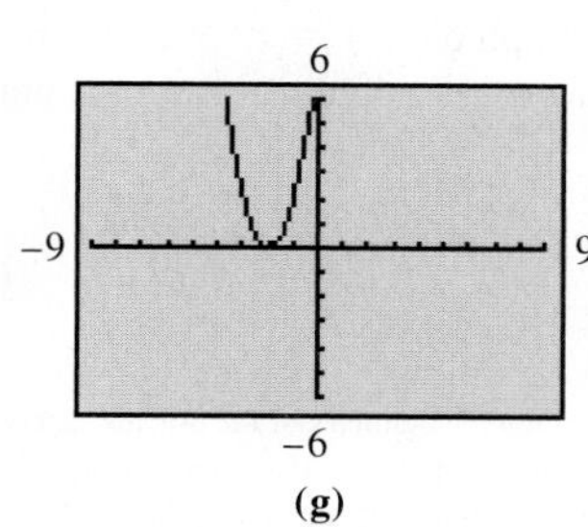

(g)

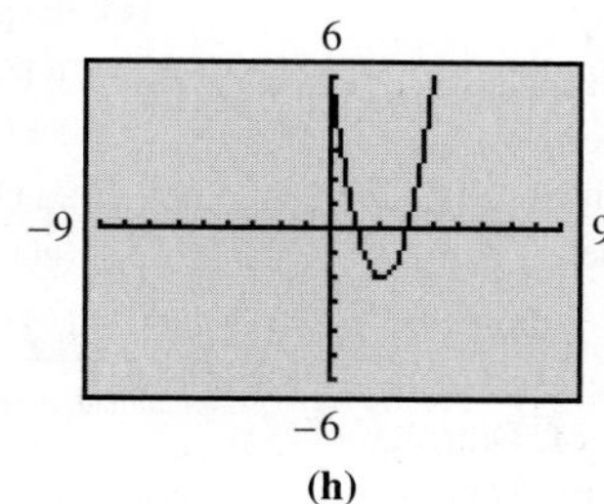

(h)

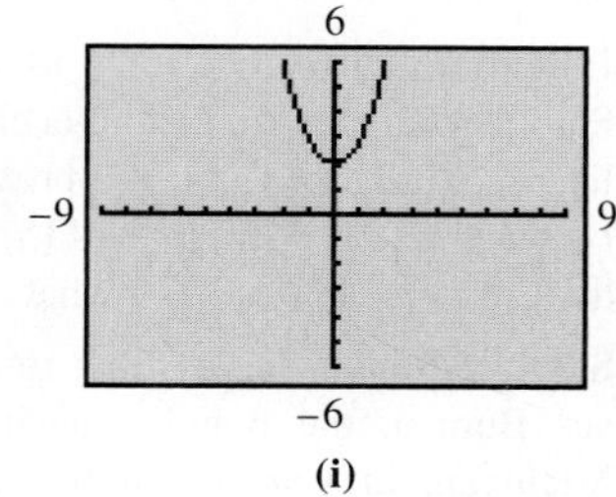

(i)

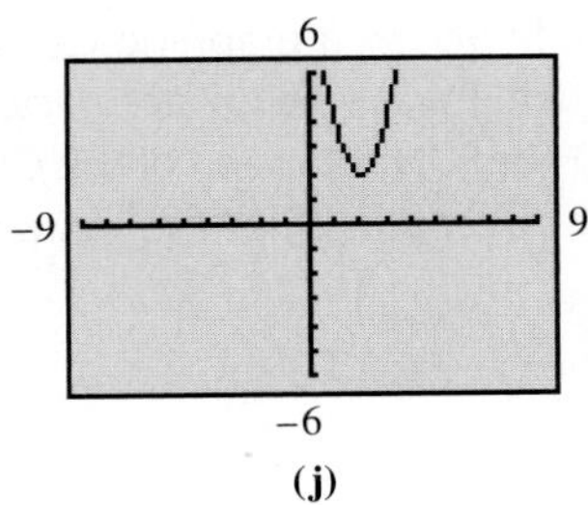

(j)

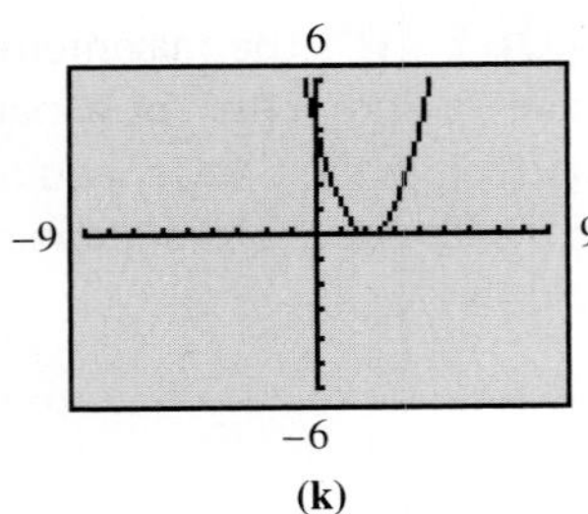

(k)

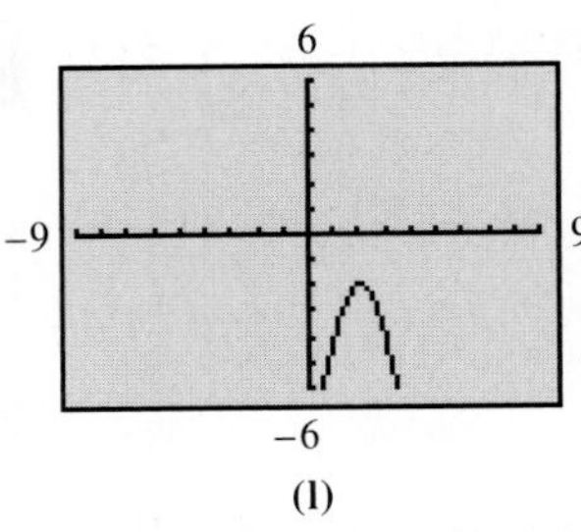

(l)

Find the rule of a quadratic function whose graph has the given vertex and passes through the given point. (See Example 4.)

21. Vertex (1, 2); point (5, 6)

22. Vertex (−3, 2); point (2, 1)

23. Vertex (−1, −2); point (1, 2)

24. Vertex (2, −4); point (5, 2)

25. Vertex (0, 0); point (2, 12)

26. Vertex (0, 0); point (−2, −5)

27. Vertex (4, −2); point (6, 6)

28. Vertex (−2, 0); point (−3, 1)

Without graphing, find the vertex of the parabola that is the graph of the given function. (See Examples 5 and 6.)

29. $f(x) = -x^2 - 6x + 3$ **30.** $g(x) = x^2 + 10x + 9$

31. $f(x) = 3x^2 - 12x + 5$ **32.** $g(x) = -4x^2 - 16x + 9$

33. $f(x) = x^2 - 8x + 25$ **34.** $g(x) = 4x^2 + 8x - 5$

Without graphing, determine the x- and y-intercepts of each of the given parabolas. (See Example 6.)

35. $f(x) = 3(x - 2)^2 - 3$ **36.** $f(x) = x^2 - 4x - 1$

37. $g(x) = 2x^2 + 8x + 6$ **38.** $g(x) = x^2 - 10x + 20$

Graph each parabola and find its vertex and axis of symmetry. (See Examples 1–6.)

39. $f(x) = (x + 2)^2$ **40.** $f(x) = -(x + 5)^2$

41. $f(x) = (x - 1)^2 - 3$ **42.** $f(x) = (x - 2)^2 + 1$

43. $f(x) = x^2 - 4x + 6$ **44.** $f(x) = x^2 + 6x + 3$

45. $f(x) = 2x^2 - 4x + 5$ **46.** $f(x) = -3x^2 + 24x - 46$

Use a calculator to work these problems.

47. **Transportation** According to data from the National Safety Council, the fatal-accident rate per 100,000 licensed drivers can be approximated by the function $f(x) = .0328x^2 - 3.55x + 115$, where x is the age of the driver ($16 \le x \le 88$). At what age is the rate the lowest?

48. **Social Science** According to data from the U.S. Census Bureau, the population (in thousands) of Detroit, Michigan, in year x can be approximated by $g(x) = -.417x^2 + 48.6x + 250$, where $x = 0$ corresponds to 1900. In what year did Detroit have its largest population?

49. **Health** Blood flow to the fetal spleen is of research interest because several diseases are associated with increased resistance in the splenic artery (the artery that goes to the spleen). Researchers have found that the index of splenic artery resistance in the fetus can be described by the function

$$y = .057x - .001x^2,$$

where x is the number of weeks of gestation.*

(a) At how many weeks is splenic artery resistance at a maximum?

(b) What is the maximum splenic artery resistance?

(c) At how many weeks is splenic artery resistance equal to 0, according to the given formula? Is your answer reasonable for this function? Explain.

50. **Physical Science** If an object is thrown upward with an initial velocity of 80 ft/second, then its height after t seconds is given by

$$h = 80t - 16t^2.$$

(a) Find the maximum height attained by the object.

(b) Find the number of seconds it takes the object to hit the ground.

In Exercises 51–53, graph the functions in parts (a)–(d) on the same set of axes; then answer part (e).

51. **(a)** $k(x) = x^2$ **(b)** $f(x) = x^2 + 2$
(c) $g(x) = x^2 + 3$ **(d)** $h(x) = x^2 + 5$
(e) Explain how the graph of $f(x) = x^2 + c$ (where c is a positive constant) can be obtained from the graph of $k(x) = x^2$.

52. **(a)** $k(x) = x^2$ **(b)** $f(x) = x^2 - 1$
(c) $g(x) = x^2 - 2$ **(d)** $h(x) = x^2 - 4$
(e) Explain how the graph of $f(x) = x^2 - c$ (where c is a positive constant) can be obtained from the graph of $k(x) = x^2$.

53. **(a)** $k(x) = x^2$ **(b)** $f(x) = (x + 2)^2$
(c) $g(x) = (x - 2)^2$ **(d)** $h(x) = (x - 4)^2$

*Abuhamad, A. Z. et. al., "Doppler Flow Velocimetry of the Splenic Artery in the Human Fetus: Is It a Marker of Chronic Hypoxia?" *American Journal of Obstetrics and Gynecology*, Vol. 172, No. 3, March 1995, pp. 820–825.

(e) Explain how the graphs of $f(x) = (x + c)^2$ and $f(x) = (x - c)^2$ (where c is a positive constant) can be obtained from the graph of $k(x) = x^2$.

54. Verify that the right side of the equation

$$ax^2 + bx + c = a\left(x - \left(\frac{-b}{2a}\right)\right)^2 + \left(c - \frac{b^2}{4a}\right)$$

equals the left side. Since the right side of the equation is in the form $a(x - h)^2 + k$, we conclude that the vertex of the parabola $f(x) = ax^2 + bx + c$ has x-coordinate $h = -b/2a$.

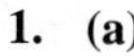

Checkpoint Answers

1. (a)

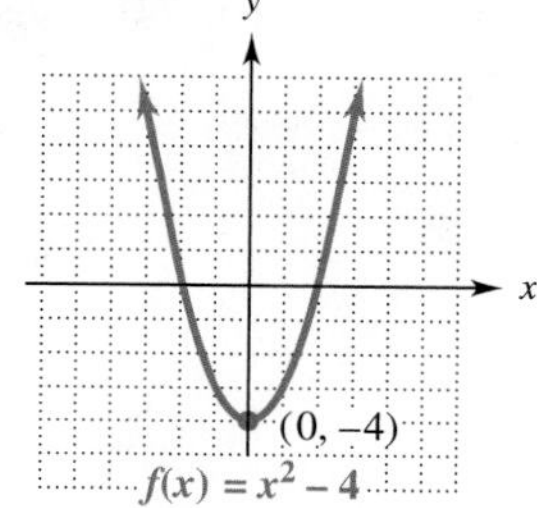

$f(x) = x^2 - 4$

(b)

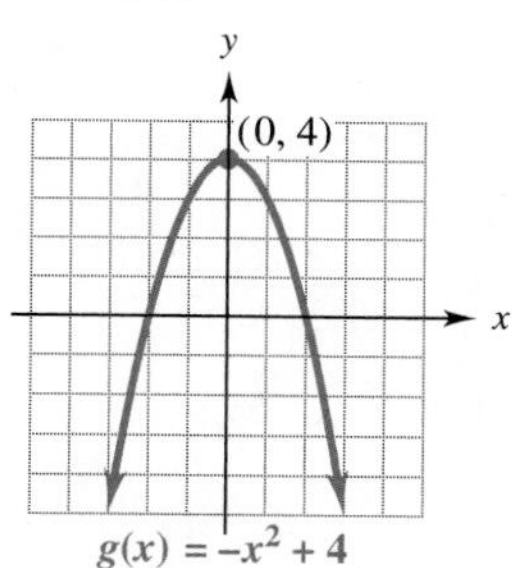

$g(x) = -x^2 + 4$

2. (a)

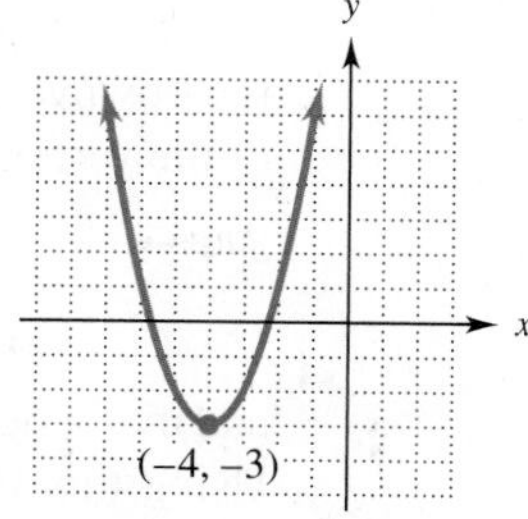

$f(x) = (x + 4)^2 - 3$

(b)

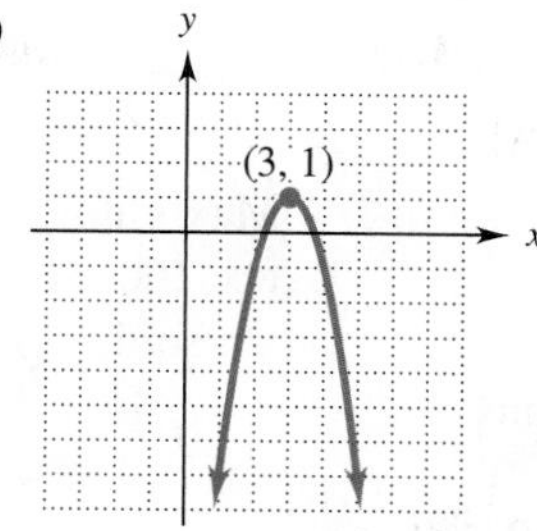

$f(x) = -2(x - 3)^2 + 1$

3. (a) $f(x) = (x + 3)^2 - 4$; $(-3, -4)$
(b) $g(x) = (x - 6)^2 - 3$; $(6, -3)$

4. (a) $\left(\frac{5}{4}, \frac{55}{8}\right)$
(b) $(3, 9.7)$

3.5 ▶ Applications of Quadratic Functions

The fact that the vertex of a parabola $y = ax^2 + bx + c$ is the highest or lowest point on the graph can be used in applications to find a maximum or a minimum value.

Example 1 Anne Kelly owns and operates Aunt Emma's Blueberry Pies. She has hired a consultant to analyze her business operations. The consultant tells her that her profits $P(x)$ from the sale of x cases of pies are given by

$$P(x) = 120x - x^2.$$

How many cases of pies should she sell in order to maximize profit? What is the maximum profit?

Solution The profit function can be rewritten as $P(x) = -x^2 + 120x$. Its graph is a downward-opening parabola. According to the box on page 172, the x-coordinate of the vertex is

$$x = \frac{-b}{2a} = \frac{-120}{2(-1)} = 60$$

and the y-coordinate is $P(60) = -60^2 + 120(60) = 3600$. So the vertex is (60, 3600). Using this fact and plotting some points, we obtain the graph in Figure 3.27. For each point on the graph,

the x-coordinate is the number of cases;

the y-coordinate is the profit on that number of cases.

Maximum profit occurs at the point with the largest y-coordinate, namely, the vertex, (60, 3600). So the maximum profit of \$3600 is obtained when 60 cases of pies are sold. 1 ✓

✓ **Checkpoint 1**

When a company sells x units of a product, its profit is $P(x) = -2x^2 + 40x + 280$. Find

(a) the number of units which should be sold so that maximum profit is received;

(b) the maximum profit.

Answers: *See page 184.*

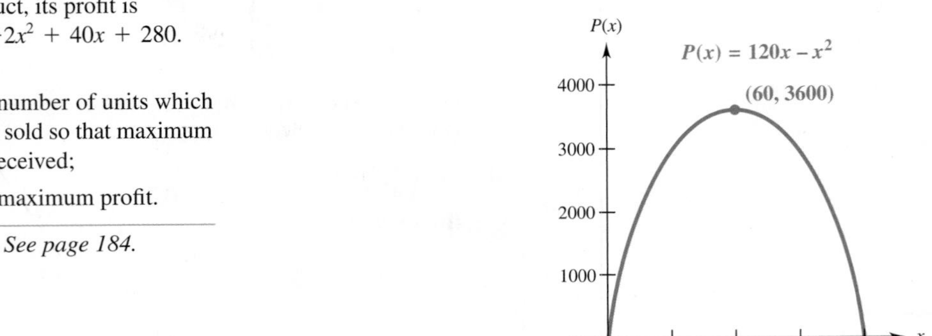

FIGURE 3.27

Supply and demand curves were introduced in Section 3.3. Here is a quadratic example.

Example 2 Suppose that the price of and demand for an item are related by

$$p = 150 - 6q^2, \quad \text{Demand function}$$

where p is the price (in dollars) and q is the number of items demanded (in hundreds). Suppose also that the price and supply are related by

$$p = 10q^2 + 2q, \quad \text{Supply function}$$

where q is the number of items supplied (in hundreds). Find the equilibrium quantity and the equilibrium price.

Solution The graphs of both of these equations are parabolas (Figure 3.28). Only those portions of the graphs which lie in the first quadrant are included, because none of supply, demand, or price can be negative.

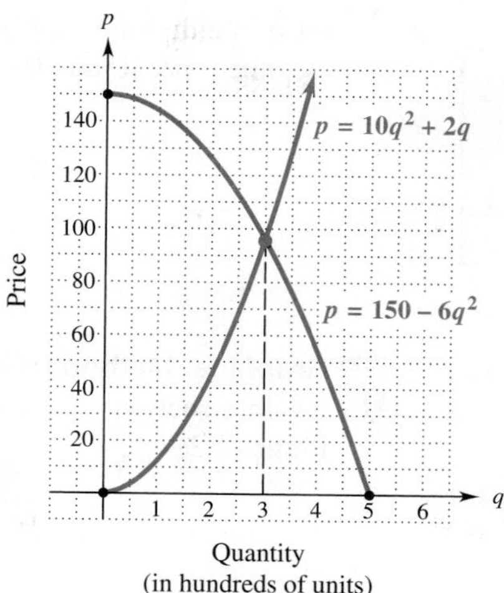

FIGURE 3.28

The point where the demand and supply curves intersect is the equilibrium point. Its first coordinate is the equilibrium quantity, and its second coordinate is the equilibrium price. These coordinates may be found in two ways.

Algebraic Method At the equilibrium point, the second coordinate of the demand curve must be the same as the second coordinate of the supply curve, so that

$$150 - 6q^2 = 10q^2 + 2q.$$

Write this quadratic equation in standard form as follows:

$$0 = 16q^2 + 2q - 150 \qquad \text{Add } -150 \text{ and } 6q^2 \text{ to both sides.}$$

$$0 = 8q^2 + q - 75. \qquad \text{Multiply both sides by } \frac{1}{2}.$$

This equation can be solved by the quadratic formula, given in Section 1.7. Here, $a = 8$, $b = 1$, and $c = -75$:

$$q = \frac{-1 \pm \sqrt{1 - 4(8)(-75)}}{2(8)}$$

$$= \frac{-1 \pm \sqrt{1 + 2400}}{16} \qquad -4(8)(-75) = 2400$$

$$= \frac{-1 \pm 49}{16} \qquad \sqrt{1 + 2400} = \sqrt{2401} = 49$$

$$q = \frac{-1 + 49}{16} = \frac{48}{16} = 3 \quad \text{or} \quad q = \frac{-1 - 49}{16} = -\frac{50}{16} = -\frac{25}{8}.$$

It is not possible to make $-25/8$ units, so discard that answer and use only $q = 3$. Hence, the equilibrium quantity is 300. Find the equilibrium price by substituting 3

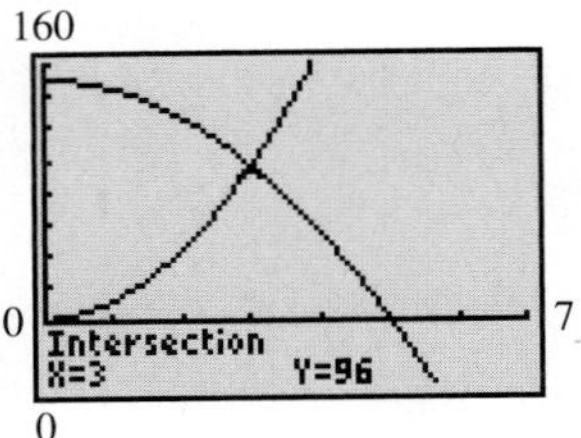

FIGURE 3.29

for q in either the supply or the demand function (and check your answer by using the other one). Using the supply function gives

$$p = 10q^2 + 2q$$
$$p = 10 \cdot 3^2 + 2 \cdot 3 \quad \text{Let } q = 3.$$
$$= 10 \cdot 9 + 6$$
$$p = \$96.$$

Graphical Method Graph the two functions on a graphing calculator, and use the intersection finder to determine that the equilibrium point is (3, 96), as in Figure 3.29. 2✓

✓Checkpoint 2

The price and demand for an item are related by $p = 32 - x^2$, while price and supply are related by $p = x^2$. Find

(a) the equilibrium quantity;

(b) the equilibrium price.

Answers: See page 184.

Example 3 The rental manager of a small apartment complex with 16 units has found from experience that each \$40 increase in the monthly rent results in an empty apartment. All 16 apartments will be rented at a monthly rent of \$500. How many \$40 increases will produce maximum monthly income for the complex?

Solution Let x represent the number of \$40 increases. Then the number of apartments rented will be $16 - x$. Also, the monthly rent per apartment will be $500 + 40x$. (There are x increases of \$40, for a total increase of $40x$.) The monthly income, $I(x)$, is given by the number of apartments rented times the rent per apartment, so

$$I(x) = (16 - x)(500 + 40x)$$
$$= 8000 + 640x - 500x - 40x^2$$
$$= 8000 + 140x - 40x^2.$$

Since x represents the number of \$40 increases and each \$40 increase causes one empty apartment, x must be a whole number. Because there are only 16 apartments, $0 \le x \le 16$. Since there is a small number of possibilities, the value of x that produces maximum income may be found in several ways.

Brute Force Method Use a scientific calculator or the table feature of a graphing calculator (as in Figure 3.30) to evaluate $I(x)$ when $x = 1, 2, \ldots, 16$ and find the largest value.

X	Y1
0	8000
1	8100
2	8120
3	8060
4	7920
5	7700
6	7400

X=0

X	Y1
7	7020
8	6560
9	6020
10	5400
11	4700
12	3920
13	3060

X=7

X	Y1
14	2120
15	1100
16	0
17	-1180
18	-2440
19	-3780
20	-5200

X=14

FIGURE 3.30

The tables show that a maximum income of \$8120 occurs when $x = 2$. So the manager should charge rent of $500 + 2(40) = \$580$, leaving two apartments vacant.

Algebraic Method The graph of $I(x) = 8000 + 140x - 40x^2$ is a downward-opening parabola (why?), and the value of x that produces maximum income occurs at the vertex. The methods of Section 3.4 show that the vertex is (1.75, 8122.50). Since x must be a whole number, evaluate $I(x)$ at $x = 1$ and $x = 2$ to see which one gives the best result:

$$\text{If } x = 1, \text{ then } I(1) = -40(1)^2 + 140(1) + 8000 = 8100.$$
$$\text{If } x = 2, \text{ then } I(2) = -40(2)^2 + 140(2) + 8000 = 8120.$$

So maximum income occurs when $x = 2$. The manager should charge a rent of $500 + 2(40) = \$580$, leaving two apartments vacant.

QUADRATIC MODELS

Real-world data can sometimes be used to construct a quadratic function that approximates the data. Such **quadratic models** can then be used (subject to limitations) to predict future behavior.

Example 4 **Business** The annual revenues (in billions of dollars) as reported by Morningstar.com for Verizon Communications, Inc., are given in the following table for the years 2000–2007:

Year	2000	2001	2002	2003	2004	2005	2006	2007
Revenue	64.7	67.2	67.6	67.8	71.3	75.1	88.1	93.5

(a) Display the information graphically.

Solution Let $x = 0$ correspond to 2000. Plot the points given by the table—(0, 64.7), (1, 67.2), and so on—as in Figure 3.31.

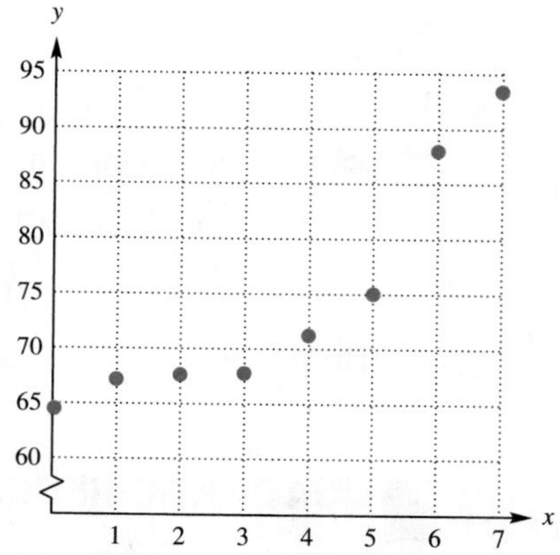

FIGURE 3.31

(b) The shape of the data points in Figure 3.31 resembles the right half of an upward-opening parabola. Using the year 2000 as the minimum, find a quadratic model $f(x) = a(x - h)^2 + k$ for this data.

Solution Recall that when a quadratic function is written in this form, the vertex of its graph is (h, k). On the basis of Figure 3.31, let (0, 64.7) be the vertex, so that

$$f(x) = a(x - 0)^2 + 64.7.$$

To find a, choose another data point, say, (7, 93.5). Assume this point lies on the parabola so that

$$f(x) = a(x - 0)^2 + 64.7$$

$$93.5 = a(7 - 0)^2 + 64.7 \quad \text{Substitute 7 for } x \text{ and 93.5 for } f(x).$$

$$93.5 = 49a + 64.7 \quad \text{Square inside parentheses.}$$

$$28.8 = 49a \quad \text{Subtract 64.7 from both sides.}$$

$$a = \frac{28.8}{49} \approx .588. \quad \text{Divide both sides by 49.}$$

Therefore, $f(x) = .588x^2 + 64.7$ is a quadratic model for this data. The graph of $f(x)$ in Figure 3.32 appears to fit the data reasonably well.

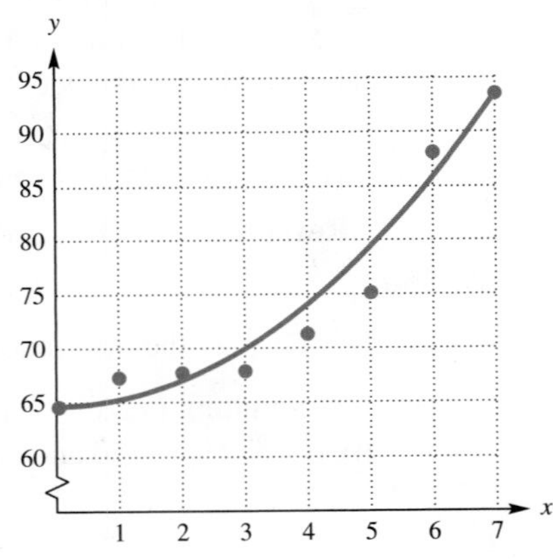

FIGURE 3.32

(c) Use the quadratic model in part (b) to estimate the revenue for 2010.

Solution The year 2010 corresponds to $x = 10$, so the number is approximately

$$f(10) = .588(10)^2 + 64.7 = 123.5.$$

The revenue for 2010 is $123.5 billion dollars. 3✓

✓Checkpoint 3

Find another quadratic model in Example 4(b) by using (0, 64.7) as the vertex and (6, 88.1) as the other point.

Answer: See page 184.

QUADRATIC REGRESSION

Linear regression was used in Section 2.3 to construct a linear function that modeled a set of data points. When the data points appear to lie on a parabola rather than on a straight line (as in Example 4), a similar least-squares regression procedure is available on most graphing calculators and spreadsheet programs to construct a quadratic model for the data. Simply follow the same steps as in linear regression, with one exception: Choose quadratic, rather than linear, regression. (Both options are on the same menu.)

Example 5 Use a graphing calculator to complete each part.

(a) Find a quadratic-regression model for the data in Example 4.

Solution Enter the first coordinates of the data points as list L_1 and the second coordinates as list L_2. Performing the quadratic regression, as in Figure 3.33, leads to the model

$$g(x) = .828x^2 - 1.84x + 66.4.$$

The number R^2 in Figure 3.33 is fairly close to 1, which indicates that this model fits the data fairly well. Figure 3.34 shows the data with the quadratic-regression line.

FIGURE 3.33

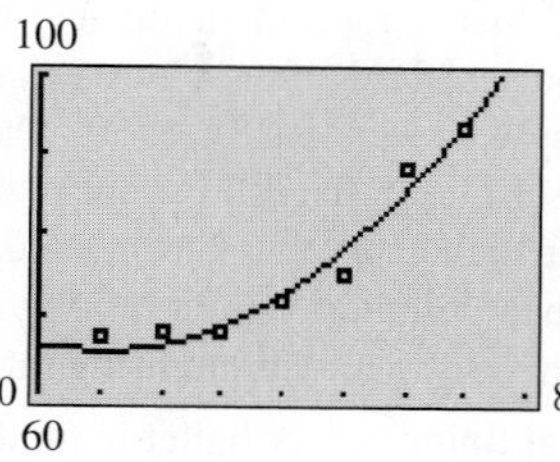

FIGURE 3.34

(b) Use the regression model to estimate the revenue for Verizon in 2010.

Solution Evaluate $g(x)$ when $x = 10$:

$$g(x) = .828(10)^2 - 1.84(10) + 66.4 = 130.8$$

This estimate corresponds to \$130.8 billion, which is \$7.3 billion higher than the estimate made in Example 4.

3.5 ▸ Exercises

Work these problems. (See Example 1.)

1. **Business** Carol Bey makes and sells candy. She has found that the cost per box for making x boxes of candy is given by

$$C(x) = x^2 - 40x + 405.$$

(a) How much does it cost per box to make 15 boxes? 18 boxes? 30 boxes?

(b) Graph the cost function $C(x)$, and mark the points corresponding to 15, 18, and 30 boxes.

(c) What point on the graph corresponds to the number of boxes that will make the cost per box as small as possible?

(d) How many boxes should she make in order to keep the cost per box at a minimum? What is the minimum cost per box?

2. **Business** Greg Tobin sells bottled water. He has found that the average amount of time he spends with each customer is related to his weekly sales volume by the function

$$f(x) = x(60 - x),$$

where x is the number of minutes per customer and $f(x)$ is the number of cases sold per week.
   (a) How many cases does he sell if he spends 10 minutes with each customer? 20 minutes? 45 minutes?
   (b) Choose an appropriate scale for the axes and sketch the graph of $f(x)$. Mark the points on the graph corresponding to 10, 20, and 45 minutes.
   (c) Explain what the vertex of the graph represents.
   (d) How long should Greg spend with each customer in order to sell as many cases per week as possible? If he does, how many cases will he sell?

3. **Natural Science** A researcher in physiology has decided that a good mathematical model for the number of impulses fired after a nerve has been stimulated is given by $y = -x^2 + 20x - 60$, where y is the number of responses per millisecond and x is the number of milliseconds since the nerve was stimulated.
   (a) When will the maximum firing rate be reached?
   (b) What is the maximum firing rate?

4. **Physical Science** A bullet is fired upward from ground level. Its height above the ground (in feet) at time t seconds is given by

$$H = -16t^2 + 1000t.$$

Find the maximum height of the bullet and the time at which it hits the ground.

5. **Business** Pat Kan owns a factory that manufactures souvenir key chains. Her weekly profit (in hundreds of dollars) is given by $P(x) = -2x^2 + 60x - 120$, where x is the number of cases of key chains sold.
   (a) What is the largest number of cases she can sell and still make a profit?
   (b) Explain how it is possible for her to lose money if she sells more cases than your answer in part (a).
   (c) How many cases should she make and sell in order to maximize her profits?

6. **Business** The manager of a bicycle shop has found that, at a price (in dollars) of $p(x) = 150 - \frac{x}{4}$ per bicycle, x bicycles will be sold.
   (a) Find an expression for the total revenue from the sale of x bicycles. (*Hint*: Revenue = Demand × Price.)
   (b) Find the number of bicycle sales that leads to maximum revenue.
   (c) Find the maximum revenue.

Work the following problems. (See Example 2.)

7. **Business** Suppose the supply of and demand for a certain textbook are given by

$$\text{supply: } p = \frac{1}{5}q^2 \quad \text{and} \quad \text{demand: } p = -\frac{1}{5}q^2 + 40,$$

where p is price and q is quantity. How many books are demanded at a price of
   (a) 10? (b) 20? (c) 30? (d) 40?
   How many books are supplied at a price of
   (e) 5? (f) 10? (g) 20? (h) 30?
   (i) Graph the supply and demand functions on the same axes.

8. **Business** Find the equilibrium quantity and the equilibrium price in Exercise 7.

9. **Business** Suppose the price p of widgets is related to the quantity q that is demanded by

$$p = 640 - 5q^2,$$

where q is measured in hundreds of widgets. Find the price when the number of widgets demanded is
   (a) 0; (b) 5; (c) 10.
   Suppose the supply function for widgets is given by $p = 5q^2$, where q is the number of widgets (in hundreds) that are supplied at price p.
   (d) Graph the demand function $p = 640 - 5q^2$ and the supply function $p = 5q^2$ on the same axes.
   (e) Find the equilibrium quantity.
   (f) Find the equilibrium price.

10. **Business** The supply function for a commodity is given by $p = q^2 + 200$, and the demand function is given by $p = -10q + 3200$
   (a) Graph the supply and demand functions on the same axes.
   (b) Find the equilibrium point.
   (c) What is the equilibrium quantity? the equilibrium price?

Business *Find the equilibrium quantity and equilibrium price for the commodity whose supply and demand functions are given.*

11. Supply: $p = 45q$; demand: $p = -q^2 + 10{,}000$.
12. Supply: $p = q^2 + q + 10$; demand: $p = -10q + 3060$.
13. Supply: $p = q^2 + 20q$; demand: $p = -2q^2 + 10q + 3000$.
14. Supply: $p = .2q + 51$; demand: $p = \dfrac{3000}{q + 5}$.

Business *The revenue function R(x) and the cost function C(x) for a particular product are given. These functions are valid only for the specified domain of values. Find the number of units that must be produced to break even.*

15. $R(x) = 200x - x^2$; $C(x) = 70x + 2200$; $0 \le x \le 100$
16. $R(x) = 300x - x^2$; $C(x) = 65x + 7000$; $0 \le x \le 150$
17. $R(x) = 400x - 2x^2$; $C(x) = -x^2 + 200x + 1900$; $0 \le x \le 100$
18. $R(x) = 500x - 2x^2$; $C(x) = -x^2 + 270x + 5125$; $0 \le x \le 125$

Business *Work each problem. (See Example 3)*

19. A charter flight charges a fare of \$200 per person, plus \$4 per person for each unsold seat on the plane. If the plane holds 100 passengers and if x represents the number of unsold seats, find the following:
   (a) an expression for the total revenue received for the flight. (*Hint*: Multiply the number of people flying, $100 - x$, by the price per ticket);
   (b) the graph for the expression in part (a);
   (c) the number of unsold seats that will produce the maximum revenue;
   (d) the maximum revenue.

20. The revenue of a charter bus company depends on the number of unsold seats. If 100 seats are sold, the price is \$50 per seat. Each unsold seat increases the price per seat by \$1. Let x represent the number of unsold seats.
   (a) Write an expression for the number of seats that are sold.
   (b) Write an expression for the price per seat.
   (c) Write an expression for the revenue.
   (d) Find the number of unsold seats that will produce the maximum revenue.
   (e) Find the maximum revenue.

21. Farmer Linton wants to find the best time to take her hogs to market. The current price is 88 cents per pound, and her hogs weigh an average of 90 pounds. The hogs gain 5 pounds per week, and the market price for hogs is falling each week by 2 cents per pound. How many weeks should Ms. Linton wait before taking her hogs to market in order to receive as much money as possible? At that time, how much money (per hog) will she get?

22. The manager of a peach orchard is trying to decide when to arrange for picking the peaches. If they are picked now, the average yield per tree will be 100 pounds, which can be sold for 40¢ per pound. Past experience shows that the yield per tree will increase about 5 pounds per week, while the price will decrease about 2¢ per pound per week.
   (a) Let x represent the number of weeks that the manager should wait. Find the price per pound.
   (b) Find the number of pounds per tree.
   (c) Find the total revenue from a tree.
   (d) When should the peaches be picked in order to produce the maximum revenue?
   (e) What is the maximum revenue?

Work these exercises. (See Example 4.)

23. **Sports** The National Center for Catastrophic Sport Injury Research keeps data on the number of fatalities directly related to football each year. These deaths occur in sandlot, pro and semipro, high school, and college-level playing. The following table gives selected years and the number of deaths*:

Year	Deaths
1970	29
1975	15
1980	9
1985	7
1990	0
1995	4
2000	3
2005	3
2007	4

   (a) Let $x = 0$ correspond to 1970. Use (20, 0) as the vertex and the data from 2005 to find a quadratic function $g(x) = a(x - h)^2 + k$ that models the data.
   (b) Use the model to estimate the number of deaths in 2006.

24. **Business** According to the Recording Industry Association of America (RIAA), the number of CDs (in millions) shipped by manufacturers for various years is given in the accompanying table. (The CD count does not include digital downloads.)

Year	CDs
1997	753
1998	847
1999	939
2000	943
2001	882
2002	803
2003	746
2004	767
2005	705
2006	620
2007	511

*www.unc.edu/depts/nccsi/FootballAnnual.pdf

(a) Let $x = 7$ correspond to 1997. Use (10, 943) as the vertex and the data for 2006 to find a quadratic function $f(x) = a(x - h)^2 + k$ that models the data.
(b) Use the model to estimate the number of CDs shipped for 2010.

25. **Business** The following table shows the People's Republic of China's gross domestic expenditures on research and development (in billions of U.S. dollars) for select years from 1996 to 2005*:

Year	Expenditures
1996	5.9
1998	8.1
2000	13.1
2002	18.8
2003	22.5
2004	28.7
2005	35.8

(a) Let $x = 6$ correspond to 1996. Use (6, 5.9) as the vertex and the data from 2004 to find a quadratic function $f(x) = a(x - h)^2 + k$ that models this data.
(b) Use the model to estimate the expenditures on research and development in 2010 and 2012.

26. **Health** The amount of personal health care expenditures (in billions of dollars) for various years is given in the table†:

Year	Expenditures
1990	607.0
1995	863.7
2000	1139.9
2001	1239.0
2002	1341.0
2003	1446.0
2004	1551.0
2005	1661.0

(a) Let $x = 0$ correspond to 1990. Use (0, 607) as the vertex and the data from 2003 to find a quadratic function $f(x) = a(x - h)^2 + k$ that models this data.
(b) Use the model to estimate the expenditures on personal health care in 2010 and 2014.

*National Science Foundation, Division of Science Resources Statistics, *National Patterns of R&D Resources* (annual series).

†U.S. Centers for Medicare and Medical Services.

In Exercises 27–30, plot the data points and use quadratic regression to find a function that models the data. (See Example 5.)

27. **Sports** Use the data in Exercise 23, with $x = 0$ corresponding to 1970. What number of injuries does this model estimate for 2006? Compare your answer with that for Exercise 23.

28. **Business** Use the data in Exercise 24, with $x = 7$ corresponding to 1997. What number of CDs does this model estimate for 2010? Compare your answer with that for Exercise 24.

29. **Business** Use the data in Exercise 25, with $x = 6$ corresponding to 1996. What spending amounts on research and development does this model estimate for 2010 and 2012? Compare your answer with those for Exercise 25.

30. **Health** Use the data in Exercise 26, with $x = 0$ corresponding to 1990. What are the amounts the model estimates for personal health care expenditures in 2010 and 2014? Compare your answer with those for Exercise 26.

Work these problems.

31. **Geometry** A field bounded on one side by a river is to be fenced on three sides to form a rectangular enclosure. There are 320 ft of fencing available. What should the dimensions be to have an enclosure with the maximum possible area?

32. **Geometry** A rectangular garden bounded on one side by a river is to be fenced on the other three sides. Fencing material for the side parallel to the river costs $30 per foot, and material for the other two sides costs $10 per foot. What are the dimensions of the garden of largest possible area if $1200 is to be spent for fencing material?

Business *Recall that profit equals revenue minus cost. In Exercises 33 and 34, find the following:*

(a) *The break-even point (to the nearest tenth)*
(b) *The x-value that makes profit a maximum*
(c) *The maximum profit*
(d) *For what x-values will a loss occur?*
(e) *For what x-values will a profit occur?*

33. $R(x) = 400x - 2x^2$ and $C(x) = 200x + 2000$, with $0 \le x \le 100$

34. $R(x) = 900x - 3x^2$ and $C(x) = 450x + 5000$, with $20 \le x \le 150$

✓ Checkpoint Answers

1. (a) 10 units (b) $480
2. (a) 4 (b) 16
3. $f(x) = .65x^2 + 64.7$

3.6 ▶ Polynomial Functions

A **polynomial function of degree *n*** is a function whose rule is given by a polynomial of degree n.* For example

$f(x) = 3x - 2$ polynomial function of degree 1;
$g(x) = 3x^2 + 4x - 6$ polynomial function of degree 2;
$h(x) = x^4 + 5x^3 - 6x^2 + x - 3$ polynomial function of degree 4.

BASIC GRAPHS

The simplest polynomial functions are those whose rules are of the form $f(x) = ax^n$ (where a is a constant).

Example 1 Graph $f(x) = x^3$.

Solution First, find several ordered pairs belonging to the graph. Be sure to choose some negative x-values, $x = 0$, and some positive x-values in order to get representative ordered pairs. Find as many ordered pairs as you need in order to see the shape of the graph. Then plot the ordered pairs and draw a smooth curve through them to obtain the graph in Figure 3.35. ✓1

Graph $f(x) = -\frac{1}{2}x^3$

Answer: See page 195.

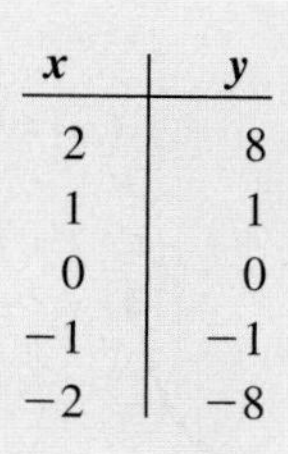

x	y
2	8
1	1
0	0
−1	−1
−2	−8

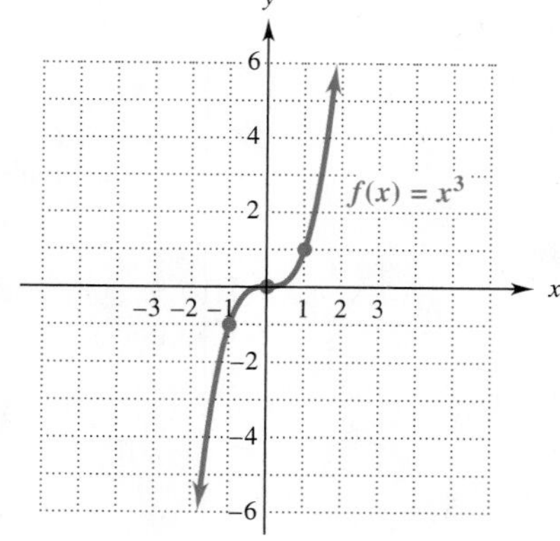

FIGURE 3.35

Example 2 Graph $f(x) = \frac{3}{2}x^4$.

Solution The following table gives some typical ordered pairs and leads to the graph in Figure 3.36 on the next page. ✓2

✓**Checkpoint 2**

Graph $g(x) = -2x^4$

Answer: See page 196.

*The degree of a polynomial was defined on page 15.

x	$f(x)$
−2	24
−1	3/2
0	0
1	3/2
2	24

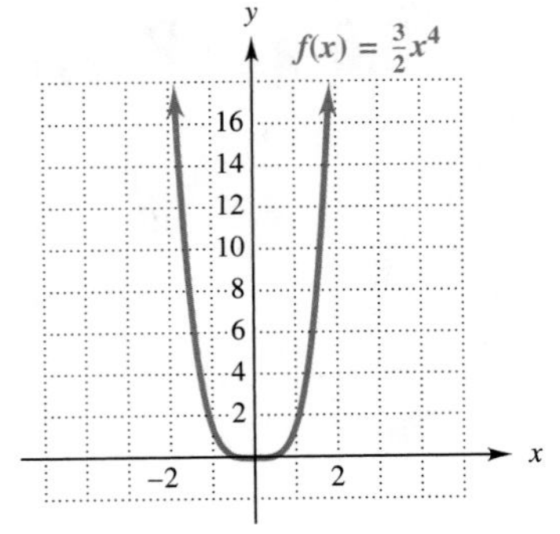

FIGURE 3.36

The graph of $f(x) = ax^n$ has one of the four basic shapes illustrated in Examples 1 and 2 and in Checkpoints 1 and 2, and summarized here.

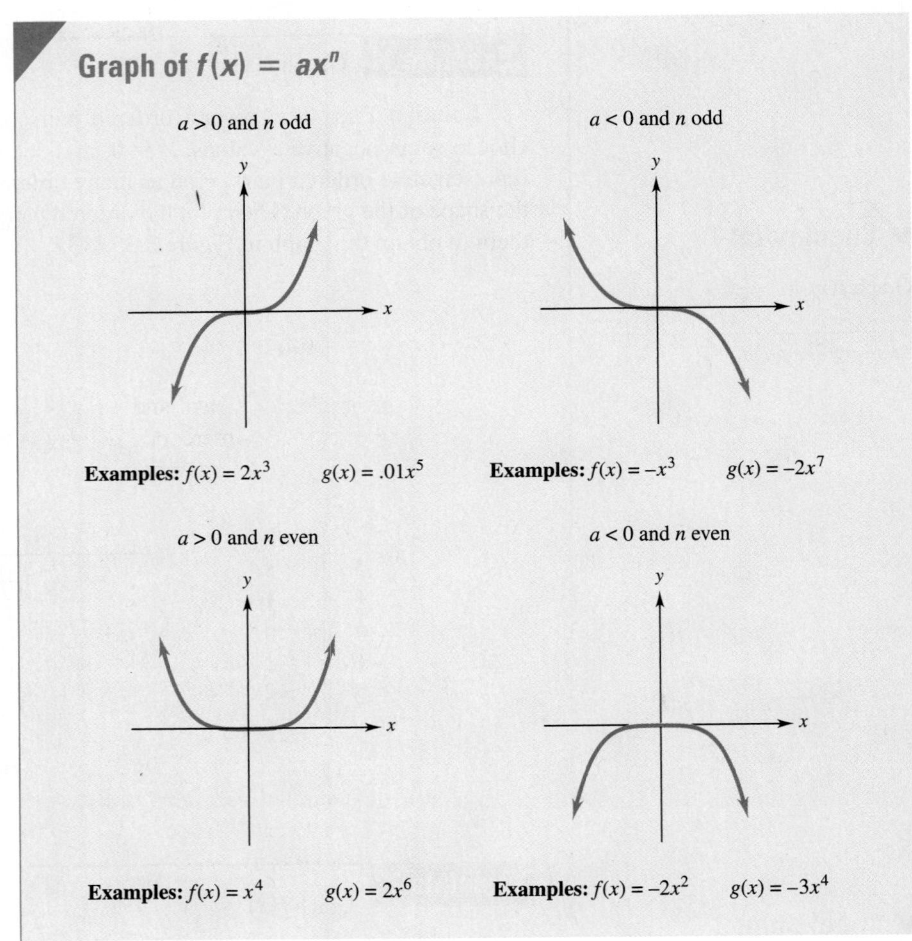

PROPERTIES OF POLYNOMIAL GRAPHS

Unlike the graphs in the preceding figures, the graphs of more complicated polynomial functions may have several "peaks" and "valleys," as illustrated in Figure 3.37. The locations of the peaks and valleys can be accurately approximated by a maximum or minimum finder on a graphing calculator. Calculus is needed to determine their exact location.

(a) $f(x) = x^3 - 4x + 2$
1 peak, 1 valley } total 2
3 x-intercepts

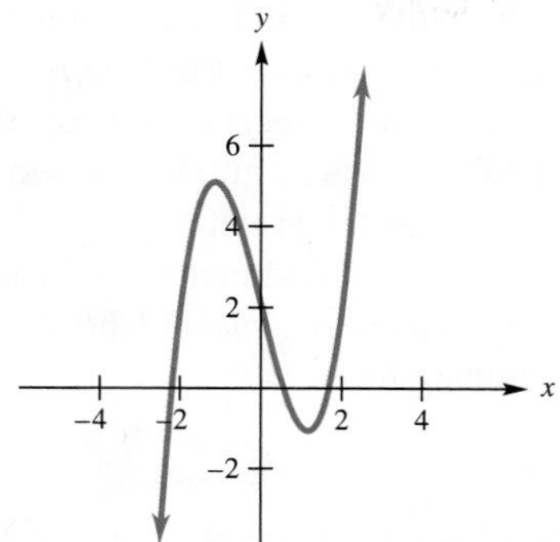

(b) $f(x) = x^5 - 5x^3 + 4x$
2 peaks, 2 valleys } total 4
5 x-intercepts

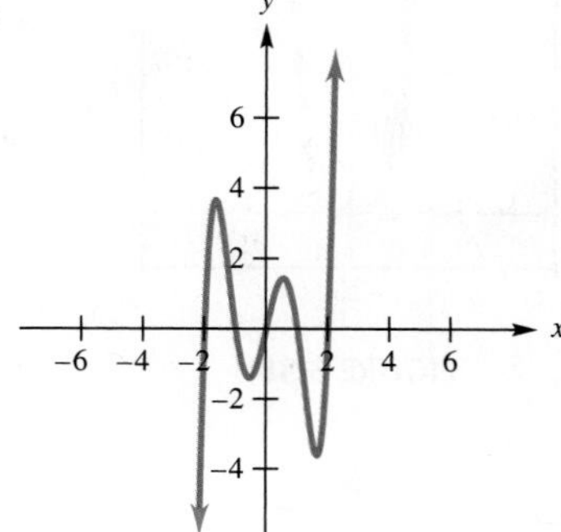

(c) $f(x) = 1.5x^4 + x^3 - 4x^2 - 3x + 4$
1 peak, 2 valleys } total 3
2 x-intercepts

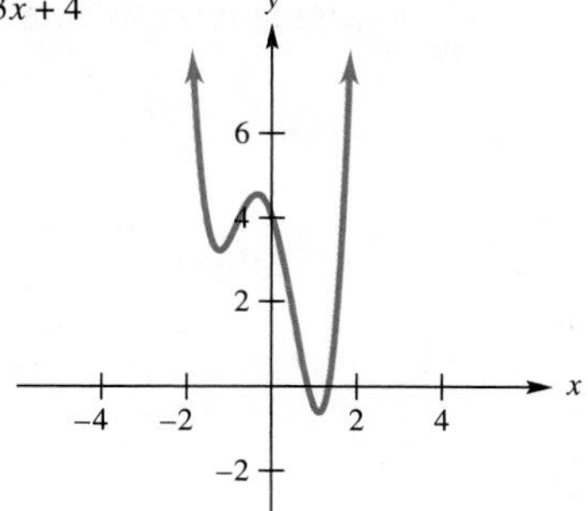

(d) $f(x) = -x^6 + x^5 + 2x^4 + 1$
2 peaks, 1 (shallow) valley } total 3
2 x-intercepts

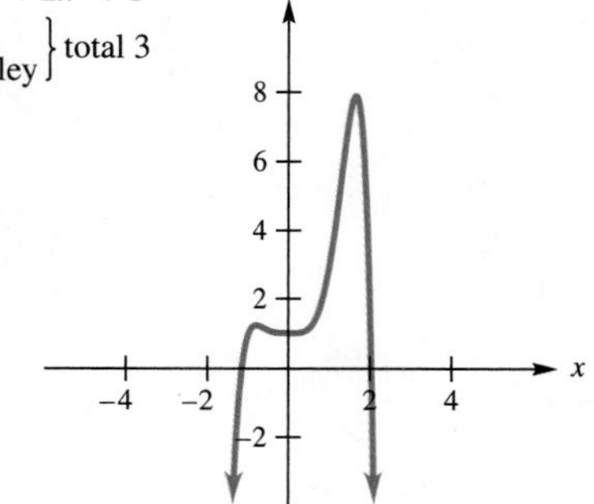

FIGURE 3.37

The total number of peaks and valleys in a polynomial graph, as well as the number of the graph's x-intercepts, depends on the degree of the polynomial, as shown in Figure 3.37 and summarized here.

Polynomial	Degree	Number of peaks & valleys	Number of x-intercepts
$f(x) = x^3 - 4x + 2$	3	2	3
$f(x) = x^5 - 5x^3 + 4x$	5	4	5
$f(x) = 1.5x^4 + x^3 - 4x^2 - 3x + 4$	4	3	2
$f(x) = -x^6 + x^5 + 2x^4 + 1$	6	3	2

In each case, the number of x-intercepts is *at most* the degree of the polynomial. The total number of peaks and valleys is at most *one less than* the degree of the polynomial. The same thing is true in every case.

> 1. The total number of peaks and valleys on the graph of a polynomial function of degree n is at most $n - 1$.
> 2. The number of x-intercepts on the graph of a polynomial function of degree n is at most n.

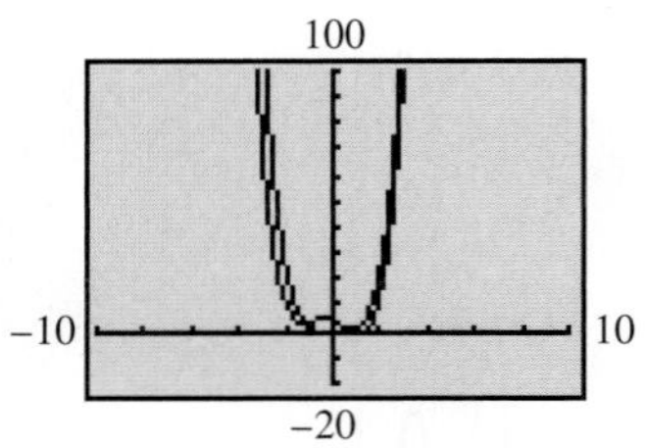

FIGURE 3.38

The domain of every polynomial function is the set of all real numbers, which means that its graph extends forever to the left and right. We indicate this by the arrows on the ends of polynomial graphs.

Although there may be peaks, valleys, and bends in a polynomial graph, the far ends of the graph are easy to describe: *they look like the graph of the highest-degree term of the polynomial.* Consider, for example, $f(x) = 1.5x^4 + x^3 - 4x^2 - 3x + 4$, whose highest-degree term is $1.5x^4$ and whose graph is shown in Figure 3.37(c). The ends of the graph shoot upward, just as the graph of $y = 1.5x^4$ does in Figure 3.36. When $f(x)$ and $y = 1.5x^4$ are graphed in the same large viewing window of a graphing calculator (Figure 3.38), the graphs look almost identical, except near the origin. This is an illustration of the following facts.

> The graph of a polynomial function is a smooth, unbroken curve that extends forever to the left and right. When $|x|$ is large, the graph resembles the graph of its highest-degree term and moves sharply away from the x-axis.

Example 3 Let $g(x) = x^3 - 11x^2 - 32x + 24$, and consider the graph in Figure 3.39.

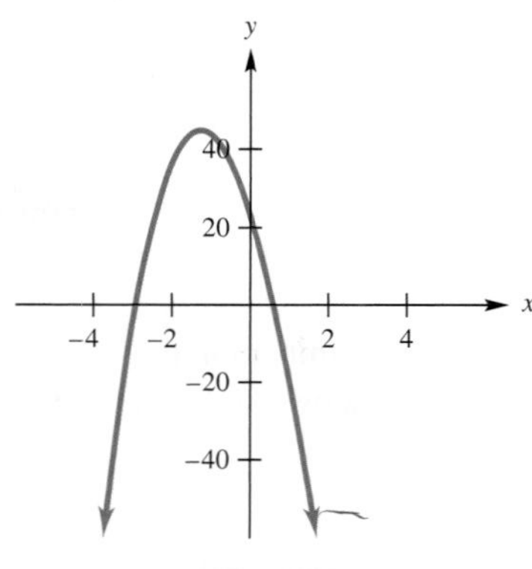

FIGURE 3.39

(a) Is Figure 3.39 a complete graph of $g(x)$; that is, does it show all the important features of the graph?

Solution The far ends of the graph of $g(x)$ should resemble the graph of its highest-degree term x^3. The graph of $f(x) = x^3$ in Figure 3.35 moves upward at the far right, but the graph in Figure 3.39 does not. So Figure 3.39 is *not* a complete graph.

(b) Use a graphing calculator to find a complete graph of $g(x)$.

Solution Since the graph of $g(x)$ must eventually start rising on the right side (as does the graph of x^3), a viewing window that shows a complete graph must extend beyond $x = 4$. By experimenting with various windows, we obtain Figure 3.40. This graph shows a total of two peaks and valleys and three x-intercepts (the maximum possible for a polynomial of degree 3). At the far ends, the graph of $g(x)$ resembles the graph of $f(x) = x^3$. Therefore, Figure 3.40 is a complete graph of $g(x)$. ✓3

✓Checkpoint 3

Find a viewing window on a graphing calculator that shows a complete graph of $f(x) = -.7x^4 + 119x^2 + 400$. (*Hint*: The graph has two x-intercepts and the maximum possible number of peaks and valleys.)

Answer: See page 196.

FIGURE 3.40

GRAPHING TECHNIQUES

Accurate graphs of first- and second-degree polynomial functions (lines and parabolas) are easily found algebraically, as we saw in Sections 2.2 and 3.4. All polynomial functions of degree 3, and some of higher degree, can be accurately graphed by hand by using calculus and algebra to locate the peaks and valleys. When a polynomial can be completely factored, the general shape of its graph can be determined algebraically by using the basic properties of polynomial graphs, as illustrated in Example 4. Obtaining accurate graphs of other polynomial functions generally requires the use of technology.

Multiply out the expression for $f(x)$ in Example 4 and determine its degree.

Answer: See page 196.

Example 4 Graph $f(x) = (2x + 3)(x - 1)(x + 2)$.

Solution Note that $f(x)$ is a polynomial of degree 3. (If you don't see why, do Checkpoint 4.) Begin by finding any x-intercepts. Set $f(x) = 0$ and solve for x: ✓4

$$f(x) = 0$$
$$(2x + 3)(x - 1)(x + 2) = 0.$$

Solve this equation by setting each of the three factors equal to 0:

$$2x + 3 = 0 \quad \text{or} \quad x - 1 = 0 \quad \text{or} \quad x + 2 = 0$$
$$x = -\frac{3}{2} \qquad x = 1 \qquad x = -2.$$

The three numbers $-3/2$, 1, and -2 divide the x-axis into four intervals:

$$x < -2, \quad -2 < x < -\frac{3}{2}, \quad -\frac{3}{2} < x < 1, \quad \text{and} \quad 1 < x.$$

These intervals are shown in Figure 3.41.

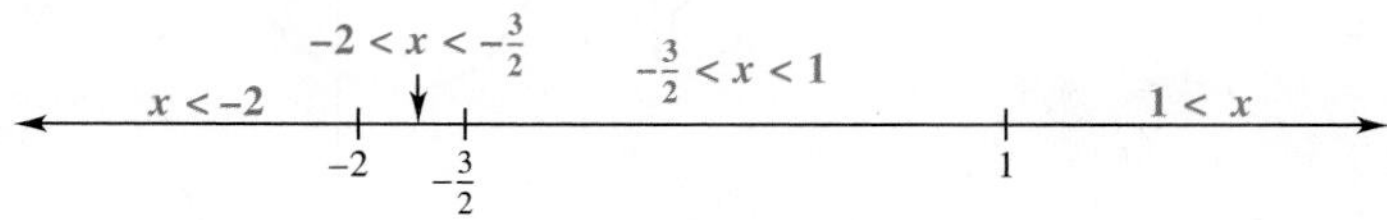

FIGURE 3.41

Since the graph is an unbroken curve, it can change from above the x-axis to below it only by passing through the x-axis. As we have seen, this occurs only at the x-intercepts: $x = -2, -3/2$, and 1. Consequently, in the interval between two intercepts (or to the left of $x = -2$ or to the right of $x = 1$), the graph of $f(x)$ must lie entirely above or entirely below the x-axis.

We can determine where the graph lies over an interval by evaluating $f(x) = (2x + 3)(x - 1)(x + 2)$ at a number in that interval. For example, $x = -3$ is in the interval where $x < -2$, and

$$f(-3) = (2(-3) + 3)(-3 - 1)(-3 + 2)$$
$$= -12.$$

Therefore, $(-3, -12)$ is on the graph. Since this point lies below the x-axis, all points in this interval (that is, all points with $x < -2$) must lie below the x-axis. By testing numbers in the other intervals, we obtain the following table.

Interval	$x < -2$	$-2 < x < -3/2$	$-3/2 < x < 1$	$x > 1$
Test Number	-3	$-7/4$	0	2
Value of f(x)	-12	11/32	-6	28
Sign of f(x)	Negative	Positive	Negative	Positive
Graph	Below x-axis	Above x-axis	Below x-axis	Above x-axis

Since the graph intersects the x-axis at the intercepts and is above the x-axis between these intercepts, there must be at least one peak there. Similarly, there must be at least one valley between $x = -3/2$ and $x = 1$, because the graph is below the x-axis there. However, a polynomial function of degree 3 can have a total of at most $3 - 1 = 2$ peaks and valleys. So there must be exactly one peak and exactly one valley on this graph.

Furthermore, when $|x|$ is large, the graph must resemble the graph of $y = 2x^3$ (the highest-degree term). The graph of $y = 2x^3$, like the graph of $y = x^3$ in Figure 3.35, moves upward to the right and downward to the left. Using these facts and plotting the x-intercepts shows that the graph must have the general shape shown in Figure 3.42. Plotting additional points leads to the reasonably accurate graph in Figure 3.43. We say "reasonably accurate" because we cannot be sure of the exact locations of the peaks and valleys on the graph without using calculus. 5

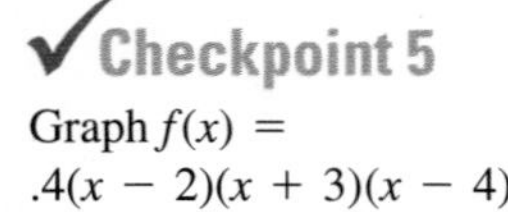

Checkpoint 5

Graph $f(x) = .4(x - 2)(x + 3)(x - 4)$.

Answer: See page 196.

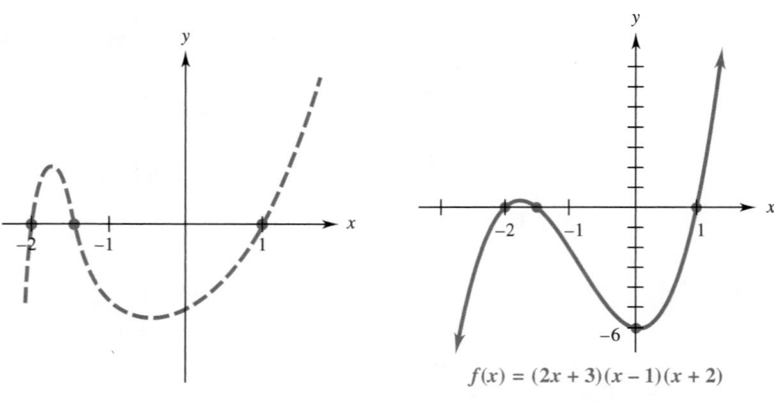

FIGURE 3.42

FIGURE 3.43

POLYNOMIAL MODELS

Regression procedures similar to those presented for linear regression in Section 2.3 and quadratic regression in Section 3.5 can be used to find cubic and quartic (degree 4) polynomial models for appropriate data.

Example 5 **Social Science** The following table shows the population of San Francisco in selected years:

Year	1950	1960	1970	1980	1990	2000	2007
Population	775,357	740,316	715,674	678,974	723,959	776,733	764,976

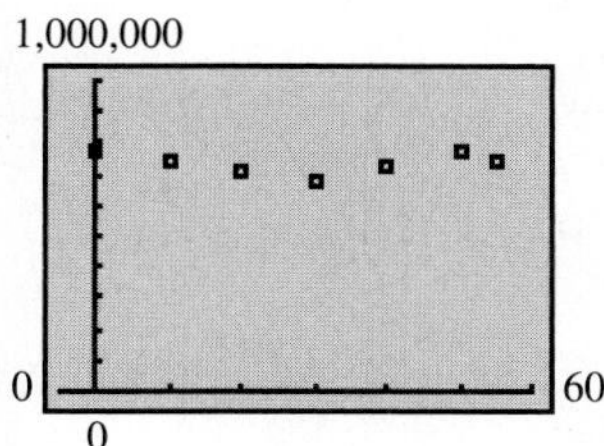

FIGURE 3.44

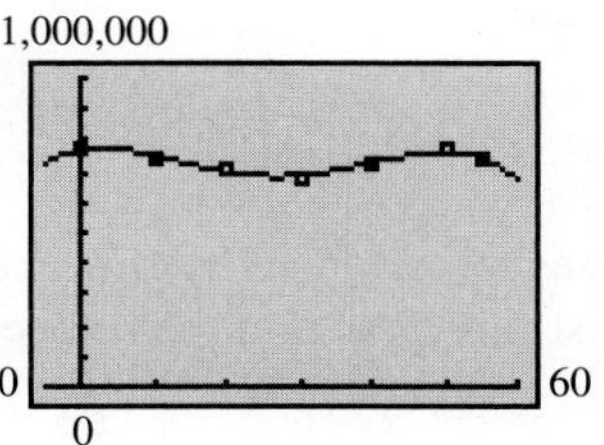

FIGURE 3.45

(a) Plot the data on a graphing calculator, with $x = 0$ corresponding to 1950.

Solution The points in Figure 3.44 suggest the general shape of a fourth-degree (quartic) polynomial.

(b) Use quartic regression to obtain a model for this data.

Solution The procedure is the same as for linear regression; just choose "quartic" in place of "linear." It produces the function

$$f(x) = -.206x^4 + 22.532x^3 - 655.074x^2 + 2100.848x + 773{,}355.255.$$

Its graph, shown in Figure 3.45, appears to fit the data well.

(c) Use the model to estimate the population of San Francisco in 1985 and 2005.

Solution 1985 and 2005 correspond to $x = 35$ and $x = 55$, respectively. Verify that

$$f(35) = 701{,}350 \quad \text{and} \quad f(55) = 771{,}036$$

Example 6 **Business** The following table shows the revenue and costs (in millions of dollars) for Continental Airlines for the years 1999–2008*:

Year	Revenue	Costs
1999	8639	8184
2000	9899	9557
2001	8969	9064
2002	8402	8853
2003	8870	8832
2004	9744	10,107
2005	11,208	11,276
2006	13,128	12,785
2007	14,232	13,773
2008	15,241	15,826

*www.Morningstar.com

(a) Let $x = 9$ corresponds to 1999. Use cubic regression to obtain models for the revenue data $R(x)$ and the costs $C(x)$.

Solution The functions obtained using cubic regression from a calculator are

$$R(x) = -5.132x^3 + 344.792x^2 - 5702.276x + 36{,}471.053;$$
$$C(x) = 8.021x^3 - 200.880x^2 + 1681.017x + 3984.969.$$

(b) Graph $R(x)$ and $C(x)$ on the same set of axes. Did costs ever exceed revenues?

Solution The graph is shown in Figure 3.46. Since the lines cross several times, we can say that costs exceeded revenues in various periods.

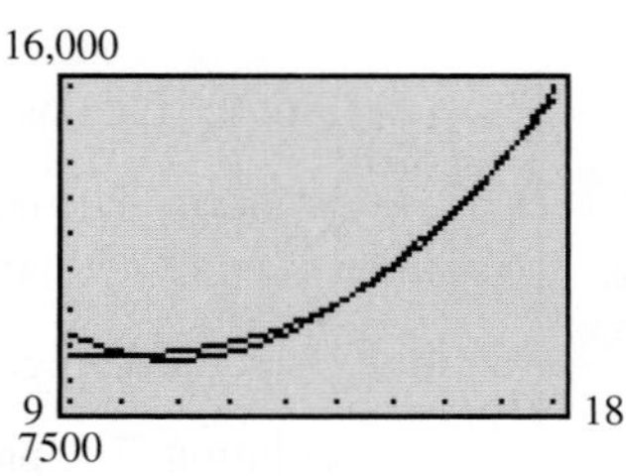

FIGURE 3.46

(c) Find the profit function $P(x)$ and show its graph.

Solution The profit function is the difference between the revenue function and the cost function. We subtract the coefficients of the cost function from the respective coefficients of the revenue function:

$$\begin{aligned} P(x) &= R(x) - C(x) \\ &= (-5.132 - 8.021)x^3 + (344.792 + 200.880)x^2 \\ &\quad + (-5702.276 - 1681.017)x + (36{,}471.053 - 3984.969) \\ &= -13.153x^3 + 545.672x^2 - 7383.293x + 32{,}486.084. \end{aligned}$$

This result leads to the graph in Figure 3.47.

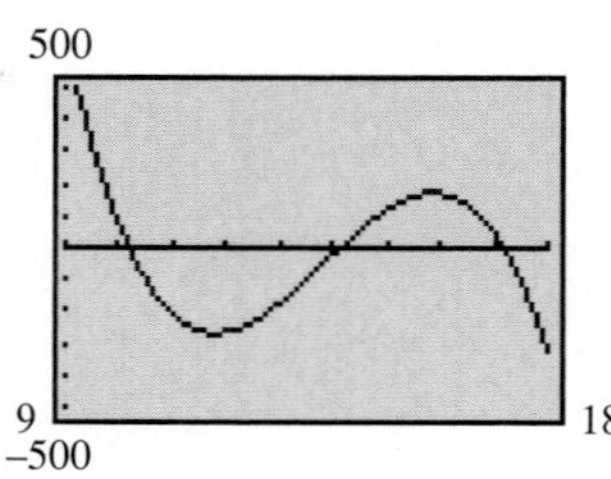

FIGURE 3.47

(d) According to the model of the profit function $P(x)$, in what years was Continental Airlines profitable?

Solution The airline was profitable in 1999, and early 2000, and from early 2004 to 2007.

3.6 ▸ Exercises

Graph each of the given polynomial functions. (See Examples 1 and 2.)

1. $f(x) = x^4$
2. $g(x) = -.5x^6$
3. $h(x) = -.2x^5$
4. $f(x) = x^7$

In Exercises 5–8, state whether the graph could possibly be the graph of (a) some polynomial function; (b) a polynomial function of degree 3; (c) a polynomial function of degree 4; (d) a polynomial function of degree 5. (See Example 3.)

5.

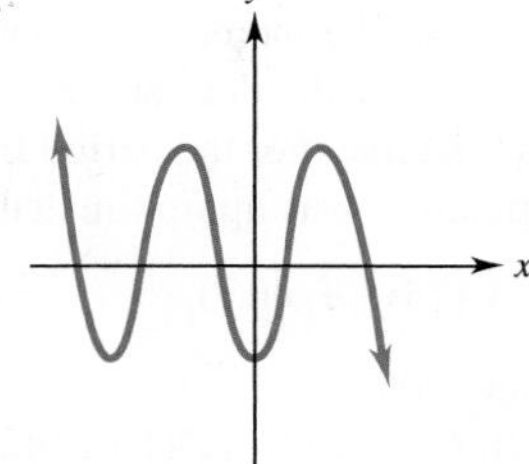

6.

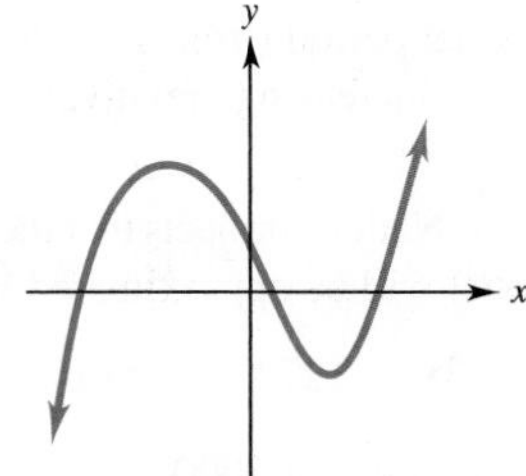

7.

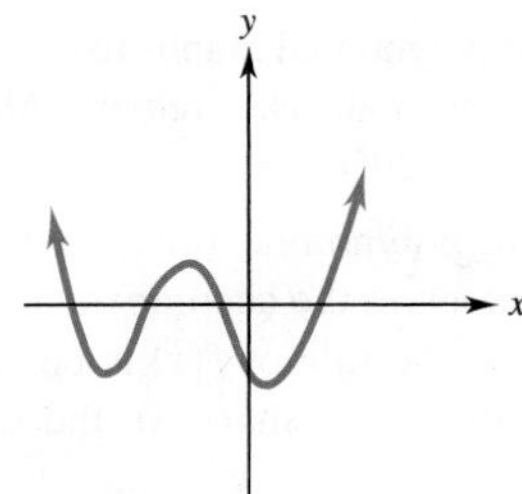

8.

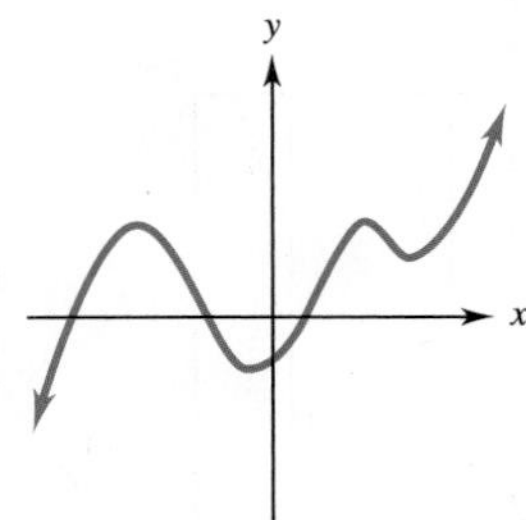

In Exercises 9–14, match the given polynomial function to its graph [(a)–(f)], without using a graphing calculator. (See Example 3 and the two boxes preceding it.)

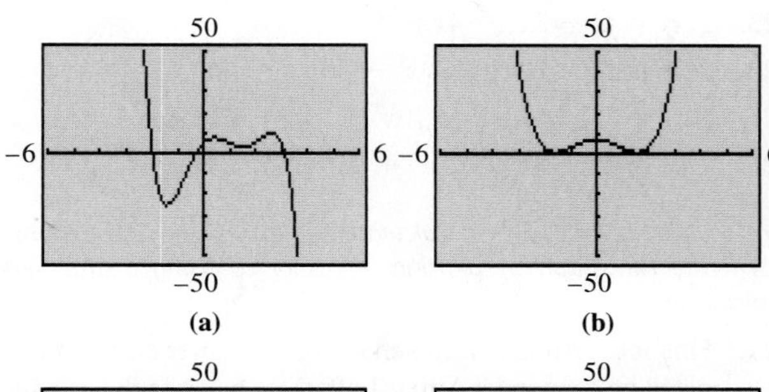

(a) (b)

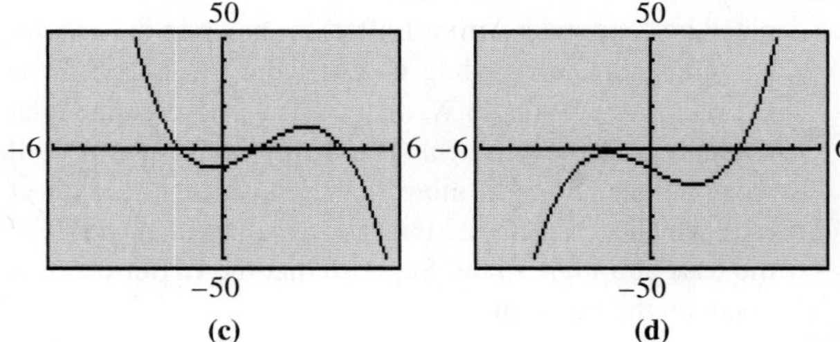

(c) (d)

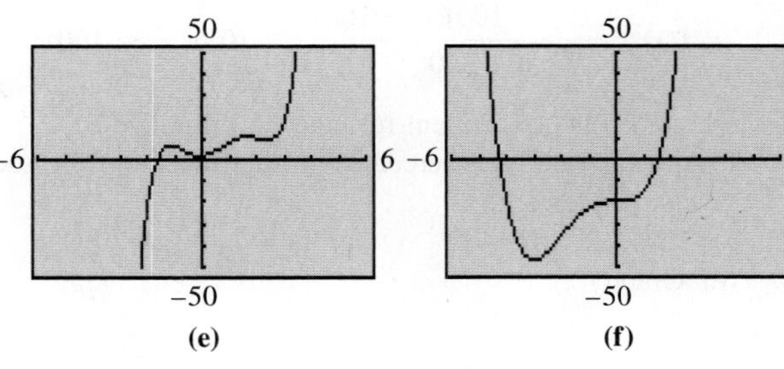

(e) (f)

9. $f(x) = x^3 - 7x - 9$
10. $f(x) = -x^3 + 4x^2 + 3x - 8$
11. $f(x) = x^4 - 5x^2 + 7$
12. $f(x) = x^4 + 4x^3 - 20$
13. $f(x) = .7x^5 - 2.5x^4 - x^3 + 8x^2 + x + 2$
14. $f(x) = -x^5 + 4x^4 + x^3 - 16x^2 + 12x + 5$

Graph each of the given polynomial functions. (See Example 4.)

15. $f(x) = (x + 3)(x - 4)(x + 1)$
16. $f(x) = (x - 5)(x - 1)(x + 1)$
17. $f(x) = x^2(x + 3)(x - 1)$
18. $f(x) = x^2(x + 2)(x - 2)$
19. $f(x) = x^3 - x^2 - 20x$
20. $f(x) = x^3 + 2x^2 - 10x$
21. $f(x) = x^3 + 4x^2 - 7x$
22. $f(x) = x^4 - 6x^2$

Exercises 23–26 require a graphing calculator. Find a viewing window that shows a complete graph of the polynomial function (that is, a graph that includes all the peaks and valleys and that indicates how the curve moves away from the x-axis at the far left and far right). There are many correct answers. Consider your answer correct if it shows all the features that appear in the window given in the answers. (See Example 3.)

23. $g(x) = x^3 - 3x^2 - 4x - 5$

24. $f(x) = x^4 - 10x^3 + 35x^2 - 50x + 24$

25. $f(x) = 2x^5 - 3.5x^4 - 10x^3 + 5x^2 + 12x + 6$

26. $g(x) = x^5 + 8x^4 + 20x^3 + 9x^2 - 27x - 7$

In Exercises 27–31, use a calculator to evaluate the functions. Generate the graph by plotting points or by using a graphing calculator.

27. Finance An idealized version of the Laffer curve (originated by economist Arthur Laffer) is shown in the accompanying graph. According to this theory, decreasing the tax rate, say, from x_2 to x_1, may actually increase the total revenue to the government. The theory is that people will work harder and earn more if they are taxed at a lower rate, which means higher total tax revenues than would be the case at a higher rate. Suppose that the Laffer curve is given by the function

$$f(x) = \frac{x(x-100)(x-160)}{240} \quad (0 \le x \le 100),$$

where $f(x)$ is government revenue (in billions of dollars) from a tax rate of x percent. Find the revenue from the given tax rates.

(a) 20% **(b)** 40% **(c)** 50% **(d)** 70%

(e) Graph $f(x)$.

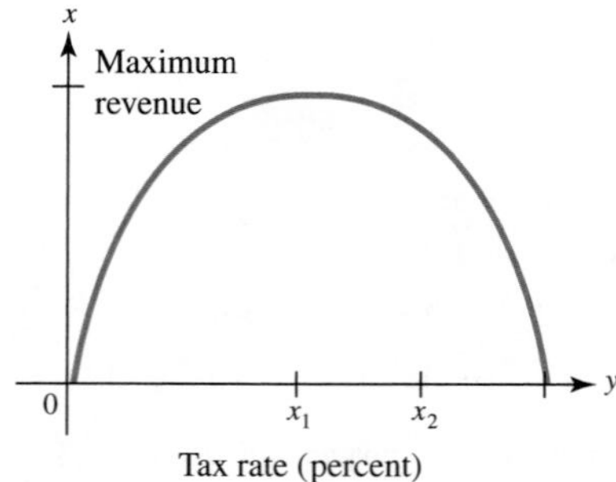

28. Health A technique for measuring cardiac output depends on the concentration of a dye after a known amount is injected into a vein near the heart. In a normal heart, the concentration of the dye at time x (in seconds) is given by the function defined by

$$g(x) = -.006x^4 + .140x^3 - .053x^2 + 1.79x.$$

(a) Find the following: $g(0)$; $g(1)$; $g(2)$; $g(3)$.

(b) Graph $g(x)$ for $x \ge 0$.

29. Physical Science The pressure of the oil in a reservoir tends to drop with time. By taking sample pressure readings for a particular oil reservoir, petroleum engineers have found that the change in pressure is given by

$$P(t) = t^3 - 18t^2 + 81t,$$

where t is time in years from the date of the first reading.

(a) Find the following: $P(0)$; $P(3)$; $P(7)$; $P(10)$.

(b) Graph $P(t)$.

(c) Over what period is the change (drop) in pressure increasing? decreasing?

30. Natural Science During the early part of the 20th century, the deer population of the Kaibab Plateau in Arizona increased rapidly because hunters had reduced the number of natural predators. The increase in population depleted the food resources of the deer and eventually caused the population to decline. For the period from 1905 to 1930, the deer population was approximated by

$$D(x) = -.125x^5 + 3.125x^4 + 4000,$$

where x is time in years from 1905.

(a) Find the following: $D(0)$; $D(5)$; $D(10)$; $D(15)$; $D(20)$; $D(25)$.

(b) Graph $D(x)$.

(c) From the graph, over what period (from 1905 to 1930) was the population increasing? relatively stable? decreasing?

31. Social Science The United Nations projects that the population of China (in millions) will be approximated by

$$g(x) = -.00096x^3 - .1x^2 + 11.3x + 1274 \quad (0 \le x \le 50),$$

where $x = 0$ corresponds to the year 2000.

(a) What populations does the model predict for 2008 and 2020?

(b) Use your knowledge of polynomial graphs to explain, without graphing, why this model is extremely unlikely to be accurate well beyond 2050.

32. A partial graph of a cubic polynomial function $f(x)$ is shown on the calculator screen in the accompanying figure. The coefficient of x^3 in the rule of $f(x)$ is negative. Sketch a graph that has the same shape as the entire graph of $f(x)$.

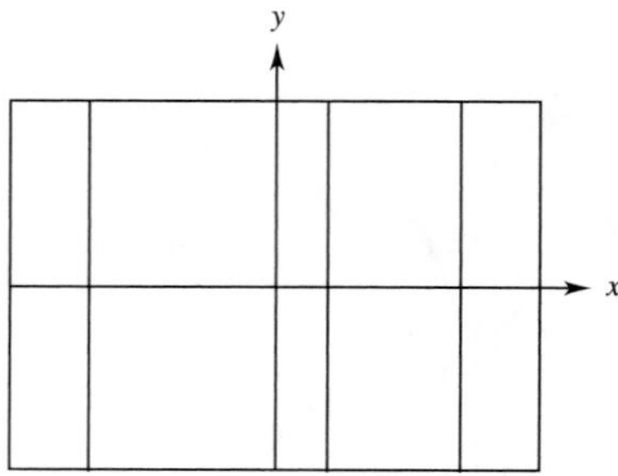

Use a graphing calculator to do the following problems. (See Example 5.)

33. **Social Science** The following table shows the actual and projected enrollment (in millions) in public high schools in selected years*:

Year	Enrollment	Year	Enrollment
1975	14.3	1995	12.5
1980	13.2	2000	13.5
1985	12.4	2005	14.9
1990	11.3	2010	14.7

(a) Plot the data points, with $x = 0$ corresponding to 1975.
(b) Use quartic regression to find a fourth-degree polynomial function $f(x)$ that models these data.
(c) Graph $f(x)$ on the same screen as the data points. Does the graph appear to fit the data well?
(d) According to this model, what is enrollment in 2010 and 2012?
(e) Use the minimum finder to estimate the year between 1975 and 2005 in which enrollment was the lowest.

34. **Business** The following table shows sales generated from lottery games (in millions) in the United States†:

Year	1990	1995	2000	2003	2004	2006
Sales	8563	10,594	9160	9655	10,472	11,015

(a) Plot the data points, with $x = 0$ corresponding to 1990, in the viewing window with $-2 \le x \le 20$ and $8000 \le y \le 12{,}000$.
(b) Use cubic regression to find a third-degree polynomial function $g(x)$ that models this data.
(c) Graph $g(x)$ on the same screen as the data points. Does the graph appear to fit the data well?
(d) Estimate lottery sales in 2005.
(e) Explain why this model is unlikely to be accurate beyond 2007.

35. **Business** The accompanying table gives the annual revenue for International Business Machines Corp. (IBM) in billions of dollars for the years 1997–2007.‡

Year	Revenue	Year	Revenue
1997	81.7	2003	96.3
1998	87.5	2004	91.1
1999	88.4	2005	91.4
2000	85.9	2006	98.8
2001	81.2	2007	104.0
2002	89.1		

(a) Let $x = 7$ correspond to 1997. Use cubic regression to find a polynomial function $R(x)$ that models this data.
(b) Graph $R(x)$ on the same screen as the data points. Does the graph appear to fit the data well?

36. **Business** The accompanying table gives the annual costs for IBM in billions of dollars for the years 1997–2007.*

Year	Cost	Year	Cost
1997	75.4	2003	87.9
1998	79.8	2004	83.2
1999	80.3	2005	81.9
2000	78.2	2006	88.4
2001	77.6	2007	92.9
2002	81.5		

(a) Let $x = 7$ correspond to 1997. Use cubic regression to find a polynomial function $C(x)$ that models these data.
(b) Graph $C(x)$ on the same screen as the data points. Does the graph appear to fit the data well?

Business *For Exercises 37 and 38, use the functions R(x) and C(x) from the answers to Exercises 35(a) and 36(a).*

37. Find the profit function $P(x)$ for IBM for the years 1997–2007.

38. In what year were profits at a minimum?

✓ Checkpoint Answers

1.

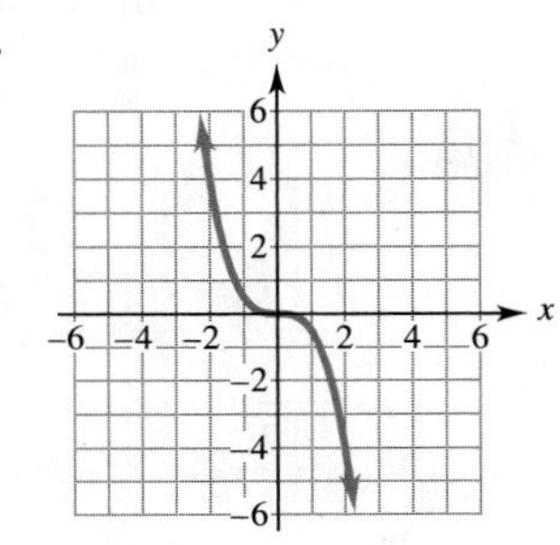

*U.S. National Center for Educational Statistics.

†2007 World Lottery Almanac.

‡Morningstar.com.

*Morningstar.com.

2.

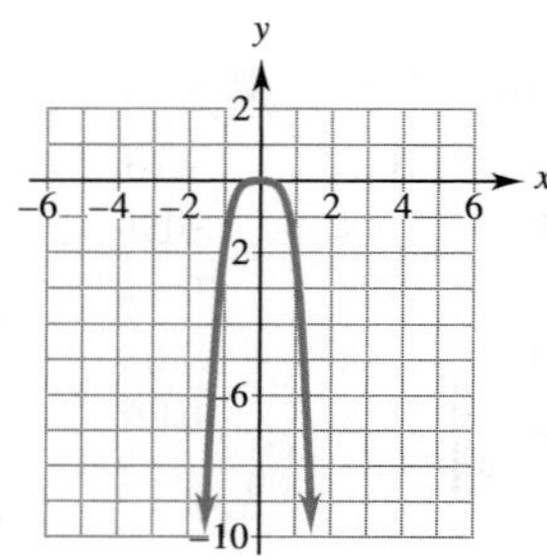

3. Many correct answers, including $-15 \le x \le 15$ and $-2000 \le y \le 6000$.

4. $f(x) = 2x^3 + 5x^2 - x - 6$; degree 3.

5.

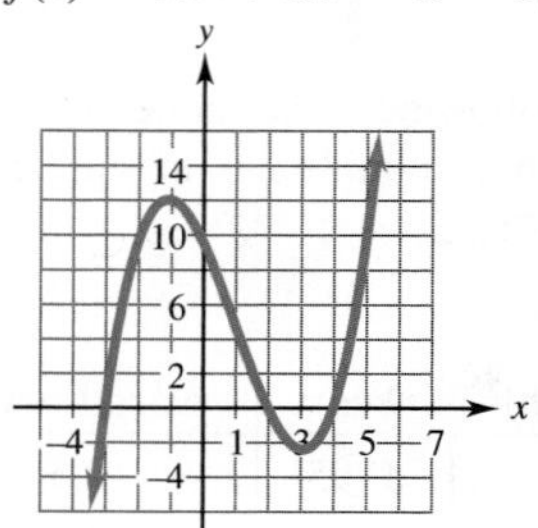

3.7 ▶ Rational Functions

A **rational function** is a function whose rule is the quotient of two polynomials, such as

$$f(x) = \frac{2}{1 + x}, \qquad g(x) = \frac{3x + 2}{2x + 4}, \qquad \text{and} \qquad h(x) = \frac{x^2 - 2x - 4}{x^3 - 2x^2 + x}.$$

Thus, a rational function is a function whose rule can be written in the form

$$f(x) = \frac{P(x)}{Q(x)},$$

where $P(x)$ and $Q(x)$ are polynomials, with $Q(x) \neq 0$. The function is undefined for any values of x that make $Q(x) = 0$, so there are breaks in the graph at these numbers.

LINEAR RATIONAL FUNCTIONS

We begin with rational functions in which both numerator and denominator are first-degree or constant polynomials. Such functions are sometimes called **linear rational functions.**

Example 1 Graph the rational function defined by $y = \dfrac{2}{1 + x}$.

Solution This function is undefined for $x = -1$, since -1 leads to a 0 denominator. For that reason, the graph of this function will not intersect the vertical line $x = -1$. Since x can take on any value except -1, the values of x can approach -1 as closely as desired from either side of -1, as shown in the following table of values.

x approaches -1 ↓ (at $x = -1.01$ and $x = -.99$)

x	-1.5	-1.2	-1.1	-1.01	$-.99$	$-.9$	$-.8$	$-.5$
$1 + x$	$-.5$	$-.2$	$-.1$	$-.01$	$.01$	$.1$	$.2$	$.5$
$\dfrac{2}{1 + x}$	-4	-10	-20	-200	200	20	10	4

↑ $|f(x)|$ gets larger and larger

The preceding table suggests that as x gets closer and closer to -1 from either side, $|f(x)|$ gets larger and larger. The part of the graph near $x = -1$ in Figure 3.48 shows this behavior. The vertical line $x = -1$ that is approached by the curve is called a *vertical asymptote.* For convenience, the vertical asymptote is indicated by a dashed line in Figure 3.48, but this line is *not* part of the graph of the function.

As $|x|$ gets larger and larger, so does the absolute value of the denominator $1 + x$. Hence, $y = 2/(1 + x)$ gets closer and closer to 0, as shown in the following table.

x	-101	-11	-2	0	9	99
$1 + x$	-100	-10	-1	1	10	100
$\frac{2}{1+x}$	$-.02$	$-.2$	-2	2	.2	.02

The horizontal line $y = 0$ is called a *horizontal asymptote* for this graph. Using the asymptotes and plotting the intercept and other points gives the graph of Figure 3.48. 1✓

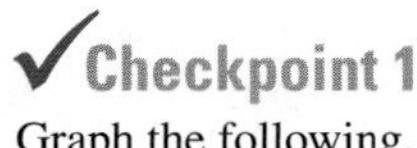

Checkpoint 1

Graph the following.

(a) $f(x) = \frac{3}{5 - x}$

(b) $f(x) = \frac{-4}{x + 4}$

Answers: See page 205.

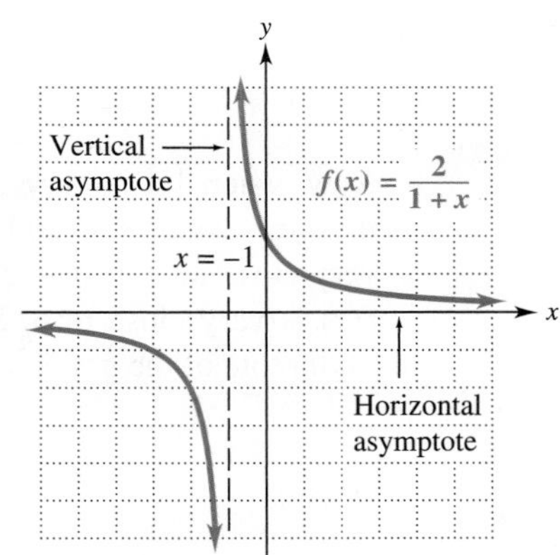

FIGURE 3.48

Example 1 suggests the following conclusion, which applies to all rational functions.

> If a number c makes the denominator zero, but the numerator nonzero, in the expression defining a rational function, then the line $x = c$ is a **vertical asymptote** for the graph of the function.

If the graph of a function approaches a horizontal line very closely when x is very large or very small, we say that this line is a **horizontal asymptote** of the graph.

In Example 1, the horizontal asymptote was the x-axis. This is not always the case, however, as the next example illustrates.

Example 2 Graph $f(x) = \dfrac{3x + 2}{2x + 4}$.

Solution Find the vertical asymptote by setting the denominator equal to 0 and then solving for x:

$$2x + 4 = 0$$
$$x = -2.$$

In order to see what the graph looks like when $|x|$ is very large, we rewrite the rule of the function. When $x \neq 0$, dividing both numerator and denominator by x does not change the value of the function:

$$f(x) = \frac{3x + 2}{2x + 4} = \frac{\dfrac{3x + 2}{x}}{\dfrac{2x + 4}{x}}$$

$$= \frac{\dfrac{3x}{x} + \dfrac{2}{x}}{\dfrac{2x}{x} + \dfrac{4}{x}} = \frac{3 + \dfrac{2}{x}}{2 + \dfrac{4}{x}}.$$

Now, when $|x|$ is very large, the fractions $2/x$ and $4/x$ are very close to 0. (For instance, when $x = 200, 4/x = 4/200 = .02$.) Therefore, the numerator of $f(x)$ is very close to $3 + 0 = 3$ and the denominator is very close to $2 + 0 = 2$. Hence, $f(x)$ is very close to $3/2$ when $|x|$ is large, so the line $y = 3/2$ is the horizontal asymptote of the graph, as shown in Figure 3.49. 2✓

✓Checkpoint 2

Graph the following.

(a) $f(x) = \dfrac{2x - 5}{x - 2}$

(b) $f(x) = \dfrac{3 - x}{x + 1}$

Answers: See page 205.

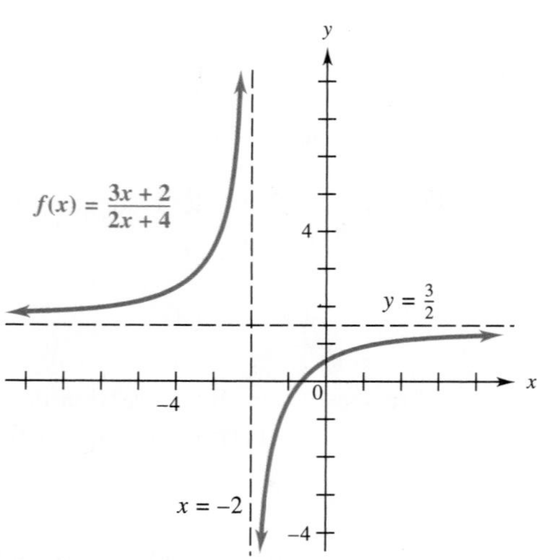

FIGURE 3.49

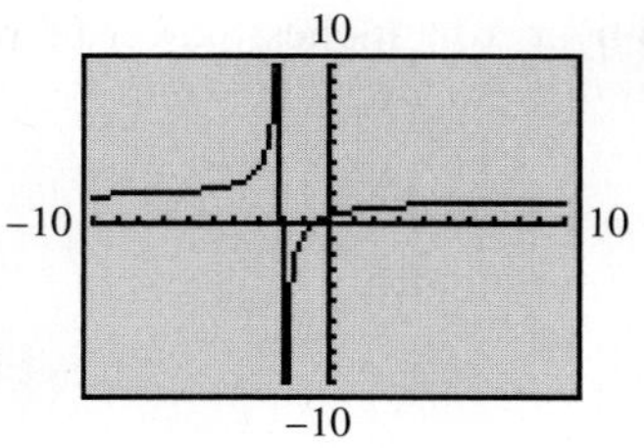

FIGURE 3.50

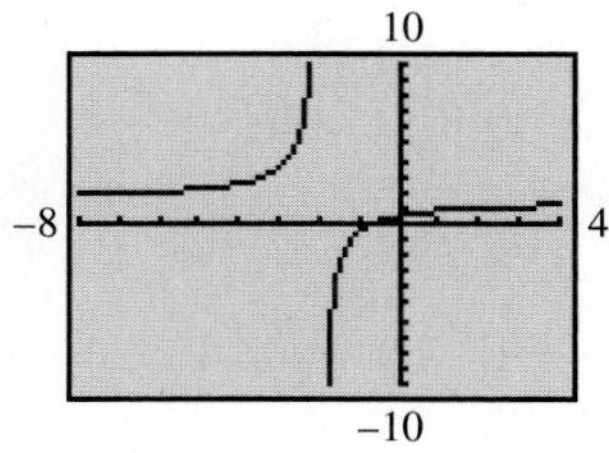

FIGURE 3.51

TECHNOLOGY TIP Depending on the viewing window, a graphing calculator may not accurately represent the graph of a rational function. For example, the graph of $f(x) = \dfrac{3x + 2}{2x + 4}$ in Figure 3.50, which should look like Figure 3.49, has an erroneous vertical line at the place where the graph has a vertical asymptote. This problem can usually be avoided by using a window that has the vertical asymptote at the center of the x-axis, as in Figure 3.51.

The horizontal asymptotes of a linear rational function are closely related to the coefficients of the x-terms of the numerator and denominator, as illustrated in Examples 1 and 2:

	Function	Horizontal Asymptote
Example 2:	$f(x) = \dfrac{3x + 2}{2x + 4}$	$y = \dfrac{3}{2}$
Example 1:	$f(x) = \dfrac{2}{1 + x} = \dfrac{0x + 2}{1x + 1}$	$y = \dfrac{0}{1} = 0$ (the x-axis)

The same pattern holds in the general case.

The graph of $f(x) = \dfrac{ax + b}{cx + d}$ (where $c \neq 0$ and $ad \neq bc$) has a vertical asymptote at the root of the denominator and has horizontal asymptote $y = \dfrac{a}{c}$.

OTHER RATIONAL FUNCTIONS

When the numerator or denominator of a rational function has degree greater than 1, the graph of the function can be more complicated than those in Examples 1 and 2. The graph may have several vertical asymptotes, as well as peaks and valleys.

Example 3 Graph $f(x) = \dfrac{2x^2}{x^2 - 4}$.

Solution Find the vertical asymptotes by setting the denominator equal to 0 and solving for x:

$$\begin{aligned} x^2 - 4 &= 0 \\ (x + 2)(x - 2) &= 0 \\ x + 2 = 0 \quad \text{or} \quad x - 2 &= 0 \\ x = -2 \qquad x &= 2. \end{aligned}$$

Since neither of these numbers makes the numerator 0, the lines $x = -2$ and $x = 2$ are vertical asymptotes of the graph. The horizontal asymptote can be determined by dividing both the numerator and denominator of $f(x)$ by x^2 (the highest power of x that appears in either one):

$$f(x) = \frac{2x^2}{x^2 - 4} = \frac{\dfrac{2x^2}{x^2}}{\dfrac{x^2 - 4}{x^2}} = \frac{\dfrac{2x^2}{x^2}}{\dfrac{x^2}{x^2} - \dfrac{4}{x^2}} = \frac{2}{1 - \dfrac{4}{x^2}}.$$

When $|x|$ is very large, the fraction $4/x^2$ is very close to 0, so the denominator is very close to 1 and $f(x)$ is very close to 2. Hence, the line $y = 2$ is the horizontal asymptote of the graph. Using this information and plotting several points in each of the three regions determined by the vertical asymptotes, we obtain Figure 3.52. 3✓

✓Checkpoint 3

List the vertical and horizontal asymptotes of the given function.

(a) $f(x) = \dfrac{3x + 5}{x + 5}$

(b) $g(x) = \dfrac{2 - x^2}{x^2 - 4}$

Answers: See page 205.

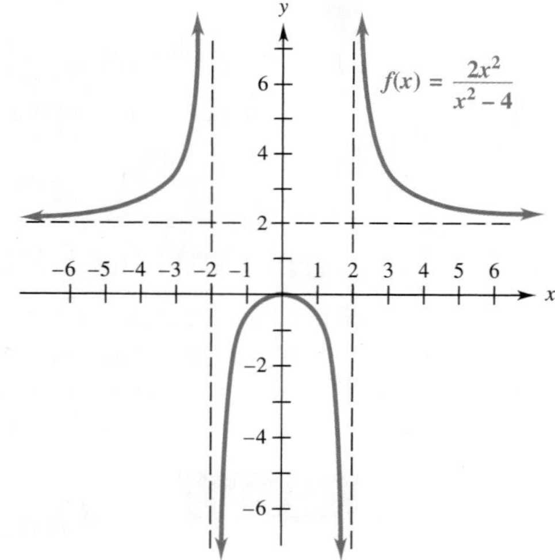

FIGURE 3.52

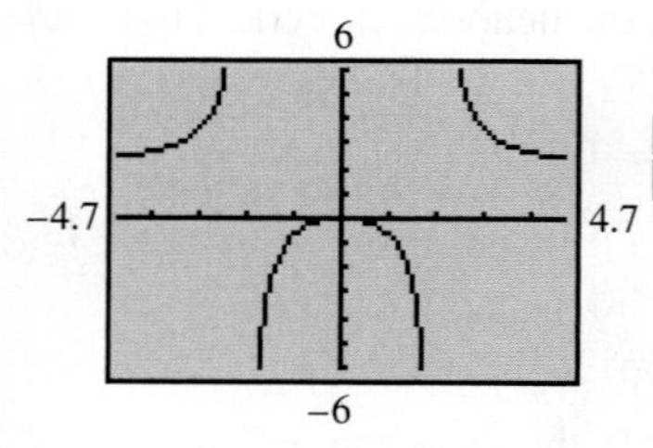

FIGURE 3.53

TECHNOLOGY TIP When a function whose graph has more than one vertical asymptote (as in Example 3) is graphed on a graphing calculator, erroneous vertical lines can sometimes be avoided by using a *decimal window* (with the *y*-range adjusted to show the graph). On TI, use (Z)DECIMAL in the ZOOM or VIEWS menu. On Casio, use INIT in the V-WINDOW menu. Figure 3.53 shows the function of Example 3 graphed in a decimal window on a TI-84+. (The *x*-range may be different on other calculators.)

The arguments used to find the horizontal asymptotes in Examples 1–3 work in the general case and lead to the following conclusion.

If the numerator of a rational function $f(x)$ is of *smaller* degree than the denominator, then the x-axis (the line $y = 0$) is the horizontal asymptote of the graph. If the numerator and denominator are of the *same* degree, say, $f(x) = \frac{ax^n + \cdots}{cx^n + \cdots}$, then the line $y = \frac{a}{c}$ is the horizontal asymptote.*

APPLICATIONS

Rational functions have a variety of applications, some of which are explored next.

Example 4 **Natural Science** In many situations involving environmental pollution, much of the pollutant can be removed from the air or water at a fairly reasonable cost, but the minute amounts of the pollutant that remain can be very expensive to remove.

Cost as a function of the percentage of pollutant removed from the environment can be calculated for various percentages of removal, with a curve fitted through the resulting data points. This curve then leads to a function that approximates the situation. Rational functions often are a good choice for these **cost–benefit functions**.

For example, suppose a cost–benefit function is given by

$$f(x) = \frac{18x}{106 - x},$$

where $f(x)$, or y, is the cost (in thousands of dollars) of removing x percent of a certain pollutant. The domain of x is the set of all numbers from 0 to 100, inclusive; any amount of pollutant from 0% to 100% can be removed. To remove 100% of the pollutant here would cost

$$y = \frac{18(100)}{106 - 100} = 300,$$

or \$300,000. Check that 95% of the pollutant can be removed for about \$155,000, 90% for about \$101,000, and 80% for about \$55,000, as shown in Figure 3.54 on the next page (in which the displayed y-coordinates are rounded to the nearest integer). ✓4

✓Checkpoint 4

Using the function in Example 4, find the cost to remove the following percentages of pollutants.

(a) 70%

(b) 85%

(c) 98%

Answers: See page 205.

*When the numerator is of larger degree than the denominator, the graph has no horizontal asymptote, but may have nonhorizontal lines or other curves as asymptotes; see Exercises 32 and 33 at the end of this section for examples.

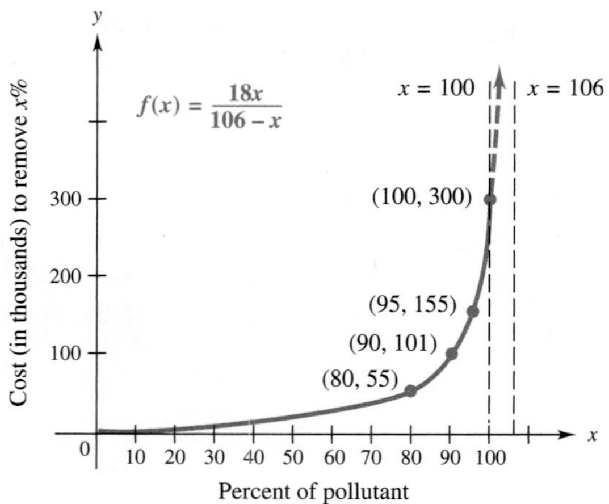

FIGURE 3.54

In management, **product-exchange functions** give the relationship between quantities of two items that can be produced by the same machine or factory. For example, an oil refinery can produce gasoline, heating oil, or a combination of the two; a winery can produce red wine, white wine, or a combination of the two. The next example discusses a product-exchange function.

Example 5 **Business** The product-exchange function for the Fruits of the Earth Winery for red wine x and white wine y, in number of cases, is

$$y = \frac{150{,}000 - 75x}{1200 + x}.$$

Graph the function and find the maximum quantity of each kind of wine that can be produced.

Solution Only nonnegative values of x and y make sense in this situation, so we graph the function in the first quadrant (Figure 3.55). Note that the y-intercept of the graph (found by setting $x = 0$) is 125 and the x-intercept (found by setting $y = 0$ and solving for x) is 2000. Since we are interested in only in the portion of the graph in Quadrant I, we can find a few more points in that quadrant and complete the graph as shown in Figure 3.55.

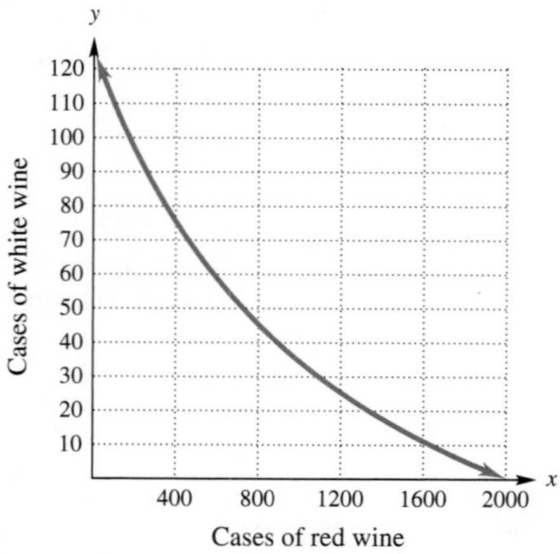

FIGURE 3.55

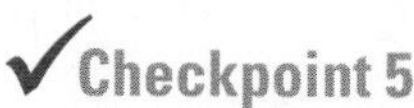

Checkpoint 5

Rework Example 5 with the product-exchange function

$$y = \frac{70{,}000 - 10x}{70 + x}$$

to find the maximum amount of each wine that can be produced.

Answer: See page 205.

The maximum value of y occurs when $x = 0$, so the maximum amount of white wine that can be produced is 125 cases, as given by the y-intercept. The x-intercept gives the maximum amount of red wine that can be produced: 2000 cases. 5

Example 6 A retailer buys 2500 specialty lightbulbs from a distributor each year. In addition to the cost of each bulb, there is a fee for each order, so she wants to order as few times as possible. However, storage costs are higher when there are fewer orders (and hence more bulbs per order to store). Past experience shows that the total annual cost (for the bulbs, ordering fees, and storage costs) is given by the rational function.

$$C(x) = \frac{.98x^2 + 1200x + 22{,}000}{x},$$

where x is the number of bulbs ordered each time. How many bulbs should be ordered each time in order to have the smallest possible cost?

Solution Graph the cost function $C(x)$ in a window with $0 \le x \le 2500$ (because the retailer cannot order a negative number of bulbs and needs only 2500 for the year).

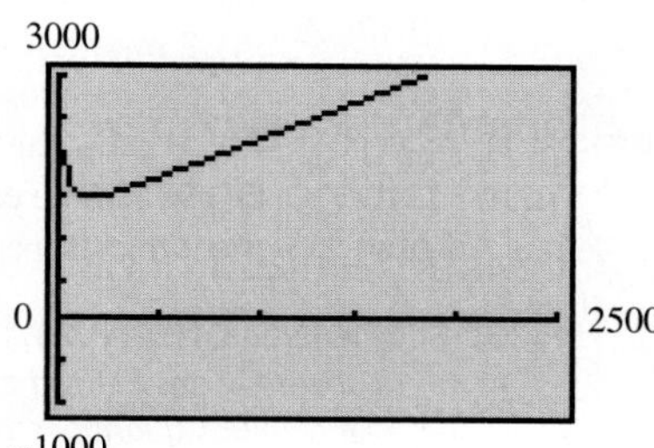

FIGURE 3.56

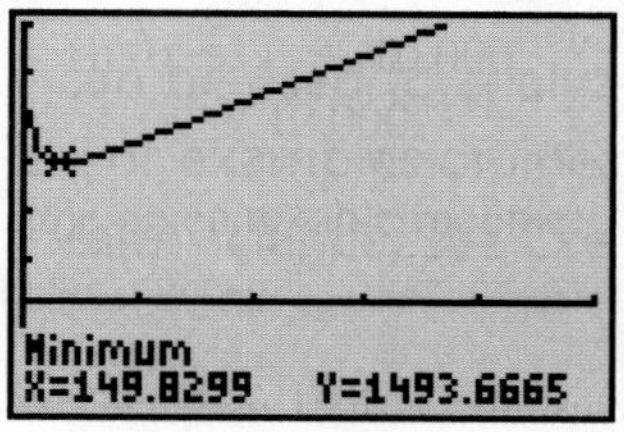

FIGURE 3.57

For each point on the graph in Figure 3.56,

the x-coordinate is the number of bulbs ordered each time;
the y-coordinate is the annual cost when x bulbs are ordered each time.

Use the minimum finder on a graphing calculator to find the point with the smallest y-coordinate, which is approximately (149.83, 1493.67), as shown in Figure 3.57. Since the retailer cannot order part of a lightbulb, she should order 150 bulbs each time, for an approximate annual cost of $1494.

3.7 ▸ Exercises

Graph each function. Give the equations of the vertical and horizontal asymptotes. (See Examples 1–3.)

1. $f(x) = \frac{1}{x + 5}$
2. $g(x) = \frac{-7}{x - 6}$
3. $f(x) = \frac{-3}{2x + 5}$
4. $h(x) = \frac{-4}{2 - x}$
5. $f(x) = \frac{3x}{x - 1}$
6. $g(x) = \frac{x - 2}{x}$
7. $f(x) = \frac{x + 1}{x - 4}$
8. $f(x) = \frac{x - 3}{x + 5}$
9. $f(x) = \frac{2 - x}{x - 3}$
10. $g(x) = \frac{3x - 2}{x + 3}$
11. $f(x) = \frac{3x + 2}{2x + 4}$
12. $f(x) = \frac{4x - 8}{8x + 1}$
13. $h(x) = \frac{x + 1}{x^2 + 2x - 8}$
14. $g(x) = \frac{1}{x(x + 2)^2}$

15. $f(x) = \dfrac{x^2 + 4}{x^2 - 4}$

16. $f(x) = \dfrac{x - 1}{x^2 - 2x - 3}$

Find the equations of the vertical asymptotes of each of the given rational functions.

17. $f(x) = \dfrac{x - 3}{x^2 + x - 2}$

18. $g(x) = \dfrac{x + 2}{x^2 - 1}$

19. $g(x) = \dfrac{x^2 + 2x}{x^2 - 4x - 5}$

20. $f(x) = \dfrac{x^2 - 2x - 4}{x^3 - 2x^2 + x}$

Work these problems. (See Example 4.)

21. **Natural Science** Suppose a cost–benefit model is given by

$$f(x) = \frac{4.3x}{100 - x},$$

where $f(x)$ is the cost, in thousands of dollars, of removing x percent of a given pollutant. Find the cost of removing each of the given percentages of pollutants.

(a) 50% **(b)** 70% **(c)** 80%
(d) 90% **(e)** 95% **(f)** 98%
(g) 99%
(h) Is it possible, according to this model, to remove *all* the pollutant?
(i) Graph the function.

22. **Natural Science** Suppose a cost–benefit model is given by

$$f(x) = \frac{6.2x}{112 - x},$$

where $f(x)$ is the cost, in thousands of dollars, of removing x percent of a certain pollutant. Find the cost of removing the given percentages of pollutants.

(a) 0% **(b)** 50% **(c)** 80%
(d) 90% **(e)** 95% **(f)** 99%
(g) 100% **(h)** Graph the function.

23. **Natural Science** The function

$$f(x) = \frac{\lambda x}{1 + (ax)^b}$$

is used in population models to give the size of the next generation $f(x)$ in terms of the current generation x.*
(a) What is a reasonable domain for this function, considering what x represents?
(b) Graph the function for $\lambda = a = b = 1$ and $x \geq 0$.
(c) Graph the function for $\lambda = a = 1$ and $b = 2$ and $x \geq 0$.
(d) What is the effect of making b larger?

24. **Natural Science** The function

$$f(x) = \frac{Kx}{A + x}$$

is used in biology to give the growth rate of a population in the presence of a quantity x of food. This concept is called Michaelis–Menten kinetics.*
(a) What is a reasonable domain for this function, considering what x represents?
(b) Graph the function for $K = 5$, $A = 2$, and $x \geq 0$.
(c) Show that $y = K$ is a horizontal asymptote.
(d) What do you think K represents?
(e) Show that A represents the quantity of food for which the growth rate is half of its maximum.

25. **Social Science** The average waiting time in a line (or queue) before getting served is given by

$$W = \frac{S(S - A)}{A},$$

where A is the average rate at which people arrive at the line and S is the average service time. At a certain fast-food restaurant, the average service time is 3 minutes. Find W for each of the given average arrival times.
(a) 1 minute
(b) 2 minutes
(c) 2.5 minutes
(d) What is the vertical asymptote?
(e) Graph the equation on the interval (0, 3).
(f) What happens to W when $A > 3$? What does this mean?

Business *Sketch the portion of the graph in Quadrant I of each of the functions defined in Exercises 26 and 27, and then estimate the maximum quantities of each product that can be produced. (See Example 5.)*

26. The product-exchange function for gasoline x and heating oil y, in hundreds of gallons per day, is

$$y = \frac{125{,}000 - 25x}{125 + 2x}.$$

27. A drug factory found that the product-exchange function for a red tranquilizer x and a blue tranquilizer y is

$$y = \frac{900{,}000{,}000 - 30{,}000x}{x + 90{,}000}.$$

28. **Physical Science** The failure of several O-rings in field joints was the cause of the fatal crash of the *Challenger* space shuttle in 1986. NASA data from 24 successful launches prior to *Challenger* suggested that O-ring failure was related to launch temperature by a function similar to

$$N(t) = \frac{600 - 7t}{4t - 100} \quad (50 \leq t \leq 85),$$

where t is the temperature (in °F) at launch and N is the approximate number of O-rings that fail. Assume that this

*See J. Maynard Smith, *Models in Ecology* (Cambridge University Press, 1974).

*See Leah Edelstein-Keshet, *Mathematical Models in Biology* (Random House, 1988).

function accurately models the number of O-ring failures that would occur at lower launch temperatures (an assumption NASA did not make).

(a) Does $N(t)$ have a vertical asymptote? At what value of t does it occur?

(b) Without actually graphing the function, what would you conjecture that the graph would look like just to the right of the vertical asymptote? What does this suggest about the number of O-ring failures that might be expected near that temperature? (The temperature at the *Challenger* launching was 31°.)

(c) Confirm your conjecture by graphing $N(t)$ between the vertical asymptote and $t = 85$.

29. Business A company has fixed costs of \$40,000 and a marginal cost of \$2.60 per unit.

(a) Find the linear cost function.

(b) Find the average cost function. (Average cost was defined in Section 3.3.)

(c) Find the horizontal asymptote of the graph of the average cost function. Explain what the asymptote means in this situation. (How low can the average cost be?)

Use a graphing calculator to do Exercises 30–33. (See Example 6.)

30. Finance Another model of a Laffer curve (see Exercise 27 of Section 3.6) is given by

$$f(x) = \frac{300x - 3x^2}{10x + 200},$$

where $f(x)$ is government revenue (in billions of dollars) from a tax rate of x percent. Find the revenue from the given tax rates.

(a) 16% (b) 25%

(c) 40% (d) 55%

(e) Graph $f(x)$.

(f) What tax rate produces maximum revenue? What is the maximum revenue?

31. Business When no more than 110 units are produced, the cost of producing x units is given by

$$C(x) = .2x^3 - 25x^2 + 1531x + 25{,}000.$$

How many units should be produced in order to have the lowest possible average cost?

32. (a) Graph $f(x) = \dfrac{x^3 + 3x^2 + x + 1}{x^2 + 2x + 1}$.

(b) Does the graph appear to have a horizontal asymptote? Does the graph appear to have some nonhorizontal straight line as an asymptote?

(c) Graph $f(x)$ and the line $y = x + 1$ on the same screen. Does this line appear to be an asymptote of the graph of $f(x)$?

33. (a) Graph $g(x) = \dfrac{x^3 - 2}{x - 1}$ in the window with $-5 \le x \le 5$ and $-6 \le y \le 12$.

(b) Graph $g(x)$ and the parabola $y = x^2 + x + 1$ on the same screen. How do the two graphs compare when $|x| \ge 2$?

✓ Checkpoint Answers

1. (a)

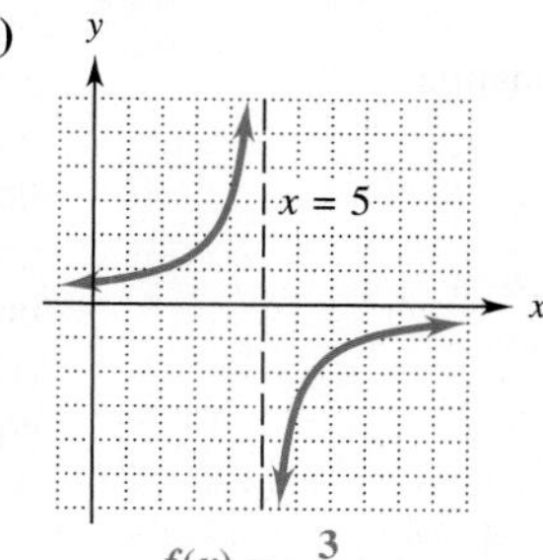

$f(x) = \dfrac{3}{5 - x}$

(b)

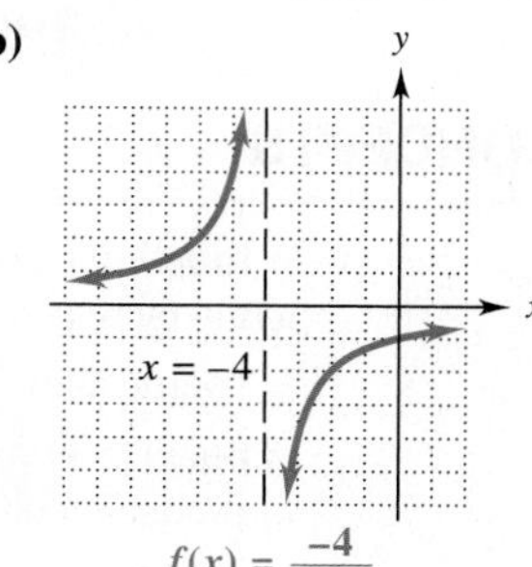

$f(x) = \dfrac{-4}{x + 4}$

2. (a)

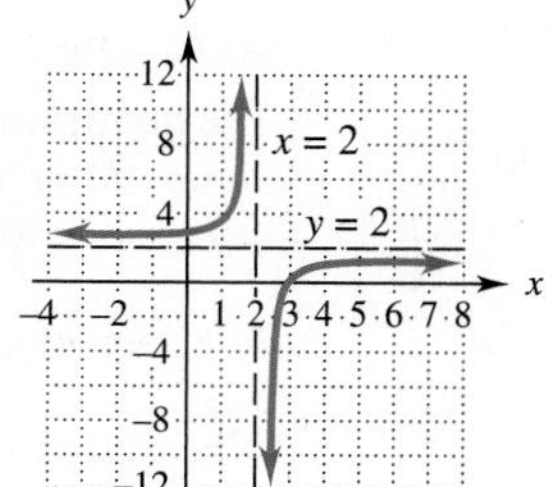

(b)

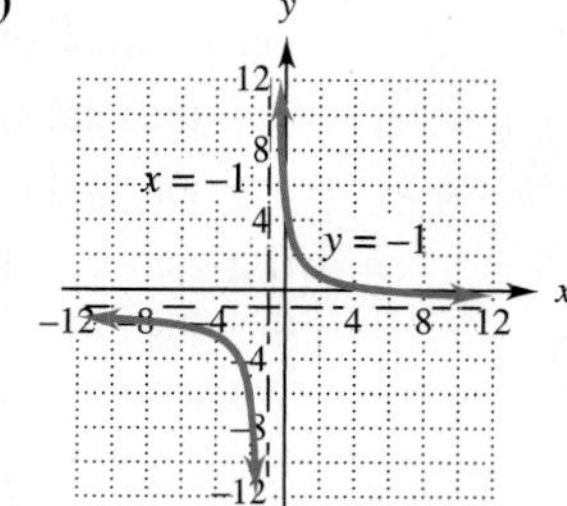

3. (a) Vertical, $x = -5$; horizontal, $y = 3$
(b) Vertical, $x = -2$ and $x = 2$; horizontal, $y = -1$

4. (a) \$35,000 (b) About \$73,000 (c) About \$221,000

5. 7000 cases of red, 1000 cases of white

CHAPTER 3 ▸ Summary

KEY TERMS AND SYMBOLS

3.1 ▸ function
domain
range
functional notation
piecewise-defined function

3.2 ▸ graph
linear function
piecewise linear function
absolute value function
greatest integer function
step function
graph reading
vertical line test

3.3 ▸ fixed costs
variable cost
average cost
linear depreciation
average rate of change
marginal cost
linear cost function
linear revenue function
break-even point
supply and demand curves
equilibrium point
equilibrium price
equilibrium quantity

3.4 ▸ quadratic function
parabola
vertex
axis

3.5 ▸ quadratic model

3.6 ▸ polynomial function
graph of $f(x) = ax^n$
properties of polynomial graphs
polynomial models

3.7 ▸ rational function
linear rational function
vertical asymptote
horizontal asymptote

CHAPTER 3 KEY CONCEPTS

A **function** consists of a set of inputs called the **domain**, a set of outputs called the **range**, and a rule by which each input determines exactly one output.

If a vertical line intersects a graph in more than one point, the graph is not that of a function.

A **linear cost function** has equation $C(x) = mx + b$, where m is the **marginal cost** (the cost of producing one more item) and b is the **fixed cost**.

If $p = f(q)$ gives the price per unit when q units can be supplied and $p = g(q)$ gives the price per unit when q units are demanded, then the **equilibrium price** and **equilibrium quantity** occur at the q-value such that $f(q) = g(q)$.

The **quadratic function** defined by $f(x) = a(x - h)^2 + k$ has a graph that is a **parabola** with vertex (h, k) and axis of symmetry $x = h$. The parabola opens upward if $a > 0$ and downward if $a < 0$.

If the equation is in the form $f(x) = ax^2 + bx + c$, the vertex is $\left(-\frac{b}{2a}, f\left(-\frac{b}{2a}\right)\right)$.

When $|x|$ is large, the graph of a **polynomial function** resembles the graph of its highest-degree term ax^n. The graph of $f(x) = ax^n$ is described on page 186.

On the graph of a polynomial function of degree n,

the total number of peaks and valleys is at most $n - 1$;

the total number of x-intercepts is at most n.

If a number c makes the denominator of a **rational function** 0, but the numerator nonzero, then the line $x = c$ is a **vertical asymptote** of the graph.

Whenever the values of y approach, but do not equal, some number k as $|x|$ gets larger and larger, the line $y = k$ is a **horizontal asymptote** of the graph.

If the numerator of a rational function is of *smaller* degree than the denominator, then the x-axis is the horizontal asymptote of the graph.

If the numerator and denominator of a rational function are of the *same* degree, say, $f(x) = \frac{ax^n + \cdots}{cx^n + \cdots}$, then the line $y = \frac{a}{c}$ is the horizontal asymptote of the graph.

CHAPTER 3 REVIEW EXERCISES

In Exercises 1–6, state whether the given rule defines a function or not.

1.

x	3	2	1	0	1	2
y	8	5	2	0	−2	−5

2.

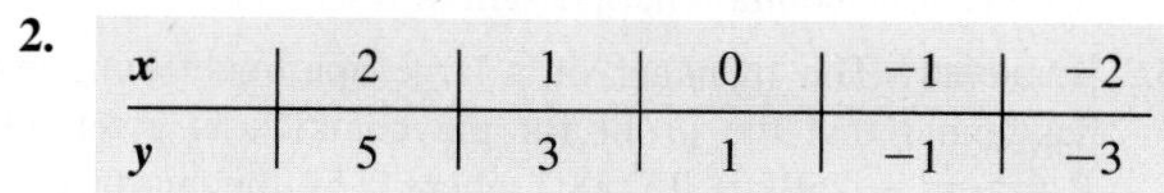

x	2	1	0	−1	−2
y	5	3	1	−1	−3

3. $y = \sqrt{x}$
4. $x = |y|$
5. $x = y^2 + 1$
6. $y = 5x - 2$

For the functions in Exercises 7–10, find

(a) $f(6)$; **(b)** $f(-2)$; **(c)** $f(p)$; **(d)** $f(r + 1)$.

7. $f(x) = 4x - 1$
8. $f(x) = 3 - 4x$
9. $f(x) = -x^2 + 2x - 4$
10. $f(x) = 8 - x - x^2$
11. Let $f(x) = 5x - 3$ and $g(x) = -x^2 + 4x$. Find each of the following:
    (a) $f(-2)$ **(b)** $g(3)$ **(c)** $g(-k)$
    (d) $g(3m)$ **(e)** $g(k - 5)$ **(f)** $f(3 - p)$
12. Let $f(x) = x^2 + x + 1$. Find each of the following:
    (a) $f(3)$ **(b)** $f(1)$ **(c)** $f(4)$
    (d) Based on your answers in parts (a)–(c), is it true that $f(a + b) = f(a) + f(b)$ for all real numbers a and b?

Graph the functions in Exercises 13–24.

13. $f(x) = |x| - 3$
14. $f(x) = -|x| - 2$
15. $f(x) = -|x + 1| + 3$
16. $f(x) = 2|x - 3| - 4$
17. $f(x) = [x - 3]$
18. $f(x) = \left[\frac{1}{2}x - 2\right]$
19. $f(x) = \begin{cases} -4x + 2 & \text{if } x \le 1 \\ 3x - 5 & \text{if } x > 1 \end{cases}$
20. $f(x) = \begin{cases} 3x + 1 & \text{if } x < 2 \\ -x + 4 & \text{if } x \ge 2 \end{cases}$
21. $f(x) = \begin{cases} |x| & \text{if } x < 3 \\ 6 - x & \text{if } x \ge 3 \end{cases}$
22. $f(x) = \sqrt{x^2}$
23. $g(x) = x^2/8 - 3$
24. $h(x) = \sqrt{x} + 2$

25. **Business** Let f be a function that gives the cost to rent a power washer for x hours. The cost is a flat \$45 for renting the washer, plus \$20 per day or fraction of a day for using the washer.
    (a) Graph f.
    (b) Give the domain and range of f.
    (c) John McDonough wants to rent the washer, but he can spend no more than \$90. What is the maximum number of days he can use it?

26. **Business** A tree removal service assesses a \$400 fee and then charges \$80 per hour for the time on an owner's property.
    (a) Is \$750 enough for 5 hours work?
    (b) Graph the ordered pairs (hours, cost).
    (c) Give the domain and range.

27. **Social Science** The percentage of children born to unmarried mothers since 1970 can be approximated by

$$f(x) = \begin{cases} .9x + 11 & \text{from 1970 to 1994} \\ 32.6 & \text{from 1994 to 2010} \end{cases}.$$

Let $x = 0$ correspond to 1970, and graph the function. What does the graph suggest about births to unmarried mothers?

28. **Health** The following graph, which tracks the percentage of children under age 18 with some kind of food allergy over a period of 10 years, appeared on the Web site for the National Center for Health Statistics:

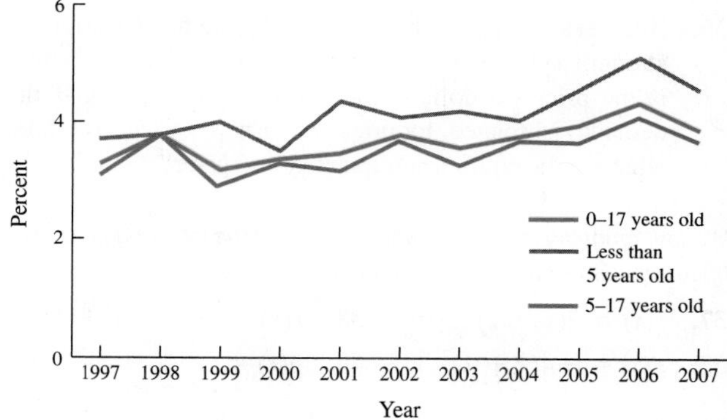

(a) What is the general trend for children less than 5 years old?
(b) Let $x = 7$ correspond to 1997. Approximate a linear function to fit the data for the percentage of children less than 5 years old with food allergies, using 3.9% in 1997 and 4.7% in 2007.
(c) If the trend continues, estimate the percentage of children less than 5 years old with food allergies in 2014.

Business *In Exercises 29–32, find the following:*

(a) *the linear cost function;*
(b) *the marginal cost;*
(c) *the average cost per unit to produce 100 units.*

29. Eight units cost \$300; fixed cost is \$60.
30. Fixed cost is \$2000; 36 units cost \$8480.
31. Twelve units cost \$445; 50 units cost \$1585.
32. Thirty units cost \$1500; 120 units cost \$5640.

33. **Business** The cost of producing x ink cartridges for a printer is given by $C(x) = 24x + 18{,}000$. Each cartridge can be sold for $28.
(a) What are the fixed costs?
(b) Find the revenue function.
(c) Find the break-even point.
(d) If the company sells exactly the number of cartridges needed to break even, what is its revenue?

34. **Business** The net amount of electricity produced in the United States (in billions of kilowatt hours) is approximated by the function

$$f(x) = 68.6x + 3009,$$

where $x = 1$ corresponds to 1991.* Using the model, find the year in which energy consumption first exceeds 4300 billion kilowatt hours.

35. **Business** Suppose the demand and price for the HBO cable channel are related by $p = -.5q + 30.95$, where p is the monthly price in dollars and q is measured in millions of subscribers. If the price and supply are related by $p = .3q + 2.15$, what are the equilibrium quantity and price?

36. **Business** Suppose the supply and price for prescription-strength Tylenol are related by $p = .0015q + 1$, where p is the price (in dollars) of a 30-day prescription. If the demand is related to price by $p = -.0025q + 64.36$, what are the equilibrium quantity and price?

Without graphing, determine whether each of the following parabolas opens upward or downward, and find its vertex.

37. $f(x) = 3(x - 2)^2 + 6$
38. $f(x) = 2(x + 3)^2 - 5$
39. $g(x) = -4(x + 1)^2 + 8$
40. $g(x) = -5(x - 4)^2 - 6$

Graph each of the following quadratic functions, and label its vertex.

41. $f(x) = x^2 - 9$
42. $f(x) = 5 - 2x^2$
43. $f(x) = x^2 + 2x - 6$
44. $f(x) = -x^2 + 8x - 1$
45. $f(x) = -x^2 - 6x + 5$
46. $f(x) = 5x^2 + 20x - 2$
47. $f(x) = 2x^2 - 12x + 10$
48. $f(x) = -3x^2 - 12x - 2$

Determine whether the functions in Exercises 47–52 have a minimum or a maximum value, and find that value.

49. $f(x) = x^2 + 6x - 2$
50. $f(x) = x^2 + 4x + 5$
51. $g(x) = -4x^2 + 8x + 3$
52. $g(x) = -3x^2 - 6x + 3$

Solve each problem.

53. **Business** The commodity market is very unstable; money can be made or lost quickly when investing in soybeans, wheat, pork bellies, and the like. Suppose that an investor kept track of her total profit P (measured in thousands of dollars) at time t; measured in months, after she began investing and found that $P = -3t^2 + 18t - 15$. At what time is her profit largest? (*Hint*: $t > 0$ in this case.)

54. **Physical Science** The height h (in feet) of a rocket at time t seconds after liftoff is given by $h = -16t^2 + 800t$.
(a) How long does it take the rocket to reach 3200 feet?
(b) What is the maximum height of the rocket?

55. **Business** The manager of a large apartment complex has found that the profit for the complex is given by $P = -x^2 + 250x - 15{,}000$, where x is the number of units rented. For what value of x does the complex produce the largest profit?

56. **Business** A rectangular enclosure is to be built with three sides made out of redwood fencing, at a cost of $15 per running foot, and the fourth side made out of cement blocks, at a cost of $30 per running foot. $900 is available for the project. What are the dimensions of the enclosure with maximum possible area, and what is this area?

57. **Business** The following table shows the average cost of tuition and fees, in dollars, at private colleges in various years*:

Year	Cost
1977	2700
1982	4639
1987	7048
1992	10,448
1997	13,785
2002	18,060
2007	23,712

(a) Let $x = 7$ correspond to 1977. Find a quadratic function $f(x) = a(x - h)^2 + k$ that models this data, using (7, 2700) as the vertex and the data for 2007.
(b) Estimate average tuition and fees at private colleges in 2012.

58. **Social Science** The following table shows the percentage of U.S. students who carried a weapon to school in various years†:

Year	Percentage
1993	22.1
1995	20.0
1997	18.3
1999	17.3
2001	17.4
2003	17.1
2005	18.5

*U.S. Energy Information Administration.

*U.S. National Center for Education Statistics.

†The U.S. Department of Justice.

(a) Sketch a scatter plot for this data, with $x = 3$ corresponding to 1993.
(b) Use quadratic regression to find a function g that models this data.
(c) Assuming the trend continues, estimate the percentage of students carrying a weapon school in 2009.

Graph each of the following polynomial functions.

59. $f(x) = x^4 - 5$
60. $g(x) = x^3 - 4x$
61. $f(x) = x(x - 4)(x + 1)$
62. $f(x) = (x - 1)(x + 2)(x - 3)$
63. $f(x) = 3x(3x + 2)(x - 1)$
64. $f(x) = x^3 - 3x^2 - 4x$
65. $f(x) = x^4 - 5x^2 - 6$
66. $f(x) = x^4 - 7x^2 - 8$

Use a graphing calculator to do Exercises 67–70.

67. **Business** The demand equation for automobile oil filters is

$$p = -.000012q^3 - .00498q^2 + .1264q + 1508,$$

where p is in dollars and q is in thousands of items. The supply equation is

$$p = -.000001q^3 + .00097q^2 + 2q.$$

Find the equilibrium quantity and the equilibrium price.

68. **Business** The average cost (in dollars) per item of manufacturing x thousand cans of spray paint is given by

$$A(x) = -.000006x^4 + .0017x^3 + .03x^2 - 24x + 1110.$$

How many cans should be manufactured if the average cost is to be as low as possible? What is the average cost in that case?

69. **Business** Plastic racks for holding compact discs sell for \$23 each. The cost of manufacturing x racks is given by $C(x) = -.000006x^3 + .07x^2 + 2x + 1200$. A maximum of 600 racks can be produced.
(a) Find the revenue and profit functions.
(b) What is the break-even point?
(c) What is the maximum number of racks that can be made without losing money?
(d) How many racks should be made in order to have as large a profit as possible? What is that profit?

70. **Health** The following table shows the average remaining life expectancy (in years) for a person in the United States at selected ages*:

Current Age	Birth	20	40	60	80	90	100
Life Expectancy	77.8	58.8	39.9	22.6	9.2	5.0	2.6

(a) Let birth = 0, and plot the data points.
(b) Use quartic regression to find a fourth-degree polynomial $f(x)$ that models the data.
(c) What is the remaining life expectancy of a person of age 25? age 35? age 50?
(d) What is the life expectancy of a person who is exactly your age?

*National Center for Health Statistics.

List the vertical and horizontal asymptotes of each function, and sketch its graph.

71. $f(x) = \dfrac{1}{x - 3}$
72. $f(x) = \dfrac{-2}{x + 4}$
73. $f(x) = \dfrac{-3}{2x - 4}$
74. $f(x) = \dfrac{5}{3x + 7}$
75. $g(x) = \dfrac{5x - 2}{4x^2 - 4x - 3}$
76. $g(x) = \dfrac{x^2}{x^2 - 1}$

77. **Business** The average cost per carton of producing x cartons of cocoa is given by

$$A(x) = \frac{650}{2x + 40}.$$

Find the average cost per carton to make the given number of cartons.
(a) 10 cartons (b) 50 cartons (c) 70 cartons
(d) 100 cartons (e) Graph $C(x)$.

78. **Business** The cost and revenue functions (in dollars) for a frozen-yogurt shop are given by

$$C(x) = \frac{400x + 400}{x + 4} \quad \text{and} \quad R(x) = 100x,$$

where x is measured in hundreds of units.
(a) Graph $C(x)$ and $R(x)$ on the same set of axes.
(b) What is the break-even point for this shop?
(c) If the profit function is given by $P(x)$, does $P(1)$ represent a profit or a loss?
(d) Does $P(4)$ represent a profit or a loss?

79. **Business** The supply and demand functions for the yogurt shop in Exercise 78 are

$$\text{supply: } p = \frac{q^2}{4} + 25 \quad \text{and} \quad \text{demand: } p = \frac{500}{q},$$

where p is the price in dollars for q hundred units of yogurt.
(a) Graph both functions on the same axes, and from the graph, estimate the equilibrium point.
(b) Give the q-intervals where supply exceeds demand.
(c) Give the q-intervals where demand exceeds supply.

80. **Business** A cost–benefit curve for pollution control is given by

$$y = \frac{9.2x}{106 - x},$$

where y is the cost, in thousands of dollars, of removing x percent of a specific industrial pollutant. Find y for each of the given values of x.
(a) $x = 50$ (b) $x = 98$
(c) What percent of the pollutant can be removed for \$22,000?

CASE 3

Architectural Arches

From ancient Roman bridges to medieval cathedrals, modern highway tunnels, and fast-food restaurants, arches are everywhere. For centuries builders and architects have used them for both structural and aesthetic reasons. There are arches of almost every material, from stone to steel. Some of the most common arch shapes are the parabolic arch, the semicircular arch, and the Norman arch (a semicircular arch set atop a rectangle, as shown here).

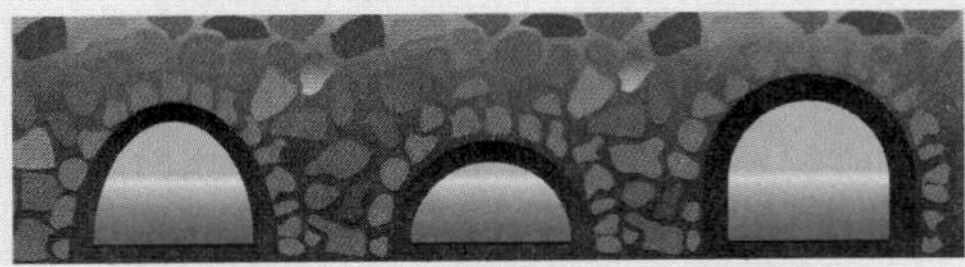

Parabolic Semicircular Norman

Note that every arch is symmetric around a vertical line through its center. The part of the arch on the left side of the line is mirror image of the part on the right side, with the line being the mirror. To describe these arches mathematically, suppose that each one is located on a coordinate plane with the origin at the intersection of the verticle symmetry line and a horizontal line at the base of the arch. For a parabolic arch, the situation looks like this, for some numbers k and c:

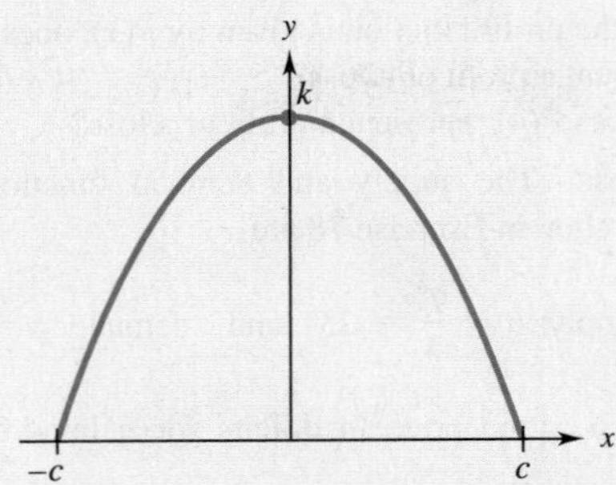

Since $(0, k)$ is the vertex, the arch is the graph of a function of the form

$$f(x) = a(x - 0)^2 + k,$$

or

$$f(x) = ax^2 + k.$$

Note that the point $(c, 0)$ is on the arch, which means that $f(c) = 0$. Therefore, we can make the following derivation:

$$f(x) = ax^2 + k$$

$$f(c) = ac^2 + k \qquad \text{Let } x = c.$$

$$0 = ac^2 + k \qquad f(c) = 0.$$

$$-k = ac^2 \qquad \text{Subtract } k \text{ from both sides.}$$

$$a = \frac{-k}{c^2} \qquad \text{Divide both sides by } c^2.$$

So the function whose graph is the shape of the parabolic arch is

$$f(x) = \frac{-k}{c^2}x^2 + k \qquad (-c \le x \le c),$$

where k is the height of the arch at its highest point and $2c$ (the distance from $-c$ to c) is the width of the arch at its base. For example, a 12-foot-high arch that is 14 feet wide at its base has $k = 12$ and $c = 7$, so it is the graph of

$$f(x) = \frac{-12}{49}x^2 + 12.$$

In order to describe semicircular and Norman arches, we must first find the equation of a circle of radius r with center at the origin. Consider a point (x, y) on the graph and the right triangle it determines*:

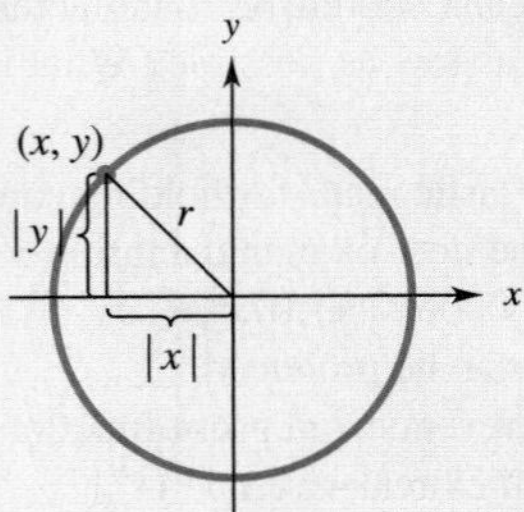

The horizontal side of the triangle has length $|x|$ (the distance from x to 0 on the x-axis), and the vertical side has length $|y|$ (the distance from y to 0 on the y-axis). The hypotenuse has length r (the radius of the circle). By the Pythagorean theorem,

$$|x|^2 + |y|^2 = r^2,$$

which is equivalent to

$$x^2 + y^2 = r^2,$$

*The figure shows a point in the second quadrant, where x is negative and y is positive, but the same argument will work for points in other quadrants.

because the absolute value of x is either x or $-x$ (see the definition on page 9), so $|x|^2 = (\pm x)^2 = x^2$, and similarly for y. Solving this equation for y shows that

$$y^2 = r^2 - x^2$$
$$y = \sqrt{r^2 - x^2} \quad \text{or} \quad y = -\sqrt{r^2 - x^2}.$$

In the first equation, y is always positive or 0 (because square roots are nonnegative), so its graph is the top half of the circle. Similarly, the second equation gives the bottom half of the circle.

Now consider a semicircle arch of radius r:

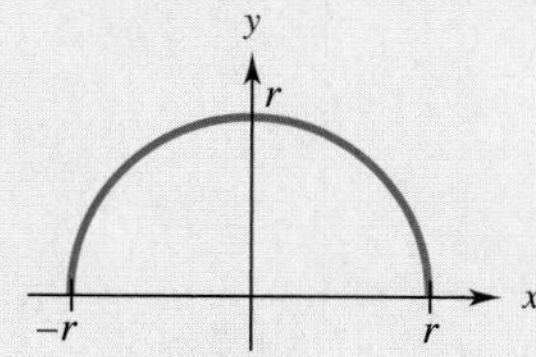

The arch is the top half of a circle with center at the origin. By the preceding paragraph, it is the graph of the function

$$g(x) = \sqrt{r^2 - x^2} \quad (-r \le x \le r).$$

For instance, a semicircular arch that is 8 feet high has $r = 8$ and is the graph of

$$g(x) = \sqrt{8^2 - x^2} = \sqrt{64 - x^2} \quad (-8 \le x \le 8).$$

A Norman arch is not the graph of a function (since its sides are vertical lines), but we can describe the semicircular top of the arch. For example, consider a Norman arch whose top has radius 8 and whose sides are 10 feet high. If the top of the arch were at ground level, then its equation would be $g(x) = \sqrt{64 - x^2}$, as we just saw. But in the actual arch, this semicircular part is raised 10 feet, so it is the graph of

$$h(x) = g(x) + 10 = \sqrt{64 - x^2} + 10 \quad (-8 \le x \le 8),$$

as shown below.

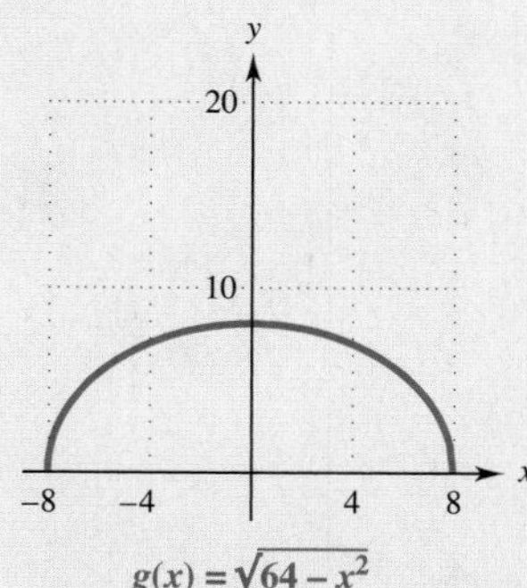

$g(x) = \sqrt{64 - x^2}$

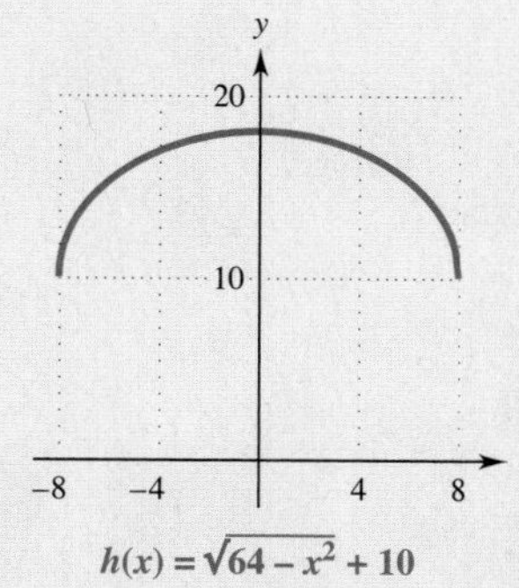

$h(x) = \sqrt{64 - x^2} + 10$

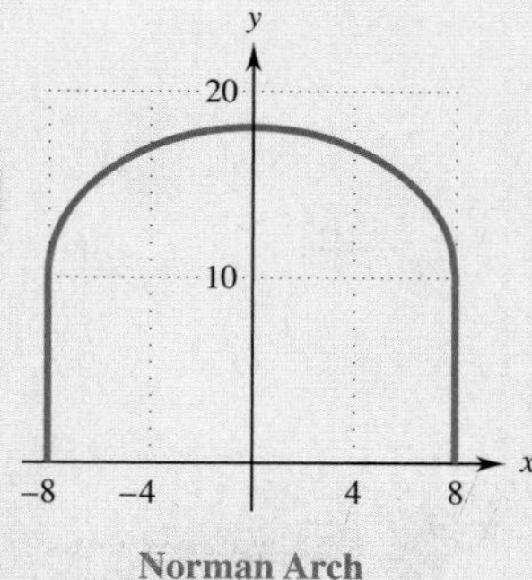

Norman Arch

EXERCISES

1. Write a function $f(x)$ that describes a parabolic arch which is 20 feet tall and 14 feet wide at the base.
2. Write a function $g(x)$ that describes a semicircular arch which is 20 feet tall. How wide is the arch at its base?
3. Write a function $h(x)$ that describes the top part of a Norman arch which is 20 feet tall and 24 feet wide at the base. How high are the vertical sides of this arch?
4. Would a truck that is 12 feet tall and 9 feet wide fit through all of the arches in Exercises 1–3? How could you fix any of the arches that are too small so that the truck would fit through?

CHAPTER

4

Exponential and Logarithmic Functions

Population growth (of humans, fish, bacteria, etc.), compound interest, radioactive decay, and a host of other phenomena can be described by exponential functions. See Exercises 40–46 on page 222. Archeologists sometimes use carbon-14 dating to determine the approximate age of an artifact (such as a dinosaur skeleton or a mummy). This procedure involves using logarithms to solve an exponential equation. See Exercises 70–72 on page 252. The Richter scale for measuring the magnitude of an earthquake is a logarithmic function. See Exercises 73–74 on page 252.

Exponential and logarithmic functions play a key role in management, economics, the social and physical sciences, and engineering. We begin with exponential growth and exponential decay functions.

4.1 ► Exponential Functions

In polynomial functions such as $f(x) = x^2 + 5x - 4$, the variable is raised to various constant exponents. In **exponential functions**, such as

$$f(x) = 10^x, \quad g(x) = 750(1.05^x), \quad h(x) = 3^{.6x}, \quad \text{and} \quad k(x) = 2^{-x^2},$$

the variable is in the exponent and the **base** is a positive constant. We begin with the simplest type of exponential function, whose rule is of the form $f(x) = a^x$, with $a > 0$.

Example 1 Graph $f(x) = 2^x$, and estimate the height of the graph when $x = 50$.

Solution Either use a graphing calculator or graph by hand: Make a table of values, plot the corresponding points, and join them by a smooth curve, as in Figure 4.1. The graph has y-intercept 1 and rises steeply to the right. Note that the graph gets very close to the x-axis on the left, but always lies *above* the axis (because *every* power of 2 is positive).

x	y
−3	1/8
−2	1/4
−1	1/2
0	1
1	2
2	4
3	8

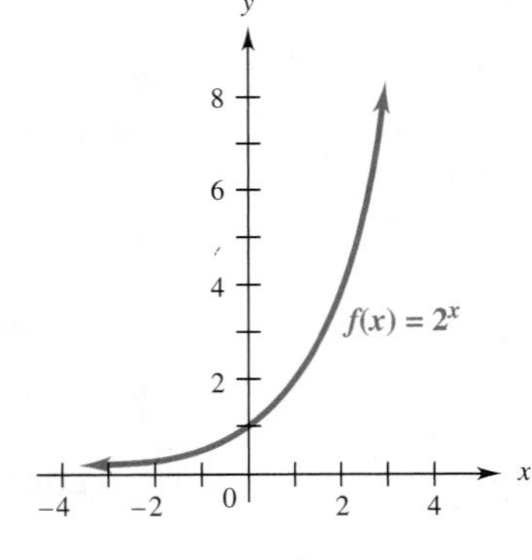

FIGURE 4.1

The graph illustrates **exponential growth**, which is far more explosive than polynomial growth. At $x = 50$, the graph would be 2^{50} units high. Since there are approximately 6 units to the inch in Figure 4.1, and since there are 12 inches to the foot and 5280 feet to the mile, the height of the graph at $x = 50$ would be approximately

$$\frac{2^{50}}{6 \times 12 \times 5280} \approx 2{,}961{,}647{,}482 \textit{ miles!}$$ ✓1

✓ **Checkpoint 1**

(a) Fill in this table:

x	$g(x) = 3^x$
−3	
−2	
−1	
0	
1	
2	
3	

(b) Sketch the graph of $g(x) = 3^x$.

Answers: See page 223.

When $a > 1$, the graph of the exponential function $h(x) = a^x$ has the same basic shape as the graph of $f(x) = 2^x$, as illustrated in Figure 4.2 and summarized in the next box.

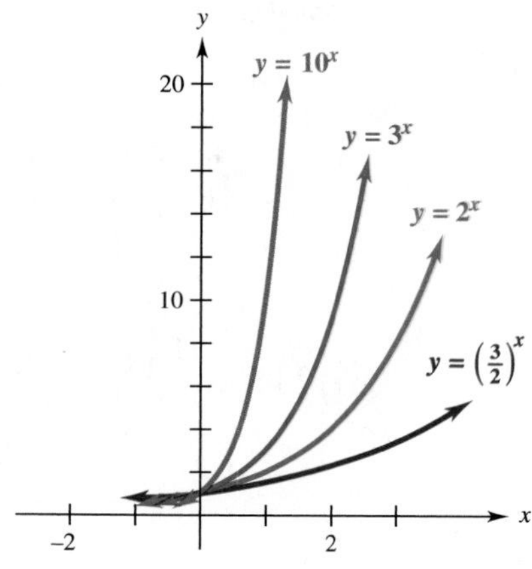

FIGURE 4.2

When $a > 1$, the function $f(x) = a^x$ has the set of all real numbers as its domain. Its graph has the shape shown here and all five of the properties listed below.

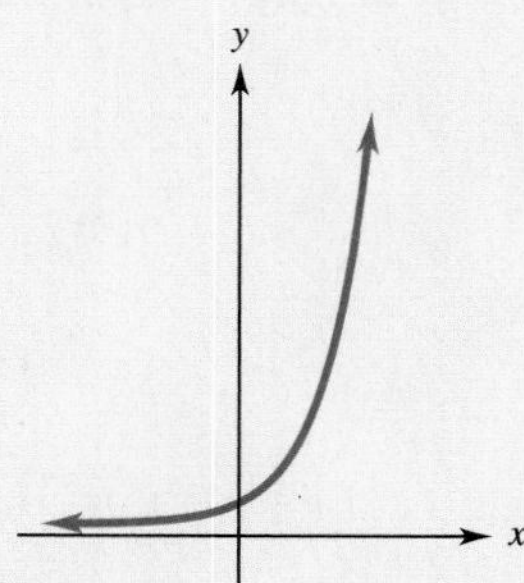

1. The graph is above the x-axis.
2. The y-intercept is 1.
3. The graph climbs steeply to the right.
4. The negative x-axis is a horizontal asymptote.
5. The larger the base a, the more steeply the graph rises to the right.

Example 2 Consider the function $g(x) = 2^{-x}$.

(a) Rewrite the rule of g so that no minus signs appear in it.

Solution By the definition of negative exponents,

$$g(x) = 2^{-x} = \frac{1}{2^x} = \left(\frac{1}{2}\right)^x.$$

(b) Graph $g(x)$.

Solution Either use a graphing calculator or graph by hand in the usual way, as shown in Figure 4.3.

x	$y = 2^{-x}$
−3	8
−2	4
−1	2
0	1
1	1/2
2	1/4
3	1/8

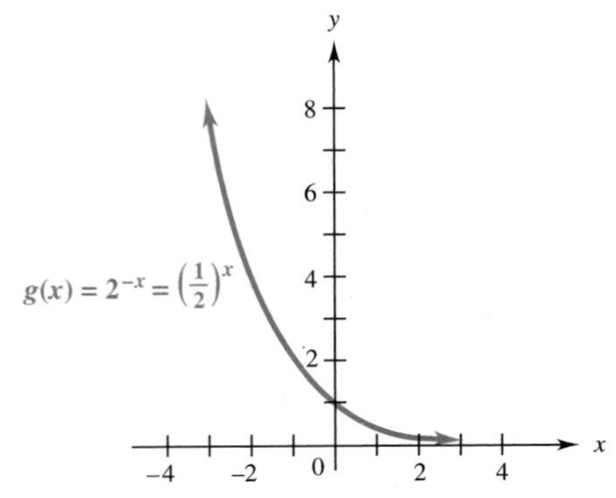

FIGURE 4.3

✓Checkpoint 2

Graph $h(x) = (1/3)^x$.

Answer: See page 223.

The graph falls sharply to the right, but never touches the x-axis, because every power of $\frac{1}{2}$ is positive. This is an example of **exponential decay**. 2✓

When $0 < a < 1$, the graph of $k(x) = a^x$ has the same basic shape as the graph of $g(x) = (1/2)^x$, as illustrated in Figure 4.4 and summarized in the next box.

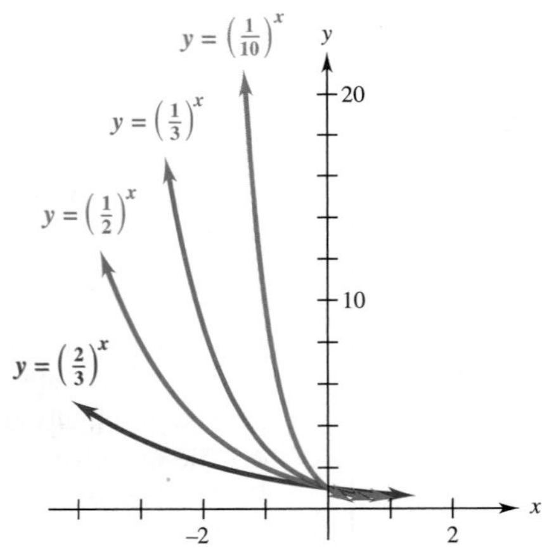

FIGURE 4.4

When $0 < a < 1$, the function $f(x) = a^x$ has the set of all real numbers as its domain. Its graph has the shape shown here and all five of the properties listed on the next page.

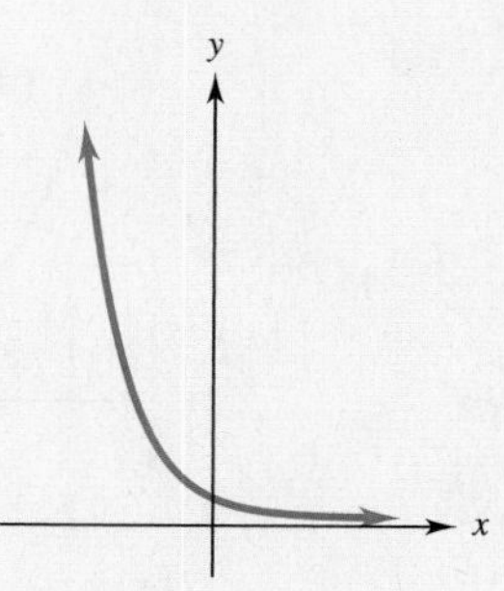

1. The graph is above the x-axis.
2. The y-intercept is 1.
3. The graph falls sharply to the right.
4. The positive x-axis is a horizontal asymptote.
5. The smaller the base a, the more steeply the graph falls to the right.

Example 3 In each case, graph $f(x)$ and $g(x)$ on the same set of axes and explain how the graphs are related.

(a) $f(x) = 2^x$ and $g(x) = (1/2)^x$

Solution The graphs of f and g are shown in Figures 4.1 and 4.3, above. Placing them on the same set of axes, we obtain Figure 4.5. It shows that the graph of $g(x) = (1/2)^x$ is the mirror image of the graph of $f(x) = 2^x$, with the y-axis as the mirror. ✓3

✓ Checkpoint 3

Use a graphing calculator to graph $f(x) = 4^x$ and $g(x) = \left(\frac{1}{4}\right)^x$ on the same screen.

Answer: See page 224.

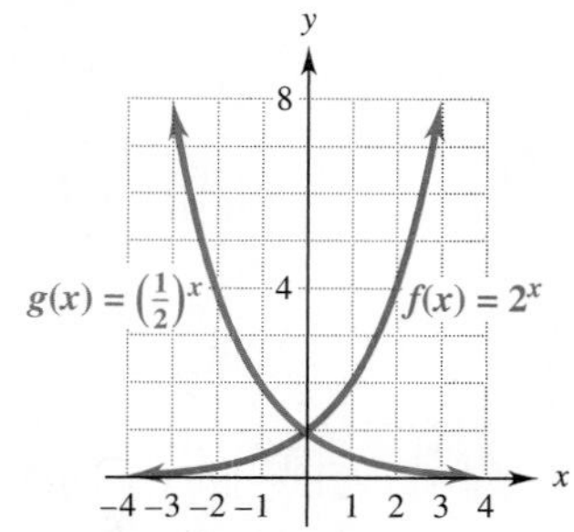

FIGURE 4.5

(b) $f(x) = 3^{1-x}$ and $g(x) = 3^{-x}$

Solution Choose values of x that make the exponent positive, zero, and negative, and plot the corresponding points. The graphs are shown in Figure 4.6 on the next page. The graph of $f(x) = 3^{1-x}$ has the same shape as the graph of $g(x) = 3^{-x}$, but is shifted 1 unit to the right, making the y-intercept (0, 3) rather than (0, 1).

(c) $f(x) = 2^{.6x}$ and $g(x) = 2^x$

Solution Comparing the graphs of $f(x) = 2^{.6x}$ and $g(x) = 2^x$ in Figure 4.7, we see that the graphs are both increasing, but the graph of $f(x)$ rises at a slower rate. This happens because of the .6 in the exponent. If the coefficient of x were greater than 1, the graph would rise at a faster rate than the graph of $g(x) = 2^x$. ✓4

✓ Checkpoint 4

Graph $f(x) = 2^{x+1}$.

Answer: See page 224.

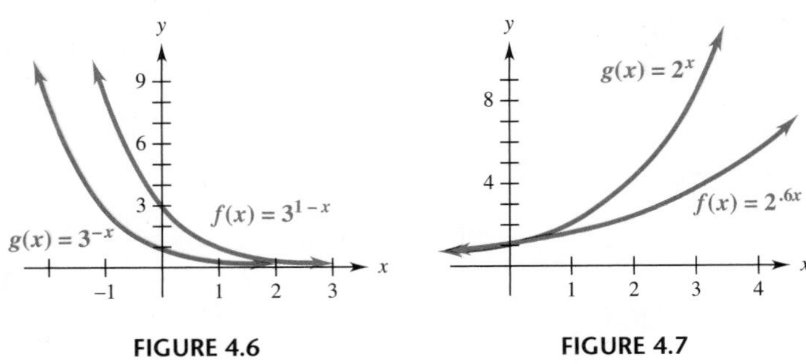

FIGURE 4.6 FIGURE 4.7

When the exponent involves a nonlinear expression in x, the graph of an exponential function may have a much different shape than the preceding ones have.

Example 4 Graph $f(x) = 2^{-x^2}$.

Solution Either use a graphing calculator or plot points and connect them with a smooth curve, as in Figure 4.8. The graph is symmetric about the y-axis; that is, if the figure were folded on the y-axis, the two halves would match. This graph has the x-axis as a horizontal asymptote. The domain is still all real numbers, but here the range is $0 < y \leq 1$. Graphs such as this are important in probability, where the normal curve has an equation similar to $f(x)$ in this example. 5

✓ Checkpoint 5

Graph $f(x) = \left(\frac{1}{2}\right)^{-x^2}$.

Answer: *See page 224.*

x	y
−2	1/16
−1.5	.21
−1	1/2
−.5	.84
0	1
.5	.84
1	1/2
1.5	.21
2	1/16

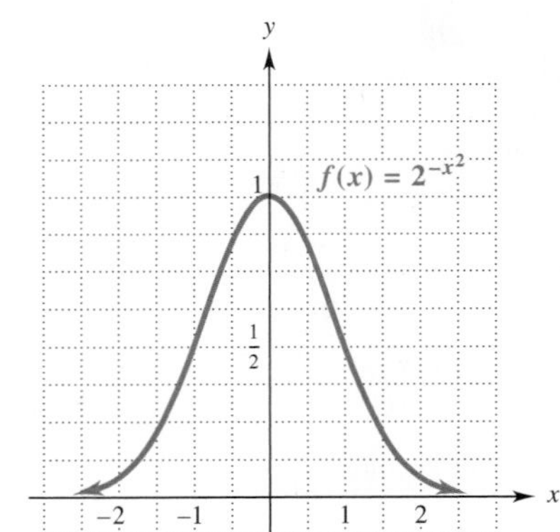

FIGURE 4.8

THE NUMBER *e*

In Case 5, we shall see that a certain irrational number, denoted e, plays an important role in the compounding of interest. This number e also arises naturally in a variety of other mathematical and scientific contexts. To 12 decimal places,

$$\mathbf{e \approx 2.718281828459.}$$

Perhaps the single most useful exponential function is the function defined by $f(x) = e^x$.

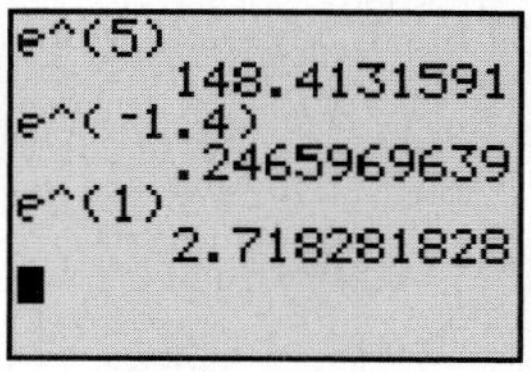

e^(5)
148.4131591
e^(-1.4)
.2465969639
e^(1)
2.718281828

FIGURE 4.9

TECHNOLOGY TIP To evaluate powers of e with a calculator, use the e^x key, as in Figure 4.9. The figure also shows how to display the decimal expansion of e by calculating e^1. 6

In Figure 4.10, the functions defined by

$$g(x) = 2^x, \quad f(x) = e^x, \quad \text{and} \quad h(x) = 3^x$$

are graphed for comparison.

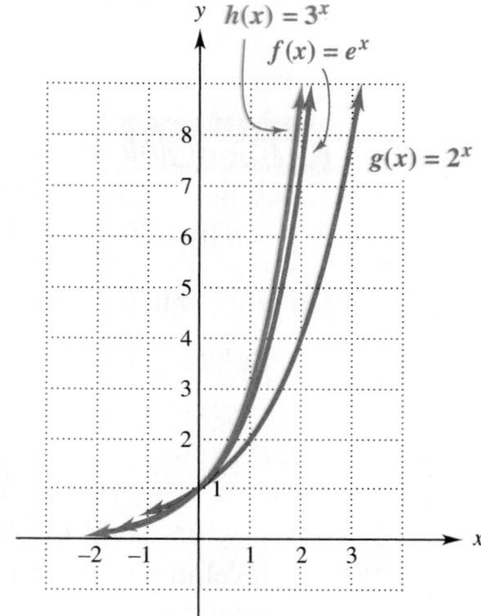

y
h(x) = 3^x
f(x) = e^x
g(x) = 2^x
8
7
6
5
4
3
2
1
−2
−1
1
2
3
x

FIGURE 4.10

Checkpoint 6

Evaluate the following powers of e.

(a) $e^{.06}$

(b) $e^{-.06}$

(c) $e^{2.30}$

(d) $e^{-2.30}$

Answers: *See page 224.*

Example 5 **Business** According to the *Wall Street Journal*, OLED* panels are increasingly being used for display screens on cell phones, digital cameras, and handheld TVs because they are thinner, provide crisper images, and use less energy than regular LED panels. The sales of OLED panels (in billions of dollars) are approximated by

$$f(x) = .07363e^{.30447x},$$

where $x = 6$ corresponds to the year 2006.

(a) What were sales in 2006?

Solution Since 2006 corresponds to $x = 6$, we evaluate $f(6)$:

$$f(6) = .07363e^{.30447*6} \approx \$.45754 \text{ billion.}$$

So sales were about $457,540,000.

(b) What were sales in 2009?

Solution 2009 corresponds to $x = 9$, so we have

$$f(9) = .07363e^{.30447*9} = \$1.1406 \text{ billion.}$$

So sales were about $1,140,600,000.

---

*Organic light-emitting diode.

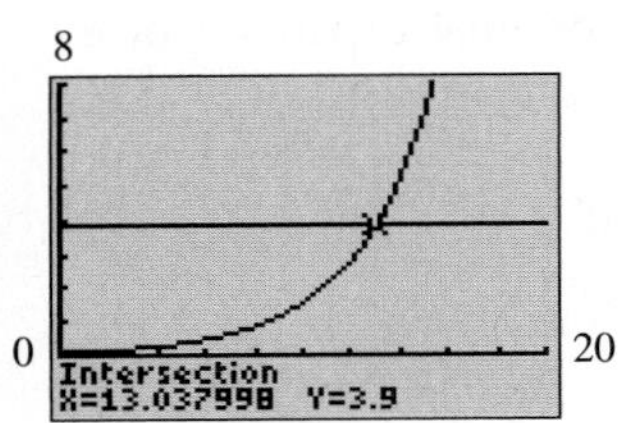

FIGURE 4.11

(c) Use a graphing calculator to determine when sales will reach \$3,900,000,000.

Solution Since f measures sales in billions of dollars, we must solve the equation $f(x) = 3.9$, that is

$$.07363e^{.30447x} = 3.9.$$

One way to do this is to find the intersection point of the graphs of $y = .07363e^{.30447x}$ and $y = 3.9$. (The calculator's intersection finder is in the same menu as its root (or zero) finder.) Figure 4.11 shows that this point is approximately (13.038, 3.9). Hence, sales will reach \$3.9 billion when $x \approx 13.038$, that is, in 2013.

Example 6 **Health** When a patient is given a 300-mg dose of the drug cimetidine intravenously, the amount C of the drug in the bloodstream t hours later is given by $C(t) = 300e^{-.3466t}$.

(a) How much of the drug is in the bloodstream after 3 hours and after 10 hours?

Solution Evaluate the function at $t = 3$ and $t = 10$:

$$C(3) = 300e^{-.3466*3} \approx 106.058 \text{ mg};$$
$$C(10) = 300e^{-.3466*10} \approx 9.37 \text{ mg}.$$

(b) Doctors want to give a patient a second 300-mg dose of cimetidine when its level in her bloodstream decreases to 75 mg. Use graphing technology to determine when this should be done.

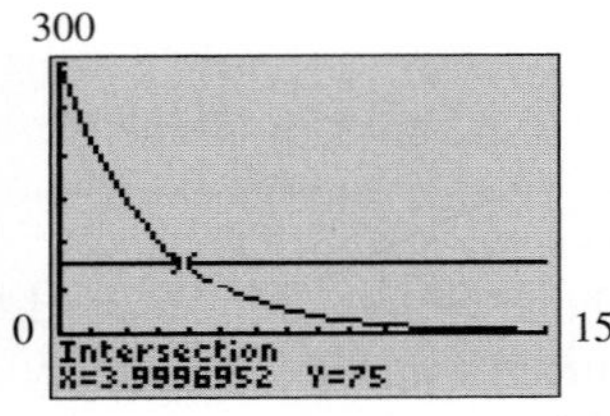

FIGURE 4.12

Solution The second dose should be given when $C(t) = 75$, so we must solve the equation

$$300e^{-.3466t} = 75.$$

Using the method of Example 5(c), we graph $y = 300e^{-.3466t}$ and $y = 75$ and find their intersection point. Figure 4.12 shows that this point is approximately (4, 75). So the second dose should be given 4 hours after the first dose.

4.1 ▶ Exercises

Classify each function as linear, quadratic, or exponential.

1. $f(x) = 6^x$ **2.** $g(x) = -5x$

3. $h(x) = 4x^2 - x + 5$ **4.** $k(x) = 4^{x+3}$

5. $f(x) = 675(1.055^x)$ **6.** $g(x) = 12e^{x^2+1}$

Without graphing,

(a) *describe the shape of the graph of each function;*
(b) *find the second coordinates of the points with first coordinates 0 and 1. (See Examples 1–3.)*

7. $f(x) = .6^x$ **8.** $g(x) = 4^{-x}$

9. $h(x) = 2^{.5x}$ **10.** $k(x) = 5(3^x)$

11. $f(x) = e^{-x}$ **12.** $g(x) = 3(16^{x/4})$

Graph each function. (See Examples 1–3.)

13. $f(x) = 3^x$ **14.** $g(x) = 3^{.5x}$

15. $f(x) = 2^{x/2}$ **16.** $g(x) = e^{x/4}$

17. $f(x) = (1/5)^x$ **18.** $g(x) = 2^{3x}$

19. Graph these functions on the same axes.
(a) $f(x) = 2^x$
(b) $g(x) = 2^{x+3}$
(c) $h(x) = 2^{x-4}$
(d) If c is a positive constant, explain how the graphs of $y = 2^{x+c}$ and $y = 2^{x-c}$ are related to the graph of $f(x) = 2^x$.

20. Graph these functions on the same axes.
 (a) $f(x) = 3^x$
 (b) $g(x) = 3^x + 2$
 (c) $h(x) = 3^x - 4$
 (d) If c is a positive constant, explain how the graphs of $y = 3^x + c$ and $y = 3^x - c$ are related to the graph of $f(x) = 3^x$.

The accompanying figure shows the graphs of $y = a^x$ for $a = 1.8, 2.3, 3.2, .4, .75$, and $.31$. They are identified by letter, but not necessarily in the same order as the values just given. Use your knowledge of how the exponential function behaves for various powers of a to match each lettered graph with the correct value of a.

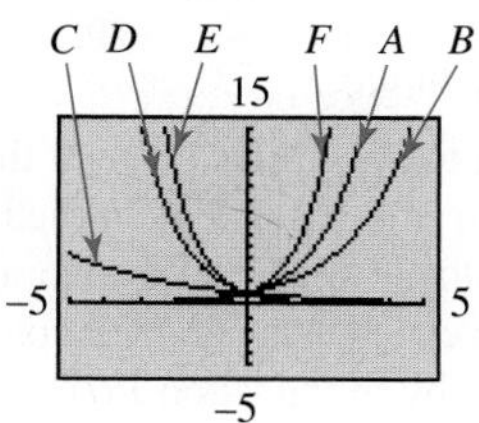

21. A	22. B	23. C
24. D	25. E	26. F

In Exercises 27 and 28, the graph of an exponential function with base a is given. Follow the directions in parts (a)–(f) in each exercise.

27.

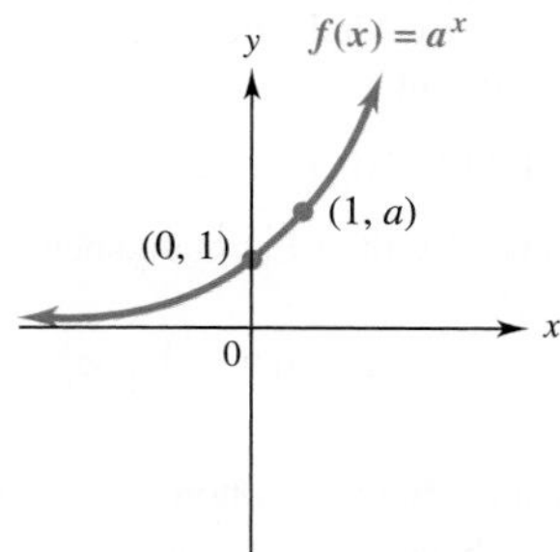

 (a) Is $a > 1$ or is $0 < a < 1$?
 (b) Give the domain and range of f.
 (c) Sketch the graph of $g(x) = -a^x$.
 (d) Give the domain and range of g.
 (e) Sketch the graph of $h(x) = a^{-x}$.
 (f) Give the domain and range of h.

28.

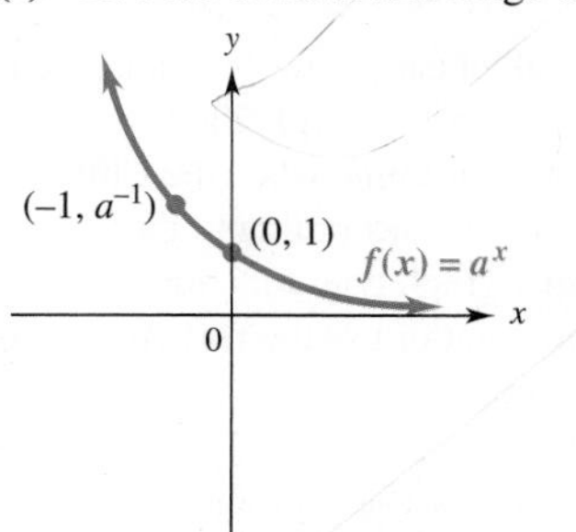

 (a) Is $a > 1$ or is $0 < a < 1$?
 (b) Give the domain and range of f.
 (c) Sketch the graph of $g(x) = a^x + 2$.
 (d) Give the domain and range of g.
 (e) Sketch the graph of $h(x) = a^{x+2}$.
 (f) Give the domain and range of h.

29. If $f(x) = a^x$ and $f(3) = 27$, find the following values of $f(x)$.
 (a) $f(1)$ (b) $f(-1)$ (c) $f(2)$ (d) $f(0)$

30. Give a rule of the form $f(x) = a^x$ to define the exponential function whose graph contains the given point.
 (a) $(3, 8)$ (b) $(-3, 64)$

Graph each function. (See Example 4.)

31. $f(x) = 2^{-x^2+2}$
32. $g(x) = 2^{x^2-2}$
33. $f(x) = x \cdot 2^x$
34. $f(x) = x^2 \cdot 2^x$

Work the following exercises.

35. **Finance** If \$1 is deposited into an account paying 6% per year compounded annually, then after t years the account will contain

$$y = (1 + .06)^t = (1.06)^t$$

dollars.
 (a) Use a calculator to complete the following table:

t	0	1	2	3	4	5	6	7	8	9	10
y	1					1.34					1.79

 (b) Graph $y = (1.06)^t$.

36. **Finance** If money loses value at the rate of 3% per year, the value of \$1 in t years is given by

$$y = (1 - .03)^t = (.97)^t.$$

 (a) Use a calculator to complete the following table:

t	0	1	2	3	4	5	6	7	8	9	10
y	1					.86					.74

 (b) Graph $y = (.97)^t$.

Work these problems. (See Example 5.)

37. **Finance** If money loses value, then as time passes, it takes more dollars to buy the same item. Use the results of Exercise 36(a) to answer the following questions.
 (a) Suppose a house costs \$105,000 today. Estimate the cost of the same house in 10 years. (*Hint*: Solve the equation $.74t = \$105,000$.)
 (b) Estimate the cost of a \$50 textbook in 8 years.

38. **Natural Science** Biologists have found that the oxygen consumption of yearling salmon is given by $g(x) = 100e^{.7x}$, where x is the speed in feet per second. Find each of the following.
(a) the oxygen consumption when the fish are still;
(b) the oxygen consumption when the fish are swimming at a speed of 2 feet per second.

39. **Business** Cell phone usage has been increasing exponentially in recent years. The number of U.S. cell phone accounts (in millions) is approximated by

$$f(x) = 110.6e^{.125x} \quad (0 \le x \le 7),$$

where $x = 0$ corresponds to the year 2000.* Estimate the number of cell phone accounts in
(a) 2002 **(b)** 2004 **(c)** 2006 **(d)** 2010
(e) Explain why your answer to part (d) is not likely to be accurate.

40. **Business** The monthly payment on a car loan at 12% interest per year on the unpaid balance is given by

$$f(n) = \frac{P}{\dfrac{1 - 1.01^{-n}}{.01}},$$

where P is the amount borrowed and n is the number of months over which the loan is paid back. Find the monthly payment for each of the following loans.
(a) \$8000 for 48 months **(b)** \$8000 for 24 months
(c) \$6500 for 36 months **(d)** \$6500 for 60 months

41. **Natural Science** The amount of plutonium remaining from 1 kilogram after x years is given by the function $W(x) = 2^{-x/24,360}$. How much will be left after
(a) 1000 years?
(b) 10,000 years?
(c) 15,000 years?
(d) Estimate how long it will take for the 1 kilogram to decay to half its original weight. Your answer may help to explain why nuclear waste disposal is a serious problem.

Business *The scrap value of a machine is the value of the machine at the end of its useful life. By one method of calculating scrap value, where it is assumed that a constant percentage of value is lost annually, the scrap value is given by*

$$S = C(1 - r)^n,$$

where C is the original cost, n is the useful life of the machine in years, and r is the constant annual percentage of value lost. Find the scrap value for each of the following machines.

42. Original cost, \$68,000; life, 10 years; annual rate of value loss, 8%

43. Original cost, \$244,000; life, 12 years; annual rate of value loss, 15%

44. Use the graphs of $f(x) = 2^x$ and $g(x) = 2^{-x}$ (not a calculator) to explain why $2^x + 2^{-x}$ is approximately equal to 2^x when x is very large.

45. **Social Science** There were fewer than a billion people on Earth when Thomas Jefferson died in 1826, and there are now more than 6 billion. If the world population continues to grow as expected, the population (in billions) in year t will be given by the function $P(t) = 4.834(1.01^{(t-1980)})$.* Estimate the world population in the following years.
(a) 2005 **(b)** 2010 **(c)** 2030
(d) What will the world population be when you reach 65 years old?

Work the following problems. (See Examples 5 and 6.)

46. **Health** The amount spent by Medicare on home health care services (in billions of dollars) is approximated by

$$g(x) = 7.88*1.112^x \quad (0 \le x \le 13),$$

where $x = 0$ corresponds to 2000.† Estimate spending in
(a) 2009. **(b)** 2010.
(c) Use technology to find the year in which spending reaches \$29,000,000,000.

47. **Business** The size of the Chinese economy is predicted to surpass the size of the U.S. economy at some point in the next 40 years. The size of an economy is measured by its gross domestic product (GDP). Projections of the GDPs of China and the United States (in trillions of dollars) are as follows, where $x = 0$ corresponds to the year 2000:‡

China: $f(x) = 1.35*1.076^x$
United States: $g(x) = 10.4*1.024^x \quad (0 \le x \le 50)$.

Find the projected GDP of each country in the given years.
(a) 2011 **(b)** 2025 **(c)** 2047
 **(d)** Use technology to determine when the Chinese GDP surpasses the U.S. GDP according to these projections. (*Hint*: Either graph both functions and find their intersection point or use the table feature of a graphing calculator.)

*Based on data from the CTIA Semiannual Wireless Survey.

*Based on data and projections by the U.S. Census Bureau.

†Based on data and projections from the Centers for Medicare and Medicaid Services, and the Congressional Budget Office.

‡Based on data from Goldman Sachs.

48. Business Construction of expressways in China has expanded significantly in recent decades. The approximate number of kilometers of expressways in China (in thousands) is given by

$$f(x) = 1.2e^{.228x} \quad (0 \le x \le 20),$$

where $x = 0$ corresponds to the year 1990. How many kilometers of expressways did China have in the given years?
(a) 2006 **(b)** 2007

(c) Assuming that this function remains valid, use technology to determine when there are 500,000 kilometers of expressways in China.

49. Health The United States has been experiencing a shortage of registered nurses, and the situation is expected to worsen in coming years. The projected shortage of full-time equivalent registered nurses (in thousands) can be approximated by

$$N(x) = 100.5*1.11^x,$$

where $x = 0$ corresponds to the year 2000.* Find the shortage in the given years.
(a) 2010 **(b)** 2013 **(c)** 2015
(d) Use technology to find the year in which the shortage will exceed 600,000 nurses.

50. Business The amount spent per person per year on cable and satellite TV can be approximated by

$$g(x) = \frac{497}{1 + 1.884e^{-.183x}},$$

where $x = 0$ corresponds to the year 2000.† Find the amount per person spent in the given years.
(a) 2005 **(b)** 2010

(c) When will per-person spending reach \$425?

51. Business Sales of music CDs (in millions of dollars) in the United States are approximated by

$$f(t) = 13.82*.935^t,$$

where $t = 0$ corresponds to the year 2000.† How much is spent on CDs in the given years?
(a) 2002 **(b)** 2006 **(c)** 2010
(d) Graph $f(t)$ for the period from 2000 to 2015.

52. Health According to projections by the Alzheimer's Association, the total number of people in the United States with Alzheimer's disease (in millions) is approximated by

$$A(x) = 4.23*1.023^x,$$

where $x = 0$ corresponds to the year 2000. Find the number of people who have Alzheimer's disease in the given years.
(a) 2010 **(b)** 2016
(c) the year in which you turn 60

*Based on data from the U.S. Department of Health and Human Services.

†Based on data and projections in the *Statistical Abstract of the United States: 2008.*

53. Health When a patient is given a 20-mg dose of aminophylline intravenously, the amount C of the drug in the bloodstream t hours later is given by $C(t) = 20e^{-.1155t}$. How much aminophylline remains in the bloodstream after the given numbers of hours?
(a) 4 hours
(b) 8 hours
(c) Approximately when does the amount of aminophylline decrease to 3.5 mg?

54. The accompanying figure shows the graph of an exponential growth function $f(x) = Pa^x$.
(a) Find P. [*Hint*: What is $f(0)$?]
(b) Find a. [*Hint*: What is $f(2)$?]
(c) Find $f(5)$.

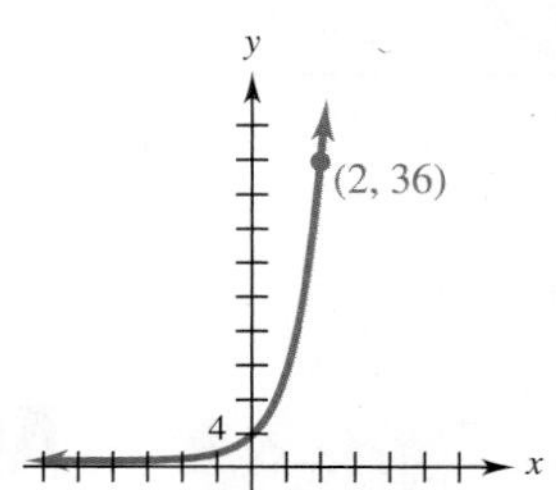

✓Checkpoint Answers

1. (a) The entries in the second column are 1/27, 1/9, 1/3, 1, 3, 9, 27, respectively.

(b)

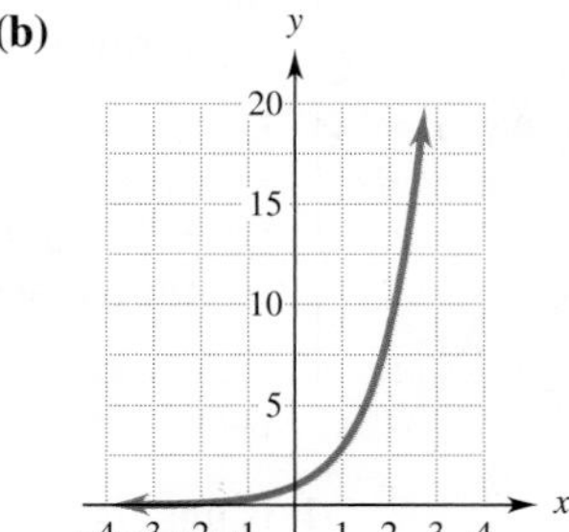

2.

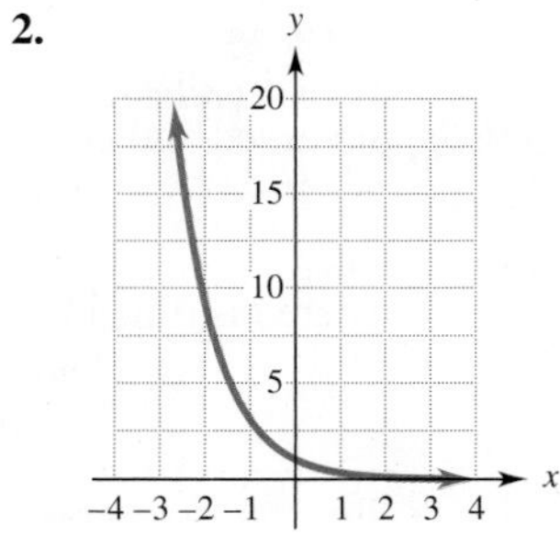

3.

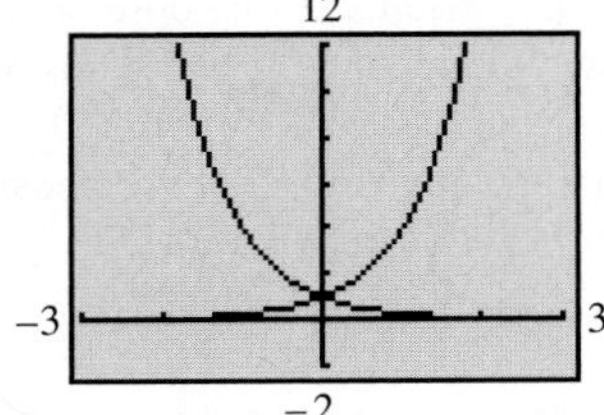

4.

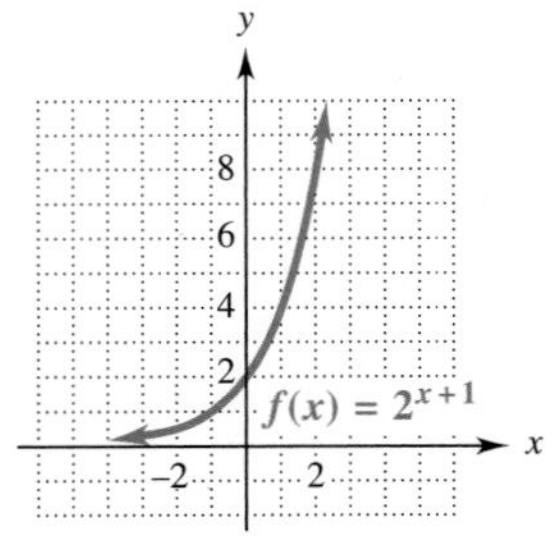

5.

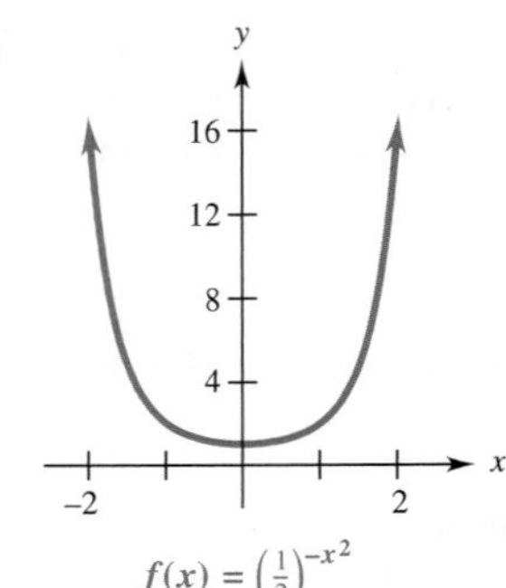

$f(x) = \left(\frac{1}{2}\right)^{-x^2}$

6. **(a)** 1.06184 **(b)** .94176
(c) 9.97418 **(d)** .10026

4.2 ▶ Applications of Exponential Functions

In many situations in biology, economics, and the social sciences, a quantity changes at a rate proportional to the quantity present. For example, a country's population might be increasing at a rate of 1.3% a year. In such cases, the amount present at time t is given by an **exponential growth function**.

Exponential Growth Function

Under normal conditions, growth can be described by a function of the form

$$f(t) = y_0 e^{kt} \quad \text{or} \quad f(t) = y_0 b^t,$$

where $f(t)$ is the amount present at time t, y_0 is the amount present at time $t = 0$, and k and b are constants that depend on the rate of growth.

It is understood that growth can involve either growing larger or growing smaller.

Example 1 **Business** In the early 2000s, online retail spending steadily increased and could be approximated by the exponential function

$$f(t) = y_0 e^{.218t},$$

where t is time in years, $t = 0$ corresponds to 2000, and $f(t)$ is in billions of dollars.*

*Based on data in the *Statistical Abstract of the United States: 2008*.

(a) If online retail sales were $25,000,000,000 in 2000, find the sales in 2007.

Solution Since y_0 represents the sales when $t = 0$ (that is, in 2000) we have $y_0 = 25$ (because sales are measured in billions of dollars). So the sales function is $f(t) = 25e^{.218t}$. To find the sales in 2007, evaluate $f(t)$ at $t = 7$ (which corresponds to 2007):

$$f(t) = 25e^{.218t}$$
$$f(7) = 25e^{.218*7} \approx 114.99.$$

Hence, sales in 2007 were about $115 billion.

(b) If this model remains accurate, what is the amount of online retail sales in 2012?

Solution Since 2012 corresponds to $t = 12$, evaluate the function at $t = 12$:

$$f(12) = 25e^{.218*12} \approx \$342.02 \text{ billion.}$$ ✓1

✓Checkpoint 1

Suppose the number of bacteria in a culture at time t is

$$y = 500e^{.4t},$$

where t is measured in hours.

(a) How many bacteria are present initially?

(b) How many bacteria are present after 10 hours?

Answers: See page 232.

Example 2 **Health** Cigarette consumption in the United States has been decreasing for some time. Based on data from the U.S. Department of Agriculture, the number of cigarettes produced since 1995 can be approximated by the function

$$g(x) = 879e^{-.04x} \quad (x \geq 5),$$

where $x = 5$ corresponds to 1995 and $g(x)$ is measured in billions.

(a) Find the number of cigarettes produced in 2005 and 2010.

Solution 2005 and 2010 correspond to $x = 15$ and $x = 20$, respectively.

So we evaluate the function at these numbers:

$$g(15) = 879^{-.04*15} \approx 482.41 \text{ billion in 2005;}$$
$$g(20) = 879^{-.04*20} \approx 394.96 \text{ billion in 2010.}$$

(b) If this model remains accurate, when will cigarette production fall to 298.5 billion?

Solution Graph $g(x) = 879e^{-.04x}$ and $y = 298.5$ on the same screen and find the x-coordinate of their intersection point. Figure 4.13 shows that production is 298.5 billion in 2017 ($x = 27$).

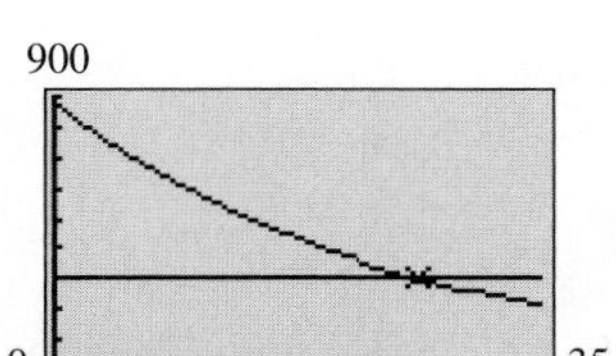

FIGURE 4.13

When a quantity is known to grow exponentially, it is sometimes possible to find a function that models its growth from a small amount of data.

Example 3 **Finance** When money is placed in a bank account that pays compound interest, the amount in the account grows exponentially, as we shall see in Chapter 5. Suppose such an account grows from $1000 to $1316 in 7 years.

(a) Find a growth function of the form $f(t) = y_0 b^t$ that gives the amount in the account at time t years.

Solution The values of the account at time $t = 0$ and $t = 7$ are given; that is, $f(0) = 1000$ and $f(7) = 1316$. Solve the first of these equations for y_0:

$$f(0) = 1000$$
$$y_0 b^0 = 1000 \quad \text{Rule of } f$$
$$y_0 = 1000. \quad b^0 = 1$$

So the rule of f has the form $f(t) = 1000b^t$. Now solve the equation $f(7) = 1316$ for b:

$$f(7) = 1316$$
$$1000b^7 = 1316 \quad \text{Rule of } f$$
$$b^7 = 1.316 \quad \text{Divide both sides by 1000.}$$
$$b = (1.316)^{1/7} \approx 1.04. \quad \text{Take the seventh root of each side.}$$

So the rule of the function is $f(t) = 1000(1.04)^t$.

(b) How much is in the account after 12 years?

Solution $f(12) = 1000(1.04)^{12} = \$1601.03.$ 2

✓Checkpoint 2

Suppose an investment grows exponentially from \$500 to \$587.12 in three years.

(a) Find a function of the form $f(t) = y_0 b^t$ that gives the value of the investment after t years.

(b) How much is the investment worth after 10 years?

Answers: See page 232.

Example 4 **Health** Infant mortality rates in the United States are shown in the following table*:

Year	Rate	Year	Rate
1920	76.7	1980	12.6
1930	60.4	1985	10.6
1940	47.0	1990	9.2
1950	29.2	1995	7.6
1960	26.0	2000	6.9
1970	20.0	2005	6.9

(a) Let $t = 0$ correspond to 1920. Use the data for 1920 and 2005 to find a function of the form $f(t) = y_0 b^t$ that models this data.

Solution Since the rate is 76.7 when $t = 0$, we have $y_0 = 76.7$. Hence, $f(t) = 76.7b^t$. Because 2005 corresponds to $t = 85$, we have $f(85) = 6.9$; that is,

$$76.7b^{85} = 6.9 \quad \text{Rule of } f$$
$$b^{85} = \frac{6.9}{76.7} \quad \text{Divide both sides by 76.7.}$$
$$b = \left(\frac{6.9}{76.7}\right)^{\frac{1}{85}} \approx .97206 \quad \text{Take } 85^{th} \text{ roots on both sides.}$$

Therefore, the function is $f(t) = 76.7(.97206^t)$.

*Infants less than 1 year old, deaths per 1000 live births. U.S. National Center for Health Statistics.

(b) Use exponential regression on a graphing calculator to find another model for the data.

Solution The procedure for entering the data and finding the function is the same as that for linear regression (just choose "exponential" instead of "linear"), as explained in Section 2.3. Depending on the calculator, one of the following functions will be produced:

$$g(t) = 80.5178 * .97013^t \quad \text{or} \quad h(t) = 80.5178e^{-.03032t}.$$

Both functions give the same values (except for slight differences due to rounding of the coefficients. They fit the data reasonably well, as shown in Figure 4.14.

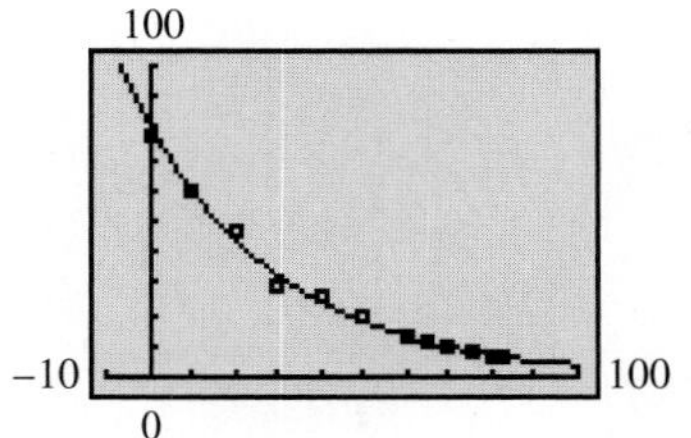

FIGURE 4.14

(c) Use the preceding models to estimate the mortality rate in 2006 and 2012.

Solution Evaluating each of the models in parts (a) and (b) at $t = 85$ and $t = 92$ shows that the models give slightly different results.

t	$f(t)$	$g(t)$
85	6.7	5.9
92	5.7	4.9

OTHER EXPONENTIAL MODELS

When a quantity changes exponentially, but does not either grow very large or decrease practically to 0, as in Examples 1–3, different functions are needed.

Example 5 **Business** Sales of a new product often grow rapidly at first and then begin to level off with time. Suppose the annual sales of an inexpensive can opener are given by

$$S(x) = 10{,}000(1 - e^{-.5x}),$$

where $x = 0$ corresponds to the time the can opener went on the market.

(a) What were the sales in each of the first three years?

Solution At the end of one year ($x = 1$), sales were

$$S(1) = 10{,}000(1 - e^{-.5(1)}) \approx 3935.$$

Sales in the next two years were

$$S(2) = 10{,}000(1 - e^{-.5(2)}) \approx 6321 \quad \text{and} \quad S(3) = 10{,}000(1 - e^{-.5(3)}) \approx 7769.$$

(b) What were the sales at the end of the 10th year?

Solution $S(10) = 10{,}000(1 - e^{-.5(10)}) \approx 9933.$

(c) Graph the function S. What does it suggest?

Solution The graph can be obtained by plotting points and connecting them with a smooth curve or by using a graphing calculator, as in Figure 4.15. The graph indicates that sales will level off after the 12th year, to around 10,000 can openers per year. 3

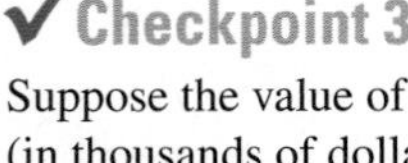

Checkpoint 3

Suppose the value of the assets (in thousands of dollars) of a certain company after t years is given by

$$V(t) = 100 - 75e^{-.2t}.$$

(a) What is the initial value of the assets?

(b) What is the limiting value of the assets?

(c) Find the value after 10 years.

(d) Graph $V(t)$.

Answers: See page 232.

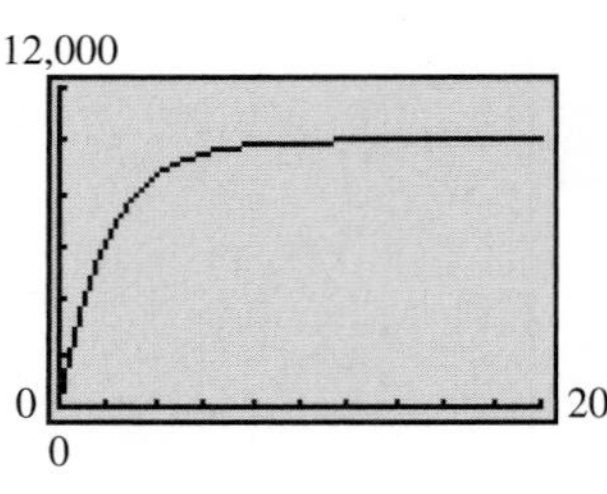

FIGURE 4.15

A variety of activities can be modeled by **logistic functions**, whose rules are of the form

$$f(x) = \frac{c}{1 + ae^{kx}}.$$

The logistic function in the next example is sometimes called a **forgetting curve**.

Example 6 **Social Science** Psychologists have measured people's ability to remember facts that they have memorized. In one such experiment, it was found that the number of facts, $N(t)$, remembered after t days was given by

$$N(t) = \frac{10.003}{1 + .0003e^{.8t}}.$$

(a) How many facts were remembered at the beginning of the experiment?

Solution When $t = 0$,

$$N(0) = \frac{10.003}{1 + .0003e^{.8(0)}} = \frac{10.003}{1.0003} = 10.$$

So 10 facts were remembered at the beginning.

(b) How many facts were remembered after one week? after two weeks?

Solution One and two weeks respectively correspond to $t = 7$ and $t = 14$:

$$N(7) = \frac{10.003}{1 + .0003e^{.8(7)}} = \frac{10.003}{1.0811} \approx 9.25;$$

$$N(14) = \frac{10.003}{1 + .0003e^{.8(14)}} = \frac{10.003}{22.9391} \approx .44.$$

So 9 facts were remembered after one week, but effectively none were remembered after two weeks (because .44 is less than "half a fact"). The graph of the function in Figure 4.16 gives a picture of this forgetting process. 4✓

✓Checkpoint 4

In Example 6,

(a) find the number of facts remembered after 10 days;

(b) use the graph to estimate when just one fact will be remembered.

Answers: See page 232.

TECHNOLOGY TIP

Many graphing calculators can find a logistic model for appropriate data.

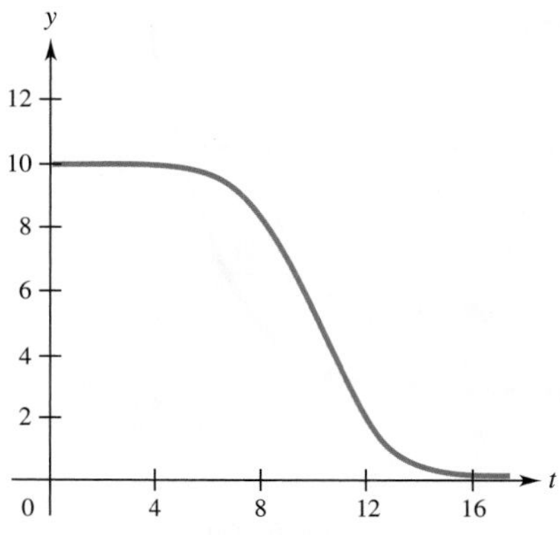

FIGURE 4.16

4.2 ▸ Exercises

1. **Finance** Suppose you owe \$800 on your credit card and you decide to make no new purchases and to make the minimum monthly payment on the account. Assuming that the interest rate on your card is 1% per month on the unpaid balance and that the minimum payment is 2% of the total (balance plus interest), your balance after t months is given by

$$B(t) = 800(.9898^t).$$

Find your balance at each of the given times.
   (a) six months
   (b) one year (remember that t is in months)
   (c) five years
   (d) eight years
   (e) On the basis of your answers to parts (a)–(d), what advice would you give to your friends about minimum payments?

2. **Health** The amount spent per person on health care in the United States is approximated by

$$H(t) = h_0{*}1.065^t,$$

where $t = 0$ corresponds to the year 2000 and $H(t)$ is in thousands of dollars.*
   (a) The amount spent per person in 2000 was \$4624. Find h_0.
   (b) To the nearest dollar, estimate the per-person costs in 2008, 2010, and 2012.

3. **Business** Sales of consumer electronics in recent years are approximated by

$$f(t) = 79.4{*}1.1^t \quad (t \geq 3),$$

where $t = 3$ corresponds to the year 2003 and $f(t)$ is in billions of dollars.† Find the sales in the given years.
   (a) 2006 **(b)** 2009
   (c) If the function remains accurate, what will sales be in 2014? Does this result seem plausible?

*Based on data from the Centers for Medicare and Medicaid Services.

†Based on data in the *Statistical Abstract of the United States: 2008.*

4. **Health** The graph shows how the risk of chromosomal abnormality in a child rises with the age of the mother.*
   (a) From the graph, read the risk of chromosomal abnormality (per 1000) at ages 20, 35, 42, and 49.
   (b) Verify by substitution that the exponential equation $y = .590e^{.061t}$ "fits" the graph for ages 20 and 35.
   (c) Does the equation in part (b) also fit the graph for ages 42 and 49? What does this mean?

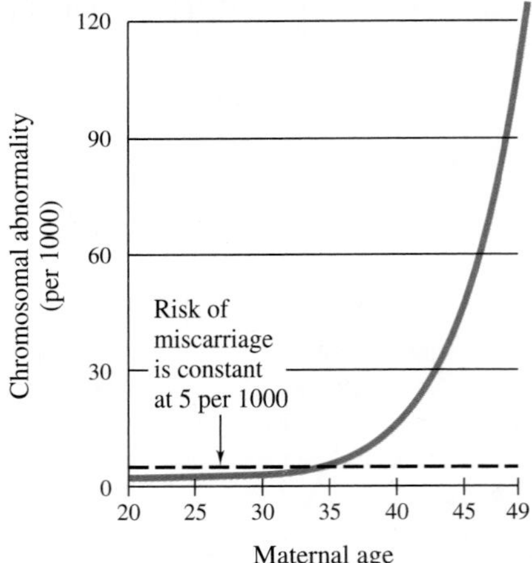

Source: American College of Obstetricians and Gynecologists.

In each of the following problems, find an exponential function of the form $f(t) = y_0b^t$ to model the data. (See Examples 3 and 4.)

5. **Business** There were about 1.5 million Hispanic-owned businesses in 2000 and 3.2 million in 2010.
   (a) Find a model for this data in which $t = 0$ corresponds to 2000 and the number of businesses is measured in millions.
   (b) If this model remains accurate, how many Hispanic-owned businesses will there be in 2014?
   (c) By experimenting with different values (or using a graphing calculator to solve an appropriate equation), estimate the year in which the number of such businesses will reach 6 million.

6. **Social Science** The U.S. Census Bureau predicts that the African-American population will increase from 35.3 million in 2000 to 59.2 million in 2050.†
   (a) Find a model for this data, in which $t = 0$ corresponds to 2000.
   (b) What is the projected African-American population in 2014? in 2030?
   (c) By experimenting with different values of t (or by using a graphing calculator to solve an appropriate equation), estimate the year in which the African-American population will reach 55 million.

7. **Business** Sales of plasma TVs in the United States were about $1.6 billion in 2003 and $6 billion in 2007.*
   (a) Let $t = 0$ correspond to 2003 and find a model for this data.
   (b) According to the model, what are sales in 2008 and 2010?

8. **Health** Medicare expenditures were $111 billion in 1990 and increased to $408.3 billion in 2006.*
   (a) Find a model for this data, in which $t = 0$ corresponds to 1990 and expenditures are in billions of dollars.
   (b) Estimate Medicare expenditures in 2010 and 2014.

In the following exercises, find the exponential model as follows: If you do not have suitable technology, use the first and last data points to find a function. (See Examples 3 and 4.) If you have a graphing calculator or other suitable technology, use exponential regression to find a function. (See Example 4.)

9. **Finance** The table shows the purchasing power of a dollar in recent years, with 2000 being the base year.† For example, the entry .88 for 2005 means that a dollar in 2005 bought what 88 cents did in 2000.

Year	Purchasing Power of $1
2000	$1.00
2001	.97
2002	.96
2003	.94
2004	.91
2005	.88
2006	.85
2007	.83
2008	.79

   (a) Find an exponential model for this data, where $t = 0$ corresponds to 2000.
   (b) Find the purchasing power of a dollar in 2010, 2012, and 2014.
   (c) Use a graphing calculator (or trial and error) to determine the year in which the purchasing power of the 2000 dollar will drop to 40 cents.

10. **Physical Science** The table shows the atmospheric pressure (in millibars) at various altitudes (in meters).

**New York Times,* February 5, 1994, p. 24. © 1994, The New York Times Company.
†*Statistical Abstract of the United States: 2008.*

**Statistical Abstract of the United States: 2008.*
†U.S. Bureau of Labor Statistics.

Altitude	Pressure
0	1,013
1000	899
2000	795
3000	701
4000	617
5000	541
6000	472
7000	411
8000	357
9000	308
10,000	265

(a) Find an exponential model for this data, in which t is measured in thousands. (For instance, $t = 2$ is 2000 meters.)

(b) Use the function in part (a) to estimate the atmospheric pressure at 1500 meters and 11,000 meters. Compare your results with the actual values of 846 millibars and 227 millibars, respectively.

11. **Health** The table shows the age-adjusted death rates per 100,000 Americans for heart disease.*

Year	Death Rate
2000	257.6
2001	247.8
2002	240.8
2003	232.3
2004	217.0
2005	211.1
2006	210.2

(a) Find an exponential model for this data, where $t = 0$ corresponds to 2000.

(b) Assuming the model remains accurate, estimate the death rate in 2010 and 2015.

(c) Use a graphing calculator (or trial and error) to determine the year in which the death rate will fall to 100.

12. **Business** The table shows outstanding consumer credit (in billions of dollars) at the beginning of various years.†

Year	Credit
1980	350.5
1985	524.0
1990	797.7
1995	1010.4
2000	1543.7
2005	2200.1
2006	2295.4
2008	2535.6

(a) Find an exponential model for this data, where $t = 0$ corresponds to 1980.

(b) If this model remains accurate, what will the outstanding consumer credit be in 2013 and 2016?

(c) In what year will consumer credit reach \$6000 billion?

Work the following problems. (See Example 5.)

13. **Business** Assembly-line operations tend to have a high turnover of employees, forcing the companies involved to spend much time and effort in training new workers. It has been found that a worker who is new to the operation of a certain task on the assembly line will produce $P(t)$ items on day t, where

$$P(t) = 25 - 25e^{-.3t}.$$

(a) How many items will be produced on the first day?

(b) How many items will be produced on the eighth day?

(c) According to the function, what is the maximum number of items the worker can produce?

14. **Social Science** The number of words per minute that an average person can type is given by

$$W(t) = 60 - 30e^{-.5t},$$

where t is time in months after the beginning of a typing class. Find each of the following.

(a) $W(0)$ **(b)** $W(1)$ **(c)** $W(4)$ **(d)** $W(6)$

Natural Science *Newton's law of cooling says that the rate at which a body cools is proportional to the difference in temperature between the body and an environment into which it is introduced. Using calculus, we can find that the temperature F(t) of the body at time t after being introduced into an environment having constant temperature T_0 is*

$$F(t) = T_0 + Cb^t,$$

where C and b are constants. Use this result in Exercises 15 and 16.

15. Boiling water at 100° Celsius is placed in a freezer at −18° Celsius. The temperature of the water is 50° Celsius after 24 minutes. Find the temperature of the water after 76 minutes.

16. Paisley refuses to drink coffee cooler than 95°F. She makes coffee with a temperature of 170°F in a room with

*U.S. Department of Health and Human Services.

†U.S. Federal Reserve.

a temperature of 70°F. The coffee cools to 120°F in 10 minutes. What is the longest amount of time she can let the coffee sit before she drinks it?

17. **Social Science** A sociologist has shown that the fraction $y(t)$ of people in a group who have heard a rumor after t days is approximated by

$$y(t) = \frac{y_0 e^{kt}}{1 - y_0(1 - e^{kt})},$$

where y_0 is the fraction of people who heard the rumor at time $t = 0$ and k is a constant. A graph of $y(t)$ for a particular value of k is shown in the figure.

(a) If $k = .1$ and $y_0 = .05$, find $y(10)$.

(b) If $k = .2$ and $y_0 = .10$, find $y(5)$.

(c) Assume the situation in part (b). How many *weeks* will it take for 65% of the people to have heard the rumor?

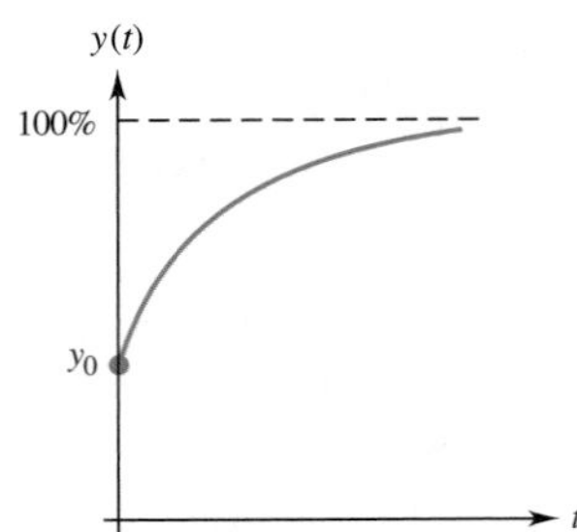

18. **Social Science** Data from the National Highway Traffic Safety Administration indicate that, in year t, the approximate percentage of people in the United States who wear seat belts when driving is given by

$$g(t) = \frac{97}{1 + .823e^{-.1401t}},$$

where $t = 0$ corresponds to the year 1994. What percentage used seat belts in

(a) 2000? (b) 2003? (c) 2004?

Assuming that this function is accurate after 2004, what percentage of people will use seat belts in

(d) 2011? (e) 2013? (f) 2015?

(g) Graph the function. Does the graph suggest that seat belt usage will ever reach 100%?

Use a graphing calculator or other technology to do the following problems. (See Example 6.)

19. **Business** The revenues for cable companies from basic cable TV (in billions of dollars) can be approximated by

$$f(x) = \frac{38.34}{1 + 17.23e^{-.17243x}},$$

where $x = 0$ corresponds to the year 1980.*

(a) Estimate the revenues in 2005 and 2010.

(b) Graph $f(x)$ for the period 1980–2030.

(c) Use the graph to determine the year in which revenues will reach \$37 billion.

(d) Based on the graph, are revenues likely to surpass \$40 billion in the foreseeable future?

20. **Natural Science** The population of fish in a certain lake at time t months is given by the function

$$p(t) = \frac{20{,}000}{1 + 24(2^{-.36t})}.$$

(a) Graph the population function from $t = 0$ to $t = 48$ (a four-year period).

(b) What was the population at the beginning of the period?

(c) Use the graph to estimate the one-year period in which the population grew most rapidly.

(d) When do you think the population will reach 25,000? What factors in nature might explain your answer?

21. **Finance** Since 2000, the national debt can be approximated by the logistic model

$$g(x) = \frac{40.35}{1 + 6.39e^{-.0866x}},$$

where $x = 0$ corresponds to the year 2000 and $g(x)$ is measured in trillions of dollars.*

(a) Estimate the national debt in 2000, 2010, and 2015.

(b) Graph $g(x)$ for the period 2000–2050.

(c) Determine when the debt will reach 20 trillion dollars.

(d) According to this model, will the national debt level off at any point in the future?

22. **Social Science** The probability P percent of having an automobile accident is related to the alcohol level t of the driver's blood by the function $P(t) = e^{21.459t}$.

(a) Graph $P(t)$ in a viewing window with $0 \le t \le .2$ and $0 \le P(t) \le 100$.

(b) At what blood alcohol level is the probability of an accident at least 50%? What is the legal blood alcohol level in your state?

✓ Checkpoint Answers

1. (a) 500 (b) About 27,300
2. (a) $f(t) = 500(1.055)^t$ (b) \$854.07
3. (a) \$25,000 (b) \$100,000
   (c) \$89,850
   (d)

y
100
75
50
25
$V(t) = 100 - 75e^{-.2t}$
0 5 10 15 20 25 x

4. (a) 5 (b) After about 12 days

*Based on data in the *Statistical Abstract of the United States: 2008.*

*Based on data from the U.S. Office of Management and Budget.

4.3 ▸ Logarithmic Functions

Until the development of computers and calculators, logarithms were the only effective tool for large-scale numerical computations. They are no longer needed for this purpose, but logarithmic functions still play a crucial role in calculus and in many applications.

COMMON LOGARITHMS

Logarithms are simply *a new language for old ideas*—essentially, a special case of exponents.

Definition of Common (Base 10) Logarithms

$$y = \log x \qquad \text{means} \qquad 10^y = x.$$

"**Log** x," which is read "the logarithm of x," is the answer to the question,

To what exponent must 10 be raised to produce x?

Example 1 To find log 10,000, ask yourself, "To what exponent must 10 be raised to produce 10,000?" Since $10^4 = 10{,}000$, we see that $\log 10{,}000 = 4$. Similarly,

$$\log 1 = 0 \qquad \text{because} \qquad 10^0 = 1;$$

$$\log .01 = -2 \qquad \text{because} \qquad 10^{-2} = \frac{1}{10^2} = \frac{1}{100} = .01;$$

$$\log \sqrt{10} = 1/2 \qquad \text{because} \qquad 10^{1/2} = \sqrt{10}.$$

✓1

✓Checkpoint 1

Find each common logarithm.

(a) log 100

(b) log 1000

(c) log .1

Answers: See page 243.

Example 2 Log(−25) is the exponent to which 10 must be raised to produce −25. But every power of 10 is positive! So there is no exponent that will produce −25. *Logarithms of negative numbers and 0 are not defined.*

Example 3 **(a)** We know that log 359 must be a number between 2 and 3 because $10^2 = 100$ and $10^3 = 1000$. By using the "log" key on a calculator, we find that log 359 (to four decimal places) is 2.5551. You can verify this statement by computing $10^{2.5551}$; the result (rounded) is 359. See the first two lines in Figure 4.17.

(b) When 10 is raised to a negative exponent, the result is a number less than 1. Consequently, the logarithms of numbers between 0 and 1 are negative. For instance, $\log .026 = -1.5850$, as shown in the third line in Figure 4.17. ✓2

✓Checkpoint 2

Find each common logarithm.

(a) log 27

(b) log 1089

(c) log .00426

Answers: See page 243.

FIGURE 4.17

NOTE On most scientific calculators, enter the number followed by the log key. On graphing calculators, press the log key followed by the number, as in Figure 4.17.

NATURAL LOGARITHMS

Although common logarithms still have some uses (one of which is discussed in Section 4.4), the most widely used logarithms today are defined in terms of the number e (whose decimal expansion begins 2.71828 . . .) rather than 10. They have a special name and notation.

Definition of Natural (Base e) Logarithms

$$y = \ln x \quad \text{means} \quad e^y = x.$$

Thus, the number $\ln x$ (which is sometimes read "el-en x") is the exponent to which e must be raised to produce the number x. For instance, $\ln 1 = 0$ because $e^0 = 1$. Although logarithms to the base e may not seem as "natural" as common logarithms, there are several reasons for using them, some of which are discussed in Section 4.4.

Example 4 **(a)** To find ln 85, use the LN key of your calculator, as in Figure 4.18. The result is 4.4427. Thus, 4.4427 is the exponent (to four decimal places) to which e must be raised to produce 85. You can verify this result by computing $e^{4.4427}$; the answer (rounded) is 85 Figure 4.18.

(b) A calculator shows that $\ln .38 = -.9676$ (rounded), which means that $e^{-.9676} \approx .38$. See Figure 4.18. 3

Checkpoint 3

Find the following.

(a) ln 6.1

(b) ln 20

(c) ln .8

(d) ln .1

Answers: See page 243.

FIGURE 4.18

Example 5 You don't need a calculator to find $\ln e^8$. Just ask yourself, "To what exponent must e be raised to produce e^8?" The answer, obviously, is 8. So $\ln e^8 = 8$.

OTHER LOGARITHMS

The procedure used to define common and natural logarithms can be carried out with any positive number $a \neq 1$ as the base in place of 10 or e.

Definition of Logarithms to the Base a

$$y = \log_a x \quad \text{means} \quad a^y = x.$$

Read $y = \log_a x$ as "y is the logarithm of x to the base a." As was the case with common and natural logarithms, $\log_a x$ is an *exponent*; it is the answer to the question,

To what power must a be raised to produce x?

For example, suppose $a = 2$ and $x = 16$. Then $\log_2 16$ is the answer to the question, To what power must 2 be raised to produce 16? It is easy to see that $2^4 = 16$, so $\log_2 16 = 4$. In other words, the exponential statement $2^4 = 16$ is equivalent to the logarithmic statement $4 = \log_2 16$.

In the definition of a logarithm to base a, note carefully the relationship between the base and the exponent:

Exponent
↓
Logarithmic form: $y = \log_a x$
↑
Base

Exponent
↓
Exponential form: $a^y = x$
↑
Base

Common and natural logarithms are the special cases when $a = 10$ and when $a = e$, respectively. Both $\log u$ and $\log_{10} u$ mean the same thing. Similarly, $\ln u$ and $\log_e u$ mean the same thing.

Example 6 This example shows several statements written in both exponential and logarithmic form.

Exponential Form	Logarithmic Form
(a) $3^2 = 9$	$\log_3 9 = 2$
(b) $(1/5)^{-2} = 25$	$\log_{1/5} 25 = -2$
(c) $10^5 = 100{,}000$	$\log_{10} 100{,}000$ (or $\log 100{,}000$) $= 5$
(d) $4^{-3} = 1/64$	$\log_4 (1/64) = -3$
(e) $2^{-4} = 1/16$	$\log_2 (1/16) = -4$
(f) $e^0 = 1$	$\log_e 1$ (or $\ln 1$) $= 0$

✓4 ✓5

✓Checkpoint 4

Write the logarithmic form of

(a) $5^3 = 125$;

(b) $3^{-4} = 1/81$;

(c) $8^{2/3} = 4$.

Answers: See page 243.

✓Checkpoint 5

Write the exponential form of

(a) $\log_{16} 4 = 1/2$;

(b) $\log_3 (1/9) = -2$;

(c) $\log_{16} 8 = 3/4$.

Answers: See page 243.

PROPERTIES OF LOGARITHMS

Some of the important properties of logarithms arise directly from their definition.

Let x and a be any positive real numbers, with $a \neq 1$, and r be any real number. Then

(a) $\log_a 1 = 0$; **(b)** $\log_a a = 1$;

(c) $\log_a a^r = r$; **(d)** $a^{\log_a x} = x$.

Property (a) was discussed in Example 1 (with $a = 10$). Property (c) was illustrated in Example 5 (with $a = e$ and $r = 8$). Property (b) is property (c) with $r = 1$. To

FIGURE 4.19

understand property (d), recall that $\log_a x$ is the exponent to which a must be raised to produce x. Consequently, when you raise a to this exponent, the result is x, as illustrated for common and natural logarithms in Figure 4.19.

The following properties are part of the reason that logarithms are so useful. They will be used in the next section to solve exponential and logarithmic equations.

The Product, Quotient, and Power Properties

Let x, y, and a be any positive real numbers, with $a \neq 1$. Let r be any real number. Then

$$\log_a xy = \log_a x + \log_a y \qquad \text{Product property}$$

$$\log_a \frac{x}{y} = \log_a x - \log_a y \qquad \text{Quotient property}$$

$$\log_a x^r = r \log_a x \qquad \text{Power property}$$

Each of these three properties is illustrated on a calculator in Figure 4.20.

```
ln(50/9)
1.714798428
ln(50)-ln(9)
1.714798428
```

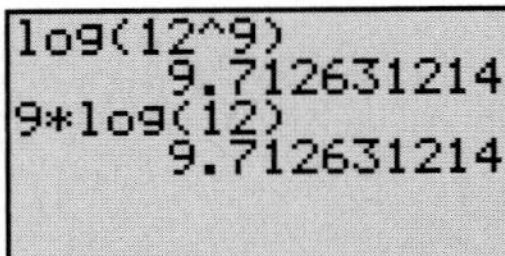

FIGURE 4.20

To prove the product property, let

$$m = \log_a x \qquad \text{and} \qquad n = \log_a y.$$

Then, by the definition of logarithm,

$$a^m = x \qquad \text{and} \qquad a^n = y.$$

Multiply to get

$$a^m \cdot a^n = x \cdot y,$$

or, by a property of exponents,

$$a^{m+n} = xy.$$

Use the definition of logarithm to rewrite this last statement as

$$\log_a xy = m + n.$$

Replace m with $\log_a x$ and n with $\log_a y$ to get

$$\log_a xy = \log_a x + \log_a y.$$

The quotient and power properties can be proved similarly.

Example 7 Using the properties of logarithms, we can write each of the following as a single logarithm:*

(a) $\log_a x + \log_a (x - 1) = \log_a x(x - 1);$ Product property

(b) $\log_a (x^2 + 4) - \log_a (x + 6) = \log_a \dfrac{x^2 + 4}{x + 6};$ Quotient property

(c) $\log_a 9 + 5 \log_a x = \log_a 9 + \log_a x^5 = \log_a 9x^5.$ Product and power properties

✓ **Checkpoint 6**

Write each expression as a single logarithm, using the properties of logarithms.

(a) $\log_a 5x + \log_a 3x^4$

(b) $\log_a 3p - \log_a 5q$

(c) $4 \log_a k - 3 \log_a m$

Answers: See page 243.

CAUTION There is no logarithm property that allows you to simplify the logarithm of a sum, such as $\log_a (x^2 + 4)$ In particular, $\log_a (x^2 + 4)$ is *not* equal to $\log_a x^2 + \log_a 4$. The product property of logarithms shows that $\log_a x^2 + \log_a 4 = \log_a 4x^2$.

Example 8 Assume that $\log_6 7 \approx 1.09$ and $\log_6 5 \approx .90$. Use the properties of logarithms to find each of the following:

(a) $\log_6 35 = \log_6 (7 \cdot 5) = \log_6 7 + \log_6 5 \approx 1.09 + .90 = 1.99;$

(b) $\log_6 5/7 = \log_6 5 - \log_6 7 \approx .90 - 1.09 = -.19;$

(c) $\log_6 5^3 = 3 \log_6 5 \approx 3(.90) = 2.70;$

(d) $\log_6 6 = 1;$

(e) $\log_6 1 = 0.$

✓ **Checkpoint 7**

Use the properties of logarithms to rewrite and evaluate each expression, given that $\log_3 7 \approx 1.77$ and $\log_3 5 \approx 1.46$.

(a) $\log_3 35$

(b) $\log_3 7/5$

(c) $\log_3 25$

(d) $\log_3 3$

(e) $\log_3 1$

Answers: See page 243.

In Example 8, several logarithms to the base 6 were given. However, they could have been found by using a calculator and the following formula.

Change-of-Base Theorem

For any positive numbers a and x (with $a \neq 1$),

$$\log_a x = \frac{\ln x}{\ln a}.$$

Example 9 To find $\log_7 3$, use the theorem with $a = 7$ and $x = 3$:

$$\log_7 3 = \frac{\ln 3}{\ln 7} \approx \frac{1.0986}{1.9459} \approx .5646.$$

You can check this result on your calculator by verifying that $7^{.5646} \approx 3$.

Example 10 **Natural Science** Environmental scientists who study the diversity of species in an ecological community use the *Shannon index* to measure diversity. If there are k different species, with n_1 individuals of species 1, n_2 individuals of species 2, and so on, then the Shannon index H is defined as

$$H = \frac{N \log_2 N - [n_1 \log_2 n_1 + n_2 \log_2 n_2 + \cdots + n_k \log_2 n_k]}{N},$$

*Here and elsewhere, we assume that variable expressions represent positive numbers and that the base a is positive, with $a \neq 1$.

where $N = n_1 + n_2 + n_3 + \cdots + n_k$. A study of the species that barn owls in a particular region typically eat yielded the following data:

Species	Number
Rats	143
Mice	1405
Birds	452

Find the index of diversity of this community.

Solution In this case, $n_1 = 143,\ n_2 = 1405,\ n_3 = 452$, and

$$N = n_1 + n_2 + n_3 = 143 + 1405 + 452 = 2000.$$

So the index of diversity is

$$\begin{aligned} H &= \frac{N \log_2 N - [n_1 \log_2 n_1 + n_2 \log_2 n_2 + \cdots + n_k \log_2 n_k]}{N} \\ &= \frac{2000 \log_2 2000 - [143 \log_2 143 + 1405 \log_2 1405 + 452 \log_2 452]}{2000}. \end{aligned}$$

To compute H, we use the change-of-base theorem:

$$\begin{aligned} H &= \frac{2000 \dfrac{\ln 2000}{\ln 2} - \left[143 \dfrac{\ln 143}{\ln 2} + 1405 \dfrac{\ln 1405}{\ln 2} + 452 \dfrac{\ln 452}{\ln 2}\right]}{2000} \\ &\approx 1.1149. \end{aligned}$$

LOGARITHMIC FUNCTIONS

For a given *positive* value of x, the definition of logarithm leads to exactly one value of y, so that $y = \log_a x$ defines a function.

> If $a > 0$ and $a \neq 1$, the **logarithmic function** with base a is defined as
>
> $$f(x) = \log_a x.$$

The most important logarithmic function is the natural logarithmic function.

Example 11 Graph $f(x) = \ln x$ and $g(x) = e^x$ on the same axes.

Solution For each function, use a calculator to compute some ordered pairs. Then plot the corresponding points and connect them with a curve to obtain the graphs in Figure 4.21.

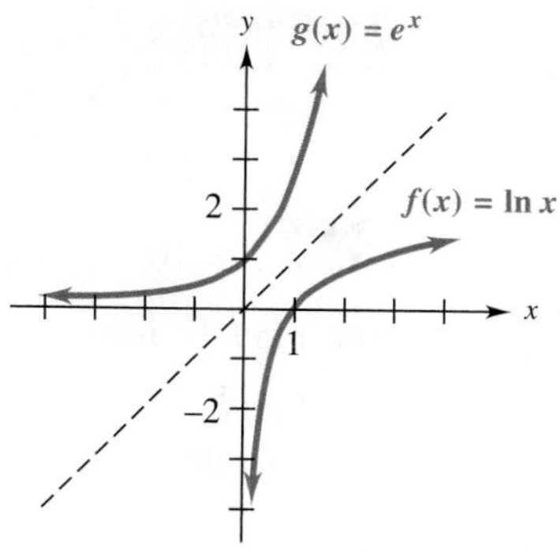

FIGURE 4.21

The dashed line in Figure 4.21 is the graph of $y = x$. Observe that the graph of $f(x) = \ln x$ is the mirror image of the graph of $g(x) = e^x$, with the line $y = x$ being the mirror. A pair of functions whose graphs are related in this way are said to be **inverses** of each other. A more complete discussion of inverse functions is given in many college algebra books. ✓8

Checkpoint 8

Verify that $f(x) = \log x$ and $g(x) = 10^x$ are inverses of each other by graphing $f(x)$ and $g(x)$ on the same axes.

Answer: See page 243.

When the base $a > 1$, the graph of $f(x) = \log_a x$ has the same basic shape as the graph of the natural logarithmic function in Figure 4.21, as summarized below.

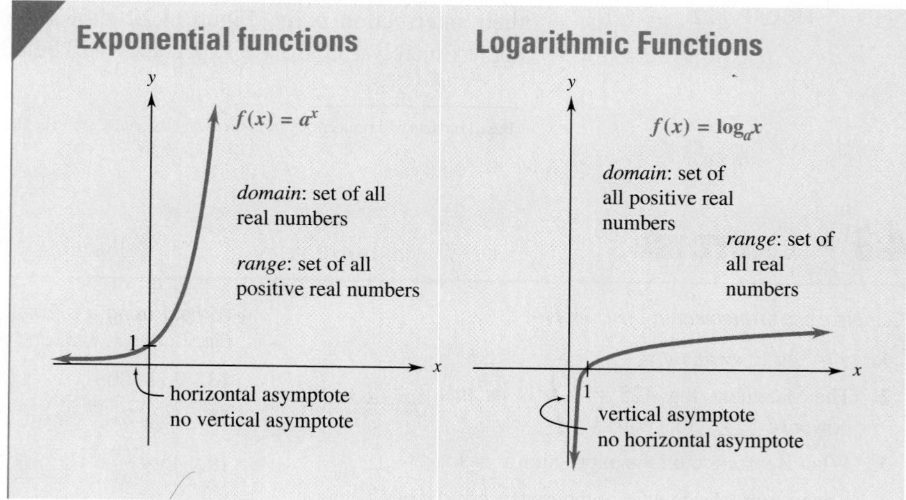

As the information in the box suggests, the functions $f(x) = \log_a x$ and $g(x) = a^x$ are inverses of each other. (Their graphs are mirror images of each other, with the line $y = x$ being the mirror.)

APPLICATIONS

Logarithmic functions are useful for, among other things, describing quantities that grow, but do so at a slower rate as time goes on.

Example 12 **Health** The life expectancy at birth of a person born in year x is approximated by the function

$$f(x) = 17.6 + 12.8 \ln x,$$

where $x = 10$ corresponds to 1910.*

(a) Find the life expectancy of persons born in 1910, 1960, and 2010.

Solution Since these years correspond to $x = 10$, $x = 60$, and $x = 110$, respectively, use a calculator to evaluate $f(x)$ at these numbers:

$$f(10) = 17.6 + 12.8 \ln(10) \approx 47.073;$$
$$f(60) = 17.6 + 12.8 \ln(60) \approx 70.008;$$
$$f(110) = 17.6 + 12.8 \ln(110) \approx 77.766.$$

So in the half-century from 1910 to 1960, life expectancy at birth increased from about 47 to 70 years, an increase of 23 years. But in the half-century from 1960 to 2010, it increased less than 8 years, from about 70 to 77.8 years.

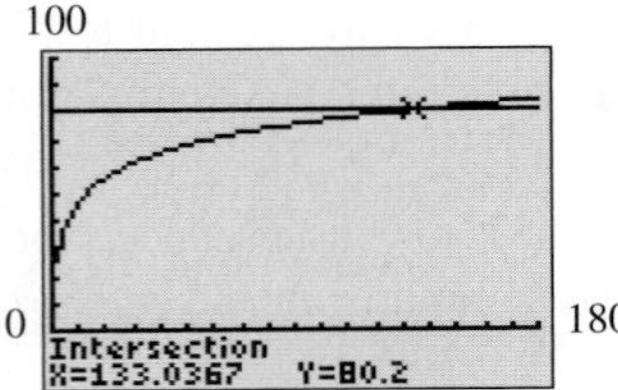

FIGURE 4.22

(b) If this function remains accurate, when will life expectancy at birth be 80.2 years?

Solution We must solve the equation $f(x) = 80.2$, that is,

$$17.6 + 12.8 \ln x = 80.2$$

In the next section, we shall see how to do this algebraically. For now, we solve the equation graphically by graphing $f(x)$ and $y = 80.2$ on the same screen and finding their intersection point. Figure 4.22 shows that the x-coordinate of this point is approximately 133. So life expectancy will be 80.2 years in 2033.

*Based on data and projections from the U.S. National Center for Health Statistics.

4.3 ▶ Exercises

Complete each statement in Exercises 1–4.

1. $y = \log_a x$ means $x =$ _____.
2. The statement $\log_5 125 = 3$ tells us that _____ is the power of _____ that equals _____.
3. What is wrong with the expression $y = \log_b$?
4. Logarithms of negative numbers are not defined because _____.

Translate each logarithmic statement into an equivalent exponential statement. (See Examples 1, 5, and 6.)

5. $\log 100{,}000 = 5$
6. $\log .001 = -3$
7. $\log_9 81 = 2$
8. $\log_2 (1/8) = -3$

Translate each exponential statement into an equivalent logarithmic statement. (See Examples 5 and 6.)

9. $10^{1.9823} = 96$
10. $e^{3.2189} = 25$
11. $3^{-2} = 1/9$
12. $16^{1/2} = 4$

Without using a calculator, evaluate each of the given expressions. (See Examples 1, 5, and 6.)

13. $\log 1000$
14. $\log .0001$
15. $\log_6 36$
16. $\log_3 81$
17. $\log_4 64$
18. $\log_5 125$
19. $\log_2 \frac{1}{4}$
20. $\log_3 \frac{1}{27}$
21. $\ln \sqrt{e}$
22. $\ln(1/e)$
23. $\ln e^{8.77}$
24. $\log 10^{74.3}$

Use a calculator to evaluate each logarithm to three decimal places. (See Examples 3 and 4.)

25. $\log 53$
26. $\log .005$
27. $\ln .0068$
28. $\ln 354$
29. Why does $\log_a 1$ always equal 0 for any valid base a?

Write each expression as the logarithm of a single number or expression. Assume that all variables represent positive numbers. (See Example 7.)

30. $\log 20 - \log 5$
31. $\log 6 + \log 8 - \log 2$
32. $3 \ln 2 + 2 \ln 3$
33. $2 \ln 5 - \frac{1}{2} \ln 25$

34. $5 \log x - 2 \log y$

35. $2 \log u + 3 \log w - 6 \log v$

36. $\ln(3x + 2) + \ln(x + 4)$ **37.** $2 \ln(x + 2) - \ln(x + 3)$

Write each expression as a sum and/or a difference of logarithms, with all variables to the first degree.

38. $\log 5x^2y^3$ **39.** $\ln \sqrt{6m^4n^2}$

40. $\ln \dfrac{3x}{5y}$ **41.** $\log \dfrac{\sqrt{xz}}{z^3}$

42. The calculator-generated table in the figure is for $y_1 = \log(4 - x)$. Why do the values in the y_1 column show ERROR for $x \geq 4$?

X	Y1
0	.60206
1	.47712
2	.30103
3	0
4	ERROR
5	ERROR
6	ERROR

X=4

Express each expression in terms of u and v, where u = ln x and v = ln y. For example, $\ln x^3 = 3(\ln x) = 3u$.

43. $\ln(x^2y^5)$ **44.** $\ln(\sqrt{x} \cdot y^2)$

45. $\ln(x^3/y^2)$ **46.** $\ln(\sqrt{x}/y)$

Evaluate each expression. (See Example 9.)

47. $\log_6 384$ **48.** $\log_{30} 78$

49. $\log_{35} 5646$ **50.** $\log_6 60 - \log_{60} 6$

Find numerical values for b and c for which the given statement is FALSE.

51. $\log(b + c) = \log b + \log c$

52. $\dfrac{\ln b}{\ln c} = \ln\left(\dfrac{b}{c}\right)$

Graph each function. (See Example 11.)

53. $y = \ln(x + 2)$ **54.** $y = \ln x + 2$

55. $y = \log(x - 3)$ **56.** $y = \log x - 3$

57. Graph $f(x) = \log x$ and $g(x) = \log(x/4)$ for $-2 \leq x \leq 8$. How are these graphs related? How does the quotient rule support your answer?

In Exercises 58 and 59, the coordinates of a point on the graph of the indicated function are displayed at the bottom of the screen. Write the logarithmic and exponential equations associated with the display.

58. $f(x) = \log x$

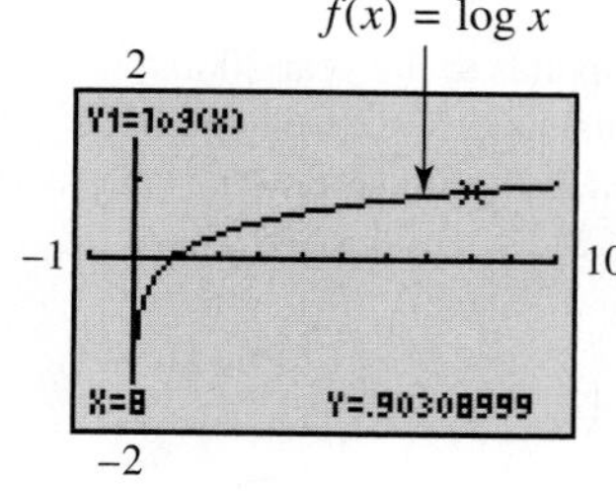

59. $g(x) = \ln x$

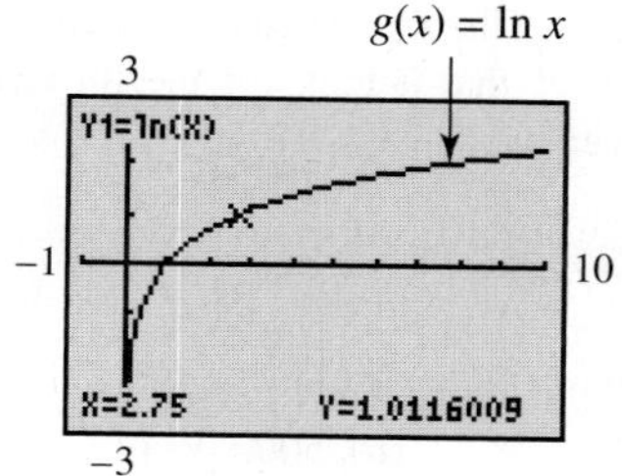

60. Match each equation with its graph. Each tick mark represents one unit.

(a) $y = \log x$ **(b)** $y = 10^x$ **(c)** $y = \ln x$ **(d)** $y = e^x$

(A)

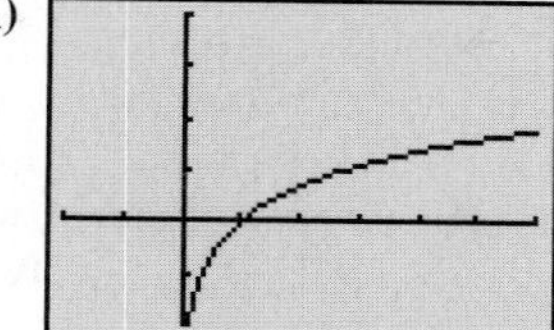

(B)

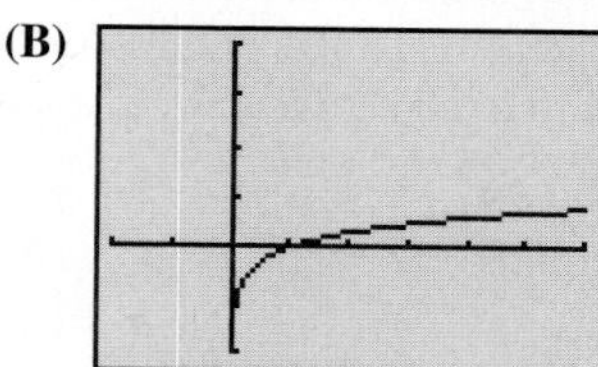

(C)

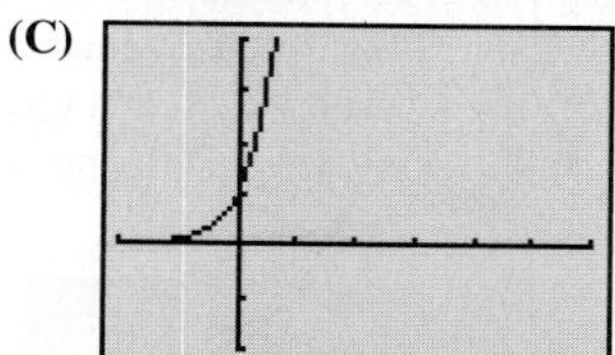

(D)

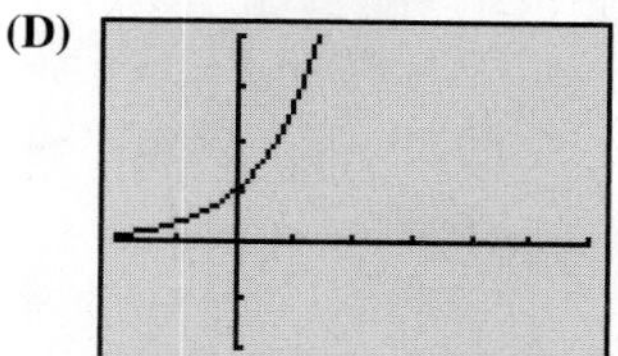

61. Finance The doubling function

$$D(r) = \frac{\ln 2}{\ln(1 + r)}$$

gives the number of years required to double your money when it is invested at interest rate r (expressed as a decimal), compounded annually. How long does it take to double your money at each of the following rates?

(a) 4% **(b)** 8% **(c)** 18% **(d)** 36%

(e) Round each of your answers in parts (a)–(d) to the nearest year, and compare them with these numbers: 72/4, 72/8, 72/18, and 72/36. Use this evidence to state a "rule of thumb" for determining approximate doubling time without employing the function D. This rule, which has long been used by bankers, is called the *rule of 72*.

62. **Health** Two people with the flu visited a college campus. The number of days, T, that it took for the flu virus to infect n people is given by

$$T = -1.43 \ln\left(\frac{10,000 - n}{4998n}\right).$$

How many days will it take for the virus to infect
(a) 500 people? (b) 5000 people?

63. **Business** Domestic sales of DVDs and videotapes (in billions of dollars) are approximated by

$$g(x) = 10.155 + 3.62 \ln x,$$

where $x = 1$ corresponds to the year 2001.*
(a) Estimate the sales in 2004 and 2008.
(b) Graph the function g for the period from 2001 to 2025.
(c) Assuming that this model remains accurate, what does the shape of the graph suggest about DVD and videotape sales?

64. **Physical Science** The barometric pressure p (in inches of mercury) is related to the height h above sea level (in miles) by the equation

$$h = -5 \ln\left(\frac{p}{29.92}\right).$$

The pressure readings given in parts (a)–(c) were made by a weather balloon. At what heights were they made?
(a) 29.92 in. (b) 20.05 in. (c) 11.92 in.
(d) Use a graphing calculator to determine the pressure at a height of 3 miles.

65. **Social Science** The expected percentage of the U.S. population age 60 and older is given by

$$g(x) = 15.93 + 2.174 \ln x \qquad (x \ge 1),$$

where $x = 1$ corresponds to the year 2001.†
(a) What is the expected percentage of people who are 60 or older in 2007, 2015, 2030, and 2050?
(b) Graph the function g.
(c) What does the shape of the graph suggest about the percentage of older people as time goes on?

*Based on data and projections from Adams Media Research.

†Based on projections by the U.S. Census Bureau.

Natural Science *These exercises deal with the Shannon index of diversity. (See Example 10.) Note that in two communities with the same number of species, a larger index indicates greater diversity.*

66. A study of barn owl prey in a particular area produced the following data:

Species	Number
Rats	662
Mice	907
Birds	531

Find the index of diversity of this community. Is this community more or less diverse than the one in Example 10?

67. An eastern forest is composed of the following trees:

Species	Number
Beech	2754
Birch	689
Hemlock	4428
Maple	629

What is the index of diversity of this community?

68. A community has high diversity when all of its species have approximately the same number of individuals. It has low diversity when a few of its species account for most of the total population. Illustrate this fact for the following two communities:

Community 1	Number
Species A	1000
Species B	1000
Species C	1000

Community 2	Number
Species A	2500
Species B	200
Species C	300

69. **Business** The total assets of bond mutual funds (in trillions of dollars) are approximated by

$$f(x) = -2.43 + 1.414 \ln x \qquad (x \ge 10),$$

where $x = 10$ corresponds to the year 2000.*
(a) What were the total assets in 2006 and 2010?
(b) If this model remains accurate, find the year in which total assets reach \$2.25 trillion.

*Based on data in the *Statistical Abstract of the United States: 2008*.

70. **Finance** Small Business Administration loans to minority-owned small businesses (in billions of dollars) are approximated by

$$h(x) = .3 + 3.55 \ln x \quad (x \geq 3),$$

where $x = 3$ corresponds to the year 2003.*

(a) What was the volume of loans in 2006 and 2009?

(b) If this model remains accurate, when will loan volume reach \$10.2 billion?

71. **Business** Total net shipments of aluminum (in billions of pounds) are approximated by

$$g(x) = 16.8 + 2.924 \ln(x + 1),$$

where $x = 0$ corresponds to the year 1990.†

(a) How many pounds of aluminum were shipped in 2004 and 2008?

(b) Assuming the continued accuracy of the model, when will aluminum shipments reach 27,100,000,000 pounds?

72. **Finance** The median family income is the "middle income"—half the families have this income or more, and half have a lesser income. The median family income can be approximated by

$$f(x) = 18{,}724.85 + 13{,}530.74 \ln x \quad (x \geq 5),$$

where $x = 5$ corresponds to the year 1995.*

(a) What was the median family income in 2009? What is it this year?

(b) If this model remains accurate, when will the median family income reach \$62,500?

*Based on data from the U.S. Small Business Administration.

†Based on data from the Aluminum Association, Inc.

✓ Checkpoint Answers

1. **(a)** 2 **(b)** 3 **(c)** −1
2. **(a)** 1.4314 **(b)** 3.0370 **(c)** −2.3706
3. **(a)** 1.8083 **(b)** 2.9957 **(c)** −.2231 **(d)** −2.3026
4. **(a)** $\log_5 125 = 3$ **(b)** $\log_3 (1/81) = -4$ **(c)** $\log_8 4 = 2/3$
5. **(a)** $16^{1/2} = 4$ **(b)** $3^{-2} = 1/9$ **(c)** $16^{3/4} = 8$
6. **(a)** $\log_a 15x^5$ **(b)** $\log_a (3p/5q)$ **(c)** $\log_a (k^4/m^3)$
7. **(a)** 3.23 **(b)** .31 **(c)** 2.92 **(d)** 1 **(e)** 0
8.

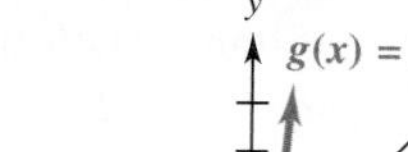

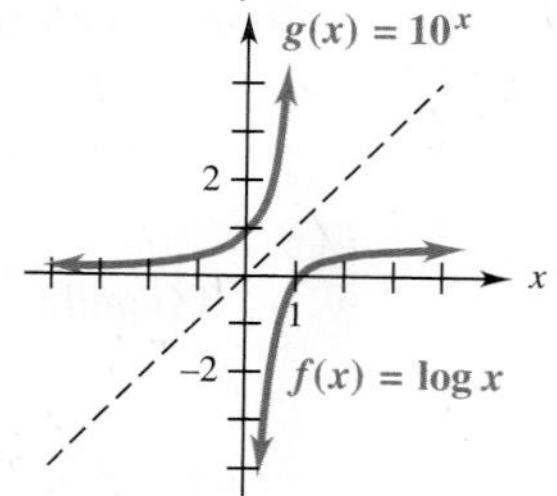

*Based on data from the U.S. Census Bureau.

4.4 ▸ Logarithmic and Exponential Equations

Many applications involve solving logarithmic and exponential equations, so we begin with solution methods for such equations.

LOGARITHMIC EQUATIONS

When an equation involves only logarithmic terms, use the logarithm properties to write each side as a single logarithm. Then use this fact.

> Let u, v, and a be positive real numbers, with $a \neq 1$.
>
> If $\log_a u = \log_a v$, then $u = v$.

Example 1 Solve $\log x = \log(x + 3) - \log(x - 1)$.

Solution First, use the quotient property of logarithms to write the right side as a single logarithm:

$$\log x = \log(x + 3) - \log(x - 1)$$

$$\log x = \log\left(\frac{x + 3}{x - 1}\right).$$

The fact in the preceding box now shows that

$$x = \frac{x+3}{x-1}$$
$$x(x-1) = x+3$$
$$x^2 - x = x+3$$
$$x^2 - 2x - 3 = 0$$
$$(x-3)(x+1) = 0$$
$$x = 3 \quad \text{or} \quad x = -1.$$

Since $\log x$ is not defined when $x = -1$, the only possible solution is $x = 3$. Use a calculator to verify that 3 actually is a solution. ✓1

✓Checkpoint 1

Solve each equation.

(a) $\log_2 (p+9) - \log_2 p = \log_2 (p+1)$

(b) $\log_3 (m+1) - \log_3 (m-1) = \log_3 m$

Answers: See page 253.

When an equation involves constants and logarithmic terms, use algebra and the logarithm properties to write one side as a single logarithm and the other as a constant. Then use the following property of logarithms, which was discussed on pages 235–236.

> If a and u are positive real numbers, with $a \neq 1$, then
> $$a^{\log_a u} = u.$$

Example 2 Solve each equation.

(a) $\log_5 (2x - 3) = 2$

Solution Since the base of the logarithm is 5, raise 5 to the exponents given by the equation:

$$5^{\log_5 (2x-3)} = 5^2$$

On the left side, use the fact in the preceding box (with $a = 5$ and $u = 2x - 3$) to conclude that

$$2x - 3 = 25$$
$$2x = 28$$
$$x = 14.$$

Verify that 14 is a solution of the original equation.

(b) $\log(x - 16) = 2 - \log(x - 1)$

Solution First rearrange the terms to obtain a single logarithm on the left side:

$$\log(x-16) + \log(x-1) = 2$$
$$\log[(x-16)(x-1)] = 2 \qquad \text{Product property of logarithms}$$
$$\log(x^2 - 17x + 16) = 2.$$

Since the base of the logarithm is 10, raise 10 to the given powers:

$$10^{\log(x^2-17x+16)} = 10^2.$$

On the left side, apply the logarithm property in the preceding box (with $a = 10$ and $u = x^2 - 17x + 16$):

$$x^2 - 17x + 16 = 100$$
$$x^2 - 17x - 84 = 0$$
$$(x + 4)(x - 21) = 0$$
$$x = -4 \quad \text{or} \quad x = 21.$$

In the original equation, when $x = -4$, $\log(x - 16) = \log(-20)$, which is not defined. So -4 is not a solution. You should verify that 21 is a solution of the original equation.

(c) $\log_2 x - \log_2 (x - 1) = 1$

Solution Proceed as before, with 2 as the base of the logarithm:

$$\log_2 \frac{x}{x-1} = 1 \qquad \text{Quotient property of logarithms}$$
$$2^{\log_2 x/(x-1)} = 2^1 \qquad \text{Exponentiate to the base 2.}$$
$$\frac{x}{x-1} = 2 \qquad \text{Use the fact in the preceding box.}$$
$$x = 2(x - 1) \qquad \text{Multiply both sides by } x - 1.$$
$$x = 2x - 2$$
$$-x = -2$$
$$x = 2.$$

Verify that 2 is a solution of the original equation. 2✓

✓Checkpoint 2

Solve each equation.

(a) $\log_5 x + 2\log_5 x = 3$

(b) $\log_6 (a + 2) - \log_6 \frac{a - 7}{5} = 1$

Answers: See page 253.

EXPONENTIAL EQUATIONS

An equation in which all the variables are exponents is called an exponential equation. When such an equation can be written as two powers of the same base, it can be solved by using the following fact.

> Let a be a positive real number, with $a \neq 1$.
>
> If $a^u = a^v$, then $u = v$.

Example 3 Solve $9^x = 27$.

Solution First, write both sides as powers of the same base. Since $9 = 3^2$ and $27 = 3^3$, we have

$$9^x = 27$$
$$(3^2)^x = 3^3$$
$$3^{2x} = 3^3.$$

Apply the fact in the preceding box (with $a = 3$, $u = 2x$, and $v = 3$):

$$2x = 3$$
$$x = \frac{3}{2}.$$

Verify that $x = 3/2$ is a solution of the original equation. 3✓

✓Checkpoint 3

Solve each equation.

(a) $8^{2x} = 4$

(b) $5^{3x} = 25^4$

(c) $36^{-2x} = 6$

Answers: See page 253.

Exponential equations involving different bases can often be solved by using the power property of logarithms, which is repeated here for natural logarithms.

If u and r are real numbers, with u positive, then

$$\ln u^r = r \ln u.$$

Although natural logarithms are used in the following examples, logarithms to any base will produce the same solutions.

Example 4 Solve $3^x = 5$.

Solution Take natural logarithms on both sides:

$$\ln 3^x = \ln 5.$$

Apply the power property of logarithms in the preceding box (with $r = x$) to the left side:

$$x \ln 3 = \ln 5$$

$$x = \frac{\ln 5}{\ln 3} \approx 1.465. \quad \text{Divide both sides by the constant ln 3.}$$

To check, evaluate $3^{1.465}$; the answer should be approximately 5, which verifies that the solution of the given equation is 1.465 (to the nearest thousandth). 4✓

✓Checkpoint 4

Solve each equation. Round solutions to the nearest thousandth.

(a) $2^x = 7$

(b) $5^m = 50$

(c) $3^y = 17$

Answers: See page 253.

CAUTION Be careful: $\frac{\ln 5}{\ln 3}$ is *not* equal to $\ln\left(\frac{5}{3}\right)$ or $\ln 5 - \ln 3$.

Example 5 Solve $3^{2x-1} = 4^{x+2}$.

Solution Taking natural logarithms on both sides gives

$$\ln 3^{2x-1} = \ln 4^{x+2}.$$

Now use the power property of logarithms and the fact that ln 3 and ln 4 are constants to rewrite the equation:

$(2x - 1)(\ln 3) = (x + 2)(\ln 4)$	Power property
$2x(\ln 3) - 1(\ln 3) = x(\ln 4) + 2(\ln 4)$	Distributive property
$2x(\ln 3) - x(\ln 4) = 2(\ln 4) + 1(\ln 3).$	Collect terms with x on one side.

Factor out x on the left side to get

$$[2(\ln 3) - \ln 4]x = 2(\ln 4) + \ln 3.$$

Divide both sides by the coefficient of x:

$$x = \frac{2(\ln 4) + \ln 3}{2(\ln 3) - \ln 4}.$$

Using a calculator to evaluate this last expression, we find that

$$x = \frac{2 \ln 4 + \ln 3}{2 \ln 3 - \ln 4} \approx 4.774.$$ 5

Checkpoint 5

Solve each equation. Round solutions to the nearest thousandth.

(a) $6^m = 3^{2m-1}$

(b) $5^{6a-3} = 2^{4a+1}$

Answers: See page 253.

Recall that $\ln e = 1$ (because 1 is the exponent to which e must be raised to produce e). This fact simplifies the solution of equations involving powers of e.

Example 6 Solve $3e^{x^2} = 600$.

Solution First divide each side by 3 to get

$$e^{x^2} = 200.$$

Now take natural logarithms on both sides; then use the power property of logarithms:

$$\begin{aligned} e^{x^2} &= 200 \\ \ln e^{x^2} &= \ln 200 \\ x^2 \ln e &= \ln 200 && \text{Power property} \\ x^2 &= \ln 200 && \ln e = 1. \\ x &= \pm\sqrt{\ln 200} \\ x &\approx \pm 2.302. \end{aligned}$$

Verify that the solutions are ± 2.302, rounded to the nearest thousandth. (The symbol $\pm$ is used as a shortcut for writing the two solutions 2.302 and -2.302.) 6

Checkpoint 6

Solve each equation. Round solutions to the nearest thousandth.

(a) $e^{.1x} = 11$

(b) $e^{3+x} = .893$

(c) $e^{2x^2-3} = 9$

Answers: See page 253.

TECHNOLOGY TIP Logarithmic and exponential equations can be solved on a graphing calculator. To solve $3^x = 5^{2x-1}$, for example, graph $y = 3^x$ and $y = 5^{2x-1}$ on the same screen. Then use the intersection finder to determine the x-coordinates of their intersection points. Alternatively, graph $y = 3^x - 5^{2x-1}$ and use the root finder to determine the x-intercepts of the graph.

APPLICATIONS

Some of the most important applications of exponential and logarithmic functions arise in banking and finance. They will be thoroughly discussed in Chapter 5. The applications here are from other fields.

Example 7 **Social Science** According to projections by the U.S. Census Bureau, the world population (in billions) in year x is approximated by the function $P(x) = 4.834(1.011^{x-1980})$. When will the population reach 7 billion?

Solution You are being asked to find the value of x for which $P(x) = 7$—that is, to solve the following equation:

$$4.834(1.011^{x-1980}) = 7$$

$$1.011^{x-1980} = \frac{7}{4.834} \quad \text{Divide both sides by 4.834.}$$

$$\ln 1.011^{x-1980} = \ln\left(\frac{7}{4.834}\right) \quad \text{Take logarithms on both sides.}$$

$$(x - 1980)\ln 1.011 = \ln\left(\frac{7}{4.834}\right) \quad \text{Power property of logarithms}$$

$$x - 1980 = \frac{\ln(7/4.834)}{\ln 1.011} \quad \text{Divide both sides by ln 1.011.}$$

$$x = \frac{\ln(7/4.834)}{\ln 1.011} + 1980 \approx 2013.8.$$

Hence, the world's population will reach 7 billion in late 2013. ✓7

✓Checkpoint 7

Use the function in Example 7 to determine when the earth's population will be 10 billion.

Answer: See page 253.

The **half-life** of a radioactive substance is the time it takes for a given quantity of the substance to decay to one-half its original mass. The half-life depends only on the substance, not on the size of the sample. It can be shown that the amount of a radioactive substance at time t is given by the function

$$f(t) = y_0\left(\frac{1}{2}\right)^{t/h} = y_0(.5^{t/h}),$$

where y_0 is the initial amount (at time $t = 0$) and h is the half-life of the substance.

Radioactive carbon-14 is found in every living plant and animal. After the plant or animal dies, its carbon-14 decays exponentially, with a half-life of 5730 years. This fact is the basis for a technique called *carbon dating* for determining the age of fossils.

Example 8 **Natural Science** A round wooden table hanging in Winchester Castle (England) was alleged to have belonged to King Arthur, who lived in the fifth century. A recent chemical analysis showed that the wood had lost 9% of its carbon-14.* How old is the table?

Solution The decay function for carbon-14 is

$$f(t) = y_0(.5^{t/5730}),$$

where $t = 0$ corresponds to the time the wood was cut to make the table. (That is when the tree died.) Since the wood has lost 9% of its carbon-14, the amount in the

*This is done by measuring the ratio of carbon-14 to nonradioactive carbon-12 in the table (a ratio that is approximately constant over long periods) and comparing it with the ratio in living wood.

table now is 91% of the initial amount y_0 (that is, $.91y_0$). We must find the value of t for which $f(t) = .91y_0$. So we must solve the equation

$$y_0(.5^{t/5730}) = .91y_0 \qquad \text{Definition of } f(t)$$

$$.5^{t/5730} = .91 \qquad \text{Divide both sides by } y_0.$$

$$\ln .5^{t/5730} = \ln .91 \qquad \text{Take logarithms on both sides.}$$

$$\left(\frac{t}{5730}\right)\ln .5 = \ln .91 \qquad \text{Power property of logarithms}$$

$$t \ln .5 = 5730 \ln .91 \qquad \text{Multiply both sides by 5730.}$$

$$t = \frac{5730 \ln .91}{\ln .5} \approx 779.63. \qquad \text{Divide both sides by ln 5.}$$

The table is about 780 years old and therefore could not have belonged to King Arthur. 8✓

✓Checkpoint 8

How old is a skeleton that has lost 65% of its carbon-14?

Answer: See page 253.

Earthquakes are often in the news. The standard method of measuring their size, the **Richter scale**, is a logarithmic function (base 10).

Example 9 **Physical Science** The intensity $R(i)$ of an earthquake, measured on the Richter scale, is given by

$$R(i) = \log\left(\frac{i}{i_0}\right),$$

where i is the intensity of the ground motion of the earthquake and i_0 is the intensity of the ground motion of the so-called *zero earthquake* (the smallest detectable earthquake, against which others are measured). The underwater earthquake that caused the disastrous 2004 tsunami in Southeast Asia measured 9.1 on the Richter scale.

(a) How did the ground motion of this tsunami compare with that of the zero earthquake?

Solution In this case,

$$R(i) = 9.1$$

$$\log\left(\frac{i}{i_0}\right) = 9.1.$$

By the definition of logarithms, 9.1 is the exponent to which 10 must be raised to produce i/i_0, which means that

$$10^{9.1} = \frac{i}{i_0}, \qquad \text{or equivalently,} \qquad i = 10^{9.1}i_0.$$

So the earthquake that produced the tsunami had $10^{9.1}$ (about 1.26 *billion*) times more ground motion than the zero earthquake.

(b) What is the Richter-scale intensity of an earthquake with 10 times as much ground motion as the 2004 tsunami earthquake?

Solution From (a), the ground motion of the tsunami quake was $10^{9.1}\,i_0$. So a quake with 10 times that motion would satisfy

$$i = 10(10^{9.1}\,i_0) = 10^1 \cdot 10^{9.1}\,i_0 = 10^{10.1}\,i_0\,.$$

Therefore, its Richter scale measure would be

$$R(i) = \log\left(\frac{i}{i_0}\right) = \log\left(\frac{10^{10.1}\, i_0}{i_0}\right) = \log 10^{10.1} = 10.1.$$

Thus, a tenfold increase in ground motion increases the Richter scale measure by only 1. 9✓

✓Checkpoint 9

Find the Richter-scale intensity of an earthquake whose ground motion is 100 times greater than the ground motion of the 2004 tsunami earthquake discussed in Example 9.

Answer: See page 253.

Example 10 **Social Science** One action that government could take to reduce carbon emissions into the atmosphere is to place a tax on fossil fuels. This tax would be based on the amount of carbon dioxide that is emitted into the air when such a fuel is burned. The *cost–benefit* equation $\ln(1 - P) = -.0034 - .0053T$ describes the approximate relationship between a tax of T dollars per ton of carbon dioxide and the corresponding percent reduction P (in decimals) in emissions of carbon dioxide.*

(a) Write P as a function of T.

Solution We begin by writing the cost–benefit equation in exponential form:

$$\ln(1 - P) = -.0034 - .0053T$$
$$1 - P = e^{-.0034-.0053T}$$
$$P = P(T) = 1 - e^{-.0034-.0053T}.$$

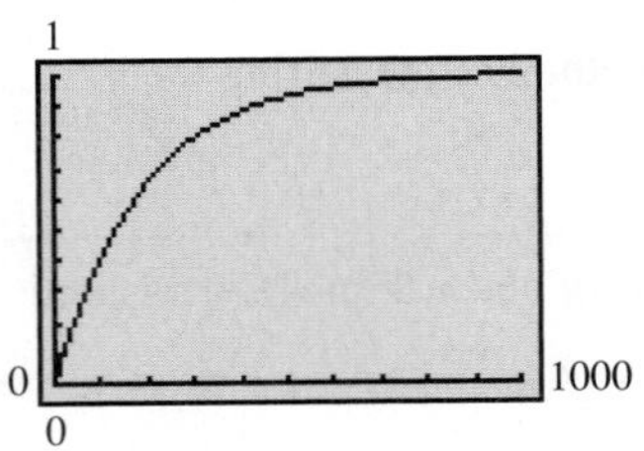

FIGURE 4.23

A calculator-generated graph of $P(T)$ is shown in Figure 4.23.

(b) Discuss the benefit of continuing to raise taxes on carbon dioxide emissions.

Solution From the graph, we see that initially there is a rapid reduction in carbon dioxide emissions. However, after a while, there is little benefit in raising taxes further.

*W. Nordhause, "To Slow or Not to Slow: The Economics of the Greenhouse Effect." New Haven, CT: Yale University.

4.4 ▶ Exercises

Solve each logarithmic equation. (See Example 1.)

1. $\ln(x + 3) = \ln(2x - 5)$
2. $\ln(8k - 7) - \ln(3 + 4k) = \ln(9/11)$
3. $\ln(3x + 1) - \ln(5 + x) = \ln 2$
4. $\ln(5x + 2) = \ln 4 + \ln(x + 3)$
5. $2\ln(x - 3) = \ln(x + 5) + \ln 4$
6. $\ln(k + 5) + \ln(k + 2) = \ln 18k$

Solve each logarithmic equation. (See Example 2.)

7. $\log_3(6x - 2) = 2$
8. $\log_5(3x - 4) = 1$
9. $\log x - \log(x + 4) = -1$
10. $\log m - \log(m + 4) = -2$
11. $\log_3(y + 2) = \log_3(y - 7) + \log_3 4$
12. $\log_8(z - 6) = 2 - \log_8(z + 15)$
13. $\ln(x + 9) - \ln x = 1$
14. $\ln(2x + 1) - 1 = \ln(x - 2)$
15. $\log x + \log(x - 9) = 1$
16. $\log(x - 1) + \log(x + 2) = 1$

Solve each equation for c.

17. $\log(3 + b) = \log(4c - 1)$
18. $\ln(b + 7) = \ln(6c + 8)$
19. $2 - b = \log(6c + 5)$
20. $8b + 6 = \ln(2c) + \ln c$

21. Suppose you overhear the following statement: "I must reject any negative answer when I solve an equation involving logarithms." Is this correct? Write an explanation of why it is or is not correct.

22. What values of x cannot be solutions of the following equation?

$$\log_a (4x - 7) + \log_a (x^2 + 4) = 0.$$

Solve these exponential equations without using logarithms. (See Example 3.)

23. $2^{x-1} = 8$
24. $16^{-x+2} = 8$
25. $25^{-3x} = 3125$
26. $81^{-2x} = 3^{x-1}$
27. $6^{-x} = 36^{x+6}$
28. $16^x = 64$
29. $\left(\frac{3}{4}\right)^x = \frac{16}{9}$
30. $2^{x^2-4x} = \frac{1}{16}$

Use logarithms to solve these exponential equations. (See Examples 4–6.)

31. $2^x = 5$
32. $5^x = 8$
33. $2^x = 3^{x-1}$
34. $4^{x+2} = 5^{x-1}$
35. $3^{1-2x} = 5^{x+5}$
36. $4^{3x-1} = 3^{x-2}$
37. $e^{3x} = 6$
38. $e^{-3x} = 5$
39. $2e^{5a+2} = 8$
40. $10e^{3z-7} = 5$

Solve each equation for c.

41. $10^{4c-3} = d$
42. $3 \cdot 10^{2c+1} = 4d$
43. $e^{2c-1} = b$
44. $3e^{5c-7} = b$

Solve these equations. (See Examples 1–6.)

45. $\log_7 (r + 3) + \log_7 (r - 3) = 1$
46. $\log_4 (z + 3) + \log_4 (z - 3) = 2$
47. $\log_3 (a - 3) = 1 + \log_3 (a + 1)$
48. $\log w + \log(3w - 13) = 1$
49. $\log_2 \sqrt{2y^2} - 1 = 3/2$
50. $\log_2 (\log_2 x) = 1$
51. $\log_2 (\log_3 x) = 1$
52. $\frac{\ln (2x + 1)}{\ln (3x - 1)} = 2$
53. $5^{-2x} = \frac{1}{25}$
54. $5^{x^2+x} = 1$
55. $2^{|x|} = 16$
56. $5^{-|x|} = \frac{1}{25}$
57. $2^{x^2-1} = 10$
58. $3^{2-x^2} = 8$
59. $2(e^x + 1) = 10$
60. $5(e^{2x} - 2) = 15$

61. Explain why the equation $4^{x^2+1} = 2$ has no solutions.

62. Explain why the equation $\log(-x) = -4$ does have a solution, and find that solution.

Work these problems. (See Example 7.)

63. **Business** According to data from the U.S. Federal Reserve, the amount of outstanding consumer debt (in trillions of dollars) is approximated by

$$g(t) = .365478*1.073306^t,$$

where $t = 0$ corresponds to the year 1980. Find the year in which consumer debt is
(a) \$4 trillion; **(b)** \$5 trillion.

64. **Business** The U.S. gross domestic product (GDP) (in trillions of dollars) is approximated by

$$f(x) = 10.4*1.024^x,$$

where $x = 0$ corresponds to the year 2000.* When will the GDP reach the following levels?
(a) \$16 trillion **(b)** \$30 trillion

65. **Health** As we saw in Example 12 of Section 4.3, the life expectancy at birth of a person born in year x is approximately

$$f(x) = 17.6 + 12.8 \ln x,$$

where $x = 10$ corresponds to the year 1910. Find the birth year of a person whose life expectancy at birth was
(a) 75.5 years; **(b)** 77.5 years; **(c)** 81 years.

66. **Health** A drug's effectiveness decreases over time. If, each hour, a drug is only 90% as effective as during the previous hour, at some point the patient will not be receiving enough medication and must receive another dose. This situation can be modeled by the exponential function $y = y_0(.90)^{t-1}$. In this equation, y_0 is the amount of the initial dose and y is the amount of medication still available t hours after the drug was administered. Suppose 200 mg of the drug is administered. How long will it take for this initial dose to reach the dangerously low level of 50 mg?

67. **Business** The Consumer Price Index (CPI) measures prices paid by urban consumers for a representative basket of goods and services. In this index, 100 corresponds to 1982–1984. The recent CPI can be approximated by

$$C(x) = 170.63e^{.03x}$$

where $x = 0$ corresponds to the year 2000.†

(a) When did the CPI reach 220?
(b) If this model remains accurate, when will the CPI reach 260?

68. **Health** The probability P percent of having an accident while driving a car is related to the alcohol level t of the driver's blood by the equation $P = e^{kt}$, where k is a constant. Accident statistics show that the probability

*Based on data and projections by Goldman Sachs.

†Based on data from the U.S. Bureau of Labor Statistics.

of an accident is 25% when the blood alcohol level is $t = .15$.

(a) Use the equation and the preceding information to find k (to three decimal places). *Note*: Use $P = 25$, and not .25.

(b) Using the value of k from part (a), find the blood alcohol level at which the probability of having an accident is 50%.

Work these exercises. (See Example 8.)

69. **Natural Science** The amount of cobalt-60 (in grams) in a storage facility at time t is given by

$$C(t) = 25e^{-.14t},$$

where time is measured in years.

(a) How much cobalt-60 was present initially?

(b) What is the half-life of cobalt-60? (*Hint*: For what value of t is $C(t) = 12.5$?)

70. **Natural Science** A Native American mummy was found recently. It had 73.6% of the amount of radiocarbon present in living beings. Approximately how long ago did this person die?

71. **Natural Science** How old is a piece of ivory that has lost 36% of its radiocarbon?

72. **Natural Science** A sample from a refuse deposit near the Strait of Magellan had 60% of the carbon-14 of a contemporary living sample. How old was the sample?

Natural Science *Do these problems. (See Example 9.)*

73. In May 2008, Sichuan province, China suffered an earthquake that measured 7.9 on the Richter scale.

(a) Express the intensity of this earthquake in terms of i_0.

(b) In July of the same year, a quake measuring 5.4 on the Richter scale struck Los Angeles. Express the intensity of this earthquake in terms of i_0.

(c) How many times more intense was the China earthquake than the one in Los Angeles?

74. Find the Richter-scale intensity of earthquakes whose ground motion is

(a) $1000i_0$; (b) $100{,}000i_0$; (c) $10{,}000{,}000i_0$.

(d) Fill in the blank in this statement: Increasing the ground motion by a factor of 10^k increases the Richter intensity by _____ units.

75. The loudness of sound is measured in units called decibels. The decibel rating of a sound is given by

$$D(i) = 10 \cdot \log\left(\frac{i}{i_0}\right),$$

where i is the intensity of the sound and i_0 is the minimum intensity detectable by the human ear (the so-called *threshold sound*). Find the decibel rating of each of the sounds with the given intensities. Round answers to the nearest whole number.

(a) Whisper, $115i_0$

(b) Average sound level in the movie *Godzilla*, $10^{10}i_0$

(c) Jackhammer, $31{,}600{,}000{,}000i_0$

(d) Rock music, $895{,}000{,}000{,}000i_0$

(e) Jetliner at takeoff, $109{,}000{,}000{,}000{,}000i_0$

76. (a) How much more intense is a sound that measures 100 decibels than the threshold sound?

(b) How much more intense is a sound that measures 50 decibels than the threshold sound?

(c) How much more intense is a sound measuring 100 decibels than one measuring 50 decibels?

77. **Natural Science** Refer to Example 10.

(a) Determine the percent reduction in carbon dioxide when the tax is $60.

(b) What tax will cause a 50% reduction in carbon dioxide emissions?

78. **Social Science** The number of years, $N(r)$, since two independently evolving languages split off from a common ancestral language is approximated by

$$N(r) = -5000 \ln r,$$

where r is the proportion of the words from the ancestral language that is common to both languages now. Find each of the following.

(a) $N(.9)$ (b) $N(.5)$ (c) $N(.3)$

(d) How many years have elapsed since the split of two languages if 70% of the words of the ancestral language are common to both languages today?

(e) If two languages split off from a common ancestral language about 1000 years ago, find r.

79. **Natural Science** In the central Sierra Nevada mountains of California, the percent of moisture that falls as snow rather than rain is approximated reasonably well by

$$p = 86.3 \ln h - 680,$$

where p is the percent of moisture as snow at an altitude of h feet (with $3000 \le h < 8500$).

(a) Graph p.

(b) At what altitude is 50 percent of the moisture snow?

80. **Physical Science** The table gives some of the planets' average distances D from the sun and their period P of revolution around the sun in years. The distances have been normalized so that Earth is one unit from the sun. Thus, Jupiter's distance of 5.2 means that Jupiter's distance from the sun is 5.2 times farther than Earth's.*

Planet	D	P
Earth	1	1
Jupiter	5.2	11.9
Saturn	9.54	29.5
Uranus	19.2	84.0

*C. Ronan, *The Natural History of the Universe*. New York: Macmillan Publishing Co., 1991.

(a) Plot the points (D, P) for these planets. Would a straight line or an exponential curve fit these points best?

(b) Plot the points $(\ln D, \ln P)$ for these planets. Do these points appear to lie on a line?

(c) Determine a linear equation that approximates the data points, with $x = \ln D$ and $y = \ln P$. Use the first and last data points (rounded to 2 decimal places). Graph your line and the data on the same coordinate axes.

(d) Use the linear equation to predict the period of the planet Pluto if its distance is 39.5. Compare your answer with the true value of 248.5 years.

✓ Checkpoint Answers

1.	**(a)** 3	**(b)** $1 + \sqrt{2} \approx 2.414$	
2.	**(a)** 5	**(b)** 52	
3.	**(a)** 1/3	**(b)** 8/3	**(c)** $-1/4$
4.	**(a)** 2.807	**(b)** 2.431	**(c)** 2.579
5.	**(a)** 2.710	**(b)** .802	
6.	**(a)** 23.979	**(b)** -3.113	**(c)** ± 1.612

7. In 2046.

8. About 8679 years

9. 11.1

CHAPTER 4 ▸ Summary

KEY TERMS AND SYMBOLS

4.1 ▸ exponential function
exponential growth and decay
the number $e \approx 2.71828\ldots$

4.2 ▸ exponential growth function
logistic function

4.3 ▸ $\log x$ common (base-10 logarithm) of x
$\ln x$ natural (base-e logarithm) of x
$\log_a x$ base-a logarithm of x
product, quotient, and power properties of logarithms
change-of-base theorem
logarithmic function
inverses

4.4 ▸ logarithmic equations
exponential equations
half-life
Richter scale

CHAPTER 4 KEY CONCEPTS

An important application of exponents is the **exponential growth function**, defined as $f(t) = y_0 e^{kt}$ or $f(t) = y_0 b^t$, where y_0 is the amount of a quantity present at time $t = 0$, $e \approx 2.71828$, and k and b are constants.

The **logarithm** of x to the base a is defined as follows: For $a > 0$ and $a \neq 1$, $y = \log_a x$ means $a^y = x$. Thus, $\log_a x$ is an *exponent*, the power to which a must be raised to produce x.

Properties of Logarithms ▶ Let x, y, and a be positive real numbers, with $a \neq 1$, and let r be any real number. Then

$\log_a 1 = 0;$ $\quad \log_a a = 1;$

$\log_a a^r = r;$ $\quad a^{\log_a x} = x.$

Product property $\log_a xy = \log_a x + \log_a y$

Quotient property $\log_a \frac{x}{y} = \log_a x - \log_a y$

Power property $\log_a x^r = r \log_a x$

Solving Exponential and Logarithmic Equations ▶ Let $a > 0$, with $a \neq 1$.

If $\log_a u = \log_a v$, then $u = v$.

If $a^u = a^v$, then $u = v$.

CHAPTER 4 REVIEW EXERCISES

Match each equation with the letter of the graph that most closely resembles the graph of the equation. Assume that $a > 1$.

1. $y = a^{x+2}$ **2.** $y = a^x + 2$

3. $y = -a^x + 2$ **4.** $y = a^{-x} + 2$

(a)

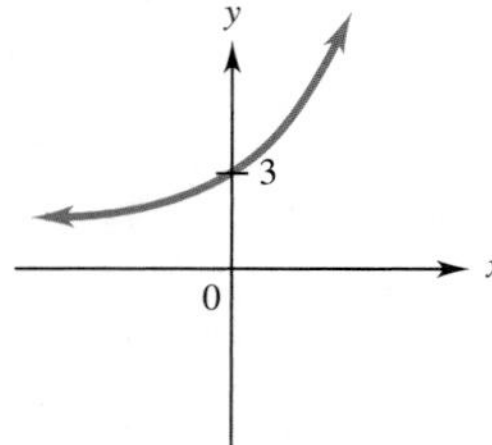

(b)

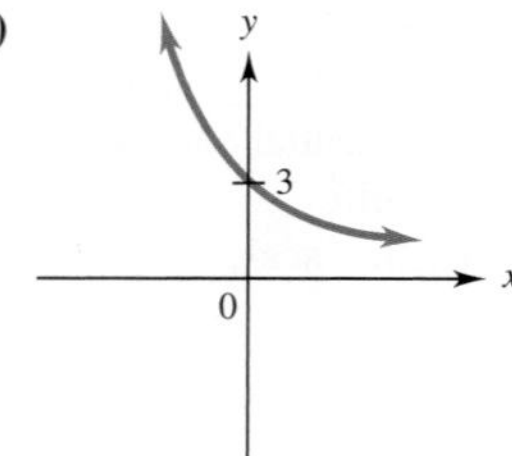

(c)

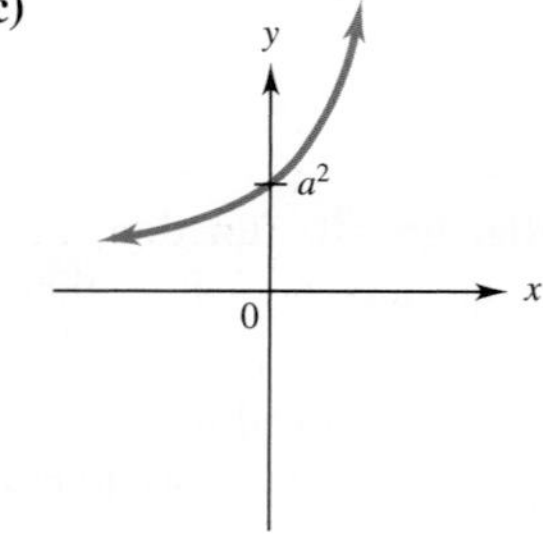

(d)

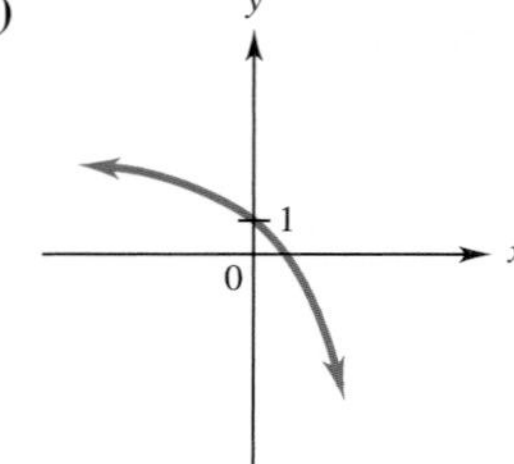

Consider the exponential function $y = f(x) = a^x$ graphed here. Answer each question on the basis of the graph.

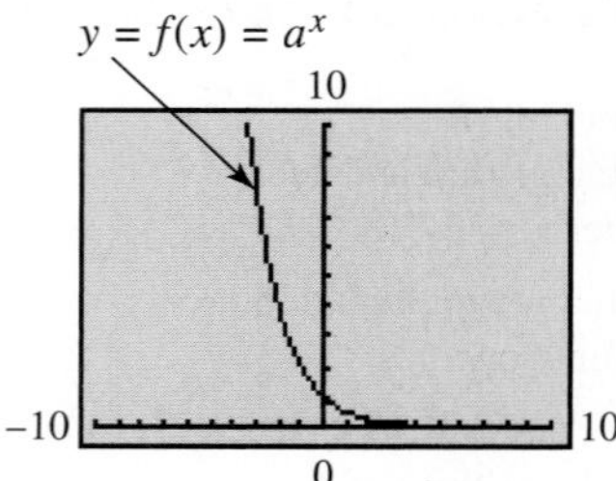

5. What is true about the value of a in comparison to 1?

6. What is the domain of f?

7. What is the range of f?

8. What is the value of $f(0)$?

Graph each function.

9. $f(x) = 4^x$ **10.** $g(x) = 4^{-x}$

11. $f(x) = \ln x + 5$ **12.** $g(x) = \log x - 3$

Work these problems.

13. Business Cigarette usage has been declining in recent years. The number of cigarettes (in billions) produced in year x is approximated by

$$f(x) = 564.13 * .9713^x,$$

where $x = 0$ corresponds to 2000.*

(a) Estimate the number of cigarettes produced in 2009 and 2011.

(b) According to this model, in what year will cigarette production drop below 100 billion per year? Do you think this is likely?

14. Business A person learning certain skills involving repetition tends to learn quickly at first. Then learning tapers off and approaches some upper limit. Suppose the number of symbols per minute a textbook typesetter can produce is given by $p(t) = 250 - 120(2.8)^{-.5t}$, where t is the number of months the typesetter has been in training. Find each of the following.

(a) $p(2)$ **(b)** $p(4)$

(c) $p(10)$ **(d)** Graph $y = p(t)$.

Translate each exponential statement into an equivalent logarithmic one.

15. $10^{2.53148} = 340$ **16.** $5^4 = 625$

17. $e^{3.8067} = 45$ **18.** $7^{1/2} = \sqrt{7}$

Translate each logarithmic statement into an equivalent exponential one.

19. $\log 10{,}000 = 4$ **20.** $\log 26.3 = 1.4200$

21. $\ln 81.1 = 4.3957$ **22.** $\log_2 4096 = 12$

Evaluate these expressions without using a calculator.

23. $\ln e^5$ **24.** $\log \sqrt[3]{10}$

25. $10^{\log 8.9}$ **26.** $\ln e^{3t^2}$

27. $\log_8 2$ **28.** $\log_8 32$

Write these expressions as a single logarithm. Assume all variables represent positive quantities.

29. $\log 4x + \log 5x^5$ **30.** $4 \log u - 5 \log u^6$

31. $3 \log b - 2 \log c$ **32.** $7 \ln x - 3(\ln x^3 + 5 \ln x)$

Solve each equation. Round to the nearest thousandth.

33. $\ln(m + 8) - \ln m = \ln 3$

34. $2 \ln(y + 1) = \ln(y^2 - 1) + \ln 5$

35. $\log(m + 3) = 2$ **36.** $\log x^3 = 2$

37. $\log_2 (3k + 1) = 4$ **38.** $\log_5 \left(\frac{5z}{z - 2}\right) = 2$

39. $\log x + \log(x - 3) = 1$

40. $\log_2 r + \log_2 (r - 2) = 3$

41. $2^{3x} = \frac{1}{64}$ **42.** $\left(\frac{9}{16}\right)^x = \frac{3}{4}$

43. $9^{2y+1} = 27^y$ **44.** $\frac{1}{2} = \left(\frac{b}{4}\right)^{1/4}$

45. $8^p = 19$ **46.** $3^z = 11$

47. $5 \cdot 2^{-m} = 35$ **48.** $2 \cdot 15^{-k} = 18$

49. $e^{-5-2x} = 5$ **50.** $e^{3x-1} = 12$

51. $6^{2-m} = 2^{3m+1}$ **52.** $5^{3r-1} = 6^{2r+5}$

53. $(1 + .003)^k = 1.089$ **54.** $(1 + .094)^z = 2.387$

Work these problems.

55. Business Expenditures by the National Park System (in billions of dollars) are approximated by

$$g(x) = -41.96 + 9.54 \ln x,$$

where $x = 90$ corresponds to the year 1990.* Find the expenditures in the given years.

(a) 2005 **(b)** 2010

(c) In what year will expenditures reach \$3.5 billion?

56. Natural Science A population is increasing according to the growth law $y = 2e^{.02t}$, where y is in millions and t is in years. Match each of the questions (a), (b), (c), and (d) with one of the solutions (A), (B), (C), or (D).

(a) How long will it take for the population to triple? **(A)** Evaluate $2e^{.02(1/3)}$.

(b) When will the population reach 3 million? **(B)** Solve $2e^{.02t} = 3 \cdot 2$ for t.

(c) How large will the population be in 3 years? **(C)** Evaluate $2e^{.02(3)}$.

(d) How large will the population be in 4 months? **(D)** Solve $2e^{.02t} = 3$ for t.

57. Natural Science The amount of polonium (in grams) present after t days is given by

$$A(t) = 10e^{-.00495t}.$$

(a) How much polonium was present initially?

(b) What is the half-life of polonium?

(c) How long will it take for the polonium to decay to 3 grams?

58. Health According to projections by the Alzheimer's Association, the number of people suffering from Alzheimer's disease in year x is approximated by

$$A(x) = 4.23 * 1.023^x,$$

where $x = 0$ corresponds to 2000. When will the number of people with Alzheimer's disease be

(a) 8 million? **(b)** 13 million?

*Based on data from the U.S. Department of Agriculture.

*Based on data from the National Park Service.

59. Health Deaths per 100,000 people in motor vehicle accidents are approximated by

$$h(x) = 21.03 * .9843^x,$$

where $x = 0$ corresponds to the year 1980.* Find the number of deaths per 100,000 people in the given years.
(a) 2002 **(b)** 2008
(c) If this model remains accurate, when will the rate drop below 12 per 100,000?

60. Natural Science One earthquake measures 4.6 on the Richter scale. A second earthquake has ground motion 1000 times greater than the first. What does the second one measure on the Richter scale?

Natural Science *Another form of Newton's law of cooling (see Section 4.2, Exercises 15 and 16) is* $F(t) = T_0 + Ce^{-kt}$, *where C and k are constants.*

61. A piece of metal is heated to 300° Celsius and then placed in a cooling liquid at 50° Celsius. After 4 minutes, the metal has cooled to 175° Celsius. Find its temperature after 12 minutes.

62. A frozen pizza has a temperature of 3.4° Celsius when it is taken from the freezer and left out in a room at 18° Celsius. After half an hour, its temperature is 7.2° Celsius. How long will it take for the pizza to thaw to 10° Celsius?

In Exercises 63–66, do part (a) and skip part (b) if you do not have a graphing calculator. If you have a graphing calculator, then skip part (a) and do part (b).

63. Business The table shows projected online retail sales (in billions of dollars) in selected years:†

Year	2006	2007	2008	2009	2010	2011
Sales	\$132.1	\$157.4	\$184.4	\$212.6	\$241.6	\$271.1

(a) Let $x = 0$ correspond to 2006. Use the data points from 2006 and 2011 to find a function of the form $f(x) = a * b^x$ that models this data.
(b) Use exponential regression to find a function $g(x)$ that models the data, with $x = 0$ corresponding to 2006.
(c) Estimate online sales in 2013.
(d) According to this model, when will online sales reach \$500 billion per year?

64. Physical Science The atmospheric pressure (in millibars) at a given altitude (in thousands of meters) is listed in the table:

Altitude	0	2000	4000	6000	8000	10,000
Pressure	1013	795	617	472	357	265

(a) Use the data points for altitudes 0 and 10,000 to find a function of the form $f(x) = a(b^x)$ that models this data, where altitude x is measured in thousands of meters. (For example, $x = 4$ means 4000 meters.)
(b) Use exponential regression to find a function $g(x)$ that models the data, with x measured in thousands of meters.
(c) Estimate the pressure at 1500 m and 11,000 m, and compare the results with the actual values of 846 and 227 millibars, respectively.
(d) At what height is the pressure 500 millibars?

65. Business The table shows the number of Southwest Airlines passengers (in millions) in selected years:*

Year	2002	2003	2004	2005	2006	2007
Passengers	63	65.7	70.9	77.7	83.8	88.7

(a) Let $x = 1$ correspond to 2002. Use the data points for 2002 and 2007 to find a function of the form $f(x) = a + b \ln x$ that models this data. [*Hint*: Use (1, 63) to find a; then use (6, 88.7) to find b.]
(b) Use logarithmic regression to find a function $g(x)$ that models the data, with $x = 1$ corresponding to 2002.
(c) Estimate the number of passengers in 2010.
(d) According to this model, when will the number of passengers reach 100 million?

66. Health The number of kidney transplants (in thousands) in selected years is shown in the table:†

Year	1991	1995	2000	2005	2006	2007
Transplants	9.678	11.083	13.613	16.481	17.091	16.626

(a) Let $x = 1$ correspond to 1991. Use the data points for 1991 and 2007 to find a function of the form $f(x) = a + b \ln x$ that models this data. [See Exercise 65(a).]
(b) Use logarithmic regression to find a function $g(x)$ that models this data, with $x = 1$ corresponding to 1991.
(c) Estimate the number of kidney transplants in 2010.
(d) If the model remains accurate, when will the number of kidney transplants reach 18,000?

*Based on data from the National Safety Council.
†*Statistical Abstract of the United States: 2008.*

*Southwest Airlines annual reports, 2002–2008.
†National Network for Organ Sharing.

CASE 4

Characteristics of the Monkeyface Prickleback*

The monkeyface prickleback (*Cebidichthys violaceus*), known to anglers as the "monkeyface eel," is found in rocky intertidal and subtidal habitats ranging from San Quintin Bay, Baja California, to Brookings, Oregon. Pricklebacks are prime targets of the few sports anglers who "poke pole" in the rocky intertidal zone at low tide. Little is known about the life history of this species. The results of a study of the length, weight, and age of this species are discussed in this case.

Data on standard length (*SL*) and total length (*TL*) were collected. Early in the study only *TL* was measured, so a conversion to *SL* was necessary. The equation relating the two lengths, calculated from 177 observations for which both lengths had been measured, is

$$SL = TL(.931) + 1.416.$$

Ages (determined by standard aging techniques) were used to estimate parameters of the von Bertanfany growth model

$$L_t = L_x(1 - e^{-kt}), \tag{1}$$

where

L_t = length at age t,
L_x = asymptotic age of the species,
k = growth completion rate, and
t_0 = theoretical age at zero length.

The constants a and b in the model

$$W = aL^b, \tag{2}$$

where

W = weight in g and
L = standard length in cm,

were determined using 139 fish ranging from 27 cm and 145 g to 60 cm and 195 g.

Growth curves giving length as a function of age are shown in the next figure. For the data marked OPERCLE, the lengths were computed from the ages by using equation (1).

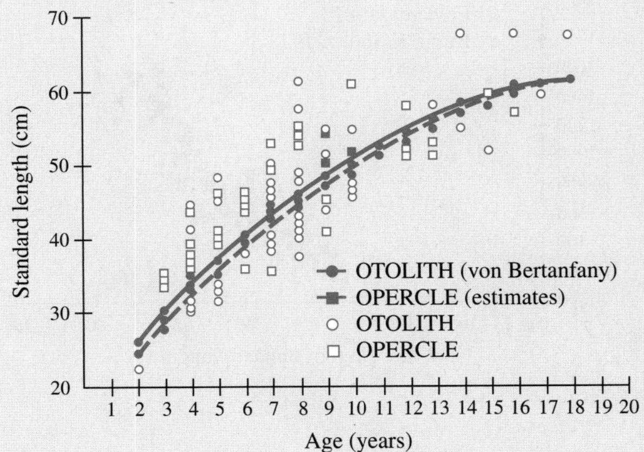

Estimated length from equation (1) at a given age was larger for males than females after age eight. See the following table:

Structure/ Sex	Age (yr)	Length (cm)	L_x	k	t_0	n
Otolith						
Est.	2–18	23–67	72	.10	−1.89	91
S.D.			8	.03	1.08	
Opercle						
Est.	2–18	23–67	71	.10	−2.63	91
S.D.			8	.04	1.31	
Opercle–Females						
Est.	0–18	15–62	62	.14	−1.95	115
S.D.			2	.02	.28	
Opercle–Males						
Est.	0–18	13–67	70	.12	−1.91	74
S.D.			5	.02	.29	

*"Characteristics of the Monkeyface Prickleback," by William H. Marshall and Tina Wyllie Echeverria. *California Fish & Game*, Vol 78, No. 2, Spring 1992.

Weight–length relationships found with equation (2) are shown in the next figure, along with data from other studies.

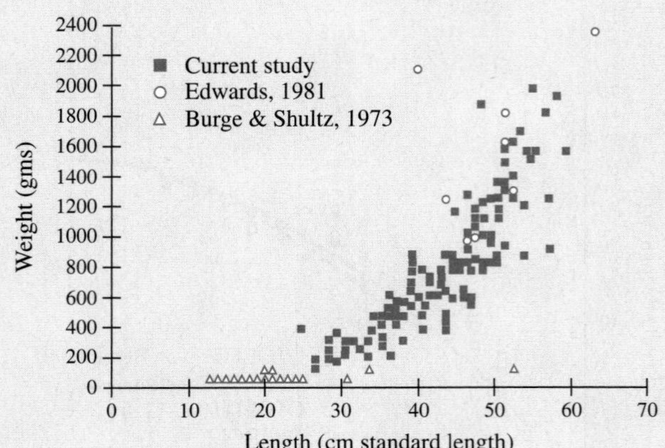

EXERCISES

1. Use equation (1) to estimate the lengths at ages 4, 11, and 17. Let $L_x = 71.5$ and $k = .1$. Compare your answers with the results in the first figure. What do you find?

2. Use equation (2) with $a = .01289$ and $b = 2.9$ to estimate the weights for lengths of 25 cm, 40 cm, and 60 cm. Compare your answers with the results in the second figure. Are your answers reasonable compared with the curve?

CHAPTER

5

Mathematics of Finance

Most people must take out a loan for a big purchase, such as a car, a major appliance, or a house. People who carry a balance on their credit cards are, in effect, also borrowing money. Loan payments must be accurately determined, and it may take some work to find the "best deal." See Exercise 54 on page 301 and Exercise 59 on page 278. We must all plan for eventual retirement, which usually involves savings accounts and investments in stocks, bonds, and annuities to fund 401K accounts or individual retirement accounts (IRAs). See Exercises 40 and 41 on page 288.

It is important for both businesspersons and consumers to understand the mathematics of finance in order to make sound financial decisions. Interest formulas for borrowing and investing money are introduced in this chapter.

NOTE We try to present realistic, up-to-date applications in this book. Because interest rates change so frequently, however, it is very unlikely that the rates in effect when this chapter was written are the same as the rates today when you are reading it. Fortunately, the mathematics of finance is the same regardless of the level of interest rates. So we have used a variety of rates in the examples and exercises. Some will be realistic and some won't by the time you see them—but all of them have occurred in the past several decades.

5.1 Simple Interest and Discount

Interest is the fee paid to use someone else's money. Interest on loans of a year or less is frequently calculated as **simple interest**, which is paid only on the amount borrowed or invested and not on past interest. The amount borrowed or deposited is called the **principal**. The **rate** of interest is given as a percent per year, expressed as a decimal. For example, $6\% = .06$ and $11\frac{1}{2}\% = .115$. The **time** during which the money is accruing interest is calculated in years. Simple interest is the product of the principal, rate, and time.

Simple Interest

The simple interest I on P dollars at a rate of interest r per year for t years is

$$I = Prt.$$

It is customary in financial problems to round interest to the nearest cent.

Example 1 To furnish her new apartment, Eleanor Chin borrowed $5000 at 9.2% interest for 11 months. How much interest will she pay?

Solution Use the formula $I = Prt$, with $P = 5000$, $r = .092$, and $t = 11/12$ years:

$$I = Prt$$

$$I = 5000 * .092 * \frac{11}{12} = 421.6667.$$

Rounding this result to the nearest penny shows that Eleanor pays $421.67 in interest. 1✓

✓Checkpoint 1

Find the simple interest for each loan.

(a) $2000 at 8.5% for 10 months

(b) $3500 at 10.5% for $1\frac{1}{2}$ years

Answers: See page 267.

Simple interest is normally used only for loans with a term of a year or less. A significant exception is the case of **corporate bonds** and similar financial instruments. A typical bond pays simple interest twice a year for a specified length of time, at the end of which the bond **matures**. At maturity, the company returns your initial investment to you.

Example 2 The XYZ Corporation is issuing 10-year bonds at an annual simple interest rate of 6%, with interest paid twice a year. Leslie Lahr buys a $10,000 bond.

(a) How much interest will she earn every six months?

Solution Use the interest formula, $I = Prt$, with $P = 10{,}000$, $r = .06$, and $t = \frac{1}{2}$:

$$I = Prt = 10{,}000 * .06 * \frac{1}{2} = \$300.$$

(b) How much interest will she earn over the 10-year life of the bond?

Solution Either use the interest formula with $t = 10$, that is,

$$I = Prt = 10{,}000 * .06 * 10 = \$6000,$$

or do it in your head: She earns \$300 every six months, which is \$600 per year, so the 10-year total is \$600 * 10 = \$6000. 2✓

✓Checkpoint 2

For the given bonds, find the semiannual interest payment and the total interest paid over the life of the bond.

(a) \$6000 GE Capital 5-year bond at 5.1% annual interest.

(b) \$12,000 Wal-Mart 8-year bond at 5.8% annual interest

Answers: *See page 267.*

FUTURE VALUE

If you deposit P dollars at simple interest rate r for t years, then the **future value** (or **maturity value**) A of this investment is the sum of the principal P and the interest I it has earned:

$$\begin{aligned} A &= \text{Principal} + \text{Interest} \\ &= P + I \\ &= P + Prt && I = Prt. \\ &= P(1 + rt). && \text{Factor out } P. \end{aligned}$$

The following box summarizes this result.

Future Value (or Maturity Value) for Simple Interest

The future value (maturity value) A of P dollars for t years at interest rate r per year is

$$\mathbf{A = P + I}, \quad \text{or} \quad \mathbf{A = P(1 + rt)}.$$

Example 3 Find each maturity value and the amount of interest paid.

(a) Rick borrows \$20,000 from his parents at 5.25% to add a room on his house. He plans to repay the loan in 9 months with a bonus he expects to receive at that time.

Solution The loan is for 9 months, or 9/12 of a year, so $t = .75$, $P = 20{,}000$, and $r = .0525$. Use the formula to obtain

$$\begin{aligned} A &= P(1 + rt) \\ &= 20{,}000[1 + .0525(.75)] \\ &\approx 20{,}787.5, && \text{Use a calculator.} \end{aligned}$$

or \$20,787.50. The maturity value A is the sum of the principal P and the interest I, that is, $A = P + I$. To find the amount of interest paid, rearrange this equation:

$$I = A - P$$

$$I = \$20{,}787.50 - \$20{,}000 = \$787.50.$$

(b) A loan of $11,280 for 85 days at 9% interest.

Solution Use the formula $A = P(1 + rt)$, with $P = 11{,}280$ and $r = .09$. Unless stated otherwise, we assume a 365-day year, so the period in years is $t = 85/365$. The maturity value is

$$A = P(1 + rt)$$
$$A = 11{,}280\left(1 + .09 * \frac{85}{365}\right)$$
$$\approx 11{,}280(1.020958904) \approx \$11{,}516.42.$$

As in part (a), the interest is

$$I = A - P = \$11{,}516.42 - \$11{,}280 = \$236.42.$$ 3

Checkpoint 3

Find each future value.

(a) $1000 at 4.6% for 6 months

(b) $8970 at 11% for 9 months

(c) $95,106 at 9.8% for 76 days

Answers: See page 267.

Example 4 Suppose you borrow $15,000 and are required to pay $15,315 in 4 months to pay off the loan and interest. What is the simple interest rate?

Solution One way to find the rate is to solve for r in the future-value formula when $P = 15{,}000$, $A = 15{,}315$, and $t = 4/12 = 1/3$:

$$P(1 + rt) = A$$
$$15{,}000\left(1 + r * \frac{1}{3}\right) = 15{,}315$$
$$15{,}000 + \frac{15{,}000r}{3} = 15{,}315 \qquad \text{Multiply out left side.}$$
$$\frac{15{,}000r}{3} = 315 \qquad \text{Subtract 15,000 from both sides.}$$
$$15{,}000r = 945 \qquad \text{Multiply both sides by 3.}$$
$$r = \frac{945}{15{,}000} = .063. \qquad \text{Divide both sides by 15,000.}$$

Therefore, the interest rate is 6.3%. 4

Checkpoint 4

You lend a friend $500. She agrees to pay you $520 in 6 months. What is the interest rate?

Answer: See page 267.

PRESENT VALUE

A sum of money that can be deposited today to yield some larger amount in the future is called the **present value** of that future amount. Present value refers to the principal to be invested or loaned, so we use the same variable P as we did for principal. In interest problems, P always represents the amount at the beginning of the period, and A always represents the amount at the end of the period. To find a formula for P, we begin with the future-value formula:

$$A = P(1 + rt).$$

Dividing each side by $1 + rt$ gives the following formula for the present value.

Present Value for Simple Interest

The **present value** P of a future amount of A dollars at a simple interest rate r for t years is

$$P = \frac{A}{1 + rt}.$$

Example 5 Find the present value of \$32,000 in 4 months at 9% interest.

Solution

$$P = \frac{A}{1 + rt} = \frac{32{,}000}{1 + (.09)\left(\frac{4}{12}\right)} = \frac{32{,}000}{1.03} = 31{,}067.96.$$

A deposit of \$31,067.96 today at 9% interest would produce \$32,000 in 4 months. These two sums, \$31,067.96 today and \$32,000.00 in 4 months, are equivalent (at 9%) because the first amount becomes the second amount in 4 months. 5✓

✓Checkpoint 5

Find the present value of the given future amounts. Assume 6% interest.

(a) \$7500 in 1 year

(b) \$89,000 in 5 months

(c) \$164,200 in 125 days

Answers: See page 267.

Example 6 Because of a court settlement, Jeff Weidenaar owes \$5000 to Chuck Synovec. The money must be paid in 10 months, with no interest. Suppose Weidenaar wants to pay the money today and that Synovec can invest it at an annual rate of 5%. What amount should Synovec be willing to accept to settle the debt?

Solution The \$5000 is the future value in 10 months. So Synovec should be willing to accept an amount which will grow to \$5000 in 10 months at 5% interest. In other words, he should accept the present value of \$5000 under these circumstances. Use the present-value formula with $A = 5000$, $r = .05$, and $t = 10/12 = 5/6$:

$$P = \frac{A}{1 + rt} = \frac{5000}{1 + .05 * \frac{5}{6}} = 4800.$$

Synovec should be willing to accept \$4800 today in settlement of the debt. 6✓

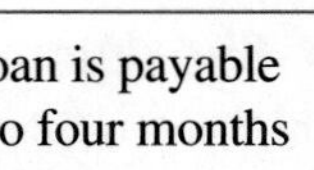

✓Checkpoint 6

Jerrell Davis is owed \$19,500 by Christine O'Brien. The money will be paid in 11 months, with no interest. If the current interest rate is 10%, how much should Davis be willing to accept today in settlement of the debt?

Answer: See page 267.

Example 7 Larry Parks owes \$6500 to Virginia Donovan. The loan is payable in one year at 6% interest. Donovan needs cash to pay medical bills, so four months before the loan is due, she sells the note (loan) to the bank. If the bank wants a return of 9% on its investment, how much should it pay Donovan for the note?

Solution First find the maturity value of the loan—the amount (with interest) that Parks must pay Donovan:

$$\begin{aligned} A &= P(1 + rt) && \text{Maturity-value formula} \\ &= 6500(1 + .06 * 1) && \text{Let } P = 6500,\ r = .06, \text{ and } t = 1. \\ &= 6500(1.06) = \$6890. \end{aligned}$$

In four months, the bank will receive \$6890. Since the bank wants a 9% return, compute the present value of this amount at 9% for four months:

$$P = \frac{A}{1 + rt} \qquad \text{Present-value formula}$$

$$= \frac{6890}{1 + .09\left(\frac{4}{12}\right)} = \$6689.32. \qquad \text{Let } A = 6890,\ r = .09, \text{ and } t = 4/12.$$

The bank pays Donovan \$6689.32 and four months later collects \$6890 from Parks. 7

Checkpoint 7

A firm accepts a \$21,000 note due in 8 months, with interest of 10.5%. Two months before it is due, the firm sells the note to a broker. If the broker wants a 12.5% return on his investment, how much should he pay for the note?

Answer: See page 267.

DISCOUNT

The preceding examples dealt with loans in which money is borrowed and simple interest is charged. For most loans, both the principal (amount borrowed) and the interest are paid at the end of the loan period. With a corporate bond (which is a loan to a company by the investor who buys the bond), interest is paid during the life of the bond and the principal is paid back at maturity. In both cases,

the borrower receives the principal,

but pays back the principal *plus* the interest.

In a **simple discount loan**, however, the interest is deducted in advance from the amount of the loan and the *balance* is given to the borrower. The *full value* of the loan must be paid back at maturity. Thus,

the borrower receives the principal *less* the interest,

but pays back the principal.

The most common examples of simple discount loans are U.S. Treasury bills (T-bills), which are essentially short-term loans to the U.S. government by investors. T-bills are sold at a **discount** from their face value and the Treasury pays back the face value of the T-bill at maturity. The discount amount is the interest deducted in advance from the face value. The Treasury receives the face value less the discount, but pays back the full face value.

Example 8 Greg Tobin buys a six-month \$6000 treasury bill that sells at a discount rate of 4.5%. What is the amount of the discount? What is the price of the T-bills?

Solution The discount rate on a T-bill is always a simple *annual* interest rate. Consequently, the discount (interest) is found with the simple interest formula, using $P = 6000$ (face value), $r = .045$ (discount rate), and $t = .5$ (because 6 months is half a year):

$$\text{Discount} = Prt = 6000 * .045 * .5 = \$135.$$

So the price of the T-bill is

$$\text{Face Value} - \text{Discount} = 6000 - 135 = \$5865.$$ 8

Checkpoint 8

The maturity times and discount rates for \$10,000 T-bills sold on September 25, 2008, are given. Find the discount amount and the price of each T-bills.

(a) one year; 1.955%

(b) three months; 1.420%

(c) six months; 1.790%

Answers: See page 267.

In a simple discount loan, such as a T-bill, the discount rate is not the actual interest rate the borrower pays. In Example 8, the discount rate 4.5% was applied to the face value of \$6000, rather than the \$5865 that the Treasury (the borrower) received.

Example 9 Find the actual interest rate paid by the Treasury in Example 8.

Solution Use the formula for simple interest, $I = Prt$, with r as the unknown. Here, $P = 5865$ (the amount the Treasury received) and $I = 135$ (the discount amount). Since this is a six-month T-bill, $t = .5$, and we have

$$I = Prt$$

$$135 = 5865(r)(.5)$$

$$135 = 2932.5r \quad \text{Multiply out right side.}$$

$$r = \frac{135}{2932.5} \approx .0460. \quad \text{Divide both sides by 2932.5.}$$

So the actual interest rate is about 4.6%. 9 ✓

✓Checkpoint 9

Find the actual interest rate paid by the Treasury for each T-bill in Checkpoint 8.

Answer: See page 267.

As illustrated in Example 9, the discount rate is always lower than the actual interest rate.

5.1 ▶ Exercises

Unless stated otherwise, "interest" means simple interest, and "interest rate" and "discount rate" refer to annual rates. Assume 365 days in a year.

1. What factors determine the amount of interest earned on a fixed principal?

Find the interest on each of these loans. (See Example 1.)

2. \$35,000 at 6% for 9 months

3. \$2850 at 7% for 8 months

4. \$1875 at 5.3% for 7 months

5. \$3650 at 6.5% for 11 months

6. \$5160 at 7.1% for 58 days

7. \$2830 at 8.9% for 125 days

8. \$8940 at 9%; loan made on May 7 and due September 19

9. \$5328 at 8%; loan made on August 16 and due December 30

10. \$7900 at 7%; loan made on July 7 and due October 25

For each of the given bonds, whose interest rates are provided, find the semiannual interest payment and the total interest earned over the life of the bond. (See Example 2.)

11. \$10,000 GE Capital Corporation 15-year bond at 7.5%

12. \$7000 Wal-Mart Stores 5-year bond at 4.75%

13. \$15,000 Principal Life Income Fund 3-year bond at 5.15%

14. \$24,000 Proctor and Gamble 10-year bond at 6.875%

Find the future value of each of these loans. (See Example 3.)

15. \$12,000 loan at 3.5% for 3 months

16. \$3475 loan at 7.5% for 6 months

17. \$6500 loan at 5.25% for 8 months

18. \$24,500 loan at 9.6% for 10 months

19. What is meant by the *present value* of money?

20. In your own words, describe the *maturity value* of a loan.

Find the present value of each future amount. (See Examples 5 and 6.)

21. \$15,000 for 9 months; money earns 6%

22. \$48,000 for 8 months; money earns 5%

23. \$15,402 for 120 days; money earns 6.3%

24. \$29,764 for 310 days; money earns 7.2%

The given Treasury bills were sold in August 2008. Find (a) the price of the T-bill, and (b) the actual Interest rate paid by the Treasury. (See Examples 8 and 9.)

25. Three-months \$8000 T-bill with discount rate 1.870%

26. Six-month \$18,000 T-bill with discount rate 1.925%

27. Six-month \$12,000 T-bill with discount rate 2.020%

28. One-year \$6000 T-bill with discount rate 2.140%

Finance *Work the given, applied problems.*

29. In March 1868, Winston Churchill's grandfather, L.W. Jerome, issued \$1000 bonds (to pay for a road to a race track he owned in what is now the Bronx). The bonds carried a 7% annual interest rate payable semiannually. Mr. Jerome paid the interest until March 1874, at which time New York City assumed responsibility for the bonds (and the road they financed).*

(a) The first of these bonds matured in March 2009. At that time, how much interest had New York City paid on this bond?

(b) Another of these bonds will not mature until March 2147! At that time, how much interest will New York City have paid on it?

30. An accountant for a corporation forgot to pay the firm's income tax of \$725,896.15 on time. The government charged a penalty of 9.8% interest for the 34 days the money was late. Find the total amount (tax and penalty) that was paid.

31. Mike Branson invested his summer earnings of \$3000 in a savings account for college. The account pays 2.5% interest. How much will this amount to in 9 months?

32. To pay for textbooks, a student borrows \$450 dollars from a credit union at 6.5% simple interest. He will repay the loan in 38 days, when he expects to be paid for tutoring. How much interest will he pay?

33. An account invested in a money market fund grew from \$67,081.20 to \$67,359.39 in a month. What was the interest rate, to the nearest tenth?

34. A \$100,000 certificate of deposit held for 60 days is worth \$101,133.33. To the nearest tenth of a percent, what interest rate was earned?

35. Dave took out a \$7500 loan at 7% and eventually repaid \$7675 (principal and interest). What was the time period of the loan?

36. What is the time period of a \$10,000 loan at 6.75%, in which the total amount of interest paid was \$618.75?

37. Tuition of \$1769 will be due when the spring term begins in 4 months. What amount should a student deposit today, at 3.25%, to have enough to pay the tuition?

38. A firm of accountants has ordered 7 new computers at a cost of \$5104 each. The machines will not be delivered for 7 months. What amount could the firm deposit in an account paying 6.42% to have enough to pay for the machines?

39. John Sun Yee needs \$6000 to pay for remodeling work on his house. A contractor agrees to do the work in 10 months. How much should Yee deposit at 3.6% to accumulate the \$6000 at that time?

40. Lorie Reilly decides to go back to college. For transportation, she borrows money from her parents to buy a small car for \$7200. She plans to repay the loan in 7 months. What amount can she deposit today at 5.25% to have enough to pay off the loan?

41. A six-month \$4000 Treasury bill sold for \$3930. What was the discount rate?

42. A three-month \$7600 Treasury bill carries a discount of \$80.75. What is the discount rate for this T-bill?

Finance *Work the next set of problems, in which you are to find the annual simple interest rate. Consider any fees, dividends, or profits as part of the total interest.*

43. A stock that sold for \$22 at the beginning of the year was selling for \$24 at the end of the year. If the stock paid a dividend of \$.50 per share, what is the simple interest rate on an investment in this stock? (*Hint*: Consider the interest to be the increase in value plus the dividend.)

44. Jerry Ryan borrowed \$8000 for nine months at an interest rate of 7%. The bank also charges a \$100 processing fee. What is the actual interest rate for this loan?

45. You are due a tax refund of \$760. Your tax preparer offers you a no-interest loan to be repaid by your refund check, which will arrive in four weeks. She charges a \$60 fee for this service. What actual interest rate will you pay for this loan? (*Hint*: The time period of this loan is not 4/52, because a 365-day year in 52 weeks and 1 day. So use days in your computations.)

46. Your cousin is due a tax refund of \$400 in six weeks. His tax preparer has an arrangement with a bank to get him the \$400 now. The bank charges an administrative fee of \$29 plus interest at 6.5%. What is the actual interest rate for this loan? (See the hint for Exercise 45.)

Finance *Work these problems. (See Example 7.)*

47. A building contractor gives a \$13,500 promissory note to a plumber who has loaned him \$13,500. The note is due in nine months with interest at 9%. Three months after the note is signed, the plumber sells it to a bank. If the bank gets a 10% return on its investment, how much will the plumber receive? Will it be enough to pay a bill for \$13,650?

48. Shalia Johnson owes \$7200 to the Eastside Music Shop. She has agreed to pay the amount in seven months at an interest rate of 10%. Two months before the loan is due, the store needs \$7550 to pay a wholesaler's bill. The bank will buy the note, provided that its return on the investment is 11%. How much will the store receive? Is it enough to pay the bill?

49. Let y_1 be the future value after t years of \$100 invested at 8% annual simple interest. Let y_2 be the future value after t years of \$200 invested at 3% annual simple interest.

(a) Think of y_1 and y_2 as functions of t and write the rules of these functions.

(b) Without graphing, describe the graphs of y_1 and y_2.

**New York Times*, February 13, 2009.

(c) Verify your answer to part (b) by graphing y_1 and y_2 in the first quadrant.

(d) What do the slopes and y-intercepts of the graphs represent (in terms of the investment situation that they describe)?

50. If $y = 16.25t + 250$ and y is the future value after t years of P dollars at interest rate r, what are P and r? (*Hint*: See Exercise 49.)

✓Checkpoint Answers

1. (a) \$141.67 (b) \$551.25
2. (a) \$153; \$1530 (b) \$348; \$5568
3. (a) \$1023 (b) \$9710.03 (c) \$97,046.68
4. 8%
5. (a) \$7075.47 (b) \$86,829.27 (c) \$160,893.96
6. \$17,862.60
7. \$22,011.43
8. (a) \$195.50; \$9804.50 (b) \$35.50; \$9964.50 (c) \$89.50; \$9910.50
9. (a) About 1.994% (b) About 1.425% (c) About 1.806%

5.2 ▶ Compound Interest

With annual simple interest, you earn interest each year on your original investment. With annual **compound interest**, however, you earn interest both on your original investment *and* on any previously earned interest. To see how this process works, suppose you deposit \$1000 at 5% annual interest. The following chart shows how your account would grow with both simple and compound interest:

	SIMPLE INTEREST		COMPOUND INTEREST	
End of Year	**Interest Earned**	**Balance**	**Interest Earned**	**Balance**
	Original Investment: \$1000		*Original Investment*: \$1000	
1	1000(.05) = \$50	\$1050	1000(.05) = \$50	\$1050
2	1000(.05) = \$50	\$1100	1050(.05) = \$52.50	\$1102.50
3	1000(.05) = \$50	\$1150	1102.50(.05) = \$55.13*	\$1157.63

As the chart shows, simple interest is computed each year on the original investment, but compound interest is computed on the entire balance at the end of the preceding year. So simple interest always produces \$50 per year in interest, whereas compound interest produces \$50 interest in the first year and increasingly larger amounts in later years (because you earn interest on your interest). 1✓

✓Checkpoint 1

Extend the chart in the text by finding the interest earned and the balance at the end of years 4 and 5 for **(a)** simple interest and **(b)** compound interest.

Answer: See page 279.

Example 1 If \$7000 is deposited in an account that pays 4% interest compounded annually, how much money is in the account after nine years?

Solution After one year, the account balance is

$$\begin{aligned} 7000 + 4\% \text{ of } 7000 &= 7000 + (.04)7000 \\ &= 7000(1 + .04) \quad \text{Distributive property} \\ &= 7000(1.04) = \$7280. \end{aligned}$$

*Rounded to the nearest cent.

The initial balance has grown by a factor of 1.04. At the end of the second year, the balance is

$$\begin{aligned}7280 + 4\% \text{ of } 7280 &= 7280 + (.04)7280 \\ &= 7280(1 + .04) \qquad \text{Distributive property} \\ &= 7280(1.04) = 7571.20.\end{aligned}$$

Once again, the balance at the beginning of the year has grown by a factor of 1.04. This is true in general: If the balance at the beginning of a year is P dollars, then the balance at the end of the year is

$$\begin{aligned}P + 4\% \text{ of } P &= P + .04P \\ &= P(1 + .04) \\ &= P(1.04).\end{aligned}$$

So the account balance grows like this:

$$7000 \xrightarrow{\textbf{Year 1}} 7000(1.04) \xrightarrow{\textbf{Year 2}} \underbrace{[7000(1.04)](1.04)}_{7000(1.04)^2} \xrightarrow{\textbf{Year 3}} \underbrace{[7000(1.04)(1.040)](1.04)}_{7000(1.04)^3} \rightarrow \cdots.$$

At the end of nine years, the balance is

$$7000(1.04)^9 = \$9963.18 \qquad \text{(rounded to the nearest penny).}$$

The argument used in Example 1 applies in the general case and leads to this conclusion.

Compound Interest

If P dollars are invested at interest rate i per period, then the **compound amount** (future value) A after n compounding periods is

$$A = P(1 + i)^n.$$

In Example 1, for instance, we had $P = 7000$, $n = 9$, and $i = .04$ (so that $1 + i = 1 + .04 = 1.04$).

NOTE Compare this future value formula for compound interest with the one for simple interest from the previous section, using t as the number of years:

Compound interest	$A = P(1 + r)^t$;
Simple interest	$A = P(1 + rt)$.

The important distinction between the two formulas is that, in the compound interest formula, the number of years, t, is an *exponent*, so that money grows much more rapidly when interest is compounded.

Example 2 Suppose $1000 is deposited for six years in an account paying 8.31% per year compounded annually.

(a) Find the compound amount.

Solution In the formula above, $P = 1000$, $i = .0831$, and $n = 6$. The compound amount is

$$A = P(1 + i)^n$$
$$A = 1000(1.0831)^6$$
$$A = \$1614.40.$$

(b) Find the amount of interest earned.

Solution Subtract the initial deposit from the compound amount:

$$\text{Amount of interest} = \$1614.40 - \$1000 = \$614.40.$$ 2

✓Checkpoint 2

Suppose $17,000 is deposited at 4% compounded annually for 11 years.

(a) Find the compound amount.

(b) Find the amount of interest earned.

Answers: See page 279.

TECHNOLOGY TIP Spreadsheets are ideal for performing financial calculations. Figure 5.1 shows a Microsoft Excel spreadsheet with the formulas for compound and simple interest used to create columns *B* and *C*, respectively, when $1000 is invested at an annual rate of 10%. Notice how rapidly the compound amount increases compared with the maturity value with simple interest. For more details on the use of spreadsheets in the mathematics of finance, see the *Spreadsheet Manual* that is available with this book.

	A	B	C
1	period	compound	simple
2	1	1100	1100
3	2	1210	1200
4	3	1331	1300
5	4	1464.1	1400
6	5	1610.51	1500
7	6	1771.561	1600
8	7	1948.7171	1700
9	8	2143.58881	1800
10	9	2357.947691	1900
11	10	2593.74246	2000
12	11	2853.116706	2100
13	12	3138.428377	2200
14	13	3452.271214	2300
15	14	3797.498336	2400
16	15	4177.248169	2500
17	16	4594.972986	2600
18	17	5054.470285	2700
19	18	5559.917313	2800
20	19	6115.909045	2900
21	20	6727.499949	3000

FIGURE 5.1

Example 3 If a \$16,000 investment grows to \$50,000 in 18 years, what is the interest rate (assuming annual compounding)?

Solution Use the compound interest formula, with $P = 16{,}000$, $A = 50{,}000$, and $n = 18$, and solve for i:

$$P(1+i)^n = A$$

$$16{,}000(1+i)^{18} = 50{,}000$$

$$(1+i)^{18} = \frac{50{,}000}{16{,}000} = 3.125 \qquad \text{Divide both sides by 16,000.}$$

$$\sqrt[18]{(1+i)^{18}} = \sqrt[18]{3.125} \qquad \text{Take 18th roots on both sides.}$$

$$1 + i = \sqrt[18]{3.125}$$

$$i = \sqrt[18]{3.125} - 1 = .06535. \qquad \text{Subtract 1 from both sides.}$$

So the interest rate is about 6.535%.

Interest can be compounded more than once a year. Common **compounding periods** include

semiannually (2 periods per year),
quarterly (4 periods per year),
monthly (12 periods per year), and
daily (usually 365 periods per year).

When the annual interest i is compounded m times per year, the interest rate per period is understood to be i/m.

Example 4 In October 2008, an online advertisement by Pacific Mercantile bank offered a \$10,000 certificate of deposit (CD) for one year at 4.3% interest. Find the value of the CD (compound amount) after one year if interest is compounded according to the given periods.

(a) Annually

Solution Apply the formula $A = P(1+i)^n$ with $P = 10{,}000$, $i = .043$, and $n = 1$:

$$A = P(1+i)^n = 10{,}000(1+.043)^1 = 10{,}000(1.043) = \$10{,}430.$$

(b) Semiannually

Solution Use the same formula and value of P. Here, interest is compounded twice a year, so the number of periods is $n = 2$ and the interest rate per period is $i = .043/2$:

$$A = P(1+i)^n = 10{,}000\left(1 + \frac{.043}{2}\right)^2 = \$10{,}434.62.$$

(c) Quarterly

Solution Proceed as in part (b), but now interest is compounded 4 times a year, so $n = 4$ and the interest rate per period is $i = .043/4$:

$$A = P(1+i)^n = 10{,}000\left(1 + \frac{.043}{4}\right)^4 = \$10{,}436.98.$$

TECHNOLOGY TIP TI-84+ and most Casios have a "TVM solver" for financial computations (in the TI APPS/financial menu or the Casio main menu); a similar one can be downloaded for TI-86/89. Figure 5.2 shows the solution of Example 4(e) on such a solver (FV means future value). The use of these solvers is explained in the next section. Most of the problems in this section can be solved just as quickly with an ordinary calculator.

FIGURE 5.2

(d) Monthly

Solution Interest is compounded 12 times a year, so $n = 12$ and $i = .043/12$:

$$A = P(1 + i)^n = 10{,}000\left(1 + \frac{.043}{12}\right)^{12} = \$10{,}438.58.$$

(e) Daily*

Solution Interest is compounded 365 times a year, so $n = 365$ and $i = .043/365$:

$$A = P(1 + i)^n = 10{,}000\left(1 + \frac{.043}{365}\right)^{365} = \$10{,}439.35.$$

Example 4 shows that the more often interest is compounded, the larger is the amount of interest earned. Since interest is rounded to the nearest penny, however, there is a limit on how much you can earn. In Example 4, for instance, the compound amount will grow to \$10,439.38, but no larger, no matter how many times per year interest is compounded. Nevertheless, the idea of compounding more and more frequently leads to a method of computing interest, called **continuous compounding**, that is used in certain financial situations. See Case 5 at the end of this chapter for details.

Example 5 The given CDs were advertised online by various banks in October 2008. Find the future value of each one.

(a) Triad Bank: \$100,000 for 1.5 years at 2.88% compounded quarterly.

Solution Use the compound interest formula with $P = 100{,}000$. Interest is compounded 4 times a year, so the interest rate per period is $i = .0288/4$. Since there are 4 periods per year, the number of periods in 1.5 years is $n = 4(1.5) = 6$. The future value is

$$A = P(1 + i)^n = 100{,}000\left(1 + \frac{.0288}{4}\right)^{6} = \$104{,}398.51.$$

(b) Discover Bank: \$2500 for 2.5 years at 4.37% compounded daily.

Solution Here, $P = 2500$ and $i = .0437/365$ (because interest is compounded 365 times a year). The number of periods in 2.5 years is $n = 365(2.5) = 912.5$. The future value is

$$A = P(1 + i)^n = 2500\left(1 + \frac{.0437}{365}\right)^{912.5} = \$2788.58.$$ 3

Checkpoint 3

Find the compound amount.

(a) \$10,000 at 8% compounded quarterly for 7 years

(b) \$36,000 at 6% compounded monthly for 3.5 years

Answers: See page 279.

Ordinary corporate or municipal bonds usually make semiannual simple interest payments. With a **zero-coupon bond**, however, there are no interest payments during the life of the bond. The investor receives a single payment when the bond matures, consisting of his original investment and the interest (compounded semiannually)

*This is the compounding frequency actually used by Pacific Mercantile Bank.

that it has earned. Zero-coupon bonds are sold at a substantial discount from their face value, and the buyer receives the face value of the bond when it matures. The difference between the face value and the price of the bond is the interest earned.

Example 6 Doug Payne bought a 15-year zero-coupon bond paying 4.5% interest (compounded semiannually) for \$12,824.50. What is the face value of the bond?

Solution Use the compound interest formula with $P = 12{,}824.50$. Interest is paid twice a year, so the rate per period is $i = .045/2$, and the number of periods in 15 years is $n = 30$. The compound amount will be the face value:

$$A = P(1+i)^n = 12{,}824.50(1+.045/2)^{30} = 24{,}999.99618.$$

Rounding to the nearest cent, we see that the face value of the bond in \$25,000. 

✓Checkpoint 4

Find the face value of the zero coupon.

(a) 30-year bond at 6% sold for \$2546

(b) 15-year bond at 5% sold for \$16,686

Answers: See page 279.

Example 7 Suppose that the inflation rate is 3.5% (which means that the overall level of prices is rising 3.5% a year). How many years will it take for the overall level of prices to double?

Solution We want to find the number of years it will take for \$1 worth of goods or services to cost \$2. Think of \$1 as the present value and \$2 as the future value, with an interest rate of 3.5%, compounded annually. Then the compound amount formula becomes

$$P(1+i)^n = A$$
$$1(1+.035)^n = 2,$$

which simplifies as

$$1.035^n = 2.$$

We must solve this equation for n. There are several ways to do this.

Graphical Use a graphing calculator (with x in place of n) to find the intersection point of the graphs of $y_1 = 1.035^x$ and $y_2 = 2$. Figure 5.3 shows that the intersection point has (approximate) x-coordinate 20.14879. So it will take about 20.15 years for prices to double.

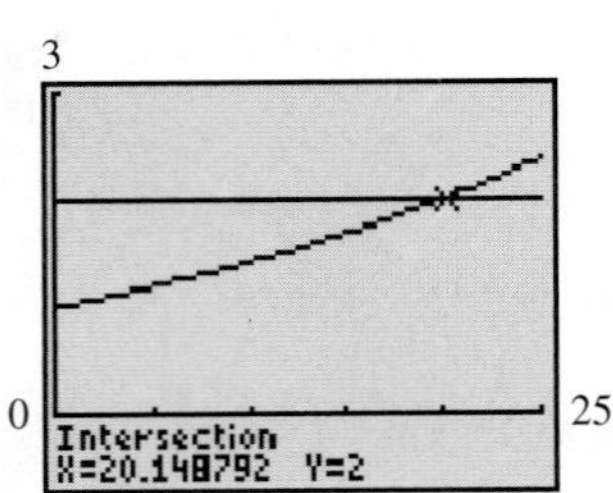

FIGURE 5.3

Algebraic The same answer can be obtained by using natural logarithms, as in Section 4.4:

$$1.035^n = 2$$
$$\ln 1.035^n = \ln 2 \quad \text{Take the logarithm of each side.}$$
$$n \ln 1.035 = \ln 2 \quad \text{Power property of logarithms}$$
$$n = \frac{\ln 2}{\ln 1.035} \quad \text{Divide both sides by ln 1.035.}$$
$$n \approx 20.14879. \quad \text{Use a calculator.}$$

✓5

✓Checkpoint 5

Using a calculator, find the number of years it will take for \$500 to increase to \$750 in an account paying 6% interest compounded semiannually.

Answer: See page 279.

EFFECTIVE RATE (APY)

If you invest \$100 at 9%, compounded monthly, then your balance at the end of one year is

$$A = P(1 + i)^n = 100\left(1 + \frac{.09}{12}\right)^{12} = \$109.38.$$

You have earned \$9.38 in interest, which is 9.38% of your original \$100. In other words, \$100 invested at 9.38% compounded *annually* will produce the same amount of interest (namely, \$100 * .0938 = \$9.38) as does 9% compounded monthly. In this situation, 9% is called the **nominal** or **stated rate**, while 9.38% is called the **effective rate** or **annual percentage yield (APY)**.

In the discussion that follows, the nominal rate is denoted r and the APY (effective rate) is denoted r_E. As illustrated in the preceding paragraph,

> *the APY r_E is the annual compounding rate needed to produce the same amount of interest in one year, as the nominal rate does with more frequent compounding.*

Example 8 In October 2008, National City Bank in Cleveland offered its customers a 4-year \$25,000 CD at 5.12% interest, compounded daily. Find the APY.

Solution The italicized statement given previously means that we must have the following:

\$25,000 at rate r_E, compounded annually = \$25,000 at 5.12%, compounded daily

$$25{,}000\left(1 + r_E\right)^1 = 25{,}000\left(1 + \frac{.0512}{365}\right)^{365} \quad \text{Compound interest formula}$$

$$(1 + r_E) = \left(1 + \frac{.0512}{365}\right)^{365} \quad \text{Divide both sides by 25,000.}$$

$$r_E = \left(1 + \frac{.0512}{365}\right)^{365} - 1 \quad \text{Subtract 1 from both sides.}$$

$$r_E \approx .0525.$$

So the APY is about 5.25%.

The argument in Example 8 can be carried out with 25,000 replaced by P, .0512 by r, and 365 by m. The result is the effective-rate formula.

Effective Rate (APY)

The effective rate (APY) corresponding to a stated rate of interest, r, compounded m times per year is

$$r_E = \left(1 + \frac{r}{m}\right)^m - 1.$$

Example 9 Find the APY for each of the given money market checking accounts (with balances between \$50,000 and \$100,000), which were advertised in October 2008.

(a) Imperial Capital Bank: 3.35% compounded monthly.

Solution Use the effective-rate formula with $r = .0335$ and $m = 12$:

$$r_E = \left(1 + \frac{r}{m}\right)^m - 1 = \left(1 + \frac{.0335}{12}\right)^{12} - 1 = .034019.$$

So the APY is about 3.40%, a slight increase over the nominal rate of 3.35%.

(b) U.S. Bank: 2.33% compounded daily.

Solution Use the formula with $r = .0233$ and $m = 365$:

$$r_E = \left(1 + \frac{r}{m}\right)^m - 1 = \left(1 + \frac{.0233}{365}\right)^{365} - 1 = .023572.$$

The APY is about 2.36%. 6

✓Checkpoint 6

Find the APY corresponding to a nominal rate of

(a) 12% compounded monthly;

(b) 8% compounded quarterly.

Answers: See page 279.

TECHNOLOGY TIP Effective rates (APYs) can be computed on TI-84+ by using "Eff" in the APPS financial menu, as shown in Figure 5.4 for Example 10.

FIGURE 5.4

Example 10 Bank A is now lending money at 10% interest compounded annually. The rate at Bank B is 9.6% compounded monthly, and the rate at Bank C is 9.7% compounded quarterly. If you need to borrow money, at which bank will you pay the least interest?

Solution Compare the APYs:

$$\text{Bank } A: \quad \left(1 + \frac{.10}{1}\right)^1 - 1 = .10 = 10\%;$$

$$\text{Bank } B: \quad \left(1 + \frac{.096}{12}\right)^{12} - 1 \approx .10034 = 10.034\%;$$

$$\text{Bank } C: \quad \left(1 + \frac{.097}{4}\right)^4 - 1 \approx .10059 = 10.059\%.$$

The lowest APY is at Bank A, which has the highest nominal rate. 

Checkpoint 7

Find the APY corresponding to a nominal rate of

(a) 4% compounded quarterly;

(b) 7.9% compounded daily.

Answers: See page 279.

NOTE Although you can find both the stated interest rate and the APY for most certificates of deposit and other interest-bearing accounts, most bank advertisements mention only the APY.

PRESENT VALUE FOR COMPOUND INTEREST

The formula for compound interest, $A = P(1 + i)^n$, has four variables: A, P, i, and n. Given the values of any three of these variables, the value of the fourth can be found. In particular, if A (the future amount), i, and n are known, then P can be found. Here, P is the amount that should be deposited today to produce A dollars in n periods.

Example 11 Keisha Jones must pay a lump sum of \$6000 in 5 years. What amount deposited today at 6.2% compounded annually will amount to \$6000 in 5 years?

Solution Here, $A = 6000$, $i = .062$, $n = 5$, and P is unknown. Substituting these values into the formula for the compound amount gives

$$6000 = P(1.062)^5$$

$$P = \frac{6000}{(1.062)^5} = 4441.49,$$

or \$4441.49. If Jones leaves \$4441.49 for 5 years in an account paying 6.2% compounded annually, she will have \$6000 when she needs it. To check your work, use the compound interest formula with $P = \$4441.49$, $i = .062$, and $n = 5$. You should get $A = \$6000.00$. 8✓

✓Checkpoint 8

Find P in Example 11 if the interest rate is

(a) 6%;

(b) 10%.

Answers: See page 279.

As Example 11 shows, \$6000 in 5 years is (approximately) the same as \$4441.49 today (if money can be deposited at 6.2% annual interest). An amount that can be deposited today to yield a given amount in the future is called the *present value* of the future amount. By solving $A = P(1 + i)^n$ for P, we get the following general formula for present value.

Present Value for Compound Interest

The **present value** of A dollars compounded at an interest rate i per period for n periods is

$$P = \frac{A}{(1 + i)^n}, \qquad \text{or} \qquad P = A(1 + i)^{-n}.$$

Example 12 A zero-coupon bond with face value \$15,000 and a 6% interest rate (compounded semiannually) will mature in 9 years. What is a fair price to pay for the bond today?

Solution Think of the bond as a 9-year investment paying 6%, compounded semiannually, whose future value is \$15,000. Its present value (what it is worth today) would be a fair price. So use the present value formula with $A = 15{,}000$. Since interest is compounded twice a year, the interest rate per period is $i = .06/2 = .03$ and the number of periods in nine years is $n = 9(2) = 18$. Hence,

$$P = \frac{A}{(1 + i)^n} = \frac{15{,}000}{(1 + .03)^{18}} \approx 8810.919114.$$

So a fair price would be the present value of \$8810.92. 9✓

✓Checkpoint 9

Find the fair price (present value) in Example 12 if the interest rate is 7.5%.

Answer: See page 279.

Example 13 Assuming an annual inflation rate of 3.5%, how much did an item that sells for $100 today cost four years ago?

Solution Think of the price four years ago as the present value P and $100 as the future value A. Then $i = .035$, $n = 4$, and the present value is

$$P = \frac{A}{(1+i)^n} = \frac{100}{(1+.035)^4} = 87.1442.$$

So the item cost $87.14 four years ago. 10

✓Checkpoint 10

What did a $100 item sell for 3 years ago if the annual inflation rate has been 4%?

Answer: *See page 279.*

SUMMARY

At this point, it seems helpful to summarize the notation and the most important formulas for simple and compound interest. We use the following variables:

P = principal or present value;

A = future or maturity value;

r = annual (stated or nominal) interest rate;

t = number of years;

m = number of compounding periods per year;

i = interest rate per period;

n = total number of compounding periods;

r_E = effective rate (APY).

Simple Interest	**Compound Interest**
$A = P(1 + rt)$	$A = P(1 + i)^n$
$P = \dfrac{A}{1 + rt}$	$P = \dfrac{A}{(1 + i)^n} = A(1 + i)^{-n}$
	$r_E = \left(1 + \dfrac{r}{m}\right)^m - 1$

5.2 ▸ Exercises

Interest on the zero-coupon bonds here is compounded semiannually.

1. In the preceding summary what is the difference between r and i? between t and n?
2. Explain the difference between simple interest and compound interest.
3. What factors determine the amount of interest earned on a fixed principal?
4. In your own words, describe the *maturity value* of a loan.
5. What is meant by the *present value* of money?
6. If interest is compounded more than once per year, which rate is higher, the stated rate or the effective rate?

Find the compound amount for each of the following deposits. (See Examples 1, 2, 4, and 5.)

7. $1000 at 4% compounded annually for 6 years
8. $1000 at 6% compounded annually for 10 years
9. $470 at 8% compounded semiannually for 12 years
10. $15,000 at 4.6% compounded semiannually for 11 years

11. \$6500 at 4.5% compounded quarterly for 8 years
12. \$9100 at 6.1% compounded quarterly for 4 years

Find the amount of interest earned by each of the following deposits. (See Examples 2, 4, and 5.)

13. \$26,000 at 6% compounded annually for 5 years
14. \$22,000 at 5% compounded annually for 8 years
15. \$6000 at 4% compounded semiannually for 6.4 years
16. \$2500 at 4.5% compounded semiannually for 8 years
17. \$5124.98 at 6.3% compounded quarterly for 5.2 years
18. \$27,630.35 at 4.6% compounded quarterly for 3.9 years

Find the interest rate (with annual compounding) that makes the statement true. (See Example 3.)

19. \$3000 grows to \$3606 in 5 years
20. \$2550 grows to \$3905 in 11 years
21. \$8500 grows to \$12,161 in 7 years
22. \$9000 grows to \$17,118 in 16 years

Find the face value (to the nearest dollar) of the zero-coupon bond. (See Example 6.)

23. 15-year bond at 5.2%; price \$4630
24. 10-year bond at 4.1%; price \$13,328
25. 20-year bond at 3.5%; price \$9992
26. 25-year bond at 4.4%; price \$10,106
27. 30-year bond at 6.3%; price \$3888.50
28. How do the nominal, or stated, interest rate and the effective interest rate (APY) differ?

Find the APY corresponding to the given nominal rates. (See Examples 8–10).

29. 4% compounded semiannually
30. 6% compounded quarterly
31. 5% compounded quarterly
32. 4.7% compounded semiannually
33. 5.2% compounded monthly
34. 6.2% compounded monthly

Find the present value of the given future amounts. (See Example 11.)

35. \$12,000 at 5% compounded annually for 6 years
36. \$8500 at 6% compounded annually for 9 years
37. \$17,230 at 4% compounded quarterly for 10 years
38. \$5240 at 6% compounded quarterly for 8 years

What price should you be willing to pay for each of these zero-coupon bonds? (See Example 12.)

39. 5-year \$5000 bond; interest at 3.5%
40. 10-year \$10,000 bond; interest at 4%
41. 15-year \$20,000 bond; interest at 4.7%
42. 20-year \$15,000 bond; interest at 5.3%
43. If money can be invested at 8% compounded quarterly, which is larger, \$1000 now or \$1210 in 5 years? Use present value to decide.
44. If money can be invested at 6% compounded annually, which is larger, \$10,000 now or \$15,000 in 6 years? Use present value to decide.

Finance *Work the following applied problems.*

45. A small business borrows \$50,000 for expansion at 9% compounded monthly. The loan is due in 4 years. How much interest will the business pay?
46. A developer needs \$80,000 to buy land. He is able to borrow the money at 10% per year compounded quarterly. How much will the interest amount to if he pays off the loan in 5 years?
47. Lora Reilly has inherited \$10,000 from her uncle's estate. She will invest the money for 2 years. She is considering two investments: a money market fund that pays a guaranteed 5.8% interest compounded daily and a 2-year Treasury note at 6% annual interest. Which investment pays the most interest over the 2-year period?
48. Which of these 20-year zero-coupon bonds will pay more interest at maturity: one that sells for \$4510, with a 6.1% interest rate, or one that sells for \$5809, with a 4.8% interest rate?
49. Suppose \$10,000 is invested at an annual rate of 5% for 10 years. Find the future value if interest is compounded as follows.
    (a) annually
    (b) quarterly
    (c) monthly
    (d) daily
50. In the New Testament, Jesus commends a widow who contributed 2 mites to the temple treasury (Mark 12:42–44). A mite was worth roughly 1/8 of a cent. Suppose the temple had invested those 2 mites at 4% interest compounded quarterly. How much would the money be worth 2000 years later?
51. As the prize in a contest, you are offered \$1000 now or \$1210 in 5 years. If money can be invested at 6% compounded annually, which is larger?
52. Two partners agree to invest equal amounts in their business. One will contribute \$10,000 immediately. The other plans to contribute an equivalent amount in 3 years, when she expects to acquire a large sum of money. How much should she contribute at that time to match her partner's investment now, assuming an interest rate of 6% compounded semiannually?

53. The pie graph shows the percent of baby boomers ages 46 to 49 who said they had investments with a total value as shown in each category.* Note that 30% say they have saved less than \$10,000 and 28% don't know or gave no answer. Assume that the money is invested at an average rate of 8% compounded quarterly for 20 years, when this age group will be ready for retirement. Find the range of amounts each group (except the "Don't know or no answer" group) in the graph will have saved for retirement if no more is added.

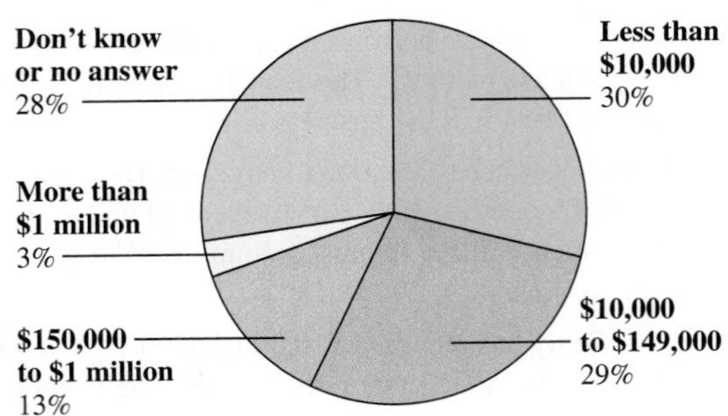

Note: Figures add to more than 100% because of rounding.

Sources: Census Bureau (age distribution); Merrill Lynch Baby Boom Retirement Index (investments); William M. Mercer, Inc. (life expectancy).

54. Scott Silva borrowed \$5200 from his friend Joe Vetere to buy computer equipment. He repaid the loan 10 months later with simple interest at 7%. Vetere then invested the proceeds in a 5-year certificate of deposit paying 6.3% compounded quarterly. How much will he have at the end of 5 years? (*Hint*: You need to use both simple and compound interest.)

55. In 1995, O. G. McClain of Houston, Texas, mailed a \$100 check to a descendant of Texas independence hero Sam Houston to repay a \$100 debt of McClain's great-great-grandfather, who died in 1835, to Sam Houston.† A bank estimated the interest on the loan to be \$420 million for the 160 years it was due. Find the interest rate the bank was using, assuming that interest is compounded annually.

56. In the Capital Appreciation Fund, a mutual fund from T. Rowe Price, a \$10,000 investment grew to \$24,828 over the 10-year period 1998–2008. Find the annual interest rate, compounded yearly, that this investment earned.

57. A \$10,000 investment in the Developing Technology Fund, also from T. Rowe Price, decreased in value to \$3901 over the 8-year period 2000–2008. Find the annual interest rate, compounded yearly, for this investment. (Your answer should be negative.)

New York Times, December 31, 1995, Sec, 3, p. 5.
†*New York Times*, March 30, 1995.

58. An investor buys a 20-year \$20,000 zero-coupon bond for \$7020. What annual interest rate will she earn if she holds the bond to maturity? (*Hint*: Let r be the annual interest rate; solve the compound amount formula with $i = r/2$.)

59. The Flagstar Bank in Michigan offered a 5-year certificate of deposit (CD) at 4.38% interest compounded quarterly.* On the same day on the Internet, Principal Bank offered a 5-year CD at 4.37% interest compounded monthly. Find the APY for each CD. Which bank paid a higher APY?

60. The Westfield Bank in Ohio offered the CD rates shown in the accompanying table in October 2008. The APY rates shown assume monthly compounding. Find the corresponding nominal rates to the nearest hundredth. (*Hint*: Solve the effective-rate equation for r.)

Term	6 mo	1 yr	2 yr	3 yr	5 yr
APY (%)	2.25	2.50	3.00	3.25	3.75

61. A company has agreed to pay \$2.9 million in 5 years to settle a lawsuit. How much must it invest now in an account paying 5% interest compounded monthly to have that amount when it is due?

62. Bill Poole wants to have \$20,000 available in 5 years for a down payment on a house. He has inherited \$16,000. How much of the inheritance should he invest now to accumulate the \$20,000 if he can get an interest rate of 5.5% compounded quarterly?

63. If inflation has been running at 3.75% per year and a new car costs \$23,500 today, what would it have cost three years ago?

64. If inflation is 2.4% per year and a washing machine costs \$345 today, what did a similar model cost five years ago?

Use the approach in Example 7 to find the time it would take for the general level of prices in the economy to double at the average annual inflation rates in Exercises 65–67.

65. 3%   66. 4%   67. 5%

68. The consumption of electricity has increased historically at 6% per year. If it continues to increase at this rate indefinitely, find the number of years before the electric utility companies will need to double their generating capacity.

69. Suppose a conservation campaign coupled with higher rates causes the demand for electricity to increase at only 2% per year, as it has recently. Find the number of years before the utility companies will need to double their generating capacity.

*June 1, 2005.

70. You decide to invest a \$16,000 bonus in a money market fund that guarantees a 5.5% annual interest rate compounded monthly for 7 years. A one-time fee of \$30 is charged to set up the account. In addition, there is an annual administrative charge of 1.25% of the balance in the account at the end of each year.
  (a) How much is in the account at the end of the first year?
  (b) How much is in the account at the end of the seventh year?

The following exercises are from professional examinations.

71. On January 1, 2002, Jack deposited \$1000 into Bank X to earn interest at the rate of j per annum compounded semi-annually. On January 1, 2007, he transferred his account to Bank Y to earn interest at the rate of k per annum compounded quarterly. On January 1, 2010, the balance at Bank Y was \$1990.76. If Jack could have earned interest at the rate of k per annum compounded quarterly from January 1, 2002, through January 1, 2010, his balance would have been \$2203.76. Which of the following represents the ratio k/j?*
  (a) 1.25
  (b) 1.30
  (c) 1.35
  (d) 1.40
  (e) 1.45

*Problem adapted from "Course 140 Examination, Mathematics of Compound Interest" of the Education and Examination Committee of the Society of Actuaries. Reprinted by permission of the Society of Actuaries.

72. On January 1, 2009, Tone Company exchanged equipment for a \$200,000 non-interest-bearing note due on January 1, 2012. The prevailing rate of interest for a note of this type on January 1, 2009, was 10%. The present value of \$1 at 10% for three periods is 0.75. What amount of interest revenue should be included in Tone's 2010 income statement?*
  (a) \$7500
  (b) \$15,000
  (c) \$16,500
  (d) \$20,000

✓ Checkpoint Answers

1.

	Year	Interest	Balance
(a)	4	\$50	\$1200
	5	\$50	\$1250
(b)	4	\$57.88	\$1215.51
	5	\$60.78	\$1276.29

2. (a) \$26,170.72 (b) \$9170.72
3. (a) \$17,410.24 (b) \$44,389.18
4. (a) \$15,000 (b) \$35,000
5. About 7 years ($n = 6.86$)
6. (a) 12.68% (b) 8.24%
7. (a) 4.06% (b) 8.219%
8. (a) \$4483.55 (b) \$3725.53
9. \$7732.24 **10.** \$88.90

*Adapted from the Uniform CPA Examination, American Institute of Certified Public Accountants.

5.3 ▶ Annuities, Future Value, and Sinking Funds

So far in this chapter, only lump-sum deposits and payments have been discussed. Many financial situations, however, involve a sequence of payments at regular intervals, such as weekly deposits in a savings account or monthly payments on a mortgage or car loan. Such periodic payments are the subject of this section and the next.

The analysis of periodic payments will require an algebraic technique that we now develop. Suppose x is a real number. For reasons that will become clear later, we want to find the product

$$(x - 1)(1 + x + x^2 + x^3 + \cdots + x^{11}).$$

Using the distributive property to multiply this expression out, we see that all but two of the terms cancel:

$$\begin{aligned} &x(1 + x + x^2 + x^3 + \cdots + x^{11}) - 1(1 + x + x^2 + x^3 + \cdots + x^{11}) \\ &= (x + x^2 + x^3 + \cdots + x^{11} + x^{12}) - 1 - x - x^2 - x^3 - \cdots - x^{11} \\ &= x^{12} - 1. \end{aligned}$$

Hence, $(x - 1)(1 + x + x^2 + x^3 + \cdots + x^{11}) = x^{12} - 1$. Dividing both sides by $x - 1$, we have

$$1 + x + x^2 + x^3 + \cdots + x^{11} = \frac{x^{12} - 1}{x - 1}.$$

The same argument, with any positive integer n in place of 12 and $n - 1$ in place of 11, produces the following result:

If x is a real number and n is a positive integer, then

$$1 + x + x^2 + x^3 + \cdots + x^{n-1} = \frac{x^n - 1}{x - 1}.$$

```
1+5+5^2+5^3+5^4+
5^5+5^6
           19531
(5^7-1)/(5-1)
           19531
```

FIGURE 5.5

For example, when $x = 5$ and $n = 7$, we see that

$$1 + 5 + 5^2 + 5^3 + 5^4 + 5^5 + 5^6 = \frac{5^7 - 1}{5 - 1} = \frac{78{,}124}{4} = 19{,}531.$$

A calculator can easily add up the terms on the left side, but it is faster to use the formula (Figure 5.5).

ORDINARY ANNUITIES

A sequence of equal payments made at equal periods of time is called an **annuity**. The time between payments is the **payment period**, and the time from the beginning of the first payment period to the end of the last period is called the **term of the annuity**. Annuities can be used to accumulate funds—for example, when you make regular deposits in a savings account. Or they can be used to pay out funds—as when you receive regular payments from a pension plan after you retire.

Annuities that pay out funds are considered in the next section. This section deals with annuities in which funds are accumulated by regular payments into an account or investment that earns compound interest. The **future value** of such an annuity is the final sum on deposit—that is, the total amount of all deposits and all interest earned by them.

We begin with **ordinary annuities**—ones where the payments are made at the *end* of each period and the frequency of payments is the same as the frequency of compounding the interest.

Example 1 \$1500 is deposited at the end of each year for the next 6 years in an account paying 8% interest compounded annually. Find the future value of this annuity.

Solution Figure 5.6 shows the situation schematically.

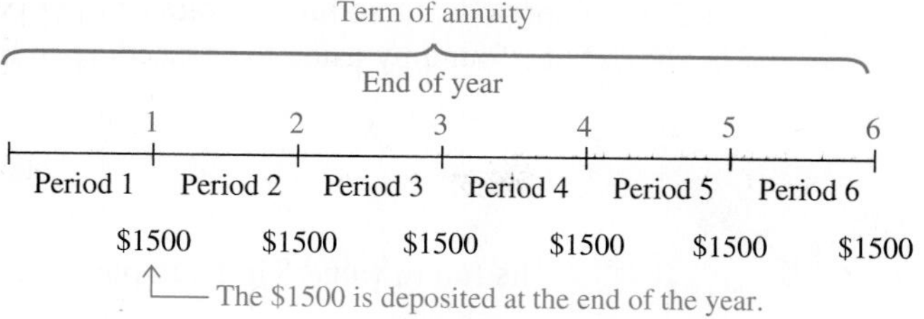

FIGURE 5.6

To find the future value of this annuity, look separately at each of the $1500 payments. The first $1500 is deposited at the end of period 1 and earns interest for the remaining 5 periods. From the formula in the box on page 268, the compound amount produced by this payment is

$$1500(1 + .08)^5 = 1500(1.08)^5.$$

The second $1500 payment is deposited at the end of period 2 and earns interest for the remaining 4 periods. So the compound amount produced by the second payment is

$$1500(1 + .08)^4 = 1500(1.08^4).$$

Continue to compute the compound amount for each subsequent payment, as shown in Figure 5.7. Note that the last payment earns no interest.

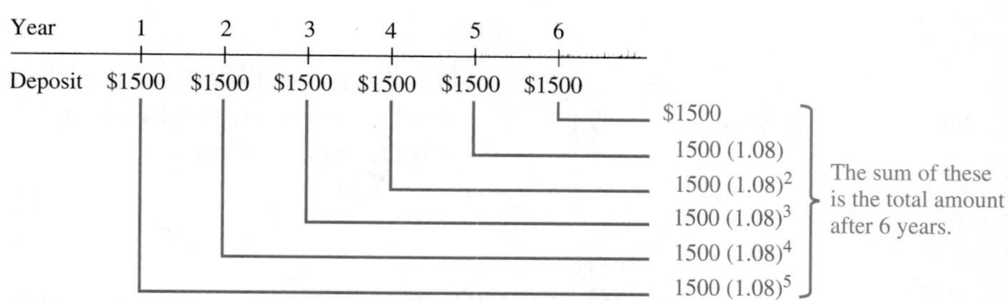

FIGURE 5.7

✓Checkpoint 1

Complete these steps for an annuity of $2000 at the end of each year for 3 years. Assume interest of 6% compounded annually.

(a) The first deposit of $2000 produces a total of _____.

(b) The second deposit becomes _____.

(c) No interest is earned on the third deposit, so the total in the account is _____.

Answers: See page 289.

The last column of Figure 5.7 shows that the total amount after 6 years is the sum

$$1500 + 1500 \cdot 1.08 + 1500 \cdot 1.08^2 + 1500 \cdot 1.08^3 + 1500 \cdot 1.08^4 + 1500 \cdot 1.08^5$$
$$= 1500(1 + 1.08 + 1.08^2 + 1.08^3 + 1.08^4 + 1.08^5). \quad (1)$$

Now apply the algebraic fact in the box on page 280 to the expression in parentheses (with $x = 1.08$ and $n = 6$). It shows that the sum (the future value of the annuity) is

$$1500 \cdot \frac{1.08^6 - 1}{1.08 - 1} = \$11{,}003.89. \quad ✓1$$

Example 1 is the model for finding a formula for the future value of any annuity. Suppose that a payment of R dollars is deposited at the end of each period for n

periods, at an interest rate of i per period. Then the future value of this annuity can be found by using the procedure in Example 1, with these replacements:

1500	.08	1.08	6	5
↓	↓	↓	↓	↓
R	i	$1 + i$	n	$n - 1$

The future value S in Example 1—call it S—is the sum **(1)**, which now becomes

$$S = R[1 + (1 + i) + (1 + i)^2 + \cdots + (1 + i)^{n-2} + (1 + i)^{n-1}].$$

Apply the algebraic fact in the box on page 280 to the expression in brackets (with $x = 1 + i$). Then we have

$$S = R\left[\frac{(1 + i)^n - 1}{(1 + i) - 1}\right] = R\left[\frac{(1 + i)^n - 1}{i}\right].$$

The quantity in brackets in the right-hand part of the preceding equation is sometimes written $s_{\overline{n}|i}$ (read "s-angle-n at i"). So we can summarize as follows.*

Future Value of an Ordinary Annuity

The future value S of an ordinary annuity used to accumulate funds is given by

$$S = R\left[\frac{(1 + i)^n - 1}{i}\right], \quad \text{or} \quad S = R \cdot s_{\overline{n}|i},$$

where

R is the payment at the end of each period,
i is the interest rate per period, and
n is the number of periods.

TECHNOLOGY TIP Most computations with annuities can be done quickly with a spreadsheet program or a graphing calculator. On a calculator, use the TVM solver if there is one (see the Tip on page 271); otherwise, use the programs in the Program Appendix at www.pearsonhighered.com/mwa10e.

Figure 5.8 shows how to do Example 1 on a TI-84+ TVM solver. First, enter the known quantities: N = number of payments, I% = annual interest rate, PV = present value, PMT = payment per period (entered as a negative amount), P/Y = number of payments per year, and C/Y = number of compoundings per year. At the bottom of the screen, set PMT: to "END" for ordinary annuities. Then put the cursor next to the unknown amount FV (future value), and press SOLVE.

Note: P/Y and C/Y should always be the same for problems in this book. If you use the solver for ordinary compound interest problems, set PMT = 0 and enter either PV or FV (whichever is known) as a negative amount.

```
N=6
I%=8
PV=0
PMT=-1500
•FV=11003.89356
P/Y=1
C/Y=1
PMT:END BEGIN
```

FIGURE 5.8

*We use S for the future value here instead of A, as in the compound interest formula, to help avoid confusing the two formulas.

Example 2 Chris Webber is an athlete who feels that his playing career will last 7 years. To prepare for his future, he deposits \$122,000 at the end of each year for 7 years in an account paying 6% compounded annually. How much will he have on deposit after 7 years?

Solution His payments form an ordinary annuity with $R = 122{,}000$, $n = 7$, and $i = .06$. The future value of this annuity (by the previous formula) is

$$S = 122{,}000\left[\frac{(1.06)^7 - 1}{.06}\right] = \$1{,}024{,}048.19.$$

✓2

✓Checkpoint 2

Johnson Building Materials deposits \$2500 at the end of each year into an account paying 8% per year compounded annually. Find the total amount on deposit after

(a) 6 years;

(b) 10 years.

Answers: See page 289.

Example 3 Anne teaches at a public university in North Carolina. She contributes \$625 at the end of each month (including some matching funds from the university) to the CREF Stock Fund.* For the past 10 years, this fund has returned 3.84% a year, compounded monthly.

(a) Assuming that the 3.84% rate continues, how much will she have in her account after 15 years?

Solution Anne's payments form an ordinary annuity, with monthly payment $R = 625$. The interest per month is $i = .0384/12$, and the number of months in 15 years is $n = 12 * 5 = 180$. The future value of this annuity is

$$S = R\left[\frac{(1+i)^n - 1}{i}\right] = 625\left[\frac{(1 + .0384/12)^{180} - 1}{.0384/12}\right] = \$151{,}811.20.$$

(b) Assume the economy has gotten better and that the fund now has a return of 7.72%, compounded monthly. Since Anne's salary has risen over the first 15 years, she can now contribute \$1000 per month. At the end of the next 15 years, how much is her account worth?

Solution Deal separately with the two parts of her account (the \$1000 contributions in the future and the \$151,811.20 already in the account). The contributions form an ordinary annuity as in part (a). Now we have $R = 1000$, $i = .0772/12$, and $n = 180$. So the future value is

$$S = R\left[\frac{(1+i)^n - 1}{i}\right] = 1000\left[\frac{(1 + .0772/12)^{180} - 1}{.0772/12}\right] = \$337{,}581.42.$$

Meanwhile, the \$151,811.20 from the first 15 years is also earning interest at 7.72%, compounded monthly. By the compound amount formula (Section 5.2), the future value of this money is

$$151{,}811.20(1 + .0772/12)^{180} = \$481{,}510.79$$

So the total amount in Anne's account after 30 years is the sum

$$\$337{,}581.42 + \$481{,}510.79 = \$810{,}092.21.$$

✓3

✓Checkpoint 3

Find the total value of the account in part (b) of Example 3 if the fund's return for the last 15 years is 9.78%, compounded monthly.

Answer: See page 289.

*TIAA-CREF provides retirement accounts for many colleges and universities. This is one of the funds available to participants (but not to the general public).

SINKING FUNDS

A **sinking fund** is a fund set up to receive periodic payments. Corporations and municipalities use sinking funds to repay bond issues, to retire preferred stock, to provide for replacement of fixed assets, and for other purposes. If the payments are equal and are made at the end of regular periods, they form an ordinary annuity.

Example 4 A business sets up a sinking fund so that it will be able to pay off bonds it has issued when they mature. If it deposits \$12,000 at the end of each quarter in an account that earns 5.2% interest, compounded quarterly, how much will be in the sinking fund after 10 years?

Solution The sinking fund is an annuity, with $R = 12{,}000$, $i = .052/4$, and $n = 4(10) = 40$. The future value is

$$S = R\left[\frac{(1+i)^n - 1}{i}\right] = 12{,}000\left[\frac{(1+.052/4)^{40} - 1}{.052/4}\right] = \$624{,}369.81.$$

So there will be about \$624,370 in the sinking fund.

Example 5 A firm borrows 6 million dollars to build a small factory. The bank requires it to set up a \$200,000 sinking fund to replace the roof after 15 years. If the firm's deposits earn 6% interest, compounded annually, find the payment it should make at the end of each year into the sinking fund.

Solution This situation is an annuity with future value $S = 200{,}000$, interest rate $i = .06$, and $n = 15$. Solve the future-value formula for R:

$$S = R\left[\frac{(1+i)^n - 1}{i}\right]$$

$$200{,}000 = R\left[\frac{(1+.06)^{15} - 1}{.06}\right]$$ Let $S = 200{,}000$, $i = .06$, and $n = 15$.

$$200{,}000 = R[23.27597]$$ Compute the quantity in brackets.

$$R = \frac{200{,}000}{23.27597} = \$8592.55.$$ Divide both sides by 23.27597.

So the annual payment is about \$8593. ✓4

✓Checkpoint 4

Francisco Arce needs \$8000 in 6 years so that he can go on an archaeological dig. He wants to deposit equal payments at the end of each quarter so that he will have enough to go on the dig. Find the amount of each payment if the bank pays

(a) 12% interest compounded quarterly;

(b) 8% interest compounded quarterly.

Answers: See page 289.

Example 6 As an incentive for a valued employee to remain on the job, a company plans to offer her a \$100,000 bonus, payable when she retires in 20 years. If the company deposits \$200 a month in a sinking fund, what interest rate must it earn, with monthly compounding, in order to guarantee that the fund will be worth \$100,000 in 20 years?

Solution The sinking fund is an annuity with $R = 200$, $n = 12(20) = 240$, and future value $S = 100{,}000$. We must find the interest rate. If x is the annual interest

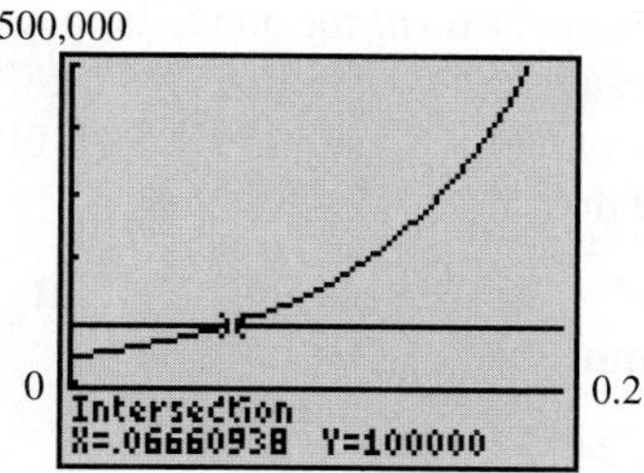

FIGURE 5.9

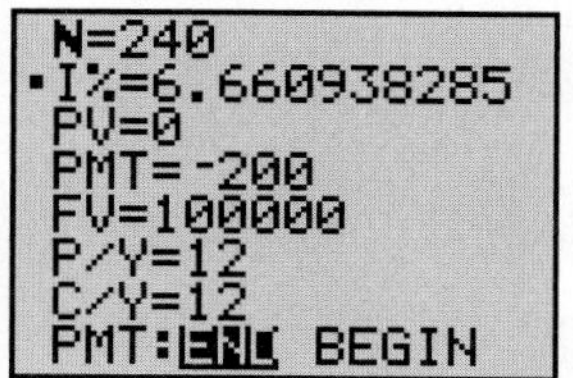

FIGURE 5.10

Checkpoint 5

Pete's Pizza deposits $5800 at the end of each quarter for 4 years.

(a) Find the final amount on deposit if the money earns 6.4% compounded quarterly.

(b) Pete wants to accumulate $110,000 in the 4-year period. What interest rate (to the nearest tenth) will be required?

Answers: See page 289.

rate in decimal form, then the interest rate per month is $i = x/12$. Inserting these values into the future-value formula, we have

$$R\left[\frac{(1+i)^n - 1}{i}\right] = S$$

$$200\left[\frac{(1+x/12)^{240} - 1}{x/12}\right] = 100{,}000.$$

This equation is hard to solve algebraically. You can get a rough approximation by using a calculator and trying different values for x. With a graphing calculator, you can get an accurate solution by graphing

$$y_1 = 200\left[\frac{(1+x/12)^{240} - 1}{x/12}\right] \quad \text{and} \quad y_2 = 100{,}000$$

and finding the x-coordinate of the point where the graphs intersect. Figure 5.9 shows that the company needs an interest rate of about 6.661%. The same answer can be obtained on a TVM solver (Figure 5.10). 5

ANNUITIES DUE

The formula developed previously is for *ordinary annuities*—annuities with payments at the *end* of each period. The results can be modified slightly to apply to **annuities due**—annuities where payments are made at the *beginning* of each period.

An example will illustrate how this is done. Consider an annuity due in which payments of $100 are made for 3 years, and an ordinary annuity in which payments of $100 are made for 4 years, both with 5% interest, compounded annually. Figure 5.11 computes the growth of each payment separately (as was done in Example 1).

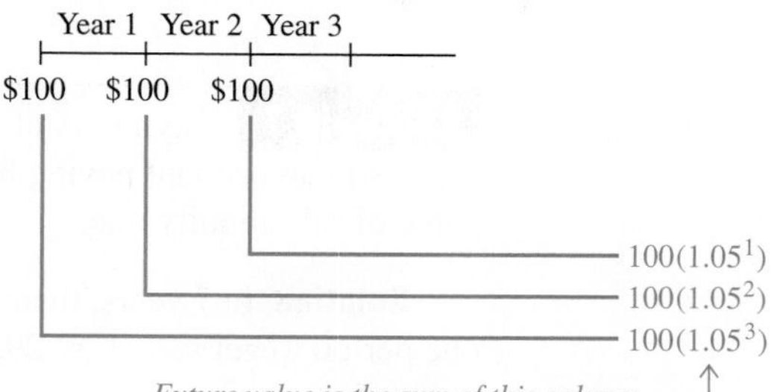

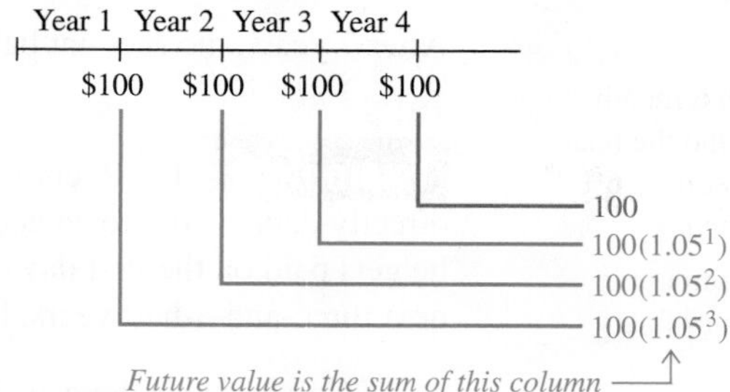

FIGURE 5.11

Figure 5.11 shows that the future values are the same, *except* for one \$100 payment on the ordinary annuity (shown in red). So we can use the formula on page 282 to find the future value of the 4-year ordinary annuity and then subtract one \$100 payment to get the future value of the 3-year annuity due:

Future value of 3-year annuity due = **Future value of 4-year ordinary annuity** − **One payment**

$$S = 100\left[\frac{1.05^4 - 1}{.05}\right] - 100 = \$331.01.$$

Essentially the same argument works in the general case.

Future Value of an Annuity Due

The future value S of an annuity due used to accumulate funds is given by

$$S = R\left[\frac{(1+i)^{n+1} - 1}{i}\right] - R$$

S = Future value of an ordinary annuity of $n + 1$ payments − One payment,

where

R is the payment at the beginning of each period,

i is the interest rate per period, and

n is the number of periods.

Example 7 Payments of \$500 are made at the beginning of each quarter for 7 years in an account paying 8% interest, compounded quarterly. Find the future value of this annuity due.

Solution In 7 years, there are $n = 28$ quarterly periods. For an annuity due, add one period to get $n + 1 = 29$, and use the formula with $i = .08/4 = .02$:

$$S = R\left[\frac{(1+i)^{n+1} - 1}{i}\right] - R = 500\left[\frac{(1+.02)^{29} - 1}{.02}\right] - 500 = \$18{,}896.12.$$

After 7 years, the account balance will be \$18,896.12. ✓6

✓Checkpoint 6

(a) Ms. Black deposits \$800 at the beginning of each 6-month period for 5 years. Find the final amount if the account pays 6% compounded semiannually.

(b) Find the final amount if this account were an ordinary annuity.

Answers: See page 289.

Example 8 Jay Rechtien plans to have a fixed amount from his paycheck directly deposited into an account that pays 5.5% interest, compounded monthly. If he gets paid on the first day of the month and wants to accumulate \$13,000 in the next three-and-a-half years, how much should he deposit each month?

Solution Jay's deposits form an annuity due whose future value is $S = 13{,}000$. The interest rate is $i = .055/12$. There are 42 months in three-and-a-half years.

TECHNOLOGY TIP When a TVM solver is used for annuities due, the PMT: setting at the bottom of the screen should be "BEGIN". See Figure 5.12, which shows the solution of Example 8.

N=42
I%=5.5
PV=0
PMT=-280.110136
FV=13000
P/Y=12
C/Y=12
PMT:END BEGIN

FIGURE 5.12

Since this is an annuity due, add one period, so that $n + 1 = 43$. Then solve the future-value formula for the payment R:

$$R\left[\frac{(1 + i)^{n+1} - 1}{i}\right] - R = S$$

$$R\left[\frac{(1 + .055/12)^{43} - 1}{.055/12}\right] - R = 13{,}000 \qquad \text{Let } i = .055/12,\ n = 43, \text{ and } S = 13{,}000.$$

$$R\left(\left[\frac{(1 + .055/12)^{43} - 1}{.055/12}\right] - 1\right) = 13{,}000 \qquad \text{Factor out } R \text{ on left side.}$$

$$R(46.4103) = 13{,}000 \qquad \text{Compute left side.}$$

$$R = \frac{13{,}000}{46.4103} = 280.110. \qquad \text{Divide both sides by } 46.4103.$$

Jay should have $280.11 deposited from each paycheck.

5.3 ▶ Exercises

Note: Unless stated otherwise, all payments are made at the end of the period.

Find each of these sums (to 4 decimal places).

1. $1 + 1.05 + 1.05^2 + 1.05^3 + \cdots + 1.05^{14}$

2. $1 + 1.046 + 1.046^2 + 1.046^3 + \cdots + 1.046^{21}$

Find the future value of the ordinary annuities with the given payments and interest rates. (See Examples 1, 2, 3(a), and 4.)

3. $R = \$12{,}000$, 6.2% interest compounded annually for 8 years

4. $R = \$20{,}000$, 4.5% interest compounded annually for 12 years

5. $R = \$865$, 6% interest compounded semiannually for 10 years

6. $R = \$7300$, 9% interest compounded semiannually for 6 years

7. $R = \$1200$, 8% interest compounded quarterly for 10 years

8. $R = \$20{,}000$, 6% interest compounded quarterly for 12 years

Find the final amount (rounded to the nearest dollar) in each of these retirement accounts, in which the rate of return on the account and the regular contribution change over time. (See Example 3.)

9. $400 per month invested at 4%, compounded monthly, for 10 years; then $600 per month invested at 6%, compounded monthly, for 10 years.

10. $500 per month invested at 5%, compounded monthly, for 20 years; then $1000 per month invested at 8%, compounded monthly, for 20 years.

11. $1000 per quarter invested at 4.2%, compounded quarterly, for 10 years; then $1500 per quarter invested at 7.4%, compounded quarterly, for 15 years.

12. $1500 per quarter invested at 7.4%, compounded quarterly, for 15 years; then $1000 per quarter invested at 4.2%, compounded quarterly, for 10 years. (Compare with Exercise 11.)

Find the amount of each payment to be made into a sinking fund to accumulate the given amounts. Payments are made at the end of each period. (See Example 5.)

13. $11,000; money earns 5% compounded semiannually for 6 years

14. $65,000; money earns 6% compounded semiannually for $4\frac{1}{2}$ years

15. $50,000; money earns 8% compounded quarterly for $2\frac{1}{2}$ years

16. $25,000; money earns 9% compounded quarterly for $3\frac{1}{2}$ years

17. $6000; money earns 6% compounded monthly for 3 years

18. $9000; money earns 7% compounded monthly for $2\frac{1}{2}$ years

Find the interest rate needed for the sinking fund to reach the required amount. Assume that the compounding period is the same as the payment period. (See Example 6.)

19. $50,000 to be accumulated in 10 years; annual payments of $3940.

20. $100,000 to be accumulated in 15 years; quarterly payments of $1200.

21. $38,000 to be accumulated in 5 years; quarterly payments of $1675.

22. \$77,000 to be accumulated in 20 years; monthly payments of \$195.

23. What is meant by a sinking fund? List some reasons for establishing a sinking fund.

24. Explain the difference between an ordinary annuity and an annuity due.

Find the future value of each annuity due. (See Example 7.)

25. Payments of \$500 for 10 years at 5% compounded annually

26. Payments of \$1050 for 8 years at 3.5% compounded annually

27. Payments of \$16,000 for 11 years at 4.7% compounded annually

28. Payments of \$25,000 for 12 years at 6% compounded annually

29. Payments of \$1000 for 9 years at 8% compounded semiannually

30. Payments of \$750 for 15 years at 6% compounded semiannually

31. Payments of \$100 for 7 years at 9% compounded quarterly

32. Payments of \$1500 for 11 years at 7% compounded quarterly

Find the payment that should be used for the annuity due whose future value is given. Assume that the compounding period is the same as the payment period. (See Example 8.)

33. \$8000; quarterly payments for 3 years; interest rate 4.4%

34. \$12,000; annual payments for 6 years; interest rate 5.1%.

35. \$55,000; monthly payments for 12 years; interest rate 5.7%.

36. \$125,000; monthly payments for 9 years; interest rate 6%.

Finance *Work the following applied problems.*

37. A typical pack-a-day smoker in Cleveland, OH, spends about \$150 per month on cigarettes. Suppose the smoker invests that amount at the end of each month in a savings account at 4.8% compounded monthly. What would the account be worth after 40 years?

38. For the past 10 years, the class A shares of the Europacific Growth Fund (a mutual fund from American Funds) have increased in value at the rate of 8.1%, compounded monthly. If you had invested \$300 a month in this fund for those 10 years, how much would be in your account now?

39. Becky Anderson deposits \$12,000 at the end of each year for 9 years in an account paying 6% interest compounded annually.
    (a) Find the final amount she will have on deposit.
    (b) Becky's brother-in-law works in a bank that pays 5% compounded annually. If she deposits money in this bank instead of the other one, how much will she have in her account?
    (c) How much would Becky lose over 9 years by using her brother-in-law's bank?

40. Raul Vasquez, a 25-year-old professional, puts \$750 in a retirement fund at the end of each quarter until he reaches age 60. The account pays 8% interest compounded quarterly.
    (a) How much will be in the account when he is 60?
    (b) If Raul makes no further deposits after age 60, how much will he have for retirement at age 65?

41. At the end of each quarter, a 50-year-old woman puts \$1200 in a retirement account that pays 7% interest compounded quarterly. When she reaches age 60, she withdraws the entire amount and places it in a mutual fund that pays 9% interest compounded monthly. From then on, she deposits \$300 in the mutual fund at the end of each month. How much is in the account when she reaches age 65?

42. Hassi is paid on the first day of the month, and \$80 is automatically deducted from his pay and deposited in a savings account. If the account pays 7.5% interest compounded monthly, how much will be in the account after 3 years and 9 months?

43. A father opened a savings account for his daughter on the day she was born, depositing \$1000. Each year on her birthday, he deposits another \$1000, making the last deposit on her 21st birthday. If the account pays 6.5% interest compounded annually, how much is in the account at the end of the day on the daughter's 21st birthday?

44. Jasspreet Kaur deposits \$2435 at the beginning of each semiannual period for 8 years in an account paying 6% compounded semiannually. She then leaves that money alone, with no further deposits, for an additional 5 years. Find the final amount on deposit after the entire 13-year period.

45. Chuck Hickman deposits \$10,000 at the beginning of each year for 12 years in an account paying 5% compounded annually. He then puts the total amount on deposit in another account paying 6% compounded semiannually for another 9 years. Find the final amount on deposit after the entire 21-year period.

46. Suppose that the best rate that the company in Example 6 can find is 6.3%, compounded monthly (rather than the 6.661% it wants). Then the company must deposit more in the sinking fund each month. What monthly deposit will guarantee that the fund will be worth \$100,000 in 20 years?

47. David Horwitz needs \$10,000 in 8 years.
    (a) What amount should he deposit at the end of each quarter at 5% compounded quarterly so that he will have his \$10,000?
    (b) Find Horwitz's quarterly deposit if the money is deposited at 5.8% compounded quarterly.

48. Harv's Meats knows that it must buy a new deboner machine in 4 years. The machine costs $12,000. In order to accumulate enough money to pay for the machine, Harv decides to deposit a sum of money at the end of each 6 months in an account paying 6% compounded semiannually. How much should each payment be?

49. Barbara Margolius wants to buy a $24,000 car in 6 years. How much money must she deposit at the end of each quarter in an account paying 5% compounded quarterly so that she will have enough to pay for her car?

50. The Chinns agree to sell an antique vase to a local museum for $19,000. They want to defer the receipt of this money until they retire in 5 years (and are in a lower tax bracket). If the museum can earn 5.8%, compounded annually, find the amount of each annual payment it should make into a sinking fund so that it will have the necessary $19,000 in 5 years.

51. Diane Gray sells some land in Nevada. She will be paid a lump sum of $60,000 in 7 years. Until then, the buyer pays 8% simple interest quarterly.
   (a) Find the amount of each quarterly interest payment.
   (b) The buyer sets up a sinking fund so that enough money will be present to pay off the $60,000. The buyer wants to make semiannual payments into the sinking fund; the account pays 6% compounded semiannually. Find the amount of each payment into the fund.

52. Joe Seniw bought a rare stamp for his collection. He agreed to pay a lump sum of $4000 after 5 years. Until then, he pays 6% simple interest semiannually.
   (a) Find the amount of each semiannual interest payment.
   (b) Seniw sets up a sinking fund so that enough money will be present to pay off the $4000. He wants to make annual payments into the fund. The account pays 8% compounded annually. Find the amount of each payment.

53. To save for retirement, Karla Harby put $300 each month into an ordinary annuity for 20 years. Interest was compounded monthly. At the end of the 20 years, the annuity was worth $147,126. What annual interest rate did she receive?

54. Jennifer Wall made payments of $250 per month at the end of each month to purchase a piece of property. After 30 years, she owned the property, which she sold for $330,000. What annual interest rate would she need to earn on an ordinary annuity for a comparable rate of return?

55. When Joe and Sarah graduate from college, each expects to work a total of 45 years. Joe begins saving for retirement immediately. He plans to deposit $600 at the end of each quarter into an account paying 8.1% interest, compounded quarterly, for 10 years. He will then leave his balance in the account, earning the same interest rate, but make no further deposits for 35 years. Sarah plans to save nothing during the first 10 years and then begin depositing $600 at the end of each quarter in an account paying 8.1% interest, compounded quarterly, for 35 years.
   (a) Without doing any calculations, predict which one will have the most in his or her retirement account after 45 years. Then test your prediction by answering the following questions (calculation required to the nearest dollar).
   (b) How much will Joe contribute to his retirement account?
   (c) How much will be in Joe's account after 45 years?
   (d) How much will Sarah contribute to her retirement account?
   (e) How much will be in Sarah's account after 45 years?

56. In a 1992 Virginia lottery, the jackpot was $27 million. An Australian investment firm tried to buy all possible combinations of numbers, which would have cost $7 million. In fact, the firm ran out of time and was unable to buy all combinations, but ended up with the only winning ticket anyway. The firm received the jackpot in 20 equal annual payments of $1.35 million.* Assume these payments meet the conditions of an ordinary annuity.
   (a) Suppose the firm can invest money at 8% interest compounded annually. How many years would it take until the investors would be further ahead than if they had simply invested the $7 million at the same rate? (*Hint*: Experiment with different values of n, the number of years, or use a graphing calculator to plot the value of both investments as a function of the number of years.)
   (b) How many years would it take in part (a) at an interest rate of 12%?

✓ Checkpoint Answers

1. **(a)** $2247.20 **(b)** $2120.00 **(c)** $6367.20
2. **(a)** $18,339.82 **(b)** $36,216.41
3. $1,060,578.92
4. **(a)** $232.38 **(b)** $262.97
5. **(a)** $104,812.44 **(b)** 8.9%
6. **(a)** $9446.24 **(b)** $9171.10

**Washington Post*, March 10, 1992, p. A1.

5.4 Annuities, Present Value, and Amortization

In the annuities studied previously, regular deposits were made into an interest-bearing account and the value of the annuity increased from 0 at the beginning to some larger amount at the end (the future value). Now we expand the discussion to include annuities that begin with an amount of money and make regular payments each period until the value of the annuity decreases to 0. Examples of such annuities are lottery jackpots, structured settlements imposed by a court in which the party at fault (or his or her insurance company) makes regular payments to the injured party, and trust funds that pay the recipients a fixed amount at regular intervals.

In order to develop the essential formula for dealing with "payout annuities," we need another useful algebraic fact. If x is a nonzero number and n is a positive integer, verify the following equality by multiplying out the right-hand side:*

$$x^{-1} + x^{-2} + x^{-3} + \cdots + x^{-(n-1)} + x^{-n} = x^{-n}(x^{n-1} + x^{n-2} + x^{n-3} + \cdots + x^{1} + 1).$$

Now use the sum formula in the box on page 280 to rewrite the expression in parentheses on the right-hand side:

$$x^{-1} + x^{-2} + x^{-3} + \cdots + x^{-(n-1)} + x^{-n} = x^{-n}\left(\frac{x^n - 1}{x - 1}\right)$$
$$= \frac{x^{-n}(x^n - 1)}{x - 1} = \frac{x^0 - x^{-n}}{x - 1} = \frac{1 - x^{-n}}{x - 1}.$$

We have proved the following result:

> If x is a nonzero real number and n is a positive integer, then
>
> $$x^{-1} + x^{-2} + x^{-3} + \cdots + x^{-n} = \frac{1 - x^{-n}}{x - 1}.$$

PRESENT VALUE

In Section 5.2, we saw that the present value of A dollars at interest rate i per period for n periods is the amount that must be deposited today (at the same interest rate) in order to produce A dollars in n periods. Similarly, the **present value of an annuity** is the amount that must be deposited today (at the same compound interest rate as the annuity) to provide all the payments for the term of the annuity. It does not matter whether the payments are invested to accumulate funds or are paid out to disperse funds; the amount needed to provide the payments is the same in either case. We begin with ordinary annuities.

Example 1 Your rich aunt has funded an annuity that will pay you \$1500 at the end of each year for six years. If the interest rate is 8%, compounded annually, find the present value of this annuity.

*Remember that powers of x are multiplied by *adding* exponents and that $x^n x^{-n} = x^{n-n} = x^0 = 1$.

Solution Look separately at each payment you will receive. Then find the present value of each payment—the amount needed now in order to make the payment in the future. The sum of these present values will be the present value of the annuity, since it will provide all of the payments.

To find the first \$1500 payment (due in one year), the present value of \$1500 at 8% annual interest is needed now. According to the present-value formula for compound interest on page 275 (with $A = 1500$, $i = .08$, and $n = 1$), this present value is

$$\frac{1500}{1 + .08} = \frac{1500}{1.08} = 1500(1.08^{-1}) \approx \$1388.89.$$

This amount will grow to \$1500 in one year.

For the second \$1500 payment (due in two years), we need the present value of \$1500 at 8% interest, compounded annually for two years. The present-value formula for compound interest (with $A = 1500$, $i = .08$, and $n = 2$) shows that this present value is

$$\frac{1500}{(1 + .08)^2} = \frac{1500}{1.08^2} = 1500(1.08^{-2}) \approx \$1286.01.$$

Less money is needed for the second payment because it will grow over two years instead of one.

A similar calculation shows that the third payment (due in three years) has present value $\$1500(1.08^{-3})$. Continue in this manner to find the present value of each of the remaining payments, as summarized in Figure 5.13.

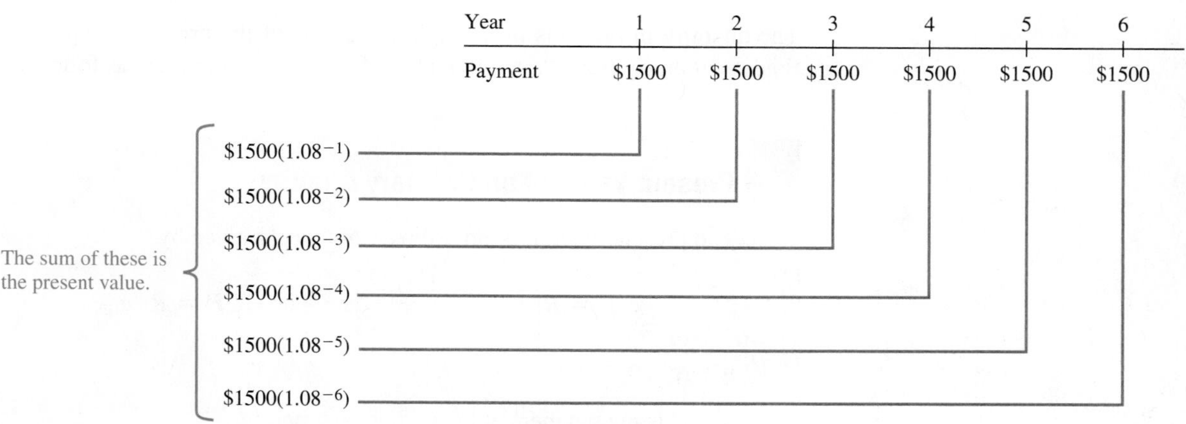

FIGURE 5.13

The left-hand column of Figure 5.13 shows that the present value is

$$1500 \cdot 1.08^{-1} + 1500 \cdot 1.08^{-2} + 1500 \cdot 1.08^{-3} + 1500 \cdot 1.08^{-4} + 1500 \cdot 1.08^{-5} + 1500 \cdot 1.08^{-6}$$
$$= 1500(1.08^{-1} + 1.08^{-2} + 1.08^{-3} + 1.08^{-4} + 1.08^{-5} + 1.08^{-6}). \qquad \textbf{(1)}$$

Now apply the algebraic fact in the box on page 290 to the expression in parentheses (with $x = 1.08$ and $n = 6$). It shows that the sum (the present value of the annuity) is

$$1500\left[\frac{1 - 1.08^{-6}}{1.08 - 1}\right] = 1500\left[\frac{1 - 1.08^{-6}}{.08}\right] = \$6934.32.$$

This amount will provide for all six payments and leave a zero balance at the end of six years (give or take a few cents due to rounding to the nearest penny at each step). 1 ✓

✓Checkpoint 1

Show that \$6934.32 will provide all the payments in Example 1 as follows:

(a) Find the balance at the end of the first year after the interest has been added and the \$1500 payment subtracted.

(b) Repeat part (a) to find the balances at the ends of years 2 through 6.

Answers: See page 302.

Example 1 is the model for finding a formula for the future value of any ordinary annuity. Suppose that a payment of R dollars is made at the end of each period for n periods, at interest rate i per period. Then the present value of this annuity can be found by using the procedure in Example 1, with these replacements:

$$\begin{array}{cccc} 1500 & .08 & 1.08 & 6 \\ \downarrow & \downarrow & \downarrow & \downarrow \\ R & i & 1+i & n \end{array}.$$

The future value in Example 1 is the sum in equation **(1)**, which now becomes

$$P = R[(1 + i)^{-1} + (1 + i)^{-2} + (1 + i)^{-3} + \cdots + (1 + i)^{-n}].$$

Apply the algebraic fact in the box on page 290 to the expression in brackets (with $x = 1 + i$). Then we have

$$P = R\left[\frac{1 - (1 + i)^{-n}}{(1 + i) - 1}\right] = R\left[\frac{1 - (1 + i)^{-n}}{i}\right].$$

The quantity in brackets in the right-hand part of the preceding equation is sometimes written $a_{\overline{n}|i}$ (read "a-angle-n at i"). So we can summarize as follows.

Present Value of an Ordinary Annuity

The present value P of an ordinary annuity is given by

$$P = R\left[\frac{1 - (1 + i)^{-n}}{i}\right], \quad \text{or} \quad P = R \cdot a_{\overline{n}|i},$$

where

R is the payment at the end of each period,
i is the interest rate per period, and
n is the number of periods.

CAUTION Do not confuse the formula for the present value of an annuity with the one for the future value of an annuity. Notice the difference: The numerator of the fraction in the present-value formula is $1 - (1 + i)^{-n}$, but in the future-value formula, it is $(1 + i)^n - 1$.

Example 2 Jim Riles was in an auto accident. He sued the person at fault and was awarded a structured settlement in which an insurance company will pay him \$600 at the end of each month for the next seven years. How much money should the insurance company invest now at 4.7%, compounded monthly, to guarantee that all the payments can be made?

Solution The payments form an ordinary annuity. The amount needed to fund all the payments is the present value of the annuity. Apply the present-value formula with $R = 600$, $n = 7 \cdot 12 = 84$, and $i = .047/12$ (the interest rate per month). The insurance company should invest

$$P = R\left[\frac{1-(1+i)^{-n}}{i}\right] = 600\left[\frac{1-(1+.047/12)^{-84}}{.047/12}\right] = \$42{,}877.44.$$ 2✓

✓Checkpoint 2

An insurance company offers to pay Jane Parks an ordinary annuity of \$1200 per quarter for five years *or* the present value of the annuity now. If the interest rate is 6%, find the present value.

Answer: See page 302.

Example 3 To supplement his pension in the early years of his retirement, Ralph Taylor plans to use \$124,500 of his savings as an ordinary annuity that will make monthly payments to him for 20 years. If the interest rate is 5.2%, how much will each payment be?

Solution The present value of the annuity is $P = \$124{,}500$, the monthly interest rate is $i = .052/12$, and $n = 12 \cdot 20 = 240$ (the number of months in 20 years). Solve the present-value formula for the monthly payment R:

$$P = R\left[\frac{1-(1+i)^{-n}}{i}\right]$$

$$124{,}500 = R\left[\frac{1-(1+.052/12)^{-240}}{.052/12}\right]$$

$$R = \frac{124{,}500}{\left[\frac{1-(1+.052/12)^{-240}}{.052/12}\right]} = \$835.46.$$

Taylor will receive \$835.46 a month (about \$10,026 per year) for 20 years. 3✓

✓Checkpoint 3

Carl Dehne has \$80,000 in an account paying 4.8% interest, compounded monthly. He plans to use up all the money by making equal monthly withdrawals for 15 years. If the interest rate is 4.8%, find the amount of each withdrawal.

Answer: See page 302.

Example 4 Surinder Sinah and Maria Gonzalez are graduates of Kenyon College. They both agree to contribute to an endowment fund at the college. Sinah says he will give \$500 at the end of each year for 9 years. Gonzalez prefers to give a single donation today. How much should she give to equal the value of Sinah's gift, assuming that the endowment fund earns 7.5% interest, compounded annually.

Solution Sinah's gift is an ordinary annuity with annual payments of \$500 for 9 years. Its *future* value at 7.5% annual compound interest is

$$S = R\left[\frac{(1+i)^n - 1}{i}\right] = 500\left[\frac{(1+.075)^9 - 1}{.075}\right] = 500\left[\frac{1.075^9 - 1}{.075}\right] = \$6114.92.$$

We claim that for Gonzalez to equal this contribution, she should today contribute an amount equal to the *present* value of this annuity, namely,

$$P = R\left[\frac{1-(1+i)^{-n}}{i}\right] = 500\left[\frac{1-(1+.075)^{-9}}{.075}\right] = 500\left[\frac{1-1.075^{-9}}{.075}\right] = \$3189.44.$$

To confirm this claim, suppose the present value $P = \$3189.44$ is deposited today at 7.5% interest, compounded annually for 9 years. According to the compound interest formula on page 268, P will grow to

$$3189.44(1 + .075)^9 = \$6114.92,$$

the future value of Sinah's annuity. So at the end of 9 years, Gonzalez and Sinah will have made identical gifts.

Example 4 illustrates the following alternative description of the present value of an "accumulation annuity":

> The present value of an annuity for accumulating funds is the single deposit that would have to be made today to produce the future value of the annuity (assuming the same interest rate and period of time.) 4 ✓

✓Checkpoint 4

What lump sum deposited today would be equivalent to equal payments of

(a) $650 at the end of each year for 9 years at 4% compounded annually?

(b) $1000 at the end of each quarter for 4 years at 4% compounded quarterly?

Answers: See page 302.

Corporate bonds, which were introduced in Section 5.1, are routinely bought and sold in financial markets. In most cases, interest rates when a bond is sold differ from the interest rate paid by the bond (known as the **coupon rate**). In such cases, the price of a bond will not be its face value, but will instead be based on current interest rates. The next example shows how this is done.

Example 5 A 15-year $10,000 bond with a 5% coupon rate was issued five years ago and is now being sold. If the current interest rate for similar bonds is 7%, what price should a purchaser be willing to pay for this bond?

Solution According to the simple interest formula (page 260), the interest paid by the bond each half-year is

$$I = Prt = 10{,}000 \cdot .05 \cdot \frac{1}{2} = \$250.$$

Think of the bond as a two-part investment: The first is an annuity that pays $250 every six months for the next 10 years; the second is the $10,000 face value of the bond, which will be paid when the bond matures, 10 years from now. The purchaser should be willing to pay the present value of each part of the investment, assuming 7% interest, compounded semiannually.* The interest rate per period is $i = .07/2$, and the number of six-month periods in 10 years is $n = 20$. So we have:

Present value of annuity

$$P = R\left[\frac{1 - (1 + i)^{-n}}{i}\right]$$
$$= 250\left[\frac{1 - (1 + .07/2)^{-20}}{.07/2}\right]$$
$$= \$3553.10$$

Present value of $10,000 in 10 years

$$P = A(1 + i)^{-n}$$
$$= 10{,}000(1 + .07/2)^{-20}$$
$$= \$5025.66.$$

So the purchaser should be willing to pay the sum of these two present values:

$$\$3553.10 + \$5025.66 = \$8578.76.$$ 5 ✓

✓Checkpoint 5

Suppose the current interest rate for bonds is 4% instead of 7% when the bond in Example 5 is sold. What price should a purchaser be willing to pay for it?

Answer: See page 302.

*The analysis here does not include any commissions or fees charged by the financial institution that handles the bond sale.

NOTE: Example 5 and Checkpoint 5 illustrate the inverse relation between interest rates and bond prices: If interest rates rise, bond prices fall, and if interest rates fall, bond prices rise.

LOANS AND AMORTIZATION

If you take out a car loan or a home mortgage, you repay it by making regular payments to the bank. From the bank's point of view, your payments are an annuity that is paying it a fixed amount each month. The present value of this annuity is the amount you borrowed.

Example 6 A car costs \$22,000. After a down payment of \$4000, the balance will be paid off in 48 monthly payments, with interest of 7% per year on the unpaid balance. Find the amount of each payment.

Solution After a \$4000 down payment, the loan amount is \$18,000. Use the present-value formula for an annuity, with $P = 18{,}000$, $n = 48$, and $i = .07/12$ (the monthly interest rate). Then solve for the payment R.

$$P = R\left[\frac{1-(1+i)^{-n}}{i}\right]$$

$$18{,}000 = R\left[\frac{1-(1+.07/12)^{-48}}{.07/12}\right]$$

$$R = \frac{18{,}000}{\left[\frac{1-(1+.07/12)^{-48}}{.07/12}\right]} = \$431.03.$$ 6

Checkpoint 6

Kelly Erin buys a small business for \$174,000. She agrees to pay off the cost in payments at the end of each semiannual period for 7 years, with interest of 8% compounded semiannually on the unpaid balance. Find the amount of each payment.

Answer: See page 302.

A loan is **amortized** if both the principal and interest are paid by a sequence of equal periodic payments. The periodic payment needed to amortize a loan may be found, as in Example 6, by solving the present-value formula for R.

Amortization Payments

A loan of P dollars at interest rate i per period may be amortized in n equal periodic payments of R dollars made at the end of each period, where

$$R = \frac{P}{\left[\frac{1-(1+i)^{-n}}{i}\right]} = \frac{Pi}{1-(1+i)^{-n}}.$$

Example 7 The Newells buy a home for \$140,000, with a down payment of \$30,000. They take out a 30-year \$110,000 mortgage at an annual interest rate of 6.7%.

(a) Find the monthly payment needed to amortize this loan.

TECHNOLOGY TIP A TVM solver on a graphing calculator can find the present value of an annuity or the payment on a loan: Fill in the known information, put the cursor next to the unknown item (PV or PMT), and press SOLVE. Figure 5.14 shows the solution to Example 7(a) on a TVM solver. Alternatively, you can use the program in the Program Appendix at www.pearsonhighered.com/mwa10e.

FIGURE 5.14

Find the Newells' remaining balance after 20 years.

Answer: See page 302.

Solution Apply the formula in the preceding box, with $P = 110{,}000$, $n = 12 \cdot 30 = 360$ (the number of monthly payments in 30 years), and monthly interest rate $i = .067/12$:*

$$R = \frac{Pi}{1 - (1 + i)^{-n}} = \frac{110{,}000(.067/12)}{1 - (1 + .067/12)^{-360}} = 709.8057758.$$

Monthly payments of \$709.81 are required to amortize the loan.

(b) After 10 years, approximately how much do the Newells owe on their mortgage?

Solution You may be tempted to say that after 10 years of payments on a 30-year mortgage, the balance will be reduced by a third. However, a significant portion of each payment goes to pay interest. So much less than a third of the mortgage is paid off in the first 10 years, as we now see.

After 10 years (120 payments), the 240 remaining payments can be thought of as an annuity. The present value of this annuity is the (approximate) remaining balance on the mortgage. Hence, we use the present-value formula with $R = 709.81$, $i = .067/12$, and $n = 240$:

$$P = 709.81\left[\frac{1 - (1 + .067/12)^{-240}}{.067/12}\right] = \$93{,}717.37.$$

So the remaining balance is about \$93,717.37. The actual balance probably differs slightly from this figure because payments and interest amounts are rounded to the nearest penny. ✓7

Example 7(b) illustrates an important fact: Even though equal *payments* are made to amortize a loan, the loan *balance* does not decrease in equal steps. The method used to estimate the remaining balance in Example 7(b) works in the general case. If n payments are needed to amortize a loan and x payments have been made, then the remaining payments form an annuity of $n - x$ payments. So we apply the present-value formula with $n - x$ in place of n to obtain this result.

Remaining Balance

If a loan can be amortized by n payments of R dollars each at an interest rate i per period, then the *approximate* remaining balance B after x payments is

$$B = R\left[\frac{1 - (1 + i)^{-(n-x)}}{i}\right].$$

AMORTIZATION SCHEDULES

The remaining-balance formula is a quick and convenient way to get a reasonable estimate of the remaining balance on a loan, but it is not accurate enough for a bank or business, which must keep its books exactly. To determine the exact remaining balance after each loan payment, financial institutions normally use an **amortization**

*Mortgage rates are quoted in terms of annual interest, but it is always understood that the monthly rate is $\frac{1}{12}$ of the annual rate and that interest is compounded monthly.

schedule, which lists how much of each payment is interest, how much goes to reduce the balance, and how much is still owed after each payment.

Example 8 Beth Hill borrows $1000 for one year at 12% annual interest, compounded monthly.

(a) Find her monthly payment.

Solution Apply the amortization payment formula with $P = 1000$, $n = 12$, and monthly interest rate $i = .12/12 = .01$. Her payment is

$$R = \frac{Pi}{1 - (1 + i)^{-n}} = \frac{1000(.01)}{1 - (1 + .01)^{-12}} = \$88.85.$$

(b) After making five payments, Hill decides to pay off the remaining balance. Approximately how much must she pay?

Solution Apply the remaining-balance formula just given, with $R = 88.85$, $i = .01$, and $n - x = 12 - 5 = 7$. Her approximate remaining balance is

$$B = R\left[\frac{1 - (1 + i)^{-(n-x)}}{i}\right] = 88.85\left[\frac{1 - (1 + .01)^{-7}}{.01}\right] = \$597.80.$$

(c) Construct an amortization schedule for Hill's loan.

Solution An amortization schedule for the loan is shown in the accompanying table. It was obtained as follows: The annual interest rate is 12% compounded monthly, so the interest rate per month is $12\%/12 = 1\% = .01$. When the first payment is made, one month's interest, namely, $.01(1000) = \$10$, is owed. Subtracting this from the $88.85 payment leaves $78.85 to be applied to repayment. Hence, the principal at the end of the first payment period is $1000 - 78.85 = \$921.15$, as shown in the "payment 1" line of the table.

When payment 2 is made, one month's interest on the new balance of $921.15 is owed, namely, $.01(921.15) = \$9.21$. Continue as in the preceding paragraph to compute the entries in this line of the table. The remaining lines of the table are found in a similar fashion.

Payment Number	Amount of Payment	Interest for Period	Portion to Principal	Principal at End of Period
0	—	—	—	$1000.00
1	$88.85	$10.00	$78.85	921.15
2	88.85	9.21	79.64	841.51
3	88.85	8.42	80.43	761.08
4	88.85	7.61	81.24	679.84
5	88.85	6.80	82.05	597.79
6	88.85	5.98	82.87	514.92
7	88.85	5.15	83.70	431.22
8	88.85	4.31	84.54	346.68
9	88.85	3.47	85.38	261.30
10	88.85	2.61	86.24	175.06
11	88.85	1.75	87.10	87.96
12	88.84	.88	87.96	0

Note that Hill's remaining balance after five payments differs slightly from the estimate made in part (b).

The final payment in the amortization schedule in Example 8(c) differs from the other payments. It often happens that the last payment needed to amortize a loan must be adjusted to account for rounding earlier and to ensure that the final balance will be exactly 0.

TECHNOLOGY TIP Most Casio graphing calculators can produce amortization schedules. For other calculators, use the amortization table program in the Program Appendix (www.pearsonhighered.com/mwa10e). Spreadsheets are another useful tool for creating amortization tables. Microsoft Excel has a built-in feature for calculating monthly payments. Figure 5.15 shows an Excel amortization table for Example 6. For more details, see the *Spreadsheet Manual*, also available with this book.

	A	B	C	D	E	F
1	Prnt#	Payment	Interest	Principal	End Principal	
2	0				1000	
3	1	88.85	10.00	78.85	921.15	
4	2	88.85	9.21	79.64	841.51	
5	3	88.85	8.42	80.43	761.08	
6	4	88.85	7.61	81.24	679.84	
7	5	88.85	6.80	82.05	597.79	
8	6	88.85	5.98	82.87	514.92	
9	7	88.85	5.15	83.70	431.22	
10	8	88.85	4.31	84.54	346.68	
11	9	88.85	3.47	85.38	261.30	
12	10	88.85	2.61	86.24	175.06	
13	11	88.85	1.75	87.10	87.96	
14	12	88.85	0.88	87.97	-0.01	

FIGURE 5.15

ANNUITIES DUE

We want to find the present value of an annuity due in which 6 payments of R dollars are made at the *beginning* of each period, with interest rate i per period, as shown schematically in Figure 5.16.

```
1     2     3     4     5     6
|-----|-----|-----|-----|-----|-----
R     R     R     R     R     R
```

FIGURE 5.16

The present value is the amount needed to fund all 6 payments. Since the first payment earns no interest, R dollars are needed to fund it. Now look at the last 5 payments by themselves in Figure 5.17.

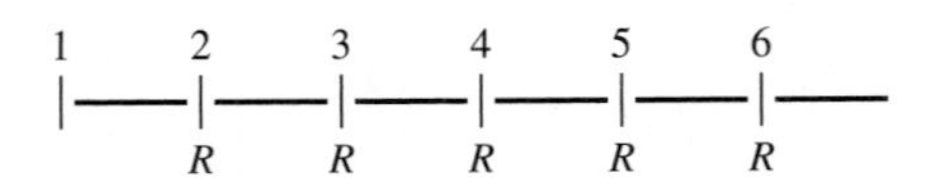

```
1     2     3     4     5     6
|-----|-----|-----|-----|-----|-----
      R     R     R     R     R
```

FIGURE 5.17

If you think of these 5 payments as being made at the end of each period, you see that they form an ordinary annuity. The money needed to fund them is the present value of this ordinary annuity. So the present value of the annuity due is given by

$$R + \text{Present value of the ordinary annuity of 5 payments}$$

$$R + R\left[\frac{1-(1+i)^{-5}}{i}\right].$$

Replacing 6 by n and 5 by $n - 1$, and using the argument just given, produces the general result that follows.

Present Value of an Annuity Due

The present value P of an annuity due is given by

$$P = R + R\left[\frac{1-(1+i)^{-(n-1)}}{i}\right],$$

P = One payment + Present value of an ordinary annuity of $n - 1$ payments

where

R is the payment at the beginning of each period,
i is the interest rate per period, and
n is the number of periods.

Example 9 In a recent Powerball lottery, a couple won about \$58.6 million. They could choose to receive 30 yearly payments of \$1,953,333, beginning immediately, or an equivalent single lump-sum payment now (which is often called the "cash value"). If the Powerball Lottery Association can earn 4.8% annual interest, how much is the cash value?

Solution The yearly payments form a 30-payment annuity due. An equivalent amount now is the present value of this annuity. Apply the present-value formula with $R = 1{,}953{,}333$, $i = .048$, and $n = 30$:

$$P = R + R\left[\frac{1-(1+i)^{-(n-1)}}{i}\right] = 1{,}953{,}333 + 1{,}953{,}333\left[\frac{1-(1+.048)^{-29}}{.048}\right]$$

$$= \$32{,}199{,}176.53.$$

The couple chose the cash value of \$32,199,176.53.

```
N=30
I%=4.8
▪PV=32199176.53
PMT=-1953333
FV=0
P/Y=1
C/Y=1
PMT:END BEGIN
```

FIGURE 5.18

TECHNOLOGY TIP Figure 5.18 shows the solution of Example 9 on a TVM solver. Since this is an annuity due, the PMT: setting at the bottom of the screen is "BEGIN".

5.4 ▶ Exercises

Unless noted otherwise, all payments and withdrawals are made at the end of the period.

1. Explain the difference between the present value of an annuity and the future value of an annuity.

Find the present value of each ordinary annuity. (See Examples 1, 2, and 4.)

2. Payments of \$890 each year for 16 years at 6% compounded annually
3. Payments of \$1400 each year for 8 years at 6% compounded annually
4. Payments of \$10,000 semiannually for 15 years at 7.5% compounded semiannually
5. Payments of \$50,000 quarterly for 10 years at 5% compounded quarterly
6. Payments of \$15,806 quarterly for 3 years at 6.8% compounded quarterly

Find the amount necessary to fund the given withdrawals. (See Examples 1 and 2.)

7. Quarterly withdrawals of \$650 for 5 years; interest rate is 4.9%, compounded quarterly.
8. Yearly withdrawals of \$1200 for 14 years; interest rate is 5.6%, compounded annually.
9. Monthly withdrawals of \$425 for 10 years; interest rate is 6.1%, compounded monthly.
10. Semiannual withdrawals of \$3500 for 7 years; interest rate is 5.2%, compounded semiannually.

Find the payment made by the ordinary annuity with the given present value. (See Example 3.)

11. \$90,000; monthly payments for 22 years; interest rate is 4.9%, compounded monthly.
12. \$45,000; monthly payments for 11 years; interest rate is 5.3%, compounded monthly.
13. \$275,000; quarterly payments for 18 years; interest rate is 6%, compounded quarterly.
14. \$330,000; quarterly payments for 30 years; interest rate is 6.1% compounded quarterly.

Find the lump sum deposited today that will yield the same total amount as payments of \$10,000 at the end of each year for 15 years at each of the given interest rates. (See Example 4 and the box following it.)

15. 3% compounded annually
16. 4% compounded annually
17. 6% compounded annually
18. What sum deposited today at 5% compounded annually for 8 years will provide the same amount as \$1000 deposited at the end of each year for 8 years at 6% compounded annually?
19. What lump sum deposited today at 8% compounded quarterly for 10 years will yield the same final amount as deposits of \$4000 at the end of each 6-month period for 10 years at 6% compounded semiannually?

Find the price a purchaser should be willing to pay for the given bond. Assume that the coupon interest is paid twice a year. (See Example 5.)

20. \$20,000 bond with coupon rate 4.5% that matures in 8 years; current interest rate is 5.9%.
21. \$15,000 bond with coupon rate 6% that matures in 4 years; current interest rate is 5%.
22. \$25,000 bond with coupon rate 7% that matures in 10 years; current interest rate is 6%.
23. \$10,000 bond with coupon rate 5.4% that matures in 12 years; current interest rate is 6.5%.
24. What does it mean to amortize a loan?

Find the payment necessary to amortize each of the given loans. (See Examples 6, 7(a), and 8(a).)

25. \$2500; 8% compounded quarterly; 6 quarterly payments
26. \$41,000; 9% compounded semiannually; 10 semiannual payments
27. \$90,000; 7% compounded annually; 12 annual payments
28. \$140,000; 12% compounded quarterly; 15 quarterly payments
29. \$7400; 8.2% compounded semiannually; 18 semiannual payments
30. \$5500; 9.5% compounded monthly; 24 monthly payments

Find the monthly house payment necessary to amortize the given loans. (See Example 7(a).)

31. \$149,560 at 6.75% for 25 years
32. \$170,892 at 7.11% for 30 years
33. \$153,762 at 5.45% for 30 years
34. \$96,511 at 8.57% for 25 years

Find the monthly payment and estimate the remaining balance (to the nearest dollar). Assume interest is on the unpaid balance. (See Examples 7 and 8.)

35. 3-year car loan for \$8500 at 6.6%; remaining balance after 2 years.

36. 2-year loan for $3000 at 8%; remaining balance after 18 months.

37. 30-year mortgage for $130,000 at 6.8%; remaining balance after 12 years.

38. 15-year mortgage for $205,000 at 5%; remaining balance after 4.5 years.

Use the amortization table in Example 8(c) to answer the questions in Exercises 39–42.

39. How much of the 5th payment is interest?

40. How much of the 10th payment is used to reduce the debt?

41. How much interest is paid in the first 5 months of the loan?

42. How much interest is paid in the last 5 months of the loan?

Find the cash value of the lottery jackpot (to the nearest dollar). Yearly jackpot payments begin immediately (26 for Mega Millions and 30 for Powerball). Assume the lottery can invest at the given interest rate. (See Example 9.)

43. Powerball: $57.6 million; 5.1% interest

44. Powerball: $207 million; 5.78% interest

45. Mega Millions: $41.6 million; 4.735% interest

46. Mega Millions: $23.4 million; 4.23% interest

Finance *Work the following applied problems.*

47. An auto stereo dealer sells a stereo system for $600 down and monthly payments of $30 for the next 3 years. If the interest rate is 1.25% per month on the unpaid balance, find
(a) the cost of the stereo system;
(b) the total amount of interest paid.

48. John Kushida buys a used car costing $6000. He agrees to make payments at the end of each monthly period for 4 years. He pays 12% interest, compounded monthly.
(a) What is the amount of each payment?
(b) Find the total amount of interest Kushida will pay.

49. A speculator agrees to pay $15,000 for a parcel of land; this amount, with interest, will be paid over 4 years with semiannual payments at an interest rate of 10% compounded semiannually. Find the amount of each payment.

Finance *A student education loan has two repayment options. The standard plan repays the loan in 10 years with equal monthly payments. The extended plan allows from 12 to 30 years to repay the loan. A student borrows $35,000 at 7.43% compounded monthly.*

50. Find the monthly payment and total interest paid under the standard plan.

51. Find the monthly payment and total interest paid under the extended plan with 20 years to pay off the loan.

52. If the lottery in Example 9 can invest at a 5.5% annual rate, what is the cash value of the jackpot (to the nearest dollar)?

Finance *Use the formula for the approximate remaining balance to work each problem. (See Examples 7(b) and 8(b).)*

53. When Teresa Flores opened her law office, she bought $14,000 worth of law books and $7200 worth of office furniture. She paid $1200 down and agreed to amortize the balance with semiannual payments for 5 years at 12% compounded semiannually.
(a) Find the amount of each payment.
(b) When her loan had been reduced below $5000, Flores received a large tax refund and decided to pay off the loan. How many payments were left at this time?

54. Kareem Adams buys a house for $285,000. He pays $60,000 down and takes out a mortgage at 6.9% on the balance. Find his monthly payment and the total amount of interest he will pay if the length of the mortgage is
(a) 15 years;
(b) 20 years;
(c) 25 years.
(d) When will half the 20-year loan be paid off?

55. The Beyes plan to purchase a home for $212,000. They will pay 20% down and finance the remainder for 30 years at 7.2% interest, compounded monthly.*
(a) How large are their monthly payments?
(b) What will be their approximate loan balance right after they have made their 96th payment?
(c) How much interest will they pay during the 7th year of the loan?
(d) If they were to increase their monthly payments by $150, how long would it take to pay off the loan?

Finance *Work each problem.*

56. Sandi Goldstein has inherited $25,000 from her grandfather's estate. She deposits the money in an account offering

*Exercises 55–56 and 59–61 were supplied by Norman Lindquist at Western Washington University.

6% interest compounded annually. She wants to make equal annual withdrawals from the account so that the money (principal and interest) lasts exactly 8 years.

(a) Find the amount of each withdrawal.

(b) Find the amount of each withdrawal if the money must last 12 years.

57. Elizabeth Bernardi and her employer contribute $400 at the end of each month to her retirement account, which earns 7% interest, compounded monthly. When she retires after 45 years, she plans to make monthly withdrawals for 30 years. If her account earns 5% interest, compounded monthly, then when she retires, what is her maximum possible monthly withdrawal (without running out of money)?

58. Jim Milliken won a $15,000 prize. On March 1, he deposited it in an account earning 5.2% interest, compounded monthly. On March 1 one year later, he begins to withdraw the same amount at the beginning of each month for a year. Assuming that he uses up all the money in the account, find the amount of each monthly withdrawal.

59. Jeni Ramirez plans to retire in 20 years. She will make 240 equal monthly contributions to her retirement account. One month after her last contribution, she will begin the first of 120 monthly withdrawals from the account. She expects to withdraw $3500 per month. How large must her monthly contributions be in order to accomplish her goal if her account is assumed to earn interest of 10.5% compounded monthly for the duration of her contributions and the 120 months of withdrawals?

60. Ron Okimura also plans to retire in 20 years. Ron will make 120 equal monthly contributions to his account. Ten years (120 months) after his last contribution, he will begin the first of 120 monthly withdrawals from the account. Ron also expects to withdraw $3500 per month. His account also earns interest of 10.5% compounded monthly for the duration of his contributions and the 120 months of withdrawals. How large must Ron's monthly contributions be in order to accomplish his goals?

61. Madeline and Nick Swenson took out a 30-year mortgage for $160,000 at 9.8% interest compounded monthly. After they had made 12 years of payments (144 payments), they decided to refinance the remaining loan balance for 25 years at 7.2% interest, compounded monthly. What will be the balance on their loan 5 years after they refinance?

62. Four years after it was issued, a 10-year $5000 bond with a 4% coupon rate is being sold for $4670. What rate of return will the buyer get? (*Hint*: Let r be the annual rate, so that the rate per 6-month period is $i = r/2$. As in Example 5, the price of the bond is the sum of the present value of an annuity and the present value of $5000 in 6 years. Set up the price equation and use technology to solve for r.)

Finance *In Exercises 63–66, prepare an amortization schedule showing the first four payments for each loan. (See Example 8(c).)*

63. An insurance firm pays $4000 for a new printer for its computer. It amortizes the loan for the printer in 4 annual payments at 8% compounded annually.

64. Large semitrailer trucks cost $72,000 each. Ace Trucking buys such a truck and agrees to pay for it by a loan that will be amortized with 9 semiannual payments at 6% compounded semiannually.

65. One retailer charges $1048 for a certain computer. A firm of tax accountants buys 8 of these computers. It makes a down payment of $1200 and agrees to amortize the balance with monthly payments at 12% compounded monthly for 4 years.

66. Joan Varozza plans to borrow $20,000 to stock her small boutique. She will repay the loan with semiannual payments for 5 years at 7% compounded semiannually.

✓ Checkpoint Answers

1. **(a)** $5989.07
   (b) $4968.20; $3865.66; $2674.91; $1388.90; $0.01
2. $20,602.37 **3.** $624.33
4. **(a)** $4832.97 **(b)** $14,717.87
5. $10,817.57 **6.** $16,472.40
7. About $61,955.22 (more than half of the original mortgage)

CHAPTER 5 ▶ Summary

KEY TERMS AND SYMBOLS

5.1 ▶ simple interest
principal
rate
time
future value (maturity value)
present value
discount and T-bills

5.2 ▶ compound interest
compound amount
compounding period
nominal rate (stated rate)

effective rate (APY)
present value

5.3 annuity
payment period
term of an annuity
ordinary annuity
future value of an ordinary annuity
sinking fund
annuity due
future value of an annuity due

5.4 present value of an ordinary annuity
amortization payments
remaining balance
amortization schedule
present value of an annuity due

CHAPTER 5 KEY CONCEPTS

A Strategy for Solving Finance Problems We have presented a lot of new formulas in this chapter. By answering the following questions, you can decide which formula to use for a particular problem.

1. Is simple or compound interest involved?

Simple interest is normally used for investments or loans of a year or less; compound interest is normally used in all other cases.

2. If simple interest is being used, what is being sought—interest amount, future value, present value, or discount?

3. If compound interest is being used, does it involve a lump sum (single payment) or an annuity (sequence of payments)?

(a) For a lump sum,

(i) is ordinary compound interest involved?

(ii) what is being sought—present value, future value, number of periods, or effective rate (APY)?

(b) For an annuity,

(i) is it an ordinary annuity (payment at the end of each period) or an annuity due (payment at the beginning of each period)?

(ii) what is being sought—present value, future value, or payment amount?

Once you have answered these questions, choose the appropriate formula and work the problem. As a final step, consider whether the answer you get makes sense. For instance, the amount of interest or the payments in an annuity should be fairly small compared with the total future value.

Key Formulas

List of Variables

r is the annual interest rate.

m is the number of periods per year.

i is the interest rate per period. $i = \frac{r}{m}$

t is the number of years.

n is the number of periods. $n = tm$

P is the principal or present value.

A is the future value of a lump sum.

S is the future value of an annuity.

R is the periodic payment in an annuity.

B is the remaining balance on a loan.

Interest

	Simple Interest	Compound Interest
Interest	$I = Prt$	$I = A - P$
Future value	$A = P(1 + rt)$	$A = P(1 + i)^n$
Present value	$P = \frac{A}{1 + rt}$	$P = \frac{A}{(1 + i)^n} = A(1 + i)^{-n}$
		Effective rate (or APY) $r_E = \left(1 + \frac{r}{m}\right)^m - 1$

Discount If D is the **discount** on a T-bill with face value P at simple interest rate r for t years, then $D = Prt$.

Annuities

Ordinary annuity — Future value: $S = R\left[\frac{(1+i)^n - 1}{i}\right] = R \cdot s_{\overline{n}|i}$

Present value: $P = R\left[\frac{1-(1+i)^{-n}}{i}\right] = R \cdot a_{\overline{n}|i}$

Annuity due — Future value: $S = R\left[\frac{(1+i)^{n+1} - 1}{i}\right] - R$

Present value: $P = R + R\left[\frac{1-(1+i)^{-(n-1)}}{i}\right]$

CHAPTER 5 REVIEW EXERCISES

Find the simple interest for the following loans.

1. $4902 at 6.5% for 11 months
2. $42,368 at 9.22% for 5 months
3. $3478 at 7.4% for 88 days
4. $2390 at 8.7% from May 3 to July 28

Find the semiannual (simple) interest payment and the total interest earned over the life of the bond.

5. $12,000 Merck Company 6-year bond at 4.75% annual interest
6. $20,000 General Electric 9-year bond at 5.25% annual interest

Find the maturity value for each simple interest loan.

7. $7750 at 6.8% for 4 months
8. $15,600 at 8.2% for 9 months
9. What is meant by the present value of an amount A?

Find the present value of the given future amounts; use simple interest.

10. $459.57 in 7 months; money earns 5.5%
11. $80,612 in 128 days; money earns 6.77%
12. A 9-month $7000 Treasury bill sells at a discount rate of 3.5%. Find the amount of the discount and the price of the T-bill.
13. A 6-month $10,000 T-bill sold at a 4% discount. Find the actual rate of interest paid by the Treasury.
14. For a given amount of money at a given interest rate for a given period greater than 1 year, does simple interest or compound interest produce more interest? Explain.

Find the compound amount and the amount of interest earned in each of the given scenarios.

15. $2800 at 6% compounded annually for 12 years
16. $57,809.34 at 4% compounded quarterly for 6 years
17. $12,903.45 at 6.37% compounded quarterly for 29 quarters
18. $4677.23 at 4.57% compounded monthly for 32 months

Find the amount of compound interest earned by each deposit.

19. $22,000 at 5.5%, compounded quarterly for 6 years
20. $2975 at 4.7%, compounded monthly for 4 years

Find the face value (to the nearest dollar) of the zero-coupon bond.

21. 5-year bond at 3.9%; price $12,366
22. 15-year bond at 5.2%; price $11,575

Find the APY corresponding to the given nominal rate.

23. 5% compounded semiannually
24. 6.5% compounded daily

Find the present value of the given amounts at the given interest rate.

25. $42,000 in 7 years; 12% compounded monthly
26. $17,650 in 4 years; 8% compounded quarterly
27. $1347.89 in 3.5 years; 6.2% compounded semiannually
28. $2388.90 in 44 months; 5.75% compounded monthly

Find the price that a purchaser should be willing to pay for these zero-coupon bonds.

29. 10-year $15,000 bond; interest at 4.4%
30. 25-year $30,000 bond; interest at 6.2%
31. What is meant by the future value of an annuity?

Find the future value of each annuity.

32. $1288 deposited at the end of each year for 14 years; money earns 7% compounded annually
33. $4000 deposited at the end of each quarter for 8 years; money earns 6% compounded quarterly
34. $233 deposited at the end of each month for 4 years; money earns 6% compounded monthly

35. \$672 deposited at the beginning of each quarter for 7 years; money earns 5% compounded quarterly

36. \$11,900 deposited at the beginning of each month for 13 months; money earns 7% compounded monthly

37. What is the purpose of a sinking fund?

Find the amount of each payment that must be made into a sinking fund to accumulate the given amounts. Assume payments are made at the end of each period.

38. \$6500; money earns 5% compounded annually; 6 annual payments

39. \$57,000; money earns 6% compounded semiannually for $8\frac{1}{2}$ years

40. \$233,188; money earns 5.7% compounded quarterly for $7\frac{3}{4}$ years

41. \$56,788; money earns 6.12% compounded monthly for $4\frac{1}{2}$ years

Find the present value of each ordinary annuity.

42. Payments of \$850 annually for 4 years at 5% compounded annually

43. Payments of \$1500 quarterly for 7 years at 8% compounded quarterly

44. Payments of \$4210 semiannually for 8 years at 5.6% compounded semiannually

45. Payments of \$877.34 monthly for 17 months at 6.4% compounded monthly

Find the amount necessary to fund the given withdrawals (which are made at the end of each period).

46. Quarterly withdrawals of \$800 for 4 years with interest rate 4.6%, compounded quarterly

47. Monthly withdrawals of \$1500 for 10 years with interest rate 5.8%, compounded monthly

48. Yearly withdrawals of \$3000 for 15 years with interest rate 6.2%, compounded annually

Find the payment for the ordinary annuity with the given present value.

49. \$150,000; monthly payments for 15 years, with interest rate 5.1%, compounded monthly

50. \$25,000; quarterly payments for 8 years, with interest rate 4.9%, compounded quarterly

51. Find the lump-sum deposit today that will produce the same total amount as payments of \$4200 at the end of each year for 12 years. The interest rate in both cases is 4.5%, compounded annually.

52. If the current interest rate is 6.5%, find the price (to the nearest dollar) that a purchaser should be willing to pay for a \$24,000 bond with coupon rate 5% that matures in 6 years.

Find the amount of the payment necessary to amortize each of the given loans.

53. \$32,000 at 8.4% compounded quarterly; 10 quarterly payments

54. \$5607 at 7.6% compounded monthly; 32 monthly payments

Find the monthly house payments for the given mortgages.

55. \$56,890 at 6.74% for 25 years

56. \$77,110 at 8.45% for 30 years

57. Find the approximate remaining balance after five payments have been made on the loan in Exercise 53.

58. Find the approximate remaining balance after payments have been made for 15 years on the mortgage in Exercise 55.

A portion of an amortization table is given here for a \$127,000 loan at 8.5% interest compounded monthly for 25 years.

Payment Number	Amount of Payment	Interest for Period	Portion to Principal	Principal at End of Period
0	—	—	—	\$127,000.00
1	\$1022.64	\$899.58	\$123.06	126,876.94
2	1022.64	898.71	123.93	126,753.01
3	1022.64	897.83	124.81	126,628.20
4	1022.64	896.95	125.69	126,502.51
5	1022.64	896.06	126.58	126,375.93
6	1022.64	895.16	127.48	126,248.45

Use the table to answer the following questions.

59. How much of the fifth payment is interest?

60. How much of the sixth payment is used to reduce the debt?

61. How much interest is paid in the first 3 months of the loan?

62. How much has the debt been reduced at the end of the first 6 months?

Finance *Work the following applied problems.*

63. In August 2008, a Pennsylvania family won a Powerball lottery prize of \$86,280,000.
   (a) If they had chosen to receive the money in 30 yearly payments, beginning immediately, what would have been the amount of each payment?
   (b) They actually chose the one-time lump-sum cash option. If the interest rate is 5.87%, approximately how much did they receive?

64. A firm of attorneys deposits \$15,000 of profit-sharing money in an account at 6% compounded semiannually for $7\frac{1}{2}$ years. Find the amount of interest earned.

65. Tom, a graduate student, is considering investing \$500 now, when he is 23, or waiting until he is 40 to invest \$500. How much more money will he have at age 65 if he invests now, given that he can earn 5% interest compounded quarterly?

66. According to a financial Web site, on June 15, 2005, Frontenac Bank of Earth City, Missouri, paid 3.94% interest, compounded quarterly, on a 2-year CD, while E*TRADE Bank of Arlington, Virginia, paid 3.93% compounded daily.* What was the effective rate for the two CDs, and which bank paid a higher effective rate?

67. Chalon Bridges deposits semiannual payments of \$3200, received in payment of a debt, in an ordinary annuity at 6.8% compounded semiannually. Find the final amount in the account and the interest earned at the end of 3.5 years.

68. Each year, a firm must set aside enough funds to provide employee retirement benefits of \$52,000 in 20 years. If the firm can invest money at 7.5% compounded monthly, what amount must be invested at the end of each month for this purpose?

69. In 1995, Oseola McCarty donated \$150,000 to the University of Southern Mississippi to establish a scholarship fund.† What is unusual about this donation is that the entire amount came from what she was able to save each month from her work as a washerwoman, a job she began in 1916 at the age of 8, when she dropped out of school. How much would Ms. McCarty have had to put into her savings account at the end of every 3 months to accumulate \$150,000 over 79 years? Assume that she received an interest rate of 5.25% compounded quarterly.

70. Some pension experts recommend that you start drawing at least 40% of your full pension as early as possible.‡ Suppose you have built up a pension with \$12,000 annual payments by working 10 years for a company when you leave to accept a better job. The company gives you the option of collecting half the full pension when you reach age 55 or the full pension at age 65. Assume an interest rate of 8% compounded annually. By age 75, how much will each plan produce? Which plan would produce the larger amount?

71. In 3 years, Ms. Nguyen must pay a pledge of \$7500 to her favorite charity. What lump sum can she deposit today at 10% compounded semiannually so that she will have enough to pay the pledge?

72. To finance the \$15,000 cost of their kitchen remodeling, the Chews will make equal payments at the end of each month for 36 months. They will pay interest at the rate of 7.2% compounded monthly. Find the amount of each payment.

73. To expand her business, the owner of a small restaurant borrows \$40,000. She will repay the money in equal payments at the end of each semiannual period for 8 years at 9% interest compounded semiannually. What payments must she make?

74. The Taggart family bought a house for \$91,000. They paid \$20,000 down and took out a 30-year mortgage for the balance at 9%.

(a) Find their monthly payment.

(b) How much of the first payment is interest?

After 180 payments, the family sold their house for \$136,000. They paid closing costs of \$3700 plus 2.5% of the sale price.

(c) Estimate the current mortgage balance at the time of the sale. (See Example 8 in Section 5.4.)

(d) Find the total closing costs.

(e) Find the amount of money they received from the sale after paying off the mortgage.

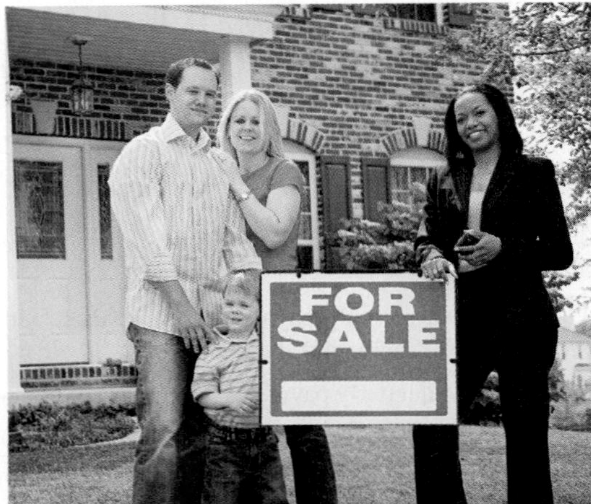

75. Over a 20-year period, the class A shares of the Davis New York Venture mutual fund increased in value at the rate of 11.2%, compounded monthly. If you had invested \$250 at the end of each month in this fund, what would have the value of your account been at the end of those 20 years?

*www.bankrate.com.

†*New York Times*, November 12, 1996, pp. A1, A22.

‡*Smart Money*, October 1994, "Pocket That Pension," p. 33.

76. Gene deposits $500 per quarter to his nest-egg account. The account earns interest of 9%, compounded quarterly. Right after Gene makes his 16th deposit, he loses his job and cannot make any deposits for the next 5 years (20 quarters). Eventually, Gene gets another job and again begins making deposits to his account. Since he missed so many deposits while he was out of work, Gene now deposits $750 per quarter. His first $750 deposit comes exactly 20 quarters after his last $500 deposit. What will be Gene's account balance right after he has made his 32nd $750 deposit?*

77. The proceeds of a $10,000 death benefit are left on deposit with an insurance company for 7 years at an annual effective interest rate of 5%.† The balance at the end of 7 years is paid to the beneficiary in 120 equal monthly payments of X, with the first payment made immediately. During the payout period, interest is credited at an annual effective interest rate of 3%. Which of the following is the correct value of X?
(a) 117 **(b)** 118 **(c)** 129 **(d)** 135 **(e)** 158

78. The *New York Times* posed a scenario with two individuals, Sue and Joe, who each have $1200 a month to spend on housing and investing. Each takes out a mortgage for $140,000. Sue gets a 30-year mortgage at a rate of 6.625%. Joe gets a 15-year mortgage at a rate of 6.25%. Whatever money is left after the mortgage payment is invested in a mutual fund with a return of 10% annually.*

(a) What annual interest rate, compounded monthly, gives an effective annual rate of 10%?
(b) What is Sue's monthly payment?
(c) If, after the payment in part (b), Sue invests the remainder of her $1200 each month in a mutual fund with the interest rate found in part (a), how much money will she have in the fund at the end of 30 years?
(d) What is Joe's monthly payment?
(e) You found in part (d) that Joe has nothing left to invest until his mortgage is paid off. If he then invests the entire $1200 monthly in a mutual fund with the interest rate in part (a), how much money will he have at the end of 30 years (i.e., after 15 years of paying the mortgage and 15 years of investing)?
(f) Who is ahead at the end of the 30 years and by how much?
(g) Discuss to what extent the difference found in part (f) is due to the different interest rates or to the different amounts of time.

79. Cathy wants to retire on $55,000 per year for her life expectancy of 20 years. She estimates that she will be able to earn interest of 9%, compounded annually, throughout her lifetime. To reach her retirement goal, Cathy will make annual contributions to her account for the next 25 years. One year after making her last contribution, she will take her first retirement check. How large must her yearly contributions be?

*Exercises 76 and 79 supplied by Norman Lindquist, Western Washington University 76.

†Problem from "Course 140 Examination, Mathematics of Compound Interest" of the Education and Examination Committee of the Society of Actuaries. Reprinted by permission of the Society of Actuaries.

**New York Times*, September 27, 1998, p. BU 10.

CASE 5
Continuous Compounding

Informally, you can think of *continuous compounding* as a process in which interest is compounded *very* frequently (for instance, every nanosecond). You will occasionally see an ad for a certificate of deposit in which interest is compounded continuously. That's pretty much a gimmick in most cases, because it produces only a few more cents than daily compounding. However, continuous compounding does play a serious role in certain financial situations, notably in the pricing of derivatives.* So let's see what is involved in continuous compounding.

*Derivatives are complicated financial instruments. But investors have learned the hard way that they can sometimes cause serious problems—as was the case in the recession that began in 2008, which was blamed in part on the misuse of derivatives.

As a general rule, the more often interest is compounded, the better off you are as an investor. (See Example 4 of Section 5.2.) But there is, alas, a limit on the amount of interest, no matter how often it is compounded. To see why this is so, suppose you have \$1 to invest. The Exponential Bank offers to pay 100% annual interest, compounded n times per year and rounded to the nearest penny. Furthermore, you may pick any value for n that you want. Can you choose n so large that your \$1 will grow to \$5 in a year? We will test several values of n in the formula for the compound amount, with $P = 1$. In this case, the annual interest rate (in decimal form) is also 1. If there are n periods in the year, the interest rate per period is $i = 1/n$. So the amount that your dollar grows to is:

$$A = P(1 + i)^n = 1\left(1 + \frac{1}{n}\right)^n.$$

A computer gives the following results for various values of n:

Interest Is Compounded . . .	n	$\left(1 + \frac{1}{n}\right)^n$
Annually	1	$\left(1 + \frac{1}{1}\right)^1 = 2$
Semiannually	2	$\left(1 + \frac{1}{2}\right)^2 = 2.25$
Quarterly	4	$\left(1 + \frac{1}{4}\right)^4 \approx 2.4414$
Monthly	12	$\left(1 + \frac{1}{12}\right)^{12} \approx 2.6130$
Daily	365	$\left(1 + \frac{1}{365}\right)^{365} \approx 2.71457$
Hourly	8760	$\left(1 + \frac{1}{8760}\right)^{8760} \approx 2.718127$
Every minute	525,600	$\left(1 + \frac{1}{525{,}600}\right)^{525{,}600} \approx 2.7182792$
Every second	31,536,000	$\left(1 + \frac{1}{31{,}53{,}600}\right)^{31{,}536{,}000} \approx 2.7182818$

Because interest is rounded to the nearest penny, the compound amount never exceeds \$2.72, no matter how big n is. (A large computer was used to develop the table, and the figures in it are accurate. If you try these computations with your calculator, however, your answers may not agree exactly with those in the table because of round-off error in the calculator.)

The preceding table suggests that as n takes larger and larger values, the corresponding values of $\left(1 + \frac{1}{n}\right)^n$ get closer and closer to a specific real number whose decimal expansion begins $2.71828\cdots$. This is indeed the case, as is shown in calculus, and the number $2.71828\cdots$ is denoted e. This fact is sometimes expressed by writing

$$\lim_{n\to\infty}\left(1+\frac{1}{n}\right)^n = e,$$

which is read "the limit of $\left(1+\frac{1}{n}\right)^n$ as n approaches infinity is e."

The preceding example is typical of what happens when interest is compounded n times per year, with larger and larger values of n. It can be shown that no matter what interest rate or principal is used, there is always an upper limit (involving the number e) on the compound amount, which is called the compound amount from **continuous compounding**.

Continuous Compounding

The compound amount A for a deposit of P dollars at an interest rate r per year compounded continuously for t years is given by

$$A = Pe^{rt}.$$

Most calculators have an e^x key for computing powers of e. See the Technology Tip on page 219 for details on using a calculator to evaluate e^x.

Example 1 Suppose $5000 is invested at an annual rate of 4% compounded continuously for 5 years. Find the compound amount.

Solution In the formula for continuous compounding, let $P = 5000$, $r = .04$, and $t = 5$. Then a calculator with an e^x key shows that

$$A = 5000e^{(.04)5} = 5000e^{.2} \approx \$6107.01.$$

You can readily verify that daily compounding would have produced a compound amount about 6¢ less.

EXERCISES

1. Find the compound amount when $20,000 is invested at 6% compounded continuously for
   (a) 2 years; **(b)** 10 years; **(c)** 20 years.
2. **(a)** Find the compound amount when $2500 is invested at 5.5%, compounded monthly for two years.
   (b) Do part (a) when the interest rate is 5.5% compounded continuously.
3. How much more does a deposit of $25,000 earn in 5 years when it is invested at 5% compounded continuously, as opposed to 5% compounded daily?
4. It can be shown that if interest is compounded continuously at nominal rate r, then the effective rate r_E is $e^r - 1$. Find the effective rate of continuously compounded interest if the nominal rate is
   (a) 4.5%; **(b)** 5.7%; **(c)** 7.4%.
5. Suppose you win a court case and the defendant has up to 8 years to pay you $5000 in damages. Assume that the defendant will wait until the last possible minute to pay you.
   (a) If you can get an interest rate of 3.75% compounded continuously, what is the present value of the $5000? [*Hint*: Solve the continuous-compounding formula for P.]
   (b) If the defendant offers you $4000 immediately to settle his debt, should you take the deal?

CHAPTER

6

Systems of Linear Equations and Matrices

The structure of some crystals can be described by a large system of linear equations. A variety of resource allocation problems involving many variables can be handled by solving an appropriate system of linear equations. Technology (such as computers and graphing calculators) is very helpful for handling large systems. Smaller ones can easily be solved by hand. See Exercises 22 and 23 on page 341.

This chapter deals with **linear** (or **first-degree**) **equations** such as

$$2x + 3y = 14 \quad \text{linear equation in two variables,}$$
$$4x - 2y + 5z = 8 \quad \text{linear equation in three variables,}$$

and so on. A **solution** of such an equation is an ordered list of numbers that, when substituted for the variables in the order they appear, produces a true statement. For instance, (1, 4) is a solution of the equation $2x + 3y = 14$ because substituting $x = 1$ and $y = 4$ produces the true statement $2(1) + 3(4) = 14$.

Many applications involve **systems of linear equations**, such as these two:

Two equations in two variables	Three equations in four variables
$\begin{aligned} 5x - 3y &= 7 \\ 2x + 4y &= 8 \end{aligned}$	$\begin{aligned} 2x + y + z &= 3 \\ x + y + z + w &= 5 \\ -4x + z + w &= 0. \end{aligned}$

A **solution of a system** is a solution that satisfies *all* the equations in the system. For instance, in the right-hand system of equations above, (1, 0, 1, 3) is a solution of all three equations (check it) and hence is a solution of the system. By contrast, (1, 1, 0, 3) is a solution of the first two equations, but not the third. Hence, (1, 1, 0, 3) is not a solution of the system.

This chapter presents methods for solving such systems, including matrix methods. Matrix algebra and other applications of matrices are also discussed.

6.1 Systems of Two Linear Equations in Two Variables

The graph of a linear equation in two variables is a straight line. The coordinates of each point on the graph represent a solution of the equation (Section 2.1). Thus, the solution of a system of two such equations is represented by the point or points where the two lines intersect. There are exactly three geometric possibilities for two lines: They intersect at a single point, or they coincide, or they are distinct and parallel. As illustrated in Figure 6.1, each of these geometric possibilities leads to a different number of solutions for the system.

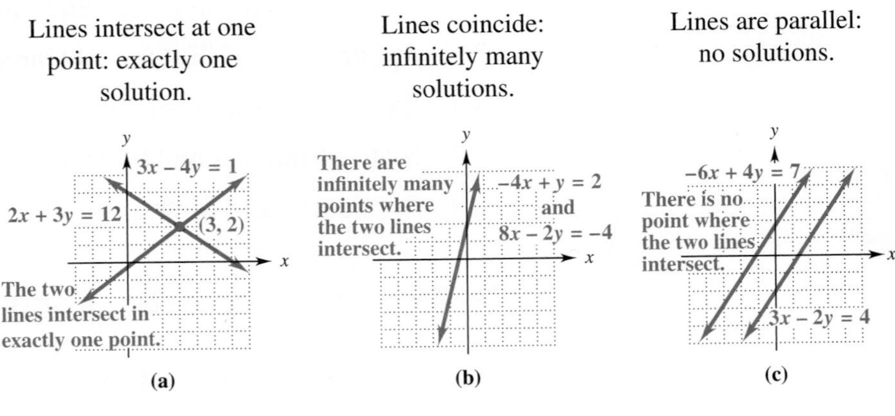

FIGURE 6.1

THE SUBSTITUTION METHOD

Example 1 illustrates the substitution method for solving a system of two equations in two variables.

Example 1 Solve the system

$$\begin{aligned} 2x - y &= 1 \\ 3x + 2y &= 4. \end{aligned}$$

Solution Begin by solving the first equation for y:

$$\begin{aligned} 2x - y &= 1 \\ -y &= -2x + 1 && \text{Subtract } 2x \text{ from both sides.} \\ y &= 2x - 1. && \text{Multiply both sides by } -1. \end{aligned}$$

Substitute this expression for y in the second equation and solve for x:

$$\begin{aligned} 3x + 2y &= 4 \\ 3x + 2(2x - 1) &= 4 && \text{Substitute } 2x - 1 \text{ for } y. \\ 3x + 4x - 2 &= 4 && \text{Multiply out the left side.} \\ 7x - 2 &= 4 && \text{Combine like terms.} \\ 7x &= 6 && \text{Add 2 to both sides.} \\ x &= 6/7. && \text{Divide both sides by 7.} \end{aligned}$$

Therefore, every solution of the system must have $x = 6/7$. To find the corresponding solution for y, substitute $x = 6/7$ in one of the two original equations and solve for y. We shall use the first equation.

$$\begin{aligned} 2x - y &= 1 \\ 2\left(\frac{6}{7}\right) - y &= 1 && \text{Substitute } 6/7 \text{ for } x. \\ \frac{12}{7} - y &= 1 && \text{Multiply out the left side.} \\ -y &= -\frac{12}{7} + 1 && \text{Subtract } 12/7 \text{ from both sides.} \\ y &= \frac{12}{7} - 1 = \frac{5}{7}. && \text{Multiply both sides by } -1. \end{aligned}$$

Hence, the solution of the original system is $x = 6/7$ and $y = 5/7$. We would have obtained the same solution if we had substituted $x = 6/7$ in the second equation of the original system, as you can easily verify. 1 ✓

✓ Checkpoint 1

Use the substitution method to solve this system:

$$\begin{aligned} x - 2y &= 3 \\ 2x + 3y &= 13. \end{aligned}$$

Answer: See page 318.

The substitution method is useful when at least one of the equations has a variable with coefficient 1. That is why we solved for y in the first equation of Example 1. If we had solved for x in the first equation or for x or y in the second, we would have had a lot more fractions to deal with.

TECHNOLOGY TIP A graphing calculator can be used to solve systems of two equations in two variables. Solve each of the given equations for y and graph them on the same screen. In Example 1, for instance we would graph

$$y_1 = 2x - 1 \quad \text{and} \quad y_2 = \frac{-3x + 4}{2}.$$

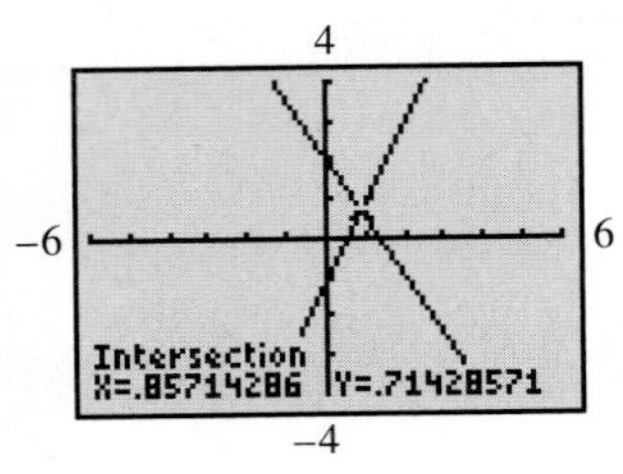

FIGURE 6.2

Use the intersection finder to determine the point where the graphs intersect (the solution of the system), as shown in Figure 6.2. Note that this is an *approximate* solution, rather than the exact solution found algebraically in Example 1.

THE ELIMINATION METHOD

The elimination method of solving systems of linear equations is often more convenient than substitution, as illustrated in the next four examples.

Example 2 Solve the system

$$5x + y = 4$$
$$3x + 2y = 1.$$

Solution Multiply the first equation by -2, so that the coefficients of y in the two equations are negatives of each other:

$$-10x - 2y = -8 \quad \text{First equation multiplied by } -2.$$
$$3x + 2y = 1.$$

Any solution of this system of equations must also be a solution of the sum of the two equations:

$$\begin{array}{rcll} -10x - 2y &=& -8 & \\ 3x + 2y &=& 1 & \\ \hline -7x &=& -7 & \text{Sum; variable } y \text{ is eliminated.} \\ x &=& 1. & \text{Divide both sides by } -7. \end{array}$$

To find the corresponding value of y, substitute $x = 1$ in one of the original equations, say the first one:

$$5x + y = 4$$
$$5(1) + y = 4 \quad \text{Substitute 1 for } x.$$
$$y = -1. \quad \text{Subtract 5 from both sides.}$$

Therefore, the solution of the original system is $(1, -1)$. ✓2

✓Checkpoint 2

Use the elimination method to solve this system:

$$x + 2y = 4$$
$$3x - 4y = -8.$$

Answer: See page 318.

Example 3 Solve the system

$$3x - 4y = 1$$
$$2x + 3y = 12.$$

Solution Multiply the first equation by 2 and the second equation by -3 to get

$$6x - 8y = 2$$
$$-6x - 9y = -36.$$

The multipliers 2 and -3 were chosen so that the coefficients of x in the two equations would be negatives of each other. Any solution of both these equations must also be a solution of their sum:

$$\begin{aligned} 6x - 8y &= 2 \\ -6x - 9y &= -36 \\ \hline -17y &= -34 \quad \text{Sum; variable } x \text{ is eliminated.} \\ y &= 2. \end{aligned}$$

To find the corresponding value of x, substitute 2 for y in either of the original equations. We choose the first equation:

$$\begin{aligned} 3x - 4(2) &= 1 \\ 3x - 8 &= 1 \\ 3x &= 9 \\ x &= 3. \end{aligned}$$

Therefore, the solution of the system is (3, 2). The graphs of both equations of the system are shown in Figure 6.1(a). They intersect at the point (3, 2), the solution of the system. 3✓

✓Checkpoint 3

Solve the system of equations

$$\begin{aligned} 3x + 2y &= -1 \\ 5x - 3y &= 11. \end{aligned}$$

Draw the graph of each equation on the same set of axes.

Answer: See page 318.

DEPENDENT AND INCONSISTENT SYSTEMS

A system of equations that has a unique solution, such as those in Example 1–3, is called an **independent system**. Now we consider systems that have many solutions or no solutions at all.

Example 4 Solve the system

$$\begin{aligned} -4x + y &= 2 \\ 8x - 2y &= -4. \end{aligned}$$

Solution If you solve each equation in the system for y, you see that the two equations are actually the same:

$$\begin{aligned} -4x + y &= 2 \\ y &= 4x + 2 \end{aligned} \qquad\qquad \begin{aligned} 8x - 2y &= -4 \\ -2y &= -8x - 4 \\ y &= 4x + 2. \end{aligned}$$

So the two equations have the same graph, as shown in Figure 6.1(b), and the system has infinitely many solutions (namely, all solutions of $y = 4x + 2$). A system such as this is said to be **dependent**.

You do not have to analyze a system as was just done in order to find out that the system is dependent. The elimination method will warn you. For instance, if you attempt to eliminate x in the original system by multiplying both sides of the first equation by 2 and adding the results to the second equation, you obtain

$$\begin{aligned} -8x + 2y &= 4 \\ 8x - 2y &= -4 \\ \hline 0 &= 0. \quad \text{Both variables are eliminated.} \end{aligned}$$

The equation "$0 = 0$" is an algebraic indication that the system is dependent and has infinitely many solutions. 4✓

✓Checkpoint 4

Solve the following system:

$$\begin{aligned} 3x - 4y &= 13 \\ 12x - 16y &= 52. \end{aligned}$$

Answer: See page 319.

Example 5 Solve the system

$$\begin{aligned} 3x - 2y &= 4 \\ -6x + 4y &= 7. \end{aligned}$$

Solution The graphs of these equations are parallel lines (each has slope 3/2), as shown in Figure 6.1(c). Therefore, the system has no solution. However, you do not need the graphs to discover this fact. If you try to solve the system algebraically by multiplying both sides of the first equation by 2 and adding the results to the second equation, you obtain

$$\begin{aligned} 6x - 4y &= 8 \\ -6x + 4y &= 7 \\ \hline 0 &= 15. \end{aligned}$$

The false statement "0 = 15" is the algebraic signal that the system has no solution. A system with no solutions is said to be **inconsistent**. 5✓

✓Checkpoint 5

Solve the system

$$\begin{aligned} x - y &= 4 \\ 2x - 2y &= 3. \end{aligned}$$

Draw the graph of each equation on the same set of axes.

Answer: See page 319.

APPLICATIONS

In many applications, the answer is the solution of a system of equations. Solving the system is the easy part—*finding* the system, however, may take some thought.

Example 6 Eight hundred people attend a basketball game, and total ticket sales are \$3102. If adult tickets are \$6 and student tickets are \$3, how many adults and how many students attended the game?

Solution Let x be the number of adults and y the number of students. Then

Number of adults + Number of students = Total attendance.

$$x + y = 800.$$

A second equation can be found by considering ticket sales:

Adult ticket sales + Student ticket sales = Total ticket sales

$$\begin{pmatrix}\text{Price}\\\text{per}\\\text{ticket}\end{pmatrix} \times \begin{pmatrix}\text{Number}\\\text{of}\\\text{adults}\end{pmatrix} + \begin{pmatrix}\text{Price}\\\text{per}\\\text{ticket}\end{pmatrix} \times \begin{pmatrix}\text{Number}\\\text{of}\\\text{students}\end{pmatrix} = 3102.$$

$$6x + 3y = 3102.$$

To find x and y we must solve this system of equations:

$$\begin{aligned} x + y &= 800 \\ 6x + 3y &= 3102. \end{aligned}$$

The system can readily be solved by hand or by using technology. See Checkpoint 6, which shows that 234 adults and 566 students attended the game. 6✓

✓Checkpoint 6

(a) Use substitution or elimination to solve the system of equations in Example 6.

(b) Use technology to solve the system. (*Hint*: Solve each equation for y, graph both equations on the same screen, and use the intersection finder.)

Answers: See page 319.

6.1 ▸ Exercises

Determine whether the given ordered list of numbers is a solution of the system of equations.

1. $(-1, 3)$

$$\begin{aligned} 2x + y &= 1 \\ -3x + 2y &= 9 \end{aligned}$$

2. $(2, -.5)$

$$\begin{aligned} .5x + 8y &= -3 \\ x + 5y &= -5 \end{aligned}$$

Use substitution to solve each system. (See Example 1.)

3. $\begin{aligned} 3x - y &= 1 \\ x + 2y &= -9 \end{aligned}$

4. $\begin{aligned} x + y &= 7 \\ x - 2y &= -5 \end{aligned}$

5. $\begin{aligned} 3x - 2y &= 4 \\ 2x + y &= -1 \end{aligned}$

6. $\begin{aligned} 5x - 3y &= -2 \\ -x - 2y &= 3 \end{aligned}$

Use elimination to solve each system. (See Examples 2–5.)

7. $\begin{aligned} x - 2y &= 5 \\ 2x + y &= 3 \end{aligned}$

8. $\begin{aligned} 3x - y &= 1 \\ -x + 2y &= 4 \end{aligned}$

9. $\begin{aligned} 2x - 2y &= 12 \\ -2x + 3y &= 10 \end{aligned}$

10. $\begin{aligned} 3x + 2y &= -4 \\ 4x - 2y &= -10 \end{aligned}$

11. $\begin{aligned} x + 3y &= -1 \\ 2x - y &= 5 \end{aligned}$

12. $\begin{aligned} 4x - 3y &= -1 \\ x + 2y &= 19 \end{aligned}$

13. $\begin{aligned} 2x + 3y &= 15 \\ 8x + 12y &= 40 \end{aligned}$

14. $\begin{aligned} 2x + 5y &= 8 \\ 6x + 15y &= 18 \end{aligned}$

15. $\begin{aligned} 2x - 8y &= 2 \\ 3x - 12y &= 3 \end{aligned}$

16. $\begin{aligned} 3x - 2y &= 4 \\ 6x - 4y &= 8 \end{aligned}$

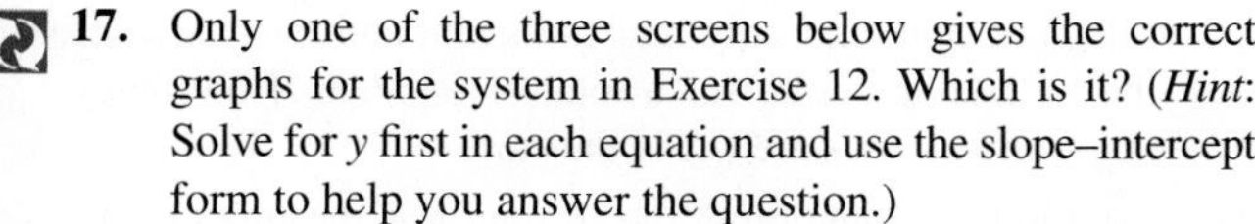

17. Only one of the three screens below gives the correct graphs for the system in Exercise 12. Which is it? (*Hint*: Solve for y first in each equation and use the slope–intercept form to help you answer the question.)

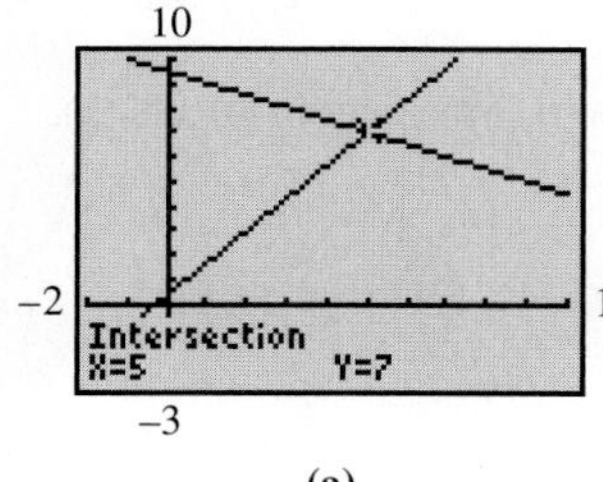

(a)

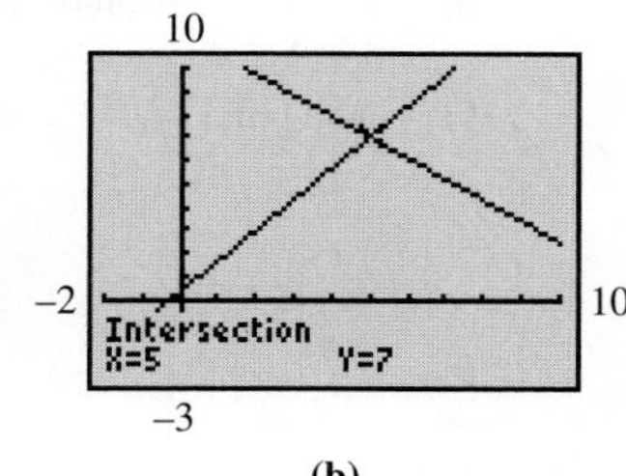

(b)

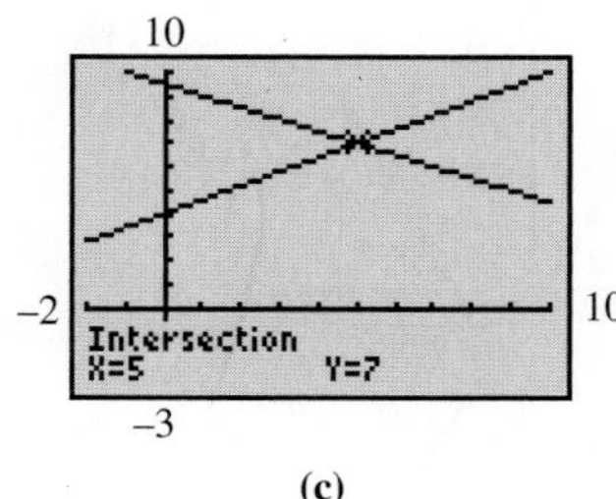

(c)

In Exercises 18 and 19, multiply both sides of each equation by a common denominator to eliminate the fractions. Then solve the system.

18. $\begin{aligned} \frac{x}{2} + \frac{y}{3} &= 8 \\ \frac{2x}{3} + \frac{3y}{2} &= 17 \end{aligned}$

19. $\begin{aligned} \frac{x}{5} + 3y &= 31 \\ 2x - \frac{y}{5} &= 8 \end{aligned}$

20. Business When Neil Simon opens a new play, he has to decide whether to open the show on Broadway or off Broadway. For example, he decided to open his play *London Suite* off Broadway. From information provided by Emanuel Azenberg, his producer, the following equations were developed:

$$\begin{aligned} 43{,}500x - y &= 1{,}295{,}000 \\ 27{,}000x - y &= 440{,}000, \end{aligned}$$

where x represents the number of weeks that the show has run and y represents the profit or loss from the show. The first equation is for Broadway, and the second equation is for off Broadway.*

(a) Solve this system of equations to determine when the profit or loss from the show will be equal for each venue. What is the amount of that profit or loss?

(b) Discuss which venue is favorable for the show.

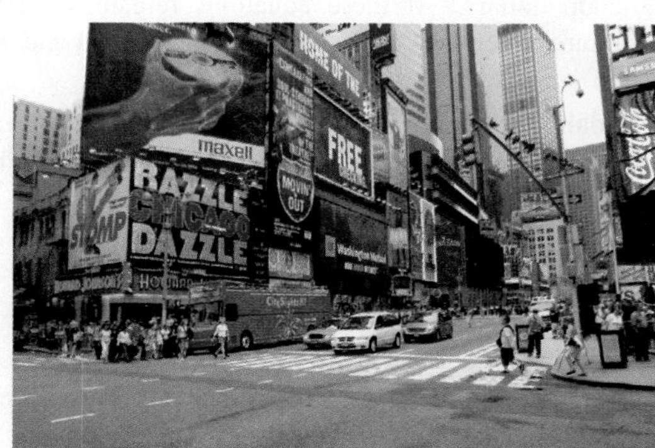

21. Social Science One of the factors that contribute to the success or failure of a particular army during war is its ability to get new troops ready for service. It is possible to analyze the rate of change in the number of troops of two hypothetical armies with the simplified model

$$\begin{aligned} \text{Rate of increase (Red Army)} &= 200{,}000 - .5r - .3b \\ \text{Rate of increase (Blue Army)} &= 350{,}000 - .5r - .7b, \end{aligned}$$

*Albert Goetz, "Basic Economics: Calculating against Theatrical Disaster," *Mathematics Teacher* 89, no. 1 (January 1996): 30–32. Reprinted with permission, ©1996 by the National Council of Teachers of Mathematics. All rights reserved.

where r is the number of soldiers in the Red Army at a given time and b is the number of soldiers in the Blue Army at the same time. The factors .5 and .7 represent each army's efficiency at bringing new soldiers into the fight.*

(a) Solve this system of equations to determine the number of soldiers in each army when the rate of increase for each is zero.

(b) Describe what might be going on in a war when the rate of increase is zero.

22. **Social Science** The population y in year x of Birmingham, AL, and Lexington, KY, is approximated by the equations

$$\begin{aligned} &\textit{Lexington:} && -2.85x + y = 202 \\ &\textit{Birmingham:} && 4.14x + 2y = 570, \end{aligned}$$

where $x = 0$ corresponds to 1980 and y is in thousands.† In what year do the two cities have the same population?

23. **Finance** On the basis of data from 1990 to 2006, the median income y in year x for men and women is approximated by the equations

$$\begin{aligned} &\textit{Men:} && -247x + 2y = 61{,}552 \\ &\textit{Women:} && -854x + 3y = 45{,}765, \end{aligned}$$

where $x = 0$ corresponds to 1990 and y is in constant 2006 dollars.‡ If these equations remain valid in the future, when will the median income of men and women be the same?

24. **Health** The death rate per 100,000 people y in year x from heart disease and cancer is approximated by the equations

$$\begin{aligned} &\textit{Heart Disease:} && 13.6x + 2y = 644 \\ &\textit{Cancer:} && 1.8x + y = 216, \end{aligned}$$

where $x = 0$ corresponds to 1990.§ If these equations remain accurate, when will the death rates for heart disease and cancer be the same?

25. **Business** In November 2008, HBO released the complete series of *The Sopranos* on DVD. According to an ad in the *New York Times*, there were 86 episodes on 33 discs. If 5 discs were soundtracks and bonus material and 6 discs each had 4 episodes, how many discs had 2 episodes and how many had 3?

26. **Business** A 200-seat theater charges \$8 for adults and \$5 for children. If all seats were filled and the total ticket income was \$1435, how many adults and how many children were in the audience?

27. **Transportation** A plane flies 3000 miles from San Francisco to Boston in 5 hours, with a tailwind all the way. The return trip on the same route, now with a headwind, takes 6 hours. Assuming both remain constant, find the speed of the plane and the speed of the wind. [*Hint*: If x is the plane's speed and y the wind speed (in mph), then the plane travels to Boston at $x + y$ mph because the plane and the wind go in the same direction; on the return trip, the plane travels at $x - y$ mph. (Why?)]

28. **Transportation** A plane flying into a headwind travels 2200 miles in 5 hours. The return flight along the same route with a tailwind takes 4 hours. Find the wind speed and the plane's speed (assuming both are constant). (See the hint for Exercise 27.)

29. **Finance** Shirley Cicero has \$16,000 invested in Boeing and GE stock. The Boeing stock currently sells for \$30 a share and the GE stock for \$70 a share. If GE stock triples in value and Boeing stock goes up 50%, her stock will be worth \$34,500. How many shares of each stock does she own?

30. **Business** An apparel shop sells skirts for \$45 and blouses for \$35. Its entire stock is worth \$51,750, but sales are slow and only half the skirts and two-thirds of the blouses are sold, for a total of \$30,600. How many skirts and blouses are left in the store?

31. (a) Find the equation of the straight line through (1, 2) and (3, 4).

(b) Find the equation of the line through (−1, 1) with slope 3.

(c) Find a point that lies on both of the lines in (a) and (b).

✓ Checkpoint Answers

1. (5, 1)
2. (0, 2)
3. (1, −2)

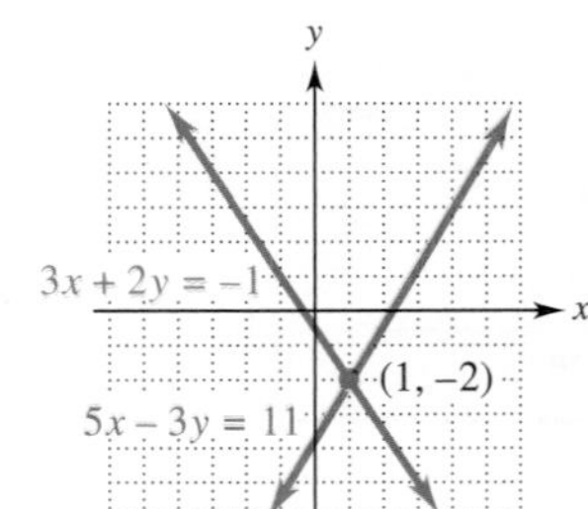

*Ian Bellamy, "Modeling War," *Journal of Peace Research* 36, no. 6 (1999): 729–739. ©1999 Sage Publications, Ltd.

†www.city-data.com

‡U.S. Census Bureau.

§U.S. Department of Health and Human Services.

4. All ordered pairs that satisfy the equation $3x - 4y = 13$ (or $12x - 16y = 52$)
5. No solution

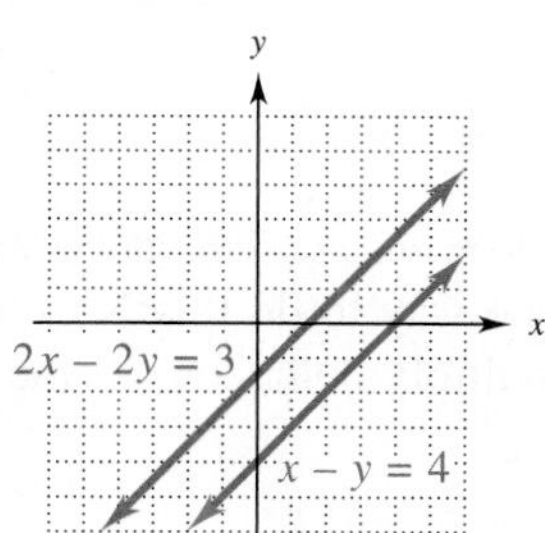

6. (a) $x = 234$ and $y = 566$
(b) 800

6.2 ▸ Larger Systems of Linear Equations

Two systems of equations are said to be **equivalent** if they have the same solutions. The basic procedure for solving a large system of equations is to transform the system into a simpler, equivalent system and then solve this simpler system.

Three operations, called **elementary operations**, are used to transform a system into an equivalent one:

1. **Interchange any two equations in the system.**
Changing the order of the equations obviously does not affect the solutions of the equations or the system.
2. **Multiply an equation in the system by a nonzero constant.**
Multiplying an equation by a nonzero constant does not change its solutions. So it does not change the solutions of the system.
3. **Replace an equation in the system by the sum of itself and a constant multiple of another equation.**
We are not saying here that the replacement equation has the same solutions as the one it replaces—it doesn't—only that the new *system* has the same solutions as the original system (that is, the systems are equivalent). No formal proof will be given, but Checkpoint 1 illustrates this fact.* 1 ✓

✓ Checkpoint 1

Verify that $x = 2$, $y = 1$ is the solution of the system

$$x - 3y = -1$$
$$3x + 2y = 8.$$

(a) Replace the second equation by the sum of itself and -3 times the first equation.

(b) What is the solution of the system in part (a)?

Answers: See page 332.

Example 1 shows how to use elementary operations on a system to eliminate certain variables and produce an equivalent system that is easily solved. The italicized statements provide an outline of the procedure. Here and later on, we use R_1 to denote the first equation in a system, R_2 the second equation, and so on.

*This operation was used in a slightly different form in Section 6.1, when the sum of an equation and a constant multiple of another equation was found in order to eliminate one of the variables. See, for instance, Examples 2 and 3 of that section.

Example 1 Solve the system

$$\begin{aligned} 2x + y - z &= 2 \\ x + 3y + 2z &= 1 \\ x + y + z &= 2. \end{aligned}$$

Solution First, *use elementary operations to produce an equivalent system in which 1 is the coefficient of x in the first equation.* One way to do this is to interchange the first two equations (another would be to multiply both sides of the first equation by $\frac{1}{2}$:

$$\begin{aligned} x + 3y + 2z &= 1 \quad \text{Interchange } R_1 \text{ and } R_2. \\ 2x + y - z &= 2 \\ x + y + z &= 2. \end{aligned}$$

Next, *use elementary operations to produce an equivalent system in which the x-term has been eliminated from the second and third equations.* To eliminate the x-term from the second equation, replace the second equation by the sum of itself and -2 times the first equation:

$$\begin{array}{rr} -2R_1 & -2x - 6y - 4z = -2 \\ R_2 & 2x + y - z = 2 \\ \hline -2R_1 + R_2 & -5y - 5z = 0 \end{array}$$

$$\begin{aligned} x + 3y + 2z &= 1 \\ -5y - 5z &= 0 \quad -2R_1 + R_2 \\ x + y + z &= 2. \end{aligned}$$

To eliminate the x-term from the third equation of this last system, replace the third equation by the sum of itself and -1 times the first equation:

$$\begin{array}{rr} -1R_1 & -x - 3y - 2z = -1 \\ R_3 & x + y + z = 2 \\ \hline -1R_1 + R_3 & -2y - z = 1 \end{array}$$

$$\begin{aligned} x + 3y + 2z &= 1 \\ -5y - 5z &= 0 \\ -2y - z &= 1. \quad -1R_1 + R_3 \end{aligned}$$

Now that x has been eliminated from all but the first equation, we ignore the first equation and work on the remaining ones. *Use elementary operations to produce an equivalent system in which 1 is the coefficient of y in the second equation.* This can be done by multiplying the second equation in the system by $-\frac{1}{5}$:

$$\begin{aligned} x + 3y + 2z &= 1 \\ y + z &= 0 \quad -\tfrac{1}{5}R_2 \\ -2y - z &= 1. \end{aligned}$$

Then *use elementary operations to obtain an equivalent system in which y has been eliminated from the third equation.* Replace the third equation by the sum of itself and 2 times the second equation:

$$\begin{aligned} x + 3y + 2z &= 1 \\ y + z &= 0 \\ z &= 1. \quad 2R_2 + R_3 \end{aligned}$$

The solution of the third equation is obvious: $z = 1$. Now work backward in the system. Substitute 1 for z in the second equation and solve for y, obtaining $y = -1$. Finally, substitute 1 for z and -1 for y in the first equation and solve for x, obtaining $x = 2$. This process is known as **back substitution**. When it is finished, we have the solution of the original system, namely, $(2, -1, 1)$. It is always wise to check the solution by substituting the values for x, y, and z in *all* equations of the original system.

The procedure used in Example 1 to eliminate variables and produce a system in which back substitution works can be carried out with any system, as summarized below. In this summary, the first variable that appears in an equation with nonzero coefficient is called the **leading variable** of that equation, and its nonzero coefficient is called the **leading coefficient**.

The Elimination Method for Solving Large Systems of Linear Equations

Use elementary operations to transform the given system into an equivalent one as follows:

1. Make the leading coefficient of the first equation 1 either by interchanging equations or by multiplying the first equation by a suitable constant.
2. Eliminate the leading variable of the first equation from each later equation by replacing the later equation by the sum of itself and a suitable multiple of the first equation.
3. Repeat Steps 1 and 2 for the second equation: Make its leading coefficient 1 and eliminate its leading variable from each later equation by replacing the later equation by the sum of itself and a suitable multiple of the second equation.
4. Repeat Steps 1 and 2 for the third equation, fourth equation, and so on, until it is not possible to go any further.

Then solve the resulting system by back substitution.

At various stages in the elimination process, you may have a choice of elementary operations that can be used. As long as the final result is a system in which back substitution can be used, the choice does not matter. To avoid unnecessary errors, choose elementary operations that minimize the amount of computation and, as far as possible, avoid complicated fractions. 2✓

✓Checkpoint 2

Use the elimination method to solve each system.

(a) $$\begin{aligned} 2x + y &= -1 \\ x + 3y &= 2 \end{aligned}$$

(b) $$\begin{aligned} 2x - y + 3z &= 2 \\ x + 2y - z &= 6 \\ -x - y + z &= -5 \end{aligned}$$

Answers: See page 332.

MATRIX METHODS

You may have noticed that the variables in a system of equations remain unchanged during the solution process. So we need to keep track of only the coefficients and the constants. For instance, consider the system in Example 1:

$$\begin{aligned} 2x + y - z &= 2 \\ x + 3y + 2z &= 1 \\ x + y + z &= 2. \end{aligned}$$

This system can be written in an abbreviated form without listing the variables as

$$\begin{bmatrix} 2 & 1 & -1 & 2 \\ 1 & 3 & 2 & 1 \\ 1 & 1 & 1 & 2 \end{bmatrix}.$$

Such a rectangular array of numbers, consisting of horizontal **rows** and vertical **columns**, is called a **matrix** (plural, **matrices**). Each number in the array is an **element**, or **entry**. To separate the constants in the last column of the matrix from the coefficients of the variables, we sometimes use a vertical line, producing the following **augmented matrix**:

$$\left[\begin{array}{ccc|c} 2 & 1 & -1 & 2 \\ 1 & 3 & 2 & 1 \\ 1 & 1 & 1 & 2 \end{array}\right].$$ 3

✓Checkpoint 3

(a) Write the augmented matrix of the following system:

$$\begin{aligned} 4x - 2y + 3z &= 4 \\ 3x + 5y + z &= -7 \\ 5x - y + 4z &= 6. \end{aligned}$$

(b) Write the system of equations associated with the following augmented matrix:

$$\left[\begin{array}{cc|c} 2 & -2 & -2 \\ 1 & 1 & 4 \\ 3 & 5 & 8 \end{array}\right].$$

Answers: See page 332.

The rows of the augmented matrix can be transformed in the same way as the equations of the system, since the matrix is just a shortened form of the system. The following **row operations** on the augmented matrix correspond to the elementary operations used on systems of equations.

> Performing any one of the following **row operations** on the augmented matrix of a system of linear equations produces the augmented matrix of an equivalent system:
>
> **1.** Interchange any two rows.
> **2.** Multiply each element of a row by a nonzero constant.
> **3.** Replace a row by the sum of itself and a constant multiple of another row of the matrix.

Row operations on a matrix are indicated by the same notation we used for elementary operations on a system of equations. For example, $2R_3 + R_1$ indicates the sum of 2 times row 3 and row 1. 4

✓Checkpoint 4

Perform the given row operations on the matrix

$$\begin{bmatrix} -1 & 5 \\ 3 & -2 \end{bmatrix}.$$

(a) Interchange R_1 and R_2.

(b) Find $2R_1$.

(c) Replace R_2 by $-3R_1 + R_2$.

(d) Replace R_1 by $2R_2 + R_1$.

Answers: See page 332.

Example 2 Use matrices to solve the system

$$\begin{aligned} x - 2y &= 6 - 4z \\ x + 13z &= 6 - y \\ -2x + 6y - z &= -10. \end{aligned}$$

Solution First, put the system in the required form, with the constants on the right side of the equals sign and the terms with variables *in the same order* in each equation on the left side of the equals sign. Then write the augmented matrix of the system.

$$\begin{aligned} x - 2y + 4z &= 6 \\ x + y + 13z &= 6 \\ -2x + 6y - z &= -10 \end{aligned} \qquad \left[\begin{array}{ccc|c} 1 & -2 & 4 & 6 \\ 1 & 1 & 13 & 6 \\ -2 & 6 & -1 & -10 \end{array}\right].$$

The matrix method is the same as the elimination method, except that row operations are used on the augmented matrix instead of elementary operations on the corresponding system of equations, as shown in the following side-by-side comparison:

Equation Method		**Matrix Method**
Replace the second equation by the sum of itself and -1 times the first equation: $$\begin{aligned} x - 2y + 4z &= 6 \\ 3y + 9z &= 0 \\ -2x + 6y - z &= -10. \end{aligned}$$	$\leftarrow -1R_1 + R_2 \rightarrow$	Replace the second row by the sum of itself and -1 times the first row: $$\left[\begin{array}{ccc\|c} 1 & -2 & 4 & 6 \\ 0 & 3 & 9 & 0 \\ -2 & 6 & -1 & -10 \end{array}\right].$$
Replace the third equation by the sum of itself and 2 times the first equation: $$\begin{aligned} x - 2y + 4z &= 6 \\ 3y + 9z &= 0 \\ 2y + 7z &= 2. \end{aligned}$$	$\leftarrow 2R_1 + R_3 \rightarrow$	Replace the third row by the sum of itself and 2 times the first row: $$\left[\begin{array}{ccc\|c} 1 & -2 & 4 & 6 \\ 0 & 3 & 9 & 0 \\ 0 & 2 & 7 & 2 \end{array}\right].$$
Multiply both sides of the second equation by $\frac{1}{3}$: $$\begin{aligned} x - 2y + 4z &= 6 \\ y + 3z &= 0 \\ 2y + 7z &= 2. \end{aligned}$$	$\leftarrow \frac{1}{3}R_2 \rightarrow$	Multiply each element of row 2 by $\frac{1}{3}$: $$\left[\begin{array}{ccc\|c} 1 & -2 & 4 & 6 \\ 0 & 1 & 3 & 0 \\ 0 & 2 & 7 & 2 \end{array}\right].$$
Replace the third equation by the sum of itself and -2 times the second equation: $$\begin{aligned} x - 2y + 4z &= 6 \\ y + 3z &= 0 \\ z &= 2. \end{aligned}$$	$\leftarrow -2R_2 + R_3 \rightarrow$	Replace the third row by the sum of itself and -2 times the second row: $$\left[\begin{array}{ccc\|c} 1 & -2 & 4 & 6 \\ 0 & 1 & 3 & 0 \\ 0 & 0 & 1 & 2 \end{array}\right].$$

Now use back substitution:

$$z = 2 \qquad \begin{aligned} y + 3(2) &= 0 \\ y &= -6 \end{aligned} \qquad \begin{aligned} x - 2(-6) + 4(2) &= 6 \\ x + 20 &= 6 \\ x &= -14. \end{aligned}$$

The solution of the system is $(-14, -6, 2)$. ✓5

✓Checkpoint 5

Complete the matrix solution of the system with this augmented matrix:

$$\left[\begin{array}{ccc|c} 1 & 1 & 1 & 2 \\ 1 & -2 & 1 & -1 \\ 0 & 3 & 1 & 5 \end{array}\right].$$

Answer: See page 332.

A matrix, such as the last one in Example 2, is said to be in **row echelon form** when

all rows consisting entirely of zeros (if any) are at the bottom;

the first nonzero entry in each row is 1 (called a *leading* 1); and

each leading 1 appears to the right of the leading 1s in any preceding rows.

When a matrix in row echelon form is the augmented matrix of a system of equations, as in Example 2, the system can readily be solved by back substitution. So the matrix solution method amounts to transforming the augmented matrix of a system of equations into a matrix in row echelon form.

THE GAUSS–JORDAN METHOD

The **Gauss–Jordan method** is a variation on the matrix elimination method used in Example 2. It replaces the back substitution used there with additional elimination of variables, as illustrated in the next example.

Example 3 Use the Gauss–Jordan method to solve the system in Example 2.

Solution First, set up the augmented matrix. Then apply the same row operations used in Example 2 until you reach the final row echelon matrix, which is

$$\left[\begin{array}{ccc|c} 1 & -2 & 4 & 6 \\ 0 & 1 & 3 & 0 \\ 0 & 0 & 1 & 2 \end{array}\right].$$

Now use additional row operations to make the entries (other than the 1s) in columns two and three into zeros. Make the first two entries in column three 0 as follows:

$$\left[\begin{array}{ccc|c} 1 & -2 & 4 & 6 \\ 0 & 1 & 0 & -6 \\ 0 & 0 & 1 & 2 \end{array}\right] \quad -3R_3 + R_2$$

$$\left[\begin{array}{ccc|c} 1 & -2 & 0 & -2 \\ 0 & 1 & 0 & -6 \\ 0 & 0 & 1 & 2 \end{array}\right] \quad -4R_3 + R_1$$

Now make the first entry in column two 0.

$$\left[\begin{array}{ccc|c} 1 & 0 & 0 & -14 \\ 0 & 1 & 0 & -6 \\ 0 & 0 & 1 & 2 \end{array}\right]. \quad 2R_2 + R_1$$

The last matrix corresponds to the system $x = -14$, $y = -6$, $z = 2$. So the solution of the original system is $(-14, -6, 2)$.

The final matrix in the Gauss–Jordan method is said to be in **reduced row echelon form**, meaning that it is in row echelon form *and* every column containing a leading 1 has zeros in all its other entries, as in Example 3. As we saw there, the solution of the system can be read directly from the reduced row echelon matrix.

In the Gauss–Jordan method, row operations may be performed in any order, but it is best to transform the matrix systematically. Either follow the procedure in Example 3 (which first puts the system into a form in which back substitution can be used and then eliminates additional variables) or work column by column from left to right, as in Checkpoint 6. ✓6

✓Checkpoint 6

Use the Gauss–Jordan method to solve the system

$$\begin{aligned} x + 2y &= 11 \\ -4x + y &= -8 \\ 5x + y &= 19 \end{aligned}$$

as instructed in parts (a)–(g). Give the shorthand notation for the required row operations in parts (b)–(f) and the new matrix in each step.

(a) Set up the augmented matrix.

(b) Get 0 in row 2, column 1.

(c) Get 0 in row 3, column 1.

(d) Get 1 in row 2, column 2.

(e) Get 0 in row 1, column 2.

(f) Get 0 in row 3, column 2.

(g) List the solution of the system.

Answers: See page 332.

TECHNOLOGY AND SYSTEMS OF EQUATIONS

When working by hand, it is usually better to use the matrix elimination method (Example 2), because errors with back substitution are less likely to occur than errors in performing the additional row operations needed in the Gauss–Jordan method. When you are using technology, however, the Gauss–Jordan method (Example 3) is more efficient: The solution can be read directly from the final reduced row echelon matrix without any "hand work," as illustrated in the next example.

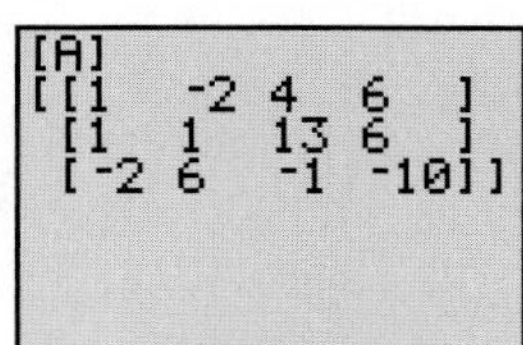

FIGURE 6.3

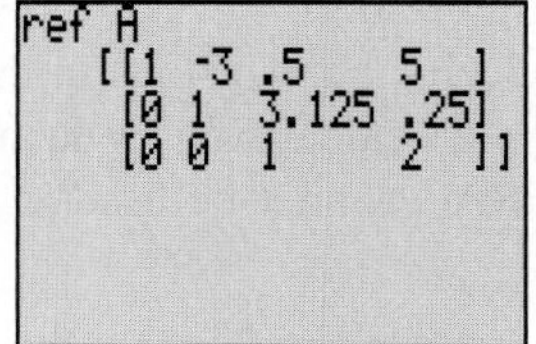

FIGURE 6.4

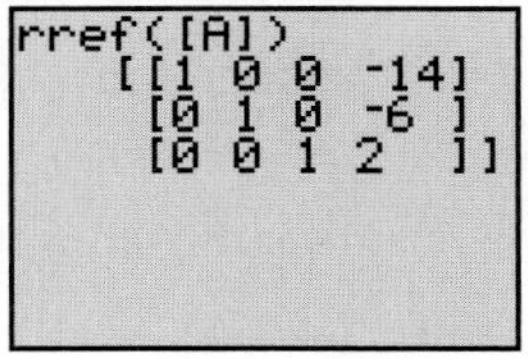

FIGURE 6.5

Example 4 Use a graphing calculator to solve the system in Example 2.

Solution There are three ways to proceed.

Matrix Elimination Method Enter the augmented matrix of the system into the calculator (Figure 6.3). Use REF (in the MATH or OPS submenu of the MATRIX menu) to put this matrix in row echelon form (Figure 6.4*). The system corresponding to this matrix is

$$\begin{aligned} x - 3y + \quad .5z &= 5 \\ y + 3.125z &= .25 \\ z &= 2. \end{aligned}$$

Because the calculator used a different sequence of row operations than was used in Example 2, it produced a different row echelon matrix (and corresponding system). However, back substitution shows that the solutions of this system are the same ones found in Example 2:

$$z = 2 \qquad \begin{aligned} y + 3.125(2) &= .25 \\ y + 6.25 &= .25 \\ y &= -6 \end{aligned} \qquad \begin{aligned} x - 3(-6) + .5(2) &= 5 \\ x + 18 + 1 &= 5 \\ x &= -14. \end{aligned}$$

As you can see, a significant amount of hand work is involved in this method.

Gauss–Jordan Method Enter the augmented matrix (Figure 6.3). Use RREF (in same menu as REF) to produce the same reduced row echelon matrix obtained in Example 3 (Figure 6.5). The solution of the system can be read directly from the matrix: $x = -14$, $y = -6$, $z = 2$.

System Solver Method This is essentially the same as the Gauss–Jordan method, with an extra step at the beginning. Call up the solver (see the Technology Tip below) and enter the number of variables and equations. The solver will display an appropriately sized matrix, with all entries 0. Change the entries so that the matrix becomes the augmented matrix of the system (Figure 6.6). Press SOLVE and the solution will appear (Figure 6.7). 7✓

On a graphing calculator, use the Gauss–Jordan method or the system solver to solve

$$\begin{aligned} x - \quad y + 5z &= -6 \\ 4x + 2y + 4z &= \quad 4 \\ x - 5y + 8z &= -17 \\ x + 3y + 2z &= \quad 5. \end{aligned}$$

Answer: See page 332.

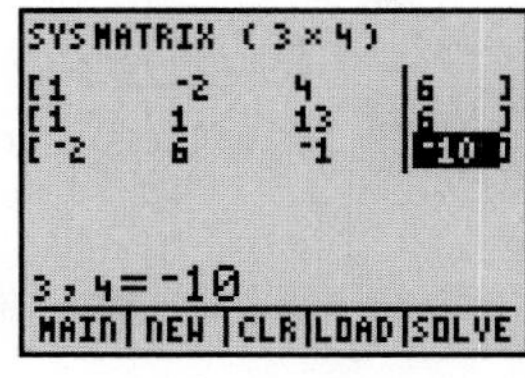

FIGURE 6.6

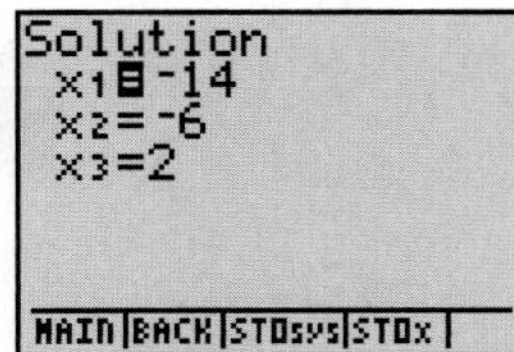

FIGURE 6.7

TECHNOLOGY TIP Many graphing calculators have system solvers. Select POLYSMLT in the APPS menu of TI-84+, SIMULT on the TI-86 keyboard, or EQUA in the Casio main menu. *Note*: The TI-86 and Casio solvers work only for systems that have the same number of variables as equations *and* have a unique solution (so they cannot be used in Checkpoint 7 or in Examples 5–8).

*A TI-86 was used for Figure 6.4.

Example 5 Use a graphing calculator to solve the system

$$\begin{aligned} x + 2y - z &= 0 \\ 3x - y + z &= 6 \\ 7x + 28y - 8z &= -5 \\ 5x + 3y - z &= 6. \end{aligned}$$

Solution Enter the augmented matrix into the calculator (Figure 6.8). Then use RREF to put this matrix in reduced row echelon form (Figure 6.9). In Figure 6.9, you must use the arrow key to scroll over to see the full decimal expansions in the right-hand column. This inconvenience can be avoided on TI calculators by using FRAC (in the MATH menu), as in Figure 6.10. (*Note*: FRAC cannot be used in a system solver.)

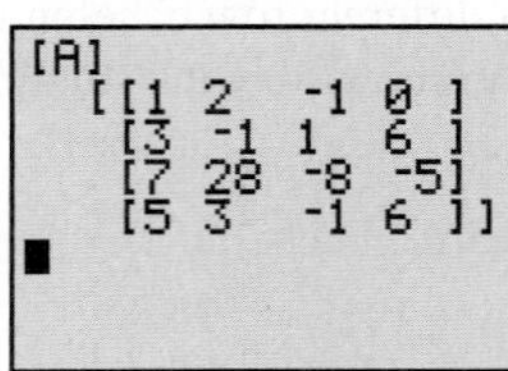

FIGURE 6.8

```
rref([A])
[[1 0 0 1.57142...
[0 1 0 -.28571...
[0 0 1 1      ...
[0 0 0 0      ...
```

FIGURE 6.9

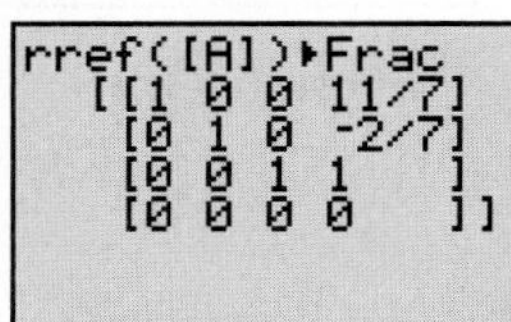

FIGURE 6.10

The answers can now be read from the last column of the matrix in Figure 6.10: $x = 11/7, y = -2/7, z = 1$. 8✓

✓Checkpoint 8

Use a graphing calculator to solve the following system:

$$\begin{aligned} x + 3y &= 4 \\ 4x + 8y &= 4 \\ 6x + 12y &= 6. \end{aligned}$$

Answer: See page 332.

INCONSISTENT AND DEPENDENT SYSTEMS

Recall that a system of equations in two variables may have exactly one solution, an infinite number of solutions, or no solutions at all. This fact was illustrated geometrically in Figure 6.1. The same thing is true for systems with three or more variables (and the same terminology is used): A system has exactly one solution (an **independent system**), an infinite number of solutions (a **dependent system**), or no solutions at all (an **inconsistent system**).

Example 6 Solve the system

$$\begin{aligned} 2x + 4y &= 4 \\ 3x + 6y &= 8 \\ 2x + y &= 7. \end{aligned}$$

Solution There are several possible ways to proceed.

Manual Method Write the augmented matrix and perform row operations to obtain a first column whose entries (from top to bottom) are 1, 0, 0:

$$\left[\begin{array}{cc|c} 2 & 4 & 4 \\ 3 & 6 & 8 \\ 2 & 1 & 7 \end{array}\right]$$

$$\begin{bmatrix} 1 & 2 & | & 2 \\ 3 & 6 & | & 8 \\ 2 & 1 & | & 7 \end{bmatrix} \quad \tfrac{1}{2}R_1$$

$$\begin{bmatrix} 1 & 2 & | & 2 \\ 0 & 0 & | & 2 \\ 2 & 1 & | & 7 \end{bmatrix}. \quad -3R_1 + R_2$$

Stop! The second row of the matrix denotes the equation $0x + 0y = 2$. Since the left side of this equation is always 0 and the right side is 2, it has no solution. Therefore, the original system has no solution.

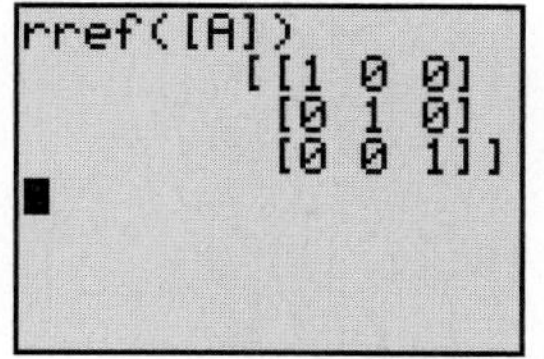

FIGURE 6.11

Calculator Method Enter the augmented matrix into a graphing calculator and use RREF to put it into reduced row echelon form, as in Figure 6.11. The last row of that matrix corresponds to $0x + 0y = 1$, which has no solution. Hence, the original system has no solution. 9 ✓

Checkpoint 9

Solve each system.

(a) $x - y = 4$
$-2x + 2y = 1$

(b) $3x - 4y = 0$
$2x + y = 0$

Answers: See page 332.

Whenever the solution process produces a row whose elements are all 0 *except* the last one, as in Example 6, the system is inconsistent and has no solutions. However, if a row with a 0 for *every* entry is produced, it corresponds to an equation such as $0x + 0y + 0z = 0$, which has infinitely many solutions. So the system may have solutions, as in the next example.

Example 7 Solve the system

$$\begin{aligned} 2x - 3y + 4z &= 6 \\ x - 2y + z &= 9 \\ y + 2z &= -12. \end{aligned}$$

Solution Use matrix elimination as far as possible, beginning with the augmented matrix of the system:

$$\begin{bmatrix} 2 & -3 & 4 & | & 6 \\ 1 & -2 & 1 & | & 9 \\ 0 & 1 & 2 & | & -12 \end{bmatrix}$$

$$\begin{bmatrix} 1 & -2 & 1 & | & 9 \\ 2 & -3 & 4 & | & 6 \\ 0 & 1 & 2 & | & -12 \end{bmatrix} \quad \text{Interchange } R_1 \text{ and } R_2.$$

$$\begin{bmatrix} 1 & -2 & 1 & | & 9 \\ 0 & 1 & 2 & | & -12 \\ 0 & 1 & 2 & | & -12 \end{bmatrix} \quad -2R_1 + R_2$$

$$\begin{bmatrix} 1 & -2 & 1 & | & 9 \\ 0 & 1 & 2 & | & -12 \\ 0 & 0 & 0 & | & 0 \end{bmatrix}. \quad -R_2 + R_3 \qquad \begin{aligned} x - 2y + z &= 9 \\ y + 2z &= -12. \end{aligned}$$

The last augmented matrix above represents the system shown to its right. Since there are only two rows in the matrix, it is not possible to continue the process. The fact that the corresponding system has one variable (namely, z) that is not the leading variable of an equation indicates a dependent system. To find its solutions, first solve the second equation for y:

$$y = -2z - 12.$$

Now substitute the result for y in the first equation and solve for x:

$$\begin{aligned} x - 2y + z &= 9 \\ x - 2(-2z - 12) + z &= 9 \\ x + 4z + 24 + z &= 9 \\ x + 5z &= -15 \\ x &= -5z - 15. \end{aligned}$$

Each choice of a value for z leads to values for x and y. For example,

$$\begin{aligned} &\text{if } z = 1, && \text{then } x = -20 && \text{and} && y = -14; \\ &\text{if } z = -6, && \text{then } x = 15 && \text{and} && y = 0; \\ &\text{if } z = 0, && \text{then } x = -15 && \text{and} && y = -12. \end{aligned}$$

There are infinitely many solutions of the original system, since z can take on infinitely many values. The solutions are all ordered triples of the form

$$(-5z - 15, -2z - 12, z),$$

where z is any real number. 10✓

✓Checkpoint 10

Use the following values of z to find additional solutions for the system of Example 7.

(a) $z = 7$

(b) $z = -14$

(c) $z = 5$

Answers: See page 332.

Since both x and y in Example 7 were expressed in terms of z, the variable z is called a **parameter**. If we had solved the system in a different way, x or y could have been the parameter. The system in Example 7 had one more variable than equations. If there are two more variables than equations, there usually will be two parameters, and so on.

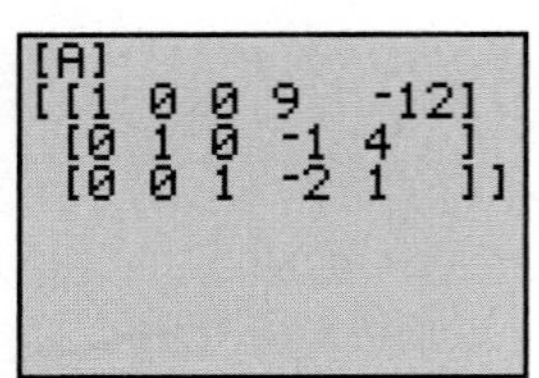

FIGURE 6.12

Example 8 Row operations were used to reduce the augmented matrix of a system of three equations in four variables (x, y, z, and w) to the reduced row echelon matrix in Figure 6.12. Solve the system.

Solution First write out the system represented by the matrix:

$$\begin{aligned} x + \qquad\qquad 9w &= -12 \\ y - \qquad w &= 4 \\ z - 2w &= 1. \end{aligned}$$

Let w be the parameter. Solve the first equation for x, the second for y, and the third for z:

$$x = -9w - 12; \qquad y = w + 4; \qquad z = 2w + 1.$$

The solutions are given by $(-9w - 12, w + 4, 2w + 1, w)$, where w is any real number.

TECHNOLOGY TIP When the system solvers on TI-86 and Casio produce an error message, the system might be inconsistent (no solutions) or dependent (infinitely many solutions). You must use REF or REFF or manual methods to solve the system.

The TI-84+ solver usually solves dependent systems directly; Figure 6.13 shows its solutions for Example 7. When the message "no solution found" is displayed, as in Figure 6.14, select RREF at the bottom of the screen to display the reduced row echelon matrix of the system. From that, you can determine the solutions (if there are any) or that no solutions are possible.

FIGURE 6.13

FIGURE 6.14

6.2 ▸ Exercises

Obtain an equivalent system by performing the stated elementary operation on the system. (See Example 1.)

1. Interchange equations 1 and 2.

$$\begin{aligned} 2x - 4y + 5z &= 1 \\ x \qquad\quad - 3z &= 2 \\ 5x - 8y + 7z &= 6 \\ 3x - 4y + 2z &= 3 \end{aligned}$$

2. Interchange equations 1 and 3.

$$\begin{aligned} 2x - 2y + z &= -6 \\ 3x + y + 2z &= 2 \\ x + y - 2z &= 0 \end{aligned}$$

3. Multiply the second equation by -1.

$$\begin{aligned} 3x \qquad + z + 2w + 18v &= 0 \\ 4x - y + \qquad w + 24v &= 0 \\ 7x - y + z + 3w + 42v &= 0 \\ 4x \qquad + z + 2w + 24v &= 0 \end{aligned}$$

4. Multiply the third equation by $1/2$.

$$\begin{aligned} x + 2y + 4z &= 3 \\ x \qquad\quad + 2z &= 0 \\ 2x + 4y + z &= 3 \end{aligned}$$

5. Replace the second equation by the sum of itself and -2 times the first equation.

$$\begin{aligned} x + y + 2z + 3w &= 1 \\ 2x + y + 3z + 4w &= 1 \\ 3x + y + 4z + 5w &= 2 \end{aligned}$$

6. Replace the third equation by the sum of itself and -1 times the first equation.

$$\begin{aligned} x + 2y + 4z &= 6 \\ y + z &= 1 \\ x + 3y + 5z &= 10 \end{aligned}$$

7. Replace the third equation by the sum of itself and -2 times the second equation.

$$\begin{aligned} x + 12y - 3z + 4w &= 10 \\ 2y + 3z + w &= 4 \\ 4y + 5z + 2w &= 1 \\ 6y - 2z - 3w &= 0 \end{aligned}$$

8. Replace the third equation by the sum of itself and 3 times the second equation.

$$\begin{aligned} 2x + 2y - 4z + w &= -5 \\ 2y + 4z - w &= 2 \\ -6y - 4z + 2w &= 6 \\ 2y + 5z - 3w &= 7 \end{aligned}$$

Solve the system by back substitution.

9. $\begin{aligned} x + 3y - 4z + 2w &= 1 \\ y + z - w &= 4 \\ 2z + 2w &= -6 \\ 3w &= 9 \end{aligned}$

10. $\begin{aligned} x + 5z + 6w &= 10 \\ y + 3z - 2w &= 4 \\ z - 4w &= -6 \\ 2w &= 4 \end{aligned}$

11. $\begin{aligned} 2x + 2y - 4z + w &= -5 \\ 3y + 4z - w &= 0 \\ 2z - 7w &= -6 \\ 5w &= 15 \end{aligned}$

12. $\begin{aligned} 3x - 2y - 4z + 2w &= 6 \\ 2y + 5z - 3w &= 7 \\ 3z + 4w &= 0 \\ 3w &= 15 \end{aligned}$

Write the augmented matrix of each of the given systems. Do not solve the systems. (See Examples 2–5.)

13. $\begin{aligned} 2x + y + z &= 3 \\ 3x - 4y + 2z &= -5 \\ x + y + z &= 2 \end{aligned}$

14. $\begin{aligned} 3x + 4y - 2z - 3w &= 0 \\ x - 3y + 7z + 4w &= 9 \\ 2x + 5z - 6w &= 0 \end{aligned}$

Write the system of equations associated with each of the given augmented matrices. Do not solve the systems.

15. $\left[\begin{array}{ccc|c} 2 & 3 & 8 & 20 \\ 1 & 4 & 6 & 12 \\ 0 & 3 & 5 & 10 \end{array}\right]$

16. $\left[\begin{array}{ccc|c} 3 & 2 & 6 & 18 \\ 2 & -2 & 5 & 7 \\ 1 & 0 & 5 & 20 \end{array}\right]$

Use the indicated row operation to transform each matrix. (See Examples 2 and 3.)

17. Interchange R_2 and R_3.

$\left[\begin{array}{ccc|c} 1 & 2 & 3 & -1 \\ 6 & 5 & 4 & 6 \\ 2 & 0 & 7 & -4 \end{array}\right]$

18. Replace R_3 by $-3R_1 + R_3$.

$\left[\begin{array}{cccc|c} 1 & 5 & 2 & 0 & -1 \\ 8 & 5 & 4 & 6 & 6 \\ 3 & 0 & 7 & 1 & -4 \end{array}\right]$

19. Replace R_2 by $2R_1 + R_2$.

$\left[\begin{array}{cccc|c} -4 & -3 & 1 & -1 & 2 \\ 8 & 2 & 5 & 0 & 6 \\ 0 & -2 & 9 & 4 & 5 \end{array}\right]$

20. Replace R_3 by $\frac{1}{4}R_3$.

$\left[\begin{array}{ccc|c} 2 & 5 & 1 & -1 \\ -4 & 0 & 4 & 6 \\ 6 & 0 & 8 & -4 \end{array}\right]$

In Exercises 21–24, the reduced row echelon form of the augmented matrix of a system of equations is given. Find the solutions of the system. (See Example 3.)

21. $\left[\begin{array}{cccc|c} 1 & 0 & 0 & 0 & 3/2 \\ 0 & 1 & 0 & 0 & 17 \\ 0 & 0 & 1 & 0 & -5 \\ 0 & 0 & 0 & 1 & 0 \end{array}\right]$

22. $\left[\begin{array}{ccccc|c} 1 & 0 & 0 & 0 & 0 & 6 \\ 0 & 1 & 0 & 0 & 0 & 4 \\ 0 & 0 & 1 & 0 & 0 & 5 \\ 0 & 0 & 0 & 0 & 1 & 2 \\ 0 & 0 & 0 & 0 & 0 & 1 \end{array}\right]$

23. $\left[\begin{array}{cccc|c} 1 & 0 & 0 & 1 & 12 \\ 0 & 1 & 0 & 2 & -3 \\ 0 & 0 & 1 & 0 & -5 \\ 0 & 0 & 0 & 0 & 0 \end{array}\right]$

24. $\left[\begin{array}{cccc|c} 1 & 0 & 0 & 0 & 7 \\ 0 & 1 & 0 & 0 & 2 \\ 0 & 0 & 1 & 0 & -5 \\ 0 & 0 & 0 & 1 & 3 \\ 0 & 0 & 0 & 0 & 0 \\ 0 & 0 & 0 & 0 & 0 \end{array}\right]$

In Exercises 25–30, perform row operations on the augmented matrix as far as necessary to determine whether the system is independent, dependent, or inconsistent. (See Examples 6–8.)

25. $\begin{aligned} x + 2y &= 0 \\ y - z &= 2 \\ x + y + z &= -2 \end{aligned}$

26. $\begin{aligned} x + 2y + z &= 0 \\ y + 2z &= 0 \\ x + y - z &= 0 \end{aligned}$

27. $\begin{aligned} x + 2y + 4z &= 6 \\ y + z &= 1 \\ x + 3y + 5z &= 10 \end{aligned}$

28. $\begin{aligned} x + y + 2z + 3w &= 1 \\ 2x + y + 3z + 4w &= 1 \\ 3x + y + 4z + 5w &= 2 \end{aligned}$

29. $\begin{aligned} a - 3b - 2c &= -3 \\ 3a + 2b - c &= 12 \\ -a - b + 4c &= 3 \end{aligned}$

30. $\begin{aligned} 2x + 2y + 2z &= 6 \\ 3x - 3y - 4z &= -1 \\ x + y + 3z &= 11 \end{aligned}$

Write the augmented matrix of the system and use the matrix method to solve the system. (See Examples 2 and 4.)

31. $\begin{aligned} -x + 3y + 2z &= 0 \\ 2x - y - z &= 3 \\ x + 2y + 3z &= 0 \end{aligned}$

32. $\begin{aligned} 3x + 7y + 9z &= 0 \\ x + 2y + 3z &= 2 \\ x + 4y + z &= 2 \end{aligned}$

33. $\begin{aligned} x - 2y + 4z &= 6 \\ x + 2y + 13z &= 6 \\ -2x + 6y - z &= -10 \end{aligned}$

34. $\begin{aligned} x - 2y + 5z &= -6 \\ x + 2y + 3z &= 0 \\ x + 3y + 2z &= 5 \end{aligned}$

35. $\begin{aligned} x + y + z &= 200 \\ x - 2y &= 0 \\ 2x + 3y + 5z &= 600 \\ 2x - y + z &= 200 \end{aligned}$

36. $\begin{aligned} 2x - y + 2z &= 3 \\ -x + 2y - z &= 0 \\ 3y - 2z &= 1 \\ x + y - z &= 1 \end{aligned}$

37. $\begin{aligned} x + y + z &= 5 \\ 2x + y - z &= 2 \\ x - y + z &= -2 \end{aligned}$

38. $\begin{aligned} 2x + y + 3z &= 9 \\ -x - y + z &= 1 \\ 3x - y + z &= 9 \end{aligned}$

Use the Gauss–Jordan method to solve each of the given systems of equations. (See Examples 3–5.)

39. $\begin{aligned} x + 2y + z &= 5 \\ 2x + y - 3z &= -2 \\ 3x + y + 4z &= -5 \end{aligned}$

40. $\begin{aligned} 3x - 2y + z &= 6 \\ 3x + y - z &= -4 \\ -x + 2y - 2z &= -8 \end{aligned}$

41. $\begin{aligned} x + 3y - 6z &= 7 \\ 2x - y + 2z &= 0 \\ x + y + 2z &= -1 \end{aligned}$

42. $\begin{aligned} x &= 1 - y \\ 2x &= z \\ 2z &= -2 - y \end{aligned}$

43. $\begin{aligned} x - 2y + 4z &= 9 \\ x + y + 13z &= 6 \\ -2x + 6y - z &= -10 \end{aligned}$

44. $\begin{aligned} x - y + 5z &= -6 \\ 3x + 3y - z &= 10 \\ x + 2y + 3z &= 5 \end{aligned}$

Solve the system by any method.

45. $\begin{aligned} x + 3y + 4z &= 14 \\ 2x - 3y + 2z &= 10 \\ 3x - y + z &= 9 \\ 4x + 2y + 5z &= 9 \end{aligned}$

46. $\begin{aligned} 4x - y + 3z &= -2 \\ 3x + 5y - z &= 15 \\ -2x + y + 4z &= 14 \\ x + 6y + 3z &= 29 \end{aligned}$

47. $\begin{aligned} x + 8y + 8z &= 8 \\ 3x - y + 3z &= 5 \\ -2x - 4y - 6z &= 5 \end{aligned}$

48. $\begin{aligned} 3x - 2y - 8z &= 1 \\ 9x - 6y - 24z &= -2 \\ x - y + z &= 1 \end{aligned}$

49. $\begin{aligned} 5x + 3y + 4z &= 19 \\ 3x - y + z &= -4 \end{aligned}$

50. $\begin{aligned} 3x + y - z &= 0 \\ 2x - y + 3z &= -7 \end{aligned}$

51. $\begin{aligned} x - 2y + z &= 5 \\ 2x + y - z &= 2 \\ -2x + 4y - 2z &= 2 \end{aligned}$

52. $\begin{aligned} 2x + 3y + z &= 9 \\ 4x + y - 3z &= -7 \\ 6x + 2y - 4z &= -8 \end{aligned}$

53. $\begin{aligned} -8x - 9y &= 11 \\ 24x + 34y &= 2 \\ 16x + 11y &= -57 \end{aligned}$

54. $\begin{aligned} 2x + y &= 7 \\ x - y &= 3 \\ x + 3y &= 4 \end{aligned}$

55. $\begin{aligned} x + 2y &= 3 \\ 2x + 3y &= 4 \\ 3x + 4y &= 5 \\ 4x + 5y &= 6 \end{aligned}$

56. $\begin{aligned} x - y &= 2 \\ x + y &= 4 \\ 2x + 3y &= 9 \\ 3x - 2y &= 6 \end{aligned}$

57. $\begin{aligned} x + y - z &= -20 \\ 2x - y + z &= 11 \end{aligned}$

58. $\begin{aligned} 4x + 3y + z &= 1 \\ -2x - y + 2z &= 0 \end{aligned}$

59. $\begin{aligned} 2x + y + 3z - 2w &= -6 \\ 4x + 3y + z - w &= -2 \\ x + y + z + w &= -5 \\ -2x - 2y - 2z + 2w &= -10 \end{aligned}$

60. $\begin{aligned} x + y + z + w &= -1 \\ -x + 4y + z - w &= 0 \\ x - 2y + z - 2w &= 11 \\ -x - 2y + z + 2w &= -9 \end{aligned}$

61. $\begin{aligned} x + 2y - z &= 3 \\ 3x + y + w &= 4 \\ 2x - y + z + w &= 2 \end{aligned}$

62. $\begin{aligned} x - 2y - z - 3w &= -3 \\ -x + y + z &= 2 \\ 4y + 3z - 6w &= -2 \end{aligned}$

63. $\begin{aligned} \frac{3}{x} - \frac{1}{y} + \frac{4}{z} &= -13 \\ \frac{1}{x} + \frac{2}{y} - \frac{1}{z} &= 12 \\ \frac{4}{x} - \frac{1}{y} + \frac{3}{z} &= -7 \end{aligned}$

[*Hint*: Let $u = 1/x$, $v = 1/y$, and $w = 1/z$, and solve the resulting system.]

64. $\begin{aligned} \frac{1}{x+1} - \frac{2}{y-3} + \frac{3}{z-2} &= 4 \\ \frac{5}{y-3} - \frac{10}{z-2} &= -5 \\ \frac{-3}{x+1} + \frac{4}{y-3} - \frac{1}{z-2} &= -2 \end{aligned}$

[*Hint*: Let $u = 1/(x+1)$, $v = 1/(y-3)$, and $w = 1/(z-2)$.]

Technology is recommended for these exercises.

65. **Social Science** The population y of the listed city in year x is approximated by the given equations, in which $x = 0$ corresponds to 1970 and y is in thousands:

Sacramento, CA: $-5.7x + y = 242.9$;
Pittsburgh, PA: $5.4x + y = 498.2$;
Oakland, CA: $-1.6x + y = 337.2$.

In what year did all three cities have the same population? What was that population?

66. **Business** The owner of a small business borrows money on three separate credit cards: x dollars on his Mastercard, y dollars on his Visa, and z dollars on his American Express card. These amounts satisfy the following equations:

$$\begin{aligned} 1.18x + 1.15y + 1.09z &= 11{,}244.25 \\ 3.54x - .55y + .27z &= 3{,}732.75 \\ .06x + .05y + .03z &= 414.75. \end{aligned}$$

How much did the owner borrow on each card?

67. **Business** A band concert is attended by x adults, y teenagers, and z preteen children. These numbers satisfied the following equations:

$$\begin{aligned} x + 1.25y + .25z &= 457.5 \\ x + .6y + .4z &= 390 \\ 3.16x + 3.48y + .4z &= 1297.2. \end{aligned}$$

How many adults, teenagers, and children were present?

Work these problems.

68. Find constants a, b, and c such that the points $(2, 3)$, $(-1, 0)$, and $(-2, 2)$ lie on the graph of the equation $y = ax^2 + bx + c$. (*Hint*: Since $(2, 3)$ is on the graph, we must have $3 = a(2^2) + b(2) + c$; that is, $4a + 2b + c = 3$. Similarly, the other two points lead to two more equations. Solve the resulting system for a, b, and c.)

69. Graph the equations in the given system. Then explain why the graphs show that the system is inconsistent.

$$\begin{aligned} 2x + 3y &= 8 \\ x - y &= 4 \\ 5x + y &= 7 \end{aligned}$$

70. Explain why a system with more variables than equations cannot have a unique solution (that is, be an independent system). (*Hint*: When you apply the elimination method to such a system, what must happen?)

✓ Checkpoint Answers

1. **(a)** The system becomes $\begin{aligned} x - 3y &= -1 \\ 11y &= 11. \end{aligned}$ **(b)** $x = 2, y = 1$
2. **(a)** $(-1, 1)$ **(b)** $(3, 1, -1)$
3. **(a)** $\left[\begin{array}{ccc|c} 4 & -2 & 3 & 4 \\ 3 & 5 & 1 & -7 \\ 5 & -1 & 4 & 6 \end{array}\right]$ **(b)** $\begin{aligned} 2x - 2y &= -2 \\ x + y &= 4 \\ 3x + 5y &= 8 \end{aligned}$
4. **(a)** $\begin{bmatrix} 3 & -2 \\ -1 & 5 \end{bmatrix}$ **(b)** $\begin{bmatrix} -2 & 10 \\ 3 & -2 \end{bmatrix}$ **(c)** $\begin{bmatrix} -1 & 5 \\ 6 & -17 \end{bmatrix}$ **(d)** $\begin{bmatrix} 5 & 1 \\ 3 & -2 \end{bmatrix}$
5. $(-1, 1, 2)$
6. **(a)** $\left[\begin{array}{cc|c} 1 & 2 & 11 \\ -4 & 1 & -8 \\ 5 & 1 & 19 \end{array}\right]$ **(b)** $4R_1 + R_2 \left[\begin{array}{cc|c} 1 & 2 & 11 \\ 0 & 9 & 36 \\ 5 & 1 & 19 \end{array}\right]$
   (c) $-5R_1 + R_3 \left[\begin{array}{cc|c} 1 & 2 & 11 \\ 0 & 9 & 36 \\ 0 & -9 & -36 \end{array}\right]$
   (d) $\frac{1}{9}R_2 \left[\begin{array}{cc|c} 1 & 2 & 11 \\ 0 & 1 & 4 \\ 0 & -9 & -36 \end{array}\right]$
   (e) $-2R_2 + R_1 \left[\begin{array}{cc|c} 1 & 0 & 3 \\ 0 & 1 & 4 \\ 0 & -9 & -36 \end{array}\right]$
   (f) $9R_2 + R_3 \left[\begin{array}{cc|c} 1 & 0 & 3 \\ 0 & 1 & 4 \\ 0 & 0 & 0 \end{array}\right]$
   (g) $(3, 4)$
7. $(1, 2, -1)$
8. $(-5, 3)$
9. **(a)** No solution **(b)** $(0, 0)$
10. **(a)** $(-50, -26, 7)$ **(b)** $(55, 16, -14)$ **(c)** $(-40, -22, 5)$

6.3 ▸ Applications of Systems of Linear Equations

There are no hard and fast rules for solving applied problems, but it is usually best to begin by identifying the unknown quantities and letting each be represented by a variable. Then look at the given data to find one or more relationships among the unknown quantities that lead to equations. If the equations are all linear, use the techniques of the preceding sections to solve the system.

Example 1 The U-Drive Rent-a-Truck Company plans to spend \$3 million on 200 new vehicles. Each van will cost \$10,000, each small truck \$15,000, and each large truck \$25,000. Past experience shows that U-Drive needs twice as many vans as small trucks. How many of each kind of vehicle can the company buy?

Solution Let x be the number of vans, y the number of small trucks, and z the number of large trucks. Then

$$\begin{pmatrix}\text{Number of}\\ \text{vans}\end{pmatrix} + \begin{pmatrix}\text{Number of}\\ \text{small trucks}\end{pmatrix} + \begin{pmatrix}\text{Number of}\\ \text{large trucks}\end{pmatrix} = \text{Total number of vehicles}$$

$$x + y + z = 200. \qquad \textbf{(1)}$$

Similarly,

$$\begin{pmatrix}\text{Cost of } x\\ \text{vans}\end{pmatrix} + \begin{pmatrix}\text{Cost of } y\\ \text{small trucks}\end{pmatrix} + \begin{pmatrix}\text{Cost of } z\\ \text{large trucks}\end{pmatrix} = \text{Total cost}$$

$$10{,}000x + 15{,}000y + 25{,}000z = 3{,}000{,}000.$$

Dividing both sides by 5000 produces the equivalent equation

$$2x + 3y + 5z = 600. \qquad \textbf{(2)}$$

Finally, the number of vans is twice the number of small trucks; that is, $x = 2y$, or equivalently,

$$x - 2y = 0. \qquad \textbf{(3)}$$

We must solve the system given by equations (1)–(3):

$$\begin{aligned} x + y + z &= 200\\ 2x + 3y + 5z &= 600\\ x - 2y \quad\;\; &= 0. \end{aligned}$$

Manual Method Form the augmented matrix and transform it into row echelon form:

$$\left[\begin{array}{ccc|c} 1 & 1 & 1 & 200\\ 2 & 3 & 5 & 600\\ 1 & -2 & 0 & 0 \end{array}\right]$$

$$\left[\begin{array}{ccc|c} 1 & 1 & 1 & 200\\ 0 & 1 & 3 & 200\\ 0 & -3 & -1 & -200 \end{array}\right] \quad \begin{array}{l} \\ -2R_1 + R_2\\ -R_1 + R_3 \end{array}$$

$$\left[\begin{array}{ccc|c} 1 & 1 & 1 & 200\\ 0 & 1 & 3 & 200\\ 0 & 0 & 8 & 400 \end{array}\right] \quad \begin{array}{l} \\ \\ 3R_2 + R_3 \end{array}$$

$$\left[\begin{array}{ccc|c} 1 & 1 & 1 & 200\\ 0 & 1 & 3 & 200\\ 0 & 0 & 1 & 50 \end{array}\right]. \quad \begin{array}{l} \\ \\ \frac{1}{8}R_3 \end{array}$$

This row echelon matrix corresponds to the system

$$\begin{aligned} x + y + z &= 200 \\ y + 3z &= 200 \\ z &= 50. \end{aligned}$$

Use back substitution to solve this system:

$$z = 50 \qquad \begin{aligned} y + 3z &= 200 \\ y + 3(50) &= 200 \\ y + 150 &= 200 \\ y &= 50 \end{aligned} \qquad \begin{aligned} x + y + z &= 200 \\ x + 50 + 50 &= 200 \\ x + 100 &= 200 \\ x &= 100. \end{aligned}$$

Therefore, U-Drive should buy 100 vans, 50 small trucks, and 50 large trucks.

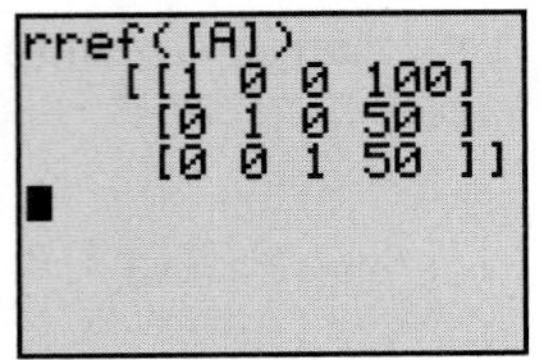

FIGURE 6.15

Calculator Method Enter the augmented matrix of the system into the calculator. Use RREF to change it into reduced row echelon form, as in Figure 6.15. The answers may be read directly from the matrix: $x = 100$ vans, $y = 50$ small trucks, and $z = 50$ large trucks. 1 ✓

Checkpoint 1

In Example 1, suppose that U-Drive can spend only \$2 million on 150 new vehicles and the company needs three times as many vans as small trucks. Write a system of equations to express these conditions.

Answer: See page 342.

Example 2 Ellen McGillicuddy plans to invest a total of \$100,000 in a money market account, a bond fund, an international stock fund, and a domestic stock fund. She wants 60% of her investment to be conservative (money market and bonds). She wants the amount in international stocks to be one-fourth of the amount in domestic stocks. Finally, she needs an annual return of \$4000. Assuming she gets annual returns of 2.5% on the money market account, 3.5% on the bond fund, 5% on the international stock fund, and 6% on the domestic stock fund, how much should she put in each investment?

Solution Let x be the amount invested in the money market account, y the amount in the bond fund, z the amount in the international stock fund, and w the amount in the domestic stock fund. Then

$$x + y + z + w = \text{total amount invested} = 100{,}000. \tag{4}$$

Use her annual return to get a second equation:

$$\begin{pmatrix}\text{2.5\% return}\\ \text{on money}\\ \text{market}\end{pmatrix} + \begin{pmatrix}\text{3.5\% return}\\ \text{on bond}\\ \text{fund}\end{pmatrix} + \begin{pmatrix}\text{5\% return on}\\ \text{international}\\ \text{stock fund}\end{pmatrix} + \begin{pmatrix}\text{6\% return}\\ \text{on domestic}\\ \text{stock fund}\end{pmatrix} = 4000$$

$$.025x + .035y + .05z + .06w = 4000. \tag{5}$$

Since Ellen wants the amount in international stocks to be one-fourth of the amount in domestic stocks, we have

$$z = \frac{1}{4}w, \qquad \text{or equivalently,} \qquad z - .25w = 0. \tag{6}$$

Finally, the amount in conservative investments is $x + y$, and this quantity should be equal to 60% of \$100,000—that is,

$$x + y = 60{,}000. \tag{7}$$

Now solve the system given by equations (4)–(7):

$$\begin{aligned} x + \quad y + \quad z + \quad w &= 100{,}000 \\ .025x + .035y + .05z + .06w &= 4{,}000 \\ z - .25w &= 0 \\ x + \quad y \qquad\qquad &= 60{,}000. \end{aligned}$$

Manual Method Write the augmented matrix of the system and transform it into row echelon form, as in Checkpoint 2. ✓2

✓Checkpoint 2

(a) Write the augmented matrix of the system given by equation (4)–(7) of Example 2.

(b) List a sequence of row operations that transforms this matrix into row echelon form.

(c) Display the final row echelon form matrix.

Answers: See page 342.

The final matrix in Checkpoint 2 represented the system

$$\begin{aligned} x + y + z + w &= 100{,}000 \\ y + 2.5z + 3.5w &= 150{,}000 \\ z - .25w &= 0 \\ w &= 32{,}000. \end{aligned}$$

Back substitution shows that

$$w = 32{,}000 \qquad z = .25w = .25(32{,}000) = 8000$$

$$y = -2.5z - 3.5w + 150{,}000 = -2.5(8000) - 3.5(32{,}000) + 150{,}000 = 18{,}000$$

$$x = -y - z - w + 100{,}000 = -18{,}000 - 8000 - 32{,}000 + 100{,}000 = 42{,}000.$$

Therefore, Ellen should put \$42,000 in the money market account, \$18,000 in the bond fund, \$8000 in the international stock fund, and \$32,000 in the domestic stock fund.

```
rref([A])
[[1 0 0 0 42000...
 [0 1 0 0 18000...
 [0 0 1 0 8000  ...
 [0 0 0 1 32000...
```

FIGURE 6.16

Calculator Method Enter the augmented matrix of the system into the calculator. Use RREF to change it into reduced row echelon form, as in Figure 6.16. The answers can easily be read from this matrix: $x = 42{,}000$, $y = 18{,}000$, $z = 8000$, and $w = 32{,}000$.

Example 3 An animal feed is to be made from corn, soybeans, and cottonseed. Determine how many units of each ingredient are needed to make a feed that supplies 1800 units of fiber, 2800 units of fat, and 2200 units of protein, given that 1 unit of each ingredient provides the numbers of units shown in the table below. The table states, for example, that a unit of corn provides 10 units of fiber, 30 units of fat, and 20 units of protein.

	Corn	Soybeans	Cottonseed	Totals
Units of Fiber	10	20	30	1800
Units of Fat	30	20	40	2800
Units of Protein	20	40	25	2200

Solution Let x represent the required number of units of corn, y the number of units of soybeans, and z the number of units of cottonseed. Since the total amount of fiber is to be 1800, we have

$$10x + 20y + 30z = 1800.$$

The feed must supply 2800 units of fat, so

$$30x + 20y + 40z = 2800.$$

Finally, since 2200 units of protein are required, we have

$$20x + 40y + 25z = 2200.$$

Thus, we must solve this system of equations:

$$\begin{aligned} 10x + 20y + 30z &= 1800 \\ 30x + 20y + 40z &= 2800 \\ 20x + 40y + 25z &= 2200. \end{aligned} \qquad \textbf{(8)}$$

Now solve the system, either manually or with technology.

Manual Method Write the augmented matrix and use row operations to transform it into row echelon form, as in Checkpoint 3. The resulting matrix represents the following system: 3✓

$$\begin{aligned} x + 2y + 3z &= 180 \\ y + \frac{5}{4}z &= 65 \\ z &= 40. \end{aligned} \qquad \textbf{(9)}$$

(a) Write the augmented matrix of system (8) of Example 3.

(b) List a sequence of row operations that transforms this matrix into row echelon form.

(c) Display the final row echelon form matrix.

Answers: See page 342.

Back substitution now shows that

$$z = 40, \quad y = 65 - \frac{5}{4}(40) = 15, \quad \text{and} \quad x = 180 - 2(15) - 3(40) = 30.$$

Thus, the feed should contain 30 units of corn, 15 of soybeans, and 40 of cottonseed.

Calculator Method Enter the augmented matrix of the system into the calculator (top of Figure 6.17). Use RREF to transform it into reduced row echelon form (bottom of Figure 6.17), which shows that $x = 30$, $y = 15$, and $z = 40$.

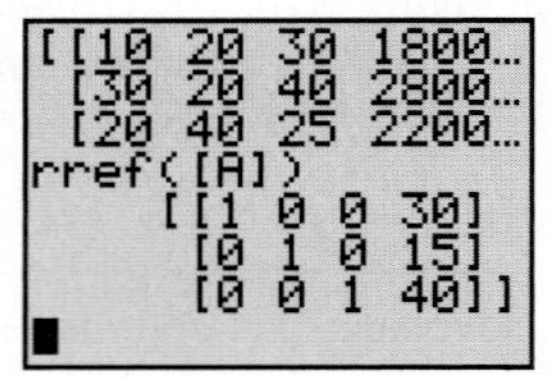

FIGURE 6.17

Example 4 The concentrations (in parts per million) of carbon dioxide (a greenhouse gas) have been measured at Mauna Loa, Hawaii, since 1959. The concentrations are known to have increased quadratically. The following table lists readings for three years*:

Year	1964	1984	2004
Carbon Dioxide	319	344	377

(a) Use the given data to construct a quadratic function that gives the concentration in year x.

Solution Let $x = 0$ correspond to 1959. Then the table is represented by the data points (5, 319), (25, 344), and (45, 377). We must find a function of the form

$$f(x) = ax^2 + bx + c$$

*C. D. Keeling and T. P. Whorf, Scripps Institution of Oceanography.

whose graph contains these three points. If (5, 319) is to be on the graph, we must have $f(5) = 319$; that is,

$$a(5^2) + b(5) + c = 319$$
$$25a + 5b + c = 319.$$

The other two points lead to these equations:

$$f(25) = 344 \qquad\qquad f(45) = 377$$
$$a(25^2) + b(25) + c = 344 \qquad\qquad a(45^2) + b(45) + c = 377$$
$$625a + 25b + c = 344 \qquad\qquad 2025a + 45b + c = 377.$$

Now work by hand or use technology to solve the following system:

$$25a + 5b + c = 319$$
$$625a + 25b + c = 344$$
$$2025a + 45b + c = 377.$$

```
[[25   5  1 319...
 [625  25 1 344...
 [2025 45 1 377...
rref([A])
    [[1 0 0 .01]
     [0 1 0 .95]
     [0 0 1 314]]
```

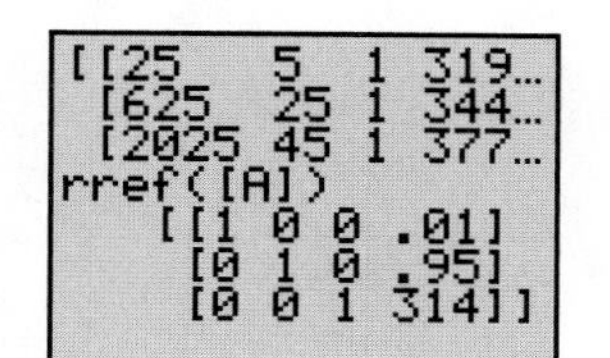

FIGURE 6.18

The reduced row echelon form of the augmented matrix in Figure 6.18 shows that the solution is $a = .01$, $b = .95$, and $c = 314$. So the function is

$$f(x) = .01x^2 + .95x + 314.$$

(b) Use this model to estimate the carbon dioxide concentrations in 2010 and 2014.

Solution The year 2010 corresponds to $x = 51$, so the concentration is

$$f(51) = .01(51^2) + .95(51) + 314 = 388.46.$$

Similarly, the concentration in 2014 ($x = 55$) is

$$f(55) = .01(55^2) + .95(55) + 314 = 396.5.$$

Example 5 Kelly Karpet Kleaners sells rug-cleaning machines. The EZ model weighs 10 pounds and comes in a 10-cubic-foot box. The compact model weighs 20 pounds and comes in an 8-cubic-foot box. The commercial model weighs 60 pounds and comes in a 28-cubic-foot box. Each of Kelly's delivery vans has 248 cubic feet of space and can hold a maximum of 440 pounds. In order for a van to be fully loaded, how many of each model should it carry?

Solution Let x be the number of EZ, y the number of compact, and z the number of commercial models carried by a van. Then we can summarize the given information in this table.

Model	Number	Weight	Volume
EZ	x	10	10
Compact	y	20	8
Commercial	z	60	28
Total for a load		440	248

Since a fully loaded van can carry 440 pounds and 248 cubic feet, we must solve this system of equations:

$$\begin{aligned} 10x + 20y + 60z &= 440 \quad \text{Weight equation} \\ 10x + 8y + 28z &= 248. \quad \text{Volume equation} \end{aligned}$$

The augmented matrix of the system is

$$\left[\begin{array}{ccc|c} 10 & 20 & 60 & 440 \\ 10 & 8 & 28 & 248 \end{array}\right].$$

FIGURE 6.19

If you are working by hand, transform the matrix into row echelon form and use back substitution. If you are using a graphing calculator, use RREF to transform the matrix into reduced row echelon form, as in Figure 6.19 (which also uses FRAC in the MATH menu to eliminate long decimals).

The system corresponding to Figure 6.19 is

$$\begin{aligned} x + \quad \frac{2}{3}z &= 12 \\ y + \frac{8}{3}z &= 16, \end{aligned}$$

which is easily solved: $x = 12 - \frac{2}{3}z$ and $y = 16 - \frac{8}{3}z$.

Hence, all solutions of the system are given by $\left(12 - \frac{2}{3}z, 16 - \frac{8}{3}z, z\right)$. The only solutions that apply in this situation, however, are those given by $z = 0, 3$, and 6, because all other values of z lead to fractions or negative numbers. (You can't deliver part of a box or a negative number of boxes). Hence, there are three ways to have a fully loaded van:

Solution	Van Load
(12, 16, 0)	12 EZ, 16 compact, 0 commercial
(10, 8, 3)	10 EZ, 8 compact, 3 commercial
(8, 0, 6)	8 EZ, 0 compact, 6 commercial

6.3 ▶ Exercises

Use systems of equations to work these applied problems. (See Examples 1–5.)

1. **Business** The U-Drive Company in Example 1 learns that each van now costs \$13,000 and each small truck \$18,000. The company decides to buy only 182 new vehicles. How many of each kind should it buy?

2. **Finance** Suppose that Ellen McGillicuddy in Example 2 finds that her annual return on the international stock fund will be only 4%. Now how much should she put in each investment?

3. **Health** To meet consumer demand, the animal feed in Example 3 now supplies only 2400 units of fat. How many units of corn, soybeans, and cottonseed are now needed?

4. **Business** Suppose that Kelly Karpet Kleaners in Example 5 finds a way to pack the EZ model in an 8-cubic-foot box. Now how many of each model should a fully loaded van carry?

5. **Business** Matt took clothes to the cleaners three times last month. First, he brought 3 shirts and 1 pair of slacks and paid \$10.96. Then he brought 7 shirts, 2 pairs of

slacks, and a sports coat and paid \$30.40. Finally, he brought 4 shirts and 1 sports coat and paid \$14.45. How much was he charged for each shirt, each pair of slacks, and each sports coat?

6. **Business** A minor league baseball park has 7000 seats. Box seats cost \$6, grandstand seats cost \$4, and bleacher seats cost \$2. When all seats are sold, the revenue is \$26,400. If the number of box seats is one-third the number of bleacher seats, how many seats of each type are there?

7. **Business** Tickets to a band concert cost \$5 for adults, \$3 for teenagers, and \$2 for preteens. There were 570 people at the concert, and total ticket receipts were \$1950. Three-fourths as many teenagers as preteens attended. How many adults, teenagers, and preteens attended?

8. **Business** Shipping charges at an online bookstore are \$4 for one book, \$6 for two books, and \$7 for three to five books. Last week, there were 6400 orders of five or fewer books, and total shipping charges for these orders were \$33,600. The number of shipments with \$7 charges was 1000 less than the number with \$6 charges. How many shipments were made in each category (one book, two books, three-to-five books)?

Work the following problems by writing and solving a system of equations. (See Examples 1–5.)

9. **Finance** An investor wants to invest \$30,000 in corporate bonds that are rated AAA, A, and B. The lower rated ones pay higher interest, but pose a higher risk as well. The average yield is 5% on AAA bonds, 6% on A bonds, and 10% on B bonds. Being conservative, the investor wants to have twice as much in AAA bonds as in B bonds. How much should she invest in each type of bond to have an interest income of \$2000?

10. **Finance** Kate borrows \$10,000. Some is from her friend at 8% annual interest, twice as much as that from her bank at 9%, and the remainder from her insurance company at 5%. She pays a total of \$830 in interest for the first year. How much did she borrow from each source?

11. **Business** Pretzels cost \$3 per pound, dried fruit \$4 per pound, and nuts \$8 per pound. How many pounds of each should be used to produce 140 pounds of trail mix costing \$6 per pound in which there are twice as many pretzels (by weight) as dried fruit?

12. **Business** An auto manufacturer sends cars from two plants, I and II, to dealerships A and B, located in a midwestern city. Plant I has a total of 28 cars to send, and plant II has 8. Dealer A needs 20 cars, and dealer B needs 16. Transportation costs based on the distance of each dealership from each plant are \$220 from I to A, \$300 from I to B, \$400 from II to A, and \$180 from II to B. The manufacturer wants to limit transportation costs to \$10,640. How many cars should be sent from each plant to each of the two dealerships?

13. **Physical Science** The stopping distance for a car traveling 25 mph is 61.7 feet, and for a car traveling 35 mph it is 106 feet.* The stopping distance in feet can be described by the equation $y = ax^2 + bx$, where x is the speed in mph.
    (a) Find the values of a and b.
    (b) Use your answers from part (a) to find the stopping distance for a car traveling 55 mph.

14. **Natural Science** An animal breeder can buy four types of tiger food. Each case of Brand A contains 25 units of fiber, 30 units of protein, and 30 units of fat. Each case of Brand B contains 50 units of fiber, 30 units of protein, and 20 units of fat. Each case of Brand C contains 75 units of fiber, 30 units of protein, and 20 units of fat. Each case of Brand D contains 100 units of fiber, 60 units of protein, and 30 units of fat. How many cases of each brand should the breeder mix together to obtain a food that provides 1200 units of fiber, 600 units of protein, and 400 units of fat?

15. **Finance** An investor plans to invest \$70,000 in a mutual fund, corporate bonds, and a fast-food franchise. She plans to put twice as much in bonds as in the mutual fund. On the basis of past performance, she expects the mutual fund to pay a 2% dividend, the bonds 10%, and the franchise 6%. She would like a dividend income of \$4800. How much should she put in each of three investments?

16. **Business** According to data from a Texas agricultural report, the amounts of nitrogen (lb/acre), phosphate (lb/acre), and labor (hr/acre) needed to grow honeydews, yellow onions, and lettuce are given by the following table†:

	Honeydews	Yellow Onions	Lettuce
Nitrogen	120	150	180
Phosphate	180	80	80
Labor	4.97	4.45	4.65

    (a) If a farmer has 220 acres, 29,100 pounds of nitrogen, 32,600 pounds of phosphate, and 480 hours of labor, can he use all of his resources completely? If so, how many acres should he allot for each crop?
    (b) Suppose everything is the same as in part (a), except that 1061 hours of labor are available. Is it possible to use all of his resources completely? If so, how many acres should he allot for each crop?

17. **Health** Computer-aided tomography (CAT) scanners take X-rays of a part of the body from different directions and put the information together to create a picture of a cross-section

**National Traffic Safety Institute Student Workbook,* 1993, p. 7.

†Miguel Paredes, Mohammad Fatehi, and Richard Hinthorn, "The Transformation of an Inconsistent Linear System into a Consistent System," *AMATYC Review*, 13, no. 2 (spring 1992).

of the body.* The amount by which the energy of the X-ray decreases, measured in linear-attenuation units, tells whether the X-ray has passed through healthy tissue, tumorous tissue, or bone, on the basis of the following table:

Type of Tissue	Linear-Attenuation Values
Healthy tissue	.1625–.2977
Tumorous tissue	.2679–.3930
Bone	.3857–.5108

The part of the body to be scanned is divided into cells. If an X-ray passes through more than one cell, the total linear-attenuation value is the sum of the values for the cells. For example, in the accompanying figure, let a, b, and c be the values for cells A, B, and C, respectively. Then the attenuation value for beam 1 is $a + b$ and for beam 2 is $a + c$.

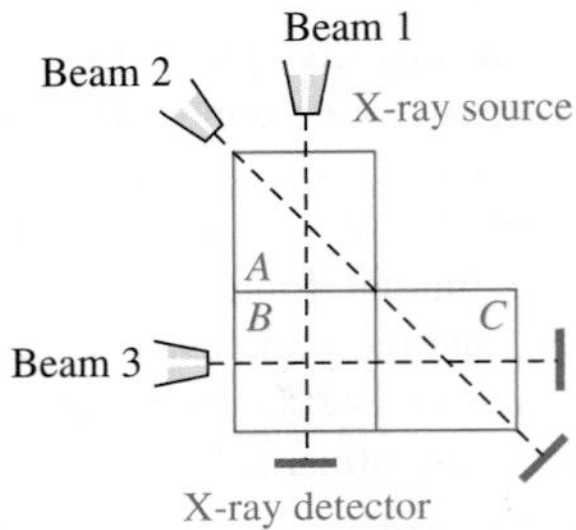

(a) Find the attenuation value for beam 3.

(b) Suppose that the attenuation values are .8, .55, and .65 for beams 1, 2, and 3, respectively. Set up and solve the system of three equations for a, b, and c. What can you conclude about cells A, B, and C?

18. Health (Refer to Exercise 17.) Four X-ray beams are aimed at four cells, as shown in the following figure:

Beam 1 Beam 2

Beam 3

A C

B D

Beam 4

(a) Suppose the attenuation values for beams 1, 2, 3, and 4 are .60, .75, .65, and .70, respectively. Do we have enough information to determine the values of a, b, c, and d? Explain.

(b) Suppose we have the data from part (a), as well as the following values for d. Find the values for a, b, and c, and make conclusions about cells A, B, C, and D in each case.

(i) .33 **(ii)** .43

(c) Two X-ray beams are added as shown in the figure. In addition to the data in part (a), we now have attenuation values of .85 and .50 for beams 5 and 6, respectively. Find the values for a, b, c, and d, and make conclusions about cells A, B, C, and D.

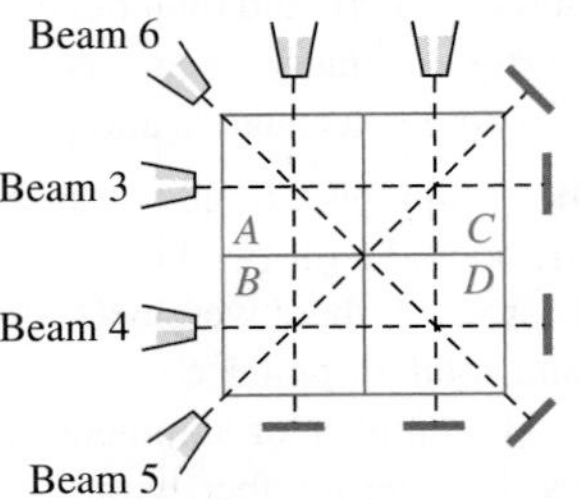

19. Transportation During rush hours, substantial traffic congestion is encountered at the intersections shown in the figure. (The arrows indicate one-way streets.)

The city wishes to improve the signals at these corners to speed the flow of traffic. The traffic engineers first gather data. As the figure shows, 700 cars per hour come down M Street to intersection A, and 300 cars per hour come down 10th Street to intersection A. x_1 of these cars leave A on M Street, and x_4 cars leave A on 10th Street.

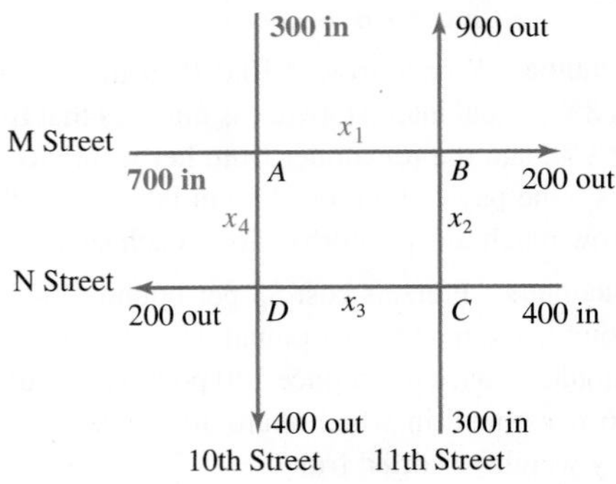

The number of cars entering A must equal the number leaving, so that

$$x_1 + x_4 = 700 + 300,$$

or

$$x_1 + x_4 = 1000.$$

For intersection B, x_1 cars enter on M Street and x_2 on 11th Street. The figure shows that 900 cars leave B on

*Exercises 17 and 18 are based on the article "Medical Applications of Linear Equations," by David Jabon, Gail Nord, Bryce W. Wilson, and Penny Coffman, *Mathematics Teacher* 89, no. 5 (May 1996) p. 98: 398.

11th Street and 200 cars leave B on M Street. So, we have

$$x_1 + x_2 = 900 + 200$$
$$x_1 + x_2 = 1100.$$

(a) Write two equations representing the traffic entering and leaving intersections C and D.
(b) Solve the system of four equations, using x_4 as the parameter.
(c) On the basis of your solution to part (b), what are the largest and smallest possible values for the number of cars leaving intersection A on 10th Street?
(d) Answer the question in part (c) for the other three variables.
(e) Verify that you could have discarded any one of the four original equations without changing the solution. What does this tell you about the original problem?

20. Transportation The diagram shows the traffic flow at four intersections during rush hour, as in Exercise 19.

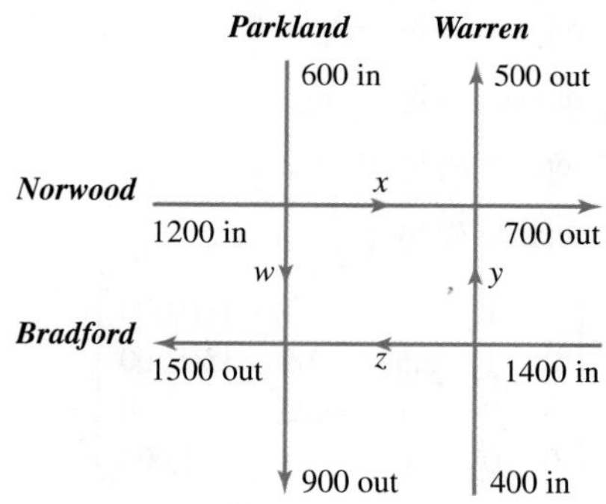

(a) What are the possible values of x, y, z, and w that will avoid any congestion? (Use w as the parameter.)
(b) What are the possible values of w?

A graphing calculator or other technology is recommended for the following exercises.

21. Health The table shows the calories, sodium, and protein in one cup of various kinds of soup.

	Progresso™ Hearty Chicken Rotini	Healthy Choice™ Hearty Chicken	Campbell's™ Chunky Chicken Noodle
Calories	100	130	110
Sodium (mg)	960	480	890
Protein (g)	7	9	8

How many cups of each kind of soup should be mixed together to produce 10 servings of soup, each of which provides 171 calories, 1158 milligrams of sodium, and 12.1 grams of protein? What is the serving size (in cups)? (*Hint*: In 10 servings, there must be 1710 calories, 11,580 milligrams of sodium, and 121 grams of protein.)

22. Health The table shows the calories, sodium, and fat in 1 ounce of various snack foods (all produced by Planters™):

	Sweet N' Crunchy Peanuts	Dry Roasted Honey Peanuts	Kettle Roasted Honey BBQ Peanuts
Calories	140	160	180
Sodium (mg)	20	110	55
Fat (g)	8	13	15

How many ounces of each kind of snack should be combined to produce 10 servings, each of which provides 284 calories, 93 milligrams of sodium, and 20.6 grams of fat? What is the serving size?

23. Finance An investment firm recommends that a client invest in bonds rated AAA, A, and B. The average yield on AAA bonds is 6%, on A bonds 7%, and on B bonds 10%. The client wants to invest twice as much in AAA bonds as in B bonds. How much should be invested in each type of bond under the following conditions?
(a) The total investment is \$25,000, and the investor wants an annual return of \$1810 on the three investments.
(b) The values in part (a) are changed to \$30,000 and \$2150, respectively.
(c) The values in part (a) are changed to \$40,000 and \$2900, respectively.

24. Business An electronics company produces transistors, resistors, and computer chips. Each transistor requires 3 units of copper, 1 unit of zinc, and 2 units of glass. Each resistor requires 3, 2, and 1 unit of the three materials, and each computer chip requires 2, 1, and 2 units of these materials, respectively. How many of each product can be made with the following amounts of materials?
(a) 810 units of copper, 410 of zinc, and 490 of glass
(b) 765 units of copper, 385 of zinc, and 470 of glass
(c) 1010 units of copper, 500 of zinc, and 610 of glass

25. Business At a pottery factory, fuel consumption for heating the kilns varies with the size of the order being fired. In the past, the company recorded the figures in the table.

x = Number of Platters	y = Fuel Cost per Platter
6	\$2.80
8	2.48
10	2.24

(a) Find an equation of the form $y = ax^2 + bx + c$ whose graph contains the three points corresponding to the data in the table.

(b) How many platters should be fired at one time in order to minimize the fuel cost per platter? What is the minimum fuel cost per platter?

26. Social Science The table shows Census Bureau projections for the population of the United States (in millions):

Year	2010	2030	2050
Population	310.2	373.5	439.0

(a) Find a quadratic function $f(x) = ax^2 + bx + c$ that gives the U.S. population (in millions) in year x, where $x = 0$ corresponds to 2000. (*Hint*: See Example 4.)

(b) Estimate the U.S. population in 2015 and 2020.

(c) In what year will the U.S. population reach 400 million?

27. Business The gross domestic product (GDP) of the United States was \$11 trillion in 2003 and is projected to be \$20 trillion in 2028 and \$30 trillion in 2044.*

(a) Let $x = 0$ correspond to 2000. Find a quadratic function $f(x) = ax^2 + bx + c$ that gives the GDP (in trillions of dollars) in year x.

(b) Estimate the GDP in 2009 and 2015.

(c) In what year will the GDP reach \$25 trillion?

28. Health The number of Alzheimer's cases in people 85 and older was 2 million in 2004 and is projected to be 3 million in 2025 and 8 million in 2050.†

(a) Let $x = 0$ correspond to 2000. Find a quadratic function that models the given data.

(b) How many people 85 or older will have Alzheimer's disease in 2020 and in 2034?

(c) In the year you turn 85, how many people your age or older are expected to have Alzheimer's disease?

29. Physical Science For certain aircraft, there exists a quadratic relationship between an airplane's maximum speed S (in knots) and its ceiling C—its highest altitude possible (in thousands of feet).‡ The following table lists three airplanes that conform to this relationship:

Airplane	Maximum Speed	Ceiling
Hawkeye	320	33
Corsair	600	40
Tomcat	1283	50

(a) If the relationship between C and S is written as $C = aS^2 + bS + c$, use a linear system of equations to determine the constants a, b, and c.

(b) A new aircraft of this type has a ceiling of 45,000 feet. Predict its top speed.

✓ Checkpoint Answers

1.
$$\begin{aligned} x + y + z &= 150 \\ 2x + 3y + 5z &= 400 \\ x - 3y \quad &= 0 \end{aligned}$$

2. (a)
$$\begin{bmatrix} 1 & 1 & 1 & 1 & 100{,}000 \\ .025 & .035 & .05 & .06 & 4000 \\ 0 & 0 & 1 & -.25 & 0 \\ 1 & 1 & 0 & 0 & 60{,}000 \end{bmatrix}$$

(b) Many sequences are possible, including this one: Replace R_2 by $-.025R_1 + R_2$; replace R_4 by $-R_1 + R_4$; replace R_2 by $\frac{1}{.01}R_2$; replace R_4 by $R_3 + R_4$; replace R_4 by $\frac{-1}{1.25}R_4$.

(c)
$$\begin{bmatrix} 1 & 1 & 1 & 1 & 100{,}000 \\ 0 & 1 & 2.5 & 3.5 & 150{,}000 \\ 0 & 0 & 1 & -.25 & 0 \\ 0 & 0 & 0 & 1 & 32{,}000 \end{bmatrix}$$

3. (a)
$$\begin{bmatrix} 10 & 20 & 30 & 1800 \\ 30 & 20 & 40 & 2800 \\ 20 & 40 & 25 & 2200 \end{bmatrix}$$

(b) Many sequences are possible, including this one: Replace R_1 by $\frac{1}{10}R_1$; replace R_2 by $\frac{1}{10}R_2$; replace R_3 by $\frac{1}{5}R_3$; replace R_2 by $-3R_1 + R_2$; replace R_3 by $-4R_1 + R_3$; replace R_2 by $-\frac{1}{4}R_2$; replace R_3 by $-\frac{1}{7}R_3$.

(c)
$$\begin{bmatrix} 1 & 2 & 3 & 180 \\ 0 & 1 & \frac{5}{4} & 65 \\ 0 & 0 & 1 & 40 \end{bmatrix}$$

---

*Goldman Sachs.

†Alzheimer's Association.

‡D. Sanders, *Statistics: A First Course*, Fifth Edition (McGraw Hill, 1995).

6.4 ▶ Basic Matrix Operations

Until now, we have used matrices only as a convenient shorthand to solve systems of equations. However, matrices are also important in the fields of management, natural science, engineering, and social science as a way to organize data, as Example 1 demonstrates.

Example 1 The EZ Life Company manufactures sofas and armchairs in three models: A, B, and C. The company has regional warehouses in New York, Chicago, and San Francisco. In its August shipment, the company sends 10 model A sofas, 12 model B sofas, 5 model C sofas, 15 model A chairs, 20 model B chairs, and 8 model C chairs to each warehouse.

This data might be organized by first listing it as follows:

Sofas	10 model A	12 model B	5 model C;
Chairs	15 model A	20 model B	8 model C.

Alternatively, we might tabulate the data:

		MODEL		
		A	B	C
FURNITURE	Sofa	10	12	5
	Chair	15	20	8

With the understanding that the numbers in each row refer to the type of furniture (sofa or chair) and the numbers in each column refer to the model (A, B, or C), the same information can be given by a matrix as follows:

$$M = \begin{bmatrix} 10 & 12 & 5 \\ 15 & 20 & 8 \end{bmatrix}$$ ✓1

✓Checkpoint 1

Rewrite matrix M in Example 1 in a matrix with three rows and two columns.

Answer: See page 351.

A matrix with m horizontal rows and n vertical columns has dimension, or size, $m \times n$. The number of rows is always given first.

Example 2

(a) The matrix $\begin{bmatrix} 6 & 5 \\ 3 & 4 \\ 5 & -1 \end{bmatrix}$ is a 3×2 matrix.

(b) $\begin{bmatrix} 5 & 8 & 9 \\ 0 & 5 & -3 \\ -4 & 0 & 5 \end{bmatrix}$ is a 3×3 matrix.

(c) $[1 \quad 6 \quad 5 \quad -2 \quad 5]$ is a 1×5 matrix.

(d) A graphing calculator displays a 4 × 1 matrix like this:

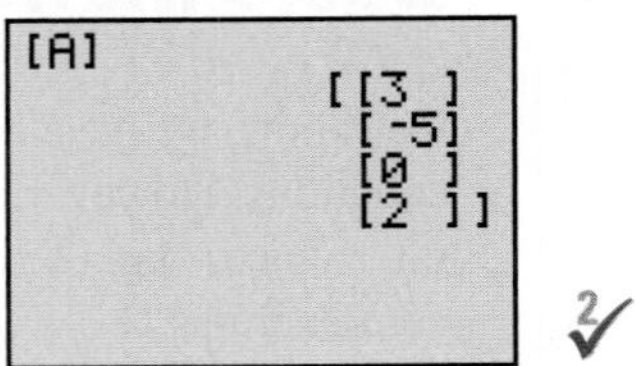

✓ 2

✓ Checkpoint 2

Give the size of each of the following matrices.

(a) $\begin{bmatrix} 2 & 1 & -5 & 6 \\ 3 & 0 & 7 & -4 \end{bmatrix}$

(b) $\begin{bmatrix} 1 & 2 & 3 \\ 4 & 5 & 6 \\ 9 & 8 & 7 \end{bmatrix}$

Answers: *See page 351.*

A matrix with only one row, as in Example 2(c), is called a **row matrix**, or **row vector**. A matrix with only one column, as in Example 2(d), is called a **column matrix**, or **column vector**. A matrix with the same number of rows as columns is called a **square matrix**. The matrix in Example 2(b) is a square matrix, as are

$$A = \begin{bmatrix} -5 & 6 \\ 8 & 3 \end{bmatrix} \quad \text{and} \quad B = \begin{bmatrix} 0 & 0 & 0 & 0 \\ -2 & 4 & 1 & 3 \\ 0 & 0 & 0 & 0 \\ -5 & -4 & 1 & 8 \end{bmatrix}.$$

✓ 3

✓ Checkpoint 3

Use the numbers 2, 5, −8, and 4 to write

(a) a row matrix;

(b) a column matrix;

(c) a square matrix.

Answers: *See page 351.*

When a matrix is denoted by a single letter, such as the matrix A above, the element in row i and column j is denoted a_{ij}. For example, $a_{21} = 8$ (the element in row 2, column 1). Similarly, in matrix B, $b_{42} = -4$ (the element in row 4, column 2).

ADDITION

The matrix given in Example 1,

$$M = \begin{bmatrix} 10 & 12 & 5 \\ 15 & 20 & 8 \end{bmatrix},$$

shows the August shipment from the EZ Life plant to each of its warehouses. If matrix N below gives the September shipment to the New York warehouse, what is the total shipment for each item of furniture to the New York warehouse for the two months?

$$N = \begin{bmatrix} 45 & 35 & 20 \\ 65 & 40 & 35 \end{bmatrix}$$

If 10 model A sofas were shipped in August and 45 in September, then altogether 55 model A sofas were shipped in the two months. Adding the other corresponding entries gives a new matrix, Q, that represents the total shipment to the New York warehouse for the two months:

$$Q = \begin{bmatrix} 55 & 47 & 25 \\ 80 & 60 & 43 \end{bmatrix}$$

It is convenient to refer to Q as the *sum* of M and N.

The way these two matrices were added illustrates the following definition of addition of matrices:

> The **sum** of two $m \times n$ matrices X and Y is the $m \times n$ matrix $X + Y$ in which each element is the sum of the corresponding elements of X and Y.

It is important to remember that only matrices that are the same size can be added.

Example 3 Find each sum if possible.

(a) $\begin{bmatrix} 5 & -6 \\ 8 & 9 \end{bmatrix} + \begin{bmatrix} -4 & 6 \\ 8 & -3 \end{bmatrix} = \begin{bmatrix} 5+(-4) & -6+6 \\ 8+8 & 9+(-3) \end{bmatrix} = \begin{bmatrix} 1 & 0 \\ 16 & 6 \end{bmatrix}.$

(b) The matrices

$$A = \begin{bmatrix} 5 & 8 \\ 6 & 2 \end{bmatrix} \quad \text{and} \quad B = \begin{bmatrix} 3 & 9 & 1 \\ 4 & 2 & 5 \end{bmatrix}$$

are of different sizes, so it is not possible to find the sum $A + B$. 4✓

✓Checkpoint 4

Find each sum when possible.

(a) $\begin{bmatrix} 2 & 5 & 7 \\ 3 & -1 & 4 \end{bmatrix} + \begin{bmatrix} -1 & 2 & 0 \\ 10 & -4 & 5 \end{bmatrix}$

(b) $\begin{bmatrix} 1 \\ 2 \\ 3 \end{bmatrix} + \begin{bmatrix} 2 & -1 \\ 4 & 5 \\ 6 & 0 \end{bmatrix}$

(c) $[5 \quad 4 \quad -1] + [-5 \quad 2 \quad 3]$

Answers: See page 351.

TECHNOLOGY TIP Spreadsheets and graphing calculators can find matrix sums, as illustrated in Figure 6.20.

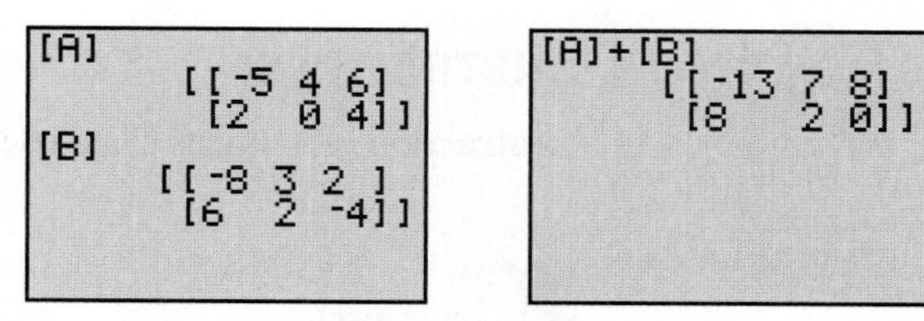

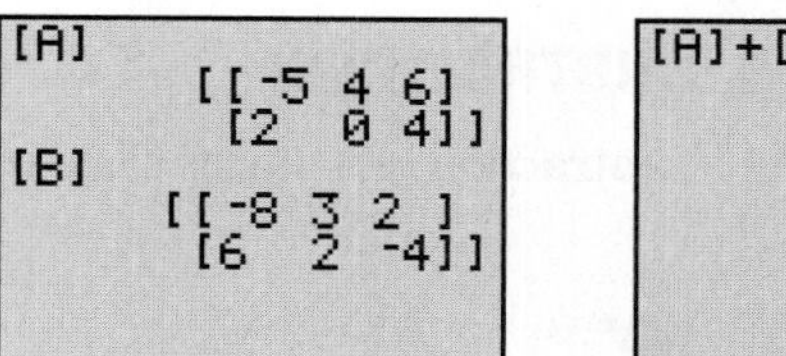

FIGURE 6.20

Example 4 The September shipments of the three models of sofas and chairs from the EZ Life Company to the New York, San Francisco, and Chicago warehouses are given respectively in matrices N, S, and C as follows:

$$N = \begin{bmatrix} 45 & 35 & 20 \\ 65 & 40 & 35 \end{bmatrix}; \quad S = \begin{bmatrix} 30 & 32 & 28 \\ 43 & 47 & 30 \end{bmatrix}; \quad C = \begin{bmatrix} 22 & 25 & 38 \\ 31 & 34 & 35 \end{bmatrix}.$$

What was the total amount shipped to the three warehouses in September?

Solution The total of the September shipments is represented by the sum of the three matrices N, S, and C:

$$N + S + C = \begin{bmatrix} 45 & 35 & 20 \\ 65 & 40 & 35 \end{bmatrix} + \begin{bmatrix} 30 & 32 & 28 \\ 43 & 47 & 30 \end{bmatrix} + \begin{bmatrix} 22 & 25 & 38 \\ 31 & 34 & 35 \end{bmatrix}$$

$$= \begin{bmatrix} 97 & 92 & 86 \\ 139 & 121 & 100 \end{bmatrix}.$$

For example, from this sum, the total number of model C sofas shipped to the three warehouses in September was 86. 5✓

✓Checkpoint 5

From the result of Example 4, find the total number of the following shipped to the three warehouses.

(a) Model A chairs

(b) Model B sofas

(c) Model C chairs

Answers: See page 351.

Example 5 A drug company is testing 200 patients to see if Painoff (a new headache medicine) is effective. Half the patients receive Painoff and half receive a placebo. The data on the first 50 patients is summarized in this matrix:

	Pain Relief Obtained	
	Yes	No
Patient took Painoff	22	3
Patient took placebo	8	17

For example, row 2 shows that, of the people who took the placebo, 8 got relief, but 17 did not. The test was repeated on three more groups of 50 patients each, with the results summarized by these matrices:

$$\begin{bmatrix} 21 & 4 \\ 6 & 19 \end{bmatrix}; \quad \begin{bmatrix} 19 & 6 \\ 10 & 15 \end{bmatrix}; \quad \begin{bmatrix} 23 & 2 \\ 3 & 22 \end{bmatrix}.$$

The total results of the test can be obtained by adding these four matrices:

$$\begin{bmatrix} 22 & 3 \\ 8 & 17 \end{bmatrix} + \begin{bmatrix} 21 & 4 \\ 6 & 19 \end{bmatrix} + \begin{bmatrix} 19 & 6 \\ 10 & 15 \end{bmatrix} + \begin{bmatrix} 23 & 2 \\ 3 & 22 \end{bmatrix} = \begin{bmatrix} 85 & 15 \\ 27 & 73 \end{bmatrix}.$$

Because 85 of 100 patients got relief with Painoff and only 27 of 100 did so with the placebo, it appears that Painoff is effective. 6 ✓

✓ Checkpoint 6

Later, it was discovered that the data in the last group of 50 patients in Example 5 was invalid. Use a matrix to represent the total test results after that data was eliminated.

Answer: See page 351.

SUBTRACTION

Subtraction of matrices can be defined in a manner similar to matrix addition.

> The **difference** of two $m \times n$ matrices X and Y, is the $m \times n$ matrix $X - Y$ in which each element is the difference of the corresponding elements of X and Y.

Example 6 Find the following.

(a) $\begin{bmatrix} 1 & 2 & 3 \\ 0 & -1 & 5 \end{bmatrix} - \begin{bmatrix} -2 & 3 & 0 \\ 1 & -7 & 2 \end{bmatrix}$

Solution

$$\begin{bmatrix} 1 & 2 & 3 \\ 0 & -1 & 5 \end{bmatrix} - \begin{bmatrix} -2 & 3 & 0 \\ 1 & -7 & 2 \end{bmatrix} = \begin{bmatrix} 1-(-2) & 2-3 & 3-0 \\ 0-1 & -1-(-7) & 5-2 \end{bmatrix}$$
$$= \begin{bmatrix} 3 & -1 & 3 \\ -1 & 6 & 3 \end{bmatrix}.$$

(b) $[8 \quad 6 \quad -4] - [3 \quad 5 \quad -8]$

Solution

$$[8 \quad 6 \quad -4] - [3 \quad 5 \quad -8] = [8-3 \quad 6-5 \quad -4-(-8)]$$
$$= [5 \quad 1 \quad 4].$$

(c) $\begin{bmatrix} -2 & 5 \\ 0 & 1 \end{bmatrix} - \begin{bmatrix} 3 \\ 5 \end{bmatrix}$

Solution The matrices are of different sizes and thus cannot be subtracted. 7 ✓

✓ Checkpoint 7

Find each of the following differences when possible.

(a) $\begin{bmatrix} 2 & 5 \\ -1 & 0 \end{bmatrix} - \begin{bmatrix} 6 & 4 \\ 3 & -2 \end{bmatrix}$

(b) $\begin{bmatrix} 1 & 5 & 6 \\ 2 & 4 & 8 \end{bmatrix} - \begin{bmatrix} 2 & 1 \\ 10 & 3 \end{bmatrix}$

(c) $[5 \quad -4 \quad 1] - [6 \quad 0 \quad -3]$

Answers: See page 351.

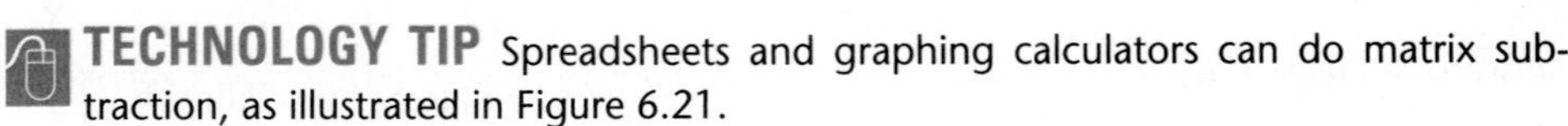

TECHNOLOGY TIP Spreadsheets and graphing calculators can do matrix subtraction, as illustrated in Figure 6.21.

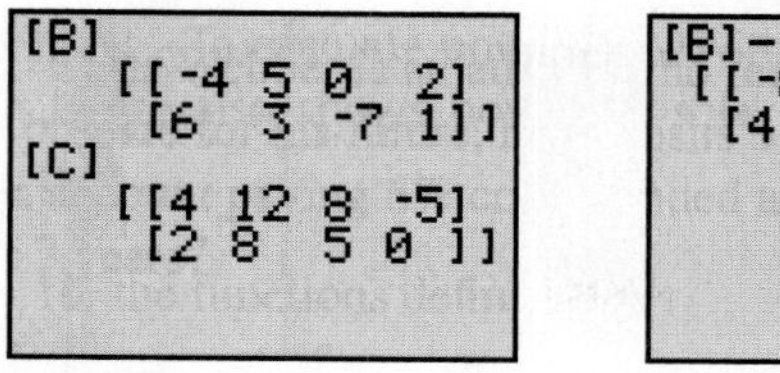

```
[B]
   [[-4 5 0  2]
    [6  3 -7 1]]
[C]
   [[4 12 8 -5]
    [2 8  5 0 ]]
```

```
[B]-[C]
   [[-8 -7 -8  7]
    [4  -5 -12 1]]
```

FIGURE 6.21

Example 7 During September, the Chicago warehouse of the EZ Life Company shipped out the following numbers of each model, where the entries in the matrix have the same meaning as given earlier:

$$K = \begin{bmatrix} 5 & 10 & 8 \\ 11 & 14 & 15 \end{bmatrix}.$$

What was the Chicago warehouse's inventory on October 1, taking into account only the number of items received and sent out during the month?

Solution The number of each kind of item received during September is given by matrix C from Example 4; the number of each model sent out during September is given by matrix K. The October 1 inventory is thus represented by the matrix $C - K$:

$$C - K = \begin{bmatrix} 22 & 25 & 38 \\ 31 & 34 & 35 \end{bmatrix} - \begin{bmatrix} 5 & 10 & 8 \\ 11 & 14 & 15 \end{bmatrix} = \begin{bmatrix} 17 & 15 & 30 \\ 20 & 20 & 20 \end{bmatrix}.$$

SCALAR MULTIPLICATION

Suppose one of the EZ Life Company warehouses receives the following order, written in matrix form, where the entries have the same meaning as given earlier:

$$\begin{bmatrix} 5 & 4 & 1 \\ 3 & 2 & 3 \end{bmatrix}.$$

Later, the store that sent the order asks the warehouse to send six more of the same order. The six new orders can be written as one matrix by multiplying each element in the matrix by 6, giving the product

$$6\begin{bmatrix} 5 & 4 & 1 \\ 3 & 2 & 3 \end{bmatrix} = \begin{bmatrix} 30 & 24 & 6 \\ 18 & 12 & 18 \end{bmatrix}.$$

In work with matrices, a real number, like the 6 in the preceding multiplication, is called a **scalar**.

> The **product** of a scalar k and a matrix X is the matrix kX, each of whose elements is k times the corresponding element of X.

Example 8

(a) $(-3)\begin{bmatrix} 2 & -5 \\ 1 & 7 \\ 4 & -6 \end{bmatrix} = \begin{bmatrix} -6 & 15 \\ -3 & -21 \\ -12 & 18 \end{bmatrix}.$

✓Checkpoint 8

Find each product.

(a) $-3\begin{bmatrix} 4 & -2 \\ 1 & 5 \end{bmatrix}$

(b) $4\begin{bmatrix} 2 & 4 & 7 \\ 8 & 2 & 1 \\ 5 & 7 & 3 \end{bmatrix}$

Answers: See page 351.

(b) Graphing calculators can also do scalar multiplication (Figure 6.22). ✓8

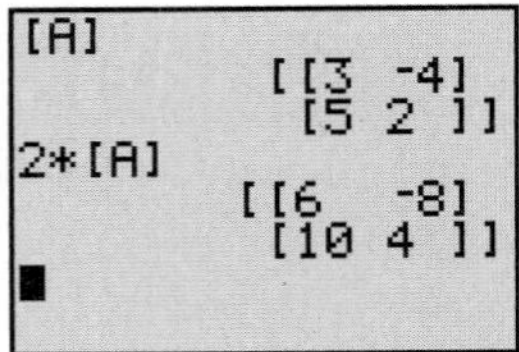

FIGURE 6.22

TECHNOLOGY TIP To compute the negative of matrix A on a calculator, use the "negative" key $(-)$, as shown in Figure 6.23.

```
[A]
      [[2 3   ]
       [1 1.5]]
-[A]
    [[-2 -3   ]
     [-1 -1.5]]
```

FIGURE 6.23

Recall that the *negative* of a real number a is the number $-a = (-1)a$. The negative of a matrix is defined similarly.

> The **negative** (or *additive inverse*) of a matrix A is the matrix $(-1)A$—that is, the matrix obtained by multiplying each element of A by -1. It is denoted $-A$.

Example 9 Find $-A$ and $-B$ when

$$A = \begin{bmatrix} 1 & 2 & 3 \\ 0 & -1 & 5 \end{bmatrix} \quad \text{and} \quad B = \begin{bmatrix} -2 & 3 & 0 \\ 1 & -7 & 2 \end{bmatrix}.$$

Solution By the preceding definition,

$$-A = \begin{bmatrix} -1 & -2 & -3 \\ 0 & 1 & -5 \end{bmatrix} \quad \text{and} \quad -B = \begin{bmatrix} 2 & -3 & 0 \\ -1 & 7 & -2 \end{bmatrix}.$$ ✓9

✓Checkpoint 9

Let A and B be the matrices in Example 9. Find

(a) $A + (-B)$;

(b) $A - B$.

(c) What can you conclude from parts (a) and (b)?

Answers: See page 351.

A matrix consisting only of zeros is called a **zero matrix** and is denoted O. There is an $m \times n$ zero matrix for each pair of values of m and n—for instance,

$$\begin{bmatrix} 0 & 0 \\ 0 & 0 \end{bmatrix}; \qquad \begin{bmatrix} 0 & 0 & 0 & 0 \\ 0 & 0 & 0 & 0 \end{bmatrix}.$$

2 × 2 zero matrix   2 × 4 zero matrix

The negative of a matrix and zero matrices have the following properties, as illustrated in Checkpoint 9 and Exercises 26–28.

> Let A and B be any $m \times n$ matrices and let O be the $m \times n$ zero matrix. Then
>
> $$A + (-B) = A - B;$$
> $$A + (-A) = O = A - A;$$
> $$A + O = A = O + A.$$

6.4 ▸ Exercises

Find the size of each of the given matrices. Identify any square, column, or row matrices. (See Example 2.) Give the additive inverse of each matrix.

1. $\begin{bmatrix} 7 & -8 & 4 \\ 0 & 13 & 9 \end{bmatrix}$

2. $\begin{bmatrix} -7 & 23 \\ 5 & -6 \end{bmatrix}$

3. $\begin{bmatrix} -3 & 0 & 11 \\ 1 & \frac{1}{4} & -7 \\ 5 & -3 & 9 \end{bmatrix}$

4. $[6 \quad -4 \quad \frac{2}{3} \quad 12 \quad 2]$

5. $\begin{bmatrix} 7 \\ 11 \end{bmatrix}$

6. $[-5]$

7. If A is a 5×3 matrix and $A + B = A$, what do you know about B?

8. If C is a 3×3 matrix and D is a 3×4 matrix, then $C + D$ is _____.

Perform the indicated operations where possible. (See Examples 3–6.)

9. $\begin{bmatrix} 1 & 2 & 7 & -1 \\ 8 & 0 & 2 & -4 \end{bmatrix} + \begin{bmatrix} -8 & 12 & -5 & 5 \\ -2 & -3 & 0 & 0 \end{bmatrix}$

10. $\begin{bmatrix} 1 & 7 \\ 2 & -3 \\ 3 & 7 \end{bmatrix} + \begin{bmatrix} 2 & 8 \\ 6 & 8 \\ -1 & 9 \end{bmatrix}$

11. $\begin{bmatrix} -1 & -5 & 9 \\ 2 & 2 & 3 \end{bmatrix} + \begin{bmatrix} 4 & 4 & -7 \\ 1 & -1 & 2 \end{bmatrix}$

12. $\begin{bmatrix} 2 & 4 \\ -8 & 2 \end{bmatrix} + \begin{bmatrix} 9 & -5 \\ 8 & 5 \end{bmatrix}$

13. $\begin{bmatrix} -3 & -2 & 5 \\ 3 & 9 & 0 \end{bmatrix} - \begin{bmatrix} 1 & 5 & -2 \\ -3 & 6 & 8 \end{bmatrix}$

14. $\begin{bmatrix} 9 & 1 \\ 0 & -3 \\ 4 & 10 \end{bmatrix} - \begin{bmatrix} 1 & 9 & -4 \\ -1 & 1 & 0 \end{bmatrix}$

Let $A = \begin{bmatrix} -2 & 0 \\ 5 & 3 \end{bmatrix}$ *and* $B = \begin{bmatrix} 0 & 2 \\ 4 & -6 \end{bmatrix}$. *Find each of the following.*

15. $2A$
16. $-3B$
17. $-4B$
18. $5A$
19. $-4A + 5B$
20. $3A - 10B$

Let $A = \begin{bmatrix} 1 & -2 \\ 4 & 3 \end{bmatrix}$ *and* $B = \begin{bmatrix} 2 & -1 \\ 0 & 5 \end{bmatrix}$. *Find a matrix X satisfying the given equation.*

21. $2X = 2A + 3B$
22. $3X = A - 3B$

Using matrices

$$O = \begin{bmatrix} 0 & 0 \\ 0 & 0 \end{bmatrix}, P = \begin{bmatrix} m & n \\ p & q \end{bmatrix}, T = \begin{bmatrix} r & s \\ t & u \end{bmatrix}, \text{ and } X = \begin{bmatrix} x & y \\ z & w \end{bmatrix},$$

verify that the statements in Exercises 23–28 are true.

23. $X + T$ is a 2×2 matrix.

24. $X + T = T + X$ (commutative property of addition of matrices).

25. $X + (T + P) = (X + T) + P$ (associative property of addition of matrices).

26. $X + (-X) = O$ (inverse property of addition of matrices).

27. $P + O = P$ (identity property of addition of matrices).

28. Which of the preceding properties are valid for matrices that are not square?

Work the following exercises. (See Examples 1 and 7.)

29. **Business** When ticket holders fail to attend, major league sports teams lose the money these fans would have spent on refreshments, souvenirs, etc. The percentage of fans who don't show up is 16% in basketball and hockey, 20% in football, and 18% in baseball. The lost revenue per fan is \$18.20 in basketball, \$18.25 in hockey, \$19 in football, and \$15.40 in baseball. The total annual lost revenue is \$22.7 million in basketball, \$35.8 million in hockey, \$51.9 million in football, and \$96.3 million in baseball.* Express this information in matrix form; specify what the rows and columns represent.

30. **Finance** 83% of undergraduate students had credit cards in 2001 and 76% in 2004. Average credit card debt was \$2327 in 2001 and \$2169 in 2004. The percentage of students with balances of \$7000 or more was 6% in 2001 and 7% in 2004. The average number of credit cards per

*American Demographics.

student was 4.25 in 2001 and 4.09 in 2004.* Write this information first as a 4 × 2 matrix and then as a 2 × 4 matrix.

31. **Health** The shortage of organs for transplants is a continuing problem in the United States. At the end of 2000, there were 3929 people waiting for a heart transplant, 3514 for a lung transplant, 16,095 for a liver transplant, and 44,589 for a kidney transplant. Corresponding figures for 2003 were 3444, 3800, 16,927, and 53,563, respectively. In 2006, the figures were 2814, 2857, 16,861, and 66,961, respectively.† Express this information as a matrix, labeling rows and columns.

Work these exercises. (See Examples 1, 4, 6, and 7.)

32. **Management** There are three convenience stores in Gambier. This week, Store I sold 88 loaves of bread, 48 quarts of milk, 16 jars of peanut butter, and 112 pounds of cold cuts. Store II sold 105 loaves of bread, 72 quarts of milk, 21 jars of peanut butter, and 147 pounds of cold cuts. Store III sold 60 loaves of bread, 40 quarts of milk, no peanut butter, and 50 pounds of cold cuts.
    (a) Use a 3 × 4 matrix to express the sales information for the three stores.
    (b) During the following week, sales on these products at Store I increased by 25%, sales at Store II increased by one-third, and sales at Store III increased by 10%. Write the sales matrix for that week.
    (c) Write a matrix that represents total sales over the two-week period.

33. **Health** Heart disease death rates (per 100,000 people) vary in different age groups. The rates for people 15–24 years old were 2.8 in 1995, 2.6 in 2000, and 2.7 in 2003 and 2005. In 1995, the death rates were 109.6 for people 45–54 years old and 795.4 for people 65–74 years old. The death rates for people 45–54 years old were 94.2 in 2000, 92.5 in 2003, and 89.7 in 2005. For people 65–74, the death rates were 665.6 in 2000, 585.0 in 2003, and 518.9 in 2005.‡ Write this information in matrix form, labling rows and columns.

34. **Social Science** The following table gives the educational attainment of the U.S. population 25 years and older in various years§:

	Male		Female	
	Four Years of High School or More	Four Years of College or More	Four Years of High School or More	Four Years of College or More
1960	39.5	9.7	42.5	5.8
1970	51.9	13.5	52.8	8.1
1980	67.3	20.1	65.8	12.8
1990	77.7	24.4	77.5	18.4
1995	81.7	26.0	81.6	20.0
2000	84.2	27.8	84.0	23.6
2007	85.0	29.5	86.4	28.0

    (a) Write a 2 × 7 matrix for the educational attainment of males.
    (b) Write a 2 × 7 matrix for the educational attainment of females.
    (c) Use the matrices from parts (a) and (b) to write a matrix showing how much more (or less) education males have attained than females.

35. **Transportation** The tables give the death rates (per million person-trips) for male and female drivers for various ages and numbers of passengers*:

MALE DRIVERS

Age	Number of Passengers			
	0	1	2	≥3
16	2.61	4.39	6.29	9.08
17	1.63	2.77	4.61	6.92
30–59	.92	.75	.62	.54

FEMALE DRIVERS

Age	Number of Passengers			
	0	1	2	≥3
16	1.38	1.72	1.94	3.31
17	1.26	1.48	2.82	2.28
30–59	.41	.33	.27	.40

    (a) Write a matrix A for the death rate of male drivers.
    (b) Write a matrix B for the death rate of female drivers.
    (c) Use the matrices from parts (a) and (b) to write a matrix showing the difference between the death rates of males and females.

*Nellie Mae.

†United Network of Organ Sharing and National Organ Procurement and Transplantation Network.

‡U.S. Department of Health and Human Services, Center for Health Statistics.

§U.S. Census Bureau.

*Li-Hui Chen, Susan Baker, Elisa Braver, and Guohua Li, "Carrying Passengers as a Risk Factor for Crashes Fatal to 16- and 17-Year-Old Drivers," *JAMA* 283, no. 12 (March 22/29, 2000): 1578–1582.

36. **Transportation** Use matrix operations on the matrices found in Exercise 35(a) and (b) to obtain one matrix that gives the combined death rates for males and females (per million person-trips) of drivers of various ages, with varying numbers of passengers. $\left[\textit{Hint}: \text{Consider } \frac{1}{2}(A + B).\right]$

✓ Checkpoint Answers

1. $\begin{bmatrix} 10 & 15 \\ 12 & 20 \\ 5 & 8 \end{bmatrix}$

2. **(a)** 2×4 **(b)** 3×3

3. **(a)** $[2 \quad 5 \quad -8 \quad 4]$

   (b) $\begin{bmatrix} 2 \\ 5 \\ 8 \\ -4 \end{bmatrix}$ **(c)** $\begin{bmatrix} 2 & 5 \\ -8 & 4 \end{bmatrix}$ or $\begin{bmatrix} 2 & -8 \\ 5 & 4 \end{bmatrix}$ (Other answers are possible.)

4. **(a)** $\begin{bmatrix} 1 & 7 & 7 \\ 13 & -5 & 9 \end{bmatrix}$ **(b)** Not possible

   (c) $[0 \quad 6 \quad 2]$

5. **(a)** 139 **(b)** 92 **(c)** 100

6. $\begin{bmatrix} 62 & 13 \\ 24 & 51 \end{bmatrix}$

7. **(a)** $\begin{bmatrix} -4 & 1 \\ -4 & 2 \end{bmatrix}$ **(b)** Not possible

   (c) $[-1 \quad -4 \quad 4]$

8. **(a)** $\begin{bmatrix} -12 & 6 \\ -3 & -15 \end{bmatrix}$ **(b)** $\begin{bmatrix} 8 & 16 & 28 \\ 32 & 8 & 4 \\ 20 & 28 & 12 \end{bmatrix}$

9. **(a)** $\begin{bmatrix} 3 & -1 & 3 \\ -1 & 6 & 3 \end{bmatrix}$ **(b)** $\begin{bmatrix} 3 & -1 & 3 \\ -1 & 6 & 3 \end{bmatrix}$

   (c) $A + (-B) = A - B$.

6.5 ▶ Matrix Products and Inverses

To understand the reasoning behind the definition of matrix multiplication, look again at the EZ Life Company. Suppose sofas and chairs of the same model are often sold as sets, with matrix W showing the number of each model set in each warehouse:

$$\begin{array}{l} \\ \text{New York} \\ \text{Chicago} \\ \text{San Francisco} \end{array} \begin{array}{c} \begin{array}{ccc} \text{A} & \text{B} & \text{C} \end{array} \\ \begin{bmatrix} 10 & 7 & 3 \\ 5 & 9 & 6 \\ 4 & 8 & 2 \end{bmatrix} \end{array} = W.$$

If the selling price of a model A set is \$800, of a model B set is \$1000, and of a model C set is \$1200, find the total value of the sets in the New York warehouse as follows:

Type	Number of Sets		Price of Set		Total
A	10	×	\$ 800	=	\$ 8,000
B	7	×	1000	=	7,000
C	3	×	1200	=	3,600
			Total for New York		\$18,600

The total value of the three kinds of sets in New York is \$18,600. ✓1

The work done in the preceding table is summarized as

$$10(\$800) + 7(\$1000) + 3(\$1200) = \$18,600.$$

✓ Checkpoint 1

In this example of the EZ Life Company, find the total value of the New York sets if model A sets sell for \$1200, model B for \$1600, and model C for \$1300.

Answer: See page 365.

In the same way, the Chicago sets have a total value of

$$5(\$800) + 9(\$1000) + 6(\$1200) = \$20{,}200,$$

and in San Francisco, the total value of the sets is

$$4(\$800) + 8(\$1000) + 2(\$1200) = \$13{,}600.$$

The selling prices can be written as a column matrix P and the total value in each location as a column matrix V:

$$P = \begin{bmatrix} 800 \\ 1000 \\ 1200 \end{bmatrix} \quad \text{and} \quad V = \begin{bmatrix} 18{,}600 \\ 20{,}200 \\ 13{,}600 \end{bmatrix}.$$

Consider how the first row of the matrix W and the single column P lead to the first entry of V:

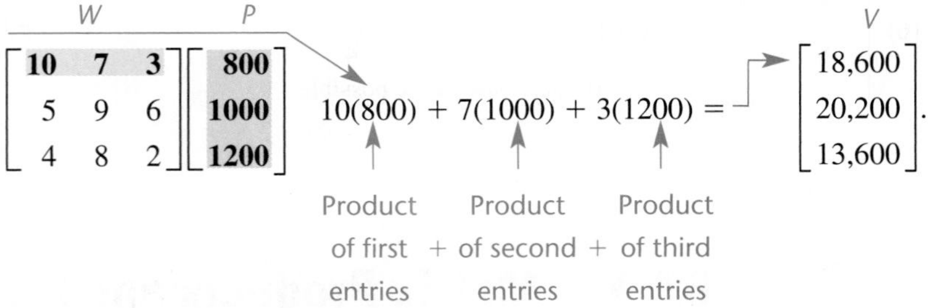

Similarly, adding the products of corresponding entries in the second row of W and the column P produces the second entry in V. The third entry in V is obtained in the same way by using the third row of W and column P. This suggests that it is reasonable to *define* the product WP to be V:

$$WP = \begin{bmatrix} 10 & 7 & 3 \\ 5 & 9 & 6 \\ 4 & 8 & 2 \end{bmatrix} \begin{bmatrix} 800 \\ 1000 \\ 1200 \end{bmatrix} = \begin{bmatrix} 18{,}600 \\ 20{,}200 \\ 13{,}600 \end{bmatrix} = V$$

Note the sizes of the matrices here: The product of a 3×3 matrix and a 3×1 matrix is a 3×1 matrix.

MULTIPLYING MATRICES

In order to define matrix multiplication in the general case, we first define the **product of a row of a matrix and a column of a matrix** (with the same number of entries in each) to be the *number* obtained by multiplying the corresponding entries (first by first, second by second, etc.) and adding the results. For instance,

$$[3 \quad -2 \quad 1] \cdot \begin{bmatrix} 4 \\ 5 \\ 0 \end{bmatrix} = 3 \cdot 4 + (-2) \cdot 5 + 1 \cdot 0 = 12 - 10 + 0 = 2.$$

Now **matrix multiplication** is defined as follows.

Let A be an $m \times n$ matrix and let B be an $n \times k$ matrix. The **product matrix** AB is the $m \times k$ matrix whose entry in the ith row and jth column is

the product of the ith row of A and the jth column of B.

CAUTION Be careful when multiplying matrices. Remember that the number of *columns* of A must equal the number of *rows* of B in order to get the product matrix AB. The final product will have as many rows as A and as many columns as B.

Example 1 Suppose matrix A is 2×2 and matrix B is 2×4. Can the product AB be calculated? If so, what is the size of the product?

Solution The following diagram helps decide the answers to these questions:

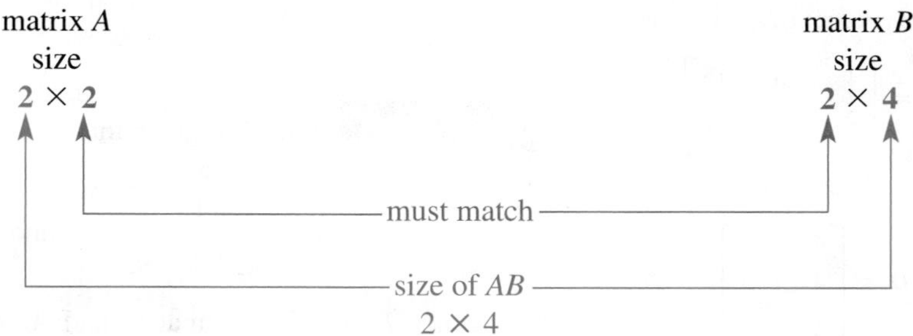

The product AB can be calculated because A has two columns and B has two rows. The product will be a 2×4 matrix. ✓2

✓Checkpoint 2

Matrix A is 4×6 and matrix B is 2×4.

(a) Can AB be found? If so, give its size.

(b) Can BA be found? If so, give its size.

Answers: See page 365.

Example 2 Find the product CD when

$$C = \begin{bmatrix} -3 & 4 & 2 \\ 5 & 0 & 4 \end{bmatrix} \quad \text{and} \quad D = \begin{bmatrix} -6 & 4 \\ 2 & 3 \\ 3 & -2 \end{bmatrix}.$$

Solution Here, matrix C is 2×3 and matrix D is 3×2, so matrix CD can be found and will be 2×2.

Step 1 row 1, column 1

$$\begin{bmatrix} \mathbf{-3} & \mathbf{4} & \mathbf{2} \\ 5 & 0 & 4 \end{bmatrix}\begin{bmatrix} \mathbf{-6} & 4 \\ \mathbf{2} & 3 \\ \mathbf{3} & -2 \end{bmatrix} \quad (-3) \cdot (-6) + 4 \cdot 2 + 2 \cdot 3 = 32.$$

Hence, 32 is the entry in row 1, column 1, of CD, as shown in Step 5 below.

Step 2 row 1, column 2

$$\begin{bmatrix} \mathbf{-3} & \mathbf{4} & \mathbf{2} \\ 5 & 0 & 4 \end{bmatrix}\begin{bmatrix} -6 & \mathbf{4} \\ 2 & \mathbf{3} \\ 3 & \mathbf{-2} \end{bmatrix} \quad (-3) \cdot 4 + 4 \cdot 3 + 2 \cdot (-2) = -4.$$

So -4 is the entry in row 1, column 2, of CD, as shown in Step 5. Continue in this manner to find the remaining entries of CD.

Step 3 row 2, column 1

$$\begin{bmatrix} -3 & 4 & 2 \\ \mathbf{5} & \mathbf{0} & \mathbf{4} \end{bmatrix} \begin{bmatrix} \mathbf{-6} & 4 \\ \mathbf{2} & 3 \\ \mathbf{3} & -2 \end{bmatrix} \quad \mathbf{5} \cdot (\mathbf{-6}) + \mathbf{0} \cdot \mathbf{2} + \mathbf{4} \cdot \mathbf{3} = -18.$$

Step 4 row 2, column 2

$$\begin{bmatrix} -3 & 4 & 2 \\ \mathbf{5} & \mathbf{0} & \mathbf{4} \end{bmatrix} \begin{bmatrix} -6 & \mathbf{4} \\ 2 & \mathbf{3} \\ 3 & \mathbf{-2} \end{bmatrix} \quad \mathbf{5} \cdot \mathbf{4} + \mathbf{0} \cdot \mathbf{3} + \mathbf{4} \cdot (\mathbf{-2}) = 12.$$

Step 5 The product is

$$CD = \begin{bmatrix} -3 & 4 & 2 \\ 5 & 0 & 4 \end{bmatrix} \begin{bmatrix} -6 & 4 \\ 2 & 3 \\ 3 & -2 \end{bmatrix} = \begin{bmatrix} 32 & -4 \\ -18 & 12 \end{bmatrix}.$$ [3]

✓Checkpoint 3

Find the product CD, given that

$$C = \begin{bmatrix} 1 & 3 & 5 \\ 2 & -4 & -1 \end{bmatrix}$$

and

$$D = \begin{bmatrix} 2 & -1 \\ 4 & 3 \\ 1 & -2 \end{bmatrix}.$$

Answer: See page 365.

Example 3 Find BA, given that

$$A = \begin{bmatrix} 1 & 7 \\ -3 & 2 \end{bmatrix} \quad \text{and} \quad B = \begin{bmatrix} 1 & 0 & -1 \\ 3 & 1 & 4 \end{bmatrix}.$$

Since B is a 2×3 matrix and A is a 2×2 matrix, the product BA is not defined. [4]

✓Checkpoint 4

Give the size of each of the following products, if the product can be found.

(a) $\begin{bmatrix} 2 & 4 \\ 6 & 8 \end{bmatrix} \begin{bmatrix} 1 & 2 & 3 \\ 0 & -1 & 2 \end{bmatrix}$

(b) $\begin{bmatrix} 1 & 2 \\ 5 & 10 \\ 12 & 7 \end{bmatrix} \begin{bmatrix} 2 & 4 \\ 3 & 6 \\ 9 & 1 \end{bmatrix}$

(c) $\begin{bmatrix} 5 \\ 2 \\ 4 \end{bmatrix} [1 \quad 0 \quad 6]$

Answers: See page 365.

TECHNOLOGY TIP Graphing calculators can find matrix products. However, if you use a graphing calculator to try to find the product in Example 3, the calculator will display an error message.

Matrix multiplication has some similarities to the multiplication of numbers.

> For any matrices A, B, and C such that all the indicated sums and products exist, matrix multiplication is associative and distributive:
>
> $\mathbf{A(BC) = (AB)C;\ A(B + C) = AB + AC;\ (B + C)A = BA + CA.}$

However, there are important differences between matrix multiplication and the multiplication of numbers. (See Exercises 19–22 at the end of this section.) In particular, matrix multiplication is *not* commutative.

> If A and B are matrices such that the products AB and BA exist,
>
> AB may not equal BA.

Figure 6.24 shows an example of this situation.

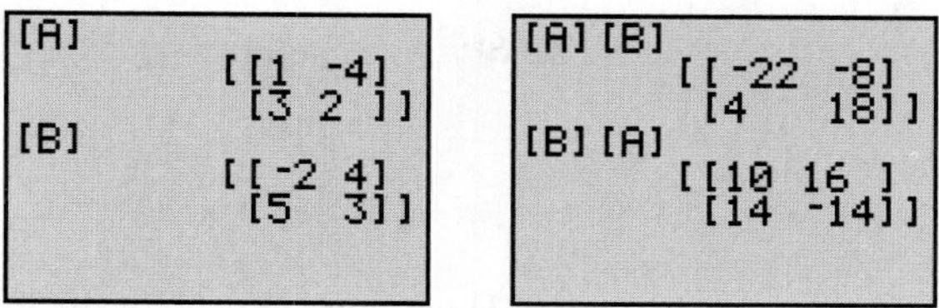

FIGURE 6.24

Example 4 A contractor builds three kinds of houses, models *A*, *B*, and *C*, with a choice of two styles, Spanish or contemporary. Matrix *P* shows the number of each kind of house planned for a new 100-home subdivision:

$$\begin{array}{l} \\ \text{Model } A \\ \text{Model } B \\ \text{Model } C \end{array} \begin{array}{c} \begin{array}{cc} \text{Spanish} & \text{Contemporary} \end{array} \\ \begin{bmatrix} 0 & 30 \\ 10 & 20 \\ 20 & 20 \end{bmatrix} \end{array} = P.$$

The amounts for each of the exterior materials used depend primarily on the style of the house. These amounts are shown in matrix Q (concrete is measured in cubic yards, lumber in units of 1000 board feet, brick in thousands, and shingles in units of 100 square feet):

$$\begin{array}{l} \\ \text{Spanish} \\ \text{Contemporary} \end{array} \begin{array}{c} \begin{array}{cccc} \text{Concrete} & \text{Lumber} & \text{Brick} & \text{Shingles} \end{array} \\ \begin{bmatrix} 10 & 2 & 0 & 2 \\ 50 & 1 & 20 & 2 \end{bmatrix} \end{array} = Q.$$

Matrix R gives the cost for each kind of material:

$$\begin{array}{l} \\ \text{Concrete} \\ \text{Lumber} \\ \text{Brick} \\ \text{Shingles} \end{array} \begin{array}{c} \text{Cost per Unit} \\ \begin{bmatrix} 20 \\ 180 \\ 60 \\ 25 \end{bmatrix} \end{array} = R.$$

(a) What is the total cost for each model of house?

Solution First find the product PQ, which shows the amount of each material needed for each model of house:

$$PQ = \begin{bmatrix} 0 & 30 \\ 10 & 20 \\ 20 & 20 \end{bmatrix} \begin{bmatrix} 10 & 2 & 0 & 2 \\ 50 & 1 & 20 & 2 \end{bmatrix}$$

$$PQ = \begin{array}{c} \begin{array}{cccc} \text{Concrete} & \text{Lumber} & \text{Brick} & \text{Shingles} \end{array} \\ \begin{bmatrix} 1500 & 30 & 600 & 60 \\ 1100 & 40 & 400 & 60 \\ 1200 & 60 & 400 & 80 \end{bmatrix} \end{array} \begin{array}{l} \\ \text{Model } A \\ \text{Model } B. \\ \text{Model } C \end{array}$$

Now multiply PQ and R, the cost matrix, to get the total cost for each model of house:

$$\begin{bmatrix} 1500 & 30 & 600 & 60 \\ 1100 & 40 & 400 & 60 \\ 1200 & 60 & 400 & 80 \end{bmatrix} \begin{bmatrix} 20 \\ 180 \\ 60 \\ 25 \end{bmatrix} = \overset{\text{Cost}}{\begin{bmatrix} 72{,}900 \\ 54{,}700 \\ 60{,}800 \end{bmatrix}} \begin{matrix} \text{Model } A \\ \text{Model } B. \\ \text{Model } C \end{matrix}$$

(b) How much of each of the four kinds of material must be ordered?

Solution The totals of the columns of matrix PQ will give a matrix whose elements represent the total amounts of each material needed for the subdivision. Call this matrix T and write it as a row matrix:

$$T = [3800 \quad 130 \quad 1400 \quad 200].$$

(c) What is the total cost for material?

Solution Find the total cost of all the materials by taking the product of matrix T, the matrix showing the total amounts of each material, and matrix R, the cost matrix. [To multiply these and get a 1×1 matrix representing total cost, we must multiply a 1×4 matrix by a 4×1 matrix. This is why T was written as a row matrix in (b).] So, we have

$$TR = [3800 \quad 130 \quad 1400 \quad 200] \begin{bmatrix} 20 \\ 180 \\ 60 \\ 25 \end{bmatrix} = [188{,}400].$$

(d) Suppose the contractor builds the same number of homes in five subdivisions. What is the total amount of each material needed in this case?

Solution Determine the total amount of each material for each model for all five subdivisions. Multiply PQ by the scalar 5 as follows:

$$5\begin{bmatrix} 1500 & 30 & 600 & 60 \\ 1100 & 40 & 400 & 60 \\ 1200 & 60 & 400 & 80 \end{bmatrix} = \begin{bmatrix} 7500 & 150 & 3000 & 300 \\ 5500 & 200 & 2000 & 300 \\ 6000 & 300 & 2000 & 400 \end{bmatrix}.$$

We can introduce notation to help keep track of the quantities a matrix represents. For example, we can say that matrix P from Example 4 represents models/styles, matrix Q represents styles/materials, and matrix R represents materials/cost. In each case, the meaning of the rows is written first and the columns second. When we found the product PQ in Example 4, the rows of the matrix represented models and the columns represented materials. Therefore, we can say that the matrix product PQ represents models/materials. The common quantity, styles, in both P and Q was eliminated in the product PQ. Do you see that the product $(PQ)R$ represents models/cost?

In practical problems, this notation helps decide in what order to multiply two matrices so that the results are meaningful. In Example 4(c), we could have found either product RT or product TR. However, since T represents subdivisions/materials and R represents materials/cost, the product TR gives subdivisions/cost. ✓5

✓Checkpoint 5

Let matrix A be

$$\text{Brand} \begin{matrix} X \\ Y \end{matrix} \overset{\begin{matrix}\text{Vitamin} \\ C \quad E \quad K\end{matrix}}{\begin{bmatrix} 2 & 7 & 5 \\ 4 & 6 & 9 \end{bmatrix}}$$

and matrix B be

$$\text{Vitamin} \begin{matrix} C \\ E \\ K \end{matrix} \overset{\begin{matrix}\text{Cost} \\ X \quad Y\end{matrix}}{\begin{bmatrix} 12 & 14 \\ 18 & 15 \\ 9 & 10 \end{bmatrix}}.$$

(a) What quantities do matrices A and B represent?

(b) What quantities does the product AB represent?

(c) What quantities does the product BA represent?

Answers: See page 365.

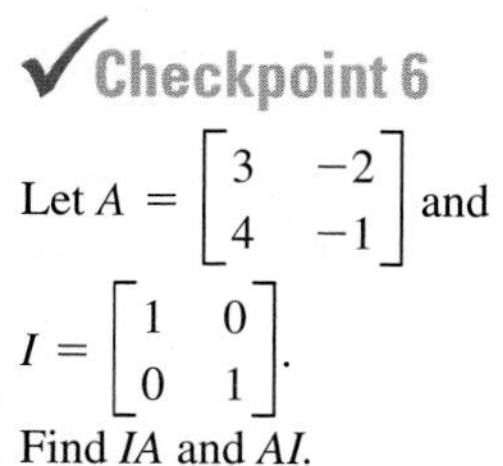

Checkpoint 6

Let $A = \begin{bmatrix} 3 & -2 \\ 4 & -1 \end{bmatrix}$ and

$I = \begin{bmatrix} 1 & 0 \\ 0 & 1 \end{bmatrix}$.

Find IA and AI.

Answer: *See page 365.*

IDENTITY MATRICES

Recall from Section 1.1 that the real number 1 is the identity element for multiplication of real numbers: For any real number a, $a \cdot 1 = 1 \cdot a = a$. In this section, an **identity matrix I** is defined that has properties similar to those of the number 1.

If I is to be the identity matrix, the products AI and IA must both equal A. The 2×2 identity matrix that satisfies these conditions is

$$I = \begin{bmatrix} 1 & 0 \\ 0 & 1 \end{bmatrix}. \quad 6✓$$

To check that I is really the 2×2 identity matrix, let

$$A = \begin{bmatrix} a & b \\ c & d \end{bmatrix}.$$

Then AI and IA should both equal A:

$$AI = \begin{bmatrix} a & b \\ c & d \end{bmatrix}\begin{bmatrix} 1 & 0 \\ 0 & 1 \end{bmatrix} = \begin{bmatrix} a(1) + b(0) & a(0) + b(1) \\ c(1) + d(0) & c(0) + d(1) \end{bmatrix} = \begin{bmatrix} a & b \\ c & d \end{bmatrix} = A;$$

$$IA = \begin{bmatrix} 1 & 0 \\ 0 & 1 \end{bmatrix}\begin{bmatrix} a & b \\ c & d \end{bmatrix} = \begin{bmatrix} 1(a) + 0(c) & 1(b) + 0(d) \\ 0(a) + 1(c) & 0(b) + 1(d) \end{bmatrix} = \begin{bmatrix} a & b \\ c & d \end{bmatrix} = A.$$

This verifies that I has been defined correctly. (It can also be shown that I is the only 2×2 identity matrix.)

The identity matrices for 3×3 matrices and 4×4 matrices are, respectively,

$$I = \begin{bmatrix} 1 & 0 & 0 \\ 0 & 1 & 0 \\ 0 & 0 & 1 \end{bmatrix} \quad \text{and} \quad I = \begin{bmatrix} 1 & 0 & 0 & 0 \\ 0 & 1 & 0 & 0 \\ 0 & 0 & 1 & 0 \\ 0 & 0 & 0 & 1 \end{bmatrix}.$$

By generalizing these findings, an identity matrix can be found for any n by n matrix. This identity matrix will have 1s on the main diagonal from upper left to lower right, with all other entries equal to 0.

TECHNOLOGY TIP An $n \times n$ identity matrix can be displayed on most graphing calculators by using IDENTITY n or IDENT n or IDENMAT(n). Look in the MATH or OPS submenu of the TI MATRIX menu, or the OPTN MAT menu of Casio.

INVERSE MATRICES

Recall that for every nonzero real number a, the equation $ax = 1$ has a solution, namely, $x = 1/a = a^{-1}$. Similarly, for a square matrix A, we consider the matrix equation $AX = I$. This equation does not always have a solution, but when it does, we use special terminology. If there is a matrix A^{-1} satisfying

$$AA^{-1} = I,$$

(that is, A^{-1} is a solution of $AX = I$), then A^{-1} is called the **inverse matrix** of A. In this case, it can be proved that $A^{-1}A = I$ and that A^{-1} is unique (that is, a square matrix has no more than one inverse). When a matrix has an inverse, it can be found by using the row operations given in Section 6.2, as we shall see later.

CAUTION Only square matrices have inverses, but not every square matrix has one. A matrix that does not have an inverse is called a **singular matrix**. Note that the symbol A^{-1} (read "A-inverse") does *not* mean $1/A$ or $1/A$; the symbol A^{-1} is just the notation for the inverse of matrix A. There is no such thing as matrix division.

Example 5 Given matrices A and B as follows, determine whether B is the inverse of A:

$$A = \begin{bmatrix} 1 & 2 \\ 4 & 6 \end{bmatrix}; \quad B = \begin{bmatrix} -3 & 1 \\ 2 & -\frac{1}{2} \end{bmatrix}.$$

Solution B is the inverse of A if $AB = I$ and $BA = I$, so we find those products:

$$AB = \begin{bmatrix} 1 & 2 \\ 4 & 6 \end{bmatrix}\begin{bmatrix} -3 & 1 \\ 2 & -\frac{1}{2} \end{bmatrix} = \begin{bmatrix} 1 & 0 \\ 0 & 1 \end{bmatrix} = I;$$

$$BA = \begin{bmatrix} -3 & 1 \\ 2 & -\frac{1}{2} \end{bmatrix}\begin{bmatrix} 1 & 2 \\ 4 & 6 \end{bmatrix} = \begin{bmatrix} 1 & 0 \\ 0 & 1 \end{bmatrix} = I.$$

Therefore, B is the inverse of A; that is, $A^{-1} = B$. (It is also true that A is the inverse of B, or $B^{-1} = A$. 7✓

✓Checkpoint 7

Given $A = \begin{bmatrix} 2 & 1 \\ 3 & 8 \end{bmatrix}$ and $B = \begin{bmatrix} -1 & 3 \\ 1 & -2 \end{bmatrix}$, determine whether they are inverses.

Answer: *See page 365.*

Example 6 Find the multiplicative inverse of

$$A = \begin{bmatrix} 2 & 4 \\ 1 & -1 \end{bmatrix}.$$

Solution Let the unknown inverse matrix be

$$A^{-1} = \begin{bmatrix} x & y \\ z & w \end{bmatrix}.$$

By the definition of matrix inverse, $AA^{-1} = I$, or

$$AA^{-1} = \begin{bmatrix} 2 & 4 \\ 1 & -1 \end{bmatrix}\begin{bmatrix} x & y \\ z & w \end{bmatrix} = \begin{bmatrix} 1 & 0 \\ 0 & 1 \end{bmatrix}.$$

Use matrix multiplication to get

$$\begin{bmatrix} 2x + 4z & 2y + 4w \\ x - z & y - w \end{bmatrix} = \begin{bmatrix} 1 & 0 \\ 0 & 1 \end{bmatrix}.$$

Setting corresponding elements equal to each other gives the system of equations

$$2x + 4z = 1 \qquad \textbf{(1)}$$
$$2y + 4w = 0 \qquad \textbf{(2)}$$
$$x - z = 0 \qquad \textbf{(3)}$$
$$y - w = 1. \qquad \textbf{(4)}$$

Since equations (1) and (3) involve only x and z, while equations (2) and (4) involve only y and w, these four equations lead to two systems of equations:

$$\begin{aligned} 2x + 4z &= 1 \\ x - z &= 0 \end{aligned} \quad \text{and} \quad \begin{aligned} 2y + 4w &= 0 \\ y - w &= 1. \end{aligned}$$

Writing the two systems as augmented matrices gives

$$\left[\begin{array}{cc|c} 2 & 4 & 1 \\ 1 & -1 & 0 \end{array}\right] \quad \text{and} \quad \left[\begin{array}{cc|c} 2 & 4 & 0 \\ 1 & -1 & 1 \end{array}\right].$$

Note that the row operations needed to transform both matrices are the same because the first two columns of both matrices are identical. Consequently, we can save time by combining these matrices into the single matrix

$$\left[\begin{array}{cc|cc} 2 & 4 & 1 & 0 \\ 1 & -1 & 0 & 1 \end{array}\right]. \tag{5}$$

Columns 1–3 represent the first system and columns 1, 2, and 4 represent the second system. Now use row operations as follows:

$$\left[\begin{array}{cc|cc} 1 & -1 & 0 & 1 \\ 2 & 4 & 1 & 0 \end{array}\right] \quad \text{Interchange } R_1 \text{ and } R_2.$$

$$\left[\begin{array}{cc|cc} 1 & -1 & 0 & 1 \\ 0 & 6 & 1 & -2 \end{array}\right] \quad -2R_1 + R_2$$

$$\left[\begin{array}{cc|cc} 1 & -1 & 0 & 1 \\ 0 & 1 & \frac{1}{6} & -\frac{1}{3} \end{array}\right] \quad \tfrac{1}{6}R_2$$

$$\left[\begin{array}{cc|cc} 1 & 0 & \frac{1}{6} & \frac{2}{3} \\ 0 & 1 & \frac{1}{6} & -\frac{1}{3} \end{array}\right]. \quad R_2 + R_1 \tag{6}$$

The left half of the augmented matrix in equation (6) is the identity matrix, so the Gauss–Jordan process is finished and the solutions can be read from the right half of the augmented matrix. The numbers in the first column to the right of the vertical bar give the values of x and z. The second column to the right of the bar gives the values of y and w. That is,

$$\left[\begin{array}{cc|cc} 1 & 0 & x & y \\ 0 & 1 & z & w \end{array}\right] = \left[\begin{array}{cc|cc} 1 & 0 & \frac{1}{6} & \frac{2}{3} \\ 0 & 1 & \frac{1}{6} & -\frac{1}{3} \end{array}\right],$$

so that

$$A^{-1} = \begin{bmatrix} x & y \\ z & w \end{bmatrix} = \begin{bmatrix} \frac{1}{6} & \frac{2}{3} \\ \frac{1}{6} & -\frac{1}{3} \end{bmatrix}.$$

Check by multiplying A and A^{-1}. The result should be I:

$$AA^{-1} = \begin{bmatrix} 2 & 4 \\ 1 & -1 \end{bmatrix}\begin{bmatrix} \frac{1}{6} & \frac{2}{3} \\ \frac{1}{6} & -\frac{1}{3} \end{bmatrix} = \begin{bmatrix} \frac{1}{3} + \frac{2}{3} & \frac{4}{3} - \frac{4}{3} \\ \frac{1}{6} - \frac{1}{6} & \frac{2}{3} + \frac{1}{3} \end{bmatrix} = \begin{bmatrix} 1 & 0 \\ 0 & 1 \end{bmatrix} = I.$$

Thus, the original augmented matrix in equation (5) has A as its left half and the identity matrix as its right half, while the final augmented matrix in equation (6), at the end of the Gauss–Jordan process, has the identity matrix as its left half and the inverse matrix A^{-1} as its right half:

$$[A \mid I] \rightarrow [I \mid A^{-1}].$$ 8✓

✓Checkpoint 8

Carry out the process in Example 6 on a graphing calculator as follows: Enter matrix (5) in the example as matrix [B] on the calculator. Then find RREF [B]. What is the result?

Answer: See page 365.

The procedure in Example 6 can be generalized as follows.

To obtain an **inverse matrix** A^{-1} for any $n \times n$ matrix A for which A^{-1} exists, follow these steps:

1. Form the augmented matrix $[A \mid I]$, where I is the $n \times n$ identity matrix.
2. Perform row operations on $[A \mid I]$ to get a matrix of the form $[I \mid B]$.
3. Matrix B is A^{-1}.

Example 7

Find A^{-1} if $A = \begin{bmatrix} 1 & 0 & 1 \\ 2 & -2 & -1 \\ 3 & 0 & 0 \end{bmatrix}$.

Solution First write the augmented matrix $[A \mid I]$:

$$[A \mid I] = \left[\begin{array}{ccc|ccc} 1 & 0 & 1 & 1 & 0 & 0 \\ 2 & -2 & -1 & 0 & 1 & 0 \\ 3 & 0 & 0 & 0 & 0 & 1 \end{array}\right].$$

You can either use technology (as in Checkpoint 8) or work by hand (as shown next) to transform the left side of this matrix into the 3×3 identity matrix:

$$\left[\begin{array}{ccc|ccc} 1 & 0 & 1 & 1 & 0 & 0 \\ 0 & -2 & -3 & -2 & 1 & 0 \\ 0 & 0 & -3 & -3 & 0 & 1 \end{array}\right] \quad \begin{array}{l} \\ -2R_1 + R_2 \\ -3R_1 + R_3 \end{array}$$

$$\left[\begin{array}{ccc|ccc} 1 & 0 & 1 & 1 & 0 & 0 \\ 0 & 1 & \frac{3}{2} & 1 & -\frac{1}{2} & 0 \\ 0 & 0 & 1 & 1 & 0 & -\frac{1}{3} \end{array}\right] \quad \begin{array}{l} \\ -\frac{1}{2}R_2 \\ -\frac{1}{3}R_3 \end{array}$$

$$\left[\begin{array}{ccc|ccc} 1 & 0 & 0 & 0 & 0 & \frac{1}{3} \\ 0 & 1 & 0 & -\frac{1}{2} & -\frac{1}{2} & \frac{1}{2} \\ 0 & 0 & 1 & 1 & 0 & -\frac{1}{3} \end{array}\right]. \quad \begin{array}{l} -1R_3 + R_1 \\ -\frac{3}{2}R_3 + R_2 \\ \end{array}$$

Looking at the right half of the preceding matrix, we see that

$$A^{-1} = \begin{bmatrix} 0 & 0 & \frac{1}{3} \\ -\frac{1}{2} & -\frac{1}{2} & \frac{1}{2} \\ 1 & 0 & -\frac{1}{3} \end{bmatrix}.$$

Verify that AA^{-1} is I.

The best way to find matrix inverses on a graphing calculator is illustrated in the next example.

Example 8 Use a graphing calculator to find the inverse of the following matrices (if they have inverses):

$$A = \begin{bmatrix} 2 & 1 & -1 \\ 1 & 3 & 2 \\ 1 & 1 & 1 \end{bmatrix} \quad \text{and} \quad B = \begin{bmatrix} 2 & 4 \\ 3 & 6 \end{bmatrix}.$$

Solution Enter matrix A into the calculator [Figure 6.25(a)]. Then use the x^{-1} key to find the inverse matrix, as in Figure 6.25(b). (Using ∧ and -1 for the inverse results in an error message on most calculators.)

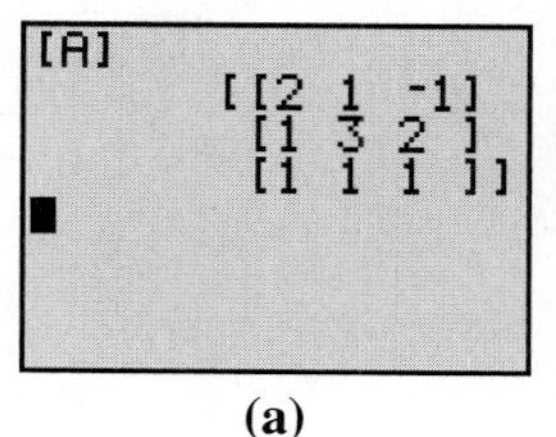

(a)

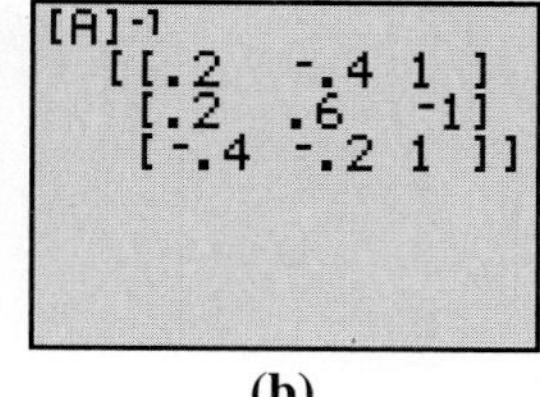

(b)

FIGURE 6.25

Now enter matrix B into the calculator and use the x^{-1} key. The result is an error message (Figure 6.26), which indicates that the matrix is singular; it does not have an inverse. (If you were working by hand, you would have found that the appropriate system of equations has no solution.) 9✓

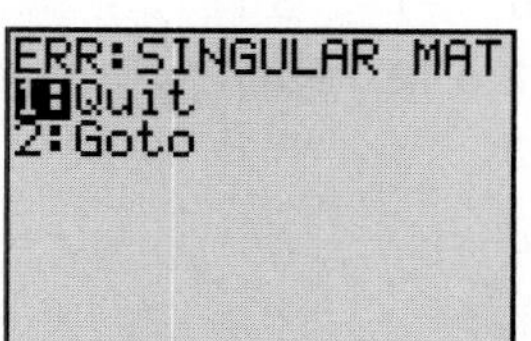

FIGURE 6.26

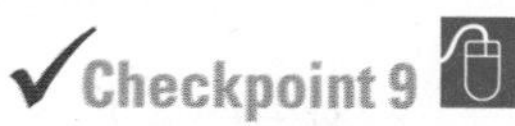

✓Checkpoint 9

Use a graphing calculator to find the inverses of these matrices (if they exist).

(a) $A = \begin{bmatrix} 1 & 2 & 3 \\ 4 & -1 & 0 \\ 5 & 1 & 3 \end{bmatrix}$

(b) $B = \begin{bmatrix} 2 & 3 \\ -1 & 4 \end{bmatrix}$

Answers: See page 365.

TECHNOLOGY TIP Because of round-off error, a graphing calculator may sometimes display an "inverse" for a matrix that doesn't actually have one. So always verify your results by multiplying A and A^{-1}. If the product is not the identity matrix, then A does not have an inverse.

6.5 ▸ Exercises

In Exercises 1–6, the sizes of two matrices A and B are given. Find the sizes of the product AB and the product BA whenever these products exist. (See Example 1.)

1. A is 2×2 and B is 2×2.
2. A is 3×3 and B is 3×2.
3. A is 3×5 and B is 5×3.
4. A is 4×3 and B is 3×6.
5. A is 4×2 and B is 3×4.
6. A is 7×3 and B is 2×7.
7. To find the product matrix AB, the number of _____ of A must be the same as the number of _____ of B.
8. The product matrix AB has the same number of _____ as A and the same number of _____ as B.

Find each of the following matrix products. (See Examples 2–4.)

9. $\begin{bmatrix} 1 & 2 \\ 3 & 4 \end{bmatrix}\begin{bmatrix} -1 \\ 3 \end{bmatrix}$

10. $\begin{bmatrix} -2 & 5 \\ 7 & 0 \end{bmatrix}\begin{bmatrix} 2 \\ 6 \end{bmatrix}$

11. $\begin{bmatrix} 2 & 2 & -1 \\ 5 & 0 & 1 \end{bmatrix}\begin{bmatrix} 0 & -2 \\ -1 & 5 \\ 0 & 2 \end{bmatrix}$

12. $\begin{bmatrix} -9 & 3 & 1 \\ 3 & 0 & 0 \end{bmatrix}\begin{bmatrix} 2 \\ -1 \\ 4 \end{bmatrix}$

13. $\begin{bmatrix} -4 & 1 \\ 2 & -3 \end{bmatrix}\begin{bmatrix} 1 & 0 \\ 0 & 1 \end{bmatrix}$

14. $\begin{bmatrix} 1 & 0 \\ 0 & 1 \end{bmatrix}\begin{bmatrix} 3 & -2 \\ 1 & -5 \end{bmatrix}$

15. $\begin{bmatrix} 1 & 0 & 0 \\ 0 & 1 & 0 \\ 0 & 0 & 1 \end{bmatrix}\begin{bmatrix} 3 & -5 & 7 \\ -2 & 1 & 6 \\ 0 & -3 & 4 \end{bmatrix}$

16. $\begin{bmatrix} -8 & 9 \\ 3 & -4 \\ -1 & 6 \end{bmatrix}\begin{bmatrix} 1 & 0 & 0 \\ 0 & 1 & 0 \end{bmatrix}$

17. $\begin{bmatrix} 1 & 2 & 3 \\ 4 & 0 & 6 \\ 7 & 8 & 9 \end{bmatrix}\begin{bmatrix} -1 & 4 \\ 7 & 0 \\ 1 & 2 \end{bmatrix}$

18. $\begin{bmatrix} -2 & 0 & 3 \\ 4 & -3 & -1 \end{bmatrix}\begin{bmatrix} 2 & 0 & -1 & 5 \\ 0 & 1 & 0 & -1 \\ 4 & 2 & 5 & -4 \end{bmatrix}$

In Exercises 19–21, use the matrices

$$A = \begin{bmatrix} -3 & -9 \\ 2 & 6 \end{bmatrix} \quad \textit{and} \quad B = \begin{bmatrix} 4 & 6 \\ 2 & 3 \end{bmatrix}.$$

19. Show that $AB \neq BA$. Hence, matrix multiplication is not commutative.

20. Show that $(A + B)^2 \neq A^2 + 2AB + B^2$.

21. Show that $(A + B)(A - B) \neq A^2 - B^2$.

22. Show that $D^2 = D$, where

$$D = \begin{bmatrix} 1 & 0 & 0 \\ \frac{1}{2} & 0 & \frac{1}{2} \\ 0 & 0 & 1 \end{bmatrix}.$$

Given matrices

$$P = \begin{bmatrix} m & n \\ p & q \end{bmatrix}, \quad X = \begin{bmatrix} x & y \\ z & w \end{bmatrix}, \quad \textit{and} \quad T = \begin{bmatrix} r & s \\ t & u \end{bmatrix},$$

verify that the statements in Exercises 23–26 are true.

23. $(PX)T = P(XT)$ (associative property)

24. $P(X + T) = PX + PT$ (distributive property)

25. $k(X + T) = kX + kT$ for any real number k

26. $(k + h)P = kP + hP$ for any real numbers k and h

Determine whether the given matrices are inverses of each other by computing their product. (See Example 5.)

27. $\begin{bmatrix} 5 & 2 \\ 3 & -1 \end{bmatrix}$ and $\begin{bmatrix} -1 & 2 \\ 3 & -4 \end{bmatrix}$

28. $\begin{bmatrix} 3 & 5 \\ 7 & 9 \end{bmatrix}$ and $\begin{bmatrix} -\frac{9}{8} & \frac{5}{8} \\ \frac{7}{8} & -\frac{3}{8} \end{bmatrix}$

29. $\begin{bmatrix} 3 & -1 \\ -4 & 2 \end{bmatrix}$ and $\begin{bmatrix} 1 & \frac{1}{2} \\ 2 & \frac{3}{2} \end{bmatrix}$

30. $\begin{bmatrix} 1 & 1 \\ .1 & .2 \end{bmatrix}$ and $\begin{bmatrix} 2 & -10 \\ -1 & 10 \end{bmatrix}$

31. $\begin{bmatrix} 1 & 1 & 1 \\ 2 & 3 & 0 \\ 1 & 2 & 1 \end{bmatrix}$ and $\begin{bmatrix} 1.5 & .5 & -1.5 \\ -1 & 0 & 1 \\ .5 & -.5 & .5 \end{bmatrix}$

32. $\begin{bmatrix} 2 & 5 & 4 \\ 1 & 4 & 3 \\ 1 & 3 & 2 \end{bmatrix}$ and $\begin{bmatrix} 1 & 2 & 1 \\ -5 & 8 & 2 \\ 7 & -11 & -3 \end{bmatrix}$

Find the inverse, if it exists, for each of the given matrices. (See Example 6.)

33. $\begin{bmatrix} 2 & -3 \\ -1 & 2 \end{bmatrix}$

34. $\begin{bmatrix} 2 & 1 \\ 1 & 1 \end{bmatrix}$

35. $\begin{bmatrix} -1 & 2 \\ 1 & -1 \end{bmatrix}$

36. $\begin{bmatrix} -3 & -5 \\ 6 & 10 \end{bmatrix}$

37. $\begin{bmatrix} 1 & 2 \\ 3 & 6 \end{bmatrix}$

38. $\begin{bmatrix} 1 & 2 \\ 3 & 4 \end{bmatrix}$

39. $\begin{bmatrix} 1 & -1 & 0 \\ -1 & 2 & 3 \\ 1 & 0 & 2 \end{bmatrix}$

40. $\begin{bmatrix} 0 & 1 & -1 \\ 2 & -2 & -1 \\ -1 & 1 & 1 \end{bmatrix}$

41. $\begin{bmatrix} 1 & 4 & 3 \\ 1 & -3 & -2 \\ 2 & 5 & 4 \end{bmatrix}$

42. $\begin{bmatrix} 1 & 2 & 0 \\ 3 & -1 & 2 \\ -2 & 3 & -2 \end{bmatrix}$

43. $\begin{bmatrix} 1 & -1 & 4 \\ 0 & 1 & 3 \\ 2 & -3 & 4 \end{bmatrix}$

44. $\begin{bmatrix} 5 & 0 & 2 \\ 2 & 2 & 1 \\ -3 & 1 & -1 \end{bmatrix}$

Use a graphing calculator to find the inverse of each matrix.

45. $\begin{bmatrix} 1 & 2 & 3 \\ 1 & 4 & 2 \\ 0 & 1 & -1 \end{bmatrix}$

46. $\begin{bmatrix} 2 & 2 & -4 \\ 2 & 6 & 0 \\ -3 & -3 & 5 \end{bmatrix}$

47. $\begin{bmatrix} 1 & 0 & -2 & 0 \\ -2 & 1 & 2 & 2 \\ 3 & -1 & -2 & -3 \\ 0 & 1 & 4 & 1 \end{bmatrix}$

48. $\begin{bmatrix} 1 & 1 & 0 & 2 \\ 2 & -1 & 1 & -1 \\ 3 & 3 & 2 & -2 \\ 1 & 2 & 1 & 0 \end{bmatrix}$

A graphing calculator or other technology is recommended for part (c) of Exercises 49–51.

49. **Social Science** The average birth and death rates per million people for several regions of the world and the populations (in millions) of those regions are shown in the following tables*:

Region	Births	Deaths
Asia	.024	.008
Latin America	.025	.007
North America	.015	.009
Europe	.011	.011

Year	Asia	Latin America	North America	Europe
1970	1996	286	226	460
1980	2440	365	252	484
1990	2906	455	277	499
2000	3683	519	310	729
2025 (projected)	4723	697	364	702

(a) Write the information in the first table as a 4×2 matrix R.

(b) Write the information in the second table as a 5×4 matrix P.

(c) Find the product PR.

(d) Explain what PR represents.

(e) From matrix PR, what was the total number of births in 2000? What total number of deaths is projected for 2025?

50. **Business** The first table shows the number of employees (in millions) in various sectors of the economy over a three-year period, and the second table gives the average weekly wage per employee (in dollars) in each sector over the same period†:

Year	2004	2005	2006
Private	108.5	110.6	112.7
State Government	4.5	4.5	4.6
Local Government	13.6	13.7	13.8
Federal Government	2.7	2.7	2.7

Year	2004	2005	2006
Private	753	779	816
State Government	791	812	844
Local Government	708	725	753
Federal Government	1111	1151	1198

(a) Write the information in the first table as a 4×3 matrix A.

(b) Write the information in the second table as a 3×4 matrix B.

(c) Find the product AB.

(d) Explain what each of these entries in AB represent: row 1, column 1; row 2, column 2; row 3, column 3; row 4, column 4. Do the other entries represent anything meaningful?

(e) From matrix AB, what was the total weekly payroll (in dollars) of local government over the three-year period?

51. **Social Science** The following table shows the population (in millions) of four countries in various years*:

	2001	2010
United States	278.1	300.1
Canada	31.6	34.3
Argentina	37.4	41.1
Japan	126.8	127.3

The next table gives the average birth and death rates for these countries*:

	Birth Rate	Death Rate
United States	.01425	.00865
Canada	.01145	.00775
Argentina	.0175	.00755
Japan	.00945	.00925

(a) Write the information in the first table as a 2×4 matrix A.

(b) Write the information in the second table as a 4×2 matrix B.

(c) Find the product AB.

(d) Explain what AB represents.

(e) According to matrix AB, what is the total number of people born in these four countries combined in 2001?

*U.S. Census Bureau and the United Nations Population Fund.

†U.S. Bureau of Labor Statistics.

*U.S. Census Bureau, U.S. Department of Commerce.

52. **Health** The first table shows the number of live births (in thousands). The second shows the infant mortality rates (deaths per 1000 live births) in those years.*

Year	2002	2003	2004	2005
Black	594	600	616	633
White	3175	3226	3223	3229

Year	2002	2003	2004	2005
Black	14.4	14.0	13.8	13.7
White	5.8	5.7	5.7	5.7

(a) Write the information in the first table as a 4×2 matrix C.
(b) Write the information in the second table as a 2×4 matrix D.
(c) Find the product DC.
(d) Which entry in DC gives the total number of black infant deaths from 2002 to 2005?
(e) Which entry in DC gives the total number of white infant deaths from 2002 to 2005?

53. **Business** The first table shows the per-capita charges (in dollars) over a four-year period for physician services and the per-capita charges for prescription drugs over the same period, and the second table shows the percentage of these charges that were covered by private insurance†:

Year	2003	2004	2005	2006
Physician Services	1259	1339	1424	1493
Prescription Drugs	600	642	673	723

Year	2003	2004	2005	2006
Physician Services	82.6	82.7	82.7	82.6
Prescription Drugs	65.4	66.1	66.2	66.6

(a) Write the information in the first table as a 4×2 matrix A.
(b) Write the information in the second table as a 2×4 matrix B.
(c) Find AB and round each entry to two decimal places.
(d) Find BA and round each entry to two decimal places.
(e) Explain the meaning (if any) of each entry in AB and in BA.

54. **Business** Burger Barn's three locations sell hamburgers, fries, and soft drinks. Barn I sells 900 burgers, 600 orders of fries, and 750 soft drinks each day. Barn II sells 1500 burgers a day and Barn III sells 1150. Soft drink sales number 900 a day at Barn II and 825 a day at Barn III. Barn II sells 950 orders of fries per day and Barn III sells 800.
(a) Write a 3×3 matrix S that displays daily sales figures for all locations.
(b) Burgers cost $1.50 each, fries $.90 an order, and soft drinks $.60 each. Write a 1×3 matrix P that displays the prices.
(c) What matrix product displays the daily revenue at each of the three locations?
(d) What is the total daily revenue from all locations?

55. **Business** The four departments of Stagg Enterprises need to order the following amounts of the same products:

	Paper	Tape	Ink Cartridges	Memo Pads	Pens
Department 1	10	4	3	5	6
Department 2	7	2	2	3	8
Department 3	4	5	1	0	10
Department 4	0	3	4	5	5

The unit price (in dollars) of each product is as follows for two suppliers:

Product	Supplier A	Supplier B
Paper	9	12
Tape	6	6
Ink Cartridges	24	18
Memo Pads	4	4
Pens	8	12

(a) Use matrix multiplication to get a matrix showing the comparative costs for each department for the products from the two suppliers.
(b) Find the total cost to buy products from each supplier. From which supplier should the company make the purchase?

56. **Business** The Perulli Candy Company makes three types of chocolate candy: Cheery Cherry, Mucho Mocha, and Almond Delight. The company produces its products in San Diego, Mexico City, and Managua, using two main ingredients: chocolate and sugar.
(a) Each kilogram of Cheery Cherry requires .5 kilogram of sugar and .2 kilogram of chocolate; each kilogram of Mucho Mocha requires .4 kilogram of sugar and .3 kilogram of chocolate; and each kilogram of Almond Delight requires .3 kilogram of sugar and .3 kilogram of chocolate. Put this information into a 2×3 matrix, labeling the rows and columns.
(b) The cost of 1 kilogram of sugar is $3 in San Diego, $2 in Mexico City, and $1 in Managua. The cost of 1 kilogram of chocolate is $3 in San Diego, $3 in Mexico City, and $4 in Managua. Put this information into a matrix in such a way that when you

*U.S. National Center for Health Statistics.
†U.S. Centers for Medicare and Medicaid Services.

multiply it with your matrix from part (a), you get a matrix representing the cost by ingredient of producing each type of candy in each city.

(c) Multiply the matrices in parts (a) and (b), labeling the product matrix.

(d) From part (c), what is the combined cost of sugar and chocolate required to produce 1 kilogram of Mucho Mocha in Managua?

(e) Perulli Candy needs to produce quickly a special shipment of 100 kilograms of Cheery Cherry, 200 kilograms of Mucho Mocha, and 500 kilograms of Almond Delight, and it decides to select one factory to fill the entire order. Use matrix multiplication to determine in which city the total cost of sugar and chocolate combined required to produce the order is the smallest.

✓ Checkpoint Answers

1. \$27,100
2. **(a)** No
   (b) Yes; 2×6
3. $CD = \begin{bmatrix} 19 & -2 \\ -13 & -12 \end{bmatrix}$
4. **(a)** 2×3
   (b) Not possible
   (c) 3×3
5. **(a)** A = brand/vitamin;
   B = vitamin/cost.
   (b) AB = brand/cost.
   (c) Not meaningful, although the product BA can be found.
6. $IA = \begin{bmatrix} 3 & -2 \\ 4 & -1 \end{bmatrix} = A$, and $AI = \begin{bmatrix} 3 & -2 \\ 4 & -1 \end{bmatrix} = A$.
7. No, because $AB \neq I$.
8. RREF [B] is matrix (6) in the example.
9. **(a)** No inverse
   (b) Use the FRAC key (if you have one) to simplify the answer, which is
   $$B^{-1} = \begin{bmatrix} 4/11 & -3/11 \\ 1/11 & 2/11 \end{bmatrix}.$$

6.6 ▶ Applications of Matrices

This section gives a variety of applications of matrices.

SOLVING SYSTEMS WITH MATRICES

Consider this system of linear equations:

$$2x - 3y = 4$$
$$x + 5y = 2.$$

Let

$$A = \begin{bmatrix} 2 & -3 \\ 1 & 5 \end{bmatrix}, \quad X = \begin{bmatrix} x \\ y \end{bmatrix}, \quad \text{and} \quad B = \begin{bmatrix} 4 \\ 2 \end{bmatrix}.$$

Since

$$AX = \begin{bmatrix} 2 & -3 \\ 1 & 5 \end{bmatrix}\begin{bmatrix} x \\ y \end{bmatrix} = \begin{bmatrix} 2x - 3y \\ x + 5y \end{bmatrix} \quad \text{and} \quad B = \begin{bmatrix} 4 \\ 2 \end{bmatrix},$$

the original system is equivalent to the single matrix equation $AX = B$. Similarly, any system of linear equations can be written as a matrix equation $AX = B$. The matrix A is called the **coefficient matrix**. ✓1

✓ Checkpoint 1

Write the matrix of coefficients, the matrix of variables, and the matrix of constants for the system

$$2x + 6y = -14$$
$$-x - 2y = 3.$$

Answer: See page 378.

A matrix equation $AX = B$ can be solved if A^{-1} exists. Assuming that A^{-1} exists and using the facts that $A^{-1}A = I$ and $IX = X$, along with the associative property of multiplication of matrices, gives

$$\begin{aligned} AX &= B \\ A^{-1}(AX) &= A^{-1}B && \text{Multiply both sides by } A^{-1}. \\ (A^{-1}A)X &= A^{-1}B && \text{Associative property} \\ IX &= A^{-1}B && \text{Inverse property} \\ X &= A^{-1}B. && \text{Identity property} \end{aligned}$$

When multiplying by matrices on both sides of a matrix equation, be careful to multiply in the same order on both sides, since multiplication of matrices is not commutative (unlike the multiplication of real numbers). This discussion is summarized below.

Suppose that a system of equations with the same number of equations as variables is written in matrix form as $AX = B$, where A is the square matrix of coefficients, X is the column matrix of variables, and B is the column matrix of constants. If A has an inverse, then the unique solution of the system is $X = A^{-1}B$.*

Example 1 Consider this system of equations:

$$\begin{aligned} x + y + z &= 2 \\ 2x + 3y &= 5 \\ x + 2y + z &= -1. \end{aligned}$$

(a) Write the system as a matrix equation.

Solution We have these three matrices:

Coefficient Matrix	*Matrix of Variables*		*Matrix of Constants*
$A = \begin{bmatrix} 1 & 1 & 1 \\ 2 & 3 & 0 \\ 1 & 2 & 1 \end{bmatrix}$,	$X = \begin{bmatrix} x \\ y \\ z \end{bmatrix}$,	and	$B = \begin{bmatrix} 2 \\ 5 \\ -1 \end{bmatrix}$.

So the matrix equation is

$$AX = B$$

$$\begin{bmatrix} 1 & 1 & 1 \\ 2 & 3 & 0 \\ 1 & 2 & 1 \end{bmatrix}\begin{bmatrix} x \\ y \\ z \end{bmatrix} = \begin{bmatrix} 2 \\ 5 \\ -1 \end{bmatrix}.$$

*If A does not have an inverse, then the system either has no solution or has an infinite number of solutions. Use the methods of Sections 6.1 or 6.2.

(b) Find A^{-1} and solve the equation.

Solution Use Exercise 31 of Section 6.5, technology, or the method of Section 6.5 to find that

$$A^{-1} = \begin{bmatrix} 1.5 & .5 & -1.5 \\ -1 & 0 & 1 \\ .5 & -.5 & .5 \end{bmatrix}.$$

Hence,

$$X = A^{-1}B = \begin{bmatrix} 1.5 & .5 & -1.5 \\ -1 & 0 & 1 \\ .5 & -.5 & .5 \end{bmatrix}\begin{bmatrix} 2 \\ 5 \\ -1 \end{bmatrix} = \begin{bmatrix} 7 \\ -3 \\ -2 \end{bmatrix}.$$

Thus, the solution of the original system is (7, −3, −2). 2✓

✓Checkpoint 2

Use the inverse matrix to solve the system in Example 1 if the constants for the three equations are 12, 0, and 8, respectively.

Answer: See page 378.

Example 2 Use the inverse of the coefficient matrix to solve the system

$$\begin{aligned} x + 1.5y &= 8 \\ 2x + \quad 3y &= 10. \end{aligned}$$

Solution The coefficient matrix is $A = \begin{bmatrix} 1 & 1.5 \\ 2 & 3 \end{bmatrix}$. A graphing calculator will indicate that A^{-1} does not exist. If we try to carry out the row operations, we see why:

$$\left[\begin{array}{cc|cc} 1 & 1.5 & 1 & 0 \\ 2 & 3 & 0 & 1 \end{array}\right]$$

$$\left[\begin{array}{cc|cc} 1 & 1.5 & 1 & 0 \\ 0 & 0 & -2 & 1 \end{array}\right]. \qquad -2R_1 + R_2$$

The next step cannot be performed because of the zero in the second row, second column. Verify that the original system has no solution. 3✓

✓Checkpoint 3

Solve the system in Example 2 if the constants are, respectively, 3 and 6.

Answer: See page 378.

INPUT–OUTPUT ANALYSIS

An interesting application of matrix theory to economics was developed by Nobel Prize winner Wassily Leontief. His application of matrices to the interdependencies in an economy is called **input–output analysis**. In practice, input–output analysis is very complicated, with many variables. We shall discuss only simple examples with just a few variables.

Input–output models are concerned with the production and flow of goods and services. A typical economy is composed of a number of different sectors (such as manufacturing, energy, transportation, agriculture, etc.). Each sector requires input from other sectors (and possibly from itself) to produce its output. For instance, manufacturing output requires energy, transportation, and manufactured items (such as tools and machinery). If an economy has n sectors, then the inputs required by the various sectors from each other to produce their outputs can be described by an $n \times n$ matrix called the **input–output matrix** (or the **technological matrix**).

Example 3 Suppose a simplified economy involves just three sectors—agriculture, manufacturing, and transportation—all in appropriate units. The production of 1 unit of agriculture requires $\frac{1}{2}$ unit of manufacturing and $\frac{1}{4}$ unit of transportation. The production of 1 unit of manufacturing requires $\frac{1}{4}$ unit of agriculture and $\frac{1}{4}$ unit of transportation. The production of 1 unit of transportation requires $\frac{1}{3}$ unit of agriculture and $\frac{1}{4}$ unit of manufacturing. Write the input–output matrix of this economy.

Solution Since there are three sectors in the economy, the input–output matrix A is 3×3. Each row and each column is labeled by a sector of the economy, as shown below. The first column lists the units from each sector of the economy that are required to produce one unit of agriculture. The second column lists the units required from each sector to produce 1 unit of manufacturing, and the last column lists the units required from each sector to produce 1 unit of transportation.

$$\begin{array}{ll} & \textit{Output} \\ & \begin{array}{ccc} \text{Agriculture} & \text{Manufacturing} & \text{Transporting} \end{array} \\ \textit{Input} \begin{array}{l} \text{Agriculture} \\ \text{Manufacturing} \\ \text{Transportation} \end{array} & \begin{bmatrix} 0 & \frac{1}{4} & \frac{1}{3} \\ \frac{1}{2} & 0 & \frac{1}{4} \\ \frac{1}{4} & \frac{1}{4} & 0 \end{bmatrix} = A. \end{array}$$

✓**Checkpoint 4**

Write a 2×2 input–output matrix in which 1 unit of electricity requires $\frac{1}{2}$ unit of water and $\frac{1}{3}$ unit of electricity, while 1 unit of water requires no water but $\frac{1}{4}$ unit of electricity.

Answer: See page 378.

Example 3 is a bit unrealistic in that no sector of the economy requires any input from itself. In an actual economy, most sectors require input from themselves as well as from other sectors to produce their output (as discussed in the paragraph preceding Example 3). Nevertheless, it is easier to learn the basic concepts from simplified examples, so we shall continue to use them.

The input–output matrix gives only a partial picture of an economy. We also need to know the amount produced by each sector and the amount of the economy's output that is used up by the sectors themselves in the production process. The remainder of the total output is available to satisfy the needs of consumers and others outside the production system.

Example 4 Consider the economy whose input–output matrix A was found in Example 3.

(a) Suppose this economy produces 60 units of agriculture, 52 units of manufacturing, and 48 units of transportation. Write this information as a column matrix.

Solution Listing the sectors in the same order as the rows of input–output matrix, we have

$$X = \begin{bmatrix} 60 \\ 52 \\ 48 \end{bmatrix}.$$

The matrix X is called the **production matrix**.

(b) How much from each sector is used up in the production process?

Solution Since $\frac{1}{4}$ unit of agriculture is used to produce each unit of manufacturing and there are 52 units of manufacturing output, the amount of agriculture used up by manufacturing is $\frac{1}{4} \times 52 = 13$ units. Similarly $\frac{1}{3}$ unit of agriculture is used to produce a unit of transportation, so $\frac{1}{3} \times 48 = 16$ units of agriculture are used up by transportation. Therefore $13 + 16 = 29$ units of agriculture are used up in the economy's production process.

A similar analysis shows that the economy's production process uses up

$$\underset{\text{agricul-ture}}{\tfrac{1}{2} \times 60} + \underset{\text{transpor-tation}}{\tfrac{1}{4} \times 48} = 30 + 12 = 42 \text{ units of manufacturing}$$

and

$$\underset{\text{agricul-ture}}{\tfrac{1}{4} \times 60} + \underset{\text{manufac-turing}}{\tfrac{1}{4} \times 52} = 15 + 13 = 28 \text{ units of transportation.}$$

(c) Describe the conclusions of part (b) in terms of the input–output matrix A and the production matrix X.

Solution The matrix product AX gives the amount from each sector that is used up in the production process, as shown here [with selected entries color coded as in part (b)]:

$$AX = \begin{bmatrix} 0 & \frac{1}{4} & \frac{1}{3} \\ \frac{1}{2} & 0 & \frac{1}{4} \\ \frac{1}{4} & \frac{1}{4} & 0 \end{bmatrix} \begin{bmatrix} 60 \\ 52 \\ 48 \end{bmatrix} = \begin{bmatrix} 0 \cdot 60 + \frac{1}{4} \cdot 52 + \frac{1}{3} \cdot 48 \\ \frac{1}{2} \cdot 60 + 0 \cdot 52 + \frac{1}{4} \cdot 48 \\ \frac{1}{4} \cdot 60 + \frac{1}{4} \cdot 52 + 0 \cdot 48 \end{bmatrix}$$

$$= \begin{bmatrix} \frac{1}{4} \cdot 52 + \frac{1}{3} \cdot 48 \\ \frac{1}{2} \cdot 60 + \frac{1}{4} \cdot 48 \\ \frac{1}{4} \cdot 60 + \frac{1}{4} \cdot 52 \end{bmatrix} = \begin{bmatrix} 29 \\ 42 \\ 28 \end{bmatrix}.$$

(d) Find the matrix $D = X - AX$ and explain what its entries represent.

Solution From parts (a) and (c), we have

$$D = X - AX = \begin{bmatrix} 60 \\ 52 \\ 48 \end{bmatrix} - \begin{bmatrix} 29 \\ 42 \\ 28 \end{bmatrix} = \begin{bmatrix} 31 \\ 10 \\ 20 \end{bmatrix}.$$

The matrix D lists the amount of each sector that is *not* used up in the production process and hence is available to groups outside the production process (such as consumers). For example, 60 units of agriculture are produced and 29 units are used up in the process, so the difference $60 - 29 = 31$ is the amount of agriculture that is available to groups outside the production process. Similar remarks apply to manufacturing and transportation. The matrix D is called the **demand matrix**. Matrices A and D show that the production of 60 units of agriculture, 52 units of manufacturing, and 48 units of transportation would satisfy an outside demand of 31 units of agriculture, 10 units of manufacturing, and 20 units of transportation. 5✓

✓Checkpoint 5

(a) Write a 2×1 matrix X to represent the gross production of 9000 units of electricity and 12,000 units of water.

(b) Find AX, using A from Checkpoint 4.

(c) Find D, using $D = X - AX$.

Answers: See page 378.

Example 4 illustrates the general situation. In an economy with n sectors, the input–output matrix A is $n \times n$. The production matrix X is a column matrix whose

n entries are the outputs of each sector of the economy. The demand matrix D is also a column matrix with n entries. This matrix is defined by

$$\mathbf{D = X - AX.}$$

D lists the amount from each sector that is available to meet the demands of consumers and other groups outside the production process.

In Example 4, we knew the input–output matrix A and the production matrix X and used them to find the demand matrix D. In practice, however, this process is reversed: The input–output matrix A and the demand matrix D are known, and we must find the production matrix X needed to satisfy the required demands. Matrix algebra can be used to solve the equation $D = X - AX$ for X:

$$D = X - AX$$

$$D = IX - AX \quad \text{Identity property}$$

$$D = (I - A)X. \quad \text{Distributive property}$$

If the matrix $I - A$ has an inverse, then

$$\mathbf{X = (I - A)^{-1}D.}$$

Example 5 Suppose, in the three-sector economy of Examples 3 and 4, there is a demand for 516 units of agriculture, 258 units of manufacturing, and 129 units of transportation. What should production be for each sector?

Solution The demand matrix is

$$D = \begin{bmatrix} 516 \\ 258 \\ 129 \end{bmatrix}.$$

Find the production matrix by first calculating $I - A$:

$$I - A = \begin{bmatrix} 1 & 0 & 0 \\ 0 & 1 & 0 \\ 0 & 0 & 1 \end{bmatrix} - \begin{bmatrix} 0 & \frac{1}{4} & \frac{1}{3} \\ \frac{1}{2} & 0 & \frac{1}{4} \\ \frac{1}{4} & \frac{1}{4} & 0 \end{bmatrix} = \begin{bmatrix} 1 & -\frac{1}{4} & -\frac{1}{3} \\ -\frac{1}{2} & 1 & -\frac{1}{4} \\ -\frac{1}{4} & -\frac{1}{4} & 1 \end{bmatrix}.$$

Using a calculator with matrix capability or row operations, find the inverse of $I - A$:

$$(I - A)^{-1} = \begin{bmatrix} 1.3953 & .4961 & .5891 \\ .8372 & 1.3643 & .6202 \\ .5581 & .4651 & 1.3023 \end{bmatrix}$$

(The entries are rounded to four decimal places.*) Since $X = (I - A)^{-1}D$,

$$X = \begin{bmatrix} 1.3953 & .4961 & .5891 \\ .8372 & 1.3643 & .6202 \\ .5581 & .4651 & 1.3023 \end{bmatrix} \begin{bmatrix} 516 \\ 258 \\ 129 \end{bmatrix} = \begin{bmatrix} 924 \\ 864 \\ 576 \end{bmatrix}$$

(rounded to the nearest whole numbers).

*Although we show the matrix $(I - A)^{-1}$ with entries rounded to four decimal places, we did not round off in calculating $(I - A)^{-1}D$. If the rounded figures are used, the numbers in the product may vary slightly in the last digit.

From the last result, we see that the production of 924 units of agriculture, 864 units of manufacturing, and 576 units of transportation is required to satisfy demands of 516, 258, and 129 units, respectively.

Example 6 An economy depends on two basic products: wheat and oil. To produce 1 metric ton of wheat requires .25 metric ton of wheat and .33 metric ton of oil. The production of 1 metric ton of oil consumes .08 metric ton of wheat and .11 metric ton of oil. Find the production that will satisfy a demand of 500 metric tons of wheat and 1000 metric tons of oil.

Solution The input–output matrix is

$$A = \begin{array}{c} \\ \end{array}\begin{matrix} \text{Wheat} & \text{Oil} \\ \end{matrix}$$

$$A = \begin{bmatrix} .25 & .08 \\ .33 & .11 \end{bmatrix} \begin{matrix} \text{Wheat} \\ \text{Oil} \end{matrix},$$

(columns: Wheat, Oil)

and we also have

$$I - A = \begin{bmatrix} 1 & 0 \\ 0 & 1 \end{bmatrix} - \begin{bmatrix} .25 & .08 \\ .33 & .11 \end{bmatrix} = \begin{bmatrix} .75 & -.08 \\ -.33 & .89 \end{bmatrix}.$$

Next, use technology or the methods of Section 6.5 to calculate $(I - A)^{-1}$:

$$(I - A)^{-1} = \begin{bmatrix} 1.3882 & .1248 \\ .5147 & 1.1699 \end{bmatrix} \quad \text{(rounded)}.$$

The demand matrix is

$$D = \begin{bmatrix} 500 \\ 1000 \end{bmatrix}.$$

Consequently, the production matrix is

$$X = (I - A)^{-1}D = \begin{bmatrix} 1.3882 & .1248 \\ .5147 & 1.1699 \end{bmatrix}\begin{bmatrix} 500 \\ 1000 \end{bmatrix} = \begin{bmatrix} 819 \\ 1427 \end{bmatrix},$$

where the production numbers have been rounded to the nearest whole numbers. The production of 819 metric tons of wheat and 1427 metric tons of oil is required to satisfy the indicated demand. 6✓

TECHNOLOGY TIP If you are using a graphing calculator to determine X, you can calculate $(I - A)^{-1}D$ in one step without finding the intermediate matrices $I - A$ and $(I - A)^{-1}$.

✓Checkpoint 6

A simple economy depends on just two products: beer and pretzels.

(a) Suppose $\frac{1}{2}$ unit of beer and $\frac{1}{2}$ unit of pretzels are needed to make 1 unit of beer, and $\frac{3}{4}$ unit of beer is needed to make 1 unit of pretzels. Write the technological matrix A for the economy.

(b) Find $I - A$.

(c) Find $(I - A)^{-1}$.

(d) Find the gross production X that will be needed to get a net production of

$$D = \begin{bmatrix} 100 \\ 1000 \end{bmatrix}.$$

Answers: See page 378.

CODE THEORY

Governments need sophisticated methods of coding and decoding messages. One example of such an advanced code uses matrix theory. This code takes the letters in the words and divides them into groups. (Each space between words is treated as a letter; punctuation is disregarded.) Then, numbers are assigned to the letters of the alphabet. For our purposes, let the letter *a* correspond to 1, *b* to 2, and so on. Let the number 27 correspond to a space.

For example, the message

mathematics is for the birds

can be divided into groups of three letters each:

mat hem ati cs– is– for –th e–b ird s––

(We used − to represent a space.) We now write a column matrix for each group of three symbols, using the corresponding numbers instead of letters. For example, the first four groups can be written as

$$\overset{mat}{\begin{bmatrix} 13 \\ 1 \\ 20 \end{bmatrix}}, \quad \overset{hem}{\begin{bmatrix} 8 \\ 5 \\ 13 \end{bmatrix}}, \quad \overset{ati}{\begin{bmatrix} 1 \\ 20 \\ 9 \end{bmatrix}}, \quad \overset{cs-}{\begin{bmatrix} 3 \\ 19 \\ 27 \end{bmatrix}}.$$

The entire message consists of ten 3 × 1 column matrices:

$$\begin{bmatrix} 13 \\ 1 \\ 20 \end{bmatrix}, \begin{bmatrix} 8 \\ 5 \\ 13 \end{bmatrix}, \begin{bmatrix} 1 \\ 20 \\ 9 \end{bmatrix}, \begin{bmatrix} 3 \\ 19 \\ 27 \end{bmatrix}, \begin{bmatrix} 9 \\ 19 \\ 27 \end{bmatrix}, \begin{bmatrix} 6 \\ 15 \\ 18 \end{bmatrix}, \begin{bmatrix} 27 \\ 20 \\ 8 \end{bmatrix}, \begin{bmatrix} 5 \\ 27 \\ 2 \end{bmatrix}, \begin{bmatrix} 9 \\ 18 \\ 4 \end{bmatrix}, \begin{bmatrix} 19 \\ 27 \\ 27 \end{bmatrix}.$$ ✓7

✓Checkpoint 7

Write the message "*when*" using 2 × 1 matrices.

Answer: See page 378.

Although you could transmit these matrices, a simple substitution code such as this is very easy to break.

To get a more reliable code, we choose a 3 × 3 matrix M that has an inverse. Suppose we choose

$$M = \begin{bmatrix} 1 & 3 & 3 \\ 1 & 4 & 3 \\ 1 & 3 & 4 \end{bmatrix}.$$

Then encode each message group by multiplying by M—that is,

$$\begin{bmatrix} 1 & 3 & 3 \\ 1 & 4 & 3 \\ 1 & 3 & 4 \end{bmatrix}\begin{bmatrix} 13 \\ 1 \\ 20 \end{bmatrix} = \begin{bmatrix} 76 \\ 77 \\ 96 \end{bmatrix}, \quad \begin{bmatrix} 1 & 3 & 3 \\ 1 & 4 & 3 \\ 1 & 3 & 4 \end{bmatrix}\begin{bmatrix} 8 \\ 5 \\ 13 \end{bmatrix} = \begin{bmatrix} 62 \\ 67 \\ 75 \end{bmatrix},$$

$$\begin{bmatrix} 1 & 3 & 3 \\ 1 & 4 & 3 \\ 1 & 3 & 4 \end{bmatrix}\begin{bmatrix} 1 \\ 20 \\ 9 \end{bmatrix} = \begin{bmatrix} 88 \\ 108 \\ 97 \end{bmatrix},$$

and so on. The coded message consists of the ten 3 × 1 column matrices

$$\begin{bmatrix} \mathbf{76} \\ \mathbf{77} \\ \mathbf{96} \end{bmatrix}, \begin{bmatrix} \mathbf{62} \\ \mathbf{67} \\ \mathbf{75} \end{bmatrix}, \begin{bmatrix} \mathbf{88} \\ \mathbf{108} \\ \mathbf{97} \end{bmatrix}, \ldots, \begin{bmatrix} \mathbf{181} \\ \mathbf{208} \\ \mathbf{208} \end{bmatrix}.$$

The message would be sent as a string of numbers:

76, 77, 96, 62, 67, 75, 88, 108, 97, . . . , 181, 208, 208.

Note that the same letter may be encoded by different numbers. For instance, the first a in "mathematics" is 77 and the second a is 88. This makes the code harder to break. ✓8

✓Checkpoint 8

Use the following matrix to find the 2 × 1 matrices to be transmitted for the message in Checkpoint 7:

$$\begin{bmatrix} 2 & 1 \\ 5 & 0 \end{bmatrix}.$$

Answer: See page 378.

The receiving agent rewrites the message as the ten 3 × 1 column matrices shown in color previously. The agent then decodes the message by multiplying each of these column matrices by the matrix M^{-1}. Verify that

$$M^{-1} = \begin{bmatrix} 7 & -3 & -3 \\ -1 & 1 & 0 \\ -1 & 0 & 1 \end{bmatrix}.$$

So the first two matrices of the coded message are decoded as

$$\begin{bmatrix} 7 & -3 & -3 \\ -1 & 1 & 0 \\ -1 & 0 & 1 \end{bmatrix}\begin{bmatrix} 76 \\ 77 \\ 96 \end{bmatrix} = \begin{bmatrix} 13 \\ 1 \\ 20 \end{bmatrix}\begin{matrix} m \\ a \\ t \end{matrix}$$

and

$$\begin{bmatrix} 7 & -3 & -3 \\ -1 & 1 & 0 \\ -1 & 0 & 1 \end{bmatrix}\begin{bmatrix} 62 \\ 67 \\ 75 \end{bmatrix} = \begin{bmatrix} 8 \\ 5 \\ 13 \end{bmatrix}\begin{matrix} h \\ e \\ m \end{matrix}.$$

The other blocks are decoded similarly.

ROUTING

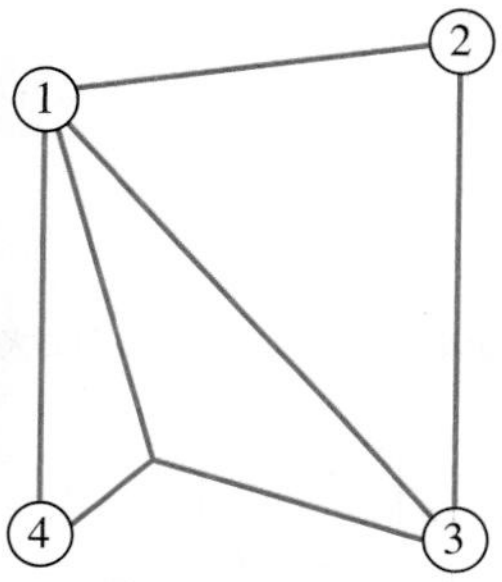

FIGURE 6.27

The diagram in Figure 6.27 shows the roads connecting four cities. Another way of representing this information is via matrix A, where the entries represent the number of roads connecting two cities without passing through another city.* For example, from the diagram, we see that there are two roads connecting city 1 to city 4 without passing through either city 2 or 3. This information is entered in row one, column four, and again in row four, column one, of matrix A:

$$A = \begin{bmatrix} 0 & 1 & 2 & 2 \\ 1 & 0 & 1 & 0 \\ 2 & 1 & 0 & 1 \\ 2 & 0 & 1 & 0 \end{bmatrix}.$$

Note that there are no roads connecting each city to itself. Also, there is one road connecting cities 3 and 2.

How many ways are there to go from city 1 to city 2 by going, for example, through exactly one other city? Because we must go through one other city, we must go through either city 3 or city 4. In the diagram in Figure 6.27, we see that we can go from city 1 to city 2 through city 3 in two ways. We can go from city 1 to city 3 in two ways and then from city 3 to city 2 in one way, so there are $2 \cdot 1 = 2$ ways to get from city 1 to city 2 through city 3. It is not possible to go from city 1 to city 2 through city 4, because there is no direct route between cities 4 and 2.

The matrix A^2 gives the number of ways to travel between any two cities by passing through exactly one other city. Multiply matrix A by itself to get A^2. Let the entry in the first row, second column, of A^2 be b_{12}. (Remember, we use a_{ij} to denote the entry in the ith row and jth column of matrix A.) The entry b_{12} is found as follows:

$$\begin{aligned} b_{12} &= a_{11}a_{12} + a_{12}a_{22} + a_{13}a_{32} + a_{14}a_{42} \\ &= 0 \cdot 1 + 1 \cdot 0 + 2 \cdot 1 + 2 \cdot 0 \\ &= 2. \end{aligned}$$

The first product, $0 \cdot 1$, in this calculation represents the number of ways to go from city 1 to city 1 (that is, 0) and then from city 1 to city 2 (that is, 1). The 0 result indicates

*Campbell, *Matrices with Applications*, 1st ed., section 2.3, example 5, pp. 50–51. © 1968. Electronically reproduced by permission of Pearson Education, Inc., Upper Saddle River, New Jersey.

that such a trip does not involve a third city. The only nonzero product $(2 \cdot 1)$ represents the two routes from city 1 to city 3 and the one route from city 3 to city 2, which result in the $2 \cdot 1$, or 2, routes from city 1 to city 2 by going through city 3.

Similarly, A^3 gives the number of ways to travel between any two cities by passing through exactly two cities. Also, $A + A^2$ represents the total number of ways to travel between two cities with at most one intermediate city. 9

Checkpoint 9

Use a graphing calculator to find the following.

(a) A^3

(b) $A + A^2$

Answers: See page 378.

The diagram can be given many other interpretations. For example, the lines could represent lines of mutual influence between people or nations, or they could represent communication lines, such as telephone lines.

6.6 ▶ Exercises

Solve the matrix equation AX = B for X. (See Example 1.)

1. $A = \begin{bmatrix} 1 & -1 \\ 5 & 6 \end{bmatrix}, B = \begin{bmatrix} 4 \\ -2 \end{bmatrix}$

2. $A = \begin{bmatrix} 1 & 2 \\ 1 & 5 \end{bmatrix}, B = \begin{bmatrix} -3 \\ 5 \end{bmatrix}$

3. $A = \begin{bmatrix} 3 & 1 \\ 4 & 2 \end{bmatrix}, B = \begin{bmatrix} 3 & 4 \\ 5 & 6 \end{bmatrix}$

4. $A = \begin{bmatrix} 1 & 4 \\ 2 & 7 \end{bmatrix}, B = \begin{bmatrix} 0 & 8 \\ 4 & 1 \end{bmatrix}$

5. $A = \begin{bmatrix} 2 & 1 & 0 \\ -4 & -1 & 3 \\ 3 & 1 & -2 \end{bmatrix}, B = \begin{bmatrix} 1 \\ 4 \\ 0 \end{bmatrix}$

6. $A = \begin{bmatrix} 3 & -1 & 0 \\ 0 & 1 & 2 \\ 9 & 0 & 5 \end{bmatrix}, B = \begin{bmatrix} -3 \\ 6 \\ 12 \end{bmatrix}$

Use the inverse of the coefficient matrix to solve each system of equations. (The inverses for Exercises 9–14 were found in Exercises 41 and 44–48 of Section 6.5.) (See Example 1.)

7. $\begin{aligned} x + 2y + 3z &= 10 \\ 2x + 3y + 2z &= 6 \\ -x - 2y - 4z &= -1 \end{aligned}$

8. $\begin{aligned} -x + y - 3z &= 3 \\ 2x + 4y - 4z &= 6 \\ -x + y + 4z &= -1 \end{aligned}$

9. $\begin{aligned} x + 4y + 3z &= -12 \\ x - 3y - 2z &= 0 \\ 2x + 5y + 4z &= 7 \end{aligned}$

10. $\begin{aligned} 5x \qquad + 2z &= 3 \\ 2x + 2y + z &= 4 \\ -3x + y - z &= 5 \end{aligned}$

11. $\begin{aligned} x + 2y + 3z &= 4 \\ x + 4y + 2z &= 8 \\ y - z &= -4 \end{aligned}$

12. $\begin{aligned} 2x + 2y - 4z &= 12 \\ 2x + 6y \qquad &= 16 \\ -3x - 3y + 5z &= -20 \end{aligned}$

13. $\begin{aligned} x \qquad - 2z \qquad &= 4 \\ -2x + y + 2z + 2w &= -8 \\ 3x - y - 2z - 3w &= 12 \\ y + 4z + w &= -4 \end{aligned}$

14. $\begin{aligned} x + y \qquad + 2w &= 3 \\ 2x - y + z - w &= 3 \\ 3x + 3y + 2z - 2w &= 5 \\ x + 2y + z \qquad &= 3 \end{aligned}$

Use matrix algebra to solve the given matrix equations for X. Check your work.

15. $N = X - MX$; $N = \begin{bmatrix} 8 \\ -12 \end{bmatrix}, M = \begin{bmatrix} 0 & 1 \\ -2 & 1 \end{bmatrix}$

16. $A = BX + X$; $A = \begin{bmatrix} 4 & 6 \\ -2 & 2 \end{bmatrix}, B = \begin{bmatrix} -2 & -2 \\ 3 & 3 \end{bmatrix}$

To complete these exercises, write a system of equations and use the inverse of the coefficient matrix to solve the system.

17. Business Donovan's Dandy Furniture makes dining room furniture. A buffet requires 15 hours for cutting, 20 hours for assembly, and 5 hours for finishing. A chair requires 5 hours for cutting, 8 hours for assembly, and 5 hours for finishing. A table requires 10 hours for cutting, 6 hours for assembly, and 6 hours for finishing. The cutting department has 4900 hours of labor available each week, the assembly department has 6600 hours available, and the finishing department has 3900 hours available. How many pieces of each type of furniture should be produced each week if the factory is to run at full capacity?

18. Natural Science A hospital dietician is planning a special diet for a certain patient. The total amount per meal of food groups *A*, *B*, and *C* must equal 400 grams. The diet should include one-third as much of group *A* as of group *B*, and the sum of the amounts of group *A* and group *C* should equal twice the amount of group *B*.

(a) How many grams of each food group should be included?

(b) Suppose we drop the requirement that the diet include one-third as much of group *A* as of group *B*. Describe the set of all possible solutions.

(c) Suppose that, in addition to the conditions given in part (a), foods A and B cost 2 cents per gram, food C costs 3 cents per gram, and a meal must cost \$8. Is a solution possible?

19. Natural Science Three species of bacteria are fed three foods: I, II, and III. A bacterium of the first species consumes 1.3 units each of foods I and II and 2.3 units of food III each day. A bacterium of the second species consumes 1.1 units of food I, 2.4 units of food II, and 3.7 units of food III each day. A bacterium of the third species consumes 8.1 units of I, 2.9 units of II, and 5.1 units of III each day. If 16,000 units of I, 28,000 units of II, and 44,000 units of III are supplied each day, how many of each species can be maintained in this environment?

20. Business A company produces three combinations of mixed vegetables, which sell in 1-kilogram packages. Italian style combines .3 kilogram of zucchini, .3 of broccoli, and .4 of carrots. French style combines .6 kilogram of broccoli and .4 of carrots. Oriental style combines .2 kilogram of zucchini, .5 of broccoli, and .3 of carrots. The company has a stock of 16,200 kilograms of zucchini, 41,400 kilograms of broccoli, and 29,400 kilograms of carrots. How many packages of each style should the company prepare in order to use up its supplies?

21. Business A national chain of casual clothing stores recently sent shipments of jeans, jackets, sweaters, and shirts to its stores in various cities. The number of items shipped to each city and their total wholesale cost are shown in the table. Find the wholesale price of one pair of jeans, one jacket, one sweater, and one shirt.

City	Jeans	Jackets	Sweaters	Shirts	Total Cost
Cleveland	3000	3000	2200	4200	\$507,650
St. Louis	2700	2500	2100	4300	459,075
Seattle	5000	2000	1400	7500	541,225
Phoenix	7000	1800	600	8000	571,500

22. Health A 100-bed nursing home provides two levels of long-term care: regular and maximum. Patients at each level have a choice of a private room or a less expensive semiprivate room. The tables show the number of patients in each category at various times last year. The total daily costs for all patients were \$24,040 in January, \$23,926 in April, \$23,760 in July, and \$24,042 in October. Find the daily cost of each of the following: a private room (regular care), a private room (maximum care), a semiprivate room (regular care), and a semiprivate room (maximum care).

	REGULAR-CARE PATIENTS	
Month	**Semiprivate**	**Private**
January	22	8
April	26	8
July	24	14
October	20	10

	MAXIMUM-CARE PATIENTS	
Month	**Semiprivate**	**Private**
January	60	10
April	54	12
July	56	6
October	62	8

Find the production matrix for the given input–output and demand matrices. (See Examples 3–6.).

23. $A = \begin{bmatrix} \frac{1}{2} & \frac{2}{5} \\ \frac{1}{4} & \frac{1}{5} \end{bmatrix}, D = \begin{bmatrix} 2 \\ 4 \end{bmatrix}$

24. $A = \begin{bmatrix} .1 & .03 \\ .07 & .6 \end{bmatrix}, D = \begin{bmatrix} 5 \\ 10 \end{bmatrix}$

Exercises 25 and 26 refer to Example 6.

25. Business If the demand is changed to 690 metric tons of wheat and 920 metric tons of oil, how many units of each commodity should be produced?

26. Business Change the technological matrix so that the production of 1 metric ton of wheat requires $\frac{1}{5}$ metric ton of oil (and no wheat) and the production of 1 metric ton of oil requires $\frac{1}{3}$ metric ton of wheat (and no oil). To satisfy the same demand matrix, how many units of each commodity should be produced?

Work these problems. (See Examples 3–6.)

27. Business A simplified economy has only two industries: the electric company and the gas company. Each dollar's worth of the electric company's output requires \$.40 of its own output and \$.50 of the gas company's output. Each dollar's worth of the gas company's output requires \$.25 of its own output and \$.60 of the electric company's output. What should the production of electricity and gas be (in dollars) if there is a \$12 million demand for gas and a \$15 million demand for electricity?

28. Business A two-segment economy consists of manufacturing and agriculture. To produce 1 unit of manufacturing output requires .40 unit of its own output and .20 unit of agricultural output. To produce 1 unit of agricultural output requires .30 unit of its own output and .40 unit of manufacturing output. If there is a demand of 240 units of manufacturing and 90 units of agriculture, what should be the output of each segment?

29. **Business** A primitive economy depends on two basic goods: yams and pork. The production of 1 bushel of yams requires $\frac{1}{4}$ bushel of yams and $\frac{1}{2}$ of a pig. To produce 1 pig requires $\frac{1}{6}$ bushel of yams. Find the amount of each commodity that should be produced to get

(a) 1 bushel of yams and 1 pig;

(b) 100 bushels of yams and 70 pigs.

30. **Business** A simplified economy is based on agriculture, manufacturing, and transportation. Each unit of agricultural output requires .4 unit of its own output, .3 unit of manufacturing output, and .2 unit of transportation output. One unit of manufacturing output requires .4 unit of its own output, .2 unit of agricultural output, and .3 unit of transportation output. One unit of transportation output requires .4 unit of its own output, .1 unit of agricultural output, and .2 unit of manufacturing output. There is demand for 35 units of agricultural, 90 units of manufacturing, and 20 units of transportation output. How many units should each segment of the economy produce?

31. **Business** In his work *Input–Output Economics*, Leontief provides an example of a simplified economy with just three sectors: agriculture, manufacturing, and households (that is, the sector of the economy that produces labor.)* It has the following input–output matrix:

	Agriculture	Manufacturing	Households
Agriculture	.25	.40	.133
Manufacturing	.14	.12	.100
Households	.80	3.60	.133

(a) How many units from each sector does the manufacturing sector require to produce 1 unit?

(b) What production levels are needed to meet a demand of 35 units of agriculture, 38 units of manufacturing, and 40 units of households?

(c) How many units of agriculture are used up in the economy's production process?

32. **Business** A much simplified version of Leontief's 42-sector analysis of the 1947 American economy has the following input–output matrix†:

	Agriculture	Manufacturing	Households
Agriculture	.245	.102	.051
Manufacturing	.099	.291	.279
Households	.433	.372	.011

(a) What information about the needs of agricultural production is given by column 1 of the matrix?

(b) Suppose the demand matrix (in billions of dollars) is

$$D = \begin{bmatrix} 2.88 \\ 31.45 \\ 30.91 \end{bmatrix}.$$

Find the amount of each commodity that should be produced.

33. **Business** An analysis of the 1958 Israeli economy is simplified here by grouping the economy into three sectors, with the following input–output matrix*:

	Agriculture	Manufacturing	Energy
Agriculture	.293	0	0
Manufacturing	.014	.207	.017
Energy	.044	.010	.216

(a) How many units from each sector does the energy sector require to produce one unit?

(b) If the economy's production (in thousands of Israeli pounds) is 175,000 of agriculture, 22,000 of manufacturing, and 12,000 of energy, how much is available from each sector to satisfy the demand from consumers and others outside the production process?

(c) The actual 1958 demand matrix is

$$D = \begin{bmatrix} 138{,}213 \\ 17{,}597 \\ 1786 \end{bmatrix}.$$

How much must each sector produce to meet this demand?

34. **Business** The 1981 Chinese economy can be simplified to three sectors: agriculture, industry and construction, and transportation and commerce.† The input–output matrix is

	Agri.	Industry/Constr.	Trans./Commerce
Agri.	.158	.156	.009
Industry/Constr.	.136	.432	.071
Trans./Commerce	.013	.041	.011

The demand [in 100,000 *renminbi* (RMB), the unit of money in China] is

$$D = \begin{bmatrix} 106{,}674 \\ 144{,}739 \\ 26{,}725 \end{bmatrix}.$$

*Wassily Leontief, *Input–Output Economics*, 2d ed. (Oxford University Press, 1986), pp. 19–27.

†Ibid., pp. 6–9.

*Ibid., pp. 174–175.

†*Input–Output Tables of China: 1981*, China Statistical Information and Consultancy Service Centre, 1987, pp. 17–19.

(a) Find the amount each sector should produce.
(b) Interpret the economic value of an increase in demand of 1 RMB in agriculture exports.

35. **Business** The 1987 economy of the state of Washington has been simplified to four sectors: natural resource, manufacturing, trade and services, and personal consumption. The input–output matrix is as follows*:

	Natural Resources	Manufacturing	Trade and Services	Personal Consumption
Natural Resources	.1045	.0428	.0029	.0031
Manufacturing	.0826	.1087	.0584	.0321
Trade and Services	.0867	.1019	.2032	.3555
Personal Consumption	.6253	.3448	.6106	.0798

Suppose the demand (in millions of dollars) is

$$D = \begin{bmatrix} 450 \\ 300 \\ 125 \\ 100 \end{bmatrix}.$$

Find the amount each sector should produce.

36. **Business** The 1963 economy of the state of Nebraska has been condensed to six sectors; livestock, crops, food products, mining and manufacturing, households, and other. The input–output matrix is as follows†:

$$\begin{bmatrix} .178 & .018 & .411 & 0 & .005 & 0 \\ .143 & .018 & .088 & 0 & .001 & 0 \\ .089 & 0 & .035 & 0 & .060 & .003 \\ .001 & .010 & .012 & .063 & .007 & .014 \\ .141 & .252 & .088 & .089 & .402 & .124 \\ .188 & .156 & .103 & .255 & .008 & .474 \end{bmatrix}.$$

(a) Find the matrix $(I - A)^{-1}$ and interpret the value in row two, column one, of this matrix.

(b) Suppose the demand (in millions of dollars) is

$$D = \begin{bmatrix} 1980 \\ 650 \\ 1750 \\ 1000 \\ 2500 \\ 3750 \end{bmatrix}.$$

Find the dollar amount each sector should produce.

37. **Business** Input–output analysis can also be used to model how changes in one city can affect cities that are connected with it in some way.* For example, if a large manufacturing company shuts down in one city, it is very likely that the economic welfare of all of the cities around it will suffer. Consider three Pennsylvania communities: Sharon, Farrell, and Hermitage. Due to their proximity to each other, residents of these three communities regularly spend time and money in the other communities. Suppose that we have gathered information in the form of an input–output matrix

$$\begin{bmatrix} .2 & .1 & .1 \\ .1 & .1 & 0 \\ .5 & .6 & .7 \end{bmatrix}.$$

This matrix can be thought of as the likelihood that a person from a particular community will spend money in each of the communities.

(a) Treat this matrix like an input–output matrix and calculate $(I - A)^{-1}$, where A is the given input–output matrix.

(b) Interpret the entries of this inverse matrix.

38. **Social Science** Use the method discussed in the text to encode the message

Anne is home.

Break the message into groups of two letters and use the matrix

$$M = \begin{bmatrix} 1 & 3 \\ 2 & 7 \end{bmatrix}.$$

39. **Social Science** Use the matrix of Exercise 38 to encode the message

Head for the hills!

*Robert Chase, Philip Bourque, and Richard Conway, Jr., *The 1987 Washington State Input–Output Study*, Report to the Graduate School of Business Administration, University of Washington, September 1993.

†F. Charles Lamphear and Theodore Roesler, "1970 Nebraska Input–Output Tables," *Nebraska Economic and Business Report No. 10*, Bureau of Business Research, University of Nebraska, Lincoln.

*The idea for this problem came from an example created by Thayer Watkins, Department of Economics, San Jose State University.

40. **Social Science** Decode the following message, which was encoded by using the matrix M of Exercise 38:

$$\begin{bmatrix} 90 \\ 207 \end{bmatrix}, \begin{bmatrix} 39 \\ 87 \end{bmatrix}, \begin{bmatrix} 26 \\ 57 \end{bmatrix}, \begin{bmatrix} 66 \\ 145 \end{bmatrix}, \begin{bmatrix} 61 \\ 142 \end{bmatrix}, \begin{bmatrix} 89 \\ 205 \end{bmatrix}.$$

Work these routing problems.

41. **Social Science** Use matrix A in the discussion on routing (pages 373–374) to find A^2. Then answer the following questions: How many ways are there to travel from
   (a) city 1 to city 3 by passing through exactly one city?
   (b) city 2 to city 4 by passing through exactly one city?
   (c) city 1 to city 3 by passing through at most one city?
   (d) city 2 to city 4 by passing through at most one city?

42. **Social Science** The matrix A^3 (see Exercise 41) was found in Checkpoint 9 of the text. Use it to answer the following questions.
   (a) How many ways are there to travel between cities 1 and 4 by passing through exactly two cities?
   (b) How many ways are there to travel between cities 1 and 4 by passing through at most two cities?

43. **Business** A small telephone system connects three cities. There are four lines between cities 3 and 2, three lines connecting city 3 with city 1, and two lines between cities 1 and 2.
   (a) Write a matrix B to represent this information.
   (b) Find B^2.
   (c) How many lines that connect cities 1 and 2 go through exactly one other city (city 3)?
   (d) How many lines that connect cities 1 and 2 go through at most one other city?

44. **Transportation** The figure shows four southern cities served by Supersouth Airlines.

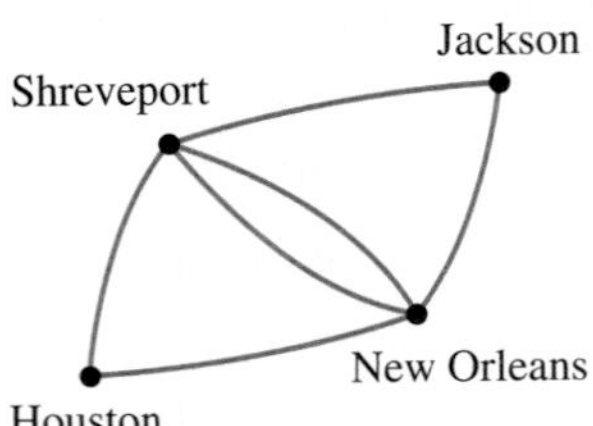

   (a) Write a matrix to represent the number of nonstop routes between cities.
   (b) Find the number of one-stop flights between Houston and Jackson.
   (c) Find the number of flights between Houston and Shreveport that require at most one stop.
   (d) Find the number of one-stop flights between New Orleans and Houston.

45. **Natural Science** The figure shows a food web. The arrows indicate the food sources of each population. For example, cats feed on rats and on mice.

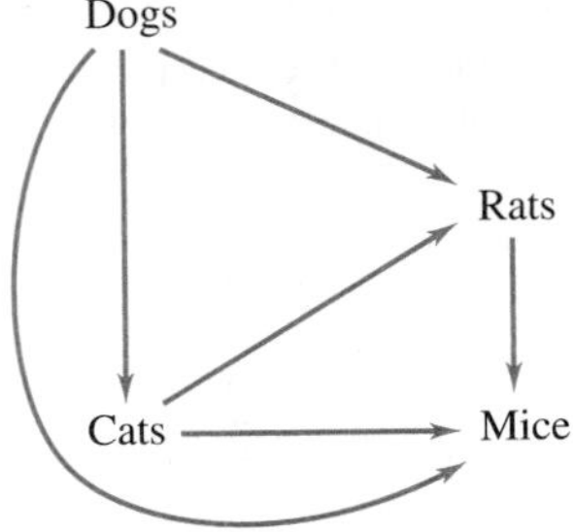

   (a) Write a matrix C in which each row and corresponding column represent a population in the food chain. Enter a 1 when the population in a given row feeds on the population in the given column.
   (b) Calculate and interpret C^2.

✓ Checkpoint Answers

1. $A = \begin{bmatrix} 2 & 6 \\ -1 & -2 \end{bmatrix}, \quad X = \begin{bmatrix} x \\ y \end{bmatrix}, \quad \text{and} \quad B = \begin{bmatrix} -14 \\ 3 \end{bmatrix}.$

2. (6, −4, 10)   3. $(3 - 1.5y, y)$ for all real numbers y

4.

	Elec.	Water
Elec.	$\frac{1}{3}$	$\frac{1}{4}$
Water	$\frac{1}{2}$	0

5. (a) $\begin{bmatrix} 9000 \\ 12{,}000 \end{bmatrix}$ (b) $\begin{bmatrix} 6000 \\ 4500 \end{bmatrix}$ (c) $\begin{bmatrix} 3000 \\ 7500 \end{bmatrix}$

6. (a) $\begin{bmatrix} \frac{1}{2} & \frac{3}{4} \\ \frac{1}{2} & 0 \end{bmatrix}$ (b) $\begin{bmatrix} \frac{1}{2} & -\frac{3}{4} \\ -\frac{1}{2} & 1 \end{bmatrix}$

   (c) $\begin{bmatrix} 8 & 6 \\ 4 & 4 \end{bmatrix}$ (d) $\begin{bmatrix} 6800 \\ 4400 \end{bmatrix}$

7. $\begin{bmatrix} 23 \\ 8 \end{bmatrix}, \begin{bmatrix} 5 \\ 14 \end{bmatrix}$ 8. $\begin{bmatrix} 54 \\ 115 \end{bmatrix}, \begin{bmatrix} 24 \\ 25 \end{bmatrix}$

9. (a)

```
[A]^3
 [[12 12 22 21]
  [12 4  9  6 ]
  [22 9  12 12]
  [21 6  12 8 ]]
```

   (b)

```
[A]+[A]²
      [[9 3 5 4]
       [3 2 3 3]
       [5 3 6 5]
       [4 3 5 5]]
```

CHAPTER 6 ▸ Summary

KEY TERMS AND SYMBOLS

6.1 ▸ linear equation
system of linear equations
solution of a system
substitution method
elimination method
independent system
dependent system
inconsistent system

6.2 ▸ equivalent systems
elementary operations
elimination method
row
column
matrix (matrices)
element (entry)
augmented matrix
row operations
row echelon form
Gauss–Jordan method
reduced row echelon form
independent system
inconsistent system
dependent system
parameter

6.3 ▸ applications of systems of linear equations

6.4 ▸ row matrix (row vector)
column matrix (column vector)
square matrix
additive inverse of a matrix
zero matrix
scalar
product of a scalar and a matrix

6.5 ▸ product matrix
identity matrix
inverse matrix
singular matrix

6.6 ▸ coefficient matrix
input–output model
input–output matrix
production matrix
demand matrix
code theory
routing theory

CHAPTER 6 KEY CONCEPTS

Solving Systems of Equations ▸ The following **elementary operations** are used to transform a system of equations into a simpler equivalent system:

1. Interchange any two equations.
2. Multiply both sides of an equation by a nonzero constant.
3. Replace an equation by the sum of itself and a constant multiple of another equation in the system.

The **elimination method** is a systematic way of using elementary operations to transform a system into an equivalent one that can be solved by **back substitution**. See Section 6.2 for details.

The matrix version of the elimination method uses the following **matrix row operations**, which correspond to using elementary row operations with back substitution on a system of equations:

1. Interchange any two rows.
2. Multiply each element of a row by a nonzero constant.
3. Replace a row by the sum of itself and a constant multiple of another row in the matrix.

The **Gauss–Jordan method** is an extension of the elimination method for solving a system of linear equations. It uses row operations on the augmented matrix of the system. See Section 6.2 for details.

Operations on Matrices ▸ The **sum** of two $m \times n$ matrices X and Y is the $m \times n$ matrix $X + Y$ in which each element is the sum of the corresponding elements of X and Y. The **difference** of two $m \times n$ matrices X and Y is the $m \times n$ matrix $X - Y$ in which each element is the difference of the corresponding elements of X and Y.

The **product** of a scalar k and a matrix X is the matrix kX, with each element being k times the corresponding element of X.

The **product matrix** AB of an $m \times n$ matrix A and an $n \times k$ matrix B is the $m \times k$ matrix whose entry in the ith row and jth column is the product of the ith row of A and the jth column of B.

The **inverse matrix** A^{-1} for any $n \times n$ matrix A for which A^{-1} exists is found as follows: Form the augmented matrix $[A|I]$, and perform row operations on $[A|I]$ to get the matrix $[I|A^{-1}]$.

CHAPTER 6 REVIEW EXERCISES

Solve each of the following systems.

1. $\begin{aligned} -5x - 3y &= -3 \\ 2x + y &= 4 \end{aligned}$

2. $\begin{aligned} 3x - y &= 8 \\ 2x + 3y &= 6 \end{aligned}$

3. $\begin{aligned} 3x - 5y &= 16 \\ 2x + 3y &= -2 \end{aligned}$

4. $\begin{aligned} \frac{1}{4}x - \frac{1}{3}y &= -\frac{1}{4} \\ \frac{1}{10}x + \frac{2}{5}y &= \frac{2}{5} \end{aligned}$

5. **Business** An office supply manufacturer makes two kinds of paper clips: standard and extra large. To make 1000 standard paper clips requires $\frac{1}{4}$ hour on a cutting machine and $\frac{1}{2}$ hour on a machine that shapes the clips. Making 1000 extra-large paper clips requires $\frac{1}{3}$ hour on each machine. The manager of paper clip production has 4 hours per day available on the cutting machine and 6 hours per day on the shaping machine. How many of each kind of clip can he make?

6. **Business** Gretchen Schmidt plans to buy shares of two stocks. One costs \$32 per share and pays dividends of \$1.20 per share. The other costs \$23 per share and pays dividends of \$1.40 per share. She has \$10,100 to spend and wants to earn dividends of \$540. How many shares of each stock should she buy?

7. **Business** Joyce Pluth has money in two investment funds. Last year, the first fund paid a dividend of 8% and the second a dividend of 2%, and Joyce received a total of \$780. This year, the first fund paid a 10% dividend and the second only 1%, and Joyce received \$810. How much does she have invested in each fund?

Solve each of the following systems.

8. $\begin{aligned} x - 2y &= 1 \\ 4x + 4y &= 2 \\ 10x + 8y &= 4 \end{aligned}$

9. $\begin{aligned} x + y - 4z &= 0 \\ 2x + y - 3z &= 2 \end{aligned}$

10. $\begin{aligned} 3x + y - z &= 3 \\ x \qquad + 2z &= 6 \\ -3x - y + 2z &= 9 \end{aligned}$

11. $\begin{aligned} 4x - y - 2z &= 4 \\ x - y - \frac{1}{2}z &= 1 \\ 2x - y - z &= 8 \end{aligned}$

Solve each of the following systems.

12. $\begin{aligned} x + z &= -3 \\ y - z &= 6 \\ 2x + 3z &= 5 \end{aligned}$

13. $\begin{aligned} 2x + 3y + 4z &= 8 \\ -x + y - 2z &= -9 \\ 2x + 2y + 6z &= 16 \end{aligned}$

14. $\begin{aligned} 5x - 8y + z &= 1 \\ 3x - 2y + 4z &= 3 \\ 10x - 16y + 2z &= 3 \end{aligned}$

15. $\begin{aligned} x - 2y + 3z &= 4 \\ 2x + y - 4z &= 3 \\ -3x + 4y - z &= -2 \end{aligned}$

16. $\begin{aligned} 3x + 2y - 6z &= 3 \\ x + y + 2z &= 2 \\ 2x + 2y + 5z &= 0 \end{aligned}$

17. **Finance** You are given \$144 in one-, five-, and ten-dollar bills. There are 35 bills. There are two more ten-dollar bills than five-dollar bills. How many bills of each type are there?

18. **Social Science** A social service agency provides counseling, meals, and shelter to clients referred by sources I, II, and III. Clients from source I require an average of \$100 for food, \$250 for shelter, and no counseling. Source II clients require an average of \$100 for counseling, \$200 for food, and nothing for shelter. Source III clients require an average of \$100 for counseling, \$150 for food, and \$200 for shelter. The agency has funding of \$25,000 for counseling, \$50,000 for food, and \$32,500 for shelter. How many clients from each source can be served?

19. **Business** The Waputi Indians make woven blankets, rugs, and skirts. Each blanket requires 24 hours for spinning the yarn, 4 hours for dying the yarn, and 15 hours for weaving. Rugs require 30, 5, and 18 hours, and skirts require 12, 3, and 9 hours, respectively. If there are 306, 59, and 201 hours available for spinning, dying, and weaving, respectively, how many of each item can be made? (*Hint*: Simplify the equations you write, if possible, before solving the system.)

20. **Business** Each week at a furniture factory, 2000 work hours are available in the construction department, 1400 work hours in the painting department, and 1300 work hours in the packing department. Producing a chair requires 2 hours of construction, 1 hour of painting, and 2 hours for packing. Producing a table requires 4 hours of construction, 3 hours of painting, and 3 hours for packing. Producing a chest requires 8 hours of construction, 6 hours of painting, and 4 hours for packing. If all available time is used in every department, how many of each item are produced each week?

For each of the following, find the dimensions of the matrix and identify any square, row, or column matrices.

21. $\begin{bmatrix} 2 & 3 \\ 5 & 9 \end{bmatrix}$

22. $\begin{bmatrix} 2 & -1 \\ 4 & 6 \\ 5 & 7 \end{bmatrix}$

23. $[12 \quad 4 \quad -8 \quad -1]$

24. $\begin{bmatrix} -7 & 5 & 6 & 4 \\ 3 & 2 & -1 & 2 \\ -1 & 12 & 8 & -1 \end{bmatrix}$

25. $\begin{bmatrix} 6 & 8 & 10 \\ 5 & 3 & -2 \end{bmatrix}$

26. $\begin{bmatrix} -9 \\ 15 \\ 4 \end{bmatrix}$

27. **Natural Science** The activities of a grazing animal can be classified roughly into three categories: grazing, moving, and resting. Suppose horses spend 8 hours grazing, 8 moving, and 8 resting; cattle spend 10 grazing, 5 moving, and 9 resting; sheep spend 7 grazing, 10 moving, and 7 resting; and goats spend 8 grazing, 9 moving, and 7 resting. Write this information as a 4×3 matrix.

28. **Business** The New York Stock Exchange reports in the daily newspapers give the dividend, price-to-earnings ratio, sales (in hundreds of shares), last price, and change in price for each company. Write the following stock reports as a 4×5 matrix: American Telephone & Telegraph, 5, 7, 2532, $52\frac{3}{8}$, $-\frac{1}{4}$; General Electric, 3, 9, 1464, 56, $+\frac{1}{8}$; Gulf Oil, 2.50, 5, 4974, 41, $-1\frac{1}{2}$; Sears, 1.36, 10, 1754, 18, $+\frac{1}{2}$.

Given the matrices

$$A = \begin{bmatrix} 4 & 6 \\ -2 & -2 \\ 5 & 9 \end{bmatrix}, \quad B = \begin{bmatrix} 1 & 2 & -3 \\ 2 & 3 & 0 \\ 0 & 1 & 4 \end{bmatrix}, \quad C = \begin{bmatrix} 5 & 0 \\ -1 & 3 \\ 4 & 7 \end{bmatrix},$$

$$D = \begin{bmatrix} 6 \\ 1 \\ 0 \end{bmatrix}, \quad E = [1 \quad 3 \quad -4], \quad F = \begin{bmatrix} -1 & 2 \\ 6 & 7 \end{bmatrix}, \text{ and}$$

$$G = \begin{bmatrix} 2 & 5 \\ 1 & 6 \end{bmatrix},$$

find each of the following (if possible).

29. $-B$

30. $-D$

31. $3A - 2C$

32. $F + 3G$

33. $2B - 5C$

34. $G - 2F$

35. **Business** Refer to Exercise 28. Write a 4×2 matrix using the sales and price changes for the four companies. The next day's sales and price changes for the same four companies were 2310, 1258, 5061, and 1812 and $-\frac{1}{4}$, $-\frac{1}{4}$, and $+\frac{1}{2}$, and $+\frac{1}{2}$, respectively. Write a 4×2 matrix using these new sales and price change figures. Use matrix addition to find the total sales and price changes for the two days.

36. **Business** An oil refinery in Tulsa sent 110,000 gallons of oil to a Chicago distributor, 73,000 to a Dallas distributor, and 95,000 to an Atlanta distributor. Another refinery in New Orleans sent the following respective amounts to the same three distributors: 85,000, 108,000, and 69,000. The next month, the two refineries sent the same distributors new shipments of oil as follows: from Tulsa, 58,000 to Chicago, 33,000 to Dallas, and 80,000 to Atlanta; from New Orleans, 40,000, 52,000, and 30,000, respectively.

(a) Write the monthly shipments from the two distributors to the three refineries as 3×2 matrices.

(b) Use matrix addition to find the total amounts sent to the refineries from each distributor.

Use the matrices shown before Exercise 29 to find each of the following (if possible).

37. AG

38. EB

39. GF

40. CA

41. AGF

42. **Health** In a study, the numbers of head and neck injuries among hockey players wearing full face shields and half face shields were compared. The following table provides the rates per 1000 athlete exposures for specific injuries that caused a player wearing either shield to miss one or more events*:

	Half Shield	Full Shield
Head and Face Injuries, Excluding Concussions	3.54	1.41
Concussions	1.53	1.57
Neck Injuries	.34	.29
Other	7.53	6.21

If an equal number of players in a large league wear each type of shield and the total number of athlete exposures for the league in a season is 8000, use matrix operations to estimate the total number of injuries of each type.

43. **Business** An office supply manufacturer makes two kinds of paper clips: standard and extra large. To make a unit of standard paper clips requires $\frac{1}{4}$ hour on a cutting machine and $\frac{1}{2}$ hour on a machine that shapes the clips. A unit of extra-large paper clips requires $\frac{1}{3}$ hour on each machine.

(a) Write this information as a 2×2 matrix (size/machine).

(b) If 48 units of standard and 66 units of extra-large clips are to be produced, use matrix multiplication to find out how many hours each machine will operate. (*Hint*: Write the units as a 1×2 matrix.)

---

*Brian Benson, Nicholas Nohtadi, Sarah Rose, and Willem Meeuwisse, "Head and Neck Injuries among Ice Hockey Players Wearing Full Face Shields vs. Half Face Shields," *JAMA*, 282, no. 24, (December 22/29, 1999): 2328–2332.

44. **Business** Theresa DePalo buys shares of three stocks. Their cost per share and dividend earnings per share are \$32, \$23, and \$54, and \$1.20, \$1.49, and \$2.10, respectively. She buys 50 shares of the first stock, 20 shares of the second, and 15 shares of the third.
   (a) Write the cost per share and earnings per share of the stocks as a 3×2 matrix.
   (b) Write the number of shares of each stock as a 1×3 matrix.
   (c) Use matrix multiplication to find the total cost and total dividend earnings of these stocks.

45. If $A = \begin{bmatrix} 3 & 0 \\ 2 & 1 \end{bmatrix}$, find a matrix B such that both AB and BA are defined and $AB \neq BA$.

46. Is it possible to do Exercise 45 if $A = \begin{bmatrix} 4 & 0 \\ 0 & 4 \end{bmatrix}$? Explain.

Find the inverse of each of the following matrices, if an inverse exists for that matrix.

47. $\begin{bmatrix} -2 & 2 \\ 0 & 5 \end{bmatrix}$

48. $\begin{bmatrix} 3 & -1 \\ -5 & 2 \end{bmatrix}$

49. $\begin{bmatrix} 6 & 4 \\ 3 & 2 \end{bmatrix}$

50. $\begin{bmatrix} 3 & 0 \\ -1 & 4 \end{bmatrix}$

51. $\begin{bmatrix} 2 & 0 & 6 \\ 1 & -1 & 0 \\ 0 & 1 & -3 \end{bmatrix}$

52. $\begin{bmatrix} 2 & -1 & 0 \\ 1 & 0 & 2 \\ 1 & -4 & 0 \end{bmatrix}$

53. $\begin{bmatrix} 2 & 3 & 5 \\ -2 & -3 & -5 \\ 1 & 4 & 2 \end{bmatrix}$

54. $\begin{bmatrix} 1 & 3 & 6 \\ 4 & 0 & 9 \\ 5 & 15 & 30 \end{bmatrix}$

55. $\begin{bmatrix} 1 & 3 & -2 & -1 \\ 0 & 1 & 1 & 2 \\ -1 & -1 & 1 & -1 \\ 1 & -1 & -3 & -2 \end{bmatrix}$

56. $\begin{bmatrix} 3 & 2 & 0 & -1 \\ 2 & 0 & 1 & 2 \\ 1 & 2 & -1 & 0 \\ 2 & -1 & 1 & 1 \end{bmatrix}$

Refer again to the matrices shown before Exercise 29 to find each of the following (if possible).

57. F^{-1}

58. G^{-1}

59. $(G - F)^{-1}$

60. $(F + G)^{-1}$

61. B^{-1}

62. Explain why the matrix $\begin{bmatrix} a & 0 \\ c & 0 \end{bmatrix}$, where a and c are nonzero constants, cannot have an inverse.

Solve each of the following matrix equations $AX = B$ for X.

63. $A = \begin{bmatrix} -3 & 4 \\ -1 & 2 \end{bmatrix}, \quad B = \begin{bmatrix} 3 \\ -1 \end{bmatrix}$

64. $A = \begin{bmatrix} 1 & 3 \\ -2 & 4 \end{bmatrix}, \quad B = \begin{bmatrix} 9 \\ 6 \end{bmatrix}$

65. $A = \begin{bmatrix} 1 & 0 & 2 \\ -1 & 1 & 0 \\ 3 & 0 & 4 \end{bmatrix}, \quad B = \begin{bmatrix} 8 \\ 4 \\ -6 \end{bmatrix}$

66. $A = \begin{bmatrix} 2 & 4 & 0 \\ 1 & -2 & 0 \\ 0 & 0 & 3 \end{bmatrix}, \quad B = \begin{bmatrix} 72 \\ -24 \\ 48 \end{bmatrix}$

Use the method of matrix inverses to solve each of the following systems.

67. $\begin{aligned} x + y &= -2 \\ 2x + 5y &= 2 \end{aligned}$

68. $\begin{aligned} 5x - 3y &= -2 \\ 2x + 7y &= -9 \end{aligned}$

69. $\begin{aligned} 2x + y &= 10 \\ 3x - 2y &= 8 \end{aligned}$

70. $\begin{aligned} x - 2y &= 7 \\ 3x + y &= 7 \end{aligned}$

71. $\begin{aligned} x + y + z &= 1 \\ 2x - y &= -2 \\ 3y + z &= 2 \end{aligned}$

72. $\begin{aligned} x &= -3 \\ y + z &= 6 \\ 2x - 3z &= -9 \end{aligned}$

73. $\begin{aligned} 3x - 2y + 4z &= 4 \\ 4x + y - 5z &= 2 \\ -6x + 4y - 8z &= -2 \end{aligned}$

74. $\begin{aligned} x + 2y &= -1 \\ 3y - z &= -5 \\ x + 2y - z &= -3 \end{aligned}$

Solve each of the following problems by any method.

75. **Business** A wine maker has two large casks of wine. One is 9% alcohol and the other is 14% alcohol. How many liters of each wine should be mixed to produce 40 liters of wine that is 12% alcohol?

76. **Business** A gold merchant has some 12-carat gold (12/24 pure gold), and some 22-carat gold (22/24 pure). How many grams of each could be mixed to get 25 grams of 15-carat gold?

77. **Natural Science** A chemist has a 40% acid solution and a 60% solution. How many liters of each should be used to get 40 liters of a 45% solution?

78. **Business** How many pounds of tea worth \$4.60 a pound should be mixed with tea worth \$6.50 a pound to get 10 pounds of a mixture worth \$5.74 a pound?

79. **Business** A machine in a pottery factory takes 3 minutes to form a bowl and 2 minutes to form a plate. The material for a bowl costs \$.25, and the material for a plate costs \$.20. If the machine runs for 8 hours and exactly \$44 is spent for material, how many bowls and plates can be produced?

80. **Transportation** A boat travels at a constant speed a distance of 57 kilometers downstream in 3 hours and then turns around and travels 55 kilometers upstream in 5 hours. What are the speeds of the boat and the current?

81. **Business** Ms. Tham invests \$50,000 three ways—at 8%, $8\frac{1}{2}$%, and 11%. In total, she receives \$4436.25 per year in interest. The interest on the 11% investment is \$80

more than the interest on the 8% investment. Find the amount she has invested at each rate.

82. **Business** Tickets to the homecoming football game cost \$4 for students, \$10 for alumni, \$12 for other adults, and \$6 for children. The total attendance was 3750, and the ticket receipts were \$29,100. Six times more students than children attended. The number of alumni was $\frac{4}{5}$ of the number of students. How many students, alumni, other adults, and children were at the game?

83. Given the input–output matrix $A = \begin{bmatrix} 0 & \frac{1}{4} \\ \frac{1}{2} & 0 \end{bmatrix}$ and the demand matrix $D = \begin{bmatrix} 2100 \\ 1400 \end{bmatrix}$, find each of the following.

(a) $I - A$
(b) $(I - A)^{-1}$
(c) the production matrix X

84. **Business** An economy depends on two commodities: goats and cheese. It takes $\frac{2}{3}$ unit of goats to produce 1 unit of cheese and $\frac{1}{2}$ unit of cheese to produce 1 unit of goats.
(a) Write the input–output matrix for this economy.
(b) Find the production required to satisfy a demand of 400 units of cheese and 800 units of goats.

Use technology to do Exercises 85–87.

85. **Business** In a simple economic model, a country has two industries: agriculture and manufacturing. To produce \$1 of agricultural output requires \$.10 of agricultural output and \$.40 of manufacturing output. To produce \$1 of manufacturing output requires \$.70 of agricultural output and \$.20 of manufacturing output. If agricultural demand is \$60,000 and manufacturing demand is \$20,000, what must each industry produce? (Round answers to the nearest whole number.)

86. **Business** Here is the input–output matrix for a small economy:

	Agriculture	Services	Mining	Manufacturing
Agriculture	.02	.9	0	.001
Services	0	.4	0	.06
Mining	.01	.02	.06	.07
Manufacturing	.25	.9	.9	.4

(a) How many units from each sector does the service sector require to produce 1 unit?
(b) What production levels are needed to meet a demand for 760 units of agriculture, 1600 units of services, 1000 units of mining, and 2000 units of manufacturing?
(c) How many units of manufacturing production are used up in the economy's production process?

87. Use this input–output matrix to answer the questions below.

	Agriculture	Construction	Energy	Manufacturing	Transportation
Agriculture	.18	.017	.4	.005	0
Construction	.14	.018	.09	.001	0
Energy	.9	0	.4	.06	.002
Manufacturing	.19	.16	.1	.008	.5
Transportation	.14	.25	.9	.4	.12

(a) How many units from each sector does the energy sector require to produce 1 unit?
(b) If the economy produces 28,067 units of agriculture, 9383 units of construction, 51,372 units of energy, 61,364 units of manufacturing, and 90,403 units of transportation, how much is available from each sector to satisfy the demand from consumers and others outside the production system?
(c) A new demand matrix for the economy is

$$\begin{bmatrix} 2400 \\ 850 \\ 1400 \\ 3200 \\ 1800 \end{bmatrix}.$$

How much must each sector produce to meet this demand?

88. **Business** The following matrix represents the number of direct flights between four cities:

	A	B	C	D
A	0	1	0	1
B	1	0	0	1
C	0	0	0	1
D	1	1	1	0

(a) Find the number of one-stop flights between cities A and C.
(b) Find the total number of flights between cities B and C that are either direct or one stop.
(c) Find the matrix that gives the number of two-stop flights between these cities.

89. **Social Science** **(a)** Use the matrix $M = \begin{bmatrix} 2 & 6 \\ 1 & 4 \end{bmatrix}$ to encode the message "leave now."
(b) What matrix should be used to decode this message?

CASE 6

Matrix Operations and Airline Route Maps

Airline route maps are usually published on airline Web sites, as well as in in-flight magazines. The purpose of these maps is to show what cities are connected to each other by nonstop flights provided by the airline. We can think of these maps as another type of **graph**, and we can use matrix operations to answer questions of interest about the graph. In order to study these graphs, a bit of terminology will be helpful. In this context, a **graph** is a set of points called **vertices** or **nodes** and a set of lines called **edges** connecting some pairs of vertices. Two vertices connected by an edge are said to be **adjacent**. Consider, for example, the Massachusetts and Rhode Island route map for Cape Air from 2009 (Figure 1).* Here, the vertices are the cities to which Cape Air flies, and two vertices are connected if there is a nonstop flight between them.

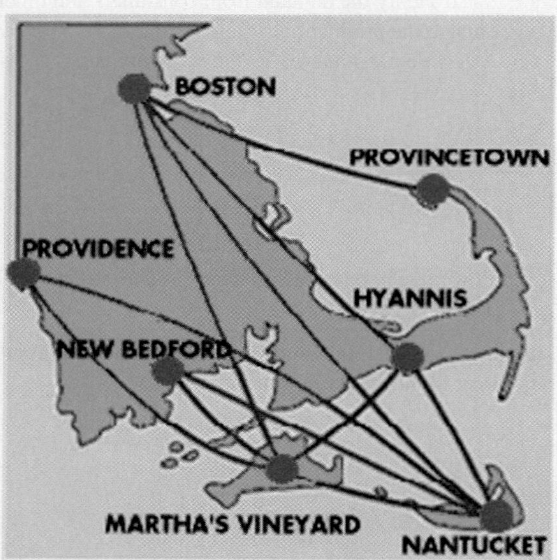

FIGURE 1 Cape Air route map, 2009.

Some natural questions arise about graphs. It might be important to know if two vertices are connected by a sequence of two edges, even if they are not connected by a single edge. In the route map, Provincetown and Hyannis are connected by a two-edge sequence, meaning that a passenger would have to stop in Boston while flying between those cities on Cape Air. It might be important to know if it is possible to get from a vertex to another vertex in a given number of flights. In the example, a passenger on Cape Air can get from any city in the company's network to any other city, given enough flights. But how many flights are enough? This is another issue of interest: What is the minimum number of steps required to get from one vertex to another? What is the minimum number of steps required to get from any vertex on the graph to any other? While these questions are relatively easy to answer for a small graph, as the number of vertices and edges grows, it becomes harder to keep track of all the different ways the vertices are connected. Matrix notation and computation can help to answer these questions.

The **adjacency matrix** for a graph with n vertices is an $n \times n$ matrix whose (i, j) entry is 1 if the ith and jth vertices are connected and 0 if they are not. If the vertices in the Cape Air graph respectively correspond to Boston (B), Hyannis (H), Martha's Vineyard (M), Nantucket (N), New Bedford (NB), Providence (P), and Provincetown (PT), then the adjacency matrix for Cape Air is as follows.

$$A = \begin{array}{c} \begin{matrix} \text{B} & \text{H} & \text{M} & \text{N} & \text{NB} & \text{P} & \text{PT} \end{matrix} \\ \begin{bmatrix} 0 & 1 & 1 & 1 & 0 & 0 & 1 \\ 1 & 0 & 1 & 1 & 0 & 0 & 0 \\ 1 & 1 & 0 & 1 & 1 & 1 & 0 \\ 1 & 1 & 1 & 0 & 1 & 1 & 0 \\ 0 & 0 & 1 & 1 & 0 & 0 & 0 \\ 0 & 0 & 1 & 1 & 0 & 0 & 0 \\ 1 & 0 & 0 & 0 & 0 & 0 & 0 \end{bmatrix} \end{array} \begin{matrix} \text{B} \\ \text{H} \\ \text{M} \\ \text{N} \\ \text{NB} \\ \text{P} \\ \text{PT} \end{matrix} .$$

Adjacency matrices can be used to address the questions about graphs raised earlier. Which vertices are connected by a two-edge sequence? How many different two-edge sequences connect each pair of vertices? Consider the matrix A^2, which is A multiplied by itself. For example, let the (6, 2) entry in the matrix A^2 be named b_{62}. This entry in A^2 is the product of the 6th row of A and the 2nd column of A, or

$$\begin{aligned} b_{62} &= a_{61}a_{12} + a_{62}a_{22} + a_{63}a_{32} + a_{64}a_{42} + a_{65}a_{52} + a_{66}a_{62} + a_{67}a_{72} \\ &= 0\cdot 1 + 0\cdot 0 + 1\cdot 1 + 1\cdot 1 + 0\cdot 0 + 0\cdot 0 + 0\cdot 0 \\ &= 2, \end{aligned}$$

which happens to be the number of two-flight sequences that connect city 6 (Providence) and city 2 (Hyannis). A careful look at Figure 1 confirms this fact. This calculation works because, in order for a two-flight sequence to occur between Providence and Hyannis, Providence and Hyannis must each connect to an intermediate city. Since Providence connects to Martha's Vineyard (city 3) and Martha's Vineyard connects to Hyannis, $a_{63}a_{32} = 1 \cdot 1 = 1$. Thus, there is 1 two-flight sequence from Providence to Hyannis that passes through Martha's Vineyard.

*Compliments of Cape Air. See www.flycapeair.com.

Since Providence does not connect to Boston (city 1), but Boston does connect with Hyannis, $a_{61}a_{12} = 0 \cdot 1 = 0$. Hence, there is no two-flight sequence from Providence to Hyannis that passes through Boston. To find the total number of two-flight sequences between Providence and Hyannis, simply sum over all intermediate points. Notice that this sum, which is

$$a_{61}a_{12} + a_{62}a_{22} + a_{63}a_{32} + a_{64}a_{42} + a_{65}a_{52} + a_{66}a_{62} + a_{67}a_{72},$$

is just b_{62}, the (6, 2) entry in the matrix A^2. So we see that the number of two-step sequences between vertex i and vertex j in a graph with adjacency matrix A is the (i, j) entry in A^2. A more general result is the following:

The number of k-step sequences between vertex i and vertex j in a graph with adjacency matrix A is the (i,j) entry in A^k.

If A is the adjacency matrix for Figure 1, then

$$A^2 = \begin{array}{c} \begin{array}{ccccccc} \text{B} & \text{H} & \text{M} & \text{N} & \text{NB} & \text{P} & \text{PT} \end{array} \\ \begin{bmatrix} 4 & 2 & 2 & 2 & 2 & 2 & 0 \\ 2 & 3 & 2 & 2 & 2 & 2 & 1 \\ 2 & 2 & 5 & 4 & 1 & 1 & 1 \\ 2 & 2 & 4 & 5 & 1 & 1 & 1 \\ 2 & 2 & 1 & 1 & 2 & 2 & 0 \\ 2 & 2 & 1 & 1 & 2 & 2 & 0 \\ 0 & 1 & 1 & 1 & 0 & 0 & 1 \end{bmatrix} \end{array} \begin{array}{l} \text{B} \\ \text{H} \\ \text{M} \\ \text{N} \\ \text{NB} \\ \text{P} \\ \text{PT} \end{array}$$

and

$$A^3 = \begin{array}{c} \begin{array}{ccccccc} \text{B} & \text{H} & \text{M} & \text{N} & \text{NB} & \text{P} & \text{PT} \end{array} \\ \begin{bmatrix} 6 & 8 & 12 & 12 & 4 & 4 & 4 \\ 8 & 6 & 11 & 11 & 4 & 4 & 2 \\ 12 & 11 & 10 & 11 & 9 & 9 & 2 \\ 12 & 11 & 11 & 10 & 9 & 9 & 2 \\ 4 & 4 & 9 & 9 & 2 & 2 & 2 \\ 4 & 4 & 9 & 9 & 2 & 2 & 2 \\ 4 & 2 & 2 & 2 & 2 & 2 & 0 \end{bmatrix} \end{array} \begin{array}{l} \text{B} \\ \text{H} \\ \text{M} \\ \text{N}\,. \\ \text{NB} \\ \text{P} \\ \text{PT} \end{array}$$

Since the (6, 3) entry in A^2 is 1, there is one two-step sequence from Providence to Martha's Vineyard. Likewise, there are four three-step sequences between Hyannis and New Bedford, since the (2, 5) entry in A^3 is 4.

In observing the figure, note that some two-step or three-step sequences may not be meaningful. On the Cape Air route map, Nantucket is reachable in two steps from Boston (via Hyannis or Martha's Vineyard), but in reality this does not matter, since there is a nonstop flight between the two cities. A better question to ask of a graph might be, "What is the least number of edges that must be traversed to go from vertex i to vertex j?"

To answer this question, consider the matrix $S_k = A + A^2 + \cdots + A^k$. The (i, j) entry in this matrix tallies the number of ways to get from vertex i to vertex j in k steps or less. If such a trip is impossible, this entry will be zero. Thus, to find the shortest number of steps between the vertices, continue to compute S_k as k increases; the first k for which the (i, j) entry in S_k is nonzero is the shortest number of steps between i and j. Note that although the shortest number of steps may be computed, the method does not determine what those steps are.

If you are interested in other airlines' route maps, visit the Web site dir.yahoo.com/Business_and_Economy/Shopping_and_Services/Travel_and_Transportation/Airlines/. This site includes links to many obscure and smaller airlines, as well as to the more well-known carriers.

EXERCISES

1. Which Cape Air cities may be reached by a two-flight sequence from New Bedford? Which may be reached by a three-flight sequence?
2. It was shown previously that there are four three-step sequences between Hyannis and New Bedford. Describe these three-step sequences.
3. Which trips in the Cape Air network take the greatest number of flights?
4. The route map for the northern routes of Big Sky Airlines for May 2001 is given in Figure 2.* Produce an adjacency matrix for this map.

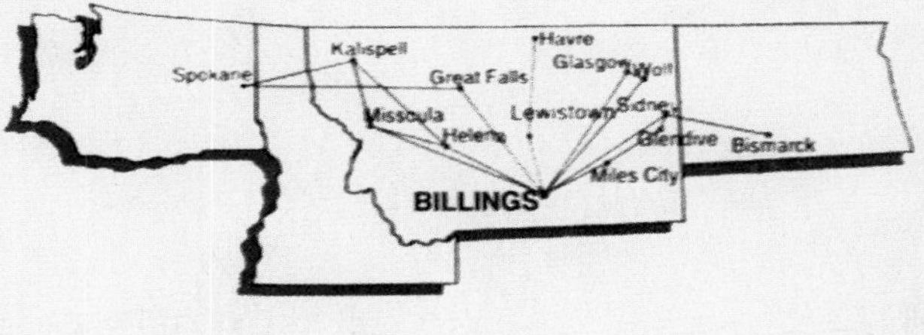

FIGURE 2 Big Sky Airlines northern route map, May 2001.

5. Which Big Sky cities could be reached by a three-flight sequence from Helena? For which cities did it take *at least* three flights to get from them to Helena?
6. Which trips in the Big Sky network took the largest number of flights? How many flights did these trips take?

*Airline route map courtesy of Big Sky Airlines, Billings, MT. Big Sky ceased operations in March 2008.

CHAPTER

7

Linear Programming

Linear programming is one of the most remarkable (and useful) mathematical techniques developed in the last 65 years. It is used to deal with a variety of issues faced by businesses, financial planners, medical personnel, sports leagues, and others. Typical applications include maximizing company profits by adjusting production schedules, minimizing shipping costs by locating warehouses efficiently, and maximizing pension income by choosing the best mix of financial products. See Exercise 17 on page 433, Exercise 33 on page 457, and Exercise 23 on page 411.

Many real-world problems involve inequalities. For example, a factory may have no more than 200 workers on a shift and must manufacture at least 3000 units at a cost of no more than \$35 each. How many workers should it have per shift in order to produce the required number of units at minimal cost? *Linear programming* is a method for finding the optimal (best possible) solution for such problems—if there is one.

In this chapter, we shall study two methods of solving linear programming problems: the graphical method and the simplex method. The graphical method requires a knowledge of **linear inequalities**, which are inequalities involving only first-degree polynomials in x and y. So we begin with a study of such inequalities.

7.1 Graphing Linear Inequalities in Two Variables

Examples of linear inequalities in two variables include

$$x + 2y < 4, \qquad 3x + 2y > 6, \qquad \text{and} \qquad 2x - 5y \geq 10.$$

A solution of a linear inequality is an ordered pair that satisfies the inequality. For example (4, 4) is a solution of

$$3x - 2y \leq 6.$$

(Check by substituting 4 for x and 4 for y.) A linear inequality has an infinite number of solutions, one for every choice of a value for x. The best way to show these solutions is to sketch the **graph of the inequality**, which consists of all points in the plane whose coordinates satisfy the inequality.

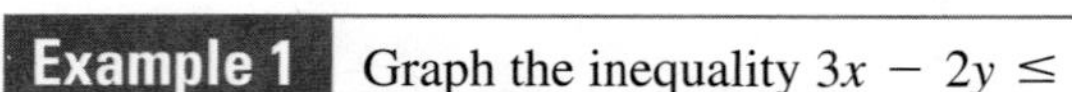

Example 1 Graph the inequality $3x - 2y \leq 6$.

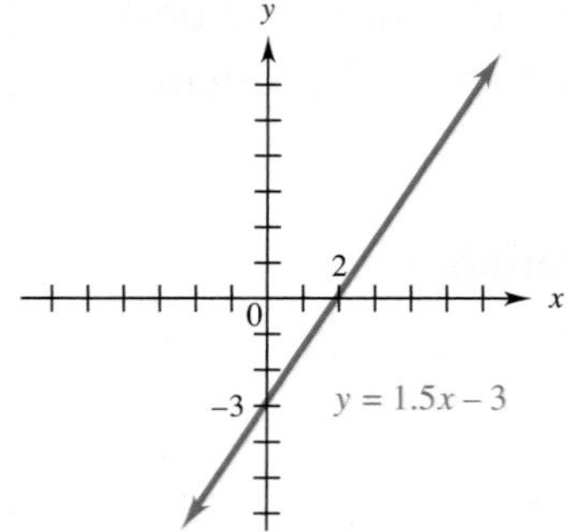

FIGURE 7.1

Solution First, solve the inequality for y:

$$3x - 2y \leq 6$$
$$-2y \leq -3x + 6$$
$$y \geq \frac{3}{2}x - 3 \qquad \text{Multiply by } -\tfrac{1}{2}\text{; reverse the inequality.}$$
$$y \geq 1.5x - 3.$$

This inequality has the same solutions as the original one. To solve it, note that the points on the line $y = 1.5x - 3$ certainly satisfy $y \geq 1.5x - 3$. Plot some points and graph this line, as in Figure 7.1.

The points on the line satisfy "y *equals* $1.5x - 3$." The points satisfying "y is *greater than* $1.5x - 3$" are the points *above* the line (because they have larger second coordinates than the points on the line; see Figure 7.2). Similarly, the points satisfying

$$y < 1.5x - 3$$

lie below the line (because they have smaller second coordinates), as shown in Figure 7.2. The line $y = 1.5x - 3$ is the **boundary line**.

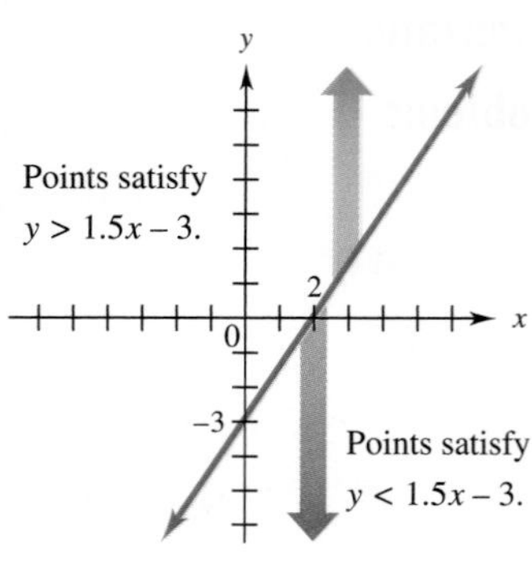

FIGURE 7.2

Thus, the solutions of $y \geq 1.5x - 3$ are all points *on or above* the line $y = 1.5x - 3$. The line and the shaded region of Figure 7.3 make up the graph of the inequality $y \geq 1.5x - 3$. 1

Checkpoint 1

Graph the given inequalities.

(a) $2x + 5y \leq 10$

(b) $x - y \geq 4$

Answers: See page 397.

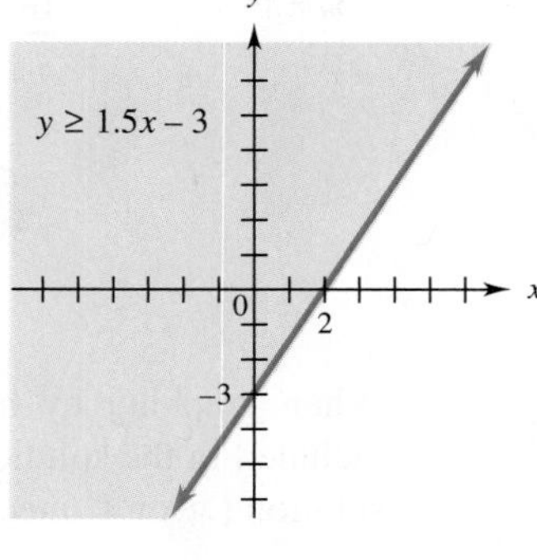

FIGURE 7.3

Example 2 Graph $x + 4y < 4$.

Solution First obtain an equivalent inequality by solving for y:

$$4y < -x + 4$$

$$y < -\frac{1}{4}x + 1$$

$$y < -.25x + 1.$$

The boundary line is $y = -.25x + 1$, but it is *not* part of the solution, since points *on* the line do not satisfy $y < -.25x + 1$. To indicate this, the line is drawn dashed in Figure 7.4. The points *below* the boundary line are the solutions of $y < -.25x + 1$, because they have smaller second coordinates than the points on the line $y = -.25x + 1$. The shaded region in Figure 7.4 (excluding the dashed line) is the graph of the inequality $y < -.25x + 1$. 2

Checkpoint 2

Graph the given inequalities.

(a) $2x + 3y > 12$

(b) $3x - 2y < 6$

Answers: See page 397.

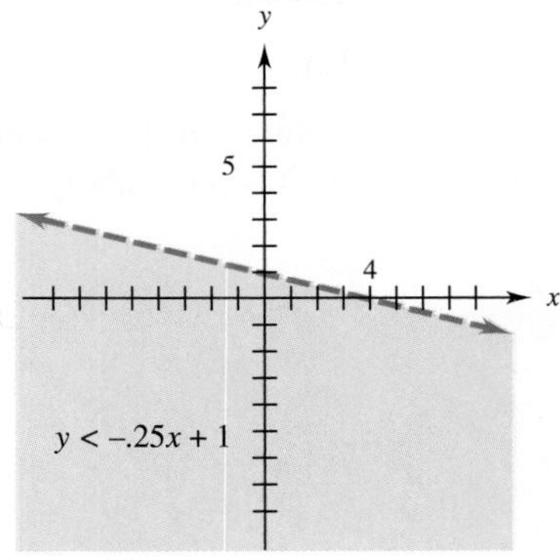

FIGURE 7.4

Examples 1 and 2 show that the solutions of a linear inequality form a **half-plane** consisting of all points on one side of the boundary line (and possibly the line itself). When an inequality is solved for y, the inequality symbol immediately tells whether the points above (>), on (=), or below (<) the boundary line satisfy the inequality, as summarized here.

Inequality	Solution Consists of All Points
$y \geq mx + b$	*on or above* the line $y = mx + b$
$y > mx + b$	*above* the line $y = mx + b$
$y \leq mx + b$	*on or below* the line $y = mx + b$
$y < mx + b$	*below* the line $y = mx + b$

When graphing by hand, draw the boundary line $y = mx + b$ solid when it is included in the solution ($\geq$ or $\leq$ inequalities) and dashed when it is not part of the solution (> or < inequalities).

TECHNOLOGY TIP To shade the area above or below the graph of Y_1 on TI-84+, go to the Y = menu and move the cursor to the left of Y_1. Press ENTER until the correct shading pattern appears (◥ for above the line and ◣ for below the line). Then press GRAPH. On TI-86/89, use the STYLE key in the Y= menu instead of the ENTER key. For other calculators, consult your instruction manual.

Example 3 Graph $5y - 2x \leq 10$.

Solution Solve the inequality for y:

$$5y \leq 2x + 10$$
$$y \leq \frac{2}{5}x + 2$$
$$y \leq .4x + 2.$$

The graph consists of all points on or below the boundary line $y \leq .4x + 2$, as shown in Figure 7.5. (See the Technology Tip.) 3✓

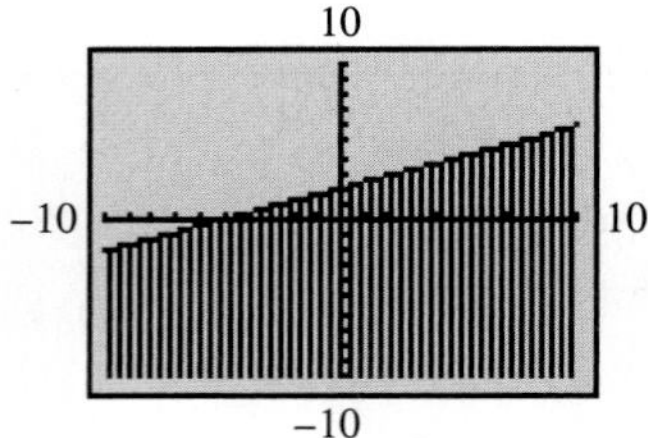

FIGURE 7.5

✓Checkpoint 3

Use a graphing calculator to graph $2x < y$.

Answer: See page 397.

CAUTION You cannot tell from a calculator-produced graph whether the boundary line is included. It is included in Figure 7.5, but not in the answer to Checkpoint 3.

Example 4 Graph each of the given inequalities.

(a) $y \geq 2$

Solution The boundary line is the horizontal line $y = 2$. The graph consists of all points on or above this line (Figure 7.6).

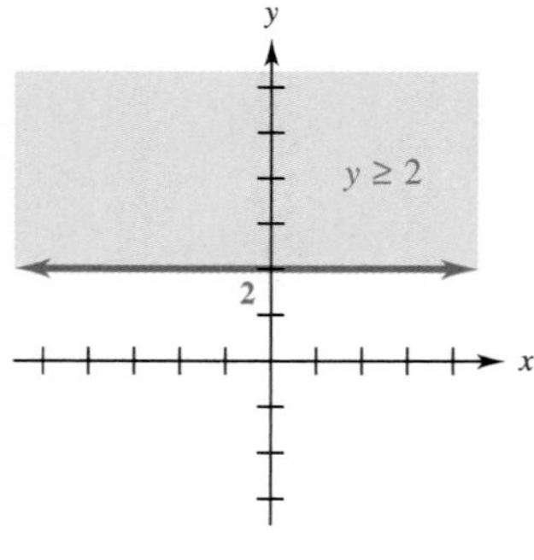

FIGURE 7.6

(b) $x \leq -1$

Solution This inequality does not fit the pattern discussed earlier, but it can be solved by a similar technique. Here, the boundary line is the vertical line $x = -1$, and it is included in the solution. The points satisfying $x < -1$ are all points to the left of this line (because they have x-coordinates smaller than -1). So the graph consists of the points that are *on or to the left of* the vertical line $x = -1$, as shown in Figure 7.7. [4]

Checkpoint 4

Graph each of the following.

(a) $x \geq 3$

(b) $y - 3 \leq 0$

Answers: See page 397.

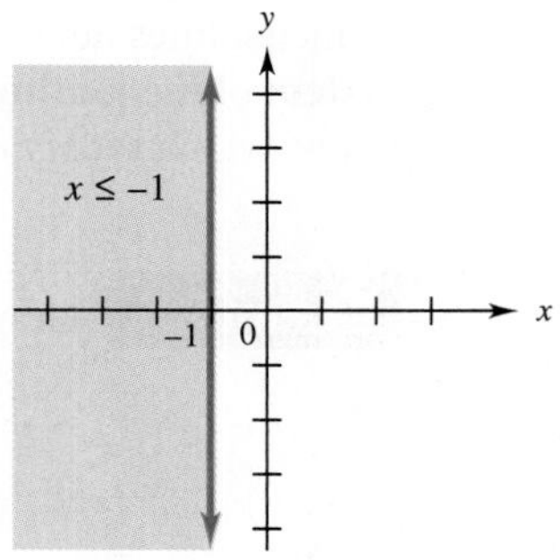

FIGURE 7.7

An alternative technique for solving inequalities that does not require solving for y is illustrated in the next example. Feel free to use it if you find it easier than the technique presented in Examples 1–4.

Example 5 Graph $4y - 2x \geq 6$.

Solution The boundary line is $4y - 2x = 6$, which can be graphed by finding its intercepts:

x-intercept: Let $y = 0$.	y-intercept: Let $x = 0$.
$4(0) - 2x = 6$	$4y - 2(0) = 6$
$x = -3.$	$y = \frac{6}{4} = 1.5.$

The graph contains the half-plane above or below this line. To determine which, choose a test point—any point not on the boundary line, say, (0, 0). Letting $x = 0$ and $y = 0$ in the inequality produces

$$4(0) - 2(0) \geq 6, \qquad \text{a } \textit{false} \text{ statement.}$$

Therefore, (0, 0) is not in the solution. So the solution is the half-plane that does *not* include (0, 0), as shown in Figure 7.8. If a different test point is used, say, (3, 5), then substituting $x = 3$ and $y = 5$ in the inequality produces

$$4(5) - 2(3) \geq 6, \qquad \text{a } \textit{true} \text{ statement.}$$

Therefore, the solution of the inequality is the half-plane containing (3, 5), as shown in Figure 7.8.

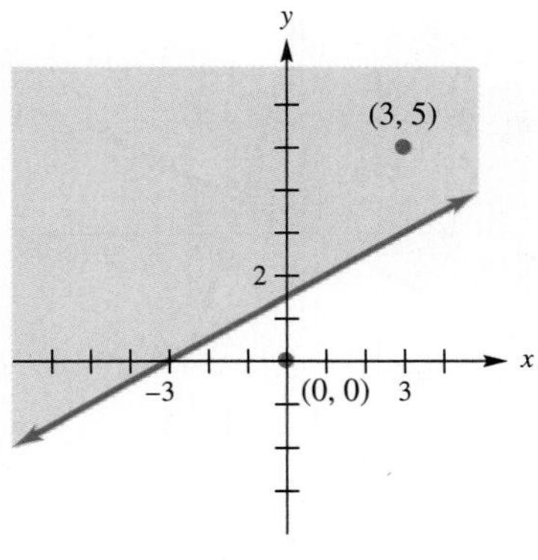

FIGURE 7.8

NOTE When using the method of Example 5, (0, 0) is the best choice for the test point because it makes the calculation very easy. The only time that (0, 0) cannot be used is for inequalities of the form $ax + by \geq 0$ (or $>$ or $<$ or $\leq$); in such cases, (0, 0) is on the line $ax + by = 0$.

SYSTEMS OF INEQUALITIES

Real-world problems often involve many inequalities. For example, a manufacturing problem might produce inequalities resulting from production requirements, as well as inequalities about cost requirements. A set of at least two inequalities is called a **system of inequalities**. The **graph** of a system of inequalities is made up of all those points which satisfy *all* the inequalities of the system.

Example 6 Graph the system

$$3x + y \leq 12$$
$$x \leq 2y.$$

Solution First, solve each inequality for y:

$$3x + y \leq 12 \qquad\qquad x \leq 2y$$
$$y \leq -3x + 12 \qquad\qquad y \geq \frac{x}{2}.$$

Then the original system is equivalent to this one:

$$y \leq -3x + 12$$
$$y \geq \frac{x}{2}.$$

Checkpoint 5

Graph the system

$$x + y \leq 6$$
$$2x + y \geq 4.$$

Answer: See page 397.

The solutions of the first inequality are the points *on or below* the line $y = -3x + 12$ (Figure 7.9). The solutions of the second inequality are the points *on or above* the line $y = x/2$ (Figure 7.10). So the solutions of the *system* are the points that satisfy both of these conditions, as shown in Figure 7.11. 5

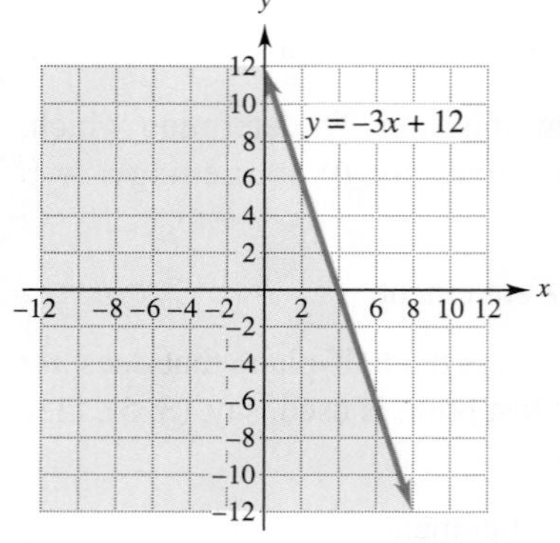

FIGURE 7.9

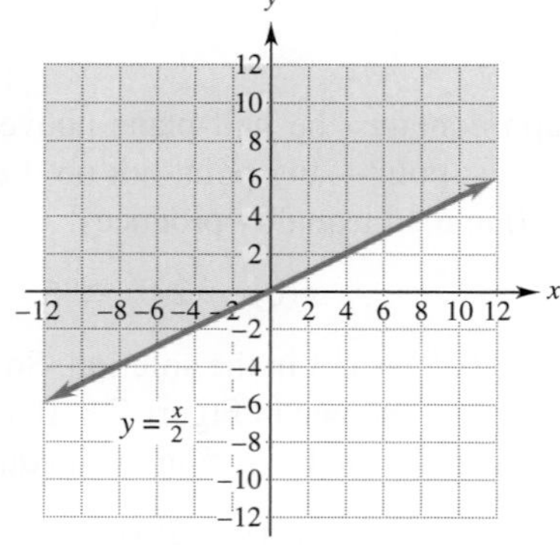

FIGURE 7.10

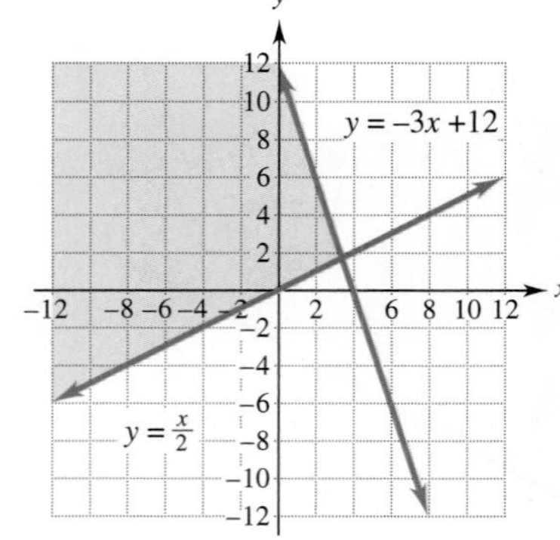

FIGURE 7.11

The shaded region in Figure 7.11 is sometimes called the **region of feasible solutions**, or just the **feasible region**, since it consists of all the points that satisfy (are feasible for) every inequality of the system.

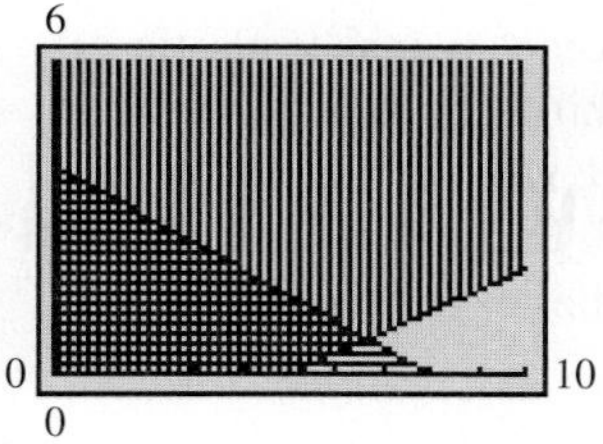

FIGURE 7.12

Example 7 Graph the feasible region for the system

$$2x - 5y \le 10$$
$$x + 2y \le 8$$
$$x \ge 0, y \ge 0.$$

Solution Begin by solving the first two inequalities for y:

$$\begin{aligned} 2x - 5y &\le 10 \\ -5y &\le -2x + 10 \\ y &\ge .4x - 2 \end{aligned} \qquad\qquad \begin{aligned} x + 2y &\le 8 \\ 2y &\le -x + 8 \\ y &\le -.5x + 4. \end{aligned}$$

Then the original system is equivalent to this one:

$$y \ge .4x - 2$$
$$y \le -.5x + 4$$
$$x \ge 0, y \ge 0.$$

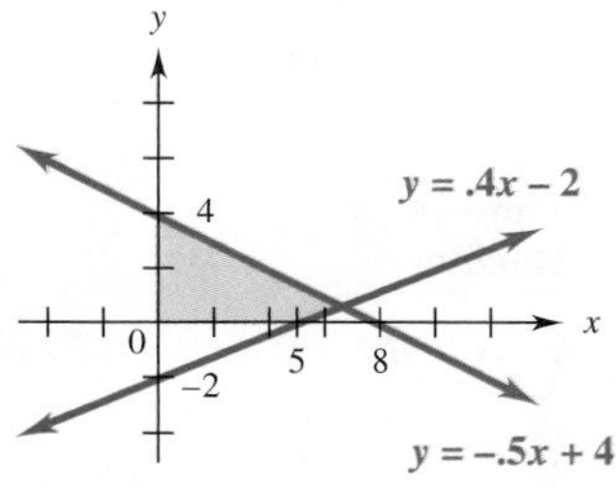

FIGURE 7.13

The inequalities $x \ge 0$ and $y \ge 0$ restrict the graph to the first quadrant. So the feasible region consists of all points in the first quadrant that are on or above the line $y = .4x - 2$ *and* on or below the line $y = -.5x + 4$. In the calculator-generated graph of Figure 7.12, the feasible region is the region with both vertical and horizontal shading. This is confirmed by the hand-drawn graph of the feasible region in Figure 7.13. 6✓

✓**Checkpoint 6**

Graph the feasible region of the system

$$x + 4y \le 8$$
$$x - y \ge 3$$
$$x \ge 0, y \ge 0.$$

Answer: See page 397.

Example 8 Graph the feasible region for the system

$$5y - 2x \le 10$$
$$x \ge 3, \quad y \ge 2.$$

Solution Solve the first inequality for y (as in Example 3) to obtain an equivalent system:

$$y \le .4x + 2$$
$$x \ge 3, \quad y \ge 2.$$

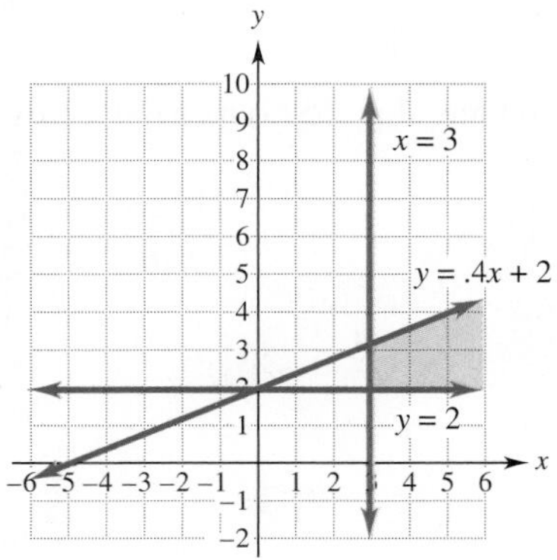

FIGURE 7.14

As shown in Example 3, Checkpoint 4, and Example 4(a), the feasible region consists of all points that lie

on or below the line $y \le .4x + 2$ *and*
on or to the right of the vertical line $x = 3$ *and*
on or above the horizontal line $y = 2$,

as shown in Figure 7.14.

APPLICATIONS

As we shall see in the rest of this chapter, many realistic problems lead to systems of linear inequalities. The next example is typical of such problems.

Example 9 **Business** Midtown Manufacturing Company makes plastic plates and cups, both of which require time on two machines. Producing a unit of plates requires 1 hour on machine A and 2 on machine B, while producing a unit of cups requires 3 hours on machine A and 1 on machine B. Each machine is operated for at most 15 hours per day. Write a system of inequalities expressing these conditions, and graph the feasible region.

Solution Let x represent the number of units of plates to be made and y represent the number of units of cups. Then make a chart that summarizes the given information.

		Time on Machine	
	Number of Units	***A***	***B***
Plates	x	1	2
Cups	y	3	1
Maximum Time Available		15	15

We must have $x \geq 0$ and $y \geq 0$, because the company cannot produce a negative number of cups or plates. On machine A, producing x units of plates requires a total of $1 \cdot x = x$ hours, while producing y units of cups requires $3 \cdot y = 3y$ hours. Since machine A is available no more than 15 hours a day,

$$x + 3y \leq 15$$

$$y \leq -\frac{x}{3} + 5.$$

Similarly, the requirement that machine B be used no more than 15 hours a day gives

$$2x + y \leq 15$$

$$y \leq -2x + 15.$$

So we must solve the system

$$y \leq -\frac{x}{3} + 5$$

$$y \leq -2x + 15$$

$$x \geq 0, y \geq 0.$$

The feasible region is shown in Figure 7.15.

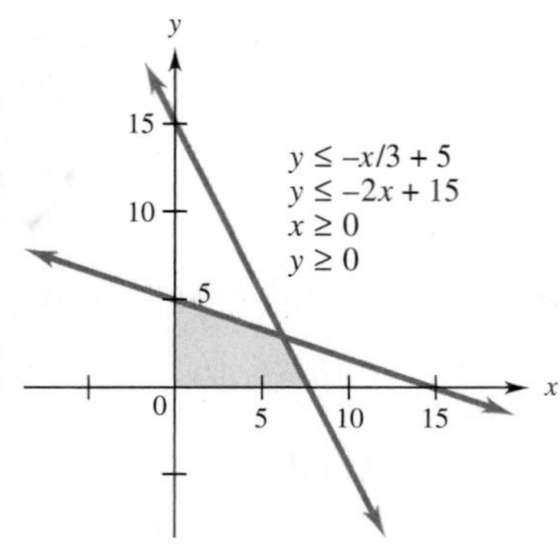

FIGURE 7.15

7.1 ▸ Exercises

Match the inequality with its graph, which is one of the ones shown.

1. $y \geq -x - 2$ **2.** $y \leq 2x - 2$ **3.** $y \leq x + 2$

4. $y \geq x + 1$ **5.** $6x + 4y \geq -12$ **6.** $3x - 2y \geq -4$

A.

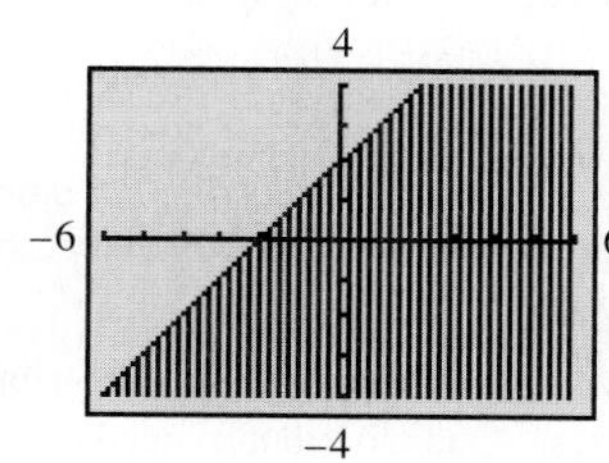

B.

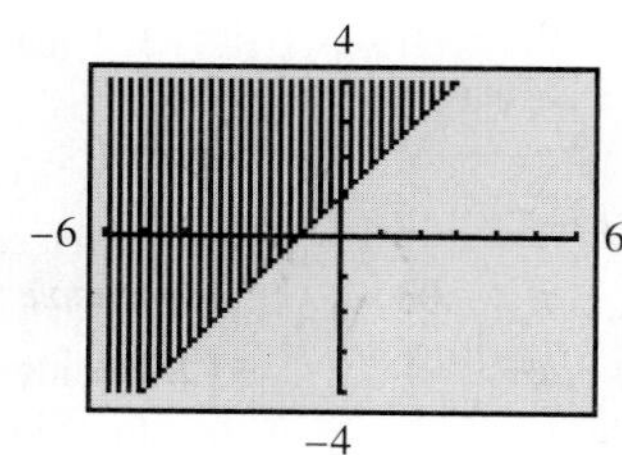

C.

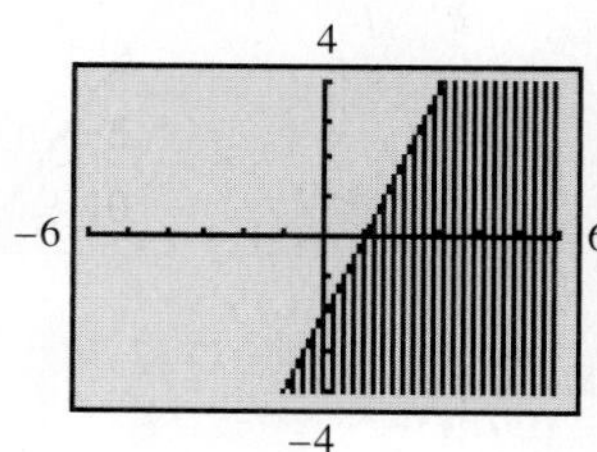

D.

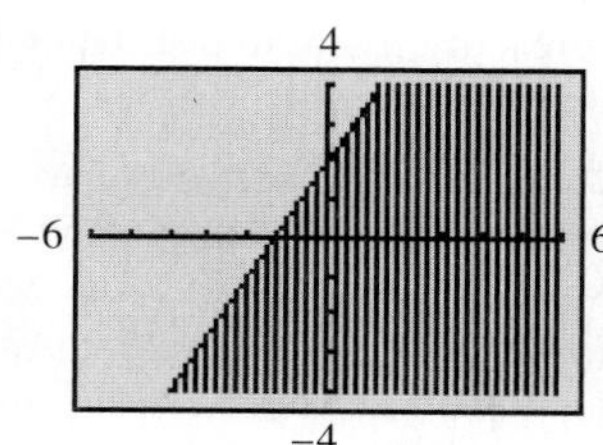

E.

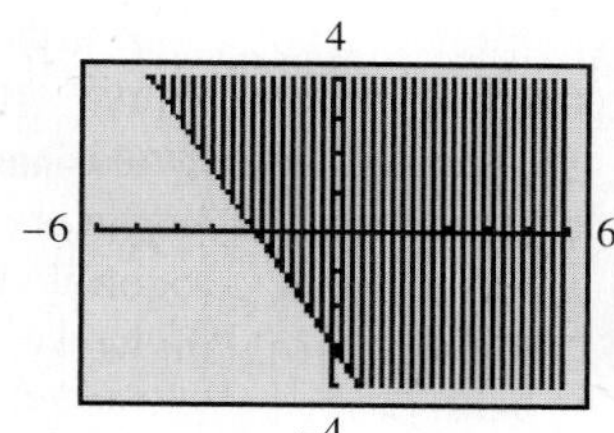

F.

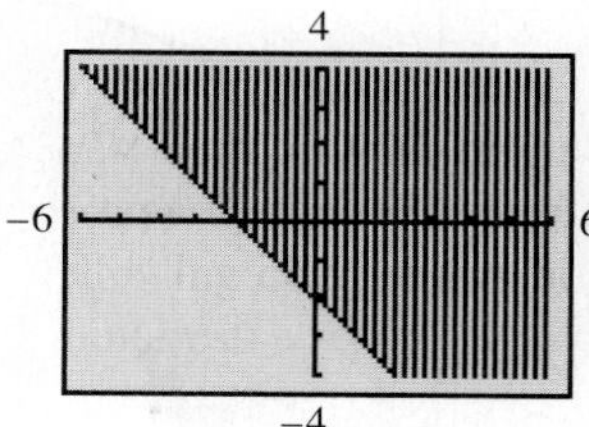

Graph each of the given linear inequalities. (See Examples 1–5.)

7. $y < 5 - 2x$

8. $y < x + 3$

9. $3x - 2y \geq 18$

10. $2x + 5y \geq 10$

11. $2x - y \leq 4$

12. $4x - 3y \leq 24$

13. $y \leq -4$

14. $x \geq -2$

15. $3x - 2y \geq 18$

16. $3x + 2y \geq -4$

17. $3x + 4y \geq 12$

18. $4x - 3y > 9$

19. $2x - 4y \leq 3$

20. $4x - 3y < 12$

21. $x \leq 5y$

22. $2x \geq y$

23. $-3x \leq y$

24. $-x \geq 6y$

25. $y \leq x$

26. $y > -2x$

27. In your own words, explain how to determine whether the boundary line of an inequality should be solid or dashed.

28. When graphing $y \leq 3x - 6$, would you shade above or below the line $y = 3x - 6$? Explain your answer.

Graph the feasible region for the given systems of inequalities. (See Examples 6 and 7.)

29. $y \geq 3x - 6$
$y \geq -x + 1$

30. $x + y \leq 4$
$x - y \geq 2$

31. $2x + y \leq 5$
$x + 2y \leq 5$

32. $x - y \geq 1$
$x \leq 3$

33. $2x + y \geq 8$
$4x - y \leq 3$

34. $4x + y \geq 9$
$2x + 3y \leq 7$

35. $2x - y \leq 1$
$3x + y \leq 6$

36. $x + 3y \leq 6$
$2x + 4y \geq 7$

37. $-x - y \leq 5$
$2x - y \leq 4$

38. $6x - 4y \geq 8$
$3x + 2y \geq 4$

39. $3x + y \geq 6$
$x + 2y \geq 7$
$x \geq 0$
$y \geq 0$

40. $2x + 3y \geq 12$
$x + y \geq 4$
$x \geq 0$
$y \geq 0$

41. $-2 \leq x \leq 3$
$-1 \leq y \leq 5$
$2x + y \leq 6$

42. $-2 \leq x \leq 2$
$y \geq 1$
$x - y \geq 0$

43. $2y - x \geq -5$
$y \leq 3 + x$
$x \geq 0$
$y \geq 0$

44. $2x + 3y \leq 12$
$2x + 3y \geq -6$
$3x + y \leq 4$
$x \geq 0$
$y \geq 0$

45. $3x + 4y \geq 12$
$2x - 3y \leq 6$
$0 \leq y \leq 2$
$x \geq 0$

46. $0 \leq x \leq 9$
$x - 2y \geq 4$
$3x + 5y \leq 30$
$y \geq 0$

Find a system of inequalities that has the given graph.

47.

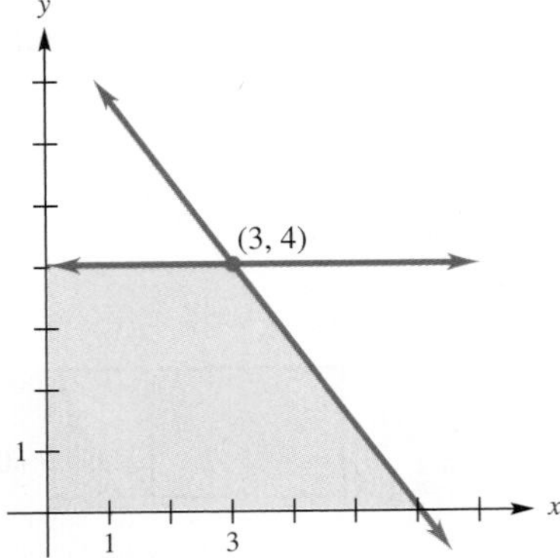

48.

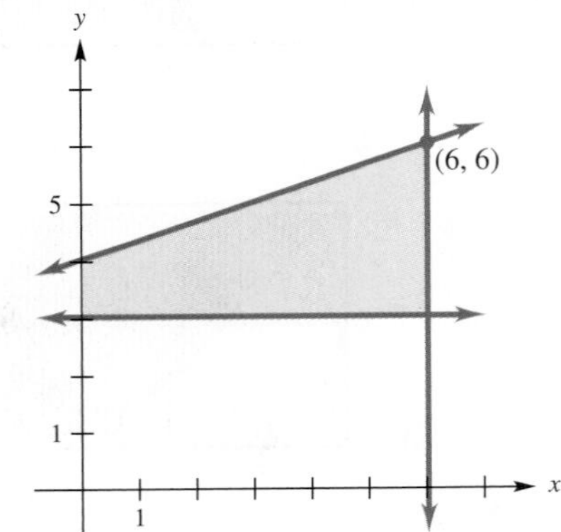

In Exercises 49 and 50, find a system of inequalities whose feasible region is the interior of the given polygon.

49. Rectangle with vertices $(2, 3)$, $(2, -1)$, $(7, 3)$, and $(7, -1)$

50. Triangle with vertices $(2, 4)$, $(-4, 0)$, and $(2, -1)$

51. **Business** Cindi Herring and Kent Merrill produce handmade shawls and afghans. They spin the yarn, dye it, and then weave it. A shawl requires 1 hour of spinning, 1 hour of dyeing, and 1 hour of weaving. An afghan needs 2 hours of spinning, 1 of dyeing, and 4 of weaving. Together, they spend at most 8 hours spinning, 6 hours dyeing, and 14 hours weaving.

(a) Complete the following table.

	Number	Hours Spinning	Hours Dyeing	Hours Weaving
Shawls	x			
Afghans	y			
Maximum Number of Hours Available		8	6	14

(b) Use the table to write a system of inequalities that describes the situation.

(c) Graph the feasible region of this system of inequalities.

52. **Business** A manufacturer of electric shavers makes two models: the regular and the flex. Because of demand, the number of regular shavers made is never more than half the number of flex shavers. The factory's production cannot exceed 1200 shavers per week.

(a) Write a system of inequalities that describes the possibilities for making x regular and y flex shavers per week.

(b) Graph the feasible region of this system of inequalities.

In each of the following, write a system of inequalities that describes all the given conditions, and graph the feasible region of the system. (See Example 9.)

53. **Business** Southwestern Oil supplies two distributors located in the northwest. One distributor needs at least 3000 barrels of oil, and the other at least 5000 barrels. Southwestern can send out at most 10,000 barrels. Let x = the number of barrels of oil sent to distributor 1 and y = the number sent to distributor 2.

54. **Business** The California Almond Growers have 2400 boxes of almonds to be shipped from their plant in Sacramento to Des Moines and San Antonio. The Des Moines market needs at least 1000 boxes, while the San Antonio market must have at least 800 boxes. Let x = the number of boxes to be shipped to Des Moines and y = the number of boxes to be shipped to San Antonio.

55. **Business** A cement manufacturer produces at least 3.2 million barrels of cement annually. He is told by the Environmental Protection Agency that his operation emits 2.5 pounds of dust for each barrel produced. The EPA has ruled that annual emissions must be reduced to 1.8 million pounds. To do this, the manufacturer plans to replace the present dust collectors with two types of electronic precipitators. One type would reduce emissions to .5 pound per barrel and would cost 16¢ per barrel. The other would reduce the dust to .3 pound per barrel and would cost 20¢ per barrel. The manufacturer does not want to spend more than .8 million dollars on the precipitators. He needs to know how many barrels he should produce with each type. Let x = the number of barrels (in millions produced) with the first type and y = the number of barrels (in millions produced) with the second type.

56. **Health** A dietician is planning a snack package of fruit and nuts. Each ounce of fruit will supply 1 unit of protein, 2 units of carbohydrates, and 1 unit of fat. Each ounce of nuts will supply 1 unit of protein, 1 unit of carbohydrates, and 1 unit of fat. Every package must provide at least 7 units of protein, at least 10 units of carbohydrates, and no more than 9 units of fat. Let x be the number of ounces of fruit and y the number of ounces of nuts to be used in each package.

Checkpoint Answers

1. (a)

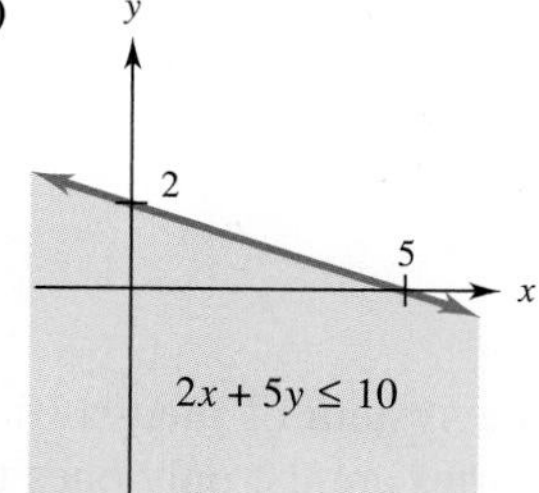

(b)

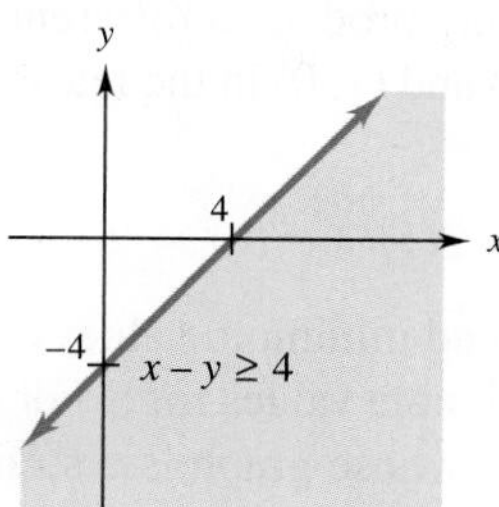

2. (a)

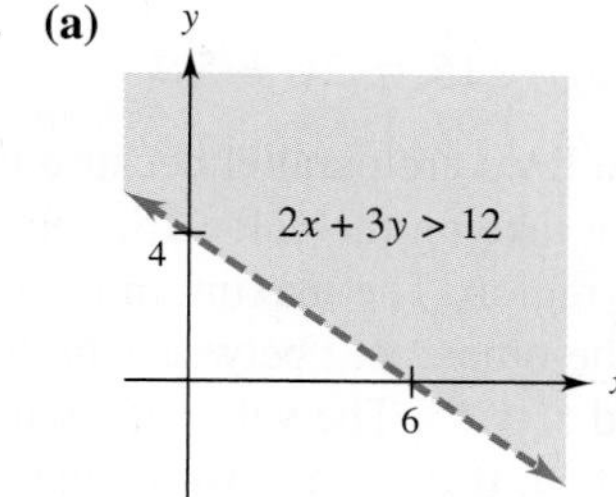

(b)

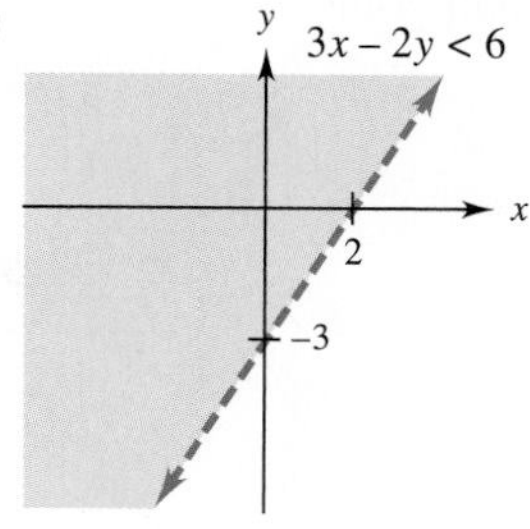

3.

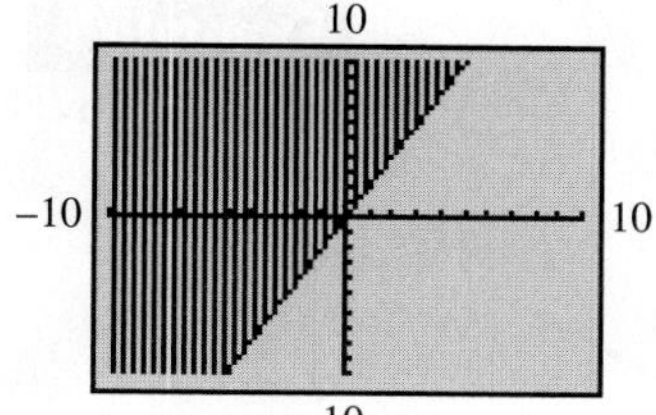

4. (a)

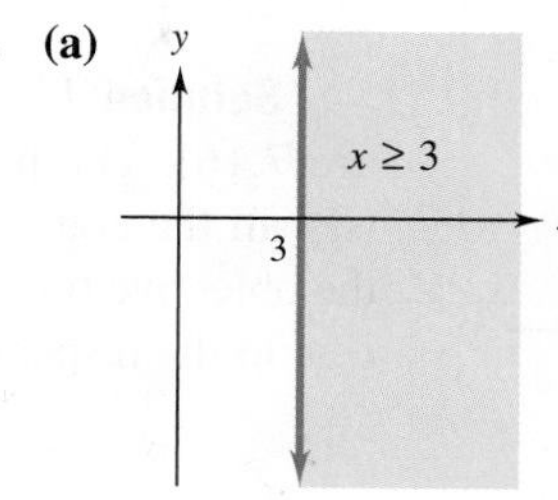

(b)

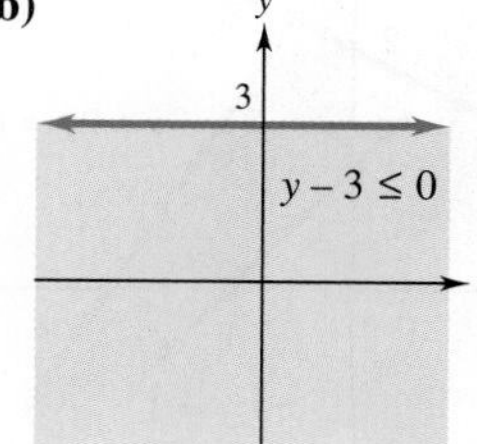

5.

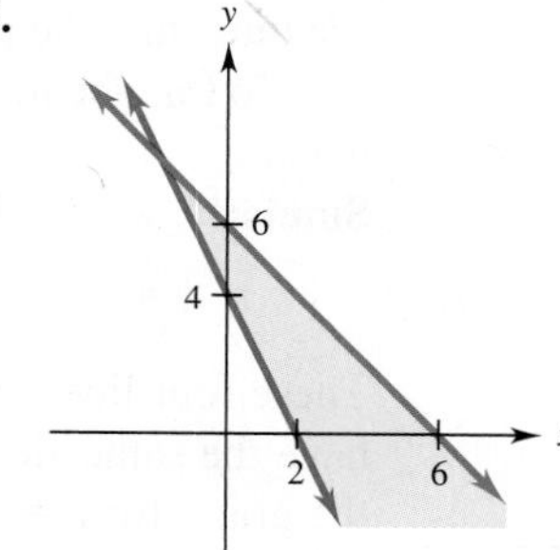

6.

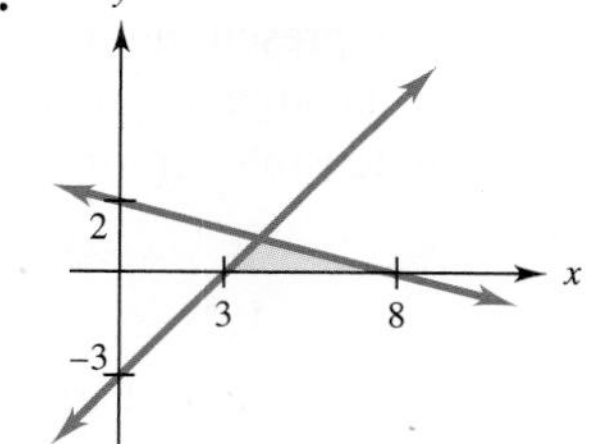

7.2 ▶ Linear Programming: The Graphical Method

Many problems in business, science, and economics involve finding the optimal value of a function (for instance, the maximum value of the profit function or the minimum value of the cost function), subject to various **constraints** (such as transportation costs, environmental protection laws, availability of parts, and interest rates). **Linear programming** deals with such situations. In linear programming, the function to be optimized, called the **objective function**, is linear and the constraints are given by linear inequalities. Linear programming problems that involve only two variables can be solved by the graphical method, explained in Example 1.

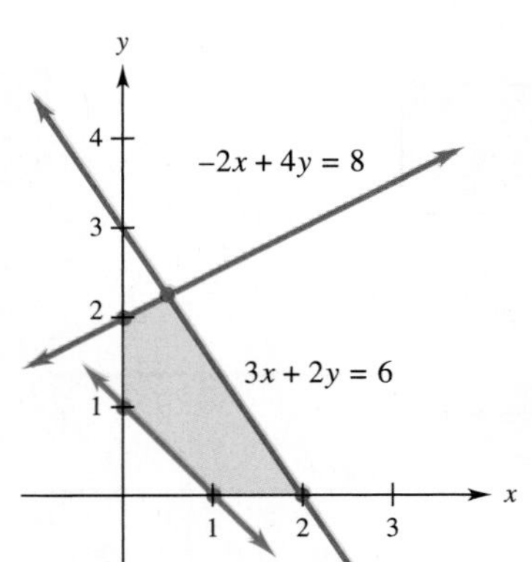

FIGURE 7.16

Example 1 Find the maximum and minimum values of the objective function $z = 2x + 5y$, subject to the following constraints:

$$3x + 2y \leq 6$$
$$-2x + 4y \leq 8$$
$$x + y \geq 1$$
$$x \geq 0, y \geq 0.$$

Solution First, graph the feasible region of the system of inequalities (Figure 7.16). The points in this region or on its boundaries are the only ones that satisfy all the constraints. However, each such point may produce a different value of the objective function. For instance, the points (.5, 1) and (1, 0) in the feasible region lead to the respective values

$$z = 2(.5) + 5(1) = 6 \quad \text{and} \quad z = 2(1) + 5(0) = 2.$$

We must find the points that produce the maximum and minimum values of z.

To find the maximum value, consider various possible values for z. For instance, when $z = 0$, the objective function is $0 = 2x + 5y$, whose graph is a straight line. Similarly, when z is 5, 10, and 15, the objective function becomes (in turn)

$$5 = 2x + 5y, \quad 10 = 2x + 5y, \quad \text{and} \quad 15 = 2x + 5y.$$

These four lines are graphed in Figure 7.17. (All the lines are parallel because they have the same slope.) The figure shows that z cannot take on the value 15, because the graph for $z = 15$ is entirely outside the feasible region. The maximum possible value of z will be obtained from a line parallel to the others and between the lines representing the objective function when $z = 10$ and $z = 15$. The value of z will be as large as possible, and all constraints will be satisfied, if this line just touches the feasible region. This occurs with the green line through point A.

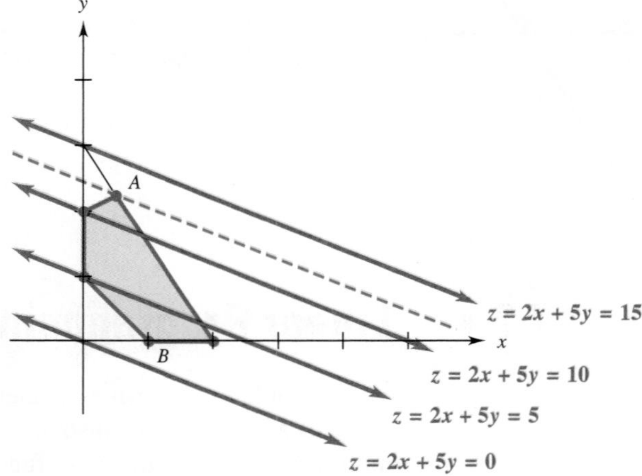

FIGURE 7.17

The point A is the intersection of the graphs of $3x + 2y = 6$ and $-2x + 4y = 8$. (See Figure 7.16.) Its coordinates can be found either algebraically or graphically (using a graphing calculator).

Algebraic Method

Solve the system

$$3x + 2y = 6$$
$$-2x + 4y = 8,$$

as in Section 6.1, to get $x = \frac{1}{2}$ and $y = \frac{9}{4}$. Hence, A has coordinates $\left(\frac{1}{2}, \frac{9}{4}\right) = (.5, 2.25)$.

Graphical Method

Solve the two equations for y:

$$y = -1.5x + 3$$
$$y = .5x + 2.$$

Graph both equations on the same screen and use the intersection finder to find that the coordinates of the intersection point A are $(.5, 2.25)$.

The value of z at point A is

$$z = 2x + 5y = 2(.5) + 5(2.25) = 12.25.$$

Thus, the maximum possible value of z is 12.25. Similarly, the minimum value of z occurs at point B, which has coordinates $(1, 0)$. The minimum value of z is $2(1) + 5(0) = 2$. 1✓

Checkpoint 1

Suppose the objective function in Example 1 is changed to $z = 5x + 2y$.

(a) Sketch the graphs of the objective function when $z = 0$, $z = 5$, and $z = 10$ on the region of feasible solutions given in Figure 7.16.

(b) From the graph, decide what values of x and y will maximize the objective function.

Answers: See page 404.

Points such as A and B in Example 1 are called corner points. A **corner point** is a point in the feasible region where the boundary lines of two constraints cross. The feasible region in Figure 7.16 is **bounded** because the region is enclosed by boundary lines on all sides. Linear programming problems with bounded regions always have solutions. However, if Example 1 did not include the constraint $3x + 2y \leq 6$, the feasible region would be **unbounded**, and there would be no way to *maximize* the value of the objective function.

Some general conclusions can be drawn from the method of solution used in Example 1. Figure 7.18 shows various feasible regions and the lines that result from various values of z. (Figure 7.18 shows the situation in which the lines are in order from left to right as z increases.) In part (a) of Figure 7.18, the objective function takes on its minimum value at corner point Q and its maximum value at P. The minimum is again at Q in part (b), but the maximum occurs at P_1 or P_2, or any point on the line segment connecting them. Finally, in part (c), the minimum value occurs at Q, but the objective function has no maximum value because the feasible region is unbounded.

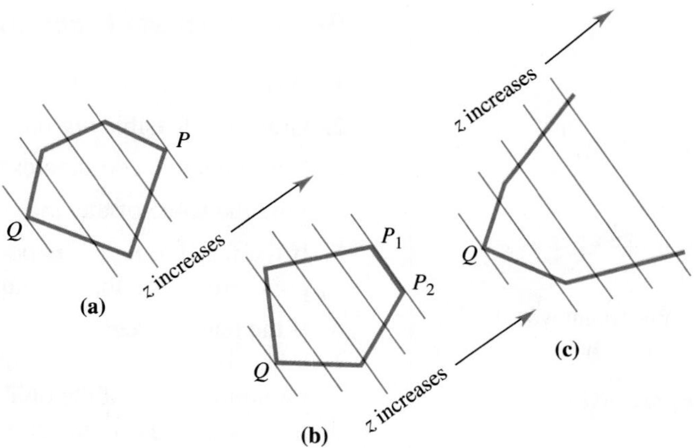

FIGURE 7.18

The preceding discussion suggests the **corner point theorem.**

Corner Point Theorem

If the feasible region is bounded, then the objective function has both a maximum and a minimum value, and each occurs at one or more corner points.

If the feasible region is unbounded, the objective function may not have a maximum or minimum. But if a maximum or minimum value exists, it will occur at one or more corner points.

This theorem simplifies the job of finding an optimum value. First, graph the feasible region and find all corner points. Then test each point in the objective function. Finally, identify the corner point producing the optimum solution.

With the theorem, the problem in Example 1 could have been solved by identifying the five corner points of Figure 7.16: (0, 1), (0, 2), (.5, 2.25), (2, 0), and (1, 0). Then, substituting each of these points into the objective function $z = 2x + 5y$ would identify the corner points that produce the maximum and minimum values of z.

Corner Point	Value of $z = 2x + 5y$	
(0, 1)	$2(0) + 5(1) = 5$	
(0, 2)	$2(0) + 5(2) = 10$	
(.5, 2.25)	$2(.5) + 5(2.25) = 12.25$	**(maximum)**
(2, 0)	$2(2) + 5(0) = 4$	
(1, 0)	$2(1) + 5(0) = 2$	**(minimum)**

From these results, the corner point (.5, 2.25) yields the maximum value of 12.25 and the corner point (1, 0) gives the minimum value of 2. These are the same values found earlier. ✓2

A summary of the steps for solving a linear programming problem by the graphical method is given here.

✓Checkpoint 2

(a) Identify the corner points in the given graph.

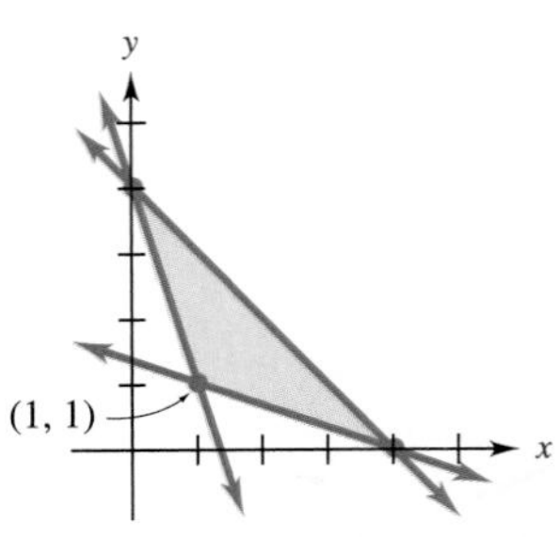

(b) Which corner point would minimize $z = 2x + 3y$?

Answers: See page 405.

Solving a Linear Programming Problem Graphically

1. Write the objective function and all necessary constraints.
2. Graph the feasible region.
3. Determine the coordinates of each of the corner points.
4. Find the value of the objective function at each corner point.
5. If the feasible region is bounded, the solution is given by the corner point producing the optimum value of the objective function.
6. If the feasible region is an unbounded region in the first quadrant and both coefficients of the objective function are positive,* then the minimum value of the objective function occurs at a corner point and there is no maximum value.

*This is the only case of an unbounded region that occurs in the applications considered here.

Example 2 Sketch the feasible region for the following set of constraints:

$$3y - 2x \geq 0$$
$$y + 8x \leq 52$$
$$y - 2x \leq 2$$
$$x \geq 3.$$

Then find the maximum and minimum values of the objective function $z = 5x + 2y$.

Solution Graph the feasible region, as in Figure 7.19. To find the corner points, you must solve these four systems of equations:

A	B	C	D
$y - 2x = 2$ $x = 3$	$3y - 2x = 0$ $x = 3$	$3y - 2x = 0$ $y + 8x = 52$	$y - 2x = 2$ $y + 8x = 52$

The first two systems are easily solved by substitution, which shows that $A = (3, 8)$ and $B = (3, 2)$. The other two systems can be solved either graphically (as in Figure 7.20) or algebraically (see Checkpoint 3). Hence, $C = (6, 4)$ and $D = (5, 12)$. 3✓

✓Checkpoint 3

Use the elimination method (see Section 6.1) to solve the last system and find the coordinates of D in Example 2.

Answer: See page 405.

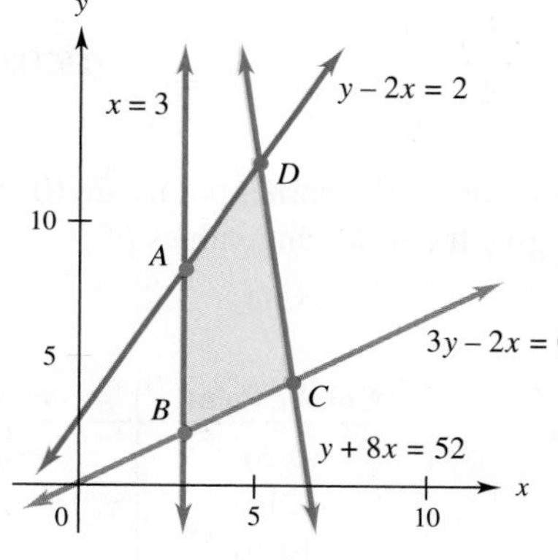

FIGURE 7.19

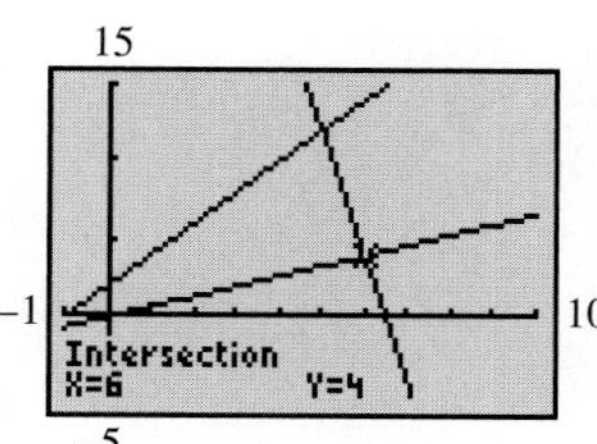

FIGURE 7.20

Use the corner points from the graph to find the maximum and minimum values of the objective function.

Corner Point	Value of $z = 5x + 2y$	
(3, 8)	$5(3) + 2(8) = 31$	
(3, 2)	$5(3) + 2(2) = 19$	(minimum)
(6, 4)	$5(6) + 2(4) = 38$	
(5, 12)	$5(5) + 2(12) = 49$	(maximum)

The minimum value of $z = 5x + 2y$ is 19, at the corner point (3, 2). The maximum value is 49, at (5, 12). 4✓

✓Checkpoint 4

Use the region of feasible solutions in the accompanying sketch to find the given values.

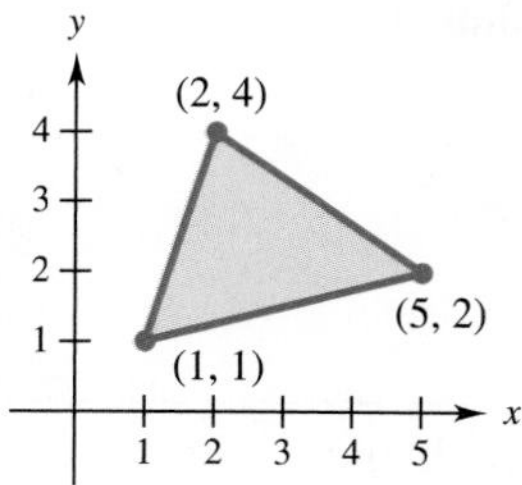

(a) The values of x and y that maximize $z = 2x - y$

(b) The maximum value of $z = 2x - y$

(c) The values of x and y that minimize $z = 4x + 3y$

(d) The minimum value of $z = 4x + 3y$

Answers: See page 405.

Example 3 Solve the following linear programming problem:

$$\begin{aligned} \text{Minimize} \quad & z = x + 2y \\ \text{subject to} \quad & x + y \le 10 \\ & 3x + 2y \ge 6 \\ & x \ge 0, y \ge 0. \end{aligned}$$

Solution The feasible region is shown in Figure 7.21.

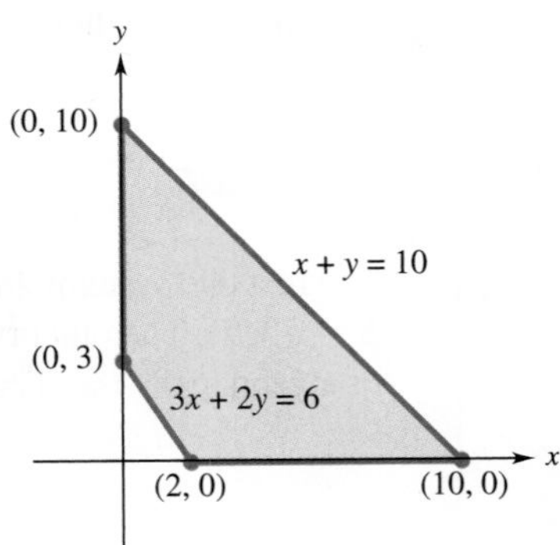

FIGURE 7.21

From the figure, the corner points are (0, 3), (0, 10), (10, 0), and (2, 0). These corner points give the following values of z.

Corner Point	Value of $z = x + 2y$	
(0, 3)	$0 + 2(3) = 6$	
(0, 10)	$0 + 2(10) = 20$	
(10, 0)	$10 + 2(0) = 10$	
(2, 0)	$2 + 2(0) = 2$	(minimum)

The minimum value of z is 2; it occurs at (2, 0). ✓5

✓Checkpoint 5

The given sketch shows a feasible region. Let $z = x + 3y$. Use the sketch to find the values of x and y that

(a) minimize z;

(b) maximize z.

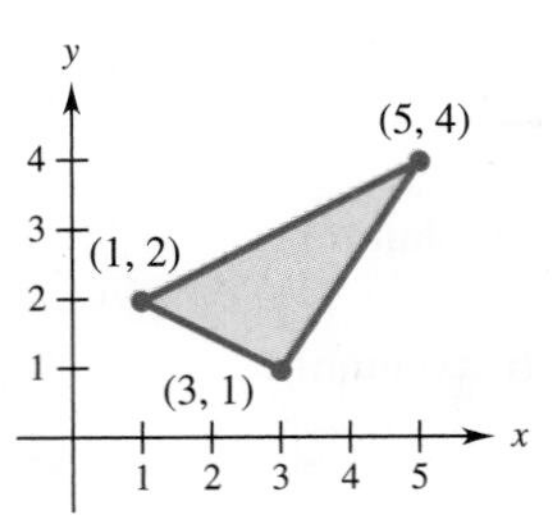

Answers: See page 405.

Example 4 Solve the following linear programming problem:

$$\begin{aligned} \text{Minimize} \quad & z = 2x + 4y \\ \text{subject to} \quad & x + 2y \ge 10 \\ & 3x + y \ge 10 \\ & x \ge 0, y \ge 0. \end{aligned}$$

Solution Figure 7.22 shows the hand-drawn graph with corner points (0, 10), (2, 4), and (10, 0), as well as the calculator graph with the corner point (2, 4). Find the value of z for each point.

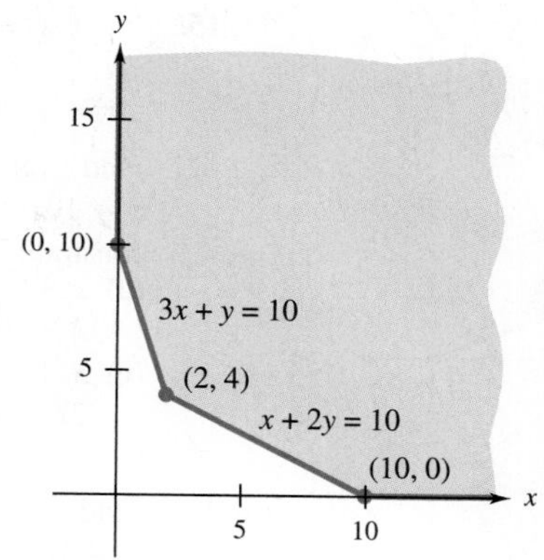

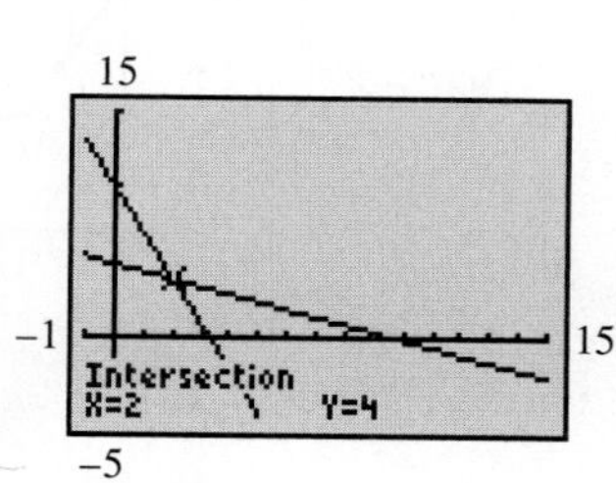

FIGURE 7.22

The sketch shows a region of feasible solutions. From the sketch, decide what ordered pair would minimize $z = 2x + 4y$.

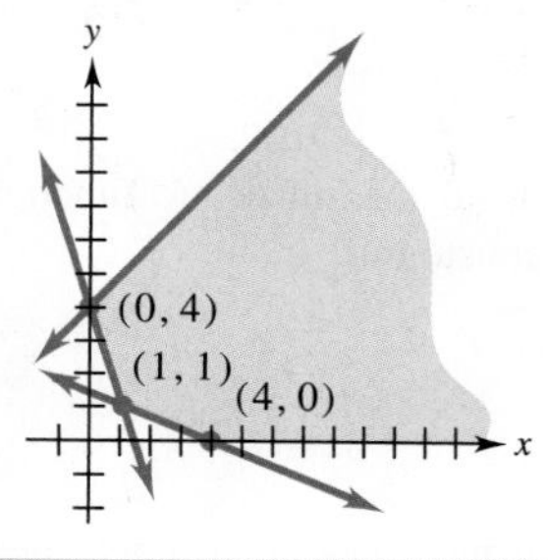

Answer: See page 405.

Corner Point	Value of $z = 2x + 4y$	
(0, 10)	$2(0) + 4(10) = 40$	
(2, 4)	$2(2) + 4(4) = 20$	(minimum)
(10, 0)	$2(10) + 4(0) = 20$	(minimum)

In this case, both (2, 4) and (10, 0), as well as all the points on the boundary line between them, give the same optimum value of z. So there is an infinite number of equally "good" values of x and y that give the same minimum value of the objective function $z = 2x + 4y$. The minimum value is 20. 6 ✓

7.2 ▶ Exercises

Exercises 1–6 show regions of feasible solutions. Use these regions to find maximum and minimum values of each given objective function. (See Examples 1 and 2.)

1. $z = 6x + y$

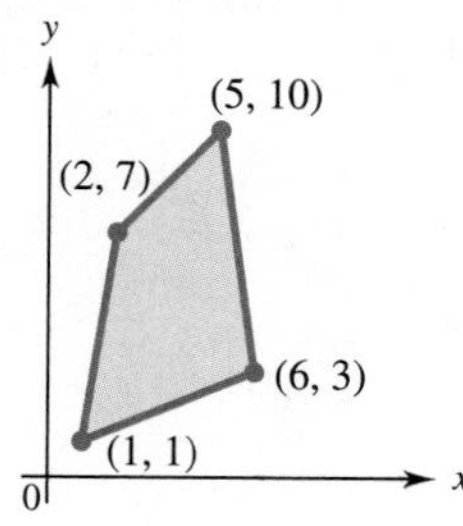

2. $z = 4x + y$

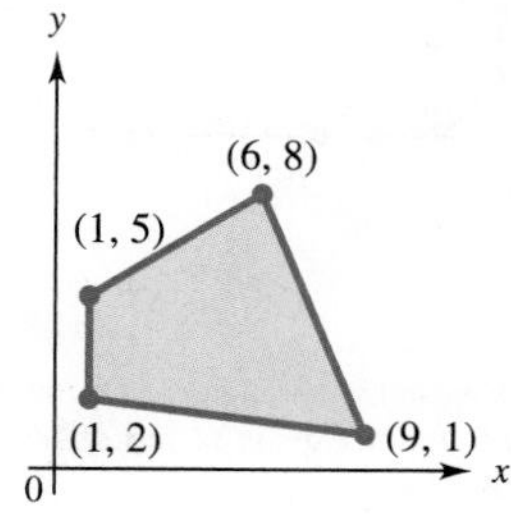

3. $z = .3x + .5y$

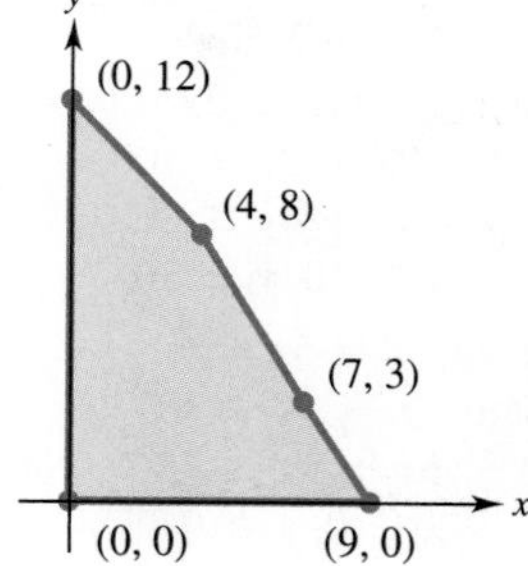

4. $z = .35x + 1.25y$

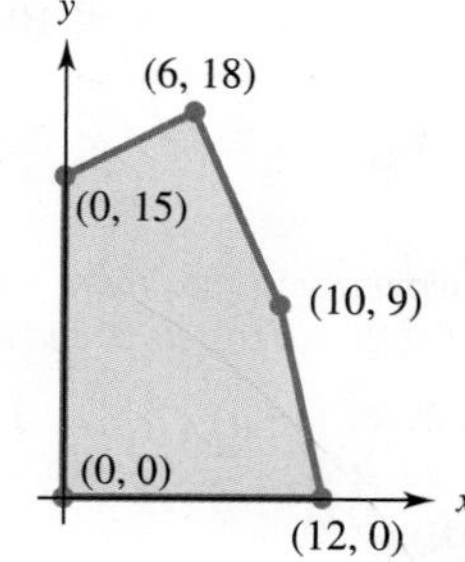

5. (a) $z = x + 5y$
(b) $z = 2x + 3y$
(c) $z = 2x + 4y$
(d) $z = 4x + y$

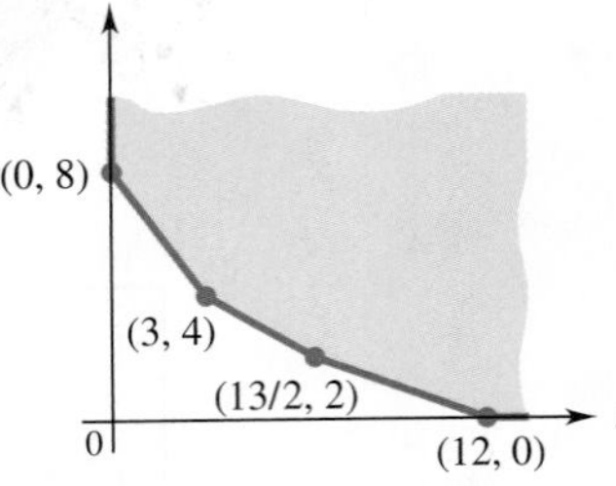

6. (a) $z = 5x + 2y$
(b) $z = 5x + 6y$
(c) $z = x + 2y$
(d) $z = x + y$

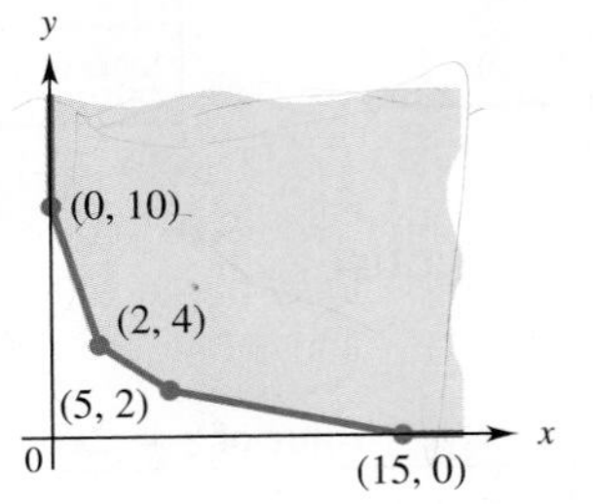

Use graphical methods to solve Exercises 7–12. (See Examples 2–4.)

7. Maximize $z = 4x + 3y$
subject to $2x + 3y \le 6$
$4x + y \le 6$
$x \ge 0, y \ge 0.$

8. Minimize $z = x + 3y$
subject to $2x + y \le 10$
$5x + 2y \ge 20$
$-x + 2y \ge 0$
$x \ge 0, y \ge 0.$

9. Minimize $z = 2x + y$
subject to $3x - y \ge 12$
$x + y \le 15$
$x \ge 2, y \ge 3.$

10. Maximize $z = x + 3y$
subject to $2x + 3y \le 100$
$5x + 4y \le 200$
$x \ge 10, y \ge 20.$

11. Maximize $z = 5x + y$
subject to $x - y \le 10$
$5x + 3y \le 75$
$x \ge 0, y \ge 0.$

12. Maximize $z = 4x + 5y$
subject to $10x - 5y \le 100$
$20x + 10y \ge 150$
$x \ge 0, y \ge 0.$

Find the minimum and maximum values of $z = 3x + 4y$ (if possible) for each of the given sets of constraints. (See Examples 2–4.)

13. $3x + 2y \ge 6$
$x + 2y \ge 4$
$x \ge 0, y \ge 0$

14. $2x + y \le 20$
$10x + y \ge 36$
$2x + 5y \ge 36$

15. $x + y \le 6$
$-x + y \le 2$
$2x - y \le 8$

16. $-x + 2y \le 6$
$3x + y \ge 3$
$x \ge 0, y \ge 0$

17. Find values of $x \ge 0$ and $y \ge 0$ that maximize $z = 10x + 12y$, subject to each of the following sets of constraints.
(a) $x + y \le 20$
$x + 3y \le 24$
(b) $3x + y \le 15$
$x + 2y \le 18$
(c) $x + 2y \ge 10$
$2x + y \ge 12$
$x - y \le 8$

18. Find values of $x \ge 0$ and $y \ge 0$ that minimize $z = 3x + 2y$, subject to each of the following sets of constraints.
(a) $10x + 7y \le 42$
$4x + 10y \ge 35$
(b) $6x + 5y \ge 25$
$2x + 6y \ge 15$
(c) $2x + 5y \ge 22$
$4x + 3y \le 28$
$2x + 2y \le 17$

19. Explain why it is impossible to maximize the function $z = 3x + 4y$ subject to the constraints
$x + y \ge 8$
$2x + y \le 10$
$x + 2y \le 8$
$x \ge 0, y \ge 0.$

20. You are given the following multiple-choice linear programming problem.*

Maximize $z = c_1x_1 + c_2x_2$
subject to $2x_1 + x_2 \le 11$
$-x_1 + 2x_2 \le 2$
$x_1 \ge 0, x_2 \ge 0.$

If $c_2 > 0$, determine the range of c_1/c_2 for which $(x_1, x_2) = (4, 3)$ is an optimal solution.
(a) $[-2, 1/2]$ (b) $[-1/2, 2]$ (c) $[-11, -1]$
(d) $[1, 11]$ (e) $[-11, 11]$

✓ Checkpoint Answers

1. (a)

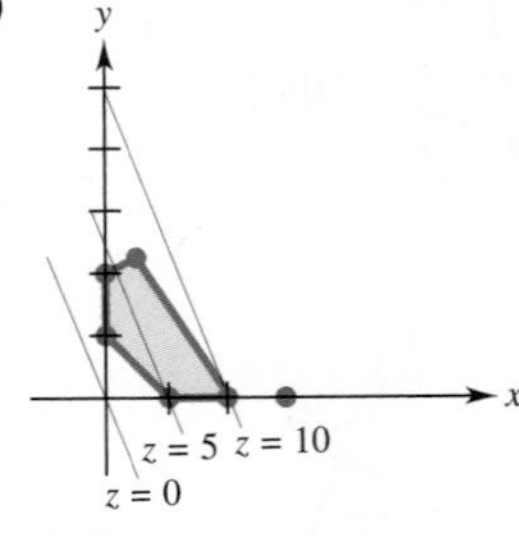

(b) (2, 0)

*Problem from "Course 130 Examination Operations Research" of the *Education and Examination Committee of the Society of Actuaries*. Reprinted by permission of the Society of Actuaries.

2. **(a)** (0, 4), (1, 1), (4, 0) **(b)** (1, 1)
3. $D = (5, 12)$
4. **(a)** (5, 2) **(b)** 8 **(c)** (1, 1) **(d)** 7
5. **(a)** (3, 1) **(b)** (5, 4)
6. (1, 1)

7.3 ▶ Applications of Linear Programming

In this section, we show several applications of linear programming with two variables.

Example 1 **Business** A 4-H Club member raises goats and pigs. She wants to raise no more than 16 animals, including no more than 10 goats. She spends \$25 to raise a goat and \$75 to raise a pig, and she has \$900 available for the project. Find the maximum profit she can make if each goat produces a profit of \$14 and each pig a profit of \$40.

Solution The total profit is determined by the number of goats and pigs. So let x be the number of goats to be produced and let y be the number of pigs. Then summarize the information of the problem in a table.

	Number	Cost to Raise	Profit Each
Goats	x	\$25	\$14
Pigs	y	\$75	\$40
Maximum Available	16	\$900	

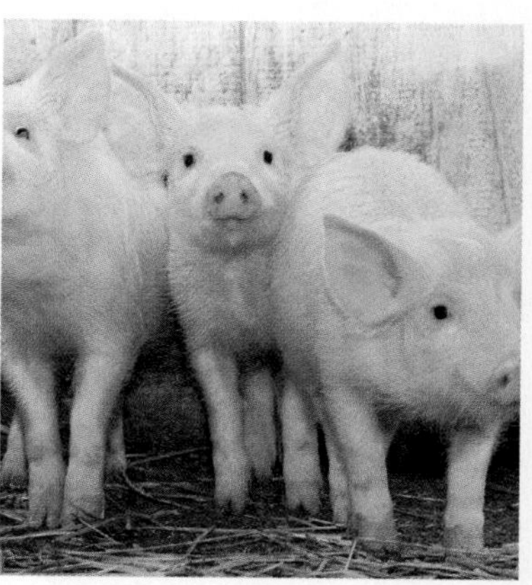

Use this table to write the necessary constraints. Since the total number of animals cannot exceed 16, the first constraint is

$$x + y \leq 16.$$

"No more than 10 goats" means that

$$x \leq 10.$$

The cost to raise x goats at \$25 per goat is $25x$ dollars, while the cost for y pigs at \$75 each is $75y$ dollars. Only \$900 is available, so

$$25x + 75y \leq 900.$$

Dividing both sides by 25 gives the equivalent inequality

$$x + 3y \leq 36.$$

The number of goats and pigs cannot be negative, so

$$x \geq 0 \quad \text{and} \quad y \geq 0.$$

The 4-H Club member wants to know the number of goats and the number of pigs that should be raised for maximum profit. Each goat produces a profit of \$14, and each pig produces a profit of \$40. If z represents total profit, then

$$z = 14x + 40y$$

is the objective function that is to be maximized.

We must solve the following linear programming problem:

$$\begin{aligned} &\text{Maximize} \quad & z = 14x + 40y & \quad \text{Objective function} \\ &\text{subject to} \quad & \left.\begin{aligned} x + y &\le 16 \\ x &\le 10 \\ x + 3y &\le 36 \\ x \ge 0, y &\ge 0. \end{aligned}\right\} & \quad \text{Constraints} \end{aligned}$$

Using the methods of the previous section, graph the feasible region for the system of inequalities given by the constraints, as in Figure 7.23.

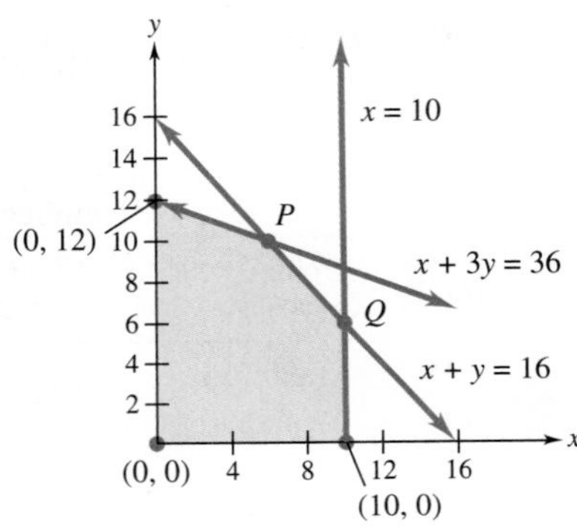

FIGURE 7.23

The corner points (0, 12), (0, 0), and (10, 0) can be read directly from the graph. Find the coordinates of the other corner points by solving a system of equations or with a graphing calculator. 1✓

Test each corner point in the objective function to find the maximum profit.

✓Checkpoint 1

Find the corner points P and Q in Figure 7.23.

Answer: See page 412.

Corner Point	$z = 14x + 40y$
(0, 12)	$14(0) + 40(12) = 480$
(6, 10)	$14(6) + 40(10) = 484$ (maximum)
(10, 6)	$14(10) + 40(6) = 380$
(10, 0)	$14(10) + 40(0) = 140$
(0, 0)	$14(0) + 40(0) = 0$

The maximum value for z of 484 occurs at (6, 10). Thus, 6 goats and 10 pigs will produce a maximum profit of \$484.

Example 2 **Business** An office manager needs to purchase new filing cabinets. He knows that Ace cabinets cost \$40 each, require 6 square feet of floor space, and hold 8 cubic feet of files. On the other hand, each Excello cabinet costs \$80, requires

8 square feet of floor space, and holds 12 cubic feet. His budget permits him to spend no more than \$560 on files, while the office has room for no more than 72 square feet of cabinets. The manager desires the greatest storage capacity within the limitations imposed by funds and space. How many of each type of cabinet should he buy?

Solution Let x represent the number of Ace cabinets to be bought, and let y represent the number of Excello cabinets. The information given in the problem can be summarized as follows.

	Number	Cost of Each	Space Required	Storage Capacity
Ace	x	\$40	6 sq ft	8 cu ft
Excello	y	\$80	8 sq ft	12 cu ft
Maximum Available		\$560	72 sq ft	

The constraints imposed by cost and space are

$$\begin{aligned} 40x + 80y &\leq 560 \quad \text{Cost} \\ 6x + 8y &\leq 72. \quad \text{Floor space} \end{aligned}$$

The number of cabinets cannot be negative, so $x \geq 0$ and $y \geq 0$. The objective function to be maximized gives the amount of storage capacity provided by some combination of Ace and Excello cabinets. From the information in the chart, the objective function is

$$z = \text{Storage capacity} = 8x + 12y.$$

In sum, the given problem has produced the following linear programming problem:

$$\begin{aligned} \text{Maximize} \quad & z = 8x + 12y \\ \text{subject to} \quad & 40x + 80y \leq 560 \\ & 6x + 8y \leq 72 \\ & x \geq 0, y \geq 0. \end{aligned}$$

A graph of the feasible region is shown in Figure 7.24. Three of the corner points can be identified from the graph as (0, 0), (0, 7), and (12, 0). The fourth corner point, labeled Q in the figure, can be found algebraically or with a graphing calculator to be (8, 3). 2 

✓**Checkpoint 2**

Find the corner point labeled P on the region of feasible solutions in the given graph.

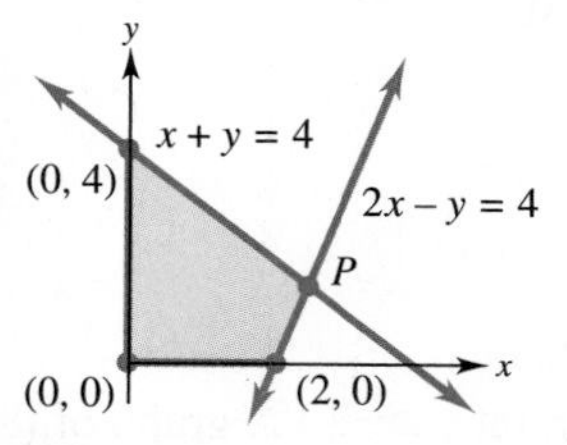

Answer: See page 412.

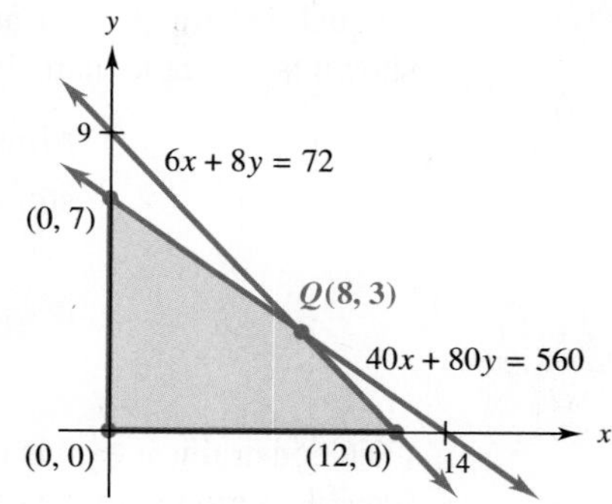

FIGURE 7.24

Use the corner point theorem to find the maximum value of z.

Corner Point	Value of $z = 8x + 12y$
(0, 0)	0
(0, 7)	84
(8, 3)	100 (maximum)
(12, 0)	96

The objective function, which represents storage space, is maximized when $x = 8$ and $y = 3$. The manager should buy 8 Ace cabinets and 3 Excello cabinets. ✓3

✓Checkpoint 3

A popular cereal combines oats and corn. At least 27 tons of the cereal are to be made. For the best flavor, the amount of corn should be no more than twice the amount of oats. Oats cost \$300 per ton, and corn costs \$200 per ton. How much of each grain should be used to minimize the cost?

(a) Make a chart to organize the information given in the problem.

(b) Write an equation for the objective function.

(c) Write four inequalities for the constraints.

Answers: See page 412.

Example 3 **Health** Certain laboratory animals must have at least 30 grams of protein and at least 20 grams of fat per feeding period. These nutrients come from food *A*, which costs 18¢ per unit and supplies 2 grams of protein and 4 of fat, and food *B*, with 6 grams of protein and 2 of fat, costing 12¢ per unit. Food *B* is bought under a long-term contract requiring that at least 2 units of *B* be used per serving. How much of each food must be bought to produce the minimum cost per serving?

Solution Let x represent the amount of food *A* needed and y the amount of food *B*. Use the given information to produce the following table.

Food	Number of Units	Grams of Protein	Grams of Fat	Cost
A	x	2	4	18¢
B	y	6	2	12¢
Minimum Required		30	20	

Use the table to develop the linear programming problem. Since the animals must have *at least* 30 grams of protein and 20 grams of fat, use $\geq$ in the constraint inequalities for protein and fat. The long-term contract provides a constraint not shown in the table, namely, $y \geq 2$. So we have the following problem:

$$\begin{aligned}
\text{Minimize} \quad & z = .18x + .12y && \text{Cost} \\
\text{subject to} \quad & 2x + 6y \geq 30 && \text{Protein} \\
& 4x + 2y \geq 20 && \text{Fat} \\
& y \geq 2 && \text{Contract} \\
& x \geq 0, y \geq 0.
\end{aligned}$$

(The constraint $y \geq 0$ is redundant because of the constraint $y \geq 2$.) A graph of the feasible region with the corner points identified is shown in Figure 7.25.

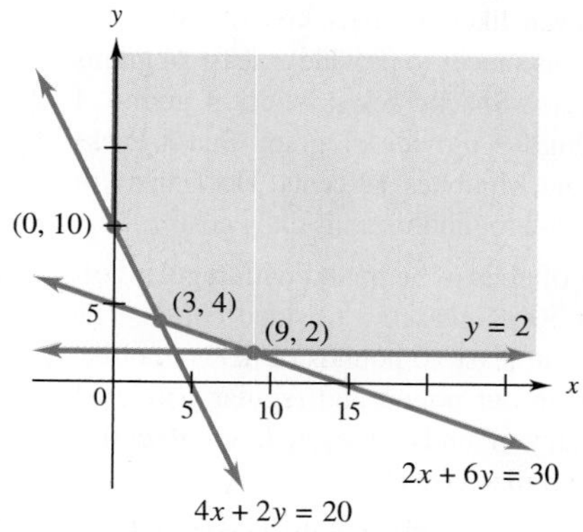

FIGURE 7.25

Use the corner point theorem to find the minimum value of z as shown in the table.

Corner Points	$z = .18x + .12y$	
(0, 10)	$.18(0) + .12(10) = 1.20$	
(3, 4)	$.18(3) + .12(4) = 1.02$	(minimum)
(9, 2)	$.18(9) + .12(2) = 1.86$	

The minimum value of 1.02 occurs at (3, 4). Thus, 3 units of food A and 4 units of food B will produce a minimum cost of $1.02 per serving. 4 ✓

✓ Checkpoint 4

Use the information in Checkpoint 3 to do the following.

(a) Graph the feasible region and find the corner points.

(b) Determine the minimum value of the objective function and the point where it occurs.

(c) Is there a maximum cost?

Answers: See page 412.

The feasible region in Figure 7.25 is an unbounded one; The region extends indefinitely to the upper right. With this region, it would not be possible to *maximize* the objective function, because the total cost of the food could always be increased by encouraging the animals to eat more.

7.3 ▶ Exercises

Write the constraints in Exercises 1–4 as linear inequalities and identify all variables used. In some instances, not all of the information is needed to write the constraints. (See Examples 1–3.)

1. A canoe requires 8 hours of fabrication and a rowboat 5 hours. The fabrication department has at most 110 hours of labor available each week.
2. Doug Gilbert needs at least 2800 miligrams of vitamin C per day. Each Supervite pill provides 250 milligrams, and each Vitahealth pill provides 350 milligrams.
3. A candidate can afford to spend no more than $9500 on radio and TV advertising. Each radio spot costs $250, and each TV ad costs $750.
4. A hospital dietician has two meal choices: one for patients on solid food that costs $2.75 and one for patients on liquids that costs $3.75. There is a maximum of 600 patients in the hospital.

Solve these linear programming problems, which are somewhat simpler than the examples in the text.

5. **Business** A chain saw requires 4 hours of assembly and a wood chipper 6 hours. A maximum of 48 hours of assembly time is available. The profit is $150 on a chain saw and $220 on a chipper. How many of each should be assembled for maximum profit?

6. **Health** Mark Donovan likes to snack frequently during the day, but he wants his snacks to provide at least 24 grams of protein per day. Each Snack-Pack provides 4 grams of protein, and each Minibite provides 1 gram. Snack-Packs cost 50 cents each and Minibites 12 cents. How many of each snack should he use to minimize his daily cost?

7. **Business** Deluxe coffee is to be mixed with regular coffee to make at least 50 pounds of a blended coffee. The mixture must contain at least 10 pounds of deluxe coffee. Deluxe coffee costs \$6 per pound and regular coffee \$5 per pound. How many pounds of each kind of coffee should be used to minimize costs?

8. **Business** Pauline Wong sells computers on commission. She spends 2 hours on selling a laptop and 1 hour on selling a desktop computer. She works no more than 40 hours per week and routinely sells at least 1 laptop and at least 4 desktops each week. To maximize her income, how many of each type should she sell if her commission is
   (a) \$30 per laptop and \$14 per desktop;
   (b) \$30 per laptop and \$16 per desktop?

9. **Business** The company in Exercise 1 cannot sell more than 10 canoes each week and always sells at least 6 rowboats. The profit on a canoe is \$400, and the profit on a rowboat is \$225. Assuming the same situation as in Exercise 1, how many of each should be made per week to maximize profits?

10. **Health** Doug Gilbert of Exercise 2 pays 3 cents for each Supervite pill and 4 cents for each Vitahealth pill. Because of its other ingredients, he cannot take more than 7 Supervite pills per day. Assuming the same conditions as in Example 2, how many of each pill should he take to provide the desired level of vitamin C at minimum cost?

11. **Social Science** The candidate in Exercise 3 wants to have at least 8 radio spots and at least 3 TV ads. A radio spot reaches 600 people, and a TV ad reaches 2000 people. Assuming the monetary facts given in Exercise 3, how many of each kind should be used to reach the largest number of people?

12. **Health** The hospital in Exercise 4 always has at least 100 patients on solid foods and at least 100 on liquids. Assuming the facts in Exercise 4, what number of each type of patient would minimize food costs?

Solve the following linear programming problems. (See Examples 1–3.)

13. **Business** A company is considering two insurance plans with the types of coverage and premiums shown in the table:

	Policy A	*Policy B*
Fire/Theft	\$10,000	\$15,000
Liability	\$180,000	\$120,000
Premium per unit	\$50	\$40

(For example, \$50 buys one unit of plan A, consisting of \$10,000 of fire and theft insurance and \$180,000 of liability insurance.)
   (a) The company wants at least \$300,000 of fire/theft insurance and at least \$3,000,000 of liability insurance from these plans. How many units should be purchased from each plan to minimize the cost of the premiums? What is the minimum premium?
   (b) Suppose the premium for policy A is reduced to \$25. Now how many units should be purchased from each plan to minimize the cost of the premiums? What is the minimum premium?

14. **Business** A manufacturer of refrigerators must ship at least 100 refrigerators to its two West Coast warehouses. Each warehouse holds a maximum of 100 refrigerators. Warehouse A holds 25 refrigerators already, and warehouse B has 20 on hand. It costs \$12 to ship a refrigerator to warehouse A and \$10 to ship one to warehouse B. Union rules require that at least 300 workers be hired. Shipping a refrigerator to Warehouse A requires 4 workers, while shipping a refrigerator to Warehouse B requires 2 workers. How many refrigerators should be shipped to each warehouse to minimize costs? What is the minimum cost?

15. **Health** Mike May has been told that each day he needs at least 16 units of vitamin A, at least 5 units of vitamin B-1, and at least 20 units of vitamin C. Each Brand X pill contain 8 units of vitamin A, 1 of vitamin B-1, and 2 of vitamin C, while each Brand Z pill contains 2 units of vitamin A, 1 of vitamin B-1, and 7 of vitamin C. A Brand X pill costs 15 cents, and a Brand Z pill costs 30 cents. How many pills of each brand should he buy to minimize his daily cost? What is the minimum cost?

16. **Business** The manufacturing process requires that oil refineries manufacture at least 2 gallons of gasoline for every gallon of fuel oil. To meet the winter demand for fuel oil, at least 3 million gallons a day must be produced. The demand for gasoline is no more than 12 million gallons per day. It takes .25 hour to ship each million gallons of gasoline and 1 hour to ship each million gallons of fuel oil out of the warehouse. No more than 6.6 hours are available for shipping. If the refinery sells gasoline for \$1.25 per gallon and fuel oil for \$1 per gallon, how much of each should be produced to maximize revenue? Find the maximum revenue.

17. **Business** A machine shop manufactures two types of bolts. The bolts require time on each of three groups of machines, but the time required on each group differs, as shown in the table:

		Machine Group		
		I	II	III
Bolts	Type 1	.1 min	.1 min	.1 min
	Type 2	.1 min	.4 min	.02 min

Production schedules are made up one day at a time. In a day, there are 240, 720, and 160 minutes available, respectively, on these machines. Type 1 bolts sell for 10¢ and type 2 bolts for 12¢. How many of each type of bolt should be manufactured per day to maximize revenue? What is the maximum revenue?

18. **Health** Kim Walrath has a nutritional deficiency and is told to take at least 2400 mg of iron, 2100 mg of vitamin B-1, and 1500 mg of vitamin B-2. One Maxivite pill contains 40 mg of iron, 10 mg of vitamin B-1, and 5 mg of vitamin B-2 and costs 6¢. One Healthovite pill provides 10 mg of iron, 15 mg of vitamin B-1, and 15 mg of vitamin B-2 and costs 8¢.
   (a) What combination of Maxivite and Healthovite pills will meet Kim's requirements at lowest cost? What is the lowest cost?
   (b) In your solution for part (a), does Kim receive more than the minimum amount she needs of any vitamin? If so, which vitamin is it?
   (c) Is there any way that Kim can avoid receiving more than the minimum she needs and still meet the other constraints and minimize the cost? Explain.

19. **Social Science** An anthropology article recounts a hypothetical situation that could be described by a linear programming model.* Suppose a population gathers plants and animals for survival. Its members need at least 360 units of energy, 300 units of protein, and 8 hides during some time period. One unit of plants provides 30 units of energy, 10 units of protein, and no hides. One animal provides 20 units of energy, 25 units of protein, and 1 hide. Only 25 units of plants and 25 animals are available. It costs the population 30 hours of labor to gather one unit of a plant and 15 hours for an animal. Find how many units of plants and how many animals should be gathered to meet the requirements with a minimum number of hours of labor.

20. **Business** The Miers Company produces small engines for several manufacturers. The company receives orders from two assembly plants for its engine. Plant I needs at least 50 engines, and plant II needs at least 27 engines. The company can send at most 105 engines to these two assembly plants. It costs $20 per engine to ship to plant I and $35 per engine to ship to plant II. Plant I gives Miers $15 in rebates toward its products for each engine Miers buys, while plant II gives similar $10 rebates. Miers estimates that it needs at least $1200 in rebates to cover products it plans to buy from the two plants. How many engines should be shipped to each plant to minimize shipping costs? What is the minimum cost?

21. **Business** A greeting card manufacturer has 500 boxes of a particular card in warehouse I and 290 boxes of the same card in warehouse II. A greeting card shop in San Jose orders 350 boxes of the card, and another shop in Memphis orders 250 boxes. The shipping costs per box to these shops from the two warehouses are shown in the following table:

		DESTINATION	
		San Jose	Memphis
Warehouse	I	$.25	$.22
	II	$.23	$.21

How many boxes should be shipped to each city from each warehouse to minimize shipping costs? What is the minimum cost? (*Hint*: Use x, $350 - x$, y, and $250 - y$ as the variables.)

22. **Business** *Hotnews Magazine* publishes a U.S. and a Canadian edition each week. There are 30,000 subscribers in the United States and 20,000 in Canada. Other copies are sold at newsstands. Postage and shipping costs average $80 per thousand copies for the United States and $60 per thousand copies for Canada. Surveys show that no more than 120,000 copies of each issue can be sold (including subscriptions) and that the number of copies of the Canadian edition should not exceed twice the number of copies of the U.S. edition. The publisher can spend at most $8400 a month on postage and shipping. If the profit is $200 for each thousand copies of the U.S. edition and $150 for each thousand copies of the Canadian edition, how many copies of each version should be printed to earn as large a profit as possible? What is that profit?

23. **Finance** A pension fund manager decides to invest at most $50 million in U.S. Treasury Bonds paying 4% annual interest and in mutual funds paying 6% annual interest. He plans to invest at least $20 million in bonds and at least $6 million in mutual funds. Bonds have an initial fee of $300 per million dollars, while the fee for mutual funds is $100 per million. The fund manager is allowed to spend no more than $8400 on fees. How much should be invested in each to maximize annual interest? What is the maximum annual interest?

24. **Natural Science** A certain predator requires at least 10 units of protein and 8 units of fat per day. One prey of Species I provides 5 units of protein and 2 units of fat; one prey of Species II provides 3 units of protein and 4 units of fat. Capturing and digesting each Species II prey requires 3 units of energy, and capturing and digesting each Species I prey requires 2 units of energy. How many of each prey would meet the predator's daily food requirements with the least expenditure of energy? Are the answers reasonable? How could they be interpreted?

*Van A. Reidhead, "Linear Programming Models in Archaeology," *Annual Review of Anthropology* 8 (1979): 543–578.

25. **Social Science** Students at Upscale U. are required to take at least 4 humanities and 4 science courses. The maximum allowable number of science courses is 12. Each humanities course carries 4 credits and each science course 5 credits. The total number of credits in science and humanities cannot exceed 92. Quality points for each course are assigned in the usual way: the number of credit hours times 4 for an A grade, times 3 for a B grade, and times 2 for a C grade. Susan Katz expects to get B's in all her science courses. She expects to get C's in half her humanities courses, B's in one-fourth of them, and A's in the rest. Under these assumptions, how many courses of each kind should she take in order to earn the maximum possible number of quality points?

26. **Social Science** In Exercise 25, find Susan's grade point average (the total number of quality points divided by the total number of credit hours) at each corner point of the feasible region. Does the distribution of courses that produces the highest number of quality points also yield the highest grade point average? Is this a contradiction?

*The importance of linear programming is shown by the inclusion of linear programming problems on most qualification examinations for Certified Public Accountants. Exercises 27–29 are reprinted from one such examination.**

The Random Company manufactures two products: Zeta and Beta. Each product must pass through two processing operations. All materials are introduced at the start of Process No. 1. There are no work-in-process inventories. Random may produce either one product exclusively or various combinations of both products, subject to the following constraints:

	Process No. 1	Process No. 2	Contribution Margin per Unit
Hours required to produce 1 unit of:			
Zeta	1 hour	1 hour	\$4.00
Beta	2 hours	3 hours	\$5.25
Total capacity in hours per day	1000 hours	1275 hours	

A shortage of technical labor has limited Beta production to 400 units per day. There are no constraints on the production of Zeta other than the hour constraints shown in the schedule. Assume that all the relationships between capacity and production are linear.

27. Given the objective to maximize total contribution margin, what is the production constraint for Process No. 1?
(a) Zeta + Beta ≤ 1000 (b) Zeta + 2 Beta ≤ 1000
(c) Zeta + Beta ≥ 1000 (d) Zeta + 2 Beta ≥ 1000

28. Given the objective to maximize total contribution margin, what is the labor constraint for production of Beta?
(a) Beta ≤ 400 (b) Beta ≥ 400
(c) Beta ≤ 425 (d) Beta ≥ 425

29. What is the objective function of the data presented?
(a) Zeta + 2 Beta = \$9.25
(b) \$4.00 Zeta + 3(\$5.25) Beta = total contribution margin
(c) \$4.00 Zeta + \$5.25 Beta = total contribution margin
(d) 2(\$4.00) Zeta + 3(\$5.25) Beta = total contribution margin

*Material from *Uniform CPA Examinations and Unofficial Answers*, copyright © 1973, 1974, 1975 by the American Institute of Certified Public Accountants, Inc., is reprinted with permission.

✓ Checkpoint Answers

1. $P = (6, 10)$
$Q = (10, 6)$
2. $\left(\frac{8}{3}, \frac{4}{3}\right)$
3. (a)

	Number of Tons	Cost/Ton
Oats	x	\$300
Corn	y	\$200
	27	

(b) $z = 300x + 200y$
(c) $x + y \geq 27$
$y \leq 2x$
$x \geq 0$
$y \geq 0$

4. (a)
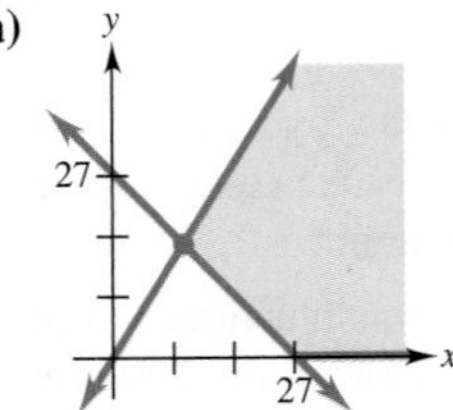

Corner points: (27, 0), (9, 18)
(b) \$6300 at (9, 18)
(c) No

7.4 ▶ The Simplex Method: Maximization

For linear programming problems with more than two variables or with two variables and many constraints, the graphical method is usually inefficient or impossible, so the **simplex method** is used. This method, which is introduced in this chapter, was developed for the U.S. Air Force by George B. Danzig in 1947. It is now used in industrial planning, factory design, product distribution networks, sports scheduling, truck routing, resource allocation, and a variety of other ways.

Because the simplex method is used for problems with many variables, it usually is not convenient to use letters such as x, y, z, or w as variable names. Instead, the symbols x_1 (read "x-sub-one"), x_2, x_3, and so on, are used. In the simplex method, all constraints must be expressed in the linear form

$$a_1x_1 + a_2x_2 + a_3x_3 + \cdots \le b,$$

where $x_1, x_2, x_3, \ldots$ are variables, $a_1, a_2, a_3, \ldots$ are coefficients, and b is a constant.

We first discuss the simplex method for linear programming problems such as the following:

$$\begin{aligned}
\text{Maximize} \quad & z = 2x_1 + 3x_2 + x_3 \\
\text{subject to} \quad & x_1 + x_2 + 4x_3 \le 100 \\
& x_1 + 2x_2 + x_3 \le 150 \\
& 3x_1 + 2x_2 + x_3 \le 320, \\
\text{with } & x_1 \ge 0, x_2 \ge 0, x_3 \ge 0.
\end{aligned}$$

This example illustrates *standard maximum form*, which is defined as follows.

Standard Maximum Form

A linear programming problem is in **standard maximum form** if

1. the objective function is to be maximized;
2. all variables are nonnegative ($x_i \ge 0$, $i = 1, 2, 3, \ldots$);
3. all constraints involve $\le$;
4. the constants on the right side in the constraints are all nonnegative ($b \ge 0$).

Problems that do not meet all of these conditions are considered in Sections 7.6 and 7.7.

The "mechanics" of the simplex method are demonstrated in Examples 1–5. Although the procedures to be followed will be made clear, as will the fact that they result in an optimal solution, the reasons these procedures are used may not be immediately apparent. Examples 6 and 7 will supply these reasons and explain the connection between the simplex method and the graphical method used in Section 7.3.

SETTING UP THE PROBLEM

The first step is to convert each constraint, a linear inequality, into a linear equation. This is done by adding a nonnegative variable, called a **slack variable**, to each constraint. For example, convert the inequality $x_1 + x_2 \leq 10$ into an equation by adding the slack variable s_1, to get

$$x_1 + x_2 + s_1 = 10, \qquad \text{where } s_1 \geq 0.$$

The inequality $x_1 + x_2 \leq 10$ says that the sum $x_1 + x_2$ is less than or equal to 10. The variable s_1 "takes up any slack" and represents the amount by which $x_1 + x_2$ fails to equal 10. For example, if $x_1 + x_2$ equals 8, then s_1 is 2. If $x_1 + x_2 = 10$, the value of s_1 is 0.

CAUTION A different slack variable must be used for each constraint.

Example 1 Restate the following linear programming problem by introducing slack variables:

$$\begin{aligned}
\text{Maximize} \quad & z = 2x_1 + 3x_2 + x_3 \\
\text{subject to} \quad & x_1 + x_2 + 4x_3 \leq 100 \\
& x_1 + 2x_2 + x_3 \leq 150 \\
& 3x_1 + 2x_2 + x_3 \leq 320, \\
\text{with } & x_1 \geq 0, x_2 \geq 0, x_3 \geq 0.
\end{aligned}$$

Solution Rewrite the three constraints as equations by introducing nonnegative slack variables s_1, s_2, and s_3, one for each constraint. Then the problem can be restated as

$$\begin{aligned}
\text{Maximize} \quad & z = 2x_1 + 3x_2 + x_3 \\
\text{subject to} \quad & x_1 + x_2 + 4x_3 + s_1 = 100 && \text{Constraint 1} \\
& x_1 + 2x_2 + x_3 + s_2 = 150 && \text{Constraint 2} \\
& 3x_1 + 2x_2 + x_3 + s_3 = 320, && \text{Constraint 3} \\
\text{with} \quad & x_1 \geq 0, x_2 \geq 0, x_3 \geq 0, s_1 \geq 0, s_2 \geq 0, s_3 \geq 0.
\end{aligned}$$

✓1

✓Checkpoint 1

Rewrite the following set of constraints as equations by adding nonnegative slack variables:

$$\begin{aligned}
x_1 + x_2 + x_3 &\leq 12 \\
2x_1 + 4x_2 &\leq 15 \\
x_2 + 3x_3 &\leq 10.
\end{aligned}$$

Answer: See page 427.

Adding slack variables to the constraints converts a linear programming problem into a system of linear equations. These equations should have all variables on the left of the equals sign and all constants on the right. All the equations of Example 1 satisfy this condition except for the objective function, $z = 2x_1 + 3x_2 + x_3$, which may be written with all variables on the left as

$$-2x_1 - 3x_2 - x_3 + z = 0. \qquad \text{Objective Function}$$

Now the equations of Example 1 (with the constraints listed first and the objective function last) can be written as the following augmented matrix.

$$\begin{array}{ccccccc|c} x_1 & x_2 & x_3 & s_1 & s_2 & s_3 & z & \\ \hline 1 & 1 & 4 & 1 & 0 & 0 & 0 & 100 \\ 1 & 2 & 1 & 0 & 1 & 0 & 0 & 150 \\ 3 & 2 & 1 & 0 & 0 & 1 & 0 & 320 \\ \hline -2 & -3 & -1 & 0 & 0 & 0 & 1 & 0 \end{array}.$$

Constraint 1
Constraint 2
Constraint 3
Objective Function

Indicators

This matrix is the initial **simplex tableau**. Except for the last entries—the 1 and 0 on the right end—the numbers in the bottom row of a simplex tableau are called **indicators**. 2 ✓

This simplex tableau represents a system of four linear equations in seven variables. Since there are more variables than equations, the system is dependent and has infinitely many solutions. Our goal is to find a solution in which all the variables are nonnegative and z is as large as possible. This will be done by using row operations to replace the given system by an equivalent one in which certain variables are eliminated from some of the equations. The process will be repeated until the optimum solution can be read from the matrix, as explained next.

✓ Checkpoint 2

Set up the initial simplex tableau for the following linear programming problem:

Maximize $z = 2x_1 + 3x_2$

subject to $x_1 + 2x_2 \le 85$

$2x_1 + x_2 \le 92$

$x_1 + 4x_2 \le 104,$

with $x_1 \ge 0$ and $x_2 \ge 0$.

Locate and label the indicators.

Answer: See page 427.

SELECTING THE PIVOT

Recall how row operations are used to eliminate variables in the Gauss–Jordan method: A particular nonzero entry in the matrix is chosen and changed to a 1; then all other entries in that column are changed to zeros. A similar process is used in the simplex method. The chosen entry is called the **pivot**. If we were interested only in solving the system, we could choose the various pivots in many different ways, as in Chapter 6. Here, however, it is not enough just to find a solution. We must find one that is nonnegative, satisfies all the constraints, *and* makes z as a large as possible. Consequently, the pivot must be chosen carefully, as explained in the next example. The reasons this procedure is used and why it works are discussed in Example 7.

Example 2 Determine the pivot in the simplex tableau for the problem in Example 1.

Solution Look at the indicators (the last row of the tableau) and choose the most negative one:

$$\begin{array}{ccccccc|c} x_1 & x_2 & x_3 & s_1 & s_2 & s_3 & z & \\ \hline 1 & 1 & 4 & 1 & 0 & 0 & 0 & 100 \\ 1 & 2 & 1 & 0 & 1 & 0 & 0 & 150 \\ 3 & 2 & 1 & 0 & 0 & 1 & 0 & 320 \\ \hline -2 & \mathbf{-3} & -1 & 0 & 0 & 0 & 1 & 0 \end{array}.$$

Most negative indicator

The most negative indicator identifies the variable that is to be eliminated from all but one of the equations (rows)—in this case, x_2. The column containing the most negative indicator is called the **pivot column**. Now, for each *positive* entry in the pivot column, divide the number in the far right column of the same row by the positive number in the pivot column:

x_1	x_2	x_3	s_1	s_2	s_3	z		Quotients
1	**1**	4	1	0	0	0	**100**	100/1 = 100
1	**2**	1	0	1	0	0	**150**	150/2 = 75 ← Smallest
3	**2**	1	0	0	1	0	**320**	320/2 = 160
−2	−3	−1	0	0	0	1	0	

The row with the smallest quotient (in this case, the second row) is called the **pivot row**. The entry in the pivot row and pivot column is the pivot:

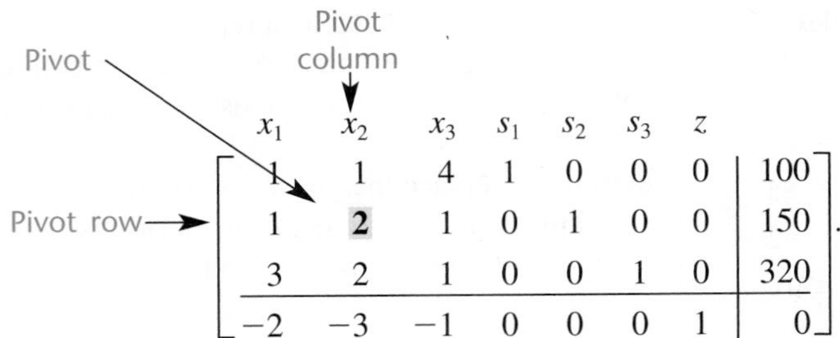

x_1	x_2	x_3	s_1	s_2	s_3	z	
1	1	4	1	0	0	0	100
1	**2**	1	0	1	0	0	150
3	2	1	0	0	1	0	320
−2	−3	−1	0	0	0	1	0

CAUTION In some simplex tableaus, the pivot column may contain zeros or negative entries. Only the positive entries in the pivot column should be used to form the quotients and determine the pivot row. If there are no positive entries in the pivot column (so that a pivot row cannot be chosen), then no maximum solution exists. ✓3

✓Checkpoint 3

Find the pivot for the following tableau:

x_1	x_2	s_1	s_2	s_3	z	
0	1	1	0	0	0	50
−2	3	0	1	0	0	78
2	4	0	0	1	0	65
−5	−3	0	0	0	1	0

Answer: See page 427.

PIVOTING

Once the pivot has been selected, row operations are used to replace the initial simplex tableau by another simplex tableau in which the pivot column variable is eliminated from all but one of the equations. Since this new tableau is obtained by row operations, it represents an equivalent system of equations (that is, a system with the same solutions as the original system). This process, which is called **pivoting**, is explained in the next example.

Example 3 Use the indicated pivot, 2, to perform the pivoting on the simplex tableau of Example 2:

x_1	x_2	x_3	s_1	s_2	s_3	z	
1	1	4	1	0	0	0	100
1	**2**	1	0	1	0	0	150
3	2	1	0	0	1	0	320
−2	−3	−1	0	0	0	1	0

Solution Start by multiplying each entry of row 2 by $\frac{1}{2}$ in order to change the pivot to 1:

x_1	x_2	x_3	s_1	s_2	s_3	z		
1	1	4	1	0	0	0	100	
$\frac{1}{2}$	**1**	$\frac{1}{2}$	0	$\frac{1}{2}$	0	0	75	$\frac{1}{2}R_2$
3	2	1	0	0	1	0	320	
−2	−3	−1	0	0	0	1	0	

Now use row operations to make the entry in row 1, column 2 a 0:

$$\begin{array}{c} \begin{matrix} x_1 & x_2 & x_3 & s_1 & s_2 & s_3 & z \end{matrix} \\ \left[\begin{array}{ccccccc|c} \frac{1}{2} & 0 & \frac{7}{2} & 1 & -\frac{1}{2} & 0 & 0 & 25 \\ \frac{1}{2} & 1 & \frac{1}{2} & 0 & \frac{1}{2} & 0 & 0 & 75 \\ 3 & 2 & 1 & 0 & 0 & 1 & 0 & 320 \\ \hline -2 & -3 & -1 & 0 & 0 & 0 & 1 & 0 \end{array}\right]. \end{array} \quad -R_2 + R_1$$

Change the 2 in row three, column two, to a 0 by a similar process:

$$\begin{array}{c} \begin{matrix} x_1 & x_2 & x_3 & s_1 & s_2 & s_3 & z \end{matrix} \\ \left[\begin{array}{ccccccc|c} \frac{1}{2} & 0 & \frac{7}{2} & 1 & -\frac{1}{2} & 0 & 0 & 25 \\ \frac{1}{2} & 1 & \frac{1}{2} & 0 & \frac{1}{2} & 0 & 0 & 75 \\ 2 & 0 & 0 & 0 & -1 & 1 & 0 & 170 \\ \hline -2 & -3 & -1 & 0 & 0 & 0 & 1 & 0 \end{array}\right]. \end{array} \quad -2R_2 + R_3$$

Finally, add 3 times row 2 to the last row in order to change the indicator −3 to 0:

$$\begin{array}{c} \begin{matrix} x_1 & x_2 & x_3 & s_1 & s_2 & s_3 & z \end{matrix} \\ \left[\begin{array}{ccccccc|c} \frac{1}{2} & 0 & \frac{7}{2} & 1 & -\frac{1}{2} & 0 & 0 & 25 \\ \frac{1}{2} & 1 & \frac{1}{2} & 0 & \frac{1}{2} & 0 & 0 & 75 \\ 2 & 0 & 0 & 0 & -1 & 1 & 0 & 170 \\ \hline -\frac{1}{2} & 0 & \frac{1}{2} & 0 & \frac{3}{2} & 0 & 1 & 225 \end{array}\right]. \end{array} \quad 3R_2 + R_4$$

The pivoting is now complete, because the pivot column variable x_2 has been eliminated from all equations except the one represented by the pivot row. The initial simplex tableau has been replaced by a new simplex tableau, which represents an equivalent system of equations.

CAUTION During pivoting, do not interchange rows of the matrix. Make the pivot entry 1 by multiplying the pivot row by an appropriate constant, as in Example 3. 4 ✓

✓Checkpoint 4

For the given simplex tableau,

(a) find the pivot;

(b) perform the pivoting and write the new tableau.

$$\begin{array}{c} \begin{matrix} x_1 & x_2 & x_3 & s_1 & s_2 & z \end{matrix} \\ \left[\begin{array}{cccccc|c} 1 & 2 & 6 & 1 & 0 & 0 & 16 \\ 1 & 3 & 0 & 0 & 1 & 0 & 25 \\ \hline -1 & -4 & -3 & 0 & 0 & 1 & 0 \end{array}\right] \end{array}$$

Answers: See page 427.

When at least one of the indicators in the last row of a simplex tableau is negative (as is the case with the tableau obtained in Example 3), the simplex method requires that a new pivot be selected and the pivoting be performed again. This procedure is repeated until a simplex tableau with no negative indicators in the last row is obtained or a tableau is reached in which no pivot row can be chosen.

Example 4 In the simplex tableau obtained in Example 3, select a new pivot and perform the pivoting.

Solution First, locate the pivot column by finding the most negative indicator in the last row. Then locate the pivot row by computing the necessary quotients and finding the smallest one, as shown here:

$$\text{Pivot row} \rightarrow \begin{bmatrix} x_1 & x_2 & x_3 & s_1 & s_2 & s_3 & z & \\ \boxed{\tfrac{1}{2}} & 0 & \tfrac{7}{2} & 1 & -\tfrac{1}{2} & 0 & 0 & 25 \\ \tfrac{1}{2} & 1 & \tfrac{1}{2} & 0 & \tfrac{1}{2} & 0 & 0 & 75 \\ 2 & 0 & 0 & 0 & -1 & 1 & 0 & 170 \\ \hline -\tfrac{1}{2} & 0 & \tfrac{1}{2} & 0 & \tfrac{3}{2} & 0 & 1 & 225 \end{bmatrix}. \quad \begin{matrix} \text{Quotients} \\ \frac{25}{\frac{1}{2}} = 50 \quad \text{Smallest} \\ \frac{75}{\frac{1}{2}} = 150 \\ 170/2 = 85 \end{matrix}$$

Pivot column (column x_1)

So the pivot is the number $\frac{1}{2}$ in row 1, column 1. Begin the pivoting by multiplying every entry in row 1 by 2. Then continue as indicated to obtain the following simplex tableau:

$$\begin{bmatrix} x_1 & x_2 & x_3 & s_1 & s_2 & s_3 & z & \\ 1 & 0 & 7 & 2 & -1 & 0 & 0 & 50 \\ 0 & 1 & -3 & -1 & 1 & 0 & 0 & 50 \\ 0 & 0 & -14 & -4 & 1 & 1 & 0 & 70 \\ \hline 0 & 0 & 4 & 1 & 1 & 0 & 1 & 250 \end{bmatrix}. \quad \begin{matrix} 2R_1 \\ -\frac{1}{2}R_1 + R_2 \\ -2R_1 + R_3 \\ \frac{1}{2}R_1 + R_4 \end{matrix}$$

Since there are no negative indicators in the last row, no further pivoting is necessary, and we call this the **final simplex tableau.**

READING THE SOLUTION

The next example shows how to read an optimal solution of the original linear programming problem from the final simplex tableau.

Example 5 Solve the linear programming problem introduced in Example 1.

Solution Look at the final simplex tableau for this problem, which was obtained in Example 4:

$$\begin{bmatrix} x_1 & x_2 & x_3 & s_1 & s_2 & s_3 & z & \\ 1 & 0 & 7 & 2 & -1 & 0 & 0 & 50 \\ 0 & 1 & -3 & -1 & 1 & 0 & 0 & 50 \\ 0 & 0 & -14 & -4 & 1 & 1 & 0 & 70 \\ \hline 0 & 0 & 4 & 1 & 1 & 0 & 1 & 250 \end{bmatrix}.$$

The last row of this matrix represents the equation

$$4x_3 + s_1 + s_2 + z = 250, \qquad \text{or equivalently,} \qquad z = 250 - 4x_3 - s_1 - s_2.$$

If x_3, s_1, and s_2 are all 0, then the value of z is 250. If any one of x_3, s_1, or s_2 is positive, then z will have a smaller value than 250. (Why?) Consequently, since we want a solution for this system in which all the variables are nonnegative and z is as large as possible, we must have

$$x_3 = 0, \qquad s_1 = 0, \qquad s_2 = 0.$$

When these values are substituted into the first equation (represented by the first row of the final simplex tableau), the result is

$$x_1 + 7\cdot 0 + 2\cdot 0 - 1\cdot 0 = 50; \qquad \text{that is,} \qquad x_1 = 50.$$

Similarly, substituting 0 for x_3, s_1, and s_2 in the last three equations represented by the final simplex tableau shows that

$$x_2 = 50, \qquad s_3 = 70, \qquad \text{and} \qquad z = 250.$$

Therefore, the maximum value of $z = 2x_1 + 3x_2 + x_3$ occurs when

$$x_1 = 50, \qquad x_2 = 50, \qquad \text{and} \qquad x_3 = 0,$$

in which case $z = 2 \cdot 50 + 3 \cdot 50 + 0 = 250$. (The values of the slack variables are irrelevant in stating the solution of the original problem.)

In any simplex tableau, some columns look like columns of an identity matrix (one entry is 1 and the rest are 0). The variables corresponding to these columns are called **basic variables** and the variables corresponding to the other columns are referred to as **nonbasic variables**. In the tableau of Example 5, for instance, the basic variables are x_1, x_2, s_3, and z (shown in blue), and the nonbasic variables are x_3, s_1, and s_2:

$$\begin{array}{c} \begin{matrix} x_1 & x_2 & x_3 & s_1 & s_2 & s_3 & z & \end{matrix} \\ \left[\begin{array}{ccccccc|c} 1 & 0 & 7 & 2 & -1 & 0 & 0 & 50 \\ 0 & 1 & -3 & -1 & 1 & 0 & 0 & 50 \\ 0 & 0 & -14 & -4 & 1 & 1 & 0 & 70 \\ \hline 0 & 0 & 4 & 1 & 1 & 0 & 1 & 250 \end{array}\right]. \end{array}$$

The optimal solution in Example 5 was obtained from the final simplex tableau by setting the nonbasic variables equal to 0 and solving for the basic variables. Furthermore, the values of the basic variables are easy to read from the matrix: Find the 1 in the column representing a basic variable; the last entry in that row is the value of that basic variable in the optimal solution. In particular, *the entry in the lower right-hand corner of the final simplex tableau is the maximum value of z.* 5✓

Checkpoint 5

A linear programming problem with slack variables s_1 and s_2 has the following final simplex tableau:

$$\begin{array}{c} \begin{matrix} x_1 & x_2 & x_3 & s_1 & s_2 & z & \end{matrix} \\ \left[\begin{array}{cccccc|c} 0 & 3 & 1 & 5 & 2 & 0 & 9 \\ 1 & -2 & 0 & 4 & 1 & 0 & 6 \\ \hline 0 & 5 & 0 & 1 & 0 & 1 & 21 \end{array}\right]. \end{array}$$

What is the optimal solution?

Answer: *See page 427.*

CAUTION If there are two identical columns in a tableau, each of which is a column in an identity matrix, only one of the variables corresponding to these columns can be a basic variable. The other is treated as a nonbasic variable. You may choose either one to be the basic variable, unless one of them is z, in which case z must be the basic variable.

The steps involved in solving a standard maximum linear programming problem by the simplex method have been illustrated in Examples 1–5 and are summarized here.

Simplex Method

1. Determine the objective function.
2. Write down all necessary constraints.
3. Convert each constraint into an equation by adding a slack variable.
4. Set up the initial simplex tableau.
5. Locate the most negative indicator. If there are two such indicators, choose one. This indicator determines the pivot column.

(Continued)

6. Use the positive entries in the pivot column to form the quotients necessary for determining the pivot. If there are no positive entries in the pivot column, no maximum solution exists. If two quotients are equally the smallest, let either determine the pivot.*
7. Multiply every entry in the pivot row by the reciprocal of the pivot to change the pivot to 1. Then use row operations to change all other entries in the pivot column to 0 by adding suitable multiples of the pivot row to the other rows.
8. If the indicators are all positive or 0, you have found the final tableau. If not, go back to Step 5 and repeat the process until a tableau with no negative indicators is obtained.†
9. In the final tableau, the *basic* variables correspond to the columns that have one entry of 1 and the rest 0. The *nonbasic* variables correspond to the other columns. Set each nonbasic variable equal to 0 and solve the system for the basic variables. The maximum value of the objective function is the number in the lower right-hand corner of the final tableau.

The solution found by the simplex method may not be unique, especially when choices are possible in steps 5, 6, or 9. There may be other solutions that produce the same maximum value of the objective function. (See Exercises 37 and 38 at the end of this section.) 6

Checkpoint 6

A linear programming problem has the following initial tableau:

$$\begin{array}{c} \begin{matrix} x_1 & x_2 & s_1 & s_2 & z \end{matrix} \\ \left[\begin{array}{ccccc|c} 1 & 1 & 1 & 0 & 0 & 40 \\ 2 & 1 & 0 & 1 & 0 & 24 \\ \hline -300 & -200 & 0 & 0 & 1 & 0 \end{array}\right]. \end{array}$$

Use the simplex method to solve the problem.

Answer: See page 427.

THE SIMPLEX METHOD WITH TECHNOLOGY

Unless indicated otherwise, the simplex method is carried out by hand in the examples and exercises of this chapter, so you can see how and why it works. Once you are familiar with the method, however,

we strongly recommend that you use technology to apply the simplex method.

Doing so will eliminate errors that occur in manual computations. It will also give you a better idea of how the simplex method is used in the real world, where applications involve so many variables and constraints that the manual approach is impractical. Readily available technology includes the following:

Graphing Calculators As noted in the Graphing Calculator Appendix of this book, a simplex program can be downloaded from www.pearsonhighered.com/mwa10e. It pauses after each round of pivoting, so you can examine the intermediate simplex tableau.

Spreadsheets Most spreadsheets have a built-in simplex method program. Figure 7.26 shows the Solver of Microsoft Excel. Spreadsheets also provide a sensitivity analysis, which allows you to see how much the constraints can be varied without changing the maximal solution.

*It may be that the first choice of a pivot does not produce a solution. In that case, try the other choice.
†Some special circumstances are noted at the end of Section 7.7.

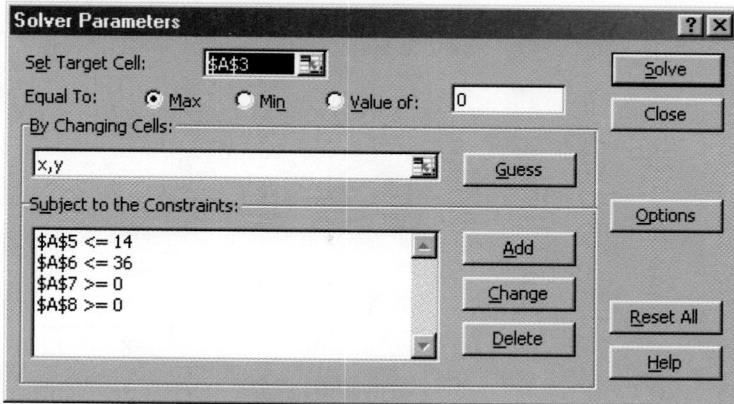

FIGURE 7.26

Other Computer Programs A variety of simplex method programs, many of which are free, can be downloaded on the Internet. Google "simplex method program" for some possibilities.

GEOMETRIC INTERPRETATION OF THE SIMPLEX METHOD

Although it may not be immediately apparent, the simplex method is based on the same geometrical considerations as the graphical method. This can be seen by looking at a problem that can be readily solved by both methods.

Example 6 In Example 2 of Section 7.3, the following problem was solved graphically (using x and y instead of x_1 and x_2, respectively):

$$\begin{aligned} \text{Maximize} \quad & z = 8x_1 + 12x_2 \\ \text{subject to} \quad & 40x_1 + 80x_2 \le 560 \\ & 6x_1 + 8x_2 \le 72 \\ & x_1 \ge 0, x_2 \ge 0. \end{aligned}$$

Graphing the feasible region (Figure 7.27) and evaluating z at each corner point shows that the maximum value of z occurs at (8, 3).

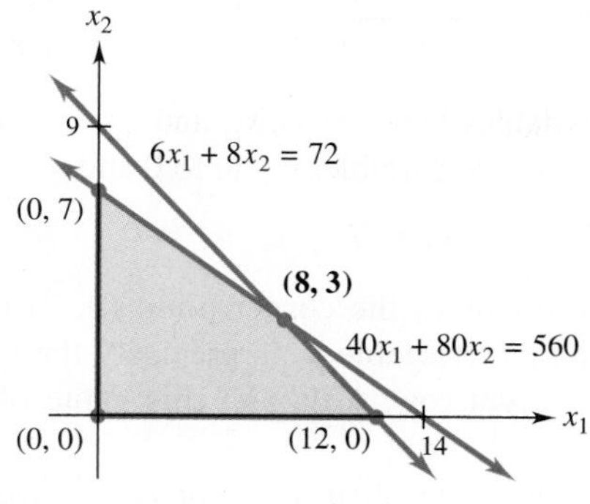

FIGURE 7.27

Corner Point	Value of $z = 8x_1 + 12x_2$	
(0, 0)	0	
(0, 7)	84	
(8, 3)	100	(maximum)
(12, 0)	96	

To solve the same problem by the simplex method, add a slack variable to each constraint:

$$\begin{aligned} 40x_1 + 80x_2 + s_1 \quad &= 560 \\ 6x_1 + 8x_2 \quad + s_2 &= 72. \end{aligned}$$

Then write the initial simplex tableau:

$$\begin{array}{c} \begin{matrix} x_1 & x_2 & s_1 & s_2 & z \end{matrix} \\ \left[\begin{array}{rrrrr|r} 40 & 80 & 1 & 0 & 0 & 560 \\ 6 & 8 & 0 & 1 & 0 & 72 \\ \hline -8 & -12 & 0 & 0 & 1 & 0 \end{array}\right]. \end{array}$$

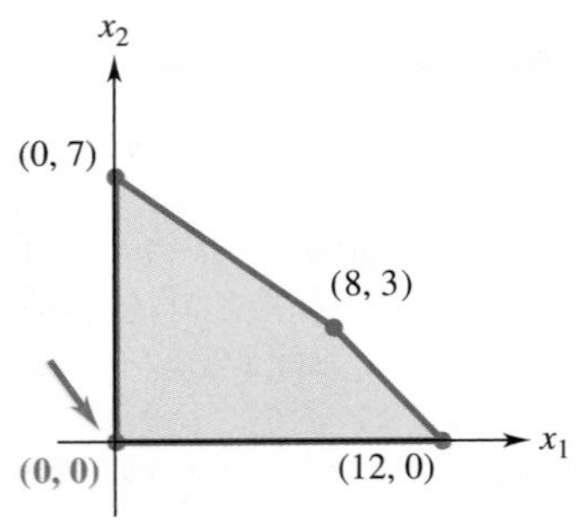

FIGURE 7.28

In this tableau, the basic variables are s_1, s_2, and z. (Why?) By setting the nonbasic variables (namely, x_1 and x_2) equal to 0 and solving for the basic variables, we obtain the following solution (which will be called a **basic feasible solution**):

$$x_1 = 0, \quad x_2 = 0, \quad s_1 = 560, \quad s_2 = 72, \quad \text{and} \quad z = 0.$$

Since $x_1 = 0$ and $x_2 = 0$, this solution corresponds to the corner point at the origin in the graphical solution (Figure 7.28).

The basic feasible solution (0, 0) given by the initial simplex tableau has $z = 0$, which is obviously not maximal. Each round of pivoting in the simplex method will produce another corner point, with a larger value of z, until we reach a corner point that provides the maximum solution.

The most negative indicator in the initial tableau is -12, and it determines the pivot column. Then we form the necessary quotients and determine the pivot row:

$$\text{Pivot row} \rightarrow \begin{array}{c} \begin{matrix} x_1 & x_2 & s_1 & s_2 & z \end{matrix} \\ \left[\begin{array}{rrrrr|r} 40 & \mathbf{80} & 1 & 0 & 0 & 560 \\ 6 & 8 & 0 & 1 & 0 & 72 \\ \hline -8 & -12 & 0 & 0 & 1 & 0 \end{array}\right]. \end{array} \quad \begin{array}{l} \text{Quotients} \\ 560/80 = 7 \quad \text{smallest} \\ 72/8 = 9 \\ \\ \end{array}$$

Pivot column (x_2)

Thus, the pivot is 80 in row 1, column 2. Performing the pivoting leads to this tableau:

$$\begin{array}{c} \begin{matrix} x_1 & x_2 & s_1 & s_2 & z \end{matrix} \\ \left[\begin{array}{rrrrr|r} \frac{1}{2} & 1 & \frac{1}{80} & 0 & 0 & 7 \\ 2 & 0 & -\frac{1}{10} & 1 & 0 & 16 \\ \hline -2 & 0 & \frac{3}{20} & 0 & 1 & 84 \end{array}\right]. \end{array} \quad \begin{array}{l} \frac{1}{80}R_1 \\ -8R_1 + R_2 \\ 12R_1 + R_3 \end{array}$$

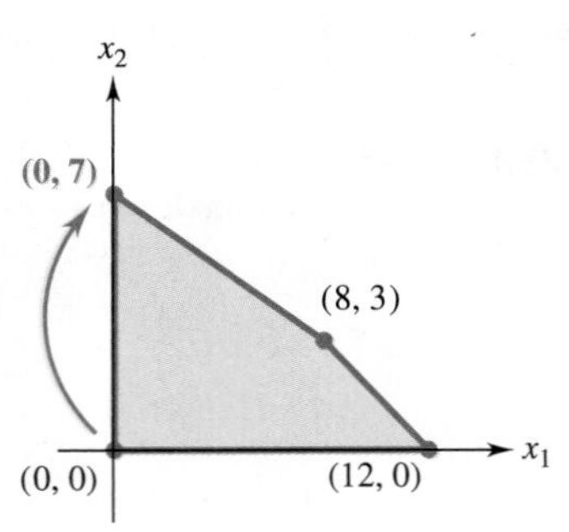

FIGURE 7.29

The basic variables here are x_2, s_2, and z, and the basic feasible solution (found by setting the nonbasic variables equal to 0 and solving for the basic variables) is

$$x_1 = 0, \quad x_2 = 7, \quad s_1 = 0, \quad s_2 = 16, \quad \text{and} \quad z = 84,$$

which corresponds to the corner point (0, 7) in Figure 7.29. Note that the new value of the pivot variable x_2 is precisely the smallest quotient, 7, that was used to select the pivot row. Although this value of z is better, further improvement is possible.

Now the most negative indicator is -2. We form the necessary quotients and determine the pivot as usual:

$$\begin{array}{c} \\ \text{Pivot row} \rightarrow \\ \\ \end{array} \begin{array}{c} x_1 \quad x_2 \quad s_1 \quad s_2 \quad z \\ \left[\begin{array}{ccccc|c} \frac{1}{2} & 1 & \frac{1}{80} & 0 & 0 & 7 \\ \boxed{2} & 0 & -\frac{1}{10} & 1 & 0 & 16 \\ \hline -2 & 0 & \frac{3}{20} & 0 & 1 & 84 \end{array}\right]. \end{array} \quad \begin{array}{l} \text{Quotients} \\ \frac{7}{1/2} = 14 \\ \frac{16}{2} = 8 \quad \text{smallest} \end{array}$$

↑ Pivot column

The pivot is 2 in row 2, column 1. Pivoting now produces the final tableau:

$$\begin{array}{c} x_1 \quad x_2 \quad s_1 \quad s_2 \quad z \\ \left[\begin{array}{ccccc|c} 0 & 1 & \frac{3}{80} & -\frac{1}{4} & 0 & 3 \\ 1 & 0 & -\frac{1}{20} & \frac{1}{2} & 0 & 8 \\ \hline 0 & 0 & \frac{1}{20} & 1 & 1 & 100 \end{array}\right]. \end{array} \quad \begin{array}{l} -\frac{1}{2}R_2 + R_1 \\ \frac{1}{2}R_2 \\ 2R_2 + R_3 \end{array}$$

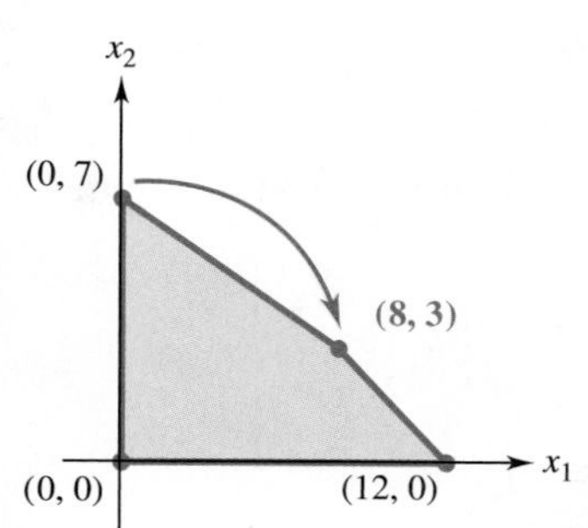

FIGURE 7.30

Here, the basic feasible solution is

$$x_1 = 8, \qquad x_2 = 3, \qquad s_1 = 0, \qquad s_2 = 0, \qquad \text{and} \qquad z = 100,$$

which corresponds to the corner point (8, 3) in Figure 7.30. Once again, the new value of the pivot variable x_1 is the smallest quotient, 8, that was used to select the pivot. Because all the indicators in the last row of the final tableau are nonnegative, (8, 3) is the maximum solution according to the simplex method. We know that this is the case, since this is the maximum solution found by the graphical method. An algebraic argument similar to the one in Example 5 could also be made.

As illustrated in Example 6, the basic feasible solution obtained from a simplex tableau corresponds to a corner point of the feasible region. Pivoting, which replaces one tableau with another, is a systematic way of moving from one corner point to another, each time improving the value of the objective function. The simplex method ends when a corner point that produces the maximum value of the objective function is reached (or when it becomes clear that the problem has no maximum solution).

When there are three or more variables in a linear programming problem, it may be difficult or impossible to draw a picture, but it can be proved that the optimal value of the objective function occurs at a basic feasible solution (corresponding to a corner point in the two-variable case). The simplex method provides a means of moving from one basic feasible solution to another until one that produces the optimal value of the objective function is reached.

EXPLANATION OF PIVOTING

The rules for selecting the pivot in the simplex method can be understood by examining how the first pivot was chosen in Example 6.

Example 7 The initial simplex tableau of Example 6 provides a basic feasible solution with $x_1 = 0$ and $x_2 = 0$:

$$\begin{array}{c} x_1 \quad x_2 \quad s_1 \quad s_2 \quad z \\ \left[\begin{array}{ccccc|c} 40 & 80 & 1 & 0 & 0 & 560 \\ 6 & 8 & 0 & 1 & 0 & 72 \\ \hline -8 & -12 & 0 & 0 & 1 & 0 \end{array}\right]. \end{array}$$

This solution certainly does not give a maximum value for the objective function $z = 8x_1 + 12x_2$. Since x_2 has the largest coefficient, z will be increased most if x_2 is increased. In other words, the most negative indicator in the tableau (which corresponds to the largest coefficient in the objective function) identifies the variable that will provide the greatest change in the value of z.

To determine how much x_2 can be increased without leaving the feasible region, look at the first two equations,

$$\begin{aligned} 40x_1 + 80x_2 + s_1 \qquad &= 560 \\ 6x_1 + 8x_2 \qquad + s_2 &= 72, \end{aligned}$$

and solve for the basic variables s_1 and s_2:

$$\begin{aligned} s_1 &= 560 - 40x_1 - 80x_2 \\ s_2 &= 72 - 6x_1 - 8x_2. \end{aligned}$$

Now x_2 is to be increased while x_1 is to keep the value 0. Hence,

$$\begin{aligned} s_1 &= 560 - 80x_2 \\ s_2 &= 72 - 8x_2. \end{aligned}$$

Since $s_1 \geq 0$ and $s_2 \geq 0$, we must have

$$\begin{aligned} 0 &\leq s_1 & 0 &\leq s_2 \\ 0 &\leq 560 - 80x_2 \quad \text{and} & 0 &\leq 72 - 8x_2 \\ 80x_2 &\leq 560 & 8x_2 &\leq 72 \\ x_2 &\leq \frac{560}{80} = 7 & x_2 &\leq \frac{72}{8} = 9. \end{aligned}$$

The right sides of these last inequalities are the quotients used to select the pivot row. Since x_2 must satisfy both inequalities, x_2 can be at most 7. In other words, the smallest quotient formed from positive entries in the pivot column identifies the value of x_2 that produces the largest change in z while remaining in the feasible region. By pivoting with the pivot determined in this way, we obtain the second tableau and a basic feasible solution in which $x_2 = 7$, as was shown in Example 6.

An analysis similar to that in Example 7 applies to each occurrence of pivoting in the simplex method. The idea is to improve the value of the objective function by adjusting one variable at a time. The most negative indicator identifies the variable that will account for the largest increase in z. The smallest quotient determines the largest value of that variable which will produce a feasible solution. Pivoting leads to a solution in which the selected variable has this largest value.

7.4 ▶ Exercises

In Exercises 1–4, (a) determine the number of slack variables needed; (b) name them; (c) use the slack variables to convert each constraint into a linear equation. (See Example 1.)

1. Maximize $z = 32x_1 + 9x_2$
subject to
$$\begin{aligned} 4x_1 + 2x_2 &\leq 20 \\ 5x_1 + x_2 &\leq 50 \\ 2x_1 + 3x_2 &\leq 25 \\ x_1 \geq 0, x_2 &\geq 0. \end{aligned}$$

2. Maximize $z = 3.7x_1 + 4.3x_2$
subject to
$$\begin{aligned} 2.4x_1 + 1.5x_2 &\leq 10 \\ 1.7x_1 + 1.9x_2 &\leq 15 \\ x_1 \geq 0, x_2 &\geq 0. \end{aligned}$$

3. Maximize $z = 8x_1 + 3x_2 + x_3$
subject to
$$\begin{aligned} 3x_1 - x_2 + 4x_3 &\leq 95 \\ 7x_1 + 6x_2 + 8x_3 &\leq 118 \\ 4x_1 + 5x_2 + 10x_3 &\leq 220 \\ x_1 \geq 0, x_2 \geq 0, x_3 &\geq 0. \end{aligned}$$

4. Maximize $z = 12x_1 + 15x_2 + 10x_3$
subject to
$$2x_1 + 2x_2 + x_3 \le 8$$
$$x_1 + 4x_2 + 3x_3 \le 12$$
$$x_1 \ge 0, x_2 \ge 0, x_3 \ge 0.$$

Introduce slack variables as necessary and then write the initial simplex tableau for each of these linear programming problems.

5. Maximize $z = 5x_1 + x_2$
subject to
$$2x_1 + 5x_2 \le 6$$
$$4x_1 + x_2 \le 6$$
$$5x_1 + 3x_2 \le 15$$
$$x_1 \ge 0, x_2 \ge 0.$$

6. Maximize $z = 5x_1 + 3x_2 + 7x_3$
subject to
$$4x_1 + 3x_2 + 2x_3 \le 60$$
$$3x_1 + 4x_2 + x_3 \le 24$$
$$x_1 \ge 0, x_2 \ge 0, x_3 \ge 0.$$

7. Maximize $z = x_1 + 5x_2 + 10x_3$
subject to
$$x_1 + 2x_2 + 3x_3 \le 10$$
$$2x_1 + x_2 + x_3 \le 8$$
$$3x_1 + 4x_3 \le 6$$
$$x_1 \ge 0, x_2 \ge 0, x_3 \ge 0.$$

8. Maximize $z = 5x_1 - x_2 + 3x_3$
subject to
$$3x_1 + 2x_2 + x_3 \le 36$$
$$x_1 + 6x_2 + x_3 \le 24$$
$$x_1 - x_2 - x_3 \le 32$$
$$x_1 \ge 0, x_2 \ge 0, x_3 \ge 0.$$

Find the pivot in each of the given simplex tableaus. (See Example 2.)

9.

x_1	x_2	x_3	s_1	s_2	z	
2	2	0	3	1	0	15
3	4	1	6	0	0	20
−2	−3	0	1	0	1	10

10.

x_1	x_2	x_3	s_1	s_2	z	
0	2	1	1	3	0	5
1	−5	0	1	2	0	8
0	−2	0	−3	1	1	10

11.

x_1	x_2	x_3	s_1	s_2	s_3	z	
6	2	1	3	0	0	0	8
0	2	0	1	0	1	0	7
6	1	0	3	1	0	0	6
−3	−2	0	2	0	0	1	12

12.

x_1	x_2	x_3	s_1	s_2	s_3	z	
0	2	0	1	2	2	0	3
0	3	1	0	1	2	0	4
1	4	0	0	3	5	0	5
0	−4	0	0	4	3	1	20

In Exercises 13–16, use the indicated entry as the pivot and perform the pivoting. (See Examples 3 and 4.)

13.

x_1	x_2	x_3	s_1	s_2	z	
1	2	4	1	0	0	56
2	**2**	1	0	1	0	40
−1	−3	−2	0	0	1	0

14.

x_1	x_2	x_3	s_1	s_2	s_3	z	
2	2	**1**	1	0	0	0	12
1	2	3	0	1	0	0	45
3	1	1	0	0	1	0	20
−2	−1	−3	0	0	0	1	0

15.

x_1	x_2	x_3	s_1	s_2	s_3	z	
1	1	1	1	0	0	0	60
3	1	**2**	0	1	0	0	100
1	2	3	0	0	1	0	200
−1	−1	−2	0	0	0	1	0

16.

x_1	x_2	x_3	s_1	s_2	s_3	z	
4	2	3	1	0	0	0	22
2	2	**5**	0	1	0	0	28
1	3	2	0	0	1	0	45
−3	−2	−4	0	0	0	1	0

For each simplex tableau in Exercises 17–20, (a) list the basic and the nonbasic variables, (b) find the basic feasible solution determined by setting the nonbasic variables equal to 0, and (c) decide whether this is a maximum solution. (See Examples 5 and 6.)

17.

x_1	x_2	x_3	s_1	s_2	z	
3	2	0	−3	1	0	29
4	0	1	−2	0	0	16
−5	0	0	−1	0	1	11

18.

x_1	x_2	x_3	s_1	s_2	s_3	z	
−3	0	$\frac{1}{2}$	1	−2	0	0	22
2	0	−3	0	1	1	0	10
4	1	4	0	$\frac{3}{4}$	0	0	17
−1	0	0	0	1	0	1	120

19.

x_1	x_2	x_3	s_1	s_2	s_3	z	
1	0	2	$\frac{1}{2}$	0	$\frac{1}{3}$	0	6
0	1	−1	5	0	−1	0	13
0	0	1	$\frac{3}{2}$	1	$-\frac{1}{3}$	0	21
0	0	2	$\frac{1}{2}$	0	3	1	18

20.

x_1	x_2	x_3	x_4	s_1	s_2	s_3	z	
−1	0	0	1	0	3	−2	0	47
2	0	1	0	0	2	$-\frac{1}{2}$	0	37
3	5	0	0	1	−1	6	0	43
4	1	0	0	0	6	0	1	86

Use the simplex method to solve Exercises 21–36.

21. Maximize $z = x_1 + 3x_2$
subject to
$$\begin{aligned} x_1 + x_2 &\le 10 \\ 5x_1 + 2x_2 &\le 20 \\ x_1 + 2x_2 &\le 36 \\ x_1 \ge 0, x_2 &\ge 0. \end{aligned}$$

22. Maximize $z = 5x_1 + x_2$
subject to
$$\begin{aligned} 2x_1 + 3x_2 &\le 8 \\ 4x_1 + 8x_2 &\le 12 \\ 5x_1 + 2x_2 &\le 30 \\ x_1 \ge 0, x_2 &\ge 0. \end{aligned}$$

23. Maximize $z = 2x_1 + x_2$
subject to
$$\begin{aligned} x_1 + 3x_2 &\le 12 \\ 2x_1 + x_2 &\le 10 \\ x_1 + x_2 &\le 4 \\ x_1 \ge 0, x_2 &\ge 0. \end{aligned}$$

24. Maximize $z = 4x_1 + 2x_2$
subject to
$$\begin{aligned} -x_1 - x_2 &\le 12 \\ 3x_1 - x_2 &\le 15 \\ x_1 \ge 0, x_2 &\ge 0. \end{aligned}$$

25. Maximize $z = 5x_1 + 4x_2 + x_3$
subject to
$$\begin{aligned} -2x_1 + x_2 + 2x_3 &\le 3 \\ x_1 - x_2 + x_3 &\le 1 \\ x_1 \ge 0, x_2 \ge 0, x_3 &\ge 0. \end{aligned}$$

26. Maximize $z = 3x_1 + 2x_2 + x_3$
subject to
$$\begin{aligned} 2x_1 + 2x_2 + x_3 &\le 10 \\ x_1 + 2x_2 + 3x_3 &\le 15 \\ x_1 \ge 0, x_2 \ge 0, x_3 &\ge 0. \end{aligned}$$

27. Maximize $z = 2x_1 + x_2 + x_3$
subject to
$$\begin{aligned} x_1 - 3x_2 + x_3 &\le 3 \\ x_1 - 2x_2 + 2x_3 &\le 12 \\ x_1 \ge 0, x_2 \ge 0, x_3 &\ge 0. \end{aligned}$$

28. Maximize $z = 4x_1 + 5x_2 + x_3$
subject to
$$\begin{aligned} x_1 + 2x_2 + 4x_3 &\le 10 \\ 2x_1 + 2x_2 + x_3 &\le 10 \\ x_1 \ge 0, x_2 \ge 0, x_3 &\ge 0. \end{aligned}$$

29. Maximize $z = 2x_1 + 2x_2 - 4x_3$
subject to
$$\begin{aligned} 3x_1 + 3x_2 - 6x_3 &\le 51 \\ 5x_1 + 5x_2 + 10x_3 &\le 99 \\ x_1 \ge 0, x_2 \ge 0, x_3 &\ge 0. \end{aligned}$$

30. Maximize $z = 4x_1 + x_2 + 3x_3$
subject to
$$\begin{aligned} x_1 + 3x_3 &\le 6 \\ 6x_1 + 3x_2 + 12x_3 &\le 40 \\ x_1 \ge 0, x_2 \ge 0, x_3 &\ge 0. \end{aligned}$$

31. Maximize $z = 300x_1 + 200x_2 + 100x_3$
subject to
$$\begin{aligned} x_1 + x_2 + x_3 &\le 100 \\ 2x_1 + 3x_2 + 4x_3 &\le 320 \\ 2x_1 + x_2 + x_3 &\le 160 \\ x_1 \ge 0, x_2 \ge 0, x_3 &\ge 0. \end{aligned}$$

32. Maximize $z = x_1 + 5x_2 - 10x_3$
subject to
$$\begin{aligned} 8x_1 + 4x_2 + 12x_3 &\le 18 \\ x_1 + 6x_2 + 2x_3 &\le 45 \\ 5x_1 + 7x_2 + 3x_3 &\le 60 \\ x_1 \ge 0, x_2 \ge 0, x_3 &\ge 0. \end{aligned}$$

33. Maximize $z = 4x_1 - 3x_2 + 2x_3$
subject to
$$\begin{aligned} 2x_1 - x_2 + 8x_3 &\le 40 \\ 4x_1 - 5x_2 + 6x_3 &\le 60 \\ 2x_1 - 2x_2 + 6x_3 &\le 24 \\ x_1 \ge 0, x_2 \ge 0, x_3 &\ge 0. \end{aligned}$$

34. Maximize $z = 3x_1 + 2x_2 - 4x_3$
subject to
$$\begin{aligned} x_1 - x_2 + x_3 &\le 10 \\ 2x_1 - x_2 + 2x_3 &\le 30 \\ -3x_1 + x_2 + 3x_3 &\le 40 \\ x_1 \ge 0, x_2 \ge 0, x_3 &\ge 0. \end{aligned}$$

35. Maximize $z = x_1 + 2x_2 + x_3 + 5x_4$
subject to
$$\begin{aligned} x_1 + 2x_2 + x_3 + x_4 &\le 50 \\ 3x_1 + x_2 + 2x_3 + x_4 &\le 100 \\ x_1 \ge 0, x_2 \ge 0, x_3 \ge 0, x_4 &\ge 0. \end{aligned}$$

36. Maximize $z = x_1 + x_2 + 4x_3 + 5x_4$
subject to
$$\begin{aligned} x_1 + 2x_2 + 3x_3 + x_4 &\le 115 \\ 2x_1 + x_2 + 8x_3 + 5x_4 &\le 200 \\ x_1 + x_3 &\le 50 \\ x_1 \ge 0, x_2 \ge 0, x_3 \ge 0, x_4 &\ge 0. \end{aligned}$$

37. The initial simplex tableau of a linear programming problem is

$$\begin{array}{cccccc|c} x_1 & x_2 & x_3 & s_1 & s_2 & z & \\ \hline 1 & 1 & 1 & 1 & 0 & 0 & 12 \\ 2 & 1 & 2 & 0 & 1 & 0 & 30 \\ \hline -2 & -2 & -1 & 0 & 0 & 1 & 0 \end{array}.$$

(a) Use the simplex method to solve the problem with column 1 as the first pivot column.
(b) Now use the simplex method to solve the problem with column 2 as the first pivot column.
(c) Does this problem have a unique maximum solution? Why?

38. The final simplex tableau of a linear programming problem is

$$\begin{array}{ccccc|c} x_1 & x_2 & s_1 & s_2 & z & \\ \hline 1 & 1 & 2 & 0 & 0 & 24 \\ 2 & 0 & 2 & 1 & 0 & 8 \\ \hline 4 & 0 & 0 & 0 & 1 & 40 \end{array}.$$

(a) What is the solution given by this tableau?
(b) Even though all the indicators are nonnegative, perform one more round of pivoting on this tableau, using column 3 as the pivot column and choosing the pivot row by forming quotients in the usual way.
(c) Show that there is more than one solution to the linear programming problem by comparing your answer in part (a) with the basic feasible solution given by the tableau found in part (b). Does it give the same value of z as the solution in part (a)?

✓ Checkpoint Answers

1. $$\begin{aligned} x_1 + x_2 + x_3 + s_1 &= 12 \\ 2x_1 + 4x_2 \qquad + s_2 &= 15 \\ x_2 + 3x_3 + s_3 &= 10 \end{aligned}$$

2. $$\begin{array}{cccccc|c} x_1 & x_2 & s_1 & s_2 & s_3 & z & \\ \hline 1 & 2 & 1 & 0 & 0 & 0 & 85 \\ 2 & 1 & 0 & 1 & 0 & 0 & 92 \\ 1 & 4 & 0 & 0 & 1 & 0 & 104 \\ \hline -2 & -3 & 0 & 0 & 0 & 1 & 0 \end{array}$$

   Indicators (under -2, -3)

3. 2 (in first column)

4. **(a)** 2

   (b) $$\begin{array}{cccccc|c} x_1 & x_2 & x_3 & s_1 & s_2 & z & \\ \hline \frac{1}{2} & 1 & 3 & \frac{1}{2} & 0 & 0 & 8 \\ -\frac{1}{2} & 0 & -9 & -\frac{3}{2} & 1 & 0 & 1 \\ \hline 1 & 0 & 9 & 2 & 0 & 1 & 32 \end{array}$$

5. $z = 21$ when $x_1 = 6, x_2 = 0$, and $x_3 = 9$.

6. $x_1 = 0, x_2 = 24, s_1 = 16, s_2 = 0, z = 4800$

7.5 ▶ Maximization Applications

Applications of the simplex method are considered in this section. First, however, we make a slight change in notation. You have noticed that the column representing the variable z in a simplex tableau never changes during pivoting. (Since all the entries except the last one in this column are 0, performing row operations has no effect on these entries—they remain 0.) Consequently, this column is unnecessary and can be omitted without causing any difficulty.

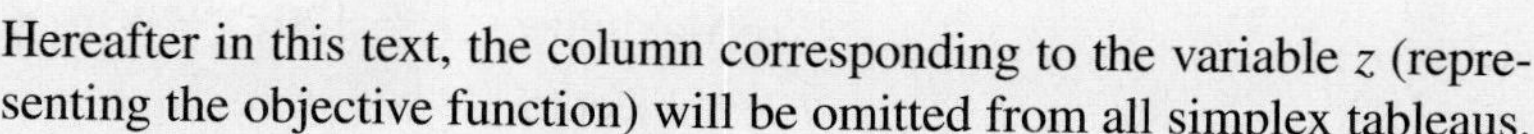

Hereafter in this text, the column corresponding to the variable z (representing the objective function) will be omitted from all simplex tableaus.

Example 1 A farmer has 110 acres of available land he wishes to plant with a mixture of potatoes, corn, and cabbage. It costs him \$400 to produce an acre of potatoes, \$160 to produce an acre of corn, and \$280 to produce an acre of cabbage. He has a maximum of \$20,000 to spend. He makes a profit of \$120 per acre of potatoes, \$40 per acre of corn, and \$60 per acre of cabbage.

(a) How many acres of each crop should he plant to maximize his profit?

Solution Let the number of acres alloted to each of potatoes, corn, and cabbage be x_1, x_2, and x_3, respectively. Then summarize the given information as follows:

Crop	Number of Acres	Cost per Acre	Profit per Acre
Potatoes	x_1	\$400	\$120
Corn	x_2	\$160	\$ 40
Cabbage	x_3	\$280	\$ 60
Maximum Available	110	\$20,000	

The constraints can be expressed as

$$\begin{aligned} x_1 + x_2 + x_3 &\le 110 && \text{Number of acres} \\ 400x_1 + 160x_2 + 280x_3 &\le 20{,}000, && \text{Production costs} \end{aligned}$$

where x_1, x_2, and x_3 are all nonnegative. The first of these constraints says that $x_1 + x_2 + x_3$ is less than or perhaps equal to 110. Use s_1 as the slack variable, giving the equation

$$x_1 + x_2 + x_3 + s_1 = 110.$$

Here, s_1 represents the amount of the farmer's 110 acres that will not be used. (s_1 may be 0 or any value up to 110.)

In the same way, the constraint $400x_1 + 160x_2 + 280x_3 \le 20{,}000$ can be converted into an equation by adding a slack variable s_2:

$$400x_1 + 160x_2 + 280x_3 + s_2 = 20{,}000.$$

The slack variable s_2 represents any unused portion of the farmer's \$20,000 capital. (Again, s_2 may have any value from 0 to 20,000.)

The farmer's profit on potatoes is the product of the profit per acre (\$120) and the number x_1 of acres, that is, $120x_1$. His profits on corn and cabbage are computed similarly. Hence, his total profit is given by

$$z = \text{profit on potatoes} + \text{profit on corn} + \text{profit on cabbage}$$
$$z = 120x_1 + 40x_2 + 60x_3.$$

The linear programming problem can now be stated as follows:

$$\begin{aligned} &\text{Maximize} && z = 120x_1 + 40x_2 + 60x_3 \\ &\text{subject to} && x_1 + x_2 + x_3 + s_1 = 110 \\ & && 400x_1 + 160x_2 + 280x_3 + s_2 = 20{,}000, \\ &\text{with} && x_1 \ge 0, x_2 \ge 0, x_3 \ge 0, s_1 \ge 0, s_2 \ge 0. \end{aligned}$$

The initial simplex tableau (without the z column) is

$$\begin{array}{c} \begin{matrix} x_1 & x_2 & x_3 & s_1 & s_2 & \end{matrix} \\ \left[\begin{array}{ccccc|c} 1 & 1 & 1 & 1 & 0 & 110 \\ 400 & 160 & 280 & 0 & 1 & 20{,}000 \\ \hline -120 & -40 & -60 & 0 & 0 & 0 \end{array}\right]. \end{array}$$

The most negative indicator is -120; column 1 is the pivot column. The quotients needed to determine the pivot row are $110/1 = 110$ and $20{,}000/400 = 50$. So the pivot is 400 in row 2, column 1. Multiplying row 2 by $1/400$ and completing the pivoting leads to the final simplex tableau:

$$\begin{array}{c} \begin{matrix} x_1 & x_2 & x_3 & s_1 & s_2 & \end{matrix} \\ \left[\begin{array}{ccccc|c} 0 & .6 & .3 & 1 & -.0025 & 60 \\ 1 & .4 & .7 & 0 & .0025 & 50 \\ \hline 0 & 8 & 24 & 0 & .3 & \mathbf{6000} \end{array}\right]. \end{array} \quad \begin{array}{l} -1R_2 + R_1 \\ \frac{1}{400}R_2 \\ 120R_2 + R_3 \end{array}$$

Setting the nonbasic variables x_2, x_3, and s_2 equal to 0, solving for the basic variables x_1 and s_1, and remembering that the value of z is in the lower right-hand corner leads to this maximum solution:

$$x_1 = 50, \quad x_2 = 0, \quad x_3 = 0, \quad s_1 = 60, \quad s_2 = 0, \quad \text{and} \quad z = 6000.$$

Therefore, the farmer will make a maximum profit of \$6000 by planting 50 acres of potatoes and no corn or cabbage.

(b) If the farmer maximizes his profit, how much land will remain unplanted? What is the explanation for this?

Solution Since 50 of 110 acres are planted, 60 acres will remain unplanted. Alternatively, note that the unplanted acres of land are represented by s_1, the slack variable in the "number of acres" constraint. In the maximal solution found in part (a), $s_1 = 60$, which means that 60 acres are left unplanted.

The amount of unused cash is represented by s_2, the slack variable in the "production costs" constraint. Since $s_2 = 0$, all the available money has been used. By using the maximal solution in part (a), the farmer has used his \$20,000 most effectively. If he had more cash, he would plant more crops and make a larger profit.

Example 2 Ana Pott, who is a candidate for the state legislature, has \$96,000 to buy TV advertising time. Ads cost \$400 per minute on a local cable channel, \$4000 per minute on a regional independent channel, and \$12,000 per minute on a national network channel. Because of existing contracts, the TV stations can provide at most 30 minutes of advertising time, with a maximum of 6 minutes on the national network channel. At any given time during the evening, approximately 100,000 people watch the cable channel, 200,000 the independent channel, and 600,000 the network channel. To get maximum exposure, how much time should Ana buy from each station?

(a) Set up the initial simplex tableau for this problem.

Solution Let x_1 be the number of minutes of ads on the cable channel, x_2 the number of minutes on the independent channel, and x_3 the number of minutes on the network channel. Exposure is measured in viewer-minutes. For instance, 100,000 people watching x_1 minutes of ads on the cable channel produces $100{,}000x_1$ viewer-minutes. The amount of exposure is given by the total number of viewer-minutes for all three channels, namely,

$$100{,}000x_1 + 200{,}000x_2 + 600{,}000x_3.$$

Since 30 minutes are available,

$$x_1 + x_2 + x_3 \leq 30.$$

The fact that only 6 minutes can be used on the network channel means that

$$x_3 \leq 6.$$

Expenditures are limited to \$96,000, so

$$\begin{aligned} \text{Cable cost} + \text{independent cost} + \text{network cost} &\leq 96{,}000 \\ 400x_1 \quad + \quad 4000x_2 \quad + \quad 12{,}000x_3 &\leq 96{,}000. \end{aligned}$$

Therefore, Ana must solve the following linear programming problem:

$$\begin{aligned} \text{Maximize} \quad & z = 100{,}000x_1 + 200{,}000x_2 + 600{,}000x_3 \\ \text{subject to} \quad & x_1 + x_2 + x_3 \leq 30 \\ & x_3 \leq 6 \\ & 400x_1 + 4000x_2 + 12{,}000x_3 \leq 96{,}000, \\ & \text{with } x_1 \geq 0, x_2 \geq 0, x_3 \geq 0. \end{aligned}$$

Introducing slack variables s_1, s_2, and s_3 (one for each constraint), rewriting the constraints as equations, and expressing the objective function as

$$-100{,}000x_1 - 200{,}000x_2 - 600{,}000x_3 + z = 0$$

leads to the initial simplex tableau:

$$\begin{array}{cccccc|c} x_1 & x_2 & x_3 & s_1 & s_2 & s_3 & \\ 1 & 1 & 1 & 1 & 0 & 0 & 30 \\ 0 & 0 & 1 & 0 & 1 & 0 & 6 \\ 400 & 4000 & 12{,}000 & 0 & 0 & 1 & 96{,}000 \\ \hline -100{,}000 & -200{,}000 & -600{,}000 & 0 & 0 & 0 & 0 \end{array}.$$

(b) Use the simplex method to find the final simplex tableau.

Solution Work by hand, or use a graphing calculator's simplex program or a spreadsheet, to obtain this final tableau:

$$\begin{array}{cccccc|c} x_1 & x_2 & x_3 & s_1 & s_2 & s_3 & \\ 1 & 0 & 0 & \frac{10}{9} & \frac{20}{9} & -\frac{25}{90{,}000} & 20 \\ 0 & 0 & 1 & 0 & 1 & 0 & 6 \\ 0 & 1 & 0 & -\frac{1}{9} & -\frac{29}{9} & \frac{25}{90{,}000} & 4 \\ \hline 0 & 0 & 0 & \frac{800{,}000}{9} & \frac{1{,}600{,}000}{9} & \frac{250}{9} & 6{,}400{,}000 \end{array}.$$

Therefore, the optimal solution is

$$x_1 = 20, \quad x_2 = 4, \quad x_3 = 6, \quad s_1 = 0, \quad s_2 = 0, \quad \text{and} \quad s_3 = 0.$$

Ana should buy 20 minutes of time on the cable channel, 4 minutes on the independent channel, and 6 minutes on the network channel.

(c) What do the values of the slack variables in the optimal solution tell you?

Solution All three slack variables are 0. This means that all the available minutes have been used ($s_1 = 0$ in the first constraint), the maximum possible 6 minutes on the national network have been used ($s_2 = 0$ in the second constraint), and all of the \$96,000 has been spent ($s_3 = 0$ in the third constraint). 1 ✓

✓Checkpoint 1

In Example 2, what is the number of viewer-minutes in the optimal solution?

Answer: *See page 434.*

Example 3 A chemical plant makes three products—glaze, solvent, and clay—each of which brings in different revenue per truckload. Production is limited, first by the number of air pollution units the plant is allowed to produce each day and second by the time available in the evaporation tank. The plant manager wants to maximize the daily revenue. Using information not given here, he sets up an initial simplex tableau and uses the simplex method to produce the following final simplex tableau:

$$\begin{array}{ccccc|c} x_1 & x_2 & x_3 & s_1 & s_2 & \\ -10 & -25 & 0 & 1 & -1 & 60 \\ 3 & 4 & 1 & 0 & .1 & 24 \\ \hline 7 & 13 & 0 & 0 & .4 & 96 \end{array}.$$

The three variables represent the number of truckloads of glaze, solvent, and clay, respectively. The first slack variable comes from the air pollution constraint and the second slack variable from the time constraint on the evaporation tank. The revenue function is given in hundreds of dollars.

(a) What is the optimal solution?

Solution

$x_1 = 0, \quad x_2 = 0, \quad x_3 = 24, \quad s_1 = 60, \quad s_2 = 0, \quad \text{and} \quad z = 96.$

(b) Interpret this solution. What do the variables represent, and what does the solution mean?

Solution The variable x_1 is the number of truckloads of glaze, x_2 the number of truckloads of solvent, x_3 the number of truckloads of clay to be produced, and z the revenue produced (in hundreds of dollars). The plant should produce 24 truckloads of clay and no glaze or solvent, for a maximum revenue of \$9600. The first slack variable, s_1, represents the number of air pollution units below the maximum number allowed. Since $s_1 = 60$, the number of air pollution units will be 60 less than the allowable maximum. The second slack variable, s_2, represents the unused time in the evaporation tank. Since $s_2 = 0$, the evaporation tank is fully used.

7.5 ▶ Exercises

Set up the initial simplex tableau for each of the given problems. You will be asked to solve these problems in Exercises 19–22.

1. **Business** A cat breeder has the following amounts of cat food: 90 units of tuna, 80 units of liver, and 50 units of chicken. To raise a Siamese cat, the breeder must use 2 units of tuna, 1 of liver, and 1 of chicken per day, while raising a Persian cat requires 1, 2, and 1 units, respectively, per day. If a Siamese cat sells for \$12 while a Persian cat sells for \$10, how many of each should be raised in order to obtain maximum gross income? What is the maximum gross income?

2. **Business** Banal, Inc., produces art for motel rooms. Its painters can turn out mountain scenes, seascapes, and pictures of clowns. Each painting is worked on by three different artists: T, D, and H. Artist T works only 25 hours per week, while D and H work 45 and 40 hours per week, respectively. Artist T spends 1 hour on a mountain scene, 2 hours on a seascape, and 1 hour on a clown. Corresponding times for D and H are 3, 2, and 2 hours and 2, 1, and 4 hours, respectively. Banal makes \$20 on a mountain scene, \$18 on a seascape, and \$22 on a clown. The head painting packer can't stand clowns, so that no more than 4 clown paintings may be done in a week. Find the number of each type of painting that should be made weekly in order to maximize profit. Find the maximum possible profit.

3. **Health** A biologist has 500 kilograms of nutrient A, 600 kilograms of nutrient B, and 300 kilograms of nutrient C. These nutrients will be used to make 4 types of food—P, Q, R, and S—whose contents (in percent of nutrient per kilogram of food) and whose "growth values" are as shown in the following table:

	P	Q	R	S
A	0	0	37.5	62.5
B	0	75	50	37.5
C	100	25	12.5	0
Growth Value	90	70	60	50

How many kilograms of each food should be produced in order to maximize total growth value? Find the maximum growth value.

4. **Natural Science** A lake is stocked each spring with three species of fish: A, B, and C. The average weights of the fish are 1.62, 2.12, and 3.01 kilograms for the three

species, respectively. Three foods—I, II, and III—are available in the lake. Each fish of species A requires 1.32 units of food I, 2.9 units of food II, and 1.75 units of food III, on the average, each day. Species B fish require 2.1 units of food I, .95 units of food II, and .6 units of food III daily. Species C fish require .86, 1.52, and 2.01 units of I, II, and III per day, respectively. If 490 units of food I, 897 units of food II, and 653 units of food III are available daily, how should the lake be stocked to maximize the weight of the fish it supports?

In each of the given exercises, (a) use the simplex method to solve the problem and (b) explain what the values of the slack variables in the optimal solution mean in the context of the problem. (See Examples 1–3).

5. **Business** A manufacturer of bicycles builds 1-, 3-, and 10-speed models. The bicycles are made of both aluminum and steel. The company has available 91,800 units of steel and 42,000 units of aluminum. The 1-, 3-, and 10-speed models need, respectively, 20, 30, and 40 units of steel and 12, 21, and 16 units of aluminum. How many of each type of bicycle should be made in order to maximize profit if the company makes \$8 per 1-speed bike, \$12 per 3-speed bike, and \$24 per 10-speed bike? What is the maximum possible profit?

6. **Social Science** Jayanta is working to raise money for the homeless by sending informational letters and making follow-up calls to local labor organizations and church groups. She discovered that each church group requires 2 hours of letter writing and 1 hour of follow-up, while, for each labor union, she needs 2 hours of letter writing and 3 hours of follow-up. Jayanta can raise \$100 from each church group and \$200 from each union local, and she has a maximum of 16 hours of letter-writing time and a maximum of 12 hours of follow-up time available per month. Determine the most profitable mixture of groups she should contact and the most money she can raise in a month.

7. **Social Science** A political party is planning a half-hour television show. The show will have 3 minutes of direct requests for money from viewers. The remaining time will feature three of the party's politicians: a senator, a congresswoman, and a governor. The senator, a party "elder statesman," demands that he be on screen at least twice as long as the governor. The total time taken by the senator and the governor must be at least twice the time taken by the congresswoman. On the basis of a preshow survey, it is believed that 40, 60, and 50 (in thousands) viewers will watch the program for each minute the senator, congresswoman, and governor, respectively, are on the air. Find the time that should be allotted to each politician in order to get the maximum number of viewers. Find the maximum number of viewers.

8. **Business** The Cut-Right Company sells sets of kitchen knives. The Basic Set consists of 2 utility knives and 1 chef's knife. The Regular Set consists of 2 utility knives, 1 chef's knife, and 1 slicer. The Deluxe Set consists of 3 utility knives, 1 chef's knife, and 1 slicer. The profit is \$30 on a Basic Set, \$40 on a Regular Set, and \$60 on a Deluxe Set. The factory has on hand 800 utility knives, 400 chef's knives, and 200 slicers. Assuming that all sets will be sold, how many of each type should be made up in order to maximize profit? What is the maximum profit?

9. **Business** The Muro Manufacturing Company makes two kinds of plasma screen television sets. It produces the Flexscan set, which makes a \$350 profit, and the Panoramic I, which makes a \$500 profit. On the assembly line, the Flexscan requires 5 hours, and the Panoramic I takes 7 hours. The cabinet shop spends 1 hour on the cabinet for the Flexscan and 2 hours on the cabinet for the Panoramic I. Both sets require 4 hours for testing and packing. On a particular production run, the Muro Company has available 3600 work-hours on the assembly line, 900 work-hours in the cabinet shop, and 2600 work-hours in the testing and packing department. How many sets of each type should it produce to make a maximum profit? What is the maximum profit?

10. **Business** The Texas Poker Company assembles three different poker sets. Each Royal Flush poker set contains 1000 poker chips, 4 decks of cards, 10 dice, and 2 dealer buttons. Each Deluxe Diamond poker set contains 600 poker chips, 2 decks of cards, 5 dice, and 1 dealer button. The Full House poker set contains 300 poker chips, 2 decks of cards, 5 dice, and 1 dealer button. The Texas Poker Company has 2,800,000 poker chips, 10,000 decks of cards, 25,000 dice, and 6000 dealer buttons in stock. It earns a profit of \$38 for each Royal Flush poker set, \$22 for each Deluxe Diamond poker set, and \$12 for each Full House poker set. How many of each type of poker set should it assemble to maximize profit? What is the maximum profit?

Use the simplex method to solve the given problems. (See Examples 1–3.)

11. **Business** The Fancy Fashions Store has \$8000 available each month for advertising. Newspaper ads cost \$400 each, and no more than 20 can be run per month. Radio ads cost \$200 each, and no more than 30 can run per month. TV ads cost \$1200 each, with a maximum of 6 available each month. Approximately 2000 women will see each newspaper ad, 1200 will hear each radio commercial, and 10,000 will see each TV ad. How much of each type of advertising should be used if the store wants to maximize its ad exposure?

12. **Business** Caroline's Quality Candy Confectionery is famous for fudge, chocolate cremes, and pralines. Its candy-making equipment is set up to make 100-pound

batches at a time. Currently there is a chocolate shortage, and the company can get only 120 pounds of chocolate in the next shipment. On a week's run, the confectionery's cooking and processing equipment is available for a total of 42 machine hours. During the same period, the employees have a total of 56 work hours available for packaging. A batch of fudge requires 20 pounds of chocolate, while a batch of cremes uses 25 pounds of chocolate. The cooking and processing take 120 minutes for fudge, 150 minutes for chocolate cremes, and 200 minutes for pralines. The packaging times, measured in minutes per 1-pound box, are 1, 2, and 3, respectively, for fudge, cremes, and pralines. Determine how many batches of each type of candy the confectionery should make, assuming that the profit per 1-pound box is 50¢ on fudge, 40¢ on chocolate cremes, and 45¢ on pralines. Also, find the maximum profit for the week.

13. **Finance** A political party is planning its fund-raising activities for a coming election. It plans to raise money through large fund-raising parties, letters requesting funds, and dinner parties where people can meet the candidate personally. Each large fund-raising party costs \$3000, each mailing costs \$1000, and each dinner party costs \$12,000. The party can spend up to \$102,000 for these activities. From experience, it is known that each large party will raise \$200,000, each letter campaign will raise \$100,000, and each dinner party will raise \$600,000. The party is able to carry out as many as 25 of these activities.
   (a) How many of each should the party plan in order to raise the maximum amount of money? What is the maximum amount?
   (b) Dinner parties are more expensive than letter campaigns, yet the optimum solution found in part (a) includes dinner parties, but no letter campaigns. Explain how this is possible.

14. **Business** A baker has 60 units of flour, 132 units of sugar, and 102 units of raisins. A loaf of raisin bread requires 1 unit of flour, 1 unit of sugar, and 2 units of raisins, while a raisin cake needs 2, 4, and 1 units, respectively. If raisin bread sells for \$3 a loaf and a raisin cake for \$4, how many of each should be baked so that the gross income is maximized? What is the maximum gross income?

15. **Health** Rachel Reeve, a fitness trainer, has an exercise regimen that includes running, biking, and walking. She has no more than 15 hours per week to devote to exercise, including at most 3 hours for running. She wants to walk at least twice as many hours as she bikes. A 130-pound person like Rachel will burn on average 531 calories per hour running, 472 calories per hour biking, and 354 calories per hour walking. How many hours per week should Rachel spend on each exercise to maximize the number of calories she burns? What is the maximum number of calories she will burn? (*Hint*: Write the constraint involving walking and biking in the form ≤ 0.)

16. **Health** Joe Vetere's exercise regimen includes light calisthenics, swimming, and playing the drums. He has at most 10 hours per week to devote to these activities. He wants the total time he does calisthenics and plays the drums to be at least twice as long as he swims. His neighbors, however, will tolerate no more than 4 hours per week on the drums. A 190-pound person like Joe will burn an average of 388 calories per hour doing calisthenics, 518 calories per hour swimming, and 345 calories per hour playing the drums. How many hours per week should Joe spend on each exercise to maximize the number of calories he burns? What is the maximum number of calories he will burn?

Business *The next two problems come from past CPA examinations.* Select the appropriate answer for each question.*

17. The Ball Company manufactures three types of lamps, labeled *A*, *B*, and *C*. Each lamp is processed in two departments: I and II. Total available person-hours per day for departments I and II are 400 and 600, respectively. No additional labor is available. Time requirements and profit per unit for each type of lamp are as follows:

	A	*B*	*C*
Person-Hours in I	2	3	1
Person-Hours in II	4	2	3
Profit per Unit	\$5	\$4	\$3

The company has assigned you as the accounting member of its profit-planning committee to determine the numbers of types of *A*, *B*, and *C* lamps that it should produce in order to maximize its total profit from the sale of lamps. The following questions relate to a linear programming model that your group has developed:
   (a) The coefficients of the objective function would be
      (1) 4, 2, 3;
      (2) 2, 3, 1;
      (3) 5, 4, 3;
      (4) 400,600.
   (b) The constraints in the model would be
      (1) 2, 3, 1;
      (2) 5, 4, 3;
      (3) 4, 2, 3;
      (4) 400,600.

*Material from *Uniform CPA Examination Questions and Unofficial Answers*, copyright © by the American Institute of Certified Public Accountants, Inc., is reprinted with permission.

(c) The constraint imposed by the available number of person-hours in department I could be expressed as
(1) $4X_1 + 2X_2 + 3X_3 \leq 400$;
(2) $4X_1 + 2X_2 + 3X_3 \geq 400$;
(3) $2X_1 + 3X_2 + 1X_3 \leq 400$;
(4) $2X_1 + 3X_2 + 1X_3 \geq 400$.

18. The Golden Hawk Manufacturing Company wants to maximize the profits on products *A*, *B*, and *C*. The contribution margin for each product is as follows:

Product	**Contribution Margin**
A	$2
B	5
C	4

The production requirements and the departmental capacities are as follows:

	Production Requirements by Product (Hours)			**Departmental Capacity (Total Hours)**
Department	***A***	***B***	***C***	
Assembling	2	3	2	30,000
Painting	1	2	2	38,000
Finishing	2	3	1	28,000

(a) What is the profit-maximization formula for the Golden Hawk Company?
(1) $2A + $5B + $4C = *X* (where *X* = profit)
(2) 5A + 8B + 5C ≤ 96,000
(3) $2A + $5B + $4C ≤ *X*
(4) $2A + $5B + $4C = 96,000

(b) What is the constraint for the Painting Department of the Golden Hawk Company?
(1) 1A + 2B + 2C ≥ 38,000
(2) $2A + $5B + $4C ≥ 38,000
(3) 1A + 2B + 2C ≤ 38,000
(4) 2A + 3B + 2C ≤ 30,000

19. Solve the problem in Exercise 1.

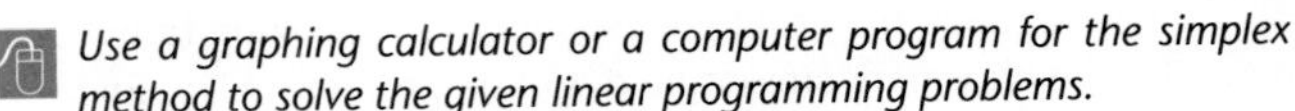

Use a graphing calculator or a computer program for the simplex method to solve the given linear programming problems.

20. Exercise 2. Your final answer should consist of whole numbers (because Banal can't sell half a painting).

21. Exercise 3

22. Exercise 4

✓ Checkpoint Answer

1. $z = 6{,}400{,}000$

7.6 ▶ The Simplex Method: Duality and Minimization

NOTE Sections 7.6 and 7.7 are independent of each other and may be read in either order.

Here, we present a method of solving *minimization* problems in which all constraints involve ≥ and all coefficients of the objective function are positive. When it applies, this method may be more efficient than the method discussed in Section 7.7. However, the method in Section 7.7 applies to a wider variety of problems—both minimization and maximization problems, even those that involve *mixed constraints* (≤, =, or ≥)—and it has no restrictions on the objective function.

We begin with a necessary tool from matrix algebra: if *A* is a matrix, then the **transpose** of *A* is the matrix obtained by interchanging the rows and columns of *A*.

Example 1 Find the transpose of each matrix.

(a) $A = \begin{bmatrix} 2 & -1 & 5 \\ 6 & 8 & 0 \\ -3 & 7 & -1 \end{bmatrix}$

Solution Write the rows of matrix A as the columns of the transpose:

$$\text{Transpose of } A = \begin{bmatrix} 2 & 6 & -3 \\ -1 & 8 & 7 \\ 5 & 0 & -1 \end{bmatrix}.$$

(b) $A = \begin{bmatrix} 1 & 2 & 4 & 0 \\ 2 & 1 & 7 & 6 \end{bmatrix}.$

Solution The transpose of $\begin{bmatrix} 1 & 2 & 4 & 0 \\ 2 & 1 & 7 & 6 \end{bmatrix}$ is $\begin{bmatrix} 1 & 2 \\ 2 & 1 \\ 4 & 7 \\ 0 & 6 \end{bmatrix}$. ✓1

✓Checkpoint 1

Give the transpose of each matrix.

(a) $\begin{bmatrix} 2 & 4 \\ 6 & 3 \\ 1 & 5 \end{bmatrix}$

(b) $\begin{bmatrix} 4 & 7 & 10 \\ 3 & 2 & 6 \\ 5 & 8 & 12 \end{bmatrix}$

Answers: See page 445.

TECHNOLOGY TIP Most graphing calculators can find the transpose of a matrix. Look for this feature in the MATRIX MATH menu (TI) or the OPTN MAT menu (Casio). The transpose of matrix A from Example 1(a) is shown in Figure 7.31.

```
            [[2  -1 5 ]
             [6   8 0 ]
             [-3 7  -1]]
[A]T
             [[2  6 -3]
              [-1 8 7 ]
              [5  0 -1]]
```

FIGURE 7.31

We now consider linear programming problems satisfying the following conditions:

1. The objective function is to be minimized.
2. All the coefficients of the objective function are nonnegative.
3. All constraints involve $\geq$.
4. All variables are nonnegative.

The method of solving minimization problems presented here is based on an interesting connection between maximization and minimization problems: Any solution of a maximizing problem produces the solution of an associated minimizing problem, and vice versa. Each of the associated problems is called the **dual** of the other. Thus, duals enable us to solve minimization problems of the type just described by the simplex method introduced in Section 7.4.

When dealing with minimization problems, we use y_1, y_2, y_3, etc., as variables and denote the objective function by w. The next two examples show how to construct the dual problem. Later examples will show how to solve both the dual problem and the original one.

Example 2 Construct the dual of this problem:

$$\begin{aligned} \text{Minimize} \quad & w = 8y_1 + 16y_2 \\ \text{subject to} \quad & y_1 + 5y_2 \geq 9 \\ & 2y_1 + 2y_2 \geq 10 \\ & y_1 \geq 0, y_2 \geq 0. \end{aligned}$$

Solution Write the augmented matrix of the system of inequalities *and* include the coefficients of the objective function (not their negatives) as the last row of the matrix:

Constants

Objective function →
$$\left[\begin{array}{cc|c} 1 & 5 & 9 \\ 2 & 2 & 10 \\ \hline 8 & 16 & 0 \end{array}\right].$$

Now form the transpose of the preceding matrix:

$$\left[\begin{array}{cc|c} 1 & 2 & 8 \\ 5 & 2 & 16 \\ \hline 9 & 10 & 0 \end{array}\right].$$

In this last matrix, think of the first two rows as constraints and the last row as the objective function. Then the dual maximization problem is as follows:

$$\begin{aligned} \text{Maximize} \quad & z = 9x_1 + 10x_2 \\ \text{subject to} \quad & x_1 + 2x_2 \le 8 \\ & 5x_1 + 2x_2 \le 16 \\ & x_1 \ge 0, x_2 \ge 0. \end{aligned}$$

Example 3 Write the duals of the given minimization linear programming problems.

(a) Minimize $w = 10y_1 + 8y_2$
subject to
$$\begin{aligned} y_1 + 2y_2 &\ge 2 \\ y_1 + y_2 &\ge 5 \\ y_1 \ge 0, y_2 &\ge 0. \end{aligned}$$

Solution Begin by writing the augmented matrix for the given problem:

$$\left[\begin{array}{cc|c} 1 & 2 & 2 \\ 1 & 1 & 5 \\ \hline 10 & 8 & 0 \end{array}\right].$$

Form the transpose of this matrix to get

$$\left[\begin{array}{cc|c} 1 & 1 & 10 \\ 2 & 1 & 8 \\ \hline 2 & 5 & 0 \end{array}\right].$$

The dual problem is stated from this second matrix as follows (using x instead of y):

$$\begin{aligned} \text{Maximize} \quad & z = 2x_1 + 5x_2 \\ \text{subject to} \quad & x_1 + x_2 \le 10 \\ & 2x_1 + x_2 \le 8 \\ & x_1 \ge 0, x_2 \ge 0. \end{aligned}$$

(b) Minimize $w = 7y_1 + 5y_2 + 8y_3$

subject to

$$\begin{aligned} 3y_1 + 2y_2 + y_3 &\geq 10 \\ y_1 + y_2 + y_3 &\geq 8 \\ 4y_1 + 5y_2 \quad &\geq 25 \\ y_1 \geq 0, y_2 \geq 0, y_3 &\geq 0. \end{aligned}$$

Solution Find the augmented matrix for this problem, as in part(a). Form the dual matrix, which represents the following problem:

Maximize $z = 10x_1 + 8x_2 + 25x_3$

subject to

$$\begin{aligned} 3x_1 + x_2 + 4x_3 &\leq 7 \\ 2x_1 + x_2 + 5x_3 &\leq 5 \\ x_1 + x_2 \quad &\leq 8 \\ x_1 \geq 0, x_2 \geq 0, x_3 &\geq 0. \end{aligned}$$ ✓2

✓Checkpoint 2

Write the dual of the following linear programming problem:

Minimize

$w = 2y_1 + 5y_2 + 6y_3$

subject to

$$\begin{aligned} 2y_1 + 3y_2 + y_3 &\geq 15 \\ y_1 + y_2 + 2y_3 &\geq 12 \\ 5y_1 + 3y_2 \quad &\geq 10 \\ y_1 \geq 0, y_2 \geq 0, y_3 &\geq 0. \end{aligned}$$

Answer: See page 445.

In Example 3, all the constraints of the minimization problems were ≥ inequalities, while all those in the dual maximization problems were ≤ inequalities. This is generally the case; inequalities are reversed when the dual problem is stated.

The following table shows the close connection between a problem and its dual.

Given Problem	Dual Problem
m variables	n variables
n constraints	m constraints (m slack variables)
Coefficients from objective function	Constraint constants
Constraint constants	Coefficients from objective function

Now that you know how to construct the dual problem, we examine how it is related to the original problem and how both may be solved.

Example 4 Solve this problem and its dual:

Minimize $w = 8y_1 + 16y_2$

subject to

$$\begin{aligned} y_1 + 5y_2 &\geq 9 \\ 2y_1 + 2y_2 &\geq 10 \\ y_1 \geq 0, y_2 &\geq 0. \end{aligned}$$

Solution In Example 2, we saw that the dual problem is

Maximize $z = 9x_1 + 10x_2$

subject to

$$\begin{aligned} x_1 + 2x_2 &\leq 8 \\ 5x_1 + 2x_2 &\leq 16 \\ x_1 \geq 0, x_2 &\geq 0. \end{aligned}$$

✓Checkpoint 3

Use the corner points in Figure 7.32(a) to find the minimum value of $w = 8y_1 + 16y_2$ and where it occurs.

Answer: See page 445.

In this case, both the original problem and the dual may be solved geometrically, as in Section 7.2. Figure 7.32(a) on the next page shows the region of feasible solutions for the original minimization problem, and Checkpoint 3 shows that

the minimum value of w is 48 at the vertex (4, 1). ✓3

✓Checkpoint 4

Use Figure 7.32(b) to find the maximum value of $z = 9x_1 + 10x_2$ and where it occurs.

Answer: See page 445.

Figure 7.32(b) shows the region of feasible solutions for the dual maximization problem, and Checkpoint 4 shows that

the maximum value of z is 48 at the vertex (2, 3). ✓4

Even though the regions and the corner points are different, the minimization problem and its dual have the same solution, 48.

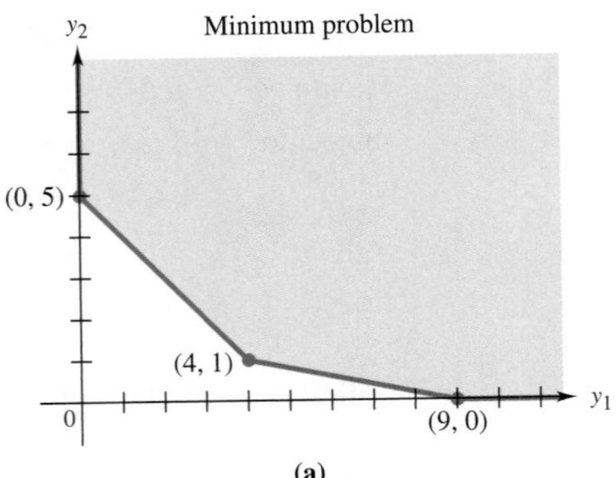

(a)

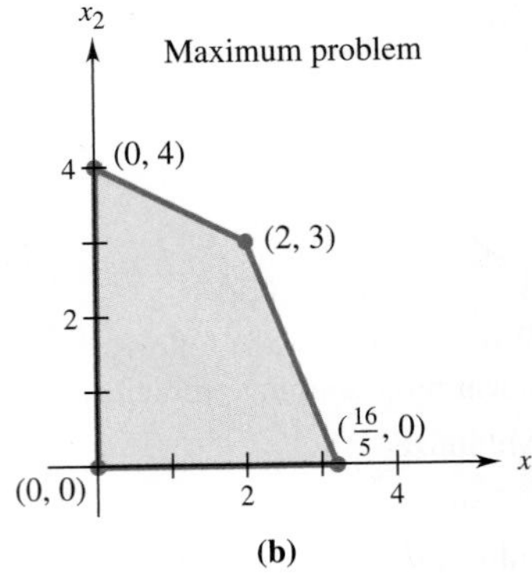

(b)

FIGURE 7.32

The next theorem, whose proof requires advanced methods, guarantees that what happened in Example 4 happens in the general case as well.

Theorem of Duality

The objective function w of a minimizing linear programming problem takes on a minimum value if, and only if, the objective function z of the corresponding dual maximizing problem takes on a maximum value. The maximum value of z equals the minimum value of w.

Geometric solution methods were used in Example 4, but the simplex method can also be used. In fact, the final simplex tableau shows the solutions for both the original minimization problem and the dual maximization problem, as illustrated in the next example.

Example 5 Use the simplex method to solve the minimization problem in Example 4.

Solution First, set up the dual problem, as in Example 4:

$$\begin{aligned} \text{Maximize} \quad & z = 9x_1 + 10x_2 \\ \text{subject to} \quad & x_1 + 2x_2 \le 8 \\ & 5x_1 + 2x_2 \le 16 \\ & x_1 \ge 0, x_2 \ge 0. \end{aligned}$$

This is a maximization problem in standard form, so it can be solved by the simplex method. Use slack variables to write the inequalities and the objective function as equations:

$$\begin{aligned} x_1 + 2x_2 + s_1 \quad\quad &= 8 \\ 5x_1 + 2x_2 \quad + s_2 \quad &= 16 \\ -9x_1 - 10x_2 \quad\quad + z &= 0, \end{aligned}$$

with $x_1 \geq 0$, $x_2 \geq 0$, $s_1 \geq 0$, and $s_2 \geq 0$. So the initial tableau is

$$\begin{array}{c} \begin{matrix} x_1 & x_2 & s_1 & s_2 & \end{matrix} \\ \left[\begin{array}{cccc|c} 1 & \mathbf{2} & 1 & 0 & 8 \\ 5 & 2 & 0 & 1 & 16 \\ \hline -9 & -10 & 0 & 0 & 0 \end{array}\right]. \end{array} \quad \begin{array}{l} \text{Quotients} \\ 8/2 = 4 \\ 16/2 = 8 \end{array}$$

The quotients show that the first pivot is the 2 shaded in blue. Pivoting is done as follows:

$$\begin{array}{c} \begin{matrix} x_1 & x_2 & s_1 & s_2 & \end{matrix} \\ \left[\begin{array}{cccc|c} \frac{1}{2} & 1 & \frac{1}{2} & 0 & 4 \\ 4 & 0 & -1 & 1 & 8 \\ \hline -4 & 0 & 5 & 0 & 40 \end{array}\right]. \end{array} \quad \begin{array}{l} \frac{1}{2}R_1 \\ -2R_1 + R_2 \\ 10R_1 + R_3 \end{array}$$

✓Checkpoint 5

In the second tableau, find the next pivot.

Answer: See page 445.

Checkpoint 5 shows that the new pivot is the 4 in row 2, column 1. ✓5 Pivoting leads to the final simplex tableau:

$$\begin{array}{c} \begin{matrix} x_1 & x_2 & s_1 & s_2 & \end{matrix} \\ \left[\begin{array}{cccc|c} 0 & 1 & \frac{5}{8} & -\frac{1}{8} & 3 \\ 1 & 0 & -\frac{1}{4} & \frac{1}{4} & 2 \\ \hline 0 & 0 & 4 & 1 & 48 \end{array}\right]. \end{array} \quad \begin{array}{l} -\frac{1}{2}R_2 + R_1 \\ \frac{1}{4}R_2 \\ 4R_2 + R_3 \end{array}$$

The final simplex tableau shows that the maximum value of 48 occurs when $x_1 = 2$ and $x_2 = 3$. In Example 4, we saw that the minimum value of 48 occurs when $y_1 = 4$ and $y_2 = 1$. Note that this information appears in the last row (shown in blue). The minimum value of 48 is in the lower right-hand corner, and the values where this occurs ($y_1 = 4$ and $y_2 = 1$) are in the last row at the bottom of the slack-variable columns.

A minimization problem that meets the conditions listed after Example 1 can be solved by the method used in Example 5 and summarized here.

Solving Minimization Problems with Duals

1. Find the dual standard maximization problem.*
2. Use the simplex method to solve the dual maximization problem.
3. Read the optimal solution of the original minimization problem from the final simplex tableau:

(Continued)

*The coefficients of the objective function in the minimization problem are the constants on the right side of the constraints in the dual maximization problem. So when all these coefficients are nonnegative (condition 2), the dual problem is in standard maximum form.

y_1 is the last entry in the column corresponding to the first slack variable;
y_2 is the last entry in the column corresponding to the second slack variable; and so on.

These values of y_1, y_2, y_3, etc., produce the minimum value of w, which is the entry in the lower right-hand corner of the tableau.

Example 6

$$\begin{aligned} \text{Minimize} \quad & w = 3y_1 + 2y_2 \\ \text{subject to} \quad & y_1 + 3y_2 \geq 6 \\ & 2y_1 + y_2 \geq 3 \\ & y_1 \geq 0, y_2 \geq 0. \end{aligned}$$

Solution Use the given information to write the matrix:

$$\left[\begin{array}{cc|c} 1 & 3 & 6 \\ 2 & 1 & 3 \\ \hline 3 & 2 & 0 \end{array}\right].$$

Transpose to get the following matrix for the dual problem:

$$\left[\begin{array}{cc|c} 1 & 2 & 3 \\ 3 & 1 & 2 \\ \hline 6 & 3 & 0 \end{array}\right].$$

Write the dual problem from this matrix as follows:

$$\begin{aligned} \text{Maximize} \quad & z = 6x_1 + 3x_2 \\ \text{subject to} \quad & x_1 + 2x_2 \leq 3 \\ & 3x_1 + x_2 \leq 2 \\ & x_1 \geq 0, x_2 \geq 0. \end{aligned}$$

Solve this standard maximization problem by the simplex method. Start by introducing slack variables, giving the system

$$\begin{aligned} x_1 + 2x_2 + s_1 \qquad\qquad &= 3 \\ 3x_1 + x_2 \qquad + s_2 \qquad &= 2 \\ -6x_1 - 3x_2 - 0s_1 - 0s_2 + z &= 0, \end{aligned}$$

with $x_1 \geq 0$, $x_2 \geq 0$, $s_1 \geq 0$, and $s_2 \geq 0$.

The initial tableau for this system is

$$\begin{array}{c} \begin{array}{cccc} x_1 & x_2 & s_1 & s_2 \end{array} \\ \left[\begin{array}{cccc|c} 1 & 2 & 1 & 0 & 3 \\ \boxed{3} & 1 & 0 & 1 & 2 \\ \hline -6 & -3 & 0 & 0 & 0 \end{array}\right], \end{array} \qquad \begin{array}{l} \text{Quotients} \\ 3/1 = 3 \\ 2/3 \end{array}$$

with the pivot as indicated. Two rounds of pivoting produce the following final tableau:

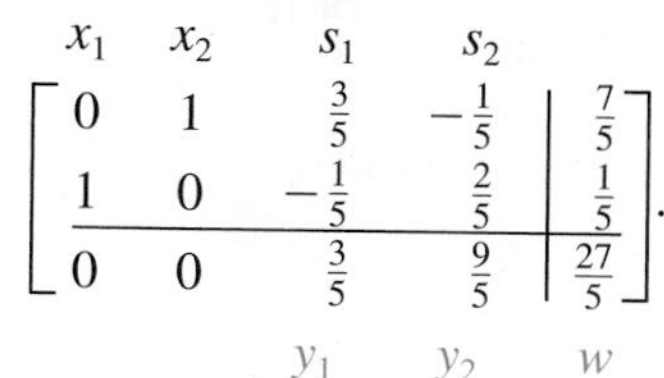

$$\begin{array}{c} \\ \\ \\ \\ \end{array}\begin{matrix} x_1 & x_2 & s_1 & s_2 & \\ \end{matrix}$$

$$\left[\begin{array}{cccc|c} 0 & 1 & \frac{3}{5} & -\frac{1}{5} & \frac{7}{5} \\ 1 & 0 & -\frac{1}{5} & \frac{2}{5} & \frac{1}{5} \\ \hline 0 & 0 & \frac{3}{5} & \frac{9}{5} & \frac{27}{5} \end{array}\right].$$

$$\begin{matrix} & & y_1 & y_2 & w \end{matrix}$$

As indicated in blue below the final tableau, the last entries in the columns corresponding to the slack variables (s_1 and s_2) give the values of the original variables y_1 and y_2 that produce the minimal value of w. This minimal value of w appears in the lower right-hand corner (and is the same as the maximal value of z in the dual problem). So the solution of the given minimization problem is as follows:

The minimum value of $w = 3y_1 + 2y_2$, subject to the given constraints, is $\frac{27}{5}$ and occurs when $y_1 = \frac{3}{5}$ and $y_2 = \frac{9}{5}$. 6 ✓

✓Checkpoint 6

Minimize $w = 10y_1 + 8y_2$
subject to

$$y_1 + 2y_2 \geq 2$$
$$y_1 + y_2 \geq 5$$
$$y_1 \geq 0, y_2 \geq 0.$$

Answer: See page 445.

Example 7 A minimization problem in three variables was solved by the use of duals. The final simplex tableau for the dual maximization problem is shown here:

$$\begin{matrix} x_1 & x_2 & x_3 & s_1 & s_2 & s_3 & \end{matrix}$$

$$\left[\begin{array}{cccccc|c} 3 & 1 & 1 & 0 & 9 & 0 & 1 \\ 13 & -1 & 0 & 1 & -2 & 0 & 10 \\ 9 & 10 & 0 & 0 & 7 & 1 & 7 \\ \hline 5 & 1 & 0 & 4 & 1 & 7 & 28 \end{array}\right].$$

(a) What is the optimal solution of the dual minimization problem?

Solution Looking at the bottom of the columns corresponding to the slack variables s_1, s_2, and s_3, we see that the solution of the minimization problem is

$y_1 = 4$, $\quad y_2 = 1$, $\quad$ and $\quad y_3 = 7$, $\quad$ with a minimal value of $w = 28$.

(b) What is the optimal solution of the dual maximization problem?

Solution Since s_1, s_2, and s_3 are slack variables by part (a), the variables in the dual problem are x_1, x_2, and x_3. Read the solution from the final tableau, as in Sections 7.4 and 7.5:

$x_1 = 0$, $\quad x_2 = 0$, $\quad$ and $\quad x_3 = 1$, $\quad$ with a maximal value of $z = 28$.

FURTHER USES OF THE DUAL

The dual is useful not only in solving minimization problems, but also in seeing how small changes in one variable will affect the value of the objective function. For example, suppose an animal breeder needs at least 6 units per day of nutrient *A* and at least 3 units of nutrient *B* and that the breeder can choose between two different feeds: feed 1 and feed 2. Find the minimum cost for the breeder if each bag of feed 1 costs \$3 and provides 1 unit of nutrient *A* and 2 units of *B*, while each bag of feed 2 costs \$2 and provides 3 units of nutrient *A* and 1 of *B*.

If y_1 represents the number of bags of feed 1 and y_2 represents the number of bags of feed 2, the given information leads to the following minimization problem:

$$\begin{aligned} \text{Minimize} \quad & w = 3y_1 + 2y_2 \\ \text{subject to} \quad & y_1 + 3y_2 \geq 6 \\ & 2y_1 + y_2 \geq 3 \\ & y_1 \geq 0, y_2 \geq 0. \end{aligned}$$

This minimization linear programming problem is the one we solved in Example 6 of this section. In that example, we formed the dual and reached the following final tableau:

$$\begin{array}{c} \begin{array}{cccc} x_1 & x_2 & s_1 & s_2 \end{array} \\ \left[\begin{array}{cccc|c} 0 & 1 & \frac{3}{5} & -\frac{1}{5} & \frac{7}{5} \\ 1 & 0 & -\frac{1}{5} & \frac{2}{5} & \frac{1}{5} \\ \hline 0 & 0 & \frac{3}{5} & \frac{9}{5} & \frac{27}{5} \end{array}\right]. \end{array}$$

This final tableau shows that the breeder will obtain minimum feed costs by using $\frac{3}{5}$ bag of feed 1 and $\frac{9}{5}$ bags of feed 2 per day, for a daily cost of $\frac{27}{5} = 5.40$ dollars.

Now look at the data from the feed problem shown in the following table:

	Units of Nutrient (Per Bag)		Cost per Bag
	A	*B*	
Feed 1	1	2	\$3
Feed 2	3	1	\$2
Minimum Nutrient Needed	6	3	

If x_1 and x_2 are the cost *per unit* of nutrients *A* and *B*, the constraints of the dual problem can be stated as follows (see page 440–441):

$$\begin{aligned} \text{Cost of feed 1:} \quad & x_1 + 2x_2 \leq 3 \\ \text{Cost of feed 2:} \quad & 3x_1 + x_2 \leq 2. \end{aligned}$$

The solution of the dual problem, which maximizes nutrients, also can be read from the final tableau above:

$$x_1 = \frac{1}{5} = .20 \quad \text{and} \quad x_2 = \frac{7}{5} = 1.40.$$

This means that a unit of nutrient *A* costs $\frac{1}{5}$ of a dollar = \$.20, while a unit of nutrient B costs $\frac{7}{5}$ dollars = \$1.40. The minimum daily cost, \$5.40, is found as follows.

$$\begin{aligned} (\$.20 \text{ per unit of } A) \times (6 \text{ units of } A) &= \$1.20 \\ +\ (\$1.40 \text{ per unit of } B) \times (3 \text{ units of } B) &= \$4.20 \\ \hline \text{Minimum daily cost} &= \$5.40. \end{aligned}$$

The numbers .20 and 1.40 are called the **shadow costs** of the nutrients. These two numbers from the dual, \$.20 and \$1.40, also allow the breeder to estimate feed costs for "small" changes in nutrient requirements. For example, an increase of 1 unit in the requirement for each nutrient would produce a total cost as follows:

\$5.40	6 units of *A*, 3 of *B*
.20	1 extra unit of *A*
1.40	1 extra unit of *B*
\$7.00.	Total cost per day

7.6 ▸ Exercises

Find the transpose of each matrix. (See Example 1.)

1. $\begin{bmatrix} 3 & -4 & 5 \\ 1 & 10 & 7 \\ 0 & 3 & 6 \end{bmatrix}$

2. $\begin{bmatrix} 3 & -5 & 9 & 4 \\ 1 & 6 & -7 & 0 \\ 4 & 18 & 11 & 9 \end{bmatrix}$

3. $\begin{bmatrix} 3 & 0 & 14 & -5 & 3 \\ 4 & 17 & 8 & -6 & 1 \end{bmatrix}$

4. $\begin{bmatrix} 15 & -6 & -2 \\ 13 & -1 & 11 \\ 10 & 12 & -3 \\ 24 & 1 & 0 \end{bmatrix}$

State the dual problem for each of the given problems, but do not solve it. (See Examples 2 and 3.)

5. Minimize $w = 3y_1 + 5y_2$
subject to
$3y_1 + y_2 \geq 4$
$-y_1 + 2y_2 \geq 6$
$y_1 \geq 0, y_2 \geq 0.$

6. Minimize $w = 4y_1 + 7y_2$
subject to
$y_1 + y_2 \geq 17$
$3y_1 + 6y_2 \geq 21$
$2y_1 + 4y_2 \geq 19$
$y_1 \geq 0, y_2 \geq 0.$

7. Minimize $w = 2y_1 + 8y_2$
subject to
$y_1 + 7y_2 \geq 18$
$4y_1 + y_2 \geq 15$
$5y_1 + 3y_2 \geq 20$
$y_1 \geq 0, y_2 \geq 0.$

8. Minimize $w = y_1 + 2y_2 + 6y_3$
subject to
$3y_1 + 4y_2 + 6y_3 \geq 8$
$y_1 + 5y_2 + 2y_3 \geq 12$
$y_1 \geq 0, y_2 \geq 0, y_3 \geq 0.$

9. Minimize $w = 5y_1 + y_2 + 3y_3$
subject to
$7y_1 + 6y_2 + 8y_3 \geq 18$
$4y_1 + 5y_2 + 10y_3 \geq 20$
$y_1 \geq 0, y_2 \geq 0, y_3 \geq 0.$

10. Minimize $w = 4y_1 + 3y_2 + y_3$
subject to
$y_1 + 2y_2 + 3y_3 \geq 115$
$2y_1 + y_2 + 8y_3 \geq 200$
$y_1 - y_3 \geq 50$
$y_1 \geq 0, y_2 \geq 0, y_3 \geq 0.$

11. Minimize $w = 8y_1 + 9y_2 + 3y_3$
subject to
$y_1 + y_2 + y_3 \geq 5$
$y_1 + y_2 \geq 4$
$2y_1 + y_2 + 3y_3 \geq 15$
$y_1 \geq 0, y_2 \geq 0, y_3 \geq 0.$

12. Minimize $w = y_1 + 2y_2 + y_3 + 5y_4$
subject to
$y_1 + y_2 + y_3 + y_4 \geq 50$
$3y_1 + y_2 + 2y_3 + y_4 \geq 100$
$y_1 \geq 0, y_2 \geq 0, y_3 \geq 0, y_4 \geq 0.$

Use duality to solve the problem that was set up in the given exercise.

13. Exercise 9

14. Exercise 8

15. Exercise 11

16. Exercise 12

Use duality to solve the given problems. (See Examples 5 and 6.)

17. Minimize $w = 2y_1 + y_2 + 3y_3$
subject to
$y_1 + y_2 + y_3 \geq 100$
$2y_1 + y_2 \geq 50$
$y_1 \geq 0, y_2 \geq 0, y_3 \geq 0.$

18. Minimize $w = 2y_1 + 4y_2$
subject to
$4y_1 + 2y_2 \geq 10$
$4y_1 + y_2 \geq 8$
$2y_1 + y_2 \geq 12$
$y_1 \geq 0, y_2 \geq 0.$

19. Minimize $w = 3y_1 + y_2 + 4y_3$
subject to
$2y_1 + y_2 + y_3 \geq 6$
$y_1 + 2y_2 + y_3 \geq 8$
$2y_1 + y_2 + 2y_3 \geq 12$
$y_1 \geq 0, y_2 \geq 0, y_3 \geq 0.$

20. Minimize $w = y_1 + y_2 + 3y_3$
subject to
$$2y_1 + 6y_2 + y_3 \geq 8$$
$$y_1 + 2y_2 + 4y_3 \geq 12$$
$$y_1 \geq 0, y_2 \geq 0, y_3 \geq 0.$$

21. Minimize $w = 6y_1 + 4y_2 + 2y_3$
subject to
$$2y_1 + 2y_2 + y_3 \geq 2$$
$$y_1 + 3y_2 + 2y_3 \geq 3$$
$$y_1 + y_2 + 2y_3 \geq 4$$
$$y_1 \geq 0, y_2 \geq 0, y_3 \geq 0.$$

22. Minimize $w = 12y_1 + 10y_2 + 7y_3$
subject to
$$2y_1 + y_2 + y_3 \geq 7$$
$$y_1 + 2y_2 + y_3 \geq 4$$
$$y_1 \geq 0, y_2 \geq 0, y_3 \geq 0.$$

23. Minimize $w = 20y_1 + 12y_2 + 40y_3$
subject to
$$y_1 + y_2 + 5y_3 \geq 20$$
$$2y_1 + y_2 + y_3 \geq 30$$
$$y_1 \geq 0, y_2 \geq 0, y_3 \geq 0.$$

24. Minimize $w = 4y_1 + 5y_2$
subject to
$$10y_1 + 5y_2 \geq 100$$
$$20y_1 + 10y_2 \geq 150$$
$$y_1 \geq 0, y_2 \geq 0.$$

25. Minimize $w = 4y_1 + 2y_2 + y_3$
subject to
$$y_1 + y_2 + y_3 \geq 4$$
$$3y_1 + y_2 + 3y_3 \geq 6$$
$$y_1 + y_2 + 3y_3 \geq 5$$
$$y_1 \geq 0, y_2 \geq 0, y_3 \geq 0.$$

26. Minimize $w = 3y_1 + 2y_2$
subject to
$$2y_1 + 3y_2 \geq 60$$
$$y_1 + 4y_2 \geq 40$$
$$y_1 \geq 0, y_2 \geq 0.$$

27. **Health** Glenn Russell, who is dieting, requires two food supplements: I and II. He can get these supplements from two different products—*A* and *B*—as shown in the following table:

		Supplement (Grams per Serving)	
		I	II
Product	*A*	4	2
	B	2	5

Glenn's physician has recommended that he include at least 20 grams of supplement I and 18 grams of supplement II in his diet. If product *A* costs 24¢ per serving and product *B* costs 40¢ per serving, how can he satisfy these requirements most economically?

28. **Business** An animal food must provide at least 54 units of vitamins and 60 calories per serving. One gram of soybean meal provides at least 2.5 units of vitamins and 5 calories. One gram of meat by-products provides at least 4.5 units of vitamins and 3 calories. One gram of grain provides at least 5 units of vitamins and 10 calories. If a gram of soybean meal costs 8¢, a gram of meat by-products 9¢, and a gram of grain 10¢, what mixture of these three ingredients will provide the required vitamins and calories at minimum cost?

29. **Business** A brewery produces regular beer and a lower-carbohydrate "light" beer. Steady customers of the brewery buy 12 units of regular beer and 10 units of light beer monthly. While setting up the brewery to produce the beers, the management decides to produce extra beer, beyond the need to satisfy the steady customers. The cost per unit of regular beer is \$36,000, and the cost per unit of light beer is \$48,000. Every unit of regular beer brings in \$100,000 in revenue, while every unit of light beer brings in \$300,000. The brewery wants at least \$7,000,000 in revenue. At least 20 additional units of beer can be sold. How much of each type of beer should be made so as to minimize total production costs?

30. **Business** Joan McKee has a part-time job conducting public-opinion interviews. She has found that a political interview takes 45 minutes and a market interview takes 55 minutes. To allow more time for her full-time job, she needs to minimize the time she spends doing interviews. Unfortunately, to keep her part-time job, she must complete at least 8 interviews each week. Also, she must earn at least \$60 per week at this job, at which she earns \$8 for each political interview and \$10 for each market interview. Finally, to stay in good standing with her supervisor, she must earn at least 40 bonus points per week; she receives 6 bonus points for each political interview and 5 points for each market interview. How many of each interview should she do each week to minimize the time spent?

31. You are given the following linear programming problem (P):*

Maximize $z = x_1 + 2x_2$
subject to
$$-2x_1 + x_2 \geq 1$$
$$x_1 - 2x_2 \geq 1$$
$$x_1 \geq 0, x_2 \geq 0.$$

The dual of (P) is (D). Which of the following statements is true?

(a) (P) has no feasible solution and the objective function of (D) is unbounded.

(b) (D) has no feasible solution and the objective function of (P) is unbounded.

(c) The objective functions of both (P) and (D) are unbounded.

---

*Problem 2 from "November 1989 course 130 Examination Operations Research" of the *Education and Examination Committee of the Society of Actuaries*. Reprinted by permission of the Society of Actuaries.

(d) Both (P) and (D) have optimal solutions.
(e) Neither (P) nor (D) has a feasible solution.

32. **Business** Refer to the end of this section in the text, on minimizing the daily cost of feeds.
(a) Find a combination of feeds that will cost \$7.00 and give 7 units of A and 4 units of B.
(b) Use the dual variables to predict the daily cost of feed if the requirements change to 5 units of A and 4 units of B. Find a combination of feeds to meet these requirements at the predicted price.

33. **Business** A small toy-manufacturing firm has 200 squares of felt, 600 ounces of stuffing, and 90 feet of trim available to make two types of toys: a small bear and a monkey. The bear requires 1 square of felt and 4 ounces of stuffing. The monkey requires 2 squares of felt, 3 ounces of stuffing, and 1 foot of trim. The firm makes \$1 profit on each bear and \$1.50 profit on each monkey. The linear programming problem to maximize profit is

$$\begin{aligned} \text{Maximize} \quad & z = x_1 + 1.5x_2 \\ \text{subject to} \quad & x_1 + 2x_2 \le 200 \\ & 4x_1 + 3x_2 \le 600 \\ & x_2 \le 90 \\ & x_1 \ge 0, x_2 \ge 0. \end{aligned}$$

The final simplex tableau is

$$\left[\begin{array}{ccccc|c} 1 & 0 & -.6 & .4 & 0 & 120 \\ 0 & 0 & -.8 & .2 & 1 & 50 \\ 0 & 1 & .8 & -.2 & 0 & 40 \\ \hline 0 & 0 & .6 & .1 & 0 & 180 \end{array}\right].$$

(a) What is the corresponding dual problem?
(b) What is the optimal solution to the dual problem?

(c) Use the shadow values to estimate the profit the firm will make if its supply of felt increases to 210 squares.
(d) How much profit will the firm make if its supply of stuffing is cut to 590 ounces and its supply of trim is cut to 80 feet?

34. Refer to Example 1 in Section 7.5.
(a) Give the dual problem.
(b) Use the shadow values to estimate the farmer's profit if land is cut to 90 acres, but capital increases to \$21,000.
(c) Suppose the farmer has 110 acres, but only \$19,000. Find the optimal profit and the planting strategy that will produce this profit.

✓ Checkpoint Answers

1. **(a)** $\begin{bmatrix} 2 & 6 & 1 \\ 4 & 3 & 5 \end{bmatrix}$ **(b)** $\begin{bmatrix} 4 & 3 & 5 \\ 7 & 2 & 8 \\ 10 & 6 & 12 \end{bmatrix}$

2. $$\begin{aligned} \text{Maximize} \quad & z = 15x_1 + 12x_2 + 10x_3 \\ \text{subject to} \quad & 2x_1 + x_2 + 5x_3 \le 2 \\ & 3x_1 + x_2 + 3x_3 \le 5 \\ & x_1 + 2x_2 \le 6 \\ & x_1 \ge 0,\ x_2 \ge 0,\ x_3 \ge 0. \end{aligned}$$

3. 48 when $y_1 = 4$ and $y_2 = 1$
4. 48 when $x_1 = 2$ and $x_2 = 3$
5. The 4 in row 2, column 1
6. $y_1 = 0$ and $y_2 = 5$, for a minimum of 40

7.7 ▸ The Simplex Method: Nonstandard Problems

NOTE Section 7.7 is independent of Section 7.6 and may be read first, if desired.

So far, the simplex method has been used to solve problems in which the variables are nonnegative and all the other constraints are of one type (either all $\le$ or all $\ge$). Now we extend the simplex method to linear programming problems with nonnegative variables and mixed constraints ($\le$, $=$, and $\ge$).

The solution method to be used here requires that all inequality constraints be written so that the constant on the right side is nonnegative. For instance, the inequality

$$4x_1 + 5x_2 - 12x_3 \le -30$$

can be replaced by the equivalent one obtained by multiplying both sides by -1 and reversing the direction of the inequality sign:

$$-4x_1 - 5x_2 + 12x_3 \ge 30.$$

MAXIMIZATION WITH ≤ AND ≥ CONSTRAINTS

As is always the case when the simplex method is involved, each inequality constraint must be written as an equation. Constraints involving ≤ are converted to equations by adding a nonnegative slack variable, as in Section 7.4. Similarly, constraints involving ≥ are converted to equations by *subtracting* a nonnegative **surplus variable**. For example, the inequality $2x_1 - x_2 + 5x_3 \geq 12$ is written as

$$2x_1 - x_2 + 5x_3 - s_1 = 12,$$

where $s_1 \geq 0$. The surplus variable s_1 represents the amount by which $2x_1 - x_2 + 5x_3$ exceeds 12.

Example 1 Restate the following problem in terms of equations, and write its initial simplex tableau:

$$\begin{aligned} \text{Maximize} \quad & z = 4x_1 + 10x_2 + 6x_3 \\ \text{subject to} \quad & x_1 + 4x_2 + 4x_3 \geq 8 \\ & x_1 + 3x_2 + 2x_3 \leq 6 \\ & 3x_1 + 4x_2 + 8x_3 \leq 22 \\ & x_1 \geq 0, x_2 \geq 0, x_3 \geq 0. \end{aligned}$$

Solution In order to write the constraints as equations, subtract a surplus variable from the ≥ constraint and add a slack variable to each ≤ constraint. So the problem becomes

$$\begin{aligned} \text{Maximize} \quad & z = 4x_1 + 10x_2 + 6x_3 \\ \text{subject to} \quad & x_1 + 4x_2 + 4x_3 - s_1 = 8 \\ & x_1 + 3x_2 + 2x_3 + s_2 = 6 \\ & 3x_1 + 4x_2 + 8x_3 + s_3 = 22 \\ & x_1 \geq 0, x_2 \geq 0, x_3 \geq 0, s_1 \geq 0, s_2 \geq 0, s_3 \geq 0. \end{aligned}$$

Write the objective function as $-4x_1 - 10x_2 - 6x_3 + z = 0$ and use the coefficients of the four equations to write the initial simplex tableau (omitting the z column):

$$\begin{array}{cccccc|c} x_1 & x_2 & x_3 & s_1 & s_2 & s_3 & \\ 1 & 4 & 4 & -1 & 0 & 0 & 8 \\ 1 & 3 & 2 & 0 & 1 & 0 & 6 \\ \hline 3 & 4 & 8 & 0 & 0 & 1 & 22 \\ -4 & -10 & -6 & 0 & 0 & 0 & 0 \end{array}. \checkmark 1$$

✓Checkpoint 1

(a) Restate this problem in terms of equations:

$$\begin{aligned} \text{Maximize} \quad & z = 3x_1 - 2x_2 \\ \text{subject to} \quad & 2x_1 + 3x_2 \leq 8 \\ & 6x_1 - 2x_2 \geq 3 \\ & x_1 + 4x_2 \geq 1 \\ & x_1 \geq 0, x_2 \geq 0. \end{aligned}$$

(b) Write the initial simplex tableau.

Answers: See page 458.

The tableau in Example 1 resembles those which have appeared previously, and similar terminology is used. The variables whose columns have one entry that is ±1 and the rest that are 0 will be called **basic variables**; the other variables are nonbasic. A solution obtained by setting the nonbasic variables equal to 0 and solving for the basic variables (by looking at the constants in the right-hand column) will be called a **basic solution**. A basic solution that is feasible is called a **basic feasible**

solution. In the tableau of Example 1, for instance, the basic variables are s_1, s_2, and s_3, and the basic solution is

$$x_1 = 0, \quad x_2 = 0, \quad x_3 = 0, \quad s_1 = -8, \quad s_2 = 6, \quad \text{and} \quad s_3 = 22.$$

However, because one variable is negative, this solution is not feasible. 2✓

✓Checkpoint 2

State the basic solution given by each tableau. Is it feasible?

(a)

$$\begin{array}{c} \begin{matrix} x_1 & x_2 & s_1 & s_2 & s_3 & \end{matrix} \\ \left[\begin{array}{rrrrr|r} 3 & -5 & 1 & 0 & 0 & 12 \\ 4 & 7 & 0 & 1 & 0 & 6 \\ 1 & 3 & 0 & 0 & -1 & 5 \\ \hline -7 & 4 & 0 & 0 & 0 & 0 \end{array}\right] \end{array}$$

(b)

$$\begin{array}{c} \begin{matrix} x_1 & x_2 & x_3 & s_1 & s_2 & \end{matrix} \\ \left[\begin{array}{rrrrr|r} 9 & 8 & -1 & 1 & 0 & 12 \\ -5 & 3 & 0 & 0 & 1 & 7 \\ \hline 4 & 2 & 3 & 0 & 0 & 0 \end{array}\right] \end{array}$$

Answers: See page 458.

The solution method for problems such as the one in Example 1 consists of two stages. **Stage I** consists of finding a basic *feasible* solution that can be used as the starting point for the simplex method. (This stage is unnecessary in a standard maximization problem, because the solution given by the initial tableau is always feasible.) There are many systematic ways of finding a feasible solution, all of which depend on the fact that row operations produce a tableau that represents a system with the same solutions as the original one. One such technique is explained in the next example. Since the immediate goal is to find a feasible solution, not necessarily an optimal one, the procedures for choosing pivots differ from those in the ordinary simplex method.

Example 2 Find a basic feasible solution for the problem in Example 1, whose initial tableau is

$$\begin{array}{c} \begin{matrix} x_1 & x_2 & x_3 & s_1 & s_2 & s_3 & \end{matrix} \\ \left[\begin{array}{rrrrrr|r} 1 & 4 & 4 & -1 & 0 & 0 & 8 \\ 1 & 3 & 2 & 0 & 1 & 0 & 6 \\ 3 & 4 & 8 & 0 & 0 & 1 & 22 \\ \hline -4 & -10 & -6 & 0 & 0 & 0 & 0 \end{array}\right] \end{array}.$$

Solution In the basic solution given by this tableau, s_1 has a negative value. The only nonzero entry in its column is the -1 in row 1. Choose any *positive* entry in row 1 except the entry on the far right. The column that the chosen entry is in will be the pivot column. We choose the first positive entry in row 1: the 1 in column 1. The pivot row is determined in the usual way by considering quotients (the constant at the right end of the row, divided by the positive entry in the pivot column) in each row except the objective row:

$$\frac{8}{1} = 8, \qquad \frac{6}{1} = 6, \qquad \frac{22}{3} = 7\frac{1}{3}.$$

The smallest quotient is 6, so the pivot is the 1 in row 2, column 1. Pivoting in the usual way leads to the tableau

$$\begin{array}{c} \begin{matrix} x_1 & x_2 & x_3 & s_1 & s_2 & s_3 & \end{matrix} \\ \left[\begin{array}{rrrrrr|r} 0 & 1 & 2 & -1 & -1 & 0 & 2 \\ 1 & 3 & 2 & 0 & 1 & 0 & 6 \\ 0 & -5 & 2 & 0 & -3 & 1 & 4 \\ \hline 0 & 2 & 2 & 0 & 4 & 0 & 24 \end{array}\right] \end{array} \begin{array}{l} -R_2 + R_1 \\ \\ -3R_2 + R_3 \\ 4R_2 + R_4 \end{array}$$

and the basic solution

$$x_1 = 6, \quad x_2 = 0, \quad x_3 = 0, \quad s_1 = -2, \quad s_2 = 0, \quad \text{and} \quad s_3 = 4.$$

Since the basic variable s_1 is negative, this solution is not feasible. So we repeat the pivoting process. The s_1 column has a -1 in row 1, so we choose a positive entry in that row, namely, the 1 in row 1, column 2. This choice makes column 2 the pivot column. The pivot row is determined by the quotients $\frac{2}{1} = 2$ and $\frac{6}{3} = 2$. (Negative entries in the pivot column and the entry in the objective row are not used.) Since there is a tie, we can choose either row 1 or row 2. We choose row 1 and use the 1 in row 1, column 2, as the pivot. Pivoting produces the tableau

$$\begin{array}{cccccc|c} x_1 & x_2 & x_3 & s_1 & s_2 & s_3 & \\ 0 & 1 & 2 & -1 & -1 & 0 & 2 \\ 1 & 0 & -4 & 3 & 4 & 0 & 0 \\ 0 & 0 & 12 & -5 & -8 & 1 & 14 \\ \hline 0 & 0 & -2 & 2 & 6 & 0 & 20 \end{array} \quad \begin{array}{l} \\ \\ -3R_1 + R_2 \\ 5R_1 + R_3 \\ -2R_1 + R_4 \end{array}$$

and the basic *feasible* solution

$$x_1 = 0, \quad x_2 = 2, \quad x_3 = 0, \quad s_1 = 0, \quad s_2 = 0, \quad \text{and} \quad s_3 = 14.$$

Once a basic feasible solution has been found, Stage I is ended. The procedures used in Stage I are summarized here.*

Finding a Basic Feasible Solution

1. If any basic variable has a negative value, locate the -1 in that variable's column and note the row it is in.
2. In the row determined in Step 1, choose a positive entry (other than the one at the far right) and note the column it is in. This is the pivot column.
3. Use the positive entries in the pivot column (except in the objective row) to form quotients and select the pivot.
4. Pivot as usual, which results in the pivot column's having one entry that is 1 and the rest that are 0's.
5. Repeat Steps 1–4 until every basic variable is nonnegative, so that the basic solution given by the tableau is feasible. If it ever becomes impossible to continue, then the problem has no feasible solution.

One way to make the required choices systematically is to choose the first possibility in each case (going from the top for rows and from the left for columns). However, any choice meeting the required conditions may be used. For maximum efficiency, it is usually best to choose the pivot column in Step 2, so that the pivot is in the same row chosen in Step 1, if this is possible. 3

Checkpoint 3

The initial tableau of a maximization problem is shown. Use column 1 as the pivot column for carrying out Stage I, and state the basic feasible solution that results.

$$\begin{array}{cccc|c} x_1 & x_2 & s_1 & s_2 & \\ 1 & 3 & 1 & 0 & 70 \\ 2 & 4 & 0 & -1 & 50 \\ \hline -8 & -10 & 0 & 0 & 0 \end{array}$$

Answer: See page 458.

In **Stage II**, the simplex method is applied as usual to the tableau that produced the basic feasible solution in Stage I. Just as in Section 7.4, each round of pivoting replaces the basic feasible solution of one tableau with the basic feasible solution of a new tableau in such a way that the value of the objective function is increased, until an optimal value is obtained (or it becomes clear that no optimal solution exists).

*Except in rare cases that do not occur in this book, this method either eventually produces a basic feasible solution or shows that one does not exist. The *two-phase method* using artificial variables, which is discussed in more advanced texts, works in all cases and often is more efficient.

Example 3 Solve the linear programming problem in Example 1 of this section.

Solution A basic feasible solution for this problem was found in Example 2 by using the tableau shown below. However, this solution is not maximal, because there is a negative indicator in the objective row. So we use the simplex method. The most negative indicator determines the pivot column, and the usual quotients determine that the number 2 in row 1, column 3, is the pivot:

$$\begin{array}{c} \begin{matrix} x_1 & x_2 & x_3 & s_1 & s_2 & s_3 & \end{matrix} \\ \left[\begin{array}{cccccc|c} 0 & 1 & \boxed{2} & -1 & -1 & 0 & 2 \\ 1 & 0 & -4 & 3 & 4 & 0 & 0 \\ 0 & 0 & 12 & -5 & -8 & 1 & 14 \\ \hline 0 & 0 & -2 & 2 & 6 & 0 & 20 \end{array}\right]. \end{array} \quad \begin{array}{l} \text{Quotients} \\ 2/2 \leftarrow \text{Smallest} \\ \\ 14/12 \\ \\ \end{array}$$

↑ Most negative indicator

Pivoting leads to the final tableau:

$$\begin{array}{c} \begin{matrix} x_1 & x_2 & x_3 & s_1 & s_2 & s_3 & \end{matrix} \\ \left[\begin{array}{cccccc|c} 0 & \frac{1}{2} & 1 & -\frac{1}{2} & -\frac{1}{2} & 0 & 1 \\ 1 & 0 & -4 & 3 & 4 & 0 & 0 \\ 0 & 0 & 12 & -5 & -8 & 1 & 14 \\ \hline 0 & 0 & -2 & 2 & 6 & 0 & 20 \end{array}\right] \end{array} \quad \begin{array}{l} \frac{1}{2}R_1 \\ \\ \\ \\ \end{array}$$

$$\begin{array}{c} \begin{matrix} x_1 & x_2 & x_3 & s_1 & s_2 & s_3 & \end{matrix} \\ \left[\begin{array}{cccccc|c} 0 & \frac{1}{2} & 1 & -\frac{1}{2} & -\frac{1}{2} & 0 & 1 \\ 1 & 2 & 0 & 1 & 2 & 0 & 4 \\ 0 & -6 & 0 & 1 & -2 & 1 & 2 \\ \hline 0 & 1 & 0 & 1 & 5 & 0 & 22 \end{array}\right]. \end{array} \quad \begin{array}{l} \\ 4R_1 + R_2 \\ -12R_1 + R_3 \\ 2R_1 + R_4 \end{array}$$

Therefore, the maximum value of z occurs when $x_1 = 4$, $x_2 = 0$, and $x_3 = 1$, in which case $z = 22$. ✓4

✓Checkpoint 4

Complete Stage II and find an optimal solution for Checkpoint 3 on page 448. What is the optimal value of the objective function z?

Answer: See page 458.

MINIMIZATION PROBLEMS

When dealing with minimization problems, we use y_1, y_2, y_3, etc., as variables and denote the objective function by w. The two-stage method for maximization problems illustrated in Examples 1–3 also provides a means of solving minimization problems. To see why, consider this simple fact: When a number t gets smaller, $-t$ gets larger, and vice versa. For instance, if t goes from 6 down to -8, then $-t$ goes from -6 up to 8. Thus, if w is the objective function of a linear programming problem, the feasible solution that produces the minimum value of w also produces the maximum value of $-w$, and vice versa. Therefore, to solve a minimization problem with objective function w, we need only solve the maximization problem with the same constraints and objective function $z = -w$.

Example 4

$$\begin{aligned} \text{Minimize} \quad & w = 2y_1 + y_2 - y_3 \\ \text{subject to} \quad & -y_1 - y_2 + y_3 \le -4 \\ & y_1 + 3y_2 + 3y_3 \ge 6 \\ & y_1 \ge 0, y_2 \ge 0, y_3 \ge 0. \end{aligned}$$

Solution Make the constant in the first constraint positive by multiplying both sides by -1. Then solve this maximization problem:

$$\begin{aligned} &\text{Maximize} \quad z = -w = -2y_1 - y_2 + y_3 \\ &\text{subject to} \quad y_1 + y_2 - y_3 \geq 4 \\ &\qquad\qquad\quad y_1 + 3y_2 + 3y_3 \geq 6 \\ &\qquad\qquad\quad y_1 \geq 0, y_2 \geq 0, y_3 \geq 0. \end{aligned}$$

Convert the constraints to equations by subtracting surplus variables, and set up the first tableau:

$$\begin{array}{c} \begin{matrix} y_1 & y_2 & y_3 & s_1 & s_2 & \end{matrix} \\ \left[\begin{array}{ccccc|c} 1 & 1 & -1 & -1 & 0 & 4 \\ 1 & 3 & 3 & 0 & -1 & 6 \\ \hline 2 & 1 & -1 & 0 & 0 & 0 \end{array}\right]. \end{array}$$

The basic solution given by this tableau, namely, $y_1 = 0, y_2 = 0, y_3 = 0, s_1 = -4$, and $s_2 = -6$, is not feasible, so the procedures of Stage I must be used to find a basic feasible solution. In the column of the negative basic variable s_1, there is a -1 in row 1; we choose the first positive entry in that row, so that column 1 will be the pivot column. The quotients $\frac{4}{1} = 4$ and $\frac{6}{1} = 6$ show that the pivot is the 1 in row 1, column 1. Pivoting produces this tableau:

$$\begin{array}{c} \begin{matrix} y_1 & y_2 & y_3 & s_1 & s_2 & \end{matrix} \\ \left[\begin{array}{ccccc|c} 1 & 1 & -1 & -1 & 0 & 4 \\ 0 & 2 & 4 & 1 & -1 & 2 \\ \hline 0 & -1 & 1 & 2 & 0 & -8 \end{array}\right]. \end{array} \quad \begin{array}{l} \\ -R_1 + R_2 \\ -2R_1 + R_3 \end{array}$$

The basic solution $y_1 = 4, y_2 = 0, y_3 = 0, s_1 = 0$, and $s_2 = -2$ is not feasible because s_2 is negative, so we repeat the process. We choose the first positive entry in row 2 (the row containing the -1 in the s_2 column), which is in column 2, so that column 2 is the pivot column. The relevant quotients are $\frac{4}{1} = 4$ and $\frac{2}{2} = 1$, so the pivot is the 2 in row 2, column 2. Pivoting produces a new tableau:

$$\begin{array}{c} \begin{matrix} y_1 & y_2 & y_3 & s_1 & s_2 & \end{matrix} \\ \left[\begin{array}{ccccc|c} 1 & 1 & -1 & -1 & 0 & 4 \\ 0 & 1 & 2 & \frac{1}{2} & -\frac{1}{2} & 1 \\ \hline 0 & -1 & 1 & 2 & 0 & -8 \end{array}\right] \end{array} \quad \begin{array}{l} \\ \frac{1}{2}R_2 \\ \\ \end{array}$$

$$\begin{array}{c} \begin{matrix} y_1 & y_2 & y_3 & s_1 & s_2 & \end{matrix} \\ \left[\begin{array}{ccccc|c} 1 & 0 & -3 & -\frac{3}{2} & \frac{1}{2} & 3 \\ 0 & 1 & 2 & \frac{1}{2} & -\frac{1}{2} & 1 \\ \hline 0 & 0 & 3 & \frac{5}{2} & -\frac{1}{2} & -7 \end{array}\right]. \end{array} \quad \begin{array}{l} -R_2 + R_1 \\ \\ R_2 + R_3 \end{array}$$

The basic solution $y_1 = 3, y_2 = 1, y_3 = 0, s_1 = 0$, and $s_2 = 0$ is feasible, so Stage I is complete. However, this solution is not optimal, because the objective row contains the negative indicator $-\frac{1}{2}$ in column 5. According to the simplex method, column 5 is the next pivot column. The only positive ratio, $3/\frac{1}{2} = 6$, is in row 1, so the pivot is $\frac{1}{2}$ in row 1, column 5. Pivoting produces the final tableau:

$$\begin{array}{c} \\ \\ \end{array}\begin{array}{ccccc|c} y_1 & y_2 & y_3 & s_1 & s_2 & \\ 2 & 0 & -6 & -3 & 1 & 6 \\ 0 & 1 & 2 & \frac{1}{2} & -\frac{1}{2} & 1 \\ \hline 0 & 0 & 3 & \frac{5}{2} & -\frac{1}{2} & -7 \end{array} \quad 2R_1$$

$$\begin{array}{ccccc|c} y_1 & y_2 & y_3 & s_1 & s_2 & \\ 2 & 0 & -6 & -3 & 1 & 6 \\ 1 & 1 & -1 & -1 & 0 & 4 \\ \hline 1 & 0 & 0 & 1 & 0 & -4 \end{array}. \quad \begin{array}{l} \\ \frac{1}{2}R_1 + R_2 \\ \\ \frac{1}{2}R_1 + R_3 \end{array}$$

Since there are no negative indicators, the solution given by this tableau ($y_1 = 0$, $y_2 = 4$, $y_3 = 0$, $s_1 = 0$, and $s_2 = 6$) is optimal. The maximum value of $z = -w$ is -4. Therefore, the minimum value of the original objective function w is $-(-4) = 4$, which occurs when $y_1 = 0$, $y_2 = 4$, and $y_3 = 0$. 5✓

✓Checkpoint 5

Minimize $w = 2y_1 + 3y_2$
subject to $y_1 + y_2 \geq 10$
$2y_1 + y_2 \geq 16$
$y_1 \geq 0, y_2 \geq 0.$

Answer: See page 458.

EQUATION CONSTRAINTS

Recall that, for any real numbers a and b,

$$a = b \qquad \text{exactly when } a \geq b \text{ and simultaneously } a \leq b.$$

Thus, an equation such as $y_1 + 3y_2 + 3y_3 = 6$ is equivalent to this pair of inequalities:

$$y_1 + 3y_2 + 3y_3 \geq 6$$
$$y_1 + 3y_2 + 3y_3 \leq 6.$$

In a linear programming problem, each equation constraint should be replaced in this way by a pair of inequality constraints. Then the problem can be solved by the two-stage method.

Example 5

$$\begin{aligned} \text{Minimize} \quad & w = 2y_1 + y_2 - y_3 \\ \text{subject to} \quad & -y_1 - y_2 + y_3 \leq -4 \\ & y_1 + 3y_2 + 3y_3 = 6 \\ & y_1 \geq 0, y_2 \geq 0, y_3 \geq 0. \end{aligned}$$

Solution Multiply the first inequality by -1 and replace the equation by an equivalent pair of inequalities, as just explained, to obtain this problem:

$$\begin{aligned} \text{Maximize} \quad & z = -w = -2y_1 - y_2 + y_3 \\ \text{subject to} \quad & y_1 + y_2 - y_3 \geq 4 \\ & y_1 + 3y_2 + 3y_3 \geq 6 \\ & y_1 + 3y_2 + 3y_3 \leq 6 \\ & y_1 \geq 0, y_2 \geq 0, y_3 \geq 0. \end{aligned}$$

Convert the constraints to equations by subtracting surplus variables s_1 and s_2 from the first two inequalities and adding a slack variable s_3 to the third. Then the first tableau is

$$\begin{array}{cccccc|c} y_1 & y_2 & y_3 & s_1 & s_2 & s_3 & \\ 1 & 1 & -1 & -1 & 0 & 0 & 4 \\ 1 & 3 & 3 & 0 & -1 & 0 & 6 \\ 1 & 3 & 3 & 0 & 0 & 1 & 6 \\ \hline 2 & 1 & -1 & 0 & 0 & 0 & 0 \end{array}.$$

The basic solution given by this tableau is $y_1 = 0, y_2 = 0, y_3 = 0, s_1 = -4, s_2 = -6$, and $s_3 = 6$, which is not feasible. So we begin Stage I. The basic variable s_1 has a negative value because of the -1 in row 1. We choose the first positive entry in that row. So column 1 will be the pivot column. The quotients are $\frac{4}{1} = 4$ in row 1 and $\frac{6}{1} = 6$ in rows 2 and 3, which means that the pivot is the 1 in row 1, column 1. Pivoting produces this tableau:

$$\begin{array}{cccccc|c} y_1 & y_2 & y_3 & s_1 & s_2 & s_3 & \\ \hline 1 & 1 & -1 & -1 & 0 & 0 & 4 \\ 0 & 2 & 4 & 1 & -1 & 0 & 2 \\ 0 & 2 & 4 & 1 & 0 & 1 & 2 \\ \hline 0 & -1 & 1 & 2 & 0 & 0 & -8 \end{array}.$$

Now the basic variable s_2 has a negative value because of the -1 in row 2. We choose the first nonzero entry in this row and form the quotients $\frac{4}{1} = 4$ in row 1 and $\frac{2}{2} = 1$ in rows 2 and 3. Thus, there are two choices for the pivot, and we take the 2 in row 2, column 2. Pivoting produces the tableau

$$\begin{array}{cccccc|c} y_1 & y_2 & y_3 & s_1 & s_2 & s_3 & \\ \hline 1 & 0 & -3 & -1.5 & .5 & 0 & 3 \\ 0 & 1 & 2 & .5 & -.5 & 0 & 1 \\ 0 & 0 & 0 & 0 & 1 & 1 & 0 \\ \hline 0 & 0 & 3 & 2.5 & -.5 & 0 & -7 \end{array}.$$

This tableau gives the basic feasible solution $y_1 = 3, y_2 = 1, y_3 = 0, s_1 = 0, s_2 = 0$, and $s_3 = 0$, so Stage I is complete. Now apply the simplex method. One round of pivoting produces the final tableau:

$$\begin{array}{cccccc|c} y_1 & y_2 & y_3 & s_1 & s_2 & s_3 & \\ \hline 1 & 0 & -3 & -1.5 & 0 & -.5 & 3 \\ 0 & 1 & 2 & .5 & 0 & .5 & 1 \\ 0 & 0 & 0 & 0 & 1 & 1 & 0 \\ \hline 0 & 0 & 3 & 2.5 & 0 & .5 & -7 \end{array}.$$

Therefore, the minimum value of $w = -z$ is $w = -(-7) = 7$, which occurs when $y_1 = 3, y_2 = 1$, and $y_3 = 0$.

You may have noticed that Example 5 is just Example 4 with the last inequality constraint replaced by an equation constraint. Note, however, that the optimal solutions are different in the two examples. The minimal value of w found in Example 4 is smaller than the one found in Example 5, but does not satisfy the equation constraint in Example 5.

The two-stage method used in Examples 1–5 is summarized here.

Solving Nonstandard Problems

1. Replace each equation constraint by an equivalent pair of inequality constraints.
2. If necessary, write each constraint with a positive constant.

3. Convert a minimization problem to a maximization problem by letting $z = -w$.
4. Add slack variables and subtract surplus variables as needed to convert the constraints into equations.
5. Write the initial simplex tableau.
6. Find a basic feasible solution for the problem if such a solution exists (Stage I).
7. When a basic feasible solution is found, use the simplex method to solve the problem (Stage II).

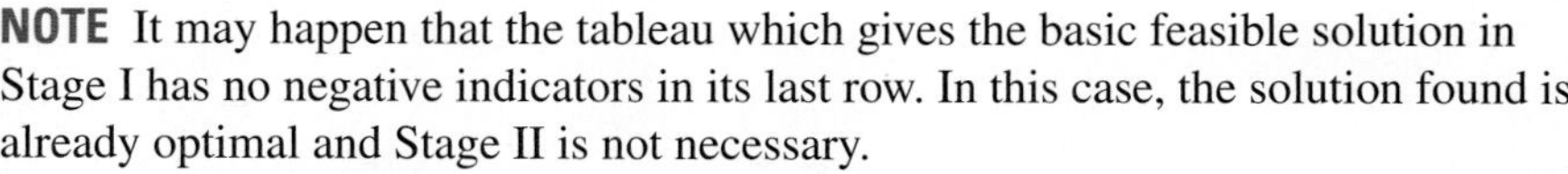

NOTE It may happen that the tableau which gives the basic feasible solution in Stage I has no negative indicators in its last row. In this case, the solution found is already optimal and Stage II is not necessary.

APPLICATIONS

Many real-world applications of linear programming involve mixed constraints. Since they typically include a large number of variables and constraints, technology is normally required to solve such problems.

Example 6 A college textbook publisher has received orders from two colleges: C_1 and C_2. C_1 needs 500 books, and C_2 needs 1000. The publisher can supply the books from either of two warehouses. Warehouse W_1 has 900 books available, and warehouse W_2 has 700. The costs to ship a book from each warehouse to each college are as follows:

		To	
		C_1	C_2
From	W_1	\$1.20	\$1.80
	W_2	\$2.10	\$1.50

How many books should be sent from each warehouse to each college to minimize the shipping costs?

Solution To begin, let

$$y_1 = \text{the number of books shipped from } W_1 \text{ to } C_1;$$
$$y_2 = \text{the number of books shipped from } W_2 \text{ to } C_1;$$
$$y_3 = \text{the number of books shipped from } W_1 \text{ to } C_2;$$
$$y_4 = \text{the number of books shipped from } W_2 \text{ to } C_2.$$

C_1 needs 500 books, so $y_1 + y_2 = 500$, which is equivalent to this pair of inequalities:

$$y_1 + y_2 \geq 500$$
$$y_1 + y_2 \leq 500.$$

Similarly, $y_3 + y_4 = 1000$, which is equivalent to

$$y_3 + y_4 \geq 1000$$
$$y_3 + y_4 \leq 1000.$$

Since W_1 has 900 books available and W_2 has 700 available,

$$y_1 + y_3 \le 900 \qquad \text{and} \qquad y_2 + y_4 \le 700.$$

The company wants to minimize shipping costs, so the objective function is

$$w = 1.20y_1 + 2.10y_2 + 1.80y_3 + 1.50y_4.$$

Now write the problem as a system of linear equations, adding slack or surplus variables as needed, and let $z = -w$:

$$\begin{aligned}
y_1 + y_2 \qquad\qquad - s_1 \qquad\qquad &= 500\\
y_1 + y_2 + \qquad\qquad s_2 \qquad\qquad &= 500\\
y_3 + y_4 \qquad - s_3 \qquad &= 1000\\
y_3 + y_4 + \qquad s_4 \qquad &= 1000\\
y_1 + y_3 + \qquad\qquad s_5 \qquad &= 900\\
y_2 \qquad y_4 + \qquad\qquad s_6 &= 700\\
1.20y_1 + 2.10y_2 + 1.80y_3 + 1.50y_4 \qquad\qquad + z &= 0.
\end{aligned}$$

Set up the initial simplex tableau:

$$\begin{array}{c}
\begin{array}{cccccccccc} y_1 & y_2 & y_3 & y_4 & s_1 & s_2 & s_3 & s_4 & s_5 & s_6 \end{array}\\
\left[\begin{array}{cccccccccc|c}
1 & 1 & 0 & 0 & -1 & 0 & 0 & 0 & 0 & 0 & 500\\
1 & 1 & 0 & 0 & 0 & 1 & 0 & 0 & 0 & 0 & 500\\
0 & 0 & 1 & 1 & 0 & 0 & -1 & 0 & 0 & 0 & 1000\\
0 & 0 & 1 & 1 & 0 & 0 & 0 & 1 & 0 & 0 & 1000\\
1 & 0 & 1 & 0 & 0 & 0 & 0 & 0 & 1 & 0 & 900\\
0 & 1 & 0 & 1 & 0 & 0 & 0 & 0 & 0 & 1 & 700\\
\hline
1.20 & 2.10 & 1.80 & 1.50 & 0 & 0 & 0 & 0 & 0 & 0 & 0
\end{array}\right].
\end{array}$$

The basic solution here is not feasible, because $s_1 = -500$ and $s_3 = -1000$. Stages I and II could be done by hand here, but because of the large size of the matrix, it is more efficient to use technology, such as the program in the Graphing Calculator Appendix. Stage I takes four rounds of pivoting and produces the feasible solution in Figure 7.33.

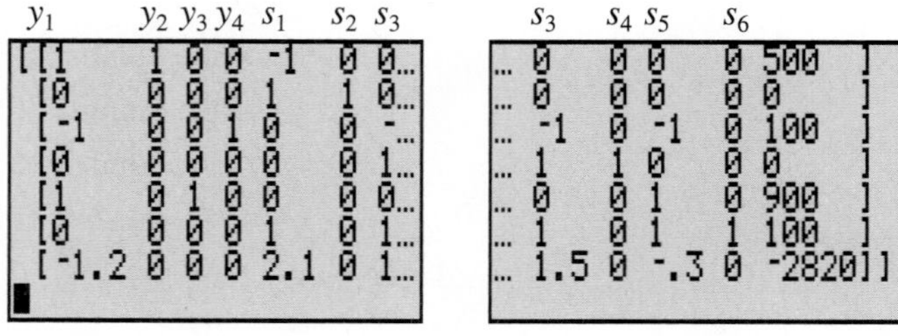

FIGURE 7.33

Because of the small size of a calculator screen, you must scroll to the right to see the entire matrix. Now Stage II begins. Two rounds of pivoting produce the final tableau (Figure 7.34).

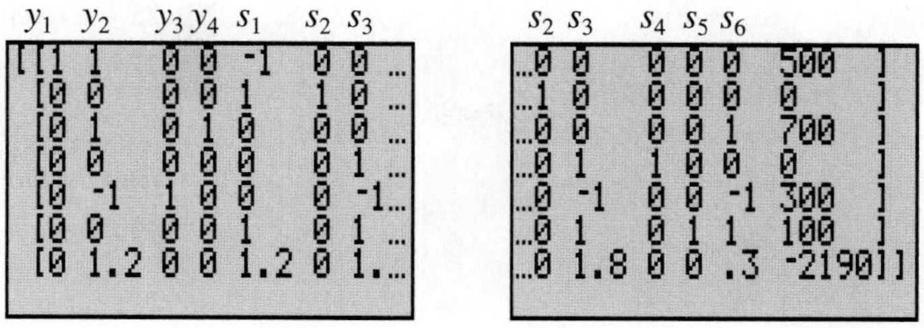

```
 y1 y2  y3 y4 s1   s2 s3          s2 s3   s4 s5 s6
[[1 1   0 0 -1    0 0 ...        ...0 0   0 0 0  500  ]
 [0 0   0 0 1     1 0 ...        ...1 0   0 0 0  0    ]
 [0 1   0 1 0     0 0 ...        ...0 0   0 0 1  700  ]
 [0 0   0 0 0     0 1 ...        ...0 1   1 0 0  0    ]
 [0 -1  1 0 0     0 -1...        ...0 -1  0 0 -1 300  ]
 [0 0   0 0 1     0 1 ...        ...0 1   0 1 1  100  ]
 [0 1.2 0 0 1.2   0 1...         ...0 1.8 0 0 .3 -2190]]
```

FIGURE 7.34

The optimal solution is $y_1 = 500$, $y_2 = 0$, $y_3 = 300$, and $y_4 = 700$, which results in a minimum shipping cost of \$2190. (Remember that the optimal value for the original minimization problem is the negative of the optimal value of the associated maximization problem.)

7.7 ▸ Exercises

In Exercises 1–4, (a) restate the problem in terms of equations by introducing slack and surplus variables and (b) write the initial simplex tableau. (See Example 1.)

1. Maximize $z = -5x_1 + 4x_2 - 2x_3$
subject to
$$-2x_2 + 5x_3 \geq 8$$
$$4x_1 - x_2 + 3x_3 \leq 12$$
$$x_1 \geq 0, x_2 \geq 0, x_3 \geq 0.$$

2. Maximize $z = -x_1 + 4x_2 - 2x_3$
subject to
$$2x_1 + 2x_2 + 6x_3 \leq 10$$
$$-x_1 + 2x_2 + 4x_3 \geq 7$$
$$x_1 \geq 0, x_2 \geq 0, x_3 \geq 0.$$

3. Maximize $z = 2x_1 - 3x_2 + 4x_3$
subject to
$$x_1 + x_2 + x_3 \leq 100$$
$$x_1 + x_2 + x_3 \geq 75$$
$$x_1 + x_2 \geq 27$$
$$x_1 \geq 0, x_2 \geq 0, x_3 \geq 0.$$

4. Maximize $z = -x_1 + 5x_2 + x_3$
subject to
$$2x_1 + x_3 \leq 40$$
$$x_1 + x_2 \geq 18$$
$$x_1 + x_3 = 20$$
$$x_1 \geq 0, x_2 \geq 0, x_3 \geq 0.$$

Convert Exercises 5–8 into maximization problems with positive constants on the right side of each constraint, and write the initial simplex tableau. (See Examples 4 and 5.)

5. Minimize $w = 2y_1 + 5y_2 - 3y_3$
subject to
$$y_1 + 2y_2 + 3y_3 \geq 115$$
$$2y_1 + y_2 + y_3 \leq 200$$
$$y_1 + y_3 \geq 50$$
$$y_1 \geq 0, y_2 \geq 0, y_3 \geq 0.$$

6. Minimize $w = 7y_1 + 6y_2 + y_3$
subject to
$$y_1 + y_2 + y_3 \geq 5$$
$$-y_1 + y_2 \leq -4$$
$$2y_1 + y_2 + 3y_3 \geq 15$$
$$y_1 \geq 0, y_2 \geq 0, y_3 \geq 0.$$

7. Minimize $w = 10y_1 + 8y_2 + 15y_3$
subject to
$$y_1 + y_2 + y_3 \geq 12$$
$$5y_1 + 4y_2 + 9y_3 \geq 48$$
$$y_1 \geq 0, y_2 \geq 0, y_3 \geq 0.$$

8. Minimize $w = y_1 + 2y_2 + y_3 + 5y_4$
subject to
$$-y_1 - y_2 + y_3 - y_4 \leq -50$$
$$3y_1 + y_2 + 2y_3 + y_4 = 100$$
$$y_1 \geq 0, y_2 \geq 0, y_3 \geq 0, y_4 \geq 0.$$

Use the two-stage method to solve Exercises 9–20. (See Examples 1–5.)

9. Maximize $z = 12x_1 + 10x_2$
subject to
$$x_1 + 2x_2 \geq 24$$
$$x_1 + x_2 \leq 40$$
$$x_1 \geq 0, x_2 \geq 0.$$

10. Find $x_1 \geq 0$, $x_2 \geq 0$, and $x_3 \geq 0$ such that
$$x_1 + x_2 + x_3 \leq 150$$
$$x_1 + x_2 + x_3 \geq 100$$
and $z = 2x_1 + 5x_2 + 3x_3$ is maximized.

11. Find $x_1 \geq 0$, $x_2 \geq 0$, and $x_3 \geq 0$ such that
$$x_1 + x_2 + 2x_3 \leq 38$$
$$2x_1 + x_2 + x_3 \geq 24$$
and $z = 3x_1 + 2x_2 + 2x_3$ is maximized.

12. Maximize $z = 6x_1 + 8x_2$
subject to
$$\begin{aligned} 3x_1 + 12x_2 &\geq 48 \\ 2x_1 + 4x_2 &\leq 60 \\ x_1 \geq 0, x_2 &\geq 0. \end{aligned}$$

13. Find $x_1 \geq 0$ and $x_2 \geq 0$ such that
$$\begin{aligned} x_1 + 2x_2 &\leq 18 \\ x_1 + 3x_2 &\geq 12 \\ 2x_1 + 2x_2 &\leq 30 \end{aligned}$$
and $z = 5x_1 + 10x_2$ is maximized.

14. Find $y_1 \geq 0$, $y_2 \geq 0$ such that
$$\begin{aligned} 10y_1 + 5y_2 &\geq 100 \\ 20y_1 + 10y_2 &\geq 160 \end{aligned}$$
and $w = 4y_1 + 5y_2$ is minimized.

15. Minimize $w = 3y_1 + 2y_2$
subject to
$$\begin{aligned} 2y_1 + 3y_2 &\geq 60 \\ y_1 + 4y_2 &\geq 40 \\ y_1 \geq 0, y_2 &\geq 0. \end{aligned}$$

16. Minimize $w = 3y_1 + 4y_2$
subject to
$$\begin{aligned} y_1 + 2y_2 &\geq 10 \\ y_1 + y_2 &\geq 8 \\ 2y_1 + y_2 &\leq 22 \\ y_1 \geq 0, y_2 &\geq 0. \end{aligned}$$

17. Maximize $z = 3x_1 + 2x_2$
subject to
$$\begin{aligned} x_1 + x_2 &= 50 \\ 4x_1 + 2x_2 &\geq 120 \\ 5x_1 + 2x_2 &\leq 200 \end{aligned}$$
with $x_1 \geq 0, x_2 \geq 0.$

18. Maximize $z = 10x_1 + 9x_2$
subject to
$$\begin{aligned} x_1 + x_2 &= 30 \\ x_1 + x_2 &\geq 25 \\ 2x_1 + x_2 &\leq 40 \end{aligned}$$
with $x_1 \geq 0, x_2 \geq 0.$

19. Minimize $w = 32y_1 + 40y_2$
subject to
$$\begin{aligned} 20y_1 + 10y_2 &= 200 \\ 25y_1 + 40y_2 &\leq 500 \\ 18y_1 + 24y_2 &\geq 300 \end{aligned}$$
with $y_1 \geq 0, y_2 \geq 0.$

20. Minimize $w = 15y_1 + 12y_2$
subject to
$$\begin{aligned} y_1 + 2y_2 &\leq 12 \\ 3y_1 + y_2 &\geq 18 \\ y_1 + y_2 &= 10 \end{aligned}$$
with $y_1 \geq 0, y_2 \geq 0.$

Use the two-stage method to solve the problem that was set up in the given exercise.

21. Exercise 1 **22.** Exercise 2

23. Exercise 3 **24.** Exercise 4

 **25.** Exercise 5 **26.** Exercise 6

 **27.** Exercise 7 **28.** Exercise 8

In Exercises 29–32, set up the initial simplex tableau, but do not solve the problem. (See Example 6.)

29. Business A company is developing a new additive for gasoline. The additive is a mixture of three liquid ingredients: I, II, and III. For proper performance, the total amount of additive must be at least 10 ounces per barrel of gasoline. However, for safety reasons, the amount of additive should not exceed 15 ounces per barrel of gasoline. At least $\frac{1}{4}$ ounce of ingredient I must be used for every ounce of ingredient II, and at least 1 ounce of ingredient III must be used for every ounce of ingredient I. If the costs of I, II, and III are \$.30, \$.09, and \$.27 per ounce, respectively, find the mixture of the three ingredients that produces the minimum cost of the additive. What is the minimum cost?

30. Business A popular soft drink called Sugarlo, which is advertised as having a sugar content of no more than 10%, is blended from five ingredients, each of which has some sugar content. Water may also be added to dilute the mixture. The sugar content of the ingredients and their costs per gallon are given in the table:

	Ingredient					
	1	**2**	**3**	**4**	**5**	**Water**
Sugar content (%)	.28	.19	.43	.57	.22	0
Cost (\$/gal.)	.48	.32	.53	.28	.43	.04

At least .01 of the content of Sugarlo must come from ingredient 3 or 4, .01 must come from ingredient 2 or 5, and .01 from ingredient 1 or 4. How much of each ingredient should be used in preparing at least 15,000 gallons of Sugarlo to minimize the cost? What is the minimum cost?

31. Business The manufacturer of a popular personal computer has orders from two dealers. Dealer D_1 wants 32 computers, and dealer D_2 wants 20 computers. The manufacturer can fill the orders from either of two warehouses, W_1 or W_2. W_1 has 25 of the computers on hand, and W_2 has 30. The costs (in dollars) to ship one computer to each dealer from each warehouse are as follows:

		To	
		D_1	D_2
From	W_1	\$14	\$22
	W_2	\$12	\$10

How should the orders be filled to minimize shipping costs? What is the minimum cost?

32. Natural Science Mark, who is ill, takes vitamin pills. Each day, he must have at least 16 units of vitamin A, 5 units of vitamin B_1, and 20 units of vitamin C. He can choose between pill 1, which costs 10¢ and contains 8 units of vitamin A, 1 unit of vitamin B_1, and 2 units of vitamin C, and pill 2, which costs 20¢ and contains 2 units of vitamin A, 1 unit of vitamin B_1, and 7 units of vitamin C. How many of each pill should he buy in order to minimize his cost?

Use the two-stage method to solve Exercises 33–40. (See Examples 5 and 6.)

33. Transportation Southwestern Oil supplies two distributors in the Northwest from two outlets: S_1 and S_2. Distributor D_1 needs at least 3000 barrels of oil, and distributor D_2 needs at least 5000 barrels. The two outlets can each furnish up to 5000 barrels of oil. The costs per barrel to ship the oil are given in the table:

		Distributors	
		D_1	D_2
Outlets	S_1	\$30	\$20
	S_2	\$25	\$22

There is also a shipping tax per barrel as given in the table below.

	D_1	D_2
S_1	\$2	\$6
S_2	\$5	\$4

Southwestern Oil is determined to spend no more than \$40,000 on shipping tax. How should the oil be supplied to minimize shipping costs?

34. Transportation Change Exercise 33 so that the two outlets each furnish exactly 5000 barrels of oil, with everything else the same. Solve the problem as in Example 5.

35. Business Topgrade Turf lawn seed mixture contains three types of seeds: bluegrass, rye, and Bermuda. The costs per pound of the three types of seed are 12¢, 15¢, and 5¢, respectively. In each batch, there must be at least 20% bluegrass seed, and the amount of Bermuda seed must be no more than two-thirds the amount of rye seed. To fill current orders, the company must make at least 5000 pounds of the mixture. How much of each kind of seed should be used to minimize costs?

36. Business Change Exercise 35 so that the company must make exactly 5000 pounds of the mixture. Solve the problem as in Example 5.

37. Finance A bank has set aside a maximum of \$25 million for commercial and home loans. Every million dollars in commercial loans requires 2 lengthy application forms, while every million dollars in home loans requires 3 lengthy application forms. The bank cannot process more than 72 application forms at this time. The bank's policy is to loan at least four times as much for home loans as for commercial loans. Because of prior commitments, at least \$10 million will be used for these two types of loans. The bank earns 12% on home loans and 10% on commercial loans. What amount of money should be allotted for each type of loan to maximize the interest income?

38. Finance Virginia Keleske has decided to invest a \$100,000 inheritance in government securities that earn 7% per year, municipal bonds that earn 6% per year, and mutual funds that earn an average of 10% per year. She will spend at least \$40,000 on government securities, and she wants at least half the inheritance to go to bonds and mutual funds. Government securities have an initial fee of 2%, municipal bonds an initial fee of 1%, and mutual funds an initial fee of 3%. Virginia has \$2400 available to pay initial fees. How much money should go into each type of investment to maximize the interest while meeting the constraints? What is the maximum interest she can earn?

39. Business A brewery produces regular beer and a lower-carbohydrate "light" beer. Steady customers of the brewery buy 12 units of regular beer and 10 units of light beer. While setting up the brewery to produce the beers, the management decides to produce extra beer, beyond that needed to satisfy the steady customers. The cost per unit of regular beer is \$36,000, and the cost per unit of light beer is \$48,000. The number of units of light beer should not exceed twice the number of units of regular beer. At least 20 additional units of beer can be sold. How much of each type of beer should be made so as to minimize total production costs?

40. Business The chemistry department at a local college decides to stock at least 800 small test tubes and 500 large test tubes. It wants to buy at least 1500 test tubes to take advantage of a special price. Since the small tubes are broken twice as often as the large, the department will order at least twice as many small tubes as large. If the small test tubes cost 15¢ each and the large ones, made of a cheaper glass, cost 12¢ each, how many of each size should the department order to minimize cost?

Business *Use technology to solve the following exercises, whose initial tableaus were set up in Exercises 29–31.*

41. Exercise 29 **42.** Exercise 30 **43.** Exercise 31

✓Checkpoint Answers

1. (a) Maximize $z = 3x_1 - 2x_2$
subject to
$$2x_1 + 3x_2 + s_1 = 8$$
$$6x_1 - 2x_2 - s_2 = 3$$
$$x_1 + 4x_2 - s_3 = 1$$
$x_1 \geq 0, x_2 \geq 0, s_1 \geq 0, s_2 \geq 0, s_3 \geq 0.$

(b)
$$\begin{array}{ccccc|c} x_1 & x_2 & s_1 & s_2 & s_3 & \\ 2 & 3 & 1 & 0 & 0 & 8 \\ 6 & -2 & 0 & -1 & 0 & 3 \\ 1 & 4 & 0 & 0 & -1 & 1 \\ \hline -3 & 2 & 0 & 0 & 0 & 0 \end{array}$$

2. (a) $x_1 = 0, x_2 = 0, s_1 = 12, s_2 = 6, s_3 = -5$; no
(b) $x_1 = 0, x_2 = 0, x_3 = 0, s_1 = 12, s_2 = 7$; yes

3.
$$\begin{array}{cccc|c} x_1 & x_2 & s_1 & s_2 & \\ 0 & 1 & 1 & \frac{1}{2} & 45 \\ 1 & 2 & 0 & -\frac{1}{2} & 25 \\ \hline 0 & 6 & 0 & -4 & 200 \end{array}$$
$x_1 = 25, x_2 = 0, s_1 = 45$, and $s_2 = 0$.

4. The optimal value $z = 560$ occurs when $x_1 = 70, x_2 = 0, s_1 = 0$, and $s_2 = 90$

5. $y_1 = 10$ and $y_2 = 0$; $w = 20$

CHAPTER 7 ▸ Summary

KEY TERMS AND SYMBOLS

7.1 ▸ linear inequality
graphs of linear inequalities
boundary
half-plane
system of inequalities
region of feasible solutions (feasible region)

7.2 ▸ linear programming
objective function
constraints
corner point
bounded feasible region
unbounded feasible region
corner point theorem

7.3 ▸ applications of linear programming

7.4 ▸ standard maximum form
slack variable
simplex tableau
indicator
pivot and pivoting
final simplex tableau
basic variables
nonbasic variables
basic feasible solution

7.6 ▸ transpose of a matrix
dual
theorem of duality
shadow costs

7.7 ▸ surplus variable
basic variables
basic solution
basic feasible solution
Stage I
Stage II

CHAPTER 7 KEY CONCEPTS

Graphing a Linear Inequality ▸ Graph the boundary line as a solid line if the inequality includes "or equal to," and as a dashed line otherwise. Shade the half-plane for which the inequality is true. The graph of a system of inequalities, called the **region of feasible solutions**, includes all points that satisfy all the inequalities of the system at the same time.

Solving Linear Programming Problems ▸ **Graphical Method:** Determine the objective function and all necessary constraints. Graph the region of feasible solutions. The maximum or minimum value will occur at one or more of the corner points of this region.

Simplex Method: Determine the objective function and all necessary constraints. Convert each constraint into an equation by adding slack variables. Set up the initial simplex tableau. Locate the most negative indicator. Form the quotients to determine the pivot. Use row operations to change the pivot to 1 and all other numbers in that column to 0. If the indicators are all positive or 0, this is the final tableau. If not, choose a new pivot and repeat the process until no indicators are negative. Read the solution from the final tableau. The optimum value of the objective function is the number in the lower right corner of the final tableau. For problems with **mixed constraints**, replace each equation constraint by a pair of inequality constraints. Then add slack variables and subtract surplus variables as needed to convert each constraint into an equation. In Stage I, use row operations to transform the matrix until the solution is feasible. In Stage II, use the simplex method as just described. For **minimization** problems, let the objective function be w and set $-w = z$. Then proceed as with mixed constraints.

Solving Minimization Problems with Duals ▸ Find the dual maximization problem. Solve the dual problem with the simplex method. The minimum value of the objective function w is the maximum value of the dual objective function z. The optimal solution is found in the entries in the bottom row of the columns corresponding to the slack variables.

CHAPTER 7 REVIEW EXERCISES

Graph each of the given linear inequalities.

1. $y \le 3x + 2$ **2.** $2x - y \ge 6$

3. $3x + 4y \ge 12$ **4.** $y \le 4$

Graph the solution of each of the given systems of inequalities.

5. $x + y \le 6$
$2x - y \ge 3$

6. $4x + y \ge 8$
$2x - 3y \le 6$

7. $2 \le x \le 5$
$1 \le y \le 6$
$x - y \le 3$

8. $x + 2y \le 4$
$2x - 3y \le 6$
$x \ge 0, y \ge 0$

Set up a system of inequalities for each of the given problems, and then graph the region of feasible solutions.

9. **Business** A bakery makes both cakes and cookies. Each batch of cakes requires 2 hours in the oven and 3 hours in the decorating room. Each batch of cookies needs $1\frac{1}{2}$ hours in the oven and $\frac{2}{3}$ of an hour in the decorating room. The oven is available no more than 15 hours a day, while the decorating room can be used no more than 13 hours a day.

10. **Business** A company makes two kinds of pizza: special and basic. The special has toppings of cheese, tomatoes, and vegetables. Basic has just cheese and tomatoes. The company sells at least 6 units a day of the special pizza and 4 units a day of the basic. The cost of the vegetables (including tomatoes) is \$2 per unit for special and \$1 per unit for basic. No more than \$32 per day can be spent on vegetables (including tomatoes). The cheese used for the special is \$5 per unit, and the cheese for the basic is \$4 per unit. The company can spend no more than \$100 per day on cheese.

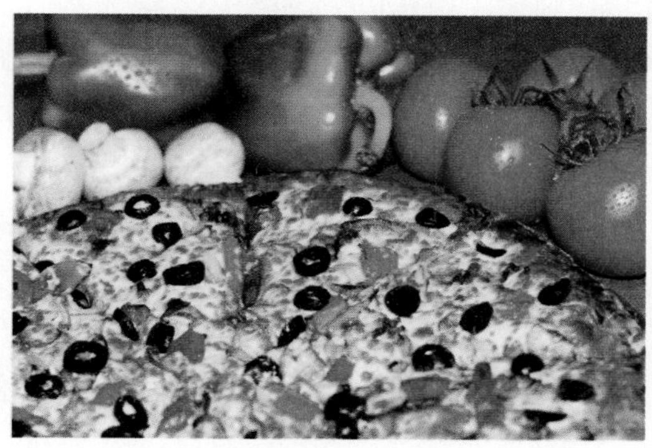

Use the given regions to find the maximum and minimum values of the objective function $z = 3x + 4y$.

11.

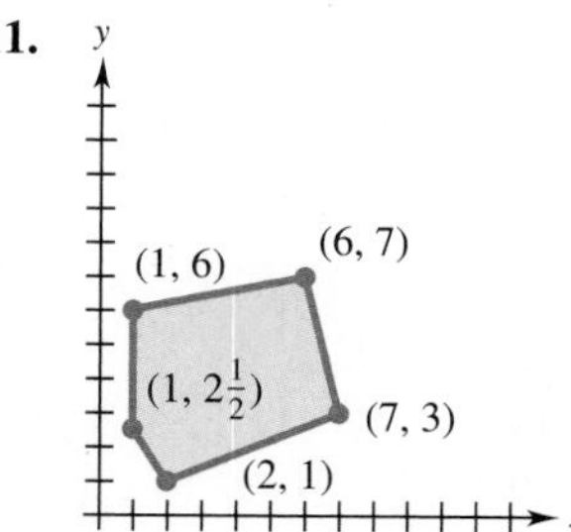

12.

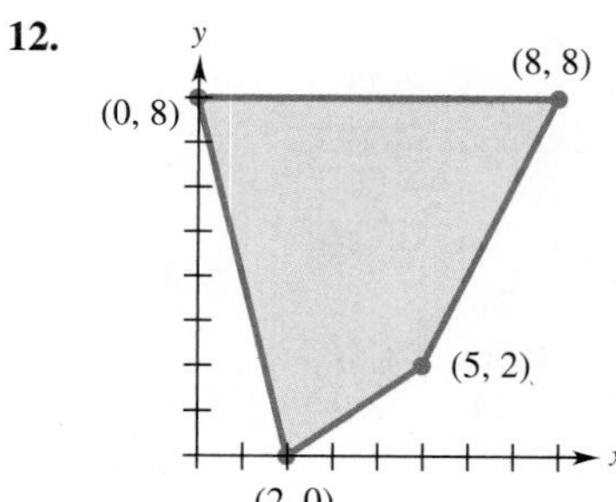

Use the graphical method to solve Exercises 13–16.

13. Maximize $z = 6x + 2y$
subject to $2x + 7y \le 14$
$2x + 3y \le 10$
$x \ge 0, y \ge 0.$

14. Find $x \ge 0$ and $y \ge 0$ such that

$$8x + 9y \ge 72$$
$$6x + 8y \ge 72$$

and $w = 2x + 10y$ is minimized.

15. Find $x \ge 0$ and $y \ge 0$ such that

$$x + y \le 50$$
$$2x + y \ge 20$$
$$x + 2y \ge 30$$

and $w = 5x + 2y$ is minimized.

16. Maximize $z = 5x - 2y$
subject to $3x + 2y \le 12$
$5x + y \ge 5$
$x \ge 0, y \ge 0.$

17. **Business** How many batches of cakes and cookies should the bakery of Exercise 9 make in order to maximize profits if cookies produce a profit of \$20 per batch and cakes produce a profit of \$30 per batch?

18. **Business** How many units of each kind of pizza should the company of Exercise 10 make in order to maximize revenue if special pizza sells for \$20 per unit and basic sells for \$18 per unit?

For each of the following problems, (a) add slack variables and (b) set up the initial simplex tableau.

19. Maximize $z = 5x_1 + 6x_2 + 3x_3$
subject to
$$\begin{aligned} x_1 + x_2 + x_3 &\le 100 \\ 2x_1 + 3x_2 &\le 500 \\ x_1 + 2x_3 &\le 350 \\ x_1 \ge 0, x_2 \ge 0, x_3 &\ge 0. \end{aligned}$$

20. Maximize $z = 2x_1 + 9x_2$
subject to
$$\begin{aligned} 3x_1 + 5x_2 &\le 47 \\ x_1 + x_2 &\le 25 \\ 5x_1 + 2x_2 &\le 35 \\ 2x_1 + x_2 &\le 30 \\ x_1 \ge 0, x_2 &\ge 0. \end{aligned}$$

21. Maximize $z = x_1 + 8x_2 + 2x_3$
subject to
$$\begin{aligned} x_1 + x_2 + x_3 &\le 90 \\ 2x_1 + 5x_2 + x_3 &\le 120 \\ x_1 + 3x_2 &\le 80 \\ x_1 \ge 0, x_2 \ge 0, x_3 &\ge 0. \end{aligned}$$

22. Maximize $z = 15x_1 + 12x_2$
subject to
$$\begin{aligned} 2x_1 + 5x_2 &\le 50 \\ x_1 + 3x_2 &\le 25 \\ 4x_1 + x_2 &\le 18 \\ x_1 + x_2 &\le 12 \\ x_1 \ge 0, x_2 &\ge 0. \end{aligned}$$

For each of the following, use the simplex method to solve the maximization linear programming problems with initial tableaus as given.

23.
$$\begin{array}{c} \begin{matrix} x_1 & x_2 & x_3 & s_1 & s_2 & \end{matrix} \\ \left[\begin{array}{ccccc|c} 1 & 2 & 3 & 1 & 0 & 28 \\ 2 & 4 & 8 & 0 & 1 & 32 \\ \hline -5 & -2 & -3 & 0 & 0 & 0 \end{array}\right] \end{array}$$

24.
$$\begin{array}{c} \begin{matrix} x_1 & x_2 & s_1 & s_2 & \end{matrix} \\ \left[\begin{array}{cccc|c} 2 & 1 & 1 & 0 & 10 \\ 9 & 3 & 0 & 1 & 15 \\ \hline -2 & -3 & 0 & 0 & 0 \end{array}\right] \end{array}$$

25.
$$\begin{array}{c} \begin{matrix} x_1 & x_2 & x_3 & s_1 & s_2 & s_3 & \end{matrix} \\ \left[\begin{array}{cccccc|c} 1 & 2 & 2 & 1 & 0 & 0 & 50 \\ 4 & 24 & 0 & 0 & 1 & 0 & 20 \\ 1 & 0 & 2 & 0 & 0 & 1 & 15 \\ \hline -5 & -3 & -2 & 0 & 0 & 0 & 0 \end{array}\right] \end{array}$$

26.
$$\begin{array}{c} \begin{matrix} x_1 & x_2 & s_1 & s_2 & s_3 & \end{matrix} \\ \left[\begin{array}{ccccc|c} 1 & -2 & 1 & 0 & 0 & 38 \\ 1 & -1 & 0 & 1 & 0 & 12 \\ 2 & 1 & 0 & 0 & 1 & 30 \\ \hline -1 & -2 & 0 & 0 & 0 & 0 \end{array}\right] \end{array}$$

Use the simplex method to solve the problem that was set up in the given exercise.

27. Exercise 19
28. Exercise 20
29. Exercise 21
30. Exercise 22

For Exercises 31–34, (a) select appropriate variables, (b) write the objective function, and (c) write the constraints as inequalities.

31. **Business** Roberta Hernandez sells three items—A, B, and C—in her gift shop. Each unit of A costs her \$2 to buy, \$1 to sell, and \$2 to deliver. For each unit of B, the costs are \$3, \$2, and \$2, respectively, and for each unit of C, the costs are \$6, \$2, and \$4, respectively. The profit on A is \$4, on B it is \$3, and on C it is \$3. How many of each item should she order to maximize her profit if she can spend \$1200 to buy, \$800 to sell, and \$500 to deliver?

32. **Business** An investor is considering three types of investment: a high-risk venture into oil leases with a potential return of 15%, a medium-risk investment in bonds with a 9% return, and a relatively safe stock investment with a 5% return. He has \$50,000 to invest. Because of the risk, he will limit his investment in oil leases and bonds to 30% and his investment in oil leases and stock to 50%. How much should he invest in each to maximize his return, assuming investment returns are as expected?

33. **Business** The Aged Wood Winery makes two white wines—Fruity and Crystal—from two kinds of grapes and sugar. The wines require the amounts of each ingredient per gallon and produce a profit per gallon as shown in the following table:

	Grape A (bushels)	Grape B (bushels)	Sugar (pounds)	Profit (dollars)
Fruity	2	2	2	12
Crystal	1	3	1	15

The winery has available 110 bushels of grape A, 125 bushels of grape B, and 90 pounds of sugar. How much of each wine should be made to maximize profit?

34. **Business** A company makes three sizes of plastic bags: 5 gallon, 10 gallon, and 20 gallon. The production time in hours for cutting, sealing, and packaging a unit of each size is as follows:

Size	Cutting	Sealing	Packaging
5 gallon	1	1	2
10 gallon	1.1	1.2	3
20 gallon	1.5	1.3	4

There are at most 8 hours available each day for each of the three operations. If the profit per unit is \$1 for 5-gallon bags, \$.90 for 10-gallon bags, and \$.95 for 20-gallon bags, how many of each size should be made per day to maximize the profit?

35. When is it necessary to use the simplex method rather than the graphical method?

36. What types of problems can be solved with the use of slack variables and surplus variables?

37. What kind of problem can be solved with the method of duals?

38. In solving a linear programming problem, you are given the following initial tableau:

$$\left[\begin{array}{rrrrr|r} 4 & 2 & 3 & 1 & 0 & 9 \\ 5 & 4 & 1 & 0 & 1 & 10 \\ \hline -6 & -7 & -5 & 0 & 0 & 0 \end{array}\right].$$

(a) What is the problem being solved?

(b) If the 1 in row 1, column 4, were a -1 rather than a 1, how would it change your answer to part (a)?

(c) After several steps of the simplex algorithm, the following tableau results:

$$\left[\begin{array}{rrrrr|r} .6 & 0 & 1 & .4 & -.2 & 1.6 \\ 1.1 & 1 & 0 & -.1 & .3 & 2.1 \\ \hline 4.7 & 0 & 0 & 1.3 & 1.1 & 22.7 \end{array}\right].$$

What is the solution? (List only the values of the original variables and the objective function. Do not include slack or surplus variables.)

(d) What is the dual of the problem you found in part (a)?

(e) What is the solution of the dual you found in part (d)? (Do not perform any steps of the simplex algorithm; just examine the tableau given in part (c).)

The tableaus in Exercises 39–41 are the final tableaus of minimization problems solved by the method of duals. State the solution and the minimum value of the objective function for each problem.

39. $$\left[\begin{array}{rrrrrr|r} 1 & 0 & 0 & 3 & 1 & 2 & 12 \\ 0 & 0 & 1 & 4 & 5 & 3 & 5 \\ 0 & 1 & 0 & -2 & 7 & -6 & 8 \\ \hline 0 & 0 & 0 & 5 & 7 & 3 & 172 \end{array}\right]$$

40. $$\left[\begin{array}{rrrrrr|r} 0 & 0 & 1 & 6 & 3 & 1 & 2 \\ 1 & 0 & 0 & 4 & -2 & 2 & 8 \\ 0 & 1 & 0 & 10 & 7 & 0 & 12 \\ \hline 0 & 0 & 0 & 9 & 5 & 8 & 62 \end{array}\right]$$

41. $$\left[\begin{array}{rrrr|r} 1 & 0 & 7 & -1 & 100 \\ 0 & 1 & 1 & 3 & 27 \\ \hline 0 & 0 & 7 & 2 & 640 \end{array}\right]$$

Use the method of duals to solve these minimization problems.

42. Minimize $w = 5y_1 + 2y_2$
subject to $2y_1 + 3y_2 \geq 6$
$2y_1 + y_2 \geq 7$
$y_1 \geq 0, y_2 \geq 0.$

43. Minimize $w = 18y_1 + 10y_2$
subject to $y_1 + y_2 \geq 17$
$5y_1 + 8y_2 \geq 42$
$y_1 \geq 0, y_2 \geq 0.$

44. Minimize $w = 4y_1 + 5y_2$
subject to $10y_1 + 5y_2 \geq 100$
$20y_1 + 10y_2 \geq 150$
$y_1 \geq 0, y_2 \geq 0.$

Write the initial simplex tableau for each of these mixed-constraint problems.

45. Maximize $z = 20x_1 + 30x_2$
subject to $5x_1 + 10x_2 \leq 120$
$10x_1 + 15x_2 \geq 200$
$x_1 \geq 0, x_2 \geq 0.$

46. Minimize $w = 4y_1 + 2y_2$
subject to $y_1 + 3y_2 \geq 6$
$2y_1 + 8y_2 \leq 21$
$y_1 \geq 0, y_2 \geq 0.$

47. Minimize $w = 12y_1 + 20y_2 - 8y_3$
subject to $y_1 + y_2 + 2y_3 \geq 48$
$y_1 + y_2 \leq 12$
$y_3 \geq 10$
$3y_1 + y_3 \geq 30$
$y_1 \geq 0, y_2 \geq 0, y_3 \geq 0.$

48. Maximize $w = 6x_1 - 3x_2 + 4x_3$
subject to $2x_1 + x_2 + x_3 \leq 112$
$x_1 + x_2 + x_3 \geq 80$
$x_1 + x_2 \leq 45$
$x_1 \geq 0, x_2 \geq 0, x_3 \geq 0.$

The given tableaus are the final tableaus of minimization problems solved by letting $w = -z$. Give the solution and the minimum value of the objective function for each problem.

49. $$\left[\begin{array}{rrrrrr|r} 0 & 1 & 0 & 2 & 5 & 0 & 17 \\ 0 & 0 & 1 & 3 & 1 & 1 & 25 \\ 1 & 0 & 0 & 4 & 2 & \frac{1}{2} & 8 \\ \hline 0 & 0 & 0 & 2 & 5 & 0 & -427 \end{array}\right]$$

50. $$\left[\begin{array}{rrrrrrr|r} 0 & 0 & 2 & 1 & 0 & 6 & 6 & 92 \\ 1 & 0 & 3 & 0 & 0 & 0 & 2 & 47 \\ 0 & 1 & 0 & 0 & 0 & 1 & 0 & 68 \\ 0 & 0 & 4 & 0 & 1 & 0 & 3 & 35 \\ \hline 0 & 0 & 5 & 0 & 0 & 2 & 9 & -1957 \end{array}\right]$$

Use the two-stage method to solve these mixed-constraint problems.

51. Exercise 45

52. Exercise 46

53. Minimize $w = 4y_1 - 8y_2$
subject to
$$\begin{aligned} y_1 + y_2 &\leq 50 \\ 2y_1 - 4y_2 &\geq 20 \\ y_1 - y_2 &\leq 22 \\ y_1 \geq 0, y_2 &\geq 0. \end{aligned}$$

54. Maximize $z = 2x_1 + 4x_2$
subject to
$$\begin{aligned} 3x_1 + 2x_2 &\leq 12 \\ 5x_1 + x_2 &\geq 5 \\ x_1 \geq 0, x_2 &\geq 0. \end{aligned}$$

Business *Solve the following maximization problems, which were begun in Exercises 31–34.*

55. Exercise 31

56. Exercise 32

57. Exercise 33

58. Exercise 34

Business *Solve the following minimization problems.*

59. Cauchy Canners produces canned corn, beans, and carrots. Demand for vegetables requires the company to produce at least 1000 cases per month. Based on past sales, it should produce at least twice as many cases of corn as of beans and at least 340 cases of carrots. It costs \$10 to produce a case of corn, \$15 to produce a case of beans, and \$25 to produce a case of carrots. How many cases of each vegetable should be produced to minimize costs? What is the minimum cost?

60. A contractor builds boathouses in two basic models: the Atlantic and the Pacific. Each Atlantic model requires 1000 feet of framing lumber, 3000 cubic feet of concrete, and \$2000 for advertising. Each Pacific model requires 2000 feet of framing lumber, 3000 cubic feet of concrete, and \$3000 for advertising. Contracts call for using at least 8000 feet of framing lumber, 18,000 cubic feet of concrete, and \$15,000 worth of advertising. If the total spent on each Atlantic model is \$3000 and the total spent on each Pacific model is \$4000, how many of each model should be built to minimize costs?

Business *Solve these mixed-constraint problems.*

61. Brand X Cannery produces canned whole tomatoes and tomato sauce. This season, the company has available 3,000,000 kilograms of tomatoes for these two products. To meet the demands of regular customers, it must produce at least 80,000 kilograms of sauce and 800,000 kilograms of whole tomatoes. The cost per kilogram is \$4 to produce canned whole tomatoes and \$3.25 to produce tomato sauce. Labor agreements require that at least 110,000 person-hours be used. Each 1-kilogram can of sauce requires 3 minutes for one worker to produce, and each 1-kilogram can of whole tomatoes requires 6 minutes for one worker. How many kilograms of tomatoes should Brand X use for each product to minimize cost? (For simplicity, assume that the production of y_1 kilograms of canned whole tomatoes and y_2 kilograms of tomato sauce requires $y_1 + y_2$ kilograms of tomatoes.)

62. A steel company produces two types of alloys. A run of type I requires 3000 pounds of molybdenum and 2000 tons of iron ore pellets, as well as \$2000 in advertising. A run of type II requires 3000 pounds of molybdenum and 1000 tons of iron ore pellets, as well as \$3000 in advertising. Total costs are \$15,000 on a run of type I and \$6000 on a run of type II. The company has on hand 18,000 pounds of molybdenum and 7000 tons of iron ore pellets and wants to use all of it. It plans to spend at least \$14,000 on advertising. How much of each type should be produced to minimize cost? What is the minimum cost?

CASE 7

Cooking with Linear Programming

Constructing a nutritious recipe can be a difficult task. The recipe must produce food that tastes good, and it must also balance the nutrients that each ingredient brings to the dish. This balancing of nutrients is very important in several diet plans that are currently popular. Many of these plans restrict the intake of certain nutrients (usually fat) while allowing for large amounts of other nutrients (protein and carbohydrates are popular choices). The number of calories in the dish is also often minimized. Linear programming can be used to help create recipes that balance nutrients.

In order to develop solutions to this type of problem, we will need to have nutritional data for the ingredients in our recipes. This data can be found in the U.S. Department of Agriculture's USDA Nutrient Database for Standard Reference, available at www.nal.usda.gov/fnic/foodcomp/search. This database contains the nutrient levels for hundreds of basic foods. The nutrient levels are given per 100 grams of food. Unfortunately, grams are not often used in recipes; instead, kitchen measures like cups, tablespoons, and fractions of vegetables are used. Table 1 shows the conversion factors from grams to more familiar kitchen units and gives serving sizes for various food.

Table 1 **Serving Sizes of Various Food**

Food	Serving Size
Beef	6 oz = 170 g
Egg	1 egg = 61 g
Feta Cheese	$\frac{1}{4}$ cup = 38 g
Lettuce	$\frac{1}{2}$ cup = 28 g
Milk	1 cup = 244 g
Oil	1 Tbsp = 13.5 g
Onion	1 onion = 110 g
Salad Dressing	1 cup = 250 g
Soy Sauce	1 Tbsp = 18 g
Spinach	1 cup = 180 g
Tomato	1 tomato = 123 g

Consider creating a recipe for a spinach omelet from eggs, milk, vegetable oil, and spinach. The nutrients of interest will be protein, fat, and carbohydrates. Calories will also be monitored. The amounts of the nutrients and calories for these ingredients are given in Table 2.

Table 2 **Nutritional Values per 100 g of Food**

Nutrient (units)	Eggs	Milk	Oil	Spinach
Calories (kcal)	152	61.44	884	23
Protein (g)	10.33	3.29	0	2.9
Fat (g)	11.44	3.34	100	.26
Carbohydrates (g)	1.04	4.66	0	3.75

Let x_1 be the number of 100-gram units of eggs to use in the recipe, x_2 be the number of 100-gram units of milk, x_3 be the number of 100-gram units of oil, and x_4 be the number of 100-gram units of spinach. We will want to minimize the number of calories in the dish while providing at least 15 grams of protein, 4 grams of carbohydrates, and 20 grams of fat. The cooking technique specifies that at least $\frac{1}{8}$ of a cup of milk (30.5 grams) must be used in the recipe. We should thus minimize the objective function (using 100-gram units of food)

$$z = 152x_1 + 61.44x_2 + 884x_3 + 23x_4$$

subject to

$$\begin{aligned} 10.33x_1 + 3.29x_2 + 0x_3 + 2.90x_4 &\geq 15 \\ 11.44x_1 + 3.34x_2 + 100x_3 + .26x_4 &\geq 20 \\ 1.04x_1 + 4.66x_2 + 0x_3 + 3.75x_4 &\geq 4 \\ 0x_1 + 1x_2 + 0x_3 + 0x_4 &\geq .305. \end{aligned}$$

Of course, all variables are subject to nonnegativity constraints:

$$x_1 \geq 0, \quad x_2 \geq 0, \quad x_3 \geq 0, \quad \text{and} \quad x_4 \geq 0.$$

Using a graphing calculator or a computer with linear programming software, we get the following solution:

$$x_1 = 1.2600, \quad x_2 = .3050, \quad x_3 = .0448, \quad \text{and} \quad x_4 = .338.$$

This recipe produces an omelet with 257.63 calories. The amounts of each ingredient are 126 grams of eggs, 30.5 grams of milk, 4.48 grams of oil, and 33.8 grams of spinach. Converting to kitchen units using Table 1, we find the recipe to be approximately 2 eggs, $\frac{1}{8}$ cup milk, 1 teaspoon oil, and $\frac{1}{4}$ cup spinach.

EXERCISES

1. Consider preparing a high-carbohydrate Greek salad using feta cheese, lettuce, salad dressing, and tomato. The amount of carbohydrates in the salad should be maximized. In

addition, the salad should have less than 260 calories, over 210 milligrams of calcium, and over 6 grams of protein. The salad should also weigh less than 400 grams and be dressed with at least 2 tablespoons ($\frac{1}{8}$ cup) of salad dressing. The amounts of the nutrients and calories for these ingredients are given in Table 3.

Table 3 **Nutritional Values per 100 g of Food**

Nutrient (units)	Feta Cheese	Lettuce	Salad Dressing	Tomato
Calories (kcal)	263	14	448.8	21
Calcium (mg)	492.5	36	0	5
Protein (g)	10.33	1.62	0	.85
Carbohydrates (g)	4.09	2.37	2.5	4.64

Use linear programming to find the number of 100-gram units of each ingredient in such a Greek salad, and convert to kitchen units by using Table 1. (*Hint:* Since the ingredients are measured in 100-gram units, the constant in the weight contraint is 4, not 400.)

2. Consider preparing a stir-fry using beef, oil, onion, and soy sauce. A low-calorie stir-fry is desired, which contains less than 10 grams of carbohydrates, more than 50 grams of protein, and more than 3.5 grams of vitamin C. In order for the wok to function correctly, at least one teaspoon (or 4.5 grams) of oil must be used in the recipe. The amounts of the nutrients and calories for these ingredients are given in Table 4.

Table 4 **Nutritional Values per 100 g of Food**

Nutrient (units)	Beef	Oil	Onion	Soy Sauce
Calories (kcal)	215	884	38	60
Protein (g)	26	0	1.16	10.51
Carbohydrates (g)	0	1	8.63	5.57
Vitamin C (g)	0	0	6.4	0

Use linear programming to find the number of 100-gram units of each ingredient to be used in the stir-fry, and convert to kitchen units by using Table 1.

Sets and Probability

We often use the relative frequency of an event from a survey to estimate unknown probabilities. For example, we can estimate the probability that a high-earning chief executive officer is in his sixties; see Example 10 in Section 8.4. Other applications of probability occur in the health and social sciences. Examples include estimating the probability of having a healthy weight, owning two cars within a household, or working full time; see Exercises 45, 52, and 57 on pages 504–505.

CASE 8: Medical Diagnosis

Federal officials cannot predict exactly how the number of traffic deaths is affected by the trend toward fewer drunken drivers and the increased use of seat belts. Economists cannot tell exactly how stricter federal regulations on bank loans affect the U.S. economy. The number of traffic deaths and the growth of the economy are subject to many factors that cannot be predicted precisely.

Probability theory enables us to deal with uncertainty. The basic concepts of probability are discussed in this chapter, and applications of probability are discussed in the next chapter. Sets and set operations are the basic tools for the study of probability, so we begin with them.

8.1 ▶ Sets

Think of a set as a well-defined collection of objects. A set of coins might include one of each type of coin now put out by the U.S. government. Another set might be made up of all the students in your English class. By contrast, a collection of young adults does not constitute a set unless the designation "young adult" is clearly defined. For example, this set might be defined as those aged 18–29.

In mathematics, sets are often made up of numbers. The set consisting of the numbers 3, 4, and 5 is written as

$$\{3, 4, 5\},$$

where **set braces**, { }, are used to enclose the numbers belonging to the set. The numbers, 3, 4, and 5 are called the **elements**, or **members**, of this set. To show that 4 is an element of the set {3, 4, 5}, we use the symbol $\in$ and write

$$4 \in \{3, 4, 5\},$$

read, "4 is an element of the set containing 3, 4, and 5."

Also, $5 \in \{3, 4, 5\}$. Place a slash through the symbol $\in$ to show that 8 is *not* an element of this set:

$$8 \notin \{3, 4, 5\}.$$

This statement is read, "8 is not an element of the set {3, 4, 5}."

Sets are often named with capital letters, so that if

$$B = \{5, 6, 7\},$$

then, for example, $6 \in B$ and $10 \notin B$. ✓1

✓Checkpoint 1

Indicate whether each statement is *true* or *false*.

(a) $9 \in \{8, 4, -3, -9, 6\}$.

(b) $4 \notin \{3, 9, 7\}$.

(c) If $M = \{0, 1, 2, 3, 4\}$, then $0 \in M$.

Answers: See page 475.

Sometimes a set has no elements. Some examples are the set of female presidents of the United States in the period 1788–2008, the set of counting numbers less than 1, and the set of men more than 10 feet tall. A set with no elements is called the **empty set**. The symbol $\varnothing$ is used to represent the empty set.

CAUTION Be careful to distinguish between the symbols 0, $\varnothing$, and {0}. The symbol 0 represents a *number*; $\varnothing$ represents a *set* with no elements; and {0} represents a *set* with one element, the number 0. Do not confuse the empty set symbol $\varnothing$ with the zero symbol 0 on a computer screen or printout.

Two sets are **equal** if they contain exactly the same elements. The sets {5, 6, 7}, {7, 6, 5}, and {6, 5, 7} all contain exactly the same elements and are equal. In symbols,

$$\{5, 6, 7\} = \{7, 6, 5\} = \{6, 5, 7\}.$$

This means that the ordering of the elements in a set is unimportant. Sets that do not contain exactly the same elements are *not equal.* For example, the sets {5, 6, 7} and {5, 6, 7, 8} do not contain exactly the same elements and are not equal. We show this by writing

$$\{5, 6, 7\} \neq \{5, 6, 7, 8\}.$$

Sometimes we describe a set by a common property of its elements rather than by a list of its elements. This common property can be expressed with **set-builder notation**; for example,

$$\{x|x \text{ has property } P\}$$

(read, "the set of all elements x such that x has property P") represents the set of all elements x having some property P.

Example 1 List the elements belonging to each of the given sets.

(a) $\{x|x$ is a natural number less than $5\}$

Solution The natural numbers less than 5 make up the set $\{1, 2, 3, 4\}$.

(b) $\{x|x$ is a state that borders Florida$\}$

Solution The states that border Florida make up the set = {Alabama, Georgia}. 

✓Checkpoint 2

List the elements in the given sets.

(a) $\{x|x$ is a counting number more than 5 and less than $8\}$

(b) $\{x|x$ is an integer, and $-3 < x \leq 1\}$

Answers: See page 475.

The **universal set** in a particular discussion is a set that contains all of the objects being discussed. In grade-school arithmetic, for example, the set of whole numbers might be the universal set, whereas in a college calculus class the universal set might be the set of all real numbers. When it is necessary to consider the universal set being used, it will be clearly specified or easily understood from the context of the problem.

Sometimes, every element of one set also belongs to another set. For example, if

$$A = \{3, 4, 5, 6\}$$

and

$$B = \{2, 3, 4, 5, 6, 7, 8\},$$

then every element of A is also an element of B. This is an example of the following definition.

A set A is a **subset** of a set B (written $A \subseteq B$) provided that every element of A is also an element of B.

Example 2 For each case, decide whether $M \subseteq N$.

(a) M is the set of all businesses making a profit in the last calendar year. N is the set of all businesses.

Solution Each business making a profit is also a business, so $M \subseteq N$.

(b) M is the set of all first-year students at a college at the end of the academic year, and N is the set of all 18-year-old students at the college at the end of the academic year.

Solution By the end of the academic year, some first-year students are older than 18, so there are elements in M that are not in N. Thus, M is not a subset of N, written $M \not\subseteq N$.

Every set A is a subset of itself, because the statement "every element of A is also an element of A" is always true. It is also true that the empty set is a subset of every set.*

> For any set A,
>
> $$\emptyset \subseteq A \quad \text{and} \quad A \subseteq A.$$

A set A is said to be a **proper subset** of a set B (written $A \subset B$) if every element of A is an element of B, but B contains at least one element that is not a member of A.

Example 3 Decide whether $E \subset F$.

(a) $E = \{2, 4, 6, 8\}$ and $F = \{1, 2, 3, 4, 5, 6, 7, 8, 9, 10\}$.

Solution Since each element of E is an element of F and F contains several elements not in E, $E \subset F$.

(b) E is the set of registered voters in Texas. F is the set of adults aged 18 years or older.

Solution To register to vote, one must be at least 18 years old. Not all adults at least 18 years old, however, are registered. Thus, every element of E is contained in F and F contains elements not in E. Therefore, $E \subset F$.

(c) E is the set of diet soda drinks. F is the set of diet soda drinks sweetened with Nutrasweet®.

Solution Some diet soda drinks are sweetened with the sugar substitute Splenda®. In this case, E is not a proper subset of F (written $E \not\subset F$), nor is it a subset of F at all ($E \not\subseteq F$). 3✓

✓Checkpoint 3

Indicate whether each statement is *true* or *false*.

(a) $\{10, 20, 30\} \subseteq \{20, 30, 40, 50\}$.

(b) $\{x|x \text{ is a minivan}\} \subseteq \{x|x \text{ is a motorvehicle}\}$.

(c) $\{a, e, i, o, u\} \subset \{a, e, i, o, u, y\}$.

(d) $\{x|x$ is a U.S. state that begins with the letter "A"$\}$ $\subset$ {Alabama, Alaska, Arizona, Arkansas}.

Answers: See page 475.

Example 4 List all possible subsets for each of the given sets.

(a) $\{7, 8\}$

Solution A good way to find the subsets of $\{7, 8\}$ is to use a **tree diagram**—a systematic way of listing all the subsets of a given set. The tree diagram in Figure 8.1(a) shows there are four subsets of $\{7, 8\}$:

$$\emptyset, \quad \{7\}, \quad \{8\}, \quad \text{and} \quad \{7, 8\}.$$

(b) {a, b, c}

Solution The tree diagram in Figure 8.1(b) shows that there are 8 subset of {a, b, c}:

$$\emptyset, \quad \{a\}, \quad \{b\}, \quad \{c\}, \quad \{a, b\}, \quad \{a, c\}, \quad \{b, c\}, \quad \text{and} \quad \{a, b, c\}.$$ 4✓

Checkpoint 4

List all subsets of {w, x, y, z,}.

Answer: See page 475.

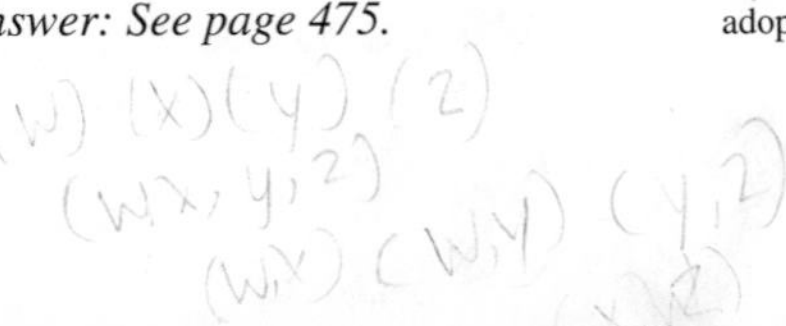

*This fact is not intuitively obvious to most people. If you wish, you can think of it as a convention that we agree to adopt in order to simplify the statements of several results later.

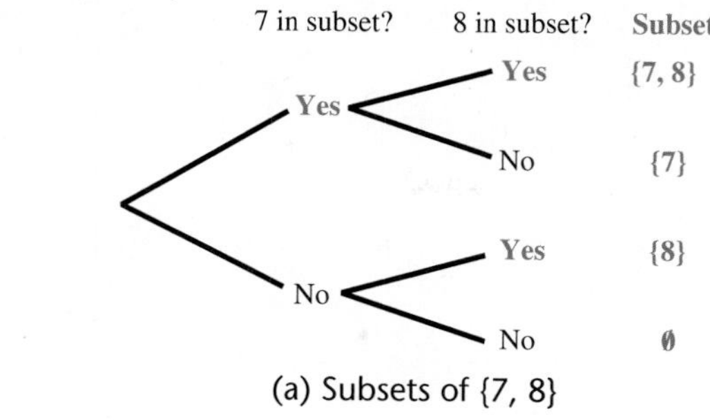

(a) Subsets of {7, 8}

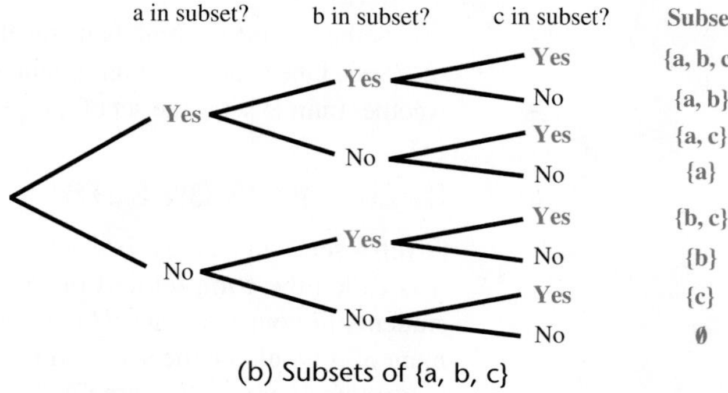

(b) Subsets of {a, b, c}

FIGURE 8.1

✓Checkpoint 5

Find the number of subsets for each of the given sets.

(a) $\{x|x \text{ is a season of the year}\}$

(b) $\{-6, -5, -4, -3, -2, -1, 0\}$

(c) $\{6\}$

Answers: See page 475.

By using the fact that there are two possibilities for each element (either it is in the subset or it is not), we have found that a set with 2 elements has 4 ($= 2^2$) subsets and a set with 3 elements has 8 ($= 2^3$) subsets. Similar arguments work for any finite set and lead to the following conclusion.

> A set of n distinct elements has 2^n subsets.

✓Checkpoint 6

Refer to sets A, B, C, and U in the diagram.

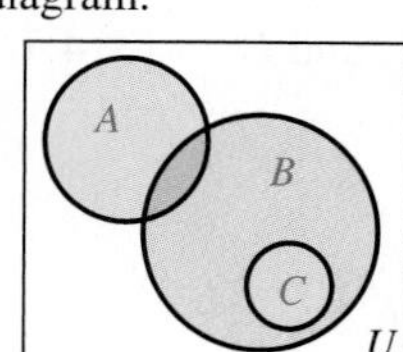

(a) Is $A \subseteq B$?

(b) Is $C \subseteq B$?

(c) Is $C \subseteq U$?

(d) Is $\varnothing \subseteq A$?

Answers: See page 475.

Example 5 Find the number of subsets for each of the given sets.

(a) {blue, brown, hazel, green}

Solution Since this set has 4 elements, it has $2^4 = 16$ subsets.

(b) $\{x|x \text{ is a month of the year}\}$

Solution This set has 12 elements and therefore has $2^{12} = 4096$ subsets.

(c) $\varnothing$

Solution Since the empty set has 0 elements, it has $2^0 = 1$ subset, $\varnothing$ itself. ✓5

Venn diagrams are sometimes used to illustrate relationships among sets. The Venn diagram in Figure 8.2, on the next page, shows a set A that is a subset of a set B, because A is entirely in B. (The areas of the regions are not meant to be proportional to the sizes of the corresponding sets.) The rectangle represents the universal set U. ✓6

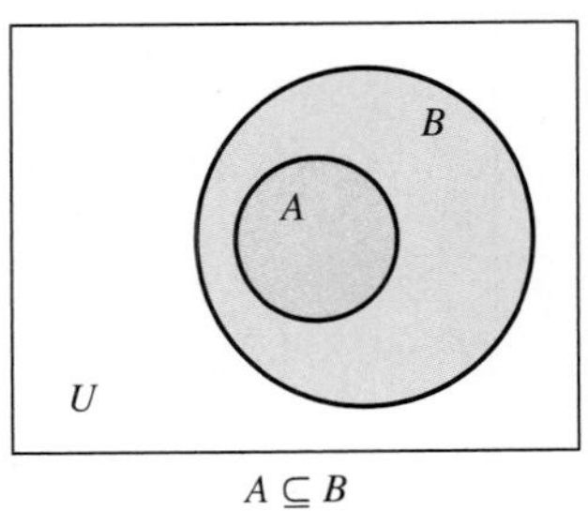

$A \subseteq B$

FIGURE 8.2

Some sets have infinitely many elements. We often use the notation "..." to indicate such sets. One example of an infinite set is the set of natural numbers, $\{1, 2, 3, 4, \ldots\}$. Another infinite set is the set of integers, $\{\ldots, -3, -2, -1, 0, 1, 2, 3, \ldots\}$.

OPERATIONS ON SETS

Given a set A and a universal set U, the set of all elements of U that do *not* belong to A is called the **complement** of set A. For example, if A is the set of all the female students in your class and U is the set of all students in the class, then the complement of A would be the set of all male students in the class. The complement of set A is written A' (read "A-prime"). The Venn diagram in Figure 8.3 shows a set B. Its complement, B', is shown in color.

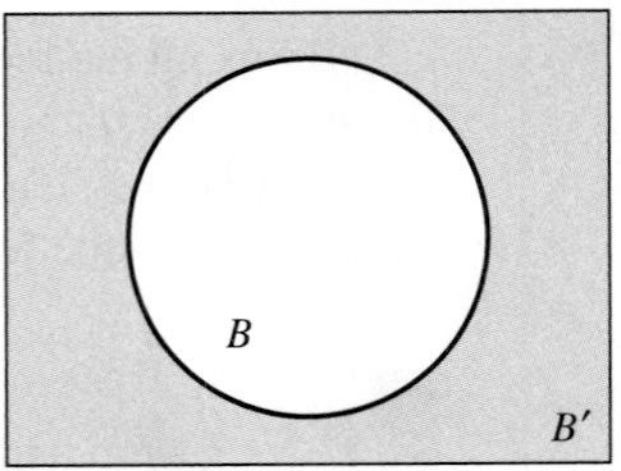

FIGURE 8.3

Some textbooks use $\overline{A}$ to denote the complement of A. This notation conveys the same meaning as A'.

Checkpoint 7

Let $U = \{a, b, c, d, e, f, g\}$, with $K = \{c, d, f, g\}$ and $R = \{a, c, d, e, g\}$. Find

(a) K';

(b) R'.

Answers: See page 475.

Example 6 Let $U = \{1, 2, 3, 4, 5, 6, 7\}$, $A = \{1, 3, 5, 7\}$, and $B = \{3, 4, 6\}$. Find the given sets.

(a) A'

Solution Set A' contains the elements of U that are not in A:

$$A' = \{2, 4, 6\}.$$

(b) B'

Solution $B' = \{1, 2, 5, 7\}$.

(c) $\varnothing'$ and U'

Solution $\varnothing' = U$ and $U' = \varnothing$. [7]

Given two sets A and B, the set of all elements belonging to *both* set A and set B is called the **intersection** of the two sets, written $A \cap B$. For example, the elements that belong to both $A = \{1, 2, 4, 5, 7\}$ and $B = \{2, 4, 5, 7, 9, 11\}$ are 2, 4, 5, and 7, so

$$\begin{aligned} \text{A and B} &= A \cap B \\ &= \{1, 2, 4, 5, 7\} \cap \{2, 4, 5, 7, 9, 11\} \\ &= \{2, 4, 5, 7\}. \end{aligned}$$

The Venn diagram in Figure 8.4 shows two sets A and B, with their intersection, $A \cap B$, shown in color.

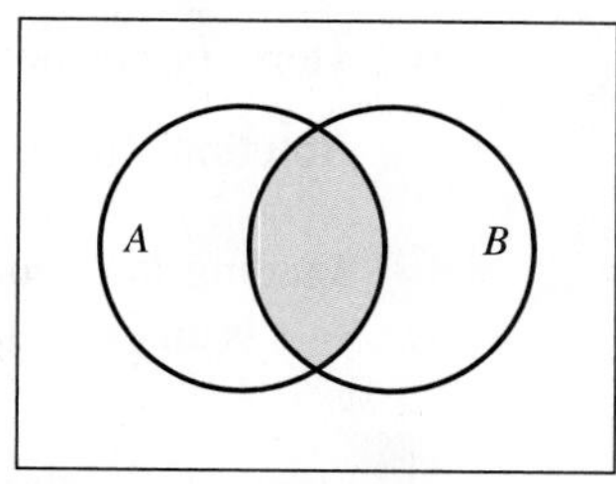

$A \cap B$

FIGURE 8.4

Find the following.

(a) $\{1, 2, 3, 4\} \cap \{3, 5, 7, 9\}$

(b) Suppose K is the set of all blue-eyed blondes in a class and J is the set of all blue-eyed brunettes in the class. Let $P = \{x | x \text{ is a brown-eyed redhead}\}$. If the class has only blondes or brunettes, find $K \cap P$.

Answers: See page 475.

Example 7 Find the given sets.

(a) $\{9, 15, 25, 36\} \cap \{15, 20, 25, 30, 35\}$

Solution $\{15, 25\}$. The elements 15 and 25 are the only ones belonging to both sets.

(b) $\{x | x \text{ is a teenager}\} \cap \{x | x \text{ is a senior citizen}\}$

Solution $\varnothing$ since no teenager is a senior citizen. 8✓

Two sets that have no elements in common are called **disjoint sets**. For example, there are no elements common to both {50, 51, 54} and {52, 53, 55, 56}, so these two sets are disjoint, and

$$\{50, 51, 54\} \cap \{52, 53, 55, 56\} = \varnothing.$$

The result of this example can be generalized as follows.

> For any sets A and B,
>
> if A and B are disjoint sets, then $A \cap B = \varnothing$.

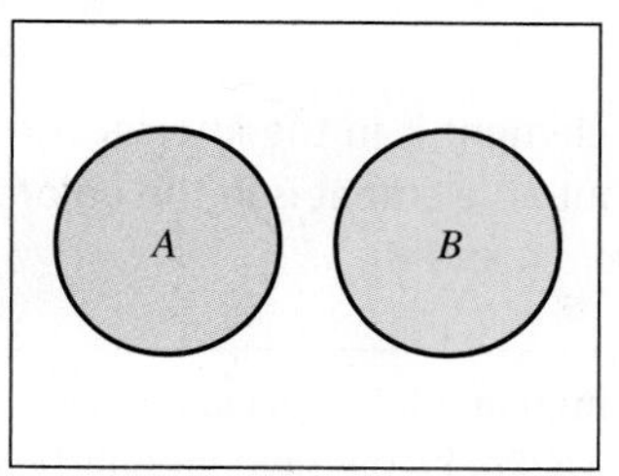

A and B are disjoint sets.

FIGURE 8.5

Figure 8.5 is a Venn diagram of disjoint sets.

The set of all elements belonging to set A or to set B, or to both sets, is called the **union** of the two sets, written $A \cup B$. For example, for sets A = {1, 3, 5} and B = {3, 5, 7, 9},

$$\begin{aligned} \text{A or B} &= \text{A} \cup \text{B} \\ &= \{1, 3, 5\} \cup \{3, 5, 7, 9\} \\ &= \{1, 3, 5, 7, 9\}. \end{aligned}$$

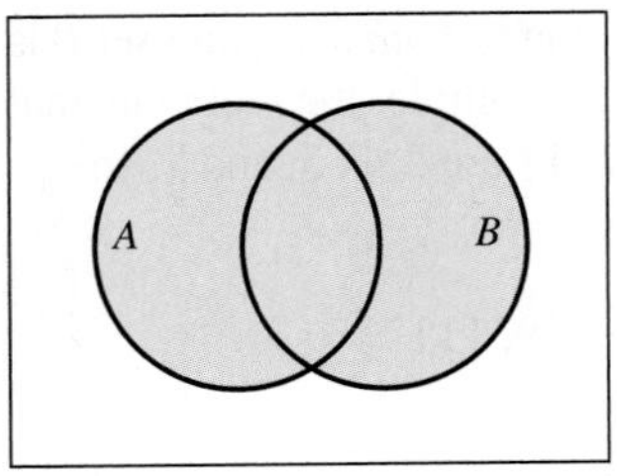

$A \cup B$

FIGURE 8.6

The Venn diagram in Figure 8.6 shows two sets A and B, with their union, $A \cup B$, shown in color.

Example 8 Find the given sets.

(a) $\{1, 2, 5, 9, 14\} \cup \{1, 3, 4, 8\}$.

Solution Begin by listing the elements of the first set, $\{1, 2, 5, 9, 14\}$. Then include any elements from the second set *that are not already listed.* Doing this gives

$$\{1, 2, 5, 9, 14\} \cup \{1, 3, 4, 8\} = \{1, 2, 3, 4, 5, 8, 9, 14\}.$$

(b) {terriers, spaniels, chows, dalmatians} ∪ {spaniels, collies, bulldogs}

Solution {terriers, spaniels, chows, dalmatians, collies, bulldogs}. 9✓

✓Checkpoint 9

Work the given union problems.

(a) Find $\{a, b, c\} \cup \{a, c, e\}$.

(b) Describe $K \cup J$ in words for the sets given in Checkpoint 8(b).

Answers: See page 475.

Finding the complement of a set, the intersection of two sets, or the union of two sets is an example of a *set operation.*

Operations on Sets

Let A and B be any sets, with U signifying the universal set. Then

the **complement** of A, written A', is

$$A' = \{x | x \notin A \text{ and } x \in U\};$$

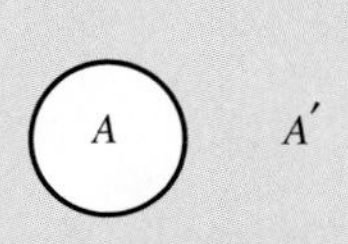

the **intersection** of A and B is

$$A \cap B = \{x | x \in A \textbf{ and } x \in B\};$$

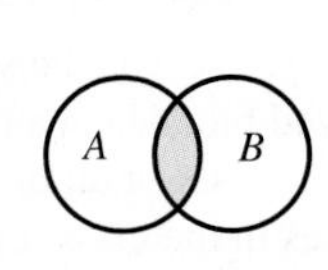

the **union** of A and B is

$$A \cup B = \{x | x \in A \textbf{ or } x \in B \textbf{ or both}\}.$$

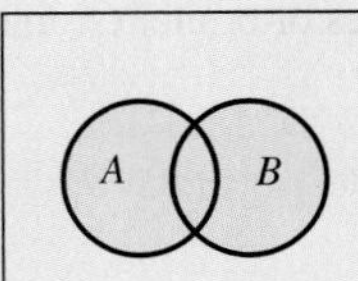

CAUTION As shown in the preceding definitions, an element is in the intersection of sets A and B if it is in *both* A and B at the same time, but an element is in the union of sets A and B if it is in *either* set A or set B, *or* in both sets A and B.

Example 9 The following table gives the 52-week high and low prices, the last price, and the change in price from the day before for five stocks on a recent day.*

*Morningstar.com, February 15, 2009.

Stock	High	Low	Last	Change
Allstate	52.16	17.12	21.23	−.30
Apple	192.24	78.20	98.99	−.11
Microsoft	32.10	16.75	19.09	−.17
Pepsi	75.25	43.78	52.57	+.57
UPS	75.08	41.40	44.84	−.06

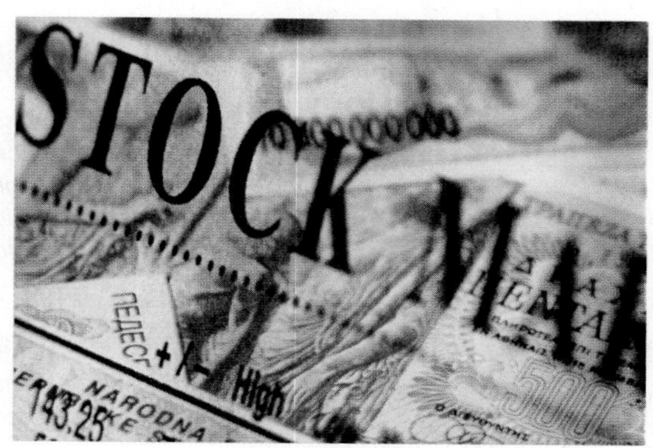

Let the universal set U consist of the five stocks listed in the table. Let A contain all stocks with a high price greater than \$60, B all stocks with a last price between \$25 and \$55, and C all stocks with a negative change. Find the results of the given set operations.

(a) B'

Solution Set B consists of Pepsi and UPS. Set B' contains all the listed stocks that are not in set B, so

$$B' = \{\text{Allstate, Apple, Microsoft}\}.$$

(b) $A \cap C$

Solution Set A consists of Apple, Pepsi, and UPS, and set C consists of Allstate, Apple, Microsoft, and UPS. Hence,

$$A \cap C = \{\text{Apple, UPS}\}.$$

(c) $A \cup B$

Solution $A \cup B = \{\text{Apple, Pepsi, UPS}\}$. 10✓

✓Checkpoint 10

In Example 9, find the given set of stocks.

(a) All stocks with a last price between \$25 and \$55 and with a negative price change

(b) All stocks with a last price between \$25 and \$55 or with a negative price change

Answers: *See page 475.*

8.1 ▶ Exercises

Write true or false for each statement.

1. $3 \in \{2, 5, 7, 9, 10\}$
2. $6 \in \{-2, 6, 9, 5\}$
3. $9 \notin \{2, 1, 5, 8\}$
4. $3 \notin \{7, 6, 5, 4\}$
5. $\{2, 5, 8, 9\} = \{2, 5, 9, 8\}$
6. $\{3, 7, 12, 14\} = \{3, 7, 12, 14, 0\}$
7. {all whole numbers greater than 7 and less than 10} = {8, 9}
8. {all counting numbers not greater than 3} = {0, 1, 2}
9. $\{x \mid x \text{ is an odd integer}, 6 \le x \le 18\} = \{7, 9, 11, 15, 17\}$

10. $\{x|x \text{ is a vowel}\} = \{a, e, i, o, u\}$

11. The elements of a set may be sets themselves, as in $\{1, \{1, 3\}, \{2\}, 4\}$. Explain why the set $\{\varnothing\}$ is not the same set as $\{0\}$.

12. What is set-builder notation? Give an example.

Let $A = \{-3, 0, 3\}$, $B = \{-2, -1, 0, 1, 2\}$, $C = \{-3, -1\}$, $D = \{0\}$, $E = \{-2\}$, and $U = \{-3, -2, -1, 0, 1, 2, 3\}$. Insert $\subseteq$ or $\not\subseteq$ to make the given statements true. (See Example 2.)

13. A_____U

14. E_____A

15. A_____E

16. B_____C

17. $\varnothing$_____A

18. $\{0, 2\}$_____D

19. D_____B

20. A_____C

Find the number of subsets of the given set. (See Example 5.)

21. $\{A, B, C\}$

22. {red, yellow, blue, black, white}

23. $\{x|x \text{ is an integer strictly between 0 and 8}\}$

24. $\{x|x \text{ is a whole number less than 4}\}$

Find the complement of each set. (See Example 6.)

25. The set in Exercise 23 if U is the set of all integers.

26. The set in Exercise 24 if U is the set of all whole numbers.

27. Describe the intersection and union of sets. How do they differ?

Insert $\cap$ or $\cup$ to make each statement true. (See Examples 7 and 8.)

28. $\{5, 7, 9, 19\}$_____$\{7, 9, 11, 15\} = \{7, 9\}$

29. $\{8, 11, 15\}$_____$\{8, 11, 19, 20\} = \{8, 11\}$

30. $\{2, 1, 7\}$_____$\{1, 5, 9\} = \{1\}$

31. $\{6, 12, 14, 16\}$_____$\{6, 14, 19\} = \{6, 14\}$

32. $\{3, 5, 9, 10\}$_____$\varnothing = \varnothing$

33. $\{3, 5, 9, 10\}$_____$\varnothing = \{3, 5, 9, 10\}$

34. $\{1, 2, 4\}$_____$\{1, 2, 4\} = \{1, 2, 4\}$

35. $\{1, 2, 4\}$_____$\{1, 2\} = \{1, 2, 4\}$

36. Is it possible for two nonempty sets to have the same intersection and union? If so, give an example.

Let $U = \{a, b, c, d, e, f, 1, 2, 3, 4, 5, 6\}$, $X = \{a, b, c, 1, 2, 3\}$, $Y = \{b, d, f, 1, 3, 5\}$, and $Z = \{b, d, 2, 3, 5\}$.

List the members of each of the given sets, using set braces. (See Examples 6–8.)

37. $X \cap Y$

38. $X \cup Y$

39. X'

40. Y'

41. $X' \cap Y'$

42. $X' \cap Z$

43. $X \cup (Y \cap Z)$

44. $Y \cap (X \cup Z)$

Let U = {all students in this school},
M = {all students taking this course},
N = {all students taking accounting}, and
P = {all students taking philosophy}.

Describe each of the following sets in words.

45. M'

46. $M \cup N$

47. $N \cap P$

48. $N' \cap P'$

49. Refer to the sets listed in the directions for Exercises 13–20. Which pairs of sets are disjoint?

50. Refer to the sets listed in the directions for Exercises 37–44. Which pairs of sets are disjoint?

Refer to Example 9 in the text. Describe each of the sets in Exercises 51–54 in words; then list the elements of each set.

51. A'

52. $B \cup C$

53. $A' \cap B'$

54. $B' \cup C$

Business *An electronics store classifies credit applicants by sex, marital status, and employment status. Let the universal set be the set of all applicants, M be the set of male applicants, S be the set of single applicants, and E be the set of employed applicants. Describe the following sets in words.*

55. $M \cap E$

56. $M' \cap S$

57. $M' \cup S'$

Business *The U.S. advertising volume (in millions of dollars) spent by certain types of media in 2006 and 2007 is shown in the following table:**

Medium	2006	2007
Newspapers	46,555	42,133
Magazines	13,168	13,787
Broadcast Television	46,880	44,521
Cable Television	25,025	26,319
Radio	19,643	19,152
Direct Mail	58,642	60,225
Internet	9,100	10,529

List the elements of each set.

58. The set of all media that collected more than \$40,000 million in both 2006 and 2007.

59. The set of all media that collected less than \$11,000 million in 2006 or 2007 (or both years).

60. The set of all media that had revenues rise from 2006 to 2007.

Business *The top seven cable television providers as of September 2008 are listed here.† Use this information for Exercises 61–66.*

Rank	Cable Provider	Subscribers
1	Comcast Cable Communications	24,406,000
2	Time Warner Cable	13,266,000
3	Cox Communications	5,382,125
4	Charter Communications	5,146,100
5	Cablevision Systems	3,112,000
6	Bright House Networks LLC	2,331,089
7	Suddenlink Communications	1,395,189

**Statistical Abstract of the United States*: 2009.

†National Cable Television Association, www.ncta.com.

List the elements of the following sets.

61. F, the set of cable providers with more than 8 million subscribers.

62. G, the set of cable providers with between 2 and 5 million subscribers.

63. H, the set of cable providers with over 2.5 million subscribers.

64. $F \cup G$ **65.** $H \cap F$ **66.** G'

Health *The following table shows some symptoms of an underactive thyroid and an overactive thyroid:*

Underactive Thyroid	Overactive Thyroid
Sleepiness, s	Insomnia, i
Dry hands, d	Moist hands, m
Intolerance of cold, c	Intolerance of heat, h
Goiter, g	Goiter, g

Let U be the smallest possible set that includes all the symptoms listed, N be the set of symptoms for an underactive thyroid, and O be the set of symptoms for an overactive thyroid. Use the lowercase letters in the table to list the elements of each set.

67. O' **68.** N'

69. $N \cap O$ **70.** $N \cup O$

✓ Checkpoint Answers

1. **(a)** False **(b)** True **(c)** True
2. **(a)** $\{6, 7\}$ **(b)** $\{-2, -1, 0, 1\}$
3. **(a)** False **(b)** True **(c)** True **(d)** False
4. $\varnothing, \{w\}, \{x\}, \{y\}, \{z\}, \{w, x\}, \{w, y\}, \{w, z\}, \{x, y\}, \{x, z\}, \{y, z\}, \{w, x, y\}, \{w, x, z\}, \{w, y, z\}, \{x, y, z\}, \{w, x, y, z\}$
5. **(a)** 16 **(b)** 128 **(c)** 2
6. **(a)** No **(b)** Yes **(c)** Yes **(d)** Yes
7. **(a)** $\{a, b, e\}$ **(b)** $\{b, f\}$
8. **(a)** $\{3\}$ **(b)** $\varnothing$
9. **(a)** $\{a, b, c, e\}$
   (b) All members of the class who have blue eyes
10. **(a)** {UPS} **(b)** {Allstate, Apple, Microsoft, Pepsi, UPS}

8.2 ▸ Applications of Venn Diagrams

We used Venn diagrams in the last section to illustrate set union and intersection. The rectangular region in a Venn diagram represents the universal set U. Including only a single set A inside the universal set, as in Figure 8.7, divides U into two nonoverlapping regions. Region 1 represents A', those elements outside set A, while region 2 represents those elements belonging to set A. (The numbering of these regions is arbitrary.)

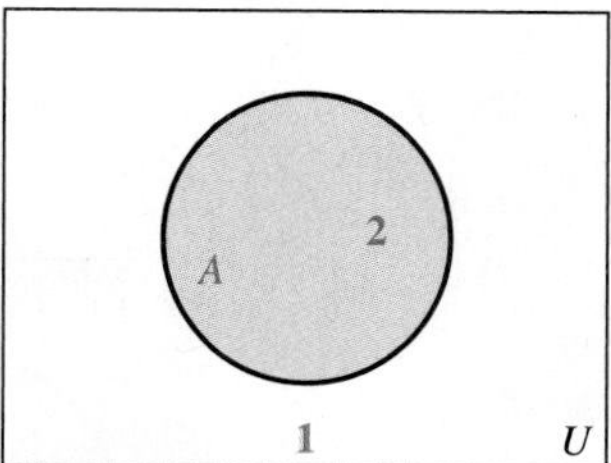

One set leads to 2 regions. (Numbering is arbitrary.)

FIGURE 8.7

Two sets lead to 4 regions. (Numbering is arbitrary.)

FIGURE 8.8

The Venn diagram of Figure 8.8 shows two sets inside U. These two sets divide the universal set into four nonoverlapping regions. As labeled in Figure 8.8, region 1 includes those elements outside both set A and set B. Region 2 includes those elements belonging to A and not to B. Region 3 includes those elements belonging to both A and B. Which elements belong to region 4? (Again, the numbering is arbitrary.)

Example 1 Draw a Venn diagram similar to Figure 8.8, and shade the regions representing the given sets.

(a) $A' \cap B$

Solution Set A' contains all the elements outside set A. As labeled in Figure 8.8, A' is represented by regions 1 and 4. Set B is represented by the elements in regions 3 and 4. The intersection of sets A' and B, the set $A' \cap B$, is given by the region common to regions 1 and 4 and regions 3 and 4. The result, region 4, is shaded in Figure 8.9.

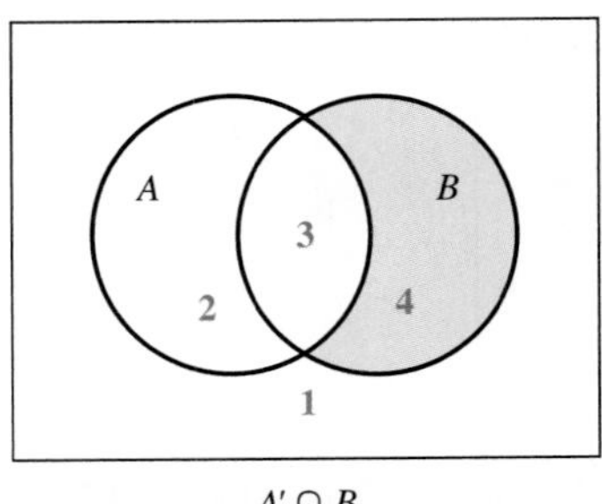

FIGURE 8.9

(b) $A' \cup B'$

Solution Again, set A' is represented by regions 1 and 4 and set B' by regions 1 and 2. To find $A' \cup B'$, identify the region that represents the set of all elements in A', B', or both. The result, which is shaded in Figure 8.10, includes regions 1, 2, and 4. ✓1

Draw Venn diagrams for the given set operations.

(a) $A \cup B'$

(b) $A' \cap B'$

Answers: See page 485.

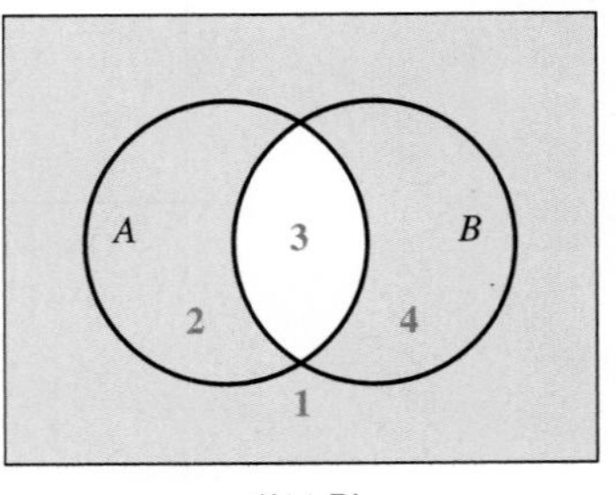

FIGURE 8.10

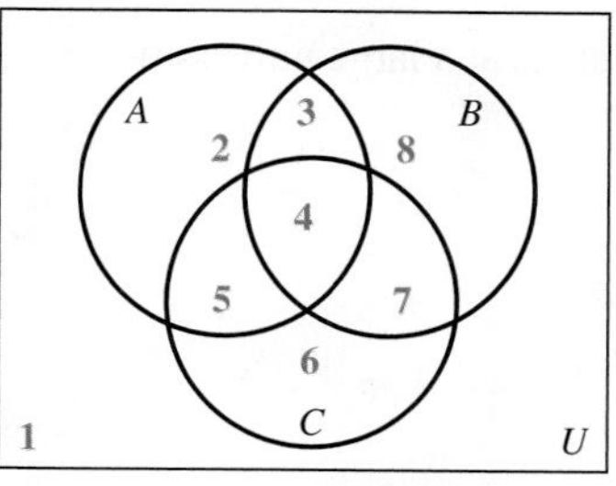

Three sets lead to 8 regions.

FIGURE 8.11

Venn diagrams also can be drawn with three sets inside U. These three sets divide the universal set into eight nonoverlapping regions that can be numbered (arbitrarily) as in Figure 8.11.

Example 2 Shade $A' \cup (B \cap C')$ in a Venn diagram.

Solution First find $B \cap C'$; see Figure 8.12. Set B is represented by regions 3, 4, 7, and 8, and set C' by regions 1, 2, 3, and 8. The overlap of these regions, regions 3 and 8, represents the set $B \cap C'$. Set A' is represented by regions 1, 6, 7, and 8. The union of regions 3 and 8 and regions 1, 6, 7, and 8 contains regions 1, 3, 6, 7, and 8, which are shaded in Figure 8.12. 2✓

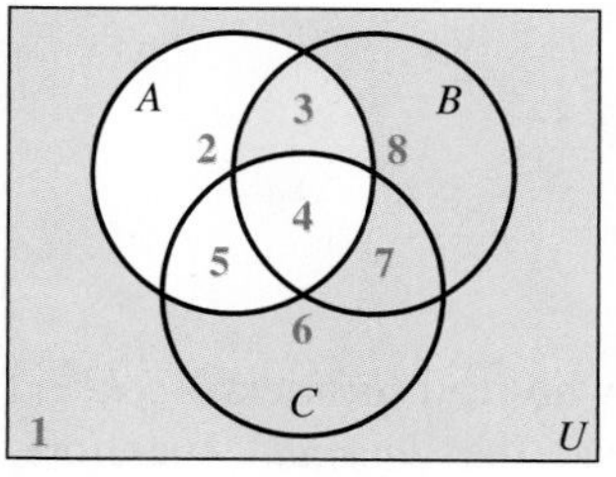

$A' \cup (B \cap C')$

FIGURE 8.12

✓Checkpoint 2

Draw Venn diagrams for the given set operations.

(a) $(B' \cap A) \cup C$

(b) $(A \cup B)' \cap C$

Answers: See page 485.

Venn diagrams can be used to solve problems that result from surveying groups of people. As an example, suppose a researcher collecting data on 100 households finds that

29 have a DVD player (DVD),

21 have a digital camera (DC), and

15 have both.

The researcher wants to answer the following questions:

(a) How many do not have a digital camera?

(b) How many have neither a DVD player nor a digital camera?

(c) How many have a DVD player, but not a digital camera?

A Venn diagram like the one in Figure 8.13 will help sort out the information. In Figure 8.13(a), we put the number 15 in the region common to both a digital camera and a DVD player, because 15 households have both. Of the 29 with a DVD player, $29 - 15 = 14$ have no digital camera, so in Figure 8.13(b) we put 14 in the region for a DVD player, but no digital camera. Similarly, $21 - 15 = 6$ households have a digital camera, but not a DVD player, so we put 6 in that region. Finally, the diagram shows that $100 - 6 - 15 - 14 = 65$ households have neither a digital camera nor a DVD player. Now we can answer the questions:

(a) $65 + 14 = 79$.

(b) 65 have neither.

(c) 14 have a DVD player, but not a digital camera. 3✓

✓Checkpoint 3

(a) Place numbers in the regions on a Venn diagram if the data on the 100 households in Example 2 showed

35 DVD players;
28 DCs;
20 with both.

(b) How many have a DC, but not a DVD player?

Answers: See page 485.

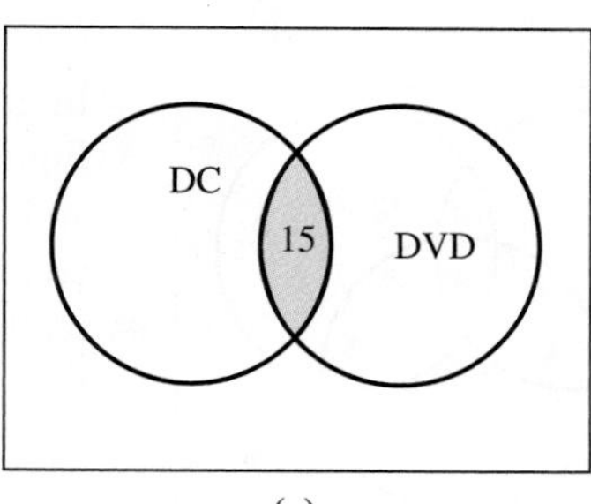

(a)

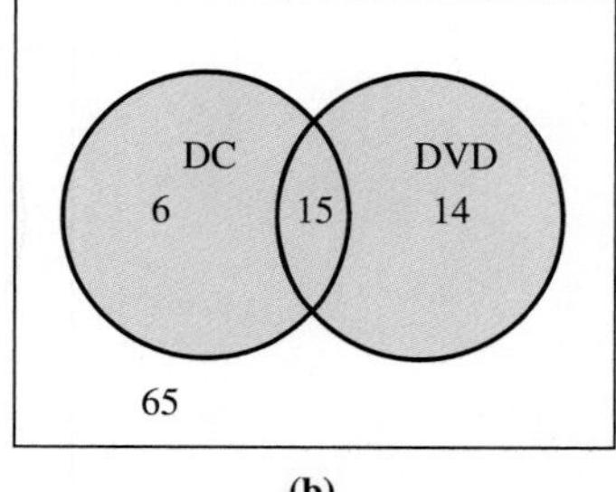

(b)

FIGURE 8.13

Example 3 A group of 60 freshman business students at a large university was surveyed, with the following results:

19 of the students read *Business Week*;
18 read the *Wall Street Journal*;
50 read *Fortune*;
13 read *Business Week* and the *Journal*;
11 read the *Journal* and *Fortune*;
13 read *Business Week* and *Fortune*;
9 read all three magazines.

Use the preceding data to answer the following questions:

(a) How many students read none of the publications?
(b) How many read only *Fortune*?
(c) How many read *Business Week* and the *Journal*, but not *Fortune*?

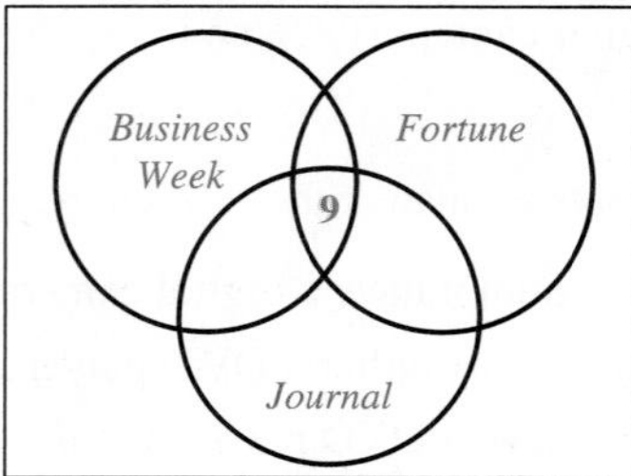

FIGURE 8.14(a)

Solution Once again, use a Venn diagram to represent the data. Since 9 students read all three publications, begin by placing 9 in the area in Figure 8.14(a) that belongs to all three regions.

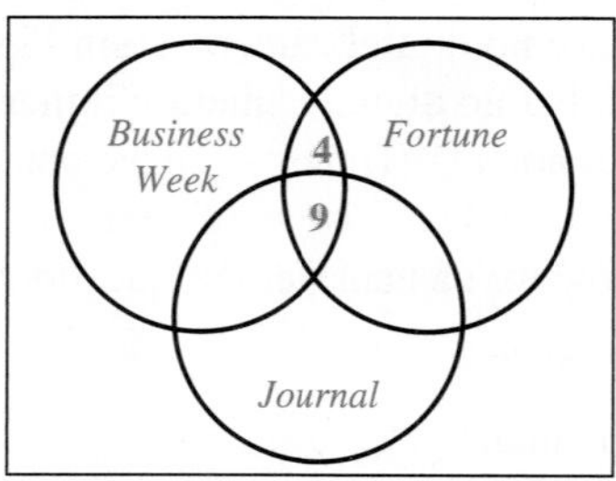

FIGURE 8.14(b)

Of the 13 students who read *Business Week* and *Fortune*, 9 also read the *Journal*. Therefore, only $13 - 9 = 4$ students read just *Business Week* and *Fortune*. So place a 4 in the region common only to *Business Week* and *Fortune* readers, as in Figure 8.14(b).

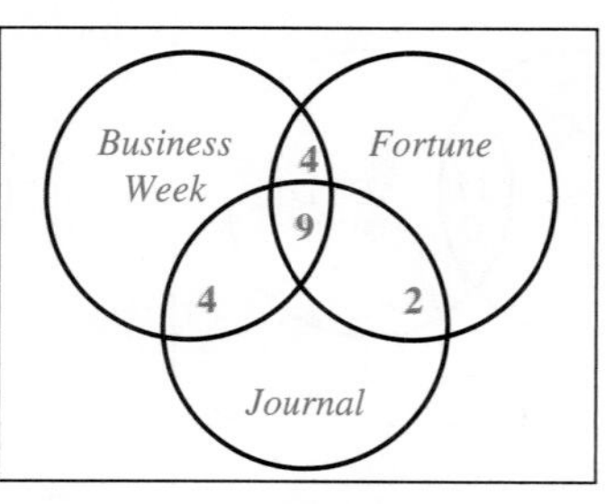

FIGURE 8.14(c)

In the same way, place a 4 in the region of Figure 8.14(c) common only to *Business Week* and the *Journal* readers, and 2 in the region common only to *Fortune* and the *Journal* readers.

2
Business Week
Fortune
4
9
4
2
Journal

FIGURE 8.14(d)

The data shows that 19 students read *Business Week.* However, 4 + 9 + 4 = 17 readers have already been placed in the *Business Week* region. The balance of this region in Figure 8.14(d) will contain only 19 − 17 = 2 students. These 2 students read *Business Week* only—not *Fortune* and not the *Journal.*

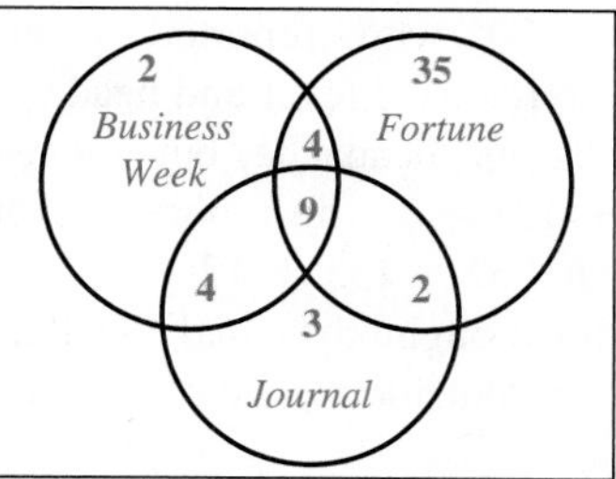

FIGURE 8.14(e)

In the same way, 3 students read only the *Journal* and 35 read only *Fortune*, as shown in Figure 8.14(e).

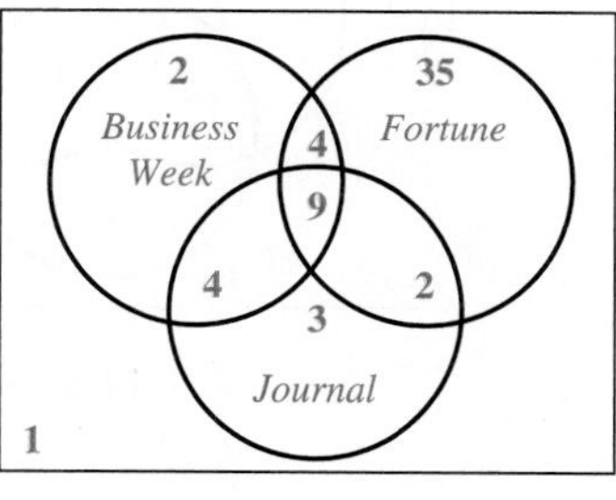

FIGURE 8.14(f)

A total of 2 + 4 + 3 + 4 + 9 + 2 + 35 = 59 students are placed in the various regions of Figure 8.14(e). Since 60 students were surveyed, 60 − 59 = 1 student reads none of the three publications, and so 1 is placed outside the other regions in Figure 8.14(f).

Figure 8.14(f) can now be used to answer the questions asked at the beginning of this example:

(a) Only 1 student reads none of the publications.

(b) There are 35 students who read only *Fortune.*

(c) The overlap of the regions representing *Business Week* and the *Journal* shows that 4 students read *Business Week* and the *Journal*, but not *Fortune.* ✓4

✓Checkpoint 4

In Example 3, how many students read exactly,

(a) 1 of the publications?

(b) 2 of the publications?

Answers: See page 485.

Example 4 **Health** Mark McCloney, M.D., saw 100 patients exhibiting flu symptoms such as fever, chills, and headache. Dr. McCloney reported the following information on patients exhibiting symptoms:

Of the 100 patients,

74 reported a fever;

72 reported chills;

67 reported a headache;

55 reported both a fever and chills;

47 reported both a fever and a headache;

49 reported both chills and a headache;

35 reported all three;

3 thought they had the flu, but did not report fever, chills, or headache.

Create a Venn Diagram to represent this data. It should show the number of people in each region.

Solution Begin with the 35 patients who reported all three symptoms. This leaves $55 - 35 = 20$ who reported fever and chills, but not headache; $47 - 35 = 12$ who reported fever and headache, but not chills; and $49 - 35 = 14$ who reported chills and headache, but not fever. With this information, we have $74 - (35 + 20 + 12) = 7$ who reported fever alone; $72 - (35 + 20 + 14) = 3$ with chills alone; and $67 - (35 + 12 + 14) = 6$ with headache alone. The remaining 3 patients who thought they had the flu, but did not report fever, chills, or headache are denoted outside the 3 circles. See Figure 8.15. ✓5

✓Checkpoint 5

In Example 4, suppose 75 patients reported a fever and only 2 thought they had the flu, but did not report fever, chills, or headache. Then how many

(a) reported only a fever?

(b) reported a fever or chills?

(c) reported a fever, chills, or headache?

Answers: See page 485.

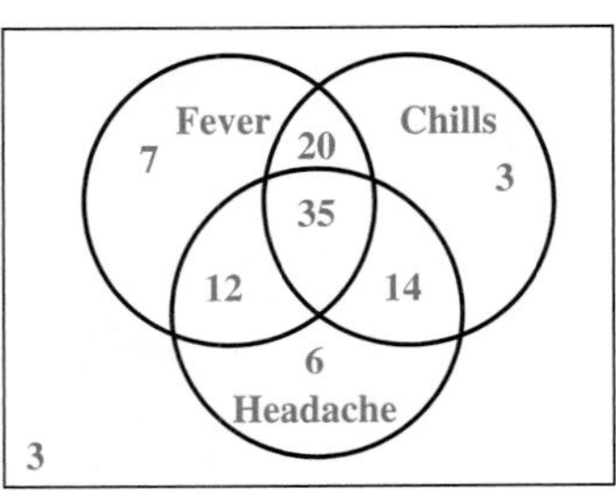

FIGURE 8.15

NOTE In all the preceding examples, we started in the innermost region with the intersection of the categories. This is usually the best way to begin solving problems of this type.

We use the symbol $n(A)$ to denote the *number* of elements in A. For instance, if $A = \{w, x, y, z\}$, then $n(A) = 4$. Next, we prove the following useful fact.

Addition Rule for Counting

$$n(A \cup B) = n(A) + n(B) - n(A \cap B).$$

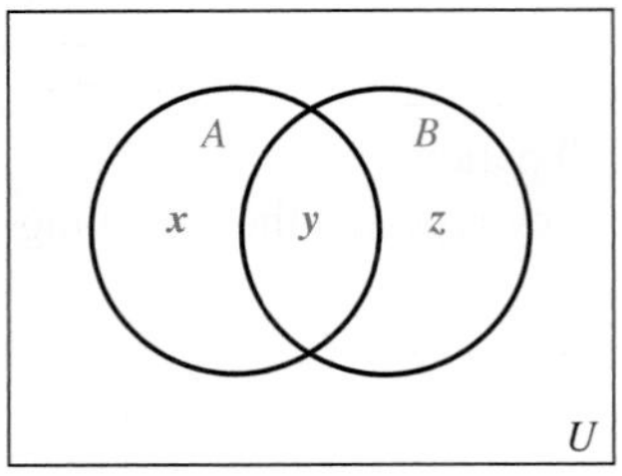

For example, if $A = \{r, s, t, u, v\}$ and $B = \{r, t, w\}$, then $A \cap B = \{r, t\}$, so that $n(A) = 5$, $n(B) = 3$, and $n(A \cap B) = 2$. By the formula in the box, $n(A \cup B) = 5 + 3 - 2 = 6$, which is certainly true, since $A \cup B = \{r, s, t, u, v, w\}$.

Here is a proof of the statement in the box: Let x be the number of elements in A that are not in B, y be the number of elements in $A \cap B$, and z be the number of elements in B that are not in A, as indicated in Figure 8.16. That diagram

FIGURE 8.16

shows that $n(A \cup B) = x + y + z$. It also shows that $n(A) = x + y$ and $n(B) = y + z$, so that

$$\begin{aligned} n(A) + n(B) - n(A \cap B) &= (x + y) + (z + y) - y \\ &= x + y + z \\ &= n(A \cup B). \end{aligned}$$

Example 5 A group of 10 students meets to plan a school function. All are majoring in accounting or economics or both. Five of the students are economics majors, and 7 are majors in accounting. How many major in both subjects?

Solution Let A represent the set of accounting majors and B represent the set of economics majors. Use the union rule, with $n(A) = 5$, $n(B) = 7$, and $n(A \cup B) = 10$. We must find $n(A \cap B)$:

$$\begin{aligned} n(A \cup B) &= n(A) + n(B) - n(A \cap B) \\ 10 &= 5 + 7 - n(A \cap B). \end{aligned}$$

So,

$$n(A \cap B) = 5 + 7 - 10 = 2.$$ 6

Checkpoint 6

If $n(A) = 10$, $n(B) = 7$, and $n(A \cap B) = 3$, find $n(A \cup B)$.

Answer: *See page 485.*

Example 6 **Natural Science** The following table gives the amounts (in thousand short tons) of air pollutant emission in 2007 for highway vehicles (cars, trucks, etc., denoted A) and off-highway machines (farm equipment, construction equipment, industrial machinery, etc., denoted B), where the air pollutants are sulfur dioxide (denoted C), volatile organic compounds (denoted D), carbon monoxide (denoted E), and nitrogen oxides (denoted F).*

		C Sulfur Dioxide	*D* Volatile Organic Compounds	*E* Carbon Monoxide	*F* Nitrogen Oxides	Total
A	**Highway**	91	3602	41,610	5563	50,866
B	**Off Highway**	396	2650	18,762	4164	25,972
	Total	487	6252	60,372	9727	76,838

Find the number of short tons of pollutants in the given sets.

(a) $A \cap E$

Solution The set $A \cap E$ consists of all pollutants that are generated by highway vehicles *and* that are carbon monoxide. From the table, we see there were 41,610 thousand such short tons.

(b) $A \cup E$

Solution The set $A \cup E$ consists of all pollutants that are generated from highway vehicles *or* that are carbon monoxide. Using the addition rule for

**Statistical Abstract of the United States: 2009.*

counting, we have $n(A \cup E) = n(A) + n(E) - n(A \cap E)$. From the table, we have $n(A) = 50{,}866$ and $n(E) = 60{,}372$. From part (a), we have $n(A \cap E) = 41{,}610$. Thus,

$$\begin{aligned} n(A \cup E) &= n(A) + n(E) - n(A \cap E) \\ &= 50{,}866 + 60{,}372 - 41{,}610 \\ &= 69{,}628 \text{ thousand short tons.} \end{aligned}$$

(c) $(E \cup F) \cap A'$

Solution Begin with the set $E \cup F$, which contains all the carbon monoxide and nitrogen oxide pollutants. Of this set, take those amounts that are *not* for highway vehicles, for a total of $18{,}762 + 4164 = 22{,}926$ thousand short tons. 7 ✓

✓ Checkpoint 7

Refer to Example 6 and find the number of short tons in each set.

(a) $B \cup C$

(b) $(B \cap D) \cup C'$

Answers: See page 485.

8.2 ▶ Exercises

Sketch a Venn diagram like the one shown, and use shading to show each of the given sets. (See Example 1.)

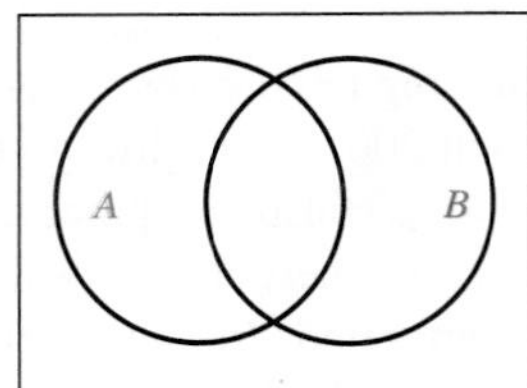

1. $A \cap B'$
2. $A \cup B'$
3. $B' \cup A'$
4. $A' \cap B'$
5. $B' \cup (A \cap B')$
6. $(A \cap B) \cup A'$
7. U'
8. $\varnothing'$
9. Three sets divide the universal set into at most _____ regions.

10. What does the notation $n(A)$ represent?

Sketch a Venn diagram like the one shown, and use shading to show each of the given sets. (See Example 2.)

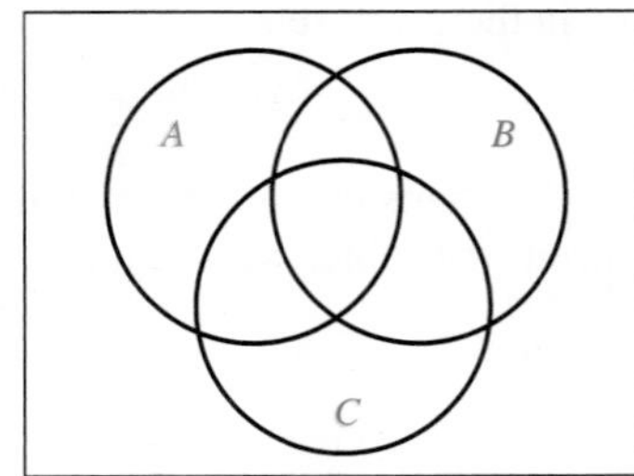

11. $(A \cap C') \cup B$
12. $A \cap (B \cup C')$
13. $A' \cap (B \cap C)$
14. $(A' \cap B') \cap C$
15. $(A \cap B') \cup C$
16. $(A \cap B') \cap C$

Use Venn diagrams to answer the given questions. (See Examples 2 and 4.)

17. **Social Science** In 2007, the percentage of children under 18 years of age who lived with both parents was 70.7, the percentage of children under 18 years of age who lived only with their mother was 22.6, and the percentage of children under 18 years of age who lived with neither parent was 3.5.* What percentage of children under age 18 lived with their father only?

18. **Business** In 2007, the total number of union members (in thousands) age 16 and above was 129,766. There were 29,409 members ages 25 to 34 years old, 30,296 ages 35 to 44 years old, 29,731 members ages 45 to 54 years old, 16,752 members ages 55 to 64 years old, and 4183 members ages 65 years and over.* How many union members were in the age group 16 to 24?

19. **Business** The human resources director for a commercial real estate company received the following numbers of applications from people with the given information:

    66 with sales experience;

    40 with a college degree;

    23 with a real estate license;

    26 with sales experience and a college degree;

    16 with sales experience and a real estate license;

    15 with a college degree and a real estate license;

    11 with sales experience, a college degree, and a real estate license;

Statistical Abstract of the United States: 2009.

22 with neither sales experience, a college degree, nor a real estate license.

(a) How many applicants were there?
(b) How many applicants did not have sales experience?
(c) How many had sales experience and a college degree, but not a real estate license?
(d) How many had only a real estate license?

20. **Business** A pet store keeps track of the purchases of customers over a four-hour period. The store manager classifies purchases as containing a dog product, a cat product, a fish product, or a product for a different kind of pet. She found that:

83 customers purchased a dog product;

101 customers purchased a cat product;

22 customers purchased a fish product;

31 customers purchased a dog and a cat product;

8 customers purchased a dog and a fish product;

10 customers purchased a cat and a fish product;

6 customers purchased a dog, a cat, and a fish product;

34 customers purchased a product for a pet other than a dog, cat, or fish.

(a) How many purchases were for a dog product only?
(b) How many purchases were for a cat product only?
(c) How many purchases were for a dog or fish product?
(d) How many purchases were there in total?

21. **Natural Science** A marine biologist surveys people who fish on Lake Erie and caught at least one fish to determine whether they had caught a walleye, a smallmouth bass, or a yellow perch in the last year. He finds:

124 caught at least one walleye;

133 caught at least one smallmouth bass;

146 caught at least one yellow perch;

75 caught at least one walleye and at least one smallmouth bass;

67 caught at least one walleye and at least one yellow perch;

79 caught at least one smallmouth bass and at least one yellow perch;

45 caught all three.

(a) Find the total number of people surveyed.
(b) How many caught at least one walleye or at least one smallmouth bass?
(c) How many caught only walleye?

22. **Health** Human blood can contain either no antigens, the A antigen, the B antigen, or both the A and B antigens. A third antigen, called the Rh antigen, is important in human reproduction and, like the A and B antigens, may or may not be present in an individual. Blood is called type A positive if the individual has the A and Rh antigens, but not the B antigen. A person having only the A and B antigens is said to have type AB-negative blood. A person having only the Rh antigen has type O-positive blood. Other blood types are defined in a similar manner. Identify the blood type of the individuals in regions (a)–(g) of the Venn diagram.

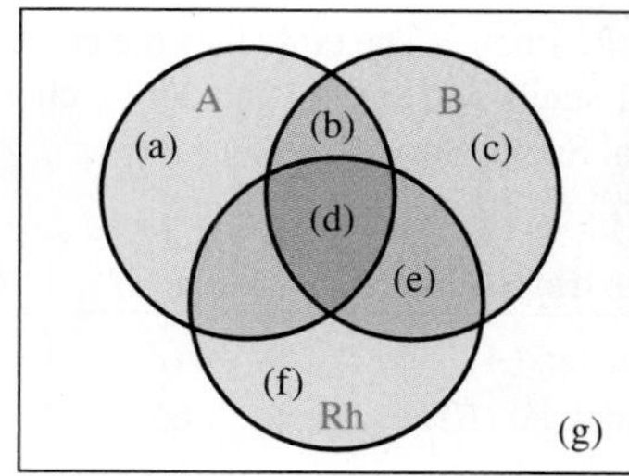

23. **Natural Science** Use the diagram from Exercise 22. In a certain hospital, the following data was recorded:

25 patients had the A antigen;

17 had the A and B antigens;

27 had the B antigen;

22 had the B and Rh antigens;

30 had the Rh antigen;

12 had none of the antigens;

16 had the A and Rh antigens;

15 had all three antigens.

How many patients

(a) were represented?
(b) had exactly one antigen?
(c) had exactly two antigens?
(d) had O-positive blood?
(e) had AB-positive blood?
(f) had B-negative blood?
(g) had O-negative blood?
(h) had A-positive blood?

24. **Business** In reviewing the portfolios of 365 of its clients, a mutual funds company categorized whether the clients were invested in international stock funds, domestic stock funds, or bond funds. It found that,

125 were invested in domestic stocks, international stocks, and bond funds;

145 were invested in domestic stocks and bond funds;

300 were invested in domestic stocks;

200 were invested in international and domestic stocks;

18 were invested in international stocks and bond funds, but not domestic stocks;

35 were invested in bonds, but not in international or domestic stocks;

87 were invested in international stocks, but not in bond funds.

(a) How many were invested in international stocks?
(b) How many were invested in bonds, but not international stocks?
(c) How many were not invested in bonds?
(d) How many were invested in international or domestic stocks?

25. **Social Science** The table lists the cross-classification of marital status and sex for the adults chosen for the 2006 General Social Survey (GSS).*

Marital Status	Male (M)	Female (F)
Married (A)	1018	1152
Widowed (B)	65	301
Divorced (C)	320	412
Separated (D)	61	95
Never Married (E)	534	546

Using the letters given in the table, find the number of respondents in each set.
(a) $A \cap M$
(b) $C \cap (F \cup M)$
(c) $D \cup F$
(d) $B' \cap E'$

26. **Social Science** The number of active-duty servicewomen in different segments of the U.S. armed forces as of September 2006 is given in the table:†

	Army (A)	Air Force (B)	Navy (C)	Marines (D)
Officers (O)	12,459	12,836	7649	1101
Enlisted (E)	57,825	54,957	42,400	10,049

Use this information and the letters given to find the number of female military personnel in each of the given sets.
(a) $A \cup B$
(b) $E \cup (C \cup D)$
(c) $O' \cap B'$

27. **Business** The table gives the number (rounded to the nearest thousand) of manufacturing firms (A) and construction firms (B) cross-classified by level of income and industry type.‡

Industry Type	Under \$1 million (C)	\$1 to \$4.9 million (D)	\$5 to \$9.9 million (E)	\$10 to \$49.9 million (F)	More than \$50 million (G)
A	165,000	70,000	18,000	18,000	7000
B	573,000	137,000	21,000	18,000	3000

Using the letters given in the table, find the number of firms in each of the given sets.
(a) $C \cup D$
(b) $B \cap G$
(c) $A \cap (E \cup F)$
(d) $(F \cup G)'$
(e) $A' \cap C'$

28. **Health** The top four causes of death in the United States for persons ages 16–34 are motor vehicle crash, homicide, suicide, and accidental poisoning. The table gives the numbers of these deaths in 2005:*

	Age Group		
Cause of Death	16–20 (A)	21–24 (B)	25–34 (C)
Motor Vehicle Crash (D)	5665	4587	7047
Homicide (E)	2571	2717	4752
Suicide (F)	1905	2120	4990
Accidental Poisoning (G)	896	1553	4386

Using the letters given in the table, find the number of deaths in each of the given sets.
(a) $A \cup B$
(b) $D \cup E$
(c) $E \cap B$
(d) $D' \cap C$

29. Restate the union rule in words.

Use Venn diagrams to answer the given questions. (See Example 5.)

30. If $n(A) = 5$, $n(B) = 8$, and $n(A \cap B) = 4$, what is $n(A \cup B)$?

31. If $n(A) = 12$, $n(B) = 27$, and $n(A \cup B) = 30$, what is $n(A \cap B)$?

32. Suppose $n(B) = 7$, $n(A \cap B) = 3$, and $n(A \cup B) = 20$. What is $n(A)$?

33. Suppose $n(A \cap B) = 5$, $n(A \cup B) = 35$, and $n(A) = 13$. What is $n(B)$?

*www.norc.org/GSS&Website/.

†www.infoplease.com.

‡*Statistical Abstract of the United States*: 2009.

*www.nhtsa.gov.

Draw a Venn diagram and use the given information to fill in the number of elements for each region.

34. $n(U) = 48, n(A) = 26, n(A \cap B) = 12, n(B') = 30$
35. $n(A) = 28, n(B) = 12, n(A \cup B) = 30, n(A') = 19$
36. $n(A \cup B) = 17, \quad n(A \cap B) = 3, \quad n(A) = 8, \quad n(A' \cup B') = 21$
37. $n(A') = 28, \quad n(B) = 25, \quad n(A' \cup B') = 45, \quad n(A \cap B) = 12$
38. $n(A) = 28, \; n(B) = 34, \; n(C) = 25, \; n(A \cap B) = 14, \; n(B \cap C) = 15, \; n(A \cap C) = 11, \; n(A \cap B \cap C) = 9, \; n(U) = 59$
39. $n(A) = 54, \quad n(A \cap B) = 22, \quad n(A \cup B) = 85, \quad n(A \cap B \cap C) = 4, \; n(A \cap C) = 15, \; n(B \cap C) = 16, \; n(C) = 44, n(B') = 63$

*In Exercises 40–43, show that the statements are true by drawing Venn diagrams and shading the regions representing the sets on each side of the equals signs.**

40. $(A \cup B)' = A' \cap B'$ 41. $(A \cap B)' = A' \cup B'$
42. $A \cap (B \cup C) = (A \cap B) \cup (A \cap C)$
43. $A \cup (B \cap C) = (A \cup B) \cap (A \cup C)$
44. Explain in words the statement about sets in question 40.
45. Explain in words the statement about sets in question 41.
46. Explain in words the statement about sets in question 42.
47. Explain in words the statement about sets in question 43.

*The statements in Exercises 40 and 41 are known as De Morgan's laws. They are named for the English mathematician Augustus De Morgan (1806–71).

✓ Checkpoint Answers

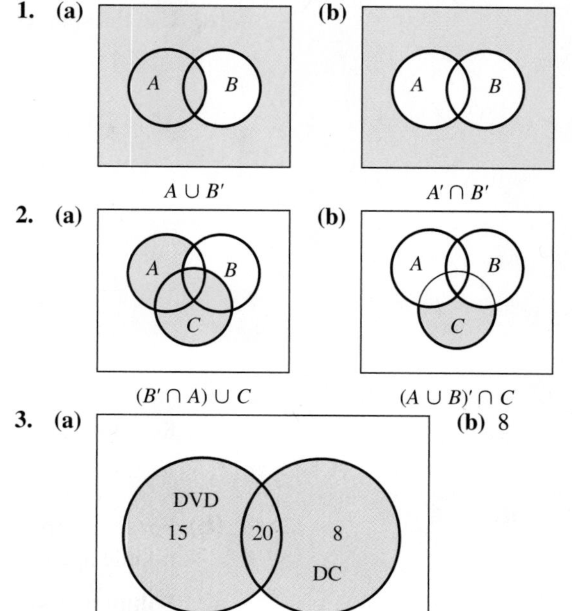

4. (a) 40 (b) 10 5. (a) 8 (b) 92 (c) 98
6. 14
7. (a) 26,063 thousand short tons
(b) 76,351 thousand short tons

8.3 ▶ Introduction to Probability

If you go to a pizzeria and order two large pizzas at \$14.99 each, you can easily find the *exact* price of your purchase: \$29.98. For the manager at the pizzeria, however, it is impossible to predict the *exact* number of pizzas to be purchased daily. The number of pizzas purchased during a day is *random*: The quantity cannot be predicted exactly. A great many problems that come up in applications of mathematics involve random phenomena—phenomena for which exact prediction is impossible. The best that we can do is determine the *probability* of the possible outcomes.

RANDOM EXPERIMENTS AND SAMPLE SPACES

A **random experiment** (sometimes called a random phenomenon) has outcomes that we cannot predict, but that nonetheless have a regular distribution in a large number of repetitions. We call a repetition from a random experiment a **trial**. The possible results of each trial are called **outcomes**. For instance, when we flip a coin, the outcomes are heads and tails. We do not know whether a particular flip will yield heads or tails, but we do know that if we flip the coin a large a number of times, about half the flips will be heads and half will be tails. Each flip of the coin is a trial. The **sample space** (denoted by S) for a random experiment is the set of all possible outcomes. For the coin flipping, the sample space is

$$S = \{\text{heads, tails}\}.$$

Example 1 Give the sample space for each random experiment.

(a) Use the spinner in Figure 8.17.

FIGURE 8.17

Solution The 7 outcomes are 1, 2, 3, . . . 7, so the sample space is

$$\{1, 2, 3, 4, 5, 6, 7\}.$$

(b) For the purposes of a public opinion poll, respondents are classified as young, middle aged, or senior and as male or female.

Solution A sample space for this poll could be written as a set of ordered pairs:

{(young, male), (young, female), (middle aged, male),
(middle aged, female), (senior, male), (senior, female)}.

(c) An experiment consists of studying the numbers of boys and girls in families with exactly 3 children. Let b represent *boy* and g represent *girl.*

Solution For this experiment, drawing a tree diagram can be helpful. First, we draw two starting branches to the left to indicate that the first child can be either a boy or a girl. From each of those outcomes, we draw two branches to indicate that the second child can be either a boy or girl. Last, we draw two branches from each of those outcomes to indicate that after the second child, the third child can be either a boy or a girl. The result is the tree in Figure 8.18.

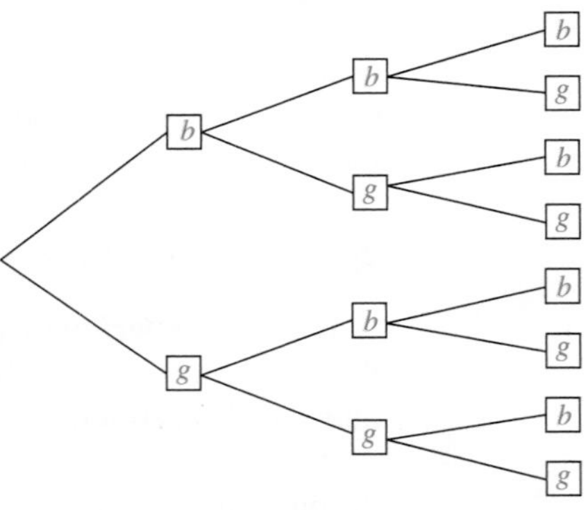

FIGURE 8.18

We can now easily list the members of the sample space S. We follow the eight paths of the branches to yield

$$S = \{bbb, bbg, bgb, bgg, gbb, gbg, ggb, ggg\}.$$ 

✓Checkpoint 1

Draw a tree diagram for the random experiment of flipping a coin two times, and determine the sample space.

Answer: See page 495.

EVENTS

An **event** is an outcome, or a set of outcomes, of a random experiment. Thus, an event is a subset of the sample space. For example, if the sample space for tossing a coin is $S = \{h, t\}$, then one event is $E = \{h\}$, which represents the outcome "heads."

An ordinary die is a cube whose six different faces show the following numbers of dots: 1, 2, 3, 4, 5, and 6. If the die is fair (not "loaded" to favor certain faces over others), then any one of the faces is equally likely to come up when the die is rolled. The sample space for the experiment of rolling a single fair die is $S = \{1, 2, 3, 4, 5, 6\}$. Some possible events are as follows:

The die shows an even number: $E_1 = \{2, 4, 6\}$.

The die shows a 1: $E_2 = \{1\}$.

The die shows a number less than 5: $E_3 = \{1, 2, 3, 4\}$.

The die shows a multiple of 3: $E_4 = \{3, 6\}$.

Example 2 For the sample space S in Example 1(c) on the previous page, write the given events in set notation.

(a) Event H: The family has exactly two girls.

Solution Families with three children can have exactly two girls with either bgg, gbg, or ggb, so that event H is

$$H = \{bgg, gbg, ggb\}.$$

(b) Event K: The three children are the same sex.

Solution Two outcomes satisfy this condition: all boys and all girls, or

$$K = \{bbb, ggg\}.$$

(c) Event J: The family has three girls.

Solution Only ggg satisfies this condition, so

$$J = \{ggg\}.$$ 2✓

✓Checkpoint 2

Suppose a die is tossed. Write the given events in set notation.

(a) The number showing is less than 3.

(b) The number showing is 5.

(c) The number showing is 8.

Answers: See page 495.

If an event E equals the sample space S, then E is a **certain event**. If event $E = \emptyset$, then E is an **impossible event**.

Example 3 Suppose a fair die is rolled. Then the sample space is $\{1, 2, 3, 4, 5, 6\}$. Find the requested events.

(a) The event "the die shows a 4."

Solution $\{4\}$.

(b) The event "the number showing is less than 10."

Solution The event is the entire sample space $\{1, 2, 3, 4, 5, 6\}$. This event is a certain event; if a die is rolled, the number showing (either 1, 2, 3, 4, 5, or 6) must be less than 10.

(c) The event "the die shows a 7."

Solution The empty set, ∅; this is an impossible event. 

✓Checkpoint 3

Which of the events listed in Checkpoint 2 is

(a) certain?

(b) impossible?

Answers: See page 495.

Since events are sets, we can use set operations to find unions, intersections, and complements of events. Here is a summary of the set operations for events.

Set Operations for Events

Let E and F be events for a sample space S. Then

$E \cap F$ occurs when both E **and** F occur;

$E \cup F$ occurs when E **or** F **or both** occur;

E' occurs when E does **not** occur.

Example 4 A study of college students grouped the students into various categories that can be interpreted as events when a student is selected at random. Consider the following events:

E: The student is under 20 years old;

F: The student is male;

G: The student is a business major.

Describe each of the following events in words.

(a) E'

Solution E' is the event that the student is 20 years old or older.

(b) $F' \cap G$

Solution $F' \cap G$ is the event that the student is not male and the student is a business major—that is, the student is a female business major.

(c) $E' \cup G$

Solution $E' \cup G$ is the event that the student is 20 or over or is a business major. Note that this event includes all students 20 or over, regardless of major. 4✓

✓Checkpoint 4

Write the set notation for the given events for the experiment of rolling a fair die if $E = \{1, 3\}$ and $F = \{2, 3, 4, 5\}$.

(a) $E \cap F$

(b) $E \cup F$

(c) E'

Answers: See page 495.

Two events that cannot both occur at the same time, such as getting both a head and a tail on the same toss of a coin, are called **disjoint events**. (Disjoint events are sometimes referred to as *mutually exclusive events.*)

Disjoint Events

Events E and F are disjoint events if $E \cap F = \varnothing$.

For any event E, E and E' are disjoint events.

Example 5 Let $S = \{1, 2, 3, 4, 5, 6\}$, the sample space for tossing a die. Let $E = \{4, 5, 6\}$, and let $G = \{1, 2\}$. Are E and G disjoint events?

Solution Yes, because they have no outcomes in common; $E \cap G = \emptyset$. See Figure 8.19. 5 ✓

✓Checkpoint 5

In Example 5, let $F = \{2, 4, 6\}$, $K = \{1, 3, 5\}$, and G remain the same. Are the given events disjoint?

(a) F and K

(b) F and G

Answers: See page 495.

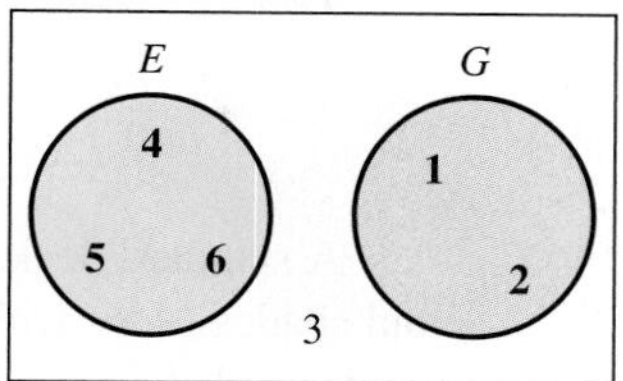

FIGURE 8.19

PROBABILITY

For sample spaces with *equally likely* outcomes, the probability of an event is defined as follows.

Basic Probability Principle

Let S be a sample space of equally likely outcomes, and let event E be a subset of S. Then the **probability that event E occurs** is

$$P(E) = \frac{n(E)}{n(S)}.$$

By this definition, the **probability of an event** is a number that indicates the relative likelihood of the event.

CAUTION The basic probability principle applies only when the outcomes are equally likely.

Example 6 Suppose a single fair die is rolled, with the sample space $S = \{1, 2, 3, 4, 5, 6\}$. Give the probability of each of the following events.

(a) E: The die shows an even number.

Solution Here, $E = \{2, 4, 6\}$, a set with three elements. Because S contains six elements,

$$P(E) = \frac{3}{6} = \frac{1}{2}.$$

(b) F: The die shows a number less than 10.

Solution Event F is a certain event, with

$$F = \{1, 2, 3, 4, 5, 6\},$$

so that

$$P(F) = \frac{6}{6} = 1.$$

(c) G: The die shows an 8.

Solution This event is impossible, so

$$P(G) = \frac{0}{6} = 0.$$ 6

Checkpoint 6

A fair die is rolled. Find the probability of rolling

(a) an odd number;

(b) 2, 4, 5, or 6;

(c) a number greater than 5;

(d) the number 7.

Answers: See page 495.

A standard deck of 52 cards has four suits—hearts (♥), clubs (♣), diamonds (♦), and spades (♠)—with 13 cards in each suit. The hearts and diamonds are red, and the spades and clubs are black. Each suit has an ace (A), a king (K), a queen (Q), a jack (J), and cards numbered from 2 to 10. The jack, queen, and king are called face cards and for many purposes can be thought of as having values 11, 12, and 13, respectively. The ace can be thought of as the low card (value 1) or the high card (value 14). See Figure 8.20. We will refer to this standard deck of cards often in our discussion of probability.

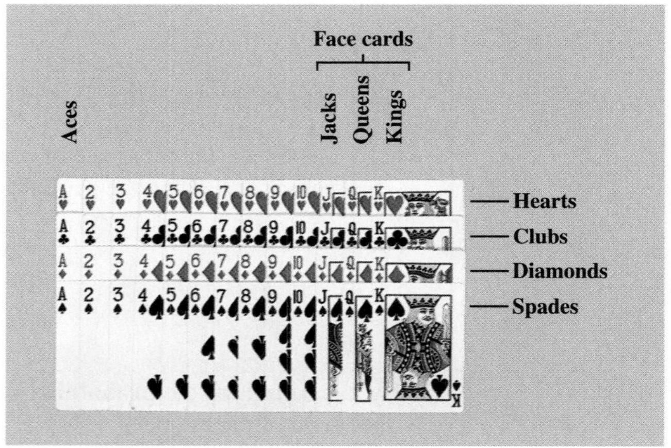

FIGURE 8.20

Example 7 If a single card is drawn at random from a standard, well-shuffled, 52-card deck, find the probability of each of the given events.

(a) Drawing an ace

Solution There are 4 aces in the deck. The event "drawing an ace" is

$$\{\text{heart ace, diamond ace, club ace, spade ace}\}.$$

Therefore,

$$P(\text{ace}) = \frac{4}{52} = \frac{1}{13}.$$

(b) Drawing a face card

Solution Since there are 12 face cards,

$$P(\text{face card}) = \frac{12}{52} = \frac{3}{13}.$$

(c) Drawing a spade

Solution The deck contains 13 spades, so

$$P(\text{spade}) = \frac{13}{52} = \frac{1}{4}.$$

(d) Drawing a spade or a heart

Solution Besides the 13 spades, the deck contains 13 hearts, so

$$P(\text{spade or heart}) = \frac{26}{52} = \frac{1}{2}. \quad 7✓$$

✓Checkpoint 7

A single playing card is drawn at random from an ordinary 52-card deck. Find the probability of drawing

(a) a queen;

(b) a diamond;

(c) a red card.

Answers: See page 495.

In the preceding examples, the probability of each event was a number between 0 and 1, inclusive. The same thing is true in general. Any event E is a subset of the sample space S, so $0 \leq n(E) \leq n(S)$. Since $P(E) = n(E)/n(S)$, it follows that $0 \leq P(E) \leq 1$.

For any event E,

$$\mathbf{0 \leq P(E) \leq 1.}$$

RELATIVE FREQUENCY PROBABILITY

In many real-life problems, it is not possible to establish exact probabilities for events. Instead, useful estimates are often found by drawing on past experience. This approach is called **relative frequency probability**. We calculate our estimate of the probability by determining the percentage of the responses with the characteristic of interest. Estimates based on relative frequency probability are sometimes called *empirical probabilities*. The next example shows one approach to finding such relative frequency probabilities.

Example 8 **Business** The table gives the frequency of households having an average monthly electric bill in various dollar amounts from a portion of the respondents of the 2007 American Community Survey conducted by the U.S. Census Bureau.*

Monthly Electric Bill (Dollars)	Frequency
0–49.99	270
50.00–99.99	253
100.00–149.99	203
150.00–199.99	112
200.00–249.99	71
250.00–299.99	34
300.00 or higher	57

*Based on data available at www.census.gov/acs.

(a) Find the relative frequency probability of having a monthly electric bill between \$0 and \$49.99.

Solution Let us define event A to be the event that a household has an average electric bill between \$0 and \$49.99. To find the relative frequency probability, we first need to find the total number of respondents. This is $270 + 253 + 203 + 112 + 71 + 34 + 57 = 1000$. We then divide the frequency in the 0–49.99 category (in this case, 270) by 1000 to obtain

$$P(A) = \frac{270}{1000} = .270.$$

(b) Find the relative frequency probability for event B of having an average monthly electric bill \$250 or higher.

Solution Here, we add together the number of respondents in the categories 250.00–299.99 and 300 or higher to obtain

$$P(B) = \frac{34 + 57}{1000} = \frac{91}{1000} = .091.$$ ✓8

✓Checkpoint 8

From the data given in Example 8, find the probability that a household has an electric bill less than \$200.

Answer: *See page 495.*

After conducting a study such as the American Community Survey, we can use the relative frequency probability estimates from the sample to make estimates for the entire population of the United States. We also usually use the term "probability" rather than "relative frequency probability." So we say the probability that a randomly chosen household in the United States has an average monthly electric bill of less than \$50 is approximately .270.

A table of frequencies, as in Example 8, sets up a probability distribution; that is, for each possible outcome of an experiment, a number, called the probability of that outcome, is assigned. This assignment may be done in any reasonable way (on a relative frequency basis, as in Example 8, or by theoretical reasoning, as in Example 6), provided that it satisfies the following conditions.

Properties of Probability

Let S be a sample space consisting of n distinct outcomes $s_1, s_2, \ldots, s_n$. An acceptable probability assignment consists of assigning to each outcome s_i a number p_i (the probability of s_i) according to the following rules:

1. The probability of each outcome is a number between 0 and 1:

$$0 \le p_1 \le 1, \quad 0 \le p_2 \le 1, \ldots, \quad 0 \le p_n \le 1.$$

2. The sum of the probabilities of all possible outcomes is 1:

$$p_1 + p_2 + p_3 + \cdots + p_n = 1.$$

8.3 ▸ Exercises

1. What is meant by a "fair" coin or die?
2. What is the sample space for a random experiment?

Write sample spaces for the random experiments in Exercises 3–9. (See Example 1.)

3. A month of the year is chosen for a wedding.
4. A day in April is selected for a bicycle race.
5. A student is asked how many points she earned on a recent 80-point test.
6. A person is asked the number of hours (to the nearest hour) he watched television yesterday.
7. The management of an oil company must decide whether to go ahead with a new oil shale plant or to cancel it.
8. A coin is tossed and a die is rolled.
9. The quarter of the year in which a company's profits were highest.
10. Define an event.
11. Define disjoint events in your own words.

Decide whether the events in Exercises 12–17 are disjoint.

12. Owning an SUV and owning a Hummer
13. Wearing a hat and wearing glasses
14. Being married and being under 30 years old
15. Being a doctor and being under 5 years old
16. Being male and being a nurse
17. Being female and being a pilot

For the random experiments in Exercises 18–20, write out an equally likely sample space, and then write the indicated events in set notation. (See Examples 2 and 3.)

18. A marble is drawn at random from a bowl containing 3 yellow, 4 white, and 8 blue marbles.
   (a) A yellow marble is drawn.
   (b) A blue marble is drawn.
   (c) A white marble is drawn.
   (d) A black marble is drawn.
19. Six people live in a dorm suite. Two are to be selected to go to the campus café to pick up a pizza. Of course, no one wants to go, so the six names (Connie, Kate, Lindsey, Jackie, Taisa, and Nicole) are placed in a hat. After the hat is shaken, two names are selected.
   (a) Taisa is selected.
   (b) The two names selected have the same number of letters.
20. An unprepared student takes a three-question true-or-false quiz in which he flips a coin to guess the answers. If the coin is heads, he guesses true, and if the coin is tails, he guesses false.
   (a) The student guesses true twice and guesses false once.
   (b) The student guesses all false.
   (c) The student guesses true once and guesses false twice.

In Exercises 21–23, write out the sample space and assume each outcome is equally likely. Then give the probability of the requested outcomes.

21. In deciding what color and style to paint a room, Greg has narrowed his choices to three colors—forest sage, evergreen whisper, and opaque emerald—and two styles—rag painting and colorwash.
   (a) Greg picks a combination with colorwash.
   (b) Greg picks a combination with opaque emerald or rag painting.
22. Tami goes shopping and sees three kinds of shoes: flats, 2″ heels, and 3″ heels. They come in two shades of beige (light and dark) and black.
   (a) The shoe selected has a heel and is black.
   (b) The shoe selected has no heel and is beige.
   (c) The shoe selected has a heel and is beige.
23. Doug Hall is shopping for a new patio umbrella. There is a 10-foot and a 12-foot model, and each is available in beige, forest green, and rust.
   (a) Doug buys a 12-foot forest green umbrella.
   (b) Doug buys a 10-foot umbrella.
   (c) Doug buys a rust-colored umbrella.

A single fair die is rolled. Find the probabilities of the given events. (See Example 6.)

24. Getting a 5
25. Getting a number less than 4
26. Getting a number greater than 4
27. Getting a 2 or a 5
28. Getting a multiple of 3
29. Getting any number except 3

David Klein wants to adopt a puppy from an animal shelter. At the shelter, he finds eight puppies that he likes: a male and female puppy from each of the four breeds of beagle, boxer, collie, and Labrador. The puppies are each so cute that Dave cannot make up his mind, so he decides to pick the dog randomly.

30. Write the sample space for the outcomes, assuming each outcome is equally likely.

Find the probability that Dave chooses the given puppy.

31. A male dog
32. A collie
33. A female Labrador
34. A beagle or a boxer
35. Anything except a Labrador

Business *The following table gives the number of fatal work injuries categorized by cause from 2007.**

Cause	Number of Fatalities
Transportation accidents	2234
Assaults and violent acts	839
Contacts with objects and equipment	916
Falls	835
Exposure to harmful substances or environments	488
Fires and explosions	159

Find the probability that a randomly chosen work fatality had the given cause. (See Example 8.)

36. A fall **37.** Fires and explosions

38. **Social Science** Respondents for the 2006 General Social Survey (GSS) indicated the following categorizations pertaining to attendance at religious services:†

Attendance	Number of Respondents
Never	1020
Less than once a year	302
Once a year	571
Several times a year	502
Once a month	308
2–3 times a month	380
Nearly every week	240
Every week	839
More than once a week	329
Don't know/no answer	19
Total	4510

Find the probability that a randomly chosen person in the United States attends religious services
(a) several times a year;
(b) 2–3 times a month;
(c) nearly every week or more frequently.

39. **Social Science** Married respondents to the 2006 General Social Survey indicated the age at which they were first married.† The table gives the results:

Age (Years)	Number of Respondents
Less than 18	80
18–19	215
20–21	250
22–23	178
24–25	143
26–30	187
31–40	90
Over 40	25

Find the probability that a person in the United States was
(a) 18 or 19 when he or she was married;
(b) less than 22 when he or she was married;
(c) over 25 when he or she was married.

40. **Health** For a medical experiment, people are classified as to whether they smoke, have a family history of heart disease, or are overweight. Define events E, F, and G as follows:

E: person smokes;

F: person has a family history of heart disease;

G: person is overweight.

Describe each of the following events in words.
(a) G'
(b) $F \cap G$
(c) $E \cup G'$

41. **Health** Refer to Exercise 40. Describe each of the events that follow in words.
(a) $E \cup F$
(b) $E' \cap F$
(c) $F' \cup G'$

42. **Health** The National Health and Nutrition Examination Study (NHANES) is conducted every several years by the U.S. Centers for Disease Control and Prevention. In the 2005–2006 survey of 2683 American women ages 18 to 85, we have the following classifications of height (in inches).*

Height (Inches)	Females
Less than 60	301
60–62.9	846
63–65.9	1040
66–68.9	437
69–71.9	56
72 or more	3

*U.S. Bureau of Labor Statistics.

†Based on data available at www.norc.org/GSS+Website.

*Based on data available at www.cdc.gov/nchs/nhanes.htm.

Find the probability that an American woman is
(a) six feet tall or taller;
(b) less than 63 inches tall;
(c) between 63 and 65.9 inches tall.

An experiment is conducted for which the sample space is $S = \{s_1, s_2, s_3, s_4, s_5\}$. Which of the probability assignments in Exercises 43–48 is possible for this experiment? If an assignment is not possible, tell why.

43.

Outcomes	s_1	s_2	s_3	s_4	s_5
Probabilities	.09	.32	.21	.25	.13

44.

Outcomes	s_1	s_2	s_3	s_4	s_5
Probabilities	.92	.03	0	.02	.03

45.

Outcomes	s_1	s_2	s_3	s_4	s_5
Probabilities	1/3	1/4	1/6	1/8	1/10

46.

Outcomes	s_1	s_2	s_3	s_4	s_5
Probabilities	1/5	1/3	1/4	1/5	1/10

47.

Outcomes	s_1	s_2	s_3	s_4	s_5
Probabilities	.64	−.08	.30	.12	.02

48.

Outcomes	s_1	s_2	s_3	s_4	s_5
Probabilities	.05	.35	.5	.2	−.3

✓ Checkpoint Answers

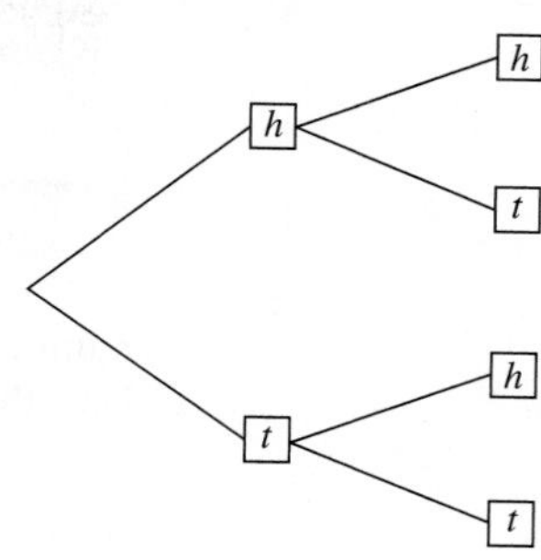

1. $\{hh, ht, th, tt\}$
2. **(a)** $\{1, 2\}$ **(b)** $\{5\}$ **(c)** $\varnothing$
3. **(a)** None **(b)** Part (c)
4. **(a)** $\{3\}$ **(b)** $\{1, 2, 3, 4, 5\}$ **(c)** $\{2, 4, 5, 6\}$
5. **(a)** Yes **(b)** No
6. **(a)** 1/2 **(b)** 2/3 **(c)** 1/6 **(d)** 0
7. **(a)** 1/13 **(b)** 1/4 **(c)** 1/2
8. .838

8.4 ▶ Basic Concepts of Probability

We determine the probability of more complex events in this section.

To find the probability of the union of two sets E and F in a sample space S, we use the union rule for counting given in Section 8.2:

$$n(E \cup F) = n(E) + n(F) - n(E \cap F).$$

Dividing both sides by $n(S)$ yields

$$\frac{n(E \cup F)}{n(S)} = \frac{n(E)}{n(S)} + \frac{n(F)}{n(S)} - \frac{n(E \cap F)}{n(S)}$$

$$P(E \cup F) = P(E) + P(F) - P(E \cap F).$$

This discussion is summarized in the next rule.

Addition Rule for Probability

For any events E and F from a sample space S,

$$P(E \cup F) = P(E) + P(F) - P(E \cap F).$$

In words, we have

$$P(E \text{ or } F) = P(E) + P(F) - P(E \text{ and } F).$$

(Although the addition rule applies to any events E and F from any sample space, the derivation we have given is valid only for sample spaces with equally likely simple events.)

Example 1 When playing American roulette, the croupier (attendant) spins a marble that lands in one of the 38 slots in a revolving turntable. The slots are numbered 1 to 36, with two additional slots labeled 0 and 00 that are painted green. Half the remaining slots are colored red, and half are black. (See Figure 8.21)

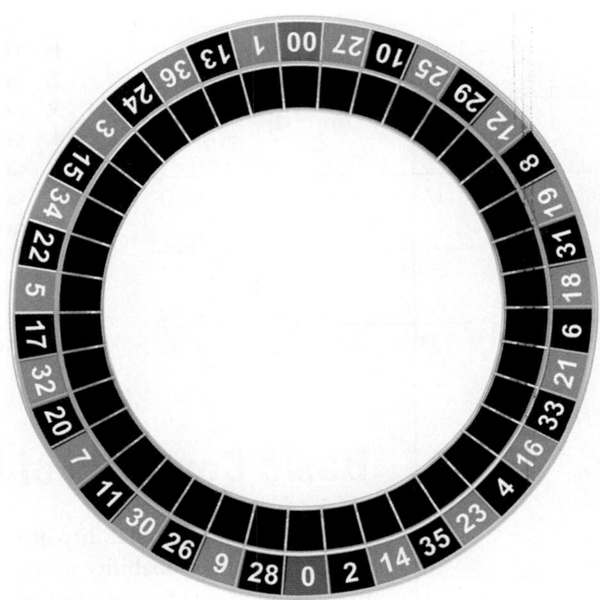

FIGURE 8.21

If we consider the numbers 0 and 00 as neither even nor odd, find the probability that the marble will land in a red or even number.

Solution Let R represent the event of the marble landing in a red slot and E the event of the marble landing in an even-numbered slot. There are 18 slots that are colored red, so $P(R) = 18/38$. There are also 18 even numbers between 1 and 36, so $P(E) = 18/38$. In order to use the addition rule, we also need to know the number of slots that are red and even numbered. Looking at Figure 8.21, we can see there are 8 such slots, which implies that $P(R \cap E) = 8/38$. Using the addition rule, we find the probability that the marble will land in a slot that is red or even numbered is

$$\begin{aligned} P(R \cup E) &= P(R) + P(E) - P(R \cap E) \\ &= \frac{18}{38} + \frac{18}{38} - \frac{8}{38} = \frac{28}{38} = \frac{14}{19}. \end{aligned}$$ ✓

✓Checkpoint 1

If an American roulette wheel is spun, find the probability of the marble landing in a black slot or a slot whose number is divisible by 3.

Answer: See page 506.

CAUTION Recall from Section 8.1 that the word "or" always indicates use of the addition rule.

Example 2 Suppose two fair dice (plural of *die*) are rolled. Find each of the given probabilities.

(a) The first die shows a 2 or the sum of the results is 6 or 7.

Solution The sample space for the throw of two dice is shown in Figure 8.22, where 1-1 represents the event "the first die shows a 1 and the second die shows a 1," 1-2 represents the event "the first die shows a 1 and the second die shows a 2," and so on. Let A represent the event "the first die shows a 2" and B represent the event "the sum of the results is 6 or 7." These events are indicated in color in Figure 8.22. From the diagram, event A has 6 elements, B has 11 elements, and the sample space has 36 elements. Thus,

$$P(A) = \frac{6}{36}, \quad P(B) = \frac{11}{36}, \quad \text{and} \quad P(A \cap B) = \frac{2}{36}.$$

By the addition rule,

$$P(A \cup B) = P(A) + P(B) - P(A \cap B),$$

$$P(A \cup B) = \frac{6}{36} + \frac{11}{36} - \frac{2}{36} = \frac{15}{36} = \frac{5}{12}.$$

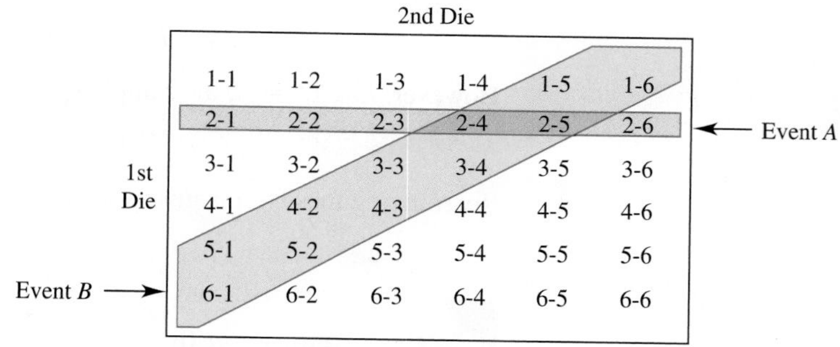

FIGURE 8.22

(b) The sum is 11 or the second die shows a 5.

Solution $P(\text{sum is } 11) = 2/36$, $P(\text{second die shows } 5) = 6/36$, and $P(\text{sum is } 11 \text{ and second die shows } 5) = 1/36$, so

$$P(\text{sum is 11 or second die shows 5}) = \frac{2}{36} + \frac{6}{36} - \frac{1}{36} = \frac{7}{36}.$$ 2

Checkpoint 2

In the random experiment of Example 2, find the given probabilities.

(a) The sum is 5 or the second die shows a 3.

(b) Both dice show the same number, or the sum is at least 11.

Answers: See page 506.

If events E and F are disjoint, then $E \cap F = \varnothing$ by definition; hence, $P(E \cap F) = 0$. Applying the addition rule yields the useful fact that follows.

Addition Rule for Disjoint Events

For disjoint events E and F,

$$P(E \cup F) = P(E) + P(F).$$

Example 3 Assume that the probability of a couple having a baby boy is the same as the probability of the couple having a baby girl. If the couple has 3 children, find the probability that at least 2 of them are girls.

Solution The event of having at least 2 girls is the union of the disjoint events $E =$ "the family has exactly 2 girls" and $F =$ "the family has exactly 3 girls." Using the equally likely sample space

$$\{ggg, ggb, gbg, bgg, gbb, bgb, bbg, bbb\},$$

where b represents a boy and g represents a girl, we see that $P(2 \text{ girls}) = 3/8$ and $P(3 \text{ girls}) = 1/8$. Therefore,

$$\begin{aligned} P(\text{at least 2 girls}) &= P(2 \text{ girls}) + P(3 \text{ girls}) \\ &= \frac{3}{8} + \frac{1}{8} = \frac{1}{2}. \end{aligned}$$ ✓3

✓Checkpoint 3

In Example 3, find the probability of having no more than 2 girls.

Answer: *See page 506.*

By definition of E', for any event E from a sample space S,

$$E \cup E' = S \quad \text{and} \quad E \cap E' = \emptyset.$$

Because $E \cap E' = \emptyset$, events E and E' are disjoint, so that

$$P(E \cup E') = P(E) + P(E').$$

However, $E \cup E' = S$, the sample space, and $P(S) = 1$. Thus,

$$P(E \cup E') = P(E) + P(E') = 1.$$

Rearranging these terms gives the following useful rule.

Complement Rule

For any event E,

$$\mathbf{P(E') = 1 - P(E)} \quad \text{and} \quad \mathbf{P(E) = 1 - P(E')}.$$

Example 4 If a fair die is rolled, what is the probability that any number but 5 will come up?

Solution If E is the event that 5 comes up, then E' is the event that any number but 5 comes up. $P(E) = 1/6$, so we have $P(E') = 1 - 1/6 = 5/6$. ✓4

✓Checkpoint 4

(a) Let $P(K) = 2/3$. Find $P(K')$.

(b) If $P(X') = 3/4$, find $P(X)$.

Answers: *See page 506.*

Example 5 If two fair dice are rolled, find the probability that the sum of the numbers showing is greater than 3.

Solution To calculate this probability directly, we must find each of the probabilities that the sum is 4, 5, 6, 7, 8, 9, 10, 11, and 12 and then add them. It is much simpler to first find the probability of the complement, the event that the sum is less than or equal to 3:

$$P(\text{sum} \le 3) = P(\text{sum is } 2) + P(\text{sum is } 3)$$
$$= \frac{1}{36} + \frac{2}{36} = \frac{3}{36} = \frac{1}{12}.$$

Now use the fact that $P(E) = 1 - P(E')$ to get

$$P(\text{sum} > 3) = 1 - P(\text{sum} \le 3) = 1 - \frac{1}{12} = \frac{11}{12}.$$ 5 ✓

✓Checkpoint 5

In Example 5, find the probability that the sum of the numbers rolled is at least 5.

Answer: See page 506.

ODDS

Sometimes probability statements are given in terms of **odds**: a comparison of $P(E)$ with $P(E')$. For example, suppose $P(E) = \frac{4}{5}$. Then $P(E') = 1 - \frac{4}{5} = \frac{1}{5}$. These probabilities predict that E will occur 4 out of 5 times and E' will occur 1 out of 5 times. Then we say that the **odds in favor** of E are 4 to 1, or 4:1.

Odds

The **odds in favor** of an event E are defined as the ratio of $P(E)$ to $P(E')$, or

$$\frac{P(E)}{P(E')}, \quad P(E') \neq 0.$$

Example 6 Suppose the weather forecaster says that the probability of rain tomorrow is 1/3. Find the odds in favor of rain tomorrow.

Solution Let E be the event "rain tomorrow." Then E' is the event "no rain tomorrow." Since $P(E) = 1/3$, $P(E') = 2/3$. By the definition of odds, the odds in favor of rain are

$$\frac{1/3}{2/3} = \frac{1}{2}, \quad \text{written 1 to 2 or 1:2.}$$

On the other hand, the odds that it will *not* rain, or the odds *against* rain, are

$$\frac{2/3}{1/3} = \frac{2}{1}, \quad \text{written 2 to 1.}$$

If the odds in favor of an event are, say, 3 to 5, then the probability of the event is 3/8, while the probability of the complement of the event is 5/8. (Odds of 3 to 5 indicate 3 outcomes in favor of the event out of a total of 8 outcomes.) The above example suggests the following generalization:

If the odds favoring event E are m to n, then

$$P(E) = \frac{m}{m+n} \quad \text{and} \quad P(E') = \frac{n}{m+n}.$$

Example 7 Often, weather forecasters give probability in terms of percentage. Suppose the weather forecaster says that there is a 40% chance that it will snow tomorrow. Find the odds of snow tomorrow.

Solution In this case, we can let E be the event "snow tomorrow." Then E' is the event "no snow tomorrow." Now, we have $P(E) = .4 = 4/10$ and $P(E') = .6 = 6/10$. By the definition of odds in favor, the odds in favor of snow are

$$\frac{4/10}{6/10} = \frac{4}{6} = \frac{2}{3}, \quad \text{written 2 to 3 or 2:3.}$$

It is important to put the final fraction into lowest terms in order to communicate the odds. ✓6

✓ Checkpoint 6

In Example 7, suppose $P(E) = 9/10$. Find the odds

(a) in favor of E;

(b) against E.

Suppose the chance of snow is 80%. Find the odds

(c) in favor of snow;

(d) against snow.

Answers: See page 506.

Example 8 The odds that a particular bid will be the low bid are 4 to 5.

(a) Find the probability that the bid will be the low bid.

Solution Odds of 4 to 5 show 4 favorable chances out of 4 + 5 = 9 chances altogether, so

$$P(\text{bid will be low bid}) = \frac{4}{4+5} = \frac{4}{9}.$$

(b) Find the odds against that bid being the low bid.

Solution There is a 5/9 chance that the bid will not be the low bid, so the odds against a low bid are

$$\frac{P(\text{bid will not be low})}{P(\text{bid will be low})} = \frac{5/9}{4/9} = \frac{5}{4},$$

or 5:4. ✓7

✓ Checkpoint 7

If the odds in favor of event E are 1 to 5, find

(a) $P(E)$;

(b) $P(E')$.

Answers: See page 506.

APPLICATIONS

Example 9 **Business** Let A represent living in a two-bedroom dwelling and B represent paying less than \$100 a month for the average electric bill. From the 2007 American Community Survey,* we have the following probabilities:

$$P(A) = .267, \quad P(B) = .510, \quad \text{and} \quad P(A \cap B) = .173.$$

(a) Find the probability that an American does *not* live in a two-bedroom dwelling and pays *\$100 or more* a month in electricity.

Solution Place the given information on a Venn diagram, starting with .173 in the intersection of the regions A and B. (See Figure 8.23.) As stated earlier, event A has probability .267. Since .173 has already been placed inside the intersection of A and B,

$$.267 - .173 = .094$$

goes inside region A, but outside the intersection of A and B. In the same way,

$$.510 - .173 = .337$$

goes inside region B, but outside the overlap.

*www.census.gov/acs.

The event we want is $A' \cap B'$. From the Venn diagram in Figure 8.23, the labeled regions have a total probability of

$$.094 + .173 + .337 = .604.$$

Since the entire region of the Venn diagram must have probability 1, the region outside A and B, namely $A' \cap B'$, has probability

$$1 - .604 = .396.$$

So the probability that an American does not live in a two-bedroom dwelling and pays $100 or more a month for electricity is .396.

(b) Find the probability that an American does not live in a two-bedroom dwelling *or* pays $100 or more a month for electricity.

Solution The corresponding region $A' \cup B'$, from Figure 8.23, has probability

$$.396 + .094 + .337 = .827.$$ 8 ✓

✓Checkpoint 8

Using the data from Example 9, find the probability that an American

(a) pays $100 or more a month for electricity;

(b) does not live in a two-bedroom dwelling and pays less than $100 a month for electricity.

Answers: See page 506.

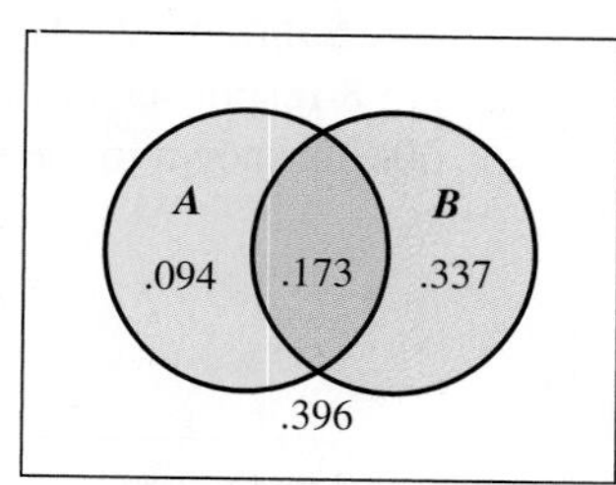

FIGURE 8.23

Example 10 **Business** Data from the 2008 *Forbes* magazine survey of the 100 highest paid chief executive officers (CEOs) is cross-classified by the CEO's age in years and the CEO's annual compensation (in millions of dollars), as shown in the table:

		Annual Compensation (Millions of dollars)			
		Less than 25	25–49	50 or more	Total
Age (Years)	40s	2	5	2	9
	50s	27	18	10	55
	60s	8	16	7	31
	70s or older	1	2	2	5
	Total	38	41	21	100

Let E be the event "the CEO earns less than $25 million" and F be the event "the CEO's age is in the 60s."

(a) Find $P(E)$.

Solution We need to find the total number of CEOs for whom the pay is less than $25 million for all of the different age groups, as shown in the shaded column in the following table:

		Annual Compensation (Millions of dollars)			
		Less than 25	25–49	50 or more	Total
Age (Years)	40s	2	5	2	9
	50s	27	18	10	55
	60s	8	16	7	31
	70s or older	1	2	2	5
	Total	38	41	21	100

Thus $2 + 27 + 8 + 1 = 38$ is the total number of CEOs who earned less than \$25 million dollars and

$$P(E) = 38/100 = .38.$$

(b) Find $P(E \cap F)$.

Solution We look to find the number that satisfy both conditions. We see there are 8 responses in the column for less than \$25 million and the row for age in the 60s, as shown in the shaded area in the following table:

		Annual Compensation (Millions of dollars)			
		Less than 25	25–49	50 or more	Total
Age (Years)	40s	2	5	2	9
	50s	27	18	10	55
	60s	8	16	7	31
	70s or older	1	2	2	5
	Total	38	41	21	100

Thus, $P(E \cap F) = 8/100 = .08$.

(c) Find $P(E \cup F)$.

Solution We can use the additive rule for probability to find $P(E \cup F)$. We know from part (a) that $P(E) = .38$. In a similar manner to part (a), we can find $P(F)$. There are $8 + 16 + 7 = 31$ CEOs whose age is in the 60s, so $P(F) = 31/100 = .31$. With the answer to (b), using the additive rule yields

$$P(E \cup F) = P(E) + P(F) - P(E \cap F) = .38 + .31 - .08 = .61.$$ ✓9

✓ Checkpoint 9

Let G be the event "the CEO earns \$50 million or more" and H be the event "the CEO's age is in the 40s." Find

(a) $P(G)$;

(b) $P(G \cap H)$;

(c) $P(G \cup H)$.

Answers: See page 506.

8.4 ▶ Exercises

Assume a single spin of the roulette wheel is made. (See Example 1.) Find the probability for the given events.

1. The marble lands on a green or black slot.
2. The marble lands on a green or even slot.
3. The marble lands on an odd or black slot.

Also with a single spin of the roulette wheel, find the probability of winning with the given bets.

4. The marble will land in a slot numbered 13–18.
5. The marble will land in a slots 0, 00, 1, 2, or 3.

6. The marble will land in a slot that is a multiple of 3.

7. The marble will land in a slot numbered 25–36.

Two dice are rolled. Find the probabilities of rolling the given sums. (See Examples 2, 4, and 5.)

8. **(a)** 2 **(b)** 4
(c) 5 **(d)** 6

9. **(a)** 8 **(b)** 9
(c) 10 **(d)** 13

10. **(a)** 9 or more
(b) Less than 7
(c) Between 5 and 8 (exclusive)

11. **(a)** Not more than 5
(b) Not less than 8
(c) Between 3 and 7 (exclusive)

Tami goes shopping and sees three kinds of shoes: flats, 2" heels, and 3" heels. The shoes come in two shades of beige (light and dark) and black. If each option has an equal chance of being selected, find the probabilities of the given events.

12. The shoes Tami buys have a heel.

13. The shoes Tami buys are black.

14. The shoes Tami buys have a 2" heel and are beige.

Ms. Elliott invites 10 relatives to a party: her mother, 3 aunts, 2 uncles, 2 sisters, 1 male cousin, and 1 female cousin. If the chances of any one guest arriving first are equally likely, find the probabilities that the given guests will arrive first.

15. **(a)** A sister or an aunt
(b) A sister or a cousin
(c) A sister or her mother

16. **(a)** An aunt or a cousin
(b) A male or an uncle
(c) A female or a cousin

Use Venn diagrams to work Exercises 17–21. See Example 9.

17. Suppose $P(E) = .30$, $P(F) = .51$, and $P(E \cap F) = .19$. Find each of the given probabilities.
(a) $P(E \cup F)$ **(b)** $P(E' \cap F)$
(c) $P(E \cap F')$ **(d)** $P(E' \cup F')$

18. Let $P(Z) = .40$, $P(Y) = .30$, and $P(Z \cup Y) = .58$. Find each of the given probabilities.
(a) $P(Z' \cap Y')$ **(b)** $P(Z' \cup Y')$
(c) $P(Z' \cup Y)$ **(d)** $P(Z \cap Y')$

19. **Health** In 2007, in the state of North Carolina, the probability that a woman giving birth was under age 20 (event A) was .116. The probability that the child weighed 2500 grams or less—a common cutoff for classification as having low birth weight (event B)—was .093. The probability that the mother was under age 20 and the child had a low birth weight was .013.* Find the probability of the given event.
(a) $A \cup B$ **(b)** $A' \cap B$
(c) $A' \cap B'$ **(d)** $A' \cup B'$

20. **Business** Data from a portion of the 2007 American Community Survey indicated that the probability a household earned $100,000 or more (event E) was 0.192. The probability a household owned three or more cars (event F) was .609. The probability a household earned $100,000 or more and owned three or more cars was .163. Find the probability of the given event.
(a) $E' \cup F'$ **(b)** $E' \cap F$
(c) $E \cup F'$

21. **Social Science** Data from the 2006 General Social Survey indicated the probability of being married is .473, the probability of being generally "very happy" is .308, and the probability of being both married and generally "very happy" is .201.† Find the probability of the given event.
(a) Not being married and not being happy
(b) Being married or not being happy
(c) Not being married and being happy

22. Define what is meant by odds.

A single fair die is rolled. Find the odds in favor of getting the results in Exercises 23–26. (See Examples 6 and 7.)

23. 2

24. 2, 3, 4

25. 2, 3, 5, or 6

26. Some number greater than 5

27. A marble is drawn from a box containing 3 yellow, 4 white, and 8 blue marbles. Find the odds in favor of drawing the given marbles.
(a) A yellow marble
(b) A blue marble
(c) A white marble

28. Find the odds of *not* drawing a white marble in Exercise 27.

29. Two dice are rolled. Find the odds of rolling a 7 or an 11.

30. In the "Ask Marilyn" column of *Parade* magazine, a reader wrote about the following game: "You and I each roll a die. If your die is higher than mine, you win. Otherwise, I win." The reader thought that the probability that each player wins is 1/2. Is this correct? If not, what is the probability that each player wins?*

---

*Based on data available at http://arc.irss.unc.edu/dvn.

†Based on data available at www.norc.edu/GSS+Website.

each player wins is 1/2. Is this correct? If not, what is the probability that each player wins?*

Social Science *For Exercises 31–33, find the odds of the event occurring from the given probability that a bachelor's degree recipient in 2006 majored in the given discipline.*†

31. Business; probability 21/100.

32. Biological or biomedical sciences; probability 1/20.

33. Education; probability 7/100.

Social Science *For Exercises 34–37, convert the given odds to the probability that the event will occur.*‡

34. The odds that an American adult is a high school graduate are 6:1.

35. The odds that an American adult is a college graduate are 2:5.

36. The odds that an American adult earned a bachelor's degree (and not a higher degree) are 1:4.

37. The odds that an American adult earned an advanced degree are 1:9.

Business *For Exercises 38–39, find the odds of the event occurring from the given probability.*§

38. In December 2008, the probability of being unemployed in the United States was 9/125.

39. In December 2008, the probability of being unemployed in Rhode Island was 1/10.

One way to solve a probability problem is to repeat the experiment many times, keeping track of the results. Then the probability can be approximated by using the basic definition of the probability of an event E, which is $P(E) = n(E)/n(S)$, where E occurs n(E) times out of n(S) trials of an experiment. This is called the ***Monte Carlo method*** *of finding probabilities. If physically repeating the experiment is too tedious, it may be simulated with the use of a random-number generator, available on most computers and scientific or graphing calculators. To simulate a coin toss or the roll of a die on a graphing calculator, change the setting to fixed decimal mode with 0 digits displayed. To simulate multiple tosses of a coin, press RAND (or RANDOM or RND#) in the PROB submenu of the MATH (or OPTN) menu, and then press ENTER repeatedly. Interpret 0 as a head and 1 as a tail. To simulate multiple rolls of a die, press RAND × 6 + .5, and then press ENTER repeatedly.*

40. Suppose two dice are rolled. Use the Monte Carlo method with at least 50 repetitions to approximate the given probabilities. Compare them with the results of Exercise 10.
(a) P(the sum is 9 or more)
(b) P(the sum is less than 7)

41. Suppose two dice are rolled. Use the Monte Carlo method with at least 50 repetitions to approximate the given probabilities. Compare them with the results of Exercise 11.
(a) P(the sum is not more than 5)
(b) P(the sum is not less than 8)

42. Suppose three dice are rolled. Use the Monte Carlo method with at least 100 repetitions to approximate the given probabilities.
(a) P(the sum is 5 or less)
(b) P(neither a 1 nor a 6 is rolled)

43. Suppose a coin is tossed 5 times. Use the Monte Carlo method with at least 50 repetitions to approximate the given probabilities.
(a) P(exactly 4 heads)
(b) P(2 heads and 3 tails)

44. Business Suppose that 8% of a certain batch of calculators have a defective case and that 11% have defective batteries. Also, 3% have both a defective case and defective batteries. A calculator is selected from the batch at random. Find the probability that the calculator has a good case and good batteries.

Health *Using the categories defined by the Centers for Disease Control and Prevention for weight status, the table cross-classifies adults by age and weight status.**

Age (Years)	Underweight	Healthy Weight	Overweight	Obese	Total
18–39	81	872	735	712	2400
40–64	48	421	601	739	1809
65 or Higher	49	264	355	309	977
Total	178	1557	1691	1760	5186

Define A as the event of being ages 18–39, B as the event of being ages 40–64, and C as the event of being age 65 or older. Let D be the event of being underweight, E be the event of having a healthy weight, F be the event of being overweight, and G be the event of being obese. Use the table for Exercises 45–50 to find the probability of the given event.

45. The event of being of a healthy weight

46. The event of being overweight or obese

47. The event of not being underweight

48. The event of being 65 or higher and being obese

49. The event of being 18–39 or of a healthy weight

50. The event of being 18–39 and overweight

Business *The table gives the number of vehicles owned cross-classified by income level for households from the 2007 American Community Survey:*

**Parade* Magazine, November 6, 1994, p. 10. Reprinted by permission of the William Morris Agency, Inc., on behalf of the author. Copyright © 1994 by Marilyn vos Savant.

†Digest of Educational Statistics.

‡www.census.gov.

§U.S. Bureau of Labor Statistics.

*Based on data available at www.cdc.gov/nchs/nhanes.htm.

Income	Number of Vehicles 0	1	2	3	4	5	Total
Less than \$100,000	13	95	234	305	100	16	763
\$100,000 or more	0	4	23	64	65	25	181
Total	13	99	257	369	165	41	944

Let A indicate the event "the household earns less than \$100,000," let B indicate the event "the number of vehicles is 1," and let C indicate "the number of vehicles is 2." Use the table to find the requested probabilities in Exercises 51–56.

51. $P(A)$

52. $P(C)$

53. $P(B \cup C)$

54. $P(A' \cap B)$

55. $P(A \cup C)$

56. $P(A' \cup B)$

Business *Data from the 2006 General Social Survey can allow us to estimate how much part-time and full-time employees work per week. Use the following table to find the probabilities of the events in Exercises 57–60:*

Labor Force Status	Hours Worked per Week 0–19	20–29	30–39	40–49	50–59	60 or more	Total
Working Full Time	22	43	251	1251	401	354	2322
Working Part Time	143	165	103	16	1	12	440
Total	165	208	354	1267	402	366	2762

57. Working full time

58. Working part time and 0–19 hours

59. Working full time and 40–49 hours

60. Working part time or working less than 30 hours

61. Natural Science Color blindness is an inherited characteristic that is more common in males than in females. If M represents male and C represents red–green color blindness, we use the relative frequencies of the incidences of males and red–green color blindness as probabilities to get

$$P(C) = .039, \quad P(M \cap C) = .035, \text{ and } P(M \cup C) = .495.*$$

Find the given probabilities.

(a) $P(C')$ **(b)** $P(M)$

(c) $P(M')$ **(d)** $P(M' \cap C')$

(e) $P(C \cap M')$ **(f)** $P(C \cup M')$

62. Natural Science Gregor Mendel, an Austrian monk, was the first to use probability in the study of genetics. In an effort to understand the mechanism of characteristic transmittal from one generation to the next in plants, he counted the number of occurrences of various characteristics. Mendel found that the flower color in certain pea plants obeyed this scheme:

Pure red crossed with pure white produces red.

From its parents, the red offspring received genes for both red (R) and white (W), but in this case red is *dominant* and white *recessive*, so the offspring exhibits the color red. However, the offspring still carries both genes, and when two such offspring are crossed, several things can happen in the third generation. The following table, called a *Punnett square*, shows the equally likely outcomes:

		Second Parent	
		R	W
First Parent	R	RR	RW
	W	WR	WW

Use the fact that red is dominant over white to find each of the given probabilities. Assume that there are an equal number of red and white genes in the population.

(a) P(a flower is red) **(b)** P(a flower is white)

63. Natural Science Mendel (see Exercise 62) found no dominance in snapdragons, with one red gene and one white gene producing pink-flowered offspring. These second-generation pinks, however, still carry one red and one white gene, and when they are crossed, the next generation still yields the Punnett square in Exercise 62. Find each of the given probabilities.

(a) P(red) **(b)** P(pink) **(c)** P(white)

(Mendel verified these probability ratios experimentally and did the same for many characteristics other than flower color. His work, published in 1866, was not recognized until 1890.)

*The probabilities of a person being male or female are from *The World Almanac and Book of Facts*, 2002. The probabilities of a male and female being color blind are from *Parsons' Diseases of the Eye* (18th ed.), by Stephen J. H. Miller (Churchill Livingston, 1990), p. 269. This reference gives a range of 3 to 4% for the probability of gross color blindness in men; we used the midpoint of that range.

64. **Social Science** Answers to a question on the legalization of marijuana in the 2006 General Social Survey indicated that there were 1044 females who answered the question, 336 females who favored making marijuana legal, and 672 men and women together who favored making marijuana legal. With the entire number of males and females who answered the question being 1828, find the probability that a person is
 (a) female and not in favor of legalization;
 (b) not in favor of legalization;
 (c) female or in favor of legalization;
 (d) male and not in favor of legalization.
 (*Hint*: Draw two circles respectively denoting females and favoring legalization that overlap in the middle.)

65. **Social Science** Answers to a question on the existence of an afterlife in the 2006 General Social Survey indicated that there were 1124 men who responded to the question, 891 men who said they believed in an afterlife, and 2177 men and women who believed in an afterlife. With a total of 2629 men and women who answered the question, find the probability that a person
 (a) believes in an afterlife;
 (b) is male and believes in an afterlife;
 (c) is female and does not believe in an afterlife.
 (*Hint*: Draw two circles respectively denoting males and believing in an afterlife that overlap in the middle.)

Finance *The National Association of College and University Business Officers researched the change in university and college endowments from 2007 to 2008. The table below shows findings for 203 colleges and universities with 2008 endowment levels above \$300 million dollars. It indicates how many schools with a particular endowment level had their endowment increase in value, how many had their endowment decrease by between 0 and 5%, and how many had their endowment decrease by more than 5%. Among these 203 colleges and universities, find the probabilities of the events in Exercises 66–70.*

	Change in Value			
Endowment Value (Dollars)	**Positive Increase**	**Between 0 and 5% Decrease**	**More than 5% Decrease**	**Total**
300–499 Million	19	27	16	62
500–999 Million	24	26	14	64
1–1.999 Billion	22	17	6	45
2 Billion or More	20	8	4	32
Total	85	78	40	203

66. The endowment had a positive increase in value.
67. The endowment was valued at \$2 billion or more and lost more than 5%.
68. The endowment did not decrease by 5% or more.
69. The endowment had between a 0 and 5% reduction or had a value of \$300–499 million.
70. The endowment was less than \$2 billion and had a positive increase.

✓ Checkpoint Answers

1. 13/19
2. (a) 1/4 (b) 2/9
3. 7/8
4. (a) 1/3 (b) 1/4
5. 5/6
6. (a) 9 to 1 (b) 1 to 9 (c) 4 to 1 (d) 1 to 4
7. (a) 1/6 (b) 5/6
8. (a) .490 (b) .337
9. (a) .21 (b) .02 (c) .28

8.5 ▶ Conditional Probability and Independent Events

Did you ever wonder what salary the president of your college or university earns per year? The *Chronicle of Higher Education* conducted a survey in November 2008 to examine that question. The following table examines whether presidents of public universities and community colleges earned less than \$250,000 or \$250,000 or more in compensation.*

*The figures in the table do not include house, car, and supplemental compensation. From http://chronicle.com/stats/990/public.htm.

	Earned less than \$250,000	Earned \$250,000 or more	Total
University President	53	101	154
Community College President	21	14	35
Total	74	115	189

Let A be the event "earned \$250,000 or more," and let B be the event "community college president." We can find $P(A), P(A'), P(B)$, and $P(B')$. For example, the table shows that a total of 115 presidents earned \$250,000 or more, so

$$P(A) = \frac{115}{189} \approx .608. \quad \checkmark^{1}$$

✓Checkpoint 1

Use the data in the table to find

(a) $P(B)$;
(b) $P(A')$;
(c) $P(B')$.

Answers: See page 519.

Suppose we want to know the probability that a community college president earns \$250,000 or more. From the table, of 35 community college presidents, 14 earned \$250,000 or more, so

$$P(\text{community college president earns \$250,000 or more}) = \frac{14}{35} = .4.$$

This is a different number from the probability of making \$250,000 or more, .608, because *we have additional information* (the president is the president of a community college) *that has reduced the sample space.* In other words, we found the probability of earning \$250,000 or more, A, given the additional information that the president is a community college president, B. This probability is called the *conditional probability* of event A, given that event B has occurred, written $P(A|B)$ and often read as "the probability of A given B."

In the preceding example,

$$P(A|B) = \frac{14}{35} = .4.$$

If we divide the numerator and denominator by 189 (the size of the sample space), this quantity can be written as

$$P(A|B) = \frac{\frac{14}{189}}{\frac{35}{189}} = \frac{P(A \cap B)}{P(B)},$$

where $P(A \cap B)$ represents, as usual, the probability that both A and B will occur.

To generalize this result, assume that E and F are two events for a particular experiment. Assume also that the sample space S for the experiment has n possible equally likely outcomes. Suppose event F has m elements and $E \cap F$ has k elements $(k \leq m)$. Then, using the fundamental principle of probability yields

$$P(F) = \frac{m}{n} \quad \text{and} \quad P(E \cap F) = \frac{k}{n}.$$

We now want to find $P(E|F)$: the probability that E occurs, given that F has occurred. Since we assume that F has occurred, we reduce the sample space to F;

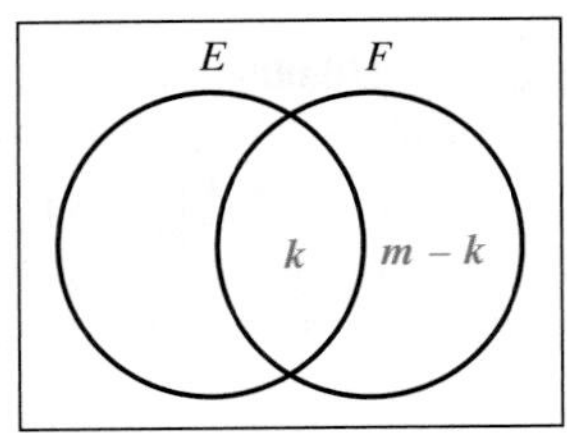

Event F has a total of m elements.

FIGURE 8.24

that is, we look only at the m elements inside F. (See Figure 8.24.) Of these m elements, there are k elements for which E also occurs, because $E \cap F$ has k elements. This yields

$$P(E|F) = \frac{k}{m}.$$

Divide numerator and denominator by n to get

$$P(E|F) = \frac{k/n}{m/n} = \frac{P(E \cap F)}{P(F)}.$$

The last result motivates the following definition of conditional probability. The **conditional probability** of an event E, given event F, written $P(E|F)$, is

$$P(E|F) = \frac{P(E \cap F)}{P(F)} = \frac{P(E \text{ and } F)}{P(F)}, \quad P(F) \neq 0.$$

This definition tells us that, for equally likely outcomes, conditional probability is found by *reducing the sample space to event* F and then finding the number of outcomes in F that are also in event E. Thus,

$$\boldsymbol{P(E|F) = \frac{n(E \cap F)}{n(F)}}.$$

Although the definition of conditional probability was motivated by an example with equally likely outcomes, it is valid in all cases. For an intuitive explanation, think of the formula as giving the probability that both E and F occur, compared with the entire probability of F occuring.

✓Checkpoint 2

The table shows the results of the 2006 General Social Survey regarding happiness for married and never married respondents.

	Very Happy	Partially Happy or Not Happy	Total
Married	600	813	1413
Never Married	144	586	730
Total	744	1399	2143

Let M represent married respondents and V represent very happy respondents. Find each of the given probabilities.

(a) $P(V|M)$

(b) $P(V|M')$

(c) $P(M|V)$

(d) $P(M'|V')$

(e) State the probability of part (d) in words.

Answers: See page 519.

Example 1 **Business** Use the information from the table on the salary of university and community college presidents (page 507) to find the given probabilities.

(a) $P(A|B')$

Solution In words, this is the probability of earning \$250,000 or more, given that the president is a university president, or

$$P(A|B') = \frac{n(A \cap B')}{n(B')} = \frac{101}{154} \approx .656.$$

(b) $P(B|A)$

Solution This represents the probability that the president is a community college president, given that he or she makes \$250,000 or more. Reduce the sample space to A. Then find $n(A \cap B)$ and $n(A)$. The result is

$$P(B|A) = \frac{n(A \cap B)}{n(A)} = \frac{14}{115} \approx .122.$$

So if a president earns \$250,000 or more, then the probability is .122 that the president is a community college president.

(c) $P(B'|A')$

Solution Here, we want the probability the president is a university president, given that he or she earns less than \$250,000:

$$P(B'|A') = \frac{n(B' \cap A')}{n(A')} = \frac{53}{74} \approx .716.$$ [2]

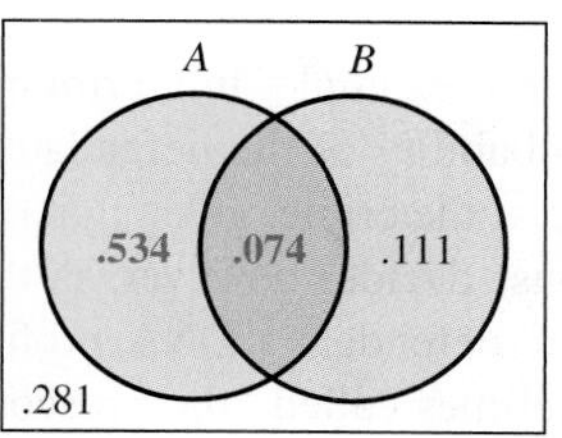

FIGURE 8.25

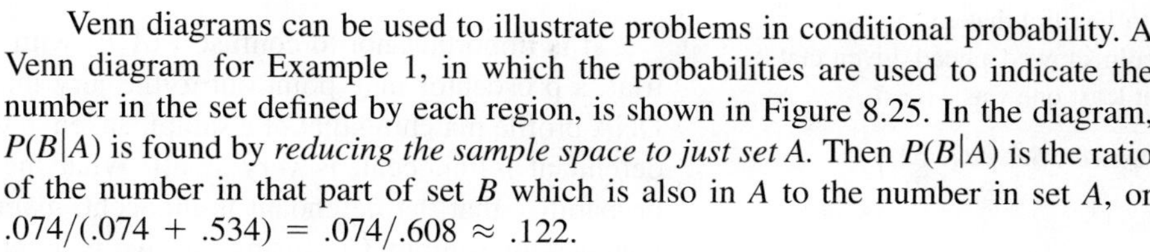

Venn diagrams can be used to illustrate problems in conditional probability. A Venn diagram for Example 1, in which the probabilities are used to indicate the number in the set defined by each region, is shown in Figure 8.25. In the diagram, $P(B|A)$ is found by *reducing the sample space to just set A*. Then $P(B|A)$ is the ratio of the number in that part of set B which is also in A to the number in set A, or $.074/(.074 + .534) = .074/.608 \approx .122$.

Example 2 Given $P(E) = .4$, $P(F) = .5$, and $P(E \cup F) = .7$, find $P(E|F)$.

Solution Find $P(E \cap F)$ first. Then use a Venn diagram to find $P(E|F)$. By the addition rule,

$$\begin{aligned} P(E \cup F) &= P(E) + P(F) - P(E \cap F) \\ .7 &= .4 + .5 - P(E \cap F) \\ P(E \cap F) &= .2. \end{aligned}$$

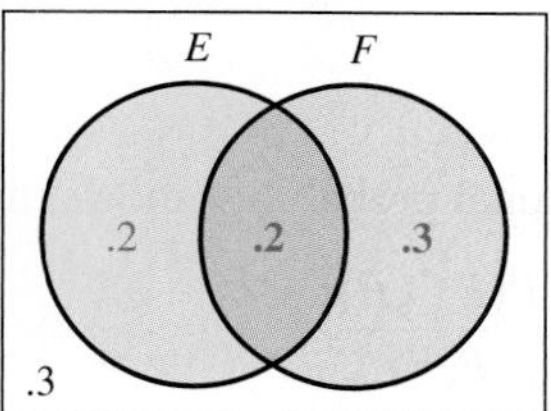

FIGURE 8.26

Now use the probabilities to indicate the number in each region of the Venn diagram in Figure 8.26. $P(E|F)$ is the ratio of the probability of that part of E which is in F to the probability of F, or

$$P(E|F) = \frac{P(E \cap F)}{P(F)} = \frac{.2}{.5} = \frac{2}{5} = .4.$$ [3]

✓Checkpoint 3

Find $P(F|E)$ if $P(E) = .3$, $P(F) = .4$, and $P(E \cup F) = .6$.

Answer: See page 519.

Example 3 Two fair coins were tossed, and it is known that at least one was a head. Find the probability that both were heads.

Solution The sample space has four equally likely outcomes: $S = \{hh, ht, th, tt.\}$ Define two events

$$E_1 = \text{at least 1 head} = \{hh, ht, th\}$$

and

$$E_2 = \text{2 heads} = \{hh\}.$$

Because there are four equally likely outcomes, $P(E_1) = 3/4$. Also, $P(E_1 \cap E_2) = 1/4$. We want the probability that both were heads, given that at least one was a head; that is, we want to find $P(E_2|E_1)$. Because of the condition that at least one coin was a head, the reduced sample space is

$$\{hh, ht, th\}.$$

Since only one outcome in this reduced sample space is two heads,

$$P(E_2|E_1) = \frac{1}{3}.$$

Alternatively, use the definition given earlier:

$$P(E_2|E_1) = \frac{P(E_2 \cap E_1)}{P(E_1)} = \frac{1/4}{3/4} = \frac{1}{3}.$$ ✓4

✓Checkpoint 4

In Example 3, find the probability that exactly one coin showed a head, given that at least one was a head.

Answer: See page 519.

It is important not to confuse $P(A|B)$ with $P(B|A)$. For example, in a criminal trial, a prosecutor may point out to the jury that the probability of the defendant's DNA profile matching that of a sample taken at the scene of the crime, given that the defendant is innocent, is very small. What the jury must decide, however, is the probability that the defendant is innocent, given that the defendant's DNA profile matches the sample. Confusing the two is an error sometimes called "the prosecutor's fallacy," and the 1990 conviction of a rape suspect in England was overturned by a panel of judges who ordered a retrial, because the fallacy made the original trial unfair.* This mistake is often called "confusion of the inverse."

In the next section, we will see how to compute $P(A|B)$ when we know $P(B|A)$.

PRODUCT RULE

If $P(E) \neq 0$ and $P(F) \neq 0$, then the definition of conditional probability shows that

$$P(E|F) = \frac{P(E \cap F)}{P(F)} \quad \text{and} \quad P(F|E) = \frac{P(F \cap E)}{P(E)}.$$

Using the fact that $P(E \cap F) = P(F \cap E)$, and solving each of these equations for $P(E \cap F)$, we obtain the following rule.

Product Rule of Probability

If E and F are events, then $P(E \cap F)$ may be found by either of these formulas:

$$\boldsymbol{P(E \cap F) = P(F) \cdot P(E|F)} \quad \text{or} \quad \boldsymbol{P(E \cap F) = P(E) \cdot P(F|E)}.$$

The **product rule** gives a method for finding the probability that events E and F both occur. Here is a simple way to remember the ordering of E and F in the probability rule:

$$P(E \cap F) = P(F) \cdot P(E|F) \quad \text{or} \quad P(E \cap F) = P(E) \cdot P(F|E).$$

Example 4 **Business** According to data from the U.S. Census Bureau, we can estimate the probability that a business is female owned as .282. We can also estimate the probability that a female-owned business has one to four employees as .504. What is the probability that a business is female owned *and* has one to four employees?

*David Pringle, "Who's the DNA Fingerprinting Pointing At?", *New Scientist*, January 29, 1994, pp. 51–52.

Solution Let F represent the event of "having a female-owned business" and E represent the event of "having one to four employees." We want to find $P(F \cap E)$. By the product rule,

$$P(F \cap E) = P(F)P(E|F).$$

From the given information, $P(F) = .282$, and the probability that a female-owned business has one to four employees is $P(E|F) = .504$. Thus,

$$P(F \cap E) = .282(.504) \approx .142.$$ 5 ✓

✓Checkpoint 5

In a litter of puppies, 3 were female and 4 were male. Half the males were black. Find the probability that a puppy chosen at random from the litter would be a black male.

Answer: See page 519.

In Section 8.1, we used a tree diagram to find the number of subsets of a given set. By including the probabilities for each branch of a tree diagram, we convert it to a **probability tree**. The following examples show how a probability tree is used with the product rule to find the probability of a sequences of events.

Example 5 A company needs to hire a new director of advertising. It has decided to try to hire either person A or person B, both of whom are assistant advertising directors for its major competitor. To decide between A and B, the company does research on the campaigns managed by A or B (none are managed by both) and finds that A is in charge of twice as many advertising campaigns as B. Also, A's campaigns have yielded satisfactory results three out of four times, while B's campaigns have yielded satisfactory results only two out of five times. Suppose one of the competitor's advertising campaigns (managed by A or B) is selected randomly.

We can represent this situation schematically as follows: Let A denote the event "Person A does the job" and B the event "Person B does the job." Let S be the event "satisfactory results" and U the event "unsatisfactory results." Then the given information can be summarized in the probability tree in Figure 8.27. Since A does twice as many jobs as B, $P(A) = 2/3$ and $P(B) = 1/3$, as noted on the first-stage branches of the tree. When A does a job, the probability of satisfactory results is $3/4$ and of unsatisfactory results $1/4$, as noted on the second-stage branches. Similarly, the probabilities when B does the job are noted on the remaining second-stage branches. The composite branches labeled 1–4 represent the four disjoint possibilities for the running and outcome of the campaign.

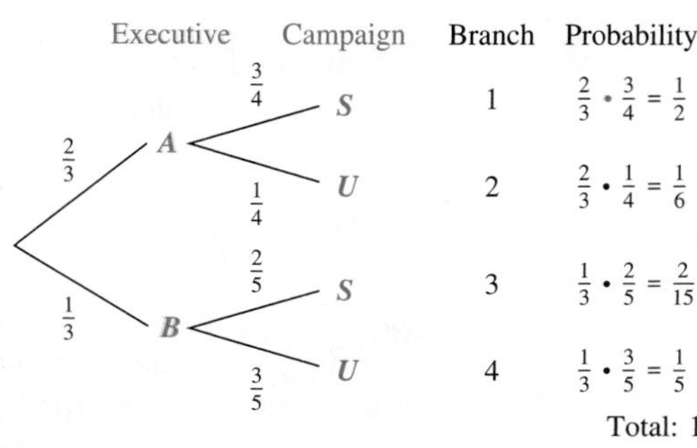

FIGURE 8.27

(a) Find the probability that A is in charge of a campaign that produces satisfactory results.

Solution We are asked to find $P(A \cap S)$. We know that when A does the job, the probability of success is 3/4; that is, $P(S|A) = 3/4$. Hence, by the product rule,

$$P(A \cap S) = P(A) \cdot P(S|A) = \frac{2}{3} \cdot \frac{3}{4} = \frac{1}{2}.$$

The event $(A \cap S)$ is represented by branch 1 of the tree, and as we have just seen, its probability is the product of the probabilities that make up that branch.

(b) Find the probability that B runs a campaign that produces satisfactory results.

Solution We must find $P(B \cap S)$. This event is represented by branch 3 of the tree, and as before, its probability is the product of the probabilities of the pieces of that branch:

$$P(B \cap S) = P(B) \cdot P(S|B) = \frac{1}{3} \cdot \frac{2}{5} = \frac{2}{15}.$$

(c) What is the probability that the selected campaign is satisfactory?

Solution The event S is the union of the disjoint events $A \cap S$ and $B \cap S$, which are represented by branches 1 and 3 of the tree diagram. By the addition rule,

$$P(S) = P(A \cap S) + P(B \cap S) = \frac{1}{2} + \frac{2}{15} = \frac{19}{30}.$$

Thus, the probability of an event that appears on several branches is the sum of the probabilities of each of these branches.

(d) What is the probability that the selected campaign is unsatisfactory?

Solution $P(U)$ can be read from branches 2 and 4 of the tree:

$$P(U) = \frac{1}{6} + \frac{1}{5} = \frac{11}{30}.$$

Alternatively, because U is the complement of S,

$$P(U) = 1 - P(S) = 1 - \frac{19}{30} = \frac{11}{30}.$$

(e) Find the probability that either A runs the campaign or the results are satisfactory (or possibly both).

Solution Event A combines branches 1 and 2, while event S combines branches 1 and 3, so use branches 1, 2, and 3:

$$P(A \cup S) = \frac{1}{2} + \frac{1}{6} + \frac{2}{15} = \frac{4}{5}.$$ ✓6

✓Checkpoint 6

Find each of the given probabilities for the Scenario in Example 5.

(a) $P(U|A)$

(b) $P(U|B)$

Answers: See page 519.

Example 6 Suppose 6 potential jurors remain in a jury pool and 2 are to be selected to sit on the jury for the trial. The races of the 6 potential jurors are 1 Hispanic, 3 Caucasian, and 2 African-American. If we select one juror at a time, find the probability that one Caucasian and one African-American are drawn.

Solution A probability tree showing the various possible outcomes is given in Figure 8.28. In this diagram, C represents the event "selecting a Caucasian juror" and A represents "selecting an African-American juror." On the first draw, $P(C$ on the 1st$) = 3/6 = 1/2$ because three of the six jurors are Caucasian. On the second draw, $P(A$ on the 2nd$|C$ on the 1st$) = 2/5$. One Caucasian juror has been removed, leaving 5, of which 2 are African-American.

We want to find the probability of selecting exactly one Caucasian and exactly one African-American. Two events satisfy this condition: selecting a Caucasian first and then selecting an African-American (branch 2 of the tree) and drawing an African-American juror first and then selecting a Caucasian juror (branch 4). For branch 2,

$$P(C \text{ on 1st}) \cdot P(A \text{ on 2nd} | C \text{ on 1st}) = \frac{1}{2} \cdot \frac{2}{5} = \frac{1}{5}.$$ 7 ✓

For branch 4, on which the African-American juror is selected first,

$$P(A \text{ first}) \cdot P(C \text{ second} | A \text{ first}) = \frac{1}{3} \cdot \frac{3}{5} = \frac{1}{5}.$$

Since these two events are disjoint, the final probability is the sum of the two probabilities.

$$P(\text{one } C, \text{ one } A) = P(C \text{ on 1st}) \cdot P(A \text{ on 2nd} | C \text{ on 1st}) + P(A \text{ on 1st}) \cdot P(C \text{ on 2nd} | A \text{ on 1st}) = \frac{2}{5}.$$ 8 ✓

✓Checkpoint 7

In Example 6, find the probability of selecting an African-American juror and then a Caucasian juror.

Answer: See page 519.

✓Checkpoint 8

In Example 6, find the probability of selecting a Caucasian juror and then a Hispanic juror.

Answer: See page 519.

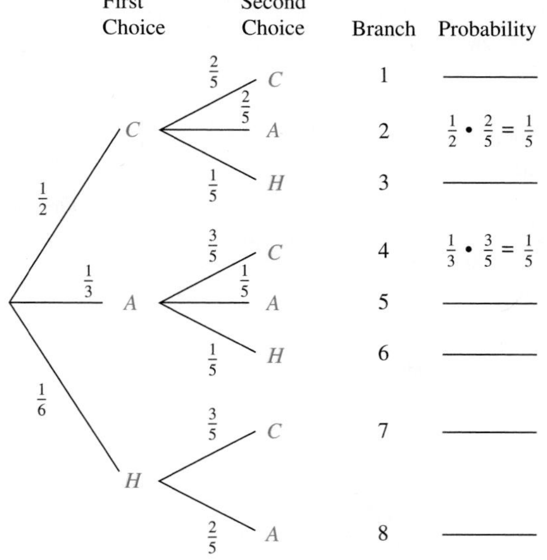

FIGURE 8.28

The product rule is often used in dealing with *stochastic processes*, which are mathematical models that evolve over time in a probabilistic manner. For example, selecting different jurors is such a process, because the probabilities change with each successive selection. (Particular stochastic processes are studied further in Section 9.5.)

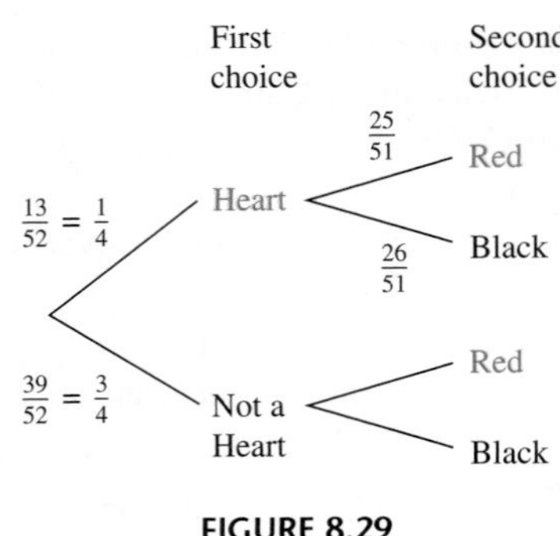

FIGURE 8.29

Example 7 Two cards are drawn without replacement from an ordinary deck (52 cards). Find the probability that the first card is a heart and the second card is red.

Solution Start with the probability tree of Figure 8.29. (You may wish to refer to the deck of cards shown on page 490.) On the first draw, since there are 13 hearts in the 52 cards, the probability of drawing a heart first is $13/52 = 1/4$. On the second draw, since a (red) heart has been drawn already, there are 25 red cards in the remaining 51 cards. Thus, the probability of drawing a red card on the second draw, given that the first is a heart, is $25/51$. By the product rule of probability,

$$\begin{aligned} &P(\text{heart on 1st and red on 2nd}) \\ &= P(\text{heart on 1st}) \cdot P(\text{red on 2nd} \mid \text{heart on 1st}) \\ &= \frac{1}{4} \cdot \frac{25}{51} = \frac{25}{204} \approx .123. \end{aligned}$$

9 ✓

✓Checkpoint 9

Find the probability of drawing a heart on the first draw and a black card on the second if two cards are drawn without replacement.

Answer: See page 519.

Example 8 Three cards are drawn, without replacement, from an ordinary deck. Find the probability that exactly 2 of the cards are red.

Solution Here, we need a probability tree with three stages, as shown in Figure 8.30. The three branches indicated with arrows produce exactly 2 red cards from the draws. Multiply the probabilities along each of these branches and then add:

$$\begin{aligned} P(\text{exactly 2 red cards}) &= \frac{26}{52} \cdot \frac{25}{51} \cdot \frac{26}{50} + \frac{26}{52} \cdot \frac{26}{51} \cdot \frac{25}{50} + \frac{26}{52} \cdot \frac{26}{51} \cdot \frac{25}{50} \\ &= \frac{50{,}700}{132{,}600} = \frac{13}{34} \approx .382. \end{aligned}$$

10 ✓

✓Checkpoint 10

Use the tree in Example 8 to find the probability that exactly one of the cards is red.

Answer: See page 519.

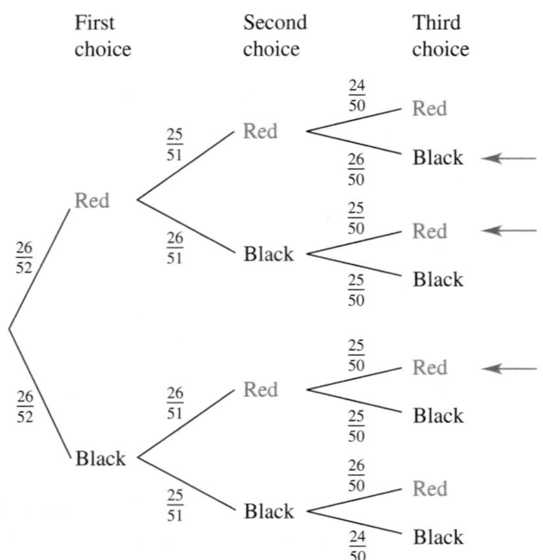

FIGURE 8.30

INDEPENDENT EVENTS

Suppose in Example 7 that we draw the two cards *with* replacement rather than without replacement. (That is, we put the first card back, shuffle them and then draw the second card.) If the first card is a heart, then the probability of drawing a red card on the second draw is $26/52$, rather than $25/51$, because there are still 52 cards in the deck, 26 of them red. In this case, $P(\text{red second}|\text{heart first})$ is the same as $P(\text{red second})$. The value of the second card is not affected by the value of the first card. We say that the event that the second card is red is *independent* of the event that the first card is a heart, since knowledge of the first card does not influence what happens to the second card. On the other hand, when we draw *without* replacement, the events that the first card is a heart and the second is red are *dependent* events. The fact that the first card is a heart means that there is one fewer red card in the deck, influencing the probability that the second card is red.

As another example, consider tossing a fair coin twice. If the first toss shows heads, the probability that the next toss is heads is still $1/2$. Coin tosses are independent events, since the outcome of one toss does not influence the outcome of the next toss. Similarly, rolls of a fair die are independent events. On the other hand, the events "the milk is old" and "the milk is sour" are dependent events: If the milk is old, there is an increased chance that it is sour. Also, in the example at the beginning of this section, the events A (the president earned \$250,000 or more) and B (the president was the head of a community college) are dependent events, because information about the institution type affects the probability of salary level. That is, $P(A|B)$ is different from $P(A)$.

If events E and F are independent, then the knowledge that E has occurred gives no (probability) information about the occurrence or nonoccurrence of event F. That is, $P(F)$ is exactly the same as $P(F|E)$, or

$$P(F|E) = P(F).$$

This, in fact, is the formal definition of independent events.

E and F are **independent events** if

$$\mathbf{P(F|E) = P(F)} \quad \text{or} \quad \mathbf{P(E|F) = P(E)}.$$

If the events are not independent, they are **dependent events**.

When E and F are independent events, $P(F|E) = P(F)$, and the product rule becomes

$$P(E \cap F) = P(E) \cdot P(F|E) = P(E) \cdot P(F).$$

Conversely, if this equation holds, it follows that $P(F) = P(F|E)$. Consequently, we have the useful rule that follows.

Product Rule for Independent Events

E and F are independent events if and only if

$$\mathbf{P(E \cap F) = P(E) \cdot P(F)}.$$

Example 9 A calculator requires a keystroke assembly and a logic circuit. Assume that 99% of the keystroke assemblies are satisfactory and 97% of the logic circuits are satisfactory. Find the probability that a finished calculator will be satisfactory.

Solution If the failure of a keystroke assembly and the failure of a logic circuit are independent events, then

$$\begin{aligned} &P(\text{satisfactory calculator}) \\ &\quad = P(\text{satisfactory keystroke assembly}) \cdot P(\text{satisfactory logic circuit}) \\ &\quad = (.99)(.97) \approx .96. \end{aligned}$$ 11

✓Checkpoint 11

Find the probability of getting 4 successive heads on 4 tosses of a fair coin.

Answer: See page 519.

CAUTION It is common for students to confuse the ideas of *disjoint* events and *independent* events. Events E and F are disjoint if $E \cap F = \emptyset$. For example, if a family has exactly one child, the only possible outcomes are $B = \{\text{boy}\}$ and $G = \{\text{girl}\}$. These two events are disjoint. However, the events are *not* independent, since $P(G|B) = 0$ (if a family with only one child has a boy, the probability that it has a girl is then 0). Since $P(G|B) \neq P(G)$, the events are not independent. Of all the families with exactly *two* children, the events $G_1 = \{\text{first child is a girl}\}$ and $G_2 = \{\text{second child is a girl}\}$ are independent, because $P(G_2|G_1)$ equals $P(G_2)$. However, G_1 and G_2 are not disjoint, since $G_1 \cap G_2 = \{\text{both children are girls}\} \neq \emptyset$.

To show that two events E and F are independent, we can show that $P(F|E) = P(F)$, that $P(E|F) = P(E)$, or that $P(E \cap F) = P(E) \cdot P(F)$. Another way is to observe that knowledge of one outcome does not influence the probability of the other outcome, as we did for coin tosses.

NOTE In some cases, it may not be apparent from the physical description of the problem whether two events are independent or not. For example, it is not obvious whether the event that a baseball player gets a hit tomorrow is independent of the event that he got a hit today. In such cases, it is necessary to use the definition and calculate whether $P(F|E) = P(F)$, or, equivalently, whether $P(E \cap F) = P(E) \cdot P(F)$.

Example 10 The probability that someone living in the United States lives in Massachusetts is .02, the probability that someone living in the United States speaks English at home is .83, and the probability that someone living in the United States lives in Massachusetts or speaks English at home is .8334.* Are the events "living in Massachusetts" and "speaking English at home" independent?

Solution Let M represent the event "living in Massachusetts" and E represent the event "speaking English at home." We must determine whether

$$P(E|M) = P(E) \quad \text{or} \quad P(M|E) = P(M).$$

*Data taken from U.S. Census population estimates and the 2007 American Community Survey. Some probabilities are rounded to the nearest hundredth.

✔ **Checkpoint 12**

The probability of living in Texas is .08, the probability of speaking English at home is .83, and the probability of living in Texas or speaking English at home is .854. Are the events "speaking English at home" and "living in Texas" independent?

Answer: See page 519.

We know that $P(E) = .83$, $P(M) = .02$, and $P(E \cup M) = .8334$. We can use the addition rule (or a Venn diagram) to find that $P(E \cap M) = .0166$, $P(E|M) = .83$, and $P(M|E) = .02$. We have

$$P(E|M) = P(E) = .83 \quad \text{and} \quad P(M|E) = P(M) = .02,$$

so the events "living in Massachusetts" and "speaking English at home" are independent. 12✔

Although we showed that $P(E|M) = P(M)$ and $P(M|E) = P(E)$ in Example 10, only one of these results is needed to establish independence. It is also important to note that independence of events does not necessarily follow our intuition; it is established from the mathematical definition of independence.

8.5 ▸ Exercises

If a single fair die is rolled, find the probability of rolling the given events. (See Examples 1 and 2.)

1. 3, given that the number rolled was odd
2. 5, given that the number rolled was even
3. An odd number, given that the number rolled was 3

If two fair dice are rolled (recall the 36-outcome sample space), find the probability of rolling the given events.

4. A sum of 8, given that the sum was greater than 7
5. A sum of 6, given that the roll was a "double" (two identical numbers)
6. A double, given that the sum was 9

If two cards are drawn without replacement from an ordinary deck, find the probabilities of the given event. (See Example 7.)

7. The second is a heart, given that the first is a heart
8. The second is black, given that the first is a spade
9. A jack and a 10 are drawn.
10. An ace and a 4 are drawn.
11. In your own words, explain how to find the conditional probability $P(E|F)$.
12. Your friend asks you to explain how the product rule for independent events differs from the product rule for dependent events. How would you respond?
13. Another friend asks you to explain how to tell whether two events are dependent or independent. How would you reply? (Use your own words.)

Decide whether the two events listed are independent.

14. S is the event that it snows tomorrow and L is the event that the instructor is late for class.
15. R is the event that four semesters of theology are required to graduate from a certain college and A is the event that the college is a religiously affiliated school.
16. R is the event that it rains in the Amazon jungle and H is the event that an instructor in New York City writes a difficult exam.
17. T is the event that Tom Cruise's next movie grosses over $200 million and R is the event that the Republicans have a majority in Congress in 2016.
18. A student reasons that the probability in Example 3 of both coins being heads is just the probability that the other coin is a head—that is, 1/2. Explain why this reasoning is wrong.
19. In a two-child family, if we assume that the probabilities of a child being male and a child being female are each .5, are the events "the children are the same sex" and "at most one child is male" independent? Are they independent for a three-child family?
20. Let A and B be independent events with $P(A) = \frac{1}{4}$ and $P(B) = \frac{1}{5}$. Find $P(A \cap B)$ and $P(A \cup B)$.

Business *Use a probability tree or Venn diagram in Exercises 21 and 22. (See Examples 2 and 5–9.) A shop that produces custom kitchen cabinets has two employees: Sitlington and Čapek. 95% of Čapek's work is satisfactory, and 10% of Sitlington's work is unsatisfactory. 60% of the shop's cabinets are made by Čapek, and the rest are made by Sitlington. Find the given probabilities.*

21. An unsatisfactory cabinet was made by Čapek. (*Hint*: Consider which event came first.)
22. A finished cabinet is unsatisfactory.
23. **Business** According to the Bureau of Labor Statistics, for 2008, 48.4% of the civilian population 16 years or older was male, 34% was not in the labor force, and 69.3% was male or not in the labor force. Find the probability of not being in the labor force, given that the person is male.

24. **Business** Using the information from Exercise 23 and the fact that 20.9% of the population is female and not in the labor force, find the probability of not being in the labor force, given that the person is female.

25. **Health** The table cross-classifies gender and height status for 5186 American adults.*

	Under 6 Feet	6 Feet or Taller	Total
Male	2108	395	2503
Female	2680	3	2683
Total	4788	398	5186

Find the probability of the given event.

(a) Being 6 feet or taller, given the person is male
(b) Being 6 feet or taller, given the person is female
(c) Being female, given the person is 6 feet or taller
(d) Being male, given the person is under 6 feet

Health *Using the categories defined by the Centers for Disease Control and Prevention for weight status, the table cross-classifies sex and weight status for the same adults in Exercise 25.**

	Underweight	Healthy Weight	Overweight	Obese	Total
Male	85	709	952	757	2503
Female	93	848	739	1003	2683
Total	178	1557	1691	1760	5186

Find the probability of the given event.

26. Being obese, given the person is male
27. Being obese, given the person is female
28. Being male, given the person is of a healthy weight
29. Being female, given the person is underweight

Natural Science *In a letter to the journal* Nature, *Robert A. J. Matthews gives the following table of outcomes of forecast and actual weather over 1000 1-hour walks, based on the United Kingdom's Meteorological Office's 83% accuracy in 24-hour forecasts.*†

	Rain	No Rain	Sum
Forecast of Rain	66	156	222
Forecast of No Rain	14	764	778
Sum	80	920	1000

30. Verify that the probability that the forecast called for rain, given that there was rain, is indeed 83%. Also, verify that the probability that the forecast called for no rain, given that there was no rain, is 83%.

31. Calculate the probability that there was rain, given that the forecast called for rain.

32. Calculate the probability that there was no rain, given that the forecast called for no rain.

33. Observe that your answer to Exercise 32 is higher than 83% and that your answer to Exercise 31 is much lower. Discuss which figure best describes the accuracy of the weather forecast in recommending whether you should carry an umbrella.

Natural Science *The following table shows frequencies for red–green color blindness, where M represents that a person is male and C represents that a person is color blind.*

	M	M'	Totals
C	.042	.007	.049
C'	.485	.466	.951
Totals	.527	.473	1.000

Use the table to find the given probabilities.

34. $P(M)$
35. $P(C)$
36. $P(M \cap C)$
37. $P(M \cup C)$
38. $P(M|C)$
39. $P(M'|C)$

40. Are the events C and M dependent? Recall that two events E and F are dependent if $P(E|F) \neq P(E)$. (See Example 10.)

41. Are the events M' and C dependent?

42. **Natural Science** A scientist wishes to determine whether there is any dependence between color blindness (C) and deafness (D). Given the probabilities listed in the table, what should his findings be (see Example 10)?

	D	D'	Totals
C	.0004	.0796	.0800
C'	.0046	.9154	.9200
Totals	.0050	.9950	1.0000

Social Science *The Motor Vehicle Department in a certain state has found that the probability of a person passing the test for a driver's license on the first try is .75. The probability that an individual who fails on the first test will pass on the second try is .80, and the probability that an individual who fails the first and second tests will pass the third time is .70. Find the probability of the given event.*

43. A person fails both the first and second tests
44. A person will fail three times in a row
45. A person will require at least two tries to pass the test

*Based on data available at www.cdc.gov/nchs/nhanes.htm.
†Robert A. J. Matthews, *Nature,* 382, August 29, 1996: p. 3.

Business *The table shows results from a subset of the 2007 American Community Survey.*

	Household Income				
Number of Vehicles	**\$0–49,999**	**\$50,000–99,999**	**\$100,000–149,999**	**\$150,000 or more**	**Total**
0–2	504	199	64	46	813
3 or more	36	80	36	35	187
Total	540	279	100	81	1000

Find the probability of the given event.

46. Earning \$150,000 or more, given that a household has 3 or more cars

47. Having 0–2 cars, given that a household earns \$0–49,999

48. Having 3 or more cars, given that a household earns between \$100,000–149,999

49. Earning \$50,000–99,999, given that the household has 0–2 cars

Transportation *The numbers of domestic and international flights for American Airlines, Continental Air Lines, and United Air Lines for 2007 are given in the table.**

Airline	Domestic	International	Total
American	618,615	147,768	766,383
Continental	321,818	87,656	409,474
United	479,660	64,161	543,821
Total	1,420,093	299,585	1,719,678

Find the probability of the given event.

50. A Continental flight was domestic

51. An international flight was American

52. A United flight was international

53. A non-American flight was international

Suppose the probability that the first record by a singing group will be a hit is .32. If the first record is a hit, so are all the group's subsequent records. If the first record is not a hit, the probability of the group's second record and all subsequent ones being hits is .16. If the first two records are not hits, the probability that the third is a hit is .08. The probability that a record is hit continues to decrease by half with each successive nonhit record. Find the probability of the given event.

54. The group will have at least one hit in its first four records

55. The group will have exactly one hit in its first three records

56. The group will have a hit in its first six records if the first three are not hits

*www.transtats.bts.gov.

Work the given problems on independent events. (See Examples 9 and 10.)

57. **Business** Corporations such as banks, where a computer is essential to day-to-day operations, often have a second, backup computer in case of failure by the main computer. Suppose that there is a .003 chance that the main computer will fail in a given period and a .005 chance that the backup computer will fail while the main computer is being repaired. Suppose these failures represent independent events, and find the fraction of the time the corporation can assume that it will have computer service. How realistic is our assumption of independence?

58. **Transportation** According to data from the U.S. Department of Transportation, Delta Airlines was on time approximately 76% of the time in 2008. Use this information, and assume that the event that a given flight takes place on time is independent of the event that another flight is on time to answer the following questions

(a) Elisabeta Gueyara plans to visit her company's branch offices; her journey requires 3 separate flights on Delta Airlines. What is the probability that all of these flights will be on time?

(b) How reasonable do you believe it is to suppose the independence of being on time from flight to flight?

59. **Natural Science** The probability that a key component of a space rocket will fail is .03.

(a) How many such components must be used as backups to ensure that the probability that at least one of the components will work is .999999?

(b) Is it reasonable to assume independence here?

60. **Natural Science** A medical experiment showed that the probability that a new medicine is effective is .75, the probability that a patient will have a certain side effect is .4, and the probability that both events will occur is .3. Decide whether these events are dependent or independent.

61. **Social Science** A teacher has found that the probability that a student studies for a test is .6, the probability that a student gets a good grade on a test is .7, and the probability that both events occur is .52. Are these events independent?

62. **Business** Refer to Exercises 46–49. Are the events of a household having 3 or more cars and a household earning \$150,000 or more independent?

✓ Checkpoint Answers

1. (a) .185 (b) .392 (c) .815
2. (a) .425 (b) .197 (c) .806 (d) .419
(e) The probability of never being married, given that the person is partially or not happy
3. 1/3 **4.** 2/3 **5.** 2/7
6. (a) 1/4 (b) 3/5
7. 1/5 **8.** 1/10 **9.** $13/102 \approx .1275$
10. $13/34 \approx .382$ **11.** 1/16 **12.** No

8.6 ▶ Bayes' Formula

Suppose the probability that a person gets lung cancer, given that the person smokes a pack or more of cigarettes daily, is known. For a research project, it might be necessary to know the probability that a person smokes a pack or more of cigarettes daily, given that the person has lung cancer. More generally, if $P(E|F)$ is known for two events E and F, can $P(F|E)$ be found? The answer is yes, we can find $P(F|E)$ by using the formula to be developed in this section. To develop this formula, we can use a probability tree to find $P(F|E)$. Since $P(E|F)$ is known, the first outcome is either F or F'. Then, for each of these outcomes, either E or E' occurs, as shown in Figure 8.31.

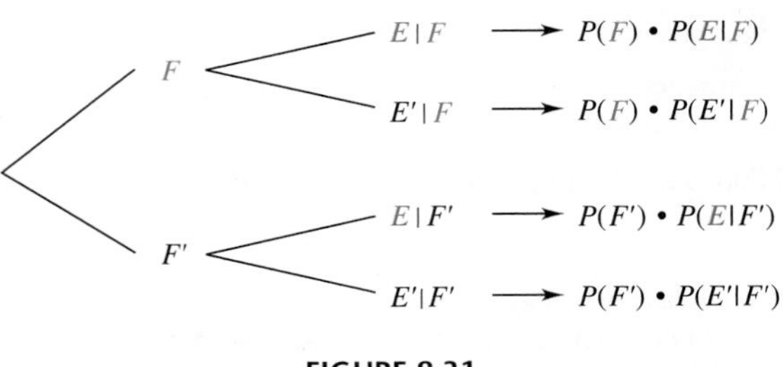

FIGURE 8.31

The four cases have the probabilities shown on the right. By the definition of conditional probability and the product rule,

$$P(E) = P(F \cap E) + P(F' \cap E),$$

$$P(F \cap E) = P(F) \cdot P(E|F), \quad \text{and} \quad P(F' \cap E) = P(F') \cdot P(E|F').$$

By substitution,

$$P(E) = P(F) \cdot P(E|F) + P(F') \cdot P(E|F')$$

and

$$P(F|E) = \frac{P(F \cap E)}{P(E)} = \frac{P(F) \cdot P(E|F)}{P(F) \cdot P(E|F) + P(F') \cdot P(E|F')}.$$

We have proved a special case of Bayes' formula, which is generalized later in this section. ✓1

✓Checkpoint 1

Use the special case of Bayes' formula to find $P(F|E)$ if $P(F) = .2$, $P(E|F) = .1$, and $P(E|F') = .3$.
[*Hint*: $P(F') = 1 - P(F)$.]

Answer: See page 526.

Bayes' Formula (Special Case)

$$P(F|E) = \frac{P(F) \cdot P(E|F)}{P(F) \cdot P(E|F) + P(F') \cdot P(E|F')}.$$

Example 1 **Business** For a fixed length of time, the probability of worker error on a certain production line is .1, the probability that an accident will occur when there is a worker error is .3, and the probability that an accident will occur when there is no worker error is .2. Find the probability of a worker error if there is an accident.

Solution Let E represent the event of an accident, and let F represent the event of a worker error. From the given information,

$$P(F) = .1, \quad P(F') = 1 - .1 = .9 \quad P(E|F) = .3, \quad \text{and} \quad P(E|F') = .2.$$

These probabilities are shown on the probability tree in Figure 8.32.

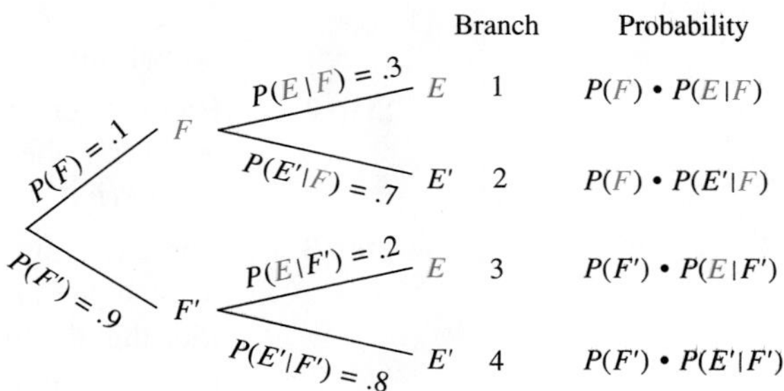

FIGURE 8.32

Applying Bayes' Formula, we find that

$$P(F|E) = \frac{P(F) \cdot P(E|F)}{P(F) \cdot P(E|F) + P(F') \cdot P(E|F')}$$

$$= \frac{(.1)(.3)}{(.1)(.3) + (.9)(.2)} \approx .143. \quad ✓2$$

✓Checkpoint 2

In Example 1, find $P(F'|E)$.

Answer: See page 526.

If we rewrite the special case of Bayes' Formula, replacing F by F_1 and F' by F_2, then it says:

$$P(F_1|E) = \frac{P(F_1) \cdot P(E|F_1)}{P(F_1) \cdot P(E|F_1) + P(F_2) \cdot P(E|F_2)}.$$

Since $F_1 = F$ and $F_2 = F'$ we see that F_1 and F_2 are disjoint and that their union is the entire sample space. The generalization of Bayes' Formula to more than two possibilities follows this same pattern.

Bayes' Formula

Suppose $F_1, F_2, \ldots F_n$ are pairwise disjoint events (meaning that any two of them are disjoint) whose union is the sample space. Then for an event E and for each i with $1 \le i \le n$,

$$\boldsymbol{P(F_i|E) = \frac{P(F_i) \cdot P(E|F_i)}{P(F_1) \cdot P(E|F_1) + \cdots + P(F_n) \cdot P(E|F_n)}}.$$

This result is known as Bayes' formula, after the Reverend Thomas Bayes (1702–61), whose paper on probability was published about 245 years ago.

The statement of Bayes' formula can be daunting. It may be easier to remember the formula by thinking of the probability tree that produced it. Go through the following steps.

Using Bayes' Formula

Step 1 Start a probability tree with branches representing events $F_1, F_2, \ldots, F_n$. Label each branch with its corresponding probability.

Step 2 From the end of each of these branches, draw a branch for event E. Label this branch with the probability of getting to it, or $P(E|F_i)$.

Step 3 There are now n different paths that result in event E. Next to each path, put its probability: the product of the probabilities that the first branch occurs, $P(F_i)$, and that the second branch occurs, $P(E|F_i)$: that is, $P(F_i) \cdot P(E|F_i)$.

Step 4 $P(F_i|E)$ is found by dividing the probability of the branch for F_i by the sum of the probabilities of all the branches producing event E.

Example 2 illustrates this process.

Example 2 **Social Science** The 2006 General Social Survey of women who are age 18 or older indicated that 86% of married women have one or more children, 31% of never married women have one or more children, and 86% of women who are divorced, separated, or widowed have one or more children. The survey also indicated that 48% of women age 18 or older were currently married, 24% had never been married, and 28% were divorced, separated, or widowed (labeled "other"). Find the probability that a woman who has one or more children is married.

Solution Let E represent the event "having one or more children," with F_1 representing "married women," F_2 representing "never married women," and F_3 "other". Then

$$P(F_1) = .48; \quad P(E|F_1) = .86;$$
$$P(F_2) = .24; \quad P(E|F_2) = .31;$$
$$P(F_3) = .28; \quad P(E|F_3) = .86.$$

We need to find $P(F_1|E)$, the probability that a woman is married, given that she has one or more children. First, draw a probability tree using the given information, as in Figure 8.33. The steps leading to event E are shown.

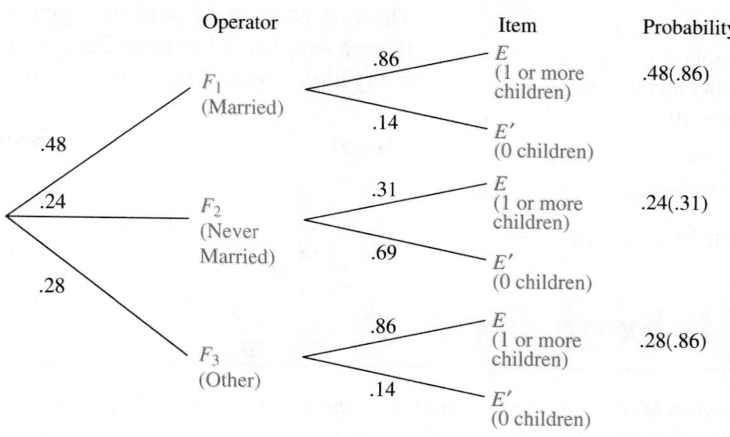

FIGURE 8.33

Find $P(F_1|E)$, using the top branch of the tree shown in Figure 8.35, by dividing the probability of this branch by the sum of the probabilities of all the branches leading to E:

$$P(F_1|E) = \frac{.48(.86)}{.48(.86) + .24(.31) + .28(.86)} = \frac{.4128}{.7280} \approx .567.$$ 3

✓Checkpoint 3

In Example 2, find

(a) $P(F_2|E)$;

(b) $P(F_3|E)$.

Answers: See page 526.

Example 3 **Business** A manufacturer buys items from six different suppliers. The fraction of the total number of items obtained from each supplier, along with the probability that an item purchased from that supplier is defective, is shown in the following table:

Supplier	Fraction of Total Supplied	Probability of Being Defective
1	.05	.04
2	.12	.02
3	.16	.07
4	.23	.01
5	.35	.03
6	.09	.05

Find the probability that a defective item came from supplier 5.

Solution Let F_1 be the event that an item came from supplier 1, with F_2, F_3, F_4, F_5, and F_6 defined in a similar manner. Let E be the event that an item is defective. We want to find $P(F_5|E)$. By Bayes' formula,

$$P(F_5|E) = \frac{(.35)(.03)}{(.05)(.04)+(.12)(.02)+(.16)(.07)+(.23)(.01)+(.35)(.03)+(.09)(.05)}$$

$$= \frac{.0105}{.0329} \approx .319.$$

✓**Checkpoint 4**

In Example 3, find the probability that the defective item came from

(a) supplier 3;

(b) supplier 6.

Answers: See page 526.

There is about a 32% chance that a defective item came from supplier 5. Even though supplier 5 has only 3% defectives, his probability of being "guilty" is relatively high, about 32%, because of the large fraction of items he supplies. ✓4

8.6 ▶ Exercises

For two events M and N, $P(M) = .4, P(N|M) = .3$, and $P(N|M') = .4$. Find each of the given probabilities. (See Example 1.)

1. $P(M|N)$ **2.** $P(M'|N)$

For disjoint events R_1, R_2, and R_3, $P(R_1) = .05$, $P(R_2) = .6$, and $P(R_3) = .35$. In addition, $P(Q|R_1) = .40$, $P(Q|R_2) = .30$, and $P(Q|R_3) = .60$. Find each of the given probabilities. (See Examples 2 and 3.)

3. $P(R_1|Q)$ **4.** $P(R_2|Q)$

5. $P(R_3|Q)$ **6.** $P(R_1'|Q)$

Suppose three jars have the following contents: 2 black balls and 1 white ball in the first; 1 black ball and 2 white balls in the second; 1 black ball and 1 white ball in the third. If the probability of selecting one of the three jars is 1/2, 1/3, and 1/6, respectively, find the probability that if a white ball is drawn, it came from the given jar.

7. The second jar **8.** The third jar

Business *According to Forbes's list of the 100 top-earning chief executive officers (CEOs) of 2008, the probability of a CEO on the list being under 60 years old was .64. The probability of earning \$50 million or more for those under age 60 was .1875. The probability of earning \$50 million or more for those age 60 or older was .25. Find the probability of the given event.*

9. A CEO who was earning \$50 million or more was under age 60.

10. A CEO who was earning less than \$50 million was age 60 or older.

Business *Data from the Bureau of Labor Statistics indicates that for December 2008, 38.2% of the labor force had a high school diploma or fewer years of education, 27.9% had some college or an associate's degree, and 33.9% had a bachelor's degree or more education. Of those with a high school diploma or fewer years of education, 8.4% were unemployed. Of those with some college or an associate's degree, 5.6% were unemployed, and of those with a bachelor's degree or more education, 3.7% were unemployed. Find the probability that a randomly chosen labor force participant has the given characteristics.*

11. Some college or an associate's degree, given that he or she is unemployed

12. A bachelors degree or more education, given that he or she is unemployed

13. A high school diploma or less education, given that he or she is employed

14. A bachelor's degree or more education, given that he or she is employed

Transportation *In 2007, the probability was .179 that in a traffic fatality, the victim was age 20 or younger. The probability that a fatal accident for someone age 20 or younger involved an alcohol-impaired driver was .25. The probability that a fatal accident for someone older than age 20 involved an alcohol-impaired driver was .326.* Suppose we know a fatality involved an alcohol-impaired driver. Find the probability that the victim was in the given age group.*

15. Age 20 or younger

16. Older than age 20

Social Science *According to data from the U.S. Census Bureau and the 2006 General Social Survey, the probability of being male in the United States was .493. The probability a male considered himself "very happy" was .303, "pretty happy" was .570, and "not too happy" was .126. The probability a female considered herself "very happy" was .31, "pretty happy" was .555, and "not too happy" was .134. Find the probability that a person selected at random had the given characteristics.*

17. Male, given that he was "not too happy"

18. Female, given that she was "very happy"

19. **Management** The following information pertains to three shipping terminals operated by Krag Corp.†

Terminal	Percentage of Cargo Handled	Percent Error
Land	50	2
Air	40	4
Sea	10	14

*www-nrd.nhtsa.dot.gov/Pubs/811016.PDF.

†Uniform CPA Examination, November 1989.

Krag's internal auditor randomly selects one set of shipping documents and ascertains that the set selected contains an error. Which of the following gives the probability that the error occurred in the Land Terminal?

(a) .02
(b) .10
(c) .25
(d) .50

Health *In a test for toxemia, a disease that affects pregnant women, a woman lies on her left side and then rolls over on her back. The test is considered positive if there is a 20-mm rise in her blood pressure within 1 minute. The results have produced the following probabilities, where T represents having toxemia at some time during the pregnancy and N represents a negative test:*

$$P(T'|N) = .90 \quad \text{and} \quad P(T|N') = .75.$$

Assume that $P(N') = .11$, *and find each of the given probabilities.*

20. $P(N|T)$ **21.** $P(N'|T)$

Health *In 2007, 9.5% of the population was ages 18–24. Of these people, 28.1% did not have health insurance. Among all others in the United States, 13.9% did not have health insurance.* Find the probability of each event.*

22. A person without health insurance is age 18–24.

23. A person with health insurance is not age 18–24.

24. **Health** The probability that a person with certain symptoms has hepatitis is .8. The blood test used to confirm this diagnosis gives positive results for 90% of people with the disease and 5% of those without the disease. What is the probability that an individual who has the symptoms and who reacts positively to the test actually has hepatitis?

Health *The 2005–2006 National Health and Nutrition Examination Survey (NHANES) collected data on age and cholesterol level. If we define high cholesterol as having a total cholesterol level of 200 or higher, the following table gives percentages and probabilities of having high cholesterol for three age categories.†*

Age Group	Percentage in Sample	Probability of High Cholesterol
18–39	46.3	.315
40–64	34.9	.504
65 and older	18.8	.405

Find the probability of the given event.

25. Being 18–39, given that the person has high cholesterol.

26. Being 65 and older, given that the person does not have high cholesterol.

27. **Social Science** A recent study by the Harvard School of Public Health reported that 78.9% of college students living in a fraternity or sorority house are binge drinkers. For students living in a regular dormitory, the rate of binge drinking is 44.5%, and for students living off campus, the rate is 43.7%.* Suppose that 10% of U.S. students live in a fraternity or sorority house, 20% live in regular dormitories, and 70% live off campus.

(a) What is the probability that a randomly selected student is a binge drinker?

(b) If a randomly selected student is a binge drinker, what is the probability that he or she lives in a fraternity or sorority house?

Finance *In 2008, credit unions whose deposits were insured by the National Credit Union Association (NCUA) were classified into three categories: those with fewer than 25,000 members, those with between 25,000 and 49,999 members, and those with 50,000 members or more. Of interest is whether the credit unions' total deposits had decreased from 2007 levels. The table summarizes the percentages of the three sizes of credit unions and the probability of decreasing deposits from 2007 to 2008.†*

Credit Union Size (Number of Members)	Percentage of Credit Unions	Probability of Decreasing Deposits
Fewer than 25,000	91%	.479
25,000–49,999	4.7%	.213
50,000 or more	4.3%	.117

Find the probability that a credit union selected at random had the given characteristics.

28. Fewer than 25,000 members, given that the total deposits decreased

29. 50,000 or more members, given that the total deposits decreased

30. Between 25,000–49,999 members, given that the total deposits did not decrease

*U.S. Census Bureau.

†Based on data from www.cdc.gov/nchs/nhanes.htm.

*Wechsler, H., Lee, J., Kuo, M., and Lee, H., "College Binge Drinking in the 1990s: A Continuing Problem, Results of the Harvard School of Public Health 1999 College Alcohol Study," online at http://www.hsph.harvard.edu/cas/rpt2000/CAS2000rpt.shtml.

†Based on data from www.ncua.gov.

Social Science *The table gives proportions of people over age 15 in the U.S. population, and proportions of people that live alone, in a recent year.* Use this table for Exercises 31–35.*

Age	Proportion in Population Age 15 or Higher	Proportion Living Alone
15–24	.177	.038
25–34	.169	.097
35–44	.179	.086
45–64	.318	.144
65 and higher	.157	.287

31. Find the probability that a randomly selected person age 15 or older who lives alone is between the ages of 45 and 64.

32. Find the probability that a randomly selected person age 15 or older who lives alone is age 65 or older.

33. Find the probability of not living alone for any person age 15 or higher.

34. Find the probability of not living alone for one person age 15–24.

35. Find the probability of not living alone for any adult age 65 or older.

**Statistical Abstract of the United States*: 2009.

✓ Checkpoint Answers

1. $1/13 \approx .077$ **2.** $6/7 \approx .857$
3. **(a)** .102 **(b)** .331
4. **(a)** .340 **(b)** .137

CHAPTER 8 ▶ Summary

KEY TERMS AND SYMBOLS

$\{\ \}$	set braces
$\in$	is an element of
$\notin$	is not an element of
$\emptyset$	empty set
$\subseteq$	is a subset of
$\not\subseteq$	is not a subset of
$\subset$	is a proper subset of
A'	complement of set A
$\cap$	set intersection
$\cup$	set union
$P(E)$	probability of event E
$P(F\mid E)$	probability of event F, given that event E has occurred

8.1 ▶ set
element (member)
empty set
set-builder notation
universal set
subset
set operations
tree diagram
Venn diagram
complement
intersection
disjoint sets
union

8.2 ▶ addition rule for counting

8.3 ▶ experiment
trial
outcome
sample space
event
certain event
impossible event
disjoint events
basic probability principle
relative frequency probability
probability distribution

8.4 ▶ addition rule for probability
complement rule
odds

8.5 ▶ conditional probability
product rule of probability
probability tree
independent events
dependent events

8.6 ▶ Bayes' formula

CHAPTER 8 KEY CONCEPTS

Sets ▶ Set A is a **subset** of set B if every element of A is also an element of B.

A set of n elements has 2^n subsets.

Let A and B be any sets with universal set U.

The **complement** of A is $A' = \{x | x \notin A \text{ and } x \in U\}$.
The **intersection** of A and B is $A \cap B = \{x | x \in A \text{ and } x \in B\}$.
The **union** of A and B is $A \cup B = \{x | x \in A \text{ or } x \in B \text{ or both}\}$.

$n(A \cup B) = n(A) + n(B) - n(A \cap B)$, where $n(X)$ is the number of elements in set X.

Probability Summary ▶

Basic Probability Principle Let S be a sample space of equally likely outcomes, and let event E be a subset of S. Then the probability that event E occurs is

$$P(E) = \frac{n(E)}{n(S)}.$$

Addition Rule For any events E and F from a sample space S,

$$P(E \cup F) = P(E) + P(F) - P(E \cap F).$$

For disjoint events E and F,

$$P(E \cup F) = P(E) + P(F).$$

Complement Rule $P(E) = 1 - P(E')$ and $P(E') = 1 - P(E)$.

Odds The odds in favor of event E are $\dfrac{P(E)}{P(E')}$, $P(E') \neq 0$.

Properties of Probability

1. For any event E in sample space S, $0 \leq P(E) \leq 1$.
2. The sum of the probabilities of all possible distinct outcomes is 1.

Conditional Probability The conditional probability of event E, given that event F has occurred, is

$$P(E|F) = \frac{P(E \cap F)}{P(F)}, \quad \text{where} \quad P(F) \neq 0.$$

For equally likely outcomes, conditional probability is found by reducing the sample space to event F; then

$$P(E|F) = \frac{n(E \cap F)}{n(F)}.$$

Product Rule of Probability If E and F are events, then $P(E \cap F)$ may be found by either of these formulas:

$$P(E \cap F) = P(F) \cdot P(E|F) \quad \text{or} \quad P(E \cap F) = P(E) \cdot P(F|E).$$

If E and F are independent events, then $P(E \cap F) = P(E) \cdot P(F)$.

Bayes' Formula

$$P(F_i|E) = \frac{P(F_i) \cdot P(E|F_i)}{P(F_1) \cdot P(E|F_1) + P(F_2) \cdot P(E|F_2) + \cdots + P(F_n) \cdot P(E|F_n)}.$$

CHAPTER 8 REVIEW EXERCISES

Write true or false for each of the given statements.

1. $9 \in \{8, 4, -3, -9, 6\}$
2. $4 \in \{3, 9, 7\}$
3. $2 \notin \{0, 1, 2, 3, 4\}$
4. $0 \notin \{0, 1, 2, 3, 4\}$
5. $\{3, 4, 5\} \subseteq \{2, 3, 4, 5, 6\}$
6. $\{1, 2, 5, 8\} \subseteq \{1, 2, 5, 10, 11\}$
7. $\{1, 5, 9\} \subset \{1, 5, 6, 9, 10\}$
8. $0 \subseteq \varnothing$

List the elements in the given sets.

9. $\{x | x \text{ is a national holiday}\}$
10. $\{x | x \text{ is an integer}, -3 \leq x < 1\}$

11. {all counting numbers less than 5}

12. $\{x | x$ is a leap year between 1989 and 2006$\}$

Let U = {Vitamins A, B_1, B_2, B_3, B_6, B_{12}, C, D, E}, M = {Vitamins A, C, D, E}, and N = {Vitamins A, B_1, B_2, C, E}. Find the given sets.

13. M' **14.** N'

15. $M \cap N$ **16.** $M \cup N$

17. $M \cup N'$ **18.** $M' \cap N$

Consider these sets:
U = {Students taking Intermediate Accounting};
A = {Females};
B = {Finance majors};
C = {Students older than 22};
D = {Students with a GPA > 3.5}.

Describe each of the following in words.

19. $A \cap C$ **20.** $B \cap D$

21. $A \cup D$ **22.** $A' \cap D$

23. $B' \cap C'$

Draw a Venn diagram and shade the given set in it.

24. $B \cup A'$ **25.** $A' \cap B$

26. $A' \cap (B' \cap C)$ **27.** $(A \cup B)' \cap C$

Business *In early 2009, reporters for* The Big Money *and* Slate *collected data on every major motion picture that grossed $75 million or more domestically for 2004–2008.* They found the following regarding the 161 movies:*

56 were action movies;
87 were rated PG-13;
44 were 2 hours or longer;
43 were action and PG-13;
27 were action and 2 hours or longer;
28 were rated PG-13 and 2 hours or longer;
20 were action, rated PG-13, and 2 hours or longer.

28. How many movies were action movies and were *not* 2 hours or longer?

29. How many movies were action or rated PG-13?

30. How many movies were neither action, PG-13, nor 2 hours or longer?

Write sample spaces for the given scenarios.

31. A die is rolled and the number of points showing is noted.

32. A card is drawn from a deck containing only 4 aces.

33. A color is selected from the set {red, blue, green}, and then a number is chosen from the set {10, 20, 30}.

Business *A student purchases a digital music player and installs 10 songs on the device to see how it works. The genres of the songs are rock (3 songs), pop (4 songs), and alternative (3 songs). She listens to the first two songs on shuffle mode.*

*www.thebigmoney.com.

34. Write the sample space for the genre if shuffle mode picks songs at random. (*Note*: Shuffle mode is allowed to play the same song twice in a row.)

35. Are the outcomes in the sample space for Exercise 34 equally likely?

Business *A customer wants to purchase a computer and printer. She has narrowed her selection among 2 Dell models, 1 Gateway model, and 2 HP models for the computer and 2 Epson models and 3 HP models for the printer.*

36. Write the sample space for the brands among which she can choose for the computer and printer.

37. Are the outcomes in the sample space for Exercise 36 equally likely?

Business *A company sells computers and copiers. Let E be the event "a customer buys a computer," and let F be the event "a customer buys a copier." In Exercises 38 and 39, write each of the given scenarios, using $\cap$, $\cup$, or $'$ as necessary.*

38. A customer buys neither a computer nor a copier.

39. A customer buys at least one computer or copier.

40. A student gives the answer to a probability problem as 6/5. Explain why this answer must be incorrect.

41. Describe what is meant by disjoint sets, and give an example.

42. Describe what is meant by mutually exclusive events, and give an example.

43. How are disjoint sets and mutually exclusive events related?

Finance *The Standard and Poor's 500 index had the following allocations, as of December 31, 2008.**

Sector	Percentage
Consumer Discretionary	8.40
Consumer Staples	12.88
Energy	13.34
Financials	13.29
Health Care	14.79
Industrials	11.08
Information Technology	15.27
Materials	2.93
Telecommunication Services	3.83
Utilities	4.19

Find the probability that a company chosen at random from the S&P 500 was from the given sectors.

44. Consumer Discretionary or Consumer Staples

45. Information Technology or Telecommunication Services

*www.standardandpoors.com.

Finance *The table gives the number of institutions insured by the Federal Deposit Insurance Corporation (FDIC), cross-classified by asset size and institution type as of June 30, 2008:**

Asset Size	Commercial Banks	Saving Institutions	Total
Less than $25 million	422	55	477
$25 million–$50 million	960	115	1075
$50 million–$100 million	1557	178	1735
$100 million–$300 million	2517	393	2910
$300 million–$500 million	691	163	854
$500 million–$1 billion	548	159	707
$1 billion–$3 billion	310	100	410
$3 billion–$10 billion	114	32	146
$10 billion or greater	84	32	116
Total	7203	1227	8430

(Note: Asset-size categories have the left endpoint included in the count. For example, asset size $50 million–$100 million includes institutions with at least $50 million and less than $100 million.)

If an FDIC-insured institution is chosen at random, find the probability of the given event.

46. $3 billion or greater in assets

47. $25 million–$50 million in assets and is a savings institution

48. $1 billion–$3 billion in assets and is a commercial bank

49. Less than $25 million in assets or is a savings institution

50. $10 billion or greater in assets or is a commercial bank

51. $500 million–$1 billion in assets, given that it is a savings institution

52. $500 million–$1 billion in assets, given that it is a commercial bank

53. Given that an institution has $10 billion or greater in assets, what is the probability it is a savings institution?

54. Given that an institution has less than $25 million in assets, what is the probability it is a commercial bank?

Health *The partial table shows the four possible (equally likely) combinations when both parents are carriers of the sickle-cell anemia trait. Each carrier parent has normal cells (N) and trait cells (T).*

		2nd Parent	
		N_2	T_2
1st Parent	N_1		N_1T_2
	T_1		

55. Complete the table.

56. If the disease occurs only when two trait cells combine, find the probability that a child born to these parents will have sickle-cell anemia.

57. Find the probability that the child will carry the trait, but not have the disease, if a normal cell combines with a trait cell.

58. Find the probability that the child neither is a carrier nor has the disease.

Find the probabilities for the given sums when two fair dice are rolled.

59. 8

60. No more than 4

61. At least 9

62. Odd and greater than 8

63. 2, given that the sum is less than 4

64. 7, given that at least one die shows a 4

Suppose $P(E) = .62$, $P(F) = .45$, and $P(E \cap F) = .28$. Find each of the given probabilities.

65. $P(E \cup F)$

66. $P(E \cap F')$

67. $P(E' \cup F)$

68. $P(E' \cap F')$

69. For the events E and F, $P(E) = .2$, $P(E|F) = .3$, and $P(F) = .4$. Find each of the given probabilities.
(a) $P(E'|F)$ **(b)** $P(E|F')$

70. Define independent events, and give an example of one.

71. Are independent events always disjoint? Are they ever disjoint? Give examples.

Business *Of the appliance repair shops listed in the phone book, 80% are competent and 20% are not. A competent shop can repair an appliance correctly 95% of the time; an incompetent shop can repair an appliance correctly 60% of the time. Suppose an appliance was repaired correctly. Find the probability that it was repaired by the given type of shop.*

72. A competent shop

73. An incompetent shop

Suppose an appliance was repaired incorrectly. Find the probability that it was repaired by the given type of shop.

74. A competent shop

75. An incompetent shop

76. **Business** A manufacturer buys items from four different suppliers. The fraction of the total number of items that is obtained from each supplier, along with the probability that an item purchased from that supplier is defective, is shown in the following table:

Supplier	Fraction of Total Supplied	Probability of Defective
1	.17	.04
2	.39	.02
3	.35	.07
4	.09	.03

(a) Find the probability that a defective item came from supplier 4.

*www.fdic.gov/sod.

(b) Find the probability that a defective item came from supplier 2.

77. Social Science The following tables list the number of passengers who were on the Titanic and the number of passengers who survived, according to class of ticket:*

	CHILDREN		WOMEN	
	On	Survived	On	Survived
First Class	6	6	144	140
Second Class	24	24	165	76
Third Class	79	27	93	80
Total	109	57	402	296

	MEN		TOTALS	
	On	Survived	On	Survived
First Class	175	57	325	203
Second Class	168	14	357	114
Third Class	462	75	634	182
Total	805	146	1316	499

Use this information to determine the given probabilities. (Round answers to two decimal places.)

(a) What is the probability that a randomly selected passenger was in second class?
(b) What is the overall probability of surviving?
(c) What is the probability of a first-class passenger surviving?
(d) What is the probability of a child who was in third class surviving?
(e) Given that a survivor is from first class, what is the probability that she was a woman?
(f) Given that a male has survived, what is the probability that he was in third class?
(g) Are the events "third-class survival" and "male survival" independent events? What does this imply?

78. Social Science The following partial table gives results of the 2006 General Social Survey question, "Are federal income taxes too high, about right, or too low?", categorized by sex:

Sex	Too High	About Right	Too Low	Total
Male	479		8	
Female		428	15	1098
Total	1134	792		

(a) Complete the table.
(b) How many were surveyed?
(c) How many men think taxes are about right?
(d) How many women think taxes are too high?
(e) How many women are in the survey?
(f) How many who think taxes are too high are male?
(g) Rewrite the event stated in part (f), using the expression "given that."
(h) Find the probability of the outcome in parts (f) and (g).
(i) Find the probability that a woman thinks that taxes are about right.

*Sandra L. Takis, "Titanic: A Statistical Exploration," *Mathematics Teacher* 92, no. 8, (November 1999): p. 660–664. Reprinted with permission.

Additional Probability Review Exercises

Use these exercises for practice, deciding which rule, principle, or formula to apply.

1. Suppose $P(E) = .4$, $P(F) = .22$, and $P(E \cup F) = .52$. Find
(a) $P(E \cup F')$; **(b)** $P(E \cap F')$; **(c)** $P(E' \cup F)$.

2. A jar contains 2 white, 3 orange, 5 yellow, and 8 black marbles. If a marble is drawn at random, find the probability that it is
(a) white; **(b)** orange;
(c) not black; **(d)** orange or yellow.

3. Finance The sector weightings for the investments in the American Century Growth Fund as of March 17, 2009 are given in the table.*

Sector	Percent
Information Technology	27.36
Health Care	18.32
Consumer Staples	12.50
Industrials	10.78
Consumer Discretionary	9.62
Energy	8.71
Financials	4.71
Materials	3.23
Utilities	2.72
Telecommunication Services	2.05

Find the probability that an investment selected at random from this fund has the given characteristics.
(a) Is in the Consumer Discretionary Sector
(b) Is in the Materials or Utilities Sector
(c) Is not in the Health Care Sector

4. Finance The American Century Growth Fund described in Exercise 3 is invested in the following geographical regions:

*www.americancentury.com.

Country	Percentage Invested
United States	95.35
Switzerland	2.36
Denmark	1.13
Bermuda	.86
Netherlands	.30

Find the probability that an investment selected at random from this fund has the given characteristics.
(a) Is in Bermuda or the United States
(b) Is in Europe
(c) Is not in the United States

5. A single fair die is rolled. Find the probability that the die shows
(a) a 2, given that the number was odd;
(b) a 4, given that the number was even;
(c) an even number, given that the number was 6.

6. **Finance** In 2007, the United States Department of Education spent the following dollar amounts on elementary and secondary education programs (where the dollar amounts are given in millions—for example, 14,842.9 in the table signifies $14,842,900,000):*

Program	Amount
Grants for the disadvantaged	14,842.9
School improvement programs	7697.0
Indian education	120.9
Special education	11,543.0
Vocational and adult education	2091.6
Education reform: Goals 2000	16.5

Find the probability that funds for a particular project
(a) came from special education;
(b) came from special education, or vocational and adult education;
(c) did not come from grants for the disadvantaged.

Social Science *The projected population of the United States (in thousands) by age groups in 2015 and 2050 are given in the table:*†

Age Group	2015	2050
0–4	22,076	28,148
5–17	56,030	73,425
18–44	116,686	150,400
45–64	83,911	98,490
65–84	40,545	69,506
85 and older	6292	19,041

Find the probability that a randomly selected person is of the given age group in the given year.

7. Age 18–44 in 2015
8. Age 18–44 in 2050
9. Age 65 or older in 2015
10. Age 65 or older in 2050

11. **Social Science** In one area, 4% of the population drives luxury cars. However, 17% of the certified public accountants (CPAs) drive luxury cars. Are the events "person drives a luxury car" and "person is a CPA" independent?

12. Suppose $P(E) = .05$, $P(F) = .1$, and $P(E \cap F) = .02$. Find **(a)** $P(E' \cap F)$; **(b)** $P(E' \cup F')$; **(c)** $P(E \cap F')$.

13. One orange and four red slips of paper are placed in a box. Two red and three orange slips are placed in a second box. A box is chosen at random, and a slip of paper is selected from it. The probability of choosing the first box is $3/8$. If the selected slip of paper is orange, what is the probability that it came from the first box?

14. Find the probability that the slip of paper in Exercise 13 came from the second box, given that it is red.

15. **Business** A manufacturing firm finds that 70% of its new hires turn out to be good workers and 30% poor workers. All current workers are given a reasoning test. Of the good workers, 80% pass it; 40% of the poor workers pass it. Assume that these figures will hold true in the future. If the company makes the test part of its hiring procedure and hires only people who meet the previous requirements and pass the test, what percentage of the new hires will turn out to be good workers?

Social Science *The 2006 General Social Survey estimates the percentage of women working full time at 42.6%. Of these women, 46.0% are married. It also estimates the percentage of women working part time as 11.7%, with 53.4% of these women married. The percentage of women not working full or part time, but who are married, is 40.9%. Find the probability that a woman has the given characteristics.*

16. Is married
17. Is married and works full time
18. Is not married and works part time
19. Is not married and is not working full or part time

Business *On a given weekend in the fall, a tire company can buy television advertising time for a college football game, a professional baseball game, or a professional football game. If the company sponsors the college football game, there is a 70% chance that the company will get a high rating. There is a 50% chance if the company sponsors a professional baseball game and a 60% chance if it sponsors a professional football game. The probability of the company sponsoring these various games is .5, .2, and .3, respectively. Suppose the company does get a high rating; find the probability that it sponsored the given type of game.*

20. A college football game
21. A professional football game

*U.S. National Center for Education Statistics.

†U.S. Census Bureau.

22. **Transportation** As reported by the National Highway Traffic Safety Administration (NHSTA), in 2006, the state of Washington had the highest rate of seat belt compliance; the odds that a driver was using a seat belt were 26:1. What is the probability that a driver in Washington in 2006 was *not* using a seat belt?

23. **Transportation** The same report as cited in Exercise 22 stated that Wyoming and New Hampshire had the lowest rates of seat belt compliance. The probability of using the seat belt in those states was .635. What are the odds that a driver in either of those two states in 2006 was using a seat belt?

Health *The table cross-classifies gender and height with data collected from the 2005–2006 National Health and Nutrition Examination Survey (NHANES).**

Gender	Less than 60 Inches	Between 61 and 66 Inches	Between 67 and 72 Inches	Greater than 72 Inches	Total
Male	38	441	1645	379	2503
Female	316	1897	467	3	2683
Total	354	2338	2112	382	5186

Find the probability that a person selected at random had the given characteristics.

24. Male and between 67 and 72 inches tall
25. Female or less than 60 inches tall
26. 72 inches tall or shorter
27. Between 61 and 66 inches tall, given that the person is a male
28. Female, given that the person is between 67 and 72 inches tall

*Table compiled by the author from data available at www.cdc.gov/nchs/nhanes.htm.

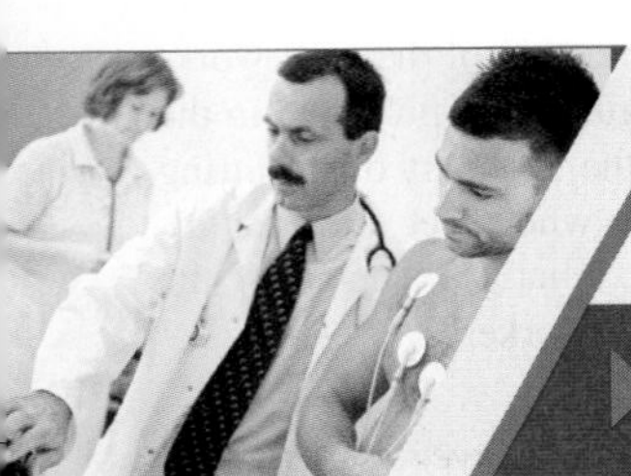

CASE 8 Medical Diagnosis

When patients undergo medical testing, a positive test result for a disease or condition can be emotionally devastating. In many cases, however, testing positive does not necessarily imply that the patient actually has the disease. Bayes' formula can be very helpful in determining the probability of actually having the disease when a patient tests positive.

Let us label the event of having the disease as D and not having the disease as D'. We will denote testing positive for the disease as T and testing negative as T'. Suppose a medical test is calibrated on patients so that we know that among patients with the disease, the test is positive 99.95% of the time. (This quantity is often called the **sensitivity** of the test.) Among patients known not to have the disease, 99.90% of the time the test gave a negative result. (This quantity is often called the **specificity** of the test.) In summary, we have

$$\text{Sensitivity} = P(T|D) = .9995 \quad \text{and} \quad \text{Specificity} = P(T'|D') = .9990.$$

Using the complement rule, we find that the probability the test will give a negative result when a patient has the disease is

$$P(T'|D) = 1 - P(T|D) = 1 - .9995 = .0005.$$

Similarly, for those patients without the disease, the probability of testing positive is .0010, calculated by

$$P(T|D') = 1 - P(T'|D') = 1 - .9990 = .0010.$$

These results do not yet answer the question of interest: If a patient tests positive for the disease, what is the probability the patient actually has the disease? Using our notation, we want to know $P(D|T)$. There are two steps to finding this probability. The first is that we need an estimate of the prevalence of the disease in the general population. Let us assume that one person in a thousand has the disease. We can then calculate that

$$P(D) = \frac{1}{1000} = .001 \quad \text{and} \quad P(D') = 1 - .001 = .999.$$

With this information, and the previous results from testing, we can now use Bayes' formula to find $P(D|T)$:

$$P(D|T) = \frac{P(D)P(T|D)}{P(D)P(T|D) + P(D')P(T|D')}.$$

Using $P(D) = .001$, $P(D') = .999$, the sensitivity $P(T|D) = .9995$ and the complement to the specificity $P(T|D') = .0010$, we have

$$P(D|T) = \frac{(.001)(.9995)}{(.001)(.9995) + (.999)(.0010)} = \frac{.0009995}{.0019985} \approx .5001.$$

Hence, the probability the patient actually has the disease after testing positive for the disease is only .5. This is approximately the same probability as guessing "heads" when flipping a coin. It seems paradoxical that a test which has such high sensitivity (in this case, .9995) and specificity (in this case, .999) could lead to a probability of merely .5 that a person who tests positive for the disease actually has the disease. This is why it is imperative to have confirmatory tests run after testing positive.

The other factor in the calculation is the prevalence of the disease among the general population. In our example, we used $P(D) = .001$. Often, it is very difficult to know how prevalent a disease is among the general population. If the disease is more prevalent, such as 1 in 100, or $P(D) = .01$, we find the probability of a patient's having the disease, given that the patient tests positive, as

$$P(D|T) = \frac{(.01)(.9995)}{(.01)(.9995) + (.99)(.0010)} = \frac{.009995}{.010985} \approx .9099.$$

So when the disease has higher prevalence, then the probability of having the disease after testing positive is also higher. If the disease has a lower prevalence (as in the case of our first example), then the probability of having the disease after testing positive could be much lower than one might otherwise think.

EXERCISES

1. Suppose the specificity of a test is 0.999. Find $P(T|D')$.
2. If the sensitivity of a test for a disease is .99 and the prevalence of the disease is .005, use your answer to Exercise 1 to find the probability of a patient's having the disease, given that the patient tested positive, or $P(D|T)$.
3. Recalculate your answer to Exercise 2 using a prevalence of disease of .0005.

Counting, Probability Distributions, and Further Topics in Probability

Probability has applications to quality control in manufacturing and to decision making in business. It plays a role in testing new medications, in evaluating DNA evidence in criminal trials, and in a host of other situations. See Exercises 1 and 2 on page 565, and Exercises 41–43 on page 545. Sophisticated counting techniques are often necessary for determining the probabilities used in these applications. See Exercises 33–35 on pages 566–567.

Probability distributions enable us to compute the "average value" or "expected outcome" when an experiment or process is repeated a number of times. These distributions are introduced in Section 9.1 and used in Sections 9.4 and 9.6. The other focus of this chapter is the development of effective ways to count the possible outcomes of an experiment without actually listing them all (which can be *very* tedious when large numbers are involved). These counting techniques are introduced in Section 9.2 and are used to find probabilities throughout the rest of the chapter.

9.1 Probability Distributions and Expected Value

Probability distributions were introduced briefly in Section 8.4. Now we take a more complete look at them. In this section, we shall see that the *expected value* of a probability distribution is a type of average. A probability distribution depends on the idea of a *random variable*, so we begin with that.

RANDOM VARIABLES

One of the questions asked in the 2006 National Health and Nutrition Examination Study (NHANES) had to do with respondents' daily hours of TV or video use.* The answer to that question, which we will label x, is one of the numbers 0 through 6 (corresponding to the numbers of hours of use). Since the value of x is random, x is called a random variable.

Random Variable

A **random variable** is a function that assigns a real number to each outcome of an experiment.

The following table gives each possible outcome of the study question on TV and video use together with the probability $P(x)$ of each outcome x.

x	0	1	2	3	4	5	6
$P(x)$	.13	.19	.27	.16	.10	.13	.02

A table that lists all the outcomes with the corresponding probabilities is called a **probability distribution**. The sum of the probabilities in a probability distribution must always equal 1. (The sum in some distributions may vary slightly from 1 because of rounding.)

Instead of writing the probability distribution as a table, we could write the same information as a set of ordered pairs:

$$\{(0, .13), (1, .19), (2, .27), (3, .16), (4, .10), (5, .13), (6, .02)\}.$$

There is just one probability for each value of the random variable.

*National Health and Nutrition Examination Study, www.cdc.gov/nchs/nhanes.htm.

The information in a probability distribution is often displayed graphically as a special kind of bar graph called a **histogram**. The bars of a histogram all have the same width, usually 1 unit. The heights of the bars are determined by the probabilities. A histogram for the data in the probability distribution on the previous page is given in Figure 9.1. A histogram shows important characteristics of a distribution that may not be readily apparent in tabular form, such as the relative sizes of the probabilities and any symmetry in the distribution.

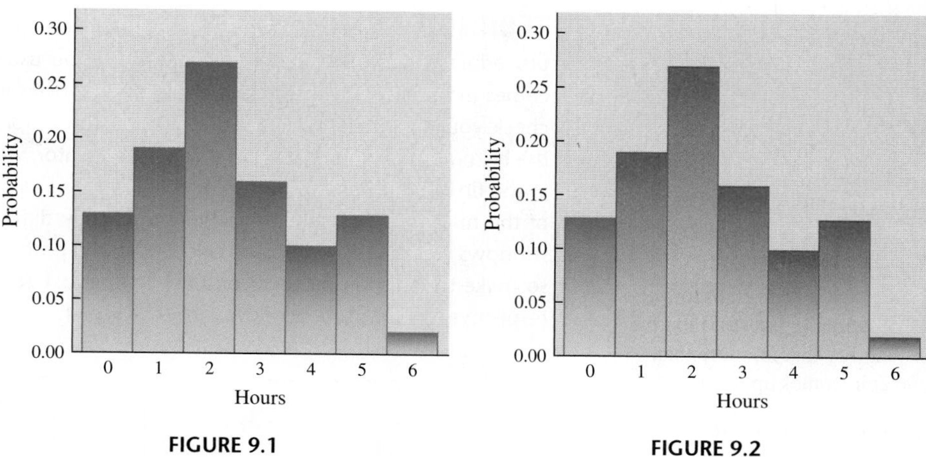

FIGURE 9.1

FIGURE 9.2

The area of the bar above $x = 0$ in Figure 9.1 is the product of 1 and .13, or $1 \cdot .13 = .13$. Since each bar has a width of 1, its area is equal to the probability that corresponds to its x-value. The probability that a particular value will occur is thus given by the area of the appropriate bar of the graph. For example, the probability that one or more hours are spent watching TV or a video is the sum of the areas for $x = 1, x = 2, x = 3, x = 4, x = 5$, and $x = 6$. This area, shown in red in Figure 9.2, corresponds to .87 of the total area, since

$$\begin{aligned} P(x \geq 1) &= P(x = 1) + P(x = 2) + P(x = 3) + P(x = 4) \\ &\qquad + P(x = 5) + P(x = 6) \\ &= .19 + .27 + .16 + .10 + .13 + .02 \\ &= .87. \end{aligned}$$

Example 1

(a) Give the probability distribution for the number of heads showing when two coins are tossed.

Solution Let x represent the random variable "number of heads." Then x can take on the value 0, 1, or 2. Now find the probability of each outcome. When two coins are tossed, the sample space is {TT, TH, HT, HH}. So the probability of getting one head is $2/4 = 1/2$. Similar analysis of the other cases produces this table.

x	0	1	2
$P(x)$	1/4	1/2	1/4

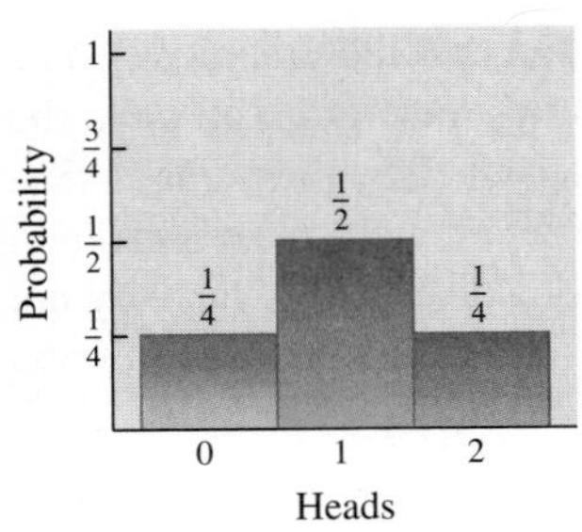

FIGURE 9.3

✓Checkpoint 1

(a) Give the probability distribution for the number of heads showing when three coins are tossed.

(b) Draw a histogram for the distribution in part (a). Find the probability that no more than one coin comes up heads.

Answers: See page 546.

(b) Draw a histogram for the distribution in the table. Find the probability that at least one coin comes up heads.

Solution The histogram is shown in Figure 9.3. The portion in red represents

$$P(x \geq 1) = P(x = 1) + P(x = 2)$$
$$= \frac{3}{4}. \quad ✓^{1}$$

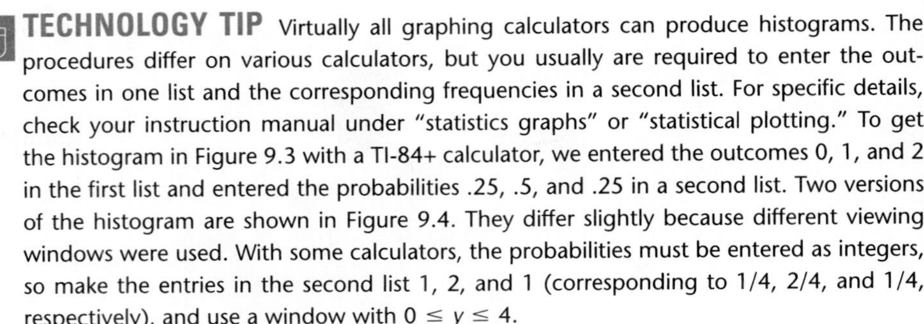

TECHNOLOGY TIP Virtually all graphing calculators can produce histograms. The procedures differ on various calculators, but you usually are required to enter the outcomes in one list and the corresponding frequencies in a second list. For specific details, check your instruction manual under "statistics graphs" or "statistical plotting." To get the histogram in Figure 9.3 with a TI-84+ calculator, we entered the outcomes 0, 1, and 2 in the first list and entered the probabilities .25, .5, and .25 in a second list. Two versions of the histogram are shown in Figure 9.4. They differ slightly because different viewing windows were used. With some calculators, the probabilities must be entered as integers, so make the entries in the second list 1, 2, and 1 (corresponding to 1/4, 2/4, and 1/4, respectively), and use a window with $0 \leq y \leq 4$.

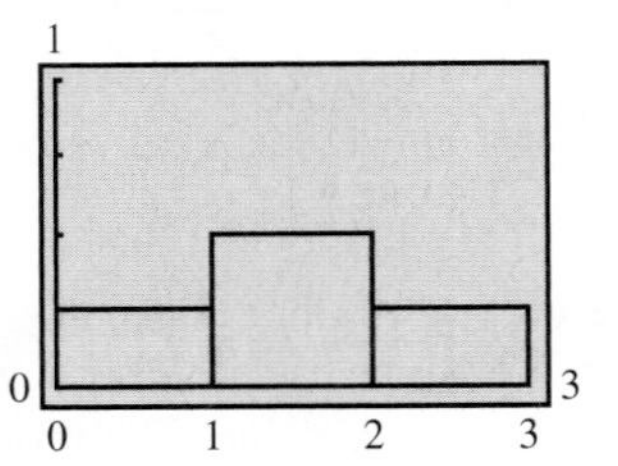

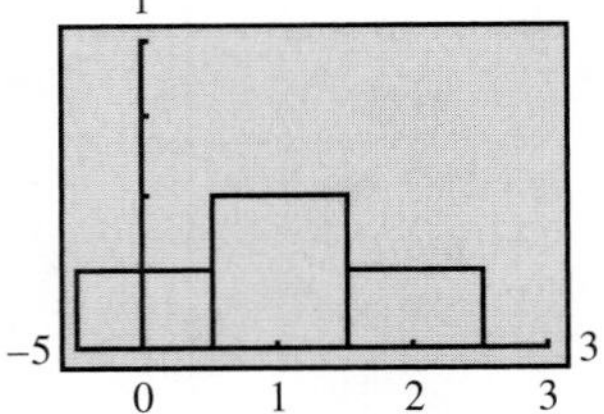

FIGURE 9.4

EXPECTED VALUE

In working with probability distributions, it is useful to have a concept of the typical or average value that the random variable takes on. In Example 1, for instance, it seems reasonable that, on the average, one head shows when two coins are tossed. This does not tell what will happen the next time we toss two coins; we may get two heads, or we may get none. If we tossed two coins many times, however, we would expect that, in the long run, we would average about one head for each toss of two coins.

A way to solve such problems in general is to imagine flipping two coins 4 times. Based on the probability distribution in Example 1, we would expect that 1 of the 4 times we would get 0 heads, 2 of the 4 times we would get 1 head, and 1 of the 4 times we would get 2 heads. The total number of heads we would get, then, is

$$0 \cdot 1 + 1 \cdot 2 + 2 \cdot 1 = 4.$$

The expected number of heads per toss is found by dividing the total number of heads by the total number of tosses:

$$\frac{0 \cdot 1 + 1 \cdot 2 + 2 \cdot 1}{4} = 0 \cdot \frac{1}{4} + 1 \cdot \frac{1}{2} + 2 \cdot \frac{1}{4} = 1.$$

Notice that the expected number of heads turns out to be the sum of the three values of the random variable x, multiplied by their corresponding probabilities. We can use this idea to define the *expected value* of a random variable as follows.

Expected Value

Suppose that the random variable x can take on the n values $x_1, x_2, x_3, \ldots, x_n$. Suppose also that the probabilities that these values occur are, respectively, $p_1, p_2, p_3, \ldots, p_n$. Then the **expected value** of the random variable is

$$E(x) = x_1p_1 + x_2p_2 + x_3p_3 + \cdots + x_np_n.$$

Example 2 **Social Science** In the example with the TV and video usage on page 536, find the expected number of hours per day of viewing.

Solution Multiply each outcome in the table on page 536 by its probability, and sum the products:

$$\begin{aligned} E(x) &= 0\cdot.13 + 1\cdot.19 + 2\cdot.27 + 3\cdot.16 + 4\cdot.10 + 5\cdot.13 + 6\cdot.02 \\ &= 2.38. \end{aligned}$$

On the average, a respondent of the survey will indicate 2.38 hours of TV or video usage. 2✓

✓Checkpoint 2

Find the expected value of the number of heads showing when four coins are tossed.

Answer: See page 546.

Physically, the expected value of a probability distribution represents a balance point. If we think of the histogram in Figure 9.1 as a series of weights with magnitudes represented by the heights of the bars, then the system would balance if supported at the point corresponding to the expected value.

Example 3 **Business** Suppose a local symphony decides to raise money by raffling off a microwave oven worth \$400, a dinner for two worth \$80, and two books worth \$20 each. A total of 2000 tickets are sold at \$1 each. Find the expected value of winning for a person who buys one ticket in the raffle.

Solution Here, the random variable represents the possible amounts of net winnings, where net winnings = amount won − cost of ticket. For example, the net winnings of the person winning the oven are \$400 (amount won) − \$1 (cost of ticket) = \$399, and the net winnings for each losing ticket are \$0 − \$1 = −\$1.

The net winnings of the various prizes, as well as their respective probabilities, are shown in the table below. The probability of winning \$19 is 2/2000, because there are 2 prizes worth \$20. (We have not reduced the fractions in order to keep all the denominators equal.) Because there are 4 winning tickets, there are 1996 losing tickets, so the probability of winning −\$1 is 1996/2000.

x	\$399	\$79	\$19	−\$1
$P(x)$	1/2000	1/2000	2/2000	1996/2000

The expected winnings for a person buying one ticket are

$$399\left(\frac{1}{2000}\right) + 79\left(\frac{1}{2000}\right) + 19\left(\frac{2}{2000}\right) + (-1)\left(\frac{1996}{2000}\right) = -\frac{1480}{2000}$$
$$= -.74.$$

On the average, a person buying one ticket in the raffle will lose \$.74, or 74¢.

It is not possible to lose 74¢ in this raffle: Either you lose \$1, or you win a prize worth \$400, \$80, or \$20, minus the \$1 you paid to play. But if you bought tickets in many such raffles over a long time, you would lose 74¢ per ticket, on the average. It is important to note that the expected value of a random variable may be a number that can never occur in any one trial of the experiment. 3

Checkpoint 3

Suppose you buy 1 of 10,000 tickets at \$1 each in a lottery where the prize is \$5,000. What are your expected net winnings? What does this answer mean?

Answer: See page 546.

NOTE An alternative way to compute expected value in this and other such examples is to calculate the expected amount won and then subtract the cost of the ticket afterward. The amount won is either \$400 (with probability 1/2000), \$80 (with probability 1/2000), \$20 (with probability 2/2000), or \$0 (with probability 1996/2000). The expected winnings for a person buying one ticket are then

$$400\left(\frac{1}{2000}\right) + 80\left(\frac{1}{2000}\right) + 20\left(\frac{2}{2000}\right) + 0\left(\frac{1996}{2000}\right) - 1 = -\frac{1480}{2000}$$
$$= -.74.$$

Example 4 Each day, Lynette and Tanisha toss a coin to see who buys coffee (at \$1.75 a cup). One tosses, while the other calls the outcome. If the person who calls the outcome is correct, the other buys the coffee; otherwise the caller pays. Find Lynette's expected winnings.

Solution Assume that an honest coin is used, that Tanisha tosses the coin, and that Lynette calls the outcome. The possible results and corresponding probabilities are shown in the following table:

	Possible Results			
Result of Toss	Heads	Heads	Tails	Tails
Call	Heads	Tails	Heads	Tails
Caller Wins?	Yes	No	No	Yes
Probability	1/4	1/4	1/4	1/4

Lynette wins a \$1.75 cup of coffee whenever the results and calls match, and she loses \$1.75 when there is no match. Her expected winnings are

$$1.75\left(\frac{1}{4}\right) + (-1.75)\left(\frac{1}{4}\right) + (-1.75)\left(\frac{1}{4}\right) + 1.75\left(\frac{1}{4}\right) = 0.$$

On the average, over the long run, Lynette breaks even. 4

Checkpoint 4

Find Tanisha's expected winnings.

Answer: See page 546.

A game with an expected value of 0 (such as the one in Example 4) is called a **fair game**. Casinos do not offer fair games. If they did, they would win (on the average) \$0 and have a hard time paying the help! Casino games have expected winnings for the house that vary from 1.5 cents per dollar to 60 cents per dollar. The next example examines the popular game of roulette.

Example 5 **Business** As we saw in Chapter 8, an American roulette wheel has 38 slots. Two of the slots are marked 0 and 00 and are colored green. The remaining slots are numbered 1–36 and are colored red and black (18 red and 18 black). One simple wager is to bet \$1 on the color red. If the marble lands in a red slot, the player gets his or her dollar back, plus a \$1 of winnings. Find the expected winnings for a \$1 bet on red.

Solution For this bet, there are only two possible outcomes: winning or losing. The random variable has outcomes +1 if the marble lands in a red slot and −1 if it does not. We need to find the probability for these two outcomes. Since there are 38 total slots, 18 of which are colored red, the probability of winning a dollar is 18/38. The player will lose if the marble lands in any of the remaining 20 slots, so the probability of losing the dollar is 20/38. Thus, the probability distribution is

x	-1	$+1$
$P(x)$	$\frac{20}{38}$	$\frac{18}{38}$

The expected winnings are

$$E(x) = -1\left(\frac{20}{38}\right) + 1\left(\frac{18}{38}\right) = -\frac{2}{38} \approx -.053.$$

The winnings on a dollar bet for red average out to losing about a nickel on every spin of the roulette wheel. In other words, a casino earns on average 5.3 cents on every dollar bet on red. 5✓

✓Checkpoint 5

A gambling game requires a \$1 bet. If the player wins, she gets \$100, but if she loses, she loses her \$1. The probability of winning is .005. What are the expected winnings of this game?

Answer: See page 546.

Exercises 17–20 at the end of the section ask you to find the expected winnings for other bets on games of chance. The idea of expected value can be very useful in decision making, as shown by the next example.

Example 6 **Finance** Suppose that, at age 50, you receive a letter from Mutual of Mauritania Insurance Company. According to the letter, you must tell the company immediately which of the following two options you will choose: Take \$50,000 at age 60 (if you are alive, and \$0 otherwise), or take \$65,000 at age 70 (again, if you are alive, and \$0 otherwise). Based *only* on the idea of expected value, which should you choose?

Solution Life insurance companies have constructed elaborate tables showing the probability of a person living a given number of years into the future. From a recent such table, the probability of living from age 50 to age 60 is .88, while the probability of living from age 50 to 70 is .64. The expected values of the two options are as follows.

First Option: (50,000)(.88) + (0)(.12) = 44,000;

Second Option: (65,000)(.64) + (0)(.36) = 41,600.

Strictly on the basis of expected value, choose the first option. 6✓

✓Checkpoint 6

After college, a person is offered two jobs. With job *A*, after five years, there is a 50% chance of making \$60,000 per year and a 50% chance of making \$45,000. With job *B*, after five years, there is a 30% chance of making \$80,000 per year and a 70% chance of making \$35,000. Based strictly on expected value, which job should be taken?

Answer: See page 546.

Example 7 **Social Science** The table gives the probability distribution for the number of children of respondents to the 2006 General Social Survey,* for those with 7 or fewer children.

x	0	1	2	3	4	5	6	7
$P(x)$	.273	.159	.257	.165	.087	.031	.019	.010

Find the expected value for the number of children.

Solution Using the formula for the expected value, we have

$$\begin{aligned} E(x) &= 0(.273) + 1(.159) + 2(.257) + 3(.165) \\ &\quad + 4(.087) + 5(.031) + 6(.019) + 7(.010) \\ &= 1.855. \end{aligned}$$

For those respondents with 7 or fewer children, the number of children, on average, is 1.855.

*www.norc.org/GSS+Website.

9.1 ▶ Exercises

For each of the experiments described, let x determine a random variable and use your knowledge of probability to prepare a probability distribution. (Hint: Use a tree diagram.)

1. Four children are born, and the number of boys is noted. (Assume an equal chance of a boy or a girl for each birth.)
2. Two dice are rolled, and the total number of points is recorded.
3. A game spinner with equal chance of obtaining values 1, 2, 3, 4 is spun 2 times. The sum of the values from 2 spins is computed.
4. Two names are drawn from a hat, signifying who should go pick up pizza. Three of the names are on the swim team and two are not. The number of swimmers selected is counted.

Draw a histogram for each of the given exercises, and shade the region that gives the indicated probability. (See Example 1.)

5. Exercise 1; $P(x \leq 2)$
6. Exercise 2; $P(x \geq 11)$
7. Exercise 3; P(at least one queen)
8. Exercise 4; P(fewer than two swimmers)

Find the expected value for each random variable. (See Example 2.)

9.

x	1	3	5	7
$P(x)$	.1	.5	.2	.2

10.

y	0	15	30	40
$P(y)$	.15	.20	.40	.25

11.

z	0	2	4	8	16
$P(z)$	.21	.24	.21	.17	.17

12.

x	5	10	15	20	25
$P(x)$	.40	.30	.15	.10	.05

Find the expected values for the random variables x whose probability functions are graphed.

13.

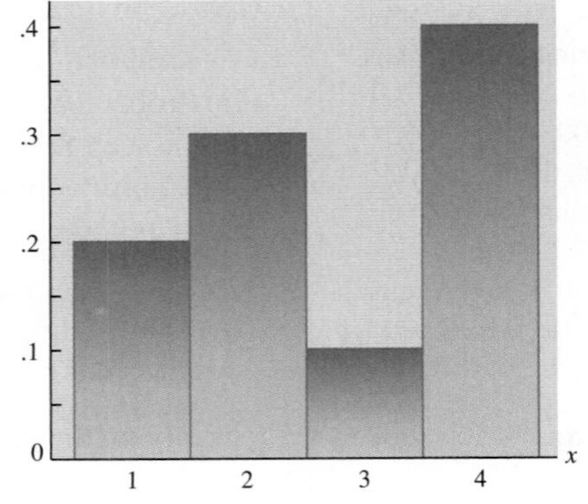

14. 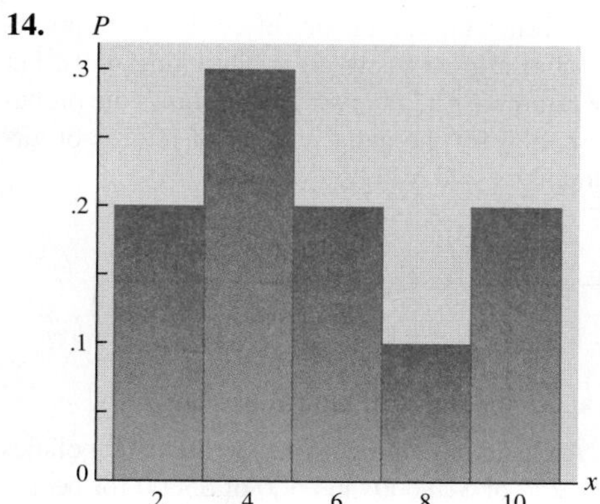

15. 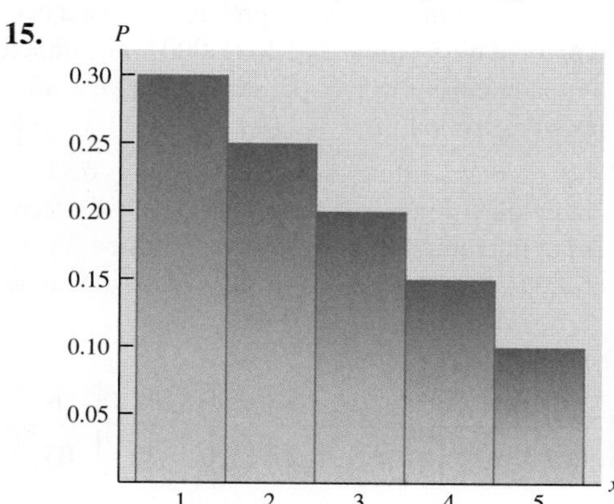

16. 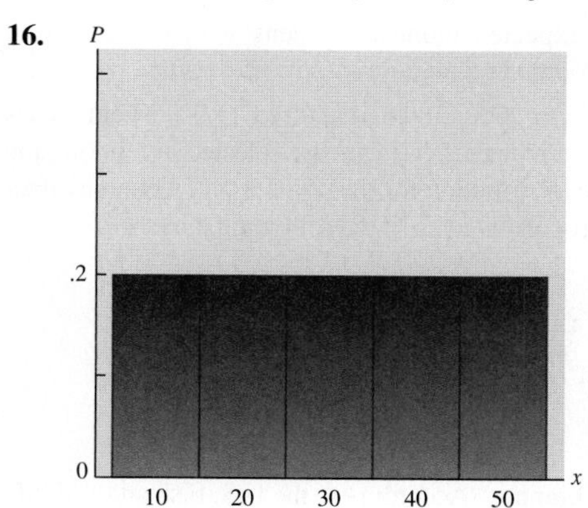

Find the expected winnings for the games of chance described in Exercises 17–20. (See Example 5.)

17. In one form of roulette, you bet \$1 on "even." If one of the 18 even numbers comes up, you get your dollar back, plus another one. If one of the 20 noneven (18 odd, 0, and 00) numbers comes up, you lose your dollar.

18. Repeat Exercise 17 if there are only 19 noneven numbers (no 00).

19. *Numbers* is a game in which you bet \$1 on any three-digit number from 000 to 999. If your number comes up, you get \$500.

20. In one form of the game Keno, the house has a pot containing 80 balls, each marked with a different number from 1 to 80. You buy a ticket for \$1 and mark one of the 80 numbers on it. The house then selects 20 numbers at random. If your number is among the 20, you get \$3.20 (for a net winning of \$2.20).

21. **Business** An online gambling site offers a first prize of \$50,000 and two second prizes of \$10,000 each for registered users when they place a bet. A random bet will be selected over a 24-hour period. Two million bets are received in the contest. Find the expected winnings if you can place one registered bet of \$1 in the given period.

Business *A contest at a fast-food restaurant offered the following cash prizes and probabilities of winning when one buys a large order of French fries for \$1.89:*

Prize	Probability
\$100,000	1/8,504,860
\$50,000	1/302,500
\$10,000	1/282,735
\$1000	1/153,560
\$100	1/104,560
\$25	1/9,540

22. Find the expected winnings if the player buys one large order of French fries.

23. Find the expected winnings if the player buys 25 large orders of French fries in multiple visits.

24. **Business** According to the Web site of Mars, the makers of M&M's Plain Chocolate Candies, 21% of the candies produced are green.* If we select 4 candies from a bag at random and record the number of green candies, the probability distribution is as follows:

x	0	1	2	3	4
$P(x)$	.3895	.4141	.1651	.0293	.0019

Find the expected value for the number of green candies.

25. **Social Science** Recently, it has been estimated that 85.7% of the U.S. population has achieved a high school degree or higher.† If four adults are selected at random, the probability distribution for the number with a high school degree or higher is given in the table:

x	0	1	2	3	4
$P(x)$	.0004	.0100	.0901	.3600	.5394

Find the expected value.

26. **Social Science** Recently, it has been estimated that 28.7% of the population has earned a college degree or

*http://us.mms.com/us/.

†www.census.gov.

higher.* If four people are chosen at random, the following table gives the probability distribution for the number with a college degree or higher:

x	0	1	2	3	4
$P(x)$	.258	.416	.251	.067	.007

Find $E(x)$.

For Exercises 27–30, determine whether the probability distributions are valid or not. If not, explain why.

27.

x	5	10	15	20	25	30	35
$P(x)$	.01	.09	.25	.45	.05	.20	−.05

28.

x	−2	−1	0	1	2	3	4
$P(x)$	.05	.10	.75	.02	.03	.04	.01

29.

x	1	3	5	7	9	11
$P(x)$	.01	.02	.03	.04	.05	.85

30.

x	−10	−5	0	5	10
$P(x)$	.50	.10	−.20	.30	.30

For Exercises 31–35, fill in the missing value(s) to make a valid probability distribution.

31.

x	5	10	15	20	25	30
$P(x)$	.01	.09	.25	.45	.05	

32.

x	−3	−2	−1	0	1	2	3
$P(x)$		.15	.15	.15	.15	.15	.15

33.

x	10	20	30	40
$P(x)$	.20		.25	.30

34.

x	−50	−40	−30	−20	−10	0	10
$P(x)$	.05	.25	.10	.10	.05		

35.

x	1	2	3	6	12	24	48
$P(x)$	.10	.10	.20	.25	.05		

*www.census.gov.

36. Business During the month of July, a home improvement store sold a great many air-conditioning units, but some were returned. The following table shows the probability distribution for the daily number of returns of air-conditioning units sold in July:

x	0	1	2	3	4	5
$P(x)$	.55	.31	.08	.04	.01	.01

Find the expected number of returns per day.

37. Finance An insurance company has written 100 policies of \$15,000, 250 of \$10,000, and 500 of \$5000 for people age 20. If experience shows that the probability that a person will die in the next year at age 20 is .0007, how much can the company expect to pay out during the year after the policies were written?

38. Social Science According to a recent study, 62% of teenagers have television sets in their bedrooms.* If 6 teens are selected at random, the probability distribution for the number of teens with televisions in their bedrooms is as follows.

x	0	1	2	3	4	5	6
$P(x)$	.003	.029	.120	.262	.320	.209	.057

Find the expected number of teens with televisions in their bedrooms in a random sample of 6 teens.

39. Business In 2008, Toyota captured 15.9% of auto sales in the United States.† If 3 cars are selected at random, the probability distribution for the number that were manufactured by Toyota is given in the following table:

x	0	1	2	3
$P(x)$	.595	.337	.064	.004

Find $E(x)$.

40. Natural Science According to the U.S. Fish and Wildlife service, 16.7% of the endangered species in the United States are birds. If we select 5 endangered species at random, the probability distribution for the number of bird species is given in the following table:

x	0	1	2	3	4	5
$P(x)$	.4011	.4020	.1612	.0323	.0032	.0001

Find $E(x)$.

*http://health.usnews.com/blogs/on-parenting/2008/4/7/tv-in-the-bedroom-bad-idea.htm.

†http://online.wsj.com.

41. **Business** Levi Strauss and Company uses expected value to help its salespeople rate their accounts.* For each account, a salesperson estimates potential additional volume and the probability of getting it. The product of these figures gives the expected value of the potential, which is added to the existing volume. The totals are then classified as *A*, *B*, or *C* as follows: $40,000 or below, class *C*; above $40,000, up to and including $55,000, class *B*; above $55,000, class *A*. Complete the chart.

Account Number	Existing Volume	Potential Additional Volume	Probability of Additional Volume	Expected Value of Potential	Existing Volume + Expected Value of Potential	Class
1	$15,000	$10,000	.25	$2,500	$17,500	*C*
2	40,000	0	—	—	40,000	*C*
3	20,000	10,000	.20			
4	50,000	10,000	.10			
5	5,000	50,000	.50			
6	0	100,000	.60			
7	30,000	20,000	.80			

42. According to Len Pasquarelli, in the first 10 games of the 2004 professional football season in the United States, two-point conversions were successful 51.2% of the time.† We can compare this rate with the historical success rate of extra-point kicks of 94%.‡
(a) Calculate the expected value of each strategy.
(b) Over the long run, which strategy will maximize the number of points scored?
(c) From this information, should a team always use only one strategy? Explain.

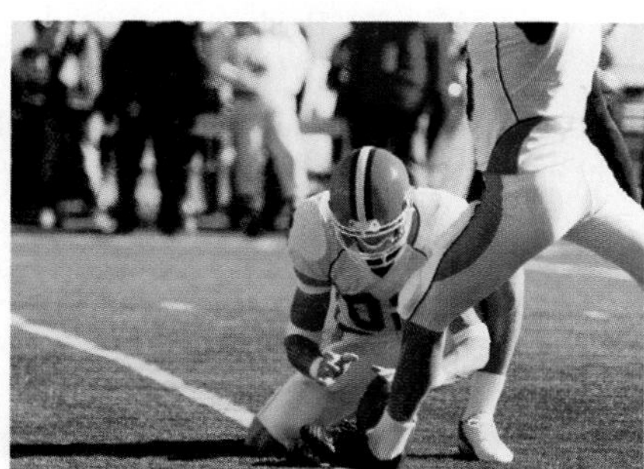

43. **Natural Science** Otitis media, or middle-ear infection, is initially treated with an antibiotic. Researchers have compared two antibiotics—amoxicillin and cefaclor—for their cost effectiveness. Amoxicillin is inexpensive, safe, and effective. Cefaclor is also safe. However, it is considerably more expensive and is generally more effective. Use the given tree diagram (in which costs are estimated as the total cost of medication, an office visit, an ear check, and hours of lost work) to complete the following tasks:*
(a) Find the expected cost of using each antibiotic to treat a middle-ear infection.
(b) To minimize the total expected cost, which antibiotic should be chosen?

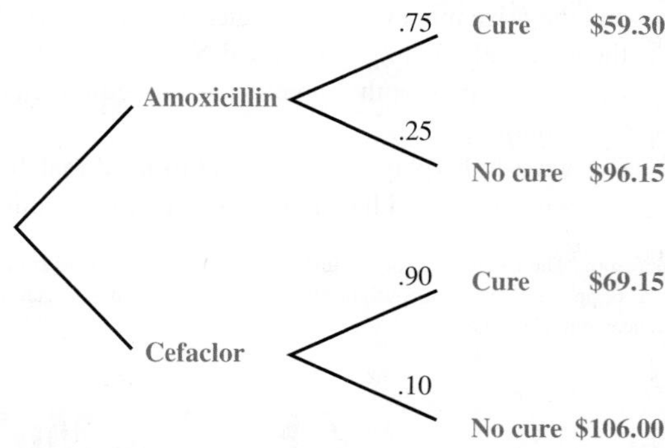

44. **Physical Science** One of the few methods that can be used in an attempt to cut the severity of a hurricane is to *seed* the storm. In this process, silver iodide crystals are dropped into the storm in order to decrease the wind speed. Unfortunately, silver iodide crystals sometimes cause the storm to *increase* its speed. Wind speeds may

*This example was supplied by James McDonald, Levi Strauss and Company, San Francisco.

†Len Pasquarelli, "Teams More Successful Going for Two," November 18, 2004,www.espn.com.

‡David Leonhardt, "In Football, 6 + 2 Often Equals 6," *New York Times*, January 16, 2000, p. 4-2.

*Jeffrey Weiss and Shoshana Melman, "Cost Effectiveness in the Choice of Antibiotics for the Initial Treatment of Otitis Media in Children: A Decision Analysis Approach." *Journal of Pediatric Infectious Disease*, vol. 7, no. 1 (1998): 23–26.

also increase or decrease even with no seeding. Use the given tree diagram to complete the following tasks.*

(a) Find the expected amount of damage under each of the options, "seed" and "do not seed."

(b) To minimize total expected damage, which option should be chosen?

Option	Probability	Change in wind speed	Property damage (millions of dollars)
Seed	0.038	+32%	335.8
	0.143	+16%	191.1
	0.392	0	100.0
	0.255	−16%	46.7
	0.172	−34%	16.3
Do not seed	0.054	+32%	335.8
	0.206	+16%	191.1
	0.480	0	100.0
	0.206	−16%	46.7
	0.054	−34%	16.3

45. In the 2008 Wimbledon Championships, Roger Federer and Rafeal Nadal played in the finals. The prize money for the winner was £750,000 (British pounds sterling), and the prize money for the runner-up was £350,000. Find the expected winnings for Rafeal Nadal if

(a) we assume both players had an equal chance of winning;

(b) we use the players' prior head-to-head match record, whereby Nadal had a .67 probability of winning.

*Data from "The Decision to Seed Hurricanes," by R. A. Howard from *Science*, Vol. 176, pp. 1191–1202, Copyright 1972 by the American Association for the Advancement of Science.

46. Bryan Miller has two cats and a dog. Each pet has a 35% probability of climbing into the chair in which Bryan is sitting, independently of how many pets are already in the chair with Bryan.

(a) Find the probability distribution for the number of pets in the chair with Bryan. (*Hint*: List the sample space.)

(b) Use the probability distribution in part (a) to find the expected number of pets in the chair with Bryan.

✓ Checkpoint Answers

1. (a)

x	$P(x)$
0	1/8
1	3/8
2	3/8
3	1/8

(b) 1/2

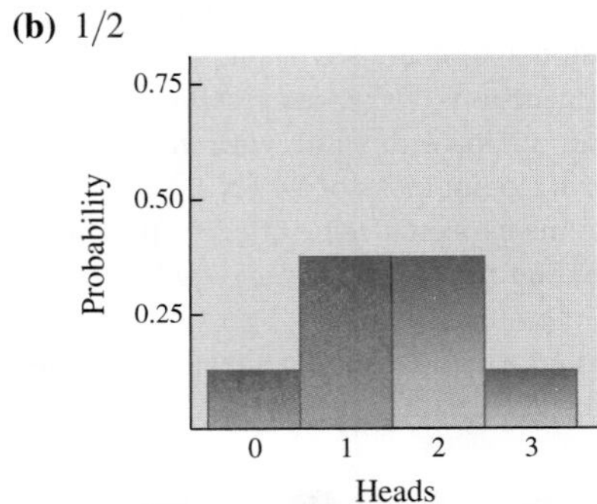

2. 2

3. −$ .50. On the average, you lose $ .50 per ticket purchased.

4. 0

5. −$.495

6. Job A has an expected salary of $52,500, and job B has an expected salary of $48,500. Take job A.

9.2 The Multiplication Principle, Permutations, and Combinations

We begin with a simple example. If there are three roads from town A to town B and two roads from town B to town C, in how many ways can someone travel from A to C by way of B? We can solve this simple problem with the help of Figure 9.5, which lists all the possible ways to go from A to C.

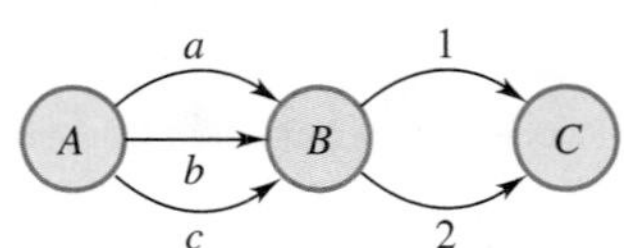

FIGURE 9.5

The possible ways to go from A through B to C are a1, a2, b1, b2, c1, and c2. So there are 6 possible ways. Note that 6 is the product of $3 \cdot 2$, 3 is the number of ways to go from A to B, and 2 is the number of ways to go from B to C.

Another way to solve this problem is to use a tree diagram, as shown in Figure 9.6. This diagram shows that, for each of the 3 roads from A, there are 2 different routes leading from B to C, making $3 \cdot 2 = 6$ different ways.

This example is an illustration of the *multiplication principle.*

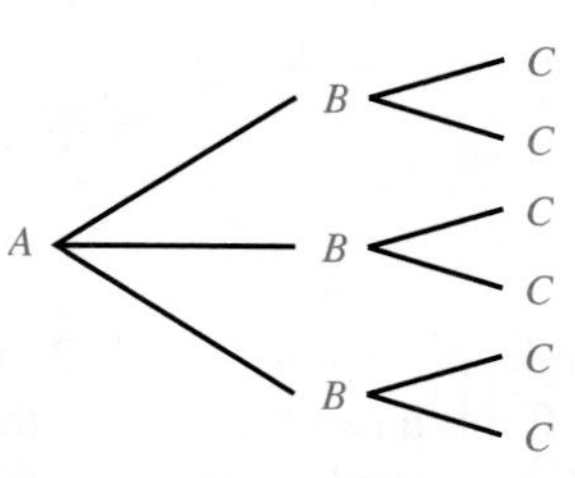

FIGURE 9.6

Multiplication Principle

Suppose n choices must be made, with

$$m_1 \text{ ways to make choice } 1,$$
$$m_2 \text{ ways to make choice } 2,$$
$$\vdots$$
$$m_n \text{ ways to make choice } n.$$

Then there are

$$m_1 \cdot m_2 \cdot \cdots \cdot m_n$$

different ways to make the entire sequence of choices.

Example 1 Suppose Angela has 9 skirts, 8 blouses, and 13 different pairs of shoes. If she is willing to wear any combination, how many different skirt–blouse–shoe choices does she have?

Solution By the multiplication principle, there are 9 skirt choices, 8 blouse choices, and 13 shoe choices, for a total of $9 \cdot 8 \cdot 13 = 936$ skirt–blouse–shoe outfits. ∎

Example 2 **Business** In March 2009, there were 881 sink faucets, 314 bath vanities, and 87 medicine cabinets available to order at the Home Depot® Web site. How many different ways could you buy one sink faucet, bath vanity, and medicine cabinet?

Solution A tree (or other diagram) would be far too complicated to use here, but the multiplication principle easily answers the question. There are

$$881 \cdot 314 \cdot 87 = 24{,}067{,}158$$

ways. ∎

Example 3 A combination lock can be set to open to any 3-letter sequence.

(a) How many sequences are possible?

Solution Since there are 26 letters of the alphabet, there are 26 choices for each of the 3 letters, and, by the multiplication principle, $26 \cdot 26 \cdot 26 = 17{,}576$ different sequences.

(b) How many sequences are possible if no letter is repeated?

Solution There are 26 choices for the first letter. It cannot be used again, so there are 25 choices for the second letter and then 24 choices for the third letter. Consequently, the number of such sequences is $26 \cdot 25 \cdot 24 = 15{,}600$. 1✓

✓Checkpoint 1

(a) In how many ways can 6 business tycoons line up their golf carts at the country club?

(b) How many ways can 4 pupils be seated in a row with 4 seats?

Answers: See page 560.

FACTORIAL NOTATION

The use of the multiplication principle often leads to products such as $5 \cdot 4 \cdot 3 \cdot 2 \cdot 1$, the product of all the natural numbers from 5 down to 1. If n is a natural number, the symbol $n!$ (read "*n factorial*") denotes the product of all the natural numbers from n down to 1. The factorial is an algebraic shorthand. For example, instead of writing $5 \cdot 4 \cdot 3 \cdot 2 \cdot 1$, we simply write $5!$. If $n = 1$, this formula is understood to give $1! = 1$.

n-Factorial

For any natural number n,

$$n! = n(n-1)(n-2)\ldots(3)(2)(1).$$

By definition, $0! = 1$.

Note that $6! = 6 \cdot 5 \cdot 4 \cdot 3 \cdot 2 \cdot 1 = 6 \cdot (5 \cdot 4 \cdot 3 \cdot 2 \cdot 1) = 6 \cdot 5!$. Similarly, the definition of $n!$ shows that

$$n! = n \cdot (n-1)!.$$

One reason that $0!$ is defined to be 1 is to make the preceding formula valid when $n = 1$, for when $n = 1$, we have $1! = 1$ and $1 \cdot (1-1)! = 1 \cdot 0! = 1 \cdot 1 = 1$, so $n! = n \cdot (n-1)!$. 2✓

✓Checkpoint 2

Evaluate:

(a) 4!

(b) 6!

(c) 1!

(d) 6!/4!

Answers: See page 560.

Almost all calculators have an $n!$ key. A calculator with a 10-digit display and scientific-notation capability will usually give the exact value of $n!$ for $n \le 13$ and approximate values of $n!$ for $14 \le n \le 69$. The value of $70!$ is approximately $1.198 \cdot 10^{100}$, which is too large for most calculators. So how would you simplify $\frac{100!}{98!}$? Depending on the type of calculator, there may be an overflow problem. The next two examples show how to avoid this problem.

TECHNOLOGY TIP The factorial key on a graphing calculator is usually located in the PRB or PROB submenu of the MATH or OPTN menu.

Example 4 Evaluate $\frac{100!}{98!}$.

Solution We use the fact that $n! = n \cdot (n-1)!$ several times:

$$\frac{100!}{98!} = \frac{100 \cdot 99!}{98!} = \frac{100 \cdot 99 \cdot 98!}{98!} = 100 \cdot 99 = 9900.$$

Example 5 Evaluate $\frac{5!}{2!\,3!}$.

Solution $\frac{5!}{2!\,3!} = \frac{5 \cdot 4!}{2!\,3!} = \frac{5 \cdot 4 \cdot 3!}{2!\,3!} = \frac{5 \cdot 4}{2 \cdot 1} = 10.$

Example 6 Morse code uses a sequence of dots and dashes to represent letters and words. How many sequences are possible with at most 3 symbols?

Solution "At most 3" means "1 or 2 or 3." Each symbol may be either a dot or a dash. Thus, the following numbers of sequences are possible in each case:

Number of Symbols	Number of Sequences
1	2
2	$2 \cdot 2 = 4$
3	$2 \cdot 2 \cdot 2 = 8$

Altogether, $2 + 4 + 8 = 14$ different sequences of at most 3 symbols are possible. Because there are 26 letters in the alphabet, some letters must be represented by sequences of 4 symbols in Morse code. 3✓

✓Checkpoint 3

How many Morse code sequences are possible with at most 4 symbols?

Answer: See page 560.

PERMUTATIONS

A **permutation** of a set of elements is an ordering of the elements. For instance, there are six permutations (orderings) of the letters *A*, *B*, and *C*, namely,

$$ABC, ACB, BAC, BCA, CAB, \text{ and } CBA,$$

as you can easily verify. As this listing shows, order counts when determining the number of permutations of a set of elements. By saying "order counts," we mean that the event *ABC* is indeed distinct from *CBA* or any other ordering of the three letters. We can use the multiplication principle to determine the number of possible permutations of any set.

Example 7 How many batting orders are possible for a 9-person baseball team?

Solution There are 9 possible choices for the first batter, 8 possible choices for the second batter, 7 for the third batter, and so on, down to the eighth batter (2 possible choices) and the ninth batter (1 possibility). So the total number of batting orders is

$$9 \cdot 8 \cdot 7 \cdot 6 \cdot 5 \cdot 4 \cdot 3 \cdot 2 \cdot 1 = 362{,}880.$$

In other words, the number of permutations of a 9-person set is 9!.

The argument in Example 7 applies to any set, leading to the conclusion that follows.

The number of permutations of an n element set is $n!$.

Sometimes we want to order only some of the elements in a set, rather than all of them.

Example 8 A teacher has 5 books and wants to display 3 of them side by side on her desk. How many arrangements of 3 books are possible?

Solution The teacher has 5 ways to fill the first space, 4 ways to fill the second space, and 3 ways to fill the third space. Because she wants only 3 books on the desk, there are only 3 spaces to fill, giving $5 \cdot 4 \cdot 3 = 60$ possible arrangements. 4 ✓

✓ **Checkpoint 4**

How many ways can a merchant with limited space display 4 fabric samples side by side from her collection of 8?

Answer: See page 560.

In Example 8, we say that the possible arrangements are *the permutations of 5 things taken 3 at a time*, and we denote the number of such permutations by ${}_5P_3$. In other words, ${}_5P_3 = 60$. More generally, an ordering of r elements from a set of n elements is called a **permutation of n things taken r at a time**, and the number of such permutations is denoted ${}_nP_r$.* To see how to compute this number, look at the answer in Example 8, which can be expressed like this:

$$ {}_5P_3 = 5 \cdot 4 \cdot 3 = 5 \cdot 4 \cdot 3 \cdot \frac{2 \cdot 1}{2 \cdot 1} = \frac{5 \cdot 4 \cdot 3 \cdot 2 \cdot 1}{2 \cdot 1} = \frac{5!}{2!} = \frac{5!}{(5-3)!}. $$

A similar analysis in the general case leads to this useful fact:

Permutations

If ${}_nP_r$ (where $r \le n$) is the number of permutations of n elements taken r at a time, then

$$ {}_nP_r = \frac{n!}{(n-r)!}. $$

TECHNOLOGY TIP The permutation function on a graphing calculator is in the same menu as the factorial key. For large values of n and r, the calculator display for ${}_nP_r$ may be an approximation.

To find ${}_nP_r$, we can either use the preceding rule or apply the multiplication principle directly, as the next example shows.

*Another notation that is sometimes used is $P(n, r)$.

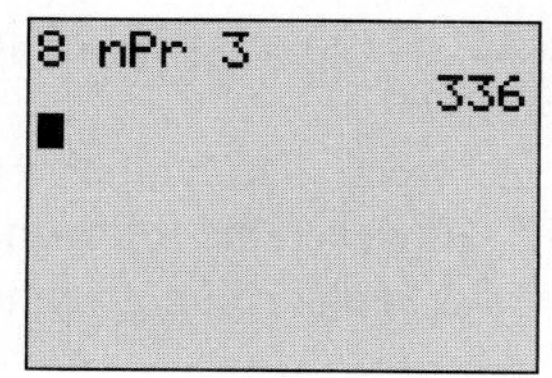

FIGURE 9.7

Example 9 Early in 2008, 8 candidates sought the Republican nomination for president at the Iowa caucus. In a poll, how many ways could voters rank their first, second, and third choices?

Solution This is the same as finding the number of permutations of 8 elements taken 3 at a time. Since there are 3 choices to be made, the multiplication principle gives ${}_8P_3 = 8 \cdot 7 \cdot 6 = 336$. Alternatively, by the formula for ${}_nP_r$,

$$ {}_8P_3 = \frac{8!}{(8-3)!} = \frac{8!}{5!} = \frac{8 \cdot 7 \cdot 6 \cdot 5 \cdot 4 \cdot 3 \cdot 2 \cdot 1}{5 \cdot 4 \cdot 3 \cdot 2 \cdot 1} = 8 \cdot 7 \cdot 6 = 336. $$

Figure 9.7 shows this result on a TI-84+ graphing calculator. 5✓

✓Checkpoint 5

Find the number of permutations of

(a) 5 things taken 2 at a time;

(b) 9 things taken 3 at a time.

Find each of the following:

(c) ${}_3P_1$;

(d) ${}_7P_3$;

(e) ${}_{12}P_2$.

Answers: See page 560.

Example 10 In a college admissions forum, 5 female and 4 male sophomore panelists discuss their college experiences with high school seniors.

(a) In how many ways can the panelists be seated in a row of 9 chairs?

Solution Find ${}_9P_9$, the total number of ways to seat 9 panelists in 9 chairs:

$$ {}_9P_9 = \frac{9!}{(9-9)!} = \frac{9!}{0!} = \frac{9!}{1} = 9 \cdot 8 \cdot 7 \cdot 6 \cdot 5 \cdot 4 \cdot 3 \cdot 2 \cdot 1 = 362{,}880. $$

So, there are 362,880 ways to seat the 9 panelists.

(b) In how many ways can the panelists be seated if the males and females are to be alternated?

Solution Use the multiplication principle. In order to alternate males and females, a female must be seated in the first chair (since there are 5 females and only 4 males), any of the males next, and so on. Thus, there are 5 ways to fill the first seat, 4 ways to fill the second seat, 4 ways to fill the third seat (with any of the 4 remaining females), and so on, or

$$ 5 \cdot 4 \cdot 4 \cdot 3 \cdot 3 \cdot 2 \cdot 2 \cdot 1 \cdot 1 = 2880. $$

So, there are 2880 ways to seat the panelists.

(c) In how many ways can the panelists be seated if the males must sit together and the females sit together?

Solution Use the multiplication principle. We first must decide how to arrange the two groups (males and females). There are 2! ways of doing this. Next, there are 5! ways of arranging the females and 4! ways of arranging the men, for a total of

$$ 2!\,5!\,4! = 2 \cdot 120 \cdot 24 = 5760 $$

ways. 6✓

✓Checkpoint 6

A collection of 3 paintings by one artist and 2 by another is to be displayed. In how many ways can the paintings be shown

(a) in a row?

(b) if the works of the artists are to be alternated?

(c) if one painting by each artist is displayed?

Answers: See page 560.

COMBINATIONS

In Example 8, we found that there are 60 ways a teacher can arrange 3 of 5 different books on a desk. That is, there are 60 permutations of 5 things taken 3 at a time. Suppose now that the teacher does not wish to arrange the books on her desk, but rather wishes to choose, at random, any 3 of the 5 books to give to a book sale to raise money for her school. In how many ways can she do this?

At first glance, we might say 60 again, but that is incorrect. The number 60 counts all possible *arrangements* of 3 books chosen from 5. However, the following arrangements, for example, would all lead to the same set of 3 books being given to the book sale:

mystery–biography–textbook	biography–textbook–mystery
mystery–textbook–biography	textbook–biography–mystery
biography–mystery–textbook	textbook–mystery–biography

The foregoing list shows 6 different *arrangements* of 3 books, but only one *subset* of 3 books. A subset of items selected *without regard to order* is called a **combination**. The number of combinations of 5 things taken 3 at a time is written ${}_5C_3$. Since they are subsets, combinations are *not ordered.*

To evaluate ${}_5C_3$, start with the $5 \cdot 4 \cdot 3$ *permutations* of 5 things taken 3 at a time. Combinations are unordered; therefore, we find the number of combinations by dividing the number of permutations by the number of ways each group of 3 can be ordered—that is, by 3!:

$$ {}_5C_3 = \frac{5 \cdot 4 \cdot 3}{3!} = \frac{5 \cdot 4 \cdot 3}{3 \cdot 2 \cdot 1} = 10. $$

There are 10 ways that the teacher can choose 3 books at random for the book sale.

Generalizing this discussion gives the formula for the number of combinations of n elements taken r at a time, written ${}_nC_r$.* In general, a set of r elements can be ordered in $r!$ ways, so we divide ${}_nP_r$ by $r!$ to get ${}_nC_r$:

$$ \begin{aligned} {}_nC_r &= \frac{{}_nP_r}{r!} \\ &= {}_nP_r \frac{1}{r!} \\ &= \frac{n!}{(n-r)!} \cdot \frac{1}{r!} \qquad \text{Definition of } {}_nP_r \\ &= \frac{n!}{(n-r)!\,r!}. \end{aligned} $$

This last form is the most useful for setting up the calculation. 7 ✓

✓ Checkpoint 7

Evaluate $\frac{{}_nP_r}{r!}$ for the given values.

(a) $n = 6, r = 2$

(b) $n = 8, r = 4$

(c) $n = 7, r = 0$

Answers: See page 560.

Combinations

The number of combinations of n elements taken r at a time, where $r \le n$, is

$$ {}_nC_r = \frac{n!}{(n-r)!\,r!}. $$

*Another notation that is sometimes used in place of ${}_nC_r$ is $\binom{n}{r}$.

Example 11 From a group of 10 students, a committee is to be chosen to meet with the dean. How many different 3-person committees are possible?

Solution A committee is not ordered, so we compute

$$_{10}C_3 = \frac{10!}{(10 - 3)!\,3!} = \frac{10!}{7!\,3!} = \frac{10 \cdot 9 \cdot 8 \cdot 7!}{7!\,3!} = \frac{10 \cdot 9 \cdot 8}{3 \cdot 2 \cdot 1} = 120.$$

TECHNOLOGY TIP The key for obtaining $_nC_r$ on a graphing calculator is located in the same menu as the key for obtaining $_nP_r$.

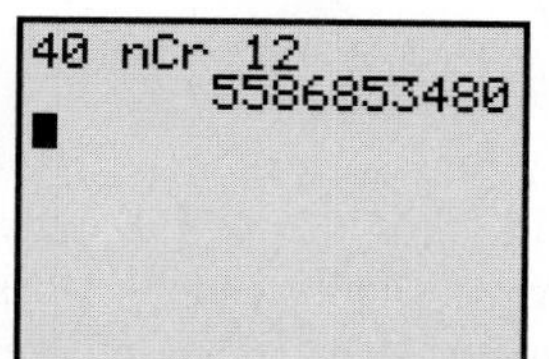

FIGURE 9.8

Example 12 In how many ways can a 12-person jury be chosen from a pool of 40 people?

Solution Since the order in which the jurors are chosen does not matter, we use combinations. The number of combinations of 40 things taken 12 at a time is

$$_{40}C_{12} = \frac{40!}{(40 - 12)!\,12!} = \frac{40!}{28!\,12!}.$$

Using a calculator to compute this number (Figure 9.8), we see that there are 5,586,853,480 possible ways to choose a jury. 8✓

✓Checkpoint 8

Use $\frac{n!}{(n - r)!\,r!}$ to evaluate $_nC_r$.

(a) $_6C_2$

(b) $_8C_4$

(c) $_7C_0$

Compare your answers with the answers to Checkpoint 7.

Answers: See page 560.

Example 13 Three managers are to be selected from a group of 30 to work on a special project.

(a) In how many different ways can the managers be selected?

Solution Here, we wish to know the number of 3-element combinations that can be formed from a set of 30 elements. (We want combinations, not permutations, since order within the group of 3 does not matter.) So, we calculate

$$_{30}C_3 = \frac{30!}{27!\,3!} = 4060.$$

There are 4060 ways to select the project group.

(b) In how many ways can the group of 3 be selected if a certain manager must work on the project?

Solution Since 1 manager has already been selected for the project, the problem is reduced to selecting 2 more from the remaining 29 managers:

$$_{29}C_2 = \frac{29!}{27!\,2!} = 406.$$

In this case, the project group can be selected in 406 ways.

(c) In how many ways can a nonempty group of at most 3 managers be selected from these 30 managers?

Solution The group is to be nonempty; therefore, "at most 3" means "1 or 2 or 3." Find the number of ways for each case:

Case	Number of Ways
1	${}_{30}C_1 = \frac{30!}{29!\,1!} = \frac{30 \cdot 29!}{29!\,1!} = 30$
2	${}_{30}C_2 = \frac{30!}{28!\,2!} = \frac{30 \cdot 29 \cdot 28!}{28! \cdot 2 \cdot 1} = 435$
3	${}_{30}C_3 = \frac{30!}{27!\,3!} = \frac{30 \cdot 29 \cdot 28 \cdot 27!}{27! \cdot 3 \cdot 2 \cdot 1} = 4060$

The total number of ways to select at most 3 managers will be the sum

$$30 + 435 + 4060 = 4525.$$ 9✓

✓Checkpoint 9

Five orchids from a collection of 20 are to be selected for a flower show.

(a) In how many ways can this be done?

(b) In how many different ways can the group of 5 be selected if 2 particular orchids must be included?

(c) In how many ways can at least 1 and at most 5 orchids be selected? (*Hint*: Use a calculator.)

Answers: See page 560.

CHOOSING A METHOD

The formulas for permutations and combinations given in this section will be very useful in solving probability problems in later sections. Any difficulty in using these formulas usually comes from being unable to differentiate among them. Both permutations and combinations give the number of ways to choose r objects from a set of n objects. The differences between permutations and combinations are outlined in the following summary.

Permutations	**Combinations**
Different orderings or arrangements of the r objects are different permutations.	Each choice or subset of r objects gives 1 combination. Order within the r objects does not matter.
${}_nP_r = \frac{n!}{(n-r)!}$	${}_nC_r = \frac{n!}{(n-r)!\,r!}$
Clue words: arrangement, schedule, order Order matters!	Clue words: group, committee, set, sample Order does not matter!

In the examples that follow, concentrate on recognizing which of the formulas should be applied.

Example 14 For each of the given problems, tell whether permutations or combinations should be used to solve the problem.

(a) How many 4-digit numbers are possible if no digits are repeated?

Solution Since changing the order of the 4 digits results in a different number, we use permutations.

(b) A sample of 3 lightbulbs is randomly selected from a batch of 15 bulbs. How many different samples are possible?

Solution The order in which the 3 lightbulbs are selected is not important. The sample is unchanged if the bulbs are rearranged, so combinations should be used.

(c) In a basketball conference with 8 teams, how many games must be played so that each team plays every other team exactly once?

Solution The selection of 2 teams for a game is an *unordered* subset of 2 from the set of 8 teams. Use combinations again.

(d) In how many ways can 4 patients be assigned to 6 hospital rooms so that each patient has a private room?

Solution The room assignments are an *ordered* selection of 4 rooms from the 6 rooms. Exchanging the rooms of any 2 patients within a selection of 4 rooms gives a different assignment, so permutations should be used. 10✓

✓Checkpoint 10

Solve the problems in Example 14.

Answers: See page 560.

Example 15 A manager must select 4 employees for promotion. Twelve employees are eligible.

(a) In how many ways can the 4 employees be chosen?

Solution Because there is no reason to consider the order in which the 4 are selected, we use combinations:

$$_{12}C_4 = \frac{12!}{4!\,8!} = 495.$$

(b) In how many ways can 4 employees be chosen (from 12) to be placed in 4 different jobs?

Solution In this case, once a group of 4 is selected, its members can be assigned in many different ways (or arrangements) to the 4 jobs. Therefore, this problem requires permutations:

$$_{12}P_4 = \frac{12!}{8!} = 11{,}880.$$ 11✓

✓Checkpoint 11

A mailman has special-delivery mail for 7 customers.

(a) In how many ways can he arrange his schedule to deliver to all 7?

(b) In how many ways can he schedule deliveries if he can deliver to only 4 of the 7?

Answers: See page 560.

Example 16 Powerball is a lottery game played in 30 states (plus the District of Columbia and the U.S. Virgin Islands). For a $1 ticket, a player selects five different numbers from 1 to 59 and one powerball number from 1 to 39 (which may be the same as one of the first five chosen). A match of all six numbers wins the jackpot. How many different selections are possible?

FIGURE 9.9

Solution The order in which the first five numbers are chosen does not matter. So we use combinations to find the number of combinations of 59 things taken 5 at a time—that is, $_{59}C_5$. There are 39 ways to choose one powerball number from 1 to 39. So, by the multiplication principle, the number of different selections is

$$_{59}C_5 \cdot 39 = \frac{59!}{(59-5)!\,5!} \cdot 39 = \frac{59! \cdot 39}{54!\,5!} = 195{,}249{,}054,$$

as shown in two ways on a graphing calculator in Figure 9.9. 12✓

✓Checkpoint 12

Under earlier Powerball rules, you had to choose five different numbers from 1 to 53 and then choose one powerball number from 1 to 42. Under those rules, how many different selections were possible?

Answer: See page 560.

Example 17 A male student going on spring break at Daytona Beach has 8 tank tops and 12 pairs of shorts. He decides he will need 5 tank tops and 6 pairs of shorts for the trip. How many ways can he choose the tank tops and the shorts?

Solution We can break this problem into two parts: finding the number of ways to choose the tank tops, and finding the number of ways to choose the shorts. For the tank tops, the order is not important, so we use combinations to obtain

$$ {}_{8}C_{5} = \frac{8!}{3!\,5!} = 56. $$

Likewise, order is not important for the shorts, so we use combinations to obtain

$$ {}_{12}C_{6} = \frac{12!}{6!\,6!} = 924. $$

We now know there are 56 ways to choose the tank tops and 924 ways to choose the shorts. The total number of ways to choose the tank tops and shorts can be found using the multiplication principle to obtain $56 \cdot 924 = 51{,}744$. 13

✓Checkpoint 13

Lacy wants to pack 4 of her 10 blouses and 2 of her 4 pairs of jeans for her trip to Europe. How many ways can she choose the blouses and jeans?

Answer: *See page 560.*

As Examples 16 and 17 show, often both combinations and the multiplication principle must be used in the same problem.

Example 18 To illustrate the differences between permutations and combinations in another way, suppose 2 cans of soup are to be selected from 4 cans on a shelf: noodle (N), bean (B), mushroom (M), and tomato (T). As shown in Figure 9.10(a), there are 12 ways to select 2 cans from the 4 cans if the order matters (if noodle first and bean second is considered different from bean and then noodle, for example). However, if order is unimportant, then there are 6 ways to choose 2 cans of soup from the 4, as illustrated in Figure 9.10(b).

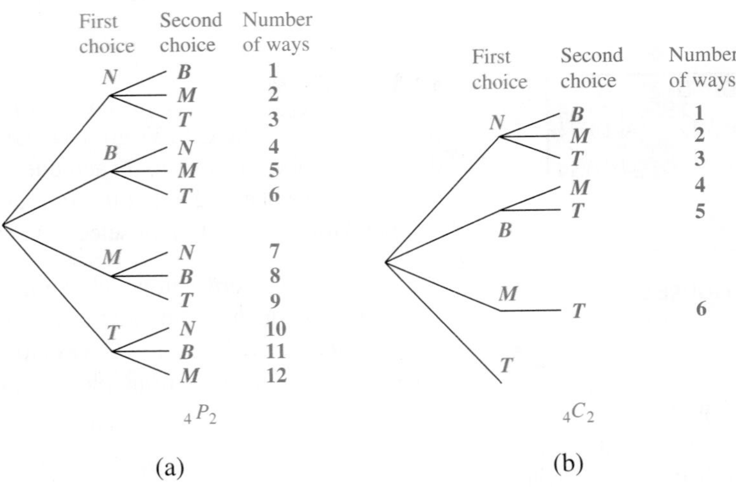

FIGURE 9.10

CAUTION It should be stressed that not all counting problems lend themselves to either permutations or combinations.

9.2 ▸ Exercises

Evaluate the given factorials, permutations, and combinations.

1. ${}_4P_2$ **2.** $3!$ **3.** ${}_8C_5$

4. $7!$ **5.** ${}_8P_1$ **6.** ${}_7C_2$

7. $4!$ **8.** ${}_4P_4$ **9.** ${}_9C_6$

10. ${}_8C_2$ **11.** ${}_{13}P_3$ **12.** ${}_9P_5$

Use a calculator to find values for Exercises 13–20.

13. ${}_{25}P_5$ **14.** ${}_{40}P_5$

15. ${}_{14}P_5$ **16.** ${}_{17}P_8$

17. ${}_{18}C_5$ **18.** ${}_{32}C_9$

19. ${}_{28}C_{14}$ **20.** ${}_{35}C_{30}$

21. Some students find it puzzling that $0! = 1$, and they think that $0!$ should equal 0. If this were true, what would be the value of ${}_4P_4$ according to the permutations formula?

22. If you already knew the value of $8!$, how could you find the value of $9!$ quickly?

Use the multiplication principle to solve the given problems. (See Examples 1–6.)

23. **Social Science** An ancient Chinese philosophical work known as the *I Ching* (*Book of Changes*) is often used as an oracle from which people can seek and obtain advice. The philosophy describes the duality of the universe in terms of two primary forces: *yin* (passive, dark, receptive) and *yang* (active, light, creative). See the accompanying figure. The yin energy is represented by a broken line (– –) and the yang by a solid line (—). These lines are written on top of one another in groups of three, known as *trigrams*. For example, the trigram ☱ is called *Tui*, the Joyous, and has the image of a lake.

(a) How many trigrams are there altogether?

(b) The trigrams are grouped together, one on top of the other, in pairs known as hexagrams. Each hexagram represents one aspect of the *I Ching* philosophy. How many hexagrams are there?

24. **Business** How many different heating–cooling units are possible if a home owner has 3 choices for the efficiency rating of the furnace, 3 options for the fan speed, and 6 options for the air condenser?

25. **Business** An auto manufacturer produces 6 models, each available in 8 different colors, with 4 different upholstery fabrics and 3 interior colors. How many varieties of the auto are available?

26. **Business** How many different 4-letter radio station call letters can be made

(a) if the first letter must be K or W and no letter may be repeated?

(b) if repeats are allowed (but the first letter still must be K or W)?

(c) How many of the 4-letter call letters (starting with K or W) with no repeats end in R?

27. **Social Science** A Social Security number has 9 digits. How many Social Security numbers are possible? The U.S. population in 2009 was approximately 307 million. Was it possible for every U.S. resident to have a unique Social Security number? (Assume no restrictions.)

28. **Social Science** The United States Postal Service currently uses 5-digit zip codes in most areas. How many zip codes are possible if there are no restrictions on the digits used? How many would be possible if the first number could not be 0?

29. **Social Science** The Postal Service is encouraging the use of 9-digit zip codes in some areas, adding 4 digits after the usual 5-digit code. How many such zip codes are possible with no restrictions?

30. **Social Science** For many years, the state of California used 3 letters followed by 3 digits on its automobile license plates.

(a) How many different license plates are possible with this arrangement?

(b) When the state ran out of new numbers, the order was reversed to 3 digits followed by 3 letters. How many new license plate numbers were then possible?

(c) Several years ago, the numbers described in part (b) were also used up. The state then issued plates with 1 letter followed by 3 digits and then 3 letters. How many new license plate numbers will this arrangement provide?

31. **Business** A recent trip to the drug store revealed 12 different kinds of Pantene® shampoo and 10 different kinds of Pantene® conditioner. How many ways can Sherri buy 1 Pantene® shampoo and 1 Pantene® conditioner?

32. **Business** A pharmaceutical salesperson has 6 doctors' offices to call on.

(a) In how many ways can she arrange her schedule if she calls on all 6 offices?

(b) In how many ways can she arrange her schedule if she decides to call on 4 of the 6 offices?

Social Science *The United States is rapidly running out of telephone numbers. In large cities, telephone companies have introduced new area codes as numbers are used up.*

33. (a) Until recently, all area codes had a 0 or a 1 as the middle digit and the first digit could not be 0 or 1. How many area codes were possible with this arrangement? How many telephone numbers does the current 7-digit sequence permit per area code? (The 3-digit sequence that follows the area code cannot start with 0 or 1. Assume that there are no other restrictions.)
    (b) The actual number of area codes under the previous system was 152. Explain the discrepancy between this number and your answer to part (a).
34. The shortage of area codes under the previous system was avoided by removing the restriction on the second digit. How many area codes are available under this new system?
35. A problem with the plan in Exercise 34 was that the second digit in the area code had been used to tell phone company equipment that a long-distance call was being made. To avoid changing all equipment, an alternative plan proposed a 4-digit area code and restricted the first and second digits as before. How many area codes would this plan have provided?
36. Still another solution to the area-code problem is to increase the local dialing sequence to 8 digits instead of 7. How many additional numbers would this plan create? (Assume the same restrictions.)
37. Define permutation in your own words.

Use permutations to solve each of the given problems. (See Examples 7–10.)

38. A baseball team has 15 players. How many 9-player batting orders are possible?
39. Tim is a huge fan of the latest album by country-music singer Kenny Chesney. If Tim has time to listen to only 5 of the 12 songs on the album, how many ways can he listen to the 5 songs?
40. **Business** From a cooler with 8 cans of different kinds of soda, 3 are selected for 3 people. In how many ways can this be done?
41. The Greek alphabet has 24 letters. How many ways can one name a fraternity using 3 Greek letters (with no repeats)?
42. **Finance** A customer speaks to his financial advisor about investment products. The advisor has 9 products available, but knows he will only have time to speak about 4. How many different ways can he give the details on 4 different investment products to the customer?
43. The student activity club at the college has 32 members. In how many different ways can the club select a president, a vice president, a treasurer, and a secretary?
44. A student can only take one class a semester, and she needs to take 4 more electives in any order. If there are 20 courses from which she can choose, how many ways can she take her 4 electives?
45. In a club with 17 members, how many ways can the club elect a president and a treasurer?

Use combinations to solve each of the given problems. (See Examples 11–13.)

46. **Business** Four items are to be randomly selected from the first 25 items on an assembly line in order to determine the defect rate. How many different samples of 4 items can be chosen?
47. **Social Science** A group of 4 students is to be selected from a group of 10 students to take part in a class in cell biology.
    (a) In how many ways can this be done?
    (b) In how many ways can the group that will *not* take part be chosen?
48. **Natural Science** From a group of 15 smokers and 21 nonsmokers, a researcher wants to randomly select 7 smokers and 6 nonsmokers for a study. In how many ways can the study group be selected?
49. The college football team has 11 seniors. The team needs to elect a group of 4 senior co-captains. How many different 4-person groups of co-captains are possible?
50. The drama department holds auditions for a play with a cast of 5 roles. If 33 students audition, how many casts of 5 people are possible?
51. Explain the difference between a permutation and a combination.
52. Padlocks with digit dials are often referred to as "combination locks." According to the mathematical definition of combination, is this an accurate description? Explain.

Exercises 53–70 are mixed problems that may require permutations, combinations, or the multiplication principle. (See Examples 14–18.)

53. Use a tree diagram to find the number of ways 2 letters can be chosen from the set $\{P, Q, R\}$ if order is important and
    (a) if repetition is allowed;
    (b) if no repeats are allowed.
    (c) Find the number of combinations of 3 elements taken 2 at a time. Does this answer differ from that to part (a) or (b)?
54. Repeat Exercise 53, using the set $\{P, Q, R, S\}$ and 4 in place of 3 in part (c).
55. **Social Science** The U.S. Senate Foreign Relations Committee in 2009 had 11 Democrats and 7 Republicans. A delegation of 5 people is to be selected to visit Iraq.
    (a) How many delegations are possible?
    (b) How many delegations would have all Republicans?

(c) How many delegations would have 3 Democrats and 2 Republicans?
(d) How many delegations would have at least one Democrat?

56. **Natural Science** In an experiment on plant hardiness, a researcher gathers 6 wheat plants, 5 barley plants, and 3 rye plants. She wishes to select 4 plants at random.
(a) In how many ways can this be done?
(b) In how many ways can this be done if exactly 2 wheat plants must be included?

57. **Business** According to the Baskin-Robbins® Web site, there are 21 "classic flavors" of ice cream.
(a) How many different double-scoop cones can be made if order does not matter (for example, putting chocolate on top of vanilla is equivalent to putting vanilla on top of chocolate)?
(b) How many different triple-scoop cones can be made if order does matter?

58. **Finance** A financial advisor offers 8 mutual funds in the high-risk category, 7 in the moderate-risk category, and 10 in the low-risk category. An investor decides to invest in 3 high-risk funds, 4 moderate-risk funds and 3 low-risk funds. How many ways can the investor do this?

59. A lottery game requires that you pick 6 different numbers from 1 to 99. If you pick all 6 winning numbers, you win $4 million.
(a) How many ways are there to choose 6 numbers if order is not important?
(b) How many ways are there to choose 6 numbers if order matters?

60. In Exercise 59, if you pick 5 of the 6 numbers correctly, you win $5,000. In how many ways can you pick exactly 5 of the 6 winning numbers without regard to order?

61. The game of Sets* consists of a special deck of cards. Each card has on it either one, two, or three shapes. The shapes on each card are all the same color, either green, purple, or red. The shapes on each card are the same style, either solid, shaded, or outline. There are three possible shapes—squiggle, diamond, and oval—and only one type of shape appears on a card. The deck consists of all possible combinations of shape, color, style, and number. How many cards are in a deck?

62. **Health** Over the course of the previous nursing shift, 16 new patients were admitted onto a hospital ward. If a nurse begins her shift by caring for 6 of the new patients, how many possible ways could the 6 patients be selected from the 16 new arrivals?

63. **Natural Science** A biologist is attempting to classify 52,000 species of insects by assigning 3 initials to each species. Is it possible to classify all the species in this way? If not, how many initials should be used?

64. One play in a state lottery consists of choosing 6 numbers from 1 to 44. If your 6 numbers are drawn (in any order), you win the jackpot.
(a) How many possible ways are there to draw the 6 numbers?
(b) If you get 2 plays for a dollar, how much would it cost to guarantee that one of your choices would be drawn?
(c) Assuming that you work alone and can fill out a betting ticket (for 2 plays) every second, and assuming that the lotto drawing will take place 3 days from now, can you place enough bets to guarantee that 1 of your choices will be drawn?

65. A cooler contains 5 cans of Pepsi®, 1 can of Diet Coke®, and 3 cans of 7UP®; you pick 3 cans at random. How many samples are possible in which the soda cans picked are
(a) only Pepsi;
(b) only Diet Coke;
(c) only 7UP;
(d) 2 Pepsi, 1 Diet Coke;
(e) 2 Pepsi, 1 7UP;
(f) 2 7UP, 1 Pepsi;
(g) 2 Diet Coke, 1 7UP?

66. A class has 9 male students and 8 female students. How many ways can the class select a committee of four people to petition the teacher not to make the final exam cumulative if the committee has to have 2 males and 2 females?

67. **Health** A hospital wants to test the viability of a new medication for attention deficit disorder. It has 35 adults volunteer for the study, but can only enroll 20 in the study. How many ways can it choose the 20 volunteers to enroll in the study?

68. Suppose a pizza shop offers 4 choices of cheese and 9 toppings. If the order of the cheeses and toppings does not matter, how many different pizza selections are possible when choosing two cheeses and 2 toppings?

69. In the game of bingo, each card has 5 columns. Column 1 has spaces for 5 numbers, chosen from 1 to 15. Column 2 similarly has 5 numbers, chosen from 16 to 30. Column 3 has a free space in the middle, plus 4 numbers chosen from 31 to 45. The 5 numbers in columns 4 and 5 are chosen from 46 to 60 and from 61 to 75, respectively. The numbers in each card can be in any order. How many different bingo cards are there?

70. A television commercial for Little Caesars® pizza announced that, with the purchase of two pizzas, one could receive free any combination of up to five toppings on each pizza. The commercial shows a young child waiting in line at one of the company's stores who calculates that there are 1,048,576 possibilities for the toppings on the two pizzas. Verify the child's calculation. Use the fact that Little Caesars has 11 toppings to choose from. Assume

*Copyright © Marsha J. Falco.

that the order of the two pizzas matters; that is, if the first pizza has combination 1 and the second pizza has combination 2, that arrangement is different from combination 2 on the first pizza and combination 1 on the second.*

If the n objects in a permutations problem are not all distinguishable—that is, if there are n_1 of type 1, n_2 of type 2, and so on, for r different types—then the number of distinguishable permutations is

$$\frac{n!}{n_1!\, n_2! \cdots n_r!}.$$

Example *In how many ways can you arrange the letters in the word Mississippi?*

This word contains 1 m, 4 i's, 4 s's, and 2 p's. To use the formula, let $n = 11$, $n_1 = 1$, $n_2 = 4$, $n_3 = 4$, and $n_4 = 2$ to get

$$\frac{11!}{1!\,4!\,4!\,2!} = 34{,}650$$

arrangements. The letters in a word with 11 different letters can be arranged in $11! = 39{,}916{,}800$ ways.

71. Find the number of distinguishable permutations of the letters in each of the given words.
(a) martini **(b)** nunnery **(c)** grinding

72. A printer has 5 W's, 4 X's, 3 Y's, and 2 Z's. How many different "words" are possible that use all these letters? (A "word" does not have to have any meaning here.)

73. Shirley is a shelf stocker at the local grocery store. She has 4 varieties of Stouffer's® frozen dinners, 3 varieties of Lean Cuisine® frozen dinners, and 5 varieties of Weight Watchers® frozen dinners. In how many distinguishable ways can she stock the shelves if
(a) the dinners can be arranged in any order?
(b) dinners from the same company are considered alike and have to be shelved together?
(c) dinners from the same company are considered alike, but do not have to be shelved together?

74. A child has a set of different-shaped plastic objects. There are 2 pyramids, 5 cubes, and 6 spheres. In how many ways can she arrange them in a row
(a) if they are all different colors?
(b) if the same shapes must be grouped?
(c) In how many distinguishable ways can they be arranged in a row if objects of the same shape are also the same color, but need not be grouped?

*Joseph F. Heiser, "Pascal and Gauss meet Little Caesars," *Mathematics Teacher*, 87 (September 1994): 389. In a letter to *Mathematics Teacher*, Heiser argued that the two combinations should be counted as the same, so the child has actually overcounted. In that case, there would be 524,800 possibilities.

✓ Checkpoint Answers

1. **(a)** 720 **(b)** 24
2. **(a)** 24 **(b)** 720 **(c)** 1 **(d)** 30
3. 30 **4.** $8 \cdot 7 \cdot 6 \cdot 5 = 1680$
5. **(a)** 20 **(b)** 504 **(c)** 3 **(d)** 210 **(e)** 132
6. **(a)** 120 **(b)** 12 **(c)** 6
7. **(a)** 15 **(b)** 70 **(c)** 1
8. **(a)** 15 **(b)** 70 **(c)** 1
9. **(a)** 15,504 **(b)** 816 **(c)** 21,699
10. **(a)** 5040 **(b)** 455 **(c)** 28 **(d)** 360
11. **(a)** 5040 **(b)** 840
12. 120,526,770 **13.** 1260

9.3 Applications of Counting

Many of the probability problems involving *dependent* events that were solved with probability trees in Chapter 8 can also be solved by using counting principles—that is, permutations and combinations. Permutations and combinations are especially helpful when the number of choices is large. The use of counting rules to solve probability problems depends on the basic probability principle introduced in Section 8.3 and repeated here.

If event E is a subset of sample space S, then the probability that event E occurs, written $P(E)$, is

$$P(E) = \frac{n(E)}{n(S)}.$$

It is also helpful to keep in mind that, in probability statements,

"and" corresponds to multiplication

and

"or" corresponds to addition.

Example 1 From a potential jury pool with 1 Hispanic, 3 Caucasian, and 2 African-American members, 2 jurors are selected one at a time without replacement. Find the probability that 1 Caucasian and 1 African-American are selected.

Solution In Example 6 of Section 8.5, it was necessary to consider the order in which the jurors were selected. With combinations, it is not necessary: Simply count the number of ways in which 1 Caucasian and 1 African-American juror can be selected. The Caucasian can be selected in ${}_3C_1$ ways, and the African-American juror can be selected in ${}_2C_1$ ways. By the multiplication principle, both results can occur in

$${}_3C_1 \cdot {}_2C_1 = 3 \cdot 2 = 6 \text{ ways,}$$

giving the numerator of the probability fraction. For the denominator, 2 jurors are selected from a total of 6 candidates. This can occur in ${}_6C_2 = 15$ ways. The required probability is

$$P(\text{1 Caucasian and 1 African American}) = \frac{{}_3C_1 \cdot {}_2C_1}{{}_6C_2} = \frac{3 \cdot 2}{15} = \frac{6}{15} = \frac{2}{5} = .40.$$

This result agrees with the answer found earlier.

Example 2 From a baseball team of 15 players, 4 are to be selected to present a list of grievances to the coach.

(a) In how many ways can this be done?

Solution Four players from a group of 15 can be selected in ${}_{15}C_4$ ways. (Use combinations, since the order in which the group of 4 is selected is unimportant.) So,

$${}_{15}C_4 = \frac{15!}{4!\,11!} = \frac{15(14)(13)(12)}{4(3)(2)(1)} = 1365.$$

There are 1365 ways to choose 4 players from 15.

(b) One of the players is Michael Branson. Find the probability that Branson will be among the 4 selected.

Solution The probability that Branson will be selected is the number of ways the chosen group includes him, divided by the total number of ways the group of 4 can be chosen. If Branson must be one of the 4 selected, the problem reduces to finding the number of ways the additional 3 players can be chosen. There are 3 chosen from 14 players; this can be done in

$${}_{14}C_3 = \frac{14!}{3!\,11!} = 364$$

ways. The number of ways 4 players can be selected from 15 is

$$n = {}_{15}C_4 = 1365.$$

The probability that Branson will be one of the 4 chosen is

$$P(\text{Branson is chosen}) = \frac{364}{1365} \approx .267.$$

(c) Find the probability that Branson will not be selected.

Solution The probability that he will not be chosen is $1 - .267 = .733$. ✓1

✓Checkpoint 1

The ski club has 8 women and 7 men. What is the probability that if the club elects 3 officers, all 3 of them will be women?

Answer: See page 567.

Example 3 **Business** A manufacturing company performs a quality-control analysis on the ceramic tile it produces. It produces the tile in batches of 24 pieces. In the quality-control analysis, the company tests 3 pieces of tile per batch. Suppose a batch of 24 tiles has 4 defective tiles.

(a) What is the probability that exactly 1 of the 3 tested tiles is defective?

Solution Let $P(1 \text{ defective})$ represent the probability of there being exactly 1 defective tile among the 3 tested tiles. To find this probability, we need to know how many ways we can select 3 tiles for testing. Since order does not matter, there are ${}_{24}C_3$ ways to choose 3 tiles:

$${}_{24}C_3 = \frac{24!}{21!\,3!} = \frac{24 \cdot 23 \cdot 22}{3 \cdot 2 \cdot 1} = 2024.$$

There are ${}_4C_1$ ways of choosing 1 defective tile from the 4 in the batch. If we choose 1 defective tile, we must then choose 2 good tiles among the 20 good tiles in the batch. We can do this in ${}_{20}C_2$ ways. By the multiplication principle, there are

$${}_4C_1 \cdot {}_{20}C_2 = \frac{4!}{3!\,1!} \cdot \frac{20!}{18!\,2!} = 4 \cdot 190 = 760$$

ways to choose exactly 1 defective tile.

Thus,

$$P(1 \text{ defective}) = \frac{760}{2024} \approx .3755.$$

(b) If at least one of the tiles in a batch is defective, the company will not ship the batch. What is the probability that the batch is not shipped?

Solution The batch will not be shipped if 1, 2, or 3 of the tiles sampled are defective. We already found the probability of there being exactly 1 defective tile in part (a). We now need to find $P(2 \text{ defective})$ and $P(3 \text{ defective})$. To find $P(2 \text{ defective})$, we need to count the number of ways to choose 2 from the 4 defective tiles in the batch and choose 1 from the 20 good tiles in the batch:

$${}_4C_2 \cdot {}_{20}C_1 = \frac{4!}{2!\,2!} \cdot \frac{20!}{19!\,1!} = 6 \cdot 20 = 120.$$

To find $P(3 \text{ defective})$, we need to count the number of ways to choose 3 from the 4 defective tiles in the batch and choose 0 from the 20 good tiles in the batch:

$${}_4C_3 \cdot {}_{20}C_0 = \frac{4!}{1!\,3!} \cdot \frac{20!}{20!\,0!} = 4 \cdot 1 = 4.$$

We now have

$$P(2 \text{ defective}) = \frac{120}{2024} \approx .0593 \text{ and } P(3 \text{ defective}) = \frac{4}{2024} \approx .0020.$$

Thus, the probability of rejecting the batch because 1, 2, or 3 tiles are defective is

$$P(1 \text{ defective}) + P(2 \text{ defective}) + P(3 \text{ defective}) \approx .3755 + .0593 + .0020 = .4368.$$

(c) Use the complement rule to find the probability the batch will be rejected.

Solution We reject the batch if at least 1 of the sampled tiles is defective. The opposite of at least 1 tile being defective is that none are defective. We can find the probability that none of the 3 sampled tiles is defective by choosing 0 from the 4 defective tiles and choosing 3 from the 20 good tiles:

$$_4C_0 \cdot {}_{20}C_3 = 1 \cdot 1140 = 1140.$$

Therefore, the probability that none of the sampled tiles is defective is

$$P(0 \text{ defective}) = \frac{1140}{2024} \approx .5632.$$

Using the complement rule, we have

$$P(\text{at least 1 defective}) \approx 1 - .5632 = .4368,$$

the same answer as in part (b). Using the complement rule can often save time when multiple probabilities need to be calculated for problems involving "at least 1." 2✓

✓Checkpoint 2

A batch of 15 granite slabs is mined, and 4 have defects. If the manager spot-checks 3 slabs at random, what is the probability that at least 1 slab is defective?

Answer: *See page 567.*

Example 4 In a common form of 5-card draw poker, a hand of 5 cards is dealt to each player from a deck of 52 cards. (For a review of a standard deck, see Figure 8.21 on page 490) There is a total of

$$_{52}C_5 = \frac{52!}{5!\,47!} = 2{,}598{,}960$$

such hands possible. Find the probability of being dealt each of the given hands.

(a) Heart-flush hand (5 hearts)

Solution There are 13 hearts in a deck; there are

$$_{13}C_5 = \frac{13!}{5!\,8!} = \frac{13(12)(11)(10)(9)}{5(4)(3)(2)(1)} = 1287$$

different hands containing only hearts. The probability of a heart flush is

$$P(\text{heart flush}) = \frac{1287}{2{,}598{,}960} \approx .000495.$$

(b) A flush of any suit (5 cards, all from 1 suit)

Solution There are 4 suits to a deck, so

$$P(\text{flush}) = 4 \cdot P(\text{heart flush}) = 4(.000495) \approx .00198.$$

(c) A full house of aces and eights (3 aces and 2 eights)

Solution There are $_4C_3$ ways to choose 3 aces from among the 4 in the deck and $_4C_2$ ways to choose 2 eights, so

$$P(3 \text{ aces, } 2 \text{ eights}) = \frac{_4C_3 \cdot {}_4C_2}{_{52}C_5} \approx .00000923.$$

(d) Any full house (3 cards of one value, 2 of another)

Solution There are 13 values in a deck, so there are 13 choices for the first value mentioned, leaving 12 choices for the second value. (Order *is* important here, since a full house of aces and eights, for example, is not the same as a full house of eights and aces.)

$$P(\text{full house}) = \frac{13 \cdot {}_4C_3 \cdot 12 \cdot {}_4C_2}{{}_{52}C_5} \approx .00144. \quad 3$$

Checkpoint 3

Find the probability of being dealt a poker hand (5 cards) with 4 kings.

Answer: See page 567.

Example 5 A cooler contains 8 different kinds of soda, among which 3 cans are Pepsi®, Classic Coke®, and Sprite®. What is the probability, when picking at random, of selecting the 3 cans in the particular order listed in the previous sentence?

Solution Use permutations to find the number of arrangements in the sample, because order matters:

$$n = {}_8P_3 = 8(7)(6) = 336.$$

Since each can is different, there is only 1 way to choose Pepsi, Classic Coke, and Sprite in that order, so the probability is

$$\frac{1}{336} = .0030. \quad 4$$

Checkpoint 4

Martha, Leonard, Calvin, and Sheila will be handling the officer duties of president, vice president, treasurer, and secretary.

(a) If the offices are assigned randomly, what is the probability that Calvin is the president?

(b) If the offices are assigned randomly, what is the probability that Sheila is president, Martha is vice president, Calvin is treasurer, and Leonard is secretary?

Answers: See page 567.

Example 6 Suppose a group of 5 people is in a room. Find the probability that at least 2 of the people have the same birthday.

Solution "Same birthday" refers to the month and the day, not necessarily the same year. Also, ignore leap years, and assume that each day in the year is equally likely as a birthday. First find the probability that *no 2 people* among 5 people have the same birthday. There are 365 different birthdays possible for the first of the 5 people, 364 for the second (so that the people have different birthdays), 363 for the third, and so on. The number of ways the 5 people can have different birthdays is thus the number of permutations of 365 things (days) taken 5 at a time, or

$${}_{365}P_5 = 365 \cdot 364 \cdot 363 \cdot 362 \cdot 361.$$

The number of ways that the 5 people can have the same or different birthdays is

$$365 \cdot 365 \cdot 365 \cdot 365 \cdot 365 = (365)^5.$$

Finally, the *probability* that none of the 5 people have the same birthday is

$$\frac{{}_{365}P_5}{(365)^5} = \frac{365 \cdot 364 \cdot 363 \cdot 362 \cdot 361}{365 \cdot 365 \cdot 365 \cdot 365 \cdot 365} \approx .973.$$

The probability that at least 2 of the 5 people *do* have the same birthday is $1 - .973 = .027$.

Example 6 can be extended for more than 5 people. In general, the probability that no 2 people among n people have the same birthday is

$$\frac{_{365}P_n}{(365)^n}.$$

The probability that at least 2 of the n people *do* have the same birthday is

$$1 - \frac{_{365}P_n}{(365)^n}.$$ 5 ✓

✓ Checkpoint 5

Evaluate $1 - \frac{_{365}P_n}{(365)^n}$ for

(a) $n = 3$;

(b) $n = 6$.

Answers: See page 567.

The following table shows this probability for various values of n:

Number of People, n	Probability That At Least 2 Have the Same Birthday
5	.027
10	.117
15	.253
20	.411
22	.476
23	**.507**
25	.569
30	.706
35	.814
40	.891
50	.970
365	1

✓ Checkpoint 6

Set up (but do not calculate) the probability that at least 2 of the 9 members of the Supreme Court have the same birthday.

Answer: See page 567.

The probability that 2 people among 23 have the same birthday is .507, a little more than half. Many people are surprised at this result; somehow it seems that a larger number of people should be required. 6 ✓

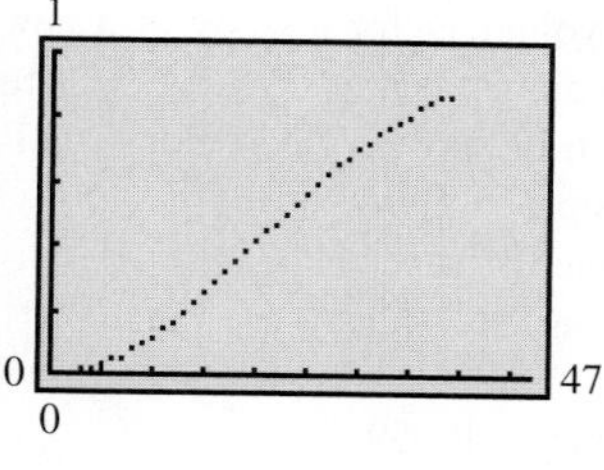

FIGURE 9.11

Using a graphing calculator, we can graph the probability formula in the previous example as a function of n, but the graphing calculator must be set to evaluate the function at integer points. Figure 9.11 was produced on a TI-84+ by letting $Y_1 = 1 - (365 \text{ nPr } X)/365^X$ on the interval $0 \le x \le 47$. (This domain ensures integer values for x.) Notice that the graph does not extend past $x = 39$. This is because $P(365, n)$ and 365^n are too large for the calculator when $n \ge 40$.

9.3 ▸ Exercises

Business *Suppose in Example 3 that the number of defective tiles in a batch of 24 is 5 rather than 4. If 3 tiles are sampled at random, the management would like to know the probability that*

1. exactly 1 of the sampled tiles is defective;

2. the batch is rejected (that is, at least 1 sampled tile is defective).

Business *A shipment of 8 computers contains 3 with defects. Find the probability that with a sample of the given size, drawn from the 8, will not contain a defective computer. (See Example 3.)*

3. 1

4. 2

5. 3

6. 5

A radio station runs a promotion at an auto show with a money box with 10 $100 tickets, 12 $50 tickets, and 20 $25 tickets. The box contains an additional 200 "dummy" tickets with no value. Three tickets are randomly drawn. Find the given probabilities. (See Examples 1 and 2).

7. All $100 tickets
8. All $50 tickets
9. Exactly two $25 tickets and no other money winners
10. One ticket of each money amount
11. No tickets with money
12. At least one money ticket

Two cards are drawn at random from an ordinary deck of 52 cards. (See Example 4.)

13. How many 2-card hands are possible?

Find the probability that the 2-card hand in Exercise 13 contains the given cards.

14. 2 kings
15. No deuces (2's)
16. 2 face cards
17. Different suits
18. At least 1 black card
19. No more than 1 diamond
20. Discuss the relative merits of using probability trees versus combinations to solve probability problems. When would each approach be most appropriate?
21. Several examples in this section used the rule $P(E') = 1 - P(E)$. Explain the advantage (especially in Example 6) of using this rule.

Natural Science *A shipment contains 8 igneous, 7 sedimentary, and 7 metamorphic rocks. If we select 5 rocks at random, find the probability that*

22. all 5 are igneous;
23. 3 are sedimentary;
24. only 1 is metamorphic.
25. In Exercise 59 in Section 9.2, we found the number of ways to pick 6 different numbers from 1 to 99 in a state lottery.
    (a) Assuming that order is unimportant, what is the probability of picking all 6 numbers correctly to win the big prize?
    (b) What is the probability if order matters?
26. In Exercise 25 (a), what is the probability of picking exactly 5 of the 6 numbers correctly?
27. Example 16 in Section 9.2 shows that the probability of winning the Powerball lottery is 1/195,249,054. If Juanita and Michelle each play Powerball on one particular evening, what is the probability that both will select the winning numbers if they make their selections independently of each other?
28. **Business** A cellular phone manufacturer randomly selects 5 phones from every batch of 50 produced. If at least one of the phones is found to be defective, then each phone in the batch is tested individually. Find the probability that the entire batch will need testing if the batch contains
    (a) 8 defective phones;
    (b) 2 defective phones.
29. **Social Science** Of the 15 members of President Barack Obama's first Cabinet, 4 were women. Suppose the president randomly selected 4 advisors from the cabinet for a meeting. Find the probability that the group of 4 would be composed as follows:
    (a) 3 woman and 1 man;
    (b) All men;
    (c) At least one woman.
30. **Business** A car dealership has 8 red, 9 silver, and 5 black cars on the lot. Ten cars are randomly chosen to be displayed in front of the dealership. Find the probability that
    (a) 4 are red and the rest are silver;
    (b) 5 are red and 5 are black;
    (c) 8 are red.

31. **Health** Twenty subjects volunteer for a study of a new cold medicine. Ten of the volunteers are ages 20–39, 8 are ages 40–59, and 2 are age 60 or older. If we select 7 volunteers at random, find the probability that
    (a) all the volunteers selected are ages 20–39;
    (b) 5 of the volunteers are ages 20–39 and 2 are age 60 or older;
    (c) 3 of the volunteers are ages 40–59.

For Exercises 32–34, refer to Example 6 in this section.

32. Set up the probability that at least 2 of the 43 men who have served as president of the United States have had the same birthday.*
33. Set up the probability that at least 2 of the 100 U.S. senators have the same birthday.
34. Set up the probability that at least 2 of the 50 U.S. governors have the same birthday.

*In fact, James Polk and Warren Harding were both born on November 2. Although Barack Obama is the 44th president, the 22nd and 24th presidents were the same man: Grover Cleveland.

*One version of the New York State lottery game Quick Draw has players selecting 4 numbers at random from the numbers 1–80. The state picks 20 winning numbers. If the player's 4 numbers are selected by the state, the player wins $55.**

35. What is the probability of winning?

36. If the state picks 3 of the player's numbers, the player wins $5. What is the probability of winning $5?

37. What is the probability of having none of your 4 numbers selected by the state?

38. During the 1988 college football season, the Big Eight Conference ended the season in a "perfect progression," as shown in the following table:†

Won	Lost	Team
7	0	Nebraska (NU)
6	1	Oklahoma (OU)
5	2	Oklahoma State (OSU)
4	3	Colorado (CU)
3	4	Iowa State (ISU)
2	5	Missouri (MU)
1	6	Kansas (KU)
0	7	Kansas State (KSU)

Someone wondered what the probability of such an outcome might be.

(a) Assuming no ties and assuming that each team had an equally likely probability of winning each game, find the probability of the perfect progression shown in the table.

(b) Under the same assumptions, find a general expression for the probability of a perfect progression in an n-team league.

39. Use a computer or a graphing calculator and the Monte Carlo method with $n = 50$ to estimate the probabilities of the given hands at poker. (See the directions for Exercises 40–43 on page 504.) Assume that aces are either high or low. Since each hand has 5 cards, you will need $50 \cdot 5 = 250$ random numbers to "look at" 50 hands. Compare these experimental results with the theoretical results.

(a) A pair of aces

(b) Any two cards of the same value

(c) Three of a kind

40. Use a computer or a graphing calculator and the Monte Carlo method with $n = 20$ to estimate the probabilities of the given 13-card bridge hands. Since each hand has 13 cards, you will need $20 \cdot 13 = 260$ random numbers to "look at" 20 hands.

(a) No aces

(b) 2 kings and 2 aces

(c) No cards of any suit—that is, only 3 suits represented

*www.nylottery.org/index.php.

†From Richard Madsen, "On the Probability of a Perfect Progression." *American Statistics* 45, no. 3 (August 1991): 214.

✓ Checkpoint Answers

1. .123
2. About .637
3. .00001847
4. (a) 1/4 (b) 1/24
5. (a) .008 (b) .040
6. $1 - {}_{365}P_9/365^9$

9.4 ▸ Binomial Probability

In Section 9.1, we learned about probability distributions where we listed each outcome and its associated probability. After learning in Sections 9.2 and 9.3 how to count the number of possible outcomes, we are now ready to understand a special probability distribution known as the *binomial distribution.* This distribution occurs when the same experiment is repeated many times and each repetition is independent of previous ones. One outcome is designated a success and any other outcome is considered a failure. For example, you might want to find the probability of rolling 8 twos in 12 rolls of a die (rolling two is a success; rolling anything else is a failure). The individual trials (rolling the die once) are called **Bernoulli trials**, or **Bernoulli processes**, after the Swiss mathematician Jakob Bernoulli (1654–1705).

If the probability of a success in a single trial is p, then the probability of failure is $1 - p$ (by the complement rule). When Bernoulli trials are repeated a fixed number of times, the resulting distribution of outcomes is called a **binomial distribution**, or a **binomial experiment**. A binomial experiment must satisfy the following conditions.

Binomial Experiment

1. The same experiment is repeated a fixed number of times.
2. There are only two possible outcomes: success and failure.
3. The probability of success for each trial is constant.
4. The repeated trials are independent.

The basic characteristics of binomial experiments are illustrated by a recent poll of small businesses (250 employees or fewer) conducted by the National Federation of Independent Businesses. The poll found that 36% of small businesses pay for the health insurance of all or almost all full-time employees.* We use Y to denote the event that a small business does this, and N to denote the event that it does not. If we sample 5 small businesses at random and use .36 as the probability for Y, we will generate a binomial experiment, since all of the requirements are satisfied:

The sampling is repeated a fixed number of times (5);

there are only two outcomes of interest (Y or N);

the probability of success is constant ($p = .36$);

"at random" guarantees that the trials are independent.

To calculate the probability that all 5 randomly chosen businesses pay for health insurance, we use the product rule for independent events (Section 8.5) and $P(Y) = .36$ to obtain

$$P(YYYYY) = P(Y) \cdot P(Y) \cdot P(Y) \cdot P(Y) \cdot P(Y) = (.36)^5 \approx .006.$$

Determining the probability that 4 out of 5 businesses chose to pay for health insurance is slightly more complicated. The business that does not pay could be the first, second, third, fourth, or fifth business surveyed. So we have the following possible outcomes:

N Y Y Y Y
Y N Y Y Y
Y Y N Y Y
Y Y Y N Y
Y Y Y Y N

So the total number of ways in which 4 successes (and 1 failure) can occur is 5, which is the number ${}_5C_4$. The probability of each of these 5 outcomes is

$$P(Y) \cdot P(Y) \cdot P(Y) \cdot P(Y) \cdot P(N) = (.36)^4(1 - .36)^1 = (.36)^4(.64).$$

Since the 5 outcomes where there are 4 Ys and one N represent disjoint events, we multiply by ${}_5C_4 = 5$:

$$P(4\ Y\text{s out of 5 trials}) = {}_5C_4(.36)^4(.64)^{5-4} = 5(.36)^4(.64)^1 \approx .054.$$

---

*www.nfib.com/.

The probability of obtaining exactly 3 *Y*s and 2 *N*s can be computed in a similar way. The probability of any one way of achieving 3 *Y*s and 2 *N*s will be

$$(.36)^3(.64)^2.$$

Again, the desired outcome can occur in more than one way. Using combinations, we find that the number of ways in which 3 *Y*s and 2 *N*s can occur is ${}_5C_3 = 10$. So, we have

$$P(3\ Ys\text{ out of 5 trials}) = {}_5C_3(.36)^3(.64)^{5-3} = 10(.36)^3(.64)^2 \approx .191.$$ ✓1

✓ Checkpoint 1

Find the probability of obtaining

(a) exactly 2 businesses that pay for health insurance;

(b) exactly 1 business that pays for health insurance;

(c) exactly no business that pays for health insurance.

Answers: See page 574.

With the probabilities just generated and the answers to Checkpoint 1, we can write the probability distribution for the number of small businesses that pay for health insurance for all or almost all their full-time employees when 5 businesses are selected at random:

x	0	1	2	3	4	5
$P(x)$	.107	.302	.340	.191	.054	.006

When the outcomes and their associated probabilities are written in this form, it is very easy to calculate answers to questions such as, What is the probability that 3 or more businesses pay for the health insurance of their full-time employees? We see from the table that

$$P(3\text{ or more }Ys) = .191 + .054 + .006 = .251.$$

Similarly, the probability of one or fewer businesses paying for health insurance is

$$P(1\text{ or fewer }Ys) = .302 + .107 = .409.$$

The example illustrates the following fact.

TECHNOLOGY TIP On the TI-84+ calculator, use "binompdf(*n*,*p*,*x*)" in the DISTR menu to compute the probability of exactly *x* successes in *n* trials (where *p* is the probability of success in a single trial). Use "binomcdf(*n*,*p*,*x*)" to compute the probability of at most *x* successes in *n* trials. Figure 9.12 shows the probability of exactly 3 successes in 5 trials and the probability of at most 3 successes in 5 trials, with the probability of success set at .36 for each case.

```
binompdf(5,.36,3
)
          .191102976
binomcdf(5,.36,3
)
        .9402056704
```

FIGURE 9.12

Binomial Probability

If p is the probability of success in a single trial of a binomial experiment, the probability of x successes and $n - x$ failures in n independent repeated trials of the experiment is

$${}_nC_x p^x (1 - p)^{n-x}.$$

Example 1 **Business** On March 26, 2009, a report from a study conducted by the Families and Work Institute found that 67% of workers age 29 or younger desire to have a job with more responsibility.* Suppose a random sample of 6 workers age 29 or younger is chosen. Find the probability of the given scenarios.

*www.familiesandwork.org.

(a) Exactly 4 of the 6 workers desire a job with more responsibility.

Solution We can think of the 6 workers sampled as independent trials, and a success occurs if a worker desires a job with more responsibility. This is a binomial experiment with $p = .67$, $n = 6$, and $x = 4$. By the binomial probability rule,

$$\begin{aligned} P(\text{exactly } 4) &= {}_6C_4(.67)^4(.33)^{6-4} \\ &= 15(.67)^4(.33)^2 \\ &\approx .329. \end{aligned}$$

(b) None of the 6 workers desires a job with more responsibility.

Solution Let $x = 0$. Then we have

$$\begin{aligned} P(\text{exactly } 0) &= {}_6C_0(.67)^0(.33)^{6-0} \\ &= 1(.67)^0(.33)^6 \\ &\approx .001. \end{aligned}$$

✓2

✓Checkpoint 2

According to the study in Example 1, 35% of workers age 29 or younger agreed with the statement, "It is better for all involved if the man earns the money and the woman takes care of the home and children." If 4 workers are selected at random, find the probability that

(a) 1 of 4 agreed with the statement;

(b) 3 out of 4 agreed with the statement.

Answers: See page 574.

Example 2 Suppose a family has 3 children.

(a) Find the probability distribution for the number of girls.

Solution Let $x =$ the number of girls in three births. According to the binomial probability rule, the probability of exactly one girl being born is

$$P(x = 1) = {}_3C_1\left(\frac{1}{2}\right)^1\left(\frac{1}{2}\right)^2 = 3\left(\frac{1}{2}\right)^3 = \frac{3}{8}.$$

The other probabilities in this distribution are found similarly, as shown in the following table:

x	0	1	2	3
$P(x)$	${}_3C_0\left(\frac{1}{2}\right)^0\left(\frac{1}{2}\right)^3 = \frac{1}{8}$	${}_3C_1\left(\frac{1}{2}\right)^1\left(\frac{1}{2}\right)^2 = \frac{3}{8}$	${}_3C_2\left(\frac{1}{2}\right)^2\left(\frac{1}{2}\right)^1 = \frac{3}{8}$	${}_3C_3\left(\frac{1}{2}\right)^3\left(\frac{1}{2}\right)^0 = \frac{1}{8}$

(b) Find the expected number of girls in a 3-child family.

Solution For a binomial distribution, we can use the following method (which is presented here with a "plausibility argument," but not a full proof): Because 50% of births are girls, it is reasonable to expect that 50% of a sample of children will be girls. Since 50% of 3 is $3(.50) = 1.5$, we conclude that the expected number of girls is 1.5. ✓3

✓Checkpoint 3

Find the probability of getting 2 fours in 8 tosses of a die.

Answer: See page 574.

The expected value in Example 2(b) was the product of the number of births and the probability of a single birth being a girl—that is, the product of the number of trials and the probability of success in a single trial. The same conclusion holds in the general case.

Expected Value for a Binomial Distribution

When an experiment meets the four conditions of a binomial experiment with n fixed trials and constant probability of success p, the expected value is

$$E(x) = np.$$

Example 3 **Health** Data from the 2005–2006 National Health and Nutrition Examination Survey indicate that 35% of adults age 18–85 answered "Yes" to the question, "Over the past 30 days, did you do any vigorous activities for at least 10 minutes that caused heavy sweating, or large increases in breathing or heart rate?"* If we select 15 adults at random, find the following probabilities,

(a) The probability that exactly 5 engaged in vigorous activity in the last 30 days

Solution The experiment is repeated 15 times, with having engaged in vigorous activity being considered a success. The probability of success is .35. Since the selection is done at random, the trials are considered independent. Thus, we have a binomial experiment, and

$$P(x = 5) = {}_{15}C_5(.35)^5(.65)^{10} \approx .212.$$

(b) The probability that at most 3 engaged in vigorous activity in the last 30 days

Solution "At most 3" means 0, 1, 2, or 3 successes. We must find the probability for each case and then use the addition rule for disjoint events:

$$P(x = 0) = {}_{15}C_0(.35)^0(.65)^{15} \approx .002;$$
$$P(x = 1) = {}_{15}C_1(.35)^1(.65)^{14} \approx .013;$$
$$P(x = 2) = {}_{15}C_2(.35)^2(.65)^{13} \approx .048;$$
$$P(x = 3) = {}_{15}C_3(.35)^3(.65)^{12} \approx .111.$$

Thus,

$$P(\text{at most } 3) = .002 + .013 + .048 + .111 = .174.$$

(c) The expected number of adults who have engaged in vigorous activity in the last 30 days.

Solution Because this is a binomial experiment, we can use the formula $E(x) = np = 15(.35) = 5.25$. In repeated samples of 15 adults, the average number who engaged in vigorous activity is 5.25. ✓4

✓Checkpoint 4

Suppose that in Example 3, 55% of adults said "Yes" to having engaged in moderate physical activity within the last 30 days. If 10 adults are selected at random, find the probability that

(a) exactly 2 said "Yes";

(b) at most 2 said "Yes."

(c) What is the expected number who say "Yes?"

Answers: See page 574.

Example 4 **Social Science** According to a study conducted by the American Veterinary Association in 2006, 37% of American households had a pet dog. If a sample of 10 random households is conducted, what is the probability that at least 1 household will have a dog?

*Based on data available at www.cdc.gov/nchs/nhanes.htm.

Solution We can treat this problem as a binomial experiment, with $n = 10$, $p = .37$, and x representing the number of households in the sample that have a dog. "At least 1 of 10" means 1 or 2 or 3, etc., up to 10. It will be simpler here to find the probability that none of the 10 selected households has a dog—that is, $P(x = 0)$—and then find the probability that at least 1 of the households has a dog, which is the value $1 - P(x = 0)$:

$$P(x = 0) = {}_{10}C_0(.37)^0(.63)^{10} \approx .0098$$

$$P(x \geq 1) = 1 - P(x = 0) \approx 1 - .0098 = .9902.$$ 5 ✓

✓Checkpoint 5

In Example 4, find the probability that

(a) at least 3 households have a dog;

(b) at most 5 of the households have a dog.

(c) What is the expected value?

Answers: See page 574.

Example 5 If each member of a 9-person jury acts independently of the other members and makes the correct determination of guilt or innocence with probability .65, find the probability that the majority of the jurors will reach a correct verdict.*

Solution Since the jurors in this particular situation act independently, we can treat the problem as a binomial experiment. Thus, the probability that the majority of the jurors will reach the correct verdict is given by

$$\begin{aligned} P(\text{at least } 5) &= {}_9C_5(.65)^5(.35)^4 + {}_9C_6(.65)^6(.35)^3 + {}_9C_7(.65)^7(.35)^2 \\ &\quad + {}_9C_8(.65)^8(.35)^1 + {}_9C_9(.65)^9 \\ &\approx .2194 + .2716 + .2162 + .1004 + .0207 \\ &= .8283. \end{aligned}$$

TECHNOLOGY TIP Some spreadsheets provide binomial probabilities. In Microsoft Excel, for example, the command "=BINOMDIST (5, 9, .65, 0)" gives .21939, which is the probability for $x = 5$ in Example 5. Alternatively, the command "= BINOMDIST (4, 9, .65, 1)" gives .17172 as the probability that 4 or fewer jurors will make the correct decision. Subtract .17172 from 1 to get .82828 as the probability that the majority of the jurors will make the correct decision. This value agrees with the value found in Example 5.

*Bernard Grofman, "A Preliminary Model of Jury Decision Making as a Function of Jury Size, Effective Jury Decision Rule, and Mean Juror Judgmental Competence," *Frontiers in Economics* (1979), pp. 98–110.

9.4 ▶ Exercises

In Exercises 1–39, see Examples 1–5.

Social Science *In 2008, the percentage of children under 18 years of age who lived with both parents was approximately 70%.* Find the probabilities that the given number of persons selected at random from 10 children under 18 years of age in 2008 lived with both parents.*

1. Exactly 6 **2.** Exactly 5

3. None **4.** All

5. At least 1 **6.** At most 4

*U.S Census Bureau, *America's Families and Living Arrangments: 2008.*

Social Science *The study in Exercise 1 also found that approximately 19% of children under 18 years of age lived with their mother only. Find the probabilities that the given number of persons selected at random from 10 children under 18 years of age in 2008 lived with their mother only.*

7. Exactly 2 **8.** Exactly 1

9. None **10.** All

11. At least 1 **12.** At most 2

A coin is tossed 5 times. Find the probability of getting

13. all heads; **14.** exactly 3 heads;

15. no more than 3 heads; **16.** at least 3 heads.

17. How do you identify a probability problem that involves a binomial experiment?

18. Why do combinations occur in the binomial probability formula?

Business *Researchers at the University of Virginia estimated that 34% of the mortgage foreclosures that occurred in the United States in 2008 were in the state of California.* If 15 foreclosed mortgages were selected at random, find the probability that*

19. exactly 3 were from California;

20. none was from California;

21. at most 2 were from California;

22. exactly 10 were not from California.

23. If 200 foreclosed mortgages were selected randomly, what would the expected number of California mortgages be?

Business *The study from the National Federation of Independent Businesses (page 568) also found that approximately 56% of small businesses choose the cost of health care as the number-one challenge facing their business. If we select 8 small businesses at random, find the probability that the given number of businesses chose the cost of health care as the number-one challenge.*

24. all 8 businesses

25. all but 1 business

26. at most 2 businesses

27. at most 7 businesses

28. What is the expected number of businesses that chose the cost of health care as the number-one challenge?

Natural Science *The probability that a birth will result in twins is .027.† Assuming independence (perhaps not a valid assumption), what are the probabilities that, out of 100 births in a hospital, there will be the given numbers of sets of twins?*

29. Exactly 2 sets of twins

30. At most 2 sets of twins

Social Science *According to the Web site Answers.com, 10–13% of Americans are left handed. Assume that the percentage is 11%. If we select 9 people at random, find the probability that the number who are left handed is*

31. exactly 2;

32. at least 2;

33. none;

34. at most 3.

35. In a class of 35 students, how many left-handed students should the instructor expect?

Social Science *Respondents to the 2006 General Social Survey (GSS) indicated that approximately 7.3% of Americans attend church services more than once a week.‡ If 16 Americans are chosen at random, find the given probabilities.*

36. Exactly 2 attend church services more than once a week

37. At most 3 attend church services more than once a week

*www.virginia.edu/uvatoday/newsRelease.php?id=7838.

†*The World Almanac and Book of Facts*, 2001, p. 873.

‡Data available at www.norc.org/GSS+Website

Health *Data from the 2005–2006 National Health and Nutrition Examination Survey (NHANES) estimates that 15.8% of American males are six feet tall or taller.**

38. If we select 9 males at random, what is the probability that at least 2 are six feet tall or taller?

39. If we select 5 males at random, will the probability that at least 2 are six feet tall or taller be higher or lower than the probability found for Exercise 38?

40. If we select 500 males at random, what is the expected number who will be six feet tall or taller?

41. **Social Science** In the "Numbers" section of *Time* magazine, it was reported that 15.2% of low-birth-weight babies graduate from high school by age 19. It was also reported that 57.5% of their normal-birth-weight siblings graduated from high school by 19.†
   (a) If 40 low-birth-weight babies are tracked through high school, what is the probability that fewer than 15 will graduate from high school by age 19?
   (b) What are some of the factors that may contribute to the wide difference in high school success between these siblings? Do you believe that low birth weight is the primary cause of the difference? What other information do you need to better answer these questions?

42. **Natural Science** In a placebo-controlled trial of Adderall XR®, a medication for attention deficit and hyperactivity disorder (ADHD), 22% of users of the drug reported loss of appetite. Only 2% of the patients taking the placebo reported loss of appetite.‡
   (a) If 100 patients who are taking Adderall XR are selected at random, what is the probability that 15 or more will experience loss of appetite?
   (b) If 100 patients who are taking the placebo are selected at random, what is the probability that 15 or more will experience loss of appetite?
   (c) Do you believe Adderall XR causes a loss of appetite? Why or why not?

43. **Natural Science** DNA evidence has become an integral part of many court cases. When DNA is extracted from cells and body fluids, genetic information is represented by bands of information, which look similar to a bar code at a grocery store. It is generally accepted that, in unrelated people, the probability of a particular band matching is 1 in 4.§
   (a) If 5 bands are compared in unrelated people, what is the probability that all 5 of the bands match? (Express your answer in terms of "1 chance in ?".)

*Based on data available at www.cdc.gov/nchs/nhanes.htm.

†"Numbers," *Time*, July 17, 2000, p. 21.

‡Advertisement in *Newsweek*, June 13, 2005, for Adderall XR®, marketed by Shire US, Inc.

§"Genetic Fingerprinting Worksheet." Centre for Innovation in Mathematics Teaching, http://www.cimt.plymouth.ac.uk

(b) If 20 bands are compared in unrelated people, what is the probability that all 20 of the bands match? (Express your answer in terms of "1 chance in ?".)

(c) If 20 bands are compared in unrelated people, what is the probability that 16 or more bands match? (Express your answer in terms of "1 chance in ?".)

(d) If you were deciding on a child's paternity and there were 16 matches out of 20 bands compared, would you believe that the person being tested was the father? Explain.

44. **Social Science** In England, a woman was found guilty of smothering her two infant children. Much of the Crown's case against her was based on the testimony from a pediatrician who indicated that the chances of 2 crib deaths occurring in both siblings was only about 1 in 73 million. This number was calculated by assuming that the probability of a single crib death is 1 in 8500 and the probability of 2 crib deaths is 1 in 8500^2 (i.e., binomial).* Why is the use of binomial probability not correct in this situation?

✓ Checkpoint Answers

1. (a) About .340 (b) About .302 (c) About .107
2. (a) About .384 (b) About .111
3. About .2605
4. (a) About .0229 (b) About .0274 (c) 5.5
5. (a) About .7794 (b) About .8795 (c) 3.7

*Stephen J. Watkins, "Conviction by Mathematical Error?" *British Medical Journal* 320, no. 7226 (January 1, 2000):2–3.

9.5 Markov Chains

In Section 8.5, we touched on **stochastic processes**—mathematical models that evolve over time in a probabilistic manner. In the current section, we study a special kind of stochastic process called a **Markov chain**, in which the outcome of an experiment depends only on the outcome of the previous experiment. In other words, the next state of the system depends only on the present state, not on preceding states. Such experiments are common enough in applications to make their study worthwhile. Markov chains are named after the Russian mathematician A. A. Markov (1856–1922), who started the theory of stochastic processes. To see how Markov chains work, we look at an example.

Example 1 **Business** A small town has only two dry cleaners: Johnson and NorthClean. Johnson's manager hopes to increase the firm's market share by an extensive advertising campaign. After the campaign, a market research firm finds that there is a probability of .8 that a Johnson customer will bring his next batch of dirty items to Johnson and a .35 chance that a NorthClean customer will switch to Johnson for his next batch. Assume that the probability that a customer comes to a given cleaner depends only on where the last load of clothes was taken. If there is a .8 chance that a Johnson customer will return to Johnson, then there must be a $1 - .8 = .2$ chance that the customer will switch to NorthClean. In the same way, there is a $1 - .35 = .65$ chance that a NorthClean customer will return to NorthClean. If an individual bringing a load to Johnson is said to be in state 1 and an individual bringing a load to NorthClean is said to be in state 2, then these probabilities of change from one cleaner to the other are as shown in the following table:

		Second Load	
	State	1	2
First Load	1	.8	.2
	2	.35	.65

The information from the table can be written in other forms. Figure 9.13 is a **transition diagram** that shows the two states and the probabilities of going from one to another.

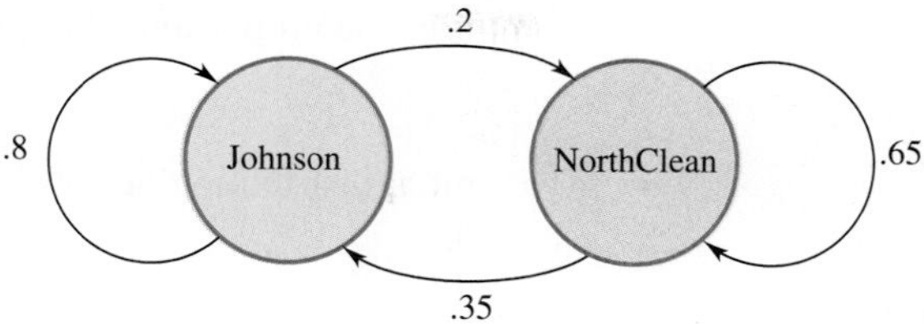

FIGURE 9.13

In a **transition matrix**, the states are indicated at the side and top, as follows:

		Second Load	
		Johnson	NorthClean
First Load	Johnson	.8	.2
	NorthClean	.35	.65

✓ 1

✓Checkpoint 1

You are given the transition matrix

		State	
		1	2
State	1	.3	.7
	2	.1	.9

(a) What is the probability of changing from state 1 to state 2?

(b) What does the number .1 represent?

(c) Draw a transition diagram for this information.

Answers: See page 586.

A **transition matrix** has the following features:

1. It is square, since all possible states must be used both as rows and as columns.
2. All entries are between 0 and 1, inclusive, because all entries represent probabilities.
3. The sum of the entries in any row must be 1, because the numbers in the row give the probability of changing from the state at the left to one of the states indicated across the top.

Example 2 **Business** Suppose that when the new promotional campaign began, Johnson had 40% of the market and NorthClean had 60%. Use the probability tree in Figure 9.14 to find how these proportions would change after another week of advertising.

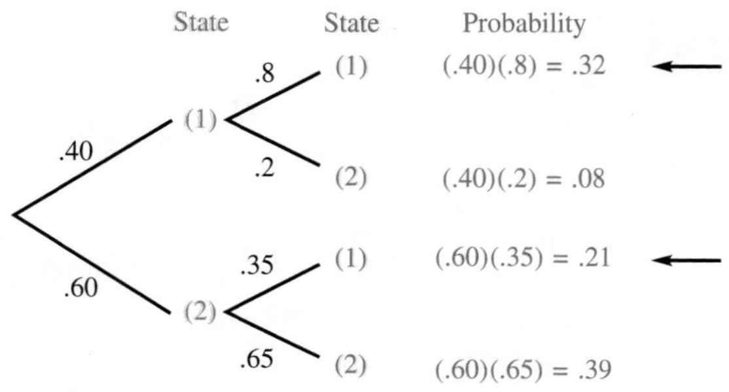

FIGURE 9.14

Solution Add the numbers indicated with arrows to find the proportion of people taking their cleaning to Johnson after one week:

$$.32 + .21 = .53$$

Similarly, the proportion taking their cleaning to NorthClean is

$$.08 + .39 = .47.$$

The initial distribution of 40% and 60% becomes 53% and 47%, respectively, after 1 week.

These distributions can be written as the *probability vectors*

$$[.40 \quad .60] \text{ and } [.53 \quad .47].$$

A **probability vector** is a one-row matrix, with nonnegative entries, in which the sum of the entries is equal to 1.

The results from the probability tree of Figure 9.14 are exactly the same as the result of multiplying the initial probability vector by the transition matrix (multiplication of matrices was discussed in Section 6.4):

$$[.4 \quad .6]\begin{bmatrix} .8 & .2 \\ .35 & .65 \end{bmatrix} = [.53 \quad .47].$$

If v denotes the original probability vector $[.4 \quad .6]$ and P denotes the transition matrix, then the market share vector after one week is $vP = [.53 \quad .47]$. To find the market share vector after two weeks, multiply the vector $vP = [.53 \quad .47]$ by P; this amounts to finding vP^2. 2✓

✓Checkpoint 2

Find the product

$$[.53 \ .47]\begin{bmatrix} .8 & .2 \\ .35 & .65 \end{bmatrix}.$$

Answer: See page 586.

Checkpoint 2 shows that after 2 weeks, the market share vector is $vP^2 = [.59 \quad .41]$. To get the market share vector after three weeks, multiply this vector by P; that is, find vP^3. Do not use the rounded answer from Checkpoint 2. 3✓

✓Checkpoint 3

Find each cleaner's market share after three weeks.

Answer: See page 586.

Continuing this process gives each cleaner's share of the market after additional weeks:

Weeks after Start	Johnson	NorthClean	
0	.4	.6	v
1	.53	.47	vP^1
2	.59	.41	vP^2
3	.61	.39	vP^3
4	.63	.37	vP^4
12	.64	.36	vP^{12}
13	.64	.36	vP^{13}

The results seem to approach the probability vector $[.64 \quad .36]$.

What happens if the initial probability vector is different from $[.4 \quad .6]$? Suppose $[.75 \quad .25]$ is used; then the same powers of the transition matrix as before give the following results:

Week after Start	Johnson	NorthClean	
0	.75	.25	v
1	.69	.31	vP^1
2	.66	.34	vP^2
3	.65	.35	vP^3
4	.64	.36	vP^4
5	.64	.36	vP^5
6	.64	.36	vP^6

The results again seem to be approaching the numbers in the probability vector [.64 .36], the same numbers approached with the initial probability vector [.4 .6]. In either case, the long-range trend is for a market share of about 64% for Johnson and 36% for NorthClean. The example suggests that this long-range trend does not depend on the initial distribution of market shares. This means that if the initial market share for Johnson was less than 64%, the advertising campaign has paid off in terms of a greater long-range market share. If the initial share was more than 64%, the campaign did not pay off.

REGULAR TRANSITION MATRICES

One of the many applications of Markov chains is in finding long-range predictions. It is not possible to make long-range predictions with all transition matrices, but for a large set of transition matrices, long-range predictions *are* possible. Such predictions are always possible with **regular transition matrices**. A transition matrix is **regular** if some power of the matrix contains all positive entries. A Markov chain is a **regular Markov chain** if its transition matrix is regular.

Example 3 Decide whether the given transition matrices are regular.

(a) $A = \begin{bmatrix} .3 & .1 & .6 \\ .0 & .2 & .8 \\ .3 & .7 & 0 \end{bmatrix}.$

Solution Square A:

$$A^2 = \begin{bmatrix} .27 & .47 & .26 \\ .24 & .60 & .16 \\ .09 & .17 & .74 \end{bmatrix}.$$

Since all entries in A^2 are positive, matrix A is regular.

(b) $B = \begin{bmatrix} .3 & 0 & .7 \\ 0 & 1 & 0 \\ 0 & 0 & 1 \end{bmatrix}.$

Solution Find various powers of B:

$$B^2 = \begin{bmatrix} .09 & 0 & .91 \\ 0 & 1 & 0 \\ 0 & 0 & 1 \end{bmatrix}; \quad B^3 = \begin{bmatrix} .027 & 0 & .973 \\ 0 & 1 & 0 \\ 0 & 0 & 1 \end{bmatrix}; \quad B^4 = \begin{bmatrix} .0081 & 0 & .9919 \\ 0 & 1 & 0 \\ 0 & 0 & 1 \end{bmatrix}.$$

Notice that all of the powers of B shown here have zeros in the same locations. Thus, further powers of B will still give the same zero entries, so that no power of matrix B contains all positive entries. For this reason, B is not regular. 

✓Checkpoint 4

Decide whether the given transition matrices are regular.

(a) $\begin{bmatrix} 0 & 1 \\ 1 & 0 \end{bmatrix}$

(b) $\begin{bmatrix} .45 & .55 \\ 1 & 0 \end{bmatrix}$

Answers: See page 586.

NOTE If a transition matrix P has some zero entries, and P^2 does as well, you may wonder how far you must compute P^n to be certain that the matrix is not regular. The answer is that if zeros occur in identical places in both P^n and P^{n+1} for any n, then they will appear in those places for all higher powers of P, so P is not regular.

Suppose that v is any probability vector. It can be shown that, for a regular Markov chain with a transition matrix P, there exists a single vector V that does not depend on v, such that $v \cdot P^n$ gets closer and closer to V as n gets larger and larger.

Equilibrium Vector of a Markov Chain

If a Markov chain with transition matrix P is regular, then there is a unique vector V such that, for any probability vector v and for large values of n,

$$v \cdot P^n \approx V.$$

Vector V is called the equilibrium vector, or the fixed vector, of the Markov chain.

In the example with Johnson Cleaners, the equilibrium vector V is approximately $[.64 \quad .36]$. Vector V can be determined by finding P^n for larger and larger values of n and then looking for a vector that the product $v \cdot P^n$ approaches. Such a strategy can be very tedious, however, and is prone to error. To find a better way, start with the fact that, for a large value of n,

$$v \cdot P^n \approx V,$$

as mentioned in the preceding box. We can multiply both sides of this result by P, $v \cdot P^n \cdot P \approx V \cdot P$, so that

$$v \cdot P^n \cdot P = v \cdot P^{n+1} \approx VP.$$

Since $v \cdot P^n \approx V$ for large values of n, it is also true that $v \cdot P^{n+1} \approx V$ for large values of n. (The product $v \cdot P^n$ approaches V, so that $v \cdot P^{n+1}$ must also approach V.) Thus, $v \cdot P^{n+1} \approx V$ and $v \cdot P^{n+1} \approx VP$, which suggests that

$$VP = V.$$

If a Markov chain with transition matrix P is regular, then the equilibrium vector V satisties

$$VP = V.$$

The equilibrium vector V can be found by solving a system of linear equations, as shown in the remaining examples.

Example 4 Find the long-range trend for the Markov chain in Examples 1 and 2, with transition matrix

$$P = \begin{bmatrix} .8 & .2 \\ .35 & .65 \end{bmatrix}.$$

Solution This matrix is regular, since all entries are positive. Let P represent this transition matrix and let V be the probability vector $[v_1 \quad v_2]$. We want to find V such that

$$VP = V,$$

or

$$[v_1 \quad v_2]\begin{bmatrix} .8 & .2 \\ .35 & .65 \end{bmatrix} = [v_1 \quad v_2].$$

Multiply on the left to get

$$[.8v_1 + .35v_2 \quad .2v_1 + .65v_2] = [v_1 \quad v_2].$$

Set corresponding entries from the two matrices equal to obtain

$$.8v_1 + .35v_2 = v_1; \quad .2v_1 + .65v_2 = v_2.$$

Simplify each of these equations:

$$-.2v_1 + .35v_2 = 0; \quad .2v_1 - .35v_2 = 0$$

These last two equations are really the same. (The equations in the system obtained from $VP = V$ are always dependent.) To find the values of v_1 and v_2, recall that $V = [v_1 \quad v_2]$ is a probability vector, so that

$$v_1 + v_2 = 1.$$

Find v_1 and v_2 by solving the system

$$\begin{aligned} -.2v_1 + .35v_2 &= 0 \\ v_1 + \quad v_2 &= 1. \end{aligned}$$

We can rewrite the second equation as $v_1 = 1 - v_2$. Now substitute for v_1 in the first equation:

$$-.2(1 - v_2) + .35v_2 = 0.$$

Solving for v_2 yields

$$\begin{aligned} -.2 + .2v_2 + .35v_2 &= 0 \\ -.2 + .55v_2 &= 0 \\ .55v_2 &= .2 \\ v_2 &= .364. \end{aligned}$$

Since $v_2 = .364$ and $v_1 = 1 - v_2$, it follows that $v_1 = 1 - .364 = .636$, and the equilibrium vector is $[.636 \quad .364] \approx [.64 \quad .36]$.

Example 5 **Business** The probability that a complex assembly line works correctly depends on whether the line worked correctly the last time it was used. The various probabilities are as given in the following transition matrix:

	Works Properly Now	Does Not
Worked Properly Before	.79	.21
Did Not	.68	.32

Find the long-range probability that the assembly line will work properly.

Solution Begin by finding the equilibrium vector $[v_1 \quad v_2]$, where

$$[v_1 \quad v_2]\begin{bmatrix} .79 & .21 \\ .68 & .32 \end{bmatrix} = [v_1 \quad v_2].$$

Multiplying on the left and setting corresponding entries equal gives the equations

$$.79v_1 + .68v_2 = v_1 \quad \text{and} \quad .21v_1 + .32v_2 = v_2,$$

or

$$-.21v_1 + .68v_2 = 0 \quad \text{and} \quad .21v_1 - .68v_2 = 0.$$

Substitute $v_1 = 1 - v_2$ in the first of these equations to get

$$\begin{aligned} -.21(1 - v_2) + .68v_2 &= 0 \\ -.21 + .21v_2 + .68v_2 &= 0 \\ -.21 + .89v_2 &= 0 \\ .89v_2 &= .21 \\ v_2 &= \frac{.21}{.89} = \frac{21}{89}, \end{aligned}$$

and $v_1 = 1 - \frac{21}{89} = \frac{68}{89}$. The equilibrium vector is $[68/89 \quad 21/89]$. In the long run, the company can expect the assembly line to run properly $\frac{68}{89} \approx 76\%$ of the time. 5

Checkpoint 5

In Example 5, suppose the company modifies the line so that the transition matrix becomes

$$\begin{bmatrix} .85 & .15 \\ .75 & .25 \end{bmatrix}$$

Find the long-range probability that the assembly line will work properly.

Answer: See page 586.

Example 6 Find the equilibrium vector for the transition matrix

$$K = \begin{bmatrix} .2 & .6 & .2 \\ .1 & .1 & .8 \\ .3 & .3 & .4 \end{bmatrix}.$$

Solution Matrix K has all positive entries and thus is regular. For this reason, an equilibrium vector V must exist such that $VK = V$. Let $V = [v_1 \quad v_2 \quad v_3]$. Then

$$[v_1 \quad v_2 \quad v_3]\begin{bmatrix} .2 & .6 & .2 \\ .1 & .1 & .8 \\ .3 & .3 & .4 \end{bmatrix} = [v_1 \quad v_2 \quad v_3].$$

Use matrix multiplication on the left:

$$[.2v_1 + .1v_2 + .3v_3 \quad .6v_1 + .1v_2 + .3v_3 \quad .2v_1 + .8v_2 + .4v_3] = [v_1 \quad v_2 \quad v_3].$$

Set corresponding entries equal:

$$\begin{aligned} .2v_1 + .1v_2 + .3v_3 &= v_1 \\ .6v_1 + .1v_2 + .3v_3 &= v_2 \\ .2v_1 + .8v_2 + .4v_3 &= v_3. \end{aligned}$$

Simplify these equations:

$$\begin{aligned} -.8v_1 + .1v_2 + .3v_3 &= 0 \\ 6v_1 - .9v_2 + .3v_3 &= 0 \\ .2v_1 + .8v_2 - .6v_3 &= 0. \end{aligned}$$

Since V is a probability vector,

$$v_1 + v_2 + v_3 = 1.$$

This gives a system of four equations in three unknowns:

$$\begin{aligned} v_1 + \quad v_2 + \quad v_3 &= 1. \\ -.8v_1 + .1v_2 + .3v_3 &= 0 \\ .6v_1 - .9v_2 + .3v_3 &= 0 \\ .2v_1 + .8v_2 - .6v_3 &= 0. \end{aligned}$$

The system can be solved with the Gauss–Jordan method set forth in Section 6.2. Start with the augmented matrix

$$\left[\begin{array}{ccc|c} 1 & 1 & 1 & 1 \\ -.8 & .1 & .3 & 0 \\ .6 & -.9 & .3 & 0 \\ .2 & .8 & -.6 & 0 \end{array}\right].$$

The solution of this system is $v_1 = 5/23$, $v_2 = 7/23$, $v_3 = 11/23$, and so

$$V = \begin{bmatrix} \frac{5}{23} & \frac{7}{23} & \frac{11}{23} \end{bmatrix} \approx [.22 \quad .30 \quad .48].$$ ✓6

✓Checkpoint 6

Find the equilibrium vector for the transition matrix

$$P = \begin{bmatrix} .3 & .7 \\ .5 & .5 \end{bmatrix}.$$

Answer: See page 586.

In Example 4, we found that [.64 .36] was the equilibrium vector for the regular transition matrix

$$P = \begin{bmatrix} .8 & .2 \\ .35 & .65 \end{bmatrix}.$$

Observe what happens when you take powers of the matrix P (the displayed entries have been rounded for easy reading, but the full decimals were used in the calculations):

$$P^2 = \begin{bmatrix} .71 & .29 \\ .51 & .49 \end{bmatrix}; \quad P^3 = \begin{bmatrix} .67 & .33 \\ .58 & .42 \end{bmatrix}; \quad P^4 = \begin{bmatrix} .65 & .35 \\ .61 & .39 \end{bmatrix};$$

$$P^5 = \begin{bmatrix} .64 & .36 \\ .62 & .38 \end{bmatrix}; \quad P^6 = \begin{bmatrix} .64 & .36 \\ .63 & .37 \end{bmatrix}; \quad P^{10} = \begin{bmatrix} .64 & .36 \\ .64 & .36 \end{bmatrix}.$$

As these results suggest, higher and higher powers of the transition matrix P approach a matrix having all identical rows—rows that have as entries the entries of the equilibrium vector V.

If you have the technology to compute matrix powers easily (such as a graphing calculator), you can approximate the equilibrium vector by taking higher and higher powers of the transition matrix until all its rows are identical. Figure 9.15 shows part of this process for the transition matrix

$$B = \begin{bmatrix} .79 & .21 \\ .68 & .32 \end{bmatrix}$$

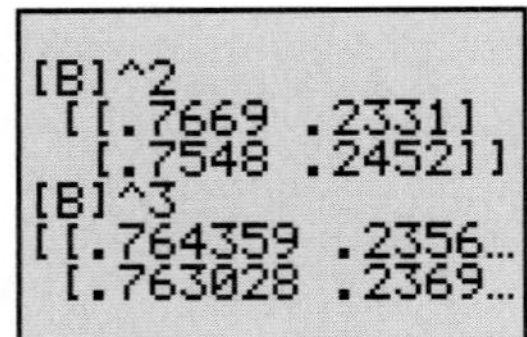

FIGURE 9.15

Figure 9.15 indicates that the equilibrium vector is [.764 .236], which is what was found algebraically in Example 5.

The results of this section can be summarized as follows.

Properties of Regular Markov Chains

Suppose a regular Markov chain has a transition matrix P.

1. As n gets larger and larger, the product $v \cdot P^n$ approaches a unique vector V for any initial probability vector v. Vector V is called the *equilibrium vector*, or *fixed vector*.
2. Vector V has the property that $VP = V$.
3. To find V, solve a system of equations obtained from the matrix equation $VP = V$ and from the fact that the sum of the entries of V is 1.
4. The powers P^n come closer and closer to a matrix whose rows are made up of the entries of the equilibrium vector V.

9.5 ▶ Exercises

Decide which of the given vectors could be a probability vector.

1. $\begin{bmatrix} \frac{1}{4} & \frac{3}{4} \end{bmatrix}$

2. $\begin{bmatrix} \frac{11}{16} & \frac{5}{16} \end{bmatrix}$

3. $[0 \quad 1]$

4. $[.3 \quad .3 \quad .3]$

5. $[.3 \quad -.1 \quad .6]$

6. $\begin{bmatrix} \frac{2}{5} & \frac{3}{10} & .3 \end{bmatrix}$

Decide which of the given matrices could be a transition matrix. Sketch a transition diagram for any transition matrices.

7. $\begin{bmatrix} .7 & .2 \\ .5 & .5 \end{bmatrix}$

8. $\begin{bmatrix} \frac{1}{4} & \frac{3}{4} \\ 0 & 1 \end{bmatrix}$

9. $\begin{bmatrix} \frac{4}{9} & \frac{1}{3} \\ \frac{1}{5} & \frac{7}{10} \end{bmatrix}$

10. $\begin{bmatrix} 0 & 1 & 0 \\ .3 & .3 & .3 \\ 1 & 0 & 0 \end{bmatrix}$

11. $\begin{bmatrix} \frac{1}{2} & \frac{1}{4} & 1 \\ \frac{2}{3} & 0 & \frac{1}{3} \\ \frac{1}{3} & 1 & 0 \end{bmatrix}$

12. $\begin{bmatrix} .2 & .3 & .5 \\ 0 & 0 & 1 \\ .1 & .9 & 0 \end{bmatrix}$

In Exercises 13–15, write any transition diagrams as transition matrices.

13.

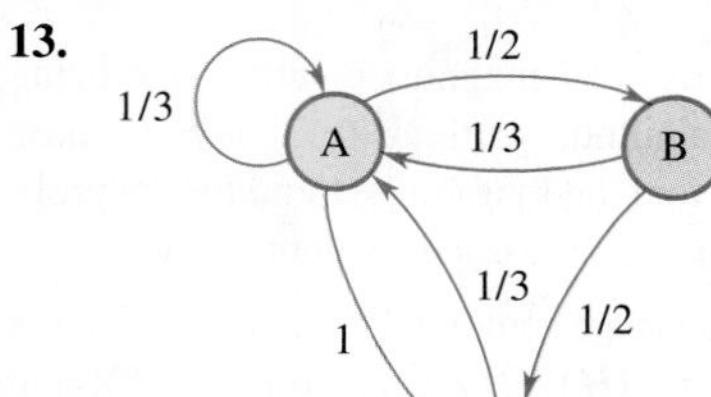

14.

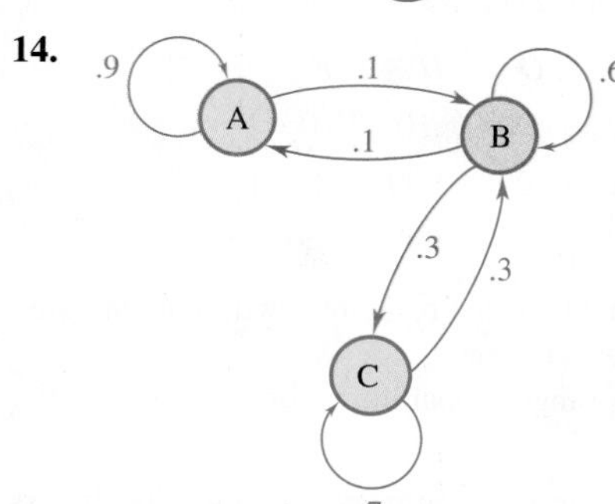

15.

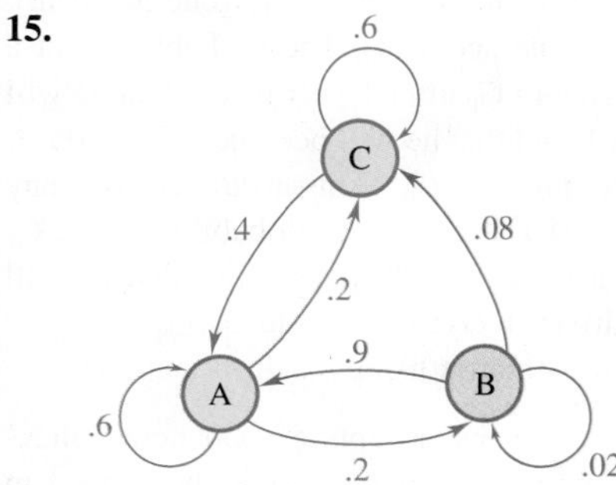

Decide whether the given transition matrices are regular. (See Example 3.)

16. $\begin{bmatrix} 1 & 0 \\ .25 & .75 \end{bmatrix}$

17. $\begin{bmatrix} .2 & .8 \\ .9 & .1 \end{bmatrix}$

18. $\begin{bmatrix} .3 & .5 & .2 \\ 1 & 0 & 0 \\ .5 & .1 & .4 \end{bmatrix}$

19. $\begin{bmatrix} 0 & 1 & 0 \\ .3 & .3 & .4 \\ 1 & 0 & 0 \end{bmatrix}$

20. $\begin{bmatrix} .25 & .40 & .30 & .05 \\ .18 & .23 & .59 & 0 \\ 0 & .15 & .36 & .49 \\ .28 & .32 & .24 & .16 \end{bmatrix}$

21. $\begin{bmatrix} .23 & .41 & 0 & .36 \\ 0 & .27 & .21 & .52 \\ 0 & 0 & 1 & 0 \\ .48 & 0 & .39 & .13 \end{bmatrix}$

Find the equilibrium vector for each of the given transition matrices. (See Examples 4 and 5.)

22. $\begin{bmatrix} .3 & .7 \\ .4 & .6 \end{bmatrix}$

23. $\begin{bmatrix} .55 & .45 \\ .19 & .81 \end{bmatrix}$

24. $\begin{bmatrix} \frac{5}{8} & \frac{3}{8} \\ \frac{7}{9} & \frac{2}{9} \end{bmatrix}$

25. $\begin{bmatrix} \frac{2}{3} & \frac{1}{3} \\ \frac{1}{8} & \frac{7}{8} \end{bmatrix}$

26. $\begin{bmatrix} .25 & .35 & .4 \\ .1 & .3 & .6 \\ .55 & .4 & .05 \end{bmatrix}$

27. $\begin{bmatrix} .16 & .28 & .56 \\ .43 & .12 & .45 \\ .86 & .05 & .09 \end{bmatrix}$

28. $\begin{bmatrix} .15 & .15 & .70 \\ .42 & .38 & .20 \\ .16 & .28 & .56 \end{bmatrix}$

29. $\begin{bmatrix} .44 & .31 & .25 \\ .80 & .11 & .09 \\ .26 & .31 & .43 \end{bmatrix}$

For each of the given transition matrices, use a graphing calculator or computer to find the first five powers of the matrix. Then find the probability that state 2 changes to state 4 after 5 repetitions of the experiment.

30. $\begin{bmatrix} .1 & .2 & .2 & .3 & .2 \\ .2 & .1 & .1 & .2 & .4 \\ .2 & .1 & .4 & .2 & .1 \\ .3 & .1 & .1 & .2 & .3 \\ .1 & .3 & .1 & .1 & .4 \end{bmatrix}$

31. $\begin{bmatrix} .3 & .2 & .3 & .1 & .1 \\ .4 & .2 & .1 & .2 & .1 \\ .1 & .3 & .2 & .2 & .2 \\ .2 & .1 & .3 & .2 & .2 \\ .1 & .1 & .4 & .2 & .2 \end{bmatrix}$

32. **Health** In a recent year, the percentage of patients at a doctor's office who received a flu shot was 26%. A campaign by the doctors and nurses was designed to increase the percentage of patients who obtain a flu shot. The doctors and nurses believed there was an 85% chance that someone who received a shot in year 1 would obtain a shot in year 2. They also believed that there was a 40% chance that a person who did not receive a shot in year 1 would receive a shot in year 2.
   (a) Give the transition matrix for this situation.
   (b) Find the percentages of patients in year 2 who received a flu shot.
   (c) Find the long-range trend for the Markov chain representing receipt of flu shots at the doctor's office.

33. **Social Science** Six months prior to an election, a poll found that only 35% of state voters planned to vote for a casino gambling initiative. After a media blitz emphasizing the new jobs that are created as a result of casinos, a new poll found that among those who did not favor it previously, 30% now favored the initiative. Among those who favored the initiative initially, 90% still favored it.
   (a) Give the transition matrix for this situation.
   (b) Find the percentage who favored the gambling initiative after the media blitz.
   (c) Find the long-term percentage who favor the initiative if the trends and media blitz continue.

34. **Business** The probability that a complex assembly line works correctly depends on whether the line worked correctly the last time it was used. There is a .91 chance that the line will work correctly if it worked correctly the time before and a .68 chance that it will work correctly if it did *not* work correctly the time before. Set up a transition matrix with this information, and find the long-run probability that the line will work correctly. (See Example 5.)

35. **Business** Suppose something unplanned occurred to the assembly line of Exercise 34, so that the transition matrix becomes

$$\begin{array}{c} \\ \text{Works} \\ \text{Doesn't Works} \end{array} \begin{array}{c} \begin{array}{cc} \text{Works} & \text{Doesn't Work} \end{array} \\ \begin{bmatrix} .81 & .19 \\ .77 & .23 \end{bmatrix} \end{array}.$$

Find the new long-run probability that the line will work properly.

36. **Natural Science** In Exercises 62 and 63 of Section 8.4 (p. 504), we discussed the effect on flower color of cross-pollinating pea plants. As shown there, since the gene for red is dominant and the gene for white is recessive, 75% of the pea plants have red flowers and 25% have white flowers, because plants with 1 red and 1 white gene appear red. If a red-flowered plant is crossed with a red-flowered plant known to have 1 red and 1 white gene, then 75% of the offspring will be red and 25% will be white. Crossing a red-flowered plant that has 1 red and 1 white gene with a white-flowered plant produces 50% red-flowered offspring and 50% white-flowered offspring.

(a) Write a transition matrix using this information.

(b) Write a probability vector for the initial distribution of colors.

(c) Find the distribution of colors after 4 generations.

(d) Find the long-range distribution of colors.

37. **Natural Science** Snapdragons with 1 red gene and 1 white gene produce pink-flowered offspring. If a red snapdragon is crossed with a pink snapdragon, the probabilities that the offspring will be red, pink, or white are 1/2, 1/2, and 0, respectively. If 2 pink snapdragons are crossed, the probabilities of red, pink, or white offspring are 1/4, 1/2, and 1/4, respectively. For a cross between a white and a pink snapdragon, the corresponding probabilities are 0, 1/2, and 1/2. Set up a transition matrix and find the long-range prediction for the fraction of red, pink, and white snapdragons.

38. **Natural Science** Markov chains can be utilized in research into earthquakes. Researchers in Italy give the following example of a transition matrix in which the rows are magnitudes of an earthquake and the columns are magnitudes of the next earthquake in the sequence.*

	2.5	**2.6**	**2.7**	**2.8**
2.5	3/7	1/7	2/7	1/7
2.6	1/2	0	1/4	1/4
2.7	1/3	1/3	0	1/3
2.8	1/4	1/2	0	1/4

Thus, the probability of a 2.5-magnitude earthquake being followed by a 2.8-magnitude earthquake is 1/7. If these trends were to persist, find the long-range trend for the probabilities of each magnitude for the subsequent earthquake.

39. **Social Science** An urban center finds that 60% of the population own a home (H), 39.5% are renters (R), and .5% are homeless (H). The study also finds the following transition probabilities per year.

$$\begin{array}{c} \\ O \\ R \\ H \end{array} \begin{array}{c} \begin{array}{ccc} O & H & R \end{array} \\ \begin{bmatrix} .90 & .10 & 0 \\ .09 & .909 & .001 \\ 0 & .34 & .66 \end{bmatrix} \end{array}$$

(a) Find the probability that residents own, rent, and are homeless after one year.

(b) Find the long-range probabilities for the three categories.

40. **Business** An insurance company classifies its drivers into three groups: G_0 (no accidents), G_1 (one accident), and G_2 (more than one accident). The probability that a driver in G_0 will stay in G_0 after 1 year is .85, that he will become a G_1 is .10, and that he will become a G_2 is .05. A driver in G_1 cannot move to G_0. (This insurance company has a long memory!) There is a .80 probability that a G_1 driver will stay in G_1 and a .20 probability that he will become a G_2. A driver in G_2 must stay in G_2.

(a) Write a transition matrix using this information.

Suppose that the company accepts 50,000 new policyholders, all of whom are in group G_0. Find the number in each group

(b) after 1 year; **(c)** after 2 years;

(d) after 3 years; **(e)** after 4 years.

(f) Find the equilibrium vector here. Interpret your result.

41. **Business** The difficulty with the mathematical model of Exercise 40 is that no "grace period" is provided; there should be a certain probability of moving from G_1 or G_2 back to G_0 (say, after 4 years with no accidents). A new

*Michele Lovallo, Vincenzo Lapenna, and Luciano Telesca, "Transition matrix analysis of earthquake magnitude sequences," *Chaos, Solitons, and Fractals* 24 (2005): 33–43.

system with this feature might produce the following transition matrix:

$$\begin{bmatrix} .85 & .10 & .05 \\ .15 & .75 & .10 \\ .10 & .30 & .60 \end{bmatrix}.$$

Suppose that when this new policy is adopted, the company has 50,000 policyholders in group G_0. Find the number in each group

(a) after 1 year;
(b) after 2 years;
(c) after 3 years.
(d) Find the equilibrium vector here. Interpret your result.

42. Suppose research on three major cell phone companies revealed the following transition matrix for the probability that a person with one cell phone carrier switches to another.

		Will Switch to	
	Company A	Company B	Company C
Now has Company A	.91	.07	.02
Now has Company B	.03	.87	.10
Now has Company C	.14	.04	.82

The current share of the market is [.26, .36, .38] for Companies A, B, and C, respectively. Find the share of the market held by each company after

(a) 1 year; **(b)** 2 years; **(c)** 3 years.
(d) What is the long-range prediction?

43. Business Data from *The Wall Street Journal* Web site indicated that the probability a new vehicle purchased in the United States in 2008 was from General Motors (GM) was .232, from Ford (F) was .150, from Toyota (T) was .159, and from other car manufacturers (O) was .459. Use the transition matrix below for market share changes from year to year to find

(a) the probability a vehicle was purchased from Ford in the next year;
(b) the long-term probability a vehicle is purchased from Toyota if the trends continue.

	GM	F	T	O
GM	.85	.04	.05	.06
F	.02	.91	.03	.04
T	.01	.01	.95	.03
O	.03	.02	.06	.89

44. Social Science At one liberal arts college, students are classified as humanities majors, science majors, or undecided. There is a 23% chance that a humanities major will change to a science major from one year to the next and a 40% chance that a humanities major will change to undecided. A science major will change to humanities with probability .12 and to undecided student with probability .38. An undecided student will switch to humanities or science with probabilities of .45 and .28, respectively. Find the long-range prediction for the fraction of students in each of these three majors.

45. Business In a queuing chain, we assume that people are queuing up to be served by, say, a bank teller. For simplicity, let us assume that once two people are in line, no one else can enter the line. Let us further assume that one person is served every minute, as long as someone is in line. Assume further that, in any minute, there is a probability of .4 that no one enters the line, a probability of .3 that exactly one person enters the line, and a probability of .3 that exactly two people enter the line, assuming that there is room. If there is not enough room for two people, then the probability that one person enters the line is .5. Let the state be given by the number of people in line.

(a) Give the transition matrix for the number of people in line:

	0	1	2
0	?	?	?
1	?	?	?
2	?	?	?

(b) Find the transition matrix for a 2-minute period.
(c) Use your result from part (b) to find the probability that a queue with no one in line has two people in line 2 minutes later.

Use a graphing calculator or computer for Exercises 46 and 47.

46. Business A company with a new training program classified each employee in one of four states: s_1, never in the program; s_2, currently in the program; s_3, discharged; s_4, completed the program. The transition matrix for this company is as follows.

	s_1	s_2	s_3	s_4
s_1	.4	.2	.05	.35
s_2	0	.45	.05	.5
s_3	0	0	1	0
s_4	0	0	0	1

(a) What percentage of employees who had never been in the program (state s_1) completed the program (state s_4) after the program had been offered five times?
(b) If the initial percentage of employees in each state was [.5 .5 0 0], find the corresponding percentages after the program had been offered four times.

47. Business Find the long-range prediction for the percentage of employees in each state for the company training program in Exercise 46.

✓ Checkpoint Answers

1. **(a)** .7
   (b) The probability of changing from state 2 to state 1
   (c)

2. [.59 .41] (rounded)
3. [.61 .39] (rounded)
4. **(a)** No **(b)** Yes
5. $5/6 \approx 83\%$
6. [5/12 7/12]

9.6 Decision Making

John F. Kennedy once remarked that he had assumed that, as president, it would be difficult to choose between distinct, opposite alternatives when a decision needed to be made. Actually, he found that such decisions were easy to make; the hard decisions came when he was faced with choices that were not as clear cut. Most decisions fall into this last category—decisions that must be made under conditions of uncertainty. In Section 9.1, we saw how to use expected values to help make a decision. Those ideas are extended in this section, where we consider decision making in the face of uncertainty. We begin with an example.

Example 1 **Business** Freezing temperatures are endangering the orange crop in central California. A farmer can protect his crop by burning smudge pots; the heat from the pots keeps the oranges from freezing. However, burning the pots is expensive, costing $20,000. The farmer knows that if he burns smudge pots, he will be able to sell his crop for a net profit (after the costs of the pots are deducted) of $50,000, provided that the freeze does develop and wipes out other orange crops in California. If he does nothing, he will either lose the $10,000 he has already invested in the crop if it does freeze or make a profit of $46,000 if it does not freeze. (If it does not freeze, there will be a large supply of oranges, and thus his profit will be lower than if there were a small supply.) What should the farmer do?

Solution He should begin by carefully defining the problem. First, he must decide on the **states of nature**—the possible alternatives over which he has no control. Here, there are two: freezing temperatures and no freezing temperatures. Next, the farmer should list the things he can control—his actions or **strategies**. He has two possible strategies: to use smudge pots or not. The consequences of each action under each state of nature, called **payoffs**, are summarized in a **payoff matrix**, as follows, where the payoffs in this case are the profits for each possible combination of events:

		States of Nature	
		Freeze	No Freeze
Strategies of Farmer	Use Smudge Pots	$50,000	$26,000
	Do Not Use Pots	−$10,000	$46,000

✓Checkpoint 1

Explain how each of the given payoffs in the matrix were obtained.

(a) $-\$10{,}000$

(b) $50,000

Answers: See page 590.

To get the $26,000 entry in the payoff matrix, use the profit if there is no freeze, namely, $46,000, and subtract the $20,000 cost of using the pots. 1✓

Once the farmer makes the payoff matrix, what then? The farmer might be an optimist (some might call him a gambler); in this case, he might assume that the best will happen and go for the biggest number of the matrix ($50,000). For that profit, he must adopt the strategy "use smudge pots."

On the other hand, if the farmer is a pessimist, he would want to minimize the worst thing that could happen. If he uses smudge pots, the worst thing that could happen to him would be a profit of $26,000, which will result if there is no freeze. If he does not use smudge pots, he might face a loss of $10,000. To minimize the worst, he once again should adopt the strategy "use smudge pots."

Suppose the farmer decides that he is neither an optimist nor a pessimist, but would like further information before choosing a strategy. For example, he might call the weather forecaster and ask for the probability of a freeze. Suppose the forecaster says that this probability is only .2. What should the farmer do? He should recall the discussion of expected value and work out the expected profit for each of his two possible strategies. If the probability of a freeze is .2, then the probability that there is no freeze is .8. This information leads to the following expected values:

$$\text{If smudge pots are used:} \quad 50{,}000(.2) + 26{,}000(.8) = 30{,}800;$$

$$\text{If no smudge pots are used:} \quad -10{,}000(.2) + 46{,}000(.8) = 34{,}800.$$

Here, the maximum expected profit, $34,800, is obtained if smudge pots are not used. 2✓

✓Checkpoint 2

What should the farmer do if the probability of a freeze is .6? What is his expected profit?

Answer: See page 590.

As the example shows, the farmer's beliefs about the probabilities of a freeze affect his choice of strategies.

Example 2 **Business** An owner of several greeting-card stores must decide in July about the type of displays to emphasize for Sweetest Day in October. He has three possible choices: emphasize chocolates, emphasize collectible gifts, or emphasize gifts that can be engraved. His success is dependent on the state of the economy in October. If the economy is strong, he will do well with the collectible gifts, while in a weak economy, the chocolates do very well. In a mixed economy, the gifts that can be engraved will do well. He first prepares a payoff matrix for all three possibilities, where the numbers in the matrix represent his profits in thousands of dollars:

		States of Nature		
		Weak Economy	Mixed	Strong Economy
Strategies	Chocolates	85	30	75
	Collectibles	45	45	110
	Engraved	60	95	85

(a) What would an optimist do?

Solution If the owner is an optimist, he should aim for the biggest number on the matrix, 110 (representing $110,000 in profit). His strategy in this case would be to display collectibles.

(b) How would a pessimist react?

Solution A pessimist wants to find the best of the worst of all bad things that can happen. If he displays collectibles, the worst that can happen is a profit of \$45,000. For displaying engravable items, the worst is a profit of \$60,000, and for displaying chocolates, the worst is a profit of \$30,000. His strategy here is to use the engravable items.

(c) Suppose the owner reads in a business magazine that leading experts believe that there is a 50% chance of a weak economy in October, a 20% chance of a mixed economy, and a 30% chance of a strong economy. How might he use this information?

Solution The owner can now find his expected profit for each possible strategy.

Chocolates	$85(.5) + 30(.20) + 75(.30)$	$= 71;$
Collectibles	$45(.5) + 45(.20) + 110(.30)$	$= 64.5;$
Engraved	$60(.5) + 95(.20) + 85(.30)$	$= 74.5.$

Here, the best strategy is to display gifts that can be engraved; the expected profit is 74.5, or \$74,500. 3 ✓

✓Checkpoint 3

Suppose the owner reads another article, which gives the following predictions: a 35% chance of a weak economy, a 25% chance of an in-between economy, and a 40% chance of a strong economy. What is the best strategy now? What is the expected profit?

Answer: See page 590.

9.6 ▶ Exercises

1. **Business** A developer has \$100,000 to invest in land. He has a choice of two parcels (at the same price): one on the highway and one on the coast. With both parcels, his ultimate profit depends on whether he faces light opposition from environmental groups or heavy opposition. He estimates that the payoff matrix is as follows (the numbers represent his profit):

	Opposition Light	Heavy
Highway	\$70,000	\$30,000
Coast	\$150,000	−\$40,000

What should the developer do if he is
(a) an optimist? **(b)** a pessimist?
(c) Suppose the probability of heavy opposition is .8. What is his best strategy? What is the expected profit?
(d) What is the best strategy if the probability of heavy opposition is only .4?

2. **Business** Mount Union College has sold out all tickets for a jazz concert to be held in the stadium. If it rains, the show will have to be moved to the gym, which has a much smaller seating capacity. The dean must decide in advance whether to set up the seats and the stage in the gym, in the stadium, or in both, just in case. The following payoff matrix shows the net profit in each case:

		States of Nature Rain	No Rain
Strategies	Set up in Stadium	−\$1550	\$1500
	Set up in Gym	\$1000	\$1000
	Set up in Both	\$750	\$1400

What strategy should the dean choose if she is
(a) an optimist?
(b) a pessimist?
(c) If the weather forecaster predicts rain with a probability of .6, what strategy should she choose to maximize the expected profit? What is the maximum expected profit?

3. **Business** An analyst must decide what fraction of the automobile tires produced at a particular manufacturing plant are defective. She has already decided that there are three possibilities for the fraction of defective items: .02, .09, and .16. She may recommend two courses of action: upgrade the equipment at the plant or make no upgrades. The following payoff matrix represents the *costs* to the company in each case, in hundreds of dollars:

		Defectives		
		.02	.09	.16
Strategies	Upgrade	130	130	130
	No Upgrade	28	180	450

What strategy should the analyst recommend if she is
(a) an optimist?
(b) a pessimist?
(c) Suppose the analyst is able to estimate probabilities for the three states of nature as follows:

Fraction of Defectives	Probability
.02	.70
.09	.20
.16	.10

Which strategy should she recommend? Find the expected cost to the company if that strategy is chosen.

4. **Business** The research department of the Allied Manufacturing Company has developed a new process that it believes will result in an improved product. Management must decide whether to go ahead and market the new product or not. The new product may be better than the old one, or it may not be better. If the new product is better and the company decides to market it, sales should increase by $50,000. If it is not better and the old product is replaced with the new product on the market, the company will lose $25,000 to competitors. If management decides not to market the new product, the company will lose $40,000 if it is better and will lose research costs of $10,000 if it is not.
   (a) Prepare a payoff matrix.
   (b) If management believes that the probability that the new product is better is .4, find the expected profits under each strategy and determine the best action.

5. **Business** A businessman is planning to ship a used machine to his plant in Nigeria. He would like to use it there for the next 4 years. He must decide whether to overhaul the machine before sending it. The cost of overhaul is $2600. If the machine fails when it is in operation in Nigeria, it will cost him $6000 in lost production and repairs. He estimates that the probability that it will fail is .3 if he does not overhaul it and .1 if he does overhaul it. Neglect the possibility that the machine might fail more than once in the 4 years.
   (a) Prepare a payoff matrix.
   (b) What should the businessman do to minimize his expected costs?

6. **Business** A contractor prepares to bid on a job. If all goes well, his bid should be $25,000, which will cover his costs plus his usual profit margin of $4000. However, if a threatened labor strike actually occurs, his bid should be $35,000 to give him the same profit. If there is a strike and he bids $25,000, he will lose $5500. If his bid is too high, he may lose the job entirely, while if it is too low, he may lose money.
   (a) Prepare a payoff matrix.
   (b) If the contractor believes that the probability of a strike is .6, how much should he bid?

7. **Business** An artist travels to craft fairs all summer long. She must book her booth at a June craft show six months in advance and decide if she wishes to rent a tent for an extra $500 in case it rains on the day of the show. If it does not rain, she believes she will earn $3000 at the show. If it rains, she believes she will earn only $2000, provided she has a tent. If she does not have a tent and it does rain, she will have to pack up and go home and will thus earn $0. Weather records over the last 10 years indicate that there is a .4 probability of rain in June.
   (a) Prepare a profit matrix.
   (b) What should the artist do to maximize her expected revenue?

8. **Business** An investor has $50,000 to invest in stocks. She has two possible strategies: buy conservative blue-chip stocks or buy highly speculative stocks. There are two states of nature: the market goes up and the market goes down. The following payoff matrix shows the net amounts she will have under the various circumstances.

	Market Up	Market Down
Buy Blue Chip	$60,000	$46,000
Buy Speculative	$80,000	$32,000

   What should the investor do if she is
   (a) an optimist?
   (b) a pessimist?
   (c) Suppose there is a .6 probability of the market going up. What is the best strategy? What is the expected profit?
   (d) What is the best strategy if the probability of a market rise is .2?

Sometimes the numbers (or payoffs) in a payoff matrix do not represent money (profits or costs, for example). Instead, they may represent utility. A utility is a number that measures the satisfaction (or lack of it) that results from a certain action. Utility numbers must be assigned by each individual, depending on how he or she feels about a situation. For example, one person might assign a utility of +20 for a week's vacation in San Francisco, with −6 being assigned if the vacation were moved to Sacramento. Work the problems that follow in the same way as the preceding ones.

9. **Social Science** A politician must plan her reelection strategy. She can emphasize jobs or she can emphasize the environment. The voters can be concerned about jobs

or about the environment. Following is a payoff matrix showing the utility of each possible outcome.

		Voters	
		Jobs	Environment
Candidate	Jobs	+40	−10
	Environment	−12	+30

The political analysts feel that there is a .35 chance that the voters will emphasize jobs. What strategy should the candidate adopt? What is its expected utility?

10. In an accounting class, the instructor permits the students to bring a calculator or a reference book (but not both) to an examination. The examination itself can emphasize either numbers or definitions. In trying to decide which aid to take to an examination, a student first decides on the utilities shown in the following payoff matrix:

		Exam Emphasizes	
		Numbers	Definition
Student Chooses	Calculator	+50	0
	Book	+15	+35

(a) What strategy should the student choose if the probability that the examination will emphasize numbers is .6? What is the expected utility in this case?

(b) Suppose the probability that the examination emphasizes numbers is .4. What strategy should the student choose?

✓Checkpoint Answers

1. (a) If the crop freezes and smudge pots are not used, the farmer's profit is −$10,000 for labor costs.
   (b) If the crop freezes and smudge pots are used, the farmer makes a profit of $50,000.
2. Use smudge pots; $40,400
3. Engravable; $78,750

CHAPTER 9 ▶ Summary

KEY TERMS AND SYMBOLS

9.1 ▶ random variable
probability distribution
histogram
expected value
fair game

9.2 ▶ ***n!*** (*n* factorial)
multiplication principle
permutations
combinations

9.4 ▶ Bernoulli trials (processes)
binomial experiment
binomial probability

9.5 ▶ stochastic processes
Markov chain
state
transition diagram
transition matrix
probability vector
regular transition matrix
regular Markov chain
equilibrium vector (fixed vector)

9.6 ▶ states of nature
strategies
payoffs
payoff matrix

CHAPTER 9 KEY CONCEPTS

Expected Value of a Probability Distribution For a random variable x with values $x_1, x_2, \ldots, x_n$ and probabilities $p_1, p_2, \ldots, p_n$, the expected value is

$$E(x) = x_1p_1 + x_2p_2 + \cdots + x_np_n.$$

Multiplication Principle If there are m_1 ways to make a first choice, m_2 ways to make a second choice, and so on, then there are $m_1m_2 \cdots m_n$ different ways to make the entire sequence of choices.

The number of **permutations** of n elements taken r at a time is ${}_nP_r = \frac{n!}{(n-r)!}$.

The number of **combinations** of n elements taken r at a time is

$${}_nC_r = \frac{n!}{(n-r)!\,r!}.$$

Binomial Experiments have the following characteristics: (1) The same experiment is repeated several times; (2) there are only *two* outcomes, labeled success and failure; (3) the probability of success is the same for each trial; and (4) the trials are independent. If the probability of success in a single trial is p, the probability of x successes in n trials is

$$_nC_x p^x(1 - p)^{n-x}.$$

Markov Chains A **transition matrix** must be square, with all entries between 0 and 1 inclusive, and the sum of the entries in any row must be 1. A Markov chain is *regular* if some power of its transition matrix P contains all positive entries. The long-range probabilities for a regular Markov chain are given by the **equilibrium**, or **fixed**, **vector** V, where, for any initial probability vector v, the products vP^n approach V as n gets larger and $VP = V$. To find V, solve the system of equations formed by $VP = V$ and the fact that the sum of the entries of V is 1.

Decision Making A **payoff matrix**, which includes all available strategies and states of nature, is used in decision making to define the problem and the possible solutions. The expected value of each strategy can help to determine the best course of action.

CHAPTER 9 REVIEW EXERCISES

In Exercises 1–3, (a) sketch the histogram of the given probability distribution, and (b) find the expected value.

1.

x	0	1	2	3
$P(x)$	.22	.54	.16	.08

2.

x	−3	−2	−1	0	1	2	3
$P(x)$	.15	.20	.25	.18	.12	.06	.04

3.

x	−10	0	10
$P(x)$	$\frac{1}{3}$	$\frac{1}{3}$	$\frac{1}{3}$

4. Health The probability distribution for the previous number of children a pregnant mother has given birth to can be estimated using data from the North Carolina Birth Registry as follows.*

x = Number of children now living	0	1	2	3	4
$P(x)$	.40898	.32788	.16616	.06281	.02110

x = Number of children now living	5	6	7	8	9
$P(x)$	.00771	.00321	.00132	.00061	.00023

Find the expected value.

*Based on data available at www.odum.unc.edu. Data modified slightly for 9 previous children or less.

5. Social Science Data from The American Community Survey yields the following probability distribution for x, the number of bedrooms for the dwellings in which Americans live.*

x	0	1	2	3	4	5
$P(x)$	.0138	.1049	.2722	.3909	.1748	.0434

Find the expected number of bedrooms.

In Exercises 6 and 7, (a) give the probability distribution, and (b) find the expected value.

6. Business A grocery store has 10 bouquets of flowers for sale, 3 of which are red rose displays. Two bouquets are selected at random, and the number of rose bouquets is noted.

7. Social Science In a class of 10 students, 3 did not do their homework. The professor selects 3 members of the class to present solutions to homework problems on the board and records how many of those selected did not do their homework.

Solve the given problems.

8. Suppose someone offers to pay you \$100 if you draw 3 cards from a standard deck of 52 cards and all the cards are hearts. What should you pay for the chance to win if it is a fair game?

9. You pay \$2 to play a game of "Over/Under," in which you will roll two dice and note the sum of the results. You can bet that the sum will be less than 7 (under), exactly 7, or greater than 7 (over). If you bet "under" and you win, you get your \$2 back, plus \$2 more. If you bet 7 and you win, you get your \$2 back, plus \$4, and if you bet "over" and

*Based on data available at www.census.gov/acs.

win, you get your $2 back, plus $2 more. What are the expected winnings for each type of bet?

10. A lottery has a first prize of $10,000, two second prizes of $1000 each, and two $100 third prizes. Ten thousand tickets are sold, at $2 each. Find the expected winnings of a person buying 1 ticket.

11. **Social Science** It can be estimated that 25% of renters pay $1140 or more a month in rent.* If we randomly select 5 households that rent, the probability distribution for x, the number of the renters that pay $1140 or more in rent, is given as follows:

x	0	1	2	3	4	5
$P(x)$	.2373	.3955	.2637	.0879	.0146	.0010

Find the expected value for x.

12. **Business** In October 2008, David Pogue of The *New York Times* estimated Apple® Computer's total worldwide market share at approximately 7.5%. If we select 3 computers at random and define x to be the number of Apple computers, the probability distribution is as follows:

x	0	1	2	3
$P(x)$	.7915	.1925	.0156	.0004

Find $E(x)$.

13. In how many ways can 8 different taxis line up at the airport?

14. How many variations are possible for gold, silver, and bronze medalists in the 50-meter swimming race if there are 8 finalists?

15. In how many ways can a sample of 3 computer monitors be taken from a batch of 12 identical monitors?

16. If 4 of the 12 monitors in Exercise 15 are broken, in how many ways can the sample of 3 include the following?
    (a) 1 broken monitor;
    (b) no broken monitors;
    (c) at least 1 broken monitor.

17. In how many ways can 6 students from a class of 30 be arranged in the first row of seats? (There are 6 seats in the first row.)

18. In how many ways can the six students in Exercise 17 be arranged in a row if a certain student must be first?

19. In how many ways can the 6 students in Exercise 17 be arranged if half the students are science majors and the other half are business majors and if
    (a) like majors must be together?
    (b) science and business majors are alternated?

*Based on data available at www.census.gov/acs.

20. Explain under what circumstances a permutation should be used in a probability problem and under what circumstances a combination should be used.

21. Discuss under what circumstances the binomial probability formula should be used in a probability problem.

Suppose 2 cards are drawn without replacement from an ordinary deck of 52 cards. Find the probabilities of the given results.

22. Both cards are black.
23. Both cards are hearts.
24. Exactly 1 is a face card.
25. At most 1 is an ace.

An ice cream stand contains 4 custard flavors, 6 ice cream flavors, and 2 frozen yogurt selections. Three customers come to the window. If each customer's selection is random, find the probability that the selections include

26. all ice cream;
27. all custard;
28. at least one frozen yogurt;
29. one custard, one ice cream, and one frozen yogurt;
30. at most one ice cream.

31. In this exercise, we study the connection between sets (from Chapter 8) and combinations.
    (a) Given a set with n elements, what is the number of subsets of size 0? of size 1? of size 2? of size n?
    (b) Using your answer from part (a), give an expression for the total number of subsets of a set with n elements.
    (c) Using your answer from part (b) and a result from Chapter 8, explain why the following equation must be true:

$$_nC_0 + {_nC_1} + {_nC_2} + \cdots + {_nC_n} = 2^n.$$

    (d) Verify the equation in part (c) for $n = 4$ and $n = 5$.

Health *According to the U.S. National Center for Health Statistics, 36% of deaths are a result of major cardiovascular disease. If 7 deaths are selected at random, find the probability that*

32. exactly 2 of the deaths were from major cardiovascular disease;

33. at least 1 of the deaths was from major cardiovascular disease.

Natural Science *Researchers studied scarring patterns on the skin of humpback whales in Alaska and estimate that 78% of the whales have been previously entangled in fishing nets.* Suppose that 6 whales are selected at random, and let x be the number of whales with scars indicating previous entanglement.*

34. Give the probability distribution for x.

35. What is the expected number of whales indicating previous entanglement?

*Janet L. Neilson, Christine M. Gabriele, and Janice M. Straley, "Humpback Whale Entanglement in Fishing Gear in Northern Southeastern Alaska," *Proceedings of the Fourth Glacier Bay Science Symposium*, 2007, pp. 204–207.

36. **Business** As of February 2009, 22% of the Janus Health Global Life Sciences Fund (a high growth investment mutual fund) was invested in foreign stocks. If 4 stocks from the fund are picked at random, find the probability that the given numbers of stocks are foreign stocks.
   (a) All 4 stocks
   (b) At least 1 stock
   (c) At most 2 stocks

37. **Business** In 2008, 15% of the credit unions insured by the National Credit Union Association (NCUA) had $100 million or more in assets.* If we select 5 credit unions at random,
   (a) give the probability distribution for x, the number of credit unions with $100 million or more in assets;
   (b) give the expected value for the number of credit unions with $100 million or more in assets.

Decide whether each matrix is a regular transition matrix.

38. $\begin{bmatrix} 0 & 1 \\ .77 & .23 \end{bmatrix}$ **39.** $\begin{bmatrix} -.2 & .4 \\ .3 & .7 \end{bmatrix}$

40. $\begin{bmatrix} .21 & .15 & .64 \\ .50 & .12 & .38 \\ 1 & 0 & 0 \end{bmatrix}$ **41.** $\begin{bmatrix} .22 & 0 & .78 \\ .40 & .33 & .27 \\ 0 & .61 & .39 \end{bmatrix}$

42. **Business** Using e-mail for professional correspondence has become a major component of a worker's day. A study classified e-mail use into 3 categories for an office day: no use, light use (1–60 minutes), and heavy use (more than 60 minutes). Researchers observed a pool of 100 office workers over a month and developed the following transition matrix of probabilities from day to day:

		Current Day		
		No Use	Light Use	Heavy Use
Prervious Day	No use	.35	.15	.50
	Light Use	.30	.35	.35
	Heavy Use	.15	.30	.55

Suppose the initial distribution for the three states is [.2, .4, .4]. Find the distribution after
   (a) 1 day;
   (b) 2 days.
   (c) What is the long-range prediction for the distribution of e-mail use?

43. **Business** An analyst at a major brokerage firm that invests in Europe, North America, and Asia has examined the investment records for a particular international stock mutual fund over several years. The analyst constructed the following transition matrix for the probability of switching the location of an equity from year to year:

		Current Year		
		Europe	North America	Asia
Previous Year	Europe	.80	.14	.06
	North America	.04	.85	.11
	Asia	.03	.13	.84

If the initial investment vector is 15% in Europe, 60% in North America, and 25% in Asia,
   (a) find the percentages in Europe, North America, and Asia after 1 year;
   (b) find the percentages in Europe, North America, and Asia after 3 years;
   (c) find the long-range percentages in Europe, North America, and Asia.

44. **Social Science** A candidate for city council can come out in favor of a new factory, be opposed to it, or waffle on the issue. The change in votes for the candidate depends on what her opponent does, with payoffs as shown in the following matrix:

		Opponent		
		Favors	Waffles	Opposes
Candidate	Favors	0	−1000	−4000
	Waffles	1000	0	−500
	Opposes	5000	2000	0

   (a) What should the candidate do if she is an optimist?
   (b) What should she do if she is a pessimist?
   (c) Suppose the candidate's campaign manager feels that there is a 40% chance that the opponent will favor the plant and a 35% chance that he will waffle. What strategy should the candidate adopt? What is the expected change in the number of votes?
   (d) The opponent conducts a new poll that shows strong opposition to the new factory. This changes the probability that he will favor the factory to 0 and the probability that he will waffle to .7. What strategy should our candidate adopt? What is the expected change in the number of votes now?

45. **Social Science** When teaching, an instructor can adopt a strategy using either active learning or lecturing to help students learn best. A class often reacts very differently to these two strategies. A class can prefer lecturing or active learning. A department chair constructs the following payoff matrix of the average point gain (out of 500 possible points) on the final exam after studying many classes that use active learning and many that use lecturing and polling students as to their preference:

		Students in class prefer	
		Lecture	Active Learning
Instructor uses	Lecture	50	−80
	Active Learning	−30	100

*Based on data available at www.ncua.org.

(a) If the department chair uses the preceding information to decide how to teach her own classes, what should she do if she is an optimist?
(b) What about if she is a pessimist?
(c) If the polling data shows that there is a 75% chance that a class will prefer the lecture format, what strategy should she adopt? What is the expected payoff?
(d) If the chair finds out that her next class has had more experience with active learning, so that there is now a 60% chance that the class will prefer active learning, what strategy should she adopt? What is the expected payoff?

*Exercises 46 and 47 are taken from actuarial examinations given by the Society of Actuaries.**

46. Business A company is considering the introduction of a new product that is believed to have probability .5 of being successful and probability .5 of being unsuccessful. Successful products pass quality control 80% of the time. Unsuccessful products pass quality control 25% of the time. If the product is successful, the net profit to the company will be \$40 million; if unsuccessful, the net loss will be \$15 million. Determine the expected net profit if the product passes quality control.
(a) \$23 million (b) \$24 million
(c) \$25 million (d) \$26 million
(e) \$27 million

47. Business A merchant buys boxes of fruit from a grower and sells them. Each box of fruit is either Good or Bad. A Good box contains 80% excellent fruit and will earn \$200 profit on the retail market. A Bad box contains 30% excellent fruit and will produce a loss of \$1000. The a priori probability of receiving a Good box of fruit is .9. Before the merchant decides to put the box on the market, he can sample one piece of fruit to test whether it is excellent. Based on that sample, he has the option of rejecting the box without paying for it. Determine the expected value of the right to sample. (*Hint*: If the merchant samples the fruit, what are the probabilities of accepting a Good box, accepting a Bad box, and not accepting the box? What are these probabilities if he does not sample the fruit?)
(a) 0 (b) \$16 (c) \$34
(d) \$72 (e) \$80

48. Business An issue of *Mathematics Teacher* included "Overbooking Airline Flights," an article by Joe Dan Austin. In this article, Austin developed a model for the expected income for an airline flight. With appropriate assumptions, the probability that exactly x of n people with reservations show up at the airport to buy a ticket is given by the binomial probability formula. Assume the following: Six reservations have been accepted for 3 seats, $p = .6$ is the probability that a person with a reservation will show up, a ticket costs \$100, and the airline must pay \$100 to anyone with a reservation who does not get a ticket. Complete the following table:

Number Who Show Up (x)	0	1	2	3	4	5	6
Airline's Income							
$P(x)$							

(a) Use the table to find $E(I)$, the expected income from the 3 seats.
(b) Find $E(I)$ for $n = 3$, $n = 4$, and $n = 5$. Compare these answers with $E(I)$ for $n = 6$. For these values of n, how many reservations should the airline book for the 3 seats in order to maximize the expected revenue?

*Problems from "Course 130 Examination, Operations Research," of the Education and Examination Committee of the Society of Actuaries. Reprinted by permission of the Society of Actuaries.

CASE 9

Quick Draw® from the New York State Lottery

At bars and restaurants in the state of New York, patrons can play an electronic lottery game called Quick Draw.* A similar game is available in many other states. There are 10 ways for a patron to play this game. Prior to the draw, a person may bet $1 on games called 10-spot, 9-spot, 8-spot, 7-spot, 6-spot, 5-spot, 4-spot, 3-spot, 2-spot, and 1-spot. Depending on the game, the player will choose numbers from 1 to 80. For the 10-spot game, the player chooses 10 numbers; for a 9-Spot game, the player chooses 9 numbers; etc. Every four minutes, the State of New York chooses 20 numbers at random from the numbers 1 to 80. For example, if a player chose the 6-Spot game, he or she will have picked 6 numbers. If 3, 4, 5, or 6 of the numbers the player picked are also numbers the state picked randomly, then the player will win money. Each game has different ways to win, with differing payoff amounts. Notice with the 10-Spot, 9-Spot, 8-Spot, and 7-Spot, a player can win by matching 0 numbers correctly. The accompanying tables show the payoffs for the different games. Notice that a player does not have to match all the numbers he or she picked in order to win.

10-Spot Game

Numbers Matched	Winnings per $1 Played
10	$100,000
9	$5,000
8	$300
7	$45
6	$10
5	$2
0	$5

9-Spot Game

Numbers Matched	Winnings per $1 Played
9	$30,000
8	$3,000
7	$125
6	$20
5	$5
0	$2

8-Spot Game

Numbers Matched	Winnings per $1 Played
8	$10,000
7	$550
6	$75
5	$6
0	$2

7-Spot Game

Numbers Matched	Winnings per $1 Played
7	$5,000
6	$100
5	$20
4	$2
0	$1

6-Spot Game

Numbers Matched	Winnings per $1 Played
6	$1,000
5	$55
4	$6
3	$1

5-Spot Game

Numbers Matched	Winnings per $1 Played
5	$300
4	$20
3	$2

*More information on Quick Draw can be found at www.nylottery.org; click on "Daily Games."

4-Spot Game

Numbers Matched	Winnings per $1 Played
4	$55
3	$5
2	$1

3-Spot Game

Numbers Matched	Winnings per $1 Played
3	$23
2	$2

2-Spot Game

Numbers Matched	Winnings per $1 Played
2	$10

1-Spot Game

Numbers Matched	Winnings per $1 Played
1	$2

With our knowledge of counting, it is possible for us to calculate the probability of winning for these different games.

Example 1 Find the probability distribution for the number of matches for the 6-spot game.

Solution Let us define x to be the number of matches when playing 6-spot. The outcomes of x are then 0, 1, 2, . . . , 6. To find the probabilities of these matches, we need to do a little thinking. First, we need to know how many ways a player can pick 6 numbers from the selection of 1 to 80. Since the order in which the player picks the numbers does not matter, the number of ways to pick 6 numbers is

$$_{80}C_6 = \frac{80!}{74!\,6!} = 300{,}500{,}200.$$

To find the probability of the outcomes of 0 to 6, we can think of the 80 choices broken into groups: 20 winning numbers the state picked and 60 losing numbers the state did not pick. If x, the number of matches, is 0, then the player picked 0 numbers from the 20 winning numbers and 6 from the 60 losing numbers. Using the multiplication principle, we find that this quantity is

$$_{20}C_0 \cdot {}_{60}C_6 = \left(\frac{20!}{20!\,0!}\right)\left(\frac{60!}{54!\,6!}\right) = (1)(50{,}063{,}860) = 50{,}063{,}860.$$

Therefore,

$$P(x = 0) = \frac{_{20}C_0 \cdot {}_{60}C_6}{_{80}C_6} = \frac{50{,}063{,}860}{300{,}500{,}200} \approx .16660.$$

Similarly for x = 1, 2, . . . , 6, and completing the probability distribution table, we have

x	$P(x)$
0	$\frac{{}_{20}C_0 \cdot {}_{60}C_6}{{}_{80}C_6} \approx .16660$
1	$\frac{{}_{20}C_1 \cdot {}_{60}C_5}{{}_{80}C_6} \approx .36349$
2	$\frac{{}_{20}C_2 \cdot {}_{60}C_4}{{}_{80}C_6} \approx .30832$
3	$\frac{{}_{20}C_3 \cdot {}_{60}C_3}{{}_{80}C_6} \approx .12982$
4	$\frac{{}_{20}C_4 \cdot {}_{60}C_2}{{}_{80}C_6} \approx .02854$
5	$\frac{{}_{20}C_5 \cdot {}_{60}C_1}{{}_{80}C_6} \approx .00310$
6	$\frac{{}_{20}C_6 \cdot {}_{60}C_0}{{}_{80}C_6} \approx .00013$

Example 2 Find the expected winnings for a $1 bet on the 6-spot game.

Solution To find the expected winnings, we take the winnings for each number of matches and subtract our $1 initial payment fee. Thus, we have the following:

x	*Net Winnings*	$P(x)$
0	−$1	0.16660
1	−$1	.36349
2	−$1	.30832
3	$0	.12982
4	$5	.02854
5	$54	.00310
6	$999	.00013

The expected winnings are

$$\begin{aligned} E(\text{winnings}) &= (-1)\cdot .16660 + (-1)\cdot .36349 \\ &\quad + (-1)\cdot .30832 + 0\cdot .12982 + 5(.02854) \\ &\quad + 54(.00310) + 999(.00013) \\ &= -.39844. \end{aligned}$$

Thus, for every $1 bet on the 6-Spot game, a player would lose about 40 cents. Put another way, the state gets about 40 cents, on average, from every $1 bet on 6-spot.

EXERCISES

1. If New York State initiates a promotion where players earn "double payoffs" for the 6-spot game, (that is, if a player matched 3 numbers, she would win $2; if she matched 4 numbers, she would win $12; etc.), find the expected winnings.
2. Would it be in the state's interest to offer such a promotion? Why or why not?
3. Find the probability distribution for the 4-spot game.
4. Find the expected winnings for the 4-spot game.
5. If the state offers double payoffs for the 4-spot game, what are the expected winnings?

CHAPTER

10

Introduction to Statistics

Statistics has applications to almost every aspect of modern life. The digital age is creating a wealth of data that needs to be summarized, visualized, and analyzed, from the earnings of major-league baseball teams to movies' box-office receipts and the sales for the soft-drink industry. See Exercises 26 and 27 on pages 615 and 616, and Exercises 54–58 on page 640.

Statistics is the science that deals with the collection and summarization of data. Methods of statistical analysis make it possible to draw conclusions about a population on the basis of data from a sample of the population. Statistical models have become increasingly useful in manufacturing, government, agriculture, medicine, and the social sciences and in all types of research. An Indianapolis race-car team, for example, is using statistics to improve its performance by gathering data on each run around the track. The team samples data 300 times a second and uses computers to process the data. In this chapter, we give a brief introduction to some of the key topics from statistical methodology.

10.1 Frequency Distributions

Researchers often wish to learn characteristics or traits of a specific **population** of individuals, objects, or units. The traits of interest are called **variables**, and it is these that we measure or label. Often, however, a population of interest is very large or constantly changing, so measuring each unit is impossible. Thus, researchers are forced to collect data on a subset of the population of interest, called a **sample**.

Sampling is a complex topic, but the universal aim of all sampling methods is to obtain a sample that "represents" the population of interest. One common way of obtaining a representative sample is to perform simple random sampling, in which every unit of the population has an equal chance to be selected to be in the sample. Suppose we wanted to study the height of students enrolled in a class. To obtain a random sample, we could place slips of paper containing the names of everyone in class in a hat, mix the papers, and draw 10 names blindly. We would then record the height (the variable of interest) for each student selected.

A simple random sample can be difficult to obtain in real life. For example, suppose you want to take a random sample of voters in your congressional district to see which candidate they prefer in the next election. If you do a telephone survey, you have a representative sample of people who are at home to answer the telephone, but those who work a lot of hours and are rarely home to answer the phone, those who have an unlisted number, those who cannot afford a telephone, and those who refuse to answer telephone surveys are underrepresented. Such people may have an opinion different from those of the people you interview.

A famous example of an inaccurate poll was made by the *Literary Digest* in 1936. Its survey indicated that Alfred Landon would win the presidential election; in fact, Franklin Roosevelt won with 62% of the popular vote. The *Digest*'s major error was mailing its surveys to a sample of those listed in telephone directories. During the Depression, many poor people did not have telephones, and the poor voted overwhelmingly for Roosevelt. Modern pollsters use sophisticated techniques to ensure that their sample is as representative as possible.

Once a sample has been collected and all data of interest is recorded, the data must be organized so that conclusions may be more easily drawn. With numeric responses, one method of organization is to group the data into intervals, usually of equal size.

Example 1 **Business** The following list gives the 2008–2009 tuition (in thousands of dollars) for a random sample of 30 private colleges that offer four-year degrees or higher:*

23	22	38	25	11	16	15	26	23	24
37	18	21	36	36	28	18	9	39	17
27	24	10	32	24	27	22	24	28	39

Identify the population and the variable, group the data into intervals, and find the frequency of each interval.

Solution The population is all private colleges that offer a four-year degree or higher. The variable of interest is the tuition (in thousands of dollars). The highest

*Data from http://chronicle.com/.

number in the list is 39, and the lowest number is 9; one convenient way to group the data is in intervals of size 5, starting with 5–9 and ending with 35–39. This grouping gives an interval for each number in the list and results in seven equal intervals of a convenient size. Too many intervals of smaller size would not simplify the data enough, while too few intervals of larger size would conceal information that the data might provide. A rule of thumb is to use from 6 to 15 intervals.

First tally the number of schools in each interval. Then total the tallies in each interval, as in the following table:

Tuition Amount	Tally	Frequency
5–9	|	1
10–14	||	2
15–19	~~||||~~	5
20–24	~~||||~~ ||||	9
25–29	~~||||~~ |	6
30–34	|	1
35–39	~~||||~~ |	6
		Total = 30

1 ✓

This table is an example of **grouped frequency distribution.**

✓Checkpoint 1

An accounting firm selected 24 complex tax returns prepared by a certain tax preparer. The number of errors per return were as follows:

8	12	0	6	10	8	0	14
8	12	14	16	4	14	7	11
9	12	7	15	11	21	22	19

Prepare a grouped frequency distribution for this data. Use intervals 0–4, 5–9, and so on.

Answer: See page 607.

The frequency distribution in Example 1 shows information about the data that might not have been noticed before. For example, the interval with the largest number of colleges is 20–24. However, some information has been lost; for example, we no longer know exactly how many colleges charged 39 (thousand dollars) in tuition.

PICTURING DATA

The information in a grouped frequency distribution can be displayed graphically with a **histogram**, which is similar to a bar graph. In a histogram, the number of observations in each interval determines the height of each bar, and the size of each interval determines the width of each bar. If equally sized intervals are used, all the bars have the same width.

A **frequency polygon** is another form of graph that illustrates a grouped frequency distribution. The polygon is formed by joining consecutive midpoints of the tops of the histogram bars with straight-line segments. Sometimes the midpoints of the first and last bars are joined to endpoints on the horizontal axis where the next midpoint would appear. (See Figure 10.1 on the next page.)

✓Checkpoint 2

Make a histogram and a frequency polygon for the distribution found in Checkpoint 1.

Answer: See page 607.

Example 2 A grouped frequency distribution of college tuition was found in Example 1. Draw a histogram and a frequency polygon for this distribution.

Solution First, draw a histogram, shown in blue in Figure 10.1. To get a frequency polygon, connect consecutive midpoints of the tops of the bars. The frequency polygon is shown in red. 2 ✓

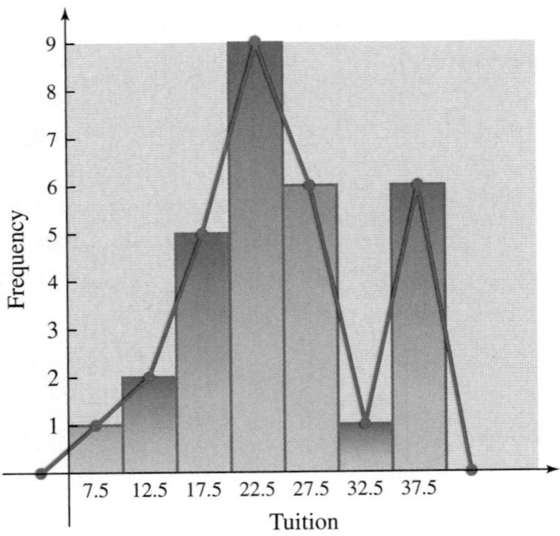

FIGURE 10.1

TECHNOLOGY TIP As noted in Section 9.1, most graphing calculators can display histograms. Many will also display frequency polygons (which are usually labeled LINE or xyLINE in calculator menus). When dealing with grouped frequency distributions, however, certain adjustments must be made on a calculator:

1. *A calculator list of outcomes must consist of single numbers, not intervals.* The table in Example 1, for instance, cannot be entered as shown. To convert the first column of the table for calculator use, choose one number in each interval—say, 7 in the interval 5–9, 12 in the interval 10–14, 17 in the interval 15–19, etc. Then use 7, 12, 17, . . . as the list of outcomes to be entered into the calculator. The frequency list (the last column of the table) remains the same.
2. *The histogram's bar width affects the shape of the graph.* If you use a bar width of 4 in Example 1, the calculator may produce a histogram with gaps in it. To avoid this, use the interval $5 \le x < 10$ in place of $5 \le x \le 9$, and similarly for the other intervals, and make 5 the bar width.

Following this procedure, we obtain the calculator-generated histogram and frequency polygon in Figure 10.2 for the data from Example 1. Note that the width of each histogram bar is 5. Some calculators cannot display both the histogram and the frequency polygon on the same screen, as is done here.

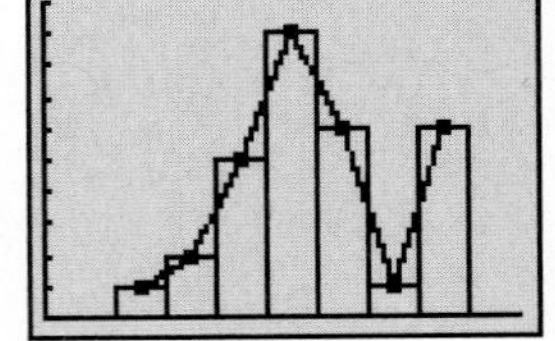

FIGURE 10.2

Stem-and-leaf plots allow us to organize the data into a distribution without the disadvantage of losing the original information. In a **stem-and-leaf plot**, we separate the digits in each data point into two parts consisting of the first one or two digits (the stem) and the remaining digit (the leaf). We also provide a key to show the reader the units of the data that was recorded.

Example 3 Construct a stem-and-leaf plot for the data in Example 1.

Solution Since the data is made up of two-digit numbers, we use the first digit for the stems: 0, 1, 2, and 3. The second digits provide the leaves. For example, if we look at the second row of the stem-and-leaf plot, we have a stem value of 1 and leaf values of 0 and 1. This corresponds to entries of 10 and 11, meaning one college had tuition of $10 (thousand) and another college had tuition of $11 (thousand). In this example, each row corresponds to an interval in the frequency table. The stems and leaves are separated by a vertical line.

Stem	Leaves
0	9
1	01
1	56788
2	122334444
2	567788
3	2
3	667899

Units: 3|9 = 39 thousand dollars

If we turn the page on its side, the distribution looks like a histogram, but still retains each of the original values. We used each stem digit twice, because, as with a histogram, using too few intervals conceals useful information about the shape of the distribution. 3✓

✓Checkpoint 3

Make a stem-and-leaf plot for the data in Example 1, using one stem each for 0, 1, 2, and 3.

Answer: See page 607.

Example 4 List the original data for the following stem-and-leaf plot of resting pulses taken on the first day of class for 36 students:

Stem	Leaves
4	8
5	278
6	034455688888
7	02222478
8	2269
9	00002289

Units: 9|0 = 90 beats per minute

The first stem and its leaf correspond to the data point 48 beats per minute. Similarly, the rest of the data are 52, 57, 58, 60, 63, 64, 64, 65, 65, 66, 68, 68, 68, 68, 68, 70, 72, 72, 72, 72, 74, 77, 78, 82, 82, 86, 89, 90, 90, 90, 90, 92, 92, 98, and 99 beats per minute. 4✓

✓Checkpoint 4

List the original data for the following heights (inches) of students:

Stem	Leaves
5	9
6	00012233334444
6	55567777799
7	0111134
7	558

Units: 7|5 = 75 inches

Answer: See page 607.

ASSESSING THE SHAPE OF A DISTRIBUTION

Histograms and stem-and-leaf plots are very useful in assessing what is called the **shape** of the distribution. One common shape of data is seen in Figure 10.3(a). When all the bars of a histogram are approximately the same height, we say the data has a **uniform** shape. In Figure 10.3(b), we see a histogram that is said to be bell shaped, or **normal**. We use the "normal" label when the frequency peaks in the middle and tapers off equally on each side. When the data does not taper off equally on each side, we say the data is **skewed**. If the data tapers off further to the left, we say the data is **left skewed** (Figure 10.3(c)). When the data tapers off further to the right, we say the data is **right skewed** (Figure 10.3(d)). (Notice that with skewed data, we say "left skewed" or "right skewed" to refer to the tail, and not the peak of the data.)

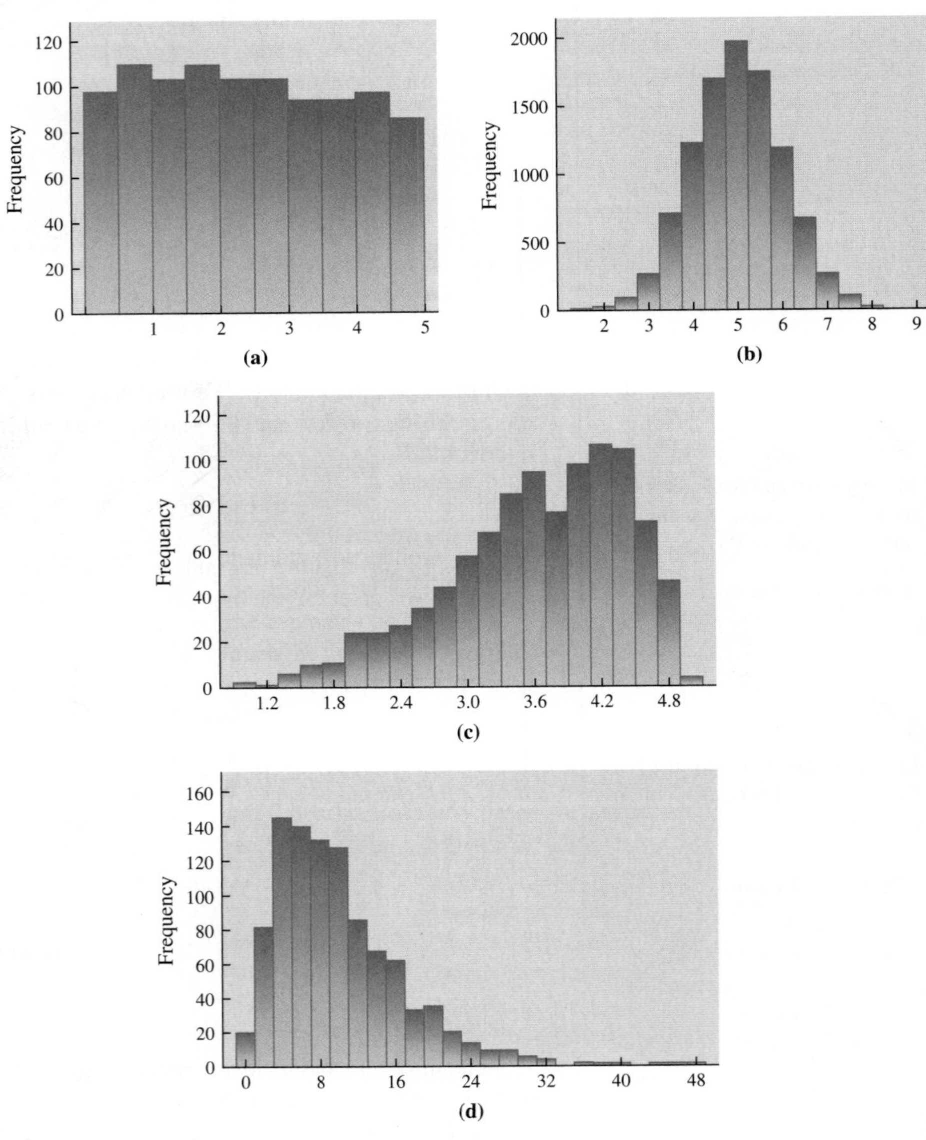

FIGURE 10.3

Example 5 **Health** Characterize the shapes of the given distributions for 1000 adult males.

(a) Height (inches); see Figure 10.4.

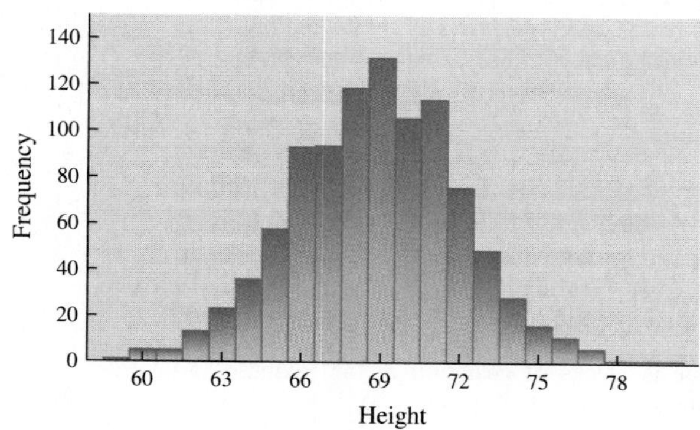

FIGURE 10.4

Solution The shape is **normal** because the shape peaks in the middle and tapers equally on each side.

(b) Body mass index (kg/m^2); see Figure 10.5.

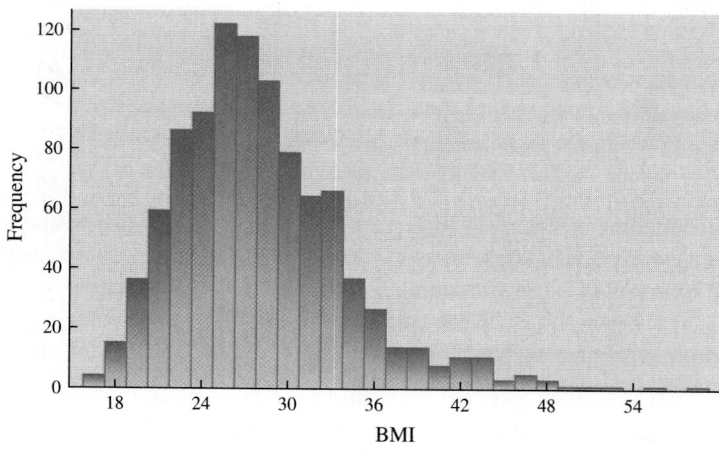

FIGURE 10.5

Solution The shape is **right skewed** because the tail is to the right. 

Checkpoint 5

Characterize the shape of the distribution from the following stem-and-leaf plot of ages (years):

Stem	Leaves
1	88888
2	22334
2	5579
3	333344
3	59
4	3344
4	667779
5	34
5	5679
6	1
6	
7	13
7	678

(*Units*: 7|8 = 78 years

Answer: See page 607.

It is important to note that most data is *not* normal, as we will see in the upcoming exercises and the next section. Using the label "normal" is a bit of a misnomer, because many important distributions, such as income, house prices, and infant birth weights, are skewed. It is also important to know that not all distributions have an easy-to-classify shape. This is especially true when samples are small.

10.1 ▶ Exercises

The data for Exercises 1–4 consists of a random sample of 50 births taken from the 2007 North Carolina Birth Registry. For each variable, (a) group the data as indicated; (b) prepare a frequency distribution with columns for intervals and frequencies; (c) construct a histogram; (d) construct a frequency polygon. (See Examples 1 and 2).*

1. The variable is the age, in years, of the mother giving birth. Use 8 intervals, starting with 14–17 (inclusive).

15	26	27	32	24	24	35	29	44	20
25	20	18	31	22	30	25	38	37	22
28	20	21	23	34	30	27	24	33	29
26	20	18	15	24	28	24	31	20	20
21	24	25	33	28	36	26	27	35	19

2. The variable is weeks of gestation of the infant. Use 9 intervals, starting with 27–28 (inclusive).

42	40	42	40	39	39	38	36	38	40
41	34	36	34	41	38	40	38	37	40
37	39	39	37	40	39	39	39	27	40
41	34	39	43	36	41	38	31	40	30
40	41	44	39	32	37	41	38	39	39

3. The variable is weight (in pounds) gained by the mother during pregnancy. Use 10 intervals, starting with 0–4 (inclusive).

8	13	15	22	25	20	35	29	5	37
49	35	49	35	32	28	42	25	25	45
47	35	45	5	30	32	16	33	19	30
40	32	39	36	7	22	28	30	0	20
10	10	35	31	41	31	25	26	20	11

4. The variable is weight (in grams) of the infant at birth. Use 8 intervals, starting with 1000–1499 (inclusive).

2608	3374	3260	3459	3204	3090	2835	3175	2778	4536
3289	2693	2693	2211	3742	3374	3799	3657	3232	3374
3459	3062	3572	2835	3600	3629	3572	3572	1106	3856
3686	3119	3033	3289	2552	3033	3005	1531	3090	1616
2523	3430	3204	2778	1644	2098	3884	3827	3941	3175

Finance *The data for Exercises 5–8 consists of random sample of 30 households from the 2007 American Community Survey.† Construct a frequency distribution and a histogram for each data set.*

5. Household income (in thousands of dollars):

159	83	15	17	80	159
127	53	102	46	79	46
149	100	99	179	27	171
14	52	33	230	49	22
39	27	9	13	86	38

6. Monthly mortgage payment (in dollars):

430	330	740	860	940	280
2400	1000	1800	300	2400	300
2000	350	3200	2100	710	1100
1300	1100	500	500	2000	800
1000	160	1400	900	700	740

7. Monthly electric bill (in dollars):

80	110	50	110	200	130
50	150	60	370	400	80
70	200	50	120	80	100
520	90	160	250	70	190
190	120	30	180	310	230

8. Annual property insurance costs (in dollars):

600	500	380	1900	370	1200
330	800	800	690	1100	100
970	600	400	600	540	3300
2400	460	2200	1300	600	1500
780	1000	1300	320	1000	4000

Construct a frequency distribution and a histogram for the data in Exercises 9 and 10.

9. **Business** The ages (in years) of the 30 highest-earning chief executive officers in 2008, according to *Forbes:**

63	43	62	65	73	55
66	56	66	70	58	61
47	53	50	55	63	62
69	57	58	58	58	49
54	59	50	62	54	53

10. **Social Science** The commuting time (in minutes) for 30 adults chosen at random:†

20	10	15	50	40	25
20	20	10	25	40	35
10	15	45	15	75	1
19	10	20	5	30	40
35	20	7	13	25	45

Construct a stem-and-leaf plot for the data in the indicated exercise. (See Example 3.)

11. Exercise 1 (use stems, 1, 1, 2, 2, 3, 3, 4)

12. Exercise 2

13. Exercise 3

14. Exercise 4 (round grams to the nearest hundred)

15. Exercise 5 (round incomes to the nearest ten thousand)

16. Exercise 6 (round dollars to the nearest hundred)

*http://arc.icss.unc.edu/dvn/dv/NCVITAL.

†Based on data from www.census.gov/acs.

*www.forbes.com.

†www.census.gov/acs.

17. Exercise 7

18. Exercise 8 (round dollars to the nearest hundred)

19. **Social Science** The following data gives the percentage of residents with a high school diploma for the 50 states and the District of Columbia in 2007:*

80	85	80	90	85	90
91	83	89	84	88	86
84	89	87	91	87	89
81	88	88	87	83	81
80	86	87	82	82	89
89	86	91	84	88	91
88	90	79	83	81	
87	89	86	89	79	
86	80	90	87	90	

Create a stem-and-leaf plot for this data.

20. **Social Science** The following data gives the percentage of residents with a bachelor's degree for the 50 states and the District of Columbia:*

21	26	20	28	23	34
26	27	27	22	28	34
25	29	35	33	26	30
19	25	38	34	30	17
30	30	25	25	24	25
35	22	31	32	25	23
35	24	19	26	22	
26	29	25	26	25	
48	20	27	24	29	

Create a stem-and-leaf plot for this data.

Describe the shape of each of the given histograms. (See Example 5.)

21.

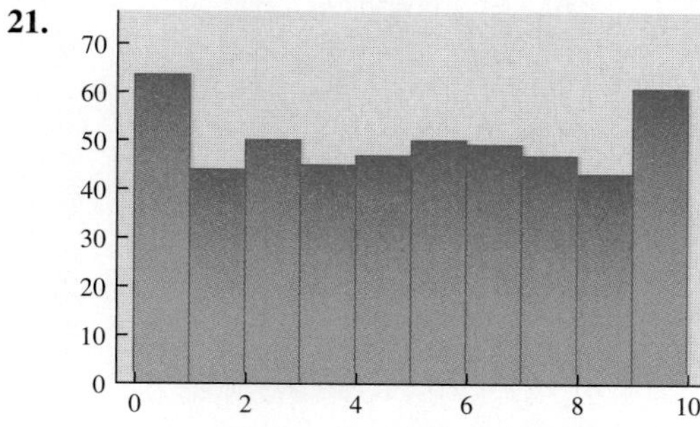

22.

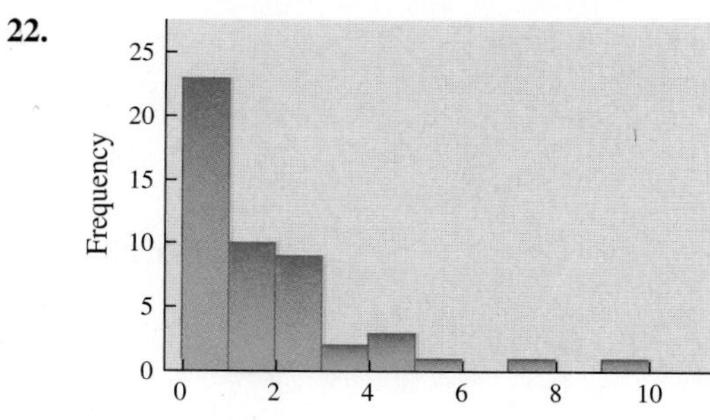

23.

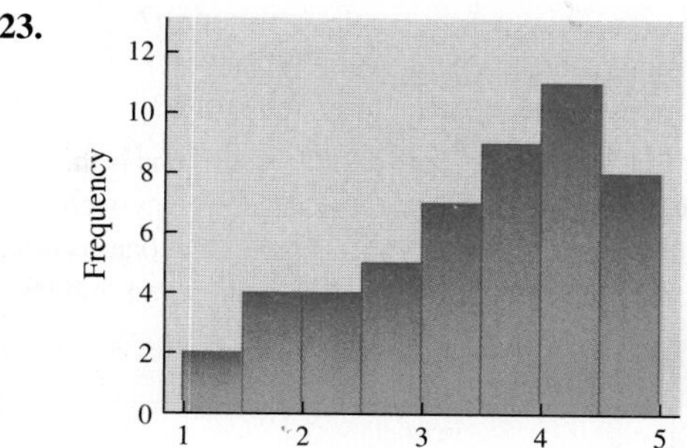

24.

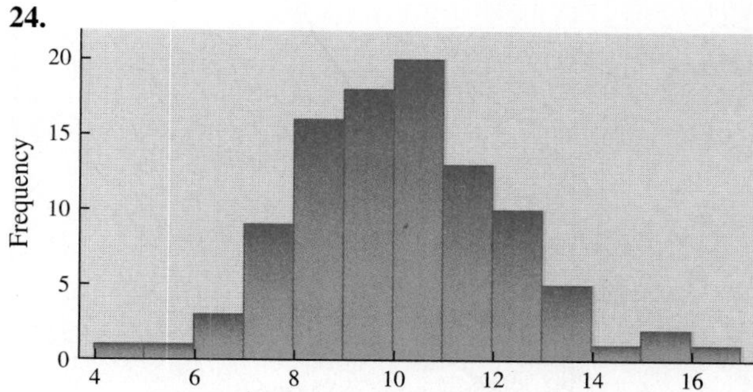

25.

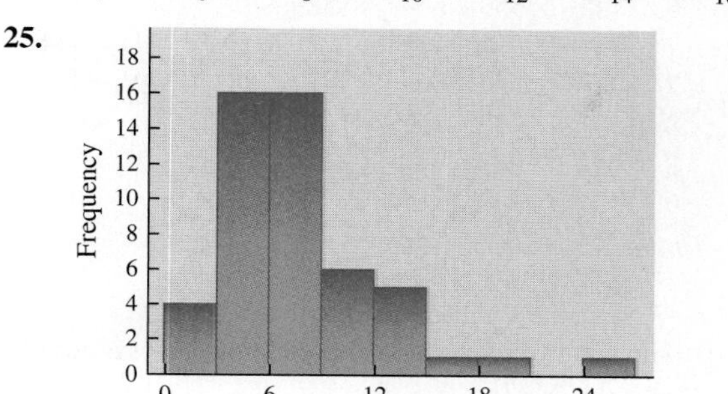

26. The grade distribution for scores on a final exam is shown in the following stem-and-leaf plot:

Stem	Leaves
2	7
3	
3	
4	01
4	899
5	4
5	5
6	122
6	58
7	00124
7	9
8	0044
8	5679
9	00223334
9	5788

Units: 9|8 = 98%

The Statistical Abstract of the United States: 2009.

(a) What is the shape of the grade distribution?
(b) How many students earned 90% or better?
(c) How many students earned less than 60%?

27. Finance The loan-to-asset ratio for credit unions is the percentage of the value of the assets that are encumbered in loans. Following is the stem-and-leaf plot of loan-to-asset ratio values for the credit unions in Massachusetts for 2008:*

Stem	Leaves
1	4
1	
2	0133
2	889
3	0334
3	556666677888899
4	00011123344444
4	55556677888899
5	0000111122222223344444
5	5666666777788889999
6	00111122222233333333444444444
6	5555556667778888999 9
7	0001111111111222233333334444444444
7	555566666667777777778888999
8	00000011223444
8	666778
9	023

Units: 9|3 = 93%

(a) What is the shape of the distribution?
(b) How many credit unions have loan-to-asset ratios of exactly 70%?
(c) Do more credit unions have ratios in the 30s or in the 40s?

28. Health The percentage of adults who smoke regularly is summarized for the 50 U.S. states and the District of Columbia in the following stem-and-leaf plot:†

Stem	Leaves
1	0
1	
1	4
1	67777777
1	8888888889999
2	000000011111
2	2222223333
2	44555
2	
2	8

Units: 2|8 = 28%

Notice that this plot has the leaves divided into five categories instead of two.
(a) Describe the shape of the distribution.
(b) How many states have a smoking percentage of 20% or higher?
(c) What is the lowest percentage? Can you guess which state that is?

29. Business The following stem-and-leaf plot gives the average insurance expenditure per insured vehicle (in tens of dollars) for the 50 states and the District of Columbia:*

Stem	Leaves
5	55689
6	012345567788999
7	0234455899
8	2444445
9	234689
10	2567
11	1288

Units: 11|8 = $1180

(a) Describe the shape of the distribution.
(b) How many states have an average expenditure of over $1000?
(c) How many states have an average expenditure of $840?

30. Business For the 22 states that produce large numbers of broiler chickens, the following stem-and-leaf plot gives the production in millions of live pounds:†

Stem	Leaves
0	0012234
0	8
1	02333
1	66
2	2
3	2
3	4
4	78
5	
5	
6	4
6	7

Units: 6|7 = 67 million pounds

(a) Describe the shape of the distribution.
(b) How many states produce fewer than 10 million pounds?

*www.ncua.org.
†www.cdc.gov/mmwr.

*www.naic.org.
†U.S. Department of Agriculture.

✓ Checkpoint Answers

1.

Interval	Frequency
0–4	3
5–9	7
10–14	9
15–19	3
20–24	2
	Total: 24

2.

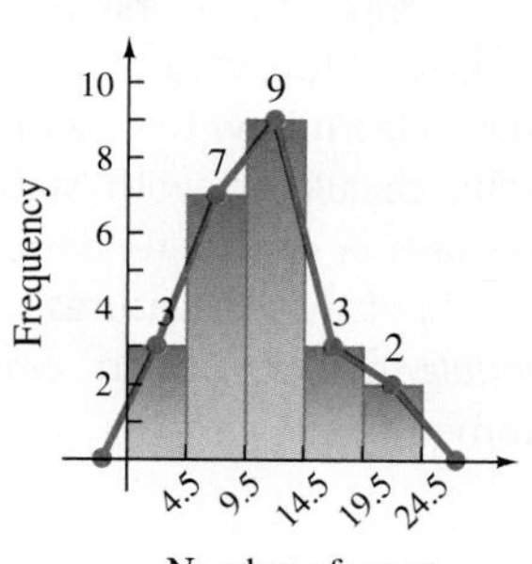

3.

Stem	Leaves
0	9
1	0156788
2	122334444567788
3	2667899

Units: 3|9 = 39 thousand dollars

4. 59, 60, 60, 60, 61, 62, 62, 63, 63, 63, 63, 64, 64, 64, 64, 65, 65, 65, 66, 67, 67, 67, 67, 67, 69, 69, 70, 71, 71, 71, 71, 73, 74, 75, 75, 78
5. Right skewed

10.2 ▶ Measures of Central Tendency

Often, we want to summarize data numerically with a measure that represents a "typical" outcome. There are several ways to do this, and we generally call such a summary a "measure of center." In this section, we learn about the three most common measures of center: the mean, median, and mode.

MEAN

The three most important measures of central tendency are the mean, the median, and the mode. The most used of these is the mean, which is similar to the expected value of a probability distribution. The **arithmetic mean** (or just the "mean") of a set of numbers is the sum of the numbers, divided by the total number of numbers. We write the sum of n numbers $x_1, x_2, x_3, \ldots, x_n$ in a compact way with **summation notation**, also called **sigma notation**. With the Greek letter Σ (sigma), the sum

$$x_1 + x_2 + x_3 + \cdots + x_n$$

is written

$$x_1 + x_2 + x_3 + \cdots + x_n = \sum_{i=1}^{n} x_i.$$

In statistics, $\Sigma_{i=1}^{n} x_i$ is often abbreviated as just Σx. The symbol $\bar{x}$ (read "x-bar") is used to represent the mean of a sample.

Mean

The mean of the n numbers $x_1, x_2, x_3, \ldots, x_n$ is

$$\bar{x} = \frac{x_1 + x_2 + \cdots + x_n}{n} = \frac{\Sigma x}{n}.$$

TECHNOLOGY TIP Computing the mean is greatly simplified by the statistical capabilities of most scientific and graphing calculators. Calculators vary considerably in how data is entered, so read your instruction manual to learn how to enter lists of data and the corresponding frequencies. On scientific calculators with statistical capabilities, there are keys for finding most of the measures of central tendency discussed in this section. On graphing calculators, most or all of these measures can be obtained with a single keystroke. (Look for a *one-variable statistics* option, which is often labeled 1-VAR, in the STAT menu or its CALC submenu.)

Example 1 **Business** The number of businesses filing for bankruptcy for the years 2004–2008 are given in the following table:*

Year	Petitions Filed
2004	34,817
2005	39,201
2006	19,695
2007	28,322
2008	43,546

Find the mean number of business bankruptcies filed annually during this period.

Solution Let $x_1 = 34{,}817$, $x_2 = 39{,}201$, and so on. Here, $n = 5$, since there are 5 numbers in the list, so

$$\bar{x} = \frac{34{,}817 + 39{,}201 + 19{,}695 + 28{,}322 + 43{,}546}{5} \approx 33{,}116.$$

The mean number of business bankruptcy petitions filed during the given years is 33,116. ✓1

The mean of data that has been arranged into a frequency distribution is found in a similar way. For example, suppose the following quiz score data is collected:

✓Checkpoint 1

Find the mean dollar amount of the following purchases of eight students selected at random at the campus bookstore during the first week of classes:

\$250.56	\$567.32
\$45.29	\$321.56
\$120.22	\$561.04
\$321.07	\$226.90

Answer: See page 617.

*www.uscourts.gov/.

Value	Frequency
84	2
87	4
88	7
93	4
99	3
	Total: 20

TECHNOLOGY TIP The mean of the five numbers in Example 1 is easily found by using the $\overline{x}$ key on a scientific calculator or the one-variable statistics key on a graphing calculator. A graphing calculator will also display additional information, which will be discussed in the next section.

The value 84 appears twice, 87 four times, and so on. To find the mean, first add 84 two times, 87 four times, and so on; or get the same result faster by multiplying 84 by 2, 87 by 4, and so on, and then adding the results. Dividing the sum by 20, the total of the frequencies, gives the mean:

$$\begin{aligned}\overline{x} &= \frac{(84\cdot 2) + (87\cdot 4) + (88\cdot 7) + (93\cdot 4) + (99\cdot 3)}{20} \\ &= \frac{168 + 348 + 616 + 372 + 297}{20} \\ &= \frac{1801}{20} \\ \overline{x} &= 90.05.\end{aligned}$$

Verify that your calculator gives the same result.

Example 2 **Social Science** An instructor of a finite-mathematics class at a small liberal-arts college collects data on the age of her students. The data is recorded in the following frequency distribution:

Age	Frequency	Age × Frequency
18	12	$18\cdot 12 = 216$
19	9	$19\cdot 9 = 171$
20	5	$20\cdot 5 = 100$
21	2	$21\cdot 2 = 42$
22	2	$22\cdot 2 = 44$
	Total: 30	Total: 573

Find the mean age.

Solution The age 18 appears 12 times, 19 nine times, and so on. To find the mean, first multiply 18 by 12, 19 by 9, and so on, to get the column "Age × Frequency," which has been added to the frequency distribution. Adding the products from this column gives a total of 573. The total from the frequency column is 30. The mean age is

$$\overline{x} = \frac{573}{30} = 19.1.$$ ✓2

✓Checkpoint 2

Find $\overline{x}$ for the following frequency distribution for the variable of years of schooling for a sample of construction workers.

Years	Frequency
7	2
9	3
11	6
13	4
15	4
16	1

Answer: See page 617.

The mean of grouped data is found in a similar way. For grouped data, intervals are used, rather than single values. To calculate the mean, it is assumed that all of the values in a given interval are located at the midpoint of the interval. The letter x is used to represent the midpoints, and f represents the frequencies, as shown in the next example.

Example 3 **Social Science** The grouped frequency distribution for annual tuition (in thousands of dollars) for the 30 private colleges described in Example 1 of Section 10.1 is as follows:

Tuition Amount	Midpoint, x	Frequency, f	Product, xf
5–9	7	1	7
10–14	12	2	24
15–19	17	5	85
20–24	22	9	198
25–29	27	6	162
30–34	32	1	32
35–39	37	6	222
		Total: 30	Total: 730

Find the mean from the grouped frequency distribution.

Solution A column for the midpoint of each interval has been added. The numbers in this column are found by adding the endpoints of each interval and dividing by 2. For the interval 5–9, the midpoint is $(5 + 9)/2 = 7$. The numbers in the product column on the right are found by multiplying each frequency by its corresponding midpoint. Finally, we divide the total of the product column by the total of the frequency column to get

$$\bar{x} = \frac{730}{30} \approx 24.3.$$

Note that information is always lost when the data is grouped. It is more accurate to use the original data, rather than the grouped frequency, when calculating the mean, but the original data might not be available. Furthermore, the mean based upon the grouped data is typically not too different from the mean based upon the original data, and there may be situations in which the extra accuracy is not worth the extra effort. ✓3

✓Checkpoint 3

Find the mean of the following grouped frequency distribution for the number of classes completed thus far in the college careers of a random sample of 52 students:

Classes	Frequency
0–5	6
6–10	10
11–20	12
21–30	15
31–40	9

Answer: See page 617.

✓Checkpoint 4

Find the mean for the college tuition data using the following intervals for the grouped frequency distribution:

Tuition Amount	Frequency
6–11	3
12–17	3
18–23	7
24–29	10
30–35	1
36–41	6

Answer: See page 617.

NOTE **1.** The midpoint of an interval in a grouped frequency distribution may be a value that none of the data assumes. For example, if we grouped the tuition data for the 30 private colleges into the intervals 6–11, 12–17, 18–23, 24–29, 30–35, 36–41, the midpoints would be 8.5, 14.5, 20.5, 26.5, 32.5, and 38.5, respectively, even though all the data as reported were whole numbers.

2. If we had used different intervals in Example 3, the mean would have come out to be a slightly different number. This is demonstrated in Checkpoint 4. ✓4

The formula for the mean of a grouped frequency distribution is as follows.

Mean of a Grouped Distribution

The mean of a distribution in which x represents the midpoints, f denotes the frequencies, and $n = \Sigma f$ is

$$\bar{x} = \frac{\Sigma(xf)}{n}.$$

The mean of a random sample is a random variable, and for this reason it is sometimes called the **sample mean**. The sample mean is a random variable because it assigns a number to the experiment of taking a random sample. If a different random sample were taken, the mean would probably have a different value, with some values being more probable than others. For example, if another set of 30 private colleges were selected in Example 3, the mean tuition amount might have been 31.4.

We saw in Section 9.1 how to calculate the expected value of a random variable when we know its probability distribution. The expected value is sometimes called the **population mean**, denoted by the Greek letter μ. In other words,

$$E(x) = \mu.$$

Furthermore, it can be shown that the expected value of $\bar{x}$ is also equal to μ; that is,

$$E(\bar{x}) = \mu.$$

For instance, consider again the 30 private colleges in Example 3. We found that $\bar{x} = 24.3$, but the value of μ, the average for all tuition amounts, is unknown. If a good estimate of μ were needed, the best guess (based on this data) is 24.3.

MEDIAN

Asked by a reporter to give the average height of the players on his team, a Little League coach lined up his 15 players by increasing height. He picked out the player in the middle and pronounced this player to be of average height. This kind of average, called the **median**, is defined as the middle entry in a set of data arranged in either increasing or decreasing order. If the number of entries is even, the median is defined to be the mean of the two middle entries. The following table shows how to find the median for the two sets of data {8, 7, 4, 3, 1} and {2, 3, 4, 7, 9, 12} after each set has been arranged in increasing order.

Odd Number of Entries	Even Number of Entries
1	2
3	3
Median = 4	4 } *Median* $= \frac{4+7}{2} = 5.5$
7	7 }
8	9
	12

NOTE As shown in the table, when the number of entries is even, the median is not always equal to one of the data entries.

Example 4 Find the median number of hours worked per week

(a) for a sample of 7 male students whose work hours were

0, 7, 10, 20, 22, 25, 30.

Solution The median is the middle number, in this case 20 hours per week. (Note that the numbers are already arranged in numerical order.) In this list, three numbers are smaller than 20 and three are larger.

(b) for a sample of 11 female students whose work hours were

20, 0, 20, 30, 35, 30, 20, 23, 16, 38, 25.

Solution First, arrange the numbers in numerical order, from smallest to largest, or vice versa:

0, 16, 20, 20, 20, 23, 25, 30, 30, 35, 38.

The middle number can now be determined; the median is 23 hours per week.

(c) for a sample of 10 students of either gender whose work hours were

25, 18, 25, 20, 16, 12, 10, 0, 35, 32.

Solution Write the numbers in numerical order:

0, 10, 12, 16, **18**, **20**, 25, 25, 32, 35.

There are 10 numbers here; the median is the mean of the two middle numbers, or

$$\text{median} = \frac{18 + 20}{2} = 19.$$

The median is 19 hours per week.

✓Checkpoint 5

Find the median for the given heights in inches.

(a) 60, 72, 64, 75, 72, 65, 68, 70

(b) 73, 58, 77, 66, 69, 69, 66, 68, 67

Answers: See page 617.

CAUTION Remember, the data must be arranged in numerical order before you locate the median. 5✓

Both the mean and the median of a sample are examples of a **statistic**, which is simply a number that gives summary information about a sample. In some situations, the median gives a truer representative or typical element of the data than the mean does. For example, suppose that in an office there are 10 salespersons, 4 secretaries, the sales manager, and Ms. Daly, who owns the business. Their annual salaries are as follows: support staff, \$30,000 each; salespersons, \$50,000 each; manager, \$70,000; and owner, \$400,000. The mean salary is

$$\bar{x} = \frac{(30{,}000)4 + (50{,}000)10 + 70{,}000 + 400{,}000}{16} = \$68{,}125.$$

However, since 14 people earn less than \$68,125 and only 2 earn more, the mean does not seem very representative. The median salary is found by ranking the salaries by size: \$30,000, \$30,000, \$30,000, \$30,000, \$50,000, \$50,000, . . . ,

TECHNOLOGY TIP Many graphing calculators (including most TI- and Casio models) display the median when doing one-variable statistics. You may have to scroll down to a second screen to find it.

$400,000. There are 16 salaries (an even number) in the list, so the mean of the 8th and 9th entries will give the value of the median. The 8th and 9th entries are both $50,000, so the median is $50,000. In this example, the median is more representative of the distribution than the mean is.

When the data includes extreme values (such as $400,000 in the preceding example), the mean may not provide an accurate picture of a typical value. So the median is often a better measure of center than the mean for data with extreme values, such as income levels and house prices. In general, the median is a better measure of center whenever we see right-skewed or left-skewed distributions.

MODE

Sue's scores on 10 class quizzes include one 7, two 8's, six 9's, and one 10. She claims that her average grade on quizzes is 9, because most of her scores are 9's. This kind of "average," found by selecting the most frequent entry, is called the **mode**.

Example 5 Find the mode for the given data sets.

(a) Ages of retirement: 55, 60, 63, 63, 70, 55, 60, 65, 68, 65, 65, 71, 65, 65

Solution The number 65 occurs more often than any other, so it is the mode. It is sometimes convenient, but not necessary, to place the numbers in numerical order when looking for the mode.

(b) Total cholesterol score: 180, 200, 220, 260, 220, 205, 255, 240, 190, 300, 240

Solution Both 220 and 240 occur twice. This list has *two* modes, so it is bimodal.

(c) Prices of new cars: $25,789, $43,231, $33,456, $19,432, $22,971, $29,876

Solution No number occurs more than once. This list has no mode. ✓6

✓Checkpoint 6

Find the mode for each of the given data sets.

(a) Highway miles per gallon of an automobile: 25, 28, 32, 19, 15, 25, 30, 25

(b) Price paid for last haircut or styling: $11, $35, $35, $10, $0, $12, $0, $35, $38, $42, $0, $25

(c) Class enrollment in six sections of calculus: 30, 35, 26, 28, 29, 19

Answers: See page 617.

The mode has the advantages of being easily found and not being influenced by data that are extreme values. It is often used in samples where the data to be "averaged" are not numerical. A major disadvantage of the mode is that there may be more than one, in case of ties, or there may be no mode at all when all entries occur with the same frequency.

The mean is the most commonly used measure of central tendency. Its advantages are that it is easy to compute, it takes all the data into consideration, and it is reliable—that is, repeated samples are likely to give similar means. A disadvantage of the mean is that it is influenced by extreme values, as illustrated in the salary example.

The median can be easy to compute and is influenced very little by extremes. A disadvantage of the median is the need to rank the data in order; this can be tedious when the number of items is large.

✓Checkpoint 7

Following is a list of the number of movies seen at a theater in the last three months by nine students selected at random:

1, 0, 2, 5, 2, 0 , 0, 1, 4

(a) Find the mean.

(b) Find the median.

(c) Find the mode.

Answers: See page 617.

Example 6 **Business** A sample of 10 working adults was asked "How many hours did you work last week?" Their responses were as follows:

40, 35, 43, 40, 30, 40, 45, 40, 55, 20.*

Find the mean, median, and mode of the data.

*www.norc.org/GSS+Website.

Solution The mean number of hours worked is

$$\bar{x} = \frac{40 + 35 + 43 + 40 + 30 + 40 + 45 + 40 + 55 + 20}{10} = 38.8 \text{ hours.}$$

After the numbers are arranged in order from smallest to largest, the middle number, or median, is 40 hours.

The number 40 occurs more often than any other, so it is the mode. 

10.2 Exercises

Find the mean for each data set. Round to the nearest tenth. (See Example 1.)

1. Secretarial salaries (U.S. dollars):

   \$21,900, \$22,850, \$24,930, \$29,710, \$28,340, \$40,000.

2. Starting teaching salaries (U.S. dollars):

   \$38,400, \$39,720, \$28,458, \$29,679, \$33,679.

3. Earthquakes on the Richter scale:

   3.5, 4.2, 5.8, 6.3, 7.1, 2.8, 3.7, 4.2, 4.2, 5.7.

4. Body temperatures of self-classified "healthy" students (degrees Fahrenheit):

   96.8, 94.1, 99.2, 97.4, 98.4, 99.9, 98.7, 98.6.

5. Lengths of foot (inches) for adult men:

   9.2, 10.4, 13.5, 8.7, 9.7.

Find the mean for each distribution. Round to the nearest tenth. (See Examples 2 and 3.)

6. Scores on a quiz, on a scale from 0 to 10:

Value	Frequency
7	4
8	6
9	7
10	11

7. Age (years) of students in an introductory accounting class:

Value	Frequency
19	3
20	5
21	25
22	8
23	2
24	1
28	1

8. Commuting distance (miles) for students at a university:

Value	Frequency
0	15
1	12
2	8
5	6
10	5
17	2
20	1
25	1

9. Estimated miles per gallon of automobiles:

Value	Frequency
9	5
11	10
15	12
17	9
20	6
28	1

10–14. *Find the median of the data in Exercises 1–5. (See Example 4.)*

Find the mode or modes for each of the given lists of numbers. (See Example 5.)

15. Ages (years) of children in a day-care facility:

    1, 2, 2, 1, 2, 2, 1, 1, 2, 2, 3, 4, 2, 3, 4, 2, 3, 2, 3.

16. Ages (years) in the intensive care unit at a local hospital:

    68, 64, 23, 68, 70, 72, 72, 68.

17. Heights (inches) of students in a statistics class:

    62, 65, 71, 74, 71, 76, 71, 63, 59, 65, 65, 64, 72, 71, 77, 63, 65.

18. Minutes of pain relief from acetaminophen after childbirth:

    60, 240, 270, 180, 240, 210, 240, 300, 330, 360, 240, 120.

19. Grade point averages for 5 students:

3.2, 2.7, 1.9, 3.7, 3.9.

20. When is the median the most appropriate measure of central tendency?

21. Under what circumstances would the mode be an appropriate measure of central tendency?

For grouped data, the modal class is the interval containing the most data values. Give the mean and modal class for each of the given collections of grouped data. (See Example 3.)

22. **Health** Weight gain (in pounds) for 50 mothers giving birth:*

Interval	Frequency
0–4	1
5–9	4
10–14	4
15–19	3
20–24	5
25–29	8
30–34	9
35–39	8
40–44	3
45–49	5

23. Weight of 50 newly born infants (in grams):

Interval	Frequency
1000–1499	1
1500–1999	3
2000–2499	2
2500–2999	9
3000–3499	21
3500–3999	13
4000–4499	0
4500–4999	1

24. To predict the outcome of the next congressional election, you take a survey of your friends. Is this a random sample of the voters in your congressional district? Explain why or why not.

*http://arc.irss.unc.edu/dvn/dv/NCVITAL.

Work each problem. (See Example 6.)

25. **Social Science** The following table shows the number of nations participating in the winter Olympic games from 1972 to 2006:*

Year	Nations Participating
1972	35
1976	37
1980	37
1984	49
1988	57
1992	64
1994	67
1998	72
2002	77
2006	80

Find the following statistics for the data:
(a) mean;
(b) median;
(c) mode.

26. **Business** The following table gives the value (in millions of dollars) of the 10 most valued baseball teams as estimated by *Forbes* in 2007:†

Rank	Team	Value
1	New York Yankees	1306
2	New York Mets	824
3	Boston Red Sox	816
4	Los Angels Dodgers	694
5	Chicago Cubs	642
6	Los Angeles Angels of Anaheim	500
7	Atlanta Braves	497
8	San Francisco Giants	494
9	St. Louis Cardinals	484
10	Philadelphia Phillies	481

(a) Find the mean value of these teams.
(b) Find the median value of these teams.
(c) What might account for the difference between these values?

The New York Times Almanac: 2008.

†www.forbes.com.

27. Business The 12 movies that have earned the most revenue (in millions of dollars) from U.S. domestic box-office receipts are given in the table:*

Rank	Title	U.S. Box-Office Receipts
1	*Titanic*	601
2	*The Dark Knight*	533
3	*Star Wars*	461
4	*Shrek 2*	436
5	*E. T.: The Extra-Terrestrial*	435
6	*Star Wars: Episode I—The Phantom Menace*	431
7	*Pirates of the Caribbean: Dead Man's Chest*	423
8	*Spider-Man*	404
9	*Star Wars: Episode III—Revenge of the Sith*	380
10	*The Lord of the Rings: The Return of the King*	377
11	*Spider-Man 2*	373
12	*The Passion of the Christ*	370

(a) Find the mean value in dollars for this group of movies.
(b) Find the median value in dollars for this group of movies.

28. Natural Science The number of recognized blood types varies by species, as indicated in the following table.†

Animal	Number of Blood Types
Pig	16
Cow	12
Chicken	11
Horse	9
Human	8
Sheep	7
Dog	7
Rhesus Monkey	6
Mink	5
Rabbit	5
Mouse	4
Rat	4
Cat	2

Find the mean, median, and mode of this data.

29. Business The revenue (in millions of dollars) for the Starbucks Corporation for 1999–2008 is given in the table:*

Year	Revenue
1999	1680.2
2000	2169.2
2001	2649.0
2002	3288.9
2003	4075.5
2004	5294.3
2005	6369.3
2006	7786.9
2007	9411.5
2008	10,383.0

(a) Calculate the mean and median for this data.
(b) What year's revenue revenue is closest to the mean?

Natural Science *The table gives the average monthly high and low temperatures, in degrees Fahrenheit, for Raleigh, NC, over the course of a year:*†

Month	High	Low
January	49	30
February	53	32
March	61	40
April	71	48
May	78	57
June	84	65
July	88	69
August	86	68
September	80	62
October	70	49
November	61	42
December	52	33

Find the mean and median for each of the given subgroups.

30. The high temperatures **31.** The low temperatures

Business *For each of Exercises 32 and 33, a frequency distribution and its histogram have been constructed from the 2009 Fan Cost index (FCI) report from Team Marketing Report®. The FCI is a measure of how much it costs a family of four to attend a major-league baseball game by taking into account ticket prices, parking, and the costs for food, drink, a program, and a cap.‡ Determine the*

*www.imdb.com as of April 18, 2009.

†*The Handy Science Answer Book*, Carnegie Library of Pittsburgh, Pennsylvania, p. 264.

*www.morningstar.com.

†www.weather.com.

‡http://www.teammarketing.com/fancost/.

shape of the distribution from the histogram, and then decide if the mean or median is a better measure of center. If the mean is the better measure, calculate the value. If the median is the better measure, give the midpoint of the interval that contains it.

32. The frequency distribution and histogram for the average ticket price:

Price (Dollars)	Frequency
10–19.99	10
20–29.99	13
30–39.99	4
40–49.99	1
50–59.99	1
60–69.99	0
70–79.99	1

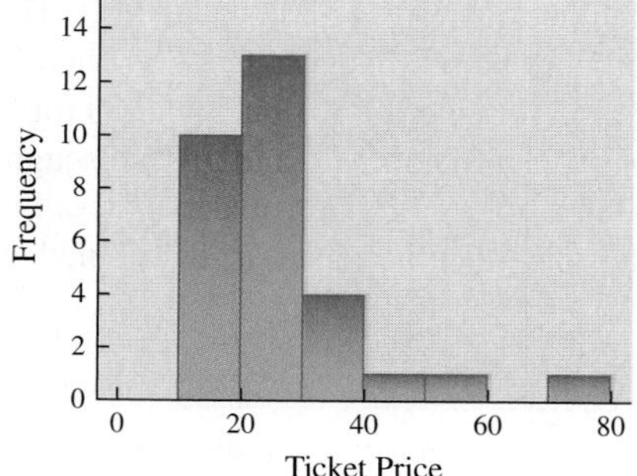

33. Business The frequency distribution and the histogram for the Fan Cost Index (FCI) are given below.

FCI (Dollars)	Frequency
100–149.99	5
150–199.99	13
200–249.99	8
250–299.99	1
300–349.99	2
350–399.99	0
400–449.99	1

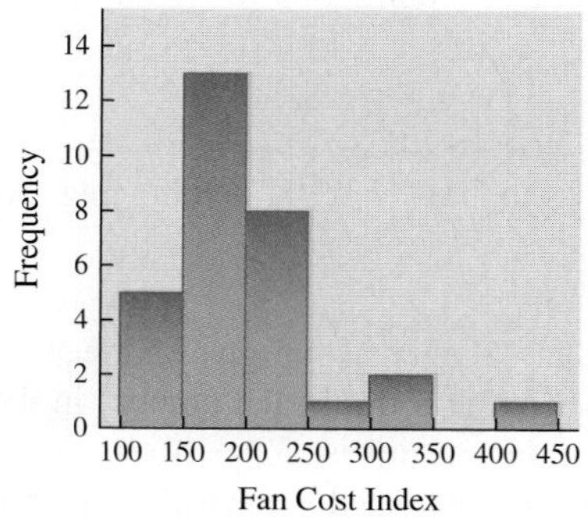

34. Health The following stem-and-leaf plot gives the distribution for the percentage of children without health insurance in 2006 (rounded to the nearest percent) for the 50 U.S. states and the District of Columbia.*

Stem	Leaves
0	4445
0	666666777777
0	888888999999
1	000001
1	222333
1	44455
1	77
1	888
2	1

Units: 2|1 = 21%

(a) Describe the shape of the distribution.
(b) Find the median percentage.

35. Health The following stem-and-leaf plot gives the percentages of all residents (rounded to the nearest percent) without health insurance for the 50 U.S. states and the District of Columbia.

Stem	Leaves
0	888999
1	00000011111
1	222233333
1	44455555
1	6777777
1	8889
2	0011
2	2
2	4

Units: 2|4 = 24%

(a) Describe the shape of the distribution.
(b) Find the median percentage.

✓Checkpoint Answers

1. \$301.75 **2.** $\bar{x} = 11.75$
3. 18.90 **4.** 24.7
5. (a) 69 inches **(b)** 68 inches
6. (a) 25 miles per gallon **(b)** \$0 and \$35
(c) No mode
7. (a) About 1.7 **(b)** 1 **(c)** 0

*www.census.gov.

10.3 Measures of Variation

The mean, median, and mode are measures of central tendency for a list of numbers, but tell nothing about the *spread* of the numbers in the list. For example, look at the following data sets of number of times per week three people ate meals at restaurants over the course of five weeks:

Jill:	3	5	6	3	3
Miguel:	4	4	4	4	4
Sharille:	10	1	0	0	9

Each of these three data sets has a mean of 4, but the amount of dispersion or variation within the lists is different. This difference may reflect different dining patterns over time. Thus, in addition to a measure of central tendency, another kind of measure is needed that describes how much the numbers vary.

The largest number of restaurant meals for Jill is 6, while the smallest is 3, a difference of 3. For Miguel, the difference is 0; for Sharille, it is 10. The difference between the largest and smallest number in a sample is called the **range**, one example of a measure of variation. The range is 3 for Jill, 0 for Miguel, and 10 for Sharille. The range has the advantage of being very easy to compute and gives a rough estimate of the variation among the data in the sample. However, it depends only on the two extremes and tells nothing about how the other data is distributed between the extremes.

TECHNOLOGY TIP Many graphing calculators show the largest and smallest numbers in a list when displaying one-variable statistics, usually on the second screen of the display.

Example 1 **Business** Find the range for each given data set for a small sample of people.

(a) Price paid for last haircut (with tip): 10, 0, 15, 30, 20, 18, 50, 120, 75, 95, 0, 5

Solution The highest number here is 120; the lowest is 0. The range is the difference of these numbers, or

$$120 - 0 = 120.$$

(b) Amount spent for last vehicle servicing: 30, 19, 125, 150, 430, 50, 225

Solution Range $= 430 - 19 = 411.$ ✓1

✓Checkpoint 1

Find the range for this sample of the number of miles from students' homes to college: 15, 378, 5, 210, 125.

Answer: See page 627.

To find another useful measure of variation, we begin by finding the **deviations from the mean**—the differences found by subtracting the mean from each number in a distribution.

Example 2 Find the deviations from the mean for the following sample of ages.

$$32, \quad 41, \quad 47, \quad 53, \quad 57.$$

Solution Adding these numbers and dividing by 5 gives a mean of 46 years. To find the deviations from the mean, subtract 46 from each number in the sample. For example, the first deviation from the mean is $32 - 46 = -14$; the last is $57 - 46 = 11$ years. All of the deviations are listed in the following table.

Age	Deviation from Mean
32	−14
41	−5
47	1
53	7
57	11

To check your work, find the sum of the deviations. It should always equal 0. (The answer is always 0 because the positive and negative deviations cancel each other out.) 2✓

✓**Checkpoint 2**

Find the deviations from the mean for the following sample of number of miles traveled by various people to a vacation location:

135, 60, 50, 425, 380.

Answer: See page 627.

To find a measure of variation, we might be tempted to use the mean of the deviations. However, as just mentioned, this number is always 0, no matter how widely the data is dispersed. To avoid the problem of the positive and negative deviations averaging to 0, we could take absolute values and find $\Sigma|x - \bar{x}|$ and then divide it by n to get the *mean deviation*. However, statisticians generally prefer to square each deviation to get nonnegative numbers and then take the square root of the mean of the squared variations in order to preserve the units of the original data (such as inches, pounds). (Using squares instead of absolute values allows us to take advantage of some algebraic properties that make other important statistical methods much easier.) The squared deviations for the data in Example 2 are shown in the following table:

Number	Deviation from Mean	Square of Deviation
32	−14	196
41	−5	25
47	1	1
53	7	49
57	11	121

In this case, the mean of the squared deviations is

$$\frac{196 + 25 + 1 + 49 + 121}{5} = \frac{392}{5} = 78.4.$$

This number is called the **population variance**, because the sum was divided by $n = 5$, the number of items in the original list.

Since the deviations from the mean must add up to 0, if we know any 4 of the 5 deviations, the 5th can be determined. That is, only $n - 1$ of the deviations are free to vary, so we really have only $n - 1$ independent pieces of information, or *degrees of freedom*. Using $n - 1$ as the divisor in the formula for the mean gives

$$\frac{196 + 25 + 1 + 49 + 121}{5 - 1} = \frac{392}{4} = 98.$$

This number, 98, is called the **sample variance** of the distribution and is denoted s^2, because it is found by averaging a list of squares. In this case, the population and sample variances differ by quite a bit. But when n is relatively large, as is the case in real-life applications, the difference between them is rather small.

Sample Variance

The variance of a sample of n numbers $x_1, x_2, x_3, \ldots, x_n$, with mean $\bar{x}$, is

$$s^2 = \frac{\Sigma(x - \bar{x})^2}{n - 1}.$$

When computing the sample variance by hand, it is often convenient to use the following shortcut formula, which can be derived algebraically from the definition in the preceding box:

$$s^2 = \frac{\Sigma x^2 - n\bar{x}^2}{n - 1}.$$

To find the sample variance, we square the deviations from the mean, so the variance is in squared units. To return to the same units as the data, we use the *square root* of the variance, called the **sample standard deviation**, denoted s.

Sample Standard Deviation

The standard deviation of a sample of n numbers $x_1, x_2, x_3, \ldots, x_n$, with mean $\bar{x}$, is

$$s = \sqrt{\frac{\Sigma(x - \bar{x})^2}{n - 1}}.$$

NOTE The **population standard deviation** is

$$\sigma = \sqrt{\frac{\Sigma(x - \bar{x})^2}{n}},$$

where n is the population size.

TECHNOLOGY TIP When a graphing calculator computes one-variable statistics for a list of data, it usually displays the following information (not necessarily in this order, and sometimes on two screens) and possibly other information as well:

Information	Notation
Number of data entries	n or $N\Sigma$
Mean	$\bar{x}$ or mean Σ
Sum of all data entries	Σx or TOT Σ
Sum of the squares of all data entries	Σx^2
Sample standard deviation	Sx or sx or $x\sigma_{n-1}$ or SSDEV
Population standard deviation	σx or $x\sigma_n$ or PSDEV
Largest/smallest data entries	maxX/minX or MAXΣ/MINΣ
Median	Med or MEDIAN

NOTE In the rest of this section, we shall deal exclusively with the sample variance and the sample standard deviation. So whenever standard deviation is mentioned, it means "sample standard deviation," not population standard deviation.

As its name indicates, the standard deviation is the most commonly used measure of variation. The standard deviation is a measure of the variation from the mean. The size of the standard deviation indicates how spread out the data is from the mean.

Example 3 Find the standard deviation for the following sample of the lengths (in minutes) of eight consecutive cell phone conversations by one person:

$$2,\quad 8,\quad 3,\quad 2,\quad 6,\quad 11,\quad 31,\quad 9.$$

Work by hand, using the shortcut variance formula on page 620.

Solution Arrange the work in columns, as shown in the table in the margin. Now use the first column to find the mean:

Time	Square of the Time
2	4
8	64
3	9
2	4
6	36
11	121
31	961
9	81
72	1280

$$\bar{x} = \frac{\sum x}{8} = \frac{72}{8} = 9 \text{ minutes.}$$

The total of the second column gives $\sum x^2 = 1280$. The variance is

$$s^2 = \frac{\sum x^2 - n\bar{x}^2}{n - 1}$$

$$= \frac{1280 - 8(9)^2}{8 - 1}$$

$$= 90.3 \text{ (rounded),}$$

and the standard deviation is

$$s \approx \sqrt{90.3} \approx 9.5 \text{ minutes.}$$ 3

✓Checkpoint 3

Find the standard deviation for a sample of the number of miles traveled by various people to a vacation location:

135, 60, 50, 425, 380.

Answer: See page 627.

TECHNOLOGY TIP The screens in Figure 10.6 show two ways to find variance and standard deviation on a TI-84+ calculator: with the LIST menu and with the STAT menu. The data points are first entered in a list—here, L_5. See your instruction book for details.

FIGURE 10.6

In a spreadsheet, enter the data in cells A1 through A8. Then, in cell A9, type "=VAR (A1..A8)" and press Enter. The standard deviation can be calculated either by taking the square root of cell A9 or by typing "=STDEV (A1..A8)" in cell A10 and pressing Enter.

CAUTION We must be careful to divide by $n - 1$, not n, when calculating the standard deviation of a sample. Many calculators are equipped with statistical keys that compute the variance and standard deviation. Some of these calculators use $n - 1$, and others use n for these computations; some may have keys for both. Check your calculator's instruction book before using a statistical calculator for the exercises.

One way to interpret the standard deviation uses the fact that, for many populations, most of the data is within three standard deviations of the mean. (See Section 10.4.) This implies that, in Example 3, most of the population data from which this sample is taken is between

$$\bar{x} - 3s = 9 - 3(9.5) = -19.5$$

and

$$\bar{x} + 3s = 9 + 3(9.5) = 37.5.$$

For Example 3, the preceding calculations imply that most phone conversations are less than 37.5 minutes long. This approach of determining whether sample observations are beyond 3 standard deviations of the mean is often employed in conducting quality control in many industries.

For data in a grouped frequency distribution, a slightly different formula for the standard deviation is used.

Standard Deviation for a Grouped Distribution

The standard deviation for a sample distribution with mean $\bar{x}$, where x is an interval midpoint with frequency f and $n = \Sigma f$, is

$$s = \sqrt{\frac{\Sigma f x^2 - n\bar{x}^2}{n - 1}}.$$

The formula indicates that the product fx^2 is to be found for each interval. Then all the products are summed, n times the square of the mean is subtracted, and the difference is divided by one less than the total frequency—that is, by $n - 1$. The square root of this result is s, the standard deviation. The standard deviation found by this formula may (and probably will) differ somewhat from the standard deviation found from the original data.

CAUTION In calculating the standard deviation for either a grouped or an ungrouped distribution, using a rounded value for the mean or variance may produce an inaccurate value.

Example 4 The following frequency distribution gives the 2008–2009 annual tuition (in hundreds of dollars) for a random sample of 30 community colleges:*

*Based on data from *The Chronicle of Higher Education*, at http://chronicle.com/.

Class	Frequency f
5–9.99	3
10–14.99	4
15–19.99	6
20–24.99	3
25–29.99	6
30–34.99	4
35–39.99	1
40–44.99	1
45–49.99	2

Find the sample standard deviation s for this data.

Solution We first need to find the mean $\overline{x}$ for this grouped data. We find the midpoint of each interval and label it x. We multiply the frequency by the midpoint x to obtain fx:

Class	Frequency f	Midpoint x	fx
5–9.99	3	7.5	22.5
10–14.99	4	12.5	50
15–19.99	6	17.5	105
20–24.99	3	22.5	67.5
25–29.99	6	27.5	165
30–34.99	4	32.5	130
35–39.99	1	37.5	37.5
40–44.99	1	42.5	42.5
45–49.99	2	47.5	95
	Total = 30		Total = 715

Therefore,

$$\overline{x} = \frac{715}{30} \approx 23.8.$$

Now that we have the mean value, we can modify our table to include columns for x^2 and fx^2. We obtain the following results:

Class	Frequency f	x^2	fx^2
5–9.99	3	56.25	168.75
10–14.99	4	156.25	625.00
15–19.99	6	306.25	1837.50
20–24.99	3	506.25	1518.75
25–29.99	6	756.25	4537.50
30–34.99	4	1056.25	4225.00
35–39.99	1	1406.25	1406.25
40–44.99	1	1806.25	1806.25
45–49.99	2	2256.25	4512.50
	Total = 30		20,637.50

We now use the formula for the standard deviation with $n = 30$ to find s:

$$s = \sqrt{\frac{\Sigma fx^2 - n\bar{x}^2}{n - 1}}$$

$$= \sqrt{\frac{20{,}637.5 - 30(23.8)^2}{30 - 1}}$$

$$\approx 11.21.$$ ✓4

✓Checkpoint 4

Find the standard deviation for the following grouped frequency distribution of the number of classes completed thus far in the college careers of a random sample of 52 students:

Classes	Frequency
0–5	6
6–10	10
11–20	12
21–30	15
31–40	9

Answer: See page 627.

NOTE A calculator is almost a necessity for finding a standard deviation. With a nongraphing calculator, a good procedure to follow is first to calculate $\bar{x}$. Then, for each x, square that number, and multiply the result by the appropriate frequency. If your calculator has a key that accumulates a sum, use it to accumulate the total in the last column of the table. With a graphing calculator, simply enter the midpoints and the frequencies, and then ask for the one-variable statistics.

10.3 ▶ Exercises

1. How are the variance and the standard deviation related?
2. Why can't we use the sum of the deviations from the mean as a measure of dispersion of a distribution?

Finance *In Exercises 3–10, expenditures (in millions of dollars) for various government services in 2005 are given for the five largest counties in the United States by population: Los Angeles, CA; Cook, IL; Harris, TX; Maricopa, AZ; and Orange, CA.* Find the range and the standard deviation for each given category.*

3. Housing: 5, 10, 3, 11, 15
4. Public Welfare: 4614, 11, 33, 695, 735
5. Health: 1770, 39, 200, 77, 322
6. Hospitals: 2916, 897, 880, 162, 0
7. Police Protection: 1118, 94, 363, 57, 270
8. Correction: 969, 389, 69, 324, 252
9. Highways: 251, 136, 372, 103, 36
10. Parks and Recreation: 227, 112, 26, 7, 1

Find the standard deviation for the grouped data in Exercises 11 and 12. (See Example 4.)

11. Number of credits for a sample of college students:

College Credits	Frequency
0–24	4
25–49	3
50–74	6
75–99	3
100–124	5
125–149	9

12. Scores on a calculus exam:

Scores	Frequency
30–39	1
40–49	6
50–59	13
60–69	22
70–79	17
80–89	13
90–99	8

13. **Natural Science** Twenty-five laboratory rats used in an experiment to test the food value of a new product made the following weight gains in grams:

**Statistical Abstract of the United States*: 2009.

5.25	5.03	4.90	4.97	5.03
5.12	5.08	5.15	5.20	4.95
4.90	5.00	5.13	5.18	5.18
5.22	5.04	5.09	5.10	5.11
5.23	5.22	5.19	4.99	4.93

Find the mean gain and the standard deviation of the gains.

14. Business An assembly-line machine turns out washers with the following thicknesses (in millimeters):

1.20	1.01	1.25	2.20	2.58	2.19	1.29	1.15
2.05	1.46	1.90	2.03	2.13	1.86	1.65	2.27
1.64	2.19	2.25	2.08	1.96	1.83	1.17	2.24

Find the mean and standard deviation of these thicknesses.

An application of standard deviation is given by Chebyshev's theorem. (P. L. Chebyshev was a Russian mathematician who lived from 1821 to 1894.) This theorem, which applies to any distribution of numerical data, states,

For any distribution of numerical data, at least $1 - 1/k^2$ of the numbers lie within k standard deviations of the mean.

Example *For any distribution, at least*

$$1 - \frac{1}{3^2} = 1 - \frac{1}{9} = \frac{8}{9}$$

of the numbers lie within 3 standard deviations of the mean. Find the fraction of all the numbers of a data set lying within the given numbers of standard deviations from the mean.

15. 2 **16.** 4 **17.** 1.5

In a certain distribution of numbers, the mean is 50, with a standard deviation of 6. Use Chebyshev's theorem to tell what percent of the numbers are

18. between 32 and 68;

19. between 26 and 74;

20. less than 38 or more than 62;

21. less than 32 or more than 68;

22. less than 26 or more than 74.

Business *The following table gives the total amounts of sales (in millions of dollars) for aerobic, basketball, and cross-training shoes in the United States from 2001–2007:**

	Aerobic	Basketball	Cross-training
2001	281	761	1476
2002	239	789	1421
2003	222	890	1407
2004	237	877	1327
2005	261	878	1437
2006	262	964	1516
2007	268	987	1561

*www.nsga.org.

23. Find the mean and standard deviation for the aerobic shoe sales.

24. Find the mean and standard deviation for the basketball shoe sales.

25. Find the mean and standard deviation for the cross-training shoe sales.

26. Which type of shoe has the most variation in its sales? Explain.

27. Natural Science The number of recognized blood types of various animal species is given in the following table:*

Animal	Number of Blood Types
Pig	16
Cow	12
Chicken	11
Horse	9
Human	8
Sheep	7
Dog	7
Rhesus Monkey	6
Mink	5
Rabbit	5
Mouse	4
Rat	4
Cat	2

In Exercise 28 of the previous section, the mean was found to be 7.38.

(a) Find the variance and the standard deviation of this data.

(b) How many of these animals have blood types that are within 1 standard deviation of the mean?

28. Social Science The table shows the salaries (in thousands of dollars) of the nine highest-paid state governors as of September 2007:†

State	Salary
CA	212
NY	179
MI	177
NJ	175
VA	175
PA	164
WA	163
TN	160
IL	151

(a) Find the mean salary of these governors. Which state has the governor with the salary closest to the mean?

**The Handy Science Answer Book*, Carnegie Library of Pittsburgh, Pennsylvania, p. 264.

†*The World Almanac and Book of Facts*: 2008.

(b) Find the standard deviation for the data.

(c) What percentage of the governors had salaries within 1 standard deviation of the mean?

(d) What percentage of the governors had salaries within 3 standard deviations of the mean?

29. Health The amounts of time that it takes for various slow-growing tumors to double in size are listed in the following table:*

Type of Cancer	Doubling Time (days)
Breast cancer	84
Rectal cancer	91
Synovioma	128
Skin cancer	131
Lip cancer	143
Testicular cancer	153
Esophageal cancer	164

(a) Find the mean and standard deviation of this data.

(b) How many of these cancers have doubling times that are within 2 standard deviations of the mean?

(c) If a person had a nonspecified tumor that was doubling every 200 days, discuss whether this particular tumor was growing at a rate that would be expected.

30. Business The Quaker Oats Company conducted a survey to determine whether a proposed premium, to be included with purchases of the firm's cereal, was appealing enough to generate new sales.† Four cities were used as test markets, where the cereal was distributed with the premium, and four cities were used as control markets, where the cereal was distributed without the premium. The eight cities were chosen on the basis of their similarity in terms of population, per-capita income, and total cereal purchase volume. The results were as follows:

		Percent Change in Average Market Shares per Month
Test Cities	1	+18
	2	+15
	3	+7
	4	+10
Control Cities	1	+1
	2	−8
	3	−5
	4	0

(a) Find the mean of the change in market share for the four test cities.

(b) Find the mean of the change in market share for the four control cities.

(c) Find the standard deviation of the change in market share for the test cities.

(d) Find the standard deviation of the change in market share for the control cities.

(e) Find the difference between the mean of part (a) and the mean of part (b). This represents the estimate of the percent change in sales due to the premium.

(f) The two standard deviations from part (c) and part (d) were used to calculate an "error" of ±7.95 for the estimate in part (e). With this amount of error, what is the smallest and largest estimate of the increase in sales? (*Hint*: Use the answer to part (e).)

On the basis of the results of this exercise, the company decided to mass-produce the premium and distribute it nationally.

31. Business The following table gives 10 samples of three measurements made during a production run:

SAMPLE NUMBER									
1	**2**	**3**	**4**	**5**	**6**	**7**	**8**	**9**	**10**
2	3	−2	−3	−1	3	0	−1	2	0
−2	−1	0	1	2	2	1	2	3	0
1	4	1	2	4	2	2	3	2	2

(a) Find the mean $\bar{x}$ for each sample of three measurements.

(b) Find the standard deviation s for each sample of three measurements.

(c) Find the mean $\bar{\bar{x}}$ of the sample means.

(d) Find the mean $\bar{s}$ of the sample standard deviations.

(e) The upper and lower control limits of the sample means here are $\bar{\bar{x}} \pm 1.954\bar{s}$. Find these limits. If any of the measurements are outside these limits, the process is out of control. Decide whether this production process is out of control.

32. Discuss what the standard deviation tells us about a distribution.

Social Science *Shown in the following table are the reading scores of a second-grade class given individualized instruction and the reading scores of a second-grade class given traditional instruction in the same school:*

Scores	Individualized Instruction	Traditional Instruction
50–59	2	5
60–69	4	8
70–79	7	8
80–89	9	7
90–99	8	6

*Vincent Collins, R. Kenneth Lodffer, and Harold Tivey, "Observations on Growth Rates of Human Tumors," *American Journal of Roentgen*, 76, no. 5 (November 1956): 988–1000.

†This example was supplied by Jeffery S. Berman, senior analyst, Marketing Information, Quaker Oats Company.

33. Find the mean and standard deviation for the individualized-instruction scores.

34. Find the mean and standard deviation for the traditional-instruction scores.

35. Discuss a possible interpretation of the differences in the means and the standard deviations in Exercises 33 and 34.

✓ Checkpoint Answers

1. 373
2. Mean is 210; deviations are −75, −150, −160, 215, and 170.
3. 179.5 miles
4. 10.92 classes

10.4 ▶ Normal Distributions and Boxplots

Suppose a bank is interested in improving its services to customers. The manager decides to begin by finding the amount of time tellers spend on each transaction, rounded to the nearest minute. The times for 75 different transactions are recorded, with the results shown in the following table, where the frequencies listed in the second column are divided by 75 to find the empirical probabilities:

Time	Frequency	Probability
1	3	$3/75 = .04$
2	5	$5/75 \approx .07$
3	9	$9/75 = .12$
4	12	$12/75 = .16$
5	15	$15/75 = .20$
6	11	$11/75 \approx .15$
7	10	$10/75 \approx .13$
8	6	$6/75 = .08$
9	3	$3/75 = .04$
10	1	$1/75 \approx .01$

Figure 10.7(a) shows a histogram and frequency polygon for the data. The heights of the bars are the empirical probabilities, rather than the frequencies. The transaction times are given to the nearest minute. Theoretically, at least, they could have been timed to the nearest tenth of a minute, or hundredth of a minute, or even more precisely. In each case, a histogram and frequency polygon could be drawn. If the times are measured with smaller and smaller units, there are more bars in the histogram and the frequency polygon begins to look more and more like the curve in Figure 10.7(b) instead of a polygon. Actually, it is possible for the transaction times to take on any real-number value greater than 0. A distribution in which the outcomes can take on any real-number value within some interval is a **continuous distribution**. The graph of a continuous distribution is a curve.

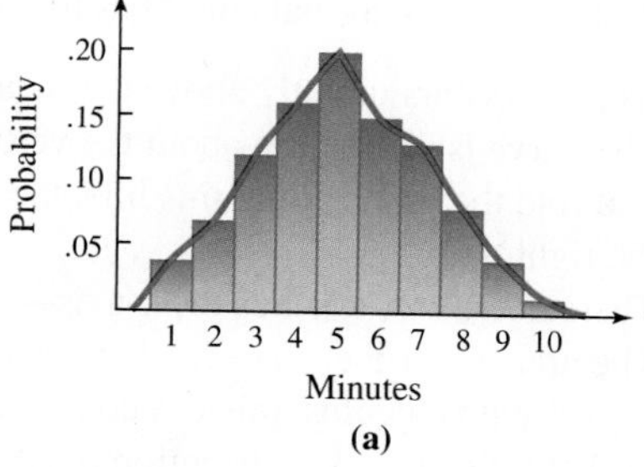

(a)

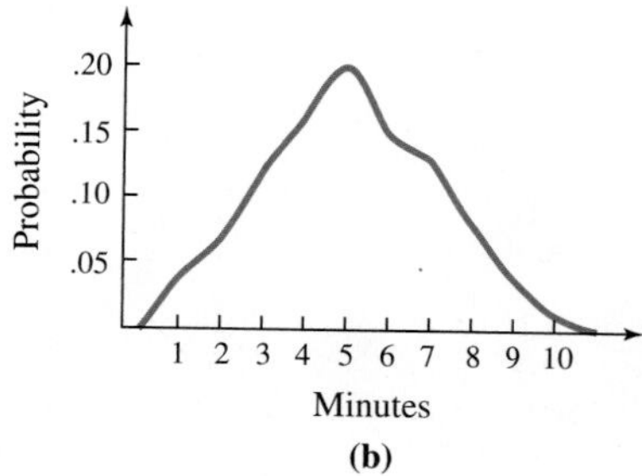

(b)

FIGURE 10.7

The distribution of heights (in inches) of college women is another example of a continuous distribution, since these heights include infinitely many possible measurements, such as 53, 58.5, 66.3, 72.666, and so on. Figure 10.8 shows the continuous distribution of heights of college women. Here, the most frequent heights occur near the center of the interval displayed.

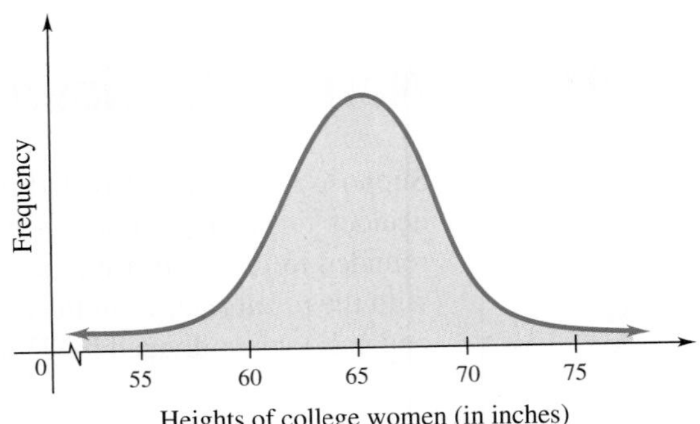

FIGURE 10.8

NORMAL DISTRIBUTIONS

As discussed on page 602, we say that data is normal (or normally distributed) when its graph is well approximated by a bell-shaped curve. (See Figure 10.9.) We call the graphs of such distributions **normal curves**. Examples of distributions that are approximately normal are the heights of college women and cholesterol levels in adults. We use the Greek letters μ (mu) to denote the mean and σ (sigma) to denote the standard deviation of a normal distribution.

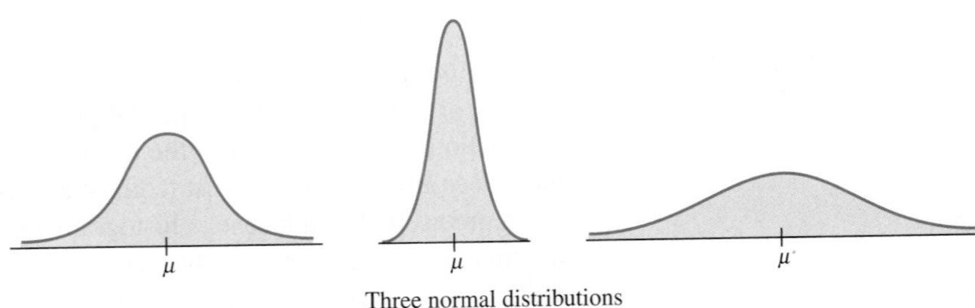

FIGURE 10.9

There are many normal distributions, depending on μ and σ. Some of the corresponding normal curves are tall and thin, and others short and wide, as shown in Figure 10.9. But every normal curve has the following properties:

1. Its peak occurs directly above the mean μ.
2. The curve is symmetric about the vertical line through the mean. (That is, if you fold the graph along this line, the left half of the graph will fit exactly on the right half).
3. The curve never touches the x-axis—it extends indefinitely in both directions.
4. The area under the curve (and above the horizontal axis) is 1. (As can be shown with calculus, this is a consequence of the fact that the sum of the probabilities in any distribution is 1.)

A normal distribution is completely determined by its mean μ and standard deviation σ.* A small standard deviation leads to a tall, narrow curve like the one in the center of Figure 10.9, because most of the data is close to the mean. A large standard deviation means the data is very spread out, producing a flat, wide curve like the one on the right in Figure 10.9.

Since the area under a normal curve is 1, parts of this area can be used to determine certain probabilities. For instance, Figure 10.10(a) is the probability distribution of the annual rainfall in a certain region. The probability that the annual rainfall will be between 25 and 35 inches is the area under the curve from 25 to 35. The general case, shown in Figure 10.10(b), can be stated as follows.

> The area of the shaded region under the normal curve from a to b is the probability that an observed data value will be between a and b.

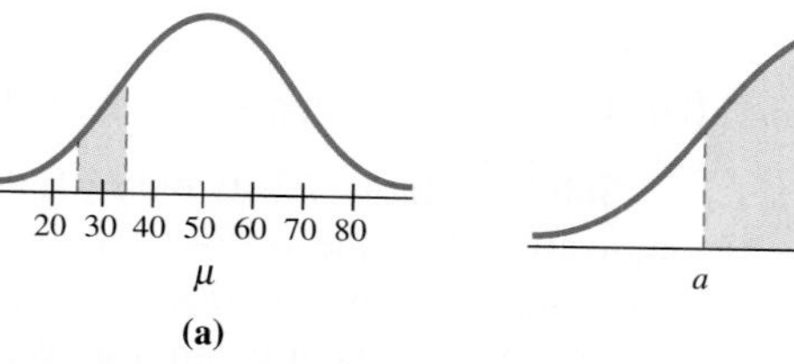

FIGURE 10.10

To use normal curves effectively, we must be able to calculate areas under portions of them. These calculations have already been done for the normal curve with mean $\mu = 0$ and standard deviation $\sigma = 1$ (which is called the **standard normal curve**) and are available in Table 2 at the back of the book. Examples 1 and 2 demonstrate how to use Table 2 to find such areas. Later, we shall see how the standard normal curve may be used to find areas under any normal curve.

The horizontal axis of the standard normal curve is usually labeled z. Since the standard deviation of the standard normal curve is 1, the numbers along the horizontal axis (the z-values) measure the number of standard deviations above or below the mean $z = 0$.

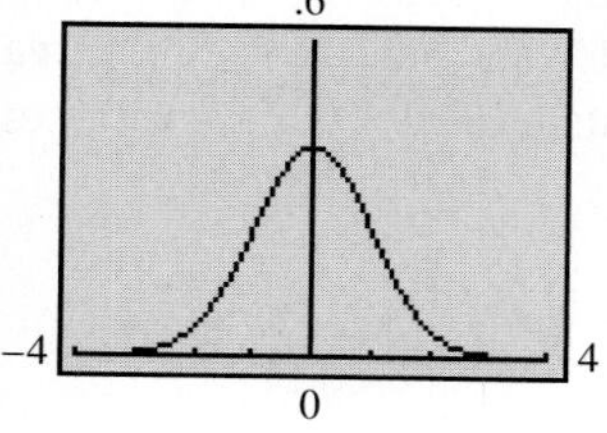

FIGURE 10.11

TECHNOLOGY TIP Some graphing calculators (such as the TI-84+ and most Casios) have the ability to graph a normal distribution, given its mean and standard deviation, and to find areas under the curve between two x-values. For an area under the curve, some calculators will give the corresponding z-value. For details, see your instruction book. (Look for "distribution" or "probability distribution.") A calculator-generated graph of the standard normal curve is shown in Figure 10.11.

---

*As shown in more advanced courses, its graph is the graph of the function

$$f(x) = \frac{1}{\sigma\sqrt{2\pi}} e^{-(x-\mu)^2/(2\sigma^2)},$$

where $e \approx 2.71828$ is the real number introduced in Section 4.1.

Example 1 Find the given areas under the standard normal curve.

(a) The area between $z = 0$ and $z = 1$, the shaded region in Figure 10.12

Solution Find the entry 1 in the z-column of Table 2. The entry next to it in the A-column is .3413, which means that the area between $z = 0$ and $z = 1$ is .3413. Since the total area under the curve is 1, the shaded area in Figure 10.12 is 34.13% of the total area under the normal curve.

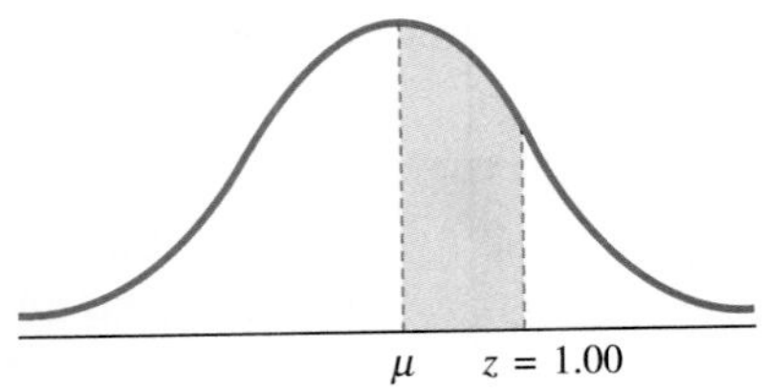

FIGURE 10.12

(b) The area between $z = -2.43$ and $z = 0$

Solution Table 2 lists only positive values of z. But the normal curve is symmetric around the mean $z = 0$, so the area between $z = 0$ and $z = -2.43$ is the same as the area between $z = 0$ and $z = 2.43$. Find 2.43 in the z-column of Table 2. The entry next to it in the A-column shows that the area is .4925. Hence, the shaded area in Figure 10.13 is 49.25% of the total area under the curve. ✓1 ✓2

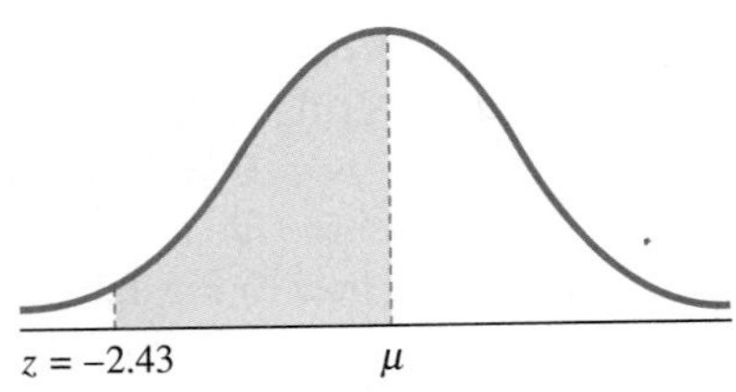

FIGURE 10.13

✓Checkpoint 1

Find the percent of the area between the mean and

(a) $z = 1.51$;

(b) $z = -2.04$.

(c) Find the percent of the area in the shaded region.

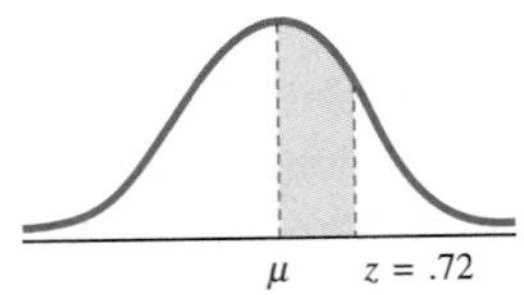

Answers: See page 640.

✓Checkpoint 2

If your calculator can graph probability distributions and find areas, use it to find the areas requested in Example 1.

Answers: See page 640.

TECHNOLOGY TIP Because of their convenience and accuracy, graphing calculators and computers have made normal-curve tables less important. Figure 10.14 shows how part (b) of Example 1 can be done on a TI-84+ calculator using a command from the DISTR menu. The second result in the calculator screen gives the area between $-\infty$ and $z = -2.43$; the entry −1E99 represents $-1 \cdot 10^{99}$, which is used to approximate $-\infty$.

FIGURE 10.14

Many statistical software packages are widely used today. All of these packages are set up in a way that is similar to a spreadsheet, and they all can be used to generate normal curve values. In addition, most spreadsheets can perform a wide range of statistical calculations.

Example 2 Use technology or Table 2 to find the percent of the total area for the given areas under the standard normal curve.

(a) The area between .88 standard deviations *below* the mean and 2.35 standard deviations *above* the mean (that is, between $z = -.88$ and $z = 2.35$)

Solution First, draw a sketch showing the desired area, as in Figure 10.15. From Table 2, the area between the mean and .88 standard deviations below the mean is .3106. Also, the area from the mean to 2.35 standard deviations above the mean is .4906. As the figure shows, the total desired area can be found by *adding* these numbers:

$$\begin{array}{r} .3106 \\ +.4906 \\ \hline .8012 \end{array}$$

The shaded area in Figure 10.15 represents 80.12% of the total area under the normal curve.

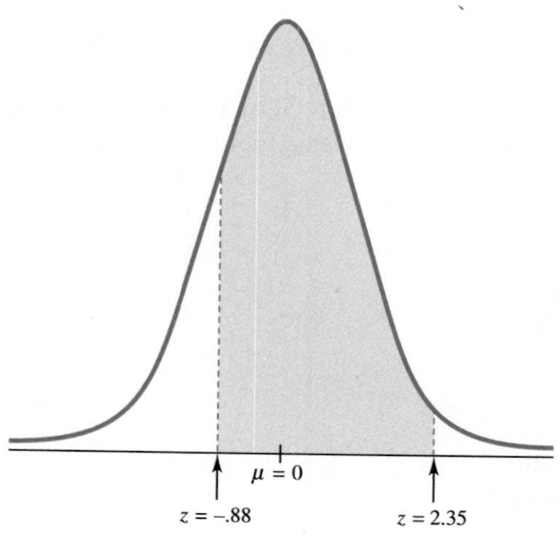

FIGURE 10.15

(b) The area between .58 standard deviations above the mean and 1.94 standard deviations above the mean

Solution Figure 10.16 on the next page shows the desired area. The area between the mean and .58 standard deviations above the mean is .2190. The area between the mean and 1.94 standard deviations above the mean is .4738. As the figure shows, the desired area is found by *subtracting* one area from the other:

$$\begin{array}{r} .4738 \\ -.2190 \\ \hline .2548 \end{array}$$

The shaded area of Figure 10.16 represents 25.48% of the total area under the normal curve.

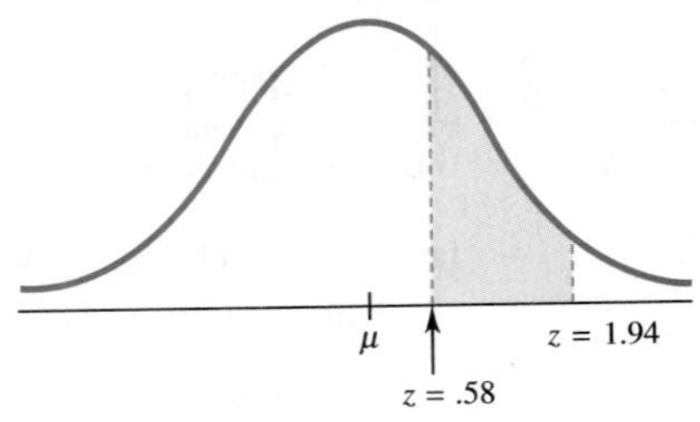

FIGURE 10.16

(c) The area to the right of 2.09 standard deviations above the mean

Solution The total area under a normal curve is 1. Thus, the total area to the right of the mean is 1/2, or .5000. From Table 2, the area from the mean to 2.09 standard deviations above the mean is .4817. The area to the right of 2.09 standard deviations is found by subtracting .4817 from .5000:

$$\begin{array}{r} .5000 \\ -.4817 \\ \hline .0183 \end{array}$$

A total of 1.83% of the total area is to the right of 2.09 standard deviations above the mean. Figure 10.17 (which is not to scale) shows the desired area. 3

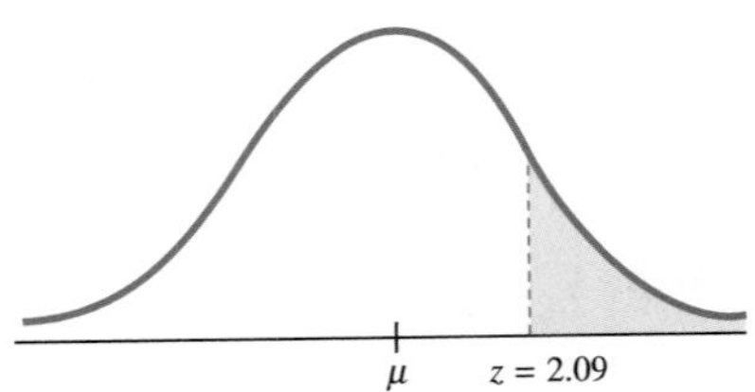

FIGURE 10.17

✓Checkpoint 3

Find the given standard normal-curve areas as percentages of the total area.

(a) Between .31 standard deviations below the mean and 1.01 standard deviations above the mean

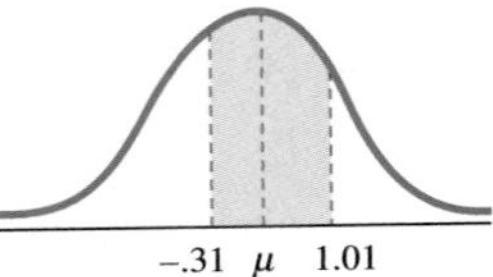

(b) Between .38 standard deviations and 1.98 standard deviations below the mean

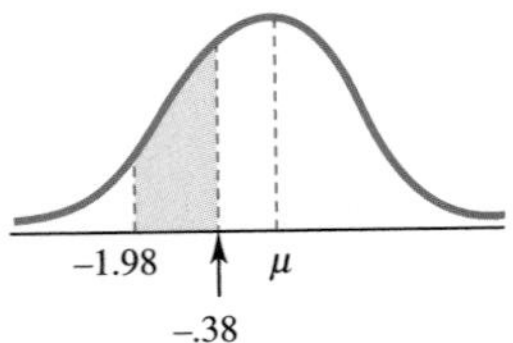

(c) To the right of 1.49 standard deviations above the mean

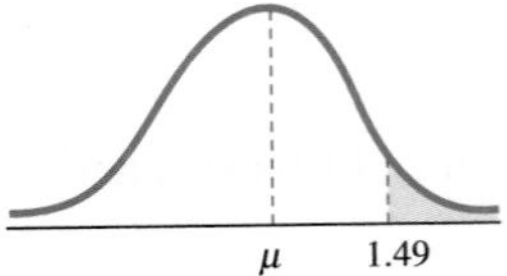

(d) What percent of the area is within 1 standard deviation of the mean? within 2 standard deviations of the mean? within 3 standard deviations of the mean? What can you conclude from the last answer?

Answers: See page 640.

The key to finding areas under *any* normal curve is to express each number x on the horizontal axis in terms of standard deviations above or below the mean. The **z-score** for x is the number of standard deviations that x lies from the mean (positive if x is above the mean, negative if x is below the mean).

Example 3 If a normal distribution has mean 60 and standard deviation 5, find the given z-scores.

(a) The z-score for $x = 65$

Solution Since 65 is 5 units above 60 and the standard deviation is 5, 65 is 1 standard deviation above the mean. So its z-score is 1.

(b) The z-score for $x = 52.5$

Solution The z-score is -1.5, because 52.5 is 7.5 units below the mean (since $52.5 - 60 = -7.5$) and 7.5 is 1.5 standard deviations (since $7.5/5 = 1.5$). 4

Checkpoint 4

Find each z-score, using the information in Example 3.

(a) $x = 36$

(b) $x = 55$

Answers: See page 640.

In Example 3(b) we found the z-score by taking the difference between 52.5 and the mean and dividing this difference by the standard deviation. The same procedure works in the general case.

If a normal distribution has mean μ and standard deviation σ, then the z-score for the number x is

$$z = \frac{x - \mu}{\sigma}.$$

The importance of z-scores is the following fact, whose proof is omitted.

Area under a Normal Curve

The area under a normal curve between $x = a$ and $x = b$ is the same as the area under the standard normal curve between the z-score for a and the z-score for b.

Therefore, by converting to z-scores and using a graphing calculator or Table 2 for the standard normal curve, we can find areas under any normal curve. Since these areas are probabilities (as explained earlier), we can now handle a variety of applications.

Graphing calculators, computer programs, and CAS programs (such as DERIVE) can be used to find areas under the normal curve and, hence, probabilities. The equation of the standard normal curve, with $\mu = 0$ and $\sigma = 1$, is

$$f(x) = (1/\sqrt{2\pi})e^{-x^2/2}.$$

A good approximation of the area under this curve (and above $y = 0$) can be found by using the x-interval $[-4, 4]$. However, calculus is needed to find such areas.

Example 4 **Business** Dixie Office Supplies finds that its sales force drives an average of 1200 miles per month per person, with a standard deviation of 150 miles. Assume that the number of miles driven by a salesperson is closely approximated by a normal distribution.

(a) Find the probability that a salesperson drives between 1200 and 1600 miles per month.

Solution Here, $\mu = 1200$ and $\sigma = 150$, and we must find the area under the normal distribution curve between $x = 1200$ and $x = 1600$. We begin by finding the z-score for $x = 1200$:

$$z = \frac{x - \mu}{\sigma} = \frac{1200 - 1200}{150} = \frac{0}{150} = 0.$$

The z-score for $x = 1600$ is

$$z = \frac{x - \mu}{\sigma} = \frac{1600 - 1200}{150} = \frac{400}{150} = 2.67.^*$$

So the area under the curve from $x = 1200$ to $x = 1600$ is the same as the area under the standard normal curve from $z = 0$ to $z = 2.67$, as indicated in Figure 10.18. A graphing calculator or Table 2 shows that this area is .4962. Therefore, the probability that a salesperson drives between 1200 and 1600 miles per month is .4962.

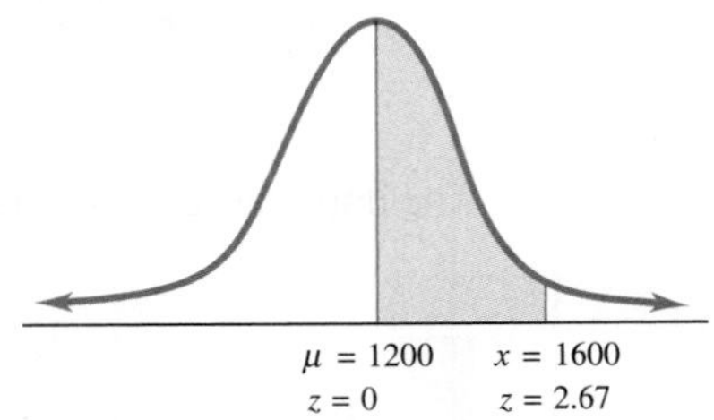

FIGURE 10.18

(b) Find the probability that a salesperson drives between 1000 and 1500 miles per month.

Solution As shown in Figure 10.19, z-scores for both $x = 1000$ and $x = 1500$ are needed.

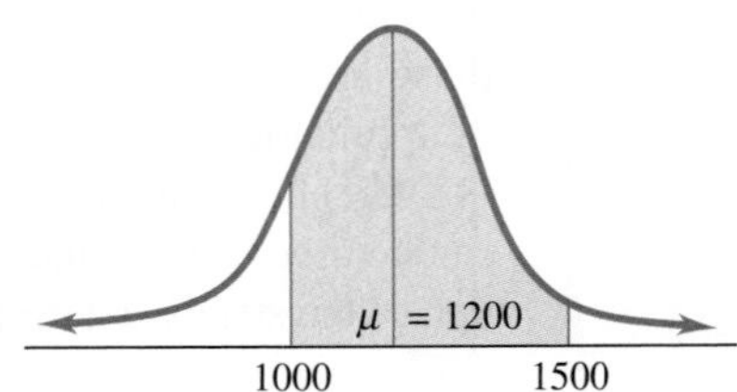

FIGURE 10.19

For $x = 1000$,

$$z = \frac{1000 - 1200}{150} = \frac{-200}{150} = -1.33.$$

For $x = 1500$,

$$z = \frac{1500 - 1200}{150} = \frac{300}{150} = 2.00.$$

*All z-scores here are rounded to two decimal places.

From Table 2, $z = 1.33$ leads to an area of .4082, while $z = 2.00$ corresponds to .4773. A total of $.4082 + .4773 = .8855$, or 88.55%, of all drivers travel between 1000 and 1500 miles per month. From this calculation, the probability that a driver travels between 1000 and 1500 miles per month is .8855. 5✓

✓Checkpoint 5

With the data from Example 4, find the probability that a salesperson drives between 1000 and 1450 miles per month.

Answer: See page 640.

Example 5 **Health** With data from the 2005–2006 National Health and Nutritional Examination Survey (NHANES), we can use 196 (mg/dL) as an estimate of the mean total cholesterol level for all Americans and 44 (mg/dL) as an estimate of the standard deviation.* Assuming total cholesterol levels to be normally distributed, what is the probability that an American chosen at random has a cholesterol level higher than 250? If 200 Americans are chosen at random, how many can we expect to have total cholesterol higher than 250?

Solution Here, $\mu = 196$ and $\sigma = 44$. The probability that a randomly chosen American has cholesterol higher than 250 is the area under the normal curve to the right of $x = 250$. The z-score for $x = 250$ is

$$z = \frac{x - \mu}{\sigma} = \frac{250 - 196}{44} = \frac{54}{44} = 1.23.$$

From Table 2, we see that the area to the right of 1.23 is $.5 - .3907 = .1093$, which is 10.93% of the total area under the curve. Therefore, the probability of a randomly chosen American having cholesterol higher than 250 is .1093.

With 10.93% of Americans having total cholesterol higher than 250, selecting 200 Americans at random yields

$$10.93\% \text{ of } 200 = .1093 \cdot 200 = 21.86.$$

Approximately 22 of these Americans can be expected to have a total cholesterol level higher than 250. 6✓

✓Checkpoint 6

Using the mean and standard deviation from Example 5, find the probability an adult selected at random has a cholesterol level below 150.

Answer: See page 640.

NOTE Notice in Example 5 that $P(z \geq 1.23) = P(z > 1.23.)$. The area under the curve is the same whether we include the endpoint or not. Notice also that $P(z = 1.23) = 0$, because no area is included.

CAUTION When calculating the normal probability, it is wise to draw a normal curve with the mean and the z-scores every time. This practice will avoid confusion as to whether you should add or subtract probabilities.

As mentioned earlier, z-scores are standard deviations, so $z = 1$ corresponds to 1 standard deviation above the mean, and so on. As found in Checkpoint 3(d) of this section, 68.26% of the area under a normal curve lies within 1 standard deviation of the mean. Also, 95.46% lies within 2 standard deviations of the mean, and 99.74% lies within 3 standard deviations of the mean. These results, summarized in Figure 10.20, can be used to get quick estimates when you work with normal curves.

*Based on data available at www.cdc.gov/nchs/nhanes.htm.

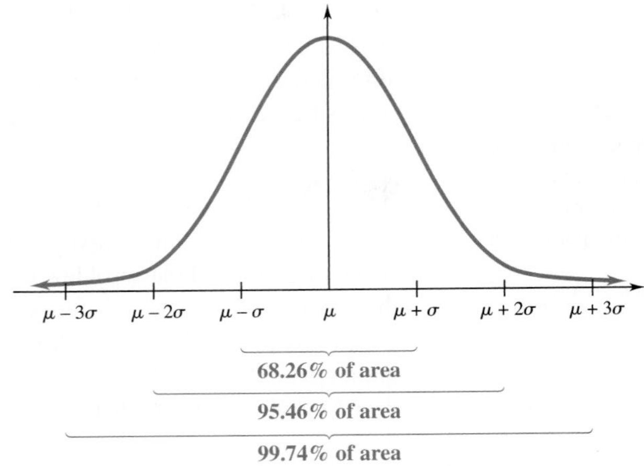

FIGURE 10.20

BOXPLOTS

The normal curve is useful because you can easily read various characteristics of the data from the picture. Boxplots are another graphical means of presenting key characteristics of a data set. The idea is to arrange the data in increasing order and choose three numbers Q_1, Q_2, and Q_3 that divide it into four equal parts, as indicated schematically in the following diagram:

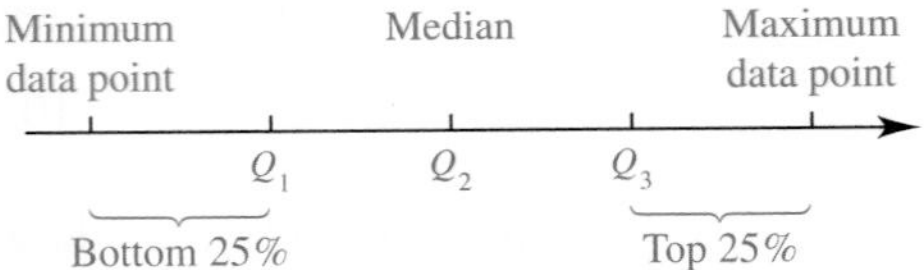

The number Q_1 is called the **first quartile**, the median Q_2 is called the **second quartile**, and Q_3 is called the **third quartile**. The minimum, Q_1, Q_2, Q_3, and the maximum are often called the five-number summary, and they are used to construct a boxplot, as illustrated in Examples 6 and 7.

Example 6 **Business** The following table gives the revenues (in billions of U.S. dollars) for Apple and Microsoft Corporations for the given years:* Construct a boxplot for the Apple revenue data.

Year	1999	2000	2001	2002	2003	2004	2005	2006	2007	2008
Apple	6.1	8.0	5.4	5.7	6.2	8.3	13.9	19.3	24.0	32.5
Microsoft	19.7	23.0	25.3	28.4	32.2	36.8	39.8	44.3	51.1	60.4

Solution We first have to order the revenues from low to high:

5.4, 5.7, 6.1, 6.2, **8.0**, **8.3**, 13.9, 19.3, 24.0, 32.5

*www.morningstar.com.

The minimum revenue is 5.4, and the maximum revenue is 32.5. Since there is an even number of revenues, the median revenue Q_2 is 8.15 (halfway between the two center entries). To find Q_1, which separates the lower 25% of the data from the rest, we first calculate

25% of n (rounded up to the nearest integer),

where n is the number of data points. Here, $n = 10$, and so $.25(10) = 2.5$, which rounds up to 3. Now count to the *third* revenue (in order) to get $Q_1 = 6.1$.

Similarly, since Q_3 separates the lower 75% of the data from the rest, we calculate

75% of n (rounded up to the nearest integer).

When $n = 10$, we have $.75(10) = 7.5$, which rounds to 8. Count to the *eighth* revenue, getting $Q_3 = 19.3$.

The five key numbers for constructing the boxplot are

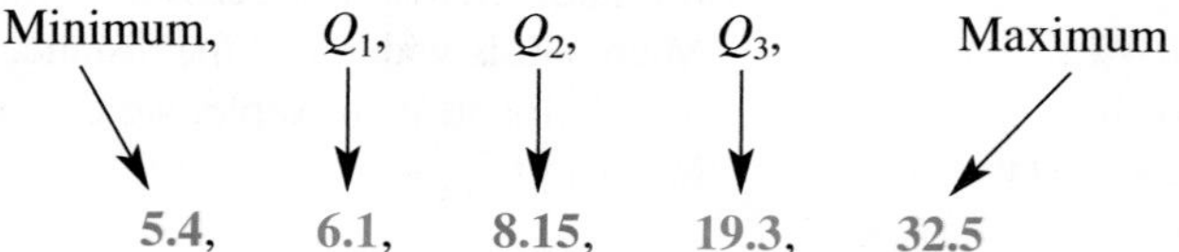

Draw a horizontal line parallel to the number axis, from the minimum to the maximum score. Around this line, construct a box whose ends are at Q_1 and Q_3, and mark the location of Q_2 by a vertical line in the box, as shown in Figure 10.21.

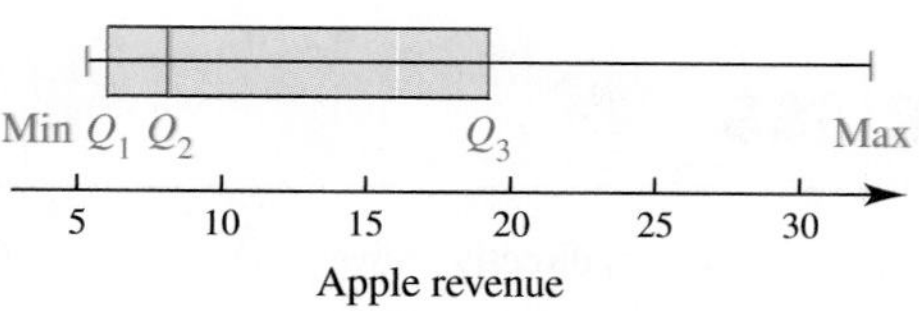

FIGURE 10.21

Boxplots are useful in showing the location of the middle 50% of the data. This is the region of the box between Q_1 and Q_3. Boxplots are also quite useful in comparing two distributions that are measured on the same scale, as we will see in Example 7.

Example 7 **Business** Construct a boxplot for the Microsoft revenue data in Example 6 with the Apple revenue boxplot on the same graph.

Solution Microsoft had increasing revenue each year, so the data is already ranked low to high:

19.7, 23.0, 25.3, 28.4, **32.2**, **36.8**, 39.8, 44.3, 51.1, 60.4.

The minimum is 19.7, the maximum is 60.4, and the median $Q_2 = 34.5$. Again, $n = 10$, and so

25% of $n = .25(10) = 2.5$, which rounds up to 3,

so that $Q_1 = 25.3$ (the *third* revenue). Similarly,

75% of $n = .75(10) = 7.5$, which rounds up to 8,

so that $Q_3 = 44.3$ (the *eighth* revenue).

We can now use the minimum, Q_1, Q_2, Q_3, and the maximum to place the boxplot of the Microsoft revenue above the Apple boxplot, as in Figure 10.22.

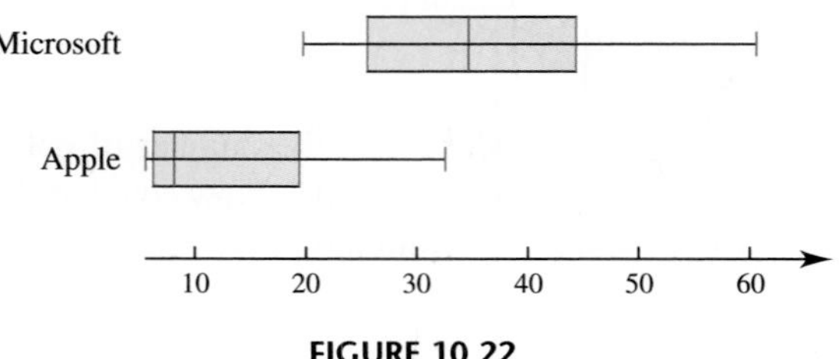

FIGURE 10.22

When we compare the revenue distributions in this way, we see how much more revenue Microsoft generated in the studied years than Apple. Because the box for Microsoft is wider, and the distance from the minimum to the maximum is also wider for Microsoft than Apple, we can also say there is a greater degree of variability for Microsoft. ✓7

✓ Checkpoint 7

Create a boxplot for the weights (in kilograms) for eight first-year students:

88, 79, 67, 69, 58, 53, 89, 57.

Answer: See page 640.

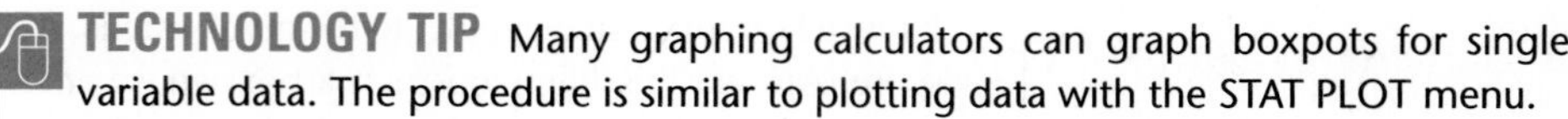

TECHNOLOGY TIP Many graphing calculators can graph boxpots for single-variable data. The procedure is similar to plotting data with the STAT PLOT menu.

10.4 ▶ Exercises

1. The peak in a normal curve occurs directly above ______.
2. The total area under a normal curve (above the horizontal axis) is ______.
3. How are z-scores found for normal distributions with $\mu \neq 0$ or $\sigma \neq 1$?
4. How is the standard normal curve used to find probabilities for normal distributions?

Find the percentage of the area under a normal curve between the mean and the given number of standard deviations from the mean. (See Example 2.)

5. 1.75
6. .26
7. −.43
8. −2.4

Find the percentage of the total area under the standard normal curve between the given z-scores. (See Examples 1 and 2.)

9. $z = 1.41$ and $z = 2.83$
10. $z = .64$ and $z = 2.11$
11. $z = -2.48$ and $z = -.05$
12. $z = -1.63$ and $z = -1.08$
13. $z = -3.05$ and $z = 1.36$
14. $z = -2.91$ and $z = -.51$

Find a z-score satisfying each of the given conditions. (Hint: Use Table 2 backward or a graphing calculator.)

15. 5% of the total area is to the right of z.
16. 1% of the total area is to the left of z.
17. 15% of the total area is to the left of z.
18. 25% of the total area is to the right of z.
19. For any normal distribution, what is the value of $P(x \leq \mu)$? of $P(x \geq \mu)$?
20. Using Chebyshev's theorem and the normal distribution, compare the probability that a number will lie within 2 standard deviations of the mean of a probability distribution. (See Exercises 15–22 of Section 10.3.) Explain what you observe.
21. Repeat Exercise 20, using 3 standard deviations.

Assume the distributions in Exercises 22–30 are all normal. (See Example 4.)

22. **Business** According to the label, a regular can of Campbell's™ soup holds an average of 305 grams, with a standard deviation of 4.2 grams. What is the probability that a can will be sold that holds more than 306 grams?
23. **Business** A jar of Adams Old Fashioned Peanut Butter contains 453 grams, with a standard deviation of 10.1 grams.

Find the probability that one of these jars contains less than 450 grams.

24. **Business** A General Electric soft-white three-way lightbulb has an average life of 1200 hours, with a standard deviation of 50 hours. Find the probability that the life of one of these bulbs will be between 1150 and 1300 hours.

25. **Business** A 100-watt lightbulb has an average brightness of 1640 lumens, with a standard deviation of 62 lumens. What is the probability that a 100-watt bulb will have a brightness between 1600 and 1700 lumens?

26. **Social Science** The scores on a standardized test in a suburban high school have a mean of 80, with a standard deviation of 12. What is the probability that a student will have a score less than 60?

27. **Health** Using the data from the same study as in Example 5, we find that the average HDL cholesterol level is 51.6 mg/dL, with a standard deviation of 14.3 mg/dL. Find the probability that an individual will have an HDL cholesterol level greater than 60 mg/dL.

28. **Business** The production of cars per day at an assembly plant has mean 120.5 and standard deviation 6.2. Find the probability that fewer than 100 cars are produced on a random day.

29. **Business** Starting salaries for accounting majors have mean \$45,000, with standard deviation \$3,200. What is the probability an individual will start at a salary above \$53,000?

30. **Social Science** The driving distance to work for residents of a certain community has mean 21 miles and standard deviation 3.6 miles. What is the probability that an individual drives between 10 and 20 miles to work?

Business *Scores on the Graduate Management Association Test (GMAT) are approximately normally distributed. The mean score for 2007–2008 was 540, with a standard deviation of 100.* For the following exercises, find the probability that a GMAT test taker selected at random earns a score in the given range, using the normal distribution as a model.*

31. Between 540 and 700
32. Between 300 and 540
33. Between 300 and 700
34. Less than 400
35. Greater than 750
36. Between 600 and 700
37. Between 300 and 400

Social Science *New studies by Federal Highway Administration traffic engineers suggest that speed limits on many thoroughfares are set arbitrarily and often are artificially low. According to traffic engineers, the ideal limit should be the "85th-percentile speed," the speed at or below which 85% of the traffic moves. Assuming that speeds are normally distributed, find the 85th-percentile speed for roads with the given conditions.*

*www.gmac.com.

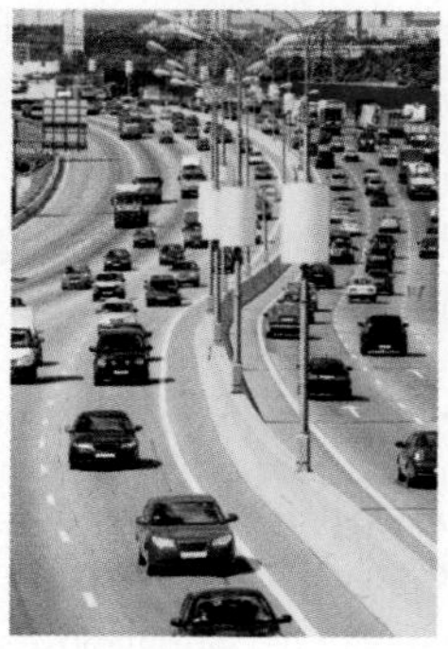

38. The mean speed is 55 mph, with a standard deviation of 10 mph.

39. The mean speed is 40 mph, with a standard deviation of 5 mph.

Social Science *One professor uses the following system for assigning letter grades in a course:*

Grade	Total Points
A	Greater than $\mu + \frac{3}{2}\sigma$
B	$\mu + \frac{1}{2}\sigma$ to $\mu + \frac{3}{2}\sigma$
C	$\mu - \frac{1}{2}\sigma$ to $\mu + \frac{1}{2}\sigma$
D	$\mu - \frac{3}{2}\sigma$ to $\mu - \frac{1}{2}\sigma$
F	Below $\mu - \frac{3}{2}\sigma$

What percentage of the students receive the given grades?

40. A
41. B
42. C

43. Do you think the system in Exercises 40–42 would be more likely to be fair in a large freshman class in psychology or in a graduate seminar of five students? Why?

Health *In nutrition, the recommended daily allowance of vitamins is a number set by the government as a guide to an individual's daily vitamin intake. Actually, vitamin needs vary drastically from person to person, but the needs are very closely approximated by a normal curve. To calculate the recommended daily allowance, the government first finds the average need for vitamins among people in the population and then the standard deviation. The recommended daily allowance is defined as the mean plus 2.5 times the standard deviation.*

44. What percentage of the population will receive adequate amounts of vitamins under this plan?

Find the recommended daily allowance for the following vitamins.

45. Mean = 550 units, standard deviation = 46 units

46. Mean = 1700 units, standard deviation = 120 units

47. Mean = 155 units, standard deviation = 14 units

48. Mean = 1080 units, standard deviation = 86 units

Social Science *The mean performance score of a large group of fifth-grade students on a math achievement test is 88. The scores are known to be normally distributed. What percentage of the students had scores as follows?*

49. More than 1 standard deviation above the mean

50. More than 2 standard deviations above the mean

Social Science *Studies have shown that women are charged an average of $500 more than men for cars.* Assume a normal distribution of overcharges with a mean of $500 and a standard deviation of $65. Find the probability of a woman's paying the given additional amounts for a car.*

51. Less than $400

52. At least $700

53. Between $350 and $600

Business *The table gives the annual revenue for Pepsi Co. and Coco-Cola Co. (in billions of U.S. dollars) for a 10-year period:†*

Year	Pepsi	Coca-Cola
1999	20.4	19.8
2000	20.4	20.5
2001	26.9	20.1
2002	25.1	19.6
2003	27.0	21.0
2004	29.3	22.0
2005	32.6	23.1
2006	35.1	24.1
2007	39.5	28.9
2008	43.3	31.9

54. What is the five-number summary for Pepsi?

55. What is the five-number summary for Coca-Cola?

56. Construct a boxplot for Pepsi.

57. Construct a boxplot for Coca-Cola.

58. Graph the boxplots for Pepsi and Coca-Cola on the same scale. Are the distributions similar? Which company has the higher median of sales over the 10-year period?

*"From repair shops to cleaners, women pay more," by Bob Dart, as appeared in *The Chicago Tribune*, May 27, 1993. Reprinted by permission of the author.

†www.morningstar.com.

Finance *The following table shows U.S. federal government payments to selected states for child nutrition programs and food stamp programs (in millions of dollars):**

State	Child Nutrition	Food Stamp
Alabama	214	33
Alaska	33	9
Arizona	254	40
Arkansas	144	26
Colorado	109	28
Connecticut	86	20
Delaware	35	9
Florida	661	82
Georgia	473	68
Hawaii	40	11
Idaho	52	11
Illinois	449	95

59. What is the five-number summary for the child nutrition programs?

60. What is the five-number summary for the food stamp programs?

61. Construct a boxplot for the child nutrition programs data.

62. Construct a boxplot for the food stamp programs data.

63. Which program has higher variability?

✓ Checkpoint Answers

1. **(a)** 43.45% **(b)** 47.93% **(c)** 26.42%
2. **(a)** 34.13% **(b)** 49.25%
3. **(a)** 46.55% **(b)** 32.82% **(c)** 6.81%
(d) 68.26%, 95.46%, 99.74%; almost all the data lies within 3 standard deviations of the mean.
4. **(a)** −4.8 **(b)** −1
5. .8607 **6.** .1469
7.

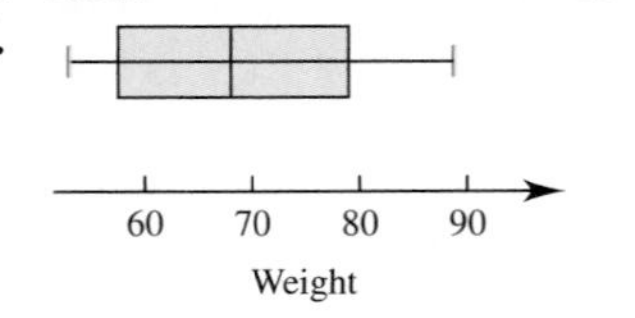

**The Statistical Abstract of the United States*: 2009.

10.5 ▶ Normal Approximation to the Binomial Distribution

As we saw in Section 9.4, many practical experiments have only two possible outcomes, sometimes referred to as success or failure. Such experiments are called Bernoulli trials or Bernoulli processes. Examples of Bernoulli trials include flipping a coin (with heads being a success, for instance, and tails a failure) and testing a computer chip coming off the assembly line to see whether it is defective. A binomial experiment consists of repeated independent Bernoulli trials, such as flipping a coin 10 times or taking a random sample of 20 computer chips from the assembly line. In Section 9.4, we found the probability distribution for several binomial experiments, such as sampling five people with bachelor's degrees in education and counting how many are women. The probability distribution for a binomial experiment is known as a **binomial distribution**.

As another example, it is reported that 40% of registered vehicles in the United States are vans, pickup trucks, or sport utility vehicles (SUVs).* Suppose an auto insurance agent wants to verify this statistic and records the type of vehicle for 10 randomly selected drivers. The agent finds that 3 out of 10, or 30%, are vans, pickups, or SUVs. How likely is this result if the figure for all vehicles is truly 40%? We can answer that question with the binomial probability formula

$$_nC_x \cdot p^x(1 - p)^{n-x},$$

where n is the sample size (10 in this case); x is the number of vans, pickups, or SUVs (3 in this case); and p is the probability that a vehicle is a van, pickup, or SUV (.40 in this case). We obtain

$$\begin{aligned} P(x = 3) &= {}_{10}C_3 \cdot (.40)^3(1 - .40)^7 \\ &= 120(.064)(.0279936) \\ &\approx .2150. \end{aligned}$$

The probability is over 20%, so this result is not unusual.

Suppose that the insurance agent takes a larger random sample, say, of 100 drivers. What is the probability that 30 or fewer vehicles are vans, pickups, or SUVs if the 40% figure is accurate? Calculating $P(x = 0) + P(x = 1) + \ldots + P(x = 30)$ is a formidable task. One solution is provided by graphing calculators or computers. There is, however, a low-tech method that has the advantage of connecting two different distributions: the normal and the binomial. The normal distribution is continuous, since the random variable can be any real number. The binomial distribution is *discrete*, because the random variable can take only integer values between 0 and n. Nevertheless, the normal distribution can be used to give a good approximation to binomial probability. We call this approximation the **normal approximation**.

In order to use the normal approximation, we first need to know the mean and standard deviation of the binomial distribution. Recall from Section 9.4 that,

---

**The Statistical Abstract of the United States*: 2009.

for the binomial distribution, $E(x) = np$. In Section 10.2, we referred to $E(x)$ as μ, and that notation will be used here. It is shown in more advanced courses in statistics that the standard deviation of the binomial distribution is given by $\sigma = \sqrt{np(1-p)}$.

Mean and Standard Deviation for the Binomial Distribution

For the binomial distribution, the mean and standard deviation are respectively given by

$$\mu = np \quad \text{and} \quad \sigma = \sqrt{np(1-p)},$$

where n is the number of trials and p is the probability of success on a single trial. ✓1

✓Checkpoint 1

Find μ and σ for a binomial distribution having $n = 120$ and $p = 1/6$.

Answer: See page 647.

Example 1 Suppose a fair coin is flipped 15 times.

(a) Find the mean and standard deviation for the number of heads.

Solution With $n = 15$ and $p = 1/2$, the mean is

$$\mu = np = 15\left(\frac{1}{2}\right) = 7.5.$$

The standard deviation is

$$\sigma = \sqrt{np(1-p)} = \sqrt{15\left(\frac{1}{2}\right)\left(1-\frac{1}{2}\right)}$$
$$= \sqrt{15\left(\frac{1}{2}\right)\left(\frac{1}{2}\right)} = \sqrt{3.75} \approx 1.94.$$

We expect, on average, to get 7.5 heads out of 15 tosses. Most of the time, the number of heads will be within three standard deviations of the mean, or between $7.5 - 3(1.94) = 1.68$ and $7.5 + 3(1.94) = 13.32$.

(b) Find the probability distribution for the number of heads, and draw a histogram of the probabilities.

Solution The probability distribution is found by putting $n = 15$ and $p = 1/2$ into the formula for binomial probability. For example, the probability of getting 9 heads is given by

$$P(x = 9) = {}_{15}C_9\left(\frac{1}{2}\right)^9\left(1-\frac{1}{2}\right)^6 \approx .15274.$$

Probabilities for the other values of x between 0 and 15, as well as a histogram of the probabilities, are shown in Figure 10.23.

x	$P(x)$
0	.00003
1	.00046
2	.00320
3	.01389
4	.04166
5	.09164
6	.15274
7	.19638
8	.19638
9	.15274
10	.09164
11	.04166
12	.01389
13	.00320
14	.00046
15	.00003

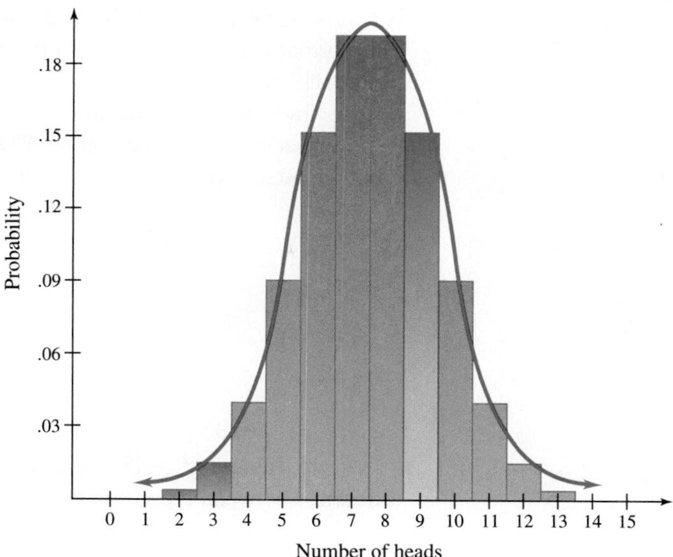

FIGURE 10.23

In Figure 10.23, we have superimposed the normal curve with $\mu = 7.5$ and $\sigma = 1.94$ over the histogram of the distribution. Notice how well the normal distribution fits the binomial distribution. This approximation was first discovered in 1718 by Abraham De Moivre (1667–1754) for the case $p = 1/2$. The result was generalized by the French mathematician Pierre-Simon Laplace (1749–1827) in a book published in 1812. As n becomes larger and larger, a histogram for the binomial distribution looks more and more like a normal curve. Figures 10.24(a) and (b), show histograms of the binomial distribution with $p = .3$, using $n = 8$ and $n = 50$, respectively.

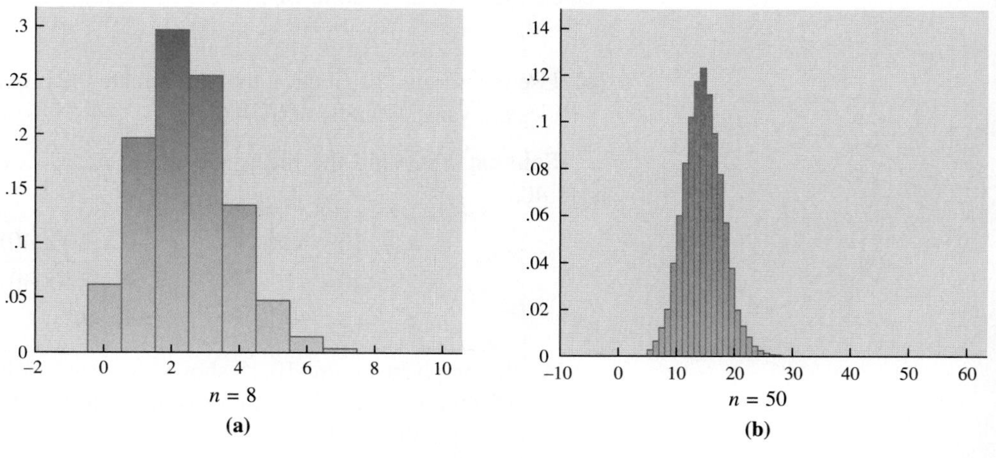

FIGURE 10.24

The probability of getting exactly 9 heads in the 15 tosses, or .15274, is the same as the area of the blue bar in Figure 10.23. As the graph suggests, the blue area is approximately equal to the area under the normal curve from $x = 8.5$ to $x = 9.5$. The normal curve is higher than the top of the bar in the left half, but lower in the right half.

To find the area under the normal curve from $x = 8.5.$ to $x = 9.5$, first find z-scores, as in the previous section. Use the mean and the standard deviation for the distribution, which we have already calculated, to get z-scores for $x = 8.5$ and $x = 9.5$:

For $x = 8.5$,

$$z = \frac{8.5 - 7.5}{1.94} = \frac{1.00}{1.94}$$
$$z \approx .52.$$

For $x = 9.5$,

$$z = \frac{9.5 - 7.5}{1.94} = \frac{2.00}{1.94}$$
$$z \approx 1.03.$$

From Table 2, $z = .52$ gives an area of .1985, and $z = 1.03$ gives .3485. The difference between these two numbers is the desired result:

$$.3485 - .1985 = .1500.$$

This answer is not far from the more accurate answer of .15274 found earlier. 

✓Checkpoint 2

Use the normal distribution to find the probability of getting exactly the given number of heads in 15 tosses of a coin.

(a) 7

(b) 10

Answers: See page 647.

CAUTION The normal-curve approximation to a binomial distribution is quite accurate, *provided that n* is large and *p* is not close to 0 or 1. As a rule of thumb, the normal-curve approximation can be used as long as both *np* and *n*(1 − *p*) are at least 5.

Example 2 **Business** Consider the previously discussed sample of 100 vehicles selected at random, where 40% of the registered vehicles are vans, pickups, or SUVs.

(a) Use the normal distribution to approximate the probability that at least 51 vehicles are vans, pickups, or SUVs.

Solution First find the mean and the standard deviation, using $n = 100$ and $p = .40$:

$$\mu = 100(.40) = 40.$$

$$\sigma = \sqrt{100(.40)(1 - .40)} = \sqrt{100(.40)(.60)} \approx 4.90.$$

As the graph in Figure 10.25 shows, we need to find the area to the right of $x = 50.5$ (since we want 51 or more vehicles to be vans, pickups, or SUVs). The z-score corresponding to $x = 50.5$ is

$$z = \frac{50.5 - 40}{4.90} \approx 2.14.$$

From Table 2, $z = 2.14$ corresponds to an area of .4838, so

$$P(z > 2.14) = .5 - .4838 = .0162.$$

This is an extremely low probability. If an insurance agent obtained 51 vehicles that were vans, pickups, or SUVs in a sample of 100, she would suspect that either her sample is not truly random or the 40% figure is too low.

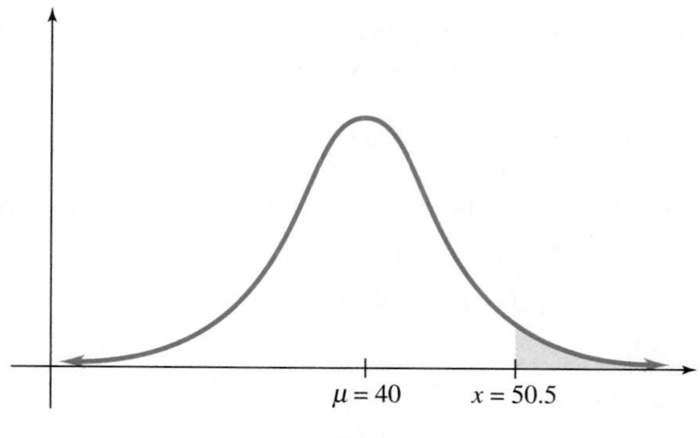

FIGURE 10.25

(b) Calculate the probability of finding between 41 and 48 vehicles that were vans, pickups, or SUVs in a random sample of 100.

Solution As Figure 10.26 shows, we need to find the area between $x = 40.5$, and $x = 48.5$:

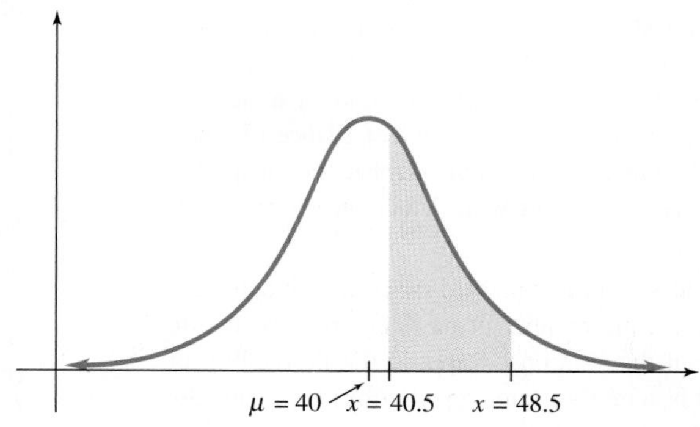

FIGURE 10.26

$$\text{If } x = 40.5, \text{ then } z = \frac{40.5 - 40}{4.90} \approx .10.$$

$$\text{If } x = 48.5, \text{ then } z = \frac{48.5 - 40}{4.90} \approx 1.73.$$

Use Table 2 to find that $z = .10$ corresponds to an area of .0398 and $z = 1.73$ yields .4582. The final answer is the difference of these numbers:

$$P(.10 \le z \le 1.73) = .4582 - .0398 = .4184.$$

The probability of finding between 41 and 48 vehicles that are vans, pickups, or SUVs is approximately .4184. ✓3

✓ Checkpoint 3

In December 2008, Michigan led the nation with the highest unemployment rate of 9.6%.* Find the approximate probability that in a random sample of 200 adults from Michigan during that month, the given numbers of people would be unemployed.

(a) Exactly 12

(b) 15 or fewer

(c) More than 10

Answers: See page 647.

*www.money.cnn.com.

10.5 Exercises

1. What must be known to find the mean and standard deviation of a binomial distribution?

2. What is the rule of thumb for using the normal distribution to approximate a binomial distribution?

Suppose 16 coins are tossed. Find the probability of getting each of the given results by using (a) the binomial probability formula and (b) the normal-curve approximation. (See Examples 1 and 2.)

3. Exactly 8 heads
4. Exactly 9 heads
5. More than 13 tails
6. Fewer than 6 tails

For the remaining exercises in this section, use the normal-curve approximation to the binomial distribution.

Suppose 1000 coins are tossed. Find the probability of getting each of the given results.

7. Exactly 500 heads
8. Exactly 510 heads
9. 475 or more heads
10. Fewer than 460 tails

A die is tossed 120 times. Find the probability of getting each of the given results.

11. Exactly twenty 5's
12. Exactly twenty-four 4's
13. More than seventeen 3's
14. Fewer than twenty-one 6's

15. A reader asked Mr. Statistics (a feature in *Fortune* magazine) about the game of 26 once played in the bars of Chicago.* In this game, the player chooses a number between 1 and 6 and then rolls a cup full of 10 dice 13 times. Out of the 130 numbers rolled, if the number chosen appears at least 26 times, the player wins. Calculate the probability of winning.

16. **Natural Science** For certain bird species, with appropriate assumptions, the number of nests escaping predation has a binomial distribution.† Suppose the probability of success (that is, a nest's escaping predation) is .3. Find the probability that at least half of 26 nests escape predation.

17. **Natural Science** Let us assume, under certain appropriate assumptions, that the probability of a young animal eating x units of food is binomially distributed, with n equal to the maximum number of food units the animal can acquire and p equal to the probability per time unit that an animal eats a unit of food. Suppose $n = 120$ and $p = .6$.
    (a) Find the probability that an animal consumes 80 units of food.
    (b) Suppose the animal must consume at least 70 units of food to survive. What is the probability that this happens?

18. **Finance** In 2007–2008, 66% of all undergraduates received some type of financial aid.* Suppose 50 undergraduates are selected at random.
    (a) Find the probability that at least 35 received financial aid.
    (b) Find the probability that at most 25 received financial aid.

19. **Finance** In 2007–2008, 27% of all undergraduates received a Federal Pell Grant.* If a random sample of 500 students is selected, find the probability that at least 150 students received a Pell Grant.

20. **Finance** In 2007–2008, 34% of all undergraduates took out a Federal Stafford loan.* If a random sample of 250 students is selected, find the probability that at most 75 students took out a Federal Stafford loan.

21. **Health** An article in the *New York Times* stated that 30% of Americans have trouble sleeping.† If 150 Americans are selected at random, find the probability that 40 or less have trouble sleeping.

22. **Health** In Exercise 21, what is the probability that more than 55 have trouble sleeping?

23. **Natural Science** A flu vaccine has a probability of 80% of preventing a person who is inoculated from getting the flu. A county health office inoculates 134 people. Find the probabilities of the given results.
    (a) Exactly 12 of the people inoculated get the flu.
    (b) No more than 12 of the people inoculated get the flu.
    (c) None of the people inoculated get the flu.

24. **Natural Science** The probability that a male will be color blind is .042. Find the probabilities that, in a group of 53 men, the given conditions will be true.
    (a) Exactly 6 are color blind.
    (b) No more than 6 are color blind.
    (c) At least 1 is color blind.

Business *In February 2008, the warranty company Square Trade estimated that 16.4% of Microsoft Xbox 360 video game systems were defective.‡*

25. If 700 systems are selected at random, find the probability that at least 100 are defective.

26. If 700 systems are selected at random, find the probability that no more than 85 are defective.

*Daniel Seligman and Patty De Llosa, "Ask Mr. Statistics," *Fortune*, May 1, 1995, p. 141.

†From G. deJong, *American Naturalist*, vol. 110.

*National Center for Education Statistics, www.nces.ed.gov.

†*The New York Times*, April 23, 2009.

‡www.squaretrade.com.

27. **Natural Science** The blood types B– and AB– are the rarest of the eight human blood types, representing 1.5% and .6% of the population, respectively.*
   (a) If the blood types of a random sample of 1000 blood donors are recorded, what is the probability that 10 or more of the samples are AB–?
   (b) If the blood types of a random sample of 1000 blood donors are recorded, what is the probability that 20 to 40, inclusive, of the samples are B–?
   (c) If a particular city had a blood drive in which 500 people gave blood and 3% of the donations were B–, would we have reason to believe that this town has a higher-than-normal number of donors who are B–? (*Hint*: Calculate the probability of 15 or more donors being B– for a random sample of 500, and then consider the probability obtained.)

Health *According to data from the American Heart Association, 49.7% of non-Hispanic white women and 42.1% of non-Hispanic black women have total cholesterol levels above 200 mg/dL.*†

28. If we select 500 non-Hispanic white women at random, what is the probability that at least 220 of the women will have a cholesterol level higher than 200 mg/dL?

29. If we select 500 non-Hispanic black women at random, what is the probability that at least 220 of the women will have a cholesterol level higher than 200 mg/dL?

30. **Health** According to *HealthDay News*, 5% of infants and young children suffer from food allergies.‡. If 250 infants are selected at random, what is the probability that between 10 and 20 inclusive have a food allergy?

31. In the 1989 U.S. Open, four golfers each made a hole in one on the same par-3 hole on the same day. *Sports Illustrated* writer R. Reilly stated the probability of getting a hole in one for a given golf pro on a given par-3 hole to be 1/3709.*
   (a) For a specific par-3 hole, use the binomial distribution to find the probability that 4 or more of the 156 golf pros in the tournament field shoot a hole in one.†
   (b) For a specific par-3 hole, use the normal approximation to the binomial distribution to find the probability that 4 or more of the 156 golf pros in the tournament field shoot a hole in one. Why must we be very cautious when using this approximation for this application?
   (c) If the probability of a hole in one remains constant and is 1/3709 for any par-3 hole, find the probability that, in 20,000 attempts by golf pros, there will be 4 or more holes in one. Discuss whether this assumption is reasonable.

✓ Checkpoint Answers

1. $\mu = 20$; $\sigma = 4.08$
2. (a) .1985 (b) .0909
3. (a) .0215 (b) .1867 (c) .9901

*National Center for Statistics and Analysis.

†www.americanheart.org.

‡www.nlm.nih.gov/medlineplus/news/fullstory_81741.html.

*R. Reilly, "King of the Hill," *Sports Illustrated*, June 1989, pp. 20–25.

†Bonnie Litwiller and David Duncan, "The Probability of a Hole in One," *School Science and Mathematics* 91, no. 1, (January 1991): 30.

CHAPTER 10 ▸ Summary

KEY TERMS AND SYMBOLS

10.1 ▸ random sample
grouped frequency distribution
histogram
frequency polygon
stem-and-leaf-plot
uniform distribution
right-skewed distribution
left-skewed distribution
normal distribution

10.2 ▸ Σ, summation (sigma) notation
$\bar{x}$, sample mean
μ, population mean
(arithmetic) mean
median
statistic
mode

10.3 ▸ s^2, sample variance
s, sample standard deviation
σ, population standard deviation
range
deviations from the mean
variance
standard deviation

10.4 ▸ μ, mean of a continuous distribution
σ, standard deviation of a normal distribution
continuous distribution
normal curves
standard normal curve
z-score
boxplot
quartile

10.5 ▸ binomial distribution

CHAPTER 10 KEY CONCEPTS

To organize the data from a sample, we use a **grouped frequency distribution**—a set of intervals with their corresponding frequencies. The same information can be displayed with a **histogram**—a type of bar graph with a bar for each interval. Each bar has width 1 and height equal to the probability of the corresponding interval. A **stem-and-leaf plot** presents the individual data in a similar form, so it can be viewed as a bar graph as well. Another way to display this information is with a **frequency polygon**, which is formed by connecting the midpoints of consecutive bars of the histogram with straight-line segments.

The **mean** $\bar{x}$ of a frequency distribution is the expected value.

For n numbers $x_1, x_2, \ldots, x_n$,

$$\bar{x} = \frac{\Sigma x}{n}.$$

For a grouped distribution,

$$\bar{x} = \frac{\Sigma(xf)}{n}.$$

The **median** is the middle entry in a set of data arranged in either increasing or decreasing order.

The **mode** is the most frequent entry in a set of numbers.

The **range** of a distribution is the difference between the largest and smallest numbers in the distribution.

The **sample standard deviation** s is the square root of the sample **variance**.

For n numbers,

$$s = \sqrt{\frac{\Sigma x^2 - n\bar{x}^2}{n - 1}}.$$

For a grouped distribution,

$$s = \sqrt{\frac{\Sigma fx^2 - n\bar{x}^2}{n - 1}}.$$

A **normal distribution** is a continuous distribution with the following properties: The highest frequency is at the mean; the graph is symmetric about a vertical line through the mean; the total area under the curve, above the x-axis, is 1. If a normal distribution has mean μ and standard deviation σ, then the z-score for the number x is $z = \frac{x - \mu}{\sigma}$.

A **boxplot** organizes a list of data using the minimum and maximum values, the median, and the first and third quartiles to give a visual overview of the distribution.

The **area under a normal curve** between $x = a$ and $x = b$ gives the probability that an observed data value will be between a and b.

The **binomial distribution** is a distribution with the following properties: For n independent repeated trials, in which the probability of success in a single trial is p, the probability of x successes is ${}_nC_x p^x(1 - p)^{n-x}$. The mean is $\mu = np$, and the standard deviation is

$$\sigma = \sqrt{np(1 - p)}.$$

CHAPTER 10 REVIEW EXERCISES

1. Discuss some reasons for organizing data into a grouped frequency distribution.
2. What is the rule of thumb for an appropriate interval in a grouped frequency distribution?

In Exercises 3 and 4, (a) write a frequency distribution; (b) draw a histogram; and (c) draw a stem-and-leaf plot.

3. The following are scores on a 100-point final exam (use intervals of 40–49, 50–59, and so on):

68	45	71	77	82	94	89	63	99	55
76	77	82	92	77	76	53	84	91	81
71	69	42	88	91	84	77	84	89	91

4. The number of units carried in one semester by the students in a business mathematics class was as follows (use intervals of 9–10, 11–12, 13–14, and 15–16):

10	9	16	12	13	15	13	16	15	11	13
12	12	15	12	14	10	12	14	15	15	13

Find the mean for each of the given data sets.

5. The data in Exercise 3
6. The data in Exercise 4
7. The following table gives the frequency counts for 44 first-year college students' waist circumference in cm:

Interval	Frequency
60–69	10
70–79	24
80–89	6
90–99	3
100–109	1

8. The following table gives the frequency counts for 44 first-year college students' caloric intake on a random day:

Interval	Frequency
0–999	1
1000–1999	12
2000–2999	14
3000–3999	11
4000–4999	5
5000–5999	1

9. What do the mean, median, and mode of a distribution have in common? How do they differ? Describe each in a sentence or two.

Find the median and the mode (or modes) for each of the given data sets.

10. Ages (years) of senior citizens tested for low calcium levels:

78, 72, 72, 73, 73, 73, 65, 68, 89, 84, 71, 80

11. Ages (years) of senior citizens tested for low calcium levels:

68, 80, 76, 66, 72, 73, 72, 74, 72, 71, 67, 77, 70

The modal class is the interval containing the most data values. Find the modal class for the distributions of the given data sets.

12. The data in Exercise 7

13. The data in Exercise 8

For the given histograms, identify the shape of the distribution.

14.

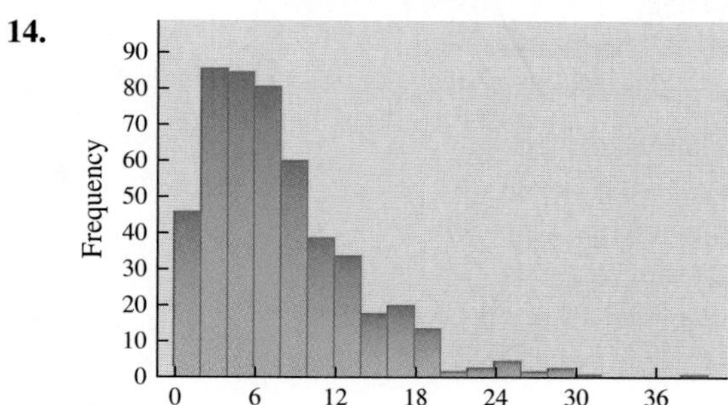

15.

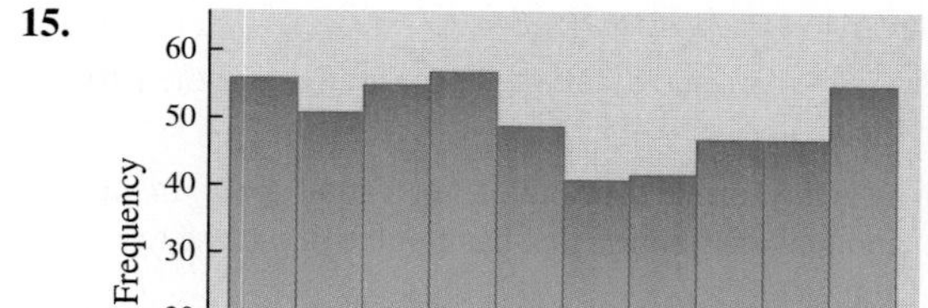

16.

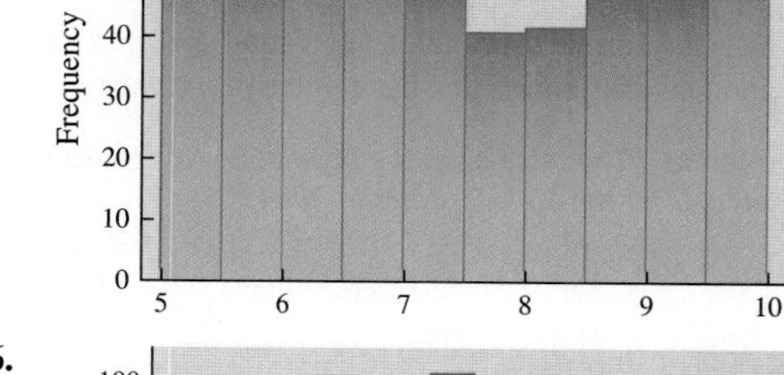

17.

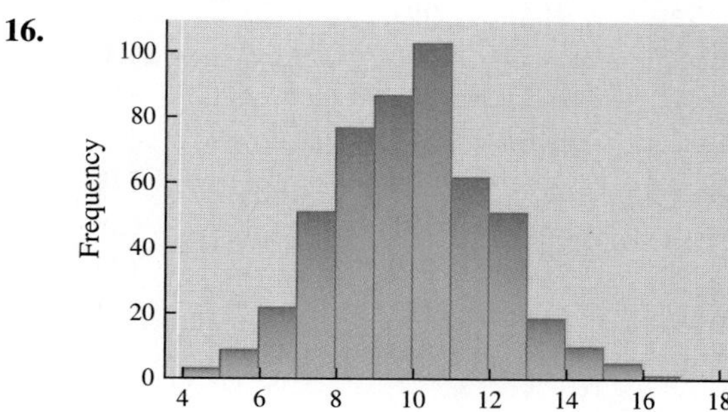

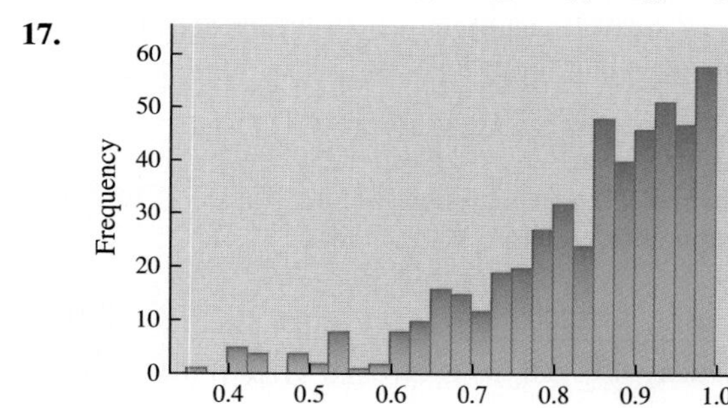

18. What is meant by the range of a distribution?

19. How are the variance and the standard deviation of a distribution related? What is measured by the standard deviation?

Find the range and standard deviation for each of the given distributions.

20. Number of days in a month of sunshine for a village:

14, 17, 18, 19, 30

21. Ages of drivers entering a fast-food restaurant:

26, 43, 17, 20, 25, 37, 54, 28, 20, 19

Find the standard deviation for the given data sets.

22. The data in Exercise 7

23. The data in Exercise 8

24. Describe the characteristics of a normal distribution.

25. What is meant by a skewed distribution?

Find the given areas under the standard normal curve.

26. Between $z = 0$ and $z = 1.35$

27. To the left of $z = .38$

28. Between $z = -1.88$ and $z = 2.41$

29. Between $z = 1.53$ and $z = 2.82$

30. Find a z-score such that 8% of the area under the curve is to the right of z.

31. Why is the normal distribution not a good approximation of a binomial distribution that has a value of p close to 0 or 1?

Social Science *The table gives the population (in thousands) of 18–24 year olds for the 50 U.S. states in 2004:**

456	126	598	206	87
74	157	997	1825	576
571	1260	531	828	2400
280	632	323	77	313
3596	316	589	1128	62
457	299	99	385	748
311	413	191	350	635
84	503	210	1185	173
1549	124	122	112	575
902	521	744	429	57

32. Using intervals that begin with 0–499, create a frequency distribution for the population data.

33. **(a)** Create a histogram for the population data.
(b) What is the shape of the distribution?

34. **Finance** The annual stock returns (percent) of Target Corporation are given in the following table:†

Year	2005	2006	2007	2008
Return	6.6	4.7	−11.7	−30.4

Find the mean and standard deviation of the return for the four-year period.

35. **Finance** The annual stock returns (percent) of Wal-Mart Stores, Inc., are given in the table:†

Year	2005	2006	2007	2008
Return	−10.3	.1	4.9	20.0

Find the mean and standard deviation of the return for the four-year period.

36. **Natural Science** The weight gains of two groups of 10 rats fed on two different experimental diets were as follows:

	Weight Gains									
Diet A	1	0	3	7	1	1	5	4	1	4
Diet B	2	1	1	2	3	2	1	0	1	0

Compute the mean and standard deviation for each group, and compare them to answer the following questions:
(a) Which diet produced the greatest mean gain?
(b) Which diet produced the most consistent gain?

37. **Natural Science** Refer to the data in Exercise 36.
(a) Construct a boxplot for each set of diet data.
(b) Use the boxplots to compare weight gains for the two diets.

38. **Business** The table gives the frequency distribution for the 161 movies that earned $75 million or more dollars in gross domestic receipts from 2004–2008:*

Interval (in Millions of Dollars)	Frequency
75–149.999999	102
150–224.999999	34
225–299.999999	13
300–374.999999	8
375–449.999999	3
450–524.999999	0
525–599.999999	1

Calculate the mean and standard deviation for the data. (*Hint*: Round the midpoint to the nearest tenth.)

Business *The table gives the number of vehicles (in thousands) sold within the United States in March 2008 and March 2009 for 12 auto manufacturers:†*

Auto Manufacturer	March 2008 Sales	March 2009 Sales
General Motors Corp.	115	68
Ford Motor Corporation	73	46
Chrysler LLC	47	24
Toyota Motor Sales USA, Inc.	130	81
American Honda Motor Co., Inc.	82	55
Nissan North America, Inc.	66	43
Hyundai Motor America	31	32
Mazda Motors of America, Inc.	22	15
Mitsubishi Motors NA, Inc.	8	3
Kia Motors America, Inc.	14	12
Volkswagen of America, Inc.	19	13
Audi of America, Inc.	7	5

**State and Metropolitan Area Data Book*: 2006.

†www.morningstar.com.

*Based on data available at www.thebigmoney.com.

†http://online.wsj.com.

39. **(a)** Find the mean and standard deviation for each set of sales.
(b) Which company is closest to the mean sales in March 2008? in March 2009?

40. Find the five-number summary for each set of sales.

41. **(a)** Construct boxplots for the two data sets on the same scale.
(b) Does it appear that sales increased or decreased? Explain.

42. **Social Science** On standard IQ tests, the mean is 100, with a standard deviation of 15. The results are very close to fitting a normal curve. Suppose an IQ test is given to a very large group of people. Find the percentage of people whose IQ score is
(a) more than 130;
(b) less than 85;
(c) between 85 and 115.

43. **Business** A machine that fills quart orange juice cartons is set to fill them with 32.1 oz. If the actual contents of the cartons vary normally, with a standard deviation of .1 oz, what percentage of the cartons contains less than a quart (32 oz)?

Social Science *The results of the 2007 National Survey on Drug Use and Health found that among young adults ages 18–25, in the last month, 16.4% had used marijuana; 1.7% had used cocaine; 6.0% had abused prescription drugs.* If we select 1000 young adults at random, approximate the probability of the given event.*

44. Less than 150 had used marijuana.

45. Between 12 and 25 had used cocaine.

46. More than 50 had abused prescription drugs.

* www.oas.samhsa.gov/nsduh/2k7nsduh/2k7Results.cfm#2.3.

CASE 10
Statistics in the Law: The *Castañeda* Decision

Statistical evidence is now routinely presented in both criminal and civil cases. In this application, we look at a famous case that established use of the binomial distribution and measurement by standard deviation as an accepted procedure.*

Defendants who are convicted in criminal cases sometimes appeal their conviction on the grounds that the jury which indicted or convicted them was drawn from a pool of jurors that does not represent the population of the district in which they live. These appeals almost always cite the Supreme Court's decision in *Castañeda v. Partida* [430 U.S. 482], a case that dealt with the selection of grand juries in the state of Texas. The decision summarizes the facts this way:

> After respondent, a Mexican-American, had been convicted of a crime in a Texas District Court and had exhausted his state remedies on his claim of discrimination in the selection of the grand jury that had indicted him, he filed a habeas corpus petition in the Federal District Court, alleging a denial of due process and equal protection under the Fourteenth Amendment, because of gross underrepresentation of Mexican-Americans on the county grand juries.

*The *Castañeda* case and many other interesting applications of statistics in law are discussed in Michael O. Finkelstein and Bruce Levin, *Statistics for Lawyers*, New York, Springer-Verlag, 1990. U.S. Supreme Court decisions are online at http://www.findlaw.com/casecode/supreme.html, and most states now have important state court decisions online.

The case went to the Appeals Court, which noted that "the county population was 79% Mexican-American, but, over an 11-year period, only 39% of those summoned for grand jury service were Mexican-American," and concluded that together with other testimony about the selection process, "the proof offered by respondent was sufficient to demonstrate a prima facie case of intentional discrimination in grand jury selection. . . ."

The state appealed to the Supreme Court, which then needed to decide whether the underrepresentation of Mexican-Americans on grand juries was indeed too extreme to be an effect of chance. To do so, they invoked the binomial distribution. Here is the argument:

> Given that 79.1% of the population is Mexican-American, the expected number of Mexican-Americans among the 870 persons summoned to serve as grand jurors over the 11-year period is approximately 688. The observed number is 339. Of course, in any given drawing some fluctuation from the expected number is predicted. The important point, however, is that the statistical model shows that the results of a random drawing are likely to fall in the vicinity of the expected value. . . .
>
> The measure of the predicted fluctuations from the expected value is the standard deviation, defined for the binomial distribution as the square root of the product of the total number in the sample (here 870) times the probability of selecting a Mexican-American (.791) times the probability of selecting a non-Mexican-American (.209). . . . Thus, in this case the standard deviation is approximately 12. As a general rule for such large samples, if the difference between the expected value and the observed number is greater than two or three standard deviations, then the hypothesis that the jury drawing was random would be suspect to a social scientist. The 11-year data here reflect a difference between the expected and observed number of Mexican-Americans of approximately 29 standard deviations. A detailed calculation reveals that the likelihood that such a substantial departure from the expected value would occur by chance is less than 1 in 10^{140}.

The Court decided that the statistical evidence supported the conclusion that jurors were not randomly selected, and that it was up to the state to show that its selection process did not discriminate against Mexican-Americans. The Court concluded the following:

> The proof offered by respondent was sufficient to demonstrate a prima facie case of discrimination in grand jury selection. Since the State failed to rebut the presumption of purposeful discrimination by competent testimony, despite two opportunities to do so, we affirm the Court of Appeals' holding of a denial of equal protection of the law in the grand jury selection process in respondent's case.

EXERCISES

1. Check the Court's calculation of 29 standard deviations as the difference between the expected number of Mexican-Americans and the number actually chosen.

2. Where do you think the Court's figure of 1 in 10^{140} came from?

3. The *Castañeda* decison also presents data from a $2\frac{1}{2}$-year period during which the state district judge supervised the selection process. During this period, 220 persons were called to serve as grand jurors, and only 100 of these were Mexican-American.

- **(a)** Considering the 220 jurors as a random selection from a large population, what is the expected number of Mexican-Americans, given the 79.1% population figure?
- **(b)** If we model the drawing of jurors as a sequence of 220 independent Bernoulli trials, what is the standard deviation of the number of Mexican-Americans?
- **(c)** About how many standard deviations is the actual number of Mexican-Americans drawn (100) from the expected number that you calculated in part (a)?
- **(d)** What does the normal-distribution table at the back of the book (Table 2) tell you about this result?

4. The following information is from a case brought by Hy-Vee stores before the Iowa Supreme Court, appealing a ruling by the Iowa Civil Rights Commission in favor of a female employee of one of its grocery stores:

> In 1985, there were 112 managerial positions in the ten Hy-Vee stores located in Cedar Rapids. Only 6 of these managers were women. During that same year there were 294 employees; 206 were men and 88 were women.

- **(a)** How far from the expected number of women in management was the actual number, assuming that gender had nothing to do with promotion? Measure the difference in standard deviations.
- **(b)** Does this look like evidence of purposeful discrimination?

CHAPTER

11

Differential Calculus

How fast is the number of subscribers to satellite radio growing? At what rate is the number of Internet users increasing? At what rate are health care expenditures increasing? These questions and many others, in fields as diverse as economics and physics, can be answered by using calculus. See Exercises 59 and 61 on page 748 and Exercise 67 on page 749.

The algebraic problems considered in earlier chapters dealt with *static* situations:

What is the revenue when x items are sold?

How much interest is earned in 2 years?

What is the equilibrium price?

Calculus, on the other hand, deals with *dynamic* situations:

At what rate is the economy growing?

How fast is a rocket going at any instant after liftoff?

How quickly can production be increased without adversely affecting profits?

The techniques of calculus will allow us to answer many questions like these that deal with rates of change.

The key idea underlying the development of calculus is the concept of limit, so we begin by studying limits.

11.1 Limits

We have often dealt with a problem like this: "Find the value of the function $f(x)$ when $x = a$." The underlying idea of "limit," however, is to examine what the function does *near* $x = a$, rather than what it does *at* $x = a$.

Example 1 The function

$$f(x) = \frac{2x^2 - 3x - 2}{x - 2}$$

is not defined when $x = 2$. (Why?) What happens to the values of $f(x)$ when x is *very close* to 2?

Solution Evaluate f at several numbers that are very close to $x = 2$, as in the following table:

x	1.99	1.999	2	2.0001	2.001
$f(x)$	4.98	4.998	—	5.0002	5.002

The table suggests that

as x gets closer and closer to 2 from either direction, the corresponding value of $f(x)$ gets closer and closer to 5.

In fact, by experimenting further, you can convince yourself that the values of $f(x)$ can be made *as close as you want* to 5 by taking values of x close enough to 2. This situation is usually described by saying "The *limit* of $f(x)$ as x approaches 2 is the number 5," which is written symbolically as

$$\lim_{x \to 2} f(x) = 5, \quad \text{or equivalently,} \quad \lim_{x \to 2} \frac{2x^2 - 3x - 2}{x - 2} = 5. \quad ✓1$$

✓Checkpoint 1

Use a calculator to estimate

$$\lim_{x \to 1} \frac{x^3 + x^2 - 2x}{x - 1}$$

by completing the following table:

x	$f(x)$
.9	
.99	
.999	
1.0001	
1.001	
1.01	
1.1	

Answer: *See page 666.*

The informal definition of "limit" that follows is similar to the situation in Example 1, but now f is any function, and a and L are fixed real numbers (in Example 1, $a = 2$ and $L = 5$).

Limit of a Function

Let f be a function, and let a and L be real numbers. Assume that $f(x)$ is defined for all x near $x = a$. Suppose that

> as x takes values very close (but not equal) to a (on both sides of a), the corresponding values of $f(x)$ are very close (and possibly equal) to L

and that

> the values of $f(x)$ can be made as close as you want to L for all values of x that are close enough to a.

Then the number L is the **limit** of the function $f(x)$ as x approaches a, which is written

$$\lim_{x \to a} f(x) = L.$$

This definition is *informal* because the expressions "near," "very close," and "as close as you want" have not been precisely defined. In particular, the tables used in Example 1 and the next set of examples provide strong intuitive evidence, but not a proof, of what the limits must be.

Example 2 If $f(x) = x^2 + x + 1$, what is $\lim_{x \to 3} f(x)$?

Solution Make a table showing the values of the function at numbers very close to 3:

	x approaches 3 from the left →			3	← x approaches 3 from the right		
x	2.9	2.99	2.9999	3	3.0001	3.01	3.1
$f(x)$	12.31	12.9301	12.9993 …		13.0007 …	13.0701	13.71

The table suggests that as x approaches 3 from either direction, $f(x)$ gets closer and closer to 13 and, hence, that

$$\lim_{x \to 3} f(x) = 13, \qquad \text{or equivalently,} \qquad \lim_{x \to 3} (x^2 + x + 1) = 13.$$

Note that the function $f(x)$ is defined when $x = 3$ and that $f(3) = 3^2 + 3 + 1 = 13$. So in this case, the limit of $f(x)$ as x approaches 3 is $f(3)$, the value of the function at 3. ■

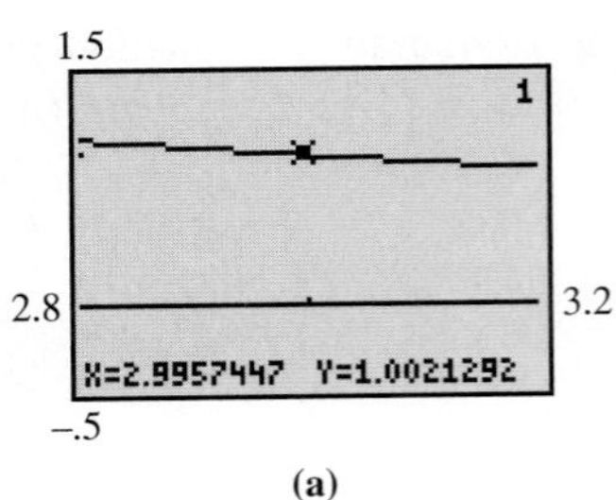

(a)

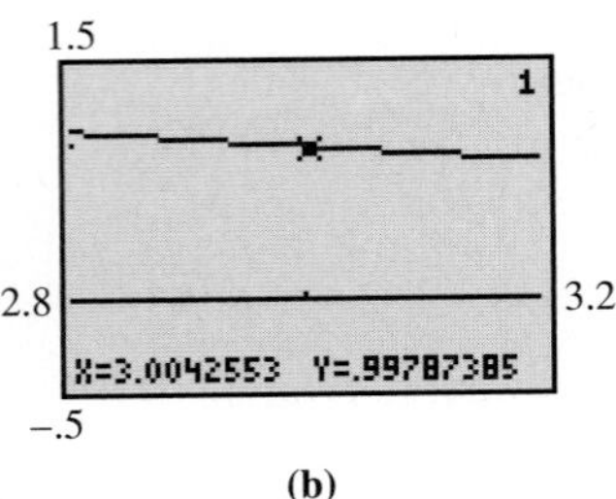

(b)

FIGURE 11.1

X	Y_1
2.99	1.005
2.999	1.0005
2.9999	1.0001
3	ERROR
3.0001	.99995
3.001	.9995
3.01	.99501

X=3

FIGURE 11.2

Example 3 Use a graphing calculator to find

$$\lim_{x\to 3}\frac{x-3}{e^{x-3}-1}.$$

Solution There are two ways to estimate the limit.

Graphical Method Graph $f(x) = \dfrac{x-3}{e^{x-3}-1}$ in a very narrow window near $x = 3$. Use the trace feature to move along the graph and observe the y-coordinates as x gets very close to 3 from either side. Figure 11.1 suggests that $\lim_{x\to 3} f(x) = 1$.

Numerical Method Use the table feature to make a table of values for $f(x)$ when x is very close to 3. Figure 11.2 shows that when x is very close to 3, $f(x)$ is very close to 1. (The table displays "error" at $x = 3$ because the function is not defined when $x = 3$.) Thus, it appears that

$$\lim_{x\to 3}\frac{x-3}{e^{x-3}-1} = 1.$$

The function has the limit 1 as x approaches 3, even though $f(3)$ is not defined.

Example 4 Find $\lim_{x\to 4} f(x)$, where f is the function whose rule is

$$f(x) = \begin{cases} 0 & \text{if } x \text{ is an integer} \\ 1 & \text{if } x \text{ is not an integer} \end{cases}$$

and whose graph is shown in Figure 11.3.

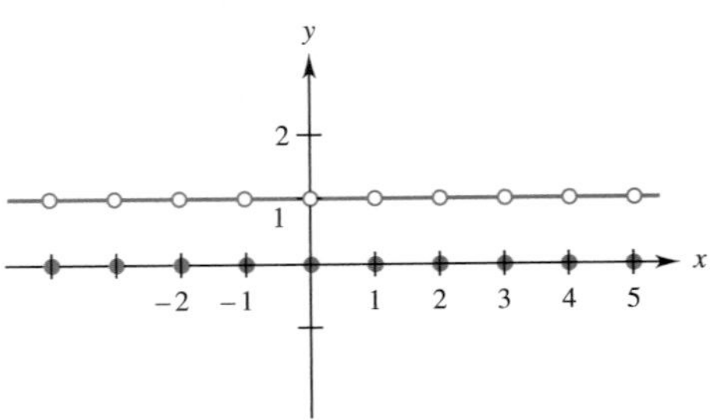

FIGURE 11.3

Solution The definition of the limit as x approaches 4 involves only values of x that are close, but not equal, to 4—corresponding to the part of the graph on either side of 4, but not at 4 itself. Now, $f(x) = 1$ for all these numbers (because the numbers very near 4, such as 3.99995 and 4.00002, are not integers). Thus, for all x very close to 4, the corresponding value of $f(x)$ is 1, so $\lim_{x\to 4} f(x) = 1$. However, since 4 is an integer, $f(4) = 0$. Therefore, $\lim_{x\to 4} f(x) \neq f(4)$.

Examples 1–4 illustrate the following facts.

Limits and Function Values

If the limit of a function $f(x)$ as x approaches a exists, then there are three possibilities:

1. $f(a)$ is not defined, but $\lim_{x \to a} f(x)$ is defined. (Examples 1 and 3)
2. $f(a)$ is defined and $\lim_{x \to a} f(x) = f(a)$. (Example 2)
3. $f(a)$ is defined, but $\lim_{x \to a} f(x) \neq f(a)$. (Example 4)

FINDING LIMITS ALGEBRAICALLY

As we have seen, tables are very useful for estimating limits. However, it is often more efficient and accurate to find limits algebraically. We begin with two simple functions.

Consider the constant function $f(x) = 5$. To compute $\lim_{x \to a} f(x)$, you must ask "When x is very close to a, what is the value of $f(x)$?" The answer is easy because no matter what x is, the value of $f(x)$ is *always* the number 5. As x gets closer and closer to a, the value of $f(x)$ is always 5. Hence,

$$\lim_{x \to a} f(x) = 5, \qquad \text{which is usually written} \qquad \lim_{x \to a} 5 = 5.$$

The same thing is true for *any* constant function.

Limit of a Constant Function

If d is a constant, then, for any real number a,

$$\lim_{x \to a} d = d.$$

Now consider the *identity function*, whose rule is $f(x) = x$. When x is very close to a number c, the corresponding value of $f(x)$ (namely, x itself) is very close to c. So we have the following conclusion.

Limit of the Identity Function

For every real number a, $\quad \lim_{x \to a} x = a.$

The facts in the two preceding boxes, together with the properties of limits that follow, will enable us to find a wide variety of limits.

Properties of Limits

Let a, r, A, and B be real numbers, and let f and g be functions such that

$$\lim_{x\to a} f(x) = A \qquad \text{and} \qquad \lim_{x\to a} g(x) = B.$$

Then the following properties hold:

1. $\lim_{x\to a}[f(x) + g(x)] = A + B = \lim_{x\to a} f(x) + \lim_{x\to a} g(x)$ *Sum Property*
(The limit of a sum is the sum of the limits.)

2. $\lim_{x\to a}[f(x) - g(x)] = A - B = \lim_{x\to a} f(x) - \lim_{x\to a} g(x)$ *Difference Property*
(The limit of a difference is the difference of the limits.)

3. $\lim_{x\to a}[f(x) \cdot g(x)] = A \cdot B = \lim_{x\to a} f(x) \cdot \lim_{x\to a} g(x)$ *Product Property*
(The limit of a product is the product of the limits.)

4. $\lim_{x\to a} \dfrac{f(x)}{g(x)} = \dfrac{A}{B} = \dfrac{\lim_{x\to a} f(x)}{\lim_{x\to a} g(x)}$ $(B \neq 0)$ *Quotient Property*
(The limit of a quotient is the quotient of the limits, provided that the limit of the denominator is nonzero.)

5. For any real number for which A^r exists,
$\lim_{x\to a}[f(x)]^r = A^r = [\lim_{x\to a} f(x)]^r.$ *Power Property*
(The limit of a power is the power of the limit.)

Although we won't prove these properties (a rigorous definition of limit is needed for that), you should find most of them plausible. For instance, if the values of $f(x)$ get very close to A and the values of $g(x)$ get very close to B when x approaches a, it is reasonable to expect that the corresponding values of $f(x) + g(x)$ will get very close to $A + B$ (Property 1) and that the corresponding values of $f(x)g(x)$ will get very close to AB (Property 3).

Example 5 Find $\lim_{x\to 3}(4x^2 - 2x + 5)$.

Solution $\lim_{x\to 3}(4x^2 - 2x + 5)$

$= \lim_{x\to 3} 4x^2 - \lim_{x\to 3} 2x + \lim_{x\to 3} 5$ Sum property

$= \lim_{x\to 3} 4 \cdot \lim_{x\to 3} \cdot x^2 - \lim_{x\to 3} 2 \cdot \lim_{x\to 3} x + \lim_{x\to 3} 5$ Product property

$= \lim_{x\to 3} 4 \cdot [\lim_{x\to 3} \cdot x]^2 - \lim_{x\to 3} 2 \cdot \lim_{x\to 3} x + \lim_{x\to 3} 5$ Power property

$= 4 \cdot 3^2 - 2 \cdot 3 + 5 = 35.$ Constant-function limit and identity-function limit

Example 5 shows that $\lim_{x\to 3} f(x) = 35$, where $f(x) = 4x^2 - 2x + 5$. Note that $f(3) = 4\cdot 3^2 - 2\cdot 3 + 5 = 35$. In other words, the limit as x approaches 3 is the value of the function at 3:

$$\lim_{x\to 3} f(x) = f(3).$$

The same analysis used in Example 5 works with any polynomial function and leads to the following conclusion.

Polynomial Limits

If $f(x)$ is a polynomial function and a is a real number, then

$$\lim_{x\to a} f(x) = f(a).$$

In other words, the limit is the value of the function at $x = a$.

✔Checkpoint 2

If $f(x) = 2x^3 - 5x^2 + 8$, find the given limits.

(a) $\lim_{x\to 2} f(x)$

(b) $\lim_{x\to -3} f(x)$

Answers: See page 666.

This property will be used frequently. 2✔

Example 6 Find each limit.

(a) $\lim_{x\to 2}[(x^2 + 1) + (x^3 - x + 3)]$

Solution $\lim_{x\to 2}[(x^2 + 1) + (x^3 - x + 3)]$

$= \lim_{x\to 2}(x^2 + 1) + \lim_{x\to 2}(x^3 - x + 3)$ Sum property

$= (2^2 + 1) + (2^3 - 2 + 3) = 5 + 9 = 14.$ Polynomial limit

(b) $\lim_{x\to -1}(x^3 + 4x)(2x^2 - 3x)$

Solution $\lim_{x\to -1}(x^3 + 4x)(2x^2 - 3x)$

$= \lim_{x\to -1}(x^3 + 4x)\cdot \lim_{x\to -1}(2x^2 - 3x)$ Product property

$= [(-1)^3 + 4(-1)]\cdot[2(-1)^2 - 3(-1)]$ Polynomial limit

$= (-1 - 4)(2 + 3) = -25.$

(c) $\lim_{x\to -1} 5(3x^2 + 2)$

Solution $\lim_{x\to -1} 5(3x^2 + 2) = \lim_{x\to -1} 5 \cdot \lim_{x\to -1}(3x^2 + 2)$ Product property

$= 5[3(-1)^2 + 2]$ Constant-function limit and polynomial limit

$= 25.$

(d) $\lim_{x\to 9}\sqrt{4x - 11}$

Solution Begin by writing the square root in exponential form.

$$\lim_{x\to 9}\sqrt{4x-11} = \lim_{x\to 9}[4x-11]^{1/2}$$
$$= [\lim_{x\to 9}(4x-11)]^{1/2} \quad \text{Power property}$$
$$= [4\cdot 9 - 11]^{1/2} \quad \text{Polynomial limit}$$
$$= [25]^{1/2} = \sqrt{25} = 5. \quad 3$$

✓Checkpoint 3

Use the limit properties to find the given limits.

(a) $\lim_{x\to 4}(3x-9)$

(b) $\lim_{x\to -1}(2x^2-4x+1)$

(c) $\lim_{x\to 2}\sqrt{3x+3}$

Answers: See page 666.

When a rational function, such as $f(x) = \frac{x}{x+2}$, is defined at $x = a$, it is easy to find the limit of $f(x)$ as x approaches a.

Example 7 Find $\lim_{x\to 5}\frac{x}{x+3}$.

Solution
$$\lim_{x\to 5}\frac{x}{x+3} = \frac{\lim_{x\to 5} x}{\lim_{x\to 5}(x+3)} \quad \text{Quotient property}$$
$$= \frac{5}{5+3} = \frac{5}{8}. \quad \text{Polynomial limit}$$

Note that $f(5) = \frac{5}{5+3} = \frac{5}{8}$. So the limit of $f(x)$ as x approaches 5 is the value of the function at 5:

$$\lim_{x\to 5} f(x) = f(5). \quad 4$$

✓Checkpoint 4

Find the given limits.

(a) $\lim_{x\to 2}\frac{2x-1}{3x+4}$

(b) $\lim_{x\to -1}\frac{x-2}{3x-1}$

Answers: See page 666.

The argument used in Example 7 works in the general case and proves the following result.

Rational Limits

If $f(x)$ is a rational function and a is a real number such that $f(a)$ is defined, then

$$\lim_{x\to a} f(x) = f(a).$$

In other words, the limit is the value of the function at $x = a$.

If a rational function is not defined at the number a, the preceding property cannot be used. Different techniques are needed in such cases to find the limit (if one exists), as we shall see in Examples 8 and 9.

The definition of the limit as x approaches a involves only the values of the function when x is *near* a, but not the value of the function *at* a. So two functions that agree for all values of x, except possibly at $x = a$, will necessarily have the same limit when x approaches a. Thus, we have the following fact.

Limit Theorem

If f and g are functions that have limits as x approaches a, and $f(x) = g(x)$ for all x near a, then

$$\lim_{x\to a} f(x) = \lim_{x\to a} g(x).$$

Example 8 Find $\lim_{x\to 2} \frac{x^2 + x - 6}{x - 2}$.

Solution The quotient property cannot be used here, because

$$\lim_{x\to 2}(x - 2) = 0.$$

We can, however, simplify the function by rewriting the fraction as

$$\frac{x^2 + x - 6}{x - 2} = \frac{(x + 3)(x - 2)}{x - 2}.$$

When $x \neq 2$, the quantity $x - 2$ is nonzero and may be cancelled, so that

$$\frac{x^2 + x - 6}{x - 2} = x + 3 \qquad \text{for all } x \neq 2.$$

Now the limit theorem can be used:

$$\lim_{x\to 2} \frac{x^2 + x - 6}{x - 2} = \lim_{x\to 2}(x + 3) = 2 + 3 = 5.$$ 5✓

✓Checkpoint 5

Find $\lim_{x\to 1} \frac{2x^2 + x - 3}{x - 1}$.

Answer: See page 666.

Example 9 Find $\lim_{x\to 4} \frac{\sqrt{x} - 2}{x - 4}$.

Solution As $x \to 4$, both the numerator and the denominator approach 0, giving the meaningless expression 0/0. To change the form of the expression, algebra can be used to rationalize the numerator by multiplying both the numerator and the denominator by $\sqrt{x} + 2$. This gives

$$\frac{\sqrt{x} - 2}{x - 4} = \frac{\sqrt{x} - 2}{x - 4} \cdot \frac{\sqrt{x} + 2}{\sqrt{x} + 2} = \frac{\sqrt{x}\cdot\sqrt{x} + 2\sqrt{x} - 2\sqrt{x} - 4}{(x - 4)(\sqrt{x} + 2)}$$

$$= \frac{x - 4}{(x - 4)(\sqrt{x} + 2)} = \frac{1}{\sqrt{x} + 2} \qquad \text{for all } x \neq 4.$$

Now use the limit theorem and the properties of limits:

$$\lim_{x\to 4} \frac{\sqrt{x} - 2}{x - 4} = \lim_{x\to 4} \frac{1}{\sqrt{x} + 2} = \frac{1}{\sqrt{4} + 2} = \frac{1}{2 + 2} = \frac{1}{4}.$$ 6✓

✓Checkpoint 6

Find the given limits.

(a) $\lim_{x\to 1} \frac{\sqrt{x} - 1}{x - 1}$

(b) $\lim_{x\to 9} \frac{\sqrt{x} - 3}{x - 9}$

Answers: See page 666.

EXISTENCE OF LIMITS

It is possible that $\lim_{x\to a} f(x)$ may not exist; that is, there may be no number L satisfying the definition of $\lim_{x\to a} f(x) = L$. This can happen in many ways, two of which are illustrated next.

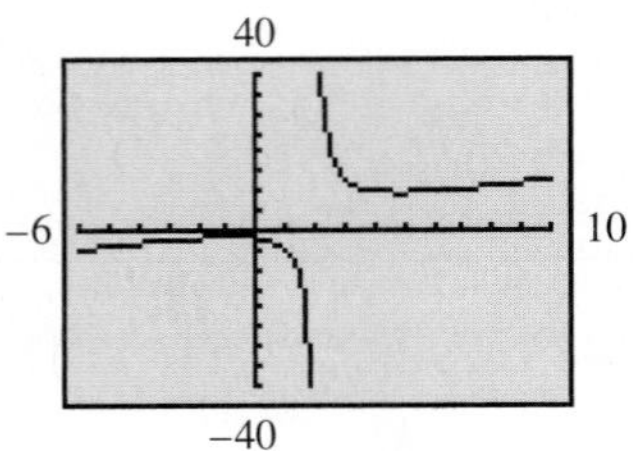

FIGURE 11.4

Example 10 If $g(x) = \frac{x^2 + 4}{x - 2}$, find $\lim_{x\to 2} g(x)$.

Solution The quotient property cannot be used, since $\lim_{x\to 2}(x - 2) = 0$, and the limit theorem does not apply because $x^2 + 4$ does not factor. So we use the following table and the graph of $g(x)$ in Figure 11.4:

	x approaches 2 from the left →				2	← x approaches 2 from the right		
x	1.8	1.9	1.99	1.999	2	2.001	2.01	2.05
$g(x)$	−36.2	−76.1	−796	−7996		8004	804	164
	$g(x)$ gets smaller and smaller →					← $g(x)$ gets larger and larger		

The table and the graph both show that as x approaches 2 from the left, $g(x)$ gets smaller and smaller, but as x approaches 2 from the right, $g(x)$ gets larger and larger. Since $g(x)$ does not get closer and closer to a single real number as x approaches 2 from either side,

$$\lim_{x\to 2} \frac{x^2 + 4}{x - 2} \text{ does not exist.}$$

7 ✓

✓ Checkpoint 7

Let $f(x) = \frac{x^2 + 9}{x - 3}$. Find the given limits.

(a) $\lim_{x\to 3} f(x)$

(b) $\lim_{x\to 0} f(x)$

Answers: See page 666.

Example 11 What is $\lim_{x\to 0} \frac{|x|}{x}$?

Solution The function $f(x) = \frac{|x|}{x}$ is not defined when $x = 0$. Recall the definition of absolute value:

$$|x| = \begin{cases} x & \text{if } x \geq 0 \\ -x & \text{if } x < 0 \end{cases}.$$

Consequently, when $x > 0$,

$$f(x) = \frac{|x|}{x} = \frac{x}{x} = 1,$$

and when $x < 0$,

$$f(x) = \frac{|x|}{x} = \frac{-x}{x} = -1.$$

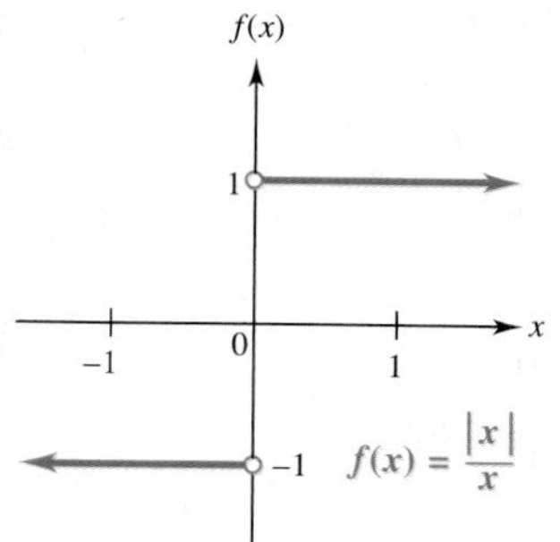

FIGURE 11.5

The graph of f is shown in Figure 11.5. As x approaches 0 from the right, x is always positive, and the corresponding value of $f(x)$ is 1. But as x approaches 0 from the left, x is always negative, and the corresponding value of $f(x)$ is -1. Thus, as x approaches 0 from *both* sides, the corresponding values of $f(x)$ do not get closer and closer to a *single* real number. Therefore, the limit does not exist.*

The function f whose graph is shown in Figure 11.6 illustrates various facts about limits that were discussed earlier in this section.

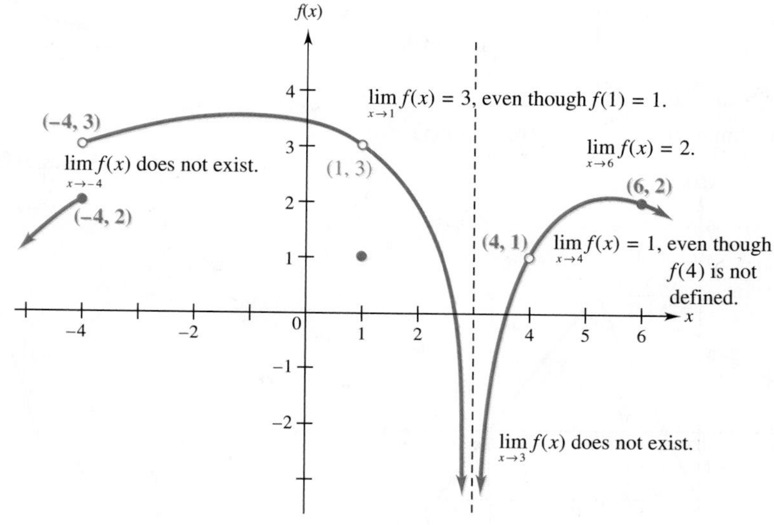

FIGURE 11.6

*However, the behavior of this function near $x = 0$ can be described by *one-sided limits*, which are discussed in the next section.

11.1 ▸ Exercises

Use the given graph to determine the value of the indicated limits. (See Examples 3, 4, 10, and 11 and Figure 11.6.)

1. **(a)** $\lim_{x\to 3} f(x)$ **(b)** $\lim_{x\to -1} f(x)$

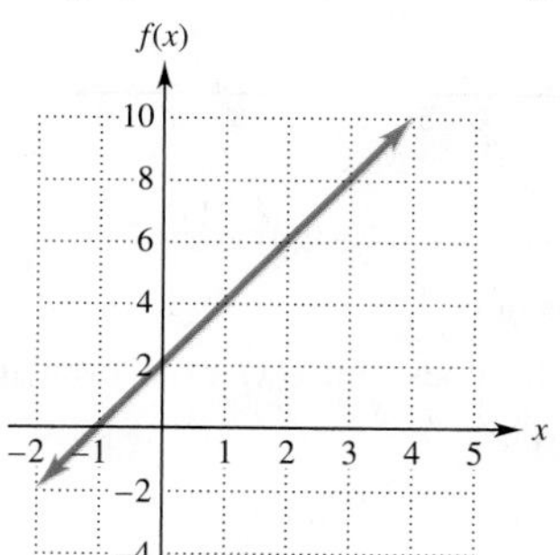

2. **(a)** $\lim_{x\to 0} F(x)$ **(b)** $\lim_{x\to 1} F(x)$

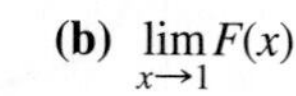

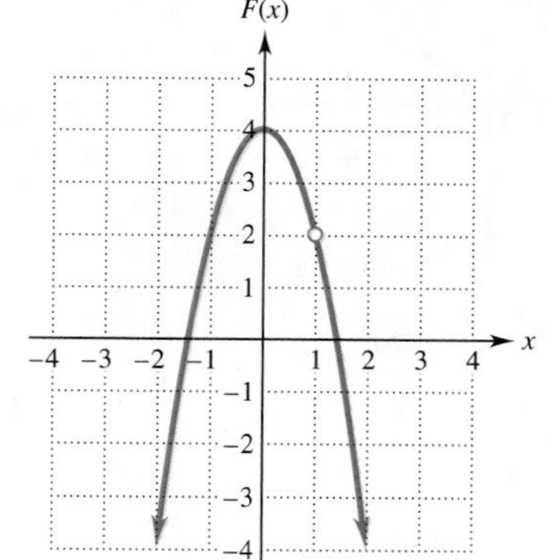

3. **(a)** $\lim_{x \to -3} f(x)$ **(b)** $\lim_{x \to 2} f(x)$

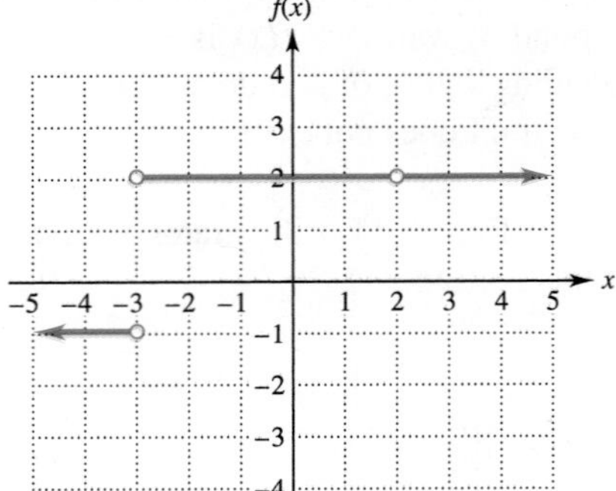

4. **(a)** $\lim_{x \to 3} g(x)$ **(b)** $\lim_{x \to 0} g(x)$

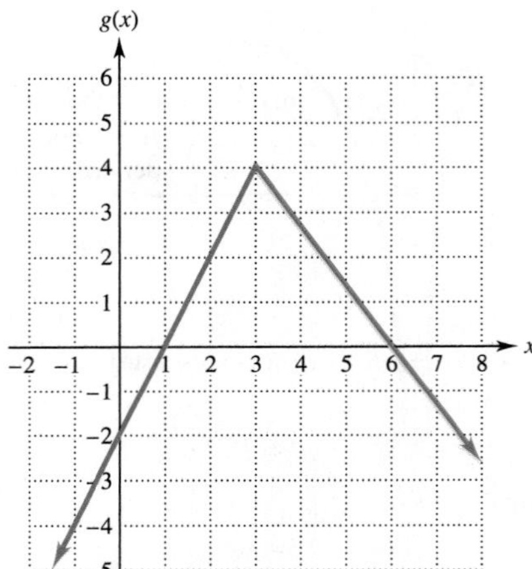

5. **(a)** $\lim_{x \to 3} f(x)$ **(b)** $\lim_{x \to 2} f(x)$

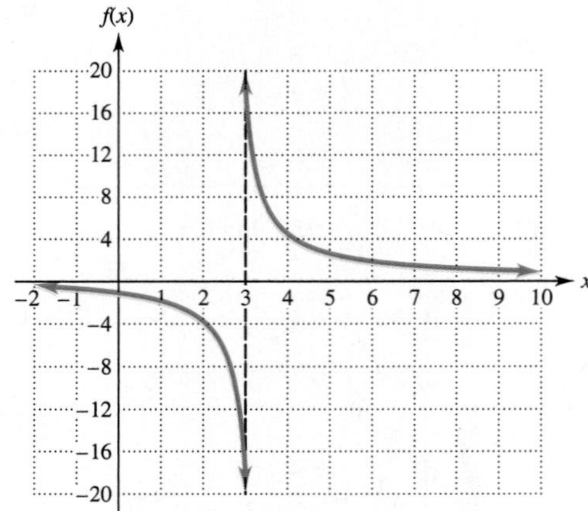

6. **(a)** $\lim_{x \to -2} h(x)$ **(b)** $\lim_{x \to 0} h(x)$

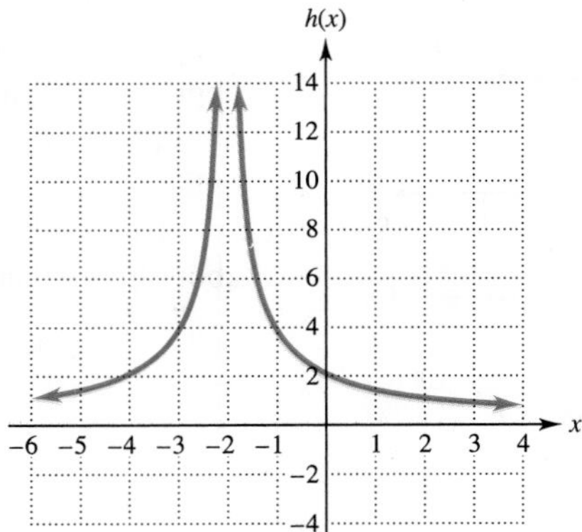

7. **(a)** $\lim_{x \to 1} g(x)$ **(b)** $\lim_{x \to 0} g(x)$

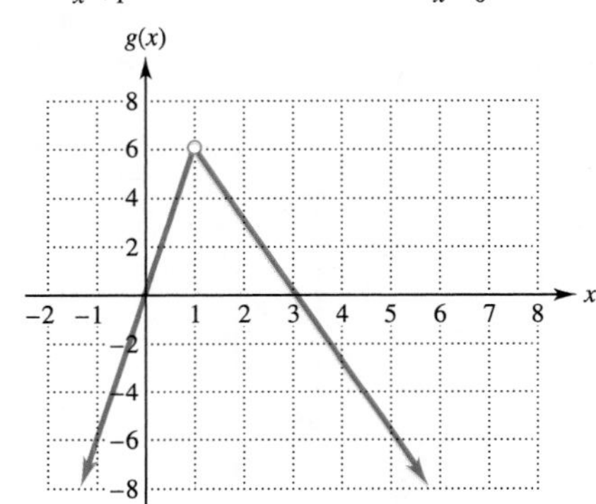

8. **(a)** $\lim_{x \to 4} f(x)$ **(b)** $\lim_{x \to 9} f(x)$

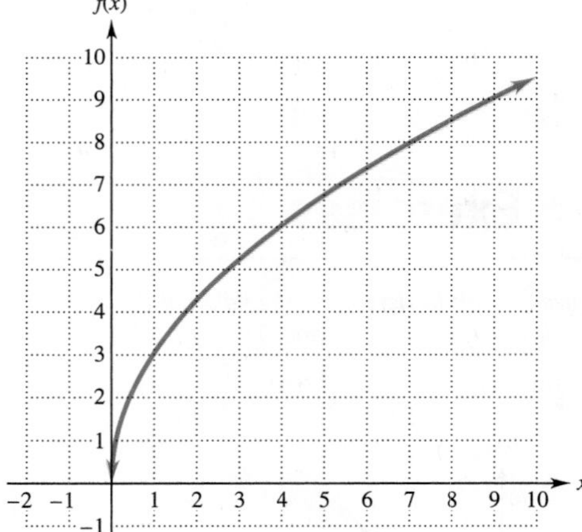

9. Explain why $\lim_{x \to 1} F(x)$ in Exercise 2(a) exists, but $\lim_{x \to -3} f(x)$ in Exercise 3(a) does not.

10. In Exercise 7(a), why does $\lim_{x \to 1} g(x)$ exist even though $g(1)$ is not defined?

Use a calculator to estimate the given limits. (See Examples 1–3.)

11. $\lim_{x\to 1}\frac{\ln x}{x-1}$

12. $\lim_{x\to 5}\frac{\ln x - \ln 5}{x-5}$

13. $\lim_{x\to 0}\frac{e^{3x}-1}{x}$

14. $\lim_{x\to 0}\frac{x}{\ln|x|}$

15. $\lim_{x\to 0}(x\ln|x|)$

16. $\lim_{x\to 0}\frac{x}{e^x-1}$

17. $\lim_{x\to 3}\frac{x^3-3x^2-4x+12}{x-3}$

18. $\lim_{x\to 4}\frac{.1x^4-.8x^3+1.6x^2+2x-8}{x-4}$

19. $\lim_{x\to -2}\frac{x^4+2x^3-x^2+3x+1}{x+2}$

20. $\lim_{x\to 0}\frac{e^{2x}+e^x-2}{e^x-1}$

Suppose $\lim_{x\to 4} f(x) = 25$ *and* $\lim_{x\to 4} g(x) = 10$. *Use the limit properties to find the given limits.*

21. $\lim_{x\to 4}[f(x)-g(x)]$

22. $\lim_{x\to 4}[g(x)\cdot f(x)]$

23. $\lim_{x\to 4}\frac{f(x)}{g(x)}$

24. $\lim_{x\to 4}[3\cdot f(x)]$

25. $\lim_{x\to 4}\sqrt{f(x)}$

26. $\lim_{x\to 4}[g(x)]^3$

27. $\lim_{x\to 4}\frac{f(x)+g(x)}{2g(x)}$

28. $\lim_{x\to 4}\frac{5g(x)+2}{1-f(x)}$

29. **(a)** Graph the function f whose rule is

$$f(x) = \begin{cases} 3-x & \text{if } x < -2 \\ x+2 & \text{if } -2 \le x < 2. \\ 1 & \text{if } x \ge 2 \end{cases}$$

Use the graph in part (a) to find the following limits:
(b) $\lim_{x\to -2} f(x)$; **(c)** $\lim_{x\to 1} f(x)$; **(d)** $\lim_{x\to 2} f(x)$.

30. **(a)** Graph the function g whose rule is

$$g(x) = \begin{cases} x^2 & \text{if } x < -1 \\ x+2 & \text{if } -1 \le x < 1. \\ 3-x & \text{if } x \ge 1 \end{cases}$$

Use the graph in part (a) to find the following limits:
(b) $\lim_{x\to -1} g(x)$; **(c)** $\lim_{x\to 0} g(x)$; **(d)** $\lim_{x\to 1} g(x)$.

Use algebra and the properties of limits as needed to find the given limits. If the limit does not exist, say so. (See Examples 5–11.)

31. $\lim_{x\to 2}(3x^3-4x^2-5x+2)$

32. $\lim_{x\to -1}(2x^3-x^2-6x+1)$

33. $\lim_{x\to 3}\frac{4x+7}{10x+1}$

34. $\lim_{x\to -2}\frac{x+6}{8x-5}$

35. $\lim_{x\to 5}\frac{x^2-25}{x-5}$

36. $\lim_{x\to -4}\frac{x^2-16}{x+4}$

37. $\lim_{x\to 4}\frac{x^2-x-12}{x-4}$

38. $\lim_{x\to 5}\frac{x^2-3x-10}{x-5}$

39. $\lim_{x\to 2}\frac{x^2-5x+6}{x^2-6x+8}$

40. $\lim_{x\to -2}\frac{x^2+3x+2}{x^2-x-6}$

41. $\lim_{x\to 4}\frac{(x+4)^2(x-5)}{(x-4)(x+4)^2}$

42. $\lim_{x\to -3}\frac{(x+3)(x-3)(x+4)}{(x+8)(x+3)(x-4)}$

43. $\lim_{x\to 3}\sqrt{x^2-3}$

44. $\lim_{x\to 3}\sqrt{x^2-7}$

45. $\lim_{x\to 4}\frac{-6}{(x-4)^2}$

46. $\lim_{x\to -2}\frac{3x}{(x+2)^3}$

47. $\lim_{x\to 0}\frac{[1/(x+3)]-1/3}{x}$

48. $\lim_{x\to 0}\frac{[-1/(x+2)]+1/2}{x}$

49. $\lim_{x\to 25}\frac{\sqrt{x}-5}{x-25}$

50. $\lim_{x\to 36}\frac{\sqrt{x}-6}{x-36}$

51. $\lim_{x\to 5}\frac{\sqrt{x}-\sqrt{5}}{x-5}$

52. **(a)** Approximate $\lim_{x\to 0}(1+x)^{1/x}$ to five decimal places. (Evaluate the function at numbers closer and closer to 0 until successive approximations agree in the first five places.)
(b) Find the decimal expansion of the number e to as many places as your calculator can manage.
(c) What do parts (a) and (b) suggest about the exact value of $\lim_{x\to 0}(1+x)^{1/x}$?

53. Business A company training program has determined that a new employee can process an average of $P(s)$ pieces of work per day after s days of on-the-job training, where

$$P(s) = \frac{105s}{s+7}.$$

Find the given quantities.
(a) $P(1)$ **(b)** $P(13)$ **(c)** $\lim_{s\to 13} P(s)$

54. Natural Science The concentration of a drug in a patient's bloodstream h hours after it was injected is given by

$$A(h) = \frac{.3h}{h^2+3}.$$

Find the given quantities.
(a) $A(.5)$ **(b)** $A(1)$ **(c)** $\lim_{h\to 1} A(h)$

55. Business The cost of manufacturing a particular DVD is

$$c(x) = 150{,}000 + 3x,$$

where x is the number of DVDs produced. The average cost per DVD, denoted by $\bar{c}(x)$, is found by dividing $c(x)$ by x. Find the given quantities.

(a) $\bar{c}(1000)$ **(b)** $\bar{c}(10{,}000)$ **(c)** $\lim_{x \to 100{,}000} \bar{c}(x)$

56. Business When the price of an essential commodity (such as gasoline) rises rapidly, consumption drops slowly at first. If the price continues to rise, however, a "tipping" point may be reached at which consumption takes a sudden, substantial drop. Suppose the accompanying graph shows the consumption of gasoline, $G(t)$, in millions of gallons, in a certain area. We assume that the price is rising rapidly. Here, t is time in months after the price began rising. Use the graph to find the given quantities.

(a) $\lim_{t \to 12} G(t)$ **(b)** $\lim_{t \to 16} G(t)$ **(c)** $G(16)$

(d) The tipping point (in months)

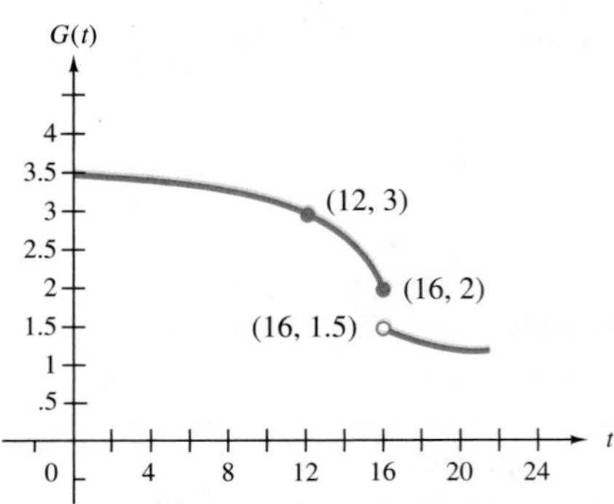

57. Social Sciences The accompanying figure shows the graphs of the functions C and I, whose rules are

$C(x)$ = population of China (in billions) in year x;

$I(x)$ = population of India (in billions) in year x.*

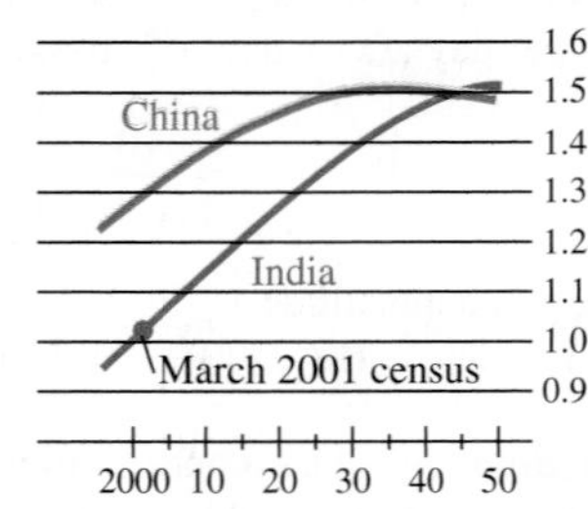

Sources: Indian Census; United Nations

Find each of the given limits.

(a) $\lim_{x \to 2030} C(x)$ **(b)** $\lim_{x \to 2015} I(x)$

(c) $\lim_{x \to 2045} C(x) - I(x)$ **(d)** $\lim_{x \to 2045} C(x) + I(x)$

58. Business The accompanying graph shows the profit from the daily production of x thousand kilograms of an industrial chemical. Use the graph to find the given limits.

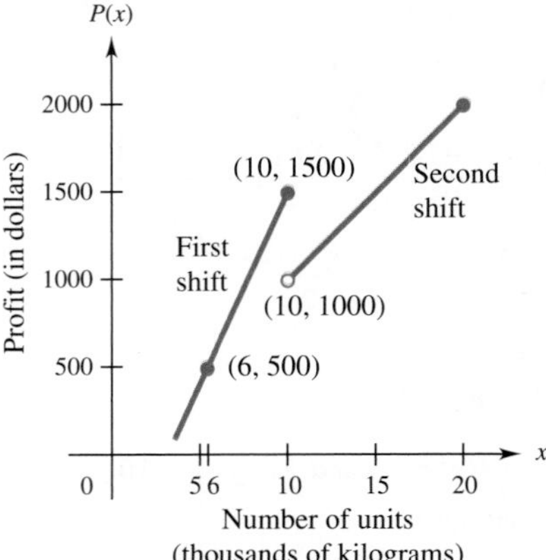

(a) $\lim_{x \to 6} P(x)$ **(b)** $\lim_{x \to 10} P(x)$ **(c)** $\lim_{x \to 15} P(x)$

(d) Use the graph to estimate the number of units of the chemical that must be produced before the second shift is beneficial.

59. Business An international long-distance phone service advertises that "All calls up to 20 minutes are 99 cents and only 7 cents per minute [or a fraction thereof] after that." So a 20-minute call costs .99, a 20.1-minute call costs \$1.06, a 21-minute call costs \$1.06, and so on. Let $P(x)$ be the price of a phone call lasting x minutes.

(a) Write the rule of the function P for $0 \le x \le 23$.

(b) Graph $P(x)$ for $0 \le x \le 23$.

Find the given limits.

(c) $\lim_{x \to 10} P(x)$ **(d)** $\lim_{x \to 20} P(x)$ **(e)** $\lim_{x \to 22.5} P(x)$

✓ Checkpoint Answers

1. 2.61; 2.9601; 2.996; 3.0004; 3.004; 3.0401; 3.41; the limit appears to be 3.
2. (a) 4 (b) −91
3. (a) 3 (b) 7 (c) 3
4. (a) $\frac{3}{10}$ (b) $\frac{3}{4}$
5. 5
6. (a) $\frac{1}{2}$ (b) $\frac{1}{6}$
7. (a) Does not exist (b) −3

*Graph appeared in the March 31, 2001, edition of the *Economist* magazine.

11.2 One-Sided Limits and Limits Involving Infinity

In addition to the limits introduced in Section 11.1, which will be used frequently, there are several other kinds of limits that will appear briefly in Sections 11.8 and 13.3. We begin with one-sided limits.

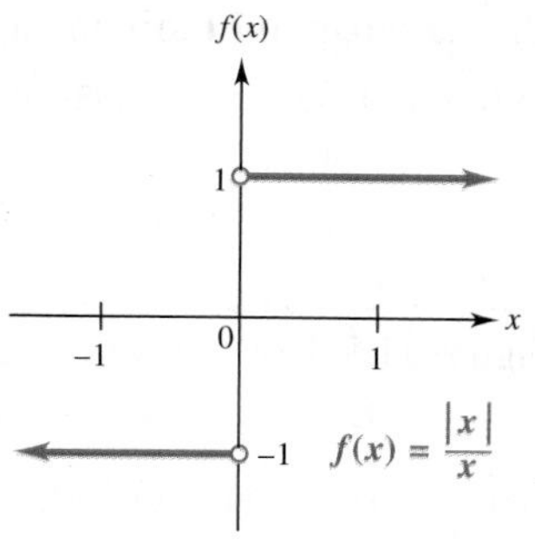

FIGURE 11.7

ONE-SIDED LIMITS

Example 11 of Section 11.1 showed that the limit as x approaches 0 of $f(x) = \frac{|x|}{x}$ does not exist. However, we can adapt the limit concept to describe the behavior of $f(x)$ near $x = 0$ as follows: First, look at the right half of the graph of $f(x)$ in Figure 11.7.

> As x takes values very close to 0 (with $x > 0$), the corresponding values of $f(x)$ are very close (in fact, equal) to 1.

We express this fact in symbols by writing

$$\lim_{x \to 0^+} f(x) = 1,$$

which is read,

> The limit of $f(x)$ as x approaches 0 from the right is 1.

Now consider the left half of the graph in Figure 11.7.

> As x takes values very close to 0 (with $x < 0$), the corresponding values of $f(x)$ are very close (in fact, equal) to -1.

We express this fact by writing

$$\lim_{x \to 0^-} f(x) = -1,$$

which is read,

> The limit of $f(x)$ as x approaches 0 from the left is -1.

The same idea carries over to the general case.

One-Sided Limits

Let f be a function, and let a, K, and M be real numbers. Assume that $f(x)$ is defined for all x near a, with $x > a$. Suppose that

> as x takes values very close (but not equal) to a (with $x > a$), the corresponding values of $f(x)$ are very close (and possibly equal) to K,

and that

> the values of $f(x)$ can be made as close as you want to K for all values of x (with $x > a$) that are close enough to a.

Then the number K is the **limit of $f(x)$ as x approaches a from the right,** which is written

$$\lim_{x \to a^+} f(x) = K.$$

The statement "M is the **limit of $f(x)$ as x approaches a from the left,**" which is written

$$\lim_{x \to a^-} f(x) = M,$$

is defined in a similar fashion. (Just replace K by M and "$x > a$" by "$x < a$" in the previous definition.)

We sometimes refer to a limit of the form $\lim_{x\to a^-} f(x)$ as a **left-hand limit** and a limit of the form $\lim_{x\to a^+} f(x)$ as a **right-hand limit**. The limit $\lim_{x\to a} f(x)$, as defined in Section 11.1, is sometimes called a **two-sided limit**.

Example 1 The graph of a function f is shown in Figure 11.8. Use it to verify that each of the following statements is true:

(a) $\lim_{x\to 6^-} f(x) = 5$; **(b)** $\lim_{x\to 6^+} f(x) = 4$; **(c)** $\lim_{x\to 6} f(x)$ is not defined;

(d) $\lim_{x\to 2^-} f(x) = 3$; **(e)** $\lim_{x\to 2^+} f(x) = 3$; **(f)** $\lim_{x\to 2} f(x) = 3$. ✓1

✓Checkpoint 1

Use Figure 11.8 to find the given limits.

(a) $\lim_{x\to 0^+} f(x)$

(b) $\lim_{x\to 0^-} f(x)$

Answers: See page 678

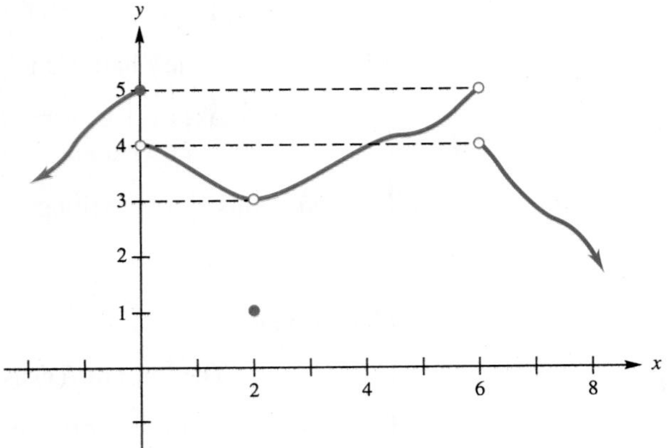

FIGURE 11.8

In parts (a)–(c) of Example 1, the left- and right-hand limits exist, but are not equal, and the two-sided limit does not exist. In parts (d)–(f), however, all three limits exist and are equal to one another. This case is an example of the following fact.

Two-Sided Limits

Let f be a function, and let a and L be real numbers. Then

$$\lim_{x\to a} f(x) = L \quad \text{exactly when} \quad \lim_{x\to a^-} f(x) = L \quad \text{and} \quad \lim_{x\to a^+} f(x) = L.$$

In other words, f has a two-sided limit at a exactly when it has both a left-hand and right-hand limit at a *and* these two limits are the same.

Although its proof is omitted, the following fact will be used frequently:

All of the properties and theorems about limits in Section 11.1 are valid for left-hand limits and for right-hand limits.

Example 2 Find each of the given limits.

(a) $\lim_{x\to 2^+} \sqrt{4 - x^2}$ and $\lim_{x\to 2^-} \sqrt{4 - x^2}$

Solution Since $f(x) = \sqrt{4 - x^2}$ is not defined when $x > 2$, the right-hand limit (which requires that $x > 2$) does not exist. For the left-hand limit, write the square root in exponential form and apply the appropriate limit properties (see pages 658 and 659).

$$\begin{aligned}\lim_{x\to 2^-} \sqrt{4 - x^2} &= \lim_{x\to 2^-} (4 - x^2)^{1/2} && \text{Exponential form}\\ &= [\lim_{x\to 2^-} (4 - x^2)]^{1/2} && \text{Power property}\\ &= 0^{1/2} = 0. && \text{Polynomial limit}\end{aligned}$$

(b) $\lim_{x\to 3^+} [\sqrt{x - 3} + x^2 + 1]$

Solution $\lim_{x\to 3^+} [\sqrt{x - 3} + x^2 + 1]$

$$\begin{aligned} &= \lim_{x\to 3^+} [(x - 3)^{1/2} + x^2 + 1] && \text{Exponential form}\\ &= \lim_{x\to 3^+} (x - 3)^{1/2} + \lim_{x\to 3^+} x^2 + \lim_{x\to 3^+} 1 && \text{Sum property}\\ &= [\lim_{x\to 3^+} (x - 3)]^{1/2} + \lim_{x\to 3^+} x^2 + \lim_{x\to 3^+} 1 && \text{Power property}\\ &= 0^{1/2} + 9 + 1 = 10. && \text{Polynomial limits}\end{aligned}$$

2✓

✓Checkpoint 2

Find the given limits.

(a) $\lim_{x\to 3^+} \sqrt{x^2 - 9}$

(b) $\lim_{x\to -2^-} [\sqrt{x^2 - 4} + x^2 + 4]$

Answers: See page 678

INFINITE LIMITS

In the next part of the discussion, it is important to remember that

the symbol ∞, which is usually read "infinity," does *not* represent a real number.

Nevertheless, the word "infinity" and the symbol ∞ are often used as a convenient shorthand to describe the behavior of some functions. Consider the function *f* whose graph is shown in Figure 11.9.

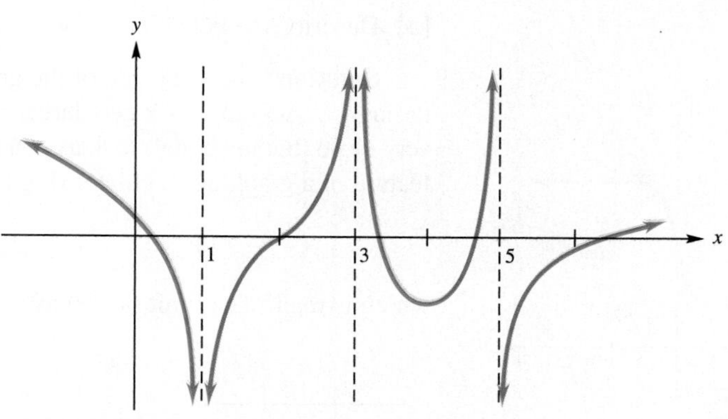

FIGURE 11.9

✓ Checkpoint 3

Use "infinite limits" to describe the behavior of the function whose graph is given, near the specified number.

(a) $f(x) = \dfrac{3}{5 - x}$ near $x = 5$

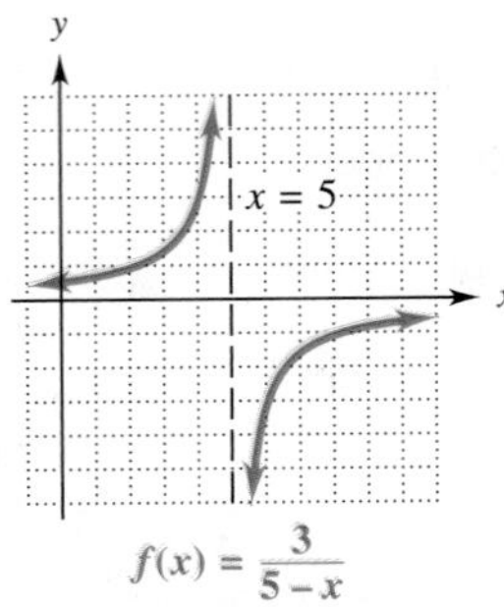

(b) $f(x) = \dfrac{2 - x}{x + 3}$ near $x = -3$

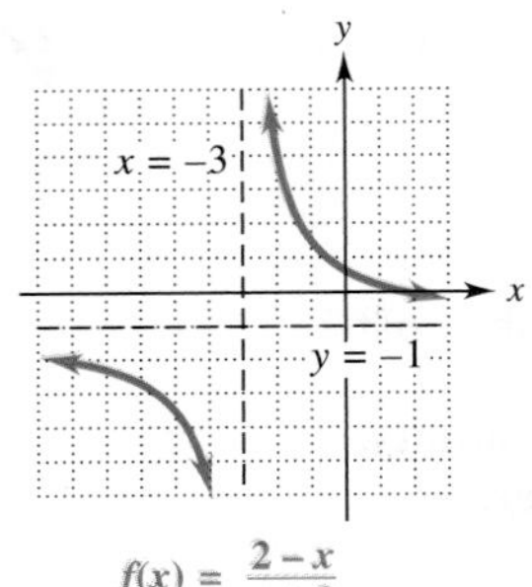

Answers: See page 678

The graph shows that as x approaches 3 (on either side), the corresponding values of $f(x)$ get larger and larger without bound. We express this behavior by writing

$$\lim_{x \to 3} f(x) = \infty,$$

which is read, "The limit of $f(x)$ as x approaches 3 is infinity."

Similarly, as x approaches 1 (on either side), the corresponding values of $f(x)$ get more and more negative (smaller and smaller) without bound. We write

$$\lim_{x \to 1} f(x) = -\infty,$$

which is read, "The limit of $f(x)$ as x approaches 1 is negative infinity."

The behavior of f near $x = 5$ can be described in a similar fashion with the use of one-sided limits. When x is close to 5, the corresponding values of $f(x)$ get very large on the left side of 5 and very small on the right side of 5, so we write

$$\lim_{x \to 5^-} f(x) = \infty \quad \text{and} \quad \lim_{x \to 5^+} f(x) = -\infty$$

and say, "The limit of $f(x)$ as x approaches 5 from the left is infinity" and "The limit of $f(x)$ as x approaches 5 from the right is negative infinity."

In all the situations just described, the language of limits and the word "infinity" are useful for describing the behavior of a function that does not have a limit in the sense discussed in Section 11.1. In particular, this language provides an algebraic way to describe precisely the fact that the function in Figure 11.9 has different types of vertical asymptotes at $x = 1$, $x = 3$, and $x = 5$. 3✓

LIMITS AT INFINITY

The word "limit" has been used thus far to describe the behavior of a function $f(x)$ when x is near a particular number a. Now we consider the behavior of a function when x is very large or very small.

Example 3 Figure 11.10 shows the graph of

$$f(x) = \frac{3}{1 + e^{-x}} + 1.$$

(a) Describe the behavior of f when x is very large.

Solution The right side of the graph appears to coincide with the horizontal line through 4. Actually, as x gets larger and larger, the corresponding values of $f(x)$ are very close (but not equal) to 4, as you can verify with a calculator or by using the trace feature of a graphing calculator (Figure 11.11).* We describe this situation by writing

$$\lim_{x \to \infty} f(x) = 4,$$

which is read "The limit of $f(x)$ as x approaches infinity is 4."

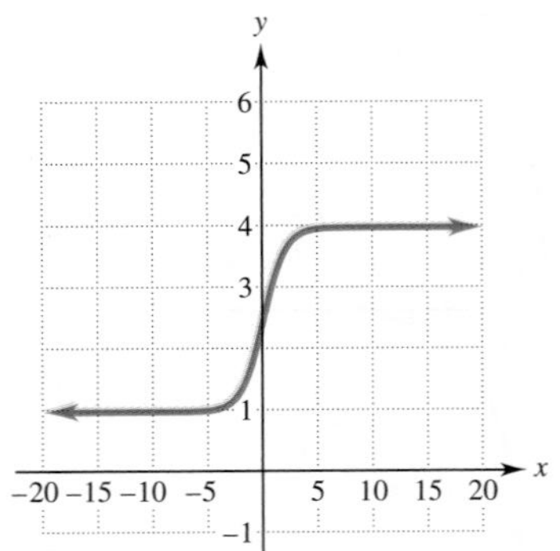

FIGURE 11.10

*Because of round-off, a TI-84+ will say that $f(x) = 4$ when $x > 28$. Actually $f(x)$ is always less than 4. Similar remarks apply to other calculators.

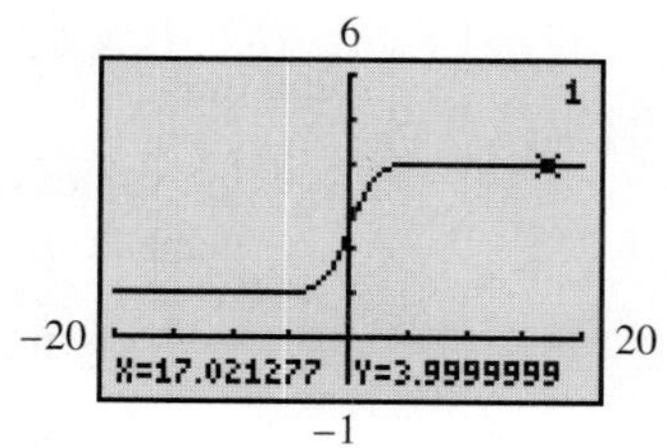

FIGURE 11.11

(b) Describe the behavior of f when x is very small.

Solution To say that x gets very small means that it moves to the left on the axis (that is, x gets more and more negative). The graph suggests that when x is very small, the corresponding values of $f(x)$ are very close to 1. You can verify this with a calculator or the trace feature on a graphing calculator (Figure 11.12).* We describe this situation by writing

$$\lim_{x \to -\infty} f(x) = 1,$$

which is read, "The limit of $f(x)$ as x approaches negative infinity is 1." 4✓

✓Checkpoint 4

The graph of a function g is shown. Use the language of limits, as in Example 3, to describe the behavior of g when

(a) x is very small;

(b) x is very large.

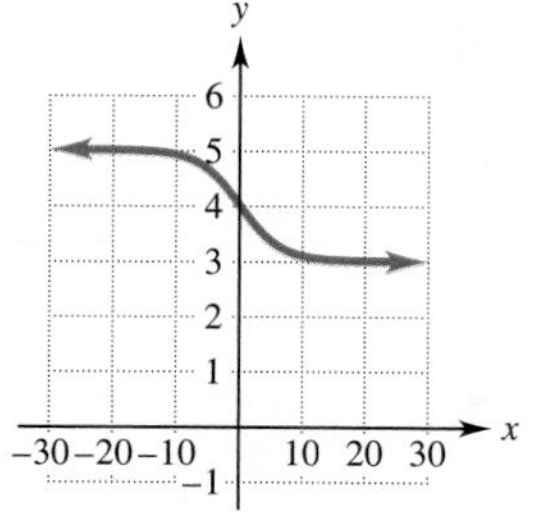

Answers: See page 678

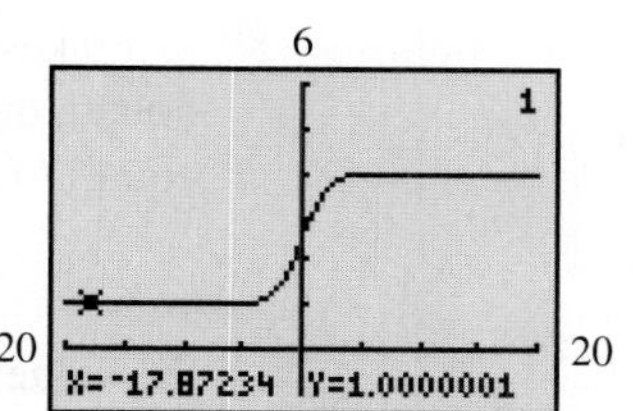

FIGURE 11.12

Once again, the words "infinity" and "negative infinity" and the symbols ∞ and $-\infty$ do *not* denote numbers. They are just a convenient shorthand to express the idea that x takes very large or very small values.

A general definition of the kind of limits illustrated in Example 3 is given next. It is an informal one, as was the definition of limit in Section 11.1.

Limits at Infinity

Let f be a function that is defined for all large positive values of x, and let L be a real number. Suppose that

as x takes larger and larger positive values, increasing without bound, the corresponding values of $f(x)$ are very close (and possibly equal) to L

(continued)

*Because of round-off, a TI-84+ will say that $f(x) = 1$ when $x < -28$. In fact, $f(x)$ is always more than 1. Similar remarks apply to other calculators.

and that

> the values of $f(x)$ can be made arbitrarily close (as close as you want) to L by taking large enough values of x.

Then we say that

the limit of $f(x)$ as x approaches infinity is L,

which is written

$$\lim_{x \to \infty} f(x) = L.$$

Similarly, if f is defined for all small negative values of x and M is a real number, then the statement

$$\lim_{x \to -\infty} f(x) = M,$$

which is read

the limit of $f(x)$ as x approaches negative infinity is M,

means that

> as x takes smaller and smaller negative values, decreasing without bound, the corresponding values of $f(x)$ are very close (and possibly equal) to M

and

> the values of $f(x)$ can be made arbitrarily close (as close as you want) to M by taking small enough values of x.

Example 4 Find the given limits.

(a) $\lim_{x \to \infty} \frac{1}{x}$.

Solution Make a table of values when x is very large:

x	50	100	1000	10,000
$1/x$	.02	.01	.001	.0001

The table shows that $1/x$ gets very close to 0 as x takes larger and larger values, which suggests that $\lim_{x \to \infty} \frac{1}{x} = 0$. You can reach the same conclusion by examining the right side of the graph of $f(x) = 1/x$ in Figure 11.13.

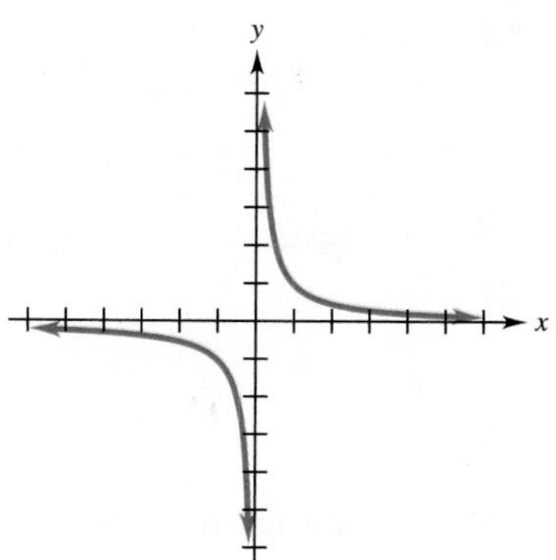

FIGURE 11.13

(b) $\lim_{x \to -\infty} \frac{1}{x}$

Solution Make a table of values when x is very small:

x	-50	-100	-1000	$-10{,}000$
$1/x$	$-.02$	$-.01$	$-.001$	$-.0001$

The table shows that $1/x$ gets very close to 0 as x gets closer and closer to 0, which suggests that $\lim_{x \to -\infty} \frac{1}{x} = 0$. The left side of the graph of $f(x) = 1/x$ in Figure 11.13 shows this result graphically.

Limits at infinity have many of the properties of ordinary limits. Consider, for instance, the constant function $f(x) = 6$. As x gets larger and larger (or smaller and smaller), the corresponding value of $f(x)$ is *always* 6. Hence, $\lim_{x \to \infty} f(x) = 6$ and $\lim_{x \to -\infty} f(x) = 6$. Similarly, for any constant d,

$$\lim_{x \to \infty} d = d \qquad \text{and} \qquad \lim_{x \to -\infty} d = d.$$

Finally, we note the following:

All of the properties of limits in Section 11.1 (page 658) are valid for limits at infinity.

Example 5 Find $\lim_{x \to \infty} \frac{4x^2 + 3x}{5x^2 - x + 2}$.

Solution Begin by writing the function in a different form. First divide both numerator and denominator by x^2 (which is the highest power of x appearing in either the numerator or denominator).* Then simplify and rewrite the result as follows:

*In Examples 5 and 6, we are interested only in nonzero values of x, and dividing both numerator and denominator by a nonzero quantity will not change the value of the function.

$$\frac{4x^2+3x}{5x^2-x+2} = \frac{\dfrac{4x^2+3x}{x^2}}{\dfrac{5x^2-x+2}{x^2}} = \frac{\dfrac{4x^2}{x^2}+\dfrac{3x}{x^2}}{\dfrac{5x^2}{x^2}-\dfrac{x}{x^2}+\dfrac{2}{x^2}} = \frac{4+\dfrac{3}{x}}{5-\dfrac{1}{x}+\dfrac{2}{x^2}}$$

$$= \frac{4+3\cdot\dfrac{1}{x}}{5-\dfrac{1}{x}+2\dfrac{1}{x}\cdot\dfrac{1}{x}}.$$

Now use the limit properties and Example 4 to compute the limit:

$$\lim_{x\to\infty}\frac{4x^2+3x}{5x^2-x+2} = \lim_{x\to\infty}\frac{4+3\cdot\dfrac{1}{x}}{5-\dfrac{1}{x}+2\cdot\dfrac{1}{x}\cdot\dfrac{1}{x}}$$ Preceding equation

$$= \frac{\lim_{x\to\infty}\left[4+3\cdot\dfrac{1}{x}\right]}{\lim_{x\to\infty}\left[5-\dfrac{1}{x}+2\cdot\dfrac{1}{x}\cdot\dfrac{1}{x}\right]}$$ Quotient property

$$= \frac{\lim_{x\to\infty}4+\lim_{x\to\infty}\left(3\cdot\dfrac{1}{x}\right)}{\lim_{x\to\infty}5-\lim_{x\to\infty}\dfrac{1}{x}+\lim_{x\to\infty}\left(2\cdot\dfrac{1}{x}\cdot\dfrac{1}{x}\right)}$$ Sum and difference properties

$$= \frac{\lim_{x\to\infty}4+\lim_{x\to\infty}3\cdot\lim_{x\to\infty}\dfrac{1}{x}}{\lim_{x\to\infty}5-\lim_{x\to\infty}\dfrac{1}{x}+\lim_{x\to\infty}2\cdot\lim_{x\to\infty}\dfrac{1}{x}\cdot\lim_{x\to\infty}\dfrac{1}{x}}$$ Product property

$$= \frac{4+3\cdot 0}{5-0+2\cdot 0\cdot 0} = \frac{4}{5}.$$ Constant-function limits and Example 4

Example 6 Find $\lim_{x\to-\infty}\dfrac{2x^2}{3-x^3}$.

Solution Divide both numerator and denominator by x^3, the highest power of x appearing in either one.* Then simplify and rewrite:

$$\frac{2x^2}{3-x^3} = \frac{\dfrac{2x^2}{x^3}}{\dfrac{3-x^3}{x^3}} = \frac{\dfrac{2x^2}{x^3}}{\dfrac{3}{x^3}-\dfrac{x^3}{x^3}} = \frac{\dfrac{2}{x}}{\dfrac{3}{x^3}-1} = \frac{2\cdot\dfrac{1}{x}}{3\cdot\dfrac{1}{x}\cdot\dfrac{1}{x}\cdot\dfrac{1}{x}-1}.$$

*In Examples 5 and 6, we are interested only in nonzero values of x, and dividing by a nonzero quantity will not change the value of the fraction.

Now proceed as in Example 5:

$$\lim_{x\to-\infty}\frac{2x^2}{3-x^3} = \lim_{x\to-\infty}\frac{2\cdot\frac{1}{x}}{3\cdot\frac{1}{x}\cdot\frac{1}{x}\cdot\frac{1}{x}-1}$$ Preceding equation

$$= \frac{\lim\limits_{x\to-\infty}\left[2\cdot\frac{1}{x}\right]}{\lim\limits_{x\to-\infty}\left[3\cdot\frac{1}{x}\cdot\frac{1}{x}\cdot\frac{1}{x}-1\right]}$$ Quotient property

$$= \frac{\lim\limits_{x\to-\infty}\left[2\cdot\frac{1}{x}\right]}{\lim\limits_{x\to-\infty}\left[3\cdot\frac{1}{x}\cdot\frac{1}{x}\cdot\frac{1}{x}\right]-\lim\limits_{x\to-\infty}1}$$ Difference property

$$= \frac{\lim\limits_{x\to-\infty}2\cdot\lim\limits_{x\to-\infty}\frac{1}{x}}{\lim\limits_{x\to-\infty}3\cdot\lim\limits_{x\to-\infty}\frac{1}{x}\cdot\lim\limits_{x\to-\infty}\frac{1}{x}\cdot\lim\limits_{x\to-\infty}\frac{1}{x}-\lim\limits_{x\to-\infty}1}$$ Product property

$$= \frac{2\cdot 0}{3\cdot 0\cdot 0\cdot 0-1} = \frac{0}{-1} = 0.$$ Constant-function limits and Example 4

✓5

✓Checkpoint 5

Find the given limits.

(a) $\lim\limits_{x\to-\infty}\frac{5x^3}{2x^3-x+4}$

(b) $\lim\limits_{x\to\infty}\frac{10x^3-9x}{28+11x^4}$

Answers: See page 678

The technique in Examples 5 and 6 carries over to any rational function $f(x) = g(x)/h(x)$, where the degree of $g(x)$ is less than or equal to the degree of $h(x)$. These limits provide a mathematical justification for the treatment of horizontal asymptotes in Section 3.7.

11.2 ▸ Exercises

The graph of the function f is shown. Use it to compute the given limits. (See Example 1.)

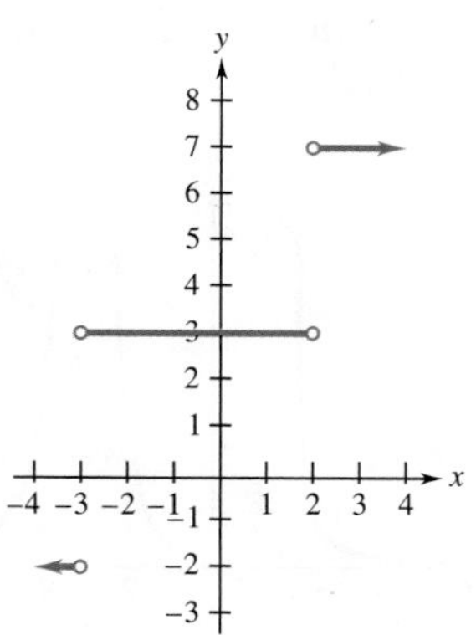

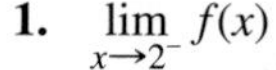

1. $\lim\limits_{x\to2^-} f(x)$

2. $\lim\limits_{x\to2^+} f(x)$

3. $\lim\limits_{x\to-3^-} f(x)$

4. $\lim\limits_{x\to-3^+} f(x)$

The graph of the function f is shown. Use it to compute the given limits. (See Example 1.)

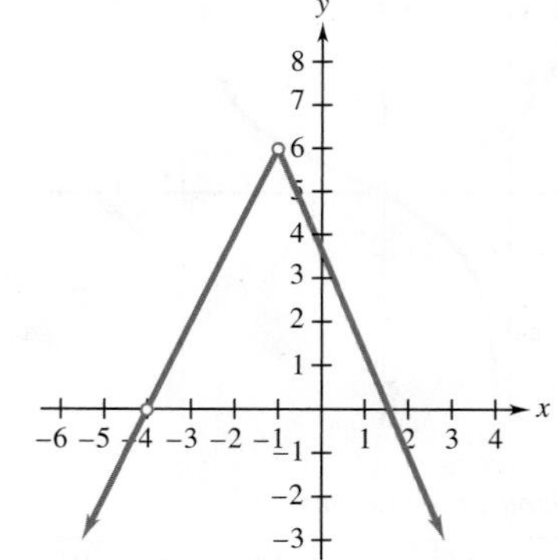

5. $\lim\limits_{x\to-1^-} f(x)$

6. $\lim\limits_{x\to-1^+} f(x)$

7. $\lim\limits_{x\to-4^-} f(x)$

8. $\lim\limits_{x\to-4^+} f(x)$

In Exercises 9–12, use the graph of the function f to determine

(a) $\lim_{x\to-2^-} f(x)$; **(b)** $\lim_{x\to0^+} f(x)$;
(c) $\lim_{x\to3^-} f(x)$; **(d)** $\lim_{x\to3^+} f(x)$.

9.

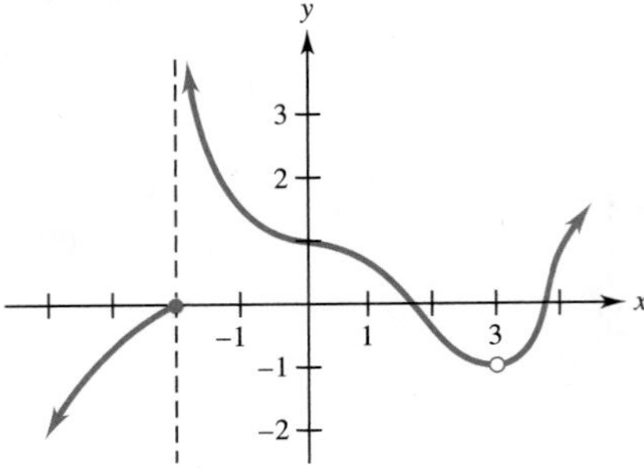

10.

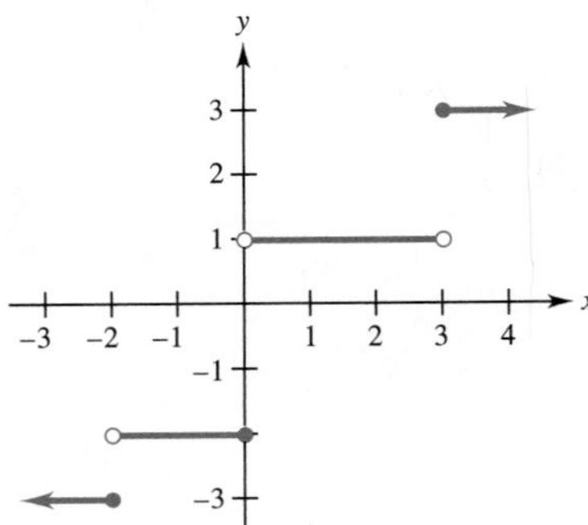

11.

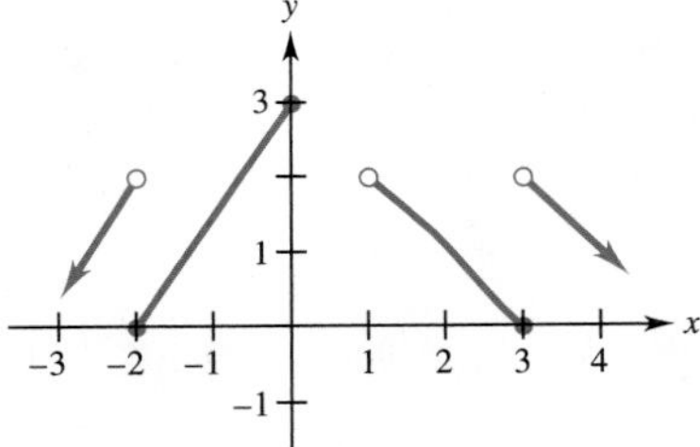

12.

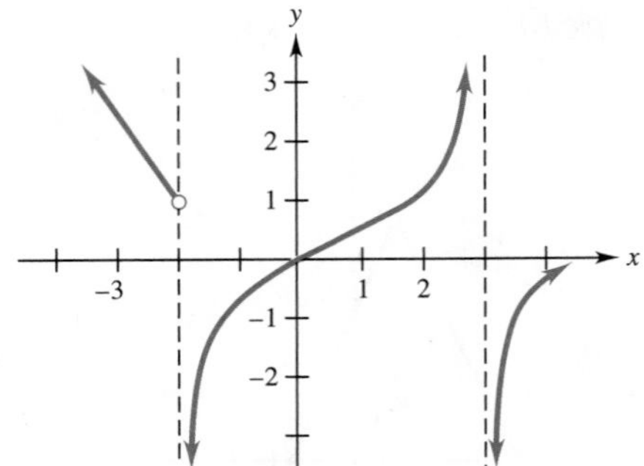

Find the given limits. (See Example 2.)

13. $\lim_{x\to2^+} \sqrt{x^2 - 4}$

14. $\lim_{x\to3^-} \sqrt{9 - x^2}$

15. $\lim_{x\to0^+} \sqrt{x} + x + 1$

16. $\lim_{x\to1^+} \sqrt{x - 1} + 1$

17. $\lim_{x\to-2^+} (x^3 - x^2 - x + 1)$

18. $\lim_{x\to-2^-} (2x^3 + 3x^2 + 5x + 2)$

19. $\lim_{x\to-5^+} \frac{\sqrt{x + 5} + 5}{x^2 - 5}$

20. $\lim_{x\to1^+} \frac{x^3 - 2}{\sqrt{x^3 - 1} + 2}$

21. Graph the function f whose rule is

$$f(x) = \begin{cases} 3 - x & \text{if } x < -2 \\ x + 2 & \text{if } -2 \le x < 2 \\ 1 & \text{if } x = 2 \\ 4 - x & \text{if } x > 2 \end{cases}$$

Use the graph to evaluate the given limits.
(a) $\lim_{x\to-2^-} f(x)$ **(b)** $\lim_{x\to-2^+} f(x)$ **(c)** $\lim_{x\to-2} f(x)$

22. Let f be the function in Exercise 21. Evaluate the given limits.
(a) $\lim_{x\to2^-} f(x)$ **(b)** $\lim_{x\to2^+} f(x)$ **(c)** $\lim_{x\to2} f(x)$

23. For the function f whose graph is shown, find
(a) $\lim_{x\to2} f(x)$; **(b)** $\lim_{x\to-2^+} f(x)$; **(c)** $\lim_{x\to-2^-} f(x)$.

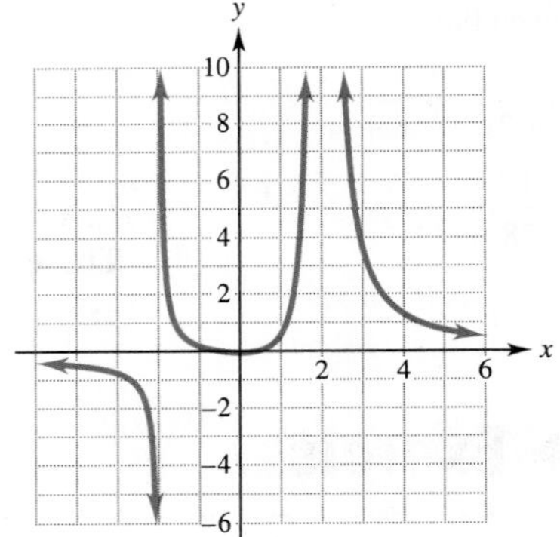

24. For the function g whose graph is shown, find
(a) $\lim_{x\to3} g(x)$; **(b)** $\lim_{x\to-2^+} g(x)$; **(c)** $\lim_{x\to-2^-} g(x)$.

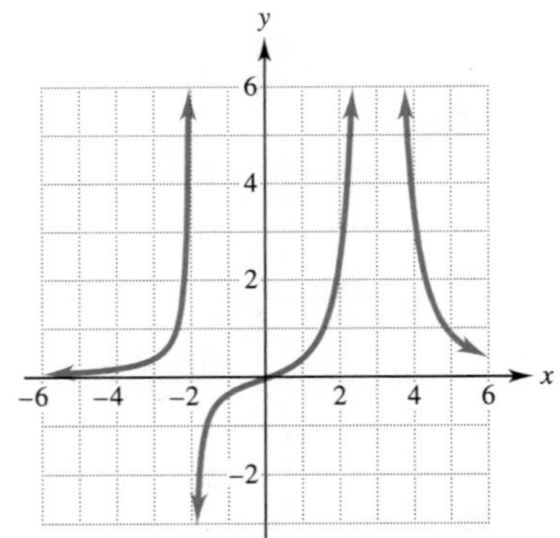

Use a calculator to estimate the given limits.

25. $\lim_{x \to 4} \dfrac{-6}{(x - 4)^2}$ **26.** $\lim_{x \to -2} \dfrac{2x}{(x + 2)^2}$

27. $\lim_{x \to -1^+} \dfrac{2}{1 + x}$ **28.** $\lim_{x \to -3^+} \dfrac{2 - x}{x + 3}$

Use the given graph of the function f to find

$$\lim_{x \to \infty} f(x) \quad \text{and} \quad \lim_{x \to -\infty} f(x).$$

(See Example 3.)

29.

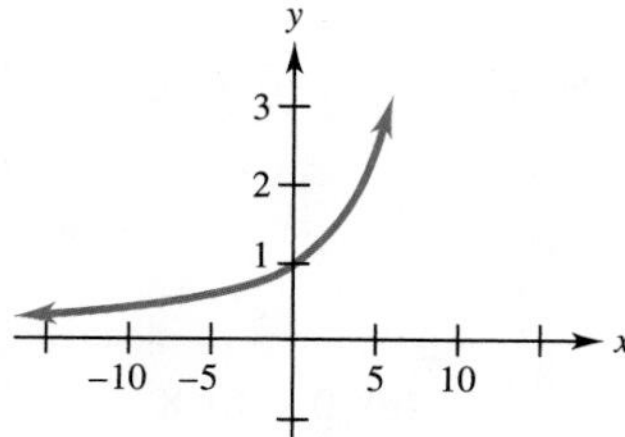

30.

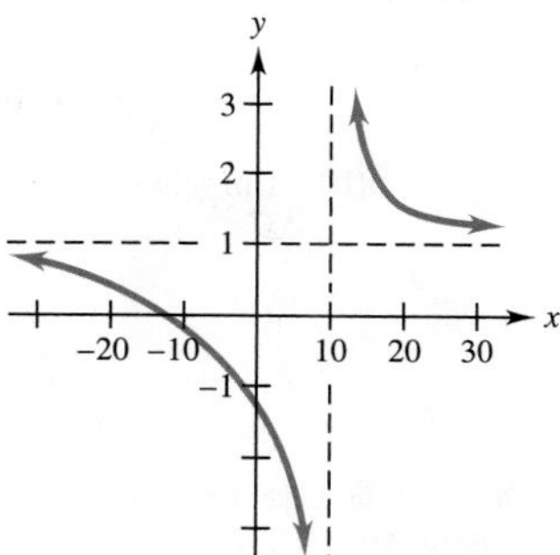

31.

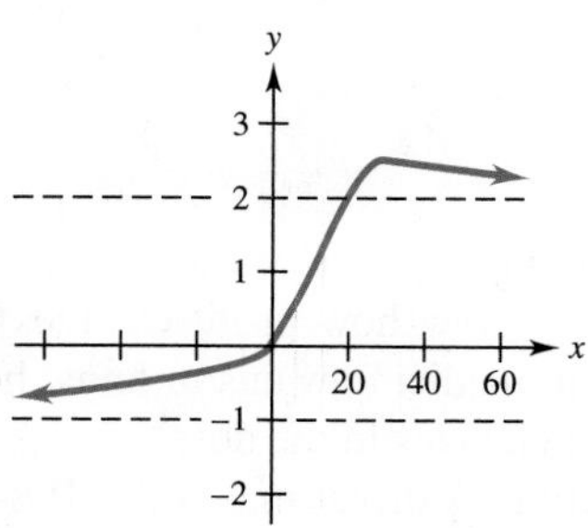

32.

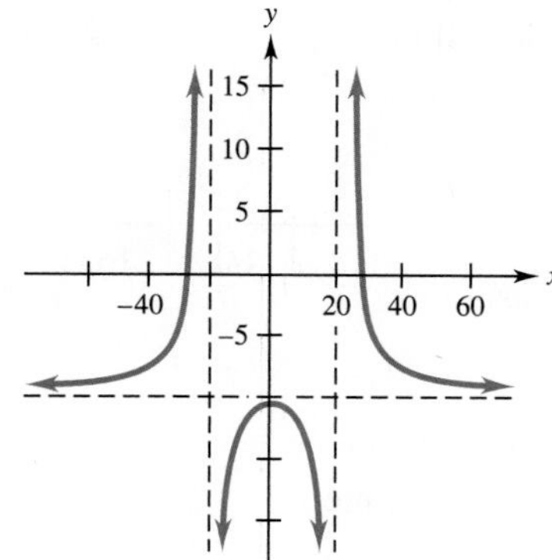

33.

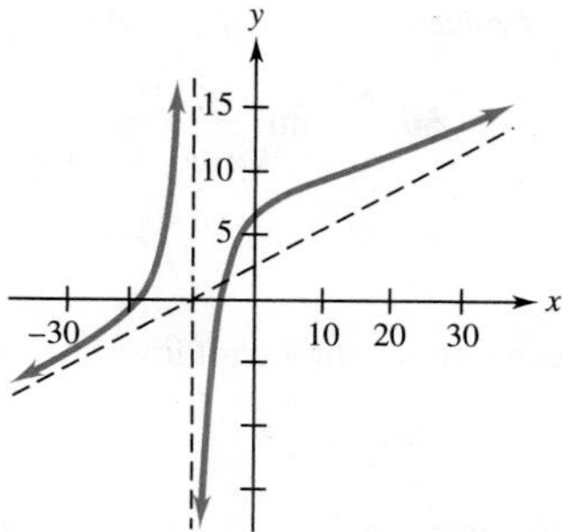

34.

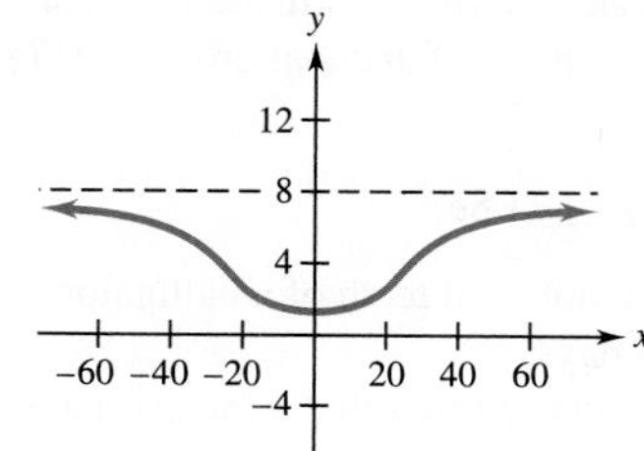

Use a calculator to estimate the limit. (See Example 4.)

35. $\lim_{x \to \infty} [\sqrt{x^2 + 1} - (x + 1)]$

36. $\lim_{x \to -\infty} [\sqrt{x^2 + x + 1} + x]$

37. $\lim_{x \to -\infty} \dfrac{x^{2/3} - x^{4/3}}{x^3}$ **38.** $\lim_{x \to \infty} \dfrac{x^{5/4} + x}{2x - x^{5/4}}$

39. $\lim_{x \to -\infty} e^{1/x}$ **40.** $\lim_{x \to \infty} \dfrac{e^{1/x}}{x}$

41. $\lim_{x \to \infty} \dfrac{\ln x}{x}$ **42.** $\lim_{x \to \infty} \dfrac{5}{1 + (1.1)^{-x/20}}$

Use the properties of limits to find the given limits. (See Examples 5 and 6.)

43. $\lim_{x \to \infty} \dfrac{10x^2 - 5x + 8}{3x^2 + 8x - 29}$ **44.** $\lim_{x \to \infty} \dfrac{44x^2 - 12x + 82}{11x^2 - 35x - 37}$

45. $\lim_{x \to -\infty} \dfrac{4x^2 + 11x + 21}{7 + 67x - x^2}$ **46.** $\lim_{x \to -\infty} \dfrac{5x + 9x^2 - 17}{21x + 5x^3 + 12x^2}$

47. $\lim_{x \to -\infty} \dfrac{8x^5 + 7x^4 - 10x^3}{25 - 4x^5}$

48. $\lim_{x \to \infty} \dfrac{7x^4 + 3x^3 - 12x^2 + x - 76}{33x^3 - 17x^2 + 6}$

49. $\lim_{x \to -\infty} \dfrac{(4x - 1)(x + 2)}{5x^2 - 3x + 1}$ **50.** $\lim_{x \to \infty} \dfrac{(3x - 1)(4x + 3)}{6x^2 - x + 7}$

51. $\lim_{x \to \infty} \left(5x - \dfrac{1}{4x^2}\right)$ **52.** $\lim_{x \to -\infty} (7x^2 + 2)^{-2}$

53. $\lim_{x \to -\infty} \left(\dfrac{18x}{x - 9} - \dfrac{3x}{x + 4}\right)$

54. $\lim_{x \to \infty} \left(\dfrac{2x}{x^2 + 2} + \dfrac{17x}{x^3 - 11}\right)$

Use the definition of absolute value to find the given limits.

55. $\lim_{x\to\infty} \frac{x}{|x|}$ **56.** $\lim_{x\to-\infty} \frac{x}{|x|}$

57. $\lim_{x\to-\infty} \frac{x}{|x| + 1}$

58. Use the change-of-base formula for logarithms (Section 4.3) to show that $\lim_{x\to\infty} \frac{\ln x}{\log x} = \ln 10$.

59. **Natural Science** Researchers have developed a mathematical model that can be used to estimate the number of teeth, $N(t)$, at time t (days of incubation) for *Alligator mississippiensis**:

$$N(t) = 71.8e^{-8.96e^{-.0685t}}.$$

(a) Find $N(65)$, the number of teeth of an alligator that hatched after 65 days.
(b) Find $\lim_{t\to\infty} N(t)$, and use this value as an estimate of the number of teeth of a newborn alligator. Does this estimate differ significantly from the estimate of part (a)?

60. **Natural Science** To develop strategies to manage water quality in polluted lakes, biologists must determine the depths of sediments and the rate of sedimentation. It has been determined that the depth of sediment $D(t)$ (in centimeters) with respect to time (in years before 1990) for Lake Coeur d'Alene, Idaho, can be estimated by the equation

$$D(t) = 155(1 - e^{-.0133t}).*$$

(a) Find $D(20)$ and interpret.
(b) Find $\lim_{t\to\infty} D(t)$ and interpret.

61. **Health** The concentration of a drug in a patient's bloodstream h hours after it was injected is given by

$$A(h) = \frac{.17h}{h^2 + 2}.$$

Find and interpret $\lim_{h\to\infty} A(h)$.

*Kulesa, P., G. Cruywagen, et al. "On a Model Mechanism for the Spatial Patterning of Teeth Primordia in the Alligator," *Journal of Theoretical Biology*, Vol. 180, 1996, pp. 287–296.

*Nord, Gail, and John Nord, "Sediment in Lake Coeur d'Alene, Idaho," *Mathematics Teacher*, Vol. 91, No. 4, April 1998, pp. 292–295.

✓ Checkpoint Answers

1. (a) 4 (b) 5
2. (a) 0 (b) 8
3. (a) $\lim_{x\to5^-} \frac{3}{5-x} = \infty$ and $\lim_{x\to5^+} \frac{3}{5-x} = -\infty$
   (b) $\lim_{x\to-3^-} \frac{2-x}{x+3} = -\infty$ and $\lim_{x\to-3^+} \frac{2-x}{x+3} = \infty$
4. (a) $\lim_{x\to-\infty} g(x) = 5$ (b) $\lim_{x\to\infty} g(x) = 3$
5. (a) $\frac{5}{2}$ (b) 0

11.3 ▶ Rates of Change

One of the main applications of calculus is determining how one variable changes in relation to another. A person in business wants to know how profit changes with respect to changes in advertising, while a person in medicine wants to know how a patient's reaction to a drug changes with respect to changes in the dose.

We begin the discussion with a familiar situation. A driver makes the 168-mile trip from Cleveland to Columbus, Ohio, in 3 hours. The following table shows how far the driver has traveled from Cleveland at various times:

Time (in hours)	0	.5	1	1.5	2	2.5	3
Distance (in miles)	0	22	52	86	118	148	168

If f is the function whose rule is

$$f(x) = \text{distance from Cleveland at time } x,$$

then the table shows, for example, that $f(2) = 118$ and $f(3) = 168$. So the distance traveled from time $x = 2$ to $x = 3$ is $168 - 118$—that is, $f(3) - f(2)$. In a similar fashion, we obtain the other entries in the following chart:

Time Interval	Distance Traveled
$x = 2$ to $x = 3$	$f(3) - f(2) = 168 - 118 = 50$
$x = 1$ to $x = 3$	$f(3) - f(1) = 168 - 52 = 116$
$x = 0$ to $x = 2.5$	$f(2.5) - f(0) = 148 - 0 = 148$
$x = .5$ to $x = 1$	$f(1) - f(.5) = 52 - 22 = 30$
$x = a$ to $x = b$	$f(b) - f(a)$

The last line of the chart shows how to find the distance traveled in any time interval $(0 \leq a < b \leq 3)$.

Since distance = average speed × time,

$$\text{average speed} = \frac{\text{distance traveled}}{\text{time interval}}.$$

In the preceding chart, you can compute the length of each time interval by taking the difference between the two times. For example, from $x = 1$ to $x = 3$ is a time interval of length $3 - 1 = 2$ hours; hence, the average speed over this interval is $116/2 = 58$ mph. Similarly, we have the following information:

Time Interval	Average Speed = $\frac{\text{Distance Traveled}}{\text{Time Interval}}$
$x = 2$ to $x = 3$	$\frac{f(3) - f(2)}{3 - 2} = \frac{168 - 118}{3 - 2} = \frac{50}{1} = 50$ mph
$x = 1$ to $x = 3$	$\frac{f(3) - f(1)}{3 - 1} = \frac{168 - 52}{3 - 1} = \frac{116}{2} = 58$ mph
$x = 0$ to $x = 2.5$	$\frac{f(2.5) - f(0)}{2.5 - 0} = \frac{148 - 0}{2.5 - 0} = \frac{148}{2.5} = 59.2$ mph
$x = .5$ to $x = 1$	$\frac{f(1) - f(.5)}{1 - .5} = \frac{52 - 22}{1 - .5} = \frac{30}{.5} = 60$ mph
$x = a$ to $x = b$	$\frac{f(b) - f(a)}{b - a}$ mph

✓Checkpoint 1

Find the average speed

(a) from $t = 1.5$ to $t = 2$;

(b) from $t = s$ to $t = r$.

Answers: See page 690.

The last line of the chart shows how to compute the average speed over any time interval $(0 \leq a < b \leq 3)$. 1✓

Now, speed (miles per hour) is simply the *rate of change* of distance with respect to time, and what was done for the distance function f in the preceding discussion can be done with any function.

Quantity	Meaning for the Distance Function	Meaning for an Arbitrary Function f
$b - a$	Time interval = change in time from $x = a$ to $x = b$	Change in x from $x = a$ to $x = b$
$f(b) - f(a)$	Distance traveled = corresponding change in distance as time changes from a to b	Corresponding change in $f(x)$ as x changes from a to b
$\dfrac{f(b) - f(a)}{b - a}$	Average speed = average rate of change of distance with respect to time as time changes from a to b	**Average rate of change** of $f(x)$ with respect to x as x changes from a to b (where $a < b$)

Example 1 If $f(x) = x^2 + 4x + 5$, find the average rate of change of $f(x)$ with respect to x as x changes from -2 to 3.

Solution This is the situation described in the last line of the preceding chart, with $a = -2$ and $b = 3$. The average rate of change is

$$\frac{f(3) - f(-2)}{3 - (-2)} = \frac{26 - 1}{5}$$
$$= \frac{25}{5} = 5. \quad 2✓$$

✓Checkpoint 2

Find the average rate of change of $f(x)$ in Example 1 when x changes from

(a) 0 to 4;

(b) 2 to 7.

Answers: See page 690.

Example 2 **Finance** Suppose $D(t)$ is the value of the Dow Jones Industrial Average (DJIA) at time t. Figure 11.14 shows the values of $D(t)$ for the end of the given year over a nine-year period.*

Year	DJIA Index
2000	10,787
2001	10,022
2002	8342
2003	10,454
2004	10,783
2005	10,718
2006	12,463
2007	13,265
2008	8776

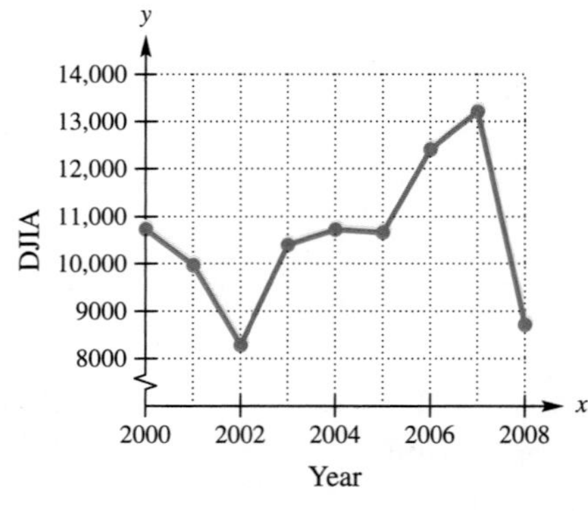

FIGURE 11.14

*The Statistical Abstract of the United States: 2009.

Approximate the average rate of change of the DJIA with respect to time over the given intervals.

(a) From the end of 2002 to the end of 2007

Solution The table says that the DJIA stood at 8342 at the end of 2002 and 13,265 at the end of 2007; that is $D(2002) = 8342$ and $D(2007) = 13{,}265$. The average rate of change over this period is

$$\frac{D(2007) - D(2002)}{2007 - 2002} = \frac{13{,}265 - 8342}{2007 - 2002} = \frac{4923}{5} = 984.6.$$

Thus, the DJIA was increasing at an average rate of 984.6 index points per year.

(b) From the end of 2000 to the end of 2002

Solution The table says that $D(2000) = 10{,}787$, and we saw earlier that $D(2002) = 8342$. The average rate of change is

$$\frac{D(2002) - D(2000)}{2002 - 2000} = \frac{8342 - 10{,}787}{2002 - 2000} = \frac{-2445}{2} = -1222.5.$$

The negative number implies that the Dow was decreasing at the average rate of 1222.5 index points per year during this period. 3✓

✓Checkpoint 3

Use Figure 11.14 to find the average rate of change of the DJIA with respect to time over the given interval.

(a) From the end of 2002 to the end of 2008

(b) From the end of 2005 to the end of 2008

Answers: See page 690.

Example 2 also illustrates the geometric interpretation of the average rate of change. In part (a), for instance,

$$\begin{array}{c}\text{average rate of change}\\ \text{from 2002 to 2007}\end{array} = \frac{D(2007) - D(2002)}{2007 - 2002} = \frac{13{,}265 - 8342}{2007 - 2002}.$$

This last number is precisely the *slope* of the line from (2002, 8342) to (2007, 13,265) on the graph of D. The same thing is true in general.

Geometric Meaning of Average Rate of Change

If f is a function, then the average rate of change of $f(x)$ with respect to x, as x changes from a to b, namely,

$$\frac{f(b) - f(a)}{b - a},$$

is the slope of the line joining the points $(a, f(a))$ and $(b, f(b))$ on the graph of f.

INSTANTANEOUS RATE OF CHANGE

Suppose a car is stopped at a traffic light. When the light turns green, the car begins to move along a straight road. Assume that the distance traveled by the car is given by the function

$$s(t) = 2t^2 \qquad (0 \le t \le 30),$$

where time t is measured in seconds and the distance $s(t)$ at time t is measured in feet. We know how to find the *average* speed of the car over any time interval, so we now turn to a different problem: determining the *exact* speed of the car at a particular instant—say, $t = 10$.*

The intuitive idea is that the exact speed at $t = 10$ is very close to the average speed over a very short time interval near $t = 10$. If we take shorter and shorter time intervals near $t = 10$, the average speeds over these intervals should get closer and closer to the exact speed at $t = 10$. In other words,

> the exact speed at $t = 10$ is the limit of the average speeds over shorter and shorter time intervals near $t = 10$.

The following chart illustrates this idea:

Intervals	Average Speed
$t = 10$ to $t = 10.1$	$\dfrac{s(10.1) - s(10)}{10.1 - 10} = \dfrac{204.02 - 200}{.1} = 40.2$
$t = 10$ to $t = 10.01$	$\dfrac{s(10.01) - s(10)}{10.01 - 10} = \dfrac{200.4002 - 200}{.01} = 40.02$
$t = 10$ to $t = 10.001$	$\dfrac{s(10.001) - s(10)}{10.001 - 10} = \dfrac{200.040002 - 200}{.001} = 40.002$

The chart suggests that the exact speed at $t = 10$ is 40 ft/sec. We can confirm this intuition by computing the average speed from $t = 10$ to $t = 10 + h$, where h is any very small nonzero number. (The chart does this for $h = .1$, $h = .01$, and $h = .001$.) The average speed from $t = 10$ to $t = 10 + h$ is

$$\begin{aligned}
\frac{s(10+h) - s(10)}{(10+h) - 10} &= \frac{s(10+h) - s(10)}{h} \\
&= \frac{2(10+h)^2 - 2\cdot 10^2}{h} \\
&= \frac{2(100 + 20h + h^2) - 200}{h} \\
&= \frac{200 + 40h + 2h^2 - 200}{h} \\
&= \frac{40h + 2h^2}{h} = \frac{h(40+2h)}{h} \qquad (h \neq 0) \\
&= 40 + 2h.
\end{aligned}$$

Saying that the time interval from 10 to $10 + h$ gets shorter and shorter is equivalent to saying that h gets closer and closer to 0. Hence, the exact speed at $t = 10$ is the limit, as h approaches 0, of the average speeds over the intervals from $t = 10$ to $t = 10 + h$; that is,

$$\begin{aligned}
\lim_{h \to 0} \frac{s(10+h) - s(10)}{h} &= \lim_{h \to 0} (40 + 2h) \\
&= 40 \text{ ft/sec.}
\end{aligned}$$

*As distance is measured in feet and time in seconds here, speed is measured in feet per second. It may help to know that 15 mph is equivalent to 22 ft/sec and 60 mph to 88 ft/sec.

In the preceding example, the car moved in one direction along a straight line. Now suppose that an object is moving back and forth along a number line, with its position at time t given by the function $s(t)$. The **velocity** of the object is the rate of motion in the direction in which it is moving. **Speed** is the absolute value of velocity. For example, a velocity of -30 ft/sec indicates a speed of 30 ft/sec in the negative direction, while a velocity of 40 ft/sec indicates a speed of 40 ft/sec in the positive direction.

Hereafter, we deal with **average velocity** and **instantaneous velocity** instead of average speed and exact speed, respectively. When the term "velocity" is used alone, it means instantaneous velocity.

Let a be a fixed number. By replacing 10 by a in the preceding discussion, we see that the average velocity of the object from time $t = a$ to time $t = a + h$ is the quotient

$$\frac{s(a+h) - s(a)}{(a+h) - a} = \frac{s(a+h) - s(a)}{h}.$$

The instantaneous velocity at time a is the limit of this quotient as h approaches 0.

Velocity

If an object moves along a straight line, with position $s(t)$ at time t, then the **velocity** of the object at $t = a$ is

$$\lim_{h \to 0} \frac{s(a+h) - s(a)}{h},$$

provided that this limit exists.

Example 3 The distance, in feet, of an object from a starting point is given by $s(t) = 2t^2 - 5t + 40$, where t is time in seconds.

(a) Find the average velocity of the object from 2 seconds to 4 seconds.

Solution The average velocity is

$$\frac{s(4) - s(2)}{4 - 2} = \frac{52 - 38}{2} = \frac{14}{2} = 7$$

feet per second.

(b) Find the instantaneous velocity at 4 seconds.

Solution For $t = 4$, the instantaneous velocity is

$$\lim_{h \to 0} \frac{s(4+h) - s(4)}{h}$$

feet per second. We have

$$\begin{aligned} s(4+h) &= 2(4+h)^2 - 5(4+h) + 40 \\ &= 2(16 + 8h + h^2) - 20 - 5h + 40 \\ &= 32 + 16h + 2h^2 - 20 - 5h + 40 \\ &= 2h^2 + 11h + 52 \end{aligned}$$

and

$$s(4) = 2(4)^2 - 5(4) + 40 = 52.$$

Thus,

$$s(4 + h) - s(4) = (2h^2 + 11h + 52) - 52 = 2h^2 + 11h,$$

and the instantaneous velocity at $t = 4$ is

$$\lim_{h \to 0} \frac{2h^2 + 11h}{h} = \lim_{h \to 0} \frac{h(2h + 11)}{h}$$
$$= \lim_{h \to 0} (2h + 11) = 11 \text{ ft/sec.}$$ 4

Checkpoint 4

In Example 3, if $s(t) = t^2 + 3$, find

(a) the average velocity from 1 second to 5 seconds;

(b) the instantaneous velocity at 5 seconds.

Answers: See page 690.

Example 4 **Health** The velocity of blood cells is of interest to physicians; a velocity slower than normal might indicate a constriction, for example. Suppose the position of a red blood cell in a capillary is given by

$$s(t) = 1.2t + 5,$$

where $s(t)$ gives the position of a cell in millimeters from some reference point and t is time in seconds. Find the velocity of this cell at time $t = a$.

Solution We have $s(a) = 1.2a + 5$. To find $s(a + h)$, substitute $a + h$ for the variable t in $s(t) = 1.2t + 5$:

$$s(a + h) = 1.2(a + h) + 5.$$

Now use the definition of velocity:

$$v(a) = \lim_{h \to 0} \frac{s(a + h) - s(a)}{h}$$
$$= \lim_{h \to 0} \frac{1.2(a + h) + 5 - (1.2a + 5)}{h}$$
$$= \lim_{h \to 0} \frac{1.2a + 1.2h + 5 - 1.2a - 5}{h} = \lim_{h \to 0} \frac{1.2h}{h} = 1.2.$$

The velocity of the blood cell at $t = a$ is 1.2 millimeters per second, regardless of the value of a. In other words, the blood velocity is a constant 1.2 millimeters per second at any time. 5

Checkpoint 5

Repeat Example 4 with $s(t) = .3t - 2$.

Answer: See page 690.

The ideas underlying the concept of the velocity of a moving object can be extended to any function $f(x)$. In place of average velocity at time t, we have the average rate of change of $f(x)$ with respect to x as x changes from one value to another. Taking limits leads to the following definition.

The **instantaneous rate of change** for a function f when $x = a$ is

$$\lim_{h \to 0} \frac{f(a + h) - f(a)}{h},$$

provided that this limit exists.

Example 5 A company determines that the cost (in hundreds of dollars) of manufacturing x cases of computer mice is

$$C(x) = -.2x^2 + 8x + 40 \qquad (0 \le x \le 20).$$

(a) Find the average rate of change of cost for manufacturing between 5 and 10 cases.

Solution Use the formula for average rate of change. The cost to manufacture 5 cases is

$$C(5) = -.2(5^2) + 8(5) + 40 = 75,$$

or \$7500. The cost to manufacture 10 cases is

$$C(10) = -.2(10^2) + 8(10) + 40 = 100,$$

or \$10,000. The average rate of change of cost is

$$\frac{C(10) - C(5)}{10 - 5} = \frac{100 - 75}{5} = 5.$$

Thus, on the average, cost is increasing at the rate of \$500 per case when production is increased from 5 to 10 cases.

(b) Find the instantaneous rate of change with respect to the number of cases produced when 5 cases are produced.

Solution The instantaneous rate of change when $x = 5$ is given by

$$\lim_{h \to 0} \frac{C(5 + h) - C(5)}{h}$$

$$= \lim_{h \to 0} \frac{[-.2(5 + h)^2 + 8(5 + h) + 40] - [-.2(5^2) + 8(5) + 40]}{h}$$

$$= \lim_{h \to 0} \frac{[-5 - 2h - .2h^2 + 40 + 8h + 40] - [75]}{h}$$

$$= \lim_{h \to 0} \frac{6h - .2h^2}{h} \qquad \text{Combine terms.}$$

$$= \lim_{h \to 0} (6 - .2h) \qquad \text{Divide by } h.$$

$$= 6. \qquad \text{Calculate the limit.}$$

When 5 cases are manufactured, the cost is increasing at the rate of \$600 per case.

The rate of change of the cost function is called the **marginal cost**.* Similarly, **marginal revenue** and **marginal profit** are the rates of change of the revenue and profit functions, respectively. Part (b) of Example 5 shows that the marginal cost when 5 cases are manufactured is \$600.

Example 6 **Business** Total revenue (billions of dollars) for year x from U.S. Aircraft shipments can be approximated by

$$f(x) = 47.7(1.04^x) \quad (5 \le x \le 17),$$

*Marginal cost for linear cost functions was discussed in Section 3.3.

where $x = 5$ corresponds to the year 1995.*

(a) Find the rate of change of revenue in 2000.

Solution Since $x = 10$ corresponds to 2000, we must find the instantaneous rate of change of $f(x)$ at $x = 10$. The algebraic techniques used in the preceding examples will not work with an exponential function, but the rate of change can be approximated by a graphing calculator or spreadsheet program. The average rate of change from 10 to $10 + h$ is

$$\frac{f(10 + h) - f(10)}{h} = \frac{47.7(1.04^{10+h}) - 47.7(1.04^{10})}{h}.$$

X	Y_1
-.001	2.7692
-1E-4	2.7693
-1E-6	2.7693
0	ERROR
1E-6	2.7693
1E-4	2.7693
.001	2.7693

X=1E-6

FIGURE 11.15

To approximate the instantaneous rate of change, we evaluate this quantity for very small values of h, as shown in Figure 11.15 (in which X is used in place of h and Y_1 is the average rate of change). The table suggests that at $x = 10$,

$$\text{rate of change} = \lim_{h \to 0} \frac{47.7(1.04^{10+h}) - 47.7(1.04^{10})}{h} \approx 2.77.$$

Thus, revenue was increasing at a rate of about 2.77 billion dollars per year in 2000.

(b) Find the rate of change in 2006.

Solution The instantaneous rate of change of $f(x)$ when $x = 16$ is

$$\lim_{h \to 0} \frac{f(16 + h) - f(16)}{h} = \lim_{h \to 0} \frac{47.7(1.04^{16+h}) - 47.7(1.04^{16})}{h} \approx 3.50,$$

X	Y_1
-.001	3.504
-1E-4	3.504
-1E-6	3.504
0	ERROR
1E-6	3.504
1E-4	3.504
.001	3.5041

X=1E-6

FIGURE 11.16

as shown in Figure 11.16. In 2006, therefore, revenue was increasing at a rate of about 3.5 billion dollars a year.

The Statistical Abstract of the United States: 2009.

11.3 ▶ Exercises

Find the average rate of change for the given functions (See Example 1.)

1. $f(x) = x^2 + 2x$ between $x = 0$ and $x = 6$
2. $f(x) = -4x^2 - 6$ between $x = 1$ and $x = 7$
3. $f(x) = 2x^3 - 4x^2 + 6$ between $x = -1$ and $x = 2$
4. $f(x) = -3x^3 + 2x^2 - 4x + 2$ between $x = 0$ and $x = 2$
5. $f(x) = \sqrt{x}$ between $x = 1$ and $x = 9$
6. $f(x) = \sqrt{3x - 2}$ between $x = 2$ and $x = 6$
7. $f(x) = \dfrac{1}{x - 1}$ between $x = -2$ and $x = 0$
8. $f(x) = .4525\, e^{1.556\sqrt{x}}$ between $x = 4$ and $x = 4.5$
9. **Business** The accompanying graph shows the total sales, in thousands of dollars, from the distribution of x thousand catalogs. Find and interpret the average rate of change of sales with respect to the number of catalogs distributed for the given changes in x.

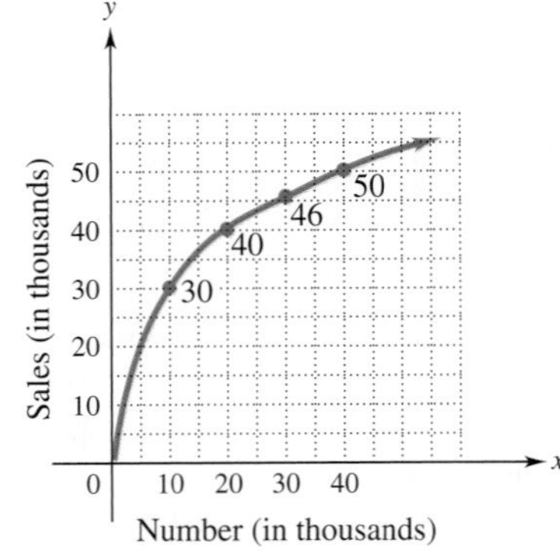

(a) 10 to 20 **(b)** 20 to 30 **(c)** 30 to 40

(d) What is happening to the average rate of change of sales as the number of catalogs distributed increases?

(e) Explain why what you have found in part (d) might happen.

10. **Social Science** The accompanying graph shows the relationship between waist circumference (in inches) and weight (in pounds) for American males (blue) and females (red).* Find the average rate of change in weight for changes in waist circumference from
   (a) 32 to 44 inches for males;
   (b) 28 to 40 inches for females.
   (c) Do we need to calculate the rate of change at any other points for males or females? Explain.

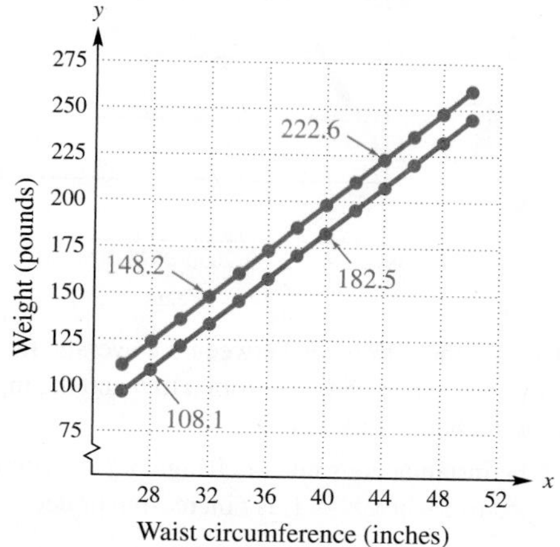

11. **Business** The accompanying graph shows the revenue (in billions of dollars) generated by the hardware industry in the United States from 1995 to 2007.† Find the approximate average change in the revenue for each period.
   (a) 1996 to 2002 (b) 2002 to 2007

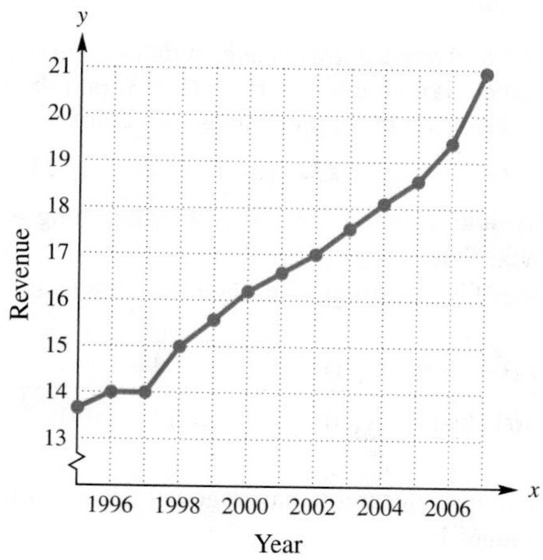

12. **Health** The accompanying graph shows the smoking rate of adults (in percentage) for the state of Massachusetts from 1998 to 2007.* Find the average rate of change in smoking percentage for each period.
   (a) 2000 to 2004 (b) 2002 to 2007

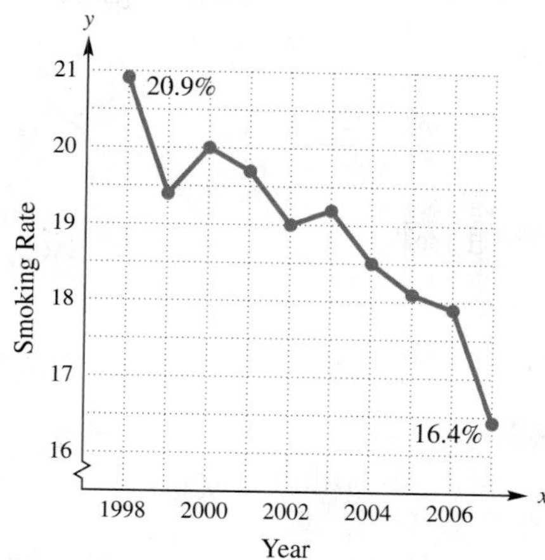

13. **Health** The Environmental Protection Agency developed the following graph to show the reduction in carbon monoxide in the United States from 1990–2007, as measured by average concentration in parts per million for 8 hours of exposure.†

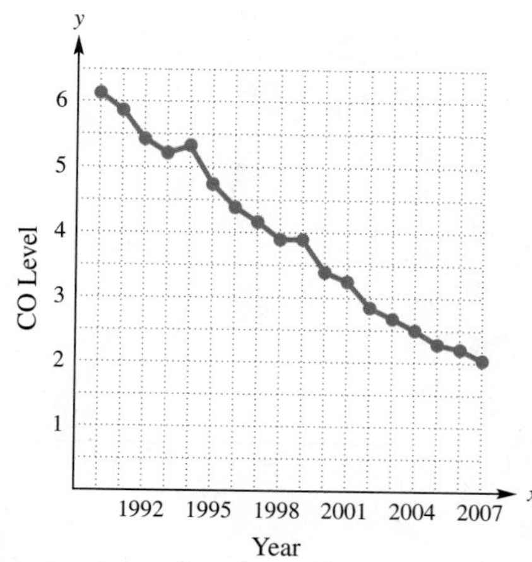

Find the average rate of change for carbon monoxide levels for each period.
   (a) 1992 to 1999 (b) 2004 to 2007

14. **Social Science** The accompanying graph, generated from data at Gallup, gives the percentage of Americans that are dissatisfied with the state of the United States.‡

*Based on data from the National Health and Nutrition Examination Study (NHANES) 2005–2006.

†*The Statistical Abstract of the United States:* 2009.

*Centers of Disease Control, www.cdc.gov.

†www.epa.gov/air/airtrends/carbon.html.

‡www.gallup.com; the values on the graph are from May of each year.

Find the average rate of change in the dissatisfaction rate for each period.
(a) 2002 to 2005 (b) 2005 to 2008 (c) 2003 to 2009

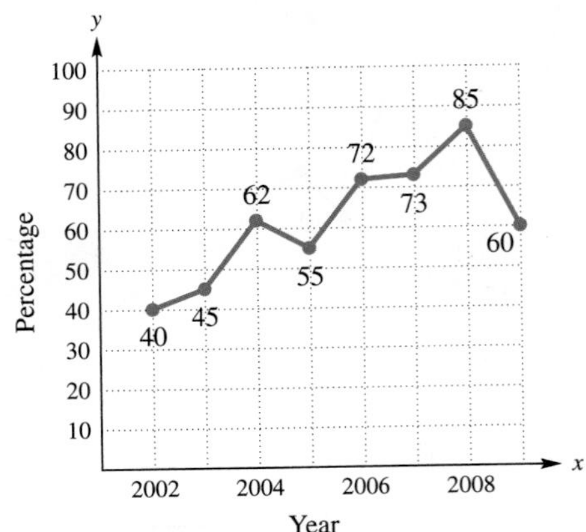

15. **Finance** The accompanying graph gives the value of the S&P 500 Index for 12 months, where $x = 1$ corresponds to June 2008.* Find the average rate of change for each period.
(a) Month 1 to month 10
(b) Month 4 to month 12
(c) Month 10 to month 12

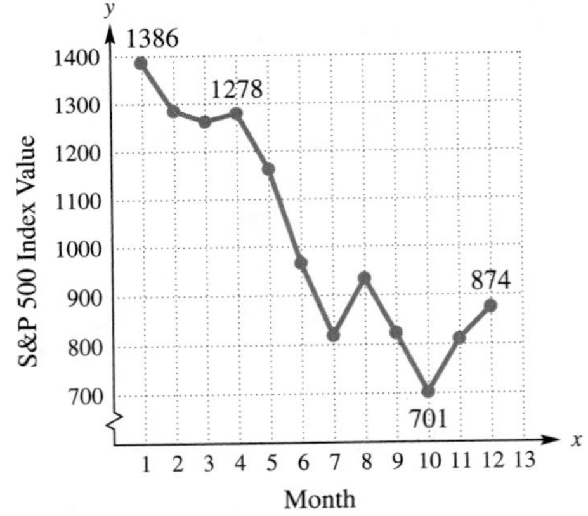

16. **Social Science** The future size of the world population depends on how soon we reach replacement-level fertility, the point at which each woman bears about 2.1 children on average. The accompanying graph shows projections for reaching that point in different years.† Estimate the average rate of change of population for each projection from 1990 to 2050. Which projection shows the smallest rate of change of world population? (Tick marks on the horizontal axis are 5 years apart.)

Ultimate World Population Size Under Different Assumptions

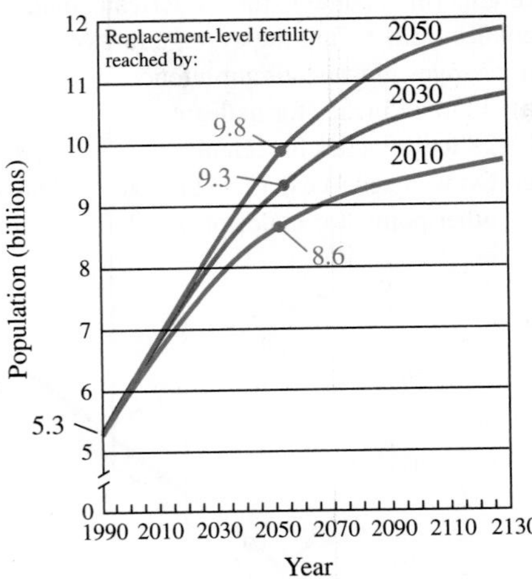

17. Explain the difference between the average rate of change of $y = f(x)$ as x changes from a to b and the instantaneous rate of change of y at $x = a$.

18. If the instantaneous rate of change of $f(x)$ with respect to x is positive when $x = 1$, is f increasing or decreasing there?

Exercises 19–21 deal with a car moving along a straight road, as discussed on pages 681–682. At time t seconds, the distance of the car (in feet) from the starting point is $s(t) = 2.2t^2$. Find the instantaneous velocity (speed) of the car at

19. $t = 5$; 20. $t = 20$.

21. What was the average speed of the car during the first 30 seconds?

An object moves along a straight line; its distance (in feet) from a fixed point at time t seconds is $s(t) = t^2 + 4t + 3$. Find the instantaneous velocity of the object at the given times. (See Example 3.)

22. $t = 6$ 23. $t = 1$ 24. $t = 10$

25. **Physical Science** A car is moving along a straight test track. The position (in feet) of the car, $s(t)$, at various times t is measured, with the following results:

t (seconds)	0	2	4	6	8	10
$s(t)$ (feet)	0	10	14	20	30	36

Find and interpret the average velocities for the following changes in t:
(a) 0 to 2 seconds; (b) 2 to 4 seconds;
(c) 4 to 6 seconds; (d) 6 to 8 seconds.
(e) Estimate the instantaneous velocity at 4 seconds
(i) by using the formula for estimating the instantaneous rate (with $h = 2$) and
(ii) by averaging the answers for the average velocity in the 2 seconds before and the 2 seconds afterwards (that is, the answers to parts (b) and (c)).

*www.morningstar.com.
†Carl Haub, Population Reference Bureau.

(f) Estimate the instantaneous velocity at 6 seconds, using the two methods in part (e).

In Exercises 26–30, find (a) $f(a + h)$; (b) $\frac{f(a + h) - f(a)}{h}$; (c) the instantaneous rate of change of f when $a = 5$. (See Examples 3–5.)

26. $f(x) = x^2 + x$

27. $f(x) = x^2 - x - 1$

28. $f(x) = x^2 + 2x + 2$

29. $f(x) = x^3$

30. $f(x) = x^3 - x$

Solve Exercises 31 and 32 by algebraic methods. (See Examples 3–5.)

31. Business The revenue (in thousands of dollars) from producing x units of an item is

$$R(x) = 10x - .002x^2.$$

(a) Find the average rate of change of revenue when production is increased from 1000 to 1001 units.
(b) Find the marginal revenue when 1000 units are produced.
(c) Find the additional revenue if production is increased from 1000 to 1001 units.
(d) Compare your answers for parts (a) and (c). What do you find?

32. Business Suppose customers in a hardware store are willing to buy $N(p)$ boxes of nails at p dollars per box, as given by

$$N(p) = 80 - 5p^2 \quad (1 \leq p \leq 4).$$

(a) Find the average rate of change of demand for a change in price from \$2 to \$3.
(b) Find the instantaneous rate of change of demand when the price is \$2.
(c) Find the instantaneous rate of change of demand when the price is \$3.
(d) As the price is increased from \$2 to \$3, how is demand changing? Is the change to be expected?

33. Health Epidemiologists in College Station, Texas, estimate that t days after the flu begins to spread in town, the percentage of the population infected by the flu is approximated by

$$p(t) = t^2 + t \quad (0 \leq t \leq 5).$$

(a) Find the average rate of change of p with respect to t over the interval from 1 to 4 days.
(b) Find the instantaneous rate of change of p with respect to t at $t = 3$.

Use technology to work Exercises 34–39. (See Example 6.)

34. Finance Outstanding consumer credit (in billions of dollars) in year x can be approximated by

$$f(x) = 762(1.0781^x),$$

where $x = 0$ corresponds to 1990.* Estimate the rate at which consumer credit was increasing in the given year.
(a) 2000 (b) 2007

*Federal Reserve Bulletin.

35. Finance Outstanding revolving consumer credit (in billions of dollars) in year x can be approximated by

$$g(x) = 268(1.0846^x),$$

where $x = 0$ corresponds to 1990.* Estimate the rate at which revolving consumer credit was increasing in the given year.
(a) 2006 (b) 2010

36. Business The number of people (in millions) who subscribe to basic cable TV can be approximated by

$$f(x) = \frac{66.7}{1 + .353(.772^x)},$$

where $x = 0$ corresponds to the year 1990.* Estimate the rate at which the number of subscribers was changing in the given year.
(a) 1995 (b) 2010
(c) What do the results of parts (a) and (b) imply?

37. Business The revenue (in billions of dollars) generated from basic cable TV can be approximated by the function

$$g(x) = 5.78 + 8.59 \ln x \quad (1 \leq x \leq 17),$$

where $x = 1$ corresponds to the year 1991.† Estimate the rate of change of revenue in the given year.
(a) 2002 (b) 2007

38. Natural Science The mesiodistal crown length (as shown in the accompanying diagram) of deciduous mandibular first molars in fetuses is related to the postconception age of the tooth as

$$L(t) = -.01t^2 + .788t - 7.048,$$

where $L(t)$ is the crown length, in millimeters, of the molar t weeks after conception.‡

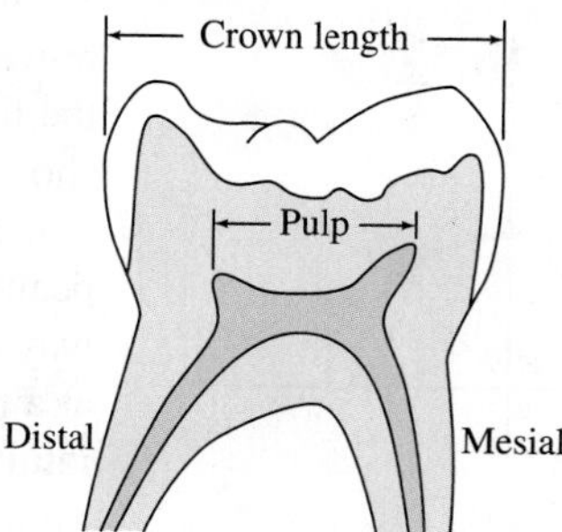

(a) Find the average rate of growth in mesiodistal crown length during weeks 22 through 28.
(b) Find the instantaneous rate of growth in mesiodistal crown length when the tooth is exactly 22 weeks of age.

*Federal Reserve Bulletin.

†SNL Kagan; www.kagan.com.

‡Harris, E. F., J. D. Hicks, and B. D. Barcroft, "Tissue Contributions to Sex and Race: Differences in Tooth Crown Size of Deciduous Molars," *American Journal of Physical Anthropology,* Vol. 115, 2001, pp. 223–237.

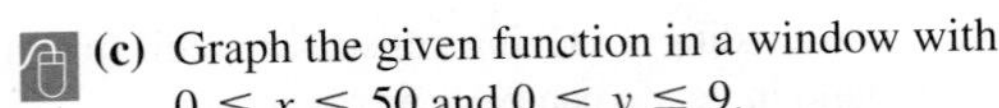

(c) Graph the given function in a window with $0 \le x \le 50$ and $0 \le y \le 9$.

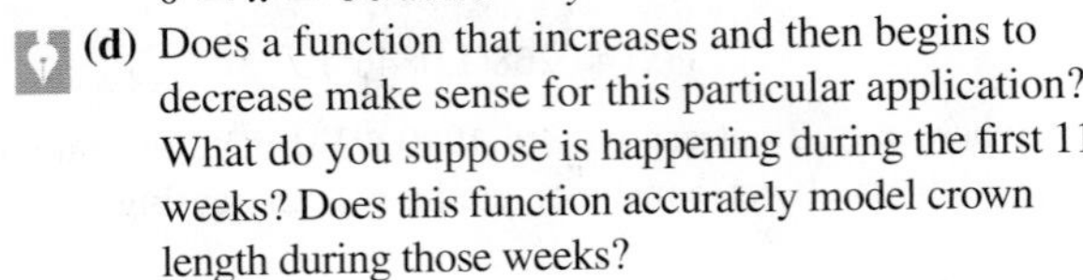

(d) Does a function that increases and then begins to decrease make sense for this particular application? What do you suppose is happening during the first 11 weeks? Does this function accurately model crown length during those weeks?

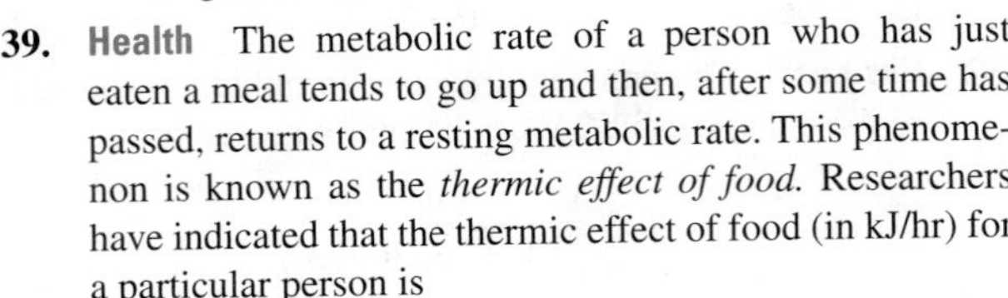

39. Health The metabolic rate of a person who has just eaten a meal tends to go up and then, after some time has passed, returns to a resting metabolic rate. This phenomenon is known as the *thermic effect of food.* Researchers have indicated that the thermic effect of food (in kJ/hr) for a particular person is

$$F(t) = -10.28 + 175.9te^{-t/1.3},$$

where t is the number of hours that have elapsed since eating a meal.*

(a) Graph the given function in a window with $0 \le x \le 6$ and $-20 \le y \le 100$.

(b) Find the average rate of change of the thermic effect of food during the first hour after eating.

(c) Use a graphing calculator to find the instantaneous rate of change of the thermic effect of food exactly 1 hour after eating.

(d) Use a graphing calculator to estimate when the function stops increasing and begins to decrease.

*Reed, G. and J. Hill, "Measuring the Thermic Effect of Food," *American Journal of Clinical Nutrition*, Vol. 63, 1996, pp. 164–169.

40. Business The number of CDs purchased annually (in millions) can be modeled by the function

$$f(x) = 69.3 + 160.81x - 7.953x^2,$$

where $x = 7$ corresponds to the year 1997.*

(a) Find the average rate of change in the number of CDs sold between 2000 and 2005.

(b) Find the instantaneous rate of change in the number of CDs sold in 2005.

(c) Which value indicates a greater rate of change? Explain.

✓ Checkpoint Answers

1. **(a)** 64 mph **(b)** $\dfrac{f(r) - f(s)}{r - s}$ mph
2. **(a)** 8 **(b)** 13
3. **(a)** Increasing at the average rate of 72.3 index points per year
   (b) Decreasing at the average rate of 647.3 index points per year
4. **(a)** 6 ft per second **(b)** 10 ft per second
5. The velocity is .3 millimeter per second.

*www.riaa.com

11.4 ▶ Tangent Lines and Derivatives

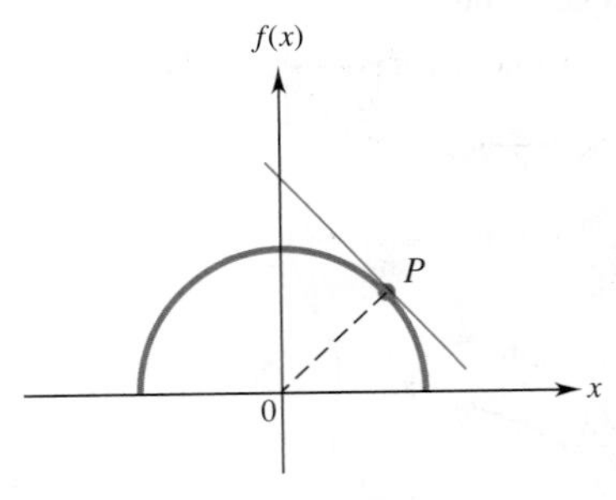

FIGURE 11.17

We now develop a geometric interpretation of the rates of change considered in the previous section. In geometry, a tangent line to a circle at a point P is defined to be the line through P that is perpendicular to the radius OP, as in Figure 11.17 (which shows only the top half of the circle). If you think of this circle as a road on which you are driving at night, then the tangent line indicates the direction of the light beam from your headlights as you pass through the point P. This analogy suggests a way of extending the idea of a tangent line to any curve: the tangent line to the curve at a point P indicates the "direction" of the curve as it passes through P. Using this intuitive idea of direction, we see, for example, that the lines through P_1 and P_3 in Figure 11.18 appear to be tangent lines, whereas the lines through P_2 and P_4 do not.

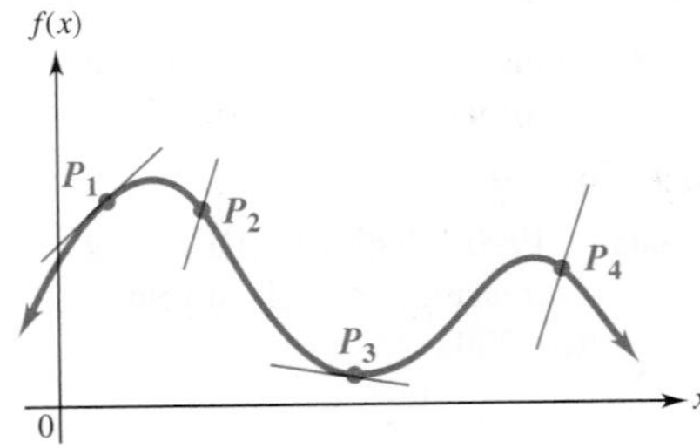

FIGURE 11.18

We can use these ideas to develop a precise definition of the tangent line to the graph of a function *f* at the point *R*. As shown in Figure 11.19, choose a second point *S* on the graph and draw the line through *R* and *S*; this line is called a **secant line**. You can think of this secant line as a rough approximation of the tangent line.

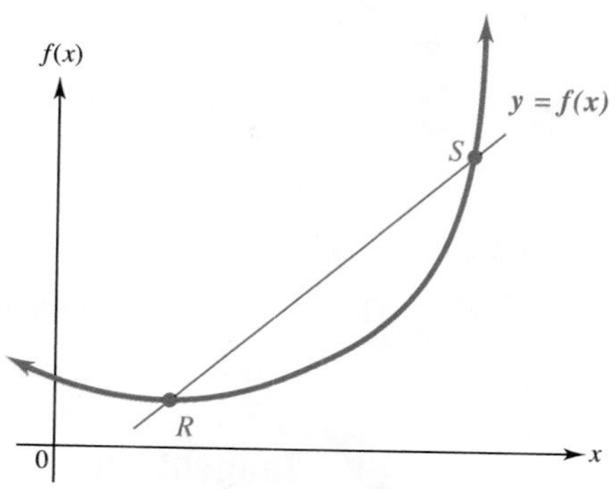

FIGURE 11.19

Now suppose that the point *S* slides down the curve closer to *R*. Figure 11.20 shows successive positions S_2, S_3, and S_4 of the point *S*. The closer *S* gets to *R*, the better the secant line *RS* approximates our intuitive idea of the tangent line at *R*.

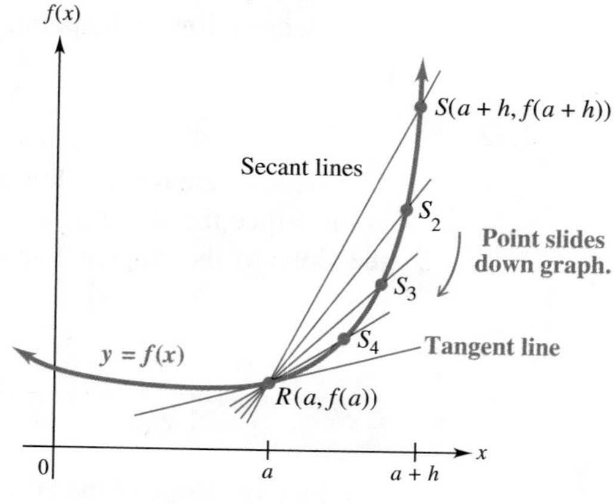

FIGURE 11.20

In particular, the closer *S* gets to *R*, the closer the slope of the secant line gets to the slope of the tangent line. Informally, we say that

the slope of tangent line at *R* = the limit of the slope of secant line *RS* as *S* gets closer and closer to *R*.

In order to make this statement more precise, suppose the first coordinate of *R* is *a*. Then the first coordinate of *S* can be written as $a + h$ for some number *h*. (In Figure 11.20, *h* is the distance on the *x*-axis between the two first coordinates.) Thus, *R* has coordinates $(a, f(a))$ and *S* has coordinates $(a + h, f(a + h))$, as shown in Figure 11.20. Consequently, the slope of the secant line *RS* is

$$\frac{f(a + h) - f(a)}{(a + h) - a} = \frac{f(a + h) - f(a)}{h}.$$

Now, as S moves closer to R, their first coordinates move closer to each other, that is, h gets smaller and smaller. Hence,

$$\begin{aligned}\text{slope of tangent line at } R &= \text{limit of slope of secant line } RS \\ &\quad \text{as } S \text{ gets closer and closer to } R \\ &= \text{limit of } \frac{f(a+h)-f(a)}{h} \\ &\quad \text{as } h \text{ gets closer and closer to } 0 \\ &= \lim_{h\to 0} \frac{f(a+h)-f(a)}{h}.\end{aligned}$$

This intuitive development suggests the following formal definition.

Tangent Line

The **tangent line** to the graph of $y = f(x)$ at the point $(a, f(a))$ is the line through this point having slope

$$\lim_{h\to 0} \frac{f(a+h)-f(a)}{h},$$

provided that this limit exists. If this limit does not exist, then there is no tangent line at the point.

The slope of the tangent line at a point is also called the **slope of the curve** at that point. Since the slope of a line indicates its direction (see the table on page 91), the slope of the tangent line at a point indicates the direction of the curve at that point.

Example 1 Consider the graph of $y = x^2 + 2$.

(a) Find the slope of the tangent line to the graph when $x = -1$.

Solution Use the preceding definition with $f(x) = x^2 + 2$ and $a = -1$. The slope of the tangent line is calculated as follows:

$$\begin{aligned}\text{Slope of tangent} &= \lim_{h\to 0} \frac{f(a+h)-f(a)}{h} \\ &= \lim_{h\to 0} \frac{[(-1+h)^2+2]-[(-1)^2+2]}{h} \\ &= \lim_{h\to 0} \frac{[1-2h+h^2+2]-[1+2]}{h} \\ &= \lim_{h\to 0} \frac{-2h+h^2}{h} = \lim_{h\to 0}(-2+h) = -2.\end{aligned}$$

So the slope of the tangent line is -2.

(b) Find the equation of the tangent line.

Solution When $x = -1$, then $y = (-1)^2 + 2 = 3$. So the tangent line passes through the point $(-1, 3)$ and has slope -2. Its equation can be found with the point–slope form of the equation of a line (see Chapter 2):

$$y - y_1 = m(x - x_1)$$
$$y - 3 = -2[x - (-1)]$$
$$y - 3 = -2(x + 1)$$
$$y - 3 = -2x - 2$$
$$y = -2x + 1.$$

Figure 11.21 shows a graph of $f(x) = x^2 + 2$ along with a graph of the tangent line at $x = -1$. 1✓

Let $f(x) = x^2 + 2$.
Find the equation of the tangent line to the graph at the point where $x = 1$.

Answer: See page 706.

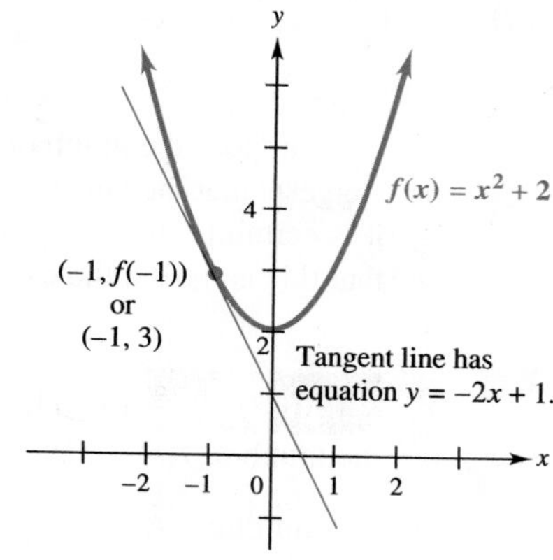

FIGURE 11.21

TECHNOLOGY TIP When finding the equation of a tangent line algebraically, you can confirm your answer with a graphing calculator by graphing both the function and the tangent line on the same screen to see if the tangent line appears to be correct. 2✓

Once a function has been graphed, many graphing calculators can draw the tangent line at any specified point. Most also display the slope of the tangent line, and some actually display its equation. Look for TANGENT or TANLN in the MATH, DRAW, or SKETCH menu.

✓Checkpoint 2

Use a graphing calculator to confirm your answer to Checkpoint 1 by graphing $f(x) = x^2 + 2$ and the tangent line at the point where $x = 1$ on the same screen.

Answer: See page 706.

In Figure 11.21, the tangent line $y = -2x + 1$ appears to coincide with the graph of $f(x) = x^2 + 2$ near the point $(-1, 3)$. This becomes more obvious when both $f(x)$ and the tangent line are graphed in a very small window on a graphing calculator (Figure 11.22, on the next page). The graph and the tangent line now appear virtually identical near $(-1, 3)$. The table of values in Figure 11.23 (in which Y_1 is $f(x)$ and Y_2 the tangent line) confirms the fact that the tangent line is a good approximation of the function when x is very close to -1.

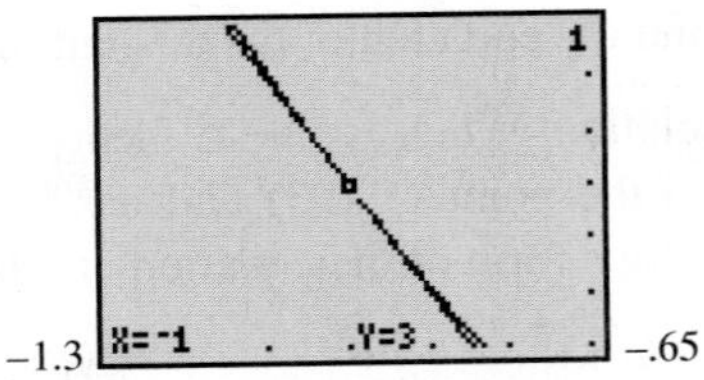

-1.3 X=-1 Y=3 -.65 1

FIGURE 11.22

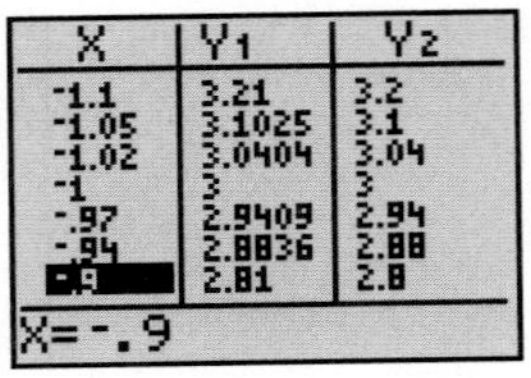

X	Y1	Y2
-1.1	3.21	3.2
-1.05	3.1025	3.1
-1.02	3.0404	3.04
-1	3	3
-.97	2.9409	2.94
-.94	2.8836	2.88
-.9	2.81	2.8

X=-.9

FIGURE 11.23

The same thing is true in the general case.

Tangent Line

If it exists, the tangent line to the graph of a function f at $x = a$ is a good approximation of the function f near $x = a$.

Suppose the graph of a function f is a straight line. The fact in the preceding box suggests that the tangent line to f at any point should be the graph of f itself (because it is certainly the best possible linear approximation of f). The next example shows that this is indeed the case.

Example 2 Let a be any real number. Find the equation of the tangent line to the graph of $f(x) = 7x + 3$ at the point where $x = a$.

Solution According to the definition, the slope of the tangent line is

$$\begin{aligned} \lim_{h\to 0} \frac{f(a+h) - f(a)}{h} &= \lim_{h\to 0} \frac{[7(a+h)+3] - [7a+3]}{h} \\ &= \lim_{h\to 0} \frac{[7a+7h+3] - 7a - 3}{h} \\ &= \lim_{h\to 0} \frac{7h}{h} = \lim_{h\to 0} 7 = 7. \end{aligned}$$

Hence, the equation of the tangent line at the point $(a, f(a))$ is

$$\begin{aligned} y - y_1 &= m(x - x_1) \\ y - f(a) &= 7(x - a) \\ y &= 7x - 7a + \overbrace{f(a)} \\ y &= 7x - 7a + 7a + 3 \\ y &= 7x + 3. \end{aligned}$$

Thus, the tangent line *is* the graph of $f(x) = 7x + 3$.

Secant lines and tangent lines (or, more precisely, their slopes) are the geometric analogues of the average and instantaneous rates of change studied in the previous section, as summarized in the following chart:

Quantity	Algebraic Interpretation	Geometric Interpretation
$\dfrac{f(a+h)-f(a)}{h}$	Average rate of change of f from $x = a$ to $x = a + h$	Slope of the secant line through $(a, f(a))$ and $(a + h, f(a + h))$
$\lim\limits_{h \to 0} \dfrac{f(a+h)-f(a)}{h}$	Instantaneous rate of change of f at $x = a$	Slope of the tangent line to the graph of f at $(a, f(a))$

THE DERIVATIVE

If $y = f(x)$ is a function and a is a number in its domain, then we shall use the symbol $f'(a)$ to denote the special limit

$$\lim_{h \to 0} \frac{f(a+h)-f(a)}{h},$$

provided that it exists. In other words, to each number a, we can assign the number $f'(a)$ obtained by calculating the preceding limit. This process defines an important new function.

Derivative

The **derivative** of the function f is the function denoted f' whose value at the number x is defined to be the number

$$f'(x) = \lim_{h \to 0} \frac{f(x+h)-f(x)}{h},$$

provided that this limit exists.

The derivative function f' has as its domain all the points at which the specified limit exists, and the value of the derivative function at the number x is the number $f'(x)$. Using x instead of a here is similar to the way that $g(x) = 2x$ denotes the function that assigns to each number a the number $2a$.

If $y = f(x)$ is a function, then its derivative is denoted either by f' or by y'. If x is a number in the domain of $y = f(x)$ such that $y' = f'(x)$ is defined, then the function f is said to be **differentiable** at x. The process that produces the function f' from the function f is called **differentiation**.

The derivative function may be interpreted in many ways, two of which were already discussed:

1. The derivative function f' gives the *instantaneous rate of change* of $y = f(x)$ with respect to x. This instantaneous rate of change can be interpreted as marginal cost, marginal revenue, or marginal profit (if the original function represents cost, revenue, or profit, respectively) or as velocity (if the original function represents displacement along a line). From now on, we will use "rate of change" to mean "instantaneous rate of change."
2. The derivative function f' gives the *slope* of the graph of f at any point. If the derivative is evaluated at $x = a$, then $f'(a)$ is the slope of the tangent line to the curve at the point $(a, f(a))$.

Example 3 Use the graph of the function $f(x)$ in Figure 11.24 to answer the given questions.

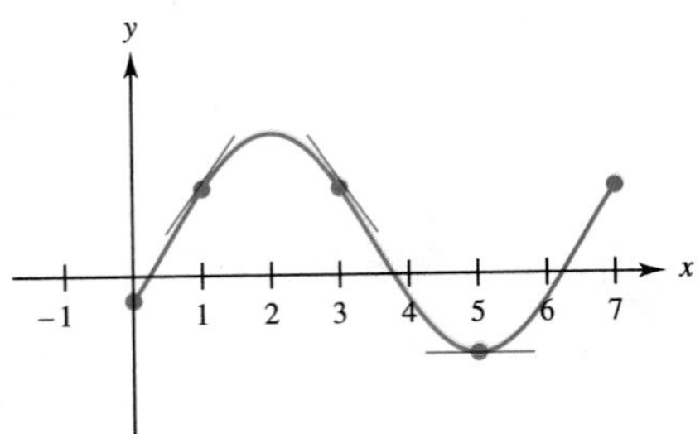

FIGURE 11.24

(a) Is $f'(3)$ positive or negative?

Solution We know that $f'(3)$ is the slope of the tangent line to the graph at the point where $x = 3$. Figure 11.24 shows that this tangent line slants downward from left to right, meaning that its slope is negative. Hence, $f'(3) < 0$.

(b) Which is larger, $f'(1)$ or $f'(5)$?

Solution Figure 11.24 shows that the tangent line to the graph at the point where $x = 1$ slants upward from left to right, meaning that its slope, $f'(1)$, is a positive number. The tangent line at the point where $x = 5$ is horizontal, so that it has slope 0. (That is, $f'(5) = 0$.) Therefore, $f'(1) > f'(5)$.

(c) For what values of x is $f'(x)$ positive?

Solution On the graph, find the points where the tangent line has positive slope (slants upward from left to right). At each such point, $f'(x) > 0$. Figure 11.24 shows that this occurs when $0 < x < 2$ and when $5 < x < 7$. ✓3

✓Checkpoint 3

The graph of a function g is shown. Determine whether the given numbers are positive, negative, or zero.

(a) $g'(0)$

(b) $g'(-1)$

(c) $g'(3)$

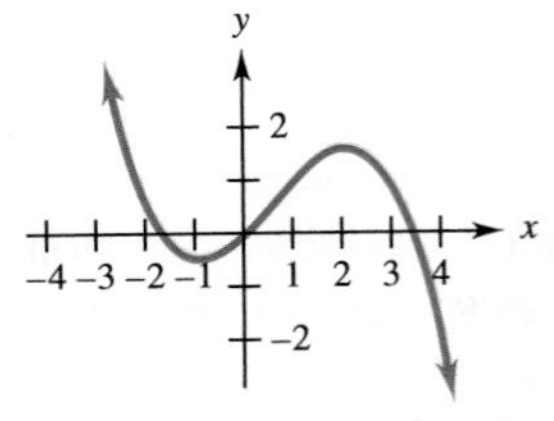

Answers: See page 706.

The rule of a derivative function can be found by using the definition of the derivative and the following four-step procedure.

Finding $f'(x)$ from the Definition of the Derivative

Step 1 Find $f(x + h)$.

Step 2 Find $f(x + h) - f(x)$.

Step 3 Divide by h to get $\dfrac{f(x + h) - f(x)}{h}$.

Step 4 Treat x as a constant and let $h \to 0$.

$$f'(x) = \lim_{h\to 0} \frac{f(x + h) - f(x)}{h} \text{ if this limit exists.}$$

Example 4 Let $f(x) = x^3 - 4x$.

(a) Find the derivative $f'(x)$.

Solution By definition,

$$f'(x) = \lim_{h \to 0} \frac{f(x + h) - f(x)}{h}.$$

Step 1 Find $f(x + h)$.
Replace x with $x + h$ in the rule of $f(x)$:

$$\begin{aligned} f(x) &= x^3 - 4x \\ f(x + h) &= (x + h)^3 - 4(x + h) \\ &= (x^3 + 3x^2h + 3xh^2 + h^3) - 4(x + h) \\ &= x^3 + 3x^2h + 3xh^2 + h^3 - 4x - 4h. \end{aligned}$$

Step 2 Find $f(x + h) - f(x)$.

Since $f(x) = x^3 - 4x$,

$$\begin{aligned} f(x + h) - f(x) &= (x^3 + 3x^2h + 3xh^2 + h^3 - 4x - 4h) - (x^3 - 4x) \\ &= x^3 + 3x^2h + 3xh^2 + h^3 - 4x - 4h - x^3 + 4x \\ &= 3x^2h + 3xh^2 + h^3 - 4h. \end{aligned}$$

Step 3 Form and simplify the quotient $\frac{f(x + h) - f(x)}{h}$:

$$\begin{aligned} \frac{f(x + h) - f(x)}{h} &= \frac{3x^2h + 3xh^2 + h^3 - 4h}{h} \\ &= \frac{h(3x^2 + 3xh + h^2 - 4)}{h} \\ &= 3x^2 + 3xh + h^2 - 4. \end{aligned}$$

Step 4 Find the limit as h approaches 0 of the result in Step 3, treating x as a constant:

$$\begin{aligned} f'(x) = \lim_{h \to 0} \frac{f(x + h) - f(x)}{h} &= \lim_{h \to 0} (3x^2 + 3xh + h^2 - 4) \\ &= 3x^2 - 4. \end{aligned}$$

Therefore, the derivative of $f(x) = x^3 - 4x$ is $f'(x) = 3x^2 - 4$.

(b) Calculate and interpret $f'(1)$.

Solution The procedure in part (a) works for *every* x and $f'(x) = 3x^2 - 4$. Hence, when $x = 1$,

$$f'(1) = 3 \cdot 1^2 - 4 = -1.$$

The number -1 is the slope of the tangent line to the graph of $f(x) = x^3 - 4x$ at the point where $x = 1$ (that is, at $(1, f(1)) = (1, -3)$).

(c) Find the equation of the tangent line to the graph of $f(x) = x^3 - 4x$ at the point where $x = 1$.

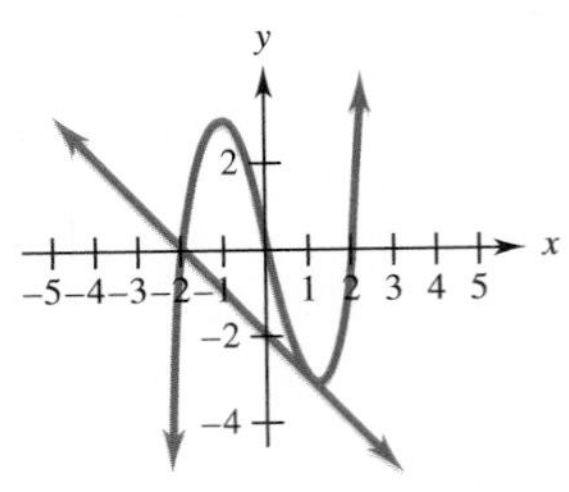

FIGURE 11.25

Solution By part (b), the point on the graph where $x = 1$ is $(1, -3)$, and the slope of the tangent line is $f'(1) = -1$. Therefore, the equation is

$$y - (-3) = (-1)(x - 1) \quad \text{Point–slope form}$$
$$y = -x - 2. \quad \text{Slope–intercept form}$$

Both $f(x)$ and the tangent line are shown in Figure 11.25. 4 ✓

✓Checkpoint 4

Let $f(x) = -5x^2 + 4$. Find the given expressions.

(a) $f(x + h)$

(b) $f(x + h) - f(x)$

(c) $\dfrac{f(x + h) - f(x)}{h}$

(d) $f'(x)$

(e) $f'(3)$

(f) $f'(0)$

Answers: See page 706.

CAUTION

1. In Example 4(a), note that $f(x + h) \neq f(x) + h$ because, by Step 1,

$$f(x + h) = x^3 + 3x^2h + 3xh^2 + h^3 - 4x - 4h,$$

but

$$f(x) + h = (x^3 - 4x) + h = x^3 - 4x + h.$$

2. In Example 4(b), do not confuse $f(1)$ and $f'(1)$. $f(1)$ is the value of the original function $f(x) = x^3 - 4x$ at $x = 1$, namely, -3, whereas $f'(1)$ is the value of the derivative function $f'(x) = 3x^2 - 4$ at $x = 1$, namely, -1.

Example 5 Let $f(x) = 1/x$. Find $f'(x)$.

Solution

Step 1 $f(x + h) = \dfrac{1}{x + h}$.

Step 2
$$f(x + h) - f(x) = \frac{1}{x + h} - \frac{1}{x}$$
$$= \frac{x - (x + h)}{x(x + h)} \quad \text{Find a common denominator.}$$
$$= \frac{x - x - h}{x(x + h)} \quad \text{Simplify the numerator.}$$
$$= \frac{-h}{x(x + h)}.$$

Step 3
$$\frac{f(x + h) - f(x)}{h} = \frac{\dfrac{-h}{x(x + h)}}{h}$$
$$= \frac{-h}{x(x + h)} \cdot \frac{1}{h} \quad \text{Invert and multiply.}$$
$$= \frac{-1}{x(x + h)}.$$

Step 4
$$f'(x) = \lim_{h \to 0} \frac{f(x + h) - f(x)}{h} = \lim_{h \to 0} \frac{-1}{x(x + h)}$$
$$= \frac{-1}{x(x + 0)} = \frac{-1}{x(x)} = \frac{-1}{x^2}.$$ 5 ✓

✓Checkpoint 5

Let $f(x) = -5/x$. Find the given expressions.

(a) $f(x + h)$

(b) $f(x + h) - f(x)$

(c) $\dfrac{f(x + h) - f(x)}{h}$

(d) $f'(x)$

(e) $f'(-1)$

Answers: See page 706.

Example 6 Let $g(x) = \sqrt{x}$. Find $g'(x)$.

Solution

Step 1 $g(x + h) = \sqrt{x + h}$.

Step 2 $g(x + h) - g(x) = \sqrt{x + h} - \sqrt{x}$.

Step 3 $\dfrac{g(x + h) - g(x)}{h} = \dfrac{\sqrt{x + h} - \sqrt{x}}{h}$.

At this point, in order to be able to divide by h, multiply both numerator and denominator by $\sqrt{x + h} + \sqrt{x}$, that is, rationalize the *numerator*:

$$\frac{g(x + h) - g(x)}{h} = \frac{\sqrt{x + h} - \sqrt{x}}{h} \cdot \frac{\sqrt{x + h} + \sqrt{x}}{\sqrt{x + h} + \sqrt{x}}$$

$$= \frac{\left(\sqrt{x + h}\right)^2 - \left(\sqrt{x}\right)^2}{h\left(\sqrt{x + h} + \sqrt{x}\right)}$$

$$= \frac{x + h - x}{h\left(\sqrt{x + h} + \sqrt{x}\right)} = \frac{1}{\sqrt{x + h} + \sqrt{x}}.$$

Step 4 $g'(x) = \lim\limits_{h \to 0} \dfrac{1}{\sqrt{x + h} + \sqrt{x}} = \dfrac{1}{\sqrt{x} + \sqrt{x}} = \dfrac{1}{2\sqrt{x}}$.

Example 7 A sales representative for a textbook-publishing company frequently makes a 4-hour drive from her home in a large city to a university in another city. If $s(t)$ represents her distance (in miles) from home t hours into the trip, then $s(t)$ is given by

$$s(t) = -5t^3 + 30t^2.$$

(a) How far from home will she be after 1 hour? after $1\frac{1}{2}$ hours?

Solution Her distance from home after 1 hour is

$$s(1) = -5(1)^3 + 30(1)^2 = 25,$$

or 25 miles. After $1\frac{1}{2}$ (or $\frac{3}{2}$) hours, it is

$$s\left(\frac{3}{2}\right) = -5\left(\frac{3}{2}\right)^3 + 30\left(\frac{3}{2}\right)^2 = \frac{405}{8} = 50.625,$$

or 50.625 miles.

(b) How far apart are the two cities?

Solution Since the trip takes 4 hours and the distance is given by $s(t)$, the university city is $s(4) = 160$ miles from her home.

(c) How fast is she driving 1 hour into the trip? $1\frac{1}{2}$ hours into the trip?

Solution Velocity (or speed) is the instantaneous rate of change in position with respect to time. We need to find the value of the derivative $s'(t)$ at $t = 1$ and $t = 1\frac{1}{2}$. 6✓

✓Checkpoint 6

Go through the four steps to find $s'(t)$, the velocity of the car at any time t, in Example 7.

Answer: See page 706.

From Checkpoint 6, $s'(t) = -15t^2 + 60t$. At $t = 1$, the velocity is

$$s'(1) = -15(1)^2 + 60(1) = 45,$$

or 45 miles per hour. At $t = 1\frac{1}{2}$, the velocity is

$$s'\left(\frac{3}{2}\right) = -15\left(\frac{3}{2}\right)^2 + 60\left(\frac{3}{2}\right) = 56.25,$$

about 56 miles per hour.

(d) Does she ever exceed the speed limit of 65 miles per hour on the trip?

Solution To find the maximum velocity, notice that the graph of the velocity function $s'(t) = -15t^2 + 60t$ is a parabola opening downward (Figure 11.26). The maximum velocity will occur at the vertex. Use graphical or algebraic methods to verify that the vertex of the parabola is (2, 60). Thus, her maximum velocity during the trip is 60 miles per hour, so she never exceeds the speed limit.

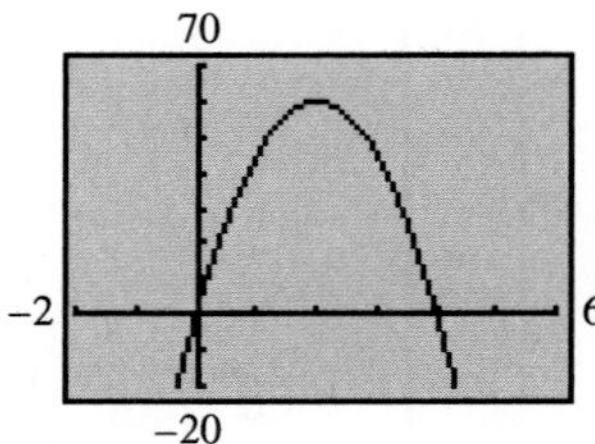

FIGURE 11.26

EXISTENCE OF THE DERIVATIVE

The definition of the derivative includes the phrase "provided that this limit exists." If the limit used to define $f'(x)$ does not exist, then, of course, the derivative does not exist at that x. For example, a derivative cannot exist at a point where the function itself is not defined. If there is no function value for a particular value of x, there can be no tangent line for that value. This was the case in Example 5: There was no tangent line (and no derivative) when $x = 0$.

Derivatives also do not exist at "corners" or "sharp points" on a graph. For example, the function graphed in Figure 11.27 is the *absolute-value function*, defined by

$$f(x) = \begin{cases} x & \text{if } x \geq 0 \\ -x & \text{if } x < 0 \end{cases}$$

and written $f(x) = |x|$. The graph has a "corner point" when $x = 0$.

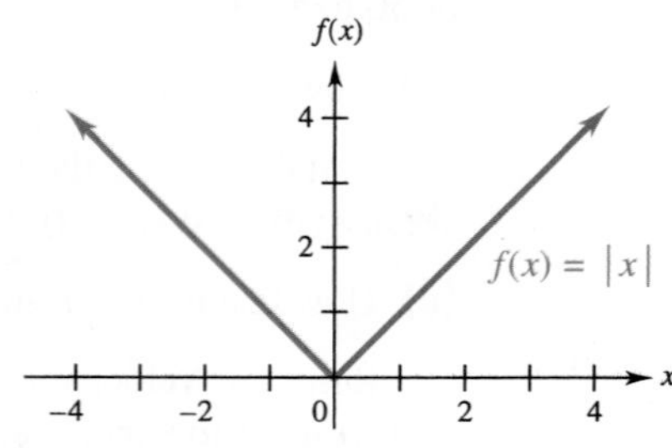

FIGURE 11.27

By the definition of derivative, the derivative at any value of x is given by

$$\lim_{h\to 0}\frac{f(x+h)-f(x)}{h} = \lim_{h\to 0}\frac{|x+h|-|x|}{h}$$

provided that this limit exists. To find the derivative at 0 for $f(x) = |x|$, replace x with 0 and $f(x)$ with $|0|$ to get

$$\lim_{h\to 0}\frac{|0+h|-|0|}{h} = \lim_{h\to 0}\frac{|h|}{h}.$$

In Example 11 of Section 11.1 (with x in place of h), we showed that

$$\lim_{h\to 0}\frac{|h|}{h}\text{ does not exist.}$$

Therefore, there is no derivative at 0. However, the derivative of $f(x) = |x|$ *does* exist for all values of x other than 0.

Since a vertical line has an undefined slope, the derivative cannot exist at any point where the tangent line is vertical, as at x_5 in Figure 11.28. This figure summarizes various ways that a derivative can fail to exist.

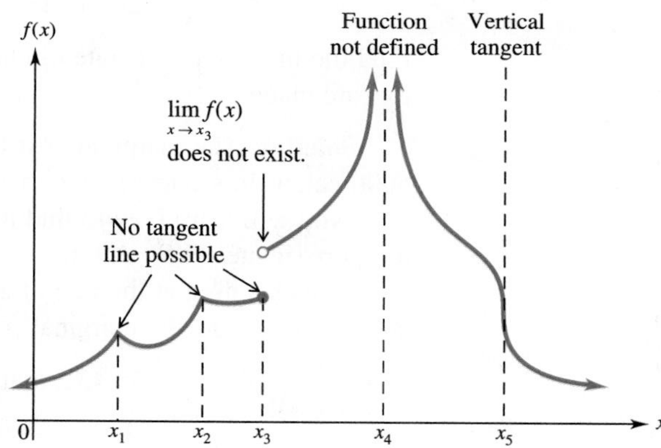

FIGURE 11.28

DERIVATIVES AND TECHNOLOGY

Many computer programs (such as Mathematica and Maple) and a few graphing calculators (such as the TI-89) can find symbolic formulas for the derivatives of most functions. Although other graphing calculators cannot find the rule of a derivative function, most of them can approximate the numerical value of the derivative function at any number where it is defined by using the **numerical derivative** feature. (See Exercise 48 at the end of this section for an explanation of computational technique used by the numerical derivative feature.)

Example 8 If $f(x) = x^3 + x^{3/2} + 6\ln x$, use a graphing calculator to find the approximate value of $f'(4)$.

Solution The numerical derivative feature is labeled nDeriv, or d/dx, or nDer and is usually in the MATH or CALC menu or one of its submenus. Check your instruction manual for the correct syntax, but on many graphing calculators, entering either

$$\text{nDeriv } (x^3 + x^{3/2} + 6 \ln x, x, 4) \qquad \text{or} \qquad d/dx(x^3 + x^{3/2} + 6 \ln x, 4)$$

produces the (approximate) value of $f'(4)$. Depending on the calculator, you are likely to get one of the following answers:

$$52.4999982, \qquad 52.5, \qquad \text{or} \qquad 52.50000102.$$

It can be shown that the exact value is $f'(4) = 52.5$. 7 ✓ ■

✓Checkpoint 7

If $f(x) = x^{2/3} - 8x^3 + 4^x$, use a graphing calculator to find the (approximate) value of

(a) $f'(2)$;

(b) $f'(3.2)$.

Answers: See page 706.

TECHNOLOGY TIP On many graphing calculators, if you have the function $f(x)$ stored in the function memory as, say, Y_1, you can use Y_1 with the nDeriv key instead of typing in the rule of $f(x)$.

Example 9 The cost (in thousands of dollars) to manufacture x thousand graphing calculators is given by

$$C(x) = .002x^4 + .1x^3 + .3x^2 - .1x + 50 \qquad (0 \le x \le 13).$$

Find the marginal cost (rate of change of cost) when 5000 and when 9000 calculators are made.

Solution The marginal-cost function is the derivative of the cost function, and 5000 calculators corresponds to $x = 5$, so the marginal cost is $C'(5)$. In later sections you will learn how to find it exactly, but for now use the numerical derivative to approximate $C'(5)$. A typical calculator shows that $C'(5) \approx 11.4$, meaning that costs are changing at the rate of \$11,400 per thousand calculators. When 9000 calculators are made, the marginal cost is

$$C'(9) \approx 35.432, \text{ that is, \$35,432 per thousand calculators.}$$ ■

CAUTION Because of the approximation methods used, the nDeriv key may display an answer at numbers where the derivative is not defined. For instance, we saw earlier that the derivative of $f(x) = |x|$ is not defined when $x = 0$, but the nDeriv key on most calculators produces 1 or 0 or -1 as $f'(0)$.

11.4 ▶ Exercises

The derivatives of each of the given functions were found in Examples 4–7. Use them to find the equation of the tangent line to the graph of the function at the given point. (See Examples 1 and 2.)

1. $f(x) = x^3 - 4x$ at $x = 2$
2. $g(x) = \sqrt{x}$ at $x = 9$
3. $f(x) = 1/x$ at $x = -2$
4. $s(t) = -5t^3 + 30t^2$ at $t = 2$

For each of the given functions, (a) find the slope of the tangent line to the graph at the given point; (b) find the equation of the tangent line. (See Examples 1 and 2.)

5. $f(x) = x^2 + 3$ at $x = 2$
6. $g(x) = 1 - 2x^2$ at $x = -1$
7. $h(x) = \dfrac{7}{x}$ at $x = 3$

8. $f(x) = \frac{-3}{x}$ at $x = -4$ **9.** $g(x) = 5\sqrt{x}$ at $x = 9$

10. $g(x) = \sqrt{x + 1}$ at $x = 15$ (*Hint*: In Step 3, multiply numerator and denominator by $\sqrt{15 + h} + \sqrt{15}$.)

Use the fact that $f'(c)$ is the slope of the tangent line to the graph of $f(x)$ at $x = c$ to work these exercises. (See Example 3.)

11. In the accompanying graph of the function f, at which of the labeled x-values is
(a) $f(x)$ the largest?
(b) $f(x)$ the smallest?
(c) $f'(x)$ the smallest?
(d) $f'(x)$ the closest to 0?

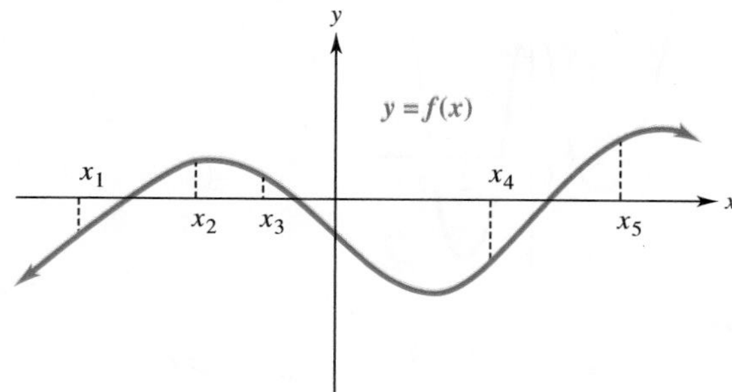

12. Sketch the graph of the *derivative* of the function g whose graph is shown. (*Hint*: Consider the slope of the tangent line at each point along the graph of g. Are there any points where there is no tangent line?)

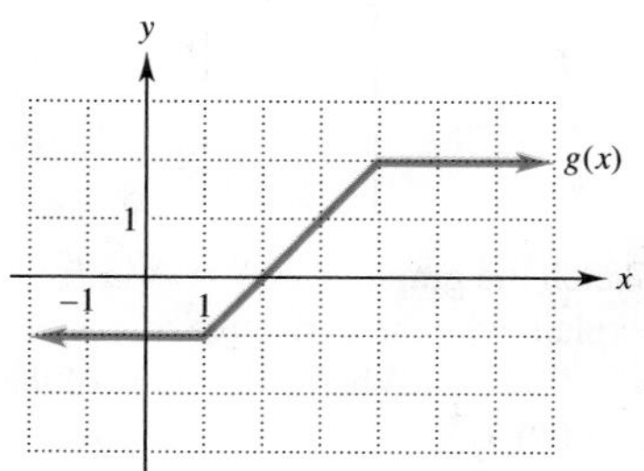

13. Sketch the graph of a function g with the property that $g'(x) > 0$ for $x < 0$ and $g'(x) < 0$ for $x > 0$. Many correct answers are possible.

14. **Physical Science** The accompanying graph shows the temperature in an oven during a self-cleaning cycle.* (The open circles on the graph are not points of discontinuity, but merely the times when the thermal door lock turns on and off.) The oven temperature is 100° when the cycle begins and 600° after half an hour. Let $T(x)$ be the temperature (in degrees Fahrenheit) after x hours.

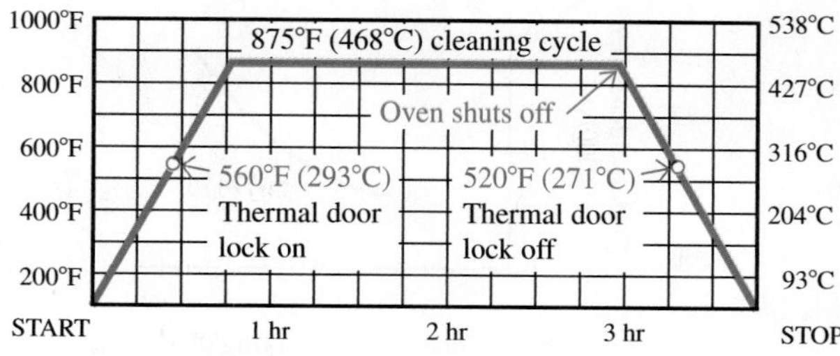

(a) Find the approximate values of x at which the derivative T' does not exist.
(b) Find and interpret $T'(.5)$.
(c) Find and interpret $T'(2)$.
(d) Find and interpret $T'(3.5)$.

15. **Physical Science** When a batter hits a baseball, the bat may not hit the center of the ball, but might instead hit over or under the center by various amounts (measured in inches). The accompanying graph shows the trajectories of balls struck by a bat swung under the ball by given amounts.* Is the derivative for a bat swung under the ball by 1.5 inches positive or negative when the ball has traveled
(a) 100 ft?  **(b)** 200 ft? 

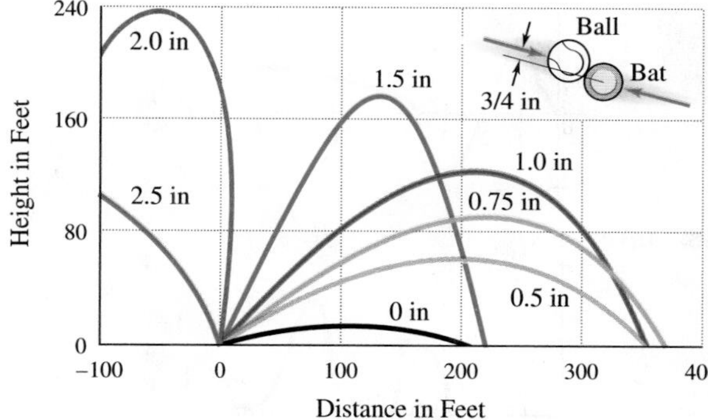

16. **Natural Science** The graph on the next page shows how the risk of coronary heart attack (CHA) rises as blood cholesterol increases.†
(a) Approximate the average rate of change of the risk of coronary heart attack as blood cholesterol goes from 100 to 300 mg/dL.
(b) Is the rate of change when blood cholesterol is 100 mg/dL higher or lower than the average rate of change in part (a)? What feature of the graph shows this?
(c) Do part (b) when blood cholesterol is 300 mg/dL.
(d) Do part (b) when blood cholesterol is 200 mg/dL.

Whirlpool Use and Care Guide, Self-Cleaning Electric Range.

*Adair, Robert K., *The Physics of Baseball*, Copyright © 1990 by HarperCollins, p. 83.

†John C. LaRosa, et al., "The Cholesterol Facts: A Joint Statement by the American Heart Association and the National Heart, Lung, and Blood Institute," from *Circulation* 81, no. 5 (May 1990): 1722.

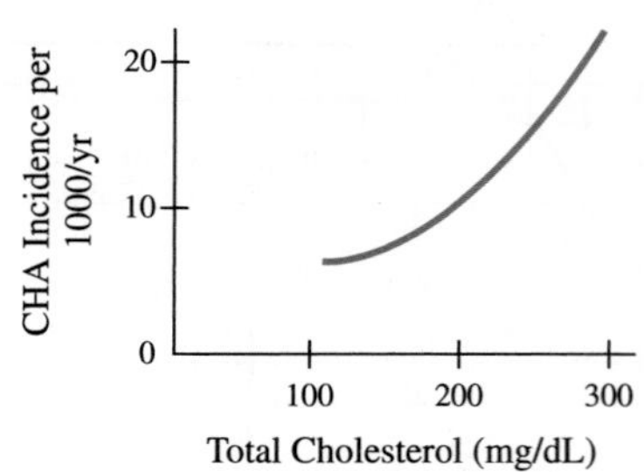

In Exercises 17 and 18, tell which graph, (a) or (b), represents velocity and which represents distance from a starting point. (Hint: Consider where the derivative is zero, positive, or negative.)

17. (a)

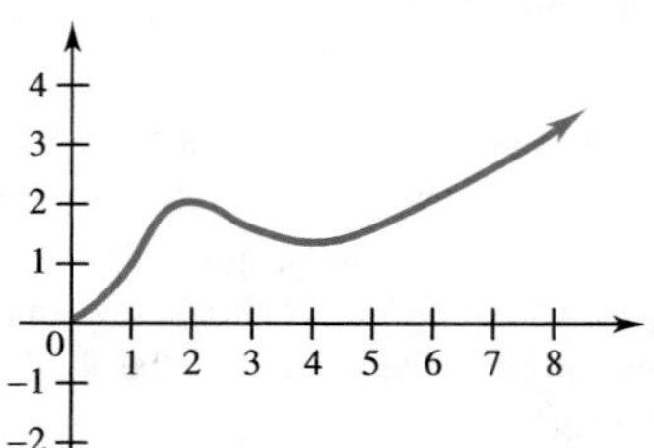

(b)

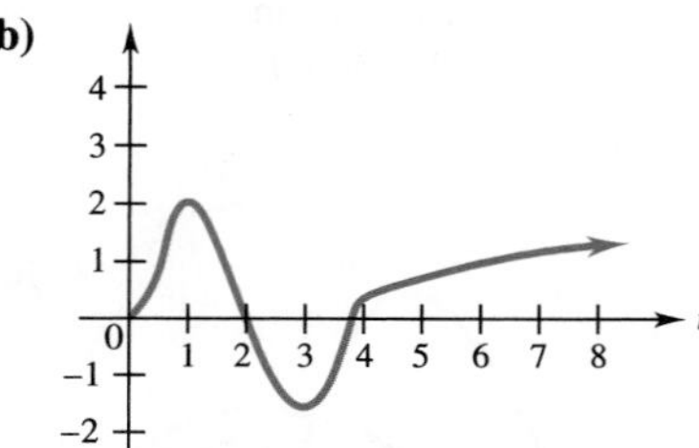

18. (a)

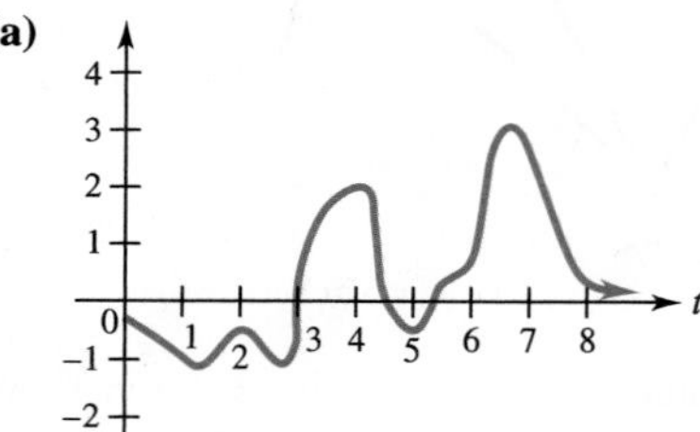

(b)

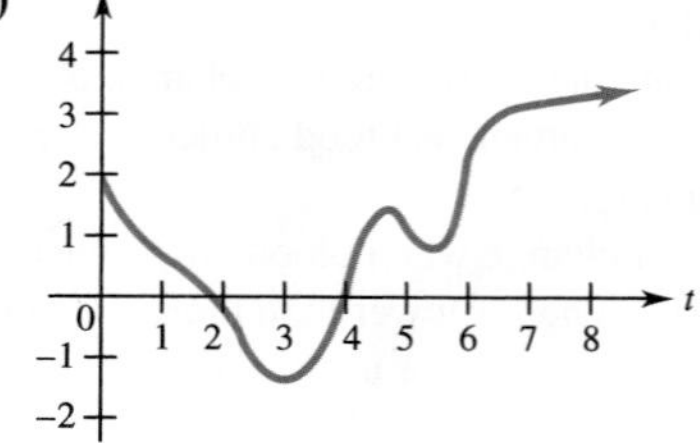

Find f′(x) for each function. Then find f′(2), f′(0), and f′(−3). (See Examples 4–6.)

19. $f(x) = -4x^2 + 11x$

20. $f(x) = 6x^2 - 4x$

21. $f(x) = 8x + 6$

22. $f(x) = x^3 + 3x$

23. $f(x) = -\dfrac{2}{x}$

24. $f(x) = \dfrac{6}{x}$

25. $f(x) = \dfrac{4}{x - 1}$

26. $f(x) = \sqrt{x}$

Find all points where the functions whose graphs are shown do not have derivatives. (See Figure 11.28 and the preceding discussion.)

27.

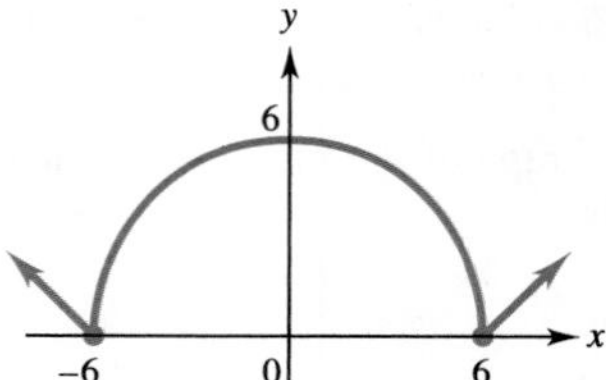

28.

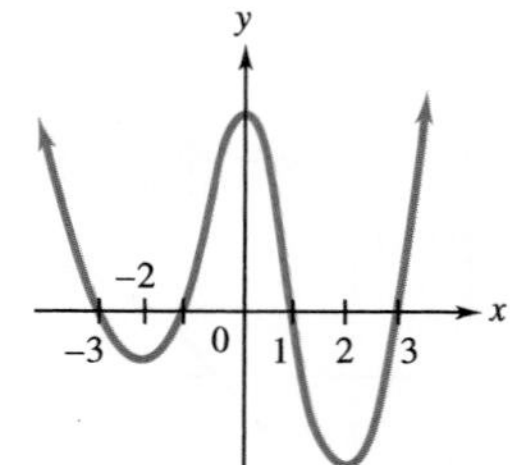

29.

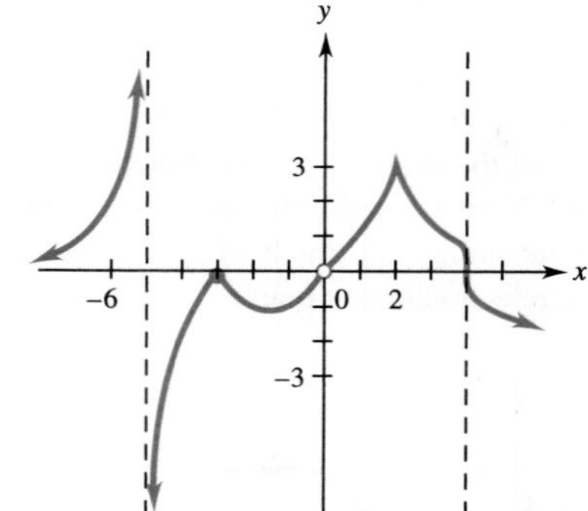

30. **(a)** Sketch the graph of $g(x) = \sqrt[3]{x}$ for $-1 \le x \le 1$.

(b) Explain why the derivative of $g(x)$ is not defined at $x = 0$. (*Hint*: What is the slope of the tangent line at $x = 0$?)

Work these exercises. (See Examples 4–7.)

31. Business The revenue generated from the sale of x picnic tables is given by

$$R(x) = 20x - \frac{x^2}{500}.$$

(a) Find the marginal revenue when $x = 1000$ units.

(b) Determine the actual revenue from the sale of the 1001st item.

(c) Compare the answers to parts (a) and (b). How are they related?

32. Business The cost of producing x items is

$$C(x) = 1000 + .24x^2 \qquad (0 \le x \le 30{,}000).$$

(a) Find the marginal cost $C'(x)$.
(b) Find $C'(100)$.
(c) Find the exact cost to produce the 101st item.
(d) Compare the answers to parts (b) and (c). How are they related?

33. **Business** The demand for a certain item is given by $D(p) = -p^2 + 5p + 1$, where p represents the price of an item in dollars.
(a) Find the rate of change of demand with respect to price.
(b) Find and interpret the rate of change of demand when the price is 12 dollars.

34. **Business** The profit (in dollars) from producing x hundred kegs of beer is given by $P(x) = 1500 + 30x - 2x^2$. Find the marginal profit at the given production levels. In each case, decide whether the firm should increase production of beer.
(a) 700 **(b)** 650 **(c)** 1200 **(d)** 1800

35. **Natural Science** In one research study, the population of a certain shellfish in an area at time t was closely approximated by the accompanying graph. Estimate and interpret the derivative at each of the marked points.

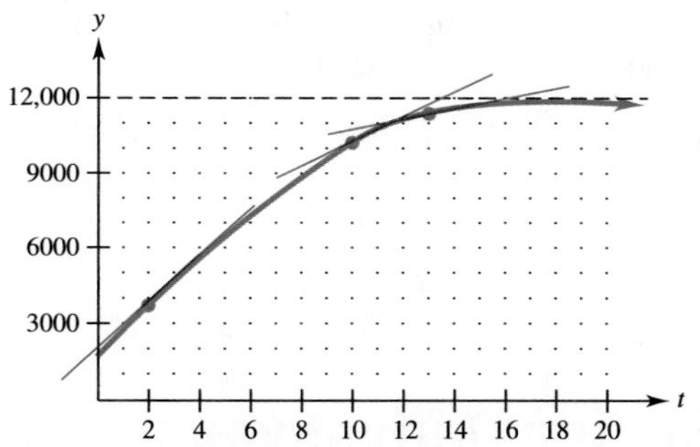

36. **Health** The eating behavior of a typical human during a meal can be described by

$$I(t) = 27 + 72t - 1.5t^2,$$

where t is the number of minutes since the meal began and $I(t)$ represents the amount (in grams) that the person has eaten at time t.*
(a) Find the rate of change of the intake of food for a person 5 minutes into a meal, and interpret it.
(b) Verify that the rate at which food is consumed is zero 24 minutes after the meal starts.
(c) Comment on the assumptions and usefulness of this function after 24 minutes. On the basis of your answer, determine a logical range for the function.

*Kissileff, H. R. and J. L. Guss, "Microstructure of Eating Behavior in Humans," *Appetite*, Vol. 36, No. 1, Feb. 2001, pp. 70–78.

Use technology to graph the (numerical) derivative of each function by graphing y = nDeriv(f(x), x). (See Example 8.)*

37. $f(x) = x^2 - 5x + 2$

38. $f(x) = .5x^3 - 3x^2 + x + 2$

39. $f(x) = \ln x + x$

40. $f(x) = e^x + 2x$

Use numerical derivatives to work these exercises. (See Examples 8 and 9.)

41. **Business** The amount of money (in billions of U.S. dollars) that China is spending on research and development can be approximated by the function $f(t) = 1.704(1.223^t)$, where $t = 6$ corresponds to the year 1996.†
(a) Estimate the value of the derivative function at 1998, 2005, and 2010.
(b) What does your answer in part (a) say about the rate at which research and development spending is increasing in China?

42. **Business** The revenue (in billions of dollars) for Southwest Airlines can be approximated by the function $R(x) = .00187x^3 + .00629x^2 - .5429x + 8.059$, where $x = 9$ corresponds to the year 1999.‡ Estimate the derivative for the years 2005 and 2010.

43. **Transportation** The number of licensed drivers in the United States (in millions) can be approximated by the function $g(x) = 145.033x^{.119}$, where $x = 6$ corresponds to the year 1996.§ Estimate and interpret the value of the derivative at 2010.

44. **Finance** A model for the value of the NASDAQ index by month is the function $f(x) = -.001234x^4 + .114805x^3 - 3.06823x^2 + 33.4445x + 1912.38$, where $x = 1$ corresponds to the value of the index on the first business day in January 2004 ($x = 2$ corresponds to February 2004, etc.)£
(a) Use the function to estimate the value of the index in July 2006 and October 2008.
(b) Use numerical derivatives to find the slope of the tangent line for April 2007.
(c) Use numerical derivatives to find the slope of the tangent line for April 2008.

Use technology for Exercise 45–48.

45. **(a)** Graph the (numerical) derivative of $f(x) = .5x^5 - 2x^3 + x^2 - 3x + 2$ for $-3 \le x \le 3$.
(b) Graph $g(x) = 2.5x^4 - 6x^2 + 2x - 3$ on the same screen.
(c) How do the graphs of $f'(x)$ and $g(x)$ compare? What does this suggest that the derivative of $f(x)$ is?

*Use nDeriv ($f(x)$, x, x) on some TI models.

†Based on data from the National Science Foundation, Division of Science Resources Statistics, *National Patterns of R&D Resources*, annual report.

‡www.morningstar.com.

§www.fhwa.dot.gov.

£www.nasdaq.com.

46. Repeat Exercise 45 for $f(x) = (x^2 + x + 1)^{1/3}$ (with $-6 \le x \le 6$) and $g(x) = \dfrac{2x + 1}{3(x^2 + x + 1)^{2/3}}$.

47. By using a graphing calculator to compare graphs, as in Exercises 45 and 46, decide which of the given functions *could* be the derivative of $y = \dfrac{4x^2 + x}{x^2 + 1}$.

(a) $f(x) = \dfrac{2x + 1}{2x}$ **(b)** $g(x) = \dfrac{x^2 + x}{2x}$

(c) $h(x) = \dfrac{2x + 1}{x^2 + 1}$ **(d)** $k(x) = \dfrac{-x^2 + 8x + 1}{(x^2 + 1)^2}$

48. If f is a function such that $f'(x)$ is defined, then it can be proved that

$$f'(x) = \lim_{h \to 0} \frac{f(x + h) - f(x - h)}{2h}.$$

Consequently, when h is very small, say, $h = .001$,

$$f'(x) \approx \frac{f(x + h) - f(x - h)}{2h} = \frac{f(x + .001) - f(x - .001)}{.002}.$$

(a) In Example 6, we saw that the derivative of $f(x) = \sqrt{x}$ is the function $f'(x) = \dfrac{1}{2\sqrt{x}}$. Make a table in which the first column lists $x = 1, 6, 11, 16,$ and 21; the second column lists the corresponding values of $f'(x)$; and the third column lists the corresponding values of

$$\frac{f(x + .001) - f(x - .001)}{.002}.$$

(b) How do the second and third columns of the table compare? (If you use the table feature on a graphing calculator, the entries in the table will be rounded off, so move the cursor over each entry to see it fully displayed at the bottom of the screen.) Your answer to this question may explain why most graphing calculators use the method in the third column to compute numerical derivatives.

✓ Checkpoint Answers

1. $y = 2x + 1$

2.

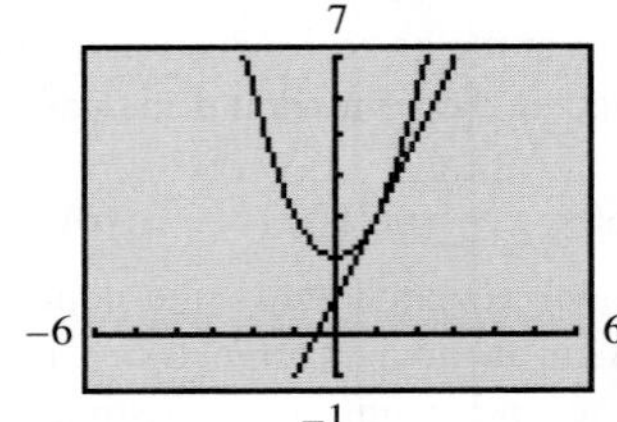

3. **(a)** Positive **(b)** Zero **(c)** Negative

4. **(a)** $-5x^2 - 10xh - 5h^2 + 4$
(b) $-10xh - 5h^2$ **(c)** $-10x - 5h$ **(d)** $-10x$
(e) -30 **(f)** 0

5. **(a)** $\dfrac{-5}{x + h}$ **(b)** $\dfrac{5h}{x(x + h)}$ **(c)** $\dfrac{5}{x(x + h)}$
(d) $\dfrac{5}{x^2}$ **(e)** 5

6. $s'(t) = -15t^2 + 60t$

7. **(a)** -73.29 **(b)** -128.2

11.5 ▶ Techniques for Finding Derivatives

In the previous section, the derivative of a function was defined as a special limit. The mathematical process of finding this limit, called *differentiation*, resulted in a new function that was interpreted in several different ways. Using the definition to calculate the derivative of a function is a very involved process, even for simple functions. In this section, we develop rules that make the calculation of derivatives much easier. Keep in mind that even though the process of finding a derivative will be greatly simplified with these rules, *the interpretation of the derivative will not change.*

In addition to y' and $f'(x)$, there are several other commonly used notations for the derivative.

Notations for the Derivative

The derivative of the function $y = f(x)$ may be denoted in any of the following ways:

$$f'(x), \quad y', \quad \frac{dy}{dx}, \quad \frac{d}{dx}[f(x)], \quad D_x y, \quad \text{or} \quad D_x[f(x)].$$

The dy/dx notation for the derivative is sometimes referred to as *Leibniz notation*, named after one of the coinventors of calculus, Gottfried Wilhelm Leibniz (1646–1716). (The other was Sir Isaac Newton (1642–1727).)

For example, the derivative of $y = x^3 - 4x$, which we found in Example 4 of the last section to be $y' = 3x^2 - 4$, can also be written in the following ways:

$$\frac{dy}{dx} = 3x^2 - 4;$$

$$\frac{d}{dx}(x^3 - 4x) = 3x^2 - 4;$$

$$D_x(x^3 - 4x) = 3x^2 - 4.$$ ✓1

✓Checkpoint 1

Use the results of some of Exercises 22–26 in the previous section to find each of the given derivatives.

(a) $\frac{d}{dx}(x^3 + 3x)$

(b) $\frac{d}{dx}\left(-\frac{2}{x}\right)$

(c) $D_x\left(\frac{4}{x-1}\right)$

(d) $D_x(\sqrt{x})$

Answers: See page 719.

A variable other than x may be used as the independent variable. For example, if $y = f(t)$ gives population growth as a function of time, then the derivative of y with respect to t could be written as

$$f'(t), \quad \frac{dy}{dt}, \quad \frac{d}{dt}[f(t)], \quad \text{or} \quad D_t[f(t)].$$

In this section, the definition of the derivative,

$$f'(x) = \lim_{h \to 0} \frac{f(x+h) - f(x)}{h},$$

is used to develop some rules for finding derivatives more easily than by the four-step process of the previous section. The first rule tells how to find the derivative of a constant function, such as $f(x) = 5$. Since $f(x + h)$ is also 5, $f'(x)$ is by definition

$$f'(x) = \lim_{h \to 0} \frac{f(x+h) - f(x)}{h}$$

$$= \lim_{h \to 0} \frac{5 - 5}{h} = \lim_{h \to 0} \frac{0}{h} = \lim_{h \to 0} 0 = 0.$$

The same argument works for $f(x) = k$, where k is a constant real number, establishing the following rule.

Constant Rule

If $f(x) = k$, where k is any real number, then

$$\mathbf{f'(x) = 0.}$$

(The derivative of a constant function is 0.)

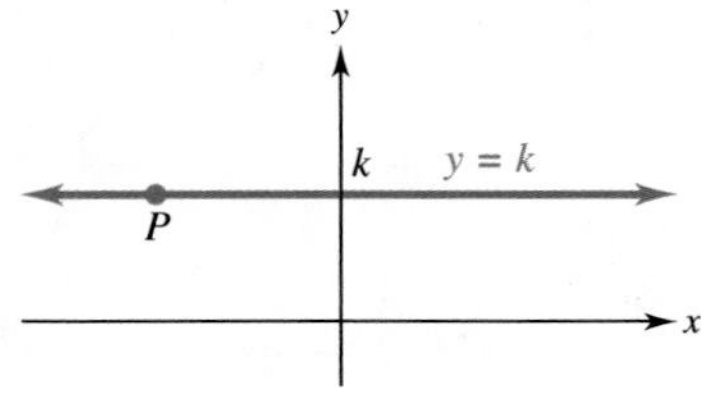

FIGURE 11.29

Figure 11.29 illustrates the constant rule; it shows a graph of the horizontal line $y = k$. At any point P on this line, the tangent line at P is the line itself. Since a horizontal line has a slope of 0, the slope of the tangent line is 0. This finding agrees with the constant rule: The derivative of a constant is 0.

Example 1 Find and label the derivative of the given function.

(a) $f(x) = 25$

Solution Since 25 is a constant, we can write $f'(x) = 0$, $\frac{d}{dx}[f(x)] = 0$, or $D_x[f(x)] = 0$.

(b) $y = \pi$

Solution Since π is a constant with value equal to 3.14159265 . . . , we have $y' = 0$, $dy/dx = 0$, or $D_x y = 0$.

(c) $y = 4^3$

Solution Since 4^3 is the constant 64, we have $y' = 0$, $dy/dx = 0$, or $D_x y = 0$. 2✓

✓Checkpoint 2

Find the derivative of the given function.

(a) $y = -4$

(b) $f(x) = \pi^3$

(c) $y = 0$

Answers: See page 719.

Functions of the form $y = x^n$, where n is a fixed real number, are common in applications. We now find the derivative functions when $n = 2$ and $n = 3$:

$$f(x) = x^2$$

$$\begin{aligned} f'(x) &= \lim_{h\to 0}\frac{f(x+h)-f(x)}{h} \\ &= \lim_{h\to 0}\frac{(x+h)^2 - x^2}{h} \\ &= \lim_{h\to 0}\frac{(x^2+2xh+h^2)-x^2}{h} \\ &= \lim_{h\to 0}\frac{2xh+h^2}{h} \\ f'(x) &= \lim_{h\to 0}(2x+h) = 2x \end{aligned}$$

$$f(x) = x^3$$

$$\begin{aligned} f'(x) &= \lim_{h\to 0}\frac{f(x+h)-f(x)}{h} \\ &= \lim_{h\to 0}\frac{(x+h)^3 - x^3}{h} \\ &= \lim_{h\to 0}\frac{(x^3+3x^2h+3xh^2+h^3)-x^3}{h} \\ &= \lim_{h\to 0}\frac{3x^2h+3xh^2+h^3}{h} \\ f'(x) &= \lim_{h\to 0}(3x^2+3xh+h^2) = 3x^2. \end{aligned}$$

Similar calculations show that

$$\frac{d}{dx}(x^4) = 4x^3 \quad \text{and} \quad \frac{d}{dx}(x^5) = 5x^4.$$

The pattern here (the derivative is the product of the exponent on the original function and a power of x that is one less) suggests that the derivative of $y = x^n$ is $y' = nx^{n-1}$, which is indeed the case. (See Exercise 67 at the end of this section.)

Furthermore, the pattern holds even when n is not a positive integer. For instance, in Example 5 of the last section we showed that

$$\text{if } f(x) = \frac{1}{x}, \quad \text{then } f'(x) = -\frac{1}{x^2}.$$

This is the same result that the preceding pattern produces:

$$\text{If } f(x) = \frac{1}{x} = x^{-1}, \quad \text{then } f'(x) = (-1)x^{-1-1} = -x^{-2} = -\frac{1}{x^2}.$$

Similarly, Example 6 in the last section showed that the derivative of $g(x) = \sqrt{x}$ is $g'(x) = \dfrac{1}{2\sqrt{x}}$, and the preceding pattern gives the same answer:

$$\text{If } g(x) = \sqrt{x} = x^{1/2}, \qquad \text{then } g'(x) = \frac{1}{2}x^{\frac{1}{2}-1} = \frac{1}{2}x^{-1/2} = \frac{1}{2}\cdot\frac{1}{x^{1/2}} = \frac{1}{2\sqrt{x}}.$$

Consequently, the following statement should be plausible.

Power Rule

If $f(x) = x^n$ for any number n, then

$$f'(x) = nx^{n-1}.$$

(The derivative of $f(x) = x^n$ is found by multiplying the exponent n on the original function by a power of x that is one less.)

Example 2 For each of the given functions, find the derivative.

(a) $y = x^8$

Solution The exponent for x is $n = 8$. Using the power rule, we multiply by 8 and decrease the power on x by 1 to obtain $y' = 8x^7$.

(b) $y = x$

Solution The exponent for x is $n = 1$. Using the power rule, we multiply by 1 and decrease the power on x by 1 to obtain $y' = 1x^0$. Recall that $x^0 = 1$ if $x \neq 0$. So we have $y' = 1$.

(c) $y = t^{3/2}$

Solution The exponent for t is $n = \frac{3}{2}$. Using the power rule, we multiply by $\frac{3}{2}$ and decrease the power on t by 1 to obtain $D_t(y) = \frac{3}{2}t^{3/2-1} = \frac{3}{2}t^{1/2}$.

(d) $y = \sqrt[3]{x}$

Solution First we rewrite $y = \sqrt[3]{x}$ as $y = x^{\frac{1}{3}}$. Using the power rule, we multiply by $\frac{1}{3}$ and decrease the power on x by 1 to obtain

$$\frac{dy}{dx} = \frac{1}{3}x^{1/3-1} = \frac{1}{3}x^{-2/3} = \frac{1}{3x^{2/3}} = \frac{1}{3\sqrt[3]{x^2}}.$$ ✓3

✓Checkpoint 3

(a) If $y = x^4$, find y'.

(b) If $y = x^{17}$, find y'.

(c) If $y = x^{-2}$, find dy/dx.

(d) If $y = t^{-5}$, find dy/dt.

Answers: See page 719.

The next rule shows how to find the derivative of the product of a constant and a function.

Constant Times a Function

Let k be a real number. If $g'(x)$ exists, then the derivative of $f(x) = k\cdot g(x)$ is

$$f'(x) = k\cdot g'(x).$$

(The derivative of a constant times a function is the constant times the derivative of the function.)

Example 3

(a) If $y = 8x^4$, find y'.

Solution Since the derivative of $g(x) = x^4$ is $g'(x) = 4x^3$ and $y = 8x^4 = 8g(x)$,

$$y' = 8g'(x) = 8(4x^3) = 32x^3.$$

(b) If $y = -\frac{3}{4}t^{12}$, find dy/dt.

Solution $$\frac{dy}{dt} = -\frac{3}{4}\left[\frac{dy}{dt}(t^{12})\right] = -\frac{3}{4}(12t^{11}) = -9t^{11}.$$

(c) Find $D_x(15x)$.

Solution $D_x(15x) = 15 \cdot D_x(x) = 15(1) = 15.$

(d) If $y = 6/x^2$, find y'.

Solution Replace $\frac{6}{x^2}$ by $6 \cdot \frac{1}{x^2}$, or $6x^{-2}$. Then

$$y' = 6(-2x^{-3}) = -12x^{-3} = -\frac{12}{x^3}.$$

X	Y_1	Y_2
5	33.541	33.541
6	36.742	36.742
7	39.686	39.686
8	42.426	42.426
9	45	45
10	47.434	47.434
11	49.749	49.749

X=11

(a)

(e) Find $D_x(10x^{3/2})$, and use a graphing calculator to confirm your answer numerically and graphically.

Solution $$D_x(10x^{3/2}) = 10\left(\frac{3}{2}x^{1/2}\right) = 15x^{1/2}.$$

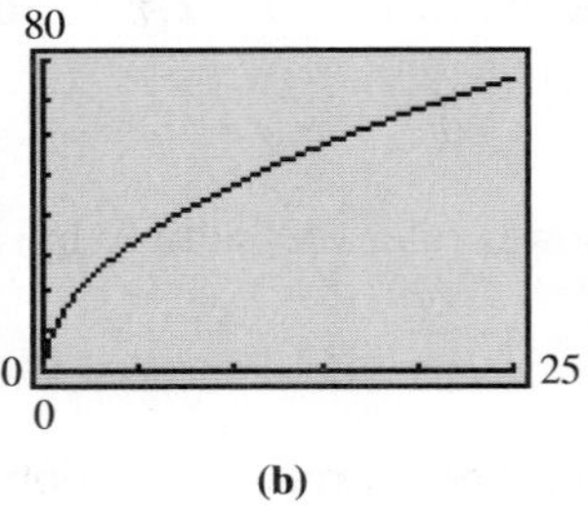

(b)

FIGURE 11.30

To confirm this result numerically, make a table of values for $y_1 = 15x^{1/2}$ and y_2 = the numerical derivative of $10x^{3/2}$; check your instruction manual for the correct syntax for y_2, which is probably one of the following:

nDeriv($10x^{3/2}$,x), Nderiv($10x^{3/2}$,x,x),
d/dx($10x^{3/2}$), or d/dx($10x^{3/2}$,x).

Figure 11.30(a) indicates that the corresponding values are identical to three decimal places. To confirm this result graphically, graph y_1 and y_2 on the same screen and verify that the graphs appear to be the same. (See Figure 11.30(b)). 4 ✓

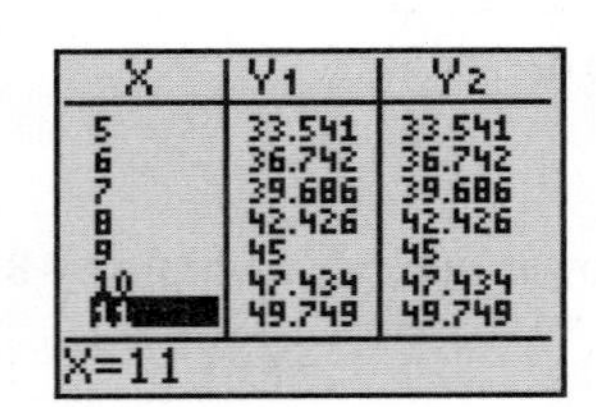

✓Checkpoint 4

Find the derivative of the given function.

(a) $y = 12x^3$

(b) $f(t) = 30t^7$

(c) $y = -35t$

(d) $y = 5\sqrt{x}$

(e) $y = -10/t$

Answers: See page 720.

Confirming your calculations numerically or graphically, as in part (e) of Example 3, is a good way to detect algebraic errors. If you compute the rule of the derivative $f'(x)$, but its graph differs from the graph of the numerical derivative of $f(x)$, then you have made a mistake. If the two graphs appear to be identical, then you are *probably* correct. (The fact that two graphs appear identical on a calculator screen does not *prove* that they really are identical.)

The final rule in this section is for the derivative of a function that is a sum or difference of functions.

Sum-or-Difference Rule

If $f(x) = u(x) + v(x)$, and if $u'(x)$ and $v'(x)$ exist, then

$$f'(x) = u'(x) + v'(x).$$

If $f(x) = u(x) - v(x)$, then

$$f'(x) = u'(x) - v'(x).$$

(The derivative of a sum or difference of two functions is the sum or difference of the derivatives of the functions.)

For a proof of this rule, see Exercise 68 at the end of this section. This rule also works for sums and differences with more than two terms.

Example 4 Find the derivatives of the given functions.

(a) $y = 6x^3 + 15x^2$

Solution Let $u(x) = 6x^3$ and $v(x) = 15x^2$. Then

$$y = u(x) + v(x) \quad \text{and}$$
$$y' = u'(x) + v'(x)$$
$$= 6(3x^2) + 15(2x) = 18x^2 + 30x.$$

(b) $p(t) = 8t^4 - 6\sqrt{t} + \dfrac{5}{t}$

Solution Rewrite $p(t)$ as $p(t) = 8t^4 - 6t^{1/2} + 5t^{-1}$; then $p'(t) = 32t^3 - 3t^{-1/2} - 5t^{-2}$, which also may be written as $p'(t) = 32t^3 - \dfrac{3}{\sqrt{t}} - \dfrac{5}{t^2}$. ✓5

(c) $f(x) = 5\sqrt[3]{x^2} + 4x^{-2} + 7$

Solution Rewrite $f(x)$ as $f(x) = 5x^{2/3} + 4x^{-2} + 7$. Then

$$D_x[f(x)] = \frac{10}{3}(x^{-1/3}) - 8x^{-3},$$

or

$$D_x[f(x)] = \frac{10}{3\sqrt[3]{x}} - \frac{8}{x^3}.$$ ✓6

The rules developed in this section make it possible to find the derivative of a function more directly than with the definition of the derivative, so that applications of the derivative can be dealt with more effectively. The examples that follow illustrate some business applications.

✓Checkpoint 5

Find the derivatives of the given functions.

(a) $y = -10x^3 - 6x + 12$

(b) $y = 5t^{8/7} + 5t^{1/2}$

(c) $f(t) = -2\sqrt{t} + 4/t$

Answers: *See page 720.*

✓Checkpoint 6

Use a graphing calculator to confirm your answer to part (c) by graphing $D_x[f(x)]$ and the numerical derivative of $f(x)$ on the same screen.

Answer: *See page 720.*

MARGINAL ANALYSIS

In business and economics, the rates of change of such variables as cost, revenue, and profit are important considerations. Economists use the word *marginal* to refer to rates of change: for example, *marginal cost* refers to the rate of change of cost with respect to the number of items produced. Since the derivative of a function gives the rate of change of the function, a marginal cost (or revenue, or profit) function is found by taking the derivative of the cost (or revenue, or profit) function. Roughly speaking, the marginal cost at some level of production x is the cost of producing the $(x + 1)$st item, as we now show. (Similar statements could be made for revenue or profit.)

Look at Figure 11.31, where $C(x)$ represents the cost of producing x units of some item. Then the cost of producing $x + 1$ units is $C(x + 1)$. The cost of the $(x + 1)$st unit is, therefore, $C(x + 1) - C(x)$. This quantity is shown on the graph in Figure 11.31.

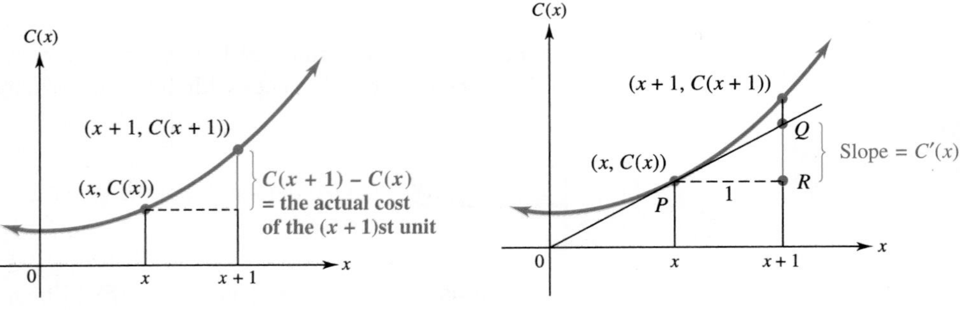

FIGURE 11.31 FIGURE 11.32

Now, if $C(x)$ is the cost function, then the marginal cost $C'(x)$ represents the slope of the tangent line at any point $(x, C(x))$. The graph in Figure 11.32 shows the cost function $C(x)$ and the tangent line at point $P = (x, C(x))$. We know that the slope of the tangent line is $C'(x)$ and that the slope can be computed using the triangle PQR in Figure 11.32:

$$C'(x) = \text{slope} = \frac{QR}{PR} = \frac{QR}{1} = QR.$$

So the length of the line segment QR is the number $C'(x)$.

Superimposing the graphs from Figures 11.31 and 11.32, as in Figure 11.33, shows that $C'(x)$ is indeed very close to $C(x + 1) - C(x)$. The two values are closest when $C'(x)$ is very large, so that 1 unit is relatively small.

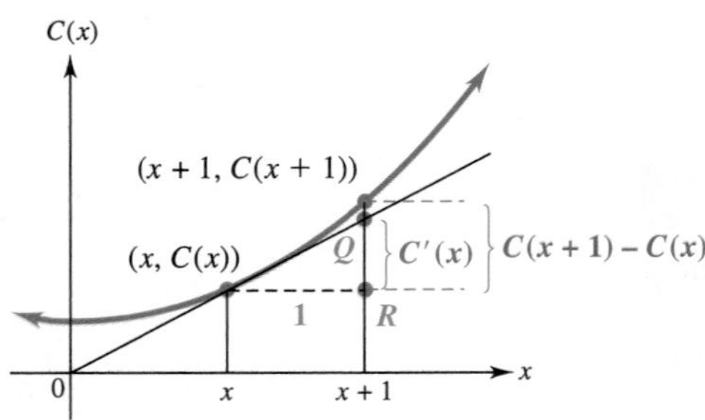

FIGURE 11.33

Therefore, we have the following conclusion.

Marginal Cost

If $C(x)$ is the cost function, then the marginal cost (rate of change of cost) is given by the derivative $C'(x)$:

$C'(x) \approx$ cost of making one more item after x items have been made.

The marginal revenue $R'(x)$ and marginal profit $P'(x)$ are interpreted similarly.

Example 5 **Business** Suppose that the total cost, in hundreds of dollars, of producing x thousand of barrels of wine is given by

$$C(x) = 4.2x^2 + 985.5x + 496 \qquad (0 \le x \le 50).$$

Find the marginal cost for the given values of x.

(a) $x = 12$

Solution To find the marginal cost, first find $C'(x)$, the derivative of the total cost function:

$$C'(x) = 8.4x + 985.5.$$

When $x = 12$,

$$C'(12) = 8.4(12) + 985.5 = 1086.3.$$

After 12 thousand barrels of wine have been produced, the cost of producing 1 thousand more barrels will be *approximately* 1083.6 hundred dollars, or \$108,360.

The *actual* cost of producing 1 thousand more barrels is $C(13) - C(12)$:

$$\begin{aligned} C(13) - C(12) &= (4.2 \cdot 13^2 + 985.5(13) + 496) - (4.2 \cdot 12^2 + 985.5(12) + 496) \\ &= 14{,}017.3 - 12{,}926.8 \\ &= 1090.5. \end{aligned}$$

The actual cost is 1090.5 hundred dollars, or \$109,050.

(b) $x = 40$

Solution After 40 thousand barrels have been produced, the cost of producing 1 thousand more barrels will be approximately

$$C'(40) = 8.4(40) + 985.5 = 1321.5,$$

or \$132,150. Compare this result with that for part (a). The cost of producing an additional 1 thousand barrels of wine is at least \$23,000 more at a production level of 40 thousand barrels than the cost of an additional 1 thousand barrels at a production level of 12 thousand barrels. Management must be careful to keep track of marginal costs. If the marginal cost of producing an extra unit exceeds the revenue received from selling it, then the company will lose money on that unit. 7 ✓

✓ Checkpoint 7

The cost in dollars to produce x units of wheat is given by

$C(x) = 5000 + 20x + 10\sqrt{x}$.

Find the marginal cost when

(a) $x = 9$;

(b) $x = 16$;

(c) $x = 25$.

(d) As more wheat is produced, what happens to the marginal cost?

Answers: See page 720.

DEMAND FUNCTIONS

The **demand function**, defined by $p = f(x)$, relates the number of units, x, of an item that consumers are willing to purchase at the price p. (Demand functions were also discussed in Section 3.3.) The total revenue $R(x)$ is related to the price per unit and the amount demanded (or sold) by the equation

$$R(x) = xp = x \cdot f(x).$$

Example 6 **Business** The demand function for a certain product is given by

$$p = \frac{50{,}000 - x}{25{,}000}.$$

Find the marginal revenue when $x = 10{,}000$ units and p is in dollars.

Solution From the function p, the revenue function is given by

$$\begin{aligned} R(x) &= xp \\ &= x\left(\frac{50{,}000 - x}{25{,}000}\right) \\ &= \frac{50{,}000x - x^2}{25{,}000} = 2x - \frac{1}{25{,}000}x^2. \end{aligned}$$

The marginal revenue is

$$R'(x) = 2 - \frac{2}{25{,}000}x.$$

When $x = 10{,}000$, the marginal revenue is

$$R'(10{,}000) = 2 - \frac{2}{25{,}000}(10{,}000) = 1.2,$$

or \$1.20 per unit. Thus, the next unit sold (at sales of 10,000) will produce additional revenue of about \$1.20. 8✓

✓**Checkpoint 8**

Suppose the demand function for x units of an item is

$$p = 5 - \frac{x}{1000},$$

where x is the price in dollars. Find

(a) the marginal revenue;

(b) the marginal revenue at $x = 500$;

(c) the marginal revenue at $x = 1000$.

Answers: See page 720.

In economics, the demand function is written in the form $p = f(x)$, as in Example 6. From the perspective of a consumer, it is probably more reasonable to think of the quantity demanded as a function of price. Mathematically, these two viewpoints are equivalent. In Example 6, the demand function could have been written from the consumer's viewpoint as

$$x = 50{,}000 - 25{,}000p.$$

Example 7 **Business** Suppose that the cost function for the product in Example 6 is given by

$$C(x) = 2100 + .25x \qquad (0 \le x \le 30{,}000).$$

Find the marginal profit from the production of the given numbers of units.

(a) 15,000

Solution From Example 6, the revenue from the sale of x units is

$$R(x) = 2x - \frac{1}{25{,}000}x^2.$$

Since profit P is given by $P = R - C$,

$$\begin{aligned} P(x) &= R(x) - C(x) \\ &= \left(2x - \frac{1}{25{,}000}x^2\right) - (2100 + .25x) \\ &= 2x - \frac{1}{25{,}000}x^2 - 2100 - .25x \\ &= 1.75x - \frac{1}{25{,}000}x^2 - 2100. \end{aligned}$$

The marginal profit from the sale of x units is

$$P'(x) = 1.75 - \frac{2}{25{,}000}x = 1.75 - \frac{1}{12{,}500}x.$$

At $x = 15{,}000$, the marginal profit is

$$P'(15{,}000) = 1.75 - \frac{1}{12{,}500}(15{,}000) = .55,$$

or \$.55 per unit.

(b) 21,875

Solution When $x = 21{,}875$, the marginal profit is

$$P'(21{,}875) = 1.75 - \frac{1}{12{,}500}(21{,}875) = 0.$$

(c) 25,000

Solution When $x = 25{,}000$, the marginal profit is

$$P'(25{,}000) = 1.75 - \frac{1}{12{,}500}(25{,}000) = -.25,$$

or $-$\$.25 per unit.

As shown by parts (b) and (c), if more than 21,875 units are sold, the marginal profit is negative. This indicates that increasing production beyond that level will *reduce* profit. 9 ✓

✓ Checkpoint 9

For a certain product, the cost is $C(x) = 1300 + .80x$ and the revenue is $R(x) = 6x - \frac{2x^2}{11{,}000}$ for x units.

(a) Find the profit $P(x)$.

(b) Find $P'(12{,}000)$.

(c) Find $P'(25{,}000)$.

(d) Interpret the results of parts (b) and (c).

Answers: See page 720.

The final example shows a medical application of the derivative as the rate of change of a function.

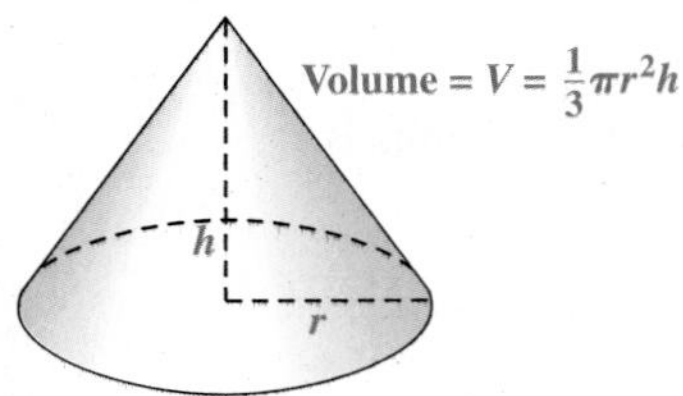

FIGURE 11.34

Example 8 **Health** A tumor has the approximate shape of a cone. (See Figure 11.34.) The radius of the tumor is fixed by the bone structure at 2 centimeters, but the tumor is growing along the height of the cone. The formula for the volume of a cone is $V = \frac{1}{3}\pi r^2 h$, where r is the radius of the base and h is the height of the cone. Find the rate of change of the volume of the tumor with respect to the height.

Solution To emphasize that the rate of change of the volume is found with respect to the height, we use the symbol dV/dh for the derivative. For this tumor, r is fixed at 2 cm. By substituting 2 for r,

$$V = \frac{1}{3}\pi r^2 h \quad \text{becomes} \quad V = \frac{1}{3}\pi \cdot 2^2 \cdot h, \quad \text{or} \quad V = \frac{4}{3}\pi h.$$

Since $4\pi/3$ is constant,

$$\frac{dV}{dh} = \frac{4\pi}{3} \approx 4.2 \text{ cu cm per cm.}$$

For each additional centimeter that the tumor grows in height, its volume will increase by approximately 4.2 cubic centimeters. 10

✓Checkpoint 10

A balloon is spherical. The formula for the volume of a sphere is $V = (4/3)\pi r^3$, where r is the radius of the sphere. Find the given quantities.

(a) dV/dr

(b) The rate of change of the volume when $r = 3$ inches

Answers: See page 720.

11.5 ▸ Exercises

Find the derivatives of the given functions. (See Examples 1–4.)

1. $f(x) = 4x^2 - 3x + 5$
2. $g(x) = 8x^2 + 2x - 12$
3. $y = 2x^3 + 3x^2 - 6x + 2$
4. $y = 4x^3 + 24x + 4$
5. $g(x) = x^4 + 3x^3 - 8x - 7$
6. $f(x) = 6x^6 - 3x^4 + x^3 - 3x + 9$
7. $f(x) = 6x^{1.5} - 4x^{.5}$
8. $f(x) = -2x^{2.5} + 8x^{.5}$
9. $y = -15x^{3/2} + 2x^{1.9} + x$
10. $y = 18x^{1.6} - 4x^{3.1} + x^4$
11. $y = 24t^{3/2} + 4t^{1/2}$
12. $y = -24t^{5/2} - 6t^{1/2}$
13. $y = 8\sqrt{x} + 6x^{3/4}$
14. $y = -100\sqrt{x} - 11x^{2/3}$
15. $g(x) = 6x^{-5} - 2x^{-1}$
16. $y = 8x^{-2} + 6x^{-1} + 1$
17. $y = 10x^{-2} + 6x^{-4} + 3x$
18. $y = 3x^{-7} + 4x^{-5} - 9x^{-4} + 21x^{-1}$
19. $f(t) = \frac{-4}{t} + \frac{8}{t^2}$
20. $f(t) = \frac{8}{t} - \frac{5}{t^2}$
21. $y = \frac{12 - 7x + 6x^3}{x^4}$
22. $y = \frac{102 - 6x + 17x^4}{x^6}$
23. $g(x) = 5x^{-1/2} - 7x^{1/2} + 101x$
24. $f(x) = -8x^{-1/2} - 8x^{1/2} + 12x$
25. $y = 7x^{-3/2} + 8x^{-1/2} + x^3 - 9$
26. $y = 2x^{-3/2} + 9x^{-1/2} + x^{-4} - x^2$
27. $y = \frac{19}{\sqrt[4]{x}}$
28. $y = \frac{-5}{\sqrt[3]{x}}$
29. $y = \frac{-8t}{\sqrt[3]{t^2}}$
30. $g(t) = \frac{10t}{\sqrt[3]{t^8}}$

Find each of the given derivatives.

31. $\frac{dy}{dx}$ if $y = 8x^{-5} - 9x^{-4} + 9x^4$
32. $\frac{dy}{dx}$ if $y = -3x^{-2} - 4x^{-5} + 5x^{-7}$
33. $D_x\left(9x^{-1/2} + \frac{2}{x^{3/2}}\right)$
34. $D_x\left(\frac{8}{\sqrt[4]{x}} - \frac{3}{\sqrt{x^3}}\right)$
35. $f'(-2)$ if $f(x) = 6x^2 - 2x$
36. $f'(3)$ if $f(x) = 9x^3 - 6x^2$
37. $f'(4)$ if $f(t) = 2\sqrt{t} - \frac{3}{\sqrt{t}}$
38. $f'(8)$ if $f(t) = -5\sqrt[3]{t} + \frac{6}{\sqrt[3]{t}}$

39. If $f(x) = -\frac{(3x^2 + x)^2}{7}$, which of the following is *closest* to $f'(1)$?

(a) -12 **(b)** -9
(c) -6 **(d)** -3
(e) 0 **(f)** 3

40. If $g(x) = -3x^{3/2} + 4x^2 - 9x$, which of the following is *closest* to $g'(4)$?

(a) 3 **(b)** 6 **(c)** 9
(d) 12 **(e)** 15 **(f)** 18

Find the slope and the equation of the tangent line to the graph of each function at the given value of x.

41. $f(x) = x^4 - 2x^2 + 1; x = 1$

42. $g(x) = -x^5 + 4x^2 - 2x + 2; x = 2$

43. $y = 4x^{1/2} + 2x^{3/2} + 1; x = 4$

44. $y = -x^{-3} + 5x^{-1} + x; x = 2$

Work these exercises. (See Examples 5 and 7.)

45. Business The profit in dollars from the sale of x expensive watches is

$$P(x) = .03x^2 - 4x + 3x^{.8} - 5000.$$

Find the marginal profit for the given values of x.

(a) $x = 100$ **(b)** $x = 1000$
(c) $x = 5000$ **(d)** $x = 10{,}000$

46. Business The total cost to produce x handcrafted weather vanes is

$$C(x) = 100 + 12x + .1x^2 + .001x^3.$$

Find the marginal cost for the given values of x.

(a) $x = 0$ **(b)** $x = 10$
(c) $x = 30$ **(d)** $x = 50$

47. Business Often, sales of a new product grow rapidly at first and then slow down with time. This is the case with the sales represented by the function

$$S(t) = 10{,}000 - 10{,}000t^{-.2} + 100t^{.1},$$

where t represents time in years. Find the rate of change of sales for the given values of t.

(a) $t = 1$ **(b)** $t = 10$

48. Business The revenue equation (in billions of dollars) for corn production in the United States is approximated by

$$R(x) = .016x^2 + 1.034x + 11.338,$$

where x is in billions of bushels.*

(a) Find the marginal-revenue equation.

(b) Find the marginal revenue for the production of 10 billion bushels.

(c) Find the marginal revenue for the production of 20 billion bushels.

49. Business The revenue equation (in hundreds of millions of dollars) for barley production in the United States is approximated by

$$R(x) = .0608x^2 + 1.4284x + 2.3418,$$

where x is in hundreds of millions of bushels.*

(a) Find the marginal-revenue equation.

(b) Find the marginal revenue for the production of 500,000,000 bushels.

(c) Find the marginal revenue for the production of 700,000,000 bushels.

50. Finance The total amount (in billions of dollars) of currency (coins and notes) for the United States can be approximated by

$$M(x) = .0052x^3 + .0943x^2 - .5983x + 33.7121,$$

where $x = 0$ corresponds to the year 1960.* Find the derivative of $M(x)$, and use it to find the rate of change of the currency in circulation in the given years.

(a) 1990 **(b)** 2010

(c) What do your answers to parts (a) and (b) tell you about the amount of currency in circulation in those years?

Work these exercises. (See Example 6.)

51. Business Assume that a demand equation is given by $x = 5000 - 100p$. Find the marginal revenue for the given production levels (values of x). (*Hint*: Solve the demand equation for p and use $R(x) = xp$.)

(a) 1000 units **(b)** 2500 units **(c)** 3000 units

52. Business Suppose that, for the situation in Exercise 51, the cost of producing x units is given by $C(x) = 3000 - 20x + .03x^2$. Find the marginal profit for each of the given production levels.

(a) 500 units **(b)** 815 units **(c)** 1000 units

Work these exercises. (See Examples 5–8).

53. Natural Science A short length of blood vessel has a cylindrical shape. The volume of a cylinder is given by $V = \pi r^2 h$. Suppose an experimental device is set up to measure the volume of blood in a blood vessel of fixed length 80 mm as the radius changes.

(a) Find dV/dr.

Suppose a drug is administered that causes the blood vessel to expand. Evaluate dV/dr for the following values of r, and interpret your answers:

(b) 4 mm; **(c)** 6 mm; **(d)** 8 mm.

---

*www.usda.gov.

*www.usda.gov.

†U.S. Federal Reserve.

54. **Social Science** According to the U.S. Census Bureau, the number of Americans (in thousands) who are expected to be over 100 years old in year x is approximated by the function

$$f(x) = .27x^2 + 3.52x + 51.78,$$

where $x = 0$ corresponds to 2000 and the formula is valid through 2045.
(a) Find a formula giving the rate of change in the number of Americans over 100 years old.
(b) What is the rate of change in the number of Americans expected to be over 100 years old in the year 2015?
(c) Is the number of Americans expected to be over 100 years old in 2015 increasing or decreasing?

55. **Finance** Living standards are defined by the total output of goods and services (gross domestic product), divided by the total population. In the United States during the 1990s and 2000s, living standards can be approximated by the function

$$f(x) = .00036x^4 - .015x^3 + .197x^2 - .257x + 28.31,$$

where $x = 0$ corresponds to the year 1990 and $f(x)$ is in millions of dollars.* Find the derivative of $f(x)$, and use it to find the rate of change in living standards in the given years.
(a) 2000 (b) 2009
(c) What do your answers to parts (a) and (b) tell you about living standards in those years?

56. **Natural Science** Workers in the insulation industry who were exposed to asbestos and employed before 1960 experienced an increased likelihood of lung cancer. If a group of insulation workers has a cumulative total of 100,000 years of work experience, with its first date of employment t years ago, then the number of lung cancer cases occurring within the group can be modeled with the function

$$N(t) = .00437t^{3.2}.†$$

Find the rate of growth of the number of workers with lung cancer in the group when the first date of employment is
(a) 5 years ago; (b) 10 years ago.

57. **Natural Science** The data in the accompanying table (from the Minneapolis *Star Tribune* of September 20, 1998) compares a dog's age with a human's. It can be modeled by the linear function

$$y_1 = 4.13x + 14.63$$

or the quadratic function

$$y_2 = -.033x^2 + 4.647x + 13.347,$$

where x is the dog's actual age and y_1 and y_2 is the dog's age in human years.*

Dog Years	Human Years
1	16
2	24
3	28
5	36
7	44
9	52
11	60
13	68
15	76

(a) Find y_1 and y_2 when $x = 5$.
(b) Find $\frac{dy_1}{dx}$ and $\frac{dy_2}{dx}$ when $x = 5$, and interpret your answers.
(c) If the first three rows are eliminated from the table, find the equation of a line that perfectly fits the reduced set of data. Interpret your findings.
(d) Of the three formulas, which do you prefer? Explain why.

58. **Health** The birth weight (in pounds) of an infant can be approximated by the function

$$g(x) = .0011x^{2.394},$$

where x represents the number of completed weeks of gestation.
(a) Find the function $g'(x)$.
(b) Evaluate $g'(32)$.
(c) Evaluate $g'(40)$.
(d) Interpret the results of parts (b) and (c).

59. **Health** To increase the velocity of the air flowing through the trachea when a human coughs, the body contracts the windpipe, producing a more effective cough. Tuchinsky determined that the velocity V of the air flowing through the trachea during a cough is given by

$$V = C(R_0 - R)R^2,$$

where C is a constant based on individual body characteristics, R_0 is the radius of the windpipe before the cough, and R is the radius of the windpipe during the cough.†
(a) Think of V as a function of R. Multiply out the rule for V and find $\frac{dV}{dR}$. (*Hint*: C and R_0 are constants here; if you cannot see what to do, it may help to consider the case when $C = 2$ and $R_0 = .15$.)

*Based on data from the U.S. Bureau of Economic Analysis and the U.S. Census Bureau.

†A. Walker, *Observation and Inference: An Introduction to the Methods of Epidemiology* (Epidemiology Resources Inc., 1991).

*Patrick Vennebush, "Media Clips: A Dog's Human Age," *Mathematics Teacher* 92 (1999): 710–712.

†Philip Tuchinsky, "The Human Cough," *UMAP Module 211* (Lexington, MA: COMAP, Inc, 1979): 1–9.

(b) It can be shown that the maximum velocity of the cough occurs when $\frac{dV}{dR} = 0$. Find the value of R that maximizes the velocity.* (Remember that R must be positive.)

60. **Health** The body mass index (BMI) is a number that can be calculated for any individual as follows: Multiply weight (in pounds) by 703 and divide by the person's height (in inches) squared. That is,

$$\text{BMI} = \frac{703w}{h^2},$$

where w is in pounds and h is in inches. The National Heart, Lung, and Blood Institute uses the BMI to determine whether a person is "overweight" ($25 \leq \text{BMI} < 30$) or "obese" ($\text{BMI} \geq 30$).

(a) Calculate the BMI for a male who weighs 220 pounds and is 6′2″ tall.

(b) How much weight would the person in part (a) have to lose until he reaches a BMI of 24.9 and is no longer "overweight"?

(c) For a 125-lb female, what is the rate of change of BMI with respect to height? (*Hint*: Take the derivative of the function $f(h) = \frac{703(125)}{h^2}$.)

(d) Calculate and interpret the meaning of $f'(65)$.

(e) Use the table function on a graphing calculator to construct a table for BMI for various weights and heights.

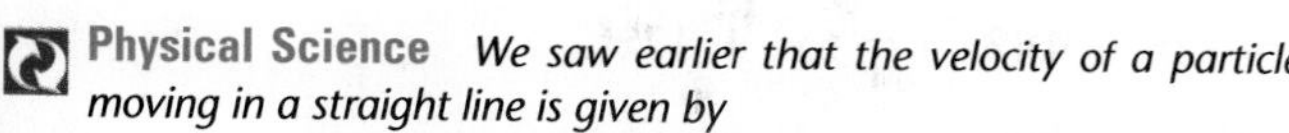
Physical Science *We saw earlier that the velocity of a particle moving in a straight line is given by*

$$\lim_{h \to 0} \frac{s(t+h) - s(t)}{h},$$

where $s(t)$ gives the position of the particle at time t. This limit is the derivative of $s(t)$, so the velocity of a particle is given by $s'(t)$. If $v(t)$ represents velocity at time t, then $v(t) = s'(t)$. For each of the given position functions, find (a) $v(t)$; (b) the velocity when $t = 0$, $t = 5$, and $t = 10$.

61. $s(t) = 8t^2 + 3t + 1$

62. $s(t) = 10t^2 - 5t + 6$

63. $s(t) = 2t^3 + 6t^2$

64. $s(t) = -t^3 + 3t^2 + t - 1$

65. **Physical Science** If a rock is dropped from a 144-foot-high building, its position (in feet above the ground) is given by $s(t) = -16t^2 + 144$, where t is the time in seconds since it was dropped.

(a) What is its velocity 1 second after being dropped? 2 seconds after being dropped?

(b) When will it hit the ground?

(c) What is its velocity upon impact?

66. **Physical Science** A ball is thrown vertically upward from the ground at a velocity of 64 feet per second. Its distance from the ground at t seconds is given by $s(t) = -16t^2 + 64t$.

(a) How fast is the ball moving 2 seconds after being thrown? 3 seconds after being thrown?

(b) How long after the ball is thrown does it reach its maximum height?

(c) How high will the ball go?

67. Perform each step and give reasons for your results in the following proof that the derivative of $y = x^n$ is $y' = n \cdot x^{n-1}$ (we prove this result only for positive integer values of n, but it is valid for all values of n):

(a) Recall the binomial theorem from algebra:

$$(p+q)^n = p^n + n \cdot p^{n-1}q + \frac{n(n-1)}{2}p^{n-2}q^2 + \cdots + q^n.$$

Evaluate $(x+h)^n$.

(b) Find the quotient $\frac{(x+h)^n - x^n}{h}$.

(c) Use the definition of the derivative to find y'.

68. Perform each step and give reasons for your results in the following proof that the derivative of $y = f(x) + g(x)$ is

$$y' = f'(x) + g'(x).$$

(a) Let $s(x) = f(x) + g(x)$. Show that

$$s'(x) = \lim_{h \to 0} \frac{[f(x+h) + g(x+h)] - [f(x) + g(x)]}{h}.$$

(b) Show that

$$s'(x) = \lim_{h \to 0} \left[\frac{f(x+h) - f(x)}{h} + \frac{g(x+h) - g(x)}{h} \right].$$

(c) Finally, show that $s'(x) = f'(x) + g'(x)$.

Use a graphing calculator or computer to graph each function and its derivative on the same screen. Determine the values of x where the derivative is (a) positive, (b) zero, and (c) negative. (d) What is true of the graph of the function in each case?

69. $g(x) = 6 - 4x + 3x^2 - x^3$

70. $k(x) = 2x^4 - 3x^3 + x$

✓ Checkpoint Answers

1. (a) $3x^2 + 3$ (b) $\frac{2}{x^2}$ (c) $\frac{-4}{(x-1)^2}$ (d) $\frac{1}{2\sqrt{x}}$

2. (a) 0 (b) 0 (c) 0

3. (a) $y' = 4x^3$ (b) $y' = 17x^{16}$
   (c) $dy/dx = -2/x^3$ (d) $dy/dt = -5/t^6$

*Interestingly, Tuchinsky also states that X-rays indicate that the body naturally contracts the windpipe to this radius during a cough.

4. (a) $36x^2$ (b) $210t^6$ (c) -35
(d) $(5/2)x^{-1/2}$, or $5/(2\sqrt{x})$ (e) $10t^{-2}$, or $10/t^2$

5. (a) $y' = -30x^2 - 6$

(b) $y' = \frac{40}{7}t^{1/7} + \frac{5}{2}t^{-1/2}$, or $y' = \frac{40}{7}t^{1/7} + \frac{5}{2t^{1/2}}$

(c) $f'(t) = \frac{-1}{\sqrt{t}} - \frac{4}{t^2}$

6. Both graphs look like this:

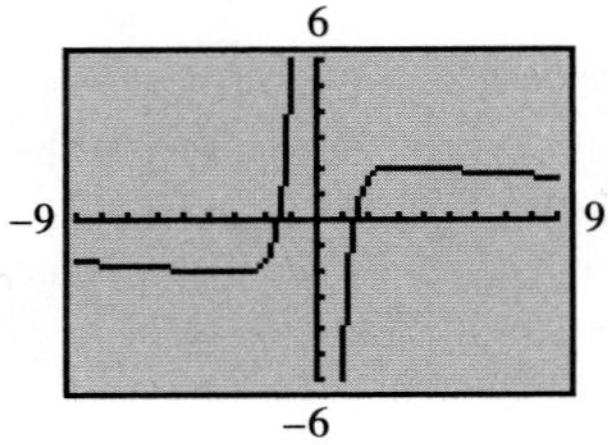

7. (a) $\$65/3 \approx \$21.67$ (b) $\$85/4 = \$21.25$
(c) \$21
(d) It decreases and approaches \$20.

8. (a) $R'(x) = 5 - \frac{x}{500}$ (b) \$4 (c) \$3

9. (a) $P(x) = 5.2x - \frac{2x^2}{11{,}000} - 1300$
(b) .84 (c) -3.89
(d) Profit is increasing by \$.84 per unit at 12,000 units in part (b) and decreasing by \$3.89 per unit at 25,000 units in part (c).

10. (a) $4\pi r^2$ (b) 36π cubic inches per inch

11.6 Derivatives of Products and Quotients

In the last section, we saw that the derivative of the sum of two functions can be obtained by taking the sum of the derivatives. What about products? Is the derivative of a product of two functions equal to the product of their derivatives? For example, if

$$u(x) = 2x + 4 \quad \text{and} \quad v(x) = 3x^2,$$

then the product of u and v is

$$f(x) = (2x + 4)(3x^2) = 6x^3 + 12x^2.$$

Using the rules of the last section, we have

$$u'(x) = 2, \quad v'(x) = 6x, \quad \text{and} \quad f'(x) = 18x^2 + 24x,$$

so that

$$u'(x) \cdot v'(x) = 12x \quad \text{and} \quad f'(x) = 18x^2 + 24x.$$

Obviously, these two functions are *not* the same, which shows that the derivative of the product is *not* equal to the product of the derivatives. For graphical confirmation of this fact, work Checkpoint 1. 1✓

✓Checkpoint 1

Let $u(x) = \ln x$ and $v(x) = \sqrt{x^2 + x}$. Using a graphing calculator and the viewing window with $0 \le x \le 5$ and $-2 \le y \le 4$, graph these two functions on the same screen:

$$\text{nDeriv}(u(x) \cdot v(x))$$

and

$$\text{nDeriv}(u(x)) \cdot \text{nDeriv}(v(x)).$$

Here, nDeriv denotes the numerical derivative. What can you conclude from the graphs?

Answer: See page 728.

The correct rule for finding the derivative of a product is as follows.

Product Rule

If $f(x) = u(x) \cdot v(x)$, and if both $u'(x)$ and $v'(x)$ exist, then

$$f'(x) = u(x) \cdot v'(x) + v(x) \cdot u'(x).$$

(The derivative of a product of two functions is the first function times the derivative of the second, plus the second function times the derivative of the first.)

To sketch the method used to prove the product rule, let

$$f(x) = u(x) \cdot v(x).$$

Then $f(x + h) = u(x + h) \cdot v(x + h)$, and, by definition,

$$\begin{aligned} f'(x) &= \lim_{h \to 0} \frac{f(x + h) - f(x)}{h} \\ &= \lim_{h \to 0} \frac{u(x + h) \cdot v(x + h) - u(x) \cdot v(x)}{h}. \end{aligned}$$

Now subtract and add $u(x + h) \cdot v(x)$ in the numerator, giving

$$f'(x) = \lim_{h \to 0} \frac{u(x + h) \cdot v(x + h) - u(x + h) \cdot v(x) + u(x + h) \cdot v(x) - u(x) \cdot v(x)}{h}.$$

Factor $u(x + h)$ from the first two terms of the numerator, and factor $v(x)$ from the last two terms. Then use the properties of limits (Section 11.1) as follows:

$$\begin{aligned} f'(x) &= \lim_{h \to 0} \frac{u(x + h)[v(x + h) - v(x)] + v(x)[u(x + h) - u(x)]}{h} \\ &= \lim_{h \to 0} u(x + h)\left[\frac{v(x + h) - v(x)}{h}\right] + \lim_{h \to 0} v(x)\left[\frac{u(x + h) - u(x)}{h}\right] \\ &= \lim_{h \to 0} u(x + h) \cdot \lim_{h \to 0} \frac{v(x + h) - v(x)}{h} + \lim_{h \to 0} v(x) \cdot \lim_{h \to 0} \frac{u(x + h) - u(x)}{h}. \quad (*) \end{aligned}$$

If u' and v' both exist, then

$$\lim_{h \to 0} \frac{u(x + h) - u(x)}{h} = u'(x) \quad \text{and} \quad \lim_{h \to 0} \frac{v(x + h) - v(x)}{h} = v'(x).$$

The fact that u' exists can be used to prove that

$$\lim_{h \to 0} u(x + h) = u(x),$$

and since no h is involved in $v(x)$,

$$\lim_{h \to 0} v(x) = v(x).$$

Substituting these results into equation $(*)$ gives

$$f'(x) = u(x) \cdot v'(x) + v(x) \cdot u'(x),$$

the desired result.

✓Checkpoint 2

Use the product rule to find the derivatives of the given functions.

(a) $f(x) = (5x^2 + 6)(3x)$

(b) $g(x) = (8x)(4x^2 + 5x)$

Answers: See page 728.

Example 1 Let $f(x) = (2x + 4)(3x^2)$. Use the product rule to find $f'(x)$.

Solution Here, f is given as the product of $u(x) = 2x + 4$ and $v(x) = 3x^2$. By the product rule and the fact that $u'(x) = 2$ and $v'(x) = 6x$,

$$\begin{aligned} f'(x) &= u(x) \cdot v'(x) + v(x) \cdot u'(x) \\ &= (2x + 4)(6x) + (3x^2)(2) \\ &= 12x^2 + 24x + 6x^2 = 18x^2 + 24x. \end{aligned}$$

This result is the same as that found at the beginning of the section. 2✓

Example 2 Find the derivative of $y = (\sqrt{x} + 3)(x^2 - 5x)$.

Solution Let $u(x) = \sqrt{x} + 3 = x^{1/2} + 3$ and $v(x) = x^2 - 5x$. Then

$$\begin{aligned} y' &= u(x) \cdot v'(x) + v(x) \cdot u'(x) \\ &= (x^{1/2} + 3)(2x - 5) + (x^2 - 5x)\left(\frac{1}{2}x^{-1/2}\right) \\ &= 2x^{3/2} + 6x - 5x^{1/2} - 15 + \frac{1}{2}x^{3/2} - \frac{5}{2}x^{1/2} \\ &= \frac{5}{2}x^{3/2} + 6x - \frac{15}{2}x^{1/2} - 15. \end{aligned}$$

3 ✓

✓ Checkpoint 3

Find the derivatives of the given functions.

(a) $f(x) = (x^2 - 3)(\sqrt{x} + 5)$

(b) $g(x) = (\sqrt{x} + 4)(5x^2 + x)$

Answers: See page 728.

We could have found the derivatives in Examples 1 and 2 by multiplying out the original functions. The product rule then would not have been needed. In the next section, however, we shall see products of functions where the product rule is essential.

What about *quotients* of functions? To find the derivative of the quotient of two functions, use the next rule.

Quotient Rule

If $f(x) = \dfrac{u(x)}{v(x)}$, if all indicated derivatives exist, and if $v(x) \neq 0$, then

$$f'(x) = \frac{v(x) \cdot u'(x) - u(x) \cdot v'(x)}{[v(x)]^2}.$$

(The derivative of a quotient is the denominator times the derivative of the numerator, minus the numerator times the derivative of the denominator, all divided by the square of the denominator.)

The proof of the quotient rule is similar to that of the product rule and is omitted here.

CAUTION Just as the derivative of a product is *not* the product of the derivatives, the derivative of a quotient is *not* the quotient of the derivatives. If you are asked to take the derivative of a product or a quotient, it is essential that you recognize that the function contains a product or quotient and then use the appropriate rule.

Example 3 Find $f'(x)$ if $f(x) = \dfrac{2x - 1}{4x + 3}$.

Solution Let $u(x) = 2x - 1$, with $u'(x) = 2$. Also, let $v(x) = 4x + 3$, with $v'(x) = 4$. Then, by the quotient rule,

$$f'(x) = \frac{v(x) \cdot u'(x) - u(x) \cdot v'(x)}{[v(x)]^2}$$

$$= \frac{(4x + 3)(2) - (2x - 1)(4)}{(4x + 3)^2}$$

$$= \frac{8x + 6 - 8x + 4}{(4x + 3)^2}$$

$$f'(x) = \frac{10}{(4x + 3)^2}.$$ 4 ✓

✓Checkpoint 4

Find the derivatives of the given functions.

(a) $f(x) = \frac{3x + 7}{5x + 8}$

(b) $g(x) = \frac{2x + 11}{5x - 1}$

Answers: See page 728.

CAUTION In the second step of Example 3, we had the expression

$$\frac{(4x + 3)(2) - (2x - 1)(4)}{(4x + 3)^2}.$$

Students often incorrectly "cancel" the $4x + 3$ in the numerator with one factor of the denominator. Because the numerator is a *difference* of two products, however, you must multiply and combine terms *before* looking for common factors in the numerator and denominator.

Example 4 Find $D_x\left(\frac{x - 2x^2}{4x^2 + 1}\right)$.

Solution Use the quotient rule:

$$D_x\left(\frac{x - 2x^2}{4x^2 + 1}\right) = \frac{(4x^2 + 1)D_x(x - 2x^2) - (x - 2x^2)D_x(4x^2 + 1)}{(4x^2 + 1)^2}$$

$$= \frac{(4x^2 + 1)(1 - 4x) - (x - 2x^2)(8x)}{(4x^2 + 1)^2}$$

$$= \frac{4x^2 - 16x^3 + 1 - 4x - 8x^2 + 16x^3}{(4x^2 + 1)^2}$$

$$= \frac{-4x^2 - 4x + 1}{(4x^2 + 1)^2}.$$ 5 ✓

✓Checkpoint 5

Find each derivative. Write your answer with positive exponents.

(a) $D_x\left(\frac{x^{-2} - 1}{x^{-1} + 2}\right)$

(b) $D_x\left(\frac{2 + x^{-1}}{x^3 + 1}\right)$

Answers: See page 728.

Example 5 Find $D_x\left(\frac{(3 - 4x)(5x + 1)}{7x - 9}\right)$.

Solution This function has a product within a quotient. Instead of multiplying the factors in the numerator first (which is an option), we can use the quotient rule together with the product rule. Use the quotient rule first to get

$$D_x\left(\frac{(3 - 4x)(5x + 1)}{7x - 9}\right)$$

$$= \frac{(7x - 9)[D_x(3 - 4x)(5x + 1)] - [(3 - 4x)(5x + 1)D_x(7x - 9)]}{(7x - 9)^2}.$$

Now use the product rule to find $D_x(3 - 4x)(5x + 1)$ in the numerator:

$$= \frac{(7x - 9)[(3 - 4x)5 + (5x + 1)(-4)] - (3 + 11x - 20x^2)(7)}{(7x - 9)^2}$$

$$= \frac{(7x - 9)(15 - 20x - 20x - 4) - (21 + 77x - 140x^2)}{(7x - 9)^2}$$

$$= \frac{(7x - 9)(11 - 40x) - 21 - 77x + 140x^2}{(7x - 9)^2}$$

$$= \frac{-280x^2 + 437x - 99 - 21 - 77x + 140x^2}{(7x - 9)^2}$$

$$= \frac{-140x^2 + 360x - 120}{(7x - 9)^2}.$$ 6✓

✓Checkpoint 6

Find each derivative.

(a) $D_x\left(\frac{(3x - 1)(4x + 2)}{2x}\right)$

(b) $D_x\left(\frac{5x^2}{(2x + 1)(x - 1)}\right)$

Answers: See page 728.

AVERAGE COST

Suppose $y = C(x)$ gives the total cost of manufacturing x items. As mentioned earlier, the average cost per item is found by dividing the total cost by the number of items. The rate of change of average cost, called the *marginal average cost*, is the derivative of the average cost.

Average Cost

If the total cost of manufacturing x items is given by $C(x)$, then the **average cost per item** is

$$\overline{C}(x) = \frac{C(x)}{x}.$$

The **marginal average cost** is the derivative of the average-cost function $\overline{C}'(x)$.

A company naturally would be interested in making the average cost as small as possible. We will see in the next chapter that this can be done by using the derivative of $C(x)/x$. The derivative often can be found with the quotient rule, as in the next example.

Example 6 **Business** The total cost (in dollars) to manufacture x mobile phones is given by

$$C(x) = \frac{50x^2 + 30x + 4}{x + 2} + 80{,}000.$$

(a) Find the average cost per phone.

Solution The average cost is given by the total cost divided by the number of items:

$$\overline{C}(x) = \frac{C(x)}{x} = \frac{1}{x}C(x) = \frac{1}{x}\left(\frac{50x^2 + 30x + 4}{x + 2} + 80{,}000\right)$$

$$= \frac{50x^2 + 30x + 4}{x^2 + 2x} + \frac{80{,}000}{x}$$

$$= \frac{50x^2 + 30x + 4}{x^2 + 2x} + 80{,}000x^{-1}.$$

(b) Find the average cost per phone for each of the following production levels:

5000; 10,000; 100,000.

Solution Evaluate $\overline{C}(x)$ at each of the numbers, either by hand or by using technology, as in Figure 11.35. Note that the average cost per phone is \$65.99 when 5000 phones are produced, and that reduces to \$50.80 when 100,000 are produced.

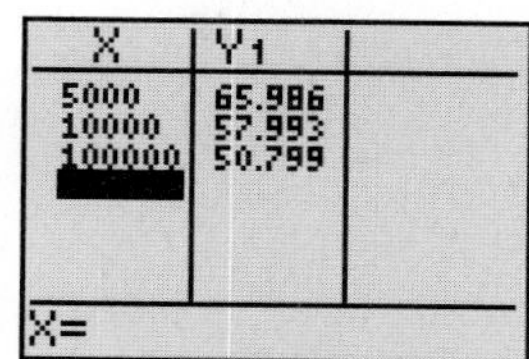

FIGURE 11.35

(c) Find the marginal average cost.

Solution The marginal average cost is the derivative of the average-cost function. Using the sum rule and the quotient rule yields

$$\overline{C}'(x) = \frac{(x^2 + 2x)(100x + 30) - (50x^2 + 30x + 4)(2x + 2)}{(x^2 + 2x)^2} + (-1)80{,}000x^{-2}$$

$$= \frac{(100x^3 + 230x^2 + 60x) - (100x^3 + 160x^2 + 68x + 8)}{(x^2 + 2x)^2} - \frac{80{,}000}{x^2}$$

$$= \frac{70x^2 - 8x - 8}{(x^2 + 2x)^2} - \frac{80{,}000}{x^2}.$$ 7

Checkpoint 7

The total cost (in dollars) to manufacture x satellite radios for automobiles is given by

$$C(x) = \frac{20x^2 + 5x + 2}{2x + 1} + 45{,}000.$$

(a) Find the average cost.

(b) Find the marginal average cost.

Answers: See page 728.

Example 7 Suppose the cost in dollars of manufacturing x hundred items is given by

$$C(x) = 4x^2 + 6x + 15.$$

(a) Find the average cost.

Solution The average cost is

$$\overline{C}(x) = \frac{C(x)}{x} = \frac{4x^2 + 6x + 15}{x} = 4x + 6 + \frac{15}{x}.$$

(b) Find the marginal average cost.

Solution The marginal average cost is

$$\frac{d}{dx}\left(\overline{C}(x)\right) = \frac{d}{dx}\left(4x + 6 + \frac{15}{x}\right) = 4 - \frac{15}{x^2}.$$

(c) Find the marginal cost.

Solution The marginal cost is

$$\frac{d}{dx}(C(x)) = \frac{d}{dx}\left(4x^2 + 6x + 15\right) = 8x + 6.$$

(d) Find the level of production at which the marginal average cost is zero.

Solution Set the derivative $\overline{C}'(x) = 0$ and solve for x:

$$4 - \frac{15}{x^2} = 0$$

$$\frac{4x^2 - 15}{x^2} = 0$$

$$4x^2 - 15 = 0$$

$$4x^2 = 15$$

$$x^2 = \frac{15}{4}$$

$$x = \pm\frac{\sqrt{15}}{2} \approx \pm 1.94.$$

You cannot make a negative number of items, so $x = 1.94$. Since x is in hundreds, the production of 194 items will yield a marginal average cost of zero dollars. 8✓

✓Checkpoint 8

If the cost function of Example 7 is given by $C(x) = x^2 + 12x + 9$, find the production level at which the marginal average cost is zero.

Answer: See page 728.

11.6 Exercises

Use the product rule to find the derivatives of the given functions. (See Examples 1 and 2.) (Hint for Exercises 6–9: Write the quantity as a product.)

1. $y = (x^2 - 2)(3x + 2)$
2. $y = (2x^2 + 3)(3x + 5)$
3. $y = (6x^3 + 2)(5x - 3)$
4. $y = (2x^2 + 4x - 3)(5x^3 + 2x + 2)$
5. $y = (x^4 - 2x^3 + 2x)(4x^2 + x - 3)$
6. $y = (3x - 4)^2$
7. $y = (5x^2 + 12x)^2$
8. $y = (3x^2 - 8)^2$
9. $y = (4x^3 + 6x^2)^2$

Use the quotient rule to find the derivatives of the given functions. (See Examples 3 and 4.)

10. $y = \frac{x + 1}{2x - 1}$
11. $y = \frac{3x - 5}{x - 3}$
12. $f(x) = \frac{7x + 1}{3x + 6}$
13. $f(t) = \frac{t^2 - 4t}{t + 3}$
14. $y = \frac{4x + 11}{x^2 - 3}$
15. $g(x) = \frac{3x^2 + x}{2x^3 - 2}$
16. $k(x) = \frac{-x^2 + 6x}{4x^3 + 3}$
17. $y = \frac{x^2 - 4x + 2}{x + 3}$
18. $y = \frac{x^2 + 7x - 2}{x - 2}$
19. $r(t) = \frac{\sqrt{t}}{3t + 4}$
20. $y = \frac{5x + 8}{\sqrt{x}}$
21. $y = \frac{9x - 8}{\sqrt{x}}$
22. $y = \frac{59}{4x + 11}$
23. $y = \frac{2 - 7x}{5 - x}$
24. $f(t) = \frac{2t^2 + t}{t - 9}$

Find the derivative of each of the given functions. (See Example 5.)

25. $f(p) = \frac{(6p - 7)(11p - 1)}{4p + 3}$
26. $f(t) = \frac{(8t - 3)(2t + 5)}{t - 7}$
27. $g(x) = \frac{x^3 - 8}{(3x + 9)(2x - 1)}$
28. $f(x) = \frac{x^3 - 1}{(4x - 1)(3x + 7)}$
29. Find the error in the following work:

$$D_x\left(\frac{2x + 5}{x^2 - 1}\right) = \frac{(2x + 5)(2x) - (x^2 - 1)2}{(x^2 - 1)^2}$$
$$= \frac{4x^2 + 10x - 2x^2 + 2}{(x^2 - 1)^2}$$
$$= \frac{2x^2 + 10x + 2}{(x^2 - 1)^2}.$$

30. Find the error in the following work:

$$D_x\left(\frac{x^2 - 4}{x^3}\right) = x^3(2x) - (x^2 - 4)(3x^2)$$
$$= 2x^4 - 3x^4 + 12x^2 = -x^4 + 12x^2.$$

Find the equation of the tangent line to the graph of f(x) at the given point.

31. $f(x) = \dfrac{x}{x-2}$ at (3, 3)

32. $f(x) = \dfrac{x}{x^2+2}$ at $\left(2, \frac{1}{3}\right)$

33. $f(x) = \dfrac{(x-1)(2x+3)}{x-5}$ at (9, 42)

34. $f(x) = \dfrac{(x+3)^2}{x+2}$ at $\left(-4, -\frac{1}{2}\right)$

Work these exercises. (See Examples 6 and 7.)

35. Business The total cost (in hundreds of dollars) to produce x toasters is

$$C(x) = \frac{4x+3}{x+5}.$$

Find the average cost for each of the following production levels.
(a) 15 toasters; **(b)** 25 toasters; **(c)** x toasters.
(d) Find the marginal average-cost function.

36. Business The total profit (in tens of dollars) from selling x cookbooks is

$$P(x) = \frac{7x-5}{3x+4}.$$

Find the average profit from each of the following sales levels:
(a) 12 books; **(b)** 20 books; **(c)** x books.
(d) Find the marginal average-profit function.

37. Business An old industrial factory site needs environmental remediation. The cost (in dollars) of removing x percent of the toxins in the soil is given by the cost–benefit function

$$C(x) = \frac{175{,}000x}{100-x}.$$

(a) What is the cost of removing 60% of the toxins? (Use $x = 60$, not .60).
(b) Find $C'(x)$.
Find the rate at which the cost of removing toxins is changing when the following percentages of toxins are removed:
(c) 80%; **(d)** 90%; **(e)** 95%.

38. Business Suppose you are the manager of a trucking firm and one of your drivers reports that, according to her calculations, her truck burns fuel at the rate of

$$G(x) = \frac{1}{200}\left(\frac{800}{x} + x\right)$$

gallons per mile when traveling at x miles per hour on a smooth, dry road.

(a) If the driver tells you that she wants to travel 20 miles per hour, what should you tell her? (*Hint*: Take the derivative of G and evaluate it for $x = 20$. Then interpret your results.)

(b) If the driver wants to go 40 miles per hour, what should you say? (*Hint*: Find $G'(40)$.)

39. Health During the course of an illness, a patient's temperature (in degrees Fahrenheit) x hours after the start of the illness is given by

$$T(x) = \frac{10x}{x^2+5} + 98.6.$$

(a) Find dT/dx.
Evaluate dT/dx at the following times, and interpret your answer:
(b) $x = 0$; **(c)** $x = 1$; **(d)** $x = 3$; **(e)** $x = 9$.

40. Physical Science The relationship between the fixed focal length F of a camera, the distance x from the object being photographed to the lens, and the distance y from the lens to the film is given by $\dfrac{1}{F} = \dfrac{1}{x} + \dfrac{1}{y}$.

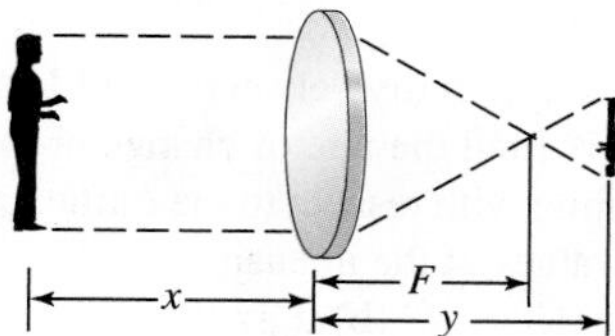

(a) Assume that the focal length is 50 mm ($F = 50$). Express y as a function of x.
(b) Find dy/dx.
Suppose the person being photographed moves away from the lens, always being kept in focus. Find the rate at which the distance y from the lens to the film is changing when the person is the following distance from the lens:
(c) 3100 mm (about 10 ft);
(d) 4900 mm (about 16 ft);
(e) 7600 mm (about 25 ft).

41. Natural Science Murrell's formula for calculating the total amount of rest, in minutes, required after performing a particular type of work activity for 30 minutes is given by the formula

$$R(w) = 30\frac{w-4}{w-1.5},$$

where w is the work expended in kilocalories per min (kcal/min).*

*Mark Sanders and Ernest McCormick, *Human Factors in Engineering and Design*, Seventh Edition (New York: McGraw-Hill, 1993), pp. 243–246.

(a) A value of 5 for w indicates light work, such as riding a bicycle on a flat surface at 10 miles per hour. Find $R(5)$.

(b) A value of 7 for w indicates moderate work, such as mowing grass with a push mower on level ground. Find $R(7)$.

(c) Find $R'(5)$ and $R'(7)$ and compare your answers. Explain whether these answers do or do not make sense.

42. Natural Science When a certain drug is introduced into a muscle, the muscle responds by contracting. The amount of contraction, s, in millimeters, is related to the concentration of the drug, x, in milliliters, by

$$s(x) = \frac{x}{m + nx},$$

where m and n are constants.

(a) Find $s'(x)$.

(b) Evaluate $s'(x)$ when $x = 50$, $m = 10$, and $n = 3$.

(c) Interpret your results for part (b).

43. Business The average number of vehicles waiting in line to enter a parking lot can be modeled by the function

$$f(x) = \frac{x^2}{2(1 - x)},$$

where x is a quantity between 0 and 1 known as the traffic intensity.* Find the rate of change of the number of vehicles waiting with respect to the traffic intensity for the following values of the intensity:

(a) $x = .1$; (b) $x = .6$.

44. Natural Science Using data collected by zoologist Reto Zach, researchers can estimate the work done by a crow to break open a whelk (a large marine snail) by the function

$$W = \left(1 + \frac{20}{H - .93}\right)H,$$

where H is the height of the whelk (in meters) when it is dropped.†

(a) Find $\frac{dW}{dH}$.

(b) The amount of work is minimized when $\frac{dW}{dH} = 0$. Find the value of H that minimizes W.

(c) Interestingly, Zach observed the crows dropping the whelks from an average height of 5.23 meters. What does this imply?

*F. Mannering and W. Kilareski, *Principles of Highway Engineering and Traffic Control*, Second Edition (John Wiley and Sons, 1997).

†Brian Kellar and Heather Thompson, "Whelk-come to Mathematics," *Mathematics Teacher* 92, no. 6 (September 1999): 475–481.

45. Social Science After t hours of instruction, a typical typing student can type

$$N(t) = \frac{70t^2}{30 + t^2}$$

words per minute (wpm).

(a) Find $N'(t)$, the rate at which the student is improving after t hours.

(b) At what rate is the student improving after 3 hours? 5 hours? 7 hours? 10 hours? 15 hours?

(c) Describe the student's progress during the first 15 hours of instruction.

46. Physical Science The intensity I of a 100-watt lightbulb (in watts per square meter) at a distance d meters from the bulb is given by

$$I(d) = \frac{7.92}{d^2}.$$

(a) Find $I'(d)$.

Find the rate at which the intensity is changing when an object is the following distance from the bulb:

(b) 3 meters; (c) 6 meters; (d) 9 meters.

✓ Checkpoint Answers

1. Since the two graphs are not the same, the derivative y' of the product function $y = (\ln x)\sqrt{x^2 + x}$ is not equal to the product of the derivatives, $u'(x) \cdot v'(x)$.

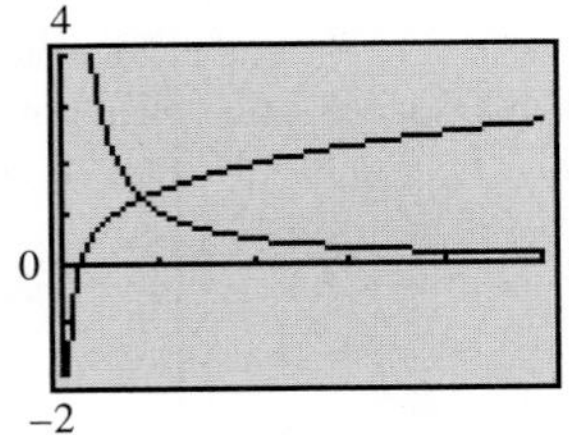

2. (a) $45x^2 + 18$ (b) $96x^2 + 80x$

3. (a) $\frac{5}{2}x^{3/2} + 10x - \frac{3}{2}x^{-1/2}$

(b) $\frac{25}{2}x^{3/2} + 40x + \frac{3}{2}x^{1/2} + 4$

4. (a) $\frac{-11}{(5x + 8)^2}$ (b) $\frac{-57}{(5x - 1)^2}$

5. (a) $\frac{-1 - 4x - x^2}{x^2 + 4x^3 + 4x^4}$ (b) $-\frac{6x^4 + 4x^3 + 1}{x^2(x^3 + 1)^2}$

6. (a) $\frac{6x^2 + 1}{x^2}$ (b) $\frac{-5x^2 - 10x}{(2x + 1)^2(x - 1)^2}$

7. (a) $\frac{20x^2 + 5x + 2}{2x^2 + x} + 45{,}000x^{-1}$

(b) $\frac{10x^2 - 8x - 2}{(2x^2 + x)^2} - \frac{45{,}000}{x^2}$

8. 300 items

11.7 ▶ The Chain Rule

Many of the most useful functions for applications are created by combining simpler functions. Viewing complex functions as combinations of simpler functions often makes them easier to understand and use.

COMPOSITION OF FUNCTIONS

Consider the function h whose rule is $h(x) = \sqrt{x^3}$. To compute $h(4)$, for example, you first find $4^3 = 64$ and then take the square root: $\sqrt{64} = 8$. So the rule of h may be rephrased as follows:

First apply the function $f(x) = x^3$.
Then apply the function $g(x) = \sqrt{x}$ to the preceding result.

The same idea can be expressed in functional notation like this:

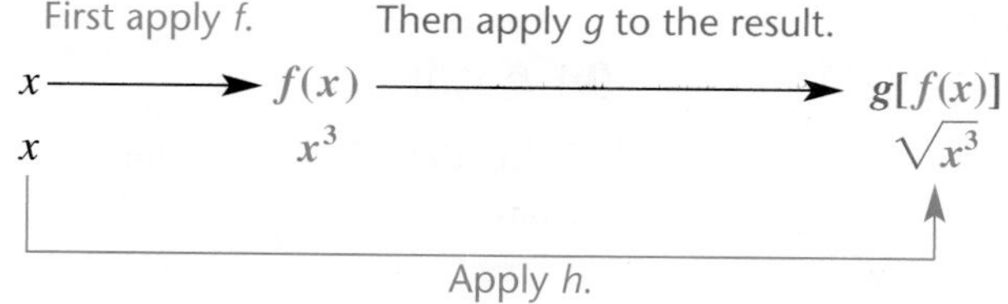

So the rule of h may be written as $h(x) = g[f(x)]$, where $f(x) = x^3$ and $g(x) = \sqrt{x}$. We can think of the functions g and f as being "composed" to create the function h. Here is a formal definition of this idea.

Composite Function

Let f and g be functions. The **composite function**, or **composition**, of g and f is the function whose values are given by $g[f(x)]$ for all x in the domain of f such that $f(x)$ is in the domain of g.

Example 1 Let $f(x) = 2x - 1$ and $g(x) = \sqrt{3x + 5}$. Find each of the following.

(a) $g[f(4)]$

Solution Find $f(4)$ first:

$$f(4) = 2 \cdot 4 - 1 = 8 - 1 = 7.$$

Then

$$g[f(4)] = g[7] = \sqrt{3 \cdot 7 + 5} = \sqrt{26}.$$

(b) $f[g(4)]$

Solution Since $g(4) = \sqrt{3 \cdot 4 + 5} = \sqrt{17}$,

$$f[g(4)] = 2 \cdot \sqrt{17} - 1 = 2\sqrt{17} - 1.$$

(c) $f[g(-2)]$

Solution This composite function does not exist, since -2 is not in the domain of g. 1 ✓

✓Checkpoint 1

Let $f(x) = 3x - 2$ and $g(x) = (x - 1)^5$. Find the following.

(a) $g[f(2)]$

(b) $f[g(2)]$

Answers: See page 740.

Example 2 Let $f(x) = 4x + 1$ and $g(x) = 2x^2 + 5x$. Find each of the following.

(a) $g[f(x)]$

Solution Use the given functions:

$$\begin{aligned} g[f(x)] &= g[4x + 1] \\ &= 2(4x + 1)^2 + 5(4x + 1) \\ &= 2(16x^2 + 8x + 1) + 20x + 5 \\ &= 32x^2 + 16x + 2 + 20x + 5 \\ &= 32x^2 + 36x + 7. \end{aligned}$$

(b) $f[g(x)]$

Solution By the definition of a composite function, with f and g interchanged, we have

$$\begin{aligned} f[g(x)] &= f[2x^2 + 5x] \\ &= 4(2x^2 + 5x) + 1 \\ &= 8x^2 + 20x + 1. \end{aligned}$$ 2 ✓

✓Checkpoint 2

Let $f(x) = \sqrt{x + 4}$ and $g(x) = x^2 + 5x + 1$. Find $f[g(x)]$ and $g[f(x)]$.

Answers: See page 740.

As Example 2 shows, $f[g(x)]$ usually is *not* equal to $g[f(x)]$. In fact, it is rare to find two functions f and g for which $f[g(x)] = g[f(x)]$.

It is often necessary to write a given function as the composite of two other functions, as is illustrated in the next example.

Example 3

(a) Express the function $h(x) = (x^3 + x^2 - 5)^4$ as the composite of two functions.

Solution One way to do this is to let $f(x) = x^3 + x^2 - 5$ and $g(x) = x^4$; then

$$g[f(x)] = g[x^3 + x^2 - 5] = (x^3 + x^2 - 5)^4 = h(x).$$

(b) Express the function $h(x) = \sqrt{4x^2 + 5}$ as the composite of two functions in two different ways.

Solution One way is to let $f(x) = 4x^2 + 5$ and $g(x) = \sqrt{x}$, so that

$$g[f(x)] = g[4x^2 + 5] = \sqrt{4x^2 + 5} = h(x).$$

Another way is to let $k(x) = 4x^2$ and $t(x) = \sqrt{x + 5}$; then

$$t[k(x)] = t[4x^2] = \sqrt{4x^2 + 5} = h(x).$$ 3 ✓

✓Checkpoint 3

Express the given function as a composite of two other functions.

(a) $h(x) = (7x^2 + 5)^4$

(b) $h(x) = \sqrt{15x^2 + 1}$

Answers: See page 740.

THE CHAIN RULE

The product and quotient rules tell how to find the derivative of fg and f/g from the derivatives of f and g. To develop a way to find the derivative of the composite function $f[g(x)]$ from the derivatives of f and g, we consider an example.

Metal expands or contracts as the temperature changes, and the temperature changes over a period of time. Think of the length y of a metal bar as a function of the temperature d, say, $y = f(d)$, and the temperature d as a function of time x, say, $d = g(x)$. Then

$$y = f(d) = f(g(x)).$$

So the composite function $f(g(x))$ gives the length y as a function of time x.

Suppose that the length of the bar is increasing at the rate of 2 mm for every degree increase in temperature and that the temperature is increasing at the rate of 4° per hour. In functional notation, this means that $f'(d) = 2$ and $g'(x) = 4$. During the course of an hour, the length of the bar will increase by $2 \cdot 4 = 8$ mm. So the rate of change of the length y with respect to time x is 8 mm/hr; that is, $y' = 8$. Consequently,

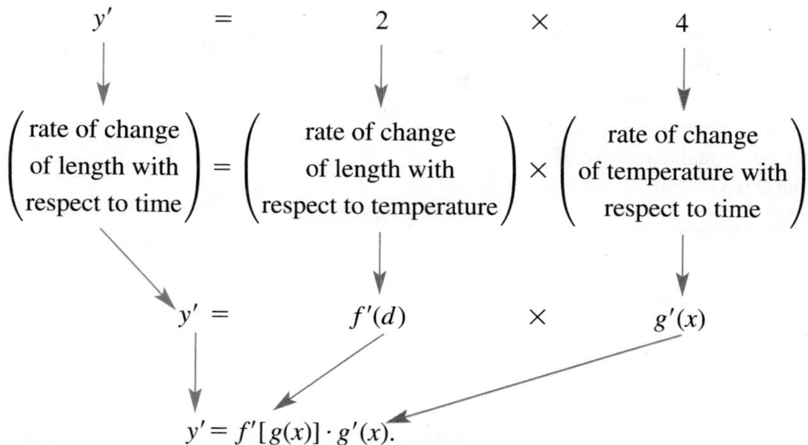

This example shows how the derivative of the composite function y is related to the derivatives of f and g. The same result holds for any composite function.

Chain Rule

If f and g are functions and $y = f[g(x)]$, then

$$y' = f'[g(x)] \cdot g'(x),$$

provided that $f'[g(x)]$ and $g'(x)$ exist.

(To find the derivative of $f[g(x)]$, find the derivative of $f(x)$, replace each x with $g(x)$, and multiply the result by the derivative of $g(x)$.)

Example 4 Use the chain rule to find $D_x\sqrt{15x^2 + 1}$.

Solution Write $\sqrt{15x^2 + 1}$ as $(15x^2 + 1)^{1/2}$. Let $f(x) = x^{1/2}$ and $g(x) = 15x^2 + 1$. Then $\sqrt{15x^2 + 1} = f[g(x)]$ and

$$D_x(15x^2 + 1)^{1/2} = f'[g(x)] \cdot g'(x).$$

Here, $f'(x) = \frac{1}{2}x^{-1/2}$, with $f'[g(x)] = \frac{1}{2}[g(x)]^{-1/2}$, and $g'(x) = 30x$ so that

$$\begin{aligned} D_x\sqrt{15x^2 + 1} &= \frac{1}{2}[g(x)]^{-1/2} \cdot g'(x) \\ &= \frac{1}{2}(15x^2 + 1)^{-1/2} \cdot (30x) \\ &= \frac{15x}{(15x^2 + 1)^{1/2}}. \end{aligned}$$

You can confirm this result graphically by graphing $y = 15x/(15x^2 + 1)^{1/2}$ and the numerical derivative of $\sqrt{15x^2 + 1}$ on the same screen; the graphs will appear identical. ✓4

Let $y = \sqrt{2x^4 + 3}$.

(a) Find dy/dx.

(b) Confirm your answer graphically by graphing dy/dx and the numerical derivative of y on the same screen.

Answers: See page 740.

The chain rule can also be stated in the Leibniz notation for derivatives. If y is a function of u, say, $y = f(u)$, and u is a function of x, say, $u = g(x)$, then

$$f'(u) = \frac{dy}{du} \quad \text{and} \quad g'(x) = \frac{du}{dx}.$$

Now y can be considered a function of x, namely, $y = f(u) = f(g(x))$. According to the chain rule, the derivative of y is

$$\frac{dy}{dx} = f'(g(x)) \cdot g'(x) = f'(u) \cdot g'(x) = \frac{dy}{du} \cdot \frac{du}{dx}.$$

Thus, we have the following alternative version of the chain rule.

The Chain Rule (Alternative Form)

If y is a function of u, say, $y = f(u)$, and if u is a function of x, say, $u = g(x)$, then $y = f(u) = f[g(x)]$ and

$$\frac{dy}{dx} = \frac{dy}{du} \cdot \frac{du}{dx},$$

provided that dy/du and du/dx exist.

One way to remember the chain rule is to *pretend* that dy/du and du/dx are fractions, with du "canceling out."

Example 5 Find dy/dx if $y = (3x^2 - 5x)^7$.

Solution Let $y = u^7$, where $u = 3x^2 - 5x$. Then $\frac{dy}{du} = 7u^6$ and $\frac{du}{dx} = 6x - 5$. Hence,

$$\begin{aligned} \frac{dy}{dx} &= \frac{dy}{du} \cdot \frac{du}{dx} \\ &= 7u^6 \cdot (6x - 5). \end{aligned}$$

Replacing u with $3x^2 - 5x$ gives

$$\frac{dy}{dx} = 7(3x^2 - 5x)^6(6x - 5).$$ 5✓

✓Checkpoint 5

Use the chain rule to find dy/dx if $y = 10(2x^2 + 1)^4$.

Answer: See page 740.

Each of the functions in Examples 4 and 5 has the form $y = u^n$, with u a function of x:.

$$y = (15x^2 + 1)^{1/2} \quad \text{and} \quad y = (3x^2 - 5x)^7.$$
$$y = u^{1/2} \qquad\qquad y = u^7.$$

Their derivatives are

$$y' = \frac{1}{2}(15x^2 + 1)^{-1/2}(30x) \quad \text{and} \quad y' = 7(3x^2 - 5x)^6(6x - 5)$$
$$y' = \frac{1}{2}u^{-1/2}\,u' \qquad\qquad y' = 7u^6\,u'.$$

Thus, these functions are examples of the following special case of the chain rule.

Generalized Power Rule

Let u be a function of x, and let $y = u^n$, for any real number n. Then

$$y' = n \cdot u^{n-1} \cdot u',$$

provided that u' exists.

(The derivative of $y = u^n$ is found by decreasing the exponent on u by 1 and multiplying the result by the exponent n and by the derivative of u with respect to x.)

Example 6

(a) Use the generalized power rule to find the derivative of $y = (3 + 5x)^2$.

Solution Let $u = 3 + 5x$ and $n = 2$. Then $u' = 5$. By the generalized power rule,

$$y' = \frac{dy}{dx} = n \cdot u^{n-1} \cdot u'$$

$$\begin{aligned} &\quad\ \ n \qquad u \quad n-1 \qquad u' \\ &= 2 \cdot (3 + 5x)^{2-1} \cdot \frac{d}{dx}(3 + 5x) \\ &= 2(3 + 5x)^{2-1} \cdot 5 \\ &= 10(3 + 5x) \\ &= 30 + 50x. \end{aligned}$$

(b) Find y' if $y = (3 + 5x)^{-3/4}$.

Solution Use the generalized power rule with $n = -\frac{3}{4}$, $u = 3 + 5x$, and $u' = 5$:

$$y' = -\frac{3}{4}(3 + 5x)^{(-3/4)-1}(5)$$
$$= -\frac{15}{4}(3 + 5x)^{-7/4}.$$ 6

Checkpoint 6

Find dy/dx for the given functions.

(a) $y = (2x + 5)^6$

(b) $y = (4x^2 - 7)^3$

(c) $f(x) = \sqrt{3x^2 - x}$

(d) $g(x) = (2 - x^4)^{-3}$

Answers: See page 740.

Example 7 Find the derivative of the given function.

(a) $y = 2(7x^2 + 5)^4$

Solution Let $u = 7x^2 + 5$. Then $u' = 14x$, and

$$y' = 2 \cdot \underset{n}{4}(\underset{u}{7x^2 + 5})^{\overset{}{4-1}} \cdot \underbrace{\frac{d}{dx}(7x^2 + 5)}_{u'}$$

(arrows label n, u, $n - 1$, u')

$$= 2 \cdot 4(7x^2 + 5)^3(14x)$$
$$= 112x(7x^2 + 5)^3.$$

(b) $y = \sqrt{9x + 2}$

Solution Write $y = \sqrt{9x + 2}$ as $y = (9x + 2)^{1/2}$. Then

$$y' = \frac{1}{2}(9x + 2)^{-1/2}(9) = \frac{9}{2}(9x + 2)^{-1/2}.$$

The derivative also can be written as

$$y' = \frac{9}{2(9x + 2)^{1/2}} \quad \text{or} \quad y' = \frac{9}{2\sqrt{9x + 2}}.$$ 7

Checkpoint 7

Find dy/dx for the given functions.

(a) $y = 12(x^2 + 6)^5$

(b) $y = 8(4x^2 + 2)^{3/2}$

Answers: See page 740.

CAUTION

(a) A common error is to forget to multiply by $g'(x)$ when using the generalized power rule. Remember, the generalized power rule is an example of the chain rule, so the derivative must involve a "chain," or product, of derivatives.

(b) Another common mistake is to write the derivative as $n[g'(x)]^{n-1}$. Remember to leave $g(x)$ unchanged and then to multiply by $g'(x)$.

Sometimes both the generalized power rule and either the product or quotient rule are needed to find a derivative, as the next two examples show.

Example 8 Find the derivative of $y = 4x(3x + 5)^5$.

Solution Write $4x(3x + 5)^5$ as the product

$$y = 4x \cdot (3x + 5)^5.$$

According to the product rule,

$$y' = 4x \cdot \frac{d}{dx}(3x + 5)^5 + \frac{d}{dx}(4x) \cdot (3x + 5)^5.$$

Use the generalized power rule to find the derivative of $(3x + 5)^5$;

$$\begin{aligned} y' &= 4x\underbrace{[5(3x + 5)^4 \cdot 3]}_{\text{Derivative of } (3x+5)^5} + \underbrace{(4)}_{\text{Derivative of } 4x}(3x + 5)^5 \\ &= 60x(3x + 5)^4 + 4(3x + 5)^5 \\ &= 4(3x + 5)^4[15x + (3x + 5)^1] \\ &= 4(3x + 5)^4(18x + 5). \end{aligned}$$

Factor out the greatest common factor, $4(3x + 5)^4$

8

✓Checkpoint 8

Find the derivatives of the given functions.

(a) $y = 9x(x + 7)^2$

(b) $y = -3x(4x^2 + 7)^3$

Answers: See page 740.

Example 9 Find the derivative of $y = \frac{(3x + 2)^7}{x - 1}$.

Solution Use the quotient rule and the generalized power rule:

$$\frac{dy}{dx} = \frac{(x - 1) \cdot \frac{d}{dx}(3x + 2)^7 - (3x + 2)^7 \frac{d}{dx}(x - 1)}{(x - 1)^2}$$ Quotient rule

$$= \frac{(x - 1)[7(3x + 2)^6 \cdot 3] - (3x + 2)^7(1)}{(x - 1)^2}$$ Generalized power rule

$$= \frac{21(x - 1)(3x + 2)^6 - (3x + 2)^7}{(x - 1)^2}$$

$$= \frac{(3x + 2)^6[21(x - 1) - (3x + 2)]}{(x - 1)^2}$$ Factor out the greatest common factor, $(3x + 2)^6$.

$$= \frac{(3x + 2)^6[21x - 21 - 3x - 2]}{(x - 1)^2}$$ Simplify inside brackets.

$$= \frac{(3x + 2)^6(18x - 23)}{(x - 1)^2}.$$

9

✓Checkpoint 9

Find the derivatives of the given functions.

(a) $y = \frac{(5x + 2)^3}{11x}$

(b) $y = \frac{(x - 12)^7}{6x + 1}$

Answers: See page 740.

APPLICATIONS

Some applications requiring the use of the chain rule or the generalized power rule are illustrated in the next three examples.

Example 10 **Business** The revenue from installing granite countertops earned by a kitchen renovation contractor is given by

$$R(n) = \frac{500(n^2 + 25)}{n + 12},$$

where n is the number of granite slabs imported from Brazil. Usually, the supplier in Brazil ships 25 slabs per week to the kitchen contractor. If the supplier in Brazil cuts back on the number of slabs shipped to the kitchen contractor at rate of 5 per week, then $dn/dt = -5$. Under this situation, how fast is revenue decreasing for the kitchen contractor after two weeks?

Solution We want to find dR/dt, the change in revenue with respect to time. By the chain rule,

$$\frac{dR}{dt} = \frac{dR}{dn} \cdot \frac{dn}{dt}.$$

First, find dR/dn as follows:

$$\frac{dR}{dn} = \frac{(n + 12)(500 \cdot 2n) - 500(n^2 + 25)}{(n + 12)^2} = \frac{500n^2 + 12{,}000n - 12{,}500}{(n + 12)^2}.$$

After two weeks of decreasing slabs at a rate of 5 per week, we have $n = 15$, so

$$\frac{dR}{dn} = \frac{500(15)^2 + 12{,}000(15) - 12{,}500}{(15 + 12)^2} = 384.09.$$

Since, $dn/dt = -5$, we have

$$\frac{dR}{dt} = (384.09)(-5) = -1920.45.$$

Revenue is being lost at a rate of \$1920.45 after two weeks. 10✓

✓Checkpoint 10

Suppose the revenue function for Example 10 is given by $R(n) = \frac{750(n^2 + 20)}{n + 10}$ and the slabs decrease at a rate of 4 per week. How fast is the revenue decreasing after one week?

Answer: See page 740.

Example 11 **Business** An elderly aunt deposits \$20,000 in an account to be used by her newly born niece to attend college. The account earns interest at the rate of r percent per year, compounded monthly. At the end of 18 years, the balance in the account is given by

$$A = 20{,}000\left(1 + \frac{r}{1200}\right)^{216}.$$

Find the rate of change of A with respect to r if $r = 1.5$, 2.5, or 3.

Solution First find dA/dr, using the generalized power rule:

$$\frac{dA}{dr} = (216)(20{,}000)\left(1 + \frac{r}{1200}\right)^{215}\left(\frac{1}{1200}\right) = 3600\left(1 + \frac{r}{1200}\right)^{215}.$$

If $r = 1.5$, we obtain

$$\frac{dA}{dr} = 3600\left(1 + \frac{1.5}{1200}\right)^{215} = 4709.19,$$

or \$4709.19 per percentage point. If $r = 2.5$, we obtain

$$\frac{dA}{dr} = 3600\left(1 + \frac{2.5}{1200}\right)^{215} = 5631.55,$$

or \$5631.55 per percentage point. If $r = 3$, we obtain

$$\frac{dA}{dr} = 3600\left(1 + \frac{3}{1200}\right)^{215} = 6158.07,$$

or \$6158.07 per percentage point. 11✓

✓Checkpoint 11

If the initial donation in Example 11 is \$25,000, find the rate of change of A with respect to r if $r = 2$.

Answer: See page 740.

The chain rule can be used to develop the formula for the **marginal-revenue product**, an economic concept that approximates the change in revenue when a manufacturer hires an additional employee. Start with $R = px$, where R is total revenue from the daily production of x units and p is the price per unit. The demand function is $p = f(x)$, as before. Also, x can be considered a function of the number of employees, n. Since $R = px$, and x—and therefore, p—depends on n, R can also be considered a function of n. To find an expression for dR/dn, use the product rule for derivatives on the function $R = px$ to get

$$\frac{dR}{dn} = p \cdot \frac{dx}{dn} + x \cdot \frac{dp}{dn}. \qquad (*)$$

By the chain rule,

$$\frac{dp}{dn} = \frac{dp}{dx} \cdot \frac{dx}{dn}.$$

Substituting for dp/dn in equation $(*)$ yields

$$\frac{dR}{dn} = p \cdot \frac{dx}{dn} + x\left(\frac{dp}{dx} \cdot \frac{dx}{dn}\right)$$

$$= \left(p + x \cdot \frac{dp}{dx}\right)\frac{dx}{dn}. \qquad \text{Factor out } \frac{dx}{dn}.$$

The expression for dR/dn gives the marginal-revenue product.

Example 12 Find the marginal-revenue product dR/dn (in dollars) when $n = 20$ if the demand function is $p = 600/\sqrt{x}$ and $x = 5n$.

Solution As shown previously,

$$\frac{dR}{dn} = \left(p + x \cdot \frac{dp}{dx}\right)\frac{dx}{dn}.$$

Find dp/dx and dx/dn. From

$$p = \frac{600}{\sqrt{x}} = 600x^{-1/2},$$

we have the derivative

$$\frac{dp}{dx} = -300x^{-3/2}.$$

Also, from $x = 5n$, we have

$$\frac{dx}{dn} = 5.$$

Then, by substitution,

$$\frac{dR}{dn} = \left[\frac{600}{\sqrt{x}} + x(-300x^{-3/2})\right]5 = \left[\frac{600}{\sqrt{x}} - \frac{300}{\sqrt{x}}\right]5 = \frac{1500}{\sqrt{x}}.$$

✓Checkpoint 12

Find the marginal-revenue product at $n = 10$ if the demand function is $p = 1000/x^2$ and $x = 8n$. Interpret your answer.

Answer: See page 740.

If $n = 20$, then $x = 100$, and

$$\frac{dR}{dn} = \frac{1500}{\sqrt{100}} = 150.$$

This means that hiring an additional employee when production is at a level of 100 items will produce an increase in revenue of $150. 12✓

11.7 ▶ Exercises

Let $f(x) = 2x^2 + 3x$ and $g(x) = 4x - 1$. Find each of the given composite functions. (See Example 1.)

1. $f[g(5)]$
2. $f[g(-4)]$
3. $g[f(5)]$
4. $g[f(-4)]$

Find $f[g(x)]$ and $g[f(x)]$ for the given functions. (See Example 2.)

5. $f(x) = 8x + 12;\ g(x) = 2x + 3$
6. $f(x) = -6x + 9;\ g(x) = 5x + 7$
7. $f(x) = -x^3 + 2;\ g(x) = 4x + 2$
8. $f(x) = 2x;\ g(x) = 6x^2 - x^3 - 1$
9. $f(x) = \frac{1}{x};\ g(x) = x^2$
10. $f(x) = \frac{2}{x^4};\ g(x) = 2 - x$
11. $f(x) = \sqrt{x + 2};\ g(x) = 8x^2 - 6x$
12. $f(x) = 9x^2 - 11x;\ g(x) = 2\sqrt{x + 2}$

Write each function as a composition of two functions. (There may be more than one way to do this.) (See Example 3.)

13. $y = (4x + 3)^5$
14. $y = (x^2 + 2)^{1/5}$
15. $y = \sqrt{6 + 3x^2}$
16. $y = \sqrt{x + 3} - \sqrt[3]{x + 3}$
17. $y = \frac{\sqrt{x} + 3}{\sqrt{x} - 3}$
18. $y = \frac{2x}{\sqrt{x} + 5}$
19. $y = (x^{1/2} - 3)^3 + (x^{1/2} - 3) + 5$
20. $y = (x^2 + 5x)^{1/3} - 2(x^2 + 5x)^{2/3} + 7$

Find the derivative of each of the given functions. (See Examples 4–7.)

21. $y = (7x - 12)^5$
22. $y = (13x - 3)^4$
23. $y = 7(4x + 3)^4$
24. $y = -4(3x - 2)^7$
25. $y = 11(3x + 5)^{3/2}$
26. $y = -3(x^4 + 7x^2)^3$
27. $y = -4(10x^2 + 8)^5$
28. $y = 13(4x - 2)^{2/3}$

Find the derivative of the function. (See Example 4.)

29. $y = -7(4x^2 + 9x)^{3/2}$
30. $y = 11(5x^2 + 6x)^{3/2}$
31. $y = 8\sqrt{4x + 7}$
32. $y = -3\sqrt{7x - 1}$
33. $y = -2\sqrt{x^2 + 4x}$
34. $y = 4\sqrt{2x^2 + 3}$

Use the product or quotient rule or the generalized power rule to find the derivative of each of the given functions. (See Examples 8 and 9.)

35. $y = (x + 1)(x - 3)^2$
36. $y = (2x + 1)^3(x - 5)$
37. $y = 5(x + 3)^2(2x - 1)^5$
38. $y = -9(x + 4)^2(2x - 3)^3$
39. $y = (3x + 1)^3\sqrt{x}$
40. $y = (3x + 5)^2\sqrt{x + 1}$
41. $y = \frac{1}{(x - 2)^3}$
42. $y = \frac{-3}{(4x + 1)^2}$
43. $y = \frac{(2x - 7)^2}{4x - 1}$
44. $y = \frac{(x - 4)^2}{5x + 2}$
45. $y = \frac{x^2 + 5x}{(3x + 1)^3}$
46. $y = \frac{4x^3 - x}{(x - 3)^2}$
47. $y = (x^{1/2} + 2)(2x + 3)$
48. $y = (4 - x^{2/3})(x + 1)^{1/2}$

Consider the following table of values of the functions f and g and their derivatives at various points:

x	1	2	3	4
$f(x)$	2	4	1	3
$f'(x)$	−6	−7	−8	−9
$g(x)$	2	3	4	1
$g'(x)$	2/7	3/7	4/7	5/7

Find each of the given derivatives.

49. **(a)** $D_x(f[g(x)])$ at $x = 1$ **(b)** $D_x(f[g(x)])$ at $x = 2$
50. **(a)** $D_x(g[f(x)])$ at $x = 1$ **(b)** $D_x(g[f(x)])$ at $x = 2$
51. If $f(x) = (2x^2 + 3x + 1)^{50}$, then which of the following is closest to $f'(0)$?
    (a) 1 **(b)** 50 **(c)** 100
    (d) 150 **(e)** 200 **(f)** 250
52. The graphs of $f(x) = 3x + 5$ and $g(x) = 4x - 1$ are straight lines.
    (a) Show that the graph of $f[g(x)]$ is also a straight line.
    (b) How are the slopes of the graphs of $f(x)$ and $g(x)$ related to the slope of the graph of $f[g(x)]$?

Work these exercises. (See Examples 10, 11, and 12.)

53. Business The cost of producing x bags of dog food is given by

$$C(x) = 600 + \sqrt{200 + 25x^2 - x} \quad (0 \le x \le 5000).$$

Find the marginal-cost function.

54. Business A vendor's weekly profit from the sales of x souvenir T-shirts at a popular tourist site is given by

$$P(x) = (x^3 + 22x - 150)^{1/3} - 600 \quad (0 \le x \le 7500).$$

(a) Use a calculator to find $P(100)$, $P(500)$, $P(2000)$, and $P(5000)$.
(b) Explain why it is reasonable that some of the numbers found in part (a) are negative.
(c) Find the marginal-profit function.

55. Business The revenue function from the sale of x ear-buds for portable music players is given by

$$R(x) = 15\sqrt{500x - 2x^2} \quad (0 \le x \le 200).$$

(a) Find the marginal-revenue function.
(b) Evaluate the marginal-revenue function at $x = 40$, 80, 120, and 160.
(c) Explain the significance of the answers found in part (b).

56. Finance An amount of \$10,000 is deposited in an account with an interest rate of r percent per year, compounded daily. At the end of 6 years, the balance in the account is given by

$$A = 10{,}000\left(1 + \frac{r}{36{,}500}\right)^{2190}.$$

Find the rate of change of A with respect to r (written as a percentage) for the given interest rates.
(a) 2% **(b)** 2.5% **(c)** 3%

57. Finance An amount of \$15,000 is deposited in an account with an interest rate of r percent per year, compounded monthly. At the end of 8 years, the balance in the account is given by

$$A = 15{,}000\left(1 + \frac{r}{1200}\right)^{96}.$$

Find the rate of change of A with respect to r (written as a percentage) for the given interest rates.
(a) 2.5% **(b)** 3.25% **(c)** 4%

58. Business Assume that the total revenue from the sale of x lawn mowers is given by

$$R(x) = 250{,}000\sqrt[3]{\frac{x}{220}}.$$

Find the marginal revenue when the given number of mowers is sold.
(a) 700 **(b)** 1200 **(c)** 1600
(d) Find the average revenue from the sale of x mowers.
(e) Find the marginal average revenue.

59. Business Find the marginal-revenue product for a manufacturer with 10 workers if the demand function is $p = 350/x^{1/2}$ and if $x = 10n$.

60. Natural Science An oil well off the Gulf Coast is leaking, spreading oil in a circle over the surface. At any time t, in minutes, after the beginning of the leak, the radius of the circular oil slick on the surface is $r(t) = t^2$ feet. Let $A(r) = \pi r^2$ represent the area of a circle of radius r. Find and interpret $A[r(t)]$.

61. Natural Science When there is a thermal inversion layer over a city (as happens often in Los Angeles), pollutants cannot rise vertically, but are trapped below the layer and must disperse horizontally. Assume that a factory smokestack begins emitting a pollutant at 8 A.M. Assume that the pollutant disperses horizontally, forming a circle. If t represents the time, in hours, since the factory began emitting pollutants ($t = 0$ represents 8 A.M.), assume that the radius of the circle of pollution is $r(t) = 2t$ miles. Let $A(r) = \pi r^2$ represent the area of a circle of radius r. Find and interpret $A[r(t)]$.

62. Business Suppose a demand function is given by

$$x = 30\left(5 - \frac{p}{\sqrt{p^2 + 1}}\right),$$

where x is the demand for a product and p is the price per item in dollars. Find the rate of change in the demand for the product (i.e., find dx/dp).

63. Business Suppose the demand for a certain brand of vacuum cleaner is given by

$$D(p) = \frac{-p^2}{100} + 500,$$

where p is the price in dollars. If the price, in terms of the cost c, is expressed as

$$p(c) = 2c - 10,$$

find the demand in terms of the cost.

64. Business A certain truck depreciates according to the formula

$$V = \frac{6000}{1 + .3t + .1t^2},$$

where t is time measured in years and $t = 0$ represents the time of purchase. Find the rate at which the value of the truck is changing at the given times.
(a) 2 years **(b)** 4 years

65. Business Suppose the cost, in dollars, of manufacturing x items is given by

$$C = 2000x + 3500$$

and the demand equation is given by

$$x = \sqrt{15{,}000 - 1.5p}.$$

In terms of the demand x,

(a) find an expression for the revenue R;
(b) find an expression for the profit P;
(c) find an expression for the marginal profit;
(d) determine the value of the marginal profit when the price is \$25.

66. Natural Science The strength of a person's reaction to a certain drug is given by

$$R(Q) = Q\left(C - \frac{Q}{3}\right)^{1/2},$$

where Q represents the quantity of the drug given to the patient and C is a constant. The derivative $R(Q)$ is called the *sensitivity* to the drug.

(a) Find $R'(Q)$.
(b) Find $R'(Q)$ if $Q = 87$ and $C = 59$.

67. Natural Science The volume and surface area of a jawbreaker candy of any radius r are given by the formulas

$$V(r) = \frac{4}{3}\pi r^3 \quad \text{and} \quad S(r) = 4\pi r^2,$$

respectively. It is estimated that the radius of a jawbreaker while in a person's mouth is

$$r(t) = 6 - \frac{3}{17}t,$$

where $r(t)$ is in mm and t is in min.*

(a) What is the life expectancy of a jawbreaker?
(b) Find $\frac{dV}{dt}$ and $\frac{dS}{dt}$ when $t = 17$, and interpret your answer.
(c) Construct an analogous experiment using some other type of food, or verify the results of this experiment.

68. Natural Science To test an individual's use of calcium, a researcher injects a small amount of radioactive calcium into the person's bloodstream. The calcium remaining in the bloodstream is measured each day for several days. Suppose the amount of the calcium remaining in the bloodstream, in milligrams per cubic centimeter, t days after the initial injection is approximated by

$$C(t) = \frac{1}{2}(2t + 1)^{-1/2}.$$

Find the rate of change of C with respect to time for each of the given number of days.

(a) $t = 0$ **(b)** $t = 4$ **(c)** $t = 6$ **(d)** $t = 7.5$

69. Social Science Studies show that after t hours on the job, the number of items a supermarket cashier can scan per minute is given by

$$F(t) = 60 - \frac{150}{\sqrt{8 + t^2}}.$$

(a) Find $F'(t)$, the rate at which the cashier's speed is increasing.
(b) At what rate is the cashier's speed increasing after 5 hours? 10 hours? 20 hours? 40 hours?
(c) Are your answers in part (b) increasing or decreasing with time? Is this reasonable? Explain.

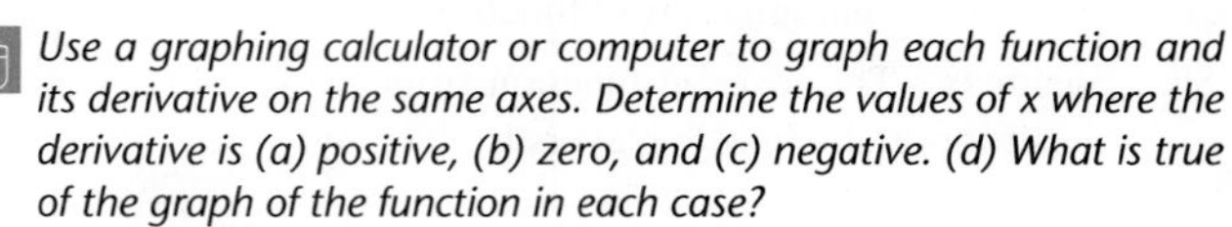

Use a graphing calculator or computer to graph each function and its derivative on the same axes. Determine the values of x where the derivative is (a) positive, (b) zero, and (c) negative. (d) What is true of the graph of the function in each case?

70. $G(x) = \dfrac{2x}{(x - 1)^2}$ **71.** $K(x) = \sqrt[3]{(2x - 1)^2}$

✓ Checkpoint Answers

1. **(a)** 243 **(b)** 1
2. $f[g(x)] = \sqrt{x^2 + 5x + 5}$;
   $g[f(x)] = (\sqrt{x + 4})^2 + 5\sqrt{x + 4} + 1$
   $= x + 5 + 5\sqrt{x + 4}$
3. There are several correct answers, including
   (a) $h(x) = f[g(x)]$, where $f(x) = x^4$ and $g(x) = 7x^2 + 5$;
   (b) $h(x) = f[g(x)]$, where $f(x) = \sqrt{x}$ and $g(x) = 15x^2 + 1$.
4. **(a)** $\dfrac{dy}{dx} = \dfrac{4x^3}{\sqrt{2x^4 + 3}}$
   (b) The two graphs appear to be identical:

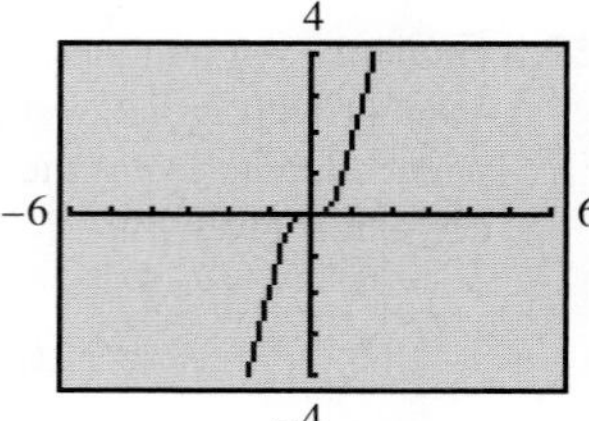

5. $160x(2x^2 + 1)^3$
6. **(a)** $12(2x + 5)^5$ **(b)** $24x(4x^2 - 7)^2$
   (c) $\dfrac{6x - 1}{2\sqrt{3x^2 - x}}$ **(d)** $\dfrac{12x^3}{(2 - x^4)^4}$
7. **(a)** $120x(x^2 + 6)^4$ **(b)** $96x(4x^2 + 2)^{1/2}$
8. **(a)** $9(x + 7)(3x + 7)$ **(b)** $-21(4x^2 + 7)^2(4x^2 + 1)$
9. **(a)** $\dfrac{11(5x + 2)^2(10x - 2)}{121x^2}$ **(b)** $\dfrac{(x - 12)^6(36x + 79)}{(6x + 1)^2}$
10. The revenue is decreasing at a rate of \$2625.39.
11. \$6437.32 per percentage point
12. −\$1.25; hiring an additional employee will produce a decrease in revenue of \$1.25.

*Roger Guffey, "The Life Expectancy of a Jawbreaker: An Application of the Composition of Functions," *Mathematics Teacher* 92, no. 2 (February 1999): 125–127.

11.8 ▸ Derivatives of Exponential and Logarithmic Functions

The exponential function $f(x) = e^x$ and the logarithmic function to the base e, $g(x) = \ln x$, were studied in Chapter 4. (Recall that $e \approx 2.71828$.) In this section, we shall find the derivatives of these functions. In order to do that, we must first find a limit that will be needed in our calculations.

We claim that

$$\lim_{h \to 0} \frac{e^h - 1}{h} = 1.$$

Although a rigorous proof of this fact is beyond the scope of this book, a graphing calculator provides both graphical and numerical support, as shown in Figure 11.36.*

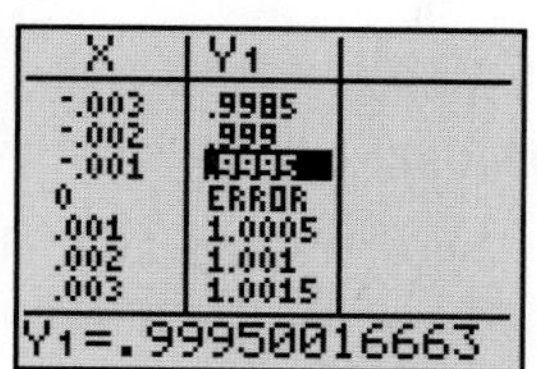

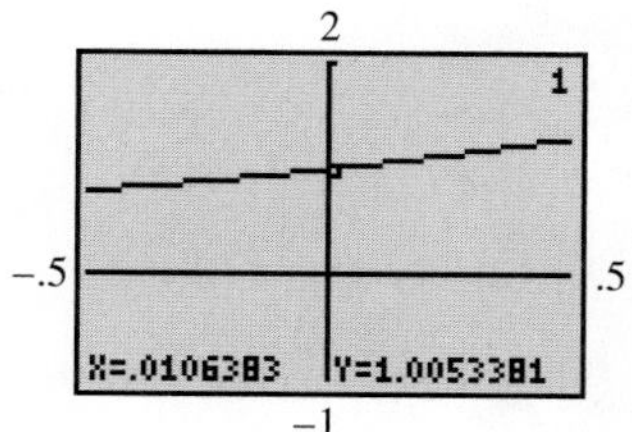

$$y_1 = \frac{e^h - 1}{h}$$

FIGURE 11.36

To find the derivative of $f(x) = e^x$, we use the definition of the derivative function,

$$f'(x) = \lim_{h \to 0} \frac{f(x + h) - f(x)}{h},$$

provided that this limit exists. (Remember that h is the variable here and x is treated as a constant.)

For $f(x) = e^x$, we see that

$$\begin{aligned} f'(x) &= \lim_{h \to 0} \frac{e^{x+h} - e^x}{h} \\ &= \lim_{h \to 0} \frac{e^x e^h - e^x}{h} && \text{Product property of exponents} \\ &= \lim_{h \to 0} \frac{e^x(e^h - 1)}{h} \\ &= \lim_{h \to 0} e^x \cdot \lim_{h \to 0} \frac{e^h - 1}{h}, && \text{Product property of limits} \end{aligned}$$

provided that the last two limits exist. But h is the variable here and x is constant; therefore,

$$\lim_{h \to 0} e^x = e^x.$$

*As usual, the variable X is used in place of h on a calculator.

Combining this fact with our previous work, we see that

$$f'(x) = \lim_{h\to 0} e^x \cdot \lim_{h\to 0} \frac{e^h - 1}{h} = e^x \cdot 1 = e^x.$$

In other words, *the exponential function* $f(x) = e^x$ *is its own derivative.*

Example 1 Find each derivative.

(a) $y = x^3e^x$

Solution The product rule and the fact that $f(x) = e^x$ is its own derivative show that

$$\begin{aligned} y' &= x^3 \cdot D_x(e^x) + D_x(x^3) \cdot e^x \\ &= x^3 \cdot e^x + 3x^2 \cdot e^x = e^x(x^3 + 3x^2). \end{aligned}$$

(b) $y = (2e^x + x)^5$

Solution By the generalized power, sum, and constant rules,

$$\begin{aligned} y' &= 5(2e^x + x)^4 \cdot D_x(2e^x + x) \\ &= 5(2e^x + x)^4 \cdot [D_x(2e^x) + D_x(x)] \\ &= 5(2e^x + x)^4 \cdot [2D_x(e^x) + 1] \\ &= 5(2e^x + x)^4(2e^x + 1). \end{aligned}$$ 1 ✓

✓Checkpoint 1

Differentiate the given expressions.

(a) $(2x^2 - 1)e^x$

(b) $(1 - e^x)^{1/2}$

Answers: See page 750.

Example 2 Find the derivative of $y = e^{x^2-3x}$.

Solution Let $f(x) = e^x$ and $g(x) = x^2 - 3x$. Then

$$y = e^{x^2-3x} = e^{g(x)} = f[g(x)],$$

and $f'(x) = e^x$ and $g'(x) = 2x - 3$. By the chain rule,

$$\begin{aligned} y' &= f'[g(x)] \cdot g'(x) \\ &= e^{g(x)} \cdot (2x - 3) \\ &= e^{x^2-3x} \cdot (2x - 3) = (2x - 3)e^{x^2-3x}. \end{aligned}$$

The argument used in Example 2 can be used to find the derivative of $y = e^{g(x)}$ for any differentiable function g. Let $f(x) = e^x$; then $y = e^{g(x)} = f[g(x)]$. By the chain rule,

$$y' = f'[g(x)] \cdot g'(x) = e^{g(x)} \cdot g'(x) = g'(x)e^{g(x)}.$$

We summarize these results as follows.

Derivatives of e^x and $e^{g(x)}$

If $y = e^x$, then $y' = e^x$.

If $y = e^{g(x)}$, then $y' = g'(x) \cdot e^{g(x)}$.

CAUTION Notice the difference between the derivative of a variable to a constant power, such as $D_x x^3 = 3x^2$, and the derivative of a constant to a variable power, like $D_x e^x = e^x$. Remember, $D_x e^x \neq xe^{x-1}$.

Example 3 Find derivatives of the given functions.

(a) $y = e^{7x}$

Solution Let $g(x) = 7x$, with $g'(x) = 7$. Then

$$y' = g'(x)e^{g(x)} = 7e^{7x}.$$

(b) $y = 5e^{-3x}$

Solution Here, let $g(x) = -3x$ and $g'(x) = -3$, so that

$$y' = 5(-3e^{-3x}) = -15e^{-3x}.$$

(c) $y = 8e^{4x^3}$

Solution Let $g(x) = 4x^3$, with $g'(x) = 12x^2$. Then

$$y' = 12x^2(8e^{4x^3}) = 96x^2e^{4x^3}.$$ 2 ✓

✓Checkpoint 2

Find the derivative of the given function.

(a) $y = 4e^{11x}$

(b) $y = -5e^{(-9x+2)}$

(c) $y = e^{-x^3}$

Answers: See page 750.

Example 4 Let $y = \dfrac{250{,}000}{2 + 15e^{-4x}}$. Find y'.

Solution Use the quotient rule:

$$y' = \frac{(2 + 15e^{-4x})(0) - 250{,}000(-60e^{-4x})}{(2 + 15e^{-4x})^2}$$

$$= \frac{15{,}000{,}000e^{-4x}}{(2 + 15e^{-4x})^2}.$$ 3 ✓

✓Checkpoint 3

Find the derivative of the given function.

(a) $y = \dfrac{e^x}{12 + x}$

(b) $y = \dfrac{9000}{15 + 3e^x}$

Answers: See page 750.

Example 5 A 100-gram sample of a radioactive substance decays exponentially. The amount left after t years is given by $A(t) = 100e^{-.12t}$. Find the rate of change of the amount after 3 years.

Solution The rate of change is given by the derivative dA/dt:

$$\frac{dA}{dt} = 100(e^{-.12t})(-.12) = -12e^{-.12t}.$$

After 3 years ($t = 3$), the rate of change is

$$\frac{dA}{dt} = -12e^{-.12(3)} = -12e^{-.36} \approx -8.4$$

grams per year.

Example 6 If $y = 10^x$, find y'.

Solution By one of the basic properties of logarithms (page 235), we have $10 = e^{\ln 10}$. Therefore,

$$y = 10^x = (e^{\ln 10})^x = e^{(\ln 10)x},$$

and

$$\begin{aligned} y' &= \frac{d}{dx}[(\ln 10)x] \cdot e^{(\ln 10)x} && \text{By the preceding box, with } g(x) = (\ln 10)x \\ &= (\ln 10)e^{(\ln 10)x} && \text{Remember that ln 10 is a constant.} \\ &= (\ln 10)(e^{\ln 10})^x \\ &= (\ln 10)10^x. && e^{\ln 10} = 10 \end{aligned}$$

✓Checkpoint 4

Adapt the solution of Example 6 to find the derivative of $y = 25^x$.

Answer: See page 750.

DERIVATIVES OF LOGARITHMIC FUNCTIONS

To find the derivative of $g(x) = \ln x$, we use the definitions and properties of natural logarithms that were developed in Section 4.3:

$$g(x) = \ln x \qquad \text{means} \qquad e^{g(x)} = x,$$

and for all $x > 0$, $y > 0$, and every real number r,

$$\ln xy = \ln x + \ln y, \qquad \ln\frac{x}{y} = \ln x - \ln y, \qquad \text{and} \qquad \ln x^r = r\ln x.$$

Note that we must have $x > 0$ and $y > 0$ because logarithms of negative numbers are not defined.

Differentiating with respect to x on each side of $e^{g(x)} = x$ shows that

$$D_x(e^{g(x)}) = D_x(x)$$
$$e^{g(x)} \cdot g'(x) = 1.$$

Because $e^{g(x)} = x$, this last equation becomes

$$x \cdot g'(x) = 1$$
$$g'(x) = \frac{1}{x}$$
$$\frac{d}{dx}(\ln x) = \frac{1}{x} \qquad \text{for all } x > 0.$$

Example 7

(a) Assume that $x > 0$, and use properties of logarithms to find the derivative of $y = \ln 6x$.

Solution

$$\begin{aligned} y' &= \frac{d}{dx}(\ln 6x) \\ &= \frac{d}{dx}(\ln 6 + \ln x) && \text{Product rule for logarithms} \\ &= \frac{d}{dx}(\ln 6) + \frac{d}{dx}(\ln x). && \text{Sum rule for derivatives} \end{aligned}$$

Be careful here: ln 6 is a *constant* ($\ln 6 \approx 1.79$), so its derivative is 0 (*not* $\frac{1}{6}$). Hence,

$$y' = \frac{d}{dx}(\ln 6) + \frac{d}{dx}(\ln x) = 0 + \frac{1}{x} = \frac{1}{x}.$$

(b) Assume that $x > 0$, and use the chain rule to find the derivative of $y = \ln 6x$.

Solution Let $f(x) = \ln x$ and $g(x) = 6x$, so that $y = \ln 6x = \ln g(x) = f(g(x))$. Then, by the chain rule,

$$y' = f'[g(x)] \cdot g'(x) = \frac{1}{g(x)} \cdot \frac{d}{dx}(6x) = \frac{1}{6x} \cdot 6 = \frac{1}{x}.$$

The argument used in Example 6(b) applies equally well in the general case. The derivative of $y = \ln g(x)$, where $g(x)$ is a function and $g(x) > 0$, can be found by letting $f(x) = \ln x$, so that $y = f[g(x)]$, and applying the chain rule:

$$y' = f'[g(x)] \cdot g'(x) = \frac{1}{g(x)} \cdot g'(x) = \frac{g'(x)}{g(x)}.$$

We summarize these results as follows.

Derivatives of ln x and ln $g(x)$

If $y = \ln x$, then $y' = \frac{1}{x}$ $\quad (x > 0)$.

If $y = \ln g(x)$, then $y' = \frac{g'(x)}{g(x)}$ $\quad (g(x) > 0)$.

Example 8 Find the derivatives of the given functions.

(a) $y = \ln(3x^2 - 4x)$

Solution Let $g(x) = 3x^2 - 4x$, so that $g'(x) = 6x - 4$. From the second formula in the preceding box,

$$y' = \frac{g'(x)}{g(x)} = \frac{6x - 4}{3x^2 - 4x}.$$

(b) $y = 3x \ln x^2$

Solution Since $3x \ln x^2$ is the product of $3x$ and $\ln x^2$, use the product rule:

$$y' = (3x)\left(\frac{d}{dx} \ln x^2\right) + (\ln x^2)\left(\frac{d}{dx} 3x\right)$$

$$= 3x\left(\frac{2x}{x^2}\right) + (\ln x^2)(3) \qquad \text{Take derivatives.}$$

$$= 6 + 3 \ln x^2$$

$$= 6 + \ln (x^2)^3 \qquad \text{Property of logarithms}$$

$$= 6 + \ln x^6. \qquad \text{Property of exponents}$$

(c) $y = \ln[(x^2 + x + 1)(4x - 3)^5]$

Solution Here, we use the properties of logarithms before taking the derivative (the same thing could have been done in part (b) by writing $\ln x^2$ as $2 \ln x$):

$$\begin{aligned} y &= \ln[(x^2 + x + 1)(4x - 3)^5] \\ &= \ln(x^2 + x + 1) + \ln(4x - 3)^5 && \text{Property of logarithms} \\ &= \ln(x^2 + x + 1) + 5\ln(4x - 3); && \text{Property of logarithms} \\ y' &= \frac{2x + 1}{x^2 + x + 1} + 5 \cdot \frac{4}{4x - 3} && \text{Take derivatives.} \\ &= \frac{2x + 1}{x^2 + x + 1} + \frac{20}{4x - 3}. \end{aligned}$$

5 ✓

✓**Checkpoint 5**

Find y' for the given functions.

(a) $y = \ln(7 + x)$

(b) $y = \ln(4x^2)$

(c) $y = \ln(8x^3 - 3x)$

(d) $y = x^2 \ln x$

Answers: See page 750.

The function $y = \ln(-x)$ is defined for all $x < 0$ (since $-x > 0$ when $x < 0$.) Its derivative can be found by applying the derivative rule for $\ln g(x)$ with $g(x) = -x$:

$$\begin{aligned} y' &= \frac{g'(x)}{g(x)} \\ &= \frac{-1}{-x} \\ &= \frac{1}{x}. \end{aligned}$$

This is the same as the derivative of $y = \ln x$, with $x > 0$. Since

$$|x| = \begin{cases} x & \text{if } x > 0 \\ -x & \text{if } x < 0 \end{cases},$$

we can combine two results into one in the manner that follows.

> If $y = \ln|x|$, then $y' = \dfrac{1}{x}$ $\quad (x \neq 0)$.

Example 9 Let $y = e^x \ln|x|$. Find y'.

Solution Use the product rule:

$$y' = e^x \cdot \frac{1}{x} + \ln|x| \cdot e^x = e^x\left(\frac{1}{x} + \ln|x|\right).$$

6 ✓

✓**Checkpoint 6**

Find the derivative of each function.

(a) $y = e^{x^2} \ln|x|$

(b) $y = x^2/\ln|x|$

Answers: See page 750.

Example 10 If $f(x) = \log x$, find $f'(x)$.

Solution By the change-of-base theorem (page 237),

$$f(x) = \log x = \frac{\ln x}{\ln 10} = \frac{1}{\ln 10} \cdot \ln x.$$

Since $1/\ln 10$ is a constant,

$$f'(x) = \frac{1}{\ln 10} \cdot \frac{d}{dx}(\ln x) = \frac{1}{\ln 10} \cdot \frac{1}{x} = \frac{1}{(\ln 10)x}.$$

Often, a population or the sales of a certain product will start growing slowly, then grow more rapidly, and then gradually level off. Such growth can frequently be approximated by a *logistic function* of the form

$$f(x) = \frac{c}{1 + ae^{kx}}$$

for appropriate constants a, c, and k. (Logistic functions were introduced in Section 4.2.)

Example 11 **Business** The number of subscribers (in millions) to cable television programming is approximated by

$$S(x) = \frac{68.0}{1 + 19.3e^{-.2x}},$$

where $x = 5$ corresponds to the year 1975.* At what rate is the number of subscribers changing in 2010?

Solution Use the quotient rule to find the derivative of $S(x)$:

$$S'(x) = \frac{(1 + 19.3e^{-.2x})(0) - 68.0(19.3(-.2)e^{-.2x})}{(1 + 19.3e^{-.2x})^2}$$

$$= \frac{262.48e^{-.2x}}{(1 + 19.3e^{-.2x})^2}.$$

The rate of change in 2010 ($x = 40$) is

$$S'(40) = \frac{262.48e^{-.2(40)}}{(1 + 19.3e^{-.2(40)})^2} \approx .086923 \text{ million per year.}$$

This means that subscribers are increasing at the rate of approximately 86,923 per year. 7 ✓

*Based on data from the *The Cable Financial Databook*.

✓Checkpoint 7

Suppose a deer population is given by

$$f(x) = \frac{10,000}{1 + 2e^{x}},$$

where x is time in years. Find the rate of change of the population when

(a) $x = 0$;

(b) $x = 5$.

(c) Is the population increasing or decreasing?

Answers: See page 750.

11.8 ▸ Exercises

Find the derivatives of the given functions. (See Examples 1–5 and 7–9.)

1. $y = e^{5x}$

2. $y = e^{-4x}$

3. $f(x) = 5e^{2x}$

4. $f(x) = 4e^{-2x}$

5. $g(x) = -4e^{-7x}$

6. $g(x) = 6e^{x/2}$

7. $y = e^{x^2}$

8. $y = e^{-x^2}$

9. $f(x) = e^{x^3/3}$

10. $y = 4e^{2x^2-4}$

11. $y = -3e^{3x^2+5}$

12. $y = xe^{3x}$

13. $y = \ln(-10x^2 + 7x)$

14. $y = \ln\sqrt{x - 8}$

15. $y = \ln\sqrt{5x + 1}$

16. $y = \ln[(4x + 1)(5x - 2)]$

17. $f(x) = \ln[(5x + 9)(x^2 + 1)]$

18. $f(x) = \ln\left(\frac{7x - 1}{12x + 5}\right)$
19. $y = x^4e^{-3x}$
20. $y = (x - 5)^2e^{3x}$
21. $y = (6x^2 - 5x)e^{-3x}$
22. $y = \ln(11 - 2x)$
23. $y = \ln\left(\frac{4 - x}{3x + 8}\right)$
24. $y = \ln(x^5 + 8x^2)^{3/2}$
25. $y = \ln(8x^3 - 3x)^{1/2}$
26. $y = -4x \ln(x + 12)$
27. $y = x \ln(9 - x^4)$
28. $y = (2x^3 - 1) \ln|x|$
29. $y = \frac{\ln|x|}{x^4}$
30. $y = \frac{3 \ln|x|}{3x + 4}$
31. $y = \frac{-5 \ln|x|}{5 - 2x}$
32. $y = \frac{3x^3}{\ln|x|}$
33. $y = \frac{x^3 - 1}{2 \ln|x|}$
34. $y = [\ln(x + 1)]^4$
35. $y = \sqrt{\ln(x - 3)}$
36. $y = \frac{e^x}{\ln|x|}$
37. $y = \frac{e^x - 1}{\ln|x|}$
38. $y = \frac{e^x + e^{-x}}{x}$
39. $y = \frac{e^x - e^{-x}}{x}$
40. $y = e^{x^3} \ln|x|$
41. $f(x) = e^{3x+2} \ln(4x - 5)$
42. $f(x) = \frac{2400}{3 + 8e^{.2x}}$
43. $y = \frac{700}{7 - 10e^{.4x}}$
44. $y = \frac{10{,}000}{9 + 4e^{-.2x}}$
45. $y = \frac{500}{12 + 5e^{-.5x}}$
46. $y = \ln(\ln|x|)$

Find the derivatives of the given functions. (See Examples 6 and 10.) The chain rule is needed in Exercises 48–52.

47. $y = 8^x$
48. $y = 8^{x^2}$
49. $y = 15^{2x}$
50. $f(x) = 2^{-x}$
51. $g(x) = \log 6x$
52. $y = \log(x^2 + 1)$

For Exercises 53–56, find the equation of the tangent line to the graph of f at the given point.

53. $f(x) = \frac{e^x}{2x + 1}$ at $(0, 1)$
54. $f(x) = \ln(1 + x^2)$ at $(0, 0)$
55. $f(x) = \frac{x^2}{e^x}$ when $x = 1$
56. $f(x) = \ln(2x^2 - 7x)$ when $x = -1$
57. If $f(x) = e^{2x}$, find $f'[\ln(1/4)]$.
58. If $g(x) = 3e \ln[\ln x]$, find $g'(e)$.

Work these exercises.

59. **Social Science** The number of Internet users (in millions) in the United States is approximated by

$$f(x) = 175e^{.014x},$$

where $x = 7$ corresponds to the year 1997.* Find the number of Internet users and the rate at which this number is changing in the given year.
(a) 2010 **(b)** 2012

60. **Business** The amount of money the average consumer spends on entertainment and reading annually in the United States can be approximated by

$$g(x) = 1176e^{.029x},$$

where $x = 5$ corresponds to the year 1985.†
(a) What is the amount of the expenditure expected to be in 2011?
(b) At what rate is the expenditure changing in 2011?

61. **Business** The amount of money the average consumer spends on broadcast and satellite radio annually in the United States can be approximated by

$$g(x) = .965e^{.287x},$$

where $x = 5$ corresponds to the year 2005.‡ Find the amount spent on broadcast and satellite radio and the rate of change in the given year.
(a) 2009 **(b)** 2010 **(c)** 2011

62. **Natural Science** The population of a bed of clams in the Great South Bay off Long Island is approximated by the logistic function

$$G(t) = \frac{52{,}000}{1 + 12e^{-.52t}},$$

where t is measured in years. Find the clam population and its rate of growth after
(a) 1 year **(b)** 4 years **(c)** 10 years.
(d) What happens to the rate of growth over time?

63. **Social Science** The number of doctorates (in thousands) awarded in the field of science in the United States is approximated by the function

$$g(x) = 17.6e^{.043x},$$

where $x = 2$ corresponds to the year 2002.§
(a) Find $g(10)$. **(b)** Find $g'(10)$.
(c) Interpret your answers for parts (a) and (b).

64. **Social Science** The number of doctorates (in thousands) awarded in the field of engineering in the United States is approximated by the function

$$h(x) = 4.16e^{.088x},$$

where $x = 2$ corresponds to the year 2002.§

*Mediamark Research, Inc.
†U.S. Bureau of Labor Statistics.
‡VSS *Communications Industry Forecast*, annual.
§www.nsf.gov/statistics/infbrief/nsf09307.

(a) Find $h(12)$. (b) Find $h'(12)$.
(c) Interpret your answers for parts (a) and (b).

65. **Natural Science** The age–weight relationship of female Arctic foxes caught in Svalbard, Norway, can be estimated by the function

$$M(t) = 3102e^{-e^{-.022(t-56)}},$$

where t is the age of the fox in days and $M(t)$ is the weight of the fox in grams.*
(a) Estimate the weight of a female fox that is 200 days old.
(b) Estimate the rate of change in weight of an Arctic fox that is 200 days old.
(c) Use a graphing calculator to graph $M(t)$, and then describe the growth pattern.
(d) Use the table function on a graphing calculator or a spreadsheet to develop a chart that shows the estimated weight and growth rate of female foxes for days 50, 100, 150, 200, 250, and 300.

66. **Health** The number of female physicians (in thousands) can be approximated by the function

$$f(x) = 55.8e^{.06x} \quad (0 \le x \le 26),$$

where $x = 0$ corresponds to the year 1980.†
(a) Find the number of female physicians for 2006.
(b) Find the rate of change for the number of female physicians in 2006.

67. **Health** Per-capita spending on health care in the United States is approximated by

$$f(x) = 1272e^{.0671x},$$

where $x = 0$ corresponds to the year 1980.‡ At what rate are health care expenditures changing in 2010?

68. **Health** One measure of whether a dialysis patient has been adequately dialyzed is the urea reduction ratio (URR). It is generally agreed that a patient has been adequately dialyzed when the URR exceeds a value of .65. The value of the URR can be calculated for a particular patient from the following formula by Gotch:

$$\text{URR} = 1 - e^{-.0056t+.04} - \frac{8t(1 - e^{-.0056t+.04})}{126t + 900}.$$

Here, t is measured in minutes.§
(a) Find the value of the URR after this patient receives dialysis for 180 minutes. Has the patient received adequate dialysis?
(b) Find the value of the URR after the patient receives dialysis for 240 minutes. Has the patient received adequate dialysis?
(c) Use the numerical derivative feature on a graphing calculator to compute the instantaneous rate of change of the URR when the time on dialysis is 240 minutes. Interpret this rate.

69. **Health** The function

$$V(t) = \frac{1100}{(1 + 1023e^{-.02415t})^4},$$

where t is in months and $V(t)$ is in cubic centimeters, models the relationship between the volume of a breast tumor and the amount of time it has been growing.*
(a) Find the tumor volume at 240 months.
(b) Assuming that the shape of a tumor is spherical, find the radius of the tumor from part (a). (*Hint*: The volume of a sphere is given by the formula $v = \frac{4}{3}\pi r^3$.)
(c) If a tumor of volume .5 cm^3 is detected, according to the formula, how long has it been growing? What does this imply?
(d) Calculate the rate of change of tumor volume at 240 months. Interpret this rate.

70. **Business** The demand function for x units of a product is

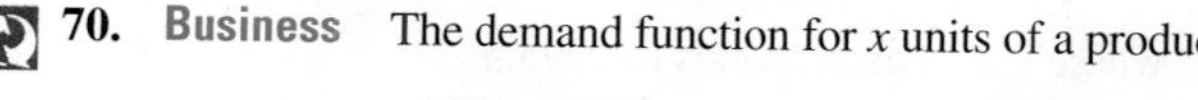

$$p = 100 - 10\ln x \quad (1 < x < 20{,}000),$$

where $x = 6n$ and n is the number of employees producing the product.
(a) Find the revenue function $R(x)$.
(b) Find the marginal-revenue product function. (See Example 12 in Section 11.7.)
(c) Evaluate and interpret the marginal-revenue product when $x = 20$.

71. **Business** Suppose the demand function for x thousand of a certain item is

$$p = 100 + \frac{50}{\ln x} \quad (x > 1),$$

where p is in dollars.
(a) Find the marginal revenue.
(b) Find the revenue from the next thousand items at a demand of 8000 ($x = 8$).

72. **Health** The amount of money (in billions of dollars) the United States federal government dedicates to health research is approximated by the function

$$f(x) = 21.1 + 4.4\ln(x) \quad (1 \le x \le 10),$$

*Pal Prestrud and Kjell Nilssen, "Growth, Size, and Sexual Dimorphism in Arctic Foxes," *Journal of Mammalogy* 76, no. 2 (May 1995): 522–530.

†American Medical Association.

‡Centers for Medicare and Medicaid Services.

§Edward Kessler, Nathan Ritchey, et al., "Urea Reduction Ratio and Urea Kinetic Modeling: A Mathematical Analysis of Changing Dialysis Parameters," *American Journal of Nephrology* 18 (1998): 471–477.

*John Spratt et al., "Decelerating Growth and Human Breast Cancer," *Cancer* 71, no. 6 (1993): 2013–2019.

where $x = 1$ corresponds to the year 2001.*

(a) Estimate the amount spent on health research in 2009 and 2010.

(b) Find the rate of change of $f(x)$ for 2009 and 2010.

73. Transportation A child is waiting at a street corner for a gap in traffic so that she can safely cross the street. A mathematical model for traffic shows that if the child's expected waiting time is to be at most 1 minute, then the maximum traffic flow (in cars per hour) is given by

$$f(x) = \frac{67{,}338 - 12{,}595 \ln x}{x},$$

where x is the width of the street in feet.† Find the maximum traffic flow and the rate of change of the maximum traffic flow with respect to street width when x is the given number.

(a) 30 feet **(b)** 40 feet

74. Business The gross domestic product (GDP) of the United States (in trillions of dollars) is approximated by

$$f(x) = -3.99 + 6.04 \ln(x) \quad (x \geq 7),$$

where $x = 7$ corresponds to the year 1997.‡ At what rate was the GDP changing in the given year?

(a) 2007 **(b)** 2010

75. Health The number of liver transplants in the United States in year x is approximated by

$$g(x) = -232.3 + 2403 \ln(x) \quad (x \geq 5),$$

where $x = 5$ corresponds to 1995.§

(a) Estimate the number of transplants for 2009.

(b) Find the rate of change for the number of transplants in 2009.

76. Social Science The population for the state of California (in millions) can be approximated by the function

$$h(x) = 34.4 + 1.05 \ln(x) \quad (x \geq 1),$$

where $x = 1$ corresponds to the year 2001.£ Find the rate of change of the population in 2010.

77. Physical Science The Richter scale provides a measure of the magnitude of an earthquake. One of the largest Richter numbers M ever recorded for an earthquake was 8.9, from the 1933 earthquake in Japan. The following formula shows a relationship between the amount of energy released, E, and the Richter number:

$$M = \frac{2 \ln E - 2 \ln .007}{3 \ln 10}.$$

Here, E is measured in kilowatt-hours (kWh).*

(a) For the 1933 earthquake in Japan, what value of E gives a Richter number $M = 8.9$?

(b) If the average household uses 247 kilowatt-hours per month, for how many months would the energy released by an earthquake of this magnitude power 10 million households?

(c) Find the rate of change of the Richter number M with respect to energy when $E = 70{,}000$ kWh.

(d) What happens to $\frac{dM}{dE}$ as E increases?

Federal R&D Funding by Budget Function, annual.

†Edward A. Bender, *An Introduction to Mathematical Modeling*, Dover Publications, Inc., 2000.

‡Bureau of Economic Analysis.

§www.unos.org.

£www.census.gov/popest/states/NST-ann-est.html.

✓ Checkpoint Answers

1. **(a)** $(2x^2 - 1)e^x + 4xe^x$ **(b)** $\frac{-e^x}{2(1 - e^x)^{1/2}}$
2. **(a)** $y' = 44e^{11x}$ **(b)** $y' = 45e^{(-9x+2)}$ **(c)** $y' = -3x^2e^{-x^3}$
3. **(a)** $y' = \frac{e^x(11 + x)}{(12 + x)^2}$ **(b)** $y' = \frac{-27{,}000e^x}{(15 + 3e^x)^2}$
4. $y' = (\ln 25)25^x$
5. **(a)** $y' = \frac{1}{7 + x}$ **(b)** $y' = \frac{2}{x}$ **(c)** $y' = \frac{24x^2 - 3}{8x^3 - 3x}$ **(d)** $y' = x(1 + 2 \ln x)$
6. **(a)** $y' = e^{x^2}\left(\frac{1}{x} + 2x \ln|x|\right)$ **(b)** $y' = \frac{2x \ln|x| - x}{(\ln|x|)^2}$
7. **(a)** About -2222 **(b)** About -33 **(c)** Decreasing

*Christopher Bradley, "Media Clips," *Mathematics Teacher* 93, no. 4 (April 2000): 300–303.

11.9 ▶ Continuity and Differentiability

Intuitively speaking, a function is **continuous** at a point if you can draw the graph of the function near that point without lifting your pencil from the paper. Conversely, a function is **discontinuous** at a point if the pencil *must* be lifted from the paper in order to draw the graph on both sides of the point.

Looking at some graphs that have points of discontinuity will clarify the idea of continuity at a point. Each of the three graphs in Figure 11.37 is continuous *except* at $x = 2$. As usual, an open circle indicates that the point is not part of the graph. You cannot draw these graphs without lifting your pencil at $x = 2$, at least for an instant.

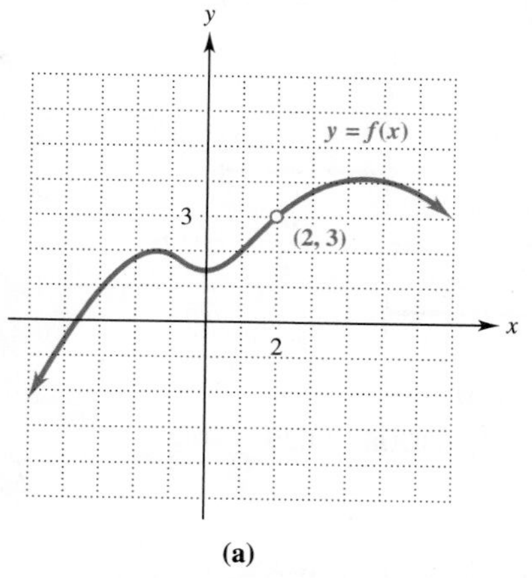

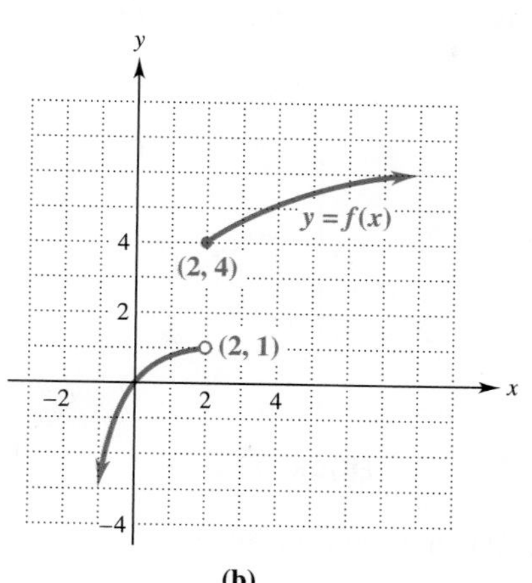

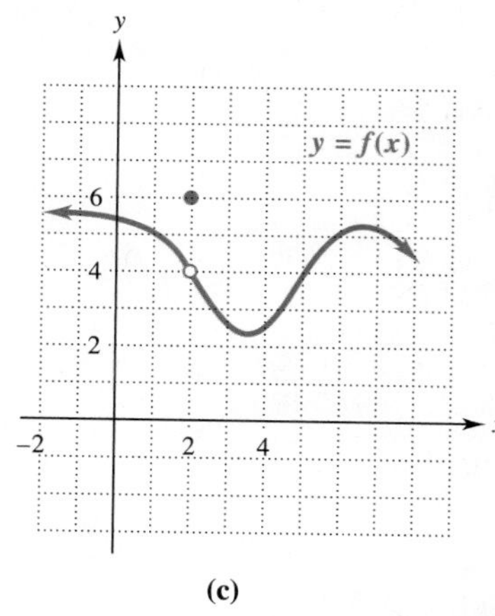

FIGURE 11.37

✔ Checkpoint 1

Find any points of discontinuity for the given functions.

(a)

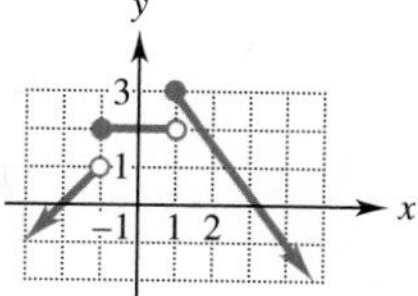

(b)

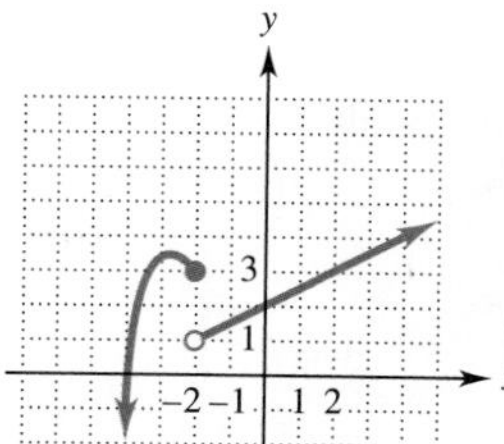

(c)

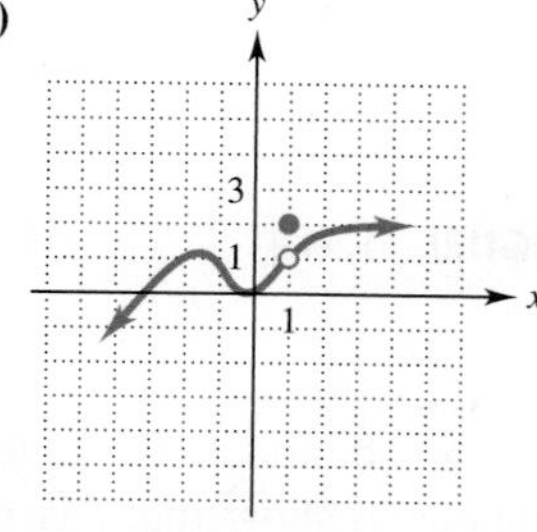

Answers: See page 760.

However, there are different reasons why each function is discontinuous at $x = 2$:

1. In graph (a), $f(2)$ is not defined, whereas $f(2) = 4$ in graph (b) and $f(2) = 6$ in graph (c).
2. In graph (b), $\lim_{x\to 2} f(x)$ does not exist (why?), but $\lim_{x\to 2} f(x) = 3$ in graph (a) and $\lim_{x\to 2} f(x) = 4$ in graph (c).
3. In graph (c), $f(2) = 6$ and $\lim_{x\to 2} f(x) = 4$, so $\lim_{x\to 2} f(x) \neq f(2)$.

By phrasing each of these considerations positively, we obtain the following definition.

Continuity at a Point

A function f is **continuous** at $x = c$ if

(a) $f(c)$ is defined;

(b) $\lim_{x\to c} f(x)$ exists;

(c) $\lim_{x\to c} f(x) = f(c)$.

If f is not continuous at $x = c$, it is **discontinuous** there.

The idea behind condition (c) is this: To draw the graph near $x = c$ without lifting your pencil, it must be the case that when x is very close to c, $f(x)$ must be very close to $f(c)$—otherwise, you would have to lift the pencil at $x = c$, as happens in Figure 11.37 at $x = 2$. 1✔

Example 1 Tell why the given functions are discontinuous at the indicated point.

(a) $f(x)$ in Figure 11.38 at $x = 3$

Solution The open circle on the graph of Figure 11.38 at the point where $x = 3$ means that $f(3)$ does not exist. Because of this, part (a) of the definition fails.

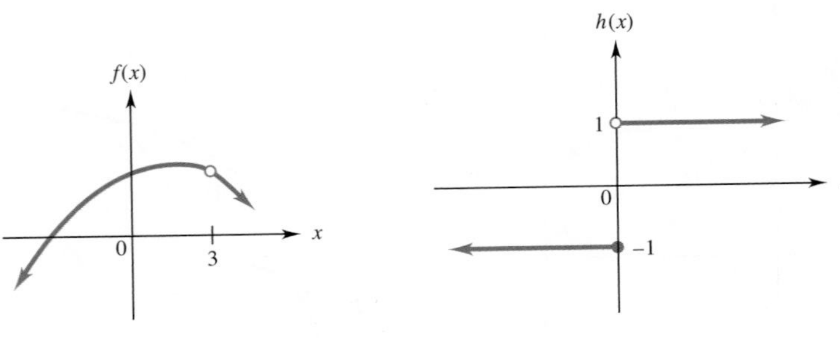

FIGURE 11.38

FIGURE 11.39

(b) $h(x)$ in Figure 11.39 at $x = 0$

Solution The graph of Figure 11.39 shows that $h(0) = -1$. Also, as x approaches 0 from the left, $h(x)$ is -1. However, as x approaches 0 from the right, $h(x)$ is 1. As mentioned in Section 11.1, for a limit to exist at a particular value of x, the values of $h(x)$ must approach a single number. Since no single number is approached by the values of $h(x)$ as x approaches 0, $\lim_{x\to 0} h(x)$ does not exist, and part (b) of the definition fails.

(c) $g(x)$ at $x = 4$ in Figure 11.40

Solution In Figure 11.40, the heavy dot above 4 shows that $g(4)$ is defined. In fact, $g(4) = 1$. However, the graph also shows that

$$\lim_{x\to 4} g(x) = -2,$$

so $\lim_{x\to 4} g(x) \neq g(4)$, and part (c) of the definition fails.

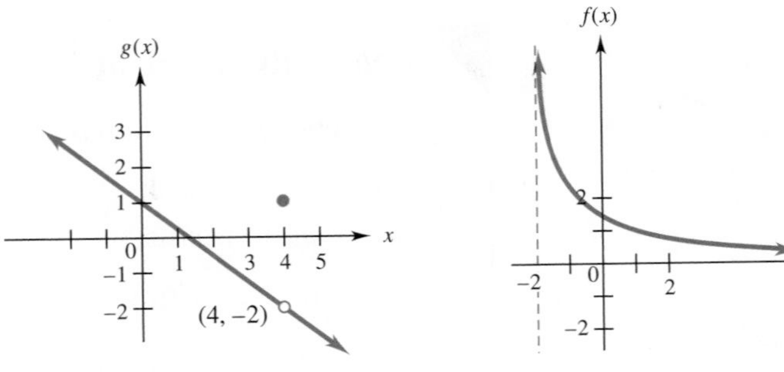

FIGURE 11.40

FIGURE 11.41

(d) $f(x)$ in Figure 11.41 at $x = -2$

Solution The function f graphed in Figure 11.41 is not defined at -2, and $\lim_{x\to -2} f(x)$ does not exist. Either of these reasons is sufficient to show that f is not continuous at -2. (Function f *is* continuous at any value of x greater than -2, however.) 2✓

✓Checkpoint 2

Tell why the given functions are discontinuous at the indicated point.

(a)

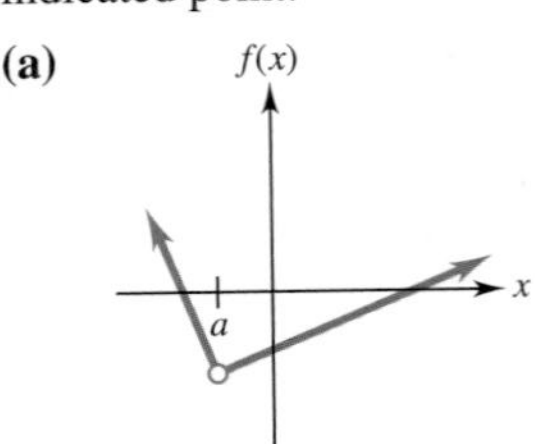

(b)

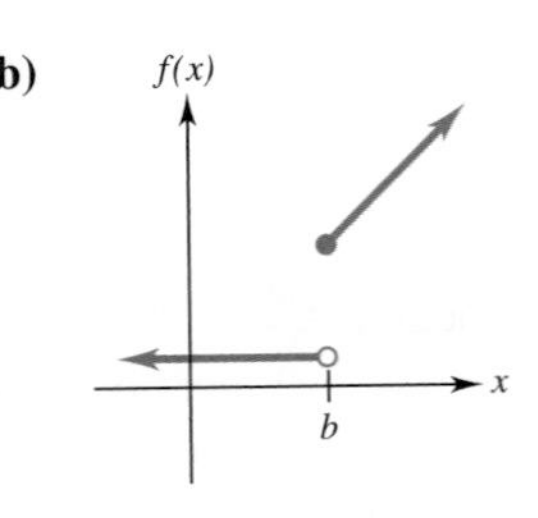

Answers: See page 760.

CONTINUITY AT ENDPOINTS

One-sided limits (see Section 11.2) can be used to define continuity at an endpoint of a graph.* Consider the endpoint $(1, 0)$ of the graph of the half-circle function $f(x) = \sqrt{1 - x^2}$ in Figure 11.42. Note that f has a left-hand limit at $x = 1$. Furthermore,

$$f(1) = 0 \qquad \text{and} \qquad \lim_{x \to 1^-} f(x) = 0, \qquad \text{so that} \qquad \lim_{x \to 1^-} f(x) = f(1).$$

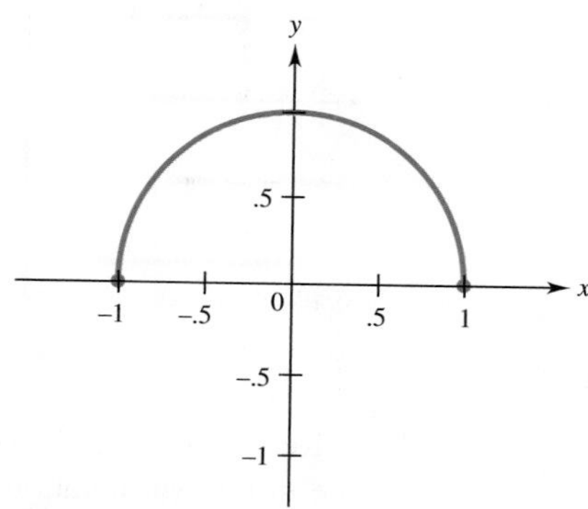

FIGURE 11.42

We say that $f(x)$ is *continuous from the left* at $x = 1$. A similar situation occurs with the other endpoint $(-1, 0)$: There is a right-hand limit at $x = -1$, and we have $\lim_{x \to -1^+} f(x) = 0 = f(-1)$. We say that $f(x)$ is *continuous from the right* at $x = -1$. This situation is an example of the following definition.

Continuity from the Left and from the Right

A function f is **continuous from the left** at $x = c$ if

(a) $f(c)$ is defined;
(b) $\lim_{x \to c^-} f(x)$ exists;
(c) $\lim_{x \to c^-} f(x) = f(c)$.

Similarly, f is **continuous from the right** at $x = c$ if

(a) $f(c)$ is defined;
(b) $\lim_{x \to c^+} f(x)$ exists;
(c) $\lim_{x \to c^+} f(x) = f(c)$.

Note that a function which is continuous at $x = c$ is automatically continuous from the left and from the right at $x = c$. 3✓

✓Checkpoint 3

State whether the given function is continuous from the left, continuous from the right, or neither at the given value of x.

(a) $f(x)$ in Figure 11.37(b) at $x = 2$

(b) $h(x)$ in Figure 11.39 at $x = 0$

(c) $g(x)$ in Figure 11.40 at $x = 4$

Answers: See page 760.

*The preceding definition of continuity does not apply to an endpoint because the two-sided limit $\lim_{x \to c} f(x)$ does not exist when $(c, f(c))$ is an endpoint.

CONTINUITY ON AN INTERVAL

When discussing the continuity of a function, it is often helpful to use interval notation, which was introduced in Chapter 1. The following chart should help you recall how it is used:

Interval	Name	Description	Interval Notation
−2 3	Open interval	$-2 < x < 3$	$(-2, 3)$
−2 3	Closed interval	$-2 \le x \le 3$	$[-2, 3]$
3	Open interval	$x < 3$	$(-\infty, 3)$
−5	Open interval	$x > -5$	$(-5, \infty)$

Remember, the symbol ∞ does not represent a number; ∞ is used for convenience in interval notation to indicate that the interval extends without bound in the positive direction. Also, $-\infty$ indicates no bound in the negative direction. ✓4

✓Checkpoint 4

Write each of the following in interval notation.

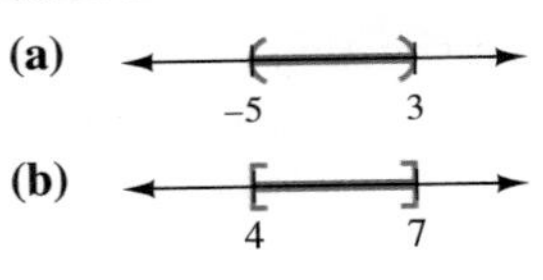

(a) −5 3

(b) 4 7

(c) −1

Answers: See page 760.

Continuity at a point was defined previously. Continuity on an interval is defined as follows.

Continuity on Open and Closed Intervals

A function f is **continuous on the open interval (a, b)** if f is continuous at every x-value in the interval (a, b).

A function f is **continuous on the closed interval $[a, b]$** if

1. f is continuous on the open interval (a, b);
2. f is continuous from the right at $x = a$;
3. f is continuous from the left at $x = b$.

Intuitively, a function is continuous on an interval if you can draw the entire graph of f over that interval without lifting your pencil from the paper.

FIGURE 11.43

Example 2 Is the function of Figure 11.43 continuous on the given x-intervals?

(a) $(0, 2)$

Solution The function g is discontinuous only at $x = -2$, 0, and 2. Hence, g is continuous at every point of the open interval $(0, 2)$, which does not include 0 or 2.

(b) $[0, 2]$

Solution The function g is not defined at $x = 0$, so it is not continuous from the right there. Therefore, it is not continuous on the closed interval $[0, 2]$.

✓Checkpoint 5

Are the functions whose graphs are shown continuous on the indicated intervals?

(a) $[-4, -1]$; $[2, 5]$

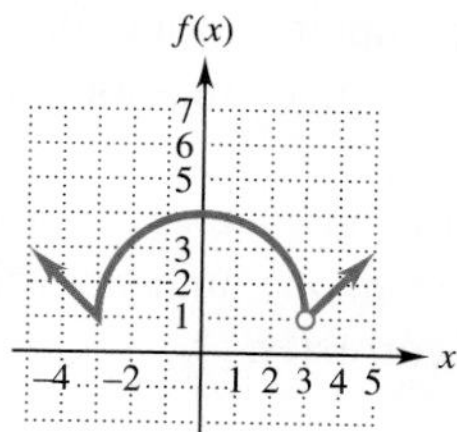

(b) $(1, 3)$; $[4, 6]$

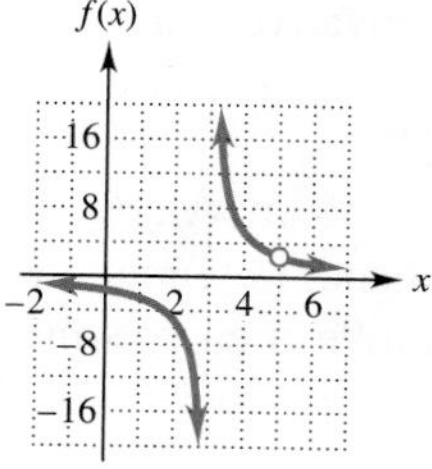

Answers: See page 760.

(c) $(1, 3)$

Solution This interval contains a point of discontinuity at $x = 2$. So g is not continuous on the open interval $(1, 3)$.

(d) $[-2, -1]$

Solution The function g is continuous on the open interval $(-2, -1)$, continuous from the right at $x = -2$, and continuous from the left at $x = -1$. Hence, g is continuous on the closed interval $[-2, -1]$. 5✓

Example 3 **Business** A tool rental firm charges an initial flat fee of \$30 to rent a power washer and then an additional \$15 an hour or fraction of an hour. Let $C(x)$ represent the cost of renting a power washer for x hours.

(a) Graph C.

Solution The charge for 1 hour of use is \$30 for the initial fee and \$15 for the hour of use, or \$45. In fact, in the interval $(0, 1]$, $C(x) = 45$. To rent the power washer for more than 1 hour, but not more than 2 hours, the charge is $30 + 2 \cdot 15 = 60$ dollars. For any value of x in the interval $(1, 2]$, $C(x) = 60$. Also in $(2, 3]$, we have $C(x) = 75$. These results lead to the graph in Figure 11.44.

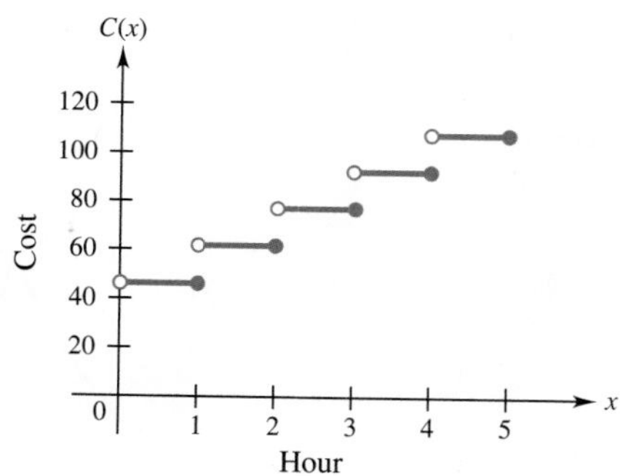

FIGURE 11.44

(b) Find any points of discontinuity for C.

Solution As the graph suggests, $C(x)$ is discontinuous at 1, 2, 3, 4, 5, and all other positive integers. However, $C(x)$ is continuous from the left at each such integer value of x. 6✓

✓Checkpoint 6

The fee structure for a legal consultation is an initial charge of \$500 and then an additional charge of \$150 per hour of service. Let $F(x)$ represent the cost of using the consultant for x hours. Find any points of discontinuity for F.

Answer: See page 760.

CONTINUITY AND TECHNOLOGY

Because of the way a graphing calculator or computer graphs functions, the graph of a continuous function may look like disconnected, closely spaced line segments (Figure 11.45). Furthermore, calculators often have trouble accurately portraying graphs at points of discontinuity. For instance, "jump discontinuities" (such as those at $x = 1$, 2, and 3 in Figure 11.44) may appear as "stair steps," with vertical line

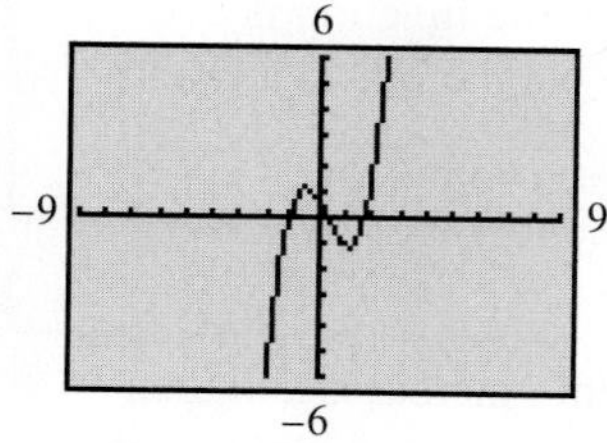

FIGURE 11.45

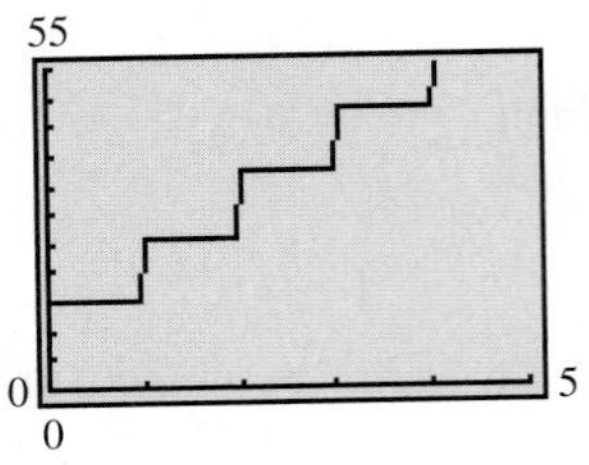

FIGURE 11.46

segments connecting separate pieces of the graph (Figure 11.46). Similarly, a hole in the graph (such as the one at $x = 4$ in Figure 11.40) may not be visible on a calculator screen, depending on what viewing window is used. Nor can a calculator indicate whether an endpoint is included in the graph.

One reason these inaccuracies appear is due to the manner in which a calculator graphs: It plots points and connects them with line segments. In effect, the calculator assumes that the function is continuous unless it actually computes a function value and determines that it is undefined. In that case, it will skip a pixel and not connect points on either side of it.

The moral of this story is that a calculator or computer is only a tool. In order to use it correctly and effectively, you must understand the mathematics involved. When you do, it is usually easy to interpret screen images correctly.

CONTINUITY AND DIFFERENTIABILITY

As shown earlier in this chapter, a function fails to have a derivative at a point where the function is not defined—where the graph of the function has a "sharp point," or where the graph has a vertical tangent line. (See Figure 11.47.)

The function graphed in Figure 11.47 is continuous on the interval (x_1, x_2) and has a derivative at each point on this interval. On the other hand, the function is also continuous on the interval $(0, x_2)$, but does *not* have a derivative at each point on the interval. (See x_1 on the graph.)

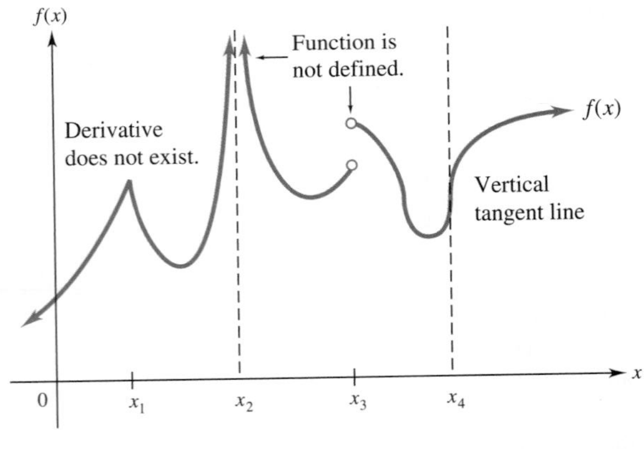

FIGURE 11.47

A similar situation holds in the general case.

> If the derivative of a function exists at a point, then the function is continuous at that point. However, a function may be continuous at a point and not have a derivative there.

Example 4 A nova is a star whose brightness suddenly increases and then gradually fades. The cause of the sudden increase in brightness is thought to be an explosion of some kind. The intensity of light emitted by a nova as a function of time is shown in Figure 11.48.* Notice that although the graph is a continuous curve, it is not differentiable at the point of the explosion.

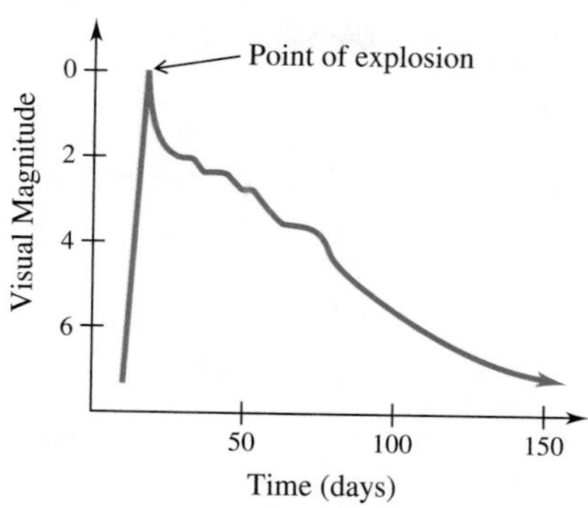

FIGURE 11.48

TECHNOLOGY TIP In some viewing windows, a calculator graph may appear to have a sharp corner when, in fact, the graph is differentiable at that point. When in doubt, try a different window to see if the corner disappears.

*Reprinted with permission of Macmillan Publishing Company from *Astronomy: The Structure of the Universe* by William J. Kaufmann, III. Copyright © 1977 by William J. Kaufmann, III.

11.9 ▸ Exercises

Find all points of discontinuity for the functions whose graphs are shown. (See Example 1.)

1.

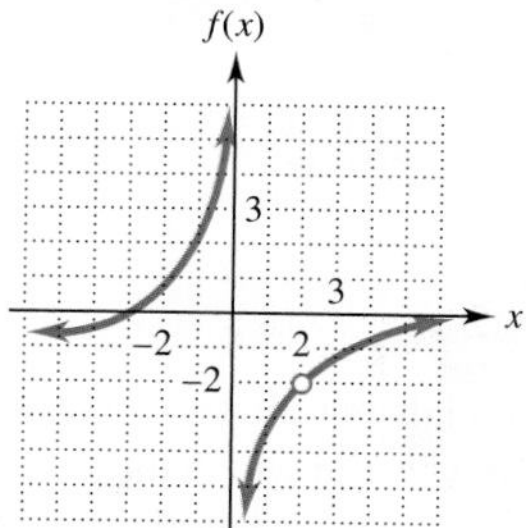

2.

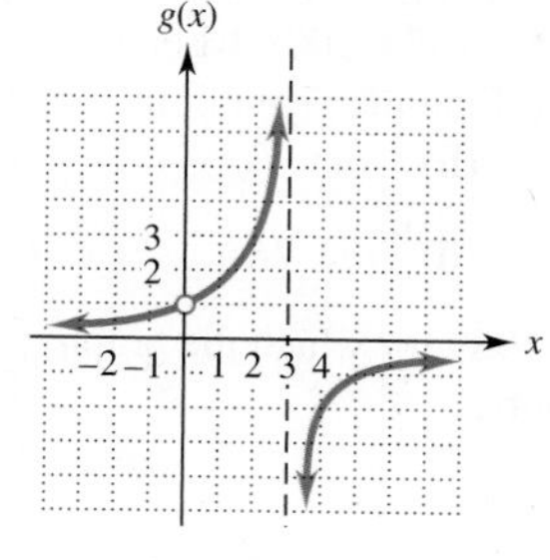

3.

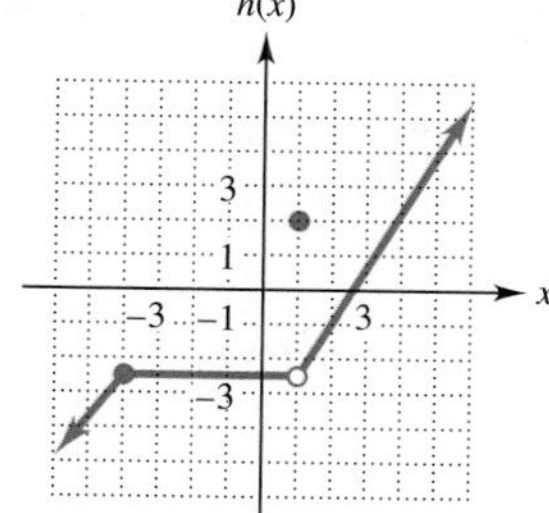

4.

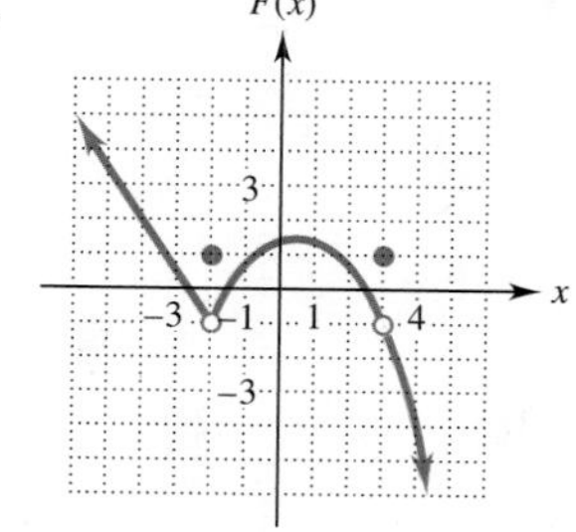

5.

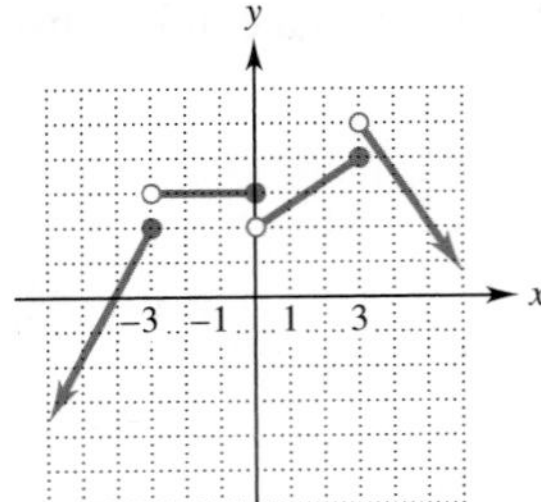

6.

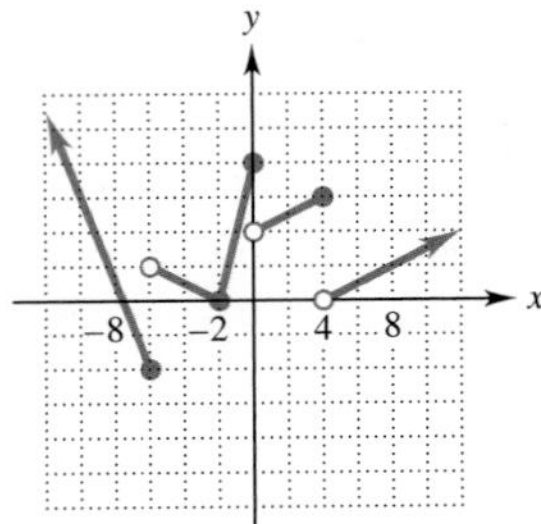

7.

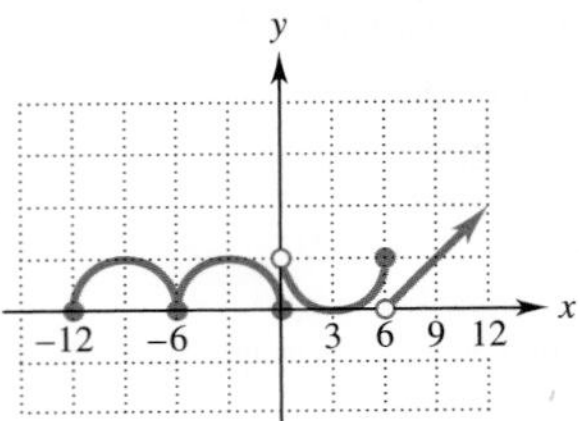

8.

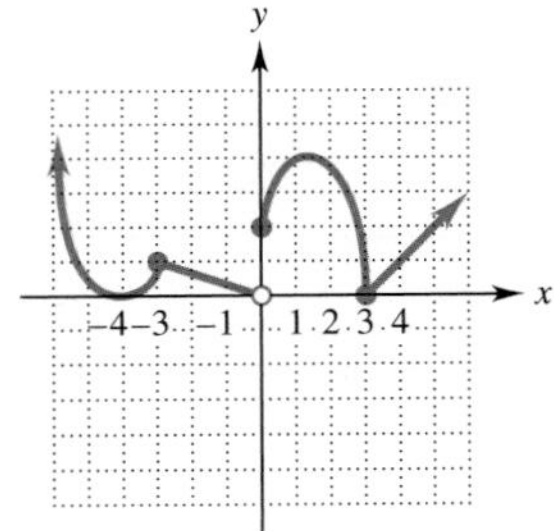

Are the given functions continuous at the given values of x?

9. $f(x) = \dfrac{4}{x - 2}$; $x = 0, x = 2$

10. $g(x) = \dfrac{5}{x + 5}$; $x = -5, x = 5$

11. $h(x) = \dfrac{1}{x(x - 3)}$; $x = 0, x = 3, x = 5$

12. $h(x) = \dfrac{-1}{(x - 2)(x + 3)}$; $x = 0, x = 2, x = 3$

13. $g(x) = \dfrac{x + 2}{x^2 - x - 2}$; $x = 1, x = 2, x = -2$

14. $h(x) = \dfrac{3x}{6x^2 + 15x + 6}$; $x = 0, x = -\dfrac{1}{2}, x = 3$

15. $g(x) = \dfrac{x^2 - 4}{x - 2}$; $x = 0, x = 2, x = -2$

16. $h(x) = \dfrac{x^2 - 25}{x + 5}$; $x = 0, x = 5, x = -5$

17. $f(x) = \begin{cases} x - 2 & \text{if } x \le 3 \\ 2 - x & \text{if } x > 3 \end{cases}$; $x = 2, x = 3$

18. $g(x) = \begin{cases} e^x & \text{if } x < 0 \\ x + 1 & \text{if } 0 \le x \le 3 \\ 2x - 3 & \text{if } x > 3 \end{cases}$; $x = 0, x = 3$

Find all x-values where the function is discontinuous.

19. $f(x) = \dfrac{7 + x}{x(x - 3)}$

20. $f(x) = \dfrac{x^2 - 9}{x + 3}$

21. $g(x) = x^2 - 5x + 12$

22. $h(x) = \dfrac{|x + 4|}{x + 4}$

23. $k(x) = e^{\sqrt{x-2}}$

24. $r(x) = \ln\left(\dfrac{x}{x - 3}\right)$

In Exercises 25–26, (a) graph the given function, and (b) find all values of x where the function is discontinuous.

25. $f(x) = \begin{cases} 2 & \text{if } x < 1 \\ x + 3 & \text{if } 1 \le x < 4 \\ 8 & \text{if } x \ge 4 \end{cases}$

26. $g(x) = \begin{cases} -10 & \text{if } x < -2 \\ x^2 + 4x - 6 & \text{if } -2 \le x < 2 \\ 10 & \text{if } x \ge 2 \end{cases}$

In Exercises 27 and 28, find the constant k that makes the given function continuous at x = 2.

27. $f(x) = \begin{cases} x + k & \text{if } x \le 2 \\ 5 - x & \text{if } x > 2 \end{cases}$

28. $g(x) = \begin{cases} x^k & \text{if } x \le 2 \\ 2x + 4 & \text{if } x > 2 \end{cases}$

29. **Finance** In past years, officials in California have tended to raise the state sales tax in years in which the state faced a budget deficit and then cut the tax when the state had a surplus. The graph on the top of the next column shows the California state sales tax for 1974–2008.* Let $T(x)$ represent the sales tax in year x. Find the given limits.

(a) $\lim_{x \to '82} T(x)$ **(b)** $\lim_{x \to '04^-} T(x)$

(c) $\lim_{x \to '04^+} T(x)$ **(d)** $\lim_{x \to '04} T(x)$

(e) List three years after 1989 for which the graph indicates a discontinuity.

*www.boe.ca.gov/sutax/taxrateshist.htm.

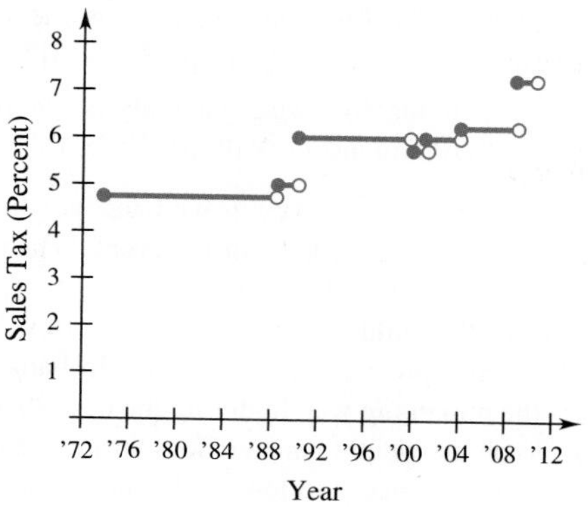

30. **Natural Science** Suppose a gram of ice is at a temperature of $-100°C$. The accompanying graph shows the temperature of the ice as an increasing number of calories of heat are applied. Where is this function discontinuous? Where is it differentiable?

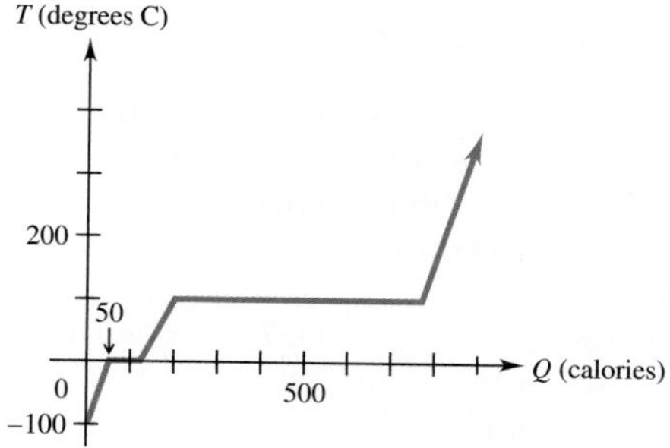

31. **Social Science** With certain skills (such as music), learning is rapid at first and then levels off. However, sudden insights may cause learning to speed up sharply. A typical graph of such learning is shown in the accompanying figure. Where is the function discontinuous? Where is it differentiable?

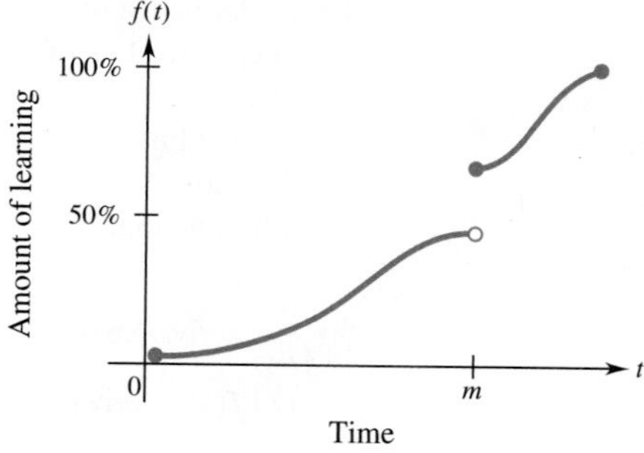

32. **Business** The average price per ton of coal was \$21.76 in 1990 and \$25.40 in 2007. The following graph approximates the price of coal over the time between the given years:*

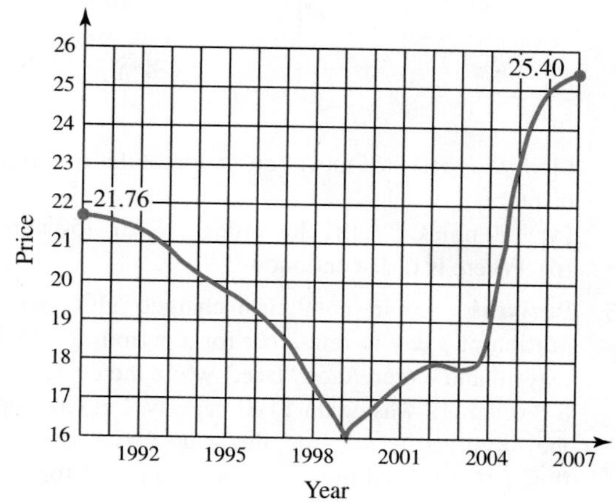

(a) What is the largest interval over which the function is continuous?
(b) Where is the function differentiable?

33. **Business** The following table gives the prices for sending a letter (weighing up to 8 oz.) via airmail from the United States to Australia as they appeared on the U.S. Postal Service Web site in 2009:†

Weight not over ... (oz)	Price
1	1.24
2	2.08
3	2.92
4	3.76
5	4.60
6	5.44
7	6.28
8	7.12

Let $C(x)$ represent the postage for a letter weighing x oz. Find the given limits.

(a) $\lim_{x\to 5^-} C(x)$ **(b)** $\lim_{x\to 5^+} C(x)$

(c) $\lim_{x\to 5} C(x)$ **(d)** $C(5)$

(e) Find all values on the interval (0, 8) where the function C is discontinuous.

(f) Sketch the graph of $y = C(x)$ on the interval (0, 8).

34. **Business** The cost to transport an automobile in a truck depends on the distance x, in miles, that the automobile is moved. Let $C(x)$ represent the cost to move an automobile x miles. One firm charges as follows:

*Based on data from the U.S. Energy Information Administration.
†www.usps.com.

Cost per Mile	Distance in Miles
5.50	$0 < x \le 200$
3.50	$200 < x \le 500$
2.50	$x > 500$

Find the cost to move an automobile the following distances:
(a) 150 miles; **(b)** 300 miles; **(c)** 750 miles.
(d) Where is C discontinuous?

35. Business A car rental firm charged \$120 per day or portion of a day to rent a car for a period of 1 to 5 days. Days 6 and 7 were then "free," while the charge for days 8 through 12 was again \$120 per day. Let $C(t)$ represent the total cost to rent the car for t days, where $0 < t \le 12$. Find the total cost of a rental for the given number of days.
(a) 5 **(b)** 6 **(c)** 7 **(d)** 8
(e) Find $\lim_{t \to 5} C(t)$. **(f)** Find $\lim_{t \to 6} C(t)$.

Write each of the given ranges in interval notation.

36.

37.

38.

39.
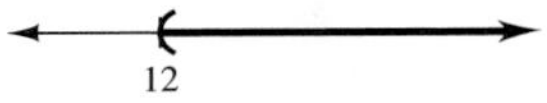

40. On which of the following intervals is the function in Exercise 5 continuous: $(-3, 0)$; $[0, 3]$; $[1, 3]$?

41. On which of the following intervals is the function in Exercise 6 continuous: $(-6, 0)$; $[0, 3]$; $(4, 8)$?

42. Health During pregnancy, a womans weight naturally increases during the course of the event. When she delivers, her weight immediately decreases by the approximate weight of the child. Suppose that a 120-lb woman gains 27 lb during pregnancy, delivers a 7-lb baby, and then, loses the remaining weight during the next 20 weeks.
(a) Graph the weight gain and loss during the pregnancy and the 20 weeks following the birth of the baby. Assume that the pregnancy lasts 40 weeks, that delivery occurs immediately after this time interval, and that the weight gain and loss before and after birth, respectively, are linear.
(b) Is this function continuous? If not, then find the value(s) of t where the function is discontinuous.

✓ Checkpoint Answers

1. **(a)** $x = -1, 1$ **(b)** $x = -2$ **(c)** $x = 1$
2. **(a)** $f(a)$ does not exist. **(b)** $\lim_{x \to b} f(x)$ does not exist.
3. **(a)** Continuous from the right
   (b) Continuous from the left
   (c) Neither
4. **(a)** $(-5, 3)$ **(b)** $[4, 7]$ **(c)** $(-\infty, -1]$
5. **(a)** Yes; no **(b)** Yes; no
6. $x = 1, 2, 3, 4, \ldots$

CHAPTER 11 ▶ Summary

KEY TERMS AND SYMBOLS

11.1 ▶ $\lim_{x \to a} f(x)$ limit of $f(x)$ as x approaches a
limit of a constant function
limit of the identity function
properties of limits
polynomial limits
limit theorem
$\lim_{x \to \infty} f(x)$ limit of $f(x)$ as x approaches infinity
$\lim_{x \to -\infty} f(x)$ limit of $f(x)$ as x approaches negative infinity
infinite limits

11.2 ▶ $\lim_{x \to a^+} f(x)$ limit of $f(x)$ as x approaches a from the right
$\lim_{x \to a^-} f(x)$ limit of $f(x)$ as x approaches a from the left

11.3 ▶ average rate of change
geometric meaning of average rate of change
velocity
instantaneous rate of change
marginal cost, revenue, profit

11.4 ▶ y', derivative of y
$f'(x)$, derivative of $f(x)$
secant line
tangent line
derivative
differentiable

11.5 ▶ $\frac{dy}{dx}$, derivative of $y = f(x)$
$D_x[f(x)]$, derivative of $f(x)$
$\frac{d}{dx}[f(x)]$, derivative of $f(x)$
derivative rules
demand function

11.6 ▸ $\overline{C}(x)$, average cost per item
product rule
quotient rule
marginal average cost

11.7 ▸ $g[f(x)]$, composite function
chain rule
generalized power rule
marginal-revenue product

11.8 ▸ derivatives of exponential functions
derivatives of logarithmic functions

11.9 ▸ continuous at a point
discontinuous
continuous from the left and from the right
interval notation
continuous on an open or closed interval

CHAPTER 11 KEY CONCEPTS

Limit of a Function ▸ Let f be a function, and let a and L be real numbers. Suppose that as x takes values very close (but not equal) to a (on both sides of a), the corresponding values of $f(x)$ are very close (and possibly equal) to L and that the values of $f(x)$ can be made arbitrarily close to L for all values of x that are close enough to a. Then L is the **limit** of f as x approaches a, written $\lim_{x \to a} f(x) = L$.

One-Sided Limits ▸ The **limit of f as x approaches a from the right**, written $\lim_{x \to a^+} f(x)$, is defined by replacing the phrase "on both sides of a" by "with $x > a$" in the definition of the limit of a function.

The **limit of f as x approaches a from the left**, written $\lim_{x \to a^-} f(x)$, is defined by replacing the phrase "on both sides of a" by "with $x < a$" in the definition of the limit of a function.

Properties of Limits ▸ Let a, k, A, and B be real numbers, and let f and g be functions such that

$$\lim_{x \to a} f(x) = A \quad \text{and} \quad \lim_{x \to a} g(x) = B.$$

Then

1. $\lim_{x \to a} k = k$ (for any constant k);

2. $\lim_{x \to a} x = a$ (for any real number a);

3. $\lim_{x \to a} [f(x) \pm g(x)] = A \pm B = \lim_{x \to a} f(x) \pm \lim_{x \to a} g(x)$;

4. $\lim_{x \to a} [f(x) \cdot g(x)] = A \cdot B = \lim_{x \to a} f(x) \cdot \lim_{x \to a} g(x)$;

5. $\lim_{x \to a} \dfrac{f(x)}{g(x)} = \dfrac{A}{B} = \dfrac{\lim_{x \to a} f(x)}{\lim_{x \to a} g(x)}$ $(B \neq 0)$;

6. for any real number r for which A^r exists,

$$\lim_{x \to a} [f(x)]^r = A^r = [\lim_{x \to a} f(x)]^r.$$

Polynomial Limits ▸ If f is a polynomial function, then $\lim_{x \to a} f(x) = f(a)$.

Limit Theorem ▸ If f and g are functions that have limits as x approaches a, and $f(x) = g(x)$ for all $x \neq a$, then $\lim_{x \to a} f(x) = \lim_{x \to a} g(x)$.

Limits at Infinity ▸ Let f be a function that is defined for all large values of x, and let L be a real number.

Suppose that as x takes larger and larger values, increasing without bound, the corresponding values of $f(x)$ are very close (and possibly equal) to L, and suppose that the values of $f(x)$ can be made arbitrarily close to L by taking large enough values of x. Then L is the **limit of f as x approaches infinity**, written $\lim_{x \to \infty} f(x) = L$.

Let f be a function that is defined for all small negative values of x, and let M be a real number. Suppose that as x takes smaller and smaller negative values, decreasing without bound, the corresponding values of $f(x)$ are very close (and possibly equal) to M, and suppose that the values of $f(x)$ can be made arbitrarily close to M by taking small enough values of x. Then L is the **limit of f as x approaches negative infinity**, written $\lim_{x \to -\infty} f(x) = M$.

Derivatives

The **instantaneous rate of change** of a function f when $x = a$ is

$$\lim_{h \to 0} \frac{f(a + h) - f(a)}{h},$$

provided that this limit exists.

The **tangent line** to the graph of $y = f(x)$ at the point $(a, f(a))$ is the line through this point and having slope $\lim_{h \to 0} \frac{f(a + h) - f(a)}{h}$, provided that this limit exists.

The **derivative** of the function f is the function denoted f' whose value at the number x is $f'(x) = \lim_{h \to 0} \frac{f(x + h) - f(x)}{h}$, provided that this limit exists.

Rules for Derivatives

(Assume that all indicated derivatives exist.)

Constant Function If $f(x) = k$, where k is any real number, then $\boldsymbol{f'(x) = 0}$.

Power Rule If $f(x) = x^n$, for any real number n, then $\boldsymbol{f'(x) = n \cdot x^{n-1}}$.

Constant Times a Function Let k be a real number. Then the derivative of $y = k \cdot f(x)$ is $\boldsymbol{y' = k \cdot f'(x)}$.

Sum-or-Difference Rule If $f(x) = u(x) \pm v(x)$, then $\boldsymbol{f'(x) = u'(x) \pm v'(x)}$.

Product Rule If $f(x) = u(x) \cdot v(x)$, then $\boldsymbol{f'(x) = u(x) \cdot v'(x) + v(x) \cdot u'(x)}$.

Quotient Rule If $f(x) = \frac{u(x)}{v(x)}$, and $v(x) \neq 0$, then $\boldsymbol{f'(x) = \frac{v(x) \cdot u'(x) - u(x) \cdot v'(x)}{[v(x)]^2}}$.

Chain Rule Let $y = f[g(x)]$. Then $\boldsymbol{y' = f'[g(x)] \cdot g'(x)}$.

Chain Rule (alternative form) If y is a function of u, say, $y = f(u)$, and if u is a function of x, say, $u = g(x)$, then $y = f(u) = f[g(x)]$, and

$$\boldsymbol{\frac{dy}{dx} = \frac{dy}{du} \cdot \frac{du}{dx}}.$$

Generalized Power Rule Let u be a function of x, and let $y = u^n$ for any real number n. Then

$$\boldsymbol{y' = n \cdot u^{n-1} \cdot u'}.$$

Exponential Function If $y = e^{g(x)}$, then $\boldsymbol{y' = g'(x) \cdot e^{g(x)}}$.

Natural-Logarithm Function If $y = \ln|x|$, then $\boldsymbol{y' = \frac{1}{x}}$.

If $y = \ln[g(x)]$, then $\boldsymbol{y' = \frac{g'(x)}{g(x)}}$.

Continuity A function f is **continuous** at $x = c$ if $f(c)$ is defined, $\lim_{x \to c} f(x)$ exists, and $\lim_{x \to c} f(x) = f(c)$.

CHAPTER 11 REVIEW EXERCISES

In Exercises 1–6, determine graphically or numerically whether the limit exists. If the limit exists, find its (approximate) value.

1. $\lim_{x \to -3} f(x)$; $\lim_{x \to -3^-} f(x)$; $\lim_{x \to -3^+} f(x)$

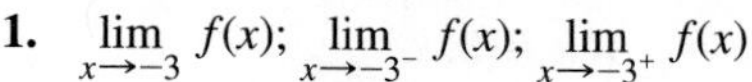

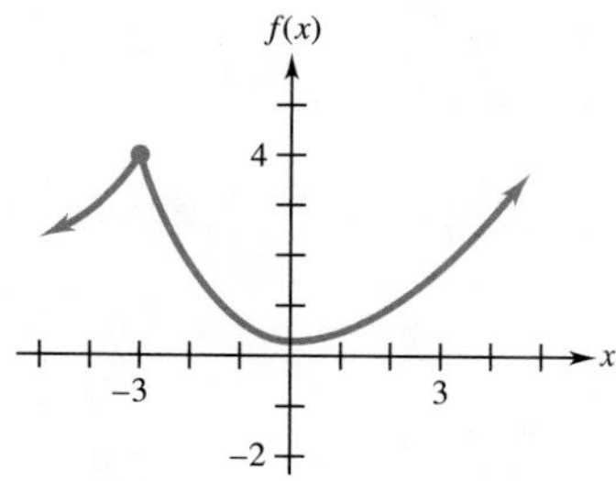

2. $\lim_{x \to -1} g(x)$; $\lim_{x \to -1^-} g(x)$; $\lim_{x \to -1^+} g(x)$

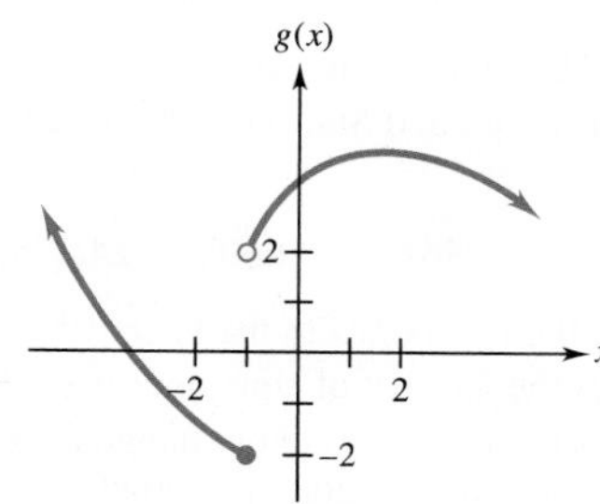

3. $\lim_{x \to 1} \dfrac{x^3 - 1.1x^2 - 2x + 2.1}{x - 1}$

4. $\lim_{x \to 2} \dfrac{x^4 + .5x^3 - 4.5x^2 - 2.5x + 3}{x - 2}$

5. $\lim_{x \to 0} \dfrac{\sqrt{2 - x} - \sqrt{2}}{x}$

6. $\lim_{x \to -1} \dfrac{10^x - .1}{x + 1}$

Find the given limit if it exists.

7. $\lim_{x \to 3} (x^2 - 5x + 2)$

8. $\lim_{x \to -2} (-3x^3 + x + 4)$

9. $\lim_{x \to 4} \dfrac{5x + 1}{x - 4}$

10. $\lim_{x \to 5} \dfrac{5x - 3}{x + 1}$

11. $\lim_{x \to 5} \dfrac{x^2 - 25}{x + 5}$

12. $\lim_{x \to 5} \dfrac{x^2 - 2x - 3}{x + 1}$

13. $\lim_{x \to -4} \dfrac{2x^2 + 3x - 20}{x + 4}$

14. $\lim_{x \to 3} \dfrac{3x^2 - 2x - 21}{x - 3}$

15. $\lim_{x \to 9} \dfrac{\sqrt{x} - 3}{x - 9}$

16. $\lim_{x \to 16} \dfrac{\sqrt{x} - 4}{x - 16}$

Find the given limit if it exists.

17. $\lim_{x \to -1^+} \sqrt{9 + 8x - x^2}$

18. $\lim_{x \to 8^+} \dfrac{x^2 - 64}{x - 8}$

19. $\lim_{x \to -5^+} \dfrac{|x + 5|}{x + 5}$

20. $\lim_{x \to 7^-} \left(\sqrt{7 - x^2 + 6x} + 2\right)$

21. $\lim_{x \to \infty} \dfrac{2x^3 - 3x^2 + 5x - 1}{4x^3 + 2x^2 - x + 10}$

22. $\lim_{x \to -\infty} \dfrac{4 - 3x - 2x^2}{x^3 + 2x + 5}$

23. $\lim_{x \to -\infty} \left(\dfrac{2x + 1}{x - 3} + \dfrac{4x - 1}{3x}\right)$

24. $\lim_{x \to \infty} \dfrac{x}{(\ln x)^2}$

Use the accompanying graph to find the average rate of change of f on the given intervals.

25. $x = 0$ to $x = 4$

26. $x = 2$ to $x = 8$

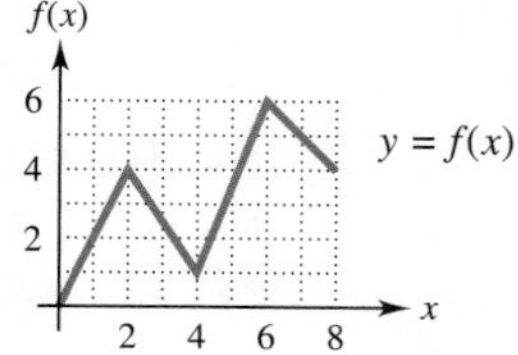

Find the average rate of change for each of the given functions.

27. $f(x) = 3x^2 - 5$, from $x = 1$ to $x = 5$

28. $g(x) = -x^3 + 2x^2 + 1$, from $x = -2$ to $x = 3$

29. $h(x) = \dfrac{6 - x}{2x + 3}$, from $x = 0$ to $x = 6$

30. $f(x) = e^{2x} + 5 \ln x$, from $x = 1$ to $x = 7$

Use the definition of the derivative to find the derivative of each of the given functions.

31. $y = 2x + 3$

32. $y = x^2 + 2x$

33. $y = 2x^2 - x - 1$

34. $y = x^3 + 5$

Find the slope of the tangent line to the given curve at the given value of x. Find the equation of each tangent line.

35. $y = x^2 - 6x$; at $x = 2$

36. $y = 8 - x^2$; at $x = 1$

37. $y = \dfrac{-2}{x + 5}$; at $x = -2$

38. $y = \sqrt{6x - 2}$; at $x = 3$

39. Business Suppose hardware store customers are willing to buy $T(p)$ boxes of nails at p dollars per box, where

$$T(p) = .06p^4 - 1.25p^3 + 6.5p^2 - 18p + 200 \quad (0 < p \le 11).$$

(a) Find the average rate of change in demand for a change in price from \$5 to \$8.

(b) Find the instantaneous rate of change in demand when the price is \$5.

(c) Find the instantaneous rate of change in demand when the price is \$8.

40. Suppose the average rate of change of a function $f(x)$ from $x = 1$ to $x = 4$ is 0. Does this mean that f is constant between $x = 1$ and $x = 4$? Explain.

Find the derivative of each of the given functions.

41. $y = 6x^2 - 7x + 3$
42. $y = x^3 - 5x^2 + 1$
43. $y = 4x^{9/2}$
44. $y = -4x^{-5}$
45. $f(x) = x^{-6} + \sqrt{x}$
46. $f(x) = 8x^{-3} - 5\sqrt{x}$
47. $y = (4t^2 + 9)(t^3 - 5)$
48. $y = (-5t + 3)(t^3 - 4t)$
49. $y = 8x^{3/4}(5x + 1)$
50. $y = 40x^{-1/4}(x^2 + 1)$
51. $f(x) = \dfrac{4x}{x^2 - 8}$
52. $g(x) = \dfrac{-5x^2}{3x + 1}$
53. $y = \dfrac{\sqrt{x} - 6}{3x + 7}$
54. $y = \dfrac{\sqrt{x} + 9}{x - 4}$
55. $y = \dfrac{x^2 - x + 1}{x - 1}$
56. $y = \dfrac{2x^3 - 5x^2}{x + 2}$
57. $f(x) = (4x - 2)^4$
58. $k(x) = (5x - 1)^6$
59. $y = \sqrt{2t - 5}$
60. $y = -3\sqrt{8t - 1}$
61. $y = 2x(3x - 4)^3$
62. $y = 5x^2(2x + 3)^5$
63. $f(u) = \dfrac{3u^2 - 4u}{(2u + 3)^3}$
64. $g(t) = \dfrac{t^3 + t - 2}{(2t - 1)^5}$
65. $y = e^{-2x^3}$
66. $y = -5e^{x^2}$
67. $y = 5x \cdot e^{2x}$
68. $y = -7x^2 \cdot e^{-3x}$
69. $y = \ln(x^2 + 4x - 1)$
70. $y = \ln(4x^3 + 2x)$
71. $y = \dfrac{\ln 6x}{x^2 - 1}$
72. $y = \dfrac{\ln(3x + 5)}{x^2 + 5x}$
73. $y = \dfrac{x^2 + 3x - 10}{x - 3}$
74. $y = \dfrac{x^2 - x - 6}{x - 2}$
75. $y = -6e^{2x}$
76. $y = 8e^{.5x}$

Find each of the given derivatives.

77. $D_x\left(\dfrac{\sqrt{x} + 1}{\sqrt{x} - 1}\right)$
78. $D_x\left(\dfrac{2x + \sqrt{x}}{1 - x}\right)$
79. $\dfrac{dy}{dt}$ if $y = \sqrt{t^{1/2} + t}$
80. $\dfrac{dy}{dx}$ if $y = \dfrac{\sqrt{x - 1}}{x}$
81. $f'(1)$ if $f(x) = \dfrac{\sqrt{8 + x}}{x + 1}$
82. $f'(-2)$ if $f(t) = \dfrac{2 - 3t}{\sqrt{2 + t}}$

Find all points of discontinuity for the given graphs.

83.

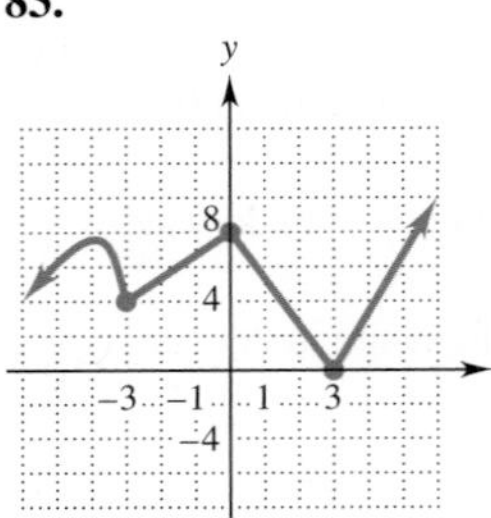

84.

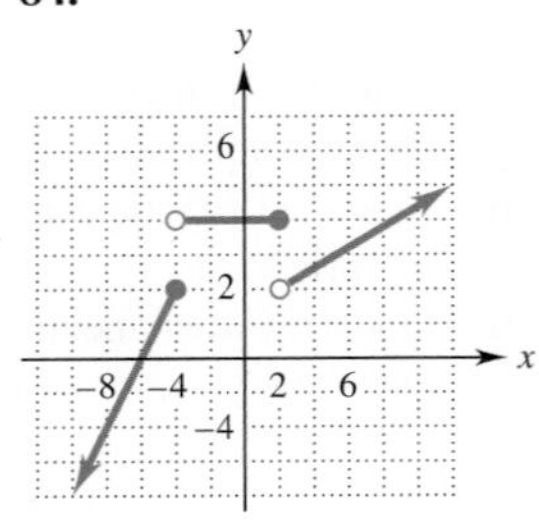

Are the given functions continuous at the given points?

85. $f(x) = \dfrac{2x - 3}{2x + 3}$; $x = -\dfrac{3}{2}, x = 0, x = \dfrac{3}{2}$
86. $g(x) = \dfrac{2x - 1}{x^3 + x^2}$; $x = -1, x = 0, x = \dfrac{1}{2}$
87. $h(x) = \dfrac{2 - 3x}{2 - x - x^2}$; $x = -2, x = \dfrac{2}{3}, x = 1$
88. $f(x) = \dfrac{x^2 - 4}{x^2 - x - 6}$; $x = 2, x = 3, x = 4$
89. $f(x) = \dfrac{x - 6}{x + 5}$; $x = 6, x = -5, x = 0$
90. $f(x) = \dfrac{x^2 - 9}{x + 3}$; $x = 3, x = -3, x = 0$

Work these problems.

91. **Finance** The amount of consumer loans (in billions of dollars) for the United States can be approximated by the function

$$L(x) = .286x^2 - 3.669x + 24.475,$$

where $x = 0$ corresponds to the year 1950.*
(a) What is the amount of consumer loans for 2005?
(b) What is the average rate of change for the amount of consumer loans for 2005 to 2009?
(c) What is the rate of change for the amount of consumer loans in 2010?

92. **Business** The total energy consumption (in quadrillion Btu's) for the United States can be approximated by the function

$$f(x) = -.000144x^3 + .014151x^2 + .1388x + 23.35,$$

where $x = 0$ corresponds to the year 1970.†
(a) Find the energy consumption for 1990, 2000, and 2008.
(b) Find the average rate of change in energy consumption between 2000 and 2008.
(c) At what rate was energy consumption changing in 2008?

93. **Business** The net summer capacity (in millions of kilowatts) for U.S. nuclear power plants can be approximated by the function

$$f(x) = 69.52e^{.018x},$$

where $x = 0$ corresponds to the year 1980.‡
(a) Find the net summer capacity in 2005.
(b) Find the rate of change of net summer capacity in 2010.

---

*http://alfred.stlouisfed.org.
†U.S. Energy Information Administration.
‡www.eia.doe.gov/emeu/aer/nuclear.html.

94. Business The net revenue for Bank of America (in billions of dollars) can be approximated by the function

$$g(x) = -.033741x^4 + 1.62176x^3 - 28.4297x^2 + 216.603x - 599.806 \quad (9 \le x \le 18),$$

where $x = 9$ corresponds to the year 1999.*

(a) Find the revenue in 2006.

(b) Find the revenue in 2008.

(c) Find the rate of change of revenue in 2007.

95. Business The amount of money (in billions of dollars) spent on high-definition television sets in the United States can be approximated by the function

$$h(x) = -13.93 + 17.25\ln(x) \quad (3 \le x \le 9),$$

where $x = 3$ corresponds to the year 2003.†

(a) How much was spent in 2007?

(b) What was the rate of change in 2008?

96. Health The number of U.S. facilities (in thousands) that have a primary focus on caring for patients with substance abuse issues can be approximated by the function

$$f(x) = .0079x^3 - .3180x^2 + 4.1636x - 3.9136 \quad (5 \le x \le 18),$$

where $x = 5$ corresponds to the year 1995.‡

(a) How many facilities were there in 2000 and 2007?

(b) At what rate was the number of facilities changing in 2007?

97. Finance The amount of foreign bonds (in billions of dollars) purchased by the United States can be approximated by the function

$$g(x) = \frac{1660}{1 + 36.08e^{-.247x}} \quad (5 \le x \le 27),$$

where $x = 5$ corresponds to the year 1985.§

(a) Find the amount of foreign bonds purchased in 2000 and 2006.

(b) What was the rate of change in 2006?

98. Business The crude-oil production (in thousands of barrels per day) for Canada can be approximated by the function

$$h(x) = 1252.3e^{.025x} \quad (1 \le x \le 26),$$

where $x = 1$ corresponds to the year 1981.£

(a) How many barrels were produced in 2005?

(b) What was the rate of change in production for 2005?

99. Physical Science The accompanying graph shows how, when a baseball bat is swung, the velocities of the hands and bat vary with the time of the swing.* Estimate and interpret the value of the derivative functions for the hands and for the bat at the time when the velocities of the two are equal. (*Note*: The rate of change of velocity is called *acceleration*.)

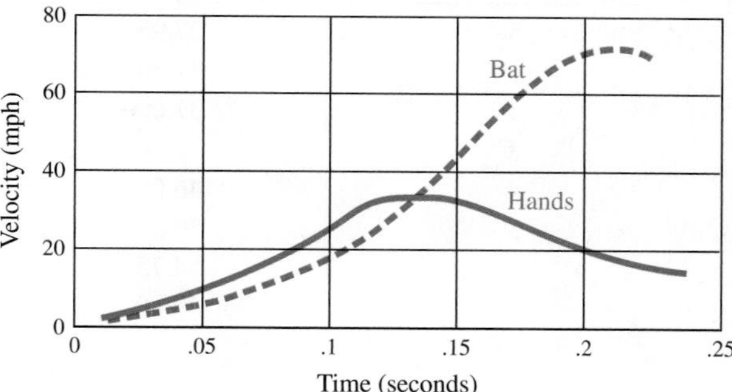

100. Natural Science Under certain conditions, the length of the monkeyface prickleback, a west-coast game fish, can be approximated by

$$L = 71.5(1 - e^{-.1t})$$

and its weight by

$$W = .01289 \cdot L^{2.9},$$

where L is the length in cm, t is the age in years, and W is the weight in grams.†

(a) Find the approximate length of a five-year-old monkeyface.

(b) Find how fast the length of a five-year-old monkeyface is increasing.

(c) Find the approximate weight of a five-year-old monkeyface. (*Hint*: Use your answer from part (a).)

(d) Find the rate of change of the weight with respect to length for a five-year-old monkeyface.

(e) Using the chain rule and your answers to parts (b) and (d), find how fast the weight of a five-year-old monkeyface is increasing.

101. Business The table on the next page gives the prices for sending a letter (weighing up to 7 pounds) via Express Mail International from the United States to France as they appeared on the U.S. Postal Service Web site in 2009. Let $C(x)$ represent the postage for a letter weighing x pounds. Find the given limits.

(a) $\lim_{x \to 6^-} C(x)$ **(b)** $\lim_{x \to 6^+} C(x)$

(c) $\lim_{x \to 6} C(x)$ **(d)** $C(6)$

*www.morningstar.com.

†Consumer Electronics Association.

‡www.samhsa.gov.

§U.S. Department of Treasury.

£U.S. Energy Information Association.

*Adair, Robert K., *The Physics of Baseball*, Second Revised Edition, HarperCollins, 1994.

†William H. Marshall and Tina Wyllie Echeverria, "Characteristics of the Monkeyface Prickleback," *California Fish and Game* 78, no. 2 (spring 1992). For more details, see Case 3 on page xxx.

(e) Find all values on the interval (0, 7) where the function C is discontinuous.
(f) Sketch the graph of $y = C(x)$ on the interval (0, 7).

Weight (lbs) not over . . .	Price (U.S. dollars)
.5	27.95
1	32.50
2	37.00
3	41.50
4	46.00
5	50.50
6	54.75
7	59.00

102. Business A power washer can be rented for a fee of $70 for the first eight hours of use, and then another $15 is added for each additional hour or fraction thereof. Let $C(x)$ represent the cost to rent a power washer for x hours.
(a) Graph $C(x)$.
(b) Find $\lim_{x\to 10^-} C(x)$.
(c) Find $\lim_{x\to 10^+} C(x)$.
(d) Find $\lim_{x\to 10} C(x)$.
(e) Find $C(10)$.

CASE 11

Price Elasticity of Demand

Any retailer who sells a product or a service is concerned with how a change in price affects demand for the article. The sensitivity of demand to price changes varies with different items. For smaller items, such as soft drinks, food staples, and lightbulbs, small percentage changes in price will not affect the demand for the item much. However, sometimes a small change in price on big-ticket items, such as cars, homes, and furniture, can have significant effects on demand.

One way to measure the sensitivity of changes in price to demand is by the ratio of percent change in demand to percent change in price. If q represents the quantity demanded and p the price of the item, this ratio can be written as

$$\frac{\Delta q/q}{\Delta p/p},$$

where Δq represents the change in q and Δp represents the change in p. The ratio is always negative, because q and p are positive, while Δq and Δp have opposite signs. (An *increase* in price causes a *decrease* in demand.) If the absolute value of this quantity is large, it shows that a small increase in price can cause a relatively large decrease in demand.

Applying some algebra, we can rewrite this ratio as

$$\frac{\Delta q/q}{\Delta p/p} = \frac{\Delta q}{q}\cdot\frac{p}{\Delta p} = \frac{p}{q}\cdot\frac{\Delta q}{\Delta p}.$$

Suppose $q = f(p)$. (Note that this is the inverse of the way our demand functions have been expressed so far; previously, we had $p = D(q)$.) Then $\Delta q = f(p + \Delta p) - f(p)$. It follows that

$$\frac{\Delta q}{\Delta p} = \frac{f(p + \Delta p) - f(p)}{\Delta p}.$$

As $\Delta p \to 0$, this quotient becomes

$$\lim_{\Delta p\to 0}\frac{\Delta q}{\Delta p} = \lim_{\Delta p\to 0}\frac{f(p + \Delta p) - f(p)}{\Delta p} = \frac{dq}{dp},$$

and

$$\lim_{\Delta p\to 0}\frac{p}{q}\cdot\frac{\Delta q}{\Delta p} = \frac{p}{q}\cdot\frac{dq}{dp}.$$

The quantity

$$E = -\frac{p}{q}\cdot\frac{dq}{dp}$$

is positive because dq/dp is negative. E is called **elasticity of demand** and measures the instantaneous responsiveness of demand to price. For example, E may be .2 for medical expenses (even though these expenses have considerable price increases each year, they still have a high demand), but may be

1.2 for stereo equipment (high-cost and nice-to-have items, but not necessities). These numbers indicate that the demand for medical services is much less responsive to price changes than the demand for stereo equipment.

If $E < 1$, the relative change in demand is less than the relative change in price, and the demand is called **inelastic**. If $E > 1$, the relative change in demand is greater than the relative change in price, and the demand is called **elastic**. When $E = 1$, the percentage changes in price and demand are relatively equal, and the demand is said to have **unit elasticity**.

Sometimes elasticity is counterintuitive. The addiction to illicit drugs is an excellent example. The quantity of the drug demanded by addicts, if anything, increases, no matter what the cost. Thus, illegal drugs are an inelastic commodity.

Example 1 The VCR market provides a good example of price elasticity of consumer electronics. In the 1990s, the demand for VCRs was expressed by the equation $q = -.025p + 20.45$, where q was the annual demand, in millions of VCRs, and p is the price of the product.*

(a) Calculate and interpret the elasticity of demand when $p = \$200$ and when $p = \$500$.

Solution Since $q = -.025p + 20.45$, we have $dq/dp = -.025$, so that

$$E = -\frac{p}{q}\cdot\frac{dq}{dp}$$

$$= -\frac{p}{-.025p + 20.45}\cdot(-.025)$$

$$= \frac{.025p}{-.025p + 20.45}.$$

Let $p = 200$ to get

$$E = \frac{.025(200)}{-.025(200) + 20.45} \approx .324.$$

Since $.324 < 1$, the demand was inelastic, and a percentage change in price resulted in a smaller percentage change in demand. For example, a 10% increase in price will cause a 3.24% decrease in demand.

If $p = 500$, then

$$E = \frac{.025(500)}{-.025(500) + 20.45} \approx 1.57.$$

Since $1.57 > 1$, the price is elastic. At this point, a percentage increase in price resulted in a greater percentage decrease in demand. A 10% increase in price resulted in a 15.7% decrease in demand.

(b) Determine the price at which demand had unit elasticity ($E = 1$). What is the significance of this price?

Solution Demand had unit elasticity at the price p that made $E = 1$, so we must solve the equation

$$E = \frac{.025p}{-.025p + 20.45} = 1$$

$$.025p = -.025p + 20.45$$

$$.05p = 20.45$$

$$p = 409.$$

Demand had unit elasticity at a price of \$409 per VCR. Unit elasticity indicates that the changes in price and demand are about the same.

The definitions from the preceding discussion can be expressed in the manner that follows.

Elasticity of Demand

Let $q = f(p)$, where q is the demand at a price p. The **elasticity of demand** is as follows:

Demand is **inelastic** if $E < 1$.

Demand is **elastic** if $E > 1$.

Demand has **unit elasticity** if $E = 1$.

EXERCISES

1. The monthly demand for beef in a given region can be expressed by the equation $q = -3.003p + 675.23$, where q is the monthly demand in tons and p is the price in dollars per 100 pounds. Determine the elasticity of demand when the price is \$70.*

2. Acme Stationery sells designer-brand pens. The demand equation for annual sales of these pens is $q = -1000p + 70{,}000$, where p is the price per pen. Normally, the pens sell for \$30 each. They are very popular, and Acme has been thinking of raising the price by one third.†
   (a) Find the elasticity of demand if $p = \$30$.
   (b) Find the elasticity of demand if the price is raised by one third. Is raising the price a good idea?

*Adapted from Todd Thibodeaux, *Pricing Plots Products' Destinies*, Consumer Electronics Vision, July/August 1998.

*Adapted from *How Demand and Supply Determine Price*, Agricultural Marketing Manual, Alberta, Canada, February 1999.

†Taken from R. Horn, *Economics 331: Warm-up Problems: Supply, Demand, Elasticity.*

3. The monthly demand for lodging in a certain city is given by $q = -2481.52p + 472{,}191.2$, where p is the nightly rate.*

(a) Find E when $p = \$100$ and when $p = \$75$.

(b) At what price is there unit elasticity?

4. Although there are other contributing factors, increases in the price of cigarettes between 1991 and 2000 showed decreasing consumption. Given that $q = -2.35p + 28.26$ represents the annual demand in billions of packs and p is the price per pack, find and interpret the demand elasticity when $p = \$3.00$.*

5. What must be true about demand if $E = 0$ everywhere?

*Extracted from Bjorn Hanson, *Price Elasticity of Lodging Demand* (PricewaterhouseCoopers, 2000).

*Obtained from Frank J. Chaloupka, *Policy Levers for the Control of Tobacco Consumption* (Impact Teen, 2000).

CHAPTER

12

Applications of the Derivative

Functions and their derivatives can be used to solve a variety of optimization problems. For example, automotive engineers want to design engines with maximum fuel efficiency. A cereal manufacturer wants to have a box that requires the smallest amount of cardboard (and hence is cheapest). A hospital or a business may want to determine the optimum intervals at which supplies should be ordered to minimize the total cost of ordering, delivery, and storage. Every business wants to maximize its profit. See Exercise 34 on page 811, Exercise 51 on page 812, and Exercise 55 on page 813.

CASE 12: A Total Cost Model for a Training Program

In the last chapter, the derivative was defined and interpreted, and formulas were developed for the derivatives of many functions. In this chapter, we investigate the connection between the derivative of a function and the graph of the function. In particular, we shall see how to use the derivative to determine algebraically the maximum and minimum values of a function, as well as the intervals on which the function is increasing or decreasing.

12.1 ▶ Derivatives and Graphs

Informally, we say that a function is **increasing** on an interval if its graph is *rising* from left to right over the interval and that a function is **decreasing** on an interval if its graph is *falling* from left to right over the interval.

Example 1 For which x-intervals is the function graphed in Figure 12.1 increasing? For which intervals is it decreasing?

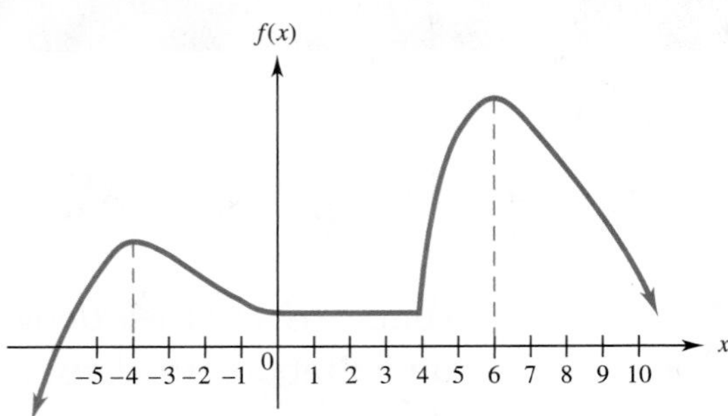

FIGURE 12.1

Solution Moving from left to right, the function is increasing up to −4, then decreasing from −4 to 0, constant (neither increasing nor decreasing) from 0 to 4, increasing from 4 to 6, and, finally, decreasing from 6 on. In interval notation, the function is increasing on $(-\infty, -4)$ and $(4, 6)$, decreasing on $(-4, 0)$ and $(6, \infty)$, and constant on $(0, 4)$. 1 ✓

✓ Checkpoint 1

For what values of x is the function whose graph is shown increasing? decreasing?

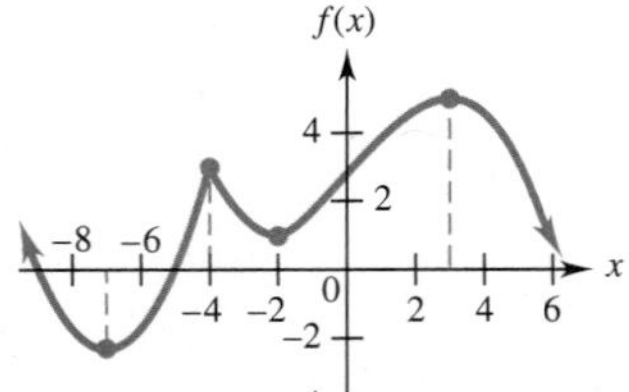

Answer: See page 785.

In order to examine the connection between the graph of a function f and the derivative of f, it is sometimes helpful to think of the graph of f as a roller-coaster track, with a roller-coaster car moving from left to right along the graph, as shown in Figure 12.2. At any point along the graph, the floor of the car (a straight-line segment) represents the tangent line to the graph at that point.

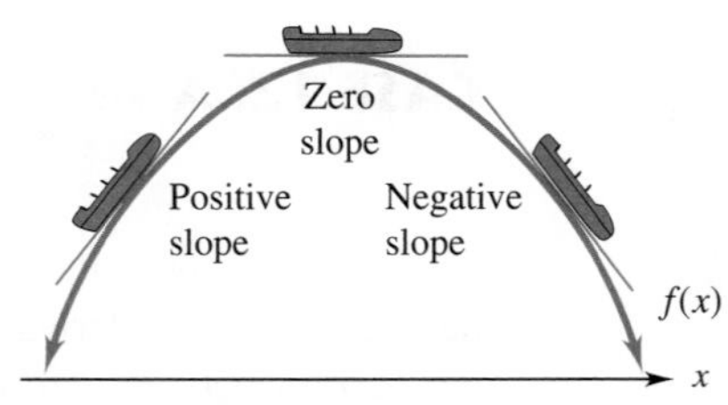

FIGURE 12.2

The slope of the tangent line is positive when the car travels uphill (the function is *increasing*), and the slope of the tangent line is negative when the car travels downhill (the function is *decreasing*). Since the slope of the tangent line at the point $(x, f(x))$ is given by the derivative $f'(x)$, we have the following useful facts.

Increasing and Decreasing Functions

Suppose a function f has a derivative at each point in an open interval. Then

1. if $f'(x) > 0$ for each x in the interval, f is *increasing* on the interval;
2. if $f'(x) < 0$ for each x in the interval, f is *decreasing* on the interval;
3. if $f'(x) = 0$ for each x in the interval, f is *constant* on the interval.

Example 2 Find the intervals on which the function

$$f(x) = x^3 + 3x^2 - 9x + 4,$$

whose graph is shown in Figure 12.3, is increasing, decreasing, and constant.

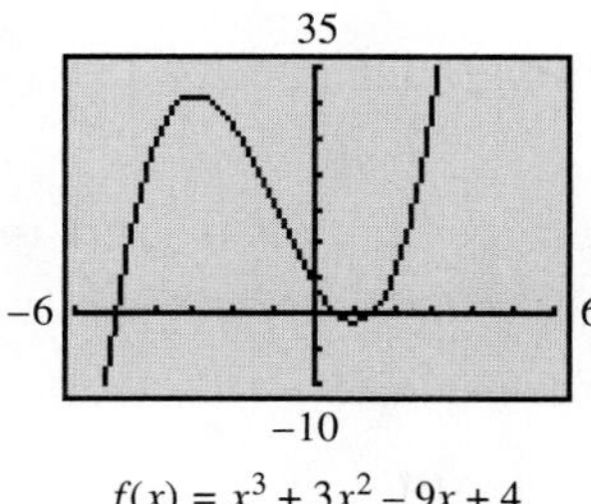

$f(x) = x^3 + 3x^2 - 9x + 4$

FIGURE 12.3

Solution The graph in Figure 12.3 does not clearly show the intervals on which f is increasing or decreasing and suggests that f is constant near $x = -3$. To get more precise information, we first find the derivative:

$$f'(x) = 3x^2 + 6x - 9.$$

According to the box preceding the example, the function f is increasing on the intervals where $f'(x) > 0$ and decreasing on the intervals where $f'(x) < 0$. So we must solve the following inequalities:

$$\begin{array}{ccc} f'(x) > 0 & \text{and} & f'(x) < 0 \\ 3x^2 + 6x - 9 > 0 & & 3x^2 + 6x - 9 < 0. \end{array}$$

As shown in Section 2.5, the first step is to solve the equation $f'(x) = 0$:

$$\begin{aligned} 3x^2 + 6x - 9 &= 0 \\ 3(x^2 + 2x - 3) &= 0 \\ 3(x + 3)(x - 1) &= 0 \\ x = -3 \quad \text{or} \quad x &= 1. \end{aligned}$$

Now use either graphical or algebraic methods to solve the inequalities.

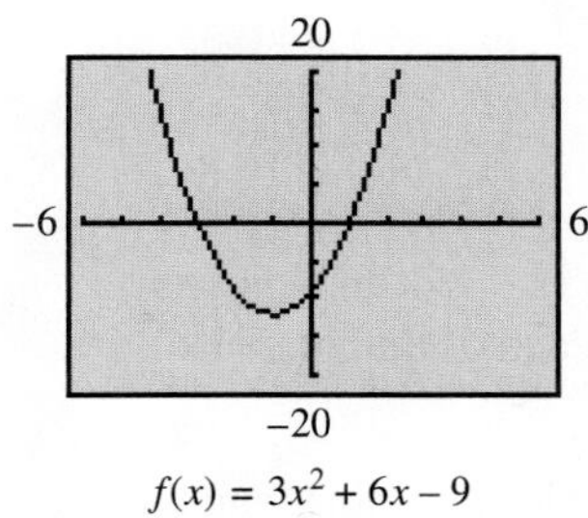

$f(x) = 3x^2 + 6x - 9$

FIGURE 12.4

Graphical The graph of the derivative $f'(x) = 3x^2 + 6x - 9$ in Figure 12.4 lies above the x-axis on the intervals $(-\infty, -3)$ and $(1, \infty)$, meaning that $f'(x) > 0$ on these intervals. Similarly, $f'(x) < 0$ on $(-3, 1)$ because the graph of $f'(x)$ is below the x-axis on $(-3, 1)$.

Algebraic The numbers -3 and 1 divide the x-axis into three intervals: $(-\infty, -3)$, $(-3, 1)$, and $(1, \infty)$. Determine the sign of $f'(x)$ on each interval by testing a number in that interval, as summarized next.

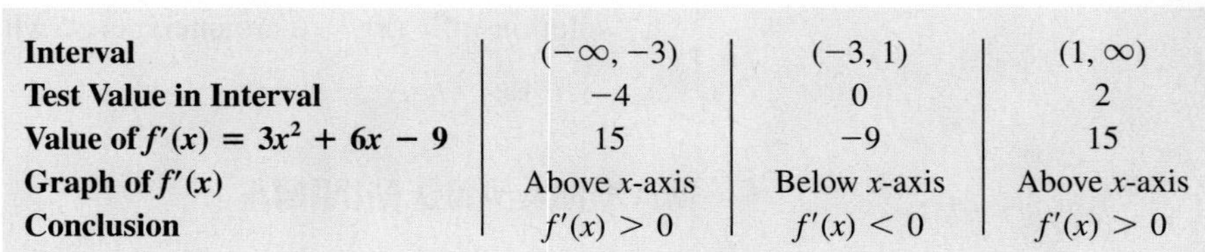

Interval	$(-\infty, -3)$	$(-3, 1)$	$(1, \infty)$
Test Value in Interval	-4	0	2
Value of $f'(x) = 3x^2 + 6x - 9$	15	-9	15
Graph of $f'(x)$	Above x-axis	Below x-axis	Above x-axis
Conclusion	$f'(x) > 0$	$f'(x) < 0$	$f'(x) > 0$

Regardless of method, the conclusion is the same: The original function f is increasing on $(-\infty, -3)$ and $(1, \infty)$, which are the intervals where $f'(x) > 0$, and decreasing on $(-3, 1)$, where $f'(x) < 0$. In particular, f is *not* constant near $x = -3$; the "flat" portion of the graph in Figure 12.3 is not actually horizontal (as you can confirm by using the trace feature of a graphing calculator). 2

Checkpoint 2

Find all intervals on which $f(x) = 4x^3 + 3x^2 - 18x + 1$ is increasing or decreasing.

Answer: *See page 785.*

CRITICAL NUMBERS

In Example 2, the numbers for which $f'(x) = 0$ were essential for determining exactly where the function was increasing and decreasing. The situation is a bit different with the absolute-value function $f(x) = |x|$, whose graph is shown in Figure 12.5. Clearly, f is decreasing on the left of $x = 0$ and increasing on the right of $x = 0$. But as we saw on pages 700–701, the derivative of $f(x) = |x|$ does not exist at $x = 0$. These examples suggest that the points where the derivative is 0 or undefined play an important role.

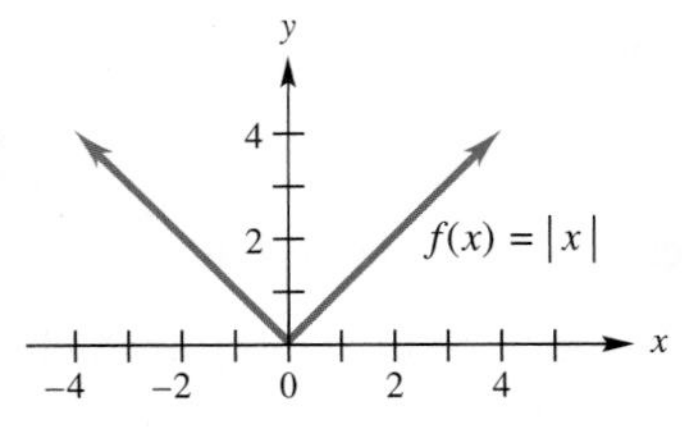

FIGURE 12.5

In view of the preceding discussion, we have the following definition.

Critical Numbers

If f is a function, then a number c for which $f(c)$ is defined and

$$\text{either } f'(c) = 0 \quad \text{or} \quad f'(c) \text{ does not exist}$$

is called a **critical number** of f. The corresponding point $(c, f(c))$ on the graph of f is called a **critical point**.

The procedure used in Example 2, which applies to all functions treated in this book, can now be summarized in the manner that follows.

Increasing/Decreasing Test

To find the intervals on which a function f is increasing or decreasing, do the following:

Step 1 Compute the derivative f'.

Step 2 Find the critical numbers of f.

Step 3 Solve the inequalities $f'(x) > 0$ and $f'(x) < 0$ graphically or algebraically (by testing a number in each of the intervals determined by the critical numbers).

The solutions of $f'(x) > 0$ are intervals on which f is increasing, and the solutions of $f'(x) < 0$ are intervals on which f is decreasing.

MAXIMA AND MINIMA

We have seen that the graph of a typical function may have "peaks" or "valleys," as illustrated in Figure 12.6.

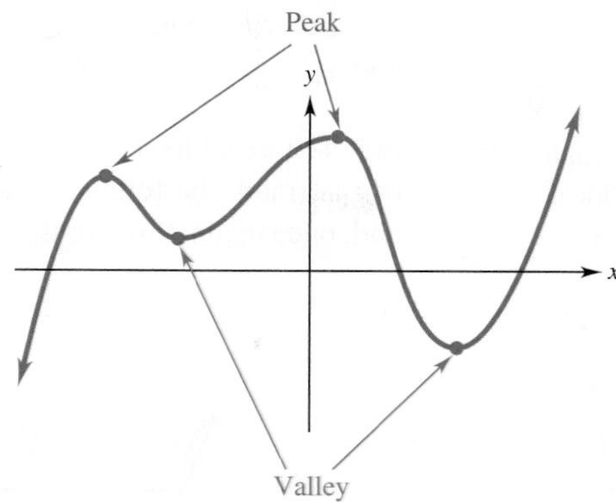

FIGURE 12.6

Peaks and valleys are sometimes called **turning points** of the graph, because as you go from left to right, the graph changes from increasing to decreasing at a peak and from decreasing to increasing at a valley.

A peak is the highest point in its neighborhood, but not necessarily the highest point on the graph. Similarly, a valley is the lowest point in its neighborhood, but not necessarily the lowest point on the graph. Consequently, a peak is called a *local maximum* and a valley a *local minimum.* More precisely, we have the following definitions.

Local Extrema

Let c be a number in the domain of a function f. Then

1. f has a **local maximum** at c if $f(x) \leq f(c)$ for all x near c;
2. f has a **local minimum** at c if $f(x) \geq f(c)$ for all x near c.

The function f is said to have a **local extremum** at c if it has a local maximum or minimum there.

 **NOTE** The plurals of *maximum*, *minimum*, and *extremum* are, respectively, *maxima*, *minima*, and *extrema.*

Example 3 Identify the local extrema of the function whose graph is shown in Figure 12.7.

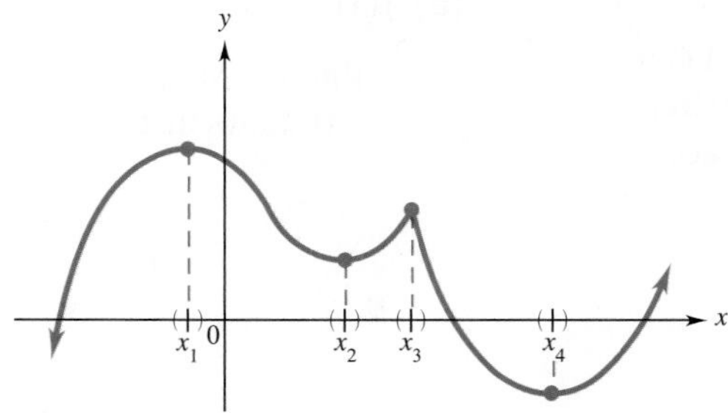

FIGURE 12.7

Solution The function has local maxima at x_1 and x_3 and local minima at x_2 and x_4. 3✓

✓Checkpoint 3

Identify the x-values of all points where the given graphs have local maxima and local minima.

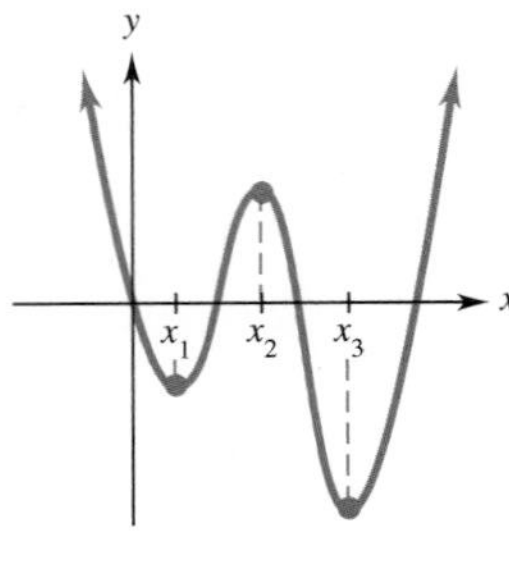

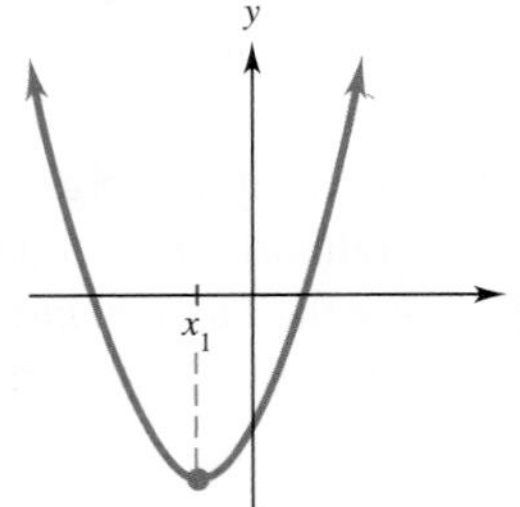

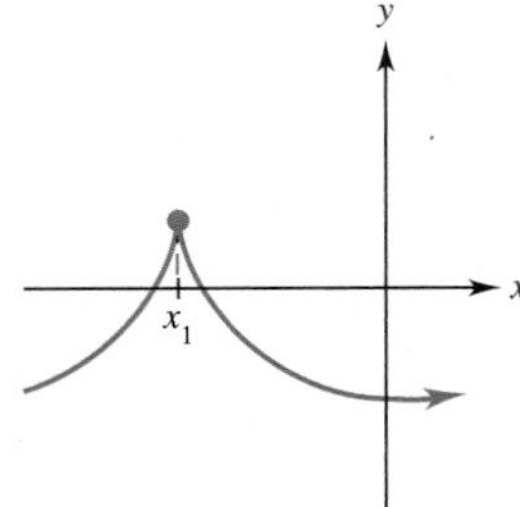

Answers: See page 785.

The *exact* location of a local extremum (rather than a calculator's approximation) can normally be found by using derivatives. To see why this is so, let f be a function and, once again, think of the graph of f as a roller-coaster track, as shown in Figure 12.8.

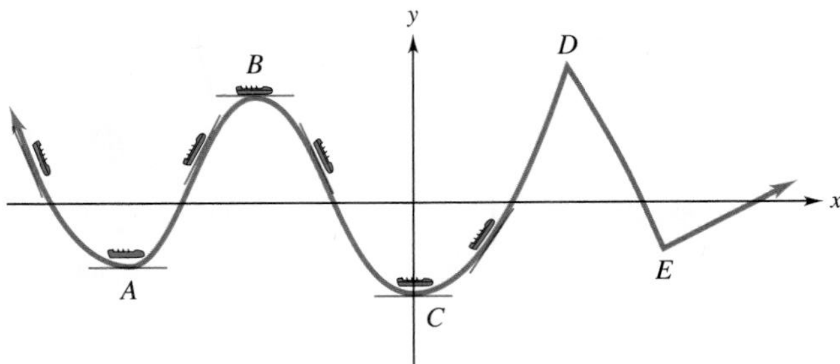

FIGURE 12.8

As the car passes through the local extrema at A, B, and C, the tangent line is horizontal and has slope 0. At D and E, however, a real roller-coaster car would have trouble. It would fly off the track at D and be unable to make the 90° change of direction at E. Notice that the graph does not have tangent lines at points D and E. (See the discussion of the existence of derivatives in Section 11.4.) Thus, a point $(c, f(c))$ where a local extremum occurs has the following property: The tangent line has slope 0, *or* there is no tangent line—that is, $f'(c) = 0$ or $f'(c)$ is not defined. In other words, c is a critical number.

Local Extrema and Critical Numbers

If f has a local extremum at $x = c$, then c is a critical number of f.

CAUTION This result says that every local extremum occurs at a critical number, *not* that every critical number produces a local extremum. Thus, the critical numbers provide a list of *possibilities*: If there is a local extremum, it must occur at a number on the list, but the list may include numbers at which there is no local extremum.

TECHNOLOGY TIP Most graphing calculators have a maximum/minimum finder that can approximate local extrema to a high degree of accuracy. Check your instruction manual.

Example 4 Find the critical numbers of the given functions.

(a) $f(x) = 2x^3 - 3x^2 - 72x + 15$

Solution We have $f'(x) = 6x^2 - 6x - 72$, so $f'(x)$ exists for every x. Setting $f'(x) = 0$ shows that

$$\begin{aligned} 6x^2 - 6x - 72 &= 0 \\ 6(x^2 - x - 12) &= 0 \\ x^2 - x - 12 &= 0 \\ (x + 3)(x - 4) &= 0 \\ x + 3 = 0 \quad \text{or} \quad x - 4 &= 0 \\ x = -3 \quad \text{or} \quad x &= 4. \end{aligned}$$

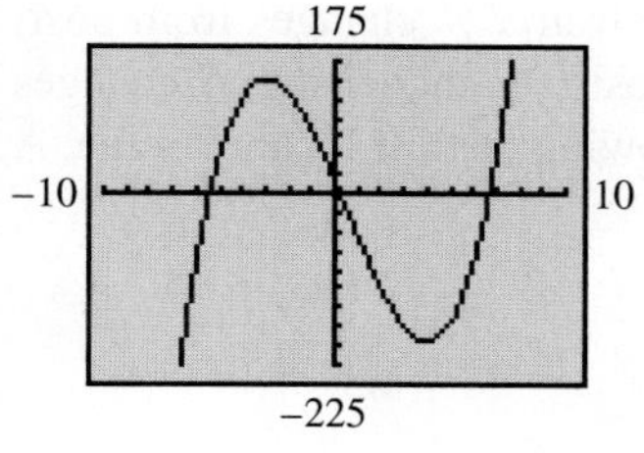

$f(x) = 2x^3 - 3x^2 - 72x + 15$

FIGURE 12.9

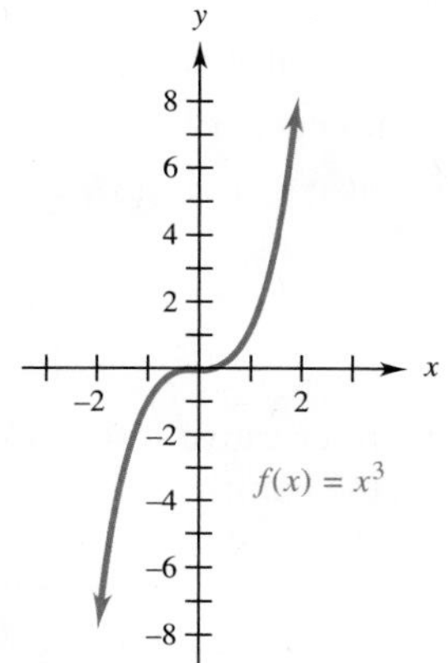

FIGURE 12.10

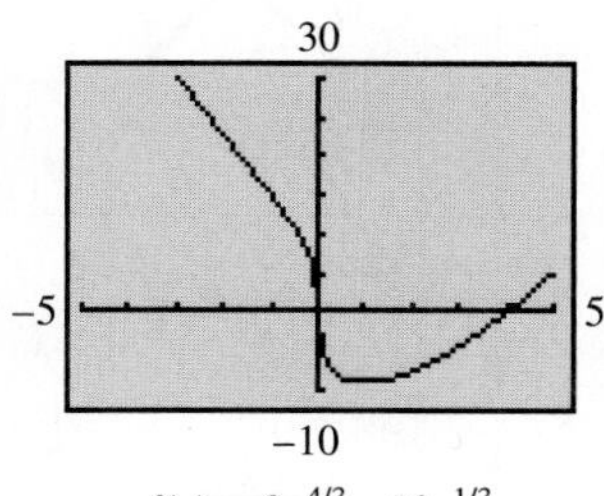

$f(x) = 3x^{4/3} - 12x^{1/3}$

FIGURE 12.11

✓Checkpoint 4

Find the critical numbers for each of the given functions.

(a) $\frac{1}{3}x^3 - x^2 - 15x + 6$

(b) $6x^{2/3} - 4x$

Answers: See page 785.

Therefore, -3 and 4 are the critical numbers of f; these are the only places where local extrema could occur. The graph of f in Figure 12.9 shows that there is a local maximum at $x = -3$ and a local minimum at $x = 4$.

(b) $f(x) = x^3$

Solution The derivative $f'(x) = 3x^2$ is 0 exactly when $x = 0$. So $x = 0$ is the only critical number of f and the only possible location for a local maximum or minimum. In this case, however, we know what the graph of f looks like. (See Figure 12.10.) The graph shows that there is no local maximum or minimum at $x = 0$.

(c) $f(x) = 3x^{4/3} - 12x^{1/3}$

Solution We first compute the derivative:

$$\begin{aligned} f'(x) &= 3 \cdot \frac{4}{3}x^{1/3} - 12 \cdot \frac{1}{3}x^{-2/3} \\ &= 4x^{1/3} - \frac{4}{x^{2/3}} \\ &= \frac{4x^{1/3}x^{2/3}}{x^{2/3}} - \frac{4}{x^{2/3}} \\ &= \frac{4x - 4}{x^{2/3}}. \end{aligned}$$

The derivative fails to exist when $x = 0$. Since the original function f is defined when $x = 0$, 0 is a critical number of f. If $x \neq 0$, then $f'(x)$ is 0 only when the numerator $4x - 4 = 0$, that is, when $x = 1$. So the critical numbers of f are 0 and 1. These numbers are the *possible* locations for local extrema. However, the graph of f in Figure 12.11 suggests that there is a local minimum at $x = 1$, but no local extremum at $x = 0$. ✓4

THE FIRST-DERIVATIVE TEST

When all the critical numbers of a function f have been found, you must then determine which ones lead to local extrema. Sometimes this can be done graphically, as in Example 4. It can also be done algebraically by using the following observation:

> At a local maximum, f changes from increasing to decreasing, and at a local minimum, f changes from decreasing to increasing, as illustrated in Figure 12.12.

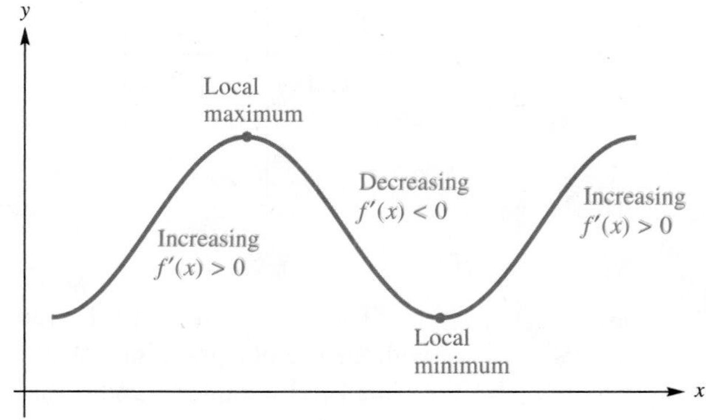

FIGURE 12.12

When f changes from increasing to decreasing, its derivative f' changes from positive to negative. Similarly, when f changes from decreasing to increasing, f' changes from negative to positive. These facts lead to the following test for local extrema; A formal proof of the test is omitted.

First-Derivative Test

Assume that $a < c < b$ and that c is the only critical number for a function f in the interval $[a, b]$. Assume that f is differentiable for all x in $[a, b]$, except possibly at $x = c$.

1. If $f'(a) > 0$ and $f'(b) < 0$, then there is a local maximum at c.
2. If $f'(a) < 0$ and $f'(b) > 0$, then there is a local minimum at c.
3. If $f'(a)$ and $f'(b)$ are both positive or both negative, then there is no local extremum at c.

The sketches in the following table show how the first-derivative test works (assume the same conditions on a, b, and c as those stated in the box):

$f(x)$ has . . .	Sign of $f'(a)$	Sign of $f'(b)$	Sketches	
Local Maximum	+	−		
Local Minimum	−	+		
No Local Extrema	+	+		
No Local Extrema	−	−		

The local extrema of the functions in Example 4(a) and 4(c) were determined with the use of a graphing calculator (once the critical numbers had been found algebraically). Examples 5 and 6 show how the first-derivative test can be used in place of graphing.

Example 5 Use the first-derivative test to find the local extrema of

$$f(x) = 2x^3 - 3x^2 - 72x + 15.$$

Solution Example 4(a) shows that the critical numbers of f are -3 and 4. To use the first-derivative test on -3, we must choose an interval $[a, b]$ containing -3, but not containing the other critical number. We shall use $a = -4$ and $b = 0$. Many other choices of a and b are possible, but we try to select ones that will make the computations easy. Since

$$f'(x) = 6x^2 - 6x - 72 = 6(x^2 - x - 12) = 6(x + 3)(x - 4),$$

it follows that

$$f'(-4) = 6(-4 + 3)(-4 - 4) = 6(-1)(-8) > 0$$

and

$$f'(0) = 6(0 + 3)(0 - 4) = 6(3)(-4) < 0.$$

(Note that it is not necessary to finish calculating the exact value of $f'(x)$ in order to determine its sign.) Thus, the value of the derivative is positive to the left of -3 and negative to the right of -3, as shown in Figure 12.13. By part 1 of the first-derivative test, there is a local maximum at $x = -3$, and it is $f(-3) = 150$.

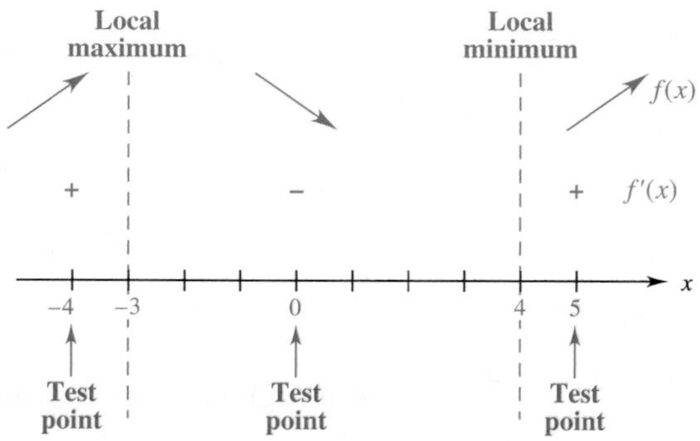

FIGURE 12.13

Similarly, we can use $a = 0$ and $b = 5$ to test the critical number $c = 4$. We just saw that $f'(0) < 0$; also,

$$f'(5) = 6(5 + 3)(5 - 4) = 6(8)(1) > 0.$$

Hence, by part 2 of the first-derivative test, there is a local minimum at $x = 4$, given by $f(4) = -193$. ✓5

✓Checkpoint 5

Find the location of all local extrema of the given functions.

(a) $f(x) = 2x^2 + 12x - 15$

(b) $g(x) = \frac{1}{3}x^3 - x^2 - 8x + 57$

Answers: See page 785.

Example 6 Use the first-derivative test to find the local extrema of

$$f(x) = 3x^{4/3} - 12x^{1/3}.$$

Solution Example 4(c) shows that the critical numbers of f are 0 and 1. We can use the intervals $[-1, 1/2]$ and $[1/2, 2]$ to test these critical numbers:

$$f'(x) = \frac{4x - 4}{x^{2/3}} = \frac{4x - 4}{\sqrt[3]{x^2}}$$

$$f'(-1) = \frac{4(-1) - 4}{\sqrt[3]{(-1)^2}} = \frac{-4 - 4}{1} < 0$$

$$f'\left(\frac{1}{2}\right) = \frac{4(1/2) - 4}{\sqrt[3]{(1/2)^2}} = \frac{-2}{\sqrt[3]{1/4}} < 0$$

$$f'(2) = \frac{4(2) - 4}{\sqrt[3]{2^2}} = \frac{4}{\sqrt[3]{4}} > 0.$$

Since $f'(x)$ is negative at both -1 and $\frac{1}{2}$, part 3 of the first-derivative test shows that there is no local extremum at $x = 0$. Since $f'(\frac{1}{2}) < 0$ and $f'(2) > 0$, there is a local minimum at $x = 1$, by part 2 of the first-derivative test. These results are shown schematically in Figure 12.14. 6✓

✓Checkpoint 6

Find any local extrema for $f(x) = x^3 + 4x^2 - 3x + 5$.

Answer: See page 785.

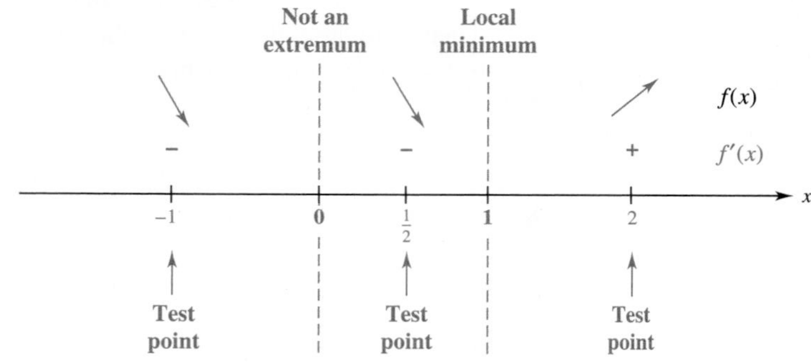

FIGURE 12.14

Although graphing technology sometimes eliminates the need to use the first-derivative test, this is not always the case, as the next example illustrates.

Example 7 Find all local extrema of $f(x) = x^2e^{5x}$.

Solution Although the graph of f in Figure 12.15 suggests that there are no local extrema, algebraic analysis shows otherwise. Using the product rule for derivatives, we have

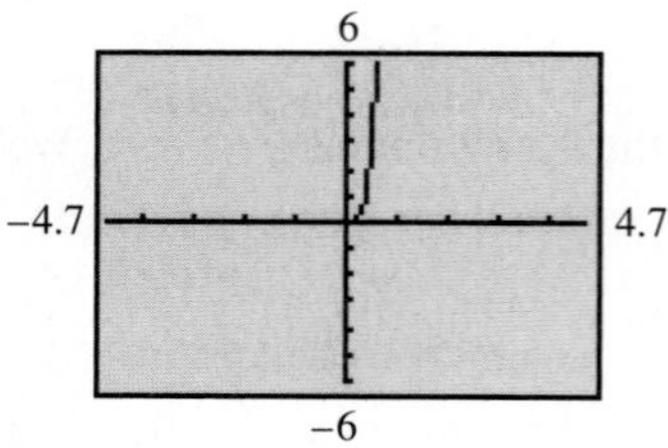

FIGURE 12.15

$$f'(x) = x^2(5e^{5x}) + 2x(e^{5x}) = 5x^2e^{5x} + 2xe^{5x}.$$

So the critical numbers are the solutions of

$$5x^2e^{5x} + 2xe^{5x} = 0$$

$$e^{5x}(5x^2 + 2x) = 0$$

$$e^{5x}(5x + 2)x = 0.$$

Since e^{5x} is never 0, this derivative can equal 0 only when

$$5x + 2 = 0 \qquad \text{or} \quad x = 0.$$

$$x = -2/5 = -.4$$

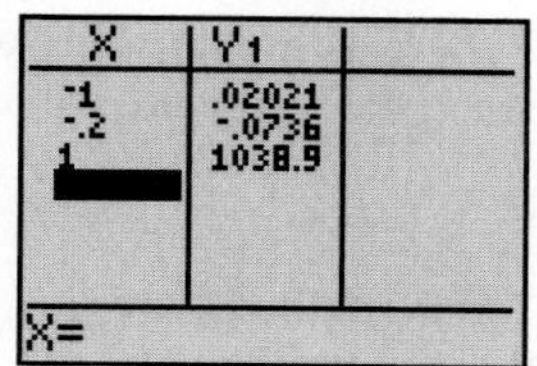

FIGURE 12.16

✓Checkpoint 7

Find a viewing window on a graphing calculator that clearly shows the local maximum and minimum of the function f in Example 7.

Answer: See page 785.

✓Checkpoint 8

Find all local extrema of $g(x) = x^3e^x$.

Answer: See page 785.

We choose the interval $[-1, -.2]$ to test $-.4$ and the interval $[-.2, 1]$ to test 0. Using a calculator to compute values of $y_1 = f'(x)$, we obtain Figure 12.16. By the first-derivative test, there is a local maximum at $x = -.4$ and there is a local minimum at $x = 0$, as indicated schematically in Figure 12.17. ✓7 ✓8

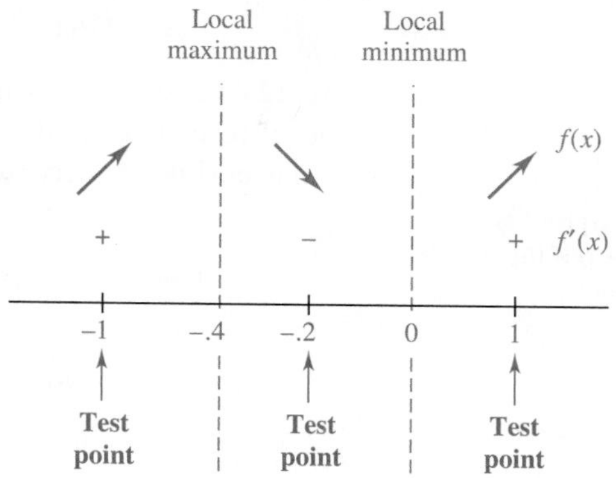

FIGURE 12.17

Example 8 **Business** The cost function $C(x)$ and the revenue function $R(x)$ (in millions of dollars) for Continental Airlines, Inc., for the years 1997–2007 are respectively approximated by

$$C(x) = 20.192x^3 - 678.052x^2 + 7794.925x - 21{,}582.1 \quad (7 \le x \le 17)$$

and

$$R(x) = 24.378x^3 - 799.425x^2 + 8790.739x - 23{,}430.8 \quad (7 \le x \le 17),$$

where $x = 7$ corresponds to 1997.* Determine where the profit function is increasing, and find the minimum profit in the period 1997–2007.

Solution The profit function is the difference between the revenue function and the cost function. We subtract the coefficients of the cost function from the respective coefficients of the revenue function:

$$\begin{aligned} P(x) &= R(x) - C(x) \\ &= (24.378 - 20.192)x^3 + (-799.425 + 678.052)x^2 \\ &\quad + (8790.739 - 7794.925)x + (-23{,}430.8 + 21{,}582.1) \\ &= 4.186x^3 - 121.373x^2 + 995.814x - 1848.7. \end{aligned}$$

To determine the critical numbers, we find $P'(x)$ and use the quadratic formula to solve $P'(x) = 0$:

$$P'(x) = 12.558x^2 - 242.746x + 995.814 = 0$$

$$x = \frac{242.746 \pm \sqrt{(-242.746)^2 - 4(12.558)(995.814)}}{2(12.558)} \approx \begin{cases} 5.9 \\ 13.4 \end{cases}.$$

Since 13.4 is the only critical number in the domain of our profit function, we apply the first-derivative test with $a = 7$ and $b = 17$. We have

$$P'(7) = -88.066 < 0 \quad \text{and} \quad P'(17) = 498.394 > 0.$$

*www.morningstar.com.

Therefore, $P(x)$ is decreasing on the (approximate) interval (7, 13.4), is increasing on (13.4, 17), and has a local minimum at approximately 13.4. Hence, the minimum profit is

$$\begin{aligned} P(x) &= 4.186x^3 - 121.373x^2 + 995.814x - 1848.7 \\ P(13.4) &= 4.186(13.4)^3 - 121.373(13.4)^2 + 995.814(13.4) - 1848.7 \\ &\approx -226.6 \text{ (that is, a loss of \$226.6 million).} \end{aligned}$$

Figure 12.18 shows $C(x)$ (in blue), $R(x)$ (in red), and $P(x)$ (in green). We see that whenever revenue exceeds cost (when red is higher than blue), the profit function (green) is positive. Otherwise, the profit function is negative. 9 ✓

✓Checkpoint 9

Find the interval where profit is increasing and $P(x) > 0$ if the profit function is defined as

$$P(x) = -600 + 70x - x^2.$$

Answer: See page 785.

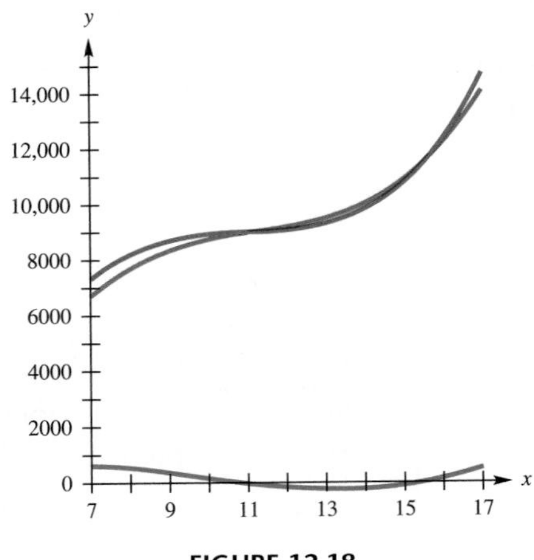

FIGURE 12.18

12.1 ▶ Exercises

For each function, list the intervals on which the function is increasing, the intervals on which it is decreasing, and the location of all local extrema. (See Examples 1 and 3.)

1.

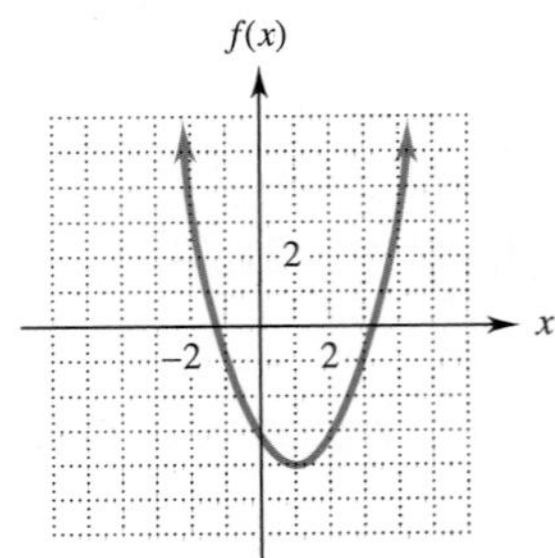

2.

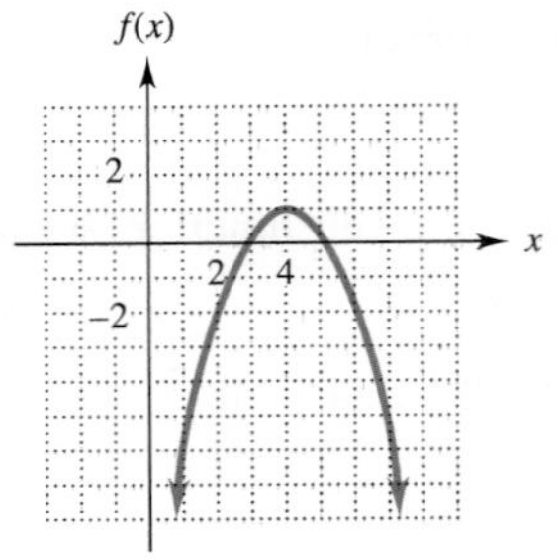

3.

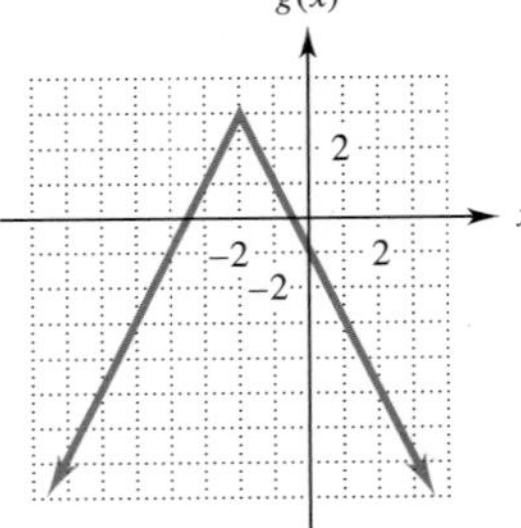

4.

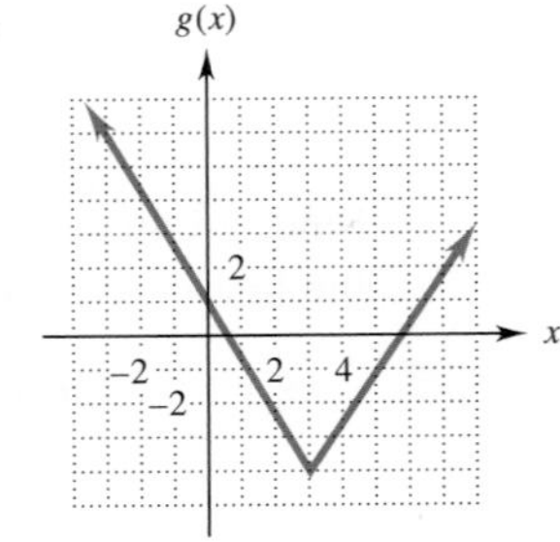

5.

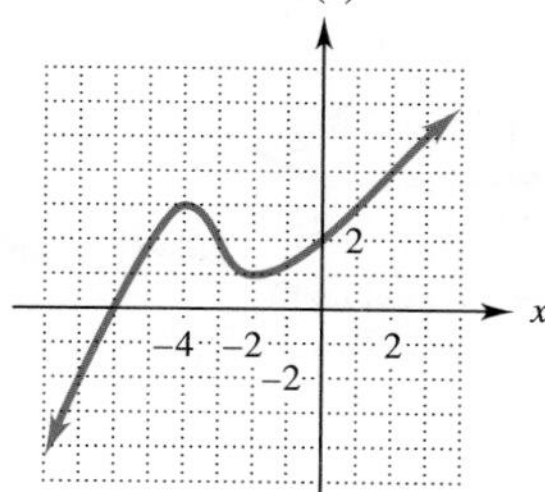

6.

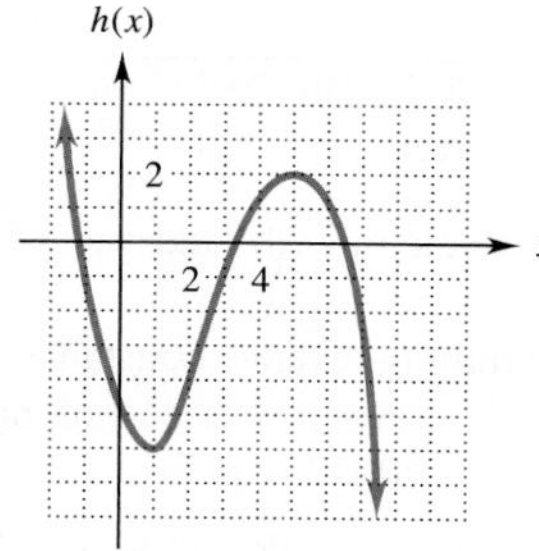

7.

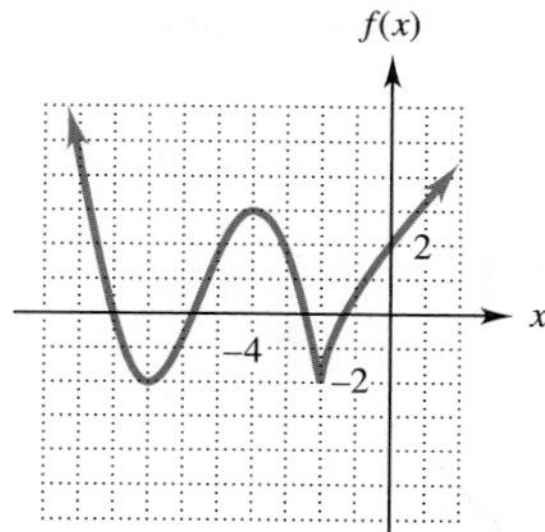

8.

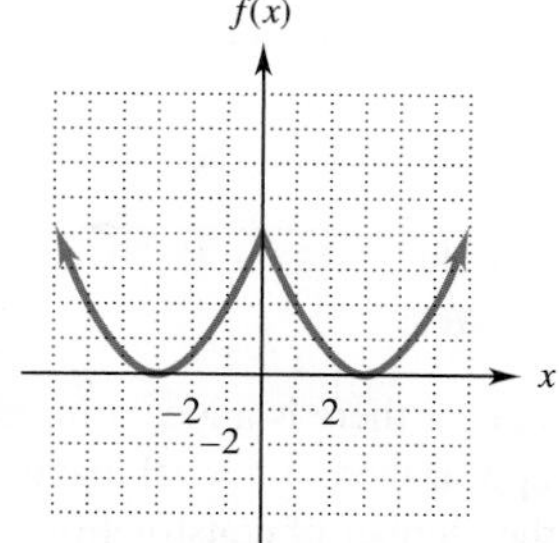

Find the intervals on which each function is increasing or decreasing. (See Example 2.)

9. $f(x) = 2x^3 - 5x^2 - 4x + 2$
10. $f(x) = 4x^3 - 9x^2 - 12x + 7$
11. $f(x) = \dfrac{x+1}{x+3}$
12. $f(x) = \dfrac{x^2+4}{x}$
13. $f(x) = \sqrt{6-x}$
14. $f(x) = \sqrt{x^2+9}$
15. $f(x) = 2x^3 + 3x^2 - 12x + 5$
16. $f(x) = 4x^3 - 15x^2 - 72x + 5$

The graph of the derivative function f' is given; list the critical numbers of the function f.

17.

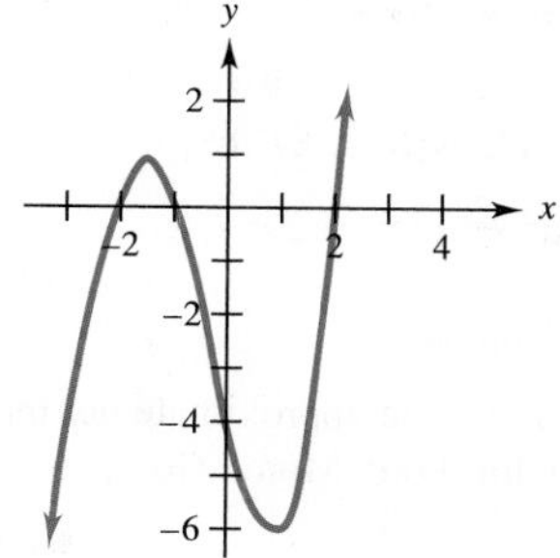

18.

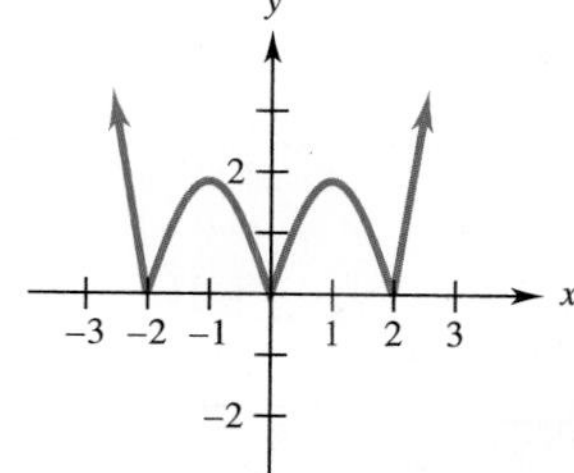

Determine the location of each local extremum of the function. (See Examples 4–7.)

19. $f(x) = x^3 + 3x^2 - 3$
20. $f(x) = x^3 - x^2 - 5x + 1$
21. $f(x) = x^3 + 6x^2 + 9x + 2$
22. $f(x) = x^3 - 3x^2 - 24x + 5$
23. $f(x) = \dfrac{4}{3}x^3 - \dfrac{21}{2}x^2 + 5x + 3$
24. $f(x) = -\dfrac{2}{3}x^3 - \dfrac{1}{2}x^2 - 3x - 4$
25. $f(x) = \dfrac{2}{3}x^3 - x^2 - 12x + 2$
26. $f(x) = \dfrac{4}{3}x^3 + 10x^2 - 24x - 3$
27. $f(x) = x^5 + 20x^2 + 8$
28. $f(x) = 3x^3 - 18.5x^2 - 4.5x - 45$

In Exercises 29–42, use the first-derivative test to determine the location of each local extremum and the value of the function at this extremum. (See Examples 5–7.)

29. $f(x) = x^{11/5} - x^{6/5} + 1$
30. $f(x) = (7 - 2x)^{2/3} - 2$
31. $f(x) = -(3 - 4x)^{2/5} + 4$
32. $f(x) = x^4 + \dfrac{1}{x}$
33. $f(x) = \dfrac{x^3}{x^3 + 1}$
34. $f(x) = \dfrac{x^2 - 2x + 1}{x - 3}$
35. $f(x) = -xe^x$
36. $f(x) = xe^{-x}$
37. $f(x) = x \cdot \ln|x|$
38. $f(x) = x - \ln|x|$
39. $f(x) = xe^{3x} - 2$
40. $f(x) = x^3e^{4x} + 1$

Use the maximum/minimum finder on a graphing calculator to determine the approximate location of all local extrema of these functions.

41. $f(x) = .2x^4 - x^3 - 12x^2 + 99x - 5$

42. $f(x) = x^5 - 12x^4 - x^3 + 232x^2 + 260x - 600$

43. $f(x) = .01x^5 + .2x^4 - x^3 - 6x^2 + 5x + 40$

44. $f(x) = .1x^5 + 3x^4 - 4x^3 - 11x^2 + 3x + 2$

Work the given exercises. (See Example 8.)

45. Business The graph shows the approximate net income (in billions of dollars) for Ford Motor Company from 1999–2008:*

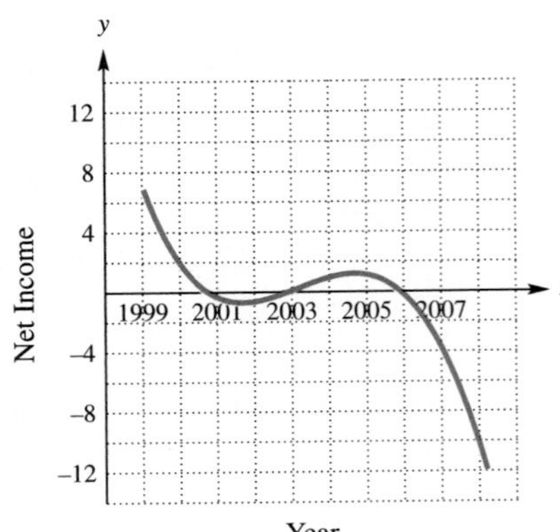

During which years is the slope 0 at some point?

46. Business The net income (in billions of dollars) for Home Depot Corp. from 2000–2009 is modeled by the following graph:*

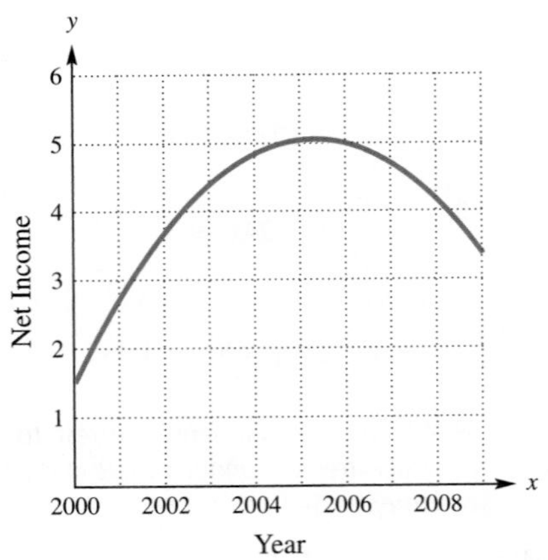

In what years is the function increasing? decreasing?

47. Business The graph shows the revenue generated (in billions of dollars) from the computer programming and computer design industries:

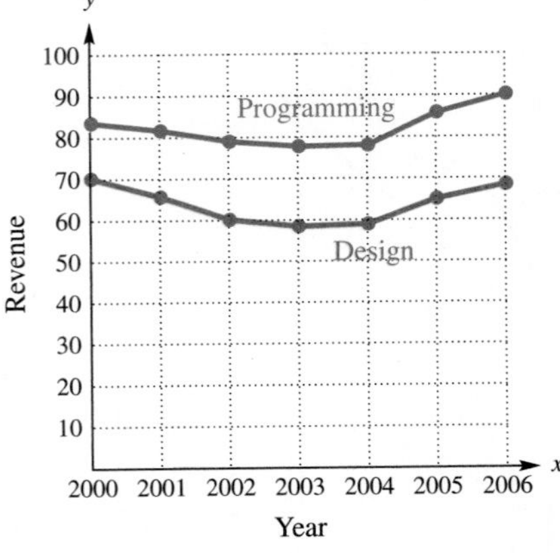

(a) In what years is the revenue of both industries increasing?

(b) In what year is programming revenue constant?

(c) Is programming or design revenue decreasing more rapidly between 2000 and 2003? Explain.

48. Business The graph shows an approximation of the number of workers (in thousands) employed in the construction industry in California, where year 0 corresponds to 1990:*

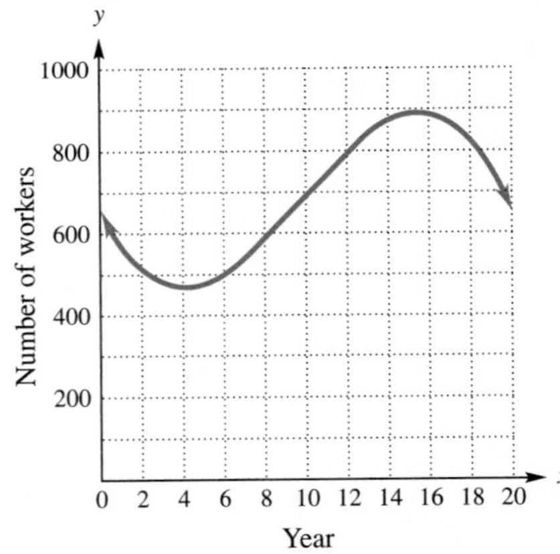

(a) In what year does it appear there is a local minimum?

(b) In what year does it appear there is a local maximum?

(c) In the year 2000, is the number of construction workers increasing or decreasing?

49. Business The function

$$f(x) = -.547x^3 + 16.072x^2 - 102.955x + 658.046$$

approximates the number of construction workers employed in California, as described in Exercise 48. Use the techniques presented in this section to find the location of all local extrema.

*www.morningstar.com.

*St. Louis Federal Reserve.

50. **Business** The profit $P(x)$ (in billions of dollars) for Dell, Inc., can be approximated by the function

$$P(x) = -.015x^3 + .167x^2 - .252x + 1.770 \quad (0 \le x \le 9),$$

where $x = 0$ corresponds to the year 2000.* Find the years in which profit is increasing and decreasing for $0 \le x \le 9$.

51. **Social Science** The standard normal probability function is used to describe many different populations. Its graph is the well-known normal curve. This function is defined by

$$f(x) = \frac{1}{\sqrt{2\pi}} e^{-x^2/2}.$$

Give the intervals where the function is increasing and decreasing.

52. **Health** The aortic pressure–diameter relation in a particular patient who underwent cardiac catheterization can be modeled by the polynomial

$$D(p) = .000002p^3 - .0008p^2 + .1141p + 16.683 \quad (55 \le p \le 130),$$

where $D(p)$ is the aortic diameter (in millimeters) and p is the aortic pressure (in mmHg).† Determine where this function is increasing and where it is decreasing within the given interval.

53. **Business** The cost and revenue for Johnson and Johnson, Inc., can be respectively approximated by

$$C(x) = -.0021x^3 + .0159x^2 + 3.1442x + 24.1502 \quad (0 \le x \le 8),$$

and

$$R(x) = -.0090x^3 + .1335x^2 + 3.8922x + 28.9465 \quad (0 \le x \le 8),$$

where $x = 0$ corresponds to the year 2000.‡ Determine the intervals within (0, 8) where the profit function is increasing.

54. **Health** The number of people $P(t)$ (in hundreds) infected t days after an epidemic begins is approximated by

$$P(t) = \frac{10 \ln(.19t + 1)}{.19t + 1}.$$

When will the number of people infected start to decline?

55. **Natural Science** The function

$$A(x) = .004x^3 - .05x^2 + .16x + .05$$

approximates the blood alcohol concentration in an average person's bloodstream x hours after drinking 8 ounces of 100-proof whiskey. The function applies only to the interval [0, 6].

(a) On what time intervals is the blood alcohol concentration increasing?

(b) On what intervals is it decreasing?

56. **Natural Science** The percent of concentration of a drug in the bloodstream x hours after the drug is administered is given by

$$K(x) = \frac{4x}{3x^2 + 27}.$$

(a) On what time intervals is the concentration of the drug increasing?

(b) On what intervals is it decreasing?

57. **Social Science** The graph shows the average number of 911 calls for police service for the Cleveland Police Department during the 24-hour day, with $x = 0$ corresponding to the hour 12:00 to 12:59 A.M.*

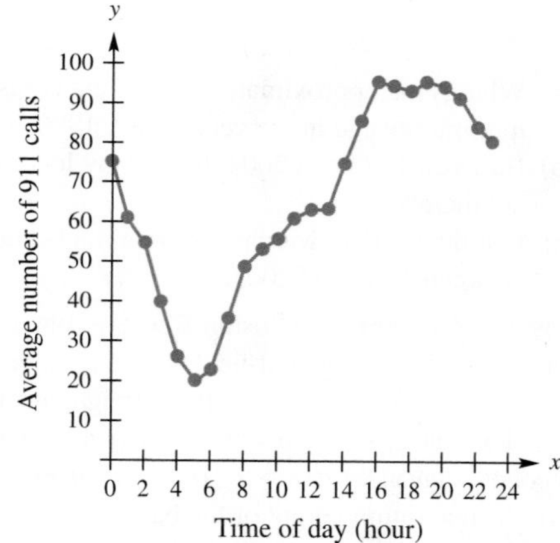

(a) List the intervals on which the average number of 911 calls is increasing.

(b) List the intervals on which the average number of 911 calls is decreasing

(c) List the times of day at which this function has a local maximum or minimum.

58. **Finance** The average interest rate for 6-month certificates of deposit (CDs) can be approximated by the function

$$f(x) = .026x^3 - .841x^2 + 8.382x - 21.47 \quad (7 \le x \le 17),$$

where $x = 7$ corresponds to the year 1997.† Find the local extrema for the interval $7 \le x \le 17$.

---

*www.morningstar.com.

†Stefanadis, C., J. Dernellis, et al., "Assessment of Aortic Line of Elasticity Using Polynomial Regression Analysis," *Circulation*, Vol. 101, No. 15, April 18, 2000, pp. 1819–1825.

‡www.morningstar.com.

*Holcomb, J., and Radke-Sharpe, N. "Forecasting Police Calls during Peak Times for the City of Cleveland," *Case Studies in Business, Industry, and Government Statistics*, 1(1), 2007, pp. 47–53.

†*Statistical Abstract of the United States:* 2009.

59. Finance The minimum wage rose from \$1.60 an hour in 1970 to \$5.85 in 2007. However, the actual purchasing power of the minimum wage has varied greatly over that period. The graph shows the value of the minimum wage in constant 2007 dollars, adjusted for inflation, from 1970 to 2007:*

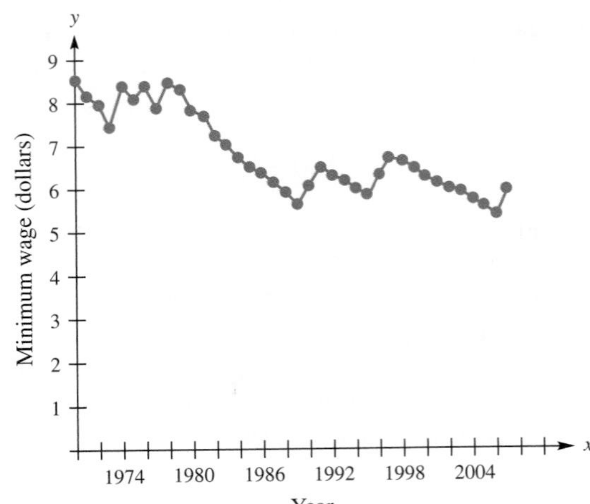

(a) What is the approximate value of the adjusted minimum wage in the years 1980, 1990, and 2000?
(b) Between 1973 and 2004, how many local maxima are there?
(c) List the years in which local minima occurred between 1986 and 2007.

60. Physical Science A Boston Red Sox pitcher stands on top of the 37-foot-high left-field wall (the "Green Monster") in Fenway Park and fires a fastball straight up. The position function that gives the height of the ball (in feet) at time t seconds is given by $s(t) = -16t^2 + 140t + 37$.† Find
(a) the maximum height of the ball;
(b) the time when the ball hits the ground;
(c) the velocity of the ball when it hits the ground.

61. Business The amount of revenue generated (in millions of dollars) from taxes on alcohol in Connecticut can be approximated by the function

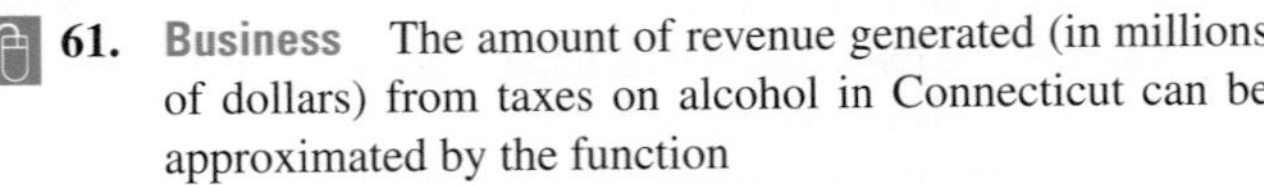

$$g(x) = 4.63 + 60.66x - 33.98x^2 + 8.686x^3 - 1.1526x^4 + .08246x^5 - .0030185x^6 + .00004435x^7 \quad (2 \le x \le 18),$$

where $x = 2$ corresponds to the year 1992.‡
(a) Using a graphing calculator or software, plot $g(x)$.
(b) How many local extrema appear in the graph?
(c) Use the maximum/minimum finder to approximate the values of the extrema.

*www.dol.gov.
†Exercise provided by Frederick Russell of Charles County Community College.
‡St. Louis Federal Reserve.

62. Natural Science A marshy region used for agricultural drainage has become contaminated with selenium. It has been determined that flushing the area with clean water will reduce the selenium for a while, but it will then begin to build up again. A biologist has found that the percentage of selenium in the soil x months after the flushing begins is given by

$$f(x) = \frac{x^2 + 36}{2x} \qquad (1 \le x \le 12).$$

When will the selenium be reduced to a minimum? What is the minimum percentage?

63. Finance The amount of revenue generated (in millions of dollars) from motor vehicle licenses in Massachusetts can be approximated by the function

$$h(x) = -.1351x^3 + 4.472x^2 - 38.864x + 332.459 \quad (2 \le x \le 18),$$

where $x = 2$ corresponds to the year 1992.*
(a) Using a graphing calculator or software, plot $h(x)$.
(b) How many local extrema appear in the graph?
(c) Find the local extrema.

64. Natural Science The number of salmon swimming upstream to spawn is approximated by

$$S(x) = -x^3 + 3x^2 + 400x + 5000 \quad (6 \le x \le 20),$$

where x represents the temperature of the water in degrees Celsius. Find the water temperature that produces the maximum number of salmon swimming upstream.

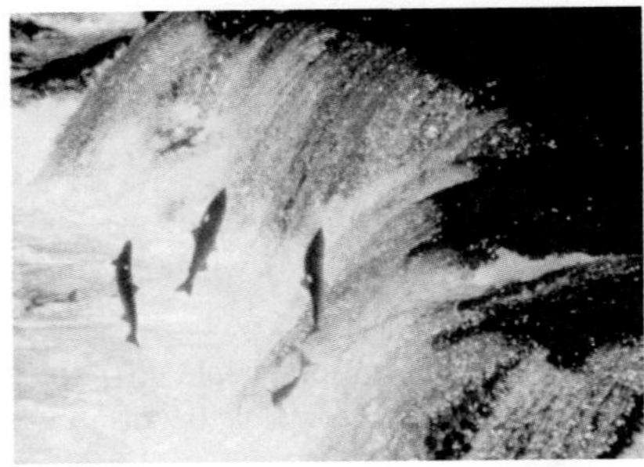

65. Social Science According to projections by the United Nations, the population of the People's Republic of China (in billions) can be approximated by the function

$$f(x) = -.000164x^2 + .014532x + 1.1385 \quad (0 \le x \le 60),$$

where $x = 0$ corresponds to the year 1990.† In what year will China reach its maximum population, and what will that population be?

*St. Louis Federal Reserve.
†http://esa.un.org/unpp.

66. **Natural Science** In the summer, the activity level of a certain type of lizard varies according to the time of day. A biologist has determined that the activity level is given by the function

$$a(t) = .008t^3 - .27t^2 + 2.02t + 7 \qquad (0 \le t \le 24),$$

where t is the number of hours after noon. When is the activity level highest? When is it lowest?

67. **Finance** The interest rate on U.S. government 6-month treasury bonds can be approximated by

$$g(x) = -.0084x^4 + .4298x^3 - 7.9062x^2 + 61.4056x - 165.102 \qquad (7 \le x \le 17),$$

where $x = 7$ corresponds to the year 1997.* Use a graphing calculator to find the local extrema of this function (within the given range of x-values). Interpret your answer.

68. **Natural Science** The microbe concentration $B(x)$, in appropriate units, of Lake Tom depends approximately on the oxygen concentration x, again in appropriate units, according to the function

$$B(x) = x^3 - 7x^2 - 160x + 1800 \qquad (0 \le x \le 20).$$

(a) Find the oxygen concentration that will lead to the minimum microbe concentration.
(b) What is the minimum microbe concentration?

69. **Social Science** A group of researchers found that people prefer training films of moderate length; shorter films contain too little information, while longer films are boring. For a training film on the care of exotic birds, the researchers determined that the ratings people gave for the film could be approximated by

$$R(t) = \frac{20t}{t^2 + 100},$$

where t is the length of the film (in minutes). Find the film length that received the highest rating.

70. Consider the function*

$$g(x) = \frac{1}{x^{12}} - 2\left(\frac{1000}{x}\right)^6.$$

(a) Using a graphing calculator, try to find any local minima, or tell why finding a local minimum is difficult for this function.
(b) Find any local minima, using the techniques of calculus.
(c) On the basis of your results in parts (a) and (b), describe circumstances under which local extrema are easier to find by using the techniques of calculus than with a graphing calculator.

Statistical Abstract of the United States: 2009.

✓Checkpoint Answers

1. Increasing on $(-7, -4)$ and $(-2, 3)$; decreasing on $(-\infty, -7)$, $(-4, -2)$, and $(3, \infty)$
2. Increasing on $(-\infty, -3/2)$ and $(1, \infty)$; decreasing on $(-3/2, 1)$
3. (a) Local maximum at x_2; local minima at x_1 and x_3
   (b) No local maximum; local minimum at x_1
   (c) Local maximum at x_1; no local minimum
4. (a) $-3, 5$ (b) $0, 1$
5. (a) Local minimum at -3
   (b) Local maximum at -2; local minimum at 4
6. Local maximum at -3: $f(-3) = 23$; local minimum at $\frac{1}{3}$: $f\left(\frac{1}{3}\right) = \frac{121}{27}$
7. There are many possibilities, including $-2 \le x \le 2$ and $-.1 \le y \le .1$.
8. Local minimum at $x = -3$; no local extremum at $x = 0$
9. $(10, 35)$

*From Ed Dubinsky, "Is Calculus Obsolete?" *Mathematics Teacher*, 88, no. 2 (February 1995): 146–148.

12.2 ▶ The Second Derivative

The first derivative of a function indicates when the function is increasing or decreasing. In many applications, however, this information isn't enough. For example, Figure 12.19 on the next page shows the prices of two different stocks over a period of months. Both stocks are worth \$5 per share at the beginning and are continually increasing in price.

It is easy to see from the graph that stock B is the better long-term investment. In real life, however, you see only *part* of the graph (the past performance of a stock). How can that help you to predict future performance? To answer this question, look at the graphs again. Both are always increasing, but after the first few months,

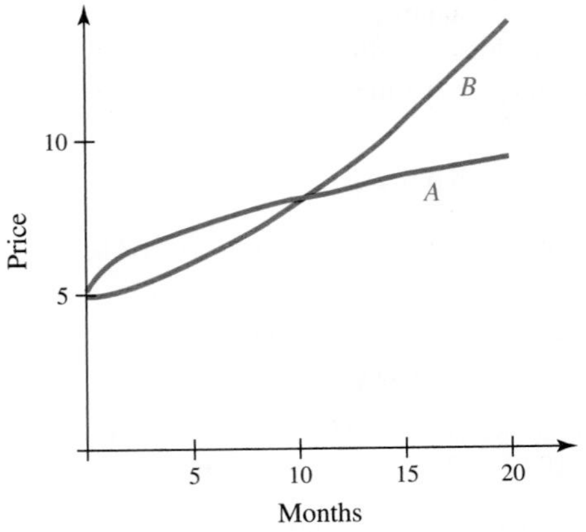

FIGURE 12.19

stock B increases at a faster rate than stock A. So the *rate* of increase plays a crucial role here.

In mathematical terms, if the price of the stock in month x is $f(x)$, then the derivative $f'(x)$ tells you when the price is increasing (or decreasing). The *rate* of increase (or decrease) of the derivative function $f'(x)$ is given by *its* derivative. In other words, the rate at which the stock price is increasing (or decreasing) is given by the derivative of the derivative of the price function. In order to deal with such situations, we need some additional terminology and notation.

HIGHER DERIVATIVES

If a function f has a derivative f', then the derivative of f', if it exists, is the **second derivative** of f, written $f''(x)$. The derivative of $f''(x)$, if it exists, is called the **third derivative** of f, and so on. By continuing this process, we can find **fourth derivatives** and other higher derivatives. For example, if $f(x) = x^4 + 2x^3 + 3x^2 - 5x + 7$, then

$$f'(x) = 4x^3 + 6x^2 + 6x - 5, \quad \text{First derivative of } f$$
$$f''(x) = 12x^2 + 12x + 6, \quad \text{Second derivative of } f$$
$$f'''(x) = 24x + 12, \quad \text{Third derivative of } f$$
$$f^{(4)}(x) = 24. \quad \text{Fourth derivative of } f$$

1 ✓

✓Checkpoint 1

Let

$f(x) = 4x^3 - 12x^2 + x - 1$.

Find

(a) $f''(0)$;
(b) $f''(4)$;
(c) $f''(-2)$.

Answers: See page 798

The second derivative of $y = f(x)$ can be written in any of the following notations:

$$f''(x), \quad y'', \quad \frac{d^2y}{dx^2}, \quad \text{or} \quad D_x^2[f(x)].$$

The third derivative can be written in a similar way. For $n \geq 4$, the nth derivative is written $f^{(n)}(x)$.

Example 1 Find the second derivative of the given functions.

(a) $f(x) = 8x^3 - 9x^2 + 6x + 4$

Solution Here, $f'(x) = 24x^2 - 18x + 6$. The second derivative is the derivative of $f'(x)$, or

$$f''(x) = 48x - 18.$$

(b) $y = \frac{4x + 2}{3x - 1}$

Solution Use the quotient rule to find y':

$$y' = \frac{(3x - 1)(4) - (4x + 2)(3)}{(3x - 1)^2} = \frac{12x - 4 - 12x - 6}{(3x - 1)^2} = \frac{-10}{(3x - 1)^2}.$$

Use the quotient rule again to find y'':

$$y'' = \frac{(3x - 1)^2(0) - (-10)(2)(3x - 1)(3)}{[(3x - 1)^2]^2}$$

$$= \frac{60(3x - 1)}{(3x - 1)^4} = \frac{60}{(3x - 1)^3}.$$

(c) $y = xe^x$

Solution Using the product rule gives

$$\frac{dy}{dx} = x \cdot e^x + e^x \cdot 1 = xe^x + e^x.$$

Differentiate this result to get $\frac{d^2y}{dx^2}$:

$$\frac{d^2y}{dx^2} = (xe^x + e^x) + e^x = xe^x + 2e^x = (x + 2)e^x.$$ 2 ✓

✓Checkpoint 2

Find the second derivatives of the given functions.

(a) $y = -9x^3 + 8x^2 + 11x - 6$

(b) $y = -2x^4 + 6x^2$

(c) $y = \frac{x + 2}{5x - 1}$

(d) $y = e^x + \ln x$

Answers: See page 798

APPLICATIONS

In the previous chapter, we saw that the first derivative of a function represents the rate of change of the function. The second derivative, then, represents the rate of change of the first derivative. This fact has a variety of applications. For one such application, we take another look at stocks *A* and *B* from the beginning of this section.

Example 2 Suppose that, over a 10-month period, the price of stock *A* is given by $f(x) = x^{1/2} + 5$ and the price of stock *B* is given by $g(x) = .1x^{3/2} + 5$.

(a) When are stocks *A* and *B* increasing in price?

Solution The first derivatives of the price functions are, respectively,

$$f'(x) = \frac{1}{2}x^{-1/2} = \frac{1}{2\sqrt{x}} \quad \text{and} \quad g'(x) = .1\left(\frac{3}{2}x^{1/2}\right) = \frac{.3}{2}\sqrt{x}.$$

Both derivatives are always positive (because $\sqrt{x}$ is positive for $x > 0$). So both price functions are increasing for all $x > 0$.

(b) At what rate are these stock prices increasing in the 10th month? What does this suggest about their future performance?

Solution The rate at which $f'(x) = \frac{1}{2}x^{-1/2}$ and $g'(x) = \frac{.3}{2}x^{1/2}$ are increasing is given by *their* derivatives (the second derivatives of f and g):

$$f''(x) = \frac{1}{2}\cdot\frac{-1}{2}x^{-3/2} = \frac{-1}{4\sqrt{x^3}} \quad \text{and} \quad g''(x) = \frac{.3}{2}\cdot\frac{1}{2}x^{-1/2} = \frac{3}{40\sqrt{x}}.$$

When $x = 10$,

$$f''(10) = \frac{-1}{4\sqrt{10^3}} \approx -.0079 \quad \text{and} \quad g''(10) = \frac{3}{40\sqrt{10}} \approx .0237.$$

The rate of increase for stock A is negative, meaning that its price is increasing at a decreasing rate. The rate of increase for stock B is positive, meaning that its price is increasing at an increasing rate. In other words, the price of stock A is increasing more and more slowly, while the price of stock B is increasing faster and faster. This result suggests that stock B is probably a better investment for the future.

The preceding discussion assumes that present trends continue (which is not guaranteed in the stock market). If they do continue for another 10 months, then Figure 12.19 shows what will happen to the stocks during this period.

Although Example 2 is much simpler than real life, second derivatives are actually used by some investors. According to an article in the *Wall Street Journal* on August 1, 2001 (when the market was in a downturn), "many investors hope it will help them get into the market ahead of any big rallies. In essence, the second-derivative approach involves looking at such things as profit warnings, analysts' earning estimates and equipment orders (including cancellations) for signs that the downward momentum is slowing."

The second derivative also plays a role in the physics of a moving particle. If a function describes the position of a moving object (along a straight line) at time t, then the first derivative gives the velocity of the object. That is, if $y = s(t)$ describes the position (along a straight line) of the object at time t, then $v(t) = s'(t)$ gives the velocity at time t.

The rate of change of velocity is called **acceleration**. Since the second derivative gives the rate of change of the first derivative, the acceleration is the derivative of the velocity. Thus, if $a(t)$ represents the acceleration at time t, then

$$\mathbf{a(t) = \frac{d}{dt}v(t) = s''(t).}$$

Example 3 Suppose that an object is moving along a straight line, with its position, in feet, at time t, in seconds, given by

$$s(t) = t^3 - 2t^2 - 7t + 9.$$

Find the given quantities.

(a) The velocity at any time t

Solution The velocity is given by

$$v(t) = s'(t) = 3t^2 - 4t - 7.$$

(b) The acceleration at any time t

Solution Acceleration is given by

$$a(t) = v'(t) = s''(t) = 6t - 4.$$

(c) When the object stops, its velocity is zero. For $t \geq 0$, when does that occur?

Solution Set $v(t) = 0$:

$$\begin{aligned} 3t^2 - 4t - 7 &= 0 \\ (3t - 7)(t + 1) &= 0 \\ 3t - 7 = 0 \quad &\text{or} \quad t + 1 = 0 \\ t = \frac{7}{3} \quad &\quad\quad t = -1. \end{aligned}$$

Since we want $t \geq 0$, only $t = \frac{7}{3}$ is acceptable here. The object will stop in $\frac{7}{3}$ seconds. 3 ✓

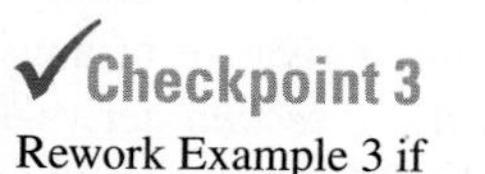
Rework Example 3 if
$s(t) = t^4 - t^3 + 10.$

Answers: See page 798

CONCAVITY

We shall now see how the second derivative provides information about how the graph of a function "bends," which is often hard to see on a calculator or computer screen. A graph is **concave upward** on an interval if it bends upward over the interval and **concave downward** if it bends downward, as shown in Figure 12.20. The graph is concave downward on the interval (a, b) and concave upward on the interval (b, c).* A point on the graph where the concavity changes (such as the point where $x = b$ in Figure 12.20) is called a **point of inflection**.

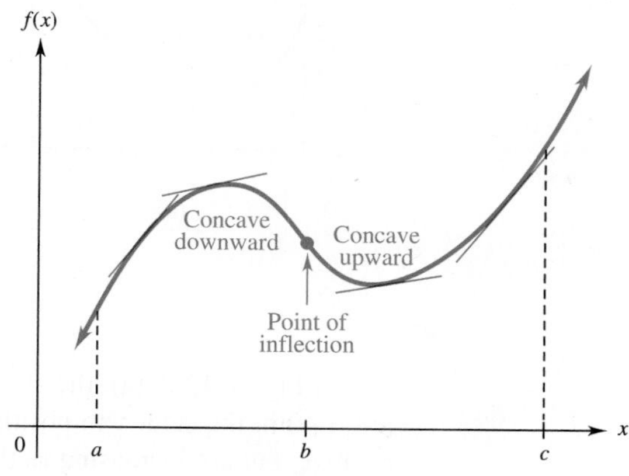

FIGURE 12.20

*Figure 12.20 also illustrates the formal definition of concavity: A function is *concave downward* on an interval if its graph lies below the tangent line at each point in the interval and *concave upward* if its graph is above the tangent line at each point in the interval.

A function that is increasing on an interval may have either kind of concavity; the same is true for a decreasing function. Some of the possibilities are illustrated in Figure 12.21.

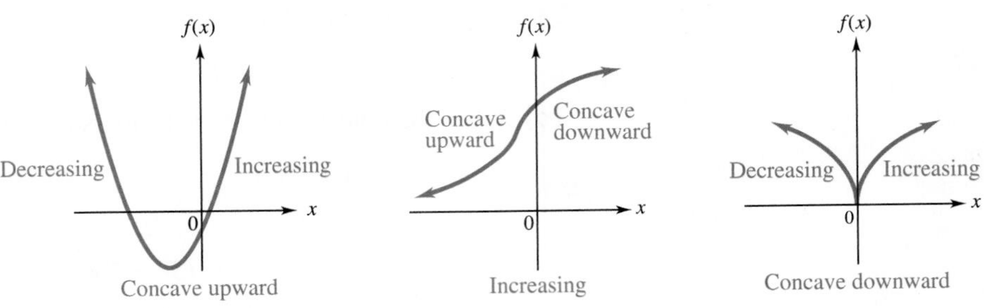

FIGURE 12.21

Next, we examine the relationship between the second derivative of a function f and the concavity of the graph of f. We have seen that when the derivative of any function is positive, that function is increasing. Consequently, if the *second* derivative of f (the derivative of the first derivative) is positive, then the *first* derivative of f is increasing. Since the first derivative gives the slope of the tangent line to the graph of f at each point, the fact that the first derivative is increasing means that the tangent-line slopes are increasing as you move from left to right along the graph of f, as illustrated in Figure 12.22.

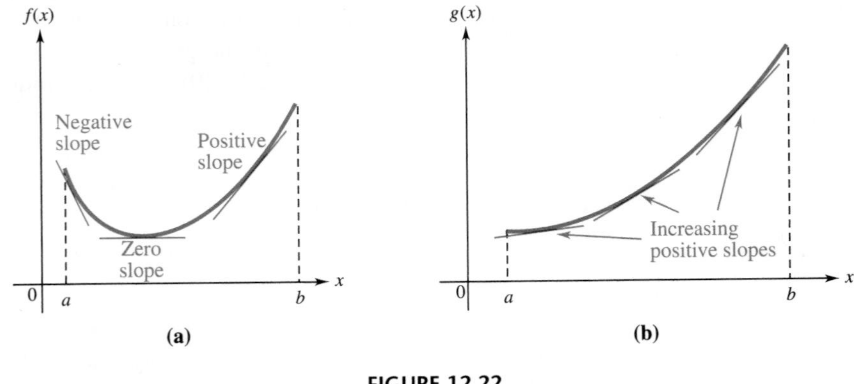

FIGURE 12.22

In Figure 12.22(a), the slopes of the tangent lines increase from negative at the left, to 0 in the center, to positive at the right. In Figure 12.22(b), the slopes are all positive, but are increasing as the tangent lines get steeper. Note that both graphs in Figure 12.22 are *concave upward.*

Similarly, when the second derivative is negative, the first derivative (slope of the tangent line) is decreasing, as illustrated in Figure 12.23. In Figure 12.23(a), the tangent-line slopes decrease from positive, to 0, to negative. In Figure 12.23(b), the slopes get more and more negative as the tangent lines drop downward more steeply. Note that both graphs are *concave downward.*

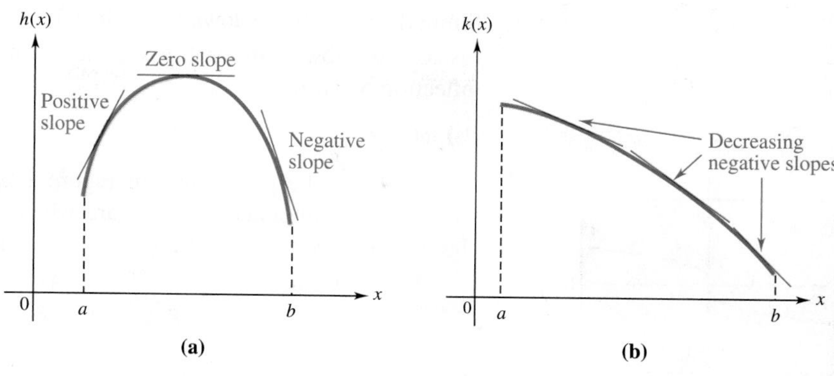

FIGURE 12.23

The preceding discussion suggests the following result.

Concavity Test

Let f be a function whose first and second derivatives exist at all points in the interval (a, b).

1. If $f''(x) > 0$ for all x in (a, b), then f is concave upward on (a, b).
2. If $f''(x) < 0$ for all x in (a, b), then f is concave downward on (a, b).

Example 4 Find the intervals over which the given function is concave upward or downward, and find any points of inflection.

(a) $f(x) = .5x^3 - 2x^2 + 5x + 1$

Solution The graph of f (Figure 12.24) suggests that f is concave upward to the right of the y-axis, but the concavity near the y-axis is not clear. Even in a window where the concavity is visible, the location of the point of inflection is not clear. (Try it!) So we use the test in the preceding box. The first derivative is $f'(x) = 1.5x^2 - 4x + 5$, and the second derivative is $f''(x) = 3x - 4$. The function is concave upward when $f''(x) > 0$—that is, when

$$3x - 4 > 0$$
$$3x > 4$$
$$x > \frac{4}{3}.$$

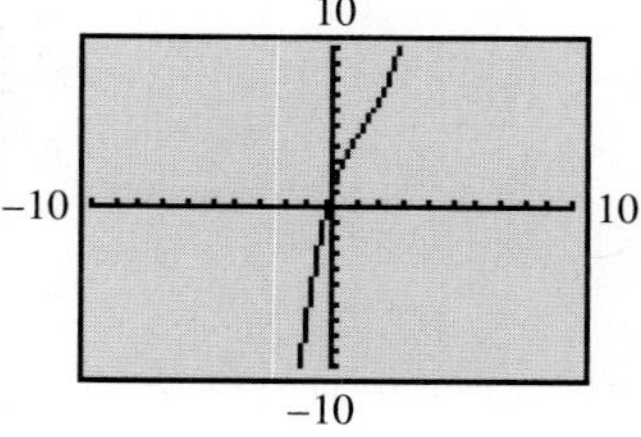

FIGURE 12.24

Similarly, f is concave downward when $3x - 4 < 0$—that is, when $x < \frac{4}{3}$. Therefore, f is concave downward on $(-\infty, \frac{4}{3})$ and concave upward on $(\frac{4}{3}, \infty)$, with a point of inflection when $x = \frac{4}{3}$.

(b) $g(x) = x^{1/3}$

Solution The graph of g in Figure 12.25 suggests that g is concave upward when $x < 0$ and concave downward when $x > 0$. We can confirm this conclusion algebraically by noting that $g'(x) = \frac{1}{3}x^{-2/3}$ and

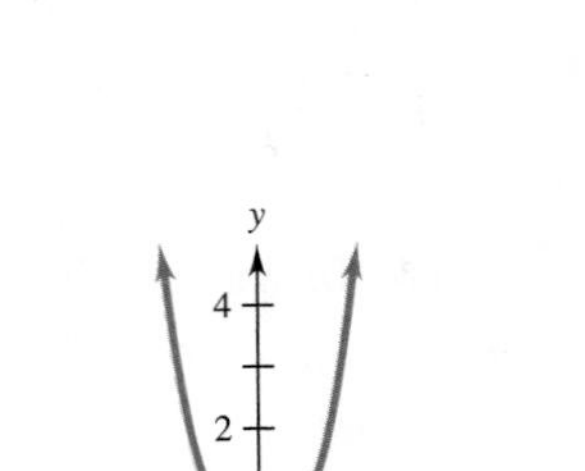

FIGURE 12.25

$$g''(x) = -\frac{2}{9}x^{-5/3} = \frac{-2}{9\sqrt[3]{x^5}}.$$

When $x < 0$, the denominator of $g''(x)$ is negative (why?), so $g''(x)$ is positive and g is concave upward on $(-\infty, 0)$. When $x > 0$, the denominator of $g''(x)$ is positive, so $g''(x)$ is negative and g is concave downward on $(0, \infty)$. The graph changes concavity at $(0, 0)$, which is a point of inflection.

In Example 4(a), the point of inflection on the graph of f occurs at $x = \frac{4}{3}$, the number at which the second derivative $f''(x) = 3x - 4$ is 0. In Example 4(b), the point of inflection occurs when $x = 0$, the number at which the second derivative $g''(x) = \dfrac{-2}{9\sqrt[3]{x^5}}$ is not defined. These facts suggest the following result.

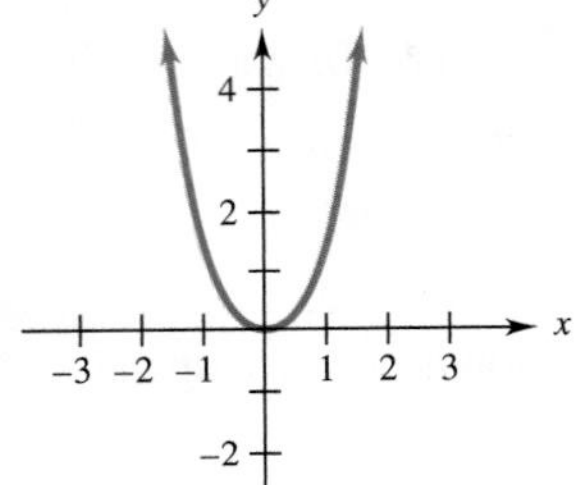

> If a function f has a point of inflection at $x = c$, then $f''(c) = 0$ or $f''(c)$ does not exist.

FIGURE 12.26

CAUTION The converse of the preceding statement is not always true: The second derivative may be 0 at a point that is not a point of inflection. For example, if $f(x) = x^4$, then $f'(x) = 4x^3$ and $f''(x) = 12x^2$. Hence, $f''(x) = 0$ when $x = 0$. However, the graph of $f(x)$ in Figure 12.26 is always concave upward, so it has no point of inflection at $x = 0$ (or anywhere else). 4

✓Checkpoint 4

Find the intervals where the given functions are concave upward. Identify any inflection points.

(a) $f(x) = 6x^3 - 24x^2 + 9x - 3$

(b) $f(x) = 2x^2 - 4x + 8$

Answers: See page 798

The **law of diminishing returns** in economics is related to the idea of concavity. The graph of the function f in Figure 12.27 shows the output y from a given input x. For instance, the input might be advertising costs and the output the corresponding revenue from sales.

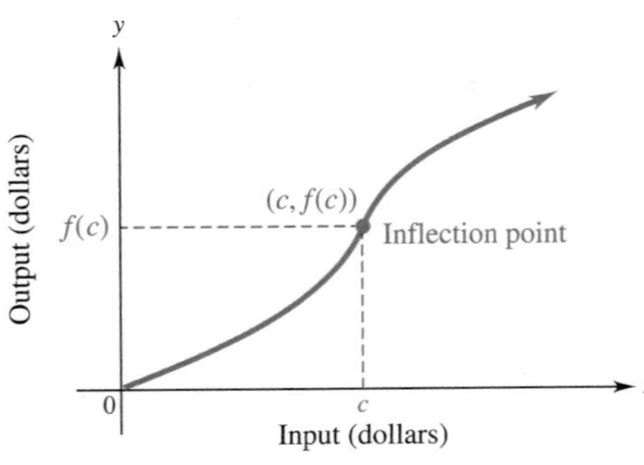

FIGURE 12.27

The graph in Figure 12.27 shows an inflection point at $(c, f(c))$. For $x < c$, the graph is concave upward, so the rate of change of the slope is increasing. This indicates that the output y is increasing at a faster rate with each additional dollar spent. When $x > c$, however, the graph is concave downward, the rate of change of the slope is decreasing, and the increase in y is smaller with each additional dollar spent. Thus, further input beyond c dollars produces diminishing returns. The point of inflection at $(c, f(c))$ is called the **point of diminishing returns**. Any investment beyond the value c is not considered a good use of capital.

Example 5 **Business** The revenue $R(x)$ generated from airline passengers is related to the number of available seat miles x by

$$R(x) = -.928x^3 + 31.492x^2 - 326.80x + 1143.88 \qquad (8 \le x \le 12),$$

where x is measured in hundreds of billions of miles and $R(x)$ is in billions of dollars.* Is there a point of diminishing returns for this function? If so, what is it?

Solution Since a point of diminishing returns occurs at an inflection point, look for an x-value that makes $R''(x) = 0$. We have

$$\begin{aligned} R'(x) &= 3(-.928)x^2 + 2(31.492)x - 326.80 \\ &= -2.784x^2 + 62.984x - 326.80 \end{aligned}$$

and

$$\begin{aligned} R''(x) &= 2(-2.784)x + 62.984 \\ &= -5.568x + 62.984. \end{aligned}$$

When we set $R''(x) = 0$ and solve for x, we obtain

$$\begin{aligned} -5.568x + 62.984 &= 0 \\ -5.568x &= -62.984 \\ x &\approx 11.31. \end{aligned}$$

Test a number in the interval (8, 11.31) to see that $R''(x)$ is positive there. Then test a number in the interval (11.31, 12) to see that $R''(x)$ is negative in that interval. Since the sign of $R''(x)$ changes from positive to negative at $x = 11.31$, the graph changes from concave upward to concave downward at that point, and there is a point of diminishing returns at the inflection point (11.31, 133.54). Increasing capacity beyond 1.131 trillion seat miles would not pay off. 5 ✓

✓Checkpoint 5

The revenue $R(x)$ generated from sales of a certain product is related to the amount x spent on advertising by

$$R(x) = \frac{1}{75{,}000}(125x^3 - x^4) \qquad (0 < x \le 100),$$

where x and $R(x)$ are in thousands of dollars. What is the point of diminishing returns?

Answer: See page 798.

MAXIMA AND MINIMA

If a function f has a local maximum at c and $f'(c)$ is defined, then the graph of f is necessarily concave downward near $x = c$. Similarly, if f has a local minimum at d and $f'(d)$ exists, then the graph is concave upward near $x = d$, as shown in Figure 12.28 on the next page.

*Based on data from the Air Transport Association of America.

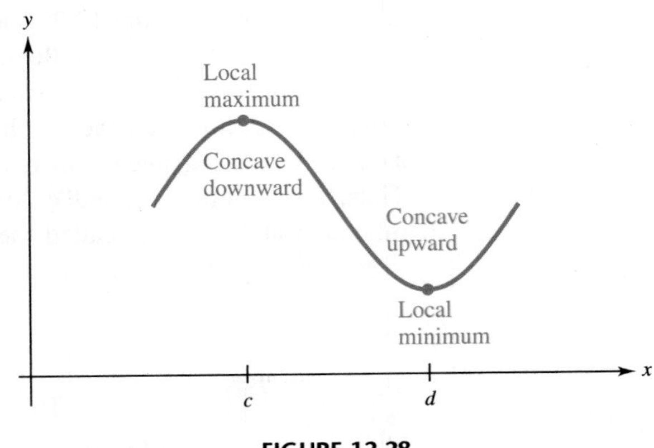

FIGURE 12.28

These facts should make the following result plausible.

The Second-Derivative Test

Let c be a critical number of the function f such that $f'(c) = 0$ and $f''(x)$ exists for all x in some open interval containing c.

1. If $f''(c) > 0$, then f has a local minimum at c.
2. If $f''(c) < 0$, then f has a local maximum at c.
3. If $f''(c) = 0$, then this test gives no information; use the first-derivative test.

Example 6 **Business** The number of passenger cars (in millions) imported into the United States in year x can be approximated by

$$g(x) = -2.527x^3 + 75.742x^2 - 602.494x + 2808.88 \qquad (0 \le x \le 16),$$

where $x = 5$ corresponds to 1995.* Find all local extrema of this function, and interpret your answers.

Solution First find the critical numbers. The first derivative is

$$\begin{aligned} g'(x) &= 3(-2.527)x^2 + 2(75.742)x - 602.494 \\ &= -7.581x^2 + 151.484x - 602.494. \end{aligned}$$

We solve the equation $g'(x) = 0$ so that $-7.581x^2 + 151.484x - 602.494 = 0$. Using the quadratic formula and a calculator yields

$$\begin{aligned} x &= \frac{-151.484 \pm \sqrt{(-151.484)^2 - 4(-7.581)(-602.494)}}{2(-7.581)} \\ &= \frac{-151.484 \pm \sqrt{4677.3742}}{-15.162} \approx \begin{cases} 5.5 \\ 14.5 \end{cases}. \end{aligned}$$

*U.S. Bureau of Transportation Statistics.

The critical numbers are 5.5 and 14.5.

Now use the second-derivative test on each critical number. The second derivative is

$$g''(x) = 2(-7.581)x + 151.484$$
$$= -15.162x + 151.484.$$

For the critical number $x = 5.5$, we have

$$g''(5.5) = -15.162(5.5) + 151.484 = 68.093 > 0,$$

which means that there is a local minimum at $x = 5.5$. Also,

$$g''(14.5) = -15.162(14.5) + 151.484 = -68.365 < 0.$$

Hence, there is a local maximum at $x = 14.5$. Thus, passenger-car imports were at a low at the midpoint of 1995 ($x = 5.5$) and at a high at the midpoint of 2004 ($x = 14.5$). 6

Checkpoint 6

Use the second-derivative test to find all local maxima and local minima for the given functions.

(a) $f(x) = 3x^2 - 24x + 1$

(b) $g(x) = x^3 - 4x^2 - 2x + 9$

Answers: See page 798.

NOTE The second-derivative test works only for those critical points c which make $f'(c) = 0$. This test does not work for those critical points c for which $f'(c)$ does not exist (since $f''(c)$ would not exist there either). Also, the second-derivative test does not work for critical points c that make $f''(c) = 0$. In both of these cases, use the first-derivative test.

12.2 ▸ Exercises

For each of these functions, find $f''(x)$, $f''(0)$, $f''(2)$, and $f''(-3)$. (See Examples 1 and 2.)

1. $f(x) = x^3 - 6x^2 + 1$
2. $f(x) = 2x^4 + x^3 - 8x^2 + 2$
3. $f(x) = (x + 3)^4$
4. $f(x) = \dfrac{2x + 5}{x - 8}$
5. $f(x) = \dfrac{x^2}{1 + x}$
6. $f(x) = \dfrac{-x}{1 - x^2}$
7. $f(x) = \sqrt{x + 4}$
8. $f(x) = \sqrt{2x + 9}$
9. $f(x) = 5x^{4/5}$
10. $f(x) = -x^{2/3}$
11. $f(x) = 2e^x$
12. $f(x) = \ln(2x - 3)$
13. $f(x) = 6e^{2x}$
14. $f(x) = 2 + e^{-x}$
15. $f(x) = \ln|x|$
16. $f(x) = \dfrac{1}{x^2}$
17. $f(x) = x\ln|x|$
18. $f(x) = \dfrac{\ln|x|}{x}$

Business *In Exercises 19 and 20, $P(t)$ is the price of a certain stock at time t during a particular day. (See Example 2.)*

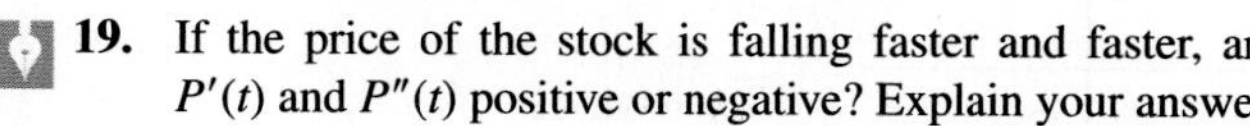

19. If the price of the stock is falling faster and faster, are $P'(t)$ and $P''(t)$ positive or negative? Explain your answer.

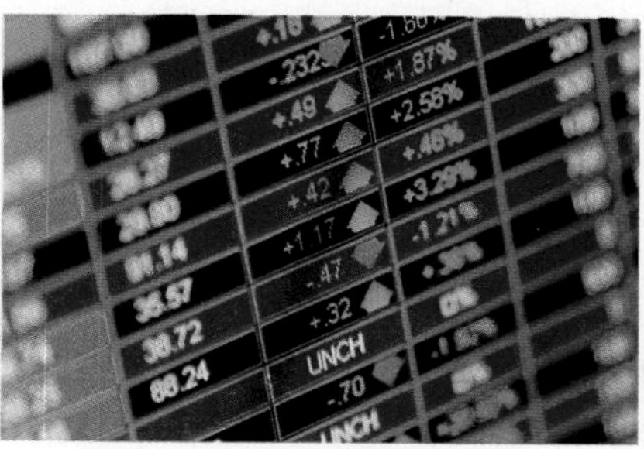

20. When the stock reaches its highest price during the day, are $P'(t)$ and $P''(t)$ positive or negative? Explain your answer.

Physical Science *Each of the functions in Exercises 21–24 gives the distance from a starting point at time t of a particle moving along a line. Find the velocity and acceleration functions. Then find the velocity and acceleration at $t = 0$ and $t = 4$. Assume that time is measured in seconds and distance is measured in centimeters. Velocity will be in centimeters per second (cm/sec) and acceleration in centimeters per second per second (cm/sec^2). (See Example 3.)*

21. $s(t) = 6t^2 + 5t$
22. $s(t) = 3t^3 - 6t^2 + 8t - 4$
23. $s(t) = 3t^3 - 4t^2 + 8t - 9$

24. $s(t) = \dfrac{-2}{3t + 4}$

Find the largest open intervals on which each function is concave upward or concave downward, and find the location of any points of inflection. (See Example 4.)

25. $f(x) = x^3 + 3x - 5$
26. $f(x) = -x^3 + 8x - 7$
27. $f(x) = x^3 + 4x^2 - 6x + 3$
28. $f(x) = 5x^3 + 12x^2 - 32x - 14$
29. $f(x) = \dfrac{2}{x - 4}$
30. $f(x) = \dfrac{-2}{x + 1}$
31. $f(x) = x^4 + 8x^3 - 30x^2 + 24x - 3$
32. $f(x) = x^4 + 8x^3 + 18x^2 + 12x - 84$

Business *In Exercises 33 and 34, find the point of diminishing returns for the given functions, where R(x) represents revenue, in thousands of dollars, and x represents the amount spent on advertising, in thousands of dollars. (See Example 5.)*

33. $R(x) = 10{,}000 - x^3 + 42x^2 + 800x;\ 0 \le x \le 20$
34. $R(x) = \dfrac{4}{27}(-x^3 + 66x^2 + 1050x - 400);\ 0 \le x \le 25$

Find all critical numbers of the functions in Exercises 35–46. Then use the second-derivative test on each critical number to determine whether it leads to a local maximum or minimum. (See Example 6.)

35. $f(x) = -2x^3 - 3x^2 - 72x + 1$
36. $f(x) = \dfrac{2}{3}x^3 + \dfrac{1}{2}x^2 - x - \dfrac{1}{4}$
37. $f(x) = x^3 + \dfrac{3}{2}x^2 - 60x + 100$
38. $f(x) = (x - 3)^5$
39. $f(x) = x^4 - 8x^2$
40. $f(x) = x^4 - 32x^2 + 7$
41. $f(x) = x + \dfrac{4}{x}$
42. $f(x) = x - \dfrac{1}{x^2}$
43. $f(x) = \dfrac{x^2 + 9}{2x}$
44. $f(x) = \dfrac{x^2 + 16}{2x}$
45. $f(x) = \dfrac{2 - x}{2 + x}$
46. $f(x) = \dfrac{x + 2}{x - 1}$

In Exercises 47–50, the rule of the derivative of a function f is given (but not the rule of f itself). Find the location of all local extrema and points of inflection of the function f.

47. $f'(x) = (x - 1)(x - 2)(x - 4)$
48. $f'(x) = (x^2 - 1)(x - 2)$
49. $f'(x) = (x - 2)^2(x - 1)$
50. $f'(x) = (x - 1)^2(x - 3)$
51. In each part, list the points (A–F) on the graph of f that satisfy the given conditions.
    (a) $f'(x) > 0$ and $f''(x) > 0$
    (b) $f'(x) < 0$ and $f''(x) > 0$
    (c) $f'(x) = 0$ and $f''(x) < 0$
    (d) $f'(x) = 0$ and $f''(x) > 0$
    (e) $f'(x) < 0$ and $f''(x) = 0$

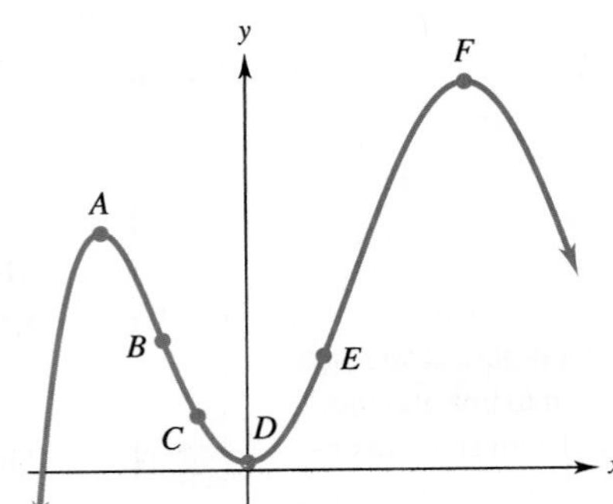

52. In each part, list the point (A–E) on the graph of f that satisfy the given conditions.
    (a) $f'(x) > 0$ and $f''(x) > 0$
    (b) $f'(x) < 0$ and $f''(x) > 0$
    (c) $f'(x) = 0$ and $f''(x) < 0$
    (d) $f'(x) = 0$ and $f''(x) > 0$
    (e) $f'(x) < 0$ and $f''(x) = 0$

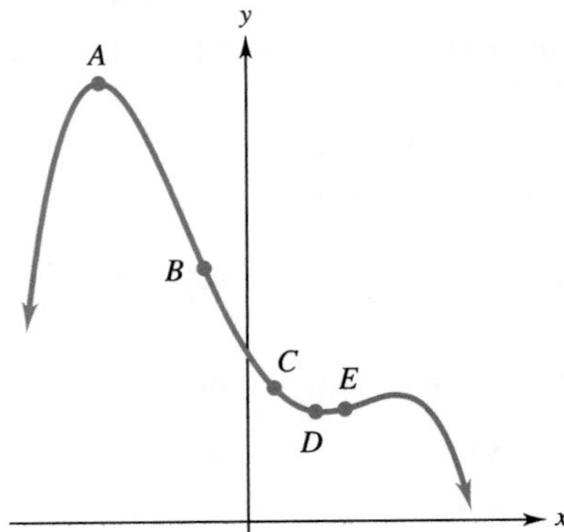

Work these problems. (See Example 2.)

53. **Finance** The price of Microsoft, Inc., stock over a 15-week period in 2009 can be approximated by the function

$$f(x) = 2.5x^{1/2} + 13 \qquad (1 \le x \le 15),$$

where $x = 1$ designates the beginning of week 1.*
    (a) When is the stock increasing in price?
    (b) At what rate is the stock increasing at the beginning of the 10th week?

*http://online.wsj.com/mdc/page/marketsdata.html.

54. Finance The price of Yahoo! stock over a 15-week period in 2009 can be approximated by the function

$$f(x) = 12.4x^{.083} \qquad (1 \le x \le 15),$$

where $x = 1$ designates the beginning of week 1.*

(a) When is the stock increasing in price?

(b) At what rate is the stock increasing at the beginning of the 12th week?

Work these problems. (See Example 3.)

55. Physical Science When an object is dropped straight down, the distance, in feet, that it falls in t seconds is given by

$$s(t) = -16t^2,$$

where negative distance (or velocity) indicates downward motion. Find the velocity at each of the following times.

(a) After 3 seconds

(b) After 5 seconds

(c) After 8 seconds

(d) Find the acceleration. (The answer here is a constant, the acceleration due to the influence of gravity alone.)

56. Physical Science If an object is thrown directly upward with a velocity of 256 ft/sec, its height above the ground after t seconds is given by $s(t) = 256t - 16t^2$. Find the velocity and the acceleration after t seconds. What is the maximum height the object reaches? When does it hit the ground?

Work these problems. (See Example 5.)

57. Business The revenue $R(x)$ (in millions of dollars) generated from cattle farming can be approximated by the function

$$R(x) = -3.21x^3 + 62.14x^2 + 172.92x + 48.65 \quad (0 \le x \le 14),$$

where x represents the number of farms (in millions).† What is the point of diminishing returns?

58. Business A local automobile dealer has found that advertising produces sales, but that too much advertising tends to "turn consumers off," resulting in reduced sales. On the basis of past experience, the dealer expects that the number $V(x)$ of vehicles sold during a month is related to the amount spent on advertising by the function

$$V(x) = -.1x^3 + 2.2x^2 + 1.2x + 15 \qquad (0 \le x \le 20),$$

where x is the amount spent on advertising (in tens of thousands of dollars). What is the point of diminishing returns?

*http://online.wsj.com/mdc/page/marketsdata.html.

†U.S. Department of Agriculture.

59. Finance The U.S. gross domestic product (GDP), in trillions of dollars, can be modeled by

$$f(x) = .0024x^3 - .0633x^2 + .994x + 3.67 \quad (5 \le x \le 17),$$

where $x = 5$ corresponds to the year 1995. Find and interpret the point of inflection.

60. Business The amount of U.S. consumer spending on television per person per year can be modeled by the function

$$g(x) = .0058x^3 - .9661x^2 + 37.426x + 117.03 \quad (5 \le x \le 11),$$

where $x = 5$ corresponds to the year 2005.* When is this function increasing?

Work these problems. You may need to use the quadratic formula to find some of the critical numbers. (See Example 6.)

61. Business U.S. national defense expenditures (in billions of constant year-2000 dollars) are approximated by

$$f(x) = .067x^3 - 2.495x^2 + 24.932x + 273.9 \quad (0 \le x \le 29),$$

where $x = 0$ corresponds to the year 1980.†

(a) Find the critical numbers of this function.

(b) In what years was defense spending at a local minimum or maximum?

(c) Find the point of inflection. What does it indicate?

62. Social Science The population (in thousands) of Baltimore, MD, is approximated by

$$g(x) = .0044x^3 - .4084x^2 + 3.572x + 947.75 \quad (0 \le x \le 57),$$

where $x = 0$ corresponds to the year 1950.‡ In what year does an inflection point for the population occur?

63. Social Science The number of aggravated assaults per 100,000 people in the United States in year x is approximated by

$$h(x) = .108x^3 - 6.135x^2 + 99.581x - 57.14 \quad (5 \le x \le 27),$$

where $x = 5$ corresponds to the year 1985.§ According to this model.

(a) In what year did assaults peak?

(b) In what year is there a point of inflection?

64. Business One store's revenue from selling x cell phones is given by

$$R(x) = (200x)e^{-.01x}.$$

**Statistical Abstract of the United States*: 2009.

†U.S. Office of Management and Budget.

‡www.census.gov.

§U.S. Department of Justice, Federal Bureau of Investigation.

(a) How many phones should be sold to maximize revenue?
(b) What is the maximum revenue?
(c) When revenue is at a maximum, what is the average price per phone?

65. **Health** The number of prescriptions (in millions) in the United States can be approximated by the function

$$f(x) = .864x^3 - 32.10x^2 + 478.04x + 429.6 \quad (5 \le x \le 17),$$

where $x = 5$ corresponds to the year 1995.*
(a) Are there any local extrema in the interval? If so, where?
(b) Is there a point of inflection? If so, what is it?

66. **Health** The birth weight (in pounds) of a newly born infant can be modeled by the function

$$f(t) = .0011t^{2.4} \qquad (17 \le t \le 45),$$

where t is the number of weeks of gestation.†
(a) Are there any local extrema in the interval? If so, where?
(b) Is there a point of inflection? If so, what is it?

67. **Health** A flu outbreak occurs in a small community. Researchers estimate that the percentage of the population infected by the flu is approximated by

$$p(t) = \frac{30t^3 - t^4}{1200} \qquad (0 \le t \le 30),$$

where t is the number of days after the flu was first observed.
(a) After how many days is the percentage of the population with the flu at a maximum?
(b) What is the maximum percentage of the population with the flu?

*www.nacds.org.

†Based on data available at www.irss.unc.edu.

✓ Checkpoint Answers

1. (a) -24 (b) 72 (c) -72
2. (a) $y'' = -54x + 16$ (b) $y'' = -24x^2 + 12$
(c) $y'' = \dfrac{110}{(5x - 1)^3}$ (d) $y'' = e^x - \dfrac{1}{x^2}$
3. (a) $v(t) = 4t^3 - 3t^2$ (b) $a(t) = 12t^2 - 6t$
(c) At 0 and $\frac{3}{4}$ second
4. (a) Concave upward on $(\frac{4}{3}, \infty)$; point of inflection is $(\frac{4}{3}, -\frac{175}{9})$
(b) $f''(x) = 4$, which is always positive; function is always concave upward, no inflection point
5. (62.5, 203.45)
6. (a) Local minimum of -47 at $x = 4$
(b) Local maximum of approximately 9.24 at $x \approx -.23$; local minimum of approximately -6.051 at $x \approx 2.90$

12.3 ▶ Optimization Applications

In most applications, the domains of the functions involved are restricted to numbers in a particular interval. For example, a factory that can produce a maximum of 40 units (because of market conditions, availability of labor, etc.) might have this cost function:

$$C(x) = -3x^3 + 135x^2 + 3600x + 12{,}000 \qquad (0 \le x \le 40).$$

Even though the rule of C is defined for all numbers x, only the numbers in the interval $[0, 40]$ are relevant, because the factory cannot produce a negative number of units or more than 40 units. In such applications, we often want to find a smallest or largest quantity—for instance, the minimum cost or the maximum profit—when x is restricted to the relevant interval. So we begin with the mathematical description of such a situation.

Let f be a function that is defined for all x in the closed interval $[a, b]$. Let c be a number in the interval. We say that f has an **absolute maximum on the interval** at c if

$$f(x) \le f(c) \qquad \textbf{for all } x \textbf{ with } a \le x \le b,$$

that is, if $(c, f(c))$ is the highest point on the graph of f over the interval $[a, b]$. Similarly, f has an **absolute minimum on the interval** at c if

$$f(x) \geq f(c) \quad \text{for all } x \text{ with } a \leq x \leq b,$$

that is, if $(c, f(c))$ is the lowest point on the graph of f over the interval $[a, b]$.

✓Checkpoint 1

Find the location of the absolute maximum and absolute minimum of the function f in Figure 12.29 on the interval $[-2, \frac{1}{2}]$.

Answer: See page 813.

Example 1 Figure 12.29 shows the graph of a function f. Consider the function f on the interval $[-2, 6]$. Since we are interested only in the interval $[-2, 6]$, the values of the function outside this interval are irrelevant. On the interval $[-2, 6]$, f has an absolute minimum at 3 (which is also a local minimum) and an absolute maximum at 6 (which is not a local maximum of the entire function). ✓1

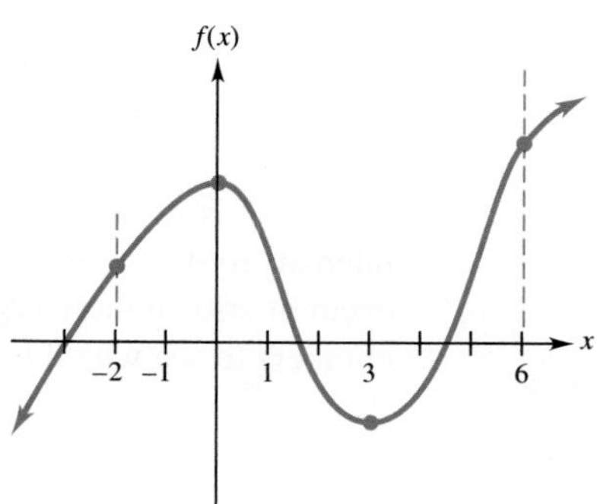

FIGURE 12.29

The absolute maximum in Example 1 occurred at $x = 6$, which is an endpoint of the interval, and the absolute minimum occurred at $x = 3$, which is a critical number of f (because f has a local minimum there). Similarly, in Checkpoint 1, the absolute maximum occurred at a critical number and the absolute minimum at an endpoint. These examples illustrate the following result, whose proof is omitted.

Extreme-Value Theorem

If a function f is continuous on a closed interval $[a, b]$, then f has both an absolute maximum and an absolute minimum on the interval. Each of these values occurs either at an endpoint of the interval or at a critical number of f.

CAUTION The extreme-value theorem may not hold on intervals that are not closed (that is, intervals that do not include one or both endpoints). For example, $f(x) = 1/x$ does not have an absolute maximum on the interval $(0, 1)$; the values of $f(x)$ get larger and larger as x approaches 0, as you can easily verify with a calculator.

Example 2 Use the extreme-value theorem to find the absolute extrema of $f(x) = 4x + \dfrac{36}{x}$ on the interval $[1, 6]$.

Solution According to the extreme-value theorem, we need consider only the critical numbers of f and the endpoints, 1 and 6, of the interval. Begin by finding the derivative and determining the critical numbers:

$$f'(x) = 4 - \frac{36}{x^2} = 0$$

$$\frac{4x^2 - 36}{x^2} = 0.$$

Since we are looking for critical numbers in [1, 6], $x \neq 0$. Because $x \neq 0, f'(x) = 0$ when

$$4x^2 - 36 = 0$$
$$4x^2 = 36$$
$$x^2 = 9$$
$$x = -3 \quad \text{or} \quad x = 3.$$

Since -3 is not in the interval [1, 6], disregard it; 3 is the only critical number of interest. By the extreme-value theorem, the absolute maximum and minimum must occur at critical numbers or endpoints—that is, at 1, 3, or 6. Evaluate f at these three numbers to see which ones give the largest and smallest values:

x-Value	Value of Function	
1	40	← Absolute maximum
3	24	← Absolute minimum
6	30	

✓ 2

✓ Checkpoint 2

Find the absolute maximum and absolute minimum values for $f(x) = -x^2 + 4x - 8$ on $[-4, 4]$.

Answer: See page 813.

A graphing calculator or other graphing technology sometimes makes it easy to find absolute extrema without taking the derivative. Technology, however, has its limits, so you should confirm your results algebraically whenever feasible.

Example 3 Find the absolute extrema of $g(x) = -.02x^3 + 600x - 20{,}000$ on [60, 135].

Solution The graph of g on [60, 135] in Figure 12.30 shows that the absolute maximum occurs at the local maximum, which is close to $x = 100$. The absolute minimum occurs at one of the endpoints, but you cannot determine which one from the graph.

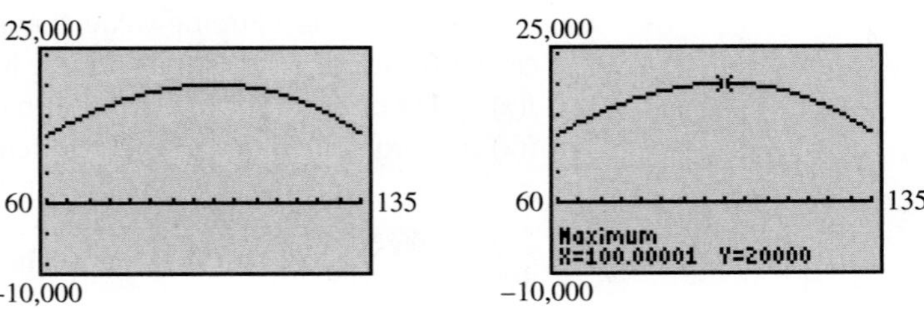

FIGURE 12.30

We first determine the critical number exactly. Since $g'(x) = -.06x^2 + 600$, the critical numbers are the solutions of

$$-.06x^2 + 600 = 0.$$

We solve as follows:

$$-.06x^2 = -600$$
$$x^2 = 10{,}000$$
$$x = -100 \quad \text{or} \quad x = 100.$$

We disregard $x = -100$ because it is not in [60, 135] and evaluate g at the endpoints and the critical number 100, as shown in the next table. The absolute maximum of 20,000 occurs at $x = 100$ (as we suspected from the graph), and the absolute minimum of 11,680 occurs at $x = 60$.

x-Value	Value of Function	
60	11,680	← Absolute minimum
100	20,000	← Absolute maximum
135	11,792.5	

✓Checkpoint 3

Find the absolute extrema of $g(x) = x^3 - 15x^2 + 48x + 50$ on [1, 6].

Answer: See page 813.

ABSOLUTE EXTREMA ON OTHER INTERVALS

The extreme-value theorem is very useful on closed intervals, but may be *false* on an open interval. For example, the function

$$f(x) = \frac{x}{(x + 2)(x - 3)}$$

has vertical asymptotes at $x = -2$ and $x = 3$. On the open interval $(-2, 3)$ between the asymptotes, the function has neither an absolute maximum nor an absolute minimum, as shown in Figure 12.31. The graph rises higher and higher near $x = -2$ and falls lower and lower near $x = 3$.

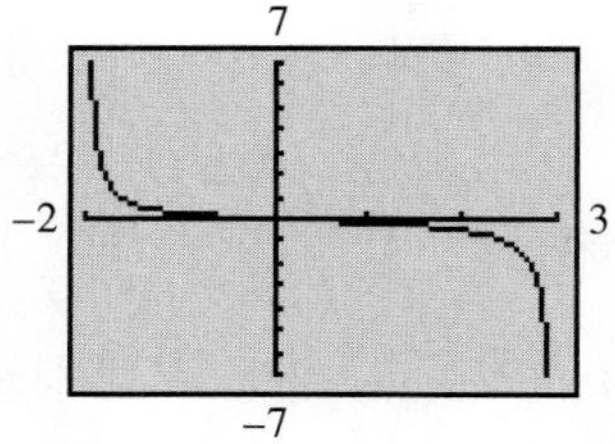

FIGURE 12.31

It is sometimes possible to identify an absolute maximum or minimum on an interval that is not closed by using the following theorem, whose proof is omitted.

Critical-Point Theorem

Suppose that a function f is continuous on an interval I and that f has exactly one critical number in the interval I, say, $x = c$.*

If f has a local maximum at $x = c$, then this local maximum is the absolute maximum of f on the interval I.

If f has a local minimum at $x = c$, then this local minimum is the absolute minimum of f on the interval I.

*I may be open, closed, or neither.

Example 4 Consider the function $f(x) = x^3 - 3x + 1$.

(a) Without graphing, show that f has an absolute minimum on the interval (0, 2).

Solution The derivative is $f'(x) = 3x^2 - 3$, which is defined everywhere, so the critical numbers are the solutions of $f'(x) = 0$:

$$\begin{aligned} 3x^2 - 3 &= 0 \\ 3x^2 &= 3 \\ x^2 &= 1 \\ x = -1 \quad \text{or} \quad x &= 1. \end{aligned}$$

The only critical number in the interval (0, 2) is $x = 1$. Use the second-derivative test to determine whether there is a local extremum at $x = 1$:

$$\begin{aligned} f'(x) &= 3x^2 - 3 \\ f''(x) &= 6x \\ f''(1) &= 6(1) = 6 > 0. \end{aligned}$$

Hence, f has a local minimum at $x = 1$. Therefore, by the critical-point theorem, the absolute minimum of f on the interval (0, 2) occurs at $x = 1$.

(b) Confirm the result of part (a) by graphing.

Solution The graph of f in Figure 12.32 shows that $(1, -1)$ is the lowest point on the graph in the interval (0, 2).

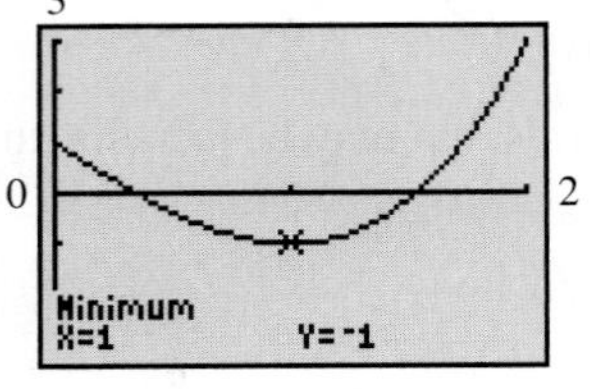

FIGURE 12.32

APPLICATIONS

When solving applied problems that involve maximum and minimum values, follow these guidelines:

Solving Applied Problems

Step 1 Read the problem carefully. Make sure you understand what is given and what is asked for.

Step 2 If possible, sketch a diagram and label the various parts.

Step 3 Decide which variable is to be maximized or minimized. Express that variable as a function of *one* other variable. Be sure to determine the domain of this function.

Step 4 Find the critical numbers for the function in Step 3.

Step 5 If the domain is a closed interval, evaluate the function at the endpoints and at each critical number to see which yields the absolute maximum or minimum. If the domain is a (not necessarily closed) interval in which there is exactly one critical number, apply the critical-point theorem, if possible, to find an absolute extremum.

CAUTION Do not skip Step 5 in the preceding box. If you are looking for a maximum and you find a single critical number in Step 4, do not automatically assume that the maximum occurs there. It may occur at an endpoint or may not exist at all.

Example 5 **Business** The cost (in thousands of dollars) of manufacturing x thousand beach towels is given by

$$C(x) = .0015x^3 - .0625x^2 + .5x + 25.$$

(a) Find the average cost function.

Solution As we saw in Section 11.6, the average cost function $\overline{C}(x)$ is given by

$$\overline{C}(x) = \frac{C(x)}{x} = \frac{.0015x^3 - .0625x^2 + .5x + 25}{x}$$

$$= .0015x^2 - .0625x + .5 + \frac{25}{x}. \quad 4$$

Checkpoint 4

Find the average cost per towel for the following production levels: 2000, 25,000, and 50,000.

Answer: See page 813.

(b) How many towels should be made in order to minimize the average cost per towel? What is the minimum average cost?

Solution The average cost function $\overline{C}(x) = .0015x^2 - .0625x + .5 + 25x^{-1}$ is defined for all $x > 0$, so endpoints play no role here. Its derivative is

$$\overline{C}'(x) = .0030x - .0625 - 25x^{-2},$$

which is defined for all $x > 0$. Hence, the critical numbers are the solutions to

$$.0030x - .0625 - 25x^{-2} = 0,$$

or

$$.0030x^3 - .0625x^2 - 25 = 0. \quad \text{Multiply both sides by } x^2.$$

Technology is needed to solve this equation. Using graphical methods, we see that the only real solution is $x \approx 30.057$ (Figure 12.33). We use the second-derivative test to determine whether this critical number is a minimum:

$$\overline{C}'(x) = .0030x - .0625 - 25x^{-2}$$

$$\overline{C}''(x) = .0030 + (-2)(-25)x^{-3} = .0030 + \frac{50}{x^3}.$$

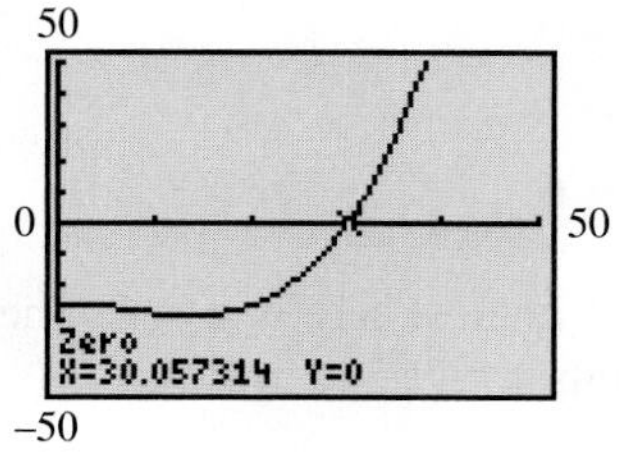

FIGURE 12.33

The second derivative is positive for all $x > 0$, so $\overline{C}'(x)$ has a local minimum at $x \approx 30.057$. By the critical-point theorem, this is an absolute minimum on the interval $(0, \infty)$. Thus, when 30,057 towels are produced, the average cost per towel is the lowest possible—namely, $\overline{C}(30.057) \approx \$.81$, or about 81 cents per towel.

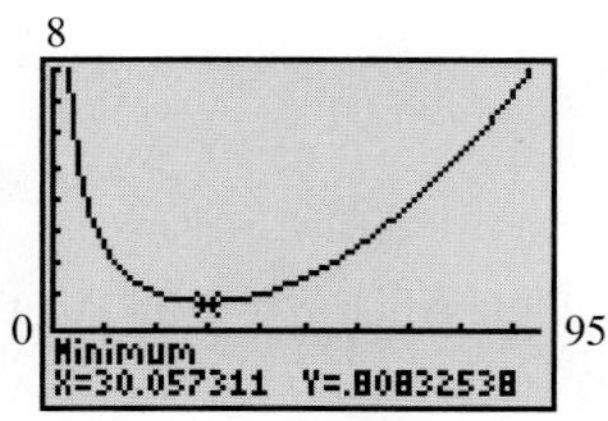

FIGURE 12.34

Alternative Solution Once we realize that technology and approximations are necessary here, we can also graph $\overline{C}(x) = .0015x^2 - .0625x + .5 + 25x^{-1}$ and use a minimum finder to determine the minimum, as in Figure 12.34. Both methods agree up to three decimal places.

Example 6 An open box is to be made by cutting a square from each corner of a 12-inch-by-12-inch piece of metal and then folding up the sides. The finished box must be at least 1.5 inches deep, but not deeper than 3 inches. What size square should be cut from each corner in order to produce a box of maximum volume?

Solution Let x represent the length of a side of the square that is cut from each corner, as shown in Figure 12.35(a). The width of the box is $12 - 2x$, as is the length. As shown in Figure 12.35(b), the depth of the box is x inches.

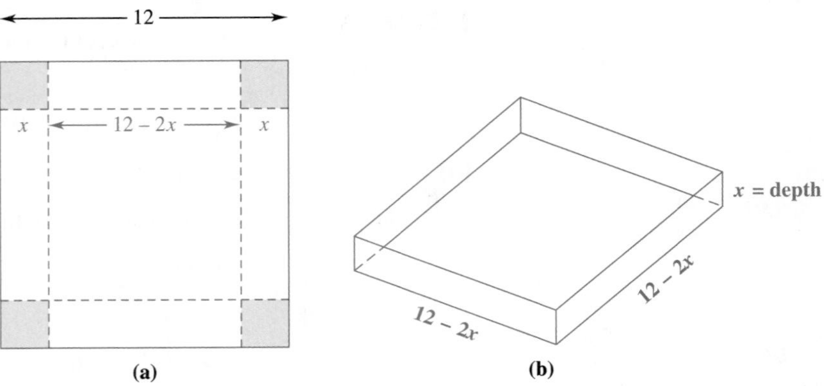

FIGURE 12.35

We must maximize the volume $V(x)$ of the box, which is given by

$$\text{Volume} = \text{length} \cdot \text{width} \cdot \text{height}$$
$$V(x) = (12 - 2x) \cdot (12 - 2x) \cdot x = 144x - 48x^2 + 4x^3.$$

Since the height x must be between 1.5 and 3 inches, the domain of this volume function is the closed interval [1.5, 3]. First find the critical numbers by setting the derivative equal to 0:

$$V'(x) = 12x^2 - 96x + 144 = 0$$
$$12(x^2 - 8x + 12) = 0$$
$$12(x - 2)(x - 6) = 0$$
$$x - 2 = 0 \quad \text{or} \quad x - 6 = 0$$
$$x = 2 \quad \text{or} \quad x = 6.$$

Since 6 is not in the domain, the only critical number of interest here is $x = 2$. The maximum volume must occur at $x = 2$ or at the endpoints $x = 1.5$ or $x = 3$.

x	$v(x)$	
1.5	121.5	
2	128	← Maximum
3	108	

The preceding table shows that the box has maximum volume when $x = 2$ and that this maximum volume is 128 cubic inches. 5 ✓

✓Checkpoint 5

An open box is to be made by cutting squares from each corner of a 20-cm-by-32-cm piece of metal and folding up the sides. Let x represent the length of the side of the square to be cut out. Find

(a) an expression for the volume $V(x)$ of the box;

(b) $V'(x)$;

(c) the value of x that leads to maximum volume (*Hint*: The solutions of the equation $V'(x) = 0$ are 4 and 40/3);

(d) the maximum volume.

Answers: See page 813.

Example 7 **Business** The U.S. Postal Service requires that boxes sent by Priority Mail have a length plus girth of no more than 108 inches, as shown in Figure 12.36. Find the dimensions of the box with largest volume that can be sent by Priority Mail, assuming that its width and height are equal.

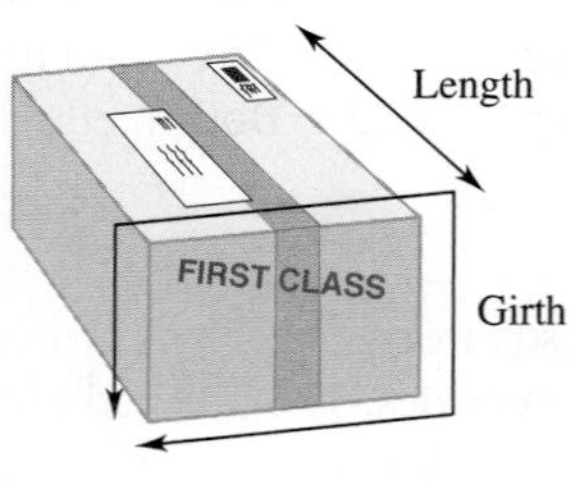

FIGURE 12.36

Solution Let x be the width and y the length of the box. Since width and height are the same, the volume of the box is

$$V = y \cdot x \cdot x = yx^2.$$

Now express V in terms of just *one* variable. Use the facts that the girth is $x + x + x + x = 4x$ and that the length plus girth is 108, so that

$$y + 4x = 108, \quad \text{or equivalently,} \quad y = 108 - 4x.$$

Substitute for y in the expression for V to get

$$V = yx^2 = (108 - 4x)x^2 = 108x^2 - 4x^3.$$

Since x and y are dimensions, we must have $x > 0$ and $y > 0$. Now, $y = 108 - 4x > 0$ implies that

$$4x < 108, \quad \text{or equivalently,} \quad x < 27.$$

Therefore, the domain of the volume function V (the set of values of x that make sense in the situation) is the open interval (0, 27). Find the critical numbers for V by setting its derivative equal to 0 and solving the equation:

$$\begin{aligned} V' = 216x - 12x^2 &= 0 \\ x(216 - 12x) &= 0 \\ x = 0 \quad \text{or} \quad 12x &= 216 \\ x &= 18. \end{aligned}$$

Use the second-derivative test to check $x = 18$, the only critical number in the domain of V. Since $V'(x) = 216x - 12x^2$,

$$V''(x) = 216 - 24x.$$

Hence, $V''(18) = 216 - 24 \cdot 18 = -216$, and V has a local maximum when $x = 18$. By the critical-point theorem, the absolute maximum of V on (0, 27) is at $x = 18$. In this case, $y = 108 - 4 \cdot 18 = 36$. Therefore, a box with dimensions 18 by 18 by 36 inches satisfies postal regulations and yields the maximum volume of $18^2 \cdot 36 = 11{,}664$ cubic inches.

Example 8 **Business** A landscape gardener wants to build an 8000-square-foot rectangular display garden along the side of a river. The garden will be fenced on three sides. (No fence is necessary along the river.) She plans to use ornamental steel fencing at \$12 per foot on the side opposite the river and chain-link fencing at \$3 per foot on the other two sides. What dimensions for the garden will minimize her costs?

Solution First draw a sketch of the situation and label the sides of the garden, as in Figure 12.37.

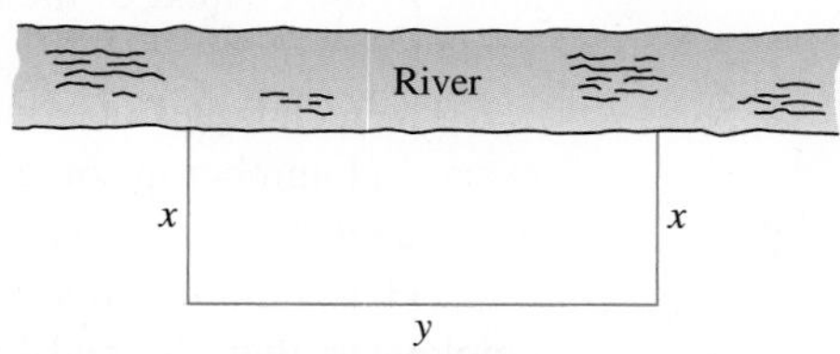

FIGURE 12.37

$2x$ feet of fencing (at \$3 per foot) are needed for the parallel sides and y feet (at \$12 per foot) for the side opposite the river. So the cost is

$$C = 2x(3) + y(12) = 6x + 12y.$$

We must find the values of x and y that make C as small as possible. First, however, we must express C in terms of a single variable. Since the garden is to have an area of 8000 square feet and the area of a rectangle is its length times its width, we have

$$xy = \text{area} = 8000$$
$$y = \frac{8000}{x}.$$

Substituting this expression for y in the cost function, we obtain

$$C = 6x + 12y = 6x + 12\left(\frac{8000}{x}\right) = 6x + \frac{96{,}000}{x}.$$

Since x is a length, it is positive. Hence, the domain of the cost function is the open interval $(0, \infty)$, and we need look only at the critical numbers. The derivative is

$$C'(x) = 6 - 96{,}000x^{-2},$$

so the critical numbers are the solutions of

$$6 - \frac{96{,}000}{x^2} = 0$$
$$6x^2 - 96{,}000 = 0$$
$$6x^2 = 96{,}000$$
$$x^2 = 16{,}000$$
$$x = \pm\sqrt{16{,}000} \approx \pm 126.5.$$

The only critical number in the domain $(0, \infty)$ is $x = \sqrt{16{,}000}$. Do Checkpoint 6 to show that this number makes C a minimum. ✓6

Therefore, the garden with the least expensive fence has dimensions

$$x = 126.5 \text{ ft} \quad \text{and} \quad y = \frac{8000}{x} \approx \frac{8000}{126.5} = 63.24 \text{ ft.}$$

✓Checkpoint 6

Show that the function C in Example 8 has a minimum at $x = \sqrt{16{,}000} \approx 126.5$ in either of the following ways:

(a) Find $C''(x)$ and $C''(\sqrt{16{,}000})$. Then use the second-derivative test and the critical-point theorem.

(b) Graph C on a graphing calculator and use the minimum finder.

Answers: See page 813.

The preceding examples illustrate some of the factors that may affect applications in the real world. First, you must be able to find a function that models the situation. The rule of this function may be defined for values of x that do not make sense in the context of the application, so the domain must be restricted to the relevant values of x.

The techniques of calculus apply to functions that are defined and continuous at every real number in some interval, so the maximum or minimum for the mathematical model (function) may not be feasible in the setting of the problem. For instance, if $C(x)$ has a minimum at $x = 80\sqrt{3}$ (≈ 138.564), where $C(x)$ is the cost of hiring x employees, then the real-life minimum occurs at either 138 or 139, whichever one leads to lower cost.

ECONOMIC LOT SIZE

Suppose that a company manufactures a constant number of units of a product per year and that the product can be manufactured in several batches of equal size during the year. On the one hand, if the company were to manufacture the item only once per year, it would minimize setup costs, but incur high warehouse costs. On the other hand, making many small batches would increase setup costs. Calculus can be used to find the number of batches per year that should be manufactured in order to minimize total cost. This number is called the **economic lot size**.

Figure 12.38 shows several of the possibilities for a product having an annual demand of 12,000 units. The top graph shows the results if only one batch of the product is made annually; in this case, an average of 6000 items will be held in a warehouse. If four batches (of 3000 each) are made at equal time intervals during a year, the average number of units in the warehouse falls to only 1500. If 12 batches are made, an average of 500 items will be in the warehouse.

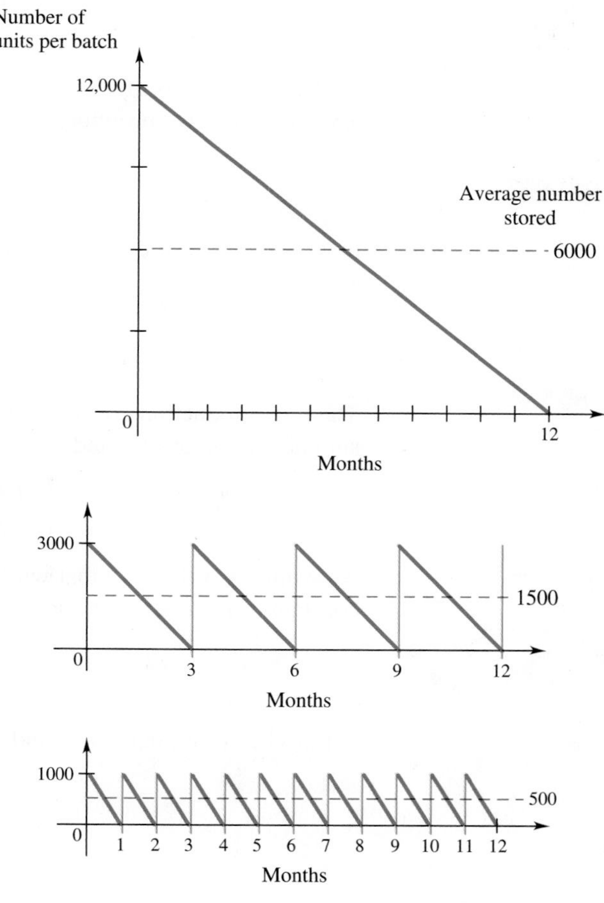

FIGURE 12.38

The following variables will be used in our discussion of economic lot size:

x = number of batches to be manufactured annually;

k = cost of storing one unit of the product for one year;

a = fixed setup cost to manufacture the product;

b = variable cost of manufacturing a single unit of the product;

M = total number of units produced annually.

The company has two types of costs associated with the production of its product: a cost associated with manufacturing the item and a cost associated with storing the finished product.

During a year, the company will produce x batches of the product, with M/x units of the product produced per batch. Each batch has a fixed cost a and a variable cost b per unit, so the manufacturing cost per batch is

$$a + b\left(\frac{M}{x}\right).$$

There are x batches per year, so the total annual manufacturing cost is

$$\left[a + b\left(\frac{M}{x}\right)\right]x. \tag{1}$$

Each batch consists of M/x units, and demand is constant; therefore, it is common to assume an average inventory of

$$\frac{1}{2}\left(\frac{M}{x}\right) = \frac{M}{2x}$$

units per year. The cost to store one unit of the product for a year is k, so the total storage cost is

$$k\left(\frac{M}{2x}\right) = \frac{kM}{2x}. \tag{2}$$

The total production cost is the sum of the manufacturing and storage costs, or the sum of expressions (1) and (2). If $T(x)$ is the total cost of producing x batches,

$$T(x) = \left[a + b\left(\frac{M}{x}\right)\right]x + \frac{kM}{2x} = ax + bM + \left(\frac{kM}{2}\right)x^{-1}.$$

Now find the value of x that will minimize $T(x)$. (Remember that a, b, k, and M are constants.) To do so, first find $T'(x)$:

$$T'(x) = a - \frac{kM}{2}x^{-2}.$$

Set this derivative equal to 0 and solve for x (remember that $x > 0$):

$$a - \frac{kM}{2}x^{-2} = 0$$

$$a = \frac{kM}{2x^2}$$

$$2ax^2 = kM$$

$$x^2 = \frac{kM}{2a}$$

$$x = \sqrt{\frac{kM}{2a}}. \tag{3}$$

The second-derivative test can be used to show that $\sqrt{kM/(2a)}$ is the annual number of batches that gives the minimum total production cost.

Example 9 **Business** A garden supplier has steady annual demand for 50,000 cubic feet of potting soil. The cost accountant for the company says that it costs $2.50 to store 1 cubic foot of potting soil for 1 year and $850 to set up the production facility to mix the ingredients for the potting soil. Find the number of batches of potting soil that should be produced for the minimum total production cost.

Solution Use equation (3), given previously:

$$x = \sqrt{\frac{kM}{2a}}$$

$$x = \sqrt{\frac{(2.50)(50{,}000)}{2(850)}} \qquad \text{Let } k = 2.50,\ M = 50{,}000, \text{ and } a = 850.$$

$$x \approx 9.$$

Nine batches of potting soil per year will lead to minimum production costs. 7 ✓

✓Checkpoint 7

In Example 9, suppose annual demand is 95,000 cubic feet, the annual cost of storage is $2.75 per cubic foot, and setup costs are $1100. Find the number of batches that should be made annually to minimize total costs.

Answer: See page 813.

12.3 ▶ Exercises

Find the location of the absolute maximum and absolute minimum of the function on the given interval. (See Example 1.)

1. $[0, 4]$

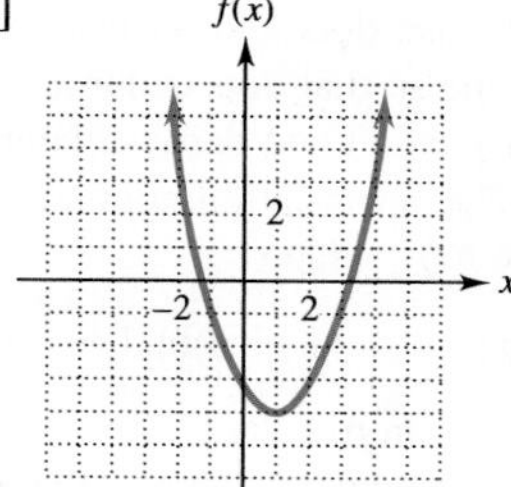

2. $[2, 5]$

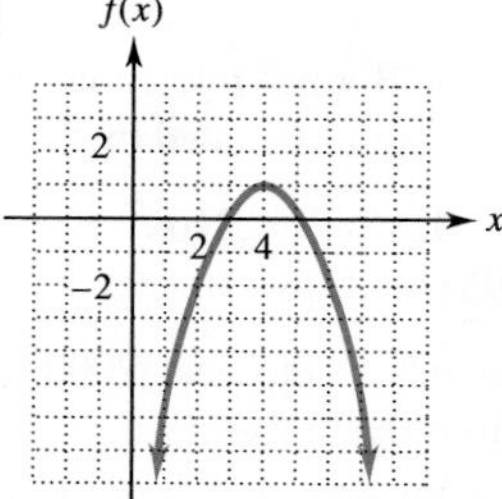

3. $[-4, 2]$

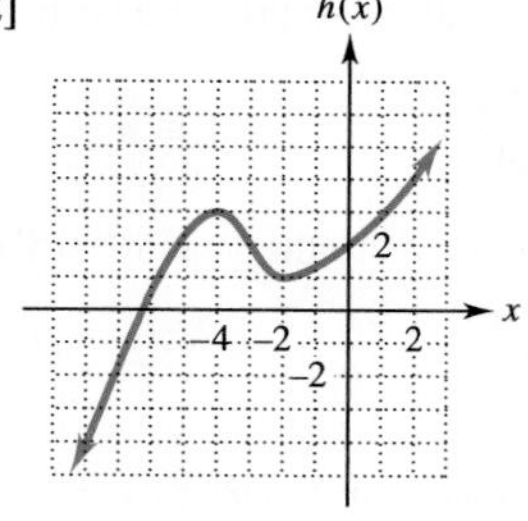

4. $[-1, 2]$

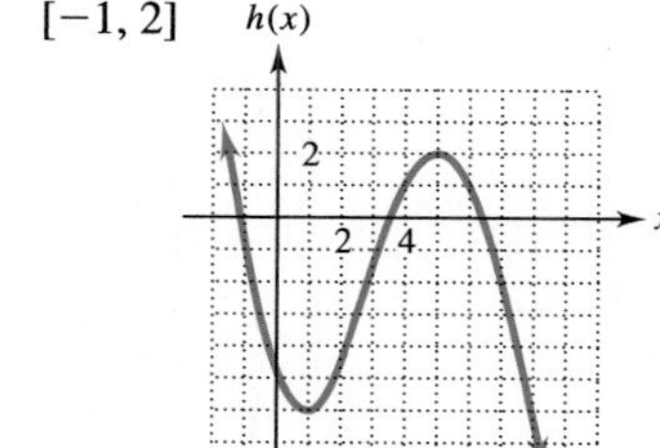

5. $[-8, 0]$

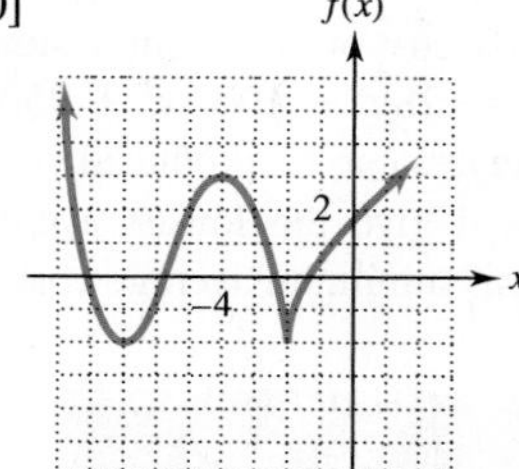

6. $[-4, 4]$

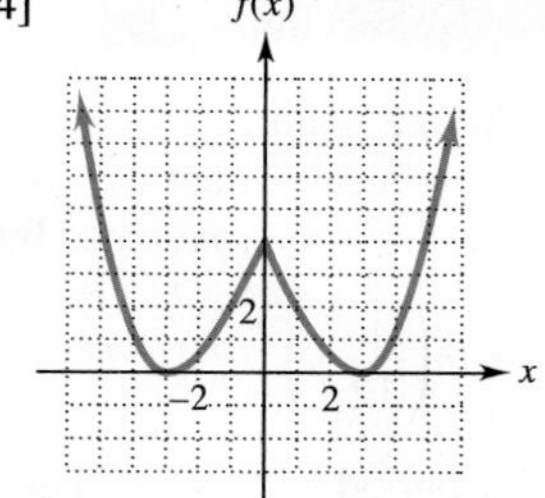

Find the absolute extrema of each function on the given interval. (See Examples 2 and 3.)

7. $f(x) = x^4 - 32x^2 - 7; [-5, 6]$

8. $f(x) = x^4 - 18x^2 + 1; [-4, 4]$

9. $f(x) = \frac{8 + x}{8 - x}; [4, 6]$

10. $f(x) = \frac{1 - x}{3 + x}; [0, 3]$

11. $f(x) = \frac{x}{x^2 + 2}; [-1, 4]$

12. $f(x) = \frac{x - 1}{x^2 + 1}; [0, 5]$

13. $f(x) = (x^2 + 18)^{2/3}; [-3, 2]$

14. $f(x) = (x^2 + 4)^{1/3}; [-1, 2]$

15. $f(x) = \frac{1}{\sqrt{x^2 + 1}}; [-1, 1]$

16. $f(x) = \frac{3}{\sqrt{x^2 + 4}}; [-2, 2]$

If possible, find an absolute extremum of each function on the given interval. (See Example 4.)

17. $g(x) = 2x^3 - 3x^2 - 12x + 1; (0, 4)$

18. $g(x) = \frac{x^2 + 1}{x}; (0, \infty)$

19. $g(x) = \frac{1}{x}; (0, \infty)$

20. $g(x) = xe^{2-x^2}; (-\infty, 0)$

21. $g(x) = 6x^{2/3} - 4x; (0, \infty)$

22. $g(x) = \frac{x}{e^x}; (0, 3)$

Work these problems. (See Example 5.)

23. **Business** The daily cost of producing x stereo receivers is given by $C(x) = .28x^3 - 100.5x^2 + 9500x$, and no more than 250 receivers can be produced each day. What production level will give the lowest average cost per receiver? What is this minimum average cost?

24. **Natural Science** A lake polluted by bacteria is treated with an antibacterial chemical. After t days, the number N of bacteria per ml of water is approximated by

$$N(t) = 20\left(\frac{t}{12} - \ln\left(\frac{t}{12}\right)\right) + 30 \quad (1 \le t \le 15).$$

(a) When during this period will the number of bacteria be a minimum?

(b) What is this minimum number of bacteria?

(c) When during the period will the number of bacteria be a maximum?

(d) What is this maximum number of bacteria?

25. **Business** A manufacturer produces gas grills that sell for \$400 each. The total cost of producing x grills is approximated by the function $C(x) = 525{,}000 - 30x + .012x^2$.

(a) Write the revenue function in this situation.

(b) Write the profit function in this situation.

(c) What number of grills should be made to guarantee maximum profit? What will that profit be?

26. **Business** Saltwater taffy can be sold wholesale for \$45 per thousand individual candies. The cost of producing x thousand candies is $C(x) = .001x^3 + .045x^2 - 1.75x$.

(a) What is the revenue function in this situation?

(b) What is the profit function in this situation?

(c) What number of candies will produce the largest possible profit?

27. **Health** A disease has hit College Station, Texas. The percentage of the population infected t days after the disease has arrived is approximated by $p(t) = 10t\,e^{-t/8}$ for $0 \le t \le 40$.

(a) After how many days is the percentage of infected people a maximum?

(b) What is this maximum percentage of the population infected?

28. **Social Science** Suppose dots and dashes are transmitted over a telegraph line so that dots occur a fraction p of the time (where $0 \le p \le 1$) and dashes occur a fraction $1 - p$ of the time. The information content of the telegraph line is given by $I(p)$, where

$$I(p) = -p\ln p - (1 - p)\ln(1 - p).$$

(a) Show that $I'(p) = -\ln p + \ln(1 - p)$.

(b) Let $I'(p) = 0$, and find the value of p that maximizes the information content.

29. **Health** The number of Botox® injections performed in the United States can be approximated by the function

$$b(x) = -.0245x^3 + .216x^2 - .095x + 1.06 \quad (0 \le x \le 7),$$

where $x = 0$ corresponds to the year 2000.* When was the number of injections the highest?

30. **Business** Sales (in billions of dollars) from Lotto games in the United States can be approximated by the function

$$l(x) = \frac{.5x}{e^{.1x}} + 8.5 \quad (0 \le x \le 17),$$

where $x = 0$ corresponds to the year 1990.† In what year were sales the highest?

**The Statistical Abstract of the United States:* 2009.

†2008 *World Lottery Almanac.*

31. **Transportation** The number (in thousands) of employees working in the railroad industry within the United States can be approximated by the function

$$f(x) = 2.96x^2 - 78.63x + 672.4 \qquad (9 \le x \le 16),$$

where $x = 9$ corresponds to the year 1999.* In what year was the number of workers at a minimum?

32. **Finance** A function modeling the federal funds effective rate (in percent) is

$$f(x) = -.00084x^3 + .0809x^2 - 2.658x + 33.72 \quad (5 \le x \le 17),$$

where $x = 5$ corresponds to the year 1995.† What is the maximum rate? What is the minimum rate?

Work these problems. (See Examples 6 and 7.)

33. **Geometry** An open box is to be made by cutting a square from each corner of a 3-foot-by-8-foot piece of cardboard and then folding up the sides. What size square should be cut from each corner in order to produce a box of maximum volume?

34. **Business** A watch manufacturing firm needs to design an open-topped box with a square base. The box must hold 32 cubic inches. Find the dimensions of the box that can be built with the minimum amount of materials.

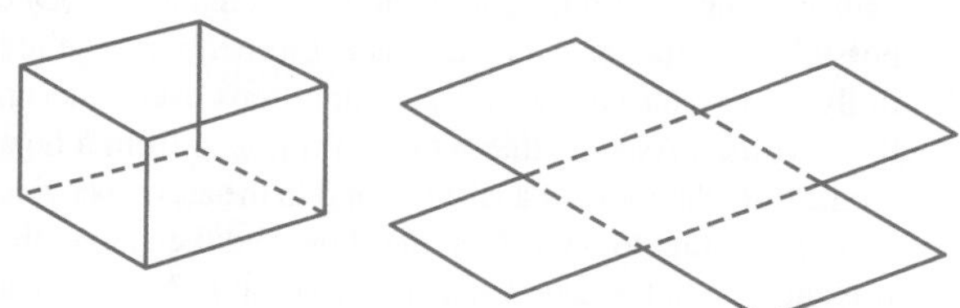

35. **Business** An artist makes a closed box with a square base. The box is to have a volume of 16,000 cubic centimeters. The material for the top and bottom of the box costs 3 cents per square centimeter, while the material for the sides costs 1.5 cents per square centimeter. Find the dimensions of the box that will lead to the minimum total cost. What is the minimum total cost?

36. **Business** A cylindrical box will be tied up with ribbon as shown in the accompanying figure. The longest piece of ribbon available is 130 cm long, and 10 cm of that is required for the bow. Find the radius and height of the box with the largest possible volume.

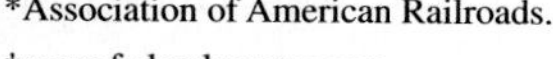

*Association of American Railroads.

†www.federalreserve.gov.

37. **Business** A company wishes to manufacture a rectangular box with a volume of 36 cubic feet that is open on top and that is twice as long as it is wide. Find the dimensions of the box produced from the minimum amount of material.

38. **Business** A cylindrical can of volume 58 cubic inches (approximately 1 quart) is to be designed. For convenient handling, it must be at least 1 inch high and 2 inches in diameter. What dimensions (radius of top, and height of can) will use the least amount of material?

Business *Work these problems. (See Example 8.)*

39. A farmer has 1200 m of fencing. He wants to enclose a rectangular field bordering a river, with no fencing needed along the river. Let x represent the width of the field.
(a) Write an expression for the length of the field.
(b) Find the area of the field.
(c) Find the value of x leading to the maximum area.
(d) Find the maximum area.

40. A rectangular field is to be enclosed with a fence. One side of the field is against an existing fence, so that no fence is needed on that side. If material for the fence costs \$2 per foot for the two ends and \$4 per foot for the side parallel to the existing fence, find the dimensions of the field of largest area that can be enclosed for \$1000.

41. A rectangular field is to be enclosed on all four sides with a fence. Fencing material costs \$3 per foot for two opposite sides and \$6 per foot for the other two sides. Find the maximum area that can be enclosed for \$2400.

42. A fence must be built to enclose a rectangular area of 20,000 ft^2. Fencing material costs \$3 per foot for the two sides facing north and south and \$6 per foot for the other two sides. Find the cost of the least expensive fence.

43. A fence must be built in a large field to enclose a rectangular area of 15,625 m^2. One side of the area is bounded by an existing fence; no fence is needed there. Material for the fence costs \$2 per meter for the two ends and \$4 per meter for the side opposite the existing fence. Find the cost of the least expensive fence.

44. **Business** A mathematics book is to contain 36 square inches of printed matter per page, with margins of 1 inch

along the sides and $1\frac{1}{2}$ inches along the top and bottom. Find the dimensions of the page that will lead to the minimum amount of paper being used for a page.

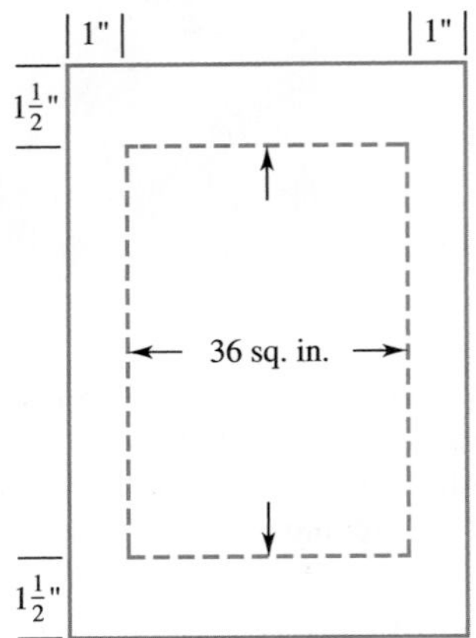

Work these problems. (See Examples 5–8.)

45. Business If the price charged for a candy bar is $p(x)$ cents, where

$$p(x) = 100 - \frac{x}{10},$$

then x thousand candy bars will be sold in a certain city.

(a) Find an expression for the total revenue from the sale of x thousand candy bars. (*Hint*: Find the product of $p(x)$, x, and 1000.)

(b) Find the value of x that leads to the maximum revenue.

(c) Find the maximum revenue.

46. Business A company makes plastic buckets for children to play with on beaches. The buckets sell for \$75 per hundred. The total cost (in dollars) of making x hundred buckets is $C(x) = 5x^2 - 20x + 12$. Assume that the company can sell all the buckets it makes. How many buckets should be made to maximize profit?

47. Business We can use the function

$$f(x) = 55.44(54.3x - 143.9)e^{-.15x}$$

to model the revenue (in millions of U.S. dollars) from the cell phone industry in India, where $x = 2$ corresponds to the year 2002.* According to this model, in what year does revenue reach its maximum?

48. Business A rock-and-roll band travels from engagement to engagement in a large bus. While traveling x miles per hour, this bus burns fuel at the rate of $G(x)$ gallons per mile, where

$$G(x) = \frac{1}{50}\left(\frac{200}{x} + \frac{x}{15}\right).$$

(a) If fuel costs \$3 per gallon, find the speed that will produce the minimum total cost for a 250-mile trip.

(b) Find the minimum total cost.

49. Business A company wishes to run a utility cable from point A on the shore to an installation at point B on the island. The island is 6 miles from the shore (at point C), and point A is 9 miles from point C. It costs \$400 per mile to run the cable on land and \$500 per mile underwater. Assume that the cable starts at A, runs along the shoreline, and then angles and runs underwater to the island. Find the point at which the line should begin to angle in order to yield the minimum total cost. (*Hint*: The length of the line underwater is $\sqrt{x^2 + 36}$.) See the accompanying figure.

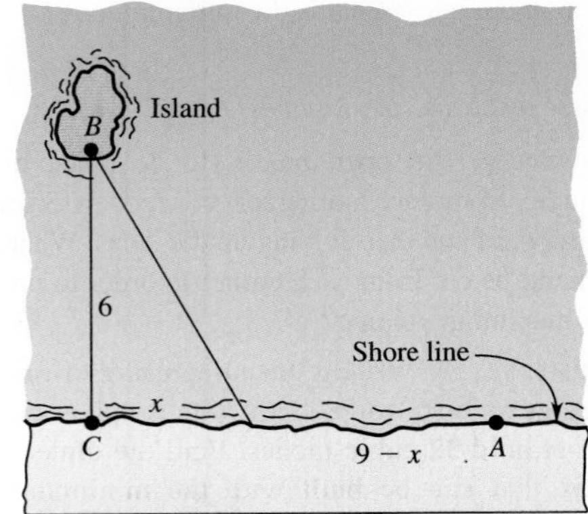

50. Natural Science Homing pigeons avoid flying over large bodies of water, preferring to fly around them instead. (One possible explanation is the fact that extra energy is required to fly over water because air pressure drops over water in the daytime.) Assume that a pigeon released from a boat 1 mile from the shore of a lake (point B in the accompanying figure) flies first to point P on the shore and then along the straight edge of the lake to reach its home at L. Assume that L is 2 miles from point A, the point on the shore closest to the boat, and that a pigeon needs $\frac{4}{3}$ as much energy to fly over water as over land. Find the location of point P if the pigeon uses the least possible amount of energy.

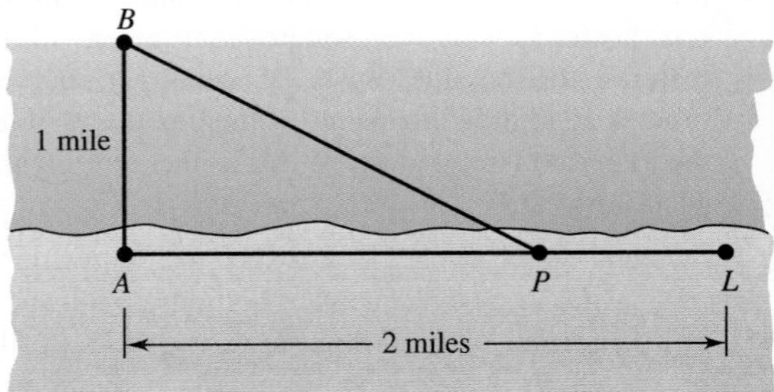

Business *The remaining exercises refer to economic lot size. (See Example 9.)*

51. A manufacturer of porcelain sinks has an annual demand of 12,300 sinks. It costs \$2.25 to store a sink for a year,

*www.researchinchina.com.

and it costs $350 to set up a factory to produce each batch. Find the number of batches of sinks that should be produced for the minimum total production cost.

52. How many sinks per batch will be produced in Exercise 51?

53. A regional market has a steady annual demand for 15,900 cases of beer. It costs $3.25 to store a case for a year. The market pays $5.50 for each order that is placed. Find the number of orders for cases of beer that should be placed each year. (*Hint*: Use the formula for economic lot size, with ordering cost in place of setup cost.)

54. Find the number of cases per order in Exercise 53.

55. A large hospital has an annual demand for 50,000 booklets on healthy eating. It costs $.75 to store one booklet for a year, and it costs $80 to place an order for a new batch of booklets. Find the optimum number of copies per order.

56. A restaurant has an annual demand for 900 bottles of a California wine. It costs $1 to store one bottle for one year, and it costs $5 to place a reorder. Find the number of orders that should be placed annually.

57. Choose the correct answer:* The economic order quantity formula assumes that
   (a) Purchase costs per unit differ due to quantity discounts.
   (b) Costs of placing an order vary with the quantity ordered.
   (c) The periodic demand for the goods is known.
   (d) Erratic usage rates are cushioned by safety stocks.

*Question from the Uniform CPA Examination of the American Institute of Certified Public Accountants, May 1991. Reprinted by permission of the Institute of Certified Public Accountants.

✓ Checkpoint Answers

1. Absolute maximum at 0; absolute minimum at -2
2. Absolute maximum of -4 at $x = 2$; absolute minimum of -40 at $x = -4$
3. Absolute maximum of 94 at $x = 2$; absolute minimum of 14 at $x = 6$
4. $12.88; $.88; $1.63
5. **(a)** $V(x) = 640x - 104x^2 + 4x^3$
   (b) $V'(x) = 640 - 208x + 12x^2$
   (c) $x = 4$
   (d) $V(4) = 1152$ cubic centimeters
6. **(a)** $C''(x) = 192{,}000x^{-3}$ and $C''(\sqrt{16{,}000}) \approx .095 > 0$, so there is an absolute minimum at $x = \sqrt{16{,}000}$ by the second-derivative test and the critical-point theorem.
   (b)

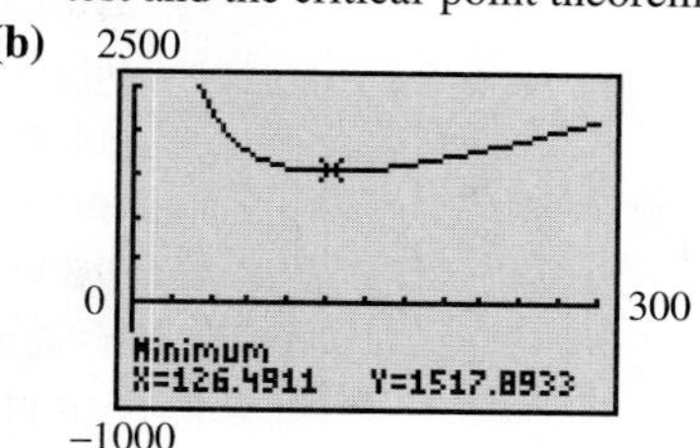

7. 11 batches

12.4 ▶ Curve Sketching

In earlier sections, we saw that the first and second derivatives of a function provide a variety of information about the graph of the function, such as the location of its local extrema, the concavity of the graph, and the intervals on which it is increasing and decreasing. This information can be very helpful for interpreting misleading screen images on a graphing calculator or computer. It also enables us to make reasonably accurate graphs of many functions by hand if graphing technology is not available.

When graphing functions by hand, you should use the following guidelines. It may not always be feasible to carry out all the steps, but you should do as many as necessary, in any convenient order, to obtain a reasonable graph.

To sketch the graph of a function $y = f(x)$, execute the following steps:

1. Find the y-intercept (if it exists) by letting $x = 0$ and computing $y = f(0)$.
2. Find the x-intercepts (if any) by letting $y = 0$ and solving the equation $f(x) = 0$ if doing so is not too difficult.

(continued)

3. If f is a rational function, find any vertical asymptotes by finding the numbers for which the denominator is 0, but the numerator is nonzero. Find any horizontal asymptotes by using the techniques of Section 3.7, as summarized in the box on page 201.
4. Find $f'(x)$ and $f''(x)$.
5. Locate any critical numbers by solving the equation $f'(x) = 0$ and determining where $f'(x)$ does not exist, but $f(x)$ does. Find the local extrema by using the first- or second-derivative test. Find the intervals where f is increasing or decreasing by solving the inequalities $f'(x) > 0$ and $f'(x) < 0$.
6. Locate potential points of inflection by solving the equation $f''(x) = 0$ and determining where $f''(x)$ does not exist, but $f(x)$ does. Find the intervals where f is concave upward or downward by solving the inequalities $f''(x) > 0$ and $f''(x) < 0$. Use this information to determine the points of inflection.
7. Use the preceding results and any other information that may be available to determine the general shape of the graph.
8. Plot the intercepts, critical points, points of inflection, and other points as needed. Connect the points with a smooth curve, using correct concavity and being careful not to draw a connected graph through points where the function is not defined.
9. If feasible, verify your results with a graphing calculator or computer. If there are significant differences between your graph and the calculator's, check for mistakes in the hand-drawn graph or adjust the calculator viewing window appropriately.

Example 1 Graph $f(x) = 2x^3 - 3x^2 - 12x + 1$.

Solution

Step 1 The y-intercept is $f(0) = 2 \cdot 0^3 - 3 \cdot 0^2 - 12 \cdot 0 + 1 = 1$.

Step 2 To find the x-intercepts, we must solve the equation

$$2x^3 - 3x^2 - 12x + 1 = 0.$$

There is no easy way to do this by hand, so skip this step. Since $f(x)$ is a polynomial function, the graph has no asymptotes, so we can also skip Step 3.

Step 4 The first derivative is $f'(x) = 6x^2 - 6x - 12$, and the second derivative is $f''(x) = 12x - 6$.

Step 5 The first derivative is defined for all x, so the only critical numbers are the solutions of $f'(x) = 0$:

$$6x^2 - 6x - 12 = 0$$

$$x^2 - x - 2 = 0 \quad \text{Divide both sides by 6.}$$

$$(x + 1)(x - 2) = 0 \quad \text{Factor.}$$

$$x = -1 \quad \text{or} \quad x = 2.$$

Using the second-derivative test on the critical number $x = -1$, we have

$$f''(-1) = 12(-1) - 6 = -18 < 0.$$

Hence, there is a local maximum when $x = -1$, that is, at the point $(-1, f(-1)) = (-1, 8)$. Similarly,

$$f''(2) = 12(2) - 6 = 18 > 0,$$

so there is a local minimum when $x = 2$ (at the point $(2, f(2)) = (2, -19)$).

Next, we determine the intervals on which f is increasing or decreasing by solving the inequalities

$$\begin{aligned} f'(x) &> 0 \quad \text{and} & f'(x) &< 0 \\ 6x^2 - 6x - 12 &> 0 & 6x^2 - 6x - 12 &< 0. \end{aligned}$$

The critical numbers divide the x-axis into three regions. Testing a number from each region, as indicated in Figure 12.39, we conclude that f is increasing on the intervals $(-\infty, -1)$ and $(2, \infty)$ and decreasing on $(-1, 2)$.

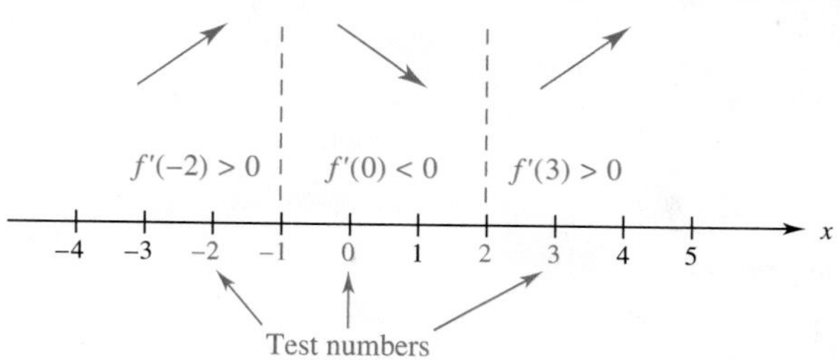

FIGURE 12.39

Step 6 The possible points of inflection are determined by the solutions of $f''(x) = 0$:

$$\begin{aligned} 12x - 6 &= 0 \\ x &= \frac{1}{2}. \end{aligned}$$

Determine the concavity of the graph by solving

$$\begin{aligned} f''(x) &> 0 \quad \text{and} & f''(x) &< 0 \\ 12x - 6 &> 0 & 12x - 6 &< 0 \\ x &> \frac{1}{2} & x &< \frac{1}{2}. \end{aligned}$$

Therefore, f is concave upward on the interval $(\frac{1}{2}, \infty)$ and concave downward on $(-\infty, \frac{1}{2})$. Consequently, the only point of inflection is $(\frac{1}{2}, f(\frac{1}{2})) = (\frac{1}{2}, -5.5)$.

Step 7 Since f is a third-degree polynomial function, we know from Section 3.6 that when x is very large in absolute value, its graph must resemble the graph of its highest-degree term, $2x^3$; that is, the graph must rise sharply on the right side and fall sharply on the left. Combining this fact with the information obtained in the preceding steps, we see that the graph of f must have the general shape shown in Figure 12.40.

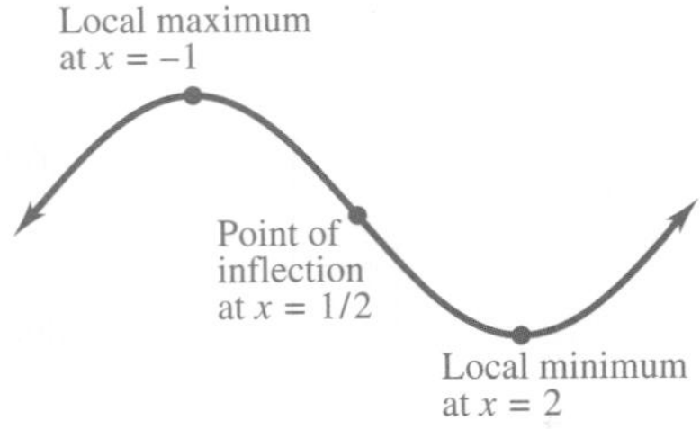

FIGURE 12.40

Step 8 Now we plot the points determined in Steps 1, 5, and 6, together with a few additional points, to obtain the graph in Figure 12.41. 1✓ 2✓

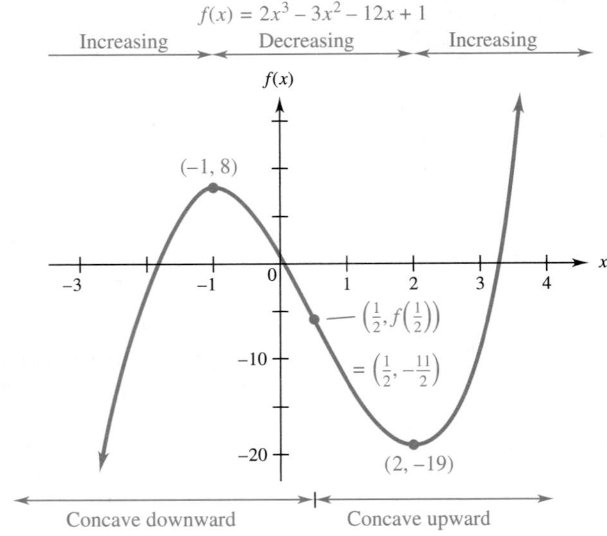

FIGURE 12.41

✓Checkpoint 1

Step 9 Use technology to verify that the graph in Figure 12.41 is correct.

Answer: See page 822.

✓Checkpoint 2

Sketch the graph of

$$f(x) = x^3 - 3x^2.$$

Answer: See page 822.

Example 2 Graph $f(x) = \dfrac{3x^2}{x^2 + 5}$.

Solution

Step 1 The y-intercept is $f(0) = 0/5 = 0$.

Step 2 To find the x-intercepts, note that $f(x)$ is always defined, because $x^2 + 5$ is always positive. Hence, $f(x) = 0$ when the numerator $3x^2 = 0$; this occurs when $x = 0$. So the point $(0, f(0)) = (0, 0)$ is both the x- and y-intercept.

Step 3 This is a rational function, but its denominator is always nonzero (why?), so there are no vertical asymptotes. Using the techniques presented in Section 3.7, we see that

$$f(x) = \frac{3x^2}{x^2 + 5} = \frac{\frac{3x^2}{x^2}}{\frac{x^2}{x^2} + \frac{5}{x^2}} = \frac{3}{1 + \frac{5}{x^2}}.$$

When x is very large in absolute value, so is x^2; thus, $5/x^2$ is very close to 0, and hence $f(x)$ is very close to $3/(1 + 0) = 3$. Consequently, the horizontal line $y = 3$ is a horizontal asymptote.

Step 4 The first derivative is

$$f'(x) = \frac{(x^2 + 5)(6x) - (3x^2)(2x)}{(x^2 + 5)^2} = \frac{30x}{(x^2 + 5)^2}.$$

The second derivative is

$$f''(x) = \frac{(x^2 + 5)^2\, 30 - (30x)(2)(x^2 + 5)(2x)}{(x^2 + 5)^4}.$$

Factor $30(x^2 + 5)$ out of the numerator:

$$f''(x) = \frac{30(x^2 + 5)[(x^2 + 5) - (x)(2)(2x)]}{(x^2 + 5)^4}.$$

Divide a factor of $(x^2 + 5)$, out of the numerator and denominator and simplify the numerator:

$$\begin{aligned} f''(x) &= \frac{30[(x^2 + 5) - (x)(2)(2x)]}{(x^2 + 5)^3} \\ &= \frac{30[(x^2 + 5) - 4x^2]}{(x^2 + 5)^3} \\ &= \frac{30(5 - 3x^2)}{(x^2 + 5)^3}. \end{aligned}$$

Step 5 Since $f'(x) = \dfrac{30x}{(x^2 + 5)^2}$ and $x^2 + 5 \neq 0$ for all x, $f'(x)$ is always defined. $f'(x) = 0$ when its numerator, $30x$, is 0; this occurs when $x = 0$. The critical number 0 divides the x-axis into two regions (Figure 12.42). Testing a number in each region shows that f is decreasing on $(-\infty, 0)$ and increasing on $(0, \infty)$. By the first-derivative test, f has a local minimum at $x = 0$.

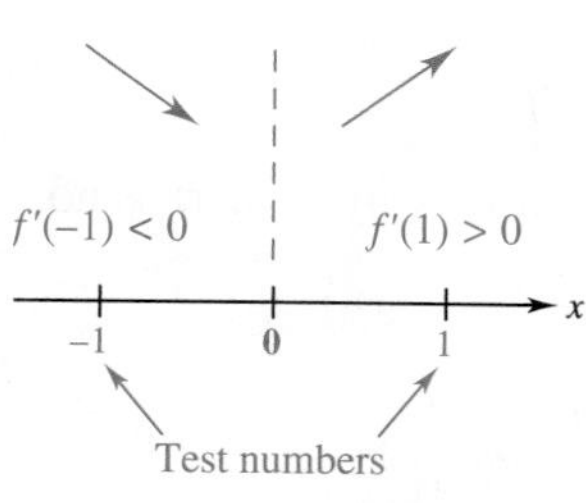

FIGURE 12.42

Step 6 $f''(x) = \dfrac{30(5 - 3x^2)}{(x^2 + 5)^3}$ is 0 when

$$\begin{aligned} 30(5 - 3x^2) &= 0 \\ 3x^2 &= 5 \\ x &= \pm\sqrt{5/3} \approx \pm 1.29. \end{aligned}$$

Testing a point in each of the three intervals defined by these points shows that f is concave downward on $(-\infty, -1.29)$ and $(1.29, \infty)$ and concave upward on $(-1.29, 1.29)$. The graph has inflection points at $(\pm\sqrt{5/3}, f(\pm\sqrt{5/3})) \approx (\pm 1.29, .75)$.

Step 7 The information about the shape of the graph obtained in Steps 4, 5, and 6 is summarized in the following chart:

Interval	$(-\infty, -1.29)$	$(-1.29, 0)$	$(0, 1.29)$	$(1.29, \infty)$
Sign of f'	−	−	+	+
Sign of f''	−	+	+	−
***f* Increasing or Decreasing**	Decreasing	Decreasing	Increasing	Increasing
Concavity of *f*	Downward	Upward	Upward	Downward
Shape of Graph	⌒	◡	◡	⌒

Step 8 Plot some points (several are needed near the origin), including the intercept at the origin, and use the fact that $y = 3$ is a horizontal asymptote to obtain the graph in Figure 12.43. 3 ✓

✓Checkpoint 3

Step 9 Use technology to verify that the graph in Figure 12.43 is correct.

Answer: See page 822.

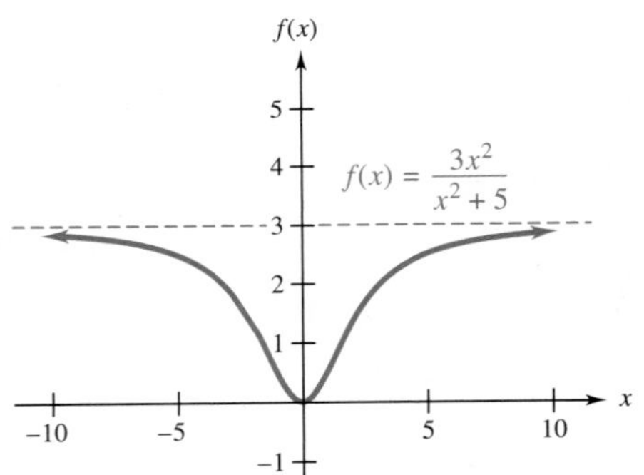

FIGURE 12.43

Example 3 Graph $f(x) = x + 1/x$.

Solution

Step 1 Since $x = 0$ is not in the domain of the function (why?), there is no y-intercept.

Step 2 To find the x-intercepts, solve $f(x) = 0$:

$$x + \frac{1}{x} = 0$$

$$x = -\frac{1}{x}$$

$$x^2 = -1.$$

Since x^2 is always positive, there is also no x-intercept.

Step 3 Note that the rule of f can be written as

$$f(x) = x + \frac{1}{x} = \frac{x^2 + 1}{x}.$$

When $x = 0$, the denominator is 0, but the numerator is nonzero, so there is a vertical asymptote at $x = 0$. Since the numerator of $f(x)$ is of a higher degree than the denominator, there is no horizontal asymptote.

Step 4 Because $f(x) = x + 1/x = x + x^{-1}$, we have

$$f'(x) = 1 - x^{-2} = 1 - \frac{1}{x^2},$$

so

$$f''(x) = 2x^{-3} = \frac{2}{x^3}.$$

Step 5 $f'(x) = 0$ when

$$\frac{1}{x^2} = 1$$
$$x^2 = 1$$
$$x = 1 \quad \text{or} \quad x = -1.$$

Hence, $x = -1$ and $x = 1$ are critical numbers. The derivative does not exist when $x = 0$, but the function is not defined there either, so $x = 0$ is not a critical number. Evaluating $f'(x)$ in each of the regions determined by the critical numbers and the asymptote shows that f is increasing on $(-\infty, -1)$ and $(1, \infty)$ and decreasing on $(-1, 0)$ and $(0, 1)$, as summarized in the chart in Step 7. By the first-derivative test, f has a relative maximum of $y = f(-1) = -2$ when $x = -1$ and a relative minimum of $y = f(1) = 2$ when $x = 1$.

Step 6 The second derivative $f''(x) = 2/x^3$ is never equal to 0 and does not exist when $x = 0$. (The function itself also does not exist at 0.) Because of this, there may be a change in concavity, but not an inflection point, when $x = 0$. The second derivative is negative when x is negative, making f concave downward on $(-\infty, 0)$. Also, $f''(x) > 0$ when $x > 0$, making f concave upward on $(0, \infty)$, as indicated in the chart in Step 7.

Step 7 The preceding information is summarized in the following chart:

Interval	$(-\infty, -1)$	$(-1, 0)$	$(0, 1)$	$(1, \infty)$
Sign of f'	+	−	−	+
Sign of f''	−	−	+	+
f Increasing or Decreasing	Increasing	Decreasing	Decreasing	Increasing
Concavity of f	Downward	Downward	Upward	Upward
Shape of Graph	 ⌒ (rising)	 ⌒ (falling)	 ◡ (falling)	 ◡ (rising)

We can determine the shape of the graph when x is very large in absolute value by noting that as x gets very large, the second term of its rule, $1/x$, gets very small, so that $f(x) = x + 1/x \approx x$. Hence, the graph gets closer and closer to the straight line $y = x$ as x becomes larger and larger. The line $y = x$ is called an **oblique**, or **slant asymptote**.

Step 8 Plot several points and use the preceding information to obtain the graph of $f(x)$ in Figure 12.44. 4 ✓

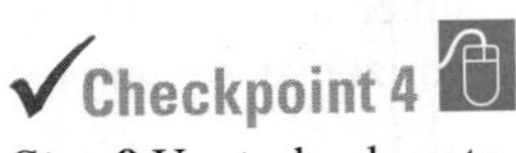

Step 9 Use technology to verify that the graph in Figure 12.44 is correct.

Answer: See page 822.

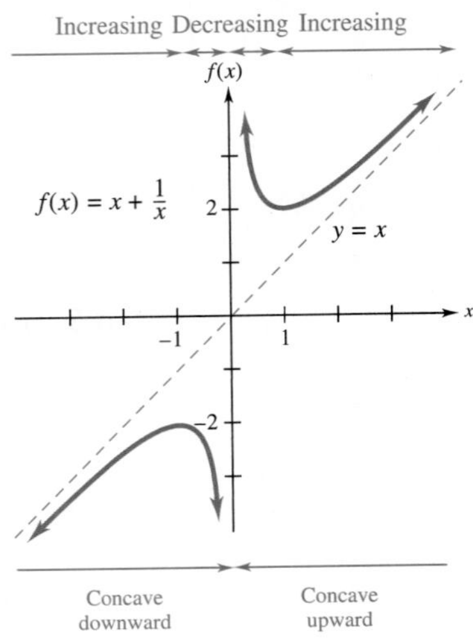

FIGURE 12.44

12.4 ▶ Exercises

Sketch the graph of the function. Identify any local extrema and points of inflection. (See Examples 1–3.)

1. $f(x) = -x^2 - 10x - 25$
2. $f(x) = x^2 - 12x + 36$
3. $f(x) = 3x^3 - 3x^2 + 1$
4. $f(x) = 2x^3 - 4x^2 + 2$
5. $f(x) = -2x^3 - 9x^2 + 108x - 10$
6. $f(x) = -2x^3 - 9x^2 + 60x - 8$
7. $f(x) = 2x^3 + \frac{7}{2}x^2 - 5x + 3$
8. $f(x) = x^3 - \frac{15}{2}x^2 - 18x - 1$
9. $f(x) = (x + 3)^4$
10. $f(x) = x^3$
11. $f(x) = x^4 - 18x^2 + 5$
12. $f(x) = x^4 - 8x^2$
13. $f(x) = x - \frac{1}{x}$
14. $f(x) = 2x + \frac{8}{x}$
15. $f(x) = \frac{x^2 + 25}{x}$
16. $f(x) = \frac{x^2 + 4}{x}$
17. $f(x) = \frac{x - 1}{x + 1}$
18. $f(x) = \frac{x}{1 + x}$

Sketch the graph of the function. Identify any local extrema and inflection points.

19. $y = x - \ln|x|$
20. $y = \frac{\ln x}{x}$
21. $y = xe^{-x}$
22. $y = x \ln|x|$

In Exercises 23–28, sketch the graph of a function f that has all of the properties listed. There are many correct answers, and your graph need not be given by an algebraic formula.

23. **(a)** The domain of f is $[0, 10]$.
    (b) $f'(x) > 0$ and $f''(x) > 0$ for all x in the domain of f.
24. **(a)** The domain of f is $[0, 10]$.
    (b) $f'(x) > 0$ and $f''(x) < 0$ for all x in the domain of f.
25. **(a)** Continuous and differentiable for all real numbers
    (b) Increasing on $(-\infty, -3)$ and $(1, 4)$
    (c) Decreasing on $(-3, 1)$ and $(4, \infty)$
    (d) Concave downward on $(-\infty, -1)$ and $(2, \infty)$
    (e) Concave upward on $(-1, 2)$

(f) $f'(-3) = f'(4) = 0$
(g) Inflection points at $(-1, 3)$ and $(2, 4)$

26. (a) Continuous for all real numbers
(b) Increasing on $(-\infty, -2)$ and $(0, 3)$
(c) Decreasing on $(-2, 0)$ and $(3, \infty)$
(d) Concave downward on $(-\infty, 0)$ and $(0, 5)$
(e) Concave upward on $(5, \infty)$
(f) $f'(-2) = f'(3) = 0$

27. (a) Continuous for all real numbers
(b) Decreasing on $(-\infty, -6)$ and $(1, 3)$
(c) Increasing on $(-6, 1)$ and $(3, \infty)$
(d) Concave upward on $(-\infty, -6)$ and $(3, \infty)$
(e) Concave downward on $(-6, 3)$
(f) y-intercept at $(0, 2)$

28. (a) Continuous and differentiable everywhere except at $x = -1$, where it has a vertical asymptote
(b) Decreasing everywhere it is defined
(c) Concave downward on $(-\infty, -1)$ and $(1, 2)$
(d) Concave upward on $(-1, 1)$ and $(2, \infty)$

29. **Natural Science** The accompanying figure shows how the risk of chromosomal abnormality in a child increases with the age of the mother.*

(a) What is the sign of the first derivative on the interval (20, 50)? Why?
(b) What is the sign of the second derivative on that interval? What does this tell you about the rate of risk?

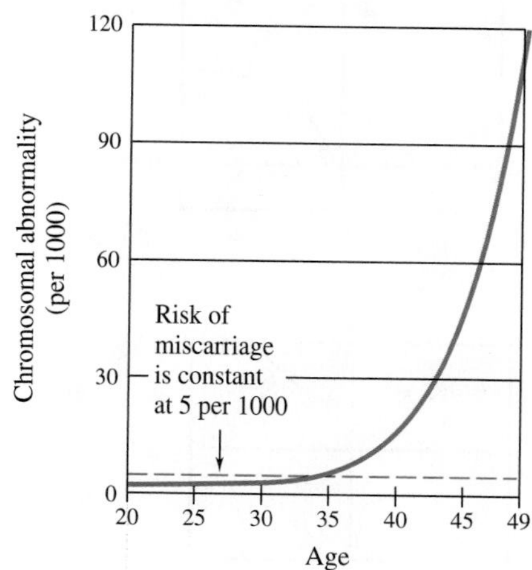

Source: American College of Obstetricians and Gynecologists.

30. **Business** The accompanying figure shows the *product life cycle* graph, with typical products marked on it. It illustrates the fact that a new product is often purchased at a faster and faster rate as people become familiar with it. In time, saturation is reached and the purchase rate stays constant until the product is made obsolete by newer products, after which it is purchased less and less often.*

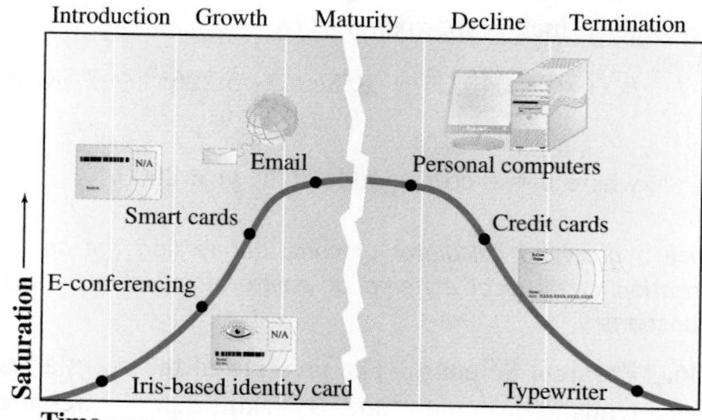

(a) Which products on the left side of the graph are closest to the left-hand point of inflection? What does the point of inflection mean here?
(b) Which product on the right side of the graph is closest to the right-hand point of inflection? What does the point of inflection mean here?
(c) Discuss where portable DVD players, fax machines, and other new technologies should be placed on the graph.

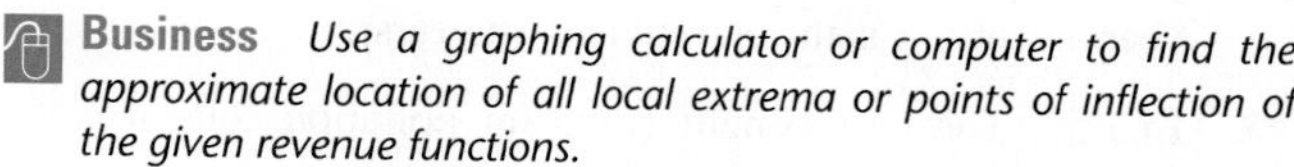
Business *Use a graphing calculator or computer to find the approximate location of all local extrema or points of inflection of the given revenue functions.*

31. The revenue (in billions of dollars) for IBM Corporation is approximated by

$$R(x) = .0725x^4 - 3.9421x^3 + 79.159x^2 - 693.2x + 2315.4 \quad (9 \le x \le 18),$$

where $x = 9$ corresponds to the year 1999.†

32. The revenue (in billions of dollars) for Eastman Kodak Corporation is approximated by

$$R(x) = -.0385x^3 + 1.461x^2 - 18.25x + 88.3 \quad (9 \le x \le 18),$$

where $x = 9$ corresponds to the year 1999.†

33. The revenue (in billions of dollars) for Polo Ralph Lauren Corporation is approximated by

$$R(x) = -.0046x^3 + .090x^2 - .101x + 2.1 \quad (0 \le x \le 9),$$

where $x = 0$ corresponds to the year 2000.†

New York Times, February 5, 1994, p. 24.

*http://www.tutor2u.net/business/marketing/products_lifecycle.asp.

†www.morningstar.com.

34. The revenue (in billions of dollars) for Coca-Cola Bottling Company is approximated by

$$R(x) = 4.102x^3 - 166.671x^2 + 2230.28x - 8536.8 \quad (9 \le x \le 18),$$

where $x = 9$ corresponds to the year 1999.*

35. The revenue (in billions of dollars) for J. C. Penney Company, Inc., is approximated by

$$R(x) = -.0432x^4 + .8586x^3 - 5.226x^2 + 7.58x + 31.4 \quad (0 \le x \le 9),$$

where $x = 0$ corresponds to the year 2000.*

Use a graphing calculator or computer to find the approximate location of all local extrema or points of inflection of the given functions.

36. Physical Science The pressure of the oil in a reservoir tends to drop with time. By taking sample pressure readings for a particular oil reservoir, petroleum engineers have found that the change in pressure is given by

$$P(t) = t^3 - 18t^2 + 81t \qquad (0 \le t \le 15),$$

where t is time in years from the date of the first reading.

37. Health The function

$$f(x) = .000000143x^4 + .0000178x^3 - .00145x^2 - .9279x + 77.8$$

gives a person's average remaining life expectancy (in years), where x is the current age (in years).†

38. Health The cost–benefit curve for pollution control is given by

$$y = \frac{9.2x}{106 - x} \qquad (0 < x \le 100),$$

where y is the cost, in thousands of dollars, or removing x percent of a specified industrial pollutant.

39. Social Science The enrollment (in millions) in U.S. public high schools is modeled by

$$g(x) = -.000039x^4 + .002393x^3 - .03311x^2 - .0845x + 14.3 \quad (0 \le x \le 35),$$

where $x = 0$ corresponds to the year 1975.‡

40. Social Science The enrollment (in millions) of female students in American institutions of higher education is approximated by

$$f(x) = .00016x^3 - .0024x^2 + .118x + 6.1 \quad (0 \le x \le 25),$$

where $x = 0$ corresponds to the year 1980.*

✓ Checkpoint Answers

1.

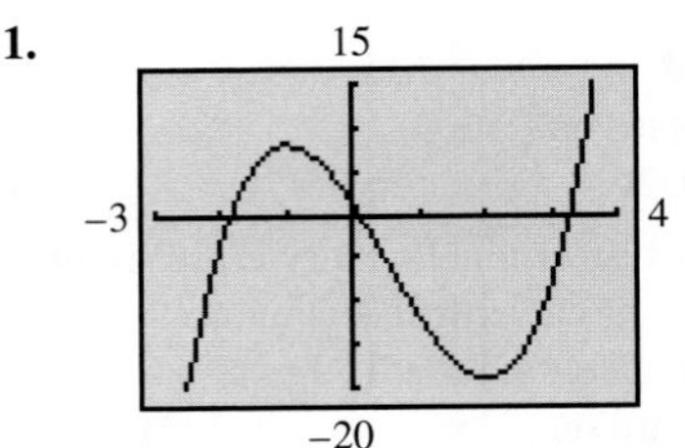

2.

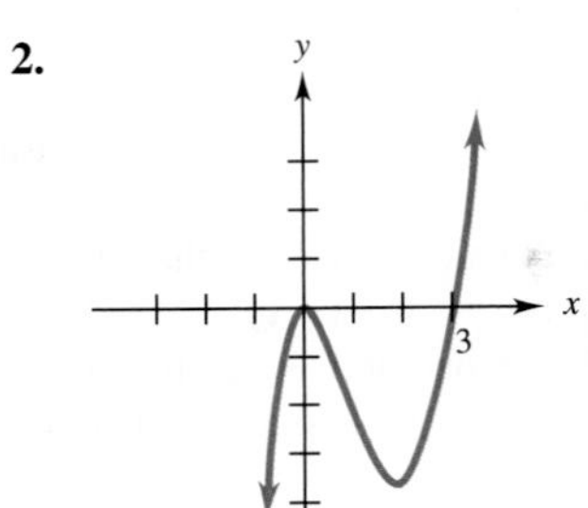

3.

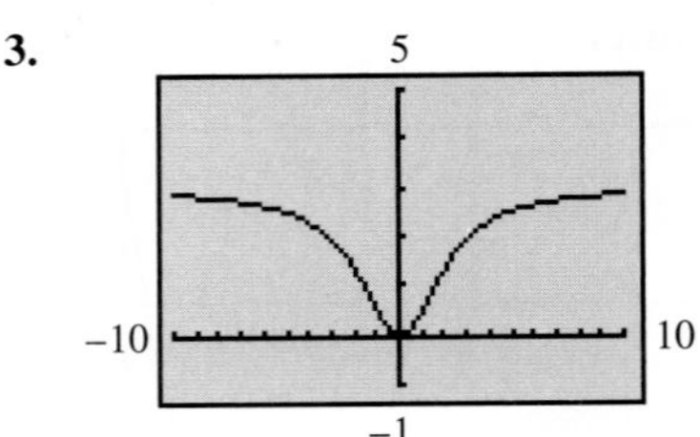

4.

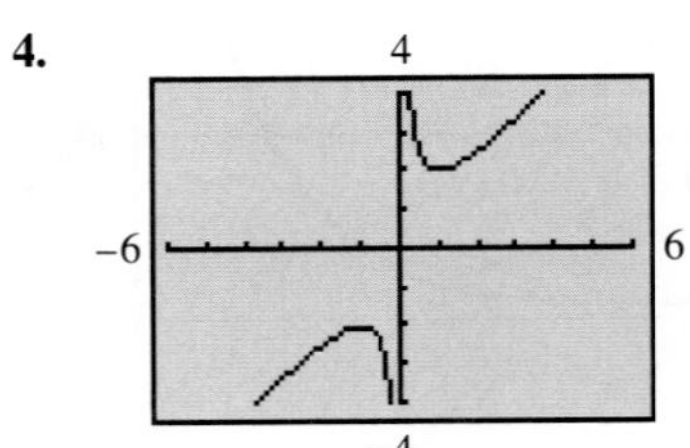

*www.morningstar.com.

†National Center for Health Statistics.

‡U.S. Center for Educational Statistics.

†U.S. Center for Educational Statistics.

CHAPTER 12 ▸ Summary

KEY TERMS AND SYMBOLS

12.1 ▸ increasing function on an interval
decreasing function on an interval
critical number
critical point
local maximum (maxima)
local minimum (minima)
local extremum (extrema)

12.2 ▸ $f''(x)$ or y'' or $\frac{d^2}{dx^2}$ or $D_x^2\,[f(x)]$, second derivative of f
$f'''(x)$, third derivative of f
$f^{(n)}(x)$, nth derivative of f
acceleration
concave upward
concave downward
point of inflection
point of diminishing returns

12.3 ▸ absolute maximum on an interval
absolute minimum on an interval
extreme-value theorem
critical-point theorem
economic lot size

12.4 ▸ curve sketching
oblique asymptote

CHAPTER 12 KEY CONCEPTS

If $f'(x) > 0$ for each x in an interval, then f is **increasing** on the interval; if $f'(x) < 0$ for each x in the interval, then f is **decreasing** on the interval; if $f'(x) = 0$ for each x in the interval, then f is **constant** on the interval.

Local Extrema ▸ Let c be a number in the domain of a function f. Then f has a **local maximum** at c if $f(x) \leq f(c)$ for all x near c, and f has a **local minimum** at c if $f(x) \geq f(c)$ for all x near c. If f has a local extremum at c, then $f'(c) = 0$ or $f'(c)$ does not exist.

First-Derivative Test ▸ Let f be a differentiable function for all x in $[a, b]$, except possibly at $x = c$. Assume that $a < c < b$ and that c is the only critical number for f in $[a, b]$. If $f'(a) > 0$ and $f'(b) < 0$, then there is a local maximum at c. If $f'(a) < 0$ and $f'(b) > 0$, then there is a local minimum at c.

Concavity ▸ Let f have derivatives f' and f'' for all x in (a, b). Then f is **concave upward** on (a, b) if $f''(x) > 0$ for all x in (a, b), f is **concave downward** on (a, b) if $f''(x) < 0$ for all x in (a, b), and f has a **point of inflection** at $x = c$ if $f''(x)$ changes sign at $x = c$.

Second-Derivative Test ▸ Let c be a critical number of f such that $f'(c) = 0$ and $f''(x)$ exists for all x in some open interval containing c. If $f''(c) > 0$, then there is a local minimum at c. If $f''(c) < 0$, then there is a local maximum at c. If $f''(c) = 0$, then the test gives no information.

Absolute Extrema ▸ Let c be in an interval $[a, b]$ where f is defined. Then f has an **absolute maximum** on the interval at c if $f(x) \leq f(c)$ for all x in $[a, b]$, and f has an **absolute minimum** on the interval at c if $f(x) \geq f(c)$ for all x in $[a, b]$.

CHAPTER 12 REVIEW EXERCISES

1. When the rule of a function is given, how can you determine where it is increasing and where it is decreasing?

2. When the rule of a function is given, how can you determine where the local extrema are located? State two algebraic ways to test whether a local extremum is a maximum or a minimum.

3. What is the difference between a local extremum and an absolute extremum? Can a local extremum be an absolute extremum? Is a local extremum necessarily an absolute extremum?

4. What information about a graph can be found from the first derivative? from the second derivative?

Find the largest open intervals on which the given functions are increasing or decreasing.

5. $f(x) = x^2 + 9x - 9$

6. $f(x) = -3x^2 - 3x + 11$

7. $g(x) = 2x^3 - x^2 - 4x + 7$

8. $g(x) = -4x^3 - 5x^2 + 8x + 1$

9. $f(x) = \dfrac{4}{x-4}$ 10. $f(x) = \dfrac{-6}{3x-5}$

Find the locations and the values of all local maxima and minima for the given functions.

11. $f(x) = 2x^3 - 3x^2 - 36x + 10$
12. $f(x) = 2x^3 - 3x^2 - 12x + 2$
13. $f(x) = x^4 - \frac{8}{3}x^3 - 6x^2 + 2$
14. $f(x) = x \cdot e^x$
15. $f(x) = 3x \cdot e^{-x}$ 16. $f(x) = \dfrac{e^x}{x-1}$

Find the second derivatives of the given functions; then find $f''(1)$ and $f''(-2)$.

17. $f(x) = 2x^5 - 5x^3 + 3x - 1$
18. $f(x) = \dfrac{3-2x}{x+3}$
19. $f(x) = -5e^{2x}$ 20. $f(x) = \ln|5x + 2|$

Sketch the graph of each of the given functions. List the location of each local extremum and point of inflection, the intervals on which the function is increasing and decreasing, and the intervals on which it is concave upward and concave downward.

21. $f(x) = -2x^3 - \frac{1}{2}x^2 - x - 3$
22. $f(x) = -\frac{4}{3}x^3 + x^2 + 30x - 7$
23. $f(x) = x^4 - \frac{4}{3}x^3 - 4x^2 + 1$
24. $f(x) = -\frac{2}{3}x^3 + \frac{9}{2}x^2 + 5x + 1$
25. $f(x) = \dfrac{x-1}{2x+1}$ 26. $f(x) = \dfrac{2x-5}{x+3}$
27. $f(x) = -4x^3 - x^2 + 4x + 5$
28. $f(x) = x^3 + \frac{5}{2}x^2 - 2x - 3$
29. $f(x) = x^4 + 2x^2$ 30. $f(x) = 6x^3 - x^4$
31. $f(x) = \dfrac{x^2+4}{x}$ 32. $f(x) = x + \dfrac{8}{x}$

Find the locations and values of all absolute maxima and absolute minima for the given functions on the given intervals.

33. $f(x) = -x^2 + 6x + 1$; [2, 4]
34. $f(x) = 4x^2 - 4x - 7$; [−1, 3]
35. $f(x) = x^3 + 2x^2 - 15x + 3$; [−.5, 3.3]
36. $f(x) = -2x^3 - x^2 + 4x - 1$; [−3, 1]

Use the extreme-value theorem of Section 12.3 to work these exercises. Round your answers to the nearest integer. You may need the quadratic formula to find some of the critical numbers.

37. **Natural Science** The number of bacteria in a culture is approximated by

$$S(x) = -x^3 + 3x^2 + 360x + 5000 \qquad (6 \le x \le 20),$$

where x represents the temperature of the culture, in degrees Celsius. Find the temperature that produces the maximum number of bacteria.

38. **Finance** The average interest rate for 6-month certificates of deposit is approximated by

$$f(x) = -.000092x^3 + .011x^2 - .537x + 11.61 \quad (5 \le x \le 27),$$

where $x = 5$ corresponds to the year 1985.* Are there local extrema in the domain of the function? If so, in what year(s) does an extremum occur?

39. **Business** The net earnings (in billions of dollars) for General Electric Corporation is approximated by

$$g(x) = -.038x^3 + 1.48x^2 - 17.6x + 78.0 \quad (9 \le x \le 18),$$

where $x = 9$ corresponds to the year 1999.† Over the years in question, what were the minimum net earnings? What were the maximum net earnings?

40. **Health** The number of heart and lung transplants (transplanted together) in recent years is approximated by

$$f(x) = .043x^3 - 1.297x^2 + 8.412x + 52.4 \quad (0 \le x \le 17),$$

where $x = 0$ corresponds to the year 1990.‡ In what year were the most heart and lung transplants performed? Approximately how many were performed?

41. **Business** The revenue (in billions of dollars) for Cisco Systems, Inc., is approximated by

$$R(x) = .0913x^3 - 3.52x^2 + 44.9x - 167.47 \quad (9 \le x \le 18),$$

where $x = 9$ corresponds to the year 1999.§
(a) Find all local extrema for $R(x)$.
(b) What is the absolute minimum revenue?
(c) What is the absolute maximum revenue?

42. **Business** Manufacturing capacity (as a percent) in the United States is approximated by

$$g(x) = -.0019x^3 + .111x^2 - 1.93x + 88.9 \quad (3 \le x \le 39),$$

---

*U.S. Federal Reserve.

†www.morningstar.com.

‡U.S. Department of Health and Human Services.

§www.morningstar.com.

where $x = 3$ corresponds to the year 1973.*

(a) Find all local and absolute extrema of the function.

(b) Locate any points of inflection.

43. **Business** The percentage of U.S. newspaper print revenues coming from classified ads is modeled by the function

$$f(x) = -.00035x^3 + .0266x^2 - .123x + 20.7 \quad (0 \le x \le 58),$$

where $x = 0$ corresponds to the year 1950.†

(a) In what year was the percentage at its maximum? What was the maximum?

(b) Where is the point of inflection?

44. **Business** The packaging department of a corporation is designing a box with a square base and top. The volume is to be 27 cubic meters. To reduce the cost, the surface area of the box is to be minimized. What dimensions (height, length, and width) should the box have?

45. **Business** A landscaper needs to design an enclosed rectangular garden with one side against an existing garage, with no fence needed there. Find the dimensions of the rectangular space of maximum area that can be enclosed with 500 meters of fence.

*St. Louis Federal Reserve.

†www.timetric.com.

46. **Business** If the garden described in Exercise 45 needs fencing on all four sides, find the dimensions of the maximum rectangular area that can be made with 500 meters of fence.

47. **Business** A company plans to package its product in a cylinder that is open at one end. The cylinder is to have a volume of 27π cubic inches. What radius should the circular bottom of the cylinder have in order to minimize the cost of the material? (*Hint*: The volume of a circular cylinder is $\pi r^2 h$, where r is the radius of the circular base and h is the height; the surface area of an open cylinder is $2\pi rh + \pi r^2$.)

48. **Business** A cell phone manufacturer produces 145,000 phones a year. It costs \$3.25 to store each phone for one year and \$230 to produce each batch. Find the number of batches that should be produced annually.

49. **Business** How many phones need to be produced for each batch in Exercise 48?

50. **Business** A chocolate maker has annual demand of 250,000 pounds per year. It costs \$1.25 to store a pound of chocolate for a year and \$350 to produce a batch.

(a) Find the number of batches to be produced annually.

(b) How many pounds should be produced in each batch?

CASE 12

A Total Cost Model for a Training Program*

In this application, we set up a mathematical model for determining the total cost required to set up a training program. Then we use calculus to find the time interval between training programs that produces the minimum total cost. The model assumes that the demand for trainees is constant and that the fixed cost of training a batch of trainees is known. Also, it is assumed that people who are trained, but for whom no job is readily available, will be paid a fixed amount per month while waiting for a job to open up.

*Based on "A Total Cost Model for a Training Program," by P. L. Goyal and S. K. Goyal, Faculty of Commerce and Administration, Concordia University. Used with permission.

The model uses the following variables:

D = demand for trainees per month;

N = number of trainees per batch;

C_1 = fixed cost of training a batch of trainees;

C_2 = variable cost of training per trainee per month;

C_3 = salary paid monthly to a trainee who has not yet been given a job after training;

m = time interval in months between successive batches of trainees;

t = length of training program in months;

$Z(m)$ = total monthly cost of program.

The total cost of training a batch of trainees is given by $C_1 + NtC_2$. However, $N = mD$, so that the total cost per batch is $C_1 + mDtC_2$.

After training, personnel are given jobs at the rate of D per month. Thus, $N - D$ of the trainees will not get a job the first month, $N - 2D$ will not get a job the second month, and so on. The $N - D$ trainees who do not get a job the first month produce total costs of $(N - D)C_3$, those not getting jobs during the second month produce costs of $(N - 2D)C_3$, and so on. Since $N = mD$, the costs during the first month can be written as

$$(N - D)C_3 = (mD - D)C_3 = (m - 1)DC_3,$$

while the costs during the second month are $(m - 2)DC_3$, and so on. The total cost for keeping the trainees without a job is thus

$$(m - 1)DC_3 + (m - 2)DC_3 + (m - 3)DC_3 + \cdots + 2DC_3 + DC_3,$$

which can be factored to give

$$DC_3[(m - 1) + (m - 2) + (m - 3) + \cdots + 2 + 1].$$

The expression in brackets is the sum of the terms of an arithmetic sequence discussed in most algebra texts. Using formulas for arithmetic sequences, we can show that the expression in brackets is equal to $m(m - 1)/2$, so we have

$$DC_3\left[\frac{m(m - 1)}{2}\right] \qquad (1)$$

as the total cost for keeping jobless trainees.

The total cost per batch is the sum of the training cost per batch, $C_1 + mDtC_2$, and the cost of keeping trainees without a proper job, given by equation (1). Since we assume that a batch of trainees is trained every m months, the total cost per month, $Z(m)$, is given by

$$Z(m) = \frac{C_1 + mDtC_2}{m} + \frac{DC_3\left[\frac{m(m - 1)}{2}\right]}{m}$$

$$= \frac{C_1}{m} + DtC_2 + DC_3\left(\frac{m - 1}{2}\right).$$

EXERCISES

1. Find $Z'(m)$.
2. Solve the equation $Z'(m) = 0$. As a practical matter, it is usually required that m be a whole number. If m does not come out to be a whole number, then m^+ and m^-, the two whole numbers closest to m, must be chosen. Calculate both $Z(m^+)$ and $Z(m^-)$; the smaller of the two provides the optimum value of Z.
3. Suppose a company finds that its demand for trainees is three per month, that a training program requires 12 months, that the fixed cost of training a batch of trainees is \$15,000, that the marginal cost per trainee per month is \$100, and that trainees are paid \$900 per month after training, but before going to work. Use your result from Exercise 2 to find m.
4. Since m is not a whole number, find m^+ and m^-.
5. Calculate $Z(m^+)$ *and* $Z(m^-)$.
6. What is the optimum time interval between successive batches of trainees? How many trainees should be in a batch?

CHAPTER

13

Integral Calculus

Social scientists use integral calculus to estimate the total population size from a rate-of-growth function. Natural scientists use similar techniques to approximate the growth of animal populations. In business and finance, integrals are used to find total revenue from marginal revenue functions. Integral calculus is also used to determine the point where profitability fails for a company. See Exercise 57 on page 837, Exercise 36 on page 882, Exercises 3 and 4 on page 880, and Exercise 17 on page 880.

CASE 13: Bounded Population Growth

The derivative and its applications, which were studied in Chapters 11 and 12, are part of what is called *differential calculus.* This chapter is devoted to the other main branch of calculus, *integral calculus.* Integrals are used to find areas, to determine the lengths of curved paths, to solve complicated probability problems, to calculate the location of an object (such as the height of a space shuttle) from its velocity, and in a variety of other ways. Such topics may seem quite different from those studied earlier, but the fundamental theorem of calculus in Section 13.4 will reveal a surprisingly close connection between differential and integral calculus.

13.1 Antiderivatives

Functions used in applications in previous chapters have provided information about the *total amount* of a quantity, such as cost, revenue, profit, temperature, gallons of oil, or distance. Derivatives of these functions provided information about the rate of change of these quantities and allowed us to answer important questions about the extrema of the functions. It is not always possible to find ready-made functions that provide information about the total amount of a quantity, but it is often possible to collect enough data to come up with a function that gives the *rate of change* of a quantity. We know that derivatives give the rate of change when the total amount is known. In this section, we shall see that this process can be reversed: When the rate of change is known, a function that gives the total amount can be obtained by a process called *antidifferentiation.* Here is the basic definition.

Antiderivatives

If $F(x)$. and $f(x)$ are functions such that $F'(x) = f(x)$, then $F(x)$ is said to be an **antiderivative** of $f(x)$.

Example 1

(a) If $F(x) = 10x$, then $F'(x) = 10$, so $F(x) = 10x$ is an antiderivative of $f(x) = 10$.

(b) For $F(x) = x^5$, $F'(x) = 5x^4$, which means that $F(x) = x^5$ is an antiderivative of $f(x) = 5x^4$.

Example 2

Find an antiderivative of $f(x) = 2x$.

Solution By remembering formulas for derivatives, it is easy to see that $F(x) = x^2$ is an antiderivative of $f(x)$ because $F'(x) = 2x = f(x)$. Note that $G(x) = x^2 + 2$ and $H(x) = x^2 - 4$ are also antiderivatives of $f(x)$, because

$$G'(x) = 2x + 0 = f(x) \quad \text{and} \quad H'(x) = 2x - 0 = f(x).$$ ✓1

✓Checkpoint 1

Find an antiderivative for each of the given expressions.

(a) $3x^2$

(b) $5x$

(c) $8x^7$

Answers: See page 838.

Any two of the antiderivatives of $f(x) = 2x$ that were found in Example 2 differ by a constant. For instance, $G(x) - F(x) = 2$ and $H(x) - G(x) = -6$. The same thing is true in the general case.

If $F(x)$ and $G(x)$ are both antiderivatives of $f(x)$, then there is a constant C such that

$$F(x) - G(x) = C.$$

(Two antiderivatives of a function can differ only by a constant.)

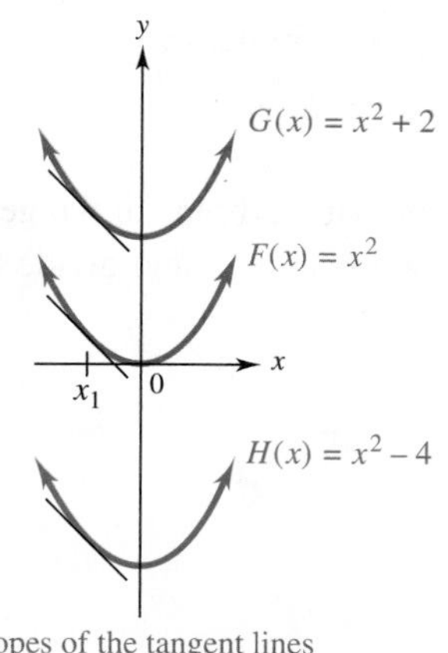

Slopes of the tangent lines at $x = x_1$ are the same.

FIGURE 13.1

The statement in the box reflects a geometric fact about derivatives, namely, that the derivative of a function gives the slope of the tangent line at any number x. For example, if you graph the three antiderivatives of $f(x) = 2x$ found in Example 2, you will see that the graphs all have the same shape because, at any x-value, their tangent lines all have the same slope, as shown in Figure 13.1.

The family of all antiderivatives of the function f is indicated by

$$\int f(x)\,dx.$$

The symbol $\int$ is the **integral sign**, $f(x)$ is the **integrand**, and $\int f(x)\,dx$ is called an **indefinite integral**. Since any two antiderivatives of $f(x)$ differ by a constant (which means that one is a constant plus the other), we can describe the indefinite integral as follows.

Indefinite Integral

If $F'(x) = f(x)$, then

$$\int f(x)\,dx = F(x) + C,$$

for any real number C.

For example, with this notation,

$$\int 2x\,dx = x^2 + C.$$

NOTE The dx in the indefinite integral $\int f(x)\,dx$ indicates that x is the variable of the function whose antiderivative is to be found, in the same way that dy/dx denotes the derivative when y is a function of the variable x. For example, in the indefinite integral $\int 2ax\,dx$, the variable of the function is x, whereas in the indefinite integral $\int 2ax\,da$, the variable is a.

The symbol $\int f(x)\,dx$ was created by G. W. Leibniz (1646–1716) in the latter part of the 17th century. The $\int$ is an elongated S from *summa*, the Latin word for *sum*. The word *integral* as a term in calculus was coined by Jakob Bernoulli (1654–1705), a Swiss mathematician who corresponded frequently with Leibniz. The relationship between sums and integrals will be clarified in Section 13.3.

Because finding an antiderivative is the inverse of finding a derivative, each formula for derivatives leads to a rule for antiderivatives. For instance, the power rule for derivatives tells us that

$$\text{if } F(x) = x^4, \quad \text{then } F'(x) = 4x^3.$$

Consequently,

$$\text{if } F(x) = \frac{1}{4}x^4, \quad \text{then } F'(x) = \frac{1}{4}(4x^3) = x^3.$$

In other words, an antiderivative of $f(x) = x^3$ is $F(x) = \frac{1}{4}x^4$. Similarly,

$$\text{if } G(x) = \tfrac{1}{8}x^8, \text{ then } G'(x) = \tfrac{1}{8}(8x^7) = x^7,$$

so $G(x) = \frac{1}{8}x^8$ is an antiderivative of $g(x) = x^7$. The same pattern holds in the general case: To find the antiderivative of x^n, increase the exponent by 1 and divide by that same number.

Power Rule for Antiderivatives

For any real number $n \neq -1$,

$$\int x^n\,dx = \frac{1}{n+1}x^{n+1} + C.$$

This result can be verified by differentiating the expression on the right in the box:

$$\frac{d}{dx}\left(\frac{1}{n+1}x^{n+1} + C\right) = \frac{n+1}{n+1}x^{(n+1)-1} + 0 = x^n.$$

(If $n = -1$, the expression in the denominator is 0, and the power rule cannot be used. We will see later how to find an antiderivative in this case.)

Example 3 Find each antiderivative.

(a) $\int x^5\,dx$

Solution Use the power rule with $n = 5$:

$$\int x^5 dx = \frac{1}{5+1}x^{5+1} + C = \frac{1}{6}x^6 + C.$$

(b) $\int \frac{1}{t^2}dt$

Solution First, write $1/t^2$ as t^{-2}. Then

$$\int \frac{1}{t^2}\,dt = \int t^{-2}\,dt = \frac{1}{-2+1}t^{-2+1} = \frac{t^{-1}}{-1} + C = \frac{-1}{t} + C.$$

(c) $\int \sqrt{u}\,du$

Solution Since $\sqrt{u} = u^{1/2}$,

$$\int \sqrt{u}\,du = \int u^{1/2}du = \frac{1}{1/2+1}u^{1/2+1} + C = \frac{1}{3/2}u^{3/2} + C = \frac{2}{3}u^{3/2} + C.$$

To check this result, differentiate $(2/3)u^{3/2} + C$; the derivative is $u^{1/2}$, the original function.

(d) $\int dx$

Solution Writing dx as $1 \cdot dx$, and using the fact that $x^0 = 1$ for any nonzero number x, we obtain

$$\int dx = \int 1\,dx = \int x^0\,dx = \frac{1}{1}x^1 + C = x + C.$$ 2

✓Checkpoint 2

Find each of the following.

(a) $\int x^6\,dx$

(b) $\int \sqrt[4]{x}\,dx$

(c) $\int 8\,dx$

Answers: See page 838.

As shown in Chapter 11, the derivative of the product of a constant and a function is the product of the constant and the derivative of the function. A similar rule applies to antiderivatives. Also, since derivatives of sums and differences are found term by term, antiderivatives can also be found term by term.

Properties of Antiderivatives

Let $f(x)$ and $g(x)$ be functions that have antiderivatives and let k be any real number.

Constant-multiple rule: $\int k \cdot f(x)\,dx = k\int f(x)\,dx.$

Sum-or-difference rule: $\int [f(x) \pm g(x)]\,dx = \int f(x)\,dx \pm \int g(x)\,dx.$

CAUTION The constant-multiple rule requires that k be a *number.* The rule does not apply to a *variable.* For example,

$$\int x\sqrt{x-1}\,dx \neq x\int \sqrt{x-1}\,dx.$$

Example 4 Find each of the given antiderivatives.

(a) $\int 4x^7\,dx$

Solution By the constant-multiple rule and the power rule,

$$\int 4x^7 = 4\int x^7\,dx = 4\left(\frac{1}{8}x^8\right) + C = \frac{1}{2}x^8 + C.$$

Since C represents any real number, it is not necessary to multiply it by 4 in the next-to-last step.

(b) $\int \frac{10}{z^6}\,dz$

Solution Convert to negative exponents and use the constant-multiple rule and the power rule:

$$\int \frac{10}{z^6} dz = \int 10z^{-6}\, dz$$

$$= 10 \int z^{-6}\, dz \qquad \text{Constant-multiple rule}$$

$$= 10\left(\frac{z^{-5}}{-5}\right) + C \qquad \text{Power rule}$$

$$= \frac{-2}{z^5} + C.$$

(c) $\int (5y^2 - 6y + 2)\, dy$

By extending the sum-or-difference rule to more than two terms, we get

$$\int (5y^2 - 6y + 2)\, dy = \int 5y^2\, dy - \int 6y\, dy + \int 2\, dy$$

$$= 5\int y^2\, dy - 6\int y\, dy + 2\int dy$$

$$= 5\left(\frac{y^3}{3}\right) - 6\left(\frac{y^2}{2}\right) + 2y + C$$

$$= \frac{5}{3}y^3 - 3y^2 + 2y + C.$$

Only one constant C is needed in the answer, because the three constants from the term-by-term antiderivatives are combined. 3 ✓

The nice thing about working with antiderivatives is that you can always check your work by taking the derivative of the result. For instance, in Example 4(c), check that $\frac{5}{3}y^3 - 3y^2 + 2y + C$ is the required antiderivative by taking the derivative:

$$\frac{d}{dy}\left(\frac{5}{3}y^3 - 3y^2 + 2y + C\right) = 5y^2 - 6y + 2.$$

The result is the original function to be integrated, so the work checks out.

✓ Checkpoint 3

Find each of the following.

(a) $\int (-6x^8)\, dx$

(b) $\int 7x^{2/3}\, dx$

(c) $\int \frac{5}{x^3}\, dx$

(d) $\int (5x^4 - 3x^2 + 8)\, dx$

(e) $\int \left(4\sqrt[3]{x} + \frac{2}{x^2}\right) dx$

Answers: See page 838.

Example 5 Find each of the given antiderivatives.

(a) $\int \frac{x^2 + 1}{\sqrt{x}} dx$

Solution First rewrite the integrand as follows:

$$\int \frac{x^2 + 1}{\sqrt{x}}\, dx = \int \left(\frac{x^2}{\sqrt{x}} + \frac{1}{\sqrt{x}}\right) dx$$

$$= \int \left(\frac{x^2}{x^{1/2}} + \frac{1}{x^{1/2}}\right) dx$$

$$= \int (x^{3/2} + x^{-1/2})\, dx. \qquad \text{Quotient rule for exponents}$$

Now find the antiderivative:

$$\int (x^{3/2} + x^{-1/2})\,dx = \int x^{3/2}\,dx + \int x^{-1/2}\,dx$$

$$= \frac{x^{5/2}}{5/2} + \frac{x^{1/2}}{1/2} + C$$

$$= \frac{2}{5}x^{5/2} + 2x^{1/2} + C.$$

(b) $\int (x^2 - 1)^2\,dx$

Solution Square the integrand first and then find the antiderivative.

$$\int (x^2 - 1)^2\,dx = \int (x^4 - 2x^2 + 1)\,dx$$

$$= \int x^4\,dx - \int 2x^2\,dx + \int 1\,dx$$

$$= \frac{x^5}{5} - \frac{2x^3}{3} + x + C$$ ✓4

✓Checkpoint 4

Find each of the following.

(a) $\int \frac{\sqrt{x}+1}{x^2}\,dx$

(b) $\int \left(\sqrt{x} + 2\right)^2\,dx$

Answers: See page 838.

As shown in Chapter 11, the derivative of $f(x) = e^x$ is $f'(x) = e^x$. Also, the derivative of $f(x) = e^{kx}$ is $f'(x) = k \cdot e^{kx}$. These results lead to the following formulas for antiderivatives of exponential functions.

Antiderivatives of Exponential Functions

If k is a real number, $k \neq 0$, then

$$\int e^x\,dx = e^x + C;$$

$$\int e^{kx}\,dx = \frac{1}{k} \cdot e^{kx} + C.$$

✓Checkpoint 5

Find each of the following.

(a) $\int (-7e^x)\,dx$

(b) $\int e^{5x}\,dx$

(c) $\int (e^{4x} - 4e^x)\,dx$

(d) $\int (-11e^{-x})\,dx$

Answers: See page 838.

Example 6 Here are some antiderivatives of exponential functions:

(a) $\int 8e^x\,dx = 8\int e^x\,dx = 8e^x + C;$

(b) $\int e^{8t}\,dt = \frac{1}{8}e^{8t} + C;$

(c) $\int 3e^{(5/6)u}\,du = 3\left(\frac{1}{5/6}e^{(5/6)u}\right) + C = 3\left(\frac{6}{5}\right)e^{(5/6)u} + C$

$$= \frac{18}{5}e^{(5/6)u} + C.$$ ✓5

The antiderivative formula $\int x^n dx = \frac{x^{n+1}}{n+1} + C$ does not hold when $n = -1$, because the denominator is 0 in that case. Nevertheless, the function $g(x) = x^{-1}$ does have an antiderivative. Recall that the derivative of $f(x) = \ln|x|$ is $f'(x) = 1/x = x^{-1}$.

Antiderivative of x^{-1}

$$\int x^{-1}\,dx = \int \frac{1}{x}\,dx = \ln|x| + C, \qquad \text{where } x \neq 0.$$

CAUTION The domain of the logarithmic function is the set of positive real numbers. However, $y = x^{-1} = 1/x$ has the set of all nonzero real numbers as its domain, so the absolute value of x *must* be used in the antiderivative.

Example 7 Here are some antiderivatives of logarithmic functions:

(a) $\int \frac{4}{x}\,dx = 4\int \frac{1}{x}\,dx = 4 \cdot \ln|x| + C;$

(b) $\int \left(-\frac{5}{x} + e^{-2x}\right) dx = -5 \cdot \ln|x| - \frac{1}{2}e^{-2x} + C.$ 6✓

✓Checkpoint 6

Find each of the following.

(a) $\int (-9/x)\,dx$

(b) $\int (8e^{4x} - 3x^{-1})\,dx$

Answers: See page 838.

APPLICATIONS

In the preceding examples, a family of antiderivative functions was found. In many applications, however, the given information allows us to determine the value of the constant C. The next three examples illustrate this idea.

Example 8 **Business** According to data from the Cellular Telecommunications Industry Association, the rate of increase in the number of cell phone subscribers (in millions) since 1990 is approximated by

$$S'(x) = 1.42x + 2.97,$$

where $x = 0$ corresponds to 1990. There were 255.4 million subscribers in 2007.* Find a function $S(x)$ that gives the number of subscribers (in millions) in year x.

Solution Since $S'(x)$ gives the rate of change of subscribers, the number of subscribers is given by

$$S(x) = \int (1.42x + 2.97)\,dx$$

$$= 1.42\frac{x^2}{2} + 2.97x + C$$

$$= .71x^2 + 2.97x + C.$$

*Cellular Telecommunications & Internet Association.

To find the value of C, use the fact that in 2007 ($x = 17$) there were 255.4 million subscribers, which says that $S(17) = 255.4$. This information gives an equation that can be solved for C:

$$\begin{aligned} S(x) &= .71x^2 + 2.97x + C \\ S(17) &= .71(17^2) + 2.97(17) + C \\ 255.4 &= .71(17^2) + 2.97(17) + C \\ 255.4 &= 255.68 + C \\ C &= -.28. \end{aligned}$$

Thus, the number of subscribers in year x is

$$S(x) = .71x^2 + 2.97x - .28.$$

Example 9 **Business** Suppose the marginal revenue from selling x units of a product is given by $40/e^{.05x} + 10$.

(a) Find the revenue function.

Solution The marginal revenue is the derivative of the revenue function, so

$$\frac{dR}{dx} = \frac{40}{e^{.05x}} + 10$$

$$R = \int \left(\frac{40}{e^{.05x}} + 10\right) dx = \int (40e^{-.05x} + 10)\, dx$$

$$= 40\left(\frac{-1}{.05}\right)e^{-.05x} + 10x + k = -800e^{-.05x} + 10x + k,$$

where k is a constant. If $x = 0$, then $R = 0$ (no items sold means no revenue), and

$$\begin{aligned} 0 &= -800e^0 + 10(0) + k \\ 800 &= k. \end{aligned}$$

Thus,

$$R = -800e^{-.05x} + 10x + 800$$

is the revenue function.

(b) Find the demand function for this product.

Solution Recall that $R = xp$, where p is the demand function. Hence,

$$-800e^{-.05x} + 10x + 800 = xp$$

$$\frac{-800e^{-.05x} + 10x + 800}{x} = p.$$

The demand function is $p = \dfrac{-800e^{-.05x} + 10x + 800}{x}$. ✓7

✓Checkpoint 7

The marginal cost at a level of production of x items is

$$C'(x) = 2x^3 + 6x - 5.$$

The fixed cost is \$800. Find the cost function $C(x)$.

Answer: See page 838.

Example 10 **Social Science** Mexico's population was 84.9 million in 1990. Its growth rate (in millions per year) is approximated by

$$f(t) = 1.1886e^{.014t} \qquad (0 \le t \le 30),$$

where $t = 0$ corresponds to 1990.*

(a) Find the rule of the population function $F(t)$ that gives the population (in millions) in year t.

Solution The derivative of the population function $F(t)$ is the rate at which the population is growing—that is, $F'(t) = 1.1886e^{.014t}$. Therefore,

$$F(t) = \int 1.1886e^{.014t}\,dt = 1.1886 \cdot \frac{1}{.014}e^{.014t} + C = 84.9e^{.014t} + C.$$

Since the population was 84.9 million in 1990 (that is, when $t = 0$), we have

$$84.9 = F(0) = 84.9e^{.014(0)} + C = 84.9e^{0} + C = 84.9 + C,$$

so that $C = 0$. Therefore, the population function is $F(t) = 84.9e^{.014t}$.

(b) What will the population be in the year 2016?

Solution Since 2016 corresponds to $t = 26$, the population will be

$$F(26) = 84.9e^{.014(26)} = 84.9e^{.364} \approx 122.18 \text{ million.}$$

*Based on data and projections from the U.S. Census Bureau.

13.1 ▶ Exercises

1. What must be true of $F(x)$ and $G(x)$ if both are antiderivatives of $f(x)$?
2. How is the antiderivative of a function related to the function?
3. In your own words, describe what is meant by an integrand.
4. Explain why the restriction $n \neq -1$ is necessary in the rule $\int x^n\,dx = \frac{1}{n+1}x^{n+1} + C$.

Find each of the given antiderivatives. (See Examples 3–7.)

5. $\int 12x\,dx$
6. $\int 25r^2 dr$
7. $\int 8p^3 dp$
8. $\int 5t^5 dt$
9. $\int 105\,dx$
10. $\int 35\,dt$
11. $\int (5z - 1)\,dz$
12. $\int (2m^2 + 3)\,dm$
13. $\int (z^2 - 4z + 6)\,dz$
14. $\int (2y^2 + 10y + 7)\,dy$
15. $\int (x^3 - 14x^2 + 22x + 8)\,dx$
16. $\int (5x^3 + 5x^2 - 10x - 6)\,dx$
17. $\int 6\sqrt{y}\,dy$
18. $\int 8z^{1/2}\,dz$
19. $\int \left(6t\sqrt{t} + 3\sqrt[4]{t}\right) dt$
20. $\int \left(12\sqrt{x} - x\sqrt{x}\right) dx$
21. $\int (56t^{1/2} + 18t^{7/2})\,dt$
22. $\int (10u^{3/2} - 8u^{5/2})\,du$
23. $\int \frac{24}{x^3}\,dx$
24. $\int \frac{-20}{x^2}\,dx$
25. $\int \left(\frac{1}{y^3} - \frac{2}{\sqrt{y}}\right) dy$
26. $\int \left(\frac{3 + 2u}{\sqrt{u}}\right) du$
27. $\int (6x^{-3} + 2x^{-1})\,dx$
28. $\int (3x^{-1} - 10x^{-2})\,dx$
29. $\int 4e^{3u}\,du$
30. $\int -e^{-3x}\,dx$
31. $\int 3e^{-.8x}\,dx$
32. $\int -4e^{.2v}\,dv$
33. $\int \left(\frac{6}{x} + 4e^{-.5x}\right) dx$
34. $\int \left(\frac{9}{x} - 3e^{-.4x}\right) dx$

35. $\int \frac{1 + 2t^4}{t} dt$

36. $\int \frac{2y^{1/2} - 3y^2}{y^3} dy$

37. $\int \left(e^{2u} + \frac{4}{u}\right) du$

38. $\int \left(\frac{2}{v} - e^{3v}\right) dv$

39. $\int (5x + 1)^2 dx$

40. $\int (2y - 1)^2 dy$

41. $\int \frac{\sqrt{x} + 1}{\sqrt[3]{x}} dx$

42. $\int \frac{1 - 2\sqrt[3]{z}}{\sqrt[3]{z}} dz$

43. The slope of the tangent line to a curve is given by

$$f'(x) = 6x^2 - 4x - 3.$$

If the point (0, 1) is on the curve, find the equation of the curve.

44. Find the equation of the curve whose tangent line has a slope of

$$f'(x) = x^{2/3}$$

if the point $(1, \frac{3}{5})$ is on the curve.

Work the given problems. (See Example 8.)

45. **Business** Imports (in billions of U.S. dollars) from Canada to the United States have changed at a rate given by $g(x) = .21x^2 + .36x + 11.38$, where x is the number of years since 2000.* The United States imported $216.63 billion worth of goods in 2001.
   (a) Find a function giving the value of imports in year x.
   (b) What was the value of imports from Canada in 2007?

46. **Business** The total U.S. sales of bicycles and bicycle-related equipment was $5.0 billion in 2000. The rate at which this number was changing since 2000 is approximated by

$$f(x) = .054x - .12,$$

where $x = 0$ corresponds to 2000.†
   (a) Find a function that gives the sales in billions of dollars in year x.
   (b) Estimate the total sales in 2008.

*U.S. Census Bureau.

†*Statistical Abstract of the United States*: 2004–2005.

Work the given problems. (See Example 9.)

47. **Health** Total health care expenditures in the United States (in billions of dollars) are increasing at a rate of $f(x) = 9.42x + 87.14$, where $x = 0$ corresponds to the year 2000.* Total health care expenditures were $1589.92 billion in 2002.
   (a) Find a function that gives the total health care expenditures in year x.
   (b) What will total health care expenditures be in 2014?

48. **Business** The marginal revenue from a product is given by $r(x) = 50 - 3x - x^2$. Find the demand function for the product.

49. **Business** The marginal profit from the sale of x hundred items of a product is $P'(x) = 4 - 6x + 3x^2$, and the "profit" when no items are sold is −$40. Find the profit function.

Business *Find the cost function for each of the given marginal cost functions.*

50. $C'(x) = x^{1/2}$; 16 units cost $70.

51. $C'(x) = x^{2/3} + 5$; 8 units cost $58.

52. $C'(x) = x^2 - 2x + 3$; 3 units cost $25.

53. $C'(x) = .2x^2 + .4x + .8$; 6 units cost $32.50.

54. $C'(x) = .0015x^3 + .033x^2 + .048x + .25$; 10 units cost $25.

55. $C'(x) = -\frac{40}{e^{.05x}} + 100$; 5 units cost $1400.

56. $C'(x) = 1.2e^{.02x}$; 2 units cost $105.

Work the given problems. (See Example 10.)

57. **Social Science** The world population was 5.282 billion in 1990. The approximate growth rate of world population is given by $f(t) = .06750396e^{.01278t}$, where $t = 0$ corresponds to 1990.†
   (a) Find a function that gives the population of the world (in billions) in year t.
   (b) Estimate the world population in 2015.

58. **Natural Science** If the rate of excretion of a biochemical compound is given by

$$f'(t) = .01e^{-.01t},$$

the total amount excreted by time t (in minutes) is $f(t)$.
   (a) Find an expression for $f(t)$.
   (b) If 0 units are excreted at time $t = 0$, how many units are excreted in 10 minutes?

*U.S. Centers for Medicare & Medicaid Services.

†U.S. Census Bureau.

59. Social Science Four years ago, there were approximately 200 million licensed drivers in the United States. Sixteen-year-olds had 20 accidents per 100 drivers. The rate at which the number of accidents per 100 drivers changed as the driver's age increased was approximated by

$$g(x) = -\frac{10.2}{x} \qquad (16 \le x \le 80),$$

where x is the age of the driver (in years).*

(a) Find a function that gives the number of accidents per 100 drivers of age x years. (Round the constant C to two decimal places.)

(b) Approximately how many accidents per 100 drivers did drivers of age 18 have? of age 20? of age 50? of age 70?

(c) What reasons might account for your answers in part (b)?

*Based on data from the National Safety Council.

✓Checkpoint Answers

1. Only one possible antiderivative is given for each.
   (a) x^3 **(b)** $\frac{5}{2}x^2$ **(c)** x^8
2. **(a)** $\frac{1}{7}x^7 + C$ **(b)** $\frac{4}{5}x^{5/4} + C$ **(c)** $8x + C$
3. **(a)** $-\frac{2}{3}x^9 + C$ **(b)** $\frac{21}{5}x^{5/3} + C$
   (c) $-\frac{5}{2}x^{-2} + C$ or $-\frac{5}{2x^2} + C$
   (d) $x^5 - x^3 + 8x + C$ **(e)** $3x^{4/3} - \frac{2}{x} + C$
4. **(a)** $-\frac{2}{\sqrt{x}} - \frac{1}{x} + C$ **(b)** $\frac{x^2}{2} + \frac{8}{3}x^{3/2} + 4x + C$
5. **(a)** $-7e^x + C$ **(b)** $\frac{1}{5}e^{5x} + C$
   (c) $\frac{1}{4}e^{4x} - 4e^x + C$ **(d)** $11e^{-x} + C$
6. **(a)** $-9 \cdot \ln|x| + C$ **(b)** $2e^{4x} - 3 \cdot \ln|x| + C$
7. $C(x) = \frac{1}{2}x^4 + 3x^2 - 5x + 800$

13.2 Integration by Substitution

In Section 13.1, we saw how to find the antiderivatives of a few simple functions. The antiderivatives of more complicated functions can sometimes be found by a process called *integration by substitution*. The technique depends on the following concept: If $u = f(x)$, then the **differential** of u, written du, is defined as

$$du = f'(x)\,dx.$$

For example, if $u = 6x^4$, then $du = 24x^3\,dx$. ✓1

✓Checkpoint 1

Find du for the given functions.

(a) $u = 9x$

(b) $u = 5x^3 + 2x^2$

(c) $u = e^{-2x}$

Answers: See page 846.

Differentials have many useful interpretations that are studied in more advanced courses. We shall use them only as a convenient notational device when finding an antiderivative such as

$$\int (3x^2 + 4)^4 6x\,dx.$$

The function $(3x^2 + 4)^4 6x$ is reminiscent of the chain rule, so we shall try to use differentials and the chain rule in *reverse* to find the antiderivative. Let $u = 3x^2 + 4$; then $du = 6x\,dx$. Now substitute u for $3x^2 + 4$ and du for $6x\,dx$ in the indefinite integral:

$$\int (3x^2 + 4)^4 6x\,dx = \int \overbrace{(3x^2 + 4)}^{u}{}^4\overbrace{(6x\,dx)}^{du}$$

$$= \int u^4\,du.$$

This last integral can now be found by the power rule:

$$\int u^4\,du = \frac{u^5}{5} + C.$$

Finally, substitute $3x^2 + 4$ for u and $6x\,dx$ for du:

$$\int (3x^2 + 4)^4\,6x\,dx = \frac{u^5}{5} + C = \frac{(3x^2 + 4)^5}{5} + C.$$

We can check the accuracy of this result by using the chain rule to take the derivative. We get

$$\frac{d}{dx}\left[\frac{(3x^2 + 4)^5}{5} + C\right] = \frac{1}{5}\cdot 5(3x^2 + 4)^4(6x) + 0$$
$$= (3x^2 + 4)^4\,6x,$$

which is the original function.

This method of integration is called **integration by substitution**. As just shown, it is simply the chain rule for derivatives in reverse. The results can always be verified by differentiation.

Example 1 Find $\int (4x + 5)^9\,dx$.

Solution We choose $4x + 5$ as u. Then $du = 4\,dx$. We are missing the constant 4. Using the fact that $4(1/4) = 1$, we can rewrite the integral:

$$\int (4x + 5)^9\,dx = \frac{1}{4}\cdot 4\int (4x + 5)^9\,dx$$

$$= \frac{1}{4}\int (4x + 5)^9(4\,dx) \qquad k\int f(x)\,dx = \int kf(x)\,dx$$

$$= \frac{1}{4}\int u^9\,du \qquad \text{Substitute.}$$

$$= \frac{1}{4}\cdot\frac{u^{10}}{10} + C = \frac{u^{10}}{40} + C \qquad \text{Find the antiderivative.}$$

$$= \frac{(4x + 5)^{10}}{40} + C. \qquad \text{Substitute.}$$ ✓2

✓Checkpoint 2

Find the given antiderivatives.

(a) $\int 8x(4x^2 - 1)^5\,dx$

(b) $\int (3x - 8)^4\,dx$

(c) $\int 18x^2(x^3 - 7)^{3/2}\,dx$

Answers: See page 846.

CAUTION When changing the x-problem to the u-problem, make sure that the change is complete—that is, that no x's are left in the u-problem.

Example 2 Find $\int x^2\sqrt{x^3 + 5}\,dx$.

Solution Rewrite the function as $x^2(x^3 + 5)^{1/2}$. An expression raised to a power is usually a good choice for u, so we let $u = x^3 + 5$; then $du = 3x^2\,dx$. The integrand does not contain the constant 3, which is part of du. One way to take care of this is to solve the differential $du = 3x^2\,dx$ for $x^2\,dx$:

$$du = 3x^2\,dx$$
$$\frac{1}{3}du = x^2\,dx.$$

Substitute $\frac{1}{3}du$ for $x^2\,dx$:

$$\int x^2\sqrt{x^3+5}\,dx = \int \sqrt{x^3+5}\,(x^2\,dx) = \int \sqrt{u}\cdot\frac{1}{3}du.$$

Now use the constant-multiple rule to bring the $\frac{1}{3}$ outside the integral sign:

$$\int x^2\sqrt{x^3+5}\,dx = \int \sqrt{u}\cdot\frac{1}{3}\,du = \frac{1}{3}\int u^{1/2}\,du$$
$$= \frac{1}{3}\cdot\frac{u^{3/2}}{3/2} + C = \frac{2}{9}u^{3/2} + C.$$

Since $u = x^3 + 5$,

$$\int x^2\sqrt{x^3+5}\,dx = \frac{2}{9}(x^3+5)^{3/2} + C. \quad 3\checkmark$$

✓Checkpoint 3

Find the given antiderivatives.

(a) $\int 4x(6x^2+3)^2\,dx$

(b) $\int x\sqrt{x^2+16}\,dx$

Answers: See page 846.

CAUTION The substitution method given in the preceding examples *will not always work.* For example, we might try to find

$$\int x^3\sqrt{x^3+1}\,dx$$

by substituting $u = x^3 + 1$, so that $du = 3x^2\,dx$. However, there is no *constant* that can be inserted inside the integral sign to give $3x^2$. This integral, and a great many others, cannot be evaluated by substitution.

With practice, choosing u will become easy if you keep two principles in mind. First, u should equal some expression in the integral that, when replaced with u, tends to make the integral simpler. Second, and more importantly, u must be an expression whose derivative is also present in the integral. The substitution should include as much of the integral as possible, so long as its derivative is still present. In Example 2, we could have chosen $u = x^3$, but $u = x^3 + 1$ is better, because it has the same derivative as x^3 and captures more of the original integral. If we carry this reasoning further, we might try $u = \sqrt{x^3+1} = (x^3+1)^{1/2}$, but this is a poor choice because $du = (1/2)(x^3+1)^{-1/2}(3x^2)\,dx$, an expression not present in the original integral.

Example 3 Find $\int \frac{x^2+5}{(x^3+15x)^2}\,dx.$

Solution Let $u = x^3 + 15x$, so that $du = (3x^2+15)\,dx = 3(x^2+5)\,dx$. The integrand is missing the 3, so multiply by $\frac{3}{3}$, putting 3 inside the integral sign and $\frac{1}{3}$ outside:

$$\int \frac{x^2+5}{(x^3+15x)^2}\,dx = \frac{1}{3}\int \frac{3(x^2+5)}{(x^3+15x)^2}\,dx$$
$$= \frac{1}{3}\int \frac{du}{u^2} = \frac{1}{3}\int u^{-2}\,du$$
$$= \frac{1}{3}\cdot\frac{u^{-1}}{-1} + C = \frac{-1}{3u} + C.$$

Substituting $x^3 + 15x$ for u gives

$$\int \frac{x^2+5}{(x^3+15x)^2}\,dx = \frac{-1}{3(x^3+15x)} + C.$$ ✓4

✓Checkpoint 4

Find the given antiderivatives.

(a) $\int z(z^2+1)^2\,dz$

(b) $\int \frac{x^2+3}{\sqrt{x^3+9x}}\,dx$

Answers: See page 846.

Recall that if $f(x)$ is a function, then by the chain rule, the derivative of the exponential function $y = e^{f(x)}$ is

$$\frac{dy}{dx} = e^{f(x)}\cdot f'(x).$$

Both the function $f(x)$ and its derivative appear on the right. This suggests that the antiderivative of a function of the form $e^{f(x)}$ can be found by letting u be the exponent.

Example 4 Find $\int 2e^{x^2}x\,dx$.

Let $u = x^2$, so that $du = 2x\,dx$. Then

$$\int 2e^{x^2}x\,dx = \int e^{x^2}2x\,dx$$
$$= \int e^u\,du$$
$$= e^u + C = e^{x^2} + C.$$

Check this answer by using the chain rule to take the derivative:

$$\frac{d}{dx}(e^{x^2} + C) = e^{x^2}(2x) + 0 = 2e^{x^2}x.$$

Example 5 Find the given antiderivatives.

(a) $\int e^{-12x}\,dx$

Solution Choose $u = -12x$, so $du = -12\,dx$. Multiply the integral by $(-\frac{1}{12})(-12)$, and use the rule for $\int e^u\,du$:

$$\int e^{-12x}\,dx = -\frac{1}{12}\cdot -12\int e^{-12x}\,dx$$
$$= -\frac{1}{12}\int e^{-12x}(-12\,dx)$$
$$= -\frac{1}{12}\int e^u\,du$$

$$= -\frac{1}{12}e^u + C$$
$$= -\frac{1}{12}e^{-12x} + C.$$

(b) $\int 6x^2 \cdot e^{x^3}\, dx$

Solution Let $u = x^3$, the exponent on e. Then $du = 3x^2\, dx$, and $(\frac{1}{3})\, du = x^2\, dx$. So

$$\int 6x^2 \cdot e^{x^3}\, dx = \int 6e^{x^3}(x^2\, dx)$$
$$= \int 6\,e^u\left(\frac{1}{3}du\right) \quad \text{Substitute.}$$
$$= 2\int e^u\, du \quad \text{Constant-multiple rule}$$
$$= 2e^u + C \quad \text{Integrate.}$$
$$= 2e^{x^3} + C. \quad \text{Substitute.}$$ 5 ✓

✓**Checkpoint 5**

Find the given antiderivatives.

(a) $\int e^{5x}\, dx$

(b) $\int 10xe^{3x^2}\, dx$

(c) $\int 2x^3 e^{x^4 - 1}\, dx$

Answers: See page 846.

Recall that the antiderivative of $f(x) = 1/x$ is $\ln|x|$. The next example uses $\int x^{-1}\, dx = \ln|x| + C$ and the method of substitution.

Example 6 Find the following.

(a) $\int \frac{dx}{9x + 6}$

Solution Choose $u = 9x + 6$, so $du = 9\, dx$. Multiply by $(\frac{1}{9})(9)$:

$$\int \frac{dx}{9x + 6} = \frac{1}{9}\cdot 9\int \frac{dx}{9x + 6} = \frac{1}{9}\int \frac{1}{9x + 6}(9\, dx)$$
$$= \frac{1}{9}\int \frac{1}{u}\, du = \frac{1}{9}\ln|u| + C = \frac{1}{9}\ln|9x + 6| + C.$$

(b) $\int \frac{(2x - 3)\, dx}{x^2 - 3x}$

Solution Let $u = x^2 - 3x$, so that $du = (2x - 3)\, dx$. Then

$$\int \frac{(2x - 3)\, dx}{x^2 - 3x} = \int \frac{du}{u} = \ln|u| + C = \ln|x^2 - 3x| + C.$$ 6 ✓

✓**Checkpoint 6**

Find the given antiderivatives.

(a) $\int \frac{4\, dx}{x - 6}$

(b) $\int \frac{(3x^2 + 5)\, dx}{x^3 + 5x + 4}$

Answers: See page 846.

The techniques used in the preceding examples can be summarized as follows.

Substitution Method

Let $u(x)$ be some function of x.

Form of the Integral	**Form of the Antiderivative**
1. $\int [u(x)]^n \cdot u'(x)\, dx,\ n \neq -1$	$\frac{[u(x)]^{n+1}}{n + 1} + C$

2. $\int e^{u(x)} \cdot u'(x)\, dx$	$e^{u(x)} + C$
3. $\int \frac{u'(x)\, dx}{u(x)}$	$\ln\lvert u(x)\rvert + C$

A slightly more complicated substitution technique is needed in some cases.

Example 7 Find $\int x\sqrt[3]{(1 - x)}\, dx$.

Solution Let $u = 1 - x$. Then $x = 1 - u$ and $dx = -du$. Now substitute:

$$\int x\sqrt[3]{(1 - x)}\, dx = \int (1 - u)\sqrt[3]{u}(-du) = \int (u - 1)u^{1/3}\, du$$

$$= \int (u^{4/3} - u^{1/3})\, du = \frac{3u^{7/3}}{7} - \frac{3u^{4/3}}{4} + C$$

$$= \frac{3}{7}(1 - x)^{7/3} - \frac{3}{4}(1 - x)^{4/3} + C. \quad 7\checkmark$$

✓Checkpoint 7

Find

$$\int x(x + 1)^{2/3}\, dx.$$

Answer: *See page 846.*

APPLICATIONS

Integration by substitution enables you to handle a wider variety of real-life applications.

Example 8 **Business** During the years 1995–2008, the approximate rate of change of revenue (in billions of dollars per year) for the commercial restaurant services industry was given by

$$R'(x) = \frac{-500{,}000(2x - 92)}{(x^2 - 92x + 2414)^2} \quad (5 \le x \le 18),$$

where $x = 5$ corresponds to 1995.* The actual revenue in 2005 was \$445 billion. Find the revenue function and estimate the revenue in 2008.

Solution The revenue function is

$$R(x) = \int R'(x)\, dx = \int \frac{-500{,}000(2x - 92)}{(x^2 - 92x + 2414)^2}\, dx.$$

Let $u = x^2 - 92x + 2414$. Then $du = (2x - 92)\, dx$ and

$$R(x) = -500{,}000\int (x^2 - 92x + 2414)^{-2}(2x - 92)\, dx$$

$$= -500{,}000\int u^{-2}\, du$$

*National Restaurant Association.

$$= -500{,}000\frac{u^{-2+1}}{-2+1} + C$$

$$= -500{,}000\frac{u^{-1}}{-1} + C$$

$$= \frac{500{,}000}{u} + C$$

$$R(x) = \frac{500{,}000}{x^2 - 92x + 2414} + C.$$

Find the value of C by using the fact that the revenue in 2005 was \$445 billion; that is, $R(15) = 445$. Thus,

$$R(x) = \frac{500{,}000}{x^2 - 92x + 2414} + C$$

$$R(15) = \frac{500{,}000}{15^2 - 92 \cdot 15 + 2414} + C$$

$$445 = 397.1 + C$$

$$C = 47.9.$$

Therefore, the revenue function is

$$R(x) = \frac{500{,}000}{x^2 - 92x + 2414} + 47.9.$$

The revenue in 2008 ($x = 18$) is

$$R(18) = \frac{500{,}000}{18^2 - 92 \cdot 18 + 2414} + 47.9 \approx \$509.9 \text{ billion.}$$ 8

✓Checkpoint 8

Sales of a new company, in thousands of dollars, are changing at a rate of

$$S'(t) = 27e^{-3t},$$

where t is time in months. Ten units were sold when $t = 0$. Find the sales function.

Answer: See page 846.

Example 9 **Social Science** To determine the top 100 popular songs of each year since 1956, Jim Quirin and Barry Cohen developed a function that represents the rate of change on the charts of *Billboard* magazine required for a song to earn a "star" on the *Billboard* "Hot 100" survey.* They developed the function

$$f(x) = \frac{A}{B + x},$$

where $f(x)$ represents the rate of change in position on the charts, x is the position on the "Hot 100" survey, and A and B are appropriate constants. The function

$$F(x) = \int f(x)\,dx$$

is defined as the "Popularity Index." Find $F(x)$.

*Formula for the "Popularity Index" from *Chartmasters' Rock 100*, Fourth Edition, by Jim Quirin and Barry Cohen. Copyright © 1987 by Chartmasters. Reprinted by permission.

Integrating $f(x)$ gives

$$F(x) = \int f(x)\,dx$$
$$= \int \frac{A}{B+x}\,dx$$
$$= A\int \frac{1}{B+x}\,dx. \quad \text{Constant-multiple rule}$$

Let $u = B + x$, so that $du = dx$. Then

$$F(x) = A\int \frac{1}{u}\,du = A\ln u + C$$
$$= A\ln(B+x) + C.$$

(The absolute value is not necessary, since $B + x$ is always positive here.)

13.2 ▸ Exercises

1. Integration by substitution is related to what differentiation method? What type of integrand suggests using integration by substitution?

2. For each of the given integrals, decide what factor should be u. Then find du.

   (a) $\int (3x^2 - 5)^4\, 2x\,dx$ **(b)** $\int \sqrt{1-x}\,dx$

   (c) $\int \frac{x^2}{2x^3+1}\,dx$ **(d)** $\int (8x-8)(4x^2-8x)\,dx$

Use substitution to find the given indefinite integrals. (See Examples 1–6 for Exercises 3–34 and Example 7 for Exercises 35–38.)

3. $\int 3(12x-1)^2\,dx$
4. $\int 5(4+9t)^3\,dt$
5. $\int \frac{5}{(3t-6)^2}\,dt$
6. $\int \frac{4}{\sqrt{5u-1}}\,du$
7. $\int \frac{x+1}{(x^2+2x-4)^{3/2}}\,dx$
8. $\int \frac{3x^2-2}{(2x^3-4x)^{5/2}}\,dx$
9. $\int r^2\sqrt{r^3+3}\,dr$
10. $\int y^3\sqrt{y^4-8}\,dy$
11. $\int (-4e^{5k})\,dk$
12. $\int (-2e^{-3z}\,dz)$
13. $\int 4w^3e^{2w^4}\,dw$
14. $\int 10ze^{-z^2}\,dz$
15. $\int (2-t)e^{4t-t^2}\,dt$
16. $\int (3-x^2)e^{9x-x^3}dx$
17. $\int \frac{e^{\sqrt{y}}}{\sqrt{y}}\,dy$
18. $\int \frac{e^{1/z^2}}{z^3}\,dz$
19. $\int \frac{-5}{12+6x}\,dx$
20. $\int \frac{9}{3-4x}\,dx$
21. $\int \frac{e^{2t}}{e^{2t}+1}\,dt$
22. $\int \frac{e^{5w+1}}{2-e^{5w+1}}\,dw$
23. $\int \frac{x+2}{(2x^2+8x)^3}\,dx$
24. $\int \frac{4y-2}{(y^2-y)^4}\,dy$
25. $\int 5\left(\frac{1}{r}+r\right)\left(1-\frac{1}{r^2}\right)dr$
26. $\int \left(\frac{2}{a}-a\right)\left(\frac{-2}{a^2}-1\right)da$
27. $\int \frac{x^2+1}{(x^3+3x)^{2/3}}\,dx$
28. $\int \frac{B^3-1}{(2B^4-8B)^{3/2}}\,dB$
29. $\int \frac{6x+7}{3x^2+7x+8}\,dx$
30. $\int \frac{x^2}{x^3+5}\,dx$
31. $\int 2x(x^2+5)^3\,dx$
32. $\int 2y^2(y^3-8)^3\,dy$
33. $\int \left(\sqrt{x^2+12x}\right)(x+6)\,dx$
34. $\int \left(\sqrt{x^2-6x}\right)(x-3)\,dx$
35. $\int \frac{(10+\ln x)^2}{x}\,dx$
36. $\int \frac{1}{x(\ln x)}\,dx$
37. $\int \frac{5u}{\sqrt{u-1}}\,du$
38. $\int \frac{2x}{(x+5)^6}\,dx$
39. $\int t\sqrt{5t-1}\,dt$
40. $\int 3r\sqrt{6-r}\,dr$

Work these problems. Round the constant C to two decimal places. (See Examples 8 and 9.)

41. **Business** A company has found that the rate of change in the cost of a new production line (in thousands of dollars) is

$$C(x) = \frac{60x}{5x^2+1},$$

where x is the number of years the line is in use.

(a) If the fixed costs are \$10,000, find the total cost function for the production line.
(b) The company will add a new production line if the total cost of this one for the first five years stays below \$40,000. Should the company add the new line?

42. **Business** The rate of expenditure for maintenance of a particular machine is given by

$$M'(x) = \sqrt{x^2 + 8x}(2x + 8),$$

where x is time measured in years. Total maintenance costs through the fourth year are \$540.
(a) Find the total maintenance function.
(b) How many years must pass before the total maintenance costs reach \$2000?

43. **Business** The rate of change of revenue from sales of digital TV sets and TV screens (in billions of dollars per year) from 2003 to 2008 is approximated by

$$g(x) = 1.1128806e^{.2483x} \qquad (3 \le x \le 8),$$

where $x = 3$ corresponds to 2003.* Revenue in 2003 was \$8.7 billion.
(a) Find the revenue function.
(b) Estimate the revenue in 2007.

44. **Business** The rate of growth of profits (in millions of dollars) in year x from a new manufacturing process is approximated by

$$f(x) = xe^{-x^2}.$$

The total profit in the third year of the new process is \$10,000.
(a) Find the total profit function. (Remember that profits are given in millions of dollars.)
(b) As time goes on, what will happen to the total profit?

45. **Business** The marginal revenue (in thousands of dollars) from the sale of x digital music players is given by

$$MR(x) = 1.8x(x^2 + 27{,}000)^{-2/3}.$$

The revenue from 150 players is \$32,000.
(a) Find the revenue function.
(b) What is the revenue from selling 250 players?
(c) How many players must be sold to produce revenue of at least \$100,000?

Statistical Abstract of the United States: 2009.

46. **Health** Per-capita consumption of cheese in the United States (in pounds) was 17.5 in 1980 and has been increasing at a rate approximated by

$$f(x) = \frac{4.33}{x + 1},$$

where $x = 0$ corresponds to the year 1980.*
(a) Find a function that gives the per-capita amount of cheese consumed in year x.
(b) Estimate the per-capita consumption of cheese in 2005 and 2010.

47. **Transportation** The rate of change in the number of urban transit vehicles such as buses and light-rail trains, in thousands, is approximated by

$$g(x) = .00040674x(x - 1970)^{.4} \qquad (x \ge 1970),$$

where x is the year.† In 1970, the number of such vehicles was 61,298.
(a) Find a function that gives the approximate number of urban transit vehicles in year x. (*Hint*: Example 7.)
(b) Estimate the number of urban transit vehicles in 2015.

✓ Checkpoint Answers

1. (a) $du = 9\,dx$ (b) $du = (15x^2 + 4x)\,dx$
(c) $du = -2e^{-2x}\,dx$
2. (a) $\frac{(4x^2 - 1)^6}{6} + C$ (b) $\frac{(3x - 8)^5}{15} + C$
(c) $\frac{12(x^3 - 7)^{5/2}}{5} + C$
3. (a) $\frac{1}{9}(6x^2 + 3)^3 + C$ (b) $\frac{1}{3}(x^2 + 16)^{3/2} + C$
4. (a) $\frac{(z^2 + 1)^3}{6} + C$ (b) $\frac{2}{3}\sqrt{x^3 + 9x} + C$
5. (a) $\frac{1}{5}e^{5x} + C$ (b) $\frac{5}{3}e^{3x^2} + C$
(c) $\frac{1}{2}e^{x^4 - 1} + C$
6. (a) $4\ln|x - 6| + C$ (b) $\ln|x^3 + 5x + 4| + C$
7. $\frac{3}{8}(x + 1)^{8/3} - \frac{3}{5}(x + 1)^{5/3} + C$
8. $S(t) = 19 - 9e^{-3t}$

*U.S. Department of Agriculture.
†U.S. Bureau of Transportation Statistics.

13.3 ▶ Area and the Definite Integral

Suppose a car travels along a straight road at a constant speed of 50 mph. The speed of the car at time t is given by the constant function $v(t) = 50$, whose graph is a horizontal straight line, as shown in Figure 13.2(a).

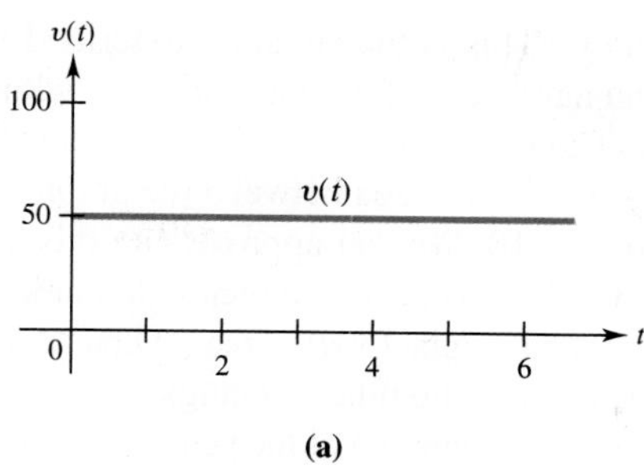

(a)

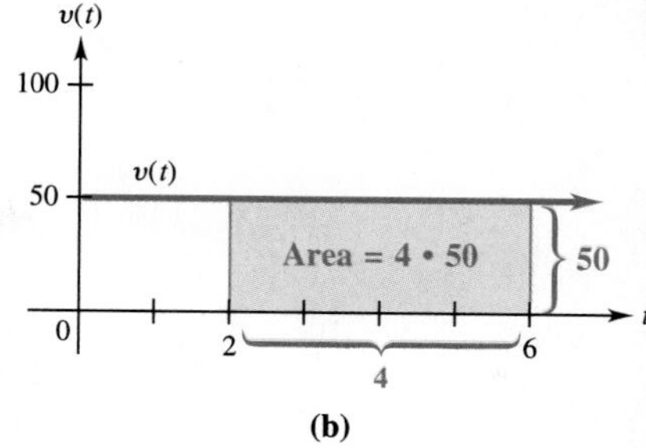

(b)

FIGURE 13.2

How far does the car travel from time $t = 2$ to $t = 6$? Since this is a 4-hour period, the answer, of course, is $4 \cdot 50 = 200$ miles. Note that 200 is precisely the *area* under the graph of the speed function $v(t)$ and above the t-axis from $t = 2$ to $t = 6$, as shown in Figure 13.2(b).

As we saw in Chapter 12, the speed function $v(t)$ is the rate of change of distance with respect to time—that is, the rate of change of the distance function $s(t)$ (which gives the distance traveled by the car at time t). Now, the distance traveled from time $t = 2$ to $t = 6$ is the amount that the distance has changed from $t = 2$ to $t = 6$. In other words, the *total change* in distance from $t = 2$ to $t = 6$ is the area under the graph of the speed (rate of change) function from $t = 2$ to $t = 6$.

A more complicated argument (omitted here) shows that a similar situation holds in the general case.

Total Change in $F(x)$

Let f be a function such that f is continuous on the interval $[a, b]$ and $f(x) \geq 0$ for all x in $[a, b]$. If $f(x)$ is the rate of change of a function $F(x)$, then the **total change in $F(x)$** as x goes from a to b is the area between the graph of $f(x)$ and the x-axis from $x = a$ to $x = b$.

Example 1 **Business** Figure 13.3 shows the graph of the function that gives the rate of change of the annual maintenance charges for a certain machine. The rate function is increasing because maintenance tends to cost more as the machine gets older. Estimate the total maintenance charges over the 10-year life of the machine.

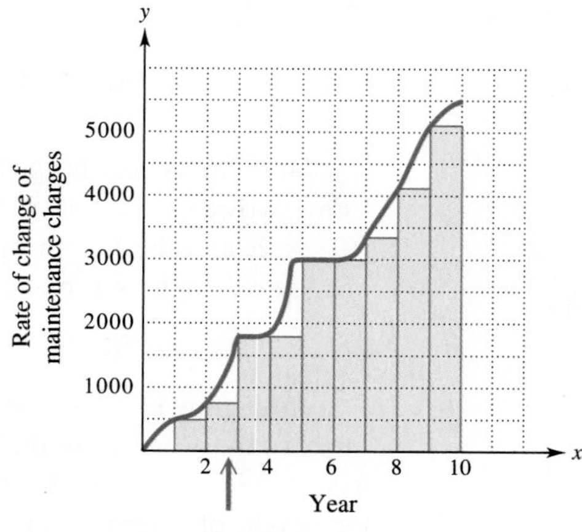

FIGURE 13.3

Solution This is the situation described in the preceding box, with $F(x)$ being the maintenance cost function and $f(x)$, whose graph is given, the rate-of-change function. The total maintenance charges are the total change in $F(x)$ from $x = 0$ to $x = 10$—that is, the area between the graph of the rate function and the x-axis from $x = 0$ to $x = 10$. We can approximate this area by using the shaded rectangles in Figure 13.3. For instance, the rectangle marked with an arrow has base 1 (from year 2 to year 3) and height 750 (the rate of change at $x = 2$, so its area is $1 \times 750 = 750$. Similarly, each of the other rectangles has base 1 and height determined by the rate of change at the beginning of the year. Consequently, we estimate the area to be the sum

$$1 \cdot 0 + 1 \cdot 500 + 1 \cdot 750 + 1 \cdot 1800 + 1 \cdot 1800 + 1 \cdot 3000 + 1 \cdot 3000 + 1 \cdot 3400 + 1 \cdot 4200 + 1 \cdot 5200 = 23{,}650.$$

Hence, the total maintenance charges over the 10 years are at least \$23,650. (The unshaded areas under the rate graph have not been accounted for in this estimate). 1 ✓

✓ Checkpoint 1

Use Figure 13.3 to estimate the maintenance charge during

(a) the first 6 years of the machine's life;

(b) the first 8 years of the machine's life.

Answers: See page 858.

AREA

The preceding examples show that the area between a graph and the x-axis has useful interpretations. In this section and the next, we develop a means of measuring such areas precisely when the function is given by an algebraic formula. The underlying idea is the same as in Example 1: Use rectangles to approximate the area under the graph.

Example 2 Find the area under the graph of $f(x) = \sqrt{4 - x^2}$ from $x = 0$ to $x = 2$, shown in Figure 13.4.

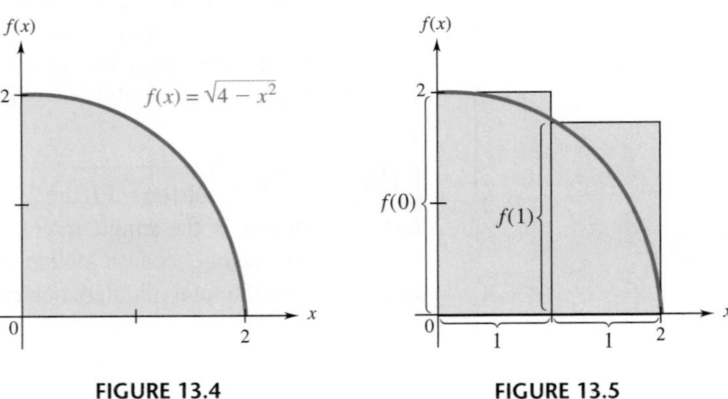

FIGURE 13.4 FIGURE 13.5

Solution A very rough approximation of the area of this region can be found by using two rectangles, as in Figure 13.5. The height of the rectangle on the left is $f(0) = 2$, and the height of the rectangle on the right is $f(1) = \sqrt{3}$. The width of each rectangle is 1, making the total area of the two rectangles

$$1 \cdot f(0) + 1 \cdot f(1) = 2 + \sqrt{3} \approx 3.7321 \text{ square units.}$$

As Figure 13.5 suggests, this approximation is greater than the actual area. To improve the accuracy of the approximation, we could divide the interval from $x = 0$ to $x = 2$ into four equal parts, each of width $\frac{1}{2}$, as shown in Figure 13.6. As before, the height of each rectangle is given by the value of f at the left-hand side of the rectangle, and its area is the width, $\frac{1}{2}$, multiplied by the height. The total area of the four rectangles is

$$\frac{1}{2}\cdot f(0) + \frac{1}{2}\cdot f\left(\frac{1}{2}\right) + \frac{1}{2}\cdot f(1) + \frac{1}{2}\cdot f\left(\frac{3}{2}\right)$$

$$= \frac{1}{2}(2) + \frac{1}{2}\left(\frac{\sqrt{15}}{2}\right) + \frac{1}{2}(\sqrt{3}) + \frac{1}{2}\left(\frac{\sqrt{7}}{2}\right)$$

$$= 1 + \frac{\sqrt{15}}{4} + \frac{\sqrt{3}}{2} + \frac{\sqrt{7}}{4} \approx 3.4957 \text{ square units.}$$

This approximation looks better, but it is still greater than the actual area desired. To improve the approximation, divide the interval from $x = 0$ to $x = 2$ into eight parts with equal widths of $\frac{1}{4}$. (See Figure 13.7.) The total area of all these rectangles is

$$\frac{1}{4}\cdot f(0) + \frac{1}{4}\cdot f\left(\frac{1}{4}\right) + \frac{1}{4}\cdot f\left(\frac{1}{2}\right) + \frac{1}{4}\cdot f\left(\frac{3}{4}\right) + \frac{1}{4}\cdot f(1) + \frac{1}{4}\cdot f\left(\frac{5}{4}\right)$$
$$+ \frac{1}{4}\cdot f\left(\frac{3}{2}\right) + \frac{1}{4}\cdot f\left(\frac{7}{4}\right).$$ ✓2

✓ Checkpoint 2

Calculate the sum

$$\frac{1}{4}\cdot f(0) + \frac{1}{4}\cdot f\left(\frac{1}{4}\right) + \cdots + \frac{1}{4}\cdot f\left(\frac{7}{4}\right),$$

using the following information:

x	$f(x)$
0	2
1/4	1.98431
1/2	1.93649
3/4	1.85405
1	1.73205
5/4	1.56125
3/2	1.32288
7/4	.96825

Answer: See page 858.

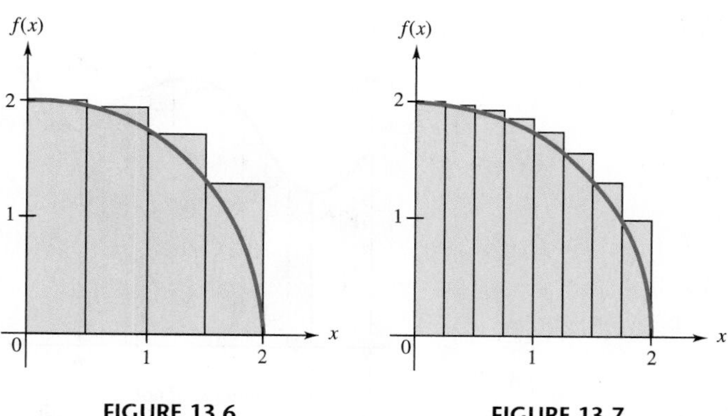

FIGURE 13.6 FIGURE 13.7

This process of approximating the area under the curve by using more and more rectangles to get a better and better approximation can be extended. To do this, divide the interval from $x = 0$ to $x = 2$ into n equal parts. Each of these n intervals has width

$$\frac{2 - 0}{n} = \frac{2}{n},$$

so each rectangle has width $2/n$ and height determined by the value of the function $f(x) = \sqrt{4 - x^2}$ at the left side of the rectangle. A computer was used to find approximations of the area for several values of n given in the table at the side.

n	Area
125	3.15675
2000	3.14257
8000	3.14184
32,000	3.14165
128,000	3.14160
512,000	3.14159

As the number n of rectangles gets larger and larger, the sum of their areas gets closer and closer to the actual area of the region. In other words, the actual area is the *limit* of these sums as n gets larger and larger without bound, which can be written

$$\text{area} = \lim_{n\to\infty} (\text{sum of areas of } n \text{ rectangles}).$$

The table suggests that this limit is a number whose decimal expansion begins 3.14159 . . . , which is the same as the beginning of the decimal approximation of π. Therefore, it seems plausible that

$$\text{area} = \lim_{n\to\infty} (\text{sum of areas of } n \text{ rectangles}) = \pi.$$

It can be shown that the region whose area was found in Example 2 is one-fourth of the interior of a circle of radius 2 with center at the origin (see Figure 13.4). Hence, its area is

$$\frac{1}{4}(\pi r^2) = \frac{1}{4}(\pi \cdot 2^2) = \pi,$$

which agrees with our answer in Example 2.

The method used in Example 2 can be generalized to find the area bounded by the curve $y = f(x)$, the x-axis, and the vertical lines $x = a$ and $x = b$, as shown in Figure 13.8. To approximate this area, we could divide the region under the curve first into 10 rectangles (Figure 13.8(a)) and then into 20 rectangles (Figure 13.8(b)). In each case, the sum of the areas of the rectangles gives an approximation of the area under the curve.

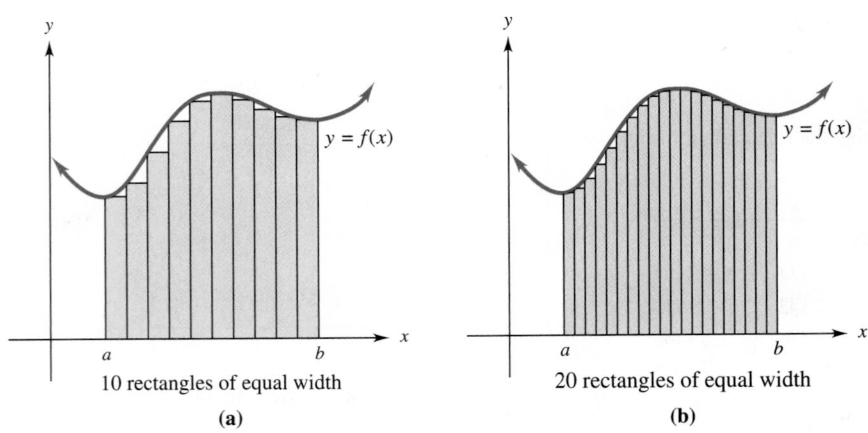

FIGURE 13.8

We can get better and better approximations by increasing the number n of rectangles. Here is a description of the general procedure: Let n be a positive integer. Divide the interval from a to b into n pieces of equal length. The symbol Δx is traditionally used to denote the length of each piece. Since the length of the entire interval is $b - a$, each of the n pieces has length

$$\Delta x = \frac{b - a}{n}.$$

Use each of these pieces as the base of a rectangle, as shown in Figure 13.9, where the endpoints of the n intervals are labeled $x_1, x_2, x_3, \ldots, x_{n+1}$. A typical rectangle, the one whose lower left corner is at x_i, has red shading. The base of this rectangle is Δx, and its height is the height of the graph over x_i, namely, $f(x_i)$, so

$$\text{Area of } i\text{th rectangle} = f(x_i) \cdot \Delta x.$$

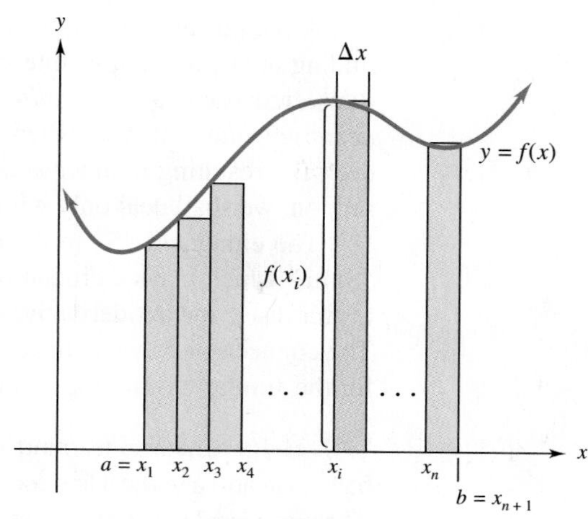

n rectangles of equal width

FIGURE 13.9

The total area under the curve is approximated by the sum of the areas of all n of the rectangles, namely,

$$f(x_1)\cdot\Delta x + f(x_2)\cdot\Delta x + f(x_3)\cdot\Delta x + \cdots + f(x_n)\cdot\Delta x.$$

The exact area is defined to be the limit of this sum (if it exists) as the number of rectangles gets larger and larger, without bound. Hence,

$$\begin{aligned}\text{Exact area} &= \lim_{n\to\infty}(f(x_1)\cdot\Delta x + f(x_2)\cdot\Delta x + f(x_3)\cdot\Delta x + \cdots + f(x_n)\cdot\Delta x)\\ &= \lim_{n\to\infty}([f(x_1) + f(x_2) + f(x_3) + \cdots + f(x_n)]\cdot\Delta x).\end{aligned}$$

This limit is called the *definite integral* of $f(x)$, from a to b and is denoted by the symbol

$$\int_a^b f(x)\,dx.$$

The preceding discussion can be summarized as follows.

The Definite Integral

If f is a continuous function on the interval $[a, b]$, then the **definite integral** of f from a to b is the number

$$\int_a^b f(x)\,dx = \lim_{n\to\infty}[(f(x_1) + f(x_2) + f(x_3) + \cdots + f(x_n))\,\Delta x],$$

where $\Delta x = (b - a)/n$ and x_i is the left-hand endpoint of the ith interval.

For instance, the area of the region in Example 2 could be written as the definite integral

$$\int_0^2 \sqrt{4 - x^2}\,dx = \pi.$$

Although the definition in the box is obviously motivated by the problem of finding areas, it is applicable to many other situations, as we shall see in later sections. In particular, *the definition of the definite integral is valid even when $f(x)$ takes negative values* (that is, when the graph goes below the x-axis). In that case, however, the resulting number is not the area between the graph and the x-axis. In this section, we shall deal only with the area interpretation of the definite integral.

The elongated "S" in the notation for the definite integral stands for the word "Sum," which plays a crucial role in the definition. This notation looks very similar to that used for antiderivatives (also called *indefinite* integrals) in earlier sections. The connection between the definite integral and antiderivatives, which is the reason for the similar terminology and notation, will be explained in the next section.

CAUTION Children learning to read sometimes confuse *b* and *d*. Both consist of a half-circle and a vertical line segment, but the location of the line segment makes all the difference. Similarly, the symbols $\int f(x)\,dx$ and $\int_a^b f(x)\,dx$ have totally different meanings—the *a* and *b* make all the difference. The indefinite integral $\int f(x)\,dx$ denotes a set of *functions* (the antiderivatives of $f(x)$), whereas the definite integral $\int_a^b f(x)\,dx$ represents a *number* (which can be interpreted as the area under the graph when $f(x) \geq 0$).

Example 3 Approximate $\int_1^4 4x\,dx$, which is the area under the graph of $f(x) = 4x$, above the x-axis, and between $x = 1$ and $x = 4$, by using six rectangles of equal width whose heights are the values of the function at the left endpoint of each rectangle.

Solution We want to find the area of the shaded region in Figure 13.10.

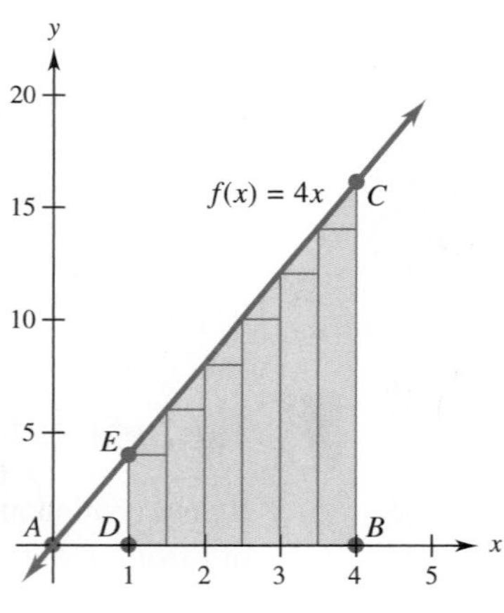

FIGURE 13.10

The heights of the six rectangles given by $f(x_i)$ are as follows:

i	x_i	$f(x_i)$
1	$x_1 = 1$	$f(1) = 4$
2	$x_2 = 1.5$	$f(1.5) = 6$
3	$x_3 = 2$	$f(2) = 8$
4	$x_4 = 2.5$	$f(2.5) = 10$
5	$x_5 = 3$	$f(3) = 12$
6	$x_6 = 3.5$	$f(3.5) = 14$

The width of each rectangle is $\Delta x = \frac{4-1}{6} = \frac{1}{2} = .5$. The sum of the areas of the six rectangles is

$$\begin{aligned} &f(x_1)\Delta x + f(x_2)\Delta x + f(x_3)\Delta x + f(x_4)\Delta x + f(x_5)\Delta x + f(x_6)\Delta x \\ &\quad = f(1)\Delta x + f(1.5)\Delta x + f(2)\Delta x + f(2.5)\Delta x + f(3)\Delta x + f(3.5)\Delta x \\ &\quad = (4)(.5) + (6)(.5) + (8)(.5) + (10)(.5) + (12)(.5) + (14)(.5) \\ &\quad = 27. \end{aligned}$$

We can check the accuracy of this approximation by noting that the area of the shaded region in Figure 13.10 is the difference of the areas of triangle ABC and triangle ADE. Triangle ABC has base 4 and height 16, and triangle ADE has base 1 and height 4. Using the formula for the area of a triangle, $A = \frac{1}{2}bh$, we see that the area of the shaded region is

$$\text{Area } ABC - \text{Area } ADE = (1/2)(4)(16) - (1/2)(1)(4) = 32 - 2 = 30.$$

So our approximation is a bit low. Using more rectangles will produce a better approximation of the area. 3✓

Divide the region of Figure 13.10 into 12 rectangles of equal width whose heights are the values of the function at the left endpoint of each rectangle.

(a) Complete this table.

i	x_i	$f(x_i)$
1	1	
2	1.25	
3	1.5	
4	1.75	
5	2	
6	2.25	
7		
8		
9		
10		
11		
12		

(b) Use the results from the table to approximate $\int_1^4 4x\,dx$.

Answers: See page 858.

TECHNOLOGY AND DEFINITE INTEGRALS

The preceding examples illustrate the difficulty of evaluating definite integrals by direct computation. In the next section, an algebraic evaluation technique that works in many cases will be developed. In all cases, technology can be used to evaluate definite integrals. Computer programs and calculators that have symbolic capabilities, such as Maple, Mathematica and the TI-89 will generally produce exact results. Other graphing calculators and spreadsheets usually provide highly accurate approximations.

TECHNOLOGY TIP There are two ways to approximate a definite integral with a graphing calculator, each of which is illustrated here for $\int_{-2}^{1}(x^3 - 5x + 6)\,dx$.

1. Approximate the limit in the definition by computing the sum for a large value of n, say, $n = 100$. Then $\Delta x = (b - a)/n = (1 - (-2))/100 = 3/100 = .03$, and the sum is

$$\begin{aligned} &[f(x_1) + f(x_2) + f(x_3) + \cdots + f(x_n)] \cdot \Delta x \\ &= [f(x_1) + f(x_2) + f(x_3) + \cdots + f(x_{100})](.03). \end{aligned}$$

To compute this sum automatically on most TI calculators, look in the submenus of the LIST menu and enter

sum(seq($x^3 - 5x + 6$, x, -2, $1 - .03$, $.03$)) × (.03).

Left endpoint   Right endpoint $-\Delta x$ Δx

On other calculators, use the RECTANGLE program in Appendix A of this text. In each case, the calculator produces the approximation 21.8393.

2. Use **numerical integration**. This feature approximates the definite integral by using more complicated summing techniques than used previously (which typically involve trapezoids or regions with parabolic boundaries). On most TI calculators, look in the MATH or CALC menu and enter

$$\text{fnInt}(x^3 - 5x + 6, x, -2, 1).$$

On the Casio, use the $\int dx$ key on the keyboard and enter

$$\int (x^3 - 5x + 6, -2, 1).$$

In each case, the calculator approximates the integral as 21.75, which actually is its exact value, as we shall see in the next section.

Example 4 Figure 13.11(a) shows the graph of

$$f(x) = x^3 - 8x^2 + 18x + 5.$$

The area under the graph from $x = 1$ to $x = 4.5$, which is shown in Figure 13.11(b), is given by

$$\int_1^{4.5} (x^3 - 8x^2 + 18x + 5)\,dx.$$

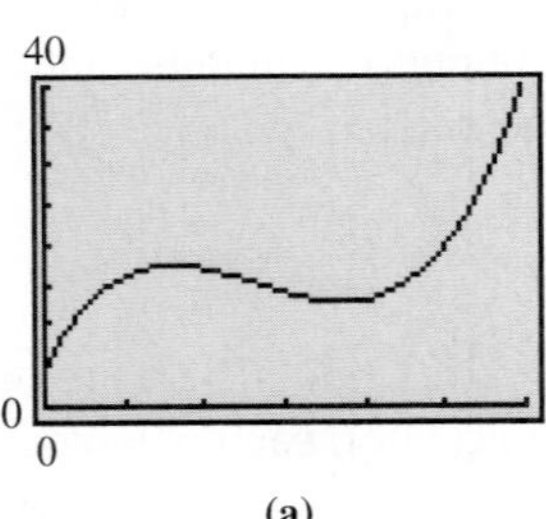

(a)

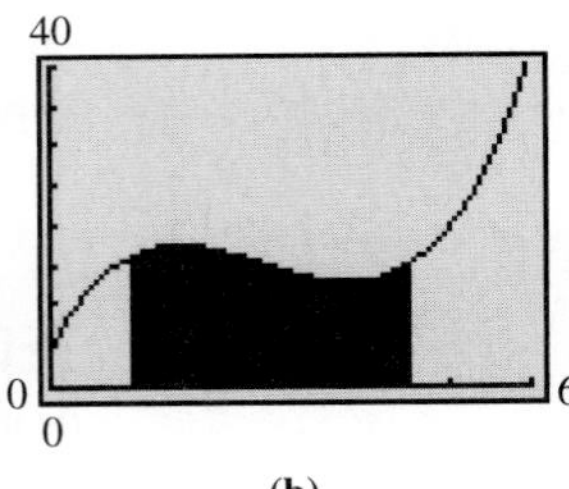

(b)

FIGURE 13.11

(a) Use a calculator and the definition presented earlier, together with 100 rectangles, to approximate this integral.

Solution For $n = 100$ rectangles, $\Delta x = (b - a)/n = (4.5 - 1)/100 = .035$. The left-hand endpoints of the bases of the rectangles are 1, 1.035, 1.07, 1.105, 1.14, . . . , 4.43, and 4.465. (The right-hand endpoint of the base of the last rectangle is 4.5.) We must compute the following sum:

$$[f(1) + f(1.035) + f(1.07) + \cdots + f(4.43) + f(4.465)](.035).$$

Compute the sum by doing Checkpoint Problem 4 in the margin. ✓4

(b) Approximate the integral by using numerical integration on a calculator.

✓ **Checkpoint 4**

If $f(x) = x^3 - 8x^2 + 18x + 5$, use the first part of the previous Technology Tip to compute the sum

$$[f(1) + f(1.035) + f(1.07) + \cdots + f(4.465)](.035).$$

Answer: *See page 858.*

FIGURE 13.12

Solution The numerical integration program on a typical calculator (see part 2 of the previous Technology Tip) produces the answer in Figure 13.12. It differs slightly from the answer in part (a) and is accurate to seven decimal places, as the techniques in the next section will show.

At the beginning of this section, we saw that if $f(x)$, (with $f(x) \geq 0$) is the rate of change of a function $F(x)$, then the total change in $F(x)$, as x goes from a to b is the area between the graph of $f(x)$, and the x-axis from $x = a$ to $x = b$. In other words,

$$\text{total change in } F(x) = \int_a^b f(x)\,dx.$$

Example 5 **Business** The marginal cost (in dollars) for manufacturing x cases of specialty souvenirs is given by

$$MC(x) = \frac{(x - 200)^2}{500}.$$

Find the amount added to the total cost when production goes from 50 cases to 300 cases.

Solution The amount added is the total change in cost from $x = 50$ to $x = 300$. Since the marginal cost function is the derivative of the cost function,

$$\text{total change} = \int_{50}^{300} \frac{(x - 200)^2}{500}\,dx.$$

Numerical integration (Figure 13.13) shows that the total change in cost from $x = 50$ to $x = 300$ is \$2916.67. ✓5

✓Checkpoint 5

If the marginal revenue from selling x computers is given by $MR(x) = .3x^3 - .5x^2 + x$, find the total revenue from selling 30 computers (that is, the total change in revenue from $x = 0$ to $x = 30$).

Answer: *See page 858.*

FIGURE 13.13

13.3 ▸ Exercises

In Exercises 1–4, estimate the required areas by using rectangles whose height is given by the value of the function at the left side of each rectangle. (See Example 1.)

1. **Business** The accompanying graph shows the rate of use of electrical energy (in kilowatt hours) in a certain city on a very hot day. Estimate the total usage of electricity on that day. Let the width of each rectangle be 2 hours.

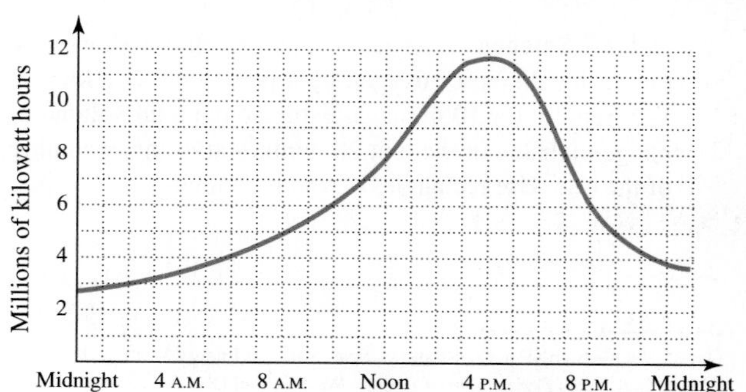

2. **Business** The accompanying graphs shows U.S. oil production and consumption rates in millions of barrels per day. The rates for the years beyond 1991 are projected on the basis of the policy in place in 1990 (labeled "Current policy base" in the graph) and on the basis of a new policy proposed by the George H. W. Bush administration in 1991 (labeled "With strategy" in the graph).* Estimate the amount of oil produced between 1990 and 2010, using the policy in place in 1990. Use rectangles with widths of 5 years. (*Hint*: There are 365 days in a year.)

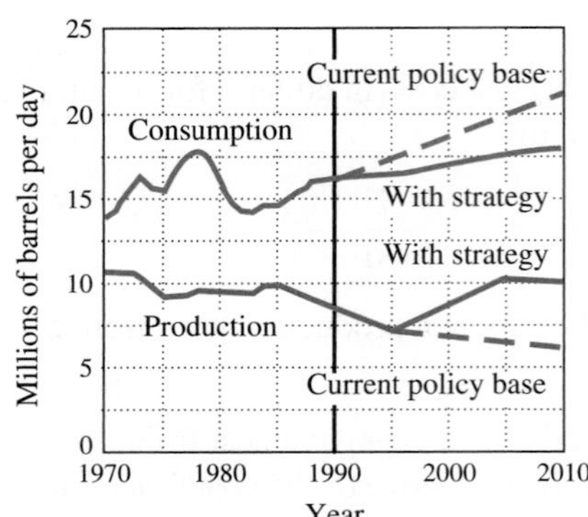

3. **Business** The accompanying graph shows coal consumption (in millions of short tons per year) in the United States in selected years.† Estimate total coal consumption for the 10-year period from 2000 to 2010, using rectangles of width 2 years.

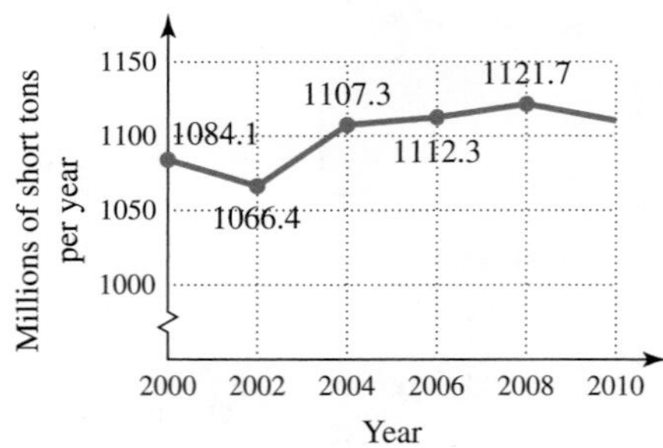

4. **Natural Science** The accompanying graph shows the rate of inhalation of oxygen by a person riding a bicycle very rapidly for 10 minutes. Estimate the total volume of oxygen inhaled in the first 20 minutes after the beginning of the ride. Use rectangles of width 1 minute.

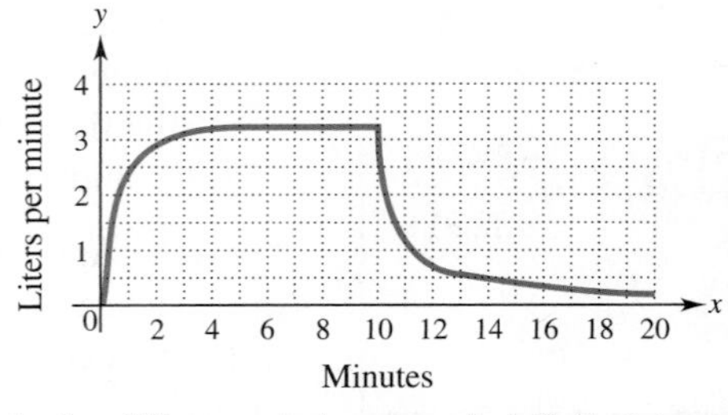

5. Explain the difference between an indefinite integral and a definite integral.

6. Complete the following statement:

$$\int_0^3 (x^2 + 2)\, dx = \lim_{n\to\infty} \underline{\qquad}, \text{ Where } \Delta x = \underline{\qquad}.$$

Approximate the area under each curve and above the x-axis on the given interval, using rectangles whose height is the value of the function at the left side of the rectangle.
(a) *Use two rectangles. (See Examples 2 and 3.)*
(b) *Use four rectangles.*
(c) *Use a graphing calculator (or other technology) and 40 rectangles. (See Example 4.)*

7. $f(x) = 3x + 8$; $[0, 4]$
8. $f(x) = 9 - 2x$; $[0, 4]$
9. $f(x) = 4 - x^2$; $[-2, 2]$
10. $f(x) = x^2 + 5$; $[-2, 2]$
11. $f(x) = e^{2x} - .5$; $[0, 2]$
12. $f(x) = \dfrac{2}{x + 1}$; $[1, 9]$

Work the given exercises. (See Examples 2 and 3.)

13. Consider the region below $f(x) = x/2$, above the x-axis, between $x = 0$ and $x = 4$. Let x_i be the left endpoint of the ith subinterval.
    (a) Use four rectangles to approximate the area of the region.
    (b) Use eight rectangles to approximate the area of the region.
    (c) Find $\int_0^4 f(x)\, dx$ by using the formula for the area of a triangle.

14. Find $\int_0^5 (5 - x)\, dx$ by using the formula for the area of a triangle.

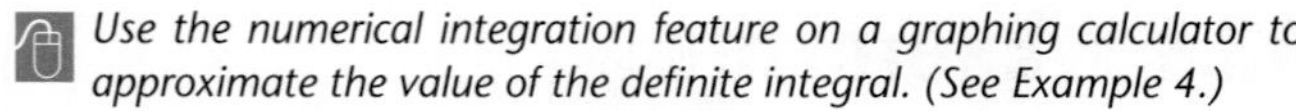

Use the numerical integration feature on a graphing calculator to approximate the value of the definite integral. (See Example 4.)

15. $\displaystyle\int_{-5}^{0} (x^3 + 6x^2 - 10x + 2)\, dx$

16. $\displaystyle\int_{-1}^{2} (-x^4 + 3x^3 - 4x^2 + 20)\, dx$

17. $\displaystyle\int_{2}^{7} 5 \ln (2x^2 + 1)\, dx$

18. $\displaystyle\int_{1}^{5} 3x \ln x\, dx$

19. $\displaystyle\int_{0}^{3} 4x^2 e^{-3x}\, dx$

20. $\displaystyle\int_{1}^{5} \frac{\ln x}{x}\, dx$

Business *A marginal revenue function MR(x) (in dollars) is given. Use numerical integration on a graphing calculator or*

*Graph, "Uncle Sam's Energy Strategy" from National Energy Strategy, February 1991. United States Department of Energy, Washington, DC.

†U.S. Department of Energy.

computer to find the total revenue over the given range. (See Example 5.)

21. $MR(x) = .04x^3 - .5x^2 + 2x \quad (0 \le x \le 99)$

22. $MR(x) = .04x^4 - .5x^3 + 2x^2 - x \quad (0 \le x \le 60)$

23. $MR(x) = .26x^4 - 6.25x^3 - 30.2x^2 + 87.5x \quad (0 \le x \le 40)$

24. $MR(x) = \dfrac{175}{1 + 3e^{-.7x}} - 20 \quad (0 \le x \le 310)$

25. Business The rate of U.S. residential electricity consumption (in billions of kilowatt hours (kWh) per year) is approximated by

$$R(x) = 927.7e^{.0244x},$$

where $x = 0$ corresponds to the year 1990.* Find the total consumption of electricity from 2004 to 2007.

26. Physical Science The velocity of a car at time x hours after starting on a trip is given by $52 - e^{-x/2}$ mph. Use the fact that velocity is the derivative of distance to find the total distance traveled by the car from time $x = 2$ to $x = 5$.

Physical Science *The graphs for Exercises 27 and 28 are from* Road & Track *magazine.*† *The curves show the velocity at time t, in seconds, when a car accelerates from a dead stop. To find the total distance traveled by the car in reaching 100 miles per hour, we must estimate the definite integral*

$$\int_0^T v(t)\,dt,$$

where T represents the number of seconds it takes for the car to reach 100 mph.

Use the graphs to estimate this distance by adding the areas of rectangles with widths of 5 seconds. The last rectangle has a width of 3. To adjust your answer to miles per hour, divide by 3600 (the number of seconds in an hour). You then have the number of miles that the car traveled in reaching 100 mph. Finally, multiply by 5280 feet per mile to convert the answers to feet.

27. Estimate the distance traveled by the Porsche 928, using the following graph:

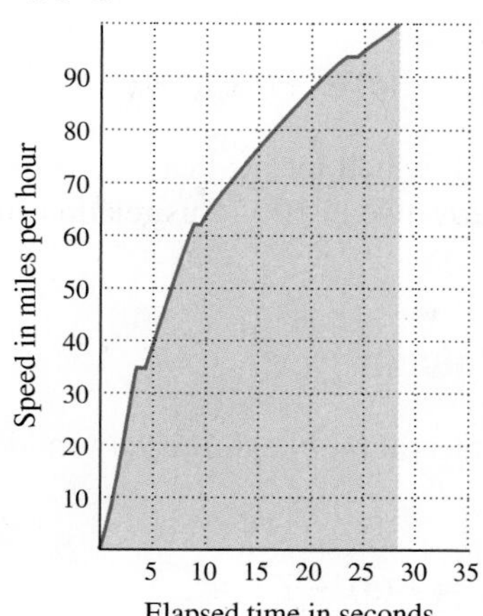

28. Estimate the distance traveled by the BMW 733i, using the following graph:

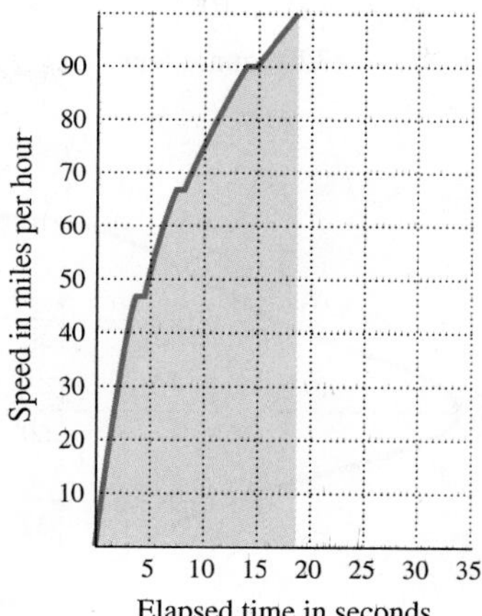

29. Two cars start from rest at a traffic light and accelerate for several minutes. The accompanying graph shows their velocities (in feet per second) as a function of time (in seconds). Car A is the one that initially has the greater velocity.*

(a) How far has car A traveled after 2 seconds? (*Hint*: Use formulas from geometry.)

(b) When is car A farthest ahead of car B?

(c) Estimate the farthest that car A gets ahead of car B. For car A, use formulas from geometry. For car B, use $n = 4$ and the value of the function at the midpoint of each interval for the height of the rectangle.

(d) Give a rough estimate of when car B catches up with car A.

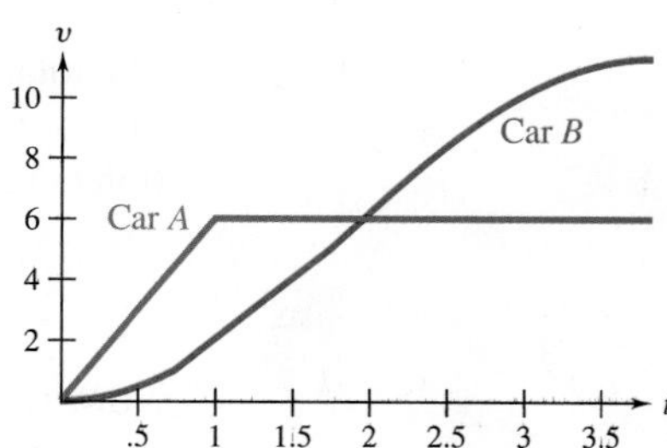

30. The booklet *All About Lawns*, published by Ortho Books, gives the following instructions for measuring the area of an irregularly shaped region.

Irregular Shapes (within 5% accuracy)
Measure a long (L) axis of the area. Every 10 feet along the length line, measure the width at right angles to the length line. Total widths and multiply by 10.

*U.S. Energy Information Administration.

†*Road & Track*, April and May 1978. Reprinted with permission of *Road & Track*.

*Based on an example given by Steve Monk of the University of Washington.

Area = ($A_1A_2 + B_1B_2 + C_1C_2$ etc.) × 10
A = (40′ + 60′ + 32′) × 10
A = 132′ × 10′
A = 1320 square feet

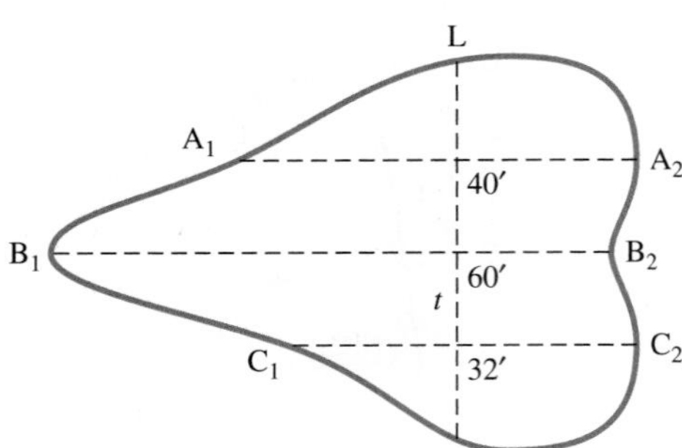

How does this method relate to the discussion in this section?

✓ Checkpoint Answers

1. **(a)** \$7850 **(b)** \$14,250
2. 3.33982 square units
3. **(a)**

i	x_i	$f(x_i)$
1	1	4
2	1.25	5
3	1.5	6
4	1.75	7
5	2	8
6	2.25	9
7	2.5	10
8	2.75	11
9	3	12
10	3.25	13
11	3.5	14
12	3.75	15

(b) 28.5

4. 52.6977
5. \$56,700

13.4 ▶ The Fundamental Theorem of Calculus

We now develop the connection between definite integrals and antiderivatives, which will explain the similar notation for these two concepts. More importantly, it provides a way to calculate definite integrals exactly.

In the last section, we saw that when $f(x)$ is the rate of change of the function $F(x)$ and $f(x) \geq 0$ on $[a, b]$, the definite integral has this interpretation:

$$\int_a^b f(x)\,dx = \text{total change in } F(x) \text{ as } x \text{ changes from } a \text{ to } b. \qquad (*)$$

The total change in $F(x)$ as x changes from a to b is the difference between the value of F at the end and the value of F at the beginning—that is, $F(b) - F(a)$. So equation $(*)$ can be stated as

$$\int_a^b f(x)\,dx = F(b) - F(a).$$

Now, $f(x)$ is the rate of change of $F(x)$, which means that $f(x)$ is the derivative of $F(x)$, or equivalently, $F(x)$ is an antiderivative of $f(x)$. This relationship is an example of the following result.

Fundamental Theorem of Calculus

Suppose f is continuous on the interval $[a, b]$ and F is *any* antiderivative of f. Then

$$\int_a^b f(x)\,dx = F(b) - F(a).$$

The proof of the fundamental theorem is discussed at the end of this section. For now, we shall concentrate on applying it. Note these facts:

1. The fundamental theorem applies to every continuous function $f(x)$. It does not require that $f(x) > 0$.
2. If the antiderivative $F(x)$ is replaced by $F(x) + C$ for any constant C, the conclusion of the fundamental theorem is the same because C is eliminated in the final answer:

$$\int_a^b f(x)\,dx = (F(b) + C) - (F(a) + C)$$

$$= F(b) - F(a).$$

3. The variable used in the integrand does not matter. Each of the following integrals represents the number $F(b) - F(a)$:

$$\int_a^b f(x)\,dx = \int_a^b f(t)\,dt = \int_a^b f(u)\,du.$$

4. The definition of $\int_a^b f(x)\,dx$ assumed that $a < b$—that is, that the lower limit of integration is the smaller number. When $a > b$, we make this definition:

$$\int_a^b f(x)\,dx = -\int_b^a f(x)\,dx.$$

For example, $\int_3^1 x^2\,dx = -\int_1^3 x^2\,dx$. Similarly, when the limits of integration are the same, we define $\int_a^a f(x)\,dx$ to be 0. The fundamental theorem is valid for such integrals.

The fundamental theorem of calculus certainly deserves its name, since it is the key connection between differential calculus and integral calculus, which were originally developed separately without knowledge of this connection between them. Most importantly for our purposes, it provides an algebraic means to evaluate definite integrals exactly.

APPLYING THE FUNDAMENTAL THEOREM

When evaluating definite integrals, the number $F(b) - F(a)$ is sometimes denoted by the symbol

$$F(x)\Big|_a^b$$

For example, if $F(x) = x^3$, then

$$x^3\Big|_1^2 \quad \text{means } F(2) - F(1) = 2^3 - 1^3 = 7.$$ ✓1

✓Checkpoint 1

Let $C(x) = x^3 + 4x^2 - x + 3$. Find the following.

(a) $C(x)\Big|_1^5$

(b) $C(x)\Big|_3^4$

Answers: See page 871.

Example 1 Evaluate each of the given integrals.

(a) $\int_1^2 4t^3\,dt.$

Solution We know that $F(t) = t^4$ is an antiderivative of $4t^3$. By the fundamental theorem,

$$\int_1^2 4t^3\,dt = t^4\Big|_1^2 = 2^4 - 1^4 = 15.$$

(b) $\int_0^5 e^{2x}\,dx$

Solution By the antiderivative rules in Section 13.1, we know that an antiderivative of e^{2x} is $\frac{1}{2}e^{2x}$. Therefore,

$$\int_0^5 e^{2x}\,dx = \frac{1}{2}e^{2x}\Big|_0^5 = \frac{1}{2}e^{10} - \frac{1}{2}e^0 = \frac{1}{2}e^{10} - \frac{1}{2}.$$ ✓2

✓Checkpoint 2

Evaluate each of the given integrals.

(a) $\int_4^6 5z\,dz$

(b) $\int_2^5 8t^3\,dt$

(c) $\int_1^9 \sqrt{z}\,dz$

Answers: See page 871.

To evaluate more complicated integrals, we need the following properties.

Properties of Definite Integrals

For any real numbers a and b for which the definite integrals exist,

1. $\int_a^b k \cdot f(x)\,dx = k \cdot \int_a^b f(x)\,dx$, for any real constant k (constant multiple of a function);

2. $\int_a^b [f(x) \pm g(x)]\,dx = \int_a^b f(x)\,dx \pm \int_a^b g(x)\,dx$ (sum or difference of functions);

3. $\int_a^b f(x)\,dx = \int_a^c f(x)\,dx + \int_c^b f(x)\,dx$, for any real number c.

Properties 1 and 2 follow directly from the similar properties of antiderivatives. Property 3 won't be proved here, but Figure 13.14 illustrates it when $f(x) > 0$ and $a < c < b$. In this case,

$$\begin{aligned}\int_a^b f(x)\,dx &= \text{area of the entire shaded region}\\ &= \text{area of blue region} + \text{area of red region}\\ &= \int_a^c f(x)\,dx + \int_c^b f(x)\,dx.\end{aligned}$$

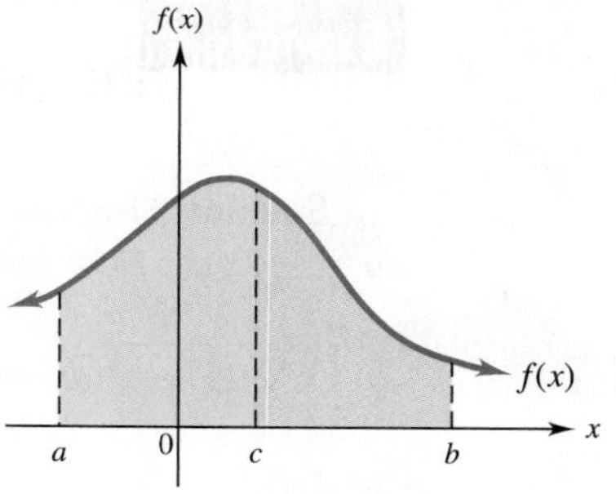

FIGURE 13.14

Example 2 Evaluate $\int_1^7 (4x^2 - 6x + 7)\,dx$.

Solution One way to evaluate this integral is to use property 2 of definite integrals to obtain

$$\int_1^7 (4x^2 - 6x + 7)\,dx = \int_1^7 4x^2\,dx - \int_1^7 6x\,dx + \int_1^7 7\,dx.$$

Next, find an antiderivative for each of the three functions on the right-hand side:

$$4x^2 \text{ has antiderivative } 4\frac{x^3}{3} = \frac{4}{3}x^3;$$

$$6x \text{ has antiderivative } 6\frac{x^2}{2} = 3x^2;$$

$$7 \text{ has antiderivative } 7x.$$

Using the fundamental theorem on the three integrals on the right-hand side above, we see that

$$\int_1^7 (4x^2 - 6x + 7)\,dx = \frac{4}{3}x^3\Big|_1^7 - 3x^2\Big|_1^7 + 7x\Big|_1^7$$

$$= \frac{4}{3}(7^3 - 1^3) - 3(7^2 - 1^2) + 7(7 - 1)$$

$$= \frac{4}{3}(343 - 1) - 3(49 - 1) + 7(6)$$

$$= 456 - 144 + 42 = 354.$$ 3 ✓

✓Checkpoint 3

Evaluate each definite integral.

(a) $\int_1^3 (4x + 3x^2)\,dx$

(b) $\int_{-2}^3 (6k^2 - 2k + 1)\,dk$

Answers: See page 871.

✓Checkpoint 4

Evaluate the given integrals.

(a) $\int_0^4 e^x\,dx$

(b) $\int_3^5 \frac{dx}{x}$

(c) $\int_2^7 \frac{4}{x}\,dx$

Answers: See page 871.

Example 3 Find $\int_1^2 \frac{dy}{y}$.

Solution In Section 13.1, we saw that an antiderivative of $\frac{1}{y}$ is $\ln|y|$. Hence,

$$\int_1^2 \frac{dy}{y} = \ln|y|\Big|_1^2 = \ln|2| - \ln|1| \approx .6931 - 0 = .6931.$$ 4 ✓

Example 4 Evaluate $\int_0^5 x\sqrt{25 - x^2}\, dx$.

Solution Use substitution to find the antiderivative. Let $u = 25 - x^2$, so that $du = -2x\, dx$. Now find the antiderivative and *express it in terms of x*:

$$\int x\sqrt{25 - x^2}\, dx = -\frac{1}{2}\int \sqrt{25 - x^2}(-2x\, dx)$$

$$= -\frac{1}{2}\int \sqrt{u}\, du \qquad \text{Substitute.}$$

$$= -\frac{1}{2}\int u^{1/2}\, du \qquad \text{Fractional exponent}$$

$$= -\frac{1}{2}\cdot\frac{u^{3/2}}{3/2} \qquad \text{Use antiderivative formula.}$$

$$= -\frac{1}{2}\cdot\frac{2}{3}u^{3/2} = -\frac{1}{3}u^{3/2} \qquad \text{Simplify.}$$

$$= -\frac{1}{3}(25 - x^2)^{3/2}. \qquad \text{Express in terms of } x.$$

Now evaluate the definite integral:

$$\int_0^5 x\sqrt{25 - x^2}\, dx = -\frac{1}{3}(25 - x^2)^{3/2}\Big|_0^5$$

$$= -\frac{1}{3}(25 - 5^2)^{3/2} - \left[-\frac{1}{3}(25 - 0^2)^{3/2}\right]$$

$$= 0 + \frac{1}{3}\cdot 25^{3/2} = \frac{125}{3}.$$ ✓5

✓Checkpoint 5

Find $\int_0^2 \frac{x}{x^2 + 1}\, dx$.

Answer: See page 871.

CAUTION When using substitution, be sure to express the antiderivative in terms of x before evaluating it. In Example 4, for instance, if you evaluate the antiderivative $-\frac{1}{3}u^{3/2}$ at 5 and 0, you get a wrong answer.

AREA

When the graph of $f(x)$ lies above the x-axis between a and b, the definite integral $\int_a^b f(x)\, dx$ gives the area between the graph of $f(x)$ and the x-axis from a to b. Now consider a function f whose graph lies below the x-axis from a to b. The shaded area in Figure 13.15 can be approximated by sums of areas of rectangles, that is, by sums of the form

$$f(x_1)\cdot\Delta x + f(x_2)\cdot\Delta x + f(x_3)\cdot\Delta x + \cdots + f(x_n)\cdot\Delta x,$$

with one difference: Since $f(x)$ is negative, the sum represents the negative of the sum of the areas of the rectangles. Consequently, the definite integral $\int_a^b f(x)\, dx$, which is the limit of such sums as n gets larger and larger, is the *negative* of the shaded area in Figure 13.15.

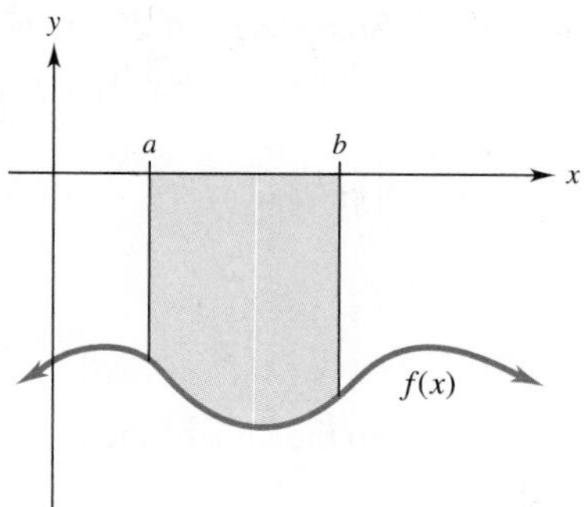

FIGURE 13.15

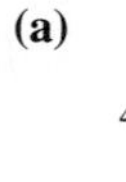

Checkpoint 6

Find each area.

(a)

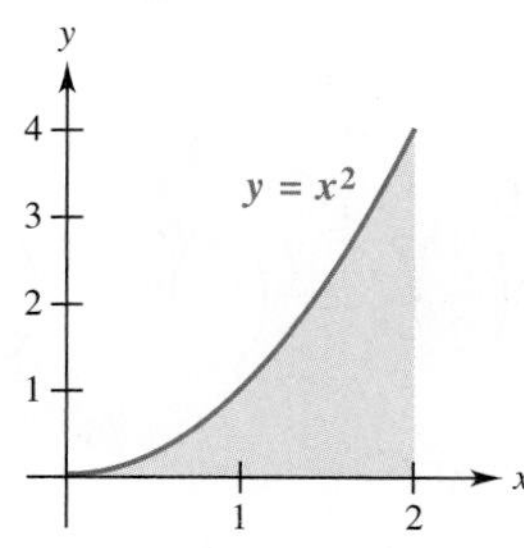

(b)

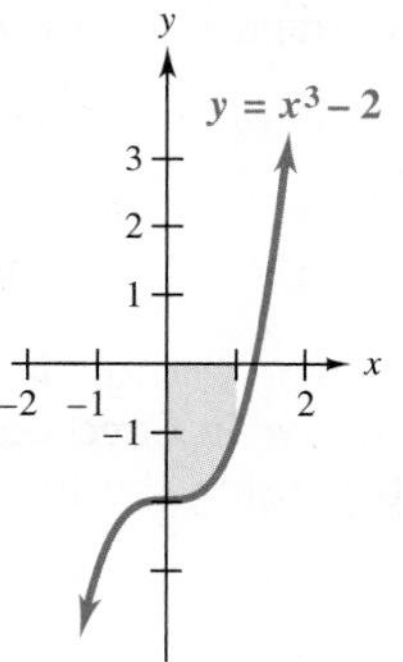

Answers: See page 871.

Example 5 Find the area between the x-axis and the graph

$$f(x) = \frac{1}{2}x^2 - 4x \qquad \text{from} \qquad x = 2 \text{ to } x = 8.$$

Solution That region, which is shaded in Figure 13.16, lies below the x-axis, and the definite integral gives the negative of its area:

$$\int_2^8 \left(\frac{1}{2}x^2 - 4x\right) dx = \left(\frac{x^3}{6} - 2x^2\right)\Bigg|_2^8 = \left(\frac{512}{6} - 128\right) - \left(\frac{8}{6} - 8\right) = -36.$$

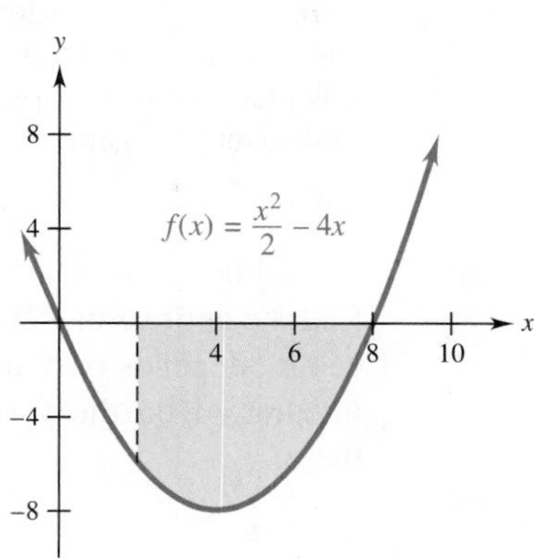

FIGURE 13.16

Therefore, the area of the region is 36.

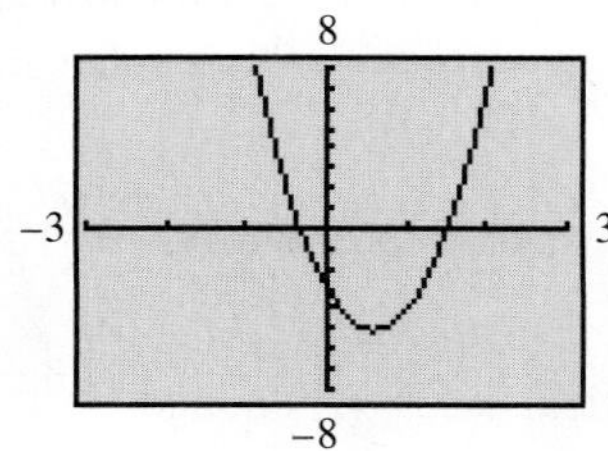

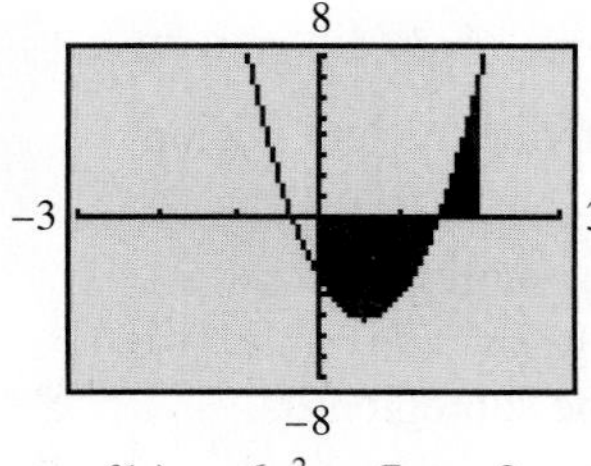

$f(x) = 6x^2 - 7x - 3$

FIGURE 13.17

Example 6 Find the area between the graph of $f(x) = 6x^2 - 7x - 3$ and the x-axis from $x = 0$ to $x = 2$, which is the shaded region in Figure 13.17.

Solution To find the total area, compute the area under the x-axis and the area over the x-axis separately. Start by finding the x-intercepts of the graph by solving

$$6x^2 - 7x - 3 = 0$$
$$(2x - 3)(3x + 1) = 0$$
$$x = 3/2 \qquad \text{or} \qquad x = -1/3.$$

Since we are concerned just with the graph between 0 and 2, the only relevant x-intercept is $\frac{3}{2}$. The area below the x-axis is the negative of

$$\int_0^{3/2} (6x^2 - 7x - 3)\,dx = \left(2x^3 - \frac{7}{2}x^2 - 3x\right)\Bigg|_0^{3/2}$$

$$= \left(2\left(\frac{3}{2}\right)^3 - \frac{7}{2}\left(\frac{3}{2}\right)^2 - 3\left(\frac{3}{2}\right)\right) - (0 - 0 - 0) = -\frac{45}{8}.$$

So the area of the lower region is $\frac{45}{8}$. The area above the x-axis is

$$\int_{3/2}^{2} (6x^2 - 7x - 3)\,dx = \left(2x^3 - \frac{7}{2}x^2 - 3x\right)\Bigg|_{3/2}^{2}$$

$$= \left(2(2^3) - \frac{7}{2}(2^2) - 3(2)\right) - \left(2\left(\frac{3}{2}\right)^3 - \frac{7}{2}\left(\frac{3}{2}\right)^2 - 3\left(\frac{3}{2}\right)\right)$$

$$= \frac{13}{8}.$$

The area of the upper region is $\frac{13}{8}$, and the total area between the graph and the x-axis is

$$\frac{45}{8} + \frac{13}{8} = \frac{58}{8} = 7.25 \text{ square units.}$$

CAUTION If you attempt to find the area in Example 6 by using a single integral over the entire interval [0, 2], you will get a wrong answer, as shown in Checkpoint 7. It is essential to compute the upper and lower areas separately, using the two integrals over $[0, \frac{3}{2}]$ and $[\frac{3}{2}, 2]$. 7 ✓

✓ Checkpoint 7

Evaluate

$$\int_0^2 (6x^2 - 7x - 3)\,dx.$$

Answer: See page 871.

Although it may be helpful, it is not necessary to graph a function in order to find the entire area between its graph and the x-axis. As Example 6 shows, the procedure depends only on finding the x-intercepts of the graph that are the limits of integration on the various integrals. Here is a summary of the procedure used there.

Finding Area

To find the area bounded by $y = f(x)$, the vertical lines $x = a$ and $x = b$, and the x-axis, use the following steps:

Step 1 Sketch a graph if convenient.

Step 2 Find any x-intercepts in $[a, b]$. These divide the total region into subregions.

Step 3 The definite integral will be *positive* for subregions above the x-axis and *negative* for subregions below the x-axis. Use separate integrals to find the areas of the subregions.

Step 4 The total area is the sum of the areas of all of the subregions.

We have seen a variety of interpretations of certain definite integrals in terms of area. We now present an interpretation that applies to all definite integrals. Consider, for example, $\int_0^2 (6x^2 - 7x - 3)\,dx$. By property 3 of definite integrals,

$$\int_0^2 (6x^2 - 7x - 3)\,dx = \int_0^{3/2} (6x^2 - 7x - 3)\,dx + \int_{3/2}^2 (6x^2 - 7x - 3)\,dx.$$

Each of the integrals on the right-hand side was evaluated in Example 6. Using those results, we see that

$$\int_0^2 (6x^2 - 7x - 3)\,dx = -\frac{45}{8} + \frac{13}{8} = \frac{13}{8} - \frac{45}{8}.$$

As Example 6 shows, this last difference is precisely

(area above the x-axis) $-$ (area below the x-axis).

The same result holds in the general case.

If f is a continuous function on $[a, b]$, then

$$\int_a^b f(x)\,dx = \begin{pmatrix}\text{area between the} \\ \text{graph and the } x\text{-axis} \\ \textit{above} \text{ the axis}\end{pmatrix} - \begin{pmatrix}\text{area between the} \\ \text{graph and the } x\text{-axis} \\ \textit{below} \text{ the axis}\end{pmatrix}$$

APPLICATIONS

In the last section, we saw that the area under the graph of a rate-of-change function $f'(x)$ from $x = a$ to $x = b$ gives the total change in $f(x)$ from a to b. We can now use the fundamental theorem to compute this change.

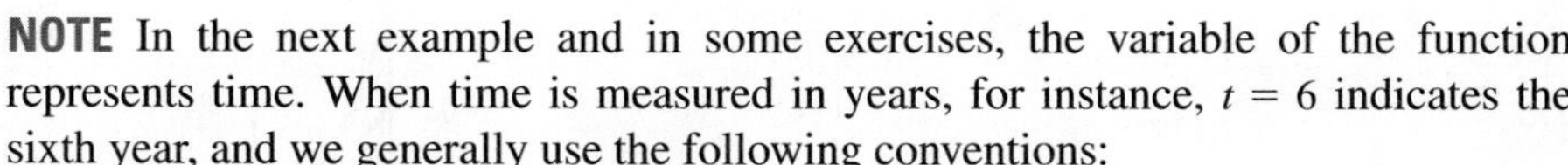

NOTE In the next example and in some exercises, the variable of the function represents time. When time is measured in years, for instance, $t = 6$ indicates the sixth year, and we generally use the following conventions:

$t = 6$ corresponds to January 1;

$t = 6.5$ corresponds to the middle of the year;

$t = 6.75$ corresponds to the end of the third quarter of the year;

and so on. Similar conventions are followed when time is measured in other units (such as minutes, hours, or days).

Example 7 **Business** Data from the U.S. Energy Information Administration was used to generate the following model for the rate of consumption of electricity (in billions of kilowatt hours per year) for the years 1990–2007, where t is measured in years and $t = 0$ corresponds to 1990:

$$C' = 2884.4e^{.0184t}.$$

At this consumption rate, what is the amount of electricity used from January 1, 2003, to January 1, 2007?

Solution The amount used is the total change in consumption from the beginning of year 13 to the beginning of year 17, so it is given by the definite integral

$$\int_{13}^{17} 2884.4e^{.0184t}\,dt = 2884.4\frac{e^{.0184t}}{.0184}\Bigg|_{13}^{17} \approx 156{,}760.9e^{.0184t}\Bigg|_{13}^{17}$$

$$= 156{,}760.9(e^{.0184(17)} - e^{.0184(13)})$$

$$\approx 15{,}208.2.$$

Therefore, approximately 15,208.2 billion kilowatt hours of electricity were used from 2003 to 2007. 8

✓Checkpoint 8

Use the function in Example 7 to find the amount of electricity used from January 1, 1995 to January 1, 2005.

Answer: *See page 871.*

In Chapter 10, you may have studied the normal probability distribution. There are many other distributions that can be used to model probability. One of these is the exponential distribution, which has the general form

$$f(x) = ae^{-ax} \qquad (0 \le x \le \infty),$$

where a is a positive constant that changes with the particular model. With this distribution—and, indeed, all probability distributions—the function is greater than or equal to zero, and the area between the curve and the x-axis equals 1.0. Figure 13.18 shows the graph of $f(x)$ for several values of a.

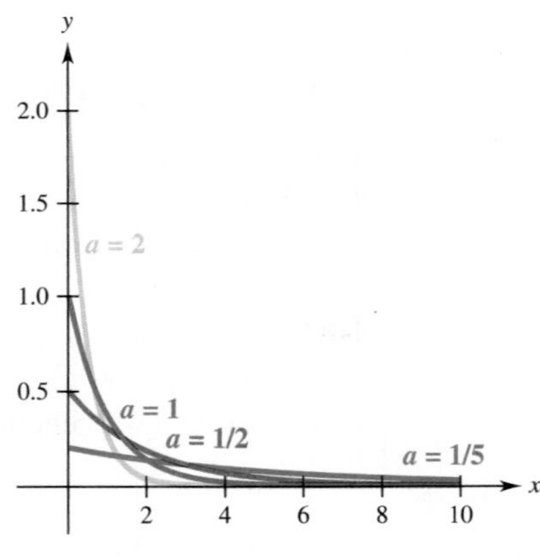

FIGURE 13.18

The next example shows how to use the exponential distribution and the fundamental theorem of calculus to estimate probabilities.

Example 8 **Natural Science** Biology students were interested in analyzing the amount of time hummingbirds spend at feeders before flying away. They found that an exponential model with $a = \frac{1}{4}$ approximated the probability of feeding time very well. Use the model to determine the probability that a hummingbird will feed for less than 5 seconds.

Solution With $a = \frac{1}{4}$, we can calculate the probability of feeding for less than 5 seconds by determining the area beneath the curve

$$f(t) = \frac{1}{4}e^{-x/4}$$

from $t = 0$ to $t = 5$ seconds. Figure 13.19 shows this area shaded in blue.

Thus, we have

$$\int_0^5 \frac{1}{4}e^{-x/4}\,dx = -e^{-x/4}\Big|_0^5 = -e^{-5/4} + e^0 = -.2865 + 1 = .7135.$$

Therefore, the probability that a hummingbird feeds for less than 5 seconds at the feeder is .7135. [9]

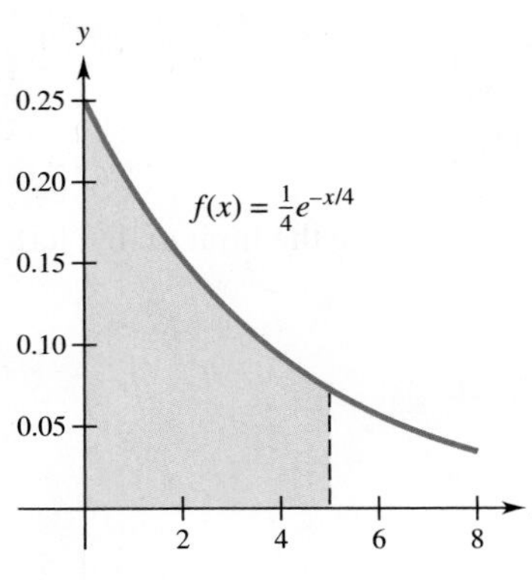

FIGURE 13.19

✓Checkpoint 9

A different biology class developed an exponential model for hummingbird feeding time with $a = .303$. With this model, find the probability that a hummingbird feeds for between 1 and 4 seconds.

Answer: *See page 871.*

PROOF OF THE FUNDAMENTAL THEOREM

Although we cannot give a rigorous proof of the fundamental theorem of calculus here, we can indicate why it is true in an important case. Suppose $f(x) > 0$ when $a \le x \le b$, so that the definite integral represents the area under the curve and above the x-axis.

Define a new function $A(x)$ by this rule:

$A(x)$ = area between the graph of $f(x)$ and the x-axis from a to x,

as shown in Figure 13.20. For instance, if the area under the graph from a to 4 is 35, then $A(4) = 35$.

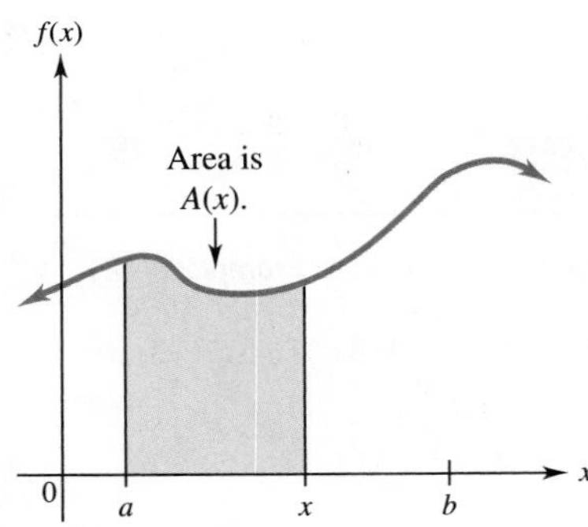

FIGURE 13.20

We now show that $A(x)$ is an antiderivative of $f(x)$. We must show that

$$A'(x) = \lim_{h\to 0} \frac{A(x + h) - A(x)}{h}$$

is actually $f(x)$. Look at the situation geometrically. When h is a small positive number, $A(x + h)$, is the area under the graph of f from a to $x + h$, and $A(x)$ is the area from a to x. Consequently, $A(x + h) - A(x)$ is the area of the shaded strip in Figure 13.21. This area can be approximated with a rectangle having base h and height $f(x)$. The area of the rectangle is $h \cdot f(x)$, and

$$A(x + h) - A(x) \approx h \cdot f(x).$$

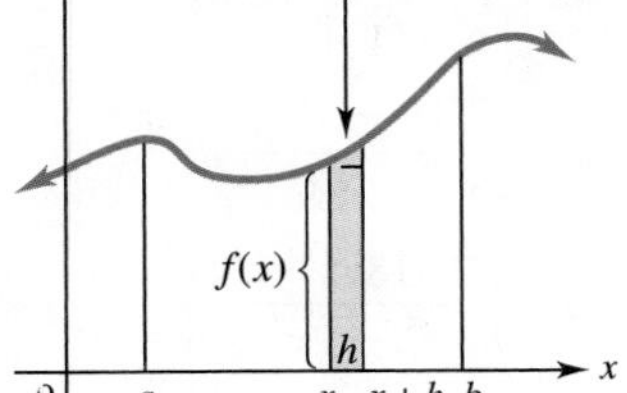

FIGURE 13.21

Dividing both sides by h gives

$$\frac{A(x + h) - A(x)}{h} \approx f(x).$$

This approximation improves as h gets smaller and smaller. Take the limit on the left as h approaches 0:

$$\lim_{h \to 0} \frac{A(x + h) - A(x)}{h} = f(x).$$

The limit is simply $A'(x)$, so

$$A'(x) = f(x).$$

This result means that A is an antiderivative of f, as we set out to show.

Since $A(x)$ is the area under the graph of f from a to x, it follows that $A(a) = 0$ and $A(b)$ is the area under the graph from a to b. But this last area is the definite integral $\int_a^b f(x)\,dx$. Putting these facts together, we have

$$\int_a^b f(x)\,dx = A(b) = A(b) - 0 = A(b) - A(a).$$

This argument suggests that the fundamental theorem is true when $f(x) > 0$ and the area function $A(x)$ is used as the antiderivative. Other arguments, which are omitted here, handle the case when $f(x)$ may not be positive and any antiderivative $F(x)$ is used.

13.4 ▶ Exercises

Evaluate each of the given definite integrals. (See Examples 1–4.)

1. $\int_{-1}^{3} (6x^2 - 7x + 8)\,dx$
2. $\int_{0}^{2} (-3x^2 + 6x + 5)\,dx$
3. $\int_{0}^{2} 3\sqrt{4u + 1}\,du$
4. $\int_{4}^{14} \sqrt{2r + 8}\,dr$
5. $\int_{0}^{1} 3(t^{1/2} + 4t)\,dt$
6. $\int_{0}^{4} -(3x^{3/2} - x^{1/2})\,dx$
7. $\int_{1}^{4} \left(10y\sqrt{y} + 3\sqrt{y}\right) dy$
8. $\int_{4}^{9} \left(4\sqrt{r} - .5r\sqrt{r}\right) dr$
9. $\int_{4}^{7} \frac{11}{(x - 3)^2}\,dx$
10. $\int_{1}^{4} \frac{-3}{(2p + 1)^2}\,dp$
11. $\int_{1}^{5} (6n^{-1} + n^{-4})\,dn$
12. $\int_{2}^{3} (3x^{-1} - 5x^{-4})\,dx$
13. $\int_{2}^{3} \left(.1e^{-.1A} + \frac{3}{A}\right) dA$
14. $\int_{1}^{2} \left(\frac{-1}{B} + 3e^{.2B}\right) dB$
15. $\int_{1}^{2} \left(e^{6u} - \frac{1}{u^2}\right) du$
16. $\int_{.5}^{1} (p^3 - e^{4p})\,dp$
17. $\int_{-1}^{0} y(2y^2 - 3)^5\,dy$
18. $\int_{0}^{2} m^2(5m^3 - 2)^3\,dm$
19. $\int_{1}^{64} \frac{\sqrt{z} - 2}{\sqrt[3]{z}}\,dz$
20. $\int_{1}^{8} \frac{9 - y^{1/3}}{y^{2/3}}\,dy$
21. $\int_{1}^{3} \frac{\ln x}{4x}\,dx$
22. $\int_{1}^{9} \frac{\sqrt{\ln x}}{x}\,dx$
23. $\int_{0}^{8} x^{1/3}\sqrt{x^{4/3} + 9}\,dx$
24. $\int_{1}^{3} \frac{4}{x(1 + \ln x)}\,dx$
25. $\int_{0}^{1} \frac{4e^t}{(3 + e^t)^2}\,dt$
26. $\int_{0}^{1} \frac{e^{2z}}{\sqrt{1 + e^{2z}}}\,dz$
27. $\int_{1}^{49} \frac{\left(1 + \sqrt{x}\right)^{4/3}}{\sqrt{x}}\,dx$
28. $\int_{2}^{8} \frac{(1 + x^{1/3})^6}{x^{2/3}}\,dx$
29. $\int_{2}^{3} \frac{4x^3 + 6}{x^4 + 6x + 9}\,dx$
30. $\int_{0}^{2} \frac{18x^2 - 24x + 6}{x^3 - 2x^2 + x + 12}\,dx$
31. Suppose the function in Example 6, $f(x) = 6x^2 - 7x - 3$ from $x = 0$ to $x = 2$, represented the annual rate of profit

of a company over a two-year period. What might the negative integral for the first year and a half indicate? What integral would represent the overall profit for the two-year period?

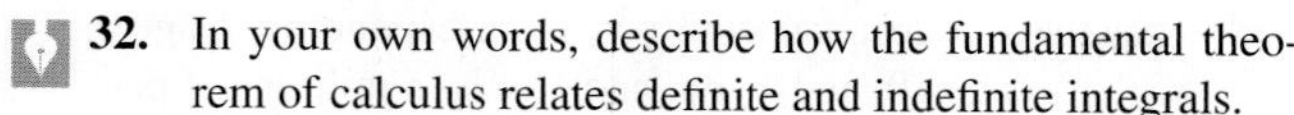

32. In your own words, describe how the fundamental theorem of calculus relates definite and indefinite integrals.

Use the definite integral to find the area between the x-axis and f(x) over the indicated interval. Check first to see if the graph crosses the x-axis in the given interval. (See Examples 5 and 6.)

33. $f(x) = 9 - x^2$; $[0, 4]$ **34.** $f(x) = x^2 - 3x - 4$; $[0, 5]$

35. $f(x) = x^3 - 1$; $[-1, 2]$ **36.** $f(x) = x^3 - 4x$; $[-2, 4]$

37. $f(x) = e^{2x} - 1$; $[-2, 1]$ **38.** $f(x) = 1 - e^{-3x}$; $[-1, 2]$

39. $f(x) = x^2e^{-x^3/2}$; $[0, 3]$ **40.** $f(x) = 5xe^{x^2}$; $[0, 1]$

41. $f(x) = \dfrac{1}{x}$; $[1, e]$ **42.** $f(x) = \dfrac{2}{x}$; $[e, e^2]$

43. $f(x) = \dfrac{12(\ln x)^3}{x}$; $[1, 4]$

44. $f(x) = \dfrac{-(6x - 4)}{(3x^2 - 4x + 4)}$; $[-1, .5]$

Find the area of each shaded region.

45.

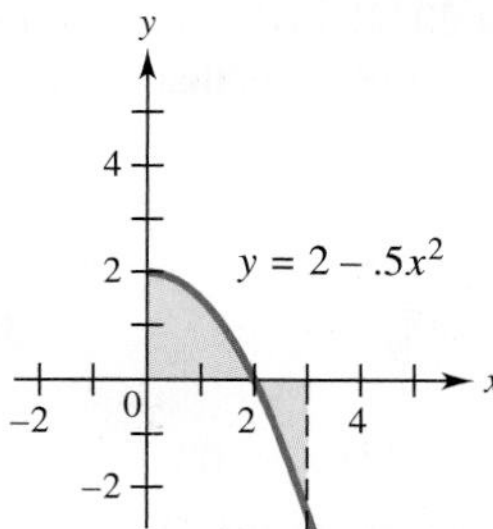

46.

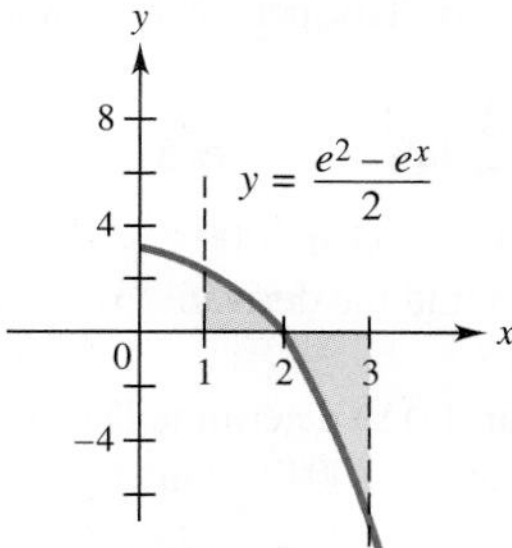

47.

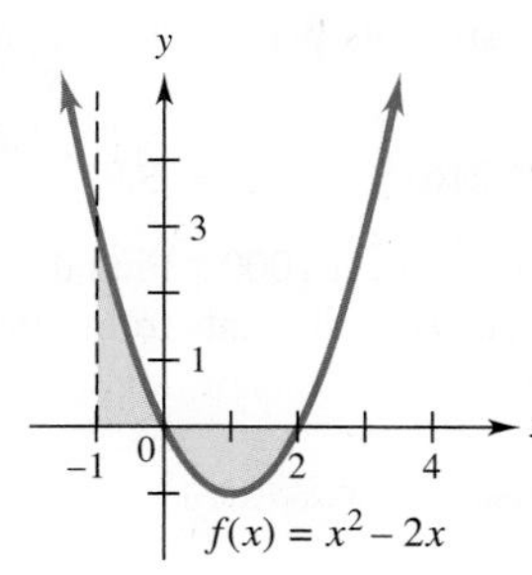

48.

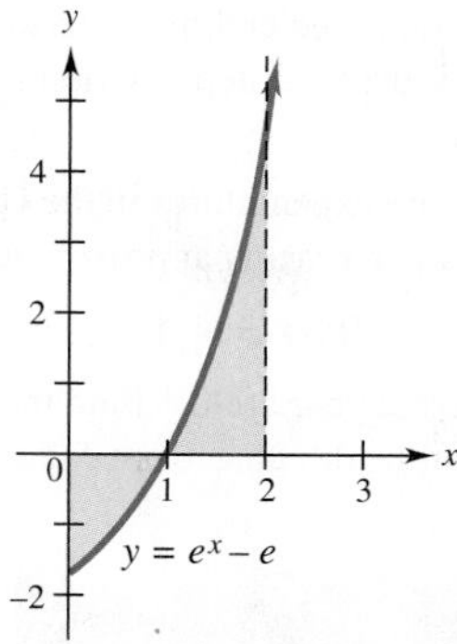

Use property 3 of definite integrals to find $\int_1^4 f(x)\,dx$ for the given functions.

49. $f(x) = \begin{cases} 2x + 3 & \text{if } x \le 2 \\ -.5x + 8 & \text{if } x > 2 \end{cases}$

50. $f(x) = \begin{cases} x^2 - 2 & \text{if } x \le 3 \\ -x^2 + 16 & \text{if } x > 3 \end{cases}$

Work these problems. (See Example 7.)

51. Business De Win Enterprises has found that its expenditure rate per day (in hundreds of dollars) on a certain type of job is given by

$$E(x) = 4x + 2,$$

where x is the number of days since the start of the job.

(a) Find the total expenditure if the job takes 10 days.

(b) How much will be spent on the job from the 10th to the 25th day?

(c) If the company wants to spend no more than $76,000 on the job, in how many days must it complete the job?

52. Business A worker new to a job will improve his efficiency with time so that it takes him fewer hours to produce an item with each day on the job, up to a certain point. Suppose the rate of change of the number of hours it takes a worker in a certain factory to produce the xth item is given by

$$H'(x) = 20 - 2x.$$

(a) What is the total number of hours required to produce the first 5 items?

(b) What is the total number of hours required to produce the first 10 items?

53. Business The rate of depreciation (in dollars per year) for a certain truck is

$$f(t) = \frac{6000(.3 + .28t)}{(1 + .3t + .14t^2)^2},$$

where t is in years and $t = 0$ is the year of purchase.

*Based on data and projections from the U.S. Centers for Medicare and Medicaid Services.

(a) Find the total depreciation at the end of 3 years.
(b) How long will it take for the total depreciation to be at least $3000?

54. **Health** The rate of health care expenditures in the United States (in trillions of dollars per year) is approximated by

$$f(x) = .0047x^2 + .087x + 1.4,$$

where $x = 0$ corresponds to the year 2000.* Find the total amount that will be spent on health care from January 1, 2009 to January 1, 2014.

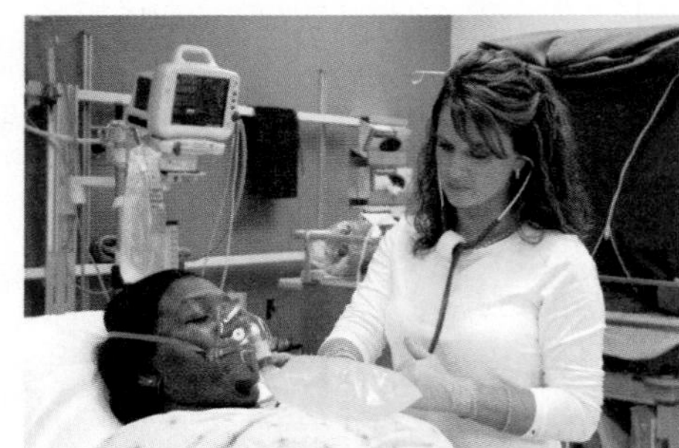

55. **Business** World energy consumption (in quadrillion BTUs) in year x is projected to be given by

$$g(x) = -.0028x^3 + .168x^2 + 7.8x + 404 \quad (0 \le x \le 20),$$

where $x = 0$ corresponds to 2000.* Find the projected total consumption from 2004 to 2020.

56. **Business** Nuclear power consumption in the United States (in quadrillion BTUs per year) over the next two decades is projected to be given by

$$f(x) = -.0000295x^4 + .0017x^3 - .035x^2 + .11x + 7.5,$$

where $x = 0$ corresponds to 2000.* How much nuclear power will be consumed from 2000 to 2010?

57. **Natural Science** An oil tanker is leaking oil at a rate given in barrels per hour by

$$L'(t) = \frac{70\ln(t + 1)}{t + 1},$$

where t is the time in hours after the tanker hits a hidden rock (when $t = 0$).
(a) Find the total number of barrels that the ship will leak on the first day.
(b) Find the total number of barrels that the ship will leak on the second day.
(c) What is happening over the long run to the amount of oil leaked per day?

58. **Health** A colony of *H. pylori* bacteria grows at the rate

$$w' = \sqrt[3]{2t + 5},$$

where w is the weight (in milligrams) after t hours. Find the change in weight of the colony from $t = 0$ to $t = 4$ hours.

*U.S. Energy Information Administration.

59. **Social Science** The 2000 U.S. Census gives an age distribution approximated by

$$f(x) = -.77x^2 + 2.5x + 40 \quad (0 \le x \le 9),$$

where x is in decades, with $x = 0$ corresponding to age 0, $x = 1$ to age 10, and so on, and $f(x)$ is in millions of people.* The population of an age group can be found by integrating this function over the interval for that age group.

(a) Find $\int_0^9 f(x)\,dx$. What does this integral represent?
(b) Baby boomers are those people born between 1945 and 1965—that is, those born in the range of 3.5 to 5.5 decades. Find the number of baby boomers.

60. **Business** Suppose that the rate of consumption of a natural resource is

$$c'(t) = ke^{rt},$$

where t is time in years, r is a constant, and k is the consumption in the year when $t = 0$. Assume that $r = .04$.
(a) In 2010, an oil company sells 1.2 billion barrels of oil. Set up a definite integral for the amount of oil that the company will sell in the 10 subsequent years.
(b) Evaluate the definite integral of part (a).
(c) The company has about 20 billion barrels of oil in reserve. To find the number of years that this amount will last, solve the equation

$$\int_0^T 1.2e^{.04t}\,dt = 20$$

for T.
(d) Rework part (c), assuming that $r = .02$.

61. **Business** The inflow of revenue to the Dell Computer Corporation (in billions of dollars per year) is approximated by

$$f(x) = 3.2 + 28\ln x \quad (x \ge 3),$$

where $x = 0$ corresponds to the year 2000.†
(a) Use the product rule to find the derivative of $y = x(\ln|x| - 1)$.
(b) Use your answer to part (a) to determine Dell's total net revenue from January 1, 2004 to January 1, 2009.

62. **Social Science** The rate at which students are receiving Pell grants (in millions of students per year) is approximately given by

$$N(x) = .015x^3 - .216x^2 + .97x + 3.9,$$

where $x = 0$ corresponds to the year 2000.‡ Find the total number of students who received Pell grants from January 1, 2005 to January 1, 2009.

*Exercise suggested by Ralph DeMarr, University of New Mexico.
†Based on data from www.dell.com.
‡Based on data from the U.S. Department of Education.

Business *A model for the amount of time it takes for the first customer to arrive at a coffee shop in the morning is modeled by the exponential distribution and the function*

$$f(x) = \frac{1}{3}e^{-x/3}.$$

(See Example 8.)

63. Find the probability that the first customer arrives in the first 5 minutes.

64. Find the probability that the first customer arrives between 1 and 3 minutes after the shop opens.

Business *A company tests batteries for cordless drills. It finds that after the first 3 hours, the life of the battery can be modeled by an exponential distribution and the function*

$$f(x) = \frac{1}{8}e^{-x/8}.$$

65. Find the probability that the battery will last between 5 and 8 hours. (*Hint*: Be sure to take into account the first three hours.)

66. Find the probability that the battery lasts more than 7 hours.

✓ Checkpoint Answers

1. **(a)** 216 **(b)** 64
2. **(a)** 50 **(b)** 1218 **(c)** 52/3
3. **(a)** 42 **(b)** 70
4. **(a)** 53.59815 **(b)** .51083 **(c)** 5.01105
5. $\frac{1}{2}\ln 5$
6. **(a)** 8/3 **(b)** 7/4
7. -4
8. Approximately 34,719.9 billion kilowatt hours
9. .4410

13.5 ▶ Applications of Integrals

If a function gives the rate of change of a quantity, then the area between the graph of the rate function and the x-axis is the total amount of the quantity. This fact has numerous applications, some of which are explored here.

Example 1 **Business** When you buy a used car, you may be offered the opportunity to buy a warranty that covers repair costs for a certain period. A company that provides such warranties uses current data and past experience to determine that the annual rate of repair costs for a particular model is given in year x by

$$r(x) = 120\,e^{.4x}.$$

What will the total repair costs be for one year and for three years?

Solution The total repair costs are represented by the area under the rate curve over the appropriate interval. For the first three years, the area is shown in Figure 13.22 and is given by the integral

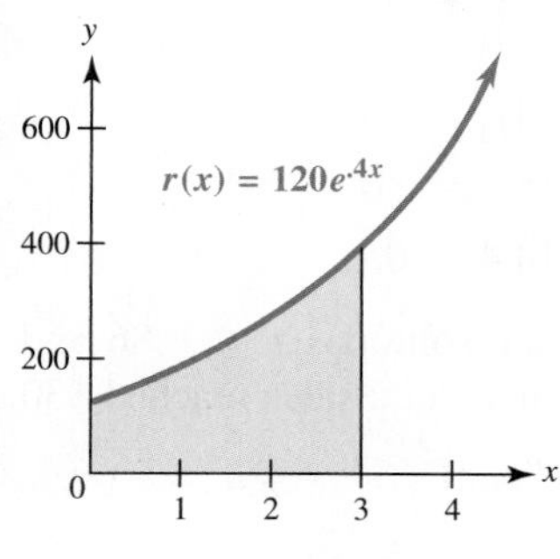

FIGURE 13.22

$$\int_0^3 120\,e^{.4x}\,dx = \frac{120}{.4}e^{.4x}\Big|_0^3 = 300\,e^{.4x}\Big|_0^3$$
$$= 300\,e^{.4(3)} - 300\,e^{.4(0)}$$
$$= 300\,e^{1.2} - 300 \approx \$696.04.$$

Similarly, the total repair costs for one year are

$$\int_0^1 120\,e^{.4x}\,dx = 300\,e^{.4x}\Big|_0^1 = 300\,e^{.4} - 300 \approx \$147.55.$$

In order to make a profit, the company must charge more than \$147.55 for a one-year warranty and more than \$696.04 for a three-year warranty. 1✓

✓ Checkpoint 1

In Example 1, find the total repair costs for

(a) two years;

(b) four years.

Answers: See page 883.

Example 2 **Business** According to data from Apple Computer, Inc., Apple's rate of net sales (in billions of dollars per year) can be approximated by

$$f(x) = 5.3x - 14.5 \qquad (x \geq 4),$$

where $x = 4$ corresponds to the year 2004.

(a) What were the total sales from the beginning of 2004 to the end of 2008?

Solution The total sales are the area under the graph of $f(x)$ above the x-axis from $x = 4$ to $x = 9$, as shown in Figure 13.23. This area is given by

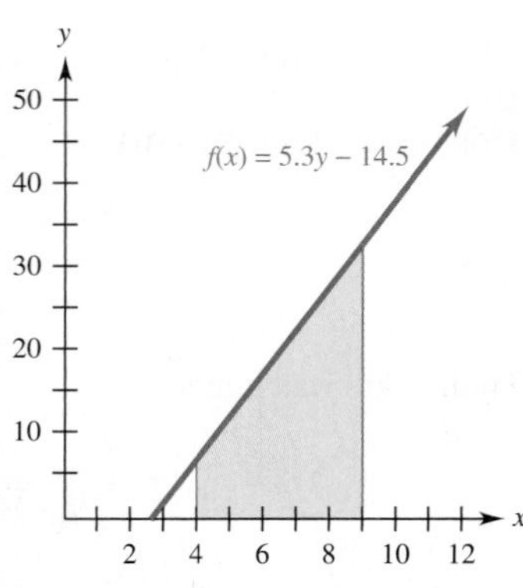

FIGURE 13.23

$$\int_4^9 (5.3x - 14.5)\,dx = \frac{5.3}{2}x^2 - 14.5x\Big|_4^9$$
$$= 2.65x^2 - 14.5x\Big|_4^9$$
$$= [2.65(9^2) - 14.5(9)] - [2.65(4^2) - 14.5(4)]$$
$$= 84.15 - (-15.6) = 84.15 + 15.6 = 99.75.$$

So the total sales in this period were about \$99.75 billion.

(b) In what year (starting in 2004) did total sales reach \$30 billion?

Solution The total sales from $x = 4$ to $x = t$ are given by $\int_4^t (5.3x - 14.5)\,dx$. So we must find t such that

$$\int_4^t (5.3x - 14.5)\,dx = 30.$$
$$\frac{5.3}{2}x^2 - 14.5x\Big|_4^t = 30$$
$$2.65x^2 - 14.5x\Big|_4^t = 30$$
$$[2.65t^2 - 14.5t] - [2.65(4^2) - 14.5(4)] = 30$$
$$2.65t^2 - 14.5t + 15.6 = 30$$
$$2.65t^2 - 14.5t - 14.4 = 0.$$

Using the quadratic formula, we obtain the approximate solutions $t \approx -.86$ and $t \approx 6.33$. Only the positive one is meaningful here. Hence, total sales reached \$30 billion when $t \approx 6.33$, that is, in early May of 2006. ✓2

✓Checkpoint 2

In Example 2, find

(a) the total sales from the beginning of 2005 to the end of 2007;

(b) the time when total sales (starting in 2004) reached \$50 billion.

Answers: See page 883.

AREA BETWEEN TWO CURVES

In some applications, it is necessary to find the area between two curves. For example, the area between the graphs of $f(x)$ and $g(x)$ from $x = a$ to $x = b$ in Figure 13.24 is shaded. This area is the area under the graph of $f(x)$ *minus* the area under the graph of $g(x)$, that is,

$$\int_a^b f(x)\,dx - \int_a^b g(x)\,dx,$$

which can be written as

$$\int_a^b [f(x) - g(x)]\, dx.$$

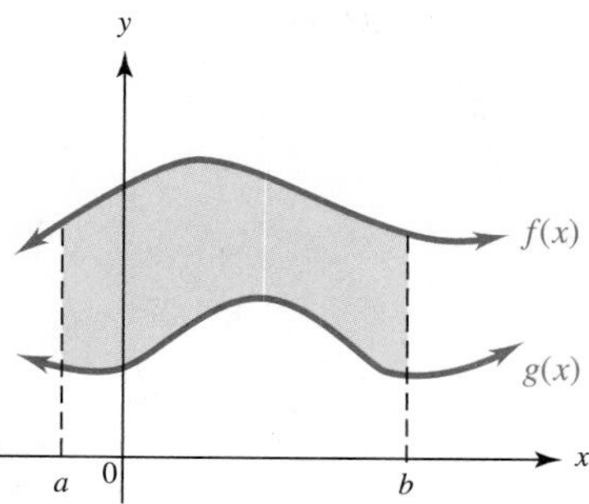

FIGURE 13.24

Similar arguments in other cases (even when the graphs are not above the x-axis) produce the following important result.

Area between Two Curves

If f and g are continuous functions and $f(x) \geq g(x)$ on the interval $[a, b]$, then the area between the graphs of $f(x)$ and $g(x)$ from $x = a$ to $x = b$ is given by

$$\int_a^b [f(x) - g(x)]\, dx.$$

Example 3 **Business** Suppose the revenue of several major investor-owned electric utilities (in billions of dollars) during the period 2003–2008 is approximated by $f(x) = 5.18x + 159$, where $x = 3$ corresponds to 2003. Their costs during the same period are approximated by $g(x) = -.16x^3 + 3.36x^2 - 16.62x + 170$. Find the profit earned from 2003 to 2008.

Solution The cost and revenue functions are graphed in Figure 13.25. Total revenue during this period is the area under the revenue curve, and total cost is the area under the cost curve. So the profit is the area between the two curves from $x = 3$ to $x = 8$, as shown in Figure 13.26.

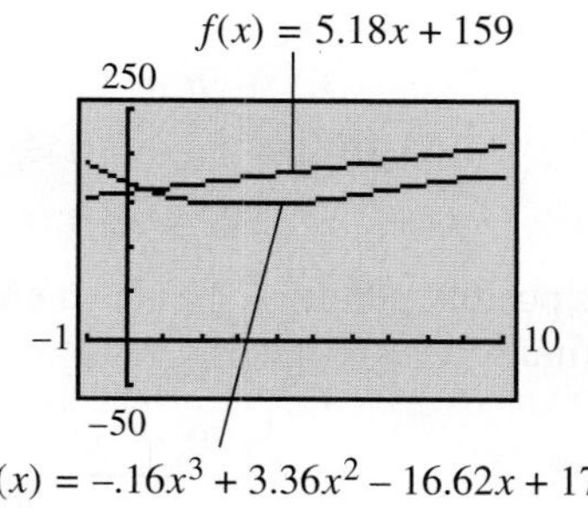

FIGURE 13.25

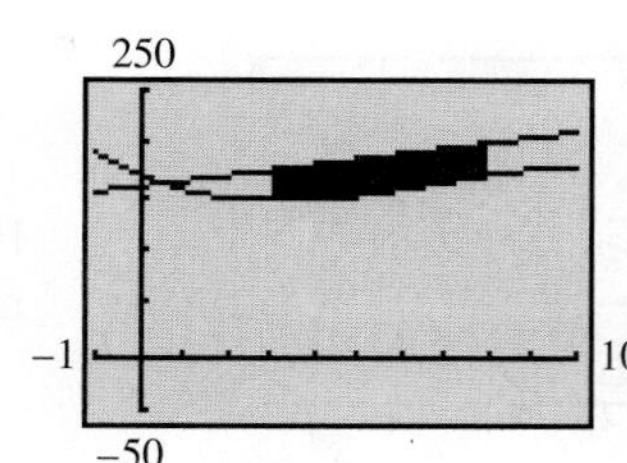

FIGURE 13.26

This area is

$$\int_3^8 [f(x) - g(x)]\,dx = \int_3^8 [(5.18x + 159) - (-.16x^3 + 3.36x^2 - 16.62x + 170)]\,dx$$

$$= \int_3^8 (.16x^3 - 3.36x^2 + 21.8x - 11)\,dx$$

$$= .16\frac{x^4}{4} - 3.36\frac{x^3}{3} + 21.8\frac{x^2}{2} - 11x\Big|_3^8$$

$$= .04x^4 - 1.12x^3 + 10.9x^2 - 11x\Big|_3^8$$

$$= [.04(8^4) - 1.12(8^3) + 10.9(8^2) - 11(8)] - [.04(3^4) - 1.12(3^3) + 10.9(3^2) - 11(3)]$$

$$= 161.9.$$

The profit was \$161,900,000,000. 3✓

✓Checkpoint 3

Find the area between the graphs of $f(x) = 8 - x^2$ and $g(x) = -x + 1$ from $x = -1$ to $x = 2$.

Answer: See page 883.

If revenue, cost, and profit are denoted by $R(x)$, $C(x)$, and $P(x)$, respectively, then

$$P(x) = R(x) - C(x),$$

so that

$$P'(x) = R'(x) - C'(x).$$

As we saw in Chapter 12, maximal profit occurs at a value of x for which $P'(x) = 0$. The previous equation shows that $P'(x) = 0$ exactly when $R'(x) = C'(x)$. In other words, maximal profit occurs when marginal revenue and marginal cost are equal.

Example 4 **Business** A manufacturer produces a fabric similar to nylon. The product has been selling well, with a marginal revenue (in hundreds of dollars) given by

$$R'(t) = -.3t^2 + 9t + 11,$$

where t is measured in hundreds of bolts of fabric. The marginal cost (also in hundreds of dollars) is given by

$$C'(t) = 2t + 6.$$

(a) To maximize profit, how many bolts should the company produce?

Solution The manufacturer should continue production until the marginal costs equal the marginal revenue. Find this point by solving the equation $R'(t) = C'(t)$ as follows:

$$-.3t^2 + 9t + 11 = 2t + 6$$

$$-3t^2 + 90t + 110 = 20t + 60 \quad \text{Multiply by 10.}$$

$$-3t^2 + 70t + 50 = 0.$$

Find the positive solution by using a graphing calculator (Figure 13.27) or the quadratic formula:

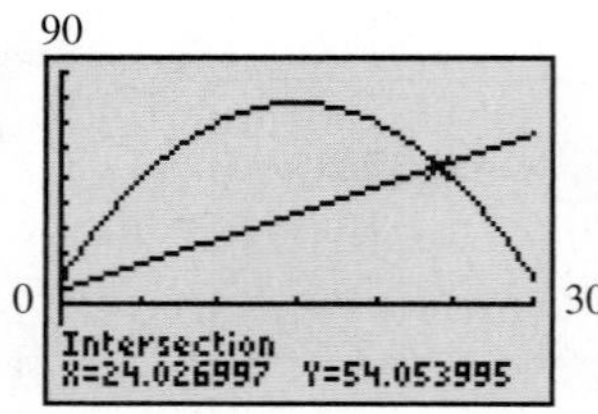

FIGURE 13.27

$$t = \frac{-70 - \sqrt{(70)^2 - 4(-3)(50)}}{2(-3)} \approx 24.0.$$

To maximize profit, the company should produce 2400 bolts of the fabric.

(b) What will the profit be on 2400 bolts?

Solution To find the profit on 2400 bolts, find the area between the graphs of the marginal revenue and marginal cost functions, shown in Figure 13.28.

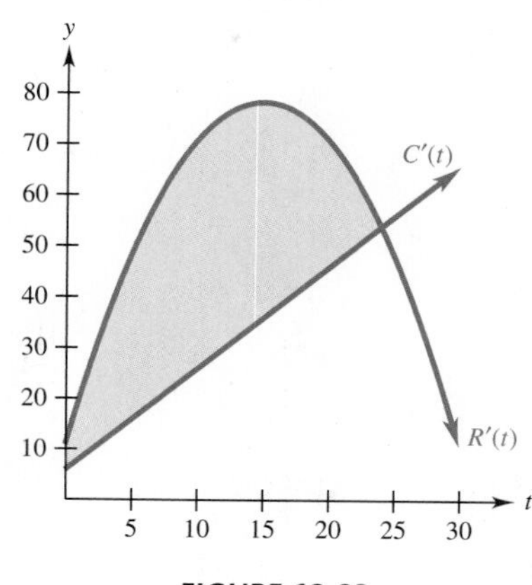

FIGURE 13.28

The area (profit) is

$$\begin{aligned}\text{Profit} &= \int_0^{24} [R'(t) - C'(t)]\,dt \\ &= \int_0^{24} [(-.3t^2 + 9t + 11) - (2t + 6)]\,dt \\ &= \int_0^{24} (-.3t^2 + 7t + 5)\,dt && \text{Combine terms.} \\ &= \left(\frac{-.3t^3}{3} + \frac{7t^2}{2} + 5t\right)\Bigg|_0^{24} && \text{Integrate.} \\ &= 753.6.\end{aligned}$$

The profit on 2400 bolts will be approximately \$75,360.

Example 5 **Business** A company is considering a new manufacturing process in one of its plants. The new process provides substantial initial savings, with the savings declining with time x according to the rate-of-savings function

$$S'(x) = 100 - x^2,$$

where x is measured in years and $S'(x)$ in thousands of dollars per year. At the same time, the cost of operating the new process increases with time x according to the rate-of-cost function (in thousands of dollars per year)

$$C'(x) = x^2 + \frac{14}{3}x.$$

(a) To maximize its savings, for how many years should the company use the new process?

Solution Figure 13.29 shows the graphs of the rate-of-savings and the rate-of-cost functions. As was the case in Example 4, maximum savings will occur when the rate of savings equals the rate of cost, that is, until the time at which these graphs intersect.

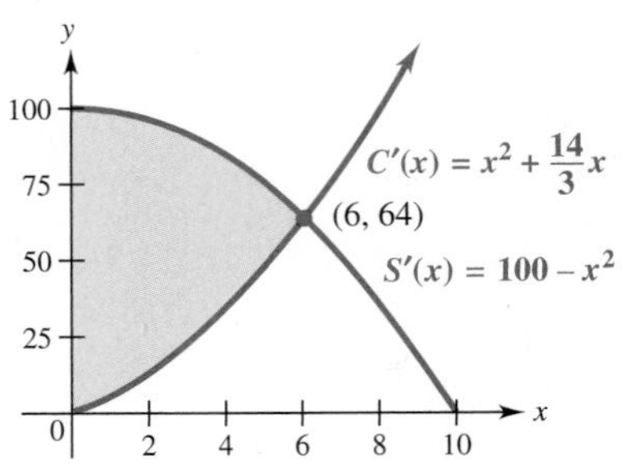

FIGURE 13.29

The graphs intersect when

$$S'(x) = C'(x),$$

or

$$100 - x^2 = x^2 + \frac{14}{3}x.$$

Solve this equation as follows:

$$0 = 2x^2 + \frac{14}{3}x - 100$$

$$0 = 3x^2 + 7x - 150 \qquad \text{Multiply by } \tfrac{3}{2}.$$

$$0 = (x - 6)(3x + 25) \qquad \text{Factor.}$$

Set each factor equal to 0 and solve both equations to get

$$x = 6 \qquad \text{or} \qquad x = -\frac{25}{3}.$$

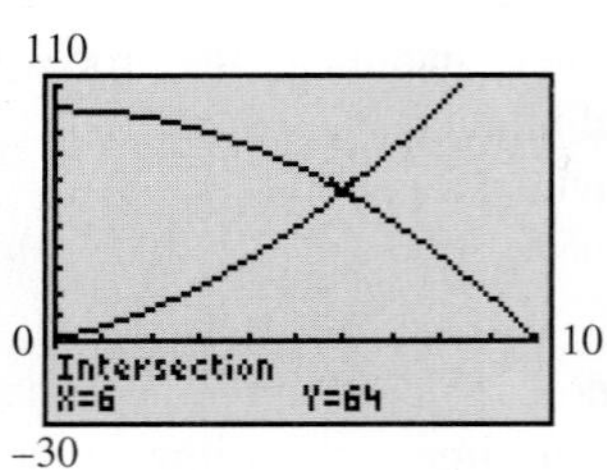

FIGURE 13.30

Only 6 is a meaningful solution here. Alternatively, the intersection point can be found by using the intersection finder on a calculator, as in Figure 13.30. The company should use the new process for 6 years.

(b) Taking the additional costs into account, how much will the company actually save during the 6-year period?

Solution The actual savings (or net savings) are the difference between the savings and the additional costs. Total savings are given by the area under the rate-of-savings curve, and total additional costs are given by the area under the rate-of-cost curve, from $x = 0$ to $x = 6$. So total net savings are given by the difference between these two areas—that is, by the shaded area in Figure 13.29. This area is given by the integral

$$\text{Net savings} = \int_0^6 [S'(x) - C'(x)]\,dx$$

$$= \int_0^6 \left[(100 - x^2) - \left(x^2 + \frac{14}{3}x\right)\right] dx$$

$$= \int_0^6 \left(100 - \frac{14}{3}x - 2x^2\right) dx \qquad \text{Combine terms.}$$

$$= 100x - \frac{7}{3}x^2 - \frac{2}{3}x^3 \Big|_0^6 \qquad \text{Integrate.}$$

$$= 100(6) - \frac{7}{3}(36) - \frac{2}{3}(216) = 372.$$

The company will save a total of $372,000 over the 6-year period. ✓4 ✓5

✓Checkpoint 4

In Example 5, find the total net savings if pollution control regulations permit the new process for only 4 years.

Answer: See page 883.

✓Checkpoint 5

If a company's rate-of-savings and rate-of-cost functions are, respectively,

$$S'(x) = 150 - 2x^2$$

and

$$C'(x) = x^2 + 15x,$$

where $S'(x)$ and $C'(x)$ give amounts in thousands of dollars per year, find the given quantities.

(a) Number of years to maximize savings

(b) Net savings during the period in part (a)

Answers: See page 883.

CONSUMERS' AND PRODUCERS' SURPLUS

The market determines the price at which a product is sold. As indicated earlier, the point of intersection of the demand curve and the supply curve for a product gives the equilibrium price. At the equilibrium price, consumers will purchase the same amount of the product that the manufacturers want to sell. Some consumers, however, will be willing to spend more for an item than the equilibrium price. The total of the differences between the equilibrium price of the item and the higher prices all those individuals would be willing to pay is called the **consumers' surplus.**

In Figure 13.31, the colored area under the demand curve is the total amount consumers are willing to spend for q_0 items. The red area under the line $y = p_0$

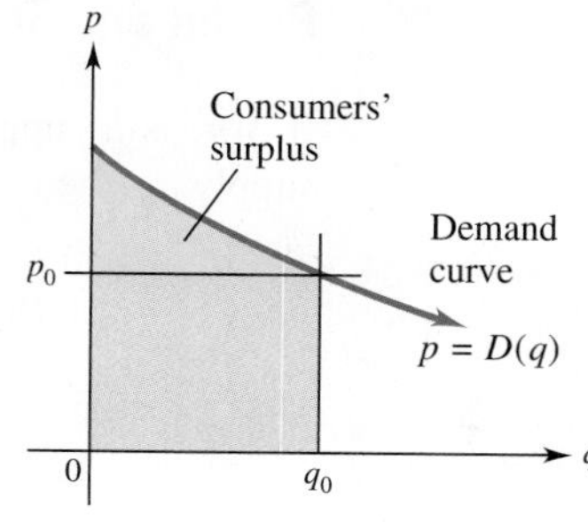

FIGURE 13.31

shows the total amount consumers actually will spend at the equilibrium price of p_0. The blue area represents the consumers' surplus. As the figure suggests, the consumers' surplus is given by the area between the two curves $p = D(q)$ and $p = p_0$, so its value can be found with a definite integral as follows.

Consumers' Surplus

If $D(q)$ is a demand function with equilibrium price p_0 and equilibrium demand q_0, then

$$\textbf{Consumers' surplus} = \int_0^{q_0} [D(q) - p_0]\, dq.$$

Similarly, if some manufacturers would be willing to supply a product at a price *lower* than the equilibrium price p_0, the total of the differences between the equilibrium price and the lower prices at which the manufacturers would sell the product is called the **producers' surplus**. If Figure 13.32 shows the (red) total area under the supply curve from $q = 0$ to $q = q_0$, which is the minimum total amount the manufacturers are willing to realize from the sale of q_0 items. The total area under the line $p = p_0$ is the amount actually realized. The difference between these two areas, the producers' surplus (blue), is also given by a definite integral.

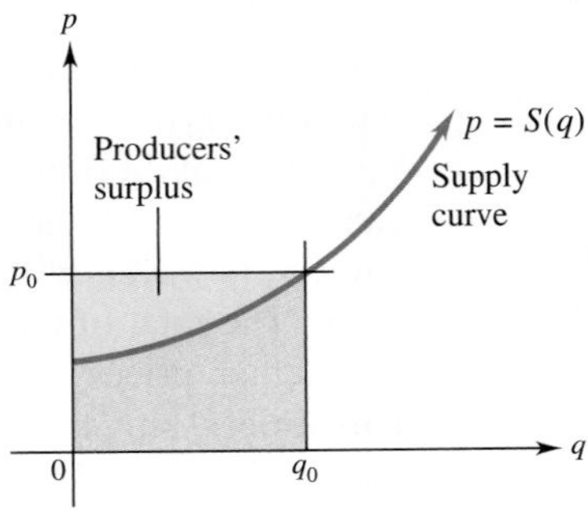

FIGURE 13.32

Producers' Surplus

If $S(q)$ is a supply function with equilibrium price p_0 and equilibrium supply q_0, then

$$\textbf{Producers' surplus} = \int_0^{q_0} [p_0 - S(q)]\, dq.$$

Example 6 **Business** Suppose the price (in dollars per ton) for oat bran is

$$D(q) = 900 - 20q - q^2$$

when the demand for the product is q tons. Also, suppose the function

$$S(q) = q^2 + 10q$$

gives the price (in dollars per ton) when the supply is q tons. Find the consumers' surplus and the producers' surplus.

Solution We begin by finding the equilibrium quantity by setting the two equations equal:

$$900 - 20q - q^2 = q^2 + 10q$$
$$0 = 2q^2 + 30q - 900$$
$$0 = q^2 + 15q - 450.$$

Use the quadratic formula or factor to see that the only positive solution of this equation is $q = 15$. At the equilibrium point where the supply and demand are both 15 tons, the price is

$$S(15) = 15^2 + 10(15) = 375,$$

or \$375. Verify that the same answer is found by computing $D(15)$. The consumers' surplus, represented by the blue area shown in Figure 13.33, is

$$\int_0^{15} [(900 - 20q - q^2) - 375]\,dq = \int_0^{15} (525 - 20q - q^2)\,dq.$$

Evaluating this definite integral gives

$$\left(525q - 10q^2 - \frac{1}{3}q^3\right)\Bigg|_0^{15} = \left[525(15) - 10(15)^2 - \frac{1}{3}(15)^3\right] - 0$$
$$= 4500.$$

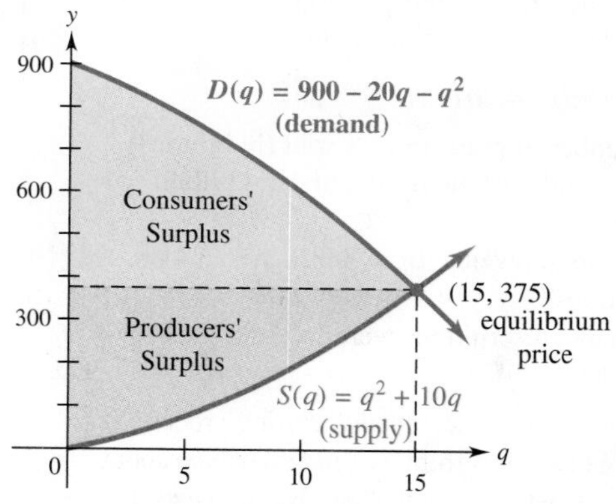

FIGURE 13.33

Here, the consumers' surplus is \$4500. The producers' surplus (the red area in Figure 13.33) is given by

$$\int_0^{15} [375 - (q^2 + 10q)]\,dq = \int_0^{15} (375 - q^2 - 10q)\,dq$$
$$= 375q - \frac{1}{3}q^3 - 5q^2\Bigg|_0^{15}$$
$$= \left[375(15) - \frac{1}{3}(15)^3 - 5(15)^2\right] - 0$$
$$= 3375.$$

The producers' surplus is \$3375. 6✓

✓Checkpoint 6

Given the demand function $D(q) = 12 - .07q$ and the supply function $S(q) = .05q$, where $D(q)$ and $S(q)$ are in dollars, find

(a) the equilibrium point;

(b) the consumers' surplus;

(c) the producers' surplus.

Answers: See page 883.

13.5 Exercises

Management *Work the given exercises. (See Examples 1 and 2.)*

1. A car-leasing firm must decide how much to charge for maintenance on the cars it leases. After careful study, the firm decides that the rate of maintenance, $M(x)$, on a new car will approximate $M(x) = 60(1 + x^2)$, where x is the number of years the car has been in use. What total maintenance cost can the company expect for a 2-year lease? What minimum amount should be added to the monthly lease payments to pay for maintenance?

2. Using the function of Exercise 1, find the total charge during the first three years. What minimum monthly charge should be added to cover a 3-year lease?

3. Over the past decade, about 13 billion tickets were sold at U.S. movie theaters. The marginal revenue from box-office receipts (in billions of dollars) was approximated by
$$MR(x) = .56x + 3.1,$$
where x is the number of tickets sold (in billions).* What was the total revenue from selling these 13 billion tickets?

4. The marginal revenue from retail prescription drug sales (in billions of dollars) is approximated by
$$MR(x) = .01151e^{3.8362x},$$
where x is the number of prescriptions sold (in billions).†
   (a) What was the total revenue for the first 3.4 billion prescriptions?
   (b) What was the total revenue from 3 billion to 3.4 billion prescriptions?
   (c) After how many prescriptions were sold did total revenue reach $900 billion?

5. A company is considering a new manufacturing process. It knows that the rate of savings from the process will be about $S(t) = 1000(t + 2)$, where t is the number of years the process has been in use. Find the total savings during the first year. Find the total savings during the first six years.

6. Assume that the new process in Exercise 5 costs $16,000. About when will it pay for itself?

7. A company is introducing a new product. Production is expected to grow slowly because of difficulties in the start-up process. It is expected that the rate of production will be approximated by $P(x) = 1000e^{.2x}$, where x is the number of years since the introduction of the product. Will the company be able to supply 20,000 units during the first four years?

8. About when will the company of Exercise 7 be able to supply its 15,000th unit?

Find the area between the two curves. (See Example 3.)

9. $y = 3x$ and $y = x^2 - 4$ from $x = -1$ to $x = 3$
10. $y = x^2 - 30$ and $y = 10 - 3x$ from $x = 1$ to $x = 5$
11. $y = x^2$ and $y = x^3$ from $x = 0$ to $x = \frac{1}{2}$
12. $y = e^x$ and $y = 3 - e^x$ from $x = -3$ to $x = 0$

Use a graphing calculator to approximate the area between the graphs of each pair of functions on the given interval.

13. $y = \ln x$ and $y = 2xe^x$; $[1, 4]$
14. $y = \ln x$ and $y = 4 - x^2$; $[2, 5]$
15. $y = \sqrt{9 - x^2}$ and $y = \sqrt{x + 1}$; $[-1, 2]$
16. $y = \sqrt{4 - 4x^2}$ and $y = \sqrt{\frac{9 - x^2}{7}}$; $[-.5, .5]$

Work these problems. (See Examples 3–5.)

17. **Business** Suppose a company wants to introduce a new machine that will produce annual savings in dollars per year at the rate given by
$$S'(x) = 150 - x^2,$$
where x is the number of years of operation of the machine, while producing annual costs in dollars per year at the rate of
$$C'(x) = x^2 + \frac{11}{4}x.$$
   (a) To maximize its net savings, for how many years should the company use this new machine?
   (b) What are the net savings during the first year of use of the machine?
   (c) What are the net savings over the period determined in part (a)?

18. **Natural Science** A new smog-control device will reduce the output of sulfur oxides from automobile exhausts. It is estimated that the rate of savings (in millions of dollars per year) to the community from the use of this device in year x will be approximated by
$$S'(x) = -x^2 + 4x + 8.$$
The new device cuts down on the production of sulfur oxides, but it causes an increase in the production of nitrous oxides. The rate of costs (in millions of dollars per year) to the community in year x is approximated by
$$C'(x) = \frac{3}{25}x^2.$$

*Based on data for 2000–2009 from www.boxofficemojo.com.

†Based on data from the *Statistical Abstract of the United States*: 2009.

(a) To maximize net savings, for how many years should the new device be used?
(b) What will be the net savings over the period found in part (a)?

19. **Business** Richardson Construction Company has an expenditure rate of $E(x) = e^{.15x}$ dollars per day on a particular paving job and an income rate of $I(x) = 120.3 - e^{.15x}$ dollars per day on the same job, where x is the number of days from the start of the job. The company's profit on that job will equal total income less total expenditures. Profit will be maximized if the job ends at the optimum time, which is the point where the two curves meet. Find the following:
(a) the optimum number of days for the job to last;
(b) the total income for the optimum number of days;
(c) the total expenditures for the optimum number of days;
(d) the maximum profit for the job.

20. **Business** A factory at Harold Levinson Industries has installed a new process that will produce an increased rate of revenue (in thousands of dollars per year) of

$$r(t) = 104 - .4e^{t/2},$$

where t is time measured in years. The new process produces additional costs (in thousands of dollars per year) at the rate of

$$c(t) = .3e^{t/2}.$$

(a) To maximize this additional profit, how long should the new process be used?
(b) Find the additional profit during this period.

21. **Business** After t years, an oil refinery is producing at the rate of

$$P(t) = \frac{20}{1.2t + 1.6}$$

trillions of gallons per year. At the same time, the oil is consumed at a rate of $C(t) = t + .8$ trillions of gallons per year.
(a) In how many years will the rate of consumption equal the rate of production?
(b) What is the total excess production before consumption and production are equal?

22. **Business** The rate of expenditure (in dollars per year) for maintenance of a certain machine is given by

$$m(x) = x^2 + 6x,$$

where x is time measured in years. The machine produces a rate of savings (in dollars per year) given by

$$s(x) = 360 - 2x^2.$$

(a) In how many years will the maintenance rate equal the savings rate?
(b) What will be the net savings over this period?

23. **Natural Science** Polluted fluid from a factory begins to enter a lake at the rate

$$f(t) = 8(1 - e^{-.25t}),$$

where t is time (in hours) and $f(t)$ is in gallons per hour. At the same time, a pollution filter begins to remove the pollution at the rate of

$$g(t) = .2t,$$

where t is in hours and $g(t)$ is in gallons per hour.
(a) How much pollution is in the lake after 20 hours?
(b) At what time will the rate at which pollution enters the lake equal the rate at which pollution is being removed?
(c) How much pollution is in the lake at the time found in part (b)?
(d) At what time will all the pollution be removed from the lake?

Business *Work the given supply-and-demand exercises, where the price is given in dollars. (See Example 6.)*

24. Find the consumers' surplus and the producers' surplus for an item having supply function

$$S(q) = 3q^2$$

and demand function

$$D(q) = 120 - \frac{q^2}{6}.$$

25. Suppose the supply function of a certain item is given by

$$S(q) = \frac{7}{5}q$$

and the demand function is given by

$$D(q) = -\frac{3}{5}q + 10.$$

(a) Graph the supply and demand curves.
(b) Find the point at which supply and demand are in equilibrium.
(c) Find the consumers' surplus.
(d) Find the producers' surplus.

26. Find the producers' surplus if the supply function for digital cameras sold by an office supply chain is given by

$$S(q) = q^{5/2} + 2.1q^{3/2} + 48.$$

Assume that supply and demand are in equilibrium at $q = 16$.

27. Suppose the supply function for concrete is given by

$$S(q) = 100 + 3q^{3/2} + q^{5/2}$$

and that supply and demand are in equilibrium at $q = 9$. Find the producers' surplus.

28. Find the consumers' surplus if the demand function for grass seed is given by

$$D(q) = \frac{80}{(3q + 1)^2},$$

assuming that supply and demand are in equilibrium at $q = 3$.

29. Find the consumers' surplus if the demand function for California chardonnay wine is given by

$$D(q) = \frac{15{,}500}{(3.2q + 7)^3}$$

and if supply and demand are in equilibrium at $q = 5$.

30. Suppose the supply function of a certain item is given by

$$S(q) = e^{q/2} - 1$$

and the demand function is given by

$$D(q) = 400 - e^{q/2}.$$

(a) Graph the supply and demand curves.
(b) Find the point at which supply and demand are in equilibrium.
(c) Find the consumers' surplus.
(d) Find the producers' surplus.

31. Repeat the four steps in Exercise 30 for the supply function

$$S(q) = q^2 + \frac{11}{4}q$$

and the demand function

$$D(q) = 150 - q^2.$$

Work the given exercises.

32. **Business** In an article in the December 1994 *Scientific American* magazine, the authors estimated future gasoline use.* Without a change in U.S. policy, auto fuel use is forecasted to rise along the projection shown at the top right in the accompanying figure. The shaded band predicts gas use if technologies for increased fuel economy are phased in by the year 2010. The moderate estimate (center curve) corresponds to an average of 46 miles per gallon for all cars on the road. Discuss the interpretation of the shaded area and other regions of the graph that pertain to the topic in this section.

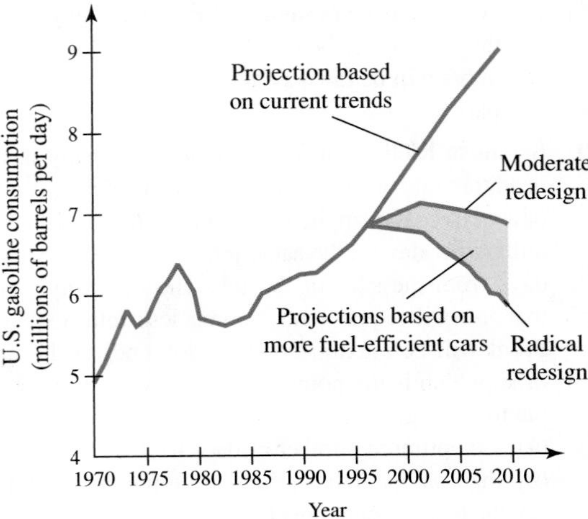

33. An artist sketches the curves $y = 3\sqrt{x}$ and $y = 2x$ on a sheet of poster board and then cuts out the area between the curves. What is that area?

34. **Business** If a large truck has been driven x thousand miles, the rate of repair costs in dollars per mile is given by

$$r(x) = .04x^{3/2}.$$

Find the total repair costs if the truck is driven
(a) 100,000 miles;
(b) 400,000 miles.

35. **Business** Data from the U.S. Bureau of Labor Statistics shows that the annual cost per worker for insurance (in dollars per year) was increasing according to the function

$$f(x) = 70.71e^{.43x},$$

where $x = 0$ corresponds to the start of 1999. Find the total increase in costs during the four years beginning in 1999.

36. **Natural Science** From 1905 to 1920, most of the predators of the Kaibab Plateau of Arizona were killed by hunters. This allowed the deer population there to grow rapidly until the deer had depleted their food sources, which caused a rapid decline in their population. The rate of change of this deer population during that time span is approximated by the function

$$D(t) = \frac{25}{2}t^3 - \frac{5}{8}t^4,$$

where t is the time in years ($0 \le t \le 25$).
(a) Find the function for the deer population if there were 4000 deer in 1905 ($t = 0$).
(b) What was the population in 1920?
(c) When was the population at a maximum?
(d) What was the maximum population?

*Adapted from "Improving Automotive Efficiency," by John DeCicco and Marc Ross (Dec. 1994).

37. **Social Science** Suppose that all the people in a country are ranked according to their incomes, starting at the bottom. Let x represent the fraction of the community making the lowest income $(0 \leq x \leq 1)$; $x = .4$ therefore represents the lower 40% of all income producers. Let $I(x)$ represent the proportion of the total income earned by the lowest x of all the people. Thus, $I(.4)$ represents the fraction of total income earned by the lowest 40% of the population. Suppose

$$I(x) = .9x^2 + .1x.$$

Find and interpret the following:
(a) $I(.1)$; **(b)** $I(.5)$; **(c)** $I(.9)$.
If income were distributed uniformly, we would have $I(x) = x$. The area under this line of complete equality is $\frac{1}{2}$. As $I(x)$ dips farther below $y = x$, there is less equality of income distribution. This inequality can be quantified by the ratio of the area between $I(x)$ and $y = x$ to $\frac{1}{2}$. This ratio is called the *coefficient of inequality* and equals $2\int_0^1 (x - I(x)\,dx$.
(d) Graph $I(x) = x$ and $I(x) = .9x^2 + .1x$ for $0 \leq x \leq 1$ on the same axes.
(e) Find the area between the curves. What does this area represent?

38. **Health** For a certain drug, the rate of release into the bloodstream of the active ingredient, in the appropriate units, is given by

$$R(t) = \frac{4}{3t} + \frac{2}{t^2},$$

where t is measured in hours after the drug is administrated. Find the total amount released
(a) from $t = 1$ to 10 hours;
(b) from $t = 10$ to 20 hours.

39. **Management** A worker new to a job will improve his efficiency with time, so that it takes him fewer hours to produce an item with each day on the job, up to a certain point. Suppose the rate of change of the number of hours it takes a worker in a certain factory to produce the xth item is given by

$$H(x) = 20 - 2x.$$

The production rate per item is a maximum when $\int_0^T H(x)\,dx$ is a maximum.
(a) How many items must be made to achieve the maximum production rate? Assume that 0 items are made in 0 hours.
(b) What is the maximum production rate per item?

✓ Checkpoint Answers

1. **(a)** \$367.66 **(b)** \$1185.91
2. **(a)** About \$59.85 billion
(b) About early April of 2007 ($t \approx 7.26$)
3. 19.5
4. \$320,000
5. **(a)** 5 **(b)** \$437,500
6. **(a)** $q = 100$ **(b)** \$350 **(c)** \$250

13.6 ▸ Tables of Integrals (Optional)

Although the fundamental theorem of calculus is a powerful tool, it cannot be used unless you can find the antiderivative of the function to be integrated. Sometimes this is difficult or even impossible. For instance,

$$\int_0^1 e^{x^2}\,dx$$

cannot be evaluated exactly. In such cases, the only option is to employ various approximation methods (such as those used by graphing calculators).

Even when a function has an antiderivative, the methods presented in earlier sections may not be adequate for finding it. Computer algebra programs and some graphing calculators, such as the TI-89, can find the antiderivatives of a wide variety of functions. For those who do not have access to such technology, tables of integrals are available. One such table is Table 3 in Appendix B at the back of this book. The following examples illustrate its use.

Example 1 Find $\int \frac{1}{\sqrt{x^2 + 16}}\,dx$.

Solution By inspecting the table, we see that if $a = 4$, this antiderivative is the same as entry 5 of the table, which is

$$\int \frac{1}{\sqrt{x^2 + a^2}}\,dx = \ln|x + \sqrt{x^2 + a^2}| + C.$$

Substituting 4 for a in this entry, we get

$$\int \frac{1}{\sqrt{x^2 + 16}}\,dx = \ln|x + \sqrt{x^2 + 16}| + C.$$

This last result could be verified by taking the derivative of the right-hand side of this last equation. 1

✓Checkpoint 1

Find the given antiderivatives.

(a) $\int \frac{4}{\sqrt{x^2 + 100}}\,dx$

(b) $\int \frac{-9}{\sqrt{x^2 - 4}}\,dx$

Answers: See page 885.

Example 2 Find $\int \frac{8}{16 - x^2}\,dx$.

Solution Convert this antiderivative into the one given in entry 7 of the table by writing the 8 in front of the integral sign (permissible only with constants) and by letting $a = 4$. Doing this gives

$$8\int \frac{1}{16 - x^2}\,dx = 8\left[\frac{1}{2\cdot 4}\ln\left|\frac{4 + x}{4 - x}\right|\right] + C$$

$$= \ln\left|\frac{4 + x}{4 - x}\right| + C. \quad 2$$

✓Checkpoint 2

Find the given antiderivatives.

(a) $\int \frac{1}{x^2 - 4}\,dx$

(b) $\int \frac{-6}{x\sqrt{25 - x^2}}\,dx$

Answers: See page 885.

Example 3 Find $\int \sqrt{4x^2 + 9}\,dx$.

Solution This antiderivative seems most similar to entry 15 of the table. However, entry 15 requires that the coefficient of the x^2 term be 1. We can satisfy that requirement here by factoring out the 4:

$$\int \sqrt{4x^2 + 9}\,dx = \int \sqrt{4\left(x^2 + \frac{9}{4}\right)}\,dx$$

$$= \int \sqrt[2]{x^2 + \frac{4}{9}}\,dx$$

$$= 2\int \sqrt{x^2 + \frac{4}{9}}\,dx.$$

Now use entry 15 with $a = \frac{2}{3}$:

$$\int \sqrt{4x^2 + 9}\,dx = 2\left[\frac{x}{2}\sqrt{x^2 + \frac{4}{9}} + \frac{(4/9)}{2}\ln\left|x + \sqrt{x^2 + \frac{4}{9}}\right|\right] + C$$

$$= x\sqrt{x^2 + \frac{4}{9}} + \frac{4}{9}\ln\left|x + \sqrt{x^2 + \frac{4}{9}}\right| + C. \quad 3$$

✓Checkpoint 3

Find the given antiderivatives.

(a) $\int \frac{3}{16x^2 - 1}\,dx$

(b) $\int \frac{-1}{9 - 100x^2}\,dx$

Answers: See page 885.

13.6 ▸ Exercises

Use the table of integrals to find each antiderivative. (See Examples 1–3.)

1. $\int \frac{-7}{\sqrt{x^2 + 36}}\,dx$
2. $\int \frac{9}{\sqrt{x^2 + 9}}\,dx$
3. $\int \frac{6}{x^2 - 9}\,dx$
4. $\int \frac{-9}{x^2 - 16}\,dx$
5. $\int \frac{-4}{x\sqrt{9 - x^2}}\,dx$
6. $\int \frac{3}{x\sqrt{121 - x^2}}\,dx$
7. $\int \frac{-5x}{3x + 1}\,dx$
8. $\int \frac{2x}{4x - 5}\,dx$
9. $\int \frac{13}{3x(3x - 5)}\,dx$
10. $\int \frac{-4}{3x(2x + 7)}\,dx$
11. $\int \frac{4}{4x^2 - 1}\,dx$
12. $\int \frac{-6}{9x^2 - 1}\,dx$
13. $\int \frac{3}{x\sqrt{1 - 9x^2}}\,dx$
14. $\int \frac{-5}{x\sqrt{1 - 16x^2}}\,dx$
15. $\int \frac{15x}{2x + 3}\,dx$
16. $\int \frac{11x}{6 - x}\,dx$
17. $\int \frac{-x}{(5x - 1)^2}\,dx$
18. $\int \frac{-6}{x(4x + 3)^2}\,dx$
19. $\int \frac{3x^4 \ln|x|}{4}\,dx$
20. $\int 4x^2 \ln|x|\,dx$
21. $\int \frac{7\ln|x|}{x^2}\,dx \left(\textit{Hint: } \frac{1}{x^2} = x^{-2}\right)$
22. $\int \frac{-3\ln|x|}{x^3}\,dx$
23. $\int xe^{-2x}\,dx$
24. $\int 6xe^{3x}\,dx$

Use Table 3 in Appendix B to solve the given problems.

25. **Business** The rate of change of revenue in dollars from the sale of x computers is

$$R'(x) = \frac{5000}{\sqrt{x^2 + 25}}.$$

Find the total revenue from the sale of the first 20 computers.

26. **Health** The rate of reaction to a drug is given by

$$r'(x) = 2x^2e^{-x},$$

where x is the number of hours since the drug was administered. Find the total reaction to the drug from $x = 1$ to $x = 5$.

27. **Natural Science** The rate of growth of a microbe population is given by

$$m'(x) = 25xe^{2x},$$

where x is time in days. What is the total accumulated growth after 3 days?

28. **Social Science** The rate (in hours per item) at which a worker in a certain job produces the xth item is

$$h'(x) = \sqrt{x^2 + 16}.$$

What is the total number of hours it will take this worker to produce the first seven items?

✓ Checkpoint Answers

1. (a) $4\ln|x + \sqrt{x^2 + 100}| + C$
   (b) $-9\ln|x + \sqrt{x^2 - 4}| + C$
2. (a) $\frac{1}{4}\ln\left|\frac{x - 2}{x + 2}\right| + C$
   (b) $\frac{6}{5}\ln\left|\frac{5 + \sqrt{25 - x^2}}{x}\right| + C$
3. (a) $\frac{3}{8}\ln\left|\frac{x - \frac{1}{4}}{x + \frac{1}{4}}\right| + C$
   (b) $-\frac{1}{60}\ln\left|\frac{\frac{3}{10} + x}{\frac{3}{10} - x}\right| + C$

13.7 ▸ Differential Equations

A **differential equation** is an equation that involves an unknown function $y = f(x)$ and a finite number of its derivatives. Solving the differential equation for y would give the unknown function. Differential equations have been important in the physical sciences and engineering for several centuries. More recently, they have been used in the social sciences, life sciences, and economics to solve a variety of problems involving population growth, biological balance, interest rates, and more.

Up to now, we have considered only equations whose solutions are *numbers*. The solutions of differential equations, however, are *functions*. For example, the solutions of the differential equation

$$\frac{dy}{dx} = 3x^2 - 2x \tag{1}$$

consist of all functions y that satisfy the equation. Since the left side of the equation is the derivative of y with respect to x, the equation says that the derivative of y is $3x^2 - 2x$, which means that y is an *antiderivative* of $3x^2 - 2x$. In other words, the solutions are all functions y such that

$$y = \int (3x^2 - 2x)\, dx = x^3 - x^2 + C. \tag{2}$$

Each different value of C in equation (2) leads to a different solution of equation (1), showing that a differential equation can have an infinite number of solutions. Equation (2) is the **general solution** of the differential equation (1). Some of the solutions of equation (1) are graphed in Figure 13.34. 1 ✓

✓ Checkpoint 1

Find the general solution of

(a) $dy/dx = 4x$;

(b) $dy/dx = -x^3$;

(c) $dy/dx = 2x^2 - 5x$.

Answers: See page 895.

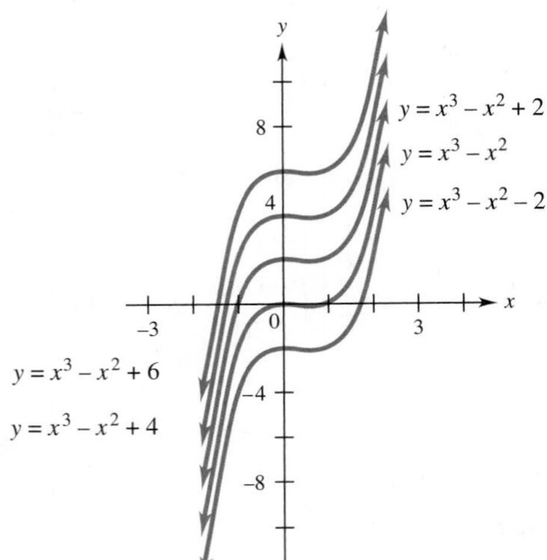

FIGURE 13.34

Equation (1) has the form

$$\frac{dy}{dx} = f(x),$$

So its solution suggests the following generalization.

General Solution of $dy/dx = f(x)$

The general solution of the differential equation $dy/dx = f(x)$ is

$$y = \int f(x)\, dx.$$

Example 1 The population P of a flock of birds is growing exponentially so that

$$\frac{dP}{dx} = 20e^{.05x},$$

where x is time in years. Find P in terms of x if there were 20 birds in the flock initially.

Solution Solve the differential equation:

$$P = \int 20e^{.05x}\,dx$$

$$= \frac{20}{.05}e^{.05x} + C$$

$$= 400e^{.05x} + C.$$

Since P is 20 when x is 0,

$$20 = 400e^{0} + C$$

$$-380 = C,$$

and

$$P = 400e^{.05x} - 380.$$

In Example 1, the given information was used to produce a solution with a specific value of C. Such a solution is called a **particular solution** of the given differential equation. The given information, $P = 20$ when $x = 0$, is called an **initial condition.** ✓2

✓Checkpoint 2

Find the particular solution in Example 1 if there were 100 birds in the flock after 2 years.

Answer: See page 895.

Sometimes a differential equation must be rewritten in the form

$$\frac{dy}{dx} = f(x)$$

before it can be solved.

Example 2 Find the particular solution of

$$\frac{dy}{dx} - 3x = 7,$$

given that $y = 2$ when $x = -2$.

Solution Add 3 x to both sides of the equation to get

$$\frac{dy}{dx} = 3x + 7.$$

The general solution is

$$y = \int (3x + 7)\,dx$$

$$y = \frac{3x^2}{2} + 7x + C.$$

Substituting 2 for y and -2 for x gives

$$2 = \frac{3(-2)^2}{2} + 7(-2) + C$$
$$C = 10.$$

The particular solution is $y = \frac{3x^2}{2} + 7x + 10.$ 3 ✓

✓Checkpoint 3

Find the particular solution of

$$2\frac{dy}{dx} - 4 = 9x^2,$$

given that $y = 4$ when $x = 1$.

Answer: See page 895.

Marginal productivity is the rate at which production changes (increases or decreases) for a unit change in investment. Thus, marginal productivity can be expressed as the first derivative of the function that gives production in terms of investment.

Example 3 **Business** Suppose the marginal productivity of a manufacturing process is given by

$$P'(x) = 3x^2 - 10, \quad (3)$$

where x is the amount of the investment in hundreds of thousands of dollars. If the process produces 100 units per month with the present investment of \$300,000 (that is, $x = 3$), by how much would production increase if the investment were increased to \$500,000?

Solution To obtain an equation for production, we proceed as above. The general solution is:

$$P(x) = \int (3x^2 - 10)\,dx$$
$$= x^3 - 10x + C.$$

To find C, use the given initial values—that is, $P(x) = 100$ when $x = 3$:

$$100 = 3^3 - 10(3) + C$$
$$C = 103.$$

Production is thus represented by

$$P(x) = x^3 - 10x + 103,$$

and if investment is increased to \$500,000, production becomes

$$P(5) = 5^3 - 10(5) + 103 = 178.$$

An increase to \$500,000 in investment will increase production from 100 units to 178 units. 4 ✓

✓Checkpoint 4

In Example 3, if marginal productivity is changed to

$$P'(x) = 3x^2 + 8x,$$

with the same initial conditions,

(a) find an equation for production;

(b) find the increase in production if investment increases to \$500,000.

Answers: See page 895.

SEPARATION OF VARIABLES

The solution method used so far in this section is essentially the same as that in Sections 13.1 and 13.2, where we studied antiderivatives. But not all differential equations can be solved so easily. Suppose, for example, that y is a function of x such that

$$\frac{dy}{dx} = \frac{x^2}{y}.$$

We cannot simply integrate, since the right side involves both variables. Nevertheless, there is a solution method that we now develop.

Consider a differential equation of the form

$$\frac{dy}{dx} = \frac{f(x)}{g(y)}. \tag{4}$$

(The example in the preceding paragraph is the case when $f(x) = x^2$ and $g(y) = y$.) Up to this point, we have used the symbol dy/dx to denote the derivative function of the function y. In advanced courses, it is shown that dy/dx can also be interpreted as a quotient of two differentials, dy divided by dx. (You may recall that differentials were used in integration by substitution.) Then we can multiply both sides of equation (4) by $g(y)\,dx$ to obtain

$$g(y)\,dy = f(x)\,dx.$$

In this form, all terms involving y (including dy) are on one side of the equation, and all terms involving x (and dx) are on the other side. Such a differential equation is said to be **separable**, since the variables x and y can be separated. A separable differential equation may be solved by integrating each side after separating the variables.

Example 4 Find the general solution of $\dfrac{dy}{dx} = -\dfrac{6x}{y}$.

Solution Separate the variables by multiplying both sides by $y\,dx$:

$$y\,dy = -6x\,dx.$$

To solve this equation, take antiderivatives on each side:

$$\int y\,dy = \int -6x\,dx$$

$$\frac{y^2}{2} + C_1 = -3x^2 + C_2$$

$$3x^2 + \frac{y^2}{2} = C_2 - C_1.$$

We can replace the constant $C_2 - C_1$ with a single constant C to obtain

$$3x^2 + \frac{y^2}{2} = C.*$$

Since powers of y are involved, it is better to leave the solution in this form rather than trying to solve for y. For each positive constant C, the graph of the solution is an ellipse, as shown in Figure 13.35 on the next page. 5✓

✓Checkpoint 5

Find the general solution of

$$\frac{dy}{dx} = \frac{5x^2}{y}.$$

Answer: See page 895.

*From now on, we will add just one constant, with the understanding that it represents the difference between the two constants obtained in the two integrations.

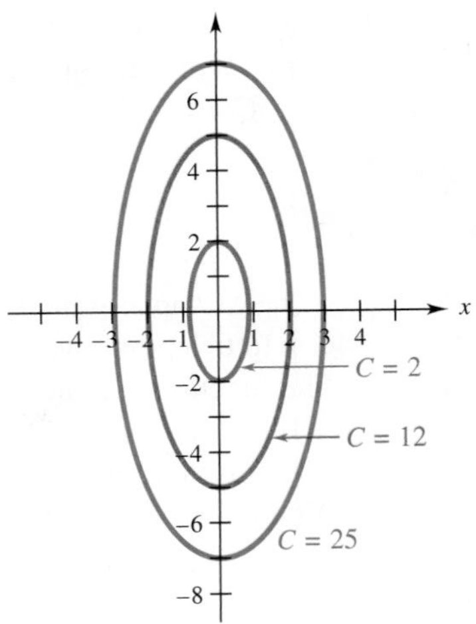

FIGURE 13.35

Example 5 **Business** Sales at a small store are now \$20,000 and are continuously increasing at an annual rate of 5%.

(a) Find a function $y = f(t)$ that gives the amount of sales, y, at time t years.

Solution To say that sales are continuously increasing at an annual rate of 5% means that the rate of increase at any time t is

$$5\% \text{ of } y = f(t), \quad \text{that is,} \quad .05y.$$

Since the rate of increase is dy/dt, we have

$$\frac{dy}{dt} = .05y.$$

To solve this equation for y, begin by separating the variables. Multiply both sides of the preceding equation by $\frac{dt}{y}$ to obtain

$$\frac{dy}{y} = .05\,dt.$$

Now take antiderivatives on both sides and use properties of logarithms and exponents to simplify the result:

$$\int \frac{dy}{y} = \int .05\,dt$$

$$\ln y = .05t + C$$

$$e^{\ln y} = e^{.05t+C}$$

$$y = e^{.05t+C} \qquad \text{Property of logarithms (page 235)}$$

$$y = e^{.05t}e^{C}. \qquad \text{Property of exponents (page 41)}$$

If we denote the constant e^C by M, the sales function has the form

$$y = Me^{.05t},$$

the same as the exponential growth functions considered in Section 4.2. To find the value of M, use the fact that $y = 20{,}000$ when $t = 0$, that is,

$$20{,}000 = Me^{.05(0)} = Me^0 = M.$$

Therefore, the sales function is $y = Me^{.05t} = 20{,}000e^{.05t}$.

(b) What was the amount of sales after 4.5 years?

Solution Evaluate y when $t = 4.5$:

$$y = 20{,}000e^{.05t} = 20{,}000e^{.05(4.5)} \approx \$25{,}046.45.$$ 6

Checkpoint 6

Suppose that current sales in Example 5 are $32,000. What will the amount of sales be in 6 years?

Answer: See page 895.

We can now explain how the exponential growth and decay functions presented in Section 4.2 were obtained. In the absence of inhibiting conditions, a population y (which might be human, animal, bacterial, etc.) grows in such a way that the rate of change of the population is proportional to the population at time x; that is, there is a constant k such that

$$\frac{dy}{dx} = ky.$$

The constant k is called the **growth-rate constant**. Example 5 is the case when $k = .05$. The same argument used there (with k in place of .05) shows that the population y at time x is given by

$$y = Me^{kx},$$

where M is the population at time $x = 0$. A positive value of k indicates growth while a negative value indicates decay.

Example 6 **Finance** In early November 2007, the value of one share of RST Corporation cost $58. Then the price of an RST share began to decrease at a continuous rate of 3% per month. What was the price of a share 16 months later, in early March 2009?

Solution If y is the price of a share at time x, then

$$y = Me^{kx},$$

where $M = 58$, $k = -.03$ (negative because the price is decreasing), and x is measured in months. So we have

$$y = 58e^{-.03x}.$$

To find the share price after 16 months, evaluate y at $x = 16$:

$$y = 58e^{-.03x} = 58e^{-.03(16)} \approx \$35.89.$$

LIMITED GROWTH

As a model of population growth, the equation $y = Me^{kx}$ is not realistic over the long run for most populations. As shown by graphs of functions of the form $y = Me^{kx}$, with both M and k positive, growth would be unbounded. Additional factors, such as space restrictions or a limited amount of food, tend to inhibit the

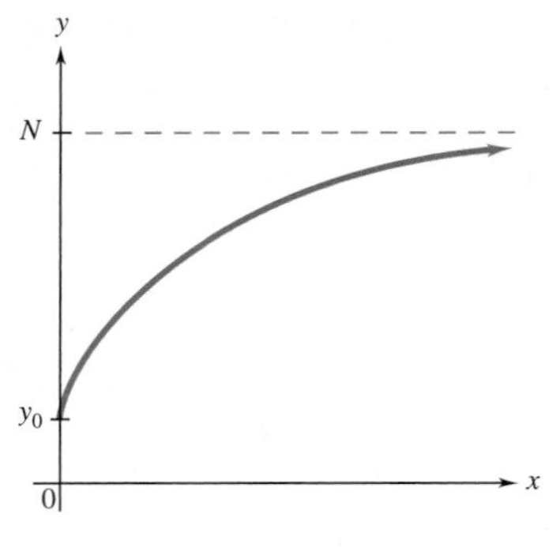

FIGURE 13.36

growth of populations as time goes on. In an alternative model that assumes a maximum population of size N, the rate of growth of a population is proportional to how close the population is to that maximum. These assumptions lead to the differential equation

$$\frac{dy}{dx} = k(N - y),$$

the limited-growth function mentioned in Chapter 4. Graphs of limited-growth functions look like the graph in Figure 13.36, where y_0 is the initial population.

Example 7 **Natural Science** Researchers believe that a certain lake can support no more than 3000 fish. There are approximately 500 fish in the lake at present, with a growth constant of .25.

(a) Write and solve a differential equation for the rate of growth of this population.

Solution Let $N = 3000$ and $k = .25$. The rate of growth of the population is given by

$$\frac{dy}{dx} = k(N - y)$$

$$\frac{dy}{dx} = .25(3000 - y).$$

To solve for y, first separate the variables.

$$\frac{dy}{3000 - y} = .25\,dx \qquad \text{Multiply both sides by } \frac{dx}{3000 - y}.$$

$$\int \frac{dy}{3000 - y} = \int .25\,dx \qquad \text{Take antiderivatives on both sides.}$$

$$-\ln(3000 - y) = .25x + C$$

$$\ln(3000 - y) = -.25x - C$$

$$3000 - y = e^{-.25x - C} = \left(e^{-.25x}\right)\left(e^{-C}\right).$$

The absolute-value bars are not needed for $\ln(3000 - y)$ because y must be less than 3000 for this population, so that $3000 - y$ is always nonnegative. Let $e^{-C} = B$. Then

$$3000 - y = Be^{-.25x}$$

$$y = 3000 - Be^{-.25x}.$$

Find B by using the fact that $y = 500$ when $x = 0$:

$$500 = 3000 - Be^0$$

$$500 = 3000 - B$$

$$B = 2500.$$

✓ Checkpoint 7

An animal population is growing at a constant rate of 4%. The habitat will support no more than 10,000 animals. There are 3000 animals present now.

(a) Write an equation giving the population y in x years.

(b) Estimate the animal population in five years.

Answers: See page 895.

Notice that the value of B is the difference between the maximum population and the initial population. Substituting 2500 for B in the equation for y gives

$$y = 3000 - 2500e^{-.25x}.$$

(b) What will the fish population be in four years?

Solution In four years, the population will be

$$y = 3000 - 2500e^{-.25(4)} = 3000 - 2500e^{-1}$$
$$= 3000 - 919.7 = 2080.3,$$

or almost 2100 fish. 7 ✓

13.7 ▸ Exercises

Find general solutions for the given differential equations. (See Examples 1–5(a).)

1. $\frac{dy}{dx} = -3x^2 + 7x$

2. $\frac{dy}{dx} = 3e^{-4x}$

3. $3x^3 - 2\frac{dy}{dx} = 0$

4. $3x^2 - 3\frac{dy}{dx} = 8$

5. $y\frac{dy}{dx} = x$

6. $y\frac{dy}{dx} = x^2 - 1$

7. $\frac{dy}{dx} = 4xy$

8. $\frac{dy}{dx} = x^2y$

9. $\frac{dy}{dx} = 4x^3y - 3x^2y$

10. $\frac{x}{3y^2} = \frac{dy}{dx}$

11. $\frac{dy}{dx} = \frac{y}{x}, x > 0$

12. $\frac{dy}{dx} = \frac{y}{x^2}$

13. $\frac{dy}{dx} = y - 7$

14. $\frac{dy}{dx} = 9 - y$

15. $\frac{dy}{dx} = y^2e^x$

16. $\frac{dy}{dx} = \frac{7e^x}{e^y}$

Find particular solutions for the given equations. (See Examples 2, 3, 5 and 6.)

17. $\frac{dy}{dx} + 2x = 3x^2$; $y = 4$ when $x = 0$

18. $\frac{dy}{dx} = 8x^3 - 3x^2 + 2x$; $y = -1$ when $x = 1$

19. $\frac{dy}{dx}(x^3 + 28) = \frac{6x^2}{y}$; $y^2 = 6$ when $x = -3$

20. $\frac{2y}{x - 3}\frac{dy}{dx} = \sqrt{x^2 - 6x}$; $y^2 = 40$ when $x = 8$

21. $\frac{dy}{dx} = \frac{x^2}{y}$; $y = 4$ when $x = 0$

22. $x^2\frac{dy}{dx} = y$; $y = -1$ when $x = 1$

23. $(5x + 3)y = \frac{dy}{dx}$; $y = 1$ when $x = 0$

24. $x\frac{dy}{dx} - y\sqrt{x} = 0$; $y = 1$ when $x = 0$

25. $\frac{dy}{dx} = \frac{7x + 1}{y - 3}$; $y = 4$ when $x = 0$

26. $\frac{dy}{dx} = \frac{x^2 + 5}{2y - 1}$; $y = 11$ when $x = 0$

27. What is the difference between a general solution and a particular solution of a differential equation?

28. What is meant by a separable differential equation?

Work the given problems. (See Example 3.)

29. Business The marginal productivity of a manufacturing process is given by

$$\frac{dy}{dx} = \frac{3x^2 - 4x}{25},$$

where x is the amount of investment (in thousands of dollars) and y is the number of units produced per month (in hundreds). Production is 2300 units when investment is \$5000. Find the production level if investment is increased to
(a) \$8000; **(b)** \$11,000.

30. Business The marginal productivity of a process is given by

$$\frac{dy}{dx} = -\frac{40}{32 - 4x},$$

where x represents the investment (in thousands of dollars). Find the productivity for each of the given investments if productivity is 100 units when the investment is \$1000.

(a) \$3000 (b) \$5000

(c) Can investments ever reach \$8000 according to this model? Why?

Work these problems. (See Examples 5 and 6.)

31. Natural Science The sell-by dating of dairy products depends on the solution of a differential equation. The rate of growth of bacteria in such products increases with time. If y is the number of bacteria (in thousands) present at a time t (in days), then the rate of growth of bacteria can be expressed as dy/dt, and we have

$$\frac{dy}{dt} = kt,$$

where k is an appropriate constant. For a certain product, $k = 8$ and $y = 50$ (in thousands) when $t = 0$.

(a) Solve the differential equation for y.

(b) Suppose the maximum allowable value for y is 550 (thousand). How should the product be dated?

32. Social Sciences A recent report by the U.S. Census Bureau predicts that the Latino-American population will increase from 26.7 million in 1995 to 96.5 million in 2050.* Assuming that the unlimited-growth model $dy/dt = ky$ fits this population growth, express the population y as a function of the year t. Let 1995 correspond to $t = 0$.

33. Social Science Suppose the rate at which a rumor spreads—that is, the number of people who have heard the rumor over a certain period of time—increases with the number of people who have heard it. If y is the number of people who have heard the rumor, then

$$\frac{dy}{dt} = ky,$$

where t is the time in days and k is a constant.

(a) If y is 1 when $t = 0$, and y is 5 when $t = 2$, find k.

Using the value of k from part (a), find y for each of the following times:

(b) $t = 3$; (c) $t = 5$; (d) $t = 10$.

34. Social Science A company has found that the rate at which a person new to the assembly line produces items is

$$\frac{dy}{dx} = 7.5e^{-.3x},$$

where x is the number of days the person has worked on the line. How many items can a new worker be expected to produce on the seventh day if he produces none when $x = 0$?

*"Population Projections of the U.S. by Age, Race, and Hispanic Origin: 1995 to 2050," U.S. Census Bureau.

35. Business Sales of a particular product have been declining continuously at the rate of 15% per year.

(a) Express the rate of sales decline as a differential equation.

(b) Find the general solution for the equation in part (a).

(c) When will sales decrease to 25% of their original level?

36. Finance If inflation grows continuously at a rate of 4%, how long will it take for \$1 to lose half its value?

37. Natural Science The amount of a tracer dye injected into the bloodstream decreases exponentially, with a decay constant of 3% per minute. If 6 cc are present initially, how many cc are present after 8 minutes? (Here, k will be negative.)

38. Finance A life insurance company invests \$5000 to fund a death benefit of \$20,000. Growth in the investment over time can be modeled by the differential equation

$$\frac{dA}{dt} = Ai,$$

where i is the interest rate and $A(t)$ is the amount invested at time t (in years). Calculate the interest rate that the investment must earn in order for the company to fund the death benefit in 24 years. (Choose one of the following.)*

(a) $\frac{-\ln 2}{12}$ (b) $\frac{-\ln 2}{24}$ (c) $\frac{\ln 2}{24}$

(d) $\frac{\ln 2}{12}$ (e) $\frac{\ln 2}{6}$

Work these problems. (See Example 7.)

39. Natural Science The rate at which the number of bacteria in a culture is changing after the introduction of a bactericide is given by

$$\frac{dy}{dx} = 30 - y,$$

where y is the number of bacteria (in thousands) present at time x. Find the number of bacteria present at each of the given times if there were 1000 thousand bacteria present at time $x = 0$.

(a) $x = 2$ (b) $x = 7$ (c) $x = 10$

40. Natural Science An isolated fish population is limited to 4000 by the amount of food available. If there are now 320 fish and the population is growing continuously at the rate of 2% per year, find the population at the end of 10 years.

Physical Science *Newton's law of cooling states that the rate of change of temperature of an object is proportional to the differ-*

*Problem 27 from the May 2003 Course 1 Examination of the *Education and Examination Committee of the Society of Actuaries*. Reprinted by permission of the Society of Actuaries.

ence in temperature between the object and the surrounding medium. Thus, if T is the temperature of the object after t hours and C is the (constant) temperature of the surrounding medium, then

$$\frac{dT}{dt} = -k(T - C),$$

where k is a constant. When a dead body is discovered within 48 hours of the death and the temperature of the medium (air or water, for example) has been fairly constant, Newton's law of cooling can be used to determine the time of death. (The medical examiner does not actually solve the equation for each case, but uses a table that is based on the formula.) Use Newton's law of cooling to work the given problems.*

41. Assume that the temperature of a body at death is 98.6°F, the temperature of the surrounding air is 68°F, and the body temperature is 90°F at the end of one hour.
(a) Find an equation that gives the body temperature T after t hours.
(b) What was the temperature of the body after two hours?
(c) When will the temperature of the body be 75°F?
(d) Approximately when will the temperature of the body be within .01°of the surrounding air?

42. Repeat Exercise 41 under these conditions: The temperature of the surrounding air is 38°F, and after one hour, the body temperature is 81°.

*Dennis Callas and David J. Hildreth, "Snapshots of Applications in Mathematics," *College Mathematics Journal* 26, No. 2 (March 1995).

Business *Elasticity of demand was discussed in Case 11 at the end of Chapter 11, where it was defined as*

$$E = -\frac{p}{q}\cdot\frac{dq}{dp}$$

for demand q and price p. Find the general demand equation q = f(p) for each of the given elasticity functions. (Hint: Set each elasticity function equal to $-\frac{p}{q}\cdot\frac{dq}{dp}$ *and then solve for q. Write the constant of integration as ln C.)*

43. $E = 2$

44. $E = \frac{4p^2}{q^2}$

✓ Checkpoint Answers

1. **(a)** $y = 2x^2 + C$ **(b)** $y = -\frac{1}{4}x^4 + C$
(c) $y = \frac{2}{3}x^3 - \frac{5}{2}x^2 + C$

2. $P = 400e^{.05x} - 342$

3. $y = \frac{3}{2}x^3 + 2x + \frac{1}{2}$

4. **(a)** $P(x) = x^3 + 4x^2 + 37$
(b) Production goes from 100 units to 262 units, an increase of 162 units.

5. $\frac{y^2}{2} - \frac{5x^3}{3} = C$

6. About $43,195.48

7. **(a)** $y = 10{,}000 - 7000e^{-.04x}$
(b) About 4269 animals

CHAPTER 13 ▸ Summary

KEY TERMS AND SYMBOLS

13.1 ▸ $\int f(x)\,dx$, indefinite integral of f
antiderivative
integral sign
integrand
power rule
constant-multiple rule
sum-or-difference rule

13.2 ▸ differential
integration by substitution

13.3 ▸ $\int_b^a f(x)\,dx$, definite integral of f
total change in $F(x)$

13.4 ▸ $F(x)\Big|_a^b = F(b) - F(a)$

13.5 ▸ consumers' surplus
producers' surplus

13.6 ▸ tables of integrals

13.7 ▸ differential equation
general solution
particular solution
initial condition
marginal productivity
separable differential equation
growth-rate constant

KEY CONCEPTS

$F(x)$ is an antiderivative of $f(x)$ if $F'(x) = f(x)$.

Indefinite Integral ▸ If $F'(x) = f(x)$, then $\int f(x)\,dx = F(x) + C$, for any real number C.

Properties of Integrals ▶ $\int k \cdot f(x)\,dx = k \cdot \int f(x)\,dx$, for any real number k.

$$\int [f(x) \pm g(x)]\,dx = \int f(x)\,dx \pm \int g(x)\,dx.$$

Rules for Integrals ▶ For $u = f(x)$ and $du = f'(x)\,dx$,

$$\int u^n du = \frac{u^{n+1}}{n+1} + C; \qquad \int e^u du = e^u + C; \qquad \int u^{-1} du = \int \frac{du}{u} = \ln|u| + C.$$

The Definite Integral ▶ If f is continuous on $[a, b]$, the definite integral of f from a to b is

$$\int_a^b f(x)\,dx = \lim_{n\to\infty} ([f(x_1) + f(x_2) + f(x_3) + \cdots + f(x_n)] \cdot \Delta x),$$

provided that this limit exists, where $\Delta x = \frac{b-a}{n}$ and x_i is the left endpoint of the ith interval.

Total Change in F(x) ▶ Let f be continuous on $[a, b]$ and $f(x) \geq 0$ for all x in $[a, b]$. If $f(x)$ is the rate of change of $F(x)$, then the total change in $F(x)$ as x goes from a to b is given by

$$\int_a^b f(x)\,dx.$$

Fundamental Theorem of Calculus ▶ Let f be continuous on $[a, b]$ and let F be any antiderivative of f. Then

$$\int_a^b f(x)\,dx = F(x)\Big|_a^b = F(b) - F(a).$$

General Solution of $\frac{dy}{dx} = f(x)$ ▶ The general solution of the differential equation $dy/dx = f(x)$ is

$$y = \int f(x)\,dx.$$

General Solution of $\frac{dy}{dx} = ky$ ▶ The general solution of the differential equation $dy/dx = ky$ is

$$y = Me^{kx}.$$

CHAPTER 13 REVIEW EXERCISES

Find each indefinite integral.

1. $\int (x^2 - 8x - 7)\,dx$
2. $\int (6 - x^2)\,dx$
3. $\int 7\sqrt{x}\,dx$
4. $\int \frac{\sqrt{x}}{8}\,dx$
5. $\int (x^{1/2} + 3x^{-2/3})\,dx$
6. $\int (8x^{4/3} + x^{-1/2})\,dx$
7. $\int \frac{-6}{x^3}\,dx$
8. $\int \frac{9}{x^4}\,dx$
9. $\int -3e^{2x}\,dx$
10. $\int 8e^{-x}\,dx$
11. $\int \frac{10}{x-3}\,dx$
12. $\int \frac{-14}{x+2}\,dx$
13. $\int 5xe^{3x^2}dx$
14. $\int 4xe^{x^2}\,dx$
15. $\int \frac{3x}{x^2-1}\,dx$
16. $\int \frac{-6x}{4-x^2}\,dx$
17. $\int \frac{12x^2\,dx}{(x^3+5)^4}$
18. $\int (x^2 - 5x)^4(2x - 5)\,dx$
19. $\int \frac{4x-5}{2x^2-5x}\,dx$
20. $\int \frac{8(2x+9)}{x^2+9x+1}\,dx$
21. $\int \frac{x^2}{e^{3x^3}}\,dx$
22. $\int 2xe^{3x^2+4}\,dx$
23. $\int -2e^{-5x}\,dx$
24. $\int 11e^{-4x}\,dx$
25. $\int \frac{3(\ln x)^5}{x}\,dx$
26. $\int \frac{10(\ln(5x+3))^2}{5x+3}\,dx$
27. $\int 25e^{-50x}\,dx$
28. $\int 4xe^{-3x^2+7}\,dx$
29. Explain how rectangles are used to approximate the area under a curve.
30. **(a)** Use five rectangles to approximate the area between the graph of $f(x) = 16x^2 - x^4 + 2$ and the x-axis from $x = -2$ to $x = 3$.
    (b) Use numerical integration to approximate this area.
31. Repeat Exercise 30 for the function $g(x) = -x^4 + 12x^2 + x + 5$ from $x = -3$ to $x = 3$.

32. Use four rectangles to approximate the area under the graph of $f(x) = 2x + 3$ and above the x-axis from $x = 0$ to $x = 4$. Let the height of each rectangle be the value of the function on the left side.

33. Find $\int_0^4 (2x + 3)\,dx$ by using the formula $A = \frac{1}{2}(B + b)h$ for the area of a trapezoid, where B and b are the lengths of the parallel sides and h is the distance between them. Compare with Exercise 32. If the answers are different, explain why.

34. Explain under what circumstances substitution is useful in integration.

Find each definite integral.

35. $\int_0^4 (x^3 - x^2)dx$

36. $\int_0^1 e^{3t}\,dt$

37. $\int_2^5 (6x^{-2} + x^{-3})\,dx$

38. $\int_2^3 (5x^{-2} + 7x^{-4})\,dx$

39. $\int_1^5 15x^{-1}\,dx$

40. $\int_1^6 \frac{8x^{-1}}{3}\,dx$

41. $\int_0^4 2e^{-5x}\,dx$

42. $\int_1^2 \frac{7}{2}e^{4x}\,dx$

43. $\int_{\sqrt{5}}^5 2x\sqrt{x^2 - 3}\,dx$

44. $\int_0^1 x\sqrt{5x^2 + 4}\,dx$

45. $\int_1^6 \frac{8x + 12}{x^2 + 3x + 9}\,dx$

46. $\int_1^4 \frac{3(\ln x)^5}{x}\,dx$

Find the area between the x-axis and f(x) over each of the given intervals.

47. $f(x) = e^x$; $[0, 2]$

48. $f(x) = 1 + e^{-x}$; $[0, 4]$

Management *Find the cost function for each of the marginal cost functions in Exercises 49–52.*

49. $C'(x) = 10 - 5x$; fixed cost is \$4.

50. $C'(x) = 2x + 3x^2$; 2 units cost \$12.

51. $C'(x) = 3\sqrt{2x - 1}$; 13 units cost \$270.

52. $C'(x) = \frac{6}{x + 1}$; fixed cost is \$18.

Work the given exercises.

53. **Business** The rate of change of sales of a new brand of tomato soup, in thousands, is given by

$$S(x) = \sqrt{x} + 3,$$

where x is the time in months that the new product has been on the market. Find the total sales after 9 months.

54. **Business** The curve shown in the accompanying figure gives the rate at which an investment accumulates income (in dollars per year). Use rectangles of width 2 units and height determined by the value of the function at the left endpoint to approximate the total income accumulated over 10 years.

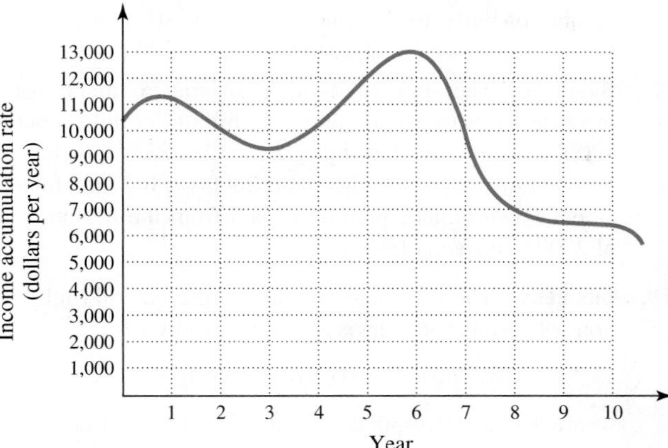

55. **Business** A manufacturer of electronic equipment requires a certain rare metal. He has a reserve supply of 4,000,000 units that he will not be able to replace. If the rate at which the metal is used is given by

$$f(t) = 100{,}000e^{.03t},$$

where t is the time in years, how long will it be before he uses up the supply? (*Hint*: Find an expression for the total amount used in t years, and set it equal to the known reserve supply.)

56. **Natural Science** The rate of change of the population of a rare species of Australian spider is given by

$$f(t) = 150 - \sqrt{3.2t + 4},$$

where $f(t)$ is the number of spiders present at time t, measured in months. Find the total number of additional spiders in the first 10 months.

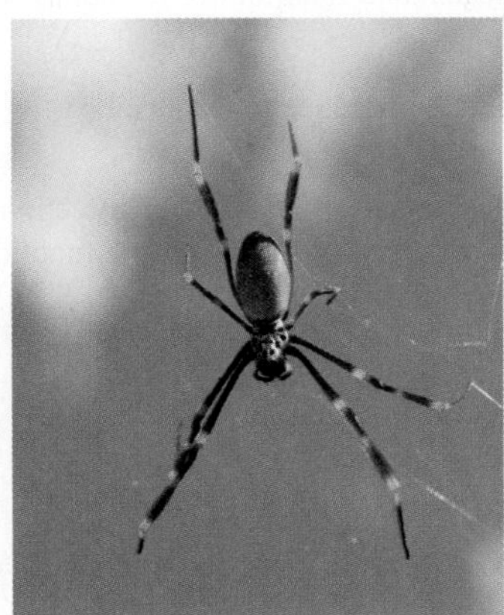

57. **Health** For the period 2000–2006, the rate of change for the number of births (in thousands per year) for mothers under the age of 20 can be approximated by

$$f(t) = .222t^3 + 2.179t^2 - 27.23t + 478.1,$$

where $t = 0$ corresponds to the year 2000.* Find the total number of births to such mothers for the six-year period from the beginning of 2000 to the end of 2005.

58. **Business** The rate at which insurance premiums increase for private automobiles (in billions of dollars per year) is approximated by $g(t) = 123.88e^{.0511t}$, where $t = 0$ corresponds to the year 2000.† Find the total amount of insurance premiums paid from the beginning of 2000 to the end of 2005.

59. **Business** The U.S. field production of crude oil (in billions of barrels per year) is approximated by

$$h(t) = -.0471t + 2.6391,$$

where $t = 5$ corresponds to the year 1995.‡ Find the total oil production from the beginning of 1995 to the end of 2008.

60. **Business** A company has installed new machinery that will produce a savings rate (in thousands of dollars) of

$$S'(x) = 225 - x^2,$$

where x is the number of years the machinery is to be used. The rate of additional costs (in thousands of dollars) to the company because of the new machinery is expected to be

$$C'(x) = x^2 + 25x + 150.$$

To maximize net savings, for how many years should the company use the new machinery? Find the net savings (in thousands of dollars) over this period.

61. Explain what consumers' surplus and producers' surplus are.

62. **Business** Suppose that the supply function of some commodity is

$$S(q) = q^2 + 5q + 100$$

and the demand function for the commodity is

$$D(q) = 350 - q^2.$$

(a) Find the producers' surplus.
(b) Find the consumers' surplus.

63. **Natural Science** The rate of infection of a disease (in people per month) is given by the function

$$I'(t) = \frac{100t}{t^2 + 2},$$

where t is the time in months since the disease broke out. Find the total number of infected people over the first four months of the disease.

64. **Business** Suppose waiting times (in minutes) in a teller line at a bank can be estimated by an exponential distribution and the function

$$f(x) = \frac{1}{2}e^{-x/2}.$$

(a) Find the probability that a customer in line waits less than 2 minutes.
(b) Find the probability that a customer in line waits more than 3 minutes.

65. **Natural Science** Suppose the amount of time between blinks of an eye for a given individual can be modeled by an exponential distribution and the function

$$f(t) = .44e^{-.44t},$$

where t is measured in seconds.
(a) Find the probability that the person will take between 1 and 2 seconds between blinks.
(b) Find the probability the person will take more than 4 seconds between blinks.

Use technology or a table of integrals to find the given antiderivatives.

66. $\displaystyle\int \frac{1}{\sqrt{x^2 - 64}}\,dx$

67. $\displaystyle\int \frac{10}{x\sqrt{25 + x^2}}\,dx$

68. $\displaystyle\int \frac{18}{x^2 - 9}\,dx$

69. $\displaystyle\int \frac{15x}{2x - 5}\,dx$

70. What is a differential equation? What is it used for?

Find general solutions for the given differential equations.

71. $\dfrac{dy}{dx} = 2x^3 + 6x + 5$

72. $\dfrac{dy}{dx} = x^2 + \dfrac{5x^4}{8}$

73. $\dfrac{dy}{dx} = \dfrac{3x + 1}{y}$

74. $\dfrac{dy}{dx} = \dfrac{e^x + x}{y - 1}$

Find particular solutions for the given differential equations.

75. $\dfrac{dy}{dx} = 5(e^{-x} - 1)$; $y = 17$ when $x = 0$

76. $\dfrac{dy}{dx} = \dfrac{x}{x^2 - 3} + 7$; $y = 52$ when $x = 2$

77. $(5 - 2x)y = \dfrac{dy}{dx}$; $y = 2$ when $x = 0$

78. $\sqrt{x}\dfrac{dy}{dx} = xy$; $y = 4$ when $x = 1$

79. **Social Science** Assume that beginning in 2000, the consumer price index (CPI) for urban consumers continuously increased at an annual rate of 2.64%.*

*U.S. National Center for Health Statistics.

†*Statistical Abstract of the United States*: 2009.

‡U.S. Energy Information Administration.

*Based on data from the U.S. Bureau of Labor Statistics.

(a) Find a differential equation that describes this situation.
(b) The CPI stood at 172.3 in 2000. Find the rule of a function that gives the CPI in year t, where $t = 0$ corresponds to the year 2000.
(c) Find the CPI in 2007.

80. **Business** The marginal profit (in hundreds of dollars) of a small bookstore is given by

$$\frac{dy}{dx} = 2.4e^{.1x},$$

where x is the number of books sold (in hundreds). Assume that profits were 0 initially.
(a) Find the profit when 800 books are sold.
(b) Find the profit when 1200 books are sold.

81. **Business** The rate at which a new worker in a certain factory produces items is given by

$$\frac{dy}{dx} = .2(125 - y),$$

where y is the number of items produced by the worker per day, x is the number of days worked, and the maximum production per day is 125 items. Assume that the worker produced 20 items the first day on the job ($x = 0$).
(a) Find the number of items the new worker will produce after 10 days.
(b) According to the function that is the solution of the differential equation, can the worker ever produce 125 items in a day?

82. **Physical Science** A roast at a temperature of 40° is put in a 300° oven. After 1 hour, the roast has reached a temperature of 150°. Newton's law of cooling states that

$$\frac{dT}{dt} = k(T - T_F),$$

where T is the temperature of an object, the surrounding medium has temperature T_F at time t, and k is a constant.
(a) Use Newton's law of cooling to find the temperature of the roast after 2.5 hours.
(b) How long does it take for the roast to reach a temperature of 250°?

83. **Natural Science** Find an equation relating x to y, given the following equations, which describe the interaction of two competing species and their growth rates:

$$\frac{dx}{dt} = .2x - .5xy;$$
$$\frac{dy}{dt} = -.3y + .4xy.$$

Find the values of x and y for which both growth rates are 0. (*Hint*: Solve the system of equations found by setting each growth rate equal to 0.)

CASE 13
Bounded Population Growth

The population growth model $\frac{dP}{dt} = kP$, whose solution is the exponential growth function $P = P_0e^{kt}$, can work well in a situation where a population is changing rapidly over a short period of time. But the model may not be useful in the long term. Exponential functions grow very large very quickly and will eventually exceed any values that the population under consideration may actually attain. For this reason, the exponential function is said to be *unbounded.* A more realistic growth model would take into account any natural bounds on the populationsize.

For example, consider a population of cattle or sheep that lives and grazes on a piece of farmland. Sheep, in particular, obtain the majority of their food intake from foraging in their environment. The sheep happily graze on the grass, plants, and weeds without exhausting the supply. This allows time for the plant life to grow back, and the eating–growing cycle continues. If the farmer populates the farmland with too many sheep, however, the plants may be eaten too quickly, failing to grow enough before some of the sheep develop health problems and possibly die. The maximum number of animals the farmland can comfortably sustain over a long period of time is called the *carrying capacity* of the land.

One model used for bounded population growth is the *logistic differential equation*

$$\frac{dP}{dt} = kP(C - P),$$

where k is a positive proportionality constant and C is the carrying capacity of the environment under consideration. When the population satisfies

$$0 < P < C,$$

the expression $kP(C - P)$ will be positive, and a positive derivative indicates that the population is increasing. If the population P exceeds the carrying capacity C, the factor $C - P$ becomes negative and

$$\frac{dP}{dt} < 0,$$

indicating that the population is decreasing, which must be the case when the carrying capacity is exceeded. The model thus agrees with the expected behavior of a bounded population.

For example, given an initial population of 10 sheep on land with carrying capacity $C = 200$ and a proportionality factor $k = .005$, the logistic differential equation becomes

$$\frac{dP}{dt} = .005P(200 - P).$$

Separating variables and integrating gives

$$\int \frac{1}{P(200 - P)}\, dP = \int .005dt. \qquad (1)$$

The right side of equation (1) is easily integrated:

$$\int .005dt = .005t + K. \qquad (2)$$

The left side of equation (1) does not match the standard integration formulas, but can be integrated by using the following identity, which you are asked to verify in Exercise 1:

$$\frac{1}{P(C - P)} = \frac{1/C}{P} + \frac{1/C}{C - P}. \qquad (3)$$

Applying this identity with $C = 200$ allows the left-hand side of equation (1) to be easily integrated:

$$\int \frac{1}{P(200 - P)}\, dP = \int \frac{1/200}{P}\, dP + \int \frac{1/200}{200 - P}\, dP$$

$$= \frac{1}{200}\ln(P) - \frac{1}{200}\ln(200 - P). \qquad (4)$$

Substituting equations (2) and (4) into equation (1) shows that

$$\frac{1}{200}\ln(P) - \frac{1}{200}\ln(200 - P) = .005t + K. \qquad (5)$$

In Exercise 2, you are asked to solve equation (5) for $P(t)$ and use the fact that $P(0) = 10$ to obtain the population function

$$P(t) = \frac{200e^t}{e^t + 19}.$$

After one year, the population would consist of $P(1) = 25$ sheep.

The graph of the population function in the accompanying figure illustrates the typical S-shaped curve for a population modeled by the logistic differential equation. The population increases rapidly at first, and then the rate of increase decreases as the population approaches the carrying capacity of its environment.

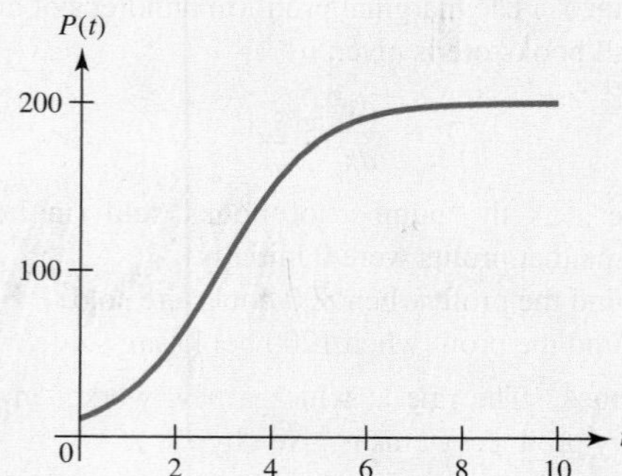

Before the logistic model can be applied, an estimate of the carrying capacity is needed. In the case of cattle and sheep, veterinary and agricultural science, along with hundreds of years of recorded experience, tell a farmer how much livestock can survive comfortably on a given plot of land. A much more difficult problem is to determine the carrying capacity of a population when many factors contribute to the carrying capacity.

Determining the carrying capacity of Earth, with respect to its human population, is one such problem. Factors such as energy supply, existing food sources, available farmland, climate, and quality of life all enter into an estimate. One estimate, assuming a relatively high standard of living for each inhabitant, puts the carrying capacity between 1.5 and 2 billion. Given the current population of approximately 6 billion, this estimate states that we have already exceeded the carrying capacity. More optimistic estimates take into account technologies that are the focus of active research. For example, some say by assuming that alternative energy sources, such as solar energy, can be harnessed fully, the carrying capacity can be increased to 23 billion. Further consideration of alternative food sources and methods provided by aquaculture and hydroponics raises that estimate to 50 billion. As you might expect, estimates of Earth's carrying capacity are a subject of great debate.

EXERCISES

1. Verify equation (3).
2. Use the techniques of Section 13.7 and the fact that $P(0) = 10$ to solve equation (5).
3. Assuming a carrying capacity of 20 billion people and a current population of 6 billion, determine when the world population will reach 18 billion. Use $k = .0011$, a value based on observed world population growth during the 1990s.
4. Under the assumptions of Exercise 3, will Earth's population ever reach 20 billion? Explain.

CHAPTER

14

Multivariate Calculus

Divers must take into account both water depths and dive times to determine safe water pressures and avoid compression sickness. Anyone who is active outside during cold weather may experience windchill (which is determined by air temperature and wind speed). The profit a business makes is dependent on a variety of factors, such as the cost of materials, wages, competitive price pressures, and interest rates. Accurate models of these and other real-life situations require functions of two or more variables. See Exercises 60 and 62 on page 925, and Exercise 30 on page 934.

Many of the ideas developed for functions of one variable also apply to functions of more than one variable. In particular, the fundamental idea of the derivative generalizes in a very natural way to functions of more than one variable.

14.1 ▶ Functions of Several Variables

Recall from Section 3.1 that a **function** consists of a set of inputs called the **domain**, a set of outputs called the **range**, and a **rule** by which each input determines exactly one output. For all of the functions studied previously in this book, both the inputs and outputs were real numbers. We now extend the concept of function to the case when the inputs are *ordered pairs* of real numbers.

Suppose, for example, that a company makes two styles of belts. It costs \$10 per belt to produce the first style and \$15 per belt to produce the second style. So the cost of producing x belts of the first style and y belts of the second style is $10x + 15y$. We can think of this cost function as a function of *two* variables whose rule is

$$C(x, y) = 10x + 15y.$$

For instance, the cost of producing 250 belts of the first style and 100 of the second style is denoted $C(250, 100)$ and is found by letting $x = 250$ and $y = 100$ in the rule of the function:

$$C(250,100) = 10 \cdot 250 + 15 \cdot 100 = 2500 + 1500 = \$4000.$$

Here is the general definition.

A **function of two variables** consists of

a set of inputs that are *ordered pairs* of real numbers (the **domain**);

a set of outputs that are real numbers (the **range**); and

a rule by which each input pair produces exactly one output number.

Essentially the same definition works for functions of three, four, or more variables—just change the inputs to ordered *triples* of real numbers for functions of three variables, to ordered *quadruples* of real numbers for functions of four variables, and so on. Functions of more than one variable are called **multivariate functions**.

Many familiar formulas can be considered as the rules of multivariate functions. For example,

- the volume V of a cone is given by a function of two variables,

$$V(r, h) = \frac{1}{3}\pi r^2 h,$$

  where r is the radius of the base and h is the height of the cone;*

- the future value A of an amount deposited at compound interest is given by a function of three variables,

$$A(P, i, n) = P(1 + i)^n,$$

*See Table 1 in Appendix B.

where P is the amount deposited, i is the interest rate per period, and n is the number of periods.*

Unless otherwise stated, the domain of a multivariate function defined by a formula or equation is assumed to be the set of all inputs for which the rule of the function produces a real number as output.

Example 1 Let $f(x, y) = 4x^2 + 2xy + 3/y$, and find each of the given quantities.

(a) $f(-1, 3)$

Solution Replace x with -1 and y with 3:

$$f(-1, 3) = 4(-1)^2 + 2(-1)(3) + \frac{3}{3} = 4 - 6 + 1 = -1.$$

(b) The domain of f

Solution Because of the quotient $3/y$, it is not possible to replace y with 0, so the domain of the function f consists of all ordered pairs (x, y) such that $y \neq 0$. ✓1

✓Checkpoint 1

Let $f(x, y) = x^3 - 4x^2 + xy$. Find

(a) $f(2, 4)$;

(b) $f(-2, 3)$.

Answers: See page 915.

Example 2 Let x represent the number of milliliters (ml) of carbon dioxide released by the lungs in 1 minute. Let y be the change in the carbon dioxide content of the blood as it leaves the lungs. (y is measured in ml of carbon dioxide per 100 ml of blood.) The total output of blood from the heart in 1 minute (measured in ml) is given by C, where C is a function of x and y such that

$$C(x, y) = \frac{100x}{y}.$$

Find $C(320, 6)$.

Solution Replace x with 320 and y with 6 to get

$$C(320, 6) = \frac{100(320)}{6}$$
$$\approx 5333 \text{ ml of blood per minute.}$$

GRAPHING EQUATIONS IN THREE VARIABLES

Equations in two variables, such as $3x + 2y = 5$, are graphed by using an x-axis and a y-axis to locate points in a plane. The plane determined by the x- and y-axes is called the ***xy*-plane**. In order to graph an equation in three variables, such as $2x + y + z = 6$, a third axis is needed—the z-axis, which goes through the origin in the xy-plane and is perpendicular to both the x-axis and the y-axis.

*See page 268.

Figure 14.1 shows one possible way to draw the three axes. In Figure 14.1, the yz-plane is in the plane of the page, with the x-axis perpendicular to the plane of the page.

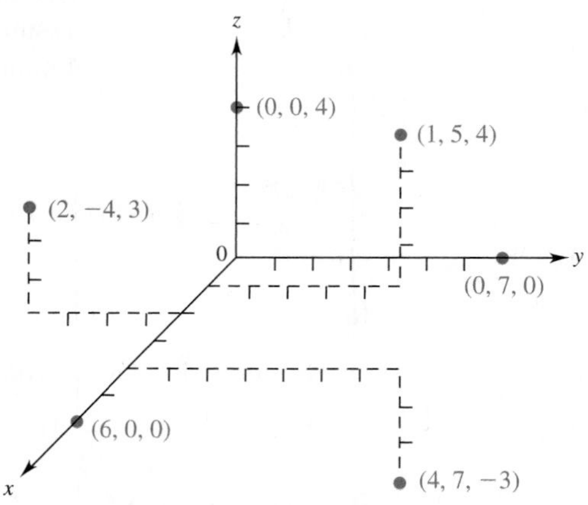

FIGURE 14.1

Just as we graphed ordered pairs earlier, we can now graph **ordered triples** of the form (x, y, z). For example, to locate the point corresponding to the ordered triple $(2, -4, 3)$, start at the origin and go 2 units along the positive x-axis. Then go 4 units in a negative direction (to the left), parallel to the y-axis. Finally, go up 3 units, parallel to the z-axis. The point representing $(2, -4, 3)$ is shown in Figure 14.1, along with several other points. The region of three-dimensional space where all coordinates are positive is called the **first octant**. 2✓

✓Checkpoint 2

Locate (3, 2, 4) on the coordinate system shown here.

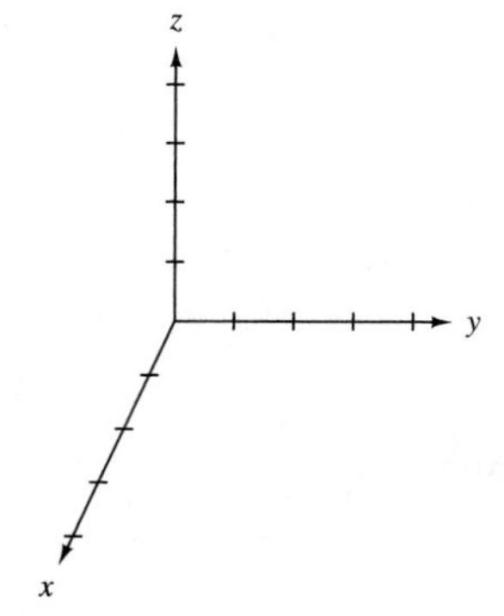

Answer: See page 915.

Some simple equations in three variables can be graphed by hand. In Chapter 2, we saw that the graph of $ax + by = c$ (where a, b, and c are constants and not both a and b are zero) is a straight line. This result generalizes to three dimensions.

Planes

If a, b, c, and d are real numbers, with a, b, and c not all zero, then the graph of

$$ax + by + cz = d$$

is a plane.

Example 3 Graph $2x + y + z = 6$.

Solution By the result shown in the preceding box, the graph of this equation is a plane. Earlier, we graphed straight lines by finding x- and y-intercepts. A similar idea helps graph a plane. To find the x-intercept—the point where the graph crosses the x-axis—let $y = 0$ and $z = 0$:

$$2x + 0 + 0 = 6$$
$$x = 3.$$

The point (3, 0, 0) is on the graph. Letting $x = 0$ and $z = 0$ gives the point (0, 6, 0), while $x = 0$ and $y = 0$ lead to (0, 0, 6). The plane through these three points includes the triangular surface shown in Figure 14.2. This region is the first-octant part of the plane that is the graph of $2x + y + z = 6$.

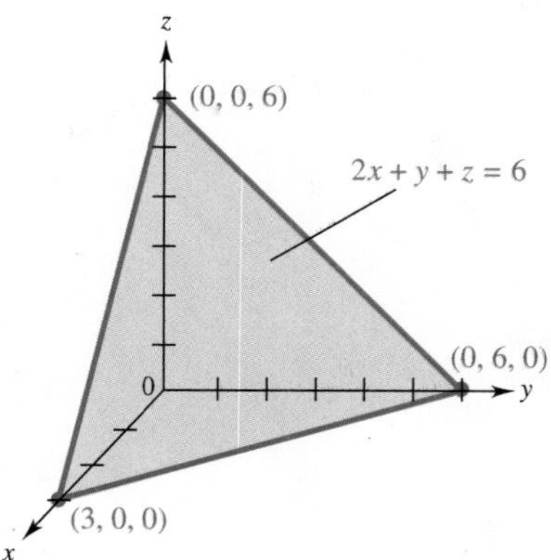

FIGURE 14.2

Throughout this discussion, we assume that all equations involve three variables. Consequently, an equation such as $x + z = 6$ is understood to have a y-term with zero coefficient: $x + 0y + z = 6$.

Example 4 Graph $x + z = 6$.

Solution To find the x-intercept, let $z = 0$, giving (6, 0, 0). If $x = 0$, we get the point (0, 0, 6). Because there is no y in the equation $x + z = 6$, there can be no y-intercept. A plane that has no y-intercept is parallel to the y-axis. The first-octant portion of the graph of $x + z = 6$ is shown in Figure 14.3.

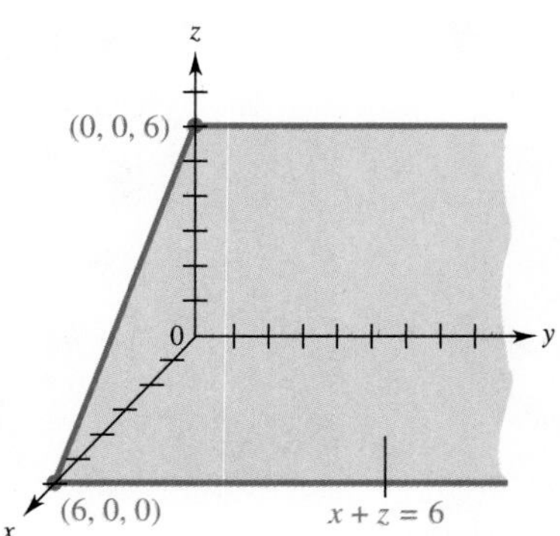

FIGURE 14.3

Example 5 Graph the three planes whose equations are respectively

$$x = 0, \quad y = 0, \quad \text{and} \quad z = 0.$$

Solution The graph of $x = 0$ consists of all points with first coordinate 0. These are the points in the ***yz*-plane** shown in Figure 14.4. Similarly, the graph of $y = 0$ consists of all points with second coordinate 0, that is, all points in the ***xz*-plane** in Figure 14.4. Finally, the graph of $z = 0$ consists of all points with third coordinate 0—the ***xy*-plane** in Figure 14.4.

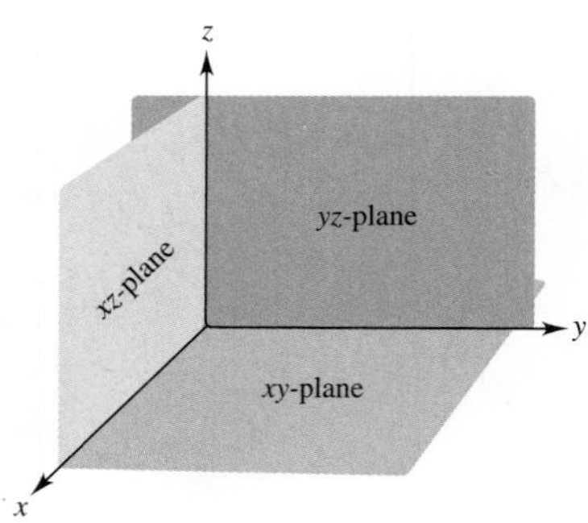

FIGURE 14.4

Example 6 Graph each of the given equations.

(a) $x = 3$

Solution This graph, which goes through (3, 0, 0), can have no y-intercept and no z-intercept. It is, therefore, a plane parallel to the y-axis and the z-axis and, therefore, to the yz-plane. The first-octant portion of the graph is shown in Figure 14.5.

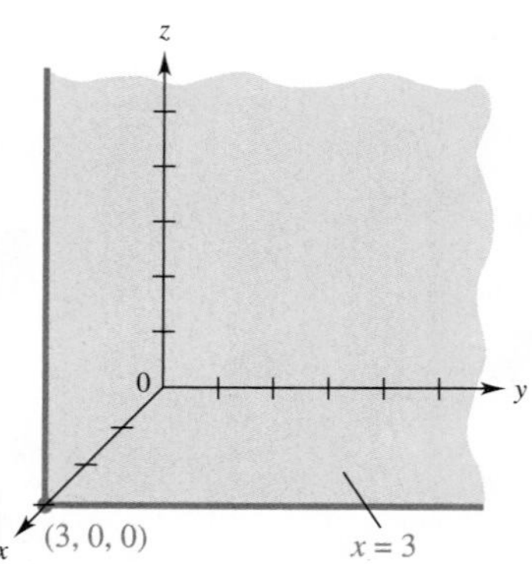

FIGURE 14.5

(b) $y = 4$

Solution This graph goes through (0, 4, 0) and is parallel to the xz-plane. The first-octant portion of the graph is shown in Figure 14.6.

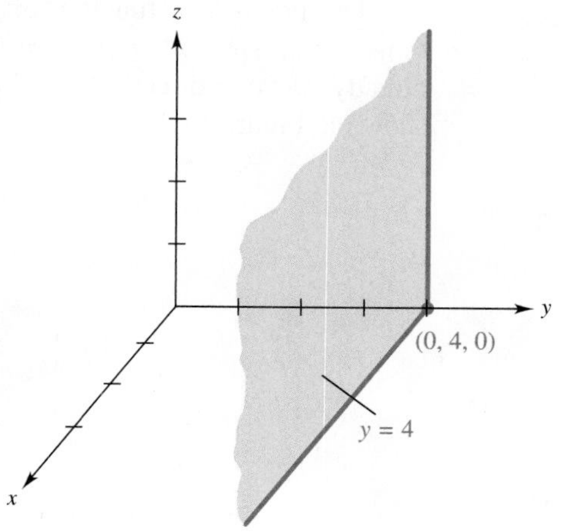

FIGURE 14.6

(c) $z = 1$.

Solution The graph is a plane parallel to the xy-plane and passing through (0, 0, 1). Its first-octant portion is shown in Figure 14.7. [3]✓

Checkpoint 3

Describe each graph and give any intercepts.

(a) $2x + 3y - z = 4$

(b) $x + y = 3$

(c) $z = 2$

Answers: See page 915.

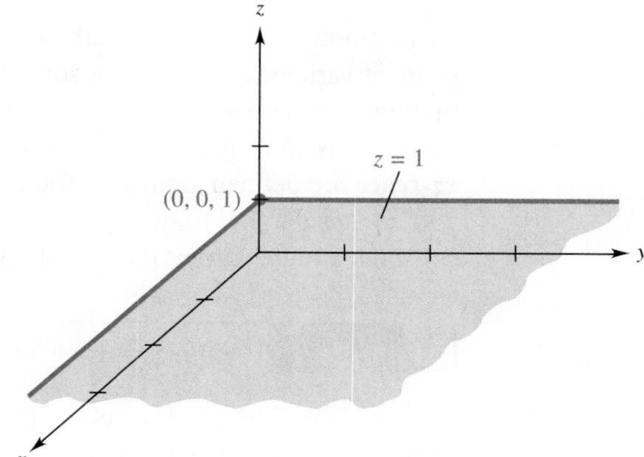

FIGURE 14.7

GRAPHING FUNCTIONS OF TWO VARIABLES

The graph of a function of one variable, $y = f(x)$, is a curve in the plane. If x_0 is in the domain of f, the point $(x_0, f(x_0))$ on the graph lies directly above or directly below (or possibly on) the number x_0 on the x-axis, as shown in Figure 14.8.

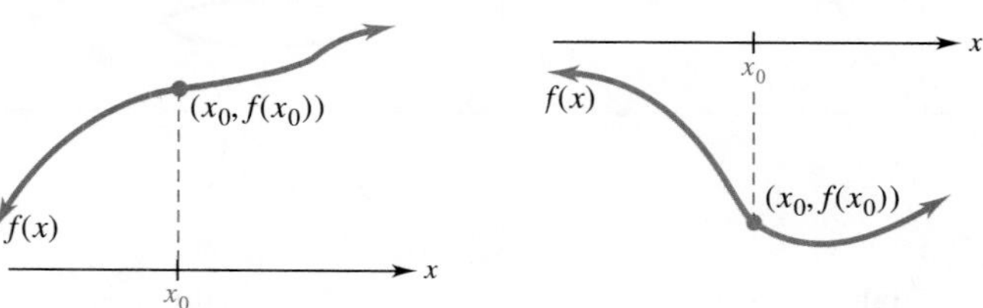

FIGURE 14.8

The graph of a function of two variables, $z = f(x, y)$, is a **surface** in three-dimensional space. If (x_0, y_0) is in the domain of f, the point $(x_0, y_0, f(x_0, y_0))$ lies directly above or directly below (or possibly on) the point (x_0, y_0) in the xy-plane, as shown in Figure 14.9.

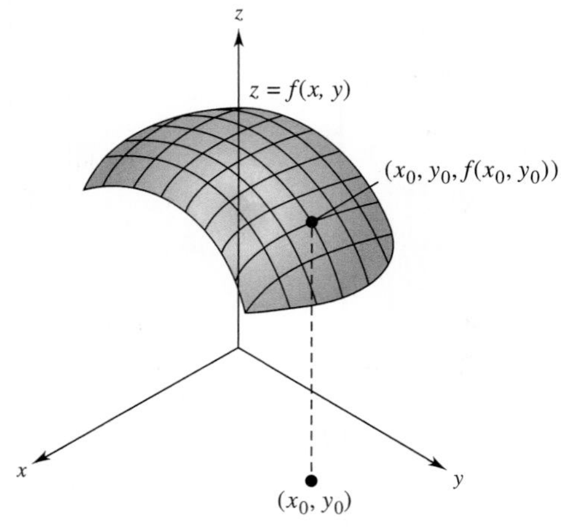

FIGURE 14.9

Computer software is available for drawing the graphs of functions of two independent variables, but you can sometimes get a good picture of a graph without it by finding various **traces**—the curves that result when a surface is cut by a plane. The ***xy*-trace** is the intersection of the surface with the xy-plane. The ***yz*-trace** and ***xz*-trace** are defined similarly. You can also determine the intersection of the surface with planes parallel to the xy-plane. Such planes are of the form $z = k$, where k is a constant, and the curves that result when they cut the surface are called **level curves**.

Example 7 Graph $z = x^2 + y^2$.

Solution The yz-plane is the plane in which every point has first coordinate 0, so its equation is $x = 0$. When $x = 0$, the equation becomes $z = y^2$, which is the equation of a parabola in the yz-plane, as shown in Figure 14.10(a). Similarly, to find

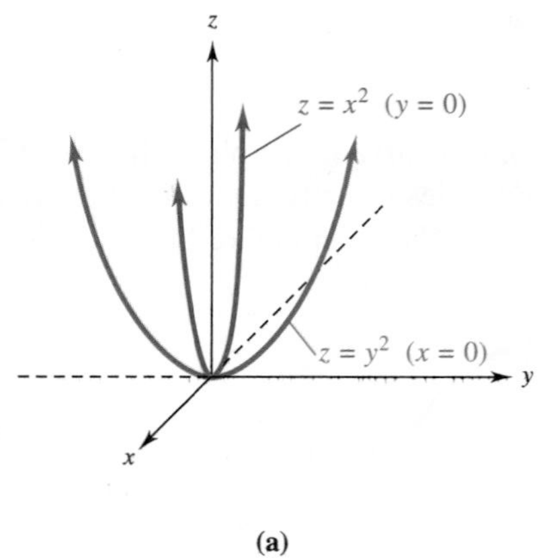

(a)

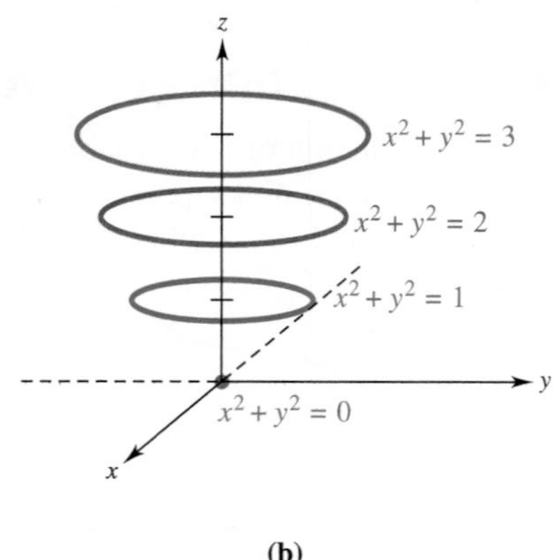

(b)

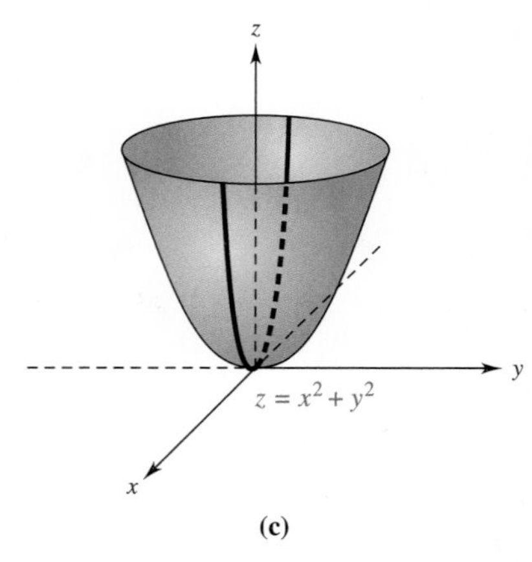

(c)

FIGURE 14.10

the intersection of the surface with the xz-plane (whose equation is $y = 0$), let $y = 0$ in the equation. It then becomes $z = x^2$, which is the equation of a parabola in the xz-plane (shown in Figure 14.10(a)). The xy-trace (the intersection of the surface with the plane $z = 0$) is the single point (0, 0, 0), because $x^2 + y^2$ is never negative and is equal to 0 only when $x = 0$ and $y = 0$.

Next, we find the level curves by intersecting the surface with the planes $z = 1$, $z = 2$, $z = 3$, etc. (all of which are parallel to the xy-plane). In each case, the result is a circle,

$$x^2 + y^2 = 1, \qquad x^2 + y^2 = 2, \qquad x^2 + y^2 = 3,$$

and so on, as shown in Figure 14.10(b). Drawing the traces and level curves on the same set of axes suggests that the graph of $z = x^2 + y^2$ is the bowl-shaped figure, called a **paraboloid**, that is shown in Figure 14.10(c).

Figure 14.11 shows the level curves from Example 7 plotted in the xy-plane. The picture can be thought of as a topographical map that describes the surface generated by $z = x^2 + y^2$, just as the topographical map in Figure 14.12 describes the surface of the land in a part of New York State.

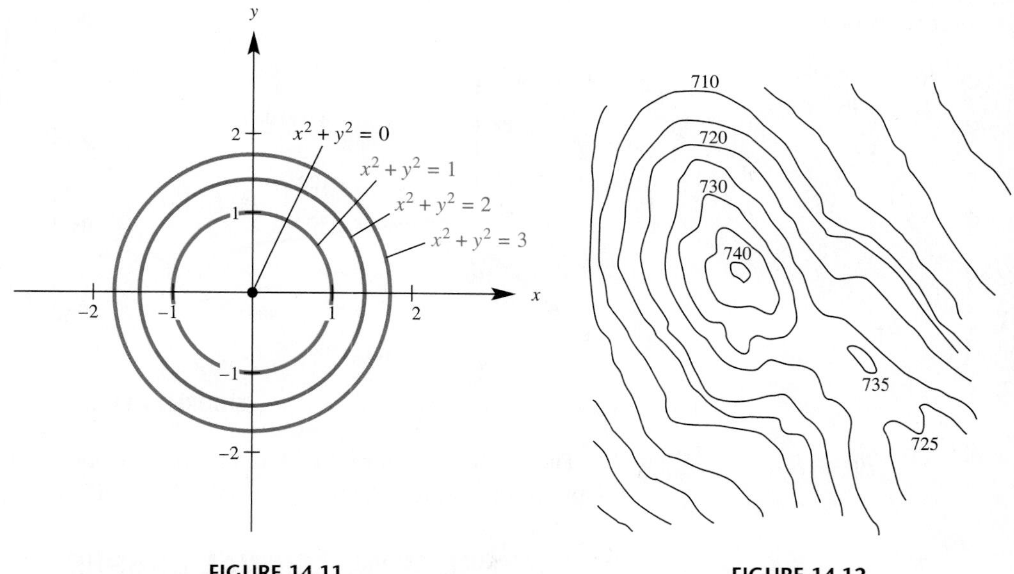

FIGURE 14.11

FIGURE 14.12

APPLICATIONS

One application of level curves in economics occurs with production functions. A **production function** $z = f(x, y)$ is a function that gives the quantity z of an item produced as a function of x and y, where x is the amount of labor and y is the amount of capital (in appropriate units) needed to produce z units. If the production function has the special form $z = P(x, y) = Ax^a y^{1-a}$, where A is a constant and $0 < a < 1$, the function is called a **Cobb–Douglas production function**.

Example 8 Find the level curve at a production of 100 items for the Cobb–Douglas production function $z = x^{2/3} y^{1/3}$.

Solution Let $z = 100$ and solve for y to get

$$100 = x^{2/3}y^{1/3}$$

$$\frac{100}{x^{2/3}} = y^{1/3}.$$

Now cube both sides to express y as a function of x:

$$y = \frac{100^3}{x^2}$$

$$y = \frac{1{,}000{,}000}{x^2}.$$

The level curve of height 100 is graphed in three dimensions in Figure 14.13(a) and on the familiar xy-plane in Figure 14.13(b). The points on the graph correspond to those values of x and y that lead to the production of 100 items. 4✓

✓Checkpoint 4

Find the equation of the level curve for the production of 100 items if the production function is $z = 5x^{1/4}y^{3/4}$.

Answer: See page 915.

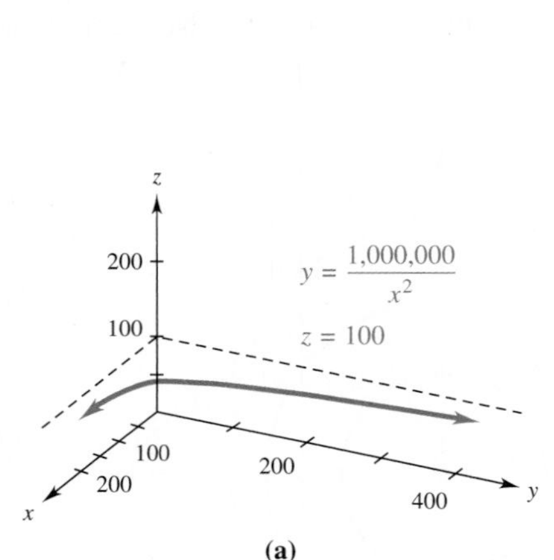

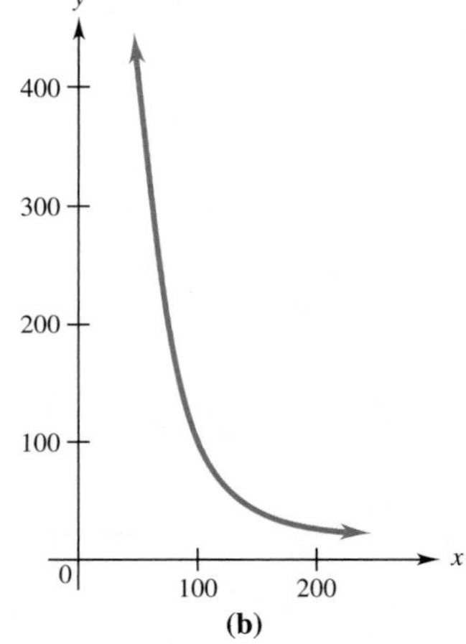

FIGURE 14.13

The curve in Figure 14.13 is called an *isoquant*, for *iso* (equal) and *quant* (amount). In Example 8, the "amounts" all "equal" 100.

OTHER THREE-DIMENSIONAL GRAPHS

The graphing techniques discussed up to now may not always be effective. So technology is normally used to graph equations in three variables. Figure 14.14 shows the graph of $z = x^2 + y^2$, which was discussed in Example 7, on a TI-89 graphing

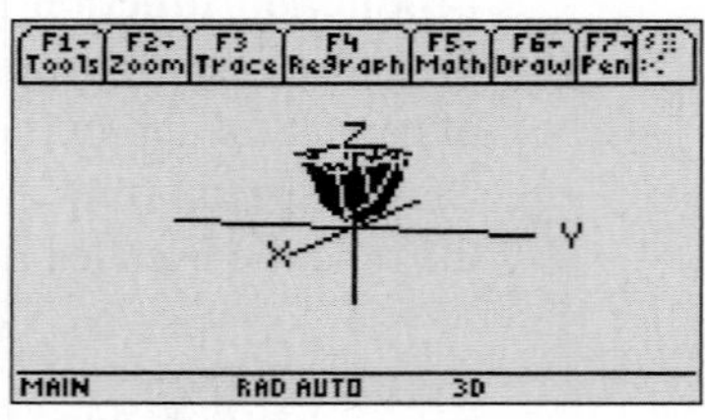

FIGURE 14.14

calculator. (Most graphing calculators cannot do three-dimensional graphing.) Figure 14.15 shows the same graph drawn by the computer program Maple™. The computer-generated graph has more detail than a graphing calculator can produce.

Paraboloid,
$z = x^2 + y^2$

xy-trace: point
yz-trace: parabola
xz-trace: parabola

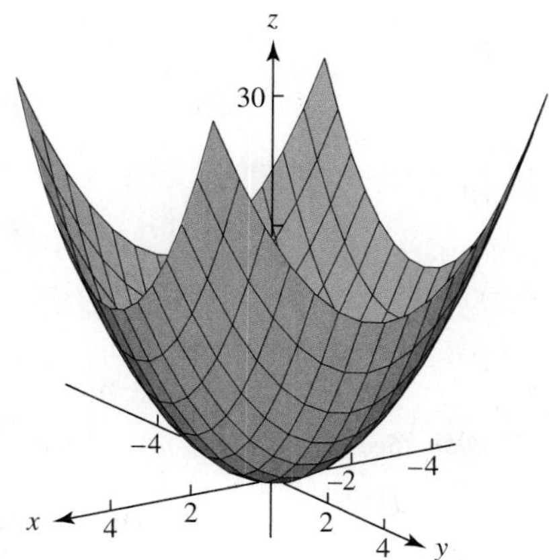

FIGURE 14.15

Figure 14.16 shows some other computer-generated graphs.

Ellipsoid,
$$\frac{x^2}{a^2} + \frac{y^2}{b^2} + \frac{z^2}{c^2} = 1$$

xy-trace: ellipse
yz-trace: ellipse
xz-trace: ellipse

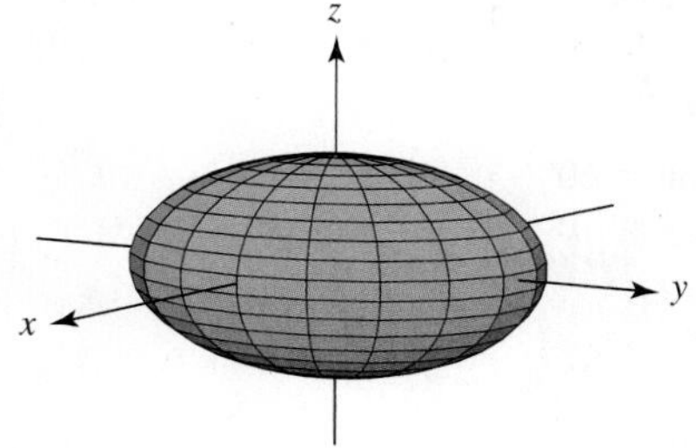

Hyperboloid of Two Sheets,
$-x^2 - y^2 + z^2 = 1$

xy-trace: none
yz-trace: hyperbola
xz-trace: hyperbola

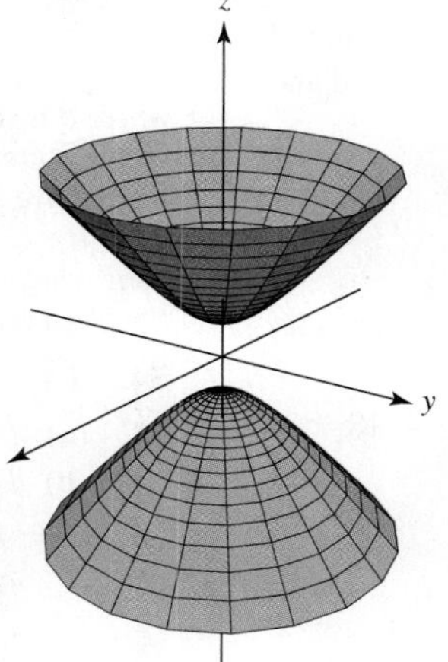

Hyperbolic Paraboloid,
$x^2 - y^2 = z^2$
(sometimes called a **saddle**)

xy-trace: two intersecting lines
yz-trace: parabola
xz-trace: parabola

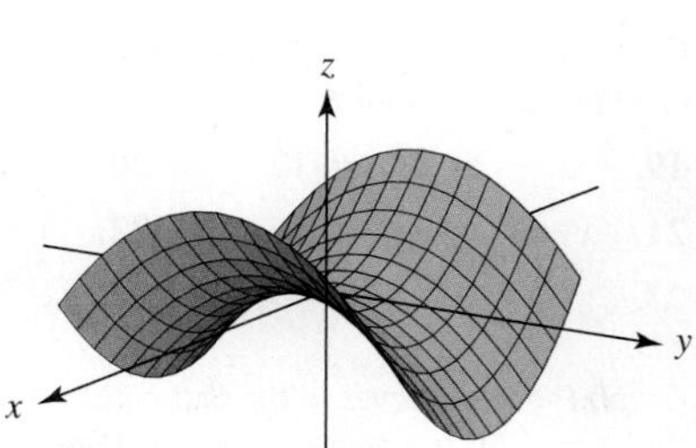

FIGURE 14.16

14.1 ▶ Exercises

For each of the given functions, find f(2, −1), f(−4, 1), f(−2, −3), and f(0, 8). (See Examples 1 and 2.)

1. $f(x, y) = 5x + 2y - 4$ **2.** $f(x, y) = 2x^2 - xy + y^2$

3. $f(x, y) = \sqrt{y^2 + 2x^2}$ **4.** $f(x, y) = \dfrac{3x + 4y}{\ln|x|}$

5. Let $g(x, y) = -x^2 - 4xy + y^3$. Find the following:
(a) $g(-2, 4)$;
(b) $g(-1, -2)$;
(c) $g(-2, 3)$;
(d) $g(5, 1)$.

6. Let $f(x, y) = \dfrac{\sqrt{9x + 5y}}{\log x}$. Find the following:
(a) $f(10, 2)$;
(b) $f(100, 1)$;
(c) $f(1000, 0)$;
(d) $f\left(\dfrac{1}{10}, 5\right)$.

Find the domain of the given function. (See Example 1.)

7. $f(x, y) = x^2 + y^2$ **8.** $f(x, y) = \sqrt{x} - \sqrt{y}$

9. $B(h, w) = \dfrac{w}{h^2}$ **10.** $K(r, s, t) = \dfrac{\sqrt{r - s}}{t + 1}$

11. $h(u, v, w) = \left|\dfrac{u + v}{v - w}\right|$ **12.** $g(x, y, z) = \dfrac{3x + 4y}{\ln z}$

13. Write the rule of a function that gives the perimeter of a rectangle of length l and width w.

14. Write the rule of a function that gives the area of a triangle of base b and height h.

15. Write the rule of a function that gives the future value of an ordinary annuity of n payments, with payment R per period and interest rate i per period.

16. Write the rule of a function that gives the present value of an annuity of n payments, with payment R per period and interest rate i per period.

17. What are the xy-, xz-, and yz-traces of a graph?

18. What is a level curve?

Graph the first-octant portion of each of the given planes. (See Examples 3, 4, and 6.)

19. $3x + 2y + z = 12$ **20.** $2x + 3y + 3z = 18$

21. $x + y = 4$ **22.** $y + z = 3$

23. $z = 4$ **24.** $y = 3$

Graph the level curves in the first octant at heights of z = 0, z = 2, and z = 4 for the given equations. (See Example 7.)

25. $3x + 2y + z = 18$ **26.** $x + 3y + 2z = 8$

27. $y^2 - x = -z$ **28.** $2y - \dfrac{x^2}{3} = z$

Business *Find the level curve at a production of 500 for each of the production functions in Exercises 29 and 30. Graph each function on the xy-plane. (See Example 8.)*

29. The production function z for the United States was once estimated as $z = x^{.7}y^{.3}$, where x stands for the amount of labor and y stands for the amount of capital.

30. A study of the connection between immigration and the fiscal problems associated with the aging of the baby-boom generation considered a production function of the form $z = x^{.6}y^{.4}$, where x represents the amount of labor and y the amount of capital.*

31. **Business** For the function in Exercise 29, what is the effect on z of doubling x? of doubling y? of doubling both x and y?

32. **Business** If labor (x) costs \$200 per unit, materials (y) cost \$100 per unit, and capital (z) costs \$50 per unit, write a function for total cost.

33. **Business** The production of a precision camera is given by

$$P(x, y) = 100\left(\frac{3}{5}x^{-2/5} + \frac{2}{5}y^{-2/5}\right)^{-5},$$

where x is the amount of labor in work-hours and y is the amount of capital.
(a) What is the production when 32 work-hours and 1 unit of capital are provided?
(b) Find the production when 1 work-hour and 32 units of capital are provided.
(c) If 32 work-hours and 243 units of capital are used, what is the production output?

Natural Science *The temperature–humidity index (THI), sometimes called the discomfort index, is given by*

$$f(d, w) = 15 + .4(d + w),$$

where d is the "dry-bulb temperature" (essentially the ambient-air temperature) and w is the "wet-bulb temperature" (which reflects the relative humidity as well as the temperature). When the THI is 75 or more, most people are uncomfortable. Find the THI and determine whether most people feel comfortable or uncomfortable under the given conditions.

34. $d = 72°$ and $w = 70°$

35. **(a)** $d = 85°$ and $w = 60°$
(b) $d = 85°$ and $w = 70°$

36. $d = 90°$ and $w = 60°$

*Storesletten, Kjetil, "Sustaining Fiscal Policy Through Immigration," *Journal of Political Economy*, Vol. 108, No. 2, April 2000, pp. 300–323.

37. Natural Science The surface area of a human (in square meters) is approximated by

$$A = .202W^{.425}H^{.725},$$

where W is the weight of the person in kilograms and H is the height in meters.* Find A for the given data.

(a) Weight = 72 kg; height = 1.78 m

(b) Weight = 65 kg; height = 1.40 m

(c) Weight = 70 kg; height = 1.60 m

(d) Using your weight and height, find your own surface area.

38. Natural Science The oxygen consumption of a well-insulated mammal that is not sweating is approximated by

$$m = \frac{2.5(T - F)}{w^{.67}},$$

where T is the internal body temperature of the animal (in °C,) F is the temperature of the outside of the animal's fur (in °C,) and w is the animal's weight in kilograms.* Find m for the given data.

(a) Internal body temperature = 38°C; outside temperature = 6°C; weight = 32 kg

(b) Internal body temperature = 40°C; outside temperature = 20°C; weight = 43 kg

Finance *The multiplier function*

$$M(n, i, t) = \frac{(1 + i)^n(1 - t) + t}{[1 + (1 - t)i]^n}$$

compares the growth of an Individual Retirement Account (IRA) with the growth of the same amount (at the same interest rate for the same period of time) in an ordinary savings account. The savings account is subject to federal income tax, but the IRA is tax free (until money is withdrawn). The three variables are the number of years, n; the annual interest rate i; and the marginal income tax rate t of the saver. When $M > 1$, the IRA grows faster than the savings account. In each of the next two exercises, find the multiplier and determine which account grows faster.

39. Money can be invested for 25 years at 5%, and the marginal income tax rate is 33%.

40. Money is invested for 40 years at 6%, and the marginal income tax rate is 28%.

41. Natural Science An article entitled "How Dinosaurs Ran" explains that the locomotion of different-sized animals can be compared when they have the same Froude number, defined as

$$F = \frac{v^2}{gl},$$

where v is the dinosaur's velocity, g is the acceleration due to gravity (9.81 m/sec^2), and l is the dinosaur's leg length.*

(a) One result described in the article is that different animals change from a trot to a gallop at the same Froude number, roughly 2.56. Find the velocity at which this change occurs for a ferret, with a leg length of .09 m, and a rhinoceros, with a leg length of 1.2 m.

(b) Ancient footprints in Texas of a sauropod, a large herbivorous dinosaur, are roughly 1 m in diameter, corresponding to a leg length of roughly 4 m. By comparing the stride divided by the leg length with that of various modern creatures, it can be determined that the Froude number for these dinosaurs is roughly .025. How fast were the sauropods traveling?

42. Natural Science Using data collected by the U.S. Forest Service, we can estimate the annual number of deer–vehicle accidents for any given county in Ohio with the function

$$A(L, T, U, C) = 53.02 + .383L + .0015T + .0028U - .0003C,$$

where A is the estimated number of accidents, L is the road length in kilometers, T is the total county land area in hundred acres (Ha), U is the urban land area in hundred acres, and C is the number of hundred acres of cropland.†

(a) Use this formula to estimate the number of deer–vehicle accidents for Mahoning County, where $L = 266$ km, $T = 107{,}484$ Ha, $U = 31{,}697$ Ha, and $C = 24{,}870$ Ha. The actual value was 396.

(b) Given the magnitude and nature of the input numbers, which of the variables have the greatest potential to influence the number of deer–vehicle accidents? Explain your answer.

43. Business Extra postage is charged for parcels sent by U.S. mail that are more than 108 inches in length and

*Exercises 37 and 38 from Clow, Duane J., and N. Scott Urquhart, *Mathematics in Biology.* Lanham, MD: Ardsley House, 1984.

*R, MacNeill Alexander, "How Dinosaurs Ran," *Scientific American* 264 (April 1991): 4.

†Aaron Iverson and Louis Iverson, "Spatial and Temporal Trends of Deer Harvest and Deer–Vehicle Accidents in Ohio," *Ohio Journal of Science* 99 (1999): 84–94.

girth combined. (Girth is the distance around the parcel perpendicular to its length. See the accompanying figure.) Express the combined length and girth as a function of L, W, and H.

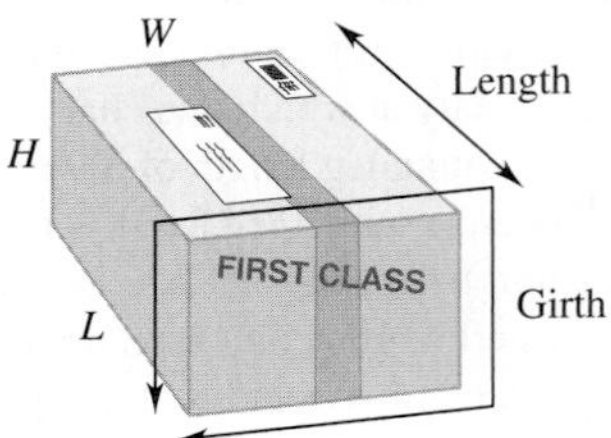

44. Natural Science Using data collected by the U.S. Forest Service, we can estimate the annual number of deer that are harvested for any given county in Ohio with the function

$$N(R, C) = 329.32 + .0377R - .0171C,$$

where N is the estimated number of harvested deer, R is the rural land area in hundred acres (Ha), and C is the number of hundred acres of cropland.* Use this formula to estimate the number of harvested deer for Tuscarawas County, where $R = 141{,}319$ Ha and $C = 37{,}960$ Ha. The actual value in 1995 was 4925 deer harvested.

By considering traces, match each equation in Exercises 45–50 with its graph in (a)–(f).

45. $z = x^2 + y^2$

46. $z^2 - y^2 - x^2 = 1$

47. $x^2 - y^2 = z$

48. $z = y^2 - x^2$

49. $\dfrac{x^2}{16} + \dfrac{y^2}{25} + \dfrac{z^2}{4} = 1$

50. $z = 5(x^2 + y^2)^{-1/2}$

(a)

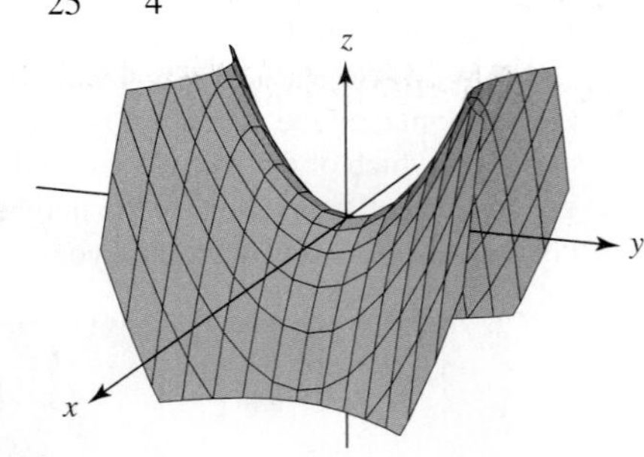

(b)

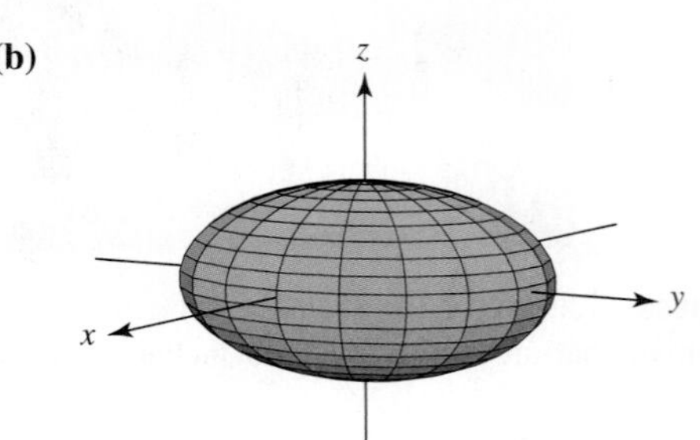

(c)

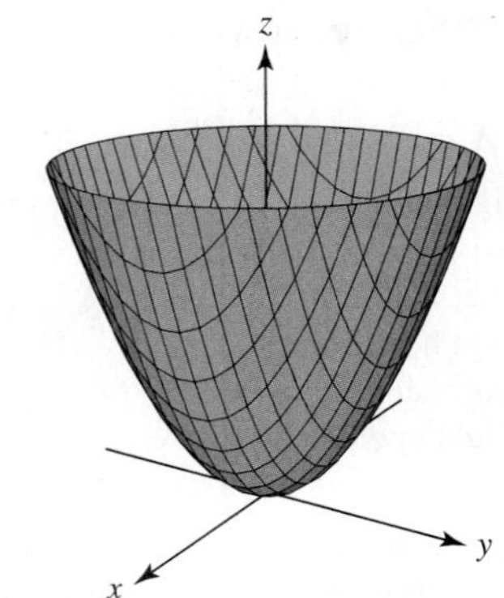

(d)

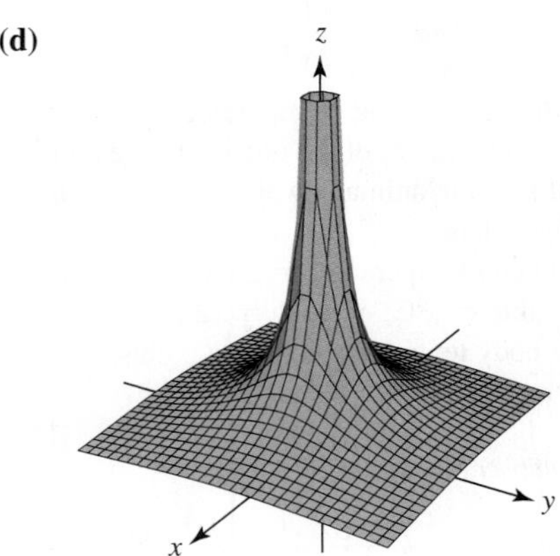

(e)

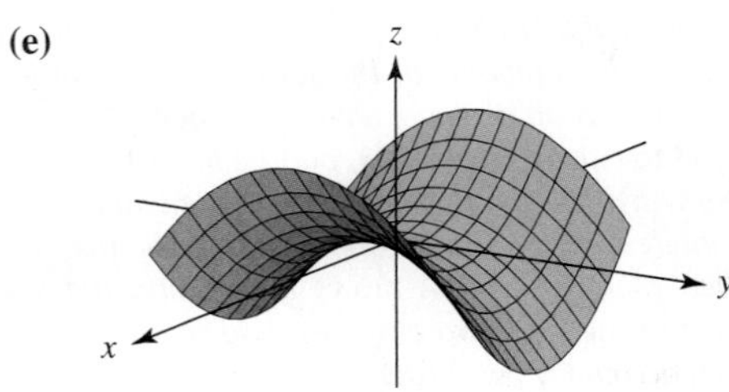

(f)

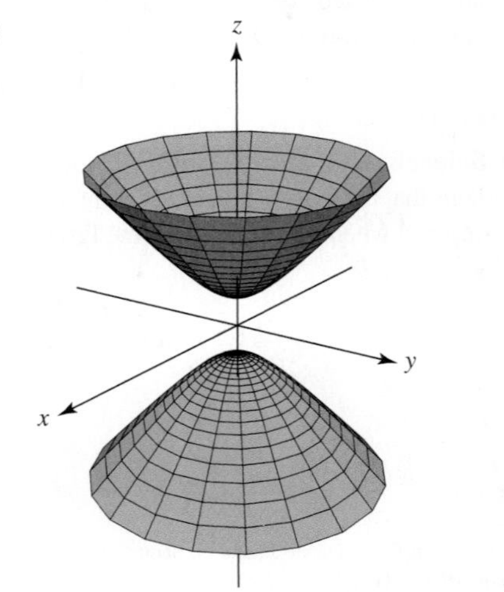

*Aaron Iverson and Louis Iverson, "Spatial and Temporal Trends of Deer Harvest and Deer–Vehicle Accidents in Ohio," *Ohio Journal of Science* 99 (1999): 84–94.

✓ Checkpoint Answers

1. (a) 0 (b) -30

2.

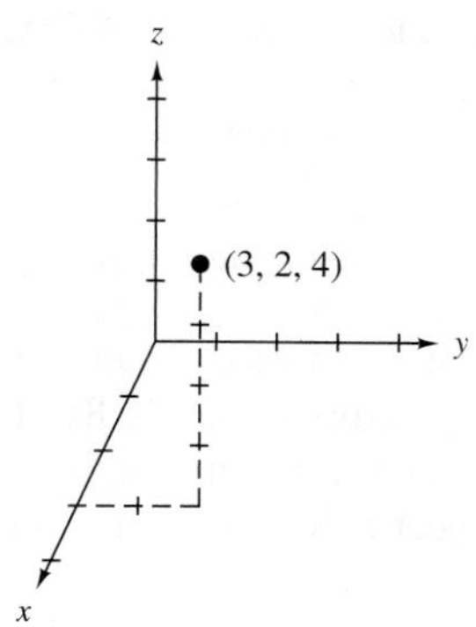

3. (a) A plane; (2, 0, 0), (0, 4/3, 0), (0, 0, -4)
   (b) A plane parallel to the z-axis; (3, 0, 0), (0, 3, 0)
   (c) A plane parallel to the x-axis and the y-axis; (0, 0, 2)

4. $y = \frac{20^{4/3}}{x^{1/3}}$

14.2 ▸ Partial Derivatives

A small firm makes only two products: emergency radios and LED crank flashlights. The profits of the firm are given by

$$P(x, y) = 40x^2 - 10xy + 5y^2 - 80,$$

where x is the number of units of radios sold and y is the number of units of flashlights sold. How will a change in x or y affect P?

Suppose that sales of radios have been steady at 10 units; only the sales of flashlights vary. The management would like to find the marginal profit with respect to y, the number of flashlights sold. Recall that marginal profit is given by the derivative of the profit function. Here, x is fixed at 10. Using this information, we begin by finding a new function, $f(y) = P(10, y)$. Let $x = 10$ to get

$$\begin{aligned} f(y) = P(10, y) &= 40(10)^2 - 10(10)y + 5y^2 - 80 \\ &= 3920 - 100y + 5y^2. \end{aligned}$$

The function $f(y)$ shows the profit from the sale of y flashlights, assuming that x is fixed at 10 units. Find the derivative df/dy to get the marginal profit with respect to y:

$$\frac{df}{dy} = -100 + 10y.$$

In this example, the derivative of the function $f(y)$ was taken with respect to y only; we assumed that x was fixed. To generalize, let $z = f(x, y)$. An intuitive definition of the *partial derivatives* of f with respect to x and y follows.

Partial Derivatives (Informal Definition)*

The **partial derivative of f with respect to x** is the derivative of f obtained by treating x as a variable and y as a constant.

The **partial derivative of f with respect to y** is the derivative of f obtained by treating y as a variable and x as a constant.

*A formal definition is given on page 922.

The symbols $f_x(x, y)$ (no prime used), $\partial z/\partial x$, and $\partial f/\partial x$ are used to represent the partial derivative of $z = f(x, y)$ with respect to x, with similar symbols used for the partial derivative with respect to y. The symbol $f_x(x, y)$ is often abbreviated as just f_x, with $f_y(x, y)$ abbreviated f_y.

Example 1 Let $f(x, y) = 4x^2 - 9xy + 6y^3$. Find f_x and f_y.

Solution To find f_x, treat x as a variable and y as a constant. Then find the derivative term by term. The derivative of $4x^2$ is $8x$. The term $-9xy$ can be written as $(-9y)x$. Since y is being treated as a constant, this term is the constant $-9y$ times x, and its derivative with respect to x is $-9y$. The last term, $6y^3$, is a constant, so its derivative is 0. In sum,

$$\begin{aligned} f(x, y) &= 4x^2 - 9xy + 6y^3 \\ f_x &= 8x - 9y + 0 \\ f_x &= 8x - 9y. \end{aligned}$$

To find f_y, treat y as a variable and x as a constant. Proceeding term by term to find the derivative, we have

$$\begin{aligned} f(x, y) &= 4x^2 - 9xy + 6y^3 \\ f_y &= 0 - 9x + 18y^2 \\ f_y &= -9x + 18y^2. \end{aligned}$$ 1✓

✓Checkpoint 1

Find f_x and f_y.

(a) $f(x, y) = -x^2y + 3xy + 2xy^2$

(b) $f(x, y) = x^3 + 2x^2y + xy$

Answers: See page 926.

Example 2 Let $f(x, y) = \ln(x^2 + y)$. Find f_x and f_y.

Solution Recall the formula for the derivative of a natural logarithmic function. If $y = \ln(g(x))$, then $y' = g'(x)/g(x)$. Using this formula and treating y as a constant, we obtain

$$f_x = \frac{D_x(x^2 + y)}{x^2 + y} = \frac{2x}{x^2 + y}.$$

Similarly, treating x as a constant leads to the following result:

$$f_y = \frac{D_y(x^2 + y)}{x^2 + y} = \frac{1}{x^2 + y}.$$ 2✓

✓Checkpoint 2

Find f_x and f_y.

(a) $f(x, y) = \ln(2x + 3y)$

(b) $f(x, y) = e^{xy}$

Answers: See page 926.

The notation

$$f_x(a, b) \quad \text{or} \quad \frac{\partial f}{\partial x}(a, b)$$

represents the value of a partial derivative when $x = a$ and $y = b$, as shown in the next example.

Example 3 Let $f(x, y) = 2x^2 + 3xy^3 + 2y + 5$. Find the following.

(a) $f_x(-1, 2)$

Solution First, find f_x by holding y constant in $f(x, y)$:

$$f(x, y) = 2x^2 + 3xy^3 + 2y + 5$$
$$f_x = 4x + 3y^3.$$

Now let $x = -1$ and $y = 2$:

$$f_x(-1, 2) = 4(-1) + 3(2)^3 = -4 + 24 = 20.$$

(b) $\dfrac{\partial f}{\partial y}(-4, -3)$

Solution Hold x constant in $f(x, y)$ to find $\dfrac{\partial f}{\partial y}$:

$$f(x, y) = 2x^2 + 3xy^3 + 2y + 5$$
$$\frac{\partial f}{\partial y} = 9xy^2 + 2.$$

Then

$$\frac{\partial f}{\partial y}(-4, -3) = 9(-4)(-3)^2 + 2 = 9(-36) + 2 = -322.$$ 3 ✓

Checkpoint 3

Let $f(x, y) = x^2 + xy^2 + 5y - 10$. Find the given partial derivatives.

(a) $f_x(2, 1)$

(b) $\dfrac{\partial f}{\partial y}(-1, 0)$

Answers: See page 926.

The derivative of a function of one variable can be interpreted as the slope of the tangent line to the graph at that point. With some modification, the same is true of partial derivatives of functions of two variables. At a point on the graph of a function of two variables, $z = f(x, y)$, there may be many tangent lines, all of which lie in the same tangent plane, as shown in Figure 14.17.

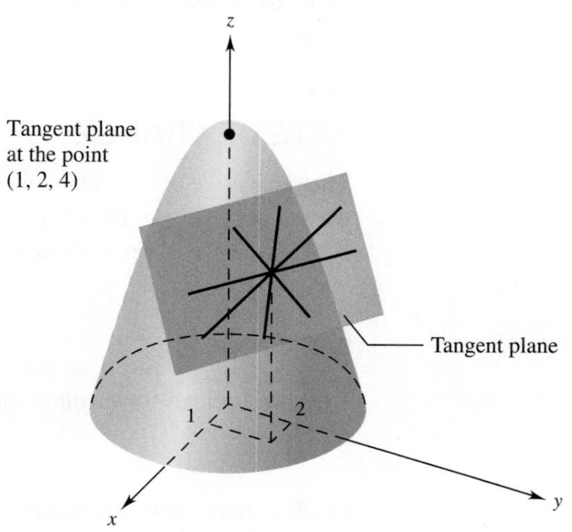

FIGURE 14.17

In any particular direction, however, there will be only one tangent line. We now demonstrate how to use partial derivatives to find the slope of the tangent lines in the x- and y-directions.

Figure 14.18 shows a surface $z = f(x, y)$ and a plane that is parallel to the xz-plane. The equation of the plane is $y = b$. (This equation corresponds to holding y fixed.)

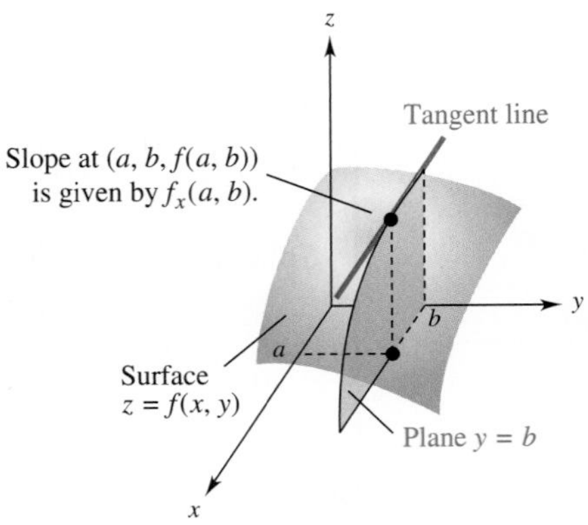

FIGURE 14.18

Every point (x, y, z) on the plane has $y = b$. Hence, every point on the curve that is the intersection of the plane and the surface must have the form $(x, b, f(x, b))$. Thus, the curve is the graph of $z = f(x, b)$. Since b is a constant, $z = f(x, b)$ is a function of the single variable x, whose derivative with respect to x is $f_x(x, b)$. When this derivative is evaluated at $x = a$, it gives the slope of the tangent line to the curve at the point $(a, b, f(a, b))$—that is, the tangent line to the surface in the x-direction at the point $(a, b, f(a, b))$. In the same way, the partial derivative with respect to y will give the slope of the tangent line to the surface in the y-direction at the point $(a, b, f(a, b))$.

RATE OF CHANGE

The derivative of $y = f(x)$ gives the rate of change of y with respect to x. In the same way, if $z = f(x, y)$, then f_x gives the rate of change of z with respect to x if y is held constant and f_y gives the rate of change of z with respect to y if x is held constant.

Example 4 Suppose that the temperature of the water at the point on a river where a nuclear power plant discharges its hot wastewater is approximated by

$$T(x, y) = 2x + 5y + xy - 40.$$

Here, x represents the temperature of the river water in degrees Celsius before it reaches the power plant and y is the number of megawatts (in hundreds) of electricity being produced by the plant.

(a) Find and interpret $T_x(9, 5)$.

Solution First, find the partial derivative T_x:

$$T_x = 2 + y.$$

This partial derivative gives the rate of change of T with respect to x. Replacing x with 9 and y with 5 gives

$$T_x(9, 5) = 2 + 5 = 7.$$

Just as the marginal cost is the approximate cost of one more item, this result, 7, is the approximate change in temperature of the output water if the input water temperature changes by 1 degree from $x = 9$ to $x = 9 + 1 = 10$ while y remains constant at 5 (500 megawatts of electricity produced).

(b) Find and interpret $T_y(9, 5)$.

Solution The partial derivative T_y is

$$T_y = 5 + x.$$

This partial derivative gives the rate of change of T with respect to y, with

$$T_y(9, 5) = 5 + 9 = 14.$$

This result, 14, is the approximate change in temperature resulting from a 1-unit increase in production of electricity from $y = 5$ to $y = 5 + 1 = 6$ (from 500 to 600 megawatts) while the input water temperature x remains constant at 9°C. 4 ✓

✓Checkpoint 4

Use the function of Example 4 to find and interpret the given partial derivatives.

(a) $T_x(5, 4)$

(b) $T_y(8, 3)$

Answers: See page 926.

As mentioned in the previous section, if $P(x, y)$ gives the output P produced by x units of labor and y units of capital, then $P(x, y)$ is a production function. The partial derivatives of this production function have practical implications. For example, $\partial P/\partial x$ gives the **marginal productivity** of labor. This represents the rate at which the output is changing with respect to changes in labor for a fixed capital investment. That is, if capital investment is held constant and labor is increased by 1 work hour, $\partial P/\partial x$ will yield the approximate change in the production level. Likewise, $\partial P/\partial y$ gives the marginal productivity of capital, which represents the rate at which the output is changing with respect to changes in capital for a fixed labor value. So if the labor force is held constant and the capital investment is increased by 1 unit, $\partial P/\partial y$ will approximate the corresponding change in the production level.

Example 5 A company that manufactures computers has determined that its production function is given by

$$P(x, y) = 500x + 800y + 3x^2y - \frac{11x^3}{15} - \frac{y^4}{4},$$

where x is the size of the labor force (in work hours per week) and y is the amount of capital (in units of \$1000) invested. Find the marginal productivity of labor and the marginal productivity of capital when $x = 50$ and $y = 20$, and interpret the results.

Solution The marginal productivity of labor is found by taking the derivative of P with respect to x:

$$\frac{\partial P}{\partial x} = 500 + 6xy - \frac{33x^2}{15}$$

$$\frac{\partial P}{\partial x}(50, 20) = 500 + 6(50)(20) - \frac{33(50^2)}{15}$$

$$= 1000.$$

Thus, if capital investment is held constant at \$20,000 and labor is increased from 50 to 51 work hours per week, production will increase by about 1000 units. In the same way, the marginal productivity of capital is $\partial P/\partial y$:

$$\frac{\partial P}{\partial y} = 800 + 3x^2 - y^3$$

$$\frac{\partial P}{\partial y}(50, 20) = 800 + 3(50)^2 - 20^3$$

$$= 300.$$

If work hours are held constant at 50 hours per week and capital investment is increased from \$20,000 to \$21,000, production will increase by about 300 units. 5✓

✓Checkpoint 5

Suppose a production function is given by $P(x, y) = 10x^2y + 100x + 400y - 5xy^2$, where x and y are defined as in Example 5. Find the marginal productivity of labor and capital when $x = 20$ and $y = 10$.

Answer: See page 926.

SECOND-ORDER PARTIAL DERIVATIVES

The second derivative of a function of one variable is very useful in determining local maxima and minima. **Second-order partial derivatives** (partial derivatives of a partial derivative) are used in a similar way for functions of two or more variables. The situation is somewhat more complicated, however, with more independent variables. For example, $f(x, y) = 4x + x^2y + 2y$ has two first-order partial derivatives:

$$f_x = 4 + 2xy \quad \text{and} \quad f_y = x^2 + 2.$$

Because each of these partial derivatives has two partial derivatives, one with respect to y and one with respect to x, there are *four* second-order partial derivatives of function f.

The same is true for any function $z = f(x, y)$. The following schematic diagram shows the partial derivatives and the second-order partial derivatives of f and the various notations used for each:

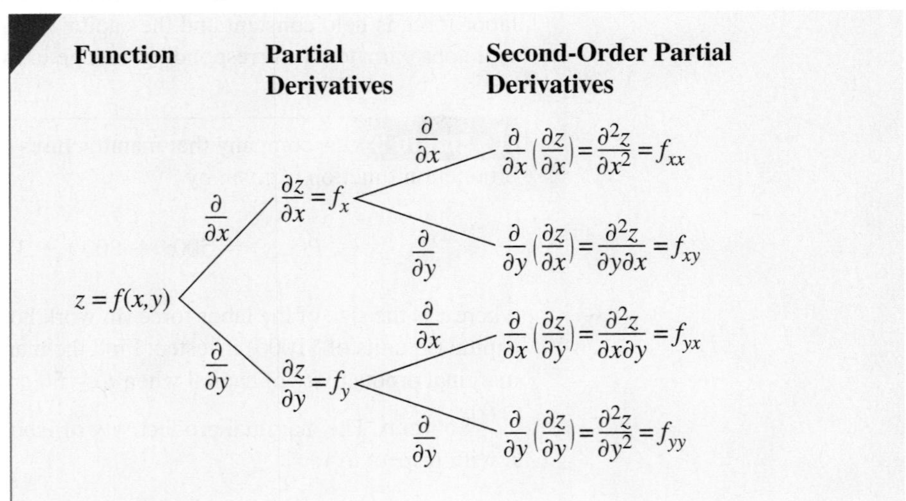

As shown in the diagram, f_{xx} is used as an abbreviation for $f_{xx}(x, y)$, with f_{yy}, f_{xy}, and f_{yx} used in a similar way. The symbol f_{xx} is read, "the partial derivative of f_x with respect to x," and f_{xy} is read, "the partial derivative of f_x with respect to y." Also, the symbol $\partial^2 z/\partial y^2$ is read, "the partial derivative of $\partial z/\partial y$ with respect to y."

NOTE For most functions found in applications, and for all the functions in this book, the second-order partial derivatives f_{xy} and f_{yx} are equal. Therefore, we need not be particular about the order in which these derivatives are found.

Example 6 Find all second-order partial derivatives of

$$f(x, y) = -4x^3 - 3x^2y^3 + 2y^2.$$

Solution First find f_x and f_y:

$$f_x = -12x^2 - 6xy^3 \quad \text{and} \quad f_y = -9x^2y^2 + 4y.$$

To find f_{xx}, take the partial derivative of f_x with respect to x:

$$f_{xx} = -24x - 6y^3.$$

Now find f_{yy} by taking the partial derivative of f_y with respect to y:

$$f_{yy} = -18x^2y + 4.$$

Next, find f_{xy} by taking the partial derivative of f_x with respect to y:

$$f_{xy} = -18xy^2.$$

Finally, find f_{yx} by taking the partial derivative of f_y with respect to x:

$$f_{yx} = -18xy^2.$$ ✓6

✓Checkpoint 6

Let $f(x, y) = 4x^2y^2 - 9xy + 8x^2 - 3y^4$. Find all second-order partial derivatives.

Answer: See page 926.

Example 7 Let $f(x, y) = 2e^x - 8x^3y^2$. Find all second-order partial derivatives.

Solution Here,

$$f_x = 2e^x - 24x^2y^2 \quad \text{and} \quad f_y = -16x^3y.$$

(Recall that if $g(x) = e^x$, then $g'(x) = e^x$.) Now find the second-order partial derivatives:

$$f_{xx} = 2e^x - 48xy^2; \qquad f_{xy} = -48x^2y;$$
$$f_{yy} = -16x^3; \qquad f_{yx} = -48x^2y.$$ ✓7

✓Checkpoint 7

Let $f(x, y) = 4e^{x+y} + 2x^3y$. Find all second-order partial derivatives.

Answer: See page 926.

Partial derivatives of multivariate functions with more than two variables are found in a way similar to that for functions with two variables. For example, to find f_x for $w = f(x, y, z)$, treat y and z as constants and differentiate with respect to x.

Example 8 Let $f(x, y, z) = xy^2z + 2x^2y - 4xz^2$. Find f_x, f_y, f_z, f_{xy}, and f_{yz}.

Solution We find the first three partial derivatives as follows:

$$f_x = y^2z + 4xy - 4z^2$$
$$f_y = 2xyz + 2x^2$$
$$f_z = xy^2 - 8xz$$

To find f_{xy}, differentiate f_x with respect to y:

$$f_{xy} = 2yz + 4x$$

In the same way, differentiate f_y with respect to z to get

$$f_{yz} = 2xy.$$ 8

✓Checkpoint 8

Let $f(x, y, z) = xyz + x^2yz + xy^2z^3$. Find f_x, f_y, f_z, and f_{xz}.

Answer: See page 926.

For those who may be interested, we close this section with the formal definition of partial derivatives. As you can see, it is very similar to the definition of the derivative of a function of one variable.

Partial Derivatives (Formal Definition)

Let $z = f(x, y)$ be a function of two variables. Then the **partial derivative of f with respect to x** is

$$f_x(x, y) = \lim_{h \to 0} \frac{f(x + h, y) - f(x, y)}{h},$$

and the **partial derivative of f with respect to y** is

$$f_y(x, y) = \lim_{h \to 0} \frac{f(x, y + h) - f(x, y)}{h},$$

provided that these limits exist.

Similar definitions could be given for functions of more than two independent variables.

14.2 ▶ Exercises

For each of the given functions, find

(a) $\frac{\partial z}{\partial x}$; **(b)** $\frac{\partial z}{\partial y}$; **(c)** $f_x(2,3)$; **(d)** $f_y(1, -2)$.

1. $z = f(x, y) = 8x^3 - 4x^2y + 9y^2$
2. $z = f(x, y) = -3x^2 - 2xy^2 + 5y^3$

In Exercises 3–14, find f_x and f_y. Then find $f_x(2, -1)$ and $f_y(-4, 3)$. Leave the answers in terms of e in Exercises 5–8 and 13 and 14. (See Examples 1–3.)

3. $f(x, y) = 6y^3 - 4xy + 5$
4. $f(x, y) = -3x^4y^3 + 10$
5. $f(x, y) = e^{2x+y}$
6. $f(x, y) = -4e^{x-y}$
7. $f(x, y) = \frac{-2}{e^{x+2y}}$
8. $f(x, y) = \frac{6}{e^{4x-y}}$
9. $f(x, y) = \frac{x + 3y^2}{x^2 + y^3}$
10. $f(x, y) = \frac{8x^2y}{x^3 - y}$
11. $f(x, y) = \ln|1 + 5x^3y^2|$
12. $f(x, y) = \ln|4x^2y + 3x|$
13. $f(x, y) = x^2e^{2xy}$
14. $f(x, y) = ye^{5x+2y}$

Find all second-order partial derivatives of the given functions. (See Examples 6 and 7.)

15. $f(x, y) = 4x^2y^2 - 16x^2 + 4y$
16. $R(x, y) = 4x^2 - 5xy^3 + 12x^2y^2$
17. $h(x, y) = -3y^2 - 4x^2y^2 + 7xy^2$
18. $P(x, y) = -16x^3 + 3xy^2 - 12x^4y^2$
19. $r(x, y) = \frac{6y}{x + y}$
20. $k(x, y) = \frac{-7x}{2x + 3y}$
21. $z = 4xe^y$
22. $z = -3ye^x$
23. $r = \ln|2x + y|$
24. $r = \ln|7x - 5y|$
25. $z = x\ln(xy)$
26. $z = (y + 1)\ln(x^3y)$

In Exercises 27 and 28, evaluate $f_{xy}(1,2)$ and $f_{yy}(3,1)$.

27. $f(x, y) = x \ln(xy)$

28. $f(x, y) = (y + 1) \ln(x^3 y)$

Find values of x and y such that both $f_x(x, y) = 0$ and $f_y(x, y) = 0$.

29. $f(x, y) = 6x^2 + 6y^2 + 6xy + 36x - 5$

30. $f(x, y) = 50 + 4x - 5y + x^2 + y^2 + xy$

31. $f(x, y) = 9xy - x^3 - y^3 - 6$

32. $f(x, y) = 2200 + 27x^3 + 72xy + 8y^2$

Find f_x, f_y, f_z and f_{yz} for the given functions. In Exercises 33 and 34, also find $f_y(3, 4, -2)$ and $f_{yz}(1, -1, 0)$ (See Example 8.)

33. $f(x, y, z) = x^2 + yz + z^4$

34. $f(x, y, z) = 3x^5 - x^2 + y^5$

35. $f(x, y, z) = \dfrac{6x - 5y}{4z + 5}$

36. $f(x, y, z) = \dfrac{2x^2 + xy}{yz - 2}$

37. $f(x, y, z) = \ln(x^2 - 5xz^2 + y^4)$

38. $f(x, y, z) = \ln(8xy + 5yz - x^3)$

39. How many partial derivatives does a function in three variables have? How many second-order partial derivatives does it have? Explain why.

40. Suppose $z = f(x, y)$ describes the cost to build a certain structure, where x represents the labor costs and y represents the cost of materials. Describe what f_x and f_y represent.

41. Business Suppose that the manufacturing cost of a calculator is approximated by

$$M(x, y) = 40x^2 + 30y^2 - 10xy + 30,$$

where x is the cost of electronic chips and y is the cost of labor. Find the given partial derivatives.

(a) $M_y(4, 2)$ **(b)** $M_x(3, 6)$
(c) $(\partial M/\partial x)(2, 5)$ **(d)** $(\partial M/\partial y)(6, 7)$

42. Business The revenue from the sale of x units of a tranquilizer and y units of an antibiotic is given by

$$R(x, y) = 5x^2 + 9y^2 - 4xy.$$

Suppose 9 units of tranquilizer and 5 units of antibiotic are sold. Use partial derivatives to answer these questions.

(a) What is the *approximate* effect on revenue if 10 units of tranquilizer and 5 units of antibiotic are sold?

(b) What is the *approximate* effect on revenue if the amount of antibiotic sold is increased to 6 units, while tranquilizer sales remain constant?

43. Business A car dealership estimates that the total weekly sales of its most popular model are a function of the car's list price p and the interest rate i offered by the manufacturer. The weekly sales are given by

$$f(p, i) = 99p - .5pi - .0025p^2.$$

(a) Find the weekly sales if the average list price is \$19,400 and the manufacturer is offering an 8% interest rate.

(b) Find and interpret $f_p(p, i)$ and $f_i(p, i)$.

(c) What would be the effect on weekly sales if the price is \$19,400 and interest rates rise from 8% to 9%?

44. Health The reaction to x units of a drug t hours after it was administered is given by

$$R(x, t) = x^2(a - x)t^2 e^{-t},$$

for $0 \le x \le a$ (where a is a constant). Find the given partial derivatives.

(a) $\dfrac{\partial R}{\partial x}$ **(b)** $\dfrac{\partial R}{\partial t}$ **(c)** $\dfrac{\partial^2 R}{\partial x^2}$ **(d)** $\dfrac{\partial^2 R}{\partial x \partial t}$

(e) Interpret your answers to parts (a) and (b).

45. Natural Science The average energy expended for an animal to walk or run 1 km can be estimated by the function

$$f(m, v) = 25.92m^{.68} + \frac{3.62m^{.75}}{v},$$

where $f(m, v)$ is the energy used (in kcal per hour), m is the mass (in g), and v is the speed of movement (in km per hour) of the animal.*

(a) Find $f(300, 10)$.

(b) Find $f_m(300, 10)$, and interpret the result.

46. Natural Science The oxygen consumption of a well-insulated mammal that is not sweating is approximated by

$$c = c(T, F, m) = \frac{2.5(T - F)}{m^{.67}} = 2.5(T - F)m^{-.67},$$

where T is the internal body temperature of the animal (in °C), F is the temperature of the outside of the animal's fur (in °C), and m is the animal's mass (in kilograms). Find the approximate change in oxygen consumption under the given conditions.

(a) The internal temperature increases from 38°C to 39°C, while the outside temperature remains at 12°C and the mass remains at 30 kg.

(b) The internal temperature is constant at 36°C, the outside temperature increases from 14°C to 15°C, and the mass remains at 25 kg.

47. Business Suppose that the cost function for the company discussed in the first paragraph of this section is given by:

$$C(x, y) = 600x + 120y + 4000,$$

where x is the number of units of emergency radios and y is the number of units of crank flashlights produced.

(a) In $C(x, y)$, what does 4000 represent?

(b) Find $\partial C/\partial x$ and $\partial C/\partial y$. What do these quantities represent?

---

*Robbins, C., *Wildlife Feeding and Nutrition*, New York: Academic Press, 1983, p. 114.

48. **Business** A bicycle manufacturer produces two models, the Basic and the Ultra. Its cost function for these products is

$$C(x, y) = 125x + 400y + 10{,}000,$$

where x is the number of Basic models and y is the number of Ultra models produced.
(a) What is the marginal cost of producing an Ultra model when the production of Basics is held constant?
(b) What is the marginal cost of producing a Basic model when the production of Ultras is held constant?

49. **Natural Science** The surface area of a human (in square meters) is approximated by

$$A(M, H) = .202M^{.425}H^{.725},$$

where M is the mass of the person (in kilograms) and H is the height (in meters). Find the *approximate* change in surface area under the given conditions.
(a) The mass changes from 72 kg to 73 kg, while the height remains 1.8 m.
(b) The mass remains stable at 70 kg, while the height changes from 1.6 m to 1.7 m.

50. **Business** Suppose the production function of a company is given by

$$P(x, y) = 100\sqrt{x^2 + y^2},$$

where x represents units of labor and y represents units of capital. (See Example 5.) Find the following when $x = 4$ and $y = 3$.
(a) The marginal productivity of labor
(b) The marginal productivity of capital

51. **Business** A manufacturer estimates that production (in hundreds of units) is a function of the respective amounts x and y of labor and capital used, as follows:

$$f(x, y) = \left[\frac{1}{3}x^{-1/3} + \frac{2}{3}y^{-1/3}\right]^{-3}.$$

(a) Find the number of units produced when 27 units of labor and 64 units of capital are utilized.
(b) Find and interpret $f_x(27, 64)$ and $f_y(27, 64)$.
(c) What would be the *approximate* effect on production of increasing labor by 1 unit?

52. **Business** A manufacturer of automobile batteries estimates that his total production in thousands of units is given by

$$f(x, y) = 3x^{1/3}y^{2/3},$$

where x is the number of units of labor and y is the number of units of capital utilized.
(a) Find and interpret $f_x(64, 125)$ and $f_y(64, 125)$ if the current level of production uses 64 units of labor and 125 units of capital.
(b) What would be the *approximate* effect on production of increasing labor to 65 units while holding capital at the current level?
(c) Suppose that sales have been good and management wants to increase either capital or labor by 1 unit. Which option would result in a larger increase in production?

53. **Business** The production function z for the United States was once estimated as

$$z = x^{.7}y^{.3},$$

where x stands for the amount of labor and y stands for the amount of capital. Find the marginal productivity of labor (that is, find $\partial z/\partial x$) and of capital (that is, find $\partial z/\partial y$). (See Example 5.)

54. **Business** A similar production function for Canada is

$$z = x^{.4}y^{.6},$$

with x, y, and z as in Exercise 53. Find the marginal productivity of labor and of capital. (See Example 5.)

55. **Natural Science** In one method of computing the quantity of blood pumped through the lungs in 1 minute, a researcher first finds each of the following (in milliliters):

b = quantity of oxygen used by body in 1 minute;
a = quantity of oxygen per liter of blood that has just gone through the lungs;
v = quantity of oxygen per liter of blood that is about to enter the lungs.

In 1 minute,

Amount of oxygen used
= amount of oxygen per liter
× number of liters of blood pumped.

If C is the number of liters pumped through the blood in 1 minute, then

$$b = (a - v) \cdot C; \quad \text{or} \quad C = \frac{b}{a - v}.$$

(a) Find C if $a = 160$, $b = 200$, and $v = 125$.
(b) Find C if $a = 180$, $b = 260$, and $v = 142$.
Find the following partial derivatives:
(c) $\partial C/\partial b$; (d) $\partial C/\partial v$.

56. **Health** A weight-loss counselor has prepared a program of diet and exercise for a client. If the client sticks to the program, the weight loss that can be expected (in pounds per week) is given by

$$\text{Weight loss} = f(n, c) = \frac{1}{8}n^2 - \frac{1}{5}c + \frac{1937}{8},$$

where c is the average daily calorie intake for the week and n is the number of 40-minute aerobic workouts per week.

(a) How many pounds can the client expect to lose by eating an average of 1200 calories per day and participating in four 40-minute workouts in a week?
(b) Find and interpret $\partial f/\partial n$.
(c) The client currently averages 1100 calories per day and does three 40-minute workouts each week. What would be the *approximate* impact on weekly weight loss of adding a fourth workout per week?

57. Social Science A developmental mathematics instructor at a large university has determined that a student's probability of success in the university's pass/fail remedial algebra course is a function of s, n, and a, where s is the student's score on the departmental placement exam, n is the number of semesters of mathematics passed in high school, and a is the student's mathematics SAT score. The instructor estimates that p, the probability of passing the course (in percent), will be

$$p = f(s, n, a) = .078a + 4(sn)^{1/2}$$

for $200 \leq a \leq 800$, $0 \leq s \leq 10$, and $0 \leq n \leq 8$. Assume that this model has some merit.
(a) If a student scores 8 on the placement exam, has taken 6 semesters of high school math, and has an SAT score of 450, what is the probability that the student will pass the course?
(b) Find p for a student with 3 semesters of high school mathematics, a placement score of 3, and an SAT score of 320.
(c) Find and interpret $f_n(3, 3, 320)$ and $f_a(3, 3, 320)$.

58. Physical Science Fitts's law is used to estimate the amount of time it takes for a person's arm to pick up a light object, move it, and then place it in a designated target area. Mathematically, Fitts's law for a particular individual is given by

$$T(s, w) = -50 + 105 \log_2\left(\frac{2s}{w}\right),$$

where s is the distance (in feet) the object is moved, w is the width of the area in which the object is being placed, and T is the time (in msec).
(a) Calculate $T(3, .5)$.
(b) Find $T_s(3, .5)$ and $T_w(3, .5)$, and interpret these values. (*Hint*: $\log_2 x = \ln x/\ln 2$.)

59. Physical Science The gravitational attraction F on a body a distance r from the center of the earth, where r is greater than the radius of the earth, is a function of the mass m of the body and the distance r. Specifically,

$$F = \frac{mgR^2}{r^2},$$

where R is the radius of the earth and g is the force of gravity—about 32 feet per second per second (ft/sec^2).
(a) Find and interpret F_m and F_r.
(b) Show that $F_m > 0$ and $F_r < 0$. Why is this reasonable?

60. Natural Science In 1908, J. Haldane constructed diving tables that give a relationship between the water pressure on body tissues for various water depths and dive times. The tables were used by divers and virtually eliminated the occurrence of decompression sickness. The pressure in atmospheres for a no-stop dive is given by the formula

$$p(l, t) = 1 + \frac{l}{33}\left(1 - e^{-.1386t}\right),$$

where t is in minutes, l is in feet, and p is in atmospheres (atm).*
(a) Find the pressure at 33 feet for a 10-minute dive.
(b) Find $p_l(33, 10)$ and $p_t(33, 10)$.
(c) Haldane estimated that decompression sickness for no-stop dives could be avoided if the diver's tissue pressure did not exceed 2.15 atmospheres. Find the maximum amount of time (including the time going down and coming back up) that a diver could stay down if he or she wants to dive to a depth of 66 feet.

61. Health The body mass index (BMI) is a number that can be calculated for any individual as follows: Multiply a person's weight by 703 and divide by the person's height squared. That is,

$$B = \frac{703w}{h^2},$$

where w is in pounds and h is in inches.† The National Heart, Lung and Blood Institute uses the BMI to determine whether a person is "overweight" ($25 \leq B < 30$) or "obese" ($B \geq 30$).
(a) Calculate the BMI for a person who weighs 220 pounds and is 74 inches tall.
(b) Calculate $\frac{\partial B}{\partial w}$ and $\frac{\partial B}{\partial h}$.

62. Natural Science In 1941, explorers Paul Siple and Charles Passel discovered that the amount of heat lost when an object is exposed to cold air depends on both the temperature of the air and the velocity of the wind. They developed the windchill index as a way to measure the danger of frostbite while performing outdoor activities. On the basis of new findings and additional data, the windchill index was recalculated in 2001. The new index is given by

$$W(V, T) = 35.74 + .6215T - 35.75V^{.16} + (.4275T)V^{.16},$$

*The estimates provided by the formula are conservative. Consult modern diving tables before making a dive. See David Westbrook, "The Mathematics of Scuba Diving," *UMAP Journal* 18, no. 2 (1997): 2–19.

†The National Institutes of Health.

where V is the wind speed in miles per hour and T is the Fahrenheit temperature for wind speeds between 4 and 45 mph.*

(a) Find the windchill for a wind speed of 20 mph and 10°F.

(b) A weather report indicates that the windchill is −22°F and the actual outdoor temperature is 5°F. Use a graphing calculator to find the corresponding wind speed to the nearest mile per hour.

(c) Find and interpret $W_V(20, 10)$ and $W_T(20, 10)$.

(d) Using the table command on a graphing calculator or a spreadsheet, develop a windchill chart for wind speeds of 5, 10, 15, and 20 mph and temperatures of 0°, 5°, 15° and 25°.

*Joint Action Group of the National Weather Service and the Meteorological Services of Canada, press release, August 2001.

✓ Checkpoint Answers

1. **(a)** $f_x = -2xy + 3y + 2y^2$; $f_y = -x^2 + 3x + 4xy$
   (b) $f_x = 3x^2 + 4xy + y$; $f_y = 2x^2 + x$
2. **(a)** $f_x = \dfrac{2}{2x + 3y}$; $f_y = \dfrac{3}{2x + 3y}$
   (b) $f_x = ye^{xy}$; $f_y = xe^{xy}$
3. **(a)** 5 **(b)** 5
4. **(a)** $T_x(5, 4) = 6$; the approximate increase in output temperature if the input temperature increases from 5 to 6 degrees
   (b) $T_y(8, 3) = 13$; the approximate increase in output temperature if the production of electricity increases from 300 to 400 megawatts
5. The marginal productivity of labor is 3600. The marginal productivity of capital is 2400.
6. $f_{xx} = 8y^2 + 16$
   $f_{yy} = 8x^2 - 36y^2$
   $f_{xy} = 16xy - 9$
   $f_{yx} = 16xy - 9$
7. $f_{xx} = 4e^{x+y} + 12xy$
   $f_{yy} = 4e^{x+y}$
   $f_{xy} = 4e^{x+y} + 6x^2$
   $f_{yx} = 4e^{x+y} + 6x^2$
8. $f_x = yz + 2xyz + y^2z^3$
   $f_y = xz + x^2z + 2xyz^3$
   $f_z = xy + x^2y + 3xy^2z^2$
   $f_{xz} = y + 2xy + 3y^2z^2$

14.3 ▶ Extrema of Functions of Several Variables

One of the most important applications of calculus is finding maxima and minima of functions. Earlier, we studied this idea extensively for functions of a single variable; now we shall see that extrema can be found for functions of two variables. In particular, an extension of the second-derivative test can be derived and used to identify maxima or minima. We begin with the definitions of local maxima and minima.

Local Maxima and Minima

Let (a, b) be in the domain of a function f.

1. f has a **local maximum** at (a, b) if there is a circular region in the xy-plane with (a, b) in its interior such that

$$f(a, b) \geq f(x, y)$$

for all points (x, y) in the circular region.

(continued)

2. f has a **local minimum** at (a, b) if there is a circular region in the xy-plane with (a, b) in its interior such that

$$f(a, b) \le f(x, y)$$

for all points (x, y) in the circular region.

As before, the term *local extremum* is used for either a local maximum or a local minimum. Examples of a local maximum and a local minimum are given in Figure 14.19 and 14.20, respectively.

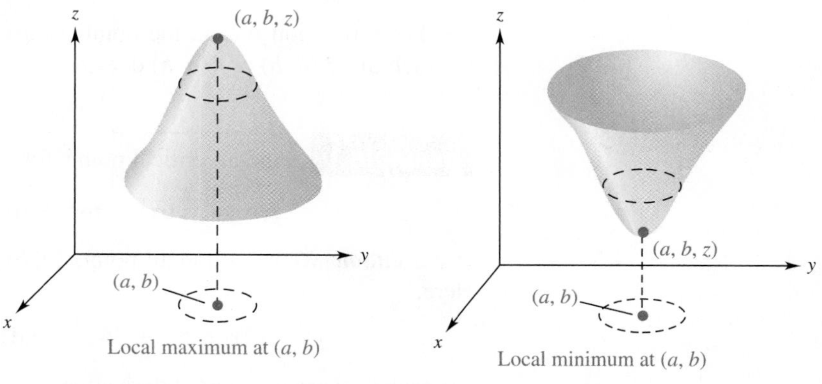

FIGURE 14.19

FIGURE 14.20

NOTE With functions of a single variable, we made a distinction between local extrema and absolute extrema. The methods for finding absolute extrema are quite involved for functions of two variables, so we will discuss only local extrema.

Tangent is horizontal.

Tangent is horizontal.

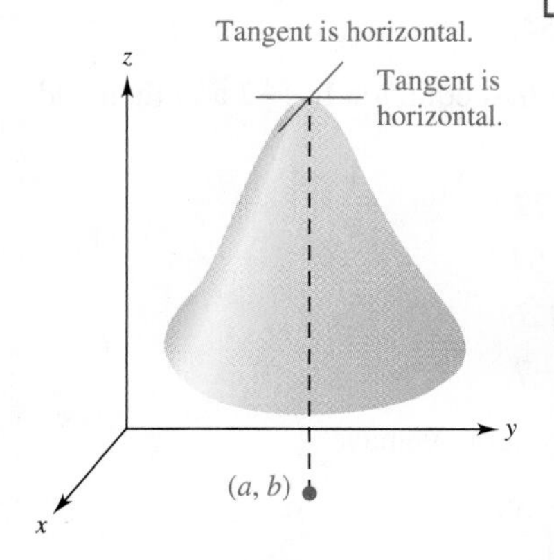

FIGURE 14.21

As suggested by Figure 14.21, at a local maximum the tangent line parallel to the x-axis has a slope of 0, as does the tangent line parallel to the y-axis. (Notice the similarity to functions of one variable.) That is, if the function $z = f(x, y)$ has a local extremum at (a, b), then $f_x(a, b) = 0$ and $f_y(a, b) = 0$, as stated in the following theorem.

Location of Extrema

If a function $z = f(x, y)$ has a local maximum or local minimum at the point (a, b), and $f_x(a, b)$ and $f_y(a, b)$ both exist, then

$$\boldsymbol{f_x(a, b) = 0} \quad \textbf{and} \quad \boldsymbol{f_y(a, b) = 0}.$$

Just as with functions of one variable, the fact that the slopes of the tangent lines are 0 is no guarantee that a local extremum has been located. For example, Figure 14.22 on the next page shows the graph of $z = f(x, y) = x^2 - y^2$. Both $f_x(0, 0) = 0$ and $f_y(0, 0) = 0$, and yet $(0, 0)$ leads to neither a local maximum nor a local minimum for the function. The point $(0, 0, 0)$ on the graph of this function is called a **saddle point**; it is a minimum when approached from one direction (in this case, the x-direction), but a maximum when approached from another direction (in this case, the y-direction). A saddle point is neither a maximum nor a minimum.

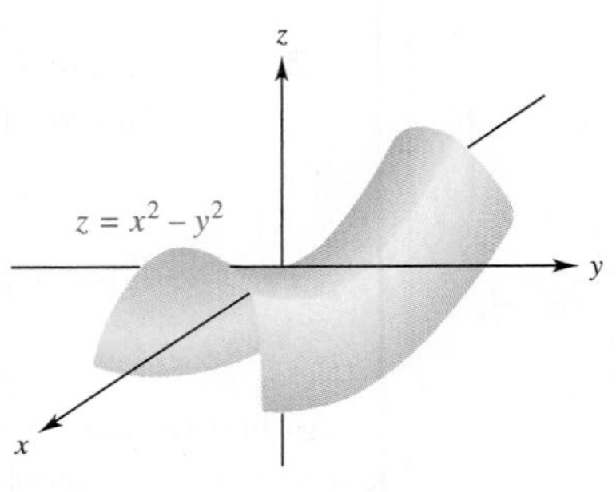

FIGURE 14.22

For a function $f(x, y)$, the points (a, b) such that $f_x(a, b) = 0$ and $f_y(a, b) = 0$ (or such that $f_x(a, b)$ or $f_y(a, b)$ does not exist) are called **critical points.**

Example 1 Find all critical points for

$$f(x, y) = 6x^2 + 6y^2 + 6xy + 36x - 54y - 5.$$

Solution We must find all points (a, b) such that $f_x(a, b) = 0$ and $f_y(a, b) = 0$. Here,

$$f_x = 12x + 6y + 36 \quad \text{and} \quad f_y = 12y + 6x - 54.$$

Set each of these two partial derivatives equal to 0:

$$12x + 6y + 36 = 0 \quad \text{and} \quad 12y + 6x - 54 = 0.$$

These two equations form a system of linear equations that we can rewrite as

$$\begin{aligned} 12x + 6y &= -36 \\ 6x + 12y &= 54. \end{aligned}$$

To solve this system by elimination, multiply the first equation by -2 and then add the equations:

$$\begin{array}{rcr} -24x - 12y &=& 72 \\ 6x + 12y &=& 54 \\ \hline -18x \qquad &=& 126 \\ x \qquad &=& -7. \end{array}$$

Substituting $x = -7$ in the first equation of the system, we have

$$\begin{aligned} 12(-7) + 6y &= -36 \\ 6y &= 48 \\ y &= 8. \end{aligned}$$

Therefore, $(-7, 8)$ is the solution of the system. Since this is the only solution, $(-7, 8)$ is the only critical point for the given function. By the previous theorem, if the function has a local extremum, it must occur at $(-7, 8)$. ✓1

✓Checkpoint 1

Find all critical points for the given functions.

(a) $f(x, y) = 4x^2 + 3xy + 2y^2 + 7x - 6y - 6$

(b) $f(x, y) = e^{-2x} + xy$

Answers: See page 935.

The results of the next theorem can be used to decide whether $(-7, 8)$ in Example 1 leads to a local maximum, a local minimum, or neither. The proof of this theorem is beyond the scope of this course.

Test for Local Extrema

Suppose $f(x, y)$ is a function such that f_{xx}, f_{yy}, and f_{xy} all exist. Let (a, b) be a critical point for which

$$f_x(a, b) = 0 \quad \text{and} \quad f_y(a, b) = 0.$$

Let M be the number defined by

$$M = f_{xx}(a, b) \cdot f_{yy}(a, b) - [f_{xy}(a, b)]^2.$$

1. If $M > 0$ and $f_{xx}(a, b) < 0$, then f has a **local maximum** at (a, b).
2. If $M > 0$ and $f_{xx}(a, b) > 0$, then f has a **local minimum** at (a, b).
3. If $M < 0$, then f has a **saddle point** at (a, b).
4. If $M = 0$, the test gives **no information**.

The following chart summarizes the conclusions of the theorem:

	$f_{xx}(a, b) < 0$	$f_{xx}(a, b) > 0$
$M > 0$	Local maximum	Local minimum
$M = 0$	No information	
$M < 0$	Saddle point	

Example 2 Example 1 showed that the only critical point of the function

$$f(x, y) = 6x^2 + 6y^2 + 6xy + 36x - 54y - 5$$

is $(-7, 8)$. Does $(-7, 8)$ lead to a local maximum, a local minimum, or neither?

Solution We can find out by using the test for local extrema. From Example 1,

$$f_x(-7, 8) = 0 \quad \text{and} \quad f_y(-7, 8) = 0.$$

Now find the various second-order partial derivatives used in finding M. From $f_x = 12x + 6y + 36$ and $f_y = 12y + 6x - 54$, we have

$$f_{xx} = 12, \quad f_{yy} = 12, \quad \text{and} \quad f_{xy} = 6.$$

(If these second-order partial derivatives had not all been constants, we would have had to evaluate them at the point $(-7, 8)$.) Now,

$$M = f_{xx}(-7, 8) \cdot f_{yy}(-7, 8) - [f_{xy}(-7, 8)]^2 = 12 \cdot 12 - 6^2 = 108.$$

Since $M > 0$ and $f_{xx}(-7, 8) = 12 > 0$, Part 2 of the theorem applies, showing that $f(x, y) = 6x^2 + 6y^2 + 6xy + 36x - 54y - 5$ has a local minimum at $(-7, 8)$. The value of this local minimum is $f(-7, 8) = -347$. ✓2

✓Checkpoint 2

Find any local maxima or minima for the functions defined in checkpoint 1.

Answer: See page 935.

Example 3 Find all points where the function

$$f(x, y) = 9xy - x^3 - y^3 - 6$$

has any local maxima or local minima.

Solution First find any critical points. Here,

$$f_x = 9y - 3x^2 \quad \text{and} \quad f_y = 9x - 3y^2.$$

Set each of these partial derivatives equal to 0:

$$\begin{aligned} f_x &= 0 \\ 9y - 3x^2 &= 0 \\ 9y &= 3x^2 \\ 3y &= x^2. \end{aligned} \qquad \begin{aligned} f_y &= 0 \\ 9x - 3y^2 &= 0 \\ 9x &= 3y^2 \\ 3x &= y^2. \end{aligned}$$

The last two equations provide some useful information. Since $3y = x^2$ and $x^2 \geq 0$, we must have $y \geq 0$. Similarly, since $3x = y^2$ and $y^2 \geq 0$, we must have $x \geq 0$. Consequently,

any critical point (x, y) of f must satisfy $x \geq 0$ and $y \geq 0$.

The substitution method can be used to solve the system of equations

$$\begin{aligned} 3y &= x^2 \\ 3x &= y^2. \end{aligned}$$

The first equation, $3y = x^2$, can be rewritten as $y = x^2/3$. Substitute this expression into the second equation to get

$$\begin{aligned} 3x &= y^2 = \left(\frac{x^2}{3}\right)^2 \\ 3x &= \frac{x^4}{9}. \end{aligned}$$

Solve this equation as follows:

$$\begin{aligned} 27x &= x^4 && \text{Multiply both sides by 9.} \\ x^4 - 27x &= 0 \\ x(x^3 - 27) &= 0. && \text{Factor} \\ x = 0 \quad \text{or} \quad x^3 - 27 &= 0 && \text{Set each factor equal to 0.} \\ x^3 &= 27 \\ x &= 3 && \text{Take the cube root on each side.} \end{aligned}$$

Use these values of x, along with the equation $3x = y^2$, to find y:

$$\begin{aligned} \text{If } x &= 0, \\ 3x &= y^2 \\ 3(0) &= y^2 \\ 0 &= y^2 \\ 0 &= y. \end{aligned} \qquad \begin{aligned} \text{If } x &= 3, \\ 3x &= y^2 \\ 3(3) &= y^2 \\ 9 &= y^2 \\ 3 = y \quad \text{or} \quad -3 &= y. \end{aligned}$$

So the solutions for the system of equations are $(0, 0)$, $(3, 3)$ and $(3, -3)$. However, as we saw earlier, a critical point here must have *both* coordinates nonnegative. Hence, the only critical points of f are $(0, 0)$ and $(3, 3)$.

To see if either critical point produces an extremum, use the test for local extrema. Here,

$$f_x = 9y - 3x^2 \quad \text{and} \quad f_y = 9x - 3y^2.$$

Hence,

$$f_{xx} = -6x, \qquad f_{yy} = -6y, \qquad \text{and} \qquad f_{xy} = 9.$$

Test each of the critical points.

For (0, 0),

$$f_{xx}(0, 0) = -6(0) = 0, \qquad f_{yy}(0, 0) = -6(0) = 0, \quad \text{and} \quad f_{xy}(0, 0) = 9,$$

so that $M = 0 \cdot 0 - 9^2 = -81$. Since $M < 0$, there is a saddle point at (0, 0).

For (3, 3),

$$f_{xx}(3, 3) = -6(3) = -18, \qquad f_{yy}(3, 3) = -6(3) = -18, \quad \text{and} \quad f_{xy}(3, 3) = 9,$$

so that $M = (-18)(-18) - 9^2 = 243$. Since $M > 0$ and $f_{xx}(3, 3) = -18 < 0$, there is a local maximum at (3, 3). If you examine Figure 14.23 carefully, you will see both this local maximum and the saddle point at (0, 0). 3✓

Find any local extrema for

$$f(x, y) = \frac{2\sqrt{2}}{3}x^3 - xy + \frac{1}{3}y^3 - 10.$$

Answer: See page 935.

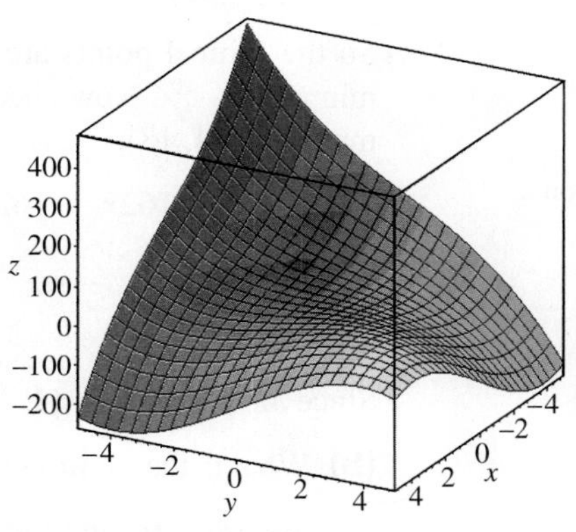

FIGURE 14.23

Example 4 A company is developing a new soft drink. The cost in dollars to produce a batch of the drink is approximated by

$$C(x, y) = 2200 + 27x^3 - 72xy + 8y^2,$$

where x is the number of kilograms of sugar per batch and y is the number of grams of flavoring per batch.

(a) Find the amounts of sugar and flavoring that result in minimum cost for a batch.

Solution Start with the following partial derivatives:

$$C_x = 81x^2 - 72y \qquad \text{and} \qquad C_y = -72x + 16y.$$

Set each of these expressions equal to 0 and solve for y:

$$\begin{aligned} 81x^2 - 72y &= 0 \\ -72y &= -81x^2 \\ y &= \frac{9}{8}x^2. \end{aligned} \qquad\qquad \begin{aligned} -72x + 16y &= 0 \\ 16y &= 72x \\ y &= \frac{9}{2}x. \end{aligned}$$

From the equation on the left, $y \geq 0$. Since $(9/8)x^2$ and $(9/2)x$ both are equal to y, they are equal to each other. Set $(9/8)x^2$ and $(9/2)x$ equal and solve the resulting equation for x:

$$\frac{9}{8}x^2 = \frac{9}{2}x$$
$$9x^2 = 36x$$
$$9x^2 - 36x = 0$$
$$9x(x - 4) = 0$$
$$9x = 0 \quad \text{or} \quad x - 4 = 0$$
$$x = 0 \quad \text{or} \quad x = 4.$$

Substitute $x = 0$ and $x = 4$ into $y = (9/2)x$ to find the corresponding y values:

$$y = \left(\frac{9}{2}\right)x = \left(\frac{9}{2}\right)(0) = 0 \quad \text{and} \quad y = \left(\frac{9}{2}\right)x = \left(\frac{9}{2}\right)(4) = 18.$$

So the critical points are (0, 0) and (4, 18). Verify that (0, 0) does not produce a minimum. ✓4 Now check to see if the critical point (4, 18) leads to a local minimum. For (4, 18),

$$C_{xx} = 162x = 162(4) = 648, \qquad C_{yy} = 16, \qquad \text{and} \qquad C_{xy} = -72.$$

Also,

$$M = (648)(16) - (-72)^2 = 5184.$$

Since $M > 0$ and $C_{xx}(4, 18) > 0$, the cost at (4, 18) is a minimum.

✓Checkpoint 4

Compute M for the critical point (0, 0) for the function in Example 4. What do you conclude?

Answer: See page 935.

(b) What is the minimum cost?

Solution To find the minimum cost, go back to the cost function and evaluate $C(4, 18)$:

$$C(x, y) = 2200 + 27x^3 - 72xy + 8y^2$$
$$C(4, 18) = 2200 + 27(4)^3 - 72(4)(18) + 8(18)^2 = 1336.$$

The minimum cost for a batch is \$1336.

TECHNOLOGY TIP Spreadsheet programs have built-in solvers that are able to optimize many functions of two or more variables. The Excel Solver is located in the Tools menu and requires that cells be identified ahead of time for each variable in the problem. It also requires that another cell be identified where the function, in terms of the variable cells, is placed. In Example 4, for instance, we could identify cells A1 and B1 to represent the variables x and y, respectively. The Solver requires that we place a guess for the answer in these cells. Thus, for our initial value or guess, we place the number 5 in each of these cells. The function must be placed in a cell in terms of cells A1 and B1. If we choose cell A3 to represent the function, then we would type "`=2200 + 27*A1^3 - 72*A1*B1 + 8*B1^2`" in cell A3.

We now click on the Tools menu and choose Solver to find a solution that either maximizes or minimizes the value of cell A3. Figure 14.24 illustrates the Solver box and the items placed in it.

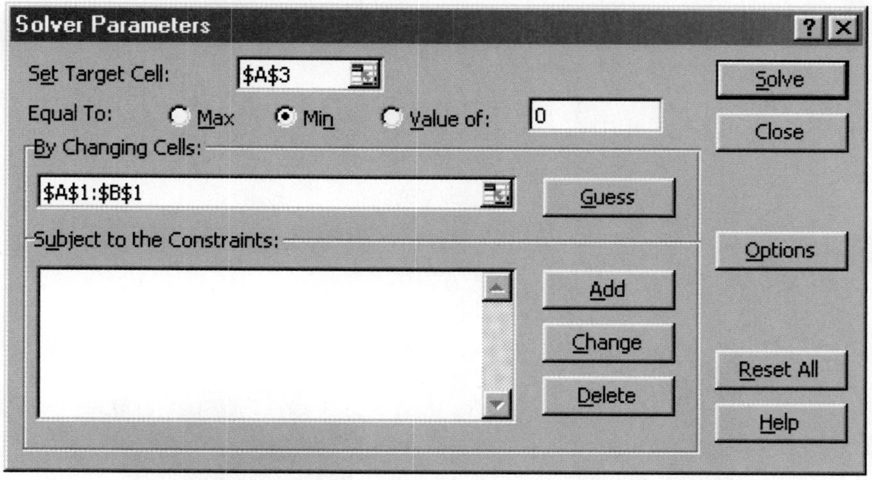

Solver Parameters
Set Target Cell: $A$3
Equal To: Max Min Value of: 0
By Changing Cells:
$A$1:$B$1
Guess
Subject to the Constraints:
Solve
Close
Options
Add
Change
Delete
Reset All
Help

FIGURE 14.24

To obtain a solution, click on Solve. The rounded solution $x = 4$ and $y = 18$ is located in cells A1 and B1, respectively. The minimum cost $C(4, 18) = 1336$ is located in cell A3.

Note: One must be careful when using the Excel Solver, because it will not find a maximizer or minimizer of a function if the initial guess is the exact place in which a saddle point occurs. For example, in the preceding problem, if our initial guess was (0, 0), the Solver would have returned the value of (0, 0) as the place where a minimum occurs. But (0, 0) is a saddle point. Thus, it is always a good idea to run the Solver for two different initial values and compare the solutions.

14.3 ▸ Exercises

1. Compare and contrast the way critical points are found for functions with one variable and for functions with more than one variable.

2. Compare and contrast the second-derivative test for $y = f(x)$ and the test for local extrema for $z = f(x, y)$.

Find all points where the given functions have any local extrema. Give the values of any local extrema. Identify any saddle points. (See Examples 1–3.)

3. $f(x, y) = 2x^2 + 4xy + 6y^2 - 8x - 10$

4. $f(x, y) = x^2 + xy + y^2 - 6x - 3$

5. $f(x, y) = x^2 - xy + y^2 + 2x + 2y + 6$

6. $f(x, y) = 2x^2 + 3xy + 2y^2 - 5x - 5y$

7. $f(x, y) = 3xy + 6y - 5x$

8. $f(x, y) = 5xy - 7x^2 - y^2 + 3x - 6y - 4$

9. $f(x, y) = 4xy - 10x^2 - 4y^2 + 8x + 8y + 9$

10. $f(x, y) = 4y^2 + 2xy + 6x + 4y - 8$

11. $f(x, y) = x^2 + xy - 2x - 2y + 2$

12. $f(x, y) = x^2 + xy + y^2 - 3x - 5$

13. $f(x, y) = x^2 - y^2 - 2x + 4y - 7$

14. $f(x, y) = 4x + 2y - x^2 + xy - y^2 + 3$

15. $f(x, y) = 2x^3 + 2y^2 - 12xy + 15$

16. $f(x, y) = 2x^2 + 4y^3 - 24xy + 18$

17. $f(x, y) = x^2 + 4y^3 - 6xy - 1$

18. $f(x, y) = 2y^3 + 5x^2 + 60xy + 25$

19. $f(x, y) = e^{xy}$

20. $f(x, y) = x^2 + e^y$

Figures (a)–(f) show the graphs of the functions defined in Exercises 21–26, respectively. Find all local extrema for each function, and then match the equation with its graph.

21. $z = -3xy + x^3 - y^3 + \frac{1}{8}$

22. $z = \frac{3}{2}y - \frac{1}{2}y^3 - x^2y + \frac{1}{16}$

23. $z = y^4 - 2y^2 + x^2 - \frac{17}{16}$

24. $z = -2x^3 - 3y^4 + 6xy^2 + \frac{1}{16}$

25. $z = -x^4 + y^4 + 2x^2 - 2y^2 + \frac{1}{16}$

26. $z = -y^4 + 4xy - 2x^2 + \frac{1}{16}$

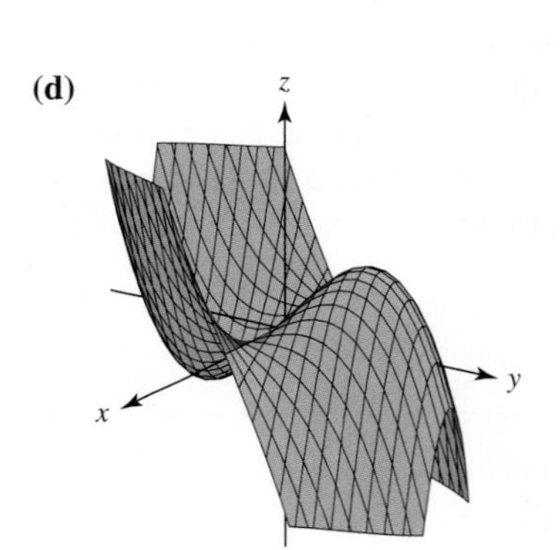

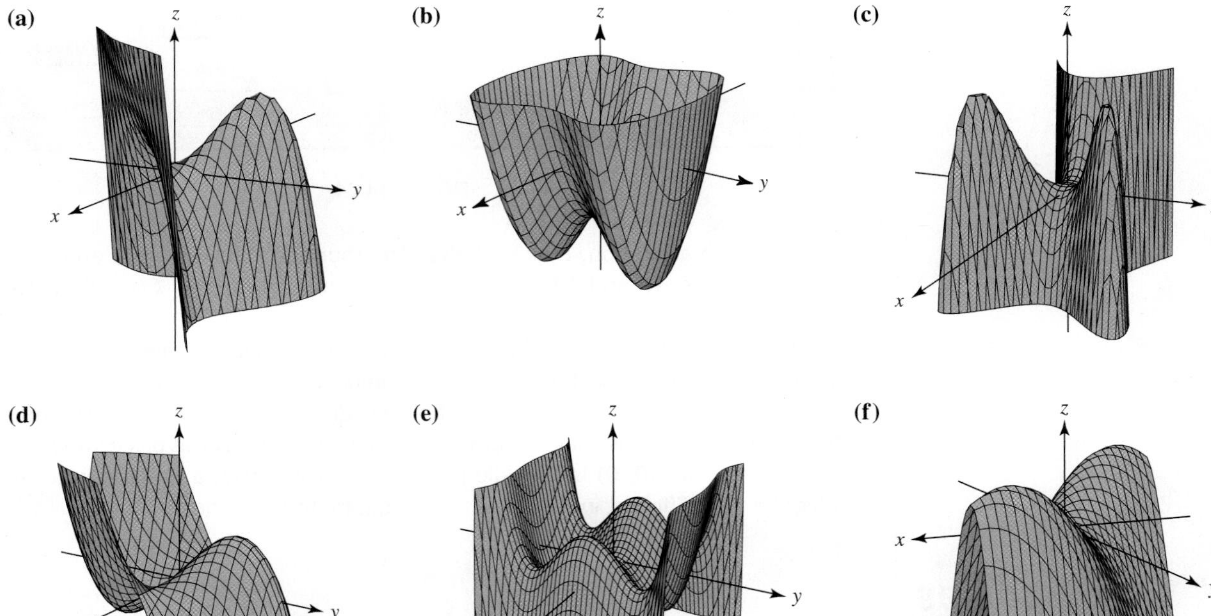

Business *Work the exercises that follow. (See Example 4.)*

27. Suppose that the profit of a certain firm is approximated by

$$P(x, y) = 1000 + 24x - x^2 + 80y - y^2,$$

where x is the cost of a unit of labor and y is the cost of a unit of goods. Find values of x and y that maximize the profit. Find the maximum profit.

28. The labor cost in dollars for manufacturing a precision camera can be approximated by

$$L(x, y) = \frac{3}{2}x^2 + y^2 - 2x - 2y - 2xy + 68,$$

where x is the number of hours required by a skilled craftsperson and y is the number of hours required by a semiskilled person. Find values of x and y that minimize the labor charge. Find the minimum labor charge.

29. The profit (in thousands of dollars) from the sales of graphing calculators is approximated by

$$P(x, y) = 800 - 2x^3 + 12xy - y^2,$$

where x is the cost of a unit of chips and y is the cost of a unit of labor. Find the maximum profit and the costs of chips and labor that produce the maximum profit.

30. The total profit from one acre of a certain crop depends on the amounts spent on fertilizer, x, and hybrid seed, y, according to the model

$$P(x, y) = -x^2 + 3xy + 160x - 5y^2 + 200y + 2{,}600{,}000.$$

Find values of x and y that lead to maximum profit. Find the maximum profit.

31. The total cost required to produce x units of electrical tape and y units of packing tape is given by

$$C(x, y) = 2x^2 + 3y^2 - 2xy + 2x - 126y + 3800.$$

Find the number of units of each kind of tape that should be produced so that the total cost is a minimum. Find the minimum total cost.

32. The total revenue, in thousands of dollars, from the sale of x spas and y solar heaters is approximated by

$$R(x, y) = 12 + 74x + 85y - 3x^2 - 5y^2 - 5xy.$$

Find the number of each that should be sold to produce maximum revenue. Find the maximum revenue.

33. A rectangular closed box is to be built at minimum cost to hold 27 cubic meters. Since the cost will depend on the surface area, find the dimensions that will minimize the surface area of the box.

34. Find the dimensions that will minimize the surface area (and hence the cost) of a rectangular fish aquarium, open on top, with a volume of 32 cubic feet.

35. The U.S. Postal Service requires that any box sent by Priority Mail have a length plus girth (distance around) totaling no more than 108 inches. (See Example 7 on page 804 for a special case.) Find the dimensions of the box with maximum volume that can be sent by Priority Mail.

36. A box that is sent through the U.S. mail by parcel post must have a length plus girth of no more than 130 inches. Find the dimensions of the box with maximum volume that can be sent by parcel post. (See Exercise 35.)

37. A piece of carry-on luggage on most airlines (such as American, Delta, and Continental) must satisfy the condition that the sum of its length, width, and height is no more than 45 inches. If no other restrictions apply, find the dimensions of the piece of carry-on luggage with maximum volume that can be brought on these airlines.*

38. Find the dimensions of the circular tube with maximum volume that can be sent by U.S. Priority Mail. See Exercise 35.

39. The monthly revenue, in hundreds of dollars, from the production of x thousand tons of grade A iron ore and y thousand tons of grade B iron ore is given by

$$R(x, y) = 2xy + 2y + 12,$$

and the corresponding cost, in hundreds of dollars, is given by

$$C(x, y) = 2x^2 + y^2.$$

*Depending on the plane and the airline, additional restrictions may apply (for instance, that the luggage fit under the seat or in the overhead compartment).

Find the amount of each grade of ore that will produce maximum profit.

40. Suppose the revenue and cost, in thousands of dollars, from the manufacture of x units of one product and y units of another are, respectively,

$$R(x, y) = 6xy + 3 - x^2 \quad \text{and} \quad C(x, y) = x^2 + 3y^3.$$

How many units of each product will produce maximum profit? What is the maximum profit?

41. The profit (in thousands of dollars) that Aunt Mildred's Metalworks earns from producing x tons of steel and y tons of aluminum can be approximated by

$$P(x, y) = 36xy - x^3 - 8y^3.$$

Find the amounts of steel and aluminum that maximize the profit, and find the value of the maximum profit.

42. The time (in hours) that a branch of Amalgamated Entities needs to spend to meet the quota set by the main office can be approximated by

$$T(x, y) = x^4 + 16y^4 - 32xy + 40,$$

where x represents how many thousands of dollars the factory spends on quality control and y represents how many thousands of dollars it spends on consulting. Find the amount of money the factory should spend on quality control and on consulting to minimize the time spent, and find the minimum number of hours.

43. Health Research has shown that the total change in color, E (measured in the form of energy as kJ/mol), when blanched potato strips are fried can be estimated by the function

$$E(t, T) = 436.16 - 10.57t - 5.46T - .02t^2 + .02T^2 + .08Tt,$$

where T is the temperature (in degrees Centigrade) and t is the frying time (in minutes).*

(a) Find the value of E prior to cooking. Assume that $T = 0$.

(b) Estimate the total change in color of a potato strip that has been cooked for 10 minutes at 180°C.

(c) Determine the critical point of this function and whether a maximum, minimum, or saddle point occurs at that point.

✓ Checkpoint Answers

1. **(a)** $(-2, 3)$ **(b)** $(0, 2)$

2. **(a)** Local minimum at $(-2, 3)$

(b) Neither a local minimum nor a local maximum at $(0, 2)$

3. Local minimum at $\left(\frac{1}{2}, \frac{\sqrt{2}}{2}\right)$; saddle point at $(0, 0)$

4. $M = -5184$, so there is a saddle point at $(0, 0)$.

*Hindra, F., and Oon-Doo Baik, "Kinetics of Quality Changes During Food Frying," *Critical Reviews in Food Science and Nutrition*, Vol. 46, 2006, pp. 239–358.

14.4 ▶ Lagrange Multipliers

It is often necessary to solve optimization problems (such as finding the maximum profit or the minimum cost). We have seen two techniques for doing this with functions of two variables:

1. Change the given problem into an equivalent one-variable problem by making a suitable substitution (as in Example 7 of Section 12.3).
2. Use the test for local extrema (as in Example 4 of Section 14.3).

The first technique works only when algebra can be used to express one variable in terms of the other. The second technique does not work when constraints are present (as is often the case in the real world). Finally, neither method applies to functions of three or more variables.

In this section, we present Lagrange's method for optimizing a function of two or more variables, subject to a constraint.* The method is used for problems of the form

$$\text{Find the optimum value of the function } f(x, y),$$
$$\text{subject to the constraint } g(x, y) = 0;$$

that is, find the point (x, y) satisfying the constraint $g(x, y) = 0$, at which $f(x, y)$ has its optimum value. The optimum value may be either a maximum value or a minimum value, depending on the problem.

Here is a typical example:

$$\text{Find the minimum value of } f(x, y) = x^2 + y^2,$$
$$\text{subject to the constrain } x + y - 4 = 0.$$

In this case, $g(x, y) = x + y - 4$ and the optimum value is the minimum value. Graphically, the point (x, y, z) satisfying $x + y - 4 = 0$ at which $z = f(x, y)$ is smallest is the lowest point on the intersection of the graph of $z = x^2 + y^2$ and the plane $x + y - 4 = 0$, as shown in Figure 14.25.† Note that this point is *not* the lowest point on the graph of $f(x, y) = x^2 + y^2$. The minimum value of $f(x, y)$ (with no constraint) occurs at $(0, 0, 0)$, which does not lie on the plane $x + y - 4 = 0$ given by the constraint.

Lagrange's method provides an algebraic way to solve problems such as this one. The key idea is to introduce an additional variable called the **Lagrange multiplier** and denoted by the lowercase Greek letter λ (lambda) and to use the *Lagrange function*

$$F(x, y, \lambda) = f(x, y) - \lambda \cdot g(x, y).$$

In our example,

$$\begin{aligned} F(x, y, \lambda) &= (x^2 + y^2) - \lambda(x + y - 4) \\ &= x^2 + y^2 - \lambda x - \lambda y + 4\lambda. \end{aligned}$$

*Named for the French mathematician Joseph Louis Lagrange (1736–1813). The method is stated here for functions of two variables, but with obvious modifications, it is valid for any number of variables.

†These graphs were found in Example 7 and Exercise 21 of Section 14.1.

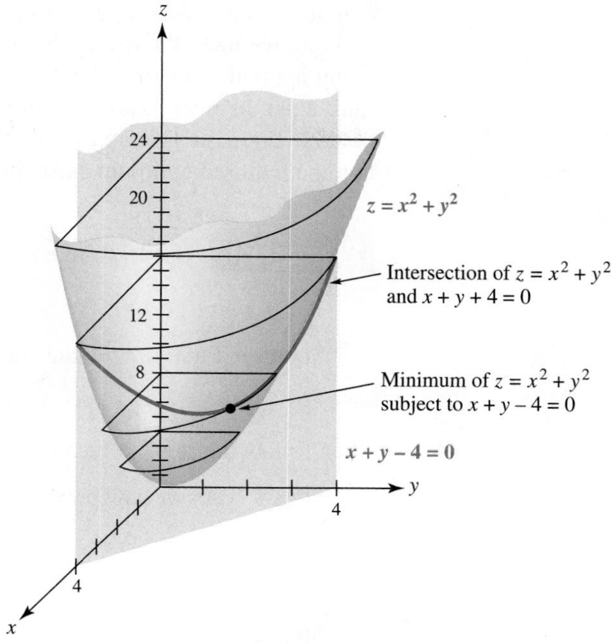

FIGURE 14.25

Lagrange's method (whose proof is beyond the scope of this book) uses the partial derivatives of the function F to find the point satisfying the constraint $g(x, y) = 0$ at which $f(x, y)$ has its optimum value (if such a point exists).

Lagrange's Method

To optimize the function $f(x, y)$ subject to the constraint $g(x, y) = 0$,

1. Form the Lagrange function

$$F(x, y, \lambda) = f(x, y) - \lambda \cdot g(x, y).$$

2. Find the partial derivatives of F: $F_x(x, y, \lambda)$, $F_y(x, y, \lambda)$, and $F_\lambda(x, y, \lambda)$.
3. Solve the following system of equations:

$$F_x(x, y, \lambda) = 0$$
$$F_y(x, y, \lambda) = 0$$
$$F_\lambda(x, y, \lambda) = 0.$$

If the original problem has a solution, it will occur at a point (x, y) for which there exists a value of λ such that (x, y, λ) is a solution of the system in Step 3.

The points found in Step 3 of Lagrange's method provide a list of *possibilities*. If the optimum value exists, it must occur at a point on the list, but the list may contain points that do not produce an optimum value.* Thus, to apply Lagrange's method, we must know that the required maximum or minimum actually exists. In many applied problems, the given conditions make it clear that an optimum value must exist. In other cases, graphical information can sometimes be used to make this determination, as Example 1 illustrates. Unless stated otherwise, you may assume that the required optimum exists in each example and exercise of this section.

Example 1 Use Lagrange's method to find the minimum value of

$$f(x, y) = x^2 + y^2, \text{ subject to the constraint } x + y = 4.$$

Solution First, rewrite the constraint in the form $g(x, y) = 0$:

$$x + y - 4 = 0.$$

Then follow the steps in the preceding box.

Step 1 As we saw previously, the Lagrange function is

$$\begin{aligned} F(x, y, \lambda) &= f(x, y) - \lambda \cdot g(x, y). \\ &= x^2 + y^2 - \lambda x - \lambda y + 4\lambda. \end{aligned}$$

Step 2 Find the partial derivatives of F:

$$\begin{aligned} F_x(x, y, \lambda) &= 2x - \lambda \\ F_y(x, y, \lambda) &= 2y - \lambda \\ F_\lambda(x, y, \lambda) &= -x - y + 4. \end{aligned}$$

Step 3 Set each partial derivative equal to 0 and solve the resulting system:

$$\begin{aligned} 2x - \lambda &= 0 \\ 2y - \lambda &= 0 \\ -x - y + 4 &= 0. \end{aligned} \qquad (1)$$

Since this is a system of *linear* equations in x, y, and λ, it could be solved by the matrix techniques of Section 6.2. However, we shall use a different technique, one that can be used even when the equations of the system are not all linear. Begin by solving the first two equations for λ:

$$\begin{aligned} 2x - \lambda &= 0 \qquad & 2y - \lambda &= 0 \\ \lambda &= 2x \qquad & \lambda &= 2y. \end{aligned}$$

Set the two expressions for λ equal to obtain

$$\begin{aligned} 2x &= 2y \\ x &= y. \end{aligned}$$

Now make the substitution $y = x$ in the third equation of the system (1):

$$\begin{aligned} -x - y + 4 &= 0 \\ -x - x + 4 &= 0 \\ -2x &= -4 \\ x &= 2. \end{aligned}$$

*This is similar to the situation with functions of one variable. Every local extremum must occur at a critical number (where the derivative is 0 or undefined), but not every critical number produces a local extremum.

Since $y = x$ and $\lambda = 2x$, we see that the only solution of system (1) is

$$x = 2, \qquad y = 2, \qquad \lambda = 2 \cdot 2 = 4.$$

Graphical considerations show that the original problem has a solution (see Figure 14.25 and the discussion preceding it), so we conclude that the minimum value of $f(x, y) = x^2 + y^2$, subject to the constraint $x + y = 4$, occurs when $x = 2$ and $y = 2$. The minimum value is

$$f(2, 2) = 2^2 + 2^2 = 8.$$

Example 2 A builder plans to construct a three-story building with a rectangular floor plan. The cost of the building, which depends on the dimensions of the floor plan, is given by

$$xy + 30x + 20y + 474{,}000,$$

where x and y are, respectively, the length and width of the rectangular floors. What length and width should be used if the building is to cost \$500,000 and have maximum area on each floor?

Solution The area of each floor is given by $A(x, y) = xy$. Since the building is to cost \$500,000, we have the following constraint:

$$xy + 30x + 20y + 474{,}000 = 500{,}000$$
$$\underbrace{xy + 30x + 20y - 26{,}000}_{g(x,\, y)} = 0.$$

So we must find the maximum value of $A(x, y)$, subject to $g(x, y) = 0$.

Step 1 The Lagrange function is

$$\begin{aligned} F(x, y, \lambda) &= A(x, y) - \lambda \cdot g(x, y). \\ &= xy - \lambda(xy + 30x + 20y - 26{,}000) \\ &= xy - \lambda xy - 30\lambda x - 20\lambda y + 26{,}000\lambda. \end{aligned}$$

Step 2 The partial derivatives of F are

$$\begin{aligned} F_x(x, y, \lambda) &= y - \lambda y - 30\lambda \\ F_y(x, y, \lambda) &= x - \lambda x - 20\lambda \\ F_\lambda(x, y, \lambda) &= -xy - 30x - 20y + 26{,}000. \end{aligned}$$

Step 3 Set each partial derivative equal to 0 and solve the system of equations

$$\begin{aligned} y - \lambda y - 30\lambda &= 0 \\ x - \lambda x - 20\lambda &= 0 \\ -xy - 30x - 20y + 26{,}000 &= 0 \end{aligned} \qquad (2)$$

as follows. Solve the first two equations for λ:

$$\begin{aligned} y - \lambda y - 30\lambda &= 0 & \qquad x - \lambda x - 20\lambda &= 0 \\ y &= \lambda y + 30\lambda & x &= \lambda x + 20\lambda \\ y &= (y + 30)\lambda & x &= (x + 20)\lambda \\ \lambda &= \frac{y}{y + 30} & \lambda &= \frac{x}{x + 20}. \end{aligned}$$

Set the two expressions for λ equal and solve the resulting equation:

$$\frac{y}{y+30} = \frac{x}{x+20}$$

$$y(x+20) = x(y+30) \quad \text{Multiply both sides by } (x+20)(y+30).$$

$$xy + 20y = xy + 30x$$

$$20y = 30x \quad \text{Subtract } xy \text{ from both sides.}$$

$$y = 1.5x. \quad \text{Divide both sides by 20.}$$

Make the substitution $y = 1.5x$ in the third equation of the system (2):

$$-xy - 30x - 20y + 26{,}000 = 0$$

$$-x(1.5x) - 30x - 20(1.5x) + 26{,}000 = 0$$

$$-1.5x^2 - 60x + 26{,}000 = 0.$$

By the quadratic formula,

$$x = \frac{-(-60) \pm \sqrt{(-60)^2 - 4(-1.5)(26{,}000)}}{2(-1.5)} \approx \begin{cases} 113.17 \\ -153.17 \end{cases}.$$

Since x is a length, the negative value does not apply. Hence,

$$y = 1.5x \approx 1.5(113.17) \approx 169.75.$$

We now have sufficient information. (There is no need to solve for λ.) Floors that measure 113.17 ft × 169.75 ft will result in the largest area per floor, which is

$$A(113.17, 169.75) = (113.17)(169.75) \approx 19{,}211 \text{ sq ft.}$$

NOTE In Examples 1 and 2, we solved the system by first solving each equation with a λ in it for λ. Then we set the expressions for λ equal and solved for one of the original variables. This is often a good approach, since we are not usually interested in the value of λ.

With a constrained function of two variables (as in the preceding examples), the Lagrange function has three variables, and a system of three equations in three variables must be solved. Next we consider constrained functions of three variables, in which case the Lagrange function has four variables and a system of four equations in four variables must be solved.

Example 3 Find three positive numbers x, y, and z whose sum is 50 and such that xyz^2 is as large as possible.

Solution We must find the maximum value of $f(x, y, z) = xyz^2$, subject to the constraint $x + y + z = 50$. Begin by writing the constraint as

$$\underbrace{x + y + z - 50}_{g(x,\,y,\,z)} = 0.$$

Step 1 The Lagrange function is

$$\begin{aligned} F(x, y, z, \lambda) &= f(x, y, z) - \lambda \cdot g(x, y, z). \\ &= xyz^2 - \lambda(x + y + z - 50) \\ &= xyz^2 - \lambda x - \lambda y - \lambda z + 50\lambda. \end{aligned}$$

Step 2 The partial derivatives of F are

$$\begin{aligned} F_x(x, y, z, \lambda) &= yz^2 - \lambda \\ F_y(x, y, z, \lambda) &= xz^2 - \lambda \\ F_z(x, y, z, \lambda) &= 2xyz - \lambda \\ F_\lambda(x, y, z, \lambda) &= -x - y - z + 50. \end{aligned}$$

Step 3 Set each partial derivative equal to 0 and solve the system of equations

$$\begin{aligned} yz^2 - \lambda &= 0 \\ xz^2 - \lambda &= 0 \\ 2xyz - \lambda &= 0 \\ -x - y - z + 50 &= 0. \end{aligned} \tag{3}$$

The first three equations are easily solved for λ:

$$\lambda = yz^2, \qquad \lambda = xz^2, \qquad \lambda = 2xyz.$$

Setting the expressions for λ equal produces three equations:

$$yz^2 = xz^2, \qquad yz^2 = 2xyz, \qquad xz^2 = 2xyz.$$

Solve the first two of these equations by doing Checkpoint 1. ✓1 The answers show that

$$y = x \qquad \text{and} \qquad z = 2x.$$

(This information is sufficient, so we need not solve the third equation.) Substitute these values into the last equation of the system (3):

$$\begin{aligned} -x - y - z + 50 &= 0 \\ -x - x - 2x + 50 &= 0 && \text{Substitute } y = x \text{ and } z = 2x. \\ -4x &= -50 && \text{Divide both sides by } -4. \\ x &= 12.5. \end{aligned}$$

✓Checkpoint 1

Assume that x, y, z are positive numbers.

(a) Solve for y:

$$yz^2 = xz^2$$

(b) Solve for z:

$$yz^2 = 2xyz.$$

Answers: See page 945.

Thus, the three numbers in the original problem are

$$x = 12.5, \qquad y = x = 12.5, \quad \text{and} \quad z = 2x = 2(12.5) = 25,$$

and the maximum value of $f(x, y, z) = xyz^2$ is $(12.5)(12.5)(25^2) = 97{,}656.25$.

Example 4 Find the dimensions of the closed rectangular box of maximum volume that can be constructed from 6 square feet of material.

Solution Let the dimensions of the box be x, y, and z, as shown in Figure 14.26 on the next page. The volume of the box is given by

$$f(x, y, z) = xyz.$$

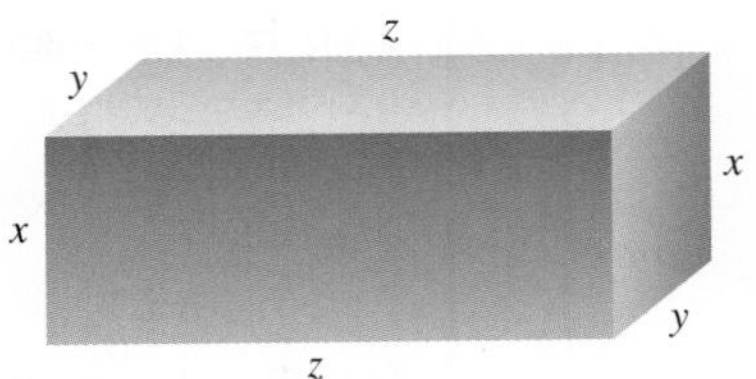

FIGURE 14.26

By considering the surface area of each side of the box in Figure 14.26, we obtain a constraint:

$$\begin{aligned} \text{Right end} + \text{left end} + \text{front} + \text{back} + \text{top} + \text{bottom} &= 6 \\ xy + xy + xz + xz + yz + yz &= 6 \\ 2xy + 2xz + 2yz &= 6 \\ xy + xz + yz &= 3. \end{aligned}$$

Divide both sides by 2.

So we must find the maximum value of $f(x, y, z) = xyz$, subject to the constraint $xy + xz + yz - 3 = 0$.

Step 1 The Lagrange function is

$$\begin{aligned} F(x, y, z, \lambda) &= f(x, y, z) - \lambda \cdot g(x, y, z). \\ &= xyz - \lambda(xy + xz + yz - 3) \\ &= xyz - \lambda xy - \lambda xz - \lambda yz + 3\lambda. \end{aligned}$$

Step 2 The partial derivatives of F are

$$\begin{aligned} F_x(x, y, z, \lambda) &= yz - \lambda y - \lambda z \\ F_y(x, y, z, \lambda) &= xz - \lambda x - \lambda z \\ F_z(x, y, z, \lambda) &= xy - \lambda x - \lambda y \\ F_\lambda(x, y, z, \lambda) &= -xy - xz - yz + 3. \end{aligned}$$

Step 3 Set each partial derivative equal to 0 and solve the system of equations

$$\begin{aligned} yz - \lambda y - \lambda z &= 0 \\ xz - \lambda x - \lambda z &= 0 \\ xy - \lambda x - \lambda y &= 0 \\ -xy - xz - yz + 3 &= 0. \end{aligned} \tag{4}$$

Begin by solving each of the first three equations for λ. ✓2

✓Checkpoint 2

Solve for λ.

(a) $yz - \lambda y - \lambda z = 0$

(b) $xz - \lambda x - \lambda z = 0$

(c) $xy - \lambda x - \lambda y = 0$

Answers: See page 945.

Checkpoint 2 shows that these solutions are

$$\lambda = \frac{yz}{y + z}, \qquad \lambda = \frac{xz}{x + z}, \qquad \text{and} \qquad \lambda = \frac{xy}{x + y}.$$

Set these expressions for λ equal, and simplify as follows (notice in the second and last steps that, since none of the dimensions of the box can be 0, we can divide both sides of each equation by x or z):

$$\frac{yz}{y + z} = \frac{xz}{x + z} \qquad \text{and} \qquad \frac{xz}{x + z} = \frac{xy}{x + y}$$

$$\frac{y}{y+z} = \frac{x}{x+z} \qquad \frac{z}{x+z} = \frac{y}{x+y}$$
$$y(x+z) = x(y+z) \qquad z(x+y) = y(x+z)$$
$$xy + yz = xy + xz \qquad zx + zy = yx + yz$$
$$yz = xz \qquad zx = yx$$
$$y = x \qquad z = y.$$

(Setting the first and third expressions equal give no additional information.) Thus, $x = y = z$. From the last equation in system (4), with $x = y$ and $z = y$,

$$-xy - xz - yz + 3 = 0$$
$$-y^2 - y^2 - y^2 + 3 = 0$$
$$-3y^2 = -3$$
$$y^2 = 1$$
$$y = \pm 1.$$

The negative solution is not applicable, so the solution of the system of equations is $x = 1, y = 1, z = 1$. In other words, the box with maximum volume under the constraint is a cube that measures 1 ft on each side.

TECHNOLOGY TIP Finding extrema of a constrained function of one or more variables can be done with a spreadsheet. In addition to the requirements stated in the Technology Tip at the end of Section 14.3, the constraint must be input into the Excel Solver. To do this, we need to input the variable part of the constraint into a designated cell. In Example 4, for instance, suppose that A5 is the designated cell. Then, in cell A5, we would type "`= A1*B1 + A1*C1 + B1*C1`".

We now click on the Tools menu and choose Solver to find a solution that either maximizes or minimizes the value of cell A3, depending on which option we choose. Figure 14.27 illustrates the Solver box and the items placed in it.

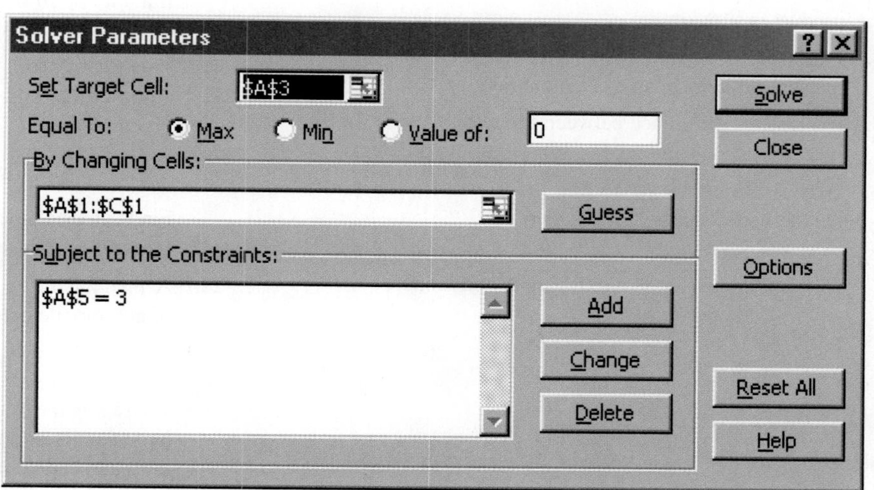

FIGURE 14.27

To obtain a solution, click on Solve. The solution $x = 1, y = 1, z = 1$ is located in cells A1, B1, and C1, respectively. The maximum volume $f(1, 1, 1) = 1$ is located in cell A3.

Note: One must be careful when using the Solver because the solution may depend on the initial value. Thus, it is always a good idea to run the Solver for two different initial values and compare the solutions.

14.4 ▶ Exercises

Find the maxima and minima in Exercises 1–10. (See Examples 1–4.)

1. Maximum of $f(x, y) = 2xy$, subject to $x + y = 20$
2. Maximum of $f(x, y) = 4xy + 8$, subject to $x + y = 16$
3. Maximum of $f(x, y) = xy^2$, subject to $x + 4y = 15$
4. Minimum of $f(x, y) = 9x^2y$, subject to $3x - y = 5$
5. Minimum of $f(x, y) = x^2 + 2y^2 - xy$, subject to $x + y = 8$
6. Minimum of $f(x, y) = 3x^2 + 4y^2 - xy - 2$, subject to $2x + y = 21$
7. Maximum of $f(x, y) = x^2 - 10y^2$, subject to $x - y = 18$
8. Maximum of $f(x, y) = 12xy - x^2 - 3y^2$, subject to $x + y = 16$
9. Maximum of $f(x, y, z) = xyz^2$, subject to $x + y + z = 6$
10. Minimum of $f(x, y, z) = xy + 2xz + 2yz$, subject to $xyz = 32$
11. Find positive numbers x and y such that $x + y = 21$ and $2xy^2$ is maximized.
12. Find positive numbers x and y such that $x + y = 24$ and $x^2y + 4$ is maximized.
13. Find three positive numbers whose sum is 102 and whose product is a maximum.
14. Find three positive numbers such that their sum is 75 and the sum of their squares is a maximum.
15. Explain the difference between the two methods we used in Sections 14.3 and 14.4 to solve extrema problems.
16. Why is it unnecessary to find the value of λ when using the method explained in this section?
17. Show that the three equations in Step 3 of the box on page 937 are equivalent to the three equations

$$f_x(x, y) = \lambda g_x(x, y),$$
$$f_y(x, y) = \lambda g_y(x, y),$$
$$g(x, y) = 0.$$

Business *Solve the given problems. (See Examples 2 and 4.)*

18. Because of terrain difficulties, two sides of a fence can be built for \$6 per ft, while the other two sides cost \$4 per ft. (See the accompanying sketch.) Find the field of maximum area that can be enclosed for \$1200.

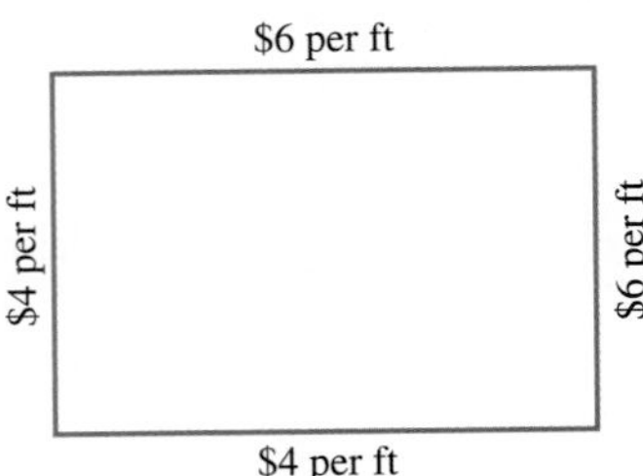

19. To enclose a yard, a fence is built against a large building, so that fencing material is used only on three sides. Material for the ends costs \$8 per ft; material for the side opposite the building costs \$6 per ft. Find the dimensions of the yard of maximum area that can be enclosed for \$1200.
20. The total cost to produce x large needlepoint kits and y small ones is given by

$$C(x, y) = 2x^2 + 6y^2 + 4xy + 10.$$

If a total of 10 kits must be made, how should production be allocated so that total cost is minimized?

21. The profit from the sale of x units of radiators for automobiles and y units of radiators for generators is given by

$$P(x, y) = -x^2 - y^2 + 4x + 8y.$$

Find values of x and y that lead to a maximum profit if the firm must produce a total of 6 units of radiators.

22. A manufacturing firm estimates that its total production of automobile batteries, in thousands of units, is

$$f(x, y) = 3x^{1/3}y^{2/3},$$

where x is the number of units of labor and y is the number of units of capital utilized. Labor costs are \$80 per unit, and capital costs are \$150 per unit. How many units each of labor and capital will maximize production if the firm can spend \$40,000 for these costs?

23. For another product, the manufacturing firm in Exercise 22 estimates that production is a function of labor x and capital y as follows:

$$f(x, y) = 12x^{3/4}y^{1/4}.$$

If \$25,200 is available for labor and capital, and if the firm's costs are \$100 and \$180 per unit, respectively, how many units of labor and capital will give maximum production?

24. A farmer has 200 m of fencing. Find the dimensions of the rectangular field of maximum area that can be enclosed by this amount of fencing.

25. Find the area of the largest rectangular field that can be enclosed with 600 m of fencing. Assume that no fencing is needed along one side of the field.

26. A cylindrical can is to be made that will hold 250π cubic inches of candy. Find the dimensions of the can with minimum surface area.

27. An ordinary 12-oz soda pop can holds about 25 cubic inches. Find the dimensions of a can with minimum surface area. Measure a can and see how close its dimensions are to the results you found.

28. A rectangular box with no top is to be built from 500 square meters of material. Find the dimensions of such a box that will enclose the maximum volume.

29. A 1-lb soda cracker box has a volume of 185 cubic inches. The end of the box is square. Find the dimensions of such a box that has minimum surface area.

30. A rectangular closed box is to be built at minimum cost to hold 27 cubic meters. Since the cost will depend on the surface area, find the dimensions that will minimize the surface area of the box.

31. Find the dimensions that will minimize the surface area (and hence the cost) of a rectangular fish aquarium, open on top, with a volume of 32 cubic feet.

32. A company needs to construct a box with an open top that will be used to transport 400 cubic yards of material, in several trips, from one place to another. Two of the sides and the bottom of the box can be made of a free, lightweight material, but only 4 square yards of the material is available. Because of the nature of the material to be transported, the two ends of the box must be made from a heavyweight material that costs \$20 per square yard. Each trip costs 10 cents.

(a) Let x, y, and z respectively denote the length, width, and height of the box. If we want to use all of the free material, show that the total cost in dollars is given by the function

$$f(x, y, z) = \frac{40}{xyz} + 40yz,$$

subject to the constraint $2xz + xy = 4$.

(b) Use the solver feature on a spreadsheet to find the dimensions of the box that minimize the transportation cost, subject to the given constraint.

33. **Social Science** The probability that the majority of a three-person jury will convict a guilty person is given by the formula

$$P(r, s, t) = rs(1 - t) + (1 - r)st + r(1 - s)t + rst$$

subject to the constraint

$$r + s + t = \alpha,$$

where r, s, and t represent each of the three jury members' probability of reaching a guilty verdict and α is some fixed constant that is generally less than or equal to the number of jurors.

(a) Form the Lagrange function.

(b) Find the values of r, s, and t that maximize the probability of convicting a guilty person when $\alpha = .75$.

(c) Find the values of r, s, and t that maximize the probability of convicting a guilty person when $\alpha = 3$.

✓ Checkpoint Answers

1. (a) $y = x$ (b) $z = 2x$

2. (a) $\lambda = \dfrac{yz}{y + z}$ (b) $\lambda = \dfrac{xz}{x + z}$ (c) $\lambda = \dfrac{xy}{x + y}$

CHAPTER 14 ▸ Summary

KEY TERMS AND SYMBOLS

14.1 ▸ $z = f(x, y)$, function of two variables; multivariate function; xy-plane; ordered triple; first octant; surface; trace; level curves; paraboloid; production function; Cobb–Douglas production function

14.2 ▶ f_x or $\frac{\partial f}{\partial x}$, partial derivative of f with respect to x

$\frac{\partial}{\partial x}\left(\frac{\partial z}{\partial x}\right)$ or $\frac{\partial^2 z}{\partial x^2}$, or f_{xx}, second-order partial derivative of $\partial z/\partial x$ (or f_x) with respect to x

$\frac{\partial}{\partial y}\left(\frac{\partial z}{\partial x}\right)$ or $\frac{\partial^2 z}{\partial y \partial x}$ or f_{xy}, second-order partial derivative of $\partial z/\partial x$ (or f_x) with respect to y

14.3 ▶ saddle point
local maximum
local minimum
critical point

14.4 ▶ constraint
Lagrange multipliers
Lagrange function
Lagrange's method

CHAPTER 14 KEY CONCEPTS

The graph of $ax + by + cz = d$ is a **plane**.

The graph of $z = f(x, y)$ is a **surface** in three-dimensional space.

The graph of $z = ax^2 + by^2$ is a **paraboloid**.

Partial Derivatives ▶ The **partial derivative of $f(x, y)$ with respect to x** is the derivative of f found by treating x as a variable and y as a constant.

The **partial derivative of $f(x, y)$ with respect to y** is the derivative of f found by treating y as a variable and x as a constant.

Second-Order Partial Derivatives ▶ For a function $z = f(x, y)$, if all indicated partial derivatives exist, then

$$\frac{\partial}{\partial x}\left(\frac{\partial z}{\partial x}\right) = \frac{\partial^2 z}{\partial x^2} = f_{xx} \qquad \frac{\partial}{\partial y}\left(\frac{\partial z}{\partial y}\right) = \frac{\partial^2 z}{\partial y^2} = f_{yy}$$

$$\frac{\partial}{\partial y}\left(\frac{\partial z}{\partial x}\right) = \frac{\partial^2 z}{\partial y \partial x} = f_{xy} \qquad \frac{\partial}{\partial x}\left(\frac{\partial z}{\partial y}\right) = \frac{\partial^2 z}{\partial x \partial y} = f_{yx}.$$

Local Extrema ▶ Let (a, b) be in the domain of a function f.

1. f has a **local maximum** at (a, b) if there is a circular region in the xy-plane with (a, b) in its interior such that

$$f(a, b) \geq f(x, y)$$

for all points (x, y) in the circular region.

2. f has a **local minimum** at (a, b) if there is a circular region in the xy-plane with (a, b) in its interior such that

$$f(a, b) \leq f(x, y)$$

for all points (x, y) in the circular region.

If $f(a, b)$ is a local extremum, then $f_x(a, b) = 0$ and $f_y(a, b) = 0$.

Test for Local Extrema ▶ Suppose $f(x, y)$ is a function such that f_{xx}, f_{yy}, and f_{xy} all exist. Let (a, b) be a critical point for which $f_x(a, b) = 0$ and $f_y(a, b) = 0$. Let $M = f_{xx}(a, b) \cdot f_{yy}(a, b) - [f_{xy}(a, b)]^2$.

If $M > 0$ and $f_{xx}(a, b) < 0$, then f has a **local maximum** at (a, b).

If $M > 0$ and $f_{xx}(a, b) > 0$, then f has a **local minimum** at (a, b).

If $M < 0$, then f has a **saddle point** at (a, b).

If $M = 0$, the test gives **no information**.

Lagrange's Method ▶ To optimize the function $f(x, y)$ subject to the constraint $g(x, y) = 0$, form the Lagrange function $F(x, y, \lambda) = f(x, y) - \lambda \cdot g(x, y)$ and find its partial derivatives $F_x(x, y, \lambda)$, $F_y(x, y, \lambda)$, and $F_\lambda(x, y, \lambda)$. Then solve the system of equations

$$F_x(x, y, \lambda) = 0, \qquad F_y(x, y, \lambda) = 0, \qquad F_\lambda(x, y, \lambda) = 0.$$

If the original problem has a solution, it will occur at a point (x, y) for which there exists a value of λ such that (x, y, λ) is a solution of the preceding system.

CHAPTER 14 REVIEW EXERCISES

Find $f(-3, 1)$ and $f(5, -2)$ for each of the given functions.

1. $f(x, y) = 6y^2 - 5xy + 2x$
2. $f(x, y) = -3x + 2x^2y^2 + 5y$
3. $f(x, y) = \dfrac{2x - 4}{x + 3y}$
4. $f(x, y) = x\sqrt{x^2 + y^2}$

5. Describe the graph of $2x + y + 4z = 12$.

6. Describe the graph of $y = 2$ on a three-dimensional grid.

Graph the first-octant portion of each plane.

7. $x + 2y + 4z = 4$
8. $3x + 2y = 6$
9. $4x + 5y = 20$
10. $x = 6$
11. Let $z = f(x, y) = -2x^2 + 5xy + y^2$. Find the following:
(a) $\dfrac{\partial z}{\partial x}$; **(b)** $\dfrac{\partial z}{\partial y}(-1, 4)$; **(c)** $f_{xy}(2, -1)$.
12. Let $z = f(x, y) = \dfrac{2y + x^2}{3y - x}$. Find the following:
(a) $\dfrac{\partial z}{\partial y}$; **(b)** $\dfrac{\partial z}{\partial x}(0, 2)$; **(c)** $f_{yy}(-1, 0)$.
13. Explain the difference between $\dfrac{\partial z}{\partial x}$ and $\dfrac{\partial z}{\partial y}$.

Find f_x and f_y.

14. $f(x, y) = 3y - 7x^2y^3$
15. $f(x, y) = 4x^3y + 12xy^3$
16. $f(x, y) = \sqrt{3x^2 + 2y^2}$
17. $f(x, y) = \dfrac{3x^2 - 2y^2}{x^2 + 4y^2}$
18. $f(x, y) = x^3e^{3y}$
19. $f(x, y) = (y + 1)^2e^{2x+y}$
20. $f(x, y) = \ln|x^2 - 4y^3|$
21. $f(x, y) = \ln|1 + x^3y^2|$
22. Explain the difference between f_{xx} and f_{xy}.

Find f_{xx} and f_{xy}.

23. $f(x, y) = 4x^3y^2 - 8xy$
24. $f(x, y) = \dfrac{2x + y}{x - 2y}$
25. $f(x, y) = -6xy^3 + 2x^2y$
26. $f(x, y) = \dfrac{3x + y}{x - 1}$
27. $f(x, y) = x^2e^y$
28. $f(x, y) = ye^{x^2}$
29. $f(x, y) = \ln(2 - x^2y)$
30. $f(x, y) = \ln(1 + 3xy^2)$
31. **Business** The cost (in dollars) of repainting a car is given by
$$C(x, y) = 4x^2 + 5y^2 - 4xy + 50,$$
where x is the number of hours of labor and y is the number of gallons of paint used. Find the cost in the given cases.
(a) 10 hours of labor and 8 gallons of paint are needed.
(b) 12 hours of labor and 10 gallons of paint are needed.
(c) 14 hours of labor and 14 gallons of paint are needed.
32. **Geometry** Write a function in terms of L, W, and H that gives the total material required to build the closed box shown in the accompanying figure.

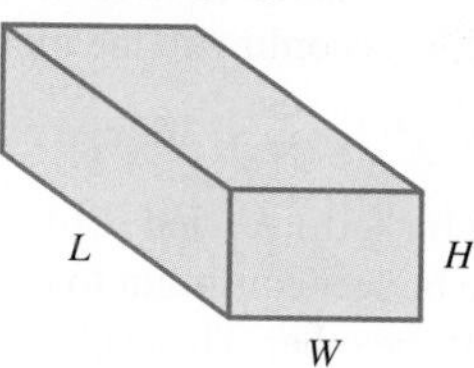

33. **Business** The manufacturing cost, in dollars, for a medium-sized business computer is given by
$$c(x, y) = 2x + y^2 + 2xy + 25,$$
where x is the memory capacity (RAM) of the computer in gigabytes and y is the number of hours of labor required. Find each of the following:
(a) $\dfrac{\partial c}{\partial x}(160, 4)$; **(b)** $\dfrac{\partial c}{\partial y}(350, 10)$.
34. **Business** The production function for a certain company is given by
$$z = x^{.7}y^{.3},$$
where x is the amount of labor (in appropriate units) and y is the amount of capital (in thousands of dollars). Find the marginal productivity of
(a) labor; **(b)** capital.

Find all points where these functions have any local extrema. Find any saddle points.

35. $z = x^2 + 2y^2 - 4y$
36. $z = x^2 + y^2 + 9x - 8y + 1$
37. $f(x, y) = x^2 + 5xy - 10x + 3y^2 - 12y$
38. $z = x^3 - 8y^2 + 6xy + 4$
39. $z = x^3 + y^2 + 2xy - 4x - 3y - 2$
40. $f(x, y) = 7x^2 + y^2 - 3x + 6y - 5xy$
41. **Business** The total cost in dollars to manufacture x solar cells and y solar collectors is
$$c(x, y) = x^2 + 5y^2 + 4xy - 70x - 164y + 1800.$$

(a) Find values of x and y that produce the minimum total cost.
(b) Find the minimum total cost.

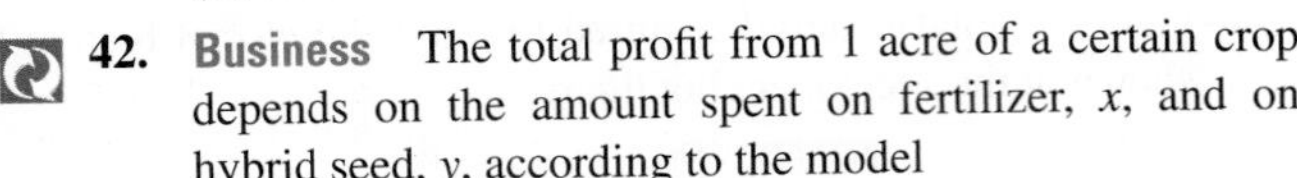

42. **Business** The total profit from 1 acre of a certain crop depends on the amount spent on fertilizer, x, and on hybrid seed, y, according to the model

$$P(x, y) = .01(-x^2 + 3xy + 160x - 5y^2 + 200y + 2600).$$

The budget for fertilizer and seed is limited to \$280.
(a) Use the budget constraint to express one variable in terms of the other. Then substitute into the profit function to get a function with one variable. Use the method shown in Chapter 12 to find the amounts spent on fertilizer and seed that will maximize profit. What is the maximum profit per acre? (*Hint*: Throughout this problem, you may ignore the coefficient of .01 until you need to find the maximum profit.)
(b) Find the amounts spent on fertilizer and seed that will maximize profit, using the method shown in Section 14.3. (*Hint*: You will not need to use the budget constraint.)
(c) Discuss any relationships between these methods.

43. Minimize $f(x, y) = x^2 + y^2$, subject to the constraint $x = y + 2$.
44. Find the maximum value of $f(x, y) = x^2y$, subject to the constraint $x + y = 4$.
45. Find positive numbers x and y whose sum is 80 such that x^2y is maximized.
46. Find positive numbers x and y whose sum is 50 such that xy^2 is maximized.
47. Notice in the previous two exercises that we specified that x and y must be positive numbers. Does a maximum exist without this requirement? Explain why or why not.
48. Use Lagrange's method to find the amounts spent on fertilizer and seeds that will maximize profit in Exercise 42.

CASE 14
Global Warming and the Method of Least Squares

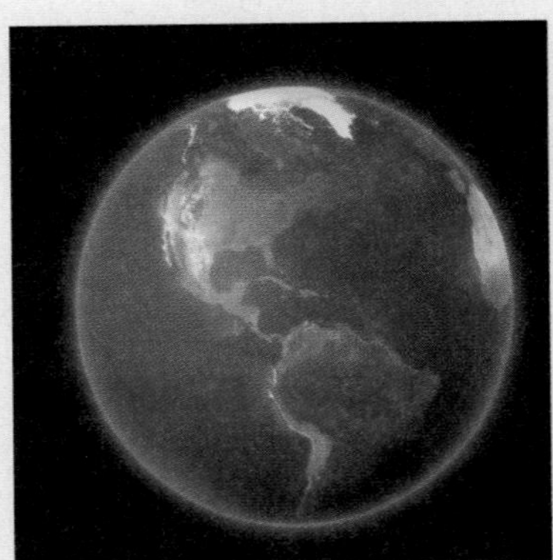

The least-squares regression line for modeling data that is approximately linear was introduced in Section 2.3. The discussion there explained why least squares are used to measure how well a line fits the data. In this case, we show how the theory of extrema of functions of several variables is used to construct the least-squares regression line.

Global warming, the idea that the atmospheric temperature of Earth is slowly increasing over time, is an issue under great debate. Some claim that global warming is a serious problem for which action must be taken; others acknowledge a warming trend, but do not feel that the trend indicates a serious problem; still others believe that there is no verifiable warming trend. In any case, it is clear that a large amount of data must be collected and analyzed before conclusions may be drawn.

The National Oceanic and Atmospheric Administration (NOAA) has compiled a great deal of historical global temperature data. Data from recent years is compiled by satellite, but a variety of methods, including the analysis of temperature-sensitive phenomena such as tree-ring density, allows for temperature estimates from centuries past. The data is often presented in terms of temperature anomalies—that is, as deviations from some norm.

For example, the NOAA reports the following temperature anomalies:

Global Land Surface Average Temperature

Year	Temperature Anomaly (°C)	Temperature (°C)
1900	−0.03	8.47
1950	−0.17	8.33
2000	0.59	9.09

The baseline temperature against which the anomalies were determined is the average global temperature over land from the years 1880 to 2000, an average computed to be 8.5°C. Adding 8.5 to each anomaly gives the actual temperature.

Figure 1 shows a plot of the adjusted temperature data.

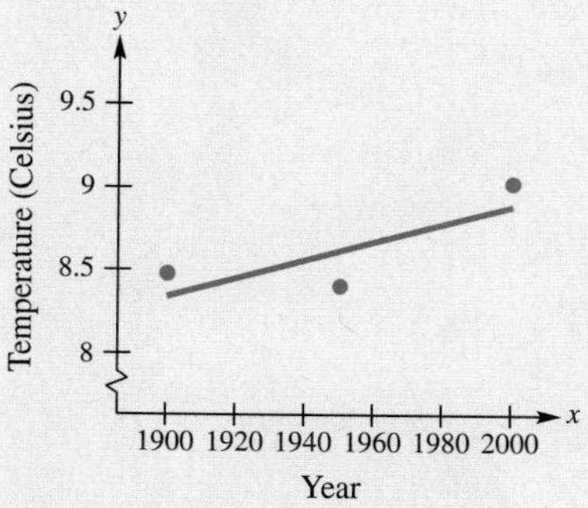

FIGURE 1 Global Land Surface Temperature

If the three points lie on a straight line, the rate at which the temperature is changing can be determined by the slope of that line. The points clearly are not on the same line, but can be approximated by a straight line as in the figure. The purpose of the least-squares method is to find the best such approximation.

Figure 2 shows the distance from each point to the approximating line.

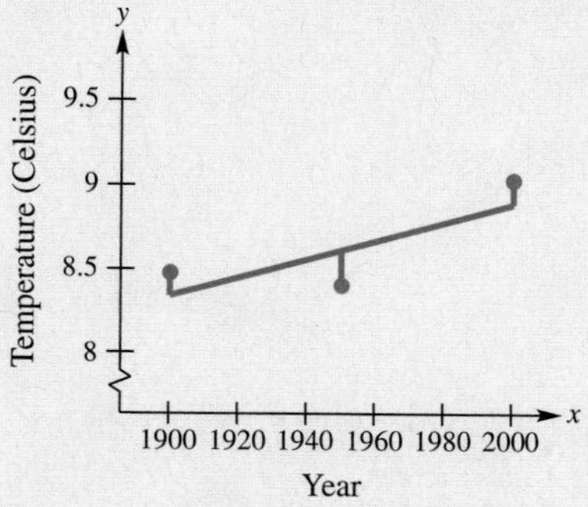

FIGURE 2 Deviations from Approximating Line

If the first point is (x_1, y_1) and the equation of the line is $y = mx + b$, then the first vertical distance is $y_1 - (mx_1 + b) = y_1 - mx_1 - b$. In the least-squares method, the values of m and b are determined so that the sum of the squares of these distances is minimized. For three points (x_1, y_1), (x_2, y_2), and (x_3, y_3), the following function of two variables must be minimized,

$$f(m, b) = (y_1 - mx_1 - b)^2 + (y_2 - mx_2 - b)^2 + (y_3 - mx_3 - b)^2. \tag{1}$$

Using the points

$$(1900, 8.47), \quad (1950, 8.33), \quad \text{and} \quad (2000, 9.09),$$

we find that equation (1) becomes

$$f(m, b) = 223.76 - 101{,}033m - 51.78b + 11{,}412{,}500m^2 + 11{,}700mb + 3b^2.$$

The partial derivatives of f with respect to m and b are

$$\frac{\partial f}{\partial m} = -101{,}033 + 22{,}825{,}000m + 11{,}700b$$

and

$$\frac{\partial f}{\partial b} = -51.78 + 11{,}700m + 6b,$$

respectively, and the resulting critical point is $m = .0062$ and $b = -3.46$. The line that best fits this data is $y = .0062x - 3.46$ and is the line drawn in Figure 1.

According to this analysis, the global land temperature increases at a rate of approximately 0.0062°C/year. It is not fair to say that this is evidence of global warming, as only three pieces of data were used. The data set was kept small to illustrate the least-squares method. This method can be generalized to use any amount of data and to fit more complicated functions, such as exponentials and sine waves, for a more detailed analysis.

EXERCISES

1. Verify that the critical point $m = .0062$ and $b = -3.46$ corresponds to a minimum of the function $f(m, b)$.

2. By differentiating equation (1) with respect to m and b without substituting the specific values for points (x_1, y_1), (x_2, y_2), and (x_3, y_3), the formulas

$$m = \frac{3(x_1y_1 + x_2y_2 + x_3y_3) - (x_1 + x_2 + x_3)(y_1 + y_2 + y_3)}{3(x_1^2 + x_2^2 + x_3^2) - (x_1 + x_2 + x_3)^2}$$

and

$$b = \frac{y_1 + y_2 + y_3}{3} - m\left(\frac{x_1 + x_2 + x_3}{3}\right)$$

can be derived. Show that these formulas give the same values for m and b given earlier.

3. The formulas in Exercise 2 generalize to any number of data points. For example, for four data points, the formulas for m and b respectively become

$$m = \frac{4(x_1y_1 + x_2y_2 + x_3y_3 + x_4y_4)}{4(x_1^2 + x_2^2 + x_3^2 + x_4^2) - (x_1 + x_2 + x_3 + x_4)^2} - \frac{(x_1 + x_2 + x_3 + x_4)(y_1 + y_2 + y_3 + y_4)}{4(x_1^2 + x_2^2 + x_3^2 + x_4^2) - (x_1 + x_2 + x_3 + x_4)^2}$$

and

$$b = \frac{y_1 + y_2 + y_3 + y_4}{4} - m\left(\frac{x_1 + x_2 + x_3 + x_4}{4}\right).$$

Using a similar formula for five data points, add the two temperature data points (1925, 8.46) and (1975, 8.55) to the preceding three and find the least-squares straight-line fit. Plot the points and the line and discuss the value of m.

APPENDIX

A Graphing Calculators

Basics

Instructions on using your graphing calculator are readily available in:

the instruction book for your calculator; and
the web site for this book (www.pearsonhighered.com/MWA10e).

In addition, you may purchase the *Graphing Calculator and Excel Spreadsheet Manual,* written specifically to accompany this textbook. It is described in the Student Supplement section of the Preface and is available at your bookstore or online at www.pearsonstore.com or at www.coursesmart.com. It is also available in MyMathLab.

Programs

The following programs are available for TI and most Casio graphing calculators. You can download them from www.pearsonhighered.com/MWA10e, and use the appropriate USB cable and software to install them in your calculator.

General

1. Fraction Conversion for Casio

Chapter 1: Algebra and Equations

2. Quadratic Formula for TI-83

Chapter 5: Mathematics of Finance

3. Present and Future Value of an Annuity
4. Loan Payment
5. Loan Balance after n Payments
6. Amortization Table for TI

Chapter 6: Systems of Linear Equations and Matrices

7. RREF Program for Casio 9750GA+, 9850, and 9860G

Chapter 7: Linear Programming

8. Simplex Method
9. Two-Stage Method

Chapter 13: Integral Calculus

10. Rectangle Approximation of $\int_a^b f(x)\,dx$ (using left endpoints)

Programs 1, 2, 6 and 7 are built into most calculators other than those mentioned. Programs 3–5 are part of the TVM Solver on TI and most Casio models, although some students may find the versions here easier to use. Programs 8–10 are not built into any calculator.

APPENDIX

B Tables

Table 1 Formulas from Geometry

CIRCLE Area: $A = \pi r^2$ Circumference: $C = 2\pi r$	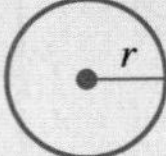
RECTANGLE Area: $A = lw$ Perimeter: $P = 2l + 2w$	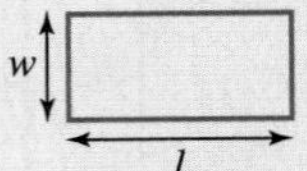
PARALLELOGRAM Area: $A = bh$ Perimeter: $P = 2a + 2b$	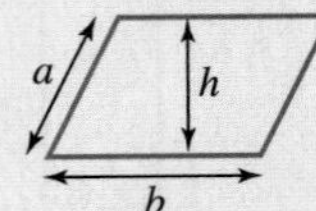
TRIANGLE Area: $A = \frac{1}{2}bh$	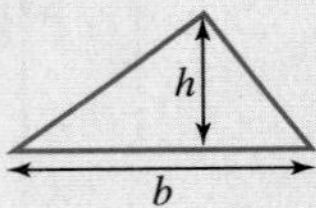
SPHERE Volume: $V = \frac{4}{3}\pi r^3$ Surface area: $A = 4\pi r^2$	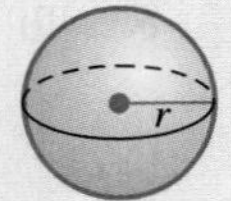
RECTANGULAR BOX Volume: $V = lwh$ Surface area: $A = 2lh + 2wh + 2lw$	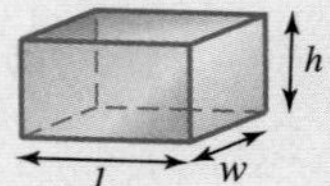
CIRCULAR CYLINDER Volume: $V = \pi r^2 h$ Surface area: $A = 2\pi r^2 + 2\pi rh$	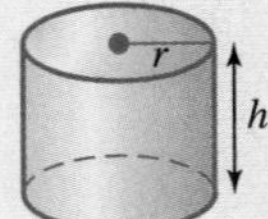
TRIANGULAR CYLINDER Volume: $V = \frac{1}{2}bhl$	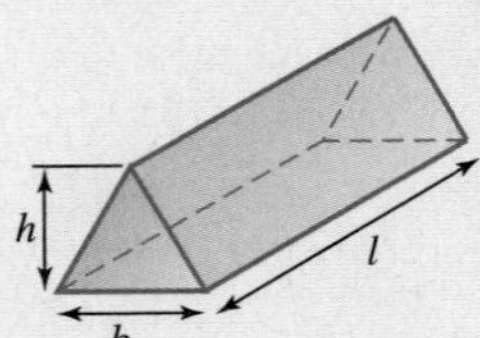
CONE Volume: $V = \frac{1}{3}\pi r^2 h$	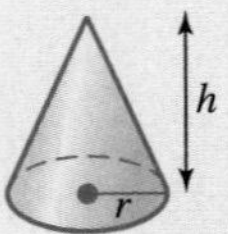

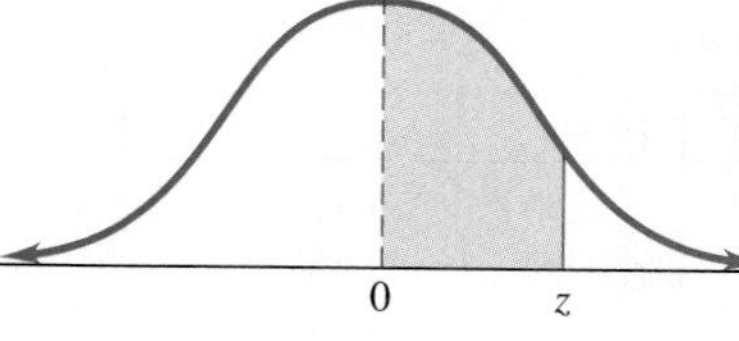

Table 2 Areas under the Normal Curve
The column under A gives the proportion of the area under the entire curve that is between $z = 0$ and a positive value of z.

z	A	z	A	z	A	z	A
.00	.0000	.48	.1844	.96	.3315	1.44	.4251
.01	.0040	.49	.1879	.97	.3340	1.45	.4265
.02	.0080	.50	.1915	.98	.3365	1.46	.4279
.03	.0120	.51	.1950	.99	.3389	1.47	.4292
.04	.0160	.52	.1985	1.00	.3413	1.48	.4306
.05	.0199	.53	.2019	1.01	.3438	1.49	.4319
.06	.0239	.54	.2054	1.02	.3461	1.50	.4332
.07	.0279	.55	.2088	1.03	.3485	1.51	.4345
.08	.0319	.56	.2123	1.04	.3508	1.52	.4357
.09	.0359	.57	.2157	1.05	.3531	1.53	.4370
.10	.0398	.58	.2190	1.06	.3554	1.54	.4382
.11	.0438	.59	.2224	1.07	.3577	1.55	.4394
.12	.0478	.60	.2258	1.08	.3599	1.56	.4406
.13	.0517	.61	.2291	1.09	.3621	1.57	.4418
.14	.0557	.62	.2324	1.10	.3643	1.58	.4430
.15	.0596	.63	.2357	1.11	.3665	1.59	.4441
.16	.0636	.64	.2389	1.12	.3686	1.60	.4452
.17	.0675	.65	.2422	1.13	.3708	1.61	.4463
.18	.0714	.66	.2454	1.14	.3729	1.62	.4474
.19	.0754	.67	.2486	1.15	.3749	1.63	.4485
.20	.0793	.68	.2518	1.16	.3770	1.64	.4495
.21	.0832	.69	.2549	1.17	.3790	1.65	.4505
.22	.0871	.70	.2580	1.18	.3810	1.66	.4515
.23	.0910	.71	.2612	1.19	.3830	1.67	.4525
.24	.0948	.72	.2642	1.20	.3849	1.68	.4535
.25	.0987	.73	.2673	1.21	.3869	1.69	.4545
.26	.1026	.74	.2704	1.22	.3888	1.70	.4554
.27	.1064	.75	.2734	1.23	.3907	1.71	.4564
.28	.1103	.76	.2764	1.24	.3925	1.72	.4573
.29	.1141	.77	.2794	1.25	.3944	1.73	.4582
.30	.1179	.78	.2823	1.26	.3962	1.74	.4591
.31	.1217	.79	.2852	1.27	.3980	1.75	.4599
.32	.1255	.80	.2881	1.28	.3997	1.76	.4608
.33	.1293	.81	.2910	1.29	.4015	1.77	.4616
.34	.1331	.82	.2939	1.30	.4032	1.78	.4625
.35	.1368	.83	.2967	1.31	.4049	1.79	.4633
.36	.1406	.84	.2996	1.32	.4066	1.80	.4641
.37	.1443	.85	.3023	1.33	.4082	1.81	.4649
.38	.1480	.86	.3051	1.34	.4099	1.82	.4656
.39	.1517	.87	.3079	1.35	.4115	1.83	.4664
.40	.1554	.88	.3106	1.36	.4131	1.84	.4671
.41	.1591	.89	.3133	1.37	.4147	1.85	.4678
.42	.1628	.90	.3159	1.38	.4162	1.86	.4686
.43	.1664	.91	.3186	1.39	.4177	1.87	.4693
.44	.1700	.92	.3212	1.40	.4192	1.88	.4700
.45	.1736	.93	.3238	1.41	.4207	1.89	.4706
.46	.1772	.94	.3264	1.42	.4222	1.90	.4713
.47	.1808	.95	.3289	1.43	.4236	1.91	.4719

(continued)

Table 2 ***(continued)***

z	A	z	A	z	A	z	A
1.92	.4726	2.42	.4922	2.92	.4983	3.42	.4997
1.93	.4732	2.43	.4925	2.93	.4983	3.43	.4997
1.94	.4738	2.44	.4927	2.94	.4984	3.44	.4997
1.95	.4744	2.45	.4929	2.95	.4984	3.45	.4997
1.96	.4750	2.46	.4931	2.96	.4985	3.46	.4997
1.97	.4756	2.47	.4932	2.97	.4985	3.47	.4997
1.98	.4762	2.48	.4934	2.98	.4986	3.48	.4998
1.99	.4767	2.49	.4936	2.99	.4986	3.49	.4998
2.00	.4773	2.50	.4938	3.00	.4987	3.50	.4998
2.01	.4778	2.51	.4940	3.01	.4987	3.51	.4998
2.02	.4783	2.52	.4941	3.02	.4987	3.52	.4998
2.03	.4788	2.53	.4943	3.03	.4988	3.53	.4998
2.04	.4793	2.54	.4945	3.04	.4988	3.54	.4998
2.05	.4798	2.55	.4946	3.05	.4989	3.55	.4998
2.06	.4803	2.56	.4948	3.06	.4989	3.56	.4998
2.07	.4808	2.57	.4949	3.07	.4989	3.57	.4998
2.08	.4812	2.58	.4951	3.08	.4990	3.58	.4998
2.09	.4817	2.59	.4952	3.09	.4990	3.59	.4998
2.10	.4821	2.60	.4953	3.10	.4990	3.60	.4998
2.11	.4826	2.61	.4955	3.11	.4991	3.61	.4999
2.12	.4830	2.62	.4956	3.12	.4991	3.62	.4999
2.13	.4834	2.63	.4957	3.13	.4991	3.63	.4999
2.14	.4838	2.64	.4959	3.14	.4992	3.64	.4999
2.15	.4842	2.65	.4960	3.15	.4992	3.65	.4999
2.16	.4846	2.66	.4961	3.16	.4992	3.66	.4999
2.17	.4850	2.67	.4962	3.17	.4992	3.67	.4999
2.18	.4854	2.68	.4963	3.18	.4993	3.68	.4999
2.19	.4857	2.69	.4964	3.19	.4993	3.69	.4999
2.20	.4861	2.70	.4965	3.20	.4993	3.70	.4999
2.21	.4865	2.71	.4966	3.21	.4993	3.71	.4999
2.22	.4868	2.72	.4967	3.22	.4994	3.72	.4999
2.23	.4871	2.73	.4968	3.23	.4994	3.73	.4999
2.24	.4875	2.74	.4969	3.24	.4994	3.74	.4999
2.25	.4878	2.75	.4970	3.25	.4994	3.75	.4999
2.26	.4881	2.76	.4971	3.26	.4994	3.76	.4999
2.27	.4884	2.77	.4972	3.27	.4995	3.77	.4999
2.28	.4887	2.78	.4973	3.28	.4995	3.78	.4999
2.29	.4890	2.79	.4974	3.29	.4995	3.79	.4999
2.30	.4893	2.80	.4974	3.30	.4995	3.80	.4999
2.31	.4896	2.81	.4975	3.31	.4995	3.81	.4999
2.32	.4898	2.82	.4976	3.32	.4996	3.82	.4999
2.33	.4901	2.83	.4977	3.33	.4996	3.83	.4999
2.34	.4904	2.84	.4977	3.34	.4996	3.84	.4999
2.35	.4906	2.85	.4978	3.35	.4996	3.85	.4999
2.36	.4909	2.86	.4979	3.36	.4996	3.86	.4999
2.37	.4911	2.87	.4980	3.37	.4996	3.87	.5000
2.38	.4913	2.88	.4980	3.38	.4996	3.88	.5000
2.39	.4916	2.89	.4981	3.39	.4997	3.89	.5000
2.40	.4918	2.90	.4981	3.40	.4997		
2.41	.4920	2.91	.4982	3.41	.4997		

Table 3 Integrals
(C is an arbitrary constant.)

1. $\int x^n \, dx = \frac{1}{n+1} x^{n+1} + C \qquad (n \neq -1)$

2. $\int e^{kx} \, dx = \frac{1}{k} e^{kx} + C$

3. $\int \frac{a}{x} \, dx = a \ln|x| + C$

4. $\int \ln|ax| \, dx = x(\ln|ax| - 1) + C$

5. $\int \frac{1}{\sqrt{x^2 + a^2}} \, dx = \ln|x + \sqrt{x^2 + a^2}| + C$

6. $\int \frac{1}{\sqrt{x^2 - a^2}} \, dx = \ln|x + \sqrt{x^2 - a^2}| + C$

7. $\int \frac{1}{a^2 - x^2} \, dx = \frac{1}{2a} \cdot \ln\left|\frac{a + x}{a - x}\right| + C \qquad (a \neq 0)$

8. $\int \frac{1}{x^2 - a^2} \, dx = \frac{1}{2a} \cdot \ln\left|\frac{x - a}{x + a}\right| + C \qquad (a \neq 0)$

9. $\int \frac{1}{x\sqrt{a^2 - x^2}} \, dx = -\frac{1}{a} \cdot \ln\left|\frac{a + \sqrt{a^2 - x^2}}{x}\right| + C \qquad (a \neq 0)$

10. $\int \frac{1}{x\sqrt{a^2 + x^2}} \, dx = -\frac{1}{a} \cdot \ln\left|\frac{a + \sqrt{a^2 + x^2}}{x}\right| + C \qquad (a \neq 0)$

11. $\int \frac{x}{ax + b} \, dx = \frac{x}{a} - \frac{b}{a^2} \cdot \ln|ax + b| + C \qquad (a \neq 0)$

12. $\int \frac{x}{(ax + b)^2} \, dx = \frac{b}{a^2(ax + b)} + \frac{1}{a^2} \cdot \ln|ax + b| + C \qquad (a \neq 0)$

13. $\int \frac{1}{x(ax + b)} \, dx = \frac{1}{b} \cdot \ln\left|\frac{x}{ax + b}\right| + C \qquad (b \neq 0)$

14. $\int \frac{1}{x(ax + b)^2} \, dx = \frac{1}{b(ax + b)} + \frac{1}{b^2} \cdot \ln\left|\frac{x}{ax + b}\right| \qquad (b \neq 0)$

15. $\int \sqrt{x^2 + a^2} \, dx = \frac{x}{2} \sqrt{x^2 + a^2} + \frac{a^2}{2} \cdot \ln|x + \sqrt{x^2 + a^2}| + C$

16. $\int x^n \cdot \ln|x| \, dx = x^{n+1}\left[\frac{\ln|x|}{n + 1} - \frac{1}{(n + 1)^2}\right] + C \qquad (n \neq -1)$

17. $\int x^n e^{ax} \, dx = \frac{x^n e^{ax}}{a} - \frac{n}{a} \cdot \int x^{n-1} e^{ax} \, dx + C \qquad (a \neq 0)$

Answers to Selected Exercises

CHAPTER 1

Section 1.1 (Page 10)

1. True **3.** Answers vary with the calculator, but 2,508,429,787/798,458,000 is the best. **5.** Distributive property **7.** Commutative property of addition **9.** Answers vary. **11.** -39 **13.** -2 **15.** 18% **17.** .75% **19.** -12 **21.** 4 **23.** -4 **25.** -1 **27.** $\frac{2040}{523}, \frac{189}{37}, \sqrt{27}, \frac{4587}{691}, 6.735, \sqrt{47}$ **29.** $12 < 18.5$ **31.** $x \geq 5.7$ **33.** $z \leq 7.5$ **35.** $<$ **37.** $<$ **39.** a lies to the right of b or is equal to b. **41.** $c < a < b$ **43.** -8 -1 **45.** -2 2 **47.** -2 **49.** 3 **51.** 7 **53.** 0 **55. (a)** 17.6 **(b)** no **57. (a)** 23.2 **(b)** yes **59.** 60° **61.** 6° **63.** 4 **65.** -19 **67.** $=$ **69.** $=$ **71.** $=$ **73.** $>$ **75.** $7 - a$ **77.** Answers vary. **79.** Answers vary. **81.** 2001, 2002, and 2007

Section 1.2 (Page 19)

1. 1,973,822.685 **3.** 289.0991339 **5.** Answers vary. **7.** 4^5 **9.** $(-6)^7$ **11.** $(5u)^{28}$ **13.** degree 4; coefficients: 6.2, -5, 4, -3, 3.7; constant term 3.7 **15.** 3 **17.** $-x^3 + x^2 - 13x$ **19.** $-6y^2 + 3y + 6$ **21.** $-6x^2 + 4x - 4$ **23.** $-18m^3 - 54m^2 + 9m$ **25.** $12z^3 + 14z^2 - 7z + 5$ **27.** $12k^2 + 16k - 3$ **29.** $6y^2 + 13y + 5$ **31.** $18k^2 - 7kq - q^2$ **33.** $4.34m^2 + 5.68m - 4.42$ **35.** $-k + 3$ **37.** $R = 5000x$; $C = 200{,}000 + 1800x$; $P = 3200x - 200{,}000$ **39.** $R = 9750x$; $C = -3x^2 + 3480x + 259{,}675$; $P = 3x^2 + 6270x - 259{,}675$ **41. (a)** \$179,000,000 **(b)** \$182,155,000 **43. (a)** \$494,000,000 **(b)** \$483,875,000 **45.** \$812,960,000 **47.** \$1,489,000,000 **49.** True **51.** False **53.** .866 **55.** .505 **57. (a)** approximately 60,501,067 cu ft **(b)** The shape becomes a rectangular box with a square base, with volume b^2h. **(c)** yes **59. (a)** 0, 1, 2, 3, or no degree (if one is the negative of the other) **(b)** 0, 1, 2, 3, or no degree (if they are equal) **(c)** 6

Section 1.3 (Page 27)

1. $12x(x - 2)$ **3.** $r(r^2 - 5r + 1)$ **5.** $6z(z^2 - 2z + 3)$ **7.** $(2y - 1)^2(14y - 4) = 2(2y - 1)^2(7y - 2)$ **9.** $(x + 5)^4(x^2 + 10x + 28)$ **11.** $(x + 1)(x + 4)$ **13.** $(x + 3)(x + 4)$ **15.** $(x + 3)(x - 2)$ **17.** $(x - 1)(x + 3)$ **19.** $(x - 4)(x + 1)$ **21.** $(z - 7)(z - 2)$ **23.** $(z + 4)(z + 6)$ **25.** $(2x - 1)(x - 4)$ **27.** $(3p - 4)(5p - 1)$ **29.** $(2z - 5)(2z - 3)$ **31.** $(2x + 1)(3x - 4)$ **33.** $(5y - 2)(2y + 5)$ **35.** $(2x - 1)(3x + 4)$ **37.** $(3a + 5)(a - 1)$ **39.** $(x + 9)(x - 9)$ **41.** $(3p - 2)^2$ **43.** $(r - 2t)(r + 5t)$ **45.** $(m - 4n)^2$ **47.** $(2u + 3)^2$ **49.** cannot be factored **51.** $(2r + 3r)(2r - 3r)$ **53.** $(x + 2y)^2$ **55.** $(3a + 5)(a - 6)$ **57.** $(7m + 2n)(3m + n)$ **59.** $(y - 7z)(y + 3z)$ **61.** $(11x + 8)(11x - 8)$ **63.** $(a - 4)(a^2 + 4a + 16)$ **65.** $(2r - 3s)(4r^2 + 6rs + 9s^2)$ **67.** $(4m + 5)(16m^2 - 20m + 25)$ **69.** $(10y - z)(100y^2 + 10yz + z^2)$ **71.** $(x^2 + 3)(x^2 + 2)$ **73.** $b^2(b + 1)(b - 1)$ **75.** $(x + 2)(x - 2)(x^2 + 3)$ **77.** $(4a^2 + 9b^2)(2a + 3b)(2a - 3b)$ **79.** $x^2(x^2 + 2)(x^4 - 2x^2 + 4)$ **81.** Answers vary. **83.** Answers vary.

Section 1.4 (Page 34)

1. $\frac{x}{7}$ **3.** $\frac{5}{7p}$ **5.** $\frac{5}{4}$ **7.** $\frac{4}{w + 6}$ **9.** $\frac{y - 4}{3y^2}$ **11.** $\frac{m - 2}{m + 3}$ **13.** $\frac{x + 3}{x + 1}$ **15.** $\frac{3}{16a}$ **17.** $\frac{3y}{x^2}$ **19.** $\frac{5}{4c}$ **21.** $\frac{3}{4}$ **23.** $\frac{3}{10}$ **25.** $\frac{2(a + 4)}{a - 3}$ **27.** $\frac{k + 2}{k + 3}$ **29.** Answers vary. **31.** $\frac{3}{35z}$ **33.** $\frac{4}{3}$ **35.** $\frac{20 + x}{5x}$ **37.** $\frac{3m - 2}{m(m - 1)}$ **39.** $\frac{37}{5(b + 2)}$ **41.** $\frac{33}{20(k - 2)}$ **43.** $\frac{7x - 1}{(x - 3)(x - 1)(x + 2)}$ **45.** $\frac{y^2}{(y + 4)(y + 3)(y + 2)}$ **47.** $\frac{x + 1}{x - 1}$ **49.** $\frac{-1}{x(x + h)}$ **51. (a)** $\frac{\pi x^2}{4x^2}$ **(b)** $\frac{\pi}{4}$ **53. (a)** $\frac{x^2}{25x^2}$ **(b)** $\frac{1}{25}$ **55. (a)** $\frac{x - 5}{x^2 - 10x}$ **(b)** 32.7 sec; 8 sec; 4.5 sec **57. (a)** \$2.4 million **(b)** No

Section 1.5 (Page 46)

1. 49 **3.** $16c^2$ **5.** $32/x^5$ **7.** $108u^{12}$ **9.** $1/7$ **11.** $1/32$ **13.** $-1/7776$ **15.** $-1/y^3$ **17.** 343 **19.** $9/16$ **21.** b^3/a **23.** 7 **25.** 1.55 **27.** 9 **29.** -16 **31.** $81/16$ **33.** $16/125$ **35.** 64 **37.** 2 **39.** $4^8 = 65{,}536$ **41.** $\frac{1}{9^{32/15}}$ **43.** z^3 **45.** $\frac{p}{9}$ **47.** $\frac{q^5}{r^3}$ **49.** $\frac{8}{25p^7}$ **51.** $2^{5/6}p^{3/2}$ **53.** $2p + 5p^{5/3}$ **55.** $\frac{1}{3y^{2/3}}$ **57.** $\frac{a^{1/2}}{49b^{5/2}}$ **59.** $x^{7/6} - x^{11/6}$ **61.** $x - y$ **63.** (f) **65.** (h) **67.** (g) **69.** (c) **71.** 5 **73.** 5 **75.** -2 **77.** 9 **79.** $\sqrt{77}$ **81.** 7 **83.** $16\sqrt{3}$ **85.** $13\sqrt{3}$ **87.** -1 **89.** $\sqrt{15} - 4\sqrt{3} + 4\sqrt{5} - 16$ **91.** $-3 - 3\sqrt{2}$ **93.** $4 + \sqrt{3}$ **95.** $\frac{7}{11 + 6\sqrt{2}}$ **97. (a)** 14 **(b)** 85 **(c)** 58.0 **99.** About \$15,996,000,000 **101.** About \$18,661,000,000 **103.** About 14,299 **105.** About 16,711 **107.** About 58,750 **109.** About 57.3 years **111.** Answers vary.

Section 1.6 (Page 57)

1. 4 **3.** 7 **5.** $-10/9$ **7.** 4 **9.** $\frac{40}{7}$ **11.** $\frac{26}{3}$
13. $-\frac{12}{5}$ **15.** $-\frac{59}{6}$ **17.** $-\frac{9}{4}$ **19.** $x = .72$ **21.** $r \approx -13.26$
23. $\frac{b - 5a}{2}$ **25.** $x = \frac{3b}{a + 5}$ **27.** $V = \frac{k}{P}$ **29.** $g = \frac{V - V_0}{t}$
31. $B = \frac{2A}{h} - b$ or $B = \frac{2A - bh}{h}$ **33.** $-2, 3$
35. $-8, 2$ **37.** $\frac{5}{2}, \frac{7}{2}$ **39.** 10 hrs **41.** 23° **43.** 71.6°
45. 2007 **47.** 2015 **49.** 2007 **51.** 2012 **53.** 1999
55. 2013 **57.** 2014 **59.** 2024 **61.** \$205.41
63. **(a)** .0352 **(b)** about .015, or 1.5% **(c)** approximately one case
65. \$8000 **67.** \$5000 **69.** about 838 mi **71.** $\frac{400}{3}$ L
73. 142 mi **75.** 70 mph **77.** 10 cm **79.** 5 cm

Section 1.7 (Page 67)

1. $-4, 14$ **3.** $0, -6$ **5.** $0, 2$ **7.** $-7, -8$ **9.** $\frac{1}{2}, 3$
11. $-\frac{1}{2}, \frac{1}{3}$ **13.** $\frac{5}{2}, 4$ **15.** $-5, -2$ **17.** $\frac{4}{3}, -\frac{4}{3}$
19. $0, 1$ **21.** $2 \pm \sqrt{7}$ **23.** $\frac{1 \pm 2\sqrt{5}}{4}$
25. $\frac{-7 \pm \sqrt{41}}{4}$; $-.1492, -3.3508$ **27.** $\frac{-1 \pm \sqrt{5}}{4}$; $.3090, -.8090$
29. $\frac{-5 \pm \sqrt{65}}{10}$; $.3062, -1.3062$ **31.** No real-number solutions
33. $-\frac{5}{2}, 1$ **35.** No real-number solutions **37.** $-5, \frac{3}{2}$ **39.** 1
41. 2 **43.** $x \approx .4701$ or 1.8240 **45.** $x \approx -1.0376$ or $.6720$
47. **(a)** 30 mph **(b)** about 35 mph **(c)** about 44 mph
49. **(a)** about 32 and 62 **(b)** about 17 and 77 **51.** **(a)** 2007 **(b)** 2010 **53.** about 1.046 ft **55.** **(a)** $x + 20$ **(b)** northbound: $5x$; eastbound: $5(x + 20)$ or $5x + 100$ **(c)** $(5x)^2 + (5x + 100)^2 = 300^2$ **(d)** about 31.23 mph and 51.23 mph
57. **(a)** $150 - x$ **(b)** $x(150 - x) = 5000$ **(c)** length 100 m; width 50 m **59.** 9 ft by 12 ft **61.** 6.25 sec **63.** **(a)** About 3.54 sec **(b)** 2.5 sec **(c)** 144 ft **65.** **(a)** 2 sec **(b)** 3/4 sec or 13/4 sec **(c)** It reaches the given height twice: once on the way up and once on the way down.
67. $t = \frac{\sqrt{2Sg}}{g}$ **69.** $h = \frac{d^2\sqrt{kL}}{L}$
71. $R = \frac{-2Pr + E^2 \pm E\sqrt{E^2 - 4Pr}}{2P}$ **73.** **(a)** $x^2 - 2x = 15$ **(b)** $x = 5$ or $x = -3$ **(c)** $z = \pm\sqrt{5}$ **75.** $\pm\frac{\sqrt{6}}{2}$
77. $\pm\sqrt{\frac{3 + \sqrt{13}}{2}}$

Chapter 1 Review (Page 70)

1. 0, 6 **3.** $-12, -6, -\frac{9}{10}, -\sqrt{4}, 0, \frac{1}{8}, 6$
5. Commutative property of multiplication **7.** Distributive property **9.** $x \geq 9$ **11.** $-7, -3, -2, 0, \pi, 8$
13. $-|3 - (-2)|, -|-2|, |6 - 4|, |8 + 1|$ **15.** -1 **17.** -1
19. 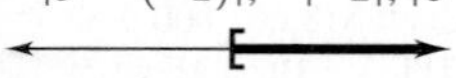

21. 

23. -18 **25.** $-\frac{7}{9}$ **27.** $4x^4 - 4x^2 + 11x$
29. $2q^4 + 14q^3 - 4q^2$ **31.** $12z^2 - 2z - 4$ **33.** $25k^2 - 4h^2$
35. $9x^2 + 24xy + 16y^2$ **37.** $k(2h^2 - 4h + 5)$
39. $a^2(5a + 2)(a + 2)$ **41.** $(3y - 2)(2y - 3)$
43. $(5a - 2)^2$ **45.** $(12p + 13q)(12p - 13q)$
47. $(3y - 1)(9y^2 + 3y + 1)$ **49.** $\frac{9x^2}{4}$
51. 4 **53.** $\frac{(y - 3)(4y - 3)}{7y}$ **55.** $\frac{(m - 1)^2}{3(m + 1)}$ **57.** $\frac{1}{6z}$
59. $\frac{64}{35q}$ **61.** $\frac{1}{5^3}$ or $\frac{1}{125}$ **63.** -1 **65.** $\frac{36}{25}$ **67.** 4^3
69. $\frac{1}{8}$ **71.** 9^3 **73.** $\frac{7}{10}$ **75.** 25
77. $\frac{1}{32}$ **79.** $\frac{1}{5^{2/3}}$ **81.** $3^{7/2}a^{5/2}$ **83.** 3 **85.** $3\sqrt{11}$
87. $3pq\sqrt[3]{2q^2}$ **89.** $\frac{n\sqrt{30m}}{6m}$ **91.** $-21\sqrt{3}$ **93.** 7
95. $\sqrt{6} - \sqrt{3}$ **97.** \$27,000,000 **99.** \$36,482,872.69
101. $-\frac{1}{3}$ **103.** -2 **105.** No solution **107.** $x = \frac{3}{8a - 2}$
109. $x = \frac{c - 3}{3a - ac - 2}$ **111.** $11, -3$ **113.** $-38, 42$
115. **(a)** 2010 **(b)** 2021 **117.** **(a)** 2005 **(b)** Answers vary.
119. \$60,000 at 8%; \$40,000 at 5% **121.** 2 **123.** 1 **125.** 0
127. $\frac{-1 \pm \sqrt{7}}{2}$ **129.** $3, -\frac{5}{2}$ **131.** $-2 \pm \sqrt{5}$
133. $\frac{-1 \pm \sqrt{3}}{3}$ **135.** $\frac{1}{2}, \frac{1}{6}$ **137.** $-\frac{8}{3}, 2$ **139.** $\frac{\pm\sqrt{3}}{3}$
141. $\frac{\pm\sqrt{3}}{3}$ **143.** $r = \frac{-Rp \pm E\sqrt{Rp}}{p}$
145. $s = \frac{a \pm \sqrt{a^2 + 4K}}{2}$ **147.** **(a)** 2116.1 pounds per sq ft; 670.5229 pounds per sq ft **(b)** 14,410 ft **149.** 50 m by 225 m or 100 m by 112.5 m **151.** 2.5 sec

Case 1 (Page 74)

1. $700 + 85x$ **3.** The \$700 refrigerator costs \$300 more.

CHAPTER 2

Section 2.1 (Page 83)

1. IV, II, I, III **3.** Yes **5.** No
7.
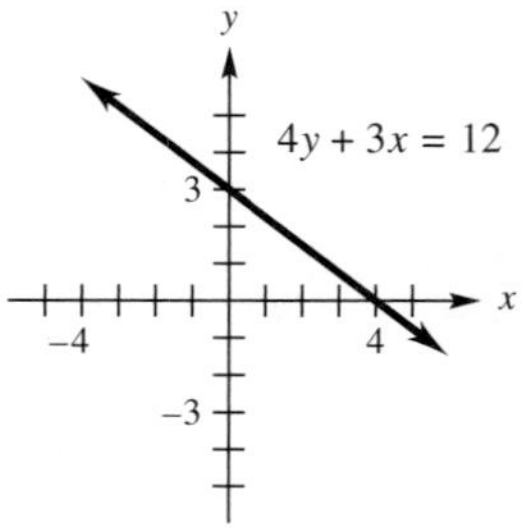

9.
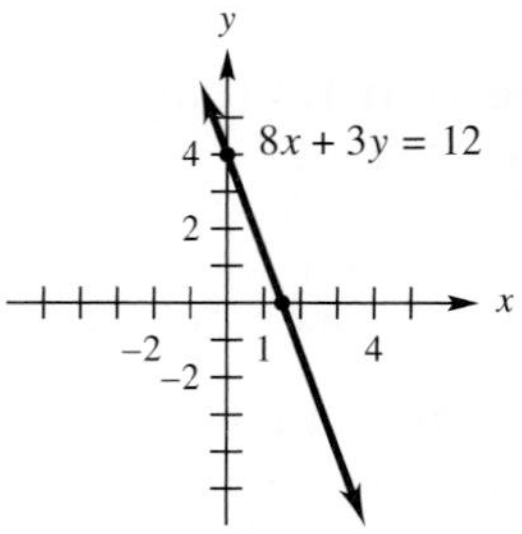

11.

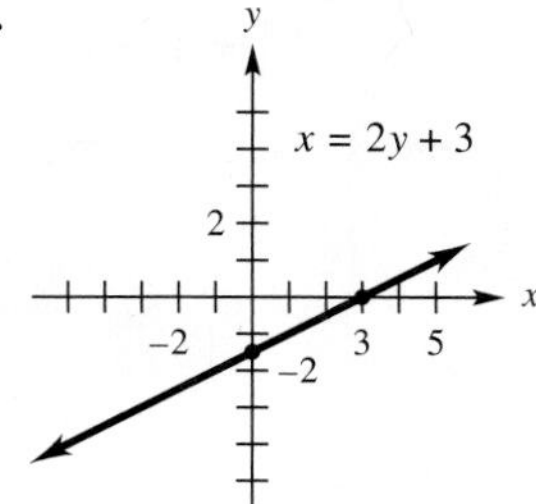

13. x-intercepts $-2.5, 3$; y-intercept 3 **15.** x-intercepts $-1, 2$; y-intercept -2 **17.** x-intercept 4; y-intercept 3 **19.** x-intercept 12; y-intercept -8 **21.** x-intercepts $3, -3$; y-intercept -9 **23.** x-intercepts $-5, 4$; y-intercept -20 **25.** no x-intercept; y-intercept 7

27.

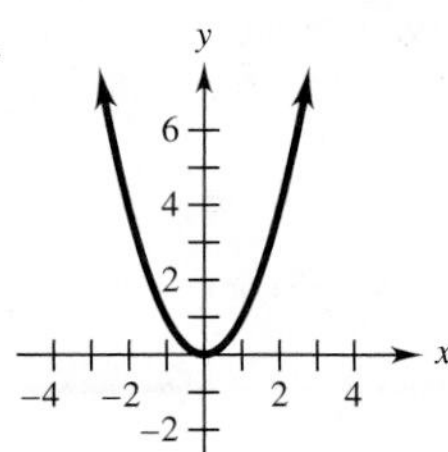

29.

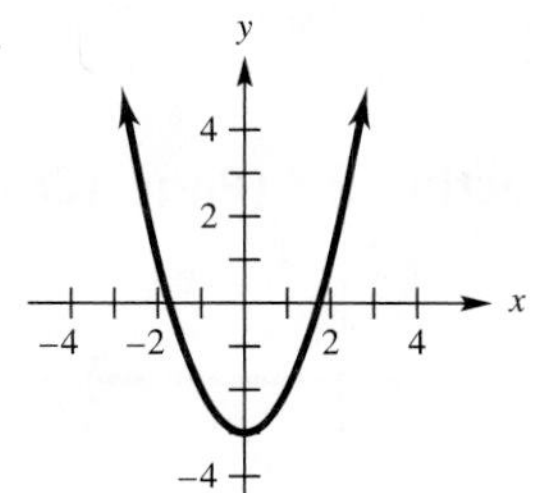

31.

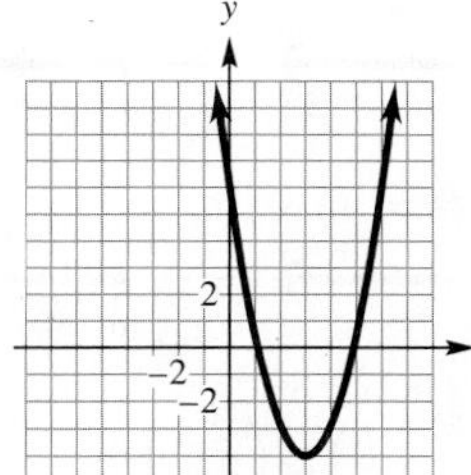

33.

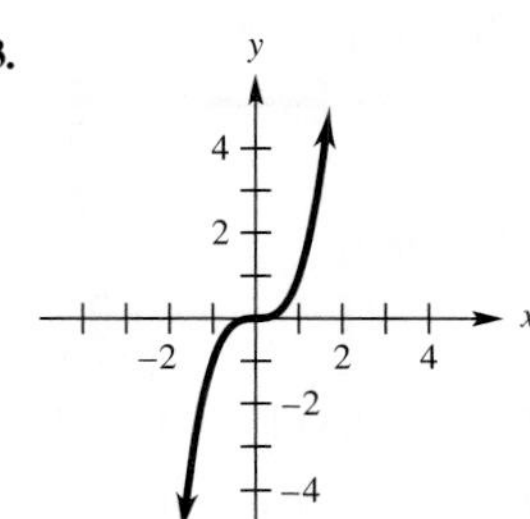

35.

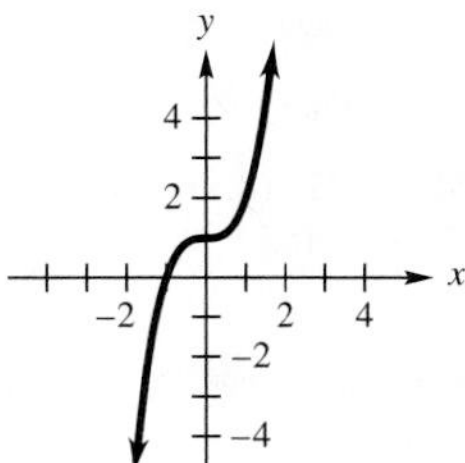

37.

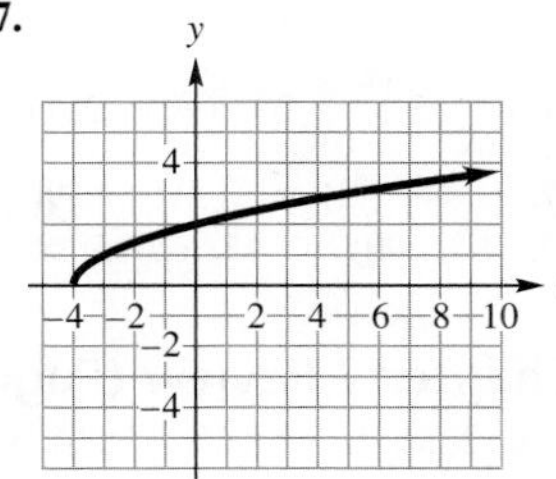

39.

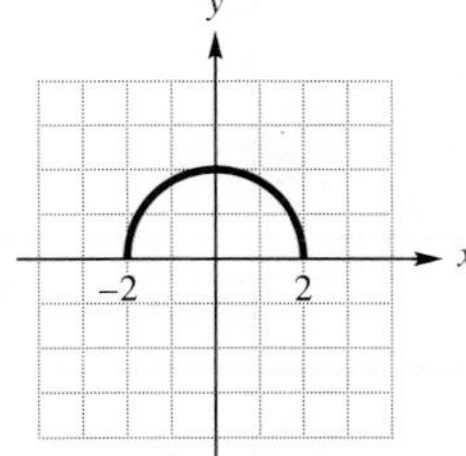

41. 9:30 AM, 8:45 AM **43.** Between 10:00 AM and 6:15 PM **45. (a)** About \$1,250,000 **(b)** About \$1,750,000 **(c)** About \$4,250,000 **47. (a)** About \$500,000 **(b)** About \$1,000,000 **(c)** About \$1,500,000 **49.** About 63, about 61, 46 **51.** 2002 **53.** About 21 **55.** 2011 and 2012 **57.** About \$740, about \$600 **59.** About \$620, 2007 **61.** No

63.

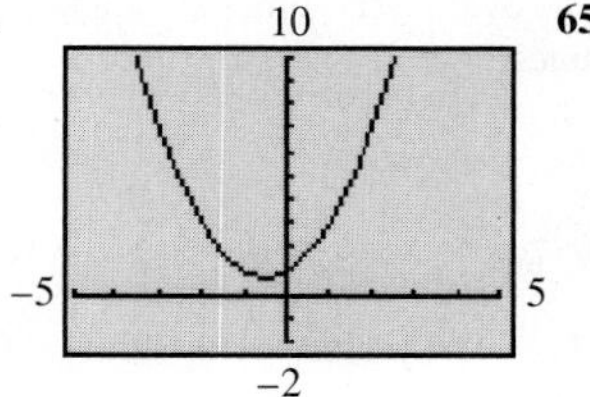

65.

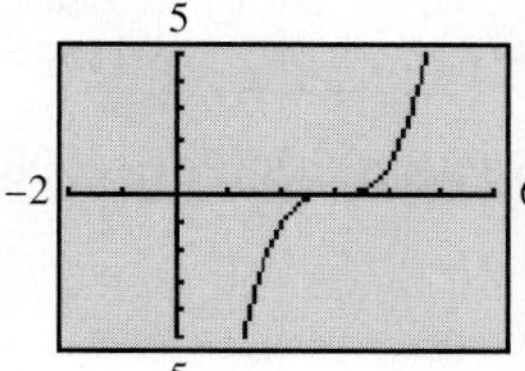

67.

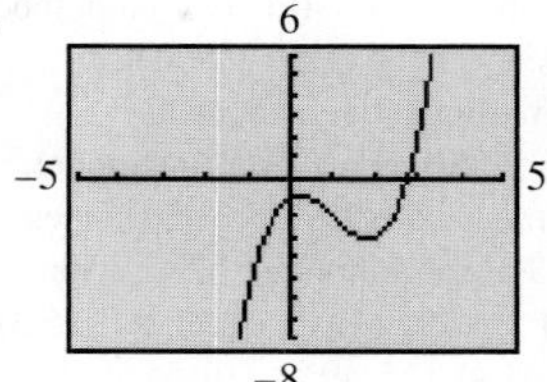

69. No;

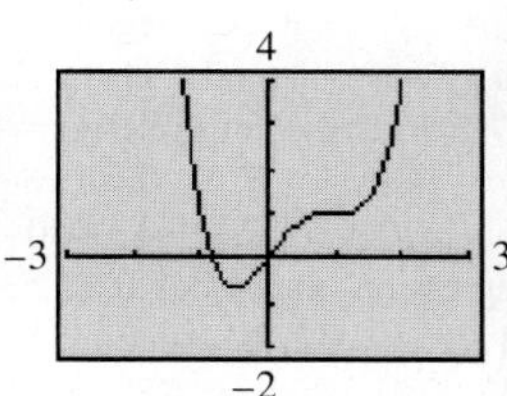

71. $x \approx -1.1038$ **73.** $x \approx 2.1017$ **75.** $x \approx -1.7521$ **77.** $r \approx 4.6580$ in. **79.** 2004 **81.** About 2.7 doctors per 1000 residents in 2006

Section 2.2 (Page 97)

1. $-\frac{3}{2}$ **3.** -2 **5.** $-\frac{5}{2}$ **7.** Not defined **9.** $y = 4x + 5$ **11.** $y = -2.3x + 1.5$ **13.** $y = -\frac{3}{4}x + 4$ **15.** $m = 2; b = -9$ **17.** $m = 3; b = -2$ **19.** $m = \frac{2}{3}; b = -\frac{16}{9}$ **21.** $m = \frac{2}{3}; b = 0$ **23.** $m = 1; b = 5$ **25. (a)** C **(b)** B **(c)** B **(d)** D

27.

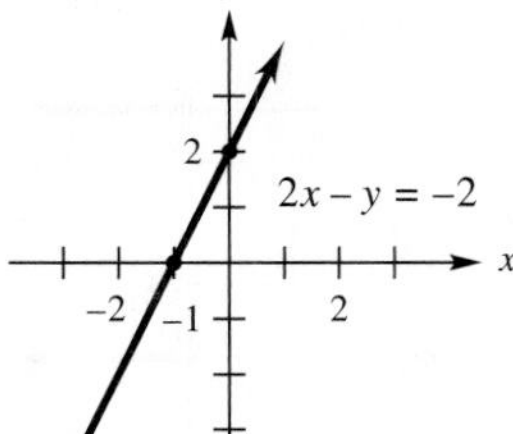

29.

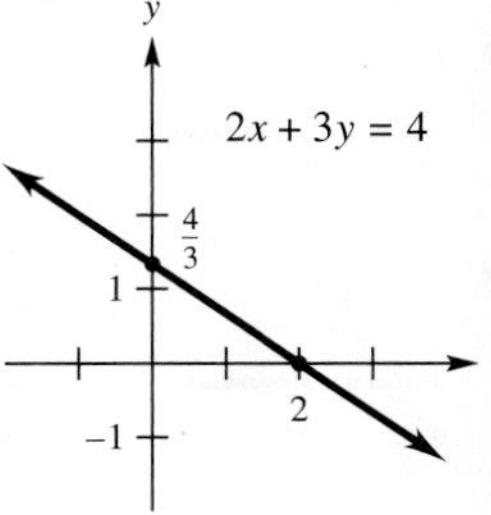

31.

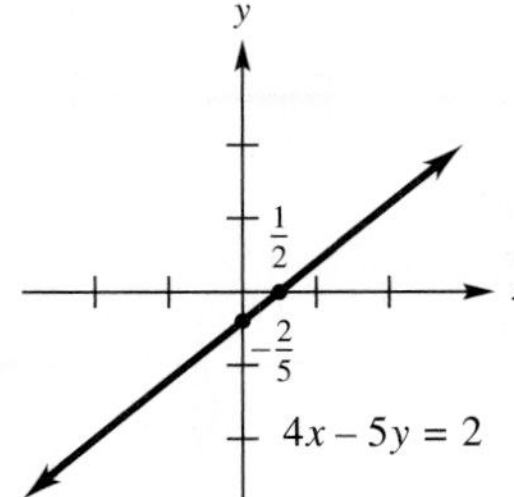

33. Perpendicular **35.** Parallel **37.** Neither **39. (a)** $\frac{2}{5}, \frac{9}{8}, -\frac{5}{2}$ **(b)** Yes **41.** $y = -\frac{2}{3}x$ **43.** $y = 3x - 3$ **45.** $y = 1$ **47.** $x = -2$ **49.** $y = 2x + 3$ **51.** $2y = 7x - 3$ **53.** $y = 5x$ **55.** $x = 6$ **57.** $y = 2x - 2$ **59.** $y = x - 6$ **61.** $y = -x + 2$ **63.** \$1330.42 **65.** \$6715.23 **67. (a)** 3255.1 million **(b)** 3672.8 million **(c)** 2021 **69. (a)** \$9,273,600,000 **(b)** 2010 **71. (a)** (0, 119.1) (2, 132.3) **(b)** $y = 6.6x + 119.1$ **(c)** 125.7 **(d)** 2007 **73. (a)** $y = 746x + 41{,}235$ **(b)** 53,917 **(c)** 2020 **75. (a)** $y = -2625x + 529{,}000$ **(b)** 2020

77. **(a)** The average decrease in time over 1 year; times are going down, in general. **(b)** 12.94 minutes

Section 2.3 (Page 108)

1. **(a)** $y = \frac{5}{9}(x - 32)$ **(b)** 10°C and 23.89°C **3.** 463.89°C

5. About 195.2, about 227.0 **7.** \$243.3 billion **9.** 4 ft
11. **(a)** 0, 0, .06, .02, −.02; sum = .06; −.2, −.1, 0, 0, 0; sum = −.3 **(b)** .0044; .05 **(c)** Model 1 **13.** No **15.** **(a)** Two point model: $y = -7.77x + 559$; regression model: $y = -7.72x + 560.45$ **(b)** Two point model: about 170; regression model: about 174
17. **(a)** Two-point model: $y = 15x + 2750$; regression-line model: $y = 14.9x + 2822$ **(b)** Two-point model: 5000, 6950, 9050; regression-line model: 5057, 6994, 9080 **(c)** Two-point model: 6275, 6500; regression-line model: 6323.5, 6500 **19.** **(a)** (9, 18.2), (10, 25.3), (11, 31.9), (12, 31.2), (13, 35.4), (14, 41.4), (15, 49.2), (16, 55.9), (17, 57.4) **(b)** $y = 4.89x - 25.13$ **(c)** About \$82.4 billion
21. **(a)** $y = .03119x + .5635$ **(b)** About .938 billion; about 1.06 billion
23. **(a)** $y = .913x + 18.3$ **(b)** Answers vary **(c)** \$45.7 thousand
25. **(a)** $y = 14.7x + 76$ **(b)** 223 billion **(c)** 2007

Section 2.4 (Page 118)

1. Answers vary.
3. $[-4, \infty)$ **5.** $(-\infty, 0)$

7. $\left(-\infty, \frac{10}{3}\right]$ **9.** $(-\infty, -8)$

11. $(-\infty, 3)$ **13.** $(-1, \infty)$

15. $(-\infty, 1]$ **17.** $\left(\frac{1}{5}, \infty\right)$

19. $(-5, 7)$ **21.** $\left[\frac{7}{3}, 5\right]$

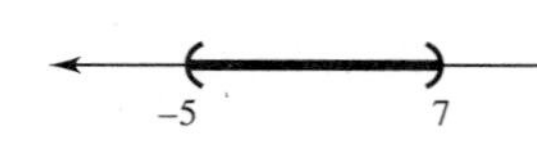

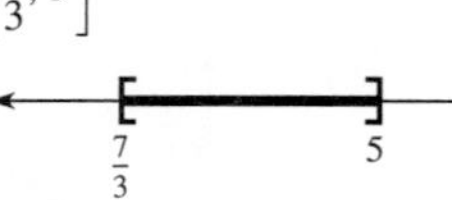

23. $\left[-\frac{11}{2}, \frac{7}{2}\right]$ **25.** $\left[-\frac{17}{7}, \infty\right)$

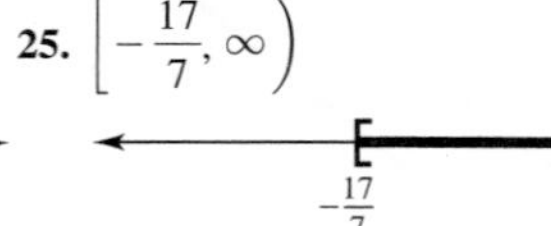

27. $x \geq 2$ **29.** $-3 < x \leq 5$ **31.** **(a)** Let x represent the number of mg per L of lead in the water. **(b)** $.038 \leq x \leq .042$ **(c)** Yes
33. **(a)** $0 < x \leq 8025$; $8025 < x \leq 32{,}550$; $32{,}550 < x \leq 78{,}850$; $78{,}850 < x \leq 164{,}550$; $164{,}550 < x \leq 357{,}700$; $x > 357{,}700$
(b) $0 < T \leq 802.50$; $802.50 < T \leq 4481.25$; $4481.25 < T \leq 16{,}056.25$; $16{,}056.25 < T \leq 40{,}052.25$; $40{,}052.25 < T \leq 103{,}791.75$; $T > 103{,}791.75$
35. $(-2, 2)$ **37.** No solution

39. $(-3, -2)$ **41.** $\left(-\infty, -\frac{5}{3}\right]$ or $[1, \infty)$

43. $\left(-\frac{3}{2}, \frac{13}{10}\right)$

45. $76 \leq T \leq 90$ **47.** $40 \leq T \leq 82$
49. **(a)** $25.33 \leq R_L \leq 28.17$; $36.58 \leq R_E \leq 40.92$
(b) $5699.25 \leq T_L \leq 6338.25$; $8230.5 \leq T_E \leq 9207$
51. 2001–2003 **53.** 2007–2010 **55.** **(a)** 2001 and earlier **(b)** 2005 and later **57.** $x \geq 400$ **59.** $x \geq 50$
61. Impossible to break even

Section 2.5 (Page 126)

1. $\left[-4, \frac{3}{2}\right]$ **3.** $(-\infty, -3)$ or $(-1, \infty)$

5. $\left[-2, \frac{1}{4}\right]$ **7.** $(-\infty, -1)$ or $\left(\frac{1}{4}, \infty\right)$

9. $[-6, 6]$ **11.** $(-\infty, 0)$ or $(16, \infty)$

13. $[-3, 0]$ or $[3, \infty)$ **15.** $[-7, -2]$ or $[2, \infty]$
17. $(-\infty, -5)$ or $(-1, 3)$ **19.** $\left(-\infty, -\frac{1}{2}\right)$ or $\left(0, \frac{4}{3}\right)$
21. No. **23.** $(-.1565, 2.5565)$ **25.** $[-2.2635, .7556]$ or $[3.5079, \infty)$ **27.** $(.5, .8393)$ **29.** $(-\infty, 1)$ or $[4, \infty)$
31. $\left(\frac{7}{2}, 5\right)$ **33.** $(-\infty, 2)$ or $(5, \infty)$ **35.** $(-\infty, -1)$
37. $(-\infty, -2)$ or $(0, 3)$ **39.** $[-1, .5]$ **41.** $(8, \infty)$
43. $(80, 200)$ **45.** $(2005, \infty)$ **47.** $x \geq 1530$

Chapter 2 Review (Page 128)

1. (−2, 3), (0, −5), (3, −2), (4, 3)
3. **5.**

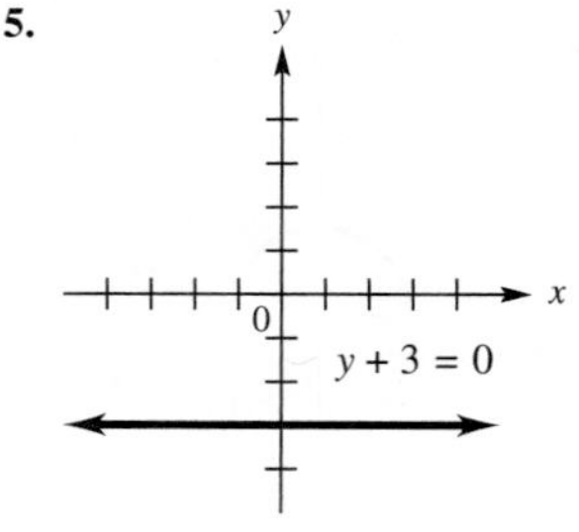

7.

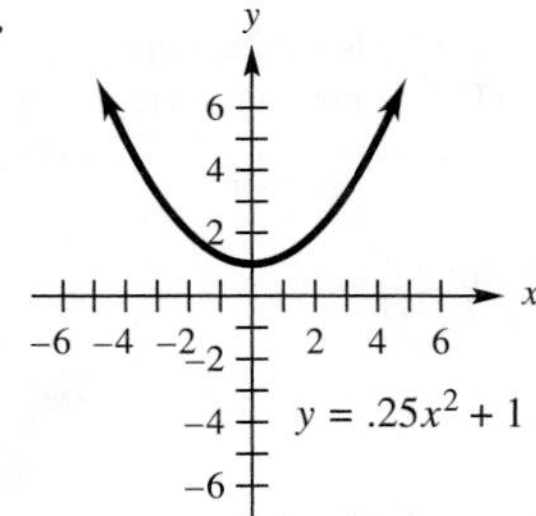

9. (a) About 11:30 AM to about 7:30 PM **(b)** From midnight until about 5 AM and after about 10:30 PM **11.** Answers vary. **13.** -3

15. $-\frac{1}{4}$ **17.** 3 **19.** 0 **21.** -3

23.

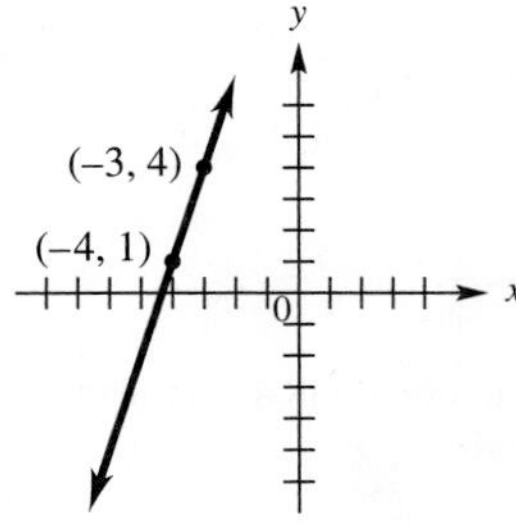

25. $3y = 2x - 13$ **27.** $4y = -5x + 17$ **29.** $x = -1$ **31.** $3y = 5x + 15$ **33. (a)** $y = -.3x + 30.2$ **(b)** Negative; answers vary. **(c)** 25.4 million metric tons **35. (a)** $y = .444x + 14.096$ **(b)** $y = .426x + 14.085$ **(c)** \$15.43, \$15.36; off by 6 cents, off by 1 cent. **(d)** \$18.54, \$18.35 **37. (a)** $y \approx 66.53x + 141.8$ **(b)** $r \approx .98$; yes **(c)** \$2,270.8 billion

39. $\left(\frac{3}{8}, \infty\right)$ **41.** $\left(-\infty, \frac{1}{4}\right]$ **43.** $\left[-\frac{1}{2}, 2\right]$ **45.** $[-8, 8]$

47. $(-\infty, 2]$ or $[5, \infty)$ **49.** $\left[-\frac{9}{5}, 1\right]$ **51.** (d)

53. (a) $y = 26.5x + 492$ **(b)** 2002 and beyond **(c)** 2007 and beyond **55.** $(-3, 2)$ **57.** $(-\infty, -5]$ or $\left[\frac{3}{2}, \infty\right)$

59. $(-\infty, -5]$ or $[-2, 3]$ **61.** $[-2, 0)$ **63.** $\left(-1, \frac{3}{2}\right)$

65. $[-19, -5)$ or $(2, \infty)$ **67.** 2004–2006

Case 2 Exercises (Page 131)

1. $y = .361x - 22.08$ **3.** \$−.42; answers vary. **5.** $y = 0$; answers vary.

CHAPTER 3

Section 3.1 (Page 140)

1. Function **3.** Function **5.** Not a function **7.** Function **9.** $(-\infty, \infty)$ **11.** $(-\infty, \infty)$ **13.** $(-\infty, 0]$ **15.** All real numbers except 2 **17.** All real numbers except 2 and -2 **19.** All real numbers such that $x > -4$ and $x \neq 3$ **21.** $(-\infty, \infty)$ **23. (a)** 8 **(b)** 8 **(c)** 8 **(d)** 8 **25. (a)** 48 **(b)** 6 **(c)** 25.38 **(d)** 28.42 **27. (a)** $\sqrt{7}$ **(b)** 0 **(c)** $\sqrt{5.7}$ **(d)** Not defined

29. (a) 12 **(b)** 23 **(c)** 12.91 **(d)** 49.41 **31. (a)** $\frac{\sqrt{3}}{15}$

(b) Not defined **(c)** $\frac{\sqrt{1.7}}{6.29}$ **(d)** Not defined **33. (a)** 13

(b) 9 **(c)** 6.5 **(d)** 24.01 **35. (a)** $6 - p$ **(b)** $6 + r$ **(c)** $3 - m$ **37. (a)** $\sqrt{4 - p}\ (p \leq 4)$ **(b)** $\sqrt{4 + r}\ (r \geq -4)$ **(c)** $\sqrt{1 - m}\ (m \leq 1)$ **39. (a)** $p^3 + 1$ **(b)** $-r^3 + 1$ **(c)** $m^3 + 9m^2 + 27m + 28$ **41. (a)** $\frac{3}{p - 1}\ (p \neq 1)$

(b) $\frac{3}{-r - 1}\ (r \neq -1)$ **(c)** $\frac{3}{m + 2}\ (m \neq -2)$

43. 2 **45.** $2x + h$ **47.**

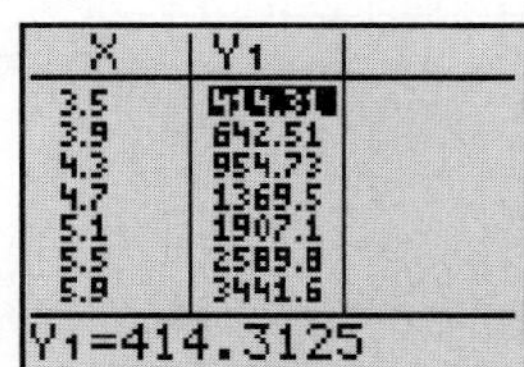

X	Y1
3.5	[illegible]
3.9	642.51
4.3	954.73
4.7	1369.5
5.1	1907.1
5.5	2589.8
5.9	3441.6

Y1=414.3125

49. (a) \$952.30 **(b)** \$3788.12 **(c)** \$8137.63 **51. (a)** \$ 34.0 billion **(b)** \$57.76 billion **53. (a)** 28.3 million **(b)** 37.0 million **55.** $2050 - 500t$ **57.** $f(x) = .03x$

59. (a) $y = x^2$ **(b)** $y = \frac{d^2}{2}$ **61.** 30,060

Section 3.2 (Page 151)

1.

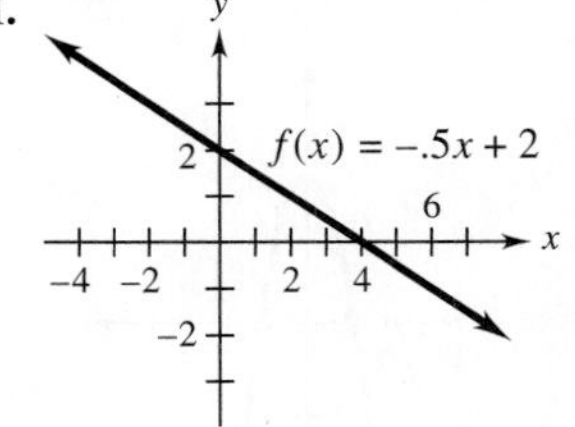

3.

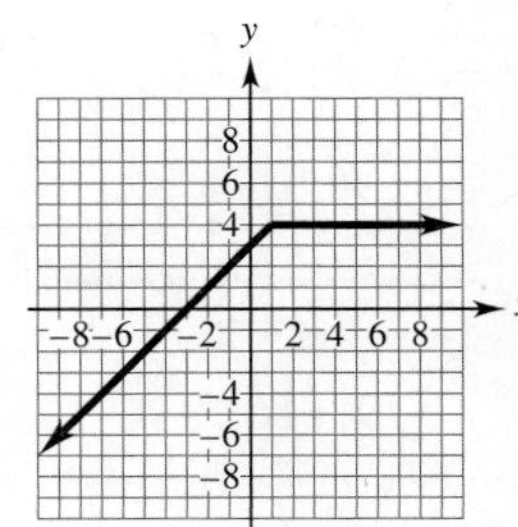

$$f(x) = \begin{cases} x + 3 & \text{if } x \leq 1 \\ 4 & \text{if } x > 1 \end{cases}$$

5.

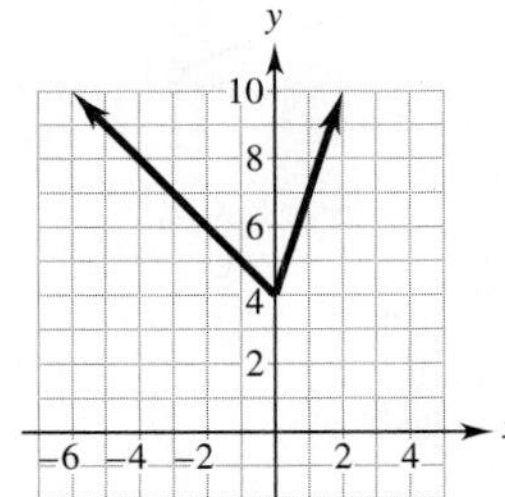

$$y = \begin{cases} 4 - x & \text{if } x \leq 0 \\ 3x + 4 & \text{if } x > 0 \end{cases}$$

7.

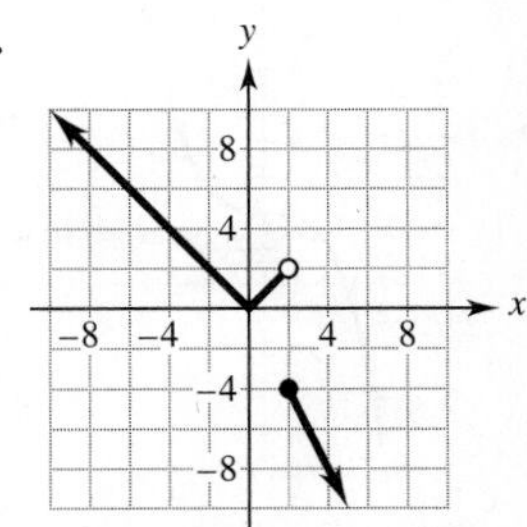

$$f(x) = \begin{cases} |x| & \text{if } x < 2 \\ -2x & \text{if } x \geq 2 \end{cases}$$

9.

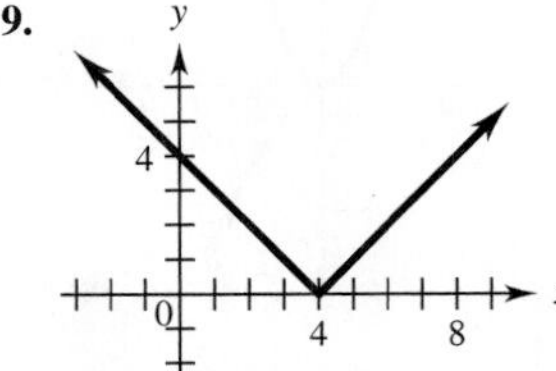

11.

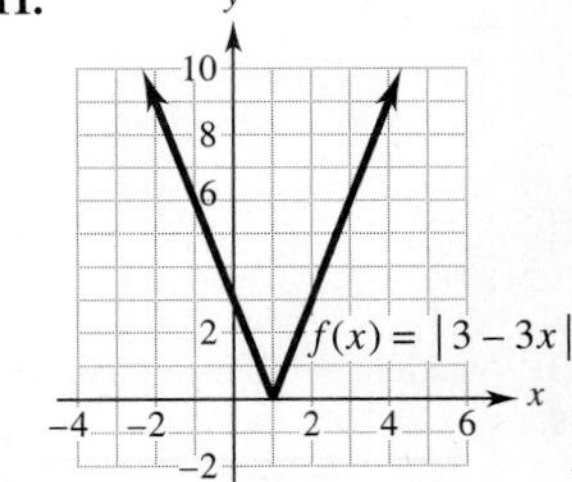

13.

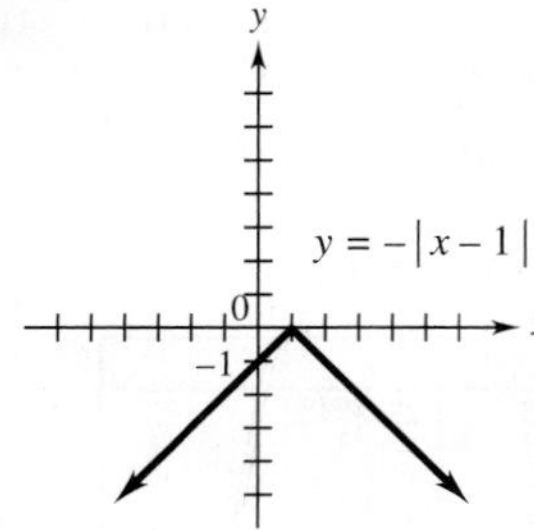

15.

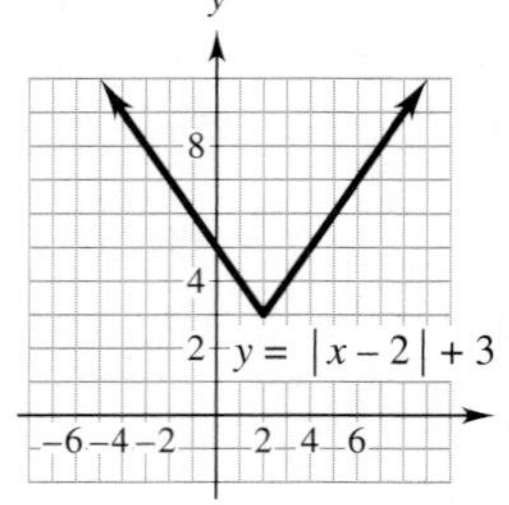

17.

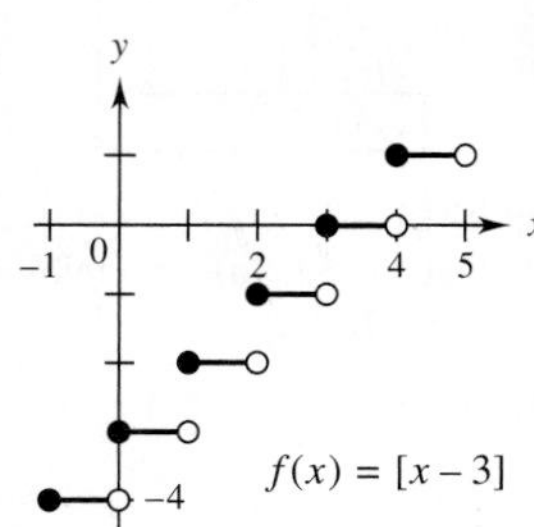

19.

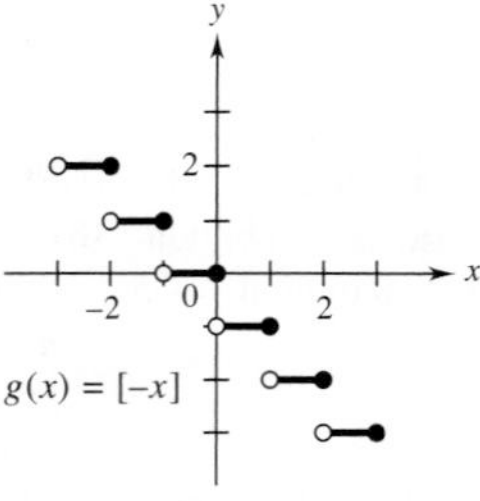

21.

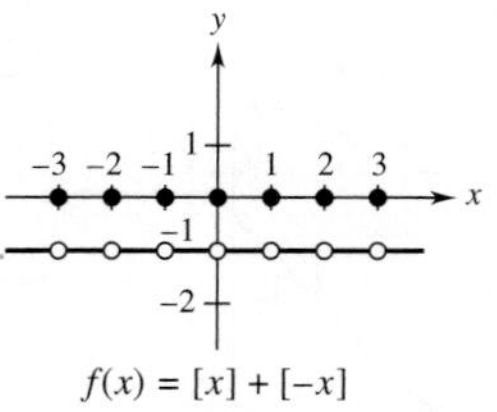

23.

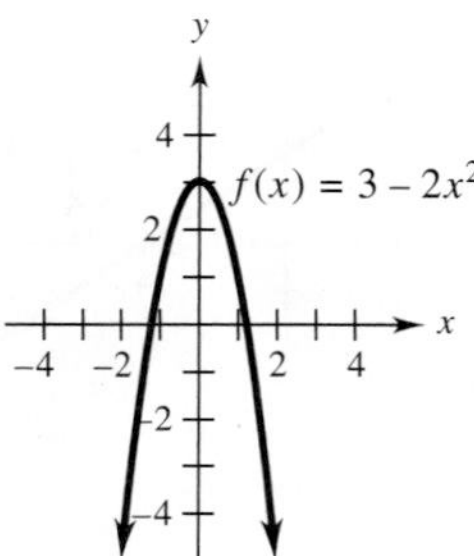

25.

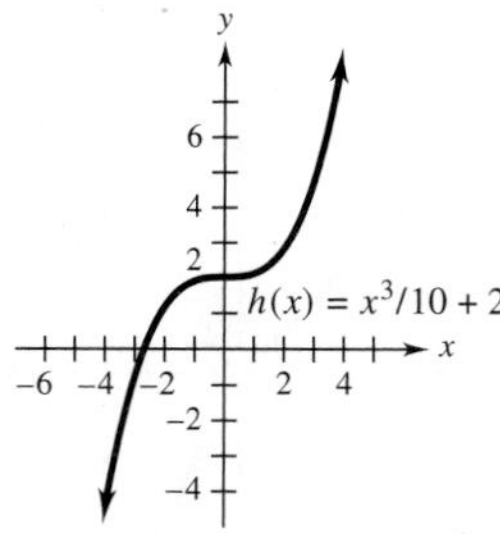

27.

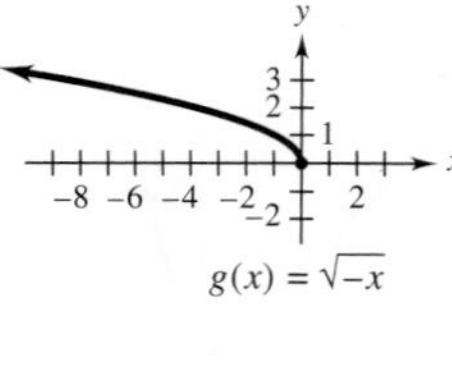

29.

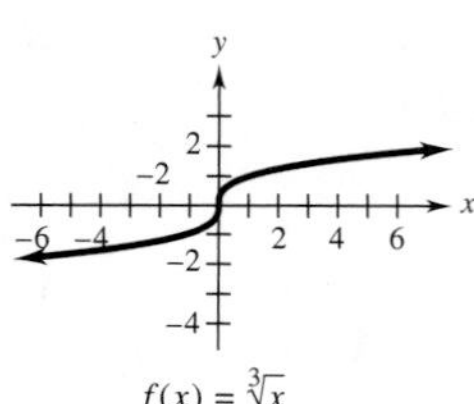

31.

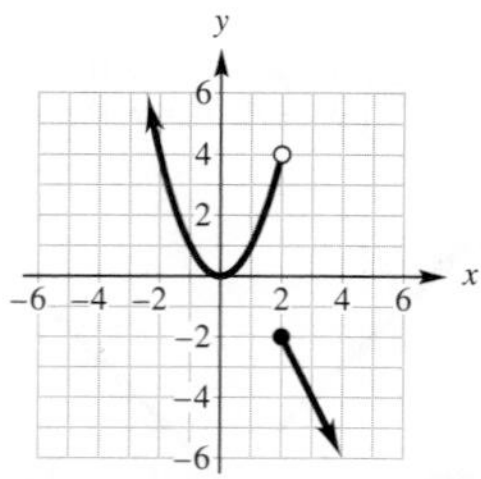

33. Function **35.** Not a function **37.** Function

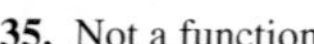

39.

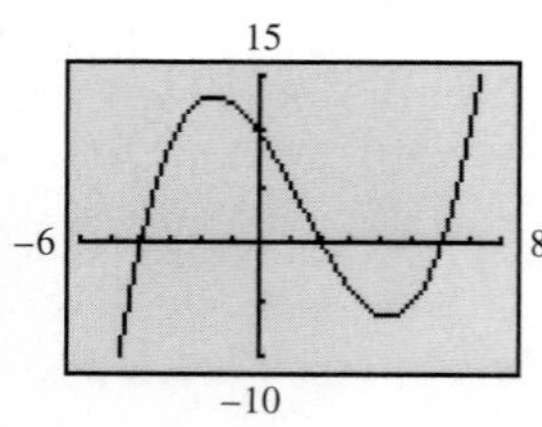

41.

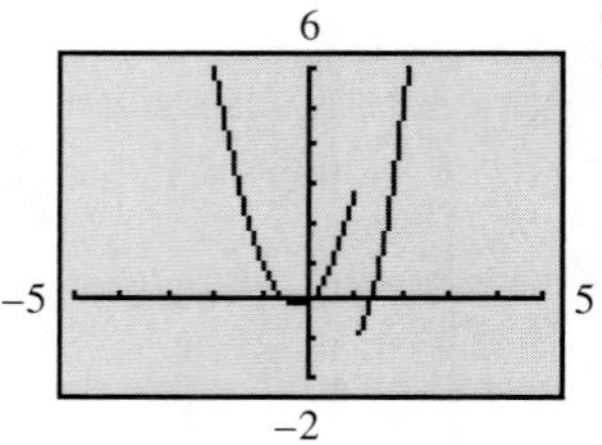

(1, −1) is on the graph; (1, 3) is not on the graph

43. $x = -4, 2, 6$ **45.** Peak at (.5078, .3938); valleys at (−1.9826, −4.2009) and (3.7248, −8.7035)

47. (a)

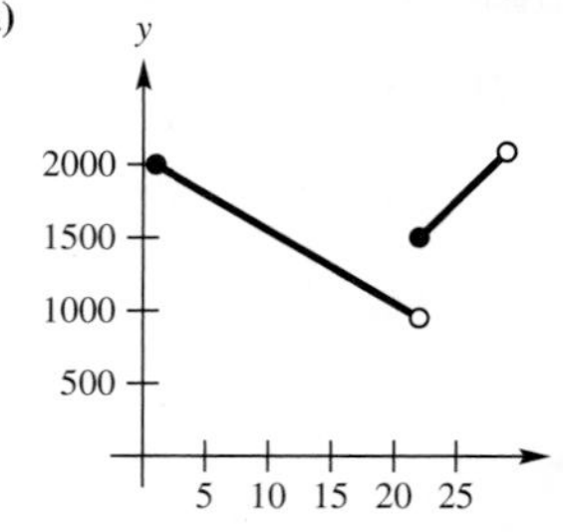

(b) Adjusted for inflation, the maximum yearly IRA contribution fell from \$2000 to \$1000 during this period.

49. (a)

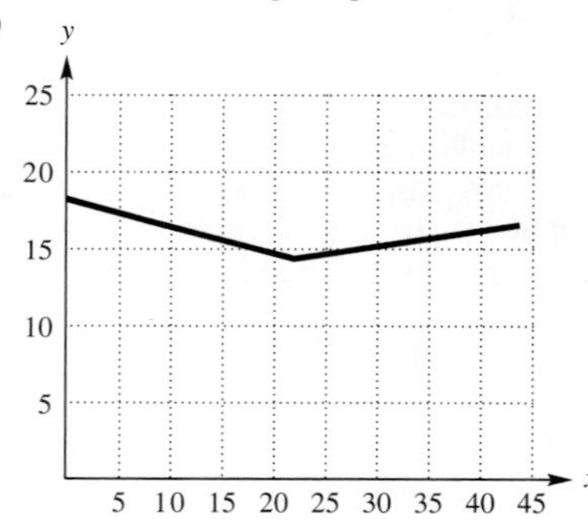

(b) 14.39 **51. (a)** $f(x) = \begin{cases} 1.3x + 15.1 & \text{if } 0 \le x \le 25 \\ 9.5x - 190 & \text{if } x > 25 \end{cases}$

(b)

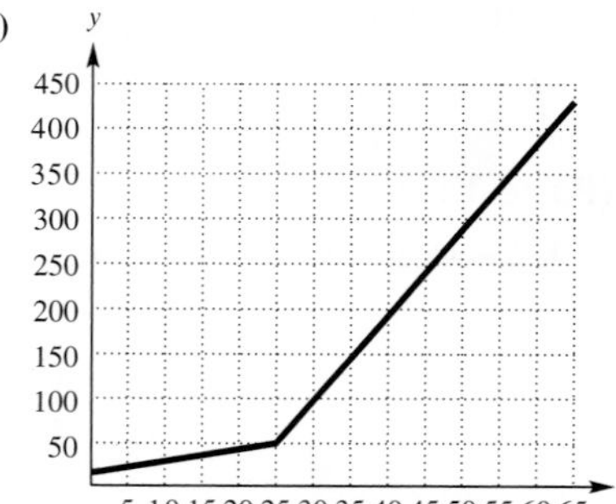

(c) 323 **(d)** 399

53. (a) No **(b)** From early 1997 to late 1998 and from early 2001 to early 2002 **(c)** From 1990 to early 1997; from late 1998 to early 2001; after early 2002

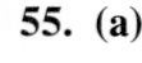
55. (a)

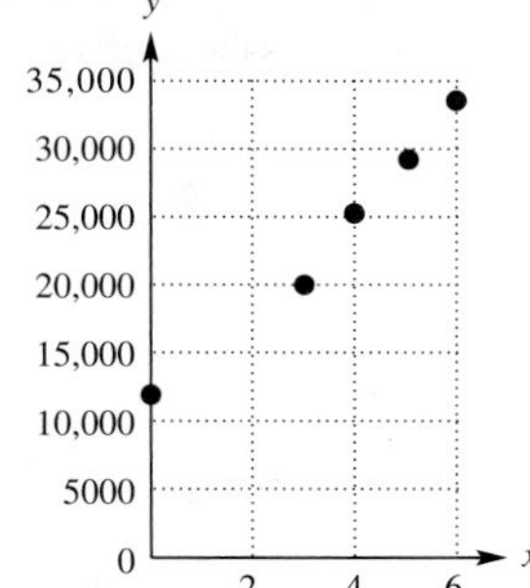

(b) $y = 3629x + 11{,}999$ **(c)** $f(x) = 3629x + 11{,}999$ **(d)** $f(4) = 26{,}515$, yes **57. (a)** 33; 39 **(b)** The figure has vertical line segments, which can't be part of the graph of a function. (Why?) To make the figure into the graph of f, delete the vertical line segments; then, for each horizontal segment of the graph, put a closed dot on the left end and an open circle on the right end (as in Figure 3.7). **59. (a)** \$34.99 **(b)** \$24.99 **(c)** \$64.99 **(d)** 74.99

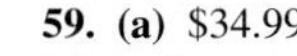
(e)

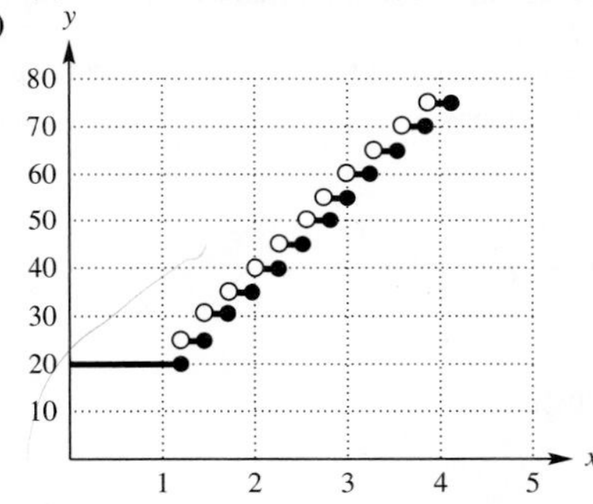

61. There are many correct answers, including

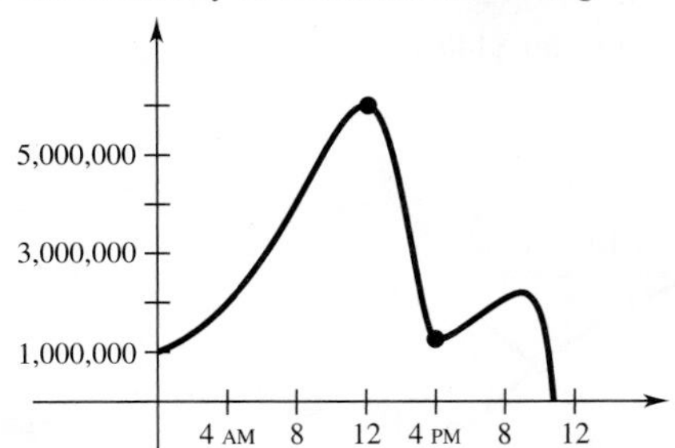

Section 3.3 (Page 164)

1. Let $C(x)$ be the cost of renting a saw for x hours; $C(x) = 25 + 5x$. **3.** Let $C(x)$ be the cost (in dollars) for x half hours; $C(x) = 8 + 2.5x$. **5.** $C(x) = 36x + 200$ **7.** $C(x) = 120x + 3800$ **9.** \$48, \$15.60, \$13.80 **11.** \$55.50, \$11.40, \$8.46 **13. (a)** $f(x) = -1916x + 16{,}615$ **(b)** 7035 **(c)** \$1916 per year **15. (a)** $f(x) = -11{,}875x + 120{,}000$ **(b)** $[0, 8]$ **(c)** \$48,750 **17. (a)** \$80,000 **(b)** \$42.50 **(c)** \$122,500; \$1,440,000 **(d)** \$122.50; \$45 **19. (a)** $C(x) = .097x + 1.32$ **(b)** \$98.32 **(c)** \$98.417 **(d)** \$.097, or 9.7¢ **(e)** \$.097, or 9.7¢ **21.** $R(x) = 525{,}000 + 1.43x$ **23. (a)** $C(x) = 10x + 750$ **(b)** $R(x) = 35x$ **(c)** $P(x) = 25x - 750$ **(d)** \$1750 **25. (a)** $C(x) = 18x + 300$ **(b)** $R(x) = 28x$ **(c)** $P(x) = 10x - 300$ **(d)** \$700 **27. (a)** $C(x) = 12.50x + 20{,}000$ **(b)** $R(x) = 30x$ **(c)** $P(x) = 17.50x - 20{,}000$ **(d)** $-\$18{,}250$ (a loss) **29.** $(3, -1)$ **31.** $\left(-\frac{11}{4}, -\frac{61}{4}\right)$ **33. (a)** 200,000 policies ($x = 200$)

(b)

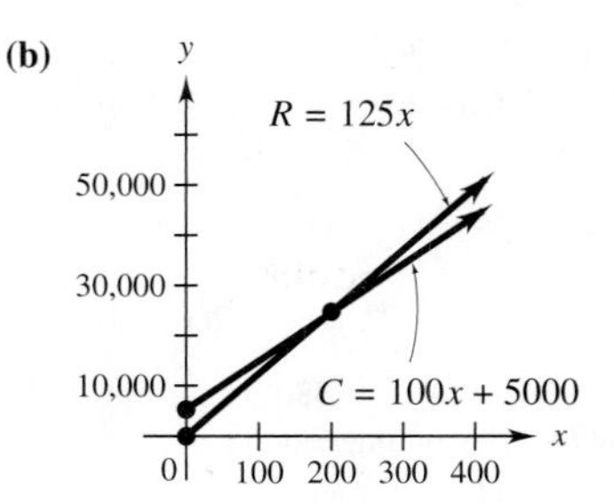

(c) Revenue: \$12,500; cost: \$15,000 **35. (a)** $C(x) = .126x + 1.5$ **(b)** \$2.382 million **(c)** About 17.857 units **37.** Break-even point is about 467 units; do not produce the item. **39.** Break-even point is about 1037 units; produce the item. **41.** The compensation of about \$20.20/hour was the same in late 2000. **43. (a)** Domestic: $f(x) = -.14x + 7.4$; Imported: $g(x) = .24x + 5.9$

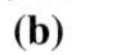
(b)

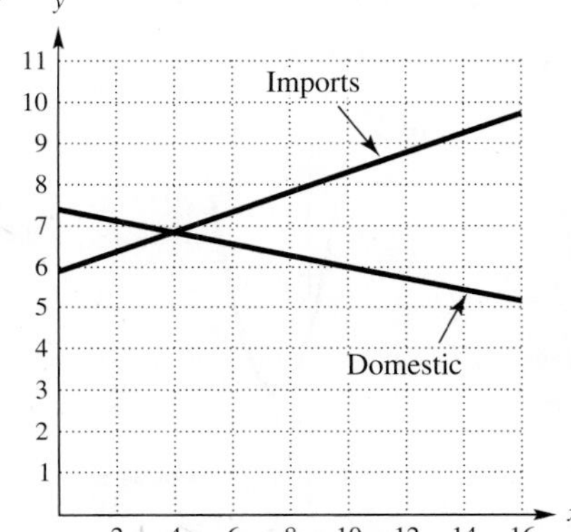

(c) In 2004, domestic production equaled imports. **45.** \$140 **47.** 10 items **49. (a)** \$16 **(b)** \$11 **(c)** \$6 **(d)** 8 units **(e)** 4 units **(f)** 0 units

(g)

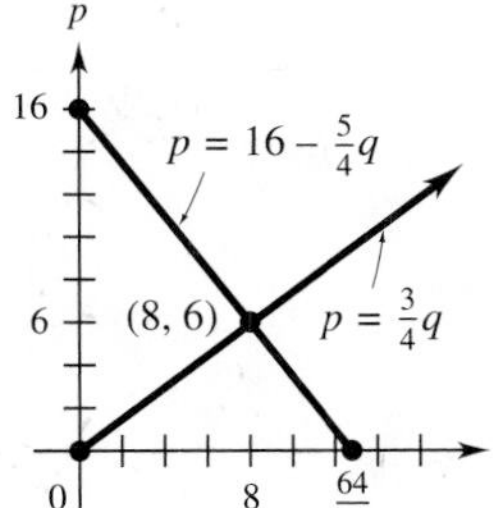

(h) 0 units **(i)** $\frac{40}{3}$ units **(j)** $\frac{80}{3}$ units **(k)** See part (g). **(l)** 8 units **(m)** \$6 **51. (a)**

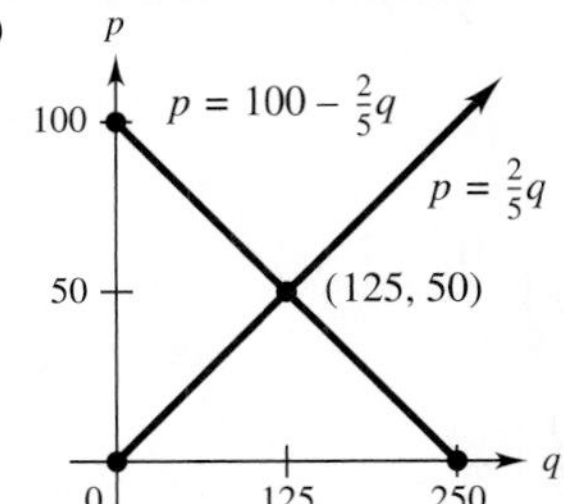

(b) 125 units **(c)** 50¢ **(d)** $[0, 125)$ **53.** Total cost increases when more items are made (because it includes the cost of all previously made items), so the graph cannot move downward. No; the average cost can decrease as more items are made, so its graph can move downward.

Section 3.4 (Page 173)

1. Upward **3.** Downward **5.** Upward **7.** (5, 7); downward **9.** (−1, −9); upward **11.** (4.5, 7.1); upward **13.** i **15.** k **17.** j **19.** f **21.** $f(x) = \frac{1}{4}(x - 1)^2 + 2$ **23.** $f(x) = (x + 1)^2 - 2$ **25.** $f(x) = 3x^2$ **27.** $f(x) = 2(x - 4)^2 - 2$ **29.** (−3, 12) **31.** (2, −7) **33.** (4, 9) **35.** x-intercepts 1, 3; y-intercept 9 **37.** x-intercepts −1, −3; y-intercept 6 **39.** (−2, 0), $x = -2$ **41.** (1, −3), $x = 1$

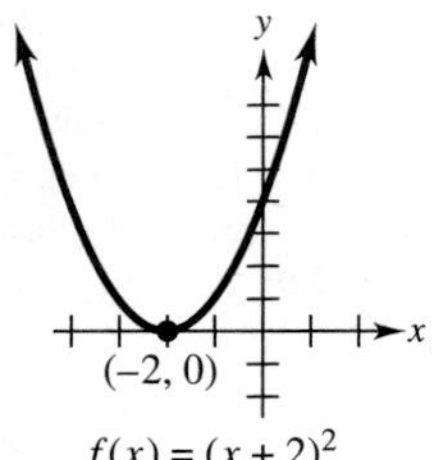

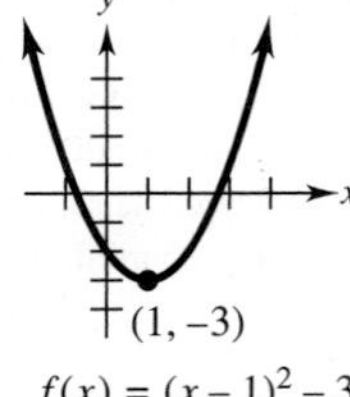

43. (2, 2), $x = 2$ **45.** (1, 3), $x = 1$

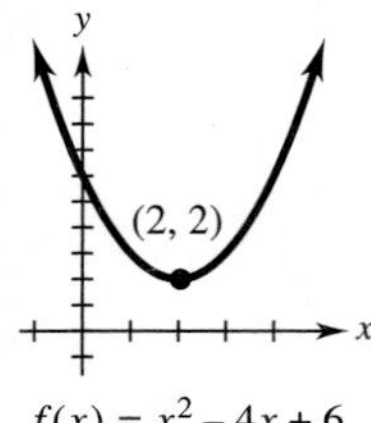

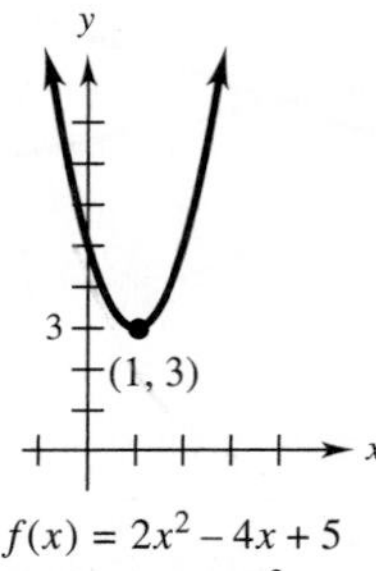

47. 54 **49. (a)** 28.5 weeks **(b)** .81225 **(c)** 57 weeks; answers vary.

51. (a)–(d)

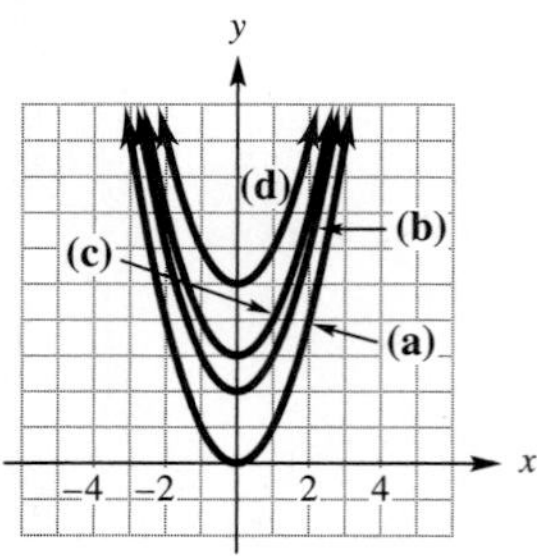

(e) The graph of $f(x) = x^2 + c$ is the graph of $k(x) = x^2$ shifted c units upward.

53. (a)–(d)

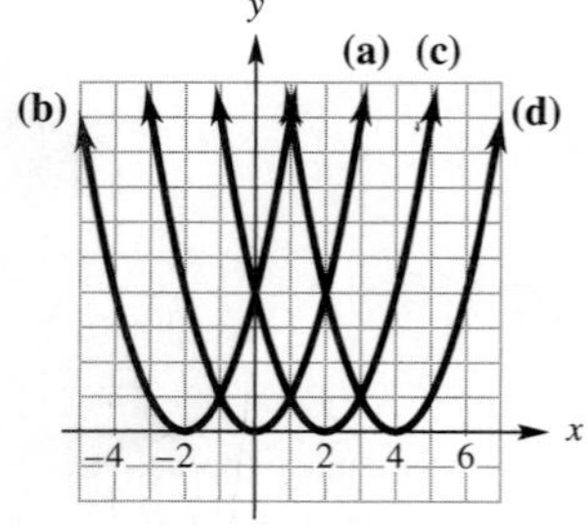

(e) The graph of $f(x) = (x + c)^2$ is the graph of $k(x) = x^2$ shifted c units to the left. The graph of $f(x) = (x - c)^2$ is the graph of $k(x) = x^2$ shifted c units to the right.

Section 3.5 (Page 181)

1. (a) \$30; \$9; \$105

(b)

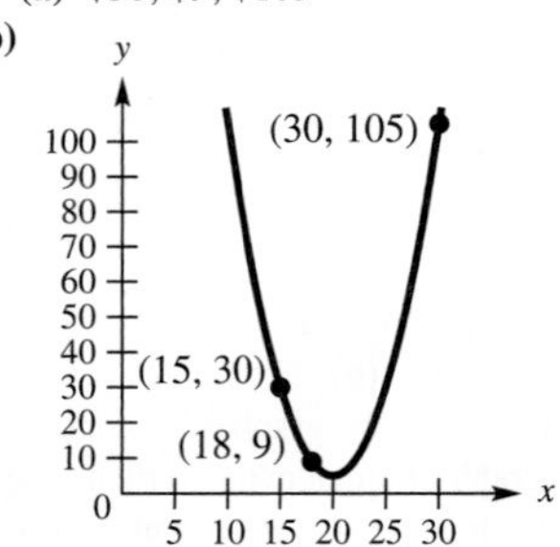

(c) (20, 5) **(d)** 20 boxes per day; \$5 per box **3. (a)** 10 milliseconds **(b)** 40 responses per millisecond **5. (a)** 27 cases **(b)** Answers vary. **(c)** 15 cases **7. (a)** About 12 books **(b)** 10 books **(c)** About 7 books **(d)** 0 books **(e)** 5 books **(f)** About 7 books **(g)** 10 books **(h)** about 12 books

(i)

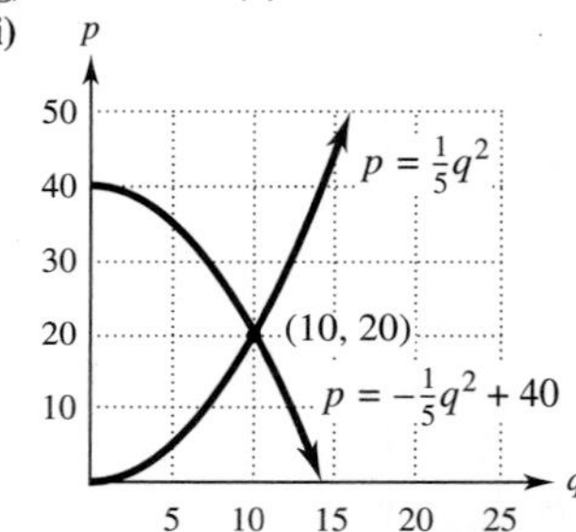

9. (a) \$640 **(b)** \$515 **(c)** \$140

(d)

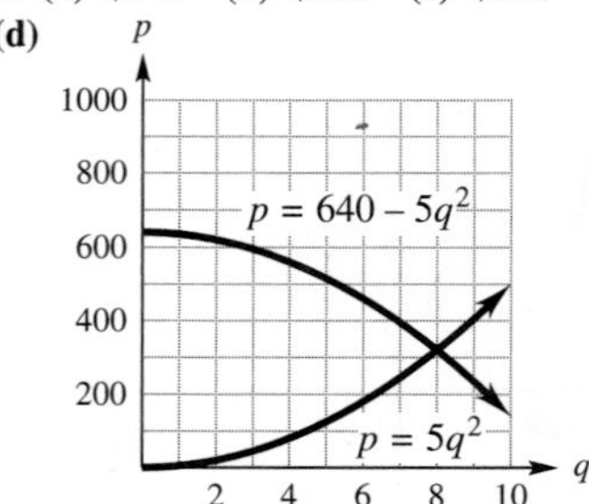

(e) 800 units **(f)** \$320 **11.** 80; \$3600 **13.** 30; \$1500 **15.** 20 **17.** 10 **19. (a)** $R(x) = (100 - x)(200 + 4x) = 20{,}000 + 200x - 4x^2$

(b)

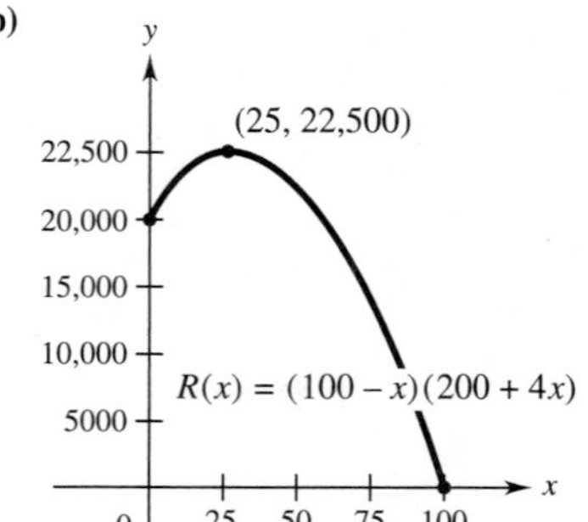

(c) 25 seats **(d)** \$22,500 **21.** 13 weeks; \$96.10/hog **23. (a)** $g(x) = .013(x - 20)^2$ **(b)** About 3

25. **(a)** $f(x) = .356(x - 6)^2 + 5.9$ **(b)** \$75.7 billion; \$97.0 billion
27. $g(x) = .037x^2 - 1.95x + 26.5$

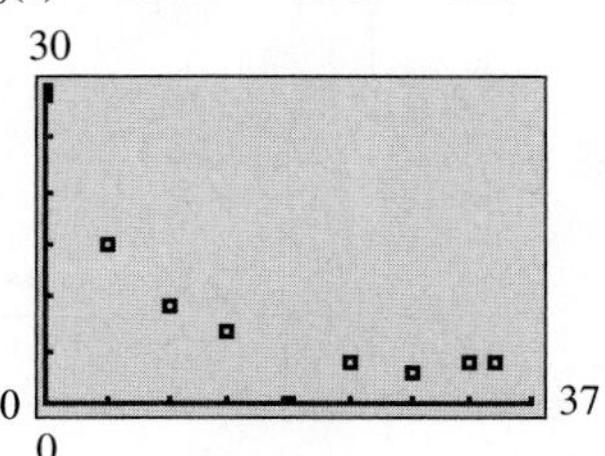

This model estimates about 4 deaths, 1 more than Exercise 23.
29. $g(x) = .348x^2 - 4.12x + 18.5$

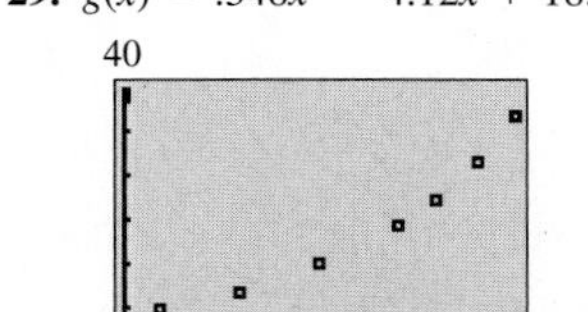

\$75.3 billion and \$96.3 billion. These values are very close to those for Exercise 25. **31.** 80 ft by 160 ft **33.** **(a)** 11.3 and 88.7 **(b)** 50 **(c)** \$3000 **(d)** $x < 11.3$ or $x > 88.7$ **(e)** $11.3 < x < 88.7$

Section 3.6 (Page 193)

1.

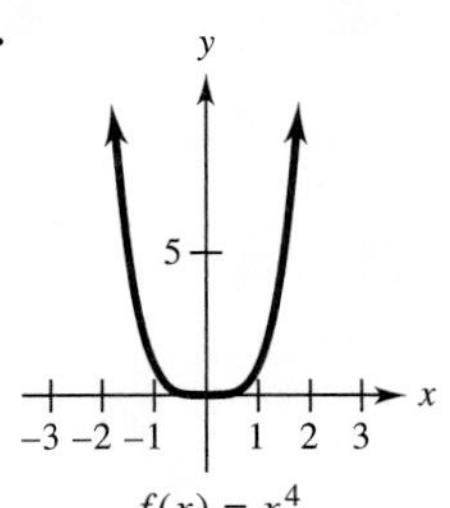

3.

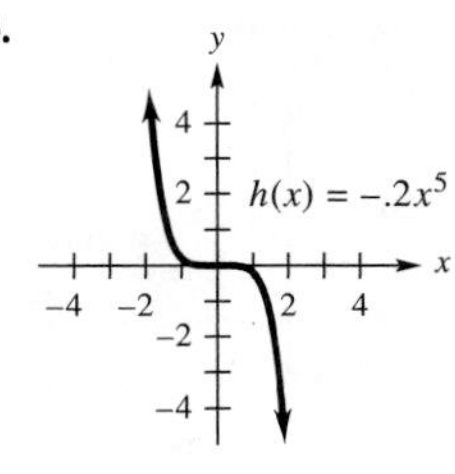

5. **(a)** Yes **(b)** No **(c)** No **(d)** Yes **7.** **(a)** Yes **(b)** No **(c)** Yes **(d)** No **9.** d **11.** b **13.** e

15.

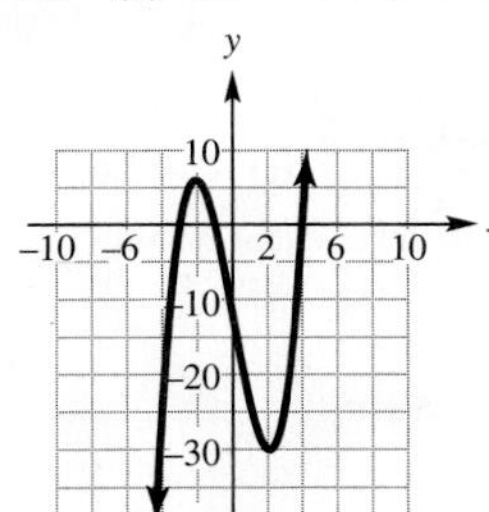

17.

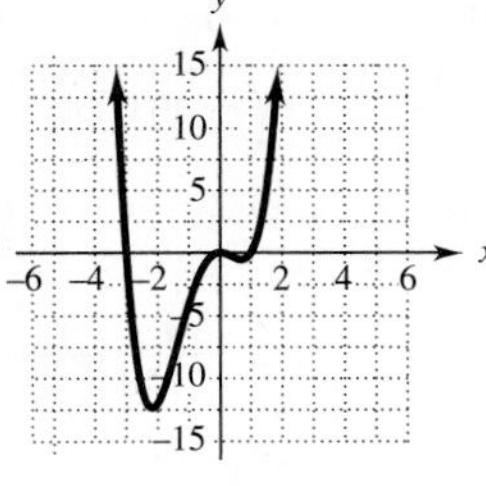

19.

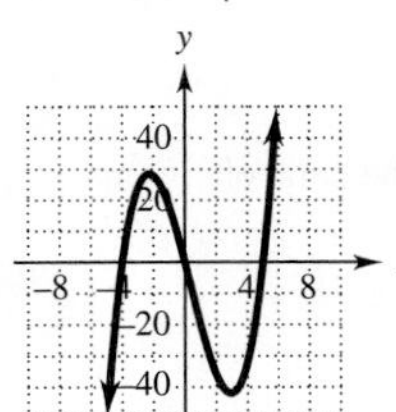

21.

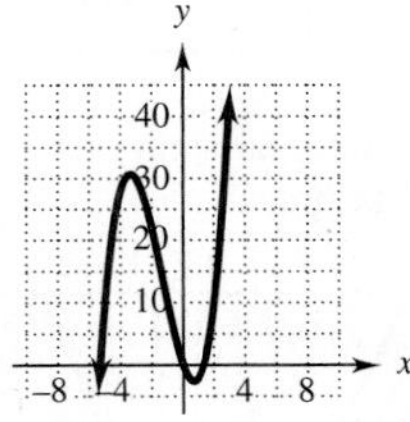

23. $-3 \leq x \leq 5$ and $-20 \leq y \leq 5$ **25.** $-3 \leq x \leq 4$ and $-35 \leq y \leq 20$ **27.** **(a)** \$933.33 billion **(b)** \$1200 billion **(c)** \$1145.8 billion **(d)** \$787.5 billion

(e)

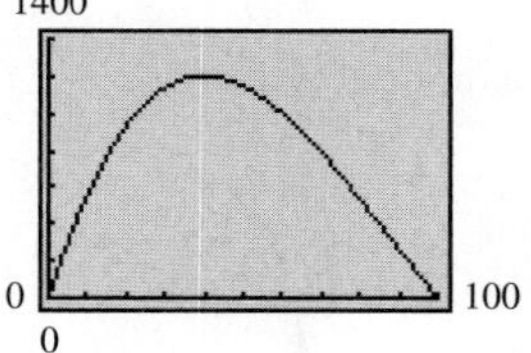

29. **(a)** 0; 108; 28; 10
(b)

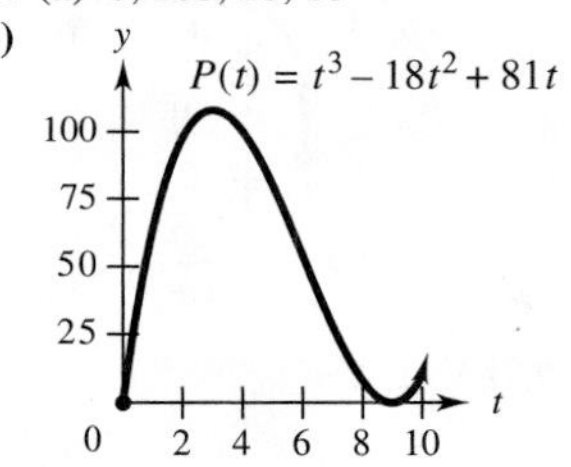

(c) Increasing for years 0 to 3 and from the 9th year on; decreasing for years 3 to 9 **31.** **(a)** 1,357,500,000; 1,452,300,000
(b) When x is large, the graph must resemble the graph of $y = -.00096x^3$, which drops down forever at the far right. (See the chart on page 186.) This would mean that China's population would become 0 at some point.
33. **(a)**

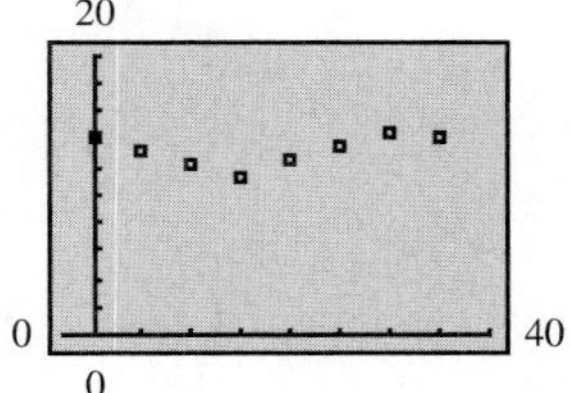

(b) $f(x) = -.000039091x^4 + .002393x^3 - .03311x^2 - .08448x + 14.28258$ **(c)** It fits reasonably well.

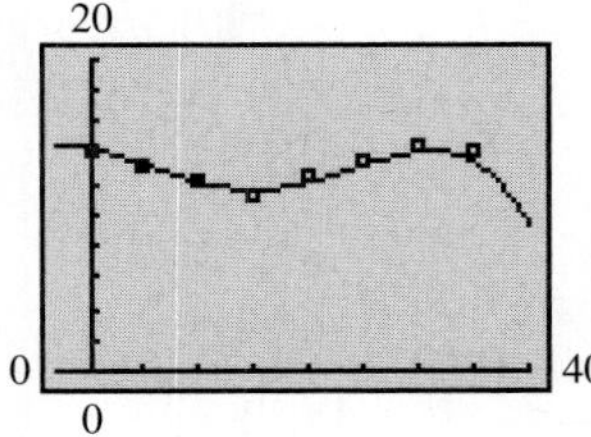

(d) 14.7 million; 13.8 million **(e)** Halfway through 1989
35. **(a)** $R(x) = .040x^3 - 1.226x^2 + 13.33x + 36.6$
(b)

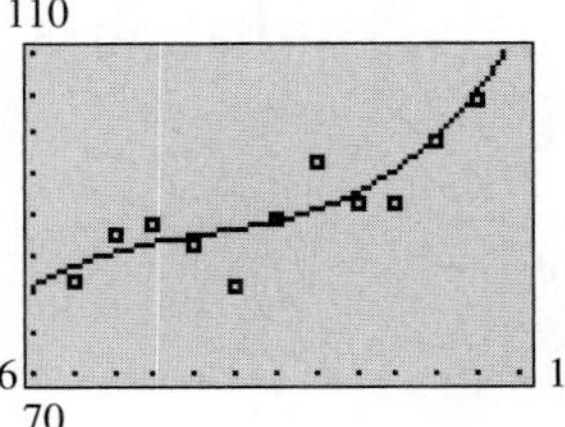

Yes, the graph fits the data reasonably well.
37. $P(x) = .005x^3 - .089x^2 + .35x + 7.4$

Section 3.7 (Page 203)

1. $x = -5, y = 0$

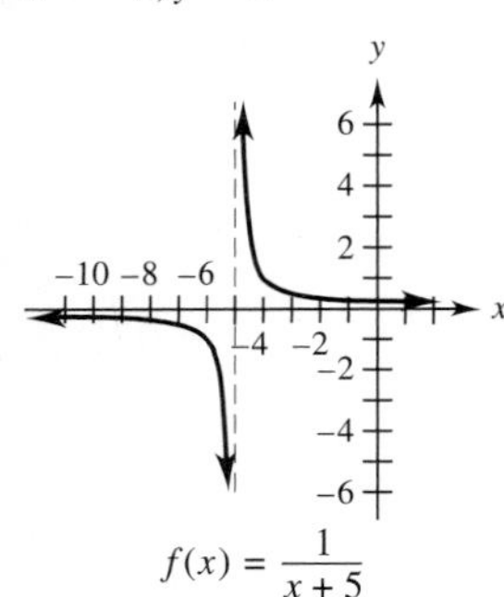

3. $x = -\frac{5}{2}, y = 0$

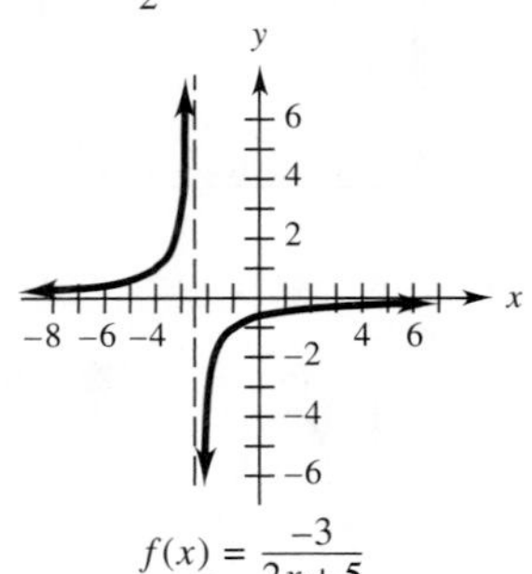

5. $x = 1, y = 3$

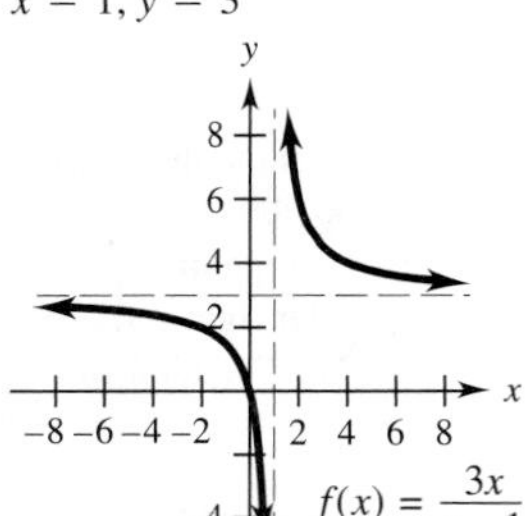

7. $x = 4, y = 1$

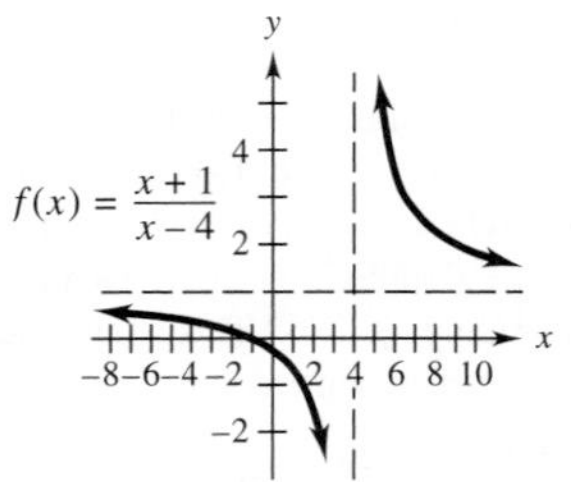

9. $x = 3; y = -1$

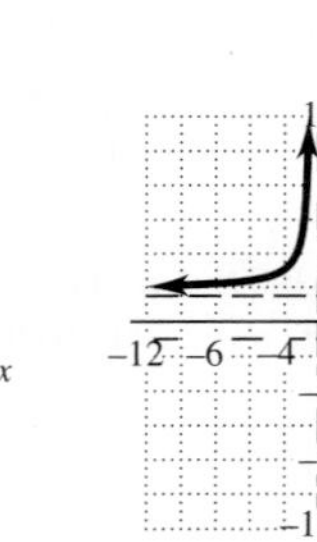

11. $x = -2, y = \frac{3}{2}$

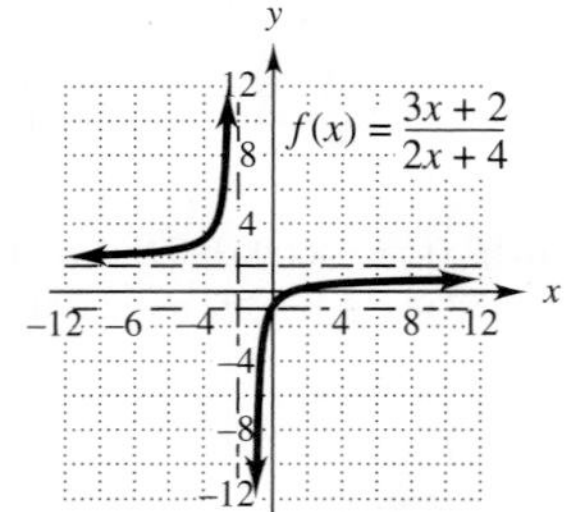

13. $x = -4, x = 2, y = 0$

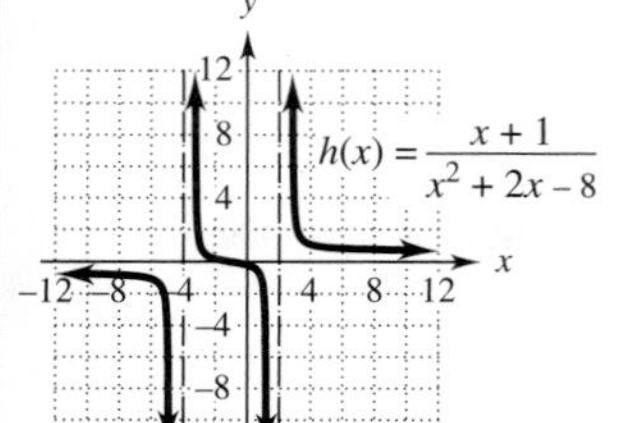

15. $x = -2, x = 2, y = 1$

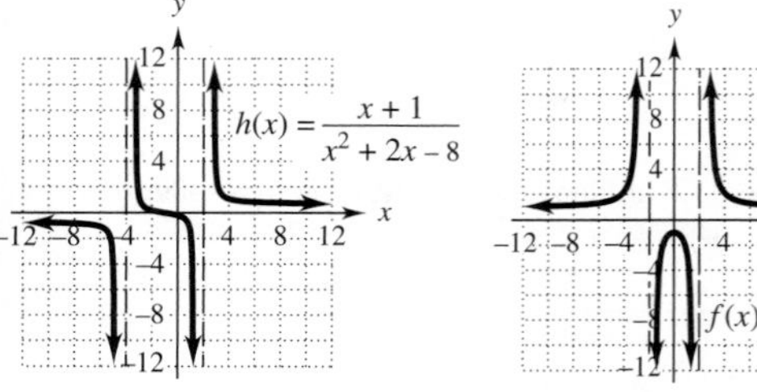

17. $x = -2, x = 1$ **19.** $x = -1, x = 5$ **21.** **(a)** \$4300 **(b)** \$10,033.33 **(c)** \$17,200 **(d)** \$38,700 **(e)** \$81,700 **(f)** \$210,700 **(g)** \$425,700 **(h)** No

(i)

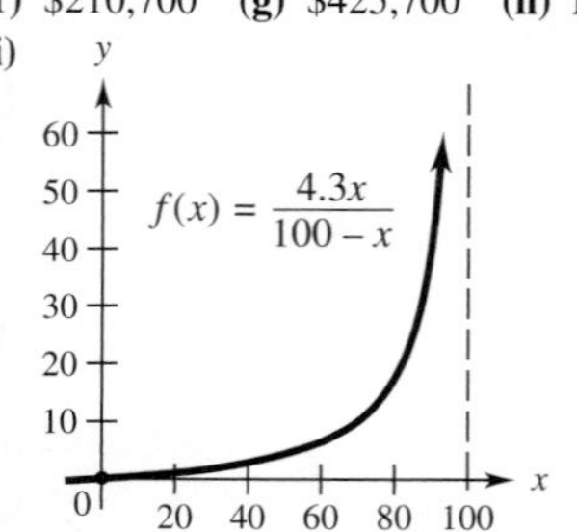

23. **(a)** $[0, \infty)$

(b)

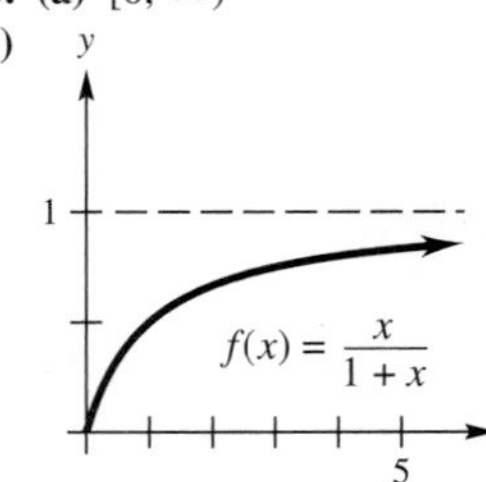

(c)

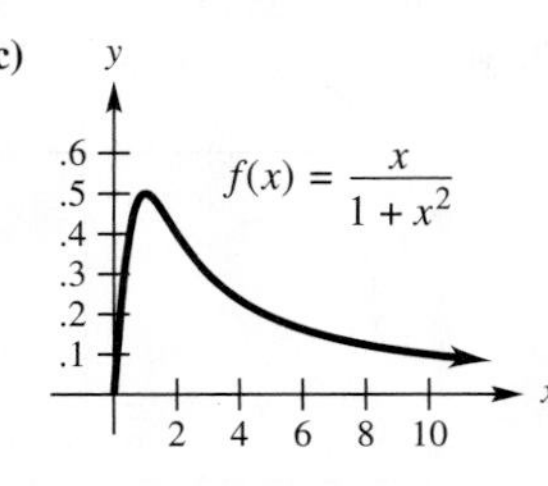

(d) Increasing b makes the next generation smaller when this generation is larger. **25.** **(a)** 6 min **(b)** 1.5 min **(c)** .6 min **(d)** $A = 0$

(e)

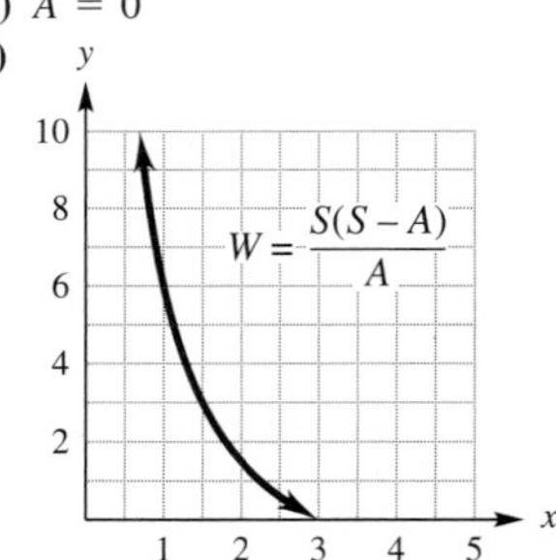

(f) W becomes negative. The waiting time approaches 0 as A approaches 3. The formula does not apply for $A > 3$ because there will be no waiting if people arrive more than 3 min apart.

27.

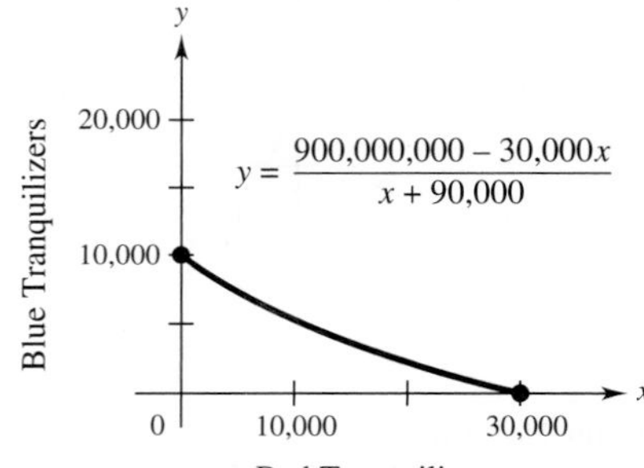

30,000 reds; 10,000 blues

29. **(a)** $C(x) = 2.6x + 40,000$

(b) $\overline{C}(x) = \frac{2.6x + 40,000}{x} = 2.6 + \frac{40,000}{x}$ **(c)** $y = 2.6$; the average cost may get close to, but will never equal, \$2.60.

31. About 73.9

33. **(a)**

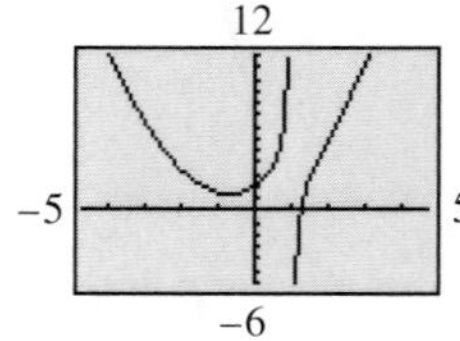

(b) They appear almost identical, because the parabola is an asymptote of the graph.

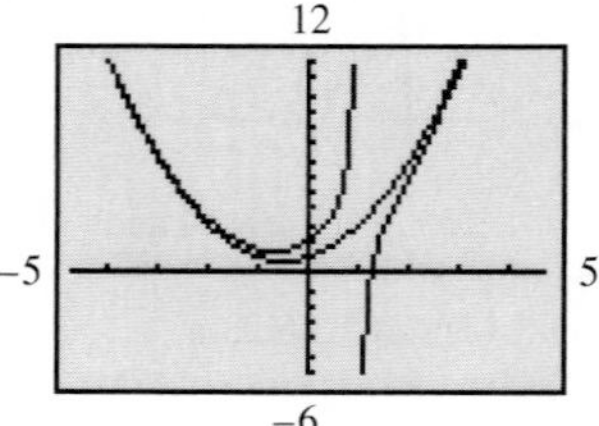

Chapter 3 Review Exercises (Page 207)

1. Not a function **3.** Function **5.** Not a function **7. (a)** 23 **(b)** -9 **(c)** $4p - 1$ **(d)** $4r + 3$ **9. (a)** -28 **(b)** -12 **(c)** $-p^2 + 2p - 4$ **(d)** $-r^2 - 3$ **11. (a)** -13 **(b)** 3 **(c)** $-k^2 - 4k$ **(d)** $-9m^2 + 12m$ **(e)** $-k^2 + 14k - 45$ **(f)** $12 - 5p$

13.

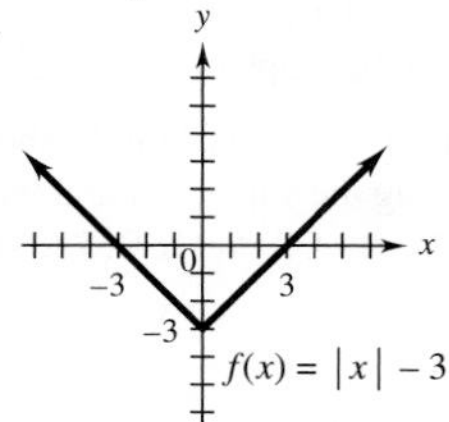

15.

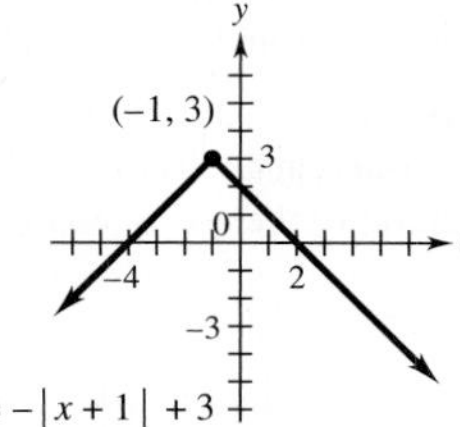

17.

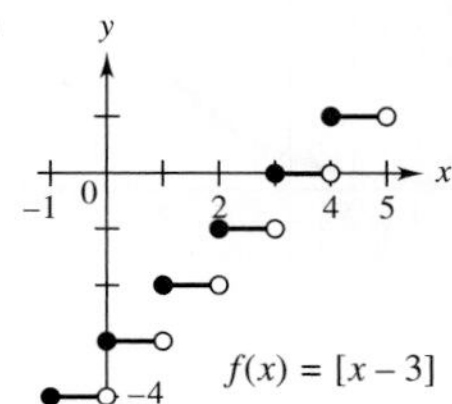

19.

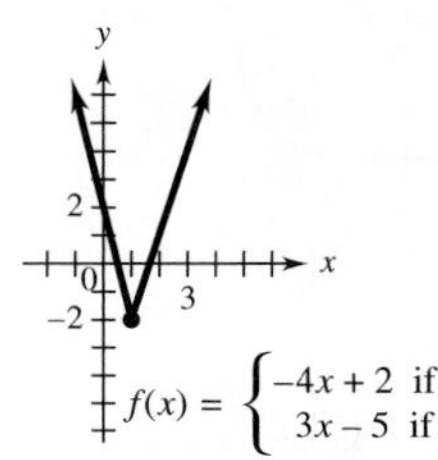

21.

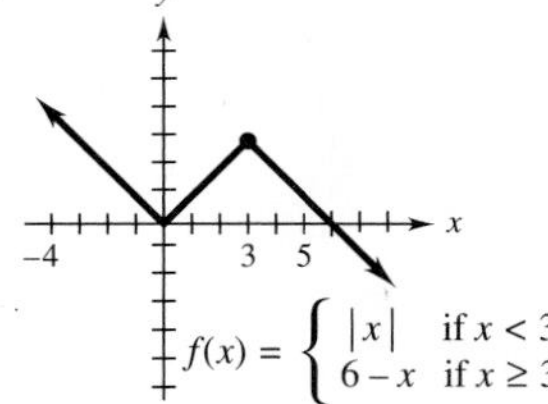

23.

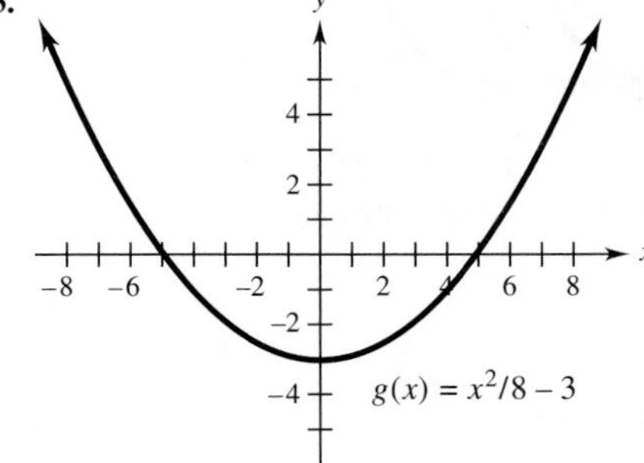

25. (a)

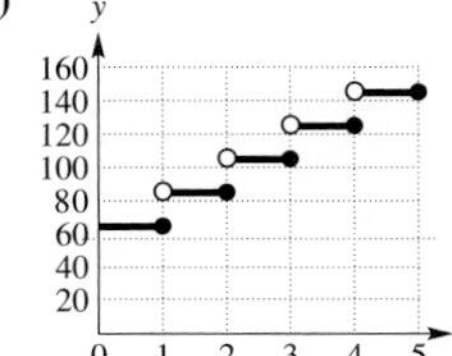

(b) Domain: $(0, \infty)$ range: $\{65, 85, 105, 125, \ldots\}$ **(c)** 2 days

27. These births appear to be leveling off.

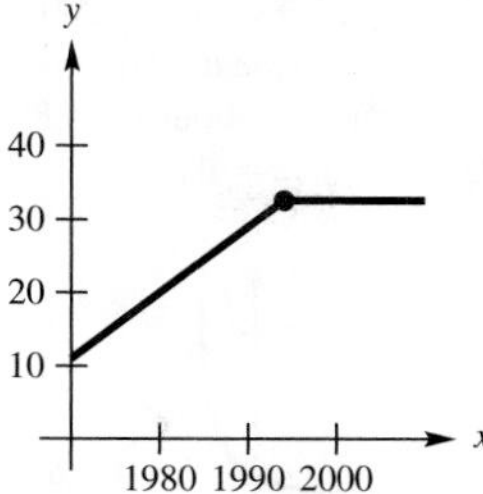

29. (a) $C(x) = 30x + 60$ **(b)** \$30 **(c)** \$30.60 **31. (a)** $C(x) = 30x + 85$ **(b)** \$30 **(c)** \$30.85 **33. (a)** \$18,000 **(b)** $R(x) = 28x$ **(c)** 4500 cartridges **(d)** \$126,000 **35.** Equilibrium quantity is 36 million subscribers at a price of \$12.95 per month. **37.** Upward; (2, 6) **39.** Downward; $(-1, 8)$

41.

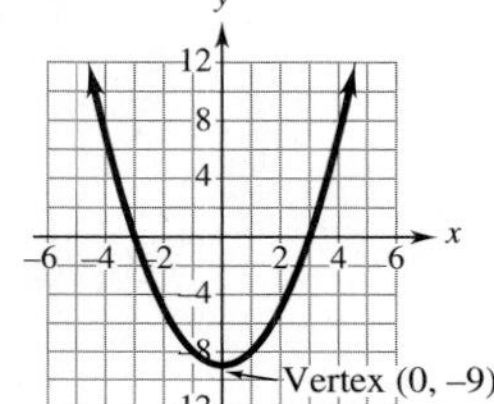

43.

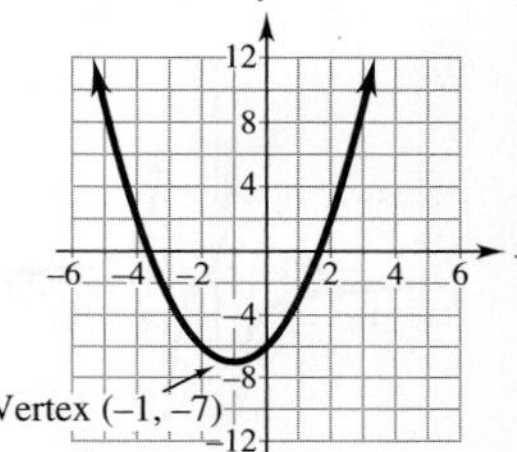

45.

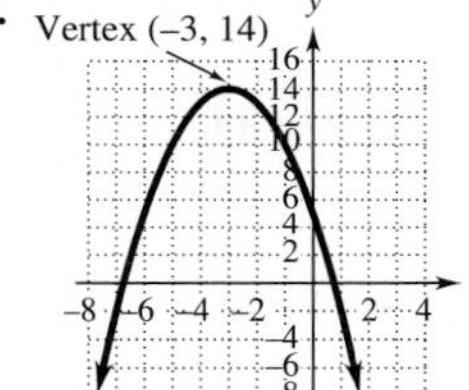

47.

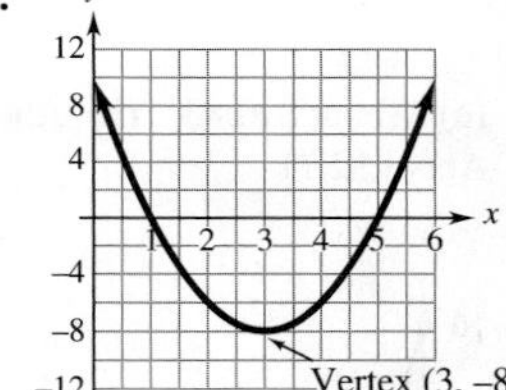

49. Minimum value; -11 **51.** Maximum value; 7 **53.** 3 months **55.** 125 units **57. (a)** $f(x) = 23.35(x - 7)^2 + 2700$ **(b)** \$31,304

59.

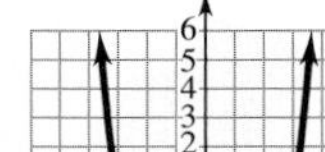
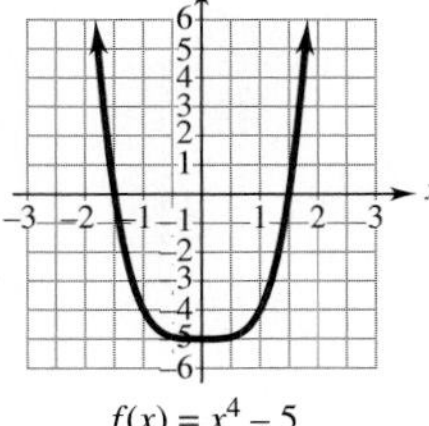

61.

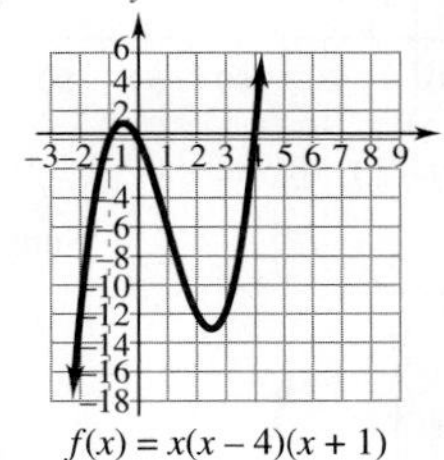

63.

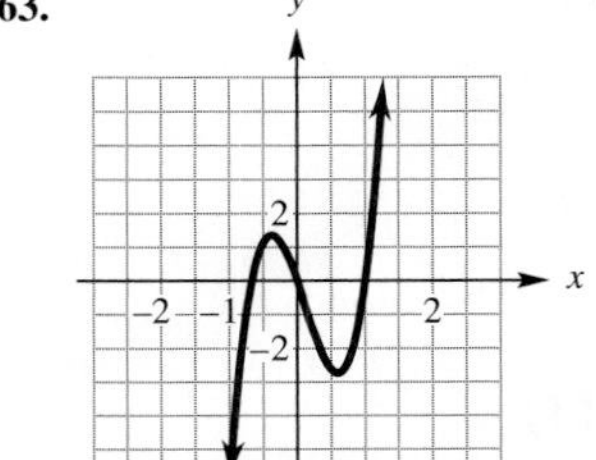

65.

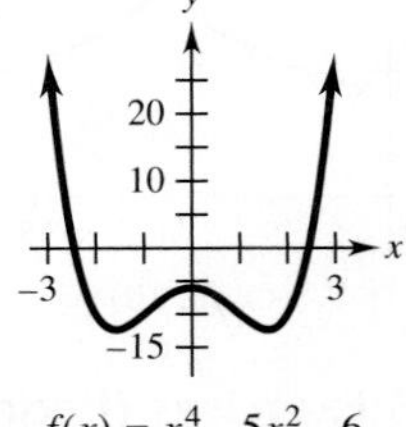

67. About 313,152; about \$690.72 per thousand
69. **(a)** $R(x) = 23x$; $P(x) = .000006x^3 - .07x^2 + 21x - 1200$
(b) About 76.54, which means that 77 must be made to earn a profit. (You can't make .54 of a rack.) **(c)** 230 **(d)** 153; about \$395.86
71. $x = 3, y = 0$ **73.** $x = 2, y = 0$

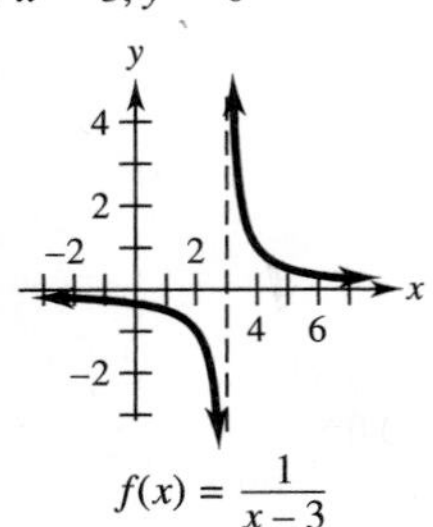

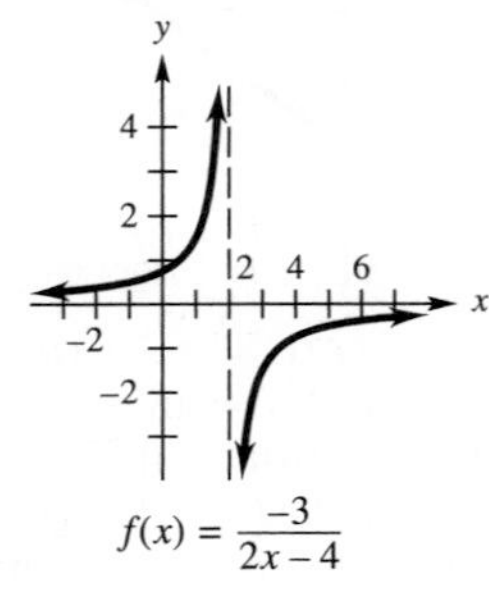

75. $x = -\frac{1}{2}, x = \frac{3}{2}, y = 0$

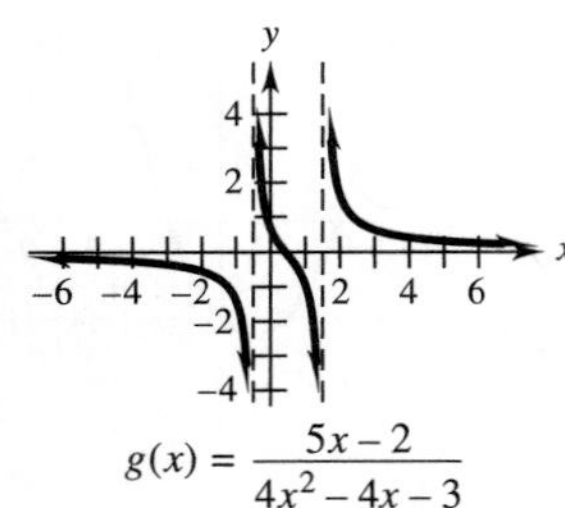

77. **(a)** About \$10.83 **(b)** About \$4.64 **(c)** About \$3.61
(d) About \$2.71
(e)

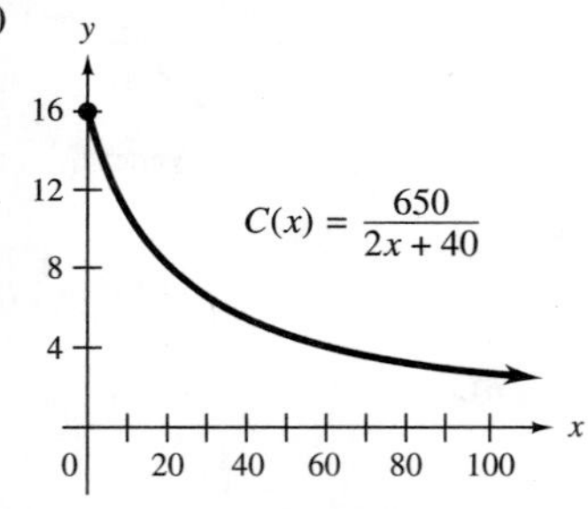

79. **(a)** (10, 50)

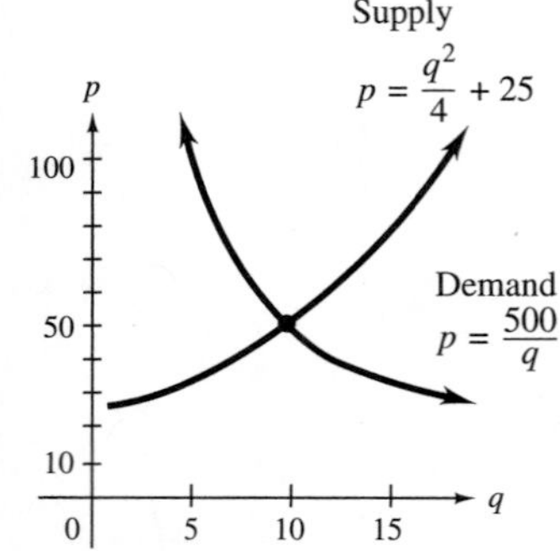

(b) $(10, \infty)$ **(c)** (0, 10)

Case 3 Exercises (Page 211)

1. $f(x) = \frac{-20}{49}x^2 + 20$ **3.** $h(x) = \sqrt{144 - x^2} + 8$; 8 ft

CHAPTER 4

Section 4.1 (Page 220)

1. Exponential **3.** Quadratic **5.** Exponential **7.** **(a)** The graph is entirely above the x-axis and falls from left to right, crossing the y-axis at 1 and then getting very close to the x-axis. **(b)** (0, 1), (1, .6) **9.** **(a)** The graph is entirely above the x-axis and rises from left to right, less steeply than the graph of $f(x) = 2^x$.
(b) $(0, 1), (1, 2^{.5}) = (1, \sqrt{2})$ **11.** **(a)** The graph is entirely above the x-axis and falls from left to right, crossing the y-axis at 1 and then getting very close to the x-axis. **(b)** $(0, 1), (1, e^{-1}) \approx (1, .367879)$

13.

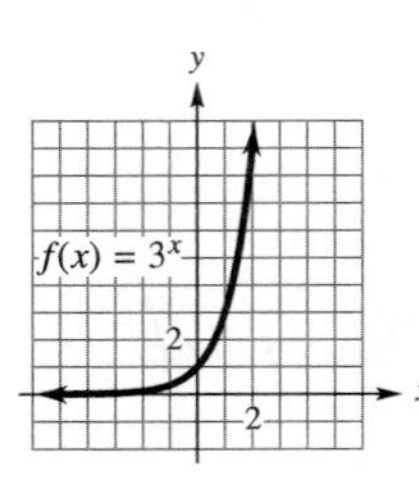

15.

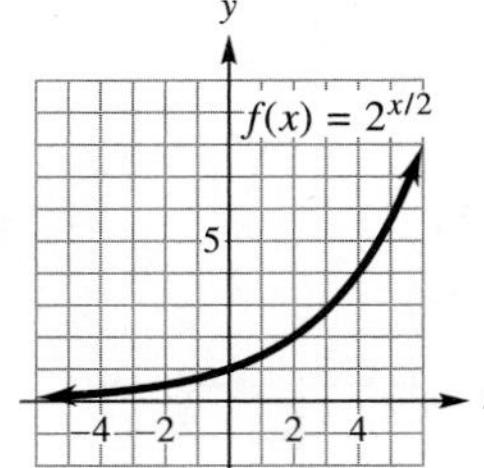

17.

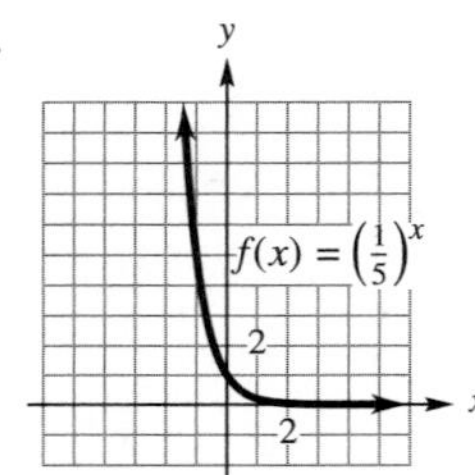

19. **(a)–(c)**

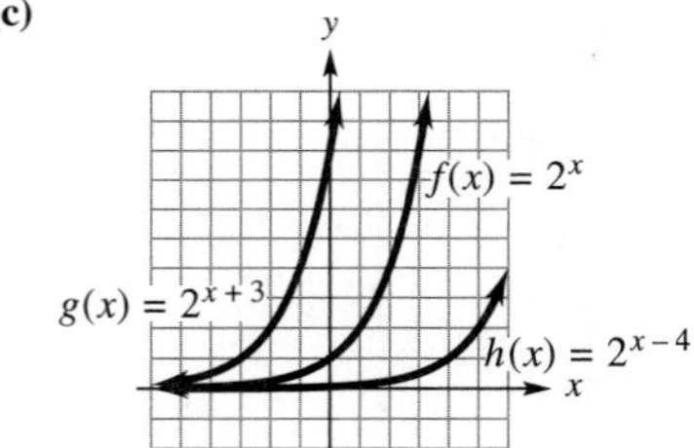

(d) Answers vary. **21.** 2.3 **23.** .75 **25.** .31
27. **(a)** $a > 1$ **(b)** Domain: $(-\infty, \infty)$; range: $(0, \infty)$
(c)

(d) Domain: $(-\infty, \infty)$; range: $(-\infty, 0)$
(e)

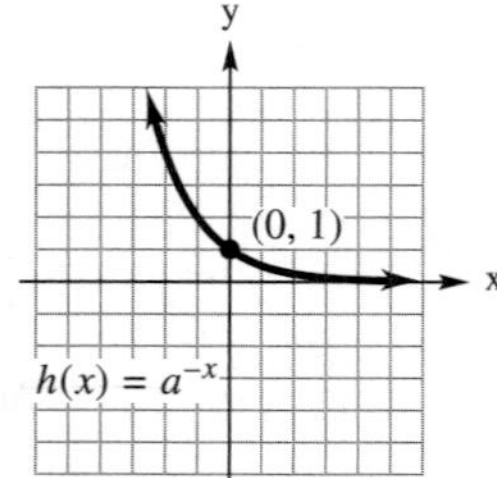

(f) Domain: $(-\infty, \infty)$; range: $(0, \infty)$ **29. (a)** 3 **(b)** $\frac{1}{3}$ **(c)** 9
(d) 1

31.

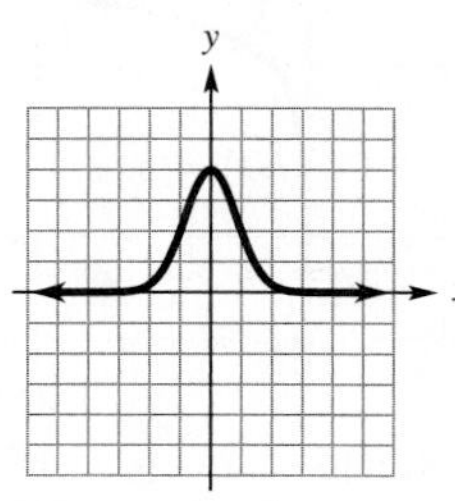

33.

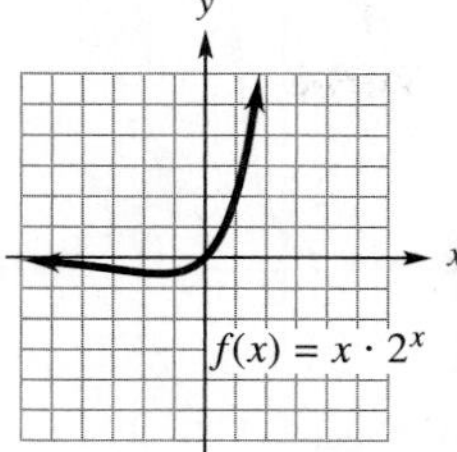

35. (a)

t	0	1	2	3	4	5	6	7	8	9	10
y	1	1.06	1.12	1.19	1.26	1.34	1.42	1.50	1.59	1.69	1.79

(b)

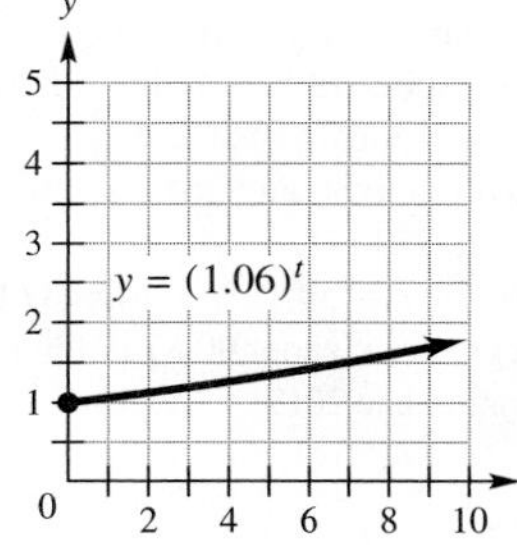

37. (a) About \$141,892 **(b)** About \$64.10 **39. (a)** About 142,000,000 **(b)** About 182,349,000 **(c)** About 234,140,000 **(d)** 386,031,000 **(e)** 386,031,000 accounts is unlikely to be accurate because the U.S. population will be about 305,000,000 in 2010. **41. (a)** About .97 kg **(b)** About .75 kg **(c)** About .65 kg **(d)** About 24,360 years **43.** \$34,706.99 **45. (a)** About 6.2 billion **(b)** About 6.5 billion **(c)** About 8 billion **(d)** Answers vary. **47. (a)** China, about \$3.022 trillion; United States, about \$13.5000 trillion **(b)** China, about \$8.426 trillion; United States, about \$18.816 trillion **(c)** China, about \$42.219 trillion; United States, about \$31.705 trillion **(d)** During 2041 **49. (a)** About 285,362 nurses **(b)** About 390,270 nurses **(c)** About 480,851 nurses **(d)** 2017 **51. (a)** About \$12,082,000 **(b)** About \$9,223,700 **(c)** About \$7,057,100
(d) 15

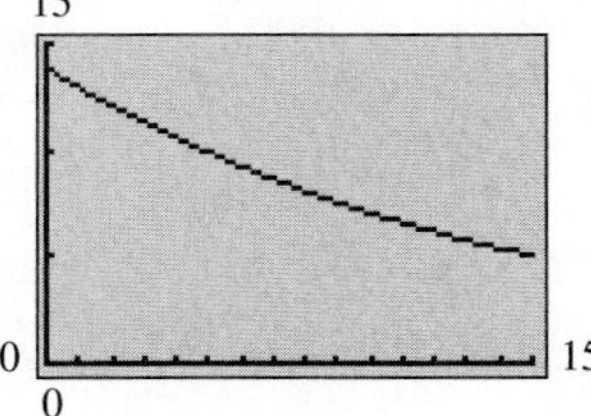

53. (a) About 12.6 mg **(b)** About 7.9 mg **(c)** In about 15.1 hours

Section 4.2 (Page 229)

1. (a) \$752.27 **(b)** \$707.39 **(c)** \$432.45 **(d)** \$298.98 **(e)** Answers vary. **3. (a)** About \$140.66 billion **(b)** About \$187.22 billion **(c)** About \$301.52 billion; answers vary. **5. (a)** $f(t) = 1.5*1.0787^t$ **(b)** About 4.33 million **(c)** 2018 **7. (a)** $f(t) = 1.6*1.3916^t$ **(b)** About \$8.35 billion; about \$16.17 billion **9. (a)** Two-point: $f(t) = .9710^t$; regression: $g(t) = 1.0103*.9717^t$ **(b)** Two-point: .745, .702, .662; regression: .758, .716, .676 **(c)** Two-point: 2031; regression: 2032 **11. (a)** Two-point: $f(t) = 257.6*.966676^t$; regression: $g(t) = 257.323*.963721^t$ **(b)** Two-point: 183.55, 154.94; regression: 177.82, 147.83 **(c)** Two-point: 2027; regression: 2025 **13. (a)** About 6 items **(b)** About 23 items **(c)** 25 items **15.** 2.6° C **17. (a)** .13 **(b)** .23 **(c)** About 2 weeks **19. (a)** \$31.138 billion; \$34.929 billion
(b) 50

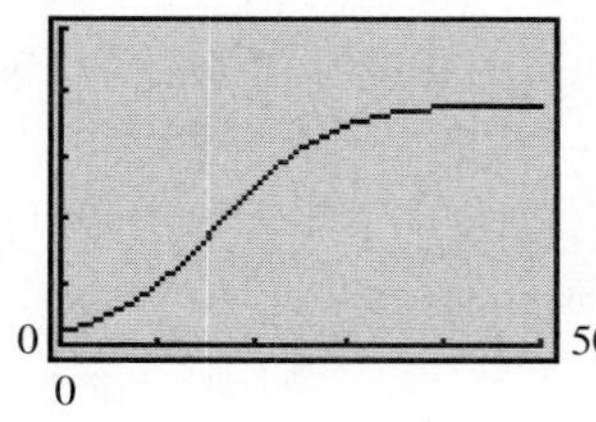

(c) In 2015 **(d)** No
21. (a) \$5.4601 trillion; \$10.941 trillion; \$14.709 trillion
(b) 50

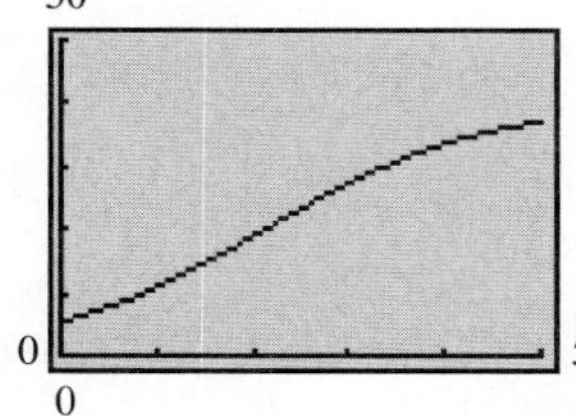

(c) In 2021 **(d)** Yes, in the last half of this century.

Section 4.3 (Page 240)

1. a^y **3.** It is missing the value that equals b^y. If that value is x, the expression should read $y = \log_b x$ **5.** $10^5 = 100,000$
7. $9^2 = 81$ **9.** $\log 96 = 1.9823$ **11.** $\log_3\left(\frac{1}{9}\right) = -2$
13. 3 **15.** 2 **17.** 3 **19.** -2 **21.** $\frac{1}{2}$ **23.** 8.77
25. 1.724 **27.** -4.991 **29.** Because $a^0 = 1$ for every valid base a. **31.** log 24 **33.** ln 5 **35.** $\log\left(\frac{u^2w^3}{v^6}\right)$
37. $\ln\left(\frac{(x+2)^2}{x+3}\right)$ **39.** $\frac{1}{2}\ln 6 + 2\ln m + \ln n$
41. $\frac{1}{2}\log x - \frac{5}{2}\log z$ **43.** $2u + 5v$ **45.** $3u - 2v$
47. 3.32112 **49.** 2.429777 **51.** Many correct answers, including, $b = 1, c = 2$.

53.

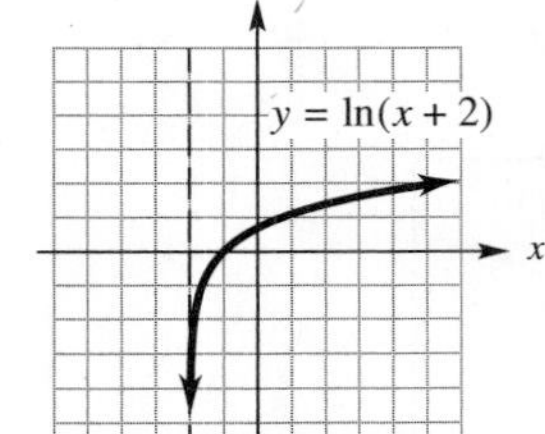

55.

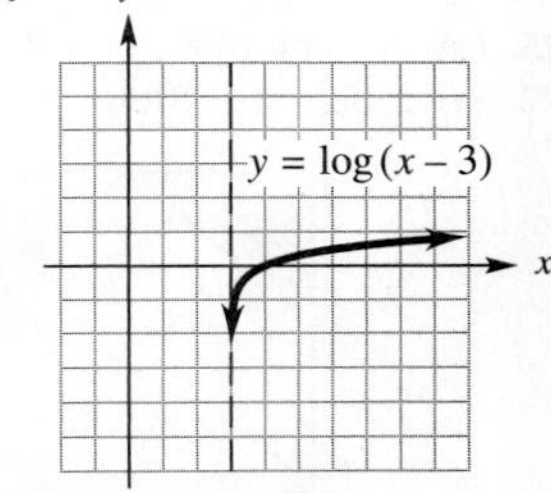

57. Answers vary. **59.** $\ln 2.75 = 1.0116009$; $e^{1.0116009} = 2.75$
61. (a) 17.67 yr **(b)** 9.01 yr **(c)** 4.19 yr **(d)** 2.25 yr
63. (a) About \$15.173 billion; about \$17.683 billion

(b)

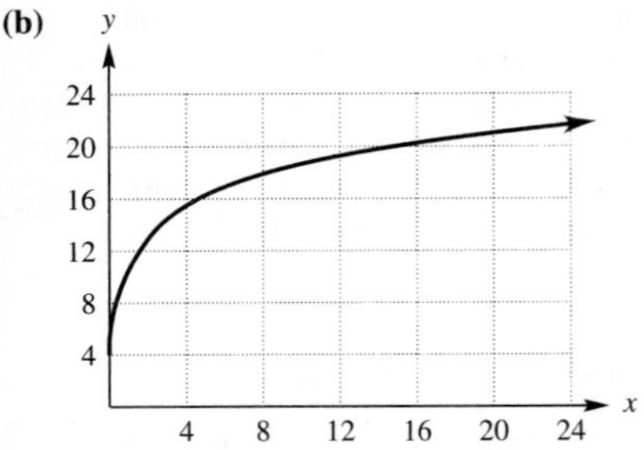

(c) Sales are increasing at a slower rate at time goes on.
65. (a) About 20.2%; about 21.8%; about 23.3%; about 24.4%
(b)

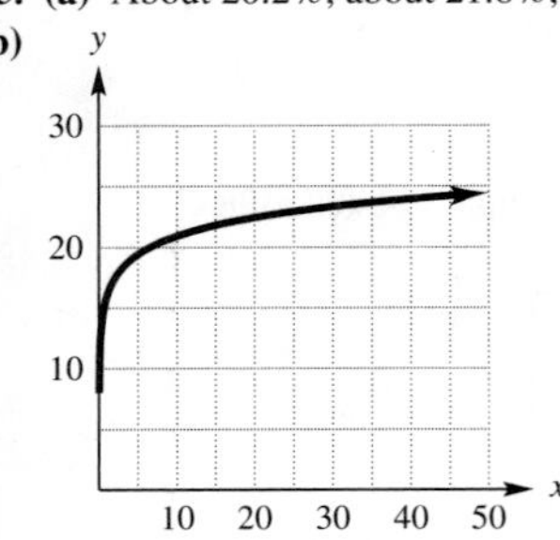

(c) It appears to be leveling off at a bit below 25%. **67.** 1.5887
69. (a) About \$1.4904 trillion; about \$1.806 trillion **(b)** In 2017
71. (a) About 24.718 billion pounds; about 25.41 pounds
(b) In 2022

Section 4.4 (Page 250)

1. 8 **3.** 9 **5.** 11 **7.** $\frac{11}{6}$ **9.** $\frac{4}{9}$ **11.** 10
13. 5.2378 **15.** 10 **17.** $\frac{4+b}{4}$ **19.** $\frac{10^{2-b}-5}{6}$
21. Answers vary. **23.** 4 **25.** $-\frac{5}{6}$ **27.** −4 **29.** −2
31. 2.3219 **33.** 2.710 **35.** −1.825 **37.** .597253
39. −.123 **41.** $\frac{\log d + 3}{4}$ **43.** $\frac{\ln b + 1}{2}$
45. 4 **47.** No solution **49.** −4, 4 **51.** 9 **53.** 1
55. 4, −4 **57.** ±2.0789 **59.** 1.386 **61.** Answers vary.
63. (a) Late 2013 ($t \approx 33.8238$) **(b)** Late 2016 ($t \approx 36.9781$)
65. (a) 1992 ($x \approx 92.1518$) **(b)** Mid-2007 ($x \approx 107.7364$)
(c) Mid-2041 ($x \approx 141.6168$) **67. (a)** Mid-2008 ($x \approx 8.471$)
(b) Early 2014 ($x \approx 14.0395$) **69. (a)** 25 g **(b)** About 4.95 yr
71. About 3689 yr old **73. (a)** Approximately $79{,}432{,}823i_0$
(b) Approximately $251{,}189i_0$ **(c)** About 316.23 times stronger
75. (a) 21 **(b)** 100 **(c)** 105 **(d)** 120 **(e)** 140
77. (a) 27.5% **(b)** \$130.14
79. (a) 120; 3000; 8000; 0
(b) About 4717 ft

Chapter 4 Review (Page 254)

1. (c) **3.** (d) **5.** $0 < a < 1$ **7.** All positive real numbers

9. $f(x) = 4^x$ **11.**

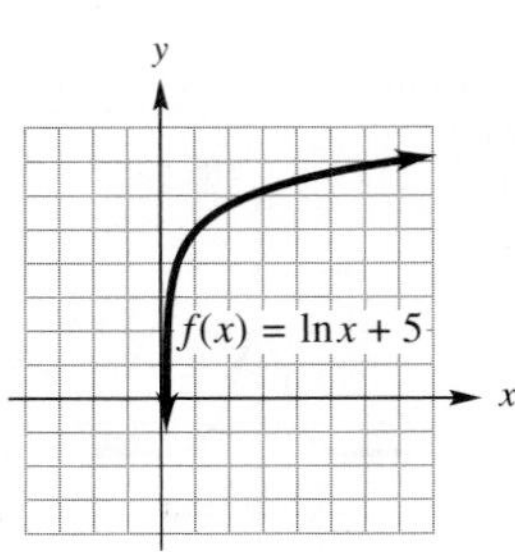

13. (a) About 434.07 billion; about 409.51 billion **(b)** In 2059; answers vary. **15.** log 340 = 2.53148 **17.** ln 45 = 3.8067
19. $10^4 = 10{,}000$ **21.** $e^{4.3957} = 81.1$ **23.** 5 **25.** 8.9
27. $\frac{1}{3}$ **29.** $\log 20x^6$ **31.** $\log\left(\frac{b^3}{c^2}\right)$ **33.** 4 **35.** 97
37. 5 **39.** 5 **41.** −2 **43.** −2 **45.** 1.416
47. −2.807 **49.** −3.305 **51.** .747 **53.** 28.463
55. (a) About \$2.439 billion **(b)** About \$2.883 billion
(c) In 2017 **57. (a)** 10 g **(b)** About 140 days **(c)** About 243 days **59. (a)** About 14.8 **(b)** About 13.5 **(c)** In mid-2015
61. 81.25° C **63. (a)** $f(x) = 132.1*1.1546^x$
(b) $g(x) = 135.553*1.1543^x$ **(c)** Two-point: about \$361.35 billion; regression: about \$370.12 billion **(d)** In 2015
65. (a) $f(x) = 63 + 14.3434 \ln x$ **(b)** $g(x) = 59.123 + 14.4487 \ln x$
(c) Two-point: about 94.52 million; regression: about 90.87 million
(d) Two-point: in early 2014; regression: in late 2017

Case 4 (Page 258)

1. about 23.6, 47.7, and 58.4; these estimates are a bit low.

CHAPTER 5

Section 5.1 (Page 265)

1. Time and interest rate **3.** \$133 **5.** \$217.48 **7.** \$86.26
9. \$158.82 **11.** \$375; \$11,250 **13.** \$386.25; \$2317.50
15. \$12,105 **17.** \$6727.50 **19.** Answers vary **21.** \$14,354.07
23. \$15,089.46 **25. (a)** \$7962.60 **(b)** About 1.879%
27. (a) \$11,878.80 **(b)** About 2.0406% **29. (a)** \$9450
(b) \$19,110 **31.** \$3056.25 **33.** 5.0% **35.** 4 months
37. \$1750.04 **39.** \$5825.24 **41.** 3.5% **43.** About 11.36%
45. About 102.91%. Ouch! **47.** \$13,725; yes
49. (a) $y_1 = 8t + 100$; $y_2 = 6\frac{3}{3}t + 200$ **(b)** The graph of y_1 is a line with slope 8 and y-intercept 100. The graph of y_2 is a line with slope $6\frac{3}{3}$ and y-intercept 200.
(c)

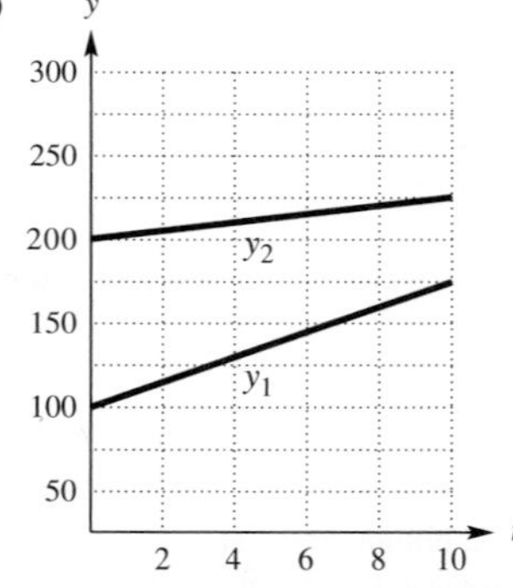

(d) The y-intercept of each graph indicates the amount invested. The slope of each graph is the annual amount of interest paid.

Section 5.2 (Page 276)

1. Answers vary. **3.** Interest rate and number of compounding periods **5.** Answers vary. **7.** \$1265.32 **9.** \$1204.75 **11.** \$9297.93 **13.** \$8793.87 **15.** \$1730.96 **17.** \$1968.48 **19.** 3.75% **21.** 5.25% **23.** \$10,000 **25.** \$20,000 **27.** \$25,000 **29.** 4.04% **31.** 5.095% **33.** 5.326% **35.** \$8954.58 **37.** \$11,572.58 **39.** \$4203.64 **41.** \$9963.10 **43.** \$1000 now **45.** \$21,570.27 **47.** Treasury note **49. (a)** \$16,288.95 **(b)** \$16,436.19 **(c)** \$16,470.09 **(d)** \$16,486.65 **51.** \$1000 now **53.** \$0–\$48,754; \$48,754–726,440; \$731,316–\$4,875,439; more than \$4,875,439 **55.** 10% **57.** −11.1% **59.** Flagstar, 4.45%; Principal, 4.46%; Principal paid a higher APY **61.** About \$2,259,696 **63.** About \$21,043 **65.** 23.4 years **67.** 14.2 years **69.** about 35 years **71.** (a)

Section 5.3 (Page 287)

1. 21.5786 **3.** \$119,625.61 **5.** \$23,242.87 **7.** \$72,482.38 **9.** About \$205,490 **11.** About \$310,831 **13.** \$797.36 **15.** \$4566.33 **17.** \$152.53 **19.** 5.19% **21.** 5.223% **23.** Answers vary. **25.** \$6603.39 **27.** \$234,295.32 **29.** \$26,671.23 **31.** \$3928.88 **33.** \$620.46 **35.** \$265.71 **37.** \$217,308.21 **39. (a)** \$137,895.79 **(b)** \$132,318.77 **(c)** \$5577.02 **41.** \$130,159.72 **43.** \$46,101.64 **45.** \$284,527.35 **47. (a)** \$256.08 **(b)** \$247.81 **49.** \$863.68 **51. (a)** \$1200 **(b)** \$3511.58 **53.** 6.5% **55. (a)** Answers vary. **(b)** \$24,000 **(c)** \$603,229 **(d)** \$84,000 **(e)** \$460,884

Section 5.4 (Page 300)

1. Answers vary. **3.** \$8693.71 **5.** \$1,566,346.66 **7.** \$11,468.10 **9.** \$38,108.61 **11.** \$557.68 **13.** \$6272.14 **15.** \$119,379.35 **17.** \$97,122.49 **19.** \$48,677.34 **21.** \$15,537.76 **23.** \$9093.14 **25.** \$446.31 **27.** \$11,331.18 **29.** \$589.31 **31.** \$1033.33 **33.** \$868.23 **35.** \$260.90; \$3022 **37.** \$847.50; \$105,429 **39.** \$6.80 **41.** \$42.04 **43.** \$30,669,881 **45.** \$24,761,633 **47. (a)** \$1465.42 **(b)** \$214.58 **49.** \$2320.83 **51.** \$280.46; \$32,310.74 **53. (a)** \$2717.36 **(b)** 2 **55. (a)** \$1151.22 **(b)** About \$152,320.58 **(c)** About \$2549.78 **(d)** About 21.22 years **57.** About \$8143.79 **59.** \$320.03 **61.** \$127,831.45

63.

Payment Number	Amount of Payment	Interest for Period	Portion to Principal	Principal at End of Period
0	–	–	–	\$4000.00
1	\$1207.68	\$320.00	\$887.68	3112.32
2	1207.68	248.99	958.69	2153.63
3	1207.68	172.29	1035.39	1118.24
4	1207.70	89.46	1118.24	0

65.

Payment Number	Amount of Payment	Interest for Period	Portion to Principal	Principal at End of Period
0	–	–	–	\$7184.00
1	\$189.18	\$71.84	117.34	7066.66
2	189.18	70.67	118.51	6948.15
3	189.18	69.48	119.70	6828.45
4	189.18	68.28	120.90	6707.55

Chapter 5 Review (Page 304)

1. \$292.08 **3.** \$62.05 **5.** \$285; \$3420 **7.** \$7925.67 **9.** Answers vary. **11.** \$78,742.54 **13.** About 4.082% **15.** \$5634.15; \$2834.15 **17.** \$20,402.98; \$7499.53 **19.** \$8532.58 **21.** \$15,000 **23.** 5.0625% **25.** \$18,207.65 **27.** \$1088.54 **29.** About \$9706.74 **31.** Answers vary. **33.** \$162,753.15 **35.** \$22,643.29 **37.** Answers vary. **39.** \$2619.29 **41.** \$916.12 **43.** \$31,921.91 **45.** \$14,222.42 **47.** \$136,340.32 **49.** \$1194.02 **51.** \$38,298.04 **53.** \$3581.11 **55.** \$392.70 **57.** About \$16,830.54 **59.** \$896.06 **61.** \$2696.12 **63. (a)** \$2,876,000 **(b)** About \$42,500,960 **65.** \$2298.58 **67.** \$24,818.76; \$2418.76 **69.** \$32.49 **71.** \$5596.62 **73.** \$3560.61 **75.** \$222,221.02 **77.** (d) **79.** \$5927.56

Case 5 (Page 309)

1. (a) \$22,549.94 **(b)** \$36,442.38 **(c)** \$66,402.34 **3.** About 55 cents **5. (a)** \$3704.09 **(b)** Yes

CHAPTER 6

Section 6.1 (Page 317)

1. Yes **3.** $(-1, -4)$ **5.** $\left(\frac{2}{7}, -\frac{11}{7}\right)$ **7.** $\left(\frac{11}{5}, -\frac{7}{5}\right)$ **9.** $(28, 22)$ **11.** $(2, -1)$ **13.** No solution **15.** $(4y + 1, y)$ for any real number y **17.** (a) **19.** $(5, 10)$ **21. (a)** $r = 175{,}000$ and $b = 375{,}000$ **(b)** Answers vary. **23.** In 2086! **25.** 4 2-episode and 18 3-episode discs **27.** Plane: 550 mph; wind: 50 mph **29.** 300 of Boeing; 100 of GE **31. (a)** $y = x + 1$ **(b)** $y = 3x + 4$ **(c)** $\left(-\frac{3}{2}, -\frac{1}{2}\right)$

Section 6.2 (Page 329)

1.
$$\begin{aligned} x \qquad - 3z &= 2 \\ 2x - 4y + 5z &= 1 \\ 5x - 8y + 7z &= 6 \\ 3x - 4y + 2z &= 3 \end{aligned}$$

3.
$$\begin{aligned} 3x \qquad + z + 2w + 18v &= 0 \\ -4x + y \qquad - w - 24v &= 0 \\ 7x - y + z + 3w + 42v &= 0 \\ 4x \qquad + z + 2w + 24v &= 0 \end{aligned}$$

5.
$$\begin{aligned} x + y + 2z + 3w &= 1 \\ -y - z - 2w &= -1 \\ 3x + y + 4z + 5w &= 2 \end{aligned}$$

7.
$$\begin{aligned} x + 12y - 3z + 4w &= 10 \\ 2y + 3z + w &= 4 \\ -z &= -7 \\ 6y - 2z - 3w &= 0 \end{aligned}$$

9. $(-68, 13, -6, 3)$ **11.** $\left(20, -9, \frac{15}{2}, 3\right)$

13. $\left[\begin{array}{ccc|c} 2 & 1 & 1 & 3 \\ 3 & -4 & 2 & -5 \\ 1 & 1 & 1 & 2 \end{array}\right]$

15.
$$\begin{aligned} 2x + 3y + 8z &= 20 \\ x + 4y + 6z &= 12 \\ 3y + 5z &= 10 \end{aligned}$$

17. $\left[\begin{array}{ccc|c} 1 & 2 & 3 & -1 \\ 2 & 0 & 7 & -4 \\ 6 & 5 & 4 & 6 \end{array}\right]$

19. $\left[\begin{array}{cccc|c} -4 & -3 & 1 & -1 & 2 \\ 0 & -4 & 7 & -2 & 10 \\ 0 & -2 & 9 & 4 & 5 \end{array}\right]$ **21.** $\left(\frac{3}{2}, 17, -5, 0\right)$

23. $(12 - w, -3 - 2w, -5, w)$ for any real number w

25. Dependent **27.** Inconsistent **29.** Independent

31. $\left(\frac{3}{2}, \frac{3}{2}, -\frac{3}{2}\right)$ **33.** $(-.8, -1.8, .8)$ **35.** $(100, 50, 50)$

37. $(0, 3.5, 1.5)$ **39.** $(-3, 4, 0)$ **41.** $(1, 0, -1)$

43. $(-93, -31, 10)$ **45.** No solution

47. $(192, 125.5, -148.5)$ **49.** $\left(\frac{1 - z}{2}, \frac{11 - z}{2}, z\right)$ for any real number z **51.** No solution **53.** $(-7, 5)$ **55.** $(-1, 2)$

57. $(-3, z - 17, z)$ for any real number z

59. $(-11.5, 13.75, -2.25, -5)$ **61.** No solution

63. $\left(\frac{1}{2}, \frac{1}{3}, -\frac{1}{4}\right)$ **65.** 1993; about 374,000

67. 220 adults; 150 teenagers; 200 children

69.

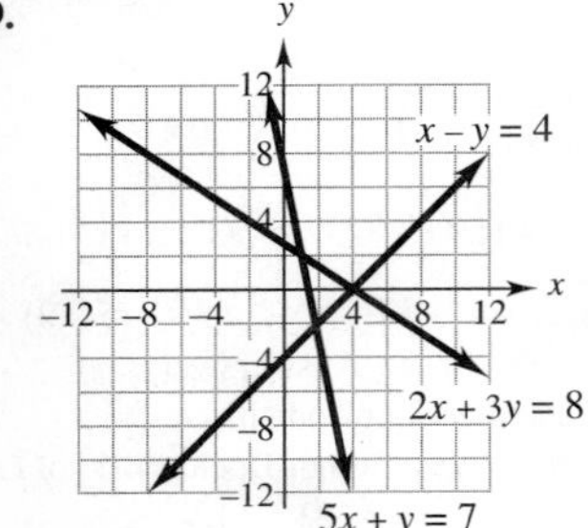

There is no point that is on all three lines.

Section 6.3 (Page 338)

1. 100 vans; 50 small trucks; 32 large trucks **3.** 10 units of corn; 25 units of soybeans; 40 units of cottonseed **5.** Shirt \$1.99; slacks \$4.99; sports coat \$6.49 **7.** 220 adults; 150 teenagers; 200 preteens **9.** \$20,000 in AAA; \$10,000 in B; none in A **11.** 40 lb pretzels, 20 lb dried fruit, 80 lb nuts **13.** **(a)** .056057, 1.06657 **(b)** About 228 ft **15.** \$15,000 in the mutual fund, \$30,000 in bonds, and \$25,000 in the food franchise **17.** **(a)** $b + c$ **(b)** .35, .45, .2; A is tumorous, B is bone, and C is healthy.

19. **(a)** $x_2 + x_3 = 700$
$x_3 + x_4 = 600$

(b) $(1000 - x_4, 100 + x_4, 600 - x_4, x_4)$ **(c)** 600; 0

(d) x_1: 1000; 400
x_2: 700; 100
x_3: 600; 0

(e) Answers vary. **21.** 4 cups of Hearty Chicken Rotini, 5 cups of Hearty Chicken, and 6 cups of Chunky Chicken Noodle; serving size 1.5 cups **23.** **(a)** \$12,000 at 6%, \$7000 at 7%, and \$6000 at 10% **(b)** \$10,000 at 6%, \$15,000 at 7%, and \$5000 at 10% **(c)** \$20,000 at 6%, \$10,000 at 7%, and \$10,000 at 10% **25.** **(a)** $y = .01x^2 - .3x + 4.24$ **(b)** 15 platters; \$1.99

27. **(a)** $f(x) = .006463x^2 + .159634x + 10.4629$ **(b)** About \$12.4 trillion; about \$14.3 trillion **(c)** In 2036

29. **(a)** $C = -.0000108S^2 + .034896S + 22.9$

(b) About 864.7 knots

Section 6.4 (Page 349)

1. 2×3; $\begin{bmatrix} -7 & 8 & -4 \\ 0 & -13 & -9 \end{bmatrix}$

3. 3×3; square matrix; $\begin{bmatrix} 3 & 0 & -11 \\ -1 & -\frac{1}{4} & 7 \\ -5 & 3 & -9 \end{bmatrix}$

5. 2×1; column matrix; $\begin{bmatrix} -7 \\ -11 \end{bmatrix}$ **7.** B is a 5×3 zero matrix.

9. $\begin{bmatrix} -7 & 14 & 2 & 4 \\ 6 & -3 & 2 & -4 \end{bmatrix}$ **11.** $\begin{bmatrix} 3 & -1 & 2 \\ 3 & 1 & 5 \end{bmatrix}$

13. $\begin{bmatrix} -4 & -7 & 7 \\ 6 & 3 & -8 \end{bmatrix}$ **15.** $\begin{bmatrix} -4 & 0 \\ 10 & 6 \end{bmatrix}$ **17.** $\begin{bmatrix} 0 & -8 \\ -16 & 24 \end{bmatrix}$

19. $\begin{bmatrix} 8 & 10 \\ 0 & -42 \end{bmatrix}$ **21.** $\begin{bmatrix} 4 & -\frac{7}{2} \\ 4 & \frac{21}{2} \end{bmatrix}$

23. $X + T = \begin{bmatrix} x & y \\ z & w \end{bmatrix} + \begin{bmatrix} r & s \\ t & u \end{bmatrix} = \begin{bmatrix} x + r & y + s \\ z + t & w + u \end{bmatrix}$; a 2×2 matrix

25.
$$X + (T + P) = \begin{bmatrix} x + (r + m) & y + (s + n) \\ z + (t + p) & w + (u + q) \end{bmatrix} = \begin{bmatrix} (x + r) + m & (y + s) + n \\ (z + t) + p & (w + u) + q \end{bmatrix} = (X + T) + P$$

27. $P + O = \begin{bmatrix} m + 0 & n + 0 \\ p + 0 & q + 0 \end{bmatrix} = \begin{bmatrix} m & n \\ p & q \end{bmatrix} = P$

29. Several possible correct answers, including

	Basketball	Hockey	Football	Baseball
Percent of no shows	16	16	20	18
Lost revenue per fan (\$)	18.20	18.25	19	15.40
Lost annual revenue (millions \$)	22.7	35.8	51.9	96.3

31. Several possible answers, including

	2000	2003	2006
Heart	3929	3444	2814
Lung	3514	3800	2857
Liver	16,095	16,927	16,861
Kidney	44,589	53,563	66,961

33. Several possible answers, including

	1995	2000	2003	2005
Ages 15–24	2.8	2.6	2.7	2.7
Ages 45–54	109.6	94.2	92.5	89.7
Ages 65–74	795.4	665.6	585.0	518.9

35. **(a)** $A = \begin{bmatrix} 2.61 & 4.39 & 6.29 & 9.08 \\ 1.63 & 2.77 & 4.61 & 6.92 \\ .92 & .75 & .62 & .54 \end{bmatrix}$

(b) $B = \begin{bmatrix} 1.38 & 1.72 & 1.94 & 3.31 \\ 1.26 & 1.48 & 2.82 & 2.28 \\ .41 & .33 & .27 & .40 \end{bmatrix}$

(c) $\begin{bmatrix} 1.23 & 2.67 & 4.35 & 5.77 \\ .37 & 1.29 & 1.79 & 4.64 \\ .51 & .42 & .35 & .14 \end{bmatrix}$

Section 6.5 (Page 361)

1. 2×2; 2×2 **3.** 3×3; 5×5 **5.** AB does not exist; 3×2

7. Columns; rows **9.** $\begin{bmatrix} 5 \\ 9 \end{bmatrix}$ **11.** $\begin{bmatrix} -2 & 4 \\ 0 & -8 \end{bmatrix}$

13. $\begin{bmatrix} -4 & 1 \\ 2 & -3 \end{bmatrix}$ **15.** $\begin{bmatrix} 3 & -5 & 7 \\ -2 & 1 & 6 \\ 0 & -3 & 4 \end{bmatrix}$

17. $\begin{bmatrix} 16 & 10 \\ 2 & 28 \\ 58 & 46 \end{bmatrix}$ **19.** $AB = \begin{bmatrix} -30 & -45 \\ 20 & 30 \end{bmatrix}$, but $BA = \begin{bmatrix} 0 & 0 \\ 0 & 0 \end{bmatrix}$.

21. $(A + B)(A - B) = \begin{bmatrix} -7 & -24 \\ -28 & -33 \end{bmatrix}$, but $A^2 - B^2 = \begin{bmatrix} -37 & -69 \\ -8 & -3 \end{bmatrix}$.

23. $(PX)T =$
$\begin{bmatrix} (mx + nz)r + (my + nw)t & (mx + nz)s + (my + nw)u \\ (px + qz)r + (py + qw)t & (px + qz)s + (py + qw)u \end{bmatrix}$.
$P(XT)$ is the same, so $(PX)T = P(XT)$.

25. $k(X + T) = k\begin{bmatrix} x + r & y + s \\ z + t & w + u \end{bmatrix}$
$= \begin{bmatrix} k(x + r) & k(y + s) \\ k(z + t) & k(w + u) \end{bmatrix}$
$= \begin{bmatrix} kx + kr & ky + ks \\ kz + kt & kw + ku \end{bmatrix}$
$= \begin{bmatrix} kx & ky \\ kz & kw \end{bmatrix} + \begin{bmatrix} kr & ks \\ kt & ku \end{bmatrix} = kX + kT$

27. No **29.** Yes **31.** Yes **33.** $\begin{bmatrix} 2 & 3 \\ 1 & 2 \end{bmatrix}$ **35.** $\begin{bmatrix} 1 & 2 \\ 1 & 1 \end{bmatrix}$

37. No inverse **39.** $\begin{bmatrix} -4 & -2 & 3 \\ -5 & -2 & 3 \\ 2 & 1 & -1 \end{bmatrix}$ **41.** $\begin{bmatrix} 2 & 1 & -1 \\ 8 & 2 & -5 \\ -11 & -3 & 7 \end{bmatrix}$

43. $\begin{bmatrix} -13 & 8 & 7 \\ -6 & 4 & 3 \\ 2 & -1 & -1 \end{bmatrix}$ **45.** $\begin{bmatrix} 6 & -5 & 8 \\ -1 & 1 & -1 \\ -1 & 1 & -2 \end{bmatrix}$

47. $\begin{bmatrix} \frac{1}{2} & -1 & -\frac{1}{2} & \frac{1}{2} \\ \frac{1}{2} & 4 & \frac{5}{2} & -\frac{1}{2} \\ -\frac{1}{4} & -\frac{1}{2} & -\frac{1}{4} & \frac{1}{4} \\ \frac{1}{2} & -2 & -\frac{3}{2} & \frac{1}{2} \end{bmatrix}$ **49. (a)** $R = \begin{bmatrix} .024 & .008 \\ .025 & .007 \\ .015 & .009 \\ .011 & .011 \end{bmatrix}$

(b) $P = \begin{bmatrix} 1996 & 286 & 226 & 460 \\ 2440 & 365 & 252 & 484 \\ 2906 & 455 & 277 & 499 \\ 3683 & 519 & 310 & 729 \\ 4723 & 697 & 364 & 702 \end{bmatrix}$

(c) $PR = \begin{bmatrix} 63.504 & 25.064 \\ 76.789 & 29.667 \\ 90.763 & 34.415 \\ 114.036 & 43.906 \\ 143.959 & 53.661 \end{bmatrix}$

(d) Rows represent the years 1970, 1980, 1990, 2000, and 2025. Column one gives the total births in those years and column two the total deaths. **(e)** 114,036,000; 53,661,000

51. (a) $\begin{bmatrix} 278.1 & 31.6 & 37.4 & 126.8 \\ 300.1 & 34.3 & 41.1 & 127.3 \end{bmatrix}$

(b) $\begin{bmatrix} .01425 & .00865 \\ .01145 & .00775 \\ .0175 & .00755 \\ .00945 & .00925 \end{bmatrix}$ **(c)** $\begin{bmatrix} 6.177505 & 4.105735 \\ 6.591395 & 4.349520 \end{bmatrix}$

(d) The rows correspond to years. The entries in each column give the total number of births and deaths, respectively, in the four countries taken together. **(e)** 6,177,505

53. (a) $A = \begin{bmatrix} 1259 & 600 \\ 1339 & 642 \\ 1424 & 673 \\ 1493 & 723 \end{bmatrix}$ **(b)** $B = \begin{bmatrix} .826 & .827 & .827 & .826 \\ .654 & .661 & .662 & .666 \end{bmatrix}$

(c) $AB = \begin{bmatrix} 1432.33 & 1437.79 & 1438.39 & 1439.53 \\ 1525.88 & 1531.72 & 1532.36 & 1533.59 \\ 1616.37 & 1622.50 & 1623.17 & 1624.44 \\ 1706.06 & 1712.61 & 1713.34 & 1714.74 \end{bmatrix}$

(d) $BA = \begin{bmatrix} 4558.15 & 2180.30 \\ 3645.49 & 1743.81 \end{bmatrix}$

(e) In AB, row 1, column 1 is the total per-capita dollar amount paid by private insurance for physician services and prescription drugs in 2003; row 2, column 2 is the same total for 2004; row 3, column 3 is the same total for 2005; row 4, column 4 is the same total for 2006. All other entries in AB are not meaningful here. In BA, row 1, column 1 is the per-capita total dollar amount paid by private insurance for physician services over the four-year period; row 2, column 2 is the per-capita total dollar amount paid by private insurance for prescription drugs over the four-year period. The other two entries in BA are not meaningful here.

55. (a)

	A	B
Dept. 1	254	290
Dept. 2	199	240
Dept. 3	170	216
Dept. 4	174	170

(b) supplier *A*: \$797; supplier *B*: \$916; buy from *A*.

Section 6.6 (Page 374)

1. $\begin{bmatrix} 2 \\ -2 \end{bmatrix}$ **3.** $\begin{bmatrix} \frac{1}{2} & 1 \\ \frac{3}{2} & 1 \end{bmatrix}$ **5.** $\begin{bmatrix} -7 \\ 15 \\ -3 \end{bmatrix}$ **7.** $(-63, 50, -9)$

9. $(-31, -131, 181)$ **11.** $(-48, 8, 12)$ **13.** $(2, 2, -1, -2)$

15. $\begin{bmatrix} -6 \\ -14 \end{bmatrix}$ **17.** 60 buffets; 600 chairs; 100 tables

19. 2340 of the first species, 10,128 of the second species, 224 of the third species **21.** jeans \$34.50; jacket \$72; sweater \$44; shirt \$21.75 **23.** $\begin{bmatrix} \frac{32}{3} \\ \frac{25}{3} \end{bmatrix}$ **25.** About 1073 metric tons of wheat, about 1431 metric tons of oil **27.** Gas \$98 million, electric \$123 million

29. (a) $\frac{7}{4}$ bushels of yams, $\frac{15}{8} \approx 2$ pigs **(b)** 167.5 bushels of yams, $153.75 \approx 154$ pigs **31. (a)** .40 unit of agriculture, .12 unit of manufacturing, and 3.60 units of households **(b)** 848 units of agriculture, 516 units of manufacturing, and 2970 units of households **(c)** About 813 units **33. (a)** .017 unit of manufacturing and .216 unit of energy **(b)** 123,725,000 pounds of agriculture, 14,792,000 pounds of manufacturing, 1,488,000 pounds of energy

(c) 195,492,000 pounds of agriculture, 25,933,000 pounds of manufacturing, 13,580,000 pounds of energy **35.** \$532 million of natural resources, \$481 million of manufacturing, \$805 million of trade and services, \$1185 million of personal consumption

37. (a) $\begin{bmatrix} 1.67 & .56 & .56 \\ .19 & 1.17 & .06 \\ 3.15 & 3.27 & 4.38 \end{bmatrix}$

(b) These multipliers imply that, if the demand for one community's output increases by \$1, then the output of the other community will increase by the amount in the row and column of that matrix. For example, if the demand for Hermitage's output increases by \$1, then output from Sharon will increase by \$.56, from Farrell by \$.06, and from Hermitage by \$4.38.

39. $\begin{bmatrix} 23 \\ 51 \end{bmatrix}, \begin{bmatrix} 13 \\ 30 \end{bmatrix}, \begin{bmatrix} 45 \\ 96 \end{bmatrix}, \begin{bmatrix} 69 \\ 156 \end{bmatrix}, \begin{bmatrix} 87 \\ 194 \end{bmatrix}, \begin{bmatrix} 23 \\ 51 \end{bmatrix}, \begin{bmatrix} 51 \\ 110 \end{bmatrix}, \begin{bmatrix} 45 \\ 102 \end{bmatrix}, \begin{bmatrix} 69 \\ 157 \end{bmatrix}$ **41. (a)** 3 **(b)** 3 **(c)** 5 **(d)** 3

43. (a) $B = \begin{bmatrix} 0 & 2 & 3 \\ 2 & 0 & 4 \\ 3 & 4 & 0 \end{bmatrix}$ **(b)** $B^2 = \begin{bmatrix} 13 & 12 & 8 \\ 12 & 20 & 6 \\ 8 & 6 & 25 \end{bmatrix}$ **(c)** 12 **(d)** 14

45. (a)

$C =$	Dogs	Rats	Cats	Mice
Dogs	0	1	1	1
Rats	0	0	0	1
Cats	0	1	0	1
Mice	0	0	0	0

(b) $\begin{bmatrix} 0 & 1 & 0 & 2 \\ 0 & 0 & 0 & 0 \\ 0 & 0 & 0 & 1 \\ 0 & 0 & 0 & 0 \end{bmatrix}$; C^2 gives the number of food sources once removed from the feeder.

47. $\begin{bmatrix} -.5 & .2 \\ 0 & .2 \end{bmatrix}$ **49.** No inverse **51.** $\begin{bmatrix} \frac{1}{4} & \frac{1}{2} & \frac{1}{2} \\ \frac{1}{4} & -\frac{1}{2} & \frac{1}{2} \\ \frac{1}{12} & -\frac{1}{6} & -\frac{1}{6} \end{bmatrix}$

53. No inverse **55.** $\begin{bmatrix} -\frac{2}{3} & -\frac{17}{3} & -\frac{14}{3} & -3 \\ \frac{1}{3} & \frac{1}{3} & \frac{1}{3} & 0 \\ -\frac{1}{3} & -\frac{10}{3} & -\frac{7}{3} & -2 \\ 0 & 2 & 1 & 1 \end{bmatrix}$ **57.** $\begin{bmatrix} -\frac{7}{19} & \frac{2}{19} \\ \frac{6}{19} & \frac{1}{19} \end{bmatrix}$

59. $\begin{bmatrix} -\frac{1}{12} & -\frac{1}{4} \\ \frac{5}{12} & \frac{1}{4} \end{bmatrix}$ **61.** $\begin{bmatrix} -1.2 & 1.1 & -.9 \\ .8 & -.4 & .6 \\ -.2 & .1 & .1 \end{bmatrix}$ **63.** $\begin{bmatrix} -5 \\ -3 \end{bmatrix}$

65. $\begin{bmatrix} -22 \\ -18 \\ 15 \end{bmatrix}$ **67.** $(-4, 2)$ **69.** $(4, 2)$ **71.** $(-1, 0, 2)$

73. No inverse; no solution for the system **75.** 16 liters of the 9%; 24 liters of the 14% **77.** 30 liters of 40% solution; 10 liters of 60% solution **79.** 80 bowls; 120 plates **81.** \$12,750 at 8%; \$27,250 at 8.5%; \$10,000 at 11%

83. (a) $\begin{bmatrix} 1 & -\frac{1}{4} \\ -\frac{1}{2} & 1 \end{bmatrix}$ **(b)** $\begin{bmatrix} \frac{8}{7} & \frac{2}{7} \\ \frac{4}{7} & \frac{8}{7} \end{bmatrix}$ **(c)** $\begin{bmatrix} 2800 \\ 2800 \end{bmatrix}$

85. Agriculture \$140,909; manufacturing \$95,455

87. (a) .4 unit agriculture; .09 unit construction; .4 unit energy; .1 unit manufacturing; .9 unit transportation **(b)** 2000 units of agriculture; 600 units of construction; 1700 units of energy; 3700 units of manufacturing; 2500 units of transportation **(c)** 29,049 units of agriculture; 9869 units of construction; 52,362 units of energy; 61,520 units of manufacturing; 90,987 units of transportation

89. (a) $\begin{bmatrix} 54 \\ 32 \end{bmatrix}, \begin{bmatrix} 134 \\ 89 \end{bmatrix}, \begin{bmatrix} 172 \\ 113 \end{bmatrix}, \begin{bmatrix} 118 \\ 74 \end{bmatrix}, \begin{bmatrix} 208 \\ 131 \end{bmatrix}$ **(b)** $\begin{bmatrix} 2 & -3 \\ -\frac{1}{2} & 1 \end{bmatrix}$

Chapter 6 Review (Page 380)

1. $(9, -14)$ **3.** $(2, -2)$ **5.** 8000 standard clips; 6000 extra-large clips **7.** \$7000 in the first fund; \$11,000 in the second fund **9.** $(2 - z, 5z - 2, z)$ for any real number z; dependent **11.** Inconsistent; no solution **13.** $(1, -2, 3)$ **15.** $(4, 3, 2)$ **17.** 19 ones; 7 fives; 9 tens **19.** 5 blankets; 3 rugs; 8 skirts **21.** 2×2; square **23.** 1×4; row **25.** 2×3

27. $\begin{bmatrix} 8 & 8 & 8 \\ 10 & 5 & 9 \\ 7 & 10 & 7 \\ 8 & 9 & 7 \end{bmatrix}$ **29.** $\begin{bmatrix} -1 & -2 & 3 \\ -2 & -3 & 0 \\ 0 & -1 & -4 \end{bmatrix}$ **31.** $\begin{bmatrix} 2 & 18 \\ -4 & -12 \\ 7 & 13 \end{bmatrix}$

33. Not defined

35. Next day; Tow-day total

$\begin{bmatrix} 2310 & -\frac{1}{4} \\ 1258 & -\frac{1}{4} \\ 5061 & \frac{1}{2} \\ 1812 & \frac{1}{2} \end{bmatrix}; \begin{bmatrix} 4842 & -\frac{1}{2} \\ 2722 & -\frac{1}{8} \\ 10{,}035 & -1 \\ 3566 & 1 \end{bmatrix}$

37. $\begin{bmatrix} 14 & 56 \\ -6 & -22 \\ 19 & 79 \end{bmatrix}$ **39.** $\begin{bmatrix} 28 & 39 \\ 35 & 44 \end{bmatrix}$ **41.** $\begin{bmatrix} 322 & 420 \\ -126 & -166 \\ 455 & 591 \end{bmatrix}$

43. (a) $\begin{bmatrix} \frac{1}{4} & \frac{1}{2} \\ \frac{1}{3} & \frac{1}{3} \end{bmatrix}$ **(b)** Cutting 34 hours; shaping 46 hours

45. Many correct answers, including $\begin{bmatrix} 1 & 2 \\ 3 & 4 \end{bmatrix}$

Case 6 (Page 385)

1. Boston, Hyannis, Martha's Vineyard, Nantucket, New Bedford, and Providence may be reached by a two-flight sequence from New Bedford; all Cape Air cities may be reached by a three-flight sequence.
3. The connection between Provincetown and Providence and the connection between Provincetown and New Bedford take three flights.
5. All Big Sky cities may be reached by a three-flight sequence from Helena. At least three flights must be used to get from Helena to Havre, Glendive, and Bismarck.

CHAPTER 7

Section 7.1 (Page 395)

1. F **3.** A **5.** E

7. **9.**

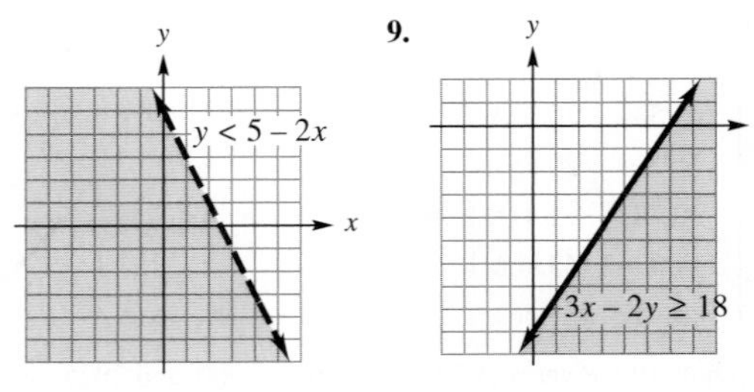

11.

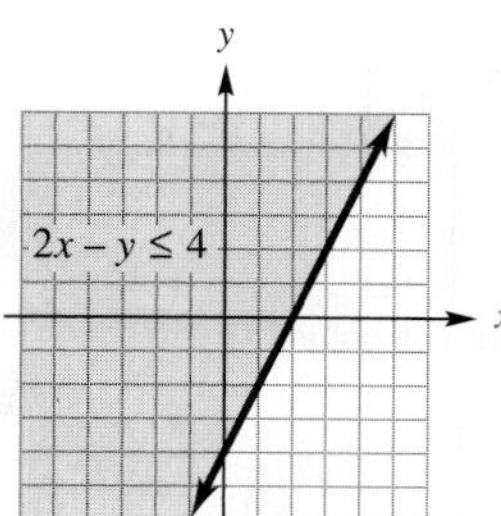

13.

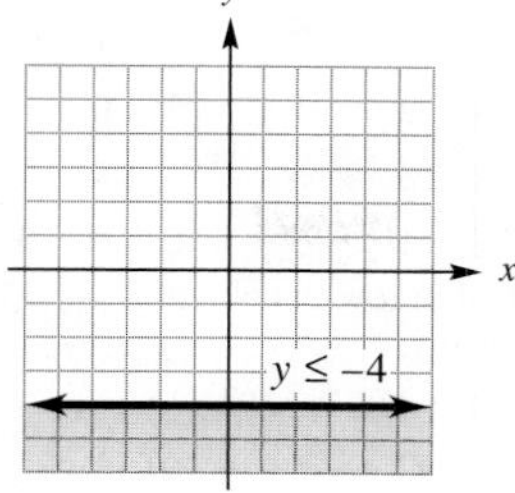

15.

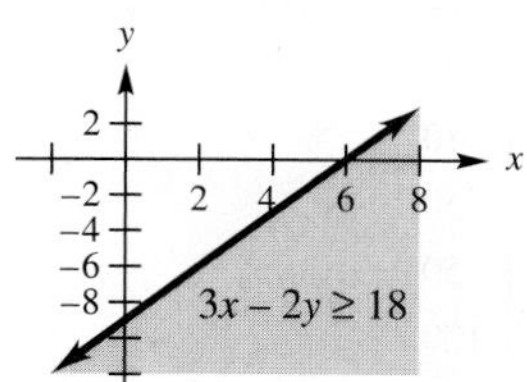

17.

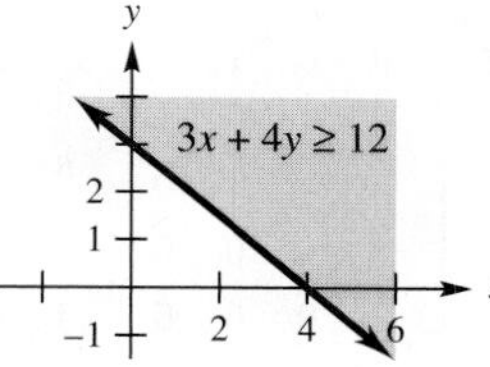

19.

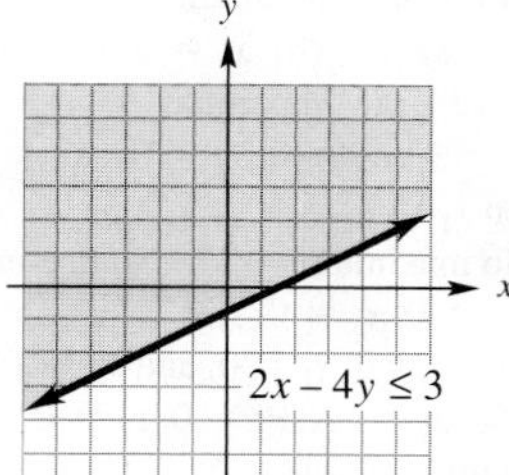

21.

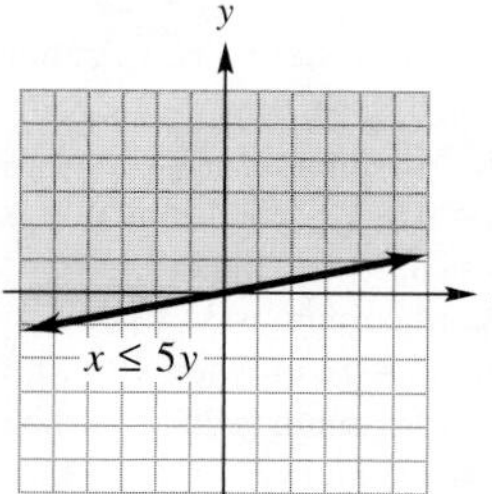

23.

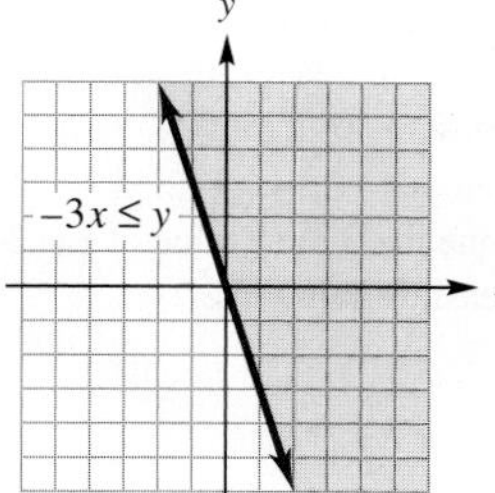

25.

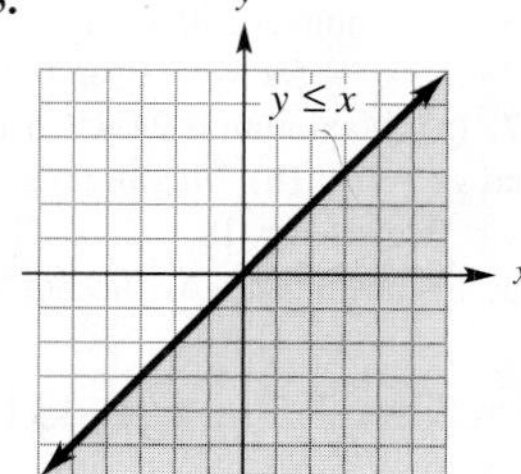

27. Answers vary.

29.

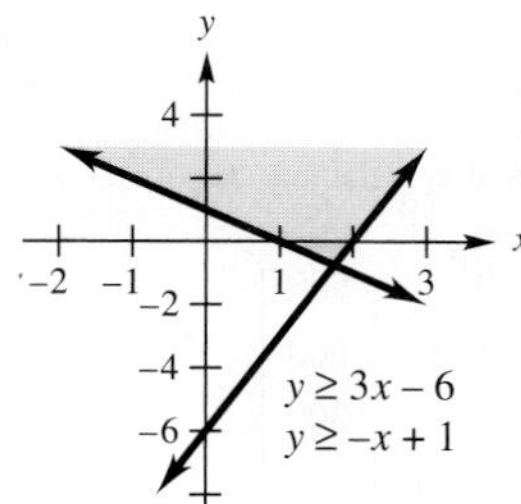

31.

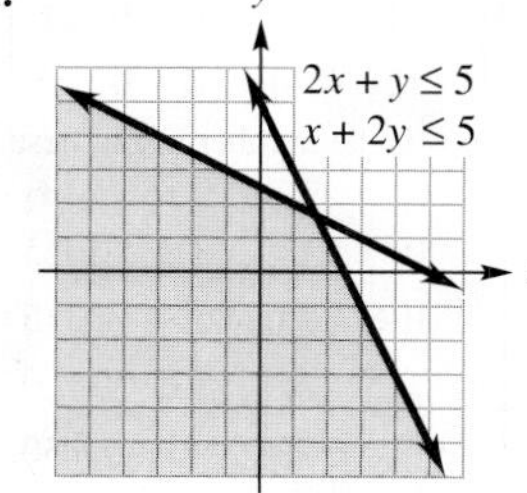

33.

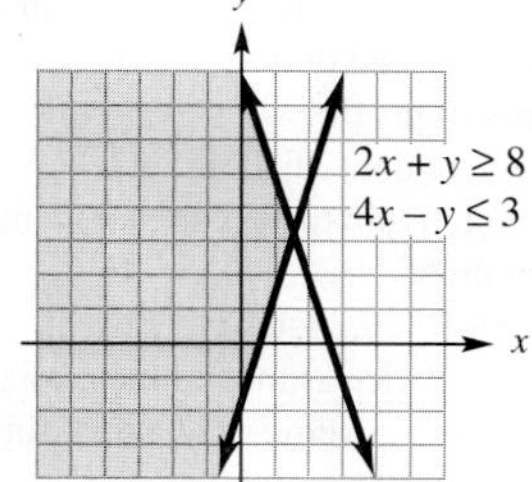

35.

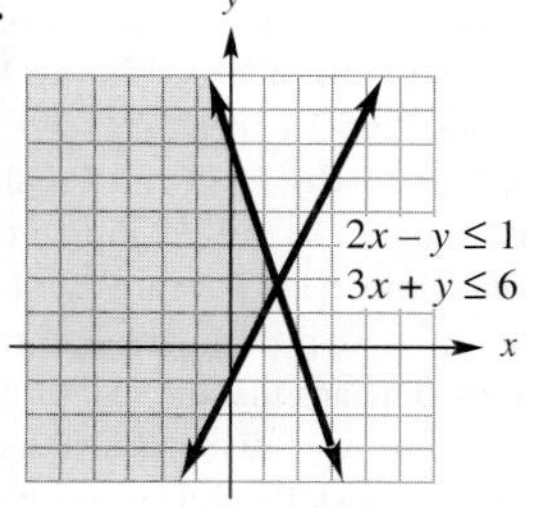

37.

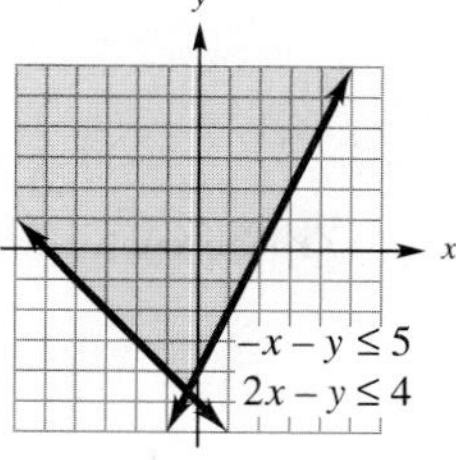

39.

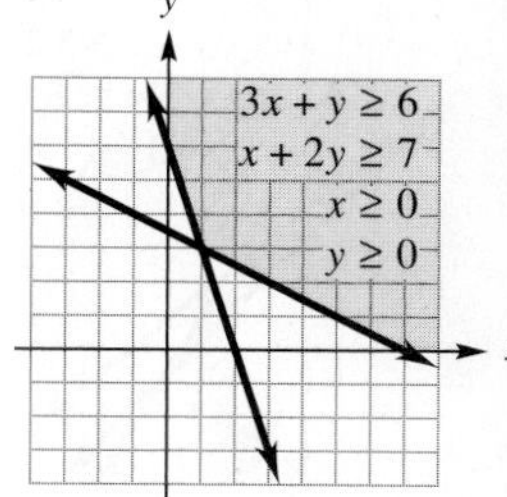

41.

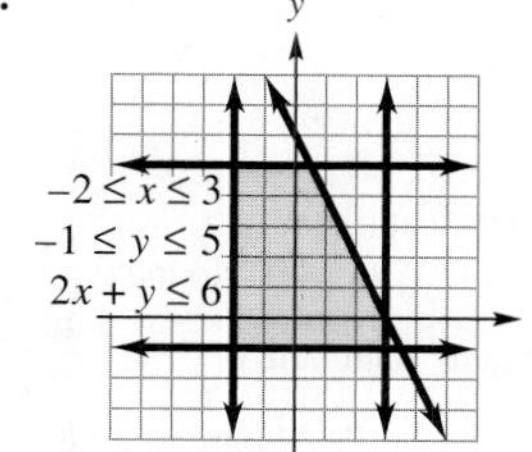

43.

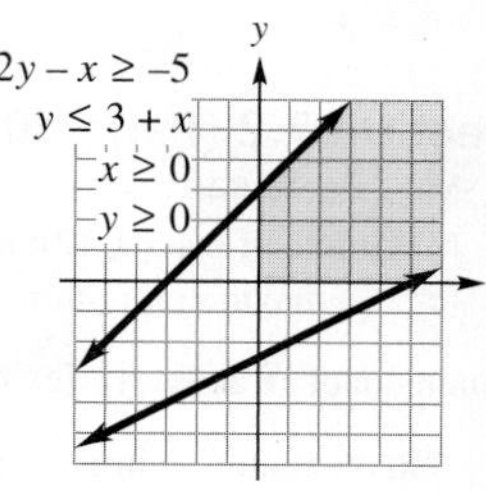

45.

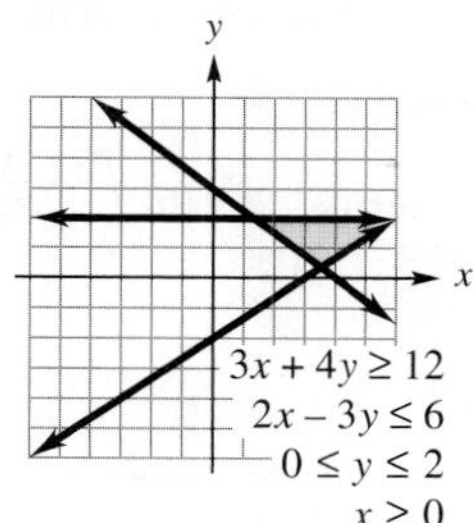

47. $x \geq 0$
$0 \leq y \leq 4$
$4x + 3y \leq 24$

49. $2 < x < 7$
$-1 < y < 3$

51. **(a)**

Number		*Hours Spinning*	*Hours Dyeing*	*Hours Weaving*
Shawls	x	1	1	1
Afghans	y	2	1	4
Maximum Number of Hours Available		8	6	14

(b) $x + 2y \leq 8$; $x + y \leq 6$; $x + 4y \leq 14$; $x \geq 0$; $y \geq 0$

(c)

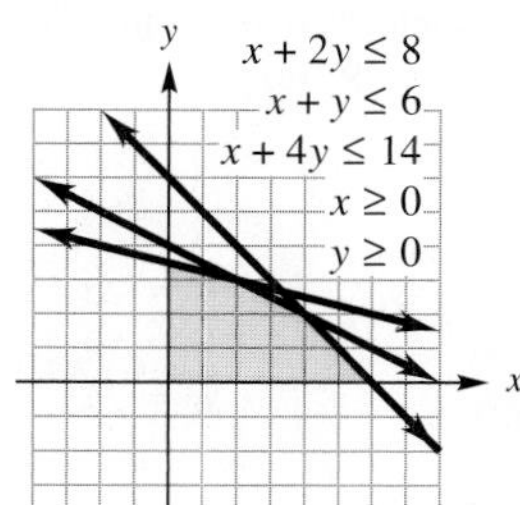

53.

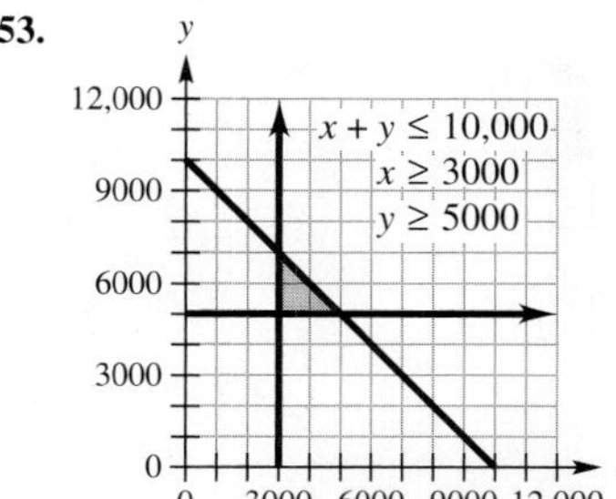

55.

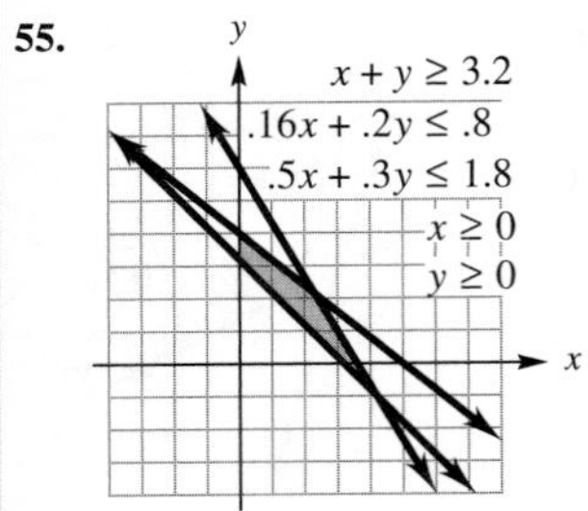

Section 7.2 (Page 403)

1. Maximum of 40 at (5, 10); minimum of 7 at (1, 1) **3.** Maximum of 6 at (0, 12); minimum of 0 at (0, 0) **5. (a)** No maximum; minimum of 12 at (12, 0) **(b)** No maximum; minimum of 18 at (3, 4) **(c)** No maximum; minimum of 21 at $\left(\frac{13}{2}, 2\right)$ **(d)** No maximum; minimum of 8 at (0, 8) **7.** Maximum of 8.4 at (1.2, 1.2) **9.** Minimum of 13 at (5, 3) **11.** Maximum of 68.75 at $\left(\frac{105}{8}, \frac{25}{8}\right)$ **13.** No maximum; minimum of 9 **15.** Maximum of 22; no minimum **17. (a)** (18, 2) **(b)** $\left(\frac{12}{5}, \frac{39}{5}\right)$ **(c)** No maximum **19.** Answers vary.

Section 7.3 (Page 409)

1. $8x + 5y \le 110$, $x \ge 0$, and $y \ge 0$, where x is the number of canoes and y is the number of rowboats **3.** $250x + 750y \le 9500$, $x \ge 0$, and $y \ge 0$, where x is the number of radio spots and y is the number of TV ads **5.** 12 chain saws; no chippers **7.** 10 lb deluxe; 40 lb regular **9.** 10 canoes; 6 rowboats **11.** 8 radio spots; 10 TV ads **13. (a)** 6 units of policy A and 16 units of policy B, for a minimum cost of \$940 **(b)** 30 units of policy A and 0 units of policy B, for a minimum cost of \$750 **15.** 3 Brand X and 2 Brand Z, for a minimum cost of \$1.05 **17.** 800 type 1 and 1600 type 2, for a maximum revenue of \$272 **19.** 0 plants and 18 animals, for a minimum of 270 hr **21.** From warehouse I, ship 60 boxes to San Jose and 250 boxes to Memphis, and from warehouse II, ship 290 boxes to San Jose and none to Memphis, for a minimum cost of \$136.70 **23.** \$20 million in bonds and \$24 million in mutual funds, for maximum interest of \$2.24 million **25.** 8 humanities, 12 science **27.** (b) **29.** (c)

Section 7.4 (Page 424)

1. (a) 3 **(b)** s_1, s_2, s_3 **(c)**
$$\begin{aligned} 4x_1 + 2x_2 + s_1 \quad\quad\quad &= 20 \\ 5x_1 + x_2 \quad + s_2 \quad &= 50 \\ 2x_1 + 3x_2 \quad\quad + s_3 &= 25 \end{aligned}$$

3. (a) 3 **(b)** s_1, s_2, s_3
(c)
$$\begin{aligned} 3x_1 - x_2 + 4x_3 + s_1 \quad\quad\quad &= 95 \\ 7x_1 + 6x_2 + 8x_3 \quad + s_2 \quad &= 118 \\ 4x_1 + 5x_2 + 10x_3 \quad\quad + s_3 &= 220 \end{aligned}$$

5.
$$\begin{array}{cccccc|c} x_1 & x_2 & s_1 & s_2 & s_3 & z & \\ 2 & 5 & 1 & 0 & 0 & 0 & 6 \\ 4 & 1 & 0 & 1 & 0 & 0 & 6 \\ 5 & 3 & 0 & 0 & 1 & 0 & 15 \\ -5 & -1 & 0 & 0 & 0 & 1 & 0 \end{array}$$

7.
$$\begin{array}{ccccccc|c} x_1 & x_2 & x_3 & s_1 & s_2 & s_3 & z & \\ 1 & 2 & 3 & 1 & 0 & 0 & 0 & 10 \\ 2 & 1 & 1 & 0 & 1 & 0 & 0 & 8 \\ 3 & 0 & 4 & 0 & 0 & 1 & 0 & 6 \\ -1 & -5 & -10 & 0 & 0 & 0 & 1 & 0 \end{array}$$

9. 4 in row 2, column 2 **11.** 6 in row 3, column 1

13.
$$\begin{array}{cccccc|c} x_1 & x_2 & x_3 & s_1 & s_2 & z & \\ -1 & 0 & 3 & 1 & -1 & 0 & 16 \\ 1 & 1 & \frac{1}{2} & 0 & \frac{1}{2} & 0 & 20 \\ 2 & 0 & -\frac{1}{2} & 0 & \frac{3}{2} & 1 & 60 \end{array}$$

15.
$$\begin{array}{ccccccc|c} x_1 & x_2 & x_3 & s_1 & s_2 & s_3 & z & \\ -\frac{1}{2} & \frac{1}{2} & 0 & 1 & -\frac{1}{2} & 0 & 0 & 10 \\ \frac{3}{2} & \frac{1}{2} & 1 & 0 & \frac{1}{2} & 0 & 0 & 50 \\ -\frac{7}{2} & \frac{1}{2} & 0 & 0 & -\frac{3}{2} & 1 & 0 & 50 \\ 2 & 0 & 0 & 0 & 1 & 0 & 1 & 100 \end{array}$$

17. (a) Basic: x_3, s_2, z; nonbasic: x_1, x_2, s_1 **(b)** $x_1 = 0, x_2 = 0, x_3 = 16, s_1 = 0, s_2 = 29, z = 11$ **(c)** Not a maximum **19. (a)** Basic: x_1, x_2, s_2, z; nonbasic: x_3, s_1, s_3 **(b)** $x_1 = 6, x_2 = 13, x_3 = 0, s_1 = 0, s_2 = 21, s_3 = 0, z = 18$ **(c)** Maximum **21.** Maximum is 30 when $x_1 = 0, x_2 = 10, s_1 = 0, s_2 = 0$, and $s_3 = 16$. **23.** Maximum is 8 when $x_1 = 4, x_2 = 0, s_1 = 8, s_2 = 2$, and $s_3 = 0$. **25.** No maximum **27.** No maximum **29.** Maximum is 34 when $x_1 = 17, x_2 = 0, x_3 = 0, s_1 = 0$, and $s_2 = 14$ or when $x_1 = 0, x_2 = 17, x_3 = 0, s_1 = 0$, and $s_2 = 14$. **31.** Maximum is 26,000 when $x_1 = 60, x_2 = 40, x_3 = 0, s_1 = 0, s_2 = 80$, and $s_3 = 0$. **33.** Maximum is 64 when $x_1 = 28, x_2 = 16, x_3 = 0, s_1 = 0, s_2 = 28$, and $s_3 = 0$. **35.** Maximum is 250 when $x_1 = 0, x_2 = 0, x_3 = 0, x_4 = 50, s_1 = 0$, and $s_2 = 50$. **37. (a)** Maximum is 24 when $x_1 = 12, x_2 = 0, x_3 = 0, s_1 = 0$, and $s_2 = 6$. **(b)** Maximum is 24 when $x_1 = 0, x_2 = 12, x_3 = 0, s_1 = 0$, and $s_2 = 18$. **(c)** The unique maximum value of z is 24, but this occurs at two different basic feasible solutions.

Section 7.5 (page 431)

1.
$$\begin{array}{ccccc|c} x_1 & x_2 & s_1 & s_2 & s_3 & \\ 2 & 1 & 1 & 0 & 0 & 90 \\ 1 & 2 & 0 & 1 & 0 & 80 \\ 1 & 1 & 0 & 0 & 1 & 50 \\ -12 & -10 & 0 & 0 & 0 & 0 \end{array},$$
where x_1 is the number of Siamese cats and x_2 is the number of Persian cats.

3.
$$\begin{array}{ccccccc|c} x_1 & x_2 & x_3 & x_4 & s_1 & s_2 & s_3 & \\ 0 & 0 & .375 & .625 & 1 & 0 & 0 & 500 \\ 0 & .75 & .5 & .375 & 0 & 1 & 0 & 600 \\ 1 & .25 & .125 & 0 & 0 & 0 & 1 & 300 \\ -90 & -70 & -60 & -50 & 0 & 0 & 0 & 0 \end{array},$$
where x_1 is the number of kilograms of P, x_2 is the number of kilograms of Q, x_3 is the number of kilograms of R, and x_4 is the number of kilograms of S. **5. (a)** Make no 1-speed or 3-speed bicycles; make 2295 10-speed bicycles; maximum profit is \$55,080 **(b)** 5280 units of aluminum are unused; all the steel is used. **7. (a)** 12 min to the senator, 9 min to the congresswoman, and 6 min to the governor, for a maximum of 1,320,000 viewers **(b)** $s_1 = 0$ means that all of the 27 available minutes were used; $s_2 = 0$ means that the senator had *exactly* twice as much time as the governor; $s_3 = 0$ means that the senator and governor had a total time *exactly* twice the time of the congresswoman.
9. (a) 300 Flexscan and 300 Panoramic I sets, for a maximum profit of \$255,000 **(b)** There are 200 unused hours in the testing and packing

department. **11.** 4 radio ads, 6 TV ads, and no newspaper ads, for a maximum exposure of 64,800 people **13. (a)** 22 fund-raising parties, no mailings, and 3 dinner parties, for a maximum of \$6,200,000 **(b)** Answers vary. **15.** 3 hours running, 4 hours biking, and 8 hours walking, for a maximum calorie expenditure of 6313 calories **17. (a)** (3) **(b)** (4) **(c)** (3) **19.** The breeder should raise 40 Siamese and 10 Persian cats, for a maximum gross income of 580. **21.** 163.6 kilograms of food P, none of Q, 1090.9 kilograms of R, 145.5 kilograms of S; maximum is 87,454.5

Section 7.6 (Page 443)

1. $\begin{bmatrix} 3 & 1 & 0 \\ -4 & 10 & 3 \\ 5 & 7 & 6 \end{bmatrix}$ **3.** $\begin{bmatrix} 3 & 4 \\ 0 & 17 \\ 14 & 8 \\ -5 & -6 \\ 3 & 1 \end{bmatrix}$

5. Maximize $z = 4x_1 + 6x_2$
subject to
$$\begin{aligned} 3x_1 - x_2 &\le 3 \\ x_1 + 2x_2 &\le 5 \\ x_1 \ge 0, x_2 &\ge 0. \end{aligned}$$

7. Maximize $z = 18x_1 + 15x_2 + 20x_3$
subject to
$$\begin{aligned} x_1 + 4x_2 + 5x_3 &\le 2 \\ 7x_1 + x_2 + 3x_3 &\le 8 \\ x_1 \ge 0, x_2 \ge 0, x_3 &\ge 0. \end{aligned}$$

9. Maximize $z = 18x_1 + 20x_2$
subject to
$$\begin{aligned} 7x_1 + 4x_2 &\le 5 \\ 6x_1 + 5x_2 &\le 1 \\ 8x_1 + 10x_2 &\le 3 \\ x_1 \ge 0, x_2 &\ge 0. \end{aligned}$$

11. Maximize $z = 5x_1 + 4x_2 + 15x_3$
subject to
$$\begin{aligned} x_1 + x_2 + 2x_3 &\le 8 \\ x_1 + x_2 + x_3 &\le 9 \\ x_1 + 3x_3 &\le 3 \\ x_1 \ge 0, x_3 \ge 0, x_3 &\ge 0. \end{aligned}$$

13. $y_1 = 0, y_2 = 4, y_3 = 0$; minimum is 4. **15.** $y_1 = 4, y_2 = 0, y_3 = \frac{7}{3}$; minimum is 39. **17.** $y_1 = 0, y_2 = 100, y_3 = 0$; minimum is 100. **19.** $y_1 = 0, y_2 = 12, y_3 = 0$; minimum is 12. **21.** $y_1 = 0, y_2 = 0, y_3 = 2$; minimum is 4. **23.** $y_1 = 10, y_2 = 10, y_3 = 0$; minimum is 320. **25.** $y_1 = 0, y_2 = 0, y_3 = 4$; minimum is 4. **27.** 4 servings of A and 2 servings of B, for a minimum cost of \$1.76 **29.** 28 units of regular beer and 14 units of light beer, for a minimum cost of \$1,680,000 **31.** (a)
33. (a) Minimize $w = 200y_1 + 600y_2 + 90y_3$
subject to
$$\begin{aligned} y_1 + 4y_2 &\ge 1 \\ 2y_1 + 3y_2 + y_3 &\ge 1.5 \\ y_1 \ge 0, y_2 \ge 0, y_3 &\ge 0. \end{aligned}$$
(b) $y_1 = .6, y_2 = .1, y_3 = 0, w = 180$ **(c)** \$186 **(d)** \$179

Section 7.7 (Page 455)

1. (a) Maximize $z = -5x_1 + 4x_2 - 2x_3$
subject to
$$\begin{aligned} -2x_2 + 5x_3 - s_1 &= 8 \\ 4x_1 - x_2 + 3x_3 + s_2 &= 12 \\ x_1 \ge 0, x_2 \ge 0, x_3 \ge 0, s_1 \ge 0, s_2 &\ge 0. \end{aligned}$$

(b)
$$\begin{array}{ccccc|c} x_1 & x_2 & x_3 & s_1 & s_2 & \\ \hline 0 & -2 & 5 & -1 & 0 & 8 \\ 4 & -1 & 3 & 0 & 1 & 12 \\ 5 & -4 & 2 & 0 & 0 & 0 \end{array}$$

3. (a) Maximize $z = 2x_1 - 3x_2 + 4x_3$
subject to
$$\begin{aligned} x_1 + x_2 + x_3 + s_1 &= 100 \\ x_1 + x_2 + x_3 - s_2 &= 75 \\ x_1 + x_2 - s_3 &= 27 \\ x_1 \ge 0, x_2 \ge 0, x_3 \ge 0, s_1 \ge 0, s_2 \ge 0, s_3 &\ge 0. \end{aligned}$$

(b)
$$\begin{array}{cccccc|c} x_1 & x_2 & x_3 & s_1 & s_2 & s_3 & \\ \hline 1 & 1 & 1 & 1 & 0 & 0 & 100 \\ 1 & 1 & 1 & 0 & -1 & 0 & 75 \\ 1 & 1 & 0 & 0 & 0 & -1 & 27 \\ -2 & 3 & -4 & 0 & 0 & 0 & 0 \end{array}$$

5. Maximize $z = -2y_1 - 5y_2 + 3y_3$
subject to
$$\begin{aligned} y_1 + 2y_2 + 3y_3 &\ge 115 \\ 2y_1 + y_2 + y_3 &\le 200 \\ y_1 + y_3 &\ge 50 \\ y_1 \ge 0, y_2 \ge 0, y_3 &\ge 0. \end{aligned}$$

$$\begin{array}{cccccc|c} y_1 & y_2 & y_3 & s_1 & s_2 & s_3 & \\ \hline 1 & 2 & 3 & -1 & 0 & 0 & 115 \\ 2 & 1 & 1 & 0 & 1 & 0 & 200 \\ 1 & 0 & 1 & 0 & 0 & -1 & 50 \\ 2 & 5 & -3 & 0 & 0 & 0 & 0 \end{array}$$

7. Maximize $z = -10y_1 - 8y_2 - 15y_3$
subject to
$$\begin{aligned} y_1 + y_2 + y_3 &\ge 12 \\ 5y_1 + 4y_2 + 9y_3 &\ge 48 \\ y_1 \ge 0, y_2 \ge 0, y_3 &\ge 0. \end{aligned}$$

$$\begin{array}{ccccc|c} y_1 & y_2 & y_3 & s_1 & s_2 & \\ \hline 1 & 1 & 1 & -1 & 0 & 12 \\ 5 & 4 & 9 & 0 & -1 & 48 \\ 10 & 8 & 15 & 0 & 0 & 0 \end{array}$$

9. Maximum is 480 when $x_1 = 40$ and $x_2 = 0$.
11. Maximum is 114 when $x_1 = 38, x_2 = 0$, and $x_3 = 0$.
13. Maximum is 90 when $x_1 = 12$ and $x_2 = 3$ or when $x_1 = 0$ and $x_2 = 9$. **15.** Minimum is 40 when $y_1 = 0$ and $y_2 = 20$.
17. Maximum is $133\frac{1}{3}$ when $x_1 = 33\frac{1}{3}$ and $x_2 = 16\frac{2}{3}$.
19. Minimum is 512 when $y_1 = 6$ and $y_2 = 8$.
21. Maximum is 112 when $x_1 = 0, x_2 = 36$, and $x_3 = 16$.
23. Maximum is 346 when $x_1 = 27, x_2 = 0$, and $x_3 = 73$.
25. Minimum is -600 when $y_1 = 0, y_2 = 0$, and $y_3 = 200$.
27. Minimum is 96 when $y_1 = 0, y_2 = 12$, and $y_3 = 0$.
29.
$$\begin{array}{ccccccc|c} y_1 & y_2 & y_3 & s_1 & s_2 & s_3 & s_4 & \\ \hline 1 & 1 & 1 & -1 & 0 & 0 & 0 & 10 \\ 1 & 1 & 1 & 0 & 1 & 0 & 0 & 15 \\ 1 & -\frac{1}{4} & 0 & 0 & 0 & -1 & 0 & 0 \\ -1 & 0 & 1 & 0 & 0 & 0 & -1 & 0 \\ .30 & .09 & .27 & 0 & 0 & 0 & 0 & 0 \end{array}$$

31.
$$\begin{array}{cccccccccc|c} y_1 & y_2 & y_3 & y_4 & s_1 & s_2 & s_3 & s_4 & s_5 & s_6 & \\ \hline 1 & 1 & 0 & 0 & -1 & 0 & 0 & 0 & 0 & 0 & 32 \\ 1 & 1 & 0 & 0 & 0 & 1 & 0 & 0 & 0 & 0 & 32 \\ 0 & 0 & 1 & 1 & 0 & 0 & -1 & 0 & 0 & 0 & 20 \\ 0 & 0 & 1 & 1 & 0 & 0 & 0 & 1 & 0 & 0 & 20 \\ 1 & 0 & 1 & 0 & 0 & 0 & 0 & 0 & 1 & 0 & 25 \\ 0 & 1 & 0 & 1 & 0 & 0 & 0 & 0 & 0 & 1 & 30 \\ 14 & 12 & 22 & 10 & 0 & 0 & 0 & 0 & 0 & 0 & 0 \end{array}$$

33. Ship 200 barrels of oil from supplier S_1 to distributor D_1; ship 2800 barrels of oil from supplier S_2 to distributor D_1; ship 2800 barrels of oil from supplier S_1 to distributor D_2; ship 2200 barrels of oil from supplier S_2 to distributor D_2. Minimum cost is \$180,400.
35. Use 1000 lb of bluegrass, 2400 lb of rye, and 1600 lb of Bermuda, for a minimum cost of \$560. **37.** Allot \$3,000,000 in commercial loans and \$22,000,000 in home loans, for a maximum return of \$2,940,000 **39.** Make 32 units of regular beer and 10 units of light beer, for a minimum cost of \$1,632,000.
41. $1\frac{2}{3}$ ounces of ingredient I, $6\frac{2}{3}$ ounces of ingredient II, and $1\frac{2}{3}$ ounces of ingredient III produce a minimum cost of \$1.55 per barrel.

43. 22 from W_1 to D_1, 10 from W_2 to D_1, none from W_1 to D_2, and 20 from W_2 to D_2, for a minimum cost of \$628.

Chapter 7 Review (Page 459)

1.

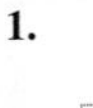

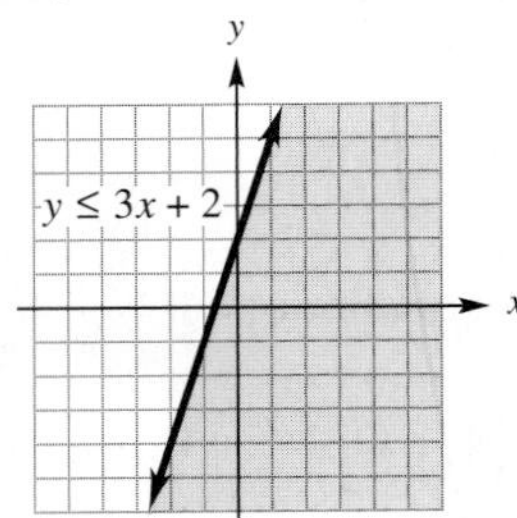

3.

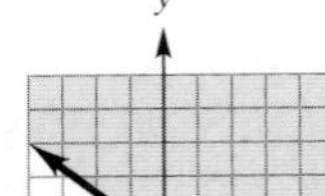

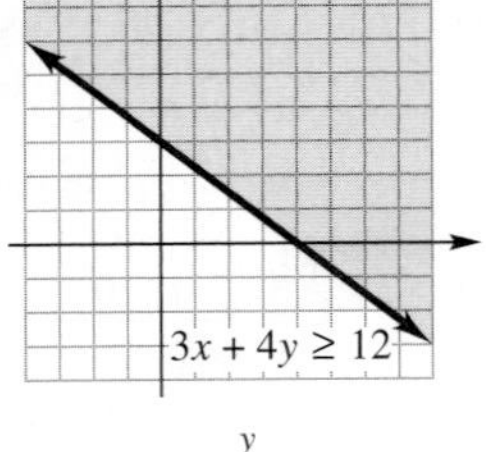

5.

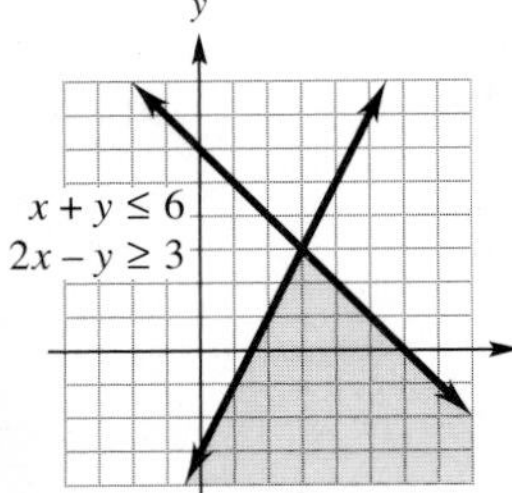

7.

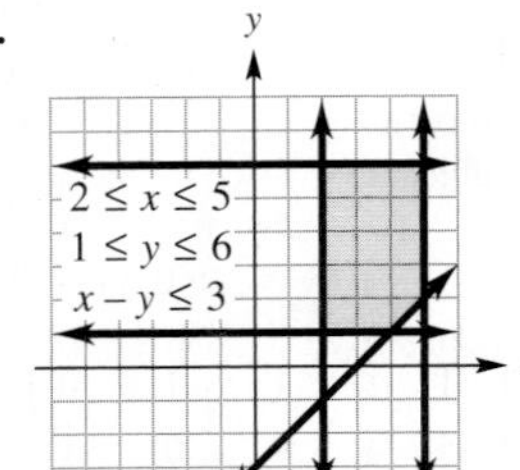

9. Let x represent the number of batches of cakes, and let y represent the number of batches of cookies. Then

$$\begin{aligned} 2x + \left(\frac{3}{2}\right)y &\le 15 \\ 3x + \left(\frac{2}{3}\right)y &\le 13 \\ x &\ge 0 \\ y &\ge 0. \end{aligned}$$

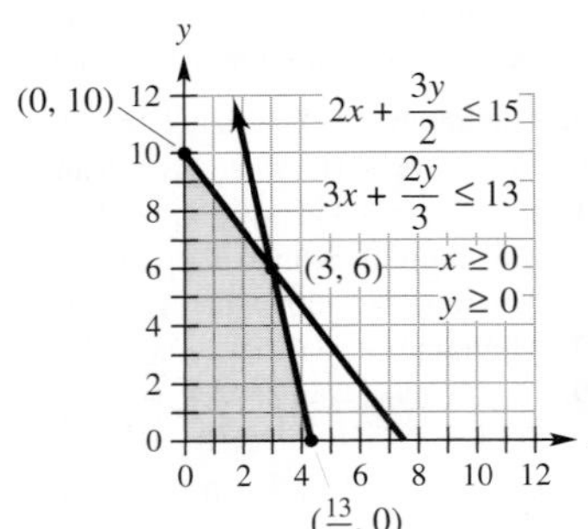

11. Maximum of 46 at (6, 7); minimum of 10 at (2, 1) **13.** Maximum of 30 when $x = 5$ and $y = 0$ **15.** Minimum of 40 when $x = 0$ and $y = 20$ **17.** Make 3 batches of cakes and 6 batches of cookies, for a maximum profit of \$210.

19. (a)
$$\begin{aligned} x_1 + x_2 + x_3 + s_1 \qquad\qquad &= 100 \\ 2x_1 + 3x_2 \qquad + s_2 \qquad &= 500 \\ x_1 \qquad + 2x_3 \qquad\qquad + s_3 &= 350 \end{aligned}$$

(b)
$$\begin{array}{cccccc|c} x_1 & x_2 & x_3 & s_1 & s_2 & s_3 & \\ 1 & 1 & 1 & 1 & 0 & 0 & 100 \\ 2 & 3 & 0 & 0 & 1 & 0 & 500 \\ 1 & 0 & 2 & 0 & 0 & 1 & 350 \\ -5 & -6 & -3 & 0 & 0 & 0 & 0 \end{array}$$

21. (a)
$$\begin{aligned} x_1 + x_2 + x_3 + s_1 \qquad\qquad &= 90 \\ 2x_1 + 5x_2 + x_3 \qquad + s_2 \qquad &= 120 \\ x_1 + 3x_2 \qquad\qquad + s_3 &= 80 \end{aligned}$$

(b)
$$\begin{array}{cccccc|c} x_1 & x_2 & x_3 & s_1 & s_2 & s_3 & \\ 1 & 1 & 1 & 1 & 0 & 0 & 90 \\ 2 & 5 & 1 & 0 & 1 & 0 & 120 \\ 1 & 3 & 0 & 0 & 0 & 1 & 80 \\ -1 & -8 & -2 & 0 & 0 & 0 & 0 \end{array}$$

23. Maximum is 80 when $x_1 = 16, x_2 = 0, x_3 = 0, s_1 = 12$, and $s_2 = 0$. **25.** Maximum is 35 when $x_1 = 5, x_2 = 0, x_3 = 5, s_1 = 35, s_2 = 0$, and $s_3 = 0$.

27. Maximum of 600 when $x_1 = 0, x_2 = 100$, and $x_3 = 0$
29. Maximum of 225 when $x_1 = 0, x_2 = \frac{15}{2}$, and $x_3 = \frac{165}{2}$
31. (a) Let x_1 = number of item A, x_2 = number of item B, and x_3 = number of item C. **(b)** $z = 4x_1 + 3x_2 + 3x_3$
(c)
$$\begin{aligned} 2x_1 + 3x_2 + 6x_3 &\le 1200 \\ x_1 + 2x_2 + 2x_3 &\le 800 \\ 2x_1 + 2x_2 + 4x_3 &\le 500 \\ x_1 \ge 0, x_2 \ge 0, x_3 &\ge 0 \end{aligned}$$

33. (a) Let x_1 = number of gallons of Fruity wine and x_2 = number of gallons of Crystal wine. **(b)** $z = 12x_1 + 15x_2$
(c)
$$\begin{aligned} 2x_1 + x_2 &\le 110 \\ 2x_1 + 3x_2 &\le 125 \\ 2x_1 + x_2 &\le 90 \\ x_1 \ge 0, x_2 &\ge 0 \end{aligned}$$

35. When there are more than 2 variables. **37.** Any standard minimization problem **39.** Minimum of 172 at (5, 7, 3, 0, 0, 0) **41.** Minimum of 640 at (7, 2, 0, 0) **43.** Minimum of 170 when $y_1 = 0$ and $y_2 = 17$

45.
$$\left[\begin{array}{cccc|c} 5 & 10 & 1 & 0 & 120 \\ 10 & 15 & 0 & -1 & 200 \\ -20 & -30 & 0 & 0 & 0 \end{array}\right]$$

47.
$$\left[\begin{array}{ccccccc|c} 1 & 1 & 2 & -1 & 0 & 0 & 0 & 48 \\ 1 & 1 & 0 & 0 & 1 & 0 & 0 & 12 \\ 0 & 0 & 1 & 0 & 0 & -1 & 0 & 10 \\ 3 & 0 & 1 & 0 & 0 & 0 & -1 & 30 \\ 12 & 20 & -8 & 0 & 0 & 0 & 0 & 0 \end{array}\right]$$

49. Minimum of 427 at (8, 17, 25, 0, 0, 0) **51.** Maximum is 480 when $x_1 = 24$ and $x_2 = 0$. **53.** Minimum is 40 when $y_1 = 10$ and $y_2 = 0$. **55.** Get 250 of A and none of B or C for maximum profit of \$1000. **57.** Make 17.5 gal of Crystal and 36.25 gal of Fruity, for a maximum profit of \$697.50. **59.** Produce 660 cases of corn, no beans, and 340 cases of carrots, for a minimum cost of \$15,100. **61.** Use 1,060,000 kilograms for whole tomatoes and 80,000 kilograms for sauce, for a minimum cost of \$4,500,000.

Case 7 (Page 463)

1. The answer in 100-gram units is 0.243037 unit of feta cheese, 2.35749 units of lettuce, 0.3125 unit of salad dressing, and 1.08698 units of tomato. Converting into kitchen units gives approximately $\frac{1}{6}$ cup feta cheese, $4\frac{1}{4}$ cups lettuce, $\frac{1}{8}$ cup salad dressing, and $\frac{7}{8}$ of a tomato.

CHAPTER 8

Section 8.1 (Page 473)

1. False **3.** True **5.** True **7.** True **9.** False **11.** Answers vary. **13.** $\subseteq$ **15.** $\not\subseteq$ **17.** $\subseteq$ **19.** $\subseteq$ **21.** 8 **23.** 128 **25.** $\{x | x$ is an integer ≤ 0 or $\ge 8\}$ **27.** Answers vary. **29.** $\cap$ **31.** $\cap$ **33.** $\cup$ **35.** $\cup$ **37.** {b, 1, 3} **39.** {d, e, f, 4, 5, 6} **41.** {e, 4, 6} **43.** {a, b, c, d, 1, 2, 3, 5} **45.** All students not taking this course. **47.** All students taking both accounting and philosophy. **49.** C and D, A and E, C and E, D and E **51.** {Allstate, Microsoft} **53.** {Allstate, Microsoft} **55.** $M \cap E$ is the set of all male employed applicants. **57.** $M' \cup S'$ is the set of all female or married applicants. **59.** {Internet} **61.** {Comcast Cable Communications, Time Warner Cable} **63.** {Comcast, Time Warner, Cox, Charter, Cablevision} **65.** {Comcast, Time Warner} **67.** $\{s, d, c\}$ **69.** $\{g\}$

Section 8.2 (Page 482)

1.

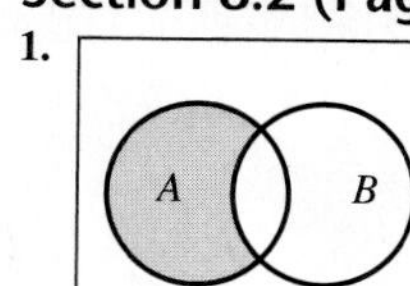

3.

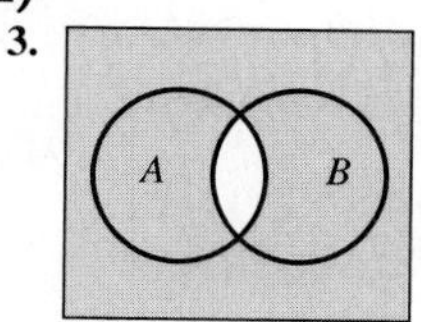

5.

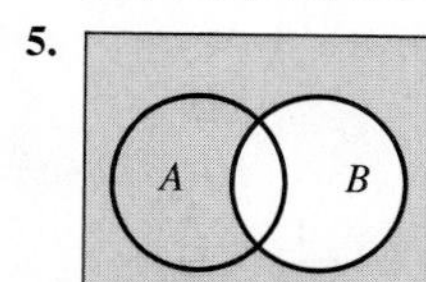

7. ∅ **9.** 8

11.

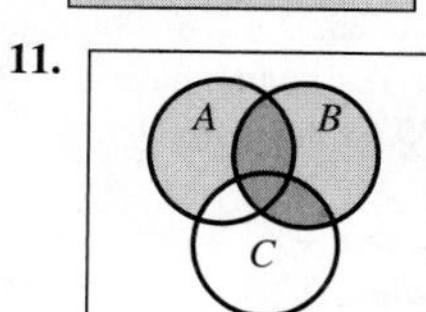

13.

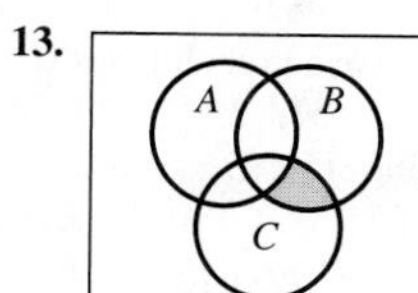

15.

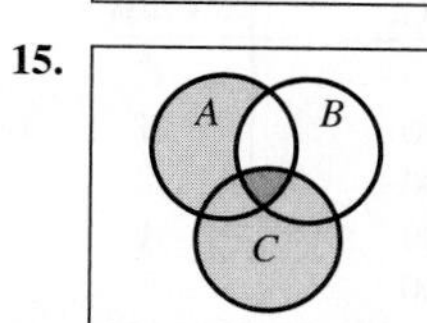

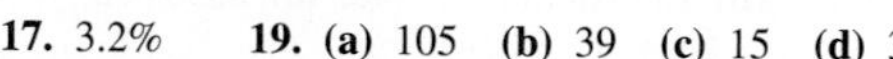

17. 3.2% **19.** **(a)** 105 **(b)** 39 **(c)** 15 **(d)** 3 **21.** **(a)** 227 **(b)** 182 **(c)** 27 **23.** **(a)** 54 **(b)** 17 **(c)** 10 **(d)** 7 **(e)** 15 **(f)** 3 **(g)** 12 **(h)** 1 **25.** **(a)** 1018 **(b)** 732 **(c)** 2567 **(d)** 3058 **27.** **(a)** 945,000 **(b)** 3000 **(c)** 36,000 **(d)** 984,000 **(e)** 179,000 **29.** Answers vary. **31.** 9 **33.** 27

35.

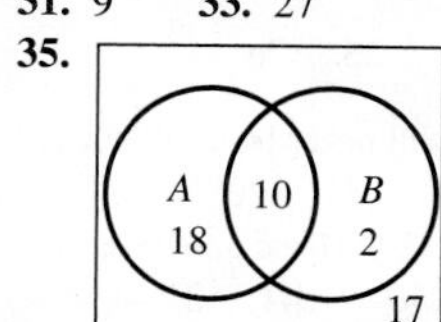

37.

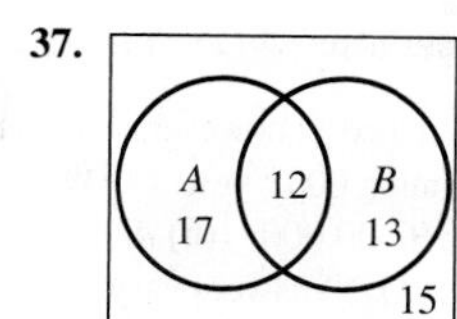

39.

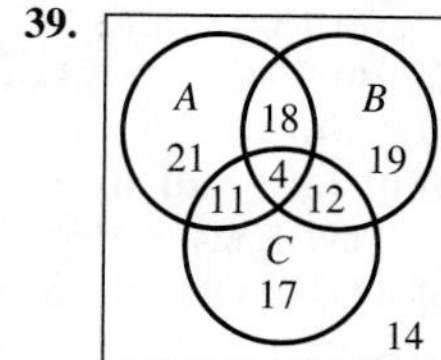

41.

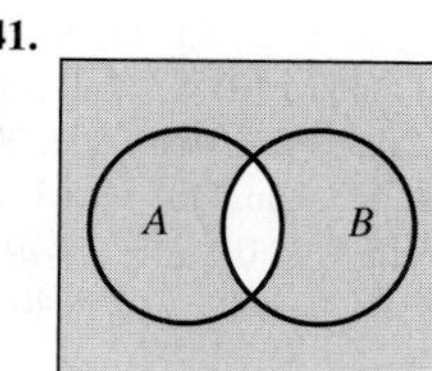

43.

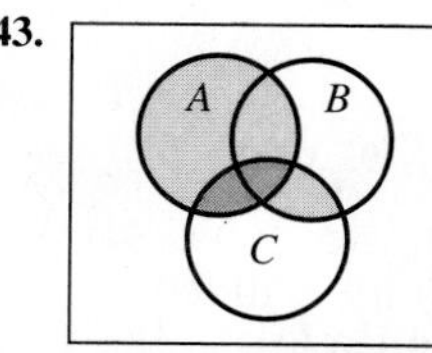

45. The complement of *A* intersect *B* equals the union of the complement of *A* and the complement of *B*. **47.** A union (*B* intersect *C*) equals (*A* union *B*) intersect (*A* union *C*).

Section 8.3 (Page 493)

1. Answers vary. **3.** {January, February, March, . . . , December} **5.** {0, 1, 2, . . . , 80} **7.** {go ahead, cancel} **9.** $\{Q_1, Q_2, Q_3, Q_4\}$ **11.** Answers vary. **13.** No **15.** Yes **17.** No **19.** S = {C&K, C&L, C&J, C&T, C&N, K&L, K&J, K&T, K&N, L&J, L&T, L&N, J&T, J&N, T&N} **(a)** {C&T, K&T, L&T, J&T, T&N} **(b)** {C&J, C&N, J&N}

21. S = {forest sage & rag painting, forest sage & colorwash, evergreen & rag painting, evergreen & colorwash, opaque & rag painting, opaque & colorwash} **(a)** $\frac{1}{2}$ **(b)** $\frac{2}{3}$ **23.** S = {10′ & beige, 10′ & forest green, 10′ & rust, 12′ & beige, 12′ & forest green, 12′ & rust} **(a)** $\frac{1}{6}$ **(b)** $\frac{1}{2}$ **(c)** $\frac{1}{3}$ **25.** $\frac{1}{2}$ **27.** $\frac{1}{3}$ **29.** $\frac{5}{6}$ **31.** $\frac{1}{2}$ **33.** $\frac{1}{8}$ **35.** $\frac{3}{4}$ **37.** $\frac{159}{5471} \approx .029$ **39.** **(a)** $\frac{215}{1168} \approx .184$ **(b)** $\frac{545}{1168} \approx .467$ **(c)** $\frac{302}{1168} \approx .259$ **41.** **(a)** The person smokes or has a family history of heart disease. **(b)** The person does not smoke and has a family history of heart disease. **(c)** The person does not have a family history of heart disease or is not overweight. **43.** Possible **45.** Not possible **47.** Not possible

Section 8.4 (Page 502)

1. $\frac{20}{38} = \frac{10}{19}$ **3.** $\frac{28}{38} = \frac{14}{19}$ **5.** $\frac{5}{38}$ **7.** $\frac{12}{38} = \frac{6}{19}$ **9.** **(a)** $\frac{5}{36}$ **(b)** $\frac{1}{9}$ **(c)** $\frac{1}{12}$ **(d)** 0 **11.** **(a)** $\frac{5}{18}$ **(b)** $\frac{5}{12}$ **(c)** $\frac{1}{3}$ **13.** $\frac{1}{3}$ **15.** **(a)** $\frac{1}{2}$ **(b)** $\frac{2}{5}$ **(c)** $\frac{3}{10}$ **17.** **(a)** .62 **(b)** .32 **(c)** .11 **(d)** .81 **19.** **(a)** .196 **(b)** .08 **(c)** .804 **(d)** .987 **21.** **(a)** .42 **(b)** .893 **(c)** .107 **23.** 1:5 **25.** 2:1 **27.** **(a)** 1:4 **(b)** 8:7 **(c)** 4:11 **29.** 2:7 **31.** 21:79 **33.** 7:93 **35.** $\frac{2}{7}$ **37.** $\frac{1}{10}$ **39.** 1:9 **41.** **(a)** .2778 **(b)** .4167 **43.** **(a)** .15625 **(b)** .3125 **45.** .300 **47.** .966 **49.** .595 **51.** .808 **53.** .377 **55.** .833 **57.** .841 **59.** .453 **61.** **(a)** .961 **(b)** .491 **(c)** .509 **(d)** .505 **(e)** .004 **(f)** .544 **63.** **(a)** $\frac{1}{4}$ **(b)** $\frac{1}{2}$ **(c)** $\frac{1}{4}$ **65.** **(a)** .828 **(b)** .339 **(c)** .083 **67.** .020 **69.** .557

Section 8.5 (Page 517)

1. $\frac{1}{3}$ **3.** 1 **5.** $\frac{1}{6}$ **7.** $\frac{4}{17}$ **9.** .012 **11.** Answers vary. **13.** Answers vary. **15.** No **17.** Yes **19.** No, yes **21.** $\frac{3}{7} = .43$ **23.** .271 **25.** **(a)** .158 **(b)** .001 **(c)** .008 **(d)** .440 **27.** .374 **29.** .522 **31.** .30 **33.** Answers vary. **35.** .049 **37.** .534 **39.** .143 **41.** Dependent **43.** .05 **45.** .25 **47.** .933 **49.** .245 **51.** .493 **53.** .159 **55.** .0457 **57.** .999985 **59.** **(a)** 3 backups **(b)** Answers vary. **61.** Dependent or "No"

Section 8.6 (Page 524)

1. $\frac{1}{3}$ **3.** .0488 **5.** .5122 **7.** .4706 **9.** .571 **11.** .259 **13.** .372 **15.** .143 **17.** .478 **19.** (c) **21.** .481 **23.** .919 **25.** .367 **27.** **(a)** .4738 **(b)** .1665 **29.** .011 **31.** .354 **33.** .871 **35.** .713

Chapter 8 Review (Page 527)

1. False **3.** False **5.** True **7.** True **9.** {New Year's Day, Martin Luther King Jr.'s Birthday, Presidents' Day, Memorial Day, Independence Day, Labor Day, Columbus Day, Veterans' Day,

Thanksgiving, Christmas} **11.** {1, 2, 3, 4} **13.** $\{B_1, B_2, B_3, B_6, B_{12}\}$ **15.** $\{A, C, E\}$ **17.** $\{A, B_3, B_6, B_{12}, C, D, E\}$ **19.** Female students older than 22 **21.** Females or students with a GPA > 3.5 **23.** Non-finance majors who are 22 or younger

25.

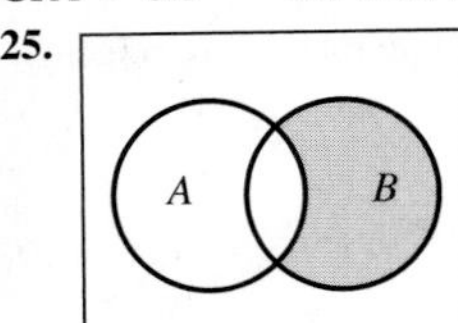

27.

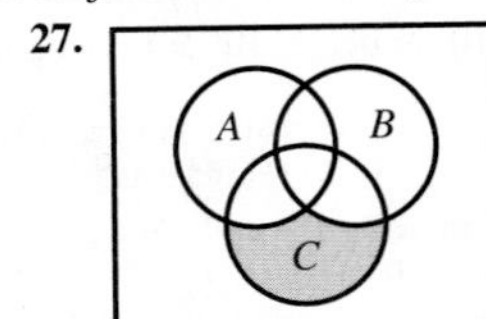

29. 100 **31.** {1, 2, 3, 4, 5, 6} **33.** {(red, 10), (red, 20), (red, 30), (blue, 10), (blue, 20), (blue, 30), (green, 10), (green, 20), (green, 30)} **35.** No. Not equal probabilities for each genre. **37.** No **39.** $E \cup F$ **41.** Answers vary. **43.** Answers vary. **45.** .1910 **47.** .014 **49.** .196 **51.** .130 **53.** .276

55.

	N_2	T_2
N_1	N_1N_2	N_1T_2
T_1	T_1N_2	T_1T_2

57. $\frac{1}{2}$ **59.** $\frac{5}{36} \approx .139$ **61.** $\frac{5}{18} \approx .278$ **63.** $\frac{1}{3}$ **65.** .79 **67.** .66 **69.** (a) .7 (b) $\frac{2}{15} \approx .1333$ **71.** Answers vary. **73.** $\frac{3}{22}$ **75.** $\frac{2}{3}$ **77.** (a) .271 (b) .379 (c) .625 (d) .342 (e) .690 (f) .514 (g) No. Answers vary.

Additional Probability Review (Page 530)

1. (a) .88 (b) .30 (c) .70 **3.** (a) .0962 (b) .0595 (c) .8168 **5.** (a) 0 (b) $\frac{1}{3}$ (c) 1 **7.** .358 **9.** .144 **11.** No **13.** $\frac{1}{6}$ **15.** 82.4% **17.** .196 **19.** .270 **21.** .286 **23.** 127:73 **25.** .525 **27.** .176

Case 8 Exercises (Page 533)

1. .001 **3.** .331

CHAPTER 9

Section 9.1 (Page 542)

1.

Number of boys	0	1	2	3	4
P(x)	.063	.25	.375	.25	.063

3.

Total	2	3	4	5	6	7	8
P(x)	$\frac{1}{16}$	$\frac{1}{8}$	$\frac{3}{16}$	$\frac{1}{4}$	$\frac{3}{16}$	$\frac{1}{8}$	$\frac{1}{16}$

5.

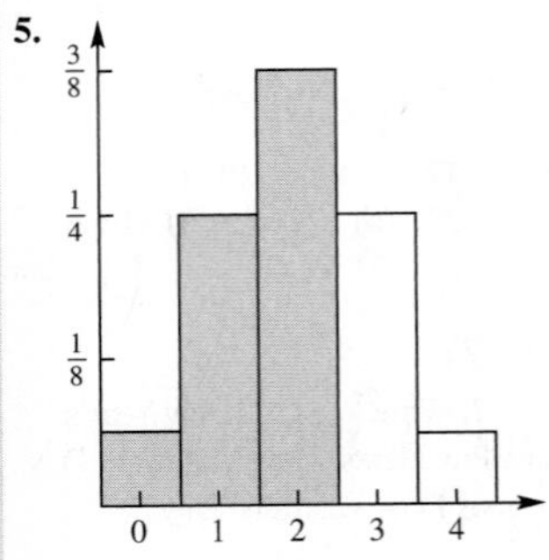

7.

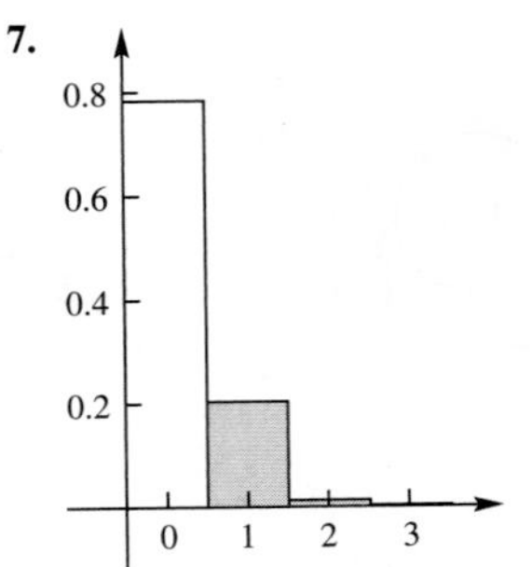

9. 4 **11.** 5.4 **13.** 2.7 **15.** 2.5 **17.** $-\frac{1}{19} \approx -.05$ **19.** −\$.50 **21.** −\$.97 **23.** −\$41.75 **25.** 3.4278 **27.** No, cannot have a probability <0. **29.** Yes, cannot have a probability <0. **31.** .15 **33.** .25 **35.** Many correct answers, including .15, .15. **37.** \$4550 **39.** .477

41.

Account Number	*Expected Value*	*Existing Volume + Expected Value of Potential*	*Class*
3	2000	22,000	*C*
4	1000	51,000	*B*
5	25,000	30,000	*C*
6	60,000	60,000	*A*
7	16,000	46,000	*B*

43. (a) \$68.51, \$72.84 (b) Amoxicillin **45.** (a) £550,000 (b) £618,000

Section 9.2 (Page 557)

1. 12 **3.** 56 **5.** 8 **7.** 24 **9.** 84 **11.** 1716 **13.** 6,375,600 **15.** 240,240 **17.** 8568 **19.** 40,116,600 **21.** $\frac{24}{0}$ is undefined. **23.** (a) 8 (b) 64 **25.** 576 **27.** 1 billion in theory; however, some numbers will never be used (such as those beginning 000); yes. **29.** 1 billion **31.** 120 **33.** (a) 160; 8,000,000 (b) Some, such as 800, 900, etc., are reserved. **35.** 1600 **37.** Answers vary. **39.** 95,040 **41.** 12,144 **43.** 863,040 **45.** 272 **47.** (a) 210 (b) 210 **49.** 330 **51.** Answers vary. **53.** (a) 9 (b) 6 (c) 3, yes **55.** (a) 8568 (b) 21 (c) 3465 (d) 8547 **57.** (a) 210 (b) 7980 **59.** (a) 1,120,529,256 (b) 806,781,064,320 **61.** 81 **63.** Not possible, 4 initials **65.** (a) 10 (b) 0 (c) 1 (d) 10 (e) 30 (f) 15 (g) 0 **67.** 3,247,943,160 **69.** 5.524×10^{26} **71.** (a) 2520 (b) 840 (c) 5040 **73.** (a) 479,001,600 (b) 6 (c) 27,720

Section 9.3 (Page 565)

1. .422 **3.** $\frac{5}{8}$ **5.** $\frac{5}{28}$ **7.** .00005 **9.** .0163 **11.** .5630 **13.** 1326 **15.** .851 **17.** .765 **19.** .941 **21.** Answers vary. **23.** .1396 **25.** (a) 8.9×10^{-10} (b) 1.2×10^{-12} **27.** 2.62×10^{-17} **29.** (a) .0322 (b) .2418 (c) .7582 **31.** (a) .0015 (b) .0033 (c) .3576 **33.** $1 - \frac{{}_{365}P_{100}}{(365)^{100}} \approx 1$ **35.** .0031 **37.** .3083 **39.** We obtained the following answers—yours should be similar: (a) .0399 (b) .5191 (c) .0226

Section 9.4 (Page 572)

1. .200 **3.** .000006 **5.** .999994 **7.** .3010 **9.** .1216 **11.** .8784 **13.** $\frac{1}{32}$ **15.** $\frac{13}{16}$ **17.** Answers vary.

19. .1222 **21.** .0719 **23.** 68 **25.** .0608 **27.** .9903
29. .247 **31.** .193 **33.** .350 **35.** 3.85 **37.** .9747
39. Lower **41.** **(a)** .9995 **(b)** Answers vary.
43. **(a)** 1 chance in 1024 **(b)** 1 chance in 1.1×10^{12} **(c)** 1 chance in 2.6×10^6 **(d)** Answers vary.

Section 9.5 (Page 582)

1. Yes **3.** Yes **5.** No **7.** No **9.** No **11.** No

13. Not a transition diagram **15.** $\begin{bmatrix} .6 & .20 & .20 \\ .9 & .02 & .08 \\ .4 & 0 & .6 \end{bmatrix}$ **17.** Yes

19. Yes **21.** No **23.** $\left[\frac{19}{64}, \frac{45}{64}\right]$ **25.** $\left[\frac{3}{11}, \frac{8}{11}\right]$

27. [.4633, .1683, .3684] **29.** [.4872, .2583, .2545]

31. $A^2 = \begin{bmatrix} .23 & .21 & .24 & .17 & .15 \\ .26 & .18 & .26 & .16 & .14 \\ .23 & .18 & .24 & .19 & .16 \\ .19 & .19 & .27 & .18 & .17 \\ .17 & .2 & .26 & .19 & .18 \end{bmatrix}$

$A^3 = \begin{bmatrix} .226 & .192 & .249 & .177 & .156 \\ .222 & .196 & .252 & .174 & .156 \\ .219 & .189 & .256 & .177 & .159 \\ .213 & .192 & .252 & .181 & .162 \\ .213 & .189 & .252 & .183 & .163 \end{bmatrix}$

$A^4 = \begin{bmatrix} .2205 & .1916 & .2523 & .1774 & .1582 \\ .2206 & .1922 & .2512 & .1778 & .1582 \\ .2182 & .1920 & .2525 & .1781 & .1592 \\ .2183 & .1909 & .2526 & .1787 & .1595 \\ .2176 & .1906 & .2533 & .1787 & .1598 \end{bmatrix}$

$A^5 = \begin{bmatrix} .21932 & .19167 & .25227 & .17795 & .15879 \\ .21956 & .19152 & .25226 & .17794 & .15872 \\ .21905 & .19152 & .25227 & .17818 & .15898 \\ .21880 & .19144 & .25251 & .17817 & .15908 \\ .21857 & .19148 & .25253 & .17824 & .15918 \end{bmatrix}$; .17794

33. **(a)** $\begin{bmatrix} .9 & .10 \\ .3 & .70 \end{bmatrix}$ **(b)** [.51, .49] **(c)** [.75, .25]

35. [.802, .198] **37.** $\left[\frac{1}{4}, \frac{1}{2}, \frac{1}{4}\right]$ **39.** **(a)** [.576, .421, .004]
(b) [.473, .526, .002] **41.** **(a)** [42,500 5000 2500]
(b) [37,125 8750 4125] **(c)** [33,281 11,513 5206]
(d) [.475 .373 .152] **43.** **(a)** .157 **(b)** .494

45. **(a)** $\begin{array}{c} \\ 0 \\ 1 \\ 2 \end{array}\begin{array}{c} \begin{array}{ccc} 0 & 1 & 2 \end{array} \\ \begin{bmatrix} .4 & .3 & .3 \\ .4 & .3 & .3 \\ 0 & .5 & .5 \end{bmatrix} \end{array}$ **(b)** $\begin{array}{c} \\ 0 \\ 1 \\ 2 \end{array}\begin{array}{c} \begin{array}{ccc} 0 & 1 & 2 \end{array} \\ \begin{bmatrix} .28 & .36 & .36 \\ .28 & .36 & .36 \\ .2 & .4 & .4 \end{bmatrix} \end{array}$

(c) .36 **47.** [0 0 .102273 .897727]

Section 9.6 (Page 588)

1. **(a)** Coast **(b)** Highway **(c)** Highway; $38,000 **(d)** Coast
3. **(a)** Do not upgrade. **(b)** Upgrade. **(c)** Do not upgrade; $10,060.

5. **(a)**

	Fails	Doesn't Fail
Overhaul	−$8600	−$2600
Don't Overhaul	−$6000	$0

(b) Don't overhaul the machine.

7. **(a)**

	No Rain	Rain
Rent tent	$2500	$1500
Do not rent tent	$3000	$0

(b) Rent tent because expected value is $2100.
9. Environment, 15.3

Chapter 9 Review (Page 591)

1. **(a)**

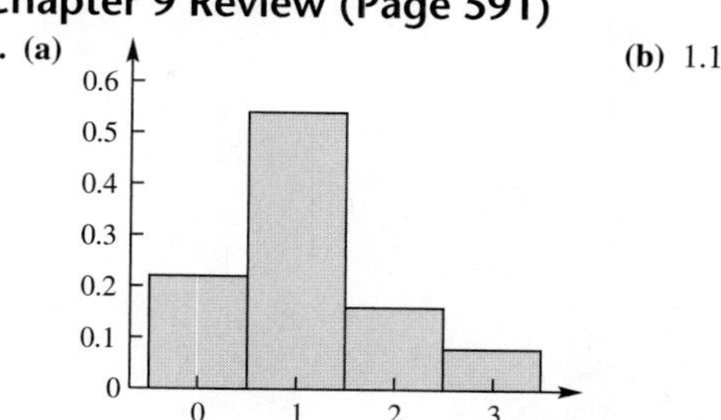

(b) 1.1

3. **(a)**

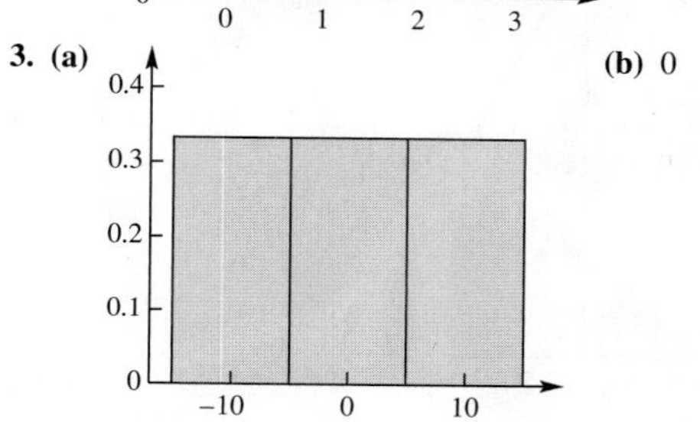

(b) 0

5. 2.7382

7. **(a)**

x	0	1	2	3
$P(x)$	.292	.525	.175	.008

(b) .889

9. Under, $E(x) = -\$33$; exactly 7, $E(x) = -\$1.00$; over, $E(x) = -\$.33$ **11.** 1.25 **13.** 40,320 **15.** 220
17. 427,518,000 **19.** **(a)** 14,905,800 **(b)** 14,905,800
21. Answers vary. **23.** .059 **25.** .995 **27.** .018
29. .218 **31.** **(a)** $1, n, \frac{n(n-1)}{2}, 1$
(b) ${}_nC_0 + {}_nC_1 + {}_nC_2 + \cdots + {}_nC_n$ **(c)** Answers vary.
33. .956 **35.** 4.68

37. **(a)**

x	0	1	2	3	4	5
$P(x)$	.4437	.3915	.1382	.0244	.0022	.0001

(b) .7504 **39.** No **41.** Yes **43.** **(a)** [.1515, .5635, .2850]
(b) [.1526, .5183, .3290] **(c)** [.1509, .4697, .3795]
45. **(a)** Active learning **(b)** Active learning **(c)** Lecture, 17.5
(d) Active learning, 48 **47.** (c)

Case 9 Exercises (Page 596)

1. +.2031

3.

x	$P(x)$
0	.30832
1	.43273
2	.21264
3	.04325
4	.00306

5. +.1947

CHAPTER 10

Section 10.1 (Page 604)

1. **(a)** and **(b)**

Interval	*Frequency*
14–17	2
18–21	11
22–25	12
26–29	11
30–33	7
34–37	5
38–41	1
42–45	1

(c) and **(d)**

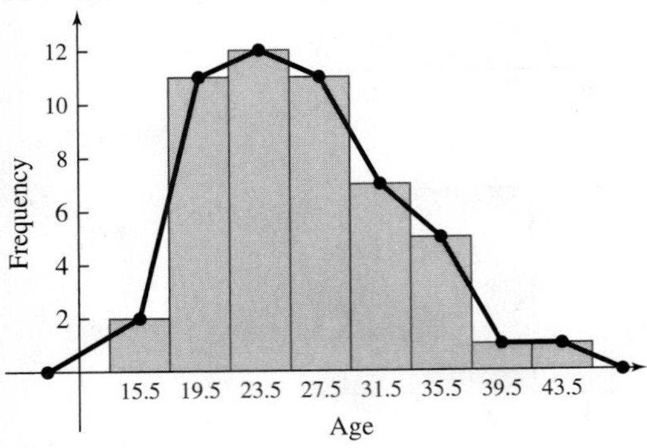

3. **(a)** and **(b)**

Interval	*Frequency*
0–4	1
5–9	4
10–14	4
15–19	3
20–24	5
25–29	8
30–34	9
35–39	8
40–44	3
45–49	5

(c) and **(d)**

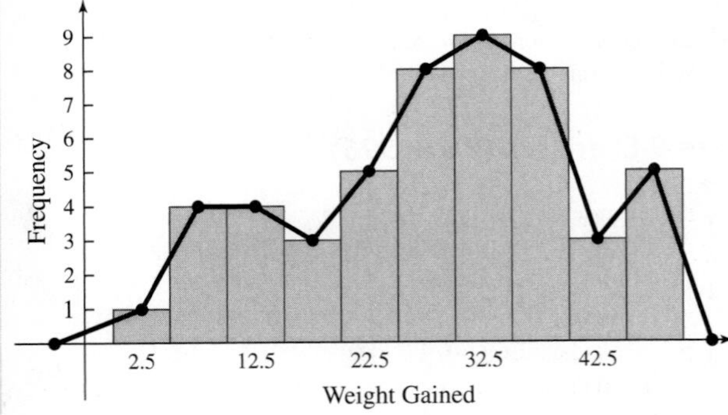

5.

Interval	*Frequency*
0–24	6
25–49	8
50–74	2
75–99	5
100–124	2
125–149	2
150–174	3
175–199	1
200–224	0
225–249	1

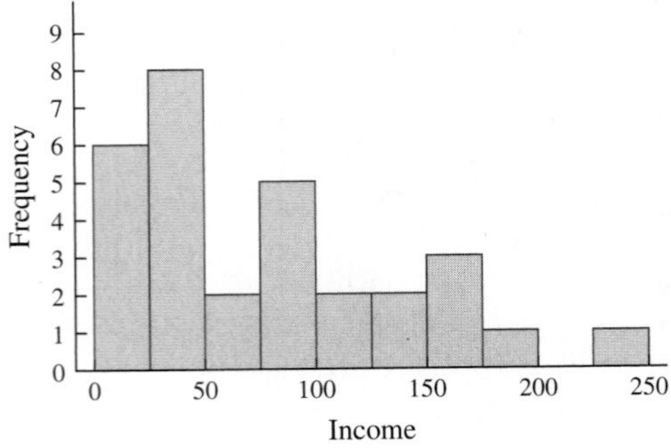

7.

Interval	*Frequency*
0–49	1
50–99	10
100–149	6
150–199	5
200–249	3
250–299	1
300–349	1
350–399	1
400–449	1
450–499	0
500–549	1

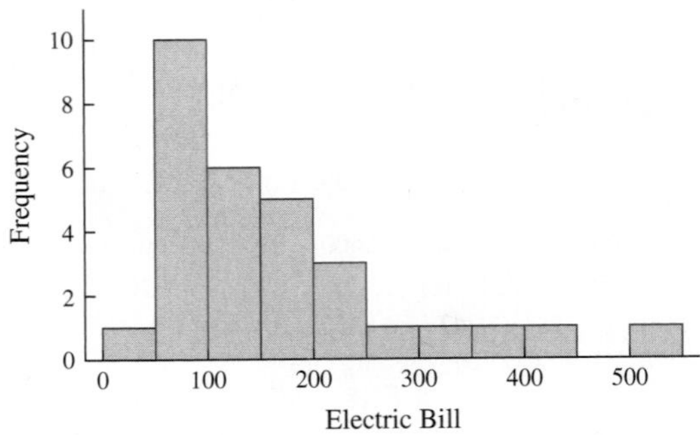

9.

Interval	*Frequency*
40–44	1
45–49	2
50–54	6
55–59	9
60–64	6
65–69	4
70–74	2

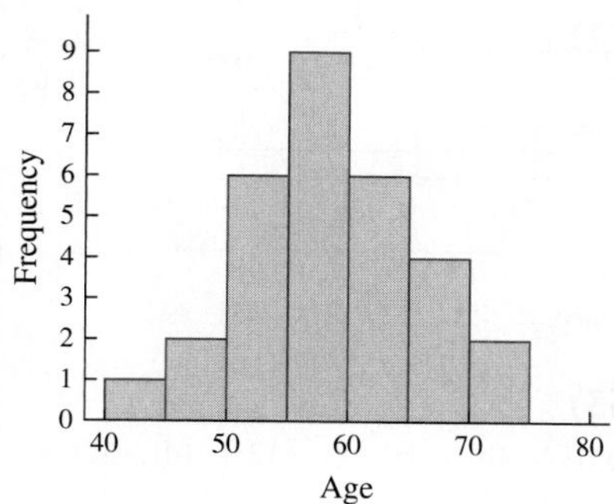

11.

Stem	*Leaves*
1	55889
2	00000011223444444
2	55566677788899
3	00112334
3	55678
4	4

Units: 4|4 = 44 years

13.

Stem	*Leaves*
0	0
0	5578
1	0013
1	569
2	00022
2	55556889
3	000112223
3	55555679
4	012
4	55799

Units: 4|9 = 49 pounds

15.

Stem	*Leaves*
0	11122233344
0	555558889
1	0003
1	56678
2	3

Units: 2|3 = 230 thousand

17.

Stem	*Leaves*
0	3
0	5556778889
1	011223
1	56899
2	003
2	5
3	1
3	7
4	0
4	
5	2

Units: 5|2 = 520 dollars

19.

Stem	*Leaves*
7	99
8	000011122333444
8	5566666777777888889999999
9	000001111

Units: 9|1 = 91%

21. Uniform **23.** Left skewed **25.** Right skewed **27.** **(a)** Normal **(b)** 3 **(c)** 40s **29.** **(a)** Right skewed **(b)** 8 **(c)** 5

Section 10.2 (Page 614)

1. \$27,955 **3.** 4.8 **5.** 10.3 **7.** 21.2 **9.** 14.8 **11.** \$33,679 **13.** 98.5 **15.** 2 **17.** 65, 71 **19.** No mode **21.** Answers vary. **23.** $\bar{x}$ = 3150 grams, 3000–3499 **25.** **(a)** 57.5 **(b)** 60.5 **(c)** 37 **27.** **(a)** \$435.333 million **(b)** \$427 million **29.** **(a)** \$5310.78 million; \$4684.9 million **(b)** 2004 **31.** $\bar{x}$ = 49.6, median = 48.5 **33.** right skewed, median = \$175 **35.** Right skewed, median = 13%

Section 10.3 (Page 624)

1. Answers vary. **3.** 12; 4.8 **5.** 1731; 728.6 **7.** 1061; 430.8 **9.** 336; 132.7 **11.** 45.2 **13.** $\bar{x} = 5.0876$, $s = .1087$ **15.** $\frac{3}{4}$ **17.** $\frac{5}{9}$ **19.** 93.75% **21.** 11.1%

23. $\bar{x}$ = \$252.8 million; s = \$20.7 million

25. $\bar{x}$ = \$1449.3 million; s = \$76.7 million

27. **(a)** $s^2 = 14.8$, $s = 3.8$ **(b)** 10 **29.** **(a)** $\bar{x}$ = 127.71 days, s = 30.16 days **(b)** All **(c)** Answers vary.

31. **(a)** $\frac{1}{3}, 2, -\frac{1}{3}, 0, \frac{5}{3}, \frac{7}{3}, 1, \frac{4}{3}, \frac{7}{3}, \frac{2}{3}$ **(b)** 2.1, 2.6, 1.5, 2.6, 2.5, .6, 1, 2.1, .6, 1.2 **(c)** 1.13 **(d)** 1.68 **(e)** −2.15, 4.41; the process is out of control. **33.** $\bar{x} = 80.17$, $s = 12.2$ **35.** Answers vary.

Section 10.4 (Page 638)

1. The mean **3.** Answers vary. **5.** 45.99% **7.** 16.64% **9.** 7.7% **11.** 47.35% **13.** 91.20% **15.** 1.64 or 1.65 **17.** −1.04 **19.** .5; .5 **21.** .889; .997 **23.** .3821 **25.** .5762 **27.** .2776 **29.** .0062 **31.** .4452 **33.** .9370 **35.** .0179 **37.** .0726 **39.** 45.2 mph **41.** 24.17% **43.** Answers vary. **45.** 665 units **47.** 190 units **49.** 15.87% **51.** .0618 **53.** .9278

55. Min = 19.6, Q_1 = 20.1, Q_2 = 21.5, Q_3 = 24.1, max = 31.9

57.

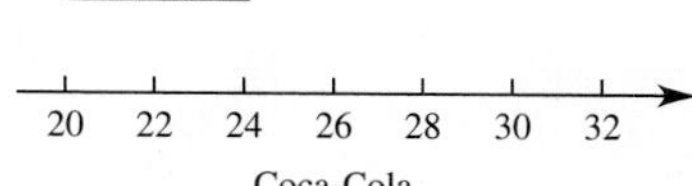

59. Min = 33, Q_1 = 40, Q_2 = 126.5, Q_3 = 254, max = 661

61.

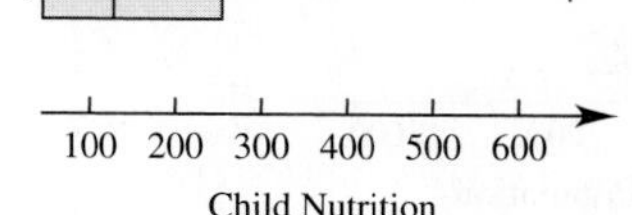

63. Child nutrition

Section 10.5 (Page 646)

1. The number of trials and the probability of success on each trial **3.** **(a)** .1964 **(b)** .1974 **5.** **(a)** .0021 **(b)** .0030 **7.** .0240 **9.** .9463 **11.** .0956 **13.** .7291 **15.** .1841 **17.** **(a)** .0238 **(b)** .6808 **19.** .0721 **21.** .2119 **23.** **(a)** .0005 **(b)** .001 **(c)** .0000 **25.** .9406 **27.** **(a)** .0764 **(b)** .121 **(c)** Yes; the probability of such a result is .0049. **29.** .2061 **31.** **(a)** 1.2139×10^{-7} **(b)** Essentially 0 **(c)** .7939

Chapter 10 Review (Page 648)

1. Answers vary.

3. (a)

Interval	Frequency
40–49	2
50–59	2
60–69	3
70–79	8
80–89	9
90–99	6

(b)

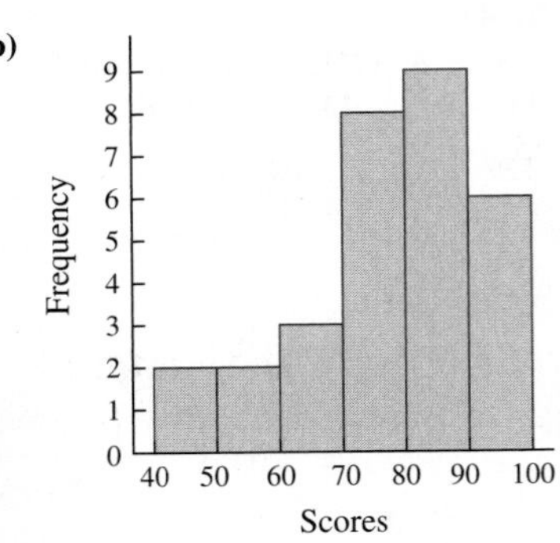

(c)

Stem	*Leaves*
4	25
5	35
6	389
7	11667777
8	122444899
9	111249

Units: 9|9 = 99 points

5. $\bar{x} = 77.27$ **7.** 75.6 cm **9.** Answers vary. **11.** 72, 72 **13.** 2000–2999 **15.** Uniform **17.** Left skewed **19.** Answers vary. **21.** Range = 37, $s = 12.10$ **23.** 1,117.7 calories **25.** Answers vary. **27.** .6480 **29.** .0606 **31.** Answers vary.

33. (a)

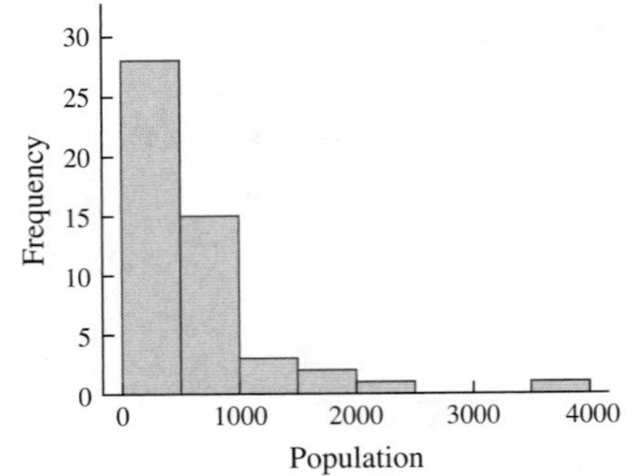

(b) Right skewed **35.** $\bar{x} = 3.675$, $s \approx 12.6$

37. (a)

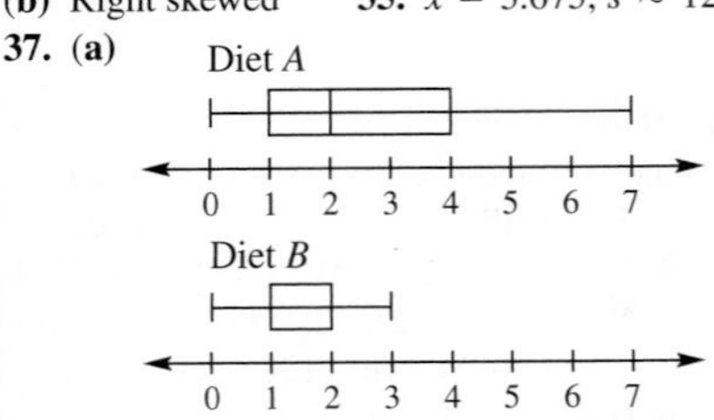

(b) Answers vary. **39. (a)** March 2008: $\bar{x} \approx 51.17$, $s \approx 42.04$; March 2009: $\bar{x} \approx 33.08$, $s \approx 25.61$ **(b)** March 2008: Chrysler; March 2009: Hyundai

41. (a)

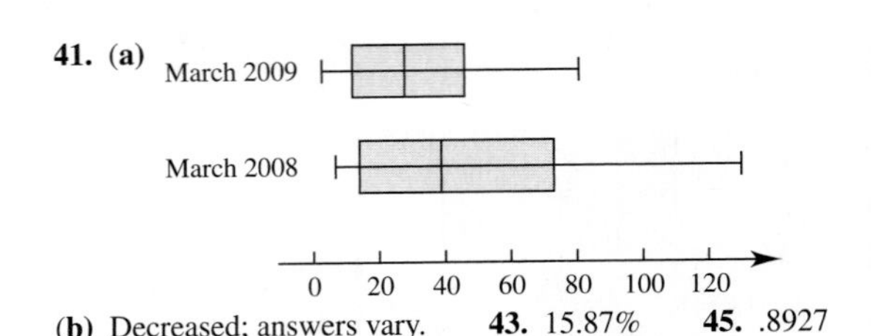

(b) Decreased; answers vary. **43.** 15.87% **45.** .8927

Case 10 (Page 663)

1. $z = -29.1$ **3. (a)** 174 **(b)** 6.03 **(c)** -12.3 **(d)** $< .004$

CHAPTER 11

Section 11.1 (Page 663)

1. (a) 8 **(b)** 0 **3. (a)** Does not exist **(b)** 2 **5. (a)** Does not exist **(b)** -4 **7. (a)** 6 **(b)** 0 **9.** Answers vary. **11.** 1 **13.** 3 **15.** 0 **17.** 5 **19.** Does not exist **21.** 15 **23.** $\frac{5}{2}$ **25.** 5 **27.** $\frac{7}{4}$

29. (a)

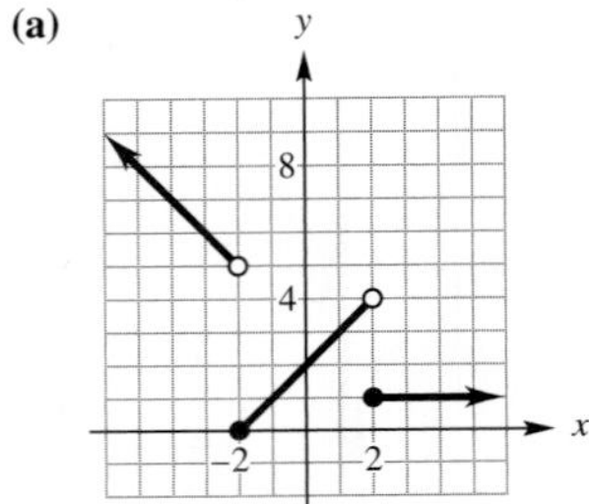

(b) Does not exist **(c)** 3 **(d)** Does not exist **31.** 0 **33.** $\frac{19}{31}$ **35.** 10 **37.** 7 **39.** $\frac{1}{2}$ **41.** Does not exist **43.** $\sqrt{6}$ **45.** Does not exist **47.** $-\frac{1}{9}$ **49.** $\frac{1}{10}$

51. $\frac{1}{2\sqrt{5}}$ or $\frac{\sqrt{5}}{10}$ **53. (a)** 13.125 **(b)** 68.25 **(c)** 68.25 **55. (a)** \$153 **(b)** \$18 **(c)** \$4.50 **57. (a)** 1.5 **(b)** 1.2 **(c)** 0 **(d)** 3

59. (a) $P(x) = \begin{cases} .99 & \text{if } 0 \le x \le 20 \\ 1.06 & \text{if } 20 < x \le 21 \\ 1.13 & \text{if } 21 < x \le 22 \\ 1.20 & \text{if } 22 < x \le 23 \end{cases}$

(b)

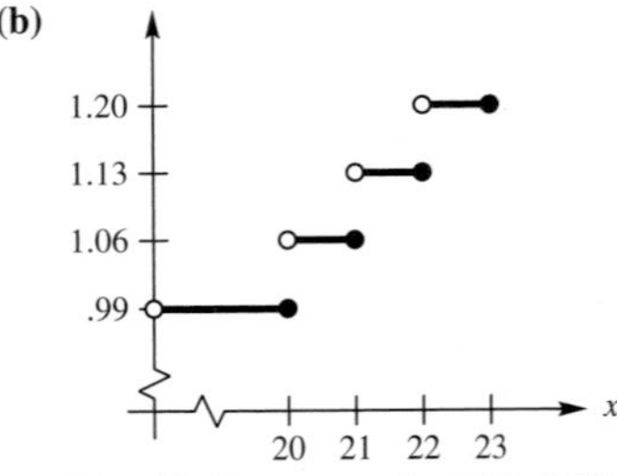

(c) .99 **(d)** Does not exist **(e)** 1.20

Section 11.2 (Page 675)

1. 3 **3.** -2 **5.** 6 **7.** 0 **9. (a)** 0 **(b)** 1 **(c)** -1 **(d)** -1 **11. (a)** 2 **(b)** Does not exist **(c)** 0 **(d)** 2 **13.** 0 **15.** 1 **17.** -9 **19.** $\frac{1}{4}$

21.

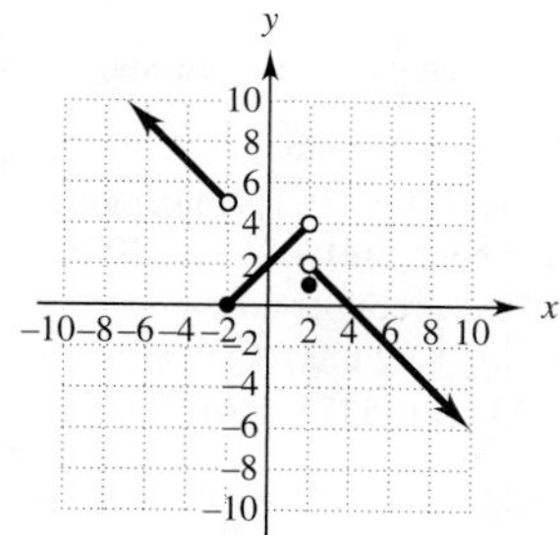

(a) 5 **(b)** 0 **(c)** Does not exist **23. (a)** ∞ **(b)** ∞ **(c)** $-\infty$ **25.** $-\infty$ **27.** ∞ **29.** ∞ and 0 **31.** 2 and -1 **33.** ∞ and $-\infty$ **35.** -1 **37.** 0 **39.** 1 **41.** 0 **43.** $\frac{10}{3}$ **45.** -4 **47.** -2 **49.** $\frac{4}{5}$ **51.** ∞ **53.** 15 **55.** 1 **57.** -1 **59. (a)** About 64.68 **(b)** 71.8; answers vary (how significant are 7 teeth?) **61.** 0; the concentration of the drug approaches 0 as time increases.

Section 11.3 (Page 686)

1. 8 **3.** 2 **5.** $\frac{1}{4}$ **7.** $-\frac{1}{3}$ **9. (a)** 1; from catalog distributions of 10,000 to 20,000, sales will have an average increase of \$1000 for each additional 1000 catalogs distributed. **(b)** $\frac{3}{5}$; from catalog distributions of 20,000 to 30,000, sales will have an average increase of \$600 for each additional 1000 catalogs distributed. **(c)** $\frac{2}{5}$; from catalog distributions of 30,000 to 40,000, sales will have an average increase of \$400 for each additional 1000 catalogs distributed. **(d)** As more catalogs are distributed, sales increase at a smaller and smaller rate. **(e)** Answers vary **11. (a)** .5 **(b)** .8 **13. (a)** $-.179$ **(b)** $-.167$ **15. (a)** -76.11 **(b)** -50.5 **(c)** 86.5 **17.** Answers vary **19.** 22 ft/sec **21.** 66 ft/sec **23.** 6 ft/sec **25. (a)** 5 ft/sec **(b)** 2 ft/sec **(c)** 3 ft/sec **(d)** 5 ft/sec **(e) (i)** 3 ft/sec; **(ii)** 2.5 ft/sec **(f) (i)** 5 ft/sec; **(ii)** 4 ft/sec **27. (a)** $a^2 + 2ah + h^2 - a - h - 1$ **(b)** $2a + h - 1$ **(c)** 9 **29. (a)** $a^3 + 3a^2h + 3ah^2 + h^3$ **(b)** $3a^2 + 3ah + h^2$ **(c)** 75 **31. (a)** \$5998 per unit **(b)** \$6000 per unit **(c)** \$5998 **(d)** Answers vary. **33. (a)** 6% per day **(b)** 7% per day **35. (a)** 79.8 **(b)** 110.44 **37. (a)** .716 **(b)** .505 **39. (a)**

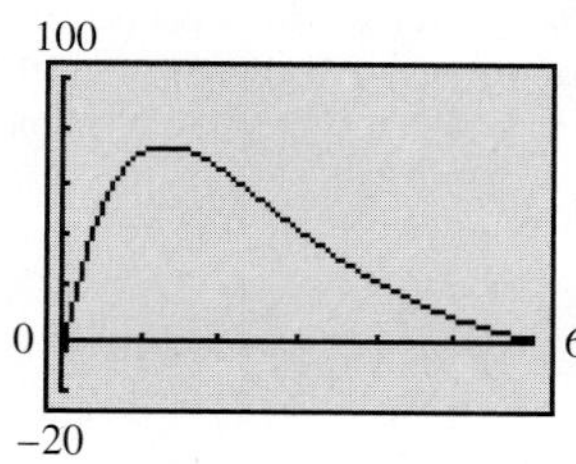

(b) About 81.51 kilojoules per hour per hour **(c)** About 18.81 kilojoules per hour per hour **(d)** At approximately $t = 1.3$ hours

Section 11.4 (Page 702)

1. $y = 8x - 16$ **3.** $y = -\frac{1}{4}x - 1$ **5. (a)** 4 **(b)** $y = 4x - 1$ **7. (a)** $-\frac{7}{9}$ **(b)** $y = -\frac{7}{9}x + \frac{14}{3}$ **9. (a)** $\frac{5}{6}$ **(b)** $y = \frac{5}{6}x + \frac{15}{2}$ **11. (a)** x_5 **(b)** x_4 **(c)** x_3 **(d)** x_2 **13.** Here is one of many possible graphs:

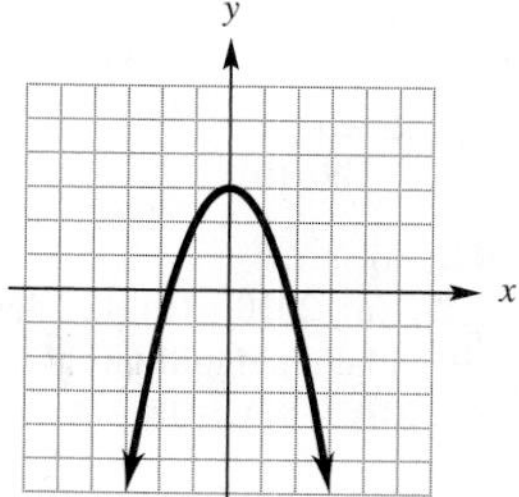

15. (a) The derivative is positive, because the tangent line is rising from left to right at $x = 100$ and thus has positive slope. **(b)** The derivative is negative, because the tangent line is falling from left to right at $x = 200$ and thus has negative slope. **17.** (a) is distance; (b) is velocity. **19.** $f'(x) = -8x + 11; -5; 11; 35$ **21.** $f'(x) = 8; 8; 8; 8$ **23.** $f'(x) = \frac{2}{x^2}, \frac{1}{2}$; does not exists; $\frac{2}{9}$ **25.** $f'(x) = \frac{-4}{(x-1)^2}; -4; -4; -\frac{1}{4}$ **27.** $-6; 6$ **29.** $-5; -3; 0; 2; 4$ **31. (a)** \$16 per table **(b)** \$15.998, or \$16 **(c)** The marginal revenue found in part (a) approximates the actual revenue from the sale of the 1001st item found in part (b). **33. (a)** $-2p + 5$ **(b)** -19; demand is decreasing at the rate of about 19 items for each increase in price of \$1. **35.** 1000; the population is increasing at a rate of 1000 shellfish per time unit. 570; the population is increasing more slowly at 570 shellfish per time unit. 250; the population is increasing at a much slower rate of 250 shellfish per time unit.

37.

39.

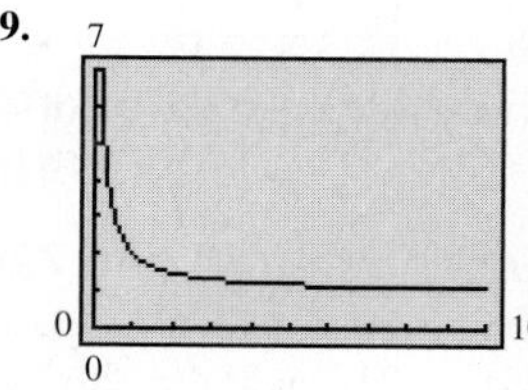

41. (a) 1.72; 7.03; 19.22 **(b)** The rate is increasing. **43.** 1.23; The rate of increase is 1.23 million drivers in 2010. **45. (a)**

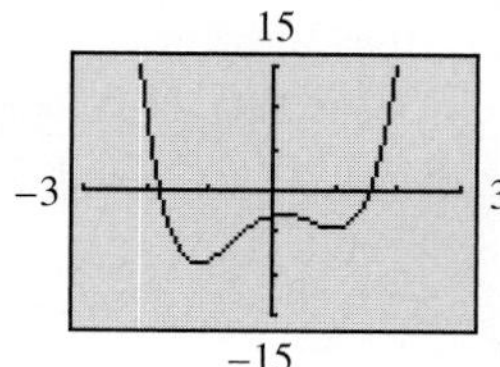

(b)

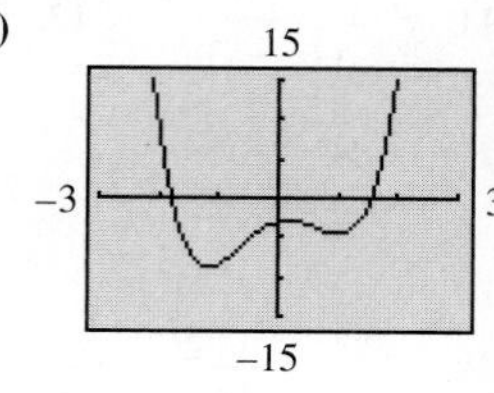

(c) The graphs appear identical, which suggests that $f'(x) = g(x)$. **47.** (d)

Section 11.5 (Page 716)

1. $f'(x) = 8x - 3$ **3.** $y' = 6x^2 + 6x - 6$ **5.** $g'(x) = 4x^3 + 9x^2 - 8$ **7.** $f'(x) = 9x^{.5} - 2x^{-.5}$ **9.** $y' = -22.5x^{1/2} + 3.8x^{.9} + 1$ **11.** $y' = 36t^{1/2} + 2t^{-1/2}$ **13.** $y' = 4x^{-1/2} + \frac{9}{2}x^{-1/4}$ **15.** $g'(x) = -30x^{-6} + 2x^{-2}$ **17.** $y' = -20x^{-3} - 24x^{-5} + 3$ **19.** $f'(t) = \frac{4}{t^2} - \frac{16}{t^3}$ **21.** $y' = -48x^{-5} + 21x^{-4} - 6x^{-2}$ **23.** $g'(x) = \frac{-5}{2}x^{-3/2} - \frac{7}{2}x^{-1/2} + 101$ **25.** $y' = \frac{-21}{2}x^{-5/2} - 4x^{-3/2} + 3x^2$

27. $y' = \frac{-19}{4}x^{-5/4}$ **29.** $y' = \frac{-8}{3}t^{-2/3}$

31. $\frac{dy}{dx} = -40x^{-6} + 36x^{-5} + 36x^3$ **33.** $-\frac{9}{2}x^{-3/2} - 3x^{-5/2}$ or $-\frac{9}{2x^{3/2}} - \frac{3}{x^{5/2}}$ **35.** -26 **37.** $\frac{11}{16}$ **39.** (b)

41. $0; y = 0$ **43.** $7; y = 7x - 3$ **45. (a)** \$2.96 **(b)** \$56.60 **(c)** \$296.44 **(d)** \$596.38 **47. (a)** 2010 **(b)** 127.45 **49. (a)** $R'(x) = .1216x + 1.4284$ **(b)** \$2.0364 hundred million **(c)** \$2.2796 hundred million **51. (a)** 30 **(b)** 0 **(c)** -10

53. (a) $\frac{dV}{dr} = 160\pi r$ **(b)** 640π; the volume of blood is increasing 640π cu mm per mm of change in the radius. **(c)** 960π; the volume of blood is increasing 960π cu mm per mm of change in the radius. **(d)** 1280π; the volume of blood is increasing 1280π cu mm per mm of change in the radius. **55. (a)** .623 **(b)** .861 **(c)** The rate of increase in living standards is higher in 2009 then in 2000.

57. (a) $y_1 = 35; y_2 = 36$ **(b)** 4.13; 4.32; these values are fairly close, and they represent a rate of change of approximately four years for a dog for one year of a human. **(c)** $y = 4x + 16$ **(d)** Answers vary. **59. (a)** $V = CR_0R^2 - CR^3$, so that $dV/dR = 2CR_0R - 3CR^2$. **(b)** $R = \frac{2}{3}R_0$ **61. (a)** $v(t) = 16t + 3$ **(b)** 3; 83; 163

63. (a) $v(t) = 6t^2 + 12t$ **(b)** 0; 210; 720

65. (a) -32 ft per sec; -64 ft per sec **(b)** In 3 sec **(c)** -96 ft per sec

67. (a) $(x + h)^n = x^n + n \cdot x^{n-1}h + \frac{n(n-1)}{2}x^{n-2}h^2 + \ldots + h^n$

(b) $\frac{(x + h)^n - x^n}{h} = n \cdot x^{n-1} + \frac{n(n-1)}{2}x^{n-2}h + \ldots + h^{n-1}$

(c) $\lim_{h \to 0} \frac{(x + h)^n - x^n}{h} = n \cdot x^{n-1} = y'$ **69. (a)–(c)** The derivative is negative on $(-\infty, \infty)$. **(d)** The graph is always falling.

Section 11.6 (Page 726)

1. $y' = 9x^2 + 4x - 6$ **3.** $y' = 120x^3 - 54x^2 + 10$

5. $y' = 24x^5 - 35x^4 - 20x^3 + 42x^2 + 4x - 6$

7. $y' = 100x^3 + 360x^2 + 288x$ **9.** $y' = 96x^5 + 240x^4 + 144x^3$

11. $y' = \frac{-4}{(x - 3)^2}$ **13.** $f'(t) = \frac{t^2 + 6t - 12}{(t + 3)^2}$

15. $g'(x) = \frac{-6x^4 - 4x^3 - 12x - 2}{(2x^3 - 2)^2}$ **17.** $y' = \frac{x^2 + 6x - 14}{(x + 3)^2}$

19. $r'(t) = \frac{-3t + 4}{2\sqrt{t}(3t + 4)^2}$ **21.** $y' = \frac{\frac{9\sqrt{x}}{2} + \frac{4}{\sqrt{x}}}{x}$ or $\frac{9x + 8}{2x\sqrt{x}}$

23. $y' = \frac{-33}{(5 - x)^2}$ **25.** $f'(p) = \frac{264p^2 + 396p - 277}{(4p + 3)^2}$

27. $g'(x) = \frac{6x^4 + 30x^3 - 27x^2 + 96x + 120}{(3x + 9)^2(2x - 1)^2}$

29. In the first step, the numerator should be $(x^2 - 1)(2) - (2x + 5)(2x)$.

31. $y = -2x + 9$ **33.** $y = -1.25x + 53.25$ **35. (a)** \$21 **(b)** \$13.73 **(c)** $\frac{4x + 3}{x^2 + 5x}$ hundred dollars per unit

(d) $\frac{-4x^2 - 6x - 15}{(x^2 + 5x)^2}$ **37. (a)** \$262,500 **(b)** $\frac{17{,}500{,}000}{(100 - x)^2}$

(c) Increasing at a rate of \$43,750 for each 1% removed

(d) Increasing at a rate of \$175,000 for each 1% removed

(e) Increasing at a rate of \$700,000 for each 1% removed

39. (a) $T'(x) = \frac{-10x^2 + 50}{(x^2 + 5)^2}$ **(b)** 2; temperature increasing at 2 degrees per hour **(c)** About 1.1111; temperature increasing at about 1.1111 degrees per hour **(d)** About $-.2041$; temperature decreasing at about .2041 degrees per hour **(e)** About $-.1028$; temperature decreasing at about .1028 degrees per hour

41. (a) 8.57 min **(b)** 16.36 min **(c)** 6.12 min/(kcal/min) and 2.48 min/(kcal/min); answers vary. **43. (a)** .1173 **(b)** 2.625

45. (a) $N'(t) = \frac{4200t}{(30 + t^2)^2}$ **(b)** 8.28 wpm; 6.94 wpm; 4.71 wpm; 2.49 wpm; .97 wpm **(c)** Answers vary.

Section 11.7 (Page 738)

1. 779 **3.** 259 **5.** $16x + 36; 16x + 27$

7. $-64x^3 - 96x^2 - 48x - 6; -4x^3 + 10$ **9.** $\frac{1}{x^2}; \frac{1}{x^2}$

11. $\sqrt{8x^2 - 6x + 2}; 8x + 16 - 6\sqrt{x + 2}$

13. If $f(x) = x^5$ and $g(x) = 4x + 3$, then $y = f[g(x)]$.

15. If $f(x) = \sqrt{x}$ and $g(x) = 6 + 3x^2$, then $y = f[g(x)]$.

17. If $f(x) = \frac{x + 3}{x - 3}$ and $g(x) = \sqrt{x}$, then $y = f[g(x)]$.

19. If $f(x) = x^3 + x + 5$ and $g(x) = x^{1/2} - 3$, then $y = f[g(x)]$.

21. $y' = 35(7x - 12)^4$ **23.** $y' = 112(4x + 3)^3$

25. $y' = \frac{99}{2}(3x + 5)^{1/2}$ **27.** $y' = -400x(10x^2 + 8)^4$

29. $y' = -\frac{21}{2}(8x + 9)(4x^2 + 9x)^{1/2}$ **31.** $y' = \frac{16}{\sqrt{4x + 7}}$

33. $y' = \frac{-2x - 4}{\sqrt{x^2 + 4x}}$ **35.** $y' = 2(x + 1)(x - 3) + (x - 3)^2$

37. $y' = 70(x + 3)(2x - 1)^4(x + 2)$

39. $y' = \frac{(3x + 1)^2(21x + 1)}{2\sqrt{x}}$ **41.** $y' = \frac{-3}{(x - 2)^4}$

43. $y' = \frac{(2x - 7)(8x + 24)}{(4x - 1)^2}$ **45.** $y' = \frac{-3x^2 - 28x + 5}{(3x + 1)^4}$

47. $y' = \frac{6x + 8\sqrt{x} + 3}{2x^{1/2}}$ **49. (a)** -2 **(b)** $-\frac{24}{7}$

51. (d) **53.** $C'(x) = \frac{(50x - 1)}{2(200 + 25x^2 - x)^{1/2}}$

55. (a) $R'(x) = \frac{15(250 - 2x)}{(500x - 2x^2)^{1/2}}$ **(b)** \$19.67; \$8.19; \$.85; $-$\$6.19 **(c)** Answers vary. **57. (a)** \$1462.33 per percentage point **(b)** \$1551.57 per percentage point **(c)** \$1646.19 per percentage point **59.** \$175 per additional worker **61.** $A[r(t)] = A(2t) = 4\pi t^2$; this function gives the area of the pollution in terms of the time since the pollutants were first emitted. **63.** $D(c) = \frac{-c^2 + 10c - 25}{25} + 500$

65. (a) $R(x) = \frac{30{,}000x - 2x^3}{3}$ **(b)** $P(x) = 8000x - \frac{2x^3}{3} - 3500$

(c) $\frac{dP}{dx} = 8000 - 2x^2$ **(d)** $-$\$21,925

67. (a) 34 min **(b)** $-\frac{108}{17}\pi$ mm^3/min; $-\frac{72}{17}\pi$ mm^2/min; Answers vary **(c)** Answers vary. **69. (a)** $F'(t) = \frac{150t}{(8 + t^2)^{3/2}}$ **(b)** About 3.96; 1.34; .36; .09 **(c)** Answers vary.

71.

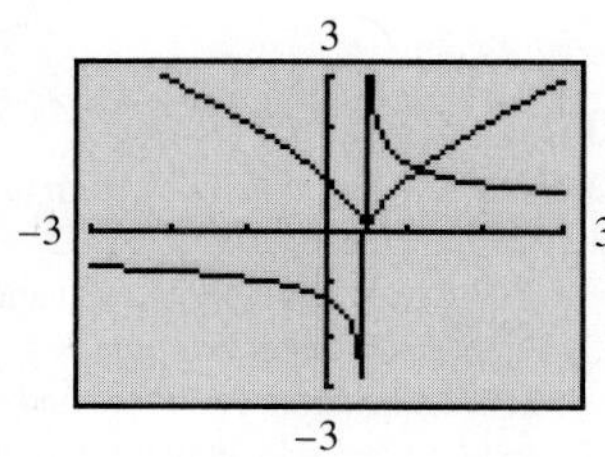

(a) $(.5, \infty)$ **(b)** None **(c)** $(-\infty, .5)$ **(d)** The derivative does not exist at $x = .5$, which corresponds to a sharp point on the graph of $K(x)$. The derivative is positive when $K(x)$ is increasing and negative when $K(x)$ is decreasing.

Section 11.8 (Page 747)

1. $y' = 5e^{5x}$ **3.** $f'(x) = 10e^{2x}$ **5.** $g'(x) = 28e^{-7x}$ **7.** $y' = 2xe^{x^2}$ **9.** $f'(x) = x^2e^{x^3/3}$ **11.** $y' = -18xe^{3x^2+5}$ **13.** $y' = \dfrac{-20x + 7}{-10x^2 + 7x}$ **15.** $y' = \dfrac{5}{2(5x + 1)}$ **17.** $f'(x) = \dfrac{15x^2 + 18x + 5}{(5x + 9)(x^2 + 1)}$ **19.** $y' = e^{-3x}(-3x^4 + 4x^3)$ **21.** $y' = e^{-3x}[-18x^2 + 27x - 5]$ **23.** $y' = \dfrac{-20}{(4 - x)(3x + 8)}$ **25.** $y' = \dfrac{3(8x^2 - 1)}{2x(8x^2 - 3)}$ **27.** $y' = \dfrac{-4x^4}{9 - x^4} + \ln(9 - x^4)$ **29.** $y' = \dfrac{1 - 4\ln|x|}{x^5}$ **31.** $y' = \dfrac{\frac{-25}{x} + 10 - 10\ln|x|}{(5 - 2x)^2}$ **33.** $y' = \dfrac{3x^3\ln|x| - (x^3 - 1)}{2x(\ln|x|)^2}$ **35.** $y' = \dfrac{1}{2(x - 3)\sqrt{\ln(x - 3)}}$ **37.** $y' = \dfrac{xe^x\ln|x| - e^x + 1}{x(\ln|x|)^2}$ **39.** $y' = \dfrac{xe^x + xe^{-x} - e^x + e^{-x}}{x^2}$ or $\dfrac{e^x(x - 1) + e^{-x}(x + 1)}{x^2}$ **41.** $f'(x) = \dfrac{4e^{3x+2}}{4x - 5} + 3e^{3x+2}\ln(4x - 5)$ **43.** $y' = \dfrac{2800e^{4x}}{(7 - 10e^{4x})^2}$ **45.** $y' = \dfrac{1250e^{-.5x}}{(12 + 5e^{-.5x})^2}$ **47.** $y' = (\ln 8)\, 8^x$ **49.** $y' = 2(\ln 15)\cdot 15^{2x}$ **51.** $g'(x) = \dfrac{1}{(\ln 10)\, x}$ **53.** $y = -x + 1$ **55.** $y = \dfrac{1}{e}x$ **57.** $\dfrac{1}{8}$ **59. (a)** About 231.5 million; increasing at a rate of about 3.24 million a year **(b)** About 238.1 million; increasing at a rate of about 3.33 million a year **61. (a)** \$12.77; increasing \$3.67 per year **(b)** \$17.02; increasing \$4.88 per year **(c)** \$22.68; increasing \$6.51 per year **63. (a)** About 27.1 thousand **(b)** About 1.2 thousand **(c)** Answers vary **65. (a)** 2974.15 g **(b)** 2.75 g/day

(c)

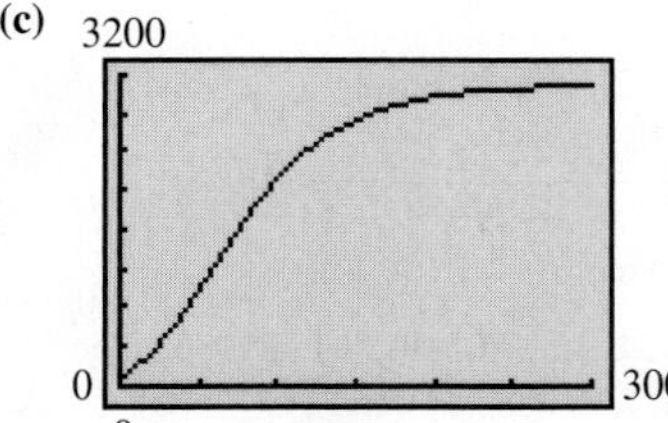

(d)

Weight	*Day*	*Rate*
990.9797	50	24.87793
2121.673	100	17.72981
2733.571	150	7.603823
2974.153	200	2.753855
3058.845	250	0.942782
3087.568	300	0.316772

67. About \$638.90 per capita **69. (a)** 3.857 cm^3 **(b)** .973 cm **(c)** About 214 mo **(d)** The tumor is 240 months old; it is increasing in volume at the instantaneous rate of .282 cm^3/mo. **71. (a)** $\dfrac{dR}{dx} = 100 + \dfrac{50(\ln x - 1)}{(\ln x)^2}$ **(b)** \$112.48 **73. (a)** About 817 cars/hr; -41.2 cars/hr per ft **(b)** About 522 cars/hr; -20.9 cars/hr per ft **75. (a)** 6843 **(b)** About 126 per year **77. (a)** $E \approx 1.567 \times 10^{11}$ kWh **(b)** About 63.4 months **(c)** About 4.14×10^{-6} **(d)** It approaches 0.

Section 11.9 (Page 757)

1. $x = 0, 2$ **3.** $x = 1$ **5.** $x = -3, 0, 3$ **7.** $x = 0, 6$ **9.** Yes, no **11.** No, no, yes **13.** Yes, no, yes **15.** Yes, no, yes **17.** Yes, no **19.** $x = 0, 3$ **21.** Continuous everywhere **23.** $x < 2$ **25. (a)**

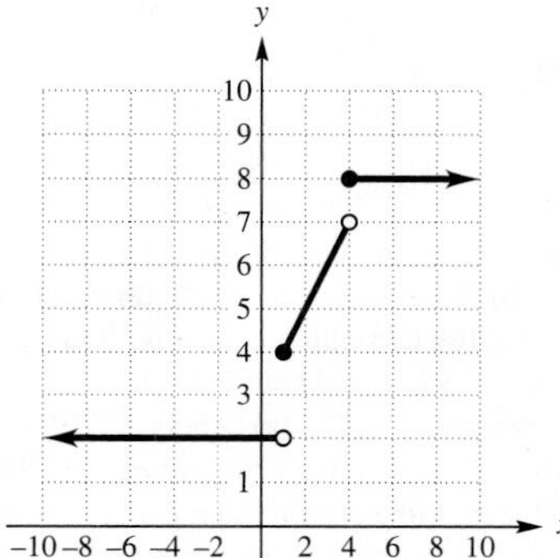

(b) $x = 1, 4$ **27.** $k = 1$ **29. (a)** 4.75 cents **(b)** 6 cents **(c)** 6.25 cents **(d)** Does not exist **(e)** Any three of the following: '91, '01, '04, '09 **31.** Discontinuous at $t = m$; differentiable everywhere *except* $t = m$ and the endpoints **33. (a)** \$4.60 **(b)** \$5.44 **(c)** Does not exist **(d)** \$4.60 **(e)** 1, 2, 3, 4, 5, 6, 7 **(f)**

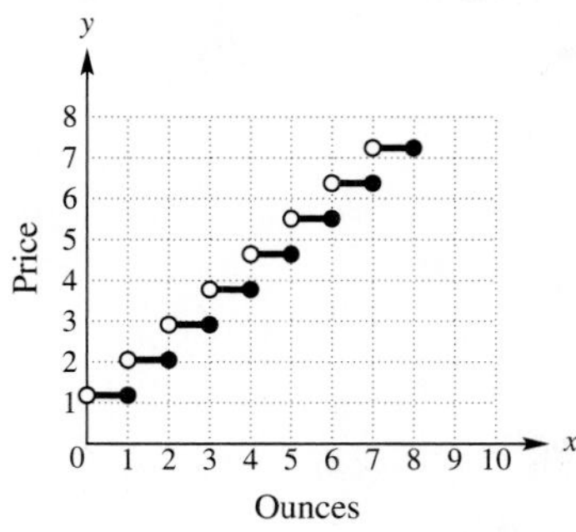

35. (a) \$600 **(b)** \$600 **(c)** \$600 **(d)** \$720 **(e)** \$600 **(f)** \$600 **37.** $[-3, 16]$ **39.** $(12, \infty)$ **41.** $(-6, 0), (4, 8)$

Chapter 11 Review (Page 763)

1. 4; 4;4 **3.** -1.2 **5.** $-.35$ **7.** -4 **9.** Does not exist **11.** 0 **13.** -13 **15.** $\dfrac{1}{6}$ **17.** 0 **19.** 1 **21.** $\dfrac{1}{2}$

23. $\frac{10}{3}$ **25.** $\frac{1}{4}$ **27.** 18 **29.** $-\frac{1}{3}$ **31.** $y' = 2$
33. $y' = 4x - 1$ **35.** $-2; y = -2x - 4$ **37.** $\frac{2}{9}; y = \frac{2}{9}x - \frac{2}{9}$
39. (a) -25.33 per dollar **(b)** -16.75 per dollar **(c)** -31.12 per dollar **41.** $y' = 12x - 7$ **43.** $y' = 18x^{7/2}$
45. $f'(x) = -6x^{-7} + \frac{1}{2}x^{-1/2}$ **47.** $y' = 20t^4 + 27t^2 - 40t$
49. $y' = 40x^{3/4} + 6(5x + 1)x^{-1/4}$ **51.** $f'(x) = \frac{-4(x^2 + 8)}{(x^2 - 8)^2}$
53. $y' = \frac{-3x + 7 + 36\sqrt{x}}{2\sqrt{x}(3x + 7)^2}$ **55.** $y' = \frac{x^2 - 2x}{(x - 1)^2}$
57. $f'(x) = 16(4x - 2)^3$ **59.** $y' = \frac{1}{\sqrt{2t - 5}}$
61. $y' = 18x(3x - 4)^2 + 2(3x - 4)^3$
63. $f'(u) = \frac{(2u + 3)(6u - 4) - 6(3u^2 - 4u)}{(2u + 3)^4}$ **65.** $y' = -6x^2e^{-2x^3}$
67. $y' = 10xe^{2x} + 5e^{2x}$ **69.** $y' = \frac{2x + 4}{x^2 + 4x - 1}$
71. $y' = \frac{x - \frac{1}{x} - 2x(\ln 6x)}{(x^2 - 1)^2}$ **73.** $y' = \frac{x^2 - 6x + 1}{(x - 3)^2}$
75. $y' = -12e^{2x}$ **77.** $y' = \frac{-1}{x^{1/2}(x^{1/2} - 1)^2}$
79. $\frac{dy}{dt} = \frac{1 + 2t^{1/2}}{4t^{1/2}(t^{1/2} + t)^{1/2}}$ **81.** $-\frac{2}{3}$ **83.** None
85. No, yes, yes **87.** No, yes, no **89.** Yes, no, yes
91. (a) \$687.83 billion **(b)** \$28.9 billion per year **(c)** \$30.651 billion per year **93. (a)** About 109.03 million kilowatts **(b)** About 2.15 million kilowatts per year **95. (a)** About \$19.6 billion **(b)** About \$2.16 billion per year **97. (a)** About \$1319.4 billion; about 1568.0 billion **(b)** About \$21.5 billion per year **99.** 0 mph per sec for the hands and approximately 640 mph per sec for the bat. These figures represent the acceleration of the hands and the bat at the moment when their velocities are equal. **101. (a)** \$54.75 **(b)** \$59.00 **(c)** Does not exist **(d)** \$54.75 **(e)** .5, 1, 2, 3, 4, 5, 6
(f)

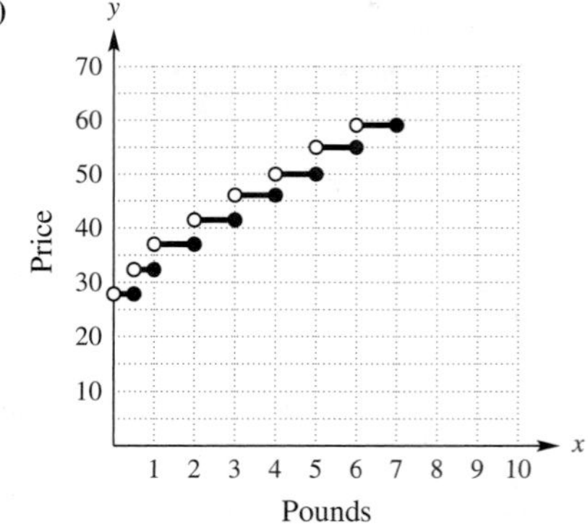

Case 11 (Page 767)

1. About .452 **3. (a)** About 1.1; about .65 **(b)** \$95.14
5. Demand is constant, no matter what the price.

CHAPTER 12

Section 12.1 (Page 780)

1. Increasing on $(1, \infty)$; decreasing on $(-\infty, 1)$; local minimum of -4 at $x = 1$ **3.** Increasing on $(-\infty, -2)$; decreasing on $(-2, \infty)$; local maximum of 3 at $x = -2$ **5.** Increasing on $(-\infty, -4)$ and $(-2, \infty)$; decreasing on $(-4, -2)$; local maximum of 3 at $x = -4$; local minimum of 1 at $x = -2$ **7.** Increasing on $(-7, -4)$ and $(-2, \infty)$; decreasing on $(-\infty, -7)$ and $(-4, -2)$; local maximum of 3 at $x = -4$; local minima of -2 at $x = -7$ and $x = -2$
9. Increasing on $\left(-\infty, -\frac{1}{3}\right)$ and $(2, \infty)$; decreasing on $\left(-\frac{1}{3}, 2\right)$
11. Increasing on $(-\infty, -3)$ and $(-3, \infty)$ **13.** Decreasing on $(-\infty, 6)$ **15.** Increasing on $(-\infty, -2)$ and $(1, \infty)$; decreasing on $(-2, 1)$ **17.** $-2, -1, 2$ **19.** Local maximum at -2; local minimum at 0 **21.** Local maximum at -3; local minimum at -1
23. Local maximum at $\frac{1}{4}$; local minimum at 5 **25.** Local maximum at -2; local minimum at 3 **27.** Local maximum at -2; local minimum at 0 **29.** Local maximum at 0; $f(0) = 1$; local minimum at $\frac{6}{11}$; $f\left(\frac{6}{11}\right) \approx .7804$ **31.** Local maximum at $\frac{3}{4}$; $f\left(\frac{3}{4}\right) = 4$
33. No local extrema **35.** Local maximum at -1; $f(-1) = \frac{1}{e}$
37. Local maximum at $-\frac{1}{e}$; $f\left(-\frac{1}{e}\right) = \frac{1}{e}$; local minimum at $\frac{1}{e}$; $f\left(\frac{1}{e}\right) = -\frac{1}{e}$ **39.** Local minimum at $-\frac{1}{3}$; $f\left(-\frac{1}{3}\right) \approx -2.1226$
41. Approximate local minimum: $(-5.5861, -563.4222)$
43. Approximate local maxima: $(-18.5239, 5982.7502)$ and $(.3837, 40.9831)$; approximate local minima: $(-2.8304, 11.4750)$ and $(4.9706, -53.7683)$ **45.** Within 2001 and 2004
47. (a) (2004, 2006] **(b)** 2003 **(c)** Design; steeper slope
49. Approximate local minimum at (4.03, 468.36); approximate local maximum at (15.55, 886.61) **51.** Increasing on $(-\infty, 0)$; decreasing on $(0, \infty)$ **53.** (0, 8) **55. (a)** (0, 2.16) **(b)** (2.16, 6) **57. (a)** (5, 16), (18, 19) **(b)** (0, 5), (16, 18), (19, 23) **(c)** 5:00 A.M., 4:00 P.M., 6:00 P.M., 7:00 P.M. **59. (a)** \$7.75, \$6.10, \$6.25 **(b)** 5 **(c)** 1989, 1995, 2006
61. (a)

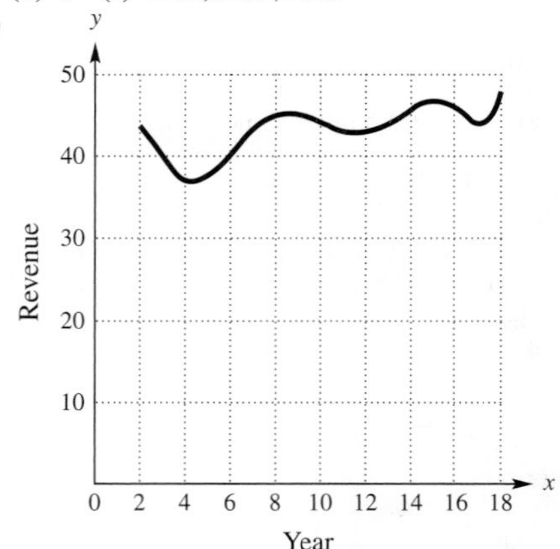

(b) 5 **(c)** (4.306, 37.072), (8.534, 45.376), (11.561, 43.033), (15.037, 46.981), (17.112, 44.285)

63. (a)

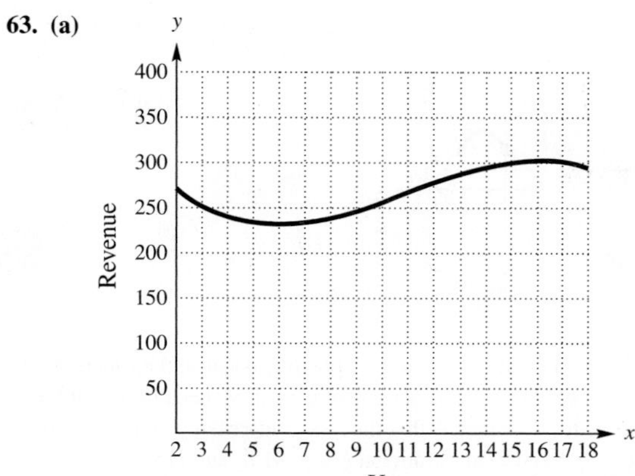

(b) 2 **(c)** Local minimum at (5.95, 231.08); local maximum at (16.12, 302.13) **65.** $x = 44.30$ (2034), with a population of approximately 1.46 billion **67.** Local maximum at (8.14, 5.81); local minimum at (13.10, 1.38) **69.** 10 min

Section 12.2 (Page 795)

1. $f''(x) = 6x - 12; -12; 0; -30$
3. $f''(x) = 12(x + 3)^2; 108; 300; 0$
5. $f''(x) = \frac{2}{(1 + x)^3}; 2; \frac{2}{27}; -\frac{1}{4}$
7. $f''(x) = -\frac{(x + 4)^{-3/2}}{4}$ or $-\frac{1}{4(x + 4)^{3/2}}; -\frac{1}{32}; -\frac{1}{4 \cdot 6^{3/2}}; -\frac{1}{4}$
9. $f''(x) = \frac{-4}{5x^{6/5}}$; undefined; $\frac{-2^{4/5}}{5}; \frac{4}{15(-3)^{1/5}}$
11. $f''(x) = 2e^x; 2; 2e^2; \frac{2}{e^3}$ **13.** $f''(x) = 24e^{2x}; 24; 24e^4; \frac{24}{e^6}$
15. $f''(x) = -\frac{1}{x^2}$; not defined; $-\frac{1}{4}; -\frac{1}{9}$
17. $f''(x) = \frac{1}{x}$; not defined; $\frac{1}{2}; -\frac{1}{3}$ **19.** Answers vary.
21. $v(t) = 12t + 5; a(t) = 12$; $v(0) = 5$ cm/sec; $v(4) = 53$ cm/sec; $a(0) = a(4) = 12$ cm/sec^2
23. $v(t) = 9t^2 - 8t + 8; a(t) = 18t - 8; v(0) = 8$ cm/sec; $v(4) = 120$ cm/sec; $a(0) = -8$ cm/sec^2; $a(4) = 64$ cm/sec^2
25. Concave upward on $(0, \infty)$; concave downward on $(-\infty, 0)$; point of inflection at $(0, -5)$ **27.** Concave upward on $\left(-\frac{4}{3}, \infty\right)$; concave downward on $\left(-\infty, -\frac{4}{3}\right)$; point of inflection at $\left(-\frac{4}{3}, \frac{425}{27}\right)$
29. Concave upward on $(4, \infty)$; concave downward on $(-\infty, 4)$; no points of inflection **31.** Concave upward on $(-\infty, -5)$ and $(1, \infty)$; concave downward on $(-5, 1)$; points of inflection at $(-5, -1248)$ and $(1, 0)$ **33.** (14, 26,688) **35.** No critical numbers; no local extrema
37. Local maximum at $x = -5$; local minimum at $x = 4$
39. Local maximum at $x = 0$; local minima at $x = 2$ and $x = -2$
41. Local maximum at $x = -2$; local minimum at $x = 2$
43. Local maximum at $x = -3$; local minimum at $x = 3$
45. No critical numbers; no local extrema **47.** Local maximum at $x = 2$; local minima at $x = 1$ and $x = 4$; points of inflection at $x = \frac{7 + \sqrt{7}}{3}$ and $x = \frac{7 - \sqrt{7}}{3}$ **49.** Local minimum at $x = 1$; points of inflection at $x = \frac{4}{3}$ and $x = 2$ **51. (a)** *E* **(b)** *C* **(c)** *A, F* **(d)** *D* **(e)** *B* **53. (a)** [1, 15] **(b)** About 40 cents per week **55. (a)** -96 ft/sec **(b)** -160 ft/sec **(c)** -256 ft/sec **(d)** -32 ft/sec^2 **57.** About 6.5 million farms **59.** (8.8, 9.2); GDP begins increasing quickly at 8.8 **61. (a)** $x = 6.93, 17.89$ **(b)** 1986, 1997 **(c)** (12.41, 327.11); the rate of defense spending stopped decreasing. **63. (a)** 1991 **(b)** 1998 **65. (a)** No **(b)** Yes, (12.38, 3067.31) **67. (a)** 22.5 days **(b)** 71.19%

Section 12.3 (Page 809)

1. Absolute maximum at $x = 4$; absolute minimum at $x = 1$
3. Absolute maximum at $x = 2$; absolute minimum at $x = -2$
5. Absolute maximum at $x = -4$; absolute minima at $x = -7$ and $x = -2$ **7.** Absolute maximum at $x = 6$; absolute minima at $x = -4$ and $x = 4$ **9.** Absolute maximum at $x = 6$; absolute minimum at $x = 4$ **11.** Absolute maximum at $x = \sqrt{2}$; absolute minimum at $x = -1$ **13.** Absolute maximum at $x = -3$; absolute minimum at $x = 0$ **15.** Absolute maximum at $x = 0$; absolute minima at $x = -1$ and $x = 1$ **17.** Absolute minimum at $x = 2$
19. No absolute extrema **21.** Absolute maximum at $x = 1$
23. 179; \$481.98 per stereo **25. (a)** $R(x) = 400x$ **(b)** $P(x) = 430x - 525{,}000 - .012x^2$ **(c)** About 17,917 grills; \$3,327,083 profit **27. (a)** 8 **(b)** 29.43% **29.** In 2005
31. 2003 **33.** 8 in. by 8 in. **35.** 20 cm by 20 cm by 40 cm; \$72
37. 3 ft by 6 ft by 2 ft **39. (a)** $1200 - 2x$ **(b)** $A(x) = 1200x - 2x^2$ **(c)** 300 m **(d)** 180,000 m^2 **41.** 20,000 ft^2 **43.** \$1000
45. (a) $R(x) = 100{,}000x - 100x^2$ **(b)** 500 **(c)** 25,000,000¢, or \$250,000 **47.** 2009 **49.** 1 mile from point *A* **51.** 6
53. 69 **55.** 3333 booklets **57.** (c)

Section 12.4 (Page 820)

1. No points of inflection **3.** Point of inflection at $x = \frac{1}{3}$

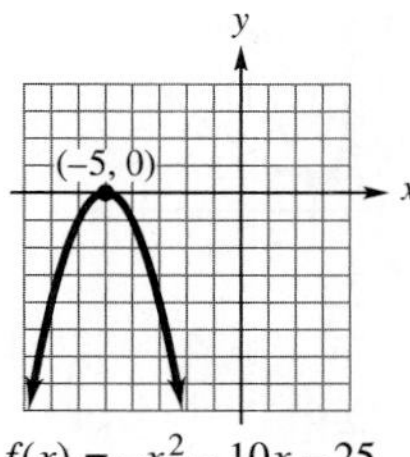

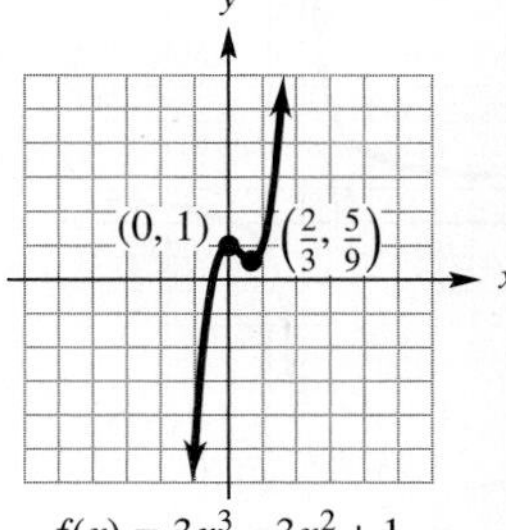

5. Point of inflection at $x = -\frac{3}{2}$ **7.** Point of inflection at $x = -\frac{7}{12}$

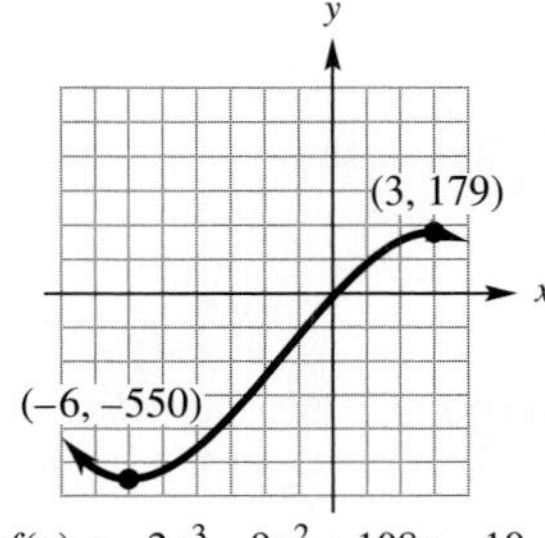

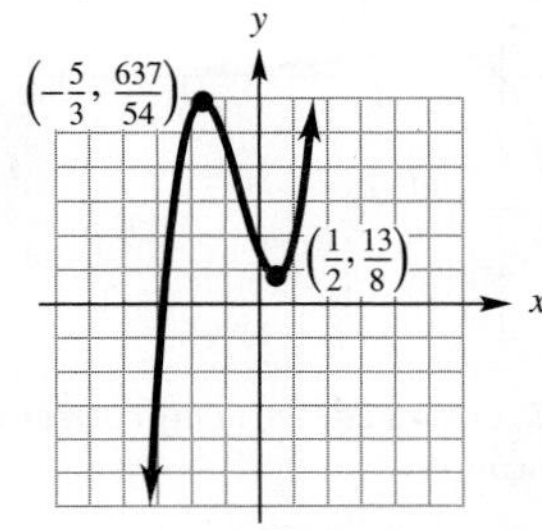

9. No points of inflection

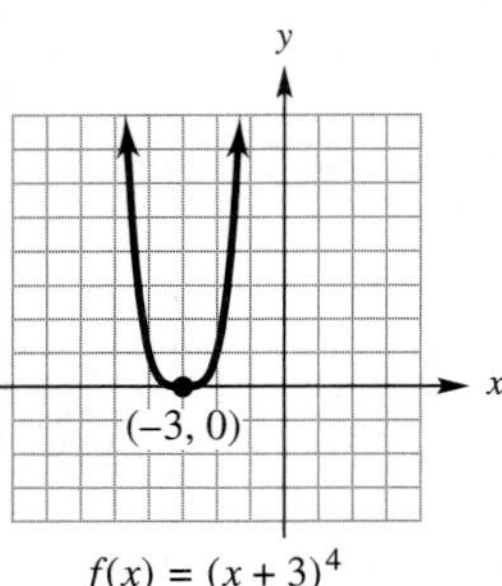

$f(x) = (x + 3)^4$

11. Points of inflection at $x = -\sqrt{3}$ and $x = \sqrt{3}$

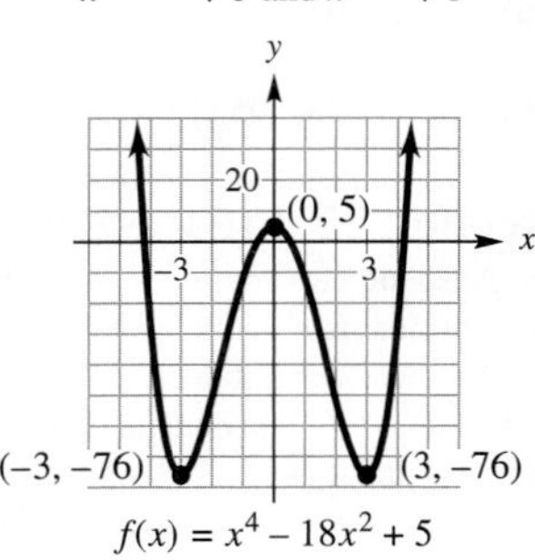

$f(x) = x^4 - 18x^2 + 5$

13. No points of inflection

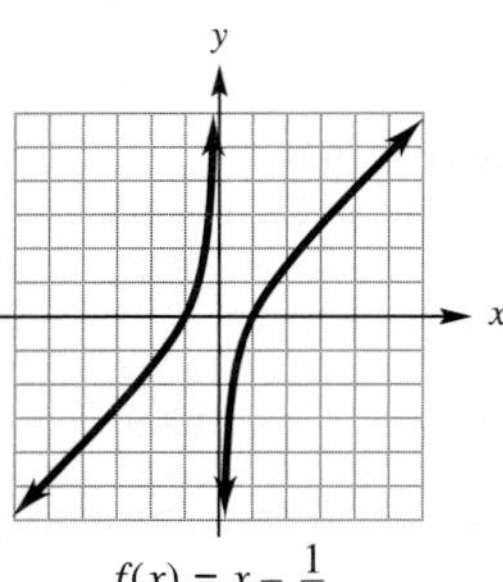

$f(x) = x - \frac{1}{x}$

15. No points of inflection

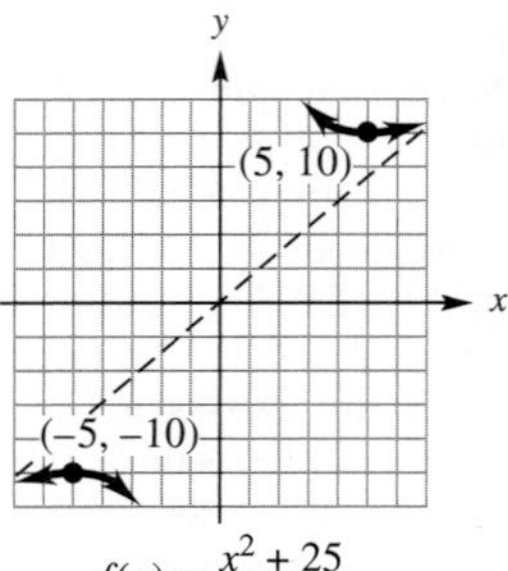

$f(x) = \frac{x^2 + 25}{x}$

17. No points of inflection

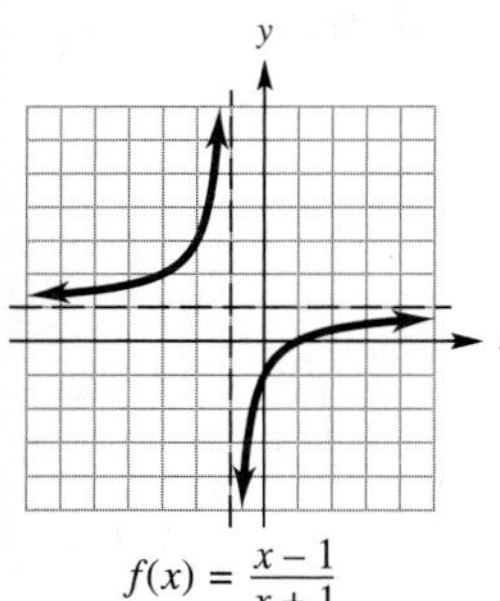

$f(x) = \frac{x - 1}{x + 1}$

19. No points of inflection

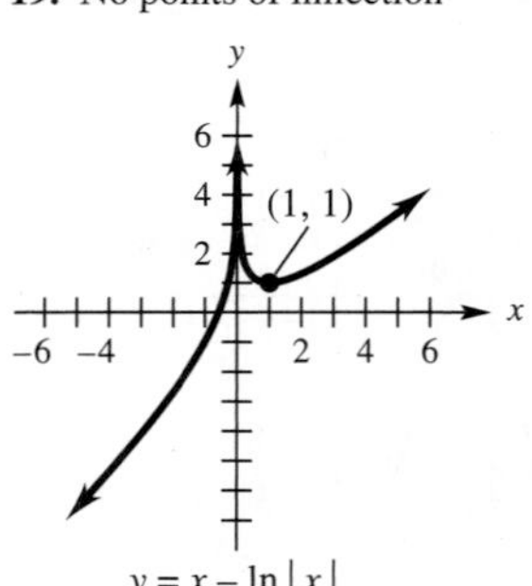

$y = x - \ln|x|$

21. Point of inflection at $x = 2$

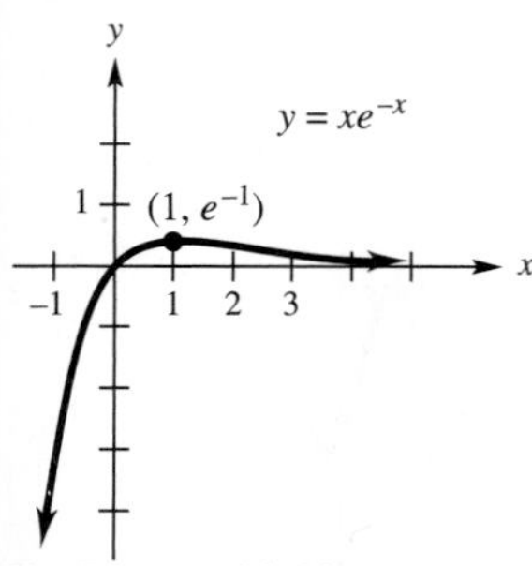

For Exercises 23–28, many correct answers are possible, including the following:

23.

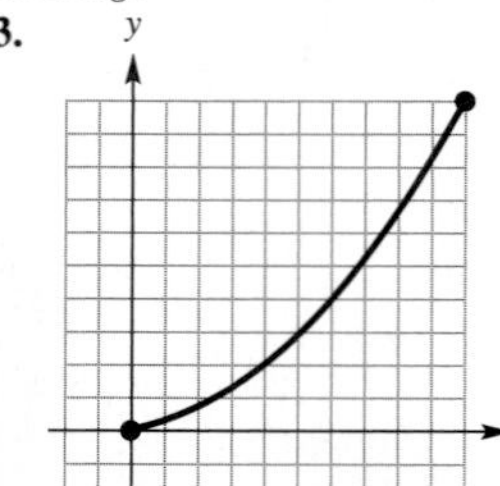

25.

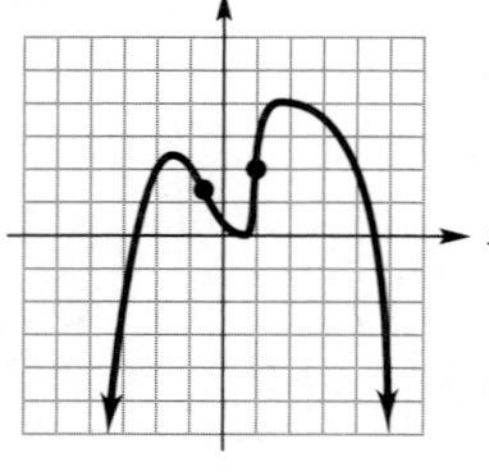

27.

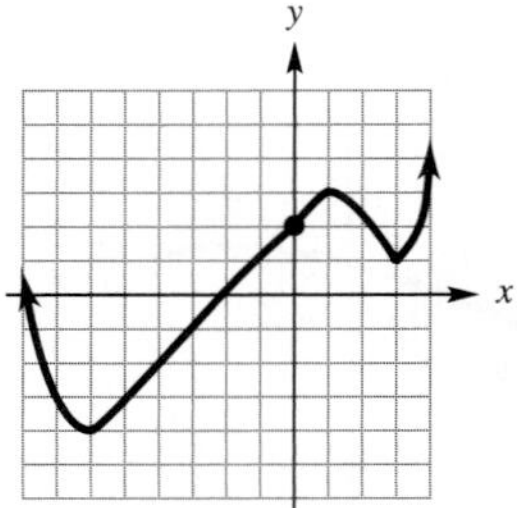

29. (a) The first derivative is always positive because the function is increasing. **(b)** The second derivative is positive because the graph is concave upward. This means that the risk is increasing at a faster and faster rate. **31.** Local minima at (10.3, 81.8) and (15.9, 93.4); local maximum at (14.5, 94.1); inflection points at (15.3, 93.7) and (11.9, 86.8) **33.** Local minimum at (.6, 2.1); inflection point at (6.5, 4.0) **35.** Local minimum at (5.9, 18.2); local maxima at (.9, 34.6) and (8.1, 20.3); inflection points at (2.8, 27.8) and (7.1, 19.3) **37.** No local maxima or minima; inflection point at (20.4, 58.4) **39.** Local minimum at (14.7, 11.7); local maximum at (32.4, 15.2); inflection points at (5.7, 13.1) and (25.0, 13.7)

Chapter 12 Review (Page 823)

1. Answers vary. **3.** Answers vary. **5.** Increasing on $\left(-\frac{9}{2}, \infty\right)$; decreasing on $\left(-\infty, -\frac{9}{2}\right)$ **7.** Increasing on $\left(-\infty, -\frac{2}{3}\right)$ and $(1, \infty)$; decreasing on $\left(-\frac{2}{3}, 1\right)$ **9.** Decreasing on $(-\infty, 4)$ and $(4, \infty)$ **11.** Local maximum of 54 at $x = -2$; local minimum of -71 at $x = 3$ **13.** Local maximum of 2 at $x = 0$; local minima of $-\frac{1}{3}$ at $x = -1$ and -43 at $x = 3$ **15.** Local maximum of $\frac{3}{e}$ at $x = 1$ **17.** $f''(x) = 40x^3 - 30x$; 10; -260 **19.** $f''(x) = -20e^{2x}$; $-20e^2$; $\frac{-20}{e^4}$ **21.** No local extrema; point of inflection at $x = -\frac{1}{12}$; decreasing on $(-\infty, \infty)$; concave upward on $\left(-\infty, -\frac{1}{12}\right)$; concave downward on $\left(-\frac{1}{12}, \infty\right)$

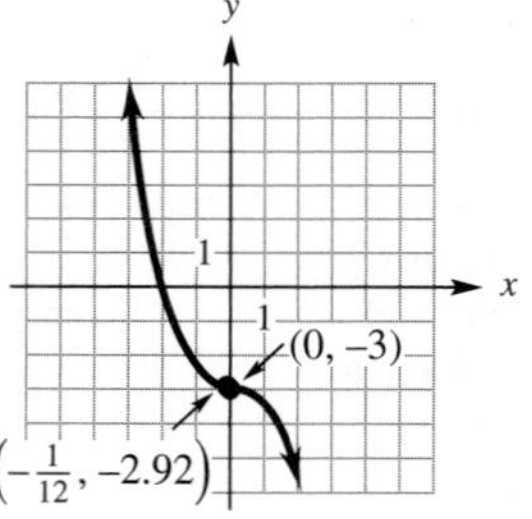

23. Local maximum at $x = 0$; local minima at $x = -1$ and $x = 2$; points of inflection at $x = \frac{1 - \sqrt{7}}{3}$ and $x = \frac{1 + \sqrt{7}}{3}$; increasing on $(-1, 0)$ and $(2, \infty)$; decreasing on $(-\infty, -1)$ and $(0, 2)$; concave upward on $\left(-\infty, \frac{1 - \sqrt{7}}{3}\right)$ and $\left(\frac{1 + \sqrt{7}}{3}, \infty\right)$; concave downward on $\left(\frac{1 - \sqrt{7}}{3}, \frac{1 + \sqrt{7}}{3}\right)$

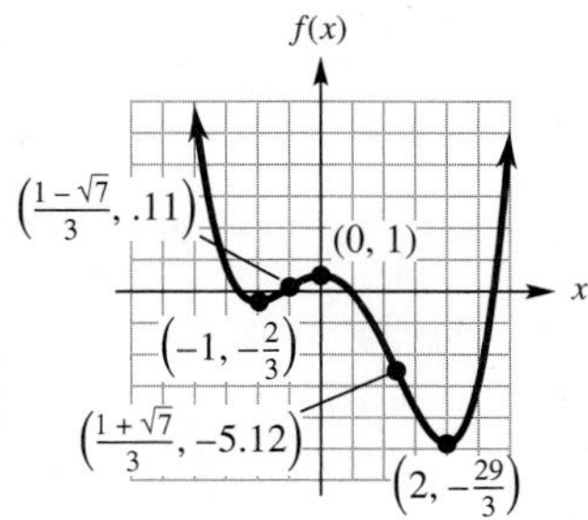

25. No local extrema or points of inflection; increasing on $\left(-\infty, -\frac{1}{2}\right)$; and $\left(-\frac{1}{2}, \infty\right)$; concave upward on $\left(-\infty, -\frac{1}{2}\right)$; concave downward on $\left(-\frac{1}{2}, \infty\right)$

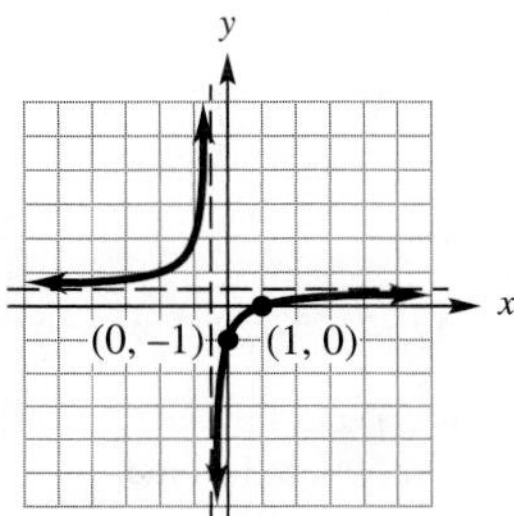

27. Local maximum at $x = \frac{1}{2}$; local minimum at $x = -\frac{2}{3}$; point of inflection at $x = -\frac{1}{12}$; increasing on $\left(-\frac{2}{3}, \frac{1}{2}\right)$; decreasing on $\left(-\infty, -\frac{2}{3}\right)$ and $\left(\frac{1}{2}, \infty\right)$; concave upward on $\left(-\infty, -\frac{1}{12}\right)$; concave downward on $\left(-\frac{1}{12}, \infty\right)$

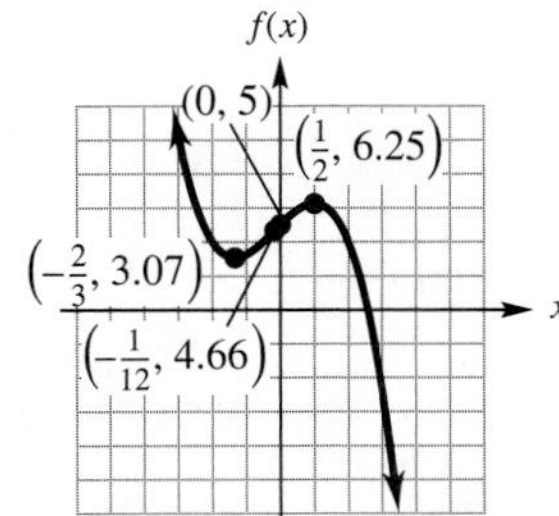

29. Local minimum at $x = 0$; increasing on $(0, \infty)$; decreasing on $(-\infty, 0)$; concave upward on $(-\infty, \infty)$. In the graph, the horizontal lines are 10 units apart.

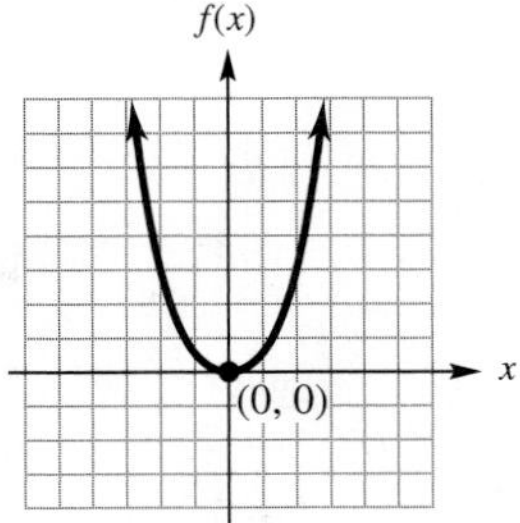

31. Local maximum at $x = -2$; local minimum at $x = 2$; no point of inflection; increasing on $(-\infty, -2)$ and $(2, \infty)$; decreasing on $(-2, 0)$ and $(0, 2)$; concave upward on $(0, \infty)$; concave downward on $(-\infty, 0)$. In the graph, the horizontal lines are 2 units apart.

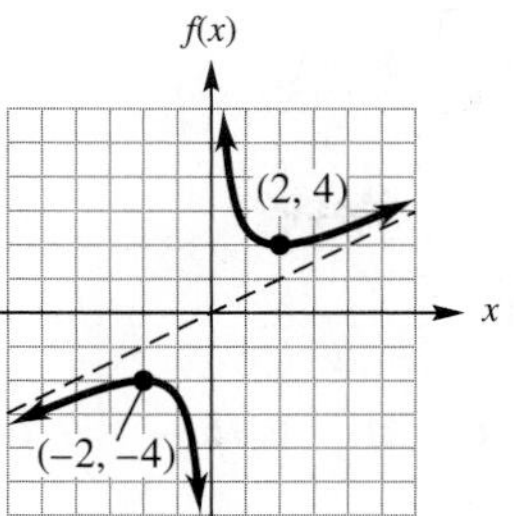

33. Absolute maximum of 10 at $x = 3$; absolute minima of 9 at $x = 2$ and $x = 4$ **35.** Absolute maximum of 11.217 at $x = 3.3$; absolute minimum of $-\frac{319}{27}$ at $x = \frac{5}{3}$ **37.** 12° **39.** Minimum at $x = 9.22$, net earnings $\approx$ \$11.76 billion; maximum at $x = 16.75$, net earnings $\approx$ \$19.85 billion **41. (a)** Local maximum at (11.74, 22.24); local minimum at (13.96, 21.74) **(b)** \$18.07 billion **(c)** \$32.71 billion **43. (a)** 1998; 37.4% **(b)** (25.33, 29.0) **45.** 125 m × 250 m **47.** 3 in. **49.** About 4531

Chapter 12 Case (Page 825)

1. $Z'(m) = -\frac{C_1}{m^2} + \frac{DC_3}{2}$ **3.** $D = 3, t = 12, C_1 = 15{,}000, C_3 = 900, m = \frac{10}{3}$

5. $Z(m^+) = Z(4) = \$11{,}400$; $Z(m^-) = Z(3) = \$11{,}300$

CHAPTER 13

Section 13.1 (Page 836)

1. $F(x)$ and $G(x)$ must differ by a constant. **3.** Answers vary. **5.** $6x^2 + C$ **7.** $2p^4 + C$ **9.** $105x + C$ **11.** $\frac{5z^2}{2} - z + C$ **13.** $\frac{z^3}{3} - 2z^2 + 6z + C$ **15.** $\frac{x^4}{4} - \frac{14x^3}{3} + 11x^2 + 8x + C$ **17.** $4y^{3/2} + C$ **19.** $\frac{12t^{5/2}}{5} + \frac{12t^{5/4}}{5} + C$ **21.** $\frac{112t^{3/2}}{3} + 4t^{9/2} + C$ **23.** $-\frac{12}{x^2} + C$ **25.** $-\frac{1}{2y^2} - 4y^{1/2} + C$ **27.** $-3x^{-2} + 2\ln|x| + C$ **29.** $\frac{4e^{3u}}{3} + C$ **31.** $-\frac{15}{4}e^{-.8x} + C$ **33.** $6\ln|x| - 8e^{-.5x} + C$ **35.** $\ln|t| + \frac{t^4}{2} + C$ **37.** $\frac{e^{2u}}{2} + 4\ln|u| + C$ **39.** $\frac{25x^3}{3} + 5x^2 + x + C$ **41.** $\frac{6x^{7/6}}{7} + \frac{3x^{2/3}}{2} + C$ **43.** $f(x) = 2x^3 - 2x^2 - 3x + 1$ **45. (a)** $G(x) = .07x^3 + .18x^3 + 11.38x + 205$ **(b)** \$317.49 billion **47. (a)** $F(x) = 4.71x^2 + 87.14x + 1396.8$ **(b)** \$3539.92 billion **49.** $x^3 - 3x^2 + 4x - 40$ **51.** $C(x) = \frac{3x^{5/3}}{5} + 5x - 1.2$ **53.** $C(x) = .067x^3 + .2x^2 + .8x + 6.1$ **55.** $C(x) = \frac{800}{e^{.05x}} + 100x + 276.96$ **57. (a)** $F(x) = 5.282e^{.01278t}$ **(b)** 7.2704 billion **59. (a)** $G(x) = 48.28 - 10.2\ln x$ **(b)** Age 18, 18.8; age 20, 17.7; age 50, 8.4; age 70, 4.9 **(c)** Answers vary.

Section 13.2 (Page 845)

1. Answers vary. **3.** $\frac{(12x-1)^3}{12}+C$ **5.** $\frac{-5}{3(3t-6)}+C$

7. $-\frac{1}{\sqrt{x^2+2x-4}}+C$ **9.** $\frac{2(r^3+3)^{3/2}}{9}+C$

11. $-\frac{4}{5}e^{5k}+C$ **13.** $\frac{e^{2w^4}}{2}+C$ **15.** $\frac{e^{4t-t^2}}{2}+C$

17. $2e^{\sqrt{y}}+C$ **19.** $-\frac{5\ln|12+6x|}{6}+C$ **21.** $\frac{1}{2}\ln|e^{2t}+1|+C$

23. $-\frac{1}{8(2x^2+8x)^2}+C$ **25.** $\frac{5\left(\frac{1}{r}+r\right)^2}{2}+C$

27. $(x^3+3x)^{1/3}+C$ **29.** $\ln|3x^2+7x+8|+C$

31. $\frac{(x^2+5)^4}{4}+C$ **33.** $\frac{(x^2+12x)^{3/2}}{3}+C$

35. $\frac{(10+\ln(x))^3}{3}+C$ **37.** $\frac{10}{3}(u-1)^{3/2}+10(u-1)^{1/2}+C$

39. $\frac{2(5t-1)^{5/2}}{125}+\frac{2(5t-1)^{3/2}}{75}+C$ **41.** **(a)** $6\ln(5x^2+1)+10$ **(b)** Yes **43.** **(a)** $G(x)=4.482e^{.2483x}-.74$ **(b)** About \$24.747 billion **45.** **(a)** $R(x)=2.7(x^2+27{,}000)^{1/3}-67.14$ **(b)** About \$53,633 **(c)** 459 **47.** **(a)** $G(x)=.00040674\,[(x-1970)^{2.4}/2.4+1970(x-1970)^{1.4}/1.4]+61.298$ **(b)** About 180,945

Section 13.3 (Page 855)

1. About 151 kWh **3.** 10,983.6 million short tons **5.** Answers vary. **7.** **(a)** 44 **(b)** 50 **(c)** 55.4 **9.** **(a)** 8 **(b)** 10 **(c)** 10.66 **11.** **(a)** 7.39 **(b)** 14.596 **(c)** 24.48 **13.** **(a)** 3 **(b)** 3.5 **(c)** 4 **15.** 228.75 **17.** 90.55 **19.** .294 **21.** \$808,680.51 **23.** \$750,533.33 **25.** About 4063.27 billion kWh **27.** About 2400 feet **29.** **(a)** 9 feet **(b)** At 2 seconds **(c)** 4 feet ahead **(d)** Between 3 and 3.5 seconds

Section 13.4 (Page 868)

1. 60 **3.** 13 **5.** 8 **7.** 138 **9.** $8\frac{1}{4}$ **11.** 9.9873

13. 1.294 **15.** 27058.06 **17.** $\frac{91}{3}$ **19.** 63.857 **21.** .1509

23. 49 **25.** .3005 **27.** 105.3946 **29.** $\ln\left(\frac{108}{37}\right)\approx 1.0712$

31. A loss, $\int_0^2(6x^2-7x-3)\,dx$ **33.** $\frac{64}{3}$ **35.** 4.75 **37.** 3.7037

39. .6667 **41.** 1 **43.** 11.08 **45.** $\frac{23}{6}$ **47.** 2.67

49. 19 **51.** **(a)** \$22,000 **(b)** \$108,000 **(c)** 19 days **53.** **(a)** \$4101.27 **(b)** 1.81 years **55.** 8294.1952 quadrillion BTUs **57.** **(a)** 363 **(b)** 167 **(c)** Decreasing to 0 **59.** **(a)** 274.14 million people age 0–90 living **(b)** 70.8 million **61.** **(a)** $y'=\ln|x|$ **(b)** About \$274.4356 billion **63.** .811 **65.** .244

Section 13.5 (Page 880)

1. \$280; \$11.67 **3.** \$87.62 billion **5.** \$2500; \$30,000 **7.** No

9. $\frac{56}{3}$ **11.** $\frac{5}{192}$ **13.** 325.04 **15.** 4.999 **17.** **(a)** 8 years **(b)** \$147.96 **(c)** About \$771 **19.** **(a)** 27 days **(b)** \$2872.12 **(c)** \$375.98 **(d)** \$2496.14 **21.** **(a)** 3.02 years **(b)** 12.72 trillion gallons **23.** **(a)** About 88.22 gal **(b)** About 40 hours **(c)** About 128 gal **(d)** About 75.78 hours

25. **(a)**

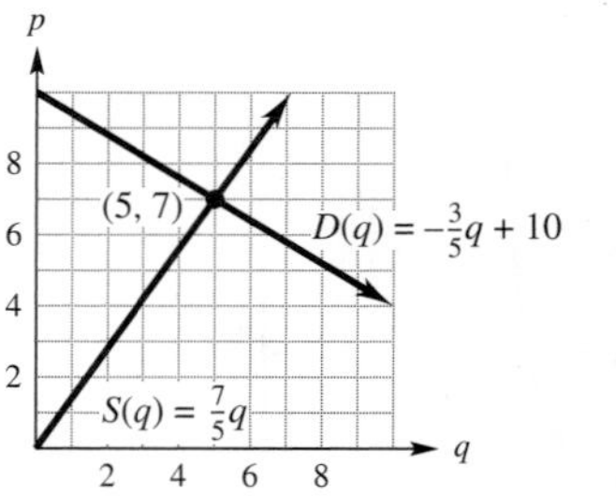

(5, 7)

$D(q)=-\frac{3}{5}q+10$

$S(q)=\frac{7}{5}q$

(b) (5, 7) **(c)** \$7.50 **(d)** \$17.50 **27.** \$1999.54 **29.** \$38.50

31. **(a)**

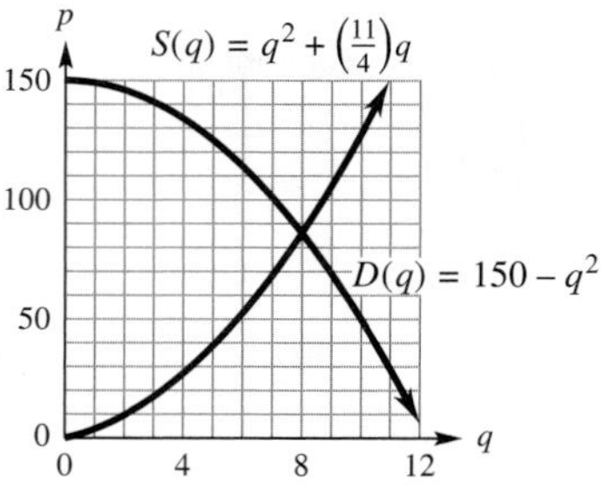

$S(q)=q^2+\left(\frac{11}{4}\right)q$

$D(q)=150-q^2$

(b) (8,86) **(c)** \$341.33 **(d)** \$429.33 **33.** $\frac{27}{16}$ **35.** \$753.89

37. **(a)** 1.9%; the lower 10% of income producers earn 1.9% of the total income of the population. **(b)** 27.5%; the lower 50% of income producers earn 27.5% of the total income of the population. **(c)** 81.9%; the lower 90% of income producers earn 81.9% of the total income of the population.

(d)

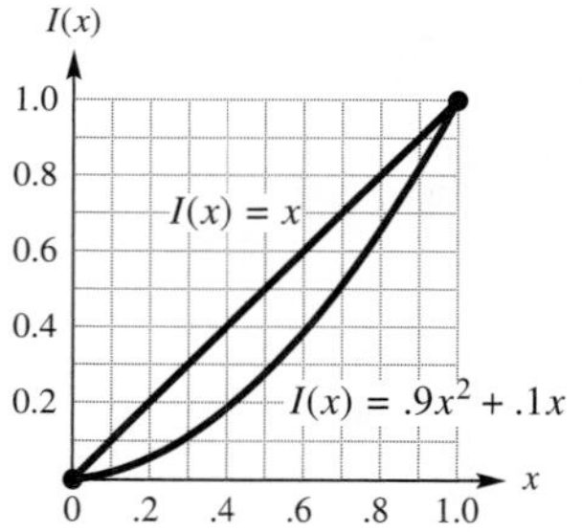

$I(x)=x$

$I(x)=.9x^2+.1x$

(e) .15 **39.** **(a)** 10 items **(b)** 10 hours per item

Section 13.6 (Page 885)

1. $-7\ln|x+\sqrt{x^2+36}|+C$ **3.** $\ln\left|\frac{x-3}{x+3}\right|+C,(x^2>9)$

5. $\frac{4}{3}\ln\left|\frac{3+\sqrt{9-x^2}}{x}\right|+C,(0<x<3)$

7. $-\frac{5x}{3}+\frac{5}{9}\ln|3x+1|+C$ **9.** $-\frac{13}{15}\ln\left|\frac{x}{3x-5}\right|+C$

11. $\ln\left|\frac{2x-1}{2x+1}\right|+C,\left(x^2>\frac{1}{4}\right)$

13. $-3\ln\left|\frac{1+\sqrt{1-9x^2}}{3x}\right|+C,\left(0<x<\frac{1}{3}\right)$

15. $\frac{15x}{2}-\frac{45}{4}\ln|2x+3|+C$ **17.** $\frac{1}{25(5x-1)}-\frac{\ln|5x-1|}{25}+C$

19. $\frac{3x^5}{4}\left(\frac{\ln|x|}{5}-\frac{1}{25}\right)+C$ **21.** $\frac{7}{x}(-\ln|x|-1)+C$

23. $-\frac{xe^{-2x}}{2}-\frac{e^{-2x}}{4}+C$ **25.** \$10,473.56 **27.** 12,613 microbes

Section 13.7 (Page 893)

1. $y = -x^3 + \frac{7x^2}{2} + C$ **3.** $y = \frac{3x^4}{8} + C$ **5.** $y^2 = x^2 + C$ **7.** $y = Me^{2x^2}$ **9.** $y = Me^{x^4 - x^3}$ **11.** $y = Mx$ **13.** $y = 7 + Me^x$ **15.** $y = -\frac{1}{e^x + C}$ **17.** $y = x^3 - x^2 + 4$ **19.** $y^2 = 4 \ln |x^3 + 28| + 6$ **21.** $y^2 = \frac{2x^3}{3} + 16$ **23.** $y = e^{2.5x^2 + 3x}$ **25.** $y^2 - 6y = 7x^2 + 2x - 8$ **27.** Answers vary. **29. (a)** 3536 units **(b)** 6356 units **31. (a)** $y = 4t^2 + 50$ **(b)** $t = 11$ days **33. (a)** $k = .8$ **(b)** 11 people **(c)** 55 people **(d)** 2981 people **35. (a)** $\frac{dy}{dt} = -.15y$ **(b)** $y = Me^{-.15t}$ **(c)** About 9.2 years **37.** 4.7 cc **39. (a)** 161.28 thousand **(b)** 30.88 thousand **(c)** 30.44 thousand **41. (a)** $T = 68 + 30.6e^{-.33t}$ **(b)** 83.8° **(c)** 4.5 hours **(d)** 24.3 hours **43.** $q = \frac{C}{p^2}$

Chapter 13 Review (Page 896)

1. $\frac{x^3}{3} - 4x^2 - 7x + C$ **3.** $\frac{14x^{3/2}}{3} + C$ **5.** $\frac{2x^{3/2}}{3} + 9x^{1/3} + C$ **7.** $3x^{-2} + C$ **9.** $-\frac{3e^{2x}}{2} + C$ **11.** $10 \ln|x - 3| + C$ **13.** $\frac{5e^{3x^2}}{6} + C$ **15.** $\frac{3 \ln|x^2 - 1|}{2} + C$ **17.** $-\frac{4}{3}(x^3 + 5)^{-3} + C$ **19.** $\ln|2x^2 - 5x| + C$ **21.** $-\frac{e^{-3x^3}}{9} + C$ **23.** $\frac{2e^{-5x}}{5} + C$ **25.** $\frac{(\ln x)^6}{2} + C$ **27.** $-\frac{e^{-50x}}{2} + C$ **29.** Answers vary. **31. (a)** 140 **(b)** 148.8 **33.** 28; answers vary. **35.** $\frac{128}{3}$ **37.** $\frac{381}{200}$ **39.** $15 \ln 5 \approx 24.1416$ **41.** .40 **43.** 66.907 **45.** 6.313 **47.** 6.3891 **49.** $C(x) = 10x - \frac{5x^2}{2} + 4$ **51.** $C(x) = (2x - 1)^{3/2} + 145$ **53.** 45,000 units **55.** 26.28 years **57.** 2,607,276 births **59.** 29,034,600,000 barrels **61.** Answers vary. **63.** About 110 people **65. (a)** .2293 **(b)** .1720 **67.** $-2 \ln\left|\frac{5 + \sqrt{25 + x^2}}{x}\right| + C$ **69.** $\frac{15x}{2} + \frac{75}{4} \ln|2x - 5| + C$ **71.** $y = \frac{x^4}{2} + 3x^2 + 5x + C$ **73.** $y^2 = 3x^2 + 2x + C$ **75.** $y = -5e^{-x} - 5x + 22$ **77.** $y = 2e^{5x - x^2}$ **79. (a)** $\frac{dy}{dt} = .0264y$ **(b)** $y = 172.3e^{.0264t}$ **(c)** 207.27 **81. (a)** 111 items **(b)** Not exactly, but for practical purposes, yes. **83.** $.2 \ln y - .5y = .3 \ln x + .4x + C, x = \frac{3}{4}$ unit and $y = \frac{2}{5}$ unit

Case 13(Page 900)

1. Answers vary. **3.** 138 years

CHAPTER 14

Section 14.1 (Page 912)

1. 4;−22; −20; 12 **3.** 3; $\sqrt{33}$; $\sqrt{17}$; 8 **5. (a)** 92 **(b)** −17 **(c)** 47 **(d)** −44 **7.** All ordered pairs (x, y), where x and y are real numbers **9.** All order pairs (h, w), where h is a nonzero real number and w is a real number **11.** All ordered triples (u, v, w), where u, v, and w are real numbers such that $v \neq w$ **13.** $P(l, w) = 2l + 2w$ **15.** $S(R, i, n) = R\frac{(1 + i)^n - 1}{i}$ **17.** Answers vary.

19.

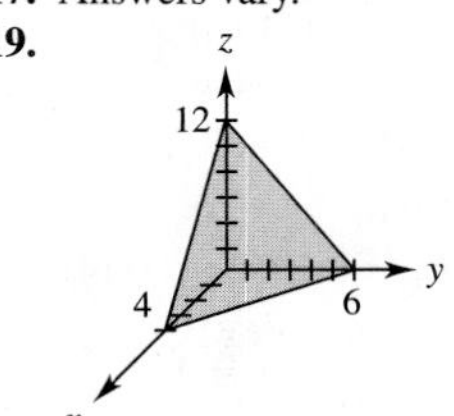

21. **23.**

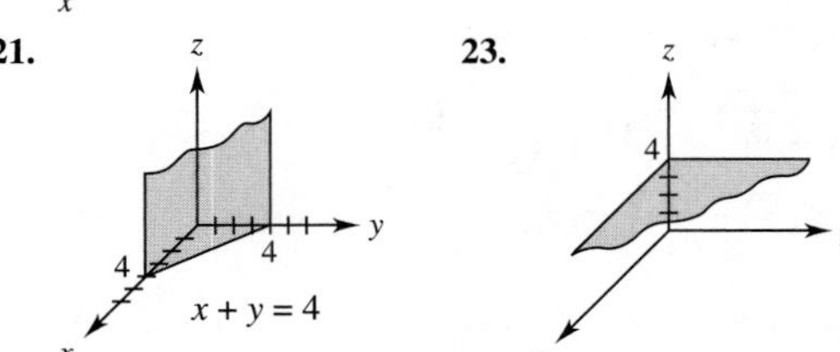

25. $3x + 2y + z = 18$

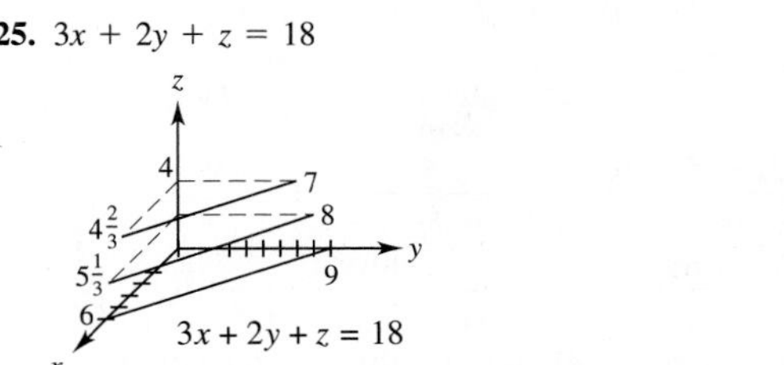

27. $y^2 - x = -z$

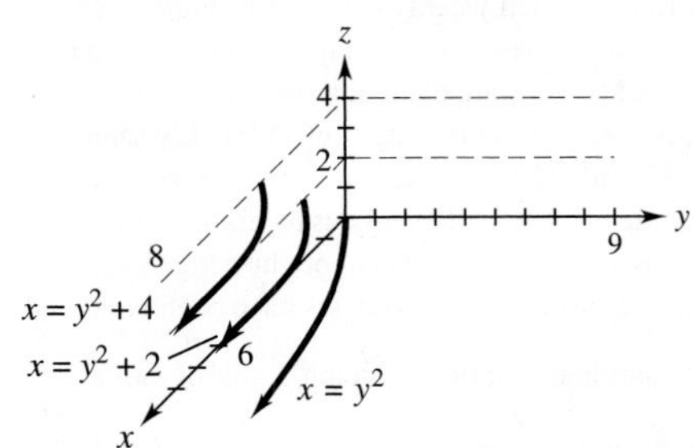

29.

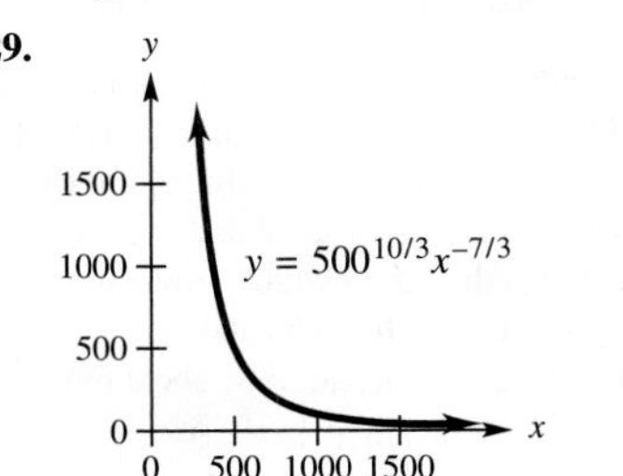

31. z is multipled by $2^{.7}$ ($\approx$ 1.6); z is multiplied by $2^{.3}$($\approx$ 1.2); z is doubled. **33. (a)** About 1987 **(b)** About 595 **(c)** About 359,768 **35. (a)** 73; comfortable **(b)** 77; uncomfortable **37. (a)** 1.8892 **(b)** 1.5198 **(c)** 1.7278 **(d)** Answers vary. **39.** 1.14; the IRA grows faster. **41. (a)** 1.5 m/sec; 5.5 m/sec **(b)** 1 m/sec **43.** $f(L, W, H) = L + 2H + 2W$ **45.** (c) **47.** (e) **49.** (b)

Section 14.2 (Page 922)

1. (a) $24x^2 - 8xy$ **(b)** $-4x^2 + 18y$ **(c)** 48 **(d)** −40 **3.** $f_x(x, y) = -4y; f_y(x, y) = -4x + 18y^2$; 4; 178

5. $f_x = 2e^{2x+y}; f_y = e^{2x+y}; 2e^3; e^{-5}$

7. $f_x = \frac{2}{e^{x+2y}}; f_y = \frac{4}{e^{x+2y}}; 2; \frac{4}{e^2}$

9. $f_x = \frac{y^3 - x^2 - 6xy^2}{(x^2+y^3)^2}; f_y = \frac{6x^2y - 3xy^2 - 3y^4}{(x^2+y^3)^2}; -\frac{17}{9}; \frac{153}{1849} \approx .083$

11. $f_x(x, y) = \frac{15x^2y^2}{1 + 5x^3y^2}; f_y(x, y) = \frac{10x^3y}{1 + 5x^3y^2}; \frac{60}{41}; \frac{1920}{2879}$

13. $f_x = 2x^2ye^{2xy} + 2xe^{2xy}; f_y = 2x^3e^{2xy}; -4e^{-4}; -128e^{-24}$

15. $f_{xx} = 8y^2 - 32; f_{yy} = 8x^2; f_{xy} = f_{yx} = 16xy$

17. $h_{xx} = -8y^2; h_{xy} = h_{yx} = -16xy + 14y; h_{yy} = -6 - 8x^2 + 14x$

19. $r_{xx} = \frac{12y}{(x+y)^3}; r_{yy} = \frac{-12x}{(x+y)^3}; r_{xy} = r_{yx} = \frac{6y - 6x}{(x+y)^3}$

21. $z_{xx} = 0; z_{yy} = 4xe^y; z_{xy} = z_{yx} = 4e^y$

23. $r_{xx} = \frac{-4}{(2x+y)^2}; r_{yy} = \frac{-1}{(2x+y)^2}; r_{xy} = r_{yx} = \frac{-2}{(2x+y)^2}$

25. $z_{xx} = \frac{1}{x}; z_{yy} = -\frac{x}{y^2}; z_{xy} = z_{yx} = \frac{1}{y}$

27. $\frac{1}{2}; -3$ **29.** $x = -4, y = 2$ **31.** $x = 0, y = 0; x = 3, y = 3$

33. $f_x = 2x; f_y = z; f_z = y + 4z^3; f_{yz} = 1; -2; 1$

35. $f_x = \frac{6}{4z+5}; f_y = -\frac{5}{4z+5}; f_z = -\frac{4(6x-5y)}{(4z+5)^2}; f_{yz} = \frac{20}{(4z+5)^2}$

37. $f_x = \frac{2x - 5z^2}{x^2 - 5xz^2 + y^4}; f_y = \frac{4y^3}{x^2 - 5xz^2 + y^4};$ $f_z = -\frac{10xz}{x^2 - 5xz^2 + y^4}; f_{yz} = \frac{40xy^3z}{(x^2 - 5xz^2 + y^4)^2}$

39. 3; 9; answers vary. **41. (a)** 80 **(b)** 180 **(c)** 110 **(d)** 360 **43. (a)** \$902,100 **(b)** $f_p(p, i) = 99 - .5i - .005p; f_i(p, i) = -.5p; f_p(p, i)$ gives the rate at which weekly sales are changing per unit change in price when the interest rate remains constant; $f_i(p, i)$ gives the rate at which weekly sales are changing per unit change in interest rate when the price remains constant. **(c)** A weekly sales decrease of \$9700 **45. (a)** 1279 kcal per hr **(b)** 2.91 kcal per hr per g; the instantaneous rate of change of energy usage for a 300-kg animal traveling at 10 km per hr is about 2.9 kcal per hr per g. **47. (a)** Fixed costs, that is, the cost when no items have been produced **(b)** $\frac{\partial C}{\partial x} = \$600$ is the marginal cost of producing 1 unit of radios when the production of flashlights remains constant. $\frac{\partial C}{\partial y} = \$120$ is the marginal cost of producing 1 unit of flashlights when the production of radios remains constant. **49. (a)** About .0112 **(b)** About .0783 **51. (a)** 4665.6 **(b)** $f_x(27, 64) = .6912$ is the rate at which production is changing when labor changes by 1 unit from 27 to 28 and capital remains constant; $f_y(27, 64) = .4374$ is the rate at which production is changing when capital changes by 1 unit from 64 to 65 and labor remains constant. **(c)** Production would be increased by about 69 units when labor is increased by 1 unit.

53. $.7x^{-.3}y^{.3}$, or $\frac{.7y^{.3}}{x^{.3}}$; $.3x^{.7}y^{-.7}$, or $\frac{.3x^{.7}}{y^{.7}}$ **55. (a)** About 5.71 **(b)** About 6.84 **(c)** $\frac{1}{a - v}$ **(d)** $\frac{b}{(a - v)^2}$ **57. (a)** 62.8% **(b)** 37.0% **(c)** $f_n(3, 3, 320) = 2$, meaning that 2% is the rate of change of the probability per additional semester of high school math. $f_a(3, 3, 320) = .078$, which means that .078% is the rate of change of the probability per unit change in SAT score. **59. (a)** $F_m = \frac{gR^2}{r^2}$; the rate of change in force per unit change in mass; $F_r = \frac{-2mgR^2}{r^3}$; the rate of change in force per unit change in distance **(b)** Answers vary. **61. (a)** 28.24 **(b)** $\frac{\partial B}{\partial w} = \frac{703}{h^2}; \frac{\partial B}{\partial h} = \frac{-1406w}{h^3}$

Section 14.3 (Page 933)

1. Answers vary. **3.** Local minimum of −22 at (3, −1) **5.** Local minimum of 2 at (−2, −2) **7.** Saddle point at $\left(-2, \frac{5}{3}\right)$ **9.** Local maximum of 17 at $\left(\frac{2}{3}, \frac{4}{3}\right)$ **11.** Saddle point at (2, −2) **13.** Saddle point at (1, 2) **15.** Saddle point at (0, 0); local minimum of −201 at (6, 18) **17.** Saddle point at (0, 0); local minimum of −7.75 at (4.5, 1.5) **19.** Saddle point at (0, 0) **21.** Local maximum of $\frac{9}{8}$ at (−1, 1); saddle point at (0, 0); (a) **23.** Local minima of $-\frac{33}{16}$ at (0, 1) and (0, −1); saddle point at (0, 0); (b) **25.** Local maxima of $\frac{17}{16}$ at (1, 0) and (−1, 0); local minima of $-\frac{15}{16}$ at (0, 1) and (0, −1); saddle points at (0, 0), (−1, 1), (1, −1), (1, 1), and (−1, −1); (e) **27.** $P(12, 40) = 2744$ **29.** $P(12, 72) = \$2,528,000$ **31.** $C(12, 25) = 2237$ **33.** 3 m × 3 m × 3 m **35.** 36 in (length) × 18 in × 18 in **37.** 15 × 15 × 15 inches **39.** 1000 tons of grade A ore and 2000 tons of grade B ore. **41.** 6 tons of steel, 3 tons of aluminum, \$216,000 **43. (a)** 436.16 kJ/mol **(b)** 137.66 kJ/mol **(c)** Saddle point at (1.75, 133°C)

Section 14.4 (Page 944)

1. 200 when $x = 10$ and $y = 10$ **3.** $\frac{125}{4}$ when $x = 5$ and $y = \frac{5}{2}$ **5.** 28 when $x = 5$ and $y = 3$ **7.** 360 when $x = 20$ and $y = 2$ **9.** $\frac{81}{4}$ when $x = \frac{3}{2}, y = \frac{3}{2}$ and $z = 3$ **11.** $x = 7, y = 14$ **13.** $x = 34, y = 34, z = 34$ **15.** Answers vary. **17.** Answers vary. **19.** 37.5 ft × 100 ft **21.** $x = 2, y = 4$ **23.** About 189 units of labor and 35 units of capital **25.** 45,000 square meters **27.** Radius about 1.58 in; height about 3.17 in **29.** About 5.70 in × 5.70 in × 5.70 in **31.** 4 ft × 4 ft × 2 ft (height) **33. (a)** $F(r, s, t, \lambda) = rs(1 - t) + (1 - r)st + r(1 - s)t + rst - \lambda(r + s + t - \alpha)$ **(b)** $r = s = t = .25$ **(c)** $r = s = t = 1.0$

Chapter 14 Review (Page 947)

1. 15; 84 **3.** Not defined; −6 **5.** Answers vary.

7. **9.**

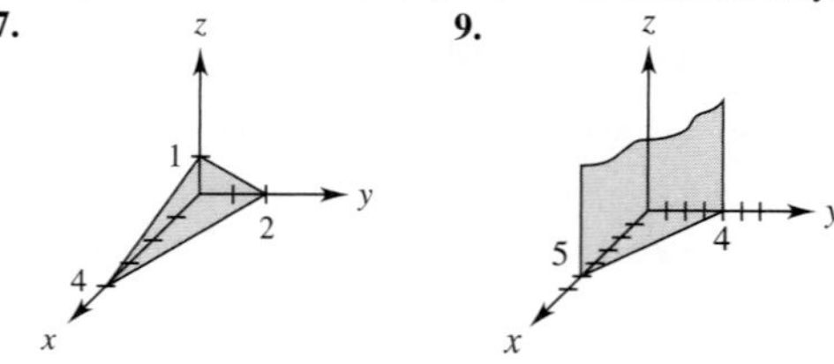

11. (a) $-4x + 5y$ **(b)** 3 **(c)** 5 **13.** Answers vary.

15. $f_x = 12x^2y + 12y^3; f_y = 4x^3 + 36xy^2$

17. $f_x = \frac{28xy^2}{(x^2 + 4y^2)^2}; f_y = \frac{-28x^2y}{(x^2 + 4y^2)^2}$

19. $f_x = 2(y + 1)^2e^{2x+y}; f_y = (y + 1)(y + 3)e^{2x+y}$

21. $f_x = \frac{3x^2y^2}{1 + x^3y^2}$; $f_y = \frac{2x^3y}{1 + x^3y^2}$ **23.** $f_{xx} = 24xy^2$; $f_{xy} = 24x^2y - 8$

25. $f_{xx} = 4y$; $f_{xy} = -18y^2 + 4x$ **27.** $f_{xx} = 2e^y$; $f_{xy} = 2xe^y$

29. $f_{xx} = \frac{-2x^2y^2 - 4y}{(2 - x^2y)^2}$; $f_{xy} = -\frac{4x}{(2 - x^2y)^2}$ **31. (a)** \$450 **(b)** \$646 **(c)** \$1030 **33. (a)** 10 **(b)** 720 **35.** Local minimum at (0, 1) **37.** Saddle point at (0, 2) **39.** Local minimum at $\left(1, \frac{1}{2}\right)$; saddle point at $\left(-\frac{1}{3}, \frac{11}{6}\right)$ **41. (a)** $x = 11, y = 12$ **(b)** \$431

43. Minimum of 2 when $x = 1$ and $y = -1$ **45.** $\frac{160}{3}$ and $\frac{80}{3}$

47. Answers vary.

Case 14 (Page 949)

1. At $m = .0062$ and $b = -3.46$, we have $f_m = f_b = 0$, $f_{mm} = 22{,}825{,}000 > 0$, and $f_{mm}f_{bb} - [f_{mb}]^2 = 60{,}000 > 0$, so f has a minimum. **3.** $m = .00532$, $b = -1.794$

Index of Applications

Business

Finance

Geometry

Health

Natural Science

Physical Science

Social Science

Transportation

Subject Index

Q

R

PERMUTATIONS

The number of permutations of n elements taken r at a time, where $r \leq n$, is

$$_nP_r = \frac{n!}{(n-r)!}.$$

COMBINATIONS

The number of combinations of n elements taken r at a time, where $r \leq n$, is

$$_nC_r = \binom{n}{r} = \frac{n!}{(n-r)!r!}.$$

BINOMIAL PROBABILITY

If p is the probability of success in a single trial of a binomial experiment, the probability of x successes and $n - x$ failures in n independent repeated trials of the experiment is

$$_nC_r p^x (1-p)^{n-x}.$$

MEAN

The mean of the n numbers, $x_1, x_2, x_3, \ldots, x_n$, is

$$\bar{x} = \frac{x_1 + x_2 + \cdots + x_n}{n} = \frac{\Sigma(x)}{n}.$$

SAMPLE STANDARD DEVIATION

The standard deviation of a sample of n numbers, $x_1, x_2, x_3, \ldots, x_n$, with mean $\bar{x}$, is

$$s = \sqrt{\frac{\Sigma(x-\bar{x})^2}{n-1}}.$$

BINOMIAL DISTRIBUTION

Suppose an experiment is a series of n independent repeated trials, where the probability of a success in a single trial is always p. Let x be the number of successes in the n trials. Then the probability that exactly x successes will occur in n trials is given by

$$_nC_r p^x (1-p)^{n-x}.$$

The mean μ and variance σ^2 of a binomial distribution are, respectively,

$$\mu = np \quad \text{and} \quad \sigma^2 = np(1-p).$$

The standard deviation is

$$\sigma = \sqrt{np(1-p)}.$$

The **derivative** of the function f is the function denoted f' whose value at the number x is defined to be the number

$$f'(x) = \lim_{h \to 0} \frac{f(x+h) - f(x)}{h},$$

provided this limit exists.

RULES FOR DERIVATIVES

Assume all indicated derivatives exist.

Constant Function If $f(x) = k$, where k is any real number, then

$$f'(x) = 0.$$

Power Rule If $f(x) = x^n$, for any real number n, then

$$f'(x) = n \cdot x^{n-1}.$$

Constant Times a Function Let k be a real number. Then the derivative of $y = k \cdot f(x)$ is

$$y' = k \cdot f'(x).$$

Sum or Difference Rule If $y = f(x) \pm g(x)$, then

$$y' = f'(x) \pm g'(x).$$

Product Rule If $f(x) = g(x) \cdot k(x)$, then

$$f'(x) = g(x) \cdot k'(x) + k(x) \cdot g'(x).$$

Quotient Rule If $f(x) = \frac{g(x)}{k(x)}$, and $k(x) \neq 0$, then

$$f'(x) = \frac{k(x) \cdot g'(x) - g(x) \cdot k'(x)}{[k(x)]^2}.$$

Chain Rule Let $y = f[g(x)]$. Then

$$y' = f'[g(x)] \cdot g'(x).$$

Chain Rule (Alternative Form) If y is a function of u, say $y = f(u)$, and if u is a function of x, say $u = g(x)$, then $y = f[g(x)]$, and

$$\frac{dy}{dx} = \frac{dy}{du} \cdot \frac{du}{dx}.$$

Generalized Power Rule Let u be a function of x, and let $y = u^n$ for any real number n. Then

$$y' = n \cdot u^{n-1} \cdot u'.$$

Exponential Function If $y = e^{g(x)}$, then

$$y' = g'(x) \cdot e^{g(x)}.$$

Natural Logarithmic Function If $y = \ln |g(x)|$, then

$$y' = \frac{g'(x)}{g(x)}.$$